Astronomical Applications of Astrometry

Ten Years of Exploitation of the Hipparcos Satellite Data

The Hipparcos satellite, developed and launched by the European Space Agency (ESA) in 1989, was the first space mission dedicated to astrometry – the accurate measurement of positions, distances, and proper motions of stars. Hipparcos pinpointed more than 100 000 stars, typically 200 times more accurately than ever before. Amongst the key achievements of its measurements are refining the cosmic distance scale, characterising the large-scale kinematic motions in the solar neighbourhood, providing precise luminosities for stellar modelling, and confirming Einstein's prediction of the effect of gravity on starlight.

This authoritative account of the Hipparcos contributions over the last decade is an outstanding reference for astronomers, astrophysicists and cosmologists. It reviews the applications of the data in different areas, describing the subject and the state-of-the-art before Hipparcos, and summarising all major contributions to the topic made by Hipparcos. It contains a detailed overview of the Hipparcos and Tycho Catalogues, their annexes and their updates. Each chapter ends with comprehensive references to relevant literature.

Michael Perryman is a Research Scientist at ESA, and a Professor in the Department of Astronomy at the University of Leiden, The Netherlands. He was the Project Scientist for the Hipparcos mission from 1981 to 1997, and has won several awards for his contributions to the field.

Astronomical Applications of Astrometry

Ten Years of Exploitation of the Hipparcos Satellite Data

Michael Perryman
European Space Agency, Noordwijk, The Netherlands
and Leiden Observatory, University of Leiden, The Netherlands

CAMBRIDGE UNIVERSITY PRESS
Cambridge, New York, Melbourne, Madrid, Cape Town, Singapore, São Paulo, Delhi

Cambridge University Press
The Edinburgh Building, Cambridge CB2 8RU, UK

Published in the United States of America by Cambridge University Press, New York

www.cambridge.org
Information on this title: www.cambridge.org/9780521514897

First published 2009

Printed in the United Kingdom at the University Press, Cambridge

A catalogue record for this publication is available from the British Library

ISBN 978-0-521-51489-7 hardback

For Julia

Contents

Preface

The context

The fundamental task of measuring stellar positions, and the derived properties of distances and space motions, has preoccupied astronomers for centuries. As one of the oldest branches of astronomy, astrometry is concerned with measurement of the positions and motions of planets and other bodies within the Solar System, of stars within our Galaxy and, at least in principle, of galaxies and clusters of galaxies within the Universe as a whole. Accurate star positions provide a celestial reference frame for representing moving objects, and for relating phenomena at different wavelengths. Determining the systematic displacement of star positions with time gives access to their motions through space. Determining their apparent annual motion as the Earth moves in its orbit around the Sun gives access to their distances through measurement of parallax. All of these quantities, and others, are accessed from high-accuracy measurements of the relative angular separation of stars. Repeated measurements over a period of time provide the pieces of a celestial jigsaw, which yield a stereoscopic map of the stars and their kinematic motions.

What follows, either directly from the observations or indirectly from modelling, are absolute physical stellar characteristics: stellar luminosities, radii, masses, and ages; and their dynamical signatures. The physical parameters are then used to understand their internal composition and structure, to disentangle their space motions and, eventually, to explain in a rigorous and consistent manner how the Galaxy was originally formed, and how it will evolve in the future. Significantly, space motions reflect dynamical perturbations of all other matter, visible or invisible.

Buried but not necessarily hidden within this fractal phase-space jigsaw are a whole host of higher-order phenomena: at the milliarcsec level, binary star signatures, General Relativistic light bending, and the dynamical consequences of dark matter are already evident; at the microarcsec level, targeted by the next generation of space astrometry missions currently under development, direct distance measurements will be extended across the Galaxy and to the Large Magellanic Cloud. Other effects will become routinely measurable at the same time. These include perspective acceleration and secular parallax evolution, more subtle metric effects, planetary perturbations to the photocentric motion, and astrometric microlensing; and at the nanoarcsec level, currently no more than an experimental concept, effects of optical interstellar scintillation, geometric cosmology, and ripples in space-time due to gravitational waves will become apparent. The bulk of this seething motion is largely below current observational capabilities, but it is there, waiting to be investigated.

Historical perspective

Measuring stellar distances, and their three-dimensional distribution and space motions, remains a difficult task, even within our solar neighbourhood. John Herschel (1792–1871) attempted to convey the unimaginable interstellar distance scales with the following analogy (quoted by Allen, 1963, p153): *'to drop a pea at the end of every mile of a voyage on a limitless ocean to the nearest fixed star, would require a fleet of 10 000 ships, each of 600 tons burthen.'*

It is useful to place milliarcsec astrometric measurements in a brief, albeit highly selective, historical context. Chapman (1990) provides further fascinating historical details of the development of angular measurements in astronomy between 1500–1850.

After the remarkable achievements of the ancient Greeks, including their first estimates of the sizes and distances of the Sun and Moon, the narrative intensifies 300–400 years ago, when three main scientific themes motivated the improvement of angular measurements: the navigational problems associated with the determination of longitude on the Earth's surface, the comprehension and acceptance of Newtonianism, and understanding the Earth's motion through space. Even before 1600, astronomers were in agreement that the crucial evidence needed to detect the Earth's motion was the measurement of trigonometric parallax, the tiny oscillation in a star's apparent position arising from the Earth's annual motion around the Sun. The early British

Astronomers Royal, for example, appreciated the importance of measuring stellar distances, and were very much preoccupied with the task. But it was to take a further 250 years until this particular piece of observational evidence could be secured.

In 1718, Edmund Halley, who had been comparing contemporary observations with those that the Greek Hipparchus and others had made, announced that three stars, Aldebaran, Sirius and Arcturus, were displaced from their expected positions by large fractions of a degree (Halley, 1718). He deduced that each star had its own distinct velocity across the line-of-sight, or proper motion: stars were moving through space.

By 1725, angular measurements had improved to a few arcsec, making it possible for James Bradley, England's third Astronomer Royal, to detect stellar aberration, as a by-product of his unsuccessful attempts to measure the distance to the bright star γ Draconis. This was an unexpected result: small positional displacements were detected, and correctly attributed to the vectorial addition of the velocity of light to that of the Earth's motion around the Sun (Bradley, 1725). His observations provided the first direct proof that the Earth was moving through space, and thus a confirmation both of Copernican theory, and Roemer's discovery of the finite velocity of light 50 years earlier. It also confirmed Newton's hypothesis of the enormity of stellar distances, and showed that the measurement of parallax would pose a technical challenge of extraordinary delicacy.

During the eighteenth century, the motions of many more stars were announced, and in 1783 William Herschel found that he could partly explain these effects by assuming that the Sun itself was moving through space. Attempts to measure parallax intensified. Nevil Maskelyne, England's fifth Astronomer Royal, spent seven months on the island of St Helena in 1761, using a zenith sector and plumb-line, in an unsuccessful attempt to measure the parallax of the bright star Sirius.

Criteria for probable proximity were developed, and after many unsuccessful attempts, the first stellar parallaxes were measured in the 1830s.

Friedrich Bessel is generally credited as being the first to publish a parallax, for 61 Cygni (Piazzi's Flying Star), from observations made between 1837–38. Thomas Henderson published a parallax for α Centauri in 1839, derived from observations made in 1832–33 at the Cape of Good Hope. In 1840, Wilhelm Struve presented his parallax for Vega from observations in 1835–1837. Confirmation that stars lay at very great but nevertheless finite distances represented a turning point in the understanding of the Universe. John Herschel, President of the Royal Astronomical Society at the time, congratulated Fellows that they had *'lived to see the day when the sounding line in the Universe of stars had at last touched bottom'* (Quoted by Hoskin, 1997, p. 219).

ESA chooses Hipparcos

THE Scientific Programme Committee of the European Space Agency decided last week to fund the astrometry mission Hipparcos as the next ESA mission after Exosat which is to be launched in 1981. Designed to improve the measurements of stellar positions by two orders of magnitude, the satellite will be launched by Ariane in mid-1986, and placed in a geostationary orbit for its lifetime of two and a half years. The total estimated cost of the project is 139.3 MAU ($185 million).

Hipparcos was the Programme Committee's final choice in spite of a recommendation by the Scientific Advisory Committee to fund a dual mission consisting of experiments to measure the Earth's magnetospheric tail and a deep space flyby of Halley's comet. By a vote of 10 votes for and one abstention, the 11 member committee, consisting of delegates from each ESA country, decided to overturn the Scientific Advisory Committee's recommendation.

Figure 1 The announcement of the selection of the Hipparcos mission by ESA's Science Programme Committee, which appeared in Nature, 1980 Vol 284, p 116, reprinted by permission from Macmillan Publishers Ltd: Nature ©1980.

The following 150 years saw enormous progress, with the development of accurate fundamental catalogues, and a huge increase in quantity and quality of astrometric data based largely on meridian circle and photographic plate measurements.

The Hipparcos mission

By the second half of the twentieth century, however, measurements from ground were running into essentially insurmountable barriers to improvements in accuracy, especially for large-angle measurements and systematic terms; a review of the instrumental status shortly in advance of the Hipparcos launch is given by Monet (1988). Problems were dominated by the effects of the Earth's atmosphere, but were compounded by complex optical terms, thermal and gravitational instrument flexures, and the absence of all-sky visibility. A proposal to make these exacting observations from space was first put forward in 1967.

Hipparcos was the result of a long process of study and lobby, and the first space experiment dedicated to astrometry. It was accepted within the European Space Agency's scientific programme in 1980 (Figure 1). It represented a major advance in physics, cost some 600 MEuro, and its execution involved some 200 European scientists and more than 2000 individuals in European industry.

The European Space Agency and its scientific advisory structure selected the Hipparcos mission based on what is referred to as the Phase A study and report (ESA, 1979). The underlying scientific motivation was to determine the physical properties of the stars through the determination of their distances, and to place theoretical studies of stellar structure and evolution, and studies of Galactic structure and kinematics, on a more secure observational footing. Observationally, the objective was to provide the positions, parallaxes, and annual proper motions for some 100 000 stars with an unprecedented accuracy of some 0.002 arcsec, a target in practice surpassed by roughly a factor of 2.

The Hipparcos satellite was launched in August 1989 and operated until 1993. The final mission results were finalised in 1996, and published by ESA in June 1997 as a compilation of 17 hardbound volumes, a celestial atlas, and six CDs, comprising the Hipparcos and Tycho Catalogues (ESA, 1997). Details of the satellite operation, and the successive steps in the data analysis, and in the validation and description of the detailed data products, are included in the published catalogue. The result have been in the scientific domain for 10 years.

The Hipparcos science

The Hipparcos results impact a very broad range of astronomical research, which can be classified into three major themes:

(a) The provision of an accurate reference frame: this has allowed the consistent and rigorous re-reduction of historical astrometric measurements, including those from Schmidt plates, meridian circles, the 100-year old Astrographic Catalogue, and 150 years of Earth-orientation measurements. These, in turn, have yielded a dense reference framework with high-accuracy long-term proper motions (the Tycho 2 Catalogue). Reduction of current state-of-the-art survey data has yielded the dense UCAC 2 Catalogue on the same reference system, and improved astrometric data from recent surveys such as SDSS and 2MASS. Implicit in the high-accuracy reference frame is the measurement of General Relativistic light bending, and the detection and characterisation of double and multiple stars.

(b) Constraints on stellar structure and evolution: the accurate distances and luminosities of 100 000 stars has provided the most comprehensive and accurate dataset of fundamental stellar parameters to date, placing constraints on internal rotation, element diffusion, convective motions, and asteroseismology. Combined with theoretical models and other data it yields evolutionary masses, radii, and ages for large numbers of stars covering a wide range of evolutionary states.

(c) Galactic kinematics and dynamics: the uniform and accurate distances and proper motions have provided a substantial advance in understanding of the kinematic and dynamical structure of the solar neighbourhood, ranging from the presence and evolution of clusters, associations and moving groups, the presence of resonance motions due to the Galaxy's central bar and spiral arms, determination of the parameters describing Galactic rotation, discrimination of the disk and halo populations, evidence for halo accretion, and the measurement of space motions of runaway stars, globular clusters, and many other types of star.

Associated with these major themes, Hipparcos has provided results in topics as diverse as Solar System science, including mass determinations of asteroids, Earth rotation and Chandler wobble, the internal structure of white dwarfs, the masses of brown dwarfs, the characterisation of exoplanets and their host stars, the height of the Sun above the Galactic mid-plane, the age of the Universe, the stellar initial mass function and star formation rates, and search strategies for extraterrestrial intelligence. The high-precision multi-epoch photometry has been used to measure variability and stellar pulsations in many classes of objects. The Hipparcos and Tycho Catalogues are now routinely used to point ground-based telescopes, navigate space missions, and drive public planetaria.

The review

The review tackles the full range of the Hipparcos scientific findings, in an analysis of the scientific literature over the 10 years since the publication of the Hipparcos and Tycho Catalogues in 1997. In this period, some 2000 or more papers have appeared in the refereed literature which directly mention the Hipparcos or Tycho Catalogues in their title or abstract. Many other papers, especially more recently, make use of the data. As the catalogues become a more routine part of the day-to-day tools of astronomy, the data are frequently used without direct attribution, or in wider-ranging applications along with other types of data. This makes more direct attribution of results to the Hipparcos mission harder and, eventually, of course, impossible. In their analysis of the productivity and impact of space-based astronomical facilities, Trimble *et al.* (2006) addressed this by *'going page by page through all the issues of 19 journals published in 2001, and identifying all the papers that reported or analysed data from any space-based astronomical facility'*.

The broad scope of the review illustrates the breadth of science touched upon by astrometric results, and makes answering the question 'what has the Hipparcos mission achieved' more tractable. It should set the new generation of advanced astrometric missions, now being developed, in a clearer context. More importantly, a broad review allows the implications of results in one area to be traced to their impact in another. It also permits a wider understanding of the strengths and limitations of the Hipparcos data in their entirety.

In illustrating the range of scientific topics, I also hope that the review inspires further detailed investigations based on the Hipparcos and Tycho Catalogue data. I have been left with the strong impression that there is much information remaining to be extracted from the catalogues, especially when taken along with data from other sources.

I have aimed for a reasonably uniform notation throughout the volume, but in certain cases considered it preferable to retain the notation of the cited article. I have modestly extended this unification to some titles of the cited references, where a wide range of abbreviations and notations can be found. References are generally restricted to refereed publications and conference proceedings, although a few less mainstream references are included where these indicated interesting work or other ideas in progress.

I stress that the work is a review, and makes marginal claim to originality. Most of the results, conclusions, and discussions are taken from the cited works. Where the authors have described their work, or their results, in clear and authoritative words I have not hesitated to borrow from them. The review has involved consulting some 5000 papers, related to the Hipparcos analyses or to the background context. It provides a snapshot of the scientific relevance of positional astronomy 10 years after the Hipparcos and Tycho Catalogue publication.

The broad approach offers numerous pitfalls in terms of the depth and expertise of the analysis presented. Relevant chapters are not aimed primarily for those already expert on the associated topic, but rather for those looking for orientation amongst the enormous literature already associated with the Hipparcos results. Accordingly, I hope that the advantages offered by a single-author survey outweigh the disadvantages imposed by such a broad review.

Conscious of the role of the individual in these large scientific endeavours, I have included in Appendix C photographs of some of the leading authors of the various Hipparcos-related papers, especially (but not exclusively) those whose figures are included. I have also included a very few individuals who were not authors of papers based on Hipparcos data, but who had some related involvement. This, I stress, is a selection: it does not necessarily signify the most important contributions, and for various reasons it is highly incomplete. Photos include some from my own collection but mostly from the authors (all included with their explicit permission), and in a few cases from non-copyrighted www sources.

Acknowledgments

The Hipparcos and Tycho Catalogues were the result of the Hipparcos space astrometry mission, undertaken by the European Space Agency, with the scientific aspects undertaken by nearly two hundred scientists within the NDAC, FAST, TDAC and INCA Scientific Consortia. The efforts of the many individuals and organisations participating in the project over many years have been an essential component of the project's successful completion. Full acknowledgments for the catalogue effort are given in the published catalogue (ESA, 1997, Volume 1).

I am most grateful to a number of colleagues who kindly reviewed the various chapters as follows: Lennart Lindegren (1); Erik Høg and Norbert Zacharias (2); Dimitri Pourbaix and Staffan Söderhjelm (3); Laurent Eyer and Carme Jordi (4); Xavier Luri and Catherine Turon (5); Jos de Bruijne (6); Corinne Charbonnel and Yveline Lebreton (7); Ulrich Bastian (8); Walter Dehnen and Michael Merrifield (9); Daniel Hestroffer (10); and Jos de Bruijne (Appendix A).

I would also like to thank the following for specific assistance: Lennart Lindegren for guidance on various aspects related to the use and transformation of astrometric parameters (Chapter 1); Catherine Turon for contributions to the section on ground-based radial velocity efforts (Chapter 1); François Ochsenbein for details of the CDS Simbad and Vizier systems (Chapter 1); Roger Sinnott and Wil Tirion for guidance on historical star atlases (Chapter 2); Tijl Verhoelst for information on unpublished results on Arcturus (Chapter 3); Laurent Eyer for discussions on variability over the HR diagram (Chapter 4); Anthony Brown on interpretation of the Lutz–Kelker bias, and for comments on luminosity calibration as a function of spectral type (Chapter 5); Nancy Houk for the current status of the Michigan Spectral Survey classification (Chapter 5); Michael Bessell for determining the *Hp* bolometric corrections (Chapter 7); Yveline Lebreton and Don VandenBerg for guidance on the compilation of stellar evolutionary models (Chapter 7); Misha Haywood for clarifying some aspects of the chemical evolution of the Galaxy (Chapter 7); Stefan Jordan and Rainer Wehrse for helpful comments on various topics (Chapter 8); Richard Branham for assistance in interpreting the effects stellar kinematics (Chapter 9); François Mignard for discussions on the Ogorodnikov–Milne model and the use of vector spherical harmonics (Chapter 9); Nicole Capitaine for clarifying various subtleties related to precession and nutation (Chapter 10);

and Jos de Bruijne for guidance on the current status of the fundamental constants (Appendix A).

I have made extensive use of figures from the cited articles to illustrate the topics covered, and I am grateful to all who authorised their use and frequently provided original figures. In addition to references to the figures made at the appropriate locations, I have used figures from the following (first) authors, almost all of whom I was able to contact to request approval:

Chapter 1: Dainis Dravins, Erik Høg, Burton Jones, Sergei Klioner, Martin Kürster, Lennart Lindegren, Leslie Morrison, Fredrik Quist, Heiner Schwan, Roland Wielen, Clifford Will, and Zi Zhu.

Chapter 2: Terry Girard, Andrew Gould, Nigel Hambly, Bob Hanson, Sebastien Lépine, Dave Monet, Quentin Parker, Jeff Pier, Yves Réquième, Roger Sinnott, Sean Urban, Bruno Viateau, Jan Vondrák, and Norbert Zacharias.

Chapter 3: Christine Allen, Frédéric Arenou, Yuri Balega, Claus Fabricius, Frank Fekel, Adam Frankoswki, Alexey Goldin, George Gontcharov, Sylvie Jancart, Alain Jorissen, Lennart Lindegren, Valeri Makarov, Christian Martin, Brian Mason, Daniel Popper‡, Dimitri Pourbaix, Fredrik Quist, Ignasi Ribas, Slavek Rucinski, Selim Selam, Staffan Söderhjelm, Anatoly Suchkov, Jocelyn Tomkin, Guillermo Torres, and Roland Wielen.

Chapter 4: Connie Aerts, Carlos Allende Prieto, Dominique Barthès, Michael Bessell, Fabien Carrier, Jørgen Christensen-Dalsgaard, Alan Cousins‡, Noel Cramer, Margarida Cunha, Albert Domingo, Jirí Dušek, Laurent Eyer, Alejandro García Gil, Gerald Handler, Erik Høg, Swetlana Hubrig, Jill Knapp, Chris Koen, Patrick de Laverny, Floor van Leeuwen, Hans-Michael Maitzen, Jaymie Matthews, Tarmo Oja, Jørgen Otzen Petersen, Alexey Pamyatnykh, John Percy, Imants Platais, Gerhard Scholz, and Christoffel Waelkens.

Chapter 5: Rodrigo Alvarez, Frédéric Arenou, Cecilia Barnbaum, Tim Bedding, Angelo Cassatella, Bing Chen, Andrei Dambis, Daniel Egret, Michael Feast, Edward Fitzpatrick, Léo Girardi, Anita Gómez, Andrew Gould, Aaron Grocholski, Frank Grundahl, Philip Keenan‡, Pavel Kroupa, Floor van Leeuwen, Gisela Maintz, John Martin, Andrzej Megier, Réné Oudmaijer, Ferhat Fikri Özeren, Giancarlo Pace, Bohdan Paczyński‡, Ernst Paunzen, Neill Reid, Michael Richmond, Jim Sowell, Krzysztof Stanek, Don VandenBerg, Walter Wegner, and Roland Wielen.

Chapter 6: Ricard Asiain, Holger Baumgardt, Jos de Bruijne, Vittorio Castellani, Bing Chen, Emmanuel Chereul, Fernando Comerón, Andrei Dambis, Thomas Dame, Benoit Famaey, Eric Feigelson, Klaus Fuhrmann, Isabelle Grenier, Jesús Hernández, Ronnie Hoogerwerf, Nina Kharchenko, Jeremy King, Yveline Lebreton, Floor van Leeuwen, Søren Madsen, Valeri Makarov, Eric Mamajek, Edmundo Moreno, Christophe Perrot, Anatoly Piskunov, Thomas Preibisch, Marilia Sartori, Jovan Skuljan, Inseok Song, Tim de Zeeuw, and Ben Zuckerman.

Chapter 7: Carlos Allende Prieto, Angel Alonso, Martin Altmann, Michaël Bazot, Gerard van Belle, Paolo Di Benedetto, Thomas Blöcker, Jos de Bruijne, Bruce Carney, Corinne Charbonnel, Giuseppe Cutispoto, Dainis Dravins, Thomas Dumm, Michele Gerbaldi, David Guenther, Swetlana Hubrig, Mikolaj Jerzykiewicz, Raul Jimenez, Karin Jonsell, Torsten Kaempf, Pierre Kervella, Henny Lamers, Yveline Lebreton, Earle Luck, Sushma Mallik, Maria Pia Di Mauro, Georges Meynet, Georges Michaud, José-Dias do Nascimento, Heidi Jo Newberg, Yuen Keong Ng, Poul Erik Nissen, Pierre North, Francesco Palla, Harald Pöhnl, Maria Sofia Randich, Bacham Eswar Reddy, Bernardo Salasnich, Klaus-Peter Schröder, Don VandenBerg, Patricia Whitelock, and Sukyoung Yi.

Chapter 8: Agnes Acker, Mario van den Ancker, Askin Ankay, Marcelo Arnal, Jacques Bergeat, David Berger, Bob Campbell, Fabien Carrier, Bing Chen, Fernando Comerón, Sophie Van Eck, Fabio Favata, Edward Fitzpatrick, Michele Gerbaldi, Patrick Guillout, Thomas Hearty, Ulrich Heber, Ronnie Hoogerwerf, Anne Marie Hubert, Fredrik Huthoff, Eric Jensen, Lex Kaper, Florian Kerber, Jill Knapp, Brigitte Konig, Howard Lanning, Jean-Louis Leroy, Jesús Maíz Apellániz, Valeri Makarov, Sergey Marchenko, John Martin, Anne-Laure Melchior, Tony Moffat, Jorge Panei, Ernst Paunzen, John Percy, Nicola Pizzolato, Judith Provencal, Klaus-Peter Schröder, Wieslaw Skórzyński, David Soderblom, Jon Sowers, Anatoly Suchkov, Peeter Tenjes, Gérard Vauclair, Jean-Luc Vergely, Sergio Vieira, Volker Weidemann, and Barry Welsh.

Chapter 9: Eugenio Carretta, Brian Chaboyer, Bing Chen, Masashi Chiba, Michel Crézé, Walter Dehnen, Wilton Dias, Dana Dinescu, Sofia Feltzing, David Fernández, Klaus Fuhrmann, Andrew Gould, Amina Helmi, Xavier Hernández, David Hogg, Johan Holmberg, Akihiko Ibukiyama, Hartmut Jahreiß, Raul Jimenez, Jacques Lépine, Jesús Maíz Apellániz, Valeri Makarov, Masanori Miyamoto, Gerhard Mühlbauer, Birgitta Nordström, Rob Olling, Alice Quillen, Neill Reid, Jerry Sellwood, Tanya Sitnik, Richard Smart, and Makoto Uemura.

Chapter 10: Conard Dahn, Thomas Dall, Daniel Hestroffer, Klaus Fuhrmann, Joan García Sanchez, Douglas Gies, Jean-Louis Halbwachs, Suzanne Hawley, Guillaume Hébrard, Paul Kalas, Chris Koen, Yuri Kolesnik, Neill Reid, Noel Robichon, Nir Shaviv, Henrik Svensmark, Guillermo Torres, Margaret Turnbull, Jan Vondrák, and Shay Zucker.

Sergey Marchenko kindly provided the extracted selection appearing as Figure 8.20, and Neill Reid kindly provided the updated version appearing as Figure 10.22.

Additionally, all figures are reproduced with the permission of the respective publishers, as follows: figures originally published in *Astronomy & Astrophysics* (and *Supplement Series*) are reproduced by permission of the Editorial Office; figures published in the *Astronomical Journal, Astrophysical Journal, Astrophysical Journal Letters,* and *Astrophysical Journal Supplement,* are reproduced by permission of the American Astronomical Society and the University of Chicago Press; figures published in the *Publications of the Astronomical Society of the Pacific* are reproduced by permission of the Astronomical Society of the Pacific and the University of Chicago Press; figures published in *Monthly Notices of the Royal Astronomical Society* are reproduced by permission of Wiley–Blackwell Publishing Ltd; figures published in the *ASP Conference Series* are reproduced by permission of the Astronomical Society of the Pacific Conference Series; figures published in *Astronomische Nachrichten* are reproduced by permission of Wiley–VCH Verlag GmbH & Co. KGaA; figures published in *Astronomy Letters, Astrophysics and Space Science, Astrophysics and Space Science Library,* and *Astronomy & Astrophysics Review* are reproduced by permission of Springer Science and Business Media; figures and text published in *Nature* are reproduced by permission of Macmillan Publishers Ltd; figures published in *New Astronomy* and *Physics Reports* are reproduced by permission of Elsevier; figures published in *International Astronomical Union Symposium Proceedings* are reproduced by permission of the IAU; figures published in ESA publications and conference proceedings are reproduced by permission of the European Space Agency; the figure from *Annual Reviews of Astronomy & Astrophysics* is reproduced by permission of Annual Reviews of Astronomy & Astrophysics; the figure from *Acta Astronomica* is reproduced by permission of the editor; the figure from *Publications of the Astronomical Society of Japan* is reproduced by permission of the Astronomical Society of Japan; the figure from *Living Reviews in Relativity* is reproduced by permission of Living Reviews in Relativity; the figure from *Monthly Notes of the Astronomical Society of South Africa* is reproduced by permission of Astronomical Society of South Africa; the figure from *Revista Mexicana de Astronomía y Astrofísica* (Serie de Conferencias) is reproduced by permission of Revista Mexicana de Astronomía y Astrofísica; the figure from *Science* is reproduced by permission of the American Association for the Advancement of Science; the figure from the *Contributions of the Astronomical Observatory of Skalnate Pleso* is reproduced by permission of the Slovak Academy of Sciences; the figure from the *12th Cambridge Workshop on Cool Stars* is reproduced by permission of the editors of the proceedings.

I thank Willie Koorts for the photo of Alan Cousins, André Heck for that of Carlos Jaschek, Frédéric Arenou for those of various colleagues, and the permission of Alistair Walker for use of the picture of Olin Eggen. The photo of Allan Sandage is reproduced courtesy of the Observatories of the Carnegie Institute of Washington.

Preparation of the review has been made feasible in its current form through the extensive use of the NASA Astrophysics Data System (ADS).

The use of LaTeX/Bibtex was indispensable. The manuscript was typeset in Adobe Utopia, using Michel Bovani's Fourier-GUTenberg math fonts, and made use of the following packages: graphicx (figures), caption (captions), natbib, bibtex and makebst (bibliographic references), chapterbib (chapter references), authorindex (author index references and compilation), and makeidx (subject indexing).

I thank José Hernández (ESAC) for providing a Java script to extract the acronyms. Gunther Thörner, Olive Buggy and Liam Gretton (ESTEC) were generous in their assistance on issues of computer infrastructure.

I acknowledge the support of the ESA Director of Science, David Southwood, and the Head of the Research and Scientific Support Department, Alvaro Giménez, during the preparation of this review.

I am grateful to Simon Mitton who, as commissioning editor, supported my proposal to undertake this review.

Finally, I would like to acknowledge four broad groups of scientists whose work is connected with this review. First, to those who participated in the development of positional astronomy from the ground over the past century or more: their creativity and meticulous skills laid the foundations and stimulated the field for the era of space astrometry to flourish. Second, to all those involved in the many aspects of the Hipparcos space astrometry mission, within ESA, industry, and the scientific community; it was a remarkable collaboration which provided the basics for a wealth of scientific investigation. Third, to those now involved in ESA's Gaia mission, which will move the frontiers of scientific understanding in this area in an even more dramatic manner; the Hipparcos discoveries should underline its importance. Fourth, to all those who used the Hipparcos and Tycho Catalogues for their own scientific research and, through their own ingenuity and questioning, produced the results presented here.

References

Allen RH, 1963, *Star Names: Their Lore and Meaning*. Dover Republication {**xv**}

Bradley J, 1725, A letter to Dr Halley giving account of a new discovered motion of the fixed stars. *Phil. Trans.*, 35, 640 {**xvi**}

Chapman A, 1990, *Dividing the Circle: the Development of Critical Angular Measurement in Astronomy 1500–1850.* Ellis Horwood, London {**xv**}

ESA, 1979, *Hipparcos Space Astrometry: Report on the Phase A Study, ESA SCI(79)10.* European Space Agency, Noordwijk {**xvii**}

ESA, 1997, *The Hipparcos and Tycho Catalogues. Astrometric and Photometric Star Catalogues derived from the ESA Hipparcos Space Astrometry Mission, ESA SP–1200 (17 volumes including 6 CDs).* European Space Agency, Noordwijk, also: VizieR Online Data Catalogue {**xvii, xviii**}

Halley E, 1718, Considerations on the change of the latitude of some of the principal fixed stars. *Phil. Trans.*, 30, 736–738 {**xvi**}

Hoskin M, 1997, *Cambridge Illustrated History of Astronomy.* Cambridge University Press {**xvi**}

Monet DG, 1988, Recent advances in optical astrometry. *ARA&A*, 26, 413–440 {**xvi**}

Trimble V, Zaich P, Bosler T, 2006, Productivity and impact of space-based astronomical facilities. *PASP*, 118, 651–655 {**xvii**}

1

The Hipparcos and Tycho Catalogues

1.1 Overview

This chapter describes various aspects of the Hipparcos and Tycho Catalogues useful for understanding the scientific exploitation considered in subsequent chapters. It describes some of the satellite measurement principles relevant for an understanding of the catalogue contents; the published intermediate astrometry data; details of the adopted reference frame; basic transformations relevant to catalogue users; details of the Tycho 2 Catalogue construction; error assessment; and details of associated data such as radial velocities and cross-identifications.

The Hipparcos Catalogue contains 118 218 entries, corresponding to an average of some three stars per square degree over the entire sky. Median precision of the five astrometric parameters ($Hp < 9$ mag) exceeded the original mission goals, and are between 0.6–1.0 mas. Some 20 000 distances were determined to better than 10%, and 50 000 to better than 20%. The inferred ratio of external to standard errors is ~ 1.0–1.2, and estimated systematic errors are below 0.1 mas. The number of solved or suspected double or multiple stars is 23 882. Photometric observations yielded multi-epoch photometry with a mean number of 110 observations per star, a median photometric precision ($Hp < 9$ mag) of 0.0015 mag, and 11 597 entries were identified as variable or possibly variable.

The Tycho Catalogue of just over 1 million stars was superseded in 2000 by the Tycho 2 Catalogue of some 2.5 million stars; both included two-colour photometry.

1.2 Observation principles

Key features Some key features of the observations were as follows: (a) through observations from space, the effects of atmospheric seeing, instrumental gravitational flexure and thermal distortions could be obviated or minimised; (b) all-sky visibility permitted a direct linking of the stars observed all over the celestial sphere; (c) the two viewing directions of the satellite, separated by a large and suitable angle, resulted in a 'rigid' connection between quasi-instantaneous one-dimensional observations in different parts of the sky. In turn, this led to parallax determinations which are absolute (rather than relative, with respect to some unknown zero-point); (d) the continuous ecliptic-based scanning of the satellite resulted in an optimum use of the available observing time, with a resulting catalogue providing reasonably homogeneous sky density and uniform astrometric accuracy; (e) the various geometrical scan configurations for each star, at multiple epochs throughout the three-year observation programme, resulted in a dense network of one-dimensional positions from which the barycentric coordinate direction (α, δ), the parallax (π), and the object's proper motion ($\mu_\alpha \cos\delta$, μ_δ) could be solved for in what was effectively a global least-squares reduction of the totality of observations. The astrometric parameters as well as their standard errors and correlation coefficients were derived in the process; (f) since the number of independent geometrical observations per object was large (typically of order 30) compared with the number of unknowns for the standard model (five astrometric unknowns per star) astrometric solutions not complying with this simple 'five-parameter' model, could be expanded to take into account the effects of double or multiple stars, or nonlinear photocentric motions ascribed to unresolved 'astrometric binaries'; (g) a somewhat larger number of actual observations per object, of order 110, provided accurate and homogeneous photometric information for each star, from which mean magnitudes, variability amplitudes, and in many cases period and variability type classification could be undertaken. Further details can be found in the published catalogue (ESA, 1997, Volume 2).

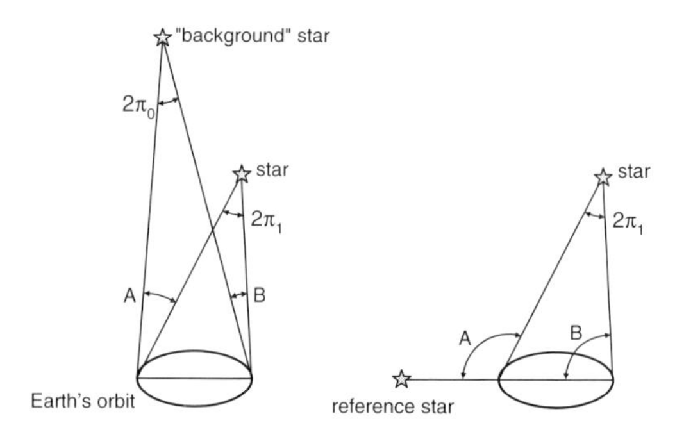

Figure 1.1 The principle of absolute parallax determination. Left: the measurement of the (small) angles A, B only allows determination of the relative parallax $\pi_1 - \pi_0 = (A - B)/2$. Right: in contrast, measurement of the large angles allows determination of the absolute parallax $\pi_1 = (A - B)/2$, independent of the distance to the reference star. From Lindegren (2005, Figure 4).

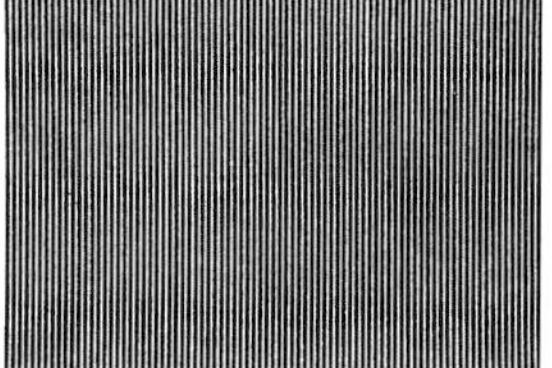

Figure 1.2 Optical micrograph of part of the main modulating grid used for the main mission (Hipparcos Catalogue) observations (left), and part of the inclined slits of the star mapper grid used for the Tycho Catalogue observations (right). From ESA (1997, Volume 2, Figure 2.14).

Understanding Hipparcos and space astrometry The basic measurement principles of Hipparcos were reasonably simple, and indeed rather elegant. A purely qualitative description of the principles are given here, to guide the understanding of the detailed operational and mathematical formulation that has been presented in detail in the published catalogue (Volume 2: the Hipparcos Satellite Operations; Volume 3: Construction of the Hipparcos Catalogue; Volume 4: Construction of the Tycho Catalogue). The space platform did not access any part of the spectral region not visible from the ground, but rather exploited the absence of the perturbing atmosphere and the essential absence of gravitational and thermal flexure, at the same time permitting all-sky visibility which is central to the concept of two simultaneous viewing directions separated by a large, fixed, angle. This, in turn, is essential for the derivation of absolute trigonometric parallaxes, independent of the assumption of distances of reference stars (Figure 1.1). Descriptions can be found in early Hipparcos studies (Høyer *et al.*, 1981), and more recently in the context of Gaia, e.g. Lindegren & Perryman (1996, Section 3.1) and Lindegren (2005, Section 3.2).

The measuring instrument consisted of two superimposed fields of view, each $0\overset{\circ}{.}9 \times 0\overset{\circ}{.}9$ in size (thus ensuring a sufficient number of stars in each field), and separated by a 'basic angle' of about 58°. Rotation about a spin axis perpendicular to the two viewing directions, and at a reasonably uniform rate, yielded modulated light signals due to a regular opaque/transparent grid at the focal surface (Figure 1.2). The signal amplitude provided the photometric intensity, while the relative signal phases yielded the instantaneous measurement of the along-scan (angular) separation between stars in the two fields of view (modulo the period of the grid, of approximately 1.208 arcsec). A slow precession of the spin axis resulted in slowly precessing great circle scans across the celestial sphere, and hence a network of one-dimensional angular measurements progressively and repeatedly covering the sky at different orientations. One great circle was scanned every 2.1 hours, resulting in approximately 14 000 great circle scans, or some 2500 'reference great circles' each of about 12-hour duration over the 3.3-year measurement period.

The measurement principle can be understood in three steps. In the first step, Figure 1.3, left illustrates a distribution of (single) stars in space with no space motions, with an observer at rest amongst them. The position of each star is fully described by just two angular coordinates α and δ. From the 100 or so repeated measurements at different orientations over the satellite's three-year lifetime, estimates of these two parameters can be extracted from the system of measurement equations, along with a formal estimate of their standard errors. As long as the instrument stability, and its associated geometrical calibration terms, varied only slowly with time, and could therefore be described by a small number of terms within the very large number of individual measurements, the system is self-calibrating, and the estimation of stellar parameters largely unaffected. The scanning 'law' was optimised, with its symmetry axis in the ecliptic plane, with a fixed angle between the spin axis and the Sun line, in order to maximise the thermal stability of the payload during its sky scanning. Small and large Sun aspect angles would have yielded poor uniformity in sky coverage or weak determinations of one or other stellar coordinate. An angle of 43° was finally adopted as compromise, still inevitably leading to asymmetries in coverage and parameter determination as a function of ecliptic latitude; details of the scanning law are given in ESA (1997, Volume 2, Chapter 8). The choice of basic angle between the two viewing directions can also be understood qualitatively, and was optimised rigorously:

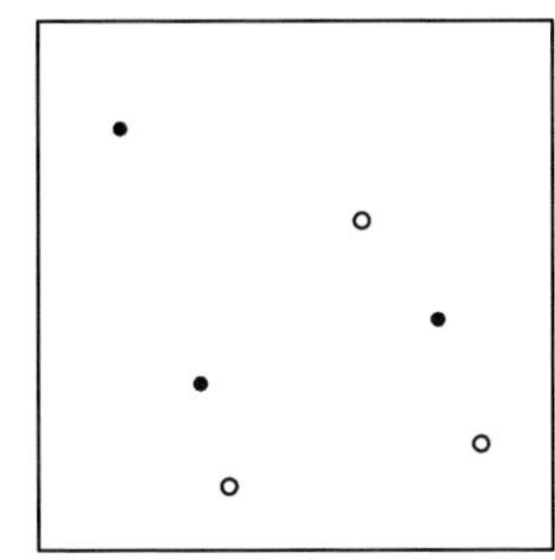
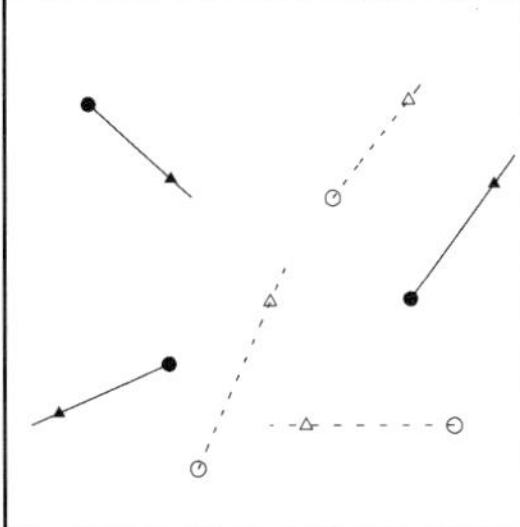
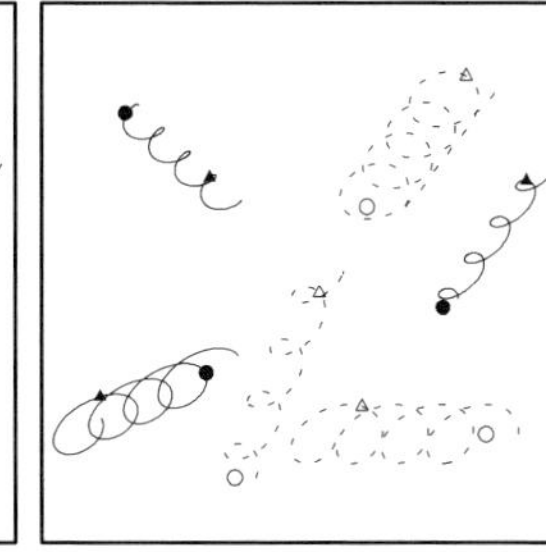

Figure 1.3 Principles of the astrometric measurements. Filled circles and solid lines show three objects from one field of view (about 1° in size), and open circles and dashed lines show three objects from a distinct sky region superimposed by virtue of the large basic angle. Left: object positions at one reference epoch. Middle: their space motions over about four years, with arbitrary proper motion vectors and scale factors; triangles show their positions at a fixed epoch near the end of the interval. Right: the total positional changes including the additional apparent motions due to annual parallax, the four 'loops' corresponding to four Earth orbits around the Sun. Again, parallaxes are of arbitrary amplitude. The parallax-induced motions are in phase for all stars in the same region of sky, so that relative measurements within one field can only provide relative parallaxes. Although the relative separations between the stars change continuously over the measurement period, they are described by just five parameters per star.

a large angle of order $\pi/2$ radians provides good interconnectivity and hence rigidity of the reference system between different parts of the sky. Yet angles near to 90°, or 60°, or indeed $(m/n)2\pi$ radians, where m and n are (small) integers, lead to weakened rigidity due to the smaller number of independent links between different parts of the sky. A basic angle of 58° was adopted, and in practice calibrated over intervals of a few hours with an accuracy of a few tens of microarcseconds (ESA, 1997, Volume 3, Figures 10.2 and 10.8).

In the second step, including a space motion of each star in the plane of the sky, orthogonal to the line-of-sight, the star's linear velocity is transformed (through its unknown distance) into an angular motion in the two angular coordinates – the star's 'proper motion' (Figure 1.3, middle). Only two additional unknowns are thereby introduced for each star, leading to a total of four unknowns for each star, compared with the 100 or so independent observations over three years. So, again, the proper motion is well determined, as are its standard errors.

In the third step, the motion of the Earth (or more strictly the orbiting satellite) around the Sun provides the periodic shift in measurement baseline necessary for the star's distance to be determined as a consequence of the apparent shift in the star's position as a function of the Earth's location (Figure 1.3, right). This apparent angular shift, the trigonometric parallax, provides a direct measurement of the star's distance through knowledge of the Earth's orbital geometry. Its inclusion adds just one unknown to the parameters describing the star's instantaneous position as a function of time. The total of five unknown parameters per star (two position components, two proper motion components, and the parallax) are the basic astrometric results provided by the measurement programme. The associated (5) standard errors and (10) covariances were determined and published, along with goodness-of-fit parameters.

Whilst of enormous astronomical importance, double and multiple stars provided considerable complications to the observations (due to the finite size and profile of the detector's sensitive field of view) and to the data analysis. Details are covered in Chapter 2. Yet the principles of analysis are easily incorporated into the qualitative framework described. If a binary star has a long orbital period such that nonlinear motions of the photocentre were insignificant over the short (three-year) measurement duration, the binary nature of the star would pass unrecognised by Hipparcos, but could show as a Hipparcos proper motion discrepant compared to those established from long temporal baseline proper motion programmes on ground. Higher-order photocentric motions could be represented by a seven-parameter, or even nine-parameter model fit (compared to the standard five-parameter model), and typically such models could be enhanced in complexity until suitable fits were obtained. A complete orbit, requiring seven elements, was determined for 45 systems. Orbital periods close to one year can become degenerate with the parallax, resulting in unreliable solutions for both. Triple or higher-order systems provided further challenges to the data processing. No valid astrometric solution could be obtained for 263 entries, of which 218 were flagged as suspected double; they are interesting cases for further ground-based and long-term astrometric monitoring.

Classical astrometry concerns only motions in the plane of the sky and ignores the star's radial velocity, its space motion along the line-of-sight. Whilst critical for an understanding of stellar kinematics, and hence population dynamics, its effect is generally imperceptible 'astrometrically', and therefore it is generally ignored in large-scale astrometric surveys. In practice, it can be measured as a Doppler shift of the spectral lines. Strictly, the radial velocity does enter a rigorous astrometric formulation (see Sections 1.4.5 and 1.13.2). Specifically, a space velocity along the line-of-sight means that the transformation from tangential linear velocity to (angular) proper motion is a function of time. The resulting effect of 'secular' or 'perspective' acceleration is the interpretation of a transverse acceleration actually arising from a purely linear space velocity with a significant radial component, with the positional effect proportional to $\pi \times \mu \times V_R$. At the accuracy levels of Hipparcos it is of (marginal) importance only for the nearest stars with the largest radial velocities and proper motions, but was accounted for in the 21 cases for which the accumulated positional effect over two years exceeds 0.1 mas (ESA, 1997, Volume 1, Section 1.2.8).

A few further complications should be mentioned. A detailed optical calibration model was included to map the transformation from sky to instrumental coordinates. Its adequacy could be verified by the detailed measurement residuals. The Earth's orbit, and the satellite's orbit with respect to the Earth, were essential for describing the location of the observer at each epoch of observation, and were supplied by an appropriate Earth ephemeris combined with accurate satellite ranging. Corrections due to special relativity (stellar aberration) made use of the corresponding satellite velocity. Modifications due to General Relativistic light bending were significant (4 mas at 90° to the ecliptic) and corrected for deterministically assuming $\gamma = 1$ in the PPN formalism. Residuals were examined to establish limits on any deviations from this General Relativistic value, as described further in Section 1.15.

1.3 Hipparcos Input Catalogue

The satellite observations relied on a pre-defined list of target stars. Stars were observed as the satellite rotated, by a sensitive region of the image dissector tube detector. This pre-defined star list formed the Hipparcos Input Catalogue: each star in the final Hipparcos Catalogue was contained in the Input Catalogue (Turon *et al.*, 1995). The Input Catalogue was compiled by the INCA Consortium (led by Catherine Turon) over the period 1982–89, finalised pre-launch, and published both in printed form (Turon *et al.*, 1992) and digitally (Turon *et al.*, 1996). Although fully superseded by the satellite results, it nevertheless includes supplemental information on multiple system components as well as compilations of radial velocities and spectral types which, not observed by the satellite, were not included in the published Hipparcos Catalogue. Its construction and properties are described in detail in Perryman *et al.* (1989, Volume II).

Figure 1.4 The third Scientific Proposal Selection Committee meeting in Paris, 6–7 April 1987. From left to right, where parentheses indicate meeting participants who were not committee members: (Jean Delhaye), Roland Wielen, Christian de Vegt, (Walter Fricke), Wilhelm Gliese, (Jean Kovalevsky). Members not in the picture: Adriaan Blaauw (chair), Jean Dommanget, Margarita Hack, Ed van den Heuvel, Carlos Jaschek, James Lequeux, Per Olof Lindblad, André Maeder, Poul Erik Nissen, Bernard Pagel, Alvio Renzini, Patrick Wayman.

Constraints on total observing time, and on the uniformity of stars across the celestial sphere for satellite operations and data analysis, led to an Input Catalogue of some 118 000 stars. It merged two components: first, a 'survey' of around 58 000 objects as complete as possible to the following limiting magnitudes

$$V \le 7.9 + 1.1 \sin|b| \quad \text{spectral type} \le \text{G5} \qquad (1.1a)$$

$$V \le 7.3 + 1.1 \sin|b| \quad \text{spectral type} > \text{G5} \qquad (1.1b)$$

where b is the Galactic latitude. If no spectral type was available, the break was taken at $B - V = 0.8$ mag. Stars constituting this survey are flagged in the Hipparcos Catalogue itself (Field H68).

The second component comprised additional stars selected according to their scientific interest, with none fainter than about $V = 13$ mag. These were selected from around 200 scientific proposals submitted on the basis of an Invitation for Proposals issued by ESA in 1982, and prioritised by the Scientific Proposal Selection Committee in consultation with the Input Catalogue Consortium (Figure 1.4). Again, this selection had to balance *a priori* scientific interest, and the observing programme's limiting magnitude, total observing time, and sky uniformity constraints. The relevant scientific proposals are listed in Perryman *et al.* (1989, Volume II, Appendix C) and in Turon *et al.* (1992, Volume 1). The

Table 1.1 Principal observational characteristics of the Hipparcos and Tycho Catalogues. ICRS is the International Celestial Reference System.

Property	Value	
Common:		
Measurement period	1989.8–1993.2	
Catalogue epoch	J1991.25	
Reference system	ICRS	
coincidence with ICRS (3 axes)	± 0.6	mas
deviation from inertial (3 axes)	± 0.25	mas yr^{-1}
Hipparcos Catalogue:		
Number of entries	118 218	
with associated astrometry	117 955	
with associated photometry	118 204	
Mean sky density	~ 3	deg^{-2}
Limiting magnitude	$V \sim 12.4$	mag
Completeness	$V = 7.3-9.0$	mag
Tycho Catalogue:		
Number of entries	1 058 332	
based on Tycho data	1 052 031	
with only Hipparcos data	6301	
Mean sky density	~ 25	deg^{-2}
Limiting magnitude	$V \sim 11.5$	mag
Completeness to 90 per cent	$V \sim 10.5$	mag
Completeness to 99.9 per cent	$V \sim 10.0$	mag
Tycho 2 Catalogue:		
Number of entries	2 539 913	
Mean sky density:		
$b = 0°$	~ 150	deg^{-2}
$b = \pm 30°$	~ 50	deg^{-2}
$b = \pm 90°$	~ 25	deg^{-2}
Completeness to 90 per cent	$V \sim 11.5$	mag
Completeness to 99 per cent	$V \sim 11.0$	mag

indication of which catalogue stars are associated with any particular proposal was maintained in the Consortium's database, but is not publicly available and is unlikely to be preserved indefinitely.

1.4 Hipparcos Catalogue and Annexes

The data reductions were carried out in parallel by two data analysis consortia, known by their acronyms FAST (led by Jean Kovalevsky) and NDAC (led initially by Erik Høg and later by Lennart Lindegren). Early descriptions of the data reduction processes were given by Kovalevsky *et al.* (1995), with details of the properties of the preliminary catalogue in comparison with ground-based stellar positions and proper motions given by Lindegren *et al.* (1995). The analyses, proceeding from nearly 1000 Gbit of satellite data, incorporated a comprehensive system of cross-checking and validation of the entire data reduction and catalogue construction process. Final results of the two independent analyses were rigorously combined into the single final Hipparcos Catalogue and associated annexes. The main features of the Hipparcos Catalogue were presented at its publication by Perryman *et al.* (1997), and the main characteristics are given in Table 1.1.

Details of the data, their reductions, the merging of the two independent astrometric solutions, and the properties of the final catalogue, are given in the published Hipparcos and Tycho Catalogues (ESA, 1997, Volume 2–4). A summary of the data fields of the Hipparcos Catalogue is given in Table 1.2.

The following basic definitions and concepts relevant to discussions of accuracy and errors were adopted throughout the project: (a) standard error: defined as the standard deviation of an estimator, providing an estimation of the random part of the total estimation error involved in estimating a population parameter from a sample (*ISO Standards Handbook 3: Statistical Methods*); (b) accuracy: defined as the uncertainty of a measured quantity, including accidental and systematic errors, often used synonymously with 'external standard error'; (c) precision: defined as the uncertainty of a measured quantity due to accidental errors, often used synonymously with 'internal (or formal) standard error' as derived, e.g. from a least-squares solution.

The notation milliarcsec, or mas, is used to denote 10^{-3} arcsec, and the notation microarcsec, or μas, is used to denote 10^{-6} arcsec.

1.4.1 Hipparcos astrometry

The standard astrometric model adopted for single stars assumes uniform rectilinear space motion relative to the Solar System barycentre. Individual positions obtained over the three-year observation period were then combined through the five-parameter model fit. The stellar motion is then described by the following five astrometric parameters (the third component of the space velocity, the radial velocity, being undetermined from the Hipparcos observations): the barycentric coordinate direction at some reference epoch, T_0, (α, δ); the rate of change of the barycentric coordinate direction expressed as proper motion components $\mu_{\alpha*} = \mu_\alpha \cos\delta$ and μ_δ, in angular measure per unit time, expressed in milliarcsec per Julian year; and the annual parallax, π, from which the coordinate distance is $(\sin\pi)^{-1}$ AU or, with sufficient approximation, π^{-1} pc if π is expressed in arcsec. These five astrometric parameters (see Table 1.4 below) are given for almost all stars in the catalogue. The $\cos\delta$ factor, signified by the asterisk in $\mu_{\alpha*}$, relates the rate of change of position in right ascension to great-circle measure. The notation was introduced in

Table 1.2 Data fields of the Hipparcos Catalogue, taken from ESA (1997). † indicates data may have been derived from satellite and/or ground-based data. Fields H71–77 appear only in the machine-readable version.

Field	Description
H1	Hipparcos Catalogue (HIP) identifier ('*' = entry is of sequence in right ascension)
H2	Proximity flag indicating objects within 10 arcsec: H = HIP, T = TYC
H3–4	Positional identifier: truncated right ascension (h m s) and declination ($\pm^\circ\ '\ ''$); epoch J1991.25, ICRS
H5†	V (Johnson) magnitude
H6	Coarse variability flag: variable or possibly variable in Hp at (mag): 1: < 0.06; 2: 0.06–0.6; 3: > 0.6
H7	Source of V magnitude in Field H5: G = ground-based, H = HIP, T = TYC, ⊔ = not available
H8–9	Right ascension, α, and declination, δ (degrees); epoch J1991.25, ICRS
H10	Reference flag for the astrometric parameters (Fields H3–4 and H8–30) of double and multiple systems: A, B, ... = specified component; * = photocentre; + = solution with respect to centre of mass
H11	Trigonometric parallax, π (mas)
H12–13	Proper motion in right ascension, $\mu_{\alpha*}$, and declination, μ_δ (mas yr^{-1}); epoch J1991.25, ICRS
H14–15	Standard error of position in right ascension, $\sigma_{\alpha*}$, and declination, σ_δ (mas); epoch J1991.25
H16	Standard error of the trigonometric parallax, σ_π (mas)
H17–18	Standard error of proper motion in right ascension, $\sigma_{\mu_{\alpha*}}$, and declination, σ_{μ_δ} (mas yr^{-1})
H19–28	Correlation coefficients: $\rho^{\delta}_{\alpha*}$, $\rho^{\pi}_{\alpha*}$, ρ^{π}_{δ}, $\rho^{\mu_{\alpha*}}_{\alpha*}$, $\rho^{\mu_{\alpha*}}_{\delta}$, $\rho^{\mu_{\alpha*}}_{\pi}$, $\rho^{\mu_\delta}_{\alpha*}$, $\rho^{\mu_\delta}_{\delta}$, $\rho^{\mu_\delta}_{\pi}$, $\rho^{\mu_\delta}_{\mu_{\alpha*}}$
H29	Percentage of rejected data, F1
H30	Goodness-of-fit statistic, F2
H31	Hipparcos Catalogue (HIP) identifier (as Field H1)
H32–33	Mean magnitude in the Tycho (star mapper) photometric system, B_T (mag), and standard error, σ_{B_T}
H34–35	Mean magnitude in the Tycho (star mapper) photometric system, V_T (mag), and standard error, σ_{V_T}
H36	Reference flag for B_T and V_T (Fields H32–35): A, B, ... = component, * = combined, – = multiple
H37–38†	Colour index in the Johnson photometric system, $B-V$ (mag), and standard error, σ_{B-V}
H39	Source of $B-V$: G = ground-based, T = Tycho, ⊔ = not available
H40–41†	Colour index in the Cousins' photometric system, $V-I$ (mag), and standard error, σ_{V-I}
H42	Source of $V-I$: A, B, C, ..., T
H43	Reference flag for colour indices (Fields H37–42) and Field H5: * = combined
H44–45	Median magnitude in the Hipparcos photometric system, Hp (mag), and standard error σ_{Hp}
H46	Scatter of the Hp observations, s (mag)
H47	Number of photometric observations for the Hp photometry, N
H48	Reference flag for Fields H44–54: A, B, ... = component, * = combined, – = multiple
H49–50	Observed magnitude at maximum/minimum luminosity (from 5th/95th distribution percentiles)
H51	Variability period from Hipparcos, P (days)
H52	Variability type: C = 'constant'; D = duplicity possibly causing variability; M = micro-variable; P = periodic; R = revised photometry; U = unsolved; ⊔ = constant or variable
H53	Variability annex flag: further data in Volume 11: 1 = periodic, 2 = 'unsolved'
H54	Variability annex flag: light-curves in Volume 12: A = folded; B = AAVSO; C = unfolded
H55	CCDM identifier
H56	Historical status of the CCDM identifier: H = Hipparcos; I = Input Catalogue; M = miscellaneous
H57	Number of separate Hipparcos Catalogue entries with the same CCDM identifier
H58	Number of components into which this entry is resolved: 1, 2, 3, 4
H59	Flag indicating details provided in the Double and Multiple Systems Annex: C, G, O, V, X
H60	Flag indicating source of absolute astrometry: F, I, L, P, S
H61	Flag indicating solution: A = good, B = fair, C = poor, D = uncertain, S = suspected non-single
H62	Component designation, brighter/fainter (in Hp), for parameters in Field H63–67
H63	Position angle of the fainter component (at epoch J1991.25), θ (degrees, rounded value)
H64	Angular separation (at epoch J1991.25), ρ (arcsec, rounded value)
H65	Standard error of the angular separation, σ_ρ (arcsec)
H66–67	Magnitude difference of components, ΔHp (mag), and standard error $\sigma_{\Delta Hp}$
H68	Flag: S = 'survey' star (i.e. complete to the magnitude limits defined in Section 1.3)
H69	Flag indicating identification chart in Volume 13: D = DSS; G = GSC
H70	Flag indicating note at end of relevant volume(s): D, G, P, W = D + P, X = D + G, Y = G + P, Z = D + G + P
H71–74	Cross-identifiers to HD/HDE/HDEC; DM: BD, CoD, CPD (see also Table 1.12)
H75	$V-I$ used for the photometric processing (see Section 4.10)
H76–77	Spectral type and source of spectral type (compilation, primarily from Hipparcos Input Catalogue)

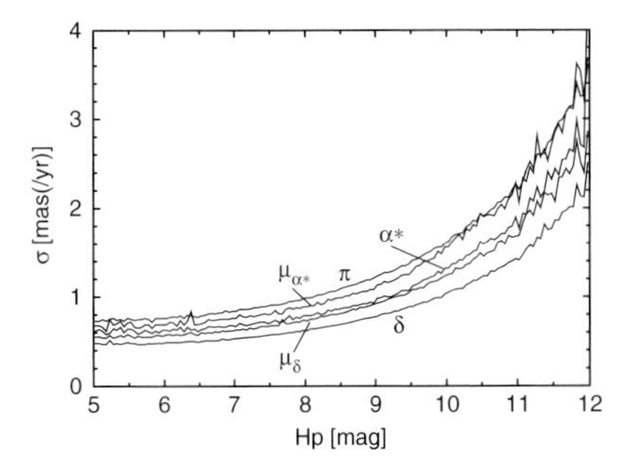

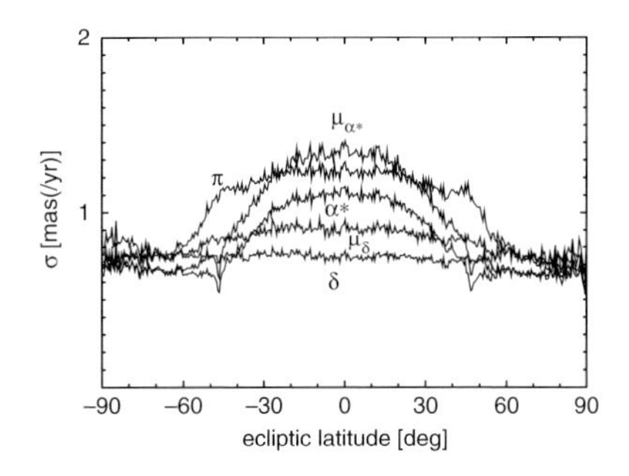

Figure 1.5 Left: median standard errors of the five astrometric parameters as a function of Hp magnitude. The unit of the standard error is milliarcsec (mas) for the positional components (α, δ) and parallax (π), and mas yr^{-1} for the proper motion components ($\mu_{\alpha*} = \mu_\alpha \cos\delta$, μ_δ). Right: median standard errors of the astrometric parameters as a function of ecliptic latitude. The dependency on ecliptic latitude is a consequence of the ecliptic-based scanning law. The errors are given at the catalogue epoch, J1991.25. From Perryman et al. (1997, Figures. 1 and 2).

the Hipparcos Catalogue publication to avoid possible confusion in the meaning of μ_α.

Median astrometric standard errors (in position, parallax, and annual proper motion) are in the range 0.7–0.9 mas for stars brighter than 9 mag at the catalogue epoch (J1991.25). The catalogue is a materialisation of the ICRS reference system, coinciding with its principal axes at the level of ± 0.6 mas, and with proper motions consistent with an inertial system at the level of ± 0.25 mas yr^{-1} (see Section 1.4.1). The 118 218 constituent stars provide a mean sky density of $\sim$ 3 stars deg^{-2}.

The 'catalogue epoch', J1991.25, corresponds to a moment in time close to the central epoch of the satellite observations. Catalogue positions are given, within the adopted reference system ICRS, at this epoch. This choice minimises the standard error of positions, but is otherwise essentially arbitrary: the provision of the correlation coefficients for each astrometric solution allows the standard errors of transformed quantities to be determined at an arbitrary epoch; including, for example, the epoch at which the standard error is minimised for each individual star, or the standard epoch J2000.0 (ESA, 1997, Volume 1, Section 1.2.7).

Figure 1.5 illustrates the median precision of each of the astrometric parameters as a function of *Hp* magnitude and ecliptic latitude, at the catalogue epoch. Detailed sky charts and histograms giving the astrometric and photometric accuracies as a function of position and magnitude are included in the Hipparcos Catalogue (ESA, 1997, Volume 1, Section 3).

The parallax determinations are trigonometric, absolute (in the sense that the parallax determination of a given star is not dependent upon either the parallaxes, or assumptions concerning the parallaxes, of other stars – including stars close by on the sky), and independent of any previous distance determinations. Analyses place a limit on the global parallax zero-point offset of less than 0.1 mas, and give confidence that the published standard errors are a reliable indication of their true external errors.

While there seems no evidence for a global zero-point shift of the Hipparcos parallaxes above an estimated 0.1 mas (ESA, 1997, Volume 3, p323), the fact that the parallax errors will likely be correlated over angular scales of 2–3° was anticipated from the measurement principle. For any astrometric parameter a, the correlation function is defined in terms of the normalised differences between the NDAC and FAST Consortium values, $\overline{\Delta a}$, as

$$R(\theta) = \frac{\langle\, \overline{\Delta a_i}\, \overline{\Delta a_j}\,\rangle}{\sqrt{\langle\, \overline{\Delta a_i^2}\,\rangle\, \langle\, \overline{\Delta a_j^2}\,\rangle}} \tag{1.2}$$

where averages are calculated over all star pairs (i, j) whose separations are in the range $\theta \pm \Delta\theta/2$. Figure 1.6, from ESA (1997, Volume 3, Figure 16.37) shows the sample correlation function for the parallax differences at a resolution of 0.°1, calculated from all $\sim 5 \times 10^9$ pairs of 100 890 stars from the Hipparcos reductions. However, the sparse mean catalogue density of about three stars deg^{-2} means that open clusters with a large local stellar concentration are probably the only regions where these small-scale systematics can be estimated, or are relevant. Van Leeuwen (2002) argued that there is no evidence for systematic or correlated errors beyond a correlation level of 0.12 and an angular scale of 1.°2.

For both the Hipparcos observations and the data reductions, an enormous simplicity arose from the absence of the atmosphere, and the decoupling of the observations from the rotation of the Earth. The

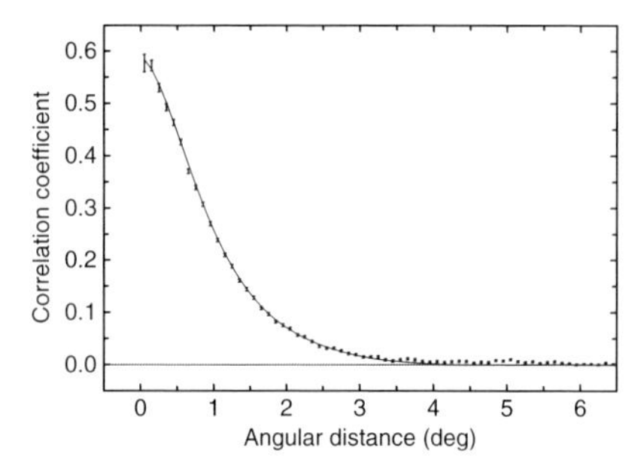

Figure 1.6 Mean sample correlation of normalised parallax difference, $R(\theta)$, as defined by Equation 1.2, out to $\theta = 6^\circ$. At angular separations less than a few degrees the correlation is strongly positive, but decreases to almost negligible values beyond about 4°. It is likely that the actual parallax errors in the published Hipparcos Catalogue exhibit a similar spatial correlation, but with some uncertain normalisation. From ESA (1997, Volume 3, Figure 16.37).

relationship between the resulting Hipparcos reference frame and the dynamical reference frame materialised by observations of Solar System objects (made from the ground and space) is considered further in Section 1.10.2 and Chapter 10.

Object directions The 'direction' to an object at each epoch of observation, as measured by a moving observer, is obtained by three successive transformations (see Murray 1983, which remains an authoritative treatment of vectorial astrometry, and ESA 1997, Volume 3, Section 12.3 for details applicable to Hipparcos): derivation of the 'coordinate direction' to the object by a translation of space-time coordinates from the adopted reference point (at the Solar System barycentre) and epoch to the observer at the time of observation; determination of the 'natural direction' to the object as measured by a hypothetical stationary observer, in a locally flat coordinate system at rest with respect to the barycentre, corresponding to the application of gravitational light deflection; and determination of the 'proper direction' taking account of the observer's motion, corresponding to the application of stellar aberration. In practice, transformations from coordinate to natural directions were carried out in space-time coordinates characterised by a spherically symmetric, heliocentric General Relativistic metric in which light-bending by the Sun (and, in NDAC only, the Earth) was taken into account. Deviations from this assumed metric were parameterised in a PPN-type formalism, from which values of $\gamma = 1.000 \pm 0.004$ (in FAST) and $\gamma = 0.992 \pm 0.005$ (in NDAC) gave further confidence in the metric formalism and global reductions (see also Section 1.15). Proper directions were computed using models of the Earth's motion with respect to the Solar System barycentre, and of the satellite motion with respect to the Earth's barycentre. The former was taken from the planetary (VSOP 82 Bretagnon, 1982) and lunar (ELP 2000 Chapront & Francou, 2003) ephemerides constructed by the Bureau des Longitudes (now the Institut de Mécanique Céleste, IMC); their use was equivalent, to within 0.01 mas, with the use of the Jet Propulsion Laboratory DE200 ephemeris. These ephemerides have since been superseded, most recently by the JPL development ephemeris solution DE414 (Konopliv *et al.*, 2006), and the numerical planetary ephemeris developed at the IMCCE–Observatoire de Paris, INPOP06 (Fienga *et al.*, 2008). The satellite position was provided by the operations centre (ESOC) with an accuracy of some 1.5 km in position and $0.2\,\mathrm{m\,s^{-1}}$ in velocity. Otherwise, the former IAU (1976) system of constants was used for the reductions (Table 1.3).

Reference system and reference frame The terminology used in the 1991 IAU resolution on reference frames and reference systems distinguishes between the use of the term 'reference system' and 'reference frame' (Wilkins, 1990). A reference system is the complete specification of how a celestial coordinate system is formed. It defines the origin and fundamental planes (or axes) of the coordinate system. It also specifies all of the constants, models, and algorithms used to transform between observable quantities and reference data that conform to the system. A reference frame consists of a set of identifiable fiducial points on the sky along with their coordinates, which serves as the practical realisation of a reference system.

While a frame represented by the directions to distant extragalactic objects intuitively complies with the requirements of an inertial frame, there are some practical considerations. The fact that such a frame is defined kinematically, through the absence of transverse motions, rather than dynamically, is why it is sometimes referred to as a 'quasi-inertial' frame. The General Relativistic definition of an inertial frame in terms of the form of the space-time metric would require practical confirmation, for example via accurate gyroscopes, which is beyond current measurement accuracies. An inertial frame can also be defined as non-rotating with respect to the Universe, related to Mach's principle in which a body's inertia is caused by the distribution of mass throughout the Universe. Finally, the Solar System may be regarded as representing a local inertial reference frame of dynamical definition, i.e. related to Newton's laws of motion. A pragmatic definition was given by Clemence (1966): 'an inertial frame of reference is defined as a frame that is free from linear and rotational accelerations'. The consistency between the adopted extragalactic reference

Table 1.3 Physical and astronomical constants used for the Hipparcos data reductions, from ESA (1997, Table 1.2.2). The unit of time is the SI second, or the Julian Year when more appropriate. The speed of light enters mainly in the computation of stellar aberration. The astronomical unit A appears in all formulae relating linear measures to the parallax; depending on the context and the units used, it is represented by a variety of numerical values, as indicated (some older texts give slightly different numerical values for A_v, usually because they assume the now-obsolete tropical year as the time unit for the proper motions); for considerations in defining A within a General Relativistic framework, see Huang et al. (1995). The heliocentric and geocentric gravitational constants are used to compute the gravitational light bending by the Sun and the Earth. The obliquity of the ecliptic has no direct significance for the data reductions, but is used as a conventional value to transform between the equatorial and ecliptic systems. The Earth ephemeris is relevant for the calculation of aberration and parallax. Note that some reference quantities have been (slightly) revised through subsequent work (see Appendix A).

Symbol	Meaning/Application	Value
	Unit of time	SI second as realised on the geoid
	Time scale	Terrestrial Time (TT)
Julian Year	Proper motion unit (mas yr^{-1})	$365.25 \times 86\,400$ s (exactly)
c	Speed of light	$299\,792\,458$ m s^{-1} (exactly)
A	Astronomical unit	$(499.004\,782$ s$) \times c$ (exactly)
		$= 1.495\,978\,701 \times 10^{11}$ m (A_m)
		$= 1000$ mas pc (A_p)
		$= 4.740\,470\,446$ km yr s^{-1} (A_v)
		$= 9.777\,922\,181 \times 10^8$ mas km yr s^{-1} (A_z)
GS	Heliocentric gravitational constant	$1.327\,124\,38 \times 10^{20}$ m^3 s^{-2}
GE	Geocentric gravitational constant	$3.986\,005 \times 10^{14}$ m^3 s^{-2}
ϵ	Obliquity of ecliptic (J2000.0)	$23^\circ\,26'\,21.448''$ (exactly)
		$= 23.^\circ439\,291\,111\,1\ldots$
	Planetary/lunar ephemeris	VSOP 82/ELP 2000

The origin of right ascension and declination: The fundamental plane of astronomical reference systems has conventionally been the extension of the Earth's equatorial plane, at some date, to infinity. The declination of a celestial object is its angular distance north or south of this plane. The right ascension is its angular distance measured eastward along the equator from some agreed reference point, traditionally the equinox, i.e. the point at which the Sun crosses the equatorial plane moving from south to north. The Sun's apparent yearly motion lies in the ecliptic, the plane of the Earth's orbit. The equinox, therefore, is a direction in space along the nodal line defined by the intersection of the ecliptic and equatorial planes. Because both of these planes are moving, the coordinate systems that they define must have a date associated with them; such a reference system must therefore be specified as the equator and equinox of some specific date, previously B1950 and more recently J2000. The previous (pre-Hipparcos) astronomical reference system was based on the equator and equinox of J2000.0 determined from observations of planetary motions, together with the IAU (1976) System of Astronomical Constants and related algorithms. The reference frame that embodied this system for practical purposes was the Fifth Fundamental Catalogue (FK5).

frame and the dynamical reference frame defined by the motion of Solar System objects is considered further in Chapter 10.

In 1995, the IAU Working Group on Reference Frames identified the International Celestial Reference System (ICRS) as the official IAU reference system, replacing FK5 (mean equator and equinox J2000) as the practical definition of celestial coordinates in the optical. The ICRS is a set of specifications defining a high-precision coordinate system with its origin at the Solar System barycentre and 'space fixed' (kinematically non-rotating) axes. The specifications include a metric tensor, a prescription for establishing and maintaining the axis directions, a list of benchmark objects with precise coordinates for each one, and standard models and algorithms that allow these coordinates to be transformed into observable quantities for any location and time. A review is given by Johnston & de Vegt (1999). The ICRS is itself materialised, with an accuracy of $\pm 30\,\mu$as, by the International Celestial Reference Frame (ICRF), a catalogue of adopted positions of 608 extragalactic radio sources observed with VLBI, all strong (> 0.1 Jy) at S and X bands, i.e. at wavelengths 13 and 3.6 cm (Ma *et al.*, 1998). Most have faint optical counterparts (typically $V > 18$) and the majority are quasars. Of these objects, 212 are defining sources that establish the orientation of the ICRS axes, with origin at the Solar System barycentre. Its construction ensured that no discontinuity larger than the uncertainty of the FK5 system occurred in the transition from FK5 to ICRS. Specifically, the ICRS axes are consistent with those of the FK5 system (mean equator and equinox J2000) within the uncertainty of the latter, $\pm$50–80 mas (Arias *et al.*, 1995), and their tie to the

Table 1.4 Astrometric characteristics of the Hipparcos and Tycho Catalogues.

Property	Value	
Hipparcos Catalogue:		
For $Hp \leq 9$ mag:		
median σ_α at J1991.25	0.77	mas
median σ_δ at J1991.25	0.64	mas
median σ_π	0.97	mas
median $\sigma_{\mu_\alpha \cos\delta}$	0.88	mas yr^{-1}
median σ_{μ_δ}	0.74	mas yr^{-1}
For each astrometric parameter:		
10 per cent better than	0.47–0.66	mas (yr^{-1})
smallest errors	0.27–0.38	mas (yr^{-1})
Fractional distance errors:		
distance < 10% ($\sigma_\pi/\pi < 0.1$)	20 853	
distance < 20% ($\sigma_\pi/\pi < 0.2$)	49 399	
Estimates of systematic errors:		
external errors/standard errors	~ 1.0–1.2	
systematic errors in astrometry	< 0.1	mas (yr^{-1})
Tycho Catalogue:		
For each astrometric parameter:		
standard errors for $V_T < 9$	7	mas (yr^{-1})
standard errors for $V_T \sim 10.5$	25	mas (yr^{-1})
smallest errors	3	mas (yr^{-1})
Estimates of systematic errors:		
external errors/standard errors	~ 1.0–1.5	
systematic errors in astrometry	~ 1	mas (yr^{-1})
Tycho 2 Catalogue (all stars):		
standard errors of positions	60	mas
standard errors of proper motions	2.5	mas yr^{-1}

best realisation of the FK5 dynamical reference system was within ±3 mas (Folkner *et al.*, 1994).

The Hipparcos reference frame The satellite observations essentially yielded highly accurate relative positions of stars with respect to each other, throughout the measurement period (1989–93). In the absence of direct observations of extragalactic sources (apart from marginal observations of 3C 273) the resulting rigid reference frame was transformed to an inertial system linked to extragalactic sources. This allows surveys at different wavelengths to be directly correlated with the Hipparcos stars, and ensures that the catalogue proper motions are, as far as possible, kinematically non-rotating. The determination of the relevant three solid-body rotation angles, and the three time-dependent rotation rates, was conducted and completed in advance of the catalogue publication by a specific working group (led by Jean Kovalevsky and Lennart Lindegren). This resulted in an accurate but indirect link to an inertial, extragalactic, reference frame.

A variety of methods to establish this reference frame link before catalogue publication were included and appropriately weighted (Lindegren & Kovalevsky, 1995; Kovalevsky *et al.*, 1997): interferometric observations of radio stars by VLBI networks, MERLIN and VLA; observations of quasars relative to Hipparcos stars using CCDs, photographic plates, and the Hubble Space Telescope; photographic programmes to determine stellar proper motions with respect to extragalactic objects (Bonn, Kiev, Lick, Potsdam, Yale/San Juan); and comparison of Earth orientation parameters obtained by VLBI and by ground-based optical observations of Hipparcos stars. Although very different in terms of instruments, observational methods and objects involved, the various techniques generally agree to within 10 mas in the orientation and 1 mas yr^{-1} in the rotation of the system. From appropriate weighting, the coordinate axes defined by the published catalogue are believed to be aligned with the extragalactic radio frame to within ± 0.6 mas at the epoch J1991.25, and non-rotating with respect to distant extragalactic objects to within ± 0.25 mas yr^{-1}. Studies by some of the groups contributing to this overall work were reported separately: the contributions by the HST Fine Guidance Sensor by Hemenway *et al.* (1997); the Bonn photographic plate observations by Geffert *et al.* (1997) and Odenkirchen *et al.* (1997); the Potsdam photographic plate observations by Hirte *et al.* (1997); and the NPM and SPM contributions by Platais *et al.* (1998).

The Hipparcos and Tycho Catalogues were constructed such that the Hipparcos Reference Frame coincides, to within observational uncertainties, with the ICRS, and representing the best estimates at the time of the catalogue completion (in 1996). The resulting Hipparcos Reference Frame is thus the materialisation of the ICRS in the optical, extending and improving the J2000(FK5) system, retaining approximately the global orientation of that system but without its regional errors.

The construction and implementation of the ICRS is supported by the International Astronomical Union (IAU). Resolution B2, passed by the 23rd General Assembly of the IAU in August 1997, states that: (a) from 1 January 1998, the IAU celestial reference system shall be the International Celestial Reference System (ICRS) as specified in the 1991 IAU Resolution on reference frames and as defined by the International Earth Rotation Service (IERS); (b) the corresponding fundamental reference frame shall be the International Celestial Reference Frame (ICRF) constructed by the IAU Working Group on Reference Frames; (c) the Hipparcos Catalogue shall be the primary realisation of the ICRS at optical wavelengths; (d) the IERS should take appropriate measures, in conjunction with the IAU Working Group on Reference Frames, to maintain the ICRF and its ties to the reference frames at other wavelengths.

At the IAU General Assembly in 2000, Resolution B1.2 restricted the number of Hipparcos stars to be considered part of the optical realisation of the ICRS.

Time scales: The rate of an atomic clock depends on the gravitational potential and its motion with respect to other clocks; thus the time scale entering the equations of motion (and its relationship with TAI) depends on the coordinate system to which the equations refer. Since 1984, *The Astronomical Almanac* referred to two such time scales: Terrestrial Dynamical Time (TDT) used for geocentric ephemerides, and Barycentric Dynamical Time (TDB) used for ephemerides referred to the Solar System barycentre. TDT differs from TAI by a constant offset, which was chosen to give continuity with ephemeris time. TDB and TDT differ by small periodic terms (arising from the transverse Doppler effect and gravitational redshift experienced by the observer) that depend on the form of the relativistic theory being used: the difference includes an annual sinusoidal term of approximately 1.66 ms amplitude, planetary terms contributing up to about 20 μs, and lunar and diurnal terms contributing up to about 2 μs. In 1991 the IAU adopted resolutions introducing new time scales which all have units of measurement consistent with the unit of time, the SI second. Terrestrial Time (TT) is the time reference for apparent geocentric ephemerides, and can be considered as equivalent to TDT. Barycentric Coordinate Time (TCB) is the coordinate time for a coordinate system with origin at the Solar System barycentre. Because of relativistic transformations, TDB, and therefore TT, differ in rate from TCB by approximately 49 seconds per century.

For civil and legal purposes it is necessary to have a time scale which approximates the diurnal rotation of the Earth relative to the Sun. Historically this has been known as Universal Time, but because the Earth's angular spin rate is variable, the Universal Time scale is non-uniform with respect to TAI. The civil time scale, which has been available through broadcast time signals since 1972, is known as Coordinated Universal Time (UTC), and differs from TAI by an integer number of seconds – it is adjusted, when judged necessary by the International Earth Rotation Service (IERS), by adding a 'leap' second at midnight on December 31, or on June 30.

The relevant part of this resolution states that: (a) Resolution B2 of the 23rd IAU General Assembly (1997) be amended by excluding from the optical realisation of the ICRS all stars flagged C, G, O, V and X in the Hipparcos Catalogue; (b) this modified Hipparcos frame be labelled the Hipparcos Celestial Reference Frame (HCRF). Effectively, this change eliminated about 18 000 of the stars in the Hipparcos Catalogue, leaving those with well determined linear proper motions. The flags referred to are given in Hipparcos Field H59 (Mason *et al.*, 2000). See, also, Brumberg & Groten (2001).

Time scale The time scale used in the Hipparcos and Tycho Catalogues, for both the astrometric and the photometric data, is Terrestrial Time, TT (Seidelmann, 1992). The practical realisation of this scale is through International Atomic Time (TAI). The basic unit of TAI and TT is the SI second, and the offset between them is conventionally 32.184 s (with deviations, attributable to the physical defects of atomic time standards, probably between the limits ±10 μs), so that the realisation of TT in terms of TAI is taken to be TT(TAI) = TAI + 32.184 s.

The standard epoch of the fundamental coordinate system J2000 corresponds to JD 2 451 545.0 terrestrial time (TT), and to the calendar date 2000 January 1, $12^{\rm h}$ = 2000 January 1.5(TT). Epoch definitions are based on the Julian year of 365.25 days. Thus the Julian epoch Jyyyy.yy corresponds to

$$\mathrm{JD} = 2\,451\,545.0 + (\mathrm{yyyy.yy} - 2000.0) \times 365.25 \quad (1.3)$$

In particular, the adopted catalogue epoch for the Hipparcos and Tycho Catalogues is

$$T_0 = \mathrm{J1991.25(TT)} = \mathrm{JD}\,2\,448\,349.0625(\mathrm{TT}) \quad (1.4)$$

which is a good approximation to the mean central epoch of the observations.

1.4.2 Hipparcos photometry

The Hipparcos Catalogue includes a variety of accurate and homogeneous photometric information for each star, in particular: the (observed) broadband Hipparcos, or *Hp*, magnitudes, in an instrument specific passband, providing the most accurate multi-epoch photometric data most suitable for variability studies (the median photometric precision for $Hp < 9$ mag is approximately 0.0015 mag); the (observed) two-colour B_T and V_T magnitudes derived from the Tycho (star mapper) observations; the (derived) Johnson V magnitude, accurate to typically 0.01 mag, and derived from a combination of satellite and ground-based photometry; and (derived) Johnson $B - V$ and Cousins' $V - I$ colour-indices, again derived from a combination of satellite and ground-based measurements.

The mean number of *Hp* photometric observations per star over the three-year observational period is 110, providing data for detailed variability classification and characterisation. The principal photometric characteristics and variability statistics are given by van Leeuwen *et al.* (1997), and summarised in Chapter 4.

Photometric results were published as arrival times at the Solar System barycentre, and expressed in terms of Barycentric Julian Date, BJD(TT).

The distinction between photometric quantities strictly estimated from the satellite observations alone, and derived photometric quantities included in the catalogue which made use of additional information, must be stressed. The *Hp* magnitudes were derived homogeneously and solely from the satellite observations. They are the most precise and accurate satellite photometry, and should be preferred for variability or other related studies whenever possible. However, a single photometric band clearly provides no colour information, e.g. for construction of the HR diagram.

The Tycho B_T and V_T are also homogeneous, but are in a non-standard system, are of lower photometric accuracy, and do not exist for all Hipparcos Catalogue entries, because a subset of the Hipparcos stars are below the magnitude completeness limit of the Tycho observations.

The derivation and inclusion of Johnson B and V, and $B-V$ and $V-I$ colour indices, provide a service to the catalogue user, but with important limitations: they were derived in different ways according to the available satellite information, ground-based photometric data, and according to the star's published spectral type. Although providing the best colour information for the catalogue as a whole, their inhomogeneity and potential construction errors means that they should be used with due caution.

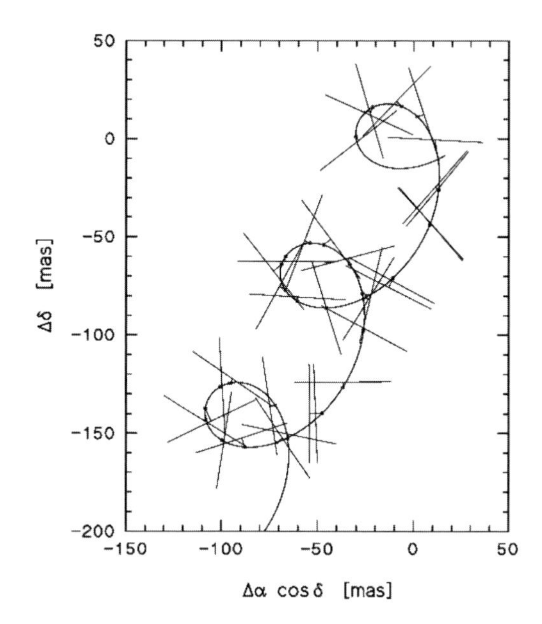

Figure 1.7 The path on the sky of one of the Hipparcos Catalogue objects, over a period of three years. Each straight line indicates the observed position of the star at a particular epoch: because the measurement is one-dimensional, the precise location along this position line is undetermined by the observation. The curve is the modelled stellar path fitted to all the measurements. The inferred position at each epoch is indicated by a dot, and the residual by a short line joining the dot to the corresponding position line. The amplitude of the oscillatory motion gives the star's parallax, with the linear component representing the star's proper motion. The intermediate astrometric data allow the quality of the model fitting to be assessed, and possibly refined. From ESA (1997, Volume 1, Figure 2.8.1).

1.4.3 Hipparcos double and multiple systems

Systems which could not be described by the standard five-parameter astrometric model were classified into five parts of the Double and Multiple Systems Annex (see Table 3.1): (C) component solutions, in which two or more components were resolved and their absolute and relative astrometry (and photometry) were reconstructed; (G) 'acceleration' solutions, in which the photocentric motion contains nonlinear time-dependent terms; (O) orbital systems, for which (partial) orbital solutions could be derived, possibly in combination with ground-based observations; (V) 'variability-induced mover' solutions, in which binarity was inferred from a time-dependent photocentric displacement; and (X) 'stochastic' solutions, in which multiplicity was evident yet uncharacterised.

Further details are given by Lindegren *et al.* (1997), and summarised in Chapter 3.

1.4.4 Intermediate astrometric and transit data

In view of the many possible alternative object models that might apply to the non-single stars, two major intermediate datasets have been included on the published catalogue CDs, and made available through the CDS: the 'intermediate astrometric data' for all Hipparcos entries, and the 'transit data' summarising the detector signals for almost a third of the entries, including all confirmed or suspected non-single stars.

The 'intermediate astrometric data' are the one-dimensional star coordinates on the individual measurement (reference) great-circles. The relevance of the data is illustrated in Figure 1.7. They allow more specialised users to retrieve and reprocess the 100 or so elementary observations made for each star during the measurement period, and to re-derive astrometric solutions based, for example, on independent information on the binary nature of the star. The data provided contain the mid-epochs and poles, referred to ICRS, of the reference great circle adopted for each orbit, by each consortium; and, for each orbit, the residuals between the observed abscissae for each star and those calculated from the set of reference astrometric parameters given in the main catalogue. Details of the relevant data reduction principles and the merging of the results from the two consortia are given in ESA (1997, Volume 3), and the data are described in Volume 1, Section 2.8. This includes, for example, a simple prescription for deriving the relevant orbit number given an observation epoch.

The transit files are the compilation of the calibrated five-parameter model data from the image dissector tube for 38 535 Hipparcos Catalogue entries, including also the detailed scanning geometry. Basically they contain the Fourier coefficients which describe the modulation of the detector signal caused by the object's motion across the modulating grid. They provide the intermediate astrometric and photometric information at the level of an individual focal plane crossing. The transit files were originally used in the NDAC double-star reductions, and were constructed for all systems recognised as double or multiple, or suspected as such during the early phases of the data reductions. As a result, the transit data are available for all double and multiple systems, plus a few thousand catalogue entries finally considered as either suspected or non-single stars. The data provide a self-contained and globally-calibrated compilation of the main detector signal parameters, and are described in ESA (1997, Volume 1, Section 2.9).

Examples of use Examples of how to use the intermediate astrometric data were presented by van Leeuwen (1997), including calibrations of a common parallax and proper motion for stars in clusters, eliminating correlations between astrometric parameters; calibrations of common luminosity properties for stars distributed over the sky, reducing the degrees of freedom and improving the homogeneity; and determination of orbital motion due to an unseen companion. Applications to RR Lyrae luminosities, period–luminosity relation for Mira stars, distance to the LMC and SMC (from the 31 and eight Hipparcos stars in these objects, respectively), distance to Pleiades and Praesepe, and use of the Solar System object data were described by van Leeuwen & Evans (1998). Lenhardt *et al.* (1999) used the intermediate data from both consortia for about 63 000 catalogue objects to estimate deviations from the PPN value of $\gamma = 1$, a zero-point correction of the catalogue parallaxes, and linear and quadratic variations of the residuals with time. Arenou (1997) discussed the application to astrometric binaries. Falin & Mignard (1999) undertook a re-analysis of the FAST consortium residuals for solutions flagged as unreliable, using additional information on multiplicity, and presenting 139 new solutions with absolute and relative astrometry. Other aspects of the re-analysis of double and multiple stars are discussed in Chapter 3.

The transit data files, and their application to re-reduction of the satellite data using more complex object models with time-variable photometric and geometric characteristics, are described by Quist & Lindegren (1999). They include the application of the two harmonics of the detector signal to aperture synthesis type imaging, equivalent to an interferometer with baselines of $\sim$0.1 and 0.2 m at 550 nm, illustrating their analysis by the case of HIP 97237, noted in the Input Catalogue as a binary with separation 0.9 arcsec and component magnitudes 12.7 and 13.6 mag (Figure 1.8). Combination with the Tycho data to improve solutions for nearby double and multiple stars with orbital motion was considered by Makarov & Fabricius (1999).

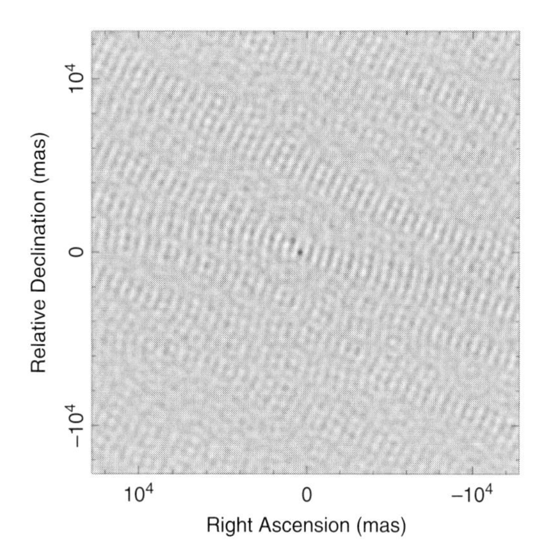

Figure 1.8 Use of the transit data for aperture synthesis imaging, applied to the 0.9 arcsec separation double, HIP 97237. The image was obtained by inverse Fourier transform of the measured complex visibilities, cleaning of the resulting image by deconvolution with the 'beam', and shifting the astrometric parameters of the reference point to that of the primary component. From Quist & Lindegren (1999, Figure 7).

In addition to the 235 astrometric binaries for which orbits were derived as part of the Hipparcos Catalogue construction, Pourbaix & Jorissen (2000) combined the intermediate astrometric data and transit data with the orbital parameters of a number of spectroscopic binaries to derive their astrometric parameters. Applied to 81 systems belonging to various families of chemically-peculiar red giants (dwarf and giant Ba stars, CH stars, and Tc-poor S stars, where the chemical peculiarity frequently arises from mass transfer across a binary system) yielded 23 reliable astrometric orbits, using a global optimization technique in 12-parameter (or, for spectroscopic binaries, nine-parameter) space. They showed that: (a) the 'cosmic error' described by Wielen (1997), namely that an unrecognised orbital motion introduces a systematic error on the proper motion, was confirmed for these objects, a comparison of the proper motion from Hipparcos with that re-derived in their work indicating that the former are indeed far off the present value for binaries with periods in the range 3–8 years; (b) Hipparcos parallaxes of unrecognised spectroscopic binaries are reliable, except for systems with periods close to one year, as expected; (c) even when a complete orbital revolution was observed by Hipparcos, the inclination is seldom precise.

The intermediate astrometric and transit data was also investigated, amongst others, for systems with a giant component (Pourbaix & Boffin, 2003); for C stars to provide improved parallaxes and luminosity calibration

(Pourbaix *et al.*, 2000); and for evolved giant stars, providing absolute magnitudes for the R-type carbon stars (Knapp *et al.*, 2001); and revised period–luminosity relationship for Galactic long-period variable stars (Knapp *et al.*, 2003, Section 4.10).

1.4.5 Transformation of astrometric data

Some of the most useful formulae for the treatment of the Hipparcos astrometric data are summarised here, taken directly from ESA (1997, Volume 1, Section 1; provided by L. Lindegren), where further details are given.

Computation of transverse motions The component of the stellar space velocity tangent to the line-of-sight, $V_{\rm T}$, is traditionally calculated as

$$V_{\rm T} = \frac{A_v \mu}{\pi} \tag{1.5}$$

where $A_v = 4.740\,47\ldots$ equals the astronomical unit, expressed in [km yr s^{-1}] (see Table 1.3). Many textbooks give slightly different values for A_v, usually because they assume the (now obsolete) tropical year as time unit for the proper motions. This expression neglects light-time effects, and should strictly include the Doppler factor $k = (1 - V_{\rm R}/c)^{-1}$, although its omission causes only a very small error on calculated space velocities of Galactic stars.

Coordinate transformations The basis vectors in the equatorial system are denoted $[\,\mathbf{x}\,\mathbf{y}\,\mathbf{z}\,]$, with $\mathbf{x}$ being the unit vector towards $(\alpha,\delta) = (0,0)$, $\mathbf{y}$ the unit vector towards $(\alpha,\delta) = (+90^\circ, 0)$, and $\mathbf{z}$ the unit vector towards $\delta = +90^\circ$. The basis vectors in the ecliptic and Galactic systems are respectively denoted $[\,\mathbf{x}_{\rm K}\,\mathbf{y}_{\rm K}\,\mathbf{z}_{\rm K}\,]$ and $[\,\mathbf{x}_{\rm G}\,\mathbf{y}_{\rm G}\,\mathbf{z}_{\rm G}\,]$. Thus, the arbitrary direction $\mathbf{u}$ may be written in terms of the equatorial, ecliptic and Galactic coordinates as

$$\mathbf{u} = [\,\mathbf{x}\,\mathbf{y}\,\mathbf{z}\,]\begin{pmatrix}\cos\delta\cos\alpha\\ \cos\delta\sin\alpha\\ \sin\delta\end{pmatrix} = [\,\mathbf{x}_{\rm K}\,\mathbf{y}_{\rm K}\,\mathbf{z}_{\rm K}\,]\begin{pmatrix}\cos\beta\cos\lambda\\ \cos\beta\sin\lambda\\ \sin\beta\end{pmatrix} \tag{1.6}$$

$$= [\,\mathbf{x}_{\rm G}\,\mathbf{y}_{\rm G}\,\mathbf{z}_{\rm G}\,]\begin{pmatrix}\cos b\cos l\\ \cos b\sin l\\ \sin b\end{pmatrix} \tag{1.7}$$

The transformation between the equatorial and ecliptic systems is given by

$$[\,\mathbf{x}_{\rm K}\,\mathbf{y}_{\rm K}\,\mathbf{z}_{\rm K}\,] = [\,\mathbf{x}\,\mathbf{y}\,\mathbf{z}\,]\,\mathbf{A}_{\rm K} \tag{1.8}$$

where

$$\mathbf{A}_{\rm K} = \begin{pmatrix}1 & 0 & 0\\ 0 & \cos\epsilon & -\sin\epsilon\\ 0 & \sin\epsilon & \cos\epsilon\end{pmatrix} \tag{1.9}$$

and $\epsilon = 23^\circ\,26'\,21.448''$ is the conventional value of the obliquity of the ecliptic (Table 1.3).

The transformation between the equatorial and Galactic systems is given by

$$[\,\mathbf{x}_{\rm G}\,\mathbf{y}_{\rm G}\,\mathbf{z}_{\rm G}\,] = [\,\mathbf{x}\,\mathbf{y}\,\mathbf{z}\,]\,\mathbf{A}_{\rm G} \tag{1.10}$$

where the matrix $\mathbf{A}_{\rm G}$ relates to the definition of the Galactic pole and centre in the ICRS system. As of 1997, no definition of this relation had been sanctioned by the IAU.[1] In order to provide an unambiguous transformation for users of the Hipparcos and Tycho Catalogues, the following definitions in the ICRS were proposed for the north Galactic pole $(\alpha_{\rm G}, \delta_{\rm G})$, and for the origin of Galactic longitude defined by the longitude of the ascending node of the Galactic plane on the equator of ICRS, l_Ω, as

$$\alpha_{\rm G} = 192.^\circ 859\,48 \tag{1.11}$$

$$\delta_{\rm G} = +27.^\circ 128\,25 \tag{1.12}$$

$$l_\Omega = 32.^\circ 931\,92 \tag{1.13}$$

Regarding the angles $\alpha_{\rm G}$, $\delta_{\rm G}$ and l_Ω as exact quantities, they are consistent with the previous (1960) definition of Galactic coordinates to a level set by the quality of optical reference frames prior to Hipparcos. The transformation matrix $A_{\rm G}$ may be computed to any desired accuracy; to 10 decimal places the result is

$$\mathbf{A}_{\rm G} = \begin{pmatrix}-0.054\,875\,560\,4 & +0.494\,109\,427\,9 & -0.867\,666\,149\,0\\ -0.873\,437\,090\,2 & -0.444\,829\,630\,0 & -0.198\,076\,373\,4\\ -0.483\,835\,015\,5 & +0.746\,982\,244\,5 & +0.455\,983\,776\,2\end{pmatrix} \tag{1.14}$$

The ecliptic longitude and latitude are thus computed from

$$\begin{pmatrix}\cos\beta\cos\lambda\\ \cos\beta\sin\lambda\\ \sin\beta\end{pmatrix} = \mathbf{A}'_{\rm K}\begin{pmatrix}\cos\delta\cos\alpha\\ \cos\delta\sin\alpha\\ \sin\delta\end{pmatrix} \tag{1.15}$$

and the Galactic longitude and latitude from

$$\begin{pmatrix}\cos b\cos l\\ \cos b\sin l\\ \sin b\end{pmatrix} = \mathbf{A}'_{\rm G}\begin{pmatrix}\cos\delta\cos\alpha\\ \cos\delta\sin\alpha\\ \sin\delta\end{pmatrix} \tag{1.16}$$

[1] The main problem arises because the transformation from B1950 to J2000 consists not only of a pure rotation of axes due to changes in precession and equinox motion, but also a change in the convention by which stellar aberration is computed in the two systems. Prior to the adoption of the J2000 system, stellar aberration was based only on the circular component of the Earth's orbital velocity. The component depending on the eccentricity of the orbit, which can amount to 0.34 arcsec, and is approximately constant for a given direction over long intervals of time, was considered to be implicitly included in the mean catalogue positions of stars. See, e.g. Aoki *et al.* (1983, 1986), Lederle & Schwan (1984), and Smith *et al.* (1989).

Epoch transformation The simplistic formulae for transforming a celestial position (α, δ) from the catalogue epoch T_0 to the arbitrary epoch T are

$$\alpha = \alpha_0 + (T - T_0)\,\mu_{\alpha*0} \sec\delta_0 \qquad (1.17)$$

$$\delta = \delta_0 + (T - T_0)\,\mu_{\delta 0} \qquad (1.18)$$

where the $\sec\delta_0$ factor compensates the $\cos\delta_0$ factor implicit in $\mu_{\alpha*}$. The resulting variances in position are given by

$$\sigma^2_{\alpha*} = [\sigma^2_{\alpha*} + 2t\rho^{\mu_{\alpha*}}_{\alpha*}\sigma_{\alpha*}\sigma_{\mu_{\alpha*}} + t^2\sigma^2_{\mu_{\alpha*}}]_0 \qquad (1.19)$$

$$\sigma^2_{\delta} = [\sigma^2_{\delta} + 2t\rho^{\mu_\delta}_{\delta}\sigma_{\delta}\sigma_{\mu_\delta} + t^2\sigma^2_{\mu_\delta}]_0 \qquad (1.20)$$

This simplified model neglects the slow changes in the proper motion components and in the parallax. A rigorous expression for transforming (α_0, δ_0, π_0, $\mu_{\alpha*0}$, $\mu_{\delta 0}$, V_{R0}) at epoch T_0 into (α, δ, π, $\mu_{\alpha*}$, μ_δ, V_R) at epoch T, along with the transformation of the 36 components of the covariances, is given in ESA (1997, Volume 1, Section 1.5.5).

Figure 1.11 below shows how the typical catalogue errors of Hipparcos, Tycho 1, and Tycho 2 propagate over the interval 1900–2050.

Space coordinates and velocity The position of a star with respect to the Solar System barycentre, **b**, measured in pc, and its barycentric space velocity, **v**, measured in km s^{-1}, are given in equatorial components by

$$\begin{pmatrix} b_x \\ b_y \\ b_z \end{pmatrix} = \mathbf{R} \begin{pmatrix} 0 \\ 0 \\ A_p/\pi \end{pmatrix} \qquad (1.21)$$

$$\begin{pmatrix} v_x \\ v_y \\ v_z \end{pmatrix} = \mathbf{R} \begin{pmatrix} k\mu_{\alpha*}A_v/\pi \\ k\mu_\delta A_v/\pi \\ kV_R \end{pmatrix} \qquad (1.22)$$

with

$$\mathbf{R} = \begin{pmatrix} -\sin\alpha & -\sin\delta\cos\alpha & \cos\delta\cos\alpha \\ \cos\alpha & -\sin\delta\sin\alpha & \cos\delta\sin\alpha \\ 0 & \cos\delta & \sin\delta \end{pmatrix} \qquad (1.23)$$

$A_p = 1000$ mas pc and $A_v = 4.74047\ldots$ km yr s^{-1} designate the astronomical unit expressed in the appropriate units (Table 1.3), and $k = (1 - V_R/c)^{-1}$ is the Doppler factor. The Galactic components of **b** and **v** are obtained through pre-multiplication by $\mathbf{A}'_G$, as in Equation 1.16.

1.5 Tycho Catalogue and Annexes

The Tycho Catalogue was compiled from the satellite's 'star mapper' measurement system (Høg *et al.*, 1997). Providing a lower accuracy in the astrometric parameters, the Tycho Catalogue nevertheless has two major attributes. First, it provided two-colour photometry to complement the astrometric data of the Hipparcos Catalogue stars. These data were included in the published Hipparcos Catalogue. Second, the Tycho measurements did not rely on a pre-determined input catalogue, and therefore provided not only a significant increase in density of the reference star grid (from the three stars per square degree of Hipparcos, to about 25 stars per square degree in the Tycho Catalogue), but a catalogue with more uniform and better-defined completeness limits. Completeness was estimated at 99.9% and 90% at $V_T = 10.0$ and $V_T = 10.5$ respectively, but with a slight dependence on sky position due to the scanning law, and on colour index due to the fact that detection was carried out in the added photometric counts from the two channels (Egret & Fabricius, 1997). The study of Holmberg *et al.* (1997), for example, also confirms that the catalogue is rather complete to $V_T \simeq 10$ and $B_T \simeq 10.5$.

With the appearance of the Tycho 2 Catalogue in 2000, the original Tycho Catalogue is now sometimes referred to as the Tycho 1 Catalogue. Tycho 2 essentially supersedes Tycho 1, with the exception of the Tycho 1 parallaxes, which are generally of limited significance. The present section describes the Tycho 1 Catalogue as published. The principles described apply equally to the Tycho 2 Catalogue.

Details of the data, the reductions, and the properties of the final catalogue, are given in the published Tycho Catalogue (ESA, 1997, Volumes 1 and 4). Other general descriptions have been given by Høg *et al.* (1995) and Großmann *et al.* (1995). Specific discussion of the Tycho star mapper background analysis, including variations due to the Van Allen radiation belts, was given by Wicenec & van Leeuwen (1995), and of the de-censoring of the faint stars in Tycho photometry by Halbwachs *et al.* (1997a).

The measurements were obtained with the Hipparcos star mapper, a system of aperiodic slits in the focal plane of the Hipparcos telescope, designed primarily for determining the satellite attitude by observation of stars with known positions (Figure 1.2, right). The light passing the slits was split by a dichroic mirror into two distinct colour channels (designated B_T and V_T) and recorded as photon counts by means of two corresponding photomultiplier tubes. The encoded photon counts were transmitted to the ground, and the subsequent reduction was undertaken by the Tycho Consortium (led by Erik Høg), within the framework of the entire mission analysis.

The basic operational principles of the Hipparcos satellite also applied to the Tycho Catalogue measurements. The various geometrical scan configurations for each star, at multiple epochs throughout the three-year observation programme and simultaneously with the

Hipparcos observations, resulted in a close connection to the Hipparcos astrometric reference frame represented by about 100 000 stars. The barycentric coordinate direction (α, δ), the parallax (π), and the object's proper motion ($\mu_\alpha \cos\delta$, μ_δ) could be solved for in least-squares reductions of the observations. The astrometric parameters as well as their standard errors and correlation coefficients were derived in the process. The large number of observations per object, of order 130, provided accurate and homogeneous photometric information for each star, from which mean magnitudes in the two passbands, B_T and V_T, were derived. Studies of variability and multiplicity were also undertaken.

The first stages of the data analysis were based upon a starting catalogue (the Tycho Input Catalogue), compiled on the basis of the Guide Star Catalogue (GSC), although subsequent steps of the Tycho analysis searched for stars, above the completeness threshold, missing from the GSC for various reasons. The final result is an astrometric and photometric reference catalogue of slightly more than one million stars: the Tycho Catalogue. It was completed in August 1996, simultaneously with the Hipparcos Catalogue, and the two were published together (ESA, 1997).

Designation of objects within the Tycho Catalogue uses the Guide Star Catalogue (GSC) numbering system (a region number, designated TYC1; and a number within the region, designated TYC2), followed by a Tycho Catalogue specific component number (TYC3); the latter was introduced to permit the parallel classification of resolved systems. As well as giving a cross-identification to the GSC, this designation system has the advantage of giving a rough indication of position on the sky. Objects contained in the GSC, TYC1 and TYC2 are identical to the identifiers defined by the GSC numbering system.

Egret & Fabricius (1997) showed that the catalogue contains about 99.9% of the stars brighter than $V = 10.0$ mag, and a large fraction of stars in the range 10.0–11.5 mag.

1.5.1 Tycho astrometry

The standard astrometric model adopted for all stars in the Tycho reductions was the same as that adopted for the Hipparcos observations: i.e. it assumes uniform rectilinear space motion relative to the Solar System barycentre. In the same way, the stellar motion on the celestial sphere is described by the following five 'astrometric' parameters: the barycentric coordinate direction (α, δ); the annual parallax, π; and the rate of change of the barycentric coordinate direction expressed as proper motion components $\mu_{\alpha*} = \mu_\alpha \cos\delta$ and μ_δ, in angular measure per unit time. The five astrometric parameters, determined from an analysis of the Tycho observations, are given for almost all stars in the catalogue. The catalogue includes astrometric quality flags, and indicators of variability and/or multiplicity. Nearly 900 000 of the catalogue entries are classified as 'recommended' reference stars, having good Tycho astrometric quality, and not recognised as double.

The principal characteristics of the astrometric results are given in Table 1.4. As in the case of the Hipparcos Catalogue, corresponding standard errors, also at the catalogue epoch (J1991.25), are given in the final catalogue, along with the associated correlation coefficients. Positions and proper motions are referred to the ICRS. The astrometric positions and their errors can be propagated to the standard epoch J2000.0, or to any other epoch, within the ICRS frame, by methods described in the published catalogue.

Median astrometric standard errors (in position, parallax, and annual proper motion) are typically around 7 mas for stars brighter than $V_T \sim 9$ mag, and approximately 25 mas for $V_T \sim 10.5$ mag, at the catalogue epoch (J1991.25). The Tycho Catalogue was adjusted so that the six global orientation and rotation components of the catalogue reference frame coincide with those of the Hipparcos Catalogue. It therefore also materialises the ICRS reference system, coinciding with its principal axes at the level of ± 0.6 mas, and with proper motions consistent with an inertial system at the level of ± 0.25 mas yr^{-1}. The 1 058 332 constituent stars provide a mean sky density of ~ 25 stars deg^{-2}.

Figure 1.9 illustrates the mean precision of each of the astrometric parameters for stars of $V_T = 9.0$ mag

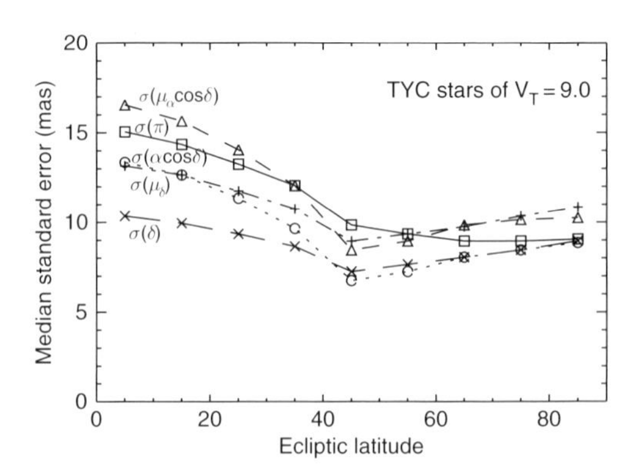

Figure 1.9 Standard errors of the five astrometric parameters as a function of ecliptic latitude for the Tycho Catalogue. The dependency on ecliptic latitude is a consequence of the ecliptic-based scanning law of the Hipparcos satellite. The errors are given at the catalogue epoch, J1991.25. The unit of the internal standard error is milliarcsec (mas) for the positional components (α, δ) and parallax (π), and mas yr^{-1} for the proper motion components ($\mu_\alpha \cos\delta, \mu_\delta$). From Høg et al. (1997, Figure 1).

as a function of ecliptic latitude. The magnitude dependence is indicated in Table 1.4. The external standard errors (i.e. accuracies) are only slightly larger for bright stars, increasing to a factor of 1.5 larger at the median magnitude $V_T = 10.5$ of the Tycho Catalogue stars.

Since for many Tycho Catalogue entries the proper motion has a standard error larger than the expected proper motion itself ($\mu \simeq 20$ mas yr^{-1} for a star of spectral type F5 and $V = 11$ mag), the accuracy of the Tycho proper motions is generally too low to calculate positions at other epochs with sufficient accuracy for reference purposes. More precise proper motions were already listed in the PPM Catalogue (Röser & Bastian, 1991, 1993) for approximately one half of the Tycho stars, although zonal errors present in the PPM Catalogue should be accounted for in any comparisons or applications. The Tycho Catalogue includes the PPM identifier where available.

1.5.2 Tycho photometry

The Tycho Catalogue includes accurate and homogeneous photometric information for each star: two-colour (B_T and V_T) magnitudes derived from the Tycho (star mapper) observations; Johnson V magnitude and $B - V$ colour index, derived from the observed $B_T - V_T$ by simplified transformations reflecting the fact that the spectral type, luminosity class, reddening, etc., are unknown for the majority of Tycho Catalogue objects; and various flags resulting from (preliminary) studies of variability based on an average of 130 transits per star during the three year observing period. The details of the Tycho B_T and V_T photometric system, and details of the corresponding transformations to V and $B - V$, are given in the published catalogue, and also described in the context of the Hipparcos main mission photometry by van Leeuwen *et al.* (1997).

Published mean photometric values were based on all transits for each star, including those transits where the star was not detected because the signal was too faint. These 'censored' observations were taken into account in a dedicated 'de-censoring' processing. In the basic Tycho data reductions, only the detections with a signal-to-noise ratio larger than 1.5 in the combined photon counts from the B_T and V_T detectors, provided astrometric and photometric estimates.

The photometric results for individual transits, including the censored observations, are provided separately in a Tycho Epoch Photometry Annex for specific subsets of stars. These individual magnitude determinations, at the specified measurements epochs, allow detection of variable stars and further studies of the variability of known variables. The principal photometric characteristics are given in Table 1.5.

Table 1.5 Mean photometric characteristics of the Tycho Catalogue. The standard errors of individual transits in the Tycho Epoch Photometry Annex are typically 10 times larger than for the mean value of a given star. Errors on B_T are typically 10 per cent higher than those quoted for V_T. Estimates of systematic errors were described in ESA (1997, Volume 4, Chapter 19).

Property	Value	
Standard errors of V_T for $V_T < 9$	0.012	mag
Standard errors of V_T for $V_T \sim 10.5$	0.06	mag
Smallest errors of V_T	0.003	mag
Estimates of systematic errors:		
external errors/standard errors	~ 1.0	
systematic errors in photometry	< 0.01	mag

The Tycho experiment also performed astrometric and photometric observations of Solar System objects. These are described in the published Hipparcos and Tycho Catalogues (ESA, 1997), and also summarised in Chapter 10.

Figure 1.10, left shows a comparison of Tycho mean magnitudes with ground-based magnitudes for the standard stars available for the reductions. Although only relatively few standard stars are fainter than $V_T =$ 11 mag, this figure and other studies undertaken as part of the catalogue construction and validation demonstrate that the systematic errors are relatively small at all magnitudes. Detailed sky distributions and histograms giving the astrometric and photometric accuracies as a function of position and magnitude are included in the documentation accompanying the published Hipparcos and Tycho Catalogues.

Figure 1.10, right shows an example of the epoch photometry results, for the Mira variable R Car (TYC 8945–1871–1). This star has an amplitude of more than 4.5 mag and a period of 302.098 days. The figure shows how the errors of a single Tycho observation depend on the magnitude itself: the error bars shown in the plots are the 1σ errors as given in the Tycho Epoch Photometry Annex A.

Dimeo *et al.* (1997) argued that the large number of stars in the Tycho Catalogue, and the large number of observations per star, make the photometric database an ideal control sample to search for the presence of microlensing events in the solar neighbourhood, to be compared with the large-scale microlensing experiments towards the LMC and the Galactic bulge. They estimated that only 0.1 event is likely to be present in the database, reducing to about 0.01 for typical events of 15 days duration once the effects of temporal sampling and photometric errors are taken into account. Such a search, either for the Tycho or Tycho 2 Catalogues, has not been reported.

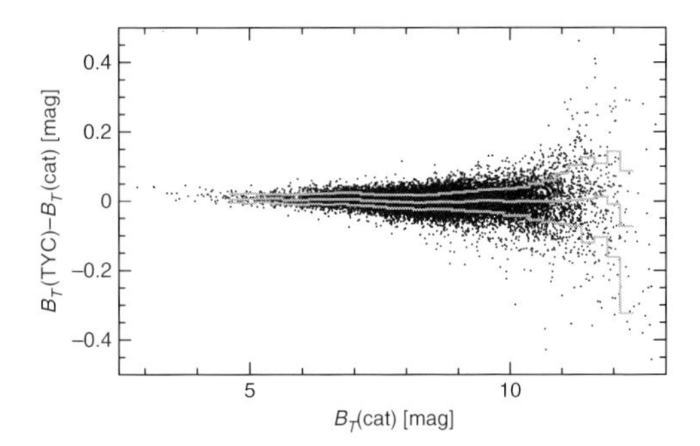

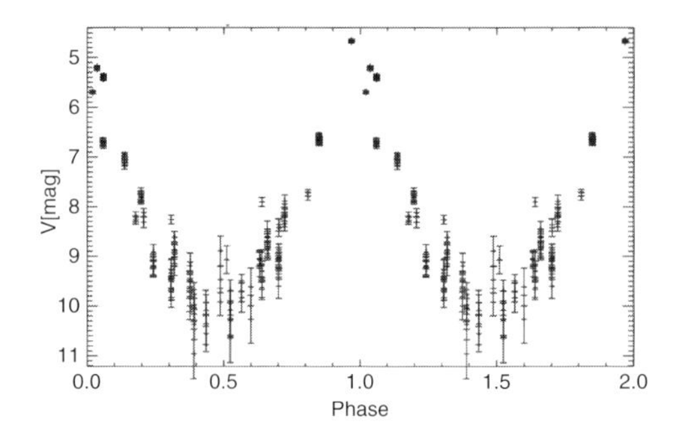

Figure 1.10 Left: Tycho mean magnitudes of 17 683 standard stars compared with ground-based magnitudes, for B_T. The three lines represent the median of the difference and the 15th and 85th percentiles, indicating the systematic errors and robust estimates of the standard errors of Tycho photometry. Right: Phase diagram of the Mira-type star R Car (TYC 8945–1871–1, period = 302.098 days) showing the increase of the errors of the Tycho single observations with magnitude. The groups of apparently 'outlying' data points are correct measurements, illustrating the typical variations of Mira light-curves from cycle to cycle. From Høg et al. (1997, Figures 2 and 3).

1.5.3 Tycho double and multiple systems

Double stars were subject to dedicated analysis, resulting in the resolution of separations down to ∼1.5 arcsec. Detection of multiplicity was effective down ∼0.5 arcsec. The Tycho Catalogue contains various flags indicating such entries.

Halbwachs *et al.* (1997b) investigated the distribution of mass ratios of main-sequence binaries using the B_T and V_T data and the mass–luminosity relation, demonstrating a form similar to that of the initial mass function with exponent ~ -2.

1.6 Post-publication Hipparcos reductions

Following the Hipparcos Catalogue publication in 1997, the data reduction teams concluded their collaborative and collective efforts. The adopted publication schedule was a delicate compromise between publishing the results in a useable, authoritative, and fully documented form as early as possible, whilst ensuring it was free from substantial random or systematic errors.

Subsequent studies have led to improved solutions in the notable case of double and multiple stars. These are discussed in Chapter 3, with a summary compilation of the affected catalogue entries given in Table 3.2.

Between 1997 and 2007, Floor van Leeuwen and colleagues (Cambridge), previously responsible for the image dissector tube and star mapper data processing within NDAC, have continued a deeper investigation into subtle effects in the satellite attitude and instrument calibration.

They identified a number of defects in the data that had not been fully accounted for, such as scan-phase discontinuities and micrometeoroid-induced attitude jumps. While confirming the stability of the basic angle, they also identified some occurrences of sub-optimal connectivity between the attitude solution for the two fields of view as a result of the distribution and weighting of reference stars. Essentially, if the bright stars used for attitude reference are not uniformly distributed on the sky, small distortions of the reference system can result. In general, the effect was adequately accounted for in the original reductions, but larger distortions seem to exist in regions with a dense concentration of bright reference stars, such as open clusters. If such dense regions happen to be poorly connected to the rest of the celestial sphere due to a paucity of bright stars around the small circle separated by the basic angle of 58°, local zonal errors can result.

The work resulted in a series of papers identifying improvements in the attitude treatment, leading to decreased standard errors in the astrometric parameters of the brightest stars (van Leeuwen & Fantino, 2003b): improved treatment of atmospheric drag during perigee passages (Dalla Torre & van Leeuwen, 2003), an improved analysis of thermal drifts (van Leeuwen & Penston, 2003), an improved calibration of solar radiation, gravity gradient, and magnetic moment torques (Fantino & van Leeuwen, 2003), and an improved, fully-dynamic, attitude reconstruction (van Leeuwen & Fantino, 2003a).

Despite these effects, van Leeuwen (2005) underlined the important conclusion that the Hipparcos Catalogue as published is generally reliable within the quoted accuracies. A resulting new and complete reduction of the Hipparcos data was eventually undertaken (van Leeuwen & Fantino, 2005). This has led to improved astrometric accuracies for stars brighter than $Hp = 9.0$ mag, reaching a factor of ~ 3 for $Hp < 4.5$ mag.

At the time of writing this review, these new data are being published (van Leeuwen, 2007a,b), and the first scientific consequences are appearing in the literature (van Leeuwen *et al.*, 2007).

1.7 Post-publication Tycho reductions

Various connected initiatives followed the publication of the Tycho Catalogue in 1997, leading to the publication of the Tycho 2 Catalogue in 2000. Before entering descriptions of the TRC and the ACT Catalogues, note that these catalogues (although not the work involved) were superseded by the Tycho 2 Catalogue.

The TRC, ACT, and AC 2000 Catalogues The Tycho Reference Catalogue (TRC) and the Astrographic Catalogue plus Tycho reference catalogue (ACT) independently combined the positional information in the Astrographic Catalogue (epoch ~1910, and discussed further in Chapter 2) with that in the Tycho Catalogue (epoch J1991.25) to obtain improved proper motions. Median accuracies of both are around 40 mas in each positional component at J1991.25, and 2.5 mas yr^{-1} in proper motions. With an average of some 24 reference stars per square degree, and homogeneous two-colour Tycho photometry, the catalogues provide a dense reference system with an accuracy of better than 100 mas maintained over several decades.

The TRC (Høg *et al.*, 1998) contains positions and proper motions for 990 182 stars of the Tycho Catalogue. Details of the construction, including the reduction of the Astrographic Catalogue onto the Hipparcos system, identification of Tycho catalogue stars in the AC, derivation of proper motions, and correction of AC systematic errors, are given in Kuzmin *et al.* (1999).

The ACT (Urban *et al.*, 1998b) contains positions and proper motions for 988 758 stars of the Tycho Catalogue. It includes cross-references to the Hipparcos and Tycho Catalogues, AC 2000, Bonner Durchmusterung (BD), Cordoba Durchmusterung (CD), Cape Durchmusterung (CPD), and Henry Draper (HD) catalogues. Urban *et al.* (1998a) also published the AC 2000, the Astrographic Catalogue on the Hipparcos system, from which the ACT was constructed. AC 2000 is a positional catalogue of 4 621 836 stars covering the entire sky, and with an average epoch of position of 1907. A revision, AC 2000.2, was subsequently published (Urban *et al.*, 2001).

A comparison of the Hipparcos, Tycho, TRC, and ACT proper motions was carried out by Hoogerwerf & Blaauw (2000) [the second author was using the Hipparcos data 61 years after his first scientific publication]. They reported that the proper motion errors in the ACT Catalogue are underestimated in all zones, sometimes by as much as 30%; that there are systematic differences, as large as 1.2 mas yr^{-1}, between the proper motions in the TRC and ACT depending on zone; and they confirmed that the proper motion errors of faint stars in the Tycho Catalogue ($B_T > 10$ mag) are underestimated by as much as 40%, consistent with the ratio of external errors to standard errors estimated in the published catalogue and summarised in Table 1.4.

The Tycho 2 Catalogue The detector used for the Hipparcos Catalogue observations, an image dissector tube, required *a priori* star positions to direct the 30 arcsec diameter sensitive region to the known star positions. In contrast, the Tycho observations covered the whole sky, and a star signal at a given location was extracted down to a given signal-to-noise completeness limit, or to a slightly fainter limit if a star was known at a given position. This made it possible to re-analyse the Tycho data stream, based on the improved end-of-mission satellite attitude and calibration data, allowing a more precise superposition of each sky region covered by different scans, and thus allowing the extraction of star information to a slightly fainter signal-to-noise limit.

This led to the construction, subsequent to the Tycho Catalogue publication in 1997, of the Tycho 2 Catalogue, also under the leadership of Erik Høg. This essentially supersedes the Tycho Catalogue (with the exception of the Tycho 1 parallaxes), and provides positions, proper motions and two-colour photometry of the 2.5 million brightest stars. The catalogue was the result of a collaboration between (primarily) the Copenhagen University Observatory (CUO) and the USNO. CUO re-analysed the Tycho data stream, which led to better positions for the Tycho stars, as well as extending the number of stars from 1 million to 2.5 million. The USNO was responsible for computing the proper motions, which was done by combining the data with that from the Astrographic Catalogue and 143 other ground-based astrometric catalogues. The result is a global reference catalogue that is 99% complete to $V = 11.0$ and 90% complete at $V = 11.5$; some evidence that the completeness limit is at slightly brighter magnitudes in some regions was reported by Sanner & Geffert (2001) in their study of open clusters (see Section 6.7). Positional accuracies range from about 10–100 mas, depending on magnitude. Proper motion accuracies are around 2.5 mas yr^{-1}. Components of double stars with separations down to 0.8 arcsec are included.

Supplement 1 lists 17 588 good quality stars from the Hipparcos and Tycho 1 Catalogues which are not in Tycho 2. Supplement 2 lists 1146 Tycho 1 stars which are probably either false or heavily disturbed.

The catalogue (Høg *et al.*, 2000c) is described in Høg *et al.* (2000b), with details of its construction and verification given in Høg *et al.* (2000a). Tycho 2 not only supersedes the 1997 Tycho Catalogue (or Tycho 1 Catalogue), except for the parallaxes, but also the ACT and

TRC catalogues, themselves based on Tycho 1. A comparison of the Tycho 2 proper motions with those of Hipparcos and ACT are given in Urban *et al.* (2000).

Details of the double star content and associated processing of the Tycho 2 Catalogue are described in Section 3.3. As a result, the Tycho Double Star Catalogue, TDSC, is a catalogue of absolute astrometry and B_T, V_T photometry for 66 219 components of 32 631 double and multiple star systems.

Galaxies in the Tycho 2 Catalogue have been identified by cross-correlation with galaxy catalogues (Metz & Geffert, 2004). This effort identified 181 galaxies in total: 116 in the Tycho 2 Catalogue, 35 uncertain galaxies in the Tycho 2 Catalogue, and 30 galaxies in the supplement. Galaxies were identified by cross-correlation with independent catalogues of galaxies and quasars (the quasar 3C 273, present in the Hipparcos Catalogue, was also detected). While all have a PGC (Principal Galaxy Catalogue, Paturel *et al.*, 1989) number, not all were found from positional correlation with the PGC, so that other as yet unidentified galaxies may still exist in Tycho 2. As these authors point out, these galaxies are not suitable for a direct link to an extragalactic reference frame since they have mean positional errors exceeding 100 mas.

To summarise the astrometric hierarchy of the Hipparcos, Tycho 1 and Tycho 2 Catalogues: users searching for maximum parallax information should search first Hipparcos, and then Tycho 1 (ignoring Tycho 2). Users searching for maximum position and proper motion information should search first Hipparcos, and then Tycho 2 (ignoring Tycho 1), but should also take into account that the Hipparcos and Tycho 2 Catalogues refer to different temporal baselines. Users searching for maximum double and multiple star information should search first Hipparcos and its associated double and multiple systems annex, and then the Tycho Double Star Catalogue (which includes the Tycho 2 doubles).

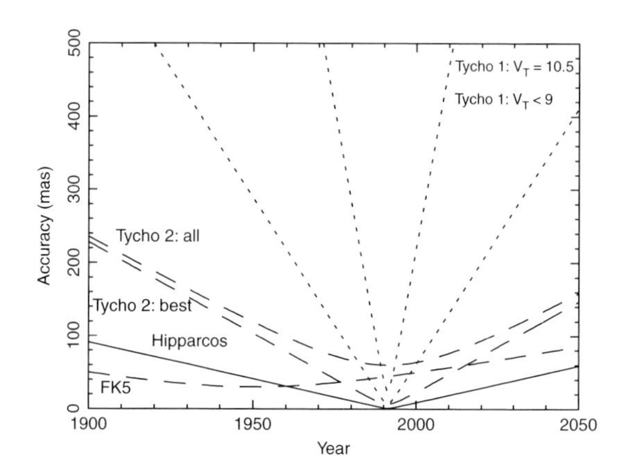

Figure 1.11 Typical accuracies of the FK5, Hipparcos, Tycho 1, and Tycho 2 Catalogues as a function of time. In each case, the position at the catalogue epoch, t_0, is propagated according to the proper motions given in Table 1.4, with $\sigma_t^2 = \sigma_0^2 + \sigma_{\rm pm}^2(t-t_0)^2$. Tycho 1 dependencies are shown for two representative magnitudes. For Tycho 2 a typical proper motion error of 2.5 mas yr^{-1} applies to both bright stars (positional error at J1991.25 of 7 mas) and faint stars (positional error at J1991.25 of 60 mas). For FK5 simplified values are: $\sigma_0 = 30$ mas, $\sigma_{\rm pm} = 0.8$ mas yr^{-1}, $t_0 = 1950$.

1.8 Catalogue products

1.8.1 Organisation

The Hipparcos and Tycho Catalogues were made available as a 17-volume publication (ESA, 1997). These are organised as shown in Table 1.6, which provides a reference for further description of the Hipparcos Catalogue contents.

The three-volume cartographic representation (Volumes 14–16) was constructed in collaboration with (and led by) Sky Publishing Corporation, and appears separately as Sky Publishing's *Millennium Sky Atlas* (Sinnott & Perryman, 1997). Nearby, variable, high proper motion, and multiple systems are indicated with the corresponding data from Hipparcos, as well as nonstellar objects added from various other sources (nebulae, open and globular star clusters, and some 8000 galaxies).

The Tycho Catalogue is available as part of the 17-volume publication, notably through Volume 1 which includes descriptions of the various datasets of the Tycho (and Hipparcos) Catalogues, and Volume 17 which includes all data catalogues on CDs.

The 70 data fields of the Hipparcos Catalogue (see Table 1.2) and the 57 data fields of the Tycho Catalogue cover the astrometric results, mean photometry and summary variability data, double star information, various flags, and cross-identifications. The two catalogues have a similar data structure, facilitating their combined use. For example, Fields H8–30 of the Hipparcos Catalogue, and Fields T8–30 of the Tycho Catalogue cover the astrometric data with the same format and meaning in the two cases, comprising astrometric parameters, standard errors and correlations, and solution flags.

1.8.2 Availability

The datasets were made available in various forms:

Hipparcos Catalogue in printed form The Hipparcos Catalogue, and a subset of its various annexes, is given in printed form within the ESA 17-volume publication (see Table 1.6). With the rapid evolution of digital media in the early 1990s, a long-life acid-free

Table 1.6 Organisation and content of the Hipparcos and Tycho Catalogues. The main sections contained in ESA (1997, Volume 1) are indicated to serve as reference for further description and mathematical formulation.

Volume	Information contained
Volume 1	Introduction and Guide to the Data Astrometric Data: reference frame, times scales, angular coordinates, epochs, variance–covariance data and correlations, stellar motion Photometric Data, Magnitudes and Variability: single stars, double stars, statistical and variability indicators, period and amplitude estimation, photometric transformations, $V - I$ colour index Double and Multiple Systems Transformation of Astrometric Data: error propagation, coordinate transformation, epoch transformation, space coordinates and velocity, relation to J2000(FK5) Description of the Catalogue and Annexes: Hipparcos Catalogue, Tycho Catalogue, double and multiple star annex, variability annex, Hipparcos and Tycho epoch photometry annexes, Solar System objects, intermediate astrometric data, transit data, identification charts and tables, machine-readable files and CDs Statistical Properties: Hipparcos Catalogue, Tycho Catalogue, astrophysics Tables: nearest, largest proper motions and transverse velocities, most luminous
Volume 2	The Hipparcos Satellite Operations
Volume 3	Construction of the Hipparcos Catalogue
Volume 4	Construction of the Tycho Catalogue
Volumes 5–9	The Hipparcos Catalogue (notes at end of each volume) Volumes 5–9: $0^h - 3^h$; $4^h - 8^h$; $9^h - 13^h$; $14^h - 18^h$; $19^h - 23^h$
Volume 10	Hipparcos Double and Multiple Systems Annex Part C: Component Solutions Part G: Acceleration Solutions Part O: Orbital Solutions Part V: VIM ('Variability-Induced Mover') Solutions Part X: Stochastic Solutions Notes on Double and Multiple Systems Solar System Objects Hipparcos: Astrometric Catalogue Hipparcos: Photometric Catalogue Tycho: Astrometric and Photometric Catalogue
Volume 11	Hipparcos Variability Annex: Tables Part 1: Periodic Variables Part 2: Unsolved Variables Photometric Notes and References; by Number; by Author Spectral Types for Hipparcos Catalogue Entries
Volume 12	Hipparcos Variability Annex: Light-Curves Part A: Folded; Part B: AAVSO; Part C: Unsolved
Volume 13	Identification Charts Part D: Charts from the STScI Digitised Sky Survey Part G: Charts from the Guide Star Catalogue Identification Tables Table 1: HIP Inconsistent with HIC Cross-Identifiers Table 2: HD (Henry Draper) Catalogue Numbers Table 3: HR (Bright Star) Catalogue Numbers Table 4: Bayer and Flamsteed Names Table 5: Variable Star Names Table 6: Common Star Names
Volumes 14–16	*Sky & Telescope*'s Millennium Star Atlas Volumes 14–16: $0^h - 7^h$; $8^h - 15^h$; $16^h - 23^h$
Volume 17	The Hipparcos and Tycho Catalogues on 6 ASCII CDs

paper was used, with a checksum at the bottom-right of each page to allow future scanning. Some 2000 catalogue sets were produced and distributed world-wide to libraries and astronomical institutes. Reviews included Strickland (1998).

Hipparcos and Tycho 1 Catalogue on CD All mission products, including the main Hipparcos and Tycho 1 Catalogues, intermediate astrometric data, and the catalogues of double and multiple stars and epoch photometry, were included on six CDs within the above publication. The six CDs included various data related to the Tycho 1 Catalogue. CD1 contains the main Tycho Catalogue. CD4 contains the Tycho Epoch Photometry Annex A with transit data for 34 446 objects. Epoch photometry for a significantly larger fraction of the Tycho Catalogue objects (481 553, including those in Annex A) is available as the Tycho Epoch Photometry Annex B through the CDS.

Catalogues via the CDS, etc. All catalogues, including the Tycho 2 and associated catalogues, are available from the on-line services and catalogue distribution facility of the Centre de Données Astronomique de Strasbourg (CDS, Egret *et al.*, 1997). Its role in the cross-identification of large surveys is described, for example, by Bonnarel *et al.* (2000).

The Catalogue Access Service of the Sternberg Astronomical Institute, which includes Tycho 2 as well as 2MASS, UCAC 2, and NOMAD, is described by Koposov & Bartunov (2006).

ESA www site www.rssd.esa.int/Hipparcos provides online access to Volume 1 of the catalogue (the detailed catalogue and annex descriptions); search facilities for the Hipparcos and Tycho 1 Catalogues, the Hipparcos intermediate astrometry data, the epoch photometry annex, and the transit data; selected statistical diagrams showing the astrometric accuracies, HR diagrams, etc. Various interactive facilities are available, notably providing access to the intermediate astrometric data, access to the epoch photometry data allowing trial variability periods to be inserted, and an animation facility to visualise stellar motions anywhere on the sky.

Celestia 2000 A single CD with access software (for both PC and Mac), Celestia 2000, was also released in 1997 containing the primary information from the Hipparcos and Tycho 1 Catalogues (Turon, 1998).

1.9 Recommended catalogues

Following the release of the Hipparcos and Tycho Catalogues, many older astrometric and related catalogues were either superseded or incorporated within other catalogues (see Chapter 2). In an attempt to put some order into a potentially confusing situation, a summary of currently recommended catalogues for astrometric and related data is provided in Table 1.7, taken directly from the compilation maintained at the USNO (Zacharias *et al.*, 2004; Zacharias, 2006).

1.10 Catalogue investigations post-publication

1.10.1 Error assessment: Internal

A number of early papers, primarily by members of the Hipparcos data processing community, examined the internal and external accuracy of the published catalogue, e.g. van Leeuwen (2002).

An important method to assess the parallax zero-point is to examine the distribution and numbers of negative parallaxes. That the catalogue should contain (non-physical) negative parallaxes at all may at first sight seem surprising or perturbing, but it is a natural result of the distribution of distances convolved with the measurement error as a function of magnitude. The distribution of negative parallaxes in fact provides a powerful test of the overall catalogue fidelity. Lindegren (1995) used the Richardson–Lucy deconvolution to derive an estimate of the distribution of true parallaxes simultaneously with the external standard error of the observations, arguing that the latter are in good agreement with the formal standard errors. Harris *et al.* (1997) compared 23 stars from the USNO ground-based CCD parallax programme having formal errors around 0.5 mas, finding excellent agreement with the Hipparcos results.

Other accuracy assessments undertaken include: a comparison of ground-based stellar positions and proper motions with provisional Hipparcos results (Lindegren *et al.*, 1995); a method for determining the individual accuracy of astrometric catalogues (Wielen, 1995); a test of Hipparcos parallaxes on multiple stars (Shatsky & Tokovinin, 1998); and a general method for catalogue accuracy estimation (Nefedjev *et al.*, 2006).

Through work done during the catalogue compilation and described in the published catalogue, and extensive studies undertaken subsequently, it is generally accepted that the quoted astrometric standard errors are reliable estimates of their uncertainties, and that global effects in the positions, parallaxes, or annual proper motions are typically below the level of $\sim$0.1 mas. The measurement process means that slightly larger zonal errors on angular scales of a few degrees, of 0.2–0.3 mas, cannot be excluded, although the low star density of about three per square degree makes this difficult to quantify.

So far there has been no publication which contradicts the statistical properties of the Hipparcos Catalogue. Discrepancies have so far only surfaced when the results are compared with indirect determinations. The

Table 1.7 Catalogue information and recommendations, based on Zacharias et al. (2004). See the referenced www page for detailed explanations and up-to-date information. Two merged datasets are also provided: the Naval Observatory Merged Astrometric Dataset, and the Washington Comprehensive Catalogue Database (WCCD). The digitised Schmidt plate catalogues (GSC 2.3, USNO B1, and SuperCOSMOS) are based on a variety of plate material, and their systematic errors are discussed in Chapter 2. This compilation is not exhaustive: other relevant astrometric catalogues re-reduced to the Hipparcos/Tycho system are discussed in Chapter 2, and include derived catalogues such as SDSS–USNO B (Munn et al., 2004); astrograph-based catalogues such as NPM (Hanson et al., 2004) and SPM (Girard et al., 2004); Carte du Ciel based catalogues (Ducourant et al., 2006; Vicente et al., 2007); and proper-motion threshold catalogues, notably rNLTT (Salim & Gould, 2003) and LSPM (Lépine & Shara, 2005).

Catalogue	No. objects	Comments
Astrometric data – currently recommended:		
Hipparcos	118 218	space observations, $V < 12$
Tycho 2	2.5 million	space + early-epoch ground, $V < 12$
GSC 2.3	997 million	digitised plates, see Section 2.5.1
USNO B1.0	1042 million	digitised plates, see Section 2.5.2
SuperCOSMOS	1900 million	digitised plates, see Section 2.5.3
UCAC2: USNO CCD Astrograph Catalogue #2	48 million	86% of sky, $R < 16$, see Section 2.7.1
UCAC2 Bright Star Supplement	430 000	provisionally completes UCAC2
Astrometry – forthcoming:		
UCAC3 (final)	80 million	expected availability 2008, $V < 16.5$
Parallaxes:		
Hipparcos Catalogue	118 218	see above
General Catalogue of Trigonometric Parallaxes	8112	half fainter than Hipparcos
Double star catalogues (see box on page 97):		
WDS: Washington Double Star Catalogue	100 000	extensive compilation
6th Orbit Catalogue	1888	determined orbits; as of June 2006
Infrared sources (see Section 2.8):		
Catalogue of Positions for IR Stellar Sources	37 700	IRAS + optical catalogues
2MASS: Two-Micron All Sky Survey	470 million	no proper motions, JHK
Spectral types:		
HD: Henry Draper Catalogue	225 300	completed in 1924
MSS: Michigan Spectral Survey	161 000	ongoing, see Section 5.4
Variable stars:		
Hipparcos and Tycho 2 Catalogues	as above	accurate, but short interval
GCVS + updated name-lists	40 000	see Section 4.1.5
Astrometric data – superseded/not recommended:		
FK5	1535	fundamental frame pre-Hipparcos
FK5 Extension	3117	provided uniformity to $V < 9$
IRS: International Reference Stars	36 027	combination of AGK3R+SRS
ACRS	320 211	replaced SAO, to $V < 10.5$
PPM North	181 731	roughly same stars as ACRS
PPM South	197 179	higher density than ACRS
Tycho 1	1 million	limited accuracy in proper motions
ACT Reference Catalogue	988 758	to improve Tycho 1 proper motions
TRC Tycho Reference Catalogue	990 182	to improve Tycho 1 proper motions
AC: Astrographic Catalogue	5 million	epoch around 1905
UCAC1: USNO CCD Astrograph Catalogue #1	27 million	southern hemisphere, $R < 16$
GSC 1.2: Guide Star Catalogue #1.2	19 million	no proper motions, $V < 16$
USNO A2.0	526 million	no proper motions, $V < 19 - 20$
USNO SA2.0	54 million	subset of USNO A2
GSC 2.2: Guide Star Catalogue #2.2	435 million	no proper motions, $V < 19$

Stellar data from the CDS: Data related to the Hipparcos and Tycho stars are, of course, continually updated in the literature. This may include updates of catalogue cross-identifiers, and data such as long-term proper motions, new radial velocities, metallicities, revised MK spectral types, revised astrometric solutions based on supplementary binary star data, etc. A centralised compilation of the 'best' available data associated with each catalogue star is in principle one of the objectives of the CDS database, accessible via the SIMBAD interface. Its role in the cross-identification of large surveys is described, for example, by Bonnarel *et al.* (2000).

Individual objects can be interrogated via the SIMBAD astronomical database, http://simbad.u-strasbg.fr/simbad. This provides basic data, cross-identifications, bibliography and measurements for astronomical objects outside the solar system. SIMBAD can be queried by identifier (for example HIP number), by coordinates, and by various criteria. Lists of objects and scripts can also be submitted. Criteria include proper motion (pm), parallax (plx), radial velocity (radvel), spectral type (sptype/sptypes), right ascension (ra), catalogue identifier (cat), and many others. For example, the qualification

dec> 60 & sptypes >= 'A0' & sptype <= 'A9' & splum = 'V' & pm >= 100

returns luminosity class V stars with $\delta > 60^\circ$, proper motion above 100 mas yr^{-1}, and spectral type in the range A0–A9. Including the criterion 'cat=HIP' would further restrict the list to the Hipparcos Catalogue stars. Note that, for example, 'sptypes' accesses the SIMBAD database, so that objects whose spectral types have been updated since the original Hipparcos Catalogue compilation will be duly located.

The VizieR service, http://vizier.u-strasbg.fr/viz-bin/VizieR, provides access to the most complete library of published astronomical catalogues and data tables available on line, organised in a self-documented database. Query tools allow the user to select relevant data tables and to extract and format records matching given criteria. Catalogues can be accessed by their usual acronym (HIP for Hipparcos), or by their CDS/ADC designation (e.g. I/239 or 1239 assigned to the Hipparcos Catalogue; with the various annexes also accessible, thus I/239/hip_main, I/239/h_dm_com, etc; I/259/tyc2 assigned to the Tycho 2 Catalogue, etc.). Multiple catalogue access and browse mode are also supported. Tables from articles published in various major journals are also available. For example, choosing 'I/239/hip_main', and modifying the parameters with 'Maximum Entries' to 'unlimited', and 'Output layout' to VOTable returns a local copy of the Hipparcos Catalogue in the VOTable format (as used in the Virtual Observatory). Similarly, entering 'radial velocity' returns all catalogues or tables including published radial velocities. This includes (in principle) all those listed in Table 1.10 below. In this way, all published radial velocity data related to the Hipparcos Catalogue entries can be accessed.

To manipulate, visualise, or cross-match two or more catalogues, various tools are provided based on the Virtual Observatory protocol: TOPCAT (http://www.star.bris.ac.uk/~mbt/topcat/) is a java application that can be downloaded to manipulate (and plot) tables locally; VOTPlot (http://vo.iucaa.ernet.in/) is another java application to plot tabular data; Aladin (http://aladin.u-strasbg.fr/) allows the manipulation of images and catalogues. All these tools can also communicate with each other according to the common standard protocol.

Thus to find the entries of the Hipparcos Catalogue also included in the 'CADARS' catalogue of stellar diameters, use TOPCAT to download the two catalogues, and load them as local copies, from where cross-matching, performing selections, and computing derived data can be carried out. Similar procedures could be used to access the white dwarfs, RR Lyrae stars, etc., included in the Hipparcos Catalogue, according to the most recently available classification criteria.

Errors related to the Hipparcos Catalogue and communicated to the CDS (mostly related to cross-identifications or spectral types) are available at http://cdsarc.u-strasbg.fr/viz-bin/getCatFile?I/239.

complex case of the Pleiades cluster is considered in detail in Section 6.4.

1.10.2 Error assessment: External

Introduction Hipparcos allowed the construction of a global astrometric reference frame from space observations, with the resulting rigid grid directly linked to an extragalactic reference frame. The Hipparcos and Tycho Catalogues thus completely avoided the complex problem faced in earlier ground-based astrometry: that of linking together observations made at different geographic locations and at different epochs, from the Earth, which is 'wobbling' due to the combined effects of precession, nutation, and short-term and unpredictable polar motion. These effects are described further in Chapter 10. Essentially, the Hipparcos Catalogue now provides an independent reference frame within which previous observations made from the Earth's surface can be analysed to study details of these effects present in observational data acquired over the last century or more.

A reference frame defined by the ephemerides (i.e. a compilation of position versus time) of one or more Solar System bodies represents a dynamical reference frame. The ephemerides incorporate theories of the Earth's motion as well other Solar System bodies, such that the dynamical reference frame embodies in a very fundamental way the moving equator and ecliptic (and hence the equinox). Dynamical reference frames are somewhat impractical for typical astronomical observations, and the ICRS no longer embraces a dynamical reference formulation.

Earlier high-accuracy astrometric catalogues were based on the simultaneous observation of stars and planets, leading to a 'fundamental catalogue' in which right ascension and declination were measured directly.

The dynamical reference system post-Hipparcos: Since the mid-1980s, astronomical measurements of the Earth's rotation, from which astronomical time is determined, have depended heavily on VLBI, with classical methods based on star transits being phased out. Thus, even without Hipparcos, the definition of the FK5 reference frame became less relevant. VLBI revealed, in addition, that the models of the Earth's precession and nutation that were part of the IAU 1976 system were inadequate for modern astrometric precision. In particular, the 'constant of precession' (the long-term rate of change of the orientation of the Earth's axis in space) had been overestimated by about 0.3 arcsec per century.

At its General Assembly in 2000 the IAU defined a system of space-time coordinates for the Solar System, and for the Earth, within the framework of General Relativity, by specifying the form of the metric tensors for each, and the four-dimensional space-time transformation between them. The former is called the Barycentric Celestial Reference System (BCRS), and the latter, the Geocentric Celestial Reference System (GCRS). The ICRS can be considered a particular implementation of the BCRS.

In 2000, the IAU also adopted new models for the description of the Earth's instantaneous orientation within the ICRS. The new models include the IAU 2000A precession–nutation model, a new definition of the celestial pole, and a new reference point, called the Celestial Ephemeris Origin, for measuring the rotational angle of the Earth around its instantaneous axis. Some aspects of the models were not finalized until late 2002. Algorithms that incorporate these models, used to transform ICRS catalogue data to observable quantities, are given in the IERS Conventions (2003). Numerical values for fundamental astronomical constants, and computer code implementing the new algorithms, are also given.

The Jet Propulsion Laboratory DE405/LE405 planetary and lunar ephemerides (which supersede DE200 and DE403) have also been aligned to the ICRS. These ephemerides provide the positions and velocities of the eight classical planets, Pluto, and the Moon with respect to the Solar System barycentre, in rectangular coordinates. The data are represented in Chebyshev series form and Fortran subroutines are provided to read and evaluate the series for any date and time. DE405 spans the years 1600 to 2200; a long version, DE406, spans the years −3000 to +3000 with lower precision.

The German series of fundamental catalogues, FK= 'Fundamental Katalog', began with the Fundamental Catalog (FC) of Auwers (1879, 1883), followed by the Neuer Fundamental Katalog (NFK) of Peters (1907), and thereafter the FK3, FK4, and FK5. Each comprised only a relatively small number of fundamental reference stars, from which higher density catalogues were contructed from the interpolation of meridian circle or photographic plate observations.

This section examines the properties of the most accurate fundamental astrometric catalogues constructed previously from the ground. They are essentially superseded by the Hipparcos Catalogue but with two caveats: old ground-based observations allow a rediscussion of the long-term effects of precession and nutation, and the short-term effects of polar motion (Section 10.4); and the early-epoch positions provide accurate long-term photocentric motions which allow the identification of otherwise unrecognisable binary systems through comparison with the quasi-instantaneous proper motions measured by Hipparcos (Section 1.11).

Error assessment with respect to FK5 Of particular interest is the relationship between the Hipparcos Catalogue and the FK5 Catalogue. The FK5 is the ground-based catalogue with the highest position and proper motion accuracies, adopted as the basic stellar reference frame by the IAU in 1976. It was constructed based on dynamical as well as kinematic considerations. For the 1535 stars of the basic FK5 (Fricke *et al.*, 1988), the average precision of positions is 30 mas at the mean catalogue epoch (1955 in right ascension and 1944 in declination), with mean proper motion errors of 0.6 mas yr^{-1} for the northern hemisphere and 1.0 mas yr^{-1} for the southern, leading to positional errors at the Hipparcos Catalogue epoch around 40–60 mas. There are 3117 stars in the FK5 Extension with somewhat lower accuracy in position and proper motion.

Since all the stars in the basic FK5 Catalogue are also contained in the Hipparcos Catalogue, the relationship between the two reference frames can be investigated by direct comparison of the positions and proper motions in the two catalogues (recall that the Hipparcos coordinate axes are considered linked to the extragalactic reference frame to within ± 0.6 mas at the epoch J1991.25, and non-rotating with respect to it to within ± 0.25 mas yr^{-1}). Results of the first such comparison were given in ESA (1997, Volume 3).

A complete characterisation of the relation between J2000(FK5) and ICRS(Hipparcos) is difficult to achieve, given the relatively small number of stars defining the FK5 reference frame, and the intricate and possibly colour- and magnitude-dependent pattern of systematic (in particular zonal) differences, combined with the perturbing effects on the proper motions of undetected astrometric binaries. Nevertheless, certain relations on a global scale may be established with relative ease, in particular the most fundamental one corresponding to a difference in the mean orientation and spin between the two catalogues. This gave, in mas and mas yr^{-1} respectively (ESA, 1997, Volume 1, Section 1.5.7)

$$\epsilon_{x,y,z} = -18.8 \pm 2.3,\ -12.3 \pm 2.3,\ +16.8 \pm 2.3 \tag{1.24}$$

$$\omega_{x,y,z} = -0.10 \pm 0.10,\ +0.43 \pm 0.10,\ +0.88 \pm 0.10 \tag{1.25}$$

where the orientation parameters refer to the epoch J1991.25. This result was based on the catalogue differences for all 1535 FK5 stars without filtering; no star was removed in the comparison. The remaining differences, once the rotation has been applied, may be as large as 150 mas, because of the large zonal differences which show up in the harmonics of higher degree. A similar decomposition based on only 1232 FK5 stars, after the double stars and the suspected astrometric binaries had been excluded, led to very similar values for the orientation and spin. Likewise various binnings in cells of 100, 200 or 400 square degrees gave comparable results. Selecting stars according to their brightness produced different solutions for the rotation and spin parameters slightly outside the above standard errors, demonstrating the difficulty of establishing a well-defined relation between J2000(FK5) and ICRS(Hipparcos). Figure 1.12 illustrates the differences reported in ESA (1997, Volume 3, Chapter 19).

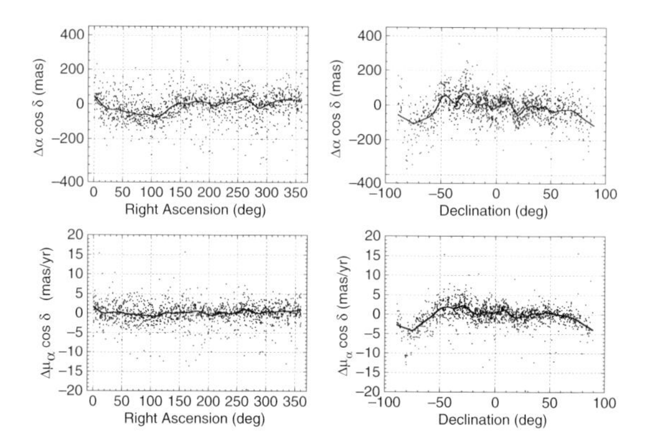

Figure 1.12 Top: differences in right ascension between the FK5 and Hipparcos Catalogues at epoch J1991.25, in the sense FK5–Hipparcos. The solid line results from a robust smoothing of the data. Bottom: differences in proper motion in right ascension. Plots of differences in declination show somewhat similar features. From ESA (1997, Volume 3, Figures 19.3 and 19.5).

The analysis was extended by Mignard & Froeschlé (2000). They decomposed the position and proper motion differences as a set of orthogonal vector harmonics, the first degree representing a pure rotation. No convenient analytic representation of the higher-order terms was found, and they gave instead zonal corrections over the sky in tabular form. Although the overall rotation is rather small, local deviations reach 100 mas in position and 2.5 mas yr^{-1} in proper motion.

A further comparison of the two proper motion systems was carried out by Zhu & Yang (1999) and Zhu (2000, 2003), who extended the analysis to include the PPM and ACRS catalogues (analysis with respect to PPM was also presented in Volume 3 of the published catalogue), each containing about 320 000 stars, reduced to the FK5 system, and thus providing higher stellar densities and fainter magnitudes for the comparisons. Known multiple systems were excluded. Examples for the case of PPM are shown in Figures 1.14 and 1.15 below. Unrecognised systematic terms in position or colour index propagate into, for example, incorrect estimates of Galactic rotation as a function of spectral class when based on these catalogues.

A difference between the two proper motion systems implies that one frame is rotating at constant angular rate with respect to the other. This can be interpreted as a correction to the precession constant, and to a combination of a non-precessional motion of the equinox and/or a correction to planetary precession (see also Section 10.4). Errors in these would lead to spin components of the form (Fricke, 1977a,b; Zhu & Yang, 1999; Mignard & Froeschlé, 2000)

$$\omega_{x,y,z} = 0, \Delta p \sin\epsilon, -\Delta p \cos\epsilon + (\Delta e + \Delta\lambda) \qquad (1.26)$$

where Δp is the correction to the 1976 IAU value of the precession used for the FK5, Δe is the 'fictitious' non-precessional motion of the equinox of FK5, $\Delta\lambda$ is a correction to planetary precession (likely to be negligible), and ϵ is the obliquity of the ecliptic. The rotational components derived by Mignard & Froeschlé (2000), similar to those derived by Zhu & Yang (1999, also subdivided by spectral type), yield $\Delta p = -1.5 \pm 0.7\,\mathrm{mas\,yr^{-1}}$ and $\Delta e + \Delta\lambda = -2.1 \pm 0.7\,\mathrm{mas\,yr^{-1}}$. A larger correction of $\Delta p = -3.0 \pm 0.2\,\mathrm{mas\,yr^{-1}}$ was derived, independently of Hipparcos data, by Charlot *et al.* (1995) from 16 years of VLBI, and by Chapront *et al.* (1999) from 24 years of lunar laser ranging observations. This value implies $\omega_x = 0$, $\omega_y = +0.43 \pm 0.10$, $\omega_z = +0.88 \pm 0.10\,\mathrm{mas\,yr^{-1}}$.

Both Mignard & Froeschlé (2000) and Zhu (2000) concluded that the FK5 proper motion system is non-rigid, with differential rotation depending on position on the sky, magnitude, and colour index. Walter & Hering (2005) also concluded that the Hipparcos proper motions are essentially free of unmodelled rotations, confirming that the Hipparcos frame is inertial at the accuracy level of the proper motion link to the ICRF, perhaps with a minor exception in the direction of the z axis. The estimated precessional corrections are then accounted for by small changes in the FK5 proper motions of the order of their errors. Similar rotational and zonal effects are also seen in the PPM and ACRS which were tied to the FK5 (Table 1.8), while the SPM 2.0, tied to an inertial system defined by faint galaxies, shows much smaller systematics and regional errors (Zhu & Yang, 1999). Studies by Zhu (2007), showing that various subsamples of proper motions from the PPM and ACRS Catalogues do not give consistent values of the precession correction when

Table 1.8 Components of the global rotation derived from FK5, PPM, ACRS, and SPM 2.0, relative to the Hipparcos proper motion system: (1) from Mignard & Froeschlé (2000), updated from ESA (1997, Volume 3); (2) from Schwan (2001a); (3) from Walter & Hering (2005), who also subdivide the data into d > 100 pc and d > 200 pc subsets; (4) from Zhu (2000, Table 1), binaries excluded.

	FK5–Hip[1]	FK5–Hip[2]	FK5–Hip[3]	PPM–Hip[4]	ACRS–Hip[4]	SPM–Hip[4]
Number of stars	1233	1151	1151	< 109 145	< 96 191	9386
ω_x	-0.30 ± 0.10	-0.34 ± 0.08	-0.30 ± 0.07	-0.67 ± 0.03	-0.42 ± 0.10	-0.10 ± 0.17
ω_y	$+0.60 \pm 0.10$	$+0.74 \pm 0.08$	$+0.65 \pm 0.07$	$+0.84 \pm 0.03$	$+0.56 \pm 0.10$	-0.48 ± 0.14
ω_z	$+0.70 \pm 0.10$	$+0.89 \pm 0.08$	$+0.77 \pm 0.08$	$+0.18 \pm 0.03$	-0.08 ± 0.10	$+0.17 \pm 0.15$

compared with Hipparcos, are discussed further in Section 10.4.

Stone (1997) made an independent check of the systematics for 689 FK5 stars observed with the Flagstaff Astrometric Scanning Transit Telescope, confirming the Hipparcos results. Schwan (2001a) made a comparison at the mean epoch of the FK5, 1949.4, rather than at the Hipparcos mean catalogue epoch, thus minimising the effect of the FK5 proper motions on the positions, and thus better separating the positional and proper motion systems. Similar results were found, and analytical representations of the systematic differences were provided. A discussion of parametric and non-parametric regression methods to evaluate the Hipparcos–FK5 residuals was made by Marco *et al.* (2004). Studies of the systematics continue to be reported (Sekowski, 2006; Martínez Usó *et al.*, 2006; Zhu, 2006).

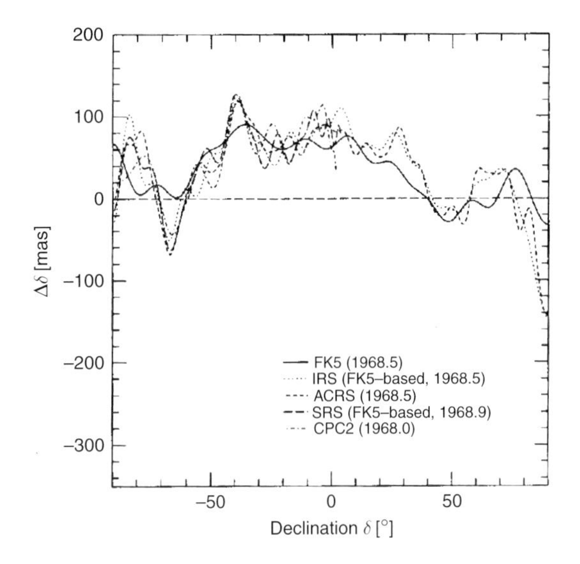

Figure 1.13 Example of the systematic differences between Hipparcos and earlier ground-based catalogues published in the FK5 system, in this case for $\Delta\delta$ as a function of δ, the comparison epoch being indicated. From Schwan (2002, Figure 2).

Other fundamental catalogues Comparisons between Hipparcos and other major fundamental catalogues of the twentieth century have been undertaken by Schwan (2001b, 2002). Since these major catalogues were generally adjusted to the fundamental catalogues FK5 (or to its predecessors FK4, FK3, etc.) their systematics largely reflects those of the associated fundamental catalogue. Comparison involved the transition from each catalogue system to the IAU (1976) system (including elimination of e-terms of aberration, precessional corrections, equinox correction, change from tropical to Julian century, and transformation to J2000), global rotation of each catalogue onto the Hipparcos system, and the determination of regional errors depending on the right ascension, declination and apparent magnitude. Various plots indicating the various systematic terms and the improvement of systematic accuracies achieved over the past 100 years are included (Figure 1.13). Schwan (2001b) treated the FK5, FK4, FK3, NFK, PGC, GC, N30, PPM, and Perth 70 catalogues; Schwan (2002) continued the analysis for the AGK3R and AGK3, SRS, CPC2, IRS, ACRS, AGK2A, AGK2, and SAO catalogues.

1.11 Catalogue combinations to reveal long-period binaries

A number of studies have compared individual Hipparcos proper motions with high-accuracy estimates from other sources, with a view to identifying unrecognised binaries from discordant proper motion estimates. The basic idea is that the Hipparcos proper motion, while extremely accurate, is a representation of the photocentric motion over a relatively short time, i.e. the 3.3-year mission duration. For some long-period binary stars, this 'instantaneous' proper motion can differ significantly from the long-term average photocentric motion determined from ground-based observations extending over several decades. Thus two proper motion estimates with comparable formal accuracy may differ by many times their combined standard errors (Figure 1.16). Wielen *et al.* (1997, 1999d) described the phenomenon

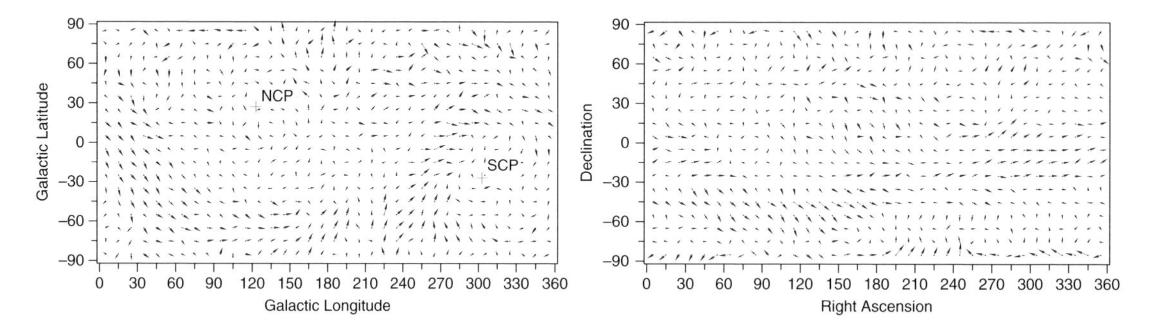

Figure 1.14 Regional proper motions differences PPM–Hipparcos, in Galactic (left) and equatorial coordinates (right). At left, NCP and SCP indicate the north and south celestial poles. From Zhu (2000, Figure 1).

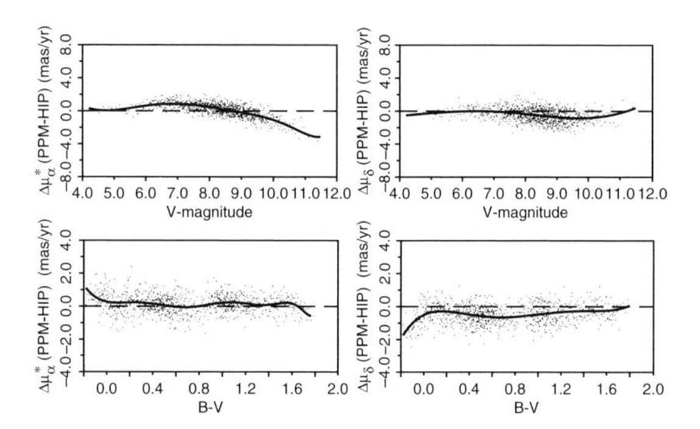

Figure 1.15 Proper motion differences PPM–Hipparcos, as a function of Hipparcos V_T (top) and Hipparcos $B-V$ (bottom). From Zhu (2000, Figures 6 and 7).

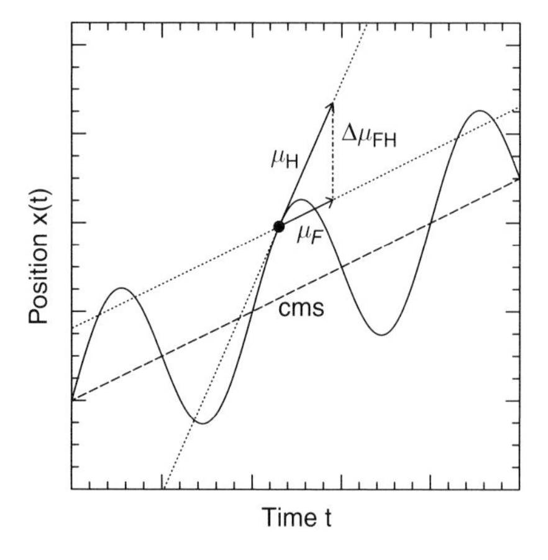

Figure 1.16 The periodic motion of an astrometric binary leads to an observable difference $\Delta\mu_{\rm FH}$ between the instantaneously measured Hipparcos proper motion $\mu_{\rm H}$ and the mean proper motion $\mu_{\rm F}$ of the photocentre. Here the orbital period of the binary is of medium length (~30 years), so that the proper motion $\mu_{\rm F}$, obtained from ground-based data (e.g. from FK5), is essentially equal to the proper motion of the centre-of-mass (cms) of the binary. From Wielen et al. (1999a, Figure 1).

as introducing 'cosmic errors' into high-precision astrometry, although the same term was used with a slightly different meaning in the context of the stochastic variables by Lindegren (1997) and Lindegren *et al.* (1997). Wielen (1997) presented methods for calculating the mean errors of stellar positions predicted from measured instantaneous data or of mean data, including this cosmic error, and extended to the comparison of astrometric catalogues.

Several studies have now been made related to this problem. Since the presence of such discordant proper motions leads to an identification of suspect or erroneous proper motions in the Hipparcos Catalogue, as well as possible new binary systems, this topic will be covered both here, in which the emphasis will be placed on the implications for proper motions, and in Section 3.4.2, where the emphasis will be placed on the detection of binary and multiple systems, the so-called $\Delta\mu$ binaries. The lists of objects whose long-term proper motions are considered suspect from this analysis corresponds to the number of possible binary detections. This is given in Table 3.2 under the category 'single stars: long-term motions'.

In this spirit, the FK6, the Sixth Catalogue of Fundamental Stars (Wielen *et al.*, 1999c), was constructed by combining the Hipparcos results with the FK5 ground-based data, measured over more than two centuries. FK6 Part I contains 878 basic fundamental stars with direct solutions, appropriate for single stars or objects which can be treated like single stars. Those classified as 'astrometrically excellent' have instantaneous and time-averaged proper motions in good agreement, and are thus best-suited to high-precision astrometry. Those with discrepant proper motions are classified as $\Delta\mu$ binaries (Wielen *et al.*, 1999a). The resulting catalogue provides the single-star mode solution (described by Wielen *et al.*, 1999b), as well as long-term and short-term prediction solutions (described by Wielen *et al.*, 2001a), depending on the epoch difference. Table 1.9

Table 1.9 Single stars and binaries resulting from the Hipparcos proper motions combined with various catalogues analysed in Heidelberg. For each combination catalogue, the table shows the number of stars classified as astrometrically excellent and solved for in single-star (SI) mode, with their resulting mean proper motions, compared with those from Hipparcos. For the $\Delta\mu$ binaries, the mean proper motion error from the long-term prediction mode is given, again compared with those from Hipparcos, in both cases taking 'cosmic errors' into account. Proper motions are in mas yr^{-1}. ARIHIP represents the best astrometric solutions from the other combined catalogues. The missing 'FK6(II)' was never published.

		Single-star mode			$\Delta\mu$ binaries		
Catalogue	Stars	Number	μ(SI)	μ(Hip)	Number	μ(LTP)	μ(Hip)
FK6(I)	878	340	0.35	0.67	199	0.50	2.21
FK6(III)	3272	1928	0.59	0.79	354	0.93	1.83
GC+HIP	20 069	14 234	0.66	0.76	1401	1.16	1.87
TYC2+HIP	89 908	72 943	0.83	1.08	2793	1.06	1.86
ARIHIP	90 842	73 023	0.89	1.13	2827	1.15	1.89

gives the typical mean proper motion errors in single-star mode, compared with those from Hipparcos, and similarly for the $\Delta\mu$ binaries in long-term prediction mode in which cosmic errors are taken into account.

Similar exercises were conducted in Heidelberg for a number of catalogue combinations, also summarised in Table 1.9. This involved the FK6 Part III (Wielen *et al.*, 2000b), which contains 3272 additional fundamental stars with high-quality ground-based data (735 from the bright extension of FK5; 1805 from the faint extension; and 732 from the remaining supplementary stars); for the astrometric catalogue GC+HIP, from a combination of the Boss General Catalogue with the Hipparcos Catalogue (Wielen *et al.*, 2001c); for TYC2+HIP from a combination of the Hipparcos Catalogue with the proper motions given in the Tycho 2 Catalogue (Wielen *et al.*, 2001d); and ARIHIP representing the best astrometric solutions from these combined catalogues: FK6, GC+HIP, and TYC2+HIP (Wielen *et al.*, 2001b). A finding list of about 4000 of the resulting $\Delta\mu$ binaries was given by Wielen *et al.* (2000a).

Odenkirchen & Brosche (1999) studied $\Delta\mu$ binaries resulting from a comparison between Hipparcos and two declination zones of the Astrographic Catalogue. From a sample of 12 000 stars they found statistical evidence for 360 astrometric binaries in the investigated zones, corresponding to about 2400 such binaries in the entire Hipparcos Catalogue, in addition to those already known. They used Monte Carlo studies to argue that the results are consistent with a binary frequency of between 70–100%.

A similar investigation was undertaken by Gontcharov *et al.* (2000, 2001). Gontcharov *et al.* (2000) considered 4638 stars in common to Hipparcos and the FK5/FK5 Extension, using 45 observational ground-based catalogues. They derived long-term proper motions, and identified a few hundred stars which showed significant nonlinear motion of the photocentres, the method being sensitive to amplitudes larger than 0.15 arcsec and periods in the range 10–100 yr. Good agreement with previously-known orbits was reported, and results for five astrometric binaries were discussed in detail. Some examples are given in Figure 3.7.

Gontcharov *et al.* (2001) constructed the PMFS Catalogue (Proper Motions of Fundamental Stars) containing long-term proper motion, over decades, of the 1535 basic FK5 stars from a combination of the Hipparcos Catalogue with six compiled and 57 observational ground-based astrometric catalogues. The proper motions from Hipparcos and ground-based compilations GC, N30, FK5, N70E, CMC9 and KSV2 (Time Service Catalogue 2) reduced to the Hipparcos system were used for the calculation of mean weighted proper motions as initial values for the first iteration of the main procedure. The nonlinear motions of 134 stars were directly separated into their proper motions of the barycentres and periodical motions of the photocentres, and the proper motions of some other 200 stars were separated from their nonlinear motions which were implicit in large differences between the PMFS and Hipparcos proper motions of the stars. The stars were classified into 760 single, 187 astrometrically wide pairs, 35 astrometrically close pairs for which the results are doubtful and 551 stellar systems with 553 brighter components in the basic FK5 for which the astrometrically observed photocentre moves, or can move, nonlinearly with an amplitude >1 mas, including almost all known visual, photocentric and spectroscopic orbital pairs. The median precision of the proper motion components is 0.5 mas yr^{-1} for proper motion in α and 0.7 mas yr^{-1} in δ.

Kaplan & Snell (2001) compared proper motions for 162 wide visual double systems with extensive measurements from the Washington Double Star Catalogue, for which nonlinear orbital motion should be negligible, to show that the mean proper motion errors from Tycho 2 are indeed at the level of about 2.5 mas yr^{-1}.

Makarov & Kaplan (2005) made an evaluation of the $\Delta\mu$ binaries resulting from a comparison of the Hipparcos and Tycho 2 proper motions, and the Hipparcos Catalogue DMSA Part G binaries, with the specific aim of estimating statistical bounds for the masses of the secondary components. Their results are discussed further in Section 3.4.2.

Frankowski *et al.* (2007) made a further analysis in parallel with the work of Makarov & Kaplan (2005). They used a χ^2 test evaluating the statistical difference between the Hipparcos and Tycho 2 proper motions for 103 134 stars in common between them. The work focuses on the detection efficiency of proper motion binaries, and is also described further in Section 3.4.2.

1.12 Reference frame studies: Optical, radio and infrared

Observations made subsequently to the Hipparcos Catalogue publication have been undertaken to confirm, maintain, or improve the link established in its construction, and have generally confirmed the fidelity of this link using longer temporal baselines, larger numbers of stars, or independent analyses.

Optical observations around Galactic radio stars In the optical, reported work has generally supported the published link between the Hipparcos frame and the ICRF. Studies have included meridian circle observations around 200 IERS sources (Kovalchuk *et al.*, 1997); observations of radio stars with astrolabes at Shanghai (Tongqi *et al.*, 1997) and Yunnan (Hu *et al.*, 1999; Hu & Wang, 2002); 689 Hipparcos stars observed by the Flagstaff Astrometric Scanning Transit Telescope confirming the orientation terms with $\epsilon_{x,y,z} = -2.2 \pm 3.3, -2.2 \pm 3.3, 3.3 \pm 2.9$ mas at epoch 1996.5 (Stone, 1998); observations of radio stars at the Valinhos CCD meridian circle (Lopes *et al.*, 1999; Assafin *et al.*, 2003a); observations of 69 radio stars with the Beijing photoelectric astrolabe at San Juan (Manrique *et al.*, 1998, 1999); positions of 315 ICRF sources from Digitised Sky Survey images (da Silva Neto *et al.*, 2000); photographic observations of 113 radio stars observed at the Engelhardt observatory (Rizvanov *et al.*, 2001); observations of 80 stars from 350 photographic plates from Torino and elsewhere (Lattanzi *et al.*, 2001); observations with the OCA photoelectric astrolabe (Martin *et al.*, 2003); positions of UCAC reference stars around 172 ICRF sources yielding an upper limit on the z-axis rotation of 0.7 mas yr^{-1} (Assafin *et al.*, 2003b; Zacharias & Zacharias, 2005); positions within 330 fields based on digitised POSS-I and POSS-II plates (Fedorov & Myznikov, 2005); MERLIN astrometry of 11 radio stars (Fey *et al.*, 2006); and VLA linked with the Pie Town VLBA observations of 46 radio stars (Boboltz *et al.*, 2007).

A non-negligible rotation of the Hipparcos proper motion system was nevertheless hinted at by Zhu (2001) based on a comparison with the SPM 2.0; by Andrei *et al.* (2003) from the Valinhos meridian circle observations; and by Bobylev (2004a,b); Bobylev & Khovirtchev (2006). The latter presented evidence for a small rotation around the Galactic y axis, of -0.37 ± 0.04 mas yr^{-1}, from proper motions of stars beyond 900 pc in the Tycho 2 and UCAC2 Catalogues, inferred from an Ogorodnikov–Milne model analysis (see Section 9.2.7).

Optical observations around extragalactic radio sources Optical observations around extragalactic radio sources, typically using a two-step technique to bridge the brighter magnitudes of the Hipparcos Catalogue, have been variously described: for 63 fields observed annually by the Carlsberg meridian circle, confirming the FK5 systematics (Argyle *et al.*, 1996); for 25–38 faint (B=20−24 mag) optical counterparts of compact extragalactic radio sources (Costa & Loyola, 1998, 1999; Costa, 2001, 2002); for 327 extragalactic sources using the Hamburg and USNO Black Birch astrographs (Zacharias *et al.*, 1999); for 23 extragalactic sources (Tang *et al.*, 2000); and for 192 fields using the Kyiv meridian axial circle (Babenko *et al.*, 2005).

Radio observations In the radio, reported work on the reference frame specifically related to the Hipparcos link includes observations of radio stars with the VLA (Lenhardt *et al.*, 1997; Boboltz *et al.*, 2003; Johnston *et al.*, 2003), and with MERLIN (Morrison *et al.*, 1997a); issues related to the transformation of early-epoch radio positions (Hering & Walter, 1998; Hering *et al.*, 1998); and a comparison of optical and radio positions of radio stars (Andrei *et al.*, 1999). As exploited during the Hipparcos reference frame link, high-precision VLBI astrometry of radio stars is of particular importance, since it provides absolute astrometry relative to the distant quasars used as VLBI phase reference calibrators. The technique yields a mean astrometric precision of 0.36 mas, reaching 0.12 mas in position, 0.05 mas in annual proper motion, and 0.10 mas in trigonometric parallax in the case of the RS CVn binary σ^2 CrB (Lestrade *et al.*, 1999). These authors also describe some associated scientific results for the sources observed.

Links to the dynamical frame The relationship between the Hipparcos reference frame and the JPL ephemerides DE403/405 have also been investigated directly, making use of the fact that these planetary ephemerides, unlike their predecessor DE200, are themselves linked to the ICRS. Krasinsky (1997) investigated the use of photoelectric observations of occultations of stars by the inner planets, thus tying Hipparcos with the frame of the DE ephemerides, and thus with the VLBI radio frame. They used 16 published timings of contacts made during the occultation of the bright Hipparcos star ϵ Gem by Mars in 1976, suggesting that the Hipparcos system needs corrections of up to 60 mas. These preliminary findings have not been confirmed.

Morrison *et al.* (1997b) used Hipparcos observations of the positions of Europa (J2) and Titan (S6), and Tycho observations of Ganymede (J3) and Callisto (J4), to investigate the JPL ephemerides of the planets Jupiter and Saturn (their reference to DExxx was to an unnumbered release, presumably close to DE403/DE405). The observed satellite positions were compared with DExxx using appropriate positions of the Galilean satellites (Arlot, 1982) and of Titan (Taylor & Shen, 1988) to calculate their offsets from the barycentres of the two planetary systems. The Hipparcos observations of J2, and the (two-dimensional) Tycho observations of J3 and J4, put tight constraints on the orbit of Jupiter, and agree closely with the series of ground-based observations made by the Carlsberg meridian telescope. Their results confirm independently the superiority of DE405 over DE200 for the ephemerides of both Jupiter and Saturn (Figure 1.17).

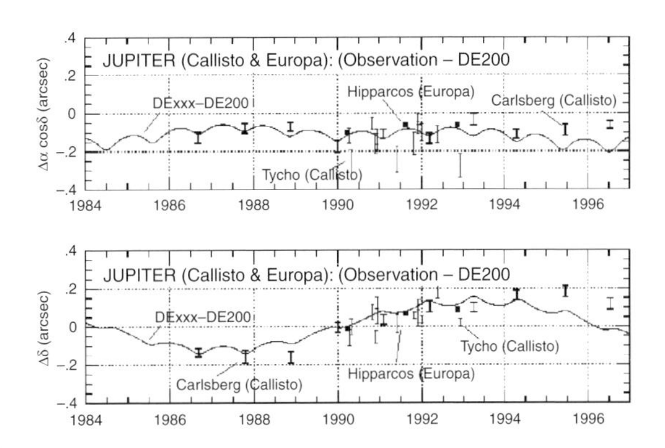

Figure 1.17 Comparison of the observed positions of Jupiter with DE200 and DExxx in α (top) and δ (bottom). Filled points: Hipparcos normal points from observations of Europa; light error bars: individual Tycho positions of Callisto; heavy error bars: mean opposition measurements from the Carlsberg telescope. From Morrison et al. (1997b, Figure 3).

Extension to fainter optical magnitudes Observations to extend and verify the Hipparcos materialisation of the ICRF to fainter optical magnitudes have been variously described: from observations at the Pulkovo Observatory extending back to 1894 (Kumkova, 2000); from observations to 16 mag with the Bordeaux and Valinhos meridian circles (Camargo *et al.*, 2003); and from more global programmes based on the UCAC, USNO B and GSC 2.3 (Urban, 2003).

Extension to the infrared This is discussed in Section 2.8.

Scientific applications of the optical–radio link Accurate linking of the optical and radio reference frames at accuracies better than 0.1 arcsec for the first time, has provided various insights into the relationship between optical and radio emission for a number of specific sources. The following are some examples. For SN 1987A, Reynolds *et al.* (1995) showed that the centres of the optical and radio emission coincide more closely than previously thought, indicating that the radio emission is due to interaction between an expanding shock front and the surrounding medium. Contreras *et al.* (1997) showed that for the unresolved contact binary Cyg OB2 No. 5, the radio and optical positions of the binary coincide, while a fainter radio component appears to result from synchrotron emission produced at the shock interaction zone between the stellar winds.

Observations of the OH 1667 MHz maser in the circumstellar shell around the Mira variable U Her were made with the VLBA over four years by van Langevelde *et al.* (2000). Using phase referencing techniques, the position of the most blue-shifted maser spot was compared with the stellar optical position measured by Hipparcos to 15 mas accuracy, confirming a model in which one of the maser spots corresponds to the stellar continuum, amplified by the maser. Vlemmings *et al.* (2002) observed the 22 GHz H_2O masers in the same object with MERLIN. The radio position of the brightest H_2O maser spot coincides with the optical position, indicating that this spot is the stellar image amplified by the maser screen in front of it, and permitting registration of the associated SiO and OH masers which form a ring around the star. Vlemmings *et al.* (2003) observed the main-line OH masers around four asymptotic giant branch stars with the VLBA over 2.5 years, yielding accurate proper motions and parallaxes for U Her, R Cas and S CrB, while additional motion of the compact blue-shifted maser of W Hya is possibly related to the stellar pulsation. From Hipparcos astrometry, the most blue-shifted maser is consistent with the amplified stellar image in the case of U Her and W Hya. Velázquez & Rodríguez (2001) studied the thermal jet associated with the young star Z CMa with the VLA, finding that the position angle of the axis of the thermal jet is coincident with the orientation of the optical jet and the orientation of the CO bipolar outflow, and using Hipparcos astrometry to establish that the jet originates from the optical component of the binary. Johnston *et al.* (2004) observed the T Tau multiple star system using MERLIN, and showed that the most intense compact radio emission may be associated with the M-type star in the binary system T Tau S, and originating from magnetic reconnection processes. Other work uses the Hipparcos parallaxes to discuss energetics of the radio emission (e.g. Stine & O'Neal, 1998).

1.13 Radial velocities

1.13.1 Data to complement the Hipparcos Catalogue

Amongst the scientific objectives identified in the ESA Phase A study (ESA, 1979), contributions to the stellar reference frame and improved distances and luminosities attracted most attention. Radial velocities as a crucial observational element (see, e.g. Latham, 2000, for an introduction to the field) began to be discussed around the same time, especially in France and Switzerland, but funding authorities were not willing to support dedicated facilities (see box on page 33). The Phase A study report (ESA, 1979, pp19–20) made specific reference to the importance of radial velocities, and the compilation of measurements was duly identified as a task for the Input Catalogue construction, as specified in the INCA Consortium proposal to ESA in 1982. With the solicitation for observing proposals on which the Hipparcos Input Catalogue would eventually be based, and the attendant activities of the Hipparcos Input Catalogue Consortium and the Proposal Selection Committee, the breadth of the Hipparcos scientific case became more evident, and the relevance of radial velocity measurements began to be discussed more widely (e.g. Fehrenbach, 1985).

Specific radial velocity observations for a subset of the Hipparcos Catalogue stars were eventually made separately for early- and late-type stars, competing for existing telescope time and instrumentation in the northern and southern hemispheres, as follows, with resulting publications summarised in Table 1.10:

(a) For early-type stars in the northern hemisphere, Charles Fehrenbach began a radial velocities survey for Hipparcos candidate stars in around 1982 (see Grenier *et al.*, 1999a). Three existing instruments were used: the Objective Prism on the Schmidt telescope (SPO, field $4 \times 4\,\mathrm{deg}^2$, for regions rich in early-type stars with $B < 12$); the Petit Prisme Objectif (PPO, field $4 \times 4\,\mathrm{deg}^2$, for regions rich in stars with $B < 8.5$); and the Marly spectrograph on the 120-cm telescope of the Observatoire de Haute Provence for isolated stars up to $B = 9$. At that time the Input Catalogue had not been fully defined, and the radial velocity observing programme was based on a very preliminary version. The aim was to target early-type stars in the range B5–F5, i.e. not observable by Coravel, with priority given to Bp–Ap and A-type stars belonging to the Hipparcos survey, taking into account radial velocities already published. Standard errors were about $3\,\mathrm{km\,s^{-1}}$. By the end of the programme, a total of 62% of the 12 286 B5–F5 stars in the northern part of the Hipparcos survey had radial velocities: 33% already had an acceptable published radial velocity, 2.6% were obtained using the SPO, 8.4% using the PPO, and 18% using the Marly spectrograph. The associated publications are listed in Table 1.10.

(b) For early-type stars in the southern Hemisphere, two ESO key-programmes for B5–F5 stars were started at the end of 1988, using the Echelec Spectrograph at the 1.52 m ESO telescope at La Silla: one for early-type stars nearer than 100 pc (Gerbaldi & Mayor, 1989), and one for early-type stars in OB and early-A associations (Hensberge *et al.*, 1990). The results of the first programme were published in three papers: radial velocities for 581 B8–F2 stars with $V < 8$ and within 100 pc (Grenier *et al.*, 1999b), masses and luminosities for 71 nearby dwarf A0 stars from the same sample (Gerbaldi *et al.*, 1999), and rotational velocities for 525 B8–F2-type stars with $V < 8$ (Royer *et al.*, 2002). Radial velocities of 186 chemically peculiar stars were also obtained with the El Leoncito slit spectrograph in a collaboration between France and Argentina (Levato *et al.*, 1996).

(c) For late-type stars in the northern hemisphere, a programme of Coravel measurements using the 1-m Swiss telescope at OHP had been ongoing since 1977. It was dedicated, independently of Hipparcos, to nearby stars, cluster stars, high proper motion stars, Cepheids, and visual binaries, many of which would eventually be included in the Hipparcos Input Catalogue (Mayor, 1985). By October 1990, 7000 F5–M stars with $B < 14$ had been observed, including about a quarter of the Hipparcos programme stars within these limits. By November 1993, a programme more focused on Hipparcos was being undertaken, to include Hipparcos 'survey' stars, and various Galactic studies (list A) which had been accepted for inclusion in the Hipparcos Input Catalogue, and proposed by individuals associated with the Coravel-OHP observations (involving Mayor, Duquennoy, Grenon, Mermilliod, Turon, Prévot, and Udry). This programme included 22 358 stars, and more than half of them had already been observed at least twice. A secondary list of Galactic programmes accepted for the Hipparcos Input Catalogue but proposed by other people (list B) included 14 234 stars, and was observed with lower priority.

(d) For late-type stars in the southern hemisphere, an ESO key-programme was accepted in January 1989 for approximately 20 000 stars later than F5, using Coravel-South on the 1.54-m Danish telescope at La Silla (Gerbaldi & Mayor, 1989), involving Mayor, Duquennoy, Grenon, Imbert, Maurice, Prévot, Andersen, Nordström, Turon, Crifo, and Udry. The overall observing programme was constructed along similar lines as for the northern part, and comprised Hipparcos 'survey' stars and list A and B Galactic study programmes, all with two observations per star. The number of stars with spectral types later than F5 in the southern programme were 13 895 for the survey, 9173 for list A, and 15 654

Steps in the acquisition of radial velocities for Hipparcos stars: Many papers using the Hipparcos data have drawn attention to the absence of radial velocities for many Hipparcos stars, and the fact that radial velocities were not acquired and published along with the final astrometric and photometric catalogues. The following summary, from unpublished notes from Catherine Turon and Michael Perryman, underlines how much work was actually invested in systematic attempts to acquire them.

Already in around 1979, Catherine Turon made an estimation of the total number of radial velocity measurements that would be required to support the Hipparcos programme. In 1980, a proposal was made by A. Bijaoui to INAG in France (the predecessor of INSU) for a new dedicated telescope and associated instrumentation, Coravel-North, to be installed at the Observatoire de Haute Provence (OHP). It would improve the existing Coravel limiting magnitude from $B < 14$ to 15.5, and would extend observations to A–F-type stars using adaptable spectral template 'masks'. At that time, the number of Hipparcos stars with no reliable radial velocity was estimated to be some 34 000 out of the expected 60 000 survey stars, and some 25 000 out of 40 000 non-survey stars with $B = 9$–13, corresponding to around 6 years of observation with a dedicated telescope assuming 3 observations per star. The cost of the project was estimated at around 5 MFrancs. This proposal, and an updated proposal in September 1981, were both unsuccessful.

In January 1982, a meeting in Marseille between Marcel Golay, Michel Mayor and Fredy Rufener from Geneva Observatory, James Lequeux, Marc Azzopardi, Louis Prévot and Yvon Georgelin from Marseille Observatory, Renaud Foy and Guy Monnet from Lyon Observatory, Albert Bijaoui from Nice Observatory, and Suzanne Grenier and Catherine Turon from Paris Observatory, continued to emphasise the importance of radial velocities being available at the time of the Hipparcos Catalogue publication, estimated then as around 1991. They decreased the acceptable number of observations per star to two, and formulated a plan for a southern hemisphere programme of 12 000 observations per year for 5–6 years, and a northern hemisphere programme of 6000 observations per year. Given that a number of observations had already been made with the OHP Coravel, and that further northern hemisphere observations could start immediately, both parts of the programme were considered achievable within 10 years. A new southern hemisphere 1.5-m telescope was proposed as a collaboration between ESO, Geneva, and INAG, with a cost of 4 MFrancs. The proposal, presented to ESO by James Lequeux and Michel Mayor in May 1982, was also unsuccessful. Nevertheless, these repeated requests probably assisted the decision to build the Elodie spectrograph for the 1.93-m telescope at OHP.

In 1987, two-years before the satellite launch, Roger Griffin, of the Institute of Astronomy, Cambridge, took up the challenge and submitted a proposal to the UK astronomical funding body, the SERC (the forerunner of PPARC, now STFC) to acquire radial velocities for the Hipparcos stars and, once these targets were exhausted, the Tycho stars. The proposal called for two identical 1-m telescopes, one in the north and one in the south, at a total cost of about £1m. As he stated, *'The Hipparcos results will be made infinitely more valuable by the provision of radial velocities to complement the transverse motions measured by the satellite.'* The proposal was supported by various members of the Hipparcos teams, including Adriaan Blaauw as chair of the Hipparcos Proposal Selection Committee, Catherine Turon as chair of the INCA Consortium, and Michael Perryman as ESA Project Scientist. However, the scientific case did not generate sufficient support within the SERC. ESA Space Science Department management also declined to consider funding a dedicated ground-based radial velocity programme for Hipparcos on the grounds that ESA had no mandate to fund ground-based astronomy.

In the event, various northern hemisphere programmes were undertaken with existing instrumentation, while southern hemisphere observations were undertaken as part of new ESO 'key programme' initiatives. Nevertheless, the systematic acquisition of radial velocities for all (possible) Hipparcos Catalogue objects, in time for inclusion in the published catalogue, was a vision which, while articulated by many, never came to fruition. This has clearly resulted in a significant limitation to the more complete and timely exploitation of the Hipparcos Catalogue data.

For ESA's Gaia mission, the importance of acquiring large-scale radial velocity data from the same space platform as that of the astrometric measurements has been fully appreciated, and a radial velocity instrument included from the start (Perryman, 2003; Munari, 2003; Katz *et al.*, 2004; Wilkinson *et al.*, 2005). Important applications include the kinematic and dynamical interpretation of the astrometric data by providing the remaining component of the six-dimensional space and velocity vector; the identification and correction for perspective acceleration; the identification of binary systems; the contribution to large-scale binary orbit determination by providing constraints on eccentricity (Pourbaix, 2002); large-scale determination of stellar rotation; and the acquisition of spectral diagnostics in the region of the adopted Ca-triplet lines for detailed physical diagnostics.

A large-scale ground-based 'counterpart' to this is the Radial Velocity Experiment (RAVE) using the Six Degree Field multiobject spectrograph on the 1.2-m UK Schmidt telescope of the Anglo–Australian Observatory. The first data release, comprising 24 748 stars drawn from the Tycho 2 and SuperCOSMOS Catalogues, has now been published (Steinmetz *et al.*, 2006).

for list B. By November 1993, 150 nights of ESO time and 150 nights of Danish time had been utilised, and by March 1994 almost all of the 23 000 stars of the core programme had been observed more than once, while about 60% had been observed more than twice. Observations finally totaling some 200 ESO and 200 Danish telescope nights continued until about mid-1995.

A status report on the Coravel measurements in 1997 indicated that more than 140 000 radial velocity measurements had been obtained for some 40 000 Hipparcos late-type stars (Udry *et al.*, 1997), the cross-correlation functions also providing rotational velocities and metallicities. The radial velocity datasets published by Nordström *et al.* (2004) and Famaey *et al.* (2005)

*Table 1.10 Radial velocity catalogues, compilations, or determinations, associated with Hipparcos stars. It includes significant datasets from publications identified during this review, and liable to be rather incomplete. * are data forming part of the Coravel database, described by Udry et al. (1997). Catalogue entries are not necessarily contained within the Hipparcos Catalogue (those after 1997 certainly are not), and entries are not necessarily mutually exclusive.*

Description	Reference	No. of stars
Hipparcos Input Catalogue:		
General Catalogue of Radial Velocities	Wilson (1953)	9616
Evans Catalogue	Evans (1978)	4866
Barbier-Brossat Catalogue	Barbier-Brossat (1989)	4094
Miscellaneous, in addition to above	Hipparcos Input Catalogue	899
Subsequent compilations:		
WEB: Wilson, Evans, Batten	Duflot *et al.* (1995b)	20 793
WEB updated to 1990	Barbier-Brossat & Figon (2000)	20 574
Stellar radial velocities 1991–98	Malaroda *et al.* (2000)	13 359
CRVAD supplement to the ASCC 2.5	Kharchenko *et al.* (2004)	34 553
Pulkovo Compilation (PCRV)	Gontcharov (2006)	35 495
CRVAD-2 supplement to the ASCC 2.5	Kharchenko *et al.* (2007)	54 907
Subsequent determinations:		
Objective prism (PPO)	Fehrenbach (1985)	734
Marly slit spectrograph	Fehrenbach *et al.* (1987a)	272
Objective prism (PPO)	Fehrenbach *et al.* (1987b)	446
Schmidt objective prism (SPO)	Fehrenbach & Burnage (1990)	391
Objective prism (PPO)	Duflot *et al.* (1990)	396
Objective prism (PPO)	Duflot *et al.* (1992)	1070
Schmidt objective prism (SPO)	Fehrenbach *et al.* (1992)	2601
Objective prism (PPO)	Duflot *et al.* (1995a)	734
Southern hemisphere, CP stars	Levato *et al.* (1996)	186
Objective prism (PPO)	Fehrenbach *et al.* (1997)	1879
Field RR Lyrae	Fernley & Barnes (1997)	56
Field RR Lyrae	Solano *et al.* (1997)	45
Elodie spectral library, F5–K7	Soubiran *et al.* (1998)	211
Northern hemisphere, B2–F5	Grenier *et al.* (1999a), includes Fehrenbach *et al.* (1987a)	2800
Southern hemisphere, B8–F2	Grenier *et al.* (1999b)	581
ROSAT extreme ultraviolet detections	Cutispoto *et al.* (1999)	51
Passage of nearby stars	García Sánchez *et al.* (1999)	77
Population II stars	Bartkevičius & Sperauskas (1999)	164
Evolved FGK stars*	de Medeiros & Mayor (1999)	2000
TW Hya association region	Zuckerman & Webb (2000)	37
Nearby M stars	Bettoni & Galletta (2001)	22
Population II binaries	Sperauskas & Bartkevičius (2002)	114
FGKM dwarfs and K giants	Nidever *et al.* (2002)	889
Late FG stars	Cutispoto *et al.* (2002)	129
Evolved FGK supergiants*	de Medeiros *et al.* (2002)	231
Overluminous F stars within 80 pc	Griffin & Suchkov (2003)	81
K giants from Tycho 2	Soubiran *et al.* (2003)	387
Perseus OB2 association, early-type	Steenbrugge *et al.* (2003)	29
Evolved double-lined binaries*	de Medeiros *et al.* (2004)	78
Nearby F and G dwarfs*	Nordström *et al.* (2004)	13 500
Nearby K and M giants*	Famaey *et al.* (2005)	6691
Population II binaries	Bartkevičius & Sperauskas (2005)	91
Scorpius–Centaurus association, B stars	Jilinski *et al.* (2006)	56
Evolved metal-poor stars	de Medeiros *et al.* (2006)	2000
Nearby solar-type binaries	Abt & Willmarth (2006)	167
RAVE measurements including Tycho 2 stars	Steinmetz *et al.* (2006)	24 748
Nearby K and M dwarfs	Upgren *et al.* (2006)	475

are part of this substantial (and otherwise unpublished) database. The former include newly-derived effective temperatures, metallicities, and isochrone ages. The latter include newly-computed $V - I$ indices.

As an end result, nearly 20 000 radial velocities were included in the Hipparcos Input Catalogue. The data included at the time of the Input Catalogue publication was rather incomplete, and did not include most radial velocities obtained by cross-correlation methods, which had either not been published, or published only in small lists. Neither did they include the ongoing observations organised in parallel with the preparation of the Input Catalogue. These data are still neither fully published nor, in terms of the overall Hipparcos stellar content, complete. Major published radial velocity datasets associated with the Hipparcos Catalogue, either included in the Input Catalogue, or obtained or compiled subsequently, are listed in Table 1.10 and commented on briefly hereafter. References to radial velocity measurements of Hipparcos/Tycho stars in the literature are probably rather incomplete.

Duflot *et al.* (1995b) gave a common version of the two catalogues of mean radial velocities by Wilson (1953) and Evans (1978) to which they have added the catalogue of spectroscopic binary systems by Batten *et al.* (1989). For each star, when possible, they give HIC/HIP and CCDM numbers. This work was supplemented by Barbier-Brossat & Figon (2000) which contains new mean velocities for 20 574 stars from observations published until December 1990.

Cross-identification between the All-Sky Compiled Catalogue of 2.5 million stars (ASCC 2.5) on the Hipparcos system, and the General Catalogue of Radial Velocities, was reported by Kharchenko *et al.* (2004), including radial velocity determination for 292 open clusters including 97 with previously unknown radial velocities, using membership from proper motions and photometry. A compilation from the literature during 1991–94 is described by Malaroda *et al.* (2000), while the VizieR online data catalogue provides an updated compilation from the literature during 1991–98. Gontcharov (2006) describes the Pulkovo Compilation of Radial Velocities (PCRV) containing mean weighted absolute radial velocities of 35 495 Hipparcos stars of various spectral types and luminosity classes over the whole celestial sphere mainly within the radius of 500 pc from the Sun. It was based on results from 203 publications. The median precision is 0.7 km s^{-1}.

Many other papers include radial velocities of individual and small samples of stars of relevance to Hipparcos stars. Amongst these are the series of photoelectric observations by Roger Griffin, which by the end of 2006 had reached Paper XVII for various source lists (see Griffin, 2006, and references therein), and Paper 191 in the case of individual spectroscopic binary orbits (Griffin & Boffin, 2006).

1.13.2 Astrometric radial velocities

The wavelength (Doppler) shift is usually understood to provide the radial velocity of a star where, conventionally, positive values indicate recession. In practice, the spectroscopic line shift represents the combination of the true velocity of the stellar centre of mass, combined with surface effects such as stellar atmospheric dynamics and gravitational redshifts. These and other aspects of the fundamental definition of radial velocity at the m s^{-1} level, including gravitational and relativistic effects, have been considered by Lindegren & Dravins (2003).

High-accuracy astrometry in principle permits the accurate determination of stellar radial velocities from geometric principles, without using spectroscopy or invoking the Doppler principle. This has two main applications, which have been explored in a series of investigations by the Lund group (Dravins *et al.*, 1997; Gullberg & Dravins, 1998; Dravins *et al.*, 1999; Lindegren *et al.*, 2000; Madsen *et al.*, 2002; Madsen, 2002). First, the method can provide fundamental radial velocity standards amongst the nearby stars. Such standards have hitherto been limited to Solar System objects, in particular asteroids, whose space motions can be derived with very high accuracy without the use of spectroscopic data. Second, a comparison between the astrometric and spectroscopic radial velocities allows certain stellar phenomena to be studied. For the Sun, the segregation of such effects has been possible because the relative Sun–Earth motion is accurately known from planetary system dynamics, and does not have to be deduced from asymmetric and shifted line profiles.

Three ways of determining astrometric radial velocities were described by Dravins *et al.* (1999), and are illustrated in Figure 1.18a–c. The first and most direct method exploits the secular change in trigonometric parallax due to the radial displacement of a star (Figure 1.18a), for which

$$v_r = -A\frac{\dot{\pi}}{\pi^2} \tag{1.27}$$

where A is the astronomical unit. An equivalent form was derived by Schlesinger (1917), who concluded that the parallax change is very small for every known star. Its detection would require extremely accurate measurements over years or decades. For Barnard's star ($\pi = 549$ mas, $V_r = -110$ km s^{-1}), the expected parallax change is 34 microarcsec yr^{-1}, which should become feasible with SIM or Gaia. Dravins *et al.* (1999) estimate the radial velocity precision that can be achieved depending on measurement accuracy and duration, showing that some 60 stars within 5 pc could be

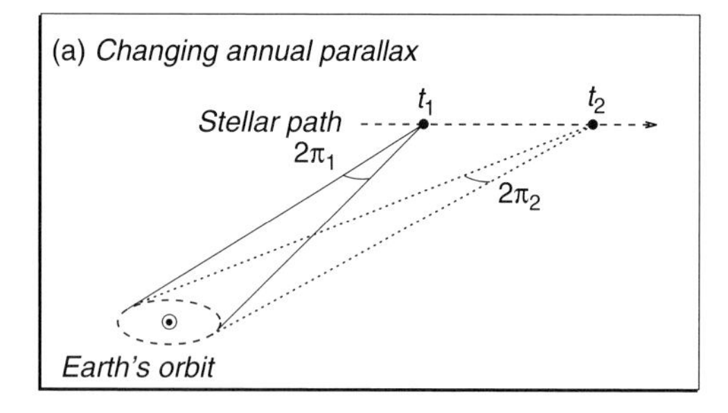

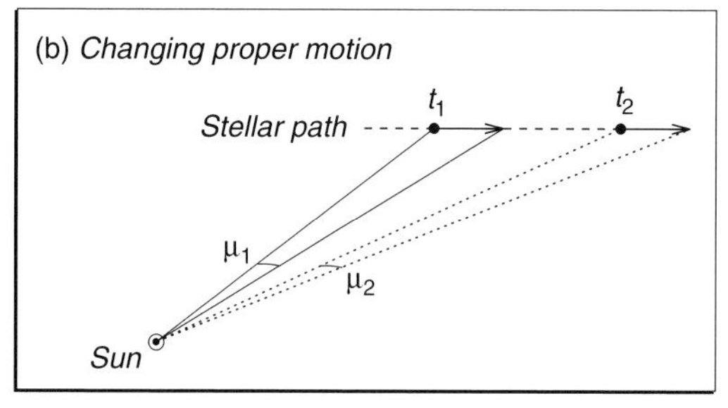

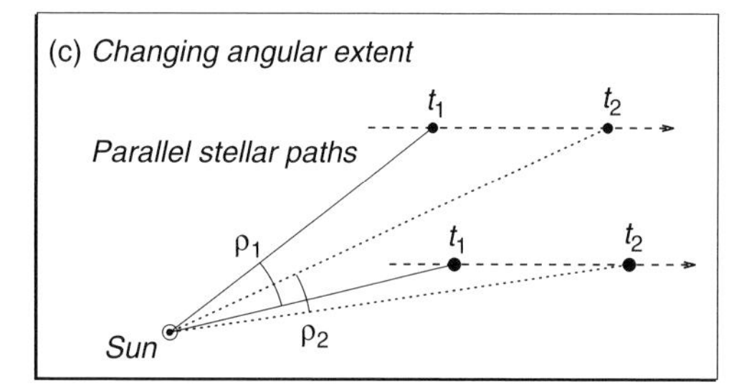

Figure 1.18 Three methods to determine stellar radial motions astrometrically: (a) changing trigonometric parallax π; (b) perspective change in the proper motion μ; (c) changing angular separation ρ of stars sharing the same space velocity, e.g. in a moving cluster. From Dravins et al. (1999, Figure 1).

measured at better than $1\,\text{km}\,\text{s}^{-1}$ from the (highly-demanding) combination of two measurements at 1 microarcsec accuracy separated by 50 years.

The second method exploits the fact that, to a good approximation, a single star moves with uniform linear velocity through space,[2] so that its space velocity can also be derived from the secular change in its proper motion (Figure 1.18b). For a given linear tangential velocity, the angular velocity (or proper motion μ), as seen from the Sun, varies inversely with the distance to the object, b. However, the tangential velocity changes due to the varying angle between the line-of-sight and the space-velocity vector. The two effects combine to produce an apparent (perspective) acceleration of the motion on the sky, or a rate of change in proper motion amounting to $\dot{\mu} = -2\mu v_r/b$ or, with $b = A/\pi$,

$$v_r = -A\frac{\dot{\mu}}{2\pi\mu} \tag{1.28}$$

The history of the application of the method is described by Dravins *et al.* (1999): formulated by Seeliger (1901), it was used by Ristenpart (1902) in an attempt to determine $\dot{\mu}$ observationally for Groombridge 1830, by Lundmark & Luyten (1922) and others for Barnard's star, and proposed as a way of confirming a hypothesised gravitational redshift of several hundred $\text{km}\,\text{s}^{-1}$ for the white dwarf van Maanen 2 (Russell & Atkinson, 1931). Photographic observations over several decades were used to determine the perspective acceleration for Barnard's star (van de Kamp, 1981, with v_r accurate to $\pm 4\,\text{km}\,\text{s}^{-1}$), for van Maanen 2 (Gatewood & Russell, 1974, with v_r accurate to $\pm 15\,\text{km}\,\text{s}^{-1}$ implying a gravitational redshift of $33 \pm 16\,\text{km}\,\text{s}^{-1}$), and for Groombridge 1830 (Beardsley *et al.*, 1974). Dravins *et al.* (1999) combined the Hipparcos results with those of the Astrographic Catalogue (AC 2000) to determine astrometric radial velocities for 16 stars with parallax–proper motion product greater than $0.5\,\text{arcsec}^2\,\text{yr}^{-1}$ (Table 1.11), and compared them with the corresponding spectroscopic values (Figure 1.19).

Figure 1.20, left illustrates the expected variation of radial velocity with time as a result of secular acceleration. In the context of high-precision radial velocity measurements to place limits on the presence of planetary systems (see Section 10.7), Kürster *et al.* (2003, 2006) observed Barnard's star for more than 5 years with UVES on the VLT (Figure 1.20, right). They measured a secular acceleration fully consistent with the predicted value of $4.50\,\text{m}\,\text{s}^{-1}\,\text{yr}^{-1}$ based on the Hipparcos proper motion and parallax combined with the absolute radial velocity of $-110.506\,\text{km}\,\text{s}^{-1}$ from Nidever *et al.* (2002). From differences between the secular and time-dependent radial

[2] The real situation is more complex: the acceleration towards the Galactic centre caused by the smoothed Galactic potential in the vicinity of the Sun is $\simeq 2 \times 10^{-13}\,\text{km}\,\text{s}^{-2}$. For a hypothetical observer near the Sun but unaffected by this acceleration, the maximum bias would be $0.06\,\text{km}\,\text{s}^{-1}$ for Barnard's star, and $0.17\,\text{km}\,\text{s}^{-1}$ for Proxima Centauri. However, since real observations are made relative to the Solar System barycentre, which itself is accelerated in the Galactic gravitational field, the observed (differential) effect will be very much smaller.

*Table 1.11 Astrometric radial velocities (in km s^{-1}), obtained by combining positions and proper motions from Hipparcos (epoch J1991.25) with old position measurements from the Astrographic Catalogue (AC 2000). Spectroscopic radial velocities are also given, from the Hipparcos Input Catalogue. Data for the binary 61 Cyg = HIP 104214+104217 = AC 1382645+1382649 refer to its centre of mass, assuming a mass ratio of 0.90. Two solutions are given: the first based on Hipparcos and AC data alone; the second (marked *) includes Bessel's visual measurements from 1838. From Dravins et al. (1999, Table 3).*

	AC 2000		Radial velocity, v_r		
HIP	No.	Epoch	Astrom.	Spectr.	Star name
439	3152964	1912.956	+7.0 ± 29.7	+22.9	
1475	1406215	1898.435	−40.8 ± 36.1	+12.0	GX And
5336	1721511	1913.868	−106.4 ± 82.7	−98.0	μ Cas
15510	3488626	1901.018	+89.6 ± 59.1	+86.8	82 Eri
19849	2125614	1892.970	−73.2 ± 32.8	−42.6	40 Eri
24186	3505363	1899.058	+249.4 ± 13.7	+245.5	Kapteyn's star
36208	282902	1908.859	−33.7 ± 38.4	+18.7	Luyten's star
54035	1340883	1930.895	+22.0 ± 36.1	−85.6	
54211	1463341	1895.620	+58.6 ± 30.5	+67.5	
57367	4195112	1924.492	+43.1 ± 106.4	–	
57939	1342199	1930.260	−139.5 ± 86.1	−98.0	Groombridge 1830
87937	146626	1905.979	−101.9 ± 6.5	−109.7	Barnard's star
104214/217	1382645/649	1921.699	−12.2 ± 34.5	−64.5	61 Cyg
104214/217	1382645/649	1921.699	−68.0 ± 11.1 *	"	"
105090	3462277	1905.316	−25.0 ± 40.9	+23.6	
108870	4384302	1901.189	−64.1 ± 23.6	−40.0	ϵ Ind
114046	3355101	1913.368	+54.0 ± 19.7	+9.5	

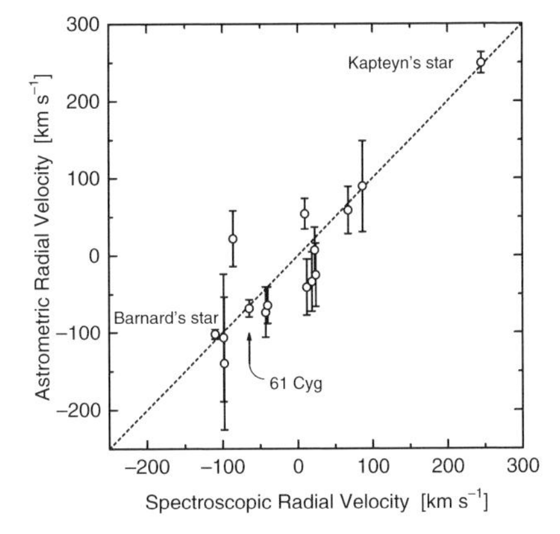

Figure 1.19 Comparison of the astrometric radial velocities in Table 1.11 with their spectroscopic counterparts. The straight line is the expected relationship v_r(astrom) $\simeq$ v_r(spectr). The three most accurate determinations are indicated. From Dravins et al. (1999, Figure 2).

velocities, they proposed that the magnetic field-free convection pattern is dominated by convective redshift, and attributed the observational scatter to stellar variability. Kürster *et al.* (2006) also measured the perspective acceleration for GJ 1, again fully consistent with the astrometric prediction.

The third astrometric method that has already been applied using data from the Hipparcos mission, concerns the secular change of the angular extent of moving star clusters (Figure 1.18c). Since all cluster stars share the same (average) velocity vector, apart from a (small) random velocity dispersion, the cluster's apparent size changes as it moves in the radial direction. This relative change, revealed by the proper motion vectors towards the cluster apex, corresponds to the relative change in distance. Since the individual stellar distances are known from parallaxes, their radial velocities follow. Details of the method are given in Lindegren *et al.* (2000). Their maximum-likelihood formulation leads to an estimate of the space velocity and internal velocity dispersion of a cluster using astrometric data only and, as a by-product, kinematically improved parallaxes are obtained for the individual cluster stars. Using the Hipparcos data for the Hyades, they derive astrometric radial velocities with a standard error of 0.47 km s^{-1} for the cluster centroid, and accurate to about 0.68 km s^{-1} for the individual stars due to their peculiar velocities, improving to 0.60 km s^{-1} if known binaries are removed. Applying this method to the Hipparcos data for the Ursa Major, Coma Berenices, Pleiades, and Praesepe clusters, and for the Scorpius–Centaurus, α Persei, and 'HIP 98321' associations, Madsen *et al.* (2002) derived astrometric radial velocities with typical accuracies of a few km s^{-1}. Taking into account ground-based spectroscopy (e.g. using the ELODIE spectrograph) reaching a precision of about 50 m s^{-1} for groups of 100 selected lines in any one star, hydrodynamic models of stellar atmospheres predicting differences of the order 1 km s^{-1} in convective line shifts between different stars,

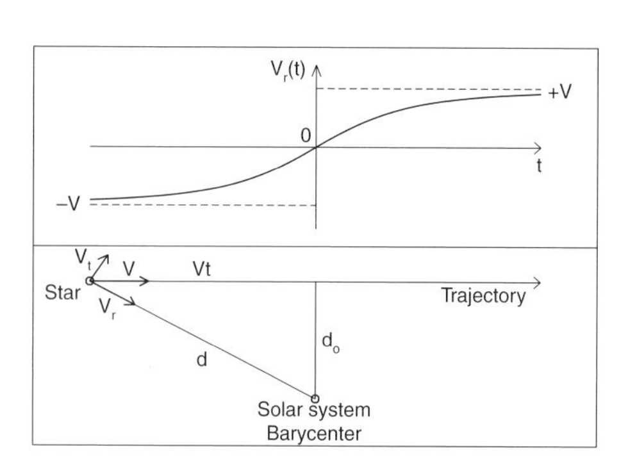

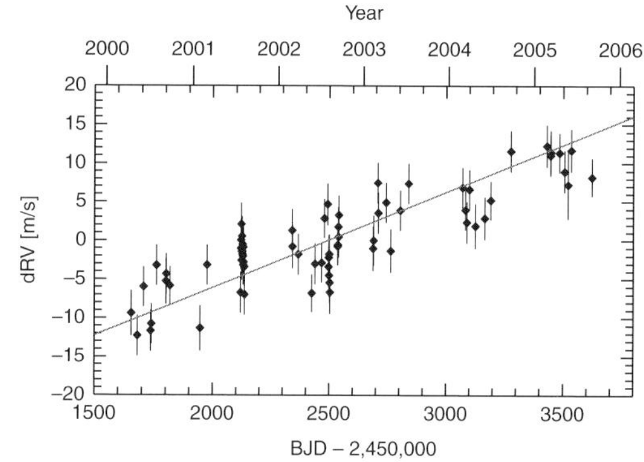

Figure 1.20 (a) Left, bottom panel: geometry of the space motion of a passing star. v_t and v_r are, respectively, the transverse and radial components of the space velocity v. The current distance is d, the minimum distance is d_0, and Vt is the trajectory from the current position to the point of nearest approach. Top panel: the absolute radial velocity as a function of time and its asymptotic limits $\pm v$. From Kürster et al. (2003, Figure 1). (b) Right: the barycentric radial velocity measurement of Barnard's star over five years, illustrating a good agreement with the predicted secular acceleration (solid line) and the fluctuations around it attributed to stellar variability. From Kürster et al. (2006, Figure 3).

and gravitational redshifts of comparable magnitude, their results provided evidence for enhanced convective blueshifts of F dwarfs, and decreased gravitational redshifts in giants. Results for the Scorpius OB2 complex indicate some expansion of its associations, albeit slower than expected from their ages. They also confirmed the gaps in the Hyades main sequence suggested by de Bruijne *et al.* (2000, see Section 6.3) as Böhm–Vitense gaps but, not being visible in other clusters, they suggest that those gaps are probably spurious.

It is useful to recall that the radial velocity strictly enters the formulae not only for space velocities, but also for the transformation of positions as a function of epoch (Section 1.4.5). Its appearance in the broader formulation of expressions accounting for relativistic effects is decribed further in Section 1.15.

1.14 Cross-identifications

For a given object, the Hipparcos Catalogue number (HIP) is identical to the Hipparcos Input Catalogue number (HIC): the use of HIP nnn rather than HIC nnn also implies the source of any associated data.

The construction of the Hipparcos Input Catalogue included a considerable effort devoted to carrying out cross-identifications of proposed stars with other catalogues. Information in the Hipparcos Input Catalogue, and in the final Hipparcos Catalogue, includes the cross-identifications shown in Table 1.12 (see the catalogues for details). All of these are preserved, for example, through the SIMBAD database of the CDS, Strasbourg.

The Tycho Catalogue provides cross-references to the Hipparcos Catalogue identifiers for 123 431 entries, with cross-identifications also given to the CCDM, PPM, HD and DM (BD/CoD/CPD) catalogues, and to the variable star catalogues GCVS and NSV. All were based on the CDS SIMBAD facility.

Other catalogue cross-indexes For double and multiple stars, and in addition to the CCDM Catalogue identifier, the Washington Catalogue of Visual Double Stars (WDS Roman, 1996) includes an updated compilation of HD, DM, and ADS identifiers, themselves partly based on the Hipparcos compilation. An updated cross-index between between the Hipparcos Catalogue number and the HD, DM, GC, HR, and Bayer–Flamsteed designations is given by Kostjuk (2004). For Tycho 2 stars, Henry Draper catalogue identifications are given for 99.8% of stars in the HD, and for 96% of the HD Extensions, by Fabricius *et al.* (2002).

MK spectral types Spectral types were assigned to 351 864 stars in Tycho 2 by cross-referencing to several catalogues of spectral types using the CDS database, resulting in the Tycho 2 Spectral Type Catalogue (Wright *et al.*, 2003). A compilation of spectral types based on six existing MK spectral type catalogues, including astrometric and photometric data mainly from Tycho 2, was published for 71 334 stars (Rybka, 2003), and subsequently updated based on 11 catalogues for 169 843 stars (Rybka, 2005). Spectral types were assigned to 584 stars in a supplement to the Bright Star Catalogue, using Hipparcos parallaxes to check their reliability (Abt, 2004).

1.15 Relativity and astrometry

Light bending As introduced already in Section 1.4.1, the reduction of the Hipparcos data necessitated the inclusion of stellar aberration up to terms in $(v/c)^2$, and

Astrometry at the milli and microarcsec level: The geometrical model for the Hipparcos reductions at the level of 1 mas is a simplification of the geometry actually applicable, illustrated in Figure 1.21. As described by Klioner (2003), and in the context of the Barycentric Celestial Reference System (BCRS) of the IAU (see Section 1.4.1), the more general model comprises five successive transformations: (i) aberration, or the effects of the observer's motion with respect to the Solar System barycentre, converts the observed source direction **s** into the BCRS coordinate direction of the light ray **n** at the point of observation; (ii) gravitational light deflection for the source at infinity, converts **n** into the direction of propagation $\boldsymbol{\sigma}$ of the light ray infinitely far from the Solar System for $t \to -\infty$; (iii) coupling of the finite distance to the source and the gravitational light deflection in the gravitational field of the Solar System, which converts $\boldsymbol{\sigma}$ into the vector **k** from the source to the observer; (iv) parallax, which converts **k** into the vector **l** from the Solar System barycentre to the source; and (v) proper motion, providing the time dependence of **l** caused by the source motion with respect to the barycentre.

At levels of 1 microarcsec, definitions of parallax, proper motion, and radial velocity have to be made compatible with General Relativity, and this introduces differences with respect to interpretation in classical Newtonian astrometry. Specifically, parallax and proper motion are no longer effects which can be considered separately, or independently of the chosen coordinate-dependent model. Defining the parallax of a source as

$$\pi(t_0) = \frac{1\,\text{AU}}{|\mathbf{x}_s(t_e)|} \tag{1.29}$$

where $\mathbf{x}_s$ is the BCRS position of the source and t_e is the BCRS coordinate time of emission of the signal by the source, Equations 78–81 of Klioner (2003) give the resulting expressions for the parallax. Second-order effects proportional to π^2 are below 3 microarcsec for $|\mathbf{x}_s| > 1$ pc. Similarly, and connected to the discussion of astrometric radial velocities in Section 1.13.2, Klioner (2003, Section 8) introduces the concepts of 'apparent proper motion', $\boldsymbol{\mu}_{\rm ap}$, and 'apparent radial velocity', $V_{\rm rad}^{\rm ap}$, and derives (his Equations 95–96)

$$\pi(t_0) = \pi_0 - \pi_0^2 \left(\frac{V_{\rm rad}^{\rm ap}}{1\,\text{AU}} \right) \Delta t_0 + \ldots \tag{1.30}$$

$$\mathbf{l}(t_0) = \mathbf{l}_0 + \boldsymbol{\mu}_{\rm ap} \Delta t_0 + \boldsymbol{\mu}_{\rm ap} c^{-1} \{[\mathbf{x}_0(t) - \mathbf{x}_0(t_0)] \cdot \mathbf{l}_0\} + \ldots \tag{1.31}$$

The amplitude of the third term in $\mathbf{l}(t_0)$ is about 150 μas for Barnard's star for which $\mu = 10.4$ arcsec yr^{-1}. For a satellite not too far from the Earth, the effect exceeds 1 μas for all stars with proper motions larger than about 60 mas yr^{-1}.

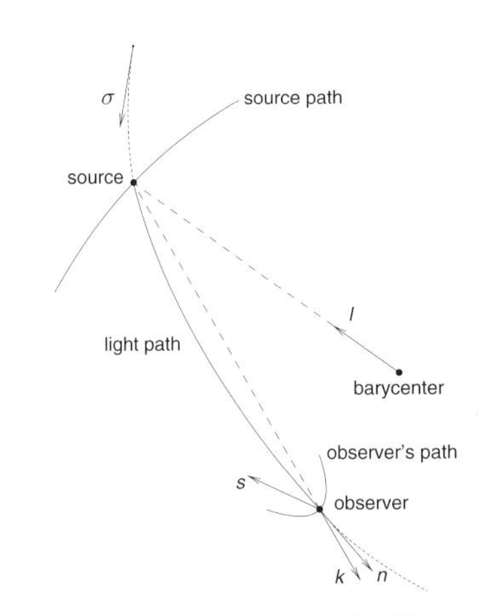

Figure 1.21 The five principal vectors used in the relativistic model of Klioner (2003): **s** *is the observed direction,* **n** *is the tangent vector to the light ray at the moment of observation,* $\boldsymbol{\sigma}$ *is the tangent vector to the light ray at* $t = -\infty$, **k** *is the coordinate vector from the source to the observer, and* **l** *the vector from the Solar System barycentre to the source. From Klioner (2003, Figure 3).*

the General Relativistic treatment of light bending due to the gravitational field of the Sun (and Earth). That the effect of relativistic light bending must be accounted for in the Hipparcos data analysis is immediately evident from the magnitude of the effect: according to General Relativity, light bending at the solar limb reaches 1.7505 arcsec, but the effect remains significant compared with the Hipparcos measurement accuracies over almost the entire celestial sphere, amounting to about 4 mas even at 90° from the Sun (i.e. orthogonal to the ecliptic). The approach adopted for Hipparcos was to introduce the predicted light bending at each measurement epoch as fully deterministic as far as effects due to the Sun (and the Earth in NDAC) are considered. The collective measurements were then assessed for any systematic departures from the prescriptions of General Relativity, with the objective of assessing whether General Relativity provides an adequate framework for the Hipparcos observations, or whether an alternative theory of gravity is indicated.

The context The comparison of metric theories of gravity, with each other and with experiment, becomes

Table 1.12 Cross-identifications to other catalogues contained in the Hipparcos Input Catalogue and/or Hipparcos Catalogue.

Type of information/Catalogue	Field/Table
Hipparcos Input Catalogue:	
Spectral type and luminosity class	Fields 20–21 (see also HIP)
Variable star name/variability type (GCVS, NSV)	Fields 22–23
Radial velocity and source	Fields 18–19
Durchmusterung Number (DM: BD, CoD, CPD)	Field 32 (see also HIP)
Henry Draper Number (HD/HDE)	Field 33 (see also HIP)
Multiple star identifier (CCDM)	Field 39
Other catalogue identifiers:	
FK5, FK5 Ext, FK4 Sup, IRS	Field 44
AGK3/CPC	Field 45
SAO	Field 46
GL, GJ, G, LHS, LTT, LP, L, BPM, CF, McC	Fields 47–48
Galactic open clusters: LMC/SMC, C*, IRC, PK, WD	Field 49
Survey star (to completeness limit)	Field 50 (see also HIP)
Hipparcos Catalogue:	
Multiple star identifier (CCDM)	Field H55
Survey star (to completeness limit)	Field H68
Henry Draper Number (HD/HDE)	Field H71
Durchmusterung Number (DM: BD, CoD, CPD)	Fields H72–75
Spectral type and source	Fields H76–77
Henry Draper Number (HD/HDE)	Identification Table 2
Bright Star Number (HR)	Identification Table 3
Bayer and Flamsteed Names	Identification Table 4
Variable Star Names	Identification Table 5
Common Star Names	Identification Table 6

particularly simple in the slow-motion, weak-field limit, known as the post-Newtonian approximation (what distinguishes one metric theory from another is the number and kind of gravitational fields it contains in addition to the metric, and the equations that determine their structure and evolution). In the Parameterized Post-Newtonian (PPN) formulation of gravitational theories, parameters are used in place of coefficients of the metric potentials. In the canonical version of the PPN formalism (Will, 1993), ten parameters are used to indicate the general metric properties: $\gamma, \beta, \xi, \alpha_{1-3}, \zeta_{1-4}$. The parameters γ (measuring space curvature per unit mass) and β (measuring nonlinearity in the superposition law for gravity) are the Eddington–Robertson–Shiff parameters used to describe the classical tests of General Relativity, and are the only non-zero parameters in General Relativity and scalar-tensor gravity.

The equations of light propagation are derived from the General Relativistic Maxwell equations, but can be derived in the limit of geometrical optics since relativistic effects depending on wavelength are much smaller than 1 microarcsec in the Solar System (Klioner, 2003, Section 6). For a spherically symmetric gravitational field, the general expression reduces to the post-Newtonian deflection angle

$$\delta\chi = \frac{2GM}{rc^2}\,\frac{(1+\gamma)}{2}\cot(\chi/2) \qquad (1.32)$$

where γ is the governing PPN parameter, equal to unity in General Relativity, c is the velocity of light, M is the mass of the deflecting body, and χ the angular distance between the deflecting body and the source. Conventionally, results are expressed in terms of $\frac{1}{2}(1+\gamma)$. The classical derivation of light bending yields only the first part (i.e. the factor 1/2) of the leading coefficient in this expression. For grazing incidence at the solar limb, the expression reduces to $\delta\chi \simeq \frac{1}{2}(1+\gamma)\,1.7505$ arcsec.

Light deflection has been observed, with various degrees of precision, on distance scales of 10^9–10^{21} m, and on mass scales from 10^{-3}–$10^{13} M_\odot$, the lower range from planetary deflection by Jupiter (Treuhaft & Lowe, 1991), and the upper range from the gravitational lensing of quasars (Dar, 1992).

The published observational confirmation based on the 1919 solar eclipse observed in Brazil (Dyson *et al.*, 1920) was of limited accuracy, while the most recent ground-based astrometric eclipse campaign undertaken in Chinguetti (Mauritania) in 1973 (Jones, 1976) yielded $\frac{1}{2}(1+\gamma) = 0.95 \pm 0.10$ (Figure 1.22).

The development of radio interferometry and radio VLBI greatly improved the measurement of light deflection: a 1995 VLBI measurement of 3C 273 and 3C 279 yielded $\frac{1}{2}(1+\gamma) = 0.9996 \pm 0.0017$ (Lebach *et al.*, 1995), and a later analysis of nearly 2 million VLBI observations of 541 radio sources from 87 VLBI sites yielded

Table 1.13 Light deflection by masses in the Solar System. The monopole effect dominates, and is summarised in the left columns for grazing incidence and for typical values of the angular separation. Columns $\chi_{\rm min}$ and $\chi_{\rm max}$ give results for the minimum and maximum angles accessible to Gaia. The value of $\chi_{\rm min}$ for the Sun corresponds to $\chi = 35°$, itself based on the original Sun aspect angle of 55° (the current value is 45°). J_2 is the quadrupole moment. The magnitude of the quadrupole effect is given for grazing incidence, and for an angle of 1°. The effect of angular momentum on quadrupole deflection is ignored, but is also significant (Epstein & Shapiro, 1980). The values are taken from Perryman et al. (2000, Table 1.17), with the grazing monople term updated according to Klioner (2003, Table 1).

	Monopole term					Quadrupole term		
Object	Grazing μas	$\chi_{\rm min}$ μas	$\chi = 45°$ μas	$\chi = 90°$ μas	$\chi_{\rm max}$ μas	J_2	Grazing μas	$\chi = 1°$ μas
Sun	1 750 000	13 000	10 000	4100	2100	$\leq 10^{-7}$	0.3	–
Earth	574	3	2.5	1.1	0	0.001	1	–
Jupiter	16 270	16 000	2.0	0.7	0	0.015	500	7×10^{-5}
Saturn	5780	6000	0.3	0.1	0	0.016	200	3×10^{-6}

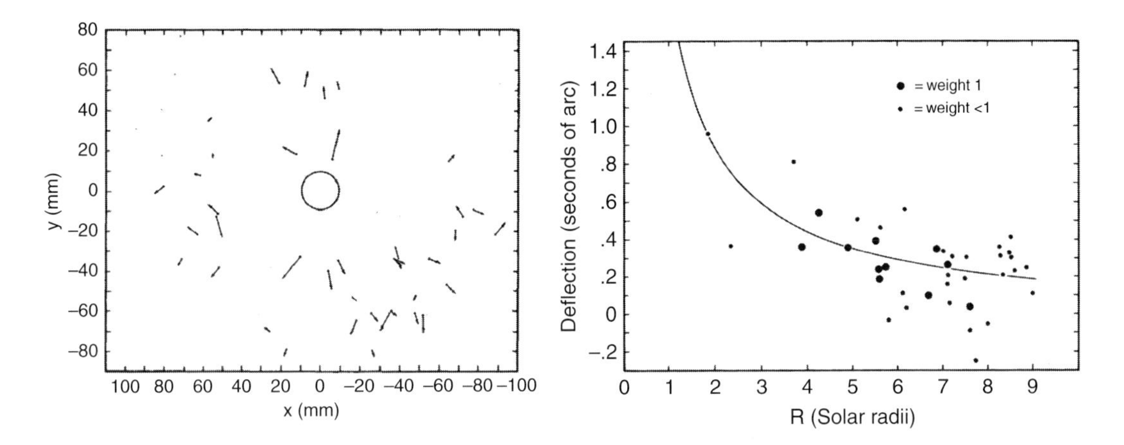

Figure 1.22 Ground-based measurements of General Relativistic light deflection by the Sun, during the 1973 total eclipse observed from Chinguetti (Mauritania). Left: the light deflection measured for individual stars, compared with non-eclipse positions. Right: the measured deflection (arcsec) as a function of radial distance from the solar limb. The figures are strictly unrelated to Hipparcos, but are included here for several reasons: their omission would have left an important astrophysical astrometric application unillustrated; they show the principle of light bending by the Sun, a significant effect for Hipparcos which had to be accounted for in the data analysis; and they illustrate the state-of-the-art in (direct) light-bending measurements before Hipparcos, demonstrating the challenging nature of Dyson & Eddington's first experimental confirmation of General Relativity. From Jones (1976, Figures 6 and 7).

$\frac{1}{2}(1+\gamma) = 0.99992 \pm 0.00023$ (Shapiro *et al.*, 2004). Fomalont & Kopeikin (2003) reported VLBA–Effelsberg observations of 'light' bending of the quasar J0842+1835 at 3.7 arcmin from Jupiter.

Future tests of light bending are being extensively discussed in view of the prospects for microarcsec space astrometry, for example, with SIM and Gaia. An accessible discussion is given by Klioner (2003) which includes, in his Table 1, the various gravitational effects on light propagation predicted for 26 Solar System bodies including the Sun, the planets, the Moon, other planetary satellites, and the largest asteroids. Other recent and related contributions are given by Kopeikin (2001), Klioner & Peip (2003), Bini *et al.* (2003), Pireaux (2004), de Felice *et al.* (2004) and de Felice *et al.* (2006).

Table 1.13 gives the deflection for the Sun and major planets, at different values of the angular separation χ, for the monopole (Equation 1.33) and quadrupole terms. For the planned Gaia mission, χ is never smaller than 45° for the Sun (assuming a Sun aspect angle of 45°), although grazing incidence is possible for the planets. With an astrometric accuracy of a few μas, the magnitude of the expected effects is considerable not only for the Sun, but also for specific observations close to planets, and notably for Jupiter (Crosta & Mignard, 2006; Souchay *et al.*, 2007). An analysis corresponding to that of the Hipparcos reductions indicates that the Gaia measurements will provide a precision of about 5×10^{-7} for γ, based on multiple observations of $\sim 10^7$ stars with $V < 13$ mag at wide angles from the Sun, with individual

Solar quadrupole and perihelion precession: The rotation of the Sun leads to a flattening of the polar regions. The solar quadrupole moment $J_2 = (C - A)/MR^2$, where C and A are the moments of inertia about the body's rotation and equatorial axes respectively, results from the rotation of the whole of the solar interior, and its mass distribution as a function of radius. Observations of the solar diameter indicate an oblateness ranging from 8.8×10^{-6} from stratospheric balloon observations (Lydon & Sofia, 1996), 1.1×10^{-5} measured from the ground (Bursa, 1986), and 9.8×10^{-6} from the MDI instrument on SOHO (Kuhn *et al.*, 1998). Improved measurements are part of the scientific case for the proposed Picard mission. These values show that the difference in diameter from equator to pole lies between 17–22 mas. Combined with a model of the solar interior and of the differential rotation constrained by helioseismology, these give estimates of $J_2 = (2.2 \pm 0.1) \times 10^{-7}$ (Mecheri *et al.*, 2004). In all of these studies, the quadrupole moment is model dependent. More direct limits on J_2 of around 3×10^{-6} are found from the effect of the solar quadrupole on lunar librations and planetary motions. A direct determination from dynamical effects will be made with Gaia, where good sampling in $a(1 - e^2)$ suggests that a limit better than 10^{-7} may be reached.

Relativistic effects, as well as the solar quadrupole moment, cause the orbital perihelion of a Solar System body to precess at the rate (Will, 1993, Equation 49, fully conservative case)

$$\Delta\omega = \frac{6\pi\lambda M_\odot}{a(1 - e^2)} + \frac{3\pi J_2 R_\odot^2}{a^2(1 - e^2)^2} \tag{1.33}$$

where $\lambda = (2\gamma - \beta + 2)/3$ is the PPN precession coefficient, M is the total mass of the two-body system, R the mean radius of the oblate body, and the rate is given in radians per revolution. While the effects are irrelevant for Hipparcos, they will be measurable for Gaia (Table 1.14). For the main belt asteroids, the precession is about seven times smaller than for Mercury in rate per revolution, although more than a hundred times in absolute rate. The three Earth-crossing asteroids indicated have perihelia precession larger than Mercury's, due to a favourable combination of distance and eccentricity. A determination of λ with an accuracy of 10^{-4} has been estimated for Gaia, with a value closer to 10^{-5} probably attainable from several tens of planets. Analysis of 24 years of lunar laser ranging also provides various constraints on γ and β (Williams *et al.*, 1996).

Table 1.14 Perihelion precession due to relativity and the solar quadrupole moment. From Perryman et al. (2000, Table 1.18).

	a		General Relativity		$J_2(=10^{-6})$
Body	AU	e	mas/rev	mas/yr	mas/rev
Mercury	0.39	0.21	102	423	0.30
Asteroid (main belt)	2.7	0.1	15	3.4	0.006
1566 Icarus	1.08	0.83	114	102	0.34
5786 Talos	1.08	0.83	114	102	0.34
3200 Phaeton	1.27	0.89	148	103	0.57

measurement accuracies better than 10 μas. Other measurements include those from Gravity Probe-B, the relativity gyroscope experiment, which is expected to yield γ to about 6×10^{-5}, while ESA's Mercury mission BepiColombo also aims at levels of $\sim 10^{-5}$.

Electromagnetic signals propagating past the Sun and returned to Earth suffer an additional non-Newtonian delay in the round-trip travel time, referred to as the Shapiro delay (Shapiro, 1964). Refraction by the solar corona is an additional complication, whose modelling can be improved by the use of dual-band observations. Active radar time delay studied using the Mariner 9 and Viking landers and orbiters resulted in $\frac{1}{2}(1+\gamma) = 1.000 \pm 0.001$ (Reasenberg *et al.*, 1979). More recent observations using the Cassini spacecraft and signal propagation passing only $1.6\,R_\odot$ from the Sun have yielded the most precise estimate of light bending to date, with $\frac{1}{2}(1+\gamma) = 1 + (1.05 \pm 1.15) \times 10^{-5}$ (Bertotti *et al.*, 2003).

The Hipparcos contribution The analysis of the astrometric residuals of the Hipparcos measurements, and the associated determination of the PPN light-bending term γ, was studied by Froeschlé *et al.* (1997). After an iterative calibration of the Hipparcos instrument parameters, each abscissa on a reference great circle had a typical precision of 3 mas for a star of 8–9 mag. Each of the two data reduction teams, NDAC and FAST, generated about 3.5 million abscissae for the full mission. In the absence of systematic errors and correlations between parameters, an unknown angular parameter in the abscissa modelling, common to all stars, could in principle be determined with a precision as high as $3/\sqrt{3.5 \times 10^6} = 0.0016$ mas. Assuming that the deflection for all stars is of the form $4 \times (1+\gamma)/2$ mas, then γ could be determined to better than 0.001.

In practice, the geometry of light deflection is similar to that of parallax displacement, although differing in direction and in their dependence on χ: varying

Brans–Dicke formulation: The Brans–Dicke theory (Brans & Dicke, 1961) remains a conceptually attractive alternative theory of gravity containing, beside the metric tensor, a scalar field ϕ and an arbitrary coupling constant ω, related by $\gamma = (1+\omega)/(2+\omega)$. The present best limit on γ from the Shapiro delay using Cassini data gives the constraint $|\omega| > 40\,000$ (Will, 2006). Other generalized scalar-tensor theories have been proposed in which $\omega = \omega(\phi)$, and which therefore changes in time with the evolution of the Universe. Because of recent developments in cosmology (e.g. inflationary models) and elementary-particle physics (e.g. string theory and Kaluza–Klein theories), these scalar-tensor theories are considered as interesting alternatives to General Relativity (Damour & Nordtvedt, 1993a,b); if this is how the Universe is evolving, then expected discrepancies of the order of $|\gamma - 1| \sim 10^{-7} - 10^{-5}$ may exist, depending on the theory. This kind of argument provides a strong motivation for any experiments able to reach these accuracies. Additional terms caused by gravitational effects on light rays at the μas level include the 'frame-dragging' effects of the motions and rotations of the Sun and the planets (Soffel, 1989). For example, the post-PPN term for a grazing ray is 11 μas for the Sun (Klioner, 2003).

as $\cot(\chi/2) = \sin\chi/(1-\cos\chi)$ for light bending (Equation 1.32), and as $\pi\sin\chi$ for the parallax effect. The difference allows the two effects to be separated in the observation equations, but they remain highly correlated. This results in an increase of the expected formal error on γ to about 0.003.

A variety of experiments were conducted to examine the robustness of the solution for γ, splitting the dataset into different groups, and solving for different combinations of parameters. The solutions were rather insensitive to these choices, and led to a result of $\gamma = 0.997 \pm 0.003$, or $\frac{1}{2}(1+\gamma) = 0.9985 \pm 0.0015$. Although this accuracy has since been superseded by both VLBI and the Cassini Shapiro-delay measurements, it is noteworthy that the results were derived from observations at large solar angles, i.e. not constrained to observations within a few $R_\odot$ of the solar limb, and with a single instrument. Figure 1.23 compares $\frac{1}{2}(1+\gamma)$ derived from the final Hipparcos data with previous determinations by other means, taken from Will (2006). Numerical values for earlier data are also listed in Soffel (1989).

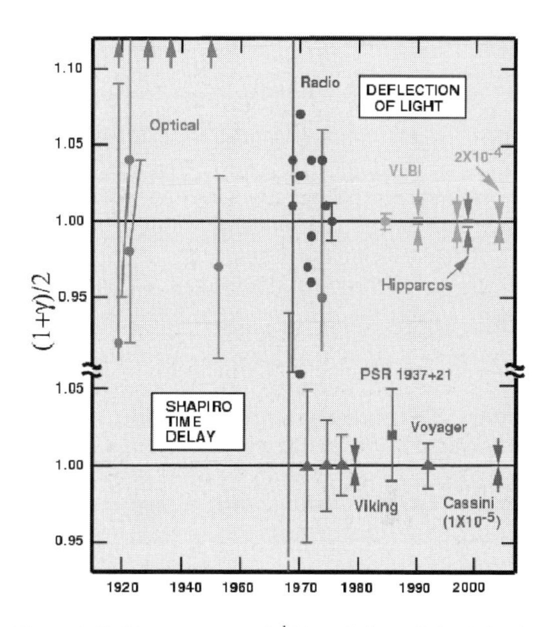

Figure 1.23 Measurements of $\frac{1}{2}(1+\gamma)$ from light-deflection (top) and time-delay measurements (bottom) versus year of experiment. The value of $\frac{1}{2}(1+\gamma)$, and γ, in General Relativity is unity. The arrows at the top denote anomalously large values from early eclipse expeditions The Hipparcos measurements yield $\frac{1}{2}(1+\gamma) = 0.9985 \pm 0.0015$ (Froeschlé et al., 1997). The latest Shapiro time-delay measurements using the Cassini spacecraft yield $\frac{1}{2}(1+\gamma) = 1 + (1.05 \pm 1.15) \times 10^{-5}$ (Bertotti et al., 2003). The latest VLBI light deflection measurements yield $\frac{1}{2}(1+\gamma) = 0.99992 \pm 0.00023$ (Shapiro et al., 2004). From Will (2006, Figure 5).

1.16 Astrometry beyond Hipparcos

Prospects from space The success of Hipparcos, and specifically the scientific interest of astrometry at the milliarcsec level, has motivated the study and acceptance of further space experiments aiming for an improvement in astrometric accuracy, limiting magnitude, and numbers of stars. A brief synopsis of the current situation is given here. Most proposals, with the notable exception of SIM, follow the same basic principles of Hipparcos, based on a continuous sky scanning with two widely-separated fields of view. Only Gaia is currently a fully approved programme, with the others noted below for completeness having been cancelled, superseded, or remaining still in the concept stage.

Roemer (Høg, 1993; Høg & Lindegren, 1994) was proposed as an ESA medium-sized mission (Lindegren *et al.*, 1993), extending the limiting magnitude and accuracy of Hipparcos to some 100 million stars, complete to $V \sim 15.5$ mag, and with accuracies in the range 0.1 mas for $V < 11$ mag to 1.5 mas at $V \sim 17$. It was rated highly, but not considered further by the external ESA advisory structure since it was considered too modest in terms of its scientific advance over Hipparcos. For the first time, the mission baselined CCD detectors for the focal plane, although CCDs had been briefly considered during the early Phase B studies of Hipparcos in around 1982, but promptly discarded due, amongst other issues, to the poor charge transfer efficiency of the early devices.

Subsequently, the more ambitious Gaia mission was proposed to ESA, originally in the form of a small optical interferometer fitting inside the Ariane 5 launch vehicle (Lindegren & Perryman, 1996), but subsequently in monolithic form (Perryman *et al.*, 2001). Gaia was accepted within the ESA scientific programme in 2000, and entered Phase B in 2006 with a target launch date at the end of 2011. The quoted references describe the astrometric capabilities, with further details given of the photometric capabilities by Jordi *et al.* (2006), and of the radial velocity measurement component by Katz *et al.* (2004) and Wilkinson *et al.* (2005). The mission targets the observation of all stars down to $V \sim 20$ mag through on-board detection, reaching astrometric accuracies of ~20 microarcsec at 15 mag, compared with 10 microarcsec at the time of mission acceptance in 2000. Comprehensive multi-colour, multi-epoch photometry will be acquired for each star (originally using filters, but more recently re-designed to be based on low-resolution spectroscopy), permitting object classification as well as determination of $T_{\rm eff}$, $\log g$, metallicity and reddening for each star. Determination of radial velocities and associated data for objects to ~17.5 mag is based on intermediate-resolution spectroscopy of the ionised Ca triplet at around 848–874 nm. There is already a rich literature associated with the scientific capabilities of Gaia, for example covering detection of up to 10–20 000 exoplanets (Casertano *et al.*, 1996), many tens of thousands of white dwarfs (Torres *et al.*, 2005), observations of asteroids and near-Earth objects (Mignard, 2002; Fienga *et al.*, 2003), microlensing events (Belokurov & Evans, 2002), supernovae (Belokurov & Evans, 2003), the determination of stellar parameters (Bailer-Jones, 2002), and of the distance scale (Tammann & Reindl, 2002). Various conferences have also been devoted to the expected results (notably Perryman & van Leeuwen, 1995; Bienaymé & Turon, 2002; Munari, 2003; Turon *et al.*, 2005).

In parallel with the early studies of Gaia, a space mission with capabilities intermediate between Hipparcos and Gaia, DIVA, was studied intensively by the German national community (Röser, 1999; Röser & Bastian, 2000). It targeted completeness to 15 mag, accuracies around 0.2 mas at $V = 9$ mag, and broad- and intermediate-band photometry. It maintained broader European support until the adoption of Gaia with an early launch date eroded its momentum and led to its termination.

The US Naval Observatory has undertaken various studies of a scanning mission following the Hipparcos/Gaia principles, and intermediate between them in terms of astrometric accuracy: in successive forms this has been known as FAME, targeting completeness for 10 million stars down to 14 mag (Seidelmann, 1995; Johnston, 2003), AMEX within NASA's Small Explorer and subsequently Mid-Explorer programmes, and OBSS reaching to 23 mag (Johnston *et al.*, 2006). A Milli-Arcsecond Pathfinder Survey (MAPS) mission, operating in a step-stare mode, has also been considered (Gaume *et al.*, 2006; Zacharias & Dorland, 2006).

Japanese teams have been studying a mission reaching accuracies of 10 microarcsec, but focusing on operation in the infrared, called JASMINE (Japan Astrometry Satellite Mission for INfrared Exploration, Gouda *et al.*, 2005); also in a smaller sky-scanning form in the z-band, reaching accuracies of around 1 mas (Nano-JASMINE, Kobayashi *et al.*, 2005, 2006a,b,c).

Russian projects in space astrometry studies over the past 10 years or more include OSIRIS and LIDA (Bagrov, 2006).

In the USA, effort has also been focused on a very different measurement concept: the Space Interferometry Mission (SIM, more recently known as SIM PlanetQuest, Shao, 2004; Unwin *et al.*, 2008). The SIM PlanetQuest study has pioneered the development of (Michelson) space interferometry, using a pointed structure and a baseline of order 10 m to measure positions with respect to a reference stellar grid (Frink *et al.*, 2001): its strength is the ability to reach high astrometric accuracy (around 1 microarcsec), even at faint magnitude (down to 20 mag), but on a relatively small number of target objects, numbering of the order of a few tens of thousands.

Space astrometry is also being continued in its more classical narrow-field form with the Fine Guidance Sensors of the Hubble Space Telescope, where parallaxes for a small sample of stars, reaching parallax accuracies of some 0.3 mas, are now obtained (Benedict & McArthur, 2005). As in all instances of narrow-field astrometry, the transformation from relative to absolute parallaxes remains a delicate challenge.

Some recent descriptions of plans for these future astrometric missions are given by Mignard & Kovalevsky (2003), Gaume (2003, 2005), and Jin *et al.* (2006).

Prospects from ground Prospects for ground-based astrometry in the post-Hipparcos era are given, for example, by Stavinschi (2001, 2003, 2006) and Kovalevsky (2003).

Specific studies of narrow-field astrometric accuracy achievable with large ground-based telescopes include work by Lazorenko & Lazorenko (2004), Lazorenko (2005, 2006), and Lazorenko *et al.* (2007).

Amongst the dedicated larger-aperture surveys, Pan-STARRS and LSST are emphasising astrometry as a valuable by-product of the photometry, and differential astrometry to about 10 mas per exposure might be feasible. Higher relative accuracy on much smaller angular scales is achieved with ground-based (speckle and longer baseline) interferometry, perhaps eventually reaching the level of 10 μas with PRIMA at the VLTI.

References

Abt HA, 2004, Spectral classification of stars in a supplement to the Bright Star Catalogue. *ApJS*, 155, 175–177 {**38**}

Abt HA, Willmarth D, 2006, The secondaries of solar-type primaries. I. The radial velocities. *ApJS*, 162, 207–226 {**34**}

Andrei AH, Assafin M, Puliaev SP, *et al.*, 1999, Early radio positions of stars. *AJ*, 117, 483–491 {**30**}

Andrei AH, Penna JL, Assafin M, *et al.*, 2003, Analysis of reference frames in the Hipparcos system. *Astronomy in Latin America* (eds. Teixeira R, Leister NV, Martin VAF, *et al.*), 101–104 {**30**}

Aoki S, Soma M, Kinoshita H, *et al.*, 1983, Conversion matrix of epoch B1950 FK4-based positions of stars to epoch J2000.0 positions in accordance with the new IAU resolutions. *A&A*, 128, 263–267 {**14**}

Aoki S, Soma M, Nakajima K, *et al.*, 1986, The conversion from the B1950 FK4-based position to the J2000 position of celestial objects. *IAU Symp. 109: Astrometric Techniques* (eds. Eichhorn HK, Leacock RJ), 123 {**14**}

Arenou F, 1997, Improvement on astrometric parameters using Hipparcos intermediate data. *The First Results of Hipparcos and Tycho, IAU Joint Discussion 14*, 21 {**13**}

Argyle RW, Einicke OH, Pilkington JDH, *et al.*, 1996, Comparison of the Carlsberg optical reference frame with the International Celestial Reference Frame. *A&A*, 312, 1031–1037 {**30**}

Arias EF, Charlot P, Feissel M, *et al.*, 1995, The extragalactic reference system of the International Earth Rotation Service, ICRS. *A&A*, 303, 604–608 {**9**}

Arlot JE, 1982, New constants for Sampson–Lieske theory of the Galilean satellites of Jupiter. *A&A*, 107, 305–310 {**31**}

Assafin M, Monken PT, da Silva Neto DN, *et al.*, 2003a, Precise optical positions for ICRF sources using the 0.6-m and 1.6-m LNA telescopes. *Bulletin of the Astronomical Society of Brazil*, 23, 49–50 {**30**}

Assafin M, Zacharias N, Rafferty TJ, *et al.*, 2003b, Optical positions of ICRF sources using UCAC reference stars. *AJ*, 125, 2728–2739 {**30**}

Babenko Y, Lazorenko PF, Karbovsky V, *et al.*, 2005, Kyiv meridian axial circle catalogue of stars in fields with extragalactic radio sources. *Kinematika i Fizika Nebesnykh Tel Supplement*, 5, 316–321 {**30**}

Bagrov AV, 2006, Russian projects of space missions for astrometry. *Astronomical Facilities of the Next Decade, IAU 26*, 1 {**44**}

Bailer-Jones CAL, 2002, Determination of stellar parameters with Gaia. *Ap&SS*, 280, 21–29 {**44**}

Barbier-Brossat M, 1989, Catalogue of mean stellar radial velocities. *A&AS*, 80, 67–71 {**34**}

Barbier-Brossat M, Figon P, 2000, Mean radial velocities catalogue of Galactic stars. *A&AS*, 142, 217–223 {**34, 35**}

Bartkevičius A, Sperauskas J, 1999, Radial velocities of Population II stars. II. *Baltic Astronomy*, 8, 325–353 {**34**}

—, 2005, Radial velocities of Population II binary stars. II. *Baltic Astronomy*, 14, 511–525 {**34**}

Batten AH, Fletcher JM, MacCarthy DG, 1989, The 8th catalogue of the orbital elements of spectroscopic binary systems. *Publications of the Dominion Astrophysical Observatory Victoria*, 17, 1 {**35**}

Beardsley WR, Gatewood G, Kamper KW, 1974, A study of an early flare, radial velocities, and parallax residuals for possible orbital motion of Groombridge 1830. *ApJ*, 194, 637–643 {**36**}

Belokurov VA, Evans NW, 2002, Astrometric microlensing with the Gaia satellite. *MNRAS*, 331, 649–665 {**44**}

—, 2003, Supernovae with 'super-Hipparcos'. *MNRAS*, 341, 569–576 {**44**}

Benedict GF, McArthur BE, 2005, High-precision stellar parallaxes from Hubble Space Telescope fine guidance sensors. *IAU Colloq. 196: Transits of Venus: New Views of the Solar System and Galaxy* (ed. Kurtz DW), 333–346 {**44**}

Bertotti B, Iess L, Tortora P, 2003, A test of General Relativity using radio links with the Cassini spacecraft. *Nature*, 425, 374–376 {**42, 43**}

Bettoni D, Galletta G, 2001, Radial velocities of giant M stars near the ecliptic. *A&A*, 368, 593–594 {**34**}

Bienaymé O, Turon C (eds.), 2002, *Gaia: A European Space Project; Les Houches, France, 14–18 May 2001*, EAS Publications Series {**44**}

Bini D, Crosta MT, de Felice F, 2003, Orbiting frames and satellite attitudes in relativistic astrometry. *Classical and Quantum Gravity*, 20, 4695–4706 {**41**}

Boboltz DA, Fey AL, Johnston KJ, *et al.*, 2003, Astrometric positions and proper motions of 19 radio stars. *AJ*, 126, 484–493 {**30**}

Boboltz DA, Fey AL, Puatua WK, *et al.*, 2007, Very Large Array plus Pie Town astrometry of 46 radio stars. *AJ*, 133, 906–916 {**30**}

Bobylev VV, 2004a, Astrometric control of the inertiality of the Hipparcos Catalogue. *Astronomy Letters*, 30, 848–853 {**30**}

—, 2004b, Kinematic control of the inertiality of ICRS catalogues. *Astronomy Letters*, 30, 251–257 {**30**}

Bobylev VV, Khovirtchev MY, 2006, Kinematic control of the inertiality of the system of Tycho 2 and UCAC 2 stellar proper motions. *Astronomy Letters*, 32, 608–621 {**30**}

Bonnarel F, Genova F, Bienaymé O, *et al.*, 2000, The role of the CDS information hub in the cross-identification of large surveys. *ASP Conf. Ser. 216: Astronomical Data Analysis Software and Systems IX*, 239 {**22, 24**}

Brans C, Dicke RH, 1961, Mach's principle and a relativistic theory of gravitation. *Physical Review*, 124, 925–935 {**43**}

Bretagnon P, 1982, Theory for the motion of all the planets: the VSOP82 solution. *A&A*, 114, 278–288 {**8**}

Brumberg VA, Groten E, 2001, IAU resolutions on reference systems and time scales in practice. *A&A*, 367, 1070–1077 {**11**}

Bursa M, 1986, The Sun's flattening and its influence on planetary orbits. *Bulletin of the Astronomical Institutes of Czechoslovakia*, 37, 312–313 {**42**}

Camargo JIB, Ducourant C, Teixeira R, *et al.*, 2003, Extension of the ICRF for selected areas down to 16 mag. II. *A&A*, 409, 361–368 {**31**}

Casertano S, Lattanzi MG, Perryman MAC, *et al.*, 1996, Astrometry from space: Gaia and planet detection. *Ap&SS*, 241, 89–104 {**44**}

Chapront J, Chapront-Touzé M, Francou G, 1999, Determination of the lunar orbital and rotational parameters and of the ecliptic reference system orientation from lunar laser ranging measurements and IERS data. *A&A*, 343, 624–633 {**26**}

Chapront J, Francou G, 2003, The lunar theory ELP2000 revisited. *Astrometry from Ground and from Space* (eds. Capitaine N, Stavinschi M), 8–14 {**8**}

Charlot P, Sovers OJ, Williams JG, *et al.*, 1995, Precession and nutation from joint analysis of radio interferometric and lunar laser ranging observations. *AJ*, 109, 418–427 {**26**}
Clemence GM, 1966, Inertial frames of reference. *QJRAS*, 7, 10–21 {**8**}
Contreras ME, Rodríguez LF, Tapia M, *et al.*, 1997, Hipparcos, VLA, and CCD observations of Cygnus OB2. V. Solving the mystery of the radio companion. *ApJ*, 488, L153 {**31**}
Costa E, Loyola P, 1998, Optical positions of compact extragalactic radio sources with respect to the Hipparcos Catalogue. *A&AS*, 131, 259–263 {**30**}
—, 1999, CCD astrometry of faint compact extragalactic radio sources. II. *A&AS*, 139, 297–304 {**30**}
Costa E, 2001, CCD astrometry of faint compact extragalactic radio sources. II. *A&A*, 367, 719–724 {**30**}
—, 2002, CCD astrometry of faint compact extragalactic radio sources. III. *A&A*, 381, 13–20 {**30**}
Crosta MT, Mignard F, 2006, Microarcsecond light bending by Jupiter. *Classical and Quantum Gravity*, 23, 4853–4871 {**41**}
Cutispoto G, Pastori L, Pasquini L, *et al.*, 2002, Fast-rotating nearby solar-type stars, Li abundances and X-ray luminosities. I. Spectral classification, $v \sin i$, Li abundances and X-ray luminosities. *A&A*, 384, 491–503 {**34**}
Cutispoto G, Pastori L, Tagliaferri G, *et al.*, 1999, Classification of extreme ultraviolet stellar sources detected by the ROSAT WFC. I. Photometric and radial velocity studies. *A&AS*, 138, 87–99 {**34**}
Dalla Torre A, van Leeuwen F, 2003, A detailed analysis of the operational orbit of the Hipparcos satellite. *Space Science Reviews*, 108, 451–470 {**18**}
Damour T, Nordtvedt K, 1993a, General Relativity as a cosmological attractor of tensor-scalar theories. *Physical Review Letters*, 70, 2217–2219 {**43**}
—, 1993b, Tensor-scalar cosmological models and their relaxation toward General Relativity. *Phys. Rev. D*, 48, 3436–3450 {**43**}
Dar A, 1992, Tests of General Relativity and Newtonian gravity at large distances and the dark matter problem. *Nuclear Physics B Proceedings Supplements*, 28, 321–326 {**40**}
da Silva Neto DN, Andrei AH, Martins RV, *et al.*, 2000, Optical positions for a sample of ICRF sources. *AJ*, 119, 1470–1479 {**30**}
de Bruijne JHJ, Hoogerwerf R, de Zeeuw PT, 2000, Two Böhm-Vitense gaps in the main sequence of the Hyades. *ApJ*, 544, L65–L67 {**38**}
de Felice F, Crosta MT, Vecchiato A, *et al.*, 2004, A General Relativistic model of light propagation in the gravitational field of the Solar System: the static case. *ApJ*, 607, 580–595 {**41**}
de Felice F, Vecchiato A, Crosta MT, *et al.*, 2006, A General Relativistic model of light propagation in the gravitational field of the Solar System: the dynamical case. *ApJ*, 653, 1552–1565 {**41**}
de Medeiros JR, Mayor M, 1999, A catalogue of rotational and radial velocities for evolved stars. *A&AS*, 139, 433–460 {**34**}
de Medeiros JR, Silva JRP, do Nascimento JD, *et al.*, 2006, A catalogue of rotational and radial velocities for evolved stars. IV. Metal-poor stars. *A&A*, 458, 895–898 {**34**}
de Medeiros JR, Udry S, Burki G, *et al.*, 2002, A catalogue of rotational and radial velocities for evolved stars. II. Ib supergiant stars. *A&A*, 395, 97–98 {**34**}
de Medeiros JR, Udry S, Mayor M, 2004, A catalogue of rotational and radial velocities for evolved stars. III. Double-lined binary systems. *A&A*, 427, 313–317 {**34**}
Dimeo T, Valls-Gabaud D, Kerins EJ, 1997, The Tycho data base as a control microlensing experiment. *Variables Stars and the Astrophysical Returns of the Microlensing Surveys*, 69–73 {**17**}
Dravins D, Lindegren L, Madsen S, *et al.*, 1997, Astrometric radial velocities from Hipparcos. *ESA SP–402: Hipparcos, Venice '97*, 733–738 {**35**}
Dravins D, Lindegren L, Madsen S, 1999, Astrometric radial velocities. I. Non-spectroscopic methods for measuring stellar radial velocity. *A&A*, 348, 1040–1051 {**35–37**}
Ducourant C, Le Campion JF, Rapaport M, *et al.*, 2006, The PM2000 Bordeaux proper motion catalogue for $11^\circ < \delta < 18^\circ$. *A&A*, 448, 1235–1245 {**23**}
Duflot M, Fehrenbach C, Mannone C, *et al.*, 1992, Radial velocity measurements. V. Ground support of the Hipparcos satellite observation programme. *A&AS*, 94, 479–517 {**34**}
—, 1995a, Radial velocities. VII. Ground-based measurements for Hipparcos. *A&AS*, 110, 177–207 {**34**}
Duflot M, Figon P, Meyssonnier N, 1995b, Radial velocities: the Wilson–Evans–Batten catalogue. *A&AS*, 114, 269 {**34, 35**}
Duflot M, Mannone C, Genty V, *et al.*, 1990, Radial velocity measurements. IV. Ground-based accompaniment to the Hipparcos observation programme. *A&AS*, 83, 251–267 {**34**}
Dyson FW, Eddington AS, Davidson CR, 1920, A determination of the deflection of light by the Sun's gravitational field, from observations made at the total eclipse of 29 May 1919. *Mem. R. Astron. Soc.*, 220, 291–333 {**40**}
Egret D, Fabricius C, 1997, The Tycho Catalogue: stellar content. *ESA SP–402: Hipparcos, Venice '97*, 31–34 {**15, 16**}
Egret D, Ochsenbein F, Genova F, 1997, On-line services at CDS for distribution of Hipparcos data. *ESA SP–402: Hipparcos, Venice '97*, 81–84 {**22**}
Epstein R, Shapiro II, 1980, Post-post-Newtonian deflection of light by the Sun. *Phys. Rev. D*, 22, 2947–2949 {**41**}
ESA, 1979, *Hipparcos Space Astrometry: Report on the Phase A Study, ESA SCI(79)10.* European Space Agency, Noordwijk {**32**}
ESA, 1997, *The Hipparcos and Tycho Catalogues. Astrometric and Photometric Star Catalogues derived from the ESA Hipparcos Space Astrometry Mission, ESA SP–1200 (17 volumes including 6 CDs).* European Space Agency, Noordwijk, also: VizieR Online Data Catalogue {**1–9, 12–17, 20, 21, 25–27**}
Evans DS, 1978, Catalogue of stellar radial velocities. *Bull. Inf. CDS*, 15, 121 {**34, 35**}
Fabricius C, Makarov VV, Knude J, *et al.*, 2002, Henry Draper catalogue identifications for Tycho 2 stars. *A&A*, 386, 709–710 {**38**}
Falin JL, Mignard F, 1999, Mining in the Hipparcos raw data. *A&AS*, 135, 231–241 {**13**}
Famaey B, Jorissen A, Luri X, *et al.*, 2005, Local kinematics of K and M giants from Coravel, Hipparcos, and Tycho 2 data. Revisiting the concept of superclusters. *A&A*, 430, 165–186 {**33, 34**}
Fantino E, van Leeuwen F, 2003, Modeling the torques affecting the Hipparcos satellite. *Space Science Reviews*, 108, 499–535 {**18**}
Fedorov PN, Myznikov AA, 2005, The X1 catalogue of positions and proper motions of faint stars around the ICRF sources. *Kinematika i Fizika Nebesnykh Tel Supplement*, 5, 322–327 {**30**}

Fehrenbach C, 1985, Les mesures de vitesses radiales au prisme objectif dans le programme Hipparcos. *IAU Colloq. 88: Stellar Radial Velocities* (eds. Philip AGD, Latham DW), 189 {**32, 34**}

Fehrenbach C, Burnage R, 1990, Radial velocity measurements. III. Ground observations accompanying the Hipparcos satellite observation programme: measurements of the radial velocities of 391 stars in 12 fields. *A&AS*, 83, 91–107 {**34**}

Fehrenbach C, Burnage R, Duflot M, *et al.*, 1987a, Radial velocity measurements. I. Ground-based observations of the programme stars for the Hipparcos satellite. *A&AS*, 71, 263–274 {**34**}

Fehrenbach C, Burnage R, Figuiere J, 1992, Radial velocity measurements. VI. Ground support of the Hipparcos satellite observation programme. *A&AS*, 95, 541–579 {**34**}

Fehrenbach C, Duflot M, Burnage R, *et al.*, 1987b, Radial velocity measurements. II. Ground-based observations of the programme stars for the Hipparcos satellite. *A&AS*, 71, 275–295 {**34**}

Fehrenbach C, Duflot M, Mannone C, *et al.*, 1997, Radial velocities. VIII. Ground-based measurements for Hipparcos. *A&AS*, 124, 255–257 {**34**}

Fernley J, Barnes TG, 1997, Radial velocities and iron abundances of field RR Lyrae stars. I. *A&AS*, 125, 313–319 {**34**}

Fey AL, Boboltz DA, Gaume RA, *et al.*, 2006, MERLIN astrometry of 11 radio stars. *AJ*, 131, 1084–1089 {**30**}

Fienga A, Bange JF, Bec-Borsenberger A, *et al.*, 2003, Close encounters of asteroids before and during the ESA Gaia mission. *A&A*, 406, 751–758 {**44**}

Fienga A, Manche H, Laskar J, *et al.*, 2008, INPOP06: a new numerical planetary ephemeris. *A&A*, 477, 315–327 {**8**}

Folkner WM, Charlot P, Finger MH, *et al.*, 1994, Determination of the extragalactic-planetary frame tie from joint analysis of radio interferometric and lunar laser ranging measurements. *A&A*, 287, 279–289 {**10**}

Fomalont EB, Kopeikin SM, 2003, The measurement of the light deflection from Jupiter: experimental results. *ApJ*, 598, 704–711 {**41**}

Frankowski A, Jancart S, Jorissen A, 2007, Proper motion binaries in the Hipparcos catalogue. Comparison with radial velocity data. *A&A*, 464, 377–392 {**30**}

Fricke W, 1977a, Arguments in favour of a change in precession. *A&A*, 54, 363–366 {**26**}

—, 1977b, Basic material for the determination of precession and of Galactic rotation and a review of methods and results. *Veroeffentlichungen des Astronomischen Rechen-Instituts Heidelberg*, 28, 1 {**26**}

Fricke W, Schwan H, Lederle T, *et al.*, 1988, Fifth fundamental catalogue (FK5). I. The basic fundamental stars. *Veroeffentlichungen des Astronomischen Rechen-Instituts Heidelberg*, 32, 1–106 {**25**}

Frink S, Quirrenbach A, Fischer DA, *et al.*, 2001, A strategy for identifying the grid stars for the Space Interferometry Mission. *PASP*, 113, 173–187 {**44**}

Froeschlé M, Mignard F, Arenou F, 1997, Determination of the PPN parameter γ with the Hipparcos data. *ESA SP–402: Hipparcos, Venice '97*, 49–52 {**42, 43**}

García Sánchez J, Preston RA, Jones DL, *et al.*, 1999, Stellar encounters with the Oort Cloud based on Hipparcos data. *AJ*, 117, 1042–1055, erratum: 118, 600 {**34**}

Gatewood G, Russell J, 1974, Astrometric determination of the gravitational redshift of van Maanen 2. *AJ*, 79, 815–818 {**36**}

Gaume RA, 2003, Status of space-based astrometric missions. *The International Celestial Reference System: Maintenance and Future Realization, IAU Joint Discussion 16*, 48 {**44**}

—, 2005, Space astrometry in the next decade. *ASP Conf. Ser. 338: Astrometry in the Age of the Next Generation of Large Telescopes* (eds. Seidelmann PK, Monet AKB), 53 {**44**}

Gaume RA, Dorland B, Makarov VV, *et al.*, 2006, Space astrometry with the Milli-Arcsecond Pathfinder Survey (MAPS): mission overview and science possibilities. *Nomenclature, Precession and New Models in Fundamental Astronomy, IAU 26, JD16*, 16 {**44**}

Geffert M, Klemola AR, Hiesgen M, *et al.*, 1997, Absolute proper motions for the calibration of the Hipparcos proper motion system. *A&AS*, 124, 157–161 {**10**}

Gerbaldi M, Faraggiana R, Burnage R, *et al.*, 1999, Search for reference A0 dwarf stars: masses and luminosities revisited with Hipparcos parallaxes. *A&AS*, 137, 273–292 {**32**}

Gerbaldi M, Mayor M, 1989, Complementary astrophysical data for Hipparcos stars: profile of two key programmes. *ESO Messenger*, 56, 12–15 {**32**}

Girard TM, Dinescu DI, van Altena WF, *et al.*, 2004, The Southern Proper Motion Programme. III. A near-complete catalogue to V=17.5. *AJ*, 127, 3060–3071 {**23**}

Gontcharov GA, 2006, Radial velocities of 35 495 Hipparcos stars in a common system. *Astronomical and Astrophysical Transactions*, 25, 145–148 {**34, 35**}

Gontcharov GA, Andronova AA, Titov OA, 2000, New astrometric binaries among Hipparcos stars. *A&A*, 355, 1164–1167 {**29**}

Gontcharov GA, Andronova AA, Titov OA, *et al.*, 2001, The proper motions of fundamental stars. I. 1535 stars from the basic FK5. *A&A*, 365, 222–227 {**29**}

Gouda N, Yano T, Kobayashi Y, *et al.*, 2005, JASMINE: Japan Astrometry Satellite Mission for INfrared Exploration. *IAU Colloq. 196: Transits of Venus: New Views of the Solar System and Galaxy* (ed. Kurtz DW), 455–468 {**44**}

Grenier S, Baylac MO, Rolland L, *et al.*, 1999a, Radial velocities. Measurements of 2800 B2–F5 stars for Hipparcos. *A&AS*, 137, 451–456 {**32, 34**}

Grenier S, Burnage R, Faraggiana R, *et al.*, 1999b, Radial velocities of Hipparcos southern B8–F2 type stars. *A&AS*, 135, 503–509 {**32, 34**}

Griffin RF, 2006, Photoelectric radial velocities. XVII. The orbits of 30 spectroscopic binaries in the southern Clube Selected Areas. *MNRAS*, 371, 1159–1172 {**35**}

Griffin RF, Boffin HMJ, 2006, Spectroscopic binary orbits from photoelectric radial velocities. Paper 191: HD 17310, HD 70645, and HD 80731. *The Observatory*, 126, 401–421 {**35**}

Griffin RF, Suchkov AA, 2003, The nature of over-luminous F stars observed in a radial velocity survey. *ApJS*, 147, 103–144 {**34**}

Großmann V, Bässgen G, Evans DW, *et al.*, 1995, Results on Tycho photometry. *A&A*, 304, 110–115 {**15**}

Gullberg D, Dravins D, 1998, Spectroscopic radial velocities: photospheric lineshifts calibrated by Hipparcos. *Highlights in Astronomy*, 11, 564 {**35**}

Halbwachs JL, di Meo T, Grenon M, *et al.*, 1997a, The decensoring of the faint stars in Tycho photometry. *A&A*, 325, 360–366 {**15**}

Halbwachs JL, Piquard S, Virelizier P, *et al.*, 1997b, A statistical study of the visual double stars in the Tycho Catalogue. *ESA SP–402: Hipparcos, Venice '97*, 263–268 {**18**}

Hanson RB, Klemola AR, Jones BF, *et al.*, 2004, Lick Northern Proper Motion Programme. III. Lick NPM2 Catalogue. *AJ*, 128, 1430–1445 {**23**}

Harris HC, Dahn CC, Monet DG, 1997, Accurate ground-based parallaxes to compare with Hipparcos. *ESA SP–402: Hipparcos, Venice '97*, 105–108 {**22**}

Hemenway PD, Duncombe RL, Bozyan EP, *et al.*, 1997, The programme to link the Hipparcos reference frame to an extragalactic reference system using the Fine Guidance Sensors of the Hubble Space Telescope. *AJ*, 114, 2796–2810 {**10**}

Hensberge H, van Dessel EL, Burger M, *et al.*, 1990, High-precision radial velocity determinations for the study of the internal kinematical and dynamical structure and evolution of young stellar groups. *ESO Messenger*, 61, 20–21 {**32**}

Hering R, Lenhardt H, Walter HG, 1998, Revised VLA positions in relation to the Hipparcos Catalogue. *Astronomische Gesellschaft Meeting Abstracts*, 26 {**30**}

Hering R, Walter HG, 1998, Updating of B1950 radio star positions by means of J2000 calibrators. *Applied and Computational Harmonic Analysis*, 3, 198–200 {**30**}

Hirte S, Schilbach E, Scholz RD, 1997, The Potsdam contribution to the extragalactic link of the Hipparcos proper motion system. *A&AS*, 126, 31–37 {**10**}

Høg E, 1993, Astrometry and photometry of 400 million stars brighter than 18 mag. *Developments in Astrometry and their Impact on Astrophysics and Geodynamics* (eds. Mueller II, Kolaczek B), IAU Symp. 156, 37–45 {**43**}

Høg E, Bässgen G, Bastian U, *et al.*, 1997, The Tycho Catalogue. *A&A*, 323, L57–L60 {**15, 16, 18**}

Høg E, Bastian U, Halbwachs JL, *et al.*, 1995, Tycho astrometry from half of the mission. *A&A*, 304, 150–159 {**15**}

Høg E, Fabricius C, Makarov VV, *et al.*, 2000a, Construction and verification of the Tycho 2 Catalogue. *A&A*, 357, 367–386 {**19**}

—, 2000b, The Tycho 2 Catalogue of the 2.5 million brightest stars. *A&A*, 355, L27–L30 {**19**}

—, 2000c, *The Tycho 2 Catalogue: Positions, Proper Motions and Two-Colour Photometry of the 2.5 Million Brightest Stars*. Copenhagen University, also: VizieR Online Data Catalogue {**19**}

Høg E, Kuzmin A, Bastian U, *et al.*, 1998, The Tycho Reference Catalogue. *A&A*, 335, L65–L68 {**19**}

Høg E, Lindegren L, 1994, The Roemer satellite project: the first high-accuracy survey of faint stars. *Galactic and Solar System Optical Astrometry* (eds. Morrison LV, Gilmore G), 246–252 {**43**}

Holmberg J, Flynn C, Lindegren L, 1997, Towards an improved model of the Galaxy. *ESA SP–402: Hipparcos, Venice '97*, 721–726 {**15**}

Hoogerwerf R, Blaauw A, 2000, The Hipparcos, Tycho, TRC, and ACT catalogues: a whole sky comparison of the proper motions. *A&A*, 360, 391–398 {**19**}

Høyer P, Poder I, Lindegren L, *et al.*, 1981, Derivation of positions and parallaxes from simulated observations with a scanning astrometry satellite. *A&A*, 101, 228–237 {**2**}

Huang TY, Han CH, Yi ZH, *et al.*, 1995, What is the astronomical unit of length? *A&A*, 298, 629–633 {**9**}

Hu H, Wang R, Li X, 1999, Optical positions of 44 radio stars from astrolabe observations. *AJ*, 117, 3066–3069 {**30**}

Hu H, Wang R, 2002, Optical positions of 55 radio stars from astrolabe observations from the Yunnan Observatory. *A&A*, 383, 1062–1066 {**30**}

Jilinski E, Daflon S, Cunha K, *et al.*, 2006, Radial velocity measurements of B stars in the Scorpius–Centaurus association. *A&A*, 448, 1001–1006 {**34**}

Jin W, Li D, Xia Y, *et al.*, 2006, Second generation astrometric satellites and post-Hipparcos advances in ground-based astrometry. *Progress in Astronomy*, 24, 100–112 {**44**}

Johnston KJ, 2003, The FAME mission. *Future EUV/UV and Visible Space Astrophysics Missions and Instrumentation. Proc. SPIE 4854* (eds. Blades JC, Siegmund OHW), 303–310 {**44**}

Johnston KJ, de Vegt C, Gaume RA, 2003, VLA radio positions of stars: 1978–1995. *AJ*, 125, 3252–3257 {**30**}

Johnston KJ, de Vegt C, 1999, Reference frames in astronomy. *ARA&A*, 37, 97–125 {**9**}

Johnston KJ, Dorland B, Gaume RA, *et al.*, 2006, The Origins Billions Star Survey: Galactic Explorer. *PASP*, 118, 1428–1442 {**44**}

Johnston KJ, Fey AL, Gaume RA, *et al.*, 2004, The enigmatic radio source T Tauri S. *ApJ*, 604, L65–L68 {**31**}

Jones BF, 1976, Gravitational deflection of light: solar eclipse of 30 June 1973. *AJ*, 81, 455–463 {**40, 41**}

Jordi C, Høg E, Brown AGA, *et al.*, 2006, The design and performance of the Gaia photometric system. *MNRAS*, 367, 290–314 {**44**}

Kaplan GH, Snell SC, 2001, An independent assessment of a subset of Hipparcos and Tycho 2 proper motions. *American Astronomical Society Meeting*, 199 {**29**}

Katz D, Munari U, Cropper M, *et al.*, 2004, Spectroscopic survey of the Galaxy with Gaia. I. Design and performance of the Radial Velocity Spectrometer. *MNRAS*, 354, 1223–1238 {**33, 44**}

Kharchenko NV, Piskunov AE, Scholz RD, 2004, Astrophysical supplements to the ASCC 2.5. I. Radial velocity data. *Astronomische Nachrichten*, 325, 439–444 {**34, 35**}

Kharchenko NV, Scholz RD, Piskunov AE, *et al.*, 2007, Astrophysical supplements to the ASCC 2.5: Ia. Radial velocities of 55 000 stars and mean radial velocities of 516 Galactic open clusters and associations. *Astronomische Nachrichten*, 328, 889–896 {**34**}

Klioner SA, 2003, A practical relativistic model for microarcsecond astrometry in space. *AJ*, 125, 1580–1597 {**39–41, 43**}

Klioner SA, Peip M, 2003, Numerical simulations of the light propagation in the gravitational field of moving bodies. *A&A*, 410, 1063–1074 {**41**}

Knapp GR, Pourbaix D, Jorissen A, 2001, Reprocessing the Hipparcos data for evolved giant stars. II. Absolute magnitudes for the R-type carbon stars. *A&A*, 371, 222–232 {**14**}

Knapp GR, Pourbaix D, Platais I, *et al.*, 2003, Reprocessing the Hipparcos data of evolved stars. III. Revised Hipparcos period–luminosity relationship for Galactic long-period variable stars. *A&A*, 403, 993–1002 {**14**}

Kobayashi Y, Gouda N, Tsujimoto T, *et al.*, 2006a, A very small astrometry satellite mission: Nano-JASMINE. *Memorie della Societa Astronomica Italiana*, 77, 1186–1188 {**44**}

—, 2006b, Nano-JASMINE: a 10 kg satellite for space astrometry. *Space Telescopes and Instrumentation I, Proc. SPIE*, 6255 {**44**}

Kobayashi Y, Gouda N, Yano T, *et al.*, 2006c, Nano-JASMINE: a 10 kg satellite for space astrometry. *36th COSPAR Scientific Assembly*, volume 36 of *COSPAR, Plenary Meeting*, 1468 {**44**}

Kobayashi Y, Yano T, Gouda N, *et al.*, 2005, Nano-JASMINE: a nano size astrometry satellite. *IAU Colloq. 196: Transits of Venus: New Views of the Solar System and Galaxy* (ed. Kurtz DW), 491–495 {**44**}

Konopliv AS, Yoder CF, Standish EM, *et al.*, 2006, A global solution for the Mars static and seasonal gravity, Mars orientation, Phobos and Deimos masses, and Mars ephemeris. *Icarus*, 182, 23–50 {**8**}

Kopeikin SM, 2001, Theory of relativistic-reference frames for high-precision astrometric space misions. *Reference Frames and Gravitomagnetism* (eds. Pascual-Sánchez JF, Floriá L, San Miguel A, *et al.*), 79–92 {**41**}

Koposov SE, Bartunov OS, 2006, SAI catalogue access services. *The Virtual Observatory in Action: New Science, New Technology, and Next Generation Facilities, IAU 26, SPS3*, 3 {**22**}

Kostjuk ND, 2004, HD-DM-GC-HR-HIP-Bayer-Flamsteed Cross-Index. *VizieR Online Data Catalogue* {**38**}

Kovalchuk A, Pinigin G, Protsyuk Y, *et al.*, 1997, Mykolayiv automatic meridian circle positions of faint stars in selected fields around extragalactic radio sources for linking the optical and radio reference frames. *The New International Celestial Reference Frame, IAU Joint Discussion 7*, 16 {**30**}

Kovalevsky J, Lindegren L, Froeschlé M, *et al.*, 1995, Construction of the intermediate Hipparcos astrometric catalogue. *A&A*, 304, 34–43 {**5**}

Kovalevsky J, Lindegren L, Perryman MAC, *et al.*, 1997, The Hipparcos Catalogue as a realization of the extragalactic reference system. *A&A*, 323, 620–633 {**10**}

Kovalevsky J, 2003, Conditions of possible programmes using small and medium size ground-based astrometric instruments. *Astrometry from Ground and from Space* (eds. Capitaine N, Stavinschi M), 209–214 {**44**}

Krasinsky GA, 1997, Linking Hipparcos with DE200/DE403 reference frame by photoelectric observations of the occultation of ϵ Gem by Mars. *The New International Celestial Reference Frame, IAU Joint Discussion 7*, 17 {**30**}

Kuhn JR, Bush RI, Scherrer P, *et al.*, 1998, The Sun's shape and brightness. *Nature*, 392, 155–157 {**42**}

Kumkova II, 2000, Densification of ICRS in the optical using old observation sets. *Models and Constants for Submicroarcsecond Astrometry, IAU Joint Discussion 2*, 11 {**31**}

Kürster M, Endl M, Rodler F, 2006, In search of terrestrial planets in the habitable zone of M dwarfs. *The Messenger*, 123, 17–20 {**36–38**}

Kürster M, Endl M, Rouesnel F, *et al.*, 2003, The low-level radial velocity variability in Barnard's star: secular acceleration, indications for convective redshift, and planet mass limits. *A&A*, 403, 1077–1087 {**36, 38**}

Kuzmin A, Høg E, Bastian U, *et al.*, 1999, Construction of the Tycho Reference Catalogue. *A&AS*, 136, 491–508 {**19**}

Latham DW, 2000, Radial velocities. *Encyclopedia of Astronomy and Astrophysics;* (ed. Murdin, P.) {**32**}

Lattanzi MG, Massone G, Poma A, *et al.*, 2001, A contribution to the link of the Hipparcos Catalogue to ICRS. *J2000, a Fundamental Epoch for Origins of Reference Systems and Astronomical Models* (ed. Capitaine N), 33–36 {**30**}

Lazorenko PF, 2005, Atmospheric limitations to astrometric detection of extra-solar planets with very large telescopes. *Kinematika i Fizika Nebesnykh Tel Supplement*, 5, 537–540 {**44**}

—, 2006, Astrometric precision of observations at VLT–FORS2. *A&A*, 449, 1271–1279 {**44**}

Lazorenko PF, Lazorenko GA, 2004, Filtration of atmospheric noise in narrow-field astrometry with very large telescopes. *A&A*, 427, 1127–1143 {**44**}

Lazorenko PF, Mayor M, Dominik M, *et al.*, 2007, High-precision astrometry on the VLT–FORS1 at time scales of few days. *A&A*, 471, 1057–1067 {**44**}

Lebach DE, Corey BE, Shapiro II, *et al.*, 1995, Measurement of the solar gravitational deflection of radio waves using Very Long Baseline Interferometry. *Physical Review Letters*, 75, 1439–1442 {**40**}

Lederle T, Schwan H, 1984, Procedure for computing the apparent places of fundamental stars from 1984 onwards. *A&A*, 134, 1–6 {**14**}

Lenhardt H, Hering R, Walter HG, 1999, Search for unmodeled effects in Hipparcos abscissa residuals. *Astronomische Gesellschaft Meeting Abstracts*, 116 {**13**}

Lenhardt H, Walter HG, Hering R, 1997, Algorithm for calculating frame rotations and its application to Hipparcos. *ESA SP–402: Hipparcos, Venice '97*, 101–104 {**30**}

Lépine S, Shara MM, 2005, A catalogue of northern stars with annual proper motions larger than 0.15 arcsec (LSPM-North Catalogue). *AJ*, 129, 1483–1522 {**23**}

Lestrade JF, Preston RA, Jones DL, *et al.*, 1999, High-precision VLBI astrometry of radio-emitting stars. *A&A*, 344, 1014–1026 {**30**}

Levato H, Malaroda S, Morrell NI, *et al.*, 1996, Radial velocities and axial rotation for a sample of chemically peculiar stars. *A&AS*, 118, 231–238 {**32, 34**}

Lindegren L, 1995, Estimating the external accuracy of Hipparcos parallaxes by deconvolution. *A&A*, 304, 61–68 {**22**}

—, 1997, The Hipparcos Catalogue Double and Multiple Systems Annex. *ESA SP–402: Hipparcos, Venice '97*, 13–16 {**28**}

—, 2005, The astrometric instrument of Gaia: principles. *ESA SP–576: The Three-Dimensional universe with Gaia* (eds. Turon C, O'Flaherty KS, Perryman MAC), 29–34 {**2**}

Lindegren L, Bastian U, Gilmore G, *et al.*, 1993, *Roemer: Proposal for the Third Medium Size ESA Mission (M3)*. Lund Observatory {**43**}

Lindegren L, Dravins D, 2003, The fundamental definition of radial velocity. *A&A*, 401, 1185–1201 {**35**}

Lindegren L, Kovalevsky J, 1995, Linking the Hipparcos Catalogue to the extragalactic reference system. *A&A*, 304, 189–201 {**10**}

Lindegren L, Madsen S, Dravins D, 2000, Astrometric radial velocities. II. Maximum-likelihood estimation of radial velocities in moving clusters. *A&A*, 356, 1119–1135 {**35, 37**}

Lindegren L, Mignard F, Söderhjelm S, *et al.*, 1997, Double star data in the Hipparcos Catalogue. *A&A*, 323, L53–L56 {**12, 28**}

Lindegren L, Perryman MAC, 1996, Gaia: Global astrometric interferometer for astrophysics. *A&AS*, 116, 579–595 {**2, 44**}

Lindegren L, Röser S, Schrijver H, *et al.*, 1995, A comparison of ground-based stellar positions and proper motions with provisional Hipparcos results. *A&A*, 304, 44 {**5, 22**}

Lopes PAA, Andrei AH, Puliaev SP, *et al.*, 1999, Observations of radio stars at the Valinhos CCD meridian circle. *A&AS*, 136, 531–537 {**30**}

Lundmark K, Luyten WJ, 1922, On the secular change in the proper motion of Barnard's star. *PASP*, 34, 126–128 {**36**}

Lydon TJ, Sofia S, 1996, A measurement of the shape of the solar disk: the solar quadrupole moment, the solar octopole moment, and the advance of perihelion of the planet Mercury. *Physical Review Letters*, 76, 177–179 {**42**}

Madsen S, Dravins D, Lindegren L, 2002, Astrometric radial velocities. III. Hipparcos measurements of nearby star clusters and associations. *A&A*, 381, 446–463 {**35, 37**}

Madsen S, 2002, Radial velocities without spectroscopy: astrometric determination of stellar radial motion. Ph.D. Thesis {**35**}

Makarov VV, Fabricius C, 1999, Revisiting Hipparcos parallaxes of nearby stars using Tycho observations. *ASP Conf. Ser. 167: Harmonizing Cosmic Distance Scales in a Post-Hipparcos Era*, 267–270 {**13**}

Makarov VV, Kaplan GH, 2005, Statistical constraints for astrometric binaries with nonlinear motion. *AJ*, 129, 2420–2427 {**30**}

Malaroda S, Levato H, Morrell NI, *et al.*, 2000, Bibliographic catalogue of stellar radial velocities: 1991–94). *A&AS*, 144, 1–4 {**34, 35**}

Manrique WT, Lizhi L, Perdomo R, *et al.*, 1999, Radio star catalogue observed in San Juan. *A&AS*, 136, 7–11 {**30**}

Manrique WT, Lu L, Podesta RC, *et al.*, 1998, San Juan radio star catalogue and comparison with Hipparcos Catalogue. *Publications of the Beijing Astronomical Observatory*, 32, 87–92 {**30**}

Marco FJ, Martínez MJ, López JA, 2004, A critical discussion on parametric and non-parametric regression methods applied to Hipparcos–FK5 residuals. *A&A*, 418, 1159–1170 {**27**}

Martínez Usó MJ, Castillo FJM, López Ortí JA, 2006, Computation of the veritable inclination between FK5 and Hipparcos equators: a critical discussion. *Nomenclature, Precession and New Models in Fundamental Astronomy, IAU 26, JD16*, 16 {**27**}

Martin VAF, Poppe PCR, Leister NV, *et al.*, 2003, Ground-based astrometry: optical-radio connection. *The International Celestial Reference System: Maintenance and Future Realization, IAU Joint Discussion 16*, 6 {**30**}

Mason BD, Corbin TE, Urban SE, 2000, The suitability of Hipparcos doubles to represent the ICRF. *AAS/Division of Dynamical Astronomy Meeting*, 31 {**11**}

Mayor M, 1985, Radial velocity measurements of late-spectral-type stars. *The European Astrometry Satellite Hipparcos: Scientific Aspects of the Input Catalogue Preparation*, 217–218 {**32**}

Ma C, Arias EF, Eubanks TM, *et al.*, 1998, The International Celestial Reference Frame as realised by Very Long Baseline Interferometry. *AJ*, 116, 516–546 {**9**}

Mecheri R, Abdelatif T, Irbah A, *et al.*, 2004, New values of gravitational moments J_2 and J_4 deduced from helioseismology. *Sol. Phys.*, 222, 191–197 {**42**}

Metz M, Geffert M, 2004, Formalism and quality of a proper motion link with extragalactic objects for astrometric satellite missions. *A&A*, 413, 771–777 {**20**}

Mignard F, 2002, Observations of Solar System objects with Gaia. I. Detection of near-Earth objects. *A&A*, 393, 727–731 {**44**}

Mignard F, Froeschlé M, 2000, Global and local bias in the FK5 from the Hipparcos data. *A&A*, 354, 732–739 {**26, 27**}

Mignard F, Kovalevsky J, 2003, Space astrometry missions: principles and objectives. *Astrometry from Ground and from Space* (eds. Capitaine N, Stavinschi M), 169–176 {**44**}

Morrison LV, Garrington ST, Argyle RW, *et al.*, 1997a, Future control of the Hipparcos frame using MERLIN. *ESA SP–402: Hipparcos, Venice '97*, 143–146 {**30**}

Morrison LV, Hestroffer D, Taylor DB, *et al.*, 1997b, Check on JPL ephemerides DExxx using Hipparcos and Tycho observations. *ESA SP–402: Hipparcos, Venice '97*, 149–152 {**31**}

Munari U (ed.), 2003, *Gaia Spectroscopy: Science and Technology*, ASP Conf. Ser. 298 {**33, 44**}

Munn JA, Monet DG, Levine SE, *et al.*, 2004, An improved proper motion catalogue combining USNO B and the Sloan Digital Sky Survey. *AJ*, 127, 3034–3042 {**23**}

Murray CA, 1983, *Vectorial Astrometry*. Adam Hilger, Bristol {**8**}

Nefedjev YA, Rakhimov LI, Rizvanov N, *et al.*, 2006, Graphical method for estimation of accuracy of astrometric catalogues. *Kinematika i Fizika Nebesnykh Tel*, 22, 219–224 {**22**}

Nidever DL, Marcy GW, Butler RP, *et al.*, 2002, Radial velocities for 889 late-type stars. *ApJS*, 141, 503–522 {**34, 36**}

Nordström B, Mayor M, Andersen J, *et al.*, 2004, The Geneva–Copenhagen survey of the solar neighbourhood: ages, metallicities, and kinematic properties of ~14 000 F and G dwarfs. *A&A*, 418, 989–1019 {**33, 34**}

Odenkirchen M, Brosche P, Borngen F, *et al.*, 1997, Absolute stellar proper motions with reference to galaxies of the M81 group. *A&AS*, 124, 189–196 {**10**}

Odenkirchen M, Brosche P, 1999, The proper motion signal of unresolved binaries in the Hipparcos Catalogue. *Astronomische Nachrichten*, 320, 397 {**29**}

Paturel G, Fouque P, Bottinelli L, *et al.*, 1989, An extragalactic data base. I. The Catalogue of Principal Galaxies. *A&AS*, 80, 299–315 {**20**}

Perryman MAC, 2003, Conference summary. *ASP Conf. Ser. 298: Gaia Spectroscopy: Science and Technology* (ed. Munari U), 391–396 {**33**}

Perryman MAC, de Boer KS, Gilmore G, *et al.*, 2001, Gaia: composition, formation and evolution of the Galaxy. *A&A*, 369, 339–363 {**44**}

Perryman MAC, Lindegren L, Kovalevsky J, *et al.*, 1997, The Hipparcos Catalogue. *A&A*, 323, L49–L52 {**5, 7**}

Perryman MAC, Pace O, de Boer KS, *et al.*, 2000, *Gaia: Report on the Concept and Technology Study*. European Space Agency ESA–SCI(2000)4 {**41, 42**}

Perryman MAC, Turon C, Lindegren L, *et al.*, 1989, *The Hipparcos Mission: Pre-Launch Status, ESA SP–1111 (3 volumes)*. European Space Agency, Noordwijk {**4**}

Perryman MAC, van Leeuwen F (eds.), 1995, *Future Possibilities for Astrometry in Space*, European Space Agency, ESA SP-379 {**44**}

Peters J, 1907, Neuer Fundamentalkatalog des Berliner Astronomischen Jahrbuchs nach den Grundlagen von A. Auwers. Für die Epochen 1875 und 1900. *Veroeffentlichungen des Koeniglichen Astronomischen Rechen-Instituts zu Berlin*, 33, 1–116 {**25**}

Pireaux S, 2004, Light deflection in Weyl gravity: constraints on the linear parameter. *Classical and Quantum Gravity*, 21, 4317–4333 {**41**}

Platais I, Kozhurina-Platais V, Girard TM, *et al.*, 1998, The Hipparcos proper motion link to the extragalactic reference system using NPM and SPM. *A&A*, 331, 1119–1129 {**10**}

Pourbaix D, 2002, Precision and accuracy of the orbital parameters derived from 2d and 1d space observations of visual or astrometric binaries. *A&A*, 385, 686–692 {**33**}

Pourbaix D, Boffin HMJ, 2003, Reprocessing the Hipparcos intermediate astrometric data of spectroscopic binaries. II. Systems with a giant component. *A&A*, 398, 1163–1177 {**13**}

Pourbaix D, Jorissen A, 2000, Re-processing the Hipparcos transit data and intermediate astrometric data of spectroscopic binaries. I. Ba, CH and Tc-poor S stars. *A&AS*, 145, 161–183 {**13**}

Pourbaix D, Knapp GR, Jorissen A, 2000, Reprocessing the Hipparcos observations of C stars: improved parallaxes and luminosity calibration. *Hipparcos and the Luminosity Calibration of the Nearer Stars, IAU Joint Discussion 13*, 39 {**14**}

Quist CF, Lindegren L, 1999, The Hipparcos transit data: what, why and how? *A&AS*, 138, 327–343 {**13**}

Reasenberg RD, Shapiro II, MacNeil PE, *et al.*, 1979, Viking relativity experiment: verification of signal retardation by solar gravity. *ApJ*, 234, L219–L221 {**42**}

Reynolds JE, Jauncey DL, Staveley-Smith L, *et al.*, 1995, Accurate registration of radio and optical images of SN 1987A. *A&A*, 304, 116–120 {**31**}

Ristenpart F, 1902, *Vierteljahrsschrift der Astron. Ges.* 37, 242 {**36**}

Rizvanov N, Dautov I, Shaimukhametov R, 2001, The comparative accuracy of photographic observations of radio stars observed at the Engelhardt Astronomical Observatory. *A&A*, 375, 670–672 {**30**}

Roman NG, 1996, WDS-DM-HD-ADS Cross-Index. *VizieR Online Data Catalogue* {**38**}

Röser S, 1999, DIVA: beyond Hipparcos and towards Gaia. *Reviews in Modern Astronomy* (ed. Schielicke RE), 97–106 {**44**}

Röser S, Bastian U, 1991, *PPM Star Catalogue North. Positions and proper motions of 181 731 stars north of $-2\overset{\circ}{.}5$ declination for equinox and epoch J2000.0. Vol 1: Zones $+80°$ to $+30°$; Vol 2: Zones $+20°$ to $-0°$*. Akademischer Verlag, Heidelberg {**17**}

—, 1993, *PPM Star Catalogue South. Positions and proper motions of 197 179 stars south of $-2\overset{\circ}{.}5$ declination for equinox and epoch J2000.0. Vol 3: Zones $-0°$ to $-20°$; Vol 4: Zones $-30°$ to $-80°$*. Akademischer Verlag, Heidelberg {**17**}

—, 2000, DIVA, the next global astrometry and photometry mission. *Astronomische Gesellschaft Meeting Abstracts* (ed. Schielicke RE), 14 {**44**}

Royer F, Gerbaldi M, Faraggiana R, *et al.*, 2002, Rotational velocities of A-type stars. I. Measurement of $v \sin i$ in the southern hemisphere. *A&A*, 381, 105–121 {**32**}

Russell HN, Atkinson RdE, 1931, *Nature*. 173, 661-663 {**36**}

Rybka SP, 2003, Compiled catalogue of MK spectral classifications, including astrometric and photometric data. *Kinematika i Fizika Nebesnykh Tel*, 19, 87–92 {**38**}

—, 2005, Compiled catalogue of the MK spectral classifications, including astrometric and photometric data and its application. *Kinematika i Fizika Nebesnykh Tel Supplement*, 5, 385–387 {**38**}

Salim S, Gould A, 2003, Improved astrometry and photometry for the Luyten Catalogue. II. Faint stars and the revised catalogue. *ApJ*, 582, 1011–1031 {**23**}

Sanner J, Geffert M, 2001, The initial mass function of open star clusters with Tycho 2. *A&A*, 370, 87–99 {**19**}

Schlesinger F, 1917, On the secular changes in the proper motion elements of certain stars. *AJ*, 30, 137–138 {**35**}

Schwan H, 2001a, An analytical representation of the systematic differences Hipparcos–FK5. *A&A*, 367, 1078–1086 {**27**}

—, 2001b, Systematic relations between the Hipparcos Catalogue and major fundamental catalogues of the 20th century. I. *A&A*, 373, 1099–1109 {**27**}

—, 2002, Systematic relations between the Hipparcos Catalogue and major fundamental catalogues of the 20th century. II. *A&A*, 387, 1123–1134 {**27**}

Seeliger H, 1901, Bemerkung über veränderliche Eigenbewegungen. *Astronomische Nachrichten*, 154, 65 {**36**}

Seidelmann PK, 1992, *Explanatory Supplement to the Astronomical Almanac*. University Science Books, New York {**11**}

—, 1995, A Fizeau optical interferometer astrometric satellite. *ESA SP-379: Future Possibilities for Astrometry in Space* (eds. Perryman MAC, van Leeuwen F), 187–189 {**44**}

Sekowski MS, 2006, The influence of choice of fundamental catalogue on calculated apparent places of stars. *Nomenclature, Precession and New Models in Fundamental Astronomy, IAU 26, JD16*, 16 {**27**}

Shao M, 2004, Science overview and status of the SIM project. *New Frontiers in Stellar Interferometry, Proc. SPIE 5491* (eds. Traub WA, Bellingham W), 328–333 {**44**}

Shapiro II, 1964, Fourth test of General Relativity. *Physical Review Letters*, 13, 789–791 {**42**}

Shapiro SS, Davis JL, Lebach DE, *et al.*, 2004, Measurement of the solar gravitational deflection of radio waves using geodetic Very Long Baseline Interferometry data, 1979–1999. *Physical Review Letters*, 92(12), 121101–1–4 {**41, 43**}

Shatsky NI, Tokovinin AA, 1998, A test of Hipparcos parallaxes on multiple stars. *Astronomy Letters*, 24, 673–676 {**22**}

Sinnott RW, Perryman MAC, 1997, *Millennium Star Atlas (3 volumes)*. Sky Publishing Corporation & European Space Agency {**20**}

Smith CA, Kaplan GH, Hughes JA, *et al.*, 1989, Mean and apparent place computations in the new IAU system. I. The transformation of astrometric catalogue systems to the equinox J2000.0. II. Transformation of mean star places from FK4 B1950.0 to FK5 J2000.0 using matrices in 6-space. *AJ*, 97, 265–279 {**14**}

Soffel MH, 1989, *Relativity in Astrometry, Celestial Mechanics and Geodesy*. Springer–Verlag, Berlin {**43**}

Solano E, Garrido R, Fernley J, *et al.*, 1997, Radial velocities and iron abundances of field RR Lyrae stars. *A&AS*, 125, 321–327 {**34**}

Soubiran C, Bienaymé O, Siebert A, 2003, Vertical distribution of Galactic disk stars. I. Kinematics and metallicity. *A&A*, 398, 141–151 {**34**}

Soubiran C, Katz D, Cayrel R, 1998, On-line determination of stellar atmospheric parameters $T_{\rm eff}$, log g, [Fe/H] from Elodie echelle spectra. II. The library of F5 to K7 stars. *A&AS*, 133, 221–226 {**34**}

Souchay J, Le Poncin-Lafitte C, Andrei AH, 2007, Close approaches between Jupiter and quasars with possible application to the scheduled Gaia mission. *A&A*, 471, 335–343 {**41**}

Sperauskas J, Bartkevičius A, 2002, Radial velocities of Population II binary stars. I. *Astronomische Nachrichten*, 323, 139–148 {**34**}

Stavinschi M, 2001, Contributions of the ground-based astrometry to the reference systems. *J2000, a Fundamental Epoch for Origins of Reference Systems and Astronomical Models* (ed. Capitaine N), 48–53 {**44**}

—, 2003, Ground-based astrometry before Gaia and DIVA. *Journal of Astronomical Data*, 9, 8–13 {**44**}

—, 2006, The small telescopes still useful for the astrometry. *Nomenclature, Precession and New Models in Fundamental Astronomy, IAU 26, JD16*, 16 {**44**}

Steenbrugge KC, de Bruijne JHJ, Hoogerwerf R, *et al.*, 2003, Radial velocities of early-type stars in the Perseus OB2 association. *A&A*, 402, 587–605 {**34**}

Steinmetz M, Zwitter T, Siebert A, *et al.*, 2006, The Radial Velocity Experiment (RAVE): first data release. *AJ*, 132, 1645–1668 {**33, 34**}

Stine PC, O'Neal D, 1998, Radio emission from young stellar objects near LkHα 101. *AJ*, 116, 890–894 {**32**}

Stone RC, 1997, Systematic errors in the FK5 Catalogue as derived from CCD observations in the extragalactic reference frame. *AJ*, 114, 850–858 {**27**}

—, 1998, New observations testing the adopted Hipparcos link to the International Celestial Reference Frame. *ApJ*, 506, L93–L96 {**30**}

Strickland D, 1998, Book Review: the Hipparcos and Tycho Catalogues. *The Observatory*, 118, 167 {**22**}

Tammann GA, Reindl B, 2002, Gaia and the extragalactic distance scale. *Ap&SS*, 280, 165–182 {**44**}

Tang Z, Wang S, Jin W, 2000, Determination of optical positions for 23 extragalactic radio sources. *MNRAS*, 319, 717–720 {**30**}

Taylor DB, Shen KX, 1988, Analysis of astrometric observations from 1967 to 1983 of the major satellites of Saturn. *A&A*, 200, 269–278 {**31**}

Tongqi X, Peizhen L, Wejing J, 1997, Comparisons of results from Hipparcos with ground-based and radio observations. *ESA SP–402: Hipparcos, Venice '97*, 117–120 {**30**}

Torres S, García-Berro E, Isern J, *et al.*, 2005, Simulating Gaia performances on white dwarfs. *MNRAS*, 360, 1381–1392 {**44**}

Treuhaft RN, Lowe ST, 1991, A measurement of planetary relativistic deflection. *AJ*, 102, 1879–1888 {**40**}

Turon C, Réquième Y, Grenon M, *et al.*, 1995, Properties of the Hipparcos Input Catalogue. *A&A*, 304, 82–93 {**4**}

Turon C, 1998, *Celestia 2000: the Hipparcos and Tycho Catalogues, SP–1200 (CD)*. European Space Agency, Noordwijk {**22**}

Turon C, Crézé M, Egret D, *et al.*, 1992, *Hipparcos Input Catalogue, ESA SP–1136 (7 volumes)*. European Space Agency, Noordwijk {**4**}

—, 1996, Hipparcos Input Catalogue, Version 2. *VizieR Online Data Catalogue*, 1196 {**4**}

Turon C, O'Flaherty KS, Perryman MAC (eds.), 2005, *The Three-Dimensional Universe with Gaia*, European Space Agency ESA SP–576 {**44**}

Udry S, Mayor M, Andersen J, *et al.*, 1997, Coravel radial velocity surveys of late-type stars of the Hipparcos mission. *ESA SP–402: Hipparcos, Venice '97*, 693–698 {**33, 34**}

Unwin SC, Shao M, Tanner AM, *et al.*, 2008, Taking the measure of the Universe: precision astrometry with SIM PlanetQuest. *PASP*, 120, 38–88 {**44**}

Upgren AR, Boyle RP, Sperauskas J, *et al.*, 2006, Kinematics of nearby K-M dwarfs: first results. *Memorie della Societa Astronomica Italiana*, 77, 1168–1171 {**34**}

Urban SE, 2003, Densification of the ICRF/HCRF in visible wavelengths. *The International Celestial Reference System: Maintenance and Future Realization, IAU Joint Discussion 16*, 41 {**31**}

Urban SE, Corbin TE, Wycoff GL, *et al.*, 1998a, The AC 2000: the Astrographic Catalogue on the system defined by the Hipparcos Catalogue. *AJ*, 115, 1212–1223 {**19**}

—, 2001, The AC 2000.2 Catalogue. *VizieR Online Data Catalogue*, 1275 {**19**}

Urban SE, Corbin TE, Wycoff GL, 1998b, The ACT Reference Catalogue. *AJ*, 115, 2161–2166 {**19**}

Urban SE, Wycoff GL, Makarov VV, 2000, Comparisons of the Tycho 2 Catalogue proper motions with Hipparcos and ACT. *AJ*, 120, 501–505 {**20**}

van de Kamp P, 1981, *Stellar Paths: Photographic Astrometry with Long-Focus Instruments*. Reidel, Astrophysics and Space Science Library Vol. 85 {**36**}

van Langevelde HJ, Vlemmings WHT, Diamond PJ, *et al.*, 2000, VLBI astrometry of the stellar image of U Her, amplified by the 1667 MHz OH maser. *A&A*, 357, 945–950 {**31**}

van Leeuwen F, 1997, Application possibilities for the Hipparcos intermediate astrometric data. *ESA SP–402: Hipparcos, Venice '97*, 203–206 {**13**}

—, 2002, Hipparcos data validation. *Highlights in Astronomy*, 12, 657 {**7, 22**}

—, 2005, Rights and wrongs of the Hipparcos data. A critical quality assessment of the Hipparcos Catalogue. *A&A*, 439, 805–822 {**18**}

—, 2007a, *Hipparcos, the New Reduction of the Raw Data*. Springer, Dordrecht {**18**}

—, 2007b, Validation of the new Hipparcos reduction. *A&A*, 474, 653–664 {**18**}

van Leeuwen F, Evans DW, Grenon M, *et al.*, 1997, The Hipparcos mission: photometric data. *A&A*, 323, L61–L64 {**11, 17**}

van Leeuwen F, Evans DW, 1998, On the use of the Hipparcos intermediate astrometric data. *A&AS*, 130, 157–172 {**13**}

van Leeuwen F, Fantino E, 2003a, Dynamic modeling of the Hipparcos attitude. *Space Science Reviews*, 108, 537–576 {**18**}

—, 2003b, Introduction to a further examination of the Hipparcos data. *Space Science Reviews*, 108, 447–449 {**18**}

—, 2005, A new reduction of the raw Hipparcos data. *A&A*, 439, 791–803 {**18**}

van Leeuwen F, Feast MW, Whitelock PA, *et al.*, 2007, Cepheid parallaxes and the Hubble constant. *MNRAS*, 379, 723–737 {**19**}

van Leeuwen F, Penston MJ, 2003, The operational environment of the Hipparcos mission. *Space Science Reviews*, 108, 471–497 {**18**}

Velázquez PF, Rodríguez LF, 2001, VLA observations of Z CMa: the orientation and origin of the thermal jet. *Revista Mexicana de Astronomia y Astrofisica*, 37, 261–267 {**31**}

Vicente B, Abad C, Garzón F, 2007, Astrometry with Carte du Ciel plates, San Fernando zone. I. Digitization and measurement using a flatbed scanner. *A&A*, 471, 1077–1089 {**23**}

Vlemmings WHT, van Langevelde HJ, Diamond PJ, *et al.*, 2003, VLBI astrometry of circumstellar OH masers: proper motions and parallaxes of four asymptotic giant branch stars. *A&A*, 407, 213–224 {**31**}

Vlemmings WHT, van Langevelde HJ, Diamond PJ, 2002, Astrometry of the stellar image of U Her amplified by the circumstellar 22 GHz water masers. *A&A*, 393, L33–L36 {**31**}

Walter HG, Hering R, 2005, Precession from Hipparcos and FK5 proper motions compared with current values: reasons for discrepancies. *A&A*, 431, 721–727 {**26, 27**}

Wicenec A, van Leeuwen F, 1995, The Tycho star mapper background analysis. *A&A*, 304, 160–167 {**15**}

Wielen R, 1995, A method for determining the individual accuracy of astrometric catalogues. *A&A*, 302, 613–622 {**22**}

—, 1997, Principles of statistical astrometry. *A&A*, 325, 367–382 {**13, 28**}

Wielen R, Dettbarn C, Jahreiß H, *et al.*, 1999a, Indications on the binary nature of individual stars derived from a comparison of their Hipparcos proper motions with ground-based data. I. Basic principles. *A&A*, 346, 675–685 {**28**}

—, 2000a, A finding list for $\Delta\mu$ binaries derived from a comparison of Hipparcos proper motions with long-term averaged data. *IAU Symp. 200*, 144 {**29**}

Wielen R, Lenhardt H, Schwan H, *et al.*, 1999b, The combination of ground-based astrometric compilation catalogues with the Hipparcos Catalogue. I. Single-star mode. *A&A*, 347, 1046–1054 {**28**}

—, 2001a, The combination of ground-based astrometric compilation catalogues with the Hipparcos Catalogue. II. Long-term predictions and short-term predictions. *A&A*, 368, 298–310 {**28**}

Wielen R, Schwan H, Dettbarn C, *et al.*, 1997, Statistical astrometry based on a comparison of individual proper motions and positions of stars in the FK5 and in the Hipparcos Catalogue. *ESA SP–402: Hipparcos, Venice '97*, 727–732 {**27**}

—, 1999c, Sixth Catalogue of Fundamental Stars (FK6). I. Basic fundamental stars with direct solutions. *Veroeffentlichungen des Astronomischen Rechen-Instituts Heidelberg*, 35 {**28**}

—, 1999d, The combination of Hipparcos data with ground-based astrometric measurement. *Modern Astrometry and Astrodynamics* (eds. Dvorak R, Haupt HF, Wodnar K), 161 {**27**}

—, 2000b, Sixth Catalogue of Fundamental Stars (FK6). III. Additional fundamental stars with direct solutions. *Veroeffentlichungen des Astronomischen Rechen-Instituts Heidelberg*, 37 {**29**}

—, 2001b, Astrometric Catalogue ARIHIP. Containing stellar data selected from the combination catalogues FK6, GC+HIP, TYC2+HIP and from the Hipparcos Catalogue. *Veroeffentlichungen des Astronomischen Rechen-Instituts Heidelberg*, 40 {**29**}

—, 2001c, Astrometric Catalogue GC+HIP. Derived from a combination of Boss' General Catalogue with the Hipparcos Catalogue. *Veroeffentlichungen des Astronomischen Rechen-Instituts Heidelberg*, 38 {**29**}

—, 2001d, Astrometric Catalogue TYC2+HIP. Derived from a combination of the Hipparcos Catalogue with the proper motions given in the Tycho 2 Catalogue. *Veroeffentlichungen des Astronomischen Rechen-Instituts Heidelberg*, 39, 1 {**29**}

Wilkinson MI, Vallenari A, Turon C, *et al.*, 2005, Spectroscopic survey of the Galaxy with Gaia. II. The expected science yield from the radial velocity spectrometer. *MNRAS*, 359, 1306–1335 {**33, 44**}

Wilkins GA, 1990, The past, present and future of reference systems for astronomy and geodesy. *IAU Symp. 141: Inertial Coordinate Systems on the Sky* (eds. Lieske JH, Abalakin VK), 39–45 {**8**}

Williams JG, Newhall XX, Dickey JO, 1996, Relativity parameters determined from lunar laser ranging. *Phys. Rev. D*, 53, 6730–6739 {**42**}

Will CM, 1993, *Theory and Experiment in Gravitational Physics*. Cambridge University Press, Second Edition {**40, 42**}

—, 2006, The confrontation between General Relativity and experiment. *Living Reviews in Relativity*, 9, 3, cited on 2007-10-01 {**43**}

Wilson RE, 1953, General Catalogue of Stellar Radial Velocities. *Carnegie Institute, Washington D.C.* {**34, 35**}

Wright CO, Egan MP, Kraemer KE, *et al.*, 2003, The Tycho 2 spectral type catalogue. *AJ*, 125, 359–363 {**38**}

Zacharias N, 2006, Accurate optical reference catalogues. *Nomenclature, Precession and New Models in Fundamental Astronomy, IAU 26, JD16*, 16 {**22**}

Zacharias MI, Zacharias N, 2005, Radio-optical reference frame link: first results using dedicated astrograph reference stars. *ASP Conf. Ser. 338: Astrometry in the Age of the Next Generation of Large Telescopes* (eds. Seidelmann PK, Monet AKB), 184–187 {**30**}

Zacharias N, Dorland B, 2006, The concept of a stare-mode astrometric space mission. *PASP*, 118, 1419–1427 {**44**}

Zacharias N, Gaume RA, Dorland B, *et al.*, 2004, *Catalogue information and recommendations*. US Naval Observatory, http://ad.usno.navy.mil/star/star_cats_rec.shtml {**22, 23**}

Zacharias N, Zacharias MI, Hall DM, *et al.*, 1999, Accurate optical positions of extragalactic radio reference frame sources. *AJ*, 118, 2511–2525 {**30**}

Zhu Z, Yang T, 1999, Overall pattern comparison of the FK5 proper motion system with Hipparcos. *AJ*, 117, 1103–1106 {**26**}

Zhu Z, 2000, Systematic differences between the FK5 proper motion system and Hipparcos. *PASP*, 112, 1103–1111 {**26–28**}

—, 2001, Hipparcos reference frame: property of inertiality. *Progress in Astronomy*, 19, 302–306 {**30**}

—, 2003, Hipparcos proper motion system with respect to FK5 and SPM 2.0 systems. *Astrometry from Ground and from Space* (eds. Capitaine N, Stavinschi M), 21–26 {**26**}

—, 2006, Precessional correction and the proper motion systems of FK5 and Hipparcos. *Acta Astronomica Sinica*, 47, 456–466 {**27**}

—, 2007, Precession constant correction and proper motion systems of FK5 and Hipparcos. *Chinese Astronomy and Astrophysics*, 31, 296–307 {**26**}

Zuckerman B, Webb RA, 2000, Identification of a nearby stellar association in the Hipparcos Catalogue: implications for recent, local star formation. *ApJ*, 535, 959–964 {**34**}

2

Derived catalogues and applications

2.1 Introduction

The use of photography to determine star positions began around 1870, flourished with the immense international cooperation of the Carte du Ciel project to map the entire celestial sphere to about 15 mag, and remained one of the most important astrometric techniques until the last decade or so of the twentieth century. Schmidt telescopes were constructed, from the 1930s onwards, with astrometry as their main objective. Such surveys have only recently been superseded by ground-based digital surveys in terms of classifying large numbers of very faint objects. In parallel, fast and accurate measuring machines and associated reduction software were developed. Kovalevsky (2002) provides details of the underlying techniques, including image formation, atmospheric effects, and plate measurements and reductions.

The development of stellar reference frames during the second half of the twentieth century has comprised both meridian and photographic observational campaigns, the former to provide reference stars with a density of about one star per square degree for the reduction of the plates obtained in the latter.

Hipparcos has allowed a re-calibration of basic meridian circle observations, of photographic plates, and of other classical astrometric instruments (Figure 2.1). Telescopes used may be classical astrographs (typified by the Carte du Ciel refracting astrograph of field $\sim 2^\circ$), Schmidt telescopes (with a larger field of view of $\sim 6^\circ$, using reflectors to minimise chromatic aberration and a correcting optical element to control spherical aberration), and long-focus instruments to obtain a larger image scale: either using refractors equipped with photographic plates, such as the Sproul in Princeton, or using reflectors as in the US Naval Observatory 1.55-m Strand telescope at Flagstaff (Figure 2.15 below).

In the reduction of photographic plates, the transformation between an area of sky on the celestial sphere, and the xy domain of a set of plate measurements, is generally a one-to-one vector transformation which must, however, take account of many factors: orientation of the plate axes, plate tilt and centre, scale factor, gnomonic projection, coma and field curvature, distortion, astigmatism, differential atmospheric refraction, chromatic aberrations, and magnitude equation (i.e. systematic positional errors as a function of magnitude). The tools for determining and applying such transformations have, of course, been thoroughly developed. Discussions of various plate-constant models can be found in, e.g. Eichhorn & Williams (1963), in the monograph by Kiselev (1989), and in the more recent introduction of treatment by orthogonal polynomials by Brosche *et al.* (1989) and Bienaymé (1993).

Before Hipparcos, a remaining obstacle was that a sufficiently accurate stellar reference frame, of adequate star density, was not available. Both attributes are necessary. A high-accuracy reference framework is required to exploit the intrinsic plate accuracy, of typically 0.1 arcsec or a little better, while a high reference star density is needed to model behaviour more complex than, say, a third-order polynomial in xy. This situation changed with the availability of the Hipparcos and Tycho Catalogues, for which accurate proper motions also ensure that a high-quality reference frame can be propagated over half a century or more, depending on application. As a result, the decade following the release of the Hipparcos and Tycho Catalogues has seen a substantial effort from groups who have re-reduced, and in many cases also re-measured, a wide variety of plate material obtained over the last 100 years.

The following sections provide a review of some of the most substantial of these investigations, and a short summary of some of the other related work.

Ground-based astrometric instruments: Astrometric measurements fall into two broad categories: narrow-field, providing relative measurements over a few degrees (photography, CCD imaging, speckle interferometry, and phase-referencing radio interferometry), and wide-field or semi-global techniques (meridian circles or transit instruments, astrolabes, optical phase interferometers, Michelson interferometry as used in NPOI or SIM, radio interferometry including VLBI, and astrometric satellites). A summary of the more specialised astrometric instruments used, now and in the recent past, is given here to provide a background for some of the comparisons and discussions made elsewhere.

* Photographic astrometry and direct CCD imaging: this is categorised, according to the instrument's focal length, image scale, and field of view, into classical astrographs (as used for the Carte du Ciel project), Schmidt telescopes to extend the field of view by dealing with spherical aberration, and long-focus instruments to obtain a larger scale (originally based on refractors to provide better mechanical stability, but more recently using reflectors). More modern astrographs, e.g. those at Lick, Yale, and the planned USNO Robotic Astrometric Telescope (URAT Zacharias *et al.*, 2006), have fields around 4.°5, while the USNO Twin and Hamburg Zone Astrographs have fields around 9° diameter;

* Meridian circles: the main problem in wide-field or semi-global astrometry is the measurement of large angles, and the provision of a fundamental system in the absence of any other reference frame. For several centuries, and still in use today, the basic instrument was the meridian circle: a combination of a transit instrument and a vertical circle, from which the right ascension is deduced from the local sidereal time and the declination from the zenith distance at the time of meridian transit. Horizontal meridian circles were developed to improve mechanical stability. The position of the Sun, when needed for the determination of the equinox, was generally obtained from classical meridian circles equipped with filters.

* Equal altitude instruments, including the astrolabe and photographic zenith tube, provide improved refraction corrections and mechanical stability. Rather than using the meridian plane as reference, they exploit the choice of a small horizontal circle, also called the parallel of altitude or almucantar, and defined by a constant apparent zenith distance. In the astrolabe, a horizontal telescope produces two images of the same star which merge into one another when the star crosses a horizontal circle of a given zenith distance. The time of this event corresponds to the moment at which the star crosses a horizontal circle with an apparent zenith distance chosen and defined by the construction of the astrolabe. Although abandoned for stellar measurements, solar astrolabes are still used for measurement of the apparent dimension of the Sun, by measuring the times at which the solar limbs are tangent to the almucantar. Astrometric observations of the Sun are difficult due to its luminosity, large angular diameter, and instrumental heating effects. The photographic zenith tube is a particular case of constant zenith distance, with $z = 0$. The measurement of the position of the zenith with respect to stars was primarily used as a contribution to the determination of the rotation of the Earth (Section 10.4.2). Comprising a vertical tube with a lens at its upper end and a mercury mirror at the bottom, each star could be photographed four times in different configurations during a transit. For several decades it was the instrument of choice for this purpose, but is now obsolete.

These and other techniques, including Hipparcos and the Hubble Fine Guidance Sensor, optical and radio interferometry, lunar occultations, laser and radar ranging, and GPS, are covered by Kovalevsky (2002). A readable account of the development of astrometric instrumentation between 1500–1850 has been given by Chapman (1990).

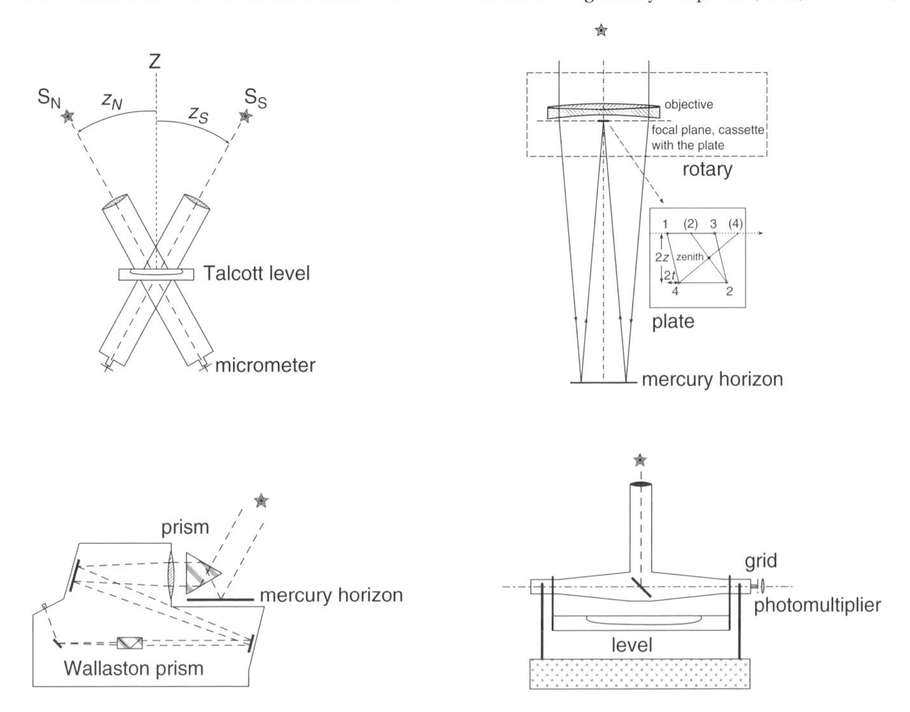

Figure 2.1 Principles of: top left: the visual zenith tube; top right: the photographic zenith tube; bottom left: the Danjon astrolabe; bottom right: the photoelectric transit instrument. The visual zenith tube and photoelectric transit instrument were used primarily in the determination of the Earth's rotation (Section 10.4.2). From Vondrák & Ron (2000, Figures 1, 2, 3, and 5).

2.2 Reference system for meridian circles

Introduction The main problem in wide-field or semi-global astrometry is the measurement of large angles. For several centuries, and still in use today, the basic instrument was the meridian circle (or transit circle): a combination of a transit instrument and a vertical circle, from which the right ascension is deduced from the local sidereal time, and the declination from the zenith distance at the time of meridian transit (Figure 2.1). But as noted by Kapteyn (1922): *'I know of no more depressing thing in the whole domain of astronomy, than to pass from the consideration of the accidental errors of our star places to that of their systematic errors. Whereas many of our meridian instruments are so perfect that by a single observation they determine the coordinates of an equatorial star with a probable error not exceeding 0.2–0.3 arcsec, the best result to be obtained from a thousand observations at all of our best observatories together may have a real error of half a second of arc and more.'* Blaauw (1998) recalls that even 50 years ago there was no unique fundamental system in which 'meridian' proper motions were defined, and researchers had the choice between the system defined by the 1000 or so stars in the FK3 and FK4, or that embodied by the General Catalogue of about 33 000 stars (Boss, 1937). The situation improved with the availability of the more accurate reference frame of the FK5, but systematics even in this latest of the ground-based fundamental catalogues left its imprints in the meridian circle catalogues, photographic plate catalogues, and others, well into the 1990s.

The central role of the 1535 bright stars of the FK5 Catalogue in defining the primary optical reference frame has already been presented in Chapter 1. It was itself the culmination of an extensive compilation of about 260 individual catalogues, observed by (mostly visual) meridian circles and more recently by some astrolabes. The FK5 defined a reference system with mean positional errors of about 0.03 arcsec at the mean catalogue epoch (1955 in α, 1944 in δ), individual mean errors of the proper motions of 0.6 mas yr^{-1} for the northern hemisphere and 1.0 mas yr^{-1} for the southern, and a mean error of the proper motion system of about 0.5 mas yr^{-1} down to $-20°$, increasing to about 1.0 mas yr^{-1} in the south (Fricke *et al.*, 1988), on top of which systematic errors in the system, principally in declination but also in right ascension south of $-50°$ had been identified (Schwan, 1993).

The pre-Hipparcos state-of-the-art in the treatment of mechanical flexure, thermal flexure, and internal refraction (or tube refraction) in meridian circle astrometry was described by Høg & Miller (1986). Even later, some 300 years after the introduction of meridian circle astrometry, a significant error source was identified and corrected by Høg & Fabricius (1988). Frequent determination of atmospheric refraction is required in fundamental observations of declination where large angles are bridged, but this had been hampered by internal refraction in the tube. The effect is as large as 1 arcsec, and quite variable, but had passed unidentified. Høg & Fabricius (1988) eliminated the cause, the temperature gradient in the tube, by a forced ventilation, resulting in an improved determination of atmospheric refraction. Ten years later the Hipparcos and Tycho Catalogues had made the measurement of large angles from the ground superfluous.

The Hipparcos contribution Réquième *et al.* (1993) and Réquième *et al.* (1995) carried out the first Hipparcos-based reductions of observations made by the Bordeaux and La Palma automatic meridian circles (see Figure 2.2), in which the positions of the FK5 reference stars were replaced by those in the Hipparcos 18-month intermediate solution (with no account taken of the poorly constrained 18-month proper motions), intended as an independent confirmation of the quality of the provisional Hipparcos reference frame.

The basic procedure for a differential reduction of meridian circle observations is the determination of a set of instrumental parameters considered as unknowns in a least-squares solution. For each night, two independent condition equations in right ascension and declination are considered, each corresponding to a reference star for which the apparent position can be constructed from the mean position given in the reference catalogue. Those used at Bordeaux were of the form

$$(\alpha - t_{\rm obs})\cos\delta = m\cos\delta + n\sin\delta + c + k(t_{\rm obs} - t_0) \tag{2.1}$$

$$\delta - \delta_{\rm obs} = P + k_1(t_{\rm obs} - t_0) + k_2\tan z + k_3\sin z \tag{2.2}$$

in right ascension and declination respectively, where $t_{\rm obs}$ is the transit time, t_0 an arbitrary origin, $\delta_{\rm obs}$ is the measured declination after applying refraction and circle division error corrections, and z is the zenith distance. Terms in $(t_{\rm obs} - t_0)$ correspond to instrument parameters, assumed to vary linearly with time. The equations yield residuals for all observed FK5 stars, comprising random and systematic instrument errors, as well as individual and systematic errors of the FK5 Catalogue. If the same long-term characteristics appear in the residuals of two different instruments, errors due to the reference system itself can reasonably be surmised.

Réquième *et al.* (1995) reported results from the Hipparcos-based re-reduction of some eight years

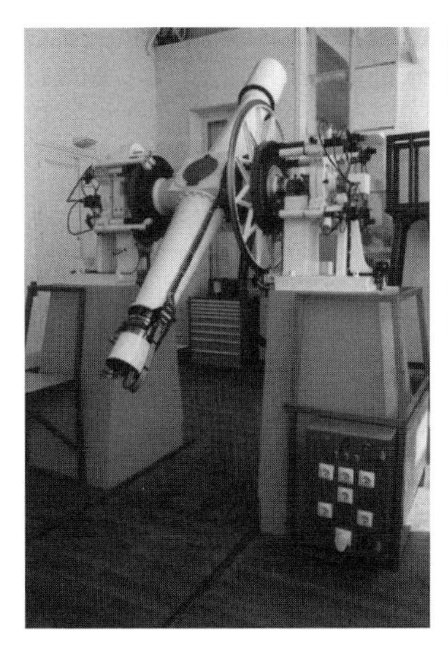

Figure 2.2 Left: the Bordeaux meridian circle was automated in 1978–79, and equipped with the first CCD detectors in 1994. Middle: the Carlsberg Meridian Telescope (formerly the Carlsberg Automatic Meridian Circle), La Palma, started observing in 1984, was upgraded to remote operation in 1997, and further upgraded with a CCD camera operating in drift-scan mode later the same year. Right: the USNO 6-inch transit circle, built in 1898 by Warner & Swasey, used continuously from then until 1995, but no longer actively used for astrometric observations. Photographs courtesy of the respective observatories (Bordeaux, La Palma, US Naval).

(about 1200 nights) of the Bordeaux automatic meridian circle, and some three years of the Carlsberg circle (1988–90), corresponding to catalogues CMC 5 and CMC 6. Results showed an excellent improvement in the post-fit residuals of the classical least-squares adjustment for both instruments (Figure 2.3). The residuals nevertheless showed small systematic errors, of order 0.03 arcsec in α and 0.05 arcsec in δ, attributed to possible defects in the measurements of the pivots, or in the calibration of the divided circle. After correction of these errors, and after incorporating an improved modelling of the instrumental parameters, a new reduction of the Bordeaux observations was made using the subsequently-available Hipparcos 30-month provisional solution, this time also making use of proper motions. Standard deviations of the residuals for 679 FK5 stars dropped to 0.024 arcsec in α and 0.040 arcsec in δ. Further tests on 173 confirmed or suspected radio stars observed at Bordeaux led to standard deviations of Bordeaux–Hipparcos (30 month) positions of 0.045 arcsec in α and 0.068 arcsec in δ, indications of the 'ultimate' accuracy to be expected in differential meridian astrometry (although still based only on the FK5 stars, which are a small subset of the Hipparcos reference stars). A detailed comparison of the Carlsberg optical reference frame with the ICRF by Argyle *et al.* (1996) showed similar systematics in the reference frame of around 0.05 arcsec in α and δ in the northern hemisphere, and around 0.07 arcsec in α and 0.10 arcsec in δ in the southern (see also Smart *et al.*, 1997).

Availability of the Hipparcos optical reference frame made classical visual meridian instruments obsolete, and several automatic circles were subsequently closed down due to their uncompetitive limiting magnitude and observing efficiency (Viateau *et al.*, 1999). However from 1988 onwards, it was demonstrated at Flagstaff (Stone, 1993; Stone *et al.*, 1996) that a CCD detector at the focus of a meridian circle, and working in drift-scan mode, could give internal errors of less than 50 mas to $V = 15-16$ mag, based on bright reference stars (down to 9 mag or so) from the Hipparcos and Tycho Catalogues, and using the dense reference system to allow for a better modelling of atmospheric refraction and other instrumental effects (Réquième *et al.*, 1995). Plans to equip other meridian circles with CCDs were followed up by Bordeaux (Viateau *et al.*, 1999, see Figure 2.4), Carlsberg at La Palma (Smart *et al.*, 1999), Mykolaiv in Ukraine (formerly Nikolayev) (Pinigin *et al.*, 1997), Valinhos at São Paulo (Lopes *et al.*, 1999), and the San Fernando (Cádiz) instrument at El Leoncito in Argentina (Muiños *et al.*, 2002).

First experiments using CCD meridian circle data reduced using Hipparcos/Tycho positions as reference, using instruments at Bordeaux (Φ = 202 mm, f = 2368 mm, declination field 28 arcmin) and Valinhos

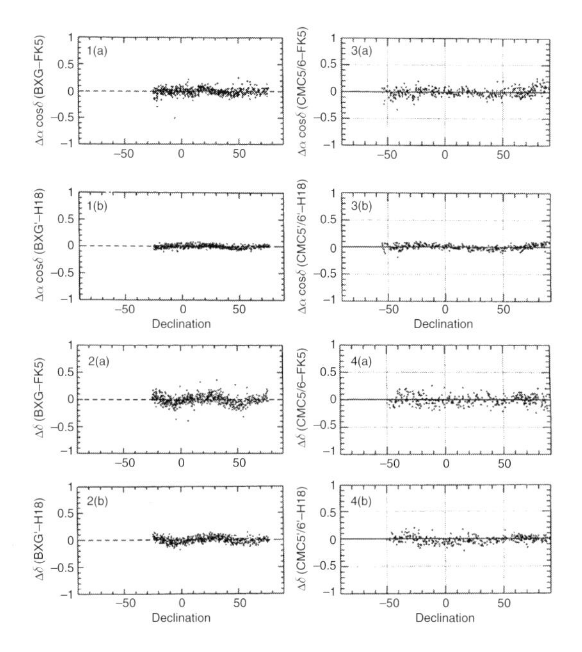

Figure 2.3 Top left pair: $\Delta\alpha$ versus δ for Bordeaux–FK5 (top) and Bordeaux (re-reduced)–Hipparcos (18 month). Bottom left pair: $\Delta\delta$ versus δ for the same instruments. Right figures: same for the Carlsberg (CMC 5/CMC 6) observations. The larger dispersions in $\Delta\delta$ arise principally from problems in modelling refraction. From Réquième et al. (1995, Figures 1–4).

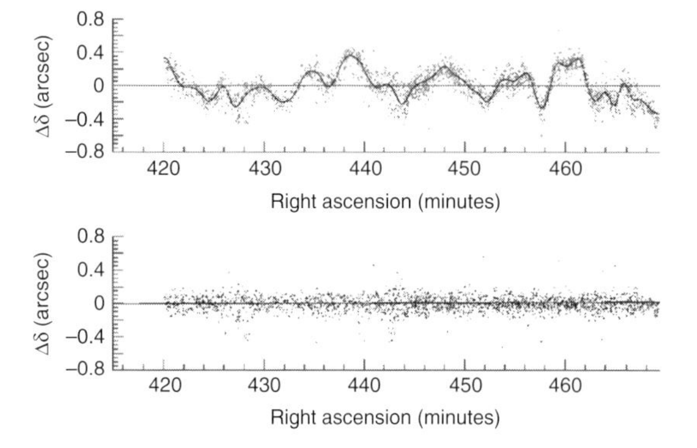

Figure 2.4 Top: residuals in declination between a preliminary reduction of a Bordeaux meridian circle strip and the mean catalogue positions (Hipparcos/Tycho densified by ACT) for a field around NGC 2384, showing residuals for stars up to V = 15 mag; σ = 0.180 arcsec. The example corresponds to poor observing conditions, at zenith distance ~ 66°, where atmospheric turbulence is very significant. A B-spline fit is shown. Bottom: after subtraction of the B-spline; σ = 0.076 arcsec. From Viateau et al. (1999, Figure 4).

São Paulo (Φ = 190 mm, f = 2590 mm, declination field 14 arcmin), were reported by Réquième *et al.* (1997), Teixeira *et al.* (1998) and Viateau *et al.* (1998, 1999). For both instruments, positional errors of about 0.05 arcsec were achieved for stars in the range 9–15 mag observed at least four times. Programmes for Bordeaux included observations of Uranus, Neptune and Pluto, some planetary satellites, selected asteroids for mass determinations, stars in open clusters, link to the extragalactic reference frame via bright quasars and radio stars, and the project 'Meridian 2000' including observations of the Bordeaux Astrographic Catalogue zone (+11° to +17°) and measurements of the Bordeaux Carte du Ciel plates intended to yield accurate proper motions to V = 15 mag (Ducourant & Rapaport, 2003).

Other re-reductions with respect to the Hipparcos and Tycho Catalogues, and plans for densification of the reference frame, were given by Kovalchuk *et al.* (1997) for the Kyiv (Ukraine) Mykolaiv Observatory's automatic axial meridian circle. This allowed characterisation of the horizontal flexure and collimator inclination as a function of temperature. Tel'nyuk-Adamchuk *et al.* (2002) reported the instrument's CCD refurbishment, leading to observations reaching $V = 16-17$ mag, with internal positional accuracies of a single observation of 0.05–0.10 arcsec. Programmes underway are observations of stars located around 209 extragalactic radio sources in the declination zone $0-30°$, and the extension of the Hipparcos–Tycho reference frame in the equatorial zone (Babenko *et al.*, 2005; Lazorenko *et al.*, 2005).

Yoshizawa & Suzuki (1997) compared reductions of the 6th annual catalogue of 6649 stars from the Tokyo photoelectric meridian circle observations between 1990–1993, made with respect to FK5 and with respect to positions from the preliminary Hipparcos H30 solution.

A larger 2k×2k CCD camera equipped with a Sloan r' filter was installed on the Carlsberg Meridian Telescope (formerly the Carlsberg Automatic Meridian Circle) in 1998. This took the magnitude limit to r' = 17 mag, and a positional accuracy in the range 0.03–0.05 arcsec. The main goal was to map the sky in the declination range $-30° < \delta < +50°$ to provide an astrometric and photometric catalogue capable of transferring the Hipparcos/Tycho reference frame to Schmidt plates (Evans, 2001; Evans *et al.*, 2002). The photometric data yield extinction data for La Palma as a valuable by-product.

The US Naval Observatory pole-to-pole catalogue $W2_{J00}$ (Rafferty & Holdenried, 2000) resulted from observations made between 1985–1996 using two transit circles, one located in Washington DC (the 6-inch circle, operated there since 1897), and the other in Blenheim, New Zealand (a 7-inch circle, originally operated since its construction in 1948 from El Leoncito, Argentina). The catalogue designation refers to the equinox of J2000.0, and extends the earlier programmes, $W5_{50}$ and $W1_{J00}$. Over 737 000 individual observations completely covering both hemispheres were made, primarily of the International Reference Stars (IRS) and FK5 stars, as well as all the classical planets (i.e. except Pluto), 13 asteroids, and some 55 000 observations of day-time objects including the Sun, Mercury, Venus, and Mars. The original objective was to form a traditional, all-sky catalogue of absolute star positions which could be firmly linked to the dynamical system. However, with the success of Hipparcos and the consequent adoption of the ICRF as the celestial reference frame, the primary focus of the pole-to-pole programme changed. The stellar positions were differentially reduced to the system of Hipparcos, and these were used to tie the planetary observations into the ICRF, resulting in a catalogue free of most of the systematic errors that would have remained if the original plan of absolute reductions had been followed. The programme resulted in a body of high-quality observational data, with an average standard deviation of a mean position of about 75 mas, that will provide input for the production of ICRF-based ephemerides, particularly for the outer planets and asteroids. As noted by the authors *'This project is the latest and largest of a long series of transit circle catalogues produced by the US Naval Observatory. It is also, because of advancing technologies, certainly the last.'*

These results refer to the some of earliest experiments making use of the Hipparcos data to reduce meridian circle observations. Subsequently, all such observations have been routinely carried out or reduced with respect to the Hipparcos reference frame in one form or another. In the process, the historic use of meridian circle or transit instruments to establish a fundamental reference frame has been made obsolete: after upgrading to use CCD detectors and with operation in drift-scan mode, instruments like the Carlsberg and FASTT now essentially operate in a narrow-field mode using many reference stars, i.e. without use of the circle readings to derive the declination, or of the transit time to derive declination.

For example, the catalogue of the Copenhagen University Observatory & Royal Greenwich Observatory (2001) supersedes the Carlsberg Meridian Catalogues 1–11 (CMC 1–11), with positions referred to the ICRS. The statistics, accordingly, merit summarising. It contains 180 812 positions and magnitudes of 176 591 stars north of declination -40°, 155 005 proper motions, and 25 848 positions and magnitudes of 184 Solar System objects obtained during the period May 1984 to May 1998, i.e. all the observations made since the instrument began operation on La Palma. The limiting magnitude is $V = 15.4$. The catalogue mainly comprises positions and proper motions for the following programmes: 36 000 International Reference Stars (IRS), 30 000 faint reference stars in a global net, 18 000 reference stars in the fields of radio sources, 17 000 stars in the Lick Northern Proper Motion catalogue, 5000 reference stars for calibration of Schmidt plates, 2600 stars in the Catalogue of Nearby Stars, 5000 stars in nearby OB associations, 10 500 F-type stars within 100 pc, 9000 G-type dwarfs and giants, K-type giants stars within 300 pc, 2200 unbiased sample of K/M-type dwarf stars, 19 400 reference stars near Veron–Cetty galaxies, 4700 variable stars (12–14 mag) in GCVS, 12 400 stars (11–14 mag) with proper motion > 0.18 arcsec yr^{-1} in NLTT, and several smaller programmes mainly aimed at Galactic kinematics. Positions and magnitudes of 12 novae and eight supernovae which occurred in the years 1991 to 1998 are included. The catalogue also contains observations of the following Solar System objects: Callisto, Ganymede, Rhea, Titan, Iapetus, Hyperion, Uranus, Oberon, Neptune, Pluto and 173 asteroids and Comet P/Wild 2. The mean error of a catalogue position at zenith is 0.09 arcsec in α and δ in CMC 1–6, improving to 0.06 arcsec in CMC 7–11. The accuracy in magnitude is 0.05 mag in CMC 1–10, improving to 0.03 mag in CMC 11. The mean error of the proper motions, derived by combining these positions with those at earlier epochs, is typically in the range 0.003–0.004 arcsec yr^{-1}. Subsequent catalogues in the series include CMC 14 (Copenhagen University Observatory *et al.*, 2006) comprising 95.9 million stars in the declination range $-30^\circ < \delta < +50^\circ$ observed between March 1999 and October 2005.

2.3 Reference system for astrolabes

Like the photoelectric meridian circles, photoelectric astrolabes are also capable of providing star positions with a precision of order 0.1 arcsec, and are therefore continuing to be used for the maintenance of the Hipparcos reference system, and the extension to fainter magnitudes. Catalogues constructed from data observed in San Juan, Argentina, since 1992, with the photoelectric astrolabe of the Beijing astronomical observatory, observing at a zenith distance of 30°, and observing in the declination zone -3° to -60°, are reported by Lu *et al.* (1996, CPASJ1), Manrique *et al.* (1999, CPASJ2) and by Lu *et al.* (2005, CPASJ3): the latter contains 6762 stars including 6156 Hipparcos stars. Systematic errors of the catalogue obtained with the Mark I photoelectric astrolabe of the Shaanxi Observatory mounted at Irkutsk, Russia, were reported by Chunlin *et al.* (1999).

The reduction of observations made over some 20 years with the photoelectric astrolabe at Calern Observatory, France, was reported by Martin *et al.* (1999). For two years, observations were made at two zenith distances, 30° and 45°, from which absolute declinations of 185 stars with a precision of 0.027 arcsec could be determined. Since 1998 the instrument has been located at the Antares Observatory in Brazil.

Table 2.1 Observatories participating in the Astrographic Catalogue/Carte du Ciel effort, and the regions, epochs, and stars measured in the context of the Astrographic Catalogue. From Urban (2006). Further details, including the number of plates and images measured, are given in Kuzmin et al. (1999, Table 1).

Zone	Decl. Range		Epoch	Number of stars
Greenwich	+90	+ 65	1892–1905	179 000
Vatican	+64	+ 55	1895–1922	256 000
Catania	+54	+ 47	1894–1932	163 000
Helsing	+46	+ 40	1892–1910	159 000
Potsdam	+39	+ 32	1893–1900	108 000
Hyderabad North	+39	+ 36	1928–1938	149 000
Uccle	+35	+ 34	1939–1950	117 000
Oxford 2	+33	+ 32	1930–1936	117 000
Oxford 1	+31	+ 25	1892–1910	277 000
Paris	+18	+ 24	1891–1927	253 000
Bordeaux	+17	+ 11	1893–1925	224 000
Toulouse	+05	+ 11	1893–1935	270 000
Algiers	+04	− 02	1891–1911	200 000
San Fernando	−03	− 09	1891–1917	225 000
Tacuba	−10	− 16	1900–1939	312 000
Hyderabad South	−17	− 23	1914–1929	293 000
Córdoba	−24	− 31	1909–1914	309 000
Perth	−32	− 37	1902–1919	229 000
Perth/Edinburgh	−38	− 40	1903–1914	139 000
Cape	−41	− 51	1897–1912	540 000
Sydney	−52	− 64	1892–1948	430 000
Melbourne	−65	− 90	1892–1940	218 000

Astrolabe observations of the Sun, based on 5924 transits observed from Calern, São Paulo, and Rio de Janeiro between 1974–1991, were reduced using Earth Orientation Parameters described in the Hipparcos reference frame by Andrei *et al.* (1997). Results showed a smaller dispersion of residuals compared to those used previously. They also reported a drop in the (unexpectedly high) correction to the obliquity of the ecliptic, from 0.43 to 0.27 arcsec.

2.4 Reference system for the Astrographic Catalogue and Carte du Ciel

A vast and unprecedented international star-mapping project was begun more than a century ago, as a result of the Astrographic Congress of more than 50 astronomers held in Paris in April 1887. Two goals were established.

For the first, the Astrographic Catalogue, the entire sky was to be photographed to 11 mag to provide a reference catalogue of star positions that would fill the magnitude gap between those previously observed by transit instrument observations down to 8 mag. Different observatories around the world were charged with surveying specific declination zones. The Astrographic Catalogue plates, of typically 6 min exposure, were in due course photographed, measured, and published in their entirety. They yielded a catalogue of positions and magnitudes down to about 11.5 mag, and completed during the first quarter of the twentieth century.

For the second goal, a second set of plates, with longer exposures but minimal overlap, was to photograph all stars to 14 mag. These plates were to be reproduced and distributed as a set of charts, the Carte du Ciel, in contrast to previous sky charts which had been constructed from the celestial coordinates of stars observed by transit instruments. Most of the Carte du Ciel plates used three exposures of 20 min, displaced to form an equilateral triangle with sides of 10 arcsec, making it easy to distinguish stars from plate flaws, and asteroids from stars. However, the charts proved to be expensive to photograph and reproduce, generally via engraved copper plates, and many zones were either not completed or properly published; the plates which were taken generally still exist, but cover only half of the sky.

The Hipparcos contribution The Astrographic Catalogue component of the project was rejuvenated with the completion of the Hipparcos and Tycho Catalogues. These allowed reduction of the old material with a dense network of accurate reference stars (even allowing for nearly 100 years of proper motions), yielding first epoch positions for stars observed by Tycho, and hence accurate two-epoch proper motions for the Tycho 2 Catalogue stars. The overall context of the Hipparcos-based reductions of the Astrographic Catalogue data are summarised at the USNO www site (Urban, 2006). Kuzmin *et al.* (1999) provide details of the

The Astrographic Catalogue and the Carte du Ciel: A fascinating account of this vast international astronomical collaboration is given by Turner (1912, the then Savilian Professor of Astronomy at Oxford University). Other aspects are covered in numerous papers in the Proceedings of IAU Symposium 133 'Mapping the Sky' (Débarbat, 1988), in the articles by Urban & Corbin (1998) expanding on the recent re-analysis of the Astrographic Catalogue, and by Jones (2000) focusing on the contribution of the instrument makers. The connection with the Hipparcos project is discussed by Arenou & Turon (2008).

For the Astrographic Catalogue, 20 observatories from around the world participated in exposing and measuring more than 22 000 glass plates (Table 2.1). Around half of the observatories ordered astrographs from the Henry brothers in France, with others coming from Howard Grubb of Dublin. Each observatory was assigned a specific declination zone to photograph. The first such plate was taken in August 1891 at the Vatican Observatory (where the exposures took more than 27 years to complete), and the last in December 1950 at the Uccle Observatory (Bruxelles), with most observations being made between 1895 and 1920. To compensate for plate defects, each area of the sky was photographed twice, using a two-fold, corner-to-centre overlap pattern, extended at the zone boundaries, such that each observatory's plates would overlap with those of the adjacent zones. The participating observatories agreed to use a standardized telescope so that all plates had a similar scale of approximately 60 arcsec mm^{-1}. The measurable areas of the plates were $2^{\circ}\!.1 \times 2^{\circ}\!.1$ (13 cm×13 cm), so the overlap pattern consisted of plates that were centred on every degree band in declination, but offset in right ascension by two degrees. Many factors, such as reference catalogue, reduction technique and print formats were left up to the individual institutions. The positional accuracy goal was 0.5 arcsec per image.

Plate measurement was a protracted affair, with measuring done by eye and recorded by hand. The original goal of 11 mag for the limiting magnitude was generally surpassed, however, with some observatories routinely measuring stars as faint as 13 mag. In total, some 4.6 million stars (8.6 million images) were observed. The brightest stars were over-exposed on the plates, not measured, and therefore missing in the resulting catalogues. The plate measurements (as rectangular coordinates), as well as the formulae to transform them to equatorial coordinates, were published in the original volumes of the Astrographic Catalogue, although the accompanying equatorial coordinates are now of only historical interest (Figure 2.5). Publication of the measurements proceeded from 1902 to 1964, and resulted in 254 printed volumes of raw data.

For decades the Astrographic Catalogue was largely ignored. The data were difficult to work with because they were available neither in machine-readable form nor in equatorial coordinates. Urban & Corbin (1998) wrote *'The history of this endeavour is one of dedicated individuals devoting tedious decades of their careers to a single goal. Some believe it is also the story of how the best European observatories of the nineteenth century lost their leadership in astronomical research by committing so many resources to this one undertaking. Long portrayed as an object lesson in overambition, the Astrographic Catalogue has more recently turned into a lesson in the way that old data can find new uses.'*

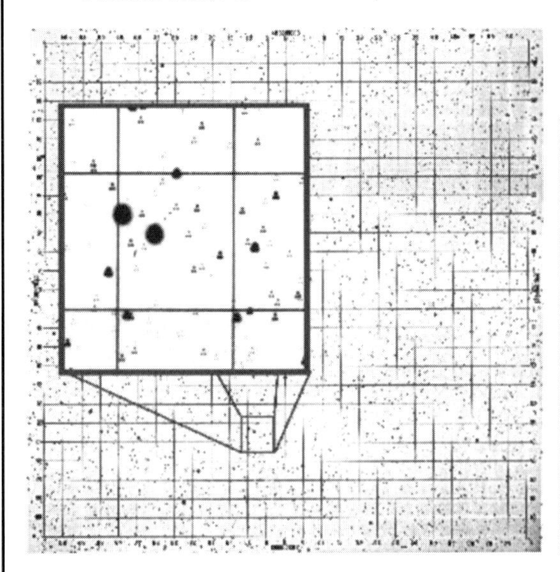

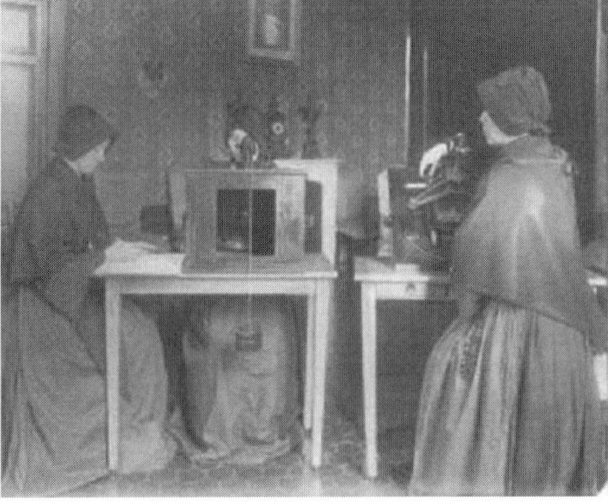

Figure 2.5 Left: contact print of a Carte du Ciel plate of the Vatican Zone: the three exposures per image are seen in the expanded inset. Right: nuns from the Congregation of the Child Mary measuring the Vatican plate collection, 1910–1921 (courtesy of the Torino Observatory and the Astrographic Catalogue/Carte du Ciel Working Group of IAU Commission 8).

re-reductions starting from the original published measurements: this included the construction of a machine-readable version of the Astrographic Catalogue from the printed volumes, undertaken at the Sternberg Astronomical Institute in Moscow between 1987–94. As noted by Kuzmin *et al.* (1999): *'It was the first and, hopefully, last time in the history of astronomy that such an amount of astrometric data was manually transformed to machine-readable form'.*

The United States Naval Observatory carried out the subsequent reductions of the Astrographic Catalogue data to a consistent system. The resulting catalogue, AC 2000 (Urban *et al.*, 1998a) and the updated AC 2000.2 (Urban *et al.*, 2001), is on the system defined by the Hipparcos Catalogue, and contains 4 621 836 stars covering the entire sky, at an average epoch of 1907. Each of the 22 zones making up the Astrographic Catalogue was reduced independently using the Astrographic Catalogue Reference Stars (ACRS, Urban *et al.*, 1998b), and are described separately in the literature (e.g. Urban & Corbin, 1995; Urban & Wycoff, 1995; Urban *et al.*, 1995; Urban & Martin, 1996; Urban *et al.*, 1996e,d,b; Urban & Jackson, 1996; Urban *et al.*, 1996c,a, etc.). Each zone was analysed for tilt, radial and tangential distortions, coma, magnitude equation and non-symmetric field distortions. Following these reductions, the data were placed on the Hipparcos system and the magnitudes were converted to be close to that of the Tycho B_T-band data (Figure 2.6). The final catalogue contains the (ICRF) positions at the mean epochs of observation, with magnitude and accuracy estimates for each star. Cross-identifications with the Hipparcos Catalogue, Tycho Catalogue and the Astrographic Catalogue Reference Stars are provided. Details of these various steps, leading also to the production of the Tycho 2 Catalogue, are given in Chapter 1.

Other related studies of the Hipparcos-based reductions of different zones of the Astrographic Catalogue have been published. Abad (1998) and Abad *et al.* (1998) used the original coordinates measured for the Astrographic Catalogue plates of the San Fernando zone, and made an independent reduction including magnitude and colour-dependent terms. Contact copies of some of the original plates were made and re-measured, with comparison of the two sets showing that no appreciable emulsion deformation was introduced during the long storage period of the original plates.

The original Carte du Ciel plates The Carte du Ciel plates, with their limiting magnitude fainter than the Astrographic Catalogue but triple exposures (the handling of which has been addressed by, e.g. Dick *et al.*, 1993; Geffert *et al.*, 1996; Ortiz Gil *et al.*, 1998), have not been used systematically for astrometry so far. A Working Group was set up under IAU Commission 8 in 1994 to consider the scientific value, the archival aspects, and measurement of the available plates (http://www.to.astro.it/astrometry/AC_CdC).

A complete digitisation, and other experiments on the astrometric measurements of the Córdoba Carte du Ciel plates, which contains the Galactic centre, were reported by Bustos Fierro & Calderón (2003) and Bustos Fierro *et al.* (2003).

Kislyuk & Yatsenko (2005) reported the construction of their FONAC 1.0 astrographic catalogue, based on re-reduction of relevant Carte du Ciel plates combined with those from the Golosiiv glass archive, taken with a Zeiss double wide-angle astrograph in 1982–1994.

Fresneau *et al.* (2007) reported measurements, reduction to the Tycho 2 system, and construction of proper motions when combined with recent observational material, for nearly 300 plates from the Sydney Observatory Carte du Ciel region.

Rapaport *et al.* (2001) observed the Bordeaux Carte du Ciel zone with the Bordeaux automatic meridian circle for some four years around the epoch 1998; the resulting catalogue M2000 includes some 2.3 million stars down to $V = 16.3$ mag. Comparison with the Hipparcos Catalogue indicates external errors below 40 mas. They also conclude that the faintest part of the Tycho Catalogue has an accuracy, at epoch 1998, some 30% larger than those inferred from the catalogue. Odenkirchen *et al.* (1998) reported experiments with plates from the Bordeaux Carte du Ciel region, scanned and digitised using the MAMA machine at the Observatoire de Paris, and with object identification using the software SExtractor. Reduction using the ACT Catalogue as reference yielded positional accuracies in the range 0.15–0.20 arcsec depending on plate position. A more comprehensive measurement of 512 plates of the Bordeaux Carte du Ciel region, scanned at the APM in Cambridge, was carried out by Ducourant *et al.* (2006) and Rapaport *et al.* (2006). In combination with the M2000 meridian circle observations, and reduced to the Hipparcos/Tycho reference system, this has led to a proper motion catalogue of 2 670 974 stars down to $V \sim 16$ mag, with proper motion accuracies in the range 1.5–6 mas yr^{-1}.

Vicente *et al.* (2007) reported the reduction of some 560 000 stars from the San Fernando zone (average epoch 1901.4), and reduced to the Tycho 2 system, yielding estimated positional accuracies of about 0.2 arcsec.

2.5 Reference system for Schmidt plates

Following Bernhard Schmidt's first 14-inch Schmidt reflector in 1934, and the 18-inch Schmidt at Palomar in 1936, larger instruments were constructed (Figure 2.7), including the Palomar 48-inch Oschin Schmidt

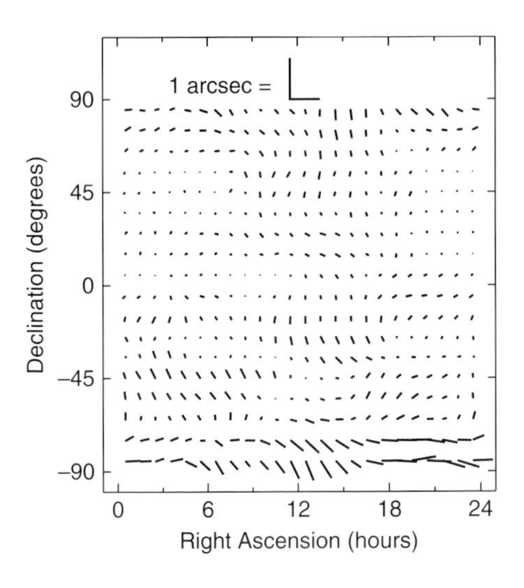

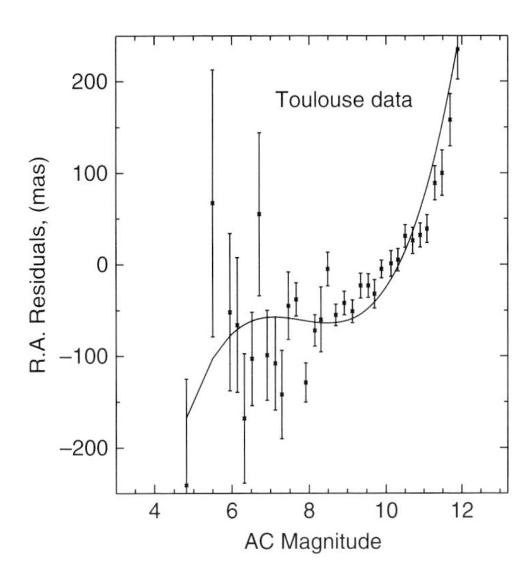

Figure 2.6 Left: systematic differences, at epoch 2000, between the AC on the system of FK5 and on the system of Hipparcos. Similar structure is seen for the difference at epoch 1900 between the ACRS and Hipparcos, demonstrating that the AC/FK5 is correctly on the system defined by the ACRS, but that a further reduction to the Hipparcos system is required. Right: differences in right ascension between the Toulouse zone of the AC 2000 and Hipparcos data as a function of magnitude, prior to the application of corrections. Each point is the average of all stars in a 0.2 mag range. The curve represents a third-order fit to the data. From Urban et al. (1998a, Figures 3a and 4a). Other examples of the systematic differences in the AC compared with Hipparcos are given by Kuzmin et al. (1999).

(operated from 1949, and named for the philanthropist Samuel Oschin), the ESO 1-m Schmidt (operated in Chile between 1973–1998), and the UK 1.2-m Schmidt (operated in Australia from 1973).

The first reductions of Schmidt plates using the Hipparcos and Tycho data were reported by Robichon *et al.* (1995). They used the unpublished 30-month Hipparcos solution to examine plate solutions using CERGA (Nice) and ESO (Chile) Schmidt plates obtained as part of a programme on Hipparcos open clusters, and scanned using the MAMA measuring machine in Paris. Other early work was reported by Lasker (1997), Morrison *et al.* (1997), and Bucciarelli *et al.* (1997, 1998). The combination of high astrometric accuracy from Hipparcos, and high sky density from Tycho, and especially when using the even higher density of Tycho 2 stars, demonstrated that the intrinsic accuracy of the plate material could be much more fully exploited. It also demonstrated that the Hipparcos/Tycho reference frame defined at 9–12 mag could be transferred to the limit of the sky survey plates at around 21 mag, taking into account the asymmetric bright star image profiles of the Schmidt plates caused by the appearance of diffraction spikes and low-light level halos due to multiple reflections in the telescope optics (Irwin *et al.*, 1997).

The most relevant Schmidt plate material referred to below includes those taken at Palomar: the first epoch Palomar Observatory Sky Survey POSS-I *O* (blue) and *E* (red) plates, the Quick *V* plates (epoch 1982), and the second epoch POSS-II *J* (blue) and *F* (red) plates; at ESO; and at the UK Schmidt Telescope: the blue plates of the Southern Sky Atlas SERC *J* and its equatorial extension SERC *EJ*, the Equatorial Red (*ER*), the second epoch red survey SES or AAO–*R*, and the infrared SERC *I* survey.

With the advent of fast high-precision measuring machines, several major plate digitisation programmes have been undertaken: they consist of scanning various subsets of the available sky survey material, in some cases glass or film copies of the original Schmidt plates. The scope of the various projects and their emphasis is different in each case: they include the programmes of the Automatic Plate Measuring machine (APM, Cambridge UK; originally using the POSS-I *O* and *E* plates, later extended to the southern hemisphere using the SERC *J*/*EJ* and SERC *ER*/AAO *R* surveys); the Automated Plate Scanner (APS, Minneapolis, based on the POSS-I and supplemented by second-epoch red plates obtained for the Luyten proper motion surveys); the COSMOS and SuperCOSMOS machines (Edinburgh); the Digitised Sky Survey (DSS) programme of the Space Telescope Science Institute; and the Precision Measuring Machine (PMM, Flagstaff USA). For a more detailed synopsis of these various surveys, see Hambly *et al.* (2001c). An introduction to the subject of digitised photographic sky surveys from Schmidt plates in particular is given by Lasker (1995), with further details being given in conference proceedings edited by MacGillivray & Thomson (1992), MacGillivray (1994), Chapman *et al.*

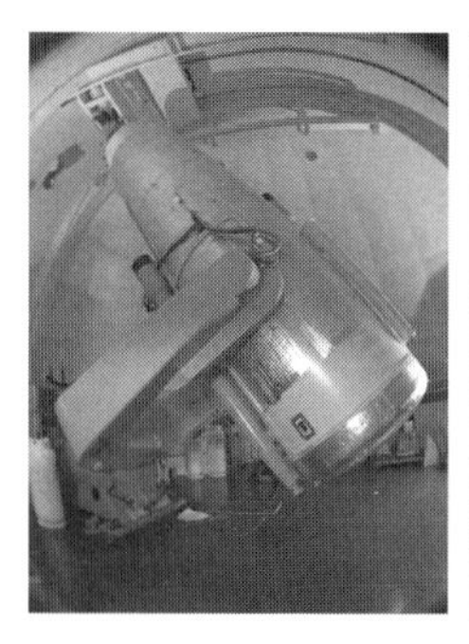

Figure 2.7 Left: the Palomar 48-inch Oschin Schmidt (courtesy Palomar Observatory and Caltech). Middle: the UK 1.2-m Schmidt (courtesy of the Anglo-Australian Observatory/David Malin Images). Right: the ESO 1-m Schmidt (courtesy of the European Southern Observatory).

Table 2.2 Properties of the earliest and most recent releases of the Guide Star Catalogue (GSC 1.1 based on FK5; McLean et al. 1998b, and GSC 2.3 based on Hipparcos/ICRF; Spagna et al. 2006) and USNO B1, also based on Hipparcos/ICRF (Monet et al., 2003). Quoted systematic errors are taken from the quoted references, but may still be underestimated in some regions.

	GSC 1.1	GSC 2.3	USNO B1
Number of objects	18×10^6	997×10^6	1042×10^6
Epochs	1	2	2
Passbands	1 (V or J)	3 (B_J, R_F, I_N)	3
Limiting magnitude	15	$B_J \sim 22$, $R_F \sim 20.5$	$V \sim 21$
Relative position error (arcsec)	0.4	0.1–0.2	0.2
Local systematic errors (arcsec)	1.0	0.10–0.15	0.10–0.15
Proper motion error (bright, mas yr^{-1})	not available	< 4	~ 4
Magnitude error (mag)	0.4	0.1–0.15	0.3

(1995), and McLean *et al.* (1998a). In the following, the contributions brought by the Hipparcos and Tycho Catalogues are emphasised.

As will become evident, the combination of Hipparcos and Tycho Catalogue data has allowed excellent progress in modern Schmidt plate reductions. Nevertheless, stars brighter than around 12 mag are still heavily over-exposed on Schmidt plates. Plate bending causes serious problems for astrometry, and deeper, denser reference star catalogues (e.g. as expected from UCAC3) are still required to fully exploit the Schmidt plate data.

2.5.1 Guide Star Catalogue

The Guide Star Catalogue (GSC) was constructed from Schmidt plates scanned by the Space Telescope Science Institute (STScI) to form the Digitised Sky Survey (DSS). Plates were scanned and digitised using modified PDS-type scanning machines at scales of 15 μm (1 arcsec) or 25 μm (1.7 arcsec) per pixel. The resulting GSC versions were constructed to support the operational requirements of the Hubble Space Telescope for off-axis guide stars (Lasker *et al.*, 1990). The Hipparcos and Tycho Catalogues contributed significantly to later versions of the GSC, in which the scanned data were re-reduced as improved astrometric reference catalogues became available during the 1990s.

Version GSC 1.0 contained some 19 million stars in the range 6–15 mag. It was constructed from digitised scans of an all-sky, single-epoch collection of Schmidt plates: using the Palomar 'Quick V' survey for $\delta > +6^\circ$ (epoch 1982), and for southern fields the UK SERC J survey (epoch 1975) and its equatorial extension (epoch 1982). Version GSC 1.1 included the missing bright stars, based on the Hipparcos Input Catalogue (Malkov & Smirnov, 1997), and with magnitudes re-calibrated using the Tycho Catalogue as photometric reference (Morrison *et al.*, 1997; Fabricius, 1997). This brought the systematic photometric errors to below 0.1 mag for stars brighter than 11.5 mag, the 1σ residuals to 0.18 mag for the Palomar Quick V plates in the north, and 0.14 mag for the SERC J plates in the south. Version GSC 1.2 was a re-reduction based on the PPM Catalogue and the Astrographic Catalogue, while Version GSC-ACT was a re-calibration using the Astrographic Catalogue/Tycho (Lasker *et al.*, 1999); see Chapter 1 for a summary of these reference catalogues. It had already been demonstrated that the Tycho Catalogue provided a sufficiently dense and precise reference catalogue to adequately map the small-scale positional distortions inherent in photographic Schmidt plates, although the fairly bright magnitude limit of $V \sim 11.5$ mag is not ideal to calibrate a possible magnitude-dependent error, especially at the fainter magnitudes (Bucciarelli *et al.*, 1997).

Version GSC 2 extended the catalogue to fainter magnitudes, and included multi-colour, multi-epoch data. It was undertaken by a larger consortium, comprising the STScI, the Italian Council for Research in Astronomy, and the Astrophysics Division of ESA-ESTEC. It added the POSS-II (*J* and *F* plates), SERC *ER*, and AAO-SES survey plates. The original POSS-I plates (epoch 1950–58) were used as the first epoch for proper motion measurements (McLean *et al.*, 1998b). Version GSC 2.2 was a preliminary version generated to support Gemini and VLT telescope operations (Space Telescope Science Institute & Osservatorio Astronomico di Torino, 2001), with magnitude limits 18.5 in photographic *F* (red) or 19.5 in photographic *J* (blue). Version GSC 2.3 is the most recent public-release version (Spagna *et al.*, 2006); properties of the original and this most recent version are summarised in Table 2.2.

2.5.2 USNO A1, A2, B1

Various substantial catalogues have been constructed or compiled at the US Naval Observatory since the early 1990s. More recently, the Hipparcos and Tycho Catalogues have been used to improve the astrometric (and photometric) reference framework of these substantial efforts. This section covers the catalogues generated by the (re-)reduction of photographic Schmidt plates by the USNO, independently of the work done at the STScI for the construction of the Guide Star Catalogue. The USNO involvement in the Astrographic Catalogue reductions was described in Section 2.4, and their contribution directly related to the construction of the Tycho 2 Catalogue (including the ACT and TRC) was described in Section 1.7. As summarised in Table 1.7 the catalogues USNO A1, and USNO A2 (as well as UJ 1.0), have already been superseded by USNO B1, and the former are noted here only for completeness.

USNO A1 was constructed around 1995–96 from Precision Measuring Machine (PMM) scans, made at the US Naval Observatory Flagstaff Station, of the first epoch POSS *O* and *E* plates for $-30^\circ < \delta < +90^\circ$, and from the SERC *J* and ESO *R* plates for $-90^\circ < \delta < -35^\circ$. The catalogue contained 491.8 million astrometric reference objects, of which 19.5 million were cross-identified with objects in the Guide Star Catalogue (GSC 1.1), which was also used as the astrometric reference. The incompleteness with respect to bright stars, and the preliminary nature of the astrometric ($\sigma \sim 0.25$ arcsec) and photometric ($\sigma \sim 0.2$–0.4 mag) calibration, meant that the catalogue had only limited distribution (Monet & Corbin, 1997).

USNO A2 resulted from various improvements undertaken with respect to USNO A1, notably the re-reduction to the ICRF reference frame through the use of the ACT Catalogue, itself derived from the AC in combination with the Hipparcos and Tycho data, and thus superseding the GSC 1.1 reference system. The resulting catalogue contains 526.2 million objects, with positions at the mean epoch of the red and blue plates, and improved (red and blue) photometry for each object. Comparison of the resulting catalogue positions with others also reduced to the ICRF, and implying remaining systematic errors in one or more of the catalogues, was reported by Assafin *et al.* (2001).

USNO B1 Version 1.0 represents a compilation of PMM scans of five complete coverages of the northern sky, and four of the southern sky, containing a mixture of colours and epochs, but which can be broadly considered as a three-colour, two-epoch catalogue (Monet *et al.*, 2003). It comprises 1.042 billion objects derived from 3.6 billion separate observations from 7435 Schmidt plates taken over the last 50 years. It is considered complete to $V = 21$, with 0.2 arcsec astrometric accuracy at epoch J2000.0, and 0.3 mag photometric accuracy in up to five colours.

The USNO A reductions used the faintest Tycho 2 stars as astrometric reference, even though these are saturated on the deep survey plates. Examples of the resulting fixed-pattern astrometric residuals for the mean first-epoch POSS *O* plate are shown in Figure 2.8. In contrast, USNO B used the less-exposed (yellow) plates from the Northern Proper Motion survey (Klemola *et al.*, 1987) and Southern Proper Motion survey (Platais *et al.*, 1998) to form the unpublished intermediate 'Yellow Sky' catalogues, YS3.0 and YS4.0, with a mean epoch of about 1975. It was this step that made use of the Tycho 2 Catalogue reference positions (see Figure 2.9). The reduction adopted implies that the proper motions in USNO B are relative rather than absolute.

Completeness with respect to high proper motion stars, a challenge because of the inhomogeneity of the underlying plate material, was assessed by Gould (2003). He concluded that the incompleteness for stars with proper motion larger than 0.18 arcsec yr^{-1} is some 6–9%, and that the reported proper motion errors are generally correct, although with an error floor of around 4 mas yr^{-1}.

2.5.3 SuperCOSMOS Sky Survey

The SuperCOSMOS Sky Survey is a third project digitising the sky atlas Schmidt plate collections, in three colours (*BRI*), with one colour (*R*) at two epochs (Hambly *et al.*, 2001b,c). It used the SuperCOSMOS measuring machine at Edinburgh, a development of the COSMOS machine which ceased operation in 1993. Hambly

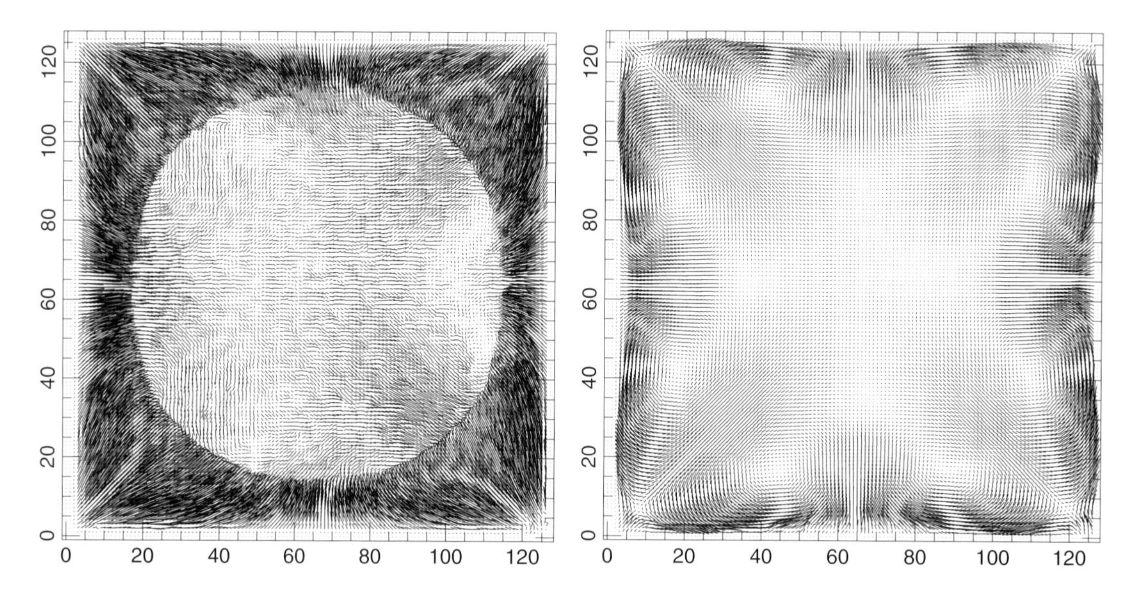

Figure 2.8 Maps of fixed pattern astrometric residuals for the mean first-epoch POSS-I O survey plate. Different structure is evident at 10 mag (left), and at 18 mag (right). The circular feature in the left plot may be due to the rapid onset of effects of asymmetry for saturated images (Monet, 2007, priv. comm.). Labelled bins are of 5 mm width, and the mean residuals are scaled such that the spacing between bins is 0.20 arcsec. From Monet et al. (2003, Figure 1).

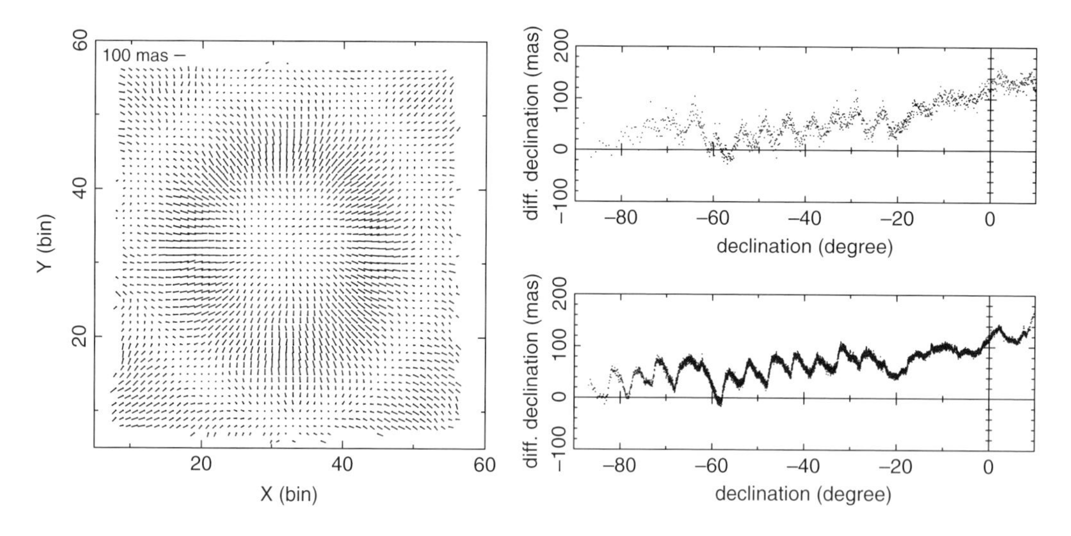

Figure 2.9 Left: field distortion pattern of the 'Yellow Sky' Catalogue YS3 (from the NPM and SPM programmes) with respect to Tycho 2 astrometry, prior to corrections, and showing the NPM data covering all Tycho 2 magnitudes. The pattern changes with magnitude, and is different for the SPM data. The largest vectors are about 100 mas. Top right: differences in YS3 declination (after field distortion pattern corrections) with respect to Tycho 2 as a function of declination. No proper motions are applied, so the data show Galactic and Solar motion. Each data point is a mean of 1000 differences. The discontinuity at −25° is the NPM/SPM boundary. Bottom right: YS3 with respect to UCAC and Tycho 2, following zero-point corrections of the SPM and NPM plates, and each data point being the mean of 4000 differences. The higher frequency oscillations in the southern data, of 30–50 mas amplitude, appear to be systematic errors at the plate level, which remain unexplained. From Zacharias et al. (2004b, Figures 9 and 10).

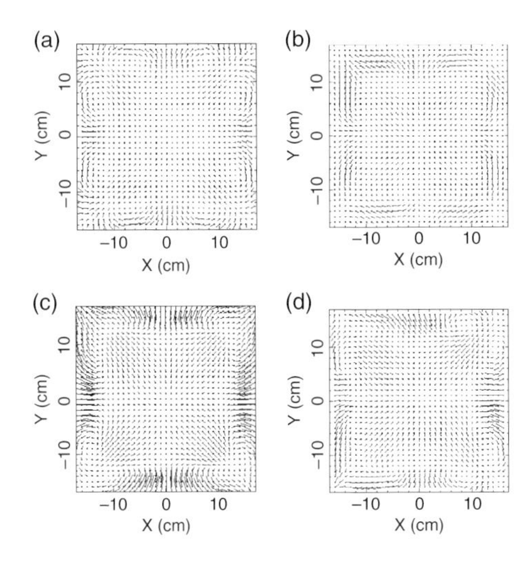

Figure 2.10 Mechanical deformation (or 'swirl') patterns for (a) UK Schmidt SERC–J/EJ plates, (b) ESO Schmidt R plates, (c) Palomar Schmidt first epoch E plates and (d) Palomar Schmidt second epoch R plates. In each case, the length of the vectors is scaled such that one tick-mark corresponds to one arcsec. The four-fold symmetry in the pattern is a characteristic of mechanical deformation of rigid glass plates. From Hambly et al. (2001a, Figure 1).

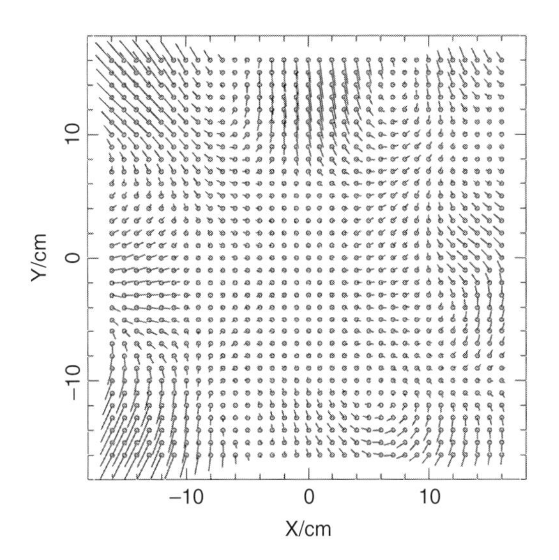

Figure 2.11 Systematic astrometric distortion pattern of SuperCOSMOS Hα survey field h67, reduced using a standard six coefficient linear fit plus a radial distortion term, and compared to the UCAC Catalogue. Systematic positional errors of more than 1 arcsec, corresponding to one tick mark on either axis, are observed in the film data, e.g. in the left-hand corners. From Parker et al. (2005, Figure 6).

et al. (2001a) describe the astrometric properties of the database. Coordinates were tied directly to the ICRF via the Hipparcos reference system (although the original South Galactic Cap survey was reduced with respect to the ACT Catalogue, all astrometry was subsequently re-reduced with respect to the Tycho 2 Catalogue following its release in 2000). The release history of the SuperCOSMOS Sky Survey is maintained on-line (http://www-wfau.roe.ac.uk/sss/history.html); July 2006 marked the final release of the full survey after inclusion of the POSS-II I and northern POSS-I E data.

The number of distinct detections from individual plates, i.e. all bands and (overlapping) Schmidt fields, is around 6×10^9; there are 1.9×10^9 distinct entries in the merged multi-colour catalogue.

The calibration procedure consisted of applying a six-coefficient linear plate model to measured positions of Tycho 2 reference stars, along with a radial distortion coefficient appropriate to Schmidt optics, and a fixed, higher-order two-dimensional correction map to account for distortion induced by mechanical deformation of the photographic material when clamped in the telescope plate holder to fit the spherical focal surface (Figure 2.10). Resulting positions are accurate to better than ± 0.2 arcsec at $J, R \sim 19, 18$ rising to ± 0.3 arcsec at $J, R \sim 22, 21$ with position-dependent systematic effects from bright to faint magnitudes at the $\sim$0.1 arcsec level. The proper motion measurements are accurate to typically $\pm$10 mas yr^{-1} at $J, R \sim 19, 18$ rising to $\pm$50 mas yr^{-1} at $J, R \sim 22, 21$ and are tied to the extragalactic reference frame. As noted previously, the Tycho 2 stars are heavily saturated on sky-limited Schmidt plates, and the SuperCOSMOS measurements reveal a systematic difference between bright and faint stars at the plate edges of up to 0.3 arcsec. The zero-point errors in the proper motions are $\leq$ 1 mas yr^{-1} for $R > 17$ and are no larger than $\sim$10 mas yr^{-1} for $R < 17$ ($R < 17$ corresponding to the onset of saturation in the scanned SuperCOSMOS images, such that the systematic centroiding becoming worse for brighter objects). A comparison with proper motions from the SPM survey was reported.

The SuperCOSMOS Sky Survey has resulted in a number of papers related to high proper motion stars and nearby stars (Hambly *et al.*, 2004), white dwarfs (Hambly *et al.*, 2005), and brown dwarfs (Scholz & Meusinger, 2002).

The AAO/UKST SuperCOSMOS Hα survey (Parker *et al.*, 2005) consists of 233 individual fields covering a swathe approximately 20° wide about the Southern Galactic Plane. Astrometric reduction also made use of the Hipparcos reference frame through the UCAC Catalogue, with the dense star grid allowing determination of the distortion maps on a film-by-film basis, films not being as mechanically stable as glass on the largest scales (Figure 2.11).

Table 2.3 Some of the major astrometric star catalogues.

Abbreviation	Catalogue	Reference
AC	Astrographic Catalogue	see Section 2.4
AC 2000	Astrographic Catalogue on the Hipparcos system	Urban *et al.* (1998a)
AC 2000.2	AC 2000.2 Catalogue	Urban *et al.* (2001)
ACRS	Astrographic Catalogue Reference Stars	Corbin & Urban (1991)
ACT	ACT Reference Catalogue	Urban *et al.* (1998b)
AGK2	Zweiter Katalog der Astronomischen Gesellschaft	Schorr & Kohlschütter (1951–57)
AGK2 Version 2	Re-measured/re-reduced using Hipparcos and Tycho Catalogues	Zacharias *et al.* (2004b)
AGK3	Catalogue of Positions and Proper Motions North of $-2\overset{\circ}{.}5$, Vols 1–8	Dieckvoss et al. (1975)
AGK3R	Catalogue of Reference Stars for the AGK3	Smith (1980)
AGK3RN	AGK3R with proper motions	Corbin (1978)
BD	Bonner Durchmusterung	Schönfeld (1886), Argelander (1903)
BPM	Bruce Proper Motion Survey	Luyten (1963)
BSC (HR)	Bright Star Catalogue (Harvard Revised)	Hoffleit & Jaschek (1982, 1991)
CMC	Carlsberg Meridian Catalogues	Carlsberg Consortium (1985–2006)
CD	Córdoba Durchmusterung	Thome (1892, 1894, 1900, 1914)
CF	Cape Catalogue of Faint Stars	Spencer-Jones & Jackson (1939)
CPC	Cape Photographic Catalogue for 1950.0	Jackson & Stoy (1954–68)
CPC2	Second Cape Photographic Catalogue	Nicholson *et al.* (1984)
CPC2 Version 2	CPC2 re-reduced using Hipparcos and Tycho Catalogues	Zacharias *et al.* (1999)
2CP50	Second Cape Catalogue for 1950.0	Stoy (1968)
CPD	Cape Photographic Durchmusterung	Gill & Kapteyn (1895–1900)
DM	Durchmusterung (BD, CD or CPD)	see BD, CD, CPD
FK4	Fourth Fundamental Catalogue	Fricke *et al.* (1963)
FK4 Sup	Preliminary Supplement to FK4	Fricke & Kopff (1963)
FK5	Fifth Fundamental Catalogue	Fricke *et al.* (1988)
FK5 Ext	FK5 Extension: new Fundamental Stars	Fricke *et al.* (1991)
GC	General Catalogue of 33 342 Stars for the Epoch 1950	Boss (1937)
GCTP	General Catalogue of Trigonometric (Stellar) Parallaxes	van Altena *et al.* (1995)
GSC	Guide Star Catalogue of the STScI	Lasker *et al.* (1990)
HD	Henry Draper Catalogue	Cannon & Pickering (1918–49)
HDE	Henry Draper Extensions	Cannon (1925–36)
HIC	Hipparcos Input Catalogue	Turon *et al.* (1992)
HIP	Hipparcos Catalogue	ESA (1997)
IRS	International Reference Stars: AGK3R and SRS catalogues	Corbin & Warren (1995)
L	Luyten Catalogue	Luyten (1942)
LHS	Luyten Half-Second Catalogue	Luyten (1979)
Lowell	Lowell Proper Motion Survey	Giclas *et al.* (1959–78)
LP	Luyten Palomar Proper Motion Catalogue	Luyten (1963–87)
LSPM North	Lépine–Shara Proper Motion Catalogue North	Lépine & Shara (2005a)
LSPM South	Lépine–Shara Proper Motion Catalogue South	Lépine & Shara (2005b)
LTT	Luyten Two-Tenth Catalogue	Luyten (1957)
NLTT	New Luyten Two-Tenth Catalogue	Luyten (1980a,b)
NPM	Lick Northern Proper Motion Programme	Hanson *et al.* (2004)
NPM Version 2	NPM re-reduced using Hipparcos and Tycho Catalogues	Hanson *et al.* (2004)
NPZT74	Northern PZT Stars Catalogue	Yasuda *et al.* (1982)
N30	Catalogue of 5268 Standard Stars, 1950.0, based on N30	Morgan (1952)
Perth 70	Catalogue of Positions of 24 900 Stars (also Perth 75, Perth 83)	Høg & von der Heide (1976)
PPM North	Positions and Proper Motions: stars north of $-2\overset{\circ}{.}5$ declination	Röser & Bastian (1991)
PPM South	Positions and Proper Motions: stars south of $-2\overset{\circ}{.}5$ declination	Röser & Bastian (1993)
PPM Bright	PPM Bright Stars Supplement	Bastian & Røser (1994)
SAO	Smithsonian Astrophysical Observatory Star Catalogue	SAO (1966)
SPM	Yale/San Juan Southern Proper Motion Catalogue	Girard *et al.* (2004)
SPM Version 3	SPM re-reduced using Hipparcos and Tycho Catalogues	Girard *et al.* (2004)
SRS	Southern Reference System Catalogue	Smith *et al.* (1990)
SSSC	Sydney Southern Star Catalogue	King & Lomb (1983)
TAC	Twin Astrographic Catalogue	Zacharias *et al.* (1996)
TAC Version 2	TAC re-reduced using Hipparcos and Tycho Catalogues	Zacharias & Zacharias (1999)
TYC	Tycho Catalogue	Høg *et al.* (1997)
TYC2	Tycho 2 Catalogue	Høg *et al.* (2000)
UCAC 1	US Naval Observatory CCD Astrograph Catalogue (UCAC 1)	Germain *et al.* (2000)
UCAC 2	US Naval Observatory CCD Astrograph Catalogue (UCAC 2)	Zacharias *et al.* (2004b)
USNO A	USNO A Catalogue	Monet (1998)
USNO B	USNO B Catalogue	Monet *et al.* (2003)
Yale	Yale Photographic Catalogues, Vols 1–32	Yale University (1926–83)

Developments in ground-based astrometric surveys: For the surveys discussed in this section, the following chronology of some of the major developments in astrometric surveys is intended to set the context and is not intended to be complete. Some other major catalogues encountered in the associated literature are listed in Table 2.3, and a convenient summary of the major attributes of these and other astrometric catalogues can be found at the www site of the Torino Observatory (www.to.astro.it/astrometry).

AGK2: between 1928 and 1931, the sky north of declination $-5°$ was photographed on 1940 glass plates each covering over $5° \times 5°$ with two dedicated astrographs located in Bonn and Hamburg, Germany. Two exposures, one of 3 minutes and one of 10 minutes, were made on each plate, and reached $B \sim 12$ mag. During the 1930s–1950s the measuring and reduction of the brighter stars were carried out, by hand, resulting in the AGK2 Catalogue.

AGK3R and AGK3: after a proposal that the AGK2 Catalogue should be observed again at Hamburg to provide proper motions, an extensive international programme of meridian observations at 10 observatories was organised, under IAU Commission 8, to provide a reference star catalogue, AGK3R, which was then used for the reduction of the photographic work carried out at Hamburg between 1956–63. This resulted in the AGK3 Catalogue, containing proper motions for all stars, which was subsequently used as the stellar reference frame in the northern hemisphere.

SAO: by the mid-1960s a high density catalogue of star positions was needed for satellite tracking. This was compiled by the Smithsonian Astrophysical Observatory for more than 250 000 stars. In each declination zone, preference was given to source catalogues with proper motions, namely the Yale Photographic Catalogues in the north, and the Cape Catalogues in the south. The resulting SAO Catalogue was limited by the generally poor quality of the first epoch material in both hemispheres (the AGK3 not yet being available in the north). Cionco *et al.* (1997) and Arias *et al.* (2000) discussed the overall orientation of SAO with respect to Hipparcos for more than 100 000 stars in common. Not surprisingly, in view of the inhomogeneous source material used in the construction of the SAO, the differences show various large distortion patterns.

SRS: the success of the AGK3R programme led to plans for a similar campaign in the southern hemisphere, formulated by the IAU in 1961. The resulting Southern Reference Star (SRS) Catalogue was constructed from observations made with 13 transit circles, with observations extending from 1961 for about two decades.

IRS: the International Reference Stars (IRS Catalogue) comprises the combination of the resulting reference stars from both hemispheres, i.e. the AGK3R in the north and the SRS in the south.

CPC2: to complement the AGK3 in the northern hemisphere, the Second Cape Photographic Catalogue, CPC2, was constructed from 5820 southern hemisphere plates taken with a new astrograph at the Cape Observatory during 1962–1972 (mean epoch 1968), and scanned with the GALAXY machine at the RGO, Herstmonceaux. This resulted in a catalogue of 276 131 stars in the range $V = 6.5-10.5$ mag (Zacharias *et al.*, 1992). Field distortions were studied by Zacharias (1995).

2.6 Other photographic surveys

2.6.1 Re-reduction of the AGK2

As part of the UCAC production, the US Naval Observatory embarked on a re-measurement, and re-reduction to the Hipparcos reference frame, of the AGK2 Catalogue plates (Zacharias *et al.*, 2004b). The AGK2 Catalogue contains about 186 000 stars with positional accuracies of about 200 mas at the observational epoch. But about 10 times more stars were in principle measurable on the plates, and with a higher accuracy than in the original analysis. The original plates were therefore loaned by the Hamburg Observatory to the US Naval Observatory for re-measurement using the USNO StarScan machine during 2001–03. Preliminary positions were obtained for over 950 000 stars from a subset of 869 of the Hamburg plates; 599 871 of these positions were used for the UCAC 2 proper motions. For well-exposed images, positional errors of about 70 mas per coordinate were obtained, based on an average of about 50 Hipparcos reference stars per plate which, combined with the $\sim$70-yr epoch difference between the UCAC observations, provides proper motions good to 1 mas yr^{-1}. This is about a factor of 2 better than previously best known (from AC 2000 minus Tycho 2) for stars in this range.

2.6.2 Re-reduction of the CPC2

Zacharias *et al.* (1997a) carried out a new set of plate reductions using Hipparcos stars as reference. The formal error of a Hipparcos star position at the mean epoch of the CPC2 is about 30 mas per coordinate, small compared to the x, y measuring precision. A standard error of unit weight of 90 mas was obtained for an average plate solution, compared to 141 mas for the earlier solution based on the Southern Reference Stars (SRS). The same systematic field distortion patterns was found in both solutions, but only the Hipparcos reference stars allowed solving for magnitude-dependent, systematic errors on a plate-by-plate basis (Figure 2.12). The Hipparcos-based reductions resulted in a Version 2 of the catalogue, containing 266 629 stars with an accuracy of about 53 mas per coordinate at the observation epoch, thus also yielding a key catalogue for proper motion determination (Zacharias *et al.*, 1999).

2.6.3 Re-reduction of the NPM and SPM

The Lick Northern Proper Motion (NPM) programme was a two-epoch (1947–1988) photographic survey set up to determine absolute proper motions (as well as positions and photometry), measured with respect to an extragalactic reference frame (based on differential

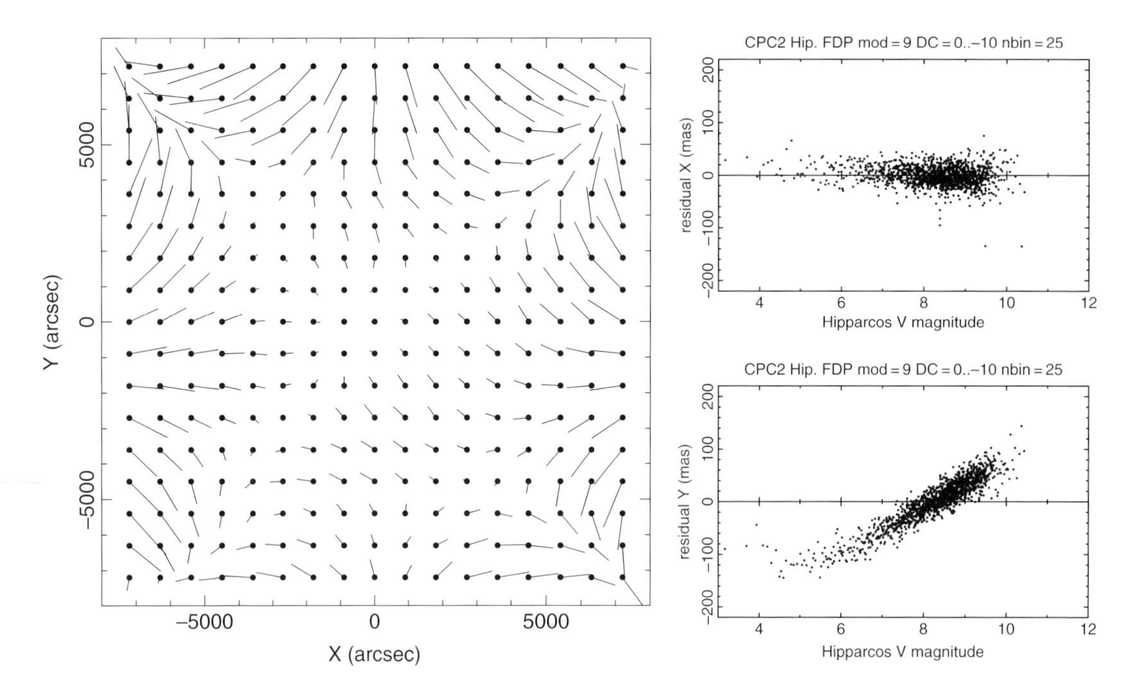

Figure 2.12 Re-reduction of the CPC2 photographic plate data using the Hipparcos and Tycho astrometric and photometric data as reference. Left: field distortion pattern obtained from a conventional plate adjustment of all CPC2 using Hipparcos reference stars. The plate size is 4°07, and the residual vectors are scaled by 10 000. About 1000 individual residuals are averaged to construct each vector. Right: systematic errors as a function of magnitude, where the residuals of a conventional plate adjustment using Hipparcos reference stars are given for the declination zone $-10^\circ < \delta < 0^\circ$; top and bottom figures show the x and y components, with each point representing the mean of 25 residuals. From Zacharias et al. (1997a, Figures 1 and 2).

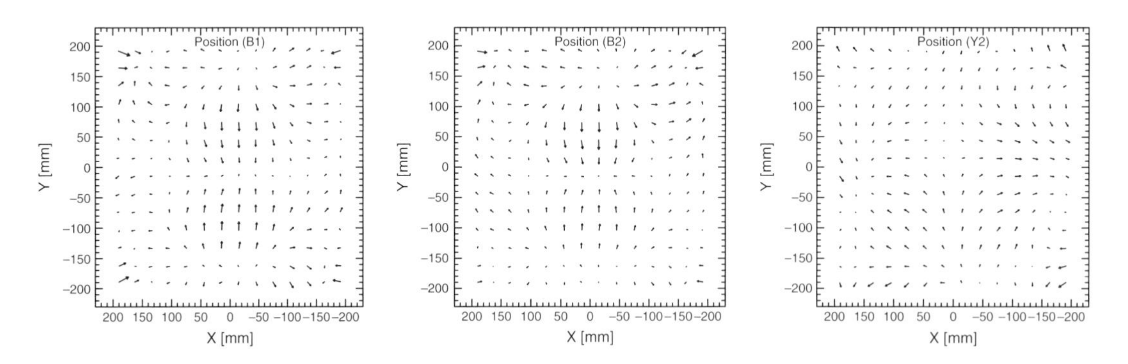

Figure 2.13 Position 'masks' for the NPM Version 2 plate reductions referred to the Hipparcos reference frame. In each case, X, Y are Lick plate coordinates, and vectors show the mean residuals for approximately 120 000 Tycho 2 stars in 14 × 14 bins, with an average of around 600 stars per bin. The vector length scale is 10 mm = 1 μm. Left: for the first epoch blue plates. The rms vector length is 0.75 μm (42 mas), and the longest is 2.10 μm (116 mas). Middle: for the second epoch blue plates. The rms vector length is 0.69 μm (38 mas), and the longest is 1.95 μm (108 mas). Right: for the second epoch yellow plates. The rms vector length is 0.62 μm (34 mas), and the longest is 1.46 μm (80 mas). From Hanson et al. (2004, Figures 2–4).

measurements with respect to some 70 000 galaxies), for 378 360 selected stars in the range $B = 8$–18, and $\delta > -23^\circ$ (Klemola *et al.*, 1987; Hanson, 1987; Hanson *et al.*, 2004). The 1993 NPM Catalogue Version 1 contains 148 940 stars in 899 fields outside the Milky Way's zone of avoidance, while the 2003 NPM Catalogue Version 2 contains 232 062 stars (122 806 faint stars for astrometry and Galactic studies, 91 648 bright $B < 14$ stars for positional reference, and 34 868 special stars selected for scientific interest) in 347 fields near the Galactic plane. The NPM Version 2 proper motions were placed on the ICRS using Tycho 2 reference stars, to an accuracy of 0.6 mas yr^{-1} in each field. The resulting rms proper motion precision is 6 mas yr^{-1}. Positional errors average 80 mas at the mean plate epoch 1968, and 200 mas at the chosen NPM Version 2 catalogue epoch 2000.

The adopted plate-constant model for the Lick NPM programme (Hanson *et al.*, 2004, Table 2) consists of 14 terms: scale, rotation, zero-point, and quadratic and cubic terms. The positional residuals based on the Tycho 2 star positions propagated to the relevant epoch then reflect higher-order positional effects present in the plates. The resulting 'masks' are shown in Figure 2.13, and indicate that: (a) all three contain systematic patterns unrepresentable by cubic terms; (b) the similarity of the two blue plate patterns testifies to the long-term stability of the Lick astrograph, important for the determination of accurate proper motions (but see below for the SPM); (c) the yellow pattern is very different, reflecting the optical characteristics of the two lens systems.

The Yale/San Juan Southern Proper Motion (SPM) programme was intended as a southern-sky complement to the Lick NPM programme, and similarly provides positions, absolute proper motions, and photographic *BV* photometry. Version 1 contained 58 880 objects at the South Galactic pole, covering about 720 deg^2 in the magnitude range $V = 5$–17.5. About 55% of all catalogue stars were randomly chosen, with the remainder representing astrophysically interesting objects drawn from various lists and databases (Girard *et al.*, 1998; Platais *et al.*, 1998; Girard *et al.*, 2004). An early study of systematic positional errors in the SPM plates was made by Platais *et al.* (1995) based on the provisional Hipparcos 30-month solution. With this limited density of reference stars they nevertheless identified a significant magnitude equation, of about 1 μm mag^{-1} over the Hipparcos magnitude range, interpreted as a lens decentering which introduces a coma-like distortion, and confirming the predictions of Conrady (1919). SPM Version 3 contains data for roughly 10.7 million objects down to $V = 17.5$, covering an irregular area of 3700 deg^2 between declinations $-20^\circ < \delta < -45^\circ$, excluding the Galactic plane. The proper motion precision for well-measured stars is estimated to be 4.0 mas yr^{-1}. For this version, the proper motions were also placed on the system of the ICRS, via Tycho 2 reference stars, and have an estimated systematic uncertainty of 0.4 mas yr^{-1}. A similar plate solution as for the NPM was adopted. Additional systematic field errors were identified in the positions, but also in the derived proper motions (Figure 2.14). The latter was unexpected since the SPM proper motions, being differential, should not suffer such effects provided the telescopes and plates remain unchanged between the two observational epochs. These proper motions residuals, corrected for in the SPM Version 3, imply that an unidentified change occurred in either the telescopes or the stored plates.

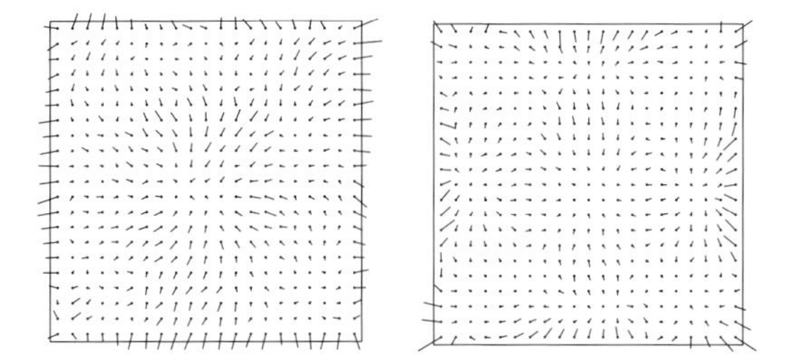

Figure 2.14 Left: mean position residuals, UCAC 0.9 minus provisional SPM Version 3, for all 156 fields. The plate dimension is 6°.4, and the maximum vector length corresponds to about 100 mas. Right: corresponding proper motion residuals, Tycho 2 minus provisional SPM Version 3. The maximum vector length corresponds to about 10 mas yr^{-1}. From Girard et al. (2004, Figure 4).

An analysis of the SPM Version 2 proper motions compared with those from Hipparcos, both considered to be quasi-inertial, was reported by Zhu (2001). They found that the two components of the spin vector difference, ω_x and ω_z, are less than the value of 0.25 mas yr^{-1} for the uncertainty of the Hipparcos inertiality, while the component ω_y is almost twice that value. They suggested that the SPM Version 2 contains a small residual colour systematic with respect to the Hipparcos proper motions.

2.6.4 Other photographic surveys

The Twin Astrographic Catalogue (TAC) is an astrometric, photographic catalogue covering most of the sky between declinations $+90^\circ$ and -18° to an average limiting magnitude of about $V = 11$ and $B = 12$. Version 1 (Zacharias *et al.*, 1996) was based on 4912 plates taken with the US Naval Observatory Twin Astrograph (blue and yellow lens) between 1977 and 1986, reduced to the system of the IRS, supplemented by proper motions obtained from a combination with the re-reduced Astrographic Catalogue, and made available as the AC zones became available. Version 2 (Zacharias & Zacharias, 1999) was based on the same plate material, but with the Hipparcos Catalogue used for a new plate-by-plate reduction, providing a significant improvement over Version 1. With an average precision of 48–120 mas per coordinate, depending on magnitude, and a higher star density than the Tycho Catalogue, the TAC is a significant catalogue for proper motion determination. Version 2 contains positions for 705 099 stars, supplemented by photographic photometry (B and V) for most.

The Pul-2 and Pul-3 Catalogues of positions and proper motions for ~59 000 stars are based on measurements of photographic plates using the Pulkovo Normal Astrograph taken between 1935–1960 for the first epoch and 1969–1986 for the second, with a mean epoch of 1963.25 (see also Kumkova, 2000). It contains stars of mainly 12–16.5 mag in nearly 150 fields with galaxies in the declination zone from −5° to +85°. The original plan, as realised by the Pul-2 Catalogue of 59 766 stars (Bobylev *et al.*, 2004), was to construct a catalogue of proper motions with respect to about 700 background galaxies. By comparing the Pul-2 and Hipparcos Catalogues, Bobylev *et al.* (2004) determined the components of the residual rotation of the Hipparcos inertial reference frame as $\omega_{x,y,z} = (-0.98 \pm 0.47, -0.03 \pm 0.38, -1.66 \pm 0.42)$. Khrutskaya *et al.* (2004) describe the construction of the Pul-3 Catalogue of 58 483 stars, now using the Tycho 2 Catalogue as the astrometric reference catalogue, yielding more than 50 000 stars fainter than 12 mag on the same reference system. The internal positional accuracy at the mean epoch of observations is 80 mas, while the accuracy of proper motions is in the range 2–12 mas yr^{-1}.

Various other observations have focused on maintaining the Hipparcos reference system through the determination of absolute proper motions of Hipparcos stars. GPM1 is a catalogue of absolute proper motions of 977 stars with respect to galaxies (Rybka & Yatsenko, 1997). Absolute proper motions for 48 Hipparcos stars in nine fields in the northern hemisphere as part of the Bonn programme for the Hipparcos reference frame link using plates from the Lick astrograph, and the ESO and Palomar Schmidt telescopes (Geffert *et al.*, 1997); and specifically in the region of M81 using plates taken between 1894–1994 (Odenkirchen *et al.*, 1997); using 3425 photographic plates obtained with the 40 cm astrograph at the Shanghai observatory from 1901–1993 (Jin *et al.*, 1997; Wang *et al.*, 1997a,b, 1998).

Tel'nyuk-Adamchuk *et al.* (2000) reported the construction of a catalogue in the polar region, north of 75°, from 14 meridian and photographic source catalogue with epochs from 1855 to the 1990s, and placed on the Hipparcos system.

The Chile–UK Quasar Survey covering 140 square degrees to $B = 20$ to investigate large-scale structure traced by quasars at $z = 0.4 - 2.2$ used astrometric calibration based on Tycho Catalogue stars to yield a positional uncertainty of 0.7 arcsec (Newman *et al.*, 1998).

Block adjustment in astrometry: As a relative measuring technique, traditional photographic astrometry determines positions of objects referred to a local plate reference frame, which is represented by a set of reference stars of a certain catalogue, rather than referred to the global reference frame represented by all stars of the reference catalogue. Compared with single plate adjustment, block adjustment avoids the contradiction of a star having several positions from various plates at a specified epoch.

Donner & Furuhjelm (1929) proposed the idea of a combined solution of neighbouring plates in the Helsingfors zone of the Carte du Ciel by considering the characteristics of overlapping plates. Eichhorn (1960) first introduced rigorous block adjustment of overlapping plates into astrometry. Further development (de Vegt, 1968; de Vegt & Ebner, 1972, 1974) led to practical applications in the re-reduction of the AGK2 (Führmann, 1979) and CPC2 (Zacharias, 1984). Since then block adjustment has acquired an extensive literature (see, e.g. Zacharias, 1988, 1992). A detailed review and bibliography is given by Eichhorn (1988).

The problem is exacerbated for CCD surveys, where the smaller field of view results in smaller numbers of primary or secondary reference stars. Yu *et al.* (2004) discuss the application of block adjustment to CCD surveys.

2.7 Reference system for CCD surveys

2.7.1 USNO catalogues: UCAC 1/2 and NOMAD

In addition to USNO's major re-reductions of existing plate material, leading to the AC 2000, USNO B, and related catalogues, a distinct effort has led to a full-sky coverage CCD astrometric programme, directly linked to the Hipparcos/Tycho Catalogues and the ICRF. The USNO CCD Astrograph Catalogue (UCAC) programme has used its 0.2 m Twin Astrograph upgraded with a 4k×4k CCD camera for an observational programme based first in the southern hemisphere, and subsequently in the northern hemisphere using the same instrument (see Figure 2.15). Observations started in the southern hemisphere at Cerro Tololo (CTIO) in Chile in January 1998 (Zacharias *et al.*, 2000a; Zacharias & Zacharias, 2003). A two-fold overlap pattern to 16 mag, covering some 40 million stars over two years, achieved accuracies of some 20 mas per coordinate between 8 and 13.5 mag, increasing to about 70 mas at the limiting magnitude, and at the mean epoch of close to 1999.0. This extension of the Hipparcos/Tycho optical reference frame beyond 12 mag was undertaken to permit the improved re-reduction of old epoch photographic data, to support continuing observational programmes with transit circle and astrograph-type instruments, and to facilitate the reduction of tertiary data such as Schmidt plate surveys and the Sloan Digital Sky Survey (Zacharias *et al.*, 1997b,c; Urban, 1998). For UCAC 1, a combination of Hipparcos, Tycho 1, and ACT stars were used for the preliminary reductions. Proper motions were then derived in two distinct ways according to magnitude: by combining the data with the ~100-year-old Astrographic Catalogue data and some 160 other photographic and transit circle catalogues, all reduced to the Hipparcos system, for stars brighter than $V = 12$;

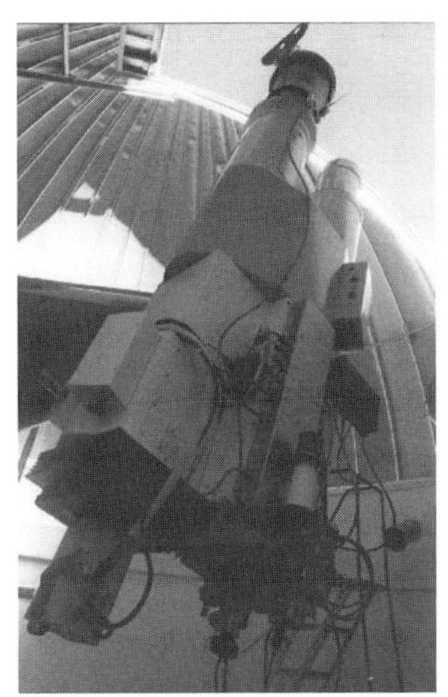

Figure 2.15 Left: the USNO 1.55-m (61-inch) Strand Telescope at Flagstaff, an astrometric reflector used to measure parallaxes since 1964, now equipped with CCDs. Middle: the US Naval Observatory Twin Astrograph at CTIO during the UCAC observations. Right: the 0.2-m (8-inch) Flagstaff Astrometric Scanning Transit Telescope (FASTT), a fully automated telescope, completed in 1981. Photographs courtesy of US Naval Observatory.

and with first and second epoch POSS, ESO, and AAO Schmidt surveys using the USNO A2 Catalogue for the fainter stars.

The preliminary catalogue, UCAC 1, the first US Naval Observatory CCD Astrograph Catalogue, comprised 27 million stars covering 80% of the southern hemisphere, yielding proper motions of around 3 mas yr^{-1} for the bright stars, and 10–15 mas yr^{-1} for the fainter stars (Zacharias *et al.*, 2000a). UCAC 1 is only some 10% as dense as USNO A2, but the positions are about a factor of 5 more precise, and of similar accuracy to those of the Tycho Catalogue (Figure 2.17). Systematic errors were estimated at the 10–15 mas level, mainly as a function of magnitude, based on comparisons with Tycho 2 (Zacharias *et al.*, 2000b).

The instrument was moved to the northern hemisphere, to Flagstaff in Arizona, in 2001, from which location sky coverage was extended to cover all northerly declinations, although the UCAC 2 Catalogue release (Zacharias *et al.*, 2004b) extended only as far north as +52°. UCAC 2 contains positions and proper motions for 48.3 million sources (mostly stars), supplemented with 2MASS photometry for 99.5% of the sources, and completely supersedes the UCAC 1 released in 2001. Reductions were based on the Tycho 2 Catalogue directly, leading to proper motion errors of about 1–3 mas yr^{-1} for stars to 12 mag, and about 4–7 mas yr^{-1} for fainter stars to 16 mag. Examples of the field distortion pattern in the two locations are given in Figure 2.16.

UCAC 2 covers over 85% of the sky; the remaining northern celestial pole area is currently being reduced. To supplement this temporarily incomplete sky coverage, and to supplement the absence of observable stars brighter than about 8 mag, the UCAC 2 Bright Star Supplement (Urban *et al.*, 2004) was created. All astrometric data were extracted from the Hipparcos Catalogue, the Hipparcos Double and Multiple System Annex, the Tycho 2 Catalogue, and the Tycho 2 Supplement 1. Photometric data were taken from the above catalogues and 2MASS. UCAC 2 BSS thus actually contains no UCAC 2 data, and contains not only bright stars, but all stars from the above-mentioned catalogues not found in the UCAC 2.

It is expected that the final catalogue, UCAC 3, will be released around mid-2008. It will be again be based on the Tycho 2 Catalogue, although with a link to extragalactic sources being part of the survey via deep CCD imaging of ICRF counterparts.

The merged dataset NOMAD Zacharias *et al.* (2004a) describe the first version of the Naval Observatory Merged Astrometric Dataset (NOMAD). The 100 GB dataset contains astrometric and photometric data for over 1 billion stars derived from the Hipparcos, Tycho 2, UCAC 2, USNO B, and the unpublished NPM- and SPM-based 'Yellow Sky' catalogues for astrometry and optical photometry, supplemented by 2MASS near-infrared photometry. For each unique star the 'best' astrometric

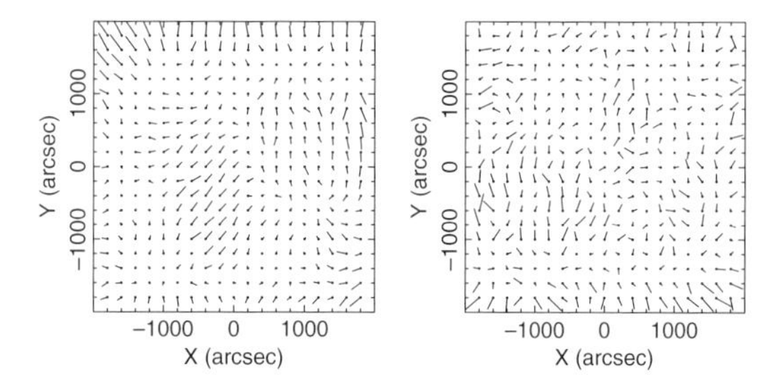

Figure 2.16 Left: field distortion pattern for the UCAC CCD astrograph data taken at CTIO. The vectors are scaled by 10 000, with the largest correction vectors about 25 mas long. Right: for the same CCD astrograph instrument data taken at Flagstaff. The much smaller number of available CCD frames cause the larger random scatter. From Zacharias et al. (2004b, Figures 2 and 3).

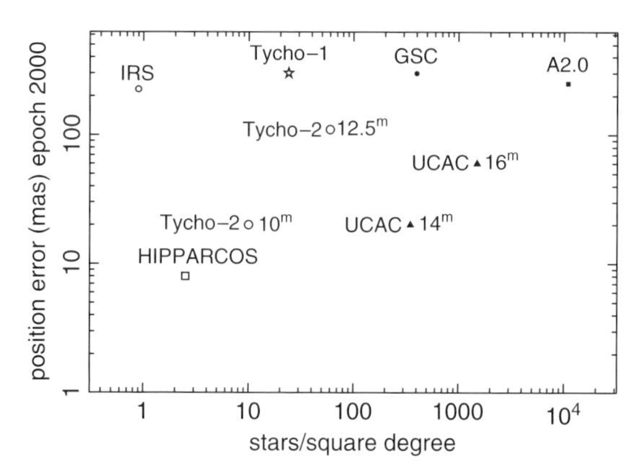

Figure 2.17 Positional accuracy versus star density for various astrometric catalogues; IRS = International Reference Stars; GSC = Guide Star Catalogue; A2.0 = USNO A2; UCAC = USNO CCD Astrograph Catalogue. For Tycho 2 and UCAC, the dependency on magnitude is indicated. From Zacharias et al. (2000a, Figure 17).

and photometric data are chosen from the source catalogues and merged into a single dataset, with flags to identify the source catalogues and cross-reference identifications. All source catalogue astrometric data are on the ICRF.

2.7.2 FASTT

Stone *et al.* (2003) describes the 0.2-m (8-inch) Flagstaff Astrometric Scanning Transit Telescope (FASTT), a fully automated telescope, completed in 1981, that takes about 41 000 CCD frames of data a year (Figure 2.15). FASTT is operated in TDI (time-delayed integration, or drift-scan mode), like the Sloan Digital Sky Survey. Astrometry is provided in support of various spacecraft missions, to predict occultation events, to calculate dynamical masses for selective asteroids, and to improve the ephemerides for thousands of asteroids, the planets Jupiter to Pluto, and 17 satellites of Jupiter through Neptune. Although most of the FASTT observing programme involves the Solar System, FASTT stellar astrometry was used to set up a number of astrometric calibration regions along the celestial equator, to verify the Hipparcos link to the ICRF (Stone *et al.*, 1996), to determine accurate positions for a large sample of radio stars, and to investigate systematic errors in the FK5 catalogue. It also produces accurate magnitudes that are being used to investigate the shapes of thousands of asteroids. By the end of 2003, FASTT had produced over 190 000 positions of Solar System objects providing a very large and homogeneous database for each object that will extend over many years and include positions accurate to ±47 to ±300 mas, depending on magnitude ($3.5 < V < 17.5$). Extensive efforts have been undertaken to improve the systematic accuracy of FASTT equatorial positions by applying corrections for differential colour refraction, distortions in the focal plane, and correcting for a positional error that is dependent on magnitude. The systematic accuracy of FASTT observations is currently about ±20 mas in both right ascension and declination.

2.7.3 Sloan Digital Sky Survey

The Sloan Digital Sky Survey (SDSS) is a survey of one-quarter of the sky in five broad optical bands; with 95% completeness limits for point sources of $u = 22.0$, $g = 22.2$, $r = 22.2$, $i = 21.3$, and $z = 20.5$. The survey will produce a database of roughly 10^8 galaxies, 10^8 stars, and 10^6 quasars with accurate photometry, astrometry, and object classification parameters (York *et al.*, 2000). Spectroscopy, covering the wavelength range 380–920 nm at $R \approx 1800$, is targeted for roughly a million galaxies, 100 000 quasars, and another 50 000 stars and serendipitous objects. The imaging camera consists of 54 CCDs in eight Dewars covering 2°.3 on the sky. The Sixth Data Release (Adelman-McCarthy, 2008) marked the completion of the imaging survey of the northern Galactic cap, comprising some 287 million objects.

The astrometric calibration is described in detail by Pier *et al.* (2003). The r CCDs serve as the astrometric reference CCDs for the SDSS, with the positions for SDSS objects being based on the r centroids and calibrations. One of two reduction strategies is employed, depending on the coverage of astrometric catalogues: whenever possible, stars detected on the r photometric CCDs are matched directly with stars in UCAC. UCAC extends down to $R = 16$, giving approximately 2–3 mag overlap with unsaturated stars on the photometric r CCDs. If UCAC coverage was not available to reduce a given imaging scan, detections of Tycho 2 stars on the astrometric CCDs are mapped onto the r CCDs (all Tycho 2 stars saturate on the r CCDs) using bright stars that have sufficient signal-to-noise ratio on the astrometric CCDs and are unsaturated on the r CCDs.

For point sources brighter than $r \sim 20$, the astrometric accuracy is 45 mas rms per coordinate when reduced against UCAC, and 75 mas rms when reduced against the brighter and lower star density Tycho 2, with an additional 20–30 mas systematic error in both cases. The relative astrometric accuracy between the r filter and each of the other filters is 25–35 mas rms. At the survey limit ($r \sim 22$), the astrometric accuracy is limited by photon statistics to approximately 100 mas rms for typical seeing. Some of the random errors were partly attributable to anomalous refraction (Section 2.9).

2.7.4 Other CCD imaging systems

The WIYN (Wisconsin, Indiana, Yale, NOAO) open cluster study makes use of the Mayall KPNO 4 m telescope CCD Mosaic Imager to obtain second-epoch positions and hence accurate proper motions of selected open clusters. This requires an astrometric calibration of the CCD Mosaic Imager, which consists of eight individual chips. To achieve this, Kozhurina-Platais *et al.* (2000) created a secondary astrometric reference system in the area of the open cluster NGC 188, based on numerous Lick astrograph plates taken in 1963–1999 and the Tycho 2 Catalogue. This allows them to determine absolute coordinates and rotation angles for each chip in the unified CCD mosaic. A major problem in calculating the various constants is the presence of unpredictable, systematic offsets in the short exposures, caused by correlated motions of stars due to atmospheric turbulence.

Bustos Fierro & Calderón (2002) carried out a detailed calibration and modelling of a small CCD field of 10 arcmin using the long-focus reflector at El Leoncito. Modelling was based on Tycho 2 and USNO A2 positions using a field in the open cluster Ruprecht 21.

2.8 Infrared reference frame

Extending the astrometric reference frame into the infrared was discussed at the time of the Hipparcos Catalogue publication by Sutton (1997), who focused on 87 bright infrared sources also appearing in the Hipparcos Catalogue.

Hindsley & Harrington (1994) had already constructed the US Naval Observatory Catalogue of Positions of Infrared Stellar Sources (CPIRSS) by identifying stars in the IRAS Point Source Catalogue (where the typical 2σ error ellipse is some 3×20 arcsec^2, but where the position angle varies across the sky) with optical positions from a hierarchy of four astrometric catalogues; yielding accuracies of about 0.2 arcsec for 33 678 infrared sources. More than 25 000 of these IRAS PSC sources were included in the Hipparcos Catalogue (Kharin, 1997). The problem of identifying the IRAS PSC stars with the optical counterparts was discussed further by Kharin (2000), and also by Knauer *et al.* (2001).

A dense infrared reference frame became a reality with the Two Micron All Sky Survey, 2MASS, released in 2003 (Cutri *et al.*, 2003; Skrutskie *et al.*, 2006). Primarily a photometric catalogue for the near infrared bands *JHK*, 2MASS also provides accurate positions for 470 million point sources at an epoch near 2000 on the ICRF, and an Extended Source Catalogue of more than 1.6 million sources. The astrometric reduction of the 2MASS data was based on the Tycho 2 Catalogue, and is described by Skrutskie *et al.* (2006, Section 4.6). It comprised three steps: point-source extraction in overlapping frames, identification of Tycho reference stars, and constraining the solution across overlapping regions of adjacent scans. Despite 2 arcsec pixels, and a best-case system point-spread function of 2.5 arcsec, typical position uncertainties for $K_S < 14$ mag are $\lesssim 100$ mas rms.

Zacharias *et al.* (2000a, 2003) compared all three-band detections of 2MASS with the USNO CCD Astrograph Catalogue (UCAC) to perform an external assessment of the accuracy of 2MASS astrometry; since both catalogues use Tycho 2 reference stars they are both nominally on the Hipparcos system. Random positional errors of UCAC are 20–70 mas, with systematic errors of about 10 mas. Based on over 51 million stars in common ($R = 8$–16) only small systematic differences (10–20 mas) between UCAC and 2MASS positions were found as a function of magnitude, colour and location in the sky and on the detector. The random errors of 2MASS positions are 60–100 mas per coordinate, depending on magnitude. The 2MASS positions are thus of higher quality than the optical positions of currently available similar sized optical catalogues like USNO B and GSC 2. Damljanović & Souchay (2003a,b) derived a similar standard error of positional differences for some 38 000 stars in common between Hipparcos–2MASS of close to 0.10 arcsec.

2.9 Atmospheric attenuation and refraction

As part of their compilation and re-reduction of Carlsberg Meridian Circle Catalogues, CMC, between 1984–1998 Copenhagen University Observatory & Royal Greenwich Observatory (2001) compiled all the meteorological data collected over the period, including the atmospheric extinction.

Cuillandre *et al.* (2004) describe the Canada–France–Hawaii Telescope (CFHT) SkyProbe system, designed to measure atmospheric attenuation by clouds. The

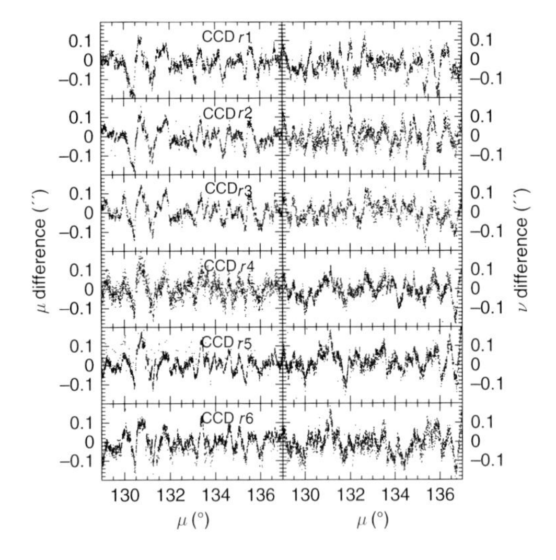

Figure 2.18 Differences in position between matched stars in two overlapping SDSS scans, reduced against Tycho 2 positions, and plotted against along-scan SDSS great-circle coordinate, μ, equivalent to time. The differences in the orthogonal great-circle coordinates μ and ν are plotted separately for each of the r CCDs. From Pier et al. (2003, Figure 14).

measurement is performed once per minute, with a crucial addition to the hardware being the full-sky Tycho photometric catalogue providing homogeneous photometry at an appropriately faint magnitude.

Langhans *et al.* (2003) reported an investigation into the effects of atmospheric refraction as a function of zenith distance using an area centred on NGC 6791, using stars with accurate positions and photometry from the Tycho Catalogue. Somewhat similar plans for refraction studies using Hipparcos stars brighter than 6 mag were reported by Tie *et al.* (2005).

Surveys operated in TDI (time-delayed integration, or drift-scan mode), like FASTT and Sloan/SDSS, suffer from atmospheric fluctuations of star positions, while the 'stare' mode operation of photographic and traditional CCD imaging cancels out much of the atmospheric turbulence effects. In their astrometric calibration of the SDSS, Pier *et al.* (2003) found that the rms errors displayed considerable coherence across the focal plane; that is, both the direction and amplitude of the residuals from CCD to CCD show a high degree of correlation (Figure 2.18). They attributed the cause of these effects to anomalous refraction (rather than to, say, telescope jitter), i.e. refraction that varies from the smooth analytical models which are functions only of zenith distance. This behaviour has been seen before in CCD mosaic calibrations (e.g. Kozhurina-Platais *et al.*, 2000), and in drift-scanning CCD observations, e.g. Stone *et al.* (1996) reported similar results with their drift-scan mode CCD observations made on the extremely stable FASTT transit telescope, although with a field of view of only about 20 arcmin. The SDSS observations, with its accurate astrometry, suggest that these effects are due to phenomena whose scale is on the order of two degrees or more on the sky. Considered to arise in atmospheric boundary layers with heights of typically a few hundred metres to two km above the Earth's surface, and horizontal wavelengths $\lambda \sim 1-10$ km, the effects would then be expected to vary with a characteristic time scale of λ/v, where v is the wind speed. Typical time signatures of a few to several tens of minutes are thus predicted, in agreement with the SDSS observations.

2.10 Proper motion surveys

2.10.1 High proper motion surveys

Surveys for stars of high proper motion are important for identifying potential stars having high space velocity, but more importantly for identifying stars possibly (but not necessarily) very close to the Sun. Ground-based parallax surveys have therefore frequently concentrated on high proper motion stars as a potential source of stars with large parallaxes, and Hipparcos followed a similar approach in the choice of stars for the Input Catalogue, notwithstanding the resulting kinematic bias that will occur in the Hipparcos sampling of nearby stars. The New Luyten Two-Tenths Catalogue (NLTT) of high proper motion stars with $\mu > 0.18$ arcsec yr^{-1} as faint as $V \sim 19$ (Luyten, 1980a,b), and the subset of the Luyten Half-Second Catalogue (LHS) with $\mu > 0.5$ arcsec yr^{-1} (Luyten, 1979), were used for the Hipparcos Input Catalogue construction, and continue to be an important source of astrometric data, for studies of nearby stars, subdwarfs, and white dwarfs.

Modern surveys for high proper motion objects is a large research field, and the following gives only an incomplete introduction. More details can be found in, e.g. Lépine & Shara (2005a).

By observing a large fraction of the brighter ($V < 11$) NLTT stars, Hipparcos obtained improved astrometry and photometry for about 13% of NLTT, although in its magnitude-limited survey ($V < 7.3-9.0$), and due to the nature of its construction based on a pre-defined observing programme, did not find any significant number of new high proper motion stars not already catalogued by Luyten. Tycho 2 includes astrometry and photometry for several thousand additional NLTT stars and, resulting from its survey-type operation, also contains several hundred previously unknown bright ($V \lesssim 11$) high proper motion stars (Gould & Salim, 2003).

Original positional uncertainties for high proper motion stars in the Luyten Half-Second Catalogue (LHS) frequently exceed 10 arcsec, and occasionally 30 arcsec.

Bakos *et al.* (2002) used cross-correlation with the Hipparcos and Tycho 2 Catalogues to derive revised coordinates for 4330 of the original 4470 stars, and revised proper motions for 4040.

Lépine *et al.* (2002) used a blink comparator to search for high proper motion stars in the Digitised Sky Survey over part of the northern sky, finding 601 stars in the range 9–20 mag with proper motions in the range 0.5–2.0 arcsec yr^{-1}. Overlap with the Hipparcos and Luyten proper motion catalogues LHS and NLTT are discussed.

Gould & Salim (2003) and Salim & Gould (2003) constructed a revised version of the New Luyten Two-Tenths (NLTT) catalogue of high proper motion stars, containing improved astrometry and photometry for the vast majority of the 59 000 stars in NLTT. The bright end was constructed by matching NLTT stars to Hipparcos, Tycho 2, and Starnet; the faint end by matching to USNO A and the Two Micron All Sky Survey (2MASS). They showed that proper motion errors for the NLTT stars with small position errors are 24 mas yr^{-1} (1σ), but deteriorate to 34 mas yr^{-1} for stars with inferior positions. They also studied the completeness of NLTT at the bright end by comparing it to Hipparcos and Tycho 2, finding that NLTT is virtually 100% complete for $V \lesssim 11.5$ and $|b| > 15°$, although completeness in this magnitude range falls to ~75% at the Galactic plane. Incompleteness near the plane is not uniform, but is rather concentrated in the interval $-80° < l < 20°$, where the Milky Way is brightest.

The Liverpool–Edinburgh high proper motion survey is based on the SuperCOSMOS Sky Survey (see Section 2.5.3), and has published a catalogue derived from the South Galactic Cap region of 6206 stars with proper motions exceeding 0.18 arcsec yr^{-1} with a magnitude limit of $R = 19.5$ mag (Pokorny *et al.*, 2003). The survey is > 90% complete within the nominal limits of the Luyten Two Tenths Catalogue of $R \leq 18.5$ mag and $\mu = 0.2$–2.5 arcsec yr^{-1}, and ≥ 80% complete for $R \leq 19.5$ mag and $\mu \leq 2.5$ arcsec yr^{-1}. Five new objects with proper motions between 1.0–2.6 arcsec yr^{-1} were reported by Hambly *et al.* (2004).

Lépine & Shara (2005a) describe the Lépine–Shara Proper Motion (LSPM) north catalogue, 61 977 stars north of the J2000 celestial equator that have proper motions larger than 0.15 arcsec yr^{-1}. They were generated primarily as a result of a systematic search for high proper motion stars in the Digitised Sky Surveys using the 'Superblink' software (and avoiding visual comparison using the more conventional blink comparator) to compare positions at the epochs of the POSS-I and POSS-II plates. Due to plate saturation at brighter magnitudes, the catalogue also incorporates stars and data from the Tycho 2 Catalogue and also, to a lesser extent, from the ASCC–2.5. LSPM expands the earlier Luyten (LHS, NLTT) catalogues, superseding them for northern declinations. Positions are given with an accuracy $\lesssim$100 mas at epoch J2000.0, and absolute proper motions are given with an accuracy of ~8 mas yr^{-1}. The absolute motions are with respect to the Tycho Catalogue, which allows the relative motions obtained with Superblink to be converted to absolute motions, thus revealing large-scale stellar motion in the Galaxy (Figure 2.19). Photometry includes B_T and V_T from Tycho 2 and ASCC 2.5, photographic $B_{\rm J}$, $R_{\rm F}$, $I_{\rm N}$ from USNO B1, infrared J, H, K_s from 2MASS, and estimated V and $V - J$ for nearly all catalogue entries, useful for initial classification. The catalogue is estimated to be over 99% complete at high Galactic latitudes ($|b| > 15°$), and over 90% complete at low Galactic latitudes ($|b| < 15°$), down to a magnitude $V = 19.0$, and has a limiting magnitude $V = 21.0$. All the northern stars listed in the LHS and NLTT catalogues were re-identified, and their positions, proper motions and magnitudes re-evaluated. The catalogue also lists a large number of new objects, which promises to expand the census of red dwarfs, subdwarfs, and white dwarfs in the vicinity of the Sun.

Lépine (2005) used LSPM-north to compile a list of 4131 dwarfs, subgiants, and giants located or suspected to be located within 33 pc of the Sun. Trigonometric parallax measurements exist for 1676 stars, with photometric and spectroscopic distance moduli available for another 783. The remaining 1672 objects are proposed as nearby star candidates for the first time, and include 539 stars suspected to be within 25 pc of the Sun, of which 63 are estimated to be within only 15 pc. They estimate that 32% (18%) of nuclear-burning stars within 33 pc (25 pc) of the Sun remain to be located. The missing systems are expected to have proper motions below the 0.15 arcsec yr^{-1} limit. A study of LSPM J2322+7847, a nearby young, low-mass proper motion companion to the young visual binary HIP 115147, discovered by the LSPM survey, was reported by Makarov *et al.* (2007).

LSPM-south is the corresponding southern catalogue of high proper motion stars generated in the same way (Lépine & Shara, 2005b), and containing over 50 000 stars with $\mu > 0.15$ arcsec yr^{-1}, again complemented at the brighter end ($V < 12$) by Tycho 2 data. Proper motions for the fainter stars are given with an accuracy ±15 mas yr^{-1}. Tests suggest that the LSPM-south is about 90% complete down to magnitude $V = 19$.

Finch *et al.* (2007) reported the discovery of 1606 proper motion systems in the range 0.18–0.40 arcsec yr^{-1} in the southern sky ($-90° < \delta < -47°$), based on SuperCOSMOS data. They infer that 31 systems are within 25 pc and two within 10 pc. The paper provides references to other recent southern high proper motion surveys.

There have been other successful surveys for high proper motion stars which have continued to employ

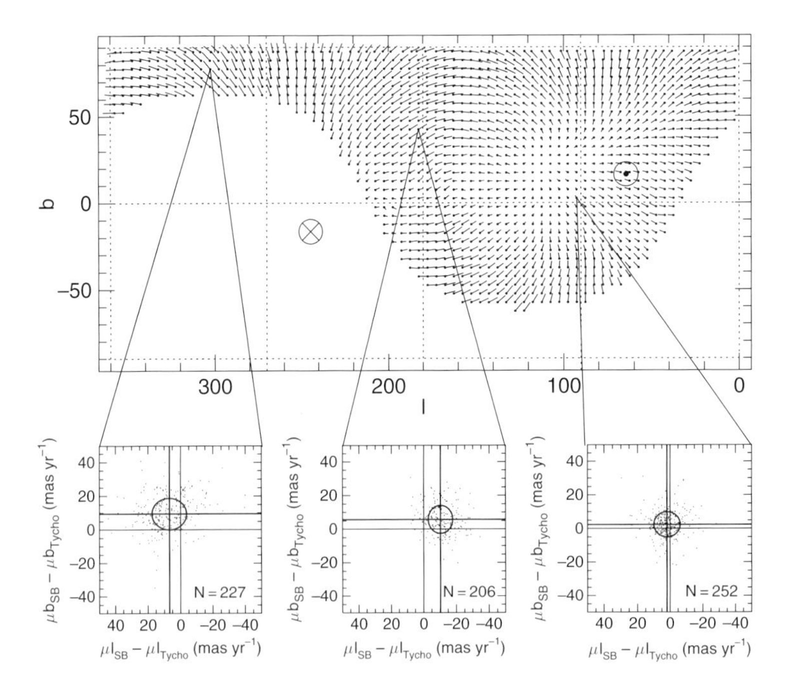

Figure 2.19 Local offsets between Superblink relative proper motions and Tycho 2 absolute proper motions, calculated using 33 300 Tycho 2 stars with proper motions in the range 0.04–0.50 arcsec yr^{-1}. The three bottom panels show local differences between the relative and absolute proper motions for stars located within 7° of the specified location. Cross-hairs mark the mean value of the offset, while the ellipse shows the mean standard deviation (1σ). The top panel plots proper motion difference vectors as a function of position on the sky, in Galactic coordinates, and also shows the apex (⊙) and antapex (⊗) of the Sun's motion. These offsets effectively map out the local mean proper motion of background field stars in the Tycho 2 (ICRS) reference system. Offsets are largest at high Galactic latitude, where there is a significant drifting motion of old disk and halo stars relative to the local standard of rest. Other effects are that of the Solar motion relative to the LSR, and the overall Galactic rotation. From Lépine & Shara (2005a, Figure 10).

plate comparisons without establishing an absolute reference system, which is not crucial if only relative motions of high proper motion stars are being searched for. These include the Calán–ESO proper motion survey using red IIIa-F plates taken 6–16 yr apart (Ruiz *et al.*, 2001), from which the discovery of Kelu 1, a free-floating brown dwarf at about 10 pc, was made (Ruiz *et al.*, 1997).

2.10.2 Other proper motions surveys

Improved proper motions resulting from a combination of the USNO B Catalogue and the Sloan Digital Sky Survey Data Release 1, covering 2099 square degrees, were independently constructed by Munn *et al.* (2004) and Gould & Kollmeier (2004), yielding some 400 000 stars with proper motions having standard errors of around 3–4 mas yr^{-1} for $r \leq 20$ mag.

Kharchenko (2001) constructed the All-Sky Compiled Catalogue of 2.5 million stars (ASCC–2.5) with a limiting magnitude $V = 12-14$ mag, a result of merging the Hipparcos and Tycho Catalogues with the ground-based catalogues PPM and CMC 11. Supplements include membership probabilities in 520 Galactic open cluster sky areas (Kharchenko *et al.*, 2004a), and radial velocity data (Kharchenko *et al.*, 2004b). Kharchenko (2004) constructed the MEGA-H proper motion catalogue of 18 169 stars by merging ASCC–2.5 with MEGA-G, a catalogue of proper motions with respect to galaxies in 47 selected areas near the main meridional section of the Galaxy (Kharchenko, 1987), also reduced to the Hipparcos system.

Fedorov & Myznikov (2005) created the X1 Catalogue of positions and proper motions for 1.3 million stars from 10–19 mag within 330 1° × 1° fields around ICRF sources in the northern hemisphere, based on a re-reduction of digitised images of the POSS-I and POSS-II Schmidt surveys obtained from USNO (Flagstaff) PMM image archive. Tycho 2 stars were used as reference: mean positional errors are 50–100 mas, and proper motions errors are about 2.5–5 mas yr^{-1} depending on magnitude.

Ivanov (2005) compiled a database of stars with proper motions exceeding 0.04 arcsec yr^{-1} using data from Hipparcos, Tycho 2, CMC 11, PPM, NPM1, NLTT, LPM, BPM and 725 other published sources. The list consists of 251 000 stars with $V < 16$ in the declination zone from -2° to $+90^\circ$. In addition to the position and proper motion, the database includes magnitudes in the *UBVRI* system, radial velocities, spectra, luminosity class, metallicity, and flags of multiplicity and variability.

Ducourant *et al.* (2005) constructed a proper motion catalogue, with positions and photometric parameters) for 1250 pre-main-sequence stars and 104 pre-main-sequence star candidates spread over all-sky major star-forming regions, and brighter than $V \sim 16.5$. Data originated from CCD meridian observations (Bordeaux and Valinhos CCD meridian circle), ESO 1.5-m and OHP 120-cm telescopes, Schmidt SERC *J*, POSS-I and POSS-II plates digitised with the MAMA measuring machine in Paris, and published catalogues including AC 2000.2, USNO A2.0, Hipparcos, Tycho 2, UCAC 2, 2MASS, and other astrometric sources. CCD meridian *V* magnitudes, and the *JHK* magnitudes from 2MASS, are included when available. Precision on proper motions vary from 2–5 mas yr^{-1}.

2.11 Parallaxes

2.11.1 Ground-based parallaxes

Hipparcos provided some 100 000 homogeneously measured absolute parallaxes of high accuracy. However, the catalogue is incomplete fainter than 8 mag and contains no objects fainter than 12 mag. Until the next space astrometry mission, the confirmation of unexpected Hipparcos parallaxes, the determination of the parallaxes of faint objects, the continued astrometric determination of stellar masses, and the detection of low-luminosity binary companions will remain the task of ground-based astrometry. Such systems now have the precision necessary to determine (relative) parallaxes with standard errors of order 1 mas. Amongst the recent ground-based parallax efforts (Dahn, 1998) the following compilations and new measurements can be noted: (a) the US Naval Observatory's Astrometric CCD programme on the Strand 1.55-m reflector: publications include parallaxes for 72 stars with $V = 15$–19.5, including the lower part of the main sequence and the cool subdwarfs by Monet *et al.* (1992); and 28 late-type dwarfs and brown dwarfs by Dahn *et al.* (2002); (b) the Multichannel Astrometric Photometer (MAP Gatewood, 1987) at the Allegheny Observatory, University of Pittsburgh (van Altena *et al.*, 1995); (c) the Navy Prototype Optical Interferometer (NPOI, a project of the US Naval Observatory, the Naval Research Laboratory, and Lowell Observatory, located near Flagstaff Arizona, from January 2002 able to track and record stellar interference fringes with six telescopes simultaneously). This superseded the Mark III stellar interferometer (a joint project of the US Naval Observatory, the Naval Research Laboratory, the Smithsonian Astrophysical Observatory, and MIT, located on Mount Wilson, and decommissioned in December 1992); (d) Cerro Tololo Inter-American Observatory parallax investigation programme, including the observation of 191 nearby stars for parallaxes as part of the RECONS nearby stars programme (Jao *et al.*, 2003); (e) the Torino Observatory parallax programme, which has determined parallaxes for six white dwarfs (Smart *et al.*, 2003), and re-examined 22 suspect objects from the Catalogue of Nearby Stars (Smart *et al.*, 2007). In addition, a discussion of parallaxes obtained with the Fine Guidance Sensors of the Hubble Space Telescope, and their comparison with Hipparcos results, is given by Benedict & McArthur (2005a), Soderblom *et al.* (2005), and Benedict & McArthur (2005b).

The Yale Trigonometric Parallax Catalogue represents the major compilation of stellar trigonometric parallaxes obtained from the ground. The Fourth Edition (van Altena *et al.*, 1995) is a revised and enlarged edition of the General Catalogue of Trigonometric Stellar Parallaxes containing 15 994 parallaxes for 8112 stars published before the end of 1995, and adding 1722 stars to those contained in the previous edition by Jenkins (1963). The mode of the parallax accuracy for the newly-added stars (standard error 0.004 arcsec) is considerably better than in the previous editions (about 0.016 arcsec). Approximately 2300 stars are not in the Hipparcos Catalogue.

Amongst other specific compilations, Bergeron *et al.* (2001) assembled a catalogue of 152 white dwarfs with parallaxes from the Yale Catalogue, and Gizis (1997) assembled almost 100 ground-based parallaxes for subdwarfs.

Comparison with Hipparcos Comparison between the Hipparcos and Allegheny parallaxes is discussed by Gatewood *et al.* (1998). For 63 stars common to both programmes, they found an average standard deviation of the difference of 2.3 mas, with no systematic difference, but an indication that the formal errors in one or both catalogues are somewhat underestimated.

Shatsky & Tokovinin (1998) compared 141 Hipparcos parallaxes with dynamical parallaxes of visual binary stars with reliable orbits, selected from the 612 physical multiple stars from the Multiple Star Catalogue of Tokovinin (1997). Data were divided into three groups according to the orbit quality and reliability of Hipparcos double-star solutions (e.g. interferometric orbits from speckle- or long-baseline interferometry, and orbits from visual measurements). Good agreement between dynamical and Hipparcos parallaxes is found for all distant systems ($\pi < 15$ mas), and for intermediate-distance systems (from 15–30 mas) unless they contain close subsystems with periods of the order of 1 year, in which case the photocentric motion perturbs the parallactic reflex motion.

2.11.2 Common proper motion systems

An interesting technique to obtain parallaxes of low-luminosity stars is via the identification of faint components of physical binaries whose primaries have measurable parallaxes, and which are then inferred to have the same parallax as the primary. The technique was used by van Biesbroeck (1944) for the low-luminosity star vB 10, by van Altena *et al.* (1995) for some common proper motion components in the Yale Catalogue, and by Oppenheimer *et al.* (2001) who applied the technique to common proper motion companions of Hipparcos stars to establish distances to some stars in their 8 pc sample.

The technique requires establishing a physical association between the two (apparent) components of a binary, to avoid inferences being based on chance proximity on the plane of the sky. This is not difficult for nearby stars, which usually have very large proper motions: the chance that two unrelated stars lying within a few arcmin of each other would have roughly similar proper motions of order 1 arcsec yr^{-1} is vanishingly small, and hence precise proper motions are not generally required.

In their systematic analysis for the common proper motion companions of Hipparcos stars, Gould & Chanamé (2004) argue that physical companions should have the same parallaxes as their primaries up to a fractional error equal to their separation in radians, which is generally small compared to the measurement error and, in any event, always less than 1%. However, the projected density of stars as a function of proper motion grows rapidly toward lower proper motions, so much higher proper motion precision is required to effectively reject spurious unrelated pairs. This has been facilitated by the publication of the revised New Luyten Two-Tenths (rNLTT) catalogue (Gould & Salim, 2003; Salim & Gould, 2003) which identifies virtually all Hipparcos counterparts of NLTT stars (Luyten, 1980b, and references therein; see also Rousseau & Perie 1997 for details of how the Luyten high proper motion stars were identified in preparation for the Hipparcos observations), and which gives new more accurate ($\sim$5.5 mas yr^{-1}) proper motions for the majority of NLTT stars in the 44% of the sky covered by the intersection of the first Palomar Observatory Sky Survey and the second 2MASS incremental release. Each rNLTT entry indicates whether the given star has an NLTT common proper motion companion, and its tabulated offset was used to search for the companion in 2MASS and USNO B1.0. Physical pairs from these NLTT binaries were identified by Chanamé & Gould (2004): from the 1147 candidate common proper motion binaries, they identified 999 physical pairs, of which 801 are disk main-sequence pairs, 116 are halo subdwarf pairs, and 82 contain at least one white dwarf.

Accordingly, Gould & Chanamé (2004) compiled a catalogue of common proper motion companions to Hipparcos stars, restricting their analysis to Hipparcos stars with accurate parallaxes that are also in the NLTT. Some of the common proper motion pairs are composed of two Hipparcos stars, each with a parallax better than 3σ. Results for the more interesting class of 424 non-Hipparcos common proper motion companions of Hipparcos stars, i.e. low-luminosity stars with new Hipparcos-based parallaxes, are shown in Figure 2.20. Of these low-luminosity components, 20 were classified as white dwarfs, either due to their position in the colour-magnitude diagram, or because they were so classified in the Villanova white dwarf database (McCook & Sion, 1999). Of these, five were not listed at Villanova as white dwarfs. Of the remainder, seven have previous parallaxes of which only two are of comparable quality to the Hipparcos parallaxes assigned by Gould & Chanamé (2004). A further 29 stars have $M_V \geq 14$, the majority of the latter being M dwarfs. The authors indicate how the technique could be applied to other datasets, for example the Luyten Double Star Catalogue or USBO B, to obtain parallaxes for additional low-luminosity stars.

2.12 Other applications

2.12.1 Celestial cartography

History Celestial cartography, or uranography, is the branch of astronomy concerned with mapping the stars and other celestial bodies. Its history is described in various books (e.g. Warner, 1979; Snyder, 1984; Whitfield, 1995; Stott, 1995), with an overview by George Lovi in the introduction to the first edition of Uranometria 2000.0 (see below). Early atlases were based on naked-eye measurements, subsequently based on telescopic, photographic and, most recently, digital observations. While frequently remarkable works of art, early star atlases were nevertheless primarily works of science, providing accurate star charts using a consistent projection scheme on which to plot the changing positions of planets, comets, and the Moon. Amongst the most notable works, including the generally-accepted 'grand' celestial atlases of Bayer, Schiller, Hevelius, Flamsteed and Bode, are: *De le Stelle Fisse* (Alessandro Piccolomini, Venice, 1540), *Uranometria* (Johann Bayer, Augsburg, 1603), *Coelum Stellatum Christianum* (Julius Schiller, Augsburg, 1627), *Harmonia Macrocosmica* (Andreas Cellarius, Amsterdam, 1661), *Globi Coelestis* (Ignace-Gaston Pardies, Paris, 1674), *Firmamentum Sobiescianum sive Uranographia* (Johann Hevelius, Gdansk, 1690), *Atlas Portatalis Coelestis* (Johann Rost, Nuremberg, 1723),

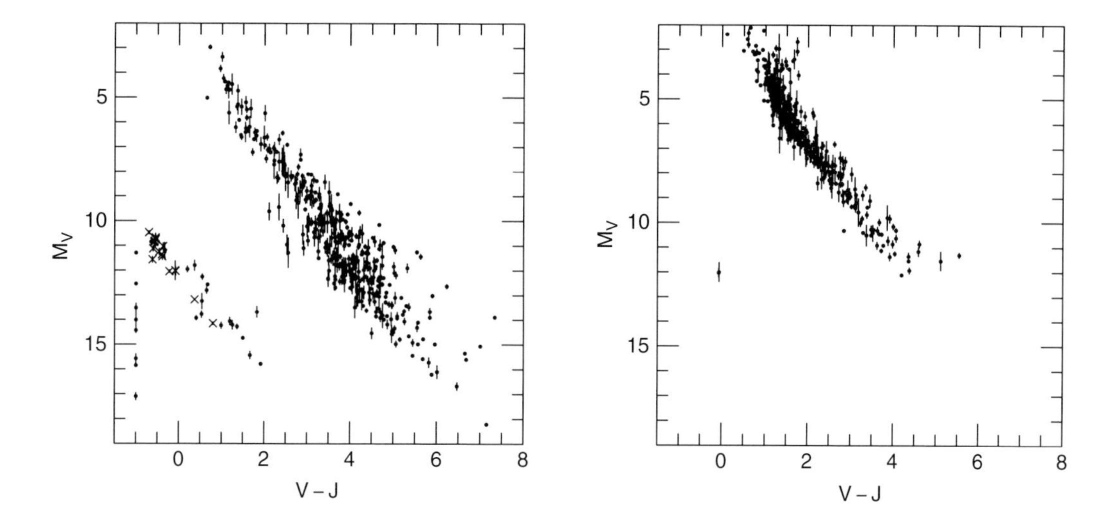

Figure 2.20 Left: colour–magnitude diagram of 424 common proper motion companions of Hipparcos stars with good (> 3σ) parallaxes from the catalogue of Gould & Chanamé (2004). Stars without J photometry are displayed at $V - J = -1$. Hipparcos white dwarfs with good parallaxes are shown as crosses; these tend to be significantly more luminous than the common proper motion white dwarfs from the catalogue. Right: colour–magnitude diagram of the Hipparcos primaries to the stars shown at left. All stars have Hipparcos photometry, which in most cases is accurate to a few percent. Importantly, the common proper motion companions in the left figure tend to fill in the region at the lower right, which is devoid of Hipparcos primaries. In both figures, error bars reflect only the parallax errors (and not the photometry errors) and so reflect the precision possible if the present, mostly photographic, photometry is replaced by CCD photometry. From Gould & Chanamé (2004, Figures 1 and 2).

Atlas Coelestis (John Flamsteed, London, 1729), *Uranographia* (Johann Bode, Berlin, 1801), *A Portraiture of the Heavens* (Francis Wollaston, London, 1811), *Neue Uranometrie* (Friedrich Argelander, Berlin, 1843), and *Atlas des Nördlichen Gestirnten Himmels (Bonner Durchmusterung)* (Friedrich Argelander *et al.*, Bonn, 1863).

Over the past century, many photographic star atlases have been created and published, including the early Franklin–Adams Charts (1914), the extensive large Schmidt telescope sky surveys starting with the National Geographic Society–Palomar Observatory Sky Atlas (1957), and smaller photographic atlases like the Falkauer Atlas (1964) and Atlas Stellarum (1970) by Hans Vehrenberg.

Modern hand-drawn or computer generated atlases include Norton's Star Atlas, which first appeared in 1910 and in its 20th edition in 2003 (6.5 mag, Ridpath), Atlas Coeli Skalnaté Pleso (7.75 mag, Bečvář, first published in Prague, 1948), the Smithsonian Astrophysical Observatory Star Atlas (9 mag, 1969; the first computer-plotted atlas), Sky Atlas 2000.0 (1st edition to 8.0 mag, Tirion, 1981), Uranometria 2000.0 (1st edition to ~9.5 mag, Tirion et al., 1987), Bright Star Atlas 2000.0 (6.5 mag, Tirion, 1990), Cambridge Star Atlas (6.5 mag, Tirion, 1991), and the Herald–Bobroff Astro Atlas (9 mag, Herald & Bobroff, 1994).

Hipparcos contributions The Millennium Star Atlas (Sinnott & Perryman, 1997) was constructed as a collaboration between the Hipparcos project, and a team at Sky & Telescope led by Roger Sinnott (Sinnott, 1997). It was the first sky atlas to include the Hipparcos and Tycho Catalogue data. It appeared as a stand-alone publication, and as three volumes of the 17-volume Hipparcos Catalogue (ESA, 1997). The atlas extended earlier undertakings in terms of completeness and uniformity to a magnitude limit of around 10–11 mag. The 1548 charts include one million stars from the Hipparcos and Tycho 1 Catalogues, three times as many as in any previous all-sky atlas; more than 8000 galaxies with their orientation; outlines of many bright and dark nebulae; the location of many open and globular clusters; and some 250 of the brightest quasars (Figure 2.21). Virtually all the non-stellar objects in the atlas are identified by type and designation. The chart scale is 100 arcsec mm^{-1}, matching that at the focus of an 8-inch f/10 Schmidt–Cassegrain. Star magnitudes are essentially Johnson V. Distance labels are given for stars within 200 light-years of the Sun. Proper motion arrows are given for stars with motions exceeding 0.2 arcsec yr^{-1}. Variable stars are indicated by amplitude and variability type. Many thousands of already known and newly-discovered double stars are depicted with tick marks indicating separation and position angle.

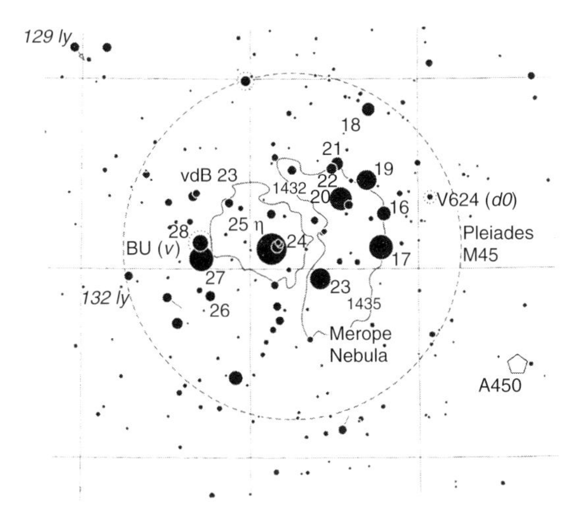

Figure 2.21 Example of part of the Millennium Star Atlas covering the Pleiades cluster. Three volumes, with 1548 charts in total, cover the entire sky. Circle sizes indicate magnitude; dashed circles indicate variable stars according to variability amplitude and type; 'dashes' indicate binary stars and their associated separation and position angle; arrows indicate high proper motion stars; distances are given for stars within 200 light-years. Other symbols indicate open and globular clusters, nebulae, galaxies, and quasars. From Sinnott & Perryman (1997).

Other major atlases since 1997 have also incorporated the Hipparcos and Tycho Catalogue data. These include Sky Atlas 2000.0 (2nd edition to 8.5 mag, Tirion & Sinnott, 1998), Cambridge Star Atlas (3rd edition to 6.5 mag, Tirion, 2001), Uranometria 2000.0 (2nd edition to 9.7 mag, Tirion, Rappaport & Remaklus, 2001), Bright Star Atlas 2000.0 (6.5 mag, Tirion & Skiff, 2001), and the Pocket Sky Atlas (7.6 mag, Sinnott, 2006). A still more ambitious project, SkyGX, comprising 3570 charts to about 12 mag, and based on the Hipparcos and Tycho 2 Catalogues via the ASCC 2.5, is currently under development (Christopher Watson, www). A full-sky visualisation of the Hipparcos and Tycho 2 Catalogues has been constructed by NASA's GSFC Scientific Visualisation Studio (Bridgman, 2007). Hipparcos and Tycho data were included in 'StarGazer', a star field visualisation system (Bogdanovski & Tsvetkov, 2006), and in 'Sky in Google Earth' in August 2007.

2.12.2 Handbooks and related compilations

The NASA SKY2000 Master Star Catalogue is a compilation of basic data for approximately 300 000 stars to a limiting magnitude of between $V = 9$ and $V = 10$. It has been used for many years for satellite acquisition and attitude determination on NASA spacecraft. Version 1 of SKY2000 was an almost complete rebuild of the old SKYMAP catalogue using the most recent data sources, while adding red magnitudes to better suit the CCD star trackers now common on NASA missions. Version 2 was significantly improved, principally by the incorporation of data from the Hipparcos and Tycho Catalogues, providing high-quality astrometric and photometric data for virtually all SKY2000 stars, and allowing considerable improvements to the data for double and variable stars.

Ochsenbein & Halbwachs (1999) updated their earlier machine-readable version of the brightest stars to include the Hipparcos data. The catalogue, of 1628 stars brighter than 5 mag (Ochsenbein & Halbwachs, 1987), was itself derived from the fourth edition of the Bright Star Catalogue (Hoffleit & Jaschek, 1982).

The fifth edition of the Monthly Sky Guide (Ridpath & Tirion, 1999) was updated to include the latest star data from Hipparcos: it provides a monthly guide to the constellations, star clusters, nebulae, galaxies, and meteor showers.

The Brightest Stars section of the Observer's Handbook, published annually by the Royal Astronomical Society of Canada, contains fundamental data for the 314 stars brighter than $V = 3.55$ mag. Magnitudes, colours, parallaxes, proper motions and radial velocities are taken from the literature, while the MK spectral types are confirmed or newly-determined. New CCD spectra are compared with existing data and types to ensure consistency with a set of primary MK standards. A few minor revisions are made. From 1998, Hipparcos data were used for the determination of absolute magnitudes and distances, resulting in several major and many minor revisions.

The Flamsteed Collection, a catalogue of the brightest stars based on John Flamsteed's original list and brought up-to-date by Hipparcos, was published by Dibon-Smith (1998).

The Astronomical Almanac is a joint publication of the US Naval Observatory (US) and Her Majesty's Nautical Almanac Office (UK). It is based to the greatest extent possible on IAU-endorsed and other internationally recognised standards. From year 2003 onwards (US Naval Observatory & Royal Greenwich Observatory, 2001), positions for bright stars and radial velocity standard stars use the Hipparcos Catalogue positions instead of FK5 positions, while positions of Solar System objects are based on the JPL DE405/LE405 ephemerides aligned with the ICRF. The various changes to the fundamental celestial reference system, changes to time scales, new models of precession and nutation, and a new method of computing Earth rotation culminated in several resolutions that were adopted by the International Astronomical Union in 1997 and 2000, and the 2006 edition of the Astronomical Almanac is the first to have the data consistent with these resolutions. Similar incorporation of the Hipparcos data into the Russian Astronomical Yearbooks is described by Glebova *et al.* (2006).

2.12.3 Satellite and telescope operations

The Hipparcos and Tycho Catalogues are now used routinely in most if not all relevant ground and space facilities. Some examples are included here where specific descriptions have been given in the literature.

The use of the Hipparcos and Tycho Catalogues in flight dynamics operations at the European Space Operations Centre, ESOC, including the improvements brought to the ISO mission operations, and the development of a generic star catalogue facility for future missions, is described by Batten *et al.* (1997).

The Tycho Catalogue was used as a primary part of the Chandra (formerly AXAF) Guide and Aspect Star Catalogue (AGASC, Green *et al.*, 1997) providing: accurate positions and magnitudes for target acquisition and guiding; the colours necessary for magnitude transformations to the Aspect Camera system; a high internal astrometric accuracy for half-arcsec *post facto* image reconstruction; and an astrometric system that ties the X-ray image astrometry into the ICRS. AGASC 1.5, incorporating Tycho 2, GSC–ACT, and 2MASS, is described by Schmidt & Green (2003). The Tycho 2 data substantially improved the photometric and astrometric measurements of stars as faint as $V = 12$, while the GSC–ACT merge decreased by about half the systematic astrometric errors down to the catalogue limit of about $V = 14.5$.

The Spitzer Space Telescope (formerly SIRTF, the Space Infrared Telescope Facility) has a visible light sensor at its focal plane to calibrate the alignment between the externally mounted star trackers and the telescope bore-sight; to establish the correspondence between the telescope coordinate system and the absolute ICRS reference frame; and to provide starting attitudes for high-accuracy absolute offset maneuvers. The Pointing Calibration and Reference Sensor (PCRS) functions as the primary absolute attitude reference for the telescope, and measures the position of Tycho Catalogue stars to an accuracy of 0.14 arcsec per axis (Mainzer *et al.*, 1998; van Bezooijen, 2003; van Bezooijen *et al.*, 2004).

Use of Hipparcos stars, in particular the ~32 000 distant K giants, as grid stars for SIM, is discussed by Frink *et al.* (1999, 2001).

Some 1000 stars from Hipparcos were used as the input catalogue for the GOMOS instrument on ESA's Earth-observation satellite Envisat (Ratier *et al.*, 1997). The medium-resolution spectrometer compares the 230–952 nm spectrum observed outside the atmosphere with the spectrum seen through the atmosphere when the star approaches the Earth's limb, to determine concentration profiles and trends in stratospheric O_3, NO_2 and NO_3. The catalogue was also used to reject double and multiple systems from the survey.

2.12.4 Education and outreach

The educational potential of the Hipparcos database, in schools and universities, was addressed by Percy (1997). He argued that science education in schools and universities is most effective when students use inquiry-based learning with real scientific data and activities. In this respect, the Hipparcos database also has significant educational potential. The astrometric data can be used for projects which illustrate important physical and astronomical concepts (as well as the role of random and systematic errors, and selection effects). The epoch photometric data can be used to develop and integrate a wide range of mathematical and science skills, at high school level and beyond. Many Hipparcos-based activities could be offered on-line, building on the success of the European Association for Astronomy Education's recent Astronomy On-Line project.

University level education in astrometry was addressed by van Altena & Stavinschi (2005) and by van Altena (2006). The former noted the role of Hipparcos in stimulating the field, but lamented that *'Astrometry is poised to enter an era of unparalleled growth and relevance due to the wealth of highly accurate data expected from the SIM and Gaia space missions. Innovative ground-based telescopes, such as the LSST, are planned which will provide less precise data, but for many more stars. The potential for studies of the structure, kinematics and dynamics of our Galaxy as well as for the physical nature of stars and the cosmological distance scale is without equal in the history of astronomy. It is therefore ironic that in two years not one course in astrometry will be taught in the US, leaving all astrometric education to Europe, China and Latin America. Who will ensure the astrometric quality control for the JWST, SIM, Gaia, LSST, to say nothing about the current large ground-based facilities, such as the VLT, Gemini, Keck, NOAO, Magellan, and LBT?'*

In terms of visualising the Hipparcos data in both the movement of stars as a function of time and their distances, Perryman used stereoscopic projection techniques using polarised light to illustrate the depth information, presenting this for the first time at the Royal Astronomical Society's George Darwin Lecture (Perryman, 1999), and subsequently as an Invited Discourse at the 2000 IAU General Assembly in Manchester (Perryman, 2002). Miyaji & Hayashi (2001) describe a 3d theatre for undergraduate astrophysics studies at Chiba University, Japan, using the Hipparcos 3d data to illustrate Galactic structure. A similar stereoscopic display tool for astronomy education and outreach,

COSMO.LAB, including a tour through the Hipparcos Catalogue data, is described by de Ruiter (2003).

Since 1998, the American Museum of Natural History and the Hayden Planetarium in New York have engaged in the 3d visualisation of the Universe. Describing itself as the most powerful virtual reality simulator in the world, it uses the Zeiss Star Projector in combination with the Digital Universe Atlas, just over 100 000 stars selected from the Hipparcos Catalogue and complemented by a further 49 stars from the Catalogue of Nearby Stars (Mendez *et al.*, 2004).

Yu & Jenkins (2004) describes Cosmic Atlas, a software program produced at the Denver Museum of Nature & Science to generate real-time digital content for the museum's Gates Planetarium, including a 3d model of the Galaxy populated with Hipparcos stars for the local Galactic neighbourhood.

Many other astronomy-related products have incorporated the Hipparcos mission data, for example, the RedShift desktop planetarium CD, which has included the Hipparcos mission data since RedShift Version 3.

References

Abad C, Vieira K, Zambrano A, 1998, Reduction of the Astrographic Catalogue, zone of San Fernando. *A&AS*, 132, 275–279 {**62**}

Abad C, 1998, Reduction and quality testing of the San Fernando Astrographic Catalogue plates. *Applied and Computational Harmonic Analysis*, 3, 146–149 {**62**}

Adelman-McCarthy JK, 2008, The sixth data release of the Sloan Digital Sky Survey. *ApJS*, 175, 297–313 {**74**}

Andrei AH, Penna JL, Benevides-Soares P, *et al.*, 1997, Astrolabe observations of the Sun on the Hipparcos reference frame. *ESA SP–402: Hipparcos, Venice '97*, 161–164 {**60**}

Arenou F, Turon C, 2008, Hipparcos, une troisième dimension pour La Carte du Ciel. *La Carte du Ciel. Histoire et actualité d'un projet scientifique international* (ed. Lamy J), Les Ulis, EDP Sciences, Paris, Observatoire de Paris {**61**}

Argyle RW, Einicke OH, Pilkington JDH, *et al.*, 1996, Comparison of the Carlsberg optical reference frame with the International Celestial Reference Frame. *A&A*, 312, 1031–1037 {**57**}

Arias EF, Cionco RG, Orellana RB, *et al.*, 2000, A comparison of the SAO–Hipparcos reference frames. *A&A*, 359, 1195–1200 {**69**}

Assafin M, Andrei AH, Martins RV, *et al.*, 2001, Investigation of USNO A2.0 Catalogue positions. *ApJ*, 552, 380–385 {**65**}

Babenko Y, Lazorenko PF, Karbovsky V, *et al.*, 2005, Kyiv meridian axial circle catalogue of stars in fields with extragalactic radio sources. *Kinematika i Fizika Nebesnykh Tel Supplement*, 5, 316–321 {**58**}

Bakos GA, Sahu KC, Németh P, 2002, Revised coordinates and proper motions of the stars in the Luyten Half-Second Catalogue. *ApJS*, 141, 187–193 {**77**}

Bastian U, Røser S, 1994, Bright Stars Supplement to PPM. *VizieR Online Data Catalogue*, 1206 {**68**}

Batten A, Marc X, McDonald A, *et al.*, 1997, The use of the Hipparcos and Tycho Catalogues in flight dynamics operations at ESOC. *ESA SP–402: Hipparcos, Venice '97*, 191–194 {**83**}

Benedict GF, McArthur BE, 2005a, High-precision stellar parallaxes from Hubble Space Telescope fine guidance sensors. *IAU Colloq. 196: Transits of Venus: New Views of the Solar System and Galaxy* (ed. Kurtz DW), 333–346 {**79**}

—, 2005b, Parallaxes with Hubble Space Telescope: how and why we do it. *ASP Conf. Ser. 338: Astrometry in the Age of the Next Generation of Large Telescopes* (eds. Seidelmann PK, Monet AKB), 58–58 {**79**}

Bergeron P, Leggett SK, Ruiz MT, 2001, Photometric and spectroscopic analysis of cool white dwarfs with trigonometric parallax measurements. *ApJS*, 133, 413–449 {**79**}

Bienaymé O, 1993, Field astrometry using orthogonal functions. *A&A*, 278, 301–106 {**54**}

Blaauw A, 1998, FK3 or GC: a difficult choice half a century ago. *Applied and Computational Harmonic Analysis*, 3, 114–122 {**56**}

Bobylev VV, Bronnikova NM, Shakht NA, 2004, Proper motions of 59 766 stars absolutized using galaxies in 149 sky fields. *Astronomy Letters*, 30, 469–479 {**72**}

Bogdanovski RG, Tsvetkov MK, 2006, StarGazer: towards the WFPDB integration. *Virtual Observatory: Plate Content Digitization, Archive Mining and Image Sequence Processing* (eds. Tsvetkov M, Golev V, Murtagh F, *et al.*), 404–408 {**82**}

Boss B, 1937, *General Catalogue of 33342 stars for the epoch 1950*. Carnegie Institution, Washington {**56, 68**}

Bridgman T, 2007, *http://svs.gsfc.nasa.gov/vis*. NASA GSFC Science Visualisation Studio {**82**}

Brosche P, Wildermann E, Geffert M, 1989, Astrometric plate reductions with orthogonal functions. *A&A*, 211, 239–244 {**54**}

Bucciarelli B, Morrison JE, McLean BJ, *et al.*, 1997, Astrometry of POSS-II plates using Tycho. *The First Results of Hipparcos and Tycho, IAU Joint Discussion 14*, 25 {**63, 65**}

—, 1998, Astrometry of POSS-II plates using Tycho. *Highlights in Astronomy*, 11, 551 {**63**}

Bustos Fierro IH, Calderón JH, 2002, Small-field CCD astrometry with a long focus reflector telescope. *Revista Mexicana de Astronomia y Astrofisica*, 38, 215–224 {**75**}

—, 2003, Measurements of Carte du Ciel plates with CCD. *Astronomy in Latin America* (eds. Teixeira R, Leister NV, Martin VAF, *et al.*), 15–22 {**62**}

Bustos Fierro IH, Melia RR, Willemoes C, *et al.*, 2003, Digitisation of Córdoba zone of Carte du Ciel plates: a complete first survey. *Astronomy in Latin America* (eds. Teixeira R, Leister NV, Martin VAF, *et al.*), 27–30 {**62**}

Chanamé J, Gould A, 2004, Disk and halo wide binaries from the revised Luyten Catalogue: probes of star formation and MACHO dark matter. *ApJ*, 601, 289–310 {**80**}

Chapman A, 1990, *Dividing the Circle: the Development of Critical Angular Measurement in Astronomy 1500–1850*. Ellis Horwood, London {**55**}

Chapman J, Cannon R, Harrison S, *et al.* (eds.), 1995, *The Future Utilisation of Schmidt Telescopes*, ASP Conf. Ser. 84 {**64**}

Chunlin L, Jiayan X, Dongming L, *et al.*, 1999, Analysis of systematic errors of the photoelectric astrolabe catalogue SIPA1. *A&AS*, 139, 43–45 {**59**}

Cionco RG, Orellana RB, Arias EF, *et al.*, 1997, Determinación preliminar de la orientación relativa SAO–Hipparcos. *Boletin de la Asociacion Argentina de Astronomia La Plata Argentina*, 41, 24–25 {**69**}

Conrady AE, 1919, Lens-systems, decentred. *MNRAS*, 79, 384–390 {**71**}
Copenhagen University Observatory, Institute of Astronomy, Cambridge, Real Instituto y Observatorio de La Armada, 2006, Carlsberg Meridian Catalogue 14 (CMC14). *VizieR Online Data Catalogue*, 1304 {**59**}
Copenhagen University Observatory, Royal Greenwich Observatory, 2001, Carlsberg Meridian Cataloguess. *VizieR Online Data Catalog*, 1256 {**59, 75**}
Corbin TE, Warren WH, 1995, International Reference Stars (IRS). *VizieR Online Data Catalogue*, 1172 {**68**}
Corbin TE, 1978, The proper motions of the AGK3R and SRS stars. *IAU Colloq. 48: Modern Astrometry* (eds. Prochazka FV, Tucker RH), 505–514 {**68**}
Cuillandre JC, Magnier EA, Isani S, *et al.*, 2004, CFHT's SkyProbe: true atmospheric attenuation measurement in the telescope field. *Scientific Detectors for Astronomy, The Beginning of a New Era* (eds. Amico P, Beletic JW, Belectic JE), 287–298 {**75**}
Cutri RM, Skrutskie MF, van Dyk S, *et al.*, 2003, *2MASS All-Sky Catalogue of Point Sources.* NASA/IPAC Infrared Science Archive {**75**}
Dahn CC, Harris HC, Vrba FJ, *et al.*, 2002, Astrometry and photometry for cool dwarfs and brown dwarfs. *AJ*, 124, 1170–1189 {**79**}
Dahn CC, 1998, Review of CCD parallax measurements. *IAU Symp. 189: Fundamental Stellar Properties*, 19–24 {**79**}
Damljanović G, Souchay J, 2003a, Cross identification of Hipparcos–2MASS second incremental data release. *Astrometry from Ground and from Space* (eds. Capitaine N, Stavinschi M), 15–20 {**75**}
—, 2003b, ICRF densification via Hipparcos-2MASS cross-identification. *The International Celestial Reference System: Maintenance and Future Realisation, IAU Joint Discussion 16*, 18 {**75**}
Débarbat S (ed.), 1988, *IAU Symp. 133: Mapping the Sky: Past Heritage and Future Directions* {**61**}
de Ruiter HR, 2003, COSMO.LAB: stereographic viewing of astronomical data. *Effective Teaching and Learning of Astronomy, 25th meeting of the IAU, Special Session 4*, Sydney, Australia, 4 {**84**}
de Vegt C, 1968, Report on overlap methods in photographic astrometry. *Highlights of Astronomy*, 1, 343–347 {**72**}
de Vegt C, Ebner H, 1972, Block adjustment methods in photographic astrometry. *A&A*, 17, 276–285 {**72**}
—, 1974, A general computer program for the application of the rigorous block adjustment solution in photographic astrometry. *MNRAS*, 167, 169–182 {**72**}
Dibon-Smith R, 1998, *The Flamsteed Collection: a catalogue of the brightest stars, based on John Flamsteed's original list and brought up to date by the Hipparcos satellite.* Clear Skies, Toronto {**82**}
Dick WR, Tucholke HJ, Brosche P, *et al.*, 1993, Hipparcos link with Carte du Ciel triple images. *A&A*, 279, 267–272 {**62**}
Donner A, Furuhjelm R, 1929, Catalogue photographique du ciel. Zone de Helsingfors entre $+39^\circ$ et $+49^\circ$. Premiere serie: Coordonnees rectilignes et equitoriales. *Univ. Helsingfors Obs.*, I/1, 52, via Yu 2004, AJ, 128, 911 {**72**}
Ducourant C, Le Campion JF, Rapaport M, *et al.*, 2006, The PM2000 Bordeaux proper motion catalogue for $11^\circ < \delta < 18^\circ$. *A&A*, 448, 1235–1245 {**62**}
Ducourant C, Rapaport M, 2003, *Astrometric projects at the Bordeaux CCD meridian circle*, 63–66. The Future of Small Telescopes In The New Millennium. Volume II: The Telescopes We Use {**58**}
Ducourant C, Teixeira R, Périé JP, *et al.*, 2005, Pre-main sequence star proper motion catalogue. *A&A*, 438, 769–778 {**79**}
Eichhorn H, 1960, Über die Reduktion von photographischen Sternpositionen und Eigenbewegungen. *Astronomische Nachrichten*, 285, 233–238 {**72**}
—, 1988, The development of the overlapping-plate method. *IAU Symp. 133: Mapping the Sky: Past Heritage and Future Directions* (ed. Débarbat S), 177–200 {**72**}
Eichhorn H, Williams CA, 1963, On the systematic accuracy of photographic astrographic data. *AJ*, 68, 221–231 {**54**}
ESA, 1997, *The Hipparcos and Tycho Catalogues. Astrometric and Photometric Star Catalogues derived from the ESA Hipparcos Space Astrometry Mission, ESA SP–1200 (17 volumes including 6 CDs).* European Space Agency, Noordwijk, also: VizieR Online Data Catalogue {**68, 81**}
Evans DW, 2001, The Carlsberg Meridian Telescope: an astrometric robotic telescope. *Astronomische Nachrichten*, 322, 347–351 {**58**}
Evans DW, Irwin MJ, Helmer L, 2002, The Carlsberg Meridian Telescope CCD drift scan survey. *A&A*, 395, 347–356 {**58**}
Fabricius C, 1997, A photometric calibration of the Guide Star Catalogue. *ESA SP–402: Hipparcos, Venice '97*, 131–134 {**64**}
Fedorov PN, Myznikov AA, 2005, The X1 catalogue of positions and proper motions of faint stars around the ICRF sources. *Kinematika i Fizika Nebesnykh Tel Supplement*, 5, 322–327 {**78**}
Finch CT, Henry TJ, Subasavage JP, *et al.*, 2007, The solar neighbourhood. XVIII. Discovery of new proper motion stars between 0.18–0.40 arcsec yr^{-1} between declinations -90° and -47°. *AJ*, 133, 2898–2907 {**77**}
Fresneau A, Vaughan AE, Argyle RW, 2007, Sydney observatory Galactic survey. Potential for detection of extreme disk-crossing stars. *A&A*, 469, 1221–1229 {**62**}
Fricke W, Kopff A, Gliese W, *et al.*, 1963, Fourth Fundamental Catalogue (FK4). *Veroeffentlichungen des Astronomischen Rechen-Instituts Heidelberg*, 10, 1–144 {**68**}
Fricke W, Kopff A, 1963, Preliminary Supplement to the Fourth Fundamental Catalogue (FK4 Sup). *Veroeffentlichungen des Astronomischen Rechen-Instituts Heidelberg*, 11, 1–47 {**68**}
Fricke W, Schwan H, Corbin TE, *et al.*, 1991, Fifth fundamental catalogue. II. The FK5 extension: new fundamental stars. *Veroeffentlichungen des Astronomischen Rechen-Instituts Heidelberg*, 33, 1–143 {**68**}
Fricke W, Schwan H, Lederle T, *et al.*, 1988, Fifth fundamental catalogue (FK5). I. The basic fundamental stars. *Veroeffentlichungen des Astronomischen Rechen-Instituts Heidelberg*, 32, 1–106 {**56, 68**}
Frink S, Quirrenbach A, Fischer DA, *et al.*, 2001, A strategy for identifying the grid stars for the Space Interferometry Mission. *PASP*, 113, 173–187 {**83**}
Frink S, Quirrenbach A, Röser S, *et al.*, 1999, Testing Hipparcos K giants as grid stars for SIM. *ASP Conf. Ser. 194: Working on the Fringe: Optical and Infrared Interferometry from Ground and Space*, 128 {**83**}
Führmann U, 1979, *Neureduction des AGK2-Sternkataloges und Untersuchung der systematischen Fehler mit Hilfe von Blockausgleichungsverfahren.* Ph.D. thesis, Hamburg {**72**}

Gatewood G, Kiewiet de Jonge J, Persinger T, 1998, Correlation of the Hipparcos and Allegheny Observatory parallax catalogues. *AJ*, 116, 1501–1503 {**79**}

Gatewood G, 1987, The multichannel astrometric photometer and atmospheric limitations in the measurement of relative positions. *AJ*, 94, 213–224 {**79**}

Geffert M, Bonnefond P, Maintz G, *et al.*, 1996, The astrometric accuracy of Carte du Ciel plates and proper motions in the field of the open cluster NGC 1647. *A&AS*, 118, 277–282 {**62**}

Geffert M, Klemola AR, Hiesgen M, *et al.*, 1997, Absolute proper motions for the calibration of the Hipparcos proper motion system. *A&AS*, 124, 157–161 {**72**}

Germain ME, Zacharias N, Urban SE, *et al.*, 2000, The first US Naval Observatory CCD Astrograph Catalogue (UCAC 1). *American Astronomical Society Meeting*, 196 {**68**}

Girard TM, Dinescu DI, van Altena WF, *et al.*, 2004, The Southern Proper Motion Programme. III. A near-complete catalogue to $V = 17.5$. *AJ*, 127, 3060–3071 {**68, 71**}

Girard TM, Platais I, Kozhurina-Platais V, *et al.*, 1998, The Southern Proper Motion Programme. I. Magnitude equation correction. *AJ*, 115, 855–867 {**71**}

Gizis JE, 1997, M-Subdwarfs: spectroscopic classification and the metallicity scale. *AJ*, 113, 806–822 {**79**}

Glebova NI, Lukashova MV, Sveshnikov ML, 2006, The Russian Astronomical Yearbooks and IAU 2000 Resolutions. *Nomenclature, Precession and New Models in Fundamental Astronomy, IAU 26, JD16*, 16 {**83**}

Gould A, 2003, Completeness of USNO B for high proper motion stars. *AJ*, 126, 472–483 {**65**}

Gould A, Chanamé J, 2004, New Hipparcos-based parallaxes for 424 faint stars. *ApJS*, 150, 455–464 {**80, 81**}

Gould A, Kollmeier JA, 2004, Proper motion catalogue from SDSS and USNO B. *ApJS*, 152, 103–111 {**78**}

Gould A, Salim S, 2003, Improved astrometry and photometry for the Luyten Catalogue. I. Bright stars. *ApJ*, 582, 1001–1010 {**76, 77, 80**}

Green PJ, Aldcroft TA, García MR, *et al.*, 1997, Using the Tycho Catalogue for AXAF. *ESA SP–402: Hipparcos, Venice '97*, 187–190 {**83**}

Hambly NC, Davenhall AC, Irwin MJ, *et al.*, 2001a, The SuperCOSMOS Sky Survey. III. Astrometry. *MNRAS*, 326, 1315–1327 {**67**}

Hambly NC, Digby AP, Oppenheimer BR, 2005, Cool white dwarfs from the SuperCOSMOS and Sloan Digital Sky Surveys. *ASP Conf. Ser. 334: 14th European Workshop on White Dwarfs* (eds. Koester D, Moehler S), 113–118 {**67**}

Hambly NC, Henry TJ, Subasavage JP, *et al.*, 2004, The solar neighbourhood. VIII. Discovery of new high proper motion nearby stars using the SuperCOSMOS Sky Survey. *AJ*, 128, 437–447 {**67, 77**}

Hambly NC, Irwin MJ, MacGillivray HT, 2001b, The SuperCOSMOS Sky Survey. II. Image detection, parametrization, classification and photometry. *MNRAS*, 326, 1295–1314 {**65**}

Hambly NC, MacGillivray HT, Read MA, *et al.*, 2001c, The SuperCOSMOS Sky Survey. I. Introduction and description. *MNRAS*, 326, 1279–1294 {**63, 65**}

Hanson RB, 1987, Lick northern proper motion programme. II. Solar motion and Galactic rotation. *AJ*, 94, 409–415 {**70**}

Hanson RB, Klemola AR, Jones BF, *et al.*, 2004, Lick Northern Proper Motion Programme. III. Lick NPM2 Catalogue. *AJ*, 128, 1430–1445 {**68, 70, 71**}

Hindsley RB, Harrington RS, 1994, The US Naval Observatory catalogue of positions of infrared stellar sources. *AJ*, 107, 280–286 {**75**}

Hoffleit D, Jaschek C, 1982, *The Bright Star Catalogue*. Yale University Observatory, 4th edition {**68**}

—, 1991, *The Bright Star Catalogue*. Yale University Observatory, 5th edition {**68, 82**}

Høg E, Bässgen G, Bastian U, *et al.*, 1997, The Tycho Catalogue. *A&A*, 323, L57–L60 {**68**}

Høg E, Fabricius C, Makarov VV, *et al.*, 2000, The Tycho 2 Catalogue of the 2.5 million brightest stars. *A&A*, 355, L27–L30 {**68**}

Høg E, Fabricius C, 1988, Atmospheric and internal refraction in meridian observations. *A&A*, 196, 301–312 {**56**}

Høg E, Miller RJ, 1986, Internal refraction in a USNO meridian circle. *AJ*, 92, 495–502 {**56**}

Høg E, von der Heide J, 1976, Perth 70. *Astronomische Abhandlungen der Hamburger Sternwarte*, 9, 1–34 {**68**}

Irwin MJ, Morrison LV, Argyle RW, 1997, Transferring Hipparcos frame to Schmidt surveys. *The New International Celestial Reference Frame, IAU Joint Discussion 7*, 12 {**63**}

Ivanov GA, 2005, Catalogue of stars with high proper motions. *Kinematika i Fizika Nebesnykh Tel*, 21, 156–158 {**78**}

Jao WC, Henry TJ, Subasavage JP, *et al.*, 2003, The solar neighbourhood. VII. Discovery and characterization of nearby multiples in the CTIO parallax investigation. *AJ*, 125, 332–342 {**79**}

Jenkins LF, 1963, *General Catalogue of Trigonometric Stellar Parallaxes*. Yale University Observatory {**79**}

Jin W, Tang Z, Li J, *et al.*, 1997, A preliminary study on the improvement of proper motion for Hipparcos stars by using photographic plates at Shanghai Observatory. *The First Results of Hipparcos and Tycho, IAU Joint Discussion 14*, 28 {**72**}

Jones D, 2000, The scientific value of the Carte du Ciel. *Astronomy and Geophysics*, 41, 16–20 {**61**}

Kapteyn JC, 1922, On the proper motions of the faint stars and the systematic errors of the Boss fundamental system. *Bull. Astron. Inst. Netherlands*, 1(8), 69–78 {**56**}

Kharchenko NV, 1987, On the general catalogue of stellar proper motions with respect to galaxies in the areas of the Galaxy main meridional section. *Kinematika i Fizika Nebesnykh Tel*, 3, 7–11 {**78**}

—, 2001, All-sky compiled catalogue of 2.5 million stars. *Kinematika i Fizika Nebesnykh Tel*, 17, 409–423 {**78**}

—, 2004, Compiled catalogue of stellar proper motions in the Hipparcos system in MEGA programme areas. *Kinematika i Fizika Nebesnykh Tel*, 20, 366–371 {**78**}

Kharchenko NV, Piskunov AE, Röser S, *et al.*, 2004a, Astrophysical supplements to the ASCC 2.5. II. Membership probabilities in 520 Galactic open cluster sky areas. *Astronomische Nachrichten*, 325, 740–748 {**78**}

Kharchenko NV, Piskunov AE, Scholz RD, 2004b, Astrophysical supplements to the ASCC 2.5. I. Radial velocity data. *Astronomische Nachrichten*, 325, 439–444 {**78**}

Kharin AS, 1997, First step of an infrared reference catalogue based on the FK5 and the creation of an infrared reference system after Hipparcos. *The New International Celestial Reference Frame, IAU Joint Discussion 7*, 14 {**75**}

—, 2000, Photometric and spectroscopic identification problems of the nearest and farthest Hipparcos infrared stars. *Hipparcos and the Luminosity Calibration of the Nearer Stars, IAU Joint Discussion 13*, 36 {**75**}

Khrutskaya EV, Khovritchev MY, Bronnikova NM, 2004, The Pul-3 catalogue of 58 483 stars in the Tycho 2 system. *A&A*, 418, 357–362 {**72**}

King DS, Lomb NR, 1983, Sydney Southern Star Catalogue. *Journal and Proceedings of the Royal Society of New South Wales*, 116, 53–70 {**68**}

Kiselev AA, 1989, Theoretical Fundamentals of Photographic Astrometry. *Moscow Izdatel Nauka* {**54**}

Kislyuk VS, Yatsenko AI, 2005, The FONAC catalogue as a result of the FON project. *Kinematika i Fizika Nebesnykh Tel Supplement*, 5, 33–39 {**62**}

Klemola AR, Jones BF, Hanson RB, 1987, Lick Northern Proper Motion programme. I. Goals, organisation, and methods. *AJ*, 94, 501–515 {**65, 70**}

Knauer TG, Ivezić Ž, Knapp GR, 2001, Analysis of stars common to the IRAS and Hipparcos surveys. *ApJ*, 552, 787–792 {**75**}

Kovalchuk A, Pinigin G, Protsyuk Y, *et al.*, 1997, First steps to re-observation of the Hipparcos/Tycho stars by ground-based automatic meridian circles. *ESA SP–402: Hipparcos, Venice '97*, 139–142 {**58**}

Kovalevsky J, 2002, *Modern Astrometry*, Second Edition. Springer–Verlag, Berlin {**54, 55**}

Kozhurina-Platais V, Platais I, Girard TM, *et al.*, 2000, WOCS: Geometry of the Mayall KPNO 4m telescope CCD mosaic. *American Astronomical Society Meeting*, 196 {**75, 76**}

Kumkova II, 2000, Densification of ICRS in the optical using old observation sets. *Models and Constants for Submicroarcsecond Astrometry*, IAU Joint Discussion 2, 11 {**72**}

Kuzmin A, Høg E, Bastian U, *et al.*, 1999, Construction of the Tycho Reference Catalogue. *A&AS*, 136, 491–508 {**60, 62, 63**}

Langhans R, Malyuto V, Potthoff H, 2003, Calculated atmospheric colour refraction and observed stellar positions. *Astronomische Nachrichten*, 324, 454–459 {**76**}

Lasker BM, 1995, Digitisation and distribution of the large photographic surveys. *PASP*, 107, 763–765 {**63**}

—, 1997, Referring Schmidt surveys to the ICRF. *The New International Celestial Reference Frame, IAU Joint Discussion 7*, 19 {**63**}

Lasker BM, Russel JN, Jenkner H, 1999, The HST Guide Star Catalogue, Version GSC-ACT. *VizieR Online Data Catalogue*, 1255 {**64**}

Lasker BM, Sturch CR, McLean BJ, *et al.*, 1990, The Guide Star Catalogue. I. Astronomical foundations and image processing. *AJ*, 99, 2019–2058 {**64, 68**}

Lazorenko PF, Babenko Y, Karbovsky V, *et al.*, 2005, The Kyiv Meridian Axial Circle catalogue of stars in fields with extragalactic radio sources. *A&A*, 438, 377–389 {**58**}

Lépine S, 2005, Nearby Stars from the LSPM-North Proper Motion Catalogue. I. Main sequence dwarfs and giants within 33 pc of the Sun. *AJ*, 130, 1680–1692 {**77**}

Lépine S, Shara MM, Rich RM, 2002, New high proper motion stars from the Digitised Sky Survey. I. Northern stars with $0.5\,\mathrm{arcsec\,yr^{-1}} < \mu < 2.0\,\mathrm{arcsec\,yr^{-1}}$ at low Galactic latitudes. *AJ*, 124, 1190–1212 {**77**}

Lépine S, Shara MM, 2005a, A catalogue of northern stars with annual proper motions larger than 0.15 arcsec (LSPM-North Catalogue). *AJ*, 129, 1483–1522 {**68, 76–78**}

—, 2005b, The LSPM-south: new catalogue of southern stars with proper motions $\mu > 0.15\,\mathrm{arcsec\,yr^{-1}}$. *American Astronomical Society Meeting Abstracts*, 207 {**68, 77**}

Lopes PAA, Andrei AH, Puliaev SP, *et al.*, 1999, Observations of radio stars at the Valinhos CCD meridian circle. *A&AS*, 136, 531–537 {**57**}

Luyten WJ, 1942, New stars with proper motion exceeding 0.5 arcsec annually. *Publications of the Astronomical Observatory, University of Minnesota*, 2, 242 {**68**}

—, 1979, *The LHS Catalogue. A Catalogue of Stars with Proper Motions Exceeding 0.5 arcsec Annually*. University of Minnesota, Minneapolis {**68, 76**}

—, 1980a, *NLTT Catalogue*. University of Minnesota, Minneapolis {**68, 76**}

—, 1980b, Proper motion survey with the forty-eight inch Schmidt telescope. LV. First supplement to the NLTT Catalogue. *Publications of the Astronomical Observatory, University of Minnesota*, 55, 1–57

Lu L, Manrique WT, Perdomo R, *et al.*, 1996, First catalogue of stars with the photoelectric astrolabe in San Juan. *A&AS*, 118, 1–5 {**59**}

—, 2005, Third San Juan photoelectric astrolabe catalogue (CPASJ3). *A&A*, 430, 327–330 {**59**} {**68, 76, 80**}

MacGillivray HT, Thomson EB (eds.), 1992, *Digitised Optical Sky Surveys*, ASSL Vol. 174 {**63**}

MacGillivray HT (ed.), 1994, *Astronomy from Wide-Field Imaging*, IAU Symp. 161 {**63**}

Mainzer AK, Young ET, Greene TP, *et al.*, 1998, Pointing calibration and reference sensor for the Space Infrared Telescope Facility. *Proc. SPIE Vol. 3356, Space Telescopes and Instruments V* (eds. Bely PY, Breckinridge JB), 1095–1101 {**83**}

Makarov VV, Zacharias N, Hennessy GS, *et al.*, 2007, The nearby young visual binary HIP 115147 and its common proper motion companion LSPM J2322+7847. *ApJ*, 668, L155–L158 {**77**}

Malkov O, Smirnov OM, 1997, Investigation of the Guide Star Catalogue. *Baltic Astronomy*, 6, 313–315 {**64**}

Manrique WT, Lizhi L, Perdomo R, *et al.*, 1999, Second San Juan photoelectric astrolabe catalogue. *A&AS*, 136, 1–5 {**59**}

Martin VAF, Leister NV, Vigouroux G, *et al.*, 1999, Absolute declinations with the photoelectric astrolabe at Calern Observatory. *A&AS*, 137, 269–272 {**59**}

McCook GP, Sion EM, 1999, A catalogue of spectroscopically identified white dwarfs. *ApJS*, 121, 1–130 {**80**}

McLean BJ, Golombek DA, Hayes JJE, *et al.* (eds.), 1998a, *New Horizons from Multi-Wavelength Sky Surveys*, IAU Symp. 179 {**64**}

McLean BJ, Hawkins C, Spagna A, *et al.*, 1998b, The Second Guide Star Catalogue. *IAU Symp. 179: New Horizons from Multi-Wavelength Sky Surveys* (eds. McLean BJ, Golombek DA, Hayes JJE, *et al.*), 431–432 {**64, 65**}

Mendez B, Craig N, Haisch BM, *et al.*, 2004, The digital universe coalition: building a prototype NVO E/PO Portal. *American Astronomical Society Meeting Abstracts*, 204 {**84**}

Miyaji S, Hayashi M, 2001, Construction of 3d theatre for undergraduate study of astronomy. *American Astronomical Society Meeting*, 199 {**83**}

Monet DG, 1998, The 526 280 881 objects in the USNO A2.0 Catalogue. *Bulletin of the American Astronomical Society*, 30, 1427 {**68**}

Monet DG, Corbin TE, 1997, USNO A and the ICRF. *The New International Celestial Reference Frame, IAU Joint Discussion 7*, 26 {**65**}

Monet DG, Dahn CC, Vrba FJ, *et al.*, 1992, US Naval Observatory CCD parallaxes of faint stars. I. Programme description and first results. *AJ*, 103, 638–665 {**79**}

Monet DG, Levine SE, Canzian B, *et al.*, 2003, The USNO B Catalogue. *AJ*, 125, 984–993 {**64–66, 68**}

Morrison JE, Lasker BM, McLean BJ, *et al.*, 1997, Some first experiments on the photometric calibration of Schmidt plates against Tycho. *ESA SP–402: Hipparcos, Venice '97*, 129–130 {**63, 64**}

Muiños JL, Bellizón F, Vallejo M, *et al.*, 2002, Observaciones meridianas con cámara CCD en el Círculo Meridiano Automático de San Fernando. *Revista Mexicana de Astronomia y Astrofisica Conf. Ser.*, volume 14, 101–101 {**57**}

Munn JA, Monet DG, Levine SE, *et al.*, 2004, An improved proper motion catalogue combining USNO B and the Sloan Digital Sky Survey. *AJ*, 127, 3034–3042 {**78**}

Newman PR, Clowes RG, Campusano LE, *et al.*, 1998, The Chile-UK quasar survey: large quasar groups. *Wide Field Surveys in Cosmology*, 408 {**72**}

Ochsenbein F, Halbwachs JL, 1987, Le Catalogue des Etoiles les Plus Brillantes. *Bulletin d'Information du Centre de Données Stellaires*, 32, 83–84 {**82**}

—, 1999, Catalogue of the brightest stars. *VizieR Online Data Catalogue*, 5053 {**82**}

Odenkirchen M, Brosche P, Borngen F, *et al.*, 1997, Absolute stellar proper motions with reference to galaxies of the M81 group. *A&AS*, 124, 189–196 {**72**}

Odenkirchen M, Soubiran C, Le Campion JF, 1998, Early epoch stellar positions from the Bordeaux Carte du Ciel. *Applied and Computational Harmonic Analysis*, 3, 145–146 {**62**}

Oppenheimer BR, Golimowski DA, Kulkarni SR, *et al.*, 2001, A coronagraphic survey for companions of stars within 8 pc. *AJ*, 121, 2189–2211 {**80**}

Ortiz Gil A, Hiesgen M, Brosche P, 1998, A new approach to the reduction of Carte du Ciel plates. *A&AS*, 128, 621–630 {**62**}

Parker QA, Phillipps S, Pierce MJ, *et al.*, 2005, The AAO/UKST SuperCOSMOS Hα survey. *MNRAS*, 362, 689–710 {**67**}

Percy JR, 1997, The educational potential of the Hipparcos database. *ESA SP–402: Hipparcos, Venice '97*, 739–742 {**83**}

Perryman MAC, 1999, A stereoscopic view of our Galaxy. *Astronomy and Geophysics*, 40, 23–29 {**83**}

—, 2002, The three-dimensional structure of our Galaxy. *Highlights of Astronomy*, 12, 3–17 {**83**}

Pier JR, Munn JA, Hindsley RB, *et al.*, 2003, Astrometric calibration of the Sloan Digital Sky Survey. *AJ*, 125, 1559–1579 {**74, 76**}

Pinigin G, Zhigang L, Zhu Z, 1997, A new role of CCD meridian circles in modern astrometry. *Astronomical and Astrophysical Transactions*, 13, 83–86 {**57**}

Platais I, Girard TM, Kozhurina-Platais V, *et al.*, 1998, The Southern Proper Motion programme. II. A catalogue at the South Galactic Pole. *AJ*, 116, 2556–2564 {**65, 71**}

Platais I, Girard TM, van Altena WF, *et al.*, 1995, A study of systematic positional errors in the SPM plates. *A&A*, 304, 141–149 {**71**}

Pokorny RS, Jones HRA, Hambly NC, 2003, The Liverpool-Edinburgh high proper motion survey. *A&A*, 397, 575–584 {**77**}

Rafferty TJ, Holdenried ER, 2000, The US Naval Observatory pole-to-pole catalogue: W2J000. *A&AS*, 141, 423–431 {**58**}

Rapaport M, Ducourant C, Le Campion JF, *et al.*, 2006, The CdC 2000 Bordeaux Carte du Ciel catalogue $+11^\circ \leq \delta \leq +18^\circ$. *A&A*, 449, 435–442 {**62**}

Rapaport M, Le Campion JF, Soubiran C, *et al.*, 2001, M2000: an astrometric catalogue in the Bordeaux Carte du Ciel zone $11^\circ < \delta < 18^\circ$. *A&A*, 376, 325–332 {**62**}

Ratier G, Bertaux JL, Langen J, *et al.*, 1997, Hipparcos as Input Catalogue for the GOMOS Envisat instrument. *ESA SP–402: Hipparcos, Venice '97*, 195–198 {**83**}

Réquième Y, Le Campion JF, Montignac G, *et al.*, 1997, CCD meridian circle reductions using Tycho positions as Reference. *ESA SP–402: Hipparcos, Venice '97*, 135–138 {**58**}

Réquième Y, Morrison LV, Helmer L, *et al.*, 1995, Meridian circle reductions using preliminary Hipparcos positions. *A&A*, 304, 121–126 {**56–58**}

Réquième Y, Morrison LV, Lindegren L, *et al.*, 1993, The impact of Hipparcos on meridian astrometry. *Positional Astronomy and Celestial Mechanics* (eds. López García A, Ortiz Gil A, Chernetenko J, *et al.*), 20–30 {**56**}

Ridpath I, Tirion W, 1999, *The Monthly Sky Guide*. Cambridge University Press {**82**}

Robichon N, Turon C, Makarov VV, *et al.*, 1995, Schmidt plate astrometric reductions using preliminary Hipparcos and Tycho data. *A&A*, 304, 132–140 {**63**}

Röser S, Bastian U, 1991, *PPM Star Catalogue North. Positions and proper motions of 181 731 stars north of $-2\overset{\circ}{.}5$ declination for equinox and epoch J2000.0. Vol 1: Zones $+80^\circ$ to $+30^\circ$; Vol 2: Zones $+20^\circ$ to -0°*. Akademischer Verlag, Heidelberg {**68**}

—, 1993, *PPM Star Catalogue South. Positions and proper motions of 197 179 stars south of $-2\overset{\circ}{.}5$ declination for equinox and epoch J2000.0. Vol 3: Zones -0° to -20°; Vol 4: Zones -30° to -80°*. Akademischer Verlag, Heidelberg {**68**}

Rousseau JM, Perie JP, 1997, Astrometric positions of stars with high proper motions in the southern hemisphere. *A&AS*, 124, 437–439 {**80**}

Ruiz MT, Leggett SK, Allard F, 1997, Kelu–1: a free-floating brown dwarf in the solar neighbourhood. *ApJ*, 491, L107–110 {**78**}

Ruiz MT, Wischnjewsky M, Rojo PM, *et al.*, 2001, Calán–ESO Proper Motion Catalogue. *ApJS*, 133, 119–160 {**78**}

Rybka SP, Yatsenko AI, 1997, GPM1: a catalogue of absolute proper motions of stars with respect to galaxies. *A&AS*, 121, 243–246 {**72**}

Salim S, Gould A, 2003, Improved astrometry and photometry for the Luyten Catalogue. II. Faint stars and the revised catalogue. *ApJ*, 582, 1011–1031 {**77, 80**}

Schmidt D, Green PJ, 2003, The AXAF (Chandra) Guide and Acquisition Star Catalogue V1.5 (AGASC 1.5). *ASP Conf. Ser. 295: Astronomical Data Analysis Software and Systems XII* (eds. Payne HE, Jedrzejewski RI, Hook RN), 81–84 {**83**}

Scholz RD, Meusinger H, 2002, SSSPM J0829–1309: a new nearby L dwarf detected in SuperCOSMOS Sky Surveys. *MNRAS*, 336, L49–L52 {**67**}

Schwan H, 1993, Completion of the FK5 Extension and a preliminary investigation of the FK5 System. *IAU Symp. 156: Developments in Astrometry and their Impact on Astrophysics and Geodynamics* (eds. Mueller II, Kolaczek B), 339–350 {**56**}

Shatsky NI, Tokovinin AA, 1998, Hipparcos versus dynamical parallaxes. *Applied and Computational Harmonic Analysis*, 3, 193–194 {**79**}

Sinnott RW, Perryman MAC, 1997, *Millennium Star Atlas (3 volumes)*. Sky Publishing Corporation & European Space Agency {**81, 82**}

Sinnott RW, 1997, The Millennium Star Atlas. *ESA SP–402: Hipparcos, Venice '97*, 79–80 {**81**}

Skrutskie MF, Cutri RM, Stiening R, *et al.*, 2006, The Two Micron All Sky Survey (2MASS). *AJ*, 131, 1163–1183 {**75**}

Smart RL, Bucciarelli B, Casalegno R, *et al.*, 1997, Linking quasars to the Carlsberg Automatic Meridan Circle frame. *The New International Celestial Reference Frame, IAU Joint Discussion 7*, 36 {**57**}

Smart RL, Bucciarelli B, Lattanzi MG, *et al.*, 1999, A direct link of the CAMC catalogue to the extragalactic frame. *A&A*, 348, 653–658 {**57**}

Smart RL, Lattanzi MG, Bucciarelli B, *et al.*, 2003, The Torino Observatory parallax programme: white dwarf candidates. *A&A*, 404, 317–323 {**79**}

Smart RL, Lattanzi MG, Jahreiß H, *et al.*, 2007, Nearby star candidates in the Torino observatory parallax program. *A&A*, 464, 787–791 {**79**}

Smith CA, Corbin TE, Hughes JA, *et al.*, 1990, The SRS Catalogue of 20 488 star positions: culmination of an international cooperative effort. *IAU Symp. 141: Inertial Coordinate System on the Sky* (eds. Lieske JH, Abalakin VK), 457–461 {**68**}

Snyder GS, 1984, *Maps of the Heavens*. Abbeville Press, New York {**80**}

Soderblom DR, Nelan E, Benedict GF, *et al.*, 2005, Confirmation of errors in Hipparcos parallaxes from Hubble Space Telescope FGS astrometry of the Pleiades. *AJ*, 129, 1616–1624 {**79**}

Space Telescope Science Institute, Osservatorio Astronomico di Torino, 2001, The Guide Star Catalogue, Version 2.2 (GSC2.2). *VizieR Online Data Catalogue*, 1271 {**65**}

Spagna A, Lattanzi MG, McLean BJ, *et al.*, 2006, The Guide Star Catalogue II: Properties of the GSC 2.3 release. *Exploiting Large Surveys for Galactic Astronomy, IAU Joint Discussion 13*, 13 {**64, 65**}

Stone RC, Monet DG, Monet AKB, *et al.*, 1996, The Flagstaff Astrometric Scanning Transit Telescope (FASTT) and star positions determined in the extragalactic reference frame. *AJ*, 111, 1721–1742 {**57, 74, 76**}

—, 2003, Upgrades to the Flagstaff Astrometric Scanning Transit Telescope: a fully automated telescope for astrometry. *AJ*, 126, 2060–2080 {**74**}

Stone RC, 1993, Recent advances with the USNO (Flagstaff) transit telescope. *IAU Symp. 156: Developments in Astrometry and their Impact on Astrophysics and Geodynamics* (eds. Mueller II, Kolaczek B), 65–70 {**57**}

Stott C, 1995, *Celestial Charts: Antique Maps of the Heavens*. Studio Editions Ltd, London {**80**}

Sutton E, 1997, Hipparcos astrometry of infrared-selected sources and the connection between optical and infrared reference frames. *PASP*, 109, 1085–1088 {**75**}

Teixeira R, Benevides-Soares P, Dominici TP, *et al.*, 1998, CCD meridian observations at Valinhos Observatory. *Applied and Computational Harmonic Analysis*, 3, 190–193 {**58**}

Tel'nyuk-Adamchuk V, Babenko Y, Lazorenko PF, *et al.*, 2002, Observing programmes of the Kyiv meridian axial circle equipped with a CCD micrometer. *A&A*, 386, 1153–1156 {**58**}

Tel'nyuk-Adamchuk V, Gregul O, Molotaj O, 2000, Proper motions of some thousands of stars from 1.5-century observations: $\delta > 75°$. *Models and Constants for Sub-microarcsecond Astrometry, IAU Joint Discussion 2*, 14–17 {**72**}

Tie QX, Yang L, Li BH, *et al.*, 2005, Determining the astronomical refraction and building the radio wave refractive delay model measured. II. Using the Hipparcos Catalogue as working catalogue. *Astronomical Research and Technology. Publications of National Astronomical Observatories of China*, 2, 246–251 {**76**}

Tokovinin AA, 1997, MSC: a catalogue of physical multiple stars. *A&AS*, 124, 75–84 {**79**}

Turner HH, 1912, *The Great Star Map; Being a Brief General Account of the International Project known as the Astrographic Chart*. John Murray, London {**61**}

Turon C, Crézé M, Egret D, *et al.*, 1992, *Hipparcos Input Catalogue, ESA SP–1136* (7 volumes). European Space Agency, Noordwijk {**68**}

Urban SE, 1998, USNO programmes to improve the optical reference frame. *Applied and Computational Harmonic Analysis*, 3, 147 {**72**}

—, 2006, *The Astrographic Catalogue and AC 2000.2*. US Naval Observatory, http://ad.usno.navy.mil/proj/AC/ {**60**}

Urban SE, Corbin TE, Wycoff GL, *et al.*, 1998a, The AC 2000: the Astrographic Catalogue on the system defined by the Hipparcos Catalogue. *AJ*, 115, 1212–1223 {**62, 63, 68**}

—, 2001, The AC 2000.2 Catalogue. *VizieR Online Data Catalogue*, 1275 {**62, 68**}

Urban SE, Corbin TE, Wycoff GL, 1998b, The ACT Reference Catalogue. *AJ*, 115, 2161–2166 {**62, 68**}

Urban SE, Corbin TE, 1995, Cape AC zone data reduced to ACRS. *VizieR Online Data Catalogue*, 1219 {**62**}

—, 1998, The Astrographic Catalogue: a century of work pays off. *S&T*, 95(6), 40 {**61**}

Urban SE, Hall DM, Corbin TE, 1996a, Toulouse AC zone data reduced to ACRS. *VizieR Online Data Catalogue*, 1232 {**62**}

Urban SE, Jackson ES, Corbin TE, 1995, Algiers AC zone data reduced to ACRS. *VizieR Online Data Catalogue*, 1223 {**62**}

—, 1996b, Paris AC zone data reduced to ACRS. *VizieR Online Data Catalogue*, 1227 {**62**}

—, 1996c, Uccle AC zone data reduced to ACRS. *VizieR Online Data Catalogue*, 1231 {**62**}

—, 1996d, Vatican AC zone data reduced to ACRS. *VizieR Online Data Catalogue*, 1226 {**62**}

Urban SE, Jackson ES, 1996, Bordeaux AC zone data reduced to ACRS. *VizieR Online Data Catalogue*, 1228 {**62**}

Urban SE, Martin JC, Corbin TE, 1996e, Oxford 2 AC zone data reduced to ACRS. *VizieR Online Data Catalogue*, 1225 {**62**}

Urban SE, Martin JC, 1996, Oxford 1 AC zone data reduced to ACRS. *VizieR Online Data Catalogue*, 1224 {**62**}

Urban SE, Wycoff GL, 1995, San Fernando AC zone data reduced to ACRS. *VizieR Online Data Catalogue*, 1222 {**62**}

Urban SE, Zacharias N, Wycoff GL, 2004, The UCAC 2 Bright Star Supplement. *VizieR Online Data Catalogue*, 1294 {**73**}

US Naval Observatory, Royal Greenwich Observatory, 2001, *The Astronomical Almanac for the year 2003*. US Government Printing Office (USGPO), Washington: and The Stationery Office, London: 2001. {**82**}

van Altena WF, 2006, Education in astrometry. *Nomenclature, Precession and New Models in Fundamental Astronomy, IAU 26, JD16*, 16 {**83**}

van Altena WF, Lee JT, Hoffleit ED, 1995, *The General Catalogue of Trigonometric Stellar Parallaxes, Fourth Edition*. Yale University Observatory, New Haven {**68, 79, 80**} van Altena WF, Stavinschi M, 2005, The future of astrometric education. *ASP Conf. Ser. 338: Astrometry in the Age of the Next Generation of Large Telescopes* (eds. Seidelmann PK, Monet AKB), 311–317 {**83**}

van Bezooijen RWH, Degen L, Nichandros H, 2004, Guide star catalogue for the Spitzer Space Telescope pointing calibration and reference sensor. *Microwave and Terahertz Photonics. Proc. SPIE, Vol. 5487* (eds. Stohr A, Jager D, Iezekiel S), 253–265 {**83**}

van Bezooijen RWH, 2003, SIRTF autonomous star tracker. *Infrared Space Telescopes and Instruments* (ed. Mather JC), 108–121 {**83**}

van Biesbroeck G, 1944, The star of lowest known luminosity. *AJ*, 51, 61–62 {**80**}

Viateau B, Réquième Y, Le Campion JF, *et al.*, 1998, Contribution of the Bordeaux CCD meridian circle to modern astrometry. *Applied and Computational Harmonic Analysis*, 3, 191–192 {**58**}

—, 1999, The Bordeaux and Valinhos CCD meridian circles. *A&AS*, 134, 173–186 {**57, 58**}

Vicente B, Abad C, Garzón F, 2007, Astrometry with Carte du Ciel plates, San Fernando zone. I. Digitization and measurement using a flatbed scanner. *A&A*, 471, 1077–1089 {**62**}

Vondrák J, Ron C, 2000, Survey of observational techniques and Hipparcos reanalysis. *ASP Conf. Ser. 208: IAU Colloq. 178: Polar Motion: Historical and Scientific Problems*, 239–250 {**55**}

Wang S, Tang Z, Jin W, *et al.*, 1997a, Determinations of high precision proper motions for 32 Hipparcos stars from photographic plates. *The New International Celestial Reference Frame, IAU Joint Discussion 7*, 40 {**72**}

—, 1997b, Improvement of proper motions for Hipparcos stars with photographic plates. *Acta Astronomica Sinica*, 38, 297 {**72**}

—, 1998, Determination of highly precise proper motions for 32 Hipparcos stars from photographic plates. *Kinematika i Fizika Nebesnykh Tel*, 14, 149–155 {**72**}

Warner DJ, 1979, *The Sky Explored. Celestial Cartography 1500–1800*. Alan R. Liss, New York {**80**}

Whitfield P, 1995, *The Mapping of the Heavens*. British Library, London {**80**}

Yasuda H, Hurukawa K, Hara H, 1982, Northern PZT stars catalog (NPZT74). *Annals of the Tokyo Astronomical Observatory*, 18, 367–427 {**68**}

York DG, Adelman SJ, Anderson JE, *et al.*, 2000, The Sloan Digital Sky Survey: technical summary. *AJ*, 120, 1579–1587 {**74**}

Yoshizawa M, Suzuki S, 1997, The Tokyo PMC catalogue 90–93: catalogue of positions of 6649 stars observed in 1990 through 1993 with Tokyo photoelectric meridian circle. *Publications of the National Astronomical Observatory of Japan*, 5, 1 {**58**}

Yu KC, Jenkins NE, 2004, Cosmic Atlas: a real-time universe simulation. *American Astronomical Society Meeting Abstracts*, 204 {**84**}

Yu Y, Tang Z, Li JL, *et al.*, 2004, Block adjustment of a group of overlapping CCD images. *AJ*, 128, 911–919 {**72**}

Zacharias MI, Zacharias N, 2003, The US Naval Observatory CCD Astrograph Catalogue project. *Astronomy in Latin America* (eds. Teixeira R, Leister NV, Martin VAF, *et al.*), 109–112 {**72**}

Zacharias N, 1984, Der CPC2-Katalog, ein photographisches Referenzsystem hoher Genauigkeit am Südhimmel. *Mitteilungen der Astronomischen Gesellschaft Hamburg*, 62, 209 {**72**}

—, 1988, Rigorous block-adjustment of the CPC2 Cape Zone. *IAU Symp. 133: Mapping the Sky: Past Heritage and Future Directions* (ed. Débarbat S), 201–206 {**72**}

—, 1992, Global block adjustment simulations using the CPC2 data structure. *A&A*, 264, 296–306 {**72**}

—, 1995, Field distortions in the CPC2 plate data. *AJ*, 109, 1880–1888 {**69**}

Zacharias N, de Vegt C, Murray CA, 1997a, CPC2 plate reductions with Hipparcos stars: first results. *ESA SP–402: Hipparcos, Venice '97*, 85–90 {**69, 70**}

Zacharias N, de Vegt C, Nicholson W, *et al.*, 1992, CPC2: the Second Cape Photographic Catalogue. II. Conventional plate adjustment and catalogue construction. *A&A*, 254, 397–421 {**69**}

Zacharias N, Germain ME, Rafferty TJ, 1997b, UCAC S: a new high precision, high density astrometric catalogue in the S. Hemisphere. *ESA SP–402: Hipparcos, Venice '97*, 177–180 {**72**}

Zacharias N, Høg E, Urban SE, *et al.*, 1997c, Comparing the Tycho Catalogue with CCD astrograph observations. *ESA SP–402: Hipparcos, Venice '97*, 121–124 {**72**}

Zacharias N, Laux U, Rakich A, *et al.*, 2006, URAT: astrometric requirements and design history. *Ground-based and Airborne Telescopes*, volume 6267 of *Proc. SPIE*, 22 {**55**}

Zacharias N, McCallon HL, Kopan E, *et al.*, 2003, Extending the ICRF into the infrared: 2MASS–UCAC astrometry. *The International Celestial Reference System: Maintenance and Future Realisation, IAU Joint Discussion 16*, 43 {**75**}

Zacharias N, Monet DG, Levine SE, *et al.*, 2004a, The Naval Observatory Merged Astrometric Dataset (NOMAD). *Bulletin of the American Astronomical Society*, 1418 {**73**}

Zacharias N, Urban SE, Zacharias MI, *et al.*, 2000a, The first US Naval Observatory CCD Astrograph Catalogue (UCAC 1). *AJ*, 120, 2131–2147 {**72–75**}

—, 2004b, The second US Naval Observatory CCD Astrograph Catalogue (UCAC 2). *AJ*, 127, 3043–3059 {**66, 68, 69, 73, 74**}

Zacharias N, Zacharias MI, de Vegt C, 1999, The Second Cape Photographic Catalogue on the Hipparcos system. *AJ*, 117, 2895–2901 {**68, 69**}

Zacharias N, Zacharias MI, Douglass GG, *et al.*, 1996, The Twin Astrographic Catalogue (TAC) Version 1.0. *AJ*, 112, 2336–2348 {**68, 71**}

Zacharias N, Zacharias MI, Urban SE, *et al.*, 2000b, Comparing Tycho 2 astrometry with UCAC 1. *AJ*, 120, 1148–1152 {**73**}

Zacharias N, Zacharias MI, 1999, The Twin Astrographic Catalogue on the Hipparcos system. *AJ*, 118, 2503–2510 {**68, 71**}

Zhu Z, 2001, Hipparcos proper motion system with respect to SPM 2.0. *PASJ*, 53, L33–L36 {**71**}

3

Double and multiple stars

3.1 Introduction

Stellar systems composed of two or more stars (binaries or multiples) exist in a wide range of configurations (see Trimble, 2002, for an introduction). They range from tight binaries with orbital periods of hours or days, and separations down to less than $10R_\odot$ ($\sim$0.1 AU), to marginally-bound wide binaries with orbital periods $>$ 10^5 yr and separations up to 20 000 AU (0.1 pc) or more. Their scientific importance is extensive, ranging from statistics of their physical properties (frequency, periods, mass ratios, eccentricities) providing constraints on theoretical models of star formation, various aspects of stellar evolution and mass transfer, and their central role for mass determination.

Binaries are extremely common in the Galaxy, with various estimates suggesting that some 50% or more of all stars occur in gravitationally bound groups having a multplicity of 2 or higher. The true binary frequency is hard to establish for any given population due to the different techniques required to probe different separation ranges, and to the inherent difficulties of detecting binaries with very small or very large angular separations. Binary frequency appears to correlate well with stellar ages: for example, low-mass pre-main-sequence stars in the star-forming region Taurus–Auriga have an (inferred) binary frequency as high as 80–100% (Leinert *et al.*, 1993; Köhler & Leinert, 1998), values confirmed by studies of other star-forming regions.

Studies of component mass ratios give different results according to observational technique, sample population, and correction for selection effects. A bimodal distribution is often reported, with one peak arising from equal-mass components, and another at a mass ratio of around 0.2–0.3. The distribution in orbital periods P has been estimated over several orders of magnitude, from days to millions of years (Duquennoy & Mayor, 1991; Fischer & Marcy, 1992). The orbital period distribution of spectroscopic binaries was also re-estimated more recently, e.g. by Goldberg *et al.* (2003). Results point to a bell-shaped distribution of $\log P$ with a maximum in the visual binary range at around 180 years. Orbital eccentricities are correlated with periods: circular orbits dominating for periods below 10 days, with longer-period systems having eccentricities from 0.1 to nearly 1.0 and an absence of circular orbits.

The formation of binary stars remains incompletely understood, but is considered as occurring in two phases: fragmentation of molecular clouds producing the condensing cores, followed by subsequent accretion and migration which will fix the final component masses and orbital parameters (see, e.g. Halbwachs *et al.*, 2003).

Double stars are studied through a range of photometric and spectroscopic methods. Traditionally, information on separations has been obtained from transit circles or astrographs, with systems studied visually using a micrometer on long-focus instruments to provide separations and position angles at the observation epoch. More recently, speckle interferometry has provided a wealth of data, with over 15 000 observations reported by the US Naval Observatory up to 2001.

The Hipparcos results have provided many new insights into the statistical distribution of binaries, the occurrence of wide binary systems, information on their Galactic dynamics for population studies, new individual stellar mass determinations, and a large number of new astrometric binaries suitable for ground-based orbit determinations.

3.2 Double and multiple stars in the Hipparcos Catalogue

A description of the double and multiple systems as detected during the Hipparcos data processing, and as

Nomenclature and classification according to method of discovery: Binary systems are often classified in terms of how they were discovered (as opposed to their actual physical or evolutionary state). These are, roughly in decreasing order of orbital period or semi-major axis:

* Common proper motion pairs: in which a pair of stars share the same proper motion, often with large angular separation and long orbital period such that orbital motion may be difficult or impossible to discern.

* Visual binaries: in which a pair of stars is resolved, usually with a small angular separation of a few arcsec. Determining a common proper motion is mostly sufficient to exclude chance alignment of unrelated stars along the line-of-sight (optical pairs). In principle, systems can be confirmed through the detection of their orbital motion, although this is impractical for orbital periods of 10^4 yr or more.

* Astrometric binaries: a physical stellar system not observed as a visual double because of its small separation and/or large magnitude difference, but inferred as double through a nonlinear proper motion of the photocentre.

* Spectrum binaries: in which the spectrum of a single image shows two sets of lines characteristic of two stars of different spectral type.

* Spectroscopic binaries: where the spectrum of a single image displays one (single-lined, SB1) or two (double-lined, SB2) sets of lines whose radial velocity signature varies periodically indicating orbital motion. The latter implies components of comparable mass and luminosity.

* Eclipsing binaries: in which the orbital plane is so close to the line-of-sight that the two stars periodically eclipse each other, giving regular minima in the combined light from the system. In close pairs, the stars are tidally deformed, giving an 'ellipsoidal' light variation even when there are no true eclipses.

Many systems show more than one of these characteristics: thus Algol is both a spectroscopic and an eclipsing binary with a period of about three days; while 70 Oph is both a visual and a spectroscopic binary with a period of 88 years.

The Hipparcos Catalogue includes common proper motion pairs, and visual binaries which form the Part C (component solutions) of the Hipparcos DMSA annex. Part O (orbital solutions), G (acceleration solutions) and Part X (stochastic solutions) typically belong to the class of astrometric binaries. The Hipparcos variability-induced movers (VIMs, recognised through photometric variability of one component), and the $\Delta\mu$ binaries (recognised by a discrepancy between the Hipparcos and long-term proper motions when combined with early epoch catalogues) are also astrometric binaries, although representing new classes when classified according to discovery method.

contained within the published Hipparcos Catalogue, is summarised by Lindegren *et al.* (1997), from which the introductory details in this section are directly taken. Some further description was given by Lindegren (1997). In addition, more details of these systems relevant to the catalogue users are given in ESA (1997, Volume 1, Section 1.4), details of the associated data processing in Volume 3, Chapter 13, and the Double and Multiple Systems Annex in printed form in Volume 10. The latter includes, for the component solution part, and for each component: Hp, B_T and V_T magnitudes, position, parallax, proper motion and associated standard errors, relative astrometry, and schematic figures illustrating the relative astrometry and photometry.

The Hipparcos Catalogue contains astrometric results for 117 955 catalogue entries, many of which are in reality double or multiple stars. For 17 917 entries, special solutions were made in order to cope with the various manifestations of multiplicity, and 6763 other entries were flagged as suspected binaries but solved as single stars. Details of the special solutions are contained in the Hipparcos Catalogue Double and Multiple Systems Annex, divided in five parts according to the type of solution: resolved systems (13 211 entries with a total of 24 588 components), astrometric binaries with curved proper motion (2622 entries) or orbital solutions (235 entries), 'variability-induced movers' (288 entries) and 'stochastic' solutions (1561 entries). Parallaxes and the positions and proper motions in the extragalactic reference system ICRS are provided for all entries.

The presence of double and multiple systems severely complicated the construction of the Hipparcos Catalogue. A global reference frame had to be established through a simultaneous solution of the astrometric parameters of a dense network of stars covering the entire sky. This process had to be restricted to observationally single stars, i.e. either real single stars or unresolved objects whose proper motions could be considered constant over the few years of satellite observations. It was thus necessary to identify the subset of stars satisfying this criterion, build the global reference frame from their observations, and afterwards attach the remaining objects to this frame using the appropriate object models. This process was further complicated by the necessity to reconcile the results of several groups treating the satellite data by different methods.

A major difficulty was that no unique alternative model could be applied to all the objects for which deviations from the 'single-star model' were detected: these had to be classified according to observational criteria which do not necessarily correspond to well-defined regions of the physical characteristics of the systems. Thus, while the astrometric processing of the 'single' stars can be characterised as quite stringent, various *ad hoc* procedures had to be adopted for the objects in the Double and Multiple Systems Annex.

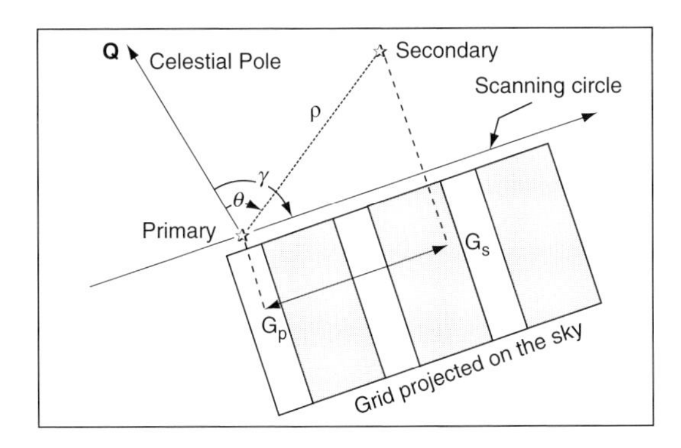

Figure 3.1 Geometry of the observation of a binary star with separation ρ and position angle θ as seen from outside the celestial sphere. The angular coordinate along the scanning circle is denoted G. The signal observed depends only on the projected phase difference $\Delta\phi$ between the phase of the secondary (at G_S) and that of the primary (at G_P). The orientation of the scanning circle is indicated by γ. From ESA (1997, Volume 3, Figure 13.5).

Their classification, as well as their published solutions, should consequently be regarded with due caution.

Observations of (suspected) multiple systems yielding additional astrometric information post catalogue publication, have led to the determination of new or revised properties of many Hipparcos objects.

3.2.1 Observational effects of multiplicity

The basic observational entity of the Hipparcos Catalogue is an 'entry', identified by a HIP number. This concept was derived from instrumental limitations and refers to whatever fell inside the ~35 arcsec diameter (FWHM) instantaneous field of view of the Hipparcos main detector system.

An entry may consist of a single star, a (resolved or unresolved) double or multiple system, or one or more components of a wide system. Unresolved objects were observed by their photocentres in the *Hp* wavelength passband of the Hipparcos main detector. The components or the photocentre may exhibit significant orbital motion about the system's centre of mass. Even if only double systems are considered, several observationally different types are distinguished (see Figure 3.3, left below) depending mainly on the separation, ρ, magnitude difference, ΔHp, and orbital period, P:

(1) Very close binaries ($\rho \lesssim 2$ mas): the photocentre practically coincides with the centre of mass, making such systems astrometrically equivalent to single stars.

(2) Unresolved systems ($2 \lesssim \rho \lesssim 100$ mas or $\Delta Hp \gtrsim 4$): (a)short-period binaries ($P \lesssim 0.1$ year) where the observations provided a quasi-random sampling of the orbit of the photocentre producing mean results usually representing the centre of mass; (b) for intermediate periods ($0.1 \lesssim P \lesssim 10$ years) the full Keplerian orbit of the photocentre could sometimes be determined, possibly with the help of ground-based elements; (c) for somewhat longer periods ($5 \lesssim P \lesssim 30$ years) the motion of the photocentre may have been significantly nonlinear, although no periodic solution was possible; (d) for moderately long periods ($10 \lesssim P \lesssim 100$ years) the motion of the photocentre may have appeared linear over the mission, although it differed significantly from the motion of the centre of mass; (e) long-period binaries ($P \gtrsim 100$ years) have no significant bias in the proper motion, although the photocentre may be offset from the centre of mass.

(3) Resolved systems ($0.1 \lesssim \rho \lesssim 10$ arcsec and $\Delta Hp \lesssim 4$ mag): the detector signal could be decomposed into the contributions of the different stellar components.

(4) Known systems with $10 < \rho < 30$ arcsec were sometimes observed as 'two-pointing doubles' by including the components as separate entries. To properly take into account their mutual influence, the components had to be treated together in the data analysis.

(5) Wide systems ($\rho \gtrsim 30$ arcsec): only one component at a time was included in the instantaneous field of view.

For multiple stars, the set of possible configurations is vastly expanded as each subpair could fall into any of the above categories. The not uncommon photometric variability of components may cause additional complications, such as spurious motion of the photocentre.

Apart from the use of ground-based information, multiplicity could basically be detected in two ways: from analysis of the detector signal (Figure 3.1), in particular the visibility expressed as a difference between the Hipparcos 'ac' and 'dc' magnitudes (van Leeuwen *et al.*, 1997), and from the residuals of the fit to a standard five-parameter astrometric model for a single star. The first method was mainly sensitive to systems of type 3 and 4, the second to type 2a–c. Type 1, 2d–e and 5 were virtually indistinguishable from single stars.

Further details of the complexities of the analysis and presentation of double and multiple systems in the context of the Hipparcos Catalogue is given in the published catalogue (ESA, 1997, Section 1.4). Systems were classified as 'double' or 'multiple', with between one to four entries with a common system identifier (CCDM number). An individual entry may also be classified as 'double' or 'multiple'. Systems may be further considered as 'single-pointing', 'two-pointing', or 'three-pointing', depending upon the number of entries considered together during the data analysis (the term 'pointing' referring to the way in which the sensitive area of the image dissector tube detector was directed to receive the light from a catalogue entry). Thus a double system may be 'single-pointing' or 'two-pointing', or may have two completely 'independent' entries. Similarly, a

Table 3.1 Number of entries in the Hipparcos Catalogue for the different categories of astrometric solutions.

Type of solution	Annex	No. of entries	Comment
Single-star solutions	–	100 038	of which 6763 flagged as suspected double
Component solutions	C	13 211	comprising 24 588 components in 12 195 solutions
Acceleration solutions	G	2622	
Orbital solutions	O	235	
Variability-induced movers	V	288	
Stochastic solutions	X	1561	
No valid astrometric solution	–	263	of which 218 flagged as suspected double
Total number of entries		118 218	
Entries with valid astrometry		117 955	

multiple system may be 'single-pointing', 'two-pointing', or 'three-pointing', or may have two to four independent entries (with a common CCDM number). The term 'multiple' is to be considered as specifically referring to a number of components equal to or larger than three. There is not always a clear distinction between independent entries and multiple entries considered as a multiple system.

3.2.2 Classification of solutions

In the limited time available for the double-star processing, it was not feasible to consider all the different situations that might apply to a given object, with due regard to available ground-based information, and to select and publish in each case the most appropriate solution. What emerged was a practical scheme dividing the astrometric solutions into the seven categories summarised in Table 3.1. The last category contains the entries without associated astrometry, e.g. because of no detectable signal, too few observations, or where the signal could not be adequately interpreted in terms of any of the attempted models. The other categories are described hereafter.

The Hipparcos Catalogue gives primarily the five astrometric parameters for all entries with associated astrometry, i.e. the position at epoch J1991.25 (α, δ), the parallax (π) and the proper motion ($\mu_{\alpha*} = \mu_\alpha \cos\delta$, μ_δ), whether the entry was treated as a single star or not. The precise meaning of the position and proper motion (e.g. component, photocentre, or centre of mass) depends on the type of solution, and the relevant flags in the catalogue must be consulted for correct interpretation. Additional information for the double and multiple systems is given in the various parts of the Annex.

Apparently single stars The vast majority of entries (85%) could be solved as single stars, characterised by their five basic astrometric parameters. Many are in reality binaries, and associated effects may be implicit in the results of some of them. For instance, binaries of type 2d may have proper motions which differ significantly from ground-based values determined with temporal baselines of several decades. The positions for types 2d–e may depend on the effective wavelength of the observations. 6763 of the entries are flagged as suspected non-single based on various criteria, although no significant or convincing non-single star solution was found. Some 4200 of the entries were listed as double or multiple in the Hipparcos Input Catalogue, most of them with large separations ($>$ 30 arcsec) or magnitude differences ($>$ 4 mag) making them effectively single for Hipparcos.

Component solutions (Part C of the Annex) These are resolved systems of type 3 and 4 for which the Annex gives the five astrometric parameters and the *Hp* magnitude of each resolved component. This allows linear (but not curved) relative motion among the components. Of the 12 195 solutions 12 005 are double, 182 triple and eight quadruple star solutions; 10 895 were solved as fixed configurations, i.e. constrained to a common parallax and proper motion; 1186 included linear relative motion, and 202 more included linear relative motion and individual parallaxes of the components. 2996 of the double stars were not previously known as such. Figure 3.2 shows the distribution of separations and magnitude differences for all pairs, and separately for the newly-detected pairs. The Annex provides a fairly complete census of binaries among the Hipparcos stars with $\Delta Hp < 3.5$ and $\rho > 0.12$–0.3 arcsec (depending on ΔHp).

The published component solutions were obtained by combining the results of the independent analyses performed by the reduction consortia FAST and NDAC. For nearly 90% of the solutions, the agreement between the two reductions was sufficiently good to permit a straightforward averaging of the data, while the differences were used to calibrate the estimated standard errors. In 1113 cases the two reductions could not be reconciled, as they typically differed by a multiple of the fundamental period of the modulating grid ($\sim$1.2 arcsec), and one of them had to be selected on

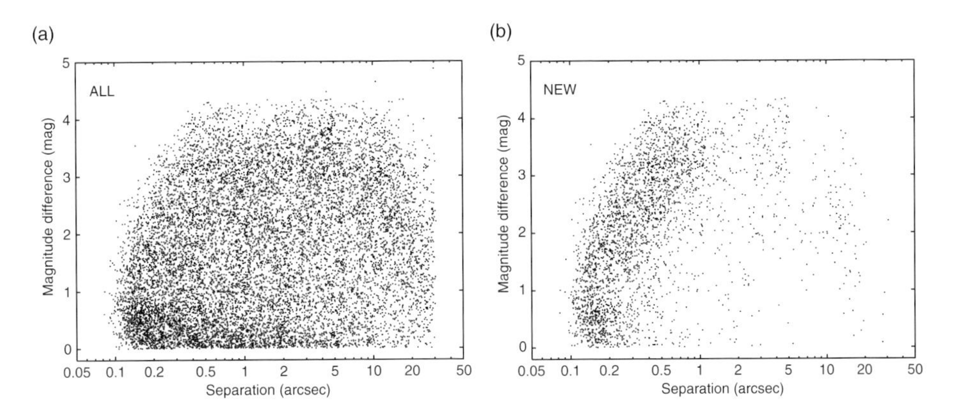

Figure 3.2 (a) Distribution of separation and magnitude difference for all 12 393 component pairs in Part C of the Annex; and (b) for the subset of 2996 pairs discovered with Hipparcos. From Lindegren et al. (1997).

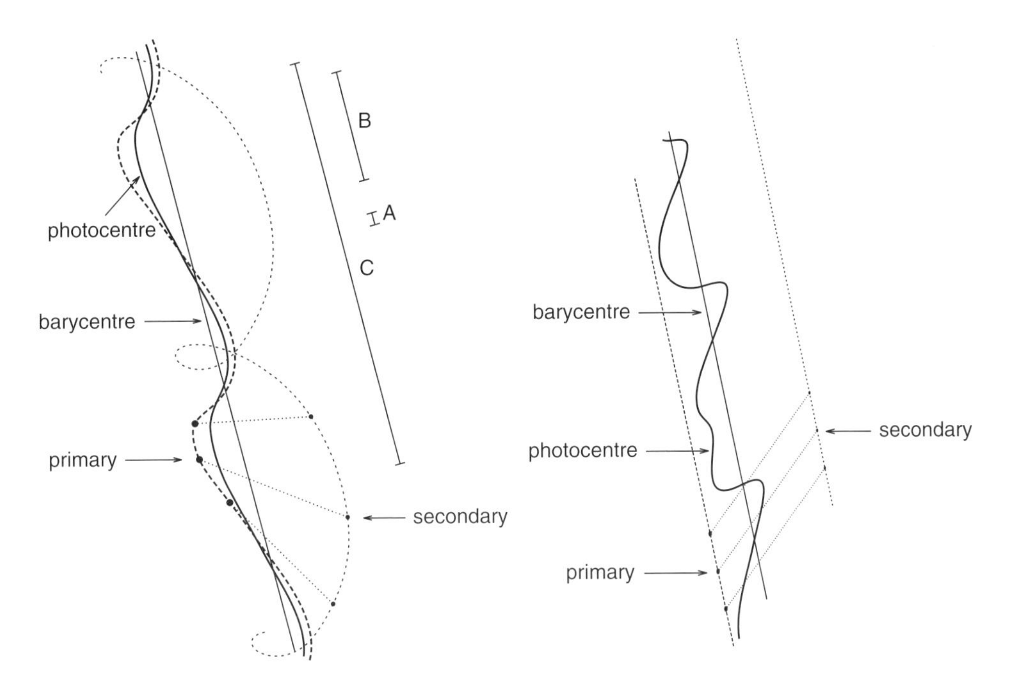

Figure 3.3 Left: for close binaries with separations below 0.1–0.3 arcsec, Hipparcos observed the system's photocentre. The orbital motions of the components around the barycentre causes the photocentre to trace a nonlinear path on the sky (thick curve). The type of solution allowed by the Hipparcos data depended mainly on the observation interval, about 3.3 years, compared with the orbital period, as illustrated by the three cases A, B, C (where the line segments represent different fractions of the orbit corresponding to the observation interval). Right: for an unresolved system in which one of the components is variable, the photocentre oscillates between the components in phase with the total intensity. Such a 'variability-induced mover', or VIM, can be detected as an astrometric binary even in the absence of significant orbital motion during the observation interval. From Lindegren (1997, Figures 2 and 3).

the basis of various criteria. Such cases are flagged, and the main parameters of alternative solutions are given as notes.

While other parts of the Hipparcos Catalogue are arranged according to the HIP number, Part C of the Annex uses the CCDM identifier from the Catalogue of Components of Double and Multiple Stars (Dommanget & Nys, 1994, and updates), sometimes corresponding to more than one HIP number. The identification of systems and components required ground-based information, often also indispensable as starting points for the highly nonlinear analysis of the space results. Such data were primarily obtained from the CCDM and a special database maintained at OATO (Torino), but in some cases dedicated observations were initiated. The updating of the ground-based information benefited from the use of a pre-release version of the Washington Catalogue of Visual Double Stars.

Acceleration solutions (Part G of the Annex) Part G lists apparently single stars with significantly nonlinear motion, for which no orbital (periodic) solution could be made, i.e. systems of type 2d. A quadratic or cubic polynomial of time was fitted to the motion in each coordinate, in addition to the standard five astrometric parameters. In tangential coordinates ($\xi \sim \Delta\alpha \cos\delta$, $\eta \sim \Delta\delta$) the apparent motion, excluding parallax, was modelled as

$$\xi(t) = \xi(0) + t\mu_{\alpha*} + \frac{1}{2}(t^2 - a)g_{\alpha*} + \frac{1}{6}(t^2 - b)t\dot{g}_{\alpha*}$$
$$\eta(t) = \eta(0) + t\mu_{\delta} + \frac{1}{2}(t^2 - a)g_{\delta} + \frac{1}{6}(t^2 - b)t\dot{g}_{\delta} \qquad (3.1)$$

where t is the time in years from J1991.25, $(g_{\alpha*}, g_\delta)$ are the accelerations in mas yr^{-2}, and $(\dot{g}_{\alpha*}, \dot{g}_\delta)$ the rates of change of the accelerations in mas yr^{-3}. The constants $a = 0.81$ yr^2 and $b = 1.69$ yr^2 were introduced to make the g and $\dot{g}$ terms approximately orthogonal to the preceding terms in Equation 3.1. Quadratic or cubic solutions were accepted depending on the statistical significance of the g and $\dot{g}$ terms. The Annex includes 2163 quadratic solutions (i.e. excluding the $\dot{g}$ terms) and 459 cubic solutions. The polynomial representation of the motion has no validity outside the mission interval (1989.9–1993.2). The proper motion $(\mu_{\alpha*}, \mu_\delta)$, representing the mean motion over the mission interval, is safer to extrapolate to other epochs, but the positional uncertainty may be considerably greater than indicated by the standard errors in proper motion.

Orbital solutions (Part O of the Annex) For some systems (type 2b) the orbital elements of the photocentre could be determined in addition to the normal five astrometric parameters for the centre of mass. In many cases some of the elements were adopted from ground-based observations of the astrometric or spectroscopic binaries, or taken as starting values for the solution. However, in nearly half of the cases the period was identified from periodogram analysis of the space data performed at the Astronomisches Rechen-Institut, Heidelberg, and refined elements subsequently determined by least-squares fitting to the Hipparcos Intermediate Astrometric Data. A complete orbit (requiring seven elements) was determined for 45 systems. In addition to details given in the published Hipparcos Catalogue, some further details of the orbital solutions made by the FAST Consortium were described by Bernstein (1994) and Emanuele *et al.* (1996).

Variability-induced movers (Part V of the Annex) Variability-induced movers, or VIMs, are unresolved binaries in which one of the components is variable. The photocentre of a VIM shows a specific motion on the sky, coupled to the variation of the total brightness of the system. Given the total magnitude for each scan across the system, a VIM solution requires two elements $(D_{\alpha*}, D_\delta)$ in addition to the five astrometric parameters. From these can be derived the position angle of the constant component with respect to the variable, and a lower limit for the separation, typically in the range 10–90 mas. The astrometric data in the main catalogue refer to the photocentre for a specific value of the total magnitude. The VIM solutions were derived at the Astronomisches Rechen-Institut, Heidelberg, by a critical examination of variable stars, using the FAST Intermediate Astrometric Data. Details are given by Wielen (1996).

Stochastic solutions (Part X of the Annex) For some objects it was not possible to find an acceptable single or double star solution in reasonable agreement with the statistical uncertainties of the individual measurements. Such objects could be unresolved binaries of type 2a–b, resolved objects where the secondary could not be located or was perturbed by variability or edge effects of the instantaneous field of view. Lacking an acceptable deterministic model for these objects, a stochastic model for the displacements relative to the centre of mass was adopted. This was achieved by quadratically increasing the standard errors of the measurements until the rms normalised residual was exactly equal to 1. The added dispersion, called the 'cosmic error', typically ranges from 3–30 mas. Solutions with a cosmic error greater than 100 mas were rejected as this was normally an indication of grid-step errors.

3.2.3 Accuracy verification

ESA (1997, Volume 2, Chapter 22) provides a comparison of the solutions with respect to ground-based observations of comparable accuracy: for astrometry based on speckle interferometry for about 1000 stars common to the Hipparcos and CHARA programmes, and for photometry based on CCD photometric observations carried out at La Palma. Relative positional errors were confirmed through comparisons with ground-based CCD measurements by Oblak *et al.* (1997a). Other comparisons between Hipparcos and ground-based observations in the separation range 1–14 arcsec were reported by Olević *et al.* (2000a, 2001).

Sinachopoulos *et al.* (1998) provided a list of very wide visual double stars in the Hipparcos Catalogue,

Double and multiple star catalogues: WDS: The Washington Double Star Catalogue (WDS, Mason *et al.*, 2001c), maintained by the US Naval Observatory, contains positions (J2000), discoverer designations, epochs, position angles, separations, magnitudes, spectral types, proper motions, and, when available, Durchmusterung numbers and notes. Now updated continuously online, it supersedes two earlier major updates (Worley & Douglass, 1984, 1997). The CD version released in 2001 included four catalogues:

(a) The Washington Double Star Summary Catalogue: it contains summary data (first and last observation dates and associated positions) for 84 489 systems based on 563 326 means, and includes cross-reference files, neglected doubles, and finder charts. It includes the 3406 systems first resolved by Hipparcos (Part C), as well as cross-references of WDS entries associated with Hipparcos Parts G, O, V and X. It also includes the 12 770 systems from the Tycho 2 Catalogue (the 11 536 known systems, and the 1234 new detections from Tycho). Hipparcos and Tycho 2 were ranked 8th and 9th respectively out of the top 25 'observers' contributing double stars to the WDS.

(b) The Fifth Catalogue of Orbits of Visual Binary Stars (Hartkopf *et al.*, 2001a): it contains 1465 orbits of 1430 systems, 58% more than the Fourth Catalogue (Worley & Heintz, 1983), and a new grading scheme. It continues the series of compilations of visual binary star orbits published by Finsen, Worley, and Heintz from the 1930s to the 1980s. The Sixth Orbit Catalogue is maintained on-line at the USNO by W.I. Hartkopf & B.D. Mason.

(c) The Third Catalogue of Interferometric Measurements of Binary Stars (Hartkopf *et al.*, 2001b): maintained at Georgia State University by CHARA for many years, the CD version includes 64 779 measures of 25 076 resolved pairs as well as unresolved pairs and measures with photometry only. It is over six times the size of the Second Catalogue, and includes double star measures from both the Hipparcos and Tycho 2 Catalogues.

(d) The Photometric Magnitude Difference Catalogue: the first publication of a previously-internal USNO product, containing reliable magnitude difference estimates for which no astrometry is quoted (and thus, inappropriate for inclusion in the WDS), containing 19 589 measures of 10 473 systems.

The WDS is the successor to the all-sky Index Catalogue of Visual Double Stars (IDS, Jeffers *et al.*, 1963). Three earlier double star catalogues of the last century, those by Burnham (1906), Innes *et al.* (1927), and Aitken & Doolittle (ADS, 1932), each covered only portions of the sky.

The Catalogue des Composantes d'Etoiles Doubles et Multiples (Catalogue of Components of Double and Multiple Stars, CCDM, Dommanget & Nys, 1994) gives identifications, bibliography, and accurate coordinates, etc. It was originally based on the Index Catalogue updated to July 1976, and developed to serve the needs of the Hipparcos mission by compiling a database of systems of which at least one component was located with an accuracy of 1 arcsec, necessary for the satellite detector pointing. The first edition (1994) contained 34 031 systems. Dommanget & Nys (2000) describes the complete list of 18 644 systems or individual components known in July 1997 which were observed by Hipparcos. Updating continued during the operational phase of the mission (Dommanget, 2000), and the second edition (2002) contains 49 325 systems.

Other compilations include:

- The Multiple Star Catalogue (MSC, Tokovinin, 1997) is a compilation of multiple systems (with 3–7 components) containing data on 612 physical multiple stars.
- The catalogue of orbits and ephemerides of visual double star systems compiled at the Ramón María Aller Observatory, Santiago de Compostela (Docobo *et al.*, 2001a), which comprised 1685 orbits for 1240 systems in 2001, and which continues the ephemerides catalogues of Muller & Couteau (1979) and Couteau *et al.* (1986).
- A compilation of spectroscopic orbits has been maintained over the past 40 years by Batten and collaborators, with the most recent versions given by Batten *et al.* (8th edition, 1989), and Pourbaix *et al.* (SB9, the 9th edition, 2004b), the latter containing 2386 systems, of which 1320 have a Hipparcos or Tycho 2 entry. At the end of 2007 it contains 2469 systems of which 2065 have a Hipparcos or Tycho 2 entry (Pourbaix 2007, priv. comm.).

between 10–100 arcsec, suitable for the astrometric calibration of CCDs (see also Sinachopoulos *et al.*, 1995).

3.3 Tycho Catalogue double stars

The original Tycho Catalogue contained astrometry and photometry for double star components with separations above about 1 arcsec, but with reliable photometry in B_T and V_T only above about 3 arcsec. The statistics, and an investigation of the mass ratios, were reported by Halbwachs *et al.* (1997). No specific double star annex was provided.

The Tycho 2 Catalogue also contained no double star annex (i.e. no specific listing of double star parameters), although it comprised a wealth of double star data in terms of separate catalogue entries for close components, from which classical double star parameters of separation, position angle, and magnitude difference can be calculated, and various associated flags. Down to a 0.8 arcsec separation limit, these included 6251 known double stars; 1234 new systems, included in the Washington Double Star (WDS) Catalogue with a 'TDS' discovery code; 4726 systems solved as single components of known WDS primaries for which the secondary was also present in the Tycho 2 Catalogue; and 1133 WDS systems seen as single in Tycho 2 for which the other component should have been measurable (Mason *et al.*, 2000b). The identification of doubles in the first analysis of Tycho 2 was based on a rather cautious search window, and several thousand probable

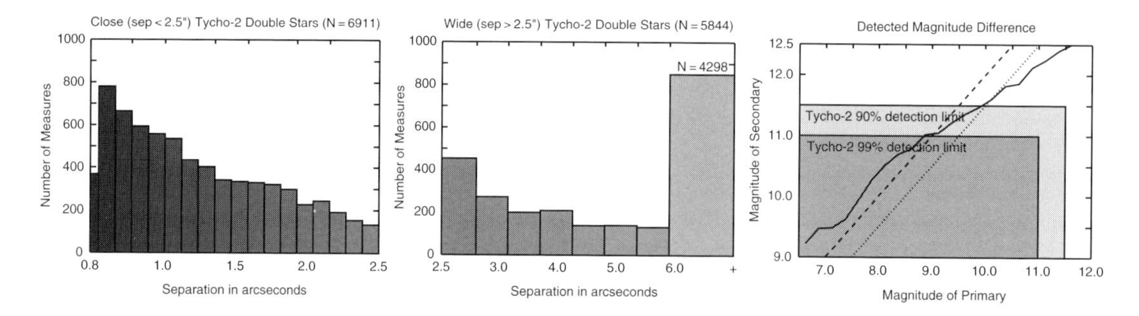

Figure 3.4 Left and middle: number of Tycho 2 measures of double stars versus separation in arcsec. Most systems with separations $\rho < 2.5$ arcsec (left) were given the 'double-star treatment' in Tycho 2. The first half-bin represents 368 systems with $\rho = 0.8$–0.85 arcsec, while other bins represents a full 0.10 arcsec. Middle: in the separation range above 2.5 arcsec, most systems were treated as single stars. For systems in the last bin, the mean separation is 28.8 arcsec. Right: Tycho 2 primary versus secondary magnitude (solid line), in 0.25 mag bins. Only bins with more than 100 members are plotted. The primary magnitude is the midpoint of the bin, while the secondary magnitude is that of the 90% limit within pairs of that bin. If the sample is cut at the 11.5 mag limit of the secondary (where Tycho 2 is 90% complete), a Δm of 1.5 is detectable (dotted line). Cutting the sample at the 11.0 mag limit of the secondary (where Tycho 2 is 99% complete), a Δm of 2.0 is detectable (dashed line). From Mason et al. (2000b, Figures 1 and 3).

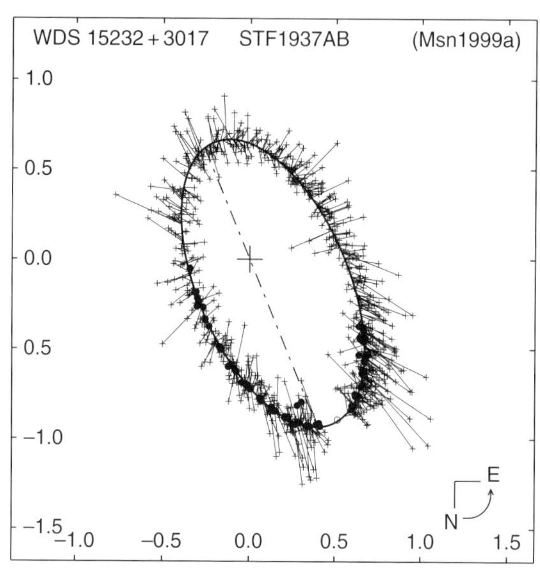

Figure 3.5 The double star measures of TDSC 39584 (WDS 15232+3017) are plotted against the definitive orbit of Mason et al. (1999a). Double star measures made with a micrometer are illustrated as plus signs while measures made by speckle interferometry are dots. The open circle at lower right is the measure of Tycho. For clarity, all other measures within 5° in position angle of the Tycho measure are omitted. Data points are connected with their predicted positions on the orbit by O–C lines. The broken line is the line of nodes and the axes of the figure are in arcsec. The direction of motion is indicated in the lower right corner. From Fabricius et al. (2002, Figure 5).

Tycho 2 doubles were omitted. Some statistical properties are illustrated in Figure 3.4.

Fabricius *et al.* (2002) re-analysed the Tycho 2 Catalogue, and resolved 17 000 more systems, including about 13 000 new discoveries. This was achieved partly by lowering the accepted separation threshold of 0.8 arcsec used in Tycho 2, at the risk of more spurious detections or larger errors in ρ, θ. They detected 13 251 visual double stars, mostly with separations between 0.3–1 arcsec, from the re-reduction. The new doubles were combined with 18 160 systems from the Washington Double Star Catalogue (WDS) identified in the Tycho 2 Catalogue, and 1220 new Tycho 2 doubles, to form the Tycho Double Star Catalogue, TDSC, a catalogue of absolute astrometry and B_T, V_T photometry for 66 219 components of 32 631 double and multiple star systems. The catalogue includes results for 32 263 single components for systems unresolved in TDSC, and a supplement gives Hipparcos and Tycho 1 data for 4777 additional components. The TDSC thus contains a total of 103 259 entries, with cross-identifications to WDS, HD, Hipparcos and Tycho 2.

For a typical double star several measurements spanning several years are required to identify non-linear, Keplerian, motion. Of the systems observed by Tycho, 13 800 of them have between 2 and 9 measures in addition to Tycho. Tycho also made measures of 6441 systems having between 10–50 historical observations, and the most suitable for orbit analysis. Of the (now) 84 486 WDS systems, only 1430 ($<$ 2%) have orbit determinations. A small number (1509) of the Tycho double star measures are of systems which are well characterised (i.e. number of measures greater than 50). These are useful in giving an independent assessment of the Tycho double star data quality, and an example is shown in Figure 3.5.

Tycho 2 colours for Hipparcos double stars The Hipparcos Double and Multiple Systems Annex gives two-colour Tycho (B_T, V_T) photometry for less than half the entries. This was due to the angular resolution of the Tycho 1 Catalogue (about 2.0 arcsec) being somewhat inferior to that of the Hipparcos main instrument (0.1 arcsec), such that the majority of close binaries were not resolved in Tycho 1, and only a photocentre solution in B_T, V_T was obtained. Additionally, some of the Hipparcos double stars were too faint for the Tycho instrument at around $V = 11.5$, while components with $\Delta m > 1.0-1.5$ mag could not be resolved in Tycho 1. The absence of two-colour photometry represented a shortcoming of the original dataset which was substantially improved by Fabricius & Makarov (2000b). They used the Identified Counts Data Base (ICDB) for more than 2.5 million stars, which was generated as part of the Tycho 2 Catalogue construction. The process of photon superposition for data from the whole mission provided improvements in limiting magnitude, astrometry, photometry, and angular resolution. As a result, they obtained two-colour photometry for Hipparcos components of 7547 double and 15 triple systems with angular separations 0.1–2.5 arcsec, of which they published 9473 components of 5173 systems with $\rho > 0.3$ arcsec, the majority without Tycho photometry in the original Hipparcos Double and Multiple Systems Annex. Verification of the resulting photometry using speckle observations has been discussed by Pluzhnik (2005).

3.4 Subsequent investigations of double and multiple stars

3.4.1 Improved solutions

The rest of this chapter describes investigations undertaken in the broad area of binary and multiple systems since the Hipparcos and Tycho Catalogues were published. Improved astrometric and orbital solutions have been published for a number of binaries, which are broadly considered according to solution type as they appear in the Hipparcos Catalogue. A rich area of research has been catalysed by the subset of apparently single stars or suspected binaries which have been revealed as $\Delta\mu$ binaries, in which their short-term Hipparcos proper motion has been found to differ from the long-term motions obtained in combination with earlier epoch ground-based observations. This section includes a treatment of the statistics and other investigations of different classes of binary, and a brief summary of studies of individual objects to which the Hipparcos data have contributed.

General considerations The statistics of the published catalogue (Table 3.1) give an indication of the number of entries which might benefit from a re-analysis on a case-by-case basis (Falin & Mignard, 1999). These include the 263 entries with no published astrometric solution (10 due to large errors in the Input Catalogue position, and the remaining 253 with simply inadequate solutions); the 2622 acceleration solutions (G), probably implying orbital periods above about 10 years; and the 1561 stochastic solutions (X). Two other indicators can and should be used to identify questionable solutions: Field H29 provides the percentage of data rejected to converge to an acceptable fit: in general this number is below 10% and a larger value is an indication that the solution should be used with caution. In addition, Field H30 quantifies the quality of the final fit, after removal of outliers, with values larger than 3–4 indicating a poor fit. Among the catalogue entries without a reliable solution, 240 appear as General Notes in the Hipparcos Catalogue for which a solution was derived after the catalogue proper was finalised.

Falin & Mignard (1999) undertook a re-analysis of the FAST consortium data for some of these entries, using additional information on multiplicity, and presented 139 new astrometric solutions. Of these, 13 new solutions were found for stars with no published solutions, and 15 were new solutions. Of the remainder, four had been flagged G (acceleration solutions) in the published catalogue, 45 had been flagged S (suspected double in Field 61), and 62 had been flagged X (stochastic solutions).

Fabricius & Makarov (2000a) presented improved astrometry for a further 257 Hipparcos entries resolved into 342 components, of which 64 systems had no published astrometry. They used the Hipparcos Transit Data files combined with results from Tycho 2 to provide better initial values for the astrometric solution. Many of their systems have separations in the range critical for the Hipparcos instrument of 13–20 arcsec, and they still advocate using the derived separations with caution.

Jorissen *et al.* (2004b) reported experiments with binary star detection using the Intermediate Astrometric Data for 163 barium stars, considered to be in binary systems from theoretical considerations. For $\pi > 5$ mas and $P < 4000$ d the binary detection rate is close to 100%, falling to 22% for the whole sample due to the presence of small parallaxes or very long periods.

A significant number of orbital binaries have also been reprocessed, based on the more recent availability of improved spectroscopic binary orbits, and these are detailed in Section 3.5.

The discovery of many more binary stars since the Hipparcos Catalogue publication, in principle means that new reductions, typically in combination with ground-based speckle data or other reliable orbital data, can be carried out, fitting the (new) model to the (original) Hipparcos observations. Re-discussion of resolved systems can make use of the published

Table 3.2 Revised astrometry, photometry or classification of Hipparcos double and multiple systems, following the Hipparcos Catalogue publication in 1997. This is a summary compilation of the associated work referred to elsewhere in this chapter; papers referring to individual objects, or only a few systems (e.g. speckle observations) are not included. The systems 'single stars: long-term motions' are those whose Hipparcos Catalogue proper motions should be considered as an inadequate representation of their long-term motions, as discussed in Section 1.11.

Reference	Hipparcos objects	Revised analysis	Systems affected	
ESA (1997)	Unreliable solutions	Notes in the catalogue	240	new solutions
Söderhjelm & Lindegren (1997)	Triple systems	Improved processing	4	improved solutions
Shatsky & Tokovinin (1998)	Orbital systems	Dynamical parallaxes	8	comparisons
Martin & Mignard (1998)	Orbital systems	Improved processing	46	improved orbits/masses
Martin *et al.* (1998)	Orbital systems	Improved processing	70	improved orbits/masses
Falin & Mignard (1999)	Unreliable solutions	Ground-based data	139	new binaries
Söderhjelm (1999)	Orbital systems	Ground-based data	205	improved orbits/masses
Arenou *et al.* (2000)	Spectroscopic binaries	Spectroscopic data	7	improved orbits/masses
Fabricius & Makarov (2000a)	Stochastic solutions	Tycho 2 data	257	new solutions
Pourbaix & Jorissen (2000)	Spectroscopic binaries	Orbital parameters	23	new orbits
Fabricius & Makarov (2000b)	Visual binaries	Revised Tycho 2 photometry	5173	new photometry
Wielen *et al.* (2001)	Single stars	Long-term motions	2827	possible binaries
Gontcharov & Kiyaeva (2002a)	Single stars	Long-term motions	11	6 new orbits
Pourbaix *et al.* (2003)	VIM solutions	Improved chromaticity	288	revised classifications
Griffin & Suchkov (2003)	Single F stars	Radial velocities yielding orbits	61	27 new binaries
Pourbaix & Boffin (2003)	Spectroscopic binaries	Improved processing	29	new orbits/masses
Gould & Chanamé (2004)	Single stars	Common proper motions: (r)NLTT	424	new companions
Jancart *et al.* (2005a)	Spectroscopic binaries	Ground-based orbits	70	improved/new fits
Balega *et al.* (2005)	New component binaries	Speckle interferometry	6	new orbits
Balega *et al.* (2006)	New component binaries	Speckle interferometry	6	new orbits
Goldin & Makarov (2006)	Stochastic solutions	Improved processing	65	new/improved orbits
Stefka & Vondrák (2006)	Single stars	Earth orientation (EOC-3)	4418	see Section 10.4.2
Gontcharov (2007)	Single stars	Long-term motions	120	new/improved orbits
Lépine & Bongiorno (2007)	Single stars	Common proper motions: LSPM	130	new companions
Frankowski *et al.* (2007)	Single stars	Long-term motions	1734	possible binaries
Goldin & Makarov (2007)	Stochastic solutions	Improved processing	81	new/improved orbits

Hipparcos Transit Data, while re-discussion of the astrometric/spectroscopic binaries can make use of the published Hipparcos Intermediate Astrometric Data (see Section 1.4.4). The latter data are available for all Hipparcos stars, while the 38 535 Transit Data files were only derived and published for known double and multiple systems, plus a few thousand suspected doubles.

Acceleration solutions Part G of the Double and Multiple Systems Annex comprises 2622 entries with 'acceleration solutions' in which a more complex model of seven or nine free parameters were demanded, corresponding to acceleration terms, $\dot{\mu}$, or second derivative of the proper motion, $\ddot{\mu}$. The number of such objects was reasonably well modelled in the study by Quist & Lindegren (2000).

Makarov & Kaplan (2005) made an evaluation of the $\Delta\mu$ binaries resulting from a comparison of the Hipparcos and Tycho 2 proper motions, and the Hipparcos Catalogue DMSA Part G binaries, with the specific aim of estimating statistical bounds for the masses of the secondary components. For 1929 $\Delta\mu$ binaries in which $\Delta\mu > 3.5\sigma$ in at least one of the coordinate components they derive, under certain assumptions,

$$\Delta\mu \leq \frac{2\pi\,\omega R_0 M_2}{M_{\rm tot}^{2/3} P^{1/3}} \tag{3.2}$$

where ω is the parallax, P the period, $M_{\rm tot}$ the total mass, and M_2 the secondary mass; the equality applies for face-on orbits. Thus, given a period estimate, a lower estimate of the secondary mass can be derived. For the acceleration solutions, they derive

$$\dot{\mu} \leq \frac{(2\pi R_1)^2 \omega M_2}{M_{\rm tot}^{2/3} P^{4/3}} \tag{3.3}$$

R_0 and R_1 are functions of the orbital eccentricity, e, and eccentric anomaly, E. From the distribution of the various solutions as a function of spectral type (Figure 3.6), they found a significant difference between the distribution of spectral types of stars with large accelerations but small proper motion differences (i.e. short periods, 1854 stars) and that of stars with large proper motion differences but insignificant accelerations (i.e. long periods $\sim$ 6–100 yr, 1161 stars). The spectral type distribution for the former sample of binaries is the same as the general distribution of all stars in the Hipparcos

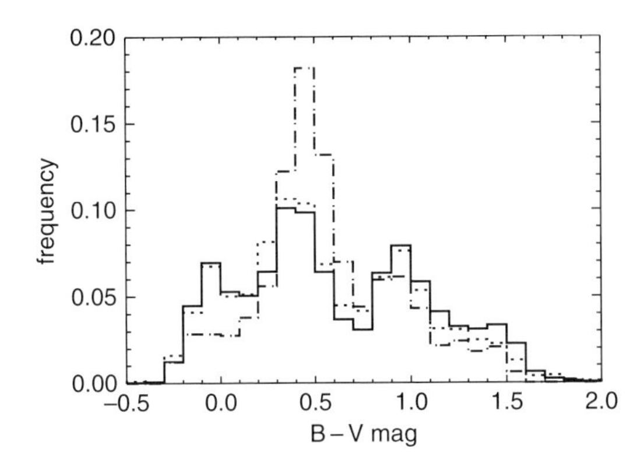

Figure 3.6 Distributions of $B-V$ colours for all 118 218 Hipparcos stars (solid line), astrometric binaries with significant accelerations but insignificant $\Delta\mu$ (dashed line; 1854 stars); and with significant $\Delta\mu$ differences but insignificant accelerations (dot-dashed line; 1161 stars). The latter sample of longer-period systems has a different spectral type distribution to that of the catalogue as a whole, perhaps indicating a difference in binary formation mechanism of the different populations. From Makarov & Kaplan (2005, Figure 2).

Catalogue, whereas the latter sample is dominated by solar-type stars ($B-V \sim 0.5$, late F and G), with an obvious dearth of blue (B and A) stars. They suggest that this difference could arise from different binary formation mechanisms, such as fragmentation versus capture, for the different populations.

Makarov (2007) extended the study to investigate statistical correlations of binarity with other basic stellar properties, e.g. mass, age, chromospheric and coronal activity, and spectral peculiarities, for various stellar categories. For nearby ($d < 100$ pc) stars that are likely to be younger than the Pleiades (~100 Myr), many of which appear to be members of sparse loosely co-moving groups, the physical conditions of the original star forming regions are imprinted in the binary statistics and the initial mass function. For the nearest late K and M dwarfs and cool white dwarfs, whose dim companions can be substellar objects or very late M dwarfs, the study probes the existence of the brown dwarf desert in the domain of long periods and low primary mass. Dwarfs and giant stars at large distances ($d > 1$ kpc) with large acceleration terms in the Hipparcos Catalogue suggest the existence of systems with massive, but dim companions. Stars with masses above $8M_\odot$ may constitute a few percent of all stars in rich open clusters, and owing to their short lifetimes, neutron stars and low-mass black holes may be fairly common in binary systems. They propose that distant accelerating binaries may contain long-period counterparts to low-mass X-ray binaries, with orbital periods of a few hours to several days.

Variability-induced movers The detection and analysis of objects classified as 'variability-induced movers' is described by Wielen (1996), and the principles are illustrated schematically in Figure 3.3, right. After publication of the Hipparcos Catalogue, which included 288 VIMs, the VIM model has been confirmed in some cases (e.g. Bertout *et al.*, 1999), although generally rather few Hipparcos VIMs have been resolved by speckle interferometry (Mason *et al.*, 1999b, 2001a). Accordingly, a re-analysis of the VIM solutions was made by Pourbaix *et al.* (2003), using the entire set of Intermediate Astrometric Data, instead of only the FAST Consortium data as used in the initial VIM analysis (see also Detournay & Pourbaix, 2002). This resulted in a decrease of 21% of the number of VIM solutions. A second step of the re-analysis involved re-processing of the reddest VIMs for which the chromaticity corrections were suspect due to the use of an inappropriate $V-I$ colour index during the initial processing (see Section 4.10). The effect is important for these objects due to their extreme red colour, their intrinsic variability on time scales comparable to the mission duration, and their time-dependent colour variations. Using the epoch-dependent colours based on Tycho 2 and Hipparcos epoch photometry as developed by Platais *et al.* (2003), they found that 89 of the 188 long-period variables flagged as VIMs in the Hipparcos Catalogue are not confirmed as VIMs. Introducing a further test based on the resulting parallax led to a rejection of 161 of the 188 long-period variables classified as VIMs in the Hipparcos Catalogue. A more detailed investigation of the resulting period–luminosity relation for long-period variables was presented by Knapp *et al.* (2003, and discussed further in Section 4.10). This more careful analysis of the reddest VIM candidates explains why many of the original Hipparcos VIM solutions were not confirmed as binaries from speckle observations.

Stochastic solutions Of the 1561 stochastic solutions, successful attempts have been made to use more accurate information from ground based and Tycho 2 data sets to reprocess the published Hipparcos Transit Data files, resulting in accurate solutions for a few hundred stars (Falin & Mignard, 1999; Fabricius & Makarov, 2000a). The remaining stars with stochastic solutions are prime suspects for yet unknown binaries.

Other analyses of objects originally published as stochastic solutions are considered in subsequent sections, and specifically within the section describing new and revised orbits for orbital binaries (Section 3.5).

3.4.2 Single stars showing evidence for binarity: The $\Delta\mu$ binaries

Although Hipparcos has provided a detailed census of binary systems to well-defined and rather stringent $\rho, \Delta m$ detection limits, unidentified binaries may still dominate the field star population, and many Hipparcos stars flagged as single may well be binaries. Systems with large Δm and periods of several decades are hard to investigate by spectroscopic, interferometric and visual methods. As described in Section 1.11 one way of detecting such objects is by comparing their Hipparcos proper motion, determined over the mission duration of about 3.3 years, with long-term proper motions determined by combining the Hipparcos results with, for example, FK5, or the Astrographic Catalogue, resulting in the so-called $\Delta\mu$ binaries. Section 1.11 summarises the implications of these studies with emphasis on the values of the Hipparcos Catalogue proper motions, while in this section the emphasis is on the consequences for binary system studies.

Wielen *et al.* (1999) show that the current catalogue accuracies permit the detection of (a) medium-period binaries with $P \sim 30$ years; (b) some nearby, long-period binaries with $P \sim 1000$ years; and (c) in some cases, short-period binaries with $P \sim 1$–3 years. Converting the proper motion difference into a (transverse) velocity difference in km s^{-1}, $\Delta v = 4.74\Delta\mu/\pi$, where $\Delta\mu$ is in mas yr^{-1} and π in mas, they derive a corresponding measurement precision of 0.043 km s^{-1} at $d = 10$ pc and 0.43 km s^{-1} at $d = 100$ pc for 847 FK5 stars, and a factor 2 lower for 11 773 stars from the General Catalogue. They give examples of confirmed single-star candidates in the case of 47 UMa (suggesting that no massive companion to this exo-planetary system exists) and δ Pav, as well as detected $\Delta\mu$ binaries γ UMa, ϵ Eri, and ι Vir. For detected $\Delta\mu$ binaries, statistical estimates of $(\rho, \Delta m)$ can be extracted from the inferred orbital period, estimates which would be significantly improved if accurate radial velocity estimates were available to provide the acceleration component. Kaplan & Makarov (2003) estimate orbit dimensions and distances at which low-mass companions and planets may be detected around main-sequence stars by SIM and Gaia using this approach.

In a study based on a sample of about 12 000 stars with $V = 7$–10 in two declination zones in the northern and equatorial sky using the Astrographic Catalogue, Odenkirchen & Brosche (1999) showed that the proper motion deviations provide statistical evidence for 360 astrometric binaries in the investigated zones, corresponding to about 2400 such binaries in the entire Hipparcos Catalogue, in addition to those already known. Their Monte Carlo simulations of orbital motion yield an acceptable approximation to the observations if a binary frequency between 70 and 100% is assumed, i.e. if most of the stars in the sample are assumed to have a companion. Thus the frequency of Hipparcos astrometric binaries appear to confirm that the frequency of non-single stars among field stars is very high.

As described in Section 1.11, Gontcharov *et al.* (2000) considered 4638 stars in common to Hipparcos and the FK5/FK5 Extension, using 45 observational ground-based catalogues. They derived long-term proper motions, summarised in their PMFS Catalogue (Proper Motions of Fundamental Stars), and identified a few hundred stars which showed significant nonlinear motion of the photocentres, the method being sensitive to amplitudes larger than 0.15 arcsec and periods in the range 10–100 yr. Good agreement with previously-known orbits was reported, and results for five astrometric binaries are discussed in detail. The analysis has been extended by Gontcharov & Kiyaeva (2002a,b), and further updated to also take account of radial velocity variations (Gontcharov, 2007) yielding some 100 updated orbits; it is discussed further in Section 3.5. Some examples of the resulting binary orbits are given in Figure 3.7.

Makarov & Kaplan (2005) made an evaluation of the $\Delta\mu$ binaries resulting from a comparison of the Hipparcos and Tycho 2 proper motions, and the Hipparcos Catalogue DMSA Part G binaries, with the specific aim of estimating statistical bounds for the masses of the secondary components. Their results are described further under 'Acceleration solutions' in Section 3.4.1.

Frankowski *et al.* (2007) made a further analysis in parallel with the work of Makarov & Kaplan (2005). They used a χ^2 test evaluating the statistical difference between the Hipparcos and Tycho 2 proper motions for 103 134 stars in common between them. The work focuses on the detection efficiency of proper motion binaries, using different kinds of control data, mostly radial velocities. The detection rate was evaluated for various star samples: for entries from the Catalogue of Spectroscopic Binary Orbits (9th edition, SB9, Pourbaix *et al.*, 2004b), for barium stars listed by Lu *et al.* (1983) which are from theoretical considerations all considered to be binaries, for spectroscopic binaries identified from radial velocity data in the Geneva–Copenhagen survey of F and G dwarfs in the solar neighbourhood from Nordström *et al.* (2004), and for stars with established radial velocity standard deviations available from the Famaey *et al.* (2005) catalogue of K and M giants in the solar neighbourhood (Figure 3.8). They show that proper motion binaries are detected efficiently for systems with $\pi \gtrsim 20$ mas, and $P \sim 1000$–30 000 d. The shortest periods in this range (1000–2000 d, i.e. one or two times the duration of the Hipparcos mission) may appear only as DMSA Part G binaries. Proper motion binaries detected among systems in the Catalogue of Spectroscopic Binary Orbits having periods shorter than about 400 d may be triple systems, in which the proper motion binary involves a component with a longer orbital period (a list of 19 candidate triple systems is given, as are binaries suspected of having low-mass, brown dwarf-like companions). Among the 37 barium stars with $\pi > 5$ mas, only seven exhibit no (spectroscopic or astrometric) evidence for binarity. Once

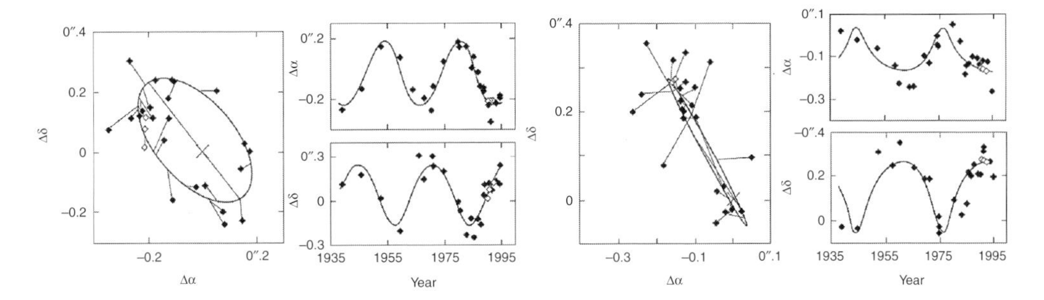

Figure 3.7 Examples of binary systems identified from the long-term proper motion analysis of Gontcharov et al. (2000), Gontcharov & Kiyaeva (2002a,b), and Gontcharov (2007). Figures show the reconstructed orbits, alongside the variations with time of the right ascension and declination differences in mas between the various individual catalogue positions and the PMFS Catalogue. Left: HIP 11072 (κ For), discovered by Hipparcos as an astrometric binary with an acceleration solution (G). The most probable period is about 26 yr. The ground-based parallax is about 0.08 arcsec, while Hipparcos gives 0. 047 ± 0. 001 arcsec. Their orbital solution implies a white dwarf secondary (or, alternatively, a red dwarf secondary if the ground-based parallax is correct). Right: HIP 45699 (83 Cnc), with a probable period of about 32 yr, and identified in FK6 as a highly-probable astrometric binary. The secondary is also probably a white dwarf. From Gontcharov & Kiyaeva (2002a, Figures 1 and 4), reproduced with kind permission of Springer Science and Business Media.

account is taken for the detection biases, the fraction of proper motion binaries shows no significant variation among the various (regular) spectral classes.

Whether the presence of 'undetected' binaries is problematic or not depends, of course, on the application. One sensitive area explored by Suchkov & McMaster (1999) relates to F stars which display strong evolutionary effects within the age range spanning those of the old halo through to young disk stars. Inferences based on star formation history or metal-enrichment in the disk rely on accurate age estimates, which are derived from isochrone fitting. Ages will be unreliable if absolute magnitudes are erroneously attributed to unidentified binaries, especially those with comparably bright components (see Figure 3.20, right). They proposed a criterion that identifies such binaries, by comparing M_V obtained from the Hipparcos parallax and Johnson V magnitude (representing the integrated flux of the binary) with the absolute magnitude derived from the dereddened Strömgren *uvby* luminosity index c_0, M_{c_0}, from the photometric compilation of Hauck & Mermilliod (1998). Their results for about 10 000 Hipparcos F stars indeed reveal the presence of a significant number of unrecognised binaries with comparably bright components (Figure 3.9). They also show that these specific binary candidates are on average older than the single stars.

A radial velocity survey of a sample of these bright ($V < 9$) 'overluminous' F stars within 80 pc was carried out by Griffin & Suchkov (2003), specifically with the aim of verifying their binary nature. Of the 111 stars selected and flagged as single by Hipparcos, 25 new binaries were discovered. Along with previously known binaries, this leads to a binary fraction of 58% for this sample, double the 29% found in the randomly selected F star sample of Nordström *et al.* (1997), and confirming the validity of the luminosity-based identification of unidentified binaries. The unconfirmed binaries are possibly of even greater interest: they show no spectroscopic signature of binarity and thus appear to be single, although their luminosity is significantly higher than normal single stars of the same surface gravity and effective temperature. Their 'overluminosity' remains an unexplained phenomenon.

Consequences for various other investigations of F stars have been reported: for example for studies of ages of single stars and various classes of binaries by Suchkov (2000); for the age–velocity relation implying prolonged main-sequence evolution for close binaries by Suchkov (2001), and for F stars with extra-solar planets by Suchkov & Schultz (2001).

3.4.3 Statistical properties

Period and eccentricity The orbital period distribution is typically rather uniform in $\log P$ over several orders of magnitude. The power law form $f(P) \sim P^{-1}$ is sometimes referred to as Öpik's relation (Öpik, 1923); an equivalent form uses the orbital separation of the components $f(s) \sim s^{-1}$.

Quist & Lindegren (2000), also reported in Quist & Lindegren (2001) and based on studies by Quist (2000), modelled the observation of double stars to examine the statistics of double star solutions in the catalogue and to set quantitative limits on binary distributions. The possibility of detecting departure from linear proper motion by a few mas opens new opportunities to study binaries in the separation range 1–10 AU, corresponding to 10–100 mas at 100 pc, and with periods up to a

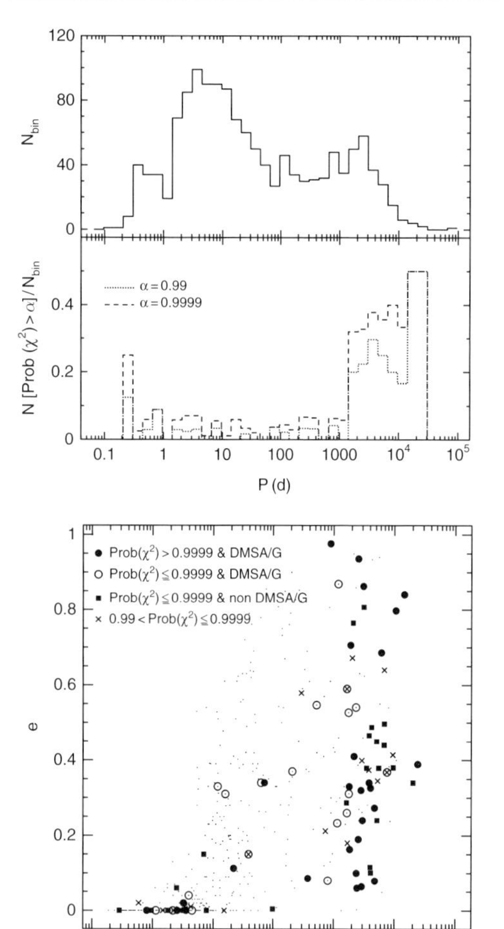

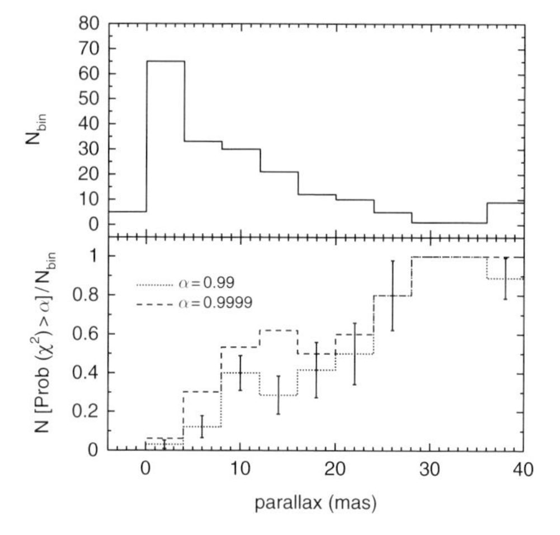

Figure 3.8 Left upper: orbital periods from the Catalogue of Spectroscopic Binary Orbits (SB9); lower: fraction for which the Hipparcos and Tycho 2 proper motions differ at confidence levels of 0.9999 (dotted) and 0.99 (dashed). Below left: eccentricity versus period for π > 10 mas. •: $\Delta\mu$ binaries detected at 0.9999 confidence and flagged as 'G'; ■ those not flagged as 'G'; ∘: systems flagged as 'G' but not fulfilling the 0.9999 confidence level; ×: not flagged as 'G' in the confidence range 0.99–0.9999. Below right, upper panel: parallaxes for P_{orb} > 1500 d. Lower panel: fraction detected as $\Delta\mu$ binaries at confidence levels of 0.99 and 0.9999. For 10 < π < 25 mas, about 50% with P_{orb} > 1500 d are detected, with more than 80% above 25 mas. From Frankowski et al. (2007, Figures 3–5).

few decades. By combining a Galaxy model (an update of that described by Holmberg *et al.*, 1997) with realistic models of the distributions of mass ratio and eccentricity, and thereafter simulating the observation process, they derived the expected numbers of the different kinds of binary solutions in the catalogue. They found that the observed number of component (C) solutions among bright (V < 7–8) main-sequence stars is not consistent with a model having the same binary frequency and distributions as found by Duquennoy & Mayor (1991) for solar-type field stars (Figure 3.10). Agreement would demand an increased multiplicity in the range a = 1–100 AU by roughly a factor 2. The number of companions per primary is estimated to be 25 ± 5% for semi-major axes from 1–10 AU, and some 43 ± 20% for the range 10–100 AU, leading to a total multiplicity of 0.9–1.2 extrapolated over all values of a. Periods from 3–30 yr often result in G solutions (filled circles in Figure 3.11) for which excellent agreement is found between the observed and calculated distributions for second- and third-order terms, while periods from 30–10^4 yr may result in undetected but significant proper motion errors (the $\Delta\mu$ binaries), of up to a few times 10 mas yr^{-1}. The fitted model predicts about three times as many orbital solutions as actually presented in the catalogue, which in turn suggests that many more orbits could be determined by analysis of the Hipparcos Intermediate Astrometric Data.

Halbwachs *et al.* (2003), discussed also in Halbwachs *et al.* (2000) and Halbwachs *et al.* (2004), combined radial velocities from Coravel (in the solar neighbourhood and in the Pleiades and Praesepe clusters) with the Hipparcos data to derive the statistical properties of main-sequence binaries with spectral types F7 to K and with periods up to 10 years. A sample of 89 spectroscopic orbits included 52 representing an unbiased selection of 405 stars (240 field stars and 165 cluster

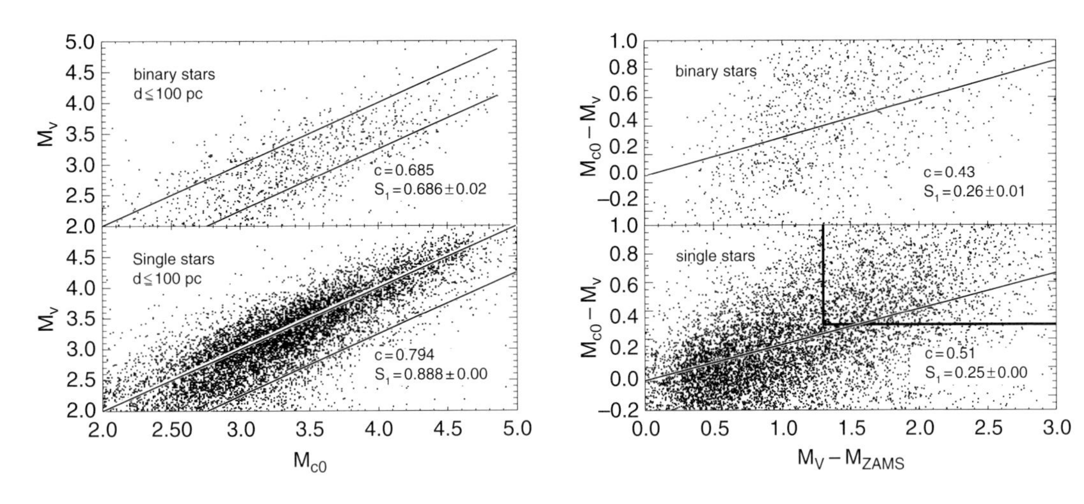

Figure 3.9 Left: M_V versus M_{c_0} for binary (top) and single (bottom) F stars. The two parallel lines are $M_V = M_{c_0}$ and $M_V = M_{c_0} - 2.5\log 2$. Most binary stars occupy the area between these lines (top), while single stars concentrate around the line $M_V = M_{c_0}$. Right: $\Delta M = M_{c_0} - M_V$ versus $(M_V - M_{\rm ZAMS})$ for known unresolved binaries (top) and single stars (bottom). The non-zero slope of the linear regression for binary stars means that the brighter the star is with respect to the ZAMS, the higher the probability that its components are of comparable brightness. For single stars, the similar non-zero slope provides evidence for the presence of unidentified binaries. Stars in the upper right corner are binary candidates with comparable components lying more than 1.3 mag above the ZAMS. From Suchkov & McMaster (1999, Figures 2 and 3).

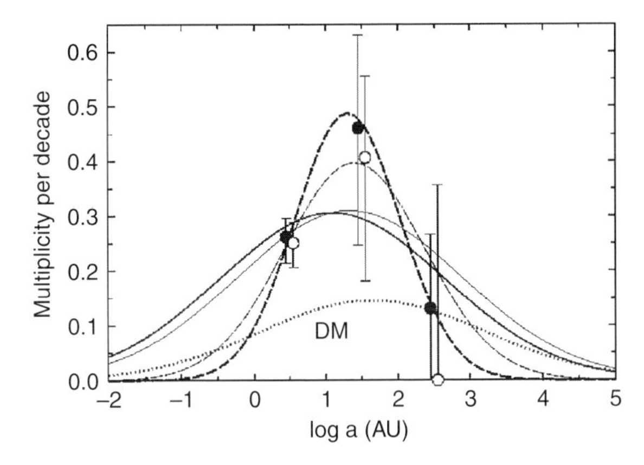

Figure 3.10 Multiplicity (companions per primary star) versus modelled semi-major axes a. Dotted curve (DM): distribution for nearby solar-type stars (Duquennoy & Mayor, 1991). Solid curves: log-normal distributions fitted to the observed counts, but constrained to the same width as the DM curve; dashed curves: without this constraint; circles: multiplicities determined by direct inversion. All assume a decreasing mass ratio distribution for MS+MS binaries, and two extreme values for the relative frequency of MS+WD binaries (thick curves and filled circles, and thin curves and open circles). From Quist & Lindegren (2000, Figure 2).

stars) based on their Hipparcos distances. Similar distributions were found for field and open cluster binaries for mass ratios, periods, the period–eccentricity relation, and binary frequencies. The distribution of mass ratios shows two maxima: a broad peak from $q \sim 0.2$–0.7, and a sharp peak for $q > 0.8$ (twins). Both are present over the full range of spectral type studied, indicating a scale-free formation process. The peak for $q > 0.8$ gradually decreases for long-period binaries. Independently of period, twins have eccentricities significantly lower than for other binaries, confirming a difference in formation processes, and they postulate twins-formation *in situ*, followed by accretion from a gaseous envelope, whereas binaries with intermediate mass ratios could be formed at wide separations, subsequently migrating closer through interactions with a circumbinary disk. The frequency of binaries with $P <$ 10 years in the solar neighbourhood is about 14%. About 0.3% of binaries are expected to appear as false positives in a planet search. Therefore, the frequency of planetary systems among stars is presently 7^{+4}_{-2}%. The extension of the distribution of mass ratios in the planetary range would result in a very sharp and very high peak, well separated from the binary stars with low mass ratios.

Bartkevičius & Gudas (2001, 2002b) used a sample of 804 Hipparcos visual binaries with known orbits, either from Hipparcos or from the catalogue of Hartkopf *et al.* (2001a). For the 59% with known radial velocities, they derived Galactic velocity components, assigning 92% of the binary sample to the thin disk, 7.6% to the thick disk, with only two objects (revised to seven in Bartkevičius & Gudas, 2002a) having halo kinematics. Acquisition of some of the missing radial velocity data has been reported subsequently, e.g. Bartkevičius & Sperauskas (2005).

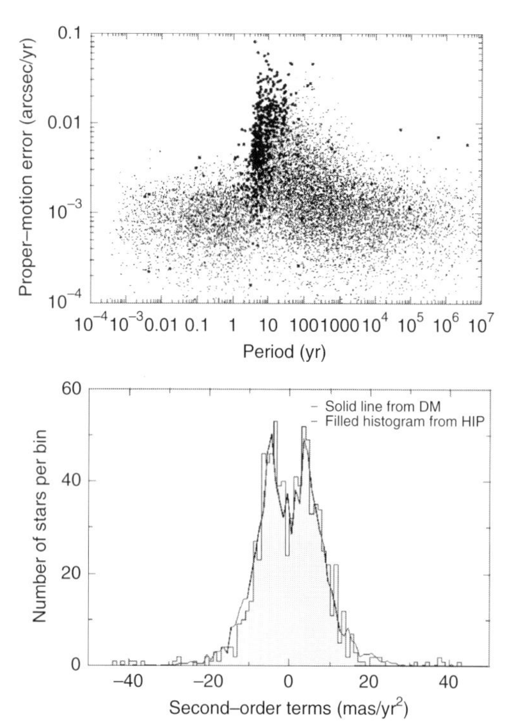

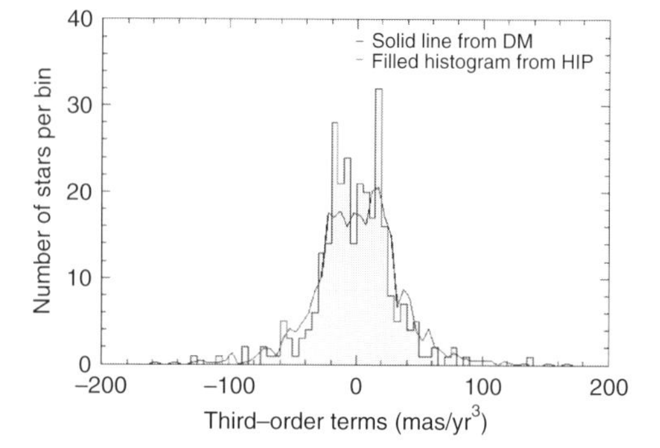

Figure 3.11 Left: proper motion errors in the synthetic catalogues versus orbital period, indicating stars with large proper motion errors due to orbital motion. • acceleration (G) solutions; most are obtained for binaries with $P \sim 3$–30 yr. Below left: quadratic acceleration solutions from Hipparcos (histogram) compared with the model distribution (curve). Below right: cubic acceleration solutions from Hipparcos (histogram) compared with the model distribution (curve). From Quist & Lindegren (2000, Figures 3–5).

Dommanget (2003) used Hipparcos trigonometric parallaxes in place of dynamical parallaxes to confirm the mass–eccentricity relation for binary systems, first noted by Doberck (1878) for visual double stars, and subsequently extended to shorter periods from spectroscopic orbits by Campbell (1910) and others.

Fisher *et al.* (2005), see also Fisher *et al.* (2004), selected a sample of spectroscopic binaries in the local solar neighbourhood ($d \leq 100$ pc and $M_V \leq 4$) to study the distributions of period, P, primary mass, m_1, and mass ratio $q(= m_2/m_1)$, as well as the initial mass function of the local population of field binaries. The sample was collated using available spectroscopic binary data and the Hipparcos Catalogue, the latter being used for distances and to refer numbers of objects to fractions of the local stellar population as a whole. They used the better-determined double-lined SB2 binaries to calibrate a Monte Carlo approach to modelling the q distribution of the single-lined SB1 binaries from their mass functions, $f(m)$, and primary masses, m_1. The total q distribution was determined by adding the observed SB2 distribution to the Monte Carlo SB1 distribution. Their results show a clear peak in the mass ratio distribution of field binaries near $q = 1$, dominated by the SB2s, but the flat distribution of the SB1s is inconsistent with their components being chosen independently at random from a steep initial mass function.

Jorissen *et al.* (2004a) searched for spectroscopic binaries from a complete sample of Hipparcos M giants, as part of their Coravel radial velocity programme. They found a spectroscopic binary frequency of about 8% for field M giants, and constructed a period–eccentricity diagram (their Figure 2), suggesting that the region occupied by spectroscopic binaries moves to larger P and smaller e moving along the sequence KIII–MIII–Ba giants (Figure 3.12). This sequence involves stars with increasingly larger orbital radii where circularisation processes are presumably operating over progressively longer time scales.

Various other discussions of the properties and statistics of binary stars in the Hipparcos Catalogue have been reported, including Arenou (1998), Urban *et al.* (2000), Söderhjelm (2001), and Arenou (2001). Raghavan *et al.* (2007) reported the start of a multiplicity survey of 454 'solar-type' stars in the solar neighbourhood ($d < 25$ pc). Soydugan *et al.* (2006) constructed a catalogue of 25 known and 197 candidate eclipsing binary systems (detached and semi-detached) where at least one component is located in the δ Scuti region of the instability strip.

Close Tycho binaries Fabricius & Makarov (2000b) used the large numbers of stars with two-colour photometry from Tycho 2 to study 'close' binary stars, especially those with separations below 1 arcsec (strictly, these are 'close' for Tycho, but not in the sense used

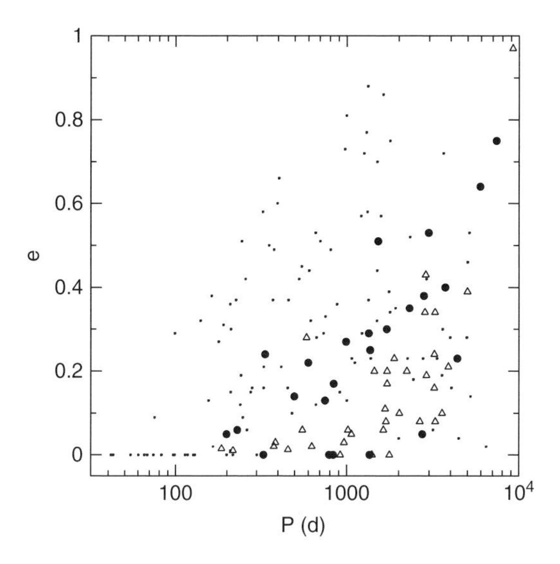

Figure 3.12 The eccentricity–period diagram for spectroscopic binaries detected from radial velocity observations of field giants selected from the Hipparcos Catalogue. Symbols show systems containing K giants (small dots), M giants (large black dots), and strong Ba stars (open triangles). The occupied region of $P-e$ space moves to the lower right of the diagram along this sequence. From Jorissen et al. (2004a, Figure 2).

more generally in double star research, where a 'close binary' typically refers to interacting systems with periods less than a few years). Photometrically-resolved components of the binaries with statistically significant trigonometric parallaxes could be placed in the HR diagram, the majority for the first time. Their results are shown in Figure 3.13, separately for the primary and secondary components. Most of the primaries belong to the main sequence, ranging from OB stars to M dwarfs. The subgiant branch, with a clear gap from the main sequence, and the red giant clump, are both well populated, extending in a broad plume to redder colours and brighter absolute magnitudes. There are two conspicuous M supergiants, and a few subdwarfs lying below the rather clear lower bound of the ZAMS. Amongst the secondary components, giants are relatively rare, as expected. An unexpected feature is the dozen or so components bluer and fainter than the main sequence. Their position indicates that they are hot subdwarfs, more likely cooler hydrogen-rich sdB stars. Two of these were already reported from this reprocessing as candidate hot subdwarfs in visual binaries (Makarov & Fabricius, 1999), with a realistic possibility to determine their masses from the orbital motion: with an angular separation of around 0.35 arcsec, their periods are estimated to be of the order of 100 years. Only two other such objects were previously known, including HD 113001B (Heber, 1992) which is also confirmed in the data.

Wide binaries There are different working definitions of what defines a wide binary, for example those with angular separations larger than 10 arcsec. Allen *et al.* (2000a) adopt a distance-independent definition as those with linear semi-major axis $a > 25$ AU. Such long-period systems appear to result from a different formation process from that of closer binaries: e.g. Abt & Levy (1976) concluded that long-period systems represent pairs of protostars which contracted separately with the secondary mass following Salpeter's law, while short-period binaries are fission systems with a tendency to equal component masses. Although the topic remains controversial, Poveda & Allen (2004) found confirmation of different formation mechanisms, with the wide binaries with $a > 40$–80 AU following Öpik's relation while the closer binaries do not.

Wide systems have small binding energies, which in turn means that they are good tracers of mass concentrations encountered through their Galactic trajectories, including individual stars, giant molecular clouds, and dark matter. This application has prompted numerous observational (e.g. Weinberg & Wasserman, 1988; Close *et al.*, 1990; Wasserman & Weinberg, 1991; Chanamé & Gould, 2004) and theoretical (e.g. Bahcall *et al.*, 1985; Wasserman & Weinberg, 1987; Poveda *et al.*, 1997) investigations, with one major obstruction being the small number of systems for which distances and space motions were known. Wide binaries are typically identified based on probability arguments, although similarity of their parallaxes, radial velocities, or proper motions provides efficient confirmation of physical association. For wide binaries, the identification of common proper motion pairs has been a valuable search technique (see Section 2.11.2).

Palasi (2000) searched for wide binaries in the Hipparcos and Tycho 2 Catalogues. F7–G9 dwarfs and subgiants were selected from the Hipparcos Catalogue according to well-defined criteria out to 50 pc, yielding 1267 stars, of which 134 have a two-component solution. They then searched for companions in both Hipparcos and Tycho 2 down to the completeness limit of $V = 11$, adopting selection criteria to avoid retaining excessive numbers of chance projections (i.e. optical, or unbound, systems). Finally 52 wide pairs were selected, of which 12 had a wide companion of the same type already in the primary sample, and six are three-component hierarchical systems. With completeness estimated above separations of about 1500 AU, they show a good fit to Öpik's relation for separations between 2000 and 15 000 AU and a sharp cut-off, with only two systems, beyond this. The resulting 5% of stars in wide systems compares to the 3% obtained in the solar neighbourhood by Close *et al.* (1990).

Ling *et al.* (2004) constructed volume-limited samples of wide Hipparcos binaries out to 100 pc and 200 pc,

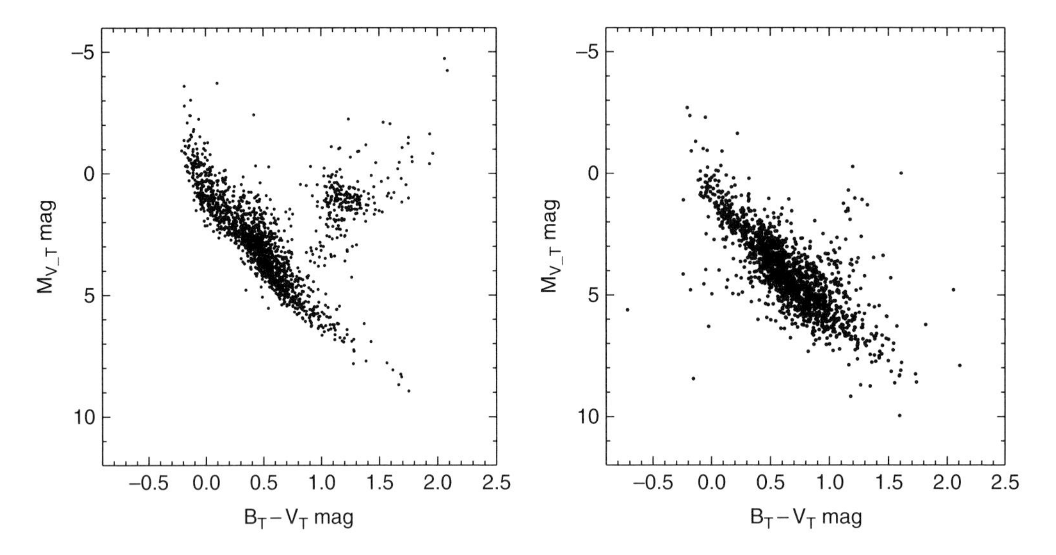

Figure 3.13 Observational HR diagrams for 1697 primary components (left) and 1699 secondary components of double and triple systems (right) with statistically significant Hipparcos parallaxes ($\sigma_\pi/\pi < 0.25$) and angular separations 0.30–2.5 arcsec. The sample is restricted to systems with max$[\sigma_{B_T}, \sigma_{V_T}] < 0.1$ mag. From Fabricius & Makarov (2000b, Figure 2).

and showed that the distribition of linear separations (using the Hipparcos parallaxes) closely follows Öpik's relation for separations $s = 10$–800 AU for the 100 pc sample, and $s = 15$–1400 AU for the 200 pc sample, rather than a Gaussian distribution (e.g. Poveda *et al.*, 2000).

Söderhjelm (1997, 2000, 2007) studied the statistics of several thousand Hipparcos visual binaries to derive the mass ratio distribution. In the first study, he used relatively small but complete samples of A–F stars, while in subsequent work, he compared the observations with a model Galaxy filtered by the instrument and observation process. Using the fact that luminosity ratios are a good indication of the mass ratios, and notwithstanding the classical Öpik (1923) bias creating an over-abundance of small-Δm systems in a magnitude-limited sample, he found a mass ratio ($q = M_2/M_1$) with a sharp peak near $q = 1$, implying an excess of very wide (50–1000 AU) binaries of equal component masses. Such systems, referred to as twins, have previously been found only at much shorter periods (e.g. Halbwachs *et al.*, 2003), where they exist with significantly lower eccentricities than is typical in other binaries. Although many of the observed characteristics of the mass distribution are reproduced by recent binary formation and evolution models (e.g. Bate, 2000; Valtonen, 2004; Bonnell & Bate, 2005; Hubber & Whitworth, 2005), the presence of a $q = 1$ peak for wide binaries would point to a specific formation process (e.g. Tokovinin, 2000), whose details however remain unclear. The $q = 1$ peak is not seen, for example, in the sample of 51 spectroscopic binaries observed in infrared spectroscopy by Mazeh *et al.* (2003).

Allen *et al.* (2000a), see also Allen *et al.* (1998) and Allen *et al.* (1999), studied the properties of wide binaries among high-velocity and metal-poor stars representative of the old disk and halo. Their occurrence as a function of kinematics or age has been a subject of investigation since Oort (1926) pointed out the paucity of visual binaries amongst high-velocity stars. More recent work has suggested that the fraction of high velocity stars that are close binaries may be around 20–30% (Carney, 1983; Stryker *et al.*, 1985), significantly lower than that for ordinary disk stars, although even the very existence of such wide systems poses interesting dynamical problems. Allen *et al.* (2000a) compiled a list of 122 wide binaries, by searching for common proper-motion companions to the more than 1200 high-velocity and metal-poor stars of Schuster *et al.* (1993). Hipparcos distances were used to convert angular into linear separations. They found 11 systems with projected separations above 10 000 AU, with expectation values for the semi-major axes estimated by means of the statistical relation (Couteau, 1960)

$$E(\log a) - E(\log s) = 0.146 \tag{3.4}$$

Furthermore, the secondaries provide a sampling of the faint end of the main sequence of old disk and halo stars. They found that the separations for the wide binaries ($\langle a \rangle > 25$ AU) follow Öpik's relation up to 10 000 AU. A subgroup with the most halo-like orbits follow Öpik's relation up to 20 000 AU. They compare these results with those found by Nigoche (2000) for wide binaries in the Orion Nebula cluster and in the Hyades with ages of a few million years and $\sim$ 625 Myr respectively, which

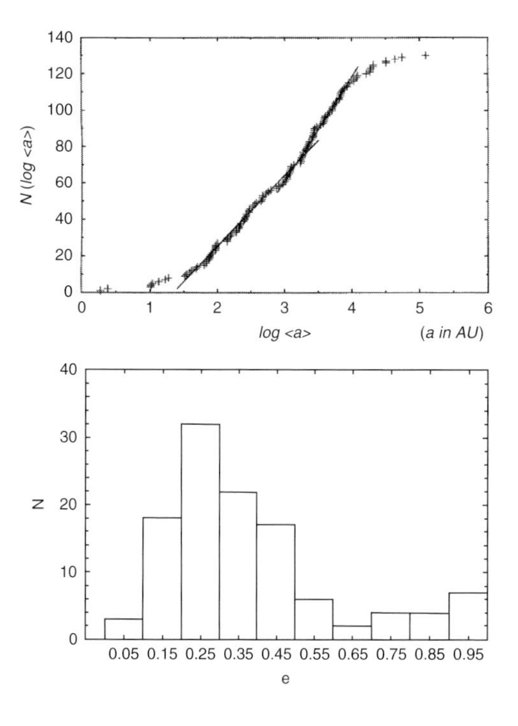

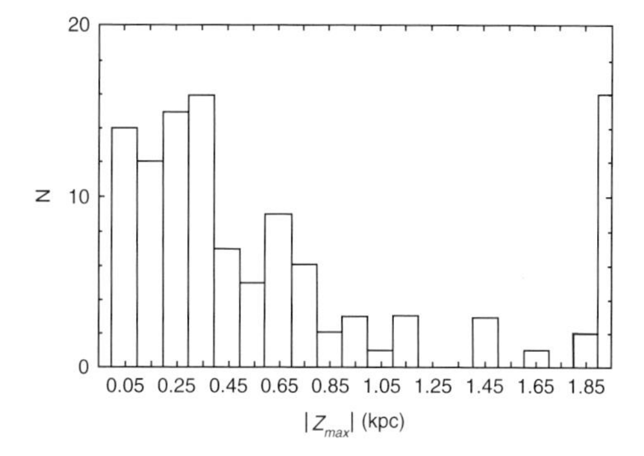

Figure 3.14 Left: cumulative distribution of the semi-major axes of the wide binaries studied by Allen et al. (2000a). Öpik's relation is represented by the straight line, with the change in slope at around 1000 AU attributed to incompleteness in the proper motion catalogues especially for large Δm. Below left: eccentricities of their Galactic orbits. Below right: distribution of $|z_{\rm max}|$, the maximum achieved distance from the Galactic plane. From Allen et al. (2000a, Figures 7–9).

obey Öpik's relation out to 30 000 AU and 18 000 AU respectively.

Valtonen (1997) has argued that Öpik's relation arises naturally for binaries formed by dynamical interaction with the disintegration of small star clusters. The departure from the power-law distribution at some upper limit of separation is then attributed to effects of dynamical interactions which tend to decrease the binding energy and hence increase the separation until its eventual dissolution. The departure sets in at much larger separations than that observed for old binaries in the solar vicinity, at around 2400 AU (Poveda *et al.*, 1997); the difference is perhaps attributable to the small time spent by these halo binaries within the Galactic thin disk where the major dynamical effects are likely to occur, combined with their large relative velocities during such encounters.

Radial velocities combined with the Hipparcos proper motions means that their Galactic orbits could be calculated, and they confirmed in the process that the majority of systems have peculiar velocities in excess of 60 km s^{-1}. Additional information on metallicity means that each could be further assigned to the old thin disk, the thick disk, or the halo population. Galactic orbits were calculated using the potential of Allen & Santillán (1993), and various results are shown in Figure 3.14. Out of 115 orbits, 48 systems have Galactic orbit eccentricities larger than 0.35, and 51 systems reach distances from the plane $|z_{\rm max}| > 500$ pc. They assign 30 binaries to the halo, 46 to the thick disk, and the remainder to the old thin disk. Most of the orbits are regular, with others having plunging orbits that reach sufficiently small pericentric distances that they are of chaotic form; the star can spend considerable time close to the plane, before energy exchange takes it several kpc from the plane (Figure 3.15). Three very metal poor extremely wide binaries were detected and discussed by Allen *et al.* (2000b), again based on the Hipparcos astrometry.

Allen & Poveda (2007) investigated membership of some of these wide binaries to phase-space groupings and moving clusters. Since both wide binaries in the halo and moving clusters are likely to be the remains of past mergers or dissolved clusters, they are expected to provide information on the dynamical and merger history of the Galaxy, and to continue to show coherent motions over times of the order of their ages. They looked for phase space groupings among the high-velocity metal-poor stars of Schuster *et al.* (2006), which itself has been updated to provide distances based on the Hipparcos astrometry, and identified a number of candidate moving clusters. In several they found a wide common proper motion binary already identified in the catalogue of wide binaries among high-velocity and metal-poor stars of Allen *et al.* (2000a). Spectroscopic follow-up studies of these stars are expected to confirm the physical reality of the group, and to discriminate between dissolved clusters or accreted systems. Other Hipparcos-based studies of wide visual binaries have been reported by Kiselev *et al.* (2007) and Sinachopoulos *et al.* (2006, 2007b,a).

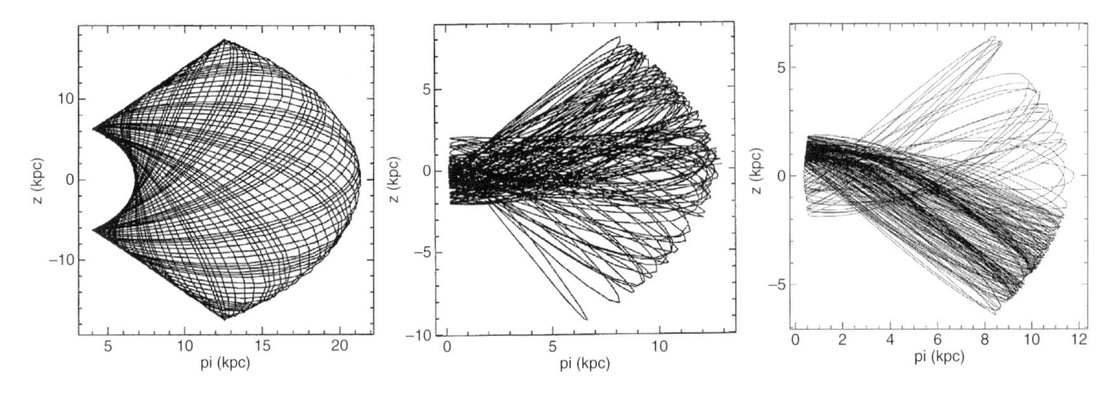

Figure 3.15 Meridional Galactic orbits for three of the wide binaries studied by Allen et al. (2000a). Left: G112–43, an extreme halo binary. Middle: G2–38, showing chaotic behaviour. Right: G106–25, also showing chaotic behaviour. From Allen et al. (2000a, Figures 10–12).

A list of 11 previously unknown halo stars from the Revised Luyten Catalogue with parallaxes that imply a wide-binary association with relevant Hipparcos stars was given by Chanamé & Gould (2004). Gould & Chanamé (2004) identified 424 common proper motion pairs by searching the NLTT and rNLTT Catalogues for companions to Hipparcos stars (see Section 2.11.2). The identification of faint Hipparcos companions yields a distance permitting the colour–magnitude relation to be constructed for these fainter stars, and allows the physical component separation to be determined from the angular separation.

Lépine & Bongiorno (2007) extended this analysis to identify proper motion companions of Hipparcos stars using the LSPM-North Catalogue with $\mu < 0.15$ arcsec yr^{-1} (Lépine & Shara, 2005). They identified 521 systems with angular separations 3–1500 arcsec, with 15 triples and 1 quadruple, of which 130 systems were newly-discovered, including 44 in which the secondary has $V > 15.0$. Completeness is claimed for secondaries with angular separations 20–300 arcsec, and apparent magnitudes $V < 19.0$. At least 9.5% of nearby ($d < 100$ pc) Hipparcos stars have distant stellar companions with projected orbital separations $s < 1000$ AU. The distribution in orbital separation is again consistent with Öpik's relation up to about 4000 AU, beyond which it follows a more steeply-decreasing power law with exponent 1.6 ± 0.1. They also report a luminosity function of the secondaries which is significantly different from that of the single star field population, showing a relative deficiency in low-luminosity ($8 < M_V < 14$) objects. The observed trends suggest either a formation mechanism biased against low-mass companions, or a progressive disruption of systems with low binding energy.

Population II binaries Zinnecker *et al.* (2004), updated from Köhler *et al.* (1998, 2000), studied the frequency, period distribution, and mass ratios for 164 halo stars in the solar neighbourhood. The sample was based on Population II stars from the list of Carney *et al.* (1994), with reliable distance estimates from Hipparcos. A visual binary survey was made using infrared speckle interferometry, adaptive optics, and direct imaging, from which they found 33 binaries, six triples, and one quadruple system. They used the Hipparcos distances to derive linear separations, finding a bimodal distribution with a peak at about 500 AU, and estimating the resulting binary frequency for projected separations above 10 AU of between 6–20%. They made a statistical estimate of the resulting binary period distribution based on assumptions of the orbital inclinations and system masses, and determined a semi-major axis distribution significantly different from that of Population I stars. Confirmation of the physical nature of some of the systems is still required, for example using second-epoch measurements to establish physical associations.

Radial velocities of Population II binary stars have been reported by Sperauskas & Bartkevičius (2002, 114 objects) and Bartkevičius & Sperauskas (2005, 91 objects); these are also listed in Table 1.10.

X-Ray binaries These are discussed in Section 8.3.

Multiplicity order Eggleton *et al.* (2007) studied the multiplicities of the 4555 stellar systems brighter than Hipparcos magnitude 6. For multiplicity from one to seven, they found frequencies of 2723, 1411, 299, 86, 22, 12 and 2. From a brighter control sample, they concluded that the higher multiplicities are significantly under-represented. They determined what observed multiplicities would be expected from a given theoretical selection, subject to observational constraints. To obtain the observed distribution, they required a theoretical sample with about twice as many triples, and five times as many sextuples. They considered a number of evolutionary scenarios involving triple and other systems, including mergers, that may mean that a system now perceived as binary may have formerly been triple.

3.5 Orbital systems

3.5.1 General properties

Methods Before detailing investigations following the Hipparcos Catalogue publication, some aspects of the orbital analysis undertaken by the FAST Consortium (as part of the Hipparcos Catalogue construction) are summarised. Bernstein (1994) described the analysis approach for double stars in general. To describe the orbital motion, they used the Thiele–Innes constants for the orientation of the orbit in space, and a Fourier series of the elliptical coordinates in the orbital plane. This leads to a linear observation equation relating the observed abscissae with the five astrometric and six orbital parameters (the seventh, the period, being fixed). Changing the trial period leads to a solution characterised by the maximum signal-to-noise ratio of the semi-major axis, or the minimum of the rms of unit weight. In many cases some of the orbital elements were adopted from ground-based observations of astrometric or spectroscopic binaries, while in around half the cases the period was identified from this periodogram analysis, with refined orbital elements then determined by least-squares fitting to the Intermediate Astrometric Data.

Emanuele *et al.* (1996) described experiments in which all seven orbital unknowns were estimated from the satellite data, using a normalisation function fitted to the great-circle satellite data. The nonlinearity of the equations leads to numerous local minima in different points of the position of the secondary component, and thus in practice demands reasonably good estimates of the seven orbital parameters to ensure convergence. They identified 355 orbital systems with suitable ground-based data, whose periods extended from a few years up to 600 years or more (their Figure 1), with 33 short-period orbits (<25 yr) and 322 long-period orbits. Orbits based on the Hipparcos data alone could be reconstructed only for the 33 short-period systems, and only some of these confirmed the ground-based results.

In general, the solutions benefit from the combination of Hipparcos data and ground-based data. For the 235 orbital pairs reported in the Hipparcos Catalogue Double and Multiple Systems Annex, Part O, all seven orbital parameters were fitted for only 45 systems, while in all other cases the value of some parameters were assumed from previous investigations, mainly spectroscopic and interferometric orbits.

As reported in Section 3.2.3, a specific comparison of 141 Hipparcos binary star parallaxes with dynamical parallaxes of visual binaries with reliable orbits was reported by Shatsky & Tokovinin (1998). The Hipparcos and dynamical parallaxes, determined according to Equation 3.11, shows generally good agreement, especially for distant systems ($\pi < 15$ mas), while at intermediate distances there are a few cases of large errors in the Hipparcos parallaxes attributable to unmodelled orbital motion in unresolved short-period (3–25 yr) orbital systems.

Stellar masses from orbital binaries Total binary system masses, $M_1 + M_2$, can be estimated when the period P and linear semi-major axis a of an orbital binary are known (Equation 3.10). Reasonably accurate total masses are therefore known for a number of systems, although the cubic dependency implies that relative mass errors are very sensitive to both parallax error, and the errors on the orbital elements a and P. Individual stellar masses, in contrast, can only be directly measured under specific circumstances or under specific assumptions (see box on page 112) and as a result, accurate masses are known for only a very small number of (main sequence) stars (see the reviews by Popper, 1980; Andersen, 1991). When stellar masses are required, for example for comparison with evolutionary models, these are usually estimated from a constructed mass–luminosity relation.

Orbital parameters for 235 systems were made available at the time of the Hipparcos Catalogue release. More careful analysis of these and other systems, frequently based on the Hipparcos Intermediate Astrometric Data, or the Hipparcos Transit Data files (essentially rectified Fourier coefficients describing the light variation at the main detector as an object transited the modulating grid at some 100 epochs during the 1990–93 mission lifetime), in combination with accurate ground-based (speckle or radial velocity) data, have led to improved orbits, improved masses for a number of systems and, under certain assumptions or specific conditions, mass estimates for individual components. Combination of the Hipparcos astrometric data with radial velocity data to improve the orbital determination and characterisation is described by Bernstein (1999), and Torres (2004).

Some early considerations of stellar mass determination for orbital binaries using Hipparcos data, and the resulting mass–luminosity relation, are given by Martin *et al.* (1997a), Lampens *et al.* (1997), Martin & Mignard (1997), and Söderhjelm *et al.* (1997). A more complete treatment is given in the series of three papers by Martin *et al.* (1997b), Martin & Mignard (1998), and Martin *et al.* (1998); and also by Söderhjelm (1999).

Martin *et al.* (1997b) studied binaries with $P <$ 20 years and $\rho > 0.3$ arcsec, in which the very specific Hipparcos (modulation grid) observing geometry (Figure 3.1) requires the introduction of a fiducial point,

Orbital binaries and astrometric orbits: Astrometric measurements describe motion on the plane of the sky. The seven classical orbital elements used to describe an astrometric orbit are $P, a'', e, i, \omega, \Omega, T_0$. P is the orbital period in years, a is the semi-major axis of the apparent orbital motion (i.e. the motion of the photocentre; here a'' indicates that the semi-major axis is in angular measure, as distinguished from the linear semi-major axis that appears as a derived quantity in spectroscopy), e is the eccentricity, i is the orbit inclination ($i = 90°$ is an edge-on orbit), ω is the periastron longitude in the plane of the orbit, Ω is the position angle of the node in the plane of projection, and T_0 is the periastron time.

The apparent motion of a binary in the plane of celestial projection is described by (Binnendijk, 1960; Heintz, 1978)

$$
\begin{aligned}
x &= A(\cos E - e) + F\sqrt{1 - e^2} \sin E \\
y &= B(\cos E - e) + G\sqrt{1 - e^2} \sin E
\end{aligned}
\tag{3.5}
$$

where x and y are the tangential coordinates, and E is the eccentric anomaly related to the mean anomaly M by

$$M = 2\pi \frac{T - T_0}{P} = E - e \sin E \tag{3.6}$$

The Thiele–Innes constants are related to the remaining orbital elements by

$$
\begin{aligned}
A &= a(+\cos\omega\cos\Omega - \sin\omega\sin\Omega\cos i) \qquad & F &= a(-\sin\omega\cos\Omega - \cos\omega\sin\Omega\cos i) \\
B &= a(+\cos\omega\sin\Omega + \sin\omega\cos\Omega\cos i) \qquad & G &= a(-\sin\omega\sin\Omega + \cos\omega\cos\Omega\cos i)
\end{aligned}
\tag{3.7}
$$

Formulation using the Hipparcos Intermediate Astrometric Data: The Intermediate Astrometric Data files give partial derivatives of the star abscissa with respect to the five parameters of the standard model, in equatorial coordinates

$$d_1 = \partial a_i/\partial\alpha_*,\ d_2 = \partial a_i/\partial\delta,\ d_3 = \partial a_i/\partial\pi,\ d_4 = \partial a_i/\partial\mu_{\alpha*},\ d_5 = \partial a_i/\partial\mu_\delta \tag{3.8}$$

where a_i is the abscissa in ith observation of a given star, α and δ are the equatorial coordinates, $\alpha_* = \alpha\cos\delta$, π is the parallax, and $\mu_{\alpha*} = \mu_\alpha\cos\delta$ and μ_δ are the orthogonal proper motion components. The abscissa is defined as a great circle arc connecting an arbitrary chosen reference zero-point and the star on a fixed great circle (close to the scan circle).

In the small-angle approximation, a linearized equation for the observed abscissa difference $\Delta a_i = a_{\rm obs} - a_{\rm calc}$, can be written as (Volume 1, Section 2.8 of ESA, 1997; Goldin & Makarov, 2006)

$$d_1\Delta x + d_2\Delta y + d_3\Delta\pi + d_4\Delta\mu_x + d_5\Delta\mu_y + d_1\sum_j \frac{\partial x}{\partial\epsilon_j}\Delta\epsilon_j + d_2\sum_j \frac{\partial y}{\partial\epsilon_j}\Delta\epsilon_j = \Delta a_i \tag{3.9}$$

where ϵ_j are the elements of the vector of seven orbital elements, $\epsilon = [P, a, e, i, \omega, \Omega, T_0]$. In this equation, the notations $\alpha*$ and δ are replaced by x and y respectively, to make them consistent with the traditionally used tangential coordinates for apparent orbits. Equation 3.9 holds only in the vicinity of a certain point in the 12-parameter space $\{\alpha, \delta, \pi, \mu_{\alpha*}, \mu_\delta, \epsilon\}$, as long as the corrections to these parameters remain small.

System mass and dynamical parallax: Application of Kepler's Law for a bound orbital system yields

$$(M_1 + M_2)\, P^2 = a^3 \tag{3.10}$$

where masses are in units of $M_\odot$, P is the orbital period in years, and a is the semi-major axis of the relative orbit in AU. Thus the combined system mass can be determined if P and a are measurable, while individual masses can only be determined if the mass ratio can be established (from the ratio of the star's distances from the barycentre, or the ratio of their speeds around it). Conversely, if masses can be estimated, e.g. from spectral types, a dynamical parallax can be calculated as

$$\pi_{\rm dyn} = a'' P^{-2/3} (M_1 + M_2)^{-1/3} \tag{3.11}$$

where a'' is in arcsec, M in units of $M_\odot$, and P in years. The method is generally not considered as providing fully independent distances because of the assumptions required to estimate the masses.

Spectroscopic orbits: Spectroscopic measurements (radial velocities) describe the motion along the line-of-sight, and the conventional spectroscopic elements for a double-lined spectroscopic binary are $P, \gamma, K_1, K_2, e, \omega, T_0$. Conventionally, ω is the longitude of periastron for the primary, whereas the convention in visual orbits is to use the longitude of periastron for the secondary: the two differing by 180°.

Combined astrometry and spectroscopy: Four orbital elements are in common between astrometric and spectroscopic solutions: P, e, ω, T_0. The combined observations therefore further constrain the three-dimensional orbit, and can yield individual component masses without the ambiguity of the orbital inclination. In the case of single-lined (SB1) systems, the mass of the primary has to be assumed, generally from the spectral type. In the case of double-lined spectroscopic binaries (SB2) the mass ratio is determined, and individual masses follow from the minimum mass derived from the spectroscopic solution, in combination with the inclination angle i provided by astrometry (e.g. Torres, 2004)

$$M_1 \sin^3 i = P(1 - e^2)^{3/2}\, (K_1 + K_2)^2\, K_2 \tag{3.12}$$

$$M_2 \sin^3 i = P(1 - e^2)^{3/2}\, (K_1 + K_2)^2\, K_1 \tag{3.13}$$

Orbital parallax: The orbital parallax is given by the ratio between the projected angular semi-major axis of the relative orbit from astrometry, $a'' \sin i$, and the projected linear semi-major axis from spectroscopy

$$a \sin i = P(K_1 + K_2)\sqrt{1 - e^2} \tag{3.14}$$

as

$$\pi_{\rm orb} = \frac{a'' \sin i}{(a_1 + a_2) \sin i} = \frac{a'' \sin i}{P(K_1 + K_2)\sqrt{1 - e^2}} \tag{3.15}$$

providing a distance estimate independent of geometric parallax. In the case of an eclipsing spectroscopic binary, $i = 90°$.

which they described as the 'hippacentre', and whose path on the sky is different to the classical relative Keplerian orbit (photocentre). Under certain conditions it is possible to derive the component mass ratio and their magnitude difference, at the same time as the astrometric parameters of the centre of mass. For smaller separations or large Δm the scale of the photocentric orbit can be recovered as a limiting case.

Martin & Mignard (1998) selected 145 potential orbital binaries in the Hipparcos Catalogue for which ground-based orbital elements gave adequate starting points. Of these, 46 eventually yielded a satisfactory solution: eight with the largest separations yielding mass ratios associated with the 'hippacentre' without further assumptions, and 38 for which the derivation of a mass ratio was only possible by including Δm from other sources. The derived parallax is then used to estimate individual masses of the components, which could be compared with ground-based mass determinations for 17 systems. Martin *et al.* (1998) continued the analysis, using the orbits updated by CHARA and others to process 70 additional orbital systems. Significant results were obtained for 22 systems, with relative accuracy better than 25% for the masses of 17 binaries. New estimates were also given for six systems previously investigated by Martin & Mignard (1998), based on reliable values of ΔHp. The mass–luminosity relation was determined for the 54 main-sequence components with $\sigma_M/M < 25\%$ and for the 23 best solutions with $\sigma_M/M < 13\%$ (Figure 3.16), with a linear regression yielding for the 23 best solutions the two equivalent forms connected through an appropriate early main-sequence bolometric correction ($BC_{Hp} \sim -0.2$)

$$\log\left(\frac{M}{M_\odot}\right) = +0.537(10) - 0.1074(20)\, M_{Hp} \tag{3.16}$$

$$\log\left(\frac{L}{L_\odot}\right) = -0.032(79) + 3.724(70)\, \log\left(\frac{M}{M_\odot}\right) \tag{3.17}$$

Söderhjelm (1999) obtained improved orbital elements and masses for 205 systems by combining the Hipparcos astrometry (using the Hipparcos Transit Data files) with ground-based observations (including old visual observations from the WDS, speckle observations from the CHARA Catalogue, and subsystems from the catalogue of spectroscopic binary orbits). The sample included well-defined systems (relative parallax error below 5%, $P < 250$ yr, and calculated separations above 0.10 arcsec during some part of the mission), supplemented with well-observed speckle targets, Hyades binaries, and other interesting multiple systems. The basic motivation for re-deriving these apparently well-known orbits was their large dependence on visual observations with large and complex systematic errors. Forcing the orbits to fit the few Hipparcos or speckle observations available makes the derived semi-major axes much more reliable, resulting in the best possible mass-determinations from the Hipparcos parallaxes. For $P > 25$–30 yr the Hipparcos Catalogue J1991.25 position alone can be combined with ground-based data, while for $P \lesssim 25$ yr, the detailed (curved) motion can be modelled only by a re-reduction using the Hipparcos Transit Data. For an undisturbed orbital binary, the model to be fitted comprises the mass ratio q, the seven orbital parameters, the five astrometric parameters for the barycentric motion, and 2 Hp magnitudes, or 15 parameters in total. For some systems, a solution for the mass ratio was attempted from the astrometric data alone, and a useful q value was obtained for 27 systems (listed in their Tables 1 and 2). Apparently anomalous solutions for which $q > 1$ signal a triple system in which the secondary is actually a close binary, while small q values may imply that a third component is in orbit with the primary. Mostly, however, the orbital curvature during the Hipparcos mission was too small to provide a q-value, but the new solutions gave improved relative orbits and sometimes improved Hipparcos parallaxes (because the original reductions could not take the orbital motion fully into account). The new orbits and the (new or original) Hipparcos parallaxes provide new system masses, and with estimated q-values derived from the Hipparcos magnitude differences, individual masses were finally derived for nearly 200 systems, of which 30 were already treated in the studies by Martin & Mignard (1998) and Martin *et al.* (1998).

From these mass determinations, mass–luminosity diagrams may be constructed. Figure 3.17 shows results

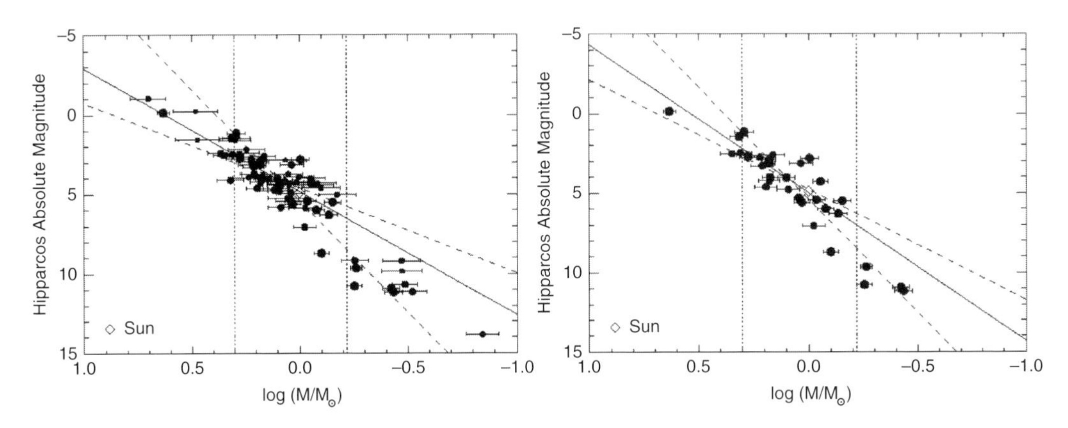

Figure 3.16 Left: mass–luminosity relation fitted over 54 main-sequence components with $\sigma_M/M < 25\%$ (75 components visible in total). Non-main-sequence stars and outliers have been removed. The dot size relates to the quality of the mass estimation. The two vertical dotted lines indicate the limits of the mass range considered for the fit, represented by the solid line. Bolometric corrections were not computed, so that the scatter in the two extreme mass ranges is not physically significant. The Sun is at the centre of the diagram (diamond). Right: restricted to mass quality: $\sigma_M/M < 13\%$. Out of 32 visible components, 23 participate in the fit. From Martin et al. (1998, Figures 4 and 5).

for estimated errors below 7.5%, 15% and 30% respectively. The highest accuracy is only achieved for a few systems which have both an accurate Hipparcos parallax, and also an accurate relative orbit defined by unbiased observations, notably speckle interferometry. The good fit to theoretical models, also over the interesting but more sparsely covered low-mass region, gives confidence both in the stellar models and in the present mass-determinations. A very narrow mass–luminosity relation is not necessarily expected because of age and abundance-differences in the sample of nearby stars. In principle, the seven or so main-sequence Hyades binaries included in the study should fall on a single mass–luminosity isochrone: the dominating error source is currently in the parallaxes, however, and a re-analysis based on the secular parallaxes making use of the proper motions (see Section 6.3, and Figure 6.8) would be appropriate. Lampens *et al.* (2004) started a programme of high-angular resolution near-infrared photometry for these 27 systems with astrometric q values, aiming for better astrophysical characterisation in terms of $T_{\rm eff}$ and chemical composition.

Arenou *et al.* (2000) revised the orbit analysis for seven double-lined spectroscopic binaries (SB2) using the Hipparcos Intermediate Astrometric Data combined with spectroscopic data (rather than the speckle data or other astrometric data used by Söderhjelm 1999). Although the SB2 binaries are not particularly favourable objects for Hipparcos due their roughly equal component masses and therefore implying only a small nonlinear motion of the photocentres, individual component masses, and their magnitude difference, could be estimated. Their mass–luminosity relationship also shows a satisfactory agreement both with theory, and with the results from Söderhjelm (1999) for main-sequence pairs with similar mass precision (Figure 3.18).

Pourbaix & Jorissen (2000) used the Hipparcos Transit Data and the Intermediate Astrometric Data, in conjunction with the orbital parameters of spectroscopic binaries, to derive revised astrometric parameters. Solutions for the five astrometric and four orbital parameters not already known from the spectroscopic orbit were applied to 81 systems for which spectroscopic orbits became available after the catalogue publication, and which belong to various families of chemically-peculiar red giants (dwarf barium stars, strong and mild barium stars, CH stars, and Tc-poor S stars). Among these, 23 yield reliable astrometric orbits, of which some examples are given in Figure 3.19. The results provided a direct test of the principles used for the detection of the $\Delta\mu$ binaries, namely that an unrecognised orbital motion introduces a systematic error on the proper motion. Comparison of the proper motion from the Hipparcos Catalogue with that re-derived by Pourbaix & Jorissen (2000) confirms that the former can be significantly displaced for binaries with periods in the range of about 3–8 yr. The Hipparcos parallaxes of unrecognised spectroscopic binaries were shown to be generally reliable, except for systems with periods close to 1 year, for which the two effects are mixed. They also showed that, even when a complete orbital revolution was observed by Hipparcos, the inclination was seldom precisely determined, and that the new solutions contribute little to the previous knowledge of masses, since

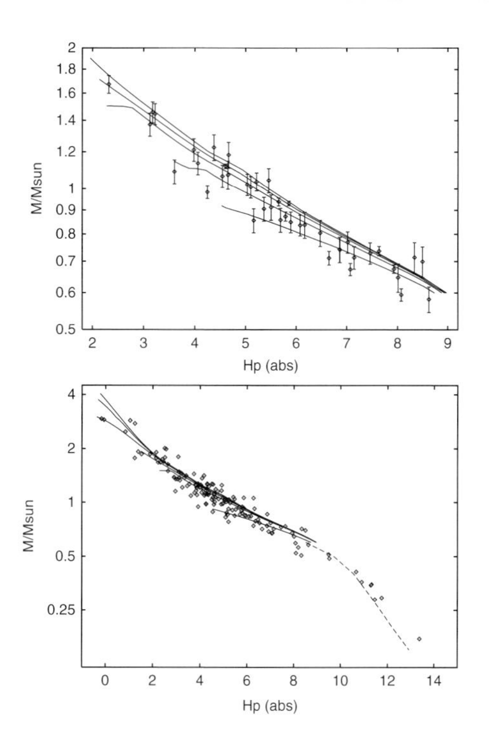

Figure 3.17 Left: mass–luminosity diagram for 42 individual main-sequence stars with $\sigma_M < 7.5\%$. Isochrones are for log age = 9.0, 9.4, 9.8, 10.2 from Bertelli et al. (1994). Below left: the 146 individual main-sequence stars with $\sigma_M < 15\%$. Isochrones are for log age = 7.5, 8.0, 8.5, 9.0, 9.4 and 10.2. The theoretical low-mass ZAMS is from Malkov et al. (1997), transformed from V to Hp magnitudes using the Hipparcos Catalogue tables and the observed Hp versus (V − I) main sequence. Below right: the 276 individual main-sequence stars with $\sigma_M < 30\%$, with the same isochrones. From Söderhjelm (1999, Figures 2–4).

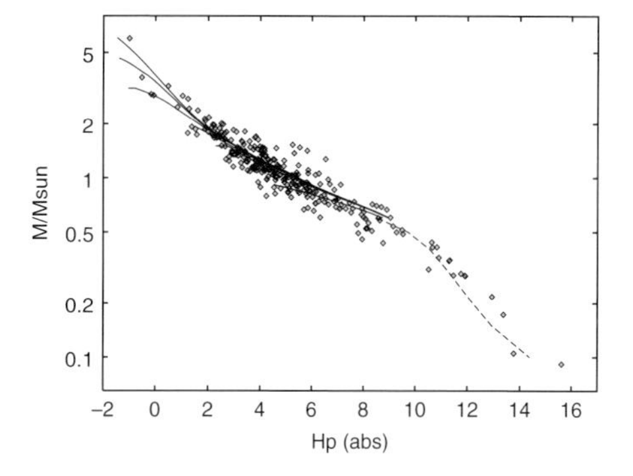

the astrometric orbit allows the inclination i to be eliminated from the mass function, but does not give access to the individual masses.

Pourbaix & Boffin (2003) updated the list of spectroscopic binaries containing a red giant compiled by Boffin *et al.* (1993), and cross-identified it with the Hipparcos Catalogue to yield a sample of 215 systems. They made a re-analysis of the Hipparcos Intermediate Astrometric Data, applying consistency tests between the Thiele–Innes and the Campbell solutions as proposed by Pourbaix & Arenou (2001), with the requirement that the most significant peak in the Hipparcos periodogram corresponds to the orbital period. They derived well-behaved solutions for 29 systems, including one double-lined spectroscopic binary, for which the newly derived astrometric solution is consistent with the spectroscopic orbit. Among these, six were new orbital solutions not present in the DMSA/O annex, while their procedure rejected 25 published DMSA/O entries. From the 29 systems, they derived the distributions of component masses as well as the mass ratios. They found that the mass of the primary is peaked around $2M_\odot$, that the secondary mass distribution is consistent with a Salpeter-like initial mass function, and that the distribution rises for smaller values of the mass ratio.

Docobo & Ling (2003) determined orbits for 20 visual doubles with separations in the range 0.1–0.5 arcsec using an analytical method not requiring prior calculation of the areal constant, and described by Docobo

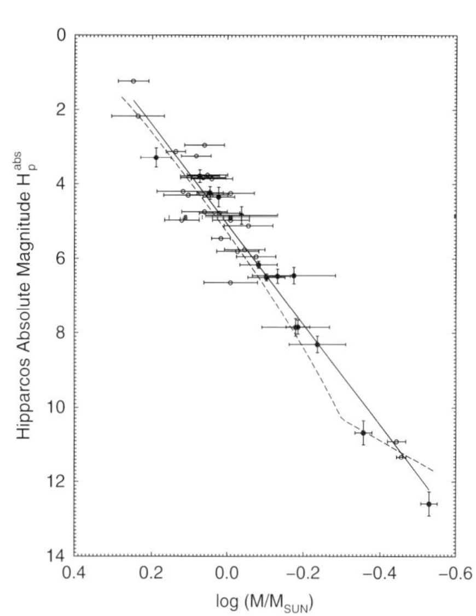

Figure 3.18 Mass–luminosity diagram determined from orbital binaries: •: results for seven SB2 mass determinations from Arenou et al. (2000); ○: results from Söderhjelm (1999). The solid line shows the linear fit $M_{Hp}^{abs} = -13.5(6)\log M + 5.07(11)$ derived by the authors; the dashed line shows the relation from Henry & McCarthy (1993). From Arenou et al. (2000, Figure 1).

(1985, 2001). Total masses were estimated from the Hipparcos parallax data for 15 systems, and for five systems not observed by Hipparcos, based on dynamical parallaxes estimated using the Baize–Romani algorithm

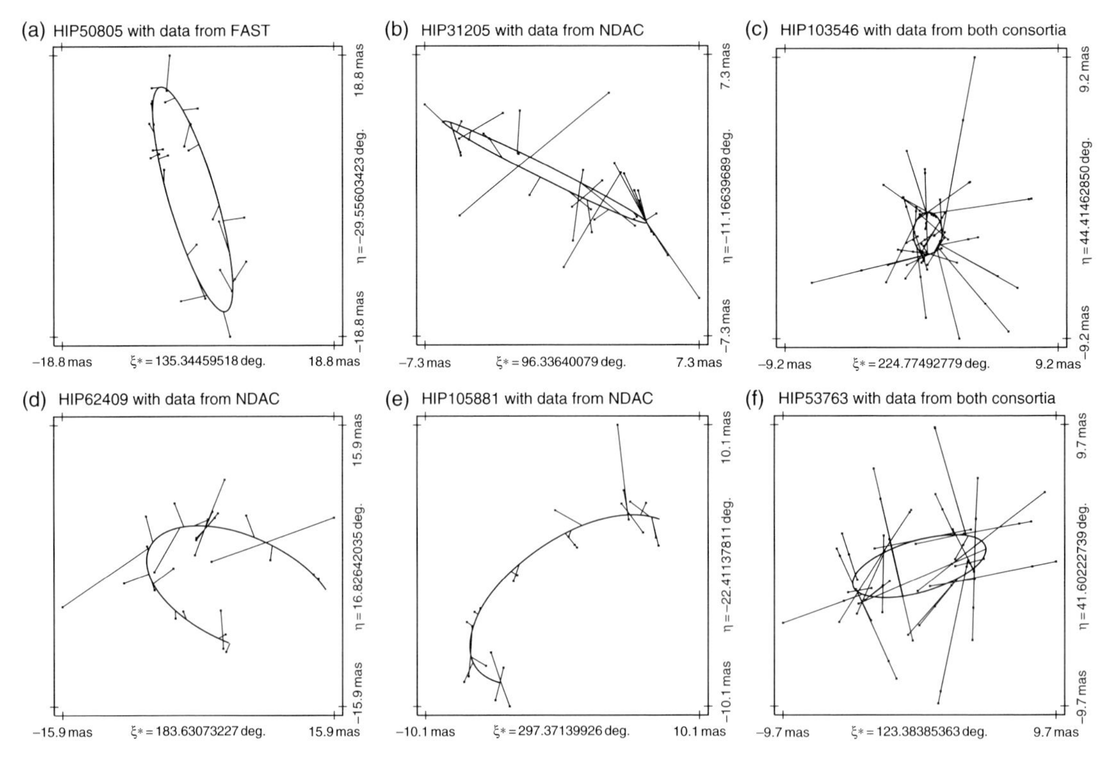

Figure 3.19 The orbital arc on the plane tangent to the line-of-sight for some representative orbits solved for by Pourbaix & Jorissen (2000). The segments connect the computed position on the orbit to the great circle (not represented, perpendicular to the segment) corresponding to the observed position (Hipparcos measurements are one-dimensional). (a) HIP 50805 is the only one of these systems for which the DMSA/O provides an orbital solution without prior information; (b) HIP 31205; (c) HIP 103546 yields an orbit at the limit of what can be extracted from the Hipparcos Intermediate Astrometric Data; (d) HIP 62409; (e) HIP 105881 is an example of an incomplete, albeit well determined, orbital arc; (f) HIP 53763 has a small parallax of about 2 mas, ill-determined since the orbital period is close to 1 yr. From Pourbaix & Jorissen (2000, Figure 1).

(Baize & Romani, 1946; Heintz, 1978). A further 14 systems were discussed by Docobo & Ling (2007).

Goldin & Makarov (2006) re-examined the Hipparcos Intermediate Astrometric Data for all 1561 stochastic solutions, employing an automated 'genetic optimisation' algorithm (instead of a more conventional grid-search algorithm) to solve the orbital fitting problem when no prior information about the orbital elements is available, e.g. from spectroscopic radial velocities. Generally, the astrometric orbital fit is a nonlinear 12-parameter adjustment problem that includes the five astrometric parameters and seven orbital parameters (see box on page 112). Obtaining a robust orbital fit becomes increasingly difficult when the orbital period P exceeds the measurement time span of about 3.3 years; and when the period exceeds nine years, the almost linear segment of the orbit is hard to distinguish from the rectilinear proper motion. In cases of long periods or highly inclined orbits the fit becomes ill-conditioned. Confidence intervals for binary detection and orbital solution were derived. The method was verified on the 235 stars with known orbital solutions, yielding a generally good agreement in the estimated periods. At a confidence level of 99% they found orbital solutions for 65 systems, most of which were previously unknown binaries, although 11 had accurate spectroscopic orbits in the literature. Reliable astrometric fits could be obtained even for $P > 3$–5 years. A few of the new probable binaries with A-type primaries with periods 444–2015 d are chemically peculiar stars, including Ap and λ Boo type, for which the anomalous spectra are explained as an admixture of the light from the unresolved, sufficiently bright and massive companions. Goldin & Makarov (2007) obtained orbital solutions for a further 81 stochastic solutions using the same genetic optimisation-based algorithm, but now combining the Intermediate Astrometric Data produced by the two data analysis consortia (NDAC and FAST) prior to running the optimisation algorithm, notwithstanding the high degree of correlation between them. As they argued: *'some of the astrometric abscissae derived from the same observations differ significantly between the two*

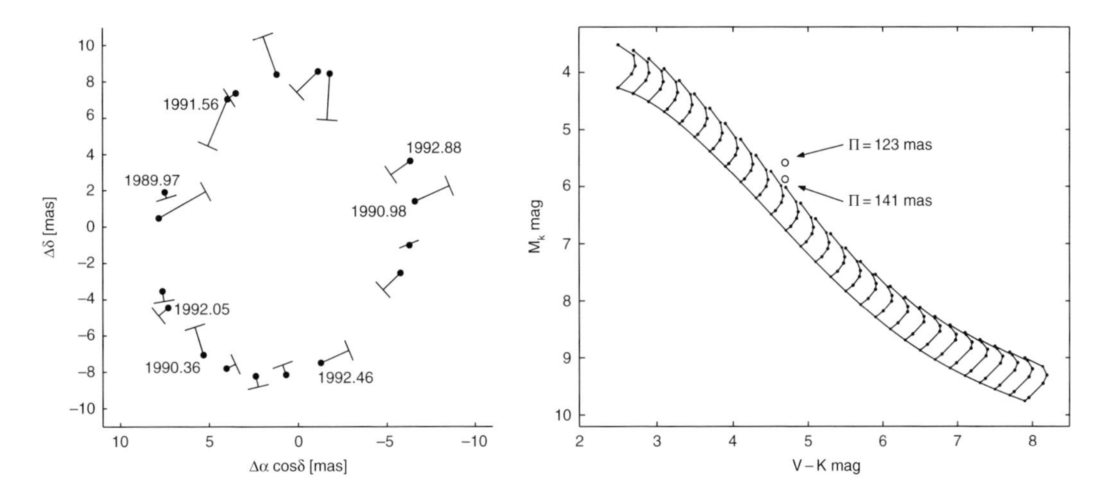

Figure 3.20 Left: apparent photocentric orbit of the new binary HIP 113699. The observed abscissa residuals are shown with straight segments connecting the estimated orbital positions (indicated with black circles) and the actual observations. The mean epoch positions of the Hipparcos grid of slits is marked with short straight segments. The χ^2-adjusted parameters are: $P = 667$ d, $e = 0.1$, $T_0 = 216$ d, $i = 36.4°$, $a_0 = 8.7$ mas, $\omega = 97°$. Right: empirical main sequence of field dwarfs from Henry et al. (2004) (lower curve) and the location of binaries with joint photometric magnitudes shown with short curves and dots for $\Delta V = 0$ (upper end points), 1, 2, 3, 6 and $+\infty$ mag. The position of the star GJ 54 are indicated with open circles for two parallaxes, 123 and 141 mas. From Goldin & Makarov (2007, Figures 2 and 3).

consortia, betraying unknown model or systematic errors. These errors can be diminished by simple averaging, but the estimated precision can only slightly improve because of the correlated noise. The safest and most conservative approach is to compute the mean values of the common data points, but to adopt the larger of the two formal errors as the expected standard deviation of the result'. An example of one of their derived orbits, and the effects of binarity on the location of a star in the HR diagram, are shown in Figure 3.20.

From the detection of $\Delta\mu$ binaries reported by Gontcharov *et al.* (2000), revealed by a difference in the Hipparcos proper motion with that determined when considering long-term proper motions from ground-based observations, Gontcharov & Kiyaeva (2002a,b) determined the photocentric orbits and component masses for 11 systems, six for the first time (α UMa, β LMi, δ And, ξ Aqr, 83 Cnc, κ For; only the first four of which are given in the former reference). Orbital determination based on Hipparcos data combined with older ground-based astrometric catalogues can be applied if the semi-major axis of the photocentric orbit exceeds 0.08 arcsec and for periods in the range 10–55 yr, being particularly efficient for components of large Δm such as white dwarfs within 50 pc.

Gontcharov (2007) further extended the analysis to also take account of radial velocity variations. Their radial velocity compilation, PRAVELO, yielding a median precision of 0.7 km s^{-1}, is then comparable to the transversal velocity precision from proper motions of 1 mas yr^{-1}, corresponding to 0.5 km s^{-1} at 100 pc. Their combined analysis yielded preliminary orbits and component masses for 100 astrometric binaries with no previous orbit calculation along with 20 previously-known systems. They found that many new orbital pairs appear to be A, F, and G stars with white dwarfs. They discuss the multiplicity of stars such as Polaris, Mizar, and Arcturus.

Ren & Fu (2007a,b) also used the Hipparcos Intermediate Astrometric Data combined with early ground-based data, obtaining local minimum solutions from the former, with the optimum solution selected in combination with the latter. They applied the method to 73 Leo using seven long-term observations from ground.

In studies of the astrometric masses for proposed extra-solar planetary candidates, Han *et al.* (2001) showed that fitting the Hipparcos observations with an orbital model when the astrometric motion caused by the companion is below the noise level can have somewhat unexpected consequences. These considerations will become crucial for Gaia, where some 500 000 orbits will need to be derived with no recourse to ground-based starting solutions. Pourbaix (2002) subsequently made extensive simulations of the precision and accuracy when deriving all seven orbital parameters from space observations of visual or astrometric binaries at different noise levels. They used two somewhat hypothetical scanning laws: a '2d' scanning (with uniform scanning over 2π rad, and uniformally over the assumed mission lifetime of 1000 or 1600 days), and a '1d' scanning (an approximation to the Hipparcos-type scanning

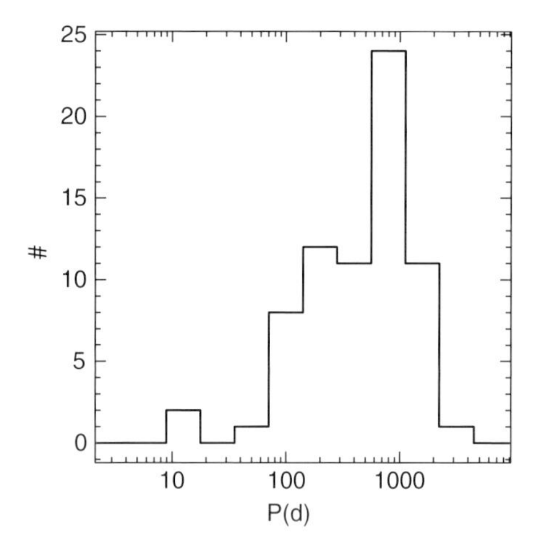

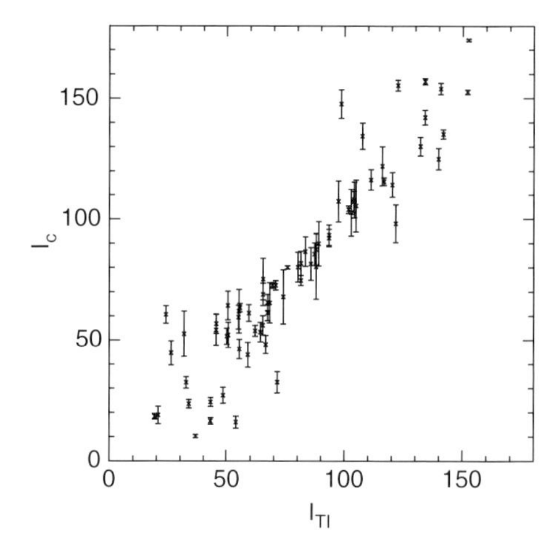

Figure 3.21 Left: distribution of the orbital periods for the 70 solutions with reliable astrometric orbital elements obtained by Jancart et al. (2005a). Right: comparison of the inclinations derived from the Thiele–Innes constants (A, B, F, G) and from the Campbell elements (a, i, ω, Ω) for these 70 systems. From Jancart et al. (2005a, Figures 8 and 9).

law). They showed the sensitivity of the estimates of P, e, a to the form of the scanning and the signal-to-noise, and also showed that for large values of e, the constants F and G in Equation 3.5 are no longer constrained, leading to unreliable estimates of e. Given that A, B, F, G appear in a linear way in the χ^2 optimisation process, and that P can be derived independently of the Keplerian nature of the orbit, they advocate a minimisation approach in which P is first estimated using a period-search technique, followed by a derivation of A, B, F, G in the two-dimensional space of e, T_0.

This approach was used by Jancart *et al.* (2005a), see also Pourbaix *et al.* (2004a), Pourbaix (2004) and Jancart *et al.* (2005b), for a reprocessing of 1374 known spectroscopic binaries from the Catalogue of Spectroscopic Binary Orbits (9th edition, Pourbaix *et al.*, 2004b) which are also contained in the Hipparcos Catalogue. The Hipparcos Intermediate Astrometric Data were used to establish whether the orbital inclination, and hence individual secondary masses, can be derived. The study was again motivated by the fact that the astrometric orbital motion is difficult to detect without prior knowledge of the spectroscopic orbital elements, and that many spectroscopic binaries were added to the Catalogue of Spectroscopic Binary Orbits since construction of the Hipparcos Catalogue. The sample covers periods from around 0.1 days to more than 30 years, and a wide range of eccentricities. It is not well-represented by SB2 systems which, because of their approximately equal masses, have generally only rather small non-linear components of photocentric motion. Statistical tests developed by Pourbaix & Arenou (2001) were used to ensure reliability of the resulting fit. This included verifying consistency between two orbit determination approaches: the first in which the Thiele–Innes constants A, B, F, G are obtained by linear χ^2 minimisation, followed by extraction of the corresponding orbital elements (a, i, ω, Ω); in the second, nonlinear approach, ω and K_1 were adopted from the spectroscopic orbit, and only two parameters of the photocentric orbit (i, Ω) were thus derived from the astrometry. They found 282 systems with detectable orbital astrometric motion, of which only 70 systems have astrometric orbit elements which are reliably determined: 29 with main-sequence primaries and 41 with giant primaries, 20 for the first time (Figure 3.21). They also suggest that many systems actually present in the original DMSA Part O have orbital solutions which would not pass their more stringent statistical tests.

Since information on masses of single-lined spectroscopic binaries (SB1) is restricted to the mass function

$$Q \sin^3 i \equiv \frac{M_2^3 \sin i}{(M_1 + M_2)^2} \tag{3.18}$$

knowledge of the inclination i provides access only to Q. For individual masses, supplementary information in the form of the mass–luminosity relation (for main-sequence stars only, there being no analogue for giants) is then required. The resulting q distribution is peaked around $q = 0.6$; the absence of 'twin' systems with $q \sim 1$ arises from bias against SB2 systems in the astrometric sample. The resulting $e - \log P$ diagram can be used to investigate the current paradigm that long-period ($P > 100$ d), low-eccentricity ($e < 0.1$) systems are never found among unevolved (i.e. pre-mass-transfer) systems. This in turn has been taken to indicate that binary systems always form in eccentric orbits, and the shortest-period systems are subsequently circularised by tidal effects. In contrast, binary systems which can be ascribed post-mass-transfer status, because they exhibit signatures of chemical pollution due to mass transfer, are often found in the avoidance region ($P > 100$ d, $e < 0.1$). Jancart *et al.* (2005a) found that, in total, eight systems fall in this region. The masses derived in their study do not offer conclusive evidence for primaries hosting a white dwarf companion, but neither do they rule out this possibility.

Triple stars Söderhjelm & Lindegren (1997) examined four triple systems each comprising an orbital binary, itself being part of a visual binary. Typically the orbital binary was treated as a single star in the published Hipparcos Catalogue solution. Söderhjelm & Lindegren (1997) showed how the Hipparcos Transit Data, combined with speckle interferometry or reliable ground-based orbits, could be used to make more complete orbital solutions. The basic model to be fitted is an orbital binary plus a third component assumed to have rectilinear motion over the Hipparcos observation period. The positions of each of the two close components are again specified by the mass ratio q, the seven orbital elements, plus the five astrometric parameters describing the barycentric motion. The third component has two positions and two proper motions offset relative to the binary centre of mass, but can normally be assumed to have the same parallax. To this are added three Hp magnitudes, together comprising 20 parameters. Two systems (Kui 83 = HIP 86221; h 3556 = HIP 14913) have a basically correct triple star solution in the Hipparcos Catalogue, and the close orbit could be studied in more detail. The other two (ϵ Hya = HIP 43109; γ And = HIP 9640) are more complex cases with only a double star solution in the Hipparcos Catalogue.

Orbital parallaxes As detailed in the box on page 112, an orbital binary which is both resolved, and is at the same time a double-lined spectroscopic binary, has a very particular importance for distance determinations. The fact that binary is resolved means that the angular separation between the two components can be measured; traditionally such a system was classified as a visual binary, but the binary may also be resolved as a result of the Hipparcos observations, or ground-based speckle or interferometry observations. An astrometric orbit yields the inclination angle i between the plane of the orbit and the plane of the sky. If it is also a double-lined spectroscopic binary, then the spectroscopic orbits of both primary and secondary can be measured, from which the linear distance $a \sin i$ can be determined. The combination of both yields both the linear separation, and the angular separation, from which the distance to the binary follows immediately, and independently of the geometric parallax. The technique is especially useful for stars in evolutionary stages not found in eclipsing binaries.

The first orbital parallax was determined for Capella by Anderson (1920) using Michelson interferometry at the Mount Wilson 100-inch telescope. His value of the semi-major axis of $a = 52.49$ mas and a corresponding distance of $d = 16.7$ pc are in reasonable agreement with more recent results by Hummel *et al.* (1994): $a = 56.47 \pm 0.05$ mas and $d = 13.3 \pm 0.1$ pc. The Hipparcos DMSA O solution (HIP 24608) used the orbital elements of Hummel *et al.* (1994) to derive $d = 12.94 \pm 0.15$ pc.

Following some other early interferometric work there were few advances in this field until the development of speckle interferometry in the 1970s (Section 3.8.4), and subsequently of improved interferometers in the 1990s (Section 3.8.6). Tomkin (2005) made a compilation of modern binary distances measured by orbital parallaxes, based on various astrometric orbits (Mark III, PTI and NPOI interferometry, and speckle interferometry) and various spectroscopic orbits (both modern accurate velocities, and old photographic radial velocities). Of the 26 systems, 23 have $d < 100$ pc, and a comparison between the resulting orbital parallax and the Hipparcos parallax is shown in Figure 3.22. The agreement is very good. For the three objects beyond 100 pc (σ Psc, ζ Aur, and 64 Ori; their Figure 3), the agreement is still close to the 1σ Hipparcos error bars.

Recent distances of two binary systems in the Pleiades do not appear in this compilation of orbital parallaxes: the eclipsing binary HD 23642 (HIP 17704 Munari *et al.*, 2004a; Southworth *et al.*, 2005a), and the double-lined spectroscopic binary Atlas (HIP 17847 Pan *et al.*, 2004). Although, in principle, the latter object fulfills the criteria for an orbital parallax determination, in practice the broad lines of the primary and secondary prevented a reliable determination of the spectroscopic orbit, so no direct measurement of the linear separation between them is available, and its distance estimate again rests on indirect arguments. Nevertheless, their pseudo-orbital parallax distance of 135 ± 2 pc, agrees to around 1σ with the Hipparcos parallax distance of $d = 117 \pm 14$ pc. In this much-debated result, there are two distinct issues: whether the mean parallax of all Hipparcos Pleiades stars is biased, a topic which is discussed in detail in Section 6.4; and whether the individual Hipparcos standard errors remain reliable even at these distances. The current evidence based on the most accurate independent distance estimates of individual stars, continues to support this assertion.

Many comparisons between Hipparcos parallaxes and dynamical parallaxes are to be found in the literature, but it should be stressed that the latter are not fundamental distance determinations, but rather are

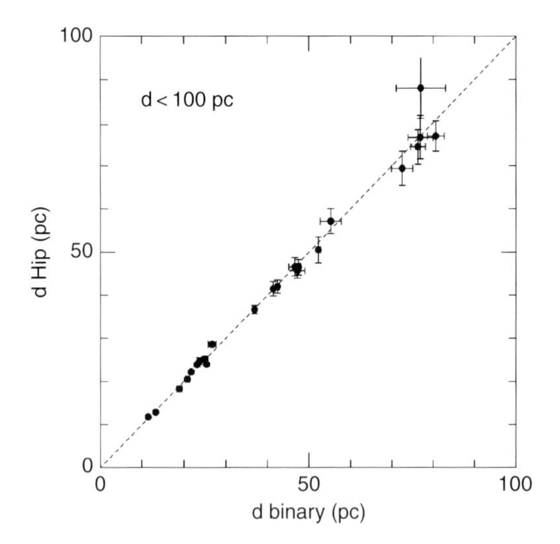

Figure 3.22 Hipparcos distances versus orbital parallax distances for 23 binary systems within 100 pc. From Tomkin (2005, Figure 2).

based on masses determined from, e.g. spectral types. For example, a comparison of 141 Hipparcos binary star parallaxes with dynamical parallaxes of visual binaries with reliable orbits was reported by Shatsky & Tokovinin (1998): it also shows generally good agreement, especially for distant systems ($\pi < 15$ mas), while at intermediate distances there are a few cases of large errors in the Hipparcos parallaxes, perhaps attributable to unmodelled orbital motion in unresolved short-period (3–25 yr) orbital systems; a comparison with the revised parallaxes derived by Söderhjelm (1999) seems not to have been made, and could be worthwhile. Tamazian *et al.* (2006) derived dynamical masses for a further 25 systems.

3.5.2 Individual orbital systems

The Hipparcos data often figure in the interpretation of spectroscopic binary orbits derived from photoelectric radial velocities in the long-running series of studies which, by the end of 2006, had reached Paper 191 (Griffin & Boffin, 2006). Many of these orbits were themselves used, for example, in the re-analysis of the Hipparcos orbital systems by Jancart *et al.* (2005a). This section focuses on a few specific double or multiple systems which have been studied with particular attention given to the Hipparcos data, including the Intermediate Astrometric Data.

Polaris Polaris (HIP 11767) is part of a multiple system comprising a single-lined spectroscopic binary with an orbital period of 29.59 yr, and more distant components. The main component of the spectroscopic binary is a low-amplitude F5 Ib supergiant, the nearest and brightest classical Cepheid, with a pulsational period of about 3.97 days. This period is increasing with time, and its pulsation amplitude (both light-curve and radial velocity) has declined significantly over the past 100 years. The close pair, α UMi A and α UMi P (also designated α UMi a), are accompanied by an 18 arcsec distant F-type dwarf, α UMi B (Herschel & Watson, 1782), and two yet more distant components, α UMi C and α UMi D at around 45 and 80 arcsec respectively (Burnham, 1894). The physical association of α UMi B has been investigated using radial velocity measurements by Kamper (1996), and supported by the Hipparcos parallax, indicating an orbital period of around 50 000 yr; while the physical association and nature of components C and D are unclear. A detailed discussion is given by Wielen *et al.* (2000).

Wielen *et al.* (2000) derived the astrometric orbit of the photocentre of the close pair α UMi AP (Figure 3.23), based on the spectroscopic orbit of α UMi A, and on the difference $\Delta\mu$ between the quasi-instantaneously measured Hipparcos proper motion and the long-term-averaged proper motion given by the FK5. While there is an ambiguity in the inclination i of the orbit, since $\Delta\mu$ cannot distinguish between a prograde ($i = 50\overset{\circ}{.}1$) or retrograde ($i = 130\overset{\circ}{.}2$) orbit, photographic observations strongly favour the latter. They determined a semi-major axis of the photocentre of AP of about 29 mas, $M_{\rm P} \sim 1.5 M_\odot$, Δm(AP) ~ 6.5 mag, and ρ(AP) ~ 160 mas. They obtained the proper motion of the centre of mass of α UMi AP with $\sigma \sim 0.45$ mas yr^{-1}, and a position of the centre of mass at J1991.31 with an accuracy of about 3.0 mas. Their ephemerides for the orbital correction from the position of the centre of mass to the instantaneous position of the photocentre of AP at an arbitrary epoch have a typical uncertainty of 5 mas. For epochs which differ from the Hipparcos epoch by more than a few years, their prediction for the position of Polaris should be significantly more accurate than using the Hipparcos data in a linear prediction, since the Hipparcos proper motion contains the instantaneous orbital motion of about 4.9 mas yr^{-1} = 3.1 km s^{-1}.

Evans *et al.* (2007) used the Hubble Space Telescope ACS camera to directly verify the predicted retrograde orbit of Wielen *et al.* (2000), and succeeded in detecting the close companion (at a separation of 176 mas and a $\Delta m \sim 6.5$ mag). Based on the predicted orbit, the Hipparcos parallax, and their measurement of the separation, they determined a mass of $4.3 \pm 1.1 M_\odot$ for the Cepheid and $1.25 \pm 0.20 M_\odot$ for the close companion.

Arcturus The K1.5 giant Arcturus (α Boo, HIP 69673) was classified by Hipparcos as a double star, with $\rho = 0.255 \pm 0.039$ arcsec, and $\Delta V = 3.33 \pm 0.31$ mag. Griffin (1998) noted that no companion had been previously

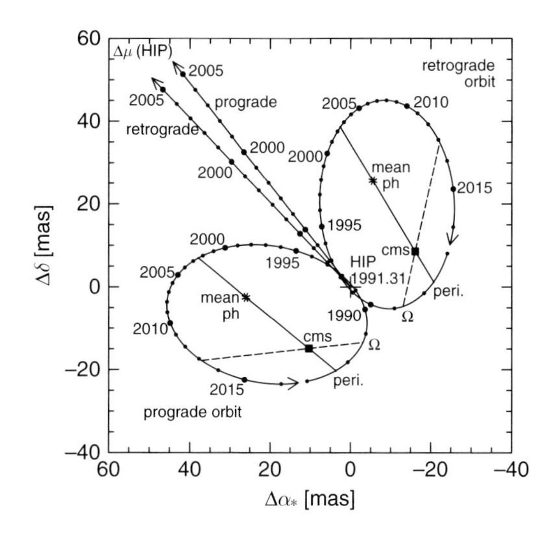

Figure 3.23 Astrometric orbit (prograde or retrograde) of the photocentre of α UMi AP. The retrograde orbit is the preferred solution. From Wielen et al. (2000, Figure 1).

reported and, in an analysis partly based on the absence of radial velocity variations, suggested that the binary classification was simply erroneous. This prompted a discussion of the differences in the solutions obtained by the FAST and NDAC data analysis teams presented by Söderhjelm & Mignard (1998), who noted that the double star solution was assigned grade 4, meaning 'doubtful, to be used with care'. They concluded that *'it is difficult to explain in detail how instrumental and random errors have conspired to mimic a close binary. In isolation, the Hipparcos data can be interpreted as due to a binary, but confronted with external evidence, such an interpretation seems impossible'.* Turner *et al.* (1999) reported observations using the natural guide star adaptive optics system on the Mount Wilson 100-inch telescope, showing the star to be single, with no companion meeting the Hipparcos Catalogue description.

Discussions of the nature of Arcturus were reopened by the study of Verhoelst *et al.* (2005). They used the IOTA interferometer with the FLUOR instrument to observe the star in four narrow filters between 2.03–2.39 μm. With a known diameter of 20.20 ± 0.08 mas (also from FLUOR/IOTA data, Perrin *et al.*, 1998), they compared the wavelength dependence of the diameter with predictions from both plane parallel and spherical atmosphere models, and found that neither could explain the observed visibilities. They showed that the data suggest the presence of a companion ($\rho = 212.7 \pm 1.5$ mas, $\theta = 157.^\circ6 \pm 1.^\circ7$, $\Delta m = 4.25 \pm 0.12$ in K), in rather good agreement with the Hipparcos data ($\rho = 255 \pm 39$ mas, $\theta = 198^\circ$, $\Delta m = 3.33 \pm 0.31$ in V). They suggested that such a companion might be difficult to detect in the configuration reported by Turner *et al.* (1999). The results nevertheless remain perplexing: Arcturus has been a spectroscopic and photometric calibrator, e.g. for ISO–SWS, but more importantly an IAU radial velocity standard (Pearce, 1955), now observed for more than 100 years. Variations of order 160 m s^{-1} and $P = 1.842 \pm 0.005$ d have been recognised more recently (Smith *et al.*, 1987), with Irwin *et al.* (1989) finding a range of 500 m s^{-1}, a long-period amplitude of 120–190 m s^{-1} and a period of 640–690 d, consistent with a companion of $M \sin i = 1.5–7.0 M_{\rm Jup}$. Intriguingly, in their catalogue of $\Delta\mu$ binaries (see Section 3.4.2) Gontcharov *et al.* (2001) identify Arcturus as an astrometric binary, with periods of 5 and 20 years present in the data. A further indication of the inability of models to match observations of Arcturus are the fact that recent LTE and NLTE models predict significantly more flux in the blue and ultraviolet bands than is observed (Short & Hauschildt, 2003). Although inconsistencies would remain, Verhoelst *et al.* (2005) postulate that the results could be explained by a fairly close binary system, with a separation of just a few AU, and seen face on, or a much wider system, with a period of centuries or longer, and seen edge on. While more recent data (as yet unpublished) from NACO and IOTA yet again appear to rule out the possibility of such a companion, long-term spectroscopic and radial velocity monitoring (Brown, 2007) continue to keep the possibility open.

V815 Her Fekel *et al.* (2005) obtained spectroscopic and radial velocity observations of the previously-known single-lined spectroscopic binary V815 Her (HIP 88848, HD 166181). Their data indicated a triple system, with a short-period circular orbit of $P_{\rm S} = 1.809$ d, and a longer-period orbit with $e = 0.76$ and $P_{\rm L} = 5.73$ yr (Figure 3.24). Hipparcos classified the solution as stochastic, implying a binary system but one lacking an acceptable deterministic model. Fekel *et al.* (2005) re-analysed the Hipparcos Intermediate Astrometric Data for this object by adopting the spectroscopic values for the orbital elements P, T, e, K, ω, in order to obtain i, Ω, and the semi-major axis of the A component. It yields an inclination $i = 78.^\circ4 \pm 1.^\circ6$, a similar parallax but with a reduced uncertainty, $\pi_{\rm revised} = 30.93 \pm 0.77$ mas compared with $\pi_{\rm Hip} = 30.69 \pm 2.08$ mas, and a revised proper motion with $\mu_{\alpha*} = 106.6 \pm 0.95$ mas yr^{-1}, $\mu_\delta = -31.0 \pm 1.1$ mas yr^{-1} (errors are from Pourbaix, priv. comm.), compared with the published value $\mu_{\alpha*} = 138.07 \pm 1.81$ mas yr^{-1}, $\mu_\delta = -18.58 \pm 1.80$ mas yr^{-1}, a result consistent with the long-term Tycho 2 Catalogue value at the 2σ rather than at the previous 30σ value. They go on to derive mass estimates suggesting that the unseen secondary in the long-term orbit may also be a binary, and discuss the possible dynamical behaviour of this possibly quadruple system.

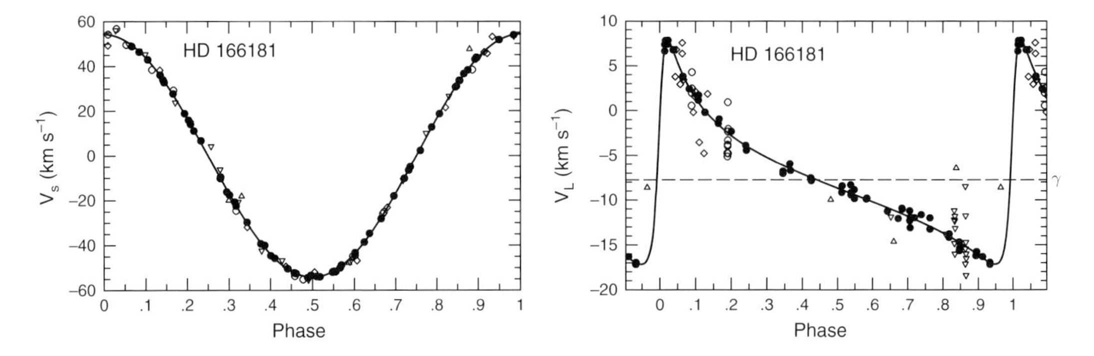

Figure 3.24 Radial velocity curve of V815 Her (HD 166181). Left: the short-period orbit (curve). Points represent the observed velocities from different observation sequences minus the velocities of the centre of mass of the spectroscopic binary in the long-period orbit, calculated from the full orbital elements from the combined spectroscopic and astrometric solution. Right: observed velocities minus the velocities in the short-period orbit, calculated from the full orbital elements. The curve represents the long-period elements. The horizontal dashed line is the centre-of-mass system velocity, γ. These curves were derived from radial velocity observations alone: they are included to illustrate the type of spectroscopic binary systems accessible to Hipparcos orbital analysis. From Fekel et al. (2005, Figures 1 and 2).

V1061 Cyg Torres *et al.* (2006), see also Torres *et al.* (2007), obtained spectroscopic and photometric observations of the eclipsing system V1061 Cyg (HIP 104263, HD 235444), previously thought to be a member of the rare class of 'cool Algols'. They show instead that it is a hierarchical triple system in which the inner eclipsing pair ($P = 2.35$ d) is composed of well-detached main-sequence stars, with the third star also being visible in the spectrum. Combining the radial velocities for the three stars, the times of eclipse, and the Hipparcos Intermediate Astrometric Data, they established the elements of the outer orbit, which is eccentric and has a period of 15.8 yr. They determined accurate values for the masses, radii, and effective temperatures of the binary components: $M_{\rm Aa} = 1.282 \pm 0.015 M_\odot$, $R_{\rm Aa} = 1.615 \pm 0.017 R_\odot$, and $T_{\rm eff}^{\rm Aa} = 6180 \pm 100$ K for the primary (star Aa); and $M_{\rm Ab} = 0.9315 \pm 0.0068 M_\odot$, $R_{\rm Ab} = 0.974 \pm 0.020 R_\odot$, and $T_{\rm eff}^{\rm Ab} = 5300 \pm 150$ K for the secondary (Ab). The masses and radii have relative errors of only 1–2%. Both stars are rotating rapidly ($v \sin i$ values are 36 ± 2 km s^{-1} and 20 ± 3 km s^{-1}) and have their rotation synchronised with the orbital motion. There are signs of activity including strong X-ray emission and possibly spots. The mass of the tertiary is determined to be $M_{\rm B} = 0.925 \pm 0.036 M_\odot$, with $T_{\rm eff}^{\rm B} = 5670 \pm 150$ K. At a distance of 166.9 ± 5.6 pc, current stellar evolution models that use a mixing length parameter $\alpha_{\rm ML}$ appropriate for the Sun agree well with the properties of the primary, but show a very large discrepancy in the radius of the secondary, in the sense that the predicted values are $\sim$10% smaller than observed. In addition, the temperature is cooler than predicted by some 200 K. These discrepancies are attributed to chromospheric activity. The magnetic fields commonly associated with stellar activity tend to inhibit convective heat transport, and the reduced convection explains why fits to models with a smaller mixing length parameter of $\alpha_{\rm ML} = 1.0$ seem to give better agreement with the observations.

HIP 50796 Torres (2006) obtained spectroscopic observations of HIP 50796, which showed it to be a single-lined binary with an orbital period of 570 d and an eccentricity of $e = 0.61$. The astrometric signature of this orbit had been detected by Hipparcos in the form of curvature in the proper-motion components, although the period was unknown at the time. Combining the new radial velocity measurements with the Hipparcos Intermediate Astrometric Data, he derived the full three-dimensional orbit, determined the dynamical mass of the unseen companion, and derived a revised trigonometric parallax that accounts for the orbital motion. Given a primary mass estimate of $0.73 M_\odot$ corresponding to a mid-K dwarf, the companion mass is determined to be $0.89 M_\odot$, or some 20% larger than the primary. The likely explanation for the larger mass without any apparent contribution to the light, is that the companion is itself a closer binary composed of M dwarfs. The near-infrared excess and X-ray emission displayed by HIP 50796 support this. Photometric modelling of the excess leads to a lower limit to the mass ratio of the close binary of $q \sim 0.8$ and individual masses of $0.44-0.48 M_\odot$ and $0.41-0.44 M_\odot$. The revised parallax ($\pi = 20.6 \pm 1.9$ mas) is more precise, and significantly smaller, than the original Hipparcos value ($\pi = 29.40 \pm 2.69$ mas).

ϕ Her Torres (2007a) studied the mercury–manganese star ϕ Her (HIP 79101, HD 145389), a well-known spectroscopic binary that has been the subject of a study by

Zavala *et al.* (2007), in which they resolved the companion using long-baseline interferometry. The total mass of the binary was consequently fairly well established, but the combination of spectroscopy with astrometry had not resulted in individual masses consistent with the spectral types of the components. The photocentric motion was clearly detected by Hipparcos. Using the Hipparcos Intermediate Astrometric Data, Torres (2007a) derived the individual masses using only astrometry. Incorporating the radial velocity measurements into the orbital solution yielded improved masses of $3.05 \pm 0.24 M_\odot$ and $1.614 \pm 0.066 M_\odot$ that are consistent with the theoretical mass–luminosity relation from the stellar evolution models of Yi *et al.* (2001). Orbital and evolutionary models are shown in Figure 3.25.

HR 6046 Torres (2007b) studied the 6-yr spectroscopic binary HR 6046 (HIP 79358, HD 145849), which in the past has been speculated to contain a compact object as the secondary. Scarfe *et al.* (2007) re-determined the orbit, and showed that the companion is an evolved but otherwise normal star of nearly identical mass as the primary, which is also a giant. The binary motion had been detected by Hipparcos. Torres (2007b) used the Hipparcos Intermediate Astrometric Data in combination with the spectroscopic results to revise the orbital solution, and to establish the orbital inclination angle, and hence the absolute masses as $M_A = 1.38^{+0.09}_{-0.03} M_\odot$ and $M_B = 1.36^{+0.07}_{-0.02} M_\odot$. Aided by other constraints, the primary star was shown to be approaching the tip of the red-giant branch, while the secondary is beginning its first ascent (Figure 3.26).

Studies determining improved orbits, component masses, and other fundamental stellar parameters for individual systems include: the chemically-peculiar system V392 Car (Debernardi & North, 2001); the M-type eclipsing binary YY Gem (Torres & Ribas, 2002); and the metallic-lined system WW Aur (Southworth *et al.*, 2005b).

Other systems Other studies of individual orbital systems, which include spectroscopic binaries, and triple and quadruple systems, and which make use of the Hipparcos data in various ways include: astrometric detection of a low-mass companion to AB Dor (Guirado *et al.*, 1997); orbit and mass of the nearby star Gliese 623AB (Barbieri *et al.*, 1997); nature of the massive O-type close binary τ CMa (van Leeuwen & van Genderen, 1997); He abundance, age and mixing-length for η Cas, ξ Boo, 70 Oph and 85 Peg (Fernandes *et al.*, 1998); multiplicity among peculiar Ap and Am stars (North *et al.*, 1998); spectroscopy of the Herbig Ae/Vega-type binary HD 35187 (Dunkin & Crawford, 1998); the active single-lined spectroscopic binary HD 6628 (Watson *et al.*, 1998); the orbit of the planet-hosting star 16 Cygni (Hauser & Marcy, 1999); dynamical versus Hipparcos parallaxes for five interferometric double stars (Olević & Jovanovic, 1999); new spectroscopic components for eight multiple systems (Tokovinin, 1999a); visual orbit for the quadruple system HD 98800 (Tokovinin, 1999b); component masses for the double-lined spectroscopic binary HIP 111170 (Halbwachs & Arenou, 1999); comparison of Hipparcos versus ground-based orbital parallaxes (Pourbaix & Lampens, 1999); fundamental stellar parameters for γ Persei (Pourbaix, 1999); nearby hot subdwarfs in HD 30187 B and HD 39927 B (Makarov & Fabricius, 1999); the 'twin' components of the chemically-peculiar SB2 system 66 Eri (Yushchenko *et al.*, 1999); revised masses of α Cen (Pourbaix *et al.*, 1999); the equal-mass binary HR 2030 (Griffin & Griffin, 2000); the orbit of the low-mass binary Gliese 600 (Tokovinin *et al.*, 2000a); the convective core of α Cen (Guenther & Demarque, 2000); the orbit and magnetic field of the Ap binary HD 81009 (Wade *et al.*, 2000); the triple system HR 7272 (Tokovinin *et al.*, 2000b); the double-lined spectroscopic binaries HD 195850 and HD 201193 (Carquillat & Ginestet, 2000); re-calculated orbits of eight double stars (Olević *et al.*, 2000b); orbits of twelve double stars (Olević & Jovanovic, 2001); double-lined spectroscopic binaries in the Am stars HD 81976 and HD 98880 (Carquillat *et al.*, 2001); seismology properties of the binary ζ Her (Morel *et al.*, 2001); masses of the triple star π Cep (Gatewood *et al.*, 2001); orbit from an astrometric study of the triple star ADS 48 (Kiyaeva *et al.*, 2001); physical properties of the massive triple system HD 135240 (δ Cir) (Penny *et al.*, 2001); double-lined eclipsing binaries as probes for stellar evolution models (Kovaleva *et al.*, 2001); masses and orbits of 16 cool magnetic Ap spectroscopic binaries (Carrier *et al.*, 2002); the chromospherically active binary system BK Psc (Gálvez *et al.*, 2002); the orbit of χ^1 Ori from astrometric and radial velocity data (Han & Gatewood, 2002); the triple system, double-lined spectroscopic binary HD 7119 (Carquillat *et al.*, 2002); the spectroscopic binaries 21 Her and γ Gem (Lehmann *et al.*, 2002); orbit and physical parameters of the quadruple system μ Ori (Fekel *et al.*, 2002); the massive compact binary in the triple system HD 36486 (δ Ori) (Harvin *et al.*, 2002); astrometric study of the low-mass binary Ross 614 (Gatewood *et al.*, 2003); the pre-main-sequence spectroscopic binary AK Sco (Alencar *et al.*, 2003); photometric studies of the triple star ER Ori (Kim *et al.*, 2003); dynamics of the quadruple system Finsen 332 triple systems (Olević & Cvetković, 2003); with luminous cool primaries and hot companions (Parsons, 2004); double-lined spectroscopic binaries in the Sco–Cen complex (Nitschelm, 2004); orbits for eight visual binaries (Ling, 2004); the

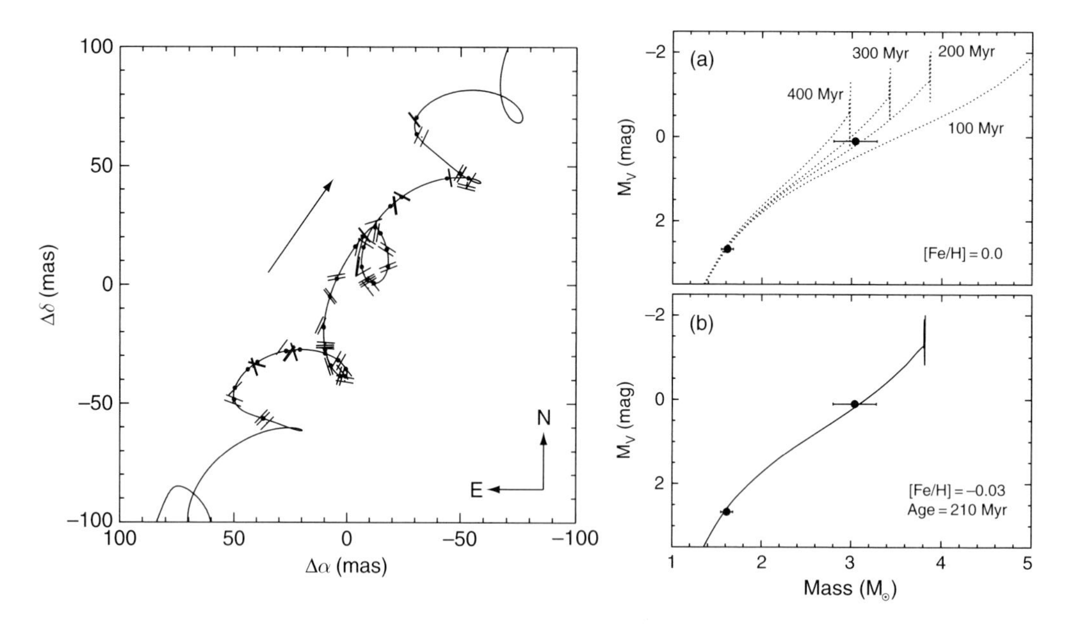

Figure 3.25 Left: path of the photocentre of ϕ Her on the plane of the sky, along with the Hipparcos observations (abscissa residuals). The irregular motion is the result of the combined effects of parallax, proper motion, and orbital motion according to the adopted global solution. The arrow indicates the direction and magnitude of the annual proper motion. Right: mass-M_V diagram for ϕ Her: top: model isochrones from the Yonsei–Yale series by Yi et al. (2001) for a range of ages, as labelled. Solar composition has been assumed; bottom: best-fitting model that reproduces all six measured properties within their uncertainties; the individual masses, the absolute visual magnitudes, and the combined $U-B$ and $B-V$ colours are simultaneously matched to better than 0.4σ. From Torres (2007a, Figures 2 and 6).

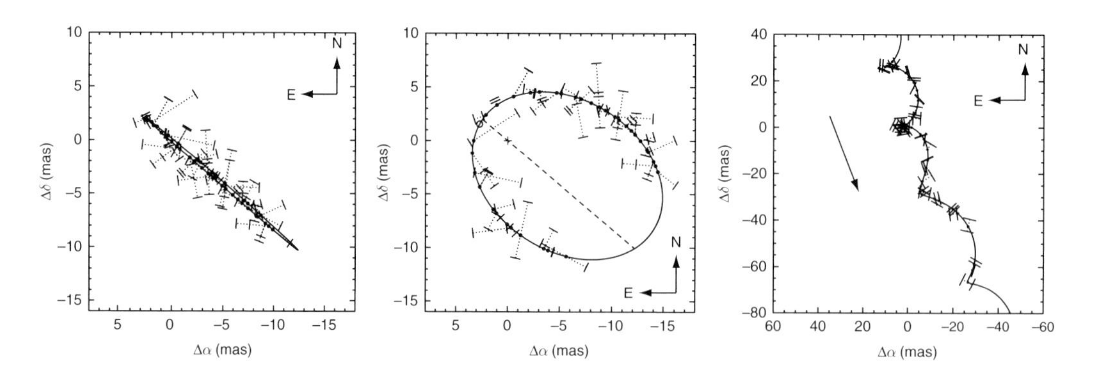

Figure 3.26 Left: photocentric motion of HR 6046 relative to the centre of mass of the binary (indicated by the plus sign) as seen by Hipparcos. The one-dimensional abscissa residuals are shown schematically with a filled circle at the predicted location, dotted lines representing the scanning direction of the satellite, and short perpendicular line segments indicating the undetermined location of the measurement on that line. Middle: as left, except that the orbit has been de-projected to appear as if it were viewed exactly face-on. The line of nodes is indicated with the dashed line. Motion on the plane of the sky in this figure is direct (counter-clockwise). The location of periastron is shown as the open circle. Right: path of the photocentre on the plane of the sky, along with the Hipparcos observations (abscissa residuals). The irregular motion is the result of the combined effects of the annual parallax, proper motion (arrow), and orbital motion. From Torres (2007b, Figures 2–4).

single-lined spectroscopic multiple system V815 Her (Fekel *et al.*, 2005); four Population I spectroscopic binary systems (Paunzen *et al.*, 2005); the close binary EE Cet (Djurašević *et al.*, 2005); the young active binary star EK Dra (König *et al.*, 2005); an astrometric study of the binary α Oph (Gatewood, 2005); orbit and dynamical mass of the visual binary Wor 2 (Tamazian *et al.*, 2005); age of the low-metallicity solar-like star 85 Peg A

(D'Antona *et al.*, 2005); the double-line spectroscopic triple system HD 131861 (Fekel *et al.*, 2006); astrometric study of the triple star ADS 9173 (Kiyaeva, 2006); a possible quadruple containing close binary stars, V899 Her (Qian *et al.*, 2006); visual orbit and dynamical mass of DG Leo (Tamazian, 2006); dynamical masses of the components in o And (Olević & Cvetković, 2006); parameters of the Algol binary V505 Sgr (Lázaro *et al.*, 2006); the double-lined binary HD 149420 (Fekel & Henry, 2006); dusty circumstellar structure of the symbiotic star CH Cyg (Biller *et al.*, 2006); orbits of red-giant spectroscopic binaries HR 1304, HR 1908, and HD 126947(Prieur *et al.*, 2006); the double-lined spectroscopic binary HD 143418 (Božić *et al.*, 2007); the triple VV Cephei-type system FR Sct (Pigulski & Michalska, 2007); orbit and masses for the visual binary A 2329 (Andrade, 2007); the spectroscopic and astrometric orbit of HR 672 (Fekel *et al.*, 2007).

3.6 Eclipsing binaries

Eclipsing binaries are important laboratories for investigating stellar parameters, providing the possibility of determining stellar masses and linear radii of the components, and therefore excellent tests of stellar evolutionary models. The subsets of spectroscopic eclipsing binaries (SB1 and SB2) and their use as mass probes and distance indicators are considered as part of Section 3.5.

The Hipparcos Catalogue lists 917 eclipsing binaries (flagged as EA, EB, EW, ...) of which 343 were newly-detected eclipsing systems identified from the epoch photometry. Among the former, 370 belonged to the catalogue of parameters for eclipsing binaries of Brancewicz & Dworak (1980).

Statistics from Hipparcos Oblak *et al.* (1997b), see also Oblak *et al.* (1998), reported the first estimation of the geometric and physical parameters of 570 systems based on the Hipparcos trigonometric parallaxes. They examined the statistical relations between Roche lobe filling, mass, radius, and separation as a function of spectral type, and made the first general confirmation of the photometric parallax determination for the furthest eclipsing and spectroscopic systems.

Oblak & Kurpinska-Winiarska (2000) updated the number of known eclipsing systems in the Hipparcos Catalogue to 993, and reported a generally good agreement between distances from Hipparcos and those based on photometric estimates, making use of the mass–luminosity relation of main-sequence binaries according to the method described by Dworak (1975).

Percy & Au-Yong (2000) made a search for short-term variability in the Hipparcos epoch photometry for B stars in eclipsing binary systems with well-determined masses. The goal was to determine accurate pulsation constants (Q values) and hence pulsation modes.

Marrese *et al.* (2005) reported the observing programme ongoing at the Asiago Observatory to obtain orbits and physical parameters of double-lined eclipsing systems by combining the Hipparcos and Tycho photometry with ground-based spectroscopy in the 850–875 nm band, simulating the photometric and spectroscopic observations to be provided by Gaia. Munari *et al.* (2004b) extrapolated the number of Hipparcos eclipsing binaries, to predict some 16 000 double-lined spectroscopic binaries that will be discovered by Gaia with $V < 13$, for which orbital solutions and physical parameters will be derived with formal accuracies better than 2%. Sarro *et al.* (2006) reported studies of an automatic classification of the light-curves of eclipsing binaries based on neural networks trained with Hipparcos data of seven different categories, including eccentric binary systems and two types of pulsating light-curve morphologies. Other classification studies have been made by Prša & Zwitter (2007).

Only limited studies have been made of the two most important observable parameters for eclipsing binaries: the period distributions (Farinella & Paolicchi, 1978; Antonello *et al.*, 1980), and the distribution of eclipse depths. The eclipse probability depends mainly on the ratio between the stellar radii and the orbit size, and appreciable numbers of eclipsing binaries are found only at periods of days or weeks. However, because of orbital evolution, these short-period systems may have started out in a much wider orbit. Söderhjelm & Dischler (2005), see also Dischler (2006), attempted to reproduce the present distributions of eclipse depth versus period, for example as observed by Hipparcos, by a large-scale population synthesis approach, starting with millions of original binaries leading to a few hundred eclipsing pairs. They used the rapid binary star evolution code (BSE, Hurley *et al.*, 2002) to follow the evolution of several hundred million binaries, starting from various simple input distributions of masses and orbit sizes. Modelling assumptions included a constant star formation rate over the last 12 Gyr, a time-independent initial mass function, an age-independent metallicity uniform in [Fe/H], various assumed mass ratio distributions, and specific distributions for the semi-major axes and orbital eccentricities. Eclipse probabilities and predicted distributions over period and eclipse depth ($P, \Delta m$) were estimated in a number of main-sequence intervals, from O-stars to brown dwarfs. Comparison between theory and the Hipparcos observations, assuming that most of the eclipsing binaries with periods in the range 0.2–20 d and eclipse depths larger than 0.1 mag were detected in the Hipparcos photometry, shows that a standard (Duquennoy & Mayor, 1991) input distribution of orbit sizes gives reasonable numbers and

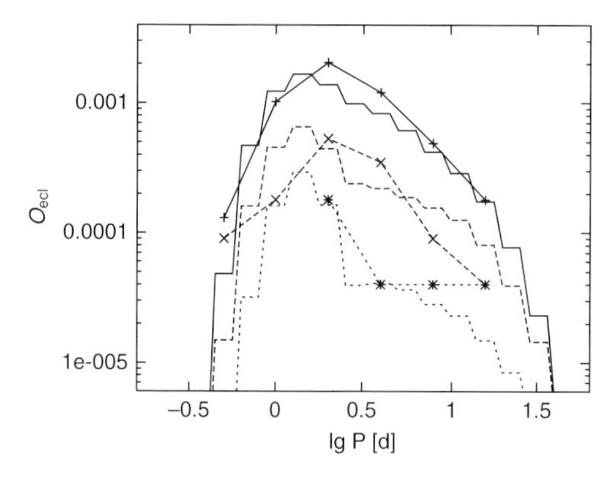

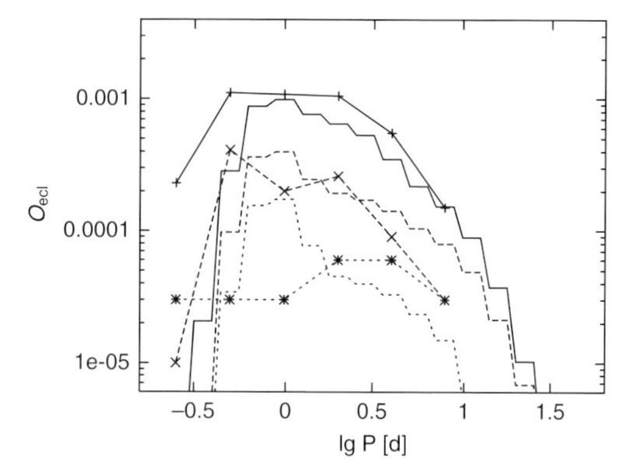

Figure 3.27 The observed numbers of binaries versus orbital period for two distance samples of the Hipparcos Catalogue compared with the theoretical predictions of Söderhjelm & Dischler (2005). Theoretical normalised distributions O_{ecl} correspond to the observed distributions for $\Delta m = 0.1$ mag (upper curves and + symbols), 0.4 mag (middle curves and × symbols) and 0.7 mag (lower curves and ⋆ symbols). Left: for the distant sample defined by $\pi = 1$–4.5 mas and comprising 22 585 Hipparcos stars, for spectral type B. Right: for the nearby sample defined by $\pi > 4.5$ mas and comprising 34 335 Hipparcos stars, for spectral type A. From Söderhjelm & Dischler (2005, Figures 14 and 15).

$(P, \Delta m)$-distributions, as long as the mass-ratio distribution is close to the observed flat ones (Figure 3.27). A random pairing model, where the primary and secondary are drawn independently from the same initial mass function, gives more than an order of magnitude too few eclipsing binaries on the upper main sequence. For a set of eclipsing OB-systems in the LMC, the observed period-distribution is different from the theoretical one, and the input orbit distributions and/or the evolutionary environment in LMC has to be different compared with the Galaxy. Dischler & Söderhjelm (2005) estimate that Gaia will detect about 500 000 eclipsing binaries, this relatively small fraction of the expected one billion star Gaia harvest arising from the fact that many eclipsing systems will escape detection, partly through the correlation between longer periods and narrower eclipses, and partly because eclipse probabilities decrease with decreasing radius, and hence mass.

Karoff *et al.* (2007) used the Hipparcos eclipsing binary statistics to estimate the completeness of the Berlin Exoplanet Search Telescope programme results on variable stars. From 43 out of 92 detected periodic variables identified as eclipsing binaries, they estimated that their survey was complete for short-period binaries, but with only a 20–30% completeness for $P > 1$ d.

Radiative flux scale and temperatures There are two hypothesis-free observational procedures for determining the radiative flux per unit area emitted from a stellar surface

$$F_i = L_i / 4\pi R^2 \tag{3.19}$$

where L_i is the stellar luminosity in band i and R is the radius.

In the angular diameter method, this follows from

$$L_i / R^2 \propto l_i / \theta^2 \tag{3.20}$$

where l_i is the flux at Earth and θ is the angular diameter. There are various observational routes to the determination of angular diameters, notably interferometry, lunar occultations, or the infrared flux method; all are complicated by the effects of limb darkening.

In the independent eclipsing binary method, L_i and R are obtained for each component of the binary from a spectroscopic and photometric analysis, with L_i from the relation $L_i \propto l_i/\pi^2$. In this case, independent values of the surface fluxes are obtained for each component. The availability of the Hipparcos parallaxes provides a significant extension of the eclipsing binary method (Popper, 1998).

Both methods require knowledge of the parallax and apparent magnitude; the eclipsing binary method also requires radial velocities of both binary components as well as a light-curve to determine the two stellar radii and the magnitude difference between the components. Both methods also require photometric indices to establish a relationship between surface brightness and effective temperature (a historical review is given by Kruszewski & Semeniuk, 1999). The working relations given by Popper (1980), based on the formula by Barnes & Evans (1976), are

$$M_V = V + 5 + 5 \log \pi \tag{3.21}$$

$$F'_V = -0.1 M_V - 0.5 \log (R/R_\odot) + 4.225 \tag{3.22}$$

where the parallax π is in arcsec, and the numerical values apply in the V band. To establish a temperature scale the additional relation is

$$\log T_{\rm eff} = F'_V - 0.1({\rm BC}) \qquad (3.23)$$

where BC is the bolometric correction, $M_V - M_{\rm bol}$.

Popper (1998) selected eclipsing binaries from the Hipparcos Catalogue within 125 pc, $\sigma_\pi/\pi < 0.10$, and with suitable spectroscopic and photometric orbital analyses providing radii, magnitude difference, and colour indices of the components. Interstellar absorption was neglected. The sample of such binaries in the solar neighbourhood for defining the surface flux and effective temperature scale is severely limited. Results for the 14 selected systems are shown in Figure 3.28. Values based on fluxes from angular diameters from the intensity interferometer measurements of Davis & Shobbrook (1977) are shown as open circles. The curve is based on the relation tabulated in Table 1 of Popper (1980). The 10 components of the five systems lying significantly below the curve have light-curves which show considerable irregularities, presumably caused by large spotted surface regions. Ignoring these systems whose fluxes probably fall below those expected from their colour indices, Popper (1998) concluded that the eclipsing binaries provide results consistent with diameters via intensity interferometry in defining and extending the radiative flux relation for stars between B6–F8.

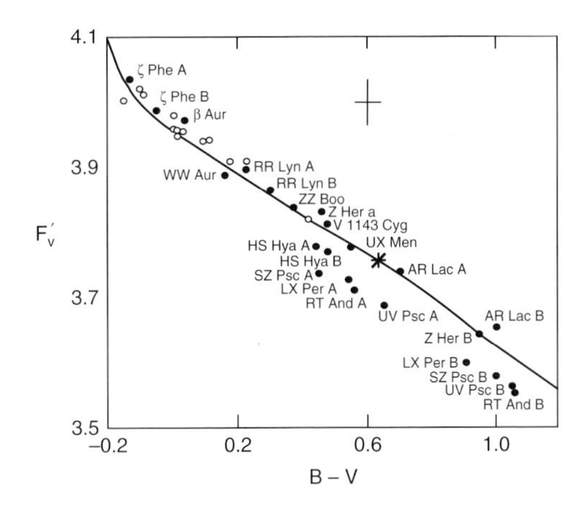

Figure 3.28 Radiative fluxes in the V band. •: components of eclipsing binaries with Hipparcos parallaxes. Each point is accompanied by the first part of its name. ○: stars with angular diameters measured by the intensity interferometer. The asterisk refers to the Sun. The curve is based primarily on angular diameters, with extrapolation to lower temperatures by various procedures including the infrared flux method. The cross shows typical uncertainties in the two coordinates. From Popper (1998, Figure 1).

Evolutionary models Masses and radii of the components of binary stars can be accurately determined allowing tests of stellar evolutionary models. Absolute dimensions of binary systems can also be used to determine distances independently of absolute magnitude calibrations or trigonometric parallax (see Section 3.5). An underlying assumption is that the components of detached binaries behave like single stars from the point of view of stellar structure and evolution, and individual effective temperatures of component stars used for the comparisons are usually estimated from observed photometric indices of the separate components obtained from an analysis of the light-curves. Ribas *et al.* (1998), see also Jordi *et al.* (1997), tested these assumptions by comparing the effective temperatures of 20 detached eclipsing binaries derived from the Hipparcos parallax (based on R, V, and BC_V), with those derived from photometric determinations obtained from standard calibrations using Strömgren or Johnson colour indices. Systems were selected according to $\sigma_\pi/\pi < 0.20$, and with radii accurate to 1–2% from Andersen (1991), and cover a temperature range from 5000–25 000 K. Only five systems are in common with those studied by Popper (1998). Effective temperatures were estimated from

$$\frac{T_{\rm eff}}{T_{{\rm eff},\odot}} = \left(\frac{R}{R_\odot}\right)^{-1/2} 10^{-0.1(M_V+{\rm BC}-M_{{\rm bol},\odot})} \qquad (3.24)$$

derived from the preceding expressions for M_V, F'_V and $T_{\rm eff}$, where M_V is the absolute visual magnitude corrected for absorption, BC is the bolometric correction, and $T_{{\rm eff},\odot}$ and $M_{{\rm bol},\odot}$ are the solar effective temperature and bolometric correction respectively. Since BC depends on effective temperature, the results were iterated to reach agreement. $M_{V,\odot}$ was taken as 4.82, and $\mathrm{BC}_\odot$ was computed for $T_{{\rm eff},\odot} = 5780$ K. The BC corrections of Flower (1996) were adopted. Comparison between the resulting effective temperature determinations using standard photometric calibrations and those derived from the Hipparcos parallaxes are shown in Figure 3.29. The photometric and Hipparcos-based effective temperatures show good agreement, with a small trend such that the Hipparcos parallax-based determinations are about 2–3% smaller than the photometric estimates, corresponding to a small systematic difference of about 0.012 dex in the temperature range covered. The large scatter of V1647 Sgr deserves further study.

Lastennet *et al.* (1999) subsequently reported the construction of Strömgren synthetic photometry based on an empirically-calibrated grid of stellar atmosphere models to derive the effective temperature of each component of double-lined spectroscopic eclipsing binaries as a function of metallicity. Their comparison with eight systems included in the Ribas *et al.* (1998) study show good agreement with the most reliable parallaxes, and add weight to the validity of the relevant BaSeL models.

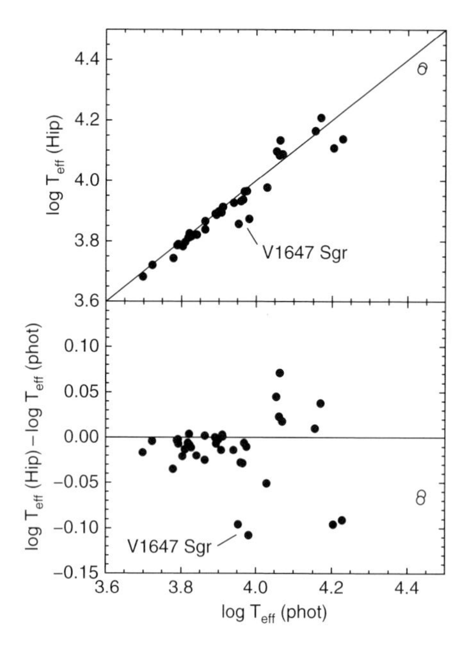

Figure 3.29 Comparison between the effective temperature determination using standard photometric calibrations and that based on the Hipparcos parallax. Residuals are shown in the lower panel. The components of the eclipsing binary CW Cep (with a particularly large error on the Hipparcos parallax, but with a distance consistent with membership of the Cepheus OB3 association) are represented with open circles, and the system V1647 Sgr (possibly affected by a visual companion) is labelled. From Ribas et al. (1998, Figure 2).

About 10 times more eclipsing binaries than used in these studies have Hipparcos trigonometric parallaxes better than 20%, many of them discovered as eclipsing variables by Hipparcos, but with the majority lacking adequate photometry or spectroscopy to be used to extend the surface brightness versus colour index relation. Kruszewski & Semeniuk (1999) compiled a list of 198 such Hipparcos eclipsing binaries within 200 pc, 156 appearing in the 'periodic variables' section, and 42 in the 'unsolved variables', candidates for further ground-based photometric and spectroscopic campaigns.

Semeniuk (2000, 2001) selected 13 Hipparcos eclipsing binaries with $\sigma_\pi/\pi < 0.25$ and homogeneous Strömgren *uvby* photometry to establish the corresponding surface brightness–colour relation in $(b-y)_0$. They obtained a relationship agreeing well with that determined by Popper (1998), although only one system was in common between the two studies.

Oblak *et al.* (2007) reported studies of 36 newly-discovered Hipparcos eclipsing binaries, of which 24 are new doubled-lined eclipsing binaries. Seven objects were found to be new spectroscopic triple systems, and they confirmed the presence of the spectroscopically visible third body in three other systems. Two triple systems, CU Cam and CN Lyn, show long-period variations of the radial velocity of the third component, as well as changes of the centre-of-mass velocity of the eclipsing system.

Individual systems Of the numerous studies of individual eclipsing binary systems based on Hipparcos data in one form or another, a particularly detailed study of the Algol-type eclipsing binary R CMa was made by Ribas *et al.* (2002), following earlier studies identifying a third body as being responsible for quasi-sinusoidal modulation of the eclipse timings (Radhakrishnan *et al.*, 1984; Demircan, 2000). They detailed a method to determine orbital properties and masses of low-mass bodies orbiting eclipsing binaries, by combining long-term eclipse timing modulations (light-travel time or LTT effect) with short-term, high-accuracy astrometry. A similar solution analysis had already been demonstrated in the case of Algol by Bachmann & Hershey (1975). The eclipses act as an accurate clock for detecting subtle variations in the distance object, analogous to the method used for discovering Earth-mass objects around radio pulsars (Wolszczan & Frail, 1992). The periodic quasi-sinusoidal variations of the eclipse arrival times have a very simple and direct physical meaning: the total path that the light has to travel varies periodically as the eclipsing pair moves around the barycentre of the triple system. The amplitude of the variation is proportional both to the mass and to the period of the third body, as well as to the sine of the orbital inclination. As discussed by Demircan (2000), nearly 60 eclipsing binaries show evidence for nearby, unseen tertiary components using LTT effects. A recent example of a brown dwarf detected around the eclipsing binary V471 Tau using this method was presented by Guinan & Ribas (2001). The method is also being employed in selected low-mass eclipsing binaries to search for extra-solar planets (Deeg *et al.*, 2000).

The LTT data alone yield only the mass function $f(M_3) = (M_3^3 \sin i_3^3)/(M_{12} + M_3)^2$ and $a_3 \sin i_3$, i.e. only upper limits to the mass and size of the orbit of the tertiary component. As for the case of simultaneous determination of spectroscopic and astrometric binary orbits, the LTT data complemented by astrometry yield the orbital inclination and hence the mass and semi-major axis of the third body. Furthermore, with the orbital elements P, e, ω known from the LTT analysis, only a small fraction of the astrometric orbit needs to be covered by high-accuracy astrometry.

A simultaneous solution of the Hipparcos astrometry combined with over a hundred years of eclipse timings (Figure 3.30) yields an orbital period of $P_{12} = 92.8 \pm 1.3$ yr (the longest period so far detected and confirmed for an eclipsing binary), an LTT semi-amplitude

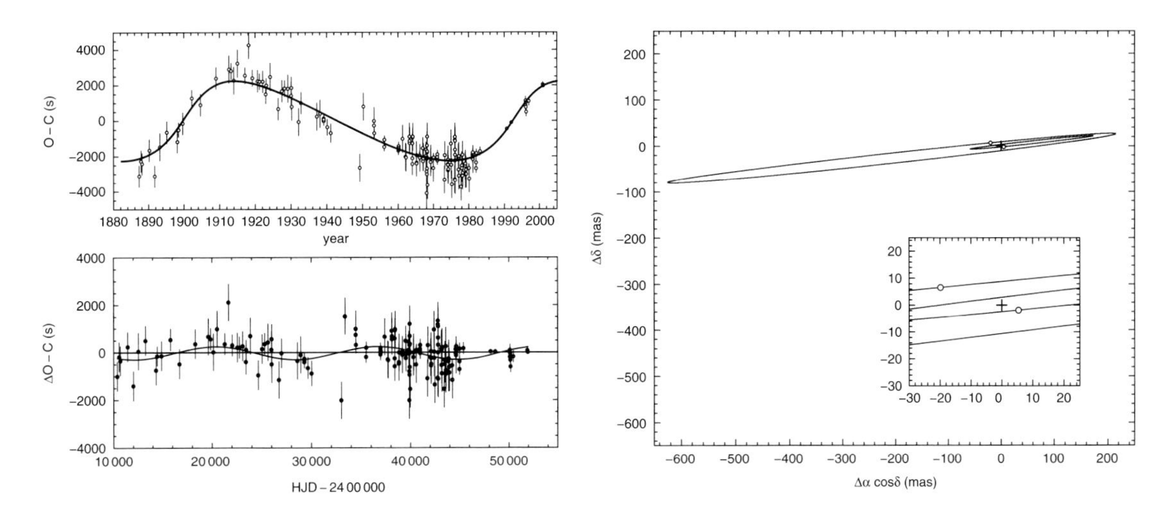

Figure 3.30 Left upper: fits to the light travel time (LTT) curve for R CMa. Left lower: residuals of the fit to the LTT orbit of the third body. The residual oscillations have been modelled with an LTT perturbation by a fourth body in a circular orbit. Right: projection on the sky of the orbits of the eclipsing pair of R CMa (small ellipse) and the tertiary component (large ellipse): +: barycentre of the triple system; ○: positions of the stars in 2002. The inset shows an enlargement of the region surrounding the barycentre. The orbital properties were derived from the simultaneous analysis of eclipse timing residuals and Hipparcos astrometry. From Ribas et al. (2002, Figures 1–3).

of 2574 ± 57 s, an angular semi-major axis of $a_{12} = 117 \pm 5$ mas, and values of the orbital eccentricity and inclination of $e_{12} = 0.49 \pm 0.05$, and $i_{12} = 91\overset{\circ}{.}7 \pm 4\overset{\circ}{.}7$, respectively; the inclination of the orbit of the third body is only made possible by the astrometric measurements. Adopting the total mass of R CMa of $M_{12} = 1.24 \pm 0.05 M_\odot$, the mass of the third body is $M_3 = 0.34 \pm 0.02 M_\odot$ and the semi-major axis of its orbit is $a_3 = 18.7 \pm 1.7$ AU. From its mass, the third body is either a dM3-4 star or, less likely, a white dwarf.

Ribas *et al.* (2002) note that the method of combining LTT analysis and astrometry complements spectroscopic searches. The LTT analysis favours the detection of long-period third bodies around eclipsing binaries because the amplitude of the time delay due to the LTT effect is proportional to $P_{12}^{2/3}$, while the spectroscopic semi-amplitude is proportional to $P_{12}^{-1/3}$. Furthermore, any strictly periodic event that can be predicted with good accuracy could be potentially useful to detect stellar or substellar companions. This might include, for example, spotted stars, pulsating stars, and transiting planets. The active RS CVn binary XY UMa (Pribulla *et al.*, 2001), for example, could be such a candidate: it displays orbital period changes explained by the mutual action of a third body, which might also explain the small discrepancy between the inferred orbital parallax (86 ± 5 pc) and Hipparcos (66 ± 6 pc) distances.

Other studies of individual orbital systems which make use of the Hipparcos data in various ways, including both photometry and astrometry, include a series of papers presenting revised elements for some 80 eclipsing binaries (Otero, 2003b; Otero & Claus, 2004; Otero, 2004; Otero & Dubovsky, 2004; Otero *et al.*, 2005; Otero & Wils, 2005a,b; Otero *et al.*, 2006a); also Shobbrook (2004), Hoogeveen (2005). New eccentric eclipsing binaries found in the Hipparcos Catalogue were reported by Otero *et al.* (2006b). Studies of other individual eclipsing systems include: the new eclipsing binaries HD 125488 and HD 126080 (Gómez Forrellad & García Melendo, 1997); the eclipsing or ellipsoidal variable V1472 Aql (Samus, 1997); the eclipsing binary system HIP 12056 (Vidal-Sainz, 1998); observations of the eclipsing variables HP Dra and V2080 Cyg (Kurpinska-Winiarska *et al.*, 2000); photometric orbital modulation in V1080 Tau (Simon *et al.*, 2000); photometry of the eclipsing variables EF Boo, VZ Psc, V417 Aql and BX And (Samec *et al.*, 2000); new eclipsing binaries in multiple systems (Kurpinska-Winiarska & Oblak, 2000); component masses of the chemically-peculiar binary system V392 Car (Debernardi & North, 2001); orbits and stellar parameters for V505 Per, V570 Per and OO Peg (Munari *et al.*, 2001); the photometric binary HD 169981 (Lehmann *et al.*, 2001); a photometric study of the eclipsing binary V899 Her (Ozdemir *et al.*, 2002); the new eclipsing system V432 Aur (Dallaporta *et al.*, 2002); radial velocities and physical parameters of HD 553 (Duemmler *et al.*, 2002); absolute dimensions of the M-Type eclipsing binary YY Gem (Torres & Ribas, 2002); the radiative flux scale defined by UV Psc (Vivekananda Rao & Radhika, 2003); the new eclipsing binary within a triple system V1154 Tau (Dallaporta *et al.*, 2003); photometric study of the binary system NN Del (Gómez Forrellad *et al.*, 2003); eclipsing binaries showing apsidal motion: V366 Pup, PT Vel and V466 Car (Otero, 2003a); orbits and stellar parameters for UW LMi, V432 Aur and CN Lyn (Marrese *et al.*, 2004); variable depths of minima of the eclipsing binary V685 Cen (Mayer *et al.*, 2004); orbital parameters for V353 Hya (Otero & Stephan, 2004); echelle spectroscopy and BV photometry of V432 Aur (Siviero *et al.*, 2004); a short-period eclipsing binary in a triple system BS Ind (Guenther *et al.*, 2005); 10 eclipsing binaries showing apsidal motion (Otero, 2005); fundamental parameters of the eclipsing binaries WW Aur and HD 23642 (Southworth *et al.*, 2004); absolute dimensions of WW Aur (Southworth *et al.*, 2005b); age and helium content of V2154 Cyg (Fernandes *et al.*, 2007); period change, spot migration and orbital solution for AR Lac (Siviero *et al.*, 2006); light-curve analysis of the eccentric eclipsing binary V744 Cas (Bulut *et al.*, 2006); apsidal motion and photometric elements of V401 Lac (Bulut & Demircan, 2006); orbit of the spectroscopic-eclipsing-interferometric triple ξ Tau (Bolton *et al.*, 2007); detection of radio emission from HIP 68718 (Anderson & Filipović, 2007); masses and radii of the components of V398 Lac (Cakırlı *et al.*, 2007); physical parameters of five southern systems (Szalai *et al.*, 2007); precise light-curve of β Aur obtained with the Wide Field Infrared Explorer (WIRE) satellite (Southworth *et al.*, 2007); astrometry and light-time effects for VW Cep, ξ Phe and HT Vir (Zasche & Wolf, 2007); apsidal motion of the eccentric eclipsing binary KL CMa (Bulut, 2008).

3.7 Contact binaries: W UMa, symbiotic, and RS CVn systems

From the morphology of their light-curves, binaries are divided into detached (EA) or contact binaries. The latter include (according to their GCVS designations) the EW or W UMa-type, the EB or β Lyr type, and the ELL or ellipsoidal type.

Classification according to evolutionary phase includes the β Lyr type (mass transfer from the more massive, lobe-filling star to its companion has just started, and is happening rapidly), Algol type or semi-detached systems (in which material is being transferred more slowly from a depleted red giant to its now more massive companion), and cataclysmic variables (comprising a closely orbiting white dwarf and red dwarf, with light from the system dominated by an accretion disk comprising material falling onto the white dwarf).

The EW, or W UMa-type, systems are eclipsing binaries with orbital periods between about 5–20 hr showing continuous light variations, and roughly equally deep eclipses. They consist of two solar-type stars surrounded by a common envelope. There is a large-scale energy transfer from the larger more massive component onto the smaller, roughly equalising the surface temperature over the entire system. This characteristic is one of the discriminating characteristics of the W UMa contact binaries, explained by the contact model of Lucy (1968). They normally show a slight reddening at both eclipses due to the combined effects of gravity and limb darkening. The components are rapidly rotating, with equatorial velocities of $\sim$ 100–200 km s^{-1}, due to spin–orbit synchronisation resulting from strong tidal interactions. Reviews of observational aspects and theoretical aspects are given by Rucinski (1992) and Eggleton (1996) respectively, with a recent catalogue of 361 field contact binaries compiled by Pribulla *et al.* (2003). This lists ephemerides for the primary minimum, the minimum and maximum visual brightness, photometric elements, $(m_1 + m_2)\sin^3 i$, spectral type, and parallax. A list of contact binaries in the Hipparcos Catalogue was given by Duerbeck (1997).

The distinction between the EW and EB or ELL is sometimes difficult, especially when colour curves are not available. The EB class is a modification of the EW class but with unequally deep eclipses, corresponding to a component temperature-difference effect. Physically, they are a mixture of contact binaries in poor thermal contact, pre-contact semi-detached systems, and semi-detached Algols, after the mass exchange and mass-ratio reversal (Rucinski, 2002b). The ellipsoidal or ELL variables are largely EW and EB binaries seen at non-eclipsing inclination angles.

Symbiotic stars are interacting binaries in which an evolved giant transfers material onto a much hotter compact companion. The binary typically comprises a red giant transferring material to a white dwarf via a stellar wind. In some systems the red giant is replaced by a yellow giant, and the white dwarf by a main sequence or neutron star. S-type symbiotic stars represent the majority and contain a non-pulsating M giant. Ellipsoidal light variations, characteristic of tidally-distorted stars, are therefore rarely observed for these systems. Classification criteria include absorption features of a late-type giant, and the presence of strong emission lines of H I and He I.

RS CVn stars, named after the prototype RS Canum Venaticorum, are active binaries in which photospheric cool spots result in distortions in the wide-band light-curves. Modelling of the effects assists estimation of the true luminosity ratio, the fractional radii, and the inclination of the orbital plane.

For the determination of the photometric elements from close binary light-curves, use is frequently made of the eclipsing binary code originally developed by Wilson & Devinney (1971). In its original form monochromatic light-curves were calculated with allowance for rotational and tidal distortion, the reflection effect, limb darkening, and gravity darkening. The inverse problem, finding the orbital elements from the observed light-curves, is accomplished by differential corrections, with probable errors obtained for all adjustable parameters.

General properties and statistics Several studies of a period–colour relation for W UMa systems have been made in the past (e.g. Rucinski, 1994, 1995, 1997b). That such a relationship might exist is based on their rather simple external structure, in which a single value for the potential replaces the two radii of detached binaries, and a single surface temperature can describe the common surface. The orbital period P and the intrinsic colour $(B-V)_0$ are correlated through the combined effects of similar geometry, Kepler's third law, and main-sequence relationships. That such a period–colour relation should apply, as for pulsating stars, rests on the fact that the orbital periods are in a simple relation to dynamical time scales, and that the radiating areas scale with the length of the period. A first revision based on Hipparcos parallaxes was made by Rucinski & Duerbeck (1997a). In a more detailed study, Rucinski & Duerbeck (1997b) used Hipparcos parallax data for 40 contact binaries of the W UMa type to construct a period–colour relation based on a wide colour range $0.26 < (B-V)_0 < 1.14$, covering $P \sim 0.24-1.15$ d, and $1.4 < M_V < 6.1$ mag, and weighted according to the relative parallax error. They derived

$$M_V(\text{Cal}) = -4.44 \log P + 3.02(B-V)_0 + 0.12 \qquad (3.25)$$

They found that the influence of mass ratio is absorbed by the colour term, that the system inclination shows a small but statistically negligible effect, and that the accuracy of the calibration is limited by the lack of reliable photometric information, where an uncertainty of 0.03 in $B-V$ contributes to a deviation of 0.1 mag in the predicted M_V.

The space density of W UMa variables has been known to be very large since the studies of Shapley (1948), since which time estimates have ranged between lower values of $\sim 10^{-6}\,\text{pc}^{-3}$ (Kraft, 1967), to as high as $11 \times 10^{-5}\,\text{pc}^{-3}$ (van't Veer, 1975), with more recently accepted values around $10^{-5}\,\text{pc}^{-3}$ (Duerbeck, 1984) and implications for theories of their evolution. These values imply a relative frequency of occurrence, amongst main-sequence stars, of around 0.1%. Hipparcos discovered a significant number of new systems, sufficient to warrant a re-determination of their space density. Rucinski (2002b) defined a sample, including the EW, EB, and EL types, complete to $V^{\max} = 7.5$ mag and estimated to include all discoverable short-period ($P < 1$ day) binaries with photometric variation larger than 0.05 mag. Of the 32 systems in the final sample, 11 were discovered by Hipparcos. The classification and selection of

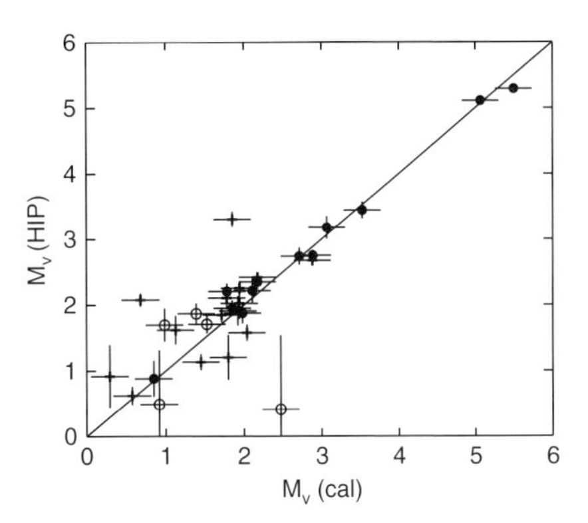

Figure 3.31 Comparison of the absolute magnitude for the sample of contact binaries determined from the Hipparcos parallaxes with those estimated from Equation 3.25 from Rucinski & Duerbeck (1997b). Symbols are: •: EQ; ∘: EB; +: ELL. Vertical error bars are from the errors in the Hipparcos parallaxes, while the errors in M_V (Cal) are assumed fixed at 0.22 mag. From Rucinski (2002b, Figure 2).

objects followed the procedure described by Rucinski (1997b,a). Separation of the EW and EB systems from the detached binaries (EA systems) was done in the a_2-a_4 plane (see Figure 3.32, bottom left), while the a_2-a_1 plane was used to separate the EW and EB systems (see Figure 3.32c, bottom right). Light-curve asymmetry (the O'Connell effect: Davidge & Milone, 1984) can be quantified in the first sine coefficient, b_1. The comparison between the Hipparcos parallax-based absolute magnitudes, and those derived from the period–colour relation of Rucinski & Duerbeck (1997b), Equation 3.25, is shown in Figure 3.31. Agreement is very good for the EW systems, and still reasonable for the EB and ELL types. The combined spatial density was estimated as $(1.02 \pm 0.24) \times 10^{-5}\,\text{pc}^{-3}$. They found that the relative frequency of occurrence, defined in relation to main-sequence stars, depends on the luminosity: at around 1/500 for $M_V > +1.5$, but rapidly decreasing for brighter binaries to a level of 1/5000 for $M_V < +1.5$ and to 1/30 000 for $M_V < +0.5$. The value of 1/130, previously determined from the deep OGLE-I sample of disk population W UMa type systems toward Baade's window, is inconsistent with these results. Possible reasons for the large discrepancy include observational effects, but also the possibility of a genuine increase in the contact-binary density in the central parts of the Galaxy. Further studies of the luminosity function of W UMa systems were presented by Rucinski (2006).

Selam (2004) studied all 64 W UMa-type systems which were newly-discovered by Hipparcos, without imposing a magnitude limit. Because of the same uncertainty in the classification of the newly-discovered Hipparcos variables noted by Rucinski (2002b), their starting list included 79 systems flagged both as EW-type

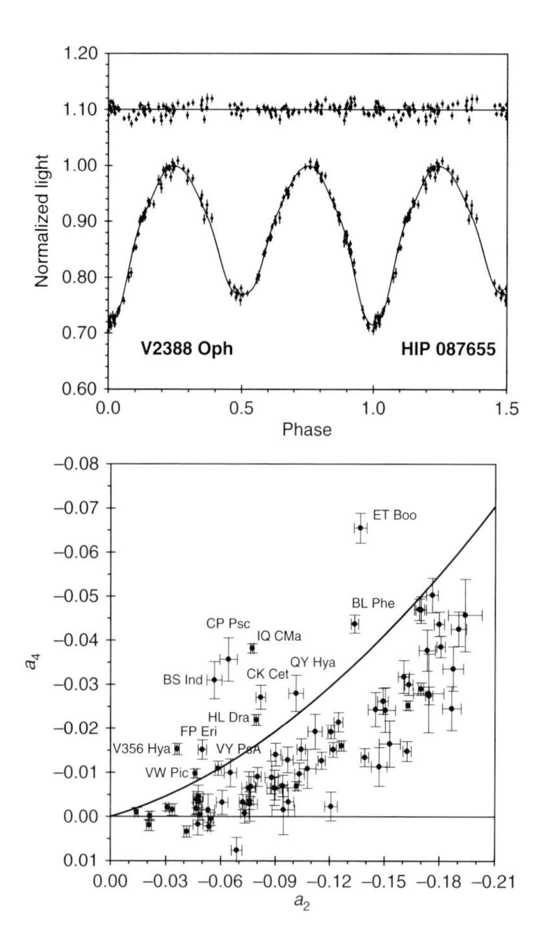

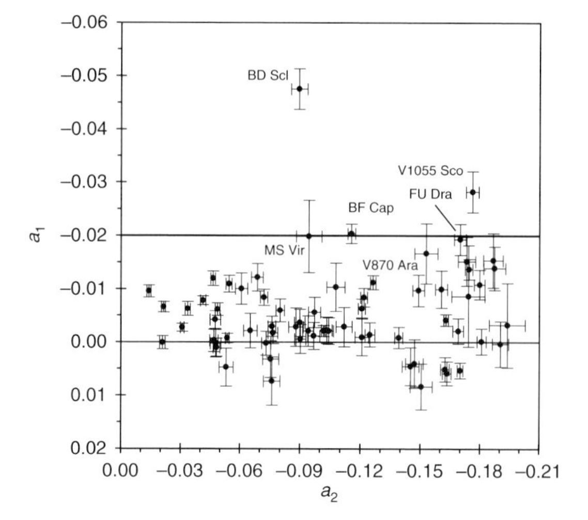

Figure 3.32 Left: Fourier fit for the Hipparcos-discovered W UMa contact binary V2388 Oph (residuals at top). Below left: Fourier coefficients a_2 versus a_4 used to separate detached binaries (above curve) from EW and EB binaries (below curve). The curve $a_4 = a_2(0.125 - a_2)$ is the theoretical position of systems at marginal contact. Below right: Fourier coefficients a_1 versus a_2 used to separate EB binaries from EW (W UMa) binaries. The separation line $a_1 = -0.02$ was suggested by Rucinski (1997a,b). From Selam (2004, Figures 1–3).

(W UMa systems) and EB-type (the β Lyr systems), with periods shorter than a day and spectral type later than A5. Their final classification was also based on Fourier analysis of the Hipparcos light-curves, and Fourier filtering again based on the coefficients a_1, a_2, and a_4, as illustrated in Figure 3.7. Physical parameters of the systems were then determined by interpolation in the nomograms determined by Rucinski (1997a), according to which the brightness variations are totally dominated by geometric causes dependent on three key parameters: the mass ratio $q = m_2/m_1$, the orbital inclination i, and the degree of contact f. Preliminary photometric solutions were obtained for 53 systems for the first time. Spectroscopically-determined mass ratios are known for 19 systems, and further high-accuracy photometry will allow more complete solutions to be obtained, notably the identification of which component is eclipsed at each light-curve minimum.

Klagyivik & Csizmadia (2004) compiled a catalogue of light-curve solutions of 159 W UMa-type systems. They determined the rate of luminosity transfer from the primary to the secondary, and established photometric distances which they compared with the Hipparcos parallaxes, reporting a good agreement between them. Fuhrmann (2004) drew attention to the blue straggler or merger candidates HD 165401 and HD 137763 & HD 137778; with other future merger candidates including σ CrB A having $P = 1.14$ days and 44 Boo B with $P = 6.4$ hr and a distance of only 12.7 pc. Another candidate is HD 220182 discussed in Gaidos *et al.* (2000), and other nearby candidates for descendants of short-lived contact systems in Rocha Pinto *et al.* (2002). Ogloza & Zakrzewski (2006) studied the spatial distribution of contact binaries, including the Hipparcos W UMa stars.

Symbiotic systems The pre-Hipparcos astrometric data for symbiotic stars was sparse: few objects had published proper motions, and few parallaxes, none of any significance, with distances obtained only by indirect methods (none were known to be associated with open or globular clusters). The Hipparcos Catalogue includes about 14 symbiotic stars, listed by Munari *et al.* (1997). Formally, the largest parallax is for R Aqr, with

$\pi = 5.07 \pm 3.15$ mas, with just one significant at the 3σ level, CH Cyg with $\pi = 3.73 \pm 0.85$ mas. They estimated the height of each star above the Galactic plane, results which tend to support the notion from ground-based infrared photometry and radial velocities that the bulk of symbiotic stars belong to the bulge or thick disk populations. Although poorly quantified, the Galactic distribution and space density is important for models which place symbiotic stars as progenitors of Type Ia supernovae. A recent catalogue of 188 confirmed and 30 suspected symbiotic stars is given by Belczyński *et al.* (2000), which includes the Hipparcos parallax although for most objects the uncertainties are rather high.

Few statistical studies of symbiotic stars based on Hipparcos data have been reported, due to the general faintness of the systems and their large distances typically at the limit of the parallax accuracies. Viotti *et al.* (1997) showed that the position of associated radio emission, generated by the ionising radiation from the hot component, is generally close to the Hipparcos optical position. In the case of multiple radio sources, the Hipparcos position is close to the central source of CH Cyg (at a Hipparcos distance of 270 ± 66 pc), to the brightest source of AG Dra, and to the SiO maser emission of R Aqr.

Individual systems Other studies of individual systems which make use of the Hipparcos data in various ways, including both photometry and astrometry, include: the characteristics of the cool component of the cataclysmic binary AE Aqr (Friedjung, 1997); the photometric versus orbital period of the symbiotic system AG Dra (Bastian, 1998); eclipsing phenomena of the symbiotic star CH Cyg (Iijima, 1998); parallaxes of cataclysmic binaries and their absolute magnitudes (Duerbeck, 1999); the early-type components of the contact binary HV UMa (Csák *et al.*, 2000); the active eclipsing binary RT And (Pribulla *et al.*, 2000); system parameters for the A-type W UMa star V1073 Cyg (Morris & Naftilan, 2000); optical, X-ray and radio observations of the RS CVn-type binary HD 61396 (Padmakar *et al.*, 2000); photometric modelling and age of RS CVn (Rodonò *et al.*, 2001); photometry of the RS CVn stars HK Lac, SZ Psc, and λ And (Percy, 2001); infrared light-curves and absolute stellar parameters of the Algol system δ Lib (Lazaro *et al.*, 2002); a multi-epoch spectrophotometric atlas of symbiotic stars (Munari & Zwitter, 2002); investigation of the new Algol-type binary HD 61273 (Royer *et al.*, 2002); photometry of eclipsing contact binaries: U Peg, YY CrB, OU Ser and EQ Tau (Pribulla & Vanko, 2002); Mg II in RS CVn stars (Cardini *et al.*, 2003); symbiotic stars in phase space: from Hipparcos to Gaia (Munari, 2003); light-curves and orbital solution for AM Leo (Hiller *et al.*, 2004); synthetic spectral analysis of the hot component in the symbiotic variable EG And (Kolb *et al.*, 2004); dimensions and distance of the contact binary V417 Aql (Lee *et al.*, 2004a); dimensions and distance of the contact binary CK Boo (Lee *et al.*, 2004b); photometric analysis of EL Aqr (Wadhwa & Zealey, 2004); photometric analysis of the contact binary systems UX Ret and CN Hyi (Wadhwa & Zealey, 2005); the Algol system XX Cep (Angione & Sievers, 2006); dense spot coverage and polar caps on the RS CVn binary SV Cam (Jeffers *et al.*, 2006); a third component in the Algol system δ Lib (Bakış *et al.*, 2006); photometric study of the W UMa eclipsing binaries VW LMi and BX Dra (Sánchez Bajo *et al.*, 2007).

Novae Novae are the visible signatures of a cataclysmic nuclear explosion caused by heating and compression of gas impacting the surface of a white dwarf as it detaches from a binary main-sequence or red giant companion as it overflows its Roche lobe. Recurrence times between successive outbursts are related to the mass of the white dwarf and the mass-transfer rate, but are typically upwards of 30 000 yr.

Hipparcos has made no significant contribution to these investigations, other than to providing a consistent set of photometric standard, notably Tycho B_T and V_T magnitudes, to establish the magnitudes of these events (e.g. Filippenko *et al.*, 1998; Hanzl, 1999; Liller & Yamamoto, 1999; Zejda *et al.*, 1999; Hanzl *et al.*, 2000; Hanzl, 2000; Platais *et al.*, 2001; Hanzl & Lehky, 2001).

Dall *et al.* (2005) argued that the common proper motion system FH Leo, a wide visual binary of separation 8.31 arcsec and observed together by Hipparcos as HIP 54268, and classified as a nova-like variable due to an outburst observed in the Hipparcos photometry cannot be a nova. This classification by Hipparcos prompted detailed follow up observations, from which their favoured explanation is a planetary accretion event. These ideas are discussed further in Section 10.7.4. More recent analysis by Vogt (2006) has proposed that the event could be explained as a dwarf nova at the same distance as the binary system, most probably a triple system with the dwarf nova orbiting one or other of the two brighter binary components, in which case it would be the first cataclysmic variable detected as a component of a multiple system.

3.8 Ground-based follow-up observations

3.8.1 Astrometry

A series of papers reporting observations of binaries with the 50 cm and 74 cm telescopes of the Nice Observatory, including discoveries by Hipparcos, and providing position angle, angular separation, and magnitude difference, include: Thorel (1998, 38 binaries); Salaman *et al.* (1999, 49 binaries); Morlet *et al.* (2000, 65 binaries); Salaman *et al.* (2001, 58 binaries); Gili & Bonneau (2001,

606 binaries); and Morlet *et al.* (2002, 167 binaries). Sinachopoulos *et al.* (2007a) measured 213 northern and 219 southern wide systems with at least one component having a Hipparcos entry.

3.8.2 Radial velocity and spectroscopy

A number of papers report radial velocity studies of Hipparcos stars, with the goal of discovering new multiple stars, or further characterising known systems. Radial velocity studies of close binary stars, including discoveries by Hipparcos, each covering 10 objects, have been reported by Lu & Rucinski (1999); Rucinski & Lu (1999); Rucinski *et al.* (2000); Lu *et al.* (2001); Rucinski *et al.* (2001); Rucinski *et al.* (2002); Rucinski (2002a); Rucinski *et al.* (2003); Pych *et al.* (2004); Rucinski *et al.* (2005); Pribulla *et al.* (2006); and Pribulla *et al.* (2007). Studies of composite spectra, which by early 2007 had reached Paper 11, have been reported by Ginestet *et al.* (1999), Carquillat *et al.* (2003), Carquillat *et al.* (2005), and Carquillat & Prieur (2007). The series of photoelectric observations by Roger Griffin had, by the end of 2006, reached Paper XVII for various source lists (see Griffin, 2006, and references therein), and Paper 191 in the case of individual spectroscopic binary orbits (Griffin & Boffin, 2006). Radial velocity studies of several hundred late-type giant stars selected from the Hipparcos Catalogue have been carried out as part of studies for the SIM PlanetQuest astrometric reference frame by Reffert *et al.* (2007). Frémat *et al.* (2006) reported radial velocity variability for 33 bright Hipparcos A-type stars. Radial velocity studies of X-ray binaries from the cross-correlation of the ROSAT All-Sky Survey–Tycho sample (see Section 8.3) have been reported by Frasca *et al.* (2006b) and Frasca *et al.* (2006a). Spectroscopic studies of hot subdwarf composite-spectrum binaries was reported by Stark & Wade (2006).

3.8.3 Photometry

Many papers report ground-based photometric and astrometric observations of binary systems, including systems observed or discovered by Hipparcos. Some of the larger-scale photometric programmes targeted at Hipparcos binaries are noted here.

Shatsky (1998) reported BVR photometry of 198 visual components in 82 multiple systems; in combination with proper motions from Hipparcos and Tycho they assess whether distant companions are indeed physical associations.

In mid-1990 a large collaborative 'Visual Double Stars' project was started, including an ESO key programme on the Dutch 91-cm telescope at La Silla introduced in 1992, with the goal of obtaining photometric and astrometric observations of visual double and multiple stars with angular separations in the range 1–15 arcsec and forming part of the Hipparcos (Input) Catalogue. The aim was to provide accurate astrometry at other epochs to assist an orbital analysis of the Hipparcos data, and to obtain complementary accurate multicolour data for systems for which high-quality astrometric data would eventually exist (Oblak *et al.*, 1999). CCD photometry and differential astrometry was published for 288 systems in the range 2–12 arcsec by Cuypers & Seggewiss (1999). The photometric data include Cousins V and I magnitudes and $V - I$ indices with internal errors of around 0.005 mag and external errors below 0.03 mag. Up to 15 measurements of each system yields internal errors of 4 mas in separations and $0.^{\circ}05$ in position angles, and providing good general agreement with the Hipparcos data. Results for a further 253 southern intermediate systems were reported by Lampens *et al.* (2001).

A programme with similar objectives was set up in the northern hemisphere using telescopes in Crete and Bulgaria, concentrating on Hipparcos and nearby stars, with results published by Lampens & Strigachev (2001, nine systems); Strigachev & Lampens (2004, 31 systems); and Lampens *et al.* (2007, 71 systems).

Lampens *et al.* (2004) obtained high-angular resolution infrared differential photometry for a sample of nearby F, G, K orbital binaries, focusing on objects that do not fit the mean mass–luminosity relation, but which have accurate Hipparcos parallaxes and high-quality orbits. The aim is to characterise each component in terms of $T_{\rm eff}$ and chemical composition. The sample includes 27 binaries with $P < 29$ yr for which new astrometric mass ratios were determined by Söderhjelm (1999). Other ground-based photometric observations of 645 Hipparcos binaries have been presented by Soulié (2006a,b).

3.8.4 Speckle interferometry

One of the most productive of the ground-based observational follow-ups to the Hipparcos mission has been in the field of speckle interferometry. A large number of close binary and multiple systems were discovered and characterised within the published Hipparcos Catalogue; more have been identified through the recognition of $\Delta\mu$ binaries, and others through the Tycho 2 Catalogue Double Star Catalogue. In addition to the measurement of over 12 000 Hipparcos double (3406 of which were new), were the 4706 entries of the combined G, O, V, or X annexes, and 6981 entries treated as single objects but classified as suspected non-single (flag S in Field H61) thus yielding 11 687 'problem stars'.

Together these provide an enormous hunting ground for observational characterisation from the ground: providing confirmation of the multiple nature, providing multi-colour photometry for component classification and, most importantly, providing additional measurements at distinct epochs for orbital determination. As described previously, accurate orbits are potentially applicable to the mass determination of the individual binary components.

Speckle interferometry is a technique which has been in use for more than 30 years, and is one which provides the most important route to characterising the dynamical properties of these objects. It allows the reconstruction, over a small field referred to as the isoplanatic patch, of diffraction-limited angular resolution using telescopes on the ground, in the visual wavebands, and over the domain most relevant to the Hipparcos and Tycho double and multiple stars: $V_{\rm lim} \sim 11$ mag, $\rho \sim 0.03-5$ arcsec, and $\Delta m \lesssim 3$ mag. The technique does not provide absolute astrometry, but rather the relative separation, position angle, and magnitude difference, at the epoch of observation. A review of speckle imaging in binary star research over the past decade has been given by Horch *et al.* (2006), who noted that *'the advent of CCD-based speckle imaging and the publication of the Hipparcos Catalogue have played important roles in the ongoing vitality of the speckle technique'.*

America: USNO, CHARA and WIYN observations In the US, speckle groups have been active at the US Naval Observatory (USNO, Washington), mainly making use of the USNO 26-inch refractor; at the Center for High Angular Resolution Astronomy (CHARA, Georgia State University, Atlanta), between 1977–1998, making use of several 2.5–4.0-m telescopes; and at the Center for Imaging Science (Rochester Institute of Technology), mostly making use of the Wisconsin–Indiana–Yale–NAOA (WIYN) 3.5-m telescope at Kitt Peak.

The USNO instrumentation, based on an intensified CCD, and data reduction methodology, has been described by Douglass *et al.* (1997); subsequently the detector was changed to an EOS-CCD, as described by Germain *et al.* (1999a), and more recently by Mason *et al.* (2006a). On the USNO 26-inch (0.66-m) refractor, where most of the recent USNO observations have been conducted, this permitted a 1 mag increase in dynamic range, to $\Delta m \sim 3.5$ mag, for $\rho \sim 2$ arcsec, and detections of secondaries as faint as 12.5 mag. Random errors are around $\sigma_\rho \sim 14$–18 mas in separation, and $\sigma_{\rm PA} \sim 0.52$–$0.57/\rho^\circ$ in position angle.

By early 2007, the US Naval Observatory had observed more than 2000 of the Hipparcos and Tycho binaries and, as reported by Brian Mason *'...the end is not yet in sight'.* Observations have been made with the CTIO 4-m, the Mount Wilson 100-inch (2.5-m), the McDonald Observatory 2.1-m, and the USNO 26-inch telescope.

The CHARA speckle interferometry programme operated at Georgia State University between 1977–1998. Mason *et al.* (1999b) reported that, at that time, over 20 000 observations had been published, and some 50 000 observations had been stored in the CHARA archives, including null and dubious detections. The relevant post-1981 observations had been made with various telescopes: the 2.5-m (the Mt Wilson 100-inch), the CFH 3.6-m, the KPNO 3.8-m, and the CTIO 4-m, corresponding to angular resolution between 0.034–0.054 arcsec at 550 nm.

The first two papers devoted to the new and problem Hipparcos binaries illustrated the various challenges to be faced. Mason *et al.* (1999b) reported USNO and CHARA observations of 848 new and problem Hipparcos binaries from archival and new dedicated observations. They found a rate of detection among the different annex types of 13% (G), 25% (V), 25% (O), 9% (X), and 8% (S, suspected). Such limited success statistics must be interpreted with some caution, since there are many parameters which affect the detectability, related to magnitude limits, changing separations of close systems, atmospheric conditions, and the content of archival target lists. But they already concluded that the relatively high proportion of S stars, for example, tended to show that these objects probably deserve such a designation in the Hipparcos Catalogue. Mason *et al.* (2001a) reported the observation of 116 new Hipparcos doubles and 469 Hipparcos problem stars from McDonald Observatory during 1998–99. Some had already been observed by Mason *et al.* (1999b): of these some were confirmed, some were detected in the first measurements and not the second, and vice versa. This was interpreted as due to the system being close to the detection limit of one or other of the critical parameters (ρ or Δm) at the time of either observation, and suggested that the verification of many of the Hipparcos double stars would require more than one attempt. As all Hipparcos stars are relatively bright, and had probably been surveyed by micrometer before (e.g. Rossiter, 1955) it is likely that the Hipparcos discoveries would indeed be of the more marginally detectable companions. Success rates were 5/226 G type, 0/34 V type, 1/43 O type, 1/49 X type, and 0/126 S type.

Speckle interferometry at the US Naval Observatory has been the subject of a series of papers, initially not focused on the Hipparcos Catalogue stars but noted here for completeness. After the low success rate on these 'problem' objects, observations of Hipparcos stars have been concentrated on the newly-resolved pairs. Results, all obtained with the 0.66-m refractor, are reported in:

∘ 2329 measurements of 467 binaries, $\rho = 0.3$–3.5 arcsec, 1990–92 (Douglass *et al.*, 1997);

∘ 2406 measurements of 547 binaries, $\rho = 0.2–3.8$ arcsec, 1993–95 (Germain *et al.*, 1999b);

∘ 2578 measurements of 590 binaries, $\rho = 0.2–4.3$ arcsec, 1995–96 (Germain *et al.*, 1999a);

∘ 1314 measurements of 625 binaries, $\rho = 0.2–5.2$ arcsec, 1997 (Douglass *et al.*, 1999);

∘ 1544 measurements of 637 binaries, $\rho = 0.2–5.2$ arcsec, 1998 (Douglass *et al.*, 2000);

∘ 1068 measurements of 815 binaries, $\rho = 0.2–6.0$ arcsec, 1999–2000 (Mason *et al.*, 2000a);

∘ 2014 measurements of 1266 mean positions, $\rho =$ 0.2–13.4 arcsec, 2000 (Mason *et al.*, 2001b);

∘ 2044 measurements of 1399 mean positions, $\rho =$ 0.2–15.0 arcsec, 2001 (Mason *et al.*, 2002);

∘ 3056 measurements of 1675 mean positions, $\rho =$ 0.2–45.2 arcsec, 2002 (Mason *et al.*, 2004a);

∘ 3047 measurements of 1572 mean positions, $\rho =$ 0.2–62.9 arcsec, 2003 (Mason *et al.*, 2004b);

∘ 1683 measurements of 805 mean positions, $\rho =$ 0.2–43.3 arcsec, 2004 (Mason *et al.*, 2006a);

∘ 1657 measurements of 1111 mean positions, $\rho =$ 0.2–17.0 arcsec, 2005 (Mason *et al.*, 2006b).

Hipparcos Catalogue objects formed only part of these observations. In the last of the above references, they also placed emphasis on observing double systems not observed recently from the ground, and the statistics again illustrate some of the limits in our knowledge of even bright double stars: 106 of the systems had not been observed in the last 50 years, 12 had not been observed in the last 100 years, and three of these were first resolved by J. Herschel in 1820.

Speckle observations of binary stars with the WIYN 3.5-m telescope at Kitt Peak started to observe the Hipparcos binaries from 1997, using both MAMA (multianode microchannel array) and CCD detectors. With the latter, they reported standard errors of 3.5 mas in separation and $1\overset{\circ}{.}2$ in position angle, with reliable astrometry obtained even on a system with $\Delta m \sim 5.3$ mag. Their results (again, only partly focused on Hipparcos binaries) have been reported in a series of papers, which include confirmation of orbital motion and consistency with the Hipparcos photometry:

∘ measurements of 154 binaries, 16 discovered by Hipparcos, 1997 (Horch *et al.*, 1999);

∘ measurements of 253 binaries, 53 discovered by Hipparcos, 1998–2000 (Horch *et al.*, 2002a);

∘ measurements of 230 binaries with spectral types A–G (Horch *et al.*, 2002b);

∘ 576 magnitude differences for 260 binaries, 1997–2000 (Horch *et al.*, 2004a).

More recent observations, including Hipparcos binaries, and using a new speckle camera RYTSI (RIT-Yale Tip–Tilt Speckle Imager) have been reported by Horch *et al.* (2004b), and Horch *et al.* (2005).

Use of the Rochester Institute CCD speckle camera at telescopes of the Cerro Tololo Inter-American Observatory (CTIO, Chile) for a series of southern hemisphere observations of binary systems, including some from the Hipparcos Catalogue, were reported by Horch *et al.* (1997, 2000, 2001, 2006). Gatewood (2005) reported an astrometric study of the binary α Oph also making use of Hipparcos, speckle, and Multichannel Astrometric Photometer (MAP) data.

In an application illustrating the use of Hipparcos data in combination with other binary information, Mason *et al.* (1998) reported a speckle interferometric survey made with the CHARA speckle camera and various 4-m class telescopes of 227 Galactic O-type stars with $V < 8$. They combined their new discovery of 15 binaries in the range $\rho \sim 0.035–1.5$ arcsec and $\Delta m < 3$ with wider pairs from the Washington Double Star Catalogue, fainter pairs from the Hipparcos Catalogue, and known spectroscopic binaries. From the resulting overall binary frequency and orbital characteristics, they concluded that binaries are common among O stars in clusters and associations, but less so among field and especially runaway stars: they found relative fractions of 72.2%, 19.4%, and 8.4% respectively, consistent with earlier indications (Gies, 1987; Garmany, 1994). They found many triple systems, suggesting a role in the ejection of stars from clusters. They found an orbital period distribution bimodal in log P, although binaries with periods of years and decades may eventually be found to fill the gap. The mass ratio distribution of visual binaries increases toward lower mass ratios, but low mass ratio companions are rare among close, spectroscopic binaries.

Russia: Special Astrophysical Observatory A series of speckle observations of Hipparcos binary and triple systems has been undertaken with an intensified CCD at the BTA 6-m telescope of the Special Astrophysical Observatory of the Russian Academy of Sciences. Balega *et al.* (1999) reported measurements for four multiple systems. Balega *et al.* (2002) reported observations made in 1998–99 of 48 Hipparcos discoveries with $\pi >$ 10 mas and $\rho < 1$ arcsec. Binaries as faint as 15 mag, with companions as close as 20 mas, can be resolved. Binarity was confirmed for all observed pairs except HIP 15597, 10 of which had fast orbital motion. Orbital elements for HIP 16602 and 21280 were improved, and a first preliminary orbit for HIP 689 was determined. A comparison between the Hipparcos and speckle component magnitude differences showed very good agreement (Figure 3.33). Absolute magnitudes and approximate spectral types were found for the components of 63 binary systems and four triples, using estimated Δm, the Hipparcos parallaxes, and the total V magnitudes. Balega *et al.* (2004) reported observations of a further

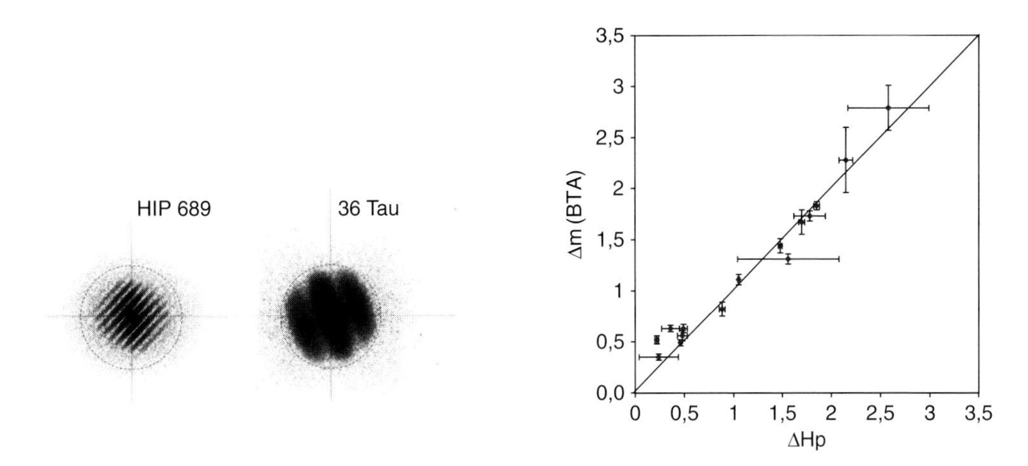

Figure 3.33 Left: ensemble average power spectra for HIP 689 and 36 Tau showing speckle fringes corresponding to their separations of 133 mas and 32 mas respectively. Right: comparison of the Hipparcos magnitude differences with the speckle magnitude difference made in the V′ filter (central wavelength 545 nm, width 30 nm) at the BTA 6-m telescope of the Special Astrophysical Observatory. From Balega et al. (2002, Figures 1 and 2).

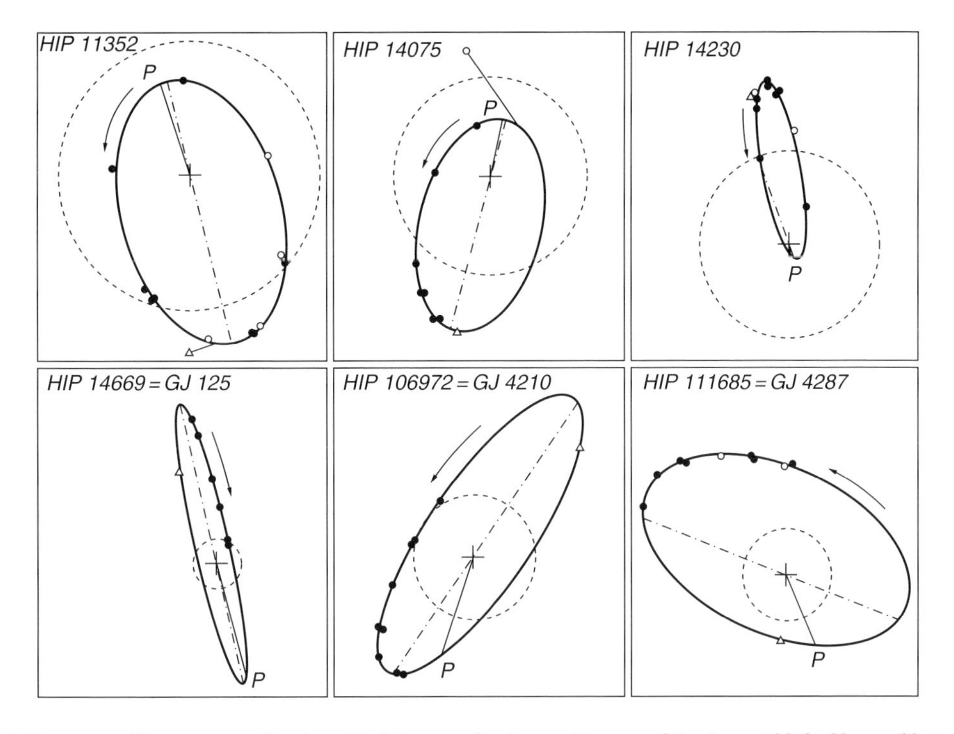

Figure 3.34 Apparent ellipses representing the orbital elements for six new Hipparcos binaries, established by speckle interferometry at the Special Astrophysical Observatory 6-m telescope between 1998–2004. Symbols indicate: •: speckle data obtained at the BTA 6-m telescope; ○: other interferometric data; △: Hipparcos first epoch measurements. Residual vectors for all measurements are plotted, but in most cases they are smaller than the points themselves. Orbital motion is indicated by an arrow. The solid line shows the periastron position, while the dot–dashed line represents the line of nodes. The dashed circle has a radius of 0.1 arcsec. From Balega et al. (2005, Figure 1).

54 new Hipparcos binaries, of which two were found to be triple systems.

Balega *et al.* (2005) reported speckle observations of six new Hipparcos binaries between 1998 and 2004, leading to definitive orbits in the range 6–28 yr for HIP 11352, HIP 14075, HIP 14230, HIP 14669, HIP 106972, and HIP 111685 (see Figure 3.34). Balega *et al.* (2006) reported a further six orbits for HIP 4809, HIP 4849, HIP 5531, HIP 19206, HIP 105947, and HIP 114922, in roughly the same period range. The total mass errors for all systems remains rather high, partly due to the high contribution from the Hipparcos parallax error. Figure 3.35 shows the location of the components of all 12 systems in the mass–luminosity diagram, in which individual masses were largely derived from mass ratios defined by the empirical mass–luminosity relation for main sequence stars from Henry & McCarthy (1993). For HIP 5531 and HIP 19206, both SB2 systems, mass ratios were obtained from spectroscopic observations. Hipparcos and dynamical parallaxes agree well, typically to within 1–2 mas, with the exception of HIP 14669 ($\pi_{\rm dyn} = 69.86$, $\pi_{\rm Hip} = 64.83$) and HIP 106972 ($\pi_{\rm dyn} = 44.85$, $\pi_{\rm Hip} = 39.78$) which both have faint M-type components. Balega *et al.* (2007) used speckle data and radial velocities to derive accurate masses for the low-mass stars GJ 765.2AB of around $0.83 M_\odot$ and $0.76 M_\odot$.

Docobo *et al.* (2006) reported speckle measurements of a further five systems, determining one new and four improved orbits, and comparing the resulting dynamical and Hipparcos parallaxes. Three further systems were analysed by Tamazian & Docobo (2006).

Other speckle observations interpreted in the context of the Hipparcos parallaxes were reported for the Mira variables R Cas (Hofmann *et al.*, 2000) and R Leo (Hofmann *et al.*, 2001).

Europe A series of speckle observations have been made at the 1.52-m telescope at Calar Alto using an intensified CCD speckle camera developed at the Ramón María Aller Observatory of the University of Santiago de Compostella in collaboration with the Special Astrophysical Observatory. Some Hipparcos binaries have been observed, and Hipparcos and dynamical parallaxes compared. Results are reported for 83 systems by Docobo *et al.* (2001b), for 101 systems by Docobo *et al.* (2004), and for 87 systems by Docobo *et al.* (2007), all in the separation range 0.1–6.6 arcsec. Woitas *et al.* (2003) reported speckle observations and a resulting orbit for the low-mass binary Gliese 22 AC.

Observations at the 2-m Bernard Lyot telescope of the Pic du Midi Observatory using the PISCO intensified CCD camera have also included some of the Hipparcos binaries. Results are reported for 43 systems observed

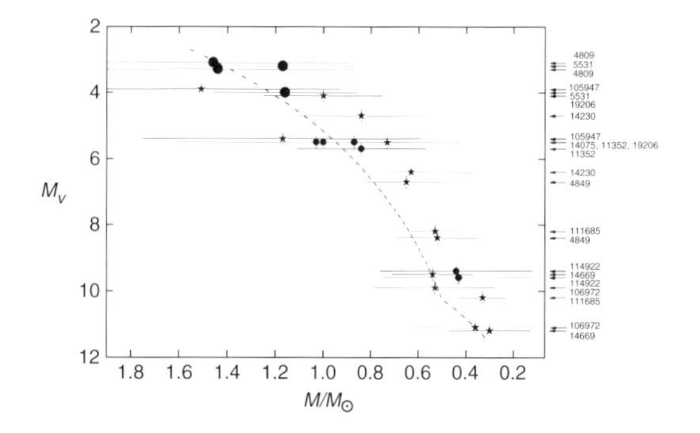

Figure 3.35 Location of the components of 12 new binaries on the mass–luminosity diagram. Symbols: •: binaries with $\Delta m \le 1$; ⋆: binaries with $\Delta m > 1$. The larger circles indicate two systems with evolved components, HIP 4809 and HIP 5531. Error bars are assumed to be the same as for the mass sums. The dashed line is the empirical mass–luminosity relation from Henry & McCarthy (1993). From Balega et al. (2006, Figure 2).

by Hipparcos by Prieur *et al.* (2001), for 47 composite spectrum stars, including the Hipparcos discovery HD 156729 (HIP 84606) with $\Delta m = 4.2$, by Prieur *et al.* (2002), and for a comparison with differential photometry for 18 close binaries by Prieur *et al.* (2003).

3.8.5 Adaptive optics

The technique of adaptive optics aims to achieve diffraction-limited angular resolution in the optical, or more technically accessible in the near infrared, by rapid adaptation of a mirror profile to compensate for the phase irregularities introduced by the atmosphere. Use of the Hipparcos and Tycho stars as references for wavefront sensors is discussed in Section 10.3. The technique also allows for follow-up of Hipparcos binary or multiple systems. Specific studies related to the nature of the Hipparcos binaries include: adaptive optics observations of Arcturus (Turner *et al.*, 1999); differential photometry using the adaptive optics system at Mt Wilson (ten Brummelaar *et al.*, 2000); a natural guide star-based multi-conjugate adaptive optics demonstrator (Bello *et al.*, 2002); star asterisms for natural guide star based multi-conjugate adaptive optics (Marchetti *et al.*, 2002); orbit, masses, and photometry for ι Cas from adaptive optics imaging (Drummond *et al.*, 2003); adaptive optic imaging of 341 Hipparcos BA-type stars within 300 pc (Ivanov *et al.*, 2006).

In an application illustrating the use of Hipparcos data in combination with other binary star information Kouwenhoven *et al.* (2005) carried out an adaptive optics survey (using the ADONIS/SHARPII+ system on the ESO 3.6-m telescope) of Hipparcos stars in the three subgroups of the nearby (130 pc) OB association Scorpius OB2: Upper Scorpius (US), Upper Centaurus

Lupus (UCL) and Lower Centaurus Crux (LCC). Membership of the associations had been defined on the basis of Hipparcos astrometry by de Zeeuw *et al.* (1999). They targeted 199 A-type and late B-type stars in the K_S band, and a subset also in the J and H band, finding 74 candidate physical companion stars, of which 41 had not been reported previously. Angular separations were in the range 0.22–12.4 arcsec, corresponding to a projected separation of 28.6–1612 AU. Absolute magnitudes for all components were derived using individual parallaxes and interstellar extinction, and masses from K_S assuming an age of 5 Myr for the US subgroup, and 20 Myr for the UCL and LCC subgroups. Companion star masses range from 0.10–$3.0 M_\odot$, with a mass ratio distribution following a power-law form proposed in a previous study by Shatsky & Tokovinin (2002), $f(q) = q^{-\Gamma}$, with $\Gamma = 0.33$. The absence of $\rho < 3.75$ arcsec companion stars in the magnitude range $K_S = 12-14$ mag may imply a lower limit on the companion mass of around $0.1 M_\odot$. Combined with visual, spectroscopic, and astrometric data on binarity in Sco OB2, they determined an overall companion star fraction of 0.52. Of the Sco OB2 Hipparcos member stars, 37 are now known to be triple: 16 consist of a primary with two visual companions, 15 consist of a primary, a visual companion, and a spectroscopic companion; and six consist of a primary, a spectroscopic companion and an astrometric companion. Five quadruple systems are known: two consist of a primary and three visual companions (HIP 69113 and HIP 81972), two consist of a primary, two visual companions and a spectroscopic companion (HIP 80112 and HIP 78384), while HIP 77820 is a quintuple system, consisting of a primary, three visual companions and one spectroscopic companion. The highest multiplicity system known in Sco OB2 is HIP 78374, which contains a primary, four visual companions and two spectroscopic companions.

3.8.6 Long-baseline interferometry

Long-baseline optical or infrared interferometry is an established technique for determining stellar angular diameters, and provides another method for investigating the geometry of individual binary systems (Monnier *et al.*, 2006; Fekel & Tomkin, 2007). Instruments used in the recent past or still in operation include the Mark III interferometer at Mount Wilson, now closed; the Palomar Testbed Interferometer at Mount Palomar (PTI, Akeson, 2006); the Infrared-Optical Telescope Array at Mount Hopkins (IOTA, Schloerb *et al.*, 2006); the US Navy Prototype Optical Interferometer (NPOI, Johnston *et al.*, 2006); the CHARA optical and near infrared interferometer of Georgia State University at Mount Wilson (ten Brummelaar *et al.*, 2003); the Sydney University Stellar Interferometer (SUSI, Davis *et al.*, 2006); the GI2T of the Observatoire de la Côte d'Azur (Mourard *et al.*, 2006); the ESO Very Large Telescope Interferometer at Cerro Paranal, Chile (VLTI, Glindemann *et al.*, 2003); the Keck Observatory Interferometer in Hawaii (Wizinowich *et al.*, 2006); the Cambridge Mullard Radio Observatory Interferometer (MROI, Buscher *et al.*, 2006); and the Mitaka Optical and Infrared Array (MIRA, Yoshizawa *et al.*, 2004).

Related studies of Hipparcos binary systems which have made use of the Hipparcos data in various ways, including both photometry and astrometry, include: astrometric infrared interferometry: sensitivity and captation field (Daigne, 1997a); optical interferometry and ground-based astrometry after Hipparcos (Daigne, 1997b); Mitaka optical and infrared array projects MIRA-I, II, and III (Nishikawa *et al.*, 1997); NPOI observations of the double stars Mizar A and Matar (Hummel *et al.*, 1998); K giants as astrometric reference stars for the Space Interferometry Mission (Frink *et al.*, 2000); near-infrared IOTA interferometry of the symbiotic star CH Cyg (Hofmann *et al.*, 2003); baseline monitoring for astrometric interferometry (Hrynevych *et al.*, 2004); calibrator stars for 200-m baseline interferometry (Mérand *et al.*, 2005); masses and distance of the triple system HD 158926 (λ Sco) from SUSI (Tango *et al.*, 2006); orbital solution and masses for σ Sco using SUSI (North *et al.*, 2007).

References

Abt HA, Levy SG, 1976, Multiplicity among solar-type stars. *ApJS*, 30, 273–306 {**107**}

Aitken RG, Doolittle E, 1932, *New general catalogue of double stars within* 120° *of the North pole*. Carnegie Institution of Washington {**97**}

Akeson RL, 2006, Recent progress at the Palomar Testbed Interferometer. *Advances in Stellar Interferometry*, Proc. SPIE 6268, 14 {**139**}

Alencar SHP, Melo CHF, Dullemond CP, *et al.*, 2003, The pre-main sequence spectroscopic binary AK Sco revisited. *A&A*, 409, 1037–1053 {**123**}

Allen C, Herrera MA, Poveda A, 1998, Wide binaries among high-velocity and metal-poor stars. *IX Latin American Regional IAU Meeting* {**108**}

—, 1999, Wide binaries among high-velocity and metal-poor stars. *Galaxy Evolution: Connecting the Distant Universe with the Local Fossil Record*, 233 {**108**}

Allen C, Poveda A, Herrera MA, 2000a, Wide binaries among high-velocity and metal-poor stars. *A&A*, 356, 529–540 {**107–110**}

Allen C, Poveda A, 2007, Halo wide binaries and moving clusters as probes of the dynamical and merger history of our Galaxy. *Binary Stars as Critical Tools and Tests in Contemporary Astrophysics, IAU Symp. 240*, 405–413 {**109**}

Allen C, Proveda A, Herrera MA, 2000b, Three very metal-poor, extremely wide binaries. *Stellar Astrophysics, Proceedings of the Pacific Rim Conference held in Hong Kong* (eds. Cheng KS, Chau HF, Chan KL, *et al.*), 439–443 {**109**}

Allen C, Santillán A, 1993, New Galactic orbits and tidal radii for globular clusters. *Revista Mexicana de Astronomia y Astrofisica*, 25, 39–50 {**109**}

Andersen J, 1991, Accurate masses and radii of normal stars. *A&A Rev.*, 3, 91–126 {**111, 127**}

Anderson JA, 1920, Application of Michelson's interferometer method to the measurement of close double stars. *ApJ*, 51, 263–275 {**119**}

Anderson MWB, Filipović MD, 2007, Detection of 6-cm radio-continuum emission from an eclipsing binary (β-Lyrae type) variable star: HIP 68718. *MNRAS*, 381, 1027–1030 {**130**}

Andrade M, 2007, Orbit, masses and spectral analysis of the visual binary A 2329. *Revista Mexicana de Astronomia y Astrofisica*, 43, 237–242 {**125**}

Angione RJ, Sievers JR, 2006, The Algol system XX Cep. *AJ*, 131, 2209–2215 {**133**}

Antonello E, Farinella P, Guerrero G, *et al.*, 1980, The period distribution of eclipsing and spectroscopic binary systems. II. *Ap&SS*, 72, 359–367 {**125**}

Arenou F, 1998, Binaries in acceleration and stochastic Hipparcos solutions. *Highlights in Astronomy*, 11, 549 {**106**}

—, 2001, Hipparcos et les binaires serrées. *Ecole de Goutelas 23, CNRS, 22–26 May 2000*, 23, 127 {**106**}

Arenou F, Halbwachs JL, Mayor M, *et al.*, 2000, Seven SB2 masses using Hipparcos intermediate astrometric data. *IAU Symp. 200*, 135–137 {**100, 114, 115**}

Bachmann PJ, Hershey JL, 1975, Orbital analysis of Algol AB, C from combined astrometric, photometric, and radial velocity data. *AJ*, 80, 836–846 {**128**}

Bahcall JN, Hut P, Tremaine S, 1985, Maximum mass of objects that constitute unseen disk material. *ApJ*, 290, 15–20 {**107**}

Baize P, Romani L, 1946, Formules nouvelles pour le calcul des parallaxes dynamiques des couples orbitaux. *Annales d'Astrophysique*, 9, 13–29 {**116**}

Bakış V, Budding E, Erdem A, *et al.*, 2006, Analysis of δ Lib including Hipparcos astrometry. *MNRAS*, 370, 1935–1945 {**133**}

Balega II, Balega YY, Hofmann KH, *et al.*, 1999, Parameters of four multiple systems from speckle interferometry. *Astronomy Letters*, 25, 797–801 {**136**}

—, 2002, Speckle interferometry of nearby multiple stars. *A&A*, 385, 87–93 {**136, 137**}

—, 2005, Orbits of new Hipparcos binaries. I. *A&A*, 433, 591–596 {**100, 137, 138**}

—, 2006, Orbits of new Hipparcos binaries. II. *A&A*, 448, 703–707 {**100, 138**}

Balega II, Balega YY, Maksimov AF, *et al.*, 2004, Speckle interferometry of nearby multiple stars. II. *A&A*, 422, 627–629 {**136**}

Balega YY, Beuzit JL, Delfosse X, *et al.*, 2007, Accurate masses of low-mass stars GJ 765.2AB. *A&A*, 464, 635–640 {**138**}

Barbieri C, Corrain G, Ragazzoni R, *et al.*, 1997, An orbit and mass of Gliese 623 AB by direct imaging with the HST–FOC. *The First Results of Hipparcos and Tycho, IAU Joint Discussion 14*, 43 {**123**}

Barnes TG, Evans DS, 1976, Stellar angular diameters and visual surface brightness. I. Late spectral types. *MNRAS*, 174, 489–502 {**126**}

Bartkevičius A, Gudas A, 2001, Kinematics of Hipparcos visual binaries. I. Stars with orbital solutions. *Baltic Astronomy*, 10, 481–587 {**105**}

—, 2002a, Gaia and Population II visual binaries. *Ap&SS*, 280, 125–128 {**105**}

—, 2002b, Kinematics of Hipparcos visual binaries. II. Stars with ground-based orbital solutions. *Baltic Astronomy*, 11, 153–203 {**105**}

Bartkevičius A, Sperauskas J, 2005, Radial velocities of Population II binary stars. II. *Baltic Astronomy*, 14, 511–525 {**105, 110**}

Bastian U, 1998, The symbiotic system AG Dra: an unexpected photometric period. *A&A*, 329, L61–L63 {**133**}

Bate MR, 2000, Predicting the properties of binary stellar systems: the evolution of accreting protobinary systems. *MNRAS*, 314, 33–53 {**108**}

Batten AH, Fletcher JM, MacCarthy DG, 1989, The 8th catalogue of the orbital elements of spectroscopic binary systems. *Publications of the Dominion Astrophysical Observatory Victoria*, 17, 1 {**97**}

Belczyński K, Mikołajewska J, Munari U, *et al.*, 2000, A catalogue of symbiotic stars. *A&AS*, 146, 407–435 {**133**}

Bello D, Leroux CR, Le Roux B, *et al.*, 2002, Performances of a NGS-based MCAO demonstrator: the NGC 3366 and NGC 2346 simulations. *Beyond Conventional Adaptive Optics. ESO Proceedings Vol. 58* (eds. Vernet E, Ragazzoni R, Esposito S, *et al.*), 231 {**138**}

Bernstein HH, 1994, Global astrometry of Hipparcos double stars within the FAST consortium. *A&A*, 283, 293–300 {**96, 111**}

—, 1999, Derivation of orbital parameters of very low-mass companions in double stars from radial velocities and observations of space-astrometry missions like Hipparcos, DIVA and Gaia. *ASP Conf. Ser. 185: IAU Colloq. 170: Precise Stellar Radial Velocities*, 410 {**111**}

Bertelli G, Bressan A, Chiosi C, *et al.*, 1994, Theoretical isochrones from models with new radiative opacities. *A&AS*, 106, 275–302 {**115**}

Bertout C, Robichon N, Arenou F, 1999, Revisiting Hipparcos data for pre-main sequence stars. *A&A*, 352, 574–586 {**101**}

Biller BA, Close LM, Li A, *et al.*, 2006, Resolving the dusty circumstellar structure of the enigmatic symbiotic star CH Cyg with the MMT adaptive optics system. *ApJ*, 647, 464–470 {**125**}

Binnendijk L, 1960, *Properties of Double Stars; a Survey of Parallaxes and Orbits*. University of Pennsylvania Press, Philadelphia {**112**}

Boffin HMJ, Cerf N, Paulus G, 1993, Statistical analysis of a sample of spectroscopic binaries containing late-type giants. *A&A*, 271, 125–138 {**115**}

Bolton CT, Grunhut JH, Hurkens R, 2007, The orbit and properties of the spectroscopic-eclipsing-interferometric triple system ξ Tau = HD 21364. *Binary Stars as Critical Tools and Tests in Contemporary Astrophysics, IAU Symp. 240*, 66–70 {**130**}

Bonnell IA, Bate MR, 2005, Binary systems and stellar mergers in massive star formation. *MNRAS*, 362, 915–920 {**108**}

Božić H, Wolf M, Harmanec P, *et al.*, 2007, HD 143418: an unusual light variable and a double-lined spectroscopic binary with a CP primary. *A&A*, 464, 263–275 {**125**}

Brancewicz HK, Dworak TZ, 1980, A catalogue of parameters for eclipsing binaries. *Acta Astronomica*, 30, 501–524 {**125**}

Brown KIT, 2007, Long-term spectroscopic and precise radial velocity monitoring of Arcturus. *PASP*, 119, 237–237 {**121**}

Bulut I, Ciček C, Erdem A, *et al.*, 2006, First ground-based photometry and light curve analysis of the eccentric eclipsing

binary V744 Cas. *Astronomische Nachrichten*, 327, 912–916 {**130**}
Bulut I, 2008, Apsidal motion and photometric elements of the eccentric eclipsing binary KL CMa. *New Astronomy*, 13, 24–27 {**130**}
Bulut I, Demircan O, 2006, Apsidal motion and photometric elements of the eccentric eclipsing binary V401 Lac. *PASJ*, 58, 159–163 {**130**}
Burnham SW, 1894, *Double star observations made with the thirty-six-inch and twelve-inch refractors of the Lick observatory, from August 1888 to June 1892*. Sacramento {**120**}
—, 1906, *A general catalogue of double stars within* 121° *of the North pole*. Carnegie Institution, Washington {**97**}
Buscher DF, Boysen RC, Dace R, *et al.*, 2006, Design and testing of an innovative delay line for the MROI. *Advances in Stellar Interferometry*, Proc. SPIE 6268, 78 {**139**}
Cakırlı Ö, Frasca A, Ibanoğlu C, *et al.*, 2007, Preliminary results on the fundamental parameters of the eclipsing binary V398 Lac. *Astronomische Nachrichten*, 328, 536–542 {**130**}
Campbell WW, 1910, Second catalogue of spectroscopic binary stars. *Lick Observatory Bulletin*, 6, 17–54 {**106**}
Cardini D, Cassatella A, Badiali M, *et al.*, 2003, A study of the Mg II 279.634 nm emission line in late-type normal and RS CVn stars. *A&A*, 408, 337–345 {**133**}
Carney BW, Latham DW, Laird JB, *et al.*, 1994, A survey of proper motion stars. XII. An expanded sample. *AJ*, 107, 2240–2289 {**110**}
Carney BW, 1983, A photometric search for halo binaries. II. Results. *AJ*, 88, 623–641 {**108**}
Carquillat JM, Ginestet N, Prieur JL, *et al.*, 2002, Contribution to the search for binaries among Am stars. III. HD 7119: a double-lined spectroscopic binary and a triple system. *MNRAS*, 336, 1043–1048 {**123**}
Carquillat JM, Ginestet N, Prieur JL, 2001, Contribution to the search of binaries among Am stars. II. HD 81976 and HD 98880, double-lined spectroscopic binaries. *A&A*, 369, 908–914 {**123**}
Carquillat JM, Ginestet N, 2000, Contribution to the study of F, G, K, M binaries. VIII. HD 195850 and HD 201193, double-lined spectroscopic binaries. *A&AS*, 144, 317–321 {**123**}
Carquillat JM, Prieur JL, Ginestet N, 2003, Contribution to the study of composite spectra: IX. Spectroscopic orbital elements of 10 systems. *MNRAS*, 342, 1271–1279 {**134**}
—, 2005, Contribution to the study of composite spectra: X. Five new spectroscopic binaries in multiple systems. *MNRAS*, 360, 718–726 {**134**}
Carquillat JM, Prieur JL, 2007, Contribution to the study of composite spectra: XI. Orbital elements of some faint systems. *Astronomische Nachrichten*, 328, 46–54 {**134**}
Carrier F, North P, Udry S, *et al.*, 2002, Multiplicity among chemically peculiar stars. II. Cool magnetic Ap stars. *A&A*, 394, 151–169 {**123**}
Chanamé J, Gould A, 2004, Disk and halo wide binaries from the revised Luyten Catalogue: probes of star formation and MACHO dark matter. *ApJ*, 601, 289–310 {**107, 110**}
Close LM, Richer HB, Crabtree DR, 1990, A complete sample of wide binaries in the solar neighbourhood. *AJ*, 100, 1968–1980 {**107**}
Couteau P, 1960, Contribution à l'étude du dénombrement des étoiles doubles visuelles. *Journal des Observateurs*, 43(3), 41–56 {**108**}
Couteau P, Morel P, Fulconis M, 1986, *Catalogue d'éphemerides d'étoiles doubles visuelles: 5*. Observatoire de Nice {**97**}
Csák B, Kiss LL, Vinkó J, *et al.*, 2000, HV UMa, a new contact binary with early-type components. *A&A*, 356, 603–611 {**133**}
Cuypers J, Seggewiss W, 1999, CCD photometry and astrometry of visual double and multiple stars of the Hipparcos Catalogue. II. CCD photometry and differential astrometry of 288 southern intermediate systems. *A&AS*, 139, 425–431 {**134**}
Daigne G, 1997a, Astrometric infrared interferometry: sensitivity and captation field. *Proc. SPIE Vol. 2871, Optical Telescopes of Today and Tomorrow* (ed. Ardeberg AL), 540–543 {**139**}
—, 1997b, Optical interferometry and ground-based astrometry after the Hipparcos mission. *NATO ASIC Proc. 501: High Angular Resolution in Astrophysics*, 337 {**139**}
Dallaporta S, Munari U, Zwitter T, 2003, V1154 Tau: a new eclipsing star within a triple system. *Informational Bulletin on Variable Stars*, 5413 {**130**}
Dallaporta S, Tomov T, Zwitter T, *et al.*, 2002, V432 Aur: a new eclipsing system. *Informational Bulletin on Variable Stars*, 5319 {**129**}
Dall TH, Schmidtobreick L, Santos NC, *et al.*, 2005, Outbursts on normal stars. FH Leo misclassified as a nova-like variable. *A&A*, 438, 317–324 {**133**}
D'Antona F, Cardini D, Di Mauro MP, *et al.*, 2005, 85 Peg A: what age for a low-metallicity solar-like star? *MNRAS*, 363, 847–856 {**125**}
Davidge TJ, Milone EF, 1984, A study of the O'Connell effect in the light curves of eclipsing binaries. *ApJS*, 55, 571–584 {**131**}
Davis J, Ireland MJ, Jacob AP, *et al.*, 2006, SUSI: an update on instrumental developments and science. *Advances in Stellar Interferometry*, Proc. SPIE 6268, 4 {**139**}
Davis J, Shobbrook RR, 1977, On uvby indices and empirical effective temperatures and bolometric corrections for B stars. *MNRAS*, 178, 651–659 {**127**}
Debernardi Y, North P, 2001, Eclipsing binaries with candidate chemically peculiar stars. II. Parameters of the system V392 Car. *A&A*, 374, 204–212 {**123, 129**}
Deeg HJ, Doyle LR, Kozhevnikov VP, *et al.*, 2000, A search for Jovian-mass planets around CM Dra using eclipse minima timing. *A&A*, 358, L5–L8 {**128**}
Demircan O, 2000, Period changes as a tool to study unseen components around eclipsing binaries. *NATO ASIC Proc. 544: Variable Stars as Essential Astrophysical Tools* (ed. Ibanoglu C), 615–622 {**128**}
Detournay S, Pourbaix D, 2002, Further processing of the Hipparcos variability induced movers. *EAS Publications, Les Houches. Gaia: A European Space Project*, 2, 367–369 {**101**}
de Zeeuw PT, Hoogerwerf R, de Bruijne JHJ, *et al.*, 1999, A Hipparcos census of the nearby OB associations. *AJ*, 117, 354–399 {**139**}
Dischler J, 2006, *Theoretical studies of binaries in astrophysics*. Ph.D. thesis, Lund University, Sweden {**125**}
Dischler J, Söderhjelm S, 2005, Predicted properties of eclipsing binaries observable by Gaia. *ESA SP–576: The Three-Dimensional Universe with Gaia* (eds. Turon C, O'Flaherty KS, Perryman MAC), 569–572 {**126**}
Djurašević G, Dimitrov D, Arbutina B, *et al.*, 2005, A study of close binary system EE Cet. *Memorie della Societa Astronomica Italiana Supplement*, 7, 168 {**124**}
Doberck W, 1878, On double star orbits. *Astronomische Nachrichten*, 91, 317–322 {**106**}

Docobo JA, 1985, On the analytic calculation of visual double star orbits. *Celestial Mechanics*, 36, 143–153 {**115**}

—, 2001, A new application mode of Docobo's analytical method for visual double star orbits calculation using the Hipparcos parallaxes. *Highlights of Spanish Astrophysics II*, 370 {**115**}

Docobo JA, Andrade M, Ling JF, *et al.*, 2004, Binary star speckle interferometry: measurements and orbits. *AJ*, 127, 1181–1186 {**138**}

Docobo JA, Andrade M, Tamazian VS, *et al.*, 2007, Binary star speckle measurements with the 1.52-m telescope at Calar Alto. *Revista Mexicana de Astronomia y Astrofisica*, 43, 141–147 {**138**}

Docobo JA, Ling JF, Prieto C, *et al.*, 2001a, Catalogue of orbits and ephemerides of visual double stars. *Acta Astronomica*, 51, 353–356 {**97**}

Docobo JA, Ling JF, 2003, Orbits and masses of twenty double stars discovered by Paul Couteau. *A&A*, 409, 989–992 {**115**}

—, 2007, Orbits and system masses of 14 visual double stars with early-type components. *AJ*, 133, 1209–1216 {**116**}

Docobo JA, Tamazian VS, Balega YY, *et al.*, 2001b, Binary star speckle measurements at Calar Alto. I. *A&A*, 366, 868–872 {**138**}

—, 2006, Speckle measurements and differential photometry of visual binaries with the 6-m telescope of the Special Astrophysical Observatory. *AJ*, 132, 994–998 {**138**}

Dommanget J, Nys O, 1994, Catalogue of the components of double and multiple stars (CCDM) first edition. *Communications de l'Observatoire Royal de Belgique*, 115, 1 {**95, 97**}

—, 2000, The visual double stars observed by the Hipparcos satellite. *A&A*, 363, 991–994, erratum: 364, 927 {**97**}

Dommanget J, 2000, The Hipparcos Catalogue and the Tycho Catalogue: analysis of the results for the visual double stars. *The Observatory*, 120, 202–210 {**97**}

—, 2003, The mass/eccentricity limit in double star astronomy. *J. Astrophys. Astron.*, 24, 99 {**106**}

Douglass GG, Hindsley RB, Worley CE, 1997, Speckle interferometry at the US Naval Observatory. I. *ApJS*, 111, 289–334 {**135**}

Douglass GG, Mason BD, Germain ME, *et al.*, 1999, Speckle interferometry at the US Naval Observatory. IV. *AJ*, 118, 1395–1405 {**136**}

Douglass GG, Mason BD, Rafferty TJ, *et al.*, 2000, Speckle interferometry at the US Naval Observatory. V. *AJ*, 119, 3071–3083 {**136**}

Drummond J, Milster S, Ryan P, *et al.*, 2003, ι Cas: orbit, masses, and photometry from adaptive optics imaging in the I and H bands. *ApJ*, 585, 1007–1014 {**138**}

Duemmler R, Iliev IK, Iliev L, 2002, The radial velocities and physical parameters of HD 553. *A&A*, 395, 885–890 {**129**}

Duerbeck HW, 1984, Constraints for cataclysmic binary evolution as derived from space distributions. *Ap&SS*, 99, 363–385 {**131**}

—, 1997, True and possible contact binaries in the Hipparcos Catalogue. *Informational Bulletin on Variable Stars*, 4513 {**130**}

—, 1999, Hipparcos parallaxes of cataclysmic binaries and the quest for their absolute magnitudes. *Informational Bulletin on Variable Stars*, 4731 {**133**}

Dunkin SK, Crawford IA, 1998, Spatially resolved optical spectroscopy of the Herbig Ae/Vega-like binary star HD 35187. *MNRAS*, 298, 275–284 {**123**}

Duquennoy A, Mayor M, 1991, Multiplicity among solar-type stars in the solar neighbourhood. II. Distribution of the orbital elements in an unbiased sample. *A&A*, 248, 485–524 {**91, 104, 105, 125**}

Dworak TZ, 1975, A catalogue of photometric parallaxes of eclipsing binaries. *Acta Astronomica*, 25, 383–416 {**125**}

Eggleton PP, Kisseleva-Eggleton L, Dearborn X, 2007, The incidence of multiplicity among bright stellar systems. *Binary Stars as Critical Tools and Tests in Contemporary Astrophysics, IAU Symp. 240*, 347–355 {**110**}

Eggleton PP, 1996, Evolution of contact binaries. *The Origins, Evolution, and Destinies of Binary Stars in Clusters* (eds. Milone EF, Mermilliod JC), ASP Conf. Ser. 90, 257–269 {**130**}

Emanuele A, Badiali M, Cardini D, *et al.*, 1996, Relative astrometry with Hipparcos: double stars with orbital motion. *A&A*, 312, 1038–1042 {**96, 111**}

ESA, 1997, *The Hipparcos and Tycho Catalogues. Astrometric and Photometric Star Catalogues derived from the ESA Hipparcos Space Astrometry Mission, ESA SP–1200 (17 volumes including 6 CDs)*. European Space Agency, Noordwijk, also: VizieR Online Data Catalogue {**92, 93, 96, 100, 112**}

Evans NR, Schaefer G, Bond HE, *et al.*, 2007, Polaris: mass and multiplicity. *Binary Stars as Critical Tools and Tests in Contemporary Astrophysics, IAU Symp. 240*, 102–104 {**120**}

Fabricius C, Høg E, Makarov VV, *et al.*, 2002, The Tycho double star catalogue. *A&A*, 384, 180–189 {**98**}

Fabricius C, Makarov VV, 2000a, Hipparcos astrometry for 257 stars using Tycho 2 data. *A&AS*, 144, 45–51 {**99–101**}

—, 2000b, Two-colour photometry for 9473 components of close Hipparcos double and multiple stars. *A&A*, 356, 141–145 {**99, 100, 106, 108**}

Falin JL, Mignard F, 1999, Mining in the Hipparcos raw data. *A&AS*, 135, 231–241 {**99–101**}

Famaey B, Jorissen A, Luri X, *et al.*, 2005, Local kinematics of K and M giants from Coravel, Hipparcos, and Tycho 2 data. Revisiting the concept of superclusters. *A&A*, 430, 165–186 {**102**}

Farinella P, Paolicchi P, 1978, The period distribution of eclipsing and spectroscopic binary systems. I. *Ap&SS*, 54, 389–406 {**125**}

Fekel FC, Barlow DJ, Scarfe CD, *et al.*, 2005, HD 166181 = V815 Her, a single-lined spectroscopic multiple system. *AJ*, 129, 1001–1007 {**121, 122, 124**}

Fekel FC, Henry GW, Barlow DJ, *et al.*, 2006, HD 131861, a double-line spectroscopic triple system. *AJ*, 132, 1910–1917 {**125**}

Fekel FC, Henry GW, 2006, Spectroscopy and photometry of the double-lined binary HD 149420. *AJ*, 131, 1724–1729 {**125**}

Fekel FC, Scarfe CD, Barlow DJ, *et al.*, 2002, The quadruple system μ Ori: three-dimensional orbit and physical parameters. *AJ*, 123, 1723–1740 {**123**}

Fekel FC, Tomkin J, 2007, Spectroscopic binary candidates for interferometers. *Binary Stars as Critical Tools and Tests in Contemporary Astrophysics, IAU Symp. 240*, 59–61 {**139**}

Fekel FC, Williamson M, Pourbaix D, 2007, The spectroscopic and astrometric orbits of HR 672. *AJ*, 133, 2431–2434 {**125**}

Fernandes J, Lebreton Y, Baglin A, *et al.*, 1998, Fundamental stellar parameters for nearby visual binary stars: η Cas, XI Boo, 70 Oph and 85 Peg. Helium abundance, age and mixing length parameter for low-mass stars. *A&A*, 338, 455–464 {**123**}

Fernandes J, Oblak E, Kurpinska-Winiarska M, 2007, The eclipsing binary system V2154 Cyg: observations and models. *Binary Stars as Critical Tools and Tests in Contemporary Astrophysics, IAU Symp. 240*, 388–392 {**130**}
Filippenko AV, Leonard DC, Modjaz M, *et al.*, 1998, Nova Ophiuchi 1998. *IAU Circ.*, 6943, 1 {**133**}
Fischer DA, Marcy GW, 1992, Multiplicity among M dwarfs. *ApJ*, 396, 178–194 {**91**}
Fisher J, Schröder KP, Smith RC, 2004, Volume-limited spectroscopic binary statistics. *Revista Mexicana de Astronomia y Astrofisica Conf. Ser.*, volume 21, 65–66 {**106**}
—, 2005, What a local sample of spectroscopic binaries can tell us about the field binary population. *MNRAS*, 361, 495–503 {**106**}
Flower PJ, 1996, Transformations from theoretical HR diagrams to colour-magnitude diagrams: effective temperatures, $B-V$ colours, and bolometric corrections. *ApJ*, 469, 355–365 {**127**}
Frankowski A, Jancart S, Jorissen A, 2007, Proper motion binaries in the Hipparcos catalogue. Comparison with radial velocity data. *A&A*, 464, 377–392 {**100, 102, 104**}
Frasca A, Guillout P, Marilli E, *et al.*, 2006a, Newly-discovered active binaries in the Rosat All-Sky Survey–Tycho sample of stellar X-ray sources. I. Orbital and physical parameters of six new binaries. *A&A*, 454, 301–309 {**134**}
Frasca A, Marilli E, Guillout P, *et al.*, 2006b, Late-type X-ray emitting binaries in the solar neighbourhood and in star-forming regions. *Ap&SS*, 304, 17–20 {**134**}
Frémat Y, Lampens P, van Cauteren P, *et al.*, 2006, Analysis of main sequence A-type stars showing radial velocity variability. *Memorie della Societa Astronomica Italiana*, 77, 174–175 {**134**}
Friedjung M, 1997, The characteristics of the cool component of the cataclysmic binary AE Aqr derived from its Hipparcos parallax. *New Astronomy*, 2, 319–322 {**133**}
Frink S, Quirrenbach A, Fischer DA, *et al.*, 2000, K giants as astrometric reference stars for the Space Interferometry Mission. *Proc. SPIE Vol. 4006, Interferometry in Optical Astronomy* (eds. Lena PJ, Quirrenbach A), 806–814 {**139**}
Fuhrmann K, 2004, Nearby stars of the Galactic disk and halo. III. *Astronomische Nachrichten*, 325, 3–80 {**132**}
Gaidos EJ, Henry GW, Henry SM, 2000, Spectroscopy and photometry of nearby young solar analogues. *AJ*, 120, 1006–1013 {**132**}
Gálvez MC, Montes D, Fernández Figueroa MJ, *et al.*, 2002, Multiwavelength optical observations of chromospherically-active binary systems. IV. The X-ray/extreme ultraviolet selected binary BK Psc (2RE J0039+103). *A&A*, 389, 524–536 {**123**}
Garmany CD, 1994, OB associations: massive stars in context. *PASP*, 106, 25–37 {**136**}
Gatewood G, 2005, An astrometric study of the binary star α Oph. *AJ*, 130, 809–814 {**124, 136**}
Gatewood G, Coban L, Han I, 2003, An astrometric study of the low-mass binary star Ross 614. *AJ*, 125, 1530–1536 {**123**}
Gatewood G, Han I, de Jonge JK, *et al.*, 2001, Hipparcos and MAP studies of the triple star π Cep. *ApJ*, 549, 1145–1150 {**123**}
Germain ME, Douglass GG, Worley CE, 1999a, Speckle interferometry at the US Naval Observatory. III. *AJ*, 117, 2511–2527 {**135, 136**}
—, 1999b, Speckle interferometry at the US Naval Observatory. II. *AJ*, 117, 1905–1920 {**136**}
Gies DR, 1987, The kinematical and binary properties of association and field O stars. *ApJS*, 64, 545–563 {**136**}
Gili R, Bonneau D, 2001, CCD measurements of visual double stars made with the 74 cm and 50 cm refractors of the Nice Observatory. *A&A*, 378, 954–957 {**134**}
Ginestet N, Griffin RF, Carquillat JM, *et al.*, 1999, Contribution to the study of composite spectra: VIII. HD 174016-7, an Ap star with a giant G . *A&AS*, 140, 279–285 {**134**}
Glindemann A, Algomedo J, Amestica R, *et al.*, 2003, The VLTI: a status report. *Interferometry for Optical Astronomy II.* (ed. Traub WA), Proc. SPIE 4838, 89–100 {**139**}
Goldberg D, Mazeh T, Latham DW, 2003, On the mass-ratio distribution of spectroscopic binaries. *ApJ*, 591, 397–405 {**91**}
Goldin A, Makarov VV, 2006, Unconstrained astrometric orbits for Hipparcos stars with stochastic solutions. *ApJS*, 166, 341–350 {**100, 112, 116**}
—, 2007, Astrometric orbits for Hipparcos stochastic binaries. *ApJS*, 173, 137–142 {**100, 116, 117**}
Gómez Forrellad JM, García Melendo E, 1997, Photometric results on three Hipparcos variables: the new eclipsing binary systems HD 125488 and HD 126080, and the star HD 341508. *Informational Bulletin on Variable Stars*, 4469 {**129**}
Gómez Forrellad JM, Sánchez Bajo F, Corbera Subirana M, *et al.*, 2003, Photometric study of the binary system NN Del. *Ap&SS*, 283, 297–304 {**130**}
Gontcharov GA, 2007, A hundred new preliminary orbits and masses from Hipparcos, ground-based astrometry and radial velocities. *Binary Stars as Critical Tools and Tests in Contemporary Astrophysics, IAU Symp. 240*, 265–269 {**100, 102, 103, 117**}
Gontcharov GA, Andronova AA, Titov OA, *et al.*, 2001, The proper motions of fundamental stars. I. 1535 stars from the basic FK5. *A&A*, 365, 222–227 {**121**}
Gontcharov GA, Andronova AA, Titov OA, 2000, New astrometric binaries among Hipparcos stars. *A&A*, 355, 1164–1167 {**102, 103, 117**}
Gontcharov GA, Kiyaeva OV, 2002a, Astrometric orbits from a direct combination of ground-based catalogues with the Hipparcos Catalogue. *Astronomy Letters*, 28, 261–271 {**100, 102, 103, 117**}
—, 2002b, Photocentric orbits from a direct combination of ground-based astrometry with Hipparcos. I. Comparison with known orbits. *A&A*, 391, 647–657 {**102, 103, 117**}
Gould A, Chanamé J, 2004, New Hipparcos-based parallaxes for 424 faint stars. *ApJS*, 150, 455–464 {**100, 110**}
Griffin REM, Griffin RF, 2000, Composite spectra. X. The equal-mass binary HR 2030. *MNRAS*, 319, 1094–1108 {**123**}
Griffin RF, 1998, Arcturus as a double star. *The Observatory*, 118, 299–301 {**120**}
—, 2006, Photoelectric radial velocities. XVII. The orbits of 30 spectroscopic binaries in the southern Clube Selected Areas. *MNRAS*, 371, 1159–1172 {**134**}
Griffin RF, Boffin HMJ, 2006, Spectroscopic binary orbits from photoelectric radial velocities. Paper 191: HD 17310, HD 70645, and HD 80731. *The Observatory*, 126, 401–421 {**120, 134**}
Griffin RF, Suchkov AA, 2003, The nature of over-luminous F stars observed in a radial velocity survey. *ApJS*, 147, 103–144 {**100, 103**}
Guenther DB, Demarque P, 2000, α Cen AB. *ApJ*, 531, 503–520 {**123**}

Guenther EW, Covino E, Alcalá JM, *et al.*, 2005, BS Indi: an enigmatic object in the Tucana association. *A&A*, 433, 629–634 {**130**}

Guinan EF, Ribas I, 2001, The best brown dwarf yet? A companion to the Hyades eclipsing binary V471 Tau. *ApJ*, 546, L43–L47 {**128**}

Guirado JC, Reynolds JE, Lestrade JF, *et al.*, 1997, Astrometric detection of a low-mass companion orbiting the star AB Dor. *ApJ*, 490, 835–839 {**123**}

Halbwachs JL, Arenou F, Mayor M, *et al.*, 2000, The nearby spectroscopic binaries re-visited with Hipparcos. *IAU Symp. 200*, 132 {**104**}

Halbwachs JL, Arenou F, 1999, On derivation of masses of the SB2 components with Gaia astrometry. *Baltic Astronomy*, 8, 301–308 {**123**}

Halbwachs JL, Mayor M, Udry S, *et al.*, 2003, Multiplicity among solar-type stars. III. Statistical properties of the F7–K binaries with periods up to 10 years. *A&A*, 397, 159–175 {**91, 104, 108**}

—, 2004, Statistical properties of solar-type close binaries. *Revista Mexicana de Astronomia y Astrofisica Conf. Ser.*, volume 21, 20–27 {**104**}

Halbwachs JL, Piquard S, Virelizier P, *et al.*, 1997, A statistical study of the visual double stars in the Tycho Catalogue. *ESA SP–402: Hipparcos, Venice '97*, 263–268 {**97**}

Hanzl D, 1999, V1493 Aql. *IAU Circ.*, 7254, 5 {**133**}

—, 2000, CI Aql. *IAU Circ.*, 7444, 3 {**133**}

Hanzl D, Lehky M, 2001, Possible nova in Puppis. *IAU Circ.*, 7557, 4 {**133**}

Hanzl D, Reszelski M, Hornoch K, *et al.*, 2000, CI Aql: probable nova in Aquila. *IAU Circ.*, 7411, 3 {**133**}

Han I, Black DC, Gatewood G, 2001, Preliminary astrometric masses for proposed extrasolar planetary companions. *ApJ*, 548, L57–L60 {**117**}

Han I, Gatewood G, 2002, A precise orbit determination of χ^1 Ori from astrometric and radial velocity data. *PASP*, 114, 224–228 {**123**}

Hartkopf WI, Mason BD, Worley CE, 2001a, The 2001 US Naval Observatory Double Star CD-ROM. II. The Fifth Catalogue of Orbits of Visual Binary Stars. *AJ*, 122, 3472–3479 {**97, 105**}

Hartkopf WI, McAlister HA, Mason BD, 2001b, The 2001 US Naval Observatory Double Star CD-ROM. III. The Third Catalogue of Interferometric Measurements of Binary Stars. *AJ*, 122, 3480–3481 {**97**}

Harvin JA, Gies DR, Bagnuolo WG, *et al.*, 2002, Tomographic separation of composite spectra. VIII. The physical properties of the massive compact binary in the triple star system HD 36486 (δ Ori A). *ApJ*, 565, 1216–1230 {**123**}

Hauck B, Mermilliod M, 1998, uvbyβ photoelectric photometric catalogue. *A&AS*, 129, 431–433 {**103**}

Hauser HM, Marcy GW, 1999, The orbit of 16 Cyg AB. *PASP*, 111, 321–334 {**123**}

Heber U, 1992, Hot sub-luminous stars. *The Atmospheres of Early-Type Stars* (eds. Heber U, Jeffery CS), volume 401 of *Lecture Notes in Physics, Springer–Verlag, Berlin*, 233–246 {**107**}

Heintz WD, 1978, *Double Stars*. Reidel, Dordrecht: Geophysics and Astrophysics Monographs {**112, 116**}

Henry TJ, McCarthy DW, 1993, The mass–luminosity relation for stars of $1.0 - 0.08\,M_\odot$. *AJ*, 106, 773–789 {**115, 138**}

Henry TJ, Subasavage JP, Brown MA, *et al.*, 2004, The solar neighbourhood. X. New nearby stars in the southern sky and accurate photometric distance estimates for red dwarfs. *AJ*, 128, 2460–2473 {**117**}

Herschel W, Watson D, 1782, Catalogue of double stars. *Royal Society of London Philosophical Transactions Series I*, 72, 112–162 {**120**}

Hiller ME, Osborn W, Terrell D, 2004, New light curves and orbital solution for AM Leo. *PASP*, 116, 337–344 {**133**}

Hofmann KH, Balega YY, Scholz M, *et al.*, 2000, Multi-wavelength bispectrum speckle interferometry of R Cas and comparison of the observations with Mira star models. *A&A*, 353, 1016–1028 {**138**}

—, 2001, Multi-wavelength bispectrum speckle interferometry of R Leo and comparison with Mira star models. *A&A*, 376, 518–531 {**138**}

Hofmann KH, Beckmann U, Berger JP, *et al.*, 2003, Near-infrared IOTA interferometry of the symbiotic star CH Cyg. *Interferometry for Optical Astronomy II. Proc. SPIE, Vol. 4838* (ed. Traub WA), 1043–1046 {**139**}

Holmberg J, Flynn C, Lindegren L, 1997, Towards an improved model of the Galaxy. *ESA SP–402: Hipparcos, Venice '97*, 721–726 {**104**}

Hoogeveen GJ, 2005, Additional data for 69 variables. *Informational Bulletin on Variable Stars*, 5652 {**129**}

Horch EP, Baptista BJ, Veillette DR, *et al.*, 2006, CCD speckle observations of binary stars from the southern hemisphere. IV. Measures during 2001. *AJ*, 131, 3008–3015 {**135, 136**}

Horch EP, Franz OG, Ninkov Z, 2000, CCD speckle observations of binary stars from the southern hemisphere. II. Measures from the Lowell–Tololo telescope during 1999. *AJ*, 120, 2638–2648 {**136**}

Horch EP, Meyer RD, van Altena WF, 2004a, Speckle observations of binary stars with the WIYN telescope. IV. Differential photometry. *AJ*, 127, 1727–1735 {**136**}

Horch EP, Ninkov Z, Franz OG, 2001, CCD speckle observations of binary stars from the southern hemisphere. III. Differential photometry. *AJ*, 121, 1583–1596 {**136**}

Horch EP, Ninkov Z, Slawson RW, 1997, CCD speckle observations of binary stars from the southern hemisphere. I. *AJ*, 114, 2117–2127 {**136**}

Horch EP, Ninkov Z, van Altena WF, *et al.*, 1999, Speckle observations of binary stars with the WIYN telescope. I. Measures during 1997. *AJ*, 117, 548–561 {**136**}

Horch EP, Robinson SE, Meyer RD, *et al.*, 2002a, Speckle observations of binary stars with the WIYN telescope. II. Relative astrometry measures during 1998–2000. *AJ*, 123, 3442–3459 {**136**}

Horch EP, Robinson SE, Ninkov Z, *et al.*, 2002b, Speckle observations of binary stars with the WIYN telescope. III. A partial survey of A, F, and G dwarfs. *AJ*, 124, 2245–2253 {**136**}

Horch EP, van Altena WF, Meyer RD, *et al.*, 2004b, Observations of Hipparcos double stars with the WIYN telescope. *Revista Mexicana de Astronomia y Astrofisica Conf. Ser.*, volume 21, 69–70 {**136**}

Horch EP, van Altena WF, Meyer RD, 2005, The WIYN speckle programme: Hipparcos binaries and beyond. *ASP Conf. Ser. 338: Astrometry in the Age of the Next Generation of Large Telescopes* (eds. Seidelmann PK, Monet AKB), 90 {**136**}

Hrynevych MA, Ligon ER, Colavita MM, 2004, Baseline monitoring for astrometric interferometry. *New Frontiers in Stellar Interferometry, Proc. SPIE Vol. 5491* (ed. Traub WA), 1649 {**139**}

Hubber DA, Whitworth AP, 2005, Binary star formation from ring fragmentation. *A&A*, 437, 113–125 {**108**}

Hummel CA, Armstrong JT, Quirrenbach A, *et al.*, 1994, Very high precision orbit of Capella by long baseline interferometry. *AJ*, 107, 1859–1867 {**119**}

Hummel CA, Mozurkewich D, Armstrong JT, *et al.*, 1998, Navy Prototype Optical Interferometer observations of the double stars Mizar A and Matar. *AJ*, 116, 2536–2548 {**139**}

Hurley JR, Tout CA, Pols OR, 2002, Evolution of binary stars and the effect of tides on binary populations. *MNRAS*, 329, 897–928 {**125**}

Iijima T, 1998, Eclipsing phenomena of the symbiotic star CH Cyg. *MNRAS*, 297, 77–83 {**133**}

Innes RTA, Dawson BH, van den Bos WH, 1927, *Southern double star catalogue* $-19°$ *to* $-90°$. Union Observatory, Johannesburg SA {**97**}

Irwin AW, Campbell B, Morbey CL, *et al.*, 1989, Long-period radial-velocity variations of Arcturus. *PASP*, 101, 147–159 {**121**}

Ivanov VD, Chauvin G, Foellmi C, *et al.*, 2006, Common proper motion search for faint companions around early-type field stars: progress report. *Ap&SS*, 304, 247–249 {**138**}

Jancart S, Jorissen A, Babusiaux C, *et al.*, 2005a, Astrometric orbits of SB9 stars. *A&A*, 442, 365–380 {**100, 118–120**}

Jancart S, Jorissen A, Pourbaix D, 2005b, Hipparcos astrometric binaries in the Ninth Catalogue of Spectroscopic Binary Orbits: a testbench for the detection of astrometric binaries with Gaia. *ESA SP–576: The Three-Dimensional Universe with Gaia* (eds. Turon C, O'Flaherty KS, Perryman MAC), 583–586 {**118**}

Jeffers HM, van Denbos WH, Greeby FM, 1963, *Index catalogue of visual double stars, 1961.0*. Publications of the Lick Observatory {**97**}

Jeffers SV, Cameron AC, Barnes JR, *et al.*, 2006, Dense spot coverage and polar caps on SV Cam. *Ap&SS*, 304, 371–373 {**133**}

Johnston KJ, Benson JA, Hutter DJ, *et al.*, 2006, The Navy Prototype Optical Interferometer (NPOI): recent developments since 2004. *Advances in Stellar Interferometry*, Proc. SPIE 6268, 6 {**139**}

Jordi C, Ribas I, Gimenez A, *et al.*, 1997, Effective temperature determination of eclipsing binaries. *ESA SP–402: Hipparcos, Venice '97*, 409–412 {**127**}

Jorissen A, Famaey B, Dedecker M, *et al.*, 2004a, Spectroscopic binaries among a complete sample of Hipparcos M giants. *Revista Mexicana de Astronomia y Astrofisica Conf. Ser.*, volume 21, 71–72 {**106, 107**}

Jorissen A, Jancart S, Pourbaix D, 2004b, Binaries in the Hipparcos data: keep digging. I. Search for binaries without a priori knowledge of their orbital elements: Application to barium stars. *ASP Conf. Ser. 318: Spectroscopically and Spatially Resolving the Components of the Close Binary Stars* (eds. Hilditch RW, Hensberge H, Pavlovski K), 141–143 {**99**}

Kamper KW, 1996, Polaris today. *JRASC*, 90, 140–157 {**120**}

Kaplan GH, Makarov VV, 2003, Astrometric detection of binary companions and planets: acceleration of proper motion. *Astronomische Nachrichten*, 324, 419–424 {**102**}

Karoff C, Rauer H, Erikson A, *et al.*, 2007, Identification of variable stars in COROT's first main observing field (LRc1). *AJ*, 134, 766–777 {**126**}

Kim CH, Lee JW, Kim HI, *et al.*, 2003, Photometric studies of the triple star ER Ori. *AJ*, 126, 1555–1562 {**123**}

Kiselev AA, Kiyaeva OV, Izmailov IS, 2007, The mass excess in the systems of wide visual double stars on the basis of apparent motion parameters method, Hipparcos parallax and WDS data. *Binary Stars as Critical Tools and Tests in Contemporary Astrophysics, IAU Symp. 240*, 129–133 {**109**}

Kiyaeva OV, 2006, Astrometric study of the triple star ADS 9173. *Astronomy Letters*, 32, 836–844 {**125**}

Kiyaeva OV, Kiselev AA, Polyakov EV, *et al.*, 2001, An astrometric study of the triple star ADS 48. *Astronomy Letters*, 27, 391–397 {**123**}

Klagyivik P, Csizmadia S, 2004, Distance to W UMa stars. *ASP Conf. Ser. 318: Spectroscopically and Spatially Resolving the Components of the Close Binary Stars* (eds. Hilditch RW, Hensberge H, Pavlovski K), 195–197 {**132**}

Knapp GR, Pourbaix D, Platais I, *et al.*, 2003, Reprocessing the Hipparcos data of evolved stars. III. Revised Hipparcos period-luminosity relationship for Galactic long-period variable stars. *A&A*, 403, 993–1002 {**101**}

Köhler R, Leinert C, 1998, Multiplicity of T Tauri stars in Taurus after ROSAT. *A&A*, 331, 977–988 {**91**}

Köhler R, Zinnecker H, Jahreiß H, 1998, Multiplicity of Population II stars. *Astronomische Gesellschaft Meeting Abstracts*, 29 {**110**}

—, 2000, Multiplicity of Population II stars. *IAU Symp. 200*, 148 {**110**}

Kolb KJ, Miller JK, Sion EM, *et al.*, 2004, Synthetic spectral analysis of the hot component in the S-type symbiotic variable EG And. *AJ*, 128, 1790–1794 {**133**}

König B, Guenther EW, Woitas J, *et al.*, 2005, The young active binary star EK Dra. *A&A*, 435, 215–223 {**124**}

Kouwenhoven MBN, Brown AGA, Zinnecker H, *et al.*, 2005, The primordial binary population. I. A near-infrared adaptive optics search for close visual companions to A star members of Scorpius OB2. *A&A*, 430, 137–154 {**138**}

Kovaleva D, Piskunov AE, Malkov O, 2001, Double-lined eclipsing binaries: probe for stellar evolution models. *Astronomische Gesellschaft Meeting Abstracts*, 501 {**123**}

Kraft RP, 1967, On the structure and evolution of W UMa stars. *PASP*, 79, 395–413 {**131**}

Kruszewski A, Semeniuk I, 1999, Nearby Hipparcos eclipsing binaries for colour-surface brightness calibration. *Acta Astronomica*, 49, 561–575 {**126, 128**}

Kurpinska-Winiarska M, Oblak E, Winiarski M, *et al.*, 2000, Observations of two Hipparcos eclipsing variables. *Informational Bulletin on Variable Stars*, 4823 {**129**}

Kurpinska-Winiarska M, Oblak E, 2000, Observations of new Hipparcos eclipsing binaries in multiple-star systems. *IAU Symp. 200*, 141 {**129**}

Lampens P, Kovalevsky J, Froeschlé M, *et al.*, 1997, On the mass–luminosity relation. *ESA SP–402: Hipparcos, Venice '97*, 421–424 {**111**}

Lampens P, Oblak E, Duval D, *et al.*, 2001, CCD photometry and astrometry for visual double and multiple stars of the Hipparcos Catalogue. III. CCD photometry and differential astrometry for 253 southern intermediate systems. *A&A*, 374, 132–150 {**134**}

Lampens P, Prieur JL, Argyle RW, 2004, Infrared differential photometry of selected orbital binaries. *Revista Mexicana de Astronomia y Astrofisica Conf. Ser.*, volume 21, 75–76 {**114, 134**}

Lampens P, Strigachev A, Duval D, 2007, Multicolour CCD measurements of visual double and multiple stars. III. *A&A*, 464, 641–645 {**134**}

Lampens P, Strigachev A, 2001, Multicolour observations of nearby visual double stars. New CCD measurements and orbits. *A&A*, 368, 572–579 {**134**}

Lastennet E, Lejeune T, Westera P, *et al.*, 1999, Metallicity-dependent effective temperature determination for eclipsing binaries from synthetic uvby Strömgren photometry. *A&A*, 341, 857–866 {**127**}

Lázaro C, Arévalo MJ, Antonopoulou E, 2006, Absolute parameters of the Algol binary V505 Sgr from infrared JK light curves. *MNRAS*, 368, 959–964 {**125**}

Lazaro C, Arévalo MJ, Claret A, 2002, Infrared light curves and absolute stellar parameters of the Algol system δ Lib: is δ Lib really an overmassive Algol binary? *MNRAS*, 334, 542–552 {**133**}

Lee JW, Kim CH, Lee CU, *et al.*, 2004a, Determinations of its absolute dimensions and distance by the analyses of light and radial velocity curves of the contact binary. I. V417 Aql. *Journal of Astronomy and Space Sciences*, 21, 73–82 {**133**}

Lee JW, Lee CU, Kim CH, *et al.*, 2004b, Determinations of its absolute dimensions and distance by the analyses of light and radial velocity curves. *Journal of Astronomy and Space Sciences*, 21, 275–282 {**133**}

Lehmann H, Andrievsky SM, Egorova I, *et al.*, 2002, The spectroscopic binaries 21 Her and γ Gem. *A&A*, 383, 558–567 {**123**}

Lehmann H, Hildebrandt G, Panov KP, *et al.*, 2001, HD 169981: an overlooked photometric binary? *A&A*, 373, 960–965 {**129**}

Leinert C, Zinnecker H, Weitzel N, *et al.*, 1993, A systematic approach for young binaries in Taurus. *A&A*, 278, 129–149 {**91**}

Lépine S, Bongiorno B, 2007, New distant companions to known nearby stars. II. Faint companions of Hipparcos stars and the frequency of wide binary systems. *AJ*, 133, 889–905 {**100, 110**}

Lépine S, Shara MM, 2005, A catalogue of northern stars with annual proper motions larger than 0.15 arcsec (LSPM-North Catalogue). *AJ*, 129, 1483–1522 {**110**}

Liller W, Yamamoto M, 1999, Nova Sagittarii 1999. *IAU Circ.*, 7153, 2 {**133**}

Lindegren L, 1997, The Hipparcos Catalogue Double and Multiple Systems Annex. *ESA SP–402: Hipparcos, Venice '97*, 13–16 {**92, 95**}

Lindegren L, Mignard F, Söderhjelm S, *et al.*, 1997, Double star data in the Hipparcos Catalogue. *A&A*, 323, L53–L56 {**92, 95**}

Ling JF, 2004, Preliminary orbits for eight visual binaries. *ApJS*, 153, 545–554 {**123**}

Ling JF, Magdalena P, Prieto C, 2004, The distribution of separations in the Hipparcos Catalogue double and multiple systems annex. *Revista Mexicana de Astronomia y Astrofisica Conf. Ser.*, volume 21, 77–78 {**107**}

Lucy LB, 1968, The structure of contact binaries. *ApJ*, 151, 1123–1135 {**130**}

Lu PK, Dawson DW, Upgren AR, *et al.*, 1983, A catalogue of spectral classification and photometry of barium stars. *ApJS*, 52, 169–181 {**102**}

Lu W, Rucinski SM, Ogłoza W, 2001, Radial velocity studies of close binary stars. IV. *AJ*, 122, 402–412 {**134**}

Lu W, Rucinski SM, 1999, Radial velocity studies of close binary stars. I. *AJ*, 118, 515–526 {**134**}

Makarov VV, 2007, Accelerating binaries near the Sun and far away. *Binary Stars as Critical Tools and Tests in Contemporary Astrophysics, IAU Symp. 240* {**101**}

Makarov VV, Fabricius C, 1999, HD 30187 B and HD 39927 B: two suspected nearby hot subdwarfs in resolved binaries. *A&A*, 349, L34–L36 {**107, 123**}

Makarov VV, Kaplan GH, 2005, Statistical constraints for astrometric binaries with non-linear motion. *AJ*, 129, 2420–2427 {**100–102**}

Malkov O, Piskunov AE, Shpil'Kina DA, 1997, Mass–luminosity relation of low-mass stars. *A&A*, 320, 79–90 {**115**}

Marchetti E, Falomo R, Bello D, *et al.*, 2002, A search for star asterisms for natural guide star based MCAO correction. *Beyond Conventional Adaptive Optics. ESO Proceedings Vol. 58* (eds. Vernet E, Ragazzoni R, Esposito S, *et al.*), 403 {**138**}

Marrese PM, Milone EF, Sordo R, *et al.*, 2005, Gaia and the fundamental stellar parameters from double-lined eclipsing binaries. *ESA SP–576: The Three-Dimensional Universe with Gaia* (eds. Turon C, O'Flaherty KS, Perryman MAC), 599–602 {**125**}

Marrese PM, Munari U, Siviero A, *et al.*, 2004, Evaluating Gaia performances on eclipsing binaries. III. Orbits and stellar parameters for UW LMi, V432 Aur and CN Lyn. *A&A*, 413, 635–642 {**130**}

Martin C, Mignard F, Froeschlé M, 1997a, Mass determination of astrometric binaries with Hipparcos. *ASSL Vol. 223: Visual Double Stars: Formation, Dynamics and Evolutionary Tracks*, 475–484 {**111**}

—, 1997b, Mass determination of astrometric binaries with Hipparcos. I. Theory and simulation. *A&AS*, 122, 571–580 {**111**}

Martin C, Mignard F, Hartkopf WI, *et al.*, 1998, Mass determination of astrometric binaries with Hipparcos. III. New results for 28 systems. *A&AS*, 133, 149–162 {**100, 111, 113, 114**}

Martin C, Mignard F, 1997, Masses of astrometric binaries. *ESA SP–402: Hipparcos, Venice '97*, 417–420 {**111**}

—, 1998, Mass determination of astrometric binaries with Hipparcos. II. Selection of candidates and results. *A&A*, 330, 585–599 {**100, 111, 113**}

Mason BD, Douglass GG, Hartkopf WI, 1999a, Binary star orbits from speckle interferometry. I. Improved orbital elements of 22 visual systems. *AJ*, 117, 1023–1036 {**98**}

Mason BD, Gies DR, Hartkopf WI, *et al.*, 1998, ICCD speckle observations of binary stars. XIX. An astrometric/spectroscopic survey of O stars. *AJ*, 115, 821 {**136**}

Mason BD, Hartkopf WI, Holdenried ER, *et al.*, 2000a, Speckle interferometry at the US Naval Observatory. VI. *AJ*, 120, 1120–1132 {**136**}

—, 2001a, Speckle interferometry of new and problem Hipparcos binaries. II. Observations obtained in 1998–1999 from McDonald Observatory. *AJ*, 121, 3224–3234 {**101, 135**}

Mason BD, Hartkopf WI, Urban SE, *et al.*, 2002, Speckle interferometry at the US Naval Observatory. VIII. *AJ*, 124, 2254–2272 {**136**}

Mason BD, Hartkopf WI, Wycoff GL, *et al.*, 2001b, Speckle interferometry at the US Naval Observatory. VII. *AJ*, 122, 1586–1601 {**136**}

—, 2004a, Speckle interferometry at the US Naval Observatory. IX. *AJ*, 127, 539–548 {**136**}

—, 2004b, Speckle interferometry at the US Naval Observatory. X. *AJ*, 128, 3012–3018 {**136**}

—, 2006a, Speckle interferometry at the US Naval Observatory. XI. *AJ*, 131, 2687–2694 {**135, 136**}

—, 2006b, Speckle interferometry at the US Naval Observatory. XII. *AJ*, 132, 2219–2230 {**136**}

Mason BD, Martin C, Hartkopf WI, *et al.*, 1999b, Speckle interferometry of new and problem Hipparcos binaries. *AJ*, 117, 1890–1904 {**101, 135**}

Mason BD, Wycoff GL, Hartkopf WI, *et al.*, 2001c, The 2001 US Naval Observatory Double Star CD-ROM. I. The Washington Double Star Catalogue. *AJ*, 122, 3466–3471 {**97**}

Mason BD, Wycoff GL, Urban SE, *et al.*, 2000b, Double stars in the Tycho 2 Catalogue. *AJ*, 120, 3244–3249 {**97, 98**}

Mayer P, Pribulla T, Chochol D, 2004, Variable depths of minima of the eclipsing binary V685 Cen. *Informational Bulletin on Variable Stars*, 5563 {**130**}

Mazeh T, Simon M, Prato L, *et al.*, 2003, The mass ratio distribution in main sequence spectroscopic binaries measured by infrared spectroscopy. *ApJ*, 599, 1344–1356 {**108**}

Mérand A, Bordé P, Coudé Du Foresto V, 2005, A catalogue of bright calibrator stars for 200-m baseline near-infrared stellar interferometry. *A&A*, 433, 1155–1162 {**139**}

Monnier JD, Schöller M, Danchi WC (eds.), 2006, *Advances in Stellar Interferometry*. Proc. SPIE 6268 {**139**}

Morel P, Berthomieu G, Provost J, *et al.*, 2001, The ζ Her binary system revisited. Calibration and seismology. *A&A*, 379, 245–256 {**123**}

Morlet G, Salaman M, Gili R, 2000, CCD measurements of visual double stars made with the 50 cm refractor of the Nice Observatory. *A&AS*, 145, 67–69 {**133**}

—, 2002, Nice Observatory CCD measurements of visual double stars (4th series). *A&A*, 396, 933–935 {**134**}

Morris SL, Naftilan SA, 2000, V1073 Cyg: a new light curve and analysis. *PASP*, 112, 852–860 {**133**}

Mourard D, Blazit A, Bonneau D, *et al.*, 2006, Recent progress and future prospects of the GI2T interferometer. *Advances in Stellar Interferometry*, Proc. SPIE 6268, 7 {**139**}

Muller P, Couteau P, 1979, *Catalogue d'éphemerides d'étoiles doubles: 4*. Observatoire de Paris {**97**}

Munari U, 2003, Symbiotic stars in phase space: from Hipparcos to Gaia. *ASP Conf. Ser. 303* (eds. Corradi RLM, Mikolajewska J, Mahoney TJ), 518 {**133**}

Munari U, Dallaporta S, Siviero A, *et al.*, 2004a, The distance to the Pleiades from orbital solution of the double-lined eclipsing binary HD 23642. *A&A*, 418, L31–L34 {**119**}

Munari U, Renzini A, Bernacca PL, 1997, Hipparcos observations of symbiotic stars. *ESA SP–402: Hipparcos, Venice '97*, 413–416 {**132**}

Munari U, Tomov T, Zwitter T, *et al.*, 2001, Evaluating Gaia performances on eclipsing binaries. I. Orbits and stellar parameters for V505 Per, V570 Per and OO Peg. *A&A*, 378, 477–486 {**129**}

Munari U, Zwitter T, Milone EF, 2004b, SB2 and eclipsing binaries with Gaia and RAVE. *ASP Conf. Ser. 318: Spectroscopically and Spatially Resolving the Components of the Close Binary Stars* (eds. Hilditch RW, Hensberge H, Pavlovski K), 422–429 {**125**}

Munari U, Zwitter T, 2002, A multi-epoch spectrophotometric atlas of symbiotic stars. *A&A*, 383, 188–196 {**133**}

Nigoche A, 2000, *Master's Thesis*. Universidad Nacional Autónoma de México {**108**}

Nishikawa J, Sato K, Fukushima T, *et al.*, 1997, Milliarcsecond astrometric interferometer: Mitaka optical and infrared array (MIRA) project. *The New International Celestial Reference Frame, IAU Joint Discussion 7*, 28 {**139**}

Nitschelm C, 2004, Discovery and confirmation of some double-lined spectroscopic binaries in the Sco–Cen complex. *ASP Conf. Ser. 318: Spectroscopically and Spatially Resolving the Components of the Close Binary Stars* (eds. Hilditch RW, Hensberge H, Pavlovski K), 291–293 {**123**}

Nordström B, Mayor M, Andersen J, *et al.*, 2004, The Geneva–Copenhagen survey of the solar neighbourhood: ages, metallicities, and kinematic properties of ~14 000 F and G dwarfs. *A&A*, 418, 989–1019 {**102**}

Nordström B, Stefanik RP, Latham DW, *et al.*, 1997, Radial velocities, rotations, and duplicity of a sample of early F-type dwarfs. *A&AS*, 126, 21–30 {**103**}

North JR, Davis J, Tuthill PG, *et al.*, 2007, Orbital solution and fundamental parameters of σ Sco. *MNRAS*, 380, 1276–1284 {**139**}

North P, Carquillat JM, Ginestet N, *et al.*, 1998, Multiplicity among peculiar A stars. I. The Ap stars HD 8441 and HD 137909, and the Am stars HD 43478 and HD 96391. *A&AS*, 130, 223–232 {**123**}

Oblak E, Cuypers J, Lampens P, *et al.*, 1997a, Accurate two-colour photometry and astrometry for Hipparcos double stars. *ESA SP–402: Hipparcos, Venice '97*, 445–448 {**96**}

Oblak E, Kurpinska-Winiarska M, Carquillat JM, 2007, Multiple stellar systems with eclipsing binaries. *Binary Stars as Critical Tools and Tests in Contemporary Astrophysics, IAU Symp. 240* {**128**}

Oblak E, Kurpinska-Winiarska M, Kundera T, *et al.*, 1997b, Statistical analysis of a sample of Hipparcos eclipsing binaries. *The First Results of Hipparcos and Tycho, IAU Joint Discussion 14*, 44 {**125**}

—, 1998, Analysis of the Hipparcos sample of eclipsing binaries. *Highlights in Astronomy*, 11, 569 {**125**}

Oblak E, Kurpinska-Winiarska M, 2000, Comparison between photometric and Hipparcos parallaxes for eclipsing binaries. *IAU Symp. 200*, 138 {**125**}

Oblak E, Lampens P, Cuypers J, *et al.*, 1999, CCD photometry and astrometry for visual double and multiple stars of the Hipparcos Catalogue. I. Presentation of the large-scale project. *A&A*, 346, 523–531 {**134**}

Odenkirchen M, Brosche P, 1999, The proper motion signal of unresolved binaries in the Hipparcos Catalogue. *Astronomische Nachrichten*, 320, 397 {**102**}

Ogloza W, Zakrzewski B, 2006, The spatial distribution of W UMa-type stars. *Ap&SS*, 304, 121–123 {**132**}

Olević D, Cvetković Z, Dačić M, 2000a, Kinematics of a sample of visual double stars. *Serbian Astronomical Journal*, 162, 101 {**96**}

—, 2001, Kinematics of a sample of visual double stars. *Balkan Meeting of Young Astronomers, eds A. Antov, R. Konstantinova-Antova, R. Bogdanovski and M. Tsvetkov, Belogradchik: Belogradchik Astronomical Observatory*, 95 {**96**}

Olević D, Cvetković Z, 2003, Dynamics of quadruple system Finsen 332. *Serbian Astronomical Journal*, 167, 63 {**123**}

—, 2006, Dynamical masses of the components in o And. *AJ*, 131, 1721–1723 {**125**}

Olević D, Jovanovic P, 1999, New orbital elements of 5 interferometric double stars. *Serbian Astronomical Journal*, 159, 87 {**123**}

—, 2001, A first and 11 recalculated orbits of double stars. *Serbian Astronomical Journal*, 163, 5 {**123**}

Olević D, Popovic G, Jovanovic P, 2000b, Recalculated orbits of 8 double stars. *Serbian Astronomical Journal*, 162, 109 {**123**}

Oort JH, 1926, The stars of high velocity. *Publs. Kapteyn Astron. Lab., Groningen*, 40, D1 {**108**}

Öpik E, 1923, *Publ. Obs. Astr. Univ. Tartu*. 26(6) {**103, 108**}

Otero SA, Claus F, 2004, New elements for 80 eclipsing binaries. II. *Informational Bulletin on Variable Stars*, 5495 {**129**}

Otero SA, 2003a, Hipparcos eclipsing binaries showing apsidal motion. *Informational Bulletin on Variable Stars*, 5482 {**130**}

—, 2003b, New elements for 80 eclipsing binaries. *Informational Bulletin on Variable Stars*, 5480 {**129**}

—, 2004, New elements for 80 eclipsing binaries. III. *Informational Bulletin on Variable Stars*, 5532 {**129**}

—, 2005, Hipparcos eclipsing binaries showing apsidal motion. II. *Informational Bulletin on Variable Stars*, 5631 {**130**}

Otero SA, Dubovsky PA, 2004, New elements for 80 eclipsing binaries. IV. *Informational Bulletin on Variable Stars*, 5557 {**129**}

Otero SA, Hoogeveen GJ, Wils P, 2006a, New elements for 80 eclipsing binaries VIII. *Informational Bulletin on Variable Stars*, 5674, 1 {**129**}

Otero SA, Stephan C, 2004, Light elements for v353 Hya, a Hipparcos eclipsing binary. *Journal of the American Association of Variable Star Observers (JAAVSO)*, 32, 105–107 {**130**}

Otero SA, Wils P, 2005a, New elements for 80 eclipsing binaries. VI. *Informational Bulletin on Variable Stars*, 5630 {**129**}

—, 2005b, New elements for 80 eclipsing binaries. VII. *Informational Bulletin on Variable Stars*, 5644 {**129**}

Otero SA, Wils P, Dubovsky PA, 2005, New elements for 80 eclipsing binaries. V. *Informational Bulletin on Variable Stars*, 5586 {**129**}

Otero SA, Wils P, Hoogeveen GJ, *et al.*, 2006b, 50 new eccentric eclipsing binaries found in the ASAS, Hipparcos and NSVS data bases. *Informational Bulletin on Variable Stars*, 5681, 1 {**129**}

Ozdemir S, Demircan O, Erdem A, *et al.*, 2002, A photometric study of the recently-discovered eclipsing binary V899 Her. *A&A*, 387, 240–243 {**129**}

Padmakar, Singh KP, Drake SA, *et al.*, 2000, Optical, X-ray and radio observations of HD 61396: a probable new RS CVn-type binary. *MNRAS*, 314, 733–742 {**133**}

Palasi J, 2000, Search for wide binaries in the Hipparcos and Tycho 2 Catalogues. *IAU Symp. 200*, 145 {**107**}

Pan XP, Shao M, Kulkarni SR, 2004, A distance of 133–137 pc to the Pleiades star cluster. *Nature*, 427, 326–328 {**119**}

Parsons SB, 2004, New and confirmed triple systems with luminous cool primaries and hot companions. *AJ*, 127, 2915–2930 {**123**}

Paunzen E, Iliev IK, Barzova IS, *et al.*, 2005, The importance of spectroscopic binary systems for the LB phenomenon. *ASP Conf. Ser. 333: Tidal Evolution and Oscillations in Binary Stars* (eds. Claret A, Giménez A, Zahn JP), 259 {**124**}

Pearce J, 1955, *IAU Trans. 9*. 441 {**121**}

Penny LR, Seyle D, Gies DR, *et al.*, 2001, Tomographic separation of composite spectra. VII. The physical properties of the massive triple system HD 135240 (δ Circini). *ApJ*, 548, 889–899 {**123**}

Percy JR, 2001, RS CVN stars in the AAVSO photoelectric photometry programme. *Journal of the American Association of Variable Star Observers*, 29, 82 {**133**}

Percy JR, Au-Yong K, 2000, Intrinsically-variable B stars in eclipsing binary systems. *Informational Bulletin on Variable Stars*, 4825 {**125**}

Perrin G, Coude Du Foresto V, Ridgway ST, *et al.*, 1998, Extension of the effective temperature scale of giants to types later than M6. *A&A*, 331, 619–626 {**121**}

Pigulski A, Michalska G, 2007, FR Sct: a triple VV Cephei-type system of particular interest. *Informational Bulletin on Variable Stars*, 5757, 1 {**125**}

Platais I, Kozhurina-Platais V, Zacharias MI, 2001, Possible nova in Puppis. *IAU Circ.*, 7556, 1 {**133**}

Platais I, Pourbaix D, Jorissen A, *et al.*, 2003, Hipparcos red stars in the $Hp\ V_T$ and VI_C systems. *A&A*, 397, 997–1010 {**101**}

Pluzhnik EA, 2005, Differential photometry of speckle-interferometric binary and multiple stars. *A&A*, 431, 587–596 {**99**}

Popper DM, 1980, Stellar masses. *ARA&A*, 18, 115–164 {**111, 126, 127**}

—, 1998, Hipparcos parallaxes of eclipsing binaries and the radiative flux scale. *PASP*, 110, 919–922 {**126–128**}

Pourbaix D, 1999, γ Per: a challenge for stellar evolution models. *A&A*, 348, 127–132 {**123**}

—, 2002, Precision and accuracy of the orbital parameters derived from 2d and 1d space observations of visual or astrometric binaries. *A&A*, 385, 686–692 {**117**}

—, 2004, Orbits from Hipparcos. *ASP Conf. Ser. 318: Spectroscopically and Spatially Resolving the Components of the Close Binary Stars* (eds. Hilditch RW, Hensberge H, Pavlovski K), 132–140 {**118**}

Pourbaix D, Arenou F, 2001, Screening the Hipparcos-based astrometric orbits of sub-stellar objects. *A&A*, 372, 935–944 {**115, 118**}

Pourbaix D, Boffin HMJ, 2003, Reprocessing the Hipparcos intermediate astrometric data of spectroscopic binaries. II. Systems with a giant component. *A&A*, 398, 1163–1177 {**100, 115**}

Pourbaix D, Jancart S, Jorissen A, 2004a, Binaries in the Hipparcos data: keep digging. II. Modeling the intermediate astrometric data of known spectroscopic systems. *ASP Conf. Ser. 318: Spectroscopically and Spatially Resolving the Components of the Close Binary Stars* (eds. Hilditch RW, Hensberge H, Pavlovski K), 144–147 {**118**}

Pourbaix D, Jorissen A, 2000, Re-processing the Hipparcos transit data and intermediate astrometric data of spectroscopic binaries. I. Ba, CH and Tc-poor S stars. *A&AS*, 145, 161–183 {**100, 114, 116**}

Pourbaix D, Lampens P, 1999, Space versus ground: a confrontation between Hipparcos and orbital parallaxes. *ASP Conf. Ser. 167: Harmonizing Cosmic Distance Scales in a Post-Hipparcos Era*, 300–304 {**123**}

Pourbaix D, Neuforge-Verheecke C, Noels A, 1999, Revised masses of α Cen. *A&A*, 344, 172–176 {**123**}

Pourbaix D, Platais I, Detournay S, *et al.*, 2003, How many Hipparcos variability-induced movers are genuine binaries? *A&A*, 399, 1167–1175 {**100, 101**}

Pourbaix D, Tokovinin AA, Batten AH, *et al.*, 2004b, SB9: the ninth catalogue of spectroscopic binary orbits. *A&A*, 424, 727–732 {**97, 102, 118**}

Poveda A, Allen C, Herrera MA, 1997, Evolutionary effects in the separations of wide binaries. *Visual Double Stars: Formation, Dynamics and Evolutionary Tracks; ASSL Vol. 223* (eds. Docobo JA, Elipe A, McAlister H), 191–198 {**107, 109**}

—, 2000, The distribution of separations of binaries. *Stellar Astrophysics, Proceedings of the Pacific Rim Conference held in Hong Kong, 1999* (eds. Cheng KS, Chau HF, Chan KL, *et al.*), 181–190 {**108**}

Poveda A, Allen C, 2004, The distribution of separations of wide binaries of different ages. *Revista Mexicana de Astronomia y Astrofisica Conf. Ser.*, volume 21, 49–57 {**107**}

Pribulla T, Chochol D, Heckert PA, *et al.*, 2001, An active binary XY UMa revisited. *A&A*, 371, 997–1011 {**129**}

Pribulla T, Chochol D, Milano L, *et al.*, 2000, Active eclipsing binary RT And revisited. *A&A*, 362, 169–188 {**133**}

Pribulla T, Kreiner JM, Tremko J, 2003, Catalogue of the field contact binary stars. *Contributions of the Astronomical Observatory Skalnate Pleso*, 33, 38–70 {**130**}

Pribulla T, Rucinski SM, Conidis G, *et al.*, 2007, Radial velocity studies of close binary stars. XII. *AJ*, 133, 1977–1987 {**134**}

Pribulla T, Rucinski SM, Lu W, *et al.*, 2006, Radial velocity studies of close binary stars. XI. *AJ*, 132, 769–780 {**134**}

Pribulla T, Vanko M, 2002, Photoelectric photometry of eclipsing contact binaries: U Peg, YY CrB, OU Ser and EQ Tau. *Contributions of the Astronomical Observatory Skalnate Pleso*, 32, 79–98 {**133**}

Prieur JL, Carquillat JM, Ginestet N, *et al.*, 2003, Speckle observations of composite spectrum stars. II. Differential photometry of the binary components. *ApJS*, 144, 263–276 {**138**}

Prieur JL, Carquillat JM, Griffin RF, 2006, Contribution to the study of binaries with spectral types F, G, K, and M XI. Orbital elements of three red-giant spectroscopic binaries: HR 1304, HR 1908, and HD 126947. *Astronomische Nachrichten*, 327, 686–690 {**125**}

Prieur JL, Koechlin L, Ginestet N, *et al.*, 2002, Speckle observations of composite spectrum stars with PISCO in 1993–1998. *ApJS*, 142, 95–104 {**138**}

Prieur JL, Oblak E, Lampens P, *et al.*, 2001, Speckle observations of binary systems measured by Hipparcos. *A&A*, 367, 865–875 {**138**}

Prša A, Zwitter T, 2007, Pipeline reduction of binary light curves from large-scale surveys. *Binary Stars as Critical Tools and Tests in Contemporary Astrophysics, IAU Symp. 240*, 217–229 {**125**}

Pych W, Rucinski SM, DeBond H, *et al.*, 2004, Radial velocity studies of close binary stars. IX. *AJ*, 127, 1712–1719 {**134**}

Qian SB, Liao WP, He JJ, *et al.*, 2006, Is V899 Her an unsolved quadruple system containing double close binary stars? *New Astronomy*, 12, 33–37 {**125**}

Quist CF, 2000, Studies of double and multiple stars using space astrometry. *Ph.D. Thesis* {**103**}

Quist CF, Lindegren L, 2000, Statistics of Hipparcos binaries: probing the 1–10 AU separation range. *A&A*, 361, 770–780 {**100, 103, 105, 106**}

—, 2001, Binarity of Hipparcos main sequence survey stars. *IAU Symp. 200*, 64 {**103**}

Radhakrishnan KR, Sarma MBK, Abhyankar KD, 1984, Photometric and spectroscopic study of R CMa. *Ap&SS*, 99, 229–236 {**128**}

Raghavan D, McAlister H, Henry TJ, 2007, A survey of stellar families: multiplicity among solar-type stars. *Binary Stars as Critical Tools and Tests in Contemporary Astrophysics, IAU Symp. 240*, 254–257 {**106**}

Reffert S, Quirrenbach A, Hekker S, *et al.*, 2007, Multiplicity in a complete sample of giant stars. *Binary Stars as Critical Tools and Tests in Contemporary Astrophysics, IAU Symp. 240* {**134**}

Ren SL, Fu YN, 2007a, A method of determining binary orbits by incorporating long-term observational data. *Acta Astronomica Sinica*, 48, 200–209 {**117**}

—, 2007b, A method to improve binary orbits by incorporating long-term observational data. *Chinese Astronomy and Astrophysics*, 31, 277–287 {**117**}

Ribas I, Arenou F, Guinan EF, 2002, Astrometric and light-travel time orbits to detect low-mass companions: a case study of the eclipsing system R CMa. *AJ*, 123, 2033–2041 {**128, 129**}

Ribas I, Gimenez A, Torra J, *et al.*, 1998, Effective temperature of detached eclipsing binaries from Hipparcos parallax. *A&A*, 330, 600–604 {**127, 128**}

Rocha Pinto HJ, Castilho BV, Maciel WJ, 2002, Chromospherically young, kinematically old stars. *A&A*, 384, 912–924 {**132**}

Rodonò M, Lanza AF, Becciani U, 2001, On the determination of the light curve parameters of detached active binaries. I. The prototype RS CVn. *A&A*, 371, 174–185 {**133**}

Rossiter RA, 1955, Catalogue of Southern Double Stars. *Publications of Michigan Observatory*, 11, 1 {**135**}

Royer F, Briot D, North P, 2002, The HD 61273 case: investigation on a new Algol. *SF2A-2002: Semaine de l'Astrophysique Française*, 563 {**133**}

Rucinski SM, 1992, Contact binaries of the W UMa type. *ASSL Vol. 177: The Realm of Interacting Binary Stars* (eds. Sahade J, McCluskey GE, Kondo Y), 111–142 {**130**}

—, 1994, M_V, $\log P$, $\log T_{\rm eff}$ calibrations for W UMa systems. *PASP*, 106, 462–471 {**131**}

—, 1995, Absolute magnitude calibration for W UMa-type systems. II. Influence of metallicity. *PASP*, 107, 648 {**131**}

—, 1997a, Eclipsing binaries in the OGLE variable star catalogue. II. Light curves of the W UMa-type systems in Baade's Window. *AJ*, 113, 1112–1121 {**131, 132**}

—, 1997b, Eclipsing binaries in the OGLE variable star catalogue. I. W Uma-type systems as distance and population tracers in Baade's Window. *AJ*, 113, 407–424 {**131, 132**}

—, 2002a, Radial velocity studies of close binary stars. VII. Methods and uncertainties. *AJ*, 124, 1746–1756 {**134**}

—, 2002b, The 7.5 mag limit sample of bright short-period binary stars. I. How many contact binaries are there? *PASP*, 114, 1124–1142 {**130, 131**}

—, 2006, Luminosity function of contact binaries based on the All Sky Automated Survey. *MNRAS*, 368, 1319–1322 {**131**}

Rucinski SM, Capobianco CC, Lu W, *et al.*, 2003, Radial velocity studies of close binary stars. VIII. *AJ*, 125, 3258–3264 {**134**}

Rucinski SM, Duerbeck HW, 1997a, Absolute magnitude calibration for the W UMa-type contact binary stars. *ESA SP–402: Hipparcos, Venice '97*, 457–460 {**131**}

—, 1997b, Absolute magnitude calibration for the W UMa-type systems based on Hipparcos data. *PASP*, 109, 1340–1350 {**131**}

Rucinski SM, Lu W, Capobianco CC, *et al.*, 2002, Radial velocity studies of close binary stars. VI. *AJ*, 124, 1738–1745 {**134**}

Rucinski SM, Lu W, Mochnacki SW, *et al.*, 2001, Radial velocity studies of close binary stars. V. *AJ*, 122, 1974–1980 {**134**}

Rucinski SM, Lu W, Mochnacki SW, 2000, Radial velocity studies of close binary stars. III. *AJ*, 120, 1133–1139 {**134**}

Rucinski SM, Lu W, 1999, Radial velocity studies of close binary stars. II. *AJ*, 118, 2451–2459 {**134**}

Rucinski SM, Pych W, Ogłoza W, *et al.*, 2005, Radial velocity studies of close binary stars. X. *AJ*, 130, 767–775 {**134**}

Salaman M, Morlet G, Gili R, 1999, CCD measurements of visual double stars made with the 50 cm refractor of the Nice Observatory. *A&AS*, 135, 499–501 {**133**}

—, 2001, Nice Observatory CCD measurements of visual double stars (3rd series). *A&A*, 369, 552–553 {**133**}

Samec RG, Tuttle J, Gray JD, *et al.*, 2000, Observations of Hipparcos eclipsing variables: EF Boo, VZ Psc, V417 Aql and BX And. *Hipparcos and the Luminosity Calibration of the Nearer Stars, IAU Joint Discussion 13*, 28 {**129**}

Samus NN, 1997, V1472 Aql: a most unusual eclipser? *Informational Bulletin on Variable Stars*, 4501 {**129**}

Sánchez Bajo F, García Melendo E, Gómez Forrellad JM, 2007, Photometric study of the W UMa eclipsing binaries VW LMi and BX Dra. *Ap&SS*, 416–422 {**133**}

Sarro LM, Sánchez Fernández C, Giménez Á, 2006, Automatic classification of eclipsing binaries light curves using neural networks. *A&A*, 446, 395–402 {**125**}

Scarfe CD, Griffin RF, Griffin REM, 2007, The double-lined spectroscopic binary HR 6046. *MNRAS*, 376, 1671–1679 {**123**}

Schloerb FP, Berger JP, Carleton NP, *et al.*, 2006, IOTA: recent science and technology. *Advances in Stellar Interferometry*, Proc. SPIE 6268, 18 {**139**}

Schuster WJ, Moitinho A, Márquez A, *et al.*, 2006, uvbyβ photometry of high-velocity and metal-poor stars. XI. Ages of halo and old disk stars. *A&A*, 445, 939–958 {**109**}

Schuster WJ, Parrao L, Contreras ME, 1993, uvbyβ photometry of high-velocity and metal-poor stars. VI. A second catalogue, and stellar populations of the Galaxy. *A&AS*, 97, 951–983 {**108**}

Selam SO, 2004, Key parameters of W UMa-type contact binaries discovered by Hipparcos. *A&A*, 416, 1097–1105 {**131, 132**}

Semeniuk I, 2000, Comparison of parallaxes from eclipsing binaries method with Hipparcos parallaxes. *Acta Astronomica*, 50, 381–386 {**128**}

—, 2001, Surface brightness–colour relation for eclipsing binaries with ubvy photometry. *Acta Astronomica*, 51, 75–80 {**128**}

Shapley H, 1948, The relative frequency of low-luminosity eclipsing binaries. *Centennial Symposia*, 249–260 {**131**}

Shatsky NI, 1998, The nature of visual components in 82 multiple systems. *Astronomy Letters*, 24, 257–269 {**134**}

Shatsky NI, Tokovinin AA, 1998, A test of Hipparcos parallaxes on multiple stars. *Astronomy Letters*, 24, 673–676 {**100, 111, 120**}

—, 2002, The mass ratio distribution of B-type visual binaries in the Sco OB2 association. *A&A*, 382, 92–103 {**139**}

Shobbrook RR, 2004, Photometry of 20 eclipsing and ellipsoidal binary systems. *Journal of Astronomical Data*, 10, 1 {**129**}

Short CI, Hauschildt PH, 2003, Atmospheric models of red giants with massive-scale non-local thermodynamic equilibrium. *ApJ*, 596, 501–508 {**121**}

Simon V, Hanzl D, Skoda P, *et al.*, 2000, Photometric orbital modulation in V1080 Tau. *A&A*, 360, 637–641 {**129**}

Sinachopoulos D, Cuypers J, Lampens P, *et al.*, 1995, VRI photometry of wide double stars with A-type primaries. *A&AS*, 112, 291–297 {**97**}

Sinachopoulos D, Gavras P, Dionatos O, *et al.*, 2007a, CCD astrometry and components instrumental magnitude difference of 432 Hipparcos wide visual double stars. *A&A*, 472, 1055–1057 {**109, 134**}

Sinachopoulos D, Gavras P, Medupe T, *et al.*, 2006, CCD measurements of Hipparcos wide visual double stars. *Recent Advances in Astronomy and Astrophysics* (ed. Solomos N), American Institute of Physics Conf. Ser. 848, 389–393 {**109**}

—, 2007b, CCD astrometry and photometry of visual double stars VI. Northern Hipparcos wide pairs measured in the years 2003–2005. *Binary Stars as Critical Tools and Tests in Contemporary Astrophysics, IAU Symp. 240*, 264–267 {**109**}

Sinachopoulos D, van Dessel EL, Geffert M, 1998, A list of Hipparcos very wide visual double stars for the astrometric calibration of imaging with CCDs. *Applied and Computational Harmonic Analysis*, 3, 195–196 {**96**}

Siviero A, Dallaporta S, Munari U, 2006, Period change, spot migration and orbital solution for the eclipsing binary AR Lac. *Baltic Astronomy*, 15, 387–394 {**130**}

Siviero A, Munari U, Sordo R, *et al.*, 2004, Asiago eclipsing binaries programme. I. V432 Aur. *A&A*, 417, 1083–1092 {**130**}

Smith PH, McMillan RS, Merline WJ, 1987, Evidence for periodic radial velocity variations in Arcturus. *ApJ*, 317, L79–L84 {**121**}

Söderhjelm S, 1997, Mass-ratio distribution from Δm statistics for nearby Hipparcos binaries. *ASSL Vol. 223: Visual Double Stars : Formation, Dynamics and Evolutionary Tracks* (eds. Docobo JA, Elipe A, McAlister H), 497–503 {**108**}

—, 1999, Visual binary orbits and masses post Hipparcos. *A&A*, 341, 121–140 {**100, 111, 113–115, 120, 134**}

—, 2000, Binary statistics from Hipparcos data: a progress report. *Astronomische Nachrichten*, 321, 165–170 {**108**}

—, 2001, Binary statistics from Hipparcos. *ASSL Vol. 264: The influence of binaries on stellar population studies*, 581 {**106**}

—, 2007, The $q = 1$ peak in the mass-ratios for Hipparcos visual binaries. *A&A*, 463, 683–691 {**108**}

Söderhjelm S, Dischler J, 2005, Eclipsing binary statistics: theory and observation. *A&A*, 442, 1003–1013 {**125, 126**}

Söderhjelm S, Lindegren L, Perryman MAC, 1997, Binary star masses from Hipparcos. *ESA SP–402: Hipparcos, Venice '97*, 251–256 {**111**}

Söderhjelm S, Lindegren L, 1997, Triple star parameters and masses. *ESA SP–402: Hipparcos, Venice '97*, 425–428 {**100, 119**}

Söderhjelm S, Mignard F, 1998, Arcturus as a double star. *The Observatory*, 118, 365–366 {**121**}

Soulié G, 2006a, Accurate equatorial coordinates of known or new components of a sixth series of some 645 double and multiple systems. I. *Observations et Travaux*, 62, 26–33 {**134**}

—, 2006b, Accurate equatorial coordinates of known or new components of a sixth series of some 645 double and multiple systems. II. *Observations et Travaux*, 63, 15–26 {**134**}

Southworth J, Bruntt H, Buzasi DL, 2007, Eclipsing binaries observed with the WIRE satellite. II. β Aur and non-linear limb darkening in light curves. *A&A*, 467, 1215–1226 {**130**}

Southworth J, Maxted PFL, Smalley B, 2005a, Eclipsing binaries as standard candles. HD 23642 and the distance to the Pleiades. *A&A*, 429, 645–655 {**119**}

Southworth J, Smalley B, Maxted PFL, *et al.*, 2004, Accurate fundamental parameters of eclipsing binary stars. *IAU Symp. 224* (eds. Zverko J, Ziznovsky J, Adelman SJ, *et al.*), 548–561 {**130**}

—, 2005b, Absolute dimensions of detached eclipsing binaries. I. The metallic-lined system WW Aur. *MNRAS*, 363, 529–542 {**123, 130**}

Soydugan E, Soydugan F, Demircan O, *et al.*, 2006, A catalogue of close binaries located in the δ Scuti region of the Cepheid instability strip. *MNRAS*, 370, 2013–2024 {**106**}

Sperauskas J, Bartkevičius A, 2002, Radial velocities of Population II binary stars. I. *Astronomische Nachrichten*, 323, 139–148 {**110**}

Stark MA, Wade RA, 2006, The nature of late-type companions in hot subdwarf composite-spectrum binaries. *Baltic Astronomy*, 15, 175–182 {**134**}

Stefka V, Vondrák J, 2006, Earth orientation catalogue EOC–3: an improved optical reference frame. *Nomenclature, Precession and New Models in Fundamental Astronomy, IAU 26, JD16*, 16 {**100**}

Strigachev A, Lampens P, 2004, Multicolour CCD measurements of nearby visual double stars. II. *A&A*, 422, 1023–1029 {**134**}

Stryker LL, Hesser JE, Hill G, *et al.*, 1985, The binary frequency of extreme subdwarfs revisited. *PASP*, 97, 247–260 {**108**}

Suchkov AA, 2000, Age difference between the populations of binary and single F stars revealed from Hipparcos data. *ApJ*, 535, L107–L110, erratum: 539, L75 {**103**}

—, 2001, Evidence for prolonged main sequence stellar evolution of F stars in close binaries. *A&A*, 369, 554–560 {**103**}

Suchkov AA, McMaster M, 1999, Evidence for a population of numerous binaries with comparably bright components among Hipparcos single F stars. *ApJ*, 524, L99–L102 {**103, 105**}

Suchkov AA, Schultz AB, 2001, Evidence for a very young age of F stars with extrasolar planets. *ApJ*, 549, L237–L240 {**103**}

Szalai T, Kiss LL, Mészáros S, *et al.*, 2007, Physical parameters and multiplicity of five southern close eclipsing binaries. *A&A*, 465, 943–952 {**130**}

Tamazian VS, 2006, Visual orbit, differential photometry, and dynamical mass of DG Leo. *AJ*, 132, 2156–2158 {**125**}

Tamazian VS, Docobo JA, Melikian ND, *et al.*, 2005, Orbit, dynamical mass, and MK type of visual binary Wor 2. *AJ*, 130, 2847–2851 {**124**}

—, 2006, MK classification and dynamical masses for late-type visual binaries. *PASP*, 118, 814–819 {**120**}

Tamazian VS, Docobo JA, 2006, Orbits and differential photometry for visual binaries A 1529, HU 610, and Cou 2031. *AJ*, 131, 2681–2686 {**138**}

Tango WJ, Davis J, Ireland MJ, *et al.*, 2006, Orbital elements, masses and distance of λ Sco A and B determined with the Sydney University Stellar Interferometer and high-resolution spectroscopy. *MNRAS*, 370, 884–890 {**139**}

ten Brummelaar TA, Mason BD, McAlister HA, *et al.*, 2000, Binary star differential photometry using the adaptive optics system at Mount Wilson Observatory. *AJ*, 119, 2403–2414 {**138**}

ten Brummelaar TA, McAlister HA, Ridgway ST, *et al.*, 2003, An update of the CHARA Array. *Interferometry for Optical Astronomy II.* (ed. Traub WA), Proc. SPIE 4838, 69–78 {**139**}

Thorel JC, 1998, Measures of visual double stars made with the 50 cm refractor at the Nice Observatory. *A&AS*, 132, 29–30 {**133**}

Tokovinin AA, 1997, MSC: a catalogue of physical multiple stars. *A&AS*, 124, 75–84 {**97**}

—, 1999a, New spectroscopic components in 8 multiple systems. *A&AS*, 136, 373–378 {**123**}

—, 1999b, The visual orbit of HD 98800. *Astronomy Letters*, 25, 669–671 {**123**}

—, 2000, On the origin of binaries with twin components. *A&A*, 360, 997–1002 {**108**}

Tokovinin AA, Balega YY, Hofmann KH, *et al.*, 2000a, The orbit of the nearby low-mass binary Gliese 600. *Astronomy Letters*, 26, 668–671 {**123**}

Tokovinin AA, Griffin RF, Balega YY, *et al.*, 2000b, The triple system HR 7272. *Astronomy Letters*, 26, 116–121 {**123**}

Tomkin J, 2005, Results and prospects for distance measurement by means of binary stars. *ASP Conf. Ser. 336: Cosmic Abundances as Records of Stellar Evolution and Nucleosynthesis* (eds. Barnes TG, Bash FN), 199 {**119, 120**}

Torres G, 2004, Combining astrometry and spectroscopy. *ASP Conf. Ser. 318: Spectroscopically and Spatially Resolving the Components of the Close Binary Stars* (eds. Hilditch RW, Hensberge H, Pavlovski K), 123–131 {**111, 112**}

—, 2006, The astrometric–spectroscopic binary system HIP 50796: an overmassive companion. *AJ*, 131, 1022–1031 {**122**}

—, 2007a, Astrometric–spectroscopic determination of the absolute masses of the Hg–Mn binary star ϕ Her. *AJ*, 133, 2684–2695 {**122–124**}

—, 2007b, Hipparcos astrometric orbit and evolutionary status of HR 6046. *AJ*, 134, 1916–1921 {**123, 124**}

Torres G, Lacy CH, Marschall LA, *et al.*, 2006, The eclipsing binary V1061 Cyg: confronting stellar evolution models for active and inactive solar-type stars. *ApJ*, 640, 1018–1038 {**122**}

—, 2007, Confronting stellar evolution models for active and inactive solar-type stars: the triple system V1061 Cyg. *Binary Stars as Critical Tools and Tests in Contemporary Astrophysics, IAU Symp. 240*, 385–389 {**122**}

Torres G, Ribas I, 2002, Absolute dimensions of the M-type eclipsing binary YY Gem (Castor C): a challenge to evolutionary models in the lower main sequence. *ApJ*, 567, 1140–1165 {**123, 129**}

Trimble V, 2002, Binary stars: overview. *Encyclopedia of Astronomy and Astrophysics* (ed. Murdin, P.), {**91**}

Turner NH, ten Brummelaar TA, Mason BD, 1999, Adaptive optics observations of Arcturus using the Mount Wilson 100-inch telescope. *PASP*, 111, 556–558 {**121, 138**}

Urban SE, Corbin TE, Wycoff GL, *et al.*, 2000, Problems of using Hipparcos of double stars. *IAU Colloq. 180: Towards Models and Constants for Sub-Microarcsecond Astrometry*, 97 {**106**}

Valtonen M, 1997, Wide binaries from few-body interactions. *Visual Double Stars: Formation, Dynamics and Evolutionary Tracks; ASSL Vol. 223* (eds. Docobo JA, Elipe A, McAlister H), 241–258 {**109**}

—, 2004, Three-body problem and multiple stellar systems. *Revista Mexicana de Astronomia y Astrofisica Conf. Ser.*, volume 21, 147–151 {**108**}

van't Veer F, 1975, The frequency of contact binaries and its consequence on their evolution. *A&A*, 40, 167–174 {**131**} van Leeuwen F, Evans DW, Grenon M, *et al.*, 1997, The Hipparcos mission: photometric data. *A&A*, 323, L61–L64 {**93**}

van Leeuwen F, van Genderen AM, 1997, The discovery of a new massive O-type close binary: τ CMa (HD 57061), based on Hipparcos and Walraven photometry. *A&A*, 327, 1070–1076 {**123**}

Verhoelst T, Bordé P, Perrin G, *et al.*, 2005, Is Arcturus a well-understood K giant?. Test of model atmospheres and potential companion detection by near-infrared interferometry. *A&A*, 435, 289–301 {**121**}

Vidal-Sainz J, 1998, HIP 12056 is an eclipsing binary system. *Informational Bulletin on Variable Stars*, 4557 {**129**}

Viotti R, Badiali M, Cardini D, *et al.*, 1997, Hipparcos astrometry and radio mapping of peculiar binary systems. *ESA SP–402: Hipparcos, Venice '97*, 405–408 {**133**}

Vivekananda Rao P, Radhika P, 2003, UV Psc and the radiative flux scale. *Ap&SS*, 283, 225–231 {**129**}

Vogt N, 2006, FH Leo, the first dwarf nova member of a multiple star system? *A&A*, 452, 985–986 {**133**}

Wade GA, Debernardi Y, Mathys G, *et al.*, 2000, An analysis of the Ap binary HD 81009. *A&A*, 361, 991–1000 {**123**}

Wadhwa SS, Zealey WJ, 2004, Photometric analysis of EL Aqr. *Ap&SS*, 291, 21–25 {**133**}

—, 2005, UX Ret and CN Hyi: Hipparcos photometry analysis. *Ap&SS*, 295, 463–472 {**133**}

Wasserman I, Weinberg MD, 1987, Theoretical implications of wide binary observations. *ApJ*, 312, 390–401 {**107**}

—, 1991, Wide binaries in the Woolley Catalogue. *ApJ*, 382, 149–167 {**107**}

Watson LC, Pollard KR, Hearnshaw JB, 1998, HD 6628: a new active, single-lined spectroscopic binary. *Informational Bulletin on Variable Stars*, 4617 {**123**}

Weinberg MD, Wasserman I, 1988, Search for wide binaries in the Yale Bright Star Catalogue. *ApJ*, 329, 253–275 {**107**}

Wielen R, 1996, Searching for variability-induced movers: an astrometric method to detect the binary nature of double stars with a variable component. *A&A*, 314, 679 {**96, 101**}

Wielen R, Dettbarn C, Jahreiß H, *et al.*, 1999, Indications on the binary nature of individual stars derived from a comparison of their Hipparcos proper motions with ground-based data. I. Basic principles. *A&A*, 346, 675–685 {**102**}

Wielen R, Jahreiß H, Dettbarn C, *et al.*, 2000, Polaris: astrometric orbit, position, and proper motion. *A&A*, 360, 399–410 {**120, 121**}

Wielen R, Schwan H, Dettbarn C, *et al.*, 2001, Astrometric Catalogue ARIHIP. Containing stellar data selected from the combination catalogues FK6, GC+HIP, TYC2+HIP and from the Hipparcos Catalogue. *Veroeffentlichungen des Astronomischen Rechen-Instituts Heidelberg*, 40 {**100**}

Wilson RE, Devinney EJ, 1971, Realisation of accurate close-binary light curves: application to MR Cyg. *ApJ*, 166, 605–619 {**130**}

Wizinowich P, Akeson RL, Colavita MM, *et al.*, 2006, Recent progress at the Keck Interferometer. *Advances in Stellar Interferometry*, Proc. SPIE 6268, 21 {**139**}

Woitas J, Tamazian VS, Docobo JA, *et al.*, 2003, Visual orbit for the low-mass binary Gliese 22 AC from speckle interferometry. *A&A*, 406, 293–298 {**138**}

Wolszczan A, Frail DA, 1992, A planetary system around the millisecond pulsar PSR 1257+12. *Nature*, 355, 145–147 {**128**}

Worley CE, Douglass GG, 1984, *The Washington Double Star Catalogue*. US Naval Observatory, Washington {**97**}

—, 1997, *The Washington Double Star Catalog* (WDS, 1996.0). *A&AS*, 125, 523–523 {**97**}

Worley CE, Heintz WD, 1983, Fourth catalogue of orbits of visual binary stars. *Publications of the US Naval Observatory*, 24, 1 {**97**}

Yi S, Demarque P, Kim YC, *et al.*, 2001, Toward better age estimates for stellar populations: the Yonsei–Yale (Y^2) isochrones for solar mixture. *ApJS*, 136, 417–437 {**123, 124**}

Yoshizawa M, Ohishi N, Suzuki S, *et al.*, 2004, *The first stellar fringes observed with Mitaka optical/infrared array (MIRA I.2) 30-m baseline*, 3–4. Annual Report of the National Astronomical Observatory of Japan, Volume 5 {**139**}

Yushchenko AV, Gopka VF, Khokhlova VL, *et al.*, 1999, Atmospheric chemical composition of the twin components of equal mass in the chemically peculiar SB2 system 66 Eri. *Astronomy Letters*, 25, 453–466 {**123**}

Zasche P, Wolf M, 2007, Combining astrometry with the light-time effect: the case of VW Cep, ζ Phe and HT Vir. *Astronomische Nachrichten*, 328, 928–937 {**130**}

Zavala RT, Adelman SJ, Hummel CA, *et al.*, 2007, The mercury–manganese binary star ϕ Her: detection and properties of the secondary and revision of the elemental abundances of the primary. *ApJ*, 655, 1046–1057 {**123**}

Zejda M, Safar J, Masi G, *et al.*, 1999, V1493 Aql = Nova Aquilae 1999. *IAU Circ.*, 7228, 4 {**133**}

Zinnecker H, Köhler R, Jahreiß H, 2004, Binary statistics among Population II stars. *Revista Mexicana de Astronomia y Astrofisica Conf. Ser.*, volume 21, 33–36 {**110**}

4

Photometry and variability

This chapter reviews the details of the photometric data acquired by the Hipparcos satellite, and the scientific results based on the large body of homogeneous photometric data, including variability, provided in the Hipparcos and Tycho Catalogues.

4.1 Hipparcos and Tycho photometric data

A description of the photometric measurements and results from Hipparcos is given by van Leeuwen *et al.* (1997a), from where the following details of the data reduction methods used, the variability analysis, and the data products, are extracted. Further descriptions of all these aspects can be found in the Hipparcos and Tycho Catalogues (ESA, 1997), in particular Volume 1, Section 1.3 (description of the photometric data), Volume 3, Chapters 14 and 21 (description of the data reduction methods and the verification of the results), and Volume 4, Chapters 8 and 9 (the Tycho photometric data analysis).

The main Hipparcos passband, *Hp*, resulted purely from an attempt to maximise the number of photons gathered in the astrometric measurements, with no consideration of astrophysical features. Photometric information was contained in both the mean intensity and the modulation amplitude of the signal as the star image passed across the main modulating grid (Figure 1.2, left). The *Hp* passband was defined by the spectral response of the S20 photocathode of the image tube detector, combined with the transmission of the optics. The large width of the *Hp* passband results in significant systematic differences between *Hp* and standard V magnitudes, depending on effective temperature (or colour), metallicity and interstellar extinction.

The high-precision *Hp* magnitudes combined with homogeneous whole-sky coverage make the *Hp* magnitudes a very important mission product, with the multi-epoch measurements providing an important database for variability studies (Table 4.3 below). The broad wavelength coverage also makes the *Hp* magnitudes an important reference for calculation of bolometric luminosities.

The Tycho astrometric and photometric data were acquired by the satellite's star trackers, a series of four aperiodic vertical and inclined slits (Figure 1.2, right). A dichroic was introduced into the optical path to provide photometric measurements in two bands, B_T and V_T, which were somewhat similar to the standard Johnson B and V. The photometric information was contained in the peak height of the signal produced by a star passing over the star mapper slits.

The Hipparcos and Tycho photometry also played an important role in the astrometric reductions, providing valuable information on stellar multiplicity and colours.

4.1.1 Magnitudes and photometric systems

The intensity of light from celestial objects is usually expressed in magnitudes, an inverse logarithmic scale. The apparent magnitude, m, is a measure of the intensity of radiation received within a particular wavelength interval. The absolute magnitude, M, is the magnitude that the object would have at a distance of 10 pc from the Sun, with

$$m - M = 5 \log d - 5 \tag{4.1}$$

where d is the distance in parsecs, and the effect of interstellar extinction along the line-of-sight is ignored. For a general discussion of the effects of extinction see, e.g. Fitzpatrick (1999).

The colour index is the difference between apparent magnitudes in different spectral regions. Usually a colour index involves only two passbands, although sometimes three play a role, e.g. m_1 and c_1 in the Strömgren system (see below). The zero-points of magnitudes and colours of many photometric systems are defined with reference to Vega (α Lyrae). The magnitude zero-points of the *UBVRI* system were set by defining Vega to have colours of zero (with $V = +0.03$ mag), and consequently the colours for Vega are approximately zero in many photometric system.

Different photometric systems employ different wavelength bands established with specific objectives, and any such system is based on a set of standard sources measured with the chosen, well-defined, passbands, and well-distributed over the sky. Temperatures and extinction can be estimated from an overall energy distribution, while certain spectral features are also used as temperature indicators, e.g. Hα and Hβ for stars cooler than 8500 K, or TiO for stars cooler than 4500 K. Other features measured at particular wavelengths can provide estimates of gravity, interstellar extinction, metallicity, α-element abundances, etc. Observations intended to replicate a specific photometric system must be transformed to the adopted standards taking into account differences in telescope reflectivity, filters, and detector response characteristics. Bessell (2000a) provides an introduction to the issues of magnitude scales and photometric systems, and Bessell (2005) provides a more comprehensive discussion of the various photometric systems.

Of the various standard systems, early (blue-sensitive) photoelectric photometry included the Johnson *UBVRI* and Kron *RI* systems. The advent of more red-sensitive detectors was accompanied by the development of the Cousins *RI* system, with R_C (~638 nm) and I_C (~797 nm) being similar to the standard photographic *R* and *I* bands. Johnson also introduced the infrared sequence *JKLMN* (approximately 1.22, 2.19, 3.45, ... μm) into which Glass introduced the additional *H* band at 1.63 μm.

Various astrophysical considerations prompted the development of the DDO, Geneva, Strömgren, Thuan–Gunn, Vilnius, Walraven, Washington, and other systems. The Strömgren *uvby* system, for example, was devised to measure the temperature, metallicity, and Balmer discontinuity, of A, B and F stars. In this system, the *u* band lies completely below the Balmer jump, *v* measures a region near 410 nm affected by metal-absorption lines, *b* is centred near 460 nm, and *y* is a narrower *V* band. The index $b - y$, like $B - V$, is a temperature indicator, while indices $m_1 = (v - b) - (b - y)$ and $c_1 = (u - v) - (v - y)$ measure metallicity and the Balmer discontinuity respectively.

Photographic emulsions, which are relevant for the photographic astrometry results discussed in Chapter 2, were originally sensitive blueward of 490 nm (O emulsions). Different chemical sensitising extended the red response to around 580 nm (G), 650 nm (D), 700 nm (F), and 880 nm (N), with further developments resulting from fine-grain emulsions and H-sensitising of the Kodak IIIaJ and IIIaF emulsions.

The Sloan Digital Sky Survey photometric system split the entire CCD sensitivity range from 300–1100 nm into five essentially non-overlapping bands, u′, g′, r′, i′, z′, related to the Thuan–Gunn system.

4.1.2 Hipparcos and Tycho photometric systems

Approximate response curves for the three passbands, *Hp*, V_T and B_T, were obtained before the start of the mission using the payload's optical and photomultiplier specifications. The passband responses were refined during the mission using the photometric calibration results. In the case of *Hp*, the passband changed over the mission due to radiation darkening of the optics, which affected the blue response of the passband more than the red. The passband was defined for an arbitrary reference time, chosen to be 1 January 1992. The reference passband was obtained through extrapolation of reduction results obtained for data up to October 1991. As a result, the actual passband for 1 January 1992 was not identical to this reference passband. Moreover, there was always a small difference in passband between the two fields of view. The reference passband defines the photometric system, and the closer it was to the actual passband, the smaller the distortions that had to be removed from the actual data by the data reductions. Magnitudes for some 22 000 photometric standard stars in the resulting photometric system, selected as described in ESA (1997, Volume 1, Section 1.3.1), were calculated and used as reference values in the data reductions. Particular emphasis was put on obtaining reliable system calibrations towards the extreme red colours, where no standard stars are available. For this purpose, ground-based photometry and AAVSO monitoring (contemporaneous with the Hipparcos observations) were obtained for a selection of Mira variables. These data were used to produce a number of reference points for different colours and at different epochs, from which the red response of *Hp* passband was reconstructed as a function of time. This reconstructed response was applied *a posteriori*, and is referred to as the 'ageing correction'. The final passbands used in the catalogue construction are shown in Figure 4.1. The magnitude scales were chosen such that $Hp = V_T = V_J$ and $B_T = B_J$ at $B - V = 0$.

4.1.3 Main mission photometric reductions

The *Hp* photometry was derived from the same signal as the astrometric data, i.e. from the measurements by the image dissector tube photomultiplier of the image signal, modulated by a grid of transparent and opaque lines in the instrument focal plane. The light was sampled at 1200 Hz, and almost eight sampling periods covered

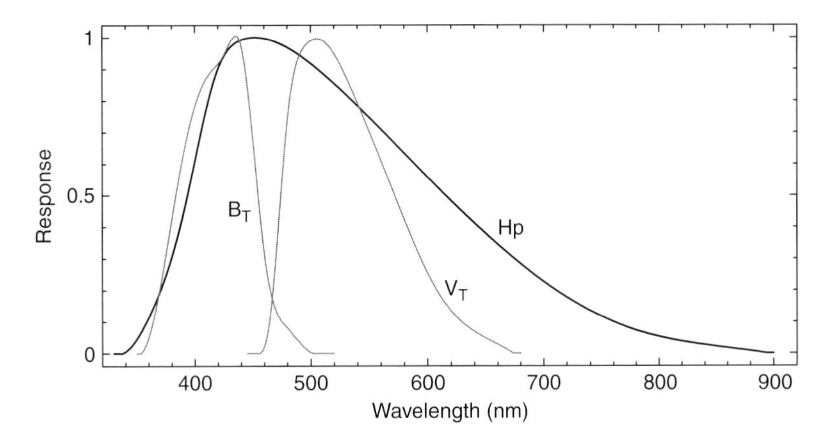

Figure 4.1 The Hipparcos (Hp) and Tycho (B_T, V_T) passbands. From van Leeuwen et al. (1997a, Figure 1).

the crossing of a single slit line. This modulated signal was accurately represented by a simple five-parameter model

$$E(N_k) = I_k = I_b + I_s[1 + M_1 \cos(p_k + g_1) + M_2 \cos 2(p_k + g_1 + g_2)] \tag{4.2}$$

where p_k is the local reference phase of the modulating grid as defined by the grid-period, the local scan-velocity, and the time of the observation. (M_1, g_1) and (M_2, g_2) are the modulation coefficients and relative signal phases for the first and second harmonic respectively and I_s and I_b are the intensities of the star and the background. The background signal was, under normal conditions (away from the radiation belt crossings), equivalent to the signal of a 14 mag star. I_k is the estimated value of N_k, the observed photon count over the integration time. The *Hp* photometry was primarily obtained from I_s, using elaborate methods for correction of the background, I_b. This part of the photometry is generally referred to as the $Hp_{\rm dc}$ photometry. The photometry obtained from $I_s M_1$ and $I_s M_2$ was also fully reduced, but is of lower accuracy (with an average value for M_1 of 0.7, the estimated errors on these photometric data are about two times larger than on the dc photometry). This part of the photometry is referred to as the $Hp_{\rm ac}$ photometry. As the modulation of the signal could be disturbed by the presence of other images in the detector field (a 30 arcsec diameter instantaneous field of view), the comparisons between the ac and dc magnitudes provide indications of such disturbances. Figure 4.2, left shows data for a star which are affected by its binary nature, where the projected component separation is dependent on the scanning geometry at the time of each transit.

Data reductions were developed and carried out independently by the FAST and NDAC consortia. The reductions involved the calibration of the detector response as a function of the position in the field of view and as a function of colour. The colour part of the calibrations was replaced *a posteriori* with the results obtained from the passband calibration, taking at the same time better care of the calibrations for the extreme red stars. This was not possible for the standard reductions due to the absence of constant stars among those with very red colours, $2 < (V - I) < 9$.

The colour-dependent coefficient evolved over the mission, reflecting the changes to the passband as a result of darkening of the main detector optics due to radiation damage. As a result, stars that were reduced using a $(V - I)$ colour different from its true value will show a drift in the *Hp* magnitude as a function of time. Similarly, variable stars with strong colour variations were reduced with a single representative colour. The *Hp* magnitudes for these stars may still be improved upon. The mechanism for corrections to the *Hp* magnitudes on the basis of colour improvements are described briefly in Volume 1, Section 1.3, and in detail in Volume 3, Chapter 14. In the final colour corrections a pseudo-colour, based on the Cousins $(V - I)$ index, was used to provide an improved treatment of stars of different luminosity classes and temperatures.

Calibrations were performed on data accumulated over 32/15 s intervals. The integration time actually spent on an individual star within that interval depended on the magnitude of the star, and on the competition for observing time from other stars in the combined field of view. Accuracies given in the epoch photometry data for different observations of the same star can therefore differ significantly, and quoted standard errors should be taken fully into account. The published epoch photometry provides the average of the measurements obtained during a field transit – it took a stellar image around 20 s to cross the field of view, and results at higher time resolution have not been preserved.

The final reduction results as obtained by FAST and NDAC showed a correlation of residuals for constant

Table 4.1 Standard errors on transit and median magnitudes. From van Leeuwen et al. (1997a, Table 1).

	⟨Transit error⟩		⟨Error on median⟩			
Hp	$Hp_{\rm dc}$	$Hp_{\rm ac}$	$Hp_{\rm dc}$	$Hp_{\rm ac}$	B_T	V_T
3	0.003	0.005	0.0004	0.0006		
5	0.005	0.009	0.0006	0.0010	0.003	0.003
7	0.008	0.019	0.0009	0.0019	0.008	0.007
9	0.015	0.037	0.0019	0.0039	0.026	0.022
11	0.033	0.072	0.0044	0.0079	0.12	0.12

Table 4.2 Mean number of transits as a function of ecliptic latitude. From van Leeuwen et al. (1997a, Table 2).

Zone	0–40°	45–50°	50–60°	60–90°
Mean	80	200	150	115

stars ranging from 0.6 for the brightest stars to 0.8 for stars of $Hp = 7$–9 down to 0.55 for the faintest stars ($Hp = 12$). These results were merged to form the Hipparcos Epoch Photometry Annex (HEPA) and Extension (HEPAE). Table 4.1 provides a summary of the photometric accuracies obtained. Table 4.2 shows how the total number of observations obtained per star varied with ecliptic latitude: the average total was 110, varying from 80 within 40° from the ecliptic, through 200 at distances of 45–50° from the ecliptic, to 115 close to the ecliptic poles.

The specific problem of red photometric standards, and the associated astrometric re-processing of red stars carried out after the Hipparcos Catalogue publication, is discussed in Section 4.10.

4.1.4 Tycho photometric reductions

The Tycho photometric data were obtained from the photon counts by the star mapper B_T and V_T photomultipliers (Figure 1.2, left). The star mapper signal consisted of four peaks at different interspacings, such that the signal provided a direct relation between position on the grid and position on the sky in one direction. By using both vertical and inclined slits, and observations from the two fields of view, the orientation of the satellite could be reconstructed, which was the main purpose of the star mapper data.

The Tycho experiment used the continuous data stream from the star mapper detectors with the aim of providing positional and photometric data for any star that could be detected. The photometric data was derived from scaling the observed signal to a 'single slit response function'. Different slit response functions were used for the various combinations of field of view, passband, slit group and upper or lower branch. The intensities were calibrated using calibration standards and a model describing the sensitivity of the detectors as a function of the position in the field of view and of star colour.

An important aspect of the Tycho photometry for fainter stars was the bias resulting from the detection of transits. By using only detected transits to derive mean magnitudes, the values that were obtained were inevitably too bright. This was corrected in a process called 'de-censoring', which also took into account the non-detections, to derive non-biased end results. The Tycho photometry has been preserved at this level of slit group crossing, but accuracies are considerably less than the main mission photometry, such that only a relatively small selection of this epoch photometry was released on CDs. The accuracies for the Tycho photometry are summarised in Table 4.1.

4.1.5 Variability analysis

All *Hp* epoch photometry was investigated for variability, and the results of this first analysis were presented in the published catalogues. First, the level of variability was determined from various measures of the spread of the observations. Then, for some 12 000 stars with a significant spread, tests on periodicity of these variations were performed. In advance of these studies, a database was established with references that could assist in this analysis. The analysis was carried out independently at the Observatoire de Genève and at the Royal Greenwich Observatory.

The time sampling, and the precision of the various epoch measurements, are critical in determining what kind of variability can be detected in the data. The basic time sampling determined by the scanning law gives intervals in the sequence 20, 108, 20, ... minutes, with this transit grouping loosely repeating after a month or more. The lower limit for periodicity detection is around 40 minutes, although lower periods can be detected in principle (Eyer & Bartholdi, 1999). The photometric precision depends strongly on magnitude, and the data are heteroscedastic (i.e. the random variables in the sequence may have different variances).

A total of 11 597 stars were found to be at least possibly variable. Of these, 2712 were found to be periodic (with 970 newly-discovered) and 5542 stars were found to be definitely variable, but no period was detected (of these 4145 are newly-discovered). The remaining 3343 stars include possible microvariables and were not further investigated. Very noticeable was the relatively large number of newly-discovered eclipsing binaries, where 343 of the 917 such systems detected were new discoveries. Over 3000 newly-discovered variables were considered well-enough determined to receive an official variable star name, a task that was carried out at short notice and in very little time by the Sternberg Institute in Moscow. Figure 4.2, right shows an example of a newly-discovered eclipsing binary.

Table 4.3 Main photometric characteristics and variability statistics of the Hipparcos Catalogue. From ESA (1997, Volume 1, p.xv).

Characteristic		N_{objects}	N_{new}
Median photometric precision (*Hp*, for $Hp < 9$ mag)	0.0015 mag		
Mean number of photometric observations per star	110		
Total number of *Hp* photometric measurements	$\sim 13 \times 10^6$		
Number of entries (possibly) variable, of which:		11597	8237
Periodic variables, of which:		2712	970
Cepheid type		273	2
RR Lyrae type		186	9
δ Scuti and SX Phoenicis type		108	35
Eclipsing binaries (e.g. EA, EB, EW, ...)		917	343
Others (e.g. M, SR, RV Tau, ...)		1238	576
Non-periodic and unsolved (e.g. RCrB, γ Cas, Z And)		5542	4145
Not investigated (including microvariables)		3343	3122

Periods given in the catalogue are mostly based on Hipparcos measurements only. However, for some eclipsing binaries with long periods the Hipparcos data alone did not contain sufficient information for a period determination. In these situations ground-based data were also incorporated. Reference epochs are given as epoch of the first primary minimum (eclipsing binaries, RV Tauri stars) or the first maximum (all other types) following BJD 2448500.0. Given the very limited time-span that was available for these investigations, they cannot be considered exhaustive, and corrections can be expected for periods and variability types, especially for newly-classified variables.

4.1.6 Data products

The Hipparcos and Tycho photometric data were published in three forms: as mean values in the astrometric catalogues, as epoch photometry, and as interpreted results of the variability analysis.

Photometric data in the Main Catalogue The main catalogue contains both mission photometry (median *Hp* and scatter values) as well as ground-based photometry ($B - V$ and $V - I$ colour indices, provided purely for the convenience of catalogue users). Variability indicators are provided, based on analysis of the *Hp* epoch photometry. When available, the Tycho B_T and V_T are also provided. Associated errors are also given. Also provided are the $(V - I)$ indices used in the reductions, which were usually, but not always, the same as the $V - I$ colour given in Field H40 of the main catalogue.

Epoch photometry files There are three epoch photometry files, two for main mission (Hipparcos Catalogue) photometry (HEPA and HEPAE), and one for Tycho Catalogue photometry. The first two provide all corresponding photometry available, while the third provides only a small selection. The HEPA and HEPAE files are constructed such that when accessing a given record number in one file, the same record number in the other file will provide the supplementary information for exactly the same observation. They therefore use the same index file. The HEPA file contains the most important data: the Hp_{dc} magnitudes, their standard errors, the epochs of observation (in TT, Terrestrial Time) and a summary flag explaining any sort of recognised problems that may exist for that observation. The HEPAE file adds to this the Hp_{ac} magnitudes and their standard errors, the coordinates of the other field of view for this particular star during the observation (for identifying accidental disturbing images), a coincidence index referring to a file of identified accidental disturbances (giving separation and magnitude of the object(s) found), and finally the background levels as detected by the two data reduction processes (important for very faint stars, where problems with background estimation could have seriously influenced the final results). Header records provide additional information on mean magnitudes, the colour index used in the data reductions, the numbers of transits available, and of transits accepted for the determination of mean values and summary information on variability.

The Tycho Epoch Photometry Annex A contains B_T and V_T epoch photometry and standard errors for 34 446 objects, mainly the brighter stars. Annex B contains data for 481 553 objects, including those in Annex A. Information on the orientation of the slits on the sky for each data point, as well as a number of flags identifying the slit group and any rejected measurements, are also included. The header records provide information on mean magnitudes, numbers of records available and used, and a number of flags identifying different problems that have been recognised for a star.

Variability results The results of the variability analysis are presented in the form of two tables, one for 'periodic variables' and one for 'unsolved', mainly non-periodic, variables.

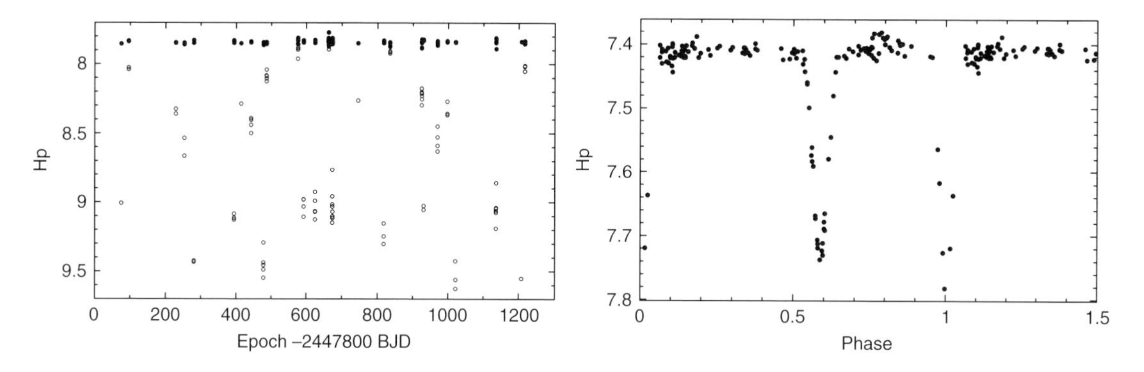

Figure 4.2 Left: Hipparcos photometry for a double star. Filled symbols refer to dc magnitudes, unaffected by multiplicity. Open symbols refer to ac magnitudes, which are affected by multiplicity. Right: Hipparcos $Hp_{\rm dc}$ photometry for a newly-discovered Algol-type eclipsing binary with eccentric orbit, HIP 270, which received the name V397 Cep. From van Leeuwen et al. (1997a, Figures 2 and 3).

The periodic variables include those for which a period and variability amplitude has been derived on the basis of the Hipparcos data. There are also cases where the Hipparcos data were insufficient to determine a period but, when folded with the period given in the literature, confirm that period – in some cases this has provided a new determination of the reference epoch. In nearly 200 cases, mainly involving Algol-type eclipsing binaries, for which the Hipparcos observations covered the full period of the light-curves inadequately, periods and reference epochs from the literature were used to fold the data. In these cases, an object may appear in the section of periodic variables even though no period is given in the main catalogue.

The 'unsolved variables' includes objects which could be classified as variable on the basis of the Hipparcos data, but for which periods could not be determined from the Hipparcos data, or where Hipparcos could not confirm a periodicity given in the literature. The majority of these are small-amplitude variables. In those cases where the Hipparcos observations have been unable to confirm a period given in the literature, an object may appear in the section of unsolved variables, even though a period may have been given in the literature. 'Microvariables' without established periods are flagged in the main Hipparcos Catalogue, but are not included in the sections on periodic or unsolved variables.

These tables provide information on periods, reference phases (if relevant), magnitude ranges and significance levels of the estimated amplitude. All periodic variables had their light-curves fitted, and from these fits accuracies of the periods were derived (van Leeuwen *et al.*, 1997b; Eyer, 1998). Pointers to notes and literature references, given in separate tables, are also provided.

In addition, there is an atlas (Volume 12 of the printed catalogue) providing three sets of light-curves: folded light-curves, showing also the fitted curves when available; not-folded light-curves, using calibrated AAVSO light-curves as background; and selected not-folded light-curves, mainly for stars with long time scale semi-regular or irregular variations. Summary variability information is provided in the main catalogue in the form of statistical indicators for variability and as references to the summary tables for periodic and unsolved variables.

Tools accessible via the ESA Hipparcos www site allow downloading the epoch photometry data points for each Hipparcos object, and the site includes Java applets for folding the data with any trial period.

4.2 Photometric properties and validation

The Hipparcos *Hp* photometry and the Tycho B_T and V_T photometry are the most substantial uniformly accurate collection of photometric data ever realised. They define a photometric reference system free from systematic errors as a function of right ascension or declination, and free from magnitude scale errors at the mmag level. At the time of the catalogue publication, the number of stars measured by Hipparcos exceeded the number measured in Johnson *UBV* over the previous 45 years (some 108 000 stars).

Ground-based observations are always made through varying amounts of the Earth's atmosphere. While the effects of refraction and extinction can be calculated rather accurately others, such as atmospheric scintillation, and the absorption and scattering by particles, are largely unpredictable, and vary on time scales ranging from minutes to seasonal changes over years. Repeated observations can average out many of these effects, but small systematic effects frequently remain.

Bessell (2000b) noted that *'Never before has such a wealth of accurately calibrated and precisely measured photometric data been obtained, data covering the whole sky, north and south of the equator. This enables intercomparison of many of the ground-based standard*

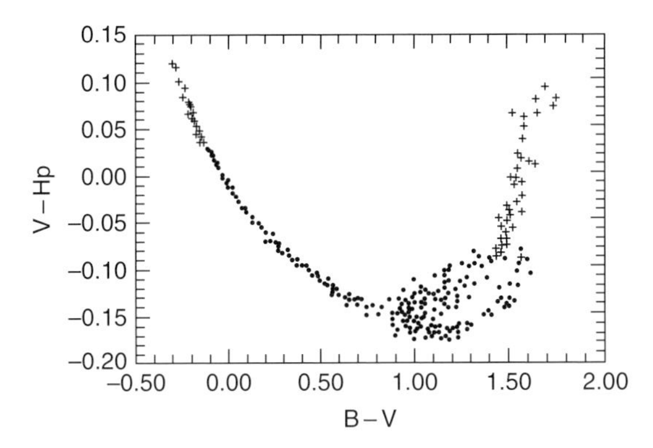

Figure 4.3 Differences between ground-based V magnitudes obtained at Sutherland, and the Hipparcos Hp magnitudes. •: second 10-year mean values for E region standard stars; ×: secondary standard stars from Kilkenny et al. (1998); ○: nearby stars from Lang (1989). B − V values are from the same sources. From Cousins (2000, Figure 1).

photometric systems and a search to be made for systematic differences'. Amongst the most precisely defined ground-based standard stars include the *UBVRI* standards of Landolt (e.g. Landolt, 1973, 1983, 2007), and the E-region standards, set up originally by Cousins (1974, 1976), continued by Menzies *et al.* (1989), and extended to very blue and red standards by Kilkenny *et al.* (1998). The stars range between $V = 4 - 9$ mag and occupy 10 regions with $\delta \sim -35°$ at roughly 3-hour intervals around the sky.

Cousins (2000) compared a long series of ground-based V magnitude measurements at Sutherland (S. Africa) with the Hipparcos observations. Figure 4.3 shows the difference $V - Hp$ as a function of $B - V$. The strong curvature shows that the Hp band is much wider than Johnson V: the negative slope at the left indicates that it extends further towards the blue, and the positive slope at the right indicates that it also extends further on the red side. The split beyond $B - V > 0.7$ shows the clear separation between dwarfs and giants due to the different atmospheric absorptions.

Bessell (2000b) used these standards to derive empirical mean relations between the Cousins–Johnson B and V magnitudes, and the Hipparcos–Tycho Hp, B_T and V_T magnitudes. The implied $V - Hp$ versus $V - I$ relation was used to derive a modification to the original Hp passband (shown in Figure 4.1), primarily affecting the blue wing of the passband. Bessell (2000b) provides tables of his proposed revision of the Hp, B_T and V_T normalised responses as a function of wavelength, as well as tables of $V - I : V - Hp : B - V : \Delta(B - V)$ and $B_T - V_T : V - V_T : \Delta(B - V) : V - Hp$, where $\Delta(B - V) = (B - V) - (B_T - V_T)$, for B–G main-sequence stars and K–M giants. Figure 4.4 shows the $V - V_T$ versus $B_T - V_T$ regression, and the $(B - V) - (B_T - V_T)$ versus $B_T - V_T$ regression, with polynomial fits, derived from these observations. Compared with Figure 4.3, the lines are straighter, with a negative slope, showing that the bandwidths for the colour indices are similar, but with shorter mean wavelengths than the corresponding Johnson bands. The straight line fits correspond to the linear approximations between Johnson V and $B - V$ and Tycho B_T and V_T given in the published catalogue over the range $-0.2 < (B_T - V_T) < 1.8$ (ESA, 1997, Vol. 1, Equation 1.3.20)

$$V = V_T - 0.090(B_T - V_T) \tag{4.3}$$

$$(B - V) = 0.850(B_T - V_T) \tag{4.4}$$

As noted in the catalogue description, these transformations apply to unreddened stars, ignore variations in luminosity class, and should not be applied to M-type stars. A more accurate transformation using linear interpolation was in practice used for the Johnson $(B - V)$ colour index given in the Tycho Catalogue.

An extensive compilation of theoretical transformations was presented by Bessell *et al.* (1998), who used synthetic spectra derived from ATLAS9 and NMARCS models to produce broadband colours and bolometric corrections for a wide range of $T_{\rm eff}$, g and [Fe/H] values. These synthetic spectra were then used by Bessell (2007, priv. comm., and available from his www pages) to construct a grid of $T_{\rm eff}$, $\log g$ and Z giving corresponding values of the following quantities (for his revised passbands): BC_V, $BC_{\rm Hip}$, $V - Hp$, $V - V_T$, $B - B_T$, $B_T - V_T$, $B - V$, $V - R$, and $V - I$. Resulting plots of various colour indices versus $T_{\rm eff}$ from the same synthetic spectra are shown in Figure 4.5. Plots of the bolometric correction $BC_{\rm Hip}$ versus $T_{\rm eff}$ are shown in Figure 7.3 below.

Maíz Apellániz (2005) used a sample of 256 Hipparcos stars with B_T and V_T magnitudes, and accurate Hubble Space Telescope spectrophotometry (using STIS) to examine the zero-points of the Tycho photometric bands. Calibration of the latter is based on a combination of Landolt *BV* photometry, ground-based spectrophotometry, and spectral energy distribution models, with four white dwarfs used as primary calibrators. The work shows that both datasets are accurate and stable enough to provide a precise cross-calibration without, for example, having to invoke modifications in the published Tycho filter response curves. The entire sample provides an estimate of the zero-point (with respect to Vega) of $(B_T - V_T)_0 = 0.020 \pm 0.001$, while a subset of seven stars provides the absolute zero-points of $(B_T)_0 = 0.078 \pm 0.009$ and $(V_T)_0 = 0.058 \pm 0.009$. An updated re-analysis by Maíz Apellániz (2006) provided a modified $(B_T - V_T)_0 = 0.033 \pm 0.001 \pm 0.005$ (random plus systematic). A corresponding analysis confirmed the published sensitivity curves for the Strömgren v, b, and y and Johnson B and V filters, but proposed revised responses for Strömgren u and Johnson U, with modified zero-points derived for all filters.

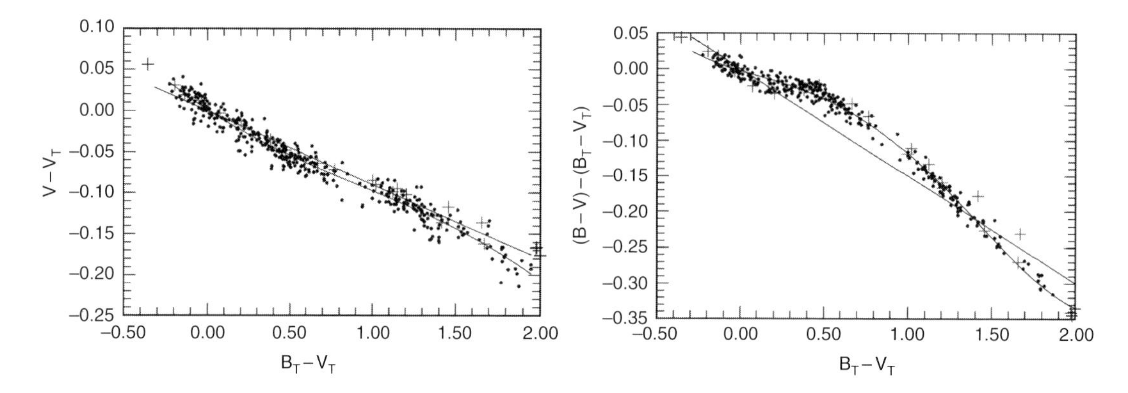

Figure 4.4 Left: the $V - V_T$ versus $B_T - V_T$ regression for the E-region stars. The straight line is the suggested approximate empirical relation given in ESA (1997); the curve is an improved polynomial fit. Right: the $(B - V) - (B_T - V_T)$ versus $B_T - V_T$ regression, with the lines having corresponding properties. In both cases the '+' signs are from synthetic photometry derived from Vilnius spectra using the revised Hipparcos passband. From Bessell (2000b, Figures 3 and 4).

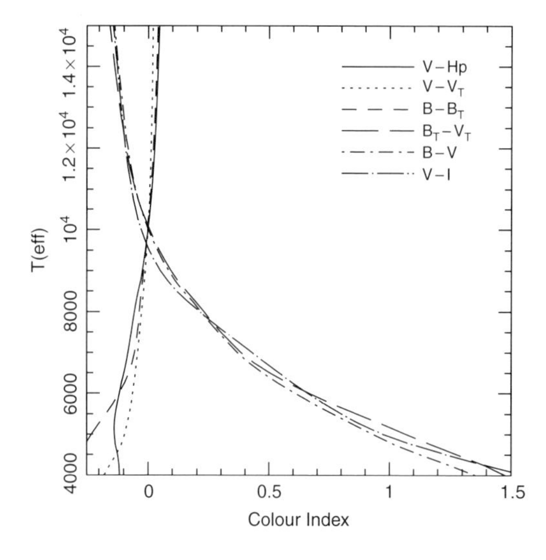

Figure 4.5 Various colour indices versus effective temperature from synthetic spectra. The determinations were made for this review by M.S. Bessell, and are available in tabular form on his www pages. The colour indices are for $\log g = 4.5$, $Z = 0.0$. *The steep relations for $V - Hp$, $V - V_T$, and $B - B_T$ reflect the similar effective wavelengths of these pairs.*

ESA (1997, Volume 1, Tables 1.3.5–1.3.6) provided $Hp - V$ versus Cousins $V - I$ for O–G5 classes II–V plus red giants of types G5III to M8III, and G, K, M dwarfs respectively. Harmanec (1998) compared Hp with UBV magnitudes for 110 stars observed at Hvar and Skalnaté Pleso (Czech Republic) to derive B and V from the catalogue values of Hp and known $B - V$ and $U - B$ colours. They proposed

$$V = Hp - 0.2964(B - V) + 0.0050(U - B) + 0.1110(B - V)^2 + 0.0157(B - V)^3 + 0.0072 \quad (4.5)$$

applicable over $-0.25 < B - V < 2.0$ and $V < 8$ mag, and providing an accuracy better than 0.01 mag.

Adelman (2001a) focused attention on those stars most photometrically stable during the Hipparcos mission, and thus most suitable as photometric or spectrophotometric standards. He identified 681 stars with amplitudes below 0.01 mag and with at least 15 accepted Hipparcos transits, covering the full Hipparcos magnitude range. Out of six G-stars considered by Lockwood *et al.* (1993) to be particularly stable, σ Dra and 70 Vir are in this most stable Hipparcos category, while ρ CrB, 64 Cet, and 31 Com belong to the next most stable group of 5956 stars with an amplitude of 0.02 mag or less. This confirms that many of the Hipparcos non-variable stars are indeed relatively stable. However, the final object in the Lockwood *et al.* (1993) list, Groombridge 1830, is noted in the Hipparcos Catalogue as an unsolved variable with an amplitude of 0.14 mag. A possible implication is that stellar variability, even amongst well-observed photometric standards, may change with time.

Kroll *et al.* (2000) reached a similar conclusion by studying 300 more-or-less random stars examined for variability on about 500 photographic plates from the Sonneberg Plate Archive taken between 1960–1996 (the total plate archive comprises some 270 000 plates), concluding that many Hipparcos constant stars reveal variability over long time scales.

Kallinger & Weiss (2002) considered whether the satellite scanning law has left some imprint in the epoch photometry calibration which shows up as spurious periodicity at specific frequencies. They selected a subset of 4863 stars considered as non-variable from Hipparcos or ground observations, subjected the Hipparcos photometry to a discrete Fourier analysis, and compared the number of peaks found in the frequency distribution with those expected in a uniform distribution.

While the amplitude spectra show generally little spurious frequency structure (their Figure 1), they did identify a $4\times$ excess in the low-frequency domain at 0–2 d^{-1}, and at around $11.55 \pm 1.77\,d^{-1}$, close to that corresponding to the spin period of the satellite of $P = 2.13$ hr, or $f = 11.25\,d^{-1}$; but with no other peaks related to other harmonics or the orbital period. They suggest that published detections of variability, based on Hipparcos data alone, with frequencies ranging from 9.78–13.32 d^{-1} and amplitudes exceeding the noise level by only a few times, should be considered as suspect. They concluded that the optimum frequency interval for detecting low-frequency variability with Hipparcos, considered by van Leeuwen (1996) to range from 0.5–11 d^{-1}, should be better restricted to 2–9 d^{-1}. They proposed this to explain the fact that Koen & Laney (2000a) reported the discovery of a period of 7.67 d (0.13 d^{-1}) for the asymptotic giant branch star RV Cam, contradicting the ground-based conclusions of Kerschbaum *et al.* (2001). They also found that 31 out of 36 A-type stars considered as low-amplitude multi-periodic variables identified by Koen (2001), fall in these spurious frequency ranges (see also Section 4.10).

Hipparcos constant stars are now frequently used as standards to verify variability in various photometric studies, e.g. Adelman *et al.* (1998); or for comparisons with photometric parallaxes, e.g. Cutispoto (1998), Cutispoto *et al.* (2001), Cutispoto *et al.* (2003). Miscellaneous considerations about the use of the Hipparcos data for the calibration of ground-based observations, and the presence of variability in the calibrating stars, are given by Adelman (1998), Collins (1999).

4.3 Photometric calibration in the optical

The validation of the photometric results made during the construction of the Hipparcos Catalogue were described in detail in ESA (1997, Volume 3, Chapter 21). The main conclusions regarding the Hipparcos photometric system homogeneity, assessed at that time through a comparison with stars from the Walraven photometric system, are also summarised here for completeness. This section includes results directly related to calibration, as well as some aspects related more to validation and/or luminosity calibration (covered further in Chapter 5).

Walraven system The internal consistency of the Hipparcos photometry, and the absence of position-dependent systematics at the level of 0.001 mag or better, was investigated during the Hipparcos Catalogue compilation, and was most convincingly demonstrated by a comparison with data obtained from the ground in the Walraven five-colour *VBLUW* photometric system (Lub & Pel, 1977). Pel (1991) had already shown that comparison of the Walraven system with the Cousins' V and $B-V$ colour index, and similarly with Strömgren y and $b-y$ versus Walraven V and $V-B$, showed no clear systematic effects as a function of magnitude, colour, right ascension and declination. Effects were encountered, however, in the comparison with Geneva V and $B2-V1$. Of the 1972 Walraven calibration standards, 1720 were also observed by Hipparcos. Comparisons between Hp and Walraven V as a function of Walraven $V-B$, and right ascension and declination, are given in Figures 21.12–21.13 of ESA (1997, Volume 3), and suggest that systematic differences are below the 0.001 mag level. Larger systematics can be expected for the few very bright stars ($Hp < 3$ mag), for stars redder than $B-V > 1.3$ (for which the Walraven system was not optimised, and for which standards are in any case problematic), and for the fainter stars ($Hp > 9-10$ mag) at low ecliptic latitudes where zodiacal light corrections were applied ESA (1997, Volume 3, Section 14.4).

Johnson system In addition to these studies carried out during the catalogue construction, Oja & Evans (1998) made a detailed examination of the systematic accuracy of ground-based photometry of various equatorial *UBV* standards using Tycho B_T and V_T. They selected three *UBV* datasets for comparison: those by Johnson (1955) supplemented by data from Oja (1996) assumed to define the *UBV* system; the equatorial standards measured by Landolt (1983); and the southern hemisphere standards in the E and F regions and in the Magellanic Clouds measured by Menzies *et al.* (1989). The relationship between V and V_T, and between $B-V$ and B_T-V_T, was first established using only the *UBV* data from Johnson (1955) and Oja (1996). They excluded stars brighter than $B = 2.2$ or $V = 2.2$ (which appear systematically in error in the Tycho data), supergiants, red late-type dwarfs, and heavily reddened stars, to derive

$$V = V_T + 0.004(1) - 0.115(3)(B_T - V_T) + 0.008(2)(B_T - V_T)^2 \tag{4.6}$$

similar to that given by Equation 1.3.33 in ESA (1997, Volume 1), which was not based purely on fundamental photometric standard stars, viz.

$$V_J = V_T + 0.0036 - 0.1284(B_T - V_T) + 0.0442(B_T - V_T)^2 - 0.015(B_T - V_T)^3 \tag{4.7}$$

The (nonlinear) relationship between $B-V$ and B_T-V_T for unreddened stars is shown in Figure 4.6. They also derived the corresponding 'reddening line' from all O and B0–B1 stars (Figure 4.6, cf. ESA 1997, Figure 1.3.6)

$$B - V = 1.024(6)(B_T - V_T) + 0.036(3) \tag{4.8}$$

Reddening for later-type stars moves the corresponding points parallel to this line. From comparison with the

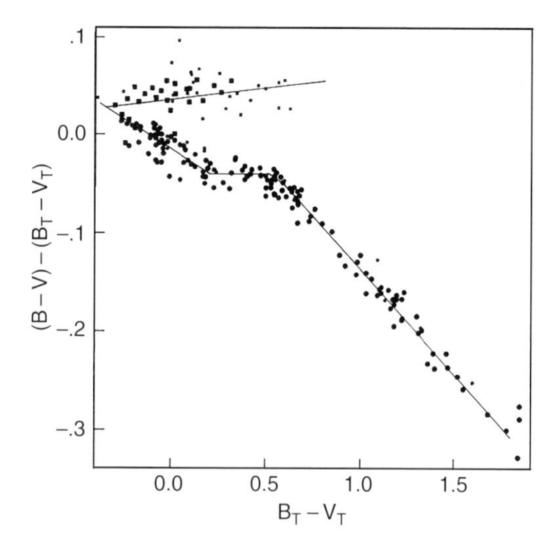

Figure 4.6 The relationship between Johnson $B-V$ and Tycho colour B_T-V_T. • = unreddened stars of spectral types later than B2; ▪ = stars of spectral types O–B1 of all reddenings, with the corresponding mean 'reddening line' indicated; reddening for later spectral types moves the corresponding points parallel to this line. Small symbols indicate points of lower accuracy, with the mean error of the colour difference exceeding 0.02 mag. A similar plot for all stars with accurate photoelectric photometry was given in ESA (1997, Figure 1.3.6). From Oja & Evans (1998, Figure 1).

other standard datasets, they concluded that the SAAO measurements of $B-V$ (for the E and F regions, for the Magellanic Cloud standards, and the equatorial standards) are generally too red by up to 0.014 mag, and that Landolt's measurements of $B-V$ of the equatorial standards are on the original Johnson system with the exception of the colour interval $-0.2 < B-V < 0.5$, where they are too red by up to 0.02 mag. Proposed correction tables are given in both cases.

Mironov & Zakharov (2002) compared the *Hp*, B_T, and V_T magnitudes with *WBVR* magnitudes from the Sternberg Astronomical Institutes Tien–Shan high altitude observatory in Kazakhstan, comprising some 13 600 northern stars with $\delta > -15^\circ$ and $V < 7.2$ mag (Cherepashchuk *et al.*, 1994; Kornilov *et al.*, 1996; Mironov, 2003). From the multiple observations of each star in both catalogues, they derived estimates of the intrinsic standard errors of each (in mag) as $\sigma_{\overline{W}} = 0.0066, \sigma_{\overline{B}} = 0.0038, \sigma_{\overline{V}} = 0.0035, \sigma_{\overline{R}} = 0.0042$ (Tien Shan), and $\sigma_{\overline{Hp}} = 0.0008, \sigma_{\overline{B_T}} = 0.0052, \sigma_{\overline{V_T}} = 0.0039$ (Hipparcos), in good agreement with the results published for each catalogue. For 6558 stars in common, they derived polynomial representation of the differences between the systems, e.g. using a second-order polynomial in V supplemented by a cubic polynomial in three colour indices to transform *WBVR* to *Hp*, B_T, and V_T, along with mappings of the form G_{WBVR}^{Hp-V} to investigate coordinate-dependent differences in α and δ. Their results indicate significant differences of amplitude ± 0.005 mag, which they infer arise from the Hipparcos ecliptic-based scanning law. However, their graphical results indicate a complex positional dependency, and in view of the consistency between Hipparcos photometry and Walraven photometry already noted, their conclusions that the ground-based observations are error-free, and that position-dependent photometric errors with an amplitude of up to 0.005 mag exist, do not appear compelling.

Strömgren system In the absence of trigonometric parallaxes, many approaches have been taken to the general problem of luminosity calibration of various stellar types. Crawford (1958) introduced a powerful index based on the photoelectric measurement of the $H\beta$ absorption line in two interference filters, one narrow and one broad, both centred on $H\beta$. This allows the determination of absolute magnitudes and temperatures of stars of early and intermediate spectral types (B–F), resulting in the M_V versus $uvby\beta$ calibrations for normal main-sequence stars given by Crawford (1975, 1978, 1979), Hilditch *et al.* (1983), and others. Open clusters and young associations were used to derive the shape of the ZAMS and the evolutionary dependence, while the zero-point was directly or indirectly fixed through the few available trigonometric parallaxes, for example using F stars in the case of Crawford (1979). Nissen & Schuster (1991) included theoretical stellar evolutionary models to derive metallicity dependence of the ZAMS. Pre-Hipparcos, the paucity of reliable trigonometric parallaxes for main-sequence A stars and peculiar stars had left open the question of whether Strömgren-based calibration is also valid for these stellar types. The availability of the Hipparcos trigonometric parallaxes has allowed these pre-Hipparcos luminosity calibrations for B–F stars to be re-assessed.

Domingo & Figueras (1999) used a Hipparcos sample of 188 main sequence normal A3–A9 stars and 197 Am stars to re-calibrate their absolute magnitudes derived from Strömgren photometry. Results of the analysis are shown in Figure 4.7. They concluded that the calibration of Crawford (1979), although reproducing rather well the absolute magnitudes of normal A-type stars, shows systematic trends and large dispersions when applied to Am stars, with differences up to 5 mas between photometric and Hipparcos parallaxes; that the corrections proposed by Guthrie (1987) for metallicity and rotation are too large; that although metallicity and rotation are highly correlated in this spectral range, a dependence of M_V on $(v \sin i)^2$ is evident for stars with $v \sin i > 100\,\mathrm{km\,s^{-1}}$; that the observational ZAMS proposed by Crawford (1979) and Mathew & Rajamohan

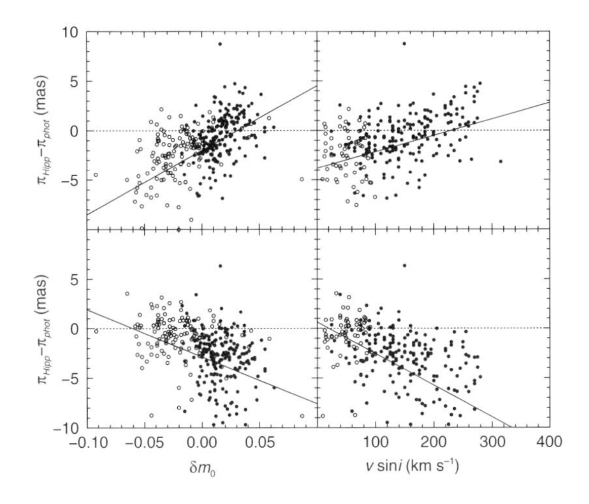

Figure 4.7 Differences between Hipparcos parallaxes and those computed using the calibrations of Crawford (1979) (upper panels) and Guthrie (1987) (lower panels) as a function of δm_0 and $v \sin i$ (• = normal A-type stars; ○ = Am stars). From Domingo & Figueras (1999, Figure 1).

(1992) are 0.2–0.3 mag below the ZAMS derived on the basis of the Hipparcos parallaxes, which are in turn in excellent agreement with the theoretical ZAMS; and that the Am stars are scattered along the whole width of the main sequence, like normal A-type stars, implying no significant differences between the evolutionary state of both stellar types.

Maitzen *et al.* (2000), following earlier results from Vogt *et al.* (1998), made a similar comparison between absolute magnitudes derived from $uvby\beta$ photometry and from Hipparcos parallaxes. Their comparisons were based on a sample of 1147 normal B–F stars with published Strömgren photometry and Hipparcos parallaxes, and a sample of 152 CP2 stars (covering the magnetic Ap stars) with $V < 8.5$ mag. Their results (Figure 4.8) show that the agreement for normal stars is acceptable although the scatter for the hottest stars is rather large. In contrast, the photometric absolute magnitudes for B-type CP2 stars are too bright by an average of 0.5 mag, while the photometric absolute magnitudes for cool A–F CP2 stars are up to 3 mag fainter than those determined directly from Hipparcos parallaxes.

Jordi *et al.* (2002) compared the ground-based and Hipparcos parallaxes for the F-type stars previously used to fix the zero-point of the Strömgren–Crawford luminosity calibration. This relied on 17 objects closer than 11 pc with $\sigma_\pi/\pi < 10.5\%$ from Woolley (1970), compared with accuracies below 1% for the same stars from Hipparcos. Combined with a large scatter on the ground-based values, the Hipparcos parallaxes are smaller by 5 ± 3 mas, and the photometric distances derived with the pre-Hipparcos calibrations underestimated by about 4%. Based on an enlarged sample of about 700 F0–G2 Hipparcos stars with $\sigma_\pi/\pi < 5\%$, Jordi

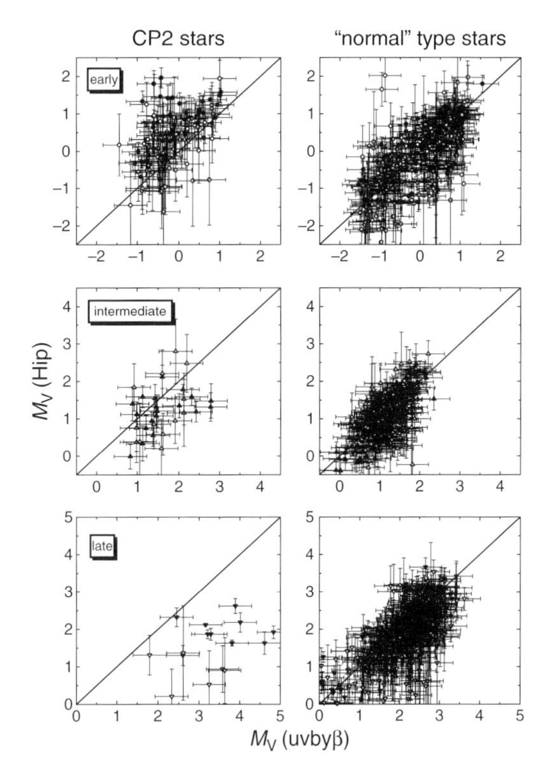

Figure 4.8 M_V(Hip) versus $M_V(uvby\beta)$ for a sample of CP2 (magnetic Ap) stars (left panels) and 'normal' type stars (right panels) divided into three subgroups (early, intermediate and late) and two accuracies of the relative Hipparcos parallax errors: filled symbols correspond to small errors; open symbols to large errors. From Maitzen et al. (2000, Figure 4).

et al. (2002) derived a revised calibration with revised evolutionary, metallicity, and ZAMS dependencies. A similar revision was attempted for B-type stars, although their scarcity in the solar neighbourhood results in only 20 Hipparcos stars with $\sigma_\pi/\pi < 5\%$.

Specific programmes to acquire Strömgren photometry for Hipparcos Catalogue stars have been reported, for example Jordi *et al.* (1996).

Vilnius system The Vilnius seven-colour photometric system (Straižys, 1992) was developed for photometric stellar classification in the presence of interstellar reddening, as well as for the identification of spectral peculiarities including emission-line, chemically- peculiar, metal-deficient, carbon-rich, and white dwarf stars. Absolute magnitude calibration as a function of spectral type and luminosity class was revised, based on the Hipparcos trigonometric parallaxes, by Malyuto *et al.* (1997) and Straižys *et al.* (1999). Photometric classification is based on the reddening-free Q parameters, which are formed from the various two-colour indices according to Straižys *et al.* (1999, Equation 4)

$$Q_{1234} = (m_1 - m_2) - E_{12}/E_{34}(m_3 - m_4) \qquad (4.9)$$

where the colour excess ratios E_{12}/E_{34} are slightly dependent on spectral type and luminosity class. Malyuto *et al.* (1997) based their calibration on the effective temperatures of 450 stars determined by the 'infrared flux method', or from angular stellar diameters and absolute magnitudes of 2100 stars determined from the Hipparcos parallaxes (see Section 7.4.2), yielding a significant improvement in calibration accuracy. For the calibration of each colour index or Q parameter, they constructed a three-dimensional diagram representing photometric quantity, effective temperature, and absolute magnitude. Such plots (e.g. Malyuto *et al.*, 1997, Figure 2 shows the $U-P$ colour index and the Q_{UPY} parameter as a function of absolute magnitude and $Y-V$ index), illustrate why particular colour indices are applicable for the classification of stars in particular ranges of temperature and luminosity.

Bartašiute *et al.* (1999) discussed application of the calibrated Vilnius photometry for 145 Hipparcos stars in the turn-off region. Lazauskaite *et al.* (2003) considered its application to the quantitative classification of metal-deficient dwarfs and Population II visual binaries.

Kazlauskas *et al.* (2003) made a similar re-calibration of the Strömvil photometric system (Straižys *et al.*, 1996). This system added the three passbands at 374, 516, and 656 nm from the Vilnius system to the *uvby* passbands from the Strömgren system, making the combined system capable of stellar classification throughout the HR diagram, also in the presence of reddening, an attribute especially useful for faint stars where the acquisition of Hβ photometry is difficult or impossible. The calibrations were based on photoelectric photometry for some 1000 stars obtained during 2000–2003, with $T_{\rm eff}$, $\log g$, and [Fe/H] obtained from the literature, and distances taken from the Hipparcos Catalogue. The resulting calibration also allows the identification of outliers due to chemical peculiarity or unrecognised binarity.

Geneva system Oja & Evans (1998, Section 4) considered possible systematics in Geneva photometry in their analysis of Tycho photometry for equatorial *UBV* standard stars. In addition to some systematic errors with right ascension and declination identified by Pel (1991), they found some evidence for a north–south zero-point error amounting to -0.007 ± 0.001 in their weighted $(B-V)$ index.

Cramer (1997) used 1580 Hipparcos trigonometric parallaxes of B to A stars, essentially absent in the immediate solar neighbourhood (and thus making a pre-Hipparcos analysis impractical), to provide a revised absolute magnitude calibration of the Geneva X and Y reddening-free parameters. The preliminary results showed that M_V was estimated correctly in the photometric temperature range 10 000–14 000 K, but that for higher temperatures, corresponding to the hotter

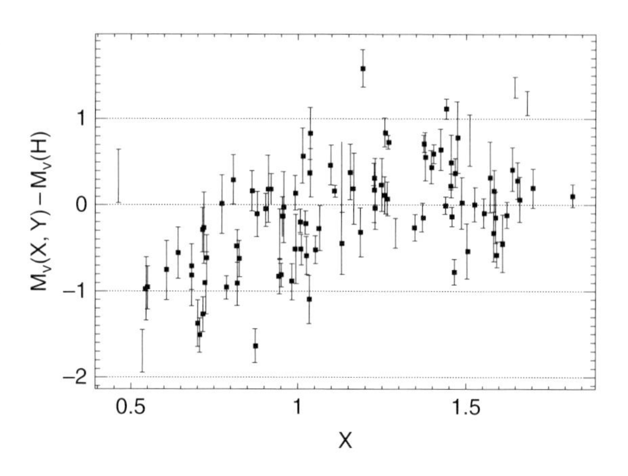

Figure 4.9 Comparison of absolute magnitudes derived from Geneva photometry, $M_V(X, Y)$, with those derived from Hipparcos, $M_V(H)$, for Ap stars. Error bars correspond to the Hipparcos magnitudes. M_V is estimated correctly for $X \lesssim 1.0$, but for the hotter stars the photometric absolute magnitude estimates are systematically overestimated. From Cramer (1997, Figure 3).

Si stars, photometric absolute magnitude estimates based on Geneva *X, Y* systematically overestimated the star's luminosity (Figure 4.9). A more extensive study of the Geneva calibrations for B-type stars, covering effects of temperature, absolute magnitude, gravity, reddening, and binarity, and drawing on the Hipparcos parallaxes for an independent luminosity calibration, is given in the review by Cramer (1999).

DDO system Høg & Flynn (1998) compared absolute magnitudes for 581 bright K giants based on Hipparcos parallaxes with absolute magnitudes determined from intermediate-band photometry in the David Dunlap Observatory (DDO) system. K giants have been used, for example, as tracers for the inner and outer disk kinematics, measuring the disk scale length, to determine properties of the bulge, and to constrain the amount of dark matter in the Galactic disk. In the disk dark matter studies of Bahcall *et al.* (1992) and Flynn & Fuchs (1994), for example, absolute magnitudes and hence distances were estimated from DDO photometry. The DDO system is an intermediate-band photometric system of six filters, four of which can be used to estimate physical parameters for late-type giants (filters 41, 42, 45 and 48 at wavelengths 416.6, 425.7, 451.7 and 488.6 nm). Three colours, C4142, C4548 and C4245 are formed from these four filters, of which C4245 is primarily sensitive to effective temperature, C4548 to luminosity, while C4142 can be used in combination with the other two colours to estimate [Fe/H] (Janes, 1979). Pre-Hipparcos, only a few giants had accurate parallax measurements, while Hipparcos observed all $V < 8.0$ giants, resulting in accurate parallaxes for around 600 objects. A number of systematic effects were identified in the DDO absolute magnitudes, in particular for the redder stars ($B - V > 1.2$) and around the clump giants, where the clump was estimated about 0.3 mag too faint by the DDO method. The revised calibration based on the Hipparcos parallaxes provides a relationship yielding photometric absolute magnitudes over the range $2 < M_V < 1$, and with $0.95 < B - V < 1.3$ and $[Fe/H] > -0.5$, with an accuracy of about 0.34 mag. The new calibration was verified against K giants with DDO photometry in 17 open clusters.

4.4 Photometric calibration in the infrared

The use of the Hipparcos and Tycho Catalogues for the extension of the ICRF astrometric reference frame into the infrared has been discussed in Section 2.8. The catalogues have also been used for photometric and related calibrations in the infrared. The use of infrared observations for the calibration of effective temperatures are discussed in Section 7.4.2.

Seitzer & Hill (1998) used JHK_S magnitudes of stars from 2MASS combined with parallaxes from Hipparcos to construct a preliminary infrared colour–magnitude diagram, M_{K_S} versus $J - K_S$, for nearby stars. Stars in common are at the faint end of the Hipparcos sample, and at the bright end of the 2MASS sample. A preliminary comparison of colour–magnitude diagrams for nearby open clusters and theoretical evolutionary tracks was noted. Some related considerations of the combination of Hipparcos with 2MASS and DeNIS survey data for kinematic and luminosity studies were noted by Alvarez & Mennessier (1998).

Martín-Luis *et al.* (2001) searched the Hipparcos and Tycho Catalogues for stars of type KIII and AV, suitable for extending the number and limiting magnitude of mid-infrared flux standards. They discuss a method for estimating fluxes in the range 1–30 μm from visible colours and spectral types, targeting the compilation of 1000 standard stars for use at ground-based telescopes.

Knauer *et al.* (2001) considered about 11 000 sources observed by both IRAS and Hipparcos, and calculated bolometric luminosities from the integrated spectral energy distributions from the ultraviolet to far-infrared wavelengths, based on distances from Hipparcos. From the 1000 sources with the best precision on luminosity, they showed that dust emission is found throughout the HR diagram, including pre-main-sequence, main-sequence, and post-main-sequence phases. They reported evidence that M giants with dust emission have luminosities 3–4 times larger than their counterparts without dust, and found a minimum threshold luminosity of about 2000 $L_\odot$ for stars to have dust, in agreement with models for radiatively driven winds. Once this threshold is achieved, the amount of circumstellar dust is independent of the stellar luminosity.

Identification of Tycho Catalogue objects with mid-infrared excess, typically attributed to the presence of circumstellar material, has been reported by Uzpen & Kobulnicky (2005), Clarke *et al.* (2005), and Uzpen *et al.* (2007). Clarke *et al.* (2005) correlated results from the Tycho 2 Catalogue and the Midcourse Experiment (MSX) Point Source Catalogue. The latter resulted from a mid-infrared survey of the Galactic plane and other areas of the sky missed by IRAS, which discovered 430 000 objects in the Galactic plane, $|b| < 6°$, some four times as many as detected by IRAS in the same region (Price *et al.*, 2001). The search yielded 1938 stars with infrared excess, of which more than half were undetected by IRAS, including young objects such as Herbig Ae/Be and Be stars, evolved objects such as OH/IR and carbon stars, and a number of B-type stars whose infrared colours could not be readily explained by known catalogued objects.

Numerous miscellaneous applications of the Hipparcos data to the physics of infrared sources have been reported, for example, to estimate mass-loss rates for the oxygen-rich asymptotic giant branch star R Cas (Truong-Bach *et al.*, 1997, 1999, using stellar parameters derived from Hipparcos); to establish positions on the HR diagram for objects with a far-infrared excess (Miroshnichenko *et al.*, 1999); to identify the infrared counterpart of the bright Galactic Z source GX 5–1 (Jonker *et al.*, 2000); and to determine for 11 G9–M2 giants, effective temperature, gravity, microturbulence, metallicity, CNO abundances, $^{12}C/^{13}C$-ratio and angular diameter from ISO SWS data, using the Hipparcos parallax to constrain the radius, luminosity and gravity-inferred mass (Decin *et al.*, 2003). An example of the combination of Hipparcos data and infrared photometry for studies of extinction is given by Knude & Fabricius (2003); this topic is covered more extensively in Section 8.5.2.

4.5 Photometric calibration in the ultraviolet

The problem of the missing opacity in the ultraviolet region of the solar spectrum (see, e.g. Allende Prieto & Lambert, 2000) was claimed to be solved by Kurucz (1992), who included millions of atomic and molecular lines previously ignored in the computation of model atmospheres, although the controversy has continued (Bell *et al.*, 1994; Balachandran & Bell, 1998). Malagnini *et al.* (1992) and Morossi *et al.* (1993) found that the theory of Kurucz (1992) underpredicted the near-ultraviolet fluxes observed in late-G and early-K stars, while others have not found such inconsistencies for late-type, metal-poor, or O, B, and A stars (Fitzpatrick & Massa, 1999).

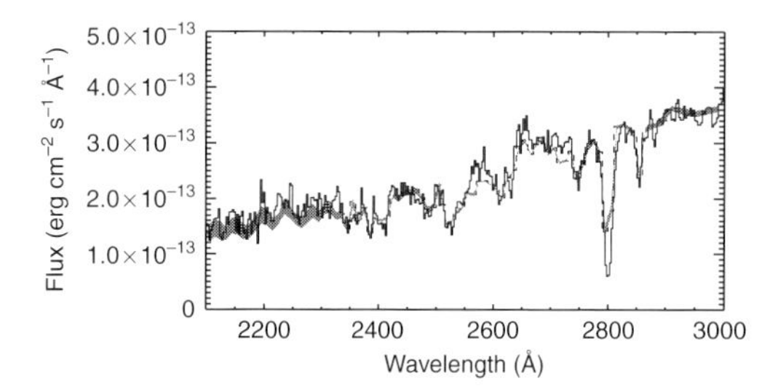

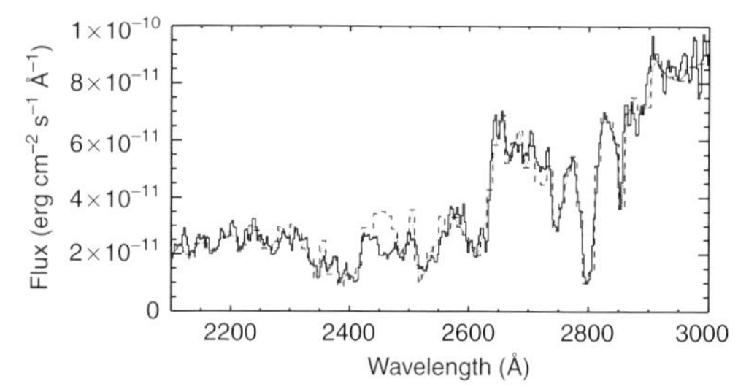

Figure 4.10 Observed (thick solid line) and theoretical (shaded and broken lines) ultraviolet fluxes at Earth for two stars from the study of Allende Prieto & Lambert (2000). The absolute fluxes are based on the Hipparcos parallaxes, and the thickness of the shaded lines represents the range of possible fits resulting from uncertainties in the derived gravity and 'dilution factor' $(\pi R)^2$, where π is the parallax and R the stellar radius. Left: G013–035, $T_{\rm eff} = 6145$ K, [Fe/H] = –1.9, $\log g = 4.5$. Right: HR 3775, $T_{\rm eff} = 5946$ K, [Fe/H] = –0.7, $\log g = 3.5$. The excellent fits are confirmation of both Hipparcos distance estimates and theoretical atmospheric models. From Allende Prieto & Lambert (2000, Figure 5).

This confusing situation was re-appraised by Allende Prieto & Lambert (2000) who used stellar near-ultraviolet fluxes measured by IUE, combined with the Hipparcos parallaxes, to yield absolute fluxes which could be compared with the predictions of theoretical atmosphere models (which yield surface flux per unit area). The absolute magnitude (calculated using the apparent visual magnitude, the parallax π, and a bolometric correction) is combined with an estimate of $T_{\rm eff}$ and theoretical evolutionary tracks to derive the stellar radius, R. Estimates of the flux emitted from the stellar surface are sensitive to estimates of R and π. The ultraviolet continuum in the spectral region between 250–350 nm is formed in the lower layers of the photosphere for late-type stars, and the ultraviolet spectra are relevant to the determination of abundances of several astrophysically interesting elements such as boron, or neutron-capture elements such as osmium, platinum, or lead. In summary, opacities and models used to compute the predicted flux can thus be checked using both the shape of the continuum, and its absolute value. Example results are shown in Figure 4.10. Allende Prieto & Lambert (2000) concluded that the Kurucz (1992) model atmospheres do reproduce the near-ultraviolet absolute continuum for stars with $4000 \leq T_{\rm eff} \leq 6000$ K, for any metallicity and gravity. Estimated values of $T_{\rm eff}$ and [Fe/H] are in excellent agreement with other reliable spectroscopic and photometric indicators, thereby implying that the average temperature stratification in the layers $0 \leq \log\tau \leq 1$ is appropriate. Together these results confirm that the fundamental hypotheses employed to construct stellar atmosphere models are adequate to interpret the near-ultraviolet continuum, and that the line and continuum opacities in the ultraviolet are indeed essentially understood.

Niemczura *et al.* (2003) described a similar procedure fitting low-resolution IUE ultraviolet spectra to theoretical stellar energy distributions from Kurucz (1993) to derive $T_{\rm eff}$, $\log g$, [m/H], $E(B-V)$, and angular diameter, θ. Combining these parameters with the Hipparcos parallax yields the stellar radius, R, and luminosity, L. Results are given for 17 standard stars from Code *et al.* (1976, for which angular diameters were measured with the stellar interferometer at Narrabri), covering spectral types O9.5–A7 and luminosity classes III–V, along with 16 stars from the LMC and 14 from the SMC. Small differences with respect to the previously-published fundamental data were reported.

García Gil *et al.* (2003) made a similar investigation of the missing opacity problem for the specific case of the standard star Vega (α Lyrae, HIP 91262), making use of more recently-compiled atomic data and revised observations. The study exploited the significant difference in $T_{\rm eff}$ between the Sun and Vega (of spectral type A0V) to help disentangle the effects of line and continuum opacity. Atomic models employed 'resonance averaged photoionisation' cross-sections from TOPBASE, and line transition probabilities and observed energy levels from the Atomic Spectroscopic Database. Observed spectra were based on IUE data in the ultraviolet with uncertainties in the absolute flux of 3–4%, and the visible spectrum of Hayes (1985) with an uncertainty of 1–2%. The Hipparcos parallax yields an accurate distance of 7.76 ± 0.03 pc, while recent interferometric observations yield an angular size of $\theta = 3.28 \pm 0.01$ mas at 2.2 μm (Ciardi *et al.*, 2001). Reasonable agreement between computed and observed fluxes are obtained for $T_{\rm eff} = 9640$ K, $\log g = 3.98$, [Fe/H] = –0.7, $\theta = 3.28$ mas, with [He/H] = –0.6 dex, [Si/H] = –0.85 dex, and [C/H] = 0.05 dex. Hydrogen plays the dominant role in shaping the ultraviolet spectrum, with H^-, He, and Fe opacities important in the visible and ultraviolet, and C I and Si II in the ultraviolet below 148 and 155 nm respectively. The 3% correction to the ultraviolet fluxes suggested by some authors appears unnecessary, although some strong lines of Fe II, Al II, and C I are still not well

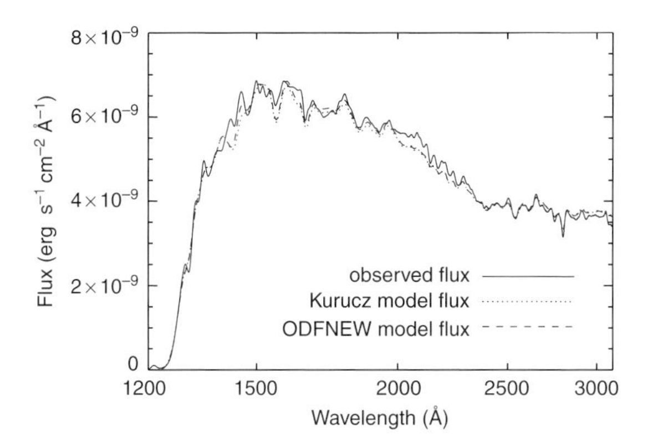

Figure 4.11 Comparison between the observed IUE spectrum of the standard star Vega, using the Hipparcos parallax to establish the absolute flux scale (solid line), and the computed spectrum using a Kurucz model atmosphere (dotted line) with $T_{\rm eff}$ = 9620 K, log g = 3.98, [M/H] = −0.7, [Si/H] = −0.90, and [C/H] = 0.03. Also shown is the computed spectrum using an ODF model atmosphere (dashed line) with $T_{\rm eff}$ = 9575 K, log g = 3.98, [M/H] = −0.7, [Si/H] = −0.90, and [C/H] = 0.03. The adopted angular diameter is 3.27 mas. From García Gil et al. (2005, Figure 5).

reproduced. More recent work is described by García Gil *et al.* (2005), and the match between theory and observation is shown in Figure 4.11. Gray (2007) summarised the problems associated with maintaining Vega as the primary photometric standard.

Mégessier (1998) used the Hipparcos results to propose for Vega $f(555.6\,\mathrm{nm}) = 3.46 \times 10^{11}\,\mathrm{W\,m^{-2}\,nm^{-1}}$ within 0.7%. Possible model inconsistencies in the infrared were noted.

Formiggini *et al.* (2002) analysed one field of the ultraviolet FAUST experiment on the ATLAS-1 Shuttle mission (Bowyer *et al.*, 1993), located in the outer region of the Ophiuchus molecular cloud, and containing 228 ultraviolet sources. Most of these were identified as normal early-type stars through correlations with catalogued objects, with an apparent dearth of 'subluminous' white dwarf or hot subdwarf stars, as found for other FAUST fields. However, correlation with Hipparcos and Tycho Catalogue parallaxes yielded 12 candidate hot subluminous stars, including six previously-unrecognised sdB stars (Section 7.5.6), and two subsequently confirmed as white dwarfs from follow-up spectroscopy. This use of parallax information suggests that other bright subluminous stars remain unrecognised in ultraviolet surveys.

4.6 Variability

One of the important by-products of the Hipparcos observations was the discovery of large numbers of new variable stars, made possible by the instrumental stability and repeated monitoring of stars over the entire sky over a period of more than three years. With no pre-selection according to variability in the case of the 'survey', it provides an important unbiased view of the occurrence of stellar variability in the solar neighbourhood. The types of variability classified in the Hipparcos Catalogue are given in Table 4.4. The location of the various intrinsic variables in the HR diagram is shown schematically in Figure 4.14 below. Studies of variability induced by multiplicity are treated in Chapter 3.

Star names and variable star names: Names of the brightest stars in the sky carry a mix of Greek (e.g. Procyon), Latin (e.g. Polaris), or Arabic (e.g. Deneb) etymology. Johann Bayer's 1603 Uranometria was a major development in celestial cartography (Section 2.12.1), and comprised 49 constellation maps, and a new ordering of star names in which Greek letters, and thereafter Roman letters, along with the Latin possessive form of the constellation name, were assigned sequentially depending on magnitude and location. John Flamsteed's 1729 Atlas Coelestis embraced the naming of more stars by assigning Arabic numerals within the constellations from west to east.

While some bright variables are accordingly identified simply by their proper names (e.g. Mira) or Bayer names, an additional naming scheme was required as more were discovered. This starts with a single capital letter assigned in order of discovery from R to Z (starting with R to avoid overlap with Bayer names which extended to Q), thereafter employing double letters AA, AB, ... to AZ, BB, BC, ... to BZ, and onwards to ZZ, avoiding only the letter J. After exhausting the 334 names of this letter-based system, variables are listed simply as V335, V336, etc. Thus V335 Ceti follows ZZ Ceti.

For Hipparcos, the variable star name was taken from the literature if the star was already known to be variable, otherwise unambiguous newly-discovered variables were assigned names by N.N. Samus and colleagues in Moscow, under the auspices of the IAU. New names assigned to Hipparcos discoveries were already included in the published Hipparcos Catalogue and annexes, including the Millennium Star Atlas. Of these, 308 are newly-identified variables in The Bright Star Catalogue (Hoffleit, 2002); see also Adelman (2001b). All 3153 assigned names appear in the 74th 'name list' of variable stars (Samus, 1997; Kazarovets *et al.*, 1999). The 78th release (Kazarovets *et al.*, 2006) lists the 1706 new variables designated in 2006, bringing the total number of named variables at that point to 40 215.

4.6.1 Variability detection methods

Of the many ways to detect periodicity, some are particularly valid only for rather strictly periodic signals, and are inefficient when there is not a full conservation of phase or amplitude: these include 'string' methods, analysis of variances, and Fourier decomposition. Methods developed for unevenly-spaced data include the Lomb–Scargle periodogram (Lomb, 1976; Scargle, 1982) and least-squares fitting by sine waves (e.g. Lehmann

Pulsations: Stellar pulsations are a widespread phenomenon, found in many post-main-sequence phases of stellar evolution, and occupying different regions of the HR diagram; see, e.g. Bradley (2000), Kurtz (2000), and reviews by Gautschy & Saio (1995, 1996).

Opacity mechanisms are responsible for many of the observed pulsations. The most typical case, as in Cepheids, RR Lyrae, and δ Scuti variables, arise as self-excited radial oscillations in partially-ionised regions in the stellar envelope in which He^{+} is ionising to He^{++} (see, e.g. Bradley, 2000). Radiant energy is transferred into ionisation energy (the γ mechanism), and leads to strong opacity gradients (the κ mechanism). As He^{+} first begins to ionise, its opacity increases with increasing temperature (contrary to the normal situation in stellar interiors), such that as the material heats through compression it traps more radiation and heats further. Expansion leads to cooling, and at maximum size the gas pressure is too low to support the envelope, and re-collapse sets in. Nonlinear effects determine the final pulsation amplitude.

In the radial pulsators, such as Cepheids and RR Lyrae, the spherical harmonic index $l = 0$. Radial oscillations can be in the fundamental or zero-order mode (node at the centre and antinode at the surface), or in overtone modes. Radial pulsations are a subset of more general non-radial pressure modes ($l > 0$), or p-modes, where pressure is the restoring force: in this case while radial motion still dominates, there is also some lateral motion. Non-radial pulsations can also be g-modes, where gravity is the restoring force.

In the HR diagram, a roughly vertical 'instability strip' includes the most regular pulsating variables, including the Cepheids and RR Lyrae variables. The red and blue borders of the instability strip are not strictly demarcated, and many non-pulsating stars lie within the strip. It extends below the main sequence to include the unevolved or mildly evolved δ Scuti stars, and continues to the pulsating white dwarfs. Light-curves of pulsating variables are typically asymmetric, and each class has characteristic properties: (a) classical Cepheids are Population I high-luminosity yellow F- and G-type pulsating giants and supergiants, mostly core He burning, with $M \sim 5$–$20M_{\odot}$. They vary by up to 1.5 mag in the visual, with periods ranging from around 1 day to over 100 days. They typically pulsate in the radial fundamental mode, and occasionally in the first overtone mode; (b) the W Virginis stars, also called Population II or type II Cepheids, are fainter low-mass analogues, $\sim 1M_{\odot}$, evolving towards or off the asymptotic giant branch, with periods ranging from 1–30 days; (c) the RR Lyrae variables are also radial mode pulsators, but are Population II objects undergoing core He burning. Luminosities are typically 40–$50L_{\odot}$, periods around 0.2–1.2 days, and amplitudes around 0.2–2 mag. Subsets pulsate in the fundamental mode (RRa and RRb), first overtone (RRc), or both (RRd).

Near the main sequence are several groups of pulsating stars, all with short periods and small amplitudes, the most obvious distinction being their pulsation periods. These include: (a) δ Scuti stars, pulsating A7–F2 stars with periods of 0.5 to several hours, and either mono-periodic or multi-periodic. Similar pulsations are seen in two classes of metal-poor stars: the SX Phoenicis class, and pulsators within the λ Bootis class; (b) rapidly-oscillating Ap (roAp) stars, a subset of Ap stars pulsating in high-overtone p-modes with amplitudes of 1–10 mmag, and periods of 4–15 minutes; (c) γ Doradus stars, non-radially pulsating F0–F5 stars with periods of about a day. They lie on or near the main sequence, on the cool side of the δ Scuti variables; (d) β Cephei stars, B0–B2 class III-V radial and non-radial p-mode pulsators, with very short pulsation periods of around 3.5–6 hr, and driven by opacity effects on iron-peak elements. About half are doubly periodic, displaying beat behaviour between the periods; (e) pulsating O stars, or α Cygni variables, are hotter than the β Cephei stars, with amplitudes of less than a few tenths of a magnitude, and periods of order 1–2 months. They are thought to be pulsating in g modes, or possibly in 'strange modes' associated with a sound-speed inversion; (f) slowly-pulsating B (SPB) stars, non-radially pulsating mid- to late-B stars with amplitudes below 0.1 mag, and periods of 0.5–3 d. They are also driven by opacity effects on iron-peak elements, but at sufficient depths to drive longer period g-modes; (g) the pulsating white dwarfs are multi-periodic g-mode pulsators of various temperatures, ranging from the hottest GW Vir type, to the coolest ZZ Ceti variables.

To the right of the instability strip and some 10 mag above the main sequence are the red variables, including the red giant branch (RGB) and asymptotic giant branch (AGB) stars. Periods range from around 25 days to over a year, frequently somewhat irregular. Bolometric magnitude variations are around 1 mag, although visual amplitudes may vary from a few hundredths up to 7–8 mag due to the presence of strong TiO absorption bands. They include the irregular or semi-regular pulsating giants and supergiants of spectral classes F to M, and the long-period variables (LPVs), of which the Mira variables are the subset with largest amplitudes.

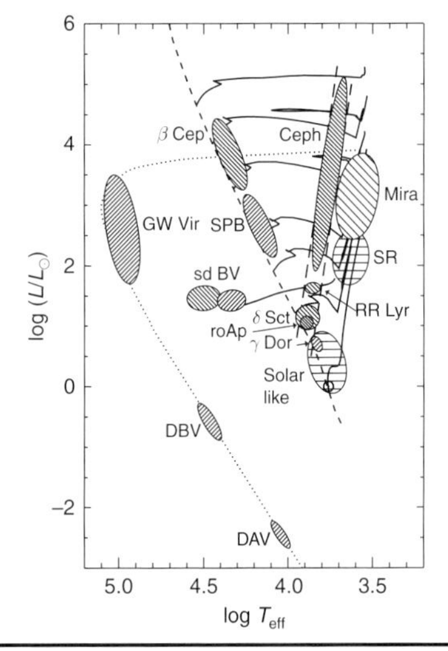

Figure 4.12 Theoretical Hertzsprung–Russell diagram showing the location of the various classes of pulsating stars in the $\log L$ *versus* $\log T_{\rm eff}$ *plane. The Ceph (Cepheids), RR Lyr, β Cep, GW Vir, δ Scu and γ Dor are named after the prototype of the class. The other acronyms are: rapidly oscillating Ap (roAp); slowly pulsating B (SPB); subdwarf B variables (sdBV); the DBV and DAV stars are variable DB (helium-rich) and DA (hydrogen-rich) white dwarfs. The parallel long-dashed lines indicate the instability strip. From Cunha et al. (2007, Figure 1).*

Table 4.4 Types of variability classified in the Hipparcos Catalogue, taken from ESA (1997, Volume 1, Table 2.4.1).

Code	Description	Class of variable
ACV	α^2 Canum Venaticorum type (including ACVO)	rotating
ACYG	α Cygni type	pulsating
BCEP	β Cephei type (including BCEPS)	pulsating
BY	BY Draconis type	rotating
CEP	Cepheids (including CEP(B))	pulsating
CST	constant stars (considered as variable by some observer(s))	–
CW	W Virginis type	pulsating
CWA	W Virginis type (periods $>$ 8 days)	pulsating
CWB	W Virginis type (periods $<$ 8 days)	pulsating
DCEP	δ Cephei type (including DCEPS)	pulsating
DSCT	δ Scuti type (including DSCTC)	pulsating
E	(E+, E/ ...)	eclipsing binary
EA	Algol type (EA+, EA/ ...)	eclipsing binary
EB	β Lyrae type (EB/ ...)	eclipsing binary
ELL	rotating ellipsoidal (ELL+... or /...)	rotating
EW	W Ursae Majoris type (EW/ ...)	eclipsing binary
FKCOM	FK Comae Berenices type	rotating
GCAS	γ Cassiopeiae type	eruptive
I	irregular (I, IA, IB, In, InT, Is)	eruptive
IN	irregular (INA, INAT, INB, INSA, INSB, INST, INT)	eruptive
IS	irregular (ISA, ISB)	eruptive
L	slow irregular (L, LB, LC)	pulsating
M	Mira Ceti type	pulsating
N	slow novae (NB, NC)	cataclysmic
NA	fast novae	cataclysmic
NL	nova-like	cataclysmic
NR	recurrent novae	cataclysmic
PVTEL	PV Telescopii type	pulsating
RCB	R Coronae Borealis type	eruptive
RR	RR Lyrae type (RR, RRAB, RRB, RRC)	pulsating
RS	RS Canum Venaticorum type	eruptive
RV	RV Tauri type (RV, RVA, RVB)	pulsating
SARV	small-amplitude red variables	pulsating/rotating
SDOR	S Doradus type	eruptive
SPB	slowly pulsating B stars	pulsating
SR	semi-regular (SR, SRA, SRB, SRC, SRD)	pulsating
SXARI	SX Arietis type	rotating
SXPHE	SX Phœnicis type	pulsating
UV	UV Ceti type	eruptive
WR	Wolf–Rayet	eruptive
XNG	X-ray nova-like system	X-ray binary
XP	X-ray pulsar	X-ray binary
ZAND	Z Andromedae type	cataclysmic

et al., 1995). Other methods are better adapted to nearly periodic signals, including wavelet (e.g. Foster, 1996), autocorrelation (e.g. Burki *et al.*, 1978; Edelson & Krolik, 1988), and the structure function or variogram method (e.g. Eyer & Genton, 1999). Some general considerations are given by Samus (2006).

The methods used for the estimation of periods and amplitudes for the published Hipparcos Catalogue are described in ESA (1997, Volume 1, Section 1.3, Appendix 3) for the approach used at the Royal Greenwich Observatory; and in Volume 3, see also Eyer (1998), Grenon (1997, 1998), and Eyer & Grenon (2000), for the method used by the Geneva Observatory. Various methods were used to estimate a period, and thereafter light-curve fitting was used to estimate amplitudes, epochs of minima, and errors.

Various subsequent treatments of the Hipparcos variability data have been published. A discussion of the equivalent Nyquist frequency for irregularly sampled data, leading to the possible detection of some very short periods, was given by Eyer & Bartholdi (1999).

Eyer & Genton (1999) described the application of a more robust 'wave variogram' analysis applied to the Hipparcos data, aiming to characterise the time scales of pseudo-periodic signals, i.e. the approximately periodic

Variability types: Stellar variability is, of course, a vast field. Already in the early 1980s, the General Catalogue of Variable Stars (GCVS, Kholopov, 1985) provided a four-volume listing of 28 211 variables of all kinds, with the New Catalogue of Suspected Variable Stars (NSV, Kholopov, 1982) comprising 14 811 stars of uncertain status. The GCVS recognises 88 types of intrinsic variables, including 34 pulsating, seven rotating, 10 X-ray, 22 eruptive and 15 cataclysmic. The following summary is included for orientation. Further introductory details can be found, for example, in Schatzman & Praderie (1993, Chapter 6), Sterken & Jaschek (1996), and Wilson (2001).

* Variability can be intrinsic (due to an internal physical origin such as pulsation), or extrinsic (e.g. in the case of eclipsing binaries).
* Non-pulsating variables near the upper main sequence include luminous blue variables (or S Dor variables); Wolf–Rayet stars where variability is caused by winds and rotation; non-periodic Oe and Be stars known as γ Cas variables; and α^2 CVn stars which are magnetic stars with spots associated with their magnetic poles.
* Other types of variable occur in binaries, including novae, dwarf novae, and cataclysmic variables. In these systems, a donor star provides a stream of gas (either from Roche-lobe overflow, or from a wind) and eruptions occur on an accreting star, powered by thermonuclear or gravitational energy. Generally, the donor is non-degenerate and the accretor is some compact degenerate object, most typically a white dwarf as in the common class of cataclysmic variables. Classical novae can brighten by around 10 mag over days, before fading over weeks or months, and are marked by only one known outburst. Recurrent novae explode like classical novae, but with frequent eruptions perhaps decades apart; accretors close to the white dwarf mass limit are inferred, which would require only small mass accretions to start a thermonuclear explosion. Dwarf novae with cyclical brightenings over weeks or months are identified with outbursts powered by gravitational energy. In symbiotic stars, ultraviolet radiation from the hot accretor interacts with the wind or atmosphere of a red giant companion.
* UV Ceti or flare stars are low-mass main-sequence stars with very high levels of chromospheric activity.

signals found in supergiant, spotted, or semi-regular variables. The method was shown to be suitable for the rejection of spurious periods, for the estimation of measurement noise, and for the identification of the pseudo-period of some typical Hipparcos variables. Considerations in the context of Gaia have also been studied (Eyer & Mignard, 2005).

Andronov (2000) discussed algorithms for the photometric classification of suspected variable stars, applied to the analysis of Tycho variables by Andronov *et al.* (2000). They include period determination using multi-harmonic or multi-frequency oscillations superimposed on a polynomial fit, and use of splines with constant and variable order, especially effective for RR Lyrae and various eclipsing binaries.

Percy *et al.* (2002) applied autocorrelation analysis to the study of Hipparcos short-period A and B-type variables, demonstrating its merits for semi-regular, or pseudo-periodic, variables. Tests to recover identified variables in the Hipparcos and Tycho datasets using autocorrelation analyses were also reported by Mironov *et al.* (2003).

Koen & Eyer (2002) made a new comprehensive search for periodic variables in the Hipparcos epoch photometry data. They used Fourier techniques and a consideration of the signal-to-noise ratio applied to 94 336 stars, ignoring those already flagged as periodic and unflagged stars fainter than $V = 10$. They identified 2675 new candidate periodic variables, of which the majority (2082) were contained in the Hipparcos 'unsolved' variables list. This number may be compared with the 2712 periodic variables already given in the Hipparcos 'periodic' variables annex. Figure 4.13 (left) shows the amplitude and period distribution of the new candidate periodic variables, while Figure 4.13 (right) shows the corresponding results from the Hipparcos periodic variables annex. The strong vertical feature in the former near $\log P \sim -1$ corresponds to the 11.25 d^{-1} rotation frequency of the satellite. Many objects classified as variables in this region are therefore likely to be spurious, although corresponding to the correct range of amplitudes and periods for δ Scuti stars. They classified the new objects according to spectral type, and noted that the high incidence of A- and F-stars at high frequencies is to be expected from the known abundant distribution of δ Scuti stars. In contrast, the large number of late-type stars with high frequencies was unexpected, and points either to aliasing problems, or a class of rapidly variable late-type stars which has been overlooked in the past.

4.6.2 Tycho variables

The contents of the Tycho Epoch Photometry Annexes have been summarised in Section 4.1.6. Briefly, Annex A is a photometric database of 34 446 bright stars, while Annex B contains photometry of nearly half a million objects. There are only a few published accounts of searches for variable stars in these Tycho databases.

Makarov *et al.* (1994) found 35 variables (21 in Annex A), of which one was a rediscovery (Bastian *et al.*, 1996). Friedrich *et al.* (1997) reported various methods to detect variables in Annex B, looking for stars with high standard deviations or scatter in their V_T magnitudes, or using different statistical moments like skewness and kurtosis to derive reasonable estimates on non-Gaussian variability in the Tycho light-curves, but without mention of the number of new variables found. Woitas (1997) reported the detection of 43 bright

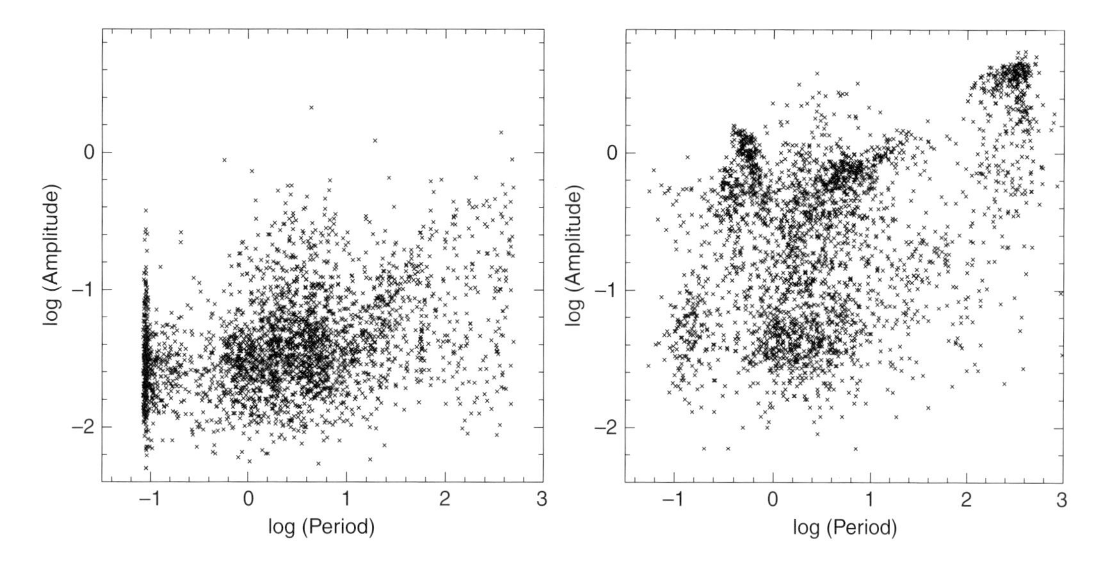

Figure 4.13 Left: amplitudes (Hp, mag) and periods (days) of the candidate variables. Right: amplitudes (Hp, mag) and periods (days) of the Hipparcos periodic variables: the most prominent clumps are the RR Lyrae (top left), Cepheid (top centre), Mira (top right), δ Scuti (bottom left), and slowly-pulsating B variables (bottom centre). From Koen & Eyer (2002, Figures 7 and 8).

variables (24 in Annex A), two of which are previously-known Hipparcos variables.

Koen & Schumann (1999) searched for periodic variables with approximately sinusoidal light-curves in Annex A. They reported 70 periodic variables not classified in the GCVS, or as Hipparcos variables, with some contained in the catalogue of suspected variables (NSV) but with unknown periods and amplitudes. In analogy with the list of M giants with relatively short periods from the Hipparcos Catalogue presented by Koen & Laney (2000a), of which those with periods shorter than about 20 days were originally considered as candidates for pulsators in high overtone modes, one of the objectives of the Tycho study was to find further examples. Some indeed were reported as M giants with periods below 20 days, with seven having periods shorter than 1 day. Conclusions regarding the existence of the Hipparcos short-period M giants were subsequently revised (see Section 4.10), and this presumably applies at least in part to the corresponding Tycho variables (see, e.g. McGough *et al.*, 2001).

Piquard *et al.* (2001) reported a more extensive search for variable stars in the Tycho observations (see also Piquard *et al.*, 2000; Andronov *et al.*, 2000). They constructed a modified Kolmogorov–Smirnov statistical test, based on a model of the satellite's parameters and thus taking into account truncated detections and censored measurements, allowing a search for variability for stars as faint as 11 mag. They identify 1091 suspected variable stars, 407 already known in Hipparcos, GCVS or NSV, and 684 of which are suspected variable, of which 496 are newly reported. Results of classification procedures applied to these variables is given by Piquard (2001).

4.6.3 Contribution of amateur astronomers

In the preparation of the Hipparcos Input Catalogue, photometric information was required for three distinct purposes: as an input to the definition of the observing programme, and specifically in defining the magnitude-limited survey; in optimising the dynamic allocation of observing time according to magnitude; and for the photometric reductions and calibrations of the satellite data. In cases where existing photometric data were inadequate, for any of a variety of reasons, new ground-based observations were acquired (Perryman *et al.*, 1989, Volume 2, Chapter 10). The particular cases of long-period and large-amplitude variable stars gave specific problems for the assignment of observing time. For about 280 stars with slow pulsations ranging from 100–800 days, future behaviour during the mission could not safely be extrapolated from their past history, and an extensive observational campaign was set up: with observations every 15 days or so over four years, corresponding to some 30 000 new observations. The collaboration of amateur observers provided a very extensive body of data for the preparation of the Hipparcos observations, specifically by the American Association of Variable Star Observers (AAVSO) and the Association Française des Observateurs d'Etoiles Variables (AFOEV), which included both visual and CCD *BVRI* observations (Mattei & Foster,

1997a; Welch, 1997). The AAVSO database now contains some 10 million observations.

Amateur observers, and students, have continued to observe specific Hipparcos variables, assisting in confirming or revising provisional periods or variability types for new variables discovered by Hipparcos (Mattei & Foster, 1997b; Ostrowski & Stencel, 1998; Mattei & Hanson, 2000).

4.7 Variability over the HR diagram

Hipparcos variability versus position in the HR diagram is shown in Figure 4.14 (left). The position of some of the main classes of intrinsically variable stars in the HR diagram is shown schematically in Figure 4.14 (right). Many such schematic plots can be found in the literature (e.g. Schatzman & Praderie, 1993, Figure 6.1), but this has been constructed to correspond directly to the variability seen in the Hipparcos M_{Hp} versus $V-I$ diagram.

Eyer & Grenon (1997) presented the variability properties in the HR diagram as a function of spectral type and luminosity class: 47% of the Hipparcos stars have complete spectral classification information, and the representation has the advantages that it is reddening free, and can be used even for very luminous stars without reliable parallaxes. Various representations of the results are shown in Figure 4.15, amongst which the following are noted. For the period distribution shown in Figure 4.15 (left), in which confirmed or suspected eclipsing binaries were removed, there is a broad period–luminosity relation for all spectral types, with the high-luminosity stars having the longest periods; for the early-type stars, β Cephei stars in luminosity classes II–IV show the shortest periods; late B and early A main-sequence stars have periods around 1 day; both G and early K dwarfs show rotationally-induced periods around 2–10 days; a period increase with increasing luminosity and decreasing temperature is conspicuous for late K and M giants and supergiants. The mean intrinsic scatter for all stars (variable, micro-variable, and constant) is shown in Figure 4.15 (middle): highly stable areas exist on both sides of the classical instability strip; the blue region narrows towards high luminosity; B8–A3 IV and V stars are the most stable, while early B are nearly all variable; the stable domain to the right of the instability strip is probably contaminated by stars with composite spectra, for which the variability is non-intrinsic; G8 II–V stars are photometrically the most stable; variability increases from G8 to M2 dwarfs with the development of activity and star spots; main-sequence stars in the instability strip are mostly microvariable with an intrinsic scatter up to 6–8 mmag; an amplitude-luminosity relation is again conspicuous for late K and M giants and supergiants; M giants show the largest intrinsic scatter. Figure 4.15 (right) shows the percentage of variables with amplitudes exceeding a given threshold: a very clear delineation of variability area is found when the threshold is set to 0.05 mag.

A sequence of investigations into the patterns of variability, with lists of specific stars recommended for further study, were reported for the following various spectral types and classes: O and B supergiants (Adelman *et al.*, 2000g); O4–B5 III–V (Adelman *et al.*, 2000f); late B III–V (Adelman *et al.*, 2000e); A0–A2 III–V (Adelman *et al.*, 2000d); A3–F0 III–V (Adelman, 2000a); A6–F9 supergiants (Adelman *et al.*, 2000a); F1–F9 III–V (Adelman *et al.*, 2000b); G0–G9 (Adelman *et al.*, 2000c); early K (Adelman, 2000b); and K5–M (Adelman, 2000c). The 2027 stars with the largest Hipparcos photometric amplitudes were considered by Adelman (2001c), with a classification according to variability type and amplitude. All of the very large amplitude variables are Miras, most of which are M stars, while a few are S and C stars. The largest amplitude of the non-Miras is the R CrB variable UW Cen with an amplitude of 2.96 mag.

Eyer & Grenon (1998) focused on low-amplitude periodic variables of interest to asteroseismology. Figure 4.16 shows the region of the HR diagram near the instability strip with all stars having $\sigma_{Hp} < 0.2$ mag, with specific symbols indicating low-amplitude periodic variables, including those specifically identified by Hipparcos. A prominent new clump of unevolved main-sequence stars lies at the red edge of the instability strip, with $M_V \sim 2.8$ and $B-V \sim 0.3$, and with periods intermediate between those of δ Scuti and γ Doradus stars. On the main sequence the blue edge of the instability strip does not extend bluewards of $B-V = 0.14$. At the time of these studies, the number of new Hipparcos discoveries (compared with previously known/suspected numbers) were, for the various types: β Cephei: 4 (59/80); slowly-pulsating B: 72–103 (11); δ Scuti: 35 (271/19); γ Doradus: 20 (11/17); roAP: 0 (29). The latter are too rapidly pulsating and have too small amplitudes to be easily detected by Hipparcos.

Wyn Evans & Belokurov (2005) and Eyer (2005) reported some early and rather promising experiments on global variability classification using 'self-organising maps' in the context of preparatory classification studies for Gaia. Self-organising maps, also known as Kohonen networks (Kohonen, 1982), are a form of neural network, providing a computational method for the visualization and analysis of high-dimensional data, based on an unsupervised learning or clustering algorithm (see Brett *et al.*, 2004, for a description based on classification of light-curves). In the analysis of the Hipparcos periodic variables reported by Eyer (2005), classification is based on period, amplitude, $V-I$ colour, and skewness of the photometric time series (Figure 4.17). A Bayesian network classifier trained on the Hipparcos Catalogue, with similar objectives, was described by López *et al.* (2006).

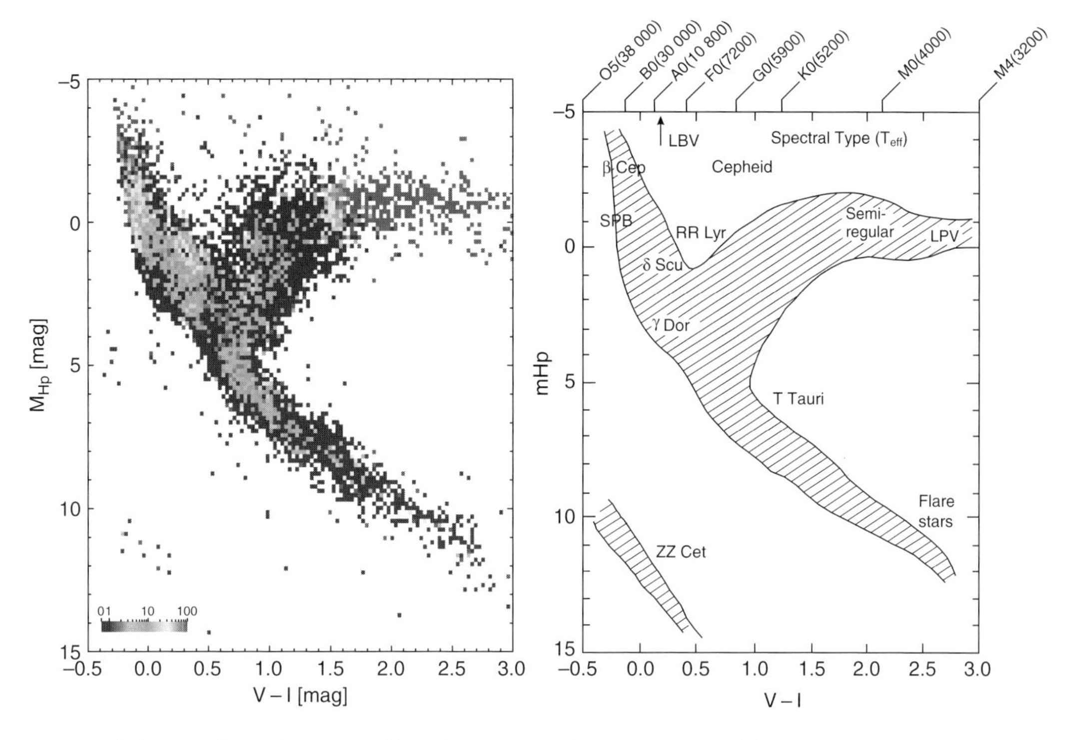

Figure 4.14 Left: fraction of stars that are variable in the M_{Hp}, $V-I$ diagram. Here, variability simply corresponds to the Hipparcos Catalogue Field H6 being non-blank. In each cell of 0.03 mag in $V-I$ and 0.15 mag in M_{Hp}, the percentage of variable stars is then coded according to the grey scale (original figure in colour). From ESA (1997, Volume 1, Figure 3.5.8). Right: schematic positions of some of the main classes of intrinsically variable stars in the HR diagram, also shown in the corresponding M_{Hp}, $V-I$ plane.

4.8 Main instability strip

4.8.1 Cepheid variables

Use of the Cepheid period–luminosity relationship as a distance indicator is discussed in Section 5.7.1, while the use of Cepheids as probes of Galactic kinematics and dynamics is discussed in Section 5.7.1. This section considers aspects of the pulsation properties where the Hipparcos data have provided some physical insight.

Of the 223 Galactic Cepheids with Hipparcos trigonometric parallaxes, most are small compared with the mean standard error, and of little individual value. In their discussion of the Cepheid period–luminosity zero-point based on Hipparcos data, Feast & Catchpole (1997) used the 26 Cepheids with the greatest weight and, of these, Polaris (α UMi, HIP 11767) is the only classical (type I) Cepheid with a parallax determined to better than 10% ($\pi = 7.56$ mas, $\sigma_\pi = 0.48$ mas), and the only Cepheid in the sample that gives a useful individual zero-point. A number of Cepheids have been identified as pulsating in the first overtone (the s-Cepheids: in the region of the instability strip where fundamental and overtone pulsators can both exist, it is not known what determines the selected pulsation behaviour). Feast & Catchpole (1997) used the relation between the fundamental and first overtone periods derived by Alcock *et al.* (1995), $P_1/P_0 = 0.720 - 0.027 \log P_0$, to derive P_0. Assuming that the overtone Cepheids indeed obey the normal period–luminosity–colour relation at their fundamental period (Beaulieu *et al.*, 1995), it is clearly important to verify that Polaris, with its pulsation period of $P = 3.97$ d, is indeed a first overtone pulsator. Evidence for this comes from the amplitude and phase lags of the radial velocity light-curves (e.g. Moskalik & Ogłoza, 2000). This Hipparcos-based demonstration that the period–luminosity zero-point given by Polaris is only consistent with that derived for the entire sample if it is a first-overtone pulsator (Feast & Catchpole, 1997, Table 2) is an independent verification of its pulsation mode.

Polaris is a particularly intriguing Cepheid (see also Section 3.5.2), whose photometric and pulsation radial velocity amplitudes have been decreasing since about the 1940s, and whose pulsation period has been increasing by about 316 s per century over the last 100 years (Arellano Ferro, 1983). Debate continues as to the cause of these changes, the evolution of pulsational instability with time, and the relationship between the photometric and radial velocity phases for overtone pulsators in

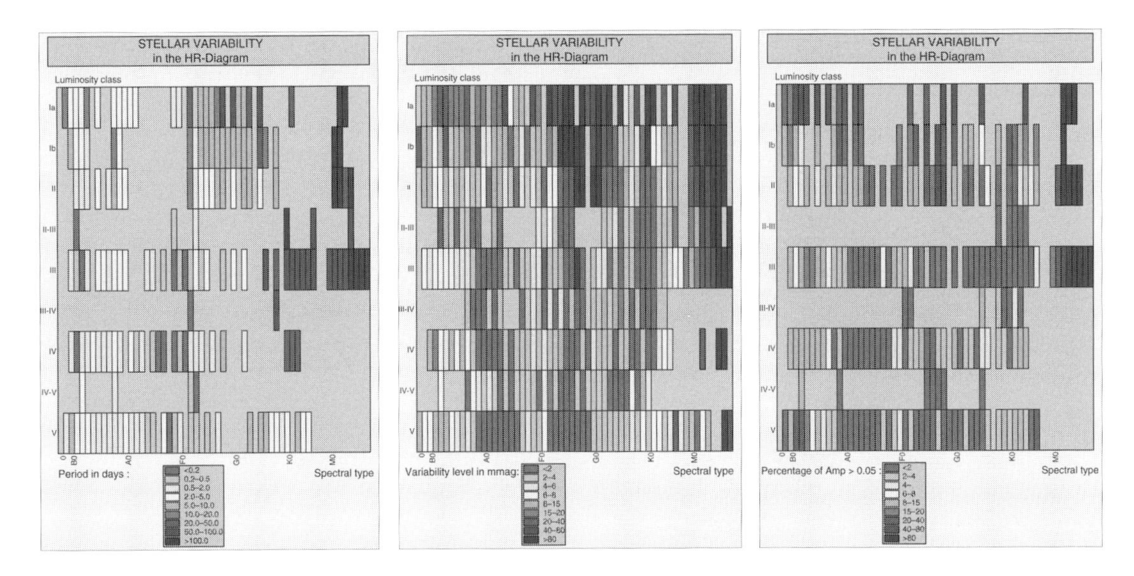

Figure 4.15 HR diagram as a function of spectral type and luminosity class, and labelled in period (left), mean intrinsic scatter (middle), and fraction of variables (right). From Eyer & Grenon (1997, Figures 2–4, originals in colour).

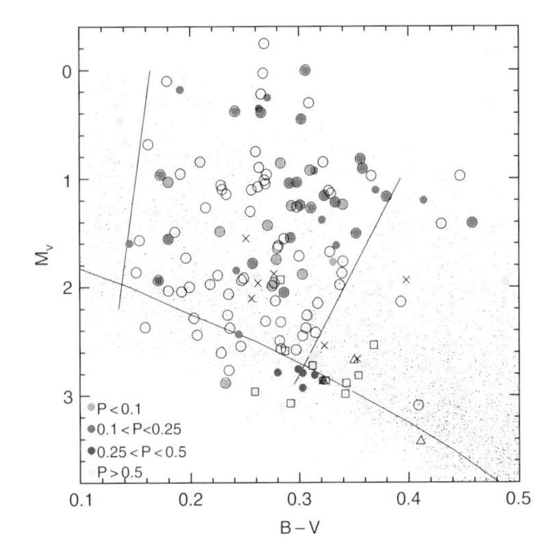

Figure 4.16 HR diagram for all Hipparcos stars having $\sigma_{Hp} < 0.2$ mag in the region of the instability strip. Symbols for the recognised variables are: □ = γ Dor; △ = suspected γ Dor; ○ = δ Scuti; × = roAp. Filled circles are the Hipparcos measurements for specified period intervals. From Eyer & Grenon (1998, Figure 2).

general (Kienzle *et al.*, 1999) and Polaris in particular (Evans *et al.*, 1998).

Other variability studies include: confirmation of the Tycho variable HD 32456 as a 3.3-day Cepheid (Bastian *et al.*, 1996), and other photometric observations (Campos Cucarella *et al.*, 1996); Fourier analysis of Hipparcos photometry of 240 Cepheid variables (Zakrzewski *et al.*, 2000); variability and multiplicity in the lower part of the Cepheid instability strip (Frémat *et al.*, 2005); and the double-mode Cepheid CO Aur (Santangelo *et al.*, 2007).

4.8.2 W Virginis variables

Variability analyses of Hipparcos W Virginis stars (Type II Cepheids) include studies of variables classified as Type II Cepheids (Berdnikov & Szabados, 1998); and period changes in TX Del and W Vir (Percy & Hoss, 2000). Reyniers & Cuypers (2005) discussed the bright supergiant HD 190390 for which classification in the post-asymptotic giant branch phase is generally assumed. Based on Geneva photometry over seven years, and supported by the Hipparcos photometry, they identified dominant frequencies of 28.6 and 11.1 days, leading to prominent beating with $P \sim 3000$ days. Based on its metal deficiency, luminosity, and pulsation period, they consider a W Vir classification more appropriate.

4.8.3 RR Lyrae variables

These are considered in Section 5.8.2.

4.9 Pulsators on or near the main sequence

An introduction to general considerations of asteroseismology, including those relevant for the δ Scuti, roAp, and β Cephei variables considered in this section, is given in Section 7.8. The main Hipparcos contributions include the unbiased identification of new examples of the different variability types, the determination of pulsation frequencies from the epoch photometry, the confirmation of pulsation models based on the parallax-based luminosities, and the derivation of physical parameters characterising the star's evolutionary phase.

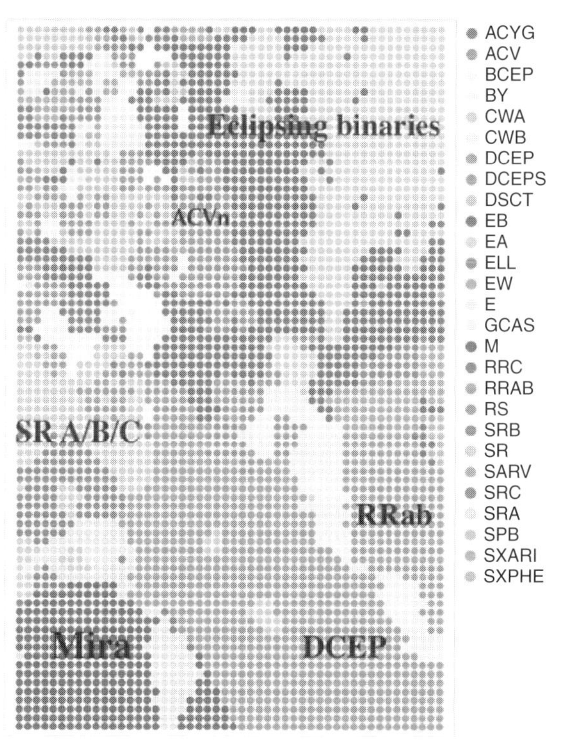

Figure 4.17 Classification of the Hipparcos periodic variable stars based on 'self-organising maps' prepared by A. Naud (Torun University). The classes of Mira, semi-regular, Cepheids, and eclipsing binaries form rather isolated groups, while the smaller amplitude variables are more scattered. From Eyer (2005, Figure 3, original in colour).

4.9.1 δ Scuti variables

General considerations The δ Scuti variables are the second most numerous group of pulsators in the Galaxy after pulsating white dwarfs. They are A7–F2 stars with periods of 0.5 to several hours. They range from about 2.5 mag above the main sequence to below the standard Population I main sequence; most stars belong to Population I while a few, designated as SX Phoenicis stars after the prototype, have low metallicities and high space velocities typical of Population II. Pulsation amplitudes in V range from a few mmag to 0.8 mag, with a typical values of 0.02 mag. Above 0.30 mag they are sometimes referred to as dwarf Cepheids, AI Vel or RRs stars (see Breger, 1979, 1980, 2000, for reviews, and further details of nomenclature, driving mechanisms, etc.).

The larger amplitude variables typically oscillate at one stable frequency, considered to be the radial fundamental mode, while the more common low-amplitude variables are generally multi-periodic, with both radial and non-radial modes inferred from the observed period ratios. Many show light-curves which are variable in phase, shape, and amplitude; interpretations have differed as to whether periods and amplitudes are unstable, or whether the apparent complexity is caused by the interaction of two or more modes with constant periods. On the main sequence, pulsation and rapid rotation can also coexist, and the average δ Scuti star rotates considerably faster than the average non-variable star (due to the fact that the classical metallic-line Am stars, which are slow rotators and lie on the main sequence, do not pulsate). Pre-Hipparcos, only the very metal-deficient star SX Phe had a significant trigonometric parallax, placing it 2 mag below the main sequence.

Pre-Hipparcos theoretical models by Petersen & Christensen-Dalsgaard (1996) used the most recent OPAL opacities and new stellar envelope models to give calibrations of the first overtone-to-fundamental mode and the second-to-first overtone period ratios in terms of the primary model parameters: metal content and mass–luminosity relation. They concluded that observed period ratios and positions in the HR diagram for the double-mode high-amplitude δ Scuti stars indicate that they are normal stars following standard evolution.

Antonello & Mantegazza (1997), see also Antonello *et al.* (1998), studied the luminosity and related parameters using the Hipparcos parallaxes, based on 103 stars with absolute magnitude errors less than 0.6 mag. They noted that: (a) stars within 50 pc appeared to separate into two groups in the colour–magnitude diagram, perhaps related to rotational velocity, although the effect tends to disappear with the inclusion of more distant stars; (b) a comparison of absolute magnitudes derived from the trigonometric parallaxes showed significant differences with those derived from photometric $uvby\beta$ indices, due to the effects of metallicity and rotational velocity; and (c) the absolute magnitudes of a few bright SX Phe stars confirm the period–luminosity relation based on globular cluster observations, although not appearing to depend on metallicity. Similar early results were reported by Liu *et al.* (1997).

Høg & Petersen (1997) used the improved Hipparcos distances to compare observations with theoretical models in the HR diagram for seven high-amplitude δ Scuti stars with accurate parallaxes. The best determined are AI Vel and SX Phe, both double-mode variables, with SX Phe belonging to Population II with $[Fe/H] \sim -1.3$, $Z \sim 0.001$, even though these objects are statistically very rare in the solar neighbourhood. It is a prototype for similar objects contributing to the blue straggler population in globular clusters, where they are probably created by mass transfer in binary systems or coalescence. Pre-Hipparcos HR diagram positions gave significant discrepancies with 'standard' stellar evolutionary models (i.e. not including rotation, mixing outside of the convection zone, or element diffusion).

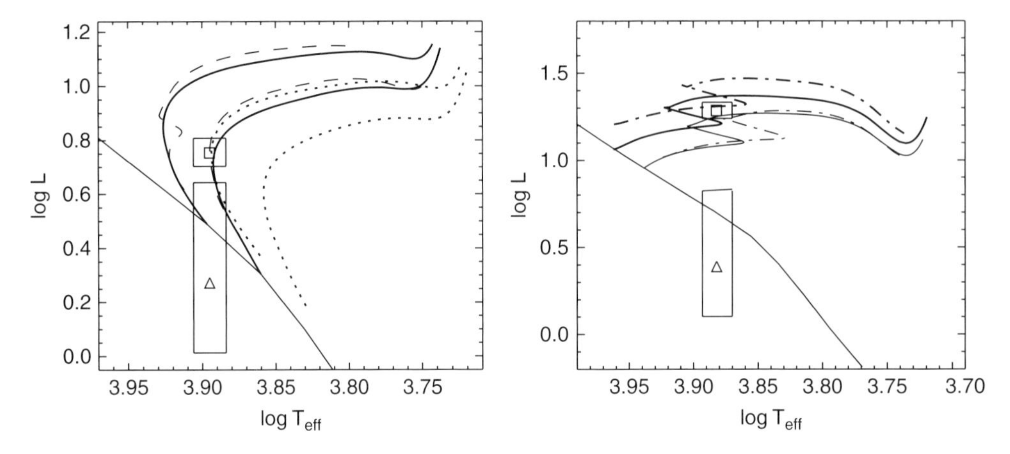

Figure 4.18 Left: comparison of standard evolution tracks in the HR diagram with observed positions of SX Phe according to the Hipparcos parallax (□) and the best earlier ground-based parallax data (△). Error boxes correspond to catalogue data and to ±200 K in effective temperature. Full curves give ZAMS and evolution tracks of mass 1.0 and 1.1 $M_\odot$ calculated for $X, Z = 0.70, 0.001$. Dashed curves are for $X, Z = 0.70, 0.002$ and masses 1.1 and 1.2 $M_\odot$, and dotted curves for $X, Z = 0.75, 0.001$ and masses 1.0 and 1.1 $M_\odot$. On three evolution tracks one model with the observed period of SX Phe is located within or close to the Hipparcos error box. Right: corresponding plots for AI Vel. Full curves give ZAMS and evolution tracks for masses 1.6 and 1.7 $M_\odot$ calculated for $X, Z = 0.70, 0.01$. Dash-dotted curves are for $X, Z = 0.70, 0.02$ and masses 1.8 and 2.0 $M_\odot$. From Høg & Petersen (1997, Figures 1 and 2).

Ground-based and Hipparcos derived luminosities are shown in Figure 4.18. For SX Phe $\pi_{\rm Hip} = 12.91 \pm 0.78$ mas compared with $\pi_{\rm ground} = 23 \pm 8$ mas. Its luminosity is now in agreement with standard stellar evolution models with mass $1.0-1.1 M_\odot$ and $Z = 0.001-0.002$, and in good agreement with the parallax predicted (via its predicted luminosity) from its oscillation frequencies, $\pi = 12 \pm 2$ mas (Petersen & Christensen-Dalsgaard, 1996). For AI Vel $\pi_{\rm Hip} = 9.99 \pm 0.53$ mas compared with $\pi_{\rm ground} = 28 \pm 11$ mas. It now fits a $1.63 M_\odot$, $Z = 0.01$ standard evolutionary model, and is in agreement with the luminosity/parallax predicted from its oscillation frequencies, $\pi = 9 \pm 2$ mas (Petersen & Christensen-Dalsgaard, 1996).

Both these objects are double-mode variables, allowing masses to be derived from pulsation theory independently of stellar evolutionary theory, from

$$P_i = Q_i M^{-1/2} R^{3/2} \tag{4.10}$$

where P_i are the normal mode periods ($i = 0$ being the fundamental mode, $i = 1$ the first overtone, etc.) and Q_i are the pulsation constants from simple pulsation models. Høg & Petersen (1997) estimated the error on the mass of SX Phe from evolution models to be around 18%. Estimates from pulsation theory are currently larger, but are dominated by errors in $T_{\rm eff}$ and photometric errors including interstellar absorption and bolometric corrections.

Solano & Pintado (1998) also derived $T_{\rm eff}$, projected rotational velocities, and surface gravities for low- and high-amplitude δ Scuti stars, also using Hipparcos parallaxes and evolutionary models.

Antonello & Pasinetti Fracassini (1998) studied the location of eight Hipparcos δ Scuti stars in the Hyades and showed that, unlike the non-variable stars, they are located both on the main sequence and above it, with the higher luminosity objects appearing to be related to high rotational velocity.

Rodríguez & Breger (2001) presented an analysis of the properties of the pulsating δ Scuti and related variables from the updated catalogue of 636 variables by Rodríguez *et al.* (2000), based on periods and amplitudes, and related contributions from Hipparcos, OGLE and MACHO. Variables in open clusters and binary/multiple systems, and the location of the δ Scuti variables in the HR diagram, are discussed on the basis of Hipparcos parallaxes and $uvby\beta$ photometry, with new borders of the classical instability presented. The Hipparcos parallaxes show that the available photometric $uvby\beta$ absolute magnitude calibrations can be applied correctly to δ Scuti variables with normal spectra, even when rotating faster than $v \sin i \sim 100$ km s^{-1}.

Handler (2002) summarised the discovery of 95 new δ Scuti variables as a by-product of the search for new γ Doradus variables in the Hipparcos 'unsolved' variables compilation (Handler, 1999). Candidates were verified from the location of known δ Scuti stars in the colour–magnitude diagram (specifically as given by Rodríguez *et al.*, 2000), and with their known period–luminosity relation (Figure 4.19). Other searches have been reported by Percy & Gilmour-Taylor (2002).

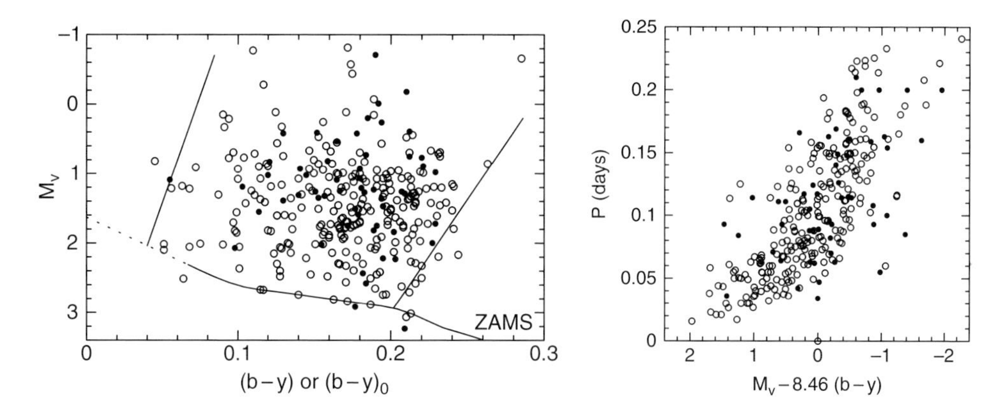

Figure 4.19 Left: colour–magnitude diagram of 252 δ Scuti stars with uvbyβ photometry (○) from the catalogue of Rodríguez et al. (2000), with 55 new candidates (•) with the same information discovered from the Hipparcos epoch photometry by Handler (2002). Blue and red edges of the δ Scuti instability strip from Rodríguez & Breger (2001) are indicated. Right: period–luminosity relation for the same 252 previously-known objects compared to 53 of the new candidates where periods could be estimated. From Handler (2002, Figure 1).

Period–luminosity relation Over the last half century, many investigators have attempted to derive a period–luminosity relation for δ Scuti stars, especially for the class of high-amplitude δ Scuti (HADS; see, e.g. McNamara, 1997a, for a summary). Høg & Petersen (1997) used their pulsation luminosities to derive a period–luminosity relation directly from the parallaxes, without recourse to photometric calibrations. They found

$$M_V = -3.74 \log P_0 - 1.91 \qquad (4.11)$$

where P_0 is the period of the fundamental mode. The star AD CMi deviates significantly from the mean relation, perhaps indicating a different evolutionary stage.

McNamara (1997a) used Strömgren *uvbyβ* photometry to determine temperatures, metallicities, and surface gravities of 26 large-amplitude δ Scuti stars, combining them with the fundamental periods to derive $M_{\rm bol}$ using the pulsation equation. Using Hipparcos parallaxes to establish the zero-point, he derived

$$M_V = -3.725 \log P_0 - 1.933 \qquad (4.12)$$

excluding outliers AD CMi and VZ Cnc, the latter probably due to the fact that it is oscillating in both first and second overtones (see Petersen & Høg, 1998). This period–luminosity relation defines a distance scale which is independent from those of the classical Cepheids and RR Lyrae stars. It can therefore be used to find M_V for RR Lyrae stars and horizontal branch stars in globular clusters, which themselves contain both SX Phe as well as RR Lyrae or horizontal branch stars. McNamara (1997a) thus derived $M_V = 0.42$ for RR Lyrae stars at [Fe/H] $= -1.9$, a value consistent with the revised Baade–Wesselink determination by McNamara (1997b), but not with the value of $M_V = 0.25$ at the same [Fe/H] determined by Feast & Catchpole (1997).

Petersen & Høg (1998), see also Petersen (1999), continued their analysis of the period–luminosity relation for 21 high-amplitude δ Scuti stars, dividing the objects into three relative parallax accuracy classes. For the six highest accuracy stars ($\sigma_\pi/\pi < 0.25$) they derived

$$\begin{aligned} M_V &= -3.73(57) \log P_0 - 1.90(59) \\ &= -3.73(57) \log (P_0/0.1) + 1.83(10) \qquad (4.13) \end{aligned}$$

where the second form with respect to a period of 0.1 day is adopted for numerical convenience. Correcting for Lutz–Kelker bias (Section 5.2.2) they determined

$$\begin{aligned} M_V &= -4.10(49) \log P_0 - 2.34(51) \\ &= -4.10(49) \log (P_0/0.1) + 1.76(9) \qquad (4.14) \end{aligned}$$

There is considerably larger scatter in the relation for the six stars of lower parallax precision, which they suggest may be due to large Lutz–Kelker type corrections (Figure 4.20). Using the same data but using the parallaxes directly in a nonlinear least-squares solution gave a similar result. Limitations in the approach include the small number of stars, the fact that the calibrating stars include both Population I and Population II, the fact that any colour term in a more comprehensive period–luminosity–colour relation is presently ignored, and possible effects of metallicity and rotation.

Petersen & Christensen-Dalsgaard (1999) revised the results of McNamara (1997a) to account for Lutz–Kelker corrections, and derived their preferred period–luminosity relation based on Hipparcos parallaxes

$$M_V = -3.725 \log P_0 - 1.969 \qquad (4.15)$$

They then developed a more detailed theoretical period–luminosity–colour–metallicity (PLCZ) relation for the Population I objects, guided by stellar evolution

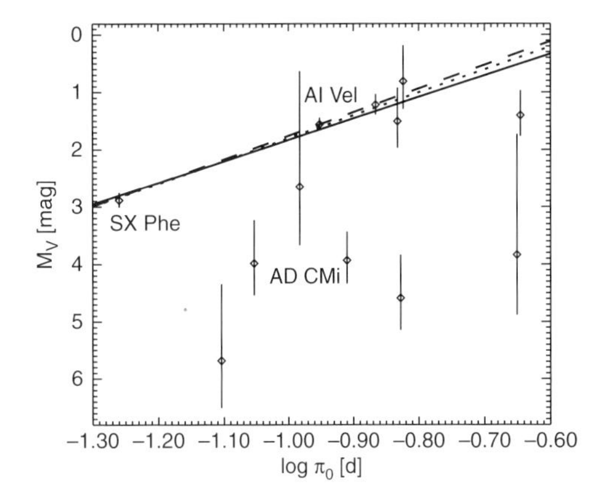

Figure 4.20 Period–luminosity diagram for the nearest 12 field high-amplitude δ Scuti stars (accuracy groups 1 and 2); Π indicates the period. Hipparcos data with 1σ error bars in π are marked by diamonds (a symmetric error in π giving asymmetric error bars in M_V). Lines gives various period–luminosity relations: full line: Equation 4.13; dashed: Equation 4.14 including Lutz–Kelker corrections, dots: a solution attributed to van Leeuwen. Six objects are well-represented by such a relation, while the poor fit of the lower points is attributed to Lutz–Kelker type errors. From Petersen & Høg (1998, Figure 2).

and pulsation theory, and working in terms of $M_{\rm bol}$ to calculate least-squares fits of the form

$$M_{\rm bol} = a(\log P_0 + 0.85) + b + c(\log T_{\rm eff} - 3.866) + d([{\rm Fe/H}] + 0.20) \qquad (4.16)$$

Of their various models reproducing results in different regions of the instability strip, they emphasise that the width of the high-amplitude part of the instability strip at constant luminosity is very small, at most ~300 K (McNamara, 1997a), which is only some 20–25% of the width of the full instability strip. Accordingly, for the high-amplitude subset, the colour term in the PLCZ relation becomes less important than in the corresponding relation for all δ Scuti stars. For this middle part of the instability strip they constructed a range of 122 theoretical models to derive

$$\begin{aligned} M_{\rm bol} = &-3.785(28)(\log P_0 + 0.85) + 1.099(3) \\ &- 12.06(49)(\log T_{\rm eff} - 3.866) \\ &- 0.271(12)([{\rm Fe/H}] + 0.20) \end{aligned} \qquad (4.17)$$

with a scatter of just 0.026 mag (see also Figure 4.21). The results suggest that the high-amplitude δ Scuti stars, and perhaps the whole δ Scuti instability strip, can be used for precise distance determinations. Its use requires rigorous mode identification to allow for the correct derivation of P_0, an approach which appears justified at least for the high-amplitude subset, for which the arguments for radial oscillations in the fundamental and first overtone modes (and the absence of non-radial modes) are now rather secure.

Individual objects In this section some of the results on individual δ Scuti objects which have relied on the Hipparcos data are noted. Without attempting a detailed introduction to the underlying theories of pulsation, analysis frequently starts from the relation

$$P\sqrt{\rho/\rho_\odot} = Q \qquad (4.18)$$

where P is the pulsation period, ρ the stellar density, and Q the pulsation constant. This can be written in various forms, e.g.

$$\begin{aligned} \log Q = &\log P + 0.5 \log(M/M_\odot) + 0.3 M_{\rm bol} \\ &+ 3 \log T_{\rm eff} - 12.7085 \quad \text{(Breger 2000)} \end{aligned} \qquad (4.19)$$

$$\begin{aligned} \log Q = &-6.454 - \log f + 0.5 \log g + 0.1 M_{\rm bol} \\ &+ \log T_{\rm eff} \quad \text{(Koen } et\ al. \text{ 2001)} \end{aligned} \qquad (4.20)$$

where f is in d^{-1}, and P and Q are in d. Using estimates of $M_{\rm bol}$, $T_{\rm eff}$, and $\log g$ or $\log M$, yields Q values which can be used with theoretical models (e.g. Stellingwerf, 1979; Fitch, 1981; Dziembowski & Pamyatnykh, 1991) to characterise pulsations in the various radial modes.

Frandsen *et al.* (1995) constructed a model fit to four observed frequencies for the binary star δ Scuti component κ^2 Boo. They derived a predicted distance of 47.9 pc, in excellent agreement with the subsequently published Hipparcos distance of 47.6 ± 1.9 pc, thus confirming their assumption that the observed modes have $m = 0$ (Bedding *et al.*, 1998a).

Templeton *et al.* (1997) presented results of five years (1983–88) of Strömgren y photometric monitoring of δ Scuti, the metal-rich ([Fe/H] = 0.49) prototype with a variability amplitude of around 0.06 mag. The data contain six distinct modes, one Fourier harmonic, and one beating mode. The dominant frequency is at $f_0 = 59.731\,129(2)$ μHz. Stellar evolutionary models were constructed to match the spectral type, radius estimate, and the Hipparcos-parallax based luminosity of $M_V = 1$. They then applied the linear non-adiabatic pulsation code of Pesnell (1990) to the shell H-burning structure, to establish the best-fit models as a function of Z. For $Z = 0.02$, for example, the best-fit model has $M = 2.1 M_\odot$, $T_{\rm eff} = 6894$ K, and $R = 4.14 R_\odot$, and the largest observed amplitude is replicated by the radial fundamental mode, with several non-radial modes being simultaneously excited. If the spectral type is correct, the Hipparcos-based luminosity specifically contradicts earlier conclusions that the strongest mode of δ Scuti is the first overtone (Balona *et al.*, 1981).

Bossi *et al.* (1998) used simultaneous photometry and spectroscopy over 11 nights to detect seven confirmed and five probable pulsation frequencies in HD 2724. The Hipparcos distance provides a luminosity

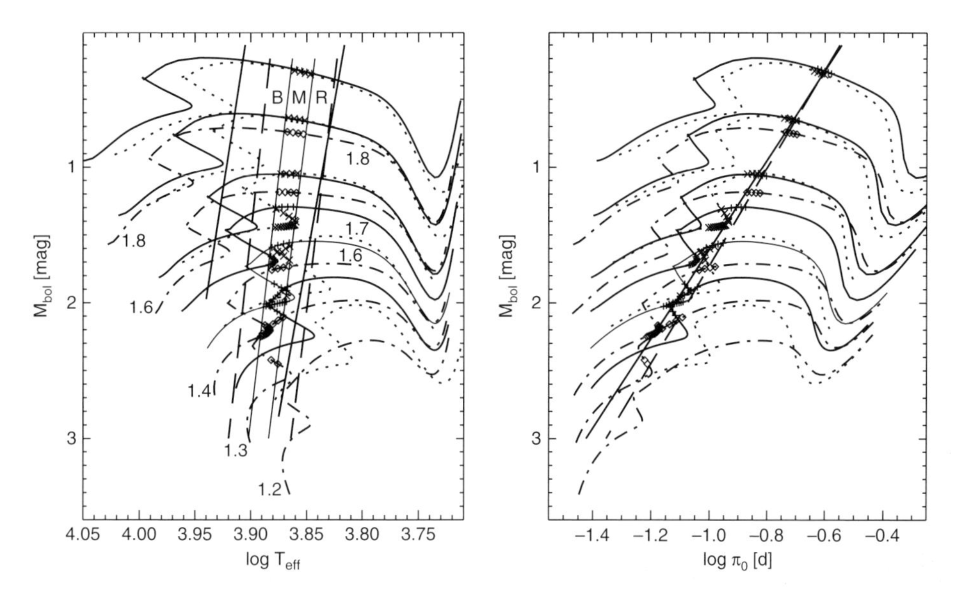

Figure 4.21 Left: HR diagram showing standard evolution tracks calculated for $Z = 0.005$ (dash-dotted curves), 0.01 (full curves) and 0.02 (dotted curves). Masses are 1.2, 1.3, 1.4, 1.6 and 1.8 $M_\odot$ for $Z = 0.005$ (all marked); 1.5, 1.6 (marked), 1.7 (marked), 2.0 and 2.2 $M_\odot$ for $Z = 0.01$; and 1.6, 1.8, 2.0, 2.2 and 2.4 $M_\odot$ for $Z = 0.02$. Each model within the high-amplitude δ Scuti instability strip (marked M and delimited by thin full lines) is given by a symbol. Adjacent strips B (blue) and R (red) were included in some PLCZ relations. Thick lines indicate the limits of the full Population I δ Scuti instability strip. Right: comparison of simple period–luminosity relations with the standard evolution tracks from the left figure, shown with the same coding (Π is the period). Each model within the high-amplitude δ Scuti instability strip is marked by a symbol. The full line is the least-squares solution for the 122 marked models. The dashed line gives the Hipparcos solution, which is in good agreement within the expected uncertainty. From Petersen & Christensen-Dalsgaard (1999, Figures 2 and 3).

indicating that the lowest frequency is the fundamental radial pulsation, thus yielding physical parameters from the associated pulsation model. This allows the surface gravity to be determined with unusual accuracy from the pulsation model, as well as yielding an estimate of the inclination of the rotation axis to the line-of-sight.

Paunzen *et al.* (1998) used photometric data for HD 84800 to establish the spectral type on the basis of the Hipparcos parallax, and to identify the observed periodicity with a high overtone mode.

Different approaches have been tried to assist mode identification (e.g. Kjeldsen & Bedding, 2001). Viskum *et al.* (1998) used new spectroscopic and photometric time-series observations of FG Vir to detect oscillations from the changes in equivalent widths of hydrogen and metal absorption lines; the absorption-line equivalent-width measurements have a strong centre-to-limb variation becoming much weaker towards the limb, while the photometric measurements have relatively small centre-to-limb variations. This difference in spatial response was used to determine modal l-values, and thus the pair of modes likely to match the observed frequency of the two $l = 0$ (radial) modes. The star is known to be near the end of its main-sequence evolution, and is multi-periodic with a main pulsation close to 1.9 hours. It is the δ Scuti star with the largest number of identified modes, with at least 24 well-determined frequencies between 100–400 μHz, and amplitudes from 0.8 mag to 22 mmag. They identified two radial modes ($l = 0$), found that the main pulsation mode at 147 μHz has $l = 1$, and that one of the radial modes (at 140 μHz) is the fundamental, implying that two modes with lower frequencies are g-modes. For the radial modes, and assuming solar metallicity, frequencies from the pulsation models used in the seismic analysis of κ^2 Boo (Frandsen *et al.*, 1995) imply a density of $0.1645 \pm 0.0005 \rho_\odot$, and a predicted distance (from the model luminosity of $L/L_\odot = 14.1 \pm 0.9$) of 84 ± 3 pc. This is in excellent agreement with the Hipparcos value of 83 ± 5 pc. They argue that if the temperature of $T_{\rm eff} = 7500 \pm 100$ K had been better known, to say ± 30 K, the distance estimate from the seismic model would be precise to within ± 1 pc. They also determine the rotation period of 3.5 days, an equatorial velocity of 33 ± 2 km s^{-1}, and an inclination angle of $40^\circ \pm 4^\circ$. Again, the importance of the Hipparcos parallax is to independently confirm the validity of the pulsation model. Similar mode identifications were confirmed by Breger *et al.* (1999), who also used the Hipparcos distance to compute new pulsation models including, for example, a revised mean density of $0.156 \pm 0.002\rho_\odot$. These agreements give confidence in deriving details of the evolutionary stage of FG Vir from the inferred $l = 1$ and $l = 2$ modes, and in deriving associated information about the convective core and convective overshooting.

Kiss *et al.* (1999a) reported radial velocity curves of the Hipparcos-discovered variable DX Ceti, and classified it as a mono-periodic high-amplitude δ Scuti star rather than as an RR Lyrae RRc variable.

Jerzykiewicz & Pamyatnykh (2000), also reported by Jerzykiewicz (2000), analysed the Hipparcos photometry of DK Vir, identifying two frequencies at 9.2095 and 7.5764 d^{-1}. Combined with ground-based data obtained in 1973 and 1980, the frequencies are refined to 9.209 45 and 7.576 41 d^{-1} and stable over this time, with corresponding V amplitudes of 8.6 ± 0.4 and 11.1 ± 0.4 mmag. Evolutionary models based on the Hipparcos parallax place the star at the end of the main-sequence phase or slightly beyond, with a mass of $2.1M_\odot$. They conclude that the two observed modes cannot both be radial, and that the higher frequency mode must be non-radial if the star is still on the main sequence.

Lampens *et al.* (2000) made a detailed study of the visual double star CCDM 23239–5349 in which the brightest component, HD 220392, is a short-period variable, probably of δ Scuti type. Hipparcos parallaxes and proper motions were used to confirm that the components are physically associated, either a common origin pair or wide binary. Both A/F-type stars are located in the δ Scuti instability strip, yet only one component is of δ Scuti type: their different evolutionary states and masses presumably underlie the reason for the observed difference in variability behaviour.

Templeton *et al.* (2000) made a detailed asteroseismology study of the multiply periodic δ Scuti component of the binary system θ Tuc. From 10 observed pulsation frequencies and mode identifications, pulsation models yield a mass in the range 1.9–$2.1M_\odot$, luminosity between 20–$25L_\odot$ and rotation velocities between 70–90 km s^{-1}. Since the model luminosity is more than a factor of two less than the Hipparcos derived value of $54L_\odot$, they infer that the secondary must be of comparable luminosity. The frequency spacing of the observed pulsation modes suggests that rotation has a strong effect on the observed spectrum. Spectroscopic observations suggest that the δ Scuti component is probably accreting mass transferred from the secondary, which is probably a $0.2M_\odot$ post-red giant branch object at $\sim$7000 K with a He core and weak or non-existent H-burning shell, which has lost most of its mass via mass transfer and winds (de Mey *et al.*, 1998). This scenario raises the possibility that tidal distortion may cause non-spherical perturbations in the δ Scuti component, whose interior may then be somewhat different from the prediction of single-star models.

Koen *et al.* (2001) analysed the Hipparcos time-series photometry of HD 21190 and HD 199434, deriving physical parameters from Strömgren photometry and Hipparcos absolute magnitudes. HD 21190 was shown to be the most evolved Ap star known. The combined Ap–δ Scuti nature makes it an important test of the models of pulsation in peculiar stars developed by Turcotte *et al.* (2000), although it is more extreme than any model they examined. Mode identifications were attempted based on amplitude ratios and phase differences from the photometric data. They used Equation 4.19 to estimate $Q = 0.019$ for HD 21190 and $Q = 0.020$ for HD 199434 which, using the models of Stellingwerf (1979), then imply pulsation in the second or third overtone for both stars assuming that the pulsations are radial.

These results are consistent with the models of Turcotte *et al.* (2000, with diffusion) in which Am stars should be stable on the main sequence, and with low-overtone modes becoming unstable as the stars evolve. However Koen *et al.* (2001) noted problems in the identification of the spherical harmonic degree of the pulsations l (using the procedure described in Koen *et al.*, 1999), possibly calling into question the photometric technique of mode identification.

Carrier *et al.* (2002) used power spectra of the Hipparcos photometry and Coravel radial velocities to analyse the multi-periodic pulsation properties of the double-lined spectroscopic binary CQ Lyn. Discovered as a δ Scuti by Hipparcos, it comprises an Am–δ Scuti star and a solar-like star with $P_{\rm orb} = 12.50736$ d. Pulsations were inferred to be $l = 2$ g-modes split by rotation (Figure 4.22). Physical properties were derived from the data combined with the Hipparcos parallax, including the rotation velocity of the primary, orbital inclination, radius, mass and age. The fact that pulsation and this metallicity can co-exist in the low instability strip implies that even for Am stars there is sufficient residual helium to drive δ Scuti pulsations.

Dallaporta *et al.* (2002) observed the 17-day period eclipsing binary GK Dra, discovered from the Hipparcos photometry. They established a period of 9.974 d, with equal depth in the primary and secondary eclipses, and with one of the components being a δ Scuti star with an amplitude of 0.04 mag and a period of about 2.7 hours.

Kiss *et al.* (2002) obtained new photometric and spectroscopic observations of V784 Cas, establishing multi-mode pulsations confirmed by the Hipparcos epoch photometry, and using the Hipparcos distance to derive evolutionary estimates of its mass and age, and providing a consistent picture of an evolved δ Scuti star with a mixture of radial plus non-radial modes.

Vidal-Sainz *et al.* (2002) observed the Hipparcos discovery V350 Peg from the ground in V band on 31 nights during 1997–1998. The period given by Hipparcos was

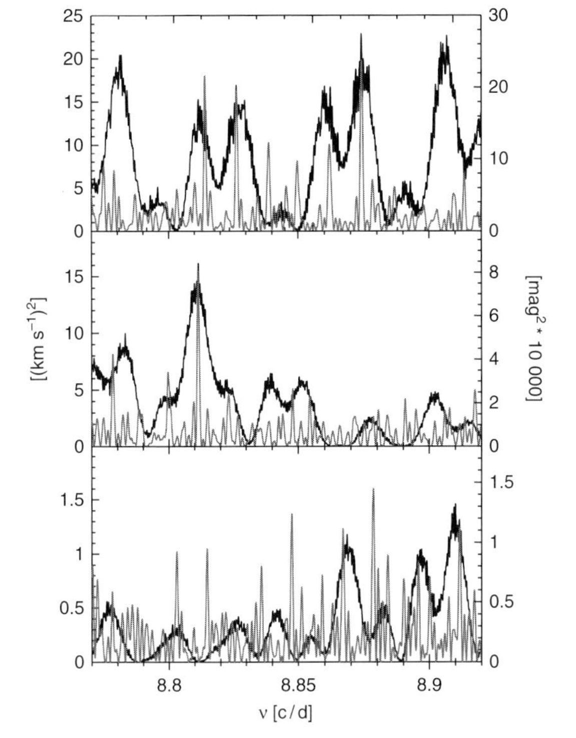

Figure 4.22 Power spectrum of the radial velocity measurements (black) and of the Hipparcos photometry (grey) of CQ Lyn. The two independent sets of measurements allow an unambiguous determination of frequencies of 8.86725 and 8.81150 d^{-1}. The first panel shows the initial Fourier transform cleaned of the binary frequency in the radial velocity case. In the second panel, only the former frequency is cleaned, and in the bottom panel both frequencies are cleaned. From Carrier et al. (2002, Figure 4).

not confirmed, but they identified two frequencies of similar amplitude, both probably non-radial modes.

Derekas *et al.* (2003) reported ground-based observations of seven bright high-amplitude δ Scuti stars, including GW UMa which was discovered through the Hipparcos photometry with a short period of 0.2032 d. The period places it on the boundary of the high-amplitude δ Scuti and the shortest period RR Lyrae stars (Poretti, 2001), while its Fourier amplitude parameter $R_{21} = 0.40$ confirms it as a high-amplitude δ Scuti. The kinematic data suggest that it is a Population II (SX Phe) star.

Another Hipparcos discovery close to the period boundary of the δ Scuti and RR Lyrae variables is V2109 Cyg, with a Hipparcos-estimated period of $P = 0.1860656$ d. The observations and analysis by Rodríguez *et al.* (2003a), see also Rodríguez *et al.* (2003b) and Gómez Forrellad (2003), obtained a very similar period from ground-based Strömgren photometry, identifying also its first harmonic, as well as a secondary frequency whose amplitude changes strongly with time and is identified with a non-radial mode. They classify it as an evolved δ Scuti with solar metal abundance, rather than as an RR Lyrae pulsator as provisionally identified in the Hipparcos analysis and supported by Kiss *et al.* (1999b).

Escolà-Sirisi *et al.* (2005) studied the Hipparcos discovery HIP 7666, classified in the Hipparcos Catalogue as an 'unsolved' system. From 23 nights of ground-based photometric data in 2000, they showed that the star is a detached eclipsing binary with an orbital period of 2.37229 d, and with one of the components, probably the primary, being a δ Scuti variable with a main pulsation period of around 25 d^{-1}. Eclipsing systems can provide accurate information on the radii and masses of binary components. The study of pulsations in such systems is also particularly interesting since they yield a near equator-on view of the pulsator, particularly favourable for the detection of non-radial modes with $l = |m|$. Additionally, the components act as a spatial filter during eclipses, causing phase and amplitude modulation of the non-radial modes, and facilitating the mode identification (Mkrtichian *et al.*, 2002).

Other photometric analyses include multiperiodicity in V929 Her (López de Coca *et al.*, 2006), studies of V966 Her and V1438 Aql (Hintz *et al.*, 2006), frequency analysis of NT Hya (Sinachopoulos & Gavras, 2007), and confirmation of the Hipparcos period in the case of QS Gem (Hintz & Brown, 2007).

Although δ Scuti variables are usually associated with the main sequence or post-main-sequence evolutionary phases, some 10 pre-main-sequence stars have been found to pulsate with time scales typical of the class. Theoretical models, based on nonlinear convective hydrodynamics, have delineated the instability strip for the first three radial modes (Marconi & Palla, 1998). Ripepi *et al.* (2003) reported multi-site observations of the pre-main-sequence star V351 Ori which, because of its photometric variations attributed to variable extinction due to circumstellar dust clouds, is considered to be on the transition from the pre-main-sequence to main-sequence phase. Five periods are identified, with suggestions that non-radial modes are present. The best-fit pulsation models are in reasonable agreement with luminosities derived from the Hipparcos parallax.

4.9.2 Rapidly-oscillating Ap (roAp) stars

Among the magnetically chemically peculiar A-type stars, the rapidly-oscillating Ap (roAp) stars are a cool subset of the SrCrEu stars: they exhibit rapid photometric and spectroscopic oscillations with periods of order minutes and amplitudes less than around 10 mmag, attributed to low spherical degree ($l \leq 3$), high overtone ($n \gg l$) p-modes. The rich oscillation spectra yield important astrophysical parameters such as the asteroseismological luminosity, the rotation period, the magnetic field strength, and the atmospheric structure. Some 30 roAp stars are known, of which 16 were observed by Hipparcos. Reviews are given by, e.g. Kurtz (1990), Martinez & Kurtz (1995), and Kurtz & Martinez (2000).

The basic behaviour of the pulsations is explained in the oblique pulsator model of Kurtz (1982). In this model, the star's pulsation axis is not aligned with its rotation axis, inclined to the line-of-sight by the angle i, but with its magnetic axis, which is itself tilted with respect to the rotation axis by an angle β, the magnetic obliquity. The pulsation modes are then seen at different aspects through the pulsation cycle which, for non-radial modes, results in amplitude modulation over the rotation period. The model then predicts that dipole ($l = 1$) pulsation modes will be split into equally-spaced triplets; while quadrupole modes ($l = 2$) will give rise to equally-spaced frequency quintuplets, where the spacing of consecutive multiplets is given by the stellar rotation frequency. More generally, the pulsation modes of some stars are further distorted by the presence of the magnetic field, and the current picture is of the pulsation axis being offset with respect to both the magnetic and rotation axes (Bigot & Dziembowski, 2002).

Gómez *et al.* (1998) used the position of Bp–Ap stars in the HR diagram to show that the roAp stars are main-sequence objects, a conclusion that also holds for the spectral subgroup of SrCrEu stars in general.

Although mode identification is in general problematic, prior mode identification is not essential in the case of pulsations of low degree l and high radial overtone, $n \gg l$, for which asymptotic pulsation theory predicts that the observed eigenfrequencies should be nearly equally spaced in frequency for the p-modes. This is the case for the roAp stars. The frequency spacing depends on the mean density, in turn most sensitively determined by the radius. Frequency measurements are then used to predict the radius, and hence evolutionary state. Independent estimates using the Hipparcos parallaxes was carried out for 12 roAp stars by Matthews *et al.* (1999), see also Matthews (2000). They derived (their Equation 4)

$$\Delta\nu = (6.64 \pm 0.36) \times 10^{-16} M^{0.5}\, T_{\rm eff}^{3}\, L^{-0.75}\ {\rm Hz} \qquad (4.21)$$

with L derived from the parallaxes, M from evolutionary tracks, and $T_{\rm eff}$ from colour indices. The Hipparcos parallaxes are generally slightly larger than the seismic predictions. Amongst objects studied by Matthews *et al.* (1999) α Cir is the brightest known roAp star, has the largest parallax, and offers the best prospect of ruling whether the Hipparcos parallax of 60.97 ± 0.58 mas for this star (and, by implication, others) is in error, or whether the underlying p-mode theory is in error, or incompletely described due to diffusion or to their high magnetic field strengths (Figure 4.23).

Handler & Paunzen (1999) used Strömgren and Geneva photometry to select candidates in the northern hemisphere, from which one or two objects were confirmed using standard Fourier techniques applied to their light-curves. They then used the Hipparcos data for 12 roAp and 54 non-oscillating Ap (noAp) stars with a reasonable precision on M_V to construct the M_V versus β diagram (Figure 4.24a). While confirming that the roAp objects are main sequence, three objects lie beyond the cool border of the δ Scuti instability strip. For some years, it was believed that the δ Scuti and roAp instability region coincides, such that a plausible driving mechanism for the roAp stars could be the same partial He^+ ionisation mechanism. However, this result points to a different excitation mechanism, supporting the model calculations of Gautschy *et al.* (1998), who suggested that pulsations are due to partial H/He ionisation in which the unstable modes arise in a chromospheric temperature-inversion layer.

Hubrig *et al.* (2000) developed this analysis. They used the proper motions and parallaxes of all 16 roAp objects observed by Hipparcos, and 30 noAp objects, to show that the kinematics of the roAp and the noAp stars are very similar, in Galactic velocity components UVW, in velocity dispersion, and in the Galactic orbit elements. They constructed the 'astrometric HR diagram' in the form proposed by Arenou & Luri (1999) in which the ordinate is constructed as $a_V = 10^{0.2M_V} = \pi\, 10^{\frac{m_V - A_V + 5}{5}}$ (see Section 5.2.4). This has the merit that the error bars, if dominated by errors on the parallax, are symmetric, and that Lutz–Kelker bias is consequently minimised. The results, shown in Figure 4.24, right, suggest that the roAp stars as a group are less luminous and less evolved than the noAp stars: the former lie some -0.47 ± 0.34 mag above the ZAMS, while the latter are some -1.20 ± 0.65 mag above the ZAMS. The evolutionary tracks indicate that the noAp stars are on average more massive, at about 1.6–$2.5 M_\odot$, than the roAp stars, at about 1.5–$2.0 M_\odot$. Finally, their radial velocity measurements indicate a real deficiency of binaries amongst roAp stars, with none known to be components of a spectroscopic binary. Similarly, no pulsating white dwarf is known to be a component of a spectroscopic binary. The question of whether binarity affects pulsation through tidal interaction is open: some authors have conjectured that tides in close binaries might drive oscillations. The question whether tidal interaction may also be efficient in damping existing pulsations has, apparently, not been addressed.

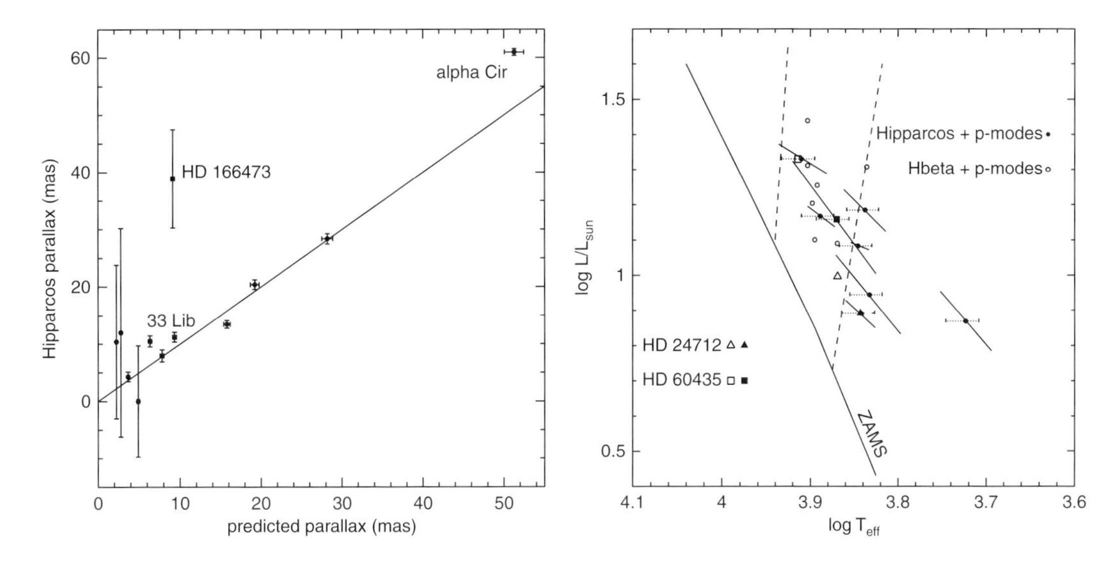

Figure 4.23 Asteroseismology for rapidly-oscillating Ap stars. Left: Hipparcos parallaxes versus asteroseismic predictions for 12 multi-periodic roAp stars. Vertical error bars are quoted parallax standard errors; horizontal error bars are due mainly to uncertainties in $T_{\rm eff}$ used to estimate luminosities from the eigenfrequency spacings. Right: the roAp stars in the HR diagram. The solid symbols are the locations based on luminosities from Hipparcos parallaxes, and on radii from $\Delta\nu$ in the eigenspectra; the dashed error bars represent uncertainty in the assumed mass ($2.0 \pm 0.5\,M_\odot$). Open symbols are the predicted locations based on radii from $\Delta\nu$, and on $T_{\rm eff}$ from $H\beta$ calibration. The dashed lines are the boundaries of the δ Scuti instability strip. The two stars indicated by the larger open and filled symbols have virtually no uncertainty in their $\Delta\nu$ values. From Matthews et al. (1999, Figures 1 and 2).

Further modelling of the roAp pulsations has used stellar models based on the Hipparcos data for estimating luminosities. This is a significant contribution because roAp have luminosities and effective temperatures which are notoriously difficult to determine. The class includes the highly peculiar Przybylski's star (HIP 56709, HD 101065, V816 Cen; Kurtz, 2002) for which temperature estimates range from less than 6000 K to over 8000 K, and whose visible spectrum is characterised by the rare-Earth elements holmium (Ho), dysprosium (Dy), neodymium (Nd), gadolinium (Gd), and samarium (Sm).

Audard *et al.* (1998) were able to reconcile some anomalous cases where the oscillation frequencies of some roAp stars were larger than the acoustic cut-off frequency obtained from earlier stellar models; their newer models have increased the cut-off frequency by about 200 μHz. Kurtz (1998) modelled the variation in pulsation amplitude which drops with increasing wavelength. Their comparison of asteroseismic with Hipparcos luminosities (Figure 4.23, left) suggests that roAp stars have lower temperatures by about 1000 K, and/or smaller radii, than the previous A star models from which the asteroseismic luminosities were calculated, or that their magnetic fields alter the frequency separations.

Kurtz *et al.* (2002) and Kurtz *et al.* (2003) discussed the case of HR 1217. Observations in 1986 showed a clear pattern of five modes with alternating frequency spacings of 33.3 μHz and 34.6 μHz, with a sixth mode at a problematic spacing of 50.0 μHz ($= 1.5 \times 33.3\mu$ Hz) to the high-frequency side. Five of the pulsation modes were either separated by 68 μHz and were alternating even and odd modes, or were separated by 34 μHz and were consecutive overtones of the same *l*. The Hipparcos parallax resolved the ambiguity in favour of the first interpretation (Matthews *et al.*, 1999), but left the puzzle of the spacing of the sixth frequency. Theoretical calculations of magneto-acoustic modes in Ap stars by Cunha (2001) subsequently suggested that there should be an additional frequency present in HR 1217 between the earlier fifth and sixth modes. Further observations of HR 1217 in 2000 using the 'Whole Earth Telescope' (WET, a world-wide network of cooperating astronomical observatories, established to obtain uninterrupted time-series measurements of variable stars, e.g. Solheim, 2003) revealed a newly-detected frequency in the pulsation spectrum matching this prediction. The possibility that different modes were excited at the various observational epochs is discussed by Kurtz *et al.* (2002). Kurtz *et al.* (2005) reported a more complete analysis of the dataset, reaching 14 μmag photometric precision, and giving a rotation period of 12.4572 d. Based on the frequency spacing and the Hipparcos parallax, specific identification of the modes is proposed.

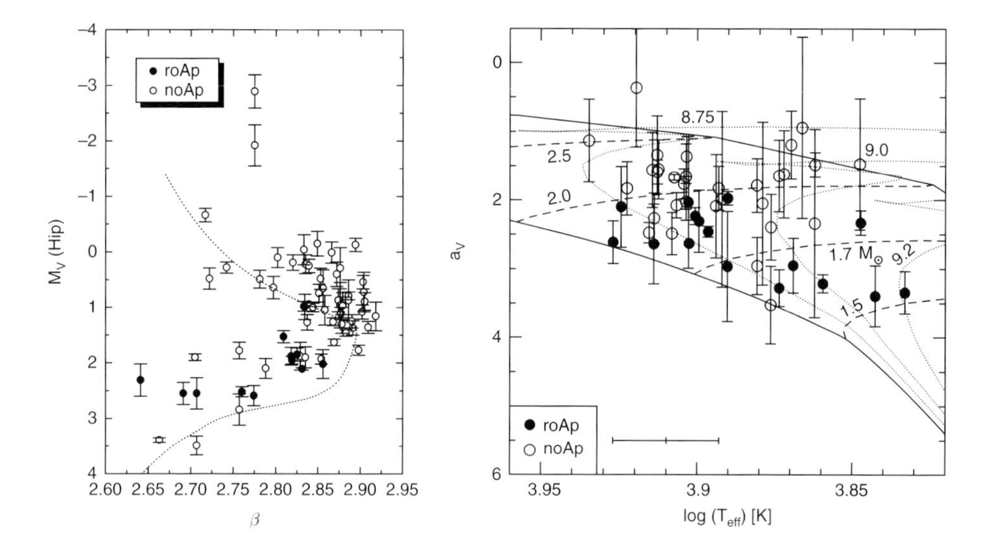

Figure 4.24 Left: M_V based on the Hipparcos parallaxes versus Strömgren β index for 12 rapidly-oscillating Ap (roAp, •) and 54 non-oscillating Ap (noAp, ∘) stars. The plot establishes that roAp stars are main-sequence objects, but the three objects signified by the filled circles at lower left are outside of the region of the δ Scuti instability strip. From Handler & Paunzen (1999, Figure 4). Right: the 'astrometric' HR diagram, with $a_V = 10^{0.2M_V}$ as ordinate, of the roAp (•) and of the noAp stars (∘). The horizontal bar at lower left indicates the typical $\pm 1\sigma$ error on $T_{\rm eff}$ (± 300 K). The lower continuous curve is the ZAMS isochrone at $\log t = 5.7$ for $Z = 0.020$, while the upper curve is the terminal-age main sequence (TAMS). Dotted curves are isochrones at the indicated $\log t$, while the dashed curves are the main-sequence evolutionary tracks for masses between 1.5 and $2.5 M_\odot$. From Hubrig et al. (2000, Figure 5).

Kochukhov *et al.* (2004) investigated the magnetic-field geometry and surface distribution of chemical elements in HR 3831. Results of the model atmosphere analysis of the spectra are combined with the Hipparcos parallax and evolutionary models to obtain accurate estimates of the fundamental stellar parameters: $T_{\rm eff} = 7650$ K, $\log L/L_\odot = 1.09$, $M/M_\odot = 1.77$ and an inclination angle of the rotation axis of $i = 68°$. They found that the variation of the longitudinal magnetic field and of the magnetic intensification of Fe I lines in the spectrum are consistent with a dipolar magnetic topology with a magnetic obliquity $\beta = 87°$ and a polar field strength $B_p = 2.5$ kG. They applied a multi-element abundance Doppler imaging inversion code for the analysis of the spectrum variability, and recovered surface distributions of 17 chemical elements, including Li, C, O, Na, Mg, Si, Ca, Ti, Cr, Mn, Fe, Co, Ba, Y, Pr, Nd, Eu.

4.9.3 γ Doradus variables

The γ Doradus stars are a class of non-radially pulsating F0–F5 stars with periods of around 8–80 hours, and amplitudes up to about 0.1 mag in V. They lie on or near the main sequence, on the cool side of the δ Scuti variables. Handler & Krisciunas (1997) give a list of 11 confirmed and 17 possible members of the group, and reviews are given by Krisciunas (1998) and Kaye *et al.* (1999). Multi-periodic variations detected in them are typically a factor of 20 longer than the period of the radial fundamental mode, such that high-order g-modes are indicated. However since convection is important for this spectral type, the usual κ mechanism instability cannot simply be invoked to explain them.

Aerts *et al.* (1998a) reported the detection of 14 new γ Doradus stars amongst the unbiased Hipparcos sample of A2–F8 (solved) variables, also making use of Geneva photometry. Their results indicate the bias towards higher temperatures of earlier ground-based surveys, with the coolest star in the Hipparcos sample having $T \sim 6000$ K (Figure 4.25, left). The objects are also shown in the context of other variability types in the same region in Figure 4.16. For most of the new candidates, more than one period is detected in the Hipparcos light-curves (Figure 4.25, right).

Handler (1999) extended the search for Hipparcos candidates to the 'unsolved' variables, following the idea that the lack of a clear periodicity could be due to unrecognised multi-periodic variations. They identified 70 candidates, most of them new, bringing the number of prime candidates to 46, with 36 additional possible members of the class. The domain of the γ Doradus instability region in the Hertzsprung–Russell diagram could be clearly delineated for the first time: it translates into $T_{\rm eff} = 7200$–7700 K on the ZAMS at $\log g \sim 4.33$, while the luminous end occurs about 1 mag above

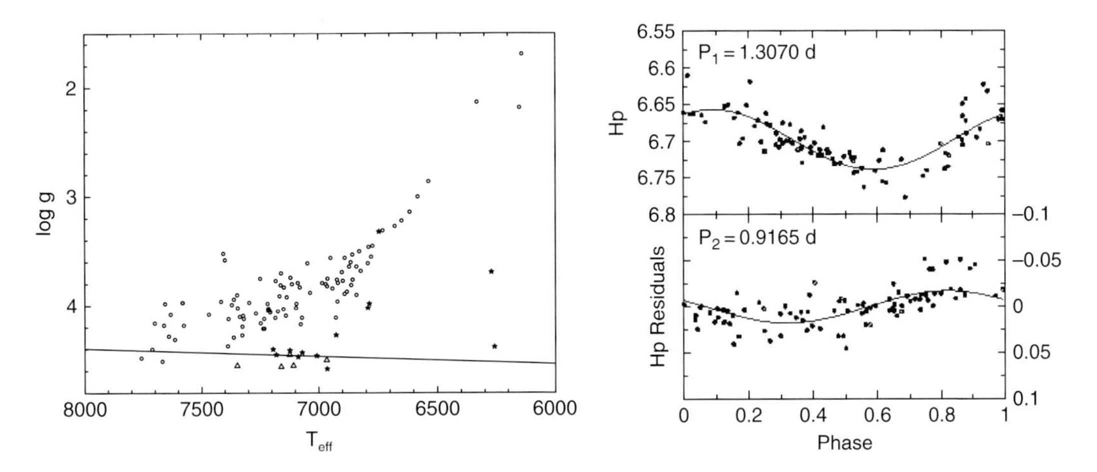

Figure 4.25 Left: the position of the new γ Doradus stars ($\star$) compared with previously known stars of this kind ($\triangle$) and with known δ Scuti stars ($\circ$) in the ($T_{\rm eff}$, $\log g$) diagram. The full line denotes the zero-age main sequence. Right: example of the phase diagrams for one of the new Hipparcos γ Doradus stars, HD 167858: the top panel shows the main period ($P_1 = 1.3070$ d); the bottom panel shows the secondary period ($P_2 = 0.9165$ d) after pre-whitening with the first. From Aerts et al. (1998a, Figures 1 and 2).

it at $\log g \sim 4.0$ between $T_{\rm eff} = 6900$–7500 K. The region partly overlaps with that of the δ Scuti stars, and the phenomenon seems to occur over a significant fraction of the main-sequence lifetime. While they remain interesting candidates for asteroseismological studies, the physical mechanism responsible for the hot and cool boundaries of the γ Doradus instability region remains unknown.

Further studies of Hipparcos candidates observed in Geneva photometry leading to a further three members were reported by Eyer & Aerts (2000a,b). Mathias *et al.* (2004) reported results of a two-year high-resolution multi-site spectroscopy campaign of 59 candidates, mostly discovered by Hipparcos. In more than 60% of cases, line profile variations can be interpreted as associated pulsations. Projected rotation velocities reach to more than 200 km s^{-1}. About 50% of the candidates are possible members of binary systems, with 20 stars being confirmed as γ Doradus variables. Jankov *et al.* (2006) obtained 230 spectra of HD 195068, carrying out a time-series analysis and confirming, amongst other results, the frequency observed in the Hipparcos photometry. Gerbaldi *et al.* (2007) studied the ultraviolet fluxes of 40 confirmed Hipparcos γ Doradus stars, and found that 29 stars (73%) have excess ultraviolet fluxes compared with atmospheric models, which could not be attributed to binarity alone.

4.9.4 β Cephei variables

The β Cephei stars are early-B type, p-mode pulsators, with very short pulsation periods of around 0.2 d, about half being doubly periodic and displaying beat behaviour between the periods. Only rather recently, and with the availability of new opacities, has it become evident that the pulsations of B-type stars, including β Cephei stars and slowly-pulsating B stars, can be explained by the κ mechanism acting in a partially ionised zone of iron lines (Cox *et al.*, 1992; Dziembowski & Pamyatnykh, 1993; Dziembowski *et al.*, 1993). One of the remaining uncertainties of current excitation models concerns the extension of the class to masses above 25 $M_\odot$.

Hipparcos observed 43 out of 63 β Cephei stars listed in Sterken & Jerzykiewicz (1993). Four were classified in the catalogue as unsolved, and six as micro-variables, with α Vir confirmed as an ellipsoidal variable. In most cases the periods agree with those already determined from ground-based observations, although three (KK Vel, κ Sco, and EN Lac) are in disagreement (Molenda-Zakowicz, 2000b). The long-term behaviour of the pulsation periods and amplitudes of EN Lac were also studied by Jerzykiewicz & Pigulski (1999). This is a $V = 5.6$ mag single-lined spectroscopic binary and an eclipsing variable with $P_{\rm orb} = 12.096\,84$ d, comprising a β Cephei variable and invisible secondary. From more than 40 years of ground-based photometry, they showed that although the period given in the Hipparcos Catalogue is spurious, the *Hp* amplitudes and epochs of maximum light of the first two of the nine largest sinusoidal terms do agree with the ground-based estimates. Time scales of the amplitude variations of these first two terms are of order 50 years, close to the growth rates of several $l \leq 2$ pulsation modes, as predicted by theories of nonlinear interaction of the pulsation modes.

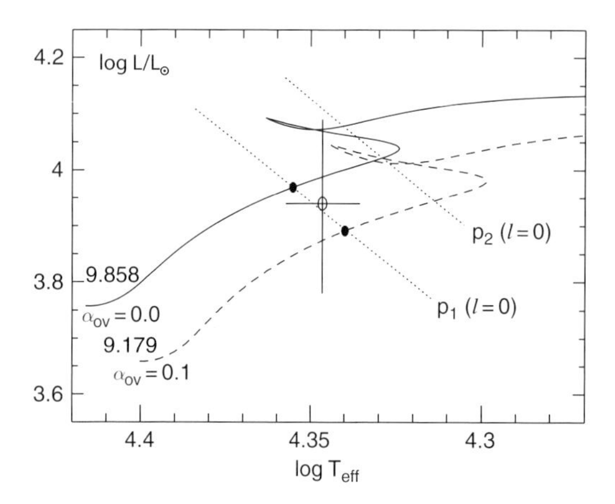

Figure 4.26 Asteroseismology for the β Cephei star ν Eri. Evolutionary tracks are shown for the models fitting the observed frequency modes, for specified values of M and α_{ov}. The two dotted lines connect models with one of the observed frequencies identified with the radial fundamental (p_1), and first overtone (p_0) mode, respectively. The values of T_{eff} and log L inferred from photometry and the Hipparcos parallax are shown with error bars. From Pamyatnykh et al. (2004, Figure 3).

Waelkens *et al.* (1998b) reported the discovery of four new candidate β Cephei stars based on their Hipparcos variability. Aerts (2000) added a further two, and for all six used ground-based follow-up photometry in the Geneva seven-colour photometric system to confirm their β Cephei-type nature, to detect the pulsation frequencies, and to identify most of the degrees of the pulsation modes. All six have frequencies in the expected range for β Cephei stars, and the degrees of the main mode range from $l = 0$ up to $l = 2$, consistent with theoretical expectations. Molenda-Zakowicz & Polubek (2005) classified one further confirmed and two new suspected β Cephei stars.

A sequence of spectroscopic and radial velocity observations, frequency analysis, mode identification, and evolutionary modelling was carried out for ν Eri by de Ridder *et al.* (2004), using the Hipparcos parallax of 5.56 ± 0.88 mas to determine the luminosity. Ausseloos *et al.* (2004) showed how the luminosity constraint allows discrimination of models with different initial hydrogen abundance (X) or metallicity (Z), while Pamyatnykh *et al.* (2004) derived characteristics of the internal rotation as well as driving mechanisms and mode instabilities (Figure 4.26).

4.9.5 Supergiants: Pulsating O and α Cyg variables

The instability mechanism for the variable supergiants has not yet been identified. One objective is therefore to compare the properties of the supergiants that show α Cyg-type variability with those of pulsators of known excitation mechanism.

Adelman & Albayrak (1997) examined the Hipparcos photometry of 26 bright early A-type supergiants, showing that they are all variable. Similarly, van Leeuwen *et al.* (1998) examined the Hipparcos photometry of 24 supergiants and hypergiants ($> 25\,M_\odot$), or α Cygni variables, 12 in the LMC, three in the SMC, and the remainder in the Galactic plane. All show microvariability, at the level of $\lesssim 0.1$ mag for the OB stars, or larger for the later types.

The 32 supergiants identified by Waelkens *et al.* (1998b) are shown with respect to other B-type variables, including other variable supergiants from the literature, in Figure 4.29, right below. They appear to extend the slowly-pulsating B and β Cephei pulsations to higher luminosity, suggesting a common cause of variability, while their periods of 1.5–24 d suggest g-mode pulsations similar to those excited in slowly-pulsating B stars (Figure 4.27). Waelkens *et al.* (1998b) concluded that the position of the Hipparcos variable supergiants is consistent with the theoretical prediction based on the κ mechanism by Pamyatnykh (1998), at least for masses up to $20M_\odot$, possibly with 'strange modes' independent of opacity (Glatzel, 1994) at masses above $40M_\odot$. Follow-up observations have been reported by Lefever *et al.* (2006, 2007). The latter observed 28 periodic B-type supergiants from Hipparcos, and from their location in the T_{eff}, log g diagram also concluded that the observed variability can be attributed to opacity-driven gravity modes.

Van Genderen *et al.* (2004) discussed variability of five LMC supergiants, including the G7 hypergiant R59 for which Hipparcos photometry is available.

4.9.6 Slowly-pulsating B stars

Slowly-pulsating B (SPB) variables are of mid-B spectral type with periods of 0.5–2 d. They are of particular interest from a pulsation point of view because, contrary to the Cepheids, RR Lyrae, and δ Scuti variables, their instability mechanism was only recently attributed to the κ mechanism acting in a partially ionised zone of metals (e.g. Gautschy & Saio, 1993). Pre-Hipparcos only 11 candidates were known, thinly populating their predicted instability domain. The slowly-pulsating B stars are the only firmly-established class of early-type main-sequence variables which pulsate in (many, high-order) non-radial g-modes, of asteroseismological importance since they penetrate deep into the stellar interior. In this respect they are the intermediate-mass main-sequence analogues of the white dwarfs.

Waelkens *et al.* (1998b), see also Waelkens *et al.* (1998a), Aerts *et al.* (1998b) and Molenda-Zakowicz (2000b), studied the 267 previously unknown B-type

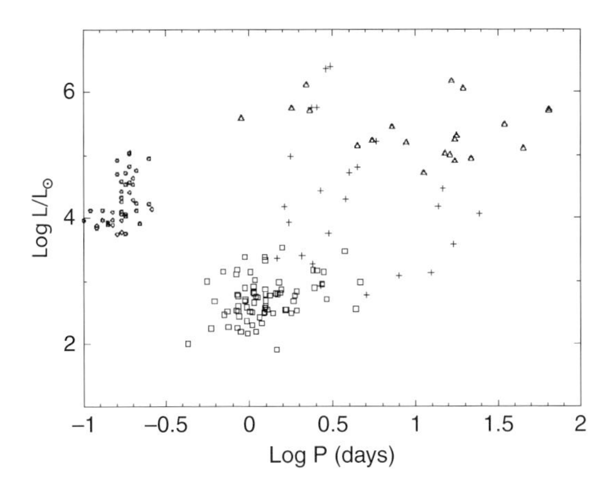

Figure 4.27 The period–luminosity relation for different types of B-type variables: ○ = β Cephei; □ = slowly pulsating B stars; + = variable supergiants discovered by Hipparcos; △ = previously-known variable supergiants from van Genderen (1985) and Burki (1978). There is a clear distinction between the location of the p-mode (β Cephei) and g-mode (slowly-pulsating B and probably supergiants) pulsators, with the new Hipparcos supergiants linking the slowly-pulsating B stars with the previously-known supergiants. From Waelkens et al. (1998b, Figure 3).

variables identified by Eyer (1998), and further classified them using multivariate discrimination based on spectral type, principal period from the Hipparcos photometry, form of the light-curve, and the 'Geneva parameters' X, Y, Z. The analysis yielded 72 new slowly-pulsating B stars, as well as four new β Cephei, 34 new variable chemically peculiar stars, 32 new supergiants with α Cyg-type variations, seven new variable Be stars, seven new eclipsing binaries, and 17 uncertain objects. Results included the few new (short-period) β Cephei stars discovered, indicating that their observational status was already rather complete, compared with the large number of new slowly-pulsating B stars discovered, as well as microvariable supergiants.

The almost tenfold increase in the discovery of objects with periods of order days also indicates the biased nature of the earlier ground-based photometric surveys. Their position in the HR diagram is shown in Figure 4.29. The slowly-pulsating B stars are homogeneously spread across the theoretical instability strip in which g-mode pulsations are predicted according to models based on the κ mechanism (Pamyatnykh, 1998). Confirmation of their nature is also provided by the presence of multi-periodicity, often immediately evident from the Hipparcos variability phase diagrams (although see Molenda-Zakowicz, 2000a), but most clearly detected through follow-up observations from ground.

Aerts *et al.* (1999b) reported long-term photometric and spectroscopic (Si II and Si III) monitoring of 17 southern slowly-pulsating B stars, including 12 of the new Hipparcos discoveries, and covering spectral types B2–B9 and the full theoretical instability strip. They found that all but one exhibit clear line-profile variability (Figure 4.28). Seven are in binary systems with five having large rotational velocities and complicated line-profile variations, suggesting that the binary environment might result in a particular spectrum of excited modes. Two of the Hipparcos discoveries are of particular interest: HD 85953, and possibly HD 131120, lie in the common part of the instability domains of the β Cephei and slowly-pulsating B stars, where theory predicts the presence of both short-period p- and long-period g-modes, although the former were not yet confirmed. HD 131120 is dominated by a single period, unusual compared with the theoretical predictions of multi-periodicity; the issue being whether the slowly-pulsating B star is truly mono-periodic, or whether it is rather a chemically peculiar star in which variation is attributed to rotation.

Subsequent work by Briquet *et al.* (2001b) and de Cat & Aerts (2002) showed that two objects, HD 55522 and HD 131120, are chemically peculiar rather than slowly-pulsating B stars, while HD 169978 turns out to be a non-pulsating, ellipsoidal variable, which was misclassified as a slowly-pulsating B star.

Mathias *et al.* (2001) similarly reported spectroscopic monitoring of 10 new northern Hipparcos slowly-pulsating B star candidates. All 10 show line-profile variability on a time scale expected for non-radial g-mode pulsations in B-type stars. For all but one they find evidence of multi-periodicity in the Hipparcos data, with only HD 28114 found to be mono-periodic. The main photometric frequency is also normally the main spectroscopic frequency, showing that each star has a dominant non-radial low-degree g-mode. Two exceptions, HD 1976 and HD 208057, are rapid rotators, leading to a poor determination of the velocity curve, and are members or suspected members of binary systems.

Further ground-based photometric follow-up of the Hipparcos discoveries were reported by Molenda-Zakowicz (2004), with related studies of multi-periodicity and classification considered by Molenda-Zakowicz & Polubek (2004) and Schoenaers & Cuypers (2004). Molenda-Zakowicz & Polubek (2005) classified four new and two suspected slowly-pulsating B stars, including the hottest-known of the type, HIP 1030. Geneva photometry over three years, with frequency analysis, was reported for 28 further objects by de Cat *et al.* (2007).

Ground-based photometry follow-up, confirming or extending the Hipparcos frequency analysis, has been reported for individual objects, notably identification of

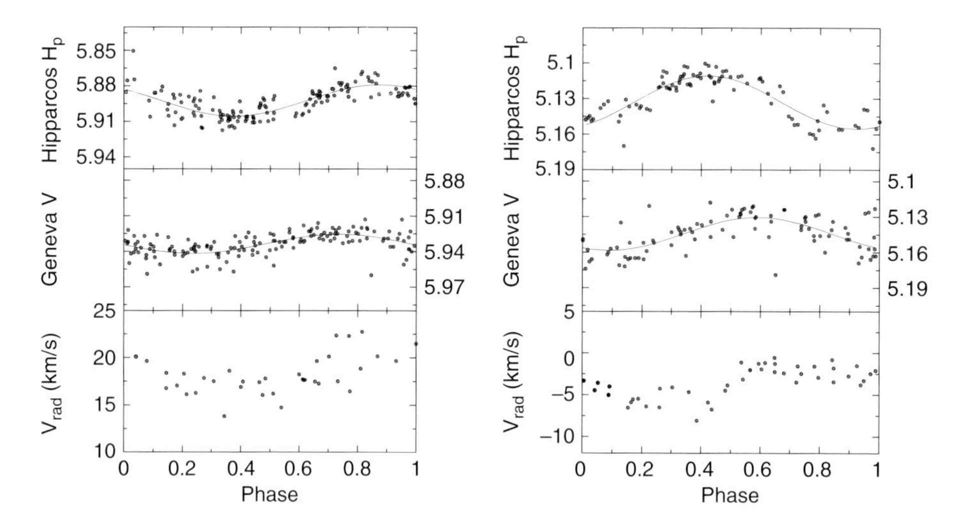

Figure 4.28 Phase diagrams of the Hipparcos photometry (top), Geneva photometry (middle), and radial-velocity (bottom) data for two slowly-pulsating B stars. Left: HD 85953 for the frequency 0.2662 d^{-1}. Right: HD 138764 for the frequency 0.7945 d^{-1}. From Aerts et al. (1999b, Figures 10 and 13).

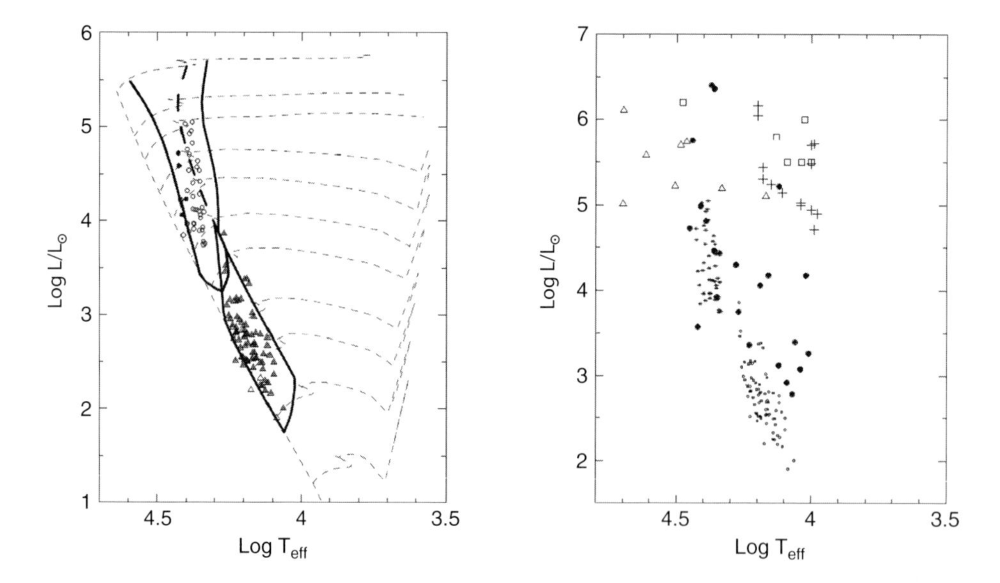

Figure 4.29 Left: position of β Cephei stars (○) and slowly-pulsating variables (△) in the HR diagram. Previously-known variables are indicated by open symbols while those discovered by Hipparcos are filled. Instability strips are those derived by Pamyatnykh (1998), based on OPAL G93/21 opacities for a composition $X = 0.70$, $Z = 0.02$. Full lines are the calculated edges of the instability domain, while the dashed line is the terminal age main sequence. Evolutionary tracks are by Schaller et al. (1992). Right: observational HR diagram for B-type variables: ⋆ = β Cephei; ○ = slowly-pulsating B star; • = variable supergiants discovered by Hipparcos; △ = variable supergiants studied by van Genderen (1985); + = variable B-type supergiants studied by Burki (1978); □ = luminous blue variables given by Lamers et al. (1998). From Waelkens et al. (1998b, Figures 1 and 2).

circumstellar dust shells for HD 42927 and HD 126341 (Aerts *et al.*, 1999a); frequency and mode analysis for 53 Per (de Ridder *et al.*, 1999); pulsation behaviour over 10 years for ι Her (Chapellier *et al.*, 2000); line-profile variation in τ Her (Masuda & Hirata, 2000); inability to fit a purely pulsation or rotation model for HD 105382 (Briquet *et al.*, 2001a); detailed spectroscopic mode identification for HD 147394 (Briquet *et al.*, 2003); and the coolest of the class known, HD 121190 (Aerts & Kolenberg, 2005).

4.9.7 Maia variables

In the region bordered by the radially pulsating δ Scuti (A7–F2) and β Cephei (B0–B3) variables, stellar pulsations seem to be largely absent. Struve (1955) suggested that, by analogy, pulsations might exist in this region, with periods of 0.1–0.3 d. He called these hypothetical variables Maia stars after the presumed prototype Maia in the Pleiades. A debate has continued about their existence since that time (McNamara, 1987).

The identification of possible Maia variables is complicated by the presence of other types of variable in the same region of the HR diagram: peculiar A (Ap) stars with periods of 0.5 d and higher; the rapidly-oscillating Ap (roAp) stars with periods of a few minutes; slowly-pulsating B stars with periods of 0.5–2 d; δ Scuti stars with periods of 0.5 to several hours; and the γ Doradus stars with periods of about 8–80 hours. Percy & Wilson (2000) accordingly defined the hypothetical Maia stars to be radial or non-radial pulsators with spectral types B7–A3, near the main sequence, which are not slowly-pulsating B, δ Scuti, or γ Doradus stars.

Scholz *et al.* (1998) used Hipparcos photometry, as well as ground-based spectroscopic and photometric observations of candidate Maia stars, including the binary γ CrB, and the frequently-quoted example of a Maia pulsator, the B9p star ET And. Hipparcos photometry results are shown in Figure 4.30. For γ CrB they reveal a period of 0.445 d, not in fact the pulsation period but nevertheless confirming the proposed pulsation model, and the Hipparcos parallax was used to confirm that the primary is indeed likely to be an A0 main-sequence star. For ET And, they reveal both the known rotation period of 1.6188 d, and an additional component of about 0.01 mag amplitude and period of 0.103 966 d (see also Lehmann *et al.*, 1999), close to the pulsation period previously proposed by Panov (1978). Of the 14 candidate stars studied, they confirm the existence of occasional pulsations in γ CrB, with evidence that pulsations exist also in γ UMi and ET And.

Percy & Wilson (2000) extended the search for Maia variables using an autocorrelation analysis of some 500 Hipparcos stars: only five showed evidence for variability periods of up to one day, and in three of these the evidence was weak. The two remaining cases are HD 29573 which may be a rotating variable, and (as in the previous study) γ CrB. Maia itself is photometrically constant. Similar results were found by de Cat *et al.* (2007).

The conclusion is that Maia variables, if they exist, are very rare and elusive.

4.10 Red variables: Long-period, Mira, and semi-regular

As stars expand and cool to become red giants or asymptotic giant branch stars, they become pulsationally unstable. Variability with $\Delta V > 0.01$ mag sets in at around spectral types K9–M0 III. As the star reaches a spectral type of late-M, the amplitude in V is several magnitudes, and it is classified as a Mira variable.

Red variables have large amplitudes and long periods, and are essentially of spectral type M (but also extending to types R, N and S). They are a mix of several types of evolved red stars which are traditionally classified empirically according to the amplitude and regularity of their light-curves: as Miras (M), semi-regular (SRa,b,c), and irregular (L), although these classes do not correspond strictly to different physical properties. They can be O-rich, C-rich, or have an excess of s-process elements (S stars, either intrinsic or extrinsic). The group includes the Mira variables, characterised by large amplitudes and periods of around 60–300 days, and semi-regular variables of smaller amplitude, of spectral types K5–M5, with periods between 30–1000 days. They occupy a key stage of stellar evolution, participate in the heavy-element enrichment of the interstellar medium and, being luminous, are good tracers of Galactic kinematics.

The irregular or semi-regular variables include several variability types among the red giants and supergiants, all generally poorly understood theoretically. RV Tau stars are of spectral type F, G, or K, with light-curves characterised by alternately deep and shallow minima which occasionally switch, and amplitudes around 1–2 mag. They are low-mass post-asymptotic giant branch stars evolving towards the white dwarf stage. The brightest is R Sct (HIP 92202) with $Hp_{\rm max} = 5.1$. The K-type semi-regular variables extend towards the region of the long-period variables, suggesting that these various types of supergiants have a common pulsational instability mechanism, but lack the properties needed for regular pulsation. A discussion of the period–luminosity relations of the various types of variable red giants derived from microlensing surveys is given by Soszynski *et al.* (2007).

Long-period variables Virtually every M giant is variable in brightness. However, details of the mechanisms causing the variability are unclear, since brightness changes vary from the strictly periodic to the completely irregular. Aside from the long-period Mira stars, which typically have periods longer than 100 d, the majority of M giants in the Hipparcos Catalogue are classified there as semi-regular (SR) pulsating stars. The GCVS defines SR variables as stars of luminosity class III and brighter, of late or intermediate spectral type, with

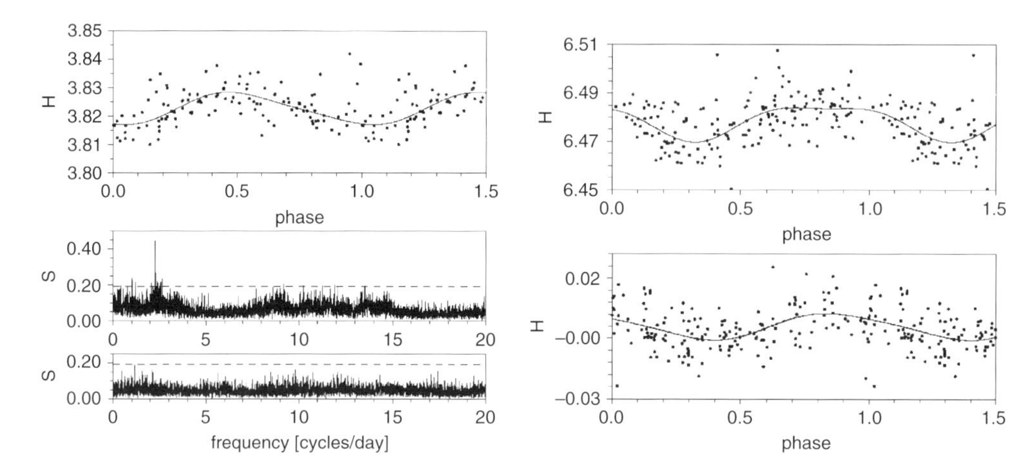

Figure 4.30 Left: Hipparcos photometry of γ CrB, folded with the period of 0.445 d (top), and showing the periodogram with the corresponding peak (middle), and with the data pre-whitened for this period (bottom). The 1% false alarm probability limit is marked by the dashed horizontal line. Right: phase diagrams of the Hipparcos photometry for ET And, folded with the rotation period of 1.6188 d (top), and residuals after subtracting the rotation period and then folded with 0.103966 d (bottom). From Scholz et al. (1998, Figures 10 and 15).

Long-period variables: Pulsation, mass, and evolutionary phase: Mass-loss rates through stellar winds are around 10^{-7}–$10^{-8} M_\odot\,\mathrm{yr}^{-1}$ for red giants, and some 10^{-6}–$10^{-7} M_\odot\,\mathrm{yr}^{-1}$ for red supergiants. These high rates imply brief stages of evolution lasting roughly a few million years. For Miras, the general correlation between period and [Fe/H] (e.g. Feast, 1981) suggests that Miras of a given period represent a homogeneous group in age and chemical composition. For stars on the asymptotic giant branch, linear theory gives the quantity $Q = P(\rho/\rho_\odot)^{1/2}$ as a function of mass for the fundamental and first harmonic (Wood, 1981). From these relations, and models of evolution along the asymptotic giant branch, it is inferred that stars evolve upwards along the asymptotic giant branch while losing mass. They become unstable at a certain luminosity, and oscillate in the first harmonic. At higher luminosities, the fundamental frequency begins to dominate leading, after a few oscillation periods, to the ejection of the envelope and the formation of a planetary nebula.

'noticeable periodicity of light changes accompanied or sometimes interrupted by different irregularities', and periods between 20–2000 d or more. For the shorter-period MIII stars, the GCVS loosely defines SRa and SRb as respectively more and less periodicity than for the SR classification. The small-amplitude red variables (SARV) are defined by Percy *et al.* (1996) as showing irregular pulsations with periods of 20–200 d or more, and amplitudes 0.1–2.5 mag. In addition, a few periodic M giants in the Hipparcos Catalogue did not receive any particular classification, notably those listed with short periods of typically less than 10 d. This nomenclature is certainly confusing, but will not be too central to the discussions in this section. What is relevant is that the K–M giants present a range of variability behaviour, and different chemical signatures partly reflecting various evolutionary or environmental conditions.

Hipparcos observed some 900 long-period variables, including Miras, semi-regular or irregular variables, O-rich, C-rich, and S stars, brighter than $V = 12.5$ mag for more than 80% of the time (Mennessier *et al.*, 1999). The Hipparcos data have contributed to diverse aspects of this important evolutionary population, including establishing certain properties as a function of Galactic population, establishing their pulsational modes, and ruling between pulsations, ellipsoidal (binary star) or rotationally-induced star spot variations, for certain anomalous variables.

The pulsation mode of the Mira variables has been debated for more than 20 years. Dynamical models (Hill & Willson, 1979) and observed velocity amplitudes favoured pulsations in the fundamental mode, while limited data indicating suspected multi-periodicity (Barthès & Tuchman, 1994) weakly favoured first overtone pulsation. New interferometric angular diameters and Hipparcos parallaxes (van Leeuwen *et al.*, 1997c) have allowed the comparison of observed and theoretical Q values (or period–luminosity relations), suggesting that most Miras pulsate in the first overtone, with a few long-period objects pulsating in the fundamental.

Mattei & Foster (1997b) reported AAVSO studies of 245 long-period variables observed by Hipparcos, showing that the periods, amplitudes, and other parameters of most Miras show discernable trends, not just random cycle-to-cycle fluctuations; that C-rich Miras fall in a significantly different range of periods and amplitudes than O-rich Miras; and that O-rich Miras of spectral type M show numerous strong relationships between period, amplitude, light-curve shape, spectral subtype, and $B-V$ and $V-I$ colour indices.

Mennessier *et al.* (1997, 1999) used the Hipparcos data for 882 long-period variables (parallaxes, proper motions, and magnitudes, supplemented by radial velocities) to provide an improved distinction between the different types of long-period variables in terms of stellar populations, absolute magnitudes, and disk scale heights. They used the 'LM method', described by Luri *et al.* (1996, see the box on page 213), to identify six discrete groupings, designated as bright disk, disk, old disk, thick disk, and extended disk, ranging in scale height from $Z_0 \sim$ 100–1200 pc respectively. As well as confirming that the mean ratio of C- to O-rich long-period variables is around 0.2–0.3, they showed that this is a strong function of initial mass, perhaps due to more efficient 'dredge-up' for more massive stars.

Barthès *et al.* (1999) developed this analysis, also based on the LM method, and applied it to a sample of about 350 O-rich long-period variables, comprising 154 Miras and 203 semi-regulars (34 SRa and 169 SRb). Using Hipparcos parallaxes and proper motions, combined with radial velocities, periods, and $V-K$ colour indices, they identified four groupings differing in kinematics and mean magnitudes: group 1, mainly composed of Miras and having the kinematics of old disk stars; group 2, mainly composed of SRb, with the same kinematics as group 1; group 3, also mainly composed of SRb, but with much younger kinematics; and group 4, corresponding to the extended disk and halo, and containing no SRb. For each group, they derived the magnitude distribution, period and de-reddened colour, as well as de-biased period–luminosity–colour relations and their various two-dimensional projections: period–luminosity, period–colour, and luminosity–colour (see Figure 4.31). They found that the SRa semi-regulars do not seem to constitute a separate class of long-period variables, while the SRb appear to comprise two populations of different ages. In a period–luminosity diagram, they constitute two evolutionary sequences towards the Mira stage. The Miras of the disk appear to pulsate in a lower-order mode. The slopes of their de-biased period–luminosity and period–colour relations are very different from those of the O-rich Miras in the LMC, suggesting that a significant number of Miras of the LMC are misclassified. Consequently, they appear not to constitute a homogeneous group, but rather include a significant proportion of metal-deficient stars, itself suggesting a relatively smooth star formation history, and that the LMC period–luminosity relation cannot simply be applied to other galaxies. More detailed aspects of the kinematic and environmental dependencies were studied by Mennessier & Luri (2001) and Mennessier *et al.* (2001).

Mennessier *et al.* (1998) made a Fourier analysis of the ground-based light-curves of the Hipparcos long-period variables to identify the phase-lag between the fundamental and first harmonic in the case of the 105 objects for which a harmonic component could be detected. They identified a dependency of these modes related to period, spectral type, and Galactic population.

De Laverny *et al.* (1998) examined the Hipparcos data for 239 long-period Mira variables for short-term photometric fluctuations. The sample included 195 M-type, 26 C-type, and 11 S-type, with periods from 100–550 days. Altogether 51 'events' in 39 M-type Miras were detected (Figure 4.32), with no similar variations observed for S and C-type Miras. Amplitude variations ranged from 0.2–1.1 mag on time scales of 2–100 hours, preferentially in late spectral types, and around minimum light phases and during the rise to maximum. They suggested that these variations may be related to molecular opacity changes and to related variations in the physical conditions leading to instabilities, or to hydrodynamic effects. Ground-based observations to investigate these variations was initiated by Ostrowski & Stencel (1999), and subsequently supported by Stencel *et al.* (2003), who suggested that these microvariations could be due to dust formation episodes in the upper atmospheres. Mais *et al.* (2006) concluded that the alleged microvariability results from undersampled, transient overtone pulsations. The variations were not supported by subsequent OGLE investigations of Woźniak *et al.* (2004). Barthès *et al.* (1998) used Fourier analysis and kinematic classification to investigate low-order pulsation resonances.

Dumm & Schild (1998) determined stellar radii of M giants in the Hipparcos Catalogue that have a parallax measured to better than 20% accuracy, using a relation between a visual surface brightness parameter and the Cousins $V-I$ colour index, calibrated with M giants with published angular diameters (see also Pijpers, 2000). The radii of (non-Mira) M giants increase from a median value of $50R_\odot$ at spectral type M0 III to $170R_\odot$ at M7/8 III, but with a large intrinsic spread for a given spectral type. They determined masses, from evolutionary tracks, in the range 0.8–$4M_\odot$. They found a close relation between stellar radius and stellar mass for a given spectral type, a linear relation between mass and radius for non-variable M giants and, with increasing amplitude of variability, larger stellar radii for a given mass (see also Hoffleit, 1999).

Of various photometric studies of Hipparcos carbon stars, e.g. near infrared photometry of 20 carbon stars by Chen *et al.* (2003), and *BV* photometry from Brno Observatory between 1979–94 for seven carbon stars by

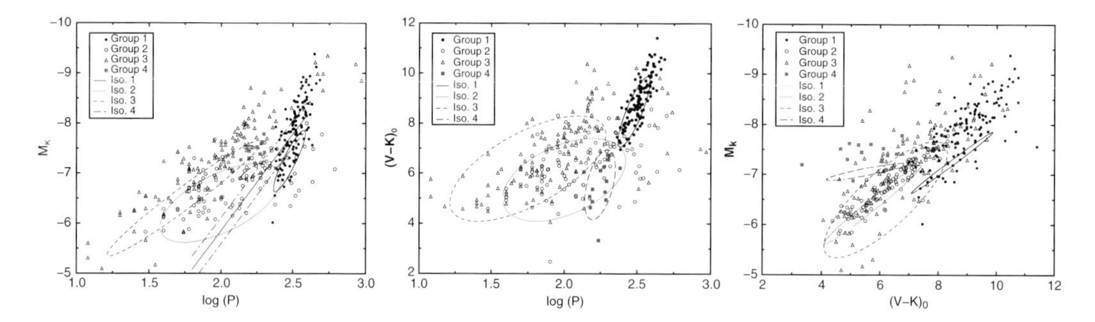

Figure 4.31 Left: period–luminosity calibrated distributions: individual data and projected model distributions (2σ probability contours in the mean period–luminosity–colour plane). The Mira strip of the LMC is also shown (thick lines). Middle: period–colour calibrated distributions: individual data and projected model distributions. Right: luminosity–colour calibrated distributions: individual data and projected model distributions. From Barthès et al. (1999, Figures 5–7).

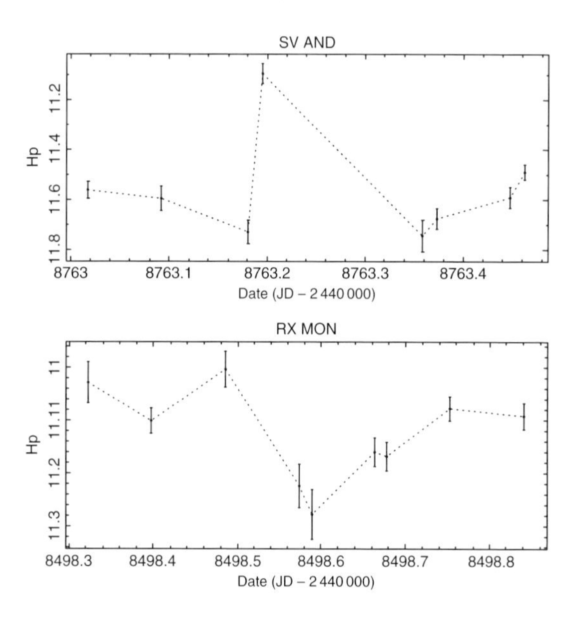

Figure 4.32 Examples of short-term variations of M-type Miras in the Hipparcos photometry. From de Laverny et al. (1998, Figure 1).

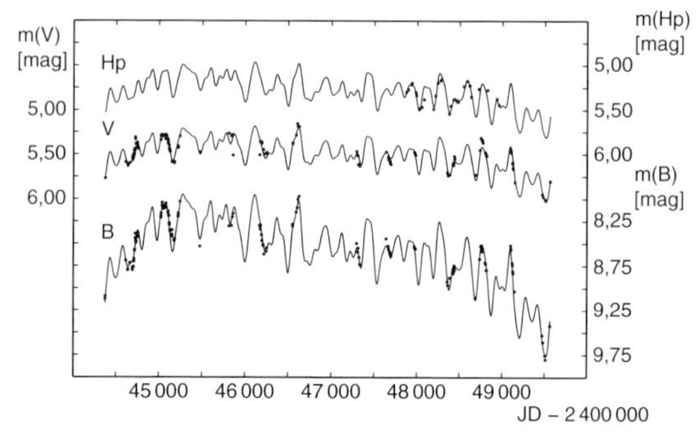

Figure 4.33 Example of ground-based photometric monitoring of carbon stars (solid curves) compared with the Hipparcos Hp, B_T and V_T epoch photometry, for the case of Y CVn. From Dušek et al. (2003, Figure 3).

Dušek *et al.* (2003), the latter authors reported that the semi-regular light-curves of these objects could be represented by a superposition of long-term changes and a set of medium-term harmonic variations (possibly pulsations) with $P \sim 50$–500 d, and also reported an almost linear correlation between a characteristic amplitude and a characteristic period (see Figure 4.33).

Lebzelter *et al.* (2005) determined the radial velocity curves of a number of bright southern long-period variables: semi-regular variables with typical Mira periods, semi-regular variables exceeding the Mira 2.5 mag amplitude limit, Miras with secondary maxima in their light-curves, and a semi-regular variable with a long secondary period. Stars with reliable Hipparcos parallaxes, plotted in a $\log P, M_K$ diagram, follow the fundamental and first overtone mode relations determined for the LMC. While all their Miras fall on the fundamental mode sequence, the semi-regular variables fall on both the first overtone and fundamental mode sequences. The semi-regular variables on the fundamental mode sequence occur at both high and low luminosities, some of them being more luminous than larger amplitude Miras. This demonstrates observationally that some parameter other than luminosity affects the stability of long-period variables, probably mass.

Various studies have been reported for individual Hipparcos long-period variables. Wing (1997) used the early Hipparcos data release to classify the nearest Miras as: R Leo (120 pc), χ Cyg (106 pc), R Cas (107 pc), etc. At mean maximum, M_V range from -2.5 to $+2.0$, with cooler stars being fainter. For 10 late-type small-amplitude variables, the nearest are L2 Pup (M5 III at 61 pc), and R Dor (M8e at 62 pc). The nearest of 23 carbon stars is U Hya (SRb at 162 pc). Lloyd & West (2001) combined Hipparcos and photoelectric V-band observations of UX Dra, to reveal a period of 175 ± 1 d which is very stable over 11 years, superimposed on a 334 ± 2 d period which shows some variation. Richwine *et al.* (2005) combined photometric data from various sources, including Hipparcos, AAVSO, and ASAS, to characterise the behaviour of 10 long-period variables in Aquila.

Various studies have been reported for individual Hipparcos semi-regular variables. Wasatonic & Guinan (1998a) combined the Hipparcos distance of V CVn (HIP 65006, $d = 375 \pm 125$ pc) with ground-based photometry to estimate the radius, luminosity, and effective temperature of this bright pulsating red giant whose properties are intermediate between the classical Mira variables and the small-amplitude red variables. The results show, for example, that the star attains a maximum radius of $\sim 400 R_\odot$ near minimum light, shrinking to $\sim 200 R_\odot$ at maximum luminosity. Bedding *et al.* (1998b) studied mode-switching in the semi-regular variable R Dor. Wasatonic & Guinan (1998b) similarly studied variations of luminosity, radius, and temperature of CE Tau. Percy & Kolin (2000) used the Hipparcos epoch photometry to study variability in several objects, reporting a mixture of confirmation of GCVS periods, and inconclusive results. For WY And, RU Cep, and SX Her, archival data were also used to study random and systematic period changes. Lebzelter & Posch (2001) studied eight small amplitude variables, all observed by Hipparcos/Tycho, six classified as semi-regular variables, and two attributed to surface spots. Pikhun *et al.* (2001) compared Hipparcos and archival photometry for UV Boo, and discussed the different periods found. Lloyd *et al.* (2002) compared the Hipparcos photometric data for V370 And, indicating periods of 228 ± 1 and 123 ± 1 d, with an absence of such periods in archival data showing comparable variability amplitudes. Thompson (2002) analysed archival data of RZ Ari over 17 years; it displays 22 clearly-defined cycles ranging from 26–71 d and a secondary period of 1100–1500 d, but with no specific period apparent from the Hipparcos data.

Glass & van Leeuwen (2007) used the revised parallaxes from the recent global re-analysis of the Hipparcos data to show that the period–luminosity sequences known to exist amongst semi-regular variables in the Magellanic Clouds and in the Galaxy bulge also appear in the solar neighbourhood M giants, with their distribution in the K versus $\log P$ diagram more closely resembling the behaviour of the Bulge stars than those in the Magellanic Clouds.

Small-amplitude red variables Small-amplitude red variables are some 100 times more numerous than Mira variables. The fact that detectable variability at around 0.02 mag sets in at type M0–1 ($T_{\rm eff} \sim 3800$ K) has been known for some time (Stebbins & Huffer, 1930; Percy & Fleming, 1992). More accurate long-term photometry from ground (Jorissen *et al.*, 1997) and from Hipparcos (Eyer & Grenon, 1997) has shown that mid- to late-K giants ($T_{\rm eff} \sim 4000$ K) are microvariable, with amplitudes of mmag. The onset of pulsations in Hipparcos data as a function of $V - I$ was investigated by Percy & Guler (1999).

Percy & Polano (1998) studied pulsation modes in the small-amplitude red variables. These are stars ascending the asymptotic giant branch; at the tip, thermal pulses occur in the He-burning shell, and mass-loss is enhanced by the large-amplitude pulsations in the Mira variables, leading quickly to the white dwarf stage. Percy & Polano (1998) determined Q-values, using temperature calibrations of colour and spectral type, a luminosity–colour relationship, and two possible values of the mass (1.0 and $1.4 M_\odot$). They concluded that the small-amplitude red variables are pulsating in a variety of modes, ranging from the fundamental to the third overtone (see also, e.g. Percy *et al.*, 2001).

Percy & Parkes (1998) investigated the pulsation modes of 13 objects with well-determined periods (some from Hipparcos), accurate Hipparcos parallaxes, and good $T_{\rm eff}$ estimates based on interferometric angular diameters and/or model atmospheres. This permits comparison of observed and theoretical Q-values. Observed values again use the relation $Q = P(\rho/\rho_\odot)^{1/2}$, requiring knowledge of the period, mass, and radius (or luminosity and temperature). Theoretical models, incorporating both dynamical and thermodynamic coupling between pulsation and convection, were from Xiong *et al.* (1998). The results (Figure 4.34) suggest that three of the 13 are pulsating in the fundamental or first overtone, nine in the first or second overtone, and one in the third overtone. Largely due to the Hipparcos parallaxes, the new observational Q-values are larger than those derived by Percy & Polano (1998). The results are consistent with those of Barthès (1998) who concluded that most of his sample of semi-regular variables in the LMC are pulsating in the second overtone. They are consistent with the models of Xiong *et al.* (1998, their Table 1) who predict that, for the coolest M giants ($\sim$ 2300–3000 K), only the fundamental mode is unstable. For slightly hotter stars ($\sim$ 2800–3800 K), the first-overtone mode is also unstable. For even hotter models, higher modes (second to fourth overtone) may also be unstable. Modes higher than fourth overtone are stable in all the models, damped by turbulent viscosity. A similar study of pulsation modes in 77 small-amplitude red giants was reported by Percy & Bakos (2003).

Short-period M giants A series of papers dealt with the identification of a few dozen M giants revealing

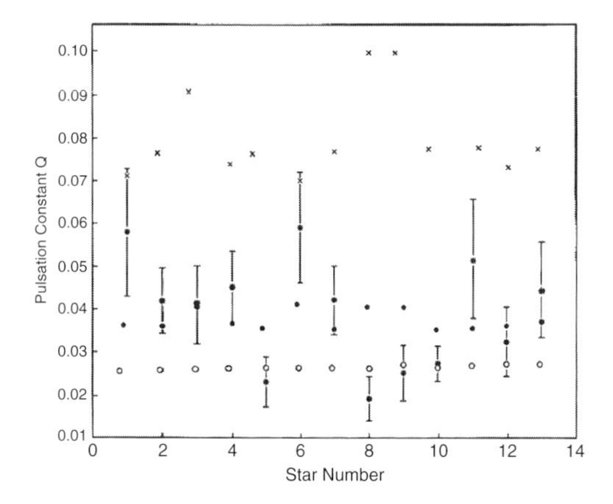

Figure 4.34 For the 13 small-amplitude red variable stars in the study of Percy & Parkes (1998), the figure shows the observational pulsation constants Q (⋆) and estimated errors, and the theoretical pulsation constants for the fundamental (×), first-overtone (•), and second-overtone (○) modes. From Percy & Parkes (1998, Figure 1).

unexpectedly rapid variability in the Hipparcos epoch photometry, showing periods as short as 1 d. Subsequent work showed that these are at least partly explicable as aliasing anomalies in the Hipparcos epoch photometry due to the specific window function noted in Section 4.6.1, perhaps combined with more complex multi-periodic behaviour. The sequence of studies is nevertheless reported.

Koen & Laney (2000b), Laney & Koen (2000) and Koen & Laney (2000a) first called attention to these M giants for which short periods were given in the Hipparcos Catalogue. Somewhat similar periodic phenomena were reported in the OGLE data, where they were considered as most likely spotted, chromospherically active stars (Udalski *et al.*, 1994). Such an interpretation would imply that star spots are not restricted to the standard families of K- and M-dwarfs (BY Dra stars), binaries with G–K class III–IV components (RS CVn stars), or rapidly rotating G–K giants (FK Com) stars. Koen & Laney (2000a) first derived the physical properties of the stars using the relationships derived for Hipparcos M giants by Dumm & Schild (1998) including

$$\log T_{\rm eff} = 3.65 - 0.035(V - I) \tag{4.22}$$

$$\log R = 2.970 - \log \pi - 0.2V + 0.365(V - I) \tag{4.23}$$

Masses were estimated from the position in the HR diagram combined with evolutionary tracks. They then considered whether the Hipparcos short-period M giants could be ellipsoidal variables, in which variability is caused by the rotation of a tidally-distorted binary component. In this case the requirement that the binary separation should be larger than the radius of the red giant, R_1, requires that the mass of the companion

$$M_2 > 3.34 \times 10^{-3}\, R^3\, P^{-2} - M_1 \tag{4.24}$$

where P is the orbital period in days. In the alternative case that the stars are spotted, the minimum variability period corresponds to rotation at the break-up speed, such that

$$P > 0.12 M_1^{-1/2}\, R^{3/2} \tag{4.25}$$

Finally, if the variability is caused by pulsation, then the pulsation constant Q can be calculated from the first form given in Equation 4.19. The resulting pulsation constants are quite small, in the range 0.002–0.03, indicative of high-overtone pulsations (e.g. Cox, 1976). The Hipparcos data provide all of the physical quantities necessary to constrain these possible models. They concluded that star spots could be ruled out in almost all cases, and ellipsoidal variations in about half. This left high-overtone pulsations as the favoured explanation.

Two subsequent studies cast doubt on the reality of many, but perhaps not all, of these putative short-period M giants (Koen *et al.*, 2002b; Percy & Hosick, 2002). Koen *et al.* (2002b) obtained time-series optical and infrared photometry and radial velocities for 20 of the objects. The optical photometry reveals variability on time scales of 15 d or more, with most showing little sign of the short-period variations claimed in the Hipparcos Catalogue. There were two notable exceptions, both of early K spectral type, although lower amplitude shorter time scale variations are also present. While thus precluding the high overtone pulsation interpretation previously proposed by Koen & Laney (2000a), they showed that pulsation remains a likely explanation for variability in at least some of the stars, albeit in low overtone or fundamental modes. They argued that invoking aliasing as a complete explanation for the short periods found by Hipparcos encounters some difficulties, and that the presence of multi-periodicity (cf. Kiss *et al.*, 1999c), as for the δ Scuti variables, along with the poor time sampling, may contribute to the spurious periods. There are other variability studies which support this interpretation (e.g. Percy *et al.*, 2001; Dušek *et al.*, 2003).

Percy & Hosick (2002) studied AG Cet, appealing to ground-based photometry, and self-correlation as well as variogram analyses of the Hipparcos light-curves, to show that at least some of the short-period red giants have spurious Hipparcos periods. As a specific example, Marinova & Percy (1999) compared more continuous AAVSO photoelectric photometry, giving a period of 77.7 d, with the Hipparcos photometry which gave additional strong but spurious periods of 1–6 d.

There is therefore a consensus that evidence for high radial overtones in these stars seems unlikely, or at least unproven.

S stars There are two kinds of cool peculiar red giant variables of spectral type S: the intrinsic S stars with high luminosity, Tc lines, and chemical peculiarities produced by intrinsic nucleosynthesis processes; and the extrinsic non-Tc S stars which occur in binary systems in which chemical enrichment is due to mass transfer from an evolved companion, as for barium stars. Van Eck *et al.* (1998) used the Hipparcos parallaxes to show that the intrinsic Tc-rich stars are thermally-pulsating asymptotic giant branch stars of low and intermediate masses, while the extrinsic Tc-poor stars are mostly low-mass (binary) stars on either the red giant branch or the early asymptotic giant branch (the two could in principle be distinguished by accurate luminosities). Adelman & Maher (1998) studied the Hipparcos variability of S stars, and suggested that all the medium- and large-amplitude S variables (> 0.3–1 mag) are intrinsic S stars, and that all the small-amplitude variables are extrinsic stars, with Miras having $P > 300$ d being intrinsic S stars.

Red standards The very high incidence of variability amongst cool stars makes the identification of red standards somewhat problematic. For example, Koen *et al.* (2002a) report that the catalogue of standard stars in use at the SAAO for $UBV(RI)_C$ photometry (mainly Harvard E- and F-region stars) has 570 entries, of which only 13 are of spectral type M, and six of those are flagged as variable or possibly variable. There are very few standards redder than about $B - V = 1.6$ or $V - I = 2.0$. Similarly, Landolt (1983) provided $UBV(RI)_C$ photometry for 223 equatorial standards, of which only 19 have $(V - I)_C > 1.7$, with eight or more known to be variable. To supplement this paucity, Koen *et al.* (2002a) used the Hipparcos database to select 546 stars with $(V - I)_C > 1.7$, $V > 7.6$, $\delta < 10°$, and no variability flag set, for which ground-based photometry was subsequently acquired at SAAO. Their results provide an assessment of the accuracy of the red star photometry given in the Hipparcos Catalogue, as well as providing a set of candidates for red standards. The calibrations are fairly robust for the dwarf stars, but still required the extrapolation of colour equations for the giants, meaning that their transformation to the standard $UBV(RI)_C$ system is less certain. Kilkenny *et al.* (2007) presented revised $UBV(RI)_CJHK$ photometry for over 100 M star standards, with *JHK* photometry for nearly 300 Hipparcos red stars not selected as standards.

Astrometric re-processing of red stars Although the Hipparcos telescope was fully reflecting, the asymmetry of the point spread function nevertheless led to a colour-dependent displacement of the image centroid. Uncorrected, this would have led to colour-dependent image displacement in the great-circle scans, and larger residuals and/or biases in the astrometric solutions for each star. This chromatic dependence was taken into account during the catalogue construction by introducing the colour of the star into the astrometric solution (ESA, 1997, Volume 3, Section 16.3). For constant stars, ground-based $B - V$, or Tycho $B_T - V_T$, were adequate to construct a Cousins' $V - I$ index used to model the broadband *Hp* energy dependency in the solution.

The large-amplitude red giant variables presented very specific problems in this regard, because the concentration of spectral energy in the red made the chromatic correction particularly sensitive to the precise $V - I$ colour index, which may also vary according to pulsational phase. Only 2989 Hipparcos stars had direct ground-based estimates of the $V - I$ index. Accordingly, a variety of estimates of variable accuracy were used to obtain $V - I$ (ESA, 1997, Volume 1, Section 1.3, Appendix 5). Furthermore, because of the iterative nature of the astrometric solution, and the lengthy numerical processing task limited by CPU power and organisational considerations relevant at the time of the catalogue completion, the Hipparcos Catalogue contains two sets of $V - I$ colour indices: the best available at the time of the catalogue release (Field H40), and that actually used for the final data processing (Field H75, which appears only in the machine-readable catalogue version). The unavailability of an adequate $V - I$ for the astrometric processing, means that some published astrometric solutions, and particularly some of the red giant star solutions, are suboptimal.

Platais *et al.* (2003a,b) provided calibrated instantaneous (epoch) Cousins $V - I$ colour indices using newly-derived *Hp* and V_{T2} (Tycho 2) photometry for Hipparcos M, S, and C spectral type stars. Three new sets of ground-based Cousins *VI* data were obtained for more than 170 carbon stars and red M giants. These datasets in combination with the published sources of *VI* photometry served to obtain the calibration curves linking Hipparcos/Tycho $Hp - V_{T2}$ with the Cousins $V - I$ index. In total, 321 carbon stars and 4464 M- and S-type stars have new $V - I$ indices. The standard error of the mean $V - I$ is about 0.1 mag or better down to $Hp \approx 9$, although it deteriorates rapidly at fainter magnitudes. These $V - I$ indices were used to verify the published Hipparcos $V - I$ colour indices. At the same time, they could be used to identify a handful of new cases where, because of poor Input Catalogue coordinates, a nearby star had been observed instead of the intended target. A considerable fraction of the DMSA C and DMSA V solutions (from the Double and Multiple Systems Annex) for

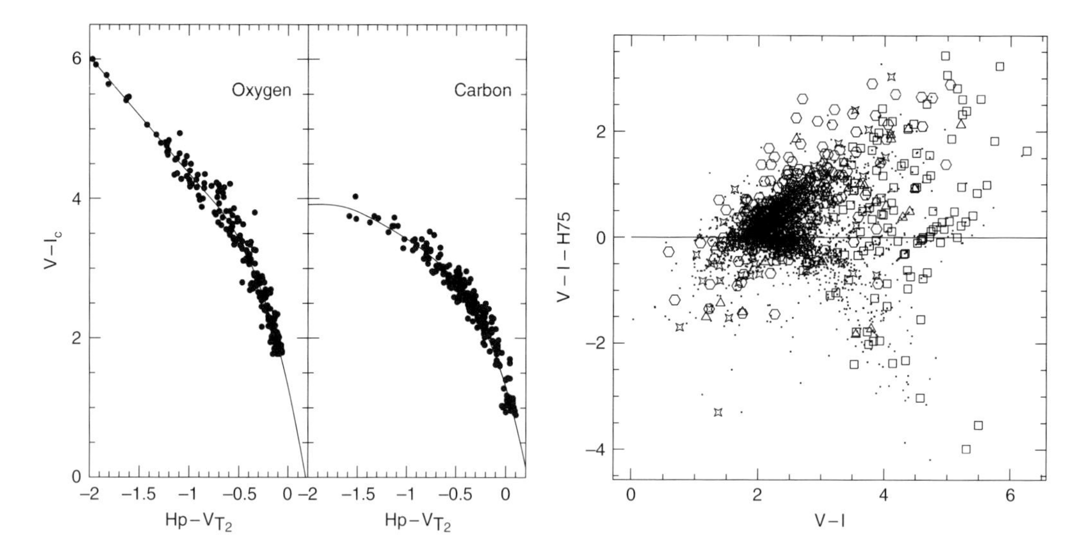

Figure 4.35 Left: colour–colour transformation for M and S stars (left panel) and carbon stars (right panel). The red end of this transformation ($Hp - V_{T2} < -1.5$) for carbon stars is uncertain due to the lack of intrinsically very red Hipparcos calibrating carbon stars. Right: comparison of the $V - I$ colour indices used in the preparation of the Hipparcos Catalogue (Field H75) with those re-derived from a colour transformation based on $Hp - V_{T2}$. Only red stars with spectral types M and S are considered. Single-star solutions are depicted by black dots; open symbols denote more complex solutions associated with the various parts of the Hipparcos Double and Multiple Systems Annex (hexagons: component solutions C; triangles: acceleration solutions G; squares: variability-induced mover (VIM) solutions V; stars: stochastic solutions X). Nearly all points in the upper right corner correspond to complex solutions, thus hinting at problems encountered in the Hipparcos data processing for these stars. From Platais et al. (2003a, Figures 6 and 9).

red stars thus appear to be spurious, most likely originating from the use of a strongly biased colour in the astrometric processing (Figure 4.35).

Knapp *et al.* (2003) analysed the K-band luminosities of a sample of Galactic long-period variables using parallaxes measured by Hipparcos, and in most cases re-computed from the Intermediate Astrometric Data using improved astrometric fits and chromaticity corrections as described by Platais *et al.* (2003a). K-band magnitudes were taken from the literature and from measurements by COBE, and corrected for interstellar and circumstellar extinction. The sample contains stars of spectral types M, S and C, and of several variability classes: Mira, semi-regular SRa, and SRb. They found that the distribution of stars in the period–luminosity plane is independent of circumstellar chemistry, but that the different variability types have different period–luminosity distributions. Both the Mira variables and the SRb variables have reasonably well-defined period–luminosity relationships, but with very different slopes (Figure 4.36). The SRa variables are distributed between the two classes, suggesting that they are a mixture of Miras and SRb, rather than a separate class of stars. New period–luminosity relationships were derived based on the revised Hipparcos parallaxes. The Miras show a similar period–luminosity relationship to that found for Large Magellanic Cloud Miras by Feast *et al.* (1989). The maximum absolute K magnitude of the sample is about -8.2 for both Miras and semi-regular stars, only slightly fainter than the expected asymptotic giant branch limit. They show that the stars with the longest periods ($P > 400$ d) have high mass-loss rates and are almost all Mira variables. Also shown in Figure 4.36 is the period–luminosity relationship for LMC Miras by Feast *et al.* (1989), with the zero-point as determined for Galactic Miras by Whitelock & Feast (2000)

$$M_K = -3.47 \log P \text{ days} + 0.85 \quad \text{Feast et al. (1989, Miras in LMC)} \tag{4.26}$$

$$M_K = -3.39(\pm 0.47) \log P \text{ days} + 0.95(\pm 3.01) \quad \text{Knapp et al. (2003, Miras)} \tag{4.27}$$

$$M_K = -1.34(\pm 0.06) \log P \text{ days} - 4.5(\pm 0.35) \quad \text{Knapp et al. (2003, semi-regulars)} \tag{4.28}$$

The period–luminosity relation for Miras agrees well with that found in the LMC, while the slope for the semi-regular variables is much shallower. They suggest that long-period variables, at least those with $P < 400$ d, may change their variability modes back and forth between Mira and semi-regular type, with different radial pulsation modes dominating in the two cases. This would

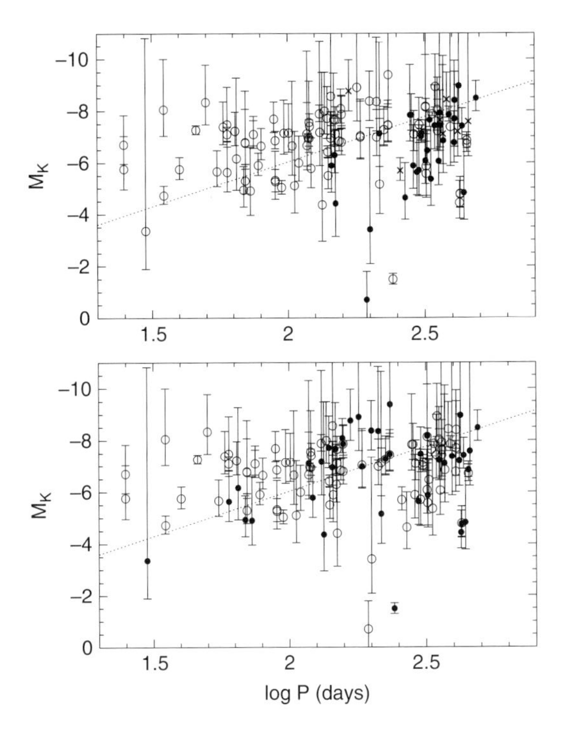

Figure 4.36 Distribution of long-period variables in the M_K – log P plane: (a) upper panel: stars indicated by variability type (•: Miras; ∘: SR/SRb; ×: SRa); (b) lower panel: stars indicated by chemical type (•: carbon stars; ∘: oxygen stars). The dashed line is the period–luminosity relationship of Equation 4.26. From Knapp et al. (2003, Figure 4).

be consistent with the different slopes of the period–luminosity relation, in which period changes by a factor of 2–4 (depending on mode) would be accompanied by only small changes in luminosity.

4.11 Individual objects

There are numerous studies of individual Hipparcos variables. They include variables discovered in the Hipparcos data, such as HD 205798 (Torres *et al.*, 1999) and various short- and medium-period variables (Kiss *et al.*, 2000), and variables discovered in the Tycho data, such as HD 143213 (a new Algol system Bastian & Born, 1997); HD 20511 (a new β Lyrae Campos Cucarella *et al.*, 1997); the new eclipsing binary systems HD 125488 and HD 126080 (Gómez Forrellad & García Melendo, 1997), and many other studies, for example: unexpected photometric period in the symbiotic system AG Dra (Bastian, 1998); the active evolved star PZ Mon (Saar, 1998); evolutionary status of HD 104237, ϵ Cha and HD 100546 (Shen & Hu, 1999); photometric orbit modulation in V1080 Tau (Simon *et al.*, 2000); orbital parameters and frequencies of ϵ Per (de Cat *et al.*, 2000); the chromospherically active binaries UX For and AG Dor (Washuettl & Strassmeier, 2001); evolutionary status of the early-type binary HD 115071 (Lloyd & Stickland, 2001); the massive close binary HD 115071 (Penny *et al.*, 2002); period and prediction for the spectrum alternator HD 191612 (Walborn *et al.*, 2004); and the close binary system SU Ind (Budding *et al.*, 2004).

References

Adelman SJ, 1998, Photometry from the Hipparcos Catalogue: constant MCP stars, comparison and check stars. *Baltic Astronomy*, 7, 427–434 {**161**}

—, 2000a, On the variability of A3–F0 luminosity class III–V stars. *Informational Bulletin on Variable Stars*, 4969 {**172**}

—, 2000b, On the variability of early K stars. *Informational Bulletin on Variable Stars*, 4958 {**172**}

—, 2000c, On the variability of K5–M stars. *Informational Bulletin on Variable Stars*, 4959 {**172**}

—, 2001a, Hipparcos photometry: the least variable stars. *A&A*, 367, 297–298 {**160**}

—, 2001b, On the variability of stars. *Informational Bulletin on Variable Stars*, 5050 {**167**}

—, 2001c, Stars with the largest Hipparcos photometric amplitudes. *Baltic Astronomy*, 10, 589–593 {**172**}

Adelman SJ, Albayrak B, 1997, On the variability of early A-type supergiants. *Informational Bulletin on Variable Stars*, 4541 {**186**}

Adelman SJ, Cay IH, Cay MT, *et al.*, 2000a, On the variability of A6 to F9 supergiants. *Informational Bulletin on Variable Stars*, 4947 {**172**}

Adelman SJ, Coursey BC, Harris EA, 2000b, On the variability of F1–F9 luminosity class III–V stars. *Informational Bulletin on Variable Stars*, 5003 {**172**}

Adelman SJ, Davis JM, Lee AS, 2000c, On the variability of G0–G9 stars. *Informational Bulletin on Variable Stars*, 4993 {**172**}

Adelman SJ, Flores RSC, Patel VJ, 2000d, On the variability of A0–A2 luminosity class III–V stars. *Informational Bulletin on Variable Stars*, 4984 {**172**}

Adelman SJ, Gentry ML, Sudiana IM, 2000e, On the variability of late BIII–V stars. *Informational Bulletin on Variable Stars*, 4968 {**172**}

Adelman SJ, Maher DW, 1998, On the variability of S stars as observed by the Hipparcos satellite. *Informational Bulletin on Variable Stars*, 4591 {**195**}

Adelman SJ, Mayer MR, Rosidivito MA, 2000f, On the variability of O4–B5 luminosity class III–V stars. *Informational Bulletin on Variable Stars*, 5008 {**172**}

Adelman SJ, Pi CM, Rayle KE, 1998, *uvby* photometry of 33 Tau, HD 50169, and HR 7786 and an assessment of FCAPT comparison stars. *A&AS*, 133, 197–200 {**161**}

Adelman SJ, Yuce K, Engin S, 2000g, On the variability of O and B supergiants. *Informational Bulletin on Variable Stars*, 4946 {**172**}

Aerts C, 2000, Follow-up photometry of six new β Cephei stars discovered from the Hipparcos mission. *A&A*, 361, 245–257 {**186**}

Aerts C, de Boeck I, Malfait K, *et al.*, 1999a, HD 42927 and HD 126341: two pulsating B stars surrounded by circumstellar dust. *A&A*, 347, 524–531 {**188**}

Aerts C, de Cat P, Peeters E, *et al.*, 1999b, Selection of a sample of bright southern slowly-pulsating B stars for long-term photometric and spectroscopic monitoring. *A&A*, 343, 872–882 {**187, 188**}

Aerts C, Eyer L, Kestens E, 1998a, The discovery of new γ Doradus stars from the Hipparcos mission. *A&A*, 337, 790–796 {**184, 185**}

Aerts C, Kolenberg K, 2005, HD 121190: a cool multi-periodic slowly-pulsating B star with moderate rotation. *A&A*, 431, 615–622 {**188**}

Aerts C, Waelkens C, de Cat P, 1998b, Slowly-pulsating B stars: new insights from Hipparcos. *IAU Symp. 185: New Eyes to See Inside the Sun and Stars*, 295 {**186**}

Alcock C, Allsman RA, Axelrod TS, *et al.*, 1995, The MACHO project LMC variable star inventory. I. Beat Cepheids-conclusive evidence for the excitation of the second overtone in classical Cepheids. *AJ*, 109, 1653–1662 {**173**}

Allende Prieto C, Lambert DL, 2000, The near-ultraviolet continuum of late-type stars. *AJ*, 119, 2445–2454 {**165, 166**}

Alvarez R, Mennessier MO, 1998, DENIS and Hipparcos: luminosity and kinematic calibrations. *ASSL Vol. 230: The Impact of Near-Infrared Sky Surveys on Galactic and Extragalactic Astronomy*, 129 {**165**}

Andronov IL, 2000, Photometric classification of the Hipparcos suspected variable stars: the mathematical aspects. *Hipparcos and the Luminosity Calibration of the Nearer Stars, IAU Joint Discussion 13*, 32 {**170**}

Andronov IL, Cuypers J, Piquard S, 2000, Light curve analysis of Tycho variable stars. *ASP Conf. Ser. 203: IAU Colloq. 176: The Impact of Large-Scale Surveys on Pulsating Star Research*, 64–65 {**170, 171**}

Antonello E, Mantegazza L, Poretti E, 1998, Luminosity of δ Scuti stars after Hipparcos satellite. *Highlights in Astronomy*, 11, 567 {**175**}

Antonello E, Mantegazza L, 1997, Luminosity and related parameters of δ Scuti stars from Hipparcos parallaxes. General properties of luminosity. *A&A*, 327, 240–244 {**175**}

Antonello E, Pasinetti Fracassini LE, 1998, Pulsating and non-pulsating stars in Hyades observed by Hipparcos satellite. *A&A*, 331, 995–1001 {**176**}

Arellano Ferro A, 1983, Period and amplitude variations of Polaris. *ApJ*, 274, 755–762 {**173**}

Arenou F, Luri X, 1999, Distances and absolute magnitudes from trigonometric parallaxes. *ASP Conf. Ser. 167: Harmonizing Cosmic Distance Scales in a Post-Hipparcos Era*, 13–32 {**182**}

Audard N, Kupka F, Morel P, *et al.*, 1998, The acoustic cut-off frequency of roAp stars. *A&A*, 335, 954–958 {**183**}

Ausseloos M, Scuflaire R, Thoul AA, *et al.*, 2004, Asteroseismology of the β Cephei star ν Eri: massive exploration of standard and non-standard stellar models to fit the oscillation data. *MNRAS*, 355, 352–358 {**186**}

Bahcall JN, Flynn C, Gould A, 1992, Local dark matter from a carefully selected sample. *ApJ*, 389, 234–250 {**164**}

Balachandran SC, Bell RA, 1998, Shallow mixing in the solar photosphere inferred from revised beryllium abundances. *Nature*, 392, 791–793 {**165**}

Balona LA, Dean JF, Stobie RS, 1981, The oscillation modes of δ Scuti. *MNRAS*, 194, 125–136 {**178**}

Bartašiute S, Ezhkova OV, Lazauskaite R, 1999, Photoelectric Vilnius photometry of Hipparcos turn-off region stars. *Baltic Astronomy*, 8, 465–482 {**164**}

Barthès D, 1998, Pulsation modes of Mira stars and questioning of linear modeling: indications from Hipparcos and the Large Magellanic Cloud. *A&A*, 333, 647–657 {**193**}

Barthès D, Luri X, Alvarez R, *et al.*, 1999, Period-luminosity-colour distribution and classification of Galactic oxygen-rich long-period variables. I. Luminosity calibrations. *A&AS*, 140, 55–67 {**191, 192**}

Barthès D, Mennessier MO, Vidal JL, *et al.*, 1998, Are low-order resonances observed in Mira pulsation? *A&A*, 334, L1–L4 {**191**}

Barthès D, Tuchman Y, 1994, Confrontation between power spectra of Mira stars and theoretical models. *A&A*, 289, 429–440 {**190**}

Bastian U, Born E, Agerer F, *et al.*, 1996, Confirmation of the classification of a new Tycho variable: HD 32456 is a 3.3-day Cepheid. *Informational Bulletin on Variable Stars*, 4306 {**170, 174**}

Bastian U, Born E, 1997, The variability type and period of HD 143213. *Informational Bulletin on Variable Stars*, 4536 {**197**}

Bastian U, 1998, The symbiotic system AG Dra: an unexpected photometric period. *A&A*, 329, L61–L63 {**197**}

Beaulieu JP, Grison P, Tobin W, *et al.*, 1995, EROS variable stars: fundamental-mode and first-overtone Cepheids in the bar of the Large Magellanic Cloud. *A&A*, 303, 137–154 {**173**}

Bedding TR, Kjeldsen H, Christensen-Dalsgaard J, 1998a, Hipparcos parallaxes for η Boo and κ Boo: two successes for asteroseismology. *ASP Conf. Ser. 154: Cool Stars, Stellar Systems, and the Sun*, 741–744 {**178**}

Bedding TR, Zijlstra AA, Jones A, *et al.*, 1998b, Mode switching in the nearby Mira-like variable R Dor. *MNRAS*, 301, 1073–1082 {**193**}

Bell RA, Paltoglou G, Tripicco MJ, 1994, The calibration of synthetic colours. *MNRAS*, 268, 771–792 {**165**}

Berdnikov LN, Szabados L, 1998, Study of neglected variable stars classified as Type II Cepheids. *Acta Astronomica*, 48, 763–774 {**174**}

Bessell MS, 2000a, Magnitude scales and photometric systems. *Encyclopedia of Astronomy and Astrophysics;* (ed. Murdin, P.) {**154**}

—, 2000b, The Hipparcos and Tycho photometric system passbands. *PASP*, 112, 961–965 {**158–160**}

—, 2005, Standard photometric systems. *ARA&A*, 43, 293–336 {**154**}

Bessell MS, Castelli F, Plez B, 1998, Model atmospheres broad-band colours, bolometric corrections and temperature calibrations for O–M stars. *A&A*, 333, 231–250 {**159**}

Bigot L, Dziembowski WA, 2002, The oblique pulsator model revisited. *A&A*, 391, 235–245 {**182**}

Bossi M, Mantegazza L, Nunez NS, 1998, Simultaneous intensive photometry and high resolution spectroscopy of δ Scuti stars. III. Mode identifications and physical calibrations in HD 2724. *A&A*, 336, 518–526 {**178**}

Bowyer S, Sasseen TP, Lampton M, *et al.*, 1993, In-flight performance and preliminary results from the Far Ultraviolet Space Telescope (Faust) flown on ATLAS-1. *ApJ*, 415, 875–881 {**167**}

Bradley PA, 2000, Stellar pulsation. *Encyclopedia of Astronomy and Astrophysics;* (ed. Murdin, P.) {**168**}

Breger M, 1979, δ Scuti and related stars. *PASP*, 91, 5–26 {**175**}

Breger M, Pamyatnykh AA, Pikall H, *et al.*, 1999, The δ Scuti star FG Vir. IV. Mode identifications and pulsation modeling. *A&A*, 341, 151–162 {**179**}

—, 1980, δ Scuti stars and dwarf Cepheids: review and pulsation modes. *Space Science Reviews*, 27, 361–368 {**175**}

—, 2000, δ Scuti stars: a review. *ASP Conf. Ser. 210: Delta Scuti and Related Stars* (eds. Breger M, Montgomery M), 3–42 {**175, 178**}

Brett DR, West RG, Wheatley PJ, 2004, The automated classification of astronomical light curves using Kohonen self-organising maps. *MNRAS*, 353, 369–376 {**172**}

Briquet M, Aerts C, de Cat P, 2001a, Optical variability of the B-type star HD 105382: pulsation or rotation? *A&A*, 366, 121–128 {**188**}

Briquet M, Aerts C, Mathias P, *et al.*, 2003, Spectroscopic mode identification for the slowly-pulsating B star HD 147394. *A&A*, 401, 281–288 {**188**}

Briquet M, de Cat P, Aerts C, *et al.*, 2001b, The B-type variable HD 131120 modeled by rotational modulation. *A&A*, 380, 177–185 {**187**}

Budding E, Rhodes M, Priestley J, *et al.*, 2004, The close binary system SU Ind. *Astronomische Nachrichten*, 325, 433–438 {**197**}

Burki G, 1978, The semi-period-luminosity–colour relation for supergiant stars. *A&A*, 65, 357–362 {**187, 188**}

Burki G, Maeder A, Rufener F, 1978, Variable stars of small amplitude. III. Semi-period of variation for seven B2 to G0 supergiant stars. *A&A*, 65, 363–367 {**169**}

Campos Cucarella F, Gómez Forrellad JM, García Melendo E, 1997, A new β Lyrae variable SAO 56342, and two new possible irregular stars: BD+32°0599 and SAO 56366. *Informational Bulletin on Variable Stars*, 4426 {**197**}

Campos Cucarella F, Guarro Flo J, Gómez Forrellad JM, *et al.*, 1996, Photometric observations of the new bright classical Cepheid HD 32456. *Informational Bulletin on Variable Stars*, 4317 {**174**}

Carrier F, Debernardi Y, Udry S, *et al.*, 2002, Analysis of a δ Scuti spectroscopic binary: CQ Lyn. *A&A*, 390, 1027–1032 {**180, 181**}

Chapellier E, Mathias P, Le Contel JM, *et al.*, 2000, The observational status of the slowly-pulsating B star ι Her. *A&A*, 362, 189–198 {**188**}

Chen P, Yang X, Wang X, 2003, Near-infrared photometry of 20 Hipparcos carbon stars. *Chinese Astronomy and Astrophysics*, 27, 285–291 {**191**}

Cherepashchuk A, Khaliullin K, Kornilov V, *et al.*, 1994, Sternberg WBVR photometric survey of bright stars. *Ap&SS*, 217, 83–85 {**162**}

Ciardi DR, van Belle GT, Akeson RL, *et al.*, 2001, On the near-infrared size of Vega. *ApJ*, 559, 1147–1154 {**166**}

Clarke AJ, Oudmaijer RD, Lumsden SL, 2005, Tycho 2 stars with infrared excess in the MSX Point Source Catalogue. *MNRAS*, 363, 1111–1124 {**165**}

Code AD, Bless RC, Davis J, *et al.*, 1976, Empirical effective temperatures and bolometric corrections for early-type stars. *ApJ*, 203, 417–434 {**166**}

Collins PL, 1999, Modeling visual photometry. I. Preliminary determination of visual bandpass. *Journal of the American Association of Variable Star Observers*, 27, 65–76 {**161**}

Cousins AWJ, 1974, Revised zero points and UBV photometry of stars in the Harvard E and F regions. *MNRAS*, 166, 711–712 {**159**}

—, 1976, VRI standards in the E regions. *Mem. Roy. Astron. Soc.*, 81, 25 {**159**}

—, 2000, Uses of the V magnitudes reobserved with the Hipparcos and Tycho photometers outside the Earth's atmosphere. *Mon. Not. Astron. Soc. S. Africa*, 59, 17 {**159**}

Cox JP, 1976, Nonradial oscillations of stars: theories and observations. *ARA&A*, 14, 247–273 {**194**}

Cox AN, Morgan SM, Rogers FJ, *et al.*, 1992, An opacity mechanism for the pulsations of OB stars. *ApJ*, 393, 272–277 {**185**}

Cramer N, 1997, Absolute magnitude calibration of Geneva photometry: B-type stars. *ESA SP–402: Hipparcos, Venice '97*, 311–314 {**164**}

—, 1999, Calibrations for B-type stars in the Geneva photometric system. *New Astronomy Review*, 43, 343–387 {**164**}

Crawford DL, 1958, Two-dimensional spectral classification by narrow-band photometry for B stars in clusters and associations. *ApJ*, 128, 185–206 {**162**}

—, 1975, Empirical calibration of the uvbyβ systems. I. The F-type stars. *AJ*, 80, 955–971 {**162**}

—, 1978, Empirical calibrations of the uvbyβ systems. II. The B-type stars. *AJ*, 83, 48–63 {**162**}

—, 1979, Empirical calibrations of the uvbyβ systems. III. The A-type stars. *AJ*, 84, 1858–1865 {**162, 163**}

Cunha MS, Aerts C, Christensen-Dalsgaard J, *et al.*, 2007, Asteroseismology and interferometry. *A&A Rev.*, 14, 217–360 {**168**}

Cunha MS, 2001, The sixth frequency of roAp star HR 1217. *MNRAS*, 325, 373–378 {**183**}

Cutispoto G, Messina S, Rodonò M, 2001, Long-term monitoring of active stars. IX. Photometry collected in 1993. *A&A*, 367, 910–930 {**161**}

—, 2003, Long-term monitoring of active stars. X. Photometry collected in 1994. *A&A*, 400, 659–670 {**161**}

Cutispoto G, 1998, Long-term monitoring of active stars. VIII. Photometry collected in February 1992. *A&AS*, 131, 321–344 {**161**}

Dallaporta S, Tomov T, Zwitter T, *et al.*, 2002, GK Dra: a δ Scuti star in a new eclipsing system discovered by Hipparcos. *Informational Bulletin on Variable Stars*, 5312 {**180**}

Decin L, Vandenbussche B, Waelkens C, *et al.*, 2003, ISO-SWS calibration and the accurate modeling of cool-star atmospheres. IV. G9 to M2 stars. *A&A*, 400, 709–727 {**165**}

Derekas A, Kiss LL, Székely P, *et al.*, 2003, A photometric monitoring of bright high-amplitude δ Scuti stars. II. Period updates for seven stars. *A&A*, 402, 733–743 {**181**}

de Cat P, Aerts C, 2002, A study of bright southern slowly-pulsating B stars. II. The intrinsic frequencies. *A&A*, 393, 965–981 {**187**}

de Cat P, Briquet M, Aerts C, *et al.*, 2007, Long-term photometric monitoring with the Mercator telescope. Frequencies and mode identification of variable O-B stars. *A&A*, 463, 243–249 {**187, 189**}

de Cat P, Telting JH, Aerts C, *et al.*, 2000, A detailed spectroscopic analysis of ε Per. I. Determination of the orbital parameters and of the frequencies. *A&A*, 359, 539–551 {**197**}

de Laverny P, Mennessier MO, Mignard F, *et al.*, 1998, Detection of short-term variations in Mira-type variables from Hipparcos photometry. *A&A*, 330, 169–174 {**191, 192**}

de Mey K, Daems K, Sterken C, 1998, θ Tucanae: a binary with a δ Scuti primary. *A&A*, 336, 527–534 {**180**}

de Ridder J, Gordon KD, Mulliss CL, *et al.*, 1999, New observations of 53 Per. *A&A*, 341, 574–578 {**188**} de Ridder J, Telting JH, Balona LA, *et al.*, 2004, Asteroseismology of the β Cephei star ν Eri. III. Extended frequency analysis and mode identification. *MNRAS*, 351, 324–332 {**186**}

Domingo A, Figueras F, 1999, Late A-type stars: new Strömgren photometric calibrations of absolute magnitudes from Hipparcos. *A&A*, 343, 446–454 {**162, 163**}

Dumm T, Schild H, 1998, Stellar radii of M giants. *New Astronomy*, 3, 137–156 {**191, 194**}

Dušek J, Mikulášek Z, Papoušek J, 2003, Preliminary analysis of light curves of seven carbon stars. *Contributions of the Astronomical Observatory Skalnate Pleso*, 33, 119–133 {**192, 194**}

Dziembowski WA, Moskalik P, Pamyatnykh AA, 1993, The opacity mechanism in B-type stars. II. Excitation of high-order g-modes in main sequence stars. *MNRAS*, 265, 588–600 {**185**}

Dziembowski WA, Pamyatnykh AA, 1991, A potential asteroseismological test for convective overshooting theories. *A&A*, 248, L11–L14 {**178**}

—, 1993, The opacity mechanism in B-type stars. I. Unstable modes in β Cephei star models. *MNRAS*, 262, 204–212 {**185**}

Edelson RA, Krolik JH, 1988, The discrete correlation function: a new method for analyzing unevenly sampled variability data. *ApJ*, 333, 646–659 {**169**}

ESA, 1997, *The Hipparcos and Tycho Catalogues. Astrometric and Photometric Star Catalogues derived from the ESA Hipparcos Space Astrometry Mission, ESA SP–1200 (17 volumes including 6 CDs)*. European Space Agency, Noordwijk, also: VizieR Online Data Catalogue {**153, 154, 157, 159–162, 169, 173, 195**}

Escolà-Sirisi E, Juan-Samsó J, Vidal-Sainz J, *et al.*, 2005, Detection of a classical δ Scuti star in the new eclipsing binary system HIP 7666. *A&A*, 434, 1063–1068 {**181**}

Evans NR, Sasselov D, Short CI, 1998, Polaris: what does a little pulsation do to an atmosphere? *ASP Conf. Ser. 154: Cool Stars, Stellar Systems, and the Sun*, 745 {**174**}

Eyer L, 1998, Geneva University, Switzerland. Ph.D. Thesis {**158, 169, 187**}

—, 2005, Variability analysis: detection and classification. *ESA SP-576: The Three-Dimensional Universe with Gaia* (eds. Turon C, O'Flaherty KS, Perryman MAC), 513–519 {**172, 175**}

Eyer L, Aerts C, 2000a, A search for new γ Doradus stars in the Geneva photometric database. *A&A*, 361, 201–206 {**185**}

—, 2000b, New γ Doradus stars from the Hipparcos mission and Geneva photometry. *ASP Conf. Ser. 203: IAU Colloq. 176: The Impact of Large-Scale Surveys on Pulsating Star Research*, 449–450 {**185**}

Eyer L, Bartholdi P, 1999, Variable stars: which Nyquist frequency? *A&AS*, 135, 1–3 {**156, 169**}

Eyer L, Genton MG, 1999, Characterization of variable stars by robust wave variograms: an application to Hipparcos mission. *A&AS*, 136, 421–428 {**169**}

Eyer L, Grenon M, 1997, Photometric variability in the HR diagram. *ESA SP–402: Hipparcos, Venice '97*, 467–472 {**172, 174, 193**}

—, 1998, Results from Hipparcos mission on stellar seismology. *IAU Symp. 185: New Eyes to See Inside the Sun and Stars*, 291–294 {**172, 174**}

—, 2000, Problems encountered in the Hipparcos variable stars analysis. *δ Scuti and Related Stars.* ASP Conf. Ser., Vol. 210 (eds. Breger M, Montgomery M), 482 {**169**}

Eyer L, Mignard F, 2005, Rate of correct detection of periodic signal with the Gaia satellite. *MNRAS*, 361, 1136–1144 {**170**}

Feast MW, 1981, Red variables of spectral class M. *ASSL Vol. 88: Physical Processes in Red Giants* (eds. Iben I, Renzini A), 193–204 {**190**}

Feast MW, Catchpole RM, 1997, The Cepheid period–luminosity zero point from Hipparcos trigonometrical parallaxes. *MNRAS*, 286, L1–L5 {**173, 177**}

Feast MW, Glass IS, Whitelock PA, *et al.*, 1989, A period–luminosity–colour relation for Mira variables. *MNRAS*, 241, 375–392 {**196**}

Fitch WS, 1981, $l = 0, 1, 2, 3$ pulsation constants for evolutionary models of δ Scuti stars. *ApJ*, 249, 218–227 {**178**}

Fitzpatrick EL, Massa D, 1999, Determining the physical properties of the B stars. I. Methodology and first results. *ApJ*, 525, 1011–1023 {**165**}

Fitzpatrick EL, 1999, Correcting for the effects of interstellar extinction. *PASP*, 111, 63–75 {**153**}

Flynn C, Fuchs B, 1994, Density of dark matter in the Galactic disk. *MNRAS*, 270, 471–479 {**164**}

Formiggini L, Brosch N, Almoznino E, *et al.*, 2002, Hidden sub-luminous stars among the FAUST ultraviolet sources towards Ophiuchus. *MNRAS*, 332, 441–455 {**167**}

Foster G, 1996, Wavelets for period analysis of unevenly sampled time series. *AJ*, 112, 1709–1729 {**169**}

Frandsen S, Jones A, Kjeldsen H, *et al.*, 1995, CCD photometry of the δ Scuti star κ^2 Boo. *A&A*, 301, 123–134 {**178, 179**}

Frémat Y, Lampens P, van Cauteren P, *et al.*, 2005, New variable and multiple stars in the lower part of the Cepheid instability strip. *Communications in Asteroseismology*, 146, 6–10 {**174**}

Friedrich S, König M, Wicenec A, 1997, Search for variables in the Tycho Epoch Photometry Annex B. *ESA SP–402: Hipparcos, Venice '97*, 441–444 {**170**}

García Gil A, Allende Prieto C, García López RJ, *et al.*, 2003, Comparisons between observed and computed visible and near-ultraviolet spectra of Vega. *ASP Conf. Ser. 288: Stellar Atmosphere Modeling* (eds. Hubeny I, Mihalas D, Werner K), 145–148 {**166**}

García Gil A, García López RJ, Allende Prieto C, *et al.*, 2005, A study of the near-ultraviolet spectrum of Vega. *ApJ*, 623, 460–471 {**167**}

Gautschy A, Saio H, Harzenmoser H, 1998, How to drive roAp stars. *MNRAS*, 301, 31–41 {**182**}

Gautschy A, Saio H, 1993, On non-radial oscillations of B-type stars. *MNRAS*, 262, 213–219 {**186**}

—, 1995, Stellar pulsations across the HR diagram: I. *ARA&A*, 33, 75–114 {**168**}

—, 1996, Stellar pulsations across the HR diagram: II. *ARA&A*, 34, 551–606 {**168**}

Gerbaldi M, Faraggiana R, Caffau E, 2007, Ultraviolet flux distributions of γ Doradus stars. *A&A*, 472, 241–246 {**185**}

Glass IS, van Leeuwen F, 2007, Semi-regular variables in the solar neighbourhood. *MNRAS*, 378, 1543–1549 {**193**}

Glatzel W, 1994, On the origin of strange modes and the mechanism of related instabilities. *MNRAS*, 271, 66–74 {**186**}

Gómez Forrellad JM, García Melendo E, 1997, Photometric results on three Hipparcos variables: the new eclipsing binary systems HD 125488 and HD 126080, and the star HD 341508. *Informational Bulletin on Variable Stars*, 4469 {**197**}

Gómez Forrellad JM, 2003, The period of V2109 Cyg revisited. *Informational Bulletin on Variable Stars*, 5377 {**181**}

Gómez AE, Luri X, Grenier S, *et al.*, 1998, The HR diagram from Hipparcos data. Absolute magnitudes and kinematics of Bp–Ap stars. *A&A*, 336, 953–959 {**182**}

Gray RO, 2007, The problems with Vega. *The Future of Photometric, Spectrophotometric and Polarimetric Standardization* (ed. Sterken C), ASP Conf. Ser. 364, 305–314 {**167**}

Grenon M, 1997, The status of stellar variability from Hipparcos photometry. *The First Results of Hipparcos and Tycho, IAU Joint Discussion 14*, 3 {**169**}

—, 1998, The stellar variability from Hipparcos photometry. *Highlights in Astronomy*, 11, 542 {**169**}

Guthrie BNG, 1987, Abundances of calcium for a sample of 57 classical Am stars. *MNRAS*, 226, 361–371 {**162, 163**}

Handler G, 1999, The domain of γ Doradus variables in the HR diagram. *MNRAS*, 309, L19–L23 {**176, 184**}

—, 2002, New bright candidate δ Scuti stars from the Hipparcos unsolved variables. *ASP Conf. Ser. 256: Observational Aspects of Pulsating B- and A Stars*, 113 {**176, 177**}

Handler G, Krisciunas K, 1997, An updated list of γ Doradus stars. *Delta Scuti Star Newsletter*, 11, 3 {**184**}

Handler G, Paunzen E, 1999, A search for rapid oscillations in chemically peculiar A-type stars. *A&AS*, 135, 57–63 {**182, 184**}

Harmanec P, 1998, A reliable transformation of Hipparcos *Hp* magnitudes into Johnson *V* and *B* magnitudes. *A&A*, 335, 173–178 {**160**}

Hayes DS, 1985, Stellar absolute fluxes and energy distributions from 0.32–4.0 μm. *Calibration of Fundamental Stellar Quantities* (eds. Hayes DS, Pasinetti LE, Philip AGD), volume 111 of *IAU Symposium*, 225–249 {**166**}

Hilditch RW, Hill G, Barnes JV, 1983, Studies of A and F stars in the region of the North Galactic Pole. V. Interstellar reddening and the uvbyβ intrinsic colour calibration. *MNRAS*, 204, 241–247 {**162**}

Hill SJ, Willson LA, 1979, Theoretical velocity structure of long-period variable star photospheres. *ApJ*, 229, 1029–1045 {**190**}

Hintz EG, Brown PJ, 2007, Revised periods for QS Gem and V367 Gem. *PASP*, 119, 274–283 {**181**}

Hintz EG, Rose MB, Bush TC, *et al.*, 2006, Establishing observational baselines for two δ Scuti variables: V966 Her and V1438 Aql. *AJ*, 132, 393–400 {**181**}

Hoffleit D, 1999, Hipparcos versus GCVS amplitudes of non-Mira M-type giants. *Journal of the American Association of Variable Star Observers*, 27, 131–140 {**191**}

—, 2002, New Hipparcos variables in the Bright Star Catalogue. *Journal of the American Association of Variable Star Observers*, 30, 139–144 {**167**}

Høg E, Flynn C, 1998, Hipparcos absolute magnitudes for metal-rich K giants and the calibration of DDO photometry. *MNRAS*, 294, 28 {**164**}

Høg E, Petersen JO, 1997, Hipparcos parallaxes and the nature of δ Scuti stars. *A&A*, 323, 827–830 {**175–177**}

Hubrig S, Kharchenko NV, Mathys G, *et al.*, 2000, Rapidly-oscillating Ap stars versus non-oscillating Ap stars. *A&A*, 355, 1031–1040 {**182, 184**}

Janes KA, 1979, Evidence for an abundance gradient in the Galactic disk. *ApJS*, 39, 135–156 {**164**}

Jankov S, Mathias P, Chapellier E, *et al.*, 2006, Non-radial pulsations in the γ Doradus star HD 195068. *A&A*, 453, 1041–1050 {**185**}

Jerzykiewicz M, 2000, Two aspects of using Hipparcos data for studying multi-periodic stellar pulsations. *ASP Conf. Ser. 203: IAU Colloq. 176: The Impact of Large-Scale Surveys on Pulsating Star Research*, 46–49 {**180**}

Jerzykiewicz M, Pamyatnykh AA, 2000, The δ Scuti star DK Vir revisited. *PASP*, 112, 1341–1349 {**180**}

Jerzykiewicz M, Pigulski A, 1999, The long-term behaviour of the pulsation periods and amplitudes of the β Cephei star EN Lac. *MNRAS*, 310, 804–810 {**185**}

Johnson HL, 1955, A photometric system. *Annales d'Astrophysique*, 18, 292–316 {**161**}

Jonker PG, Fender RP, Hambly NC, *et al.*, 2000, The infrared counterpart of the Z source GX 5-1. *MNRAS*, 315, L57–L60 {**165**}

Jordi C, Figueras F, Torra J, *et al.*, 1996, uvbyβ photometry of main sequence A-type stars. *A&AS*, 115, 401–406 {**163**}

Jordi C, Luri X, Masana E, *et al.*, 2002, The luminosity calibration of uvbyβ photometry. *Highlights in Astronomy*, 12, 684–687 {**163**}

Jorissen A, Mowlavi N, Sterken C, *et al.*, 1997, The onset of photometric variability in red giant stars. *A&A*, 324, 578–586 {**193**}

Kallinger T, Weiss WW, 2002, Detecting low-amplitude periodicities with Hipparcos. *A&A*, 385, 533–536 {**160**}

Kaye AB, Handler G, Krisciunas K, *et al.*, 1999, γ Doradus stars: defining a new class of pulsating variables. *PASP*, 111, 840–844 {**184**}

Kazarovets EV, Samus NN, Durlevich OV, *et al.*, 1999, The 74th special name-list of variable stars. *Informational Bulletin on Variable Stars*, 4659 {**167**}

—, 2006, The 78th name-list of variable stars. *Informational Bulletin on Variable Stars*, 5721 {**167**}

Kazlauskas A, Straižys V, Boyle RP, *et al.*, 2003, Calibration of the Strömvil photometric system. *Baltic Astronomy*, 12, 491–496 {**164**}

Kerschbaum F, Lebzelter T, Lazaro C, 2001, Multi-colour light variation of asymptotic giant branch stars observed with ISO. *A&A*, 375, 527–538 {**161**}

Kholopov PN, 1982, *Catalogue of Suspected Variable Stars*. Nauka Publishing House, Moscow {**170**}

—, 1985, *General Catalogue of Variable Stars*, Fourth Edition. Nauka Publishing House, Moscow {**170**}

Kienzle F, Moskalik P, Bersier D, *et al.*, 1999, Structural properties of s-Cepheid velocity curves constraining the location of the $\omega_4 = 2\omega_1$ resonance. *A&A*, 341, 818–826 {**174**}

Kilkenny D, Koen C, van Wyk F, *et al.*, 2007, Further observations of Hipparcos red stars and standards for $UBV(RI)_C$ photometry. *MNRAS*, 380, 1261–1270 {**195**}

Kilkenny D, van Wyk F, Roberts G, *et al.*, 1998, Supplementary southern standards for $UBV(RI)_C$ photometry. *MNRAS*, 294, 93–104 {**159**}

Kiss LL, Csák B, Alfaro EJ, *et al.*, 2000, Ground-based observations of short- and medium-period variables discovered by the Hipparcos satellite. *ASP Conf. Ser. 203: IAU Colloq. 176: The Impact of Large-Scale Surveys on Pulsating Star Research*, 277 {**197**}

Kiss LL, Csák B, Thomson JR, *et al.*, 1999a, DX Cet, a high-amplitude δ Scuti star. *Informational Bulletin on Variable Stars*, 4660 {**180**}

—, 1999b, V2109 Cyg, a second overtone field RR Lyrae star. *A&A*, 345, 149–155 {**181**}

Kiss LL, Derekas A, Alfaro EJ, *et al.*, 2002, The multimode pulsation of the δ Scuti star V784 Cas. *A&A*, 394, 97–106 {**180**}

Kiss LL, Szatmáry K, Cadmus RR, *et al.*, 1999c, Multi-periodicity in semi-regular variables. I. General properties. *A&A*, 346, 542–555 {**194**}

Kjeldsen H, Bedding TR, 2001, Current status of asteroseismology. *ESA SP–464: Helio- and Asteroseismology at the Dawn of the Millennium* (eds. Wilson A, Pallé PL), 361–366 {**179**}

Knapp GR, Pourbaix D, Platais I, *et al.*, 2003, Reprocessing the Hipparcos data of evolved stars. III. Revised Hipparcos period-luminosity relationship for Galactic long-period variable stars. *A&A*, 403, 993–1002 {**196, 197**}

Knauer TG, Ivezić Ž, Knapp GR, 2001, Analysis of stars common to the IRAS and Hipparcos surveys. *ApJ*, 552, 787–792 {**165**}

Knude J, Fabricius C, 2003, Luminosity and intrinsic colour calibration of main-sequence stars with 2MASS photometry: all sky local extinction. *Baltic Astronomy*, 12, 508–513 {**165**}

Kochukhov O, Drake NA, Piskunov N, *et al.*, 2004, Multi-element abundance Doppler imaging of the rapidly-oscillating Ap star HR 3831. *A&A*, 424, 935–950 {**184**}

Koen C, 2001, Multi-periodicities from the Hipparcos epoch photometry and possible pulsation in early A-type stars. *MNRAS*, 321, 44–56 {**161**}

Koen C, Eyer L, 2002, New periodic variables from the Hipparcos epoch photometry. *MNRAS*, 331, 45–59 {**170, 171**}

Koen C, Kilkenny D, van Wyk F, *et al.*, 2002a, UBV(RI)$_C$ photometry of Hipparcos red stars. *MNRAS*, 334, 20–38 {**195**}

Koen C, Kurtz DW, Gray RO, *et al.*, 2001, UBVRIJH photometry of two new luminous δ Scuti stars and the discovery of δ Scuti pulsation in the most evolved Ap star known. *MNRAS*, 326, 387–396 {**178, 180**}

Koen C, Laney D, van Wyk F, 2002b, Observations of Hipparcos short-period red giant stars. *MNRAS*, 335, 223–232 {**194**}

Koen C, Laney D, 2000a, Rapidly-oscillating M giant stars? *MNRAS*, 311, 636–648 {**161, 171, 194**}

—, 2000b, Short-period M giant stars in the Hipparcos Catalogue. *ASP Conf. Ser. 203: IAU Colloq. 176: The Impact of Large-Scale Surveys on Pulsating Star Research*, 101–104 {**194**}

Koen C, Schumann R, 1999, Sinusoidal variables from the Tycho Epoch Photometry Annex. *MNRAS*, 310, 618–628 {**171**}

Koen C, van Rooyen R, van Wyk F, *et al.*, 1999, Eight new δ Scuti stars. *MNRAS*, 309, 1051–1062 {**180**}

Kohonen T, 1982, Self-organised formation of topologically correct feature maps. *Biological Cybernetics*, 43, 59–69 {**172**}

Kornilov V, Mironov AV, Zakharov AI, 1996, The WBVR photometry of bright northern stars. *Baltic Astronomy*, 5, 379–390 {**162**}

Krisciunas K, 1998, The discovery of non-radial gravity-mode pulsations in γ Doradus-type stars. *IAU Symp. 185: New Eyes to See Inside the Sun and Stars* (eds. Deubner FL, Christensen-Dalsgaard J, Kurtz DW), 339–346 {**184**}

Kroll P, Vogt N, Braeuer HJ, *et al.*, 2000, New phenomena and statistics of stellar long-term variability. *Astronomische Gesellschaft Meeting Abstracts*, 2 {**160**}

Kurtz DW, 1982, Rapidly oscillating Ap stars. *MNRAS*, 200, 807–859 {**182**}

—, 1990, Rapidly oscillating Ap stars. *ARA&A*, 28, 607–655 {**182**}

—, 1998, Some recent results for the roAp stars. *Contributions of the Astronomical Observatory Skalnate Pleso*, 27, 264–271 {**183**}

—, 2000, Pulsating and chemically peculiar upper main sequence stars. *Encyclopedia of Astronomy and Astrophysics;* (ed. Murdin, P.) {**168**}

—, 2002, HD 101065: Przybylski's star: a most peculiar star. *ASP Conf. Ser. 279: Exotic Stars as Challenges to Evolution* (eds. Tout CA, van Hamme W), 351–364 {**183**}

Kurtz DW, Cameron C, Cunha MS, *et al.*, 2005, Pushing the ground-based limit: 14 μmag photometric precision with the definitive Whole Earth Telescope asteroseismic dataset for the rapidly-oscillating Ap star HR 1217. *MNRAS*, 358, 651–664 {**183**}

Kurtz DW, Kawaler SD, Riddle RL, *et al.*, 2002, Discovery of the 'missing' mode in HR1217 by the Whole Earth Telescope. *MNRAS*, 330, L57–L61 {**183**}

—, 2003, High precision with the Whole Earth Telescope: lessons and some results from XCov20 for the roAp Star HR 1217. *Baltic Astronomy*, 12, 105–117 {**183**}

Kurtz DW, Martinez P, 2000, Observing roAp stars with WET: a primer. *Baltic Astronomy*, 9, 253–353 {**182**}

Kurucz RL, 1992, Finding the missing solar ultraviolet opacity. *Revista Mexicana de Astronomia y Astrofisica*, 23, 181–186 {**165, 166**}

—, 1993, ATLAS9 stellar atmosphere programs and 2 km s^{-1} grid. *CD No. 13, Smithsonian Astrophysical Observatory* {**166**}

Lamers HJGLM, Bastiaanse MV, Aerts C, *et al.*, 1998, Periods, period changes and the nature of the micro-variations of luminous blue variables. *A&A*, 335, 605–621 {**188**}

Lampens P, van Camp M, Sinachopoulos D, 2000, δ Scuti stars in stellar systems: On the variability of HD 220392 and HD 220391. *A&A*, 356, 895–902 {**180**}

Landolt AU, 1973, UBV photoelectric sequences in the celestial equatorial Selected Areas 92–115. *AJ*, 78, 959–1020 {**159**}

—, 1983, UBVRI photometric standard stars around the celestial equator. *AJ*, 88, 439–460 {**159, 161, 195**}

—, 2007, UBVRI photometric standard stars around the sky at $-50°$ declination. *AJ*, 133, 2502–2523 {**159**}

Laney CD, Koen C, 2000, UBVRIJHK photometry of short-period red variables identified by the Hipparcos survey. *ASP Conf. Ser. 203: IAU Colloq. 176: The Impact of Large-Scale Surveys on Pulsating Star Research*, 133–134 {**194**}

Lang JD, 1989, *SAAO Circular*. 29 {**159**}

Lazauskaite R, Bartkevičius A, Bartašiute S, 2003, Classification of metal-deficient dwarfs in the Vilnius photometric system. *Baltic Astronomy*, 12, 547–554 {**164**}

Lebzelter T, Hinkle KH, Wood PR, *et al.*, 2005, A study of bright southern long-period variables. *A&A*, 431, 623–634 {**192**}

Lebzelter T, Posch T, 2001, Eight new small amplitude variables. *Informational Bulletin on Variable Stars*, 5089 {**193**}

Lefever K, Puls J, Aerts C, 2006, Study of a sample of periodically variable B-type supergiants. *Memorie della Societa Astronomica Italiana*, 77, 135–138 {**186**}

—, 2007, Statistical properties of a sample of periodically variable B-type supergiants. Evidence for opacity-driven gravity-mode oscillations. *A&A*, 463, 1093–1109 {**186**}

Lehmann H, Scholz G, Hildebrandt G, *et al.*, 1995, Variability investigations of possible Maia stars. *A&A*, 300, 783–790 {**169**}

—, 1999, ET And, HD 219891, or HD 219668: which one shows short-term variability? *A&A*, 351, 267–272 {**189**}

Liu YY, Baglin A, Auvergne M, *et al.*, 1997, The lower part of the instability strip: evolutionary status of δ Scuti stars. *ESA SP–402: Hipparcos, Venice '97*, 363–366 {**175**}

Lloyd C, Pickard RD, Chambers RH, 2002, The periods of the semi-regular variable V370 And. *Informational Bulletin on Variable Stars*, 5217 {**193**}

Lloyd C, Stickland DJ, 2001, The evolutionary status of the early-type binary HD 115071. *A&A*, 370, 1026–1029 {**197**}

Lloyd C, West K, 2001, Photoelectric observations of the complex low-amplitude red variable, UX Dra. *Informational Bulletin on Variable Stars*, 5202 {**192**}

Lockwood GW, Skiff BA, Thompson DT, 1993, Lessons from very long term, very high precision, photoelectric photometry. *IAU Colloq. 136: Stellar Photometry: Current Techniques and Future Developments* (eds. Butler CJ, Elliott I), 99–105 {**160**}

Lomb NR, 1976, Least-squares frequency analysis of unequally spaced data. *Ap&SS*, 39, 447–462 {**167**}

López de Coca P, Olivares I, Rolland A, *et al.*, 2006, Multiperiodicity in the δ Scuti-type star V929 Her. *Memorie della Societa Astronomica Italiana*, 77, 525–528 {**181**}

López M, Bielza C, Sarro LM, 2006, Bayesian classifiers for variable stars. *Astronomical Data Analysis Software and Systems XV* (eds. Gabriel C, Arviset C, Ponz D, *et al.*), ASP Conf. Ser. 351, 161–164 {**172**}

Lub J, Pel JW, 1977, Properties of the Walraven VBLUW photometric system. *A&A*, 54, 137–158 {**161**}

Luri X, Mennessier MO, Torra J, *et al.*, 1996, A new maximum likelihood method for luminosity calibrations. *A&AS*, 117, 405–415 {**191**}

Mais DE, Richards D, Stencel RE, 2006, Three years of Mira photometry: what has been learned? *Society for Astronomical Sciences Annual Symposium*, 25, 31 {**191**}

Maitzen HM, Paunzen E, Vogt N, *et al.*, 2000, Hβ photometry of southern chemically peculiar stars: is the uvbyβ luminosity calibration also valid for peculiar stars? *A&A*, 355, 1003–1008 {**163**}

Maíz Apellániz J, 2005, A cross-calibration between Tycho 2 photometry and Hubble Space Telescope spectrophotometry. *PASP*, 117, 615–619 {**159**}

—, 2006, A recalibration of optical photometry: Tycho 2, Strömgren, and Johnson systems. *AJ*, 131, 1184–1199 {**159**}

Makarov VV, Bastian U, Høg E, *et al.*, 1994, 35 new bright medium- and high-amplitude variables discovered by the Tycho instrument of the Hipparcos satellite. *Informational Bulletin on Variable Stars*, 4118 {**170**}

Malagnini ML, Morossi C, Buser R, *et al.*, 1992, Cool star flux spectra for population studies in galaxies. *A&A*, 261, 558–564 {**165**}

Malyuto V, Straižys V, Kazlauskas A, 1997, The Hipparcos calibration of absolute magnitudes in the Vilnius photometric system. *ESA SP–402: Hipparcos, Venice '97*, 291–294 {**163, 164**}

Marconi M, Palla F, 1998, The instability strip for pre-main sequence stars. *ApJ*, 507, L141–L144 {**181**}

Marinova MM, Percy JR, 1999, AAVSO photoelectric photometry and Hipparcos photometry of the pulsating red giant AG Cet. *Journal of the American Association of Variable Star Observers*, 27, 122–130 {**194**}

Martinez P, Kurtz DW, 1995, Observations of pulsating Ap stars in South Africa. *Ap&SS*, 230, 29–39 {**182**}

Martín-Luis F, Kidger M, Cohen M, 2001, Selection of a sample of suitable potential mid-infrared calibration stars from the Hipparcos/Tycho Catalogue. *The Calibration Legacy of the ISO Mission* {**165**}

Masuda S, Hirata R, 2000, Line-profile variation in τ Her. *A&A*, 356, 209–212 {**188**}

Mathew A, Rajamohan R, 1992, Effects of rotation on the colours and line indices of stars. V. The zero-rotation main sequence and the zero-rotation ZAMS. *Journal of Astrophysics and Astronomy*, 13, 61–107 {**163**}

Mathias P, Aerts C, Briquet M, *et al.*, 2001, Spectroscopic monitoring of 10 new northern slowly-pulsating B star candidates discovered from the Hipparcos mission. *A&A*, 379, 905–916 {**187**}

Mathias P, Le Contel JM, Chapellier E, *et al.*, 2004, Multi-site, multi-technique survey of γ Doradus candidates. I. Spectroscopic results for 59 stars. *A&A*, 417, 189–199 {**185**}

Mattei JA, Foster G, 1997a, Amateurs and the Hipparcos mission. *S&T*, 94, 30 {**172**}

—, 1997b, Studies of long-period variables. *Bulletin of the American Astronomical Society*, 29, 1284 {**172, 190**}

Mattei JA, Hanson GA, 2000, Partnership in variable star research from ground and space. *American Astronomical Society Meeting*, 196 {**172**}

Matthews JM, 2000, Asteroseismology versus astrometry testing p-mode predictions for roAp stars against Hipparcos. *NATO ASIC Proc. 544: Variable Stars as Essential Astrophysical Tools*, 387 {**182**}

Matthews JM, Kurtz DW, Martínez P, 1999, Parallaxes versus p-modes: comparing Hipparcos and asteroseismic results for pulsating Ap stars. *ApJ*, 511, 422–428 {**182, 183**}

McGough C, Samus NN, van Wyk F, 2001, CCD photometry of Tycho suspected variable stars. *American Astronomical Society Meeting*, 199 {**171**}

McNamara BJ, 1987, The Maia stars: a real class of variable stars. *Lecture Notes in Physics Vol. 274: Stellar Pulsation* (eds. Cox AN, Sparks WM, Starrfield SG), 92–95 {**189**}

McNamara DH, 1997a, Luminosities of SX Phoenicis, large-amplitude δ Scuti, and RR Lyrae stars. *PASP*, 109, 1221–1232 {**177, 178**}

—, 1997b, The absolute magnitudes of the RR Lyrae stars. *PASP*, 109, 857–867 {**177**}

Mégessier C, 1998, The visual and infrared flux calibrations. *IAU Symp. 189: Fundamental Stellar Properties*, 61 {**167**}

Mennessier MO, Alvarez R, Luri X, *et al.*, 1999, Physics and evolution of long-period variables from Hipparcos kinematics. *IAU Symp. 191: Asymptotic Giant Branch Stars*, 117 {**190, 191**}

Mennessier MO, Barthès D, Vidal JL, *et al.*, 1998, Pulsations of long-period variables and Hipparcos data. *ASP Conf. Ser. 135: A Half Century of Stellar Pulsation Interpretation* (eds. Bradley PA, Guzik JA), 395 {**191**}

Mennessier MO, Luri X, 2001, Stellar and circumstellar evolution of long-period variable stars. *A&A*, 380, 198–211 {**191**}

Mennessier MO, Mattei JA, Luri X, 1997, New aspect of long-period variable stars from Hipparcos: first results. *ESA SP–402: Hipparcos, Venice '97*, 275–278 {**191**}

Mennessier MO, Mowlavi N, Alvarez R, *et al.*, 2001, Long-period variable stars: Galactic populations and infrared luminosity calibrations. *A&A*, 374, 968–979 {**191**}

Menzies J, Cousins AWJ, Banfield RM, *et al.*, 1989, $UBV(RI)_C$ standard stars in the E- and F-regions and in the Magellanic Clouds: a revised catalogue. *South African Astronomical Observatory Circular*, 13, 1–13 {**159, 161**}

Mironov AV, 2003, The development of stellar photometry in Russia and the USSR in the twentieth century. *Baltic Astronomy*, 12, 503–507 {**162**}

Mironov AV, Zakharov AI, Nikolaev FN, 2003, On a new technique for discovering variable stars. *Baltic Astronomy*, 12, 589–594 {**170**}

Mironov AV, Zakharov AI, 2002, Systematic errors of high-precision photometric catalogues. *Ap&SS*, 280, 71–76 {**162**}

Miroshnichenko AS, Mulliss CL, Bjorkman KS, *et al.*, 1999, Six intermediate-mass stars with far-infrared excess: a search for evolutionary connections. *MNRAS*, 302, 612–624 {**165**}

Mkrtichian DE, Kusakin AV, Gamarova AY, *et al.*, 2002, Pulsating components of eclipsing binaries: new asteroseismic methods of studies and prospects. *ASP Conf. Ser. 259: IAU Colloq. 185: Radial and Non-Radial Pulsationsn as Probes of Stellar Physics* (eds. Aerts C, Bedding TR, Christensen-Dalsgaard J), 96–99 {**181**}

Molenda-Zakowicz J, 2000a, Multi-periodicity of slowly-pulsating B stars from Hipparcos photometry. *Hipparcos and the Luminosity Calibration of the Nearer Stars, IAU Joint Discussion 13*, 37 {**187**}

—, 2000b, Periods of β Cephei and slowly-pulsating B stars from Hipparcos photometry. *ASP Conf. Ser. 203: IAU Colloq. 176: The Impact of Large-Scale Surveys on Pulsating Star Research*, 434–435 {**185, 186**}

—, 2004, Short-period variables in ASAS photometry. *Communications in Asteroseismology*, 145, 50–52 {**187**}

Molenda-Zakowicz J, Polubek G, 2004, Empirical absolute magnitudes, luminosities and effective temperatures of slowly-pulsating B variables and the problem of variability classification of monoperiodic stars. *Acta Astronomica*, 54, 281–297 {**187**}

—, 2005, New β Cephei and slowly-pulsating B stars discovered in Hipparcos photometry. *Acta Astronomica*, 55, 375–388 {**186, 187**}

Morossi C, Franchini M, Malagnini ML, *et al.*, 1993, Cool stars: spectral energy distributions and model atmosphere fluxes. *A&A*, 277, 173–183 {**165**}

Moskalik P, Ogłoza W, 2000, The pulsation mode of Polaris. *ASP Conf. Ser. 203: IAU Colloq. 176: The Impact of Large-Scale Surveys on Pulsating Star Research*, 237–238 {**173**}

Niemczura E, Daszyńska J, Cugier H, 2003, The mean stellar parameters from IUE and Hipparcos data. *Advances in Space Research*, 31, 399–404 {**166**}

Nissen PE, Schuster WJ, 1991, uvbyβ photometry of high-velocity and metal-poor stars. V. Distances, kinematics and ages of halo and disk stars. *A&A*, 251, 457–468 {**162**}

Oja T, 1996, UBVRI standard stars at northern declinations. *Baltic Astronomy*, 5, 103–116 {**161**}

Oja T, Evans DW, 1998, On the systematic accuracy of the equatorial UBV standards. *A&A*, 333, 673–677 {**161, 162, 164**}

Ostrowski TA, Stencel RE, 1998, CCD photometry of Hipparcos variable stars. *Bulletin of the American Astronomical Society*, 30, 1321 {**172**}

—, 1999, Photometry of Hipparcos variable stars. *Journal of the American Association of Variable Star Observers*, 27, 37–40 {**191**}

Pamyatnykh AA, 1998, Pulsation instability domains in the upper main sequence. *ASP Conf. Ser. 135: A Half Century of Stellar Pulsation Interpretation* (eds. Bradley PA, Guzik JA), 268–269 {**186–188**}

Pamyatnykh AA, Handler G, Dziembowski WA, 2004, Asteroseismology of the β Cephei star ν Eri: interpretation and applications of the oscillation spectrum. *MNRAS*, 350, 1022–1028 {**186**}

Panov KP, 1978, *Publ. Astron. Inst. Czechoslov. Acad. Sciences*. 54, 19 {**189**}

Paunzen E, Strassmeier KG, Weiss WW, 1998, HD 84800: A new δ Scuti variable. *Informational Bulletin on Variable Stars*, 4566 {**179**}

Pel JW, 1991, A comparison of the systematic accuracy in four photometric systems. *Precision Photometry: Astrophysics of the Galaxy* (eds. Philip AGD, Upgren AR, Janes KA), 165–169 {**161, 164**}

Penny LR, Gies DR, Wise JH, *et al.*, 2002, Tomographic separation of composite spectra. IX. The massive close binary HD 115071. *ApJ*, 575, 1050–1056 {**197**}

Percy JR, Au-Yong K, Gilmour-Taylor G, *et al.*, 2002, Autocorrelation analysis of Hipparcos photometry of short-period A and B stars. *ASP Conf. Ser. 256: Observational Aspects of Pulsating B- and A Stars*, 99 {**170**}

Percy JR, Bakos AG, 2003, Pulsation modes in small amplitude red giants. *The Garrison Festschrift* (eds. Gray RO, Corbally CJ, Philip AGD), 49–56 {**193**}

Percy JR, Desjardins A, Yu L, *et al.*, 1996, Small-amplitude red variables in the AAVSO photoelectric programme: light-curves and periods. *PASP*, 108, 139–145 {**190**}

Percy JR, Fleming DEB, 1992, A photometric survey of suspected small-amplitude red variables. *PASP*, 104, 96–100 {**193**}

Percy JR, Gilmour-Taylor G, 2002, Fishing for δ Scuti stars in the Hipparcos photometric database. *Communications in Asteroseismology*, vol. 142, p. 48-49, 142, 48–49 {**176**}

Percy JR, Guler M, 1999, The onset of pulsation in red giants. *Journal of the American Association of Variable Star Observers*, 27, 1–4 {**193**}

Percy JR, Hosick J, 2002, Do red giant stars pulsate in high overtones? *MNRAS*, 334, 669–672 {**194**}

Percy JR, Hoss JX, 2000, Period changes in Population II Cepheids: TX Del and W Vir. *Journal of the American Association of Variable Star Observers*, 29, 14–18 {**174**}

Percy JR, Kolin DL, 2000, Studies of yellow semi-regular variables. *Journal of the American Association of Variable Star Observers*, 28, 1–9 {**193**}

Percy JR, Parkes M, 1998, Pulsation modes in small-amplitude red variable stars. *PASP*, 110, 1431–1433 {**193, 194**}

Percy JR, Polano S, 1998, Pulsation modes in M giants. *ASP Conf. Ser. 135: A Half Century of Stellar Pulsation Interpretation* (eds. Bradley PA, Guzik JA), 249–253 {**193**}

Percy JR, Wilson JB, Henry GW, 2001, Long-term VRI photometry of small-amplitude red variables. I. Light-curves and periods. *PASP*, 113, 983–996 {**193, 194**}

Percy JR, Wilson JB, 2000, Another search for Maia variable stars. *PASP*, 112, 846–851 {**189**}

Perryman MAC, Turon C, Lindegren L, *et al.*, 1989, *The Hipparcos Mission: Pre-Launch Status, ESA SP–1111 (3 volumes)*. European Space Agency, Noordwijk {**171**}

Pesnell WD, 1990, Non-radial, non-adiabatic stellar pulsations. *ApJ*, 363, 227–233 {**178**}

Petersen JO, 1999, δ Scuti variables as precise distance indicators. *ASP Conf. Ser. 167: Harmonizing Cosmic Distance Scales in a Post-Hipparcos Era*, 107–112 {**177**}

Petersen JO, Christensen-Dalsgaard J, 1996, Pulsation models of δ Scuti variables. I. The high-amplitude double-mode stars. *A&A*, 312, 463–474 {**175, 176**}

—, 1999, Pulsation models of δ Scuti variables. II. δ Scuti stars as precise distance indicators. *A&A*, 352, 547–554 {**177, 179**}

Petersen JO, Høg E, 1998, Hipparcos parallaxes and period–luminosity relations of high-amplitude δ Scuti stars. *A&A*, 331, 989–994 {**177, 178**}

Pijpers FP, 2000, Pulsation in giants with known radius. *The Third MONS Workshop: Science Preparation and Target Selection*, 177 {**191**}

Pikhun AI, Kudashkina LS, Brukhanov IS, 2001, The investigation of semi-regular variable UV Boo. *Odessa Astronomical Publications*, 14, 162–163 {**193**}

Piquard S, Halbwachs JL, Fabricius C, *et al.*, 2000, A search of variable stars in the Tycho observations. *ASP Conf. Ser. 203: IAU Colloq. 176: The Impact of Large-Scale Surveys on Pulsating Star Research*, 62–63 {**171**}

—, 2001, Variable stars in the Tycho photometric observations. I. Detection. *A&A*, 373, 576–588 {**171**}

Piquard S, 2001, Tycho Variable Stars. Ph.D. Thesis, Strasbourg {**171**}

Platais I, Pourbaix D, Jorissen A, *et al.*, 2003a, Hipparcos red stars in the Hp V_T and VI$_C$ systems. *A&A*, 397, 997–1010 {**195, 196**}

—, 2003b, Red variables in the Hipparcos Hp V_T system: lessons learned. *ASP Conf. Ser. 292: Interplay of Periodic, Cyclic and Stochastic Variability in Selected Areas of the HR Diagram* (ed. Sterken C), 107 {**195**}

Poretti E, 2001, Fourier decomposition and frequency analysis of the pulsating stars with $P < 1$ day in the OGLE database. I. Monoperiodic δ Scuti, RRc and RRab variables. Separation criteria and particularities. *A&A*, 371, 986–996 {**181**}

Price SD, Egan MP, Carey SJ, *et al.*, 2001, Midcourse Space Experiment survey of the Galactic plane. *AJ*, 121, 2819–2842 {**165**}

Reyniers M, Cuypers J, 2005, The evolutionary status of the bright high-latitude supergiant HD 190390. *A&A*, 432, 595–608 {**174**}

Richwine P, Bedient J, Slater T, *et al.*, 2005, The internet as a virtual observatory: new elements for ten long-period variables in Aquila. *Journal of the American Association of Variable Star Observers (JAAVSO)*, 21 {**193**}

Ripepi V, Marconi M, Bernabei S, *et al.*, 2003, Multisite observations of the pre-main sequence δ Scuti star V351 Ori. *A&A*, 408, 1047–1055 {**181**}

Rodríguez E, Arellano Ferro A, Costa V, *et al.*, 2003a, δ Scuti-type nature of the variable V2109 Cyg. *A&A*, 407, 1059–1065 {**181**}

—, 2003b, Long-period δ Scuti type variables: the case of V2109 Cyg. *ASP Conf. Ser. 292: Interplay of Periodic, Cyclic and Stochastic Variability in Selected Areas of the HR Diagram* (ed. Sterken C), 125 {**181**}

Rodríguez E, Breger M, 2001, δ Scuti and related stars: analysis of the R00 Catalogue. *A&A*, 366, 178–196 {**176, 177**}

Rodríguez E, López González MJ, López de Coca P, 2000, A revised catalogue of δ Scuti stars. *A&AS*, 144, 469–474 {**176, 177**}

Saar SH, 1998, PZ Mon: an active evolved star. *Informational Bulletin on Variable Stars*, 4580 {**197**}

Samus NN, 1997, Hipparcos stars in the GCVS. *The First Results of Hipparcos and Tycho, IAU Joint Discussion 14*, 23 {**167**}

—, 2006, Astronomical databases, space photometry, and time series analysis: an overview. *Astrophysics of Variable Stars* (eds. Sterken C, Aerts C), ASP Conf. Ser. 349, 3–7 {**169**}

Santangelo MMM, Cavalletti G, Benedetti W, 2007, UBV photoelectric photometry of the 1H/2H double mode Cepheid CO Aur. *Astronomische Nachrichten*, 328, 55–62 {**174**}

Scargle JD, 1982, Studies in astronomical time series analysis. II. Statistical aspects of spectral analysis of unevenly spaced data. *ApJ*, 263, 835–853 {**167**}

Schaller G, Schaerer D, Meynet G, *et al.*, 1992, New grids of stellar models. I. From $0.8-120\,M_\odot$ at $Z = 0.020, 0.001$. *A&AS*, 96, 269–331 {**188**}

Schatzman EL, Praderie F, 1993, *The Stars*. Springer–Verlag, Berlin {**170, 172**}

Schoenaers C, Cuypers J, 2004, Direct detection of multiple periods in variable stars. *ASP Conf. Ser. 310: IAU Colloq. 193: Variable Stars in the Local Group* (eds. Kurtz DW, Pollard KR), 283 {**187**}

Scholz G, Lehmann H, Hildebrandt G, *et al.*, 1998, Spectroscopic and photometric investigations of Maia candidate stars. *A&A*, 337, 447–459 {**189, 190**}

Seitzer P, Hill J, 1998, An infrared colour–magnitude diagram of nearby stars. *Bulletin of the American Astronomical Society*, 30, 1320 {**165**}

Shen C, Hu J, 1999, The evolutionary status of HD 104237, ϵ Cha and HD 100546. *Chinese Astronomy and Astrophysics*, 23, 493–497 {**197**}

Simon V, Hanzl D, Skoda P, *et al.*, 2000, Photometric orbital modulation in V1080 Tau. *A&A*, 360, 637–641 {**197**}

Sinachopoulos D, Gavras P, 2007, Photometric observations and frequency analysis of the δ Scuti NT Hya. *Communications in Asteroseismology*, 151, 35–38 {**181**}

Solano E, Pintado OI, 1998, Spectroscopic survey of δ Scuti stars. *Ap&SS*, 263, 267–270 {**176**}

Solheim JE, 2003, Near continuous photometry with the Whole Earth Telescope (WET). *Baltic Astronomy*, 12, 463–470 {**183**}

Soszynski I, Dziembowski WA, Udalski A, *et al.*, 2007, The Optical Gravitational Lensing Experiment: period–luminosity relations of variable red giant stars. *Acta Astronomica*, 57, 201–225 {**189**}

Stebbins J, Huffer CM, 1930, The constancy of the light of red stars. *Publications of the Washburn Observatory*, 15, 140–174 {**193**}

Stellingwerf RF, 1979, Pulsation in the lower Cepheid strip. I. Linear survey. *ApJ*, 227, 935–942 {**178, 180**}

Stencel RE, Ostrowski TA, Jurgenson CA, *et al.*, 2003, Microvariability among selected long-period variables. *The Future of Cool-Star Astrophysics: 12th Cambridge Workshop on Cool Stars, Stellar Systems, and the Sun* (eds. Brown A, Harper GM, Ayres TR), 1074–1079 {**191**}

Sterken C, Jaschek C, 1996, *Light Curves of Variable Stars, A Pictorial Atlas*. Cambridge University Press {**170**}

Sterken C, Jerzykiewicz M, 1993, β Cephei stars from a photometric point of view. *Space Science Reviews*, 62, 95–171 {**185**}

Straižys V, Crawford DL, Philip AGD, 1996, The Strömvil system: an effective combination of two medium-band photometric systems. *Baltic Astronomy*, 5, 83–101 {**164**}

Straižys V, Kazlauskas A, Bartašiute S, 1999, A post-Hipparcos calibration of the Vilnius photometric system. *ASP Conf.*

Ser. 167: Harmonizing Cosmic Distance Scales in a Post-Hipparcos Era, 324–327 {**163**}

Straižys V, 1992, *Multicolour Stellar Photometry.* Pachart Pub. House, Tucson {**163**}

Struve O, 1955, Some unusual short-period variables. *S&T*, 14, 461–463 {**189**}

Templeton MR, Bradley PA, Guzik JA, 2000, Asteroseismology of the multiply periodic δ Scuti star θ Tuc. *ApJ*, 528, 979–988 {**180**}

Templeton MR, McNamara BJ, Guzik JA, *et al.*, 1997, A new pulsation spectrum and asteroseismology of δ Scu. *AJ*, 114, 1592 {**178**}

Thompson RR, 2002, Single channel photoelectric photometry of RZ Arietis, 1983–2001. *Journal of the American Association of Variable Star Observers*, 30, 95 {**193**}

Torres G, Latham DW, Stefanik RP, *et al.*, 1999, HD 205798: a new low-amplitude variable star in Cygnus. *Informational Bulletin on Variable Stars*, 4821 {**197**}

Truong-Bach, Nguyen-Q-Rieu, Sylvester RJ, *et al.*, 1997, ISO LWS observations of H_2O from R Cas: a consistent model for its circumstellar envelope. *Ap&SS*, 255, 325–328 {**165**}

Truong-Bach, Sylvester RJ, Barlow MJ, *et al.*, 1999, H_2O from R Cas: ISO LWS-SWS observations and detailed modeling. *A&A*, 345, 925–935 {**165**}

Turcotte S, Richer J, Michaud G, *et al.*, 2000, The effect of diffusion on pulsations of stars on the upper main sequence: δ Scuti and metallic A stars. *A&A*, 360, 603–616 {**180**}

Udalski A, Kubiak M, Szymanski M, *et al.*, 1994, The Optical Gravitational Lensing Experiment. The catalogue of periodic variable stars in the Galactic bulge. I. Periodic variables in the centre of the Baade's Window. *Acta Astronomica*, 44, 317–386 {**194**}

Uzpen B, Kobulnicky HA, Monson AJ, *et al.*, 2007, The frequency of mid-infrared excess sources in Galactic surveys. *ApJ*, 658, 1264–1288 {**165**}

Uzpen B, Kobulnicky HA, 2005, Identification of mid-infrared excesses in the Tycho Catalogue. *Protostars and Planets V*, 8196 {**165**}

Van Eck S, Jorissen A, Udry S, *et al.*, 1998, The Hipparcos HR diagram of S stars: probing nucleosynthesis and dredge-up. *A&A*, 329, 971–985 {**195**}

van Genderen AM, 1985, An investigation of the micro-variations of highly luminous OBA type stars. II. *A&A*, 151, 349–360 {**187, 188**}

van Genderen AM, Sterken C, Jones A, 2004, Light variations of massive stars (α Cyg variables). XIX. The late-type supergiants R59, HDE 268822, HDE 269355, HDE 269612 and HDE 270025 in the Large Magellanic Cloud. *A&A*, 419, 667–671 {**186**}

van Leeuwen F, 1996, The Hipparcos mission. *Space Science Reviews*, 81, 201–409 {**161**}

van Leeuwen F, Evans DW, Grenon M, *et al.*, 1997a, The Hipparcos mission: photometric data. *A&A*, 323, L61–L64 {**153, 155, 156, 158**}

van Leeuwen F, Evans DW, van Leeuwen-Toczko MB, 1997b, Statistical aspects of the Hipparcos photometric data. *Statistical Challenges in Modern Astronomy II* (eds. Babu GJ, Feigelson ED), 259 {**158**}

van Leeuwen F, Feast MW, Whitelock PA, *et al.*, 1997c, First results from Hipparcos trigonometrical parallaxes of Mira-type variables. *MNRAS*, 287, 955–960 {**190**}

van Leeuwen F, van Genderen AM, Zegelaar I, 1998, Hipparcos photometry of 24 variable massive stars (α Cyg variables). *A&AS*, 128, 117–129 {**186**}

Vidal-Sainz J, Wils P, Lampens P, *et al.*, 2002, The multiple frequencies of the δ Scuti star V350 Peg. *A&A*, 394, 585–588 {**180**}

Viskum M, Kjeldsen H, Bedding TR, *et al.*, 1998, Oscillation mode identifications and models for the δ Scuti star FG Vir. *A&A*, 335, 549–560 {**179**}

Vogt N, Paunzen E, Maitzen HM, 1998, Calibration of Strömgren–Crawford photometry for Ap-stars compared to Hipparcos results. *Applied and Computational Harmonic Analysis*, 3, 204–206 {**163**}

Waelkens C, Aerts C, Grenon M, *et al.*, 1998a, Pulsating B stars discovered by Hipparcos. *ASP Conf. Ser. 135: A Half Century of Stellar Pulsation Interpretation* (eds. Bradley PA, Guzik JA), 375 {**186**}

Waelkens C, Aerts C, Kestens E, *et al.*, 1998b, Study of an unbiased sample of B stars observed with Hipparcos: the discovery of a large amount of new slowly-pulsating B stars. *A&A*, 330, 215–221 {**186–188**}

Walborn NR, Howarth ID, Rauw G, *et al.*, 2004, A period and a prediction for the spectrum alternator HD 191612. *ApJ*, 617, L61–L64 {**197**}

Wasatonic R, Guinan EF, 1998a, TiO- and V-band photometry of the pulsating red giant V CVn. *Informational Bulletin on Variable Stars*, 4579 {**193**}

—, 1998b, Variations of luminosity, radius, and temperature of the pulsating red supergiant CE Tau. *Informational Bulletin on Variable Stars*, 4629 {**193**}

Washuettl A, Strassmeier KG, 2001, A study of the chromospherically-active binaries UX For and AG Dor. *A&A*, 370, 218–229 {**197**}

Welch DL, 1997, Amateurs in the post-Hipparcos age. *S&T*, 94, 10 {**172**}

Whitelock PA, Feast MW, 2000, Hipparcos parallaxes for Mira-like long-period variables. *MNRAS*, 319, 759–770 {**196**}

Wilson R, 2001, Variable stars. *Encyclopedia of Astronomy and Astrophysics;* (ed. Murdin, P.) {**170**}

Wing RF, 1997, The distances and absolute magnitudes of some well-known red variables. *The First Results of Hipparcos and Tycho, IAU Joint Discussion 14*, 39 {**192**}

Woitas J, 1997, Detection of 43 new bright variable stars by the Tycho instrument of the Hipparcos satellite. *Informational Bulletin on Variable Stars*, 4444 {**170**}

Wood PR, 1981, Theoretical aspects of pulsation and envelope ejection in red giants. *ASSL Vol. 88: Physical Processes in Red Giants* (eds. Iben I, Renzini A), 205–223 {**190**}

Woolley RvdR, 1970, Catalogue of stars within twenty-five parsecs of the Sun. *Annals of the Royal Greenwich Observatory*, 5, 1–227 {**163**}

Woźniak PR, McGowan KE, Vestrand WT, 2004, Limits on I-band micro-variability of the Galactic bulge Mira variables. *ApJ*, 610, 1038–1044 {**191**}

Wyn Evans N, Belokurov VA, 2005, A prototype for science alerts. *ESA SP-576: The Three-Dimensional Universe with Gaia* (eds. Turon C, O'Flaherty KS, Perryman MAC), 385–392 {**172**}

Xiong DR, Deng L, Cheng Q, 1998, Turbulent convection and pulsational stability of variable stars. I. Oscillations of long-period variables. *ApJ*, 499, 355–366 {**193**}

Zakrzewski B, Ogloza W, Moskalik P, 2000, Fourier analysis of Hipparcos photometry of Cepheid variables. *Acta Astronomica*, 50, 387–397 {**174**}

5

Luminosity calibration and distance scale

5.1 Introduction

Trigonometric parallaxes provide a distance measurement (essentially) free from model assumptions, at least at levels of order 1 mas. With Hipparcos parallax accuracies of $\sigma_\pi \sim 1$ mas, distances to individual objects at 10% accuracy are achieved out to ~100 pc. In principle, the method can be applied to objects even at very large distances: the future ESA astrometric mission Gaia, with accuracies of some 10 μas at $V \sim 10$ mag, will measure distances of large numbers of sufficiently bright objects to 10% accuracy at 10 kpc. SIM PlanetQuest should exceed this by a factor of 10, reaching ~1 μas for a reasonable number of bright objects. Parallaxes a further factor of 100 more accurate still, at around 10 nanoarcsec, would still be above the effects of interstellar and interplanetary scintillation in the optical, and above stochastic gravitational wave noise, and would in principle provide direct trigonometric distance measurements out to cosmological distances. Below 1 μas, however, stellar surface structure (Eriksson & Lindegren, 2007), and relativistic modelling (Anglada Escudé, 2007), may introduce significant barriers.

At the most basic level, knowledge of the distance to an astronomical object is necessary to convert its apparent properties such as magnitude or angular radius, to absolute quantities (luminosity and linear size respectively). In the absence of direct individual measurements, distance estimates must make recourse to various creative but less direct methods. Mean distances to groups of objects can be derived with higher formal accuracy using the Hipparcos data alone, notably to the Hyades open cluster at ~45 pc and to the Pleiades open cluster at ~120–130 pc, simply by averaging (or suitably weighting) a number of individual parallaxes. Such estimates come with their own assumptions, and with a resulting confidence dependent on the amplitude and spatial form of systematic or correlated errors in the catalogue as a whole. While the accurate Hipparcos distance to the Hyades cluster now allows its use as a template for main-sequence fitting to more distant clusters (Section 6.3), such an application involves a number of additional complications or assumptions: including the extent to which the Hyades cluster typifies the properties of other clusters, and the extent to which known properties (such as chemical abundances), or less well-quantified effects (such as convection, stellar rotation, surface abundance anomalies, or extended episodes of star formation over time), restrict the applicability of main-sequence fitting due to the uncertainties in the stellar evolutionary input parameters or models.

Beyond the measurement of stars in favourable groups of clusters and associations, Hipparcos has also provided large numbers of accurate parallaxes for the full range of spectral types. In particular, it has provided improved distance estimates to high-luminosity main sequence stars which are relatively poorly represented in the solar neighbourhood, and for which pre-Hipparcos trigonometric parallaxes were therefore particularly uncertain. Provided with an estimate of the mean absolute luminosity of stars of a given spectral type, an approximate distance can then be estimated on the basis of such spectral information alone. In practice, these calibrations are still greatly complicated by a number of problems, ranging from the accuracy of the spectral classification, issues of sample bias, uncertain effects of interstellar or circumstellar reddening, and less well-modelled effects contributing to the theoretical luminosity such as stellar rotation. In short, there are many uncertainties, and often too large a spread in properties, which prevent the 'average' star from being used as an accurate 'standard candle'.

Some specific stellar types, or some specific features of the HR diagram, do however possess particularly valuable attributes as standard candles. For example, although the luminosity of most stars is reasonably constant, some high-luminosity stars exhibit regular variations of moderate to large amplitude resulting from pulsational instabilities in the stellar envelope. Best understood and most useful for Galactic structure research are the luminous Cepheid and RR Lyrae variables. The use of these and other luminosity standards relies on the hypothesis that these tracers are either true 'standard candles', with luminosities independent of metallicity or environment, or at least that these dependencies are understood and can be modelled, for example via the Cepheid period–luminosity–colour relation. To be used as a 'single step' distance calibrator requires tracers which are at the same time common enough to be represented in nearby space, and which therefore permit trigonometric distance calibration, whilst at the same time being intrinsically very bright such that they can be observed (and easily discerned) at suitably large distances. This restricts suitable tracers to luminous pulsating variables (the Cepheids, Miras, and RR Lyrae), to giant stars with well-defined properties (specifically the red 'clump' giants), or to well-defined features in the HR diagram (such as the main-sequence turn-off). Even with the Hipparcos results, however, the number of Cepheids or RR Lyrae stars with accurate trigonometric parallaxes remains small. Finally, use as constraints in cosmological models requires distances not only to nearby galaxies such as the Large Magellanic Cloud, but beyond, to galaxies participating in the general Hubble expansion flow, beyond the dynamical effects of the self-gravity of the local group of galaxies.

The approach adopted in this chapter is as follows: (1) to first focus on some of the general statistical considerations, and potential biases, relevant in applying the Hipparcos data to the problem of luminosity calibration and distance determination; (2) to examine the Hipparcos contributions to luminosity calibration across the broad range of spectral type and luminosity class; (3) to examine the Hipparcos results on each of the major tracers currently used for estimating the distance to the Large Magellanic Cloud; these sections will also cover kinematic or other information which has emerged from the Hipparcos studies, even though they may not be directly relevant to the problem of the distance scale; and (4) to summarise the resulting estimates of the distance to the Large Magellanic Cloud and beyond from these various methods.

Various workshops, meetings, and review articles have been devoted to the problem of the distance scale over the last 10 years, in large part stimulated by the results from Hipparcos, including: 'Distance Scales after Hipparcos' (Feast, 1998a); 'Hipparcos and Variable Star Distance Scales' (Feast, 1998b); 'Calibration of the Extragalactic Distance Scale' (Madore & Freedman, 1998a); 'The Distance Scale of the Universe Before and after Hipparcos' (Seggewiss, 1998b); 'Hipparcos and Primary Distance Scale Indicators' (Turon, 1998); 'The HR Diagram and the Galactic Distance Scale after Hipparcos' (Reid, 1999); 'Luminosity Calibration and Distance Scale' (Turon, 1999); 'Post-Hipparcos Galactic Distance Scales ' (Wang, 1999); 'The Distances of the Magellanic Clouds' (Walker, 1999).

5.2 Statistical biases

Users of the Hipparcos and Tycho Catalogues should be aware of a number of potential biases which might affect interpretation of the astrometric data, and more particularly in the use of the parallax data in the context of luminosity calibration.

At the most basic level are biases due to systematic errors in the measurements. In the determination of ground-based parallaxes, for example, the transformation from relative to absolute parallaxes is a potential source of such a bias. For the Hipparcos data, the available evidence indicates that for a star with true parallax π_0, the distribution of Hipparcos measurements is compatible with a Gaussian distribution around π_0, with standard error σ_π. Error correlations on a small angular scale are considered in Section 1.4.1. Other biases in luminosity, as well as parallax and proper motion themselves, might arise from undetected binary companions, as well as imperfect chromatic calibration specifically for the reddest stars (Section 4.10). A further type of bias is modelling bias, arising from fitting an incorrect model to the data. These are not considered further here.

Transformation biases result from a nonlinear transformation from a directly measured quantity with its own associated error, to some other desired quantity. A relevant example is the (superficially trivial) transformation from parallax to distance: the usual estimate of distance is $d = 1/\pi$, but this estimate is biased. The error distribution of this estimate of d is in practice obtained by multiplying the probability density function of π by the Jacobian of the transformation $\pi \rightarrow d$. The distribution function of the errors of the derived distance is skewed, with the direction of the resulting bias dependent on σ_π/π. Furthermore, for a normal distribution in π, the variance of d is infinite, making the application of a bias correction for an individual object impossible. The topic is treated extensively by Smith & Eichhorn (1996), by Kovalevsky (1998, although excluding negative parallaxes) and, in the context of the Hipparcos data, by Luri & Arenou (1997, although Figure 2 there is incorrect), Brown *et al.* (1997, see Figure 1), and Arenou & Luri (1999). An extension of the argument is applicable in determining a mean cluster distance from a number of stars, neglecting cluster depth and measurement correlations, where the estimate $1/\langle\pi_i\rangle$ is asymptotically unbiased and thus to be preferred over $\langle(1/\pi_i)\rangle$

(Arenou & Luri, 1999). A similar bias arises from the logarithmic transformation to absolute magnitude, for which $M_V = V + 5\log\pi + 5$ is a biased estimate of a star's absolute magnitude (Smith & Eichhorn, 1996; Brown *et al.*, 1997).

Sample biases may arise from selection effects and incompleteness of the Hipparcos and Tycho Catalogues. These are not at all trivial to handle in the case of the Hipparcos Catalogue, because of the way the Input Catalogue had to be constructed. A specific effort was nevertheless made to carry out part of the mission as a survey roughly complete to between $V \sim 7-8$ mag (see Section 1.3).

Truncation biases arise from the construction of a sample according to some limits on observable quantities, either because of intrinsic scatter, or because of measurement error. This is despite the fact that, provided there is no systematic error in the measurement process, an observed parallax is not in itself biased. For samples limited by parallax, the effect on derived absolute magnitudes is termed the Lutz–Kelker bias, which depends on the relative parallax error σ_π/π and the spatial distribution of the sample. For samples limited by apparent magnitude, the resulting effect is termed the Malmquist bias. The effects (specifically on absolute magnitude calibration), their precise meaning, and their inter-relationship are somewhat subtle, but potentially relevant for many Hipparcos applications.

5.2.1 Malmquist bias

For a given class of object with a true mean absolute magnitude, M_0, and dispersion, σ, an ideal volume-limited sample can give an unbiased estimate of M_0 if the distance estimate is itself unbiased. In the more common type of magnitude-limited sample, however, faint objects at larger distances are lost from the sample, while systematically brighter objects are preferentially observed as distance and volume increase, both due to intrinsic dispersion, and to observational errors. The net effect is that the mean magnitude of the sample will be brighter than that of the target population, and it cannot be used to estimate the latter without corrections being applied; the resulting Malmquist bias is $\langle M\rangle - M_0$. The presence of such a bias can be investigated through the Spaenhauer diagram (Spaenhauer, 1978), in which derived absolute magnitude is plotted against distance or parallax, from which it may be evident where the Malmquist bias starts to dominate.

Butkevich *et al.* (2005a) provides a detailed introduction to the topic, and a short historical review starting with the considerations by Eddington (1913) and Kapteyn (1914). Eddington (1914) derived the classical Malmquist bias for a homogeneous spatial distribution, $\langle M\rangle \simeq M_0 - 1.38\sigma_M^2$. Malmquist (1922) gave analytical expressions for the bias in M_0 for arbitrary spatial distributions in Euclidean space and for a Gaussian luminosity function, generalised to take interstellar absorption into account by Malmquist (1936). Jaschek & Gómez (1985) evaluated the bias using different shapes of the luminosity function and different assumptions about the space density function. Smith (1987a,b,c,d) considered various problems related to the analysis of trigonometric parallaxes. Ratnatunga & Casertano (1991) gave a numerical algorithm with emphasis on kinematic parameters. Luri *et al.* (1993) considered the influence of more realistic spatial distributions including Galactic flattening. Luri *et al.* (1996b) presented a general method allowing a simultaneous determination of luminosity, kinematic, and spatial characteristics of a given sample (see box on page 213). Teerikorpi (1997) reviewed the influence of Malmquist bias on the determination of extragalactic distances. Teerikorpi (1998) generalised the classical Malmquist bias in Euclidean space to Friedmann cosmology. Arenou & Luri (1999) included a discussion of Malmquist bias in the specific context of the Hipparcos data.

Depending on the data under consideration and the way in which mean absolute magnitudes are calculated, Butkevich *et al.* (2005a) identified three different types of Malmquist bias: integral bias (when averaging over an entire sample), magnitude-dependent bias also referred to as classical Malmquist bias or Malmquist bias of the first kind (when averaging over stars of fixed apparent magnitude), and distance-dependent bias also referred to as Malmquist bias of the second kind (when averaging over stars at the same true distance). They presented a quantitative treatment of the Spaenhauer diagram, deriving expressions for the region unaffected by distance-dependent bias.

5.2.2 Lutz–Kelker bias

What is referred to as the Lutz–Kelker bias arises in the following way. Consider a given parallax, π, with measurement error σ_π, which yields a nominal distance, d, with some upper and lower bounds. When constructing a sample based on parallax limits, the measurement error means that stars at smaller true distances can be scattered beyond the formal distance limit (out of the sample), while stars located further away can be scattered into the sample. Simply due to the different associated volumes ($V \propto d^3$), more stars from outside the nominal distance limit will be scattered into the sample than are erroneously lost from the sample. This effect causes a systematic bias in which the derived absolute magnitude is too faint. It should be stressed that the parallax of an individual star is not itself biased, and bias correction for an individual star outside of the context of a parallax-limited sample is not appropriate.

Determination of the corrections to the absolute magnitude of an individual star or a sample of stars selected according to observed trigonometric parallax has a long history, recently reviewed by Sandage & Saha (2002). These include early contributions by Eddington (1913), Dyson (1926), Jeffreys (1938), Eddington (1940), Trumpler & Weaver (1953), Ljunggren & Oja (1965), Wallerstein (1967), and West (1969). Lutz & Kelker (1973) treated the constant space density case $N(\pi)\mathrm{d}\pi \propto \pi^{-4}\,\mathrm{d}\pi$, and emphasised that the magnitude of the effect depends only on the relative parallax error σ_π/π, after which the resulting bias has entered the literature as the Lutz–Kelker effect. They tabulated, as had some of the preceding work, the magnitude of the bias as a function of σ_π/π.

For an individual object whose properties are otherwise unknown, correction for Lutz–Kelker bias, and its (typically) large associated confidence interval, may in practice provide little real knowledge of the actual bias. However, assuming a uniform distribution of stars, Lutz & Kelker (1973) established that the mean correction to the derived absolute magnitude increases with increasing relative error, reaching -0.43 mag for a 17.5% error in the parallax. Koen (1992) calculated 90% confidence intervals for the bias which, for $\sigma_\pi/\pi = 17.5\%$, ranged from $+0.33$ to -1.44 mag. Smith (1987c) showed that the Lutz–Kelker correction as formulated was valid when no additional *a priori* information is available. In the case of a Gaussian magnitude distribution, for example, Smith (1987a,c,d) showed that the correction is a function of the true absolute magnitude, the intrinsic dispersion, and σ_π/π. Examples of Lutz–Kelker corrections applied to individual objects can be found in the HST-based study of RR Lyrae by Benedict *et al.* (2002) and of the Hipparcos Cepheids by van Leeuwen *et al.* (2007).

For a magnitude-limited sample (complete in parallax to a certain limiting magnitude), this correction approaches zero because the combination of the Malmquist bias and the Lutz–Kelker bias lead to a symmetric error in magnitude, while it converges to the Lutz–Kelker value for large σ_{M_0} (Smith, 1987c). A linear approximation to the bias on the resulting mean $\langle M\rangle$, valid when $|\Delta M| < 2.17$, was derived by Smith (1987c) as

$$\delta M = \left[1 - \left(\frac{\sigma_{M_0}^2}{\sigma_{M_0}^2 + (2.17\sigma_\pi/\pi)^2}\right)\right](M_0 - \langle M\rangle) \qquad (5.1)$$

Hanson (1979) developed a formulation which has also been exploited in the use of the Hipparcos data. In the original analysis by Lutz & Kelker (1973), corrections were formulated for constant space density, for which $N(\pi) \propto \pi^{-4}$. This steep dependence on π can be partially circumvented by re-formulating the correction in terms of the observed distribution of proper motions, $N(\mu)$, which bears a simple power-law relation to $N(\pi)$ under generally valid kinematic assumptions. For a proper motion distribution of the power-law form $N(\mu) \propto \mu^{-x}$, the appropriate Lutz–Kelker correction for each individual star was given by Hanson (1979, Equation 30) as

$$\Delta M = -2.17\left[\left(n+\frac{1}{2}\right)\left(\frac{\sigma_\pi}{\pi}\right)^2 + \left(\frac{6n^2+10n+3}{4}\right) \times \left(\frac{\sigma_\pi}{\pi}\right)^4\right] \qquad (5.2)$$

where $n = x + 1$ and, again, smaller fractional errors σ_π/π result in smaller corrections. Percival *et al.* (2002) used this approach in their analysis of 43 local subdwarfs with Hipparcos parallax errors less than 13%, finding $x = 2.65 \pm 0.15$ which, with $\sigma_\pi/\pi = 0.1$, leads to $\Delta M \sim -0.09$ mag in the worst case.

Other considerations Various other studies of the correction of the Lutz–Kelker bias exist in the pre-Hipparcos literature (e.g. Turon & Crézé, 1977; Lutz, 1979, 1983, 1986). Other discussions of Malmquist and Lutz–Kelker biases affecting the Hipparcos data include Arenou & Gómez (1997); Babu & Feigelson (1997); Gómez *et al.* (1997); Arenou & Luri (1999); Smith (1999); Arenou (2000); Arenou & Luri (2002); Maíz Apellániz (2005); and Tsujimoto *et al.* (2005a,b). Numerous papers treating the Hipparcos data also provide detailed specific consideration of the effects of the Lutz–Kelker (and Malmquist) bias as part of their scientific interpretation (see, e.g. Reid, 1997; Gratton *et al.*, 1997; Bedding & Zijlstra, 1998; Lanoix *et al.*, 1999).

One of the earliest Hipparcos studies illustrates the complexities of the interpretation, and of the interrelationship between the Lutz–Kelker and Malmquist effects. Oudmaijer *et al.* (1998) used the Hipparcos data as a control sample for which distances are more-or-less precisely known, investigating the Lutz–Kelker bias in the corresponding ground-based parallax dataset. Figure 5.1, top shows the distribution of ΔM versus $(\sigma_\pi/\pi)_{\rm ground}$ for their sample of 2187 Hipparcos stars with $(\sigma_\pi/\pi)_{\rm Hip} < 0.05$. They argued that, for increasing $(\sigma_\pi/\pi)_{\rm ground}$, and up to values of $(\sigma_\pi/\pi)_{\rm ground} \sim 0.2$, the derived absolute magnitude is too faint in a manner consistent with the Lutz–Kelker predictions. Figure 5.1, bottom shows the distribution corrected according to Equation 5.1, for different values of M_0. They used this excellent fit (perhaps surprisingly so given the systematic effects likely to be present in the ground-based data) to argue that a similar effect would be present in the Hipparcos data at comparable values of $(\sigma_\pi/\pi)_{\rm Hip}$, and derived corresponding adjustments to the absolute magnitudes of the Hipparcos Cepheid sample studied by Feast & Catchpole (1997). Lutz–Kelker bias is, however, not the only effect evident in Figure 5.1, top. As subsequently emphasised by Oudmaijer *et al.* (1999), for even larger values of $(\sigma_\pi/\pi)_{\rm ground}$, derived magnitudes are too bright by up to $\sim$ 2 mag.

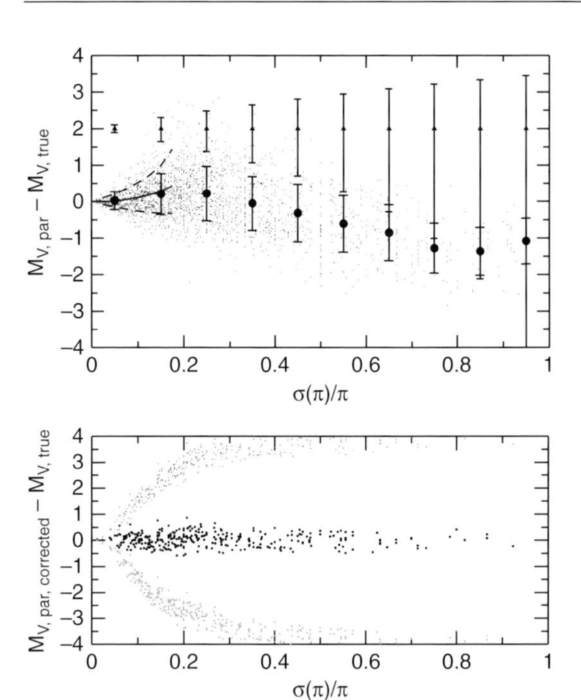

Figure 5.1 Top: ΔM as function of $(\sigma_\pi/\pi)_{\rm ground}$ for 2187 stars with $(\sigma_\pi/\pi)_{\rm Hip} < 0.05$. Triangles with error bars indicate magnitude errors propagated from parallaxes. Solid circles indicate average ΔM binned over intervals of 0.1 in σ_π/π. Thick lines indicate the Lutz–Kelker bias calculated for a uniform density of stars, with dashed lines indicating 90% confidence intervals. For $\sigma_\pi/\pi \gtrsim 0.2$ the average ΔM are systematically too bright, corresponding to the effects of the distance dependent Malmquist bias, discussed further by Oudmaijer et al. (1999). Bottom: ΔM versus $(\sigma_\pi/\pi)_{\rm ground}$ for objects with $4 < M_0 < 5$, and with $\langle M\rangle$ corrected using the approximation to the Lutz–Kelker correction given by Equation 5.1, with $M_0 = 4.49$. Upper and lower clouds of points show the solution when a different (sub-optimal) value of M_0 is used for the correction (8.5 and 0.5 respectively). From Oudmaijer et al. (1998, Figures 2 and 3).

Focusing on 200 K0V stars they identified this additional effect at larger $(\sigma_\pi/\pi)_{\rm ground}$ as the 'magnitude selection Malmquist bias'. The result was confirmed in a similar study of 166 single K0V Hipparcos stars by Butkevich *et al.* (2005b), who ignored the contribution of the Lutz–Kelker bias, but made a more rigorous attempt to exclude binary stars. For further discussion of these results, see Arenou & Luri (1999, Section 3.3).

Further detailed consideration of the Lutz–Kelker bias can be found in Sandage & Saha (2002) and Smith (2003). The former includes detailed simulations and analyses carried out in the context of the FAME astrometric mission studies.

In conclusion, and as recommended by Brown *et al.* (1997), each application should be analysed to obtain a correct estimate of any parameter of a star or sample of stars using trigonometric parallaxes, neither ignoring the possible biases, nor applying Malmquist or Lutz–Kelker corrections blindly.

5.2.3 Maximum likelihood techniques

In contrast with the method of determining mean absolute magnitudes directly from the parallaxes, where bias corrections are mandatory, the difficulties associated with the Lutz–Kelker correction can be avoided using inverse methods based on maximum likelihood (although the Malmquist bias is still present), typically with some restrictive assumptions such as uniform space density, or a truncated Gaussian luminosity function. For further details, see the discussions in Luri *et al.* (1996b), Arenou & Luri (1999), Sandage & Saha (2002), and Smith (2003).

5.2.4 Astrometry-based luminosity, or reduced parallax

In general, individual absolute magnitudes are used in the HR diagram, for example, for age determination or luminosity calibration. A common problem is how to handle individual stars with poor parallax precision, rather than simply discarding them by imposing a cut-off in σ_π/π. In particular, if a function of the absolute magnitude can be constructed that is linear in π, then averages and other statistics will behave as for a variable with normally-distributed errors. Such a method was adopted by Turon & Crézé (1977), who cited earlier use by Roman (1952) and Ljunggren & Oja (1965). In the Hipparcos literature this approach has been used by, e.g. Feast & Catchpole (1997), Koen & Laney (1998), and Arenou & Luri (1999). It is also considered in the Lutz–Kelker study by Sandage & Saha (2002).

Arenou & Luri (1999) considered the quantity which they refer to as the astrometry-based luminosity

$$a_V = 10^{0.2M_V} = \pi\, 10^{\frac{m_V - A_V + 5}{5}} \tag{5.3}$$

with π in arcsec. They refer to the corresponding HR diagram as 'the astrometric HR diagram'. By way of illustration, they simulated a sample of 1000 stars of age 10 Gyr, with [Fe/H] = −1.4 and $\sigma_{M_V} = 0.5$ mag. Figure 5.2 shows the classical HR diagram with a 30% truncation on relative parallax error, and the corresponding astrometric HR diagram. The use of the astrometry-based luminosity implies that the error bars on a_V due to parallax errors are symmetric, that there is no Lutz–Kelker type bias, that all stars may be used including those with negative parallaxes, and with the larger number of stars implying a possible gain in precision for mean values.

Groenewegen & Salaris (1999), Whitelock & Feast (2000) and Feast (2002) similarly use the 'method of reduced parallaxes', based on the construction of

$$\overline{10^{0.2M}} = \sum 0.01\,\pi\, 10^{0.2m_0} p / \sum p \tag{5.4}$$

where the parallax is in milliarcsec, m_0 is the (absorption-free) apparent magnitude, and p is a deterministic weight.

Another way to avoid the transformation bias is to transform the model expectations or simulated data to the space of the observables. An example is in the study by Schröder *et al.* (2004, Figure 4) where expected and observed parallaxes are compared.

5.2.5 Reduced proper motions

Another way of estimating distances is through the use of the 'reduced proper motion' (Luyten, 1922) , for example in the V band, to serve as a criteria for large-scale surveys in which the parallax is unknown

$$V_{\rm RPM} \equiv V + 5\log\mu = M_V + 5\log\frac{v_\perp}{47.4\,{\rm km\,s^{-1}}} \quad (5.5)$$

Here, μ is the proper motion in arcsec yr^{-1} and $v_\perp$ is the transverse velocity. The use of the observables V and μ yields a colour versus reduced proper motion diagram similar to the standard colour magnitude diagram, but with a greater scatter owing to the dispersion in $v_\perp$.

5.3 Secular and statistical parallaxes

The methods of secular parallax and classical statistical parallax are aspects of a single general method used to expand trigonometric parallax measurements to more distant objects by increasing the (temporal and spatial) baseline of the measurements. Details and the basic mathematical formulation are given in Mihalas & Binney (1981, Section 6.6), and expanded in the context of Hipparcos RR Lyrae studies by Popowski & Gould (1998b), and in the context of Hipparcos Cepheid studies by Rastorguev *et al.* (1999).

The method of secular parallax exploits the fact that the Sun moves at a speed $\sim$20 km s^{-1} relative to disk populations and $\sim$200 km s^{-1} relative to halo populations, corresponding to 4 AU and 40 AU per year respectively, so that after several decades, the accumulated linear motion is significantly larger than the Earth–Sun baseline used for the determination of trigonometric parallax. But deducing the star's distance from its secular motion transverse to the line-of-sight requires knowledge of both the solar motion and the star's transverse velocity. Since the latter is not known independently, knowledge of the solar motion contributes nothing to the distance determination for any one particular star. The method can, however, be applied to a sample of stars on the assumption of a uniform underlying velocity dispersion, in which case the solar motion can be accounted for in the average space motion.

In a related way, in the method of classical statistical parallax the distance scale to a class of star is estimated by adjusting their (linear) radial velocities and (angular) proper motions to reproduce the same velocity dispersion (e.g. Rastorguev *et al.*, 1999). Secular parallax essentially forces equality between the three space components of the velocity distribution, while classical statistical parallax forces equality of the six independent components of the velocity covariance matrix. In the modern combined version of statistical parallax, the 10 unknowns (an overall distance scaling factor plus the nine first and second velocity moments) are simultaneously estimated by maximum likelihood (Popowski & Gould, 1998b).

Table 5.1 Absolute magnitudes of the MK system as tabulated by Schmidt-Kaler (in Schaifers & Voigt, 1982, Table 13).

	Luminosity class				
Sp	V	IV	III	II	Iab
O5	−5.7	−6.0	−6.3	–	–
B0	−4.0	−4.7	−5.1	−5.7	−6.4
B5	−1.2	−1.7	−2.2	−4.0	−6.2
A0	+0.65	+0.3	+0.0	−3.0	−6.3
A5	+1.95	+1.3	+0.7	−2.8	−6.6
F0	+2.7	+2.2	+1.5	−2.5	−6.6
F5	+3.5	+2.5	+1.6	−2.3	−6.6
G0	+4.4	+3.0	+1.0	−2.3	−6.4
G5	+5.1	+3.1	+0.9	−2.3	−6.2
K0	+5.9	+3.1	+0.7	−2.3	−6.0
K5	+7.35	–	−0.2	−2.3	−5.8
M0	+8.8	–	−0.4	−2.5	−5.6
M5	+12.3	–	−0.3	–	−5.6

5.4 Absolute magnitude versus spectral type

Estimates of stellar distances are frequently based on absolute magnitudes estimated from the spectroscopic criteria of spectral type and luminosity class. The most comprehensive and widely-referenced luminosity calibration for groups of stars selected spectroscopically remains the compilation of Schmidt–Kaler (in Schaifers & Voigt, 1982, Table 13), which is also that given by Cox (2000), and reproduced in Table 5.1. However the quoted absolute magnitudes are inhomogeneous, and the associated uncertainties are seldom known.

As detailed in this section, the Hipparcos results have made a substantial impact on the absolute magnitude calibration as a function of spectral type, identifying various dependencies and subtleties in the calibrations derived previously, and showing that spectral classification is frequently insufficiently homogeneous to assign a precise mean magnitude to a given spectral type and luminosity class. There is, however, no overall synthesis of results in the form of a homogeneous revision of luminosity calibration as a function of spectral type, either in the literature or attempted here.

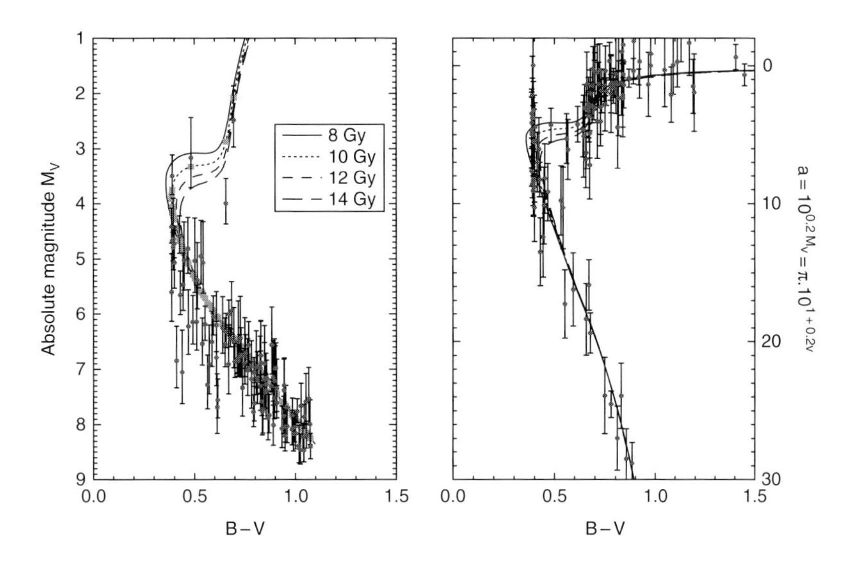

Figure 5.2 Use of the astrometry-based luminosity, using a simulated sample of 1000 stars of age 10 Gyr, with [Fe/H] = −1.4 and $\sigma_{M_V} = 0.5$ mag. Left: the classical HR diagram with $\sigma_\pi/\pi < 0.3$ (116 stars). Right: the corresponding astrometric HR diagram, using $a_V = 10^{0.2M_V}$ as ordinate, with $\sigma_{a_V} < 3$ adopted to give the same number of 116 stars, but now showing a significantly larger representation of giant stars. From Arenou & Luri (1999, Figure 8).

Maximum likelihood luminosity calibration: The LM method: Several implementations of the maximum likelihood principle to obtain luminosity calibration from kinematic data have been developed in the past (e.g. Jung, 1970; Clube & Jones, 1971). A specific algorithm designed to fully exploit the Hipparcos data and to overcome some of the limitations of earlier methods was described by Luri *et al.* (1996b), and has been used in a number of Hipparcos-based analyses.

Referred to as the LM method (for Luri–Mennessier; Mennessier *et al.*, 1997), its main characteristics are summarised by Barthès *et al.* (1999) as follows: (i) it is based on a maximum-likelihood algorithm, which is asymptotically unbiased, Gaussian, and efficient (Kendall & Stuart, 1979); (ii) it uses all available information for each star: apparent magnitude, Galactic coordinates, trigonometric parallax, proper motion, radial velocity and other relevant parameters (photometry, metallicity, period, etc.), and takes into account, as an additional constraint, the existence of mean relations between, e.g. period, luminosity and colour, whose analytical form is given *a priori*; (iii) it allows a detailed modelling of the kinematics, the spatial distribution, and also the distribution of luminosity, period and colour of the sample; (iv) it takes into account observational errors, and selection criteria relevant for the sample (i.e. the Malmquist and Lutz–Kelker bias, as discussed in Section 5.2); (v) Galactic rotation is taken into account through an Oort–Lindblad first-order differential rotation model with specified values of A, B, and R_0; (vi) the interstellar absorption is taken into account using the 3D model of Arenou *et al.* (1992); (vii) it can separate and characterise groups of stars with different properties (e.g. luminosity, kinematics, and spatial distribution), whose number is fixed in advance.

For the population corresponding to each identified group, the method provides unbiased estimates of the model parameters, e.g.: (i) the parameters of the absolute magnitude distribution, i.e. the coefficients of the mean period–luminosity–colour relation, and its dispersion; (ii) the velocity distribution: mean velocities (U_0, V_0, W_0) and dispersions $(\sigma_U, \sigma_V, \sigma_W)$; (iii) the spatial distribution defined by the scale height Z_0; (iv) the period–colour index distribution: mean of the logarithm of the period, mean de-reddened colour index, associated dispersions, and the correlation between log period and colour; (v) the percentage of the sample in each group; (vi) it also yields improved individual distance estimates, and thus improved absolute magnitude estimates, which take into account all the available information on each star: parallax and other measurements (magnitude, position, proper motion, radial velocity, period, colour). This estimation is free of any bias due to observational selection or observational errors, because both are taken into account.

A synopsis of the MK system is given in the box on page 214 to place the Hipparcos contributions into context (see also Figure 5.3).

Calibrations across the HR diagram Hipparcos represented the first large-scale capability of establishing the absolute magnitudes of the MK standards. Amongst the earliest analyses Jaschek & Gómez (1998), following from Jaschek & Gómez (1997), used only 96 standard stars from the list of MK standards collected by García (1989, from criteria recommended by W.W. Morgan) for which the absolute magnitude error was below 0.3 mag, finding that the main sequence is a wide band and that, although in general giants and dwarfs have different absolute magnitudes, the separation between

Spectral classification and the MK system: Since the early classification efforts of the Vatican Observatory astronomer Angelo Secchi around the 1860s, spectral classification schemes have been developed to provide a reference framework for stellar classification. Amongst many applications, spectral classfication has allowed the estimation of stellar distances based on spectroscopic parallaxes, a method founded on the assumption that stars with similar spectra have similar fundamental parameters, notably their luminosity. Different classification systems have been developed (Kurtz, 1984), of which only the MK system remains in widespread use today. This builds on the Harvard sequence of spectral types, OBAFGKM(SRN), preserving continuity with the HD project at Harvard College Observatory, sponsored by Henry Draper and carried out primarily by Annie Jump Cannon in the early 1900s. This classified more than 250 000 stars on a homogeneous 1d system originally based lexically on the strength of the hydrogen lines, and subsequently reordered to relate to temperature: O being the hottest, M the coolest, and SRN paralleling the temperatures of the GKM stars, but with different chemical composition (e.g. R and N stars, sometimes taken together as C stars, are carbon rich). Relevant spectral characteristics are: O = hot stars with He II absorption; B = He I absorption, H developing later; A = very strong H, decreasing later, Ca II developing; F = Ca II stronger, H weaker, metals developing; G = Ca II strong, Fe and other metals strong, H weaker; K = strong metallic lines, CH and CN bands developing; M = very red, TiO bands developing strongly.

The MKK system was introduced by Morgan *et al.* (1943) at the University of Chicago's Yerkes Observatory. They used a prismatic spectrograph covering 390–490 nm at $R = 11.5\,\mathrm{nm\,mm^{-1}}$ to introduce a framework for a 2d spectral classification, without reference to photometry, theory, or other information. The MK system was a revision by Morgan & Keenan in 1953, essentially defined by the list of standard stars of Johnson & Morgan (1953). Progressive revisions have been described by, e.g. Morgan & Keenan (1973) and Keenan (1985). As described by Morgan & Keenan (1973) *'The MK system is a phenomenology of spectral lines, blends, and bands, based on a general progression of colour index (abscissa) and luminosity (ordinate). It is defined by an array of standard stars, located on the two-dimensional spectral type versus luminosity class diagram. These standard reference points do not depend on values of any specific line intensities or ratios of intensities; they have come to be defined by the appearance of the totality of lines, blends, and bands in the ordinary photographic region. The definition of a reference point, then, is the appearance of the spectrum "as in" the standard star. For example, a star located at A2 Ia would have a spectrum having a total appearance as in the standard star α Cygni; a star of spectral type G2 V would have a spectrum whose appearance is similar to that of the Sun.'*

The MK system uses the following nomenclature (Garrison, 2000a): the Harvard HD sequence (OBA...) for the temperature axis, each letter being followed by subdivisions 0–9 (except for O-type stars for which O3 is the hottest); a luminosity class is indicated by a Roman numeral, as signified by features sensitive to surface gravity and microturbulence, thus I = supergiants, II = bright giants, III = giants, IV = subgiants, and V = main sequence (dwarfs); VI = subdwarfs, VII = white dwarfs.

Various authors have subsequently provided revised lists of proposed MK standards for various spectral ranges. The greatest precision of classification has been achieved for stars of roughly solar composition, where there are sufficient numbers in the solar neighbourhood to provide a good network of standards for all but the most luminous supergiants. Thus, giants of type G8–M3 are further subdivided into luminosity subclasses IIIa, IIIab, and IIIb. Supergiants with the highest luminosity class, 0, are defined by the four reddest of the brightest stars in the LMC. The precision of spectral types was estimated by Jaschek & Jaschek (1973) as ± 0.6 subdivisions. Keenan (1985) provides plots and tabulations of MK temperature subclasses; lists of MK types of fainter stars; published calibrations of luminosity classes for early-type stars; calibration of MK luminosity classes for types later than F8; the distribution among groups of the 426 stars in their current list of best types; and the effects of metal deficiencies on spectra of K0 III stars.

The Michigan Spectral Survey was undertaken to reclassify the HD Catalogue stars on the 2d MK system. The project uses 10° objective prism plates with a resolution of 0.2 nm, taken at the CTIO Michigan Curtis Schmidt or the KPNO Burrell Schmidt. Five of a projected seven volumes have been published (Houk & Cowley, 1975; Houk, 1978, 1982; Houk & Smith-Moore, 1988; Houk & Swift, 1999), resulting in some 161 000 re-classified stars to date (Sowell *et al.*, 2007). Results from the first four were included in the Hipparcos Input Catalogue (Fields 20–21) and also in the Hipparcos Catalogue (machine-readable only, Fields H76–77). For the luminosity class, the following designations are used in the Hipparcos Catalogue compilation: Ia0, Ia, Iab, Ib for supergiants, II for bright giants, III for giants, IV for subgiants, V for dwarfs. Subdwarfs are either noted sd followed by the spectral type, or class VI. Peculiarities of the spectra are noted in lower case letters: e for emission lines, m for enhanced metallic lines, n for nebulous lines, nn for very nebulous, p for peculiarity in the chemical composition, s for sharp lines, sh for a shell, v for variations in the spectrum, w for weak lines, etc. CN indicates stars cyanogen abundance anomaly. This description does not cover all MK designations, and a more complete description can be found in the introduction to the Michigan Spectral Survey. Plans remain in place to complete the two further projected volumes (see Houk, 1979, Table III), at Lowell and Seoul (Houk, 2007, priv. comm.).

luminosity class V and III was not evident. They also analysed the refined system of late B to early F standards defined by Garrison & Gray (1994), separating them into low and high rotational velocity stars, but finding similar effects as in the original MK system. Houk *et al.* (1997) examined the M_V distribution of 3727 early F to early K stars within 100 pc, finding the distributions non-Gaussian, but with Gaussian fits yielding estimates of M_V in good agreement with earlier luminosity estimates.

Gómez *et al.* (1997) examined the luminosity calibration of the HR diagram according to the MK classification. They identified 22 054 stars from the survey subset with known luminosity class, after eliminating

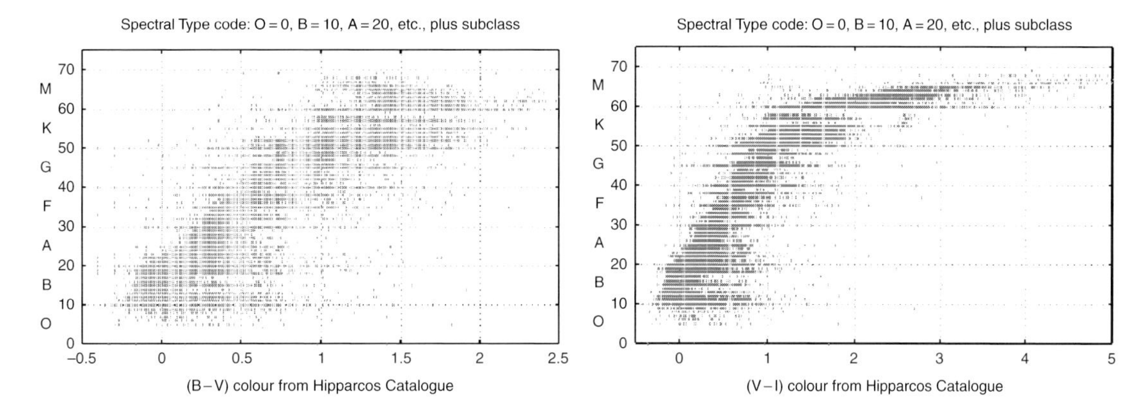

Figure 5.3 Relationship between spectral type and colour index (B − V and V − I), all taken from the Hipparcos Catalogue. The spectral type is coded in the following way: O = 0, B = 10, A = 20, etc,. with the subclass value added (e.g. an F5 star would have code 35 = 30 + 5). Figures constructed by Michael Richmond for The Amateur Sky Survey (TASS) www site.

known binaries and certain variables (Table 5.2). They applied the LM method (see page 213) to this sample to evaluate the precision of the luminosity calibration based on the MK classification (Figure 5.4a–c). The algorithm identified and separated groups differing in luminosity, kinematic, or spatial properties. The results reveal a broad main sequence (class V), reaching up to 2 mag in width, in contrast with the value of 0.3 mag given by Schmidt–Kaler. Some stars are well out of the main sequence, while for early-type stars (B to F) there is no clear separation in the HR diagram between classes III, IV, and V. Luminosity classes I, II, and III also extend over a large range in absolute magnitude, implying that a mean absolute magnitude for these groups does not provide valuable information. In agreement with the findings of Jaschek & Gómez (1997), the results showed that the relation between M_V and luminosity class has a large intrinsic dispersion, ranging from about 0.5 mag on the main sequence, 1 mag on the giant branch, to more than 1 mag for the bright giants and supergiants. Thereafter the method was applied to all survey G–K stars with and without known luminosity class, identifying seven groups in the M_V versus $B - V$ diagram (Figure 5.4d).

Sowell *et al.* (2007) also used the available MK spectral data from the Michigan Spectral Survey, along with the Hipparcos parallaxes, to construct HR diagrams (absolute V magnitude versus spectral type) for distance limits of 20, 50, 100 and 200 pc, updating previous versions of the HR diagram based on the MSC data presented for example by Houk & Fesen (1978). The HR diagram for the total of 113 286 stars is shown in Figure 5.4e.

Table 5.2 Number of stars in the Hipparcos survey with known spectral type and luminosity class, after eliminating certain binary and variable stars, according to Gómez et al. (1997, Table 1). More is known about the MK classification of Hipparcos stars than is reflected in the relevant fields of the Hipparcos Catalogue, notably as a result of ongoing work on the Michigan Spectral Survey (see box on page 214). More detailed statistics as a function of spectral type and luminosity class are presented by Sowell et al. (2007).

	I	II	III	IV	V	Total
B	177	225	802	590	1412	3206
A	54	48	541	918	2305	3866
F	57	104	451	1030	3708	5350
G	73	117	1375	548	1449	3562
K	36	122	5457	299	138	6070
Total	397	616	8626	3385	9012	22 054

O–B stars A catalogue of Galactic OB stars is maintained and described by Reed (2003, 2005), the latter (as of June 2005) comprising some 18 400 objects. He takes OB stars to mean main-sequence stars down to temperature class B2, with more luminous ones down to class B9. As he indicates, the definition of an OB star is not universal, with Vanbeveren *et al.* (1998) defining them as O–B2 V–IV, O–B3 III, O–B4 II, and all OBA Ib, Iab, Ia, and IaO stars.

Various studies of the mean absolute magnitudes of OB (and Be) stars have been made using the Hipparcos data. Briot *et al.* (1997) determined the absolute magnitudes of 213 Be stars of luminosity classes IV and V compared with those of 2077 B stars of the same spectral type. On average, they found that the Be stars are brighter than B stars of the corresponding spectral type, agreeing with some previous results. The overluminosity appears larger for later Be stars, while it is well known that emission characteristics are fainter for these stars. The very fast rotation of late Be stars, close to the break-up velocity, was considered as the most likely physical explanation. Cramer (1997) determined

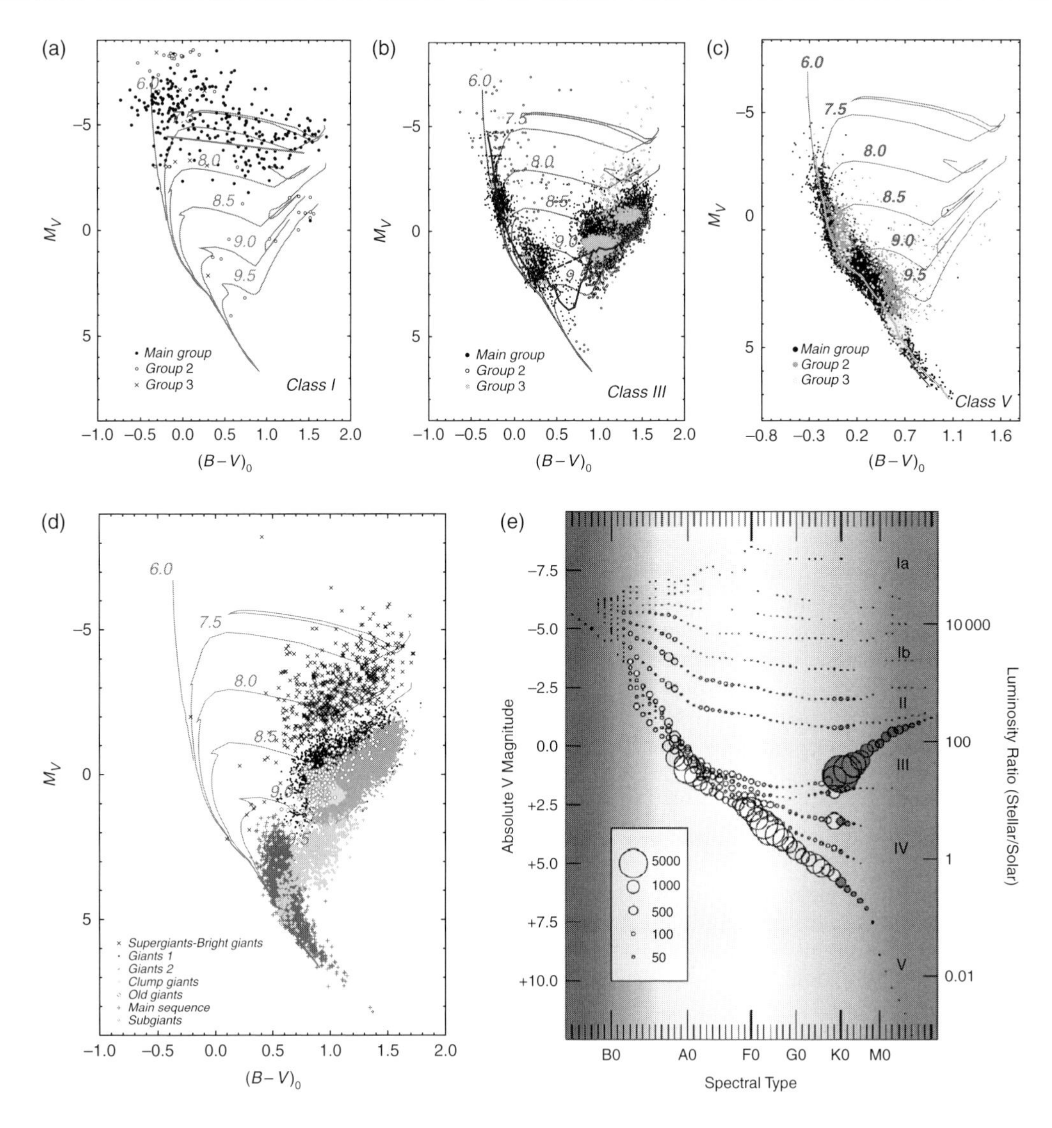

Figure 5.4 Top (a–c): calibration and individual absolute magnitudes for luminosity classes I, III, and V (classes II and IV are also included in the cited reference). Isochrones are shown for solar metallicity from Schaller et al. (1992). Groups labelled 'main group' are those with the largest numbers of stars. Lower left (d): groups resulting from the M_V versus $B-V$ calibration for all Hipparcos G–K survey stars, independent of luminosity class, and after eliminating binary and variable stars. Isochrones are again shown for solar metallicity from Schaller et al. (1992). In addition to the groups containing dwarfs and giants are five other groups on the giant branch: supergiants and bright giants, three groups of normal giants (clump stars rather concentrated in absolute magnitude and colour index, and two groups separated in absolute magnitude and colour index differing not only in luminosity but also in kinematic behaviour) and old giants belonging to the thick disk and halo. From Gómez et al. (1997, Figures 4–6, originals in colour). Lower right (e): HR diagram for 113 286 stars using Hipparcos parallaxes and spectral types from the Michigan Spectral Survey. The areas of the open circles scale to the number of stars per spectral type and luminosity class. The smallest circle represents five stars or fewer. From Sowell et al. (2007, Figure 2, original in colour).

the mean absolute magnitudes for 1580 O9–A2 stars of luminosity classes III–V in the Geneva photometric system. Lamers *et al.* (1997) and Schröder *et al.* (2004) determined the absolute magnitudes for O9V–B5III stars and found a relation between M_V from Hipparcos and from Schmidt–Kaler as a function of $v \sin i$, with slowly-rotating stars being significantly fainter by about 1 mag than rapidly rotating stars of the same spectral type (see also Section 7.7.1). Wegner (2000) determined the absolute magnitudes of OB and Be stars based on Hipparcos parallaxes, using a sample of 1207 unreddened and 441 reddened OB stars of luminosity classes

V–IV, III, Ia, and Iab–Ib–II, and a sample of 90 unreddened and 25 reddened Be stars. He derived a new calibration which differs from that of Schmidt–Kaler (in Schaifers & Voigt, 1982, Table 13) as well as from those of Briot *et al.* (1997) and Cramer (1997), attributed to an improved treatment of extinction correction. He found that both Be giants and dwarfs are brighter than normal stars of the same spectral type, but found no differences in the absolute magnitudes of fast and slow rotators. He provided tables of M_V versus spectral type and luminosity class, compared with values from Schmidt–Kaler (Figure 5.5). The study was extended to 6262 unreddened and reddened OB stars and 430 Be stars by Wegner (2006), who found that the O stars were systematically brighter than their previous estimates. Wegner (2007) extended the analysis to 30 986 unreddened and reddened A–M stars across luminosity classes I–V, finding slightly fainter absolute magnitudes than those of Schmidt–Kaler except for luminosity classes IV–V. The extent to which sample bias has been fully accounted for is, however, unclear.

Maíz Apellániz *et al.* (2004), see also Maíz Apellániz & Walborn (2003), constructed a catalogue of 378 Galactic O stars with accurate spectral classification, complete to $V < 8$ but including many fainter stars. Positions are largely taken from Tycho 2, with astrometric distances for 24 of the nearest taken from the Hipparcos Catalogue.

Fitzpatrick & Massa (2005) presented a calibration of optical (*UBV*, Strömgren $uvby\beta$, and Geneva) and near infrared (Johnson *RIJHK* and 2MASS) photometry for B and early A stars derived from Kurucz (1993) ATLAS9 model atmospheres. The B and early A stars are the most massive and luminous objects whose atmospheres are well represented by the simplifying assumptions of LTE physics, hydrostatic equilibrium, and plane-parallel geometry, and which therefore provide an important comparison with evolutionary models. They used 45 normal, nearby B and early A stars which have high-quality, low-resolution IUE spectra and accurate Hipparcos parallaxes. The calibration was based only on the ultraviolet spectral energy distributions, the absolute flux calibration of the V filter, and the Hipparcos distances to determine the appropriate model atmospheres. They used the Hipparcos distances to constrain $\log g$, using this to derive the stellar radius, R, from the spectral energy distribution matching the best-fitting ATLAS9 models. The other parameters determined as part of the fit were $T_{\rm eff}$, [m/H], the microturbulent velocity field in the atmosphere, $v_{\rm turb}$, and line-of-sight reddening $E(B-V)$. The gravities used to select the appropriate model were adjusted from the Newtonian value, GM/R^2, taking into account a correction for stellar rotation

$$g_{\rm spect} \equiv \frac{GM}{R^2} - \frac{(v \sin i)^2}{R} \tag{5.6}$$

Resulting R versus $T_{\rm eff}$ are shown in Figure 5.6, left, with $T_{\rm eff}$ versus spectral type in Figure 5.6, centre. Angular diameter measurements were available for three of their stars (γ Gem, α Leo and α Lyr) from Hanbury Brown *et al.* (1974), who obtained limb-darkened angular diameters of 1.39 ± 0.09, 1.37 ± 0.06 and 3.24 ± 0.07 mas, respectively. The radii derived by Fitzpatrick & Massa (2005), combined with the Hipparcos distances, yield angular diameters of 1.43 ± 0.02, 1.39 ± 0.02 and 3.28 ± 0.04 mas for the same stars; all are consistent to better than 1σ. Verification of the results was possible using independent estimates of $T_{\rm eff}$ and $\log g$ from Strömgren photometry. The formal fitting errors in both quantities is well-matched by estimates of the actual errors, demonstrating that the values derived from the Hipparcos fitting are free from obvious systematics. The consistency between the two determinations of $\log g$ also provides validation of the correction for stellar rotation, which becomes significant above $v_{\rm rot} = 200$ km s^{-1}. The results demonstrate that caution is needed when comparing theoretical models with stars placed in the HR diagram according to their spectroscopic-based gravities, even in the presence of moderate rotation.

A–F stars Gerbaldi *et al.* (1999) considered 71 nearby A0V stars, rejected some 30% of previously unknown binaries, defined a set of reference stars, and used the Hipparcos data compared with theoretical models to derive masses and ages. They constructed the resulting mass–luminosity relation for these stars (see Figure 7.5 below), also demonstrating good agreement with the mass–luminosity relation derived from the eclipsing binary studies of Andersen (1991).

Paunzen (1999) compared three independent lists of MK standards for B9–F2 stars: from Abt & Morrell (1995), from García (1989), and from Gray & Garrison (1987, 1989a,b) who introduced high $v \sin i$ standards into the system. They evaluated whether the different classification systems are intrinsically consistent regarding the luminosity types, whether the MK classification system is supported by Hipparcos distances, and whether the observed width of the main sequence is consistent with its theoretical value, defined as extending from the ZAMS to the TAMS (terminal-age main sequence) which has been estimated at about 1.2 mag for solar composition (Claret, 1995). Paunzen (1999) found that the three samples gave the same results, and that the inclusion of projected rotational velocities does not improve the accuracy. Ranges in M_V for luminosity classes III, IV, and V are shown in Figure 5.7. The apparent width

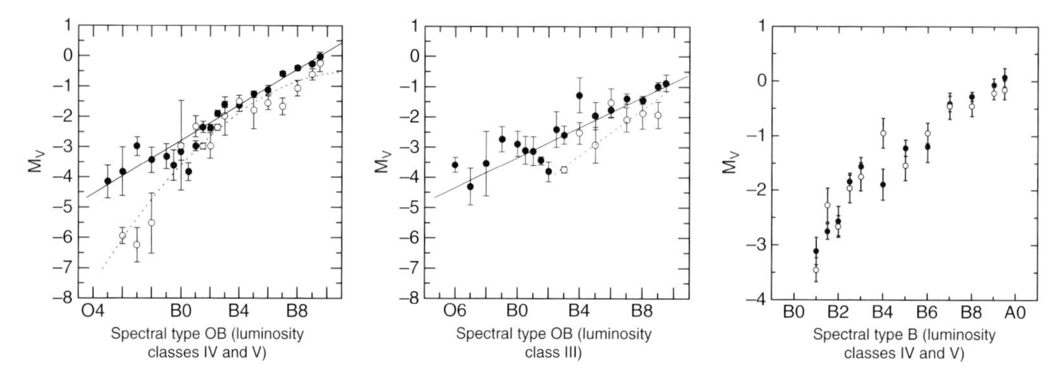

Figure 5.5 Left: absolute magnitudes of O, B (•) and Be (○) stars, of luminosity classes V–IV versus spectral type. Solid and dashed lines show the mean relation according to the number of stars and error bars for OB and OeBe stars respectively. Middle: same for luminosity class III. Right: relation between absolute magnitude and spectral type for B stars, divided into slow rotators ($v \sin i <$ 100 km s^{-1}, ○) and fast rotators ($v \sin i >$ 100 km s^{-1}, •). From Wegner (2000, Figures 1, 2 and 4).

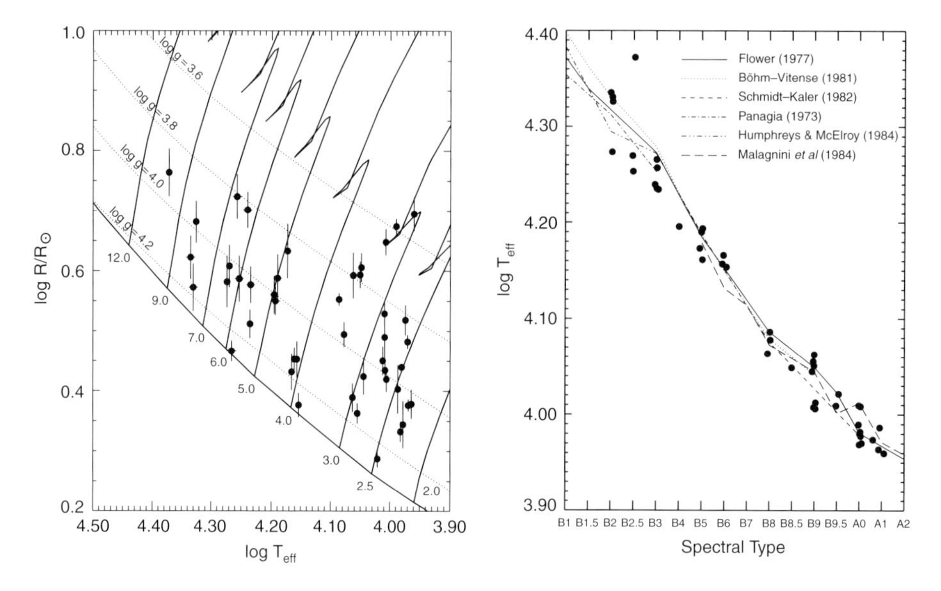

Figure 5.6 Left: theoretical HR diagram for 45 B and early A stars, in terms of $\log R/R_\odot$ *versus* $\log T_{\rm eff}$*, based on the models of Bressan et al. (1993). The thick solid curve shows the ZAMS, and the thin solid curves show the evolution tracks for the initial stellar masses listed along the ZAMS (units of $M_\odot$). Lines of constant Newtonian surface gravity, GM/R^2, are shown for several representative values. •: the 45 programme stars, where the radius, R, has been constrained by the Hipparcos distances. For most stars, the error bar for $T_{\rm eff}$ lies within the symbol. Right:* $\log T_{\rm eff}$ *as a function of spectral type for the same 45 stars. Lines show various published spectral type versus $T_{\rm eff}$ calibrations: Flower (1977); Böhm-Vitense (1981); Schmidt–Kaler (in Schaifers & Voigt, 1982, Table 13); Panagia (1973); Humphreys & McElroy (1984); and Malagnini et al. (1984). From Fitzpatrick & Massa (2005, Figures 1 and 9).*

of luminosity class V is about 2.5 mag for the whole relevant spectral range. There is no separation between luminosity class V and IV, and there is even a wide band (1.5 mag) where all three luminosity classes are found. The uncertain status of the luminosity class IV in the relevant spectral region has already been described by Keenan (1985). The width of the luminosity class III is especially broad, at about 3 mag, with only the brightest class III objects clearly separated from class V objects. The results are again consistent with those of Jaschek & Gómez (1998), but based on a much larger sample of stars. Paunzen (1999) concluded that *'the selected classification criteria for the luminosity classes are not able to distinguish between main sequence (V) and evolved giants (III), with the consequence that luminosity class IV should be rejected'*.

G–K dwarfs Two studies have estimated the absolute magnitude of K0V stars based on Hipparcos parallaxes.

Oudmaijer *et al.* (1999) used 200 objects with spectral types from the Michigan Spectral Survey, mostly within 100 pc. They accounted for Malmquist bias (an effect of order 0.2 mag in size), and estimated a mean absolute magnitude of $M_V = 5.7$, compared with 5.9 listed by Schmidt–Kaler (in Schaifers & Voigt, 1982, Table 13). Some 20% of objects were found to have a significantly brighter absolute magnitude, and they suggested that these are K0IV stars mis-classified as K0V. Butkevich *et al.* (2005b) carried out a similar analysis for 166 stars, also taking account of Malmquist bias. They found $M_V = 5.8$ with a spread of about 0.3 mag. They confirmed that some 20% of the sample are likely to be mis-classified in terms of spectral type, and that the inclusion of non-single stars may underestimate the absolute magnitude by about 0.05–1.0 mag.

Kotoneva *et al.* (2002b) showed that the luminosity of K dwarfs on the lower main sequence, $5.5 < M_V < 7.3$, is a simple function of metallicity, with $\Delta M_V = 0.04577 - 0.84375\,[\mathrm{Fe/H}]$ where ΔM_V is the luminosity difference relative to a fiducial solar metallicity isochrone. The relationship between luminosity, colour and metallicity for K dwarfs was found to be very tight, providing a new distance indicator with a range of possible applications. Alternatively, they could derive metallicities with accuracies better than 0.1 dex based on their position in the Hipparcos colour-magnitude diagram. Kotoneva *et al.* (2002a) determined metallicities for 431 single Hipparcos K dwarfs. They used isochrones to mark the stars by mass, and selected a subset of 220 stars complete within a narrow mass interval. They then fitted the data with a model of the chemical evolution of the solar cylinder based on the Galactic chemical evolution model of Chiappini *et al.* (2001), finding that only a modest cosmic scatter is required to fit the age–metallicity relation. Their model assumes two main infall episodes for the formation of the halo-thick disk and thin disk, respectively, the data confirming that the solar neighbourhood formed on a long time scale of the order of 7 Gyr.

Karataš & Schuster (2006) made a detailed calibration of absolute magnitudes for a large sample of FGK stars based on the normalised ultraviolet excess, $\delta_{0.6}$, defined and calibrated by Sandage (1969) relative to the Hyades main sequence in the two-colour $(U-B)$, $(B-V)$ diagram. For dwarf and turn-off stars, for example, a calibration based on Hipparcos parallaxes with $\sigma_\pi/\pi \le 0.1$ yields a dispersion of ± 0.24 in M_V.

G–K/M giants Egret *et al.* (1997) constructed the HR diagram for 287 G5–M3 stars on the red giant branch, classified in the MK system, and for which Hipparcos parallaxes have an accuracy of < 15%. Uncorrected for extinction but corrected for Malmquist bias (Figure 5.8), their absolute magnitudes are brighter than most other determinations.

Keenan & Barnbaum (1999) used Hipparcos parallaxes to examine the calibration of MK luminosity classes for cool giants [the first author, at age 91, was using the Hipparcos data 56 years after the publication of his original spectral analysis, and thereby gained the title of the longest publishing career in leading American journals in modern astronomy (Boeshaar, 2000)]. They constructed the HR diagram for individual giants down to $V = 6.5$ for classes IIIb and III, and to $V = 5.5$ for class IIIa, revealing for the first time the fine structure of the HR diagram for field giants (Figure 5.9). Between types G7 and K2, for example, the group of clump giants lying about 1 mag below the class III giants is particularly conspicuous.

Ginestet *et al.* (2000) analysed a sample of about 500 MK standards of cool star types (G to M), comparing them with absolute magnitudes from both Hipparcos and those from Schmidt–Kaler (in Schaifers & Voigt, 1982, Table 13). In contrast to the early-type stars results, they found that the absolute magnitude of the giants does not overlap that of the dwarfs, and that the difference between the Hipparcos and Schmidt–Kaler calibrations increases with the relative parallax error: for $\sigma_\pi/\pi \le 0.05$ only 3% of the stars show a discrepancy of one luminosity class, while this reaches 54% for $0.25 < \sigma_\pi/\pi < 0.50$.

Other studies Other studies examining the MK system in the light of the Hipparcos data include those by Stock *et al.* (1997); Stock & Stock (1999, 2001); Stock *et al.* (2002), Garrison (2000b, 2002), Grenon (2002), and Abt (2002).

Effects of binarity As discussed extensively in Chapter 3, many stars are known to be members of binary systems, and considerable evidence exists from Hipparcos astrometry, also in combination with long-term proper motions, that many stars have otherwise hidden companions. Such hidden companions commonly result in a red excess in the composite spectrum of the stellar system, and tend to increase $B-V$ and $V-I$, leading to a later spectral classification of the primary, inappropriate for its absolute magnitude (Gontcharov, 2000). The brightness anomaly of binary F stars and associated implications has been discussed by Suchkov (2001). The assessment of absolute magnitudes of the hot components of 135 stars with composite spectra has been investigated by Ginestet & Carquillat (2002).

5.5 Luminosity indicators using spectral lines

Spectral features used as indicators of spectral type and luminosity class are listed by, e.g. Jaschek & Jaschek

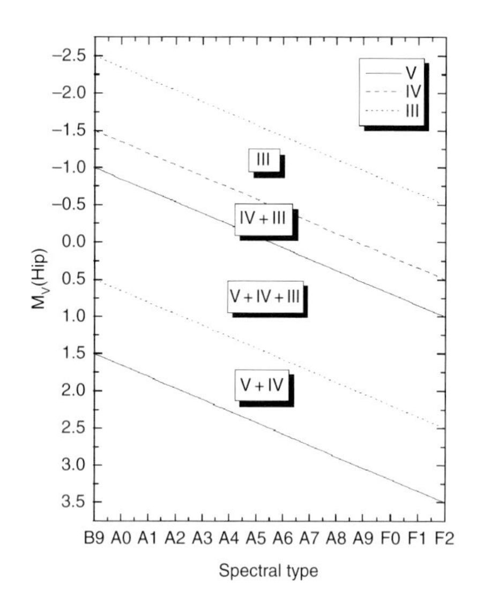

	Hipparcos, Paunzen (1999)			S–K (1982)	
	V	IV	II	V	III
B9	+1.50 : −1.00	+1.50 : −1.50	+0.50 : −2.50	+0.2	−0.6
A0	+1.65 : −0.85	+1.65 : −1.35	+0.65 : −2.35	+0.6	0.0
A1	+1.81 : −0.69	+1.81 : −1.19	+0.81 : −2.19	+1.0	+0.2
A2	+1.96 : −0.54	+1.96 : −1.04	+0.96 : −2.04	+1.3	+0.3
A3	+2.12 : −0.38	+2.12 : −0.88	+1.12 : −1.88	+1.5	+0.5
A4	+2.27 : −0.23	+2.27 : −0.73	+1.27 : −1.73		
A5	+2.42 : −0.08	+2.42 : −0.58	+1.42 : −1.58	+1.9	+0.7
A6	+2.58 : +0.08	+2.58 : −0.42	+1.58 : −1.42		
A7	+2.73 : +0.23	+2.73 : −0.27	+1.73 : −1.27	+2.2	+1.1
A8	+2.88 : +0.38	+2.88 : −0.12	+1.88 : −1.12	+2.4	+1.2
A9	+3.04 : +0.54	+3.04 : +0.04	+2.04 : −0.96		
F0	+3.19 : +0.69	+3.19 : +0.19	+2.19 : −0.81	+2.7	+1.5
F1	+3.35 : +0.85	+3.35 : +0.35	+2.35 : −0.65		
F2	+3.50 : +1.00	+3.50 : +0.50	+2.50 : −0.50	+3.6	+1.7

Figure 5.7 Left: the different bandwidths according to the table at right for the three luminosity classes, the lower limit of the luminosity class IV is the same as for class V. Right: derived upper and lower boundaries as shown in the figure for the different luminosity classes; the last two columns are the 'standard' values from Schmidt–Kaler (in Schaifers & Voigt, 1982, Table 13). From Paunzen (1999, Figure 1 and Table 3).

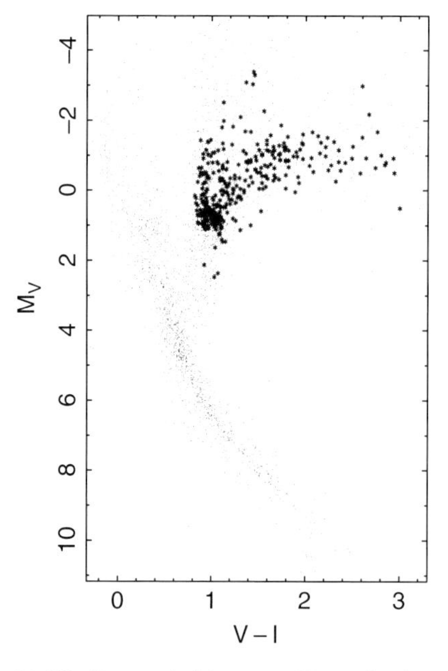

Figure 5.8 HR diagram in M_V versus $V - I$ for the Hipparcos G5–M3 stars on the giant branch with $\sigma_\pi/\pi < 0.15$ ($\star$). Dots are stars of various spectral types to illustrate the general shape of the HR diagram. From Egret et al. (1997, Figure 1).

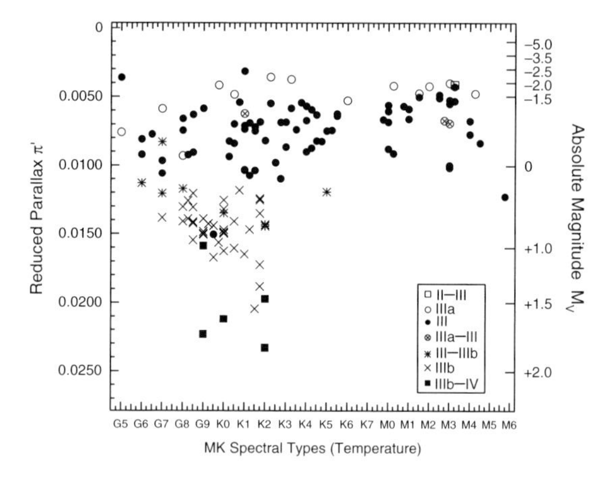

Figure 5.9 Hipparcos parallaxes and revised MK types for cool stars of normal solar composition. The diagram for stars of classes III and IIIb extends to $V = 6.5$, $V = 5.5$ for IIIa, and $V = 4.5$ for II–III. The horizontal axis shows four divisions per spectral type. For example, the halfway mark between types G0 and G1 indicates type G0.5, and the types indicated by the tick marks to the left and right of G0.5 are G0+ and G1−, respectively. From Keenan & Barnbaum (1999, Figure 1).

(1987). Amongst these, Hipparcos results have been discussed in the case of the Ca II H and K lines, and the O I triplet.

5.5.1 Wilson–Bappu effect

The Wilson–Bappu effect was a surprising and puzzling relationship between emission-line strength and stellar luminosity quantified 50 years ago by Wilson & Bappu (1957), although a correlation had been noted much earlier (Stratten, 1925).

Wilson & Bappu (1957) found that, for a sample of 185 stars of spectral types G, K, and M, and all luminosity classes, the logarithm of the width of the 'outer edges' of the Ca II H (396.8 nm) and K (393.3 nm) emission lines (see Figure 5.10), converted to a velocity width in km s^{-1}, was correlated with stellar absolute magnitude over a very large range in luminosity of $\sim$15 mag. The Sun also fits this correlation. There appeared to be no dependence on spectral type, surface temperature, or metallicity. For the next decade the phenomenon remained unexplained, but the relation was nevertheless used as an empirical tool for estimating distances to late-type stars in the absence of accurate trigonometric parallaxes (see Linsky, 1999, for a short review).

Relatively strong spectral lines in the visible and near ultraviolet, such as the H I Balmer series and the resonance lines of metals, play an important role in studying the structure of the chromosphere and transition zone (between chromosphere and corona). Explanations for the line width correlations as arising from Doppler broadening due to turbulent motion remained inadequate. An advance came in the 1970s with a picture of abundance broadening due to large optical thickness, and the corresponding inclusion of 'partial redistribution' in the treatment of radiative transfer. This describes the frequency redistribution of scattered line photons in the line core, formed at relatively shallow depths where non-radiative heating dominates, with nearly coherent scattering in the line wings, which are sampling progressively greater depths. Applied to the optically thick chromospheric resonance lines, this explained the Wilson Bappu effect for an atmosphere in hydrostatic equilibrium as a consequence of increasing chromospheric mass column density with decreasing gravity.

Wilson & Bappu (1957) and Wilson (1959) derived the following expression for the K-line width–luminosity relation

$$M_V = 27.59 - 14.94 \log W_0 \qquad (5.7)$$

Subsequent studies have re-determined this relation from different datasets, most recently using the Hipparcos distances to establish the luminosities, and thus exploring the physical origin of the relation. Wallerstein *et al.* (1999) used the Hipparcos parallaxes to derive the (weighted) relation

$$M_V = 28.83 - 15.82 \log W_0 \qquad (5.8)$$

They also found no dependence on emission strength or spectral type, except for larger widths for G stars of luminosity class Ib and II. Systematic corrections were also found for K- and M-type giants and supergiants based on Hipparcos absolute magnitudes as well as those from binary star isochrones by Parsons (2001). For a sample of 119 nearby stars with high-quality CCD spectra spanning a wide range of luminosities ($-5 < M_V < 9$), Pace *et al.* (2003) derived

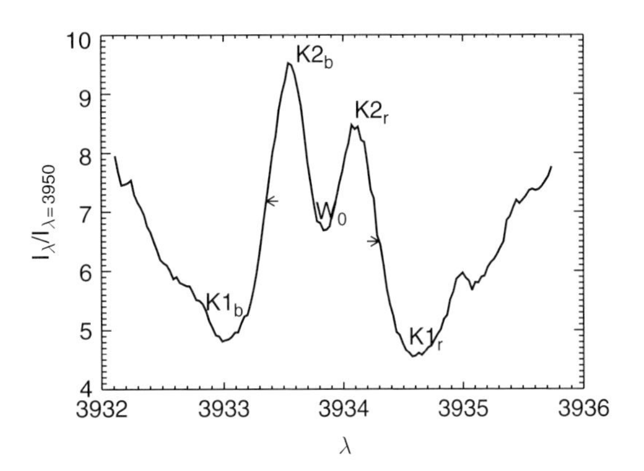

Figure 5.10 Illustration of the Ca II K emission-line profile for HD 4128, showing the emission line with central self-absorption, and the various line-width measurements including W_{base} ($=K1_r-K1_b$) employed by Scoville & Mena-Werth (1998) and described further in their Figure 1. From Pace et al. (2003, Figure 2).

$$M_V = 33.2 - 18.0 \log W_0 \qquad (5.9)$$

with a possible dependence on metallicity for [Fe/H] $\lesssim -0.4$. While the scatter of 0.6 mag is too large for the use of the correlation as a reliable distance indicator for single stars, they considered its applicability to distances of star clusters (provided they are not too metal-poor), investigating sensitivity to Lutz–Kelker bias, reddening and rotation, and concluding that the scatter arises mainly from random errors. Applied to the open cluster M 67, they derived values of 9.62–9.65 mag, in good agreement with all recent distance modulus estimates, in the range $9.55 \leq \mu_V \leq 9.85$ mag, including those based on main-sequence fitting.

Analogous relations have been found for other chromospheric emission lines with similar excitation conditions, including Mg II h and k, as well as Lyα and Hα. From IUE ultraviolet observations of the Mg II k-line (279.6 nm) 'base width' W_{base} for 94 stars Scoville & Mena-Werth (1998) used the Hipparcos parallaxes, and bolometric corrections from Böhm-Vitense (1989), to derive

$$M_V = 38.20 - 16.31 \log W_{base} \qquad (5.10)$$

with a correlation coefficient of 0.88, $\sigma = 1.17$ mag over a range of 18 mag, and no obvious dependence on metallicity. The improved absolute magnitudes did not, however, result in a reduced scatter compared with previous calibrations, rather therefore appearing to be inherent to factors affecting the Mg II k-line itself. A parallel study of IUE data by Elgarøy *et al.* (1999) updated their pre-Hipparcos study (Elgarøy *et al.*, 1997) covering 65 slowly rotating stars of spectral classes F, G, K and M, luminosity classes I–V, and of different activity levels. The independent well-defined absolute magnitudes yield a variety of detailed dependences. For the largest class of 26 K stars, they derive

$$M_V = (38.9 \pm 2.3) - (19.8 \pm 1.2) \log W_{50\%\ \mathrm{peak}} \quad (5.11)$$

The relation varies with spectral class (their Figure 3) with significant scatter around the regression lines; little evidence for a correlation for the F stars; a possible difference between dwarfs and giants of spectral class K; a magnetic activity dependency in dwarfs; and a metallicity dependence for giants and supergiants.

After excluding certain classes of stars for which the relation is expected to be more complex (specifically, chromospherically-active binaries, Mira and Cepheid variables, and rapid rotators) Cassatella *et al.* (2001) studied an enlarged sample of 230 stars, with 34 in common with those of Elgarøy *et al.* (1999), resulting in a similar dependency (Figure 5.11, left)

$$M_V = (34.56 \pm 0.29) - (16.75 \pm 0.14) \log W_0 \quad (5.12)$$

over the range $-5.4 \leq M_V \leq 9.0$. They also found a flatter relation for luminous G-type stars, but otherwise without evident dependence on effective temperature or metallicity.

RS Canum Venaticorum (RS CVn) stars are post-main-sequence stars in close binary systems, important for studies of chromospheric heating since they have the highest levels of chromospheric activity for their spectral class. They are predominantly tidally-locked, with the high activity arising from rapid rotation induced by spin–orbit coupling combined with the deepened convection zone of a post-main-sequence envelope. Özeren *et al.* (1999) studied the Mg II k-line for 41 systems observed with IUE. Although a similar overall correlation is observed, there is a dependence on luminosity class, with higher luminosity objects showing broader lines (Figure 5.11, right). Their models suggest that $\log W$ is mainly sensitive to surface gravity, and rather insensitive to effective temperature.

The availability of Hipparcos luminosities has shifted the interest of the Wilson–Bappu effect away from its use as a distance indicator, to a probe of detailed chromospheric models as a function of stellar type and luminosity. Ca II chromospheric emission continues to provide empirical indicators of stellar age, rotation, and evolutionary phase, and remains of fundamental interest for the understanding of the temperature and density structure of stellar atmospheres. The magnetic field responsible for the emission features through chromospheric heating is also responsible for a wide range of stellar activity phenomena, including coronal heating, winds, flares, and X-ray emission.

5.5.2 Equivalent width of O I

Use of the equivalent width of the O I 777.4 nm triplet for luminosity calibration has been known from the work of Merrill (1925), Keenan & Hynek (1950), Osmer (1972), and more recent work cited by Arellano Ferro *et al.* (2003). Arellano Ferro *et al.* (2003) made a revised calibration of M_V versus O I equivalent width, using improved reddening and distance estimates for a sample of 27 calibrator stars of spectral types A to G, based on accurate parallaxes and proper motions from the Hipparcos and Tycho Catalogues. Their calibration predicts absolute magnitude with accuracies of ± 0.38 mag for a sample covering a large range of M_V, from -9.5 to $+0.35$ mag. The variation of the feature in the classical Cepheid SS Sct provided a phase-dependent correction to random phase O I feature strengths in Cepheids, thus predicting mean absolute magnitudes using the above calibration. Applying such a correction led to an enlarged list of 58 calibrators by adding M_V and O I triplet strength data for 31 classical Cepheids, leading to the possibility of calculating mean Cepheid luminosities from essentially random phase observations of the O I feature.

5.5.3 Interstellar lines

The use of intensities of interstellar lines as distance indicators has a long history, with early work on the Ca II K equivalent width–distance relation reported by Struve (1928), Sanford (1937), Wilson & Merrill (1937), Evans (1941), Beals & Oke (1953), and others. The extent to which the large scatter was due to the inhomogeneous distribution of the absorbing clouds, and how much was due to distance errors, was unclear, such that use of the relation as a distance indicator remained limited. Many later studies have used the lines of Ca II, Na I, K I, and other elements to study both spatial and velocity structure of the interstellar medium, as reported in Section 8.5. However, despite the increasing quality and resolution of the spectroscopy, for nearly half a century it was difficult to improve on the early results of the equivalent width–distance relation because of problems associated with estimating stellar distances in the range of particular interest for bright early-type stars (hundreds of pc to kpc). Diffuse interstellar bands (DIB), of mostly unknown carrier origin, have also been long-known (Heger, 1921) and used as distance indicators (Herbig, 1995). In a recent survey of interstellar lines, Galazutdinov (2005) reported a tight relation between distance and strength of Ca II lines, and a poor relation with the intensities of other interstellar features.

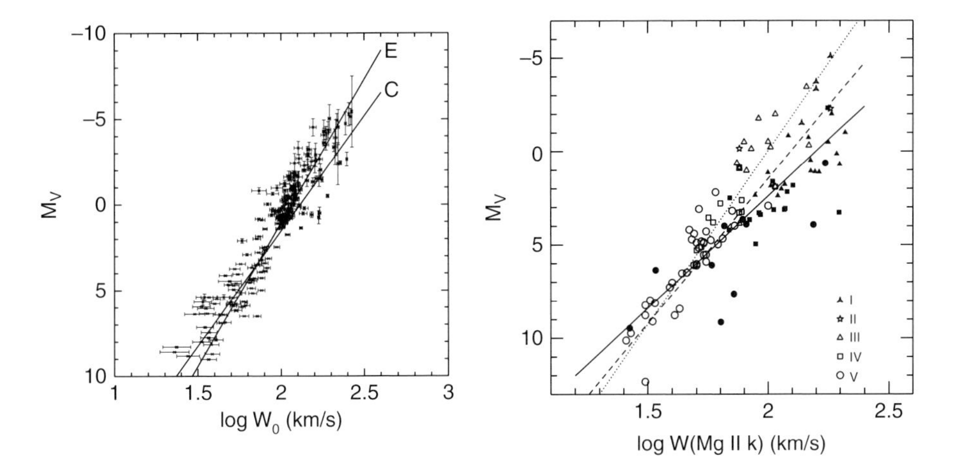

Figure 5.11 Left: the Wilson–Bappu relation in the Mg II k-line, for the 230 stars from Cassatella et al. (2001). Regression line 'E' corresponds to Equation 5.12, i.e. accounting for measurement errors on both variables. Line 'C' takes into account only the errors on M_V. From Cassatella et al. (2001, Figure 3). Right: the Wilson–Bappu relation in the Mg II k-line for RS CVn stars (solid line, and filled symbols). The dashed line shows the relation for all objects including a sample of active and quiet single stars, shown separately by open symbols and fitted by the dotted line. From Özeren et al. (1999, Figure 2).

Stimulated by these results, Megier *et al.* (2005) used the Hipparcos Catalogue as a uniform source of distance data, to investigate the distance dependence of the equivalent widths of several interstellar lines: the Ca II H and K lines (396.8468 and 393.3663 nm respectively), the K I line (769.8974 nm), and the CH line (430.0321 nm). For 147 early-type stars, they found a strong correlation between equivalent width and Hipparcos parallax, given by $\pi = 1/[2.78\,\mathrm{EW(K)} + 95]$ and $\pi - 1/[4.58\,\mathrm{EW(H)} + 102]$, where π is in arcsec and the equivalent width, EW, is in 10^{-4} nm (Figure 5.12). They thus showed that the equivalent widths of the Ca H and K lines can be used to determine distances to (for example) OB stars in the Galaxy, objects for which the absolute magnitudes are rather poorly known, and where the objects lie predominently at low Galactic latitudes appropriate to the scale height of Galactic Ca II of around 800 pc (Smoker *et al.*, 2003). Their analytical expressions, yielding a finite parallax even for zero absorption, also show that space within $\sim$ 100 pc of the Sun contains very little Ca II, in agreement with the known dimensions of the Local Bubble (Section 8.5.1).

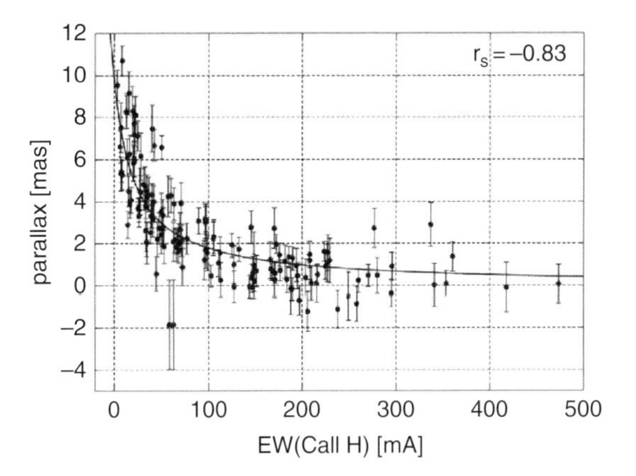

Figure 5.12 Equivalent width of the Ca II H line versus the Hipparcos parallax. The fit to the data is given by $\pi = 1/[4.58\,EW(H) + 102]$, where π is in arcsec and the equivalent width, EW, is in 10^{-4} nm. The value $r_S = -0.83$ is the Spearman rank correlation coefficient. A similar strong correlation was also found for the Ca I K line, but not for the K I or the CH lines. From Megier et al. (2005, Figure 2).

5.6 Use of standard candles

The following luminosity calibrators are currently used to estimate the distance to the Large Magellanic Cloud (LMC, Figure 5.13):

(a) Using Population I objects: these include Cepheids, red clump giants, Mira variables, eclipsing binaries, and SN 1987a, as well as main-sequence fitting (including use of the main-sequence turn-off). Estimates using Cepheids are themselves subdivided into methods based on trigonometric parallaxes, main-sequence fitting from open clusters, and the Baade–Wesselink method.

(b) Using Population II objects: these include subdwarf main-sequence fitting, horizontal branch trigonometric parallaxes, RR Lyrae stars (pulsationally-unstable horizontal branch stars), along with globular cluster dynamical models, and the use of the cooling sequence of white dwarfs in globular clusters. Estimates for the

RR Lyrae stars are themselves subdivided into methods based on statistical parallaxes, Baade–Wesselink method, and double-mode pulsators.

Hipparcos has contributed to most of these in one form or another, with the exception of methods based directly on eclipsing binaries, SN 1987a, and globular cluster dynamics. Particular insight has come from its contribution to globular cluster distance and age determination based on subdwarf main-sequence fitting, and on the theoretical understanding and practical use of the red clump giants.

Discussion of the distance scale in the literature can be confusing because of the many different 'paths' that can be followed: for example, a globular cluster distance can be determined from subdwarf main-sequence fitting, with that distance then used to derive the absolute magnitude of the RR Lyrae stars in that cluster, which are then applied as standard candles to determine the distance to the LMC. Or RR Lyrae absolute magnitudes can be derived from statistical parallaxes of relatively nearby stars, and then used to estimate the distance to the LMC. Figure 5.13 illustrates the main approaches found in the current literature and summarised in the subsequent sections. In the figure, tracers are divided into Population I and II indicators, and methods are distinguished between those (partly) based on Hipparcos data or otherwise. Anticipating the results of the following sections, some of the current resulting distances derived for the LMC according to each approach are shown at the top.

The following sections will review the methods available for these large-scale distance determination, with focus on those that have benefited from the Hipparcos data, leading to a summary of the various estimates of the LMC distance modulus. As apparent from results for open clusters (Chapter 6) and stellar evolution models (Chapter 7), open cluster main-sequence fitting will be considered as providing important constraints for developing stellar evolutionary theories once a reliable distance is known, rather than *vice versa*.

For discussions of distances to the LMC, distances will be given in terms of distance modulus, $\mu \equiv m - M = -5\log\pi - 5$, where π is the parallax in arcsec. This form is convenient for simply incorporating effects (and discussions) of reddening, and for assessing the apparent magnitude at which a particular tracer of known absolute magnitude will be observed. Where two errors for a distance estimate are given sequentially, the first refers to statistical errors, the second to estimates of the systematics. The LMC is at a distance of $\sim$ 48–55 kpc, corresponding to a trigonometric parallax of ~ 0.02 milliarcsec, and to a distance modulus of ~ 18.4–18.7 mag.

5.7 Population I distance indicators

5.7.1 Classical Cepheids

Cepheids are pulsationally-unstable stars, located in a narrow region of the HR diagram, with typical periods in the range 2–30 days, but extending to periods in the range 1–100 days. There are two subgroups. Classical Cepheids (or δ Cephei stars) are young high-mass core helium burning supergiants; they are are Population I (disk-component) objects, found in the Galactic plane, notably in spiral arms and in open clusters. Type II Cepheids, or W Virginis stars, are low-mass, Population II (metal-poor, spheroidal component) stars, found at high Galactic latitudes, in the bulge, and in globular clusters.

The immediate precursors of the classical Cepheids are massive young O and B main-sequence stars. As they evolve rapidly off the main sequence, they pass through a zone where their outer atmospheres are unstable to periodic radial oscillations, a region of the HR diagram referred to as the instability strip. In a simplified picture, high-mass stars pass through the instability region at higher luminosities (cooler temperatures) than lower-mass stars, resulting in a Cepheid instability strip which slants upwards and to the right in the HR diagram. Basic considerations lead to the prediction of a mass–luminosity relation (and hence also a radius–luminosity) for Cepheids. Since neither mass nor radius are easily observable for the majority of stars, the mass–luminosity relation cannot be used to predict luminosities and thus distances.

The importance of Cepheids as distance indicators is that there exists a correlation between period and luminosity, discovered empirically (Leavitt, 1908; Leavitt & Pickering, 1912), and subsequently explained theoretically (a historical review is given by Fernie, 1969). The relationship nevertheless shows a significant scatter about the mean line, even when corrected individually for reddening, due to the finite (temperature) width of the instability strip. If a colour-term is introduced (Martin *et al.*, 1979), the scatter is reduced to within observational uncertainties, suggesting that the basic Cepheid relation is a period–luminosity–colour one, e.g.

$$\langle M_V \rangle = \alpha \log P + \beta(\langle B_0 \rangle - \langle V_0 \rangle) + \rho_2 \tag{5.13}$$

or the equivalent in other magnitudes and colours. An accurate empirical value for the colour-term β, and whether it varies with period, has been a matter of uncertainty and debate, in part attributable to the correlation of α and β in any analysis.

Many studies fit V and I band relations independently, with the difference in resulting distance moduli assumed to be due to reddening. An alternative but equivalent approach is to fit an appropriate relation to a reddening-free 'Wesenheit index' for each Cepheid (see,

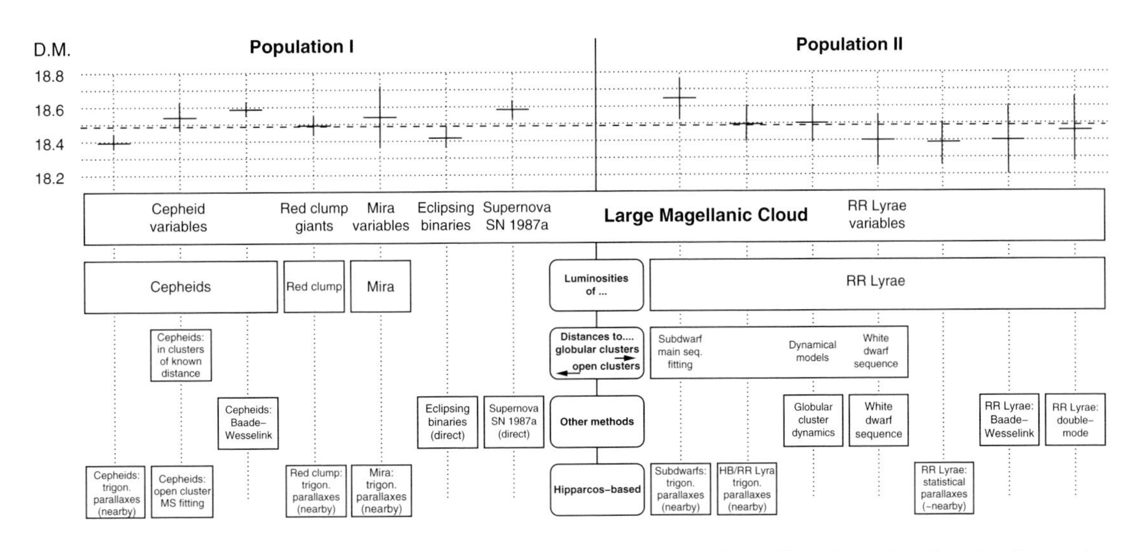

Figure 5.13 Estimated distances to the LMC summarised in this chapter. Methods are divided into those based on Population I and II indicators. Those (partly) relying on the Hipparcos data are labelled in the bottom row, and those derived independently of Hipparcos are shown in the row above. The next row indicates methods based on distance estimates to intermediate systems: open clusters (left) or globular clusters (right). The row above that indicates methods which rely on estimated luminosities of specific stellar types, which are also then observed in the LMC. The top panel indicates the resulting distance modulus and standard error derived for the LMC according to each method (see Table 5.6), with the straight mean of $(m-M)_0 = 18.49$ *indicated.*

e.g. Madore, 1982; Tanvir, 1999). For *VI* photometry, the index is defined as $W_{VI} = \langle V\rangle - R[\langle V\rangle - \langle I\rangle]$, which is extinction independent if $R = A_V/(A_V - A_I)$.

While the Cepheid period–luminosity relation has traditionally provided the most accurate method to derive distances to nearby galaxies, many complications have been encountered in pursuing such an objective to an unambiguous conclusion. Thus while the Hubble Space Telescope Key Project on the Extragalactic Distance Scale originally targeted a Cepheid distance to the Virgo cluster (some 10 times more distant than possible for Cepheid distances measured from ground), irregular velocities superimposed on the Hubble expansion component, due to gravitational perturbations from nearby galaxies, complicate the determination of H_0 (Freedman *et al.*, 2001).

There is an enormous literature on Cepheid variables, and their application to the determination of the astronomical distance scale, both within the Galaxy, and beyond. Groenewegen & Oudmaijer (2000), for example, tabulate the slope of the period–luminosity relation for 38 determinations from the literature for Cepheids in the Galaxy, the LMC and the SMC, in *V*, *I*, and *K* (mostly from a few studies in 1998–99). This section, as elsewhere in this review, concentrates on the specific contributions made by the Hipparcos observations which, of course, represent only a small part of a larger more complex picture; for references to early work see, e.g. Fernie (1969, for a historical perspective) and Feast & Walker (1987).

A main objective of Cepheid studies is to establish the slope and zero-point of the period–luminosity relation, such that an observed period directly yields the object's luminosity and thereby its distance. Until the Hipparcos results, the most accurate zero-point for the period–luminosity relation came from Cepheids in open clusters and associations through main-sequence fitting (see, e.g. Laney & Stobie, 1994).

One of the fundamental related questions is whether the period–colour and period–luminosity relations for classical Cepheids in the Galaxy, and in the Large and Small Magellanic Clouds, have the same or different slopes and zero-points; differences would greatly complicate the use of Cepheids for the extragalactic distance scale. In the review by Sandage & Tammann (2006), they conclude that the slopes and zero-points do differ, presumably due to metallicity although, despite many studies, *'it is yet to be proved or rejected that differences in metallicity are the cause'*. In contrast van Leeuwen *et al.* (2007), using Hipparcos data, conclude that the slope is the same.

Hipparcos calibration of the period–luminosity relation In general (a notable exception being van Leeuwen *et al.* 2007) a slope for the Galactic period–luminosity relation is adopted from other empirical or theoretical analyses, and the Hipparcos data have then been used to determine only the zero-point. Luminosities (and zero-points of the period–luminosity relation) can be obtained by various means (Feast, 1999): from trigonometric parallaxes, from proper motions in combination with radial velocities, from Cepheids in clusters and associations, from Baade–Wesselink luminosities, and from binary systems in which Cepheid luminosities are derived from the spectral types of hot companions detected in the far ultraviolet (Evans, 1991, 1992; Evans *et al.*, 1998a).

Groenewegen (1999) correlated the Hipparcos Catalogue with the Cepheid database of Fernie *et al.* (1995) to establish that there are 280 Cepheids in the Hipparcos Catalogue, of which 32 are either W Vir stars (Type II Cepheids), double-mode Cepheids, or Cepheids with unreliable solutions or without photometry. Of the 248 remaining classical Cepheids, 32 are first-overtone pulsators. They also compiled I-band and near-infrared photometry for these objects from the literature.

The mean standard error of the 223 Hipparcos Cepheid parallaxes considered by Feast & Catchpole (1997) is ~ 1.5 mas. The majority are distant, $d \gtrsim 500$ pc, such that the parallaxes are typically very small, and of little individual value. The closest is Polaris (α UMi = HIP 11767) with $\pi = 7.56 \pm 0.48$ mas or $d = 132 \pm 8$ pc ($\pi = 7.72 \pm 0.12$ mas in the global re-reduction used by van Leeuwen *et al.* 2007).

Trigonometric parallaxes Feast & Catchpole (1997) restricted their analysis to the 26 Cepheids with the largest astrometric weight, and obtained (their solution 6, including five objects treated as overtone pulsators)

$$\langle M_V \rangle = -2.81 \log P - 1.43(0.10) \qquad (5.14)$$

with P in days, and where the slope was adopted from that obtained by Caldwell & Laney (1991) from 88 Cepheids in the LMC. Combined with an appropriate correction for the metallicity dependence of +0.042 (from Laney & Stobie, 1994), this relation gave $\mu_{\rm LMC} = 18.70 \pm 0.10$ mag, compared with a value of 18.50 mag widely adopted previously. They argued that this 0.2 mag revision in the LMC distance modulus implies a 10% decrease in the Hubble constant, at least for those estimates of H_0 based on Cepheid observations with an adopted $\mu_{\rm LMC} = 18.50$.

The result was criticised by various authors: Szabados (1997) on the basis of unrecognised binarity; Madore & Freedman (1998b) who took account of intervening dust to argue that the Cepheids were no brighter than previously thought (see below); and Oudmaijer *et al.* (1998) who appealed to the Lutz–Kelker effect to argue that the distances had been underestimated.

Sandage & Tammann (1998) argued that the comparison of the Hipparcos re-calibration with others should be made using only local Galactic Cepheids, not based on Cepheids in the LMC that require a set of precepts not directly connected to the Hipparcos recalibration. In other words, it is possible to go directly from Equation 5.14 to other galaxies without first deriving a distance to the LMC. They estimated that the Feast & Catchpole (1997) results, using only Galactic Cepheids, gave a correction of about 4% or less to their value of $H_0 = 58 \pm 7\,{\rm km\,s^{-1}\,Mpc^{-1}}$ based on Type Ia supernovae (Saha *et al.*, 1997), which used $\langle M_V \rangle = -2.76 \log P - 1.43$, keeping all other factors and assumptions the same. Feast (1998c) in turn responded that a different reddening system was used in practice, which would still require a difference in distance scale of about 8%, compared with the 10% previously suggested.

Meanwhile, Feast & Whitelock (1997a) determined the zero-point of the Cepheid period–luminosity–colour relation, in the form established by Pont *et al.* (1994, cf. their zero-point of -2.27), based on the Hipparcos trigonometric parallaxes

$$\langle M_V \rangle = -3.80 \log P + 2.70(\langle B_0 \rangle - \langle V_0 \rangle) - 2.38(0.10) \qquad (5.15)$$

The zero-point in this equation can also be derived in a different way, based on the Hipparcos proper motions. The basic idea (described below) is that the value of the Oort constant A obtained from proper motions is essentially independent of the adopted distance scale, while that derived from radial velocities varies inversely with the distance scale, such that a comparison of the two values leads to a value for the period–luminosity–colour zero-point (cf. Equation 5.15) of -2.42 ± 0.13. Feast *et al.* (1998) similarly derived a proper-motion based value of the zero-point of the period–luminosity relation (cf. Equation 5.14) of -1.47 ± 0.13. They concluded that a combination of the radial velocity observations with the Hipparcos proper motions leads to a Cepheid period–luminosity relation in good agreement with that found using the Hipparcos parallaxes, and again confirming $\mu_{\rm LMC} = 18.70$ mag.

Pont (1999a) used appropriate Monte Carlo simulations to conclude that the Feast & Catchpole (1997) analysis is indeed unbiased, although they suggested increasing the error bar to ± 0.16 mag. Lanoix *et al.* (1999) also supported the original analysis on the basis of Monte Carlo simulations and consideration of the Lutz–Kelker bias, arguing that the Feast & Catchpole (1997) weighting gives the best zero-point and lowest dispersion, and obtaining $\langle M_V \rangle = -2.77 \log P - 1.44(0.05)$; this relation was subsequently used to determine the distances of 36 nearby galaxies by Paturel *et al.* (2002). The same conclusion was reached analytically by Koen & Laney (1998). Baumgardt *et al.* (1999) constructed a more carefully restricted sample, deriving a zero-point (in Equation 5.14) of -1.50 ± 0.17.

Madore & Freedman (1998b) went beyond the V-band period–luminosity relation, and considered the implications of the Hipparcos data for the calibration of the relation from the B-band to the near-infrared K-band, suppressing suspected overtone pulsators. They argued that the Hipparcos data alone do not allow distinguishing metallicity effects and line-of-sight reddening. Arguing that the various (Hipparcos and other) distance estimates to the LMC, at the time, gave both larger (18.65 ± 0.10 from globular clusters by Reid 1997; 18.63 ± 0.06 from RR Lyrae by Gratton *et al.* 1997) and smaller (18.37 ± 0.04 from the SN 1987A light echo by Gould & Uza 1998) they simply adopted $\mu_{\rm LMC} = 18.50 \pm 0.15$ mag, within which uncertainty the Hipparcos data confirmed the Cepheid distance scale at better than the 10% level.

Luri *et al.* (1998a), see also Luri *et al.* (1998b), used their maximum likelihood approach (LM method) to estimate the period–luminosity relation for 219 Hipparcos Galactic Cepheids, also after eliminating overtone pulsators. Assuming that the slope is as given for the LMC, their zero-point (cf. Equation 5.14) was -1.05 ± 0.17. Determining both slope and zero-point gave $\langle M_V \rangle = -2.12 \log P - 1.73$.

Groenewegen & Oudmaijer (2000) revisited the sample selection made by both Feast & Catchpole (1997) and Lanoix *et al.* (1999), making a more detailed selection of 236 objects according to completeness, flagging of overtone pulsators, and removal of double-mode pulsators, unreliable solutions, and contaminated photometry due to binarity. They established a vertical scale-height of 70 pc, as expected for a population of 3–10$M_\odot$ stars, whose flattened distribution has consequences for the precise form of the Malmquist and Lutz–Kelker type corrections. They found that in V and I the slope of the Galactic period–luminosity relation may be shallower than that for the LMC Cepheids, either for the full period range, or with a break at short periods near $\log P_0 \sim 0.7$–0.8. Taking into account possible systematic effects of slope and metallicity, they derived a best estimate of the LMC distance modulus from V and I data of $18.60 \pm 0.11(\pm 0.08 \text{ slope})(^{+0.08}_{-0.15} \text{ metallicity})$ mag.

Bono *et al.* (2002b) focused on the use of Cepheids pulsating in the first overtone. The main motivation in using these objects is that the width of their instability region is significantly smaller: specifically, current predictions suggest a width of 400 K at $\log P = 0.3$ for first overtone pulsators, compared with 900 K at $\log P = 1$ for fundamental mode pulsators. They found that zero-points predicted by Galactic Cepheid (full-amplitude nonlinear convective) models based on a mild overshooting mass–luminosity relation are in good agreement with empirical zero-points based on Hipparcos parallaxes, while those based on non-overshooting are ~ 0.2–0.3 mag brighter.

Tammann *et al.* (2003) studied 321 Galactic fundamental-mode Cepheids with *BVI* photometry from Berdnikov *et al.* (2000) and with unified colour excesses $E(B-V)$ based on data from Fernie *et al.* (1995). Distances of the 25 Cepheids in open clusters and associations from Feast (1999), and of the 28 Cepheids with Baade–Wesselink distances from Gieren *et al.* (1998) were shown to define two independent period–luminosity relations which agree very well in slope, with zero-points that agree to within 0.12 ± 0.04 mag, and which were therefore combined into a single mean Galactic period–luminosity relation. The parallax calibration by Groenewegen & Oudmaijer (2000) gave absolute magnitudes brighter by 0.21 ± 0.11 in V and 0.18 ± 0.12 mag in I at $\log P = 0.85$. They also argued that the Galactic Cepheids are redder in $(B-V)_0$ than those in LMC and SMC from Udalski *et al.* (1999a,b) respectively. They concluded that differences in the period–colour relations between the Galaxy and the Magellanic Clouds show that there is not a universal period–luminosity.

Van Leeuwen *et al.* (2007) used parallaxes for 14 Cepheids from the global re-reduction of the Hipparcos data (van Leeuwen, 2007), in 10 cases combined with parallaxes from Hubble Space Telescope observations (Benedict *et al.*, 2007). The new Hipparcos parallaxes used have been improved by up to a factor of 2 compared with the original catalogue, as used, for example, by Feast & Catchpole (1997). Overtone pulsators were included, with the fundamental period derived from the observed period, P_1 (see Section 4.8.1). Individual Lutz–Kelker corrections were also applied. They fitted the resulting data to the linear 'reddening free' relation (Madore, 1976; Ngeow *et al.*, 2005)

$$M_W = \alpha \log P + \beta(V-I) + \gamma \tag{5.16}$$

where β is an adopted ratio of total-to-selective extinction, $A_V/(A_V - A_I)$. Their finally-adopted relation based on 14 objects (Figure 5.14) is

$$M_W = -3.29 \log P + 2.45(V-I) - 2.58 \tag{5.17}$$

which can be compared with $M_W = -3.255 \log P + 2.450(V-I) - 2.724$ adopted by Freedman *et al.* (2001) in their HST key project on the Cepheid calibration of H_0, and with $M_W = -3.746 \log P + 2.563(V-I) - 2.213$ adopted by Sandage *et al.* (2006) for Galactic Cepheids in their work on a Cepheid-based H_0. Their corresponding period–luminosity and period–luminosity–colour relations in the near infrared are

$$M_K = -3.258 \log P - 2.40(0.05) \tag{5.18}$$

$$= -3.457 \log P 1.894(J-K)_0 - 3.02(0.05) \tag{5.19}$$

in which the zero-points correspond to their fits for the 220 fundamental pulsators. According to certain steps not detailed here, their results lead to the following conclusions: the slope of the reddening-free *VI* (M_W) relation is the same, within the uncertainties, in

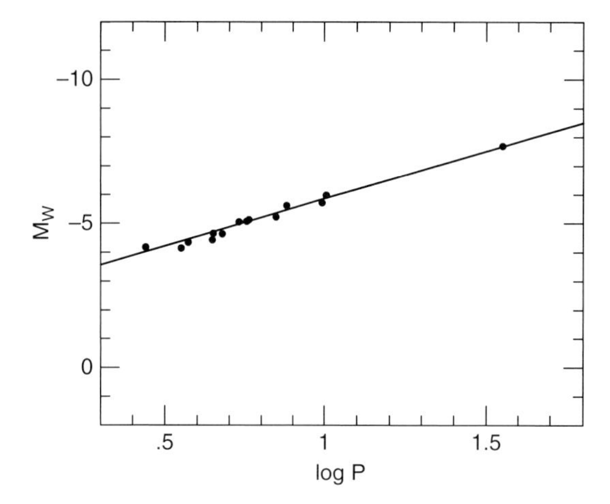

Figure 5.14 M_W*, including Lutz–Kelker correction, plotted against* $\log P$ *for the 14 Cepheids in the sample of van Leeuwen et al. (2007). The line is the relation finally adopted which has* $\alpha = -3.29$, $\beta = 2.45$ *and* $\gamma = -2.58$ *(cf. Equation 5.16). From van Leeuwen et al. (2007, Figure 1).*

the Galaxy and the LMC. This leads to LMC moduli, uncorrected for metallicity, of $\mu_{\rm LMC} = 18.52 \pm 0.03$ from the reddening-free *VI* relation; 18.47 ± 0.03 from the period–luminosity relation in *K*; and 18.45 ± 0.04 from the period–luminosity–colour relation in *JK*. Correction for metallicity leads to a true distance modulus of $\mu_{\rm LMC} = 18.39 \pm 0.05$. The revised *VI* calibration also leads to a revision of the Cepheid-based distances to the 10 galaxies on which Sandage *et al.* (2006) base their supernova Type Ia calibration and Hubble constant, and hence to a revision of their results from $H_0 = 62.3 \pm 5$ to (formally) $H_0 = 70 \pm 2\,{\rm km\,s^{-1}\,Mpc^{-1}}$. Similarly, the Freedman *et al.* (2001) result would be revised from $H_0 = 72 \pm 8$ to $H_0 = 76 \pm 6\,{\rm km\,s^{-1}\,Mpc^{-1}}$. It should be noted that the large change in the resulting LMC distance modulus compared to the results of Feast & Catchpole (1997), arises largely from the slope of the period–luminosity relation, the availability of the OGLE LMC data, and metallicity effects, rather than from the use of the Hipparcos Cepheid parallaxes which are improved but not systematically revised.

The suggested revisions to H_0 would be consistent with recent estimates from various other studies (all values in ${\rm km\,s^{-1}\,Mpc^{-1}}$): with the WMAP value of $73.2^{+3.1}_{-3.2} \pm 3$ (Spergel *et al.*, 2007); with the gravitational-lensing (B 1608+656) value of 75^{+7}_{-6} (Koopmans *et al.*, 2003); with Sunyaev–Zel'dovich and X-ray cluster results of $76^{+3.9}_{-3.4}{}^{+10.0}_{-8.0}$ (Bonamente *et al.*, 2006); and of Cepheid distances to nearby galaxies hosting Type Ia supernova giving $73 \pm 4 \pm 5$ (Riess *et al.*, 2005).

Other discussions of the Hipparcos Cepheid data have been given in the reviews by Feast (1999) and Tanvir (1999); and also by Pont (1999b); Seggewiss (1998a); Groenewegen (2000); Fouqué *et al.* (2003); Hoyle *et al.* (2003); Zhu (2003); Abrahamyan (2004); Heacox (2004); Rastorguev *et al.* (2005); and Storm (2006).

Baade–Wesselink method The Baade–Wesselink method (Baade, 1926; Wesselink, 1969), also as revised by Barnes & Evans (1976), is based on interpretation of the colour, light and radial velocity variations of a Cepheid or RR Lyrae variable during its pulsation cycle. This technique can be implemented in various ways. For example, Di Benedetto (1997) used the linear radius derived from the period according to a period–radius relation, matching it to the angular diameter inferred from the surface brightness–colour correlation using the infrared colour $V - K$ as brightness indicator. This ratio yields a distance, sometimes referred to as an 'expansion parallax'. Gieren *et al.* (1998) used the technique to derive distances to 34 Galactic Cepheids from which, by adopting the slope of the period–luminosity relation from Caldwell & Laney (1991), they derived $\mu_{\rm LMC} = 18.52 \pm 0.06$. Using calibrations from spectroscopic and interferometric techniques, Di Benedetto (1997) obtained $\mu_{\rm LMC} = 18.58 \pm 0.024$. Carretta *et al.* (2000) averaged the two results to derive a preferred Baade–Wesselink distance modulus of $\mu_{\rm LMC} = 18.55 \pm 0.10$ mag.

Di Benedetto (2002) used 219 Hipparcos Cepheids to re-calibrate four earlier Baade–Wesselink results, combining them with two period–luminosity determinations to derive $\mu_{\rm LMC} = 18.59 \pm 0.04$ mag.

Turner & Burke (2002) derived new radii for 13 bright Cepheids using a modified version of the Baade–Wesselink method using the *KHG* narrow-band spectrophotometric index. In combination with other data, they derived $\log\langle R/R_\odot\rangle = 1.064(\pm 0.0006) + 0.750(\pm 0.006)\log P_0$, which simplifies to $\langle R\rangle \propto P^{3/4}$. They found that luminosities inferred from this relation were a good match to the distance scale defined by the Hipparcos parallaxes of 34 cluster Cepheids (their Table 2 and Figures 4–5).

Other Baade–Wesselink studies with some relevance to the Hipparcos results have been reported for SU Cas (Milone *et al.*, 1999) and CK Cam (Kiss & Vinkó, 2000).

Main-sequence fitting Prior to the Hipparcos Catalogue, the most secure calibration of the Cepheid period–luminosity zero-point was considered to be that from Galactic open clusters (and, perhaps less securely due to their distance uncertainties, from associations) containing Cepheids; a compilation of relevant Cepheids is given, for example, in Feast & Walker (1987, Table 2), and in Feast (1999, Table 1). Cluster distances were fixed by main-sequence fitting, and the Cepheid luminosities followed from the cluster distances (Walker, 1998). Laney & Stobie (1994) thus obtained $\mu_{\rm LMC} = 18.49 \pm 0.04\,[\pm 0.04]$ (random + systematic). Since this distance modulus assumes a Hyades distance modulus of 3.27, slightly shorter than the precise Hipparcos determination of 3.33 ± 0.01 (Perryman *et al.*, 1998), Pont (1999a) modified this distance to $\mu_{\rm LMC} = 18.55 \pm 0.04\,[\pm 0.04]$.

While a zero-point with small internal standard error of ~ 0.05 mag can be obtained from the cluster and association data, the scatter among individual estimates is high, with a dispersion of around 0.26 mag (Feast, 1999). Much of this spread is probably attributable to uncertainties in adopted reddenings and measured metallicities. Feast (1999) concluded that *'it seems difficult at the present time to derive a definitive Cepheid zero-point from the cluster and association data'.*

Consequences of the possible membership of δ Cep with the OB association Cep OB6 (de Zeeuw *et al.*, 1999), are discussed by Feast (1999) and Di Benedetto (2002).

Lyngå & Lindegren (1998) used the Hipparcos proper motion data to confirm the long-standing assumption of membership of the Cepheids S Nor with NGC 6087, and U Sgr with M 25. Their spatial coincidence and suspected cluster membership was already well known: the history of their repeated discovery as cluster members is described by Fernie (1969).

Binarity The occurrence of binaries amongst Cepheids is above 50%, and these include many of the basic calibrators of the period–luminosity zero-point. Szabados (1997, 1999) has argued that the milliarcsec size apparent orbit of the nearby systems may have an unfavourable influence on the Hipparcos parallax determination. Most relevant work has taken care to examine the evidence for binarity, and its possible effect on the derived period–luminosity relation (e.g. Feast 1998b; and van Leeuwen *et al.* 2007, Table A1, Column 16). Information on Galactic Cepheids belonging to binary and multiple systems has been compiled by Szabados (2003).

Sensitivity of the Hipparcos parallaxes to binarity can be assessed in the case of Polaris (HIP 11767). A post-Hipparcos spectroscopic orbit of the photocentre of the close pair α UMi Aa was derived by Wielen *et al.* (2000) based on the spectroscopic orbit of α UMi A ($P_{\rm orb} = 29.59$ yr) and on the difference between the quasi-instantaneous Hipparcos proper motion and the long-term average given in the FK5 (see Section 3.5.2). These orbital elements were then used as an additional constraint on a re-reduction of the Hipparcos intermediate astrometry data for this object (Lindegren, 2000, priv. comm.). The Hipparcos parallax is barely affected, changing from 7.56 ± 0.48 mas to 7.51 ± 0.48 mas for the (preferred) retrograde orbit, and to 7.52 ± 0.48 mas for the alternative prograde orbit. Other binary Cepheids, particularly those with periods close to 1 year, may of course be more significantly affected.

Binary systems also provide a route for Cepheid luminosity calibration, via derived spectral types of hot companions detected in the far-ultraviolet (Evans, 1991, 1992; Evans *et al.*, 1998a). This approach does not make specific use of the Hipparcos data, and is not considered here further.

Variability studies Other variability properties of Cepheids are considered in Section 4.8.1.

Kinematics In addition to their use as distance indicators, the fact that Cepheids can be seen easily to large and known distances, and the fact that they trace the young (Galactic plane and spiral arm population) of the Galaxy, means that they also provide an important tracer of spiral arms and Galactic rotation. The first contribution to this study making use of the Hipparcos data was by Feast & Whitelock (1997b,a). They used 220 Galactic Cepheids with Hipparcos astrometry, together with relevant ground-based photometry. Radial velocities were take from Pont *et al.* (1994) and Metzger *et al.* (1998), with all distances reduced to a common period–luminosity–colour relation. Hipparcos proper motions were converted to components in Galactic latitude and longitude, from which the Oort constants A and B were derived from the first-order expression for Galactic rotation (see Section 9.1.3)

$$\kappa\mu_{l*} = (u_0 \sin l - v_0 \cos l)/d + (A\cos 2l + B)\cos b \quad (5.20)$$

where $\mu_{l*} = \mu_l \cos b$, $\kappa = 4.740\,47$, d was estimated from Equation 5.14, and $u_0 = 9.3\,{\rm km\,s^{-1}}$ and $v_0 = 11.2\,{\rm km\,s^{-1}}$ are the adopted Galactic components of the local solar motion (Figure 5.15). While many previous studies of Galactic rotation from proper motions have referred to a relatively small region around the Sun, the Hipparcos data cover a significant region of the Galactic disk. Although the effect of Galactic rotation on proper motions has been known from the time of Oort (1927), it does not seem to have been possible prior to the Hipparcos results to demonstrate this using individual proper motions.

From the consideration of higher-order terms in Galactic rotation, for which residuals are shown in Figure 5.16, they adopted $R_0 = 8.5 \pm 0.5$ kpc, deriving from the proper motions (with all values in kms kpc^{-1})

$$A = +14.82 \pm 0.84 \quad (5.21)$$

$$B = -12.37 \pm 0.64 \quad (5.22)$$

$$\Omega_0 = +(A - B) = +27.19 \pm 0.87 \quad (5.23)$$

$$({\rm d}\Theta/{\rm d}R)_0 = -(A + B) = -2.4 \pm 1.2 \quad (5.24)$$

and with local (Galactic) solar motion components

$$u_0 = +\ 9.32\ \mathrm{km\,s^{-1}} \tag{5.25}$$

$$v_0 = +11.18\ \mathrm{km\,s^{-1}} \tag{5.26}$$

$$w_0 = +\ 7.61 \pm 0.64\ \mathrm{km\,s^{-1}} \tag{5.27}$$

where u_0 and v_0 were adopted from the radial velocity solution, and only w_0 derived from the Hipparcos proper motion components in Galactic latitude.

Rastorguev *et al.* (1998, 1999) used the method of statistical parallaxes (Section 5.3) applied to 270 classical Cepheids with proper motions taken from Hipparcos or the Tycho Reference Catalogue, TRC (the predecessor of Tycho 2). Adjustments of 15–20% to the distance scale based on the period–luminosity relation given by Berdnikov *et al.* (1996) were proposed for those with $P < 9$ days, due to a significant fraction of first-overtone pulsators. Solar motion and Galactic rotation components were also derived from the dataset. Mishurov & Zenina (1999) reported other studies of Galactic rotation derived from Cepheid kinematics, and focused on the inferred spiral structure and spiral-arm pattern speed, described further in Section 9.7. Zhu (1999, 2000) used the Hipparcos proper motions and ground-based radial velocities to analyse the motions of classical Cepheids within the 3d Ogorodnikov–Milne model (Section 9.2.7).

Vertical distribution Dambis (2004) studied the vertical distribution of Cepheids, and its age dependence. The objective was to continue the investigations initiated by Jôeveer (1974), who found that the dispersion of vertical Galactic coordinates of classical Cepheids varies non-monotonically with age in a wave-like pattern. Dambis (2004) used classical Cepheids located at Galactocentric distances in the interval $R_0 \pm 1$ kpc, with evolutionary ages from the models of Pols *et al.* (1998) with and without overshooting. Ages were estimated from $\log t = 8.32 - 0.63 \log P$ (without overshooting), and $\log t = 8.49 - 0.66 \log P$ (with overshooting). The vertical scale height distribution is then determined as a function of mean Cepheid age (Figure 5.17). The figure is to be understood as follows: Cepheids with a very young age are found preferentially close to the Galactic plane, their assumed birth sites. Evolving in scale height with age as a result of their initial vertical velocity component, they reach their maximum Z component and return to the plane after times depending on the local mass density (Section 9.4.1). The resulting periods of vertical oscillations about the Galactic plane are found to be $P_Z = 74 \pm 2$ Myr and 104 ± 2 Myr for standard models (without overshooting) and models with overshooting, respectively. Rather than using this estimated period to constrain the local mass density, ρ_0, which is now better constrained by other Hipparcos analyses, it can instead be used to constrain stellar ages and hence evolutionary models if ρ_0 is assumed known. Thus interpreted as a consequence of vertical virial oscillations, the pattern of motions implies $\rho = 0.118 \pm 0.007 M_\odot\ \mathrm{pc^{-3}}$ and $0.060 \pm 0.004 M_\odot\ \mathrm{pc^{-3}}$ respectively. The latter value being incompatible with estimates based on Hipparcos data (see Section 9.4.1), the results provide an intriguing argument against the existence of strong convective overshooting in stars. A similar analysis was made for young open clusters (see Section 6.6). The results were also taken to favour scenarios where star formation is triggered by impacts of some massive bodies onto the Galactic plane, responsible for some excess kinetic energy in the vertical velocity component at the time of birth (Dambis, 2003a). Although this analysis only used the Hipparcos data for the constraint on ρ_0, it underlines a possible area for further investigation in the future.

5.7.2 Red clump giants

Introduction Red clump giants are low-mass core helium burning stars, so-called because of their 'clumping' in the HR diagram rather than in any spatial sense. They are the higher metallicity counterpart of the (metal-poor) horizontal branch stars (Section 7.5.6). They are very numerous, representing the dominant post-main-sequence evolutionary phase for stars with $M < 2$–$2.5 M_\odot$ (i.e. those forming an electron-degenerate core). The near constancy of the observed luminosity arises because He burning only starts when the stellar core mass reaches a critical value of $\sim 0.45 M_\odot$ (Girardi, 1999). It follows that all low-mass stars, i.e. those that develop a degenerate He core after H exhaustion, have similar core masses at the beginning of He burning, and hence similar luminosities.

Consistently observed in the colour–magnitude diagram of open clusters more than 3×10^8 yr old, Cannon (1970) speculated that a similar clump should be a prominent feature of the colour–magnitude diagram of solar neighbourhood stars. Pre-Hipparcos solar neighbourhood field giant distances, and hence luminosities, did not reach sufficient accuracies for the clump to be evident in the HR diagram, although such features had been observed for high-velocity field stars (Grenon, 1972). The age dependency of the position of the clump compared with that of the subgiant branch had also been noted (e.g. Hatzidimitriou, 1991).

Their use as distance indicators attracted further attention following a study of metal-rich globular clusters by Kuchinski *et al.* (1995), who showed that the absolute K magnitude of the horizontal branch displayed very little cluster-to-cluster variation. In the absence of nearby red clump stars with accurate parallaxes, however, no absolute distance estimates were possible.

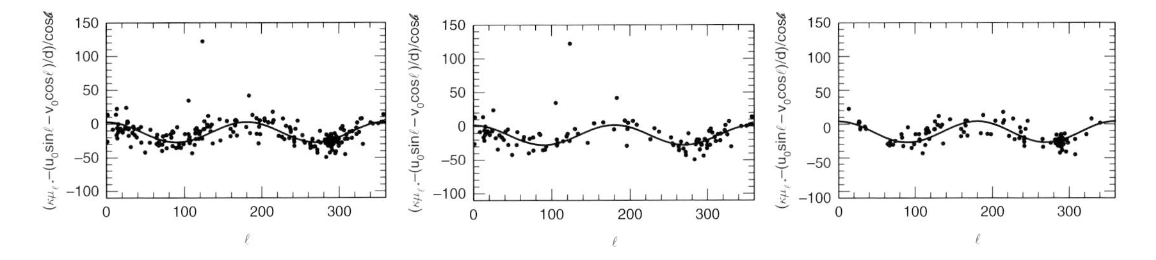

Figure 5.15 Left: proper motion in Galactic longitude multiplied by κ, and corrected for local solar motion (i.e. the combination $(A\cos 2l + B)\cos b$ in Equation 5.20), plotted against Galactic longitude (the curve corresponds to solution 2 of their Table 4). Middle: as left, but for $d < 2$ kpc (solution 5 of their Table 4). Right: as left, but for $d > 2$ kpc (solution 6 of their Table 4). The three stars which lie conspicuously above the others in the two left figures are α UMi, δ Cep and RT Aur. They are sufficiently close to the Sun that their peculiar velocities have a large effect on their proper motions. Their deviations from the mean curve can be accounted for by modest peculiar velocities, 10–20 km s^{-1}, and they have little effect on the derived Galactic rotation parameters because of the weighting system adopted. From Feast & Whitelock (1997a, Figures 2–4).

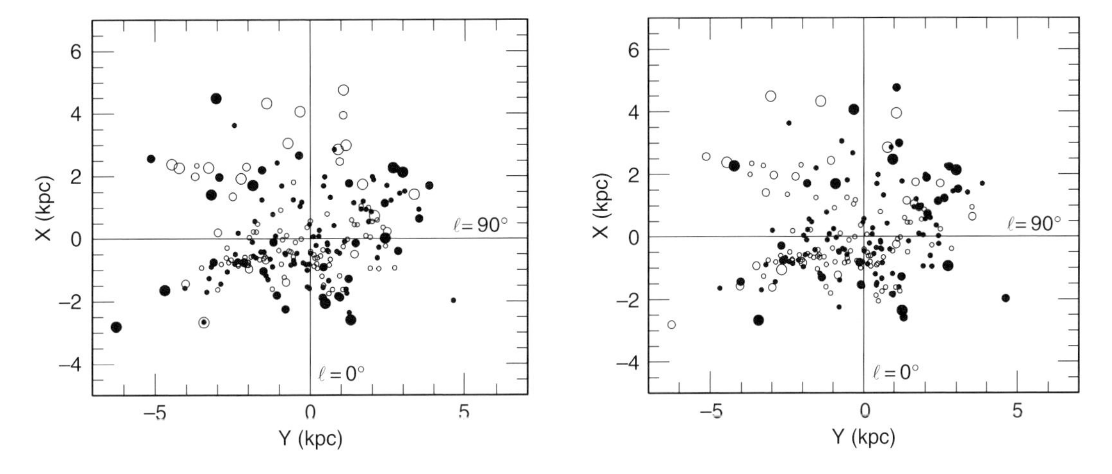

Figure 5.16 Left: location of 220 Hipparcos Cepheids in the plane of the Galaxy, showing velocity residuals for the proper motions in Galactic longitude (derived from solution 19 of their Table 5, i.e. with distances from a period–luminosity–colour calibration). Circle sizes indicate the size of the residual: large: > 40 km s^{-1}; medium: 20–40 km s^{-1}; small: 0–20 km s^{-1}. Filled circles are positive residuals and open circles are negative residuals. Right: as left, with velocity residuals from the proper motion solution in Galactic latitude (solution 1 of their Table 8) plotted on the Galactic plane. From Feast & Whitelock (1997a, Figures 5–6).

Hipparcos studies Hipparcos provided parallaxes for several hundred local red clump giants, at distances of ~100 pc. Their 'clumping' was very prominent in the first Hipparcos HR diagrams (Perryman *et al.*, 1995; Jimenez *et al.*, 1998). The Hipparcos data thus provided a very precise calibration of their mean absolute magnitude and, consequently, the possibility to use the red clump stars as absolute distance indicators both within the Galaxy and directly in external galaxies. Such a calibrator well-represented in the solar neighbourhood offers the important advantage of a more robust one-step calibration to the LMC and external galaxies, as compared with Cepheid and RR Lyrae stars where only a very small number of objects are near enough to provide luminosity calibration directly from accurate trigonometric parallaxes. No Cepheid or RR Lyrae star has a parallax accuracy better than 10%, while some 1000 Hipparcos red clump giants have such accuracies.

Paczyński & Stanek (1998) determined the absolute magnitude of some 600 red clump Hipparcos stars, fitting the I-band magnitudes to a Gaussian function superimposed on a background distribution of red giant branch stars. They found a peak value of $M_I = -0.28$, and a dispersion of ~0.2 mag, using the results to estimate a distance to the Galactic centre based on some 10 000 OGLE red clump stars in Baade's Window (Figure 5.18a,b). Stanek & Garnavich (1998) revised the value to $M_I = -0.23$, and used the results to estimate a distance to M31 using red clump stars detected from HST observations (Figure 5.18c). Stanek *et al.* (1998) also used this to make a direct comparison with the red clump in the LMC, yielding a dereddened distance modulus of

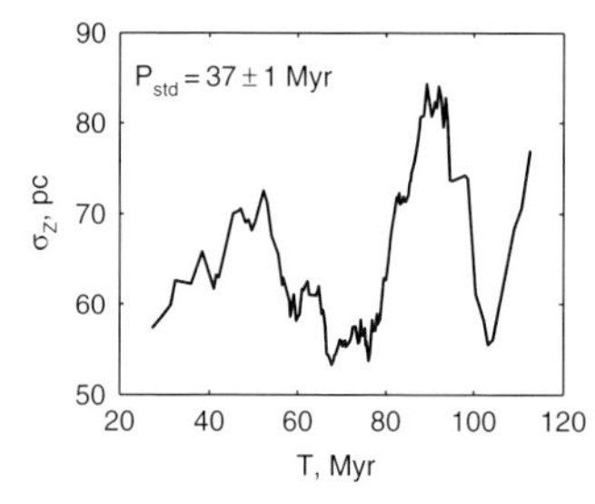

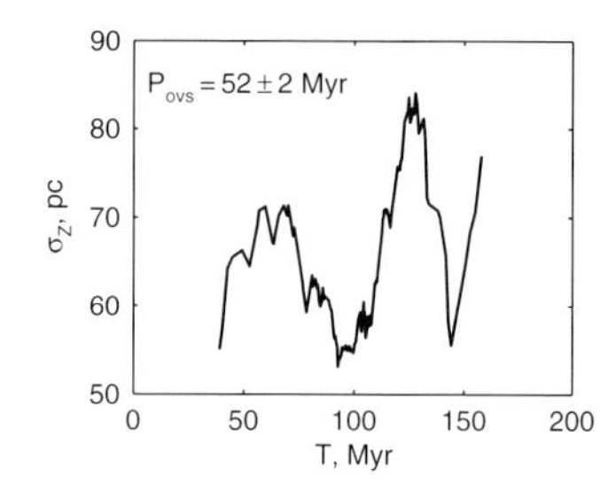

Figure 5.17 Dispersion of Galactic Cepheids perpendicular to the Galactic plane as a function of estimated age, for evolutionary models without (left) and with (right) convective overshooting. The interval between the peaks is taken to correspond to half the vertical Galactic oscillation period. From Dambis (2004, Figure 1).

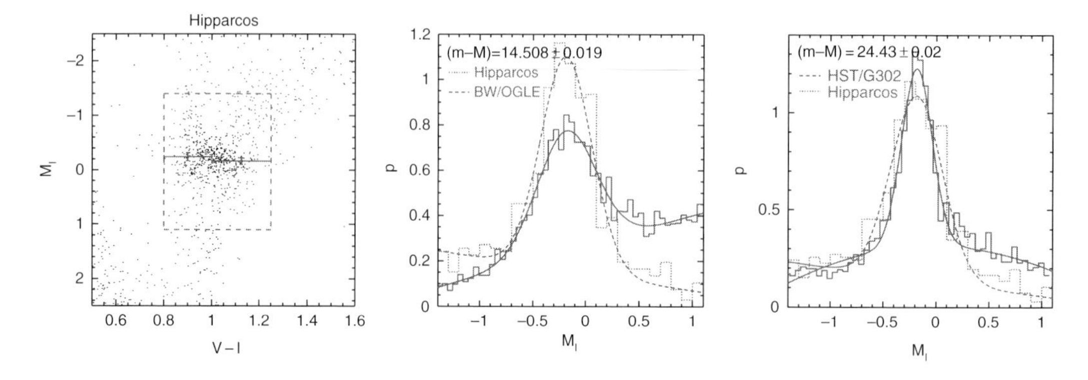

Figure 5.18 Left: colour-magnitude diagram for Hipparcos red clump stars with parallax errors smaller than 10%. The dashed rectangle surrounds the red clump region used for the comparison with Baade's Window stars. The ridge of the highest density of the red clump stars is shown for two colour bins. The average red clump colour is $\langle (V-I) \rangle \sim 1.0$, compared to the average value in Baade's Window of $\langle (V-I)_0 \rangle \sim 1.2$, indicating that the Galactic bulge stars are substantially more metal rich than the solar neighbourhood. Middle: the number of red clump stars in the solar neighbourhood, from the Hipparcos data, are shown as a function of absolute magnitude (thin solid line for both observed histogram and fit). The number of red clump stars in Baade's Window, based on the OGLE data, is shown as a thick line, adopting a preliminary distance modulus $I_0 - M_I = 14.508$. From Paczyński & Stanek (1998, Figures 2 and 3). Right: the Hipparcos data are again shown as the thin solid line. The number of red clump stars in the HST G302 field of M31 is shown as a thick line, adopting a preliminary distance modulus $I_0 - M_I = 24.43$. From Stanek & Garnavich (1998, Figure 2).

$\mu_{\rm LMC} = 18.065 \pm 0.031 \pm 0.09$ (random + systematic). Udalski *et al.* (1998) used the same method to derive a similar result of $\mu_{\rm LMC} = 18.08 \pm 0.03 \pm 0.12$.

These LMC distances were significantly shorter than any other distance determination at the time. The results prompted Girardi *et al.* (1998) and Cole (1998) to re-examine theoretical models. These analyses showed that the red clump luminosity is, in practice, dependent on both age and metallicity (being systematically brighter with decreasing metallicity). These dependencies led to differences of up to 0.6 mag in the mean value of M_I between different stellar populations, and provided an explanation for the erroneous LMC distance estimates. The results implied that population corrections, based on star formation rate and an age–metallicity relation, are needed to make precise use of the red clump as a reliable distance indicator.

The models of Girardi *et al.* (1998) showed a further unexpected result: that stars slightly heavier than the maximum mass for developing degenerate He cores would generate a secondary clumpy structure about 0.3 mag below the bluest extremity of the red clump (Figure 5.19). This secondary clump had been seen in the Hipparcos HR diagram, and had been invoked by Beaulieu & Sackett (1998) as being intrinsic to the LMC clump giant population, and not a superposition of a foreground population postulated by Zaritsky & Lin (1997) as responsible for a significant fraction of the observed microlensing towards the LMC. Girardi *et al.* (1998) accordingly revised the estimated LMC distance to $\mu_{\rm LMC} = 18.28 \pm 0.14$.

Girardi (1999, 2000) carried out further theoretical modelling of the secondary clump, leading to predictions of the clump's appearance according to age

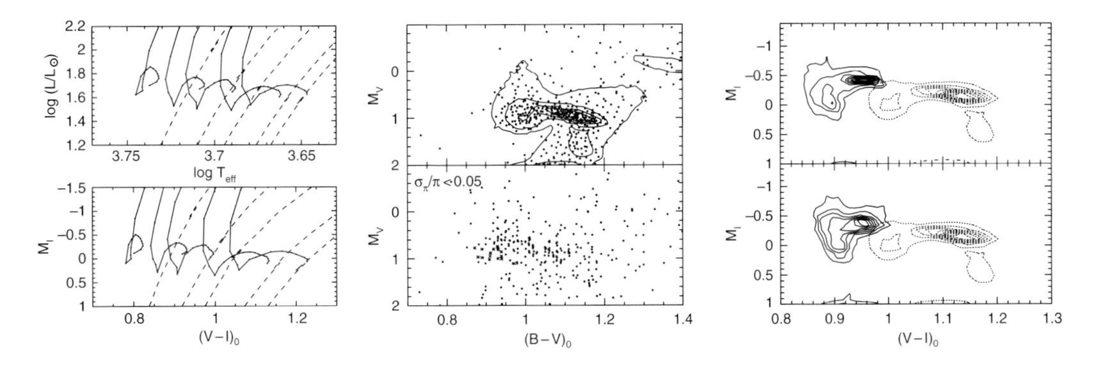

Figure 5.19 Left: position of the ZAHB (onset of quiescent He-burning) for stellar evolutionary tracks of different masses and metallicities, in the HR (top) and $V - I$ (bottom) diagrams. Dots are computed models, at typical mass intervals of $0.1M_\odot$ for $M \lesssim 2M_\odot$, and $0.2 - 0.5M_\odot$ for $M \gtrsim 2M_\odot$. Metallicities are, from left to right, $Z = 0.001, 0.004, 0.008, 0.019, 0.03$. Dashed lines correspond to the red giant branches of 4 Gyr isochrones with the same values of metallicity. Middle: distribution of clump stars in the M_V versus $B - V$ diagram, from theoretical models (upper panel), and Hipparcos data with parallax errors smaller than 5% (lower panel). Stars belonging to a secondary clump are indicated by crosses. Right: theoretical density of stars in the M_I versus $V - I$ diagram, for models representing LMC data (continuous lines). For comparison, the model for the Hipparcos data is also repeated (dotted lines). Only the contour levels corresponding to the red clump are shown. Upper panel: model with $0.004 < Z < 0.008$ and $t = 1 - 3$ Gyr; lower panel: model with star formation and chemical enrichment history similar to that suggested by Vallenari et al. (1996). From Girardi et al. (1998, Figures 1, 4 and 9).

and metallicity. Specifically, he concluded that the secondary clumping should be present in all Galactic fields containing ~1 Gyr old stars with mean metallicity above about $Z - 0.004$, and particularly strong in galaxies with increased star formation rate over the last 1 Gyr or so. Predicted masses of stars in the secondary clump were found to be sensitive to chemical composition and to the treatment of convection; these predictions could be tested by comparison with observed masses for suitable binary systems. A particularly favourable case is the Hipparcos visual binary β LMi (HD 90537), which has well-measured orbital parameters Heintz (1982) and an uncertainty in the Hipparcos parallax of 4%. The primary star, with a derived mass of $1.92 \pm 0.34M_\odot$, probably belongs to the secondary clump. Figure 5.20 shows the mean position and dispersion of the Hipparcos red clump, the position of the primary star β LMi A, and lines delimiting the lower boundary of the clump predicted by models with and without overshooting. The position of β LMi indicates a star just massive enough to ignite He in non-degenerate conditions, and would favour the case of moderate convective overshooting. Further considerations of the corresponding evolutionary models, using the Hipparcos data as observational constraints, and general luminosity predictions of horizontal branch, including RR Lyrae stars, were given by Castellani *et al.* (2000) and Girardi & Salaris (2001).

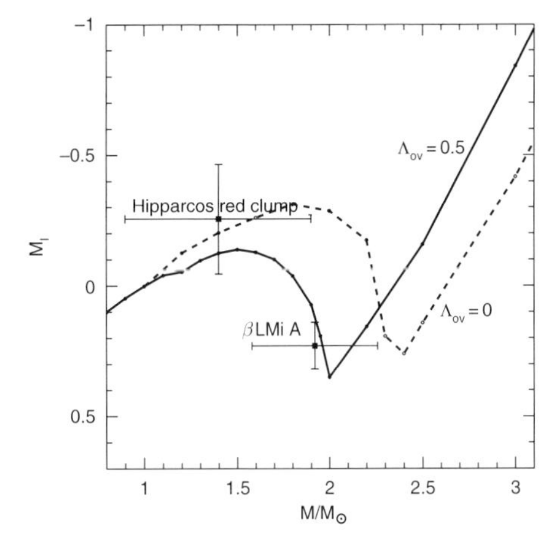

Figure 5.20 The I-band absolute magnitude of clump stars as a function of mass, for solar metallicity. Lines delimit the lower boundary of the clump as predicted by models with and without overshooting (solid and dashed lines, respectively). The mean position of the Hipparcos clump, and of the primary star of the binary β LMi, are also shown. The position of β LMi indicates a star just massive enough to ignite He in non-degenerate conditions, and would favour the case of moderate convective overshooting. From Girardi et al. (1998, Figure 11).

Udalski (2000) used observational data from the solar neighbourhood to demonstrate a weak dependence of M_I on [Fe/H] of

$$M_I = (0.13 \pm 0.07)([\mathrm{Fe/H}] + 0.25) - (0.26 \pm 0.02) \quad (5.28)$$

Predicted mean red clump magnitudes for ages in the range 0.5–12 Gyr and metallicities in the range $-1.7 \leq$ [Fe/H] $\leq +0.2$ were given by Girardi & Salaris (2001) for the V and I bands, and by Salaris & Girardi (2002) for the K band. Various studies of small samples of Galactic open clusters, each of single age and chemical composition (Sarajedini, 1999; Twarog *et al.*, 1999; Grocholski & Sarajedini, 2002) have shown broad agreement with the models in the V, I and K bands. These comparisons have confirmed that population corrections to the red clump absolute magnitude are necessary before it can be used as an accurate extragalactic distance indicator. Issues of reddening, age, chemical composition, and detailed structure of the clump, actually motivated Carretta *et al.* (2000) to discount the use of the red clump as a reliable distance indicator to the LMC.

Grocholski & Sarajedini (2002) included a comparison of M_K and [Fe/H] versus age based on theoretical models from Girardi & Salaris (2001) and K-band photometry from Alves (2000). The results (Figure 5.21) suggest that the spread in M_K results mainly from the effect of age, with both the $10^{8.8}$ and $10^{9.2}$–$10^{9.4}$ isochrones agreeing with the data. Assuming that the solar neighbourhood stars have ages near solar would imply that most of the stars have ages near the latter, i.e. 1.6–4.0 Gyr. The lack of stars around 10^9 yr may provide additional evidence for a non-constant star formation rate in the solar neighbourhood, perhaps as a result of intermittent triggering due to spiral density waves.

Popowski (2001) explored the consequences of making the RR Lyrae and red clump giant distance scales consistent in the solar neighbourhood, the Galactic bulge, and the LMC. He invoked two assumptions: that the M_V versus [Fe/H] relation for RR Lyrae stars is universal, and that M_I for clump giants in Baade's Window can be inferred from the local Hipparcos-based calibration or theoretical modelling. A comparison between the solar neighbourhood and Baade's Window sets the RR Lyrae luminosity at [Fe/H] = -1.6 in the range $M_V = 0.59$–0.70. Comparison between Baade's Window and the LMC sets red clump luminosity in the LMC in the range $M_I = -0.33$ to -0.53. The resulting distance modulus to the LMC is $\mu_{\rm LMC} = 18.24(\pm 0.08) - 18.44(\pm 0.07)$, independent of the dereddened LMC clump magnitude.

Percival & Salaris (2003) compared the effects of age and metallicity across a sample of eight Galactic open clusters and the Galactic globular cluster 47 Tuc. The open cluster distances, and hence the red clump absolute magnitudes in V, I and K, were estimated from empirical main-sequence fitting based on a large sample of local field dwarfs with accurate Hipparcos parallaxes. The nine clusters have metallicities in the range $-0.7 \leq$ [Fe/H] $\leq +0.02$ and ages from 1–11 Gyr, allowing a quantitative assessment of the age and metallicity dependences of the red clump luminosity predicted by the theoretical models of Girardi & Salaris (2001) and Salaris & Girardi (2002). They found good agreement between data and models in all three passbands, confirming the applicability of the models to single-age, single-metallicity stellar populations. They then used determinations of the star-formation rate and age–metallicity relation for the solar neighbourhood, the LMC, the SMC, and Carina, to compare the mean magnitude of the red clump in the I and K bands. This agreement between theoretical predictions and observational data provided further confirmation of the applicability of population corrections based on theoretical models.

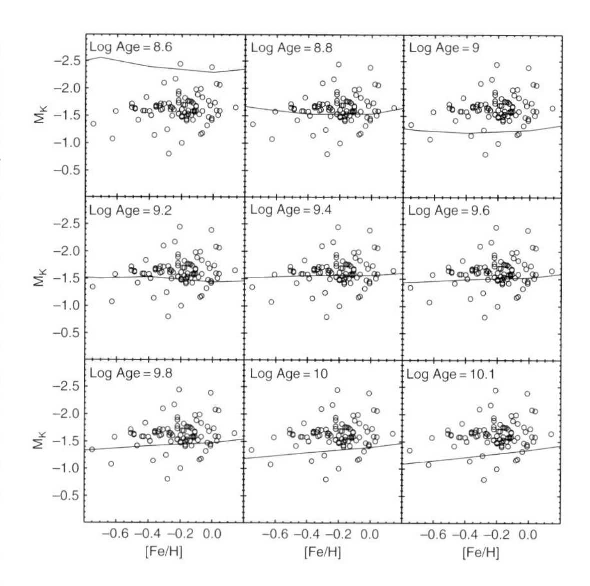

Figure 5.21 K-band absolute magnitude of solar-neighbourhood red clump stars with Hipparcos parallaxes versus metallicity. Solid lines represent the predictions of the theoretical models of Girardi & Salaris (2001). The models suggest that the vertical spread in M_K can be explained by variations in the stellar ages. From Grocholski & Sarajedini (2002, Figure 7).

Koen & Lombard (2003) discussed various statistical aspects involved in estimating the local red clump absolute magnitude.

Supporting observations of Hipparcos stars Alves (2000) assembled accurate K-band photometry for 238 Hipparcos red clump giants. Zhao *et al.* (2001) obtained high-resolution spectra for 39 Hipparcos red clump giants, and determined atmospheric parameters, [Fe/H], α-element enhancements, and masses. The majority were found to be metal-rich, [Fe/H]> 0.0, with masses around $2M_\odot$, while four objects were metal-poor, with lower masses, $M \sim 0.5-1.1M_\odot$, and lower gravities. While the typical masses coincide with the two groups predicted by Girardi & Salaris (2001), their relative proportions were taken to signify a significant peak in the star formation rate several Gyr ago. Siebert *et al.* (2003) reported high-resolution spectroscopy for 400 red clump giants selected from the Tycho 2 Catalogue. Mishenina *et al.* (2006) described a programme to provide fundamental parameters ($T_{\rm eff}$ and $\log g$) and chemical abundances for 177 Hipparcos red clump giants in the Galactic disk, contributing to studies of the Galaxy's chemical and dynamical evolution. Rybka (2007) presented a catalogue of 60 910 stars from the Tycho 2 Catalogue with a high probability (statistically estimated at 85%) of being red clump stars, based on K-band reduced proper motions. It includes astrometry from Tycho 2 and photometry from Tycho 2 and 2MASS.

Adelman (2001) examined the Hipparcos photometric variability of the red clump giants as another potential cause of errors in their use as distance calibrators, and found them to be very constant for the most part.

Application to the Galaxy Bulge Red clump stars have been detected in large numbers in the Galactic Bulge by the OGLE microlensing experiment (e.g. Udalski *et al.*, 1993; Kiraga *et al.*, 1997). Paczyński & Stanek (1998) used their early absolute magnitude calibration of $M_I = -0.28$ mag, to make a single-step determination of the distance to the Galactic centre of $R_0 = 8.4 \pm 0.4$ kpc, based on some 10 000 OGLE red clump stars detected in Baade's Window. This was revised to $R_0 = 8.2 \pm 0.15 \pm 0.15$ kpc, taking into account local interstellar extinction, by Stanek & Garnavich (1998). Paczyński *et al.* (1999) provided a list of more than 1000 additional red clump giants in Baade's Window from the OGLE-II microlensing survey, and compared them with 308 nearby clump giants with accurate Hipparcos and *UBVI* photometry.

Alves (2000) used their sample of 238 Hipparcos red clump giants with accurate K magnitudes, and determined $M_K = -1.61 \pm 0.03$ mag, from which they derived $R_0 = 8.24 \pm 0.42$ kpc. Stanek *et al.* (2000) reconsidered the photometric calibration of the bulge field, and derived $R_0 = 8.67 \pm 0.4$ kpc. Girardi & Salaris (2001) considered effects of population and α-element enhancement, showing that the combination could lead to values anywhere in the range $R_0 = 7.8 \pm 0.2$ kpc to $R_0 = 8.7 \pm 0.2$ kpc.

Application to 47 Tuc Kaluzny *et al.* (1998) estimated a distance modulus for 47 Tuc from *BVI* photometry of $\mu_{47\,\rm Tuc} = 13.32 \pm 0.03 \pm 0.036$, based on the position of its red clump giants, and assuming that the I-band brightness of the red clump is the same as for local stars observed by Hipparcos, i.e. adopting $M_I = -0.23 \pm 0.03$ as found by Stanek & Garnavich (1998). Implying a distance modulus some 0.2 mag less than the distance from subdwarf fitting, it was taken to suggest a 47 Tuc red clump of higher luminosity than in the solar neighbourhood. These conclusions would need to be revised according to the subsequent findings of age and metallicity dependencies for the clump giant luminosities.

Application to the LMC and SMC As described above, the early work by Udalski *et al.* (1998) and Stanek *et al.* (1998) led to $\mu_{\rm LMC} = 18.06$–18.08, erroneously small values subsequently attributed to effects of age and metallicity. Udalski (2000) focused on a sample of red clump stars in the Hipparcos Catalogue and LMC field giants which overlap in metallicity, to derive $\mu_{\rm LMC} = 18.24 \pm 0.08$ mag, a value also supported by Stanek *et al.* (2000).

Girardi & Salaris (2001) applied updated extinction estimates and population corrections to derive values in the range $\mu_{\rm LMC} = 18.37(\pm 0.07)$–$18.55(\pm 0.05)$ mag. They also proposed $\mu_{\rm SMC} = 18.85 \pm 0.06$ mag.

Alves *et al.* (2002) used ground-based K-band observations along with HST-based V and I observations, together combined with populations corrections, to estimate $E(B-V) = 0.089 \pm 0.015$ and $\mu_{\rm LMC} = 18.493 \pm 0.033 \pm 0.03$ mag.

Percival & Salaris (2003) used the results of their population modelling applied to eight Galactic open clusters and the Galactic globular cluster 47 Tuc to determine $\mu_{\rm LMC} = 18.49 \pm 0.06$ mag in I, and 18.46 ± 0.03 mag in K. The consistency of the distances given in the two colours also provides close agreement with the multi-colour approach of Alves *et al.* (2002). Similarly, they found $\mu_{\rm SMC} = 18.90 \pm 0.06$ mag in I, and 18.88 ± 0.03 mag in K. Pietrzyński & Gieren (2002) and Pietrzyński *et al.* (2003) reported a similar LMC distance modulus using the same Hipparcos-based K-band calibration and *JK* photometry of 18.47 ± 0.01.

Groenewegen & Salaris (2003) determined the distance to the LMC young open cluster NGC 1866 using a similar main-sequence fitting approach using the Hipparcos Hyades main sequence and corrections for metallicity based on evolutionary models to determine $\mu_{\rm LMC,\,NGC\,1866} = 18.35 \pm 0.05$. Salaris *et al.* (2003) also determined the distance to NGC 1866 using a similar main-sequence fitting approach but using nearby Hipparcos subdwarfs to determine $\mu_{\rm LMC,\,NGC\,1866} = 18.33 \pm 0.08$. In contrast, their reddening and distance determination to the LMC field stars using the multicolour red clump method yielded $\mu_{\rm LMC,\,field} = 18.53 \pm 0.07$. This perpetuates the discrepancy in the LMC distance modulus obtained by different methods.

Grocholski *et al.* (2007) measured the K-band magnitude of red clump giants in 17 LMC clusters. They then used the cluster ages and metallicities to predict each cluster's absolute K-band red-clump magnitude and thereby calculate absolute cluster distances. They found a cluster distribution in good agreement with the thick, inclined-disk geometry of the LMC, as defined by its field stars. They used the disk geometry to calculate the distance to the LMC centre, for which they found $\mu_{\rm LMC} = 18.40 \pm 0.04 \pm 0.08$ mag.

5.7.3 Mira and semi-regular variables

Mira variables, along with semi-regular variables, are a subset of the long-period variables. They are located at the tip of the asymptotic giant branch, where they experience thermal pulses. They are old, with ages of 10^9 – 10^{10} yr, but are relatively short-lived, $\sim 2\times10^5$ yr (Whitelock & Feast, 1993). They can also be studied to large distances because of their brightness. The general and photometric properties of Mira variables are summarised in Section 4.10. This section focuses on their use as distance indicators through the period–luminosity relation, and as corresponding kinematic tracers of Galactic kinematics.

Period–luminosity relation and distance scale The use of Mira variables as distance indicators was suggested, for example, by the narrow period–luminosity relation observed in the infrared for Miras in the LMC, for both M_K and $M_{\rm bol}$ (Feast *et al.*, 1989; Feast, 1996). This has been taken to imply that they all pulsate in the same mode, presumably radial, but there has been some uncertainty as to whether this is the fundamental or first overtone. The main difficulties of using Miras as universal distance indicators are the possible dependencies of the period–luminosity relation on chemical composition and mass distribution, and the dependency on pulsation mode (e.g. Kharchenko *et al.*, 2002). On account of their shorter periods, it is sometimes assumed that semi-regular variables pulsate in higher overtones than Miras; they are often considered to be Mira progenitors, in which case they would be located lower on the asymptotic giant branch.

Van Leeuwen *et al.* (1997), see also Whitelock *et al.* (1997), used Hipparcos parallaxes for 16 Mira variables which had been included in the Hipparcos Catalogue as nearby candidates for which parallaxes were expected to be significant. Linear diameters were derived for eight with known angular diameters (Figure 5.22, left). Comparison with pulsation theory showed that two, both with $P > 400$ d, are fundamental pulsators, while the others, all with $P < 400$ d, pulsate in an overtone. Adopting slopes of the period–luminosity relation from LMC data of Feast *et al.* (1989) gave relations for M_K and $M_{\rm bol}$ of the form $M = \alpha \log P + \beta$ (Figure 5.22, right). Adopting a weighting $\propto (1/\pi)^2$ gave

$$M_K = -3.47 \log P + (0.88 \pm 0.18) \tag{5.29}$$

$$M_{\rm bol} = -3.00 \log P + (2.88 \pm 0.17) \tag{5.30}$$

These result in a distance modulus for the LMC of $\mu_{\rm LMC} = 18.54 \pm 0.18$, adopted from the mean of the M_K and $M_{\rm bol}$ results separately.

Alvarez *et al.* (1997) used a maximum likelihood calibration for about 80 Hipparcos Mira variables to determine the period–luminosity relation in K, and from narrow-band measurements at 1.04 µm. They identified two significant groupings with different kinematics and scale-heights, with characteristics close to those of the thick disk/halo and old disk populations respectively. For the two most significant groups of Miras, they derived

$$M_K = -3.41 \log P + 0.976 \quad \text{85 in group 1} \tag{5.31}$$

$$= -3.18 \log P - 0.129 \quad \text{16 in group 3} \tag{5.32}$$

They concluded that the slopes of the Galactic period–luminosity relation in K are the same as for the LMC, with their two groups bracketing the relation found by van Leeuwen *et al.* (1997). They attributed the shifts of their two groups to metallicity and mass dependencies (Figure 5.23).

Pre-Hipparcos, Whitelock (1986) had argued that the Mira period–luminosity relation cannot be an evolutionary sequence. Instead, Miras and semi-regular variables within an individual globular cluster define a global sequence in the period–luminosity diagram which is shallower than the Mira period–luminosity relation. This sequence, the Whitelock track, was deemed to be a probable evolutionary track, an inference consistent with the evolutionary calculations of Vassiliadis & Wood (1993). Bedding & Zijlstra (1998) determined period–luminosity dependence for six Mira and 18 semi-regular variables with Hipparcos parallaxes better than 20% and periods $P > 50$ d (Figure 5.24). Transformed to M_K, the Whitelock track is given by

$$M_K = -(1.67 \pm 0.12) \log P - (3.05 \pm 0.25) \tag{5.33}$$

where the zero-point depends on the adopted distance scale. Bedding & Zijlstra (1998) concluded that this relation fits the Hipparcos semi-regular variables, and confirms its importance as an evolutionary track. It also implies that the semi-regular variables can be considered as Mira progenitors for those with $P > 300$ d. The separation into two sequences may be due to either a difference in pulsation mode, or to an adjustment in stellar structure.

Bergeat *et al.* (1998) used 115 Hipparcos long-period variables, including 19 Mira variables, to derive $M_K = -(3.99 \pm 0.13) \log P + (2.07 \pm 0.15)$, with a similar slope

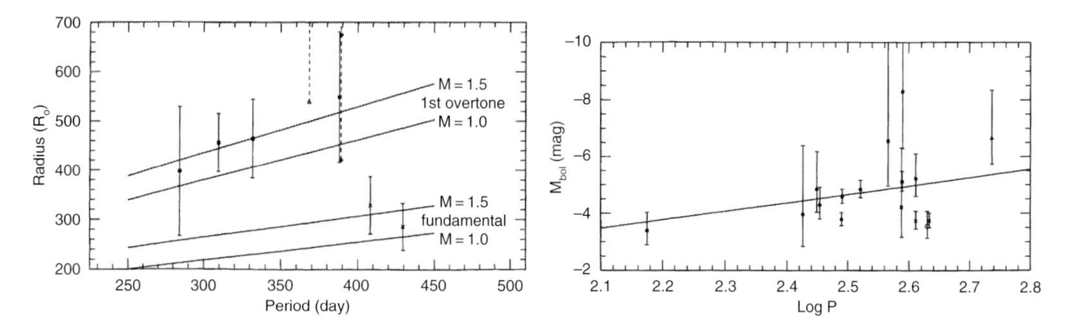

Figure 5.22 Left: the period–radius relation for Mira variables. Filled circles: R Aql, R Leo, o Cet, T Cep; crosses: χ Cyg, R Cas. The triangle is the 1σ lower limit for U Ori, while asterisks are the 1σ and 2σ lower limits for R Hya. Lines are the theoretical relations for the fundamental and first overtone pulsation. Right: the M_{bol} relation, with the same symbols. The line is for the slope of the relation determined for the LMC, with the adopted zero-point given by a distance modulus of 18.50. From van Leeuwen et al. (1997, Figures 1 and 3).

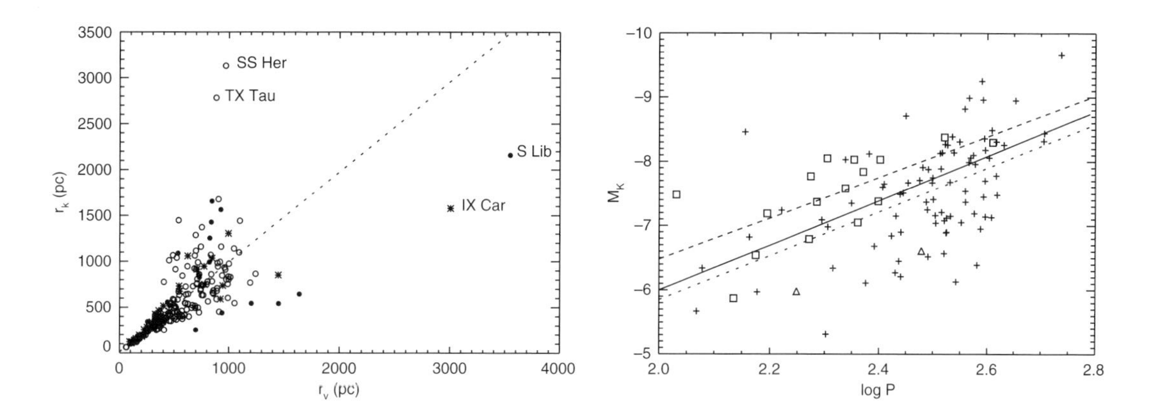

Figure 5.23 Left: Mira distance estimates using the apparent K magnitudes compared with those using the apparent V magnitudes. Asterisks represent long-period variables belonging to the young disk population, open circles to the old disk population, and filled circles to the thick disk or halo population. The regression line is also shown. Right: period–luminosity relations in the K band. Crosses represent 85 Miras belonging to the majority group 1, triangles signify a small group 2, and squares signify 16 objects of group 3. Dotted and dashed lines are the period–luminosity fit for groups 1 and 3 respectively. The solid line is the K-band relation determined by van Leeuwen et al. (1997). From Alvarez et al. (1997, Figures 1 and 3).

as found for the LMC variables, and a resulting LMC distance modulus of $\mu_{LMC} = 18.50 \pm 0.17$.

Barthès *et al.* (1999) and Barthès & Luri (2001) used the Hipparcos astrometry, ground-based radial velocities, periods, and $V - K$ colour indices to refine the maximum likelihood analysis for 350 oxygen-rich long-period variables, comprising Mira and semi-regular variables (the analysis is described further in Section 4.10). They found four groups, differing in kinematics and mean magnitudes, but with the SRa semi-regulars not obviously constituting a separate class. The SRb appear to belong to two populations of different ages. In the period–luminosity diagram, they constitute two evolutionary sequences towards the Mira stage. The Miras of the disk appear to pulsate in a lower-order mode, while the slopes of their de-biased period–luminosity and period–colour relations are very different from those of the oxygen Miras of the LMC, suggesting that a significant number of Miras of the LMC are mis-classified. This also implies that the Miras of the LMC do not constitute a homogeneous group, but include a significant proportion of metal-deficient stars, pointing to a relatively smooth star-formation history. As a consequence, they argued that trivially transposing the LMC period–luminosity relation from one galaxy to the other is not possible.

Whitelock *et al.* (2000) obtained near-infrared photometry for 193 Hipparcos Mira and semi-regular variables. They derived periods, amplitudes, and bolometric magnitudes for 92 objects, defining a Mira sample with

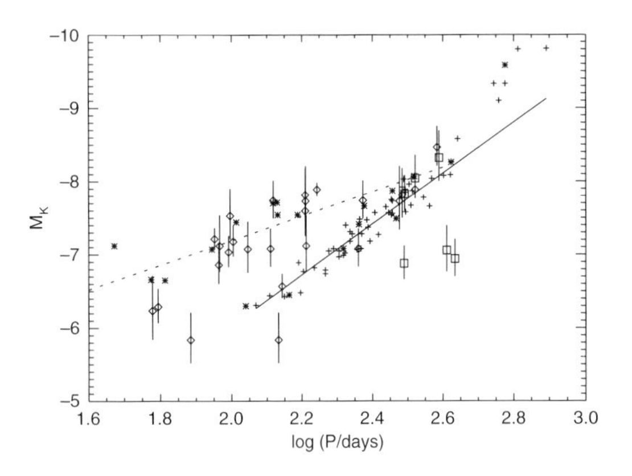

Figure 5.24 The positions of Hipparcos long-period variables in the K-band period–luminosity diagram; squares are Miras and diamonds are semi-regular variables. The solid line shows the LMC Mira relation from Feast (1996), $M_K = -3.47 \log P + 0.91$, and the crosses show the data on which it is based. Asterisks show LMC cluster long-period variables from Wood & Sebo (1996). The dotted line shows the Whitelock track (Equation 5.33) shifted upwards by 0.8 mag to pass through the semi-regular variables. From Bedding & Zijlstra (1998, Figure 1).

low mass-loss rates, $\dot{M} \lesssim 10^{-7} M_\odot\,\mathrm{yr}^{-1}$. The colour, particularly $K - L$, at a given period was found to depend on pulsation amplitude. A comparison with models suggests that this is a consequence of atmospheric extension, in the sense that large-amplitude pulsators have very extended atmospheres and redder $K - L$ and $H - K$ but bluer $J - H$ than their lower amplitude counterparts. Stars with very extended atmospheres also have higher values of $K - 12\,\mu$m. The results provide further evidence for the causal connection between pulsation and mass loss. At short periods, $\log P < 2.35$, two sequences were identified in the $Hp - K$ versus $\log P$ diagram. At a given period these two groups have, on average, the same pulsation amplitude, but different *JHKL* colours and spectral types. Whitelock & Feast (2000) used these data to eliminate the short-period red variables and low-amplitude pulsators, to construct an improved period–luminosity relation for 180 oxygen-rich Miras, yielding

$$M_K = -3.47 \log P + (0.84 \pm 0.14) \qquad (5.34)$$

They also derived linear diameters for red variables with measured angular diameters and parallaxes, to suggest that most are pulsating in the same mode which, if published model atmospheres are correct, is probably the first overtone.

Feast *et al.* (2002) used the globular cluster distance scale based on subdwarf parallaxes from Carretta *et al.* (2000) to derive estimates of M_K for globular cluster Miras, including one in the SMC globular cluster NGC 121. They derived $M_K = -3.47 \log P + (0.93 \pm 0.14)$, in close agreement with that derived from parallaxes of nearby field Miras. The mean of these two estimates, i.e. a zero-point of 0.88, together with data on LMC Miras, yields a LMC distance modulus of 18.60 ± 0.10.

Yeşilyaprak & Aslan (2004) studied the period–luminosity relationships of oxygen-rich semi-regular variables in several wavelength bands using Hipparcos parallaxes with an accuracy of better than 10%, and showed a clear dependence on period of absolute magnitude in all bands studied, from U to $100\,\mu$m. They also showed that the slope of the linear M versus $\log P$ relation is a smooth function of wavelength.

Kinematics It has long been known that the kinematics of Galactic Miras are a function of period (e.g. Feast, 1963, 2003). From a discussion of their radial velocities alone (Feast, 1963; Smak & Preston, 1965; Feast *et al.*, 1972), or in combination with proper motions in discussions of statistical and secular parallaxes (Clayton & Feast, 1969; Robertson & Feast, 1981), it is known that their asymmetric drift and their velocity dispersion become numerically larger as one moves from the longer period Miras to the shorter period ones, with the asymmetric drift reaching $\sim 100\,\mathrm{km\,s^{-1}}$ at periods of about 175 d.

As described in Section 5.7.3, Alvarez *et al.* (1997) used a maximum likelihood calibration to identify two significant groupings with different kinematics and scale heights, with characteristics close to those of the thick disk/halo and old disk populations respectively. The two groups (also seen in the pre-Hipparcos data by Luri *et al.* 1996a) exhibited different period distributions, as expected if the two groups correspond to populations of distinct initial masses, ages and metallicities.

Feast & Whitelock (2000) studied the space motion of a sample of Hipparcos Miras based on their proper motions, radial velocities, and a period–luminosity relation. They confirmed the general dependence of kinematics on pulsation period, taking into account the two colour sequences seen at shorter periods (Whitelock *et al.*, 2000). They also found that Miras with periods in the range 145–200 d in the general solar neighbourhood have a net radial outward motion from the Galactic centre of $75 \pm 18\,\mathrm{km\,s^{-1}}$. This, together with a lag behind the circular velocity of Galactic rotation of $98 \pm 19\,\mathrm{km\,s^{-1}}$, implies an elongation of their orbits, with their major axes aligned at an angle of $\sim 17^\circ$ with respect to the Sun–Galactic centre line, towards positive Galactic longitudes. This concentration was interpreted as a continuation to the solar circle and beyond of the bar-like structure of the Galactic bulge, with the orbits of some local Miras probably penetrating into the bulge. A further discussion of these results is given in Section 9.6, and a summary of the main findings was also given by Feast (2003).

Kharchenko *et al.* (2002) reached a different conclusion based on a larger sample of 724 Miras from the General Catalogue of Variable Stars, with periods between 78–612 d, and with V magnitudes, radial velocities, and proper motions reduced to the Hipparcos system. For Miras in the period range 145–200 d, no significant net radial outward motion was found. They attributed this discrepancy to the limited sample used by Feast & Whitelock (2000).

5.7.4 Other Population I distance indicators

Eclipsing binaries This technique uses eclipsing binaries detected in a number of systems (LMC, globular clusters, etc.) to derive their distances (Paczyński, 1997; Guinan, 2004). In outline, the luminosities of the component stars (from which the distances are inferred) are determined from their absolute radii and temperatures, which are themselves found from light-curves, radial velocity curves and from calibrated flux measurements. For double-lined eclipsing binaries, which also provide the basic mass–luminosity relation for a wide range of stellar masses over the interval 0.5–$20\,M_\odot$, the absolute radii of the components can be determined without any scale dependency. Different implementations and limitations of this general method are reviewed by, e.g. Clausen (2004) and Torres (2004).

Detached eclipsing double-lined spectroscopic binaries have been discovered near the main-sequence turn-off of a number of globular clusters, so that a direct measure of cluster distances via this technique is now feasible. Within the LMC, the method has been applied to at least eight systems, with six estimates for HV 2274 spanning the distance modulus range 18.20–18.49 mag reported by Clausen (2004). Guinan *et al.* (2004) derived an LMC centroid distance modulus of 18.42 ± 0.07 mag based on four systems. Estimates are again sensitive to the assumed reddening and, due to the excellent formal accuracy per object, interpretation becomes complicated by the intrinsic depth of the LMC along the line-of-sight (Fitzpatrick *et al.*, 2003).

The OGLE and MACHO microlensing programmes have revealed a new class of LMC eclipsing binaries with Cepheid components, notably S21 40876 ($V = 14.5$ mag and $P_{\rm orb} = 397$ d, containing a Cepheid with $P = 4.97$ d), and S16 119952 ($V = 17$ mag and $P_{\rm orb} = 801$ d, containing a Cepheid with $P = 2.03$ d). These hold promise as simultaneous probes of the LMC distance, as self-calibrators for the period–luminosity–colour relation, and as tests of the Baade–Wesselink method.

SN 1987a The light echo of the supernova SN 1987a has been used to estimate the LMC distance modulus, with the merit of being independent of reddening. The principle is to compare the angular size of its circumstellar ring with its absolute size derived from a light-curve analysis. The study by Panagia *et al.* (1991) implied that the observed elliptical structure is a circular ring seen at a mean inclination of $i = 42.^\circ8 \pm 2.^\circ6$, and yielded a distance modulus of 18.50 ± 0.13 mag.

A smaller value of $\mu_{\rm LMC} < 18.37 \pm 0.04$, possibly increased to 18.44 for an elliptical shape of the expansion ring, was obtained by Gould & Uza (1998). Panagia (1998) argued that it was incorrect to use lines of different excitation when comparing absolute and angular ring size, and derived instead $\mu_{\rm LMC} = 18.58 \pm 0.05$, including a small correction of 0.03 mag to take into account the position of SN 1987a within the LMC.

5.8 Population II distance indicators

The most straightforward way to derive distances to Population II stars is to measure their trigonometric parallaxes. This is possible for a few nearby subdwarfs. But other important Population II objects, the RR Lyrae and globular cluster stars, are too distant for this direct method. Accordingly, several indirect techniques have been devised, based on standard candles or other approaches (Cacciari, 1999; Gratton *et al.*, 1999). The former include the position of the red giant branch tip, the horizontal branch stars (based on theory, statistical parallaxes of RR Lyrae variables, the Baade–Wesselink method for RR Lyrae, and the position of the horizontal branch clump), main-sequence stars (including isochrone fitting and subdwarf fitting), and the white dwarf cooling sequence. Other approaches are based on detached eclipsing double-lined spectroscopic binaries, or dynamical models for globular clusters. A review of the current knowledge of the evolution of Population II stars, making use of the early Hipparcos results, was given by Caputo (1998).

5.8.1 Subdwarf main-sequence fitting

Knowledge of globular cluster distances (and kinematics) is important for a number of reasons: (a) to provide information on the formation and chemical enrichment history of the Galaxy; (b) to provide a calibration of RR Lyrae absolute magnitudes, which can then in turn be used as distance indicators for other systems such as the LMC; (c) to provide the luminosity function of globular clusters for use as distance indicators to other galaxies due to their high intrinsic luminosity; and (d) as probes of the ages of the oldest stellar systems in the Galaxy.

Globular clusters are too remote for their distances to be determined trigonometrically with Hipparcos. An important way of estimating their distances is by fitting their observed (colour–magnitude) main sequence with a corresponding sequence constructed from local stars whose luminosities have been determined directly through trigonometric parallax measurements. The technique is analogous to that of open cluster distance determination through template matching of a local main sequence. The main difference is that the template must be derived from stars similar to those found in globular clusters, i.e. the low-mass metal-poor halo population stars, or subdwarfs. These are sufficiently rare in the local stellar population, and thus generally rather distant, that the determination of an adequate template sequence has not been straightforward. The method rests on the hypothesis that the intrinsic luminosity of nearby metal-poor stars can be related to those in globular clusters, and similar complexities, due to reddening and chemical composition, apply.

Once the distance to a globular cluster is known, ages can be estimated from the position of the main-sequence turn off, represented by the turn-off luminosity or, more practically, from isochrones which provide the best representation of the photometry in the turn-off region. While the turn-off luminosity is derived from theoretical models, the predictions appear to be very robust. Absolute ages are, however, very sensitive to the adopted globular cluster distance scale: Renzini (1991) gave $\delta t/t = \delta$(distance modulus), i.e. a shift of distance modulus by 0.1 mag changes the derived ages by 10%. The main impediment to estimating accurate (absolute) ages is the continuing inability to establish the cluster distance scale unambiguously.

Relative ages of globular clusters accurate to around 1 Gyr can be derived in two main ways: the 'vertical' method using the magnitude difference between the horizontal branch at the average colour of the RR Lyrae stars and the turn-off, $\Delta V_{\rm TO}^{\rm HB}$, and the 'horizontal' method using the colour difference between the turn-off and the base of the red giant branch, $\Delta(V-I)_{\rm TO}^{\rm RGB}$. Relative rankings from these techniques agree well (Rosenberg *et al.*, 1999).

By improving the distance measurement to nearby subdwarfs, Hipparcos has made various contributions to the improvement of distance and age estimates for globular clusters, and the results will be described in detail. Many of the papers discussing Hipparcos results on globular clusters tackle the issue of ages at the same time as those of distance: Reid (1997), Gratton *et al.* (1997), Brocato *et al.* (1997), Harris *et al.* (1997), Reid (1998), Jimenez & Padoan (1998), Grundahl *et al.* (1998), Chaboyer *et al.* (1998), VandenBerg (2000), Saad & Lee (2001), Grundahl *et al.* (2002), Percival *et al.* (2002), Gratton *et al.* (2003), and Testa *et al.* (2004). Reviews of the status of the globular cluster distance scale with emphasis on the Hipparcos contributions were made by Chaboyer (1999) and Cacciari (1999).

Subdwarfs are metal-poor halo stars which, due to their low metallicity, are subluminous with respect to the disk main sequence at a given colour or, equivalently but more in line with the underlying physics, hotter for a given mass. There are a few nearby subdwarfs whose kinematics has brought them close enough to the Sun for them to have measurable trigonometric parallaxes (Figure 5.25). While there are a few M-type subdwarfs (including Kapteyn's star) within 10 pc, the nearest F and G subdwarfs lie at 40–50 pc. Before Hipparcos, few such stars were known: Sandage (1970) identified eight with known but inaccurate parallaxes; the compilation of VandenBerg *et al.* (1996) gave 15 with ground-based parallaxes and good [Fe/H], although spanning a rather narrow range of metallicity. Only seven had parallax accuracies better than 10%. Groombridge 1830, with a parallax precision of about 1%, previously carried most of the statistical weight; having been used frequently in the past to almost define the subdwarf sequence, Pont *et al.* (1998) reassuringly showed that it continues to lie on the locus of stars of similar metallicity.

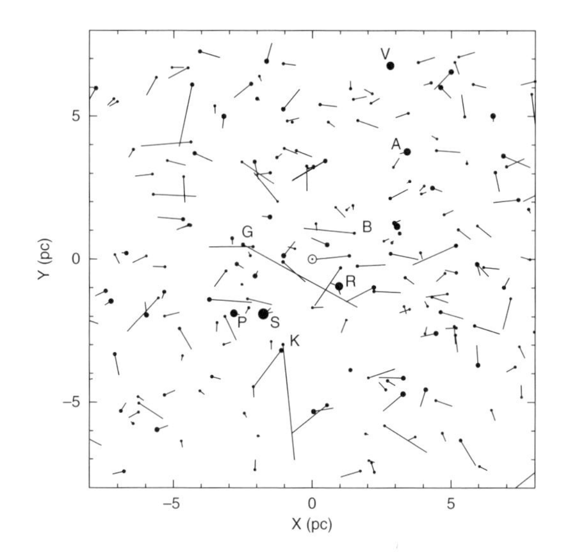

Figure 5.25 Stars within a volume of ± 8 pc in X and Y, and ± 10 pc in Z centred on the Sun, $\odot$. Space velocities, including radial velocities where known and assigned randomly otherwise, are projected onto the XY-plane, and shown as vectors with respect to the Local Standard of Rest (scaled arbitrarily). The 237 stars are shown with circles representing apparent magnitude. Bright stars: A=Altair, P=Procyon, R=Rigil Kent, S=Sirius, V=Vega. High proper motion stars: B=Barnard's Star, G=Groombridge 1830, K=Kapteyn's Star. Groombridge 1830, archetype of the halo stars near to the Sun, also has a particularly large space velocity.

The paucity of subdwarfs meant that, pre-Hipparcos, main-sequence fitting errors due to metallicity mismatches were large, and distances were very sensitive to the parts of the main sequence being fit. Although the space density of unevolved, $M_V \gtrsim 5.5$, subdwarfs at the Hipparcos Catalogue completeness limit of $V =$

Table 5.3 Main-sequence fitting distances to globular clusters based on Hipparcos subdwarfs. All are halo systems with the exception of disk clusters NGC 104 (47 Tuc) and NGC 6838 (M71). Note that other recent work continues to find higher ages, e.g. VandenBerg (2000), and the comparison with the resulting distance modulus to the LMC also suggests that at least the earlier ages given by the Hipparcos subdwarf fitting probably have to be revised upwards. Typical distance modulus errors quoted by the authors are ±0.1 mag. Some works simply quote a distance modulus uncorrected for intervening extinction, $\mu_V = (m - M)_V$, while others quote corrected values, $\mu_0 = (m - M)_0$. For ease of comparison, both values are given below: the primary results of each work are underlined, while the other is determined here by assuming a ratio of total to selective extinction given by $A_V = 3.1\,E(B - V)$ (Fitzpatrick, 1999). Values adopted for $E(B - V)$ and [Fe/H] are from the cited works. Distance moduli in the range 13–15 mag correspond to distances of 4–10 kpc. Age ranges from Gratton et al. (1997) arise from different isochrone modelling, partly arising from different assumptions on the solar absolute magnitude. Ages from Reid (1997) and Reid (1998) are indicative ranges from isochrone modelling, estimated from their Figure 13 and Figure 10 respectively. The M92 estimate from Pont et al. (1998) is from their solution excluding known binaries.

Cluster Name		$E(B-V)$	[Fe/H]	$(m-M)_V$	$(m-M)_0$	Age (Gyr)	Reference
NGC 104	47 Tuc	0.055	−0.67	13.62	13.44	9.6–10.8	Gratton *et al.* (1997)
		0.04	−0.70	13.69	13.56	9–11	Reid (1998)
		0.055	−0.67	13.55	13.38	12.5	Carretta *et al.* (2000)
		0.04	−0.70	13.33	13.21	12.0	Grundahl *et al.* (2002)
		0.04	−0.70	13.37	13.25	11.0 ± 1.4	Percival *et al.* (2002)
		0.024	−0.66	13.50	13.43	11.2 ± 1.1	Gratton *et al.* (2003)
NGC 288		0.033	−1.05	14.94	14.83	10.1–11.3	Gratton *et al.* (1997)
		0.01	−1.07	15.03	15.00	9.5–12	Reid (1998)
		0.033	−1.05	14.95	14.84	11.2	Carretta *et al.* (2000)
NGC 362		0.056	−1.12	15.04	14.86	7.8–9.3	Gratton *et al.* (1997)
		0.056	−1.12	14.98	14.80	9.9	Carretta *et al.* (2000)
NGC 4590	M68	0.05	−2.09	15.45	15.29	9–11	Reid (1997)
		0.040	−1.95	15.31	15.18	10.1–11.9	Gratton *et al.* (1997)
		0.040	−1.95	15.25	15.12	12.3	Carretta *et al.* (2000)
NGC 5904	M5	0.03	−1.40	14.54	14.45	11–13	Reid (1997)
		0.035	−1.10	14.60	14.49	9.6–11.0	Gratton *et al.* (1997)
		0.02	−1.10	14.58	14.52	9.5–12	Reid (1998)
		0.03	−1.17	14.51	14.41	8.9 ± 1.1	Chaboyer *et al.* (1998)
		0.035	−1.10	14.57	14.46	11.2	Carretta *et al.* (2000)
		0.035	−1.10	14.55	14.44	–	Testa *et al.* (2004)
NGC 6205	M13	0.02	−1.65	14.54	14.48	11–13	Reid (1997)
		0.020	−1.41	14.45	14.39	11.3–13.0	Gratton *et al.* (1997)
		0.02	−1.39	14.51	14.45	9.5–11.5	Reid (1998)
		0.021	−1.60	14.44	14.38	12.0 ± 1.5	Grundahl *et al.* (1998)
		0.02	−1.58	14.47	14.41	10.9 ± 1.4	Chaboyer *et al.* (1998)
		0.020	−1.41	14.44	14.38	12.6	Carretta *et al.* (2000)
NGC 6341	M92	0.02	−2.24	14.99	14.93	11–13	Reid (1997)
		0.025	−2.15	14.80	14.72	12.4–14.5	Gratton *et al.* (1997)
		0.02	−2.14	14.74	14.68	13–14	Pont *et al.* (1998)
		0.025	−2.15	14.72	14.64	14.8	Carretta *et al.* (2000)
NGC 6397		0.19	−1.82	12.85	12.24	10–12	Reid (1998)
		0.183	−2.03	12.58	12.01	13.9 ± 1.1	Gratton *et al.* (2003)
NGC 6752		0.02	−1.54	13.23	13.17	9–11	Reid (1997)
		0.035	−1.43	13.32	13.21	11.8–13.3	Gratton *et al.* (1997)
		0.04	−1.42	13.29	13.16	10–13	Reid (1998)
		0.04	−1.51	13.25	13.12	11.2 ± 1.3	Chaboyer *et al.* (1998)
		0.035	−1.43	13.32	13.21	12.9	Carretta *et al.* (2000)
		0.040	−1.43	13.24	13.12	13.8 ± 1.1	Gratton *et al.* (2003)
NGC 6838	M71	0.28	−0.70	14.09	13.19	7–9	Reid (1998)
		0.28	−0.70	13.71	12.84	12.0	Grundahl *et al.* (2002)
NGC 7078	M15	0.09	−2.15	15.66	15.38	9–11	Reid (1997)
NGC 7099	M30	0.05	−2.13	15.11	14.95	10–11	Reid (1997)
		0.039	−1.88	14.94	14.82	9.8–11.6	Gratton *et al.* (1997)
		0.039	−1.88	14.88	14.76	12.3	Carretta *et al.* (2000)

$7.9 + 1.1 \sin|b|$ is extremely low, such that only a few subdwarfs are present in the 'survey' part of the catalogue, the number was specifically augmented by the inclusion of all subdwarfs known at the time of construction of the Input Catalogue down to the satellite observability limit, as well as by the inclusion of high proper motion stars. High proper motion stars are not necessarily stars with high space velocities, but they comprise a significant component of high space velocity stars, and hence such surveys are higher in their representation of halo subdwarfs. As a result, Hipparcos provided parallaxes for several hundred subdwarfs between the survey completeness limit and the satellite's limiting magnitude, although for many the parallax precision remains poor. Reid *et al.* (2001) carried out a further search for previously unidentified subdwarfs in the Hipparcos Catalogue, identifying 317 candidates based on their position in the $M_V, (B-V)_T$ diagram, but eventually resulting in only nine additional candidates with accurate photometry.

The resulting main-sequence fitting procedure consists of directly comparing the observed globular cluster main sequence with a corresponding main sequence made of nearby subdwarfs with well-determined parallaxes (see, e.g. Percival *et al.*, 2002, Section 4.1). There are numerous complications in practice: the template main sequence should be constructed from nearby subdwarfs spanning a sufficient range of colour on the unevolved main sequence, and with the same metallicity as the target cluster, requiring that [Fe/H] for both samples are accurate and on a uniform scale; an error of 0.1 dex in metallicity corresponds to about 0.03 mag in distance modulus at [Fe/H] = -2.0, and 0.11 mag at [Fe/H] = -1.0, with such a metallicity error corresponding to 100 K on $T_{\rm eff}$. There remain, for example, systematic differences between the two commonly used scales of Zinn & West (1984) and Carretta & Gratton (1997). Because a fully empirical template with the required metallicity is not available, some additional steps are required: (a) observed colours of both the nearby template subdwarfs and, more crucially because of their distances, those of the globular clusters, must be corrected for interstellar reddening; (b) statistical corrections for samples selected according to parallax, parallax error, or luminosity (Lutz–Kelker or Malmquist bias) should be accounted for; and (c) corrections are needed for known and undetected binaries which have the effect of making objects redder and brighter, leading to shorter distances for affected cluster stars, and longer distances for affected field (template) subdwarfs.

Results and evolution in understanding about the Hipparcos subdwarf main-sequence fitting are outlined chronologically hereafter, and summarised in Table 5.3.

Reid (1997) used just 15 nearby Hipparcos stars with $\sigma_\pi/\pi < 0.12$ (and 20 of lower precision) to re-define the subdwarf main sequence, separating the templates and the seven globular clusters into intermediate, [Fe/H] = -1.4, and extreme, [Fe/H] = -2, subsets. Correction for Lutz–Kelker bias, and comparisons with ground-based parallaxes, were reported. Derived distances exceeded previous estimates (except for NGC 6752) due in part to the systematically smaller Hipparcos parallaxes, especially for the extreme metallicity clusters where the distance moduli are some 0.3 mag higher than previous values. He derived age estimates from three different theoretical model isochrones using different abundances, including O/Fe enhancements, and with appropriate bolometric corrections. The models of D'Antona *et al.* (1997) provided the closest match to the main-sequence, turn-off and subgiant branch data (Figure 5.26), but with none matching the exact shape of the cluster main sequences. Ages were also estimated from the bluest point of the main-sequence turn-off, where the colour–magnitude diagrams are nearly vertical. All age estimates were below 14 Gyr, with a possible age spread amongst the metal-poor halo clusters.

Gratton *et al.* (1997) used about 30 subdwarfs to calibrate nine globular clusters, with high-precision Hipparcos parallaxes ($\sigma_\pi/\pi < 0.10$), accurate chemical abundances from high-resolution spectroscopy, precise and homogeneous reddening estimates, and corrections for Lutz–Kelker bias and binary contamination. They confirmed the higher distance moduli (~0.2 mag) and younger ages (~2.8 Gyr). They estimated the mean age of the six oldest clusters as $11.8^{+2.1}_{-2.5}$ Gyr (see Table 5.4).

Pont *et al.* (1998) determined the distance of M 92 on the basis of 503 subdwarfs with Hipparcos parallaxes, correcting for Lutz–Kelker and Malmquist biases, as well as binary contamination. They found a slighty smaller distance modulus and slightly higher age of 14 ± 1.2 Gyr. From the position of the main sequence and the age of the local subdwarfs, they argue that their smaller distance and larger age is more plausible. The latter argument is illustrated by direct age dating of the local subdwarfs that have evolved onto the subgiant branch (Figure 5.27). The tightness of the main sequence is remarkable; lying nearly 1 mag below the Hyades main sequence, and with a low dispersion indicating a small age spread for the local subdwarfs, it has a turn-off sufficiently well-defined for a mean age to be determined directly, which they estimate at around 14 Gyr. The younger ages of M92 from Reid (1997) and Gratton *et al.* (1997) would imply that the extreme field subdwarfs in the solar neighbourhood are older than the cluster itself. Rather, Pont *et al.* (1998) inferred that the two populations have roughly the same age. Uncertainties in the bolometric corrections as a function of V magnitude, which are not accounted for in the age

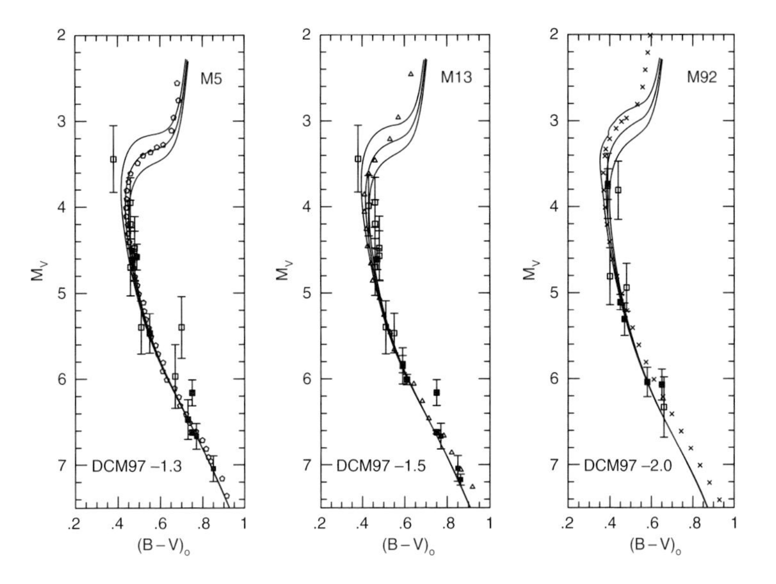

Figure 5.26 An example of the Hipparcos subdwarf main-sequence fitting for globular clusters, showing the comparison between the theoretical isochrones from D'Antona et al. (1997) and the empirical metal-poor main sequences for M5, M13 and M92. Each panel shows the relevant cluster fiducial sequence and the nearby subdwarfs of the appropriate abundance range. Isochrones are for ages of 10, 12 and 14 Gyr and for [Fe/H] = −1.3, −1.5 and −2.0. Cluster data are matched to the subdwarfs, not the isochrones. From Reid (1997, Figure 12).

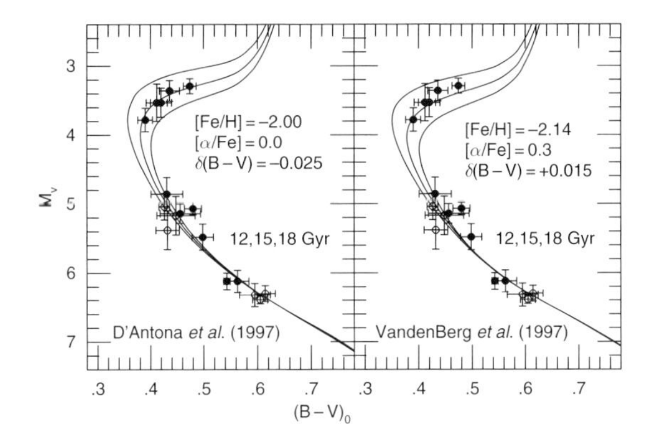

Figure 5.27 Luminosity–colour diagram and isochrones for nearby subdwarfs. The figure is for Hipparcos subdwarfs with $\sigma_\pi/\pi < 5\%$ and $-2.6 <$ [Fe/H] < -1.8, shifted in colour to [Fe/H] = −2.2, and isochrones for 12, 15, 18 Gyr from D'Antona et al. (1997) and VandenBerg et al. (2000). Isochrone colours were slightly adjusted in order to achieve the best fits to the unevolved subdwarfs. The sample also contains a few sub-giants, residing between the turn-off and the red giant branch, for which the fit remains excellent. From Pont et al. (1998, Figure 10). See also VandenBerg (2000, Figures 9 and 32).

uncertainty, could raise the age of the oldest globular clusters by up to around 2 Gyr. King *et al.* (1998) discussed possible errors due to reddening.

Reid (1998) extended his early analysis to globular clusters with higher metal abundances, based now on 91 nearby calibrating subdwarfs. As before, distances were estimated from the main-sequence fitting, and ages from isochrone fitting and the main sequence turn-off. Some of the earlier results were also revised based on improved abundance calibration. Again, younger ages were confirmed (Table 5.3).

Grundahl *et al.* (1998) estimated the distance and age of M13 using the local Hipparcos subdwarfs and Strömgren *uvbyβ* photometry, which provides rather precise estimates of $T_{\rm eff}$, [Fe/H], and $\log g$ for F and G stars, as well as reddening, for each star. The age estimate of 12 Gyr (based on [Fe/H] = −1.60, [α/Fe] = 0.3, and neglecting effects of He diffusion), is rather consistent with the other determinations, and probably less affected by systematic effects. They remarked that while the Hipparcos subdwarf calibration has placed the distance estimate on a more secure footing, it is within 0.1 mag of the pre-Hipparcos value determined by, e.g. Buckley & Longmore (1992). Reduced ages compared to the pre-Hipparcos era estimates were again attributed to improved stellar models and input physics, although model mismatch to the observed red giant branch is still evident. Two interesting features are an overluminosity of the hottest stars on the horizontal branch compared to the models (Figure 5.28a), and large scatter in c_0 on the red giant branch but not for field giants (Figure 5.28b), probably indicative of CNO abundance variations. Both features may be the signature of deep

He mixing in upper red giant branch stars, penetrating close to the H-burning shell so that some He is dredged up into the surface layers. This has been proposed to explain the observed abundance patterns of C, N, O, Na, Al and Mg, as well as the diversity of horizontal branch morphology among globular clusters having similar chemical compositions (the 'second-parameter' phenomenon). The same explanation has been put forward to explain the inconsistency in the various (field and cluster) RR Lyrae distance estimates to, for example, the Galactic centre (Section 9.2.1).

Carretta *et al.* (2000) updated the selection of the local subdwarfs and treatment of systematic biases to decrease the distance moduli of Gratton *et al.* (1997) by about 0.04 mag.

VandenBerg (2000) continued the analysis of the globular cluster distance scale and age distribution with two specific improvements: first, using stellar evolutionary models that explicitly take into account the observed abundances of α-elements; second, using the distances of field subgiants from Hipparcos parallaxes. Theoretical ZAHB loci were used to set the distance scale and absolute ages, while relative ages were determined using the $\Delta V_{\rm TO}^{\rm HB}$ method, finding the age distribution shown in Figure 5.29. Here, the oldest metal-poor clusters have an age of about 14 Gyr, with an age decreasing with increasing metallicity, a small dispersion in age at any [Fe/H] $\lesssim -1.0$, and with no more than a small dependence of age on Galactocentric distance. They proposed a reconciliation of the distance scale inferred from subdwarf main-sequence fitting and nearby RR Lyrae stars by adopting a particular metallicity scale for globular clusters, leading to a 'short' distance to the LMC of 18.30 mag or smaller, and ages $\gtrsim$ 15 Gyr.

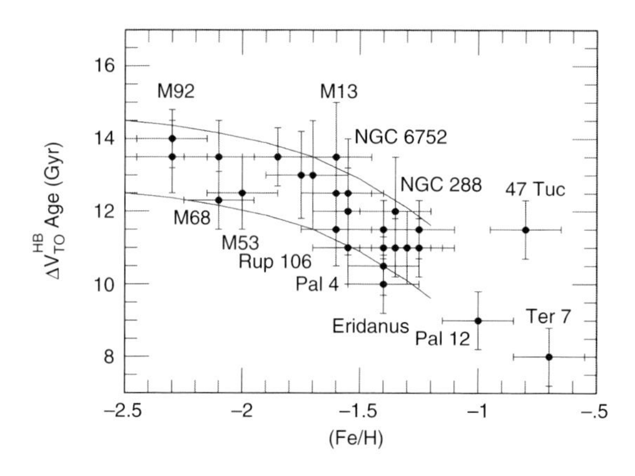

Figure 5.29 Derived $\Delta V_{\rm TO}^{\rm HB}$ ages plotted as a function of [Fe/H] for 26 globular clusters. From VandenBerg (2000, Figure 40).

Grundahl *et al.* (2002) carried out a similar analysis for M71 and 47 Tuc to that reported by Grundahl *et al.* (1998), based on local Hipparcos subdwarfs (18 stars with $\sigma_\pi/\pi < 0.08$) and Strömgren *uvby* photometry. Distances for both are systematically smaller than those previously estimated by Reid (1998) and Carretta *et al.* (2000), attributed to a different selection of calibrating subdwarfs, and reverting to the pre-Hipparcos estimates. Ages, estimated from the isochrones of VandenBerg *et al.* (2000), are around 12 Gyr, correspondingly larger than those derived by Reid (1998), with errors due to the [Fe/H] scale dominating.

Percival *et al.* (2002) found a similar result for 47 Tuc using an enlarged sample of 43 local Hipparcos subdwarfs with $M_V > 5.5$ and $\sigma_\pi/\pi < 0.13$ in the metallicity range $-1.0 <$ [Fe/H] < 0.3, for which new *BVI* photometry permitted the fitting in two colour planes, $V/(B-V)$ and $V/(V-I)$; because of the different sensitivities of these colour indices to metallicities, their consistency is a valuable test of reliability. Their revised photometry moves the main-sequence line by 0.02 mag, decreasing the distance modulus by 0.1 mag due to the steep slope of the main sequence. They found a distance modulus $\mu_0 = 13.25^{+0.06}_{-0.07}$ mag, and an age of 11.0 ± 1.4 Gyr, using the Salaris & Weiss (1998) isochrones for [Fe/H] $= -0.7$, Y $= 0.254$, $[\alpha/\mathrm{Fe}] = 0.4$. In good agreement with the Grundahl *et al.* (2002) result, they also traced the difference with that of Carretta *et al.* (2000) to the re-calibrated photometry and their preferred reddening.

Gratton *et al.* (2003) revised estimates of the distances and ages of NGC 6397, NGC 6752 and 47 Tuc. Improved reddening accurate to 0.005 mag and abundances to 0.04 dex were derived from Johnson $B-V$ and Strömgren $b-y$ colours and high-resolution spectra, which were obtained for the cluster turn-off stars and subgiants as well as for the field subdwarfs with Hipparcos parallaxes. Resulting ages for NGC 6397 and 6752 are about 13.8 ± 1.1 Gyr, with 47 Tuc about 2.6 Gyr younger; all ages are reduced by ~ 0.4 Gyr for models including He diffusion. They also derived the relation between absolute magnitude of the horizontal branch and metallicity, from these three clusters, as

$$M_V(\mathrm{HB}) = (0.22 \pm 0.05)([\mathrm{Fe/H}] + 1.5) + (0.56 \pm 0.07) \tag{5.35}$$

This value is included in the compilation of RR Lyrae absolute magnitudes given in Table 5.5.

Testa *et al.* (2004) also determined the distance to NGC 5904 (M5) via subdwarf main-sequence fitting, confirming the results of Carretta *et al.* (2000).

Butkevich *et al.* (2005a) included a synthesis of the treatment made for Malmquist and Lutz–Kelker biases (Section 5.2) in some of the Hipparcos studies of subdwarf luminosity calibration (Reid, 1997, 1998; Gratton *et al.*, 1997; Chaboyer *et al.*, 1998; Pont *et al.*, 1998), focusing on M92, a cluster common to all studies. Only Gratton *et al.* (1997) mentioned Malmquist bias; Reid (1997, 1998), and Gratton *et al.* (1997) followed the

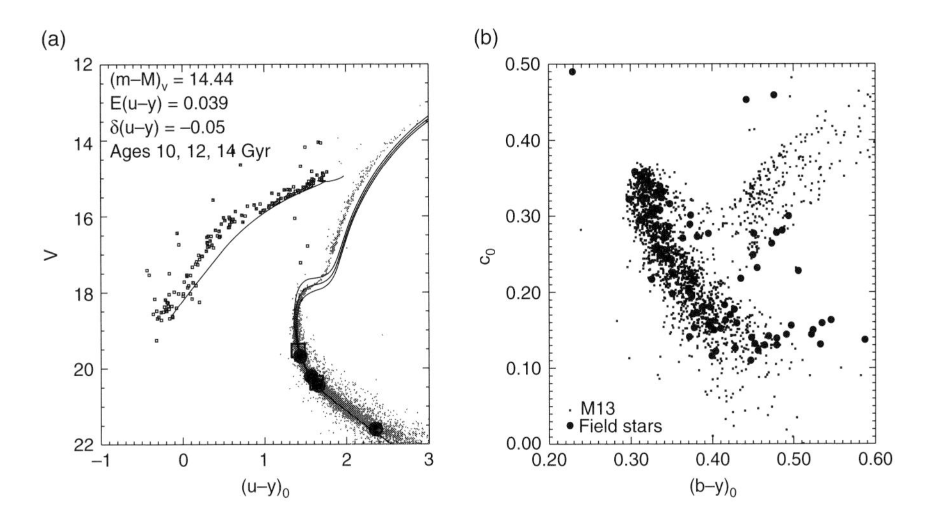

Figure 5.28 Determination of the distance and age of M13: (a) the Strömgren $(u-y)_0$, V colour–magnitude diagram, on which 10, 12, 14 Gyr isochrones (for [Fe/H] $= -1.61$ and [α/Fe] $= 0.3$), as well as local subdwarfs having Hipparcos parallaxes (filled circles), have been superposed. The authors attribute the failure of the models to match the red giant branch to problems with the adopted colour transformations; (b) the Strömgren $(b-y)_0$, c_0 diagram, with field stars having $-1.9 <$ [Fe/H] < -1.4 overplotted as filled circles. There is a large scatter in c_0 on the cluster red giant branch, but not amongst the field giants, at any given $(b-y)_0$. From Grundahl et al. (1998, Figures 1–2).

Lutz–Kelker approach of Hanson (1979); while Reid (1998), Chaboyer *et al.* (1998), and Pont *et al.* (1998) performed Monte Carlo simulations to assess the effect of biases. Butkevich *et al.* (2005a) favoured the approach of Pont *et al.* (1998), who (uniquely) found that a positive ΔM was required, essentially confirming the presence of a Malmquist bias.

Conclusions The early Hipparcos results (1997–2000) suggested systematically larger distances and smaller ages for globular clusters than the pre-Hipparcos derivations. Improved evolutionary models and the better control of systematics arising from uncertainties due to reddening and metallicity scales, subsequently led to a decrease in typical distances accompanied by an increase in typical ages. While the source of the parallaxes in all of the analyses is essentially the same (objects in the Hipparcos Catalogue) the most recent multicolour analyses of Grundahl *et al.* (2002), Percival *et al.* (2002) and Gratton *et al.* (2003) give more confidence in the overall control of these systematic errors. They confirm the suggestion by Carretta *et al.* (2000) to revise downwards the then-favoured main-sequence fitting distances by about 0.1 mag, based on a comparison between the implied distance to the LMC estimated from the resulting RR Lyrae absolute magnitudes with all other available estimates, and attributed to small inconsistencies in the reddening and metallicity scales for globular cluster and local subdwarfs.

As a consequence, age estimates for the oldest clusters consequently now lie in the range 12–14 Gyr. Some of the best constraints are provided by NGC 6397 and NGC 6752 from Gratton *et al.* (2003), i.e. an average of $13.4 \pm 0.8 \pm 0.6$ Gyr (random + systematic), where the models account for microscopic diffusion. A discussion of the consequences of these ages for cosmology, and the formation of the Galaxy, is deferred to Chapter 9.

The main source of uncertainty in the main-sequence fitting no longer rests on the absolute magnitudes of the calibrating subdwarfs, although their numbers remain limited, but in the colour domain, including line-of-sight reddening, metallicity scales and associated calibrations. An attempt to quantify these was made by Carretta *et al.* (1999); Gratton *et al.* (1999); Carretta *et al.* (2000); Gratton *et al.* (2003), who summarised typical remaining distance modulus uncertainties as due to: (a) local sample properties: parallax errors (0.01), Malmquist bias (negligible), and Lutz–Kelker effect (0.02); (b) binary contamination: field binaries (0.02), and cluster binaries (0.02–0.03); and (c) systematic differences: photometric calibrations of 0.01 mag (0.04), reddening scale of 0.015 mag (0.07), metallicity scale of 0.1 dex (0.08), and over-abundance of α-elements (negligible). Collectively these were estimated as a total uncertainty of at least ± 0.12 mag, reduced to ± 0.07 mag after the work of Gratton *et al.* (2003) in which the reddening and metallicity scale effects were collectively reduced to 0.035 mag. Gratton *et al.* (2003) show that the different assumptions about photometry, metallicities and reddening explain rather well the differences in the previous distance and age estimates for 47 Tuc, NGC 6397, and NGC 6752.

Table 5.4 Age determinations for the oldest globular clusters. The various Hipparcos subdwarf-based estimates illustrate the evolution in the understanding of various error sources with time.

Method	Reference	Value
Hipparcos subdwarfs:		
Main-sequence fitting	Reid (1997)	12 ± 1
Main-sequence fitting	Gratton *et al.* (1997)	$11.8^{+2.1}_{-2.5}$
Main-sequence fitting	Pont *et al.* (1998)	13–14
Main-sequence fitting	Carretta *et al.* (2000)	11.5 ± 2.6
Main-sequence fitting	Gratton *et al.* (2003)	13.4 ± 0.8
Other:		
Theoretical horizontal branch and MS fitting	D'Antona *et al.* (1997)	12 ± 1
Theoretical horizontal branch	Salaris *et al.* (1997)	12.2 ± 1.8
Compilations:		
Combination of independent techniques	Chaboyer *et al.* (1998)	11.5 ± 1.3
Adjustment to the LMC distance of 18.54	Carretta *et al.* (2000)	12.9 ± 2.9

Carretta *et al.* (2000) also derived a resulting mean age from local subdwarf fitting of 11.5 ± 2.6 Gyr.

Specific clusters Specific clusters have been variously discussed in the context of the Hipparcos data: **M30**: Sandquist *et al.* (1999) suggesting a high helium content relative to other clusters of similar metallicity; **M55**: Pych *et al.* (2001) derived a distance modulus of 13.86 ± 0.25 mag using the slope of the period–luminosity relation for SX Phe stars in the cluster, along with the Hipparcos parallax data for SX Phe itself; **M71**: Yim *et al.* (2004) derived a distance modulus of 13.46 ± 0.17; **M92**: Grundahl *et al.* (2000) established an age range of 12–17 Gyr on the basis of theoretical isochrones matched to Strömgren photometry. By comparison with the Hipparcos luminosity of the [Fe/H] ~ -2.5 subgiant HD 140283, they were able to tighten the constraint on age to ~16 Gyr and on distance modulus to $\mu_V < 14.60$ mag. VandenBerg (1999) has argued that the existence of such subgiants themselves provide evidence of stellar ages $\gtrsim 14$ Gyr.

5.8.2 RR Lyrae and horizontal branch stars

Globular cluster and other Population II stars having masses above solar have long since evolved into white dwarfs. In contrast, stars of approximately solar mass are well represented observationally, and their location in the HR diagram well accounted for theoretically. After ascending the giant branch, terminating in the helium flash, stars evolve rapidly onto the zero-age horizontal branch (ZAHB) with masses around 0.6–$0.8\,M_\odot$, where they basically comprise a static helium-burning core and hydrogen burning shell. Location along the ZAHB depends on metallicity, from blue in the metal-poor clusters, to red in the metal-rich clusters where they merge into the giant branch to form the region of the (red) clump giants.

RR Lyrae variables are a subset of the horizontal branch giants, occurring where the horizontal branch intersects the instability strip. They occupy a narrow colour range around spectral type A, and have pulsation periods $\lesssim 1$ day. Like the Cepheids, although less luminous but still some 5 mag above the corresponding main sequence, their distinctive light-curves allows their detection to large distances (as far as the Galactic centre in the Baade Windows of low interstellar absorption) and in crowded fields. There are two major subgroups: the RRab, most relevant to the distance scale, are metal-poor spheroidal component stars, with asymmetric light-curves, longer periods $\gtrsim 0.4$ day, large amplitudes $\Delta m \sim 0.5$–1.5 mag, and pulsating in the fundamental mode. The less numerous RRc type are old disk component stars, with symmetric almost sinusoidal light-curves, shorter periods $\lesssim 0.4$ day, smaller amplitudes $\Delta m \lesssim 0.5$ mag, and pulsating in the first overtone. There are also double-mode pulsators, denoted RRd, which pulsate simultaneously in the fundamental mode and in the first overtone.

Determination of RR Lyrae absolute magnitudes is complicated by their typically large distances which means that trigonometric parallax data is strongly limited, their dependence on metallicity, and their relatively rapid evolution away from the ZAHB. Various methods have been used for luminosity calibration: (i) direct distance estimates; (ii) using distances of RR Lyrae in Galactic globular clusters whose distance has been derived from main-sequence fitting of subdwarfs with Hipparcos parallaxes; this has the advantage of eliminating uncertainties in possible differences between RR Lyrae of the same metallicity in clusters and in the general field; (iii) the use of statistical parallaxes; (iv) Baade–Wesselink determinations; (v) using Hipparcos parallaxes of field horizontal branch stars at the centre of the instability strip at around $\log T_{\rm eff} \sim 3.85$.

Their use as calibrators for the LMC distance follows similar steps.

The RR Lyrae stars also provide tests of pulsation models, evolutionary models, even placing limits on the neutrino magnetic moment (Castellani & Degl'Innocenti, 1993), and are important kinematic tracers. Amongst many introductory references to RR Lyrae stars, a general introduction is given by Sandage (2000), and a review of the theoretical context is given by Caputo (1998). Reviews of the various Hipparcos and other results include Fernley (1998), Gratton (1998a), Popowski & Gould (1999a), Popowski (2001), and Bono (2003).

Properties of the Hipparcos RR Lyrae stars There are some 6500 Galactic RR Lyrae stars known, and listed in the GCVS, NSV, and subsequent updated name lists of variable stars. The majority are fainter than the Hipparcos and Tycho Catalogue limits. Flags indicate 179 stars classified as RR Lyrae in the Hipparcos Catalogue, of which 17 were newly-discovered in the Hipparcos epoch photometry, with some 20 others flagged in the Tycho 2 Catalogue and associated supplements (Fernley & Barnes, 1997; Mennessier & Colomé, 2002).

Determination of the absolute magnitude–metallicity relation requires uniform and well calibrated abundances. The determination of statistical parallaxes also requires radial velocities, in addition to proper motions and V magnitudes. A number of observational programmes were therefore undertaken to acquire these data for Hipparcos objects. Martínez-Delgado & Alfaro (1997) determined metallicities for 30 Hipparcos RR Lyrae using low-resolution spectroscopy and use of the Ca II line. Fernley & Barnes (1997) and Solano *et al.* (1997) determined radial velocities and iron abundances for respectively 56 and 45 Hipparcos field RR Lyrae variables. A number of new binary stars, and a number of mis-classified objects, were identified in the process.

Mennessier & Colomé (2002) correlated the positions of all known Galactic RR Lyrae stars with Tycho 2 Catalogue positions to identify, in addition to the 179 Hipparcos objects, a total of 172 Tycho 2 Catalogue RR Lyrae, thus roughly doubling the number of RR Lyrae stars with accurate proper motions. Numerous corrections to these identifications were reported by Maintz (2005).

Absolute magnitudes of RR Lyrae stars It has been known for many years that the absolute magnitude of horizontal branch in general, and RR Lyrae stars in particular, is a function of metallicity (Sandage, 1958), traditionally parameterised as $M_V(\mathrm{RR}) = \alpha[\mathrm{Fe/H}] + \beta$. There is, however, no complete agreement about the precise slope, and indeed theoretical models do not predict a simple linear relationship between luminosity and metallicity (Bono, 2003). This results in a variation of $\sim$0.25 mag resulting from a 1 dex variation in metallicity. In addition, a spread of 0.1–0.5 mag occurs through evolutionary effects, even within the same cluster. Chaboyer *et al.* (1998) and Chaboyer (1999) suggested $\alpha = 0.23 \pm 0.04$ from various (non-Hipparcos) studies; Fernley *et al.* (1998b) estimated $\alpha = 0.18 \pm 0.03$ from Baade–Wesselink absolute magnitudes; Carretta *et al.* (2000) also advocated $\alpha = 0.18 \pm 0.09$ based partly on theoretical models, although noting that it might be steeper at $[\mathrm{Fe/H}] > -1$. It is also generally assumed that the field and cluster RR Lyrae have the same absolute magnitudes.

The zero-point is normally referenced to metallicities of around $[\mathrm{Fe/H}] = -1.5$, being typical of the metal-poor globular clusters, of the horizontal branch stars with Hipparcos parallaxes considered by Gratton (1998b), of the RR Lyrae stars used in statistical parallaxes, and probably of the bar in the LMC. Having adopted a value for the slope, the zero-point, β, can then be derived in various ways, as detailed in the following. Table 5.5 is updated from Table 2 of Cacciari & Clementini (2003), where a detailed discussion of each determination can be found. Similar compilations were previously presented by, e.g. Chaboyer (1999).

Trigonometric parallaxes Hipparcos gave individually useful parallaxes for few RR Lyrae stars, of which only that for RR Lyrae itself (HIP 95497) is reasonably accurate, $\pi = 4.38 \pm 0.59$ mas (Figure 5.30). Benedict *et al.* (2002) determined an improved precision of $\pi = 3.82 \pm 0.2$ mas from HST Fine Guidance Sensor observations. Weighting and reddening corrections lead to the value in Table 5.5. At around the same time, Bono *et al.* (2002a) proposed a weighted mean of the Hipparcos, HST, and pulsational parallax for RR Lyrae of 3.87 ± 0.19 mas.

Trigonometric parallaxes of horizontal branch stars A number of other metal-poor horizontal branch stars were observed by Hipparcos, and can be used for the calibration of the RR Lyrae absolute magnitudes on the assumption that they are well approximated by field horizontal branch stars at the centre of the instability strip at around $\log T_{\mathrm{eff}} \sim 3.85$. Gratton (1998b) identified such a sample of 22 field metal-poor stars (10 blue, three RR Lyrae, nine red) with Hipparcos parallaxes. Using the cluster M5 as a template for the shape of the horizontal branch, they derived $M_V = +0.69 \pm 0.10$ at an average $[\mathrm{Fe/H}] = -1.41$ or, excluding one suspect star, $M_V = +0.60 \pm 0.12$ at an average $[\mathrm{Fe/H}] = -1.51$. Popowski & Gould (1999b) re-analysed the same sample and, after elimination of red horizontal branch stars because of possible contamination by red giant

*Table 5.5 Determination of the luminosity of RR Lyrae stars, from Cacciari & Clementini (2003, Table 2); * indicated entries added here. It is based on calibrations assuming $M_V(RR) = 0.23\,([Fe/H]+1.5)+\beta$, and evaluated at $[Fe/H]=-1.5$. The reference identifies where further details can be found (possibly a compilation of other work).*

Method	Reference	M_V(RR)
Baade–Wesselink:		
General sample*	Fernley *et al.* (1998b)	0.68±0.12
RR Ceti only	Cacciari *et al.* (2000)	0.55±0.12
Evolutionary and pulsation models:		
HB stars: evolutionary models, bright	Cacciari & Clementini (2003)	0.43±0.12
HB stars: evolutionary models, faint	"	0.56±0.12
Pulsation models: visual	"	0.58±0.12
Pulsation models: period–luminosity–colour	"	0.59±0.10
Pulsation models: double-mode	"	0.57±0.06
Fourier parameters	"	0.61±0.05
Hipparcos-based:		
Trigonometric parallaxes: RR Lyrae only	Benedict *et al.* (2002)	0.58±0.13
Trigonometric parallaxes: horizontal branch stars	Carretta *et al.* (2000)	0.62±0.11
Statistical parallaxes	Cacciari & Clementini (2003)	0.78±0.12
Weighted average of all results	Cacciari & Clementini (2003)	0.59±0.03
Hipparcos subdwarf main-sequence fitting*	Gratton *et al.* (2003)	0.56±0.07

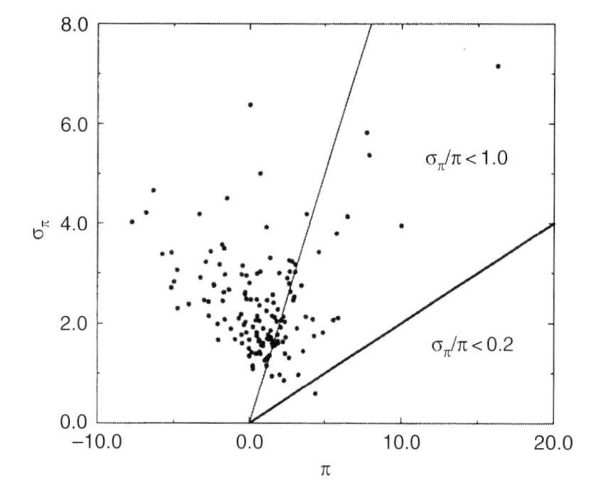

Figure 5.30 Distribution of trigonometric parallax errors σ_π versus parallax π for the RR Lyrae stars observed by Hipparcos. From Chen (1999, Figure 1).

branch stars, derived $M_V = +0.69 \pm 0.15$ at $[Fe/H] = -1.62$. Koen & Laney (1998) showed that the Gratton (1998b) distances are slightly underestimated because the intrinsic scatter in horizontal branch magnitudes was neglected when correcting for the Lutz–Kelker effect, and derived $M_V = +0.62$ at $[Fe/H] = -1.5$.

De Boer *et al.* (1997a,b), see also de Boer (1999), derived $M_V \sim 0.76 \pm 0.4$ for two cool blue horizontal branch stars. Luminosities of RR Lyrae and horizontal branch stars were also determined by McNamara (1997a). Carney *et al.* (1998) studied the red horizontal branch star HD 17072, which has the best-determined Hipparcos parallax among the metal-poor field horizontal branch stars, and supports the fainter luminosities found from statistical parallax and Baade–Wesselink analyses, in contrast to the results of main-sequence fitting of metal-poor field dwarfs to globular cluster main sequences.

Statistical parallaxes Application of the method of statistical parallax, described in Section 5.3, requires a group of stars which is dynamically homogeneous, drawn from a single velocity distribution independent of location, with all members having the same absolute magnitude. RR Lyrae stars of the Galactic spheroid constitute a good approximation to these requirements, and are rather abundant in the solar vicinity and distributed reasonably evenly over the sky. A review of the historical use of the statistical parallax technique applied to the determination of the absolute magnitudes of RR Lyrae stars is given by Dambis & Rastorguev (2001), including the earliest quoted analysis by Pavlovskaya (1953), and subsequent work by van Herk (1965), Hemenway (1975), and Clube & Dawe (1980a,b). Hawley *et al.* (1986) and Strugnell *et al.* (1986) used a rigorous algorithm making a simultaneous determination of the kinematic parameters (space velocity and velocity dispersion tensor) and the correction factor to the initial distance scale.

Using ground-based proper motions, Layden *et al.* (1996) had estimated $M_V(RR) = 0.71 \pm 0.12$ at $[Fe/H] = -1.61$. Popowski & Gould (1998a,b) and Gould & Popowski (1998) presented a mathematical analysis of the method, and re-analysed a sample of 147 RR Lyrae stars, essentially confirming these results, and giving M_V(RR) $= 0.75 \pm 0.13$ at [Fe/H] $= -1.60$, ascribing most of the difference to a correction for Malmquist bias. Popowski & Gould (1999b) provide an analytic expression for the resulting error in the distance scale.

A Hipparcos-based analysis, treating separately the halo and thick disk components, was carried out by Fernley *et al.* (1998a), Tsujimoto *et al.* (1998) and Dambis & Rastorguev (2001), all giving rather similar but not independent results. The latter used a sample of 262 RR Lyrae, separating the contributions by metallicity and kinematics using an initial distance scale to transform proper motions into space velocities, and deriving M_V(RR) $= 0.76 \pm 0.12$ for the halo population at [Fe/H] $= -1.6$. However, kinematic inhomogenities within the halo sample were found by Martin & Morrison (1998) and Borkova & Marsakov (2003) (see Section 5.8.2), which would presumably bias the results. This sensitivity to the overall model adopted for the Galactic halo, and the possibility that a significant fraction of the halo RR Lyrae stars belong to a few streams, perhaps from infalling satellite galaxies, will complicate the interpretation.

Groenewegen & Salaris (1999) used the method of 'reduced parallaxes' (Section 5.2.4) for the halo RR Lyrae stars to derive a zero-point of 0.77 ± 0.26 mag for an assumed slope of 0.18 in the M_V versus [Fe/H] relation: 0.28 mag brighter than the value from Fernley *et al.* (1998a) based on statistical parallaxes for the identical sample and using the same slope. Other analyses using the Hipparcos data have been made by Tsujimoto *et al.* (1997), Popowski (1998), Solano (1998), Solano & Barnes (1999), Tsujimoto & Yoshii (1999), Dambis (2003b), and Dambis & Vozyakova (2004).

A synthesis of various results has been given successively by Chaboyer (1999), Carretta *et al.* (2000), and by Cacciari & Clementini (2003), whose value of M_V(RR) $= 0.78 \pm 0.12$ at [Fe/H] $= -1.5$ is shown in Table 5.5.

Baade–Wesselink method Baade–Wesselink determinations of the absolute magnitudes of RR Lyrae stars follow the same principles as discussed for the Cepheid variables (Section 5.7.1), and involve similar assumptions (e.g. Jones *et al.*, 1992; McNamara, 1997b). Fernley *et al.* (1998b) summarised the method applied to 28 field RR Lyrae stars, deriving M_V(RR) $= (0.20 \pm 0.04)$([Fe/H]$+1.5) + (0.68 \pm 0.05)$ implying M_V(RR) $= 0.68 \pm 0.12$ at [Fe/H] $= -1.5$. Cacciari *et al.* (2000) made a number of modifications to the specific analysis of RR Cet, yielding M_V(RR) $= 0.55 \pm 0.12$ at [Fe/H] $= -1.5$.

Evolutionary and pulsation models There are numerous published estimates of the luminosity of horizontal branch and RR Lyrae stars based on evolutionary and pulsation models. A review is given by Cacciari & Clementini (2003), and the summary numbers are included in Table 5.5. A notably small standard error, in part due to the large numbers of objects in the class, is derived in the case of models of double-mode RR Lyrae, detected in large numbers near the bar of the LMC as part of the MACHO project. Models suggest (Caputo *et al.*, 1998) a dependency of the form

$$\log P_0 = 11.242 + 0.841 \log L - 0.679 \log M - 3.410 \log T_{\rm eff} + 0.007 \log Z \qquad (5.36)$$

$$\log P_1 = 10.845 + 0.809 \log L - 0.598 \log M - 3.323 \log T_{\rm eff} + 0.005 \log Z \qquad (5.37)$$

in which $P_{0,1}$, $T_{\rm eff}$ and Z are observed quantities, from which the mass M and luminosity L can be derived. Out of a database of 7900 RR Lyrae stars, 73 double-mode pulsators (RRd) were used to derive (theoretical) pulsation-based luminosities in this way. This led Alcock *et al.* (1997) to derive M_V(RR) $= 0.46 \pm 0.16$ mag and, with an assumed reddening of $E(B-V) = 0.086$, and a mean dereddened magnitude of their multi-mode RR Lyrae sample of $\langle V_0 \rangle = 18.94 \pm 0.03$ mag, a distance modulus of the LMC of $\mu_{\rm LMC} = 18.48 \pm 0.19$.

Bono *et al.* (2002a) used the predicted relation between the absolute K magnitude of fundamental RR Lyrae variables and their period and metal content, together with current evolutionary predictions on the mass and luminosity of horizontal branch stars, to derive a 'pulsational' parallax of $\pi = 3.858 \pm 0.131$ mas for RR Lyrae itself, including an extinction correction of $A_V = 0.12 \pm 0.01$ mag.

Summary of absolute magnitudes The weighted average of the various RR Lyrae results led Cacciari & Clementini (2003) to propose M_V(RR) $= 0.59 \pm 0.03$ at [Fe/H] $= -1.5$, in close agreement with the value of M_V(RR) $= 0.56 \pm 0.07$ at [Fe/H] $= -1.5$ proposed from subdwarf main-sequence fitting by Gratton *et al.* (2003). For the most discrepant value, based on statistical parallax, plausible reason for the discrepancies have been proposed. Collectively, these values are in agreement with those theoretical models predicting a fainter absolute magnitude for RR Lyrae stars, e.g. Straniero *et al.* (1997), Ferraro *et al.* (1999), Demarque *et al.* (2000), VandenBerg *et al.* (2000), as opposed to those predicting a brighter value, e.g. Caloi *et al.* (1997), Cassisi *et al.* (1999).

Carretta *et al.* (2000) revised the relation derived by Gratton *et al.* (1997) between the absolute magnitude of RR Lyrae, the horizontal branch, and the zero-age horizontal branch as a function of metallicity, also making use of the metal dependency of the absolute magnitude of the horizontal branch stars compared with the zero-age horizontal branch, M_V(HB) $= M_V$(ZAHB) $- (0.05 \pm$

$0.03)([\mathrm{Fe/H}] + 1.5) - (0.09 \pm 0.04)$, to propose

$$M_V(\mathrm{HB}) = (0.13 \pm 0.09)([\mathrm{Fe/H}]+1.5) + (0.54 \pm 0.07)$$
$$M_V(\mathrm{ZAHB}) = (0.18 \pm 0.09)([\mathrm{Fe/H}]+1.5) + (0.63 \pm 0.07)$$
$$M_V(\mathrm{RR}) = (0.18 \pm 0.09)([\mathrm{Fe/H}]+1.5) + (0.57 \pm 0.07)$$

Kinematics Pre-Hipparcos, Preston (1959) made a comprehensive survey of RR Lyrae stars covering a wide range of metallicity, concluding that their kinematics were consistent with both disk and halo contributions. Subsequent large-scale surveys were made by Layden (1994, 1995) and Layden *et al.* (1996), who added more stars, computed full space velocities, and concluded that the RR Lyrae stars show two chemically and kinematically distinct populations in the solar neighbourhood: the thick disk and the halo.

Wielen *et al.* (1997) derived the space velocities of 130 RR Lyrae stars using the Hipparcos proper motions and photometric distances consistent with the Hipparcos parallaxes. Mean velocities and velocity dispersions were determined as a function of [Fe/H]. The metal-poor stars showed no significant net rotation (consistent with expected properties of the halo), while the metal-rich component move like typical disk stars. Their data did not allow one to decide whether the kinematic properties of halo and disk components show a smooth transition, or whether they belong to two distinct populations (Figure 5.31).

Martin & Morrison (1998) derived space velocities for 130 nearby RR Lyrae stars, using both Hipparcos and ground-based proper motions. The velocity ellipsoids for the halo and thick-disk samples agreed with previous studies, with kinematic evidence for some thin-disk contamination in the thick-disk sample. They isolated 21 stars with $[\mathrm{Fe/H}] < -1.0$ and disk-like kinematics, representing a metal-weak tail of the thick disk that extends to $[\mathrm{Fe/H}] = -2.05$. In the halo samples, the distribution of V velocities is not Gaussian, while the U and W show unexplained structure. Systematic changes to the distance scale within the range of currently accepted values of $M_V(\mathrm{RR})$ significantly change the calculated halo kinematics. Fainter values of $M_V(\mathrm{RR})$, such as those obtained by statistical parallax of 0.6–0.7 at $[\mathrm{Fe/H}] = -1.9$, result in local halo kinematics similar to properties previously reported, while brighter values, such as those obtained via the Hipparcos subdwarf parallaxes of 0.3–0.4 at $[\mathrm{Fe/H}] = -1.9$, result in a halo with retrograde rotation and significantly enlarged velocity dispersions (Figure 5.32).

Chen (1999) used the same sample of 144 RR Lyrae stars studied by Fernley *et al.* (1998a). Pattern recognition using wavelet transforms and kernel density estimation, as well as clustering analysis, were used to search for structure in the [Fe/H] versus V velocity component (Figure 5.33). The resulting contour maps show three significant groups: the disk population at $V \sim -10\,\mathrm{km\,s^{-1}}$ and $[\mathrm{Fe/H}] = -0.4$ dex (actually comprising two subgroups with different metallicities), the halo population at $-220\,\mathrm{km\,s^{-1}}$ and -1.6 dex, and an intermediate population with $-110\,\mathrm{km\,s^{-1}}$ and -1.3 dex.

Altmann & de Boer (2000) studied the kinematics of a small sample of field horizontal branch stars, comprising 14 HBA, 2 HBB, and 5 sdB/O stars (see Section 7.5.6). The results showed a clear trend in kinematics along the horizontal branch: the HBA stars have low V component velocities with some on retrograde orbits, while the sdB/O and probably the HBB show disk-like kinematics.

Dambis & Rastorguev (2001) applied the statistical parallax technique to 262 RRab Lyrae variables, using proper motions either from Hipparcos, or from other catalogues reduced to the Hipparcos proper motion system. They determined the velocity distribution for the halo and thick disk RR Lyrae populations as $(U_0, V_0, W_0) = (-9 \pm 12, -214 \pm 10, -16 \pm 7)\,\mathrm{km\,s^{-1}}$ and $(\sigma_{UVW}) = (164 \pm 11, 105 \pm 7, 95 \pm 7)\,\mathrm{km\,s^{-1}}$ for the former, and $(U_0, V_0, W_0) = (-16 \pm 8, -41 \pm 7, -18 \pm 5)\,\mathrm{km\,s^{-1}}$ and $(\sigma_{UVW}) = (53 \pm 9, 42 \pm 8, 26 \pm 5)\,\mathrm{km\,s^{-1}}$ for the latter. From the same analysis, they derived a (short) LMC distance modulus of $\mu_{\mathrm{LMC}} = 18.22 \pm 0.11$ mag, and a distance to the Galactic centre of $7.4 \pm 0.5\,\mathrm{km\,s^{-1}}$.

Borkova & Marsakov (2003) used the same sample to compute Galactic velocity components for 209 metal-poor, $[\mathrm{Fe/H}] < -1.0$, RR Lyrae stars using proper motions, radial velocities, and photometric distances. They observed abrupt changes in the spatial and kinematical characteristics when the peculiar velocities relative to the local standard of rest cross a threshold value of $V_{\mathrm{pec}} = 280\,\mathrm{km\,s^{-1}}$. They argued that the population of metal-poor RR Lyrae stars includes two spherical subsystems occupying different volumes. They proposed that field stars with velocities below the threshold value and globular clusters with extremely blue horizontal branches form the spherical, slowly rotating subsystem of the proto-disk halo, which has a very small vertical metallicity gradient. Field stars with higher velocities and globular clusters with redder horizontal branches constitute the spheroidal subsystem of the accreted outer halo, approximately two times larger in size. It has no metallicity gradient, most of its stars have eccentric orbits, many displaying retrograde motion, and their ages are comparatively low, suggesting an extragalactic origin.

Maintz & de Boer (2005) used data for 217 RR Lyrae from the well-defined sample of 561 RR Lyrae stars to $V = 12.5$ mag constructed by Maintz (2005), to further examine the kinematics, orbits and Z-distribution. One third of the sample have orbits confined near to the

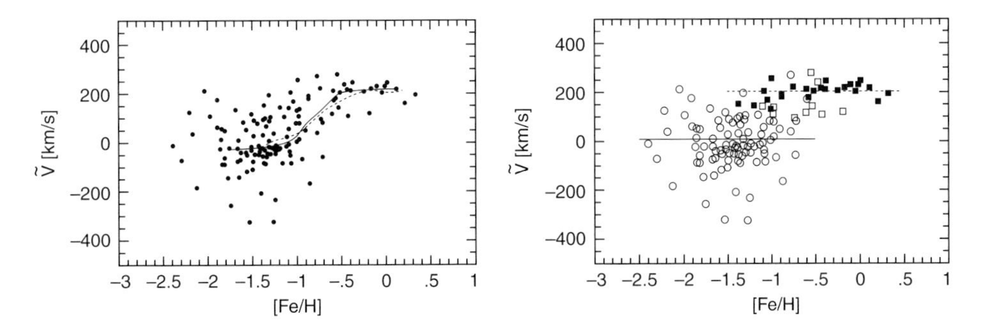

Figure 5.31 Galactic space velocity component in the direction of Galactic rotation, V, based on 130 Hipparcos RR Lyrae stars, as a function of [Fe/H]. Left: the solid curve represents a running median of V, the dashed curve represents a running mean. Right: the distribution shown separately for disk stars (filled squares) classified as having [Fe/H] > −1.4 and with peculiar velocities inside a velocity ellipsoid of UVW = 166, 100, 86 km s^{-1}, and classified as halo stars otherwise. Open squares would also be classified as disk objects if the defining velocity ellipsoid is enlarged by a factor 1.5. From Wielen et al. (1997, Figures 4 and 5).

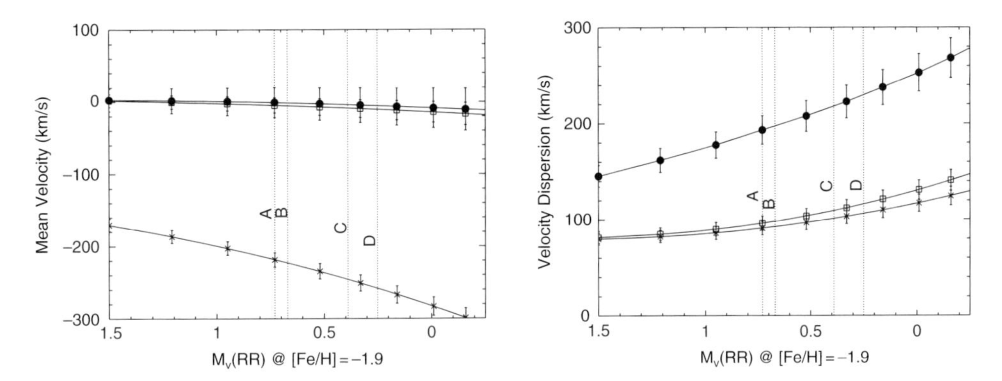

Figure 5.32 Mean U, V, and W velocities (left) and velocity dispersions (right) as functions of M_V(RR) for the halo sample of RR Lyrae stars from Martin & Morrison (1998). In both cases the symbols indicate: •: U, ×: V, and □: W. Lines A, B, C, and D mark the values of M_V(RR) adopted by Layden (1994) and by the authors, by Layden et al. (1996), by Chaboyer et al. (1998), and by Feast & Catchpole (1997), respectively. From Martin & Morrison (1998, Figures 6 and 7).

Galactic plane, while 163 have halo-like orbits fulfilling one of the following criteria: $\Theta < 100$ km s^{-1}, orbital eccentricity > 0.4, and normalised maximum orbital Z-distance > 0.45. Within their sample, roughly half of the stars were shown to have retrograde orbits.

They also inferred that the Z-distance probability distribution shows scale heights of 1.3 ± 0.1 kpc for the disk component, and 4.6 ± 0.3 kpc for the halo component (Figure 5.34). The metal-poor stars with [Fe/H] < 1.0 have a wide symmetric distribution about $\Theta = 0$, implying a halo component of old and metal poor stars with a scale height of 4–5 kpc having random orbits. The mid-plane density ratio of halo to disk stars given by this sample is ~0.16.

5.8.3 Other Population II distance indicators

Globular cluster dynamics Globular cluster distances can also be derived in a manner analogous to the use of statistical parallaxes, by adjusting the estimated distance to provide concordance between internal proper motion and radial velocity dispersions. Advantages are that only relative quantities are required, it does not make use of standard candles or evolutionary models, and it is independent of reddening (Bedin *et al.*, 2003). Rees (1996) gave distances to eight Galactic clusters, including M5, M15, M92 and NGC 6397, based on King–Michie type models. Chaboyer *et al.* (1998) restricted their analysis to six clusters to derive $M_V(\mathrm{RR}) = 0.59 \pm 0.11$ at [Fe/H] $= -1.59$, leading to $M_V(\mathrm{RR}) = (0.18 \pm 0.09)([\mathrm{Fe/H}] + 1.5) + (0.61 \pm 0.11)$ proposed by Carretta *et al.* (2000), and hence the distance to the LMC given in Table 5.6.

White dwarf cooling sequence The distance to a globular cluster can be estimated from a comparison of its DA white dwarf sequence, representing a cooling with

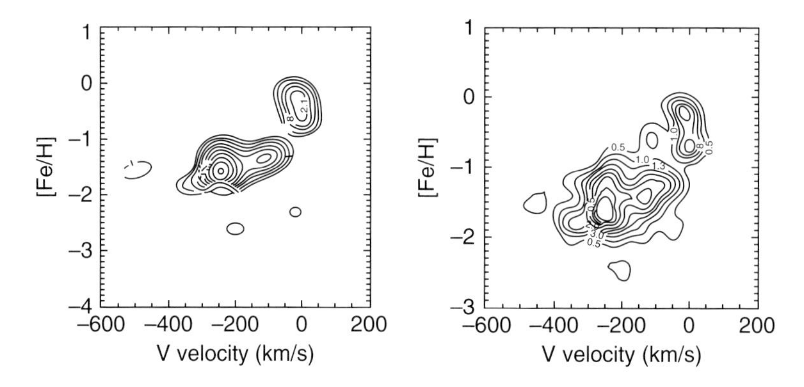

Figure 5.33 RR Lyrae kinematic analyses from Chen (1999). Left: contour map of the wavelet analysis coefficients at $\sigma = 0.5$. Right: contour map of the probability density function from the kernel density estimation for the RR Lyrae sample. From Chen (1999, Figures 2c and 3a).

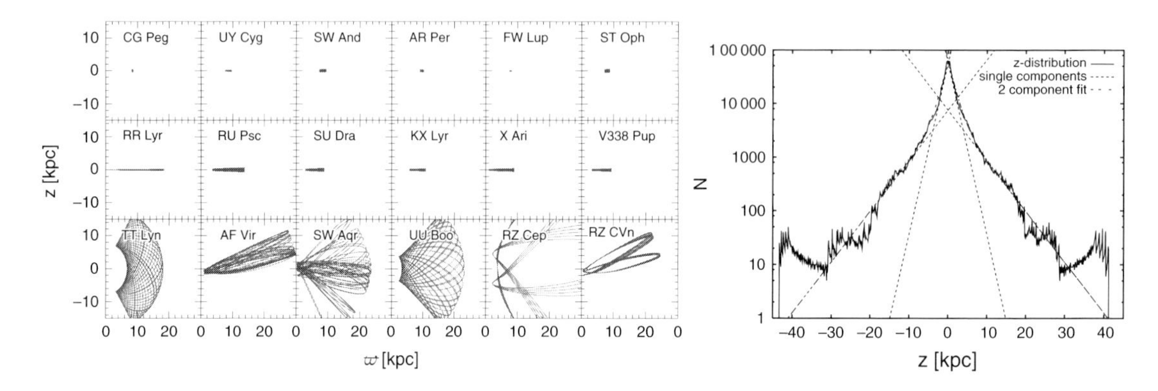

Figure 5.34 Left: meridional plots of some RR Lyrae star orbits, calculated over a time span of 10 Gyr. Top row: orbits of thin disk stars; middle row: RR Lyrae itself and five other stars with elliptical orbits in or near the plane; bottom row: orbits of stars moving far out into the halo. Right: Z-distance distribution of all sample stars, obtained by summing the individual $N(Z)$ statistics for time steps of 1 Myr. The dashed lines show the two-component fit of the logarithmic distribution based on scale heights of 1.3 ± 0.1 kpc for the disk component and 4.5 ± 0.3 kpc for the halo component. From Maintz & de Boer (2005, Figures 3 and 6).

time, with a template sequence formed from local white dwarfs with known parallaxes and masses. White dwarfs have been observed by HST in a number of clusters, and the method is considered independent of metallicity, age, and of details of convection theory.

It was first applied to NGC 6752, where Renzini *et al.* (1996) found a distance modulus of $\mu_0 = 13.05$ mag and an age of 14.5–15.5 Gyr, with and without He diffusion respectively. For the more metal-rich cluster 47 Tuc, Zoccali *et al.* (2001) updating the preliminary results from Zoccali *et al.* (1999) and using Hipparcos parallaxes for three of the local calibrators, derived $\mu_0 = 13.09 \pm 0.14$ mag and an age of 13 ± 2.2 Gyr, including effects of He diffusion. They noted that the larger distance modulus implied by the earlier subdwarf fitting results would imply implausibly low masses for the cluster white dwarfs.

Potential error sources are the difficulties of comparing HST colours for objects with very different apparent magnitudes, and the assumptions on the mass and thickness of the hydrogen envelopes of the globular cluster white dwarfs (Salaris *et al.*, 2001); the results for NGC 6791, for example, remain enigmatic (Bedin *et al.*, 2005). A comparison with the distances from Gratton *et al.* (2003) suggests good agreement in the case of NGC 6752, but poor in the case of 47 Tuc, amounting to 0.23 mag in μ_V, and degraded further in μ_0 on account of the different assumptions on $E(B-V)$. Carretta *et al.* (2000) provided the corresponding distance to the LMC given in Table 5.6.

5.9 The Magellanic Clouds

5.9.1 Distance to the Large Magellanic Cloud

As referred to throughout this chapter, the Large Magellanic Cloud (LMC) remains a basic reference point for the astronomical distance scale, at which numerous distance estimation techniques can be compared. Agreement or otherwise of distance estimates from the various methods provides an indication of the robustness, likely systematic errors, and status of the theoretical understanding associated with each of the techniques.

Its importance is underlined by the fact that the HST key project to determine H_0 assumed an LMC distance modulus of $(m - M)_0 = 18.5 \pm 0.1$ to define the fiducial Cepheid period–luminosity relation, and hence distances to external galaxies. The accuracy of their final value of $H_0 = 72 \pm 8\,\mathrm{km\,s^{-1}\,Mpc^{-1}}$ (Freedman *et al.*, 2001) was ultimately determined by the assumed LMC distance, and its assumed error constituted 6.5% of their final 9% error budget. The resulting Cepheid-based distances to external galaxies then provided the calibration of additional distance calibrators (Type Ia and Type II supernovae, the Tully–Fisher relation, surface-brightness fluctuations, and the fundamental plane) used for distance estimates to more distant galaxies and clusters.

Figure 5.13 illustrates the different paths that have been followed in this chapter. It is divided into Population I and II indicators, identifying those making direct or indirect use of the Hipparcos data, and illustrating the resulting distance derived for the LMC according to each of these, as summarised in Table 5.6.

This section will not attempt to review each of the published distance estimates to the LMC. These estimates have evolved with time, even over the past 10 years, reflecting the evolution of observational accuracy and theoretical modelling; an early post-Hipparcos review was given by Walker (1999). The topic also extends well beyond the restrictions of the Hipparcos data. Instead, the table and figure provide a selection of the latest values reported and discussed in the previous sections.

A few summary comments are nevertheless in order. A long-standing dichotomy has existed between 'long' and 'short' distance scales to the LMC. In the pre-Hipparcos literature, distance estimates via Population I indicators tended to favour the 'long' scale, $\mu_{\rm LMC} \gtrsim 18.6$. Meanwhile, estimates based RR Lyrae variables tended to support the 'short' scale, $\mu_{\rm LMC} \lesssim 18.4$. This distinction is no longer supported by the present distance estimates. Within their error bars, almost all methods now provide the same distance to the LMC. This new agreement partly arises from a consistent reddening scale in the various distance determinations, with $E(B - V) = 0.089 \pm 0.015$ proposed by Alves *et al.* (2002).

Theoretical and observational understanding of the red clump giants have progressed particularly rapidly in the last few years. The Hipparcos data have made an important contribution to this topic, providing large numbers of accurate absolute magnitudes in the solar neighbourhood, allowing careful calibration of population and age effects, and yielding $\mu_{\rm LMC} = 18.49 \pm 0.03 \pm 0.03$ (Alves *et al.*, 2002). This is identical to the straight mean of the values tabulated (ignoring the fact that not all estimates are independent, and avoiding dubious weighting schemes), and close to the value assumed by the HST key project.

There are many details remaining to be clarified, which have been noted in the preceding sections and which will not be repeated here. Amongst them might be highlighted differences found in distance estimates for the LMC cluster NGC 1866, 18.33 ± 0.08, and that for the LMC red clump field giants, 18.53 ± 0.07 (Salaris *et al.*, 2003). Depth effects in the LMC have also started to contribute to comparisons of the distance estimates, and further contributions of distance estimates to detached eclipsing double-lined spectroscopic binaries will be important in clarifying the geometry of the LMC in general, and the geometrical nature of the bar in particular.

Further in the future, one-step trigonometric parallax estimates to individual objects in the LMC, at $\pi \sim 20\,\mu$arcsec, will become accessible to space astrometry missions like Gaia and SIM. At that point, the tortuous complexity underlying the various distance estimates used to date will be relegated to historical curiosity, and still deeper insights into the physical nature of the 'standard candles' used to date will commence.

5.9.2 Dynamics of the Magellanic Clouds

Beyond our own Galaxy, Hipparcos was able to observe only a few objects because of its magnitude limit of around $V = 12$ mag. A careful pre-selection of stars satisfying magnitude and crowding constraints resulted in 36 Large Magellanic Cloud (LMC) and 11 Small Magellanic Cloud (SMC) stars contained in the final catalogue. The median standard error of the individual positions, parallaxes, and proper motions of these stars is around $1.7\,\mathrm{mas\,yr^{-1}}$, a little larger than the typical catalogue value due to their fainter visual magnitudes. A few of these were solved with only a rather poor parallax accuracy.

The LMC and SMC lie at distances of about 48 kpc and 54 kpc respectively (parallaxes around 0.02 mas), with the distance between them being around 21 kpc. Direct trigonometric distance estimates were, of course,

Table 5.6 Estimated LMC distance modulus. Most of the estimates given have been refined or influenced by the Hipparcos data. Some non-Hipparcos distance estimates are included for comparison. Where the references include separate estimates of statistical and systematic errors, these have been simply added to yield the error estimate given here. Carretta et al. (2000) reviewed the Hipparcos contributions to the use of the horizontal branch and RR Lyrae stars in distance estimates to the LMC, and their consistent treatment has been included in this compilation.

Method	Reference	Value
Population I – Hipparcos-based:		
Cepheids: trigonometric parallaxes	Van Leeuwen *et al.* (2007)	18.39 ± 0.05
Cepheids: main-sequence fitting	Pont (1999a)	18.55 ± 0.08
Cepheids: Baade–Wesselink	Di Benedetto (2002)	18.59 ± 0.04
Red clump giants	Alves *et al.* (2002)	18.49 ± 0.06
Mira variables	Van Leeuwen *et al.* (1997)	18.54 ± 0.18
Population I – other:		
Eclipsing binaries	Guinan *et al.* (2004)	18.42 ± 0.07
SN 1987a	Panagia (1998)	18.58 ± 0.05
Population II – Hipparcos-based:		
Subdwarf fitting to globular clusters	Reid (1997)	18.65 ± 0.10
HB trigonometric parallaxes	Carretta *et al.* (2000)	18.50 ± 0.11
RR Lyrae: statistical parallax	"	18.38 ± 0.12
RR Lyrae: Baade–Wesselink	"	18.40 ± 0.20
RR Lyrae: double-mode pulsation	"	18.48 ± 0.19
Population II – other:		
Globular cluster dynamical models	Carretta *et al.* (2000)	18.50 ± 0.11
Globular cluster white dwarf sequence	Carretta *et al.* (2000)	18.40 ± 0.15
Straight mean of above 14 values		18.49

impossible with Hipparcos, leading to the variety of less direct methods to estimate their distances discussed previously. From 38 single stars in both systems, the average weighted parallax is -0.1 ± 0.23 mas (not taking account of correlations between great-circle abscissae), placing marginal constraints on the zero-point of the global Hipparcos parallax system (ESA, 1997, Volume 3, Section 20.5). Some information, at the limit of the Hipparcos accuracies, has nevertheless been obtained from their proper motions.

The LMC and SMC are the largest of the dozen or so known small satellite galaxies orbiting the Milky Way. Their masses are around $2 \times 10^{10} M_\odot$ and $2 \times 10^{9} M_\odot$ respectively, and their radial velocities in the Galactic standard of rest are estimated at around $+80\,\mathrm{km\,s^{-1}}$ and $+7\,\mathrm{km\,s^{-1}}$ respectively (Gardiner *et al.*, 1994). From these small velocities, and the small orbital period of around 1 Gyr if they are currently at apogalacticon, it is generally accepted that both galaxies are currently near perigalacticon (Kroupa & Bastian, 1997a). However, whether they are gravitationally bound or unbound to the Milky Way has not been unambiguously demonstrated.

The idea that the Galaxy halo formed from the infall and dispersal of satellite galaxies after the main part of the Milky Way had collapsed is discussed in Section 9.9. Early studies of the dynamical relationship between the Milky Way and the local group galaxies were made by Kunkel & Demers (1976) and Kunkel (1979), who showed that the Galactic satellites appeared to lie within a single 'Magellanic plane', with the radial velocities of the dwarf spheroidals, some outer halo clusters, the Magellanic Clouds, and the Magellanic Stream (the H I gas trailing the Magellanic Cloud system) being consistent with Keplerian motion within this plane. Lynden-Bell (1982) proposed that the (l, b) distribution of the Galactic satellites and outer halo clusters is better described by two great streams: Ursa Minor, Draco, Carina and the Magellanic Clouds and Stream were suggested to be the debris of a former greater Magellanic galaxy, while another satellite left behind the Fornax–Leo I and II–Sculptor (FLS) stream. A considerable literature now exists on the origin, kinematics, and other properties of these satellites as cosmological substructures or interacting systems which have developed during hierarchical structure formation; see, for example, Majewski (1994), Lynden-Bell & Lynden-Bell (1995), Palma *et al.* (2002), Libeskind *et al.* (2005), Metz *et al.* (2007) and references therein.

Here, as elsewhere, attention is given to the specific contributions made by Hipparcos. While radial velocities alone have provided important insights, they provide limited information on the angular momentum vector necessary to establish a common origin. Clearly, the complete space velocity vectors are required to determine the orbits and overall mass content of these companion galaxies.

The proper motion of both Magellanic Clouds was estimated from the Hipparcos data by Kroupa & Bastian (1997a,b). Their results provide improved estimates through improved control of systematic errors compared with earlier determinations based on photographic plates spanning 14 years by Jones *et al.* (1994), and based on proper motions from the PPM Catalogue by Kroupa *et al.* (1994). Of the LMC and SMC stars observed by Hipparcos, 33 LMC and nine SMC stars were retained for the proper motion analysis. Given their proximity on the sky, Kroupa & Bastian (1997a) used the approach of van Leeuwen & Evans (1998) to take account of the correlations between the abscissae measurements. They estimated for the LMC

$$\mu_\alpha = +1.94 \pm 0.29 \text{ mas yr}^{-1}$$
$$\mu_\delta = -0.14 \pm 0.36 \text{ mas yr}^{-1} \quad (5.38)$$

and for the SMC

$$\mu_\alpha = +1.23 \pm 0.84 \text{ mas yr}^{-1}$$
$$\mu_\delta = -1.21 \pm 0.75 \text{ mas yr}^{-1} \quad (5.39)$$

with the results shown graphically in Figure 5.35. The results indicate that both galaxies are moving approximately parallel to each other on the sky, with the Magellanic Stream trailing behind. The theoretical orbit of the LMC and SMC can be further constrained by requiring them to form a bound system over a significant fraction of a Hubble time while interacting with the Galaxy. There is no unambiguous evidence for rotation of the LMC in the Hipparcos data, for which the signal of 0.25 ± 0.25 mas yr^{-1} corresponds to 58 ± 58 km s^{-1}.

Other proper motion determinations of the LMC have subsequently been made by Anguita *et al.* (2000) and Pedreros *et al.* (2002) using proper motions linked to quasars, and by Momany & Zaggia (2005) using the UCAC2 proper motions. Anguita *et al.* (2000) and Momany & Zaggia (2005) derive significantly larger values of $\mu_\delta = +2.90 \pm 0.20$ and $+4.32$ mas yr^{-1} respectively, values which would indicate that the LMC and the Milky Way are unbound. The latter work considers that the inconsistency between the UCAC2 and Hipparcos results remains unsolved, although it is perhaps attributable to systematic errors in the UCAC2 system.

The Hipparcos proper motion results were used in particle simulations of the LMC–SMC–Milky Way system by Li & Thronson (1999) taking account of gravitational and hydrodynamic interactions among gaseous and stellar components, and aiming for an improved understanding of the Magellanic Stream, the barred structure of the LMC, and star formation induced in all three galaxies due to their mutual interactions. Sawa *et al.* (1999) used the same approach and Hipparcos proper motion constraints to find orbits which reproduce the geometry and kinematics of the Magellanic Stream, as well as enhancements in star cluster formation in the LMC some 10^8 and 3×10^9 yr ago. Ružčka (2003) also made use of the Hipparcos proper motion results in a numerical model which explains the Magellanic Stream as a result of the dynamical evolution of the Milky Way with the Magellanic Cloud system.

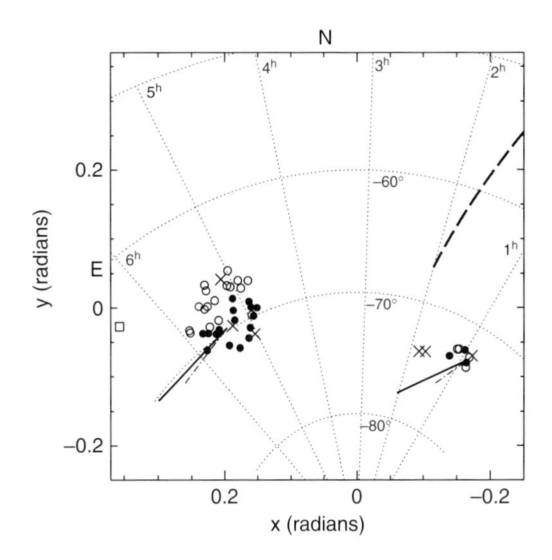

Figure 5.35 The Magellanic Cloud region shown as a gnomonic projection. Open and solid circles are Hipparcos stars with positive and negative heliocentric radial velocities, after subtraction of mean values of +274 km s^{-1} for the LMC and +148 km s^{-1} for the SMC. Crosses are stars with unknown radial velocities. Bold lines indicate the proper motion vectors of the LMC and SMC as measured by Hipparcos, with the dash–dotted lines corrected for the solar reflex motion. The bold dashed line indicates the approximate location of the Magellanic Stream. From Kroupa & Bastian (1997b, Figure 1).

Proper motion and other dynamical studies of the LMC have also made progress using other observational techniques, including use of HST observations (van der Marel *et al.*, 2002; Kallivayalil *et al.*, 2006a,b).

5.10 Other galaxies

Application of Hipparcos data to distances and space motions of other galaxies are largely indirect and at the limit of the systematic errors in the catalogue. A brief survey is given for completeness.

Paturel *et al.* (1997) and Paturel *et al.* (1998) used an early Hipparcos revision of the Cepheid period–luminosity relation to derive distances of 17 nearby galaxies. Their analysis was based on 36 Galactic Cepheids observed by Hipparcos, and 2236 multi-colour extragalactic Cepheid measurements for the 17 nearby galaxies. Their results included distance moduli of $\mu = 18.72$ for the LMC, $\mu = 24.84$ for M31, and $\mu = 28.1$ for M81 (Paturel *et al.*, 1997), later revised to $\mu = 24.6 \pm 0.2$ for M31, and $\mu = 27.6 \pm 0.2$ for M81 (Paturel *et al.*, 1998). Since the radial velocities of nearby galaxies are dominated by the local velocity field and not by the cosmological expansion field, these results cannot be used to determine H_0 directly, but they can be used for the calibration of long-range relations. Specifically they used their estimated galaxy distances to calibrate the Tully–Fisher relation between the 21-cm line width (corresponding to the amplitude of its rotation curve), W, and its absolute magnitude, M (Tully & Fisher, 1977), considered to be of the form $M = \alpha \log(W/2 \sin i) + \beta$. They corrected for Malmquist-type bias, and for inclination effects and morphological-type dependencies, by selecting galaxies similar to ('sosies' or twins, according to their nomenclature) the brightest of the nearby galaxies M31 and M81, deriving $H_0 = 53 \pm 8\,\mathrm{km\,s^{-1}\,Mpc^{-1}}$ (Paturel *et al.*, 1997), subsequently revised to $H_0 = 60 \pm 10\,\mathrm{km\,s^{-1}\,Mpc^{-1}}$ (Paturel *et al.*, 1998).

Theureau (1998) improved the robustness of the Tully–Fisher relation in the B-band, i.e. the relationship between the 21-cm line width and the optical luminosity of the galaxy in the B-band, using a sample of 5271 spiral galaxies, and showing that the zero-point of the relation is a continuous function of the galaxy's mean surface brightness (related to the relative proportion of bulge and disk luminosities). The relationship was well described by a third- or fourth-order polynomial, with the absolute calibration performed using two sets of Cepheid distance moduli; the first largely from HST, and the second based on the Hipparcos distance determinations. Control of the Malmquist-type bias is extensively discussed. The Hubble constant was then derived from a large sample of 577 galaxies, reaching velocities of $6000\,\mathrm{km\,s^{-1}}$, yielding $H_0 = 56 \pm 3\,\mathrm{km\,s^{-1}\,Mpc^{-1}}$ from HST calibrations, and $H_0 = 51 \pm 4\,\mathrm{km\,s^{-1}\,Mpc^{-1}}$ from the Hipparcos calibrations.

Stanek & Garnavich (1998) used their absolute magnitude calibration of red clump giants of $M_I = -0.23$ mag (Section 5.7.2), to make a single-step determination of the distance to M31 of $\mu_{\rm M31} = 24.471 \pm 0.35 \pm 0.045$ mag. The use of some 6300 stars in M31 led to a formal statistical error of $\lesssim 2\%$ (Figure 5.36). The same M31 distance modulus of $\mu_0 = 24.47 \pm 0.07$ mag was determined by Holland (1998) by fitting theoretical isochrones to the observed red giant branches of 14 of its globular clusters. Sarajedini & van Duyne (2001) studied red clump giants from HST observations of the outer disk of M31. They argued in the reverse sense: they assumed $\mu_{\rm M31,0} = 24.5 \pm 0.1$ from a compilation of field halo RR Lyrae, giant stars, and globular clusters by Da Costa *et al.* (2000). This distance then implies an inferred mean absolute magnitude of the red clump giants of $M_I = -0.29 \pm 0.05$.

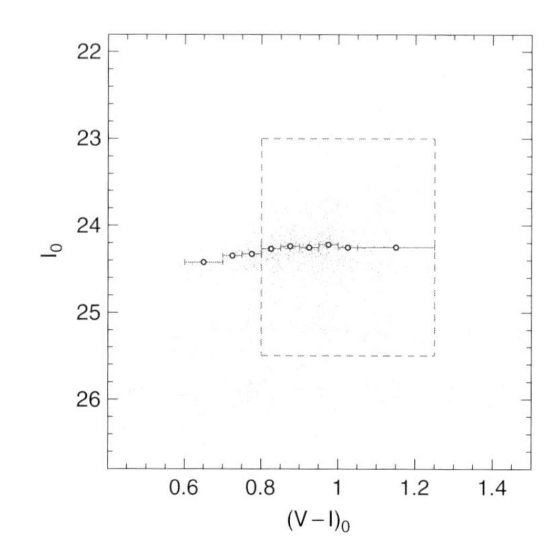

Figure 5.36 Dependence of the peak brightness of the red clump I_0 on the $(V-I)_0$ colour for the star-rich M31 field G302. The red clump exhibits a relatively sharp downturn for $(V-I)_0 < 0.8$ mag, but there is no colour dependence for $(V-I)_0 > 0.8$ mag. In the colour range $0.8 < (V-I)_0 < 1.25$, used for the comparison with the Hipparcos red clump, I_0 varies randomly between 24.22–24.27. From Stanek & Garnavich (1998, Figure 3).

Nella-Courtois *et al.* (1999) used the revised Hipparcos-based distances to 11 Galactic globular clusters derived by various workers, and available at that time (see Section 5.8.1) to re-calibrate the relationship between central velocity dispersion and luminosity for globular clusters. Such a relationship is analogous to the Faber–Jackson (luminosity–velocity dispersion) relation for elliptical galaxies, $L \propto \sigma^\gamma$, where the observed value of $\gamma \simeq 4$ can also be derived from the virial theorem using certain assumptions. From the Galactic clusters, they derived $M_V = (-3.0 \pm 0.3) \log\sigma - (5.8 \pm 0.1)$, and showed that a similar slope is seen for the 26 globular clusters studied in M31. They then derived a distance modulus for M31 of $\mu = 24.12 \pm 0.45$ mag.

Carrera *et al.* (2002) used the Hipparcos subdwarf main-sequence data to determine a distance to the Ursa Minor dwarf spheroidal galaxy of 76 ± 4 kpc, slightly higher than previous estimates.

Extending distances beyond the local group, Rejkuba (2004) determined the distance to the nearest giant elliptical NGC 5128, the central galaxy of the Centaurus group; for a review, including earlier distance determinations, see Israel (1998). A Hipparcos-independent estimate from the K-band luminosity of the tip of the red giant branch gave $(m - M)_0 = 27.87 \pm 0.16$ for a reddening of $E(B-V) = 0.11$. A reasonably consistent value of 27.96 ± 0.11 was obtained from the observation of 240 Mira long-period variables in two halo fields. The luminosity calibration is based on the K-band period–luminosity relation for the Large Magellanic Cloud, $M_K = -3.47 \log P + \beta$, for which Hipparcos determinations have given $\beta = 0.84 \pm 0.14$ based on solar neighbourhood Miras (Whitelock *et al.*, 2000), and $\beta = 0.93 \pm 0.14$ based on globular cluster Miras (Feast *et al.*, 2002). In practice, Rejkuba (2004) adopted the shorter LMC distance modulus of $(m-M)_0 = 18.50 \pm 0.04$ from Alves *et al.* (2002), adopting $\beta = 0.98 \pm 0.11$, and deriving for the Mira variables in NGC 5128 $K_0 = -3.37(\pm 0.11) \log P + 28.67(\pm 0.29)$. The similarity of the slope of the period–luminosity relation in the LMC and in NGC 5128 provides further evidence that the relationship in the K-band is rather universal.

Baum (1998) discussed a method of finding mutually consistent values for the age of the Universe, t_0, the value of H_0, and the mass density of the Universe Ω_M. The Hipparcos data for RR Lyrae stars were used to determine an age of the Universe based on the coincidence of a stellar evolution age in globular clusters, with an age estimate based on H_0 derived from globular clusters as distance indicators as calibrated in the Milky Way. Using the mean of six Hipparcos RR Lyrae absolute magnitude estimates determined until 1998, $M_V(\mathrm{RR}) = 0.61 \pm 0.15$, they derived $t_0 = 12.5 \pm 1.5$ Gyr, and $H_0 = 61 \pm 5\,\mathrm{km\,s^{-1}\,Mpc^{-1}}$, for which $\Omega_M = 0.41$ for the case $\Omega_\Lambda = 0$. Other values of Ω_M are given for different cases of a flat Universe, $\Omega_M + \Omega_\Lambda = 1$.

Using the Hipparcos reference frame to provide an absolute proper motion system, Dinescu *et al.* (2005a) determined the absolute proper motion of the Sagittarius dwarf galaxy using a cross-correlation between the 2MASS and SPM3 catalogues, yielding $\mu_l \cos b = -2.35 \pm 0.20, \mu_b = +2.07 \pm 0.20\,\mathrm{mas\,yr^{-1}}$. The Sagittarius dwarf provides an excellent possibility to study the Galaxy's dark halo potential; the information inherently present in tidal debris streams is discussed in Section 9.11.7. The most recent attempts to understand the dark halo potential from models of the interaction between Sagittarius and the Milky Way have shown that while the known orbit agrees with the location of the debris, the observed radial velocities in the leading tidal tail do not match those predicted by simple spherical halos. Helmi (2004) provided kinematic evidence from the leading stream that the dark matter halo of the Galaxy is prolate, while Johnston *et al.* (2005) found that the orbital precession and destruction favour an oblate halo. The orbit derived by Helmi (2004) is derived from a bulk proper motion having an uncertainty of $0.8\,\mathrm{mas\,yr^{-1}}$. Dinescu *et al.* (2005a) used estimates of the distance and heliocentric radial velocity of the Sagittarius dwarf (25 ± 2.5 kpc and $137 \pm 5\,\mathrm{km\,s^{-1}}$ respectively), and the Sun's peculiar velocity and local rotation rate Θ_0, to derive estimates of the galaxy's space motion, and proper motion within the Galactic rest frame. While Law *et al.* (2005) have argued that an improved proper motion will help to discriminate between models, this improved orbit has not yet been incorporated into published models.

Dinescu *et al.* (2005a) also derived a Hipparcos-referenced absolute proper motion of stars in the outer bulge of our Galaxy, yielding $\mu_l \cos b = -5.86 \pm 0.14, \mu_b = -0.59 \pm 0.14\,\mathrm{mas\,yr^{-1}}$. This results in a rotation rate for the outer bulge of $21 \pm 3\,\mathrm{km\,s^{-1}\,kpc^{-1}}$, a value in reasonable agreement with the rate of $25\,\mathrm{km\,s^{-1}\,kpc^{-1}}$ determined from radial velocities alone by Ibata & Gilmore (1995).

A stellar overdensity seen in 2MASS-selected M giants has been attributed as the remnant of a dwarf galaxy, the Canis Major dwarf (Martin *et al.*, 2004), and postulated as the core of the Monoceros tidal debris stream (Newberg *et al.*, 2002). An alternative hypothesis is that the overdensity is identified with the Milky Way warp and flare (Momany *et al.*, 2004). Proper motions of the overdensity from UCAC2, with an accuracy of $\sim 2\,\mathrm{mas\,yr^{-1}}$, may be consistent with an origin in the thin disk (Momany *et al.*, 2004) or a dwarf remnant (Peñarrubia *et al.*, 2005). Dinescu *et al.* (2005b) carried out an improved proper motion analysis for the Canis Major dwarf galaxy candidate, using SPM plates and CCD observations linked to the Hipparcos reference system, and yielding for 104 candidate stars $\mu_l \cos b = -1.47 \pm 0.37, \mu_b = -1.07 \pm 0.38\,\mathrm{mas\,yr^{-1}}$. Based on a distance to the overdensity core of 8.1 ± 0.4 kpc, a bulk radial velocity of $109 \pm 4\,\mathrm{km\,s^{-1}}$, and specific assumptions about the peculiar solar motion, R_0 and Θ_0, they determined a space velocity in the Galactic rest frame for which the Z-component is inconsistent with the known warping motion of our Galaxy (Section 9.8). Integrating the orbit in the Galactic potential model of Johnston *et al.* (1995), they derived a pericentre of 10.5 ± 0.9 kpc, an apocentre of 14 ± 0.2 kpc, a maximum distance from the Galactic plane of 2.0 ± 0.4 kpc, an orbital period $P = 342 \pm 14$ Myr, an orbital inclination $i = 15^\circ \pm 3^\circ$, and an eccentricity $e = 0.14 \pm 0.04$. With orbital characteristics close to that of the Monoceros stream's progenitor modelled by Peñarrubia *et al.* (2005), $i = 20^\circ \pm 5^\circ$ and $e = 0.10 \pm 0.05$, the derived motion makes the Canis Major dwarf a likely source of the Monoceros stream. The prograde, low-inclination orbit of low eccentricity may have led to a strong coupling with the existing disk motion, with Dinescu *et al.* (2005b) advancing the hypothesis that the interaction actually led to the present disk warp.

In the future, microarcsec astrometry will provide direct distance estimates to nearby galaxies, notably the LMC, M31 and M33, using the method of rotational parallaxes. Analogous to the orbital parallax technique, it requires individual stellar radial velocities at the level of $\sim$ 10 km s^{-1}, and proper motions at the level of a few μas yr^{-1}. Distances accurate to the 1% level have been predicted (Olling, 2007). At the most basic level, for a nearby spiral galaxy at distance D (in Mpc), and a projected rotation speed in the plane of the sky V_c (in km s^{-1}), the proper motion due to rotation is $\mu = V_c/(\kappa D)$ μarcsec yr^{-1}, where $\kappa \sim 4.74$ converts velocities in AU yr^{-1} to km s^{-1}. Resulting rotational proper motions are of order 192, 74 and 24 μas yr^{-1} for the LMC, M31, and M33 respectively (Olling, 2007). The unknown inclination angle results in a distance determined modulo tan i, which can be solved for using the principal axis method in the case of circular orbits, in which individual stars are identified along the major and minor axes at similar distances from the galaxy centre. Extension to the case of elliptical orbits, and the effects of stellar warps and spiral structure, are detailed in Olling (2007). Currently below the level at which Hipparcos can contribute, the method will be accessible to both SIM and Gaia in the future.

To conclude this section on the contribution of Hipparcos to the study of external galaxies it is recalled, as noted in Chapter 1, that a number of galaxies are actually included in the Tycho 2 Catalogue, and identified by cross-correlation with other galaxy catalogues (Metz & Geffert, 2004); the work has identified 181 galaxies in total: 116 in the Tycho 2 Catalogue, 35 uncertain galaxies in the Tycho 2 Catalogue, and 30 galaxies in the Tycho 2 supplement.

5.11 Supernovae

Very high intrinsic luminosities, and a rather uniform absolute B magnitude at maximum light (with an intrinsic dispersion below 0.2 mag, e.g. Phillips *et al.*, 1999), make the Type Ia supernovae intensively studied candidates as standard candles for determining the extragalactic distance scale. Hipparcos has made rather modest contributions to the field as follows.

Related to the estimates of H_0 reported in Section 5.10 using the calibrated Tully–Fisher relation (Paturel *et al.*, 1997, 1998), a similar value of $H_0 = 54 \pm 9$ km s^{-1} Mpc^{-1} was estimated based on a calibration of four supernova Type Ia events located in three of their calibrating galaxies (SNe 1895B and 1972E in NGC 5253, SN 1937C in IC 4182, and SN 1981B in NGC 4536), and 33 other supernovae in more distant galaxies (Paturel *et al.*, 1997); revised to $H_0 = 50 \pm 3$ km s^{-1} Mpc^{-1} for 57 external supernovae (Lanoix, 1998). These values of H_0 were still considered as upper limits due to incompleteness bias in the Cepheid period–luminosity relation.

Studies of the dispersion in absolute magnitude of Type Ia supernovae at maximum light, and the study of possible intrinsic differences between events occurring, for example, in spirals and early-type galaxies, throw light on the uniqueness of their progenitors, for which the single degenerate model (single white dwarf accreting material from a non-degenerate star) or the double degenerate model (the merger of C–O white dwarf pairs) remain contenders. Emerging evidence of diversity among Type Ia supernovae (e.g. Sahu *et al.*, 2006) make it important to assess independent distances to individual events with the highest precision. Della Valle *et al.* (1998) used the early Hipparcos re-calibration of globular cluster distances by Reid (1997) and Gratton *et al.* (1997) to determine the turnover magnitude of the luminosity function of Galaxy globular clusters, and used the fact that the absolute peak in the luminosity function is rather constant to determine the distance to the galaxy NGC 1380 and hence to the supernova observed within it, SN 1992A. They derived an absolute magnitude at maximum of $M_B(\mathrm{max}) = -18.79 \pm 0.16$, implying that the supernova is more than half a magnitude fainter than the other Type Ia supernovae for which accurate distances exist.

Accurate distances to supernova remnants are important for estimating the energy released in the explosion, and for interpreting the effects of shock waves on the structure and physical conditions of the interstellar medium. The Vela supernova remnant is one of the best-studied, and a distance of around 500 pc is often quoted (Milne, 1968; Kristian, 1970). Early pulsar dispersion measurements gave distances in the range 410–600 pc (Prentice & Ter Haar, 1969; Davies, 1969). Other works have inferred distances of around 250 pc in order to yield consistency with age estimates (Wallerstein & Silk, 1971), the transverse motion of the Vela pulsar (Oegelman *et al.*, 1989), or the total remnant energy (Jenkins & Wallerstein, 1995). Cha *et al.* (1999) determined a precise distance to the remnant by analysing high-resolution Ca II absorption spectra of 68 stars in the direction of the remnant, establishing that the absorption components with velocities above ± 25 km s^{-1} arise from the remnant rather than from the local interstellar medium, and yielding a distance of 250 ± 30 pc, consistent with the distances previously inferred from ages and energetics. This more secure distance estimate was used to infer a remnant diameter of 32 ± 4 pc (based on an angular size of $7\overset{\circ}{.}3$), a remnant age of 11 400 yr, a remnant energy of $(1-2) \times 10^{51}$ ergs, and a shock temperature of 4.5×10^6 K.

In the standard picture of stellar evolution, massive stars above $\sim 8\,M_\odot$ end their lives as supernovae, leaving behind a black hole or neutron star depending on whether they are more or less massive than some cut-off mass of around $30\,M_\odot$. However, it is conceivable that a fraction of stars in the mass range $8-30\,M_\odot$ end their lives as black hole remnants without producing a

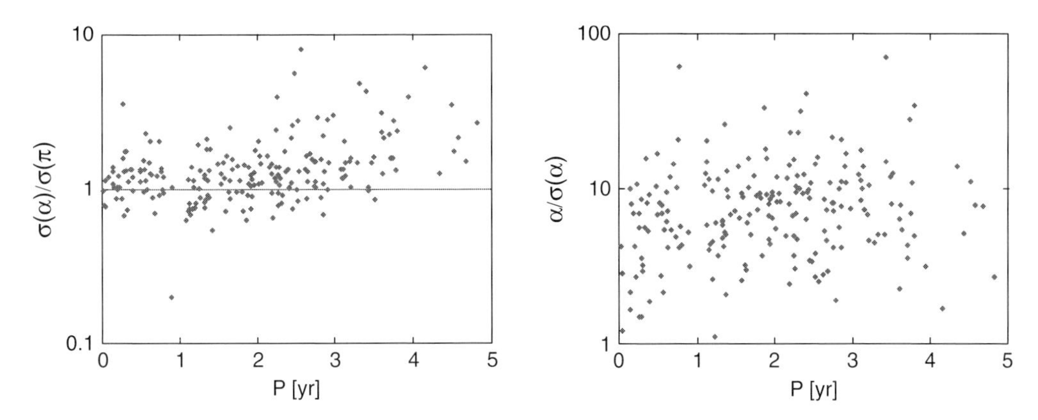

Figure 5.37 Left: the ratio of the error in the photocentric semi-major axis σ_α to the parallax error σ_π as a function of period P for astrometric binaries with orbital solutions from the Hipparcos Catalogue (from observations over 3.3 years). The plot shows 210 systems, with four lying outside of the y-axis range. As expected from general arguments, the ratio is typically unity. Right: signal-to-noise ratio α/σ_α as a function of period P for binaries detected in the Hipparcos Catalogue. From the form of the distribution, the catalogue is estimated to be complete down to roughly the 5σ detection level. From Gould & Salim (2002, Figures 1 and 3).

supernova. If such systems occurred in binaries, and if the black hole collapse occurred without disrupting the binary system, then evidence for astrometric binaries in which one component is massive but invisible should be present in the distribution of actual photocentric motions. Gould & Salim (2002) searched the Hipparcos Catalogue for the existence of black holes which might have originated from failed supernovae in astrometric binaries. For a binary whose components have masses and luminosities (M, L) and (m, l), the semi-major axis, a, is related to the period, P, by $[(m+M)/M_\odot](P/\mathrm{yr})^2 = (a/\mathrm{AU})^3$ (Kepler's third law). If the photocentric motion is fitted to a Keplerian orbit, the angular semi-major axis of the photocentre orbit, α, is related to the other parameters by

$$\frac{m^3}{M_\odot(m+M)^2}\left(\frac{L}{L+l}\right)^3 = \left(\frac{P}{\mathrm{yr}}\right)^{-2}\left(\frac{d\alpha}{\mathrm{AU}}\right)^3 \tag{5.40}$$

where d is the distance to the system. The quantities on the right can all be measured astrometrically. If the mass of the more luminous component, M, can be estimated photometrically or spectroscopically, and if the companion is dark, $l \ll L$, then the mass of the dark companion, m, can be estimated from the astrometric observations. For known astrometric binaries with orbital solutions in the Hipparcos Catalogue, the error in the photocentric semi-major axis σ_α is approximately the same as the precision on the parallax (Figure 5.37, left). The signal-to-noise ratio as a function of P for binaries in the Hipparcos Catalogue suggests that it is complete for binaries down to the roughly 5σ detection level (Figure 5.37, right). Gould & Salim (2002) went on to argue that none of the 235 Hipparcos astrometric binaries with orbital solutions, 188 of which have $P < 3.3$ yr (the observational duration of the mission), contain a clear black hole binary companion. In all cases they find a mass of the purported dark companion to be well below the black hole range, or of the order or smaller than that of the luminous star. Based on specific assumptions about the frequency of binaries as a function of period and mass, they argue that less than 10% of massive stars can end as failed supernovae. There are considerable uncertainties on this interesting number due to the uncertain fraction of companions that are black hole progenitors, and due to the unknown probability that the black hole collapse disrupts the astrometric binary. They also estimate corresponding detection probabilities for Gaia.

The large sky density and photometric uniformity has made the Tycho Catalogue photometry an important reference for magnitude estimates of supernova discoveries, for example, as reported through the IAU Central Bureau for Astronomical Telegrams (e.g. Evans *et al.*, 1998b; Hanzl, 1998; Armstrong & Hanzl, 1999; Hanzl, 1999; Yoshida & Kadota, 2000).

Finally, the next generation of scanning astrometric satellites is expected to detect large numbers of supernovae from their light-curves. For Gaia, Belokurov & Evans (2003) estimated the detection of around 21 000 supernovae during a five-year mission lifetime, of which ~14 300 are Type Ia, ~1400 are Type Ib/c and ~5700 are Type II, of which the numbers detected before maximum were predicted to be around 6300, 500, and 1700 respectively.

References

Abrahamyan HV, 2004, Absolute calibration of the period–luminosity relations for the classical Cepheids based on the

Hipparcos parallaxes and the distances of the Magellanic Clouds. *Astrophysics*, 47, 18–28 {**228**}

Abt HA, Morrell NI, 1995, The relation between rotational velocities and spectral peculiarities among A-type stars. *ApJS*, 99, 135–172 {**217**}

Abt HA, 2002, Solving Olin Wilson's mystery. *PASP*, 114, 559–562 {**219**}

Adelman SJ, 2001, On the photometric variability of red clump giants. *Baltic Astronomy*, 10, 593–597 {**235**}

Alcock C, Allsman RA, Alves DR, *et al.*, 1997, The MACHO project Large Magellanic Cloud variable star inventory. III. Multimode RR Lyrae stars, distance to the Large Magellanic Cloud, and age of the oldest stars. *ApJ*, 482, 89–97 {**249**}

Altmann M, de Boer KS, 2000, Kinematical trends among the field horizontal branch stars. *A&A*, 353, 135–146 {**250**}

Alvarez R, Mennessier MO, Barthès D, *et al.*, 1997, Oxygen-rich Mira variables: near-infrared luminosity calibrations. Populations and period–luminosity relations. *A&A*, 327, 656–661 {**236–238**}

Alves DR, 2000, K-band calibration of the red clump luminosity. *ApJ*, 539, 732–741 {**234, 235**}

Alves DR, Rejkuba M, Minniti D, *et al.*, 2002, K-band red clump distance to the Large Magellanic Cloud. *ApJ*, 573, L51–L54 {**235, 253, 254, 257**}

Andersen J, 1991, Accurate masses and radii of normal stars. *A&A Rev.*, 3, 91–126 {**217**}

Anglada Escudé G, 2007, Experiments and relativistic models for space optical astrometry: application to the Gaia mission. Ph.D. Thesis, University of Barcelona {**207**}

Anguita C, Loyola P, Pedreros MH, 2000, Proper motion of the Large Magellanic Cloud using quasars as an inertial reference system. *AJ*, 120, 845–854 {**255**}

Arellano Ferro A, Giridhar S, Rojo Arellano E, 2003, A revised calibration of the M_V versus O I 777.4 nm relationship using Hipparcos data: its application to Cepheids and evolved stars. *Revista Mexicana de Astronomia y Astrofisica*, 39, 3–15 {**222**}

Arenou F, 2000, Statistical effects on luminosity and masses from Hipparcos astrometry. *Hipparcos and the Luminosity Calibration of the Nearer Stars, IAU Joint Discussion 13*, 2–5 {**210**}

Arenou F, Gómez AE, 1997, Unbiased luminosity calibrations for Hipparcos data. *ESA SP–402: Hipparcos, Venice '97*, 287–290 {**210**}

Arenou F, Grenon M, Gómez AE, 1992, A tridimensional model of the Galactic interstellar extinction. *A&A*, 258, 104–111 {**213**}

Arenou F, Luri X, 1999, Distances and absolute magnitudes from trigonometric parallaxes. *ASP Conf. Ser. 167: Harmonizing Cosmic Distance Scales in a Post-Hipparcos Era*, 13–32 {**208–211, 213**}

Arenou F, Luri X, 2002, Statistical effects from Hipparcos astrometry. *Highlights in Astronomy*, 12, 661 {**210**}

Armstrong M, Hanzl D, 1999, Supernova 1999by in NGC 2841. *IAU Circ.*, 7157, 3 {**259**}

Baade W, 1926, Über eine Möglichkeit, die Pulsationstheorie der δ Cephei-Veränderlichen zu prüten. *Astronomische Nachrichten*, 228, 359–360 {**228**}

Babu GJ, Feigelson ED (eds.), 1997, *Statistical Challenges in Modern Astronomy II*, Springer–Verlag, New York {**210**}

Barnes TG, Evans DS, 1976, Stellar angular diameters and visual surface brightness. I. Late spectral types. *MNRAS*, 174, 489–502 {**228**}

Barthès D, Luri X, Alvarez R, *et al.*, 1999, Period–luminosity–colour distribution and classification of Galactic oxygen-rich long-period variables. I. Luminosity calibrations. *A&AS*, 140, 55–67 {**213, 237**}

Barthès D, Luri X, 2001, Period–luminosity–colour distribution and classification of Galactic oxygen-rich long-period variables. II. Confrontation with pulsation models. *A&A*, 365, 519–534 {**237**}

Baum WA, 1998, Ω, age, and H_0 implications of recent Hubble Space Telescope data in the Coma cluster. *AJ*, 116, 31–36 {**257**}

Baumgardt H, Dettbarn C, Fuchs B, *et al.*, 1999, A comparison of ground-based Cepheid period–luminosity relations with Hipparcos parallaxes. *ASP Conf. Ser. 167: Harmonizing Cosmic Distance Scales in a Post-Hipparcos Era*, 251–254 {**226**}

Beals CS, Oke JB, 1953, On the relation between distance and intensity for interstellar calcium and sodium lines. *MNRAS*, 113, 530–552 {**222**}

Beaulieu JP, Sackett PD, 1998, Red clump morphology as evidence against a new intervening stellar population as the primary source of microlensing toward the Large Magellanic Cloud. *AJ*, 116, 209–219 {**232**}

Bedding TR, Zijlstra AA, 1998, Hipparcos period–luminosity relations for Mira and semi-regular variables. *ApJ*, 506, L47–L50 {**210, 236, 238**}

Bedin LR, Piotto G, Anderson J, *et al.*, 2003, Accurate internal proper motions of globular clusters. *ASP Conf. Ser. 296: New Horizons in Globular Cluster Astronomy* (eds. Piotto G, Meylan G, Djorgovski SG, *et al.*), 360–363 {**251**}

Bedin LR, Salaris M, Piotto G, *et al.*, 2005, The white dwarf cooling sequence in NGC 6791. *ApJ*, 624, L45–L48 {**252**}

Belokurov VA, Evans NW, 2003, Supernovae with 'super-Hipparcos'. *MNRAS*, 341, 569–576 {**259**}

Benedict GF, McArthur BE, Feast MW, *et al.*, 2007, Hubble Space Telescope Fine Guidance Sensor parallaxes of Galactic Cepheid variable stars: period–luminosity relations. *AJ*, 133, 1810–1827 {**227**}

Benedict GF, McArthur BE, Fredrick LW, *et al.*, 2002, Astrometry with the Hubble Space Telescope: a parallax of the fundamental distance calibrator RR Lyrae. *AJ*, 123, 473–484 {**210, 247, 248**}

Berdnikov LN, Dambis AK, Vozyakova OV, 2000, Galactic Cepheids. Catalogue of light-curve parameters and distances. *A&AS*, 143, 211–213 {**227**}

Berdnikov LN, Vozyakova OV, Dambis AK, 1996, The BVRIJHK period–luminosity relations for Galactic classical Cepheids. *Astronomy Letters*, 22, 838–845 {**230**}

Bergeat J, Knapik A, Rutily B, 1998, The period–luminosity relation of Galactic carbon long-period variables. The distance modulus to Large Magellanic Cloud. *A&A*, 332, L53–L56 {**236**}

Boeshaar PC, 2000, Philip C. Keenan (1908–2000). *PASP*, 112, 1519–1522 {**219**}

Böhm-Vitense E, 1981, The effective temperature scale. *ARA&A*, 19, 295–317 {**218**}

—, 1989, *Introduction to Stellar Astrophysics*. Cambridge University Press {**221**}

Bono G, 2003, RR Lyrae distance scale: theory and observations. *Lecture Notes in Physics Vol. 635: Stellar Candles for the Extragalactic Distance Scale*, 635, 85–104 {**247**}

Bonamente M, Joy MK, LaRoque SJ, *et al.*, 2006, Determination of the cosmic distance scale from Sunyaev–Zel'dovich effect and Chandra X-ray measurements of high-redshift galaxy clusters. *ApJ*, 647, 25–54 {**228**}

Bono G, Caputo F, Castellani V, *et al.*, 2002a, On the pulsation parallax of the variable star RR Lyr. *MNRAS*, 332, L78–L80 {**247, 249**}

Bono G, Groenewegen MAT, Marconi M, *et al.*, 2002b, On the distance of Magellanic Clouds: first overtone Cepheids. *ApJ*, 574, L33–L37 {**227**}

Borkova TV, Marsakov VA, 2003, Two populations among the metal-poor field RR Lyrae stars. *A&A*, 398, 133–139 {**249, 250**}

Bressan A, Fagotto F, Bertelli G, *et al.*, 1993, Evolutionary sequences of stellar models with new radiative opacities. II. $Z = 0.02$. *A&AS*, 100, 647–664 {**218**}

Briot D, Robichon N, Hubert AM, 1997, Absolute magnitudes of Be stars. *ESA SP–402: Hipparcos, Venice '97*, 319–322 {**215, 217**}

Brocato E, Castellani V, Piersimoni A, 1997, The age of the globular cluster M68. *ApJ*, 491, 789–795 {**240**}

Brown AGA, Arenou F, van Leeuwen F, *et al.*, 1997, Considerations in making full use of the Hipparcos Catalogue. *ESA SP–402: Hipparcos, Venice '97*, 63–68 {**208, 209, 211**}

Buckley DRV, Longmore AJ, 1992, The distance to M13 via a subdwarf fit in the optical-infrared colour–magnitude plane. *MNRAS*, 257, 731–736 {**243**}

Butkevich AG, Berdyugin AV, Teerikorpi P, 2005a, Statistical biases in stellar astronomy: the Malmquist bias revisited. *MNRAS*, 362, 321–330 {**209, 244, 245**}

—, 2005b, The absolute magnitude of K0V stars from Hipparcos data using an analytical treatment of the Malmquist bias. *A&A*, 435, 949–954 {**211, 219**}

Cacciari C, 1999, The distance scale to globular clusters. *ASP Conf. Ser. 167: Harmonizing Cosmic Distance Scales in a Post-Hipparcos Era*, 140–160 {**239, 240**}

Cacciari C, Clementini G, Castelli F, *et al.*, 2000, Revised Baade–Wesselink analysis of RR Lyrae stars. *ASP Conf. Ser. 203: IAU Colloq. 176: The Impact of Large-Scale Surveys on Pulsating Star Research* (eds. Szabados L, Kurtz DW), 176–181 {**248, 249**}

Cacciari C, Clementini G, 2003, Globular cluster distances from RR Lyrae stars. *Lecture Notes in Physics Vol. 635: Stellar Candles for the Extragalactic Distance Scale*, 635, 105–122 {**247–249**}

Caldwell JAR, Laney CD, 1991, Cepheids in the Magellanic Clouds. *IAU Symp. 148: The Magellanic Clouds* (eds. Haynes R, Milne D), 249–257 {**226, 228**}

Caloi V, D'Antona F, Mazzitelli I, 1997, The distance scale to globular clusters through new horizontal branch models. *A&A*, 320, 823–830 {**249**}

Cannon RD, 1970, Red giants in old open clusters. *MNRAS*, 150, 111–135 {**230**}

Caputo F, 1998, Evolution of Population II stars. *A&A Rev.*, 9, 33–61 {**239, 247**}

Caputo F, Santolamazza P, Marconi M, 1998, The role of chemical composition in RR Lyrae pulsational properties. I. Periods. *MNRAS*, 293, 364–368 {**249**}

Carney BW, Lee J, Habgood MJ, 1998, The red horizontal-branch star HD 17072. *AJ*, 116, 424–428 {**248**}

Carrera R, Aparicio A, Martínez-Delgado D, *et al.*, 2002, The star-formation history and spatial distribution of stellar populations in the Ursa Minor dwarf spheroidal galaxy. *AJ*, 123, 3199–3209 {**256**}

Carretta E, Gratton RG, Clementini G, *et al.*, 1999, Biases in the main sequence fitting distances to globular clusters based on the Hipparcos Catalogue. *ASP Conf. Ser. 167: Harmonizing Cosmic Distance Scales in a Post-Hipparcos Era*, 255–258 {**245**}

—, 2000, Distances, ages, and epoch of formation of globular clusters. *ApJ*, 533, 215–235 {**228, 234, 238, 241, 244–249, 251, 252, 254**}

Carretta E, Gratton RG, 1997, Abundances for globular cluster giants. I. Homogeneous metallicities for 24 clusters. *A&AS*, 121, 95–112 {**242**}

Cassatella A, Altamore A, Badiali M, *et al.*, 2001, On the Wilson–Bappu relationship in the Mg II k line. *A&A*, 374, 1085–1091 {**222, 223**}

Cassisi S, Castellani V, Degl'Innocenti S, *et al.*, 1999, Galactic globular cluster stars: from theory to observation. *A&AS*, 134, 103–113 {**249**}

Castellani V, Degl'Innocenti S, Girardi L, *et al.*, 2000, The puzzling theoretical predictions for the luminosity of clumping He burning stars. *A&A*, 354, 150–156 {**233**}

Castellani V, Degl'Innocenti S, 1993, Stellar evolution as a probe of neutrino properties. *ApJ*, 402, 574–578 {**247**}

Chaboyer BC, 1999, Globular cluster distance determinations. *ASSL Vol. 237: Post-Hipparcos Cosmic Candles*, 111–124 {**240, 247, 249**}

Chaboyer BC, Demarque P, Kernan PJ, *et al.*, 1998, The age of globular clusters in light of Hipparcos: resolving the age problem? *ApJ*, 494, 96–110 {**240, 241, 244–247, 251**}

Cha AN, Sembach KR, Danks AC, 1999, The distance to the Vela supernova remnant. *ApJ*, 515, L25–L28 {**258**}

Chen B, 1999, The kinematics of RR Lyrae stars observed by Hipparcos. *A&A*, 344, 494–498 {**248, 250, 252**}

Chiappini C, Matteucci F, Romano D, 2001, Abundance gradients and the formation of the Milky Way. *ApJ*, 554, 1044–1058 {**219**}

Claret A, 1995, Stellar models for a wide range of initial chemical compositions until helium burning. I. From $X = 0.60$ to $X = 0.80$ for $Z = 0.02$. *A&AS*, 109, 441–446 {**217**}

Clausen JV, 2004, Eclipsing binaries as precise standard candles. *New Astronomy Review*, 48, 679–685 {**239**}

Clayton ML, Feast MW, 1969, Absolute magnitudes of Mira variables from statistical parallaxes. *MNRAS*, 146, 411–421 {**238**}

Clube SVM, Dawe JA, 1980b, Statistical parallaxes and the fundamental distance scale. II. Application of the maximum likelihood technique to RR Lyrae and Cepheid variables. *MNRAS*, 190, 591–610 {**248**}

—, 1980a, Statistical parallaxes and the fundamental distance scale. I. Description and numerical tests of maximum likelihood technique. *MNRAS*, 190, 575–610 {**248**}

Clube SVM, Jones DHP, 1971, The determination of statistical parallaxes. *MNRAS*, 151, 231–237 {**213**}

Cole AA, 1998, Age, metallicity, and the distance to the Magellanic Clouds from red clump stars. *ApJ*, 500, L137–L140 {**232**}

Cox AN, 2000, *Allen's Astrophysical Quantities*, 4th Edition. Springer, New York {**212**}

Cramer N, 1997, Absolute magnitude calibration of Geneva photometry: B-type stars. *ESA SP–402: Hipparcos, Venice '97*, 311–314 {**215, 217**}

Dambis AK, Rastorguev AS, 2001, Absolute magnitudes and kinematic parameters of the subsystem of RR Lyrae variables. *Astronomy Letters*, 27, 108–117 {**248–250**}

Dambis AK, Vozyakova OV, 2004, The kinematics and zero-point of the period–M_K relation for Galactic field RR Lyrae variables via statistical parallax. *ASP Conf. Ser. 310: IAU Colloq. 193: Variable Stars in the Local Group* (eds. Kurtz DW, Pollard KR), 128–132 {**249**}

Dambis AK, 2003a, Age dependence of the vertical distribution of young open clusters. *EAS Publications Series 10* (eds. Boily CM, Pastsis P, Portegies Zwart S, *et al.*), 147–152 {**230**}

—, 2003b, The period–M_K relation of Galactic field RR Lyrae variables. *EAS Publications Series 10* (eds. Boily CM, Pastsis P, Portegies Zwart S, *et al.*), 55–60 {**249**}

—, 2004, Age dependence of the vertical distribution of Cepheids. *ASP Conf. Ser. 310: IAU Colloq. 193: Variable Stars in the Local Group* (eds. Kurtz DW, Pollard KR), 158–161 {**230, 232**}

D'Antona F, Caloi V, Mazzitelli I, 1997, The Universe and globular clusters: an age conflict? *ApJ*, 477, 519–534 {**242, 243, 246**}

Davies RD, 1969, Distance of pulsars and the interstellar electron gas. *Nature*, 223, 355–358 {**258**}

Da Costa GS, Armandroff TE, Caldwell N, *et al.*, 2000, The dwarf spheroidal companions to M31: WFPC2 observations of Andromeda II. *AJ*, 119, 705–726 {**256**}

Della Valle M, Kissler-Patig M, Danziger J, *et al.*, 1998, Globular cluster calibration of the peak brightness of the Type Ia supernova 1992A and the value of H_0. *MNRAS*, 299, 267–276 {**258**}

Demarque P, Zinn R, Lee YW, *et al.*, 2000, The metallicity dependence of RR Lyrae absolute magnitudes from synthetic horizontal-branch models. *AJ*, 119, 1398–1404 {**249**}

de Boer KS, 1999, Horizontal-branch stars: their nature and their absolute magnitude. *ASP Conf. Ser. 167: Harmonizing Cosmic Distance Scales in a Post-Hipparcos Era*, 129–139 {**248**}

de Boer KS, Geffert M, Tucholke HJ, *et al.*, 1997a, Calculating the mass of horizontal-branch stars with Hipparcos. *ESA SP–402: Hipparcos, Venice '97*, 331–334 {**248**}

de Boer KS, Tucholke HJ, Schmidt JHK, 1997b, Calibrating horizontal-branch stars with Hipparcos. *A&A*, 317, L23–L26 {**248**}

de Zeeuw PT, Hoogerwerf R, de Bruijne JHJ, *et al.*, 1999, A Hipparcos census of the nearby OB associations. *AJ*, 117, 354–399 {**229**}

Dinescu DI, Girard TM, van Altena WF, *et al.*, 2005a, Absolute proper motion of the Sagittarius dwarf galaxy and of the outer regions of the Milky Way bulge. *ApJ*, 618, L25–L28 {**257**}

Dinescu DI, Martínez-Delgado D, Girard TM, *et al.*, 2005b, Absolute proper motion of the Canis Major dwarf galaxy candidate. *ApJ*, 631, L49–L52 {**257**}

Di Benedetto GP, 1997, Improved calibration of cosmic distance scale by Cepheid pulsation parallaxes. *ApJ*, 486, 60–74 {**228**}

—, 2002, On the absolute calibration of the Cepheid distance scale using Hipparcos parallaxes. *AJ*, 124, 1213–1220 {**228, 229, 254**}

Dyson FW, 1926, A method for correcting series of parallax observations. *MNRAS*, 86, 686–706 {**210**}

Eddington AS, 1913, On a formula for correcting statistics for the effects of a known error of observation. *MNRAS*, 73, 359–360 {**209, 210**}

—, 1914, *Stellar Movements and the Structure of the Universe.* Macmillan and Co., London {**209**}

—, 1940, The correction of statistics for accidental error. *MNRAS*, 100, 354–361 {**210**}

Egret D, Heck A, Vergely JL, *et al.*, 1997, The HR diagram of G5-M3 stars near the giant branch. *ESA SP–402: Hipparcos, Venice '97*, 335–338 {**219, 220**}

Elgarøy Ø, Engvold O, Joras P, 1997, Stellar activity and the Wilson–Bappu relation. *A&A*, 326, 165–176 {**222**}

Elgarøy Ø, Engvold O, Lund N, 1999, The Wilson–Bappu effect of the Mg II k line: dependence on stellar temperature, activity and metallicity. *A&A*, 343, 222–228 {**222**}

Eriksson U, Lindegren L, 2007, Limits of ultra-high-precision optical astrometry. Stellar surface structures. *A&A*, 476, 1389–1400 {**207**}

ESA, 1997, *The Hipparcos and Tycho Catalogues. Astrometric and Photometric Star Catalogues derived from the ESA Hipparcos Space Astrometry Mission, ESA SP–1200* (17 volumes including 6 CDs). European Space Agency, Noordwijk, also: VizieR Online Data Catalogue {**254**}

Evans JW, 1941, Interstellar line intensities and the distances of the B stars. *ApJ*, 93, 275–284 {**222**}

Evans NR, 1991, Classical Cepheid luminosities from binary companions. *ApJ*, 372, 597–609 {**226, 229**}

—, 1992, New calibrators for the Cepheid period–luminosity relation. *ApJ*, 389, 657–664 {**226, 229**}

Evans NR, Böhm-Vitense E, Carpenter K, *et al.*, 1998a, Classical Cepheid masses: U Aql. *ApJ*, 494, 768–772 {**226, 229**}

Evans R, Wild W, Hanzl D, 1998b, Supernova 1998bu in NGC 3368. *IAU Circ.*, 6921, 2 {**259**}

Feast MW, 1963, The long-period variables. *MNRAS*, 125, 367–415 {**238**}

—, 1996, The pulsation, temperatures and metallicities of Mira and semi-regular variables in different stellar systems. *MNRAS*, 278, 11–21 {**236, 238**}

—, 1998a, Distance scales after Hipparcos. *Ap&SS*, 263, 209–214 {**208**}

—, 1998b, Hipparcos and variable star distance scales. *Memorie della Societa Astronomica Italiana*, 69, 31–41 {**208, 229**}

—, 1998c, The absolute magnitudes of Cepheids and the extragalactic distance scale. *MNRAS*, 293, L27–L28 {**226**}

—, 1999, Cepheids as distance indicators. *PASP*, 111, 775–793 {**226–229**}

—, 2002, Bias in absolute magnitude determination from parallaxes. *MNRAS*, 337, 1035–1037 {**211**}

—, 2003, The Galactic kinematics of Mira variables. *ASSL Vol. 283: Mass-Losing Pulsating Stars and their Circumstellar Matter* (eds. Nakada Y, Honma M, Seki M), 83–89 {**238, 239**}

Feast MW, Catchpole RM, 1997, The Cepheid period–luminosity zero-point from Hipparcos trigonometrical parallaxes. *MNRAS*, 286, L1–L5 {**210, 211, 226–228, 251**}

Feast MW, Glass IS, Whitelock PA, *et al.*, 1989, A period–luminosity–colour relation for Mira variables. *MNRAS*, 241, 375–392 {**236**}

Feast MW, Pont F, Whitelock PA, 1998, The Cepheid period–luminosity zero-point from radial velocities and Hipparcos proper motions. *MNRAS*, 298, L43–L44 {**226**}

Feast MW, Walker AR, 1987, Cepheids as distance indicators. *ARA&A*, 25, 345–375 {**225, 228**}

Feast MW, Whitelock PA, Menzies J, 2002, Globular clusters and the Mira period–luminosity relation. *MNRAS*, 329, L7–L12 {**238, 257**}

Feast MW, Whitelock PA, 1997a, Galactic kinematics of Cepheids from Hipparcos proper motions. *MNRAS*, 291, 683–693 {**226, 229, 231**}

—, 1997b, Hipparcos parallaxes and proper motions of Cepheids and their implications. *ESA SP–402: Hipparcos, Venice '97*, 625–628 {**229**}

—, 2000, Mira kinematics from Hipparcos data: a Galactic bar to beyond the solar circle. *MNRAS*, 317, 460–487 {**238, 239**}

Feast MW, Woolley R, Yilmaz N, 1972, The kinematics of semi-regular red variables in the solar neighbourhood. *MNRAS*, 158, 23–46 {**238**}

Fernie JD, Evans NR, Beattie B, *et al.*, 1995, A database of Galactic classical Cepheids. *Informational Bulletin on Variable Stars*, 4148, 1 {**226, 227**}

Fernie JD, 1969, The period–luminosity relation: a historical review. *PASP*, 81, 707–731 {**224, 225, 229**}

Fernley J, 1998, The impact of Hipparcos on the RR Lyrae distance scale. *Memorie della Societa Astronomica Italiana*, 69, 43–47 {**247**}

Fernley J, Barnes TG, Skillen I, *et al.*, 1998a, The absolute magnitudes of RR Lyrae stars from Hipparcos parallaxes and proper motions. *A&A*, 330, 515–520 {**249, 250**}

Fernley J, Barnes TG, 1997, Radial velocities and iron abundances of field RR Lyrae stars. I. *A&AS*, 125, 313–319 {**247**}

Fernley J, Skillen I, Carney BW, *et al.*, 1998b, The slope of the RR Lyrae Mv–[Fe/H] relation. *MNRAS*, 293, L61–L64 {**247–249**}

Ferraro FR, Messineo M, Fusi Pecci F, *et al.*, 1999, The giant, horizontal, and asymptotic branches of Galactic globular clusters. I. The catalogue, photometric observables, and features. *AJ*, 118, 1738–1758 {**249**}

Fitzpatrick EL, 1999, Correcting for the effects of interstellar extinction. *PASP*, 111, 63–75 {**241**}

Fitzpatrick EL, Massa D, 2005, Determining the physical properties of B stars. II. Calibration of synthetic photometry. *AJ*, 129, 1642–1662 {**217, 218**}

Fitzpatrick EL, Ribas I, Guinan EF, *et al.*, 2003, Fundamental properties and distances of Large Magellanic Cloud eclipsing binaries. IV. HV 5936. *ApJ*, 587, 685–700 {**239**}

Flower PJ, 1977, Transformations from theoretical HR diagrams to colour–magnitude diagrams. *A&A*, 54, 31–39 {**218**}

Fouqué P, Storm J, Gieren W, 2003, Calibration of the distance scale from Cepheids. *Lecture Notes in Physics Vol. 635: Stellar Candles for the Extragalactic Distance Scale*, 635, 21–44 {**228**}

Freedman WL, Madore BF, Gibson BK, *et al.*, 2001, Final results from the Hubble Space Telescope key project to measure the Hubble Constant. *ApJ*, 553, 47–72 {**225, 227, 228, 253**}

Galazutdinov G, 2005, A survey of interstellar lines: radial velocity profiles and equivalent widths. *Journal of Korean Astronomical Society*, 38, 215–218 {**222**}

García B, 1989, A list of MK standard stars. *Bulletin d'Information du Centre de Donńees Stellaires*, 36, 27–90 {**213, 217**}

Gardiner LT, Sawa T, Fujimoto M, 1994, Numerical simulations of the Magellanic System. I. Orbits of the Magellanic Clouds and the global gas distribution. *MNRAS*, 266, 567–582 {**254**}

Garrison RF, 2000a, Classification of stellar spectra. *Encyclopedia of Astronomy and Astrophysics;* (ed. Murdin, P.) {**214**}

—, 2000b, Hipparcos and the calibration of the MK System. *Hipparcos and the Luminosity Calibration of the Nearer Stars, IAU Joint Discussion 13*, 5–7 {**219**}

—, 2002, Hipparcos and the calibration of the MK system. *Highlights in Astronomy*, 12, 673–674 {**219**}

Garrison RF, Gray RO, 1994, The late B-type stars: refined MK classification, confrontation with Strömgren photometry, and the effects of rotation. *AJ*, 107, 1556–1564 {**214**}

Gerbaldi M, Faraggiana R, Burnage R, *et al.*, 1999, Search for reference A0 dwarf stars: masses and luminosities revisited with Hipparcos parallaxes. *A&AS*, 137, 273–292 {**217**}

Gieren W, Fouque P, Gómez M, 1998, Cepheid period–radius and period–luminosity relations and the distance to the Large Magellanic Cloud. *ApJ*, 496, 17–30 {**227, 228**}

Ginestet N, Carquillat JM, Jaschek C, 2000, The absolute magnitudes of the G to M type MK standards from the Hipparcos parallaxes. *A&AS*, 142, 13–24 {**219**}

Ginestet N, Carquillat JM, 2002, Spectral classification of the hot components of a large sample of stars with composite spectra, and implication for the absolute magnitudes of the cool supergiant components. *ApJS*, 143, 513–537 {**219**}

Girardi L, 1999, A secondary clump of red giant stars: why and where. *MNRAS*, 308, 818–832 {**230, 232**}

—, 2000, Fine structure of the red clump in local group galaxies. *From Extrasolar Planets to Cosmology: VLT Opening Symposium* (eds. Bergeron J, Renzini A), 294–297 {**232**}

Girardi L, Groenewegen MAT, Weiss A, *et al.*, 1998, Fine structure of the red giant clump from Hipparcos data, and distance determinations based on its mean magnitude. *MNRAS*, 301, 149–160 {**232, 233**}

Girardi L, Salaris M, 2001, Population effects on the red giant clump absolute magnitude, and distance determinations to nearby galaxies. *MNRAS*, 323, 109–129 {**233–235**}

Gómez AE, Luri X, Mennessier MO, *et al.*, 1997, The luminosity calibration of the HR diagram revisited by Hipparcos. *ESA SP–402: Hipparcos, Venice '97*, 207–212 {**210, 214–216**}

Gontcharov GA, 2000, The influence of unseen companions on the luminosity calibration. *Hipparcos and the Luminosity Calibration of the Nearer Stars, IAU Joint Discussion 13*, 13–15 {**219**}

Gould A, Popowski P, 1998, Systematics of RR Lyrae statistical parallax. III. Apparent magnitudes and extinctions. *ApJ*, 508, 844–853 {**248**}

Gould A, Salim S, 2002, Searching for failed supernovae with astrometric binaries. *ApJ*, 572, 944–949 {**259**}

Gould A, Uza O, 1998, Upper limit to the distance to the Large Magellanic Cloud. *ApJ*, 494, 118–124 {**227, 239**}

Gratton RG, 1998a, RR Lyrae stars and Cepheids from Hipparcos. *19th Texas Symposium on Relativistic Astrophysics and Cosmology* (eds. Paul J, Montmerle T, Aubourg E), 137–137 {**247**}

—, 1998b, The absolute magnitude of field metal-poor horizontal branch stars. *MNRAS*, 296, 739–745 {**247, 248**}

Gratton RG, Bragaglia A, Carretta E, *et al.*, 2003, Distances and ages of NGC 6397, NGC 6752 and 47 Tuc. *A&A*, 408, 529–543 {**240, 241, 244–246, 248, 249, 252**}

Gratton RG, Carretta E, Clementini G, 1999, Distances and ages of globular clusters using Hipparcos parallaxes of local subdwarfs. *ASSL Vol. 237: Post-Hipparcos Cosmic Candles*, 89–110 {**239, 245**}

Gratton RG, Fusi Pecci F, Carretta E, *et al.*, 1997, Ages of globular clusters from Hipparcos parallaxes of local subdwarfs. *ApJ*, 491, 749–771 {**210, 227, 240–242, 244, 246, 249, 258**}

Gray RO, Garrison RF, 1987, The early A-type stars: refined MK classification, confrontation with Strömgren photometry, and the effects of rotation. *ApJS*, 65, 581–602 {**217**}

—, 1989a, The early F-type stars: refined classification, confrontation with Strömgren photometry, and the effects of rotation. *ApJS*, 69, 301–321 {**217**}

—, 1989b, The late A-type stars: refined MK classification, confrontation with Strömgren photometry, and the effects of rotation. *ApJS*, 70, 623–636 {**217**}

Grenon M, 1972, Photometric detection of red horizontal branch stars. *IAU Colloq. 17: Age des Etoiles* (eds. Cayrel de Strobel G, Delplace AM), 29 {**230**}

—, 2002, Photometric calibrations after Hipparcos or: the fine structure of the HR diagram revealed. *Highlights in Astronomy*, 12, 680–682 {**219**}

Grocholski AJ, Sarajedini A, Olsen KAG, *et al.*, 2007, Distances to populous clusters in the Large Magellanic Cloud via the K-band luminosity of the red clump. *AJ*, 134, 680–693 {**235**}

Grocholski AJ, Sarajedini A, 2002, WIYN open cluster study. X. The K-band magnitude of the red clump as a distance indicator. *AJ*, 123, 1603–1612 {**234**}

Groenewegen MAT, 1999, I and JHK band photometry of classical Cepheids in the Hipparcos Catalogue. *A&AS*, 139, 245–255 {**226**}

—, 2000, Numerical simulations of the Cepheid population in the Hipparcos. *ASP Conf. Ser. 203: IAU Colloq. 176: The Impact of Large-Scale Surveys on Pulsating Star Research*, 212–215 {**228**}

Groenewegen MAT, Oudmaijer RD, 2000, Multi-colour period–luminosity relations of Cepheids in the Hipparcos Catalogue and the distance to the Large Magellanic Cloud. *A&A*, 356, 849–872 {**225, 227**}

Groenewegen MAT, Salaris M, 1999, The absolute magnitudes of RR Lyrae stars from Hipparcos parallaxes. *A&A*, 348, L33–L36 {**211, 249**}

—, 2003, The distance to the Large Magellanic Cloud cluster NGC 1866; clues from the cluster Cepheid population. *A&A*, 410, 887–896 {**235**}

Grundahl F, Stetson PB, Andersen MI, 2002, The ages of the globular clusters M71 and 47 Tuc from Strömgren uvby photometry. Evidence for high ages. *A&A*, 395, 481–497 {**240, 241, 244, 245**}

Grundahl F, VandenBerg DA, Andersen MI, 1998, Strömgren photometry of globular clusters: the distance and age of M13, evidence for two populations of horizontal-branch stars. *ApJ*, 500, L179–L182 {**240, 241, 243–245**}

Grundahl F, VandenBerg DA, Bell RA, *et al.*, 2000, A distance-independent age for the globular cluster M92. *AJ*, 120, 1884–1891 {**246**}

Guinan EF, Ribas I, Fitzpatrick EL, 2004, Eclipsing binaries in local group galaxies: physical properties of the stars and calibration of the zero-point of the cosmic distance scale. *ASP Conf. Ser. 310: IAU Colloq. 193: Variable Stars in the Local Group* (eds. Kurtz DW, Pollard KR), 363–371 {**239, 254**}

Guinan EF, 2004, Seeing double in the local group: extragalactic binaries. *New Astronomy Review*, 48, 647–658 {**239**}

Hanbury Brown R, Davis J, Allen LR, 1974, The angular diameters of 32 stars. *MNRAS*, 167, 121–136 {**217**}

Hanson RB, 1979, A practical method to improve luminosity calibrations from trigonometric parallaxes. *MNRAS*, 186, 875–896 {**210, 245**}

Hanzl D, 1998, Supernova 1998S in NGC 3877. *IAU Circ.*, 6885, 2 {**259**}

—, 1999, Supernova 1999by in NGC 2841. *IAU Circ.*, 7223, 4 {**259**}

Harris WE, Bell RA, VandenBerg DA, *et al.*, 1997, NGC 2419, M92, and the age gradient in the Galactic halo. *AJ*, 114, 1030–1042 {**240**}

Hatzidimitriou D, 1991, A new age calibrator for red horizontal branch populations. *MNRAS*, 251, 545–554 {**230**}

Hawley SL, Jefferys WH, Barnes TG, *et al.*, 1986, Absolute magnitudes and kinematic properties of RR Lyrae stars. *ApJ*, 302, 626–631 {**248**}

Heacox WD, 2004, The Galactic Cepheid period–luminosity relation from Hipparcos parallaxes and proper motions. *ASP Conf. Ser. 310: IAU Colloq. 193: Variable Stars in the Local Group* (eds. Kurtz DW, Pollard KR), 547–552 {**228**}

Heger ML, 1921, *Lick Obs. Bull.* 10, 146 {**222**}

Heintz WD, 1982, Two visual-spectroscopic binary orbits: ADS 7780 and ADS 11060. *PASP*, 94, 705–707 {**233**}

Helmi A, 2004, Velocity trends in the debris of Sagittarius and the shape of the dark matter halo of our Galaxy. *ApJ*, 610, L97–L100 {**257**}

Hemenway MK, 1975, Absolute magnitudes and motions of RR Lyrae stars. *AJ*, 80, 199–207 {**248**}

Herbig GH, 1995, The diffuse interstellar bands. *ARA&A*, 33, 19–74 {**222**}

Holland S, 1998, The distance to the M31 globular cluster system. *AJ*, 115, 1916–1920 {**256**}

Houk N, 1978, *Michigan Catalogue of two-dimensional spectral types for the HD stars, Vol. 2.* Dept. of Astronomy, University of Michigan, Ann Arbour {**214**}

—, 1979, Future objective-prism spectral classification at MK dispersion. *IAU Colloq. 47: Spectral Classification of the Future* (eds. McCarthy MF, Philip AGD, Coyne GV), 51–56 {**214**}

—, 1982, *Michigan Catalogue of two-dimensional spectral types for the HD stars, Vol. 3.* Dept. of Astronomy, University of Michigan, Ann Arbour {**214**}

Houk N, Cowley AP, 1975, *Michigan Catalogue of two-dimensional spectral types for the HD stars, Vol. 1.* Dept. of Astronomy, University of Michigan, Ann Arbour {**214**}

Houk N, Fesen R, 1978, HR diagrams derived from the Michigan Spectral Catalogue. *The HR Diagram: The 100th Anniversary of Henry Norris Russell* (eds. Philip AGD, Hayes DS), IAU Symp. 80, 91–97 {**215**}

Houk N, Smith-Moore M, 1988, *Michigan Catalogue of two-dimensional spectral types for the HD stars, Vol. 4.* Dept. of Astronomy, University of Michigan, Ann Arbour {**214**}

Houk N, Swift CM, Murray CA, *et al.*, 1997, The properties of main sequence stars from Hipparcos data. *ESA SP–402: Hipparcos, Venice '97*, 279–282 {**214**}

Houk N, Swift CM, 1999, *Michigan Catalogue of two-dimensional spectral types for the HD stars, Vol. 5.* Dept. of Astronomy, University of Michigan, Ann Arbour {**214**}

Hoyle F, Shanks T, Tanvir NR, 2003, Distances to Cepheid open clusters via optical and K-band imaging. *MNRAS*, 345, 269–291 {**228**}

Humphreys RM, McElroy DB, 1984, The initial mass function for massive stars in the Galaxy and the Magellanic Clouds. *ApJ*, 284, 565–577 {**218**}

Ibata RA, Gilmore G, 1995, The outer regions of the Galactic bulge. II. Analysis. *MNRAS*, 275, 605–627 {**257**}

Israel FP, 1998, Centaurus A – NGC 5128. *A&A Rev.*, 8, 237–278 {**256**}

Jaschek C, Gómez AE, 1985, The Malmquist correction. *A&A*, 146, 387–388 {**209**}

Jaschek C, Gómez AE, 1997, The absolute magnitude of the early-type MK standards from Hipparcos parallaxes. *The First Results of Hipparcos and Tycho, IAU Joint Discussion 14*, 36 {**213, 215**}

—, 1998, The absolute magnitude of the early-type MK standards from Hipparcos parallaxes. *A&A*, 330, 619–625 {**213, 218**}

Jaschek C, Jaschek M, 1973, On the precision of the MK spectral classification system. *Spectral classification and multicolour photometry* (eds. Fehrenbach C, Westerlund BE), IAU Symp. 50, 43–51 {**214**}

—, 1987, *The Classification of Stars*. Cambridge University Press {**220**}

Jeffreys H, 1938, The correction of frequencies for a known standard error of observation. *MNRAS*, 98, 190–194 {**210**}

Jenkins EB, Wallerstein G, 1995, High-velocity, high-excitation neutral carbon in a cloud in the Vela supernova remnant. *ApJ*, 440, 227–240 {**258**}

Jimenez R, Flynn C, Kotoneva E, 1998, Hipparcos and the age of the Galactic disk. *MNRAS*, 299, 515–519 {**231**}

Jimenez R, Padoan P, 1998, The ages and distances of globular clusters with the luminosity function method: the case of M5 and M55. *ApJ*, 498, 704–709 {**240**}

Jôeveer M, 1974, Ages of δ Cephei stars and the density of gravitating masses near the Sun. *Tartu Astrofuusika Observatoorium Teated*, 46, 35–49 {**230**}

Johnson HL, Morgan WW, 1953, Fundamental stellar photometry for standards of spectral type on the revised system of the Yerkes spectral atlas. *ApJ*, 117, 313–352 {**214**}

Johnston KV, Law DR, Majewski SR, 2005, A Two Micron All-Sky Survey view of the Sagittarius dwarf galaxy. III. Constraints on the flattening of the Galactic halo. *ApJ*, 619, 800–806 {**257**}

Johnston KV, Spergel DN, Hernquist L, 1995, The disruption of the Sagittarius dwarf galaxy. *ApJ*, 451, 598–606 {**257**}

Jones BF, Klemola AR, Lin DNC, 1994, Proper motion of the Large Magellanic Cloud and the mass of the Galaxy. I. Observational results. *AJ*, 107, 1333–1337 {**255**}

Jones RV, Carney BW, Storm J, *et al.*, 1992, The Baade–Wesselink method and the distances to RR Lyrae stars. VII. The field stars SW And and DX Del and a comparison of recent Baade–Wesselink analyses. *ApJ*, 386, 646–662 {**249**}

Jung J, 1970, The derivation of absolute magnitudes from proper motions and radial velocities and the calibration of the HR diagram. *A&A*, 4, 53–69 {**213**}

Kallivayalil N, van der Marel RP, Alcock C, *et al.*, 2006a, The proper motion of the Large Magellanic Cloud using HST. *ApJ*, 638, 772–785 {**255**}

Kallivayalil N, van der Marel RP, Alcock C, 2006b, Is the SMC bound to the LMC? The Hubble Space Telescope proper motion of the SMC. *ApJ*, 652, 1213–1229 {**255**}

Kaluzny J, Wysocka A, Stanek KZ, *et al.*, 1998, BVI CCD photometry of the globular cluster 47 Tuc. *Acta Astronomica*, 48, 439–453 {**235**}

Kapteyn JC, 1914, On the individual parallaxes of the brighter Galactic helium stars in the southern hemisphere, together with considerations on the parallax of stars in general. *ApJ*, 40, 43–126 {**209**}

Karataš Y, Schuster WJ, 2006, Metallicity and absolute magnitude calibrations for UBV photometry. *MNRAS*, 371, 1793–1812 {**219**}

Keenan PC, 1985, The MK classification and its calibration. *Calibration of Fundamental Stellar Quantities* (eds. Hayes DS, Pasinetti LE, Philip AGD), IAU Symp. 111, 121–135 {**214, 218**}

Keenan PC, Barnbaum C, 1999, Revision and calibration of MK luminosity classes for cool giants by Hipparcos parallaxes. *ApJ*, 518, 859–865 {**219, 220**}

Keenan PC, Hynek JA, 1950, Neutral oxygen in stellar atmospheres. *ApJ*, 111, 1–10 {**222**}

Kendall M, Stuart A, 1979, *The Advanced Theory of Statistics. Vol. 2: Inference and Relationship*, 4th Edition. Griffin, London {**213**}

Kharchenko NV, Kilpio E, Malkov O, *et al.*, 2002, Mira kinematics in the post-Hipparcos era. *A&A*, 384, 925–936 {**236, 239**}

King JR, Stephens A, Boesgaard AM, *et al.*, 1998, Keck HIRES spectroscopy of M92 subgiants: surprising abundances near the turnoff. *AJ*, 115, 666–684 {**243**}

Kiraga M, Paczyński B, Stanek KZ, 1997, The colour–magnitude diagram in Baade's Window revisited. *ApJ*, 485, 611–617 {**235**}

Kiss LL, Vinkó J, 2000, A photometric and spectroscopic study of the brightest northern Cepheids. III. A high-resolution view of Cepheid atmospheres. *MNRAS*, 314, 420–432 {**228**}

Koen C, 1992, Confidence intervals for the Lutz–Kelker correction. *MNRAS*, 256, 65–68 {**210**}

Koen C, Laney D, 1998, On the determination of absolute magnitude zero-points from Hipparcos parallaxes. *MNRAS*, 301, 582–584 {**211, 226, 248**}

Koen C, Lombard F, 2003, Some statistical aspects of estimating the local red clump absolute magnitude. *MNRAS*, 343, 241–248 {**234**}

Koopmans LVE, Treu T, Fassnacht CD, *et al.*, 2003, The Hubble constant from the gravitational lens B 1608+656. *ApJ*, 599, 70–85 {**228**}

Kotoneva E, Flynn C, Chiappini C, *et al.*, 2002a, K dwarfs and the chemical evolution of the solar cylinder. *MNRAS*, 336, 879–891 {**219**}

Kotoneva E, Flynn C, Jimenez R, 2002b, Luminosity-metallicity relation for stars on the lower main sequence. *MNRAS*, 335, 1147–1157 {**219**}

Kovalevsky J, 1998, On the uncertainty of distances derived from parallax measurements. *A&A*, 340, L35–L38 {**208**}

Kristian J, 1970, On the optical identification of the Vela pulsar: photoelectric measurements. *ApJ*, 162, L103–L104 {**258**}

Kroupa P, Bastian U, 1997a, The Hipparcos proper motion of the Magellanic Clouds. *New Astronomy*, 2, 77–90 {**254, 255**}

—, 1997b, The motion of the Magellanic Clouds. *ESA SP–402: Hipparcos, Venice '97*, 615–616 {**255**}

Kroupa P, Röser S, Bastian U, 1994, On the motion of the Magellanic Clouds. *MNRAS*, 266, 412–420 {**255**}

Kuchinski LE, Frogel JA, Terndrup DM, *et al.*, 1995, Infrared array photometry of metal-rich globular clusters. 1. Techniques and first results. *AJ*, 109, 1131–1153 {**230**}

Kunkel WE, 1979, On the origin and dynamics of the Magellanic Stream. *ApJ*, 228, 718–733 {**254**}

Kunkel WE, Demers S, 1976, The Magellanic plane. *The Galaxy and the Local Group*, 241–247 {**254**}

Kurtz MJ, 1984, Progress in automation techniques for spectral classification. *The MK Process and Stellar Classification* (ed. Garrison RF), 133–139 {**214**}

Kurucz RL, 1993, ATLAS9 stellar atmosphere programs and 2 km s^{-1} grid. *CD No. 13, Smithsonian Astrophysical Observatory* {**217**}

Lamers HJGLM, Harzevoort JMAG, Schrijver H, *et al.*, 1997, The effect of rotation on the absolute visual magnitudes of OB stars measured with Hipparcos. *A&A*, 325, L25–L28 {**216**}

Laney CD, Stobie RS, 1994, Cepheid period–luminosity relations in K, H, J and V. *MNRAS*, 266, 441–454 {**225, 226, 228**}

Lanoix P, Paturel G, Garnier R, 1999, Direct calibration of the Cepheid period–luminosity relation. *MNRAS*, 308, 969–978 {**210, 226, 227**}

Lanoix P, 1998, Hipparcos calibration of the peak brightness of four supernovae IA and the value of H_0. *A&A*, 331, 421–427 {**258**}

Law DR, Johnston KV, Majewski SR, 2005, A Two Micron All-Sky Survey view of the Sagittarius dwarf galaxy. IV. Modeling the Sagittarius tidal tails. *ApJ*, 619, 807–823 {**257**}

Layden AC, 1994, The metallicities and kinematics of RR Lyrae variables. I. New observations of local stars. *AJ*, 108, 1016–1041 {**250, 251**}

—, 1995, The metallicities and kinematics of RR Lyrae variables. II. Galactic structure and formation from local stars. *AJ*, 110, 2288–2311 {**250**}

Layden AC, Hanson RB, Hawley SL, *et al.*, 1996, The absolute magnitude and kinematics of RR Lyrae stars via statistical parallax. *AJ*, 112, 2110–2131 {**248, 250, 251**}

Leavitt HS, Pickering EC, 1912, Periods of 25 variable stars in the Small Magellanic Cloud. *Harvard College Observatory Circular*, 173, 1–3 {**224**}

Leavitt HS, 1908, 1777 variables in the Magellanic Clouds. *Annals of Harvard College Observatory*, 60, 87–108 {**224**}

Libeskind NI, Frenk CS, Cole S, *et al.*, 2005, The distribution of satellite galaxies: the great pancake. *MNRAS*, 363, 146–152 {**254**}

Linsky JL, 1999, The Wilson–Bappu relation between Ca II emission and stellar luminosities. *ApJ*, 525, 776–780 {**221**}

Li PS, Thronson HA, 1999, New particle simulation of the Magellanic Clouds. *IAU Symp. 190: New Views of the Magellanic Clouds*, 503 {**255**}

Ljunggren B, Oja T, 1965, Photoelectric measurements of magnitudes and colours for 849 stars. *Arkiv for Astronomi*, 3, 439–465 {**210, 211**}

Luri X, Arenou F, 1997, Utilisation of Hipparcos data for distance determinations: error, bias and estimation. *ESA SP–402: Hipparcos, Venice '97*, 449–452 {**208**}

Luri X, Gómez AE, Torra J, *et al.*, 1998a, The Large Magellanic Cloud distance modulus from Hipparcos RR Lyrae and classical Cepheid data. *A&A*, 335, L81–L84 {**227**}

Luri X, Mennessier MO, Torra J, *et al.*, 1993, A new approach to the Malmquist bias. *A&A*, 267, 305–307 {**209**}

—, 1996a, Absolute magnitudes and kinematics of oxygen Mira variables. *A&A*, 314, 807–812 {**238**}

—, 1996b, A new maximum likelihood method for luminosity calibrations. *A&AS*, 117, 405–415 {**209, 211, 213**}

Luri X, Torra J, Figueras F, *et al.*, 1998b, Calibration of the classical Cepheid period–luminosity relation from Hipparcos data. *Ap&SS*, 263, 215–218 {**227**}

Lutz TE, 1979, On the use of trigonometric parallaxes for the calibration of luminosity systems. II. *MNRAS*, 189, 273–278 {**210**}

—, 1983, The calibration of absolute magnitudes from trigonometric parallaxes: sampling. *IAU Colloq. 76: Nearby Stars and the Stellar Luminosity Function* (eds. Philip AGD, Upgren AR), 41–47 {**210**}

—, 1986, Statistical problems encountered in using trigonometric parallaxes. *Astrometric Techniques* (eds. Eichhorn HK, Leacock RJ), IAU Symp. 109, 47–52 {**210**}

Lutz TE, Kelker DH, 1973, On the use of trigonometric parallaxes for the calibration of luminosity systems: theory. *PASP*, 85, 573–578 {**210**}

Luyten WJ, 1922, On the relation between parallax, proper motion, and apparent magnitude. *Lick Observatory Bulletin*, 10, 135–140 {**212**}

Lynden-Bell D, 1982, The Fornax–Leo–Sculptor system. *The Observatory*, 102, 202–208 {**254**}

Lynden-Bell D, Lynden-Bell RM, 1995, Ghostly streams from the formation of the Galaxy's halo. *MNRAS*, 275, 429–442 {**254**}

Lyngå G, Lindegren L, 1998, Hipparcos data for two open clusters containing Cepheids. *New Astronomy*, 3, 121–123 {**229**}

Madore BF, 1976, A reddening-independent formulation of the period–luminosity relation: the Wesenheit function. *The Galaxy and the Local Group* (eds. Dickens RJ, Perry JE, Smith FG, *et al.*), volume 182 of *Royal Greenwich Observatory Bulletin*, 153–159 {**227**}

—, 1982, The period–luminosity relation. IV. Intrinsic relations and reddenings for the Large Magellanic Cloud Cepheids. *ApJ*, 253, 575–579 {**225**}

Madore BF, Freedman WL, 1998a, Calibration of the extragalactic distance scale. *Stellar Astrophysics for the Local Group: VIII Canary Islands Winter School of Astrophysics*, 263–350 {**208**}

—, 1998b, Hipparcos parallaxes and the Cepheid distance scale. *ApJ*, 492, 110–115 {**226**}

Maintz G, 2005, Proper identification of RR Lyrae stars brighter than 12.5 mag. *A&A*, 442, 381–384 {**247, 250**}

Maintz G, de Boer KS, 2005, RR Lyrae stars: kinematics, orbits and z-distribution. *A&A*, 442, 229–237 {**250, 252**}

Maíz Apellániz J, Walborn NR, Galué HA, *et al.*, 2004, A Galactic O star catalogue. *ApJS*, 151, 103–148 {**217**}

Maíz Apellániz J, Walborn NR, 2003, A spectroscopic, photometric, and astrometric Galactic O-type star database. *IAU Symp. 212* (eds. van der Hucht K, Herrero A, Esteban C), 560–561 {**217**}

Maíz Apellániz J, 2005, Self-consistent distance determinations for Lutz–Kelker-limited samples. *ESA SP–576: The Three-Dimensional Universe with Gaia* (eds. Turon C, O'Flaherty KS, Perryman MAC), 179–182 {**210**}

Majewski SR, 1994, The Fornax–Leo–Sculptor stream revisited. *ApJ*, 431, L17–L21 {**254**}

Malagnini ML, Morossi C, Faraggiana R, 1984, $T_{\rm eff}$ determination from ultraviolet and visual spectrophotometry and comparison with MK classification. *The MK Process and Stellar Classification* (ed. Garrison RF), 321–325 {**218**}

Malmquist KG, 1922, *Lund Medd. Ser. I.* 100,1 {**209**}

—, 1936, *Stockholms Obs. Medd.* 26 {**209**}

Martínez-Delgado D, Alfaro EJ, 1997, Low-resolution spectroscopy of RR Lyrae stars in the Hipparcos Input Catalogue. *ESA SP–402: Hipparcos, Venice '97*, 453–456 {**247**}

Martin JC, Morrison HL, 1998, A new analysis of RR Lyrae kinematics in the solar neighbourhood. *AJ*, 116, 1724–1735 {**249–251**}

Martin NF, Ibata RA, Bellazzini M, *et al.*, 2004, A dwarf galaxy remnant in Canis Major: the fossil of an in-plane accretion on to the Milky Way. *MNRAS*, 348, 12–23 {**257**}

Martin WL, Warren PR, Feast MW, 1979, Multicolour photoelectric photometry of Magellanic Cloud Cepheids. II. An analysis of BVI observations in the LMC. *MNRAS*, 188, 139–157 {**224**}

McNamara DH, 1997a, Luminosities of SX Phoenicis, large-amplitude δ Scuti, and RR Lyrae stars. *PASP*, 109, 1221–1232 {**248**}

—, 1997b, The absolute magnitudes of the RR Lyrae stars. *PASP*, 109, 857–867 {**249**}

Megier A, Strobel A, Bondar A, *et al.*, 2005, Interstellar Ca II line intensities and the distances of the OB stars. *ApJ*, 634, 451–458 {**223**}

Mennessier MO, Colomé J, 2002, Identification of RR Lyrae stars in the Tycho 2 Catalogue. *A&A*, 390, 173–178 {**247**}

Mennessier MO, Luri X, Figueras F, *et al.*, 1997, Barium stars, Galactic populations and evolution. *A&A*, 326, 722–730 {**213**}

Merrill PW, 1925, Note on the infrared oxygen triplet in stellar spectra. *PASP*, 37, 272–275 {**222**}

Metzger MR, Caldwell JAR, Schechter PL, 1998, The shape and scale of Galactic rotation from Cepheid kinematics. *AJ*, 115, 635–647 {**229**}

Metz M, Geffert M, 2004, Formalism and quality of a proper motion link with extragalactic objects for astrometric satellite missions. *A&A*, 413, 771–777 {**258**}

Metz M, Kroupa P, Jerjen H, 2007, The spatial distribution of the Milky Way and Andromeda satellite galaxies. *MNRAS*, 374, 1125–1145 {**254**}

Mihalas D, Binney JJ, 1981, *Galactic Astronomy: Structure and Kinematics*, 2nd edition. W.H. Freeman, San Francisco {**212**}

Milne DK, 1968, Radio emission from the supernova remnant Vela-X. *Australian Journal of Physics*, 21, 201–219 {**258**}

Milone EF, Wilson WJF, Volk K, 1999, Analyses of the short-period Cepheid SU Cas. *AJ*, 118, 3016–3031 {**228**}

Mishenina TV, Bienaymé O, Gorbaneva TI, *et al.*, 2006, Elemental abundances in the atmosphere of clump giants. *A&A*, 456, 1109–1120 {**235**}

Mishurov YN, Zenina IA, 1999, Parameters of the Galactic rotation curve and spiral pattern from Cepheid kinematics. *Astronomy Reports*, 43, 487–493

Momany Y, Zaggia SR, Bonifacio P, *et al.*, 2004, Probing the Canis Major stellar over-density as due to the Galactic warp. *A&A*, 421, L29–L32 {**257**} {**230**}

Momany Y, Zaggia SR, 2005, The proper motion of the Magellanic Clouds: the UCAC 2-Hipparcos inconsistency. *A&A*, 437, 339–343 {**255**}

Morgan WW, Keenan PC, Kellman E, 1943, *An Atlas of Stellar Spectra, with an Outline of Spectral Classification*. University of Chicago Press {**214**}

Morgan WW, Keenan PC, 1973, Spectral Classification. *ARA&A*, 11, 29–50 {**214**}

Nella-Courtois HD, Lanoix P, Paturel G, 1999, Calibration of the Faber–Jackson relation for M31 globular clusters using Hipparcos data. *MNRAS*, 302, 587–592 {**256**}

Newberg HJ, Yanny B, Rockosi C, *et al.*, 2002, The ghost of Sagittarius and lumps in the halo of the Milky Way. *ApJ*, 569, 245–274 {**257**}

Ngeow CC, Kanbur SM, Nikolaev S, *et al.*, 2005, Further empirical evidence for the non-linearity of the period–luminosity relations as seen in the Large Magellanic Cloud Cepheids. *MNRAS*, 363, 831–846 {**227**}

Oegelman H, Koch-Miramond L, Auriere M, 1989, Measurement of the Vela pulsar's proper motion and detection of the optical counterpart of its compact X-ray nebula. *ApJ*, 342, L83–L86 {**258**}

Olling RP, 2007, Accurate extragalactic distances and dark energy: anchoring the distance scale with rotational parallaxes. *MNRAS*, 378, 1385–1399 {**258**}

Oort JH, 1927, Investigations concerning the rotational motion of the Galactic system together with new determinations of secular parallaxes, precession and motion of the equinox. *Bull. Astron. Inst. Netherlands*, 4, 79–89 {**229**}

Osmer PS, 1972, The atmospheres of the F-type supergiants I. Calibration of the luminosity-sensitive O I 777.4 nm line. *ApJS*, 24, 247–253 {**222**}

Oudmaijer RD, Groenewegen MAT, Schrijver H, 1998, The Lutz–Kelker bias in trigonometric parallaxes. *MNRAS*, 294, L41–L46 {**210, 211, 226**}

—, 1999, The absolute magnitude of K0V stars from Hipparcos parallaxes. *A&A*, 341, L55–L58 {**210, 211, 219**}

Özeren FF, Doyle JG, Jevremovic D, 1999, The Wilson–Bappu relation for RS CVn stars. *A&A*, 350, 635–642 {**222, 223**}

Pace G, Pasquini L, Ortolani S, 2003, The Wilson–Bappu effect: a tool to determine stellar distances. *A&A*, 401, 997–1007 {**221**}

Paczyński B, 1997, Detached eclipsing binaries as primary distance and age indicators. *The Extragalactic Distance Scale* (eds. Livio M, Donahue M, Panagia N), 273–280 {**239**}

Paczyński B, Stanek KZ, 1998, Galactocentric distance with the Optical Gravitational Lensing Experiment and Hipparcos red clump stars. *ApJ*, 494, L219–L222 {**231, 232, 235**}

Paczyński B, Udalski A, Szymanski M, *et al.*, 1999, The Optical Gravitational Lensing Experiment. UBVI photometry of stars in Baade's Window. *Acta Astronomica*, 49, 319–339 {**235**}

Palma C, Majewski SR, Johnston KV, 2002, On the distribution of orbital poles of Milky Way satellites. *ApJ*, 564, 736–761 {**254**}

Panagia N, 1973, Some physical parameters of early-type stars. *AJ*, 78, 929–934 {**218**}

—, 1998, New distance determination to the Large Magellanic Cloud. *Memorie della Societa Astronomica Italiana*, 69, 225–235 {**239, 254**}

Panagia N, Gilmozzi R, Macchetto F, *et al.*, 1991, Properties of the SN 1987a circumstellar ring and the distance to the Large Magellanic Cloud. *ApJ*, 380, L23–L26 {**239**}

Parsons SB, 2001, A large spectral class dependence of the Wilson–Bappu effect among luminous stars. *PASP*, 113, 188–194 {**221**}

Paturel G, Lanoix P, Garnier R, *et al.*, 1997, Distances for 17 nearby galaxies based on the Hipparcos calibration of the period–luminosity relation. *ESA SP–402: Hipparcos, Venice '97*, 629–634 {**255, 256, 258**}

Paturel G, Lanoix P, Teerikorpi P, *et al.*, 1998, Hubble constant from sosie galaxies and Hipparcos geometrical calibration. *A&A*, 339, 671–677 {**255, 256, 258**}

Paturel G, Teerikorpi P, Theureau G, *et al.*, 2002, Calibration of the distance scale from Galactic Cepheids. II. Use of the Hipparcos calibration. *A&A*, 389, 19–28 {**226**}

Paunzen E, 1999, A comparison of different spectral classification systems for early-type stars using Hipparcos parallaxes. *A&A*, 341, 784–788 {**217, 218, 220**}

Pavlovskaya ED, 1953, *Perem. Zvezdy.* 9, 349 {**248**}

Pedreros MH, Anguita C, Maza J, 2002, Proper motion of the Large Magellanic Cloud using quasars as an inertial reference system: the Q0459–6427 field. *AJ*, 123, 1971–1977 {**255**}

Percival SM, Salaris M, van Wyk F, *et al.*, 2002, Resolving the 47 Tucanae distance problem. *ApJ*, 573, 174–183 {**210, 240–242, 244, 245**}

Percival SM, Salaris M, 2003, An empirical test of the theoretical population corrections to the red clump absolute magnitude. *MNRAS*, 343, 539–546 {**234, 235**}

Perryman MAC, Brown AGA, Lebreton Y, *et al.*, 1998, The Hyades: distance, structure, dynamics, and age. *A&A*, 331, 81–120 {**229**}

Perryman MAC, Lindegren L, Kovalevsky J, *et al.*, 1995, Parallaxes and the HR diagram for the preliminary Hipparcos solution H30. *A&A*, 304, 69–81 {**231**}

Peñarrubia J, Martínez-Delgado D, Rix HW, *et al.*, 2005, A comprehensive model for the Monoceros tidal stream. *ApJ*, 626, 128–144 {**257**}

Phillips MM, Lira P, Suntzeff NB, *et al.*, 1999, The reddening-free decline rate versus luminosity relationship for Type Ia supernovae. *AJ*, 118, 1766–1776 {**258**}

Pietrzyński G, Gieren W, Udalski A, 2003, The Araucaria project: dependence of mean K, J, and I absolute magnitudes of red clump stars on metallicity and age. *AJ*, 125, 2494–2501 {**235**}

Pietrzyński G, Gieren W, 2002, The ARAUCARIA project: deep near-infrared survey of nearby galaxies. I. The distance to the Large Magellanic Cloud from K-band photometry of red clump stars. *AJ*, 124, 2633–2638 {**235**}

Pols OR, Schröder KP, Hurley JR, *et al.*, 1998, Stellar evolution models for $Z = 0.0001 - 0.03$. *MNRAS*, 298, 525–536 {**230**}

Pont F, Mayor M, Burki G, 1994, New radial velocities for classical Cepheids. Local galactic rotation revisited. *A&A*, 285, 415–439 {**226, 229**}

Pont F, 1999a, The Cepheid distance scale after Hipparcos. *ASP Conf. Ser. 167: Harmonizing Cosmic Distance Scales in a Post-Hipparcos Era* (eds. Egret D, Heck A), 113–128 {**226, 229, 254**}

—, 1999b, The Cepheid distance scale after Hipparcos. *ASP Conf. Ser. 167: Harmonizing Cosmic Distance Scales in a Post-Hipparcos Era*, 113–128 {**228**}

Pont F, Mayor M, Turon C, *et al.*, 1998, Hipparcos subdwarfs and globular cluster ages: the distance and age of M92. *A&A*, 329, 87–100 {**240–246**}

Popowski P, 1998, RR Lyrae stars as distance indicators. Ph.D. Thesis {**249**}

—, 2001, Harmonizing the RR Lyrae and clump distance scales: stretching the short distance scale to intermediate ranges? *MNRAS*, 321, 502–506 {**234, 247**}

Popowski P, Gould A, 1998a, Systematics of RR Lyrae statistical parallax. II. Proper motions and radial velocities. *ApJ*, 506, 271–280 {**246**}

—, 1998b, Systematics of RR Lyrae statistical parallax. I. Mathematics. *ApJ*, 506, 259–270 {**212, 248**}

—, 1999a, The RR Lyrae distance scale. *Post-Hipparcos Cosmic Candles. ASSL 237*, volume 237, 53–74 {**247**}

—, 1999b, The RR Lyrae distance scale. *ASSL Vol. 237: Post-Hipparcos Cosmic Candles*, 53–74 {**247, 249**}

Prentice AJR, Ter Haar D, 1969, H II regions and the distances to pulsars. *Nature*, 222, 964–965 {**258**}

Preston GW, 1959, A spectroscopic study of the RR Lyrae stars. *ApJ*, 130, 507–538 {**250**}

Pych W, Kaluzny J, Krzeminski W, *et al.*, 2001, Cluster ages experiment. CCD photometry of SX Phoenicis variables in the globular cluster M55. *A&A*, 367, 148–158 {**246**}

Rastorguev AS, Dambis AK, Zabolotskikh MV, 2005, Classical Cepheids and RR Lyrae stars as standard candles. *ESA SP–576: The Three-Dimensional Universe with Gaia* (eds. Turon C, O'Flaherty KS, Perryman MAC), 707–710 {**228**}

Rastorguev AS, Glushkova EV, Dambis AK, *et al.*, 1998, Open clusters, Cepheids and the problems of distance scale. *Dynamical Studies of Star Clusters and Galaxies* (eds. Kroupa P, Palouš J, Spurzem R), 195–195 {**230**}

—, 1999, Statistical parallaxes and kinematical parameters of classical Cepheids and young star clusters. *Astronomy Letters*, 25, 595–607 {**212, 230**}

Ratnatunga KU, Casertano S, 1991, Absolute magnitude calibration using trigonometric parallax: incomplete, spectroscopic samples. *AJ*, 101, 1075–1088 {**209**}

Reed BC, 2003, Catalogue of Galactic OB stars. *AJ*, 125, 2531–2533 {**215**}

—, 2005, New estimates of the solar-neighbourhood massive star birthrate and the Galactic supernova rate. *AJ*, 130, 1652–1657 {**215**}

Rees RF, 1996, Astrometric distances to globular clusters: new results. *ASP Conf. Ser. 92: Formation of the Galactic Halo...Inside and Out* (eds. Morrison HL, Sarajedini A), 289–292 {**251**}

Reid IN, 1997, Younger and brighter: new distances to globular clusters based on Hipparcos parallax measurements of local subdwarfs. *AJ*, 114, 161–179 {**210, 227, 240–244, 246, 254, 258**}

—, 1998, Hipparcos subdwarf parallaxes: metal-rich clusters and the thick disk. *AJ*, 115, 204–228 {**240, 241, 243–245**}

—, 1999, The HR diagram and the Galactic distance scale after Hipparcos. *ARA&A*, 37, 191–237 {**208**}

Reid IN, van Wyk F, Marang F, *et al.*, 2001, A search for previously unrecognized metal-poor subdwarfs in the Hipparcos astrometric catalogue. *MNRAS*, 325, 931–962 {**242**}

Rejkuba M, 2004, The distance to the giant elliptical galaxy NGC 5128. *A&A*, 413, 903–912 {**256, 257**}

Renzini A, 1991, Globular cluster ages and cosmology. *NATO ASIC Proc. 348: Observational Tests of Cosmological Inflation* (eds. Shanks T, Banday AJ, Ellis RS), 131–146 {**240**}

Renzini A, Bragaglia A, Ferraro FR, *et al.*, 1996, The white dwarf distance to the globular cluster NGC 6752 (and its age) with the Hubble Space Telescope. *ApJ*, 465, L23–L26 {**252**}

Riess AG, Li W, Stetson PB, *et al.*, 2005, Cepheid calibrations from the Hubble Space Telescope of the luminosity of two recent Type Ia supernovae and a redetermination of the Hubble constant. *ApJ*, 627, 579–607 {**228**}

Robertson BSC, Feast MW, 1981, The bolometric, infrared and visual absolute magnitudes of Mira variables. *MNRAS*, 196, 111–120 {**238**}

Roman NG, 1952, The spectra of the bright stars of types F5–K5. *ApJ*, 116, 122–143 {**211**}

Rosenberg A, Saviane I, Piotto G, *et al.*, 1999, Galactic globular cluster relative ages. *AJ*, 118, 2306–2320 {**240**}

Ružčka A, 2003, The system of the Milky Way, and Large and Small Magellanic Clouds. *Ap&SS*, 284, 519–522 {**255**}

Rybka SP, 2007, A catalogue of candidate red clump stars in Tycho 2. *Kinematika i Fizika Nebesnykh Tel*, 23, 102–133 {**235**}

Saad SM, Lee S, 2001, Distances of globular clusters based on Hipparcos parallaxes of nearby subdwarfs. *Journal of Korean Astronomical Society*, 34, 99–109 {**240**}

Saha A, Sandage AR, Labhardt L, *et al.*, 1997, Cepheid calibration of the peak brightness of Type Ia supernovae. VIII. SN 1990N in NGC 4639. *ApJ*, 486, 1–20 {**226**}

Sahu DK, Anupama GC, Prabhu TP, 2006, Photometric study of Type Ia supernova SN 2002hu. *MNRAS*, 366, 682–688 {**258**}

Salaris M, Cassisi S, García-Berro E, *et al.*, 2001, On the white dwarf distances to Galactic globular clusters. *A&A*, 371, 921–931 {**252**}

Salaris M, Degl'Innocenti S, Weiss A, 1997, The age of the oldest globular clusters. *ApJ*, 479, 665–672 {**246**}

Salaris M, Girardi L, 2002, Population effects on the red giant clump absolute magnitude: the K band. *MNRAS*, 337, 332–340 {**234**}

Salaris M, Percival SM, Brocato E, *et al.*, 2003, The distance to the Large Magellanic Cloud cluster NGC 1866 and the surrounding field. *ApJ*, 588, 801–804 {**235, 253**}

Salaris M, Weiss A, 1998, Metal-rich globular clusters in the Galactic disk: new age determinations and the relation to halo clusters. *A&A*, 335, 943–953 {**244**}

Sandage AR, 1958, The colour–magnitude diagrams of Galactic and globular clusters and their interpretation as age groups. *Ricerche Astronomiche*, 5, 41–68 {**247**}

—, 1969, New subdwarfs. II. Radial velocities, photometry, and preliminary space motions for 112 stars with large proper motion. *ApJ*, 158, 1115–1136 {**219**}

—, 1970, Main sequence photometry, colour–magnitude diagrams, and ages for the globular clusters M3, M13, M15, and M92. *ApJ*, 162, 841–870 {**240**}

—, 2000, RR Lyrae stars. *Encyclopedia of Astronomy and Astrophysics;* (ed. Murdin, P.) {**247**}

Sandage AR, Saha A, 2002, Bias properties of extragalactic distance indicators. XI. Methods to correct for observational selection bias for RR Lyrae absolute magnitudes from trigonometric parallaxes expected from the Full-Sky Astrometric Mapping Explorer Satellite. *AJ*, 123, 2047–2069 {**210, 211**}

Sandage AR, Tammann GA, Saha A, *et al.*, 2006, The Hubble Constant: a summary of the Hubble Space Telescope program for the luminosity calibration of Type Ia supernovae by means of Cepheids. *ApJ*, 653, 843–860 {**227, 228**}

Sandage AR, Tammann GA, 1998, Confirmation of previous ground-based Cepheid period–luminosity zero-points using Hipparcos trigonometric parallaxes. *MNRAS*, 293, L23–L26 {**226**}

—, 2006, Absolute magnitude calibrations of Population I and II Cepheids and other pulsating variables in the instability strip of the Hertzsprung–Russell diagram. *ARA&A*, 44, 93–140 {**225**}

Sandquist EL, Bolte M, Langer GE, *et al.*, 1999, Wide-field CCD photometry of the globular cluster M30. *ApJ*, 518, 262–283 {**246**}

Sanford RF, 1937, Regional study of the interstellar calcium lines. *ApJ*, 86, 136–152 {**222**}

Sarajedini A, van Duyne J, 2001, Deep Hubble Space Telescope WFPC2 photometry of M31's thick disk. *AJ*, 122, 2444–2457 {**256**}

Sarajedini A, 1999, WIYN open cluster study. III. The observed variation of the red clump luminosity and colour with metallicity and age. *AJ*, 118, 2321–2326 {**234**}

Sawa T, Fujimoto M, Kumai Y, 1999, The Magellanic stream and the history of the tidal interaction between the Large and Small Magellanic Clouds. *IAU Symp. 190: New Views of the Magellanic Clouds*, 499 {**255**}

Schaifers K, Voigt HH (eds.), 1982, *Landolt-Börnstein: Numerical Data and Functional Relationships in Science and Technology; New Series, Group 6 Astronomy and Astrophysics, Volume 2.* Springer–Verlag, Berlin {**212, 217–220**}

Schaller G, Schaerer D, Meynet G, *et al.*, 1992, New grids of stellar models. I. From $0.8-120\,M_\odot$ at $Z = 0.020, 0.001$. *A&AS*, 96, 269–331 {**216**}

Schröder SE, Kaper L, Lamers HJGLM, *et al.*, 2004, On the Hipparcos parallaxes of O stars. *A&A*, 428, 149–157 {**212, 216**}

Scoville F, Mena-Werth J, 1998, Recalibration of the Wilson–Bappu effect using the singly-ionised Mg k line. *PASP*, 110, 794–803 {**221**}

Seggewiss W, 1998a, A revision of the Cepheid distance scale by Hipparcos? *Supernovae and Cosmology*, 101–106 {**228**}

—, 1998b, The distance scale of the universe before and after Hipparcos. *Applied and Computational Harmonic Analysis*, 3, 150–170 {**208**}

Siebert A, Bienaymé O, Soubiran C, 2003, Spectroscopic survey of red clump stars and the vertical shape of the Galactic potential. *EAS Publications Series 10* (eds. Boily CM, Pastsis P, Portegies Zwart S, *et al.*), 49–54 {**235**}

Smak JI, Preston GW, 1965, Kinematics of the Mira variables. *ApJ*, 142, 943–963 {**238**}

Smith H, 1987a, The calibration problem. III. First-order solution for mean absolute magnitude and dispersion. *A&A*, 181, 391–393 {**209, 210**}

—, 1987b, The calibration problem. II. Trigonometric parallaxes selected according to proper motion and the problem of statistical parallaxes. *A&A*, 171, 342–347 {**209**}

—, 1987c, The calibration problem. IV. The Lutz–Kelker correction. *A&A*, 188, 233–238 {**209, 210**}

—, 1987d, The calibration problem. I. Estimation of mean absolute magnitude using trigonometric parallaxes. *A&A*, 171, 336–347 {**209, 210**}

—, 1999, The Lutz–Kelker effect in the Hipparcos era and beyond. *Modern Astrometry and Astrodynamics* (eds. Dvorak R, Haupt HF, Wodnar K), 139–150 {**210**}

—, 2003, Is there really a Lutz–Kelker bias? Reconsidering calibration with trigonometric parallaxes. *MNRAS*, 338, 891–902 {**211**}

Smith H, Eichhorn H, 1996, On the estimation of distances from trigonometric parallaxes. *MNRAS*, 281, 211–218 {**208, 209**}

Smoker JV, Rolleston WRJ, Kay HRM, *et al.*, 2003, Ca II K interstellar observations towards early-type disk and halo stars. *MNRAS*, 346, 119–134 {**223**}

Solano E, 1998, The absolute magnitude of RR Lyrae stars: from Hipparcos parallaxes and proper motions. *Ap&SS*, 263, 219–222 {**249**}

Solano E, Barnes TG, 1999, The absolute magnitude of RR Lyrae stars: from Hipparcos parallaxes and proper motions. *ASP Conf. Ser. 167: Harmonizing Cosmic Distance Scales in a Post-Hipparcos Era*, 316–319 {**249**}

Solano E, Garrido R, Fernley J, *et al.*, 1997, Radial velocities and iron abundances of field RR Lyrae stars. *A&AS*, 125, 321–327 {**247**}

Sowell JR, Trippe M, Caballero-Nieves SM, *et al.*, 2007, HR diagrams based on the HD stars in the Michigan Spectral Catalogue and the Hipparcos Catalogue. *AJ*, 134, 1089–1102 {**214–216**}

Spaenhauer AM, 1978, A systematic comparison of four methods to derive stellar space densities. *A&A*, 65, 313–321 {**209**}

Spergel DN, Bean R, Doré O, *et al.*, 2007, Three-year Wilkinson Microwave Anisotropy Probe (WMAP) observations: implications for cosmology. *ApJS*, 170, 377–408 {**228**}

Stanek KZ, Garnavich PM, 1998, Distance to M31 with the Hubble Space Telescope and Hipparcos red clump stars. *ApJ*, 503, L131–L134 {**231, 232, 235, 256**}

Stanek KZ, Kaluzny J, Wysocka A, *et al.*, 2000, UBVI colour–magnitude diagrams in Baade's Window metallicity range, implications for the red clump method, colour anomaly and the distances to the Galactic centre and the Large Magellanic Cloud. *Acta Astronomica*, 50, 191–210 {**235**}

Stanek KZ, Zaritsky D, Harris J, 1998, A short distance to the Large Magellanic Cloud with the Hipparcos-calibrated red clump stars. *ApJ*, 500, L141–L144 {**231, 235**}

Stock J, Agostinho R, Rose JA, *et al.*, 1997, Towards a multidimensional stellar classification system. *Baltic Astronomy*, 6, 41–45 {**219**}

Stock J, Stock MJ, 1999, Quantitative stellar spectral classification. *Revista Mexicana de Astronomia y Astrofisica*, 35, 143–156 {**219**}

—, 2001, Quantitative stellar spectral classification. *Revista Mexicana de Astronomia y Astrofisica Conf. Ser.*, volume 11, 83–84 {**219**}

Stock MJ, Stock J, García J, *et al.*, 2002, Quantitative stellar spectral classification. II. Early-type stars. *Revista Mexicana de Astronomia y Astrofisica*, 38, 127–140 {**219**}

Storm J, 2006, How good are RR Lyrae and Cepheids really as distance indicators? . The observational approach. *Memorie della Societa Astronomica Italiana*, 77, 188–193 {**228**}

Straniero O, Chieffi A, Limongi M, 1997, Isochrones for H-burning globular cluster stars. III. From the Sun to the globular clusters. *ApJ*, 490, 425–436 {**249**}

Stratten FJM, 1925, *Astronomical Physics*. Methuen, London {**220**}

Strugnell P, Reid IN, Murray CA, 1986, The luminosity and kinematics of RR Lyrae stars in the solar neighbourhood. I. Statistical parallaxes of RR Lyraes. *MNRAS*, 220, 413–427 {**248**}

Struve O, 1928, Further work on interstellar calcium. *ApJ*, 67, 353–390 {**222**}

Suchkov AA, 2001, Evidence for prolonged main sequence stellar evolution of F stars in close binaries. *A&A*, 369, 554–560 {**219**}

Szabados L, 1997, Hipparcos parallaxes of the nearest Cepheids – implicative of binarity. *ESA SP–402: Hipparcos, Venice '97*, 657–660 {**226, 229**}

—, 1999, Importance of precise radial velocities for Cepheid binaries. *ASP Conf. Ser. 185: IAU Colloq. 170: Precise Stellar Radial Velocities*, 211–217 {**229**}

—, 2003, Database on binaries among Galactic Classical Cepheids. *Informational Bulletin on Variable Stars*, 5394, 1 {**229**}

Tammann GA, Sandage AR, Reindl B, 2003, New period–luminosity and period–colour relations of classical Cepheids: I. Cepheids in the Galaxy. *A&A*, 404, 423–448 {**227**}

Tanvir NR, 1999, Cepheid standard candles. *ASP Conf. Ser. 167: Harmonizing Cosmic Distance Scales in a Post-Hipparcos Era*, 84–100 {**225, 228**}

Teerikorpi P, 1997, Observational selection bias affecting the determination of the extragalactic distance scale. *ARA&A*, 35, 101–136 {**209**}

—, 1998, Cosmological Malmquist bias in the Hubble diagram at high redshifts. *A&A*, 339, 647–657 {**209**}

Testa V, Chieffi A, Limongi M, *et al.*, 2004, The distance to NGC 5904 (M5) via the subdwarf main sequence fitting method. *A&A*, 421, 603–612 {**240, 241, 244**}

Theureau G, 1998, Kinematics of the local universe. VI. *B*-band Tully–Fisher relation and mean surface brightness. *A&A*, 331, 1–10 {**256**}

Torres G, 2004, Combining astrometry and spectroscopy. *ASP Conf. Ser. 318: Spectroscopically and Spatially Resolving the Components of the Close Binary Stars* (eds. Hilditch RW, Hensberge H, Pavlovski K), 123–131 {**239**}

Trumpler RJ, Weaver HF, 1953, *Statistical Astronomy*. Dover, New York {**210**}

Tsujimoto T, Miyamoto M, Yoshii Y, 1997, The absolute magnitude of RR Lyrae stars. *ESA SP–402: Hipparcos, Venice '97*, 639–642 {**249**}

—, 1998, The absolute magnitude of RR Lyrae stars derived from the Hipparcos Catalogue. *ApJ*, 492, L79–L82 {**249**}

Tsujimoto T, Yamada Y, Gouda N, 2005a, Statistical calibrations of trigonometric parallaxes. *IAU Colloq. 196: Transits of Venus: New Views of the Solar System and Galaxy* (ed. Kurtz DW), 411–419 {**210**}

—, 2005b, Statistical methods for calibrating trigonometric parallaxes. *ESA SP–576: The Three-Dimensional Universe with Gaia* (eds. Turon C, O'Flaherty KS, Perryman MAC), 719–722 {**210**}

Tsujimoto T, Yoshii Y, 1999, The absolute magnitude of RR Lyrae stars derived from the Hipparcos Catalogue. *ASP Conf. Ser. 167: Harmonizing Cosmic Distance Scales in a Post-Hipparcos Era*, 332–335 {**249**}

Tully RB, Fisher J, 1977, A new method of determining distances to galaxies. *A&A*, 54, 661–673 {**256**}

Turner DG, Burke JF, 2002, The distance scale for classical Cepheid variables. *AJ*, 124, 2931–2942 {**228**}

Turon C, 1998, Hipparcos and primary distance scale indicators. *Highlights in Astronomy*, 11, 576–577 {**208**}

—, 1999, Luminosity calibration and distance scale. *Baltic Astronomy*, 8, 181–190 {**208**}

Turon C, Crézé M, 1977, On the statistical use of trigonometric parallaxes. *A&A*, 56, 273–281 {**210, 211**}

Twarog BA, Anthony-Twarog BJ, Bricker AR, 1999, Zeroing the stellar isochrone scale: the red giant clump luminosity at intermediate metallicity. *AJ*, 117, 1816–1826 {**234**}

Udalski A, 2000, The Optical Gravitational Lensing Experiment: red clump stars as a distance indicator. *ApJ*, 531, L25–L28 {**233, 235**}

Udalski A, Soszynski I, Szymanski M, *et al.*, 1999a, The Optical Gravitational Lensing Experiment. Cepheids in the Magellanic Clouds. IV. Catalogue of Cepheids from the Large Magellanic Cloud. *Acta Astronomica*, 49, 223–317 {**227**}

—, 1999b, The Optical Gravitational Lensing Experiment. Cepheids in the Magellanic Clouds. V. Catalogue of

Cepheids from the Small Magellanic Cloud. *Acta Astronomica*, 49, 437–520 {**227**}

Udalski A, Szymanski M, Kaluzny J, *et al.*, 1993, The Optical Gravitational Lensing Experiment: colour–magnitude diagrams of the Galactic bulge. *Acta Astronomica*, 43, 69–90 {**235**}

Udalski A, Szymanski M, Kubiak M, *et al.*, 1998, Optical Gravitational Lensing Experiment. Distance to the Magellanic Clouds with the red clump stars: are the Magellanic Clouds 15% closer than generally accepted? *Acta Astronomica*, 48, 1–17 {**232, 235**}

Vallenari A, Chiosi C, Bertelli G, *et al.*, 1996, Star formation in the Large Magellanic Cloud. III. A study of the regions LMC 30, LMC 45 and LMC 61. *A&A*, 309, 367–374 {**233**}

Vanbeveren D, De Loore C, Van Rensbergen W, 1998, Massive stars. *A&A Rev.*, 9, 63–152 {**215**}

VandenBerg DA, 1999, The age structure of the halo. *ASP Conf. Ser. 165: The Third Stromlo Symposium: The Galactic Halo* (eds. Gibson BK, Axelrod RS, Putman ME), 46–58 {**246**}

—, 2000, Models for old, metal-poor stars with enhanced α-element abundances. II. Their implications for the ages of the Galaxy's globular clusters and field halo stars. *ApJS*, 129, 315–352 {**240, 241, 243, 244**}

VandenBerg DA, Stetson PB, Bolte M, 1996, The age of the Galactic globular cluster system. *ARA&A*, 34, 461–510 {**240**}

VandenBerg DA, Swenson FJ, Rogers FJ, *et al.*, 2000, Models for old, metal-poor stars with enhanced α-element abundances. I. Evolutionary tracks and ZAHB loci; observational constraints. *ApJ*, 532, 430–452 {**243, 244, 249**}

van der Marel RP, Alves DR, Hardy E, *et al.*, 2002, New understanding of Large Magellanic Cloud structure, dynamics, and orbit from carbon star kinematics. *AJ*, 124, 2639–2663 {**255**}

van Herk G, 1965, Proper motions, mean parallaxes and space velocities of RR Lyrae variables. *Bull. Astron. Inst. Netherlands*, 18, 71–105 {**248**}

van Leeuwen F, 2007, *Hipparcos, the New Reduction of the Raw Data*. Springer, Dordrecht {**227**}

van Leeuwen F, Evans DW, 1998, On the use of the Hipparcos intermediate astrometric data. *A&AS*, 130, 157–172 {**255**}

van Leeuwen F, Feast MW, Whitelock PA, *et al.*, 1997, First results from Hipparcos trigonometrical parallaxes of Mira-type variables. *MNRAS*, 287, 955–960 {**236, 237, 254**}

—, 2007, Cepheid parallaxes and the Hubble constant. *MNRAS*, 379, 723–737 {**210, 225–229, 254**}

Vassiliadis E, Wood PR, 1993, Evolution of low- and intermediate-mass stars to the end of the asymptotic giant branch with mass loss. *ApJ*, 413, 641–657 {**236**}

Walker AR, 1998, The Cepheid period–luminosity zero-point via Cepheids in Galactic open clusters. *Memorie della Societa Astronomica Italiana*, 69, 71–77 {**228**}

—, 1999, The distances of the Magellanic Clouds. *ASSL Vol. 237: Post-Hipparcos Cosmic Candles*, 125–144 {**208, 253**}

Wallerstein G, 1967, On the use of small parallaxes for calibration purposes. *PASP*, 79, 317–321 {**210**}

Wallerstein G, Machado-Pelaez L, Gonzalez G, 1999, The Ca II–M_V correlation (Wilson–Bappu effect) calibrated by Hipparcos parallaxes. *PASP*, 111, 335–341 {**221**}

Wallerstein G, Silk J, 1971, Interstellar gas in the direction of the Vela pulsar. *ApJ*, 170, 289–296 {**258**}

Wang J, 1999, Post-Hipparcos Galactic distance scales. *Progress in Astronomy*, 17, 159–167 {**208**}

Wegner W, 2000, Absolute magnitudes of OB and Be stars based on Hipparcos parallaxes. I. *MNRAS*, 319, 771–776 {**216, 218**}

—, 2006, Absolute magnitudes of OB and Be stars based on Hipparcos parallaxes. II. *MNRAS*, 371, 185–192 {**217**}

—, 2007, Absolute magnitudes of OB and Be stars based on Hipparcos parallaxes. III. *MNRAS*, 374, 1549–1556 {**217**}

Wesselink AJ, 1969, Surface brightnesses in the UBV system with applications of M_V and dimensions of stars. *MNRAS*, 144, 297–311 {**228**}

West RM, 1969, On the calibration of $M_V(K)$ for giants by means of trigonometric parallaxes. *A&A*, 3, 1–4 {**210**}

Whitelock PA, 1986, The pulsation mode and period–luminosity relationship of cool variables in globular clusters. *MNRAS*, 219, 525–536 {**236**}

Whitelock PA, Feast MW, 1993, Planetary nebulae from Miras? *Planetary Nebulae* (eds. Weinberger R, Acker A), IAU Symp. 155, 251–258 {**236**}

—, 2000, Hipparcos parallaxes for Mira-like long-period variables. *MNRAS*, 319, 759–770 {**211, 238**}

Whitelock PA, Marang F, Feast MW, 2000, Infrared colours for Mira-like long-period variables found in the Hipparcos Catalogue. *MNRAS*, 319, 728–758 {**237, 238, 257**}

Whitelock PA, van Leeuwen F, Feast MW, 1997, The luminosities and diameters of Mira variables from Hipparcos parallaxes. *ESA SP–402: Hipparcos, Venice '97*, 213–218 {**236**}

Wielen R, Fuchs B, Dettbarn C, *et al.*, 1997, Kinematics of RR Lyrae stars based on Hipparcos proper motions. *ESA SP–402: Hipparcos, Venice '97*, 599–602 {**250, 251**}

Wielen R, Jahreiß H, Dettbarn C, *et al.*, 2000, Polaris: astrometric orbit, position, and proper motion. *A&A*, 360, 399–410 {**229**}

Wilson OC, 1959, Accuracy of absolute magnitudes derived from widths of H and K emission components. *ApJ*, 130, 499–506 {**221**}

Wilson OC, Bappu MKV, 1957, H and K emission in late-type stars: dependence of line width on luminosity and related topics. *ApJ*, 125, 661–684 {**220, 221**}

Wilson OC, Merrill PW, 1937, Analysis of the intensities of the interstellar D lines. *ApJ*, 86, 44–69 {**222**}

Wood PR, Sebo KM, 1996, On the pulsation mode of Mira variables: evidence from the Large Magellanic Cloud. *MNRAS*, 282, 958–964 {**238**}

Yeşilyaprak C, Aslan Z, 2004, period–luminosity relation for M-type semi-regular variables from Hipparcos parallaxes. *MNRAS*, 355, 601–607 {**238**}

Yim HS, Chun MS, Byun YI, *et al.*, 2004, BV CCD photometry of M71: distance and age. *Journal of Astronomy and Space Sciences*, 21, 1–10 {**246**}

Yoshida S, Kadota K, 2000, Supernova 1999gh in NGC 2986. *IAU Circ.*, 7341, 4 {**259**}

Zaritsky D, Lin DNC, 1997, Evidence for an intervening stellar population toward the Large Magellanic Cloud. *AJ*, 114, 2545–2555 {**232**}

Zhao G, Qiu HM, Mao S, 2001, High-resolution spectroscopic observations of Hipparcos red clump giants: metallicity and mass determinations. *ApJ*, 551, L85–L88 {**234**}

Zhu Z, 1999, Galactic kinematics derived from classical Cepheids. *Chinese Astronomy and Astrophysics*, 23, 445–453 {**230**}

—, 2000, Local kinematics of classical Cepheids. *Ap&SS*, 271, 353–363 {**230**}

—, 2003, Distances and kinematics of classical Cepheids. *ASSL Vol. 298: Stellar Astrophysics: A Tribute to Helmut A. Abt* (eds. Cheng KS, Leung KC, Li TP), 221–227 {**228**}

Zinn R, West MJ, 1984, The globular cluster system of the Galaxy. III. Measurements of radial velocity and metallicity for 60 clusters and a compilation of metallicities for 121 clusters. *ApJS*, 55, 45–66 {**242**}

Zoccali M, Ortolani S, Renzini A, *et al.*, 1999, The distance of the globular cluster 47 Tucanae via the white dwarf cooling sequence. *ASP Conf. Ser. 167: Harmonizing Cosmic Distance Scales in a Post-Hipparcos Era* (eds. Egret D, Heck A), 336–339 {**252**}

Zoccali M, Renzini A, Ortolani S, *et al.*, 2001, The white dwarf distance to the globular cluster 47 Tucanae and its age. *ApJ*, 553, 733–743 {**252**}

6

Open clusters, groups and associations

6.1 Introduction

This chapter considers the Hipparcos contributions to the study of the Population I open clusters (in contrast with globular clusters, treated in Section 9.11), dynamical streams or moving groups, and associations. It should be recognised at the outset that the terms are potentially confusing, lack unambiguous definitions, and are occasionally used interchangeably. Broadly, clusters are reasonably dense concentrations of older stars (typically $\gtrsim$ 50–100 Myr), while associations are more extended collections of younger and more massive stars (typically $\lesssim$ 25–50 Myr). More precise definitions are given in the box on page 274.

The literature is even more confusing when it comes to the definition of a 'moving group'. The term is sometimes simply used to denote any system of stars sharing the same space motion, and thus to implicitly include open clusters, globular clusters, and associations. Even in a more restricted sense, excluding these tighter groupings, it has become clear from the Hipparcos results that the term has been used in the past to describe kinematic groups with different origins: thus 'old moving groups' (or large-scale stellar or dynamical streams, or 'superclusters') appear to include evaporating halos of open clusters, along with resonant dynamical structures in the solar neighbourhood, and systematic velocity structures perhaps imparted by spiral arm shocks. These are treated in Section 6.9. Young moving groups, with ages $\lesssim$ 25–50 Myr, appear to represent the sparser and more immediate dissipation products of the more youthful associations, and are treated in Section 6.10.

All stars are considered to have been born in dense gas clouds, and clusters and associations appear to form from massive dense cores within molecular clouds. Since it is generally considered that stars in a given cluster started their lives with a similar initial chemical composition, and at the same time, clusters also provide powerful tests for stellar evolution models.

More than 1500 open clusters are known in the Galaxy. The catalogue of Dias *et al.* (2002a) updates the previous catalogues of Lyngå (1988) and Mermilliod (1995), and comprises 1537 clusters including, for the first time, kinematic data. Dias *et al.* (2002a) already included the first of the Hipparcos discoveries: 12 from the Hipparcos Catalogue by Platais *et al.* (1998), and three probable loose clusters by Chereul *et al.* (1999), and kinematics characteristics from Hipparcos and Tycho were included for many more. Other clusters have since been discovered, including others from the Hipparcos and Tycho 2 Catalogues, detailed below, and from elsewhere (Dias *et al.*, 2006). Kharchenko *et al.* (2004, 2005c) provided a supplement to the ASCC 2.5 Catalogue (a catalogue itself based on Tycho 2), with membership probabilities for 520 Galactic open clusters for which at least three probable members (18 on average) could be identified. The compilation includes parallaxes where available, and proper motions for all objects in the Hipparcos system. A further 109 new clusters discovered from the Tycho 2 Catalogue were subsequently reported by Kharchenko *et al.* (2005a), revised to 130 new clusters by Piskunov *et al.* (2006b).

Extrapolation from the local region suggests that some 10^5 open clusters might exist throughout the entire Galaxy. Each may comprise a range of masses from the brown dwarf limit at about $0.08 M_\odot$ or below, up to about $80 M_\odot$. Since stars in any given cluster are at approximately the same distance from the Sun, their absolute magnitudes correspond to their apparent magnitudes apart from an unknown zero-point; this characteristic underpins the method of distance determination by main-sequence fitting. In this approach, a nearby main-sequence template (invariably the Hyades) is scaled to the observed main sequence of a more

Open clusters and associations: Lada & Lada (1991) define an open cluster or association empirically as a group of stars of the same physical type whose surface density significantly exceeds that of the field for stars of the same physical type.

More precisely, they consider clusters to be physically related groups of 10 or more stars whose observed stellar mass volume density would be sufficiently large, if in a state of internal virial equilibrium, to render the group stable against tidal disruption by the Galaxy ($\sim 0.1 M_\odot$ pc^{-3}; Bok, 1934) and, more stringently, by passing interstellar clouds ($\sim 1 M_\odot$ pc^{-3}; Spitzer, 1958).

They consider an association to be a loose group of 10 or more physically related stars whose stellar space density is considerably below the tidal stability limit of $\sim 1 M_\odot$ pc^{-3} (Blaauw, 1964).

Clusters with 100 or more members are designated as rich, those with less as poor. Embedded clusters are those fully or partially embedded in interstellar gas and dust (such as the Trapezium and ρ Ophiuchi), while exposed clusters have little or no associated interstellar material.

A typical cluster contain a few tens to several thousands stars, located within a typical radius of order 1–10 pc. Constituent stars progressively dissolve back into the field over time through a variety of mechanisms, notably gravitational perturbations from the disk or passing interstellar gas clouds, and supplemented by cluster ejection through gravitational encounters within the cluster. Typical lifetimes are of order 100 Myr, with the least tightly bound surviving for only a few million years, and the richest for as much as $\sim$ 1 Gyr.

This basic definition of clusters and associations embraces groups which may either be gravitationally bound (i.e. with negative total energy, including any contributions from interstellar material) or unbound. Thus, unbound clusters may have space densities exceeding $\sim 1 M_\odot$ pc^{-3}, but with large internal motions. These criteria imply, however, that the Hyades cluster, which has survived as a recognisable stellar group for some 700 Myr, but which has a present space density significantly lower than $1 M_\odot$ pc^{-3}, should therefore be classified as an association.

In practice, OB associations are characterised by low stellar densities and a large spatial extent, and survive as a recognisable group only for a short time, of order 25 Myr (T associations, concentrations of T Tauri stars, rarely include the more massive stars, and are typically found closer to star-forming regions). They are made recognisable not necessarily by their general overdensity with respect to field stars, but by their overdensity of luminous O and B stars. Their large masses and high luminosities imply that the constituent stars are young and short lived, and are therefore associated with sites of recent star formation. These characteristics do not imply that an association must have started life as a very loose grouping: Brown *et al.* (1997) showed that a group of 500 stars, with a total mass of $1000 M_\odot$ and a power-law initial mass function, initially distributed in a sphere of radius 0.5 pc, and a one-dimensional velocity dispersion of only 2 km s^{-1}, will be gravitationally unbound, and will expand to the dimensions of a typical OB association within 5–10 Myr. OB associations may contain later-type members, although these may be less-easily discernable.

remote cluster to determine its distance. This works well to first order, but there are many complications, including the effects of helium and metal composition, surface abundance anomalies, binarity, rotation, interstellar and circumstellar extinction, and cluster age, all of which make the method delicate to apply if the highest accuracies are targeted.

Membership determination, and accurate distance determination to individual objects, especially in the case of the Hyades and Pleiades, prove to be crucial in determining an accurate absolute magnitude for each star, so as to place them accurately in the HR diagram, and thereafter to compare their positions with those given by theoretical stellar evolutionary models. As will be seen in this chapter and in Chapter 7, this leads to remarkable possibilities for characterising, under certain assumptions, the cluster's age, the masses and radii of individual members, binary star parameters, the initial chemical composition of the cluster, and details of star formation and subsequent stellar and dynamical evolution. The results provide powerful diagnostics of stellar structure and evolution, including the degree of convective overshooting, insight into differential stellar rotation, internal gravitational settling, and constraints on models of late stages of stellar evolution including stars on the subgiant, red giant, and asymptotic giant branches.

This chapter first describes detection methods, largely in common for the open clusters, moving groups, and OB associations. Detection methods and membership determination have been greatly assisted by the availability of the Hipparcos and Tycho Catalogues, permitting identification of these systems based on similar distances, and common space motions. The subsequent sections present more detailed Hipparcos-based results on the nearest and richest open clusters, the Hyades and Pleiades, followed by results on other open clusters and their disruption products (the old moving groups), and OB associations and their disruption products (the young moving groups).

6.2 Detection methods

6.2.1 General considerations

With accurate distance information, compact open clusters could, in principle, be identified purely on the basis of their localisation in three-dimensional space. Spectroscopy, radial velocities, and proper motions would assist identification of possible unrelated interlopers, and would assist in identifying cluster members in progressive states of evaporation from the cluster as they disperse into the surrounding field star population. There are, however, only two reasonably rich open clusters within the relevant distance horizon of the Hipparcos parallax measurements: the Hyades at around 45 pc, and the Pleiades at around 120 pc. However, the 1σ standard errors in the Hipparcos parallaxes of $\sim$1 mas already lead to a 10% distance uncertainty on individual stars at the distance of the Pleiades. The Hipparcos proper motions, as well as radial velocities, therefore further contribute to cluster membership determination through the identification of common space motions. At the same time, the proper motions provide information on cluster dynamics and evaporation.

From simple dynamical arguments it can be shown that, for an open cluster of a few hundred stars within a volume of a few pc in radius, moving together under their mutual gravitation, the internal velocity dispersion is of the order of 1 km s^{-1} or less, and thus small compared with the typical linear velocity of the cluster as a whole relative to the Sun. The Hyades cluster, for example, has a space motion of $\sim$45 km s^{-1} with respect to the Sun, reflecting the velocity of the cloud in which the cluster formed.

Provided that the cluster is sufficiently nearby to extend over an area of, say, several degrees, the parallel motions of the stars in space yield, on the celestial sphere, directions of proper motions that appear to converge, ultimately to a unique point called the convergent-point (Figure 6.1). The cluster distance can then be determined using the well-established 'moving cluster' or 'convergent-point' method. Although the two terms are frequently used synonymously (as here), they may be formally distinguished according to whether only the gradient of the proper motion components across the cluster is measured, or whether the actual convergent-point coordinates are explicitly estimated from the directions of the proper motion vectors.

If $\mathbf{v}$ is the linear velocity of the cluster as a whole relative to the Sun, $\mathbf{b}$ the barycentric coordinate vector to a cluster member and $\langle\mathbf{b}\rangle$ its coordinate direction, and $\langle\mathbf{v}\rangle$ the unit vector to the convergent-point then, neglecting the internal velocity dispersion, the radial velocity is given by the scalar product $\rho = \langle\mathbf{b}\rangle'\mathbf{v}$. With λ denoting the angular distance between the star and the convergent-point, and μ the proper motion vector

$$\langle\mathbf{b}\rangle'\langle\mathbf{v}\rangle = \cos\lambda \tag{6.1}$$

and

$$|\mu| = \pi\,|\mathbf{v}|\sin\lambda/A_v \tag{6.2}$$

where π is the parallax of the cluster member, and $A_v = 4.740\,47$ km yr s^{-1} is the astronomical unit expressed in the appropriate form when π and μ are expressed in mas and mas yr^{-1} respectively. For the Hyades, $\lambda \simeq 33°$, the radial velocity $v_{\rm rad} \simeq 40$ km s^{-1} in the cluster centre, $|\mathbf{v}| \simeq 45$ km s^{-1}, and $|\mu| \sim 100$ mas yr^{-1}. Although $\mathbf{v}$ can in principle be determined from the radial velocity measurements alone, its resulting direction is generally not well determined because of the limited angular extent of the cluster; the usual procedure has therefore been to determine $\langle\mathbf{v}\rangle$ from proper motions, and $|\mathbf{v}|$ from radial velocities, from which λ is obtained from Equation 6.1 and π from Equation 6.2.

Although the method is conceptually simple, its application in practice is not so straightforward. Errors in the individual proper motions resulting from measurement errors, or defects in the proper motion system, lead to accidental errors in π, to an error in $\langle\mathbf{v}\rangle$ and, ultimately, to a systematic bias in λ depending on $\langle\mathbf{b}\rangle$.

6.2.2 Convergent-point method

Classical approach The convergent-point (or moving cluster) method has been implemented in practice in slightly different ways by different workers. The approach starts with a set of stars j at positions $(l, b)_j$, with proper motions $(\mu_l \cos b,\ \mu_b)_j$, and errors $(\sigma_{\mu_l \cos b}, \sigma_{\mu_b})_j$. The first step is to discard stars with insignificant proper motions, i.e. with

$$t = \frac{\mu}{\sigma_\mu} = \frac{\sqrt{\mu_l^2\cos^2 b + \mu_b^2}}{\sqrt{\sigma^2_{\mu_l \cos b} + \sigma^2_{\mu_b}}} \le t_{\rm min} \tag{6.3}$$

where $t_{\rm min}$ is typically chosen to be equal to 3–5 (e.g. Jones, 1971). The next step is to search for the maximum-likelihood coordinates $(l, b)_{\rm cp}$ of the convergent point by minimising

$$\chi^2 = \sum_{j=1}^{N} t^2_{\perp j} \tag{6.4}$$

where N is the number of stars, and $t_{\perp j}$ is the value for star j of the quantity $t_\perp$, defined as

$$t_\perp = \mu_\perp/\sigma_\perp \tag{6.5}$$

where $\mu_\perp$ is the component of the proper motion perpendicular to the direction towards the convergent-point, and $\sigma_\perp$ is its measurement error.

In case of a common space motion, the proper motion vectors will be directed towards the convergent-point, so that for all moving group members the expectation value for $\mu_\perp$ equals 0. The sum in Equation 6.4 is distributed as χ^2 with $N - 2$ degrees of freedom. If, after minimization with respect to $(l, b)_{\rm cp}$, the value of χ^2 is unacceptably high, the star with the highest value of $\mu_\perp/\sigma_\perp$ is rejected, after which minimization is repeated until a satisfactory value of χ^2 is obtained. This procedure allows for simultaneous convergent-point determination and member selection, and has been

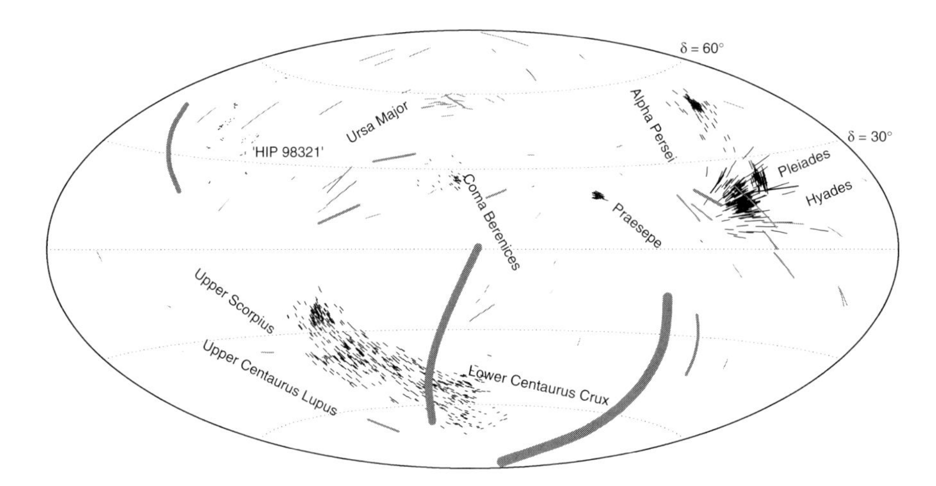

Figure 6.1 Proper motions of Hipparcos stars in some of the observed clusters and associations, over a period of 200 000 yr, showing their associated convergent-point motions. Best convergent-point accuracy is obtained in rich nearby clusters with large angular extent and large proper motion. The accuracy in the largest associations (Ursa Major, Scorpius–Centaurus) is limited by their partly unknown expansion. Stellar paths in the Ursa Major group (shaded) cover large areas of the sky. The thickness of the proper motion vectors is inversely proportional to distance: the closest star is Sirius and the two next are faint red dwarfs. Proper motions vary greatly among different clusters. From Madsen et al. (2002, Figure 2).

applied successfully to, e.g. the Hyades (van Bueren, 1952; Perryman *et al.*, 1998), and Scorpius OB2 (Jones, 1971).

Recent modifications De Bruijne (1999a) modified this implementation in three ways. First, unlike previous astrometric catalogues, the Hipparcos Catalogue gives the full covariance matrix for the measured astrometric parameters, so that it is possible to include the full error propagation. Second, even for infinitely accurate measurements, moving group members would not have proper motions directed exactly towards the convergent-point because of the intrinsic velocity dispersion in the group. As a result, selecting only stars with $t_\perp = 0$ will not identify all members. For this reason, he adapted the definition of $t_\perp$ to

$$t_\perp = \frac{\mu_\perp}{\sqrt{{\sigma_\perp}^2 + {\sigma_{\rm int}^\star}^2}} \tag{6.6}$$

where $\sigma_{\rm int}^\star$ is an estimate of the intrinsic one-dimensional velocity dispersion in the group, expressed in proper motion units. Accordingly, the definition of t is changed to

$$t = \frac{\mu}{\sqrt{{\sigma_\mu}^2 + {\sigma_{\rm int}^\star}^2}} \tag{6.7}$$

Third, the classical approach of finding the convergent-point by evaluating the sum given by Equation 6.4 on a grid, is replaced by a global direct minimization. De Bruijne (1999a) defined a membership probability $p_{\rm cp}$ as $p_{\rm cp} = 1 - p$, where p is the probability that the perpendicular component of the proper motion has a value different from zero, given the covariance matrix. He showed that

$$p_{\rm cp} = \exp\,(-\,t_\perp^2/2) \tag{6.8}$$

The above procedure is, however, biased towards inclusion of stars at larger distances. It selects members based on the absolute value of $t_\perp$, which is a function of distance: nearby stars generally have large proper motions, and are more likely to be rejected than those at larger distances, which generally have smaller proper motions. For this reason de Zeeuw *et al.* (1999), in their analysis of OB associations, combined the convergent-point method with the independent 'spaghetti' method, demanding that both criteria are satisfied for the object to be considered as a group member.

6.2.3 Other search methods

The spaghetti method This kinematic membership selection method was developed by Hoogerwerf & Aguilar (1999), and uses parallax information in addition to positions and proper motions, typically (but not necessarily) in the absence of radial velocities. The main steps are as follows.

The Hipparcos measurements constrain a star to lie on a straight line in velocity space: the proper motion and parallax determine the offset from the origin (tangential velocity), while the sky position of the star determines its direction (Figure 6.2). The covariance matrix of the astrometric parameters transforms this line into a probability distribution in velocity space. Surfaces of equal probability are elliptic cylinders, denoted as

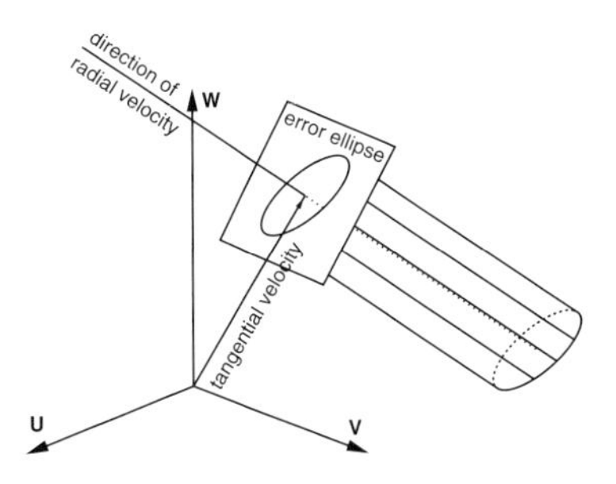

Figure 6.2 Schematic of the 'spaghetti method' for the identification of moving groups introduced by Hoogerwerf & Aguilar (1999). The five astrometric parameters of a star measured by Hipparcos define an elliptical cylinder in velocity space. The offset from the origin is determined by the tangential velocity; the orientation is set by the direction of the radial velocity. The finite thickness of the cylinder is caused by the measurement errors. All cylinders of stars in a moving group intersect. UVW are the velocity components in standard Galactic Cartesian coordinates. From Hoogerwerf & Aguilar (1999, Figure 1).

'spaghetti'. The cylinders of a set of stars with the same space motion all intersect in one point. Thus, Hoogerwerf & Aguilar (1999) identify moving groups by searching for maxima in the density of cylinders in velocity space. This 'spaghetti density' in velocity space is the sum over all stars in the sample of the individual probability distributions. Measurement of the stellar radial velocities would further reduce each elliptic cylinder to an ellipsoidal probability distribution in velocity space.

Even in the case of infinitely accurate astrometric measurements, moving group members would not necessarily have cylinders coinciding exactly at the space motion of the group because of the intrinsic velocity dispersion in the group. This broadens the associated peak in velocity space. Measurement errors broaden it further. Hoogerwerf & Aguilar (1999) therefore placed a sphere at the position of this peak, with a radius $\sigma_{\rm sp}$ given by

$$\sigma_{\rm sp}^2 = \sigma_{\rm med}^2 + \sigma_{\rm int}^2 \qquad (6.9)$$

where $\sigma_{\rm med}$ is the typical error in tangential velocity, taken as the median value of the semi-major axis of the 1σ cylinders of all stars in the range of distances where moving group members can be expected, and $\sigma_{\rm int}$ is an estimate of the group's one-dimensional velocity dispersion. At a given distance D in kpc, $\sigma_{\rm int}$ and $\sigma_{\rm int}^\star$ (Equation 6.6) are related through $\sigma_{\rm int} = A\sigma_{\rm int}^\star D$, where $A = 4.740\,47\,{\rm km\,yr\,s^{-1}}$ and $\sigma_{\rm int}$ is in units of ${\rm mas\,yr^{-1}}$. They computed for each star the integral of its probability distribution over three-dimensional velocity space restricted to the volume of the sphere with radius $\sigma_{\rm sp}$ defined in Equation 6.9, accepting those with values larger than 0.1 as members, based on Monte Carlo simulations.

The method was tested by Hoogerwerf & Aguilar (1999) on the Hyades and IC 2602 open clusters, giving good agreement in identified members with the results of Perryman *et al.* (1998) for the Hyades, and of Whiteoak (1961) for IC 2602.

Global convergence mapping Makarov & Urban (2000) introduced the technique of global convergence mapping, suitable for a general kinematic analysis of a sufficiently large sample of a few thousand or more stars, to search for moving groups and clusters in the absence of accurate radial velocity and distance measurements. Each pair of stars generates a convergent-point on the celestial sphere defined by their respective proper motion vectors. By extension, each position on the celestial sphere can be associated with an intensity of convergence, given by the total number of proper motion vectors intersecting the point, with smoothing introduced to account for measurement errors. The general velocity dispersion ellipsoid of the old stellar population, differential Galactic rotation, and the reflex solar motion, all generate a smooth background in the convergence map; for example, the apex solar motion will generate a bias in the general velocity field towards the antapex. These terms are removed by a spherical-harmonic approximation providing a smooth fit to the general background. Figure 6.3 gives examples of such convergence maps based on two different Hipparcos datasets. Application of the method to the discovery of a moving group of young stars in Carina–Vela is detailed in Section 6.10.

Epicycle correction The kinematic evolution of gravitationally unbound associations moving in the Galactic potential in the solar vicinity can be described by epicycle theory (Section 9.1.3). Blaauw (1952) used this approach to describe the effects of the Galactic potential on the trajectories of association members with initial systemic velocity equal to that of the local standard of rest. He showed that such an association stretches out quickly along the direction of Galactic rotation owing to the initial velocity dispersion. More recently, epicycle theory, and integration of the equations of motion using a realistic Galaxy gravitational potential, were used by Asiain *et al.* (1999b) to study the kinematic evolution of unbound moving groups observed by Hipparcos.

Makarov *et al.* (2004) applied the theory to a modelled association undergoing uniform expansion, showing that in about 10 Myr the shearing motions due to the Galactic force field in the planar and in the vertical directions destroys the original pattern of individual motions. As a result, the convergent-point of proper motions is substantially blurred. The same effects significantly complicate efforts to determine the past and future trajectories of clusters and associations. In line

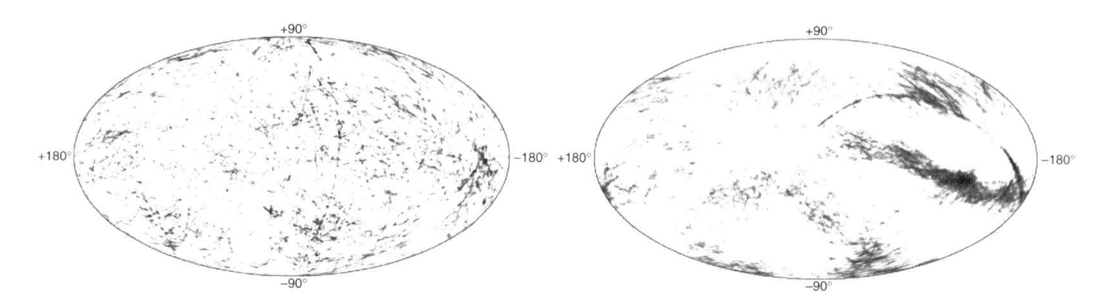

Figure 6.3 Global convergence maps, presented as Aitoff projections in Galactic coordinates. The reflex solar motion and other systematic effects have been subtracted by spherical harmonic background fits. Left: as derived from 1456 main-sequence K dwarfs in the Hipparcos Catalogue with significant proper motions, $\sigma_\mu/\mu < 0.25$. The sample represents a rather old and dynamically well-relaxed population. Sharp features correspond to the convergent-points of the Hyades and other nearby open clusters, and possibly some smaller moving groups. Right: as derived from 6334 ROSAT All-Sky Survey Bright Star Catalogue/Tycho 2 Catalogue stars with significant proper motions in Tycho 2, $\sigma_\mu/\mu < 0.25$. The extended narrow arc is produced by the Hyades cluster, elongated because of the large distance between the cluster and the convergent-point. The broad strip feature tilted by 20° to the Galactic plane is produced by the nearby Gould Belt (see Section 6.10.5). All the shaded areas should be statistically significant. From Makarov & Urban (2000, Figures 1 and 2).

with previous findings, they also predicted flattened shapes for older associations that sometimes degenerate into string- and sheet-like forms. They demonstrate how to restore the convergent properties of proper motions of an association of known age by rewriting and then inverting the resulting system of equations.

Their work quantifies why presently-known associations are all quite young, typically <30 Myr: it is simply more difficult to discover an older association, stretched in one dimension, and with a significant non-converging component in the velocity vectors. The approach may allow a systematic search for new associations using available catalogues with proper motions for millions of stars, although the corrections that have to be added to the observed proper motions to recover the convergent-point are age-dependent. A large nearby association with well-established membership could also be used to derive the local values of Oort's constants, and the vertical frequency ν tied to the local gravitational force K_z. If most stars are born in clusters and associations, the technique might also allow for a dynamical age determination of individual stars. Although the methodology has only recently been developed, and has not so far been widely applied, it represents an interesting possibility for future studies.

Orbital backtracking Connected with these detection methods is the technique of tracing back the orbits of candidate cluster members, and either confirming their physical association from the smaller spatial volume they occupied in the past, or matching the resulting intersection with the possible sites of original star formation. Orbital reconstruction by numerical integration was used by Hoogerwerf *et al.* (2001) to identify the origin of runaway stars (Section 8.4.2). Similar techniques have been used to investigate the origins of various clusters and associations discussed later in this chapter, including studies of the Pleiades moving group (Asiain *et al.*, 1999b); the β Pictoris moving group (Ortega *et al.*, 2002); the η Cha and ϵ Cha clusters (Jilinski *et al.*, 2005); the TW Hya association (de la Reza *et al.*, 2006); and various other young associations (Makarov, 2007b).

Other methods Other search techniques for identifying clusters or moving groups in the Hipparcos and Tycho Catalogues are noted here, with the associated results given in subsequent sections.

Chereul *et al.* (1997, 1998, 1999) and Figueras *et al.* (1997) used wavelet transform techniques to detect clustering in the spatial and velocity distributions in large Hipparcos samples, successfully confirming known moving groups and identifying new ones.

Platais *et al.* (1998) divided the sky into $2^\circ \times 2^\circ$ tiles, an angular scale selected to cover a cluster's core beyond $\sim$100 pc. They trimmed the catalogue entries according to certain distance, proper motion and standard error limits, then constructed the proper motion vector-point diagram to search for proper motion clusterings. The mean parallax of the candidate clusters was constructed, outliers rejected, a new scale size established from the approximate cluster distance assuming a tidal radius of 5 pc, and a new vector-point diagram constructed. Reality of 102 resulting candidate clusters was established by inspecting the colour–magnitude diagram using BV photometry from the Hipparcos Catalogue and compared with a ZAMS. The Hipparcos data were then augmented by stars from the ACT Reference Catalogue, and spectral types listed in the Hipparcos Catalogue used to verify the consistency of the colour–magnitude diagram. Mean parallaxes and proper motions were constructed from the Hipparcos Intermediate Astrometric Data using the method described by van Leeuwen & Evans (1998).

Kharchenko *et al.* (2005a) found a further 109 new clusters in the Tycho 2 Catalogue, which are on average brighter, with more members, but covering larger angular radii than those previously known. Their approach was based on selecting bright stars considered as possible 'seeds' representing new possible clusters, searching for nearby objects with similar proper motions, and confirming the reality of the cluster through visual inspection of the Digitised Sky Survey (DSS).

Vieira & Abad (2003), Abad *et al.* (2003), and Abad & Vieira (2005) described their development of Herschel's method (Trumpler & Weaver, 1953), of associating each star's proper motion with the pole of the associated great circle, and searching for clusterings of the pole positions.

Other references to cluster or moving group searches using the Hipparcos data include: improved methods for identifying moving groups (de Bruijne *et al.*, 1997); a search for moving groups in the Galactic halo (Aguilar & Hoogerwerf, 1999); searching for moving groups (Aguilar & Hoogerwerf, 2001); searches for star groups in the solar neighbourhood (Kazakevich & Orlov, 2002); and searches for stellar groups (Kazakevich *et al.*, 2003).

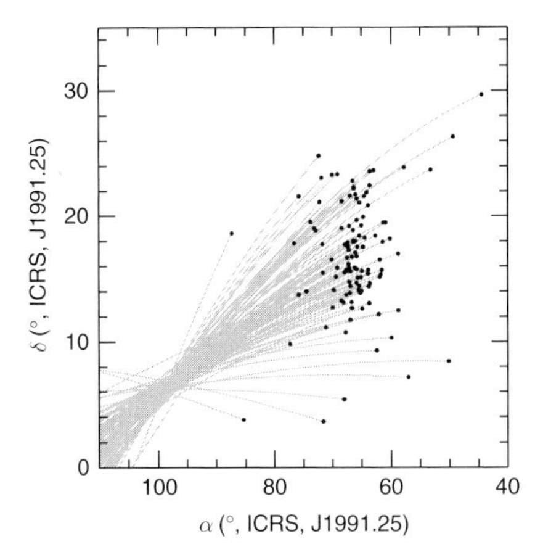

Figure 6.4 Hyades members selected from the Hipparcos data using the convergent-point method, showing positions (solid circles) and motions of the selected stars on their great circles (lines). From Perryman et al. (1998, Figure 3).

6.3 The Hyades

6.3.1 Introduction

The importance of the Hyades cluster in studies of Galactic structure, in the understanding of the chemical evolution of the Galaxy, and in the determination of the Population I distance scale, is well documented in the literature (a fairly complete review is given by Perryman *et al.*, 1998). The nearest moderately rich cluster, it has some 300–400 known members, a total mass of some 300–400 $M_\odot$, and an age of around 600–800 Myr. It has an extension in the sky of about 20°, and is a structure discernable to the naked eye from a dark location.

Although the pre-Hipparcos uncertainty in the distances of individual members limited the precise definition of the cluster's main sequence, and thereby knowledge of its helium content and corresponding evolutionary sequence, the Hyades cluster has nevertheless been used as the observational basis for several fundamental relationships in astrophysics, including the location of the main sequence in the HR diagram and the mass–luminosity relationship, as well as forming the basis for the determination of luminosities of supergiants, OB stars, and peculiar stars in clusters. Determination of the distance to the cluster has provided the zero-point for distances within our Galaxy and, indirectly through the Cepheids, one of the foundations on which the extragalactic distance scale ultimately rests.

At 40–50 pc, the Hyades cluster is somewhat beyond the distance where ground-based parallaxes of individual stars are easily measured, or generally considered as fully reliable. Over almost a century, considerable effort using a wide variety of indirect methods was therefore brought to bear on the problem of establishing the cluster distance. Distance estimates were based on a variety of geometrical manifestations of the cluster stars participating in a uniform space motion, while other estimates have been based on the average trigonometric parallax for a number of cluster stars, dynamical parallaxes for binaries, and photometric parallaxes using a variety of photometric systems. Nevertheless, the details of the HR and mass–luminosity diagrams remained imprecisely established due to limitations in the accuracy of the parallaxes of the individual members, while the distance of the cluster was open to debate: pre-Hipparcos estimates of the distance modulus ranged from 3.2 to 3.5 mag. While the last ground-based trigonometric distance estimate of van Altena *et al.* (1997a) of 3.32 ± 0.06 was in excellent agreement with the soon-to-follow Hipparcos value, a mean cluster distance was only a small part of the Hipparcos contribution: the accurate individual distances and velocities provide considerable insight into its formation, and its subsequent stellar and dynamical evolution.

6.3.2 Convergent-point analyses

For the Hyades, the streaming motion differs by only 60–70° from that of local field stars towards the solar antapex, so that observational scatter in the proper motions of member stars, and the random motions of field stars, complicates membership selection based only on proper motions.

In the basic convergent-point method it is assumed that the cluster under study is neither expanding, contracting, nor rotating, that the motion of the cluster with respect to the field is large enough to permit accurate membership discrimination, and that the system of proper motions is inertial and without systematic errors. In his review, van Altena (1974) considered that the first two criteria were adequately satisfied, but that information on the proper motion system was incomplete. Hanson (1975) considered the possibility of random motions contributing significantly to the stars' space velocities, as well as the effects of expansion, contraction, or rotation, concluding that any resulting deviations from parallel motion are insignificant at levels affecting the distance determination by the convergent-point method. Gunn *et al.* (1988) presented weak evidence for cluster rotation at the levels of $\lesssim 1\,\mathrm{km\,s^{-1}\,rad^{-1}}$ (projected), not inconsistent with these conclusions. If there is a significant velocity dispersion in the cluster, the convergent-point membership selection will lead to an artificial flattening of the distribution of candidate members in velocity space, which may in turn lead, for example, to spurious inferences of rotation. Conversely, if there is significant systematic structure in the internal velocities of a cluster, the convergent-point method may lead to a spatial bias in the selection of candidates. This would happen if the cluster possessed a significant component of rotation with the extreme internal velocities located primarily in a plane perpendicular to the great circle connecting the cluster centre and convergent-point.

The convergent-point method was first applied to the Hyades cluster by Boss (1908), using the proper motions of 41 suspected cluster members supplemented by three radial velocities. The method was further developed and discussed by Smart (1939), Brown (1950), and others. A systematic regression error arising from the quadratic form of the proper motion component coefficients in the normal equations, and leading to an upward revision of 7% in the distance to the cluster, was identified by Seares (1944, 1945). In the earliest implementations of the method, the determination of $\langle \mathbf{v} \rangle$ depends only on the directions of proper motions, and not on their absolute values. Upton (1970) derived a procedure for calculating the distance directly from the proper motion gradients across the cluster, dispensing with the intermediate step of locating the convergent-point. The cluster distance is then given by the ratio of the mean cluster radial velocity to the proper motion gradient in either coordinate.

Determination of the convergent-point from radial velocities was applied to the Hyades by Stefanik & Latham (1985), Detweiler *et al.* (1984), and Gunn *et al.* (1988), based on the methodology applied by Thackeray (1967) to Scorpius–Centaurus.

6.3.3 Hipparcos results

Perryman *et al.* (1998) used the Hipparcos absolute trigonometric parallaxes to determine individual distances to members of the Hyades cluster, from which the three-dimensional structure of the cluster was derived. Inertially-referenced proper motions were used to rediscuss distance determinations based on convergent-point analyses (Figure 6.4). A combination of parallaxes and proper motions from Hipparcos, and radial velocities from ground-based observations, were used to determine the position and velocity components of candidate members with respect to the cluster centre, providing new information on cluster membership: 13 new candidate members within 20 pc of the cluster centre were identified. Farther from the cluster centre there is a gradual merging between certain cluster members and field stars, both spatially and kinematically. Within the cluster, the kinematical structure is fully consistent with parallel space motion of the component stars with an internal velocity dispersion of about $0.3\,\mathrm{km\,s^{-1}}$. The spatial structure and mass segregation are consistent with N-body simulation results, without the need to invoke expansion, contraction, rotation, or other significant perturbations of the cluster. The distance to the observed centre of mass (a concept meaningful only in the restricted context of the cluster members contained in the Hipparcos Catalogue) is 46.34 ± 0.27 pc, corresponding to a distance modulus $m - M = 3.33 \pm 0.01$ mag for the objects within 10 pc of the cluster centre, roughly corresponding to the tidal radius.

6.3.4 Chemical composition and theoretical models

The proximity of the Hyades, and all of the attendant observational advantages this brings, means that it as an excellent laboratory for comparing observations with the predictions of evolutionary models (see Chapter 7). Greatly improved precision is seen in the HR diagrams built with the Hipparcos data combined with the best ground-based observations. Figure 6.5 shows the HR diagram in the $(M_V, B - V)$ plane for 69 cluster members. Known or suspected binaries, variable stars, and rapid rotators have been excluded (Perryman *et al.*, 1998).

Theoretical computations of internal structure show that the position of the zero-age main sequence (ZAMS, defining the locus on the HR diagram where the stars become fully supported by core hydrogen burning) depends on the initial chemical composition, described both in terms of He content, Y, and metallicity, Z. To compare observations with theoretical isochrones, the

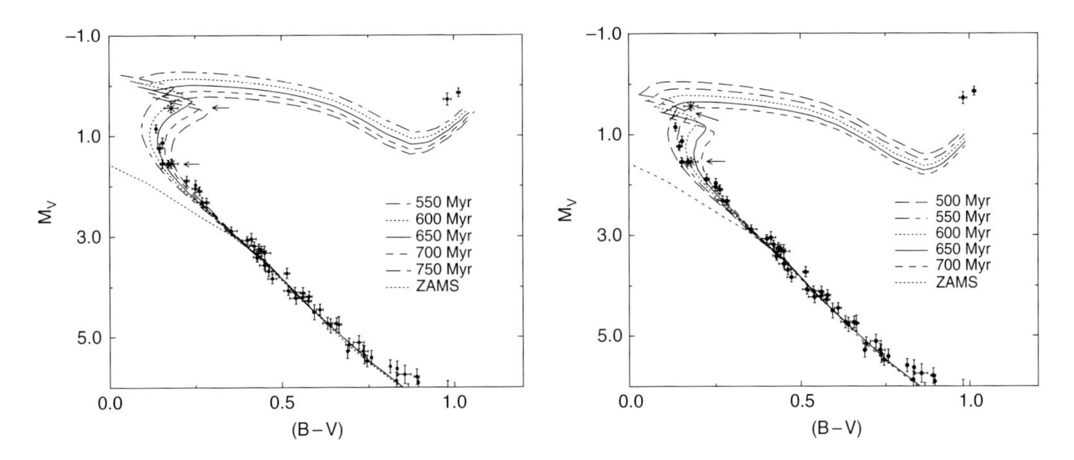

Figure 6.5 The 69 Hyades single stars measured by Hipparcos, with V and B – V taken from the catalogue, and $\sigma_{(B-V)} < 0.05$ mag. Left: the ZAMS and theoretical isochrones corresponding to ages 550–750 Myr calculated with overshooting, $\alpha_{ov} = 0.2$. Arrows indicate the components of the binary system θ^2 Tau used for the age determination. Right: the same data, but with the ZAMS and isochrones calculated without overshooting. From Perryman et al. (1998, Figures 22–23).

observables $B - V$ and M_V must first be transformed to the theoretical quantities $T_{\rm eff}$ and luminosity L. The usual procedure is to derive $T_{\rm eff}$ from $B - V$, although for the Hyades detailed spectroscopic data provide estimates of $T_{\rm eff}$ for each star to typically 50 K. Figure 6.6 shows 40 stars with $T_{\rm eff}$ and [Fe/H] = 0.14 ± 0.05 from detailed spectroscopic analysis delineating the lower part of the cluster main sequence (Cayrel de Strobel *et al.*, 1997).

Absolute magnitudes M_V can be calculated from V and π (both from Hipparcos), with suitable bolometric corrections, BC_V, derived from $T_{\rm eff}$; $\log(L/L_\odot)$ follows from $M_{\rm bol} - M_{{\rm bol},\odot}$ where $M_{{\rm bol},\odot}$ is the bolometric magnitude of the Sun (Section 7.2). These transformations depend on [Fe/H] and $\log g$. The cluster shows no reported evidence for interstellar extinction (reddening), an assumption used throughout. The mean metallicity is also determined from spectroscopic data, yielding [Fe/H] = 0.14 ± 0.05. Combination with the solar value $(Z/X)_\odot = 0.0245$ (Grevesse & Noels, 1993) yields an observational Z/X = 0.034 ± 0.007 (or $Z = 0.024 \pm 0.004$), slightly but significantly higher than the solar value.

Determination of the He abundance is more problematic. The lower part of the main sequence (Figure 6.6) is populated by low-mass stars which are only slightly evolved, and therefore rather close to their 'zero age' position. Being of intermediate or low temperature, their spectra do not show He lines, and their photospheric He abundance cannot be determined spectroscopically. Comparison with theoretical main-sequence models was therefore used to infer the He abundance from this lower part of the observed HR diagram, thus avoiding the uncertain complexities of convective core overshooting. The adopted stellar evolutionary code (CESAM: Morel, 1997) was then run with suitable input physics (equation-of-state, opacities, nuclear reaction rates, a solar mixture of heavy elements, and a value for the mixing-length parameter). Values of $Y = 0.260 \pm 0.02$, $Z = 0.024 \pm 0.04$ provided an appropriate fit to the Hyades ZAMS (Lebreton *et al.*, 1997; Perryman *et al.*, 1998; Lebreton, 2000a). Uncertainty in metallicity is the dominant source of the uncertainty on Y.

The comparison of the whole observed sequence with model isochrones yields the cluster age. Figure 6.5 shows that the optimum fit is achieved with an isochrone of 625 ± 50 Myr, $Y = 0.26$, $Z = 0.024$ and including overshooting. The five single stars in the turn-off region (which in the Hyades corresponds to the instability strip of δ Scuti stars) are also rather well represented by the 625 Myr isochrone (see also Antonello & Pasinetti Fracassini, 1998). The age uncertainty only includes the contribution from visual fitting of the isochrones. Additional errors on age result from unrecognised binaries (probably the objects apparently located 1–2σ above the resulting main sequence), stellar rotation and chromospheric activity, effective temperatures, colour calibrations and bolometric corrections, and from theoretical models through the parameterization of overshooting (Lebreton *et al.*, 1995). It is therefore reasonable to give an overall age uncertainty of at least 100 Myr.

The resulting Hyades ZAMS lies some 0.16 mag above that for solar composition ($Y = 0.2659$, $Z = 0.0175$). The fact that the He content of the Sun and Hyades are comparable appears to rule out earlier suspicions that the Hyades is He-deficient compared with field stars. On the other hand, the metallicity is higher

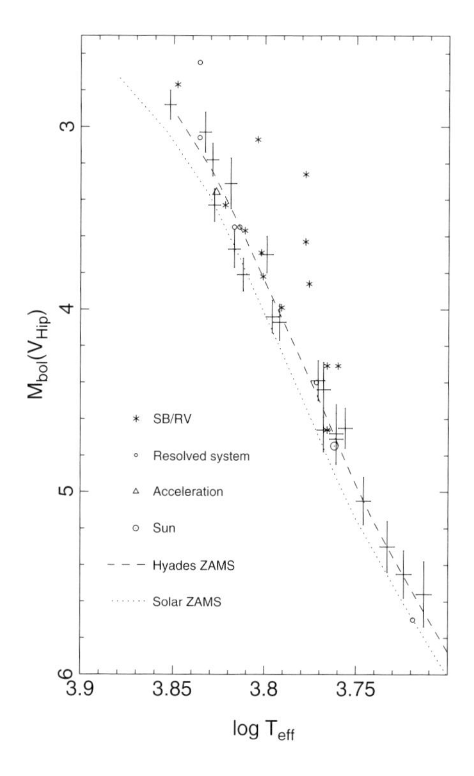

Figure 6.6 Hipparcos HR diagram for 40 selected lower main-sequence stars in the Hyades. The mean [Fe/H] is 0.14 ± 0.05. Stars with error bars are not suspected to be double or variable. Internal errors on T_{eff} are in the range 50–75 K. Double or variable stars are also indicated: objects resolved by Hipparcos or known to be double systems are shown as circles, triangles denote objects with either detected photocentric acceleration or objects possibly resolved in photometry, and ⋆ indicates a spectroscopic binary or radial velocity variable. Theoretical ZAMS loci are given for the Hyades (dashed line: Y = 0.26, Z = 0.024) and solar (dotted line: Y = 0.266, Z = 0.0175) chemical compositions. From Perryman et al. (1998, Figure 20).

for the Hyades, leading Cayrel de Strobel *et al.* (1997) to suggest that the metal enrichment of the Hyades parent cloud might have been caused by a supernova event.

Lebreton *et al.* (2001) refined the evolution modelling using binary systems with well-determined masses. For three double-lined spectroscopic binaries also resolved by speckle interferometry (51 Tau, Finsen 342 = 70 Tau, and θ^2 Tau) complete astrometric-spectroscopic orbital solutions providing individual masses and the orbital parallax, π_{orb}, were published by Torres *et al.* (1997b,c,a) respectively. For θ^1 Tau, a single-lined spectroscopic binary with astrometric information from speckle interferometry a partial solution is possible, from which individual masses can be estimated using the Hipparcos parallax. For the double-lined spectroscopic eclipsing binary system vB 22 (the vB designation numbers are from the study of van Bueren, 1952) accurate individual masses have been derived, but no orbital parallax since the system is not spatially resolved. These masses were then used with the observed luminosities, from the Hipparcos parallaxes and appropriate bolometric corrections, to derive an empirical mass–luminosity relation, to be compared with evolutionary models. Only the masses for the components of the spectroscopic eclipsing binary vB 22, with $\sigma_M/M < 1\%$, are precise enough to provide additional constraints on the He abundance (their Figure 4), with $Y = 0.26$ also providing a good fit.

Lastennet *et al.* (1999) compared the two components of each of the three binary systems 51 Tau, V818 Tau (vB 22) and θ^2 Tau with the evolutionary tracks from the Geneva and Padova groups. This aimed to establish the fidelity of the models, since the two components of each system should lie on the same isochrone (i.e. defined by the same age and chemical composition), and the predicted metallicity from all components should match the measured value. The generally good agreement is particularly notable in the case of θ^2 Tau, which comprises a main-sequence star and an evolved star, and thus testing widely different evolutionary stages. The case of V818 Tau (vB 22) is, at least at first inspection, more problematic. The masses are the most accurate known for the Hyades, and are of additional interest because the secondary ($\sim 0.77\,M_\odot$) is one of the rare stars less massive than the Sun for which a mass is known with high accuracy. Of the models studied by Lastennet *et al.* (1999), only the Padova tracks extend to suitably low masses. The distance of 50.4 ± 1.9 pc estimated by Torres *et al.* (1997b, Table 7) and used by Lastennet *et al.* (1999) (see also Lebreton *et al.*, 2001, Table 1) is, however, not from a true orbital parallax, but rather tied to that of 51 Tau through the (biased) proper motions from the PPM Catalogue. Since the binary is a double-lined eclipsing system, the radius of each component is also known with an accuracy of better than 2%, and these have been compared by Lastennet *et al.* (1999) with the radii from the Padova models computed from L and T_{eff}. While the Padova models predict a correct mass for each component, even the best-fit isochrone overestimates the radius of the more massive component by more than $0.3\,R_\odot$, and conversely for the less massive component. Lebreton *et al.* (2001) reported that models calculated with a rather high (but still acceptable) $\alpha_{MLT} > 1.8$ could fit vB 22A, while a rather low value of $\alpha_{MLT} < 1$ was required for vB 22B. Stellar radii potentially provide a powerful test of evolutionary models, however even those models generally considered the best for low-mass stars (Baraffe *et al.*, 1998) show significant discrepancies in the predicted radii for systems in this mass range (Torres & Ribas,

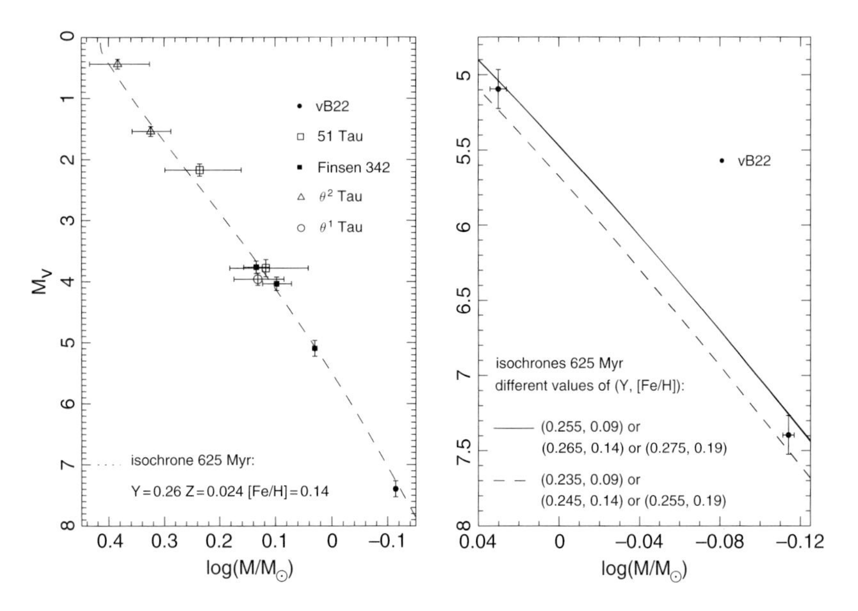

Figure 6.7 Left: the Hyades empirical and theoretical mass–luminosity relations. Masses are from Peterson & Solensky (1988) for vB 22; Torres et al. (1997a,b,c) for θ^2 Tau, Finsen 342, and θ^1 Tau respectively; and Söderhjelm (1999) for 51 Tau from Hipparcos data. The isochrone is for 625 Myr. Right: the precise positions of the members of vB 22 allow discrimination between different Y and [Fe/H] values. From Lebreton (2000b, Figure 9), reprinted with permission from the Annual Reviews of Astronomy and Astrophysics, Volume 38 ©2000 by Annual Reviews (www.annualreviews.org).

2002). Improved light-curves for vB 22 could confirm the published radii.

Figure 6.7, left shows the observed mass–luminosity relation constructed from the masses of 5 Hyades binary systems (nine components being main-sequence stars). The theoretical isochrone of 625 Myr, $Y = 0.26$, $Z = 0.024$, provide excellent agreement with the observations. The lower part of the relation is defined by the very accurate masses of the two components of vB 22. This system gives additional constraints on the Y-value derived from ZAMS calibration. Figure 6.7, right illustrates how the positions of the two vB 22 components may be used to constrain Y in the whole metallicity range allowed by observations, $[\mathrm{Fe/H}] = 0.14 \pm 0.05$ (see also Lebreton, 2000a).

6.3.5 Secular parallaxes

Improved HR diagrams have been constructed from parallaxes further refined by kinematic modelling of collective proper motions. Such model-dependent 'secular parallaxes' can provide an improvement with respect to the trigonometric parallaxes, since the Hipparcos relative proper motion accuracy is some three times larger than the Hipparcos relative parallax accuracy in the specific case of the Hyades ($\sigma_\mu/\langle\mu_{\mathrm{Hyades}}\rangle \sim 0.014$ for $\sigma_\mu = 1.5$ mas yr^{-1} and $\langle\mu_{\mathrm{Hyades}}\rangle = 111$ mas yr^{-1} compared with $\sigma_\pi/\langle\pi_{\mathrm{Hyades}}\rangle \sim 0.06$ for $\sigma_\pi = 1.3$ mas and $\langle\pi_{\mathrm{Hyades}}\rangle = 21.5$ mas). These secular parallaxes were used by Perryman *et al.* (1998) to demonstrate their statistical consistency with the trigonometric parallaxes, and by Narayanan & Gould (1999a,b) to study the possible presence and size of systematic errors in the Hipparcos data. Dravins *et al.* (1997) and Madsen (1999) developed a maximum-likelihood method to determine secular parallaxes, as part of a study of astrometric radial velocities (see also Dravins *et al.*, 1999; Lindegren *et al.*, 2000; Madsen *et al.*, 2002). In this approach, the Hipparcos Catalogue five-parameter observables for each of n cluster objects (positions, proper motions, parallaxes, and their covariances) are assumed to follow a three-dimensional Gaussian distribution with $n + 3 + 1$ model parameters: the n distances (or secular parallaxes) of the individual stars, the mean cluster centroid space motion vector $\mathbf{v}$, and the isotropic one-dimensional internal velocity dispersion σ_v. The astrometric radial velocity of an individual star can then also be determined as the scalar product

$$v_r = \mathbf{r}'\,\mathbf{v} \tag{6.10}$$

where $\mathbf{r}$ is the unit vector towards the star. Eliminating known astrometric binaries and other iteratively-rejected outliers resulted in a sample of 95 stars, with estimated parallaxes a factor of two more precise than the values published in the Hipparcos Catalogue.

De Bruijne *et al.* (2001) used the convergent-point method based on the maximum-likelihood technique described by Jones (1971), which has the merit of locating the convergent-point and, simultaneously, the corresponding cluster members. The algorithm was modified to avoid the iterative procedure leading to underestimated values of σ_v, as well as taking account of spectroscopic ground-based radial velocities which provide a further constraint on the radial component of the cluster space motion. Additional members were also selected from the Tycho 2 Catalogue. The mean accuracy on their resulting secular parallaxes is ~ 0.5 mas, a factor of 2–3 better than the published trigonometric parallaxes. They estimated an internal velocity dispersion of $\sigma_v = 0.30\,\mathrm{km\,s^{-1}}$, and confirmed the absence of significant velocity structure in the form of expansion, rotation, or shear. They found evidence for spatially correlated catalogue errors on the scale of a few degrees, consistent with the predictions of the Hipparcos Catalogue (ESA, 1997, Volume 3, Chapter 16 and Figure 16.37) with a maximum amplitude of 0.75–1.0 mas per star, a factor of ~ 2 smaller than the value reported by Narayanan & Gould (1999b).

Although a model-dependent approach, the reduced scatter in the resulting HR diagram is pronounced, and provides compelling evidence that the parallaxes have been improved, and that the majority of binaries, at least with a not too large Δm, have been identified. For 92 stars, de Bruijne *et al.* (2000, 2001) constructed the M_V versus $B-V$ diagram, which reveals two conspicuous gaps in the main sequence at around $T_{\rm eff} \sim 6400$ and ~ 7000 K (Figure 6.8). As described by de Bruijne *et al.* (2000) and references therein, these were predicted by Böhm-Vitense in the 1970s, and attributed to the use of $B-V$ as temperature indicator. Previous observational evidence for them had been sparse and disputed, and perhaps remains so (Madsen *et al.*, 2002, see also Section 7.5.1).

Narayanan & Gould (1999a) extended the moving cluster distance determination method to derive 'statistical parallaxes' for each member, taking into account individual proper motions (as in the classical convergent-point method), photometric distance estimates (from apparent magnitudes compared with absolute magnitudes based on main-sequence isochrones as a function of colour), and radial velocity gradients; the weights of these three contributions for the Hyades are in the proportion 1 : 0.33 : 0.5. They used this combination to refine the cluster membership, derive a correspondingly revised bulk motion, and hence small adjustments to the distance modulus ($m - M = 3.42 \pm 0.02$) and cluster velocity dispersion ($320 \pm 39\,\mathrm{m\,s^{-1}}$). They also used the three orbital binary systems to quantify possible systematic errors in the Hipparcos parallaxes. These potentially represent one of the most powerful independent tests of the Hipparcos parallax accuracies, and the detailed results are therefore given in Table 6.1. The table also includes the revised Hipparcos parallaxes from the global re-reduction by van Leeuwen (2007). Narayanan & Gould (1999a) derived $\langle \pi_{\rm pm} - \pi_{\rm orb} \rangle = 0.52 \pm 0.47$ (taking into account the error in the estimated cluster space velocity), thus establishing that the Hipparcos parallaxes of these three binary systems are consistent at the 1σ level with $\pi_{\rm orb}$, with any systematic error in the Hipparcos parallaxes towards the Hyades < 0.47 mas. The orbital periods are 11 yr, 6 yr, and 140 days respectively, values which should not adversely affect the astrometric parameter determination by Hipparcos.

Narayanan & Gould (1999b) compared the individual secular parallaxes predicted by the radial velocity gradient contribution to the moving cluster method with the Hipparcos trigonometric parallaxes. For both the Pleiades (their Figure 40) and Hyades (their Figure 9) they concluded that the Hipparcos parallaxes are correlated on angular scales of about 3°, with an amplitude of about 1–2 mas. Since the angular extent of the Hyades is relatively large, Narayanan & Gould (1999b) argue that this correlated spatial structure is effectively smoothed out over the cluster, with parallaxes systematically too large in some regions, and too small in others, resulting in an average parallax difference which is close to zero, and a main-sequence fitting distance in agreement with the overall parallaxes. For the smaller angular extent of the inner part of the Pleiades cluster, the effects are suggested to be more pronounced, a point considered further in Section 6.4. Their estimated correlation amplitude is somewhat larger than expected (see Section 1.4.1), and a factor 2 larger than that estimated by de Bruijne *et al.* (2001, Figure 4) using a similar approach, possible reasons for which were discussed by de Bruijne (2000a, Section 6.2). A further complication is that applying their estimates of $\pi_{\rm Hip} - \pi_{\rm pm}$ from their difference contours (their Figure 9) for the three orbital binaries results in adjusted Hipparcos parallaxes less consistent with the orbital parallaxes. The details remain confusing, perhaps implying a more complex underlying velocity distribution. The large-scale comparison of trigonometric versus photometric cluster distances by Baumgardt *et al.* (2000), described in Section 6.7, also suggests that even if this effect does occur for the Pleiades, it cannot apply generally to other Hipparcos cluster measurements.

De Bruijne *et al.* (2001) and Lebreton *et al.* (2001) carried out further modelling using their improved secular parallaxes. De Bruijne *et al.* (2001) showed that neither the Bessell *et al.* (1998) nor the Lejeune *et al.* (1998) $B-V$ versus $T_{\rm eff}$ and $T_{\rm eff}$ versus BC_V calibrations are appropriate throughout the entire mass range. Lebreton *et al.* (2001) and Castellani *et al.* (2001) showed that significant discrepancies remain with the modelling at the

Table 6.1 For the three astrometrically-resolved double-lined spectroscopic binaries in the Hyades, the table gives a comparison between the Hipparcos trigonometric parallaxes, those from the large-scale re-reduction of the Hipparcos data by van Leeuwen (2007), the secular parallaxes from de Bruijne et al. (2001) and from Narayanan & Gould (1999a), and the orbital parallaxes from Torres et al. (1997b,c,a) respectively. Although the general agreement with the orbital parallaxes is very satisfactory, the astrometrically-resolved systems are generally slightly problematic in the Hipparcos data analysis, with results quite sensitive to their detailed treatment.

Binary	HIP	$\pi_{\rm Hip}$ (mas)	$\pi'_{\rm Hip}$ (mas)	$\pi_{\rm sec,dB}$ (mas)	$\pi_{\rm sec,NG}$ (mas)	$\pi_{\rm orb}$ (mas)
51 Tau	20087	18.25 ± 0.82	18.49 ± 0.50	18.31 ± 0.69	18.45 ± 0.30	17.92 ± 0.58
70 Tau	20661	21.47 ± 0.97	21.19 ± 0.93	21.29 ± 0.37	21.35 ± 0.36	21.44 ± 0.67
θ^2 Tau	20894	21.89 ± 0.83	21.69 ± 0.46	22.24 ± 0.36	22.49 ± 0.37	21.22 ± 0.76

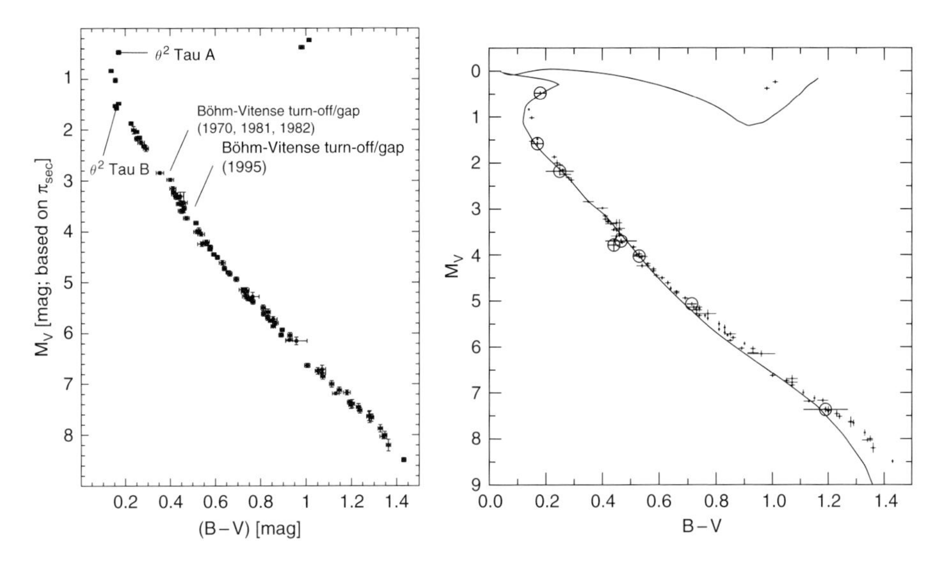

Figure 6.8 Left: colour–absolute magnitude diagram for the 92 Hyades members comprising single stars with reliable secular parallaxes. Conspicuous but artificial gaps around $B-V$ of 0.15–0.20, 0.30–0.35, and 0.95 mag are caused by the suppression of double, multiple, and peculiar stars. The (possibly spurious) first Böhm–Vitense gap (left arrow) is marked by radiative envelopes to the left, with surface convection setting in to the right: while the onset of surface convection does not alter the luminosity or effective temperature, it causes redder colours because of the lower temperatures in the deeper layers. The second gap (right arrow) coincides with the lithium dip, related to the increased depth of the surface convection zone in a narrow region around 6400±300 K; outside this region, F stars retain surface lithium for nearly their entire lifetime. From de Bruijne et al. (2001, Figure 3). Right: the Hyades HR diagram, using the data from de Bruijne et al. (2001) supplemented by the eight components of four binary systems with precise mass estimates (∘). The isochrone is for 650 Myr, $Y = 0.26$, $[Fe/H] = 0.14$, $\alpha_{\rm MLT} = \alpha_{\rm MLT,\odot}$. The mismatch between observation and theory for the low-mass stars is particularly pronounced. From Lebreton et al. (2001, Figure 5).

low-mass end, with stars systematically above the best-fitting isochrones (Figure 6.8). Improvements could be brought if the mixing-length parameter α decreases with decreasing mass (see also Prada Moroni *et al.*, 2000; Castellani *et al.*, 2001). Such a dependency was in fact reported by Schröder (1998) in his study of number counts in the HR diagram for field stars in the solar neighbourhood (see Section 7.5.1 and Figure 7.5.1).

6.3.6 Further complications

Rotation The turn-off region of the Hyades also corresponds to the δ Scuti instability strip, where stars are mainly variable and rapid rotators. Five such objects, with $v_e \sin i \sim 80$–$200\,{\rm km\,s^{-1}}$, were listed by Antonello & Pasinetti Fracassini (1998).

Rotating atmospheres affect the observational photometric data, introducing a displacement in any photometric HR diagram by amounts depending on the equatorial velocity and rotation (Maeder & Peytremann, 1972; Pérez Hernández *et al.*, 1999). This effect alone might lead to an overestimate of the age of the Hyades by ~ 50 Myr. In addition, theoretical models with uniform rotation are displaced to the right in the HR diagram (Breger & Pamyatnykh, 1998). Differential rotation induces larger effects than solid body rotation (Collins & Smith, 1985), an explanation invoked by Antonello &

Pasinetti Fracassini (1998) to explain the large displacements from the zero-age zero-rotation main sequence for the pulsating δ Scuti stars compared to the non-pulsating stars within the instability strip.

From the measurement and analysis of their oscillation frequencies and the identification of the corresponding modes by means of models (of the same age and chemical composition), it should be possible to derive the inner rotation profile, the size of their convective cores, and details of the transport processes at work in the interiors (Goupil *et al.*, 1996; Michel *et al.*, 1999). For example, the rotation profile is related to the redistribution of angular momentum by internal motions which could be generated by meridional circulation and shear turbulence in a rotating medium (Zahn, 1992). At the same time, such motions might induce internal mixing. As shown by Talon *et al.* (1997), in the HR diagram rotational effects mimic overshooting; for example, in a star of $9M_\odot$, a rotational velocity of $\sim 100\,\mathrm{km\,s^{-1}}$ is equivalent to an overshooting of $\alpha_{\rm ov} \sim 0.2$. These effects together imply that the resulting age of the Hyades might be in the range 500–650 Myr.

Surface abundance anomalies Quillen (2002) noted that, in analogy with the situation for exoplanets where an excess of short-period planets around star with enhanced metallicities has been linked to captured planetesimals, there is a possibility that observed (surface) metallicities in the Hyades might be similarly polluted. From isochrones fitted to the same secular parallaxes of de Bruijne *et al.* (2001), Quillen (2002) placed an upper limit for the metallicity scatter of $[\mathrm{Fe/H}] \lesssim 0.03$ dex, smaller than that measured spectroscopically, and suggesting that the stars formed from a gas nearly homogeneous in metallicity. Because massive stars become supernovae well before low-mass stars finish accreting, this limit also suggests that nearby supernovae are not capable of substantially enriching the interstellar medium in their own nearby star-forming regions. From the occurrence rate of stars with short-period giant planets in the solar neighbourhood, some six stars in the Hyades high-precision subset could be expected to have significantly higher metallicities. Since none are significantly displaced from the evolutionary track, no star is likely to have a metallicity enhanced by the $30\,M_\oplus$ of rocky material needed to account for the higher metallicity of the short-period exoplanet hosts, unless they happen to lie on the same isochrone as their non-polluted counterparts.

Age distribution Eggen (1998b) and Makarov (2000) have questioned the fundamental assumption of a common age for the Hyades cluster members. An age spread in open clusters was proposed by Herbig (1962) based on a comparison of the turn-off age and the gravitational contraction age. Makarov *et al.* (2000) used the Tycho 2 Catalogue proper motions in the Hyades region to confirm that their precision is similar to those of the Hipparcos Catalogue. But since they are based on observations covering up to 100 years, they are less sensitive to the effects of binary orbital motions than those of the Hipparcos observations. Makarov (2000) then studied the space velocities of 95 stars with $B - V < 0.7$ mag as a function of their X-ray luminosities from the ROSAT All Sky Survey. The internal velocity dispersion is shown to decrease with increasing X-ray luminosity, and for the highest luminosity X-ray emitters appears to be less than the expected velocity dispersion of $\sim 0.25\,\mathrm{km\,s^{-1}}$ for a system in dynamical equilibrium. The fact that the mean X-ray luminosity of main-sequence stars decreases with age was inferred from the Einstein surveys of nearby Galactic clusters and field disk stars, where the mean $\log L_{\rm X}$ for G-type stars drops from 29.7 in the α Per cluster (age $\sim$ 50 Myr) to 29.4 in the Pleiades (80 Myr) and 28.9 for the Hyades (600 Myr). While the correlation might be an indirect one related perhaps through rotational velocity, the results nevertheless suggest a spread in ages of the Hyades stars. Furthermore, half of the strongest X-ray emitters with the most coherent motions are visual, long-period ('soft') binaries, which should have been more prominently disrupted by 600 Myr of dynamical evolution. Makarov (2000) concluded that a considerable number of less massive cluster stars are not in dynamical equilibrium, most likely because they are too young, raising the possibility that the star formation history of the Hyades may have been protracted. Eggen (1998b) constructed isochrones for the cluster (as well as more extended 'supercluster') members and also concluded that both components show a larger range in age, from 600 Myr up to $\sim$ 1 Gyr, than can be easily understood as resulting from a single, prolonged epoch of star formation.

6.3.7 *N*-body analyses

Madsen (2003) used the accurate radial motions determined by purely geometric means, i.e. without using the spectroscopic Doppler effect (Dravins *et al.*, 1999; Lindegren *et al.*, 2000; Madsen *et al.*, 2002), to compare with the results of the N-body simulation code NBODY6 (Aarseth, 1999). The objective was to estimate the internal velocity dispersion of the Hyades, and to compare the results with astrometric radial velocities determined according to Equation 6.10. The assumption of a constant and isotropic velocity dispersion throughout the cluster is likely to be too simplistic; theoretically at least a variation with distance r from the cluster centre might be expected, as in the simple Plummer (1915) potential

$$\sigma_v^2 = GM/6\sqrt{r_c^2 + r^2} \tag{6.11}$$

where M is the cluster mass and r_c the core radius (~ 3 pc for the Hyades). NBODY6 incorporates algorithms to deal with stellar (including binary) encounters and stellar evolution. External perturbations are represented by a fixed, Galactic tidal field, and the cluster is assumed to move in a circular orbit around the Galactic centre. For the initial cluster configuration, stars were selected randomly from the Kroupa *et al.* (1993) initial mass function, distributed randomly in a Plummer potential with virial radius $r_v = 4$ pc, and with binaries constructed from a random pairing of the same initial mass function. The simulation was run to an age of 625 Myr, stars merging into binaries when the semi-major axis $a \leq 10 R_\odot$, and lost from the cluster when they exceed two tidal radii. Free parameters were the total particle number and the initial binary fraction. The predicted number of giants and single white dwarfs were matched to observations in order to constrain the starting conditions, eventually selected as 200 single stars and 1200 binaries, with a total initial cluster mass of 1100–1200 $M_\odot$. The models reproduce the main features of the observational HR diagram (their Figure 1) as well the distribution of stars with radial distance and magnitude. Their results suggest that the velocity dispersion decreases from $\sigma_v \sim 0.35$ km s^{-1} at the centre to nearly 0.2 km s^{-1} at 7–8 pc from the centre, increasing slightly again beyond the tidal radius of 10–11 pc, for a total current cluster mass estimated at $\sim 460 M_\odot$. Madsen (2003) noted that those stars with the smallest expected velocity dispersion are the best candidates for studies that compare astrometric and spectroscopic radial velocities to reveal astrophysical phenomena causing spectroscopic line shifts, such as convective motions and gravitational redshift.

6.3.8 Summary of uncertainties

In conclusion, while modelling of the structure and main sequence of the Hyades cluster has been remarkably successful, a number of conflicts remain with the detailed Hipparcos observations. Further accurate binary star masses and radii could improve knowledge of the He abundance, the mass–luminosity relation, and the theoretical constraints on stellar radii. Improved errors on [Fe/H] from individual abundances are desirable; C, N, and O abundances remain poorly known, while the Ne abundance is unknown. At the same time, certain abundance determinations may reflect only surface phenomena. The $B - V$ to $T_{\rm eff}$ and BC$_V$ transformations remain inadequate. Interstellar and circumstellar extinction, and even grey extinction (Section 8.5.2), require further attention. Effects of (differential) rotation and overshooting remain incompletely modelled, and the possibility that $\alpha_{\rm ov}$ is a slowly decreasing function of mass will have attendant consequences for the cluster age estimates. The comparison of non-pulsating and pulsating stars located in the instability strip should provide better insight into internal structure and effects of rotation. Model mismatch is particularly pronounced in the very low-mass region where models are systematically too blue, and where the knowledge of a few masses would provide further constraints on the equation-of-state. A non-singular epoch of star formation, and a more complex (non-relaxed) internal velocity distribution, would further complicate the modelling.

6.4 The Pleiades

6.4.1 Introduction

The Pleiades cluster is a system of about 270 known members brighter than $V = 12$, a total mass somewhere in the range 500–8000 $M_\odot$, a relatively young age of about 100 Myr, and lies at a distance of about 110–130 pc.

Independent distance determination for clusters other than the Hyades is essential for verifying the various theoretical assumptions made in the Hyades modelling. Although there exist a variety of techniques for estimating cluster distances, main-sequence fitting, and more specifically ZAMS fitting, has been the reference, the latter being based on non-evolved stars which are believed to be fairly well understood. In its basic form, the technique requires a calibrating cluster (normally the Hyades) whose distance has been derived by other methods (e.g. trigonometric parallaxes or via the proper motion convergent-point method), which is matched to the main sequence of the target cluster to determine the distance. In its simplest form, it assumes that the ZAMS for all clusters younger than the Hyades are identical, and unaffected by differences in initial chemical composition. More generally, main-sequence fitting relies on the premise that the location of a star in the HR diagram is uniquely specified by its mass, chemical composition, and age.

Progressive improvements to the technique led to Pleiades distance estimates of $d = 129$ pc ($m - M = 5.55$) by Blaauw (in Strand, 1963, p383), $d = 129$ pc ($m - M = 5.56$) when adjusted for metallicity by Turner (1979), and $d = 132$ pc ($m - M = 5.60$) by VandenBerg & Poll (1989) who used a semi-empirical fitting using computations of model stellar atmospheres, set by the He and metal abundance of the members, and constrained by the morphology of the ZAMS and further by the chemical composition of Groombridge 1830,

Table 6.2 Various distance estimates for the Pleiades: values in bold are those quoted in the cited reference, the others are derived from these values with error bars rounded (a given parallax standard error does not strictly transform to a symmetrical error on distance).

Method	Reference	Distance (pc)	π (mas)	$m - M$ (mag)
Ground-based:				
MS fitting	VandenBerg & Poll (1989)	132		**5.60**
MS fitting	Pinsonneault *et al.* (1998)	131.8 ± 2.5		$\mathbf{5.60 \pm 0.04}$
MS fitting in $(V - I)_C$	Percival *et al.* (2003)	119.7 ± 3		$\mathbf{5.39 \pm 0.06}$
MS fitting in near IR	Percival *et al.* (2005)	$\mathbf{133.8 \pm 3}$		
MS fitting	An *et al.* (2007)	135.5 ± 3		$\mathbf{5.66 \pm 0.05}$
Hipparcos:				
B stars	Cramer (1997)	130		**5.57**
Weighted mean	van Leeuwen & Hansen-Ruiz (1997)	116.1 ± 3.1	$\mathbf{8.61 \pm 0.23}$	5.32 ± 0.06
Abscissae analysis	van Leeuwen (1999a)	118.3 ± 3.5	$\mathbf{8.45 \pm 0.25}$	5.37 ± 0.06
Re-reduction	van Leeuwen (2007)	122.2 ± 2.0	$\mathbf{8.18 \pm 0.13}$	5.44 ± 0.03

the Population II subdwarf with the most reliably determined parameters. A similar distance of $d = 132 \pm 10$ pc ($m - M = 5.60 \pm 0.16$) was estimated by O'Dell *et al.* (1994) using a rather independent technique based on the knowledge of the periods, rotational velocities, and angular diameters of 10 rapidly rotating cluster stars, compared with the expected values of 'projected' cluster distance. The method uses a calibration of diameters based on their assumed dependence on V and $B - V$.

Table 6.2 summarises some of the Pleiades distance estimates relevant in the following discussion. Since different authors use different distance units (distance in pc, parallax in milliarcsec, or distance modulus in magnitudes) and while these are trivially transformed, this makes comparison more awkward. In the following, distances are quoted consistently in pc, with the original unit used by the various authors in parenthesis. The table includes a distance estimate based on the absolute magnitudes of B stars by Cramer (1997, Table 2), although this was stated as provisional, and is not discussed further here.

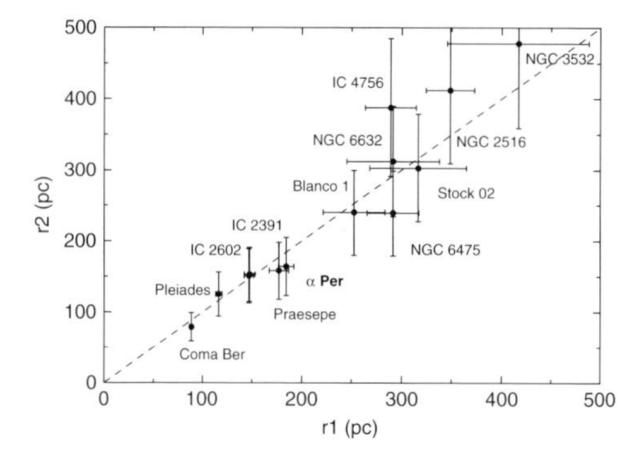

Figure 6.9 Open cluster distances from the database of Mermilliod (1995), r2, versus those derived from Hipparcos data, r1. From Chen et al. (1998, Figure 3). A similar comparison between Hipparcos and estimated ground-based parallaxes from Lyngå (1988) for the nine nearest open clusters is given by van Leeuwen (1999a, Figure 1).

6.4.2 Hipparcos distance estimates

At the Pleiades distance of about 100 pc ($\pi \sim 10$ mas) a typical Hipparcos parallax with standard error 1 mas provides an individual distance accurate to $\sim 10\%$. For the Pleiades, as well as other clusters, an improved mean cluster distance can be obtained from an average of the individual Hipparcos distance estimates.

While the Hipparcos mean distances for the nearest clusters give generally good agreement with the ground-based distance estimates taken from Lyngå (1988) or Mermilliod (1995) (see Figure 6.9), the mean Pleiades distance based on 54 members from the Hipparcos parallaxes by van Leeuwen & Hansen-Ruiz (1997, solution A) gave $d = 116.1 \pm 3.1$ pc ($\pi = 8.61 \pm 0.23$ mas), significantly smaller than the photometric estimate of $d \sim 132$ pc (see also Hansen-Ruiz & van Leeuwen, 1997).

This discrepancy has provoked an interesting controversy which has remained unresolved over the past 10 years. Despite various improved treatments of the Hipparcos data, described below, the Hipparcos distance determinations remain significantly (some 10%) nearer than indicated by main-sequence fitting. Small angular scale correlations were predicted to exist in the Hipparcos parallaxes, and have been proposed to explain the origin of such an error; but the explanation would require a correlation peak of order 1 mas in the direction of the Pleiades, a factor of some 3 higher than had been anticipated. Attempts to identify the cause of such an effect, by re-analysis of the Hipparcos Intermediate Astrometric Data, and through a large-scale re-reduction of the Hipparcos data, have also been unsuccessful.

Various efforts have also been made to identify causes for main sequence fitting discrepancies, which might arise if some of the basic properties of well-studied open clusters, such as composition (He abundance or metallicity), age, or reddening are in error, or if additional evolutionary dependencies exist. Again, these efforts have so far been unsuccessful. Independent distance estimates have also been obtained from Hubble Space Telescope and from orbital parallaxes, and many other subtleties have been identified. The following examines these various lines of investigation, which have still not been unambiguously reconciled.

Assessment of the Hipparcos data Pinsonneault *et al.* (1998) suggested that statistical correlations between right ascension and parallax ($\rho^{\pi}_{\alpha\cos\delta}$) arising from the non-uniform distribution of Hipparcos observations over time (and in turn along the parallactic ellipse) would affect all stars, including clusters. They noted that in the Pleiades the brightest stars (1) are highly concentrated near the cluster centre and are therefore subject to spatial correlations which give them nearly the same parallax, (2) have smaller σ_{π} than fainter stars which gives them more weight in the mean parallax, and (3) are those which have the highest values of $\rho^{\pi}_{\alpha\cos\delta}$ and also the highest parallaxes in the Hipparcos Catalogue. They suggested that the true parallax, close to that obtained through main sequence fitting, is obtained if the brightest stars with high $\rho^{\pi}_{\alpha\cos\delta}$ are excluded from the calculation.

In addition to the astrometric parameters, the Hipparcos Catalogue includes the Intermediate Astrometric Data (the abscissa residuals from the reference great circles) allowing a more rigorous treatment of this type of correlation (van Leeuwen, 1996; van Leeuwen & Evans, 1998; van Leeuwen, 2000). In an analysis to treat systematically the nine nearest clusters by the same determination method, and to examine the possible discrepancies due to the correlation between astrometric parameters, van Leeuwen (1999a) combined the Hipparcos Intermediate Astrometric Data for single cluster members, accounting for correlations between measurements made on the same reference great circles. In this way, the degrees of freedom are reduced with respect to the sum of the individual star solutions, which makes the astrometric parameter estimates more robust and reduces the correlations between position and parallax. Van Leeuwen (1999a) concluded that the statistical behaviour of the solutions (as demonstrated by their unit weight standard deviations) were credible, and that the revised Pleiades mean distance based on 55 stars is $d = 118.3^{+3.6}_{-3.4}$ pc (8.45 ± 0.25 mas). A similar (and clearly not independent) analysis made by Robichon *et al.* (1999b) based on 54 stars yielded $d = 118.2^{+3.2}_{-3.0}$ pc.

Other clusters were treated by the same method by both groups at the same time (Figure 6.10), and are discussed in Section 6.5. Robichon *et al.* (1999a,b) and van Leeuwen (1999a) showed that while the solution proposed by Pinsonneault *et al.* (1998) improves the situation for the Pleiades, it would introduce new difficulties for Praesepe. By means of Monte Carlo simulations of the Pleiades stars, they showed that the mean value of the Pleiades parallax does not depend on the correlations $\rho^{\pi}_{\alpha\cos\delta}$. They also examined distant stars and clusters with high $\rho^{\pi}_{\alpha\cos\delta}$. Through these tests, Robichon *et al.* (1999a,b) made the Hipparcos distances to Coma Ber and the Pleiades more secure, and did not find any obvious bias on the parallax resulting from a correlation between right ascension and parallax, either for stars within a small angular region or for the whole sky.

An independent study of the intermediate Hipparcos data treatment was undertaken by Makarov (2002) for the Pleiades, and by Makarov (2003) for Coma Ber and NGC 6231. This aimed to improve determination of the satellite attitude for the relevant transits by correcting the measured residuals by the average residual of the 700–900 reference stars from the complementary field of view, i.e. those located in the ring on the sky $58^{\circ} \pm 0\overset{\circ}{.}5$ away from the affected star. The resulting weighted mean distance for the 54 stars reduced to 129.0 ± 3.3 pc (7.75 ± 0.20 mas), rather close to the main-sequence fitting value. The plausible underlying hypothesis is that the Pleiades stars form a dense region of bright stars with undue weight assigned to them in the associated satellite attitude solution, which effectively leaves the central part of the cluster tied with only limited rigidity to the rest of the catalogue. The validity of Makarov's revised numerical adjustment was questioned by van Leeuwen (2005), and appears not to be supported by the global re-reduction of the Hipparcos data discussed hereafter.

Global re-reduction of the Hipparcos data A more comprehensive re-reduction of the Hipparcos data was

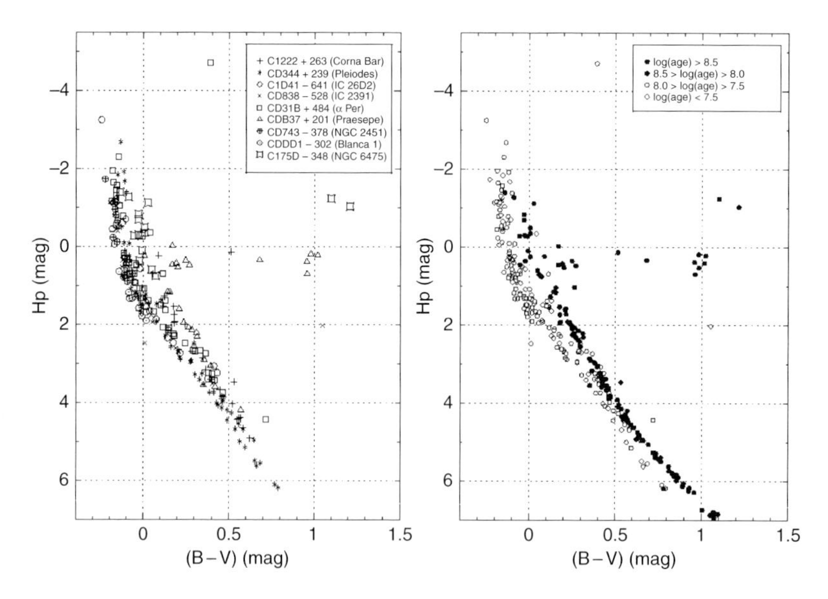

Figure 6.10 The composite HR diagram for nine open clusters. Left: showing the individual clusters; right: divided into age groups, and including the Hyades. From van Leeuwen (1999a, Figure 2).

developed in Cambridge over a period of several years (see Section 1.6), and the global results have been recently published (van Leeuwen, 2007). Although this review takes this global re-reduction as a natural dividing line in the literature not generally crossed, the Pleiades has been taken as an exception, with the following results made available in advance of publication by van Leeuwen (2007, priv. comm.).

For the Pleiades, the correlation issue has been further improved, and the mean cluster parallax accuracy has improved by a factor of about 2. The resulting value based on 53 stars (van Leeuwen, 2007) is $d = 122.2^{+2.0}_{-1.9}$ pc (8.18 ± 0.13 mas). Although shifted to a larger distance compared to the first Hipparcos distance estimate, this remains nearly 10%, and even more significant statistically, away from the main-sequence fitting value of 132 pc. Significantly, there is no discrepancy between the loci of a young open cluster like the Pleiades and the lower bound of the nearby star main sequence (Figure 6.12 below). Corresponding results for the other clusters are discussed in Section 6.5.

6.4.3 Main-sequence fitting post-Hipparcos

Mermilliod *et al.* (1997) already discussed a variety of complications related to main-sequence fitting which could provide partial explanation of the Hipparcos distance (see also Robichon *et al.*, 2000). D'Antona & Mazzitelli (1997) also drew attention to some specific problems of main-sequence fitting relevant to the Pleiades: the large errors in colour conversions below $T_{\rm eff} \sim 4000$ K, the dependence of the grey atmosphere approximation on estimates of $T_{\rm eff}$, and especially the sensitivity of the $V-I$ versus $T_{\rm eff}$ conversion: a difference of 0.1 mag at 5500 K means that a distance modulus of $m-M = 5.8$ for a 100 Myr isochrone results from the Kurucz (1993) conversion, while the Hipparcos distance modulus of $m - M = 5.3$ would be compatible with those of Bessell (1991).

Efremov *et al.* (1997) noted that the new Hipparcos distances were actually in agreement with previous estimates in the Russian literature which had pointed to a smaller distance scale, including the results of Dambis (1999) who derived a distance scale for open clusters a factor of 0.8–0.9 shorter than that generally accepted. They cited evidence for an absence of a universal correlation between He and metal abundance, $\Delta Y/\Delta Z$, attributing the Pleiades conflict in part to this effect.

Belikov *et al.* (1998) used the Hipparcos mean distance to model the observed ZAMS based on an assumed normal metallicity ($Z = 0.02$), and a He abundance, adjusted to fit the observations, of $Y = 0.34$, and an age of 90 Myr. Although there is no observational evidence to support such a high He abundance, they noted that the agreement between theoretical and observed colour-magnitude diagrams was better than with the nominal distance and normal He abundance ($Y = 0.26$–0.28). They used $M_{\rm bol}$ and $B - V$ versus $T_{\rm eff}$ relations from Schmidt–Kaler (Schaifers & Voigt, 1982, p15), combining Population I pre-main-sequence evolutionary tracks of D'Antona & Mazzitelli (1994) with post-main-sequence tracks from Schaller *et al.* (1992); an adjustment to $Y = 0.31$ was later given by Belikov *et al.* (1999b) after adopting revised tracks from D'Antona & Mazzitelli (1997). Belikov *et al.* (1998) also identified a peak in the luminosity function at $M_V = 5.5$ mag which they identified with the transition between the pre-main-sequence and main-sequence mass–luminosity relation, and argued that this implies an age spread in star formation of up to 60 Myr. In the context of the debate on whether cluster stars in general display an age difference (implying a systematic difference in age for stars of different mass), or an age spread, such an extended period of star formation would be significantly larger than previous limits of 20–30 Myr (Soderblom *et al.*, 1993).

Based on his earlier contention that the Pleiades cluster is a member of a local supercluster, or a much more extended association of young stars, with a common proper motion convergent-point, Eggen (1998d) showed that the resulting secular parallax of $\pi_{\rm clus} = 8.46$ mas would be just 3% larger than the Hipparcos parallax mean of the 14 brightest cluster members, or just 2% larger than the 8.61 ± 0.23 of van Leeuwen & Hansen-Ruiz (1997), with $Y = 0.30$ providing a fit to theoretical models.

As a test of the colour distribution of young field stars, Fischer *et al.* (1998) investigated the position of 300 nearby ($d \lesssim 20$ pc) lithium-rich G and K field stars in the colour–magnitude (M_V versus $B-V$) diagram, for which the lithium abundance indicated ages comparable to those of the Pleiades. Compared with Pleiades absolute magnitudes based on the Hipparcos distance modulus of 5.33 and an extinction of $A_V = 0.12$, and a Praesepe Hipparcos distance modulus of 6.24 with $A_V = 0.0$, they showed that while the Li-rich field stars form a sequence that generally coincides with the Praesepe stars (with the majority of Pleiades stars some 0.3–0.4 mag below), a few field stars fall on the Pleiades sequence. On the unproven assumption that simple photometric errors are not at fault, these deviant field stars may exhibit the same phenomenon observed in the Pleiades. No follow-up to this specific study has been reported (Fischer, priv. comm., 2006).

Soderblom *et al.* (1998) used similar arguments to assert that if the Pleiades stars are truly 0.3 mag fainter than the ZAMS, then comparable subluminous ZAMS stars should exist elsewhere, including in the solar neighbourhood. Of 50 field stars showing activity from Ca II H and K lines, and therefore expected to be young, they showed that they fall on or above the usually accepted ZAMS with none lying below it. Conversely, stars that do fall below the locally-defined ZAMS are old stars of low metallicity, and not young stars analogous to the Pleiades.

Pinsonneault *et al.* (1998) presented a detailed assessment of the possible problems arising from main-sequence fitting in both the metallicity-sensitive $B-V$ and metallicity-insensitive $V-I$ indices for the Pleiades, as well as the Hyades, α Per, Praesepe and Coma Ber. They examined the effect of metal abundance, demonstrating that errors on the adopted [Fe/H] = –0.03 are unlikely, but with caveats about the unknown abundances of important atmosphere opacity contributors such as Mg and Si; possible CNO anomalies; He abundance, where they arrived at a requirement of $Y = 0.37$, but detailing various lines of evidence to argue that such a value is implausible; age; and reddening, adopting $E(B-V) = 0.04$ and individual reddenings for a small number of highly reddened stars. All were considered unlikely to explain the Pleiades distance anomaly, and they concluded that the distance from main-sequence fitting is confirmed as $d = 131.8 \pm 2.5$ pc ($m - M = 5.60 \pm 0.04$). They offered some pointers for explanations based on correlations in the Hipparcos data, which subsequent analyses (see the detailed discussions in Robichon *et al.*, 1999b) or re-reductions (van Leeuwen & Fantino, 2003) do not appear to support. A possible dependence on age, as proposed by van Leeuwen (1999a), was rejected by Pinsonneault *et al.* (2000). Stello & Nissen (2001) also investigated the ZAMS location as a function of metallicity, deriving $m - M = 5.61 \pm 0.03$ and confirming the main-sequence fitting estimate of Pinsonneault *et al.* (1998), also excluding low metallicity as the cause of an observed main-sequence displacement on the basis of Strömgren photometry of F-type stars. King *et al.* (2000) measured elemental abundances in two cool Pleiades dwarfs, and while deriving Fe abundances in agreement with those for hotter stars, noted that abundances of Cr, Ca, Ti, and Al were subsolar, reflecting the pattern of the interstellar medium (see also Section 8.2.5).

It was also argued that the distances from main-sequence fitting could have larger error bars than those quoted by Pinsonneault *et al.* (1998). They depend on reddening and on transformations from the ($M_{\rm bol}$, $T_{\rm eff}$) to the colour-magnitude plane if theoretical ZAMS are used as reference (or on metallicity corrections if empirical ZAMS are compared). Robichon *et al.* (1999a) compared the solar ZAMS from Pinsonneault *et al.* (1998) with those calculated using the CESAM code both in the theoretical and in the (M_V, $B-V$) planes. They showed that while the two ZAMS are within 0.05 mag in the theoretical HR diagram, they differ by 0.15–0.20 mag in the range $B - V = 0.7$–0.8 in the (M_V, $B-V$) plane, simply because different colour-magnitude transformations have been applied. Also, main-sequence fitting often relied on rather old and inhomogeneous colour sources (in the separate Johnson and Kron–Eggen *RI* systems) requiring transformations to put all data on the same (Cousins system) scale. New photometric measurements of various cluster stars were subsequently made, as described further below.

Castellani *et al.* (2002) used the FRANEC evolutionary code and colour transformations and bolometric

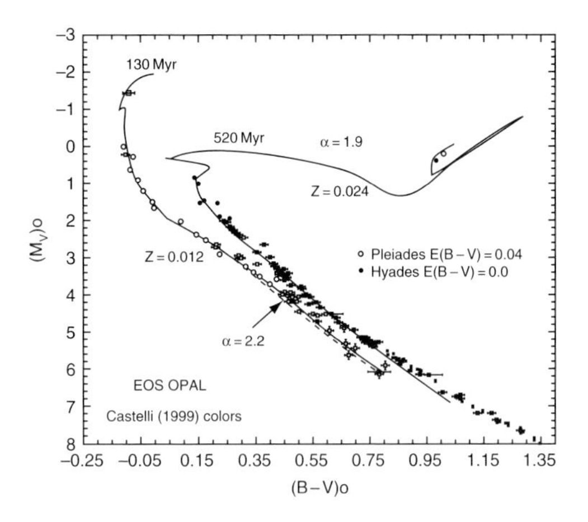

Figure 6.11 Colour-magnitude diagram of the Hyades and Pleiades, compared with the theoretical isochrones for the Hyades and Pleiades composition (Z = 0.024, Y = 0.278 and Z = 0.012, Y = 0.27, respectively). The adopted value for the mixing length is $\alpha_{\rm MLT}$ = 1.9 in both cases. The position of the Pleiades zero-age main sequence (ZAMS) for a different value of α is shown as a dashed line. From Castellani et al. (2002, Figure 2).

corrections from Castelli (1999) to fit the Hipparcos observations both for the Hyades and the Pleiades (Figure 6.11), deriving for the latter $Y = 0.27$ and $Z = 0.012$ ([Fe/H] ~ -0.15, i.e. subsolar, but still compatible with recent metallicity determinations). They demonstrated that the portion of the main sequence affected by convection can be satisfactorily explained by the same value of the mixing-length parameter ($\alpha_{\rm MLT} = 1.9$), and that the adopted stellar models can account for the location of the H-burning stars even with different metallicities, arguing that the uncertainties in chemical composition are larger than those in the Hipparcos parallaxes, and that there were consequently *'no reasons for claiming the existence of errors in the Hipparcos parallaxes or theoretical predictions'*.

Percival *et al.* (2003) considered main-sequence fitting for the Hyades, Pleiades, Praesepe and the NGC 2516 open clusters, based on a template main sequence from a sample of 54 single G-stars in the range $-0.3 \leq$ [Fe/H] $\leq +0.3$ from the local field population, acquiring homogeneous Johnson–Cousins $BV(RI)_{\rm C}$ photometry to resolve conflicts for the Pleiades and NGC 2516 between metallicities determined spectroscopically (indicating solar abundance) or photometrically (indicating subsolar values), and thus to place the metallicities of the clusters being studied on a homogeneous scale. They identified a mismatch between the Pleiades (and NGC 2516) spectroscopic and photometric metallicity which explained discrepancies between the main-sequence fitting distance moduli in $V/(B-V)$, in $V/(V-I)_{\rm C}$ and with the Hipparcos distance. Assuming [Fe/H]$= -0.4 \pm 0.1$ appropriate to the photometric (rather than spectroscopic) metallicity resulted in distance moduli of 5.46 ± 0.09 from the $(B-V)$ colour and 5.39 ± 0.06 from the $(V-I)_{\rm C}$ colour, at the same time both mutually consistent, and consistent with the Hipparcos estimate both from van Leeuwen's abscissa analysis and re-reduction (Table 6.2). They conclude that *'the widely-discussed discrepancy between main-sequence fitting and Hipparcos distances [for the Pleiades and NGC 2516] is just an artifact due to the clusters' colours which are inconsistent with their spectroscopic metallicity'*. They speculated that, since it affects the two younger clusters, fast rotation and/or chromospheric activity may be relevant factors, effects which they tentatively identify also in the data of Pinsonneault *et al.* (1998).

Percival *et al.* (2005) revised this conclusion with the release of the 2MASS All Sky Catalogue, when fitting of the infrared colours in addition to the optical bands became possible. They found that a substantial subsolar metallicity does not produce consistent results in the $V/(V-K)$ and $K/(J-K)$ colour planes. Consistent results could, however, only be found by restricting the field dwarf sample to those with $M_V \leq 6$. This corresponds to the magnitude where the Pleiades $B-V$ colours start to become anomalous according to Stauffer *et al.* (2003), a fact that the latter authors note has been known for many years, but largely ignored and certainly not explained, and which they provisionally attributed to rapid stellar rotation combined with star spots; Jones (1972) had previously postulated grey extinction (see below). Such a restriction led Percival *et al.* (2005) to a mean distance of 133.8 ± 3 pc with suspicion re-directed to the Hipparcos parallaxes, but subject to a number of caveats: the use of the field dwarfs as templates especially in view of the rapid rotation of the Pleiades members, the somewhat arbitrary exclusion of the anomalously blue lower main sequence (K dwarfs), derivation of reddening estimates, and the general issues of consistency of colour indices and the derivation of spectroscopic metallicities which themselves rely on a temperature scale, itself generally based on $B-V$ colours.

An *et al.* (2007) continued their main-sequence fitting to derive a He abundance of $Y = 0.279 \pm 0.015$, consistent with that of the Sun and of the Hyades, and a distance modulus of $(m-M)_0 = 5.66 \pm 0.01 \pm 0.05$ (internal/systematic).

6.4.4 Other distance estimates

Unlike the situation for the Hyades, the Pleiades cluster is poorly suited for a convergent-point distance determination, having only a limited angular extent on the sky of about 2°, and a small bulk radial velocity of only 5.7 km s^{-1}, giving an expansion-related proper motion at 2° from the cluster centre of only 0.23 mas yr^{-1}. Three papers refer to application of this method to the Pleiades. Madsen *et al.* (2002) showed that the expected distance error is somewhat larger than that based on parallaxes. Narayanan & Gould (1999b) quantified this by showing that the individual proper motions, photometric distance estimates, and radial velocity gradients contribute in the proportion 0.009 : 0.005 : 1.0 (cf. 1 : 0.33 : 0.5 for the Hyades, see Section 6.3), i.e. with 99% of the weight in the radial velocity gradient. They derived a mean of 133.3 pc (7.5 ± 0.6 mas), some 1σ from the other published Hipparcos estimates, but whose discrepancy from the main-sequence fitting distance they attribute to the small-angular scale correlations in the Hipparcos parallaxes discussed in Section 6.3, in which the small angular extent of the Pleiades coincides with a peak of the parallax correlation. Their resulting distance modulus has a rather large error bar ($m - M = 5.58 \pm 0.18$ mag), is in disagreement with that derived directly from Hipparcos parallaxes ($m - M = 5.36 \pm 0.07$ mag), but is in agreement with that obtained through main-sequence fitting ($m - M = 5.60 \pm 0.05$ mag). Robichon *et al.* (1999b) showed that the stars carrying the most weight in this radial velocity analysis are those with the most extreme value of the proper motion projection on the line-of-sight, i.e. the most distant members from the cluster centre parallel to the proper motion direction, with different star selection criteria yielding distances between extremes of 100 ± 16 pc to 145 ± 11 pc; if the Coravel radial velocities are free from bias, this could imply that the spatial structure of the cluster is not symmetrical, or that the internal velocity dispersion is not uniform, perhaps due to tidal distortion by the Galactic potential. A critique of the analysis of the third study by Chen & Zhao (1997), which yielded $d = 135.56 \pm 0.72$ pc, has been given by Robichon *et al.* (1999b). A similar allusion to asymmetry was given by Gatewood *et al.* (2000) who noted that the 10 stars brighter than $V = 6$ have a Hipparcos mean distance of 111.9 pc (8.94 ± 0.14 mas), while the 18 stars fainter than $V = 8$ have a mean of 126.4 pc (7.91 ± 0.68 mas). That the effect is position rather than magnitude related is indicated by the fact that even the faintest stars show a higher mean parallax near the cluster centre.

Gatewood *et al.* (2000) gave a value of 130.9 ± 7.4 pc (7.64 ± 0.43 mas) for the mean of 18 cluster members with ground-based parallaxes determined at the Allegheny Observatory using the Multichannel Astrometric Photometer. As noted by the authors, the greatest problem for narrow-field ground-based parallax work is the transformation from relative to absolute parallaxes, an iterative process which includes the estimated intrinsic luminosities of the adopted reference stars. A comparison between the ground-based and Hipparcos parallaxes for the four stars in common between the two programmes allowed no unambiguous conclusions to be drawn.

Soderblom *et al.* (2005) obtained parallaxes for three Pleiades members from Hubble Space Telescope Fine Guidance Sensor instrument, yielding a mean of $d = 134.6 \pm 3.1$ pc, from which they concluded that the Hipparcos parallaxes are in error. The FGS 1R unit used for this study is demonstrably superior to the FGS 3 unit used for earlier HST-based parallaxes, where very significant discrepancies for three stars in the Hyades observed by both FGS 3 (van Altena *et al.*, 1997b) and Hipparcos were noted by Perryman *et al.* (1998). Additional hypotheses or intermediate data are still required to derive the FGS parallaxes, notably reference star spectrophotometric absolute parallaxes, UCAC 2 proper motions, and an estimated cluster depth. None of the objects was observed by Hipparcos (at V = 10–14 mag they are fainter than the typical Hipparcos Catalogue stars), so an object-by-object comparison is not possible.

Two binary systems in the Pleiades have received specific attention: HD 23642 and HD 23850 (Table 6.3). It is useful to recall that direct distance measurements follow only for the double-lined spectroscopic binaries systems which are also astrometrically-resolved, for which both radial velocities and orbital motion can be measured, while calculations based on either orbital motion or radial velocity alone must rely on stellar models to provide a distance estimate (the more traditional method of determining distances to eclipsing binaries is to estimate the luminosity of each star from its radius and effective temperature, and apply bolometric corrections before comparing with the apparent visual magnitude to establish the distance modulus).

A determination based on the double-lined spectroscopic binary HD 23642 by Giannuzzi (1995) gave $d = 132.4$ pc ($m - M = 5.61 \pm 0.26$) for an assumed age of 1.5×10^8 yr. The discovery from the Hipparcos epoch photometry that the system is also eclipsing (the only one so far known in the cluster), with shallow eclipses (Torres, 2003), prompted a re-analysis by Munari *et al.* (2004) yielding $d = 131.9 \pm 2.1$ pc. In such analyses, there is one parameter, the temperature of the primary, which cannot be modelled and must be determined independently, although a robust estimate of $T = 9671 \pm 46$ K, with $E(B - V) = 0.012 \pm 0.004$, could be derived from existing photometry, based on [Fe/H] = 0.00 ± 0.02. The method also depends on theoretical calculations to provide bolometric corrections. Southworth *et al.* (2005), see also Southworth *et al.* (2004), analysed three methods to improve the distance estimates, based on bolometric corrections to determine absolute visual magnitudes, and on the use of empirical surface brightness relations in terms of colour indices or effective temperature (Section 7.2.2), both in the visual and in the infrared K-band, the latter yielding improved accuracy due to a smaller dependence on interstellar reddening and metal abundance. They used *JHK* photometry from 2MASS, and Tycho B_T and V_T magnitudes transformed to the Johnson system according to the updated calibration by Bessell (2000). The observed mass–radius relation was also used as a further argument against an enhanced He abundance or a metallicity departing strongly from solar. Table 6.3 summarises these distance estimates, along with the values from the Hipparcos Catalogue, and according to the large-scale re-reduction by van Leeuwen (2007).

Two new ground-based distances were given at around the same time for the bright double star Atlas (HD 23850): by Pan *et al.* (2004) and by Zwahlen *et al.* (2004). Pan *et al.* (2004) used interferometric measurements, combined with an assumed mass–luminosity relation in the absence of a spectroscopic orbit, to argue that the distance was larger than 127 pc, and probably in the range 133–137 pc. Zwahlen *et al.* (2004) subsequently published a spectroscopic orbit, and new interferometric measurements, and derived a distance of 132 ± 4 pc, although the astrometric orbit was used to constrain most of the orbital parameters for the spectroscopic orbit. Again, Table 6.3 summarises these distance estimates, along with the values from the Hipparcos Catalogue, and according to the re-reduction by van Leeuwen (2007).

A further distance constraint may ultimately be derived from asteroseismology (Section 7.8). For six δ Scuti stars in the Pleiades, none of which have a Hipparcos parallax, Fox Machado *et al.* (2006) established a grid of five metallicities, three ages, and four distance moduli, minimising the difference in observed versus computed frequencies as a function of rotation rate, and assuming a common metallicity, age and distance for all stars. They favoured $Z = 0.02$, $Y = 0.28$, an age of 70 Myr, and a distance modulus of 5.70. It does seem that further constrained by an age of 100 Myr, a distance modulus of 5.40 would be (almost) equally compatible with the $Z = 0.015$, $Y = 0.30$ model. The assumption of a single common distance of the δ Scuti stars may also be inappropriate.

Other complications An early study by Jones (1972) argued that the faintest members of the Pleiades fall almost entirely below the ZAMS, an effect he attributed to grey extinction attributable to circumstellar shells. The issue of grey extinction has recently reappeared in the context of other Hipparcos observations (Section 8.5.2). While it has not been put forward in the literature as an explanation for the Pleiades anomaly, its effects would work in the required direction.

Stello & Nissen (2001) raised the possibility that a markedly non-spherical shape of the cluster could at least in part explain the anomaly. At 6° ($\sim$ 14 pc) radius a spherical distribution would result in differences of order 1 mas. Raboud & Mermilliod (1998) estimated the core radius as $r_c = 0\overset{\circ}{.}6$ or 1.4 pc, while the tidal radius is $r_t = 7\overset{\circ}{.}4$ or 16 pc. The tidal radius of a cluster in the direction of the Galactic centre is related to the total mass through $GM_c = 4A(A - B)\, r_t^3$, where G is the gravitational constant, A and B are Oort's constants of Galactic rotation, and the equation is from King (1962) for a cluster at the same distance from the Galactic centre as the Sun. A projected ellipticity of $e = (1 - b/a) = 0.17$ was estimated by Raboud & Mermilliod (1998), consistent with the work of Wielen (1974) who predicted ratios of the three orthogonal axes of the tridimensional-ellipsoid cluster, as 2.0 : 1.4 : 1.0, with the largest axis pointing toward the Galactic centre, i.e. roughly aligned with the line-of-sight for a Galactic longitude $b = 167°$. They hypothesised that the first born bright stars (O- and B-type) formed in one part of the gas cloud, blowing it preferentially in a one-directional tail in which the fainter F- and G-type stars subsequently formed.

The reverse sequence was actually proposed by Herbig (1962), who argued as follows: '*It is proposed that ... the formation of a cluster or association is a very gradual process in which less massive stars are formed over a long interval in a massive dark cloud, from which most are unable to escape. This gradual buildup of the lower and middle parts of the luminosity function within the cloud continues until a high-luminosity O- or very early B-type star forms. The radiation of this star heats the cloud in its neighbourhood so as to ionise the hydrogen, evaporate nearby dust, and induce a degree of kinetic and turbulent activity that largely puts an end to ordinary star formation in that volume, although it may actually trigger the formation of additional large-mass stars in its immediate neighbourhood. It is approximately the time since this event that is given by conventional dating technique, not the total time since the first cluster stars. If the stars formed within a cloud and gravitationally imprisoned therein are able to escape following the dissipation of the gas that results from the formation of an O-type star, then, in the process of their diffusion outward, they might be identified as an expanding association. If the mass density due to stars alone is high enough, then the trapped stars will continue together as a group and will appear as a stable cluster.*' In this scenario, the Hipparcos mean distance has highest weight from the brightest stars, while the main-sequence fitting gives the largest weight to the fainter stars, where the slope of the main sequence is less steep.

Table 6.3 Binary star distance estimates for the Pleiades: $d_{\rm Hip}$ are Hipparcos-derived values, while $d_{\rm binary}$ are from ground-based observations. Notes: (1) the original published values; in the case of Atlas, this was already an 'orbital solution' making use of the 290-day spectroscopic orbit, but with an assumed $e = 0$; (2) the original published data taking account of the improved orbit from Zwahlen et al. (2004); (3) derived from the large-scale re-reduction of the Hipparcos data.

HD	HIP	$d_{\rm Hip}$ (pc)	$d_{\rm binary}$ (pc)	Reference
23642	17704	111 ± 12		ESA (1997)[1]
		106 ± 8		van Leeuwen (2007)[3]: $\pi = 9.43 \pm 0.75$
			131.9 ± 2.1	Munari *et al.* (2004)
			139.1 ± 3.5	Southworth *et al.* (2005)
23850 (Atlas)	17847	117 ± 14		ESA (1997)[1]
		122 ± 14		Lindegren (2006, priv. comm.)[2]
		117 ± 6		van Leeuwen (2007)[3]: $\pi = 8.52 \pm 0.40$
			135 ± 2	Pan *et al.* (2004)
			132 ± 4	Zwahlen *et al.* (2004)

In a study similar to that carried out for the Hyades (Section 6.3) Makarov & Robichon (2001) searched for additional cluster members, through a variation of the classical moving cluster method, from the long temporal baseline Tycho 2 Catalogue (and thus with proper motions less sensitive to orbital motion), and correlated the resulting 87 objects with two ROSAT X-ray surveys of the cluster. They found that strong X-ray sources exhibit a velocity dispersion in one component of only 0.20 km s^{-1}, while the low-luminosity and non-X-ray detections are consistent with a dispersion of 0.64 km s^{-1}. Although significant only at the 1.6σ level, it is in the same sense as that found for the Hyades. High X-ray luminosity was found to correlate well with visual (i.e. wide separation, $\gtrsim 10$ AU) binaries. Following the arguments of Herbig (1962), if the Pleiades were a coeval system the turn-off point should be accompanied by a distinct swing of the lower main sequence into the pre-main-sequence domain where the lowest mass stars have still not reached the ZAMS. This could in turn be caused by the presence of low-mass stars much older than the turn-off age, implying a spread in age qualitatively consistent with the distribution in internal velocity dispersion. By appealing to the relaxation time of a stellar cluster as given by $T_E \propto \overline{R}^{3/2}$ where $\overline{R} = 3.5$ pc is approximately the core radius (Chandrasekhar, 1942), a total mass of $600 M_\odot$ yields $T_E = 60$ Myr, close to the cluster age. Thus, if the star formation was extended over, say, 30 Myr (they consider an age range derived from $\log L_{\rm X}$ of 15–150 Myr), the last generation of stars would not be dynamically relaxed, and some of the solar-type stars would be considerably older than the more massive B and A stars.

6.4.5 Summary of the Pleiades distance

An examination of all of the above leaves a confusing picture. On the one hand, the original Hipparcos results, and the new reduction by van Leeuwen (2007), yield a mean distance of the observed 50 or so members of around 122 pc. Other work points to a somewhat larger mean distance of around 135 pc. Yet all other work is to a greater or lesser degree model dependent; is based on an extension of the Hyades main-sequence models (albeit permitting different chemical compositions); the Hipparcos and main-sequence samples are largely distinct, and at the same time, effects such as rapid rotation and a possible spread in age remain incompletely modelled; independent direct distance determinations either refer to different stars, or are within 1–2σ of the Hipparcos results for the same stars; and all workers ignore any depth effect in the cluster, which in the Hipparcos parallaxes extends at least from $\pi = 7$–9 mas, i.e. from $d = 111$–142 pc (with 17 stars spanning the range of $\pi = 7.5$–8.5 mas).

Two possible resolutions of this continuing paradox can be considered, which will be clarified by the next generation of space astrometry missions (SIM and Gaia) if not before: (a) the Hipparcos parallaxes remain systematically displaced by of order 1 mas in the region of the Pleiades even in the new reduction of the Hipparcos data; (b) the inconsistency lies in the different star samples (in V, and thus in $B-V$ and/or in mass) probed by the mean Hipparcos limiting magnitude (typically $B-V<0.5$) and the range probed by main-sequence fitting (typically $B-V>0.5$). This difference could involve a slightly displaced spatial distribution of these two populations, e.g. arising from a flattening and mass segregation in the viewing direction itself more-or-less aligned with the Galactic centre, or some other effect such as grey extinction.

The fact that, in both the original and re-reduced Hipparcos Catalogue data, the displacement of the Pleiades main sequence is matched by comparable displacements of some other clusters either requires a conspiracy in the patterns of systematic errors with cluster position on the sky, or implies evolutionary effects not fully accounted for in main-sequence modelling.

6.5 Distances to other nearby clusters

The Hipparcos Input Catalogue was constructed with a completeness targeted to $V \leq 7.9 + 1.1 \sin b$, where b is the Galactic latitude, with fainter stars included according to overall 'importance', but not exceeding a few stars per square degree.

The nearby clusters Hyades, Pleiades, α Per, Coma Ber, and Praesepe were given particular weight, but for other clusters, membership in the Hipparcos Catalogue is limited in surface density, and is highly incomplete. Hipparcos observed stars in all open clusters closer than 300 pc and in the richest clusters located between 300–500 pc. Each sequence covers a large range of stellar masses, and thus in principle provides further tests of the internal structure models for a wide range of initial parameters, in particular for different metallicities.

As described in Section 6.4.2, van Leeuwen (1999a) and Robichon *et al.* (1999b) used the combined abscissae method to determine the mean parallax and proper motion of several clusters using the Hipparcos Intermediate Astrometric Data for individual cluster members. Van Leeuwen (1999a) derived mean distances in this way for all nine clusters within 300 pc (except the Hyades), i.e. Coma Ber, Pleiades, IC 2602, IC 2391, α Per, Praesepe, NGC 2451, Blanco 1, and NGC 6475. Robichon *et al.* (1999b) derived mean distances for the same nine nearby clusters, nine additional rich clusters between 300–500 pc (i.e. NGC 7092, NGC 2232, IC 4756, NGC 2516, Trumpler 10, NGC 3532, Collinder 140, NGC 2547, and NGC 2422), and 32 more distant clusters with at least four Hipparcos members. For clusters closer than 500 pc, the accuracy on the mean parallax is in the range 0.2–0.5 mas and the mean proper motion accuracy is ~0.1–0.5 mas yr^{-1}. Results from the two groups are in good internal agreement.

The relative positions of the resulting main sequences in the colour–magnitude diagram (Figure 6.10) were inconsistent with the prevailing idea that differences in metallicity fully explain the relative positions of the non-evolved parts of the main sequence of different clusters. These results were discussed extensively in the literature (Mermilliod *et al.*, 1997; Robichon *et al.*, 1997a; Mermilliod, 2000; Robichon *et al.*, 2000). Some clusters have different metallicities but define the same main sequence in the $(M_V,\ B-V)$ plane (Praesepe, Coma Ber, and α Per). Coma Ber has, for example, a quasi-solar metallicity while its sequence is similar to that of the Hyades, or of the metal-rich Praesepe. Some clusters sequences (Pleiades, IC 2391 and IC 2602) are abnormally faint with respect to others, for instance Coma Ber. The metallicity of the Pleiades as determined from spectroscopy is almost solar, and similar to that of Coma Ber, but its main sequence lies ~ 0.3–0.4 mag below those of Praesepe, Coma Ber, or the Hyades. Van Leeuwen (1999a,b) suggested a possible (although unexpected) correlation between the age of a cluster and its position in the HR diagram.

The Hipparcos distances to the five closest open clusters (Hyades, Pleiades, α Per, Praesepe, and Coma Ber) was compared to those derived from main-sequence fitting by Pinsonneault *et al.* (1998); they compared theoretical isochrones, translated into the colour-magnitude plane by means of Yale colour calibrations, to observational data both in the $(M_V,\ B-V)$ and $(M_V,\ V-I)$ planes. The $B-V$ colour index is more sensitive to metallicity than $V-I$ (Alonso *et al.*, 1996), so Pinsonneault *et al.* (1998) derived as a by-product the value of the metallicity that gives the same distance modulus in the two planes and compared it to spectroscopic determinations. They judged their distance moduli to be in good agreement with Hipparcos results except for the Pleiades and Coma Ber. For Coma Ber, the problem could result from the old *VRI* colours used. For the Pleiades, the discrepancy with Hipparcos amounts to 0.24 mag, and the [Fe/H]-value derived from main sequence fitting in the two colour planes agrees with the spectroscopic determination of Boesgaard & Friel (1990), $[Fe/H] = -0.034 \pm 0.024$, although values in the range -0.03 to $+0.13$ can be found in the literature. In fact, with that metallicity the Hipparcos sequence of the Pleiades could be reproduced by classical theoretical models, provided they have a high helium content. The exact value depends on the model set and its input physics: Pinsonneault *et al.* (1998) found $Y = 0.37$, Belikov *et al.* (1998) found $Y = 0.34$ but for $[Fe/H] = 0.10$, and Lebreton (2000b) reported $Y \sim 0.31$. In any case, high helium content is only marginally supported by observations (Nissen, 1976). Pinsonneault *et al.* (1998) examined other possible origins of the discrepancy, such as erroneous metallicity, age-related effects, and reddening, and concluded that none of them is likely to be responsible for the Pleiades discrepancy.

Results of the global Hipparcos re-reduction by van Leeuwen (2007) are shown in Figure 6.12. The impression of an age-related effect responsible for the difference between the Hyades and Pleiades persists, with the close coincidence of the Hyades and Praesepe loci, and those of the Pleiades and clusters like IC 2602, IC 2391, and α Per. Coma Ber, with an age intermediate between that of the Hyades and Praesepe on the one side, and that of the Pleiades on the other, is situated between the two sequences.

Other studies and reports of distance scale and main-sequence fitting using the Hipparcos data include:

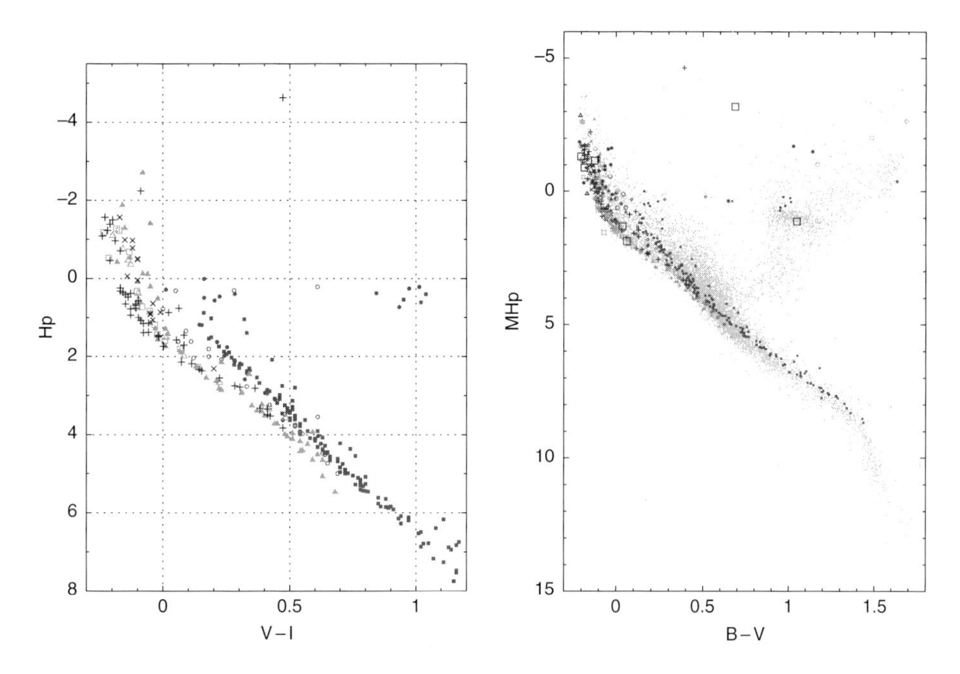

Figure 6.12 Left: the combined HR diagram for the eight nearest open clusters: ■ *Hyades;* • *Praesepe;* ○ *Coma Ber;* ▲ *Pleiades;* △ *IC 2391;* □ *IC 2602;* × *NGC 2451;* + *α Per. Right: the 9344 single stars with parallaxes better than 5% (grey background), and stars from 20 open clusters colour-coded according to age: oldest in red (Hyades, Praesepe); Pleiades and Pleiades-like ages in green; and the youngest in blue. From the re-reduction of the Hipparcos data by van Leeuwen (2007, originals in colour), reproduced with kind permission of Springer Science and Business Media.*

nearby open clusters and an improvement of the HR diagram calibrations (Robichon *et al.*, 1997b); the HR diagram of the four nearest open clusters: Hyades, Coma Ber, the Pleiades and Praesepe (Cayrel de Strobel *et al.*, 1998); open clusters, Cepheids and the problems of distance scale (Rastorguev *et al.*, 1998); sequences of nearby open clusters with Hipparcos (Robichon *et al.*, 1999c); main sequences of open clusters (Robichon *et al.*, 1999a); the WIYN open cluster study and a distance scale (Platais *et al.*, 2000); Hipparcos trigonometric parallaxes and the distance scale for open star clusters (Loktin & Beshenov, 2001); an empirical procedure to estimate distances to stellar clusters (Allende Prieto, 2001); and Hipparcos open clusters as a test for stellar evolution (Tordiglione *et al.*, 2003).

6.6 Other astrophysical applications

Chen *et al.* (1998) used cluster distances defined by the mean Hipparcos parallaxes to determine extinction as a function of distance for $|b| < 10^\circ$ (Section 8.5.2).

Dambis (1999) stressed the different distance scales on which pre-Hipparcos cluster distances have been based (different ZAMS, different theoretical grids, use of Cepheids, etc.) and constructed a homogeneous scale adopting the empirical ZAMS of Kholopov (1980), isochrones from Maeder & Meynet (1991), and calibrated using the Hipparcos distance scale of the seven highest-precision clusters. Ages and distances were derived for 203 clusters with ages $\lesssim 160$ Myr. In addition to a strong correlation between age and heliocentric distance attributable to (luminosity-dependent) selection effects, the spatial distribution in the Galactic *XY* plane shows a rather uniform radial distribution of the older clusters, but a clear delineation of the spiral arm pattern for those with ages $\lesssim 30$ Myr (Figure 6.13).

Dambis (2003) used the same 203 clusters to examine the age dependence of the vertical distribution of young open clusters. This study made limited use of the Hipparcos astrometric data, but connects to several other Hipparcos-based studies. They found that the vertical scale height of the clusters varied non-monotonically with age, instead exhibiting a wavelike pattern, similar to that reported for classical Cepheids by Jôeveer (1974), as developed in the context of Hipparcos studies by Dambis (2004, see also Section 5.7.1). Using the mass density of $\rho = 0.102 \pm 0.006\, M_\odot\,\mathrm{pc}^{-3}$ determined in the solar neighbourhood from the Hipparcos data by Holmberg & Flynn (2000, see Section 9.4.1), and cluster evolutionary ages computed with and without overshooting (Pols *et al.*, 1998), he argued that the period of the vertical oscillations, $P_z = 74 \pm 2$ Myr,

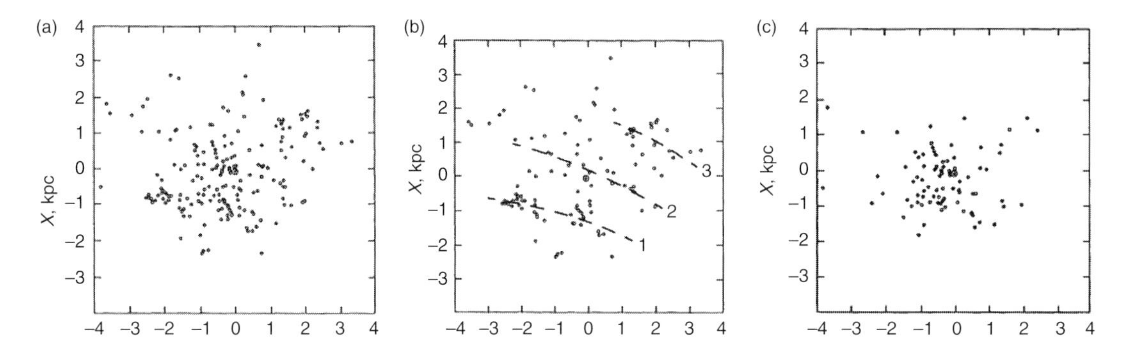

Figure 6.13 The distribution of 203 Galactic open clusters projected onto the Galactic XY plane: (a) all clusters; (b) clusters with ages $\lesssim$ 30 Myr, with dashed lines outlining the spiral arms Carina–Sagittarius (1), Cygnus–Orion (2), and Perseus (3); (c) clusters with ages $\gtrsim$ 30 Myr. From Dambis (1999, Figure 5), reproduced with kind permission of Springer Science and Business Media.

is only consistent with evolutionary models computed with no, or only mild, overshooting. As well as confirming the absence of dark matter distributed in the form of the disk, he argued that open clusters are formed with excess kinetic energy in z, with initial vertical coordinates strongly correlated with initial vertical velocity. This favours a scenario where star formation in the disk is triggered by mass infall onto the Galactic plane, or in which the vertical motions are triggered by spiral density waves (cf. Fridman *et al.*, 1998).

Loktin & Beshenov (2003) derived mean proper motions of 167 clusters from the Tycho 2 Catalogue data, using the new data to infer the Galactic rotation rate at the solar circle of $\Omega_0 = +24.6 \pm 0.8\,\mathrm{km\,s^{-1}\,kpc^{-1}}$. Frinchaboy (2006) reported multi-fibre spectrograph radial velocities of 67 open clusters to derive improved membership and bulk cluster kinematics when combined with Tycho 2 proper motions, also using these results to derive the rotational velocity of the local standard of rest.

Lépine *et al.* (2008), see also Lépine *et al.* (2006), used the observed velocity vector of open clusters in the Galactic plane (taken from Dias *et al.*, 2006) to estimate the epicycle frequency, κ (Section 9.1.3). If, for some reason, clusters are formed with non-random initial velocity perturbations, then a plot of the orientation angle of their residual velocity (with respect to the Galactic rotation curve) as a function of age reveals the epicycle frequency directly. They adopted $R_0 = 7.5$ kpc, selected clusters in the distance range $7.1 < R < 7.9$ kpc, and found $\kappa = 44 \pm 3\,\mathrm{km\,s^{-1}\,kpc^{-1}}$ and $V_0 = 230 \pm 15\,\mathrm{km\,s^{-1}}$. This method of estimating κ, and V_0/R_0, follows from the accurate distances and proper motions now available for large numbers of open clusters. The results also confirm that clusters are formed with some coherent velocity perturbations in the direction of Galactic rotation.

6.7 Searches for new clusters and members

Platais *et al.* (1998) searched for (new) star clusters and associations in the Hipparcos Catalogue using the approach described in Section 6.2.3 based on iterated vector-point diagrams tiling the celestial sphere. They found 102 suspected and confirmed open clusters, with two larger groups of clusterings perhaps constituting extended associations in Carina–Vela and Cepheus–Cygnus–Lyra–Vulpecula. They found evidence for a single new nearby cluster around the central star 'a Car' (HIP 45080), which has similar characteristics to IC 2391 and IC 2602 which flank it on the sky.

Baumgardt *et al.* (2000) focused on clusters beyond 200 pc. They used the list of Lyngå (1988), rejected clusters with limited ground-based information and those where the brightest members are fainter than 12 mag, then searched 360 clusters for possible members based on ground-based photometry and radial velocity, and Hipparcos parallaxes and proper motions. They found 630 certain and 100 possible members of 205 clusters, and derived resulting mean cluster proper motions and parallaxes. They found generally good agreement with the distance scale of Dambis (1999), but argued for a decrease of 12 ± 6% in the photometric distance scale of Loktin *et al.* (1997). They also used the test of Arenou & Luri (1999) to examine the Hipparcos parallax errors: from the absence of a broadening in the (Gaussian) distribution of $\Delta\pi/\sigma_{\rm Hip}$ they excluded errors of more than a few tenths of a mas for the vast majority of clusters (Figure 6.14), in contrast to correlation of errors over angular scales of 2–3° of up to 2 mas proposed to explain the Pleiades distance anomaly by Narayanan & Gould (1999b).

The availability of the Tycho 2 Catalogue in 2000 made further proper motion membership studies accessible. Sanner & Geffert (2001) selected nine of the well-observed clusters within 315 pc (from van Leeuwen, 1999a; Robichon *et al.*, 1999b), chosen to provide good

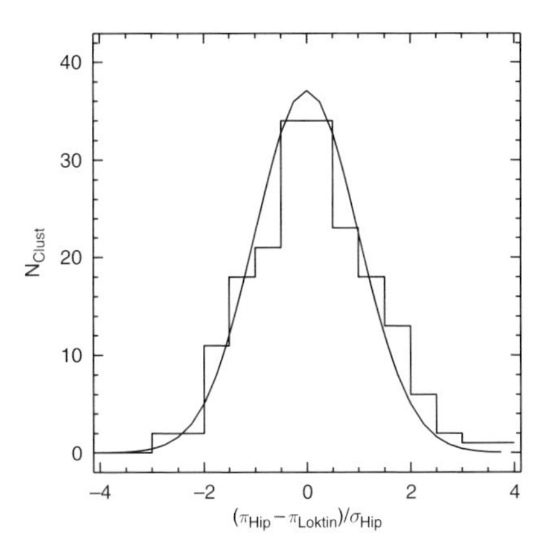

Figure 6.14 Histogram of the normalised differences $(\pi_{\rm Hip} - \pi_{\rm Loktin})/\sigma_{\rm Hip}$. A Gaussian provides a good fit to the data. Correlated errors of the order of 0.5 mas or larger would significantly broaden the observed distribution and can therefore be ruled out for the vast majority of clusters. From Baumgardt et al. (2000, Figure 3).

proper motion separation between the clusters and the field (Blanco 1, Stock 2, α Per, Pleiades, NGC 2451A, IC 2391, Praesepe, IC 2602, and NGC 7092). Common proper motion candidates for each were selected from Tycho 2, basing membership probabilities on two overlapping bivariate distributions in the vector-point diagram (Sanders, 1971). From the selected stars for each cluster (ranging between 25 for NGC 7092, 127 for the Pleiades, and up to 204 for Stock 2), they determined distances, evolutionary ages, and slope of the initial mass function. Exponents of the initial mass function ranged from $\Gamma = -0.69$ to -2.27. The study also provides an indication of the completeness limit for Tycho 2 which, they suggest, may reach to only $V_T \sim 10{-}10.5$ mag in some sky regions.

Dias *et al.* (2001, 2002b) determined mean proper motions from the Tycho 2 Catalogue for known open clusters within 1 kpc, and beyond 1 kpc respectively, selected from the database of Mermilliod (1995), and also assigning membership probabilities using the method of Sanders (1971). Dias *et al.* (2001) studied 164 clusters, from which they considered 4006 stars as members. They provided mean absolute proper motions for 112 clusters, of which 28 were determined for the first time, and found good agreement with results from Baumgardt *et al.* (2000). Dias *et al.* (2002b) determined 2021 members of 94 clusters beyond 1 kpc, deriving mean absolute proper motions, 55 for the first time.

Alessi *et al.* (2003) found 11 new clusters in the Tycho 2 Catalogue data (Alessi 1, 2, etc.). All are large, 20–110 arcmin on the sky, rather nearby ($<$ 600 pc), rather bright but poor, indicating either nearby sparse clusters or open cluster remnants. Discovery was based on similar techniques as those applied by Platais *et al.* (1998, see also Section 6.2.3), i.e. a combination of proper motion vector-point diagrams, enhancement in sky density in terms of radial density profiles, colour–magnitude diagrams, and visual inspection of the Digitised Sky Survey (DSS).

Kharchenko *et al.* (2004, 2005b,c) provided a supplement to the ASCC 2.5 Catalogue (itself based on Tycho 2), with membership probabilities for 520 Galactic open clusters for which at least three probable members (18 on average) could be identified. The compilation includes proper motions for all objects in the Hipparcos system, and parallaxes where available. A further 109 new clusters discovered from the ASCC 2.5 were subsequently reported by Kharchenko *et al.* (2005a), revised to 130 new clusters by Piskunov *et al.* (2006b). These new clusters are on average brighter, have more members, but cover larger angular radii than those previously known (Figure 6.15).

Piskunov *et al.* (2006a) used this Tycho 2-based cluster sample, considered to be complete to about 850 pc from the Sun, to draw a number of conclusions. The symmetry plane of the cluster distribution is determined to be at $Z_0 = -22 \pm 4$ pc, with a scale height of only 56 ± 3 pc (Figure 6.16). Similar figures were reported by Joshi (2005). The total surface density and volume density in the symmetry plane are $\Sigma = 114\,{\rm kpc}^{-2}$ and $D(Z_0) = 1015\,{\rm kpc}^{-3}$, respectively. They estimate the total number of open clusters in the Galactic disk to be of order of 10^5 at present. Fluctuations in the spatial and velocity distributions are attributed to the existence of four open cluster complexes of different ages containing up to a few tens of clusters, each showing the same kinematic behaviour, and a narrow age spread. The youngest cluster complex, $\log t < 7.9$, with a 19° inclination to the Galactic plane, is apparently a signature of the Gould Belt. The most abundant complex has moderate age, $\log t \sim 8.45$. The oldest, $\log t \sim 8.85$ and sparsest group was identified due to its large motion in the Galactic anticentre direction. Formation rate and lifetime of open clusters were found to be $0.23 \pm 0.03\,{\rm kpc}^{-2}\,{\rm Myr}^{-1}$ and 322 ± 31 Myr, respectively, implying a total number of cluster generations over the history of the Galaxy somewhere between 30–40.

Related work includes: seven nearby open clusters using Hipparcos data (Robichon *et al.*, 1997a); open clusters in the Hipparcos Catalogue (Baumgardt *et al.*, 1998) statistical parallaxes and kinematical parameters of classical Cepheids and young star clusters (Rastorguev *et al.*, 1999); open cluster structural parameters and proper motions (Kharchenko *et al.*, 2003); and proper motions of open star clusters from the Tycho 2 Catalogue (Beshenov & Loktin, 2004).

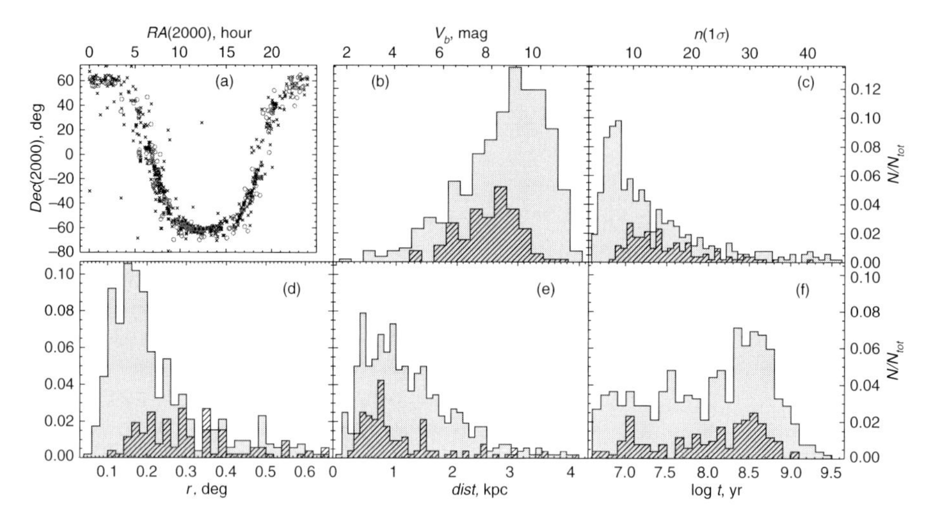

Figure 6.15 Open clusters from the ASCC 2.5 data (itself derived from the Tycho 2 Catalogue), showing previously-known and newly-discovered clusters. (a) distribution of clusters over the sky: × known clusters, ○ newly discovered. Other panels give histograms normalised to the 520 known clusters, with filled and hatched histograms indicating known and newly-discovered clusters respectively: (b) distribution of clusters versus magnitude of the brightest member; (c) distribution of the number of members; (d–f) angular radius, heliocentric distance, and age, respectively. From Kharchenko et al. (2005a, Figure 3).

6.8 Specific clusters

This section summarises some individual cluster investigations making use of the Hipparcos data, in addition to those exclusively deriving mean cluster distances from the catalogue data, including reductions based on the combined abscissae solution method described previously.

α **Per (= Per OB3):** Makarov (2006) found 139 possible cluster members, 18 new, from a combination of Tycho 2 and UCAC 2 proper motions. A convergent-point analysis gave an upper limit to the one-dimensional velocity dispersion of 1.1 km s^{-1}, and an overall contraction of the cluster is suggested. An age of 52 Myr was derived from isochrone fitting. The cluster is surrounded by an extensive sparse halo of co-moving dwarfs, found by combining the UCAC 2 proper motions with 2MASS data. Peña & Sareyan (2006) reported detailed photometry, and a resulting membership determination which they compared with that based on the Hipparcos proper motions.

Coma Ber: Odenkirchen *et al.* (1998) found 51 kinematic members from an analysis of Hipparcos, Tycho, and ACT data. They derived the cluster's distance, size, structure and an age of ~500 Myr, finding an elliptical core-halo system with major axis parallel to the Galactic motion, and a moving group of stars beyond the tidal radius, indicating cluster evaporation. Odenkirchen *et al.* (2001) extended the analysis using the Tycho 2 proper motions, which they showed are more tightly concentrated around the cluster mean motion because of the longer temporal baseline and thus lower sensitivity to binary systems. They found a one-dimensional velocity dispersion below 0.7 km s^{-1}, and confirmed at least 11 members beyond the tidal radius. These are located at the same distance as the cluster, i.e. not spread out in the vertical direction, a result attributed to the compressive tidal force of the Galactic disk perpendicular to the plane. These extra-tidal members, both Hipparcos and Tycho, are found only leading the cluster, an asymmetry which is unexplained. Abad & Vicente (1999) constructed an extended catalogue of positions and proper motions down to 14 mag, on the Hipparcos system and with comparable proper motion accuracies. Makarov (2003) used the Hipparcos Intermediate Astrometric Data to improve individual parallaxes, as well as a mean cluster parallax of 12.40 ± 0.17 mas compared to 11.49 ± 0.21 mas obtained by Robichon *et al.* (1999b), and 12.36 ± 0.23 mas from main-sequence fitting by Pinsonneault *et al.* (1998).

Collinder 21: Villanova *et al.* (2004) used Tycho 2 proper motions to show that the stellar density enhancement, previously taken as suggestive of a dissolving open cluster remnant, are chance alignments of unrelated stars.

Collinder 121: Robichon *et al.* (1999b) established a mean cluster parallax of 1.80 ± 0.24 mas (556 pc) from 13 selected Hipparcos members. De Zeeuw *et al.* (1999) identified a coherent structure of 103 stars forming a moving group at a distance of 592 ± 28 pc, with linear dimensions 100×30 pc^2, and all the characteristics

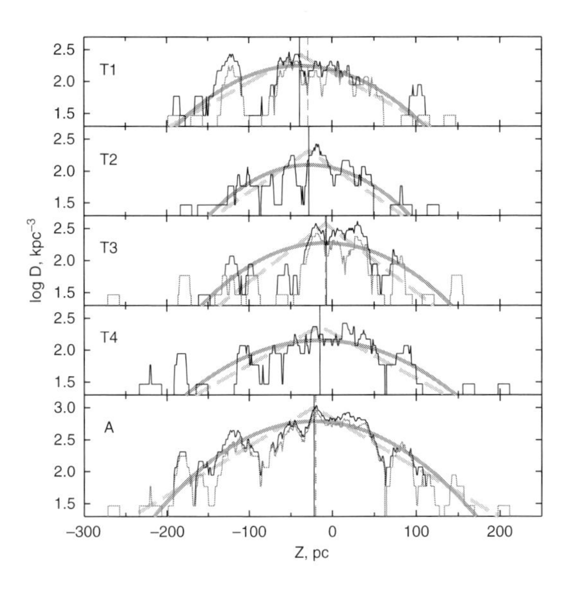

Figure 6.16 Distribution of clusters perpendicular to the Galactic plane, constructed for clusters within a completeness limit of 850 pc. Panel A: all clusters; panels T1–T4: the four age samples (from youngest, $\log t \leq 7.9$*, to oldest,* $\log t \geq 8.6$*). The thin black curves indicate smoothed spatial densities. The thin curves in panels T1, T3 and A mark 'field' subsamples of the corresponding ages. The vertical solid lines give positions (*Z_0*) of symmetry planes, and the dashed vertical lines show* Z_0 *of the field subsamples. The thick solid curves correspond to Gaussian fits, while the thick long-dashed lines give exponential fits. From Piskunov et al. (2006a, Figure 3).*

of an unbound OB association. Kaltcheva (2000) identified a more distant cluster in the same field at ~1 kpc, consistent with the original definition of the cluster by Collinder (1931), and demonstrating that the OB association is a foreground structure, designated as CMa OB2 by Eggen (1981). Similar results were reported by Zeballos *et al.* (2006). Burningham *et al.* (2003) confirmed this finding based on low-mass pre-main-sequence stars detected by XMM–Newton and ROSAT, and Kaltcheva & Makarov (2007) demonstrated agreement between the photometric parallaxes and those recomputed from the Hipparcos data taking account of correlated errors in the attitude terms. **Collinder 399:** Baumgardt (1998) used the Hipparcos and ACT proper motions to refute its existence.

IC 348: Scholz *et al.* (1999) extended membership studies to $R = 18$ mag from Schmidt plates over one square degree of the young cluster, which lies in the region of the Per OB2 association (Section 6.10). From the Hipparcos stars they derived a mean distance of 261^{+27}_{-23} pc, nearly the same as that of de Zeeuw *et al.* (1999), and confirming that IC 348 is a cluster embedded within the Per OB2 association. **IC 2391:** Platais *et al.* (2007) obtained new proper motions for 6991 stars down to $V = 13$–16, calculated membership probabilities, selected 66 stars as probable members, and obtained a distance modulus based on main-sequence fitting 0.19 mag larger than that derived from the Hipparcos parallaxes. This 'offset' between the Hipparcos and the main-sequence fitting distances for this young, ~40 Myr, cluster is about the same as that found for the Pleiades. **IC 2602:** Kaltcheva *et al.* (2005) calibrated distance estimates from Strömgren photometry and Hipparcos parallaxes, and from a database of O–B9 stars in the cluster region at 156 ± 6 pc, also identified background concentrations at 546 ± 33 pc and 1117 ± 61 pc which together are consistent with the thickness of a spiral arm. A background concentration at about 2.6 kpc could signify the interarm space between the Local Association and the more distant spiral arms. **IC 4665:** de Wit *et al.* (2006) used the Tycho 2 proper motions as part of their membership study and determination of the mass function. **M67:** Loktin (2004) presented a compiled proper motion catalogue using five catalogues of relative proper motions and the Tycho 2 Catalogue. **Melotte 227** and **Upgren 1:** Baumgardt (1998) used the Hipparcos and ACT proper motions to refute their existence.

NGC 133: Carraro (2002) used photometric data to confirm the reality of this cluster, establishing a mean distance of 630 ± 150 pc and an age < 10 Myr. **NGC 188:** Platais *et al.* (2003) used Tycho 2 data to re-calibrate old photographic plates, and to construct a catalogue of some 1050 probable members, including blue stragglers and red giants, to $V = 21$. Stetson *et al.* (2004) presented assembled photometric data for the cluster. **NGC 1252:** Baumgardt (1998) used the Hipparcos and ACT proper motions to refute its existence. **NGC 1348:** Carraro (2002) used photometric data to establish a distance of 1.9 ± 0.5 kpc and an age > 50 Myr. **NGC 1582:** Baume *et al.* (2003) combined CCD and 2MASS photometry with 53 Tycho 2 proper motions to derive a distance of 1100 ± 100 pc and an age of 300 ± 100 Myr for this heavily contaminated cluster. **NGC 1817:** Balaguer Núñez *et al.* (2004) derived proper motions from 25 photographic plates reduced to the Tycho 2 Catalogue, deriving a list of 169 probable members. **NGC 2451:** Carrier *et al.* (1999) used Hipparcos parallaxes and proper motions, combined with Geneva photometry, to confirm the reality of NGC 2451, but resolved into two clusters, A and B, with distances of 197 pc and 358 pc respectively, but both with the same age of ~ 50 Myr (see also Baumgardt, 1998). From the 12 cluster members, Robichon *et al.* (1999b) and van Leeuwen (1999a) used the combined Hipparcos abscissae solution method to derive a mean cluster parallax of 5.30 ± 0.20 mas. Platais *et al.* (2001) used an enlarged proper motion sample of

136 members to $V \sim 15$ to establish a similar distance from main-sequence fitting. **NGC 2516:** Terndrup *et al.* (2002) obtained radial velocities, rotation rates, and metallicities for a number of members. They derived a main-sequence fitting distance larger than those of Robichon *et al.* (1999b) and Sung *et al.* (2002). **NGC 2548:** Wu (2001) determined 165 probable members from 10 plates taken over 82 yr, and reduced to the Tycho 2 proper motion system. **NGC 2664** and **NGC 5385:** Villanova *et al.* (2004) used Tycho 2 proper motions to show that the stellar density enhancements, previously taken as suggestive of dissolving open cluster remnants are, in both cases, chance alignments of unrelated stars. **NGC 6231:** Makarov (2003) used the Hipparcos Intermediate Astrometric Data to improve individual parallaxes, as well as a mean cluster parallax of 1.7 ± 0.4 mas compared to -0.62 ± 0.48 mas obtained by Robichon *et al.* (1999b). All six members with negative parallaxes in the Hipparcos Catalogue yield positive parallaxes in the revised analysis. **NGC 6611:** Belikov *et al.* (1999a) constructed a compiled catalogue of 2200 stars to $V = 16.8$ mag in the region of NGC 6611, reduced to the Hipparcos system. **NGC 6791:** Chaboyer *et al.* (1999) derived age, extinction, and distance for this old, metal-rich open cluster, with the isochrones providing good agreement with the Hipparcos-based Hyades ZAMS. **NGC 6994 (M73):** Odenkirchen & Soubiran (2002) used Tycho 2 proper motions and other data to show that this controversial grouping is not a physical cluster or kinematic group.

Clusters with Cepheids Improved membership determination should assist estimates of the absolute magnitudes of Cepheids. Using Hipparcos proper motions, membership of the following Cepheids is confirmed: U Sgr in IC 4725 (M25), and S Nor in NGC 6087 (Lyngå & Lindegren, 1998; Baumgardt *et al.*, 2000); δ Cep in Cepheus OB6 (de Zeeuw *et al.*, 1999); and DL Cas in NGC 129, V Cen in NGC 5662, while rejecting membership of SZ Tau in NGC 1647 (Baumgardt *et al.*, 2000). Baumgardt *et al.* (2000) discussed various other likely and unlikely Cepheid members, including EV Sct and Y Sct in NGC 6664, BB Sgr in Collinder 394, and GH Car in Trumpler 18.

6.9 Kinematic groups

6.9.1 Introduction

In addition to clusters showing clear common spatial and kinematic properties, sparser stellar kinematic groups of 10–100 or more members have also been identified in the solar neighbourhood. Variously described as (old) moving groups, kinematic groups, dynamical streams, large-scale streams, and superclusters, these kinematically coherent groups of relatively old stars are recognisable because of the deviation of the velocities of their constituent stars from a purely circular velocity; their peculiar velocities are well above the velocity dispersion of field stars of comparable age.

The study of moving groups goes back more than a century, beginning with the work of Kapteyn (1905). Much of the recent work on kinematic groups has been by Eggen (see, e.g. Eggen, 1989). He defined a 'supercluster' as a group of gravitationally unbound stars that share the same kinematics and may occupy extended regions of the Galaxy (with an associated convergent-point), and a 'moving group' as the part of a supercluster that enters the solar neighbourhood and can be observed all over the sky (Eggen, 1994), although these definitions seem of little current relevance and will not be followed here.

Pre-Hipparcos status Considering the older and more extended moving groups of whatever nature, Table 6.4, taken from Soderblom & Mayor (1993, Table 1), lists 14 such groups in the solar neighbourhood with their Galactic space velocities. This listing includes the Hyades supercluster associated with the Hyades cluster (600 Myr), sometimes referred to as Eggen's Stream I; the Ursa Major group (or Sirius supercluster) associated with the Ursa Major cluster (300 Myr), sometimes referred to as Eggen's Stream II; the Local Association or Pleiades moving group of young stars comprising embedded clusters and associations such as the Pleiades, α Per, NGC 2516, IC 2602, and the Scorpius–Centaurus association (20–150 Myr); the IC 2391 supercluster (35–55 Myr); and the Castor moving group (200 Myr). Cluster lifetimes in the Galactic disk are typically limited to a few hundred million years as a result of internal relaxation, tidal effects of the Galactic field, and encounters with massive objects such as giant molecular clouds (Wielen, 1991), the latter also contributing to the observed increase of Galactic disk scale height with age, a process referred to as 'disk heating'. Although these relic moving groups, with distinctive kinematic and other signatures, should presumably exist at least as a continuum between bound clusters and the general field, evidence for their existence in general has been controversial, and unambiguous membership of putative groups frequently uncertain. For example, while the existence of the Hyades moving group (as opposed to the Hyades cluster) appears to be real, attempts to identify members unambiguously based on metallicity, chromospheric activity, Li abundance, or age have proved challenging.

Soderblom & Mayor (1993) argued that in order to be convincingly classified as a kinematic group, members should be moving through space at the same rate and in the same direction, with only a modest velocity dispersion; they should form a colour-magnitude diagram

Table 6.4 The 14 stellar kinematic groups in the solar neighbourhood, ordered by V, as tabulated by Soderblom & Mayor (1993, Table 1). Galactic space velocities (in km s^{-1}) are those of the defining star, and intended as indicative only. References to the groups, mostly from the work of Eggen, are given in the cited table.

Name	U	V	W
Ursa Major (Sirius) Group	+13	+1	−8
γ Leonis Group	−78	−4	−1
Hyades Group	−40	−16	−3
Local Association (Pleiades) Group	−9	−27	−12
Wolf 630 Group	+25	−33	+13
ϵ Indi Group	−78	−38	+4
ζ Herculis Group	−52	−47	−27
61 Cygni Group	−90	−53	−8
HR 1614 Group	−4	−58	−11
σ Puppis Group	−75	−88	−21
η Cephei Group	−33	−97	+10
Arcturus Group	+25	−115	−3
Groombridge 1830 Group	+277	−157	−14
Kapteyn's Star Group	+19	−288	−53

appropriate to stars of the same age and composition; and they should have the same velocity in the direction of Galactic rotation, V. This is because motions in U and W lead to oscillations of the star about the mean motion of the group, while diffusion in V removes the star from its cohort forever (Chapter 9).

The existence of old moving groups was certainly controversial before the Hipparcos results (see, as examples, Ratnatunga 1988 in the case of the Hyades and Ursa Major moving groups; Griffin 1998 in the case of the Pleiades moving group; and Taylor 2000 in the case of the HR 1614 moving group).

The debate continues, and the following summarises the Hipparcos contribution. Anticipating some of the results of this section for clarity, it now appears that these (old) moving groups in the solar neighbourhood perhaps include some evaporating halos of open clusters, while it seems rather well established that at least one of these streams (Hercules) can be identified with resonant dynamical structures due to the rotating bar, and quite possibly a number are systematic velocity structures imparted by spiral arm shocks (e.g. Gieles *et al.*, 2007). The 'young moving groups', probably representing the sparser and more immediate dissipation products of the more youthful associations, are treated separately in Section 6.10.

Cluster evaporation That at least some of these kinematically coherent groups might have resulted from cluster evaporation has been discussed by various workers, e.g. Agekyan & Belozerova (1979): within a cluster, energy exchange between members causes individual stars to acquire supercritical velocities and to leave the cluster. Tidal forces of the Galaxy accelerate the dissipation process, by diminishing the critical velocity in certain directions, beyond the 'tidal radius' of the cluster (King, 1962; Wielen, 1974). Essentially, the equipotential cluster surface becomes open at these distances, due to the effects of the Galactic tidal potential, providing an escape route for stars to evaporate from the system. Stars acquiring a velocity slightly below the critical velocity will be carried to the tidal radius, where gravitational interaction with field stars will prevent the star from returning to the central zone, instead transporting it to a halo or coronal region around the cluster from where it can orbit at a large distance, essentially quasi-stable against tidal forces (at a relative velocity of 0.5 km s^{-1}, such a star will be just 50 pc distant from the cluster after 100 Myr). For a typical open cluster, the tidal radius and the effective radius of the corona are comparable, and of order 6–10 pc. After all the stars in the central structure have evaporated, its corona can survive for several Gyr, no longer recognisable as a density enhancement in the field stars, but detectable through their common space velocities. Agekyan & Belozerova (1979) noted that Galactic disk population could well be composed of a large number of interpenetrating moving groups that originally developed as coronas around dissolving open clusters. That few moving groups have been convincingly identified to date could be simply attributable to the difficulty of detecting them.

In the case of the Hyades, $r_t \simeq 10$ pc. The existence of an extended halo formed by stars outside this tidal radius, but sharing the proper motion characteristics of the rest of the cluster members was originally noted by Pels *et al.* (1975), and has since been reported for other clusters. A striking feature in the spatial distribution of stars in the Hipparcos study of Perryman *et al.* (1998) is that about 45 stars are found between 10–20 pc, a result consistent with the simulations by Terlevich (1987), which demonstrated that since the openings of the equipotential surface are on the X-axis, stars can spend some considerable time within the cluster before they find the windows on the surface to escape through. Such N-body simulation models consistently show a halo formed by 50–80 stars in the region between 1–2 tidal radii. Some of these stars, despite having energies larger than that corresponding to the Jacobi limit, are evidently still linked to the cluster after 300–400 Myr. Clusters which might be in the late stages of dynamical evaporation have been studied by Bica *et al.* (2001) and others.

6.9.2 Detection of kinematic groups

General investigations Hipparcos studies using non-parametric methods, i.e. without *a priori* assumptions on the group's kinematic, age, or spatial struture, have found or confirmed numerous density enhancements in phase-space (i.e. in position–velocity space).

In a comprehensive search for density-velocity inhomogeneities in phase-space structure, Chereul *et al.* (1999) carried out a 3d wavelet analysis on a volume-limited (within 125 pc of the Sun) and absolute-magnitude limited ($M_V < 2.5$) sample of 2977 A–F dwarfs, using the completeness limits of the Hipparcos 'survey star' data to constrain both limits. The objective was to identify clusters and streams in either density or velocity space, without *a priori* assumptions, to estimate the fraction of clumped stars, and to compare results with previously known structures. In combination with stellar age estimates from Strömgren photometry, the state of the interstellar medium at the time of star formation, and the subsequent evaporation and mixing, could be traceable. In 3d position space they identified known clusters (Hyades, Coma Ber, and Ursa Major), three overdensity regions within the Scorpius–Centaurus association, the Hyades moving group or 'evaporation track', and three new probable loose clusters: five stars in Bootes at $d \sim 105$ pc and perhaps having a common origin with Coma Ber, Pegasus 1 with 10 stars at 97 pc, and Pegasus 2 with eight stars at 89 pc. Less than 7% of the stars belong to structures with coherent kinematics. In velocity space, streams on the largest clustering scale of $\sigma \sim 6.3\,\mathrm{km\,s^{-1}}$ correspond to four superclusters (the three of Eggen: Pleiades, Hyades, and Ursa Major/Sirius, with evidence for one previously undetected) and the Centaurus association (their Figures 15 and 17). At smaller velocity scales ($\sigma \sim 3.8\,\mathrm{km\,s^{-1}}$), these superclusters split into distinct streams with smaller velocity dispersions such as the Hyades open cluster (Figure 6.17). Despite incomplete photometry, there were good indications of the finer clumps having different ages. For example, the Hyades supercluster probably contains three groups of ages $5–6 \times 10^8$, 10^9, and $1.6–2 \times 10^9$ yr which are in an advanced stage of dispersion in the same velocity volume, and of which only part can be associated with the evaporation of known open clusters. The Ursa Major/Sirius supercluster comprises three groups of ages 10^7, 6×10^8, and 1.5×10^9 yr: the younger stream is still concentrated both kinematically and spatially while the two oldest streams are mixed in a larger volume of phase space. That these rather fragile streams originating from cluster evaporation could have survived for such long periods of $\sim 10^9$ yr has been explained either as a consequence of a heavy initial cluster mass (Portegies Zwart *et al.*, 1998) or, as discussed in more detail below, as a result of trapping into resonant orbits with the non-axisymmetric force created by the potential of the Galactic bar (Dehnen, 1998).

Asiain *et al.* (1999a) refined the approach of Chen *et al.* (1997) to pursue similar objectives in detecting moving groups in the solar neighbourhood using the four-dimensional space of the stellar velocity components and age, also avoiding *a priori* knowledge of their kinematic properties. They used the Hipparcos astrometric data for a selected sample (after removing known cluster members) of 2061 early-type (relatively young) stars within 300 pc, for which radial velocities and Strömgren photometry were known from various sources. Detection of moving group members (of which 574 were finally identified) was based on the hypothesis that they are responsible for local irregularities over an underlying Gaussian velocity distribution of field stars. Figure 6.18 shows resulting distributions of the inferred moving group candidates after subtraction of the field stars, in both the Galactic $U - V$ and the $U - \log\tau$ (velocity–age) planes. Asiain *et al.* (1999a) discussed the detailed relationship of these structures with moving groups and open clusters identified by other workers, broadly identified with: A: the Sirius group with its positive U component (including the Ursa Major cluster), B: the Pleiades group (including the Pleiades cluster and the Scorpius–Centaurus OB association), C: the Coma Ber or Cas–Tau group, D: a group possibly related to the Hyades. Two puzzling results were noted. First, the velocity dispersion and vertex deviation of the initial sample did not change significantly after elimination of the moving group candidates – the resulting values were: mean velocity components $U, V, W = -10.2, -10.8, -7.2\,\mathrm{km\,s^{-1}}$; $\sigma_U = 17.0\,\mathrm{km\,s^{-1}}$, $\sigma_U : \sigma_V : \sigma_W = 1 : 0.73 : 0.50$ and vertex deviation $\phi = 20\overset{\circ}{.}2 \pm 1\overset{\circ}{.}3$, a result on the face of it inconsistent with the suggestion that the vertex deviation in early-type stars is a consequence of the presence of stellar streams in the solar neighbourhood. Second, the velocity dispersions of the moving groups are large (up to $\sigma_U \sim 5–6\,\mathrm{km\,s^{-1}}$ and slightly smaller in σ_V) suggesting that, even under these conditions, moving groups can survive for long periods.

Dehnen (1998) studied the velocity distribution of nearby stars using a maximum-likelihood approach for a sample of 14 369 Hipparcos stars for which good positions and tangential velocities were available. In the absence of unbiased radial velocity surveys on the scale of the Hipparcos Catalogue, the use of radial velocities was explicitly excluded from this study, which rests instead on the demonstrated assumption that the velocity distribution function is independent of position on the sky, at least for nearby stars. The absence of kinematic bias in the Hipparcos Catalogue represents a significant advance over previous studies, and the sample

Kinematics of disk stars: The identification of kinematic groups in the solar neighbourhood is possible because of their specific velocity characteristics with respect to disk stars as a whole. Some relevant properties of the latter are summarised here, following the introduction given by Mihalas & Binney (1981, Section 7.1). Relative to the Local Standard of Rest, disk stars have a reasonably random distribution in all three residual velocity components U, V, W. Described in terms of a velocity ellipsoid (a formulation due to Schwarzschild (1908) based on observations in the vicinity of the Sun) it was found empirically that one axis is always oriented perpendicular to the Galactic plane, with the longest axis (the direction of maximum velocity dispersion) approximately pointed in the direction of the Galactic centre. The orientation of the velocity ellipsoid is thus determined by the Galactic longitude of the principal axis, referred to as the longitude of the vertex. The main kinematic properties are summarised by Mihalas & Binney (1981) for disk population stars as a function of stellar type as follows: (1) roughly speaking, $\langle W^2\rangle^{1/2} \simeq 0.5\,\langle U^2\rangle^{1/2}$, with the ratio $\langle V^2\rangle^{1/2}/\langle U^2\rangle^{1/2} \simeq 0.55-0.75$, and related to Oort's Constants of Galactic rotation; (2) young stars (dwarfs earlier than F, as well as A and F giants, and supergiants including classical Cepheids) show smaller velocity dispersions than older stars (dwarfs later than F5 and late-type giants), suggestive of a mechanism leading to an increase of velocity dispersion with time; (3) while the principal axis points nearly in the direction of the Galactic centre for later spectral types, it shows systematically larger departures in the vertex deviation for early spectral types; (4) the very youngest O and B stars are kinematically distinctive, presumably reflecting the velocity field of the interstellar medium from which they formed, and still largely unperturbed (due to their youth) by the general Galactic potential field or other encounters.

The asymmetric drift of a stellar population is defined as the difference between the Local Standard of Rest and the mean rotation velocity of the population. As shown in Mihalas & Binney (1981, Section 6.4), stars moving on circular orbits with $R = R_0$ and $V(R_0) = V_0$ move with the LSR; stars moving on elliptical orbits with $R < R_0$ lag the LSR; while stars moving on elliptical orbits with $R > R_0$ lead the LSR.

Wielen (1977) showed that the observed increase in velocity dispersion for disk stars as a function of age is best explained by local fluctuations of the Galaxy's gravitational field (including non-stationary spiral density waves, and the Spitzer–Schwarzschild mechanism of stellar encounters with large concentrations of the interstellar medium), which dominate the variations at birth, or those due to the global gravitational field. As a result of this diffusion, a disk star changes its space velocity randomly by more than 10 km s^{-1} per Galactic revolution, and its position by about 1.5 kpc, thus hampering the determination of stellar birthplaces. The much higher dispersion of halo objects, $\sigma_v \sim 200$ km s^{-1}, is due to the particular conditions at their time of formation during the early phase of Galactic evolution.

In the epicycle approximation in which higher-order terms of the effective potential are neglected (Chapter 9), the radial and vertical displacements in position evolve as simple harmonic motion with epicycle and vertical frequencies κ and ν, and stars would meet or 'focus' every $\Delta t = 2\pi/\kappa$.

allows the study of the nature of the velocity distribution in the solar neighbourhood for both early-type stars and for the old stellar population of the Galactic disk. The study built on that of Dehnen & Binney (1998) who used a kinematically unbiased sample of the Hipparcos Catalogue to infer, as a function of star colour, the first two moments of the velocity distribution for nearby stars, the mean and dispersion. Split into four ranges of colour indices, Dehnen (1998) constructed projections of the estimated velocity distribution in the UV, UW, and VW planes (v_x, v_y, v_z in his notation). Reassuringly, the conspicuous structures most apparent in the planar motions (UV-plane) can be identified with the moving groups discussed so far, while there is little evident structure in W. Progressively redder (and hence on average older) main-sequence stars show increased structure, confirming that some of the moving groups are old, some even older than 2–8 Gyr, and especially at large negative azimuthal velocities ($-V$) showing that the older the group is, the smaller is its (local) rotational velocity around the Galaxy. They follow an 'asymmetric drift relation', in the sense that systems with large $|U|$ lag in V with respect to the Local Standard of Rest. Underlying the moving groups is the smooth 'ellipsoidal' background distribution with axis ratios $\sigma_U : \sigma_V : \sigma_W = 1 : 0.6 : 0.35$.

Skuljan *et al.* (1999) constructed a sample of 4597 Hipparcos survey stars (with radial velocities and $B - V$) according to $\pi > 10$ mas and $\sigma_\pi/\pi < 0.1$. They constructed space velocities, and estimated the underlying probability density function $f(U, V)$ using an adaptive kernel method, with samples also subdivided by colour index into 1036 early-type stars ($B - V < 0.3$, which should contain predominantly young stars) and 3561 late-type stars ($B - V > 0.3$, which should contain a mixture of older and younger main sequence stars, young red-clump stars, and a few old red giants). Similar structure in the kinematic UV plane is seen in these two samples (the nature and statistical significance of the features at different velocity dispersion scales were assessed by wavelet analysis), and similar to that (obtained without radial velocities) reported by Dehnen (1998) (Figure 6.19). The qualitative relationship between the observed 'branch-like' structure, roughly equidistant in velocity space, and the Galactic spiral structure was also noted. The peak at $U, V = -50, 50$, noted by Skuljan *et al.* (1999) and indicated in the $B - V > 0.3$ diagram of Figure 6.19, is associated with Eggen's ζ Herculis stream.

Montes *et al.* (2001) made a specific search for young late-type stars in the five youngest kinematic groups, selecting candidates based on a variety of young-age

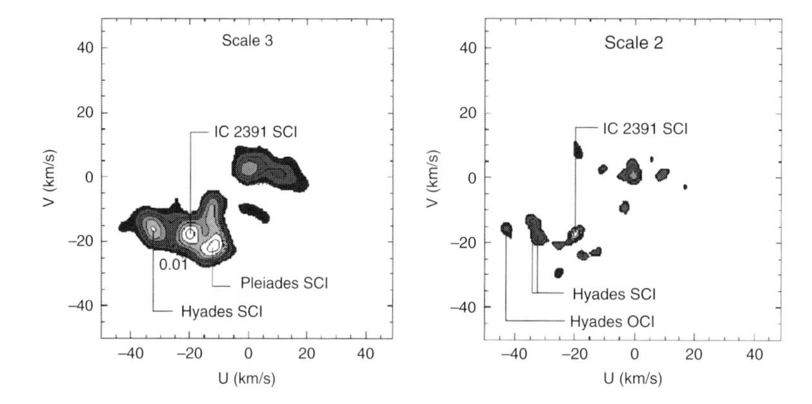

Figure 6.17 An example of velocity structure in the Galactic $U - V$ plane. The region of the Hyades supercluster showing threshholded wavelet coefficient isocontours at $W = -2.4\,\mathrm{km\,s^{-1}}$ of the velocity field (perpendicular to the Galactic plane) at their 'scale 3' at which the supercluster structure dominates ($\sigma \sim 6.3\,\mathrm{km\,s^{-1}}$, left) and at their 'scale 2' which reveals finer structure including the Hyades open cluster ($\sigma \sim 3.8\,\mathrm{km\,s^{-1}}$, right). From Chereul et al. (1999, Figure 15).

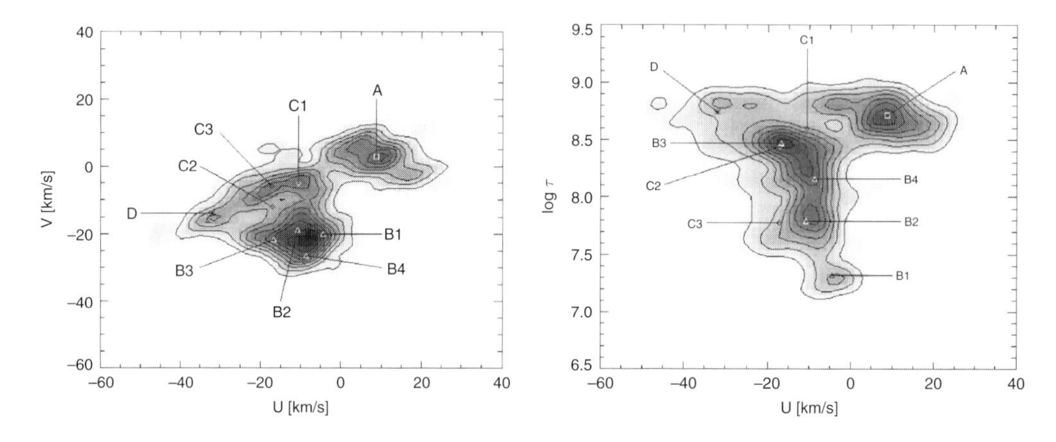

Figure 6.18 The density distributions of inferred moving group candidates, in the Galactic $U - V$ plane (left) and in the $U - \log\tau$ (velocity–age) plane (right). Not all of the structure in the left figure appears in Figure 6.17 since the latter represents one slice in the velocity component W. Structures are identified with: A: Sirius group, B: Pleiades group, C: Coma Ber or Cas-Tau, D: possibly Hyades related. From Asiain et al. (1999a, Figure 6).

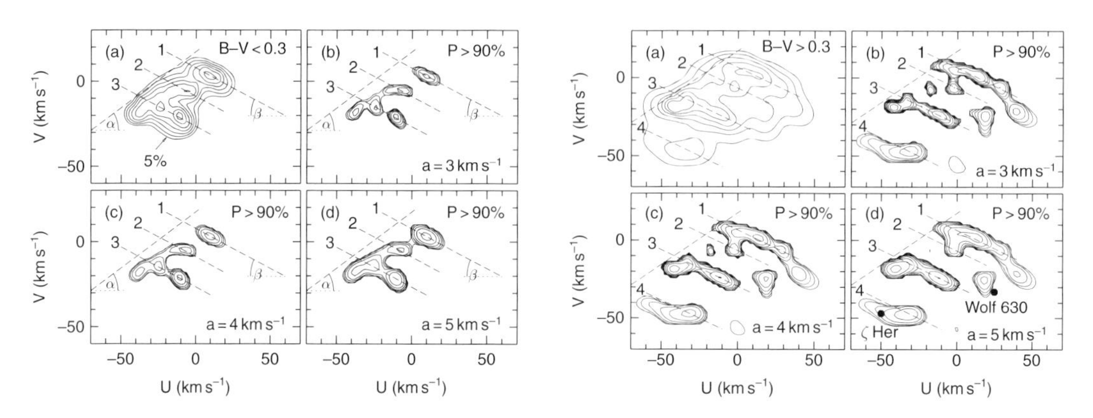

Figure 6.19 The velocity distributions of inferred moving group candidates, in the Galactic $U - V$ plane, subdivided into $B - V < 0.3$ (left) and $B - V > 0.3$ (right). In each case (a) shows the velocity distribution, and (b–d) the corresponding wavelet transforms at different velocity scales. Dashed lines show the proposed positions of 'edges' and 'branches'. From Skuljan et al. (1999, Figures 8 and 10).

indicators (high level of magnetic activity, rapid rotation, strong lithium absorption) and deriving membership probabilities based on their clustering of *UVW* velocities, their common convergent-point, and their adherence to the kinematic criteria developed by Eggen which basically treat moving groups, whose stars are extended in space, like open clusters, whose stars are concentrated in space. Imposing the most stringent set of membership criteria resulted in 45 members of the local association, 38 in the Hyades supercluster, 28 in the Ursa Major moving group, 15 in the IC 2391 supercluster, and eight in the Castor moving group.

Nordström *et al.* (2004) constructed space velocities for about 14 000 F and G dwarfs with radial velocities from the Geneva–Copenhagen Coravel survey, where the same velocity fine structure is seen (their Figure 20).

Famaey *et al.* (2005a,b) presented a kinematic analysis of 5311 K and 719 M giants in the solar neighbourhood, including radial velocity data from the Coravel database, and proper motions from the Tycho 2 Catalogue. They used the maximum-likelihood LM method (page 213) to identify distinct kinematic contributions to the overall velocity structure. The *UV*-plane again shows rich small-scale structure, with several clumps corresponding to the Hercules stream, the Sirius moving group, and the Hyades and Pleiades superclusters (Figure 6.20). The position of these streams in the *UV*-plane is then responsible for the vertex deviation of $16\overset{\circ}{.}2 \pm 5\overset{\circ}{.}6$ observed for the whole sample. The underlying velocity ellipsoid, extracted by the maximum-likelihood method after removal of the streams, is not centred on the value commonly accepted for the radial anti-solar motion, but rather on $\langle U \rangle = -2.78 \pm 1.07\,\mathrm{km\,s^{-1}}$. However, the full dataset (including the various streams) does yield the usual value for the radial solar motion of $\langle U \rangle = -10.25 \pm 0.15\,\mathrm{km\,s^{-1}}$ (see Section 9.2.4).

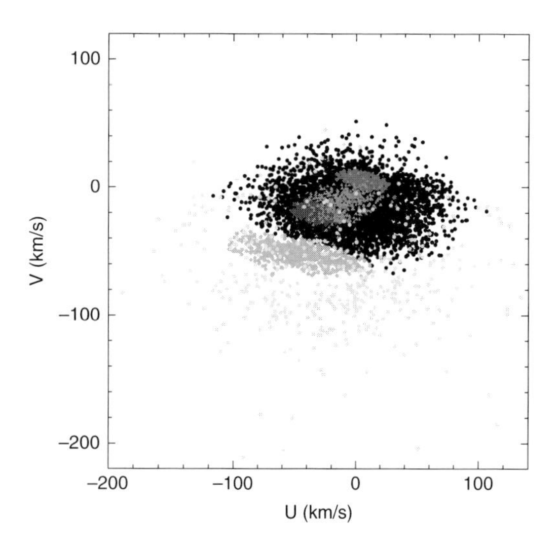

Figure 6.20 Stars from the study of Famaey et al. (2005b) plotted in the UV-plane with their values of U and V deduced from the LM method. Six different groups are represented: young giants (−10, −10), high-velocity stars (extended cloud with V ∼ −100), Hyades–Pleiades supercluster (−40, −20), Sirius group (0,+10), Hercules group (−50, −50), and stars representing a smooth background (black). From Famaey et al. (2005b, Figure 9, original in colour).

Further searches for phase structure was reported by Arifyanto & Fuchs (2006), who searched for overdensities in orbits that remain close together over time, based on the theory of Galactic orbits due to Dekker (1976). They reported three structures correlated with previously known streams (Hyades–Pleiades and Hercules in the thin disk, and the Arcturus stream in the thin disk), as well as another thick disk stream resembling the Arcturus stream.

Other searches which also identified previously-known thick disk streams, but which focused on the search for new halo star streams, were carried out by Helmi *et al.* (2006) and Dettbarn *et al.* (2007), and are described further in Section 9.9.4.

Pleiades moving group Asiain *et al.* (1999b) analysed some of the mechanisms important in the kinematic evolution of a group of unbound stars, such as the disruption effects of disk heating and differential Galactic rotation, applying these concepts to the trajectories followed by stars in each of the Pleiades moving group substructures found by Asiain *et al.* (1999a). They modelled the Galactic potential by contributions from a general axisymmetric part, the spiral arms, and the central bar. They used an axisymmetric potential, Φ, consisting of a spherical central bulge and a disk, both of the Miyamoto–Nagai form (Miyamoto & Nagai, 1975), plus a massive spherical halo, with Galactocentric distance $R_0 = 8.5\,\mathrm{kpc}$ and corresponding circular velocity $\Theta_0 = 220\,\mathrm{km\,s^{-1}}$. Spiral arm perturbations were taken as $\Phi_{\rm spiral}(R, \theta, t) = A\cos(m(\Omega_p t - \theta) + \phi(R))$, with $m = 2$ the assumed number of spiral arms, with the Sagittarius arm at $R_{\rm Sag} = 7.0\,\mathrm{kpc}$ and an inter-arm distance at the Sun's position of $\Delta R = 3.5\,\mathrm{kpc}$; $\Phi_{\rm bar}$ was taken as a triaxial ellipsoid with parameters taken from Palouš *et al.* (1993), with scale length $q_{\rm bar} = 5\,\mathrm{kpc}$, total mass $10^9 M_\odot$, and angular velocity $\Omega_{\rm bar} = 70\,\mathrm{km\,s^{-1}\,kpc^{-1}}$. The observed increase in the total stellar velocity dispersion, σ, with time was modelled by $\sigma^n(t) = \sigma_0^n + C_v t$ where σ_0 is the dispersion at birth, and C_v the apparent diffusion coefficient (Wielen, 1977).

Figure 6.21 shows the resulting disk heating law, where for the case $n = 2$, a fit yielding $\sigma_0 \sim 15\,\mathrm{km\,s^{-1}}$ and $C_v \sim 5.6 \times 10^{-7}(\mathrm{km\,s^{-1}})^2\,\mathrm{yr^{-1}}$, is in good agreement with other work. This modelled heating mechanism successfully explains the observed velocity dispersions for moving group substructures younger than

$\sim 1.5 \times 10^8$ yr, but a much smaller diffusion coefficient is required to recover the velocity dispersion of the oldest substructures, consistent with the requirement of resonance islands proposed by Dehnen (1998). In a Hipparcos study of nearly 12 000 main sequence and subgiant stars aimed at constraining star formation rates in the Solar neighbourhood, Binney *et al.* (2000) found a velocity dispersion $\sim$8 km s^{-1} at birth, increasing with time as $t^{0.33}$. The contribution to disk heating by strong transient spirals, which may also explain the presence of moving groups as a result of small-scale structure formation, is discussed by Binney (2001) and De Simone *et al.* (2004). Disk heating, and the contribution of major non-random substructures (such as clusters, streams, and resonance islands) is further discussed by Holmberg *et al.* (2007, see also their Figures 33–34).

Asiain *et al.* (1999b) used numerical integration within the adopted Galactic potential to determine the trajectories of the Pleiades subgroups B1–B4 to the time of their birth. Figure 6.22 (left) shows that the youngest group B1, composed of the Scorpius–Centaurus association members, was born in the inter-arm region, and is too young to be affected by disk heating. The birthplace of the older groups, B2–B4, lie close to minima in the spiral arm potential, consistent with their being born around these structures, conclusions that the authors note remain unchanged when considering the different spiral arm parameters derived by Mishurov & Zenina (1999). That the age of the Pleiades is close to the time elapsed since the passage of the Local Standard of Rest through the centre of the Sagittarius spiral arm has also been previously noted (e.g. Comerón *et al.*, 1994). The stellar trajectories of the individual members of subgroup B3 are shown in Figure 6.22 (right): not only do groups B3 and B4 show a maximum Galactocentric distance (corresponding to minimum kinetic energy) at the moment they were born, but clear 'focusing points' are found close to these birthplaces. However, discrepancies remain between the kinematic ages of the subgroups (at which the kinematic trajectories would yield the densest spatial concentration) and photometric ages, the latter perhaps overestimated as a result of their rapid rotation.

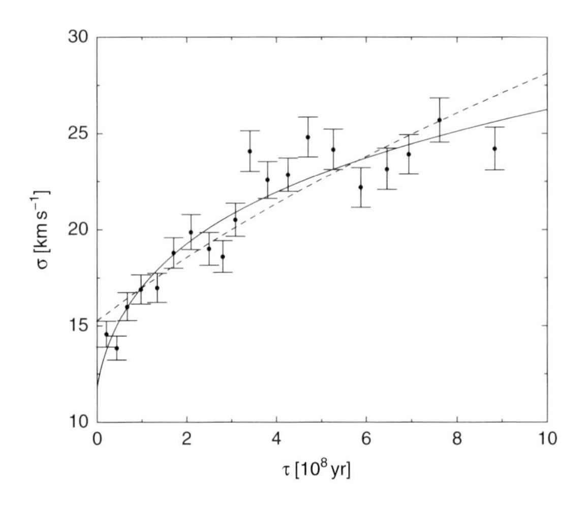

Figure 6.21 The increase of velocity dispersion of disk stars due to disk heating over the past 10^9 yr. Data are for 2061 stars closer than 300 pc from the Sun with kinematics from Hipparcos and ages from Strömgren photometry. Models are for fits to σ_0, C_v and n (solid line); and to σ_0, and C_v with $n = 2$ (dashed line). Error bars include only statistical uncertainties. From Asiain et al. (1999b, Figure 1).

Ursa Major moving group The Ursa Major kinematic group includes $\sim$100 stars across the northern and southern skies, and including the A-type stars β UMa through ζ UMa within the Ursa Major constellation (Figure 6.23) as well as Sirius (for an early review, see Roman, 1949). As for the Hyades, but in contrast to the Pleiades, its space velocity is sufficiently distinct from the Local Standard of Rest to provide good separation from the field star population. In the most comprehensive study pre-dating the Hipparcos Catalogue, Soderblom & Mayor (1993) searched for stars of appropriate age and composition in the colour-magnitude diagram, and for stars with a common space motion. They confirmed its reality, its generally adopted age of 0.3 Gyr, comparable to the typical time scale for cluster dissolution, and a mean metallicity of [Fe/H]$= -0.08 \pm 0.09$. They showed that the kinematic precision was at that time limited by parallaxes rather than radial velocities, even though most stars considered were within 25 pc.

Eggen (1998e) used the Hipparcos Catalogue members and fainter candidates with comparable space motions to argue that either the supercluster's luminosity function is very different from that of the field stars, or that only a few of the lower-mass supercluster members in the vicinity of the Sun have so far been identified.

Chupina *et al.* (2001) used the Hipparcos, Tycho, and ACT Catalogues to study the nucleus region of the Ursa Major group. They used proper motion and radial velocity data to identify a nucleus of 8–10 highly-probably members, sharing a common space motion with a velocity dispersion of 1.33 km s^{-1}, contained within a region of size $11 \times 5 \times 5\,\mathrm{pc}^3$, and elongated in the direction of the Galactic centre.

King *et al.* (2003) used the Hipparcos parallaxes, as well as additional recent radial velocity, photometry and abundance data, to establish a list of nearly 60 probable members. The velocities in Galactic *UV* coordinates reveal the non-zero vertex deviation previously established for these moving groups (Figure 6.24a–b). They revised the age upwards to

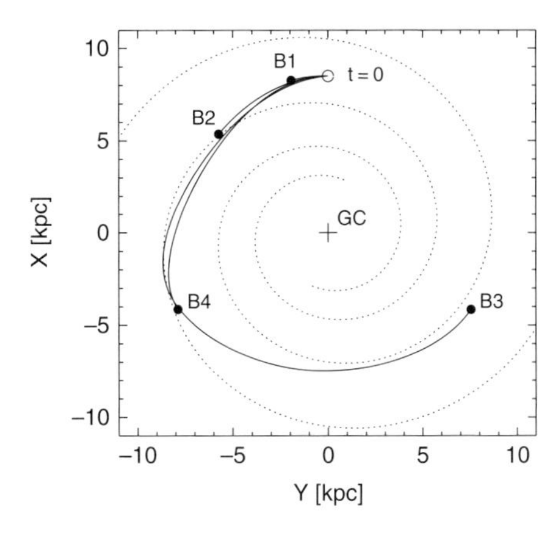

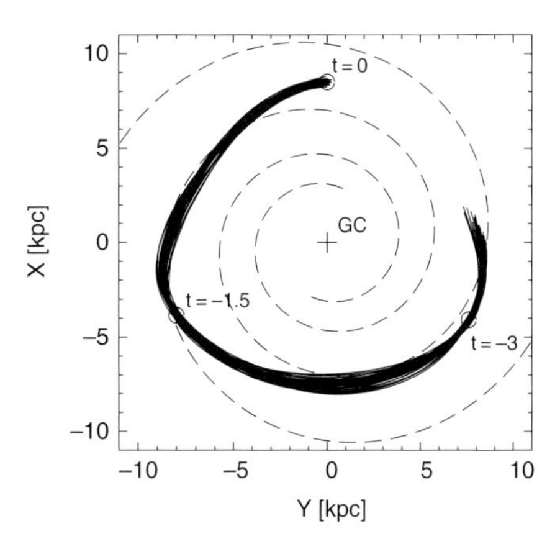

Figure 6.22 Trajectories of the nuclei of the Pleiades moving groups. Left: the four subgroups within the Pleiades moving group projected backwards in time from the present (t = 0) until their estimated mean age. Right: stellar trajectories of the B3 members, with the position of the nucleus at t = 0, −1.5, and −3 × 10^8 yr indicated by open circles. In both cases, the dashed lines represent the spiral arms according to the model described, and the reference system is rotating with the same angular velocity as the spiral arms. From Asiain et al. (1999b, Figures 3 and 8).

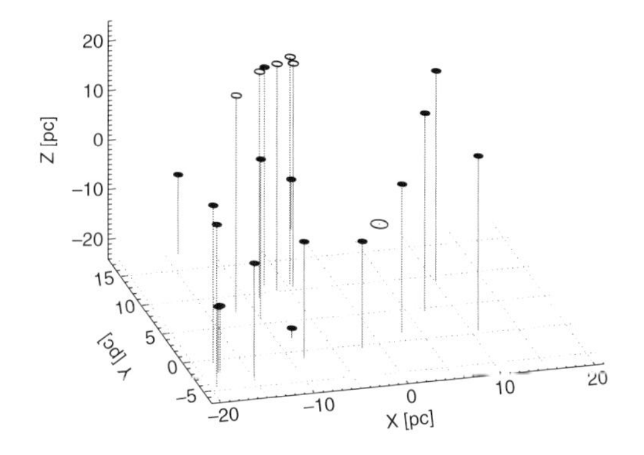

Figure 6.23 Spatial distribution of nearby bright stars of the Ursa Major group relative to the Sun (⊙). The string of open circles represents the bright stars of the Ursa Major constellation that lie within 25 pc. From Fuhrmann (2004, Figure 15). Figure copyright Wiley–VCH Verlag GmbH & Co. KGaA, and reproduced with permission.

500 ± 100 Myr (Figure 6.24c–d), in agreement with age estimates of early-type Sirius supercluster members identified by Asiain *et al.* (1999a), and of the dominant Sirius supercluster component found by Chereul *et al.* (1999), and thus identifying (if the homogeneity of the age-scale estimates is confirmed) a cluster intermediate in age between the Pleiades and Hyades clusters. Mean rotational velocities and chromospheric emission levels are comparable to those in the Hyades and smaller than those of the younger Pleiades, suggesting their rapid decline with age over 200–500 Myr.

Fuhrmann (2004) made a detailed study of a number of the individual group members based on the Hipparcos astrometry.

Castor moving group Barrado y Navascués (1998) examined the properties of 26 stars considered as candidate members of a moving group sharing the same motion and age as Castor (α Gem), originally suggested by Anosova & Orlov (1991). Based on space motions from Hipparcos, the colour–magnitude diagram, Li abundances, and activity, he argued that 16 are probably physically associated. The moving group contains several A-type stars, including Vega and Fomalhaut, two of the prototypes of the β Pic stars. He estimated an evolutionary age, from the late-type stars and the Hipparcos magnitudes, of 200 ± 100 Myr. Montes *et al.* (2001) used more stringent criteria to propose just eight Hipparcos members.

HR 1614 moving group Eggen (1978) had noted that many stars in the solar neighbourhood share the same space motion as the nearby dwarf HR 1614. Smith (1983) found that many are unusually CN-rich, amongst the richest known in the solar neighbourhood. From the 15 supercluster and 59 group members identified by Eggen (1998a, c) used Hipparcos astrometry to confirm that all but two are group and supercluster members, estimating an old age of 3–6 Gyr. He identified 3–4 of the group as blue stragglers with ages near 1 Gyr, including the contact binary AW UMa, and FK Com, the prototype of this variability class. Feltzing & Holmberg (2000) also confirmed the group's reality. They used Hipparcos astrometry, radial velocities from the literature, and Strömgren photometry to make an unbiased search in the *UV* plane, confirming 15 members of the group, and finding an age of about 2 Gyr based on isochrones from Bertelli *et al.* (1994). They note that HR 1614 stands out because it is particularly metal rich for its location in

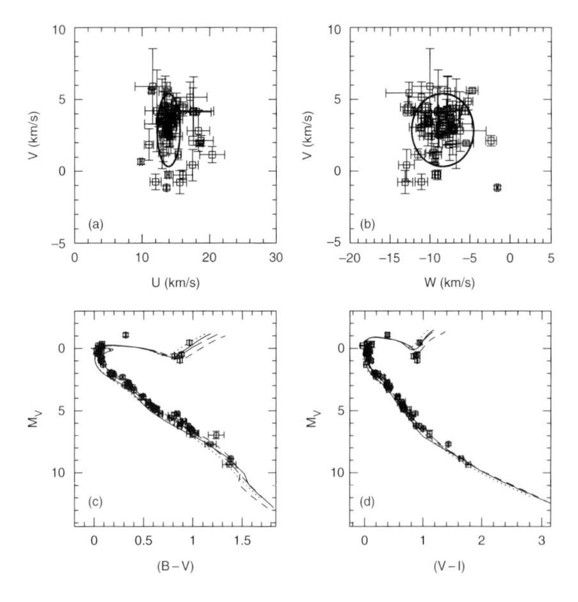

Figure 6.24 Stars in the Ursa Major moving group: (a) and (b) show the Galactic UV and VW kinematic planes, with ellipses denoting the 3σ velocity dispersion; (c) and (d) show the Hipparcos-based M_V versus $B-V$ HR diagrams for the Yonsei–Yale isochrones of 600 Myr and 400 Myr respectively (Yi et al., 2001), for various colour transformation-based isochrones for $Z = 0.01$ and $Z = 0.02$. From King et al. (2003, Figure 9).

the *UV* velocity plane; close to the LSR, for example, the group would have been completely obscured by other metal-rich stars. As described previously, Dehnen (1998) had also found numerous maxima in the velocity distribution of nearby stars, several containing exclusively (old) red stars, one of which coincides with the HR 1614 group. It is part of the stellar overdensity which Dehnen (1998) considered had been ejected from the inner disk by the Galactic bar. For a statistical analysis of the metallicities of nine old superclusters and moving groups, and its sensitivity to the starting sample, see Taylor (2000).

Others Abad & Vieira (2005), see also Vieira & Abad (2003), used their development of the approach which they attributed to Herschel, of associating each star's proper motion with the pole of the relevant great circle, and thereby identified two groups of systematic motions: the first of about 4500 stars, with $v \sim 27 \pm 1\,\mathrm{km\,s^{-1}}$ directed to $(l, b) = (177^\circ\!.8, 3^\circ\!.7)$; the second of about 4000 stars, with $v \sim 32 \pm 2\,\mathrm{km\,s^{-1}}$ directed to $(l, b) = (5^\circ\!.4, -0^\circ\!.6)$. Both are distributed rather uniformally about the Sun out to distances of ∼400 pc, and cover a wide range of spectral types.

Other studies of moving groups making use of the Hipparcos data include: late-type members of young kinematic groups (Montes *et al.*, 2000); and stellar moving groups (Mylläri *et al.*, 2001).

6.9.3 Origin of kinematic groups

The Hipparcos data have provided compelling evidence for the reality of some kinematic groups, with results confirmed by various workers, using different methods and data. With only a small possible contribution arising from the evolution and dissolution of initially gravitationally bound clusters (because of the wide range of ages in the identified subgroups), the problem is then to understand their survival over periods of $\sim 10^9$ yr, and how these old moving groups got into eccentric orbits obeying the observed asymmetric drift relation. These questions received limited attention pre-Hipparcos, probably due to their previously controversial existence.

Dehnen (1998) looked for explanations based on orbital resonances. Any non-axisymmetric force field responsible for moving clusters into progressively more eccentric orbits, and thus explaining the dependence of eccentricity with age, would need to be smooth in both space and time. He focused on the role of the Galactic bar (Section 9.6), while noting that a triaxial halo or the Magellanic Clouds could also play a role. Appealing to the mechanism proposed by Sridhar & Touma (1996), he argued that the resonant 'islands' created by the presence of the bar will sweep through phase space as the Galaxy evolves (e.g. as the bar slows down and strengthens) with stars trapped in them shifted to different orbits. Support for the role of resonances in the motion of nearby stars is given by the Feast & Whitelock (1997) redetermination of Oort's constants from the Hipparcos Cepheids (see Chapter 9), i.e. (in $\mathrm{km\,s^{-1}\,kpc^{-1}}$) $\Omega = A - B = 27.2 \pm 0.9$, $\kappa = 2(B^2 - AB)^{1/2} = 36.7 \pm 1.6$, such that $\Omega : \kappa = (2.97 \pm 0.07) : 4$. This precise 3:4 resonance between azimuthal and radial orbit frequency appears naturally in a slightly declining circular speed curve, e.g. of the form $v \propto R^{-1/9}$.

Explanation of the Hercules stream as being an unbound group of stars associated with the outer Lindblad resonance of the rotating Galactic bar was subsequently supported by the work of Fux (2001), Quillen (2003), and Mühlbauer & Dehnen (2003), and is described further in Sections 9.6 and 9.7.

Famaey *et al.* (2005b) attributed the velocity clumps corresponding to the Hyades and Pleiades superclusters, and the Sirius moving group, to additional non-axisymmetric perturbations of the Galactic potential, resulting from transient or quasi-stationary spiral waves, as modelled by De Simone *et al.* (2004). They considered but rejected the possibility that these are debris streams resulting from satellite galaxy mergers. In the case of the Hyades stream, at least part could also be the remnant of the evaporating Hyades cluster.

Famaey *et al.* (2007) proposed a simple test to distinguish these possible origins: if it is a system trapped at resonance, the mass distribution should represent the present-day mass function of the disk. If it is an evaporated cluster, the mass distribution should reflect the initial mass function of the cluster, modified by star evaporation as a function of mass. Using the Geneva–Copenhagen survey of F and G dwarfs, they found that the observed mass function is consistent with the stream being composed of trapped field stars obeying the present-day mass function, and not with evaporated cluster stars, although they propose that a small contribution from the latter probably also exists. Their hypothesis is also supported by the stream's metallicity distribution. They thus favour resonance trapping due to a spiral arm perturbation.

For the Pleiades, Hyades, and Sirius moving groups, Famaey *et al.* (2008) investigated whether all stars in the various moving groups could originate from the ongoing evaporation of their associated cluster. In each case, they found that the fraction of stars making up the velocity space overdensity is incompatible with the associated cluster isochrone, also suggesting a resonant origin for these three moving groups.

The resulting picture is thus one of dynamical streams (involving stars of different ages, not born at the same place nor at the same time) pervading the solar neighbourhood and travelling in the Galaxy with similar space velocities, arising partly from resonances with the bar, and partly from perturbations by spiral waves. In this secenario, the large number of young clusters and associations occupying a similar region of the UV plane do not occur there by chance, but could also have been deposited there by the same spiral wave perturbations associated with their formation. Comerón *et al.* (1997a) had already argued that the main moving groups presently observed in the solar neighbourhood formed in large-scale spiral shocks responsible for the observed spiral arms, with the jump in the velocity of the gas crossing the shocks being responsible for their deviation from circular motion.

This rich velocity structure in the UV plane is linked with the question of how to derive the solar motion in the presence of dynamical perturbations which clearly alter the kinematics of the solar neighbourhood: specifically, whether there exists in the solar neighbourhood a subset of stars having no net radial motion which can be used as a reference against which to measure the solar motion (Section 9.2.4).

6.10 Associations

6.10.1 Introduction

OB associations are unbound 'moving groups' of young, high-mass stars, which can be detected kinematically because of their small internal velocity dispersion (see also box on page 274). The nearby associations have a large extent on the sky, which traditionally has limited astrometric membership determination to bright stars ($V \lesssim 6$ mag), with spectral types earlier than about B5. Ambartsumian (1947) introduced the term 'association' for these groups of OB stars; he pointed out that their stellar mass density is usually less than $0.1 M_\odot$ pc^{-3}. Bok (1934) had already shown that such low-density stellar groups are unstable against Galactic tidal forces, so that OB associations must be young (Ambartsumian, 1949). This conclusion was subsequently supported by ages derived from colour-magnitude diagrams, and by the fact that such groups are usually located in or near star-forming regions. OB associations are prime sites for the study of star formation and the interaction of early-type stars with the interstellar medium (Blaauw, 1964, 1991). Ruprecht (1966) compiled the Catalogue of Star Clusters and Associations, with field boundaries, bright members, and distances, introducing an IAU-approved nomenclature, and a star list subsequently maintained for many years (Alter *et al.*, 1970; Ruprecht *et al.*, 1981).

Figure 6.25 shows the composite velocity-integrated CO survey of the Galactic plane from Dame *et al.* (2001), providing a rather complete inventory of nearby molecular clouds. The figure serves as a reference for the location of the star-forming regions and associations discussed in this section.

The Ophiuchus, Lupus and Chamaeleon molecular cloud complexes are prominent low-mass star-forming regions close to the Sun (100–200 pc), between Galactic longitudes 280 and 365°, and reaching relatively high Galactic latitudes of $+20°$ for ρ Oph and $-15°$ for Chamaeleon. They are part of an elongated structure that extends over more than 120° in longitude, from the Aquila Rift to the Vela region, which includes the Scorpius–Centaurus OB association, with its subgroups Upper Scorpius, Upper Centaurus Lupus, and Lower Centaurus Crux. Different scenarios have been proposed for the formation of this complex of stars and clouds, including sequential star-formation in which star formation begins at one extremity of a giant molecular cloud and propagates through progressive wind compression of the gas cloud (Blaauw, 1964, 1991; Preibisch & Zinnecker, 1999), the Gould Belt model which considers that a large disturbance originated in the Taurus region and is propagating in the Galactic plane (Lindblad *et al.*, 1973; Frogel & Stothers, 1977; Olano, 1982; Olano & Pöppel, 1987), the direct impact of high-velocity clouds on the disk (Lépine & Duvert, 1994), and spiral arm shocks (Sartori *et al.*, 2003).

In anticipation of the Hipparcos data, Brown *et al.* (1997) made N-body simulations to determine kinematic ages and initial sizes of OB associations. They found that tracing back the proper motions to the epoch at which the association has its smallest physical size

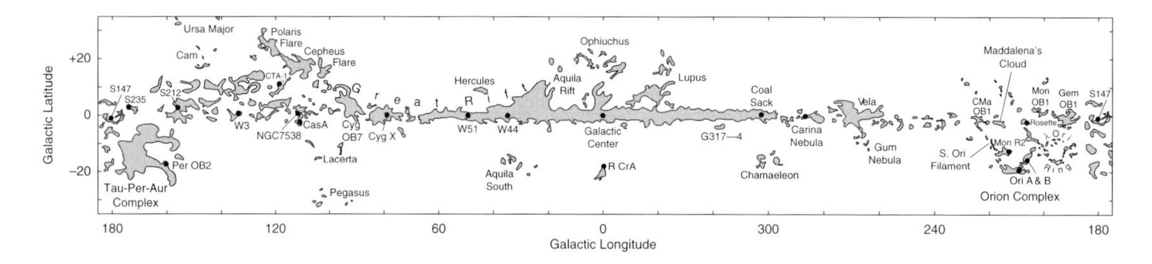

Figure 6.25 The composite velocity-integrated CO survey of the Galactic plane, considered as providing a nearly complete inventory of nearby molecular clouds. From Dame et al. (2001, Figure 2a inset).

always led to underestimated ages, and to overestimates of the initial sizes. They concluded that the long-standing discrepancy between nuclear and kinematic ages could be attributed to such underestimates of the kinematic age.

One of the main limitations of the pre-Hipparcos studies of OB associations was the determination of candidate members, as a result of the progressive dispersion of members into the field population with time, combined with largely unknown parallaxes and poorly determined space motions.

6.10.2 Large-scale studies

De Zeeuw *et al.* (1999) made a comprehensive census of the stellar content of OB associations within 1 kpc from the Sun, based on Hipparcos positions, proper motions, and parallaxes of 9150 candidate members, as part of a long-term project to study the formation, structure, and evolution of nearby young stellar groups and related star-forming regions. The Hipparcos measurements allowed a major improvement in the kinematic detection of OB associations (Figure 6.26). They identified moving groups in the Hipparcos Catalogue by combining the convergent-point implementation of de Bruijne (1999a, see also Section 6.2.2), with the 'spaghetti method' of Hoogerwerf & Aguilar (1999, see also Section 6.2.3). Astrometric members are listed for 12 young stellar groups, out to a distance of ~650 pc. These are the three subgroups of the Scorpius OB2 complex (Upper Scorpius, Upper Centaurus Lupus and Lower Centaurus Crux), as well as Vela OB2, Trumpler 10, Collinder 121, Perseus OB2, α Persei (Perseus OB3), the Cassiopeia–Taurus group, Lacerta OB1, Cepheus OB2, and a new group in Cepheus, designated as Cepheus OB6 (examples are given in Figure 6.27 for the case of Sco OB2 and α Per). The selection procedure corrected the list of previously known astrometric and photometric B- and A-type members, and identified many new members, including a significant number of F stars, as well as evolved stars, e.g. the Wolf–Rayet stars γ^2 Vel (WR 11) in Vel OB2 and EZ CMa (WR 6) in Collinder 121, and the classical Cepheid δ Cep in Cepheus OB6. Membership probabilities were given for all selected stars, and Monte Carlo simulations were used to estimate the expected number of interloper field stars. In the nearest associations, notably in Scorpius OB2, the later-type members include T Tauri objects and other stars in the final pre-main-sequence phase. This provides a firm link between the classical high-mass stellar content and ongoing low-mass star formation. Astrometric evidence for moving groups in the constellations of Corona Australis (R CrA), Canis Major (CMa OB1), Monoceros (Mon OB1), Orion (Ori OB1), Camelopardelis (Cam OB1), Cepheus (Cep OB3, Cep OB4), Cygnus (Cyg OB4, Cyg OB7), and Scutum (Sct OB2), remained inconclusive. OB associations evidently exist in many of these regions, but they are either at distances beyond ~500 pc where the Hipparcos parallaxes are of limited use, or they have unfavourable kinematics, such that the group proper motion is not distinguishable from the field stars in the Galactic disk. The mean distances of the well-established groups were found to be systematically smaller than the pre-Hipparcos photometric estimates (Figure 6.28). While part of this may be caused by the improved membership lists, a re-calibration of the upper main sequence in the HR diagram may also be called for (see also Section 5.4). The mean motions display a systematic pattern, which they discussed in relation to the Gould Belt (Section 6.10.5). Six of the 12 detected moving groups do not appear in the classical list of nearby OB associations. This is sometimes caused by the absence of O stars, but in other cases a previously known open cluster turns out to be (part of) an extended OB association. Their findings suggest that the number of unbound young stellar groups in the solar neighbourhood may be significantly larger than thought previously.

Although the Hipparcos Catalogue contains members of nearby OB associations brighter than $V = 12$, membership candidates are complete only to the catalogue completeness limit at $V = 7.3$. Hoogerwerf (2000) extended the identification of association candidates to the two catalogues generated as intermediate steps in the construction of the Tycho 2 Catalogue, the

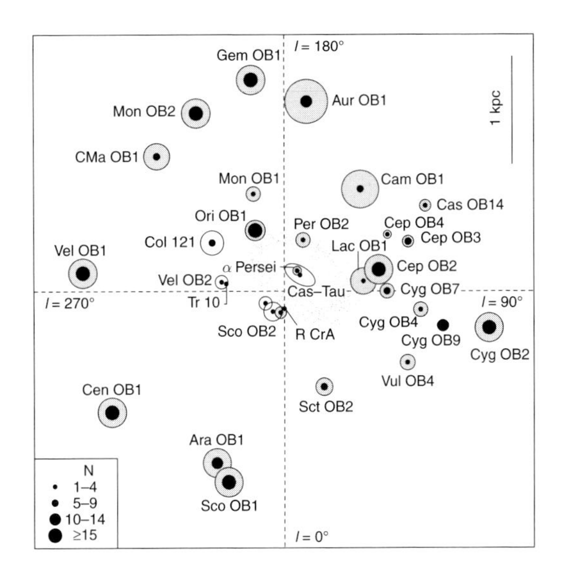

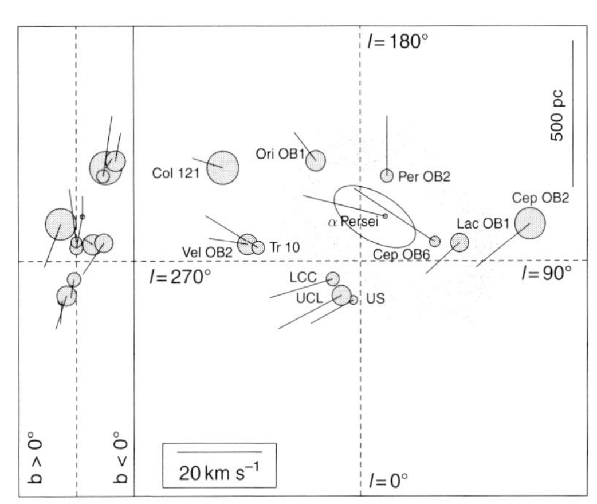

Figure 6.26 Left: Pre-Hipparcos locations of the OB associations within ~1.5 kpc, as listed by Ruprecht (1966), projected onto the Galactic plane. The Sun is at the centre of the dashed lines which give the principal directions in Galactic longitude. Circle sizes represent projected association dimensions, enlarged by a factor 2 with respect to the distance scale. The size of the central dots indicates the degree of current or recent star formation activity, as given by the number N of stars more luminous than absolute magnitude $M_V \sim -5$ (Humphreys, 1978). Associations discussed by de Zeeuw et al. (1999) which are absent from Ruprecht's list are represented as open circles. The small dots indicate the Gould Belt from the model of Olano (1982). Right: the post-Hipparcos map of nearby OB associations, showing the kinematically detected groupings projected onto the Galactic plane (right) and in cross-section (left). The grey circles indicate physical dimensions on the same scale. Lines represent streaming motions, derived from the average proper motions, mean distances and median radial velocities of the secure members, corrected for 'standard' Solar motion and Galactic rotation. The ellipse around the α Per cluster indicates the Cas–Tau association. From de Zeeuw et al. (1999, Figures 1 and 29).

Astrographic Catalogue–Tycho Catalogue (ACT) and the Tycho Reference Catalogue (TRC), which are complete to $V \sim 10.5$ and have an average proper motion accuracy of $\sim 3\,\mathrm{mas\,yr^{-1}}$ (see Chapter 1). Hoogerwerf (2000) searched for ACT/TRC stars which have proper motions consistent with the spatial velocity of the Hipparcos members of the nearby OB associations already identified by de Zeeuw *et al.* (1999), first selected using the convergent-point method, and then subjected to further constraints on the proper-motion distribution, magnitude and colour to narrow down the final number of candidate members. Monte Carlo simulations showed that the proper motion distribution, magnitude, and colour constraints removed $\sim$ 97% of the field stars, while retaining more than 90% of the cluster stars. The procedure was applied to the three subgroups of Scorpius OB2, along with Per OB3 and Cep OB6. New association candidates fainter than the completeness limit of the Hipparcos Catalogue were found in all cases except Cep OB6, although narrow-band photometry and/or radial velocities would be needed to confirm the cluster members.

Comerón *et al.* (1998) searched for expansion patterns in groups of young stars beyond the local system dominated by the Gould Belt, seeking signatures of violent formation of Galactic OB associations. Initially radially expanding motions should only become distorted by the differential rotation of the Galactic disk when the age of the structure becomes a significant fraction of the epicycle period, $2\pi/\kappa \sim 1.7 \times 10^8$ yr in the solar neighbourhood. They defined a sample of 1092 distant Hipparcos OB stars, with parallaxes < 0.003 arcsec and with $|b| < 5°$. Their proper motions are typically accurate to $0.8\,\mathrm{mas\,yr^{-1}}$ corresponding to $3\,\mathrm{km\,s^{-1}}$ at 1 kpc. Subtracting motions due to Galactic rotation, they identified, by visual inspection, two expanding structures: one in the Cygnus super bubble region and related to the associations Cyg OB1, OB3, and OB9 (and discussed in detail by Comerón *et al.*, 1993); and one related to the Canis Major OB1–R1 complex (Figure 6.29). They considered that the observed pattern of motions is unlikely to be due to spiral arm shocks. In the case of Cygnus, they favoured instead a model in which the massive stars in Cyg OB2, at the centre of the Cygnus super bubble, triggered star formation in a shell (Cyg OB1, OB3, OB7, and OB9) whose present-day motions would reflect the expansion of the shell at the time it became unstable. The space velocities are bimodal, and they

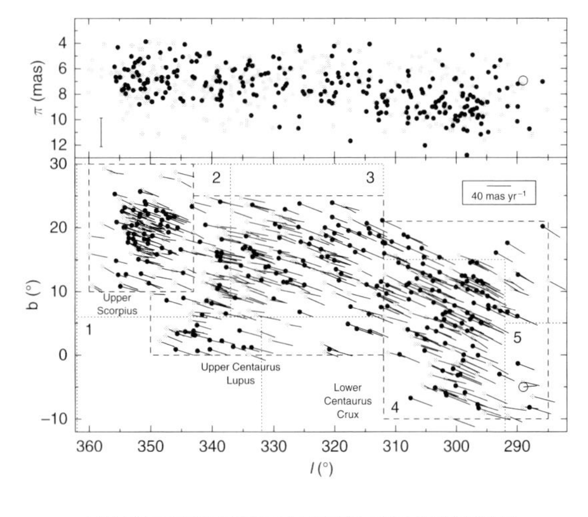

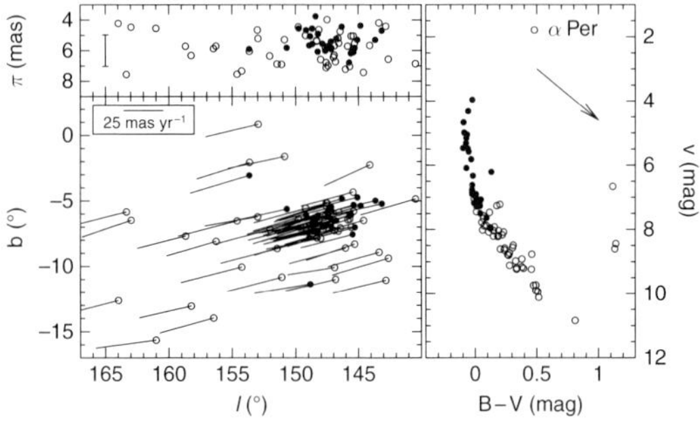

Figure 6.27 Top: positions and proper motions (bottom), and parallaxes (top), for 521 members of the Scorpius OB2 association selected from 7974 stars in the Hipparcos Catalogue in the area bounded by the dashed lines. The vertical bar in the top panel is the average $\pm 1\sigma$ parallax range for the stars shown. Black dots indicate stars with membership probability $P \geq 95\%$. Grey dots indicate remaining members. Many members near the association boundary have a low membership probability. The dotted lines are the schematic boundaries of the classical subgroups Upper Scorpius, Upper Centaurus Lupus, Lower Centaurus Crux, and the candidate subgroups (1 and 5) defined by Blaauw (1946). The large open circle represents the open cluster IC 2602. Bottom, left panel: positions and proper motions (bottom) and parallaxes (top) of the α Per (Perseus OB3) members. Filled circles indicate B-type stars, and open circles the later spectral types. Right panel: colour-magnitude diagram, not corrected for reddening, for the α Per members. The main sequence extends from B3V ($V = 4.0$ mag) to G3V ($V = 10.8$ mag). From de Zeeuw et al. (1999, Figures 9 and 18).

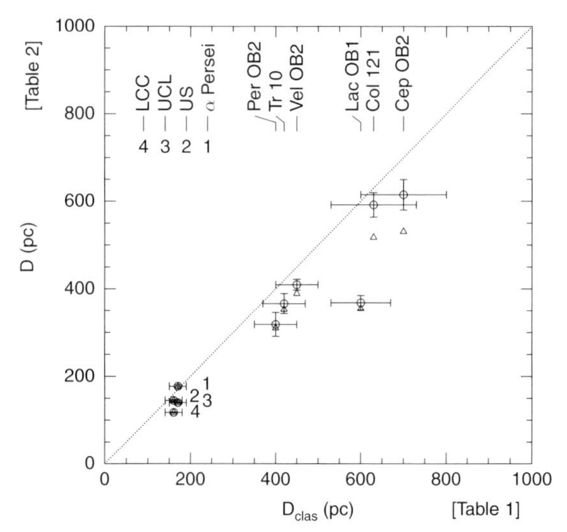

Figure 6.28 Distances D derived from the mean parallax of the Hipparcos members of the OB associations by de Zeeuw et al. (1999) versus the classical distances $D_{\rm clas}$ (Ruprecht, 1966, Table 1). For three associations $D_{\rm clas}$ was taken from another source: Collinder 121 (Feinstein, 1967), Vela OB2 (Brandt et al., 1971) and Trumpler 10 (Lyngå, 1962). ○: distances corrected for systematic effects due to selection procedures and the Hipparcos completeness limit; △: distances uncorrected for these systematic effects. From de Zeeuw et al. (1999, Figure 28).

considered stars with velocities exceeding 40 km s^{-1} to be runaways from Cyg OB2. Motions of the O and B stars in the Canis Major region were attributed to a supernova explosion, supported by the discovery of a new runaway with a large proper motion directed away from the expansion centre.

Torra *et al.* (2000b) studied the kinematics of nearby young stars from the Hipparcos Catalogue, determining the Oort constants for different samples selected by age or distance, and interpreting the systematic trends as signatures induced by the kinematic behaviour of the Gould Belt (see Section 6.10.5).

De Bruijne (1999b) used the maximum-likelihood method developed for analysis of the Hyades cluster data by Dravins *et al.* (1997) (see also Section 6.3), to simultaneously estimate the space velocity vector, the internal velocity dispersion (assumed isotropic), and an improved parallax for the individual stars, without recourse to spectroscopic data (although they refer to these as secular parallaxes, they are strictly individual 'kinematically-modelled parallaxes' as opposed to the 'moving-cluster parallax' based on the rate of change of the cluster's angular diameter, measured by proper motions and spectroscopically-determined radial velocities of its members, or the secular parallax derived in a statistical way from the observed proper motions of a group of field stars based on the solar motion with respect to the local standard of rest). Monte Carlo simulations were used to conclude that the method is robust against all systematic effects considered, including an overall expansion, except for the maximum likelihood estimate of the one-dimensional internal velocity

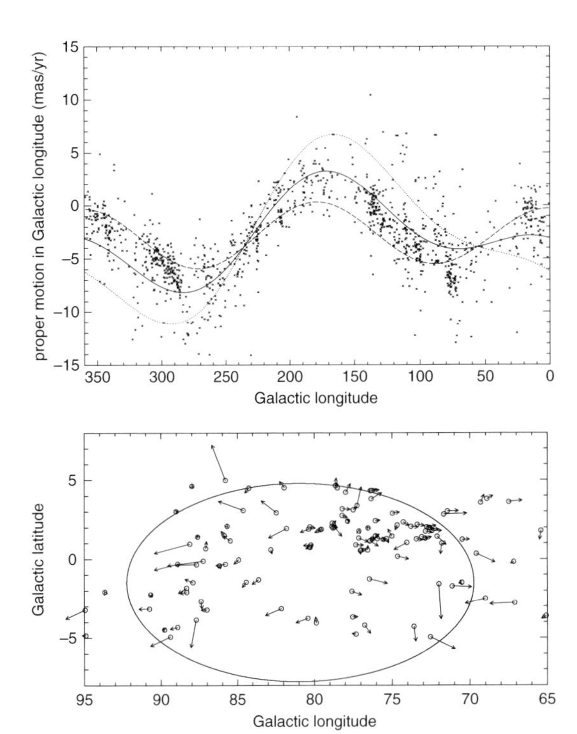

Figure 6.29 Top: distribution of stars in Galactic longitude versus proper motion in Galactic longitude. Lines represent the systematic proper motions expected from the circular Galactic rotation and the solar peculiar velocity for stars located at different distances from the Sun: r = 0.5 kpc (dotted), r = 1 kpc (solid) and r = ∞ (dashed). The latter represents the motions of distant stars for which the reflex of the solar motion (decreasing as r^{-1}) is insignificant as compared to the Galactic rotation (independent of r). It is assumed that all the stars are nearby enough so that the first order approximation to the Galactic rotation curve is valid. Bottom: positions and velocities of stars in the Cygnus region. Arrows represent residual motions in the plane of the sky, after subtraction of systematic motions due to differential Galactic rotation, assuming a distance of 1.25 kpc. The largest residual motion shown is 9.8 mas yr^{-1}. The ellipse represents the approximate outer limits of the Cygnus super bubble, as given by Cash et al. (1980). From Comerón et al. (1998, Figures 1 and 2).

dispersion σ_v, which is always underestimated. Applied to the three subgroups of Scorpius OB2, using membership established by de Zeeuw *et al.* (1999), they found that resulting modelled parallaxes are statistically consistent with, and more precise by a factor of about 2 than the Hipparcos trigonometric parallaxes, that they significantly narrow the locus of stars in the $(B-V)_0$ versus M_V colour-absolute magnitude diagram, and that they define a single-star main sequence which is consistent with the zero-age main-sequence calibration of Mermilliod (1981). They found that $\sigma_v \lesssim 1.0-1.5\,\mathrm{km\,s^{-1}}$ for all three subgroups. While the modelled parallaxes do not improve significantly the mean association distances derived by de Zeeuw *et al.* (1999) from the published Hipparcos parallaxes, they do resolve the parallax distribution, and thus the spatial structure, for Upper Centaurus Lupus and Lower Centaurus Crux.

Madsen *et al.* (2002), see also Madsen *et al.* (2000), also applied the same method to the three major subgroups of the Sco–Cen association, showing the improved HR diagrams in Madsen *et al.* (2000), and deriving astrometric radial velocities which they compared with published spectroscopic radial velocities (Figure 6.30). The figures include the loci of the 'expansion bias' in the astrometric radial velocities, assuming the inverse age of an association to be the upper limit on the relative expansion rate. Stars in Lower Centaurus Crux and in Upper Centaurus Lupus show a wide spread in the spectroscopic values (presumably largely due to measurement errors), while the mean is roughly consistent with an isotropic expansion at only about half the rate expected from the age of the association. Upper Scorpius, the youngest of the subgroups at around 5–6 Myr, does not seem to expand at all, the data rather suggesting that it contracts. There is no *a priori* reason to expect Upper Scorpius to be a bound system without expansion: the star formation appears to have dispersed the rest of the parent molecular cloud, confirming the standard picture that the removal of gas leads to a loss of binding mass of the system, such that it becomes unbound and will consequently expand (Mathieu, 1986). Madsen *et al.* (2002) also discuss results for α Persei (Per OB3), and for the 'HIP 98321' association discovered from the Hipparcos data by Platais *et al.* (1998).

Mamajek *et al.* (2002) made a spectroscopic survey of X-ray and proper motion selected samples of late-type stars in the Lower Centaurus Crux and Upper Centaurus Lupus subgroups of the nearest OB association, Scorpius–Centaurus. One sample was composed of proper motion candidates from the Astrographic Catalogue–Tycho (ACT) and Tycho Reference Catalogue (TRC) identified by Hoogerwerf (2000) with X-ray counterparts in the ROSAT All-Sky Survey (RASS) Bright Source Catalogue; the second was defined by G and K-type Hipparcos stars found to be candidate members by de Zeeuw *et al.* (1999). From optical spectra of 130 objects, they identified pre-main-sequence objects by virtue of their strong Li absorption, low surface gravities, proper motions consistent with Sco–Cen membership, and HR diagram positions consistent with being pre-main sequence. They found 93% of the RASS–ACT/TRC stars to be probable pre-main-sequence stars, compared to 73% of the Hipparcos candidates. From secular parallaxes, and Hipparcos, Tycho 2, and 2MASS photometry, they determined ages of the pre-main-sequence populations to lie between 17–23 Myr for

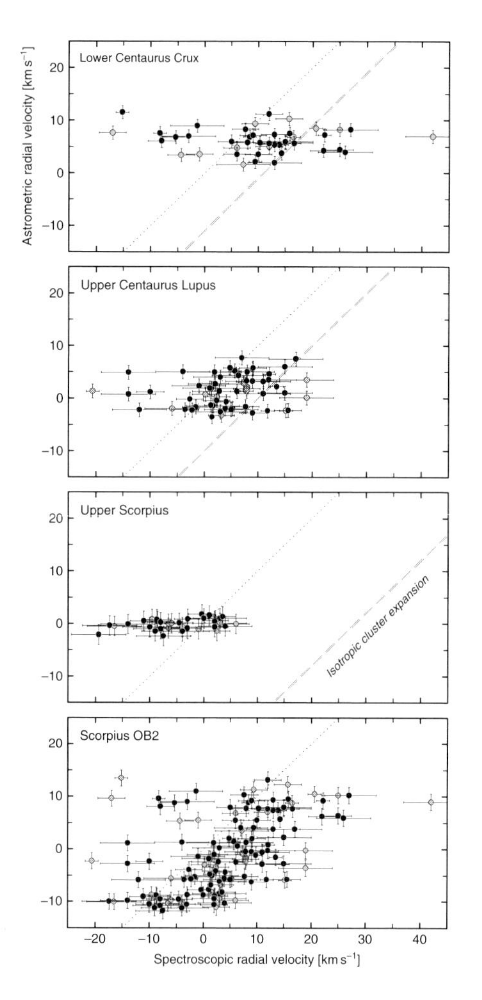

Figure 6.30 Astrometric versus spectroscopic radial velocities for stars in the Scorpius–Centaurus group of young associations, expected to undergo kinematic expansion. The top three frames show separate solutions for each subgroup. Assuming a rate of isotropic expansion equal to the inverse age of the cluster, a bias in the astrometric radial velocity would result, marked by dashed grey lines (the cluster's increasing angular size would be interpreted as an approaching motion). The assumed ages are 11, 14 and 5 Myr. The bottom frame shows the solution for all 510 stars in the groups, treated as one entity, with only stars with known spectroscopic velocities plotted. While these data do indicate some expansion of this complex of young associations, the expansion of its individual parts is significantly slower than the expected rate. From Madsen et al. (2002, Figure 7).

Lower Centaurus Crux, and 15–22 Myr for Upper Centaurus Lupus, and that 95% of the low-mass star formation in each subgroup to have occurred in less than 8 Myr and 12 Myr respectively. Using the evolutionary tracks of Bertelli *et al.* (1994), they found main-sequence turn-off ages for Hipparcos B-type members to be 16 ± 1 Myr for Lower Centaurus Crux and 17 ± 1 Myr for Upper Centaurus Lupus (Figure 6.31). Contrary to previous findings, it appears that Lower Centaurus Crux is coeval with, or slightly older than, Upper Centaurus Lupus. The secular parallaxes of the Sco–Cen pre-main-sequence stars yield distances of 85–215 pc, with 12 of the Lower Centaurus Crux members lying within 100 pc of the Sun.

Sartori *et al.* (2003) studied formation scenarios for the young stellar associations between Galactic longitudes $l = 280–360°$, based on space velocities and age distributions of the young early-type star members of the Scorpius–Centaurus OB association (i.e. the more massive stars which have already reached the main sequence or evolved away from it), and the pre-main-sequence stars belonging to the Ophiuchus, Lupus and Chamaeleon star-forming regions (i.e. the less massive stars of the same age range which have not yet reached the main sequence). All regions studied are at distances of around 100–200 pc. The study was based on Hipparcos data where available, and radial velocities from the literature. The young early-type OB association stars were based largely on the Hipparcos study of de Zeeuw *et al.* (1999), while the pre-main-sequence stars were from the compilation of 610 T Tauri and Herbig Ae/Be stars of Sartori (2000), of which 39 are in the Hipparcos Catalogue. They found that the young early-type stars and the pre-main-sequence stars of the star-forming regions follow a similar spatial distribution, with no separation between the low and the high-mass young stars, and no difference in the kinematics nor in the ages of the two populations. The pre-main-sequence stars in each star-forming region span a wide range of ages, from 1–20 Myr, while the ages of the OB subgroups are 8–10 Myr for Upper Scorpius, and 16–20 Myr both for Upper Centaurus Lupus and for Lower Centaurus Crux (Figure 6.32). They argue that, of the different scenarios for the triggering of large-scale star-formation that have been proposed, the properties are best explained by the passage of a spiral arm close to the Sun; the alignment of young stars and molecular clouds and the average velocity of the stars in the opposite direction to the Galactic rotation agree with the expected behaviour of star formation in nearby spiral arms for the situation in which the co-rotation radius lies beyond $R_\odot$ (Section 9.7). They argue that the spiral-arm interpretation explains the uniform velocity distribution over an extended region, as well as the absence of an age gradient across the three subgroups.

Hernández *et al.* (2005) also made a study of the early-type (B, A, and F) stars in a number of the nearby OB associations (Upper Scorpius, Perseus OB2, Lacerta OB1, and Orion OB1), with membership determined from the Hipparcos data, and spanning an age range of $\sim$ 3–16 Myr, with the aim of determining the

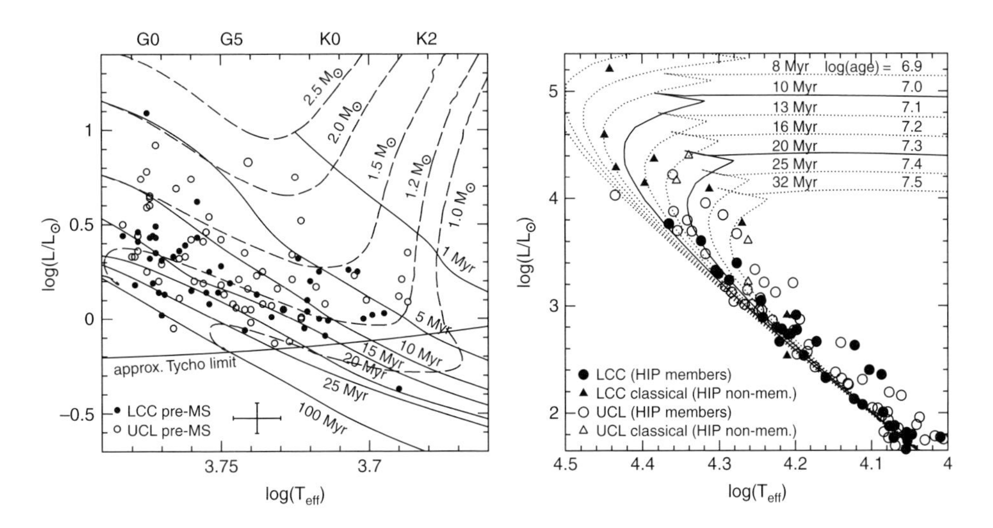

Figure 6.31 Left: theoretical HR diagram for stars identified as (possible) pre-main-sequence stars in the Upper Centaurus Lupus (○) and Lower Centaurus Crux samples (•). The pre-main-sequence evolutionary tracks of D'Antona & Mazzitelli (1997) are included. The ACT/TRC magnitude limit of $V = 11$ is shown for a distance of 150 pc, with $A_V = 0.3$ assumed. The average 1σ error bars in $\log T_{\rm eff}$, and $\log L/L_\odot$ are shown. Right: theoretical HR diagram for the B-star candidate members of Upper Centaurus Lupus and Lower Centaurus Crux using the evolutionary tracks of Bertelli et al. (1994). Only the most massive Hipparcos members were included in the age estimates. From Mamajek et al. (2002, Figures 6 and 9).

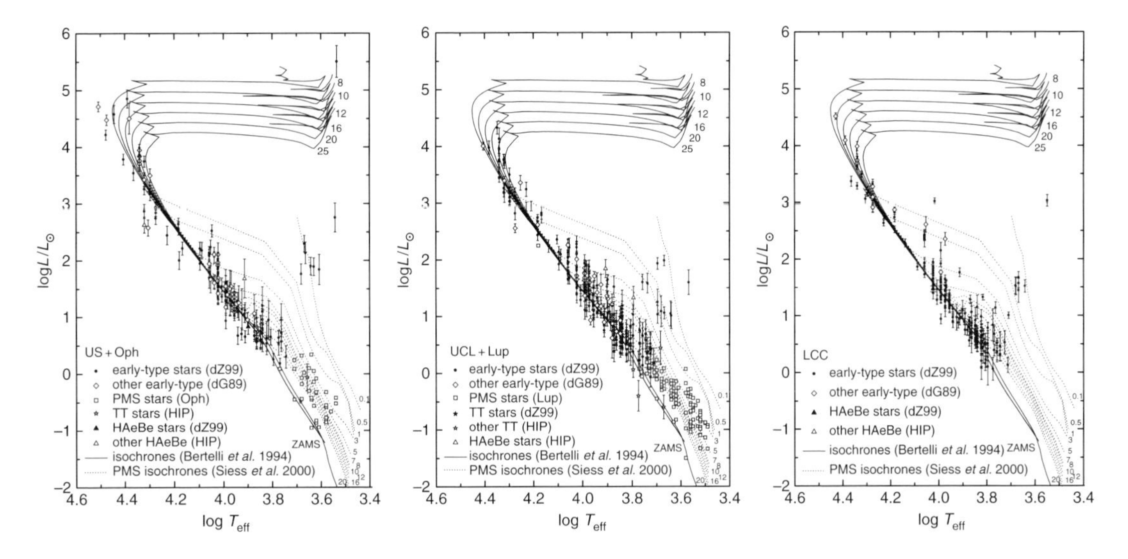

Figure 6.32 HR diagrams of the young early-type and pre-main-sequence stars in the regions studied by Sartori et al. (2003). Isochrones are for solar composition from Siess et al. (2000) for the pre-main-sequence phase, and from Bertelli et al. (1994) otherwise. Isochrone ages in Myr are indicated at the bottom of the pre-main-sequence isochrones, and at the end of the isochrones for evolution on and beyond the main sequence. Left: stars of the Upper Scorpius OB association, and the pre-main-sequence stars of Ophiuchus. Middle: stars of the Upper Centaurus Lupus OB association, and the pre-main-sequence stars of Lupus. Right: stars of the Lower Centaurus Crux OB association. From Sartori et al. (2003, Figures 11–13).

fraction of the pre-main-sequence Herbig Ae/Be stars. They obtained spectra for 440 Hipparcos stars, and determined accurate spectral types, visual extinctions, effective temperatures, luminosities and masses, using Hipparcos photometry (Figure 6.33). Using colours corrected for reddening, they found that the Herbig Ae/Be stars and the classical Be stars occupy different regions in the *JHK* diagram, and that the Herbig Ae/Be stars constitute a small fraction of the early-type stellar population even in the younger associations.

Comparing the data from associations with different ages and assuming that the near-infrared excess in the Herbig Ae/Be stars arises from optically thick dusty inner disks, they determined the evolution of the inner disk frequency with age, finding that the inner disk frequency in the age range 3–10 Myr in intermediate-mass stars is lower than that in the low-mass stars ($< 1M_\odot$). In particular, it is a factor of about 10 lower at ~3 Myr, indicating that the time scales for disk evolution are much shorter in intermediate-mass stars.

Radial velocities Verschueren *et al.* (1997) described high signal-to-noise echelle spectroscopy obtained at the ESO 1.5-m telescope for 156 early-type stars in the Sco OB2 association observed by Hipparcos, although derived radial velocities were never published. Steenbrugge *et al.* (2003) published radial velocities for 29 B and A-type stars in the Perseus OB2 association. Jilinski *et al.* (2006) published radial velocities for 56 B-type stars in the Upper Scorpius, Upper Centaurus Lupus, and Lower Centaurus Crux subgroups of the Sco–Cen association.

6.10.3 Individual associations

In addition to the general results already described, further detailed studies using the Hipparcos data have been made of various associations as follows.

Perseus OB2 Hakobyan *et al.* (2000) used the Hipparcos proper motions to identify two expanding groups of OB stars within the Per OB2 association, calculating expansion ages of 1.3 and 1.9×10^6 yr.

Belikov *et al.* (2002a) constructed a compiled astrometric and photometric catalogue of about 30 000 stars in a 10° region including the Per OB2 complex, complete to $V = 11.6$, and to $V = 18.5$ in the $1° \times 1°$ field centred on the IC 348 cluster. Positions and proper motions were reduced to the Hipparcos reference frame, with proper motion errors of ~ 1–3 mas yr^{-1}. A preliminary proper motion selection yielded more then 1000 probable Per OB2 association members. Belikov *et al.* (2002b) analysed a uniform subset of main-sequence stars earlier than A7, complete within 500 pc, to reveal an area of enhanced density with an angular size of ~ 7°, extending from its classical position towards the California nebula and the Auriga dark cloud, in total comprising 1025 main-sequence proper motion members. The shape of the association is almost spherical, with a diameter of about 40 pc, and no evidence of expansion or rotation around the line-of-sight. The distance to the association centre derived from a kinematic analysis is ~300 pc, in agreement with the Hipparcos data, but with evidence for two groups, Per OB2a and Per OB2b, located at slightly different distances. Per OB2b lacks the luminous stars observed in Per OB2a. Interpreted as an

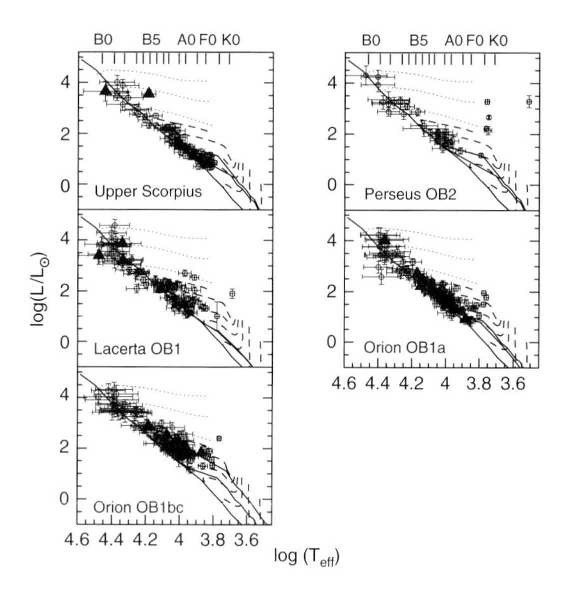

Figure 6.33 HR diagram for the Hipparcos stars in the OB associations studied by Hernández et al. (2005). ▲: emission-line stars; ○: stars without emission. Isochrones from Palla & Stahler (1993) are given for the age ranges established for each association, as well as evolutionary tracks from the same source (dashed lines) corresponding to 0.2, 0.6, 1.0, 1.5, 2.0, 3.0, and 4.0$M_\odot$ (bottom to top); tracks for 5, 9 and 15$M_\odot$ (dotted lines) are from Bernasconi (1996). The ZAMS is represented as a thick solid line. The emission-line stars and the objects without emission share the same region in these plots. From Hernández et al. (2005, Figure 4).

age difference, they suggest that the association is an example of propagating star formation, which started in Per OB2b at the edge of the Auriga clouds more than 30 Myr ago, resumed in Per OB2a after about 10 Myr, and is now in progress at the Per OB2 southern border within the Perseus cloud (IC 348).

Scorpius–Centaurus–Lupus Martín (1998) estimated that the low-mass pre-main-sequence stars identified in about 9 square degrees in Sco–Cen comprise two classical T Tauri stars, 18 weak-line T Tauri stars, and 10 post-T Tauri stars. The simultaneous presence of a mixture of T Tauri and post-T Tauri stars implies that previous isochrone-based results indicating an extremely young age for the Sco–Cen pre-main-sequence low-mass population of ~ 1 Myr (Walter *et al.*, 1994), were incorrect. A distance of about 125 pc for Sco–Cen, instead of the 160 pc used in pre-Hipparcos work, is consistent with the Hipparcos parallaxes for many of the B-type stars and would lead to older evolutionary ages. He argued that the weak-line/post-T Tauri ratio in Sco–Cen may be of order unity, suggesting that the low-mass stars of the OB association span an age range similar to the

B-type members (5–15 Myr), i.e. the low- and high-mass star populations are essentially coeval.

Preibisch *et al.* (2002) investigated the stellar population and star-formation history of the Upper Scorpius OB association over the mass range $0.1–20M_\odot$. From a spectroscopic survey, they identified 68 new (mostly M-type) pre-main-sequence stars which, combined with earlier results (Preibisch & Zinnecker, 1999) yields a total of 250 pre-main-sequence stars in the range $0.1–2M_\odot$. Including the population of 114 high-mass Hipparcos members yields a combined HR diagram for 364 high- and low-mass stars covering $0.1–20M_\odot$, with the whole stellar population characterised by a very narrow age distribution of around 5 Myr (Figure 6.34, left). From individual mass estimates, they found a power-law fit to the mass function with a slope of $\alpha = -2.6$ above $2M_\odot$, and a much flatter slope $\alpha = -0.9$ below $0.6M_\odot$ (Figure 6.34, right). The implied initial mass function is consistent with recent determinations of that in the field population.

Preibisch *et al.* (2002) argue that their results confirm earlier indications that the star formation process in Upper Scorpius was triggered, and support previous conjectures that the triggering event was a supernova shock wave originating from the nearby Upper Centaurus–Lupus association some 12 Myr ago. The structure and kinematics of the large H I loops surrounding the Scorpius–Centaurus association suggest that this shock wave passed through the former Upper Scorpius molecular cloud ~ 5–6 Myr ago (de Geus, 1992). The shock wave crossing Upper Scorpius initiated the formation of some 2500 stars with a total stellar mass of ~ 2060 $M_\odot$, including the massive stars Antares (~ 22 $M_\odot$) and its B2.5 companion (~ 8 $M_\odot$), the massive progenitor of the pulsar PSR J1932+1059 (~ 50 $M_\odot$), and ρ Oph (~ 20 $M_\odot$). When the newborn massive stars 'turned on', they immediately started to destroy the cloud by their ionising radiation and strong winds, terminating the star-formation process after $\lesssim$ 1 Myr, and explaining both the narrow age distribution, and why only about 2% of the original cloud mass was transformed into stars. About 1.5 Myr ago, the most massive star, the progenitor of the pulsar PSR J1932+1059, exploded as a supernova. This explosion led to the ejection of its companion star as the runaway star ζ Oph (Hoogerwerf *et al.*, 2001, see also Section 8.4.2), and created a strong shock wave which fully dispersed the Upper Scorpius molecular cloud and removed the remaining diffuse material. The same shock wave presumably crossed the ρ Oph cloud within the last 1 Myr (de Geus, 1992). The strong star formation activity observed now in the ρ Oph cloud might therefore have been triggered by the same shock wave (Motte *et al.*, 1998), representing the third generation of sequential triggered star formation in the Scorpius–Centaurus–Ophiuchus complex.

Kouwenhoven *et al.* (2005) addressed an improved knowledge of the mass function through a determination of the primordial binary population in Scorpius OB2 based on a near-infrared adaptive optics search for close visual companions to known A star members. Results are discussed further in Chapter 3.

Makarov (2007a) extended the kinematic analysis of pre-main-sequence stars associated with the Lupus dark cloud to a number of T Tauri stars in the UCAC 2 Catalogue (in the Hipparcos reference frame) to identify 93 stars with a streaming motion of low internal velocity dispersion, $\lesssim 1.3\,\mathrm{km\,s^{-1}}$.

Other studies Other Hipparcos-based studies of OB associations, including preliminary results or reviews, include: new results for three nearby OB associations (Hoogerwerf *et al.*, 1997); the structure and evolution of nearby OB associations (de Zeeuw *et al.*, 1997); kinematic evidence for propagating star formation induced by OB associations (Comerón *et al.*, 1997b); kinematics of Cepheus OB3 (Trullols *et al.*, 1997); spatial distribution of OB stars based on the Hipparcos Catalogue (Drimmel *et al.*, 1999); star-formation history of the Puppis–Vela region (Subramaniam & Bhatt, 1999); improved colour-magnitude diagrams for Sco OB2 (de Bruijne, 2000b); the Hipparcos view of OB associations (Brown *et al.*, 2000); OB associations and star clusters in Puppis–Vela (Subramaniam & Bhatt, 2000); bright OB stars in Canis Major–Puppis–Vela (Kaltcheva & Hilditch, 2000); a low-mass, pre-main-sequence stellar association around γ^2 Vel (Pozzo *et al.*, 2000); survey of stellar associations using proper motions (Abad *et al.*, 2001); a review of open clusters and OB associations (Brown, 2001); parallaxes and a kinematically adjusted distance scale for OB associations (Dambis *et al.*, 2001); distance to the star-forming cloud Lupus 2 (Knude & Nielsen, 2001); identification of young local associations detected by the ROSAT All-Sky Survey (Torres *et al.*, 2003); BS Ind in the Tucana association (Guenther *et al.*, 2005); new members of the Upper Scorpius association from the UKIRT infrared deep sky survey (Lodieu *et al.*, 2006); a distance of 135 ± 8 pc to the Ophiuchus star-forming region based on Hipparcos parallaxes to stars illuminating reflection nebulosity in close proximity to the embedded cluster Lynds 1688 (Mamajek, 2008); and the brightest stars of the σ Orionis cluster from a cross-correlation between sources in the Tycho 2 and 2MASS Catalogues (Caballero, 2007).

6.10.4 Young nearby streams, associations or moving groups

Until recently, the only known coeval, co-moving concentrations of stars within 70 pc or so of the Sun were the rich Hyades cluster at ~ 45 pc, and the sparse Ursa Major cluster at ~ 25 pc. Both are hundreds of Myr old.

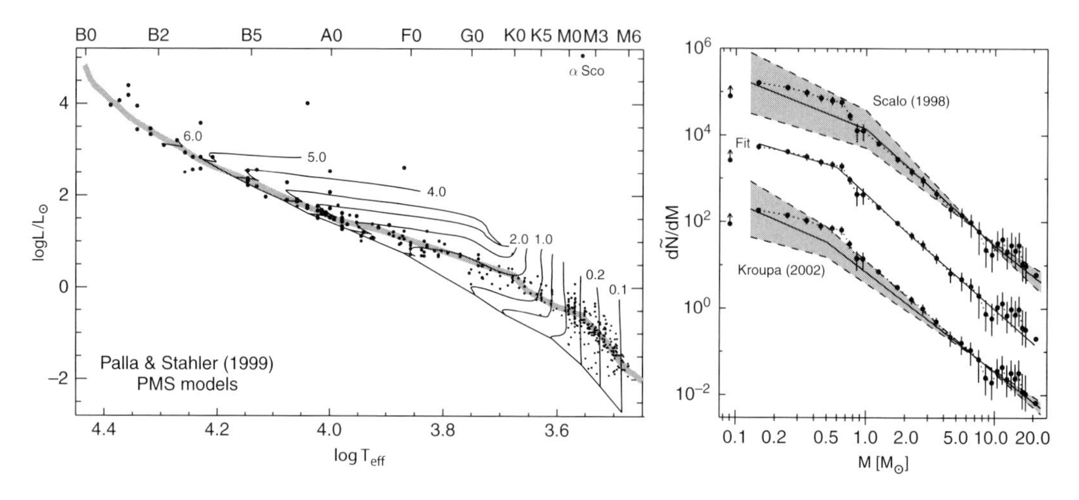

Figure 6.34 Left: HR diagram for the Upper Scorpius association members in the composite Hipparcos sample, X-ray selected sample, and spectroscopic (2dF) sample compiled by Preibisch et al. (2002). Lines show evolutionary tracks from the Palla & Stahler (1999) pre-main-sequence models, some labelled by mass. The thick solid line shows the main sequence. The 5 Myr isochrone (thick grey line) was constructed from the models of Bertelli et al. (1994) for $M = 6$–$30M_\odot$; the Palla & Stahler (1999) pre-main-sequence models for $M = 1$–$6M_\odot$; and the Baraffe et al. (1998) pre-main-sequence models for $M = 0.02$–$1M_\odot$. Right: smoothed mass function for the Upper Scorpius association compared with different models. The mass function is repeated three times, shown by filled circles connected by dotted lines, and multiplied by arbitrary factors. Middle curve: the original mass function, with the solid line showing the multi-component power-law fit. Top curve: the mass function multiplied by 30, and compared with the Scalo (1998) initial mass function (solid line); the grey-shaded area delimited by the dashed lines represents the range allowed by model errors. Bottom curve: mass function multiplied by 1/30 and compared with the Kroupa (2002) initial mass function. From Preibisch et al. (2002, Figures 5 and 7).

Starting in the late 1990s, a number of sparse moving groups, clusters, associations, or streams (different terms are used by different authors, and in this section, at least, they should be viewed as reasonably synonymous) with distances of around 50–100 pc have been discovered. Sometimes based initially on signatures of youth from far infrared and X-ray surveys, spectroscopy, and their spatial location on the sky, they are supported by the common proper motions evident in the Hipparcos and Tycho Catalogues. Derived ages fall typically in the range 5–30 Myr. Membership determination is, for some, made more difficult by the large angles that they subtend on the sky and chance coincidences. This is the case for some older (Tucana–Horologium) as well as some of the younger (β Pic and TW Hya) groups, while some very young groups (η Cha and ϵ Cha) are compact and may be bound. A number of these groups appear to be associated, and perhaps even kinematically convergent with, the Ophiuchus–Scorpius–Centaurus association. The location of some of these systems on the sky is shown in Figure 6.35. The Hipparcos data have been used to identify members, establish evolutionary ages, and trace back their places of origin. The topic is developing rapidly, and the following gives an indication of the current status of the Hipparcos contributions. A broader review is given by Zuckerman & Song (2004). The latest of the candidate groups, reported by Mamajek (2006, 2007), have ages and space motions similar to those of the Pleiades/α Per/AB Dor and the Cas–Tau/Tucana–Horologium groups.

β Pictoris An early suggestion that β Pic (Section 10.7.5) might be as young as 10 Myr was difficult to understand on the basis of its space motion and its distance from any obvious sites of recent star formation. Barrado y Navascués *et al.* (1999) found that the low-mass stars GJ 799 and GJ 803 share similar space velocities. Zuckerman *et al.* (2001a) searched for other members, amongst 22 000 stars with Hipparcos astrometry and measured radial velocities whose space motions can be calculated, restricting their search to stars within 50 pc and showing evidence for youth. They identified 17 such star systems, moving through space together with β Pic. This set of nearby 12 Myr old stars includes a 35 M_J object and various stars with dusty circumstellar disks, prime candidates for imaging programmes related to exoplanet formation. Its large spatial extent and young age imply that it did not originate from a tightly bound cluster, a conclusion supported by the existence of many wide binaries (~3000 AU) and its large velocity dispersion.

Song *et al.* (2003) identified 11 further members. Both Song *et al.* (2003) and Ortega *et al.* (2002) used the Hipparcos positions and proper motions to trace

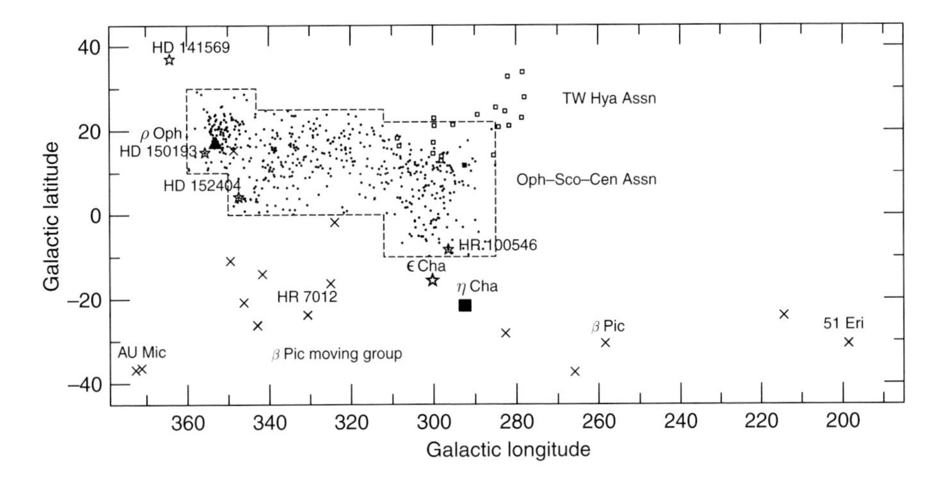

Figure 6.35 Diagram of approximately one third of the celestial sphere in Galactic coordinates showing nearby stars and pre-main-sequence groups kinematically associated with the Oph–Sco–Cen association: Upper Scorpius, Upper Centaurus Lupus and Lower Centaurus Crux subgroups (dots in dashed outline); ▲: ρ Oph embedded cluster; □: TW Hya association; ×: β Pic moving group; ▪: η Cha cluster; ⋆: ϵ Cha group (with HD 104237) and four Herbig Ae/Be systems. Mean proper motion vectors all point towards the bottom right corner (see also Figure 1 of Feigelson et al. 2003). From Feigelson et al. (2006, Figure 1).

back their space motions with time, the former using a constant velocity trajectory, the latter making use of a model Galactic potential (Figure 6.36, top left). They both arrived at a dynamical age of 11.5–12 Myr when the group occupied its minimum spatial volume. This dynamical age is encouragingly consistent with the evolutionary age determined by Zuckerman *et al.* (2001a), making different use of the Hipparcos data, of 12^{+8}_{-4} Myr. Ortega *et al.* (2002) argued that the data indicate that the group was formed gravitationally unbound, the constituent stars maintaining their original space velocities until the present time. Song *et al.* (2003) were unable to distinguish between formation in a large volume of order 30 pc in size, or in a very compact region of a few pc but with a larger velocity dispersion of up to $\sim 8\,\mathrm{km\,s^{-1}}$.

Feigelson *et al.* (2006) used Hipparcos and other data to argue that 51 Eri, at 30 pc, and the nearby (66 arcsec, ∼2000 AU) GJ 3305 are common proper motion components of a fragile wide binary also associated with the β Pic moving group.

TW Hydrae The TW Hydrae (TW Hya) association was the first of the newly-discovered nearby groups. Starting with the isolated T Tauri stars TW Hya itself, other nearby T Tauri stars were discovered as a result of their IRAS 60 μm excess. These objects, HD 98800, CD −29°8887, Hen 3–600, and CD −33°7795, were subsequently recognised as defining a young association (Kastner *et al.*, 1997); they all lie within a 10° × 10° field, at least 13° from the nearest known dark cloud, and with an age of ∼10 Myr. Further members have been recognised subsequently (e.g. Webb *et al.*, 1999).

Makarov & Fabricius (2001) used their cross-identification of the Tycho 2 Catalogue with the ROSAT Bright Star Catalogue, and a convergent-point analysis, to identify 31 candidate members, of which 23 were new (Figure 6.36, top right). The centre of mass lies at a distance of 73 pc in the direction of TW Hya, with the closest being at only 17.5 pc, a space motion similar to that of the nearby Sco–Cen associations, and a total group mass of $31M_\odot$. They estimated a lower limit to the one-dimensional velocity dispersion of $0.8\,\mathrm{km\,s^{-1}}$, and an expansion rate of $0.12\,\mathrm{km\,s^{-1}\,pc^{-1}}$ defining a dynamical age of 8.3 Myr. They considered the association as being connected with the Gould Belt structure, rather than being an isolated open cluster.

Song *et al.* (2002) used spectroscopy covering Hα and Li 670.8 nm, as well as location in the colour–magnitude diagram to confirm only three of the possible new members, eliminating most beyond 100 pc, and thus underlining the difficulty of membership identification based solely on kinematic data. Song *et al.* (2003) meanwhile identified five further association members.

Lawson & Crause (2005) determined rotation periods for 16 stars of the TW Hya association. They used the resulting bimodal distribution of the two subgroups TWA 1–13 and TWA 14–19, with median rotation periods of 4.7 d and 0.7 d respectively, as well as Hipparcos astrometry, to conclude that they formed from spatially and rotationally distinct populations: the former with a distance of ∼55 pc and an age of ∼10 Myr, and the latter at ∼90 pc, with an age of ∼17 Myr, and representing the population of low-mass stars still physically associated with the Lower Centaurus Crux subgroup.

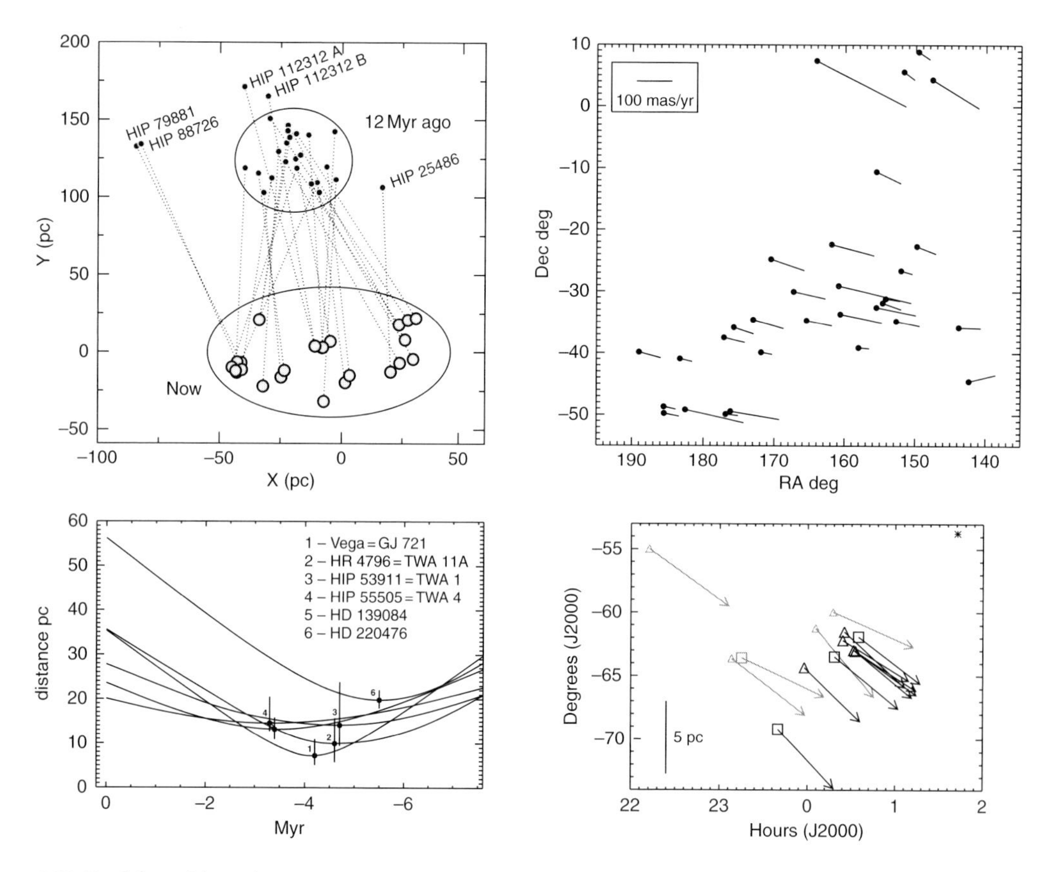

Figure 6.36 Top left: positions of β Pic moving group members traced back over 12 Myr. X, Y are distances in the co-rotating local standard of rest frame, with X positive toward the Galactic anticentre and Y positive in the direction of Galactic rotation. From Song et al. (2003, Figure 4). Top right: proper motions of the 31 candidate kinematic members of the TW Hya association selected from a convergent-point analysis of the Tycho 2 Catalogue. From Makarov & Fabricius (2001, Figure 1). Bottom left: distances of TW Hya members and Vega from the common centre of gravity in the past. The curves show the median distances derived from Monte Carlo modelling; error bars at closest approach are the 15.9 and 84.1 percentiles. From Makarov et al. (2005, Figure 1). Bottom right: candidate moving group members of the association in Tucanae, with Hipparcos proper motions extrapolated over 250 000 yr. Probable members are indicated with large symbols and dark vectors; improbable members with small symbols and light vectors. □: ROSAT X-ray sources; △: non-X-ray sources. The linear scale applies at the approximate distance of the group. From Zuckerman & Webb (2000, Figure 1).

Makarov *et al.* (2005) traced back the motions of five candidate members, using the epicycle approximation (Section 6.2.3) and Monte Carlo simulations to establish an expansion age of 4.7 ± 0.6 Myr, a characteristic initial size of 21 pc, and an initial velocity dispersion of 4–10 km s^{-1} (Figure 6.36, bottom left). The large initial size appears to preclude dynamical decay of young multiple systems (Sterzik & Durisen, 1995) as the origin of these small associations. The large initial velocity dispersion implies either a high-mass turbulent cloud as progenitor, or a violent disruption by an external cause. They suggest a scenario which accounts for the difference between the isochrone age of $\sim$10 Myr and the expansion age of 5 Myr in which star formation was stimulated in the TW Hya progenitor cloud by the near passage of the Lower Centaurus Crux OB association at a distance of 36 ± 6 pc some 11 Myr ago, but in which the newly-formed stars were not released from the cloud until a subsequent collision with one of the other molecular clouds in the North Ophiuchus region. As evident from Figure 6.36, bottom left, Vega was inside the association, and close to its centre of gravity, at the time of maximum compression 4.7 Myr ago. If this alignment was a chance encounter, the strong particle disk around Vega (Section 10.7.5) could have been enhanced by the passage through the progenitor cloud at 8 km s^{-1}.

Further related membership and dynamical evolution studies have been reported by Mamajek (2005), de la Reza *et al.* (2006), and Biller & Close (2007).

Carina–Vela Makarov & Urban (2000) discovered the sparse Carina–Vela moving group or association from ROSAT All-Sky Survey data in combination with Tycho 2 proper motions. Identification was based on global convergence mapping (Section 6.2.3). It yielded 50 candidate members, its closest only 30 pc from the Sun, strongly elongated along the Galactic plane, with three being members of the more compact young open cluster IC 2391.

Tucana–Horologium Torres *et al.* (2000) discovered a nearby association in Horologium, also based on ROSAT searches for young stars. Hipparcos parallaxes and proper motions for 11 of the 23 candidates were used to estimate its distance as ~60 pc, its size as ~50 pc, and its age as ~30 Myr. The estimated size is compatible with its expansion age for an initial velocity dispersion of $\sim 1.8\,\mathrm{km\,s^{-1}}$. It comprises at least 10 very young stars, some of which are post-T Tauri.

Zuckerman & Webb (2000) discovered a nearby association of 9–10 co-moving stars in Tucanae. They had based their all-sky search for new associations surrounding some 20 stars with a reliable IRAS 60 μm excess, and therefore with a high probability of being young. They then searched the Hipparcos Catalogue within a 6° radius of each target IRAS star for stars with the same distance and proper motion, combing the astrometric data with supporting radial velocity observations. They estimated its distance as ~45 pc (Figure 6.36, bottom right), and assigned an age of ~40 Myr based on Hα emission line strength, although this may be an overestimate based on expansion age constraints. Song *et al.* (2003) identified 11 further members.

It was subsequently realised that the two groups have the same age and space motion and occupy close positions in space, and are therefore likely to be members of one extended and dispersed stream of young stars (Zuckerman *et al.*, 2001b).

η Cha and ϵ Cha Mamajek *et al.* (1999) discovered the nearby young and compact η Cha cluster from deep ROSAT HRI data. The 12 X-ray sources have prominent stellar counterparts, with two early-type stars and 10 Li-rich, Hα emission, late-type (K3–M5). The Hipparcos data revealed that the two early-type stars (η Cha, RS Cha) and nearby HD 75505 are co-moving at ~97 pc, with ages ranging from 2–18 Myr. Mamajek *et al.* (2000) computed the cluster's Galactic motion from the Hipparcos data, and showed that the η Cha cluster, as well as members of the TW Hya association and a new group near ϵ Cha, probably originated near the giant molecular cloud complex that formed the two oldest subgroups of the Scorpius–Centaurus OB association roughly 10–15 Myr ago. Their dispersal is consistent with the velocity dispersions seen in giant molecular clouds. A large H I filament and dust lane located near η Cha was identified as part of a superbubble formed by OB winds from Scorpius–Centaurus and supernova remnants. The passage of the superbubble may have terminated its star formation and dispersed any residual associated molecular gas.

Further studies, some also making use of the Hipparcos data for membership, age, and kinematic analyses, have since been reported (Lawson *et al.*, 2001; Feigelson *et al.*, 2003; Luhman & Steeghs, 2004; Song *et al.*, 2004). The work of Lyo *et al.* (2004) summarises the cluster as comprising 17 primary and 9 secondaries within a spatial extent of ~1 pc, a distance of 97 pc, and age of 9 Myr, and with an initial mass function consistent with that of rich young clusters and field stars.

Jilinski *et al.* (2005) traced back the 3d orbits of the kinematic centres of η Cha and ϵ Cha under the action of a general Galactic potential, showing that both groups were formed in the same spatial region about 7 Myr ago. Their birthplace appears to be near to the Lower Centaurus Crux subgroup of the Sco–Cen OB association, and the epoch of minimum separation between the two in the past is consistent with the estimated evolutionary age. Ortega *et al.* (2006) included the β Pic and TW Hya groups in a similar analysis, and argued that the observed temporal sequence of group formation is consistent with the existence of a wave of star formation propagating through the Scorpius–Centaurus OB association region.

AB Dor Zuckerman *et al.* (2004) reported the discovery of a moving group of some 30 stars, with the nucleus of a dozen or so stars only some 20 pc distant, including the well-studied star AB Dor. Their estimated age is ~50 Myr.

The origin of nearby young stars Makarov (2007b) used his 'epicycle correction' approach (Makarov *et al.*, 2004, see also Section 6.2.3), to make a systematic search for close conjunctions and clusterings in the past of all known nearby young stars. The sample included 101 T Tauri, post-T Tauri, and main-sequence stars. Their 3d Galactic orbits were traced back in time using a 0.25 Myr grid extending back to 80 Myr in the past, and near approaches evaluated in time, distance, and relative velocity. Numerous clustering events were detected, and each star's orbit was also matched with those of nearby young open clusters, OB and T associations and star-forming molecular clouds, including the Ophiuchus, Lupus, Corona Australis, and Chamaeleon regions. Ejection of young stars from open clusters could be ruled out for nearly all objects, while the nearest OB associations in Scorpius–Centaurus, and especially, the dense clouds in Ophiuchus and Corona Australis have probably played a major role in the generation of the TW Hya, β Pic, and Tucana–Horologium streams

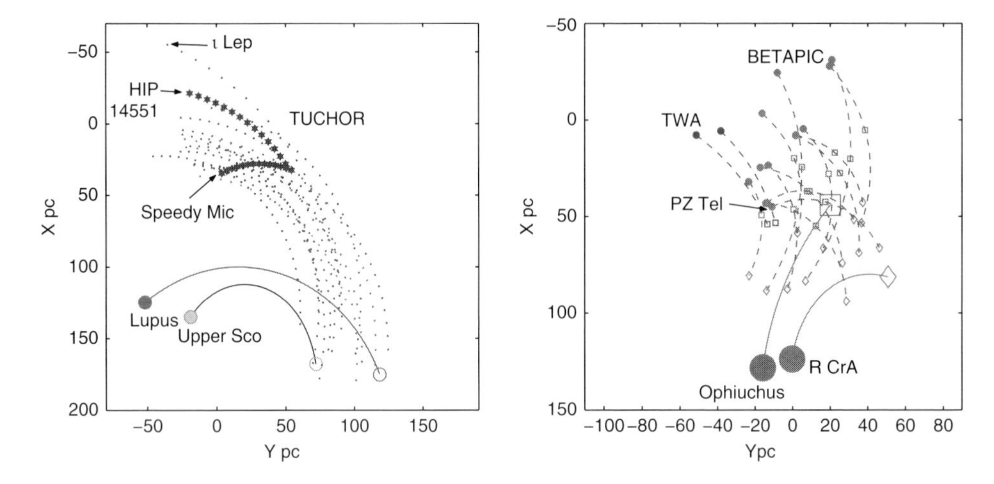

Figure 6.37 Left: trajectories of the Tucana–Horologium (TUCHOR) stars, Upper Scorpius OB association, and Lupus T-association for the past 27 Myr. Trajectories of the young ultra-rapid rotator BO Mic ('Speedy Mic', HD 197890) and HIP 14551 are shown for the past 13 Myr. Current positions of stars are in the upper left corner; •: estimated locations of the Lupus and Upper Scorpius associations today; ◦: 27 Myr ago. Tracks of the stars HIP 14551 and BO Mic are shown with asterisks. Right: trajectories of β Pic (BETAPIC) and TW Hya (TWA) stars, and the R CrA T-association for the past 20 Myr, and of the Ophiuchus star-forming region for the past 9 Myr. Positions of stars and associations are shown at various epochs: •: today; □: 9 Myr ago; ◇: 27 Myr ago. The β Pic group has an extended shape at all three epochs. From Makarov (2007b, Figures 5 and 6).

that are close to the Sun today. The core of the Tucana–Horologium association probably originated from the vicinity of the Upper Scorpius association 28 Myr ago (Figure 6.37).

These various young and nearby associations appear to represent one of the last episodes of star formation in the solar neighbourhood. A consistent picture, based on the studies described above, may now be emerging, and Zuckerman & Webb (2000) offer the following scenario. Between 10–40 Myr ago, the region through which the Sun is now passing contained an ensemble of molecular clouds forming stars at a modest rate, and responsible for the Tucanae–Horologium and TW Hya associations, related star streams, and perhaps also the β Pictoris moving group. The stars ranged primarily from A to M spectral type, but also including a few B-type stars. About 10 Myr ago, the most massive B-type star exploded as a supernova, terminating star formation (at $\sim$20 km s^{-1} with respect to the LSR, the Sun would have been separated by $\sim$200 pc from the explosion). This generated the very low density region with a radius of order 70 pc in most directions from the present position of the Sun (Section 8.5.1), and whereby molecular clouds (Taurus, Lupus, Chamaeleon, Scorpius, Ophiuchus) are seen in various directions, but all at typically 150 pc from Earth. This could also explain the fact that the β Pictoris moving group is so young, $\sim 20 \pm 10$ Myr (Barrado y Navascués *et al.*, 1999) and yet so near to Earth, $\sim$20 pc, disconnected from any obvious sites of star formation.

6.10.5 The Gould Belt

The Gould Belt is a prominent asymmetric distribution of bright stars with respect to the Galactic plane, described further in the box on page 325.

One of the major problems in all studies of the Gould Belt, and to which the Hipparcos data have contributed, is to establish reliable members. The structure crosses the Galactic plane with a small inclination angle, and therefore members are immersed in the ambient Galactic plane population along most lines-of-sight. In particular, the identification of late-type (F–M) Gould Belt stars is especially difficult; only Fresneau *et al.* (1996) have identified low-mass Gould Belt candidates on the basis of kinematic anomalies, albeit based on a small sample and only with a $<20\%$ success rate.

Lindblad *et al.* (1997a,b) selected 2440 Hipparcos stars with ages less than 30 Myr from Strömgren's 'early group', determining absolute magnitudes corrected for reddening, and estimating ages from the evolutionary models of Maeder & Meynet (1988). Stars outside the Gould Belt gave a flat rotation curve with a circular velocity at the Sun of $\Omega = 25.3 \pm 1.5$ km s^{-1} kpc^{-1}. The kinematics of the 241 Gould Belt stars differs significantly: in addition to a small outward motion of ~ 5 km s^{-1}, it rotates in the same direction as the Galactic rotation and expands, giving apparent parameters of $B = -21$ km s^{-1} kpc^{-1} and an expansion rate of $K = +12$ km s^{-1} kpc^{-1}. However, there is not an agreement between A and C from radial velocities and proper motions, suggesting that the velocity field is nonlinear.

The Gould Belt: Herschel (1847) first pointed out that the distribution of bright stars was asymmetric about the Galactic plane. Gould (1879) concluded that a prominent grouping of bright stars was aligned along a great circle crossing the Galactic plane at an angle close to 20°. This structure has been known subsequently as the Gould Belt. Campbell (1913) recognised its overall expansion. Numerous studies of the related bright stars, OB associations and young Galactic clusters have since confirmed its existence. A review of the Gould Belt system and the associated interstellar medium is given by Pöppel (1997).

Stothers & Frogel (1974) described it as follows: the O–B5 stars, supergiants, and associations within 1 kpc of the Sun populate two flat systems inclined to each other by about 20°. A more-or-less random distribution in space and age characterises the O–B5 stars of the Galactic belt, which is aligned nearly along the Milky Way. The Gould Belt is inclined to the Galactic plane (north in Sco–Oph and south in Orion), with five major features defining it. A crude diameter of the system is 750–1000 pc, and the Sun's position is eccentric, lying towards Ophichus. The nuclear age of the system is $\sim$ 30 Myr from the broad main-sequence turn-up at B2.5. Most of the O–B2 stars and youngest stellar groups near to the Sun belong to the Gould Belt, but both belts have roughly equal densities of B3–B5 stars and similar average values of interstellar extinction. Although the Gould Belt is three times as compressed vertically as the Galactic belt, each shows the same increasing concentration for stars, interstellar dust, and stellar groups. A small 'hole' around the Sun occurs in the distribution of dust and O–B5 stars.

Further work has shown that the Gould Belt is tightly correlated with many of the bright and dark nebulae as well as with large masses of neutral interstellar gas from which these young stars were probably born. The modern picture of the Gould Belt consists of a stellar component (Population I stars, OB associations and young clusters) and associated interstellar matter. It has an ellipsoidal shape with a semi-major axis of about 500 pc and a semi-minor axis of about 340 pc, the Sun being located 150–250 pc off centre. Current evidence suggests that the stars and most of the gas are in expansion, and the total mass of the gas is around $2 \times 10^6 M_\odot$ (Pöppel, 1997). Pre-Hipparcos kinematic investigations have been made by Lesh (1968), Frogel & Stothers (1977), Westin (1985), and others.

The formation of the Gould Belt is still debated. Some authors have suggested that it originated through collisions of high-velocity clouds with the Galactic disk (see, e.g. Comerón & Torra, 1992, 1994, and references). Others (e.g. Olano, 1982; Blaauw, 1991) have attributed its formation to supernova explosions, or induced star formation by strong stellar winds originating in the central Cas–Tau association. Others (e.g. Franco *et al.*, 1988; Lépine & Duvert, 1994) have pointed out that even a random spatial distribution of a few prominent OB associations would mimic an 'inclined belt' in the sky. Indeed, most of the early-type stars that make up the Gould Belt belong to the OB associations of Orion, Perseus OB2, and Scorpius–Centaurus–Lupus (Blaauw, 1991). The Lindblad Ring, an expanding neutral hydrogen feature noted by Lindblad *et al.* (1973), may have been the event giving rise to the formation of the Gould Belt (Olano, 1982), or a consequence of the activity of massive stars in the Cas–Tau association, taking place after its formation (Comerón & Torra, 1994).

Recent developments include the discovery of a population of up to 40 γ-ray sources associated with the Gould Belt (Grenier, 2000; Gehrels *et al.*, 2000). These may be relics of core-collapse supernovae resulting from massive star evolution in the belt, their presence underlining the importance of heating and enrichment of the local interstellar medium due to multiple recent supernova explosions.

They suggested that the system was born 30–40 Myr ago, possibly in a spiral arm, with an angular momentum too large to keep the system gravitationally bound. The rotation may explain why it has retained its flatness and inclination to the Galactic plane over several tens of Myr.

Guillout *et al.* (1998) reported the detection of a late-type stellar population in the direction of the Gould Belt among stars found by cross-correlating the ROSAT All-Sky Survey with the Tycho Catalogue. Regions of the Gould Belt located between $l = 195°$ and $l = 15°$ exhibit a strong density enhancement of X-ray active stars with associated X-ray luminosities typical for very young coronae. In contrast, other regions show average Galactic plane characteristics. Stars accounting for the excess extend from the solar vicinity up to about 300 pc towards a quadrant centred on $l = 240°$, but their distance distribution only extends to 180 pc from the Sun towards $l = 330°$. The structure can be understood as a disk-like arrangement of stars having the same inclination towards the Galactic plane as the Gould Belt and extending to its outer boundary. They suggested that these stars are the residuals of original associations, and interpreted them as the late-type stellar population of the Gould Belt (Figure 8.13 below). Higher sensitivity X-ray studies based on ROSAT observations of stars in

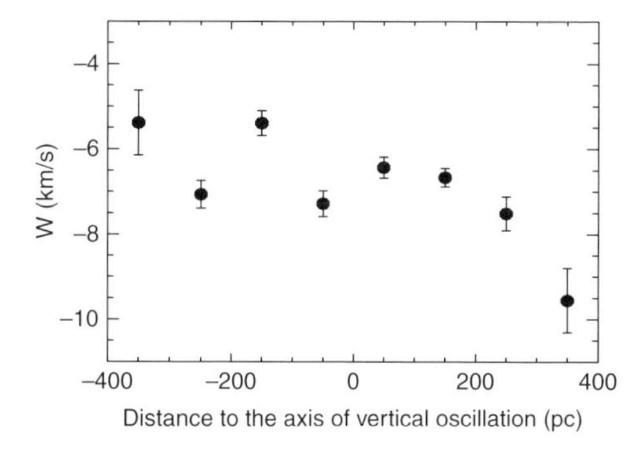

Figure 6.38 The vertical component of the space velocity, W, for stars comprising the Gould Belt, uncorrected for solar motion, as a function of the distance to the axis of vertical oscillation in pc, and averaged over a scale size of 100 pc. The cosmic dispersion of W considerably exceeds the systematic difference of velocities over the distance interval considered. From Comerón (1999, Figure 2).

the Tycho Catalogue were reported by Hempel *et al.* (2003), leading to a still larger stellar surface density at the position of the Gould Belt.

Comerón (1999) argued that the maintained arrangement of the Gould Belt stars over a considerable fraction of the vertical oscillation period of stars perpendicular to the Galactic plane, and thus implying that the coherent structure of the belt applies to the motions as well as the positions of its components, is a fact that places strong constraints on its origin. He used the Hipparcos proper motions of 323 stars not later than B2.5, to ensure the selection of only very young stars, with relative parallax errors below 30%. For the typical Gould Belt stellar distances of 400 pc, the Hipparcos proper motion errors correspond to velocity errors of about $1\,\mathrm{km\,s^{-1}}$. Proper motions were complemented by radial velocities from the literature. He assumed that the perpendicular component of the velocity, W, has systematic variations along the Galactic plane, which define a principal axis of vertical oscillation $\mathbf{G}$, such that $W = |\mathbf{G}\times\mathbf{R}|_z$, where $\mathbf{R}$ is the vector perpendicular to the axis of vertical oscillation joining it to the star. His least-squares solution demonstrated that W shows a significant trend versus distance to the vertical oscillation axis (Figure 6.38). The rather low value of $|G| = 6.5 \pm 1.8\,\mathrm{km\,s^{-1}\,kpc^{-1}}$ implies that the plane of the Gould Belt is presently near its maximum tilt. This feature was also noted by Frogel & Stothers (1977) from the absence of any significant gradient of the W-component in the direction perpendicular to the Galactic plane. The stars oscillate about the Galactic plane around an axis misaligned by $\sim 52^\circ$ with respect to the nodal line defining the intersection between the Gould Belt and the Galactic plane.

Related Hipparcos analyses imply that the Oort constant A has a smaller value than that for pure Galactic differential rotation, that B is negative and probably larger than that corresponding to pure differential rotation, and that K and probably C are both positive. Comerón (1999) used these various results to argue that expansion models in which the Gould Belt stars formed in a small volume can account for the observed values of A, C, and K, but not B. Instead, all kinematic features could be reproduced if the Gould Belt started its expansion from a plane that already had a considerable extent, and with an initial rotation around a perpendicular axis needed to explain the offset between the vertical oscillation axis and the nodal line. The data then give a rather accurate constraint on the age of the Gould Belt of 34 ± 3 Myr, consistent with the ages of the individual stars. Explanation of the Gould Belt as arising from the simple dissolution of a giant molecular cloud appears unlikely because of the required misalignment of its rotation axis with a direction perpendicular to the Galactic plane, itself apparently inconsistent with the observed rotation of giant molecular clouds (Blitz, 1993). The impact of a high-velocity cloud with the Galactic disk may provide a more adequate explanation of its formation.

Moreno *et al.* (1999) made a kinematic analysis of 252 Hipparcos stars from the Gould Belt. They derived a vertex deviation for the whole sample of $l_v \sim -64^\circ \pm 20^\circ$, a value modified to $l_v \sim 22^\circ \pm 8^\circ$ when the members of the Pleiades moving group are removed from the sample (Figure 6.39, top). This implies the existence of at least two different kinematic groups. They then modelled the evolution of a supershell in the solar neighbourhood, obtaining a fit to the shape and kinematics of the gas in the Gould Belt. Assuming that the expanding shell is also forming stars, they obtained the corresponding velocity fields for the shell and its newly-formed stars. The average vertex deviation resulting from these models for the new stars is consistent with the observed value when the Pleiades moving group members are excluded (Figure 6.39, bottom). They concluded that if the whole stellar sample is considered, a model based on a single explosion is unable to explain the main kinematic features of both the stellar and gas components. If the Pleiades moving group is removed, a better agreement between model and observations is achieved; however, the different origins of the Gould Belt and the Pleiades moving group then has to be explained.

Torra *et al.* (2000a,b), see also Torra *et al.* (1997), studied the structure and kinematics of the local system of young stars using the Hipparcos data, studying the velocity field using a classical first-order approach. Oort's constants A, B, C, K were determined for several subsamples selected by age, and the variations interpreted in terms of Galactic rotation, spiral arm structure, and the presence of the Gould Belt. They showed that the 3d picture of the stellar distribution in the solar neighbourhood reveals a distributed population of B stars that depict the Gould Belt as a disk with its members spread well outside the boundaries of the known OB associations, and that the kinematic peculiarities of the belt are preserved even when the stars belonging to the dominant associations are excluded from the analysis.

Makarov & Urban (2000) used the technique of global convergence mapping (Section 6.2.3) to construct convergence maps from 6317 ROSAT All-Sky Survey/Tycho 2 stars to reveal various sub-structure within the Gould Belt.

Maíz Apellániz (2001) constructed a self-gravitating isothermal model of the vertical structure of the Galaxy in the solar neighbourhood from Hipparcos parallaxes alone. His model predicts 12 O–B5 stars within 67 pc of the Sun. However, only three O–B5 stars are found in this volume: η UMa, α Eri, and α Pav. The difference between the expected and the measured number appears to be too large to be explained as a statistical fluctuation, and

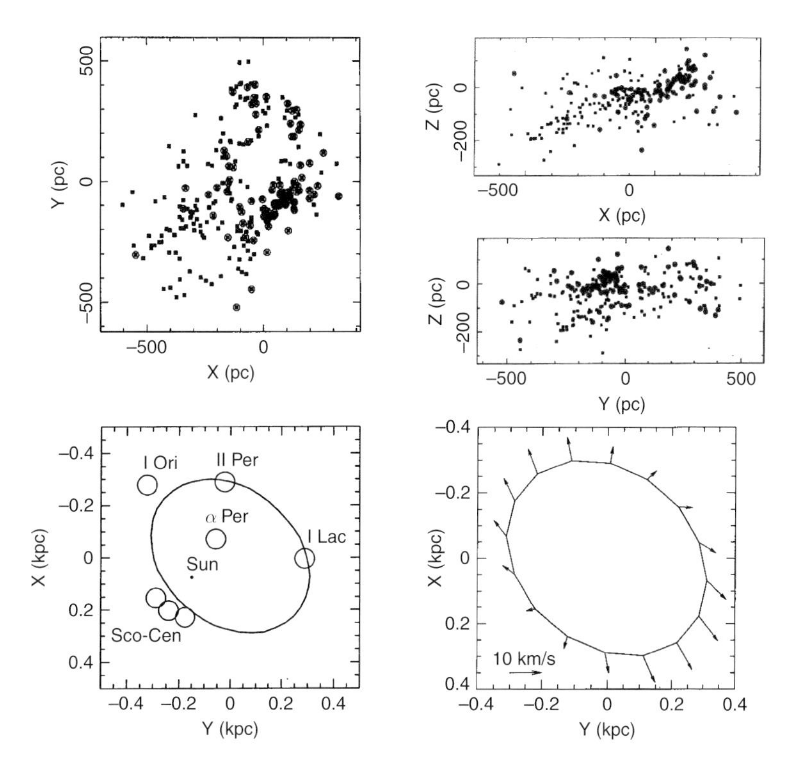

Figure 6.39 Top three panels: spatial coordinates for the stars selected as members of the Gould Belt according to the criteria of Lindblad et al. (1997b) showing projections onto the Galactic (X, Y), (X, Z) and (Y, Z) planes. Stars that share the kinematics of the Pleiades moving group are shown as circled filled squares; most are spatially related to the Sco–Cen association (the referenced work also shows the associated velocity distributions). Bottom left: present position of the expanding shell for the best-fit model with ambient pressure. The position of the Sun is also shown. Bottom right: velocity field of the model stars presently located around the shell, in the reference frame of the local standard of rest. From Moreno et al. (1999, Figures 3, 5 and 8).

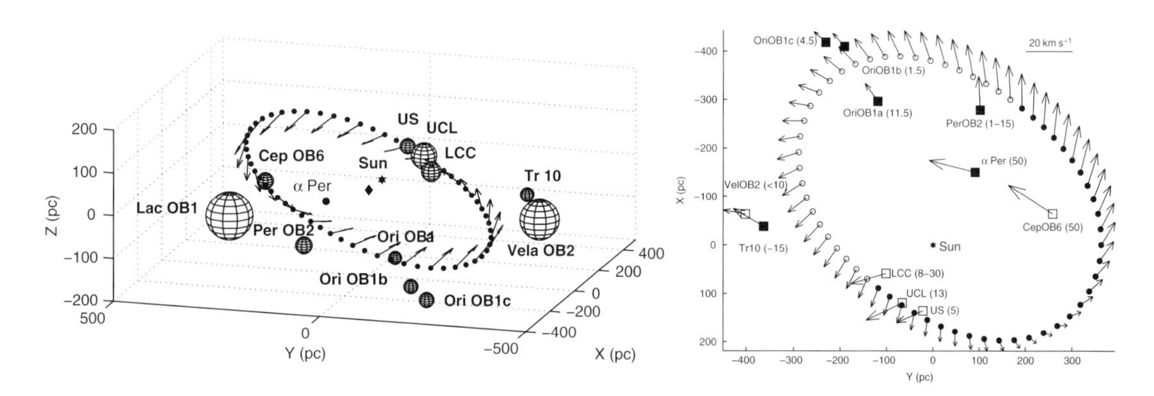

Figure 6.40 Left: 3d view of the present Gould Belt wave and its velocity field with respect to the LSR, amid the local OB associations, marked as spheres with radius proportional to their size, from de Zeeuw et al. (1999). The diamond and star denote the Gould Belt centre and the Sun, respectively. Right: present-day velocities projected on the Galactic plane of the nearby OB associations (■) (de Zeeuw et al., 1999), and of stars that would have been born in the expanding Gould Belt shell 10 Myr ago. Their present position (circles) and velocity have been computed assuming ballistic motion in the Galactic potential. All velocities are corrected for solar motion and Galactic rotation. Filled and open circles denote negative and positive vertical v_z velocities, respectively. The presumed age for each OB association is given in Myr. From Perrot & Grenier (2003, Figures 6 and 9).

they concluded that there is a local 'hole' in the distribution of early-type stars, probably associated with the Gould Belt.

Perrot & Grenier (2003) modelled the 3d dynamical evolution of the interstellar gas in the Gould Belt, comparing the model with the spatial and velocity distribution of all H I and H_2 clouds found within a few hundred parsecs from the Sun, and with the Hipparcos distances of the nearby OB associations. The model describes the expansion of a shock wave that sweeps momentum from the ambient medium, including the effects of Galactic differential rotation and its gravitational torque, interstellar density gradients within and away from the Galactic plane, possible fragmentation and drag forces in the late stages, and an initial rotation of the system. The best fit to the data yields values for the current Gould Belt semi-axes of 373 ± 5 pc and 233 ± 5 pc, and an inclination of $17.^\circ2 \pm 0.^\circ5$ (Figure 6.40). These characteristics are consistent with earlier results, but a different Belt orientation was found because of the presence of new molecular clouds and the revised distance information: the Belt centre currently lies 104 ± 4 pc away from the Sun, and the present rim coincides with most of the nearby OB associations and H_2 clouds, but its expansion bears little relation to the average association velocities; the younger ones being found farther out from the centre. A factor of two discrepancy between the dynamical age and that derived from photometric stellar ages could not be solved by adding a vertical component to the expansion, nor by adding drag forces and fragmentation, nor by introducing an initial rotation.

Other studies of the Gould Belt making use of the Hipparcos data have been reported by Palouš (1998); Cabrera Caño *et al.* (2000); Branham (2002); and Bobylev (2004).

References

Aarseth SJ, 1999, From NBODY1 to NBODY6: the growth of an industry. *PASP*, 111, 1333–1346 {**286**}

Abad C, Romero L, Bongiovanni A, 2001, Survey of stellar associations using proper motions. *Revista Mexicana de Astronomia y Astrofisica Conf. Ser.*, volume 11, 97–98 {**319**}

Abad C, Vicente B, 1999, An astrometric catalogue for the area of Coma Berenices. *A&AS*, 136, 307–312 {**300**}

Abad C, Vieira K, Bongiovanni A, *et al.*, 2003, An extension of Herschel's method for dense and extensive catalogues. Application to the determination of solar motion. *A&A*, 397, 345–351 {**279**}

Abad C, Vieira K, 2005, Systematic motions in the Galactic plane found in the Hipparcos Catalogue using Herschel's method. *A&A*, 442, 745–755 {**279, 310**}

Agekyan TA, Belozerova MA, 1979, The evaporation of stars from clusters, with the development of coronas and moving clusters. *Soviet Astronomy*, 23, 4–8 {**303**}

Aguilar LA, Hoogerwerf R, 1999, A search for moving groups in the Galactic halo. *IAU Symp. 186: Galaxy Interactions at Low and High Redshift*, 47 {**279**}

—, 2001, Searching for moving groups with Hipparcos. *Revista Mexicana de Astronomia y Astrofisica Conf. Ser.*, volume 11, 99–100 {**279**}

Alessi BS, Moitinho A, Dias WS, 2003, Searching for unknown open clusters in the Tycho 2 Catalogue. *A&A*, 410, 565–575 {**299**}

Allende Prieto C, 2001, An empirical procedure to estimate distances to stellar clusters. *ApJ*, 547, 200–206 {**297**}

Alonso A, Arribas S, Martinez-Roger C, 1996, The empirical scale of temperatures of the low main sequence (F0V–K5V). *A&A*, 313, 873–890 {**296**}

Alter G, Balazs BA, Ruprecht J, *et al.*, 1970, *Catalogue of Star Clusters and Associations,* Second Edition. Akademiai Kiado, Budapest {**311**}

Ambartsumian VA, 1947, *in Stellar Evolution and Astrophysics.* Armenian Academy of Sciences {**311**}

—, 1949, *Dokl. Akad. Nauk SSR.* 68, 22 {**311**}

An D, Terndrup DM, Pinsonneault MH, *et al.*, 2007, The distances to open clusters from main-sequence fitting. III. Improved accuracy with empirically calibrated isochrones. *ApJ*, 655, 233–260 {**288, 292**}

Anosova JP, Orlov VV, 1991, The dynamical evolution of the nearby multiple stellar systems ADS 48, ADS 6175 (α Gem = Castor), α Centauri, and ADS 9909. *A&A*, 252, 123–126 {**309**}

Antonello E, Pasinetti Fracassini LE, 1998, Pulsating and non-pulsating stars in Hyades observed by Hipparcos satellite. *A&A*, 331, 995–1001 {**281, 285, 286**}

Arenou F, Luri X, 1999, Distances and absolute magnitudes from trigonometric parallaxes. *ASP Conf. Ser. 167: Harmonizing Cosmic Distance Scales in a Post-Hipparcos Era,* 13–32 {**298**}

Arifyanto MI, Fuchs B, 2006, Fine structure in the phase space distribution of nearby subdwarfs. *A&A*, 449, 533–538 {**307**}

Asiain R, Figueras F, Torra J, *et al.*, 1999a, Detection of moving groups among early-type stars. *A&A*, 341, 427–436 {**304, 306, 307, 309**}

Asiain R, Figueras F, Torra J, 1999b, On the evolution of moving groups: an application to the Pleiades moving group. *A&A*, 350, 434–446 {**277, 278, 307–309**}

Balaguer Núñez L, Jordi C, Galadí Enríquez D, *et al.*, 2004, New membership determination and proper motions of NGC 1817. Parametric and non-parametric approach. *A&A*, 426, 819–826 {**301**}

Baraffe I, Chabrier G, Allard F, *et al.*, 1998, Evolutionary models for solar metallicity low-mass stars: mass-magnitude relationships and colour–magnitude diagrams. *A&A*, 337, 403–412 {**282, 320**}

Barrado y Navascués D, Stauffer JR, Song I, *et al.*, 1999, The age of β Pic. *ApJ*, 520, L123–L126 {**320, 324**}

Barrado y Navascués D, 1998, The Castor moving group: the age of Fomalhaut and Vega. *A&A*, 339, 831–839 {**309**}

Baume G, Villanova S, Carraro G, 2003, A study of the two northern open clusters NGC 1582 and NGC 1663. *A&A*, 407, 527–539 {**301**}

Baumgardt H, 1998, The nature of some doubtful open clusters as revealed by Hipparcos. *A&A*, 340, 402–414 {**301**}

Baumgardt H, Dettbarn C, Fuchs B, *et al.*, 1998, Open clusters in the Hipparcos Catalogue. *Dynamical Studies of Star Clusters and Galaxies* (eds. Kroupa P, Palouš J, Spurzem R), 179–183 {**299**}

Baumgardt H, Dettbarn C, Wielen R, 2000, Absolute proper motions of open clusters. I. Observational data. *A&AS*, 146, 251–258 {**284, 298, 299, 302**}

Belikov AN, Hirte S, Meusinger H, *et al.*, 1998, The fine structure of the Pleiades luminosity function and pre-main sequence evolution. *A&A*, 332, 575–585 {**290, 296**}

Belikov AN, Kharchenko NV, Piskunov AE, *et al.*, 1999a, The extremely young open cluster NGC 6611: compiled catalogue, absorption map and the HR diagram. *A&AS*, 134, 525–536 {**302**}

—, 2002a, Study of the Per OB2 star-forming complex. I. The compiled catalogue of kinematic and photometric data. *A&A*, 384, 145–154 {**318**}

—, 2002b, Study of the Per OB2 star-forming complex. II. Structure and kinematics. *A&A*, 387, 117–128 {**318**}

Belikov AN, Piskunov AE, Schilbach E, 1999b, Pleiades luminosity function: fine structure and new pre-main sequence models. *Astronomische Nachrichten*, 320, 27 {**290**}

Bernasconi PA, 1996, Grids of pre-main sequence stellar models. The accretion scenario at $Z = 0.001$ and $Z = 0.020$. *A&AS*, 120, 57–61 {**318**}

Bertelli G, Bressan A, Chiosi C, *et al.*, 1994, Theoretical isochrones from models with new radiative opacities. *A&AS*, 106, 275–302 {**309, 316, 317, 320**}

Beshenov GV, Loktin AV, 2004, Proper motions of open star clusters from Tycho 2 Catalogue data. *Astronomical and Astrophysical Transactions*, 23, 103–115 {**299**}

Bessell MS, 1991, The late-M dwarfs. *AJ*, 101, 662–676 {**290**}

—, 2000, The Hipparcos and Tycho photometric system passbands. *PASP*, 112, 961–965 {**294**}

Bessell MS, Castelli F, Plez B, 1998, Model atmospheres broadband colours, bolometric corrections and temperature calibrations for O–M stars. *A&A*, 333, 231–250 {**284**}

Bica E, Santiago BX, Dutra CM, *et al.*, 2001, Dissolving star cluster candidates. *A&A*, 366, 827–833 {**303**}

Biller BA, Close LM, 2007, A direct distance and luminosity determination for a self-luminous giant exoplanet: the trigonometric parallax to 2MASSW J1207334-393254Ab. *ApJ*, 669, L41–L44 {**322**}

Binney JJ, Dehnen W, Bertelli G, 2000, The age of the solar neighbourhood. *MNRAS*, 318, 658–664 {**308**}

Binney JJ, 2001, Secular Evolution of the Galactic Disk. *ASP Conf. Ser. 230: Galaxy Disks and Disk Galaxies*, 63–70 {**308**}

Blaauw A, 1946, A study of the Scorpius–Centaurus cluster. *Publications of the Kapteyn Astronomical Laboratory Groningen*, 52, 1–132 {**314**}

—, 1952, The evolution of expanding stellar associations; the age and origin of the Scorpius–Centaurus group. *Bull. Astron. Inst. Netherlands*, 11, 414–419 {**277**}

—, 1964, The O associations in the solar neighbourhood. *ARA&A*, 2, 213–246 {**274, 311**}

—, 1991, OB associations and the fossil record of star formation. *NATO ASIC Proc. 342: The Physics of Star Formation and Early Stellar Evolution* (eds. Lada CJ, Kylafis ND), 125–154 {**311, 325**}

Blitz L, 1993, Giant molecular clouds. *Protostars and Planets III* (eds. Levy EH, Lunine JI), 125–161 {**326**}

Bobylev VV, 2004, Determination of the rotation curve for stars of the Gould Belt using Bottlinger's formulas. *Astronomy Letters*, 30, 159–168 {**328**}

Boesgaard AM, Friel ED, 1990, Chemical composition of open clusters. I. Fe/H from high-resolution spectroscopy. *ApJ*, 351, 467–479 {**296**}

Bok BJ, 1934, The stability of moving clusters. *Harvard College Observatory Circular*, 384, 1–41 {**274, 311**}

Boss LJ, 1908, Convergent of a moving cluster in Taurus. *AJ*, 26, 31–36 {**280**}

Brandt JC, Stecher TP, Crawford DL, *et al.*, 1971, The Gum Nebula: fossil Strömgren sphere of the Vela X supernova. *ApJ*, 163, L99–L104 {**314**}

Branham RL, 2002, Kinematics of OB stars. *ApJ*, 570, 190–197 {**328**}

Breger M, Pamyatnykh AA, 1998, Period changes of δ Scuti stars and stellar evolution. *A&A*, 332, 958–968 {**285**}

Brown AGA, 2001, Open clusters and OB associations: a review. *Revista Mexicana de Astronomia y Astrofisica Conf. Ser.*, volume 11, 89–96 {**319**}

Brown AGA, Dekker G, de Zeeuw PT, 1997, Kinematic ages of OB associations. *MNRAS*, 285, 479–492 {**274, 311**}

Brown AGA, de Bruijne JHJ, Hoogerwerf R, *et al.*, 2000, OB Associations: the Hipparcos View. *Star Formation from the Small to the Large Scale* (eds. Favata F, Kaas A, Wilson A), ESA SP–445, 239–248 {**319**}

Brown A, 1950, On the determination of the convergent point of a moving cluster from proper motions. *ApJ*, 112, 225–239 {**280**}

Burningham B, Naylor T, Jeffries RD, *et al.*, 2003, On the nature of Collinder 121: insights from the low-mass pre-main sequence. *MNRAS*, 346, 1143–1150 {**301**}

Caballero JA, 2007, The brightest stars of the σ Ori cluster. *A&A*, 466, 917–930 {**319**}

Cabrera Caño J, Elias F, Alfaro EJ, 2000, Separation between Gould Belt and disk stars in the solar neighbourhood. *Ap&SS*, 272, 95–98 {**328**}

Campbell WW, 1913, *Stellar motions, with special reference to motions determined by means of the spectrograph.* Yale University Press, New Haven {**325**}

Carraro G, 2002, A photometric study of the two poorly known northern open clusters NGC 133 and NGC 1348. *A&A*, 387, 479–486 {**301**}

Carrier F, Burki G, Richard C, 1999, Geneva photometry of the open cluster NGC 2451 and its exceptional Be star HR 2968 satellite. *A&A*, 341, 469–479 {**301**}

Cash W, Charles P, Bowyer S, *et al.*, 1980, The X-ray superbubble in Cygnus. *ApJ*, 238, L71–L76 {**315**}

Castellani V, Degl'Innocenti S, Prada Moroni PG, *et al.*, 2002, Hipparcos open clusters and stellar evolution. *MNRAS*, 334, 193–197 {**291, 292**}

Castellani V, Degl'Innocenti S, Prada Moroni PG, 2001, Stellar models and Hyades: the Hipparcos test. *MNRAS*, 320, 66–72 {**284, 285**}

Castelli F, 1999, Synthetic photometry from ATLAS9 models in the UBV Johnson system. *A&A*, 346, 564–585 {**292**}

Cayrel de Strobel G, Cayrel R, Lebreton Y, 1998, Helium, [Fe/H] abundances and the HR $T_{\rm eff}$, $M_{\rm bol}$ diagram with Hipparcos data of the four nearest open clusters: Hyades, Coma Berenices, the Pleiades and Praesepe. *Highlights in Astronomy*, 11, 565–565 {**297**}

Cayrel de Strobel G, Crifo F, Lebreton Y, 1997, The impact of Hipparcos on the old problem of the helium content of the Hyades. *ESA SP–402: Hipparcos, Venice '97*, 687–688 {**281, 282**}

Chaboyer BC, Green EM, Liebert J, 1999, The age, extinction, and distance of the old, metal-rich open cluster NGC 6791. *AJ*, 117, 1360–1374 {**302**}

Chandrasekhar S, 1942, *Principles of Stellar Dynamics*. University of Chicago Press {**295**}

Chen B, Asiain R, Figueras F, *et al.*, 1997, Identification of moving groups from a sample of B, A and F type stars. *A&A*, 318, 29–36 {**304**}

Chen B, Vergely JL, Valette B, *et al.*, 1998, Comparison of two different extinction laws with Hipparcos observations. *A&A*, 336, 137–149 {**288, 297**}

Chen L, Zhao J, 1997, Maximum likelihood estimation of the mean parallax and kinematic parameters of the Pleiades. *Chinese Astronomy and Astrophysics*, 21, 433–437 {**293**}

Chereul E, Crézé M, Bienaymé O, 1997, The distribution of nearby stars in phase space mapped by Hipparcos. *ESA SP–402: Hipparcos, Venice '97*, 545–548 {**278**}

—, 1998, The distribution of nearby stars in phase space mapped by Hipparcos. II. Inhomogeneities among A-F type stars. *A&A*, 340, 384–396 {**278**}

—, 1999, The distribution of nearby stars in phase space mapped by Hipparcos: clustering and streaming among A–F type stars. *A&AS*, 135, 5–28 {**273, 278, 304, 306, 309**}

Chupina NV, Reva VG, Vereshchagin SV, 2001, The geometry of stellar motions in the nucleus region of the Ursa Major kinematic group. *A&A*, 371, 115–122 {**308**}

Collinder P, 1931, On structured properties of open galactic clusters and their spatial distribution. *Annals of the Observatory of Lund*, 2, 1 {**301**}

Collins GW, Smith RC, 1985, The photometric effect of rotation in the A stars. *MNRAS*, 213, 519–552 {**285**}

Comerón F, 1999, Vertical motion and expansion of the Gould Belt. *A&A*, 351, 506–518 {**325, 326**}

Comerón F, Torra J, Figueras F, 1997a, Understanding some moving groups in terms of a global spiral shock. *A&A*, 325, 149–158 {**311**}

Comerón F, Torra J, Gómez AE, *et al.*, 1997b, Kinematic evidence for propagating star formation induced by OB associations. *ESA SP–402: Hipparcos, Venice '97*, 479–484 {**319**}

Comerón F, Torra J, Gómez AE, 1994, On the characteristics and origin of the expansion of the local system of young objects. *A&A*, 286, 789–798 {**308**}

—, 1998, Kinematic signatures of violent formation of Galactic OB associations from Hipparcos measurements. *A&A*, 330, 975–989 {**313, 315**}

Comerón F, Torra J, Jordi C, *et al.*, 1993, Anomalous proper motions in the Cygnus super bubble region. *A&AS*, 101, 37–47 {**313**}

Comerón F, Torra J, 1992, The oblique impact of a high-velocity cloud on the Galactic disk. *A&A*, 261, 94–104 {**325**}

—, 1994, The origin of the Gould Belt by the impact of a high-velocity cloud on the Galactic disk. *A&A*, 281, 35–45 {**325**}

Cramer N, 1997, Absolute magnitude calibration of Geneva photometry: B-type stars. *ESA SP–402: Hipparcos, Venice '97*, 311–314 {**288**}

Dambis AK, 1999, Space-age distribution of young open clusters and observational selection. *Astronomy Letters*, 25, 7–13 {**290, 297, 298**}

Dambis AK, Mel'Nik AM, Rastorguev AS, 2001, Trigonometric parallaxes and a kinematically adjusted distance scale for OB associations. *Astronomy Letters*, 27, 58–64 {**319**}

—, 2003, Age dependence of the vertical distribution of young open clusters. *EAS Publications Series 10* (eds. Boily CM, Pastsis P, Portegies Zwart S, *et al.*), 147–152 {**297**}

—, 2004, Age dependence of the vertical distribution of Cepheids. *ASP Conf. Ser. 310: IAU Colloq. 193: Variable Stars in the Local Group* (eds. Kurtz DW, Pollard KR), 158–161 {**297**}

Dame TM, Hartmann D, Thaddeus P, 2001, The Milky Way in molecular clouds: a new complete CO survey. *ApJ*, 547, 792–813 {**311, 312**}

D'Antona F, Mazzitelli I, 1994, New pre-main sequence tracks for $M \leq 2.5 M_\odot$ as tests of opacities and convection model. *ApJS*, 90, 467–500 {**290**}

—, 1997, Evolution of low-mass stars. *Memorie della Societa Astronomica Italiana*, 68, 807–822 {**290, 317**}

Dehnen W, Binney JJ, 1998, Local stellar kinematics from Hipparcos data. *MNRAS*, 298, 387–394 {**305**}

Dehnen W, 1998, The distribution of nearby stars in velocity space inferred from Hipparcos data. *AJ*, 115, 2384–2396 {**304, 305, 308, 310**}

Dekker E, 1976, Spiral structure and the dynamics of galaxies. *Phys. Rep.*, 24, 315–389 {**307**}

Dettbarn C, Fuchs B, Flynn C, *et al.*, 2007, Signatures of star streams in the phase space distribution of nearby halo stars. *A&A*, 474, 857–861 {**307**}

Detweiler HL, Yoss KM, Radick RR, *et al.*, 1984, The radial velocity of the Hyades cluster. *AJ*, 89, 1038–1049 {**280**}

de Bruijne JHJ, 1999a, A refurbished convergent-point method for finding moving groups in the Hipparcos Catalogue. *MNRAS*, 306, 381–393 {**276, 312**}

—, 1999b, Structure and colour–magnitude diagrams of Scorpius OB2 based on kinematic modelling of Hipparcos data. *MNRAS*, 310, 585–617 {**314**}

—, 2000a, Astrometry from space. Ph.D. Thesis {**284**}

—, 2000b, Improved colour–magnitude diagrams for Scorpius OB2 based on kinematically modeled Hipparcos data. *ASP Conf. Ser. 198: Stellar Clusters and Associations: Convection, Rotation, and Dynamos*, 125 {**319**}

de Bruijne JHJ, Hoogerwerf R, Brown AGA, *et al.*, 1997, Improved methods for identifying moving groups. *ESA SP–402: Hipparcos, Venice '97*, 575–578 {**279**}

de Bruijne JHJ, Hoogerwerf R, de Zeeuw PT, 2000, Two Böhm-Vitense gaps in the main sequence of the Hyades. *ApJ*, 544, L65–L67 {**284**}

—, 2001, A Hipparcos study of the Hyades open cluster. Improved colour-absolute magnitude and HR diagrams. *A&A*, 367, 111–147 {**283–286**}

de Geus EJ, 1992, Interactions of stars and interstellar matter in Scorpius–Centaurus. *A&A*, 262, 258–270 {**319**}

de la Reza R, Jilinski E, Ortega VG, 2006, Dynamical evolution of the TW Hya association. *AJ*, 131, 2609–2614 {**278, 322**}

De Simone R, Wu X, Tremaine S, 2004, The stellar velocity distribution in the solar neighbourhood. *MNRAS*, 350, 627–643 {**308, 310**}

de Wit WJ, Bouvier J, Palla F, *et al.*, 2006, Exploring the lower mass function in the young open cluster IC 4665. *A&A*, 448, 189–202 {**301**}

de Zeeuw PT, Brown AGA, de Bruijne JHJ, *et al.*, 1997, Structure and evolution of nearby OB associations. *ESA SP–402: Hipparcos, Venice '97*, 495–500 {**319**}

de Zeeuw PT, Hoogerwerf R, de Bruijne JHJ, *et al.*, 1999, A Hipparcos census of the nearby OB associations. *AJ*, 117, 354–399 {**276, 300–302, 312–316, 327**}

Dias WS, Alessi BS, Moitinho A, *et al.*, 2002a, New catalogue of optically visible open clusters and candidates. *A&A*, 389, 871–873 {**273**}

Dias WS, Assafin M, Flório V, *et al.*, 2006, Proper motion determination of open clusters based on the UCAC 2 Catalogue. *A&A*, 446, 949–953 {**273, 298**}

Dias WS, Lépine JRD, Alessi BS, 2001, Proper motions of open clusters within 1 kpc based on the Tycho 2 Catalogue. *A&A*, 376, 441–447 {**299**}

—, 2002b, Proper motions of open clusters based on the Tycho 2 Catalogue. II. Clusters farther than 1 kpc. *A&A*, 388, 168–171 {**299**}

Dravins D, Lindegren L, Madsen S, *et al.*, 1997, Astrometric radial velocities from Hipparcos. *ESA SP–402: Hipparcos, Venice '97*, 733–738 {**283, 314**}

Dravins D, Lindegren L, Madsen S, 1999, Astrometric radial velocities. I. Non-spectroscopic methods for measuring stellar radial velocity. *A&A*, 348, 1040–1051 {**283, 286**}

Drimmel R, Smart RL, Lattanzi MG, 1999, Spatial distribution of OB stars based on the Hipparcos Catalogue. *Modern Astrometry and Astrodynamics* (eds. Dvorak R, Haupt HF, Wodnar K), 15–28 {**319**}

Efremov YN, Schilbach E, Zinnecker H, 1997, The Hipparcos distances of open clusters and their implication on the local variations of the $\Delta Y/\Delta Z$ ratio. *Astronomische Nachrichten*, 318, 335–338 {**290**}

Eggen OJ, 1978, Intermediate-band photometry of late-type stars. VII. The HR 1614 group of overabundant stars. *ApJ*, 222, 203–208 {**309**}

—, 1981, The region of NGC 2287 and Collinder 121. *ApJ*, 247, 507–521 {**301**}

—, 1989, Stellar superclusters, stellar groups and wide binaries. *Fundamentals of Cosmic Physics*, 13, 1–142 {**302**}

—, 1994, Stellar clusters, superclusters and groups. *Galactic and Solar System Optical Astrometry* (eds. Morrison LV, Gilmore G), 191–195 {**302**}

—, 1998a, Evolutionary oddities in old disk population clusters. *AJ*, 115, 2435–2452 {**309**}

—, 1998b, The age range of Hyades stars. *AJ*, 116, 284–292 {**286**}

—, 1998c, The HR 1614 group and Hipparcos astrometry. *AJ*, 115, 2453–2458 {**309**}

—, 1998d, The Pleiades and α Persei clusters. *AJ*, 116, 1810–1815 {**291**}

—, 1998e, The Sirius supercluster and missing mass near the Sun. *AJ*, 116, 782–788 {**308**}

ESA, 1997, *The Hipparcos and Tycho Catalogues. Astrometric and Photometric Star Catalogues derived from the ESA Hipparcos Space Astrometry Mission, ESA SP–1200 (17 volumes including 6 CDs)*. European Space Agency, Noordwijk, also: VizieR Online Data Catalogue {**284, 295**}

Famaey B, Jorissen A, Luri X, *et al.*, 2005a, Dynamical streams in the solar neighbourhood. *ESA SP–576: The Three-Dimensional Universe with Gaia* (eds. Turon C, O'Flaherty KS, Perryman MAC), 129–133 {**307**}

—, 2005b, Local kinematics of K and M giants from Coravel, Hipparcos, and Tycho 2 data. Revisiting the concept of superclusters. *A&A*, 430, 165–186 {**307, 310**}

Famaey B, Pont F, Luri X, *et al.*, 2007, The Hyades stream: an evaporated cluster or an intrusion from the inner disk? *A&A*, 461, 957–962 {**310**}

Famaey B, Siebert A, Jorissen A, 2008, On the age heterogeneity of the Pleiades, Hyades and Sirius moving groups. *A&A*, 483, 453–459 {**311**}

Feast MW, Whitelock PA, 1997, Galactic kinematics of Cepheids from Hipparcos proper motions. *MNRAS*, 291, 683–693 {**310**}

Feigelson ED, Lawson WA, Garmire GP, 2003, The ϵ Cha young stellar group and the characterization of sparse stellar clusters. *ApJ*, 599, 1207–1222, erratum: 649, 1184 {**321, 323**}

Feigelson ED, Lawson WA, Stark M, *et al.*, 2006, 51 Eri and GJ 3305: a 10–15 Myr old binary star system at 30 pc. *AJ*, 131, 1730–1739 {**321**}

Feinstein A, 1967, Collinder 121: a young southern open cluster similar to H and χ Persei. *ApJ*, 149, 107–115 {**314**}

Feltzing S, Holmberg J, 2000, The reality of old moving groups: the case of HR 1614. Age, metallicity, and a new extended sample. *A&A*, 357, 153–163 {**309**}

Figueras F, Gómez AE, Asiain R, *et al.*, 1997, Identification of moving groups in a sample of early-type main sequence stars. *ESA SP–402: Hipparcos, Venice '97*, 519–524 {**278**}

Fischer DA, Stauffer JR, Jones B, 1998, Hipparcos-based colour–magnitude diagram for Li-rich late-type field stars: a comparison with the Pleiades and Praesepe clusters. *ASP Conf. Ser. 154: Cool Stars, Stellar Systems, and the Sun*, 2097–2100 {**291**}

Fox Machado L, Pérez Hernández F, Suárez JC, *et al.*, 2006, Asteroseismic constraints on the Pleiades distance. *Memorie della Societa Astronomica Italiana*, 77, 455–457 {**294**}

Franco J, Tenorio-Tagle G, Bodenheimer P, *et al.*, 1988, On the origin of the Orion and Monoceros molecular cloud complexes. *ApJ*, 333, 826–839 {**325**}

Fresneau A, Acker A, Jasniewicz G, *et al.*, 1996, Kinematical search in the optical for low-mass stars of the Gould Belt system. *AJ*, 112, 1614–1624 {**324**}

Fridman AM, Koruzhii OV, Zasov AV, *et al.*, 1998, Vertical motions in the gaseous disk of the spiral galaxy NGC 3631. *Astronomy Letters*, 24, 764–773 {**298**}

Frinchaboy PMI, 2006, *Galactic disk dynamical tracers: open clusters and the local Milky Way rotation curve and velocity field*. Ph.D. thesis, University of Virginia {**298**}

Frogel JA, Stothers R, 1977, The local complex of O and B stars. II. Kinematics. *AJ*, 82, 890–901 {**311, 325, 326**}

Fuhrmann K, 2004, Nearby stars of the Galactic disk and halo. III. *Astronomische Nachrichten*, 325, 3–80 {**309**}

Fux R, 2001, Order and chaos in the local disk stellar kinematics induced by the Galactic bar. *A&A*, 373, 511–535 {**310**}

Gatewood G, de Jonge JK, Han I, 2000, The Pleiades, MAP-based trigonometric parallaxes of open clusters. V. *ApJ*, 533, 938–943 {**293**}

Gehrels N, Macomb DJ, Bertsch DL, *et al.*, 2000, Discovery of a new population of high-energy γ-ray sources in the Milky Way. *Nature*, 404, 363–365 {**325**}

Giannuzzi MA, 1995, The spectroscopic binary HD 23642 and the distance of the Pleiades. *A&A*, 293, 360–362 {**293**}

Gieles M, Athanassoula E, Portegies Zwart SF, 2007, The effect of spiral arm passages on the evolution of stellar clusters. *MNRAS*, 376, 809–819 {**303**}

Gould BA, 1879, Uranometria Argentina: brillantez y posición de las estrellas fijas, hasta la septima magnitud, comprendidas dentro de cien grados del polo austral, con atlas. *Resultados del Observatorio Nacional Argentino en Cordoba* {**325**}

Goupil MJ, Dziembowski WA, Goode PR, *et al.*, 1996, Can we measure the rotation rate inside stars? *A&A*, 305, 487–497 {**286**}

Grenier IA, 2000, Gamma-ray sources as relics of recent supernovae in the nearby Gould Belt. *A&A*, 364, L93–L96 {**325**}
Grevesse N, Noels A, 1993, Cosmic abundances of the elements. *Origin and Evolution of the Elements. Cambridge University Press* (eds. Prantzos N, Vangioni-Flam E, Casse M), 14–25 {**281**}
Griffin RF, 1998, 44 LMi and the 'Pleiades supercluster'. *The Observatory*, 118, 223–225 {**303**}
Guenther EW, Covino E, Alcalá JM, *et al.*, 2005, BS Indi: an enigmatic object in the Tucana association. *A&A*, 433, 629–634 {**319**}
Guillout P, Sterzik MF, Schmitt JHMM, *et al.*, 1998, Discovery of a late-type stellar population associated with the Gould Belt. *A&A*, 337, 113–124 {**325**}
Gunn JE, Griffin RF, Griffin REM, *et al.*, 1988, A new convergent point and distance modulus for the Hyades from radial velocities. *AJ*, 96, 198–210 {**280**}
Hakobyan AA, Hambaryan VV, Poghosyan AV, *et al.*, 2000, Two expanding groups of OB stars in the stellar association Per OB2. *IAU Symp. 200*, 124–125 {**318**}
Hansen-Ruiz CS, van Leeuwen F, 1997, Definition of the Pleiades main sequence in the HR diagram. *ESA SP–402: Hipparcos, Venice '97*, 295–298 {**288**}
Hanson RB, 1975, A study of the motion, membership, and distance of the Hyades cluster. *AJ*, 80, 379–401 {**280**}
Helmi A, Navarro JF, Nordström B, *et al.*, 2006, Pieces of the puzzle: ancient sub-structure in the Galactic disc. *MNRAS*, 365, 1309–1323 {**307**}
Hempel M, Berghöfer TW, Schmitt JHMM, 2003, Stellar activity in the Gould Belt. *The Future of Cool-Star Astrophysics: 12th Cambridge Workshop on Cool Stars, Stellar Systems, and the Sun* (eds. Brown A, Harper GM, Ayres TR), 805–809 {**326**}
Herbig GH, 1962, Spectral classification of faint members of the Hyades and Pleiades and the dating problem in Galactic clusters. *ApJ*, 135, 736–747 {**286, 294, 295**}
Hernández J, Calvet N, Hartmann L, *et al.*, 2005, Herbig Ae/Be stars in nearby OB associations. *AJ*, 129, 856–871 {**316, 318**}
Herschel JFWS, 1847, *Results of Astronomical Observations Made During the Years 1834–38 at the Cape of Good Hope.* Smith, Elder and Co., London {**325**}
Holmberg J, Flynn C, 2000, The local density of matter mapped by Hipparcos. *MNRAS*, 313, 209–216 {**297**}
Holmberg J, Nordström B, Andersen J, 2007, The Geneva–Copenhagen survey of the Solar neighbourhood. II. New uvby calibrations and rediscussion of stellar ages, the G dwarf problem, age–metallicity diagram, and heating mechanisms of the disk. *A&A*, 475, 519–537 {**308**}
Hoogerwerf R, 2000, OB association members in the ACT and TRC catalogues. *MNRAS*, 313, 43–65 {**312, 313, 315**}
Hoogerwerf R, Aguilar LA, 1999, Identification of moving groups and member selection using Hipparcos data. *MNRAS*, 306, 394–406 {**276, 277, 312**}
Hoogerwerf R, de Bruijne JHJ, Brown AGA, *et al.*, 1997, New results for three nearby OB associations. *ESA SP–402: Hipparcos, Venice '97*, 571–574 {**319**}
Hoogerwerf R, de Bruijne JHJ, de Zeeuw PT, 2001, On the origin of the O and B-type stars with high velocities. II. Runaway stars and pulsars ejected from the nearby young stellar groups. *A&A*, 365, 49–77 {**278, 319**}
Humphreys RM, 1978, Studies of luminous stars in nearby galaxies. I. Supergiants and O stars in the Milky Way. *ApJS*, 38, 309–350 {**313**}
Jilinski E, Daflon S, Cunha K, *et al.*, 2006, Radial velocity measurements of B stars in the Scorpius–Centaurus association. *A&A*, 448, 1001–1006 {**318**}
Jilinski E, Ortega VG, de la Reza R, 2005, On the origin of the very young groups η and ϵ Cha. *ApJ*, 619, 945–947 {**278, 323**}
Jôeveer M, 1974, Ages of δ Cephei stars and the density of gravitating masses near the Sun. *Tartu Astrofuusika Observatoorium Teated*, 46, 35–49 {**297**}
Jones BF, 1972, The Pleiades lower main sequence. *ApJ*, 171, L57–L60 {**292, 294**}
Jones DHP, 1971, The kinematics of the Scorpius–Centaurus association and Gould's belt. *MNRAS*, 152, 231–259 {**275, 276, 284**}
Joshi YC, 2005, Interstellar extinction towards open clusters and Galactic structure. *MNRAS*, 362, 1259–1266 {**299**}
Kaltcheva N, 2000, The region of Collinder 121. *MNRAS*, 318, 1023–1035 {**301**}
Kaltcheva N, Hilditch RW, 2000, The distribution of bright OB stars in the Canis Major–Puppis–Vela region of the Milky Way. *MNRAS*, 312, 753–768 {**319**}
Kaltcheva N, Jaeger S, Kaba Bah M, *et al.*, 2005, The field of IC 2602: Strömgren-Hβ photometry approach. *Astronomische Nachrichten*, 326, 738–745 {**301**}
Kaltcheva N, Makarov VV, 2007, The structure and the distance of Collinder 121 from Hipparcos and photometry: resolving the discrepancy. *ApJ*, 667, L155–L157 {**301**}
Kapteyn JC, 1905, Star streaming. *Brit. Assoc. Adv. Sci. Rep.*, 1905, 257–265 {**302**}
Kastner JH, Zuckerman B, Weintraub DA, *et al.*, 1997, X-ray and molecular emission from the nearest region of recent star formation. *Science*, 277, 67–71 {**321**}
Kazakevich E, Orlov VV, Vityazev VV, 2003, Hipparcos: search for stellar groups. *Astrometry from Ground and from Space* (eds. Capitaine N, Stavinschi M), 95–96 {**279**}
Kazakevich E, Orlov VV, 2002, Search for star groups in the solar neighbourhood. *Astrophysics*, 45, 302–312 {**279**}
Kharchenko NV, Pakulyak LK, Piskunov AE, 2003, The subsystem of open clusters in the post-Hipparcos era: cluster structural parameters and proper motions. *Astronomy Reports*, 47, 263–275 {**299**}
Kharchenko NV, Piskunov AE, Röser S, *et al.*, 2004, Astrophysical supplements to the ASCC 2.5. II. Membership probabilities in 520 Galactic open cluster sky areas. *Astronomische Nachrichten*, 325, 740–748 {**273, 299**}
—, 2005a, 109 new Galactic open clusters. *A&A*, 440, 403–408 {**273, 279, 299, 300**}
—, 2005b, All-sky census of Galactic open cluster stars. *Kinematika i Fizika Nebesnykh Tel Supplement*, 5, 381–384 {**299**}
—, 2005c, Astrophysical parameters of Galactic open clusters. *A&A*, 438, 1163–1173 {**273, 299**}
Kholopov PN, 1980, Position of zero-age main sequence for stars with high metal abundance. *AZh*, 57, 12–21 {**297**}
King I, 1962, The structure of star clusters. I. An empirical density law. *AJ*, 67, 471–485 {**294, 303**}
King JR, Soderblom DR, Fischer DA, *et al.*, 2000, Spectroscopic abundances in cool Pleiades dwarfs and NGC 2264 stars. *ApJ*, 533, 944–958 {**291**}
King JR, Villarreal AR, Soderblom DR, *et al.*, 2003, Stellar kinematic groups. II. A re-examination of the membership, activity, and age of the Ursa Major group. *AJ*, 125, 1980–2017 {**308, 310**}

Knude J, Nielsen AS, 2001, V versus $(V-I)$ distance to Lupus 2. *A&A*, 373, 714–719 {**319**}

Kouwenhoven MBN, Brown AGA, Zinnecker H, *et al.*, 2005, The primordial binary population. I. A near-infrared adaptive optics search for close visual companions to A star members of Scorpius OB2. *A&A*, 430, 137–154 {**319**}

Kroupa P, 2002, The initial mass function of stars: evidence for uniformity in variable systems. *Science*, 295, 82–91 {**320**}

Kroupa P, Tout CA, Gilmore G, 1993, The distribution of low-mass stars in the Galactic disk. *MNRAS*, 262, 545–587 {**287**}

Kurucz RL, 1993, ATLAS9 stellar atmosphere programs and 2 km s^{-1} grid. *CD No. 13, Smithsonian Astrophysical Observatory* {**290**}

Lada CJ, Lada EA, 1991, The nature, origin and evolution of embedded star clusters. *The Formation and Evolution of Star Clusters* (ed. Janes K), ASP Conf. Ser. 13, 3–22 {**274**}

Lastennet E, Valls-Gabaud D, Lejeune T, *et al.*, 1999, Consequences of Hipparcos parallaxes for stellar evolutionary models. Three Hyades binaries: V818 Tau, 51 Tau, and θ^2 Tau. *A&A*, 349, 485–494 {**282**}

Lawson WA, Crause LA, Mamajek EE, *et al.*, 2001, The η Cha cluster: photometric study of the ROSAT-detected weak-lined T Tauri stars. *MNRAS*, 321, 57–66 {**323**}

Lawson WA, Crause LA, 2005, Rotation periods for stars of the TW Hya association: the evidence for two spatially and rotationally distinct pre-main sequence populations. *MNRAS*, 357, 1399–1406 {**321**}

Lebreton Y, 2000a, Science results for stellar structure and evolution from Hipparcos. *Unsolved Problems in Stellar Evolution* (ed. Livio M), 107–125 {**281, 283**}

—, 2000b, Stellar structure and evolution: deductions from Hipparcos. *ARA&A*, 38, 35–77 {**283, 296**}

Lebreton Y, Fernandes J, Lejeune T, 2001, The helium content and age of the Hyades: constraints from five binary systems and Hipparcos parallaxes. *A&A*, 374, 540–553 {**282, 284, 285**}

Lebreton Y, Gómez AE, Mermilliod JC, *et al.*, 1997, The age and helium content of the Hyades revisited. *ESA SP–402: Hipparcos, Venice '97*, 231–236 {**281**}

Lebreton Y, Michel E, Goupil MJ, *et al.*, 1995, Accurate parallaxes and stellar ages determinations. *Astronomical and Astrophysical Objectives of Sub-Milliarcsecond Optical Astrometry* (eds. Høg E, Seidelmann PK), IAU Symp. 166, 135–142 {**281**}

Lejeune T, Cuisinier F, Buser R, 1998, A standard stellar library for evolutionary synthesis. II. The M dwarf extension. *A&AS*, 130, 65–75 {**284**}

Lépine JRD, Dias WS, Mishurov YN, *et al.*, 2006, Statistics of initial velocities of Galactic clusters. *IAU Symp. 237*, 154–158 {**298**}

Lépine JRD, Dias WS, Mishurov YN, Direct determination of the epicycle frequency in the Galactic disk, and the derived rotation velocity. *ApJ*, *MNRAS*, 386, 2081–2090 {**298**}

Lépine JRD, Duvert G, 1994, Star formation by infall of high-velocity clouds on the Galactic disk. *A&A*, 286, 60–71 {**311, 325**}

Lesh JR, 1968, The kinematics of the Gould Belt: an expanding group? *ApJS*, 17, 371–444 {**325**}

Lindblad PO, Grape K, Sandqvist A, *et al.*, 1973, On the kinematics of a local component of the interstellar hydrogen gas possibly related to the Gould Belt. *A&A*, 24, 309–312 {**311, 325**}

Lindblad PO, Loden K, Palouš J, *et al.*, 1997a, N-body models of the Gould's Belt. *Joint European and National Astronomical Meeting*, 201–201 {**324**}

Lindblad PO, Palouš J, Loden K, *et al.*, 1997b, The kinematics and nature of Gould's Belt – a 30 Myr old star-forming region. *ESA SP–402: Hipparcos, Venice '97*, 507–512 {**324, 327**}

Lindegren L, Madsen S, Dravins D, 2000, Astrometric radial velocities. II. Maximum-likelihood estimation of radial velocities in moving clusters. *A&A*, 356, 1119–1135 {**283, 286**}

Lodieu N, Hambly NC, Jameson RF, 2006, New members in the Upper Scorpius association from the UKIRT infrared deep sky survey early data release. *MNRAS*, 373, 95–104 {**319**}

Loktin AV, 2004, On putting the catalogues of the relative proper motions of open cluster stars in one system (cluster M67). *Astronomical and Astrophysical Transactions*, 23, 61–69 {**301**}

Loktin AV, Beshenov GV, 2001, Hipparcos trigonometric parallaxes and the distance scale for open star clusters. *Astronomy Letters*, 27, 386–390 {**297**}

—, 2003, Proper motions of open star clusters and the rotation rate of the Galaxy. *Astronomy Reports*, 47, 6–10 {**298**}

Loktin AV, Zakharova P, Gerasimenko T, *et al.*, 1997, The homogeneous catalogue of the main parameters of open star clusters. *Baltic Astronomy*, 6, 316–318 {**298**}

Luhman KL, Steeghs D, 2004, Spectroscopy of candidate members of the η Cha and MBM 12 young associations. *ApJ*, 609, 917–924 {**323**}

Lyngå G, 1962, On some southern Galactic clusters. *Arkiv for Astronomi*, 3, 65–92 {**314**}

—, 1988, The Lund Catalogue of open cluster data. *Astronomy from Large Databases* (eds. Murtagh F, Heck A), 379–382 {**273, 288, 298**}

Lyngå G, Lindegren L, 1998, Hipparcos data for two open clusters containing Cepheids. *New Astronomy*, 3, 121–123 {**302**}

Lyo AR, Lawson WA, Feigelson ED, *et al.*, 2004, Population and dynamical state of the η Cha sparse young open cluster. *MNRAS*, 347, 246–254 {**323**}

Madsen S, 1999, The Hyades main sequence from improved Hipparcos parallaxes. *ASP Conf. Ser. 167: Harmonizing Cosmic Distance Scales in a Post-Hipparcos Era*, 78–83 {**283**}

—, 2003, Hyades dynamics from N-body simulations: accuracy of astrometric radial velocities from Hipparcos. *A&A*, 401, 565–576 {**286, 287**}

Madsen S, Dravins D, Lindegren L, 2002, Astrometric radial velocities. III. Hipparcos measurements of nearby star clusters and associations. *A&A*, 381, 446–463 {**276, 283, 284, 286, 293, 315, 316**}

Madsen S, Lindegren L, Dravins D, 2000, Main sequences of nearby open clusters and OB associations from kinematic modeling of Hipparcos data. *ASP Conf. Ser. 198: Stellar Clusters and Associations: Convection, Rotation, and Dynamos*, 137–140 {**315**}

Maeder A, Meynet G, 1988, Tables of evolutionary star models from $0.85-120M_\odot$ with overshooting and mass loss. *A&AS*, 76, 411–425 {**324**}

—, 1991, Tables of isochrones computed from stellar models with mass loss and overshooting. *A&AS*, 89, 451–467 {**297**}

Maeder A, Peytremann E, 1972, Uniformly rotating stars with H- and metallic-line blanketed model atmospheres. *A&A*, 21, 279–284 {**285**}

Maíz Apellániz J, 2001, The spatial distribution of O–B5 stars in the solar neighbourhood as measured by Hipparcos. *AJ*, 121, 2737–2742 {**326**}

Makarov VV, Fabricius C, 2001, Internal kinematics of the TW Hya association of young stars. *A&A*, 368, 866–872 {**321, 322**}

Makarov VV, 2000, Kinematics versus X-ray luminosity segregation in the Hyades. *A&A*, 358, L63–L66 {**286**}

—, 2002, Computing the parallax of the Pleiades from the Hipparcos intermediate astrometry data: an alternative approach. *AJ*, 124, 3299–3304 {**289**}

—, 2003, Improved Hipparcos parallaxes of Coma Berenices and NGC 6231. *AJ*, 126, 2408–2410 {**289, 300, 302**}

—, 2006, Precision kinematics and related parameters of the α Persei open cluster. *AJ*, 131, 2967–2979 {**300**}

—, 2007a, The Lupus association of pre-main-sequence stars: clues to star formation scattered in space and time. *ApJ*, 658, 480–486 {**319**}

—, 2007b, Unraveling the origins of nearby young stars. *ApJS*, 169, 105–119 {**278, 323, 324**}

Makarov VV, Gaume RA, Andrievsky SM, 2005, Expansion of the TW Hya association and the encounter with Vega. *MNRAS*, 362, 1109–1113 {**322**}

Makarov VV, Odenkirchen M, Urban SE, 2000, Internal velocity dispersion in the Hyades as a test for Tycho 2 proper motions. *A&A*, 358, 923–928 {**286**}

Makarov VV, Olling RP, Teuben PJ, 2004, Kinematics of stellar associations: the epicycle approximation and the convergent point method. *MNRAS*, 352, 1199–1207 {**277, 323**}

Makarov VV, Robichon N, 2001, Internal kinematics and binarity of X-ray stars in the Pleiades open cluster. *A&A*, 368, 873–879 {**295**}

Makarov VV, Urban SE, 2000, A moving group of young stars in Carina–Vela. *MNRAS*, 317, 289–298 {**277, 278, 323, 326**}

Mamajek EE, Lawson WA, Feigelson ED, 1999, The η Chamaeleontis cluster: a remarkable new nearby young open cluster. *ApJ*, 516, L77–L80 {**323**}

—, 2000, The η Chamaeleontis cluster: origin in the Sco–Cen OB association. *ApJ*, 544, 356–374 {**323**}

Mamajek EE, Meyer MR, Liebert J, 2002, Post-T Tauri stars in the nearest OB association. *AJ*, 124, 1670–1694, erratum: 131, 2360 {**315, 317**}

Mamajek EE, 2005, A moving cluster distance to the exoplanet 2M 1207b in the TW Hya association. *ApJ*, 634, 1385–1394 {**322**}

—, 2006, A new nearby candidate star cluster in Ophiuchus at $d = 170$ pc. *AJ*, 132, 2198–2205 {**320**}

—, 2007, New nearby young star cluster candidates within 200 pc. *IAU Symp. 237* (eds. Elmegreen BG, Palouš J), 442–442 {**320**}

—, 2008, On the distance to the Ophiuchus star-forming region. *Astronomische Nachrichten*, 329, 10–14 {**319**}

Martín EL, 1998, Weak and post-T Tauri stars around B-type members of the Scorpius–Centaurus OB association. *AJ*, 115, 351–357 {**318**}

Mathieu RD, 1986, The dynamical evolution of young clusters and associations. *Highlights of Astronomy*, 7, 481–488 {**315**}

Mermilliod JC, 1981, Comparative studies of young open clusters. III. Empirical isochrones and the zero age main sequence. *A&A*, 97, 235–244 {**315**}

—, 1995, *The Database for Galactic Open Clusters (BDA)*, 127–138. Information and On-Line Data in Astronomy, ASSL 203 {**273, 288, 299**}

—, 2000, Open clusters after Hipparcos. *Very Low-Mass Stars and Brown Dwarfs* (eds. Rebolo R, Zapatero-Osorio MR), 3–16 {**296**}

Mermilliod JC, Turon C, Robichon N, *et al.*, 1997, The distance of the Pleiades and nearby clusters. *ESA SP–402: Hipparcos, Venice '97*, 643–650 {**290, 296**}

Michel E, Hernández MM, Houdek G, *et al.*, 1999, Seismology of δ Scuti stars in the Praesepe cluster. I. Ranges of unstable modes as predicted by linear analysis versus observations. *A&A*, 342, 153–166 {**286**}

Mihalas D, Binney JJ, 1981, *Galactic Astronomy: Structure and Kinematics*, 2nd edition. W.H. Freeman, San Francisco {**305**}

Mishurov YN, Zenina IA, 1999, Parameters of the Galactic rotation curve and spiral pattern from Cepheid kinematics. *Astronomy Reports*, 43, 487–493 {**308**}

Miyamoto M, Nagai R, 1975, Three-dimensional models for the distribution of mass in galaxies. *PASJ*, 27, 533–543 {**307**}

Montes D, Latorre A, Fernández Figueroa MJ, 2000, Late-type stars members of young stellar kinematic groups. *ASP Conf. Ser. 198: Stellar Clusters and Associations: Convection, Rotation, and Dynamos*, 203 {**310**}

Montes D, López-Santiago J, Gálvez MC, *et al.*, 2001, Late-type members of young stellar kinematic groups. I. Single stars. *MNRAS*, 328, 45–63 {**305, 309**}

Morel P, 1997, CESAM: a code for stellar evolution calculations. *A&AS*, 124, 597–614 {**281**}

Moreno E, Alfaro EJ, Franco J, 1999, The kinematics of stars emerging from expanding shells: an analysis of the Gould Belt. *ApJ*, 522, 276–284 {**326, 327**}

Motte F, Andre P, Neri R, 1998, The initial conditions of star formation in the ρ Oph main cloud: wide-field millimeter continuum mapping. *A&A*, 336, 150–172 {**319**}

Mühlbauer G, Dehnen W, 2003, Kinematic response of the outer stellar disk to a central bar. *A&A*, 401, 975–984 {**310**}

Munari U, Dallaporta S, Siviero A, *et al.*, 2004, The distance to the Pleiades from orbital solution of the double-lined eclipsing binary HD 23642. *A&A*, 418, L31–L34 {**293, 295**}

Mylläri A, Flynn C, Orlov VV, 2001, Stellar moving groups in Hipparcos. *ASP Conf. Ser. 228: Dynamics of Star Clusters and the Milky Way*, 329–334 {**310**}

Narayanan VK, Gould A, 1999a, A precision test of Hipparcos systematics toward the Hyades. *ApJ*, 515, 256–264 {**283–285**}

—, 1999b, Correlated errors in Hipparcos parallaxes toward the Pleiades and the Hyades. *ApJ*, 523, 328–339 {**283, 284, 293, 298**}

Nissen PE, 1976, Evidence of helium abundance differences between young groups of stars. *A&A*, 50, 343–352 {**296**}

Nordström B, Mayor M, Andersen J, *et al.*, 2004, The Geneva–Copenhagen survey of the solar neighbourhood: ages, metallicities, and kinematic properties of $\sim$14 000 F and G dwarfs. *A&A*, 418, 989–1019 {**307**}

O'Dell MA, Hendry MA, Cameron AC, 1994, New distance measurements to the Pleiades and α Persei clusters. *MNRAS*, 268, 181–193 {**288**}

Odenkirchen M, Makarov VV, Soubiran C, *et al.*, 2001, The tidal extension of the Coma star cluster revealed by Hipparcos Tycho 2 and spectroscopic data. *ASP Conf. Ser. 228: Dynamics of Star Clusters and the Milky Way*, 535–537 {**300**}

Odenkirchen M, Soubiran C, Colin J, 1998, The Coma Berenices star cluster and its moving group. *New Astronomy*, 3, 583–599 {**300**}

Odenkirchen M, Soubiran C, 2002, NGC 6994: clearly not a physical stellar ensemble. *A&A*, 383, 163–170 {**302**}

Olano CA, Pöppel WGL, 1987, Kinematical origin of the dark clouds in Taurus and of some nearby Galactic clusters. *A&A*, 179, 202–218 {**311**}

Olano CA, 1982, On a model of local gas related to Gould's belt. *A&A*, 112, 195–208 {**311, 313, 325**}

Ortega VG, de la Reza R, Jilinski E, *et al.*, 2002, The origin of the β Pictoris moving group. *ApJ*, 575, L75–L78 {**278, 320, 321**}

Ortega VG, de la Reza R, Jilinski E, 2006, Dynamical 3d evolution of nearby young moving groups. *Revista Mexicana de Astronomia y Astrofisica Conf. Ser.*, volume 26, 91–91 {**323**}

Palla F, Stahler SW, 1993, The pre-main sequence evolution of intermediate-mass stars. *ApJ*, 418, 414–425 {**318**}

—, 1999, Star formation in the Orion nebula cluster. *ApJ*, 525, 772–783 {**320**}

Palouš J, 1998, Gould's Belt: local system of young stars. *Dynamics of Galaxies and Galactic Nuclei* (eds. Duschl WJ, Einsel C), 157–162 {**328**}

Palouš J, Jungwiert B, Kopecky J, 1993, Formation of rings in weak bars: inelastic collisions and star formation. *A&A*, 274, 189–202 {**307**}

Pan XP, Shao M, Kulkarni SR, 2004, A distance of 133–137 pc to the Pleiades star cluster. *Nature*, 427, 326–328 {**294, 295**}

Pels G, Oort JH, Pels-Kluyver HA, 1975, New members of the Hyades cluster and a discussion of its structure. *A&A*, 43, 423–441 {**303**}

Percival SM, Salaris M, Groenewegen MAT, 2005, The distance to the Pleiades. Main sequence fitting in the near infrared. *A&A*, 429, 887–894 {**288, 292**}

Percival SM, Salaris M, Kilkenny D, 2003, The open cluster distance scale. A new empirical approach. *A&A*, 400, 541–552 {**288, 292**}

Pérez Hernández F, Claret A, Hernández MM, *et al.*, 1999, Photometric parameters for rotating models of A- and F-type stars. *A&A*, 346, 586–598 {**285**}

Perrot CA, Grenier IA, 2003, 3d dynamical evolution of the interstellar gas in the Gould Belt. *A&A*, 404, 519–531 {**327, 328**}

Perryman MAC, Brown AGA, Lebreton Y, *et al.*, 1998, The Hyades: distance, structure, dynamics, and age. *A&A*, 331, 81–120 {**276, 277, 279–283, 293, 303**}

Peterson DM, Solensky R, 1988, 51 Tau and the Hyades distance modulus. *ApJ*, 333, 256–266 {**283**}

Peña JH, Sareyan JP, 2006, uvbyβ photoelectric photometry of the open cluster α Per. *Revista Mexicana de Astronomia y Astrofisica*, 42, 179–194 {**300**}

Pinsonneault MH, Stauffer JR, Soderblom DR, *et al.*, 1998, The problem of Hipparcos distances to open clusters. I. Constraints from multicolour main-sequence fitting. *ApJ*, 504, 170–191 {**288, 289, 291, 292, 296, 300**}

Pinsonneault MH, Terndrup DM, Yuan Y, 2000, Main sequence fitting and the Hipparcos open cluster distance scale. *ASP Conf. Ser. 198: Stellar Clusters and Associations: Convection, Rotation, and Dynamos*, 95–104 {**291**}

Piskunov AE, Kharchenko NV, Röser S, *et al.*, 2006a, Revisiting the population of Galactic open clusters. *A&A*, 445, 545–565 {**299, 301**}

—, 2006b, The Hipparcos mission and Galactic open clusters and NGC 7538 star-forming regions. *Bulletin of the Astronomical Society of India*, 34, 129–135 {**273, 299**}

Platais I, Girard TM, van Altena WF, *et al.*, 2000, The WIYN open cluster study and a distance scale. *Hipparcos and the Luminosity Calibration of the Nearer Stars, IAU Joint Discussion 13*, 17–19 {**297**}

Platais I, Kozhurina-Platais V, Barnes S, *et al.*, 2001, WIYN open cluster study. VII. NGC 2451A and the Hipparcos distance scale. *AJ*, 122, 1486–1499 {**301**}

Platais I, Kozhurina-Platais V, Mathieu RD, *et al.*, 2003, WIYN open cluster study. XVII. Astrometry and membership to V = 21 in NGC 188. *AJ*, 126, 2922–2935 {**301**}

Platais I, Kozhurina-Platais V, van Leeuwen F, 1998, A search for star clusters from the Hipparcos data. *AJ*, 116, 2423–2430 {**273, 278, 298, 299, 315**}

Platais I, Melo C, Mermilliod JC, *et al.*, 2007, WIYN open cluster study. XXVI. Improved kinematic membership and spectroscopy of IC 2391. *A&A*, 461, 509–522 {**301**}

Plummer HC, 1915, The distribution of stars in globular clusters. *MNRAS*, 76, 107–121 {**287**}

Pols OR, Schröder KP, Hurley JR, *et al.*, 1998, Stellar evolution models for $Z = 0.0001 - 0.03$. *MNRAS*, 298, 525–536 {**297**}

Pöppel WGL, 1997, The Gould Belt system and the local interstellar medium. *Fundamentals of Cosmic Physics*, 18, 1–271 {**325**}

Portegies Zwart SF, Hut P, Makino J, *et al.*, 1998, On the dissolution of evolving star clusters. *A&A*, 337, 363–371 {**304**}

Pozzo M, Jeffries RD, Naylor T, *et al.*, 2000, The discovery of a low-mass, pre-main sequence stellar association around γ^2 Vel. *MNRAS*, 313, L23–L27 {**319**}

Prada Moroni PG, Castellani V, Degl'Innocenti S, *et al.*, 2000, The Hyades in the light of Hipparcos. *ASP Conf. Ser. 211: Massive Stellar Clusters*, 175–178 {**285**}

Preibisch T, Brown AGA, Bridges T, *et al.*, 2002, Exploring the full stellar population of the Upper Scorpius OB association. *AJ*, 124, 404–416 {**319, 320**}

Preibisch T, Zinnecker H, 1999, The history of low-mass star formation in the Upper Scorpius OB association. *AJ*, 117, 2381–2397 {**311, 319**}

Quillen AC, 2002, Using a Hipparcos-derived HR diagram to limit the metallicity scatter of stars in the Hyades: are stars polluted? *AJ*, 124, 400–403 {**286**}

—, 2003, Chaos caused by resonance overlap in the solar neighbourhood: spiral structure at the bar's outer Lindblad resonance. *AJ*, 125, 785–793 {**310**}

Raboud D, Mermilliod JC, 1998, Investigation of the Pleiades cluster. IV. The radial structure. *A&A*, 329, 101–114 {**294**}

Rastorguev AS, Glushkova EV, Dambis AK, *et al.*, 1998, Open clusters, Cepheids and the problems of distance scale. *Dynamical Studies of Star Clusters and Galaxies* (eds. Kroupa P, Palouš J, Spurzem R), 195–195 {**297**}

—, 1999, Statistical parallaxes and kinematical parameters of classical Cepheids and young star clusters. *Astronomy Letters*, 25, 595–607 {**299**}

Ratnatunga KU, 1988, A distance-independent study of extended moving groups. *AJ*, 95, 1132–1148 {**303**}

Robichon N, Arenou F, Lebreton Y, *et al.*, 1999a, Main sequences of open clusters with Hipparcos. *ASP Conf. Ser. 167: Harmonizing Cosmic Distance Scales in a Post-Hipparcos Era*, 72–77 {**289, 291, 297**}

Robichon N, Arenou F, Mermilliod JC, *et al.*, 1999b, Open clusters with Hipparcos. I. Mean astrometric parameters. *A&A*, 345, 471–484 {**289, 291, 293, 296, 298, 300–302**}

Robichon N, Arenou F, Turon C, *et al.*, 1997a, Analysis of seven nearby open clusters using Hipparcos data. *ESA SP–402: Hipparcos, Venice '97*, 567–570 {**296, 299**}

Robichon N, Lebreton Y, Arenou F, 1999c, Sequences of nearby open clusters with Hipparcos. *Ap&SS*, 265, 279–280 {**297**}

Robichon N, Lebreton Y, Turon C, *et al.*, 2000, Do Hipparcos distances of nearby open clusters really disagree with the HR diagram? *ASP Conf. Ser. 198: Stellar Clusters and Associations: Convection, Rotation, and Dynamos*, 141–144 {**290, 296**}

Robichon N, Mermilliod JC, Turon C, *et al.*, 1997b, Nearby open clusters with Hipparcos: an improvement of the HR diagram calibrations. *The First Results of Hipparcos and Tycho, IAU Joint Discussion 14*, 17–20 {**297**}

Roman NG, 1949, The Ursa Major group. *ApJ*, 110, 205–241 {**308**}

Ruprecht J, Balazs BA, White RE, 1981, *Catalogue of Star Clusters and Associations, Supplement to the Second Edition*. Akademiai Kiado, Budapest {**311**}

Ruprecht J, 1966, Classification of open star clusters. *Bulletin of the Astronomical Institutes of Czechoslovakia*, 17, 33–44 {**311, 313, 314**}

Sanders WL, 1971, An improved method for computing membership probabilities in open clusters. *A&A*, 14, 226–232 {**299**}

Sanner J, Geffert M, 2001, The initial mass function of open star clusters with Tycho 2. *A&A*, 370, 87–99 {**298**}

Sartori MJ, 2000, The star-formation scenario in the Galactic range from Ophiuchus to Chamaeleon. Ph.D. Thesis {**316**}

Sartori MJ, Lépine JRD, Dias WS, 2003, Formation scenarios for the young stellar associations between Galactic longitudes 280–360 degrees. *A&A*, 404, 913–926 {**311, 316, 317**}

Scalo JM, 1998, The inital mass function revisited: a case for variations. *ASP Conf. Ser. 142: The Stellar Initial Mass Function* (eds. Gilmore G, Howell D), 201–236 {**320**}

Schaifers K, Voigt HH (eds.), 1982, *Landolt-Börnstein: Numerical Data and Functional Relationships in Science and Technology; New Series, Group 6 Astronomy and Astrophysics*, volume 2. Springer–Verlag, Berlin {**290**}

Schaller G, Schaerer D, Meynet G, *et al.*, 1992, New grids of stellar models. I. From $0.8-120\,M_\odot$ at $Z = 0.020, 0.001$. *A&AS*, 96, 269–331 {**290**}

Scholz RD, Brunzendorf J, Ivanov GA, *et al.*, 1999, IC 348 proper motion study from digitised Schmidt plates. *A&AS*, 137, 305–321 {**301**}

Schröder KP, 1998, The solar neighbourhood HR diagram as a quantitative test for evolutionary time scales. *A&A*, 334, 901–910 {**285**}

Schwarzschild K, 1908, *Göttingen Nachrichten*. 191 {**305**}

Seares FH, 1944, Regression lines and the functional relation. *ApJ*, 100, 255–263 {**280**}

—, 1945, Regression lines and the functional relation. II. Charlier's formulae for a moving cluster. *ApJ*, 102, 366–376 {**280**}

Siess L, Dufour E, Forestini M, 2000, An internet server for pre-main sequence tracks of low- and intermediate-mass stars. *A&A*, 358, 593–599 {**317**}

Skuljan J, Hearnshaw JB, Cottrell PL, 1999, Velocity distribution of stars in the solar neighbourhood. *MNRAS*, 308, 731–740 {**305, 306**}

Smart WM, 1939, The moving cluster in Taurus. *MNRAS*, 99, 168–180 {**280**}

Smith GH, 1983, The HR 1614 moving group: evidence from DDO photometry that it contains CN-rich stars. *AJ*, 88, 1775–1783 {**309**}

Soderblom DR, King JR, Hanson RB, *et al.*, 1998, The problem of Hipparcos distances to open clusters. II. Constraints from nearby field stars. *ApJ*, 504, 192–199 {**291**}

Soderblom DR, Mayor M, 1993, Stellar kinematic groups. I. The Ursa Major group. *AJ*, 105, 226–249 {**302, 303, 308**}

Soderblom DR, Nelan E, Benedict GF, *et al.*, 2005, Confirmation of errors in Hipparcos parallaxes from Hubble Space Telescope FGS astrometry of the Pleiades. *AJ*, 129, 1616–1624 {**293**}

Soderblom DR, Stauffer JR, MacGregor KB, *et al.*, 1993, The evolution of angular momentum among zero-age main sequence solar-type stars. *ApJ*, 409, 624–634 {**291**}

Söderhjelm S, 1999, Visual binary orbits and masses post Hipparcos. *A&A*, 341, 121–140 {**283**}

Song I, Bessell MS, Zuckerman B, 2002, Additional TW Hya association members? Spectroscopic verification of kinematically selected candidates. *A&A*, 385, 862–866 {**321**}

Song I, Zuckerman B, Bessell MS, 2003, New members of the TW Hya association, β Pic moving group, and Tucana–Horologium association. *ApJ*, 599, 342–350, erratum: 603, 804-805 {**320–323**}

—, 2004, Probing the low-mass stellar end of the η Cha cluster. *ApJ*, 600, 1016–1019 {**323**}

Southworth J, Maxted PFL, Smalley B, 2005, Eclipsing binaries as standard candles. HD 23642 and the distance to the Pleiades. *A&A*, 429, 645–655 {**293, 295**}

Southworth J, Smalley B, Maxted PFL, *et al.*, 2004, Accurate fundamental parameters of eclipsing binary stars. *IAU Symp. 224* (eds. Zverko J, Ziznovsky J, Adelman SJ, *et al.*), 548–561 {**293**}

Spitzer LJ, 1958, Distribution of Galactic clusters. *ApJ*, 127, 17–27 {**274**}

Sridhar S, Touma J, 1996, Adiabatic evolution and capture into resonance: vertical heating of a growing stellar disk. *MNRAS*, 279, 1263–1273 {**310**}

Stauffer JR, Jones BF, Backman DE, *et al.*, 2003, Why are the K-dwarfs in the Pleiades so blue? *AJ*, 126, 833–847 {**292**}

Steenbrugge KC, de Bruijne JHJ, Hoogerwerf R, *et al.*, 2003, Radial velocities of early-type stars in the Perseus OB2 association. *A&A*, 402, 587–605 {**318**}

Stefanik RP, Latham DW, 1985, The Hyades: membership and convergent point from radial velocities. *Stellar Radial Velocities*, 213–222 {**280**}

Stello D, Nissen PE, 2001, The problem of the Pleiades distance. Constraints from Strömgren photometry of nearby field stars. *A&A*, 374, 105–115 {**291, 294**}

Sterzik MF, Durisen RH, 1995, Escape of T Tauri stars from young stellar systems. *A&A*, 304, L9–L12 {**322**}

Stetson PB, McClure RD, VandenBerg DA, 2004, A star catalogue for the open cluster NGC 188. *PASP*, 116, 1012–1030 {**301**}

Stothers RB, Frogel JA, 1974, The local complex of 0 and B stars. I. Distribution of stars and interstellar dust. *AJ*, 79, 456–471 {**325**}

Strand KA, 1963, *Basic Astronomical Data*. University of Chicago Press {**287**}

Subramaniam A, Bhatt HC, 1999, Star-formation history of the Puppis–Vela region using Hipparcos data. *Star Formation* (ed. Nakamoto T), 373–374 {**319**}

—, 2000, A new look at the OB associations and star clusters in the Puppis–Vela region using Hipparcos data. *Bulletin of the Astronomical Society of India*, 28, 255–256 {**319**}

Sung H, Bessell MS, Lee BW, *et al.*, 2002, The open cluster NGC 2516. I. Optical photometry. *AJ*, 123, 290–303 {**302**}

Talon S, Zahn JP, Maeder A, *et al.*, 1997, Rotational mixing in early-type stars: the main sequence evolution of a $9M_\odot$ star. *A&A*, 322, 209–217 {**286**}

Taylor BJ, 2000, A statistical analysis of the metallicities of nine old superclusters and moving groups. *A&A*, 362, 563–579 {**303, 310**}

Terlevich E, 1987, Evolution of N-body open clusters. *MNRAS*, 224, 193–225 {**303**}

Terndrup DM, Pinsonneault MH, Jeffries RD, *et al.*, 2002, Rotation and activity in the solar-metallicity open cluster NGC 2516. *ApJ*, 576, 950–962 {**302**}

Thackeray AD, 1967, Stellar radial velocities in the Scorpius–Centaurus association and IC 944. *Determination of Radial Velocities and their Applications* (eds. Batten AH, Heard JF), IAU Symp. 30, 163–166 {**280**}

Tordiglione V, Castellani V, Degl'Innocenti S, *et al.*, 2003, Hipparcos open clusters as a test for stellar evolution. *Memorie della Societa Astronomica Italiana*, 74, 520–521 {**297**}

Torra J, Fernández D, Figueras F, *et al.*, 2000a, The velocity field of young stars in the solar neighbourhood. *Ap&SS*, 272, 109–112 {**326**}

Torra J, Fernández D, Figueras F, 2000b, Kinematics of young stars. I. Local irregularities. *A&A*, 359, 82–102 {**314, 326**}

Torra J, Gómez AE, Figueras F, *et al.*, 1997, Young stars: irregularities of the velocity field. *ESA SP–402: Hipparcos, Venice '97*, 513–518 {**326**}

Torres CAO, da Silva L, Quast GR, *et al.*, 2000, A new association of post-T Tauri stars near the Sun. *AJ*, 120, 1410–1425 {**323**}

Torres CAO, Quast GR, de la Reza R, *et al.*, 2003, SACY: present status. *ASP Conf. Ser. 287: Galactic Star Formation Across the Stellar Mass Spectrum* (eds. De Buizer JM, van der Bliek NS), 439–444 {**319**}

Torres G, 2003, Discovery of a bright eclipsing binary in the Pleiades cluster. *Informational Bulletin on Variable Stars*, 5402 {**293**}

Torres G, Ribas I, 2002, Absolute dimensions of the M-type eclipsing binary YY Gem (Castor C): a challenge to evolutionary models in the lower main sequence. *ApJ*, 567, 1140–1165 {**283**}

Torres G, Stefanik RP, Latham DW, 1997a, The Hyades binaries θ^1 Tau and θ^2 Tau: the distance to the cluster and the mass–luminosity relation. *ApJ*, 485, 167–181 {**282, 283, 285**}

—, 1997b, The Hyades binary 51 Tau: spectroscopic detection of the primary, the distance to the cluster, and the mass–luminosity relation. *ApJ*, 474, 256–271 {**282, 283, 285**}

—, 1997c, The Hyades binary Finsen 342 (70 Tau): a double-lined spectroscopic orbit, the distance to the cluster, and the mass–luminosity relation. *ApJ*, 479, 268–278 {**282, 283, 285**}

Trullols E, Jordi C, Galadi Enriquez D, 1997, Kinematic analysis of Cepheus OB3. *ESA SP–402: Hipparcos, Venice '97*, 299–302 {**319**}

Trumpler RJ, Weaver HF, 1953, *Statistical Astronomy*. Dover, New York {**279**}

Turner DG, 1979, A reddening-free main sequence for the Pleiades cluster. *PASP*, 91, 642–647 {**287**}

Upton EKL, 1970, Calibration of the Hyades–Praesepe main sequence by a new treatment of the stellar motions. *AJ*, 75, 1097–1115 {**280**}

VandenBerg DA, Poll HE, 1989, On precise ZAMSs, the solar colour, and pre-main sequence lithium depletion. *AJ*, 98, 1451–1471 {**287, 288**}

van Altena WF, 1974, The distance to the Hyades cluster. *PASP*, 86, 217–222 {**280**}

van Altena WF, Lee JT, Hoffleit ED, 1997a, The trigonometric parallax of the Hyades. *Baltic Astronomy*, 6, 27–32 {**279**}

van Altena WF, Lu CL, Lee JT, *et al.*, 1997b, The distance to the Hyades cluster based on Hubble Space Telescope FGS parallaxes. *ApJ*, 486, L123–L127 {**293**}

van Bueren HG, 1952, On the structure of the Hyades cluster. *Bull. Astron. Inst. Netherlands*, 11, 385–402 {**276, 282**}

van Leeuwen F, 1996, The Hipparcos mission. *Space Science Reviews*, 81, 201–409 {**289**}

—, 1999a, Hipparcos distance calibrations for 9 open clusters. *A&A*, 341, L71–L74 {**288–291, 296, 298, 301**}

—, 1999b, Open cluster distances from Hipparcos parallaxes. *ASP Conf. Ser. 167: Harmonizing Cosmic Distance Scales in a Post-Hipparcos Era*, 52–71 {**296**}

—, 2000, Parallaxes for open clusters using the Hipparcos intermediate astrometric data. *ASP Conf. Ser. 198: Stellar Clusters and Associations: Convection, Rotation, and Dynamos*, 85–94 {**289**}

—, 2005, The Pleiades question, the definition of the zero-age main sequence, and implications. *IAU Colloq. 196: Transits of Venus: New Views of the Solar System and Galaxy* (ed. Kurtz DW), 347–360 {**289**}

—, 2007, *Hipparcos, the New Reduction of the Raw Data*. Springer, Dordrecht {**284, 285, 288, 290, 294–297**}

van Leeuwen F, Evans DW, 1998, On the use of the Hipparcos intermediate astrometric data. *A&AS*, 130, 157–172 {**279, 289**}

van Leeuwen F, Fantino E, 2003, Introduction to a further examination of the Hipparcos data. *Space Science Reviews*, 108, 447–449 {**291**}

van Leeuwen F, Hansen-Ruiz CS, 1997, The parallax of the Pleiades cluster. *ESA SP–402: Hipparcos, Venice '97*, 689–692 {**288, 291**}

Verschueren W, Brown AGA, Hensberge H, *et al.*, 1997, High signal-to-noise echelle spectroscopy in young stellar groups. I. Observations and data reduction. *PASP*, 109, 868–882 {**318**}

Vieira K, Abad C, 2003, A stellar group around the Sun detected from the Hipparcos Catalogue. *Astronomy in Latin America*, 157 {**279, 310**}

Villanova S, Carraro G, de la Fuente Marcos R, *et al.*, 2004, NGC 5385, NGC 2664 and Collinder 21: three candidate open cluster remnants. *A&A*, 428, 67–77 {**300, 302**}

Walter FM, Vrba FJ, Mathieu RD, *et al.*, 1994, X-ray sources in regions of star formation. V. The low mass stars of the Upper Scorpius association. *AJ*, 107, 692–719 {**318**}

Webb RA, Zuckerman B, Platais I, *et al.*, 1999, Discovery of seven T Tauri stars and a brown dwarf candidate in the nearby TW Hya association. *ApJ*, 512, L63–L67 {**321**}

Westin TNG, 1985, The local system of early-type stars: spatial extent and kinematics. *A&AS*, 60, 99–134 {**325**}

Whiteoak JB, 1961, A study of the Galactic cluster IC 2602. I. A photoelectric and spectroscopic investigation. *MNRAS*, 123, 245–256 {**277**}

Wielen R, 1974, The gravitational N-body problem for star clusters. *Stars and the Milky Way System* (ed. Mavridis LN), 326–354 {**294, 303**}

—, 1977, The diffusion of stellar orbits derived from the observed age-dependence of the velocity dispersion. *A&A*, 60, 263–275 {**305, 307**}

—, 1991, Dissolution of tidally limited star clusters in galaxies by passing giant molecular clouds or other massive objects. *ASP Conf. Ser. 13: The Formation and Evolution of Star Clusters* (ed. Janes K), 343–349 {**302**}

Wu Z, 2001, Measurements of the absolute proper motions for stars in the field of open cluster NGC 2548. *Shanghai Observatory Annals*, 22, 75–79 {**302**}

Yi S, Demarque P, Kim YC, *et al.*, 2001, Toward better age estimates for stellar populations: the Yonsei–Yale (Y^2) isochrones for solar mixture. *ApJS*, 136, 417–437 {**310**}

Zahn JP, 1992, Circulation and turbulence in rotating stars. *A&A*, 265, 115–132 {**286**}

Zeballos H, Barbá R, Morrell NI, *et al.*, 2006, Kinematics of stars in the line-of-sight to the open cluster Collinder 21. *Revista Mexicana de Astronomia y Astrofisica Conf. Ser.*, volume 26, 185–189 {**301**}

Zuckerman B, Song I, Bessell MS, *et al.*, 2001a, The β Pictoris moving group. *ApJ*, 562, L87–L90

Zuckerman B, Song I, Bessell MS, 2004, The AB Dor moving group. *ApJ*, 613, L65–L68 {**323**} {**320, 321**}

Zuckerman B, Song I, Webb RA, 2001b, Tucana association. *ApJ*, 559, 388–394 {**323**}

Zuckerman B, Song I, 2004, Young stars near the Sun. *ARA&A*, 42, 685–721 {**320**}

Zuckerman B, Webb RA, 2000, Identification of a nearby stellar association in the Hipparcos Catalogue: implications for recent, local star formation. *ApJ*, 535, 959–964 {**322–324**}

Zwahlen N, North P, Debernardi Y, *et al.*, 2004, A purely geometric distance to the binary star Atlas, a member of the Pleiades. *A&A*, 425, L45–L48 {**294, 295**}

7

Stellar structure and evolution

7.1 Introduction

While the chemical composition, temperatures and pressures deep in stellar interiors are out of reach of direct investigation, their observational consequences in terms of stellar luminosity, surface temperature, radii, masses, surface chemical composition, seismological oscillations, Galactic chemical enrichment, etc. are more directly accessible.

Stellar evolution is driven by changes in chemical composition caused by nuclear reactions. The development of numerical codes to calculate models of stellar structure and evolution began some 50 years ago with the pioneering works of Schwarzschild (1958) and Henyey *et al.* (1959). Matching resulting stellar models to a wide range of astronomical observations has been a hugely extensive and extremely successful field of research over many decades. Continuously improving models combined with advances in many observational areas have led to a progressively deeper understanding of the numerous physical processes that occur during the various stages of stellar formation and evolution.

Currently, the Sun provides the most stringent tests of theories of stellar structure and evolution. Its surface chemical composition is generally considered to be well-determined (Anders & Grevesse, 1989), although a significant decrease in metal content of the convection zone has recently been inferred from 3d hydrodynamical models (Asplund *et al.*, 2005; Grevesse *et al.*, 2007). Its luminosity, radius and mass are known to better than 1 part in 10^3: for their standard solar model, for example, Guenther *et al.* (1992) adopted $M_\odot = 1.989\,1 \times 10^{30}$ kg, with an uncertainty of about 0.02% directly dependent on the accuracy of G (Cohen & Taylor, 1986); $R_\odot = 6.959\,8 \times 10^8$ m, as determined from transit and eclipse measurements and corrected for an optical depth $\tau = 2/3$ as used for stellar structure calculations; $L_\odot = 3.851\,5 \times 10^{26}\,\mathrm{J\,s^{-1}}$ as an average of the solar constant measurements obtained from the Nimbus 7 and SMM satellites; and $T_{\rm eff} = 5780 \pm 30$ K. A discussion of the colours of the Sun is given by Holmberg *et al.* (2006). Its age, defined as the time it has taken to evolve from the zero-age main sequence, where nuclear reactions just begin to dominate gravitation as the primary energy source, to the present day, is known from meteorite dating to be 4.53 ± 0.04 Gyr (Guenther & Demarque, 1997, 2000); this age is adjusted for reset of the meteoritic radioactive clocks during the last high-temperature event in the primordial solar system nebula, which occurred before the Sun reached the ZAMS. Sound speed versus radius throughout most of the Sun's interior is known from inversion of the p-mode oscillations to better than 1% (Basu *et al.*, 1997), providing strong constraints on much of the input physics, including element diffusion and convective transport (Richard *et al.*, 1996; Guzik, 1998).

More generally, a stellar model is constructed by solving certain basic equations of stellar structure: conservation of mass and of energy, hydrostatic equilibrium, and energy transport via radiation, convection, and conduction (e.g. Cox & Giuli 1968 and updated by Weiss *et al.* 2004, including results from Hipparcos; Morel 1997; Chaboyer 1998). These coupled differential equations have quantifiable boundary conditions at the stellar centre and at the surface. Description of the constituent stellar plasma requires extensive inputs from nuclear and particle physics, notably opacities and nuclear reaction rates, atomic and molecular physics, thermodynamics and hydrodynamics, and radiative transfer. The complex process of convection enters as a phenomenological description involving mixing-length and overshooting, determining not only when a region is unstable to convective motion, but also the resulting heat transport efficiency.

Observational data to constrain models of stars, including the Sun, include bolometric luminosity (from magnitudes, bolometric corrections, and distance), surface chemical composition (from spectroscopy), mass (most rigorously for a modest number of favourable binary systems, or more generally from an assumed mass–luminosity relation), $T_{\rm eff}$, radius, and oscillation frequencies. Only in the case of the Sun is an independent age estimate available from the oldest meteorites (Tilton, 1988), otherwise ages are inferred indirectly through stellar evolutionary modelling.

Although these models have been enormously successful, increasingly accurate observations provide increasingly challenging constraints, and various scientific meetings have been devoted to unsolved problems in stellar structure and evolution over the last few years (e.g. Noels *et al.*, 1995; Livio, 2000, and the 2007 Cambridge conference on *Unsolved Problems in Stellar Physics*). Some of the remaining modelling uncertainties are related to transport processes in the stellar interiors, including diffusion, the treatment of convection, and the consistent inclusion of effects due to internal gravity waves together with rotation, magnetic fields, and wind losses. Others are related to atmospheric modelling, to uncertain opacities, to uncertain reactions in stable nuclei including some of the p-p chain and CNO-cycle reactions, and to fusion reactions of certain late-burning stages.

Hipparcos provided a significant improvement in precise distance determinations to a very large number of stars of varying spectral types and evolutionary stages. It has provided accurate luminosities which can be compared with the predictions of stellar evolution models. It has provided well-defined statistical samples, for example based on distance or space velocity, and the opportunity to study rather large and homogeneous samples of stars sharing similar properties, for instance, in terms of their space location or chemical composition. The contribution of more precisely determined masses, better characterised binary systems, and comprehensive photometric and variability data has also been of value.

Such information has been acquired mainly from the nearest stars observed with the highest precision, including disk and halo stars, members of open clusters, variable stars, and white dwarfs. Results to date have been successful in confirming various elements of stellar internal structure theory, yielding more precise characteristics of individual stars and clusters, and revealing some problems related to the development of stellar models. The smaller error bars on distances have also made the uncertainties on the other fundamental stellar parameters more evident, notably for effective temperatures, abundances, gravities, masses and radii.

Lebreton (2000b) presented a review of the early Hipparcos results on stellar structure and evolution, on which this chapter builds. Section 7.2 provides some background to the basic observational material entering the stellar structure and evolutionary models, and Section 7.3 provides some background to the theoretical modelling, and those aspects underpinning the microscopic physics, atmospheric models, and transport processes, focusing on the state-of-the-art at the time of the Hipparcos Catalogue publication in 1997. Section 7.4 then summarises how this basic knowledge has been augmented by Hipparcos. Later sections examine the wider results related to stellar structure and evolution where the Hipparcos data have been applied. Results related to various primary distance scale calibrators are considered further in Chapter 5, results on open clusters in Chapter 6, results for various other stages of stellar evolution in Chapter 8, and concerning the luminosity calibration of subdwarfs for the globular cluster distance scale in Chapter 9.

7.2 Observational framework and the HR diagram

This section presents an introduction to the ground-based photometric and spectroscopic data which, when combined with the Hipparcos results on astrometry and photometry, provide very homogeneous and precise sets of data necessary for comparison with theoretical models of stellar structure and evolution. The fundamental stellar parameters most relevant to the discussions of this chapter are the bolometric magnitude $M_{\rm bol}$, the effective temperature $T_{\rm eff}$, the surface gravity g, and the star's chemical composition. Although these are based on detailed photometric and/or spectroscopic analyses, the determinations also rely, to a greater or lesser extent, on model atmospheres and sometimes also on interior models.

One of the most prominent applications of an ensemble of such stellar parameters is in the comparison between the observational and theoretical Hertzsprung–Russell diagram. In its observational form the absolute visual magnitude (commonly M_V) is plotted against the colour index (commonly $B-V$); if stars are at a common distance (for example an open or globular cluster, or an external galaxy) the apparent magnitude serves in place of the absolute magnitude, with an uncertain vertical translation due to the unknown distance. In its theoretical form, in which stellar evolutionary tracks and isochrones can be plotted from theoretical models, stars are plotted in the luminosity versus $T_{\rm eff}$ plane. To compare observation with theory requires being able to transform effective temperatures to colour indices (or similar), and apparent or absolute magnitudes to bolometric magnitudes.

Masses and radii can be obtained directly for stars belonging to binary or multiple systems. Interferometry combined with distances yields stellar diameters giving direct access to $T_{\rm eff}$, but still for a very limited number of rather bright stars which then serve to calibrate other methods.

7.2.1 Bolometric magnitudes

The total energy integrated over all wavelengths is referred to as the bolometric magnitude, $M_{\rm bol}$ (see, e.g. Gray, 2000). Working in terms of the V band by way of illustration, the quantity required to transform an apparent magnitude in a particular bandpass (m_V) to an absolute magnitude in that bandpass (M_V) is the star's distance

$$m - M \equiv 5\log(d/10) = -5\log\pi - 5 \qquad (7.1)$$

where d is in pc, or the parallax π is in arcsec (and where correction due to interstellar extinction is ignored). The quantity required to transform the absolute magnitude to the bolometric magnitude, $M_{\rm bol}$, is referred to as the bolometric correction, BC, for that magnitude system

$$\mathrm{BC}(V) \equiv M_{\rm bol} - M_V = 2.5\log\frac{\int F_\nu S_\nu \mathrm{d}\nu}{\int F_\nu \mathrm{d}\nu} + C \qquad (7.2)$$

where F_ν is the flux measured at the Earth, and S_ν is the detector response for the relevant bandpass. BC is a function of the bandpass and the underlying stellar energy distribution of the star, the latter itself a function of $T_{\rm eff}$, $\log g$, metallicity, etc. The luminosity of the star is then given by

$$\log L/L_\odot = -0.4\,[M_V + \mathrm{BC} - (M_{V,\odot} + \mathrm{BC}_\odot)] \qquad (7.3)$$

Bolometric corrections are derived from empirical calibrations or from model atmospheres. The zero-point of the bolometric magnitude scale is set by reference to the Sun: Cayrel de Strobel (1996) gave $M_{\rm bol\odot} = 4.75$ and $\mathrm{BC}(V)_\odot = -0.08$ yielding $M_V = 4.83$, while Cox (2000) gives $M_{\rm bol\odot} = 4.74$, $\mathrm{BC}(V)_\odot = -0.08$, and $M_V = 4.82$ (see also Bessell *et al.*, 1998); $M_{\rm bol\odot} = 4.75$ was adopted by IAU Commission 36 at the IAU General Assembly in 1997 (Andersen, 1999, p141). Bolometric corrections are usually tabulated versus spectral type or colour index, and their dependence on $B-V$ is only moderately sensitive to luminosity class. For hot or cool stars, bolometric corrections in V are large, due to the fact that most of the flux lies outside of the V band.

As examples of the pre-Hipparcos situation, Alonso *et al.* (1995) derived bolometric fluxes, $F_{\rm bol}$, for 118 F–K dwarfs and subdwarfs, $T_{\rm eff} < 7000$ K and $-3.5 <$ [Fe/H] $<+0.5$, by integration of multi-colour photometry. For these stars, most of the energy is emitted in the optical and near infrared bands. The (small) residual flux, unmeasured because it is emitted outside of the measured bands, is then estimated from model atmospheres. The stars are sufficiently nearby so as not to be affected significantly by interstellar absorption. They provided a calibration of bolometric flux versus K, $V-K$, and [Fe/H], along with empirical bolometric corrections for main-sequence stars as a function of metallicity. Alonso *et al.* (1995) estimated that their bolometric fluxes are accurate to about 2%. Flower (1996) similarly compiled transformations between $T_{\rm eff}$, $B-V$ and BC for 335 stars spanning temperatures from 2900–52 000 K, and luminosity classes I–V.

7.2.2 Effective temperatures

Gray (2000) described the inaccuracy of the present knowledge of the effective temperature scale as one of the greatest barriers to the advancement of stellar evolutionary theory. Effective temperature is a recurrent theme in this chapter, notably through its connection with the HR diagram, so this section is included to underline the difficulties associated with its measurement. Other fundamental physical quantities also rely on an accurate knowledge of $T_{\rm eff}$: for example, and as detailed below, the radius and mass of a star can be determined given $T_{\rm eff}$, $M_{\rm bol}$ and $\log g$.

The effective temperature of a star, $T_{\rm eff}$, is defined as the temperature of a blackbody radiator with the same radius and same luminosity (total energy output), related via the Stefan–Boltzmann law

$$L = 4\pi\sigma R^2 T_{\rm eff}^4 \qquad (7.4)$$

where L is the luminosity, R is the radius, and σ is the Stefan–Boltzmann constant. Direct determination of $T_{\rm eff}$ therefore requires knowledge of both R (itself requiring knowledge of the distance d and angular diameter θ) as well as L (itself requiring knowledge of the star's bolometric magnitude, according to Equation 7.3). The diameters of 32 stars measured by intensity interferometry (Code *et al.*, 1976), six between 6500 and 10 000 K, along with the Sun formed the basis of the empirical temperature scale at that time. Angular diameters have since been measured for K and M giants using lunar occultations (e.g. Ridgway *et al.*, 1980; Richichi *et al.*, 1998; Mondal & Chandrasekhar, 2005) and, more recently, Michelson interferometry (Di Benedetto & Rabbia, 1987; Dyck *et al.*, 1996). Many individual interferometric diameters are becoming available using, for example, the VLTI, Keck, and CHARA arrays. Stellar radii can also be determined in the case of some eclipsing binary stars. Measurements of the bolometric flux may require space observations in the ultraviolet and/or infrared to quantify the contributions outside of the atmospheric transmittance, depending on temperature, or the use of bolometric corrections from other sources.

These fundamental quantities being known for only a few stars, the normal procedure is to establish a 'temperature scale' which relates $T_{\rm eff}$ determined rigorously for a set of calibrators to a more widely measurable temperature proxy, notably the spectral type or a colour index (frequently $B-V$ or $V-I$). These proxies can themselves then be used, for example, as the abscissa in an observational HR diagram, although dependency on luminosity class and metallicity complicates the picture. Various methods are used to determine effective temperatures (Bessell, 1998), and Section 7.4 will examine how the Hipparcos observations have assisted.

Many temperatures for A–K stars have been derived based on the infrared flux method, devised to measure simultaneously both $T_{\rm eff}$ and θ for stars with temperatures up to about 8000–10 000 K from a measured absolute flux distribution over the whole observable spectrum (Blackwell & Shallis, 1977; Blackwell *et al.*, 1980, 1990; Mégessier, 1995). The separation of these two parameters is possible because of the different dependence of the integrated absolute flux, F, and a monochromatic infrared flux, $F(\lambda)$, on these two parameters. Their ratio is given by $F/F(\lambda) = \sigma T^4/\phi(T, g, \lambda, A)$ where $\phi(T, g, \lambda, A)$ gives the monochromatic flux as a function of effective temperature, surface gravity, wavelength, and atomic abundances. The method is applied by calculating the right-hand side of the equation as a function of temperature for a chosen value of λ on the Rayleigh–Jeans tail (where the flux is proportional to θ^2 but only depends on the first power of T), and for selected values of $\log g$ and A, using a range of model stellar atmospheres. Iteration yields a definite value of $T_{\rm eff}$. Distances are required to correct both monochromatic and integrated fluxes for interstellar extinction. Blackwell *et al.* (1990) applied the method to 114 F–M stars, and considered that derived $T_{\rm eff}$ were accurate to better than 1%, and angular diameters to between 2 and 3%.

Alonso *et al.* (1996a,b) derived temperatures of 475 F0–K5 stars, comprising dwarfs and subdwarfs with $T_{\rm eff}$ in the range 4000–8000 K, and yielding internal accuracies of $\sim$1.5%. The zero-point of their $T_{\rm eff}$-scale is based on direct interferometric measures by Code *et al.* (1976), and the resulting systematic uncertainty is $\sim$1%.

Multi-parametric empirical calibrations of $T_{\rm eff}$ as a function of colour index, metallicity, and gravity can be derived from effective temperatures established for a representative sample of nearby stars (see, e.g. Alonso *et al.*, 1996b). In parallel, empirical calibrations serve to validate theoretical calibrations based on model atmospheres for a given set of stellar parameters. These have the advantage of covering the entire HR diagram for wide ranges of colour indices, metallicities and gravities.

For large samples of stars, the most appropriate method for the determination of $T_{\rm eff}$ and $\log g$ is based on the use of calibrated photometric indices. For early-type stars, the Strömgren $uvby\beta$ and Geneva photometric systems are widely used. Such calibrations are given for these systems, for example, by Moon & Dworetsky (1985) and Künzli *et al.* (1997) respectively.

Spectroscopic determinations of $T_{\rm eff}$ are based on the comparison of observed and model atmosphere profiles of spectral features that are sensitive to temperature. This includes the size of the Balmer discontinuity for B-type stars, the profiles of the hydrogen Balmer lines for F- and G-type stars, and the ratios of line depths of neutral metals. High-quality stellar spectra yielding a formal precision of around 50–80 K or better are commonly found in the literature (e.g. Cayrel de Strobel *et al.*, 1997b; Fuhrmann, 1998b), although the absolute temperature may be less certain.

Different techniques are required for the hottest and very coolest stars, where the bulk of the energy output is outside of the most accessible optical spectral region, where the colour indices are accordingly poor indicators of $T_{\rm eff}$, and where accurate effective temperatures and bolometric corrections are consequently less certain.

7.2.3 Surface gravities

The surface gravity of a star, g, controls the pressure in the stellar atmosphere, and affects the degree of ionisation of atoms, and hence the line and continuum absorption coefficients. The basic relations $g \propto M/R^2$ and $R^2 \propto L/T_{\rm eff}^4$ lead, on eliminating R and using Equation 7.3, to (e.g. Nissen *et al.*, 1997)

$$\log\frac{g}{g_\odot} = \log\frac{M}{M_\odot} + 4\log\frac{T_{\rm eff}}{T_{\rm eff\odot}} + 0.4(M_{\rm bol} - M_{\rm bol,\odot}) \tag{7.5}$$

which, introducing a value for $M_{\rm bol,\odot}$ gives an expression for $\log g$ based on $T_{\rm eff}$ and $M_{\rm bol}$. For example, $T_{\rm eff}$ may be estimated from Strömgren photometry, and the mass estimated from its position in a $M_V - \log T_{\rm eff}$ grid of stellar evolutionary models once the distances are accurately known.

The surface gravity can also be determined from spectroscopy. Different gravities produce different atmospheric pressures, modifying the profiles of some spectral lines. Two methods have been widely used. The first is based on the ionisation equilibrium of abundant species, such as iron, chromium, silicon or titanium in the case of F and G-type dwarfs (for metal-poor stars, $[{\rm Fe/H}] < -2$, only the iron lines may be available). By way of example, the iron abundance is determined from Fe I lines that are not sensitive to gravity, and then g is adjusted so that the analysis of Fe II lines, which are sensitive to gravity, leads to the same value of the iron abundance. The accuracy in $\log g$ is in the range 0.1–0.2 dex (Axer *et al.*, 1994). The second method uses the wings of strong lines broadened by collisional damping, exploited as a basis for stellar luminosity classification and more quantitative gravity determination (Cayrel de Strobel, 1969; Blackwell & Willis, 1977; Edvardsson, 1988). More recent use of lines such as Ca I (Cayrel *et al.*, 1996) or the Mg Ib triplet (Fuhrmann *et al.*, 1997) leads to formal uncertainties smaller than 0.15 dex. The two methods can yield systematic differences of around 0.2–0.4 dex, at least when ionisation equilibria are estimated from models in local thermodynamical equilibrium (LTE).

Thévenin & Idiart (1999) studied departure from LTE on the formation of the Fe I and Fe II lines, and found that modifications of the ionisation equilibria resulted from the over-ionisation of iron induced by significant ultraviolet fluxes. Resulting gravities inferred from iron ionisation equilibrium for 136 stars spanning a large range of metallicities were very close to those derived either from pressure-broadened strong lines or through Hipparcos parallaxes. A similar evaluation for Mg lines, using the Hipparcos data to study the dependency on surface gravity, was made by Zhao & Gehren (2000).

Rewriting of Equation 7.5 and using $M_{\rm bol} = V + 5\log\pi + 5{\rm BC} + A_V$ (where A_V accounts for effects of interstellar reddening) leads a distance estimate referred to as the spectroscopic parallax

$$\log\pi = 0.5\log\frac{g}{g_\odot} - 0.5\log\frac{M}{M_\odot} - 2\log\frac{T_{\rm eff}}{T_{\rm eff\odot}} - 0.2(V + {\rm BC} + A_V + 5 - M_{\rm bol,\odot}) \qquad (7.6)$$

The term is somewhat misleading since it is unrelated to parallax (except that they are both related to distance), but it provides a distance estimate for an object in cases where terms on the right-hand can be estimated: for example from spectroscopy for $\log g$, and from theoretical models for M and $T_{\rm eff}$.

7.2.4 Abundances

Spectroscopic determination of elemental abundances is derived from a comparison of high-resolution spectra with synthetic spectra and equivalent widths generated from model atmospheres of appropriate $T_{\rm eff}$ and $\log g$. Uncertainty in abundance determinations depends on the validity of the model atmosphere, on uncertainties in $T_{\rm eff}$ and $\log g$, and on measured oscillator strengths, and errors in the range 0.05–0.15 dex are typical (Cayrel de Strobel *et al.*, 1997b; Fuhrmann, 1998b). An additional uncertainty arises due to the solar Fe/H ratio adopted as reference; values differing by $\sim$0.15 dex exist in the literature due to difficulties in determining the solar iron abundance from Fe I or Fe II lines, because of uncertain atomic data (Axer *et al.*, 1994). Grevesse & Sauval (1999) reconciled abundances from low- and high-excitation Fe I lines as well as Fe II lines, deriving a solar photospheric abundance of $A_{\rm Fe} = 7.50 \pm 0.05$, in agreement with the meteoritic value.

7.3 Theoretical framework

7.3.1 Equation-of-state and opacities

Stellar model calculations require an equation-of-state, i.e. a description of the state of matter under a given set of physical conditions, described by two or more state functions such as its temperature, pressure, volume, or internal energy. Early stellar models were based on the ionisation equilibrium model of Saha (1920), assuming ideal gas conditions, and dealing explicitly with ions and atoms. Over a substantial region of the ρ, T plane the equation-of-state is considered to be relatively simple and can be computed efficiently using, for example, the EFF code described by Eggleton *et al.* (1973). Other important processes which enter at fairly low temperature and high density are the dissociation of molecular hydrogen, Coulomb interactions, and pressure ionisation. A didactic schematic introduction to the equation-of-state is given in Figure 1 of Pols *et al.* (1995).

Schwarzschild (1958) considered the determination of opacity to be *'by far the most bothersome factor in the entire theory [of stellar structure and evolution]'*. He took account of three contributing processes: photoionisation (bound–free transitions), inverse bremsstrahlung (free–free transitions) and electron scattering. Later work showed that spectral lines (bound–bound transitions) also contribute significantly, eventually leading to the widely-used Los Alamos opacities (Huebner *et al.*, 1977). By the 1980s, certain discrepancies between observation and theory, for example for Cepheid pulsations (Simon, 1982) and for solar oscillations (Christensen-Dalsgaard *et al.*, 1985), led to the suggestion that the Los Alamos opacities were missing important sources. Combined with the need for opacities of low-Z materials, this led to the re-examination of stellar opacities by two groups: the Opacity Project (Seaton *et al.*, 1994) and the OPAL group at Livermore (Rogers & Iglesias, 1992). They adopted independent approaches, and both eventually reported opacities generally higher than the previous Los Alamos opacities, reaching factors of 2–3 in stellar envelopes with temperatures in the range 10^5–10^6 K. With these new opacities, a number of long-standing problems in stellar evolution have been addressed and, in turn, finer tests of stellar structure have been made possible. Since opacity is very sensitive to metallicity, any errors in estimating metallicity, including differences in photospheric and interior abundances, will be problematic. Efforts have also been

Production of the elements: In the following synopsis, based on the summary by Thielemann (2002), the nomenclature $i(j, k)l$ denotes a reaction with target i, projectile j, emitted particle(s) k, and main reaction product l.

Hydrogen burning: H-burning converts ^{1}H into ^{4}He via the proton-proton (or p-p) chain, important in stars with the mass of the Sun and less, or via the CNO-cycle, important in more massive stars. The main branch of the p-p chain, for example, starts with $^1\mathrm{H}(\mathrm{p}, \mathrm{e}^+\nu)^2\mathrm{H}(\mathrm{p}, \gamma)^3\mathrm{He}$, with two ^{3}He isotopes subsequently fusing through $^3\mathrm{He}(^3\mathrm{He}, 2\mathrm{p})^4\mathrm{He}$. The net effect is to convert H into He, with energy release in the form of particles and γ-rays.

Subsequent burning stages: when C, N, O are present in subsequent generations of stars, they can participate in the CNO cycle. The dominant chain is $^{12}\mathrm{C}(\mathrm{p},\gamma)^{13}\mathrm{N}(\mathrm{e}^+\nu)^{13}\mathrm{C}(\mathrm{p},\gamma)^{14}\mathrm{N}(\mathrm{p},\gamma)^{15}\mathrm{O}(\mathrm{e}^+\nu)^{15}\mathrm{N}(\mathrm{p},\alpha)^{12}\mathrm{C}$, where the net effect is again to convert H into He, i.e. the α-particle emitted in the last step. C is necessary to initiate the sequence, but is not consumed, since the last step reproduces the ^{12}C nucleus. Main subsequent stages are He-burning, via both the triple-alpha process $^4\mathrm{He}(2\alpha,\gamma)^{12}\mathrm{C}$ and $^{12}\mathrm{C}(\alpha,\gamma)^{16}\mathrm{O}$; C-burning $^{12}\mathrm{C}(^{12}\mathrm{C},\alpha)^{20}\mathrm{Ne}$; and O-burning, $^{16}\mathrm{O}(^{16}\mathrm{O},\alpha)^{28}\mathrm{Si}$. Fusion continues beyond C and O, but now involving photodisintegration (the disintegration of nuclei by high-energy photons in the plasma), and subsequent recombination of the various products. This process of 'nuclear statistical equilibrium' is responsible for the production of Ne up to Fe. It requires higher temperatures for penetration of the increasingly higher Coulomb barriers, produces less energy as the masses increase, and ceases beyond the Fe region where the binding energy curve peaks. Many of these hydrostatic burning processes also occur under explosive conditions at higher temperatures and on shorter time scales, although their understanding also requires knowledge of nuclear reactions for unstable nuclei.

s-process: the two principal paths leading to the 'trans-Fe' elements are the s-process and the r-process. The s-process (slow neutron capture) involves neutrons which have been liberated during core and shell He-burning being captured by a nucleus, with the neutron subsequently undergoing β-decay to produce a p^+. It being easier to add the charge-less neutron to a nucleus than it is to add a proton directly, this results in the progressive build-up of elements up to Pb and Bi, starting on existing heavy nuclei around Fe. It is thought to occur in Type I supernovae and also in asymptotic giant branch stars during the thermal pulse stage. Direct evidence for the latter comes from the presence in some stellar spectra of Technetium (Tc), an s-process element with a radioactive half-life of $\sim$ 200 000 years. Since the stars are much older than this half-life, it is inferred that Tc has recently been produced in the star, and then carried to the surface via convection.

r-process: the r-process (rapid neutron capture) involves the rapid addition of many neutrons to existing nuclei, which again decay into protons, increasing the atomic number and producing the heavier elements. The r-process is a subset of explosive Si-burning, which differs strongly from its hydrostatic counterpart, and is thought to occur only in supernovae, and mostly in those of Type II (the end points of massive star evolution) rather than those of Type Ia (resulting from binary systems). Observational evidence is based on the existence of elements like Au, and the fact that in some of the oldest stars in the Galaxy, which were formed after only a small number of Type II supernovae had enriched the interstellar medium, the abundance of Fe is very low, while the abundances of r-process elements are anomalously high.

α-elements: the class of α-elements refers to those whose most abundant isotopes are integral multiples of the He nuclei or α-particle: (C, N, O), Ne, Mg, Si, S, Ar, Ca, and Ti. Type Ia supernovae predominantly produce elements of the iron peak (V, Cr, Mn, Fe, Co and Ni) as a result of normal freeze-out of charged particle reactions during cooling from statistical equilibrium. In contrast, low-density freeze-outs, most pronounced in Type II supernovae, leave a large α abundance, resulting in higher proportions of the α-elements. This includes O, so that O enhancement is well correlated with an enhancement of α-elements. C and N, as well as O, are sometimes included within the class of α-elements since they are synthesized by nuclear α-capture reactions, although the enrichment of the interstellar medium by C and N is not due to Type II supernova explosions but due to stellar winds of the more massive asymptotic giant branch stars. O (for example) is an α-element in Population II stars.

In summary, the elements are considered to have been broadly formed as follows: H in the big bang; He in the big bang and in stars; C and O in low- and high-mass stars; Ne–Fe ($Z = 10-26$) in high-mass stars; Co–Bi ($Z = 27-84$) in the s- and r-processes (asymptotic giant branch and supernovae); and Po–U ($Z = 84-92$) in the r-process in supernovae.

invested in the derivation of low-temperature opacities, including millions of molecular and atomic lines and grain absorption that are fundamental for the calculation of the envelopes and atmospheres of cool stars (Kurucz, 1991; Alexander & Ferguson, 1994).

Opacity Project and OPAL opacities have been shown to be in reasonable agreement (Seaton *et al.*, 1994; Iglesias & Rogers, 1996); and good agreement between OPAL opacities, and those of Alexander & Ferguson (1994) or Kurucz (1991), is also found in the domains where they overlap. Remaining discrepancies have been quoted as not exceeding 20%, and to be rather well understood (Iglesias & Rogers, 1996). New opacity comparisons have been made more recently using revised Opacity Project data (Badnell *et al.*, 2005).

The re-calculation of opacities required a corresponding equation-of-state. Rogers *et al.* (1996) describe the OPAL equation-of-state developed at Livermore which provides, as a function of temperature and density, tables of pressure, internal energy, entropy, and various other second-order quantities. Mihalas *et al.* (1988) described the MHD equation-of-state used in the Opacity Project (see also Däppen, 2006); this was compared with the OPAL equation-of-state by Trampedach *et al.* (2006). Other approaches used

Notations: Relevant notations used in this chapter are:
X, Y, Z = hydrogen, helium, and metal abundances by mass ('metals' embracing all elements heavier than helium)
$[X] \equiv \log(X/X_\odot)$
$[X/H] \equiv \log(N_X/N_H) - \log(N_X/N_H)_\odot$
[Fe/H] = log number abundances of Fe/H relative to solar (Fe being used as an observable proxy for metals)
$[\alpha/Fe] = \log(N_\alpha/N_{Fe}) - \log(N_\alpha/N_{Fe})_\odot$ is, similarly, the α-element abundance ratio relative to solar
$A_{el} = \log\epsilon\,(el) = \log(N_{el}/N_H) + 12$, where N_{el} is the elemental abundance by number
dex: a contraction of 'decimal exponent', with n dex meaning 10^{-n}, and 1 dex corresponding to a factor 10

to calculate stellar equations-of-state include the EFF model of Eggleton *et al.* (1973), which is relativistic, accounts for electron Fermi–Dirac statistics, but ignores Coulomb interactions; and the CEFF revision which adds a Debye–Hückel free-energy term (Christensen-Dalsgaard & Däppen, 1992; Bi *et al.*, 2000). Saumon & Chabrier (1991) developed a specific equation-of-state to interpret the first observations of very low-mass stars and brown dwarfs.

7.3.2 Atmospheres

Atmospheres are relevant at many levels in the analysis of observations, and they provide external boundary conditions for the calculation of stellar structure and necessary relations to transform theoretical (M_{bol}, T_{eff}) HR diagrams to colour–magnitude or colour–colour planes. Models have improved during the last two decades, and attention has been paid to the treatment of atomic and molecular line blanketing. The original programs MARCS (Gustafsson *et al.*, 1975) and ATLAS (Kurucz, 1979) evolved toward the ATLAS9 version appropriate for O–K stars (Kurucz, 1993b) and NMARCS for A–M stars (Brett, 1995; Bessell *et al.*, 1998). The PHOENIX code is described by, e.g. Hauschildt *et al.* (1997), Hauschildt *et al.* (2001), and references therein. Carbon (1979) and Allard *et al.* (1997) reviewed calculation details and remaining problems, such as incomplete opacity data, poor treatment of convection, neglect of non-LTE effects or assumption of plane-parallel geometry.

The external boundary conditions for interior models are commonly obtained from $T(\tau)$-laws (τ is the optical depth) derived either from theory or full atmosphere calculation. This method is suitable for low- and intermediate-mass stars, although it is not valid for masses below $\sim 0.6 M_\odot$ (Chabrier & Baraffe, 1997). Morel *et al.* (1994) and Bernkopf (1998) focused on the solar case where seismic constraints require a careful treatment of external boundary conditions. Morel *et al.* (1994) pointed out that homogeneous physics should be used in interior and atmosphere (opacities, equation-of-state, treatment of convection) and showed that the boundary level must be set deep enough, in zones where the diffusion approximation is valid. Bernkopf (1998) discussed some difficulties in reproducing Balmer lines related to the convection treatment.

7.3.3 Transport processes

Convection Classical heat transport theory suggests that stable stratified radiative transport occurs for unevolved stars with higher effective temperatures. Cooler stars have increasingly extensive convective envelopes, for which convection becomes of major importance in both energy transport and mixing. Three-dimensional numerical simulations of convection can reproduce many observational features of solar convection such as image structure, temporal development, spectra, and helioseismic properties (Stein & Nordlund, 1998). However, nearly all stellar models that have a convective envelope (including the solar model) rely on 1d phenomenological descriptions, notably the mixing-length theory (MLT) formulated by Böhm-Vitense (1953, 1958). In this approximation, convective energy is assumed to be carried by an element of fluid of fixed size that rises a specific distance adiabatically, and then is instantly absorbed by its surroundings through radiative diffusion. The distance the fluid element rises, referred to as the mixing-length distance, is assumed to be proportional to the pressure scale height; the constant of proportionality is called the mixing-length parameter α_{MLT}, which is an adjustable parameter, calibrated so that the solar model yields the observed solar radius at the present solar age, and thereafter often chosen to be constant and equal to this value. Variations of α_{MLT} in stars of various masses, metallicities, and evolutionary stages remains a matter of investigation. The theory predicts that the temperature gradient in most of the stellar convective envelope is very slightly super-adiabatic. Near the surface of the star, where convective transport efficiency drops and radiative transport efficiency rises, there is a peak in the temperature gradient that climbs well above the adiabatic temperature gradient. This region, some 0.04% $R_\odot$ thick in the Sun, is called the super-adiabatic layer. Abbett *et al.* (1997) found that the theory can reproduce the correct entropy jump across the super-adiabatic layer near the stellar surface, for $\alpha_{MLT} \approx 1.5$, but fails to describe the detailed depth structure and dynamics of convection zones. Ludwig *et al.* (1999) calibrated α_{MLT} from 2d simulations of compressible convection in solar-type stars for a broad range of T_{eff} and g-values, finding a solar α_{MLT} close to what is obtained in solar model calibration, and providing a dependence with $T_{\rm eff}$ and g which can be used to constrain the range of acceptable variations of $\alpha_{\rm MLT}$ in stellar models.

Convective penetration from unstable convection cores into the surrounding stable layers is referred to as overshooting. It extends the zone of effective mixing beyond the classical convection cores, influencing the thermal structure, and modifying the standard evolution model of stars of masses $M \gtrsim 1.2 M_\odot$, in particular extending their lifetimes (see, e.g. Maeder & Mermilliod, 1981; Bressan *et al.*, 1981). It is usually quantified by the overshooting parameter, $\alpha_{\rm ov}$, which characterises the ratio of overshooting distance to the pressure scale height. The extent of overshooting was estimated for the first time from the comparison of observed and theoretical main-sequence widths of open clusters (Maeder & Mermilliod, 1981), yielding $\alpha_{\rm ov} \sim 0.2$. Roxburgh (1997) noted that $\alpha_{\rm ov}$ is still poorly constrained, despite significant efforts made to establish the dependence of overshooting with mass, evolutionary stage, or chemical composition. Andersen (1991) noted that the simultaneous calibration of well-known binaries, where masses and radii are known to 1–2%, may provide improved constraints for $\alpha_{\rm ov}$. Models of the best-known binaries indicate a trend for $\alpha_{\rm ov}$ to increase with mass, and suggest a decrease of $\alpha_{\rm ov}$ with decreasing metallicity. Further advances are expected from asteroseismology (Brown *et al.*, 1994; Lebreton *et al.*, 1995).

Certain major convective changes during the star's evolutionary lifetime are referred to as the first, second and third 'dredge-up'. In standard models the first dredge-up occurs between the main sequence and giant branch as the surface abundances adjust to the average abundance outside the nuclear burning core (Iben, 1967). The second occurs after helium core burning on the asymptotic giant branch (Iben & Renzini, 1983). The third is a result of higher luminosity instabilities on the asymptotic giant branch (Section 7.5.7).

Other mixing processes The 'standard' models of stellar evolution were originally understood as comprising non-rotating, non-magnetic models in which convection is the only transport process responsible for mixing of the elements. These models made specific predictions about the surface abundances of stars as a function of mass, composition, and age. Evidently, any other mixing processes will affect the internal chemical composition, influencing their evolution, their age, and the relationship between initial and surface abundances. A body of evidence now exists for additional mixing processes, including the surface lithium abundance in low-mass stars, and carbon to nitrogen processing in the envelopes of red giants. In low-mass stars, microscopic diffusion due to gravitational settling carries helium and other heavy elements to the centre and modifies the evolutionary course as well as the surface abundances of Li and Be; it has been proposed to explain the low helium abundance of the solar convective zone derived from seismology (Christensen-Dalsgaard *et al.*, 1993), and now figures in the standard model for the Sun (e.g. Richard *et al.*, 1996, includes element segregation for He and 12 heavier isotopes). Diffusion acts very slowly, with time scales of order 10^9 yr (Salaris *et al.*, 2000), so that the only evolutionary stages where diffusion is relevant are the main-sequence and the white dwarf cooling sequence. The term 'standard' model is now frequently taken to include the effects of atomic diffusion not counterbalanced by any macroscopic process.

Additional turbulent mixing due to hydrodynamical instabilities related to rotation, has been invoked to explain certain depletion patterns, although such effects may inhibit microscopic diffusion (Zahn, 1992; Richard *et al.*, 1996). The various mixing and separation mechanisms (rotation, gravity waves, magnetic fields, gravitational settling, thermal diffusion, and radiative levitation), as well as associated diagnostics, and implications for clusters, tidally-locked stars, low- and high-mass stars, are discussed in reviews by Pinsonneault (1997) and Talon (2008).

7.3.4 Evolutionary tracks and isochrones

These various ingredients are used in models to compute stellar evolutionary tracks, which follow the detailed physical characteristics of a star of given initial mass and chemical composition as a function of time, typically in the theoretical HR diagram. The same models can be used to construct isochrones, which are the loci in luminosity and $T_{\rm eff}$ for any mass at a given time.

As an aid to the discussions in the remainder of this chapter, Table 7.2 below gives a compilation of the stellar evolution models found in the literature, which have been referenced in connection with the interpretation of the Hipparcos data, for various mass ranges, including an indication of principal features, and including some referenced models now superseded. Models are now frequently computed with and without overshooting, and may include α-element enhancement, rotation, diffusion, etc. Most models use OPAL opacities (Iglesias *et al.*, 1992; Rogers & Iglesias, 1992; Iglesias & Rogers, 1996), with low-temperature opacities from, e.g. Kurucz (1991) or Alexander & Ferguson (1994). Some of the more recent models (e.g. Kotoneva *et al.*, 2002b; Jimenez *et al.*, 2004; Pietrinferni *et al.*, 2004) have been specifically tested against Hipparcos observations to support the validity of the resulting evolutionary tracks.

Figure 7.2 illustrates some of the resulting evolutionary tracks for pre-main-sequence, main-sequence,

Observational features of the Hertzsprung–Russell diagram: A schematic of the HR diagram is shown in Figure 7.1, illustrating the main features relevant to the discussions in this chapter. The figure typifies the evolution of a low-mass, low-metallicity (Population II) object. Evolution for other masses and metallicities can be seen in the model isochrones in the figures elsewhere in this chapter, and further descriptive details of the various evolutionary stages can be found in, e.g. Mihalas & Binney (1981, Chapter 3) and Evans (2002).

During the pre-main-sequence phase a collapsing protostar reaches the Hayashi limit, and then evolves more slowly down the Hayashi track until thermonuclear fusion of hydrogen becomes possible as the star reaches the main sequence. The locus of points where hydrogen burning first occurs defines the zero-age main sequence. The main sequence lasts until some one tenth of the total stellar mass has been converted to helium, and the star then evolves off the main sequence as the energy generation shifts to a hydrogen burning shell surrounding a growing and inert helium core. Objects evolve rapidly through the subgiant phase and onto the giant branch. Stars with masses less than about $2M_\odot$ have degenerate helium cores, and increase substantially in luminosity, ascending the red giant branch until temperatures are high enough to initiate an explosive helium flash in the degenerate core (the triple-alpha process). The subsequent horizontal branch phase is marked by a helium burning core and a hydrogen burning shell, with the location of a star on the zero-age horizontal branch, where core helium burning starts, being determined primarily by the mass. Above about $2M_\odot$ the helium core is not degenerate, the red giant branch is missing, and the star spends most of its core helium burning life in a region to the blue of the giant branch. For Population I disk stars, the effect of higher metallicity is such that the ZAHB is moved more towards the right, merging into the giant branch and, because of the long lifetimes, resulting in the concentrated population of clump giants (see, e.g. Figure 7.21c below). As helium is exhausted in the core the star moves to a phase of shell helium burning, and the star evolves from blue to red, rising along the asymptotic giant branch which, as the name suggests, approaches the giant branch. At the tip of the asymptotic giant branch strong instabilities set in due to the proximity of the hydrogen and helium burning shells. Dense stellar winds along the asymptotic giant branch culminate in the ejection of the envelope, leaving a planetary nebula while the material is ionised by the hot stellar remnant. The end-point of the evolution of a star of low or intermediate mass is a white dwarf, which subsequently slowly cools, or a supernova explosion in the case of a more massive star.

The above is a simplified picture of the main evolutionary stages, and details depend on mass and metallicity. Within this framework, some terms merit further explanation. Subdwarfs are named after the few stars originally found in the solar neighbourhood which appear to fall distinctly below the normal main sequence in a colour–magnitude diagram, by about 0.5–0.75 mag. Their spectral lines are abnormally weak, and if the stars are sufficiently metal-poor, they can be assigned to a distinct luminosity class (VI) in the MK system; quantitative analysis shows metal abundances lower than the Sun by a factor of 10^{-2} or more, and this in turn implies that rather than being subluminous, they are instead bluer than the normal main sequence for a given L and $T_{\rm eff}$, and therefore displaced to the left in the colour–magnitude diagram. These characteristics imply that subdwarfs are identified with low-metallicity main-sequence objects belonging to the (Population II) thick disk or halo population. They can be seen, for example, in Figure 7.18 below, scattered around the lower bound of the main sequence, and merging with it due to errors in M_V as well as a spread in metallicity.

Subgiants occupy the region between the main sequence and giant branch. They were given the name by Strömberg (1930) when the few objects known at the time were not considered to fit the prevalent picture of stellar evolution. They can again be recognised spectroscopically, and are assigned luminosity class IV in the MK system, and exist as Population I or II objects. For high-mass stars, evolution off the main sequence is so rapid that few objects are found in the subgiant phase, which led to the term Hertzsprung gap being used to describe the typical absence of objects observed between the main sequence and the giant branch (see, e.g. Figure 7.21c below); boundaries in $B-V$ or $T_{\rm eff}$ depend on population and age. For the oldest disk clusters, e.g. M67 and NGC 188, with a main-sequence turn-off around type F, these ~ 1.2–$1.4M_\odot$ stars have a much slower evolution off the main sequence, and there is a resulting continuous subgiant branch with no Hertzsprung gap.

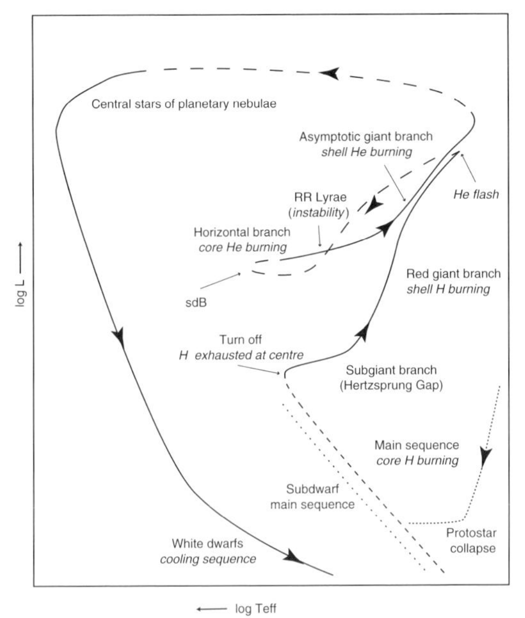

Figure 7.1 Schematic evolution track for a low-mass star from the main sequence (short-dashed line) to the degenerate white dwarf phase. Long dashed lines indicate episodes of rapid evolution and, accordingly, small numbers of observed objects. Main phases, and principal energy sources, are shown. This schematic figure should be compared with the examples of detailed models of the different phases, for different masses and chemical composition, shown in Figure 7.2. Adapted from Mihalas & Binney (1981, Figure 3.18).

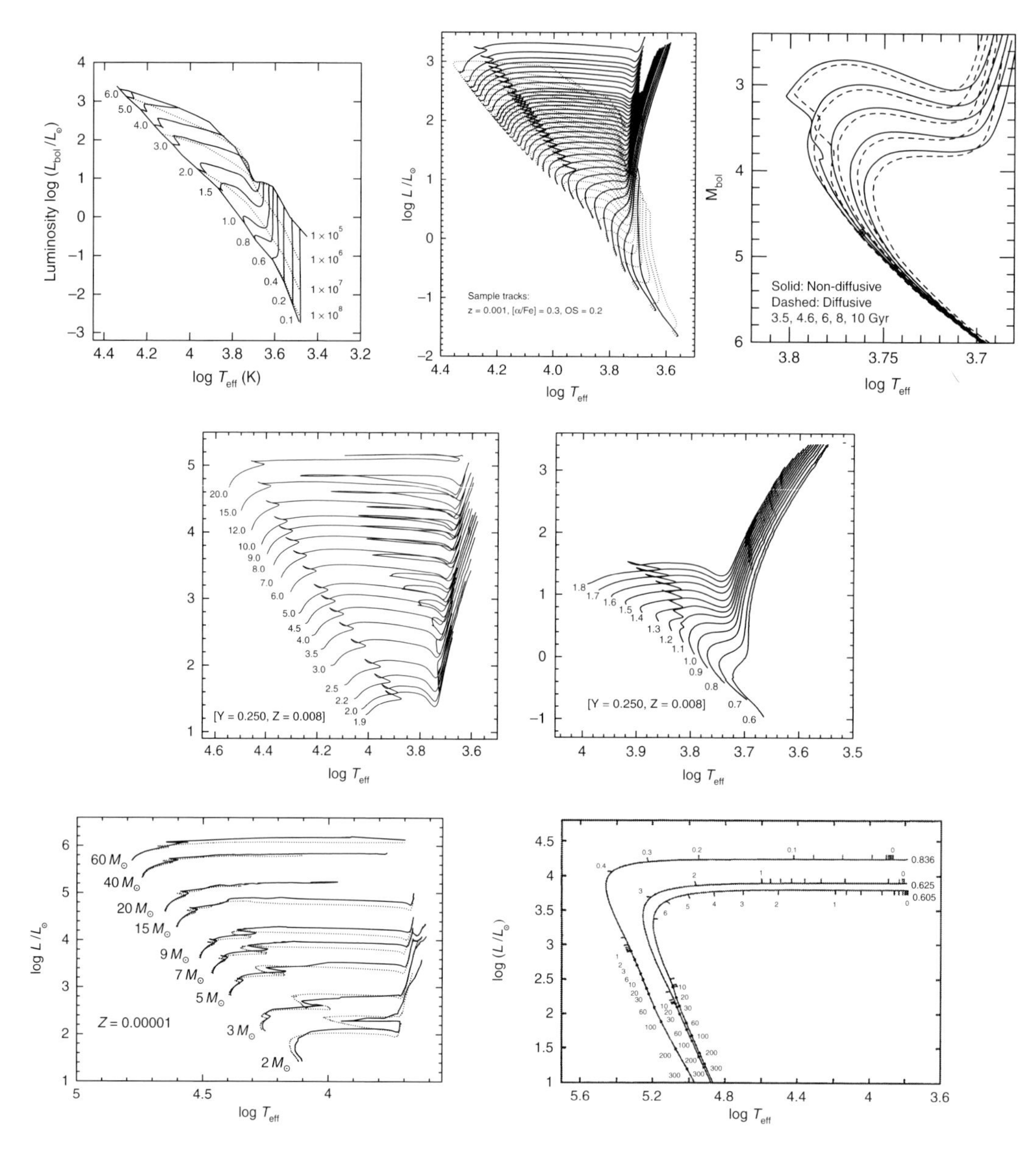

Figure 7.2 A compilation of representative theoretical evolutionary tracks and isochrones from the recent literature, illustrating paths in the theoretical HR diagram ($\log L/L_{\odot}$ versus $T_{\rm eff}$) for different stellar masses (in units of $M_{\odot}$) at different stages of their evolutionary lifetime. The plots illustrate dependencies on metallicity, α-element abundance, rotation, etc. First row, left: theoretical pre-main-sequence tracks. For each track, the evolution starts at the birthline (light solid line), and ends at the ZAMS. Selected isochrones are also shown as dotted lines. From Palla & Stahler (1999, Figure 1). First row, middle: main-sequence evolution for the mass range 0.4 – 5.0 $M_{\odot}$. Pre-main-sequence parts are shown dotted. From Yi et al. (2003, Figure 1). First row, right: non-diffusive (solid) and diffusive (dashed) isochrones for [m/H] = 0.0 and ages indicated. Sun's location is indicated $\odot$. From Michaud et al. (2004, Figure 10). Second row: (left) for the composition [$Y = 0.250, Z = 0.008$] and an α-enhanced mixture of abundances. The initial mass is indicated at the left of each curve; (right) low-mass models (same composition) up to the tip of the red giant branch. From Salasnich et al. (2000, Figures 4–5). Third row, left: non-rotating (dotted) and rotating (solid) models for metallicity $Z = 10^{-5}$. The rotating models have an initial velocity 300 km s^{-1}. From Meynet & Maeder (2002, Figure 10). Third row, right: evolution of post-AGB models with ($M_{\rm ZAMS}, M_{\rm H}$) = (3,0.605), (3,0.625), (5,0.836), where $M_{\rm H}$ = core mass $\simeq$ remnant mass. From Blöcker (1995b, Figure 7).

Table 7.1 Example of the state-of-the-art input physics and parameters for the Yonsei–Yale stellar evolutionary model set described by Yi et al. (2003).

Model property	Value or reference
Solar mixture	Grevesse & Noels (1993)
$[\alpha/\mathrm{Fe}]$ enhancement	VandenBerg *et al.* (2000)
OPAL Rosseland mean opacities	Iglesias & Rogers (1996)
Low-temperature opacities	Alexander & Ferguson (1994)
Equation-of-state	OPAL, Rogers *et al.* (1996)
Energy generation rates	Bahcall & Pinsonneault (1992)
Neutrino losses	Itoh *et al.* (1989)
Convective core overshoot	$0.2\,H_p$, with convective core
Helium diffusion	Thoul *et al.* (1994)
Mixing length parameter	$l/H_p = 1.7431$
Primordial helium abundance	$Y_0 = 0.23$
Helium enrichment parameter	$\Delta Y/\Delta Z = 2.0$

and post-main-sequence phases. State-of-the-art input physics and parameters are illustrated by the recent model set described by Yi *et al.* (2003) and listed in Table 7.1.

7.4 Fundamental parameters from Hipparcos

7.4.1 Bolometric magnitudes

There are three compelling reasons to use the Hipparcos magnitudes *Hp*, rather than say Johnson *V* magnitudes as a basis for the construction of bolometric magnitudes: their high accuracy of $\sigma \sim 0.0015$ mag; their small systematic errors and excellent uniformity independent of position, magnitude, and colour index; and the fact that the band is very wide and therefore a better approximation to the total flux.

Cayrel *et al.* (1997a) constructed bolometric corrections $\mathrm{BC}(Hp) = M_{\mathrm{bol}} - M_{Hp}$ using the Kurucz ATLAS9 code for 1500 models covering $T_{\mathrm{eff}} = 4000[250]8500$, $\log g = 0.0[0.5]5.0$, and $[\mathrm{M/H}] = -2.5[0.5]0.5$. They compared their results with bolometric magnitudes for 475 dwarfs and subdwarfs using the infrared flux method from Alonso *et al.* (1996a). Their early Hipparcos sample included few subgiant or giants stars. Bolometric corrections for giants and dwarfs are shown in Figure 7 of Haywood *et al.* (1997a), although their plans for more extensive bolometric corrections remain unpublished (Haywood, 2007, priv. comm.). A grid of metal-poor model stellar atmospheres was constructed by van't Veer-Menneret *et al.* (1999), for which they calculated bolometric corrections for the *Hp* band.

Different colour–magnitude transformations, both empirical and theoretical, have been used in the analysis of the Hipparcos data. Some of the most relevant empirical transformations have been discussed in Sections 4.2 and 4.3.

An extensive compilation of theoretical transformations were presented by Bessell *et al.* (1998), who used synthetic spectra derived from ATLAS9 and NMARCS to produce broadband colours and bolometric corrections for a wide range of T_{eff}, g and [Fe/H] values. They found fairly good agreement with empirical relations except for the coolest stars (M dwarfs and K–M giants).

These synthetic spectra were then used by Bessell (2007, priv. comm., and available from his www pages) to construct a grid of T_{eff}, $\log g$ and Z giving corresponding values of the following quantities: BC_V, $\mathrm{BC}_{\mathrm{Hip}}$, $V - Hp$, $V - V_T$, $B - B_T$, $B_T - V_T$, $B - V$, $V - R$, and $V - I$. Resulting plots of $\mathrm{BC}_{\mathrm{Hip}}$ versus T_{eff} are shown in Figure 7.3, with plots of various colour indices versus T_{eff} from the same synthetic spectra shown in Figure 4.5. Masana *et al.* (2006) adopted a T_{eff} calibration for FGK stars based on *V* and 2MASS infrared photometry, to provide T_{eff}, radius, and bolometric corrections in the *V* and *K* bands for 10 999 Hipparcos FGK dwarfs. Sensitivity of bolometric corrections to He content are considered by Girardi *et al.* (2007).

7.4.2 Effective temperatures

Blackwell & Lynas-Gray (1997, 1998) applied the infrared flux method to determine effective temperatures with accuracies of 1% for 420 stars with spectral types between A0–K3, and luminosity classes between II–V, corresponding to temperatures between 4000–10 000 K and surface gravities between $\log g = 1.0$–4.5, used as flux calibration standards for the Infrared Space Observatory (ISO). Determinations were based on narrow- and wide-band photometric data, and corrected for the effects of interstellar extinction using Hipparcos parallaxes, assuming an average of $A_V = 0.8\,\mathrm{mag\,kpc^{-1}}$ (Blackwell & Lynas-Gray, 1994). The results compare well with those of Alonso *et al.* (1996a), with differences below 0.12 ± 1.25% for the 93 stars in common (Lebreton, 2000b; Mégessier, 2000).

Effective temperatures can also be derived from the correlation between surface brightness and colour

Table 7.2 Stellar evolution models cited in the 1997–2007 literature related to the interpretation of Hipparcos data. This is neither a complete survey, nor a statement regarding their relative quality: the models are far from homogeneous, in that large progress has been made in the input physics, etc. between the earliest and latest models. Where a model has clearly been superseded according to subsequent papers, this is noted. Models treating brown dwarfs exclusively are not included.

Group	Reference	Parameter Range	Comments
Cambridge–EFF	Pols *et al.* (1998)	$M = 0.5–50$, $Z = 0.0001–0.03$	
	Pols *et al.* (1995)	$M = 0.64–125$	
	Jimenez *et al.* (2004)	$M = 0.55–80$, $Z = 0.0002–0.005$	
	Jimenez & MacDonald (1996)	$M = 0.55–1$, $Z = 0.0002–0.004$	$\alpha_{\rm MLT} = 1.0–2.0$
Firenze	Palla & Stahler (1999)	$M = 0.1–6$	pre-main sequence
	Palla & Stahler (1993)	$M = 1–6$	pre-main sequence
Genève	Meynet & Maeder (2002)	$M = 2–60$, $Z = 10^{-5}$	includes rotation
	Maeder & Meynet (2001)	$M = 9–60$, $Z = 0.004$	SMC-like; includes rotation
	Bernasconi (1996)	$M = 0.8–5$, $Z = 0.001–0.020$	pre-main sequence
	Charbonnel *et al.* (1996)	$M = 0.8–1.7$, $Z = 0.001–0.020$	ZAMS to early AGB
	Meynet *et al.* (1994)	$M = 12–120$, $Z = 0.001–0.040$	
	Schaerer *et al.* (1993a)	$M = 0.8–120$, $Z = 0.040$	
	Schaerer *et al.* (1993b)	$M = 0.8–120$, $Z = 0.008$	LMC-like
	Charbonnel *et al.* (1993)	$M = 0.8–120$, $Z = 0.004$	
	Schaller *et al.* (1992)	$M = 0.8–120$, $Z = 0.020, 0.001$	
Granada	Claret (2007)	$M = 0.8–125$, $Z = 0.04–0.10$	binary tidal evolution
	Claret (2006)	$M = 0.8–125$, $Z = 0.007–0.01$	binary tidal evolution, LMC-like
	Claret (2005)	$M = 0.8–125$, $Z = 0.002–0.004$	binary tidal evolution, SMC-like
	Claret (2004)	$M = 0.8–125$, $Z = 0.02$	binary tidal evolution
	Claret & Gimenez (1998)	Various	and references
Grenoble	Siess *et al.* (2000)	$M = 0.1–7$, $Z = 0.01–0.04$	pre-main sequence
	Siess *et al.* (1997)	$M = 0.4–5$, $Z = 0.005–0.04$	pre-main sequence/clusters
Lyon	Chabrier *et al.* (2000)	$M = 0.01–0.1$	low mass, including brown dwarfs
	Chabrier & Baraffe (1997)	$M = 0.07–0.8$	low mass
Montreal	Michaud *et al.* (2004)	$M = 0.5–1.4$	settling + radiative acceleration
Nice–Meudon	Morel (1997)	Various	CESAM code
Ohio	An *et al.* (2007)	$M = 0.2–8$, $-0.3 < [{\rm Fe/H}] < 0.20$	angular momentum evolution
	Sills *et al.* (2000)	$M = 0.1–0.5$ and $0.6–1.1$	YREC code
Padova	Marigo *et al.* (2003)	$M = 120–1000$, $Z = 0$	zero metallicity
	Marigo *et al.* (2001)	$M = 0.7–100$, $Z = 0$	zero metallicity
	Salasnich *et al.* (2000)	$M = 0.15–20$ various Y, Z	α-enhancement
	Girardi *et al.* (2000)	$M = 0.15–7$, $Z = 0.0004–0.03$	
	Girardi *et al.* (1996)		superseded
	Bertelli *et al.* (1994)	$M = 0.6–120$, $Z = 0.0004–0.05$	partly superseded
Potsdam	Herwig (2004)	$M = 4–5$, $Z = 0.0001$	AGB models, low Z
	Herwig (2000)		AGB models, overshooting
	Blöcker (1995b)	$M = 1–7$	post-AGB to white dwarfs
	Blöcker (1995a)	$M = 1–7$	
Roma	Ventura *et al.* (1998)	$M = 0.6–15$	full-spectrum turbulence
	Mazzitelli *et al.* (1995)	various	to horizontal branch
	Mazzitelli (1989)	$M = 0.7–1$	to helium flash
	D'Antona & Mazzitelli (1997)	$M = 0.02–1.5$	pre-main sequence
	D'Antona & Mazzitelli (1994)	$M = 0.015–2.5$	pre-main sequence
Teramo	Cordier *et al.* (2007)		inclusion of AGB
	Pietrinferni *et al.* (2006)	$M = 0.5–10$, $-2.60 < [{\rm Fe/H}] < 0.05$	α-enhancement
	Pietrinferni *et al.* (2004)	$M = 0.5–10$, $-2.27 < [{\rm Fe/H}] < 0.40$	
	Chieffi *et al.* (1998)	$M = 25$	to Fe core collapse
	Chieffi & Straniero (1989)	$-2.0 < [{\rm Fe/H}] < -1.0$	FRANEC code
Victoria–Regina	VandenBerg *et al.* (2006)	$M = 0.4–4.0$, $-2.31 < [{\rm Fe/H}] +0.49$	overshooting, α-enhancement
	VandenBerg *et al.* (2000)		superseded
	VandenBerg & Bell (1985)		superseded
Yale, Revised	Green & Demarque (1996)		superseded by Yonsei–Yale
Yonsei–Yale	Demarque *et al.* (2004)		improved core overshoot
	Yi *et al.* (2003)	$M = 0.4–5$, $Z = 0.00001–0.08$	supersedes Revised Yale
	Yi *et al.* (2001)	Solar	

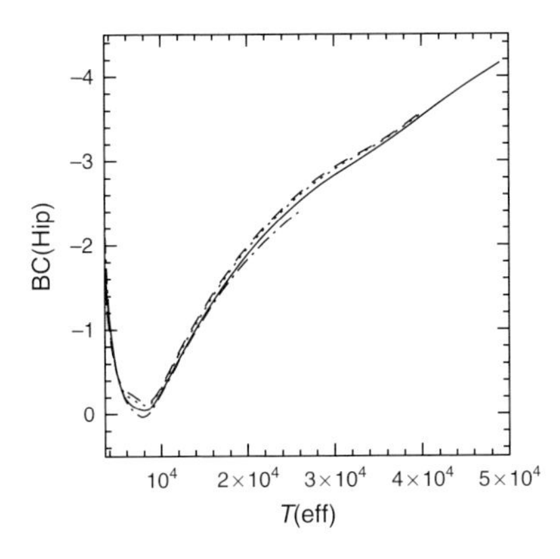

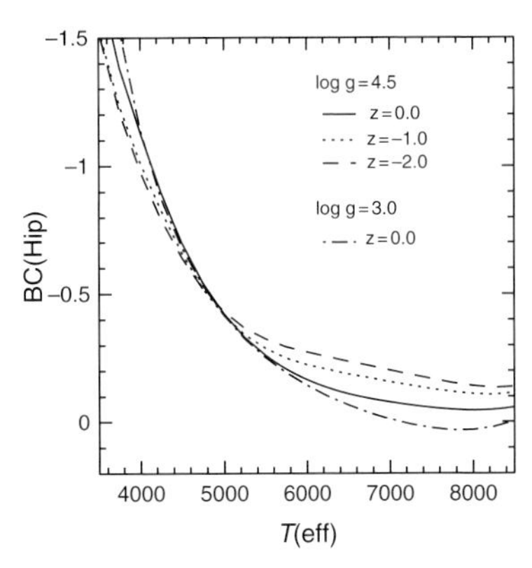

Figure 7.3 Bolometric corrections BC_{Hip} versus effective temperature from synthetic spectra. The determinations were made for this review by M.S. Bessell, and are available in tabular form on his www pages. Examples are given for a subset of the log *g*, *Z grid, and the right-hand plot is the lower T_{eff} region of the entire plot shown at left.*

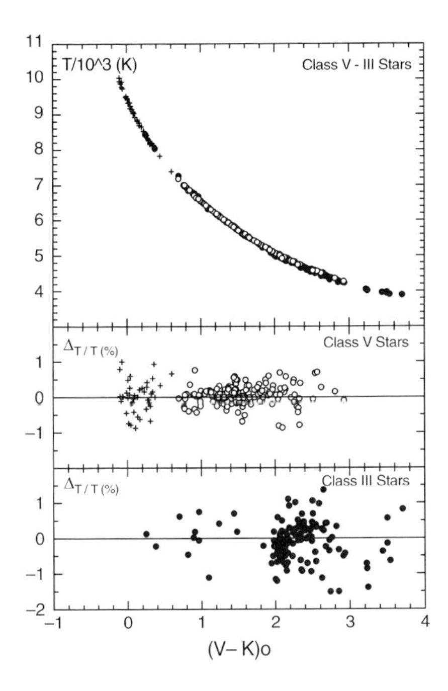

Figure 7.4 Top: individual temperatures for luminosity class III and V stars as a function of the intrinsic near-infrared colour. Middle and bottom: residuals from temperature scales represented by the best-fit second-order polynomials. Crosses indicate A-type stars. From Di Benedetto (1998, Figure 7).

index, an empirical method pioneered in the visual versus $B-V$ by Wesselink (1969) and showing a still tighter correlation in $V-R$ after the work of Barnes *et al.* (1978), who showed that the relation can be used to infer visual surface brightness for all luminosity classes and spectral types, including S and C types. The method was extended to the use of Hipparcos data by Di Benedetto (1998). They obtained a T_{eff} versus $V-K$ calibration for 327 out of 537 dwarfs and giants of types A–K selected for a flux calibration of ISO (see also Missoulis *et al.*, 1999). The work made use of high-precision K-magnitudes from ISO, bolometric fluxes from Blackwell & Lynas-Gray (1998), and Hipparcos V-magnitudes and parallaxes to correct for interstellar extinction. The visual surface brightness, defined by $S_V = V + 5\log\theta$, is first calibrated as a function of $(V-K)$ using stars with precise θ from interferometry. Then for a general star S_V is obtained from $(V-K)$, and θ is obtained from S_V and V, yielding in turn T_{eff} from F_{bol} and θ. From the resulting (T_{eff}, $V-K$) calibration, Di Benedetto (1998) derived T_{eff} values of 537 A–K dwarfs and giants with ±1% accuracy (Figure 7.4). The method produces results in good agreement with those of the infrared flux method, and is less dependent on atmosphere models. They argued that small systematic differences for A stars can be attributed to the LTE line-blanketed model atmospheres adopted by Blackwell & Lynas-Gray (1998). Otherwise, the very tight correlation for the F–K dwarfs reveals no dependency on metallicity, while gravity effects become detectable in the case of giants.

Popper (1998) used detached eclipsing binaries with rather good Hipparcos parallaxes, accurate radii, and measured V magnitude to calibrate the radiative flux as a function of $(B-V)$; he found good agreement with similar calibrations based on interferometric angular diameters. From the same data, along with appropriate bolometric corrections, Ribas *et al.* (1998) derived effective temperatures and found them to be in reasonable agreement (although systematically smaller by 2–3%) with T_{eff} derived from photometric calibrations. However the stars are rather distant, which implies rather significant internal errors on M_{bol} and T_{eff} (a parallax error of 10% is alone responsible for a T_{eff}-error of 5%). In the sample of Ribas *et al.* (1998), only a few systems

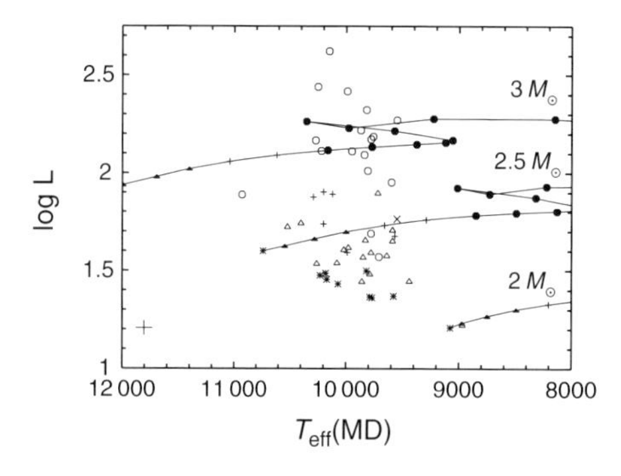

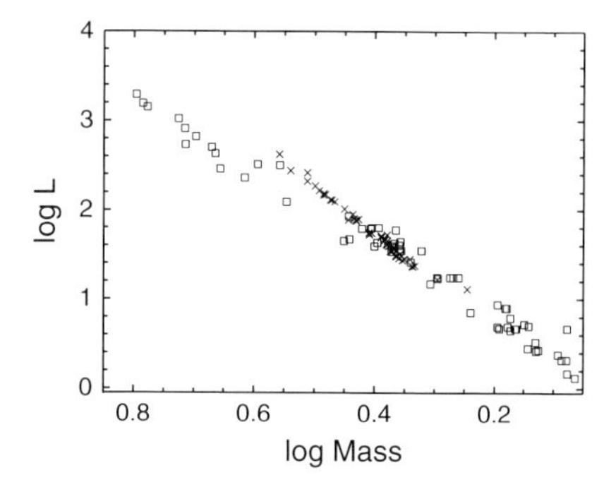

Figure 7.5 Left: HR diagram for the A0 stars from Gerbaldi et al. (1999), where the log *g ranges are indicated by:* ⋆*:* $\log g \geq 4.3$*;* △*: 4.05–4.3;* +*: 3.8–4.05;* □ ≤ *3.8. Filled triangles and squares indicate evolutionary tracks with the same intervals in* log *g, from Schaller et al. (1992).* × *indicates the position of the reference star Vega, with* $T_{\rm eff} = 9550$ *K and* $\log g = 3.95$ *according to the models of Castelli & Kurucz (1994). Right: mass versus luminosity for the stars of luminosity class V in the samples of:* ×*: Gerbaldi et al. (1999) and* □*: Andersen (1991). From Gerbaldi et al. (1999, Figures 9 and 11).*

have $\sigma_\pi/\pi < 10\%$, and because errors on radius, magnitudes, and BC also contribute, only five systems have $T_{\rm eff}$ determined to better than 3%.

The luminosity function and other properties of early-type stars allow the recent history of our solar neighbourhood to be probed back to about 1 Gyr. Gerbaldi *et al.* (1999), updated from Gerbaldi *et al.* (1998), established revised $T_{\rm eff}$, $\log g$, M, and L for 71 nearby A0 dwarf stars within about 150 pc, selecting them as representative of young objects sufficiently long-lived to be present in the solar neighbourhood in reasonable numbers. Spectroscopic observations were used to verify the accuracy of photometrically-derived atmospheric parameters $T_{\rm eff}$ and $\log g$ based on dereddened Strömgren and Geneva photometry. After rejecting known and newly-detected double stars based on their spectroscopic signatures, masses, ages, and the HR diagram were contructed for 50 stars thus considered as suitable A0 reference stars. The resulting HR diagram subdivided in ranges of $\log g$ is shown in Figure 7.5, left: $T_{\rm eff}$ was taken from calibrated Strömgren photometry, L from Hipparcos parallax-derived M_V using $M_{\rm bol\odot} = 4.75$ and bolometric corrections from Bessell *et al.* (1998). Evolutionary tracks for $Z = 0.02$ and $2.0, 2.5, 3.0 M_\odot$ were taken from Schaller *et al.* (1992). It is evident that there is a large spread in luminosity for these A0 dwarfs, and that Vega is not particularly representative of the larger sample either in L or $T_{\rm eff}$. Masses were derived based on their L and $T_{\rm eff}$ values, interpolated into the evolutionary tracks of both Schaller *et al.* (1992) and Morel (1997). The resulting mass–luminosity relation, based on (non-eclipsing) stars and using Hipparcos luminosities and evolutionary models, is shown in Figure 7.5, right. The relationship is tight, but M and L are not independent quantities, both being derived through M_V. Shown in the same figure are the $M-L$ values derived independently from eclipsing binaries by Andersen (1991), where the gap seen at around $\log M \sim 0.5$ in the latter data alone is now filled by the A stars. Ages, from the evolutionary tracks, and effects of rotation, from spectroscopy, were also discussed.

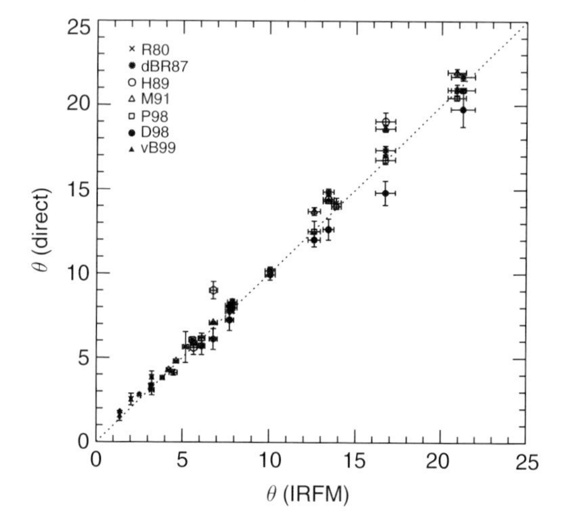

Figure 7.6 Comparison between angular radii derived by means of the infrared flux method and those directly measured by lunar occulation or Michelson interferometry. Symbols indicate the following references: R80 (Ridgway et al., 1980); dBR87 (Di Benedetto & Rabbia, 1987); H89 (Hutter et al., 1989); D98 (Dyck et al., 1998); P98 (Perrin et al., 1998); vB99 (van Belle et al., 1999). From Alonso et al. (2000, Figure 1).

Alonso *et al.* (1999) applied the infrared flux method to a sample of about 500 F0–K5 giant stars to derive their effective temperatures with an internal mean accuracy of about 1.5% and a maximum uncertainty in the zero point of the order of 0.9%, using the grid of theoretical model atmosphere fluxes distributions developed by Kurucz (1993a). Alonso *et al.* (2000) compared the angular diameters estimated by this method with those directly measured by lunar occultation or Michelson interferometry, obtaining a fairly good consistency. Although these results are based only loosely on Hipparcos data, via calibration of the infrared flux method, they illustrate the confidence that can be placed in these angular diameter estimates (Figure 7.6). Of more specific relevance to Hipparcos is the next step of their analysis, in which they used the Hipparcos distances to transform the angular diameters into linear radii: averaging individual values for about 300 stars over bins of 200 K, they derived mean linear radii of giants of solar metallicity, and tentatively extended the relation to metal-poor giants (Figure 7.7, left). They go on to discuss an application of this result to the estimation of distances of globular clusters: by comparison of the empirical $T_{\rm eff}$ versus R relation for stars on the red giant branch with the corresponding relation from theoretical isochrones. Although depending on prior knowledge of age, metallicity, helium abundance, reddening, and aspects of atmospheric convection, they argue that the method has an advantage in that fitting in the $T_{\rm eff} - R$ plane compared with fitting in the $T_{\rm eff} - M_{\rm bol}$ plane yields an error in distance in principle decreased by a factor 2. From their chosen theoretical models, they estimate that the most critical parameter to derive distances is the mixing-length, implying that the technique provides a semi-empirical test of convection efficiency in the adopted stellar models, according to how well a particular stellar model with assumed mixing length reproduces globular cluster distances obtained via the main sequence and horizontal branch fitting methods, themselves independent of convection. In the fit for the case of 47 Tuc, for example, the chosen 9 Gyr isochrone from the models of Salaris & Weiss (1998) yields a distance of 5.2 kpc (Figure 7.7, centre). Distance estimates for this and 11 other globular clusters show a generally very good agreement (Figure 7.7, right).

Ramírez & Meléndez (2005a) extended the infrared flux method of Alonso *et al.* (1999) to determine $T_{\rm eff}$ for 580 dwarfs and 470 giants covering the range $T_{\rm eff}$ = 3600–8000 K and $-4.0 <$ [Fe/H] $< +0.5$. Comparisons with direct temperature estimates from angular diameters and bolometric flux measurements, as well as those derived from Balmer line profile fitting and the surface brightness technique, show good agreement. Ramírez & Meléndez (2005b) used these results to provide $T_{\rm eff}$ versus colour relations for dwarfs and giants in 17 photometric systems, including the *UBV*, Strömgren, Vilnius, and Geneva systems, as well as the Tycho B_T and V_T bands. Polynomial fits were performed to second order in colour index and second order in [Fe/H] in all cases. Resulting $\sigma(T_{\rm eff})$ for the Tycho $(B_T - V_T)$ colour index are of modest accuracy, being 104 K based on 378 calibrating stars for dwarfs and 82 K based on 261 calibrating stars for giants, presumably due to its sensitivity to metallicity. Nevertheless, the size of the Tycho Catalogue makes the resulting $T_{\rm eff}$ estimates particularly valuable, with example applications being the search for solar analogues, or studies of the chemical evolution of the Galaxy.

Empirically constrained colour–temperature relations were constructed by VandenBerg & Clem (2003) for $BV(RI)_C$, and for *uvby* by Clem *et al.* (2004). They provided a set of transformations to Johnson $B-V$, Cousins $V-R$, and Cousins $V-I$, as well as bolometric corrections to V, for [Fe/H] = −3 to +0.3 and, in each case, values of $\log g$ from -0.5 to 5.0 for $T_{\rm eff}$ = 3000–5500 K and from 2.0 to 5.0 for $T_{\rm eff}$ = 6000–40 000 K. The transformations employed the predictions from Kurucz model atmospheres at high temperatures, $T_{\rm eff} > 8000$ K, and from MARCS model atmospheres at intermediate temperatures below 7000 K. Thus, theoretical colour–$T_{\rm eff}$ relations were used exclusively down to a minimum temperature cooler than the temperatures of turn-off stars in open and globular star clusters. For colour transformations for cooler stars down to 3000 K, corrections to the synthetic transformations were determined from observations of a few globular clusters (M92, M68, and 47 Tuc), the colour–magnitude diagrams of several open clusters (M67, the Pleiades, the Hyades, and NGC 6791), the colour–magnitude diagrams and mass–luminosity diagram for solar neighbourhood stars having good distance measurements from Hipparcos, empirical $(B-V)$ and $(V-K)$ versus $T_{\rm eff}$, and colour–colour diagrams for field giants. Application to the high precision Hipparcos secular parallaxes for the Hyades is shown in Figure 7.8.

7.4.3 Surface gravities

The method given by Equation 7.5 has been applied to metal-poor subdwarfs and subgiants with accurate distances from Hipparcos (Nissen *et al.*, 1997; Fuhrmann, 1998b; Clementini *et al.*, 1999). Nissen *et al.* (1997) showed that among the various sources of errors entering this estimate, the error on distance still dominates, but that if the distance error is lower than 20% then the error on $\log g$ may be lower than ±0.20 dex. He showed that the spectroscopic gravities are often in error by a factor 2–3, and stressed the importance of parallax-based gravities for settling the metallicity scale of metal-poor stars and for deriving accurate abundances of key elements required for nucleosynthesis studies (Figure 7.9).

Gravity effects in very metal-poor stars were also studied by Fuhrmann (1998b) who used the line profile of the Mg Ib line at 518.36 nm to determine the surface gravity and hence evolutionary stage of the halo

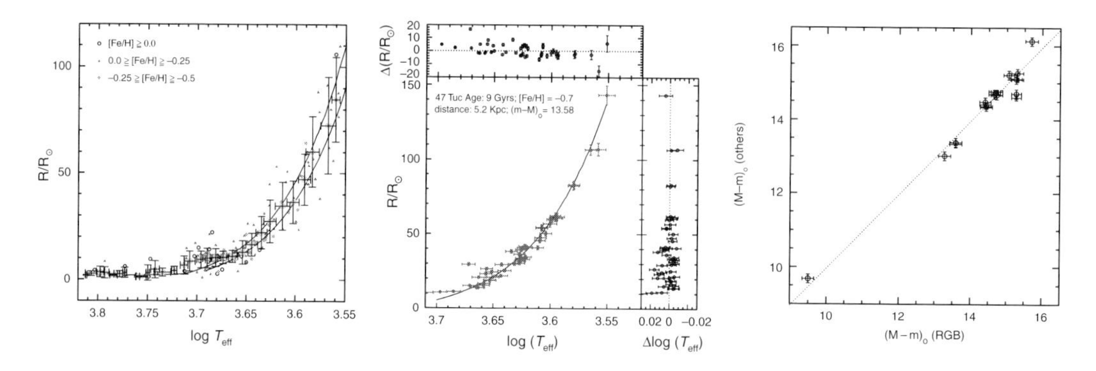

Figure 7.7 Left: squares show the mean stellar radii for [Fe/H] ≥ −0.5. Small symbols correspond to the individual stars considered in the averages, separated in metallicity groups. Theoretical isochrones, from Salaris & Weiss (1998), of the red giant branch with [Fe/H] = 0 and t = 3.5 Gyr, and of the red giant branch and subgiant branch with [Fe/H] = −0.35 and t = 4.5 Gyr are superimposed. Middle: fit of giants in 47 Tuc to the isochrone. Top and right panels show the residuals of the fit in both axes. Right: comparison between distance moduli derived by the $T_{\rm eff}$ versus R fitting to stars on the red giant branch (indicated as RGB-fitting) and those obtained by other methods: main sequence fitting (△), and horizontal-branch fitting (○). From Alonso et al. (2000, Figures 2, 6 and 9).

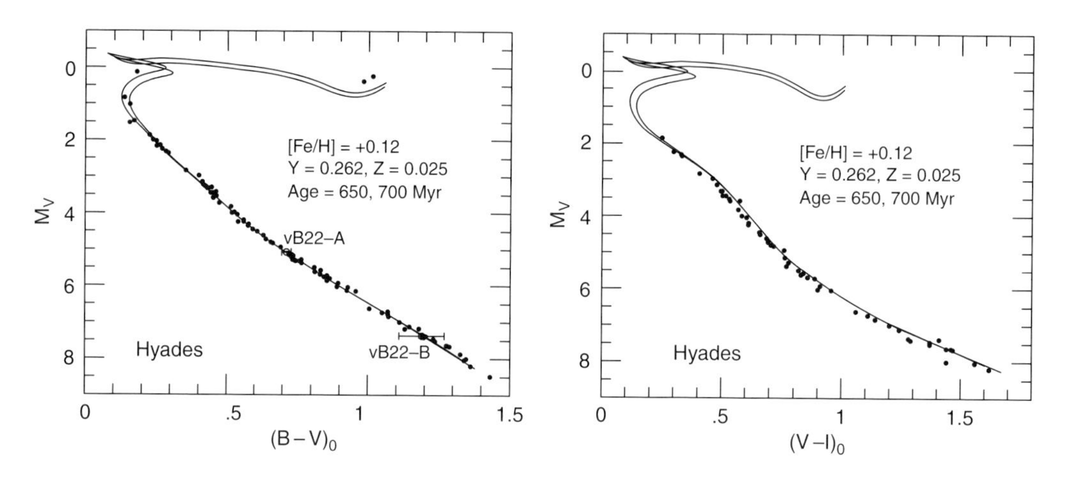

Figure 7.8 Overlay of isochrones for the indicated parameters onto the de Bruijne et al. (2001) colour–magnitude diagram for the high-fidelity sample of single Hyades members. The absolute visual magnitudes of the individual stars were derived by de Bruijne et al. (2001) from Hipparcos secular parallaxes, which have a much higher precision (by a factor of ∼3) than the trigonometric parallaxes. Left: for B − V; the positions of the components of vB 22 have also been plotted (○). Right: for the V − R colours available in the literature. From VandenBerg & Clem (2003, Figures 22 and 24).

dwarfs HD 140283 and G 84–29. The surface gravity of HD 140283 has been a long-standing enigma since a first attempted classification by Adams & Joy (1922). Large discrepancies in ground-based trigonometric parallaxes had earlier led to a wide range of estimated surface gravities extending between $\log g = 3.1$–4.8, with spectroscopic estimates ranging between $\log g = 3.1$–3.52, where the relation between g and distance was via the so-called spectroscopic parallax (Equation 7.6). The Hipparcos parallax of $\pi = 17.44 \pm 0.97$ mas ($d = 57.34^{+3.38}_{-3.01}$ pc) was used to establish $\log g = 3.69$, and revised values $M_{\rm bol} = 3.16 \pm 0.14$, $T_{\rm eff} = 5810 \pm 80$ K, $R = 2.04 \pm 0.15 R_\odot$, and [Fe/H] = −2.29. The subgiant stage is confirmed, and unidentified NLTE effects are presumably responsible for the discrepancy between the Hipparcos and spectroscopic parallaxes, even when derived from pressure-dependent line wings.

Allende Prieto *et al.* (1999) presented a consistency test of spectroscopic gravities for late-type stars, aiming to extend and verify the results of Nissen *et al.* (1997). Also based on Equation 7.5, their study was extended to many more stars, with the goal of determining if the differences between trigonometric and spectroscopic gravities depend primarily on gravity, metal content,

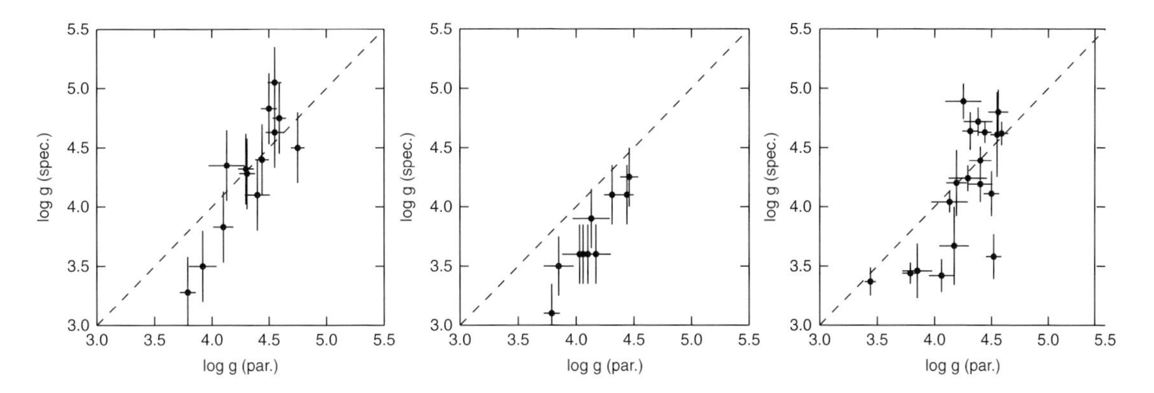

Figure 7.9 Comparison between the gravities based on Hipparcos parallaxes for a sample of metal-poor stars with various spectroscopic gravities as derived by (left to right): Tomkin et al. (1992), Magain (1989), and Axer et al. (1995). These comparisons have shown that pre-Hipparcos spectroscopic gravities were often in error by a factor of 2–3. From Nissen et al. (1997, Figures 3–5).

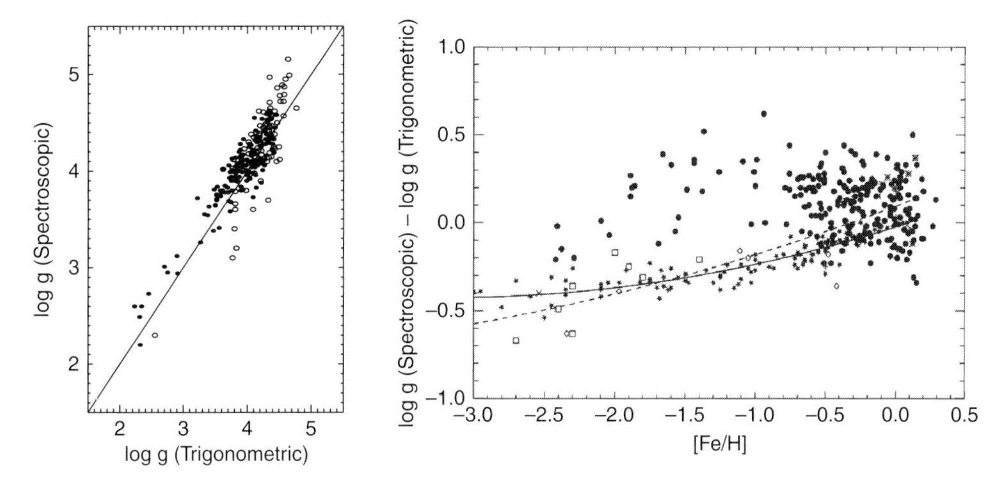

Figure 7.10 Left: comparison of spectroscopic and trigonometric gravities. •: metal-rich stars, [Fe/H] > −0.47; ∘: metal-poor stars. Solid line corresponds to the case log g (spectroscopic) = log g (trigonometric). Right: differences between spectroscopic and trigonometric gravities versus iron abundance. Different symbols identify the various sources for the spectroscopic data. The stars show the NLTE corrections to the ionisation equilibrium gravities calculated by Thévenin & Idiart (1999). Solid and dashed lines correspond to least-squares polynomial fits to the corrections of Thévenin & Idiart (1999) and the differences log g (spectroscopic) – log g (trigonometric), respectively. From Allende Prieto et al. (1999, Figures 3 and 10).

temperature, or a combination of all three parameters. They used the Hipparcos parallaxes to derive the surface gravity for a number of nearby stars, which are then used to verify gravities obtained from the photospheric iron ionisation balance (Figure 7.10). They found an approximate agreement for stars in the metallicity range $-1.0 \leq$ [Fe/H] ≤ 0, but with differences between the spectroscopic and trigonometric gravities decreasing toward lower metallicities for more metal-deficient dwarfs, $-2.5 \leq$ [Fe/H] -1.0. The results raise further doubts on the abundance analyses for extreme metal-poor stars that make use of the ionisation equilibrium to constrain the gravity.

7.4.4 Stellar radii

The relevance of stellar angular diameters for the determination of effective temperatures and related fundamental stellar data has already been described (Section 7.2.2). This section focuses on the wider application of Hipparcos data to the determination and interpretation of stellar radii.

Background There are two main problems associated with the determination of angular diameters of even the most nearby stars. The first is their very small angular sizes: numerically, $\theta = (\pi/107.5)R/R_\odot$, where the angular diameter θ and the parallax π are in the same

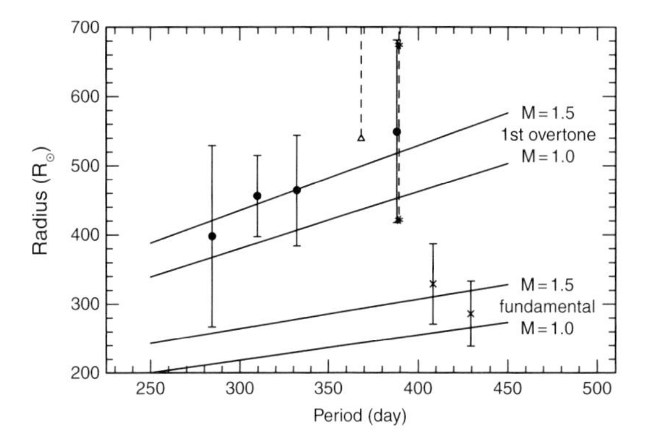

Figure 7.11 The period–radius relation for eight Mira variables, with 1σ error bars. •: R Aql, R Leo, o Cet, T Cep; ×: χ Cyg, R Cas; △: 1σ lower limit for U Ori; ⋆: 1σ and 2σ lower limits for R Hya. Lines are the predicted relations for fundamental and first overtone pulsations. From Whitelock et al. (1997, Figure 1).

units (e.g. mas). Thus even with stellar diameters reaching $500R_\odot$ or more, angular diameters range from only about 1 mas for a $1R_\odot$ main-sequence star at only 10 pc, to about 10 mas for a $100R_\odot$ giant at 100 pc. The second relates to the practical definition of the radius of a star. Since stars are gaseous, their continuum intensity, $I_\lambda(\theta)$, varies from the centre, $\theta = 0$, to the limb. For solar-type stars, Cox (2000, Section 14.7) gives $I_\lambda(\theta)/I_\lambda(0) = 1 - u_2 - v_2 + u_2 \cos\theta + v_2 \cos^2\theta$ with average values over wavelength of $\langle u_2 \rangle = +0.84$, $\langle v_2 \rangle = -0.20$. The problem is not totally facilitated by proximity, as illustrated by the associated challenges in defining the radius of the Sun (e.g. Chollet & Sinceac, 1999; Noël, 2002; Reis Neto *et al.*, 2003; Emilio & Leister, 2005).

Direct methods for determining angular diameters include interferometry, lunar occultation and, in principle, adaptive optics on large telescopes to achieve diffraction-limited resolution. Early work on direct measurements included the use of Michelson interferometry (Pease, 1931; Currie *et al.*, 1974), intensity interferometry (Hanbury Brown, 1974; Hanbury Brown *et al.*, 1974), speckle interferometry (Gezari *et al.*, 1972; Bonneau & Labeyrie, 1973; Worden, 1976) and lunar occultations (Nather & Evans, 1970).

The Stefan–Boltzmann relation provides a basic but not very accurate photometric method for determining angular diameters. Gray (1967) based his estimates on the expression for the stellar radius $R = d\sqrt{F_{E,\lambda}/F_{S,\lambda}}$ where d is the distance, $F_{S,\lambda}$ is the monochromatic flux from the star, and $E_{S,\lambda}$ is the monochromatic flux received on Earth. This forms the basis of the infrared flux method discussed in Section 7.2.2. Wesselink *et al.* (1972) provided an early catalogue of 2392 stellar radii determined photometrically and linked to 16 interferometric methods. The Baade–Wesselink method can be employed for pulsating stars such as Cepheids and RR Lyrae variables, and relies on a combination of kinematic and thermodynamic relations making use of the surface brightness, apparent magnitude, and radial velocity as a function of pulsation phase (Gautschy, 1987).

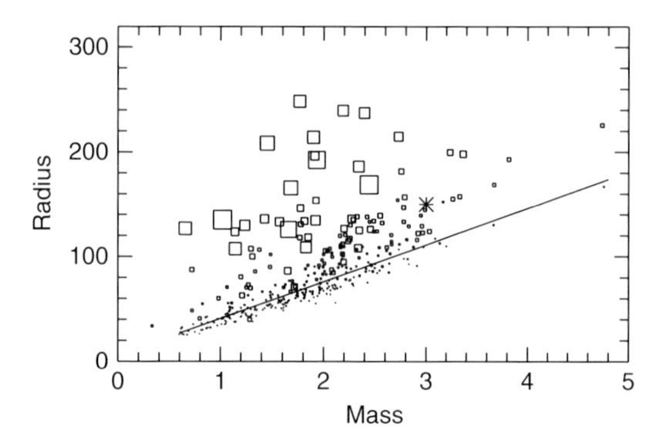

Figure 7.12 Mass–radius relation for M giants from the Hipparcos Catalogue. Masses have been derived from evolutionary tracks based on the luminosity, while linear radii have been derived from angular diameters based on the distance. Symbol size is proportional to the variability amplitude. The full line represents the relation $R/R_\odot = 6 + 35M/M_\odot$ for objects with small variability. ⋆ indicates the red giants in the symbiotic binary system BX Mon. From Dumm & Schild (1998, Figure 1), reproduced with the permission of Elsevier.

Detached eclipsing binaries offer one of the most precise methods for determining stellar diameters. If all four contacts of the two primary and two secondary eclipses are well observed, then the two stellar diameters can be determined relative to the orbital radius from the eclipse durations. Linear radii follow if radial velocities can also be measured for the two stars, i.e. in the case of double-lined detached eclipsing binaries. Roughly 45 systems are known, and they provide accuracies at the level of about 1% (Andersen, 1991). The orbital parallax based distances of these objects have been shown to be in good agreement with the more recent Hipparcos distances out to at least 100 pc (Figure 3.22), and as a result the Hipparcos results have had essentially no impact on radii already determined for these objects.

Hipparcos investigations Applications of the Hipparcos data essentially involve using the accurate trigonometric parallaxes to determine linear radii from angular diameters measured independently. This possibility has recently been dramatically enhanced by the availability of high-precision angular diameters from ground-based optical interferometry, including PTI, NPOI, IOTA, and VLTI (see Section 3.8.6).

Amongst the earliest applications of the Hipparcos data was an evaluation of a new method to derive angular diameters and brightness distributions (limb darkening) using the modulated photometric signal from the

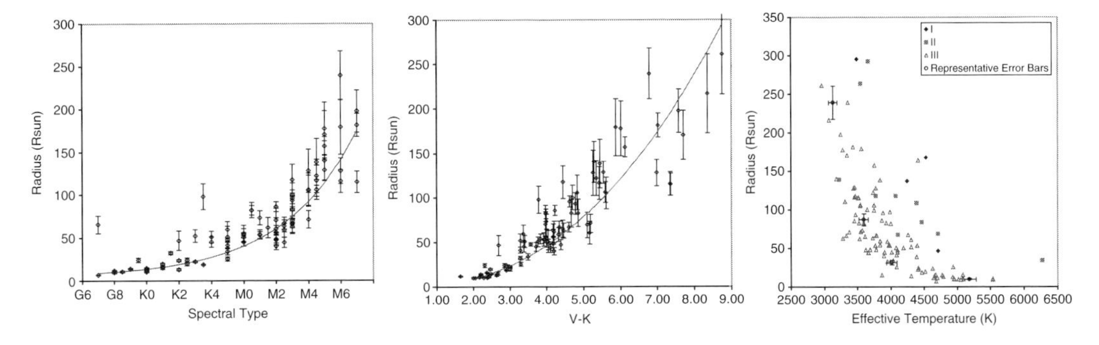

Figure 7.13 Left: radius as a function of spectral type for the sample of luminosity class I, II, and III GKM giants studied by van Belle et al. (1999). Middle: radius as a function of V − K colour for luminosity class III stars from the same sample. Right: radius as a function of effective temperature classified according to luminosity class. Representative error bars are 4% for temperature and 18% for radius. Luminosity class I and II objects are on average ~3 times the radius of class III giants for a given temperature. From van Belle et al. (1999, Figures 1, 2 and 4).

Hipparcos data itself (Hestroffer & Mignard, 1997a,b). This is based on the principle that the modulated signal of an extended object observed through a periodic grid differs from that of a point source, the same principle used for the detection of some of the Hipparcos binary systems. In practice, with a grid period of $s = 1.2074$ arcsec, only a few Solar System bodies are resolved by Hipparcos: for Ceres, whose angular diameter varies between 0.35 and 0.7 arcsec, they found that the surface optical properties depart only slightly from a uniformly bright disk. Results for the Saturnian satellite S6 Titan were also reported (see Section 3.8.6).

Whitelock *et al.* (1997) studied the period–luminosity relation for Mira variables using Hipparcos parallaxes, and also used the parallaxes to convert eight angular diameters measured interferometrically by Haniff *et al.* (1995) into linear radii. Angular diameters in the range 17–37 mas convert into photospheric radii in excess of 200–300$R_\odot$, with the dependency of R versus period (Figure 7.11) interpreted as showing that most Miras with $P < 400$ d are pulsating in an overtone, with at least some of the longer period objects pulsating in the fundamental.

A series of studies have examined linear radii and associated implications for samples of K and M giants and supergiants. Dumm & Schild (1998) started with angular diameters estimated from the relation between visual surface brightness and Cousins $V - I$, calibrated using M giants with published angular diameters. Linear radii were constructed for those with Hipparcos parallaxes better than 20%. Their results indicate that the radii of (non-Mira) M giants increase from a median value of 50$R_\odot$ at M0 III to 170$R_\odot$ at M7/8 III. They determined luminosities and, from evolutionary tracks, stellar masses, finding that M giants in the solar neighbourhood have masses in the range 0.8–$4M_\odot$. They found a close relation between radius and mass for a given spectral type, a linear relation between mass and radius for non-variable M giants given by $R/R_\odot = 6 + 35M/M_\odot$, and larger stellar radii with increasing variability amplitude for a given mass (Figure 7.12).

Dyck *et al.* (1998) reported new IOTA interferometer observations of 74 K–M giants and supergiants which, with previous observations, yielded 70 objects with estimates of $T_{\rm eff}$. These were derived from bolometric fluxes estimated from broadband photometry and Rosseland mean diameters transformed from the uniform disk diameters resulting from the interferometric visibilities (see the discussion in van Belle *et al.* 1999 for the derivation and relevance of the Rosseland diameters). For the 64 with Hipparcos parallaxes better than 30% they derived stellar radii, and showed that the objects with luminosity class II and II–III are systematically larger than luminosity class III stars at a given $T_{\rm eff}$. The sample was extended to 113 GKM giants based on angular diameters of 69 giants and supergiants obtained at the PTI by van Belle & Thompson (1998) and van Belle *et al.* (1999), with some overlap and good consistency between the results reported by Dumm & Schild (1998). Typical values are in the ranges $\theta \sim 1.5$–3 ± 0.05 mas, $T_{\rm eff} = 2500$–5500 K, $d = 50$–200 pc, and $R = 10$–$200R_\odot$. From the linear radii obtained from the Hipparcos parallaxes, and relationships between spectral type and $V - K$ colour index, they obtained

$$R = 1.76 \times (V - K)^{2.36}\, R_\odot \qquad (7.7)$$

Radius versus $T_{\rm eff}$ for luminosity classes I, II, III objects shows a tendency for the higher luminosity objects to have a greater radius at a given $T_{\rm eff}$, with classes I and II objects being some three times larger than class III giants at a given $T_{\rm eff}$ (Figure 7.13).

Jerzykiewicz & Molenda-Zakowicz (2000) re-analysed the sample of 32 O5–F8 stars with angular sizes from the Narrabri intensity interferometer determinations of

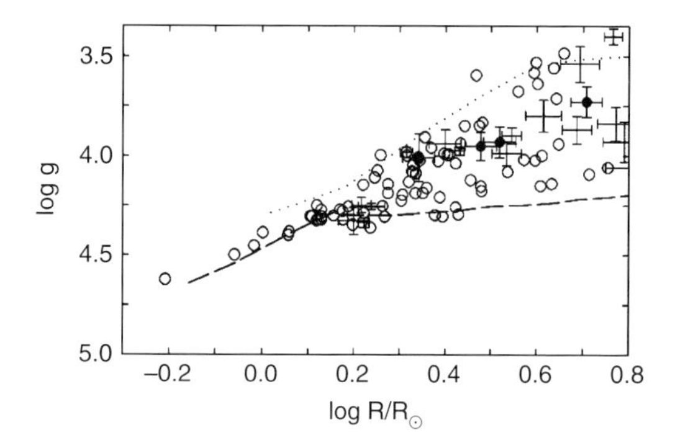

Figure 7.14 Values of log *g versus R for 25 O5–F8 stars with radii determined from stellar angular diameters and Hipparcos distances and with masses from evolutionary tracks (+), along with the same quantities for double-lined eclipsing binaries, from Andersen (1991) (○) and other sources (•). Dashed and dotted lines show ZAMS and TAMS respectively from the evolutionary tracks of Schaller et al. (1992). From Jerzykiewicz & Molenda-Zakowicz (2000, Figure 4).*

Hanbury Brown *et al.* (1974) for which total absolute fluxes were derived by Code *et al.* (1976) from OAO-2 ultraviolet spectrophotometry and ground-based visual and infrared observations. For 25 of these, Hipparcos parallaxes yield luminosities, corrected for Lutz–Kelker bias, accurate to 0.10 dex. From their position in the L versus $T_{\rm eff}$ HR diagram, they derived evolutionary masses based on the $Y = 0.30, Z = 0.02$ evolutionary tracks of Schaller *et al.* (1992). These masses and corresponding radii were used to derive log g for each star. Figure 7.14 shows these values of log g, along with those obtained directly from double-lined eclipsing binaries, mainly from Andersen (1991). That they all fall within or close to the regions of main-sequence hydrogen burning, combined with the fact that the 25 log g values from the Hipparcos distances are rather model independent, makes them good candidates for more detailed testing of stellar evolutionary models.

Various other studies have determined linear radii, and corresponding evolutionary status, for individual objects with known angular diameters using Hipparcos distances. These include lunar occultation observations of a sample of 15 M giants observed at 2.2 μm yielding two new Hipparcos-based radii (Tej & Chandrasekhar, 2000); NPOI diameter determinations of three late-type giants (Wittkowski *et al.*, 2001); IOTA observations of the Mira star T Cep (Weigelt *et al.*, 2003); VLTI–VINCI measurements of the M4 giant ψ Phe (Wittkowski *et al.*, 2004), the post-AGB binary HR 4049 (Antoniucci *et al.*, 2005), the M0 giant γ Sge (Wittkowski *et al.*, 2006b), and the M1.5 giant Menkar (Wittkowski *et al.*, 2006a); CHARA observations of the oblateness, rotational velocity, and gravity darkening of Alderamin (HD 203280) (van Belle *et al.*, 2006); and VLTI–MIDI observations of the Be star α Ara (Chesneau *et al.*, 2005).

For Cepheids, angular diameters have in the past been estimated using the Baade–Wesselink method. Interferometric determinations for Cepheids were first reported for δ Cep using the GI2T interferometer (Mourard *et al.*, 1997); and subsequently for α UMi, ζ Gem, δ Cep and η Aql using the NPOI (Nordgren *et al.*, 2000; Armstrong *et al.*, 2001); ζ Gem also using PTI (Lane *et al.*, 2000) and IOTA/FLUOR (Kervella *et al.*, 1999); and others more recently with the VLTI (Kervella *et al.*, 2004a; Kervella, 2006). Armstrong *et al.* (2001) used their angular diameters for δ Cep ($\theta = 1.520 \pm 0.014$ mas) and η Aql ($\theta = 1.69 \pm 0.04$ mas) to estimate distances from the period–radius relation, which they found to be slightly smaller than, but still consistent with, the Hipparcos distances.

VLTI–VINCI observations have yielded the angular diameters of the nearby dwarfs Sirius A (Kervella *et al.*, 2003a) and Procyon A (Kervella *et al.*, 2004b). Sirius A (α CMa, A1V) is the fifth nearest binary system, comprises a white dwarf secondary, and is the first main-sequence star whose angular diameter was measured interferometrically (Hanbury Brown *et al.*, 1967). Kervella *et al.* (2003a) determined a limb-darkened angular diameter of $\theta_{\rm LD} = 6.039 \pm 0.019$ mas and a resulting linear radius, based on the Hipparcos parallax, of $R = 1.711 \pm 0.013 R_\odot$. The results of their stellar structure and evolutionary models, based on the CESAM code of Morel (1997), provides detailed insight into its evolutionary status (Figure 7.15, left): they propose that the apparently high surface metal content of Sirius is not characteristic of the mean value of Z for the star, and is caused instead by levitation of heavy elements in the thin upper convective layer. They derive an age of 200–250 Myr, consistent with the evolutionary age estimates of Sirius B. The oldest age of Sirius A corresponds to a mass of $2.07 M_\odot$, which is the lowest mass value consistent with its error bar. Procyon A (α CMi, F5IV–V) also has a white dwarf companion, with an orbital period of 40 years. It is a prime asteroseismology target, a technique which provides strong constraints on stellar interior models. Kervella *et al.* (2004b) determined $\theta_{\rm LD} = 5.448 \pm 0.053$ mas which, combined with the Hipparcos parallax, yields $R = 2.048 \pm 0.025 R_\odot$. Their preferred model suggests that the star is close to the end of core hydrogen burning with the measured radius providing a tight constraint on the age of 2314 ± 10 Myr (Figure 7.15, right). Subtracting the cooling age of the white dwarf leads to a lifetime of 614 Myr for the white dwarf progenitor. This in turn leads to an original progenitor mass of $2.5 M_\odot$, necessary to result in a mass of $\sim 0.57 M_\odot$ for the core of the corresponding thermal-pulsating AGB star on the appropriate time scale.

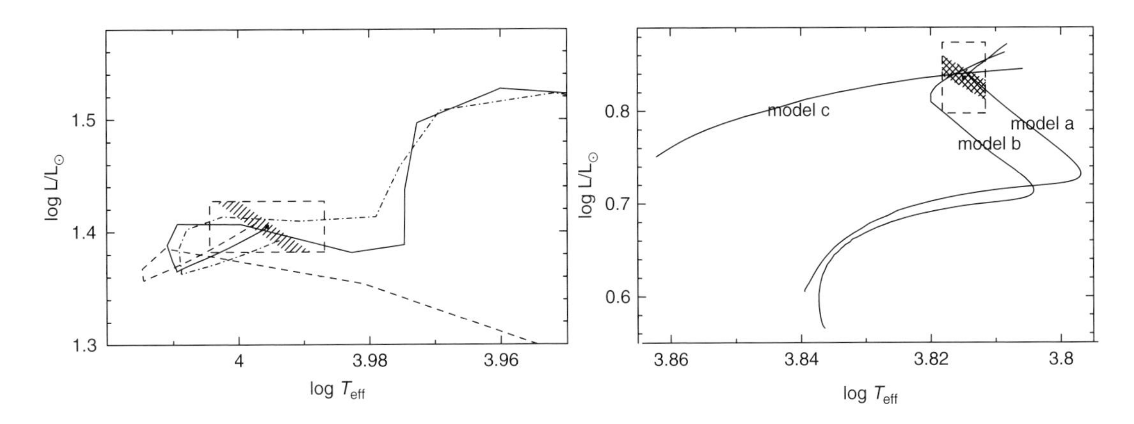

Figure 7.15 Evolutionary tracks in the HR diagram of models of Sirius A (left) from Kervella et al. (2003a, Figure 4); and for Procyon A (right) from Kervella et al. (2004b, Figure 4). In both cases, the dashed rectangle indicates the uncertainty domain in luminosity and effective temperature, while the shaded area represents the uncertainty on the interferometric radius. Left (Sirius A): the continuous line corresponds to their model 'b' with overshoot, and the dashed-dot line to their model 'a' without overshoot, both with a mass of $2.12M_\odot$. Their model 'd', corresponding to $M = 2.07M_\odot$ and an age of 243 Myr, is represented by the dashed line. Right (Procyon A): their preferred evolutionary model 'a' is compared with a model without microscopic diffusion 'b', and one without overshooting 'c', all of which affect the assumed age.

Combined analyses Soubiran *et al.* (1998) provided an echelle spectral library of 211 stars covering the full range of gravities and metallicities for F5–K7 stars ($T_{\rm eff} = 4000-6300$ K), and with associated $T_{\rm eff}$, $\log g$, and [Fe/H] values from the literature.

Allende Prieto & Lambert (1999) used the Hipparcos observations to derive estimates of $T_{\rm eff}$, M, and R for large numbers of stars in a novel analysis of the HR diagram. With mass being the key parameter determining a star's evolution, chemical composition and other factors playing a secondary role, evolutionary tracks for a range of trial masses lead to some regions being completely empty, forbidden combinations of M_V and $B-V$ where no stars are supposed to exist, while other regions are more crowded, some being crossed at different stages by stars of different masses. They used the evolutionary tracks of Bertelli *et al.* (1994) to establish 'uncertainty' regions in the HR diagrams for mass, radius, and effective temperature (Figure 7.16, left). The diagrams are then used to retrieve uncertainties in the parameters of a star lying at that location. Uncertainties in the masses are especially significant close to the position of the horizontal branch, where intermediate mass stars cross back from the giant stage and more massive stars make their way up from the main sequence, and are worst for giants and asymptotic giant branch stars. There is also significant confusion in the area where the post-AGB stars cross the upper main sequence down to the white dwarf cooling sequence, although in practice very few stars will appear in such a rapidly evolving phase. Errors in the retrieved radii and effective temperatures are small for most cases, although for the reddest evolved stars the confusion is very large. Thus R and $T_{\rm eff}$ are largely constrained by the position of a star in the colour–magnitude diagram, regardless of the existence of a wide range of possibilities for ages and metallicities. The uncertainties in R and M are quantified by comparison with the direct determinations from the eclipsing binary star sample of Andersen (1991), with $T_{\rm eff}$ from other sources (Figure 7.16, right), as well as from the infrared flux method. Equipped with estimates of these uncertainties, and discarding regions of the colour–magnitude diagram where the predictions are too vague, they applied the method to 17 219 Hipparcos stars within 100 pc, for which the results of the stellar evolutionary calculations can be used with only photometry and trigonometric parallax, and assuming roughly solar metallicity, within the limits: $R = 0.87–21R_\odot$, $M = 0.88–22.9M_\odot$, $T_{\rm eff} = 3961–33\,884$ K, and $\log g = 2.52–4.47$. Their derived luminosity function and mass function for the sample are shown in Figure 7.17.

Lyubimkov *et al.* (2002) derived $T_{\rm eff}$, $\log g$, and interstellar extinction A_V for a sample of 107 B stars from photometric indices, using the results to estimate distances which were validated by comparison with the Hipparcos parallaxes.

Takeda *et al.* (2002a) described a spectroscopic method to determine $T_{\rm eff}$, $\log g$, [Fe/H], and v_t (the micro-turbulent velocity dispersion) by finding a solution which minimises the sum of the dispersion of the Fe I abundances and the square of the Fe I–Fe II abundance difference. Takeda *et al.* (2002b) applied the

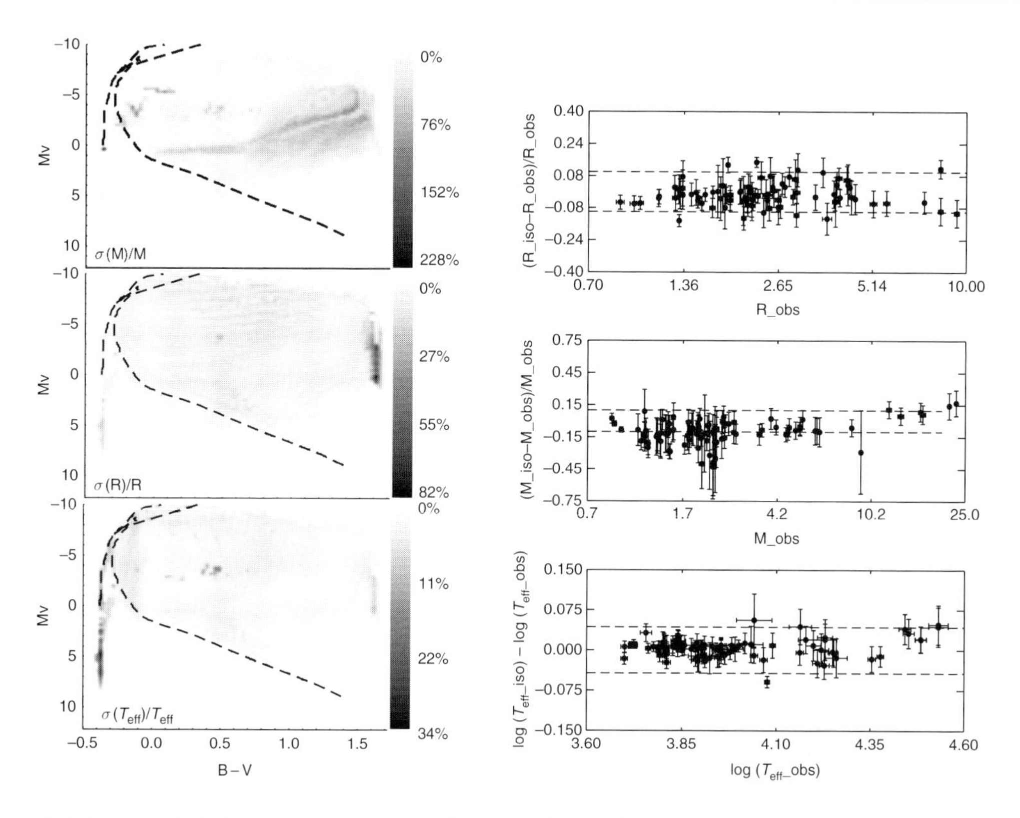

Figure 7.16 Left: uncertainties in the colour–magnitude diagram indicating the confusion for the masses, radii and effective temperatures retrieved for a given star based only on its position in the diagram. The dashed line represents an isochrone of solar metallicity and age 4 Myr. In contrast to more conventional ways of viewing the HR diagram, dark regions indicate a larger uncertainty of that parameter due to the fact that stars of different mass at different evolutionary stages can occupy it. Right: relative differences between stellar radii, masses and $T_{\rm eff}$ derived from stellar evolutionary calculations and the robust direct estimates from observations in eclipsing spectroscopic binaries compiled by Andersen (1991). From Allende Prieto & Lambert (1999, Figures 1 and 2).

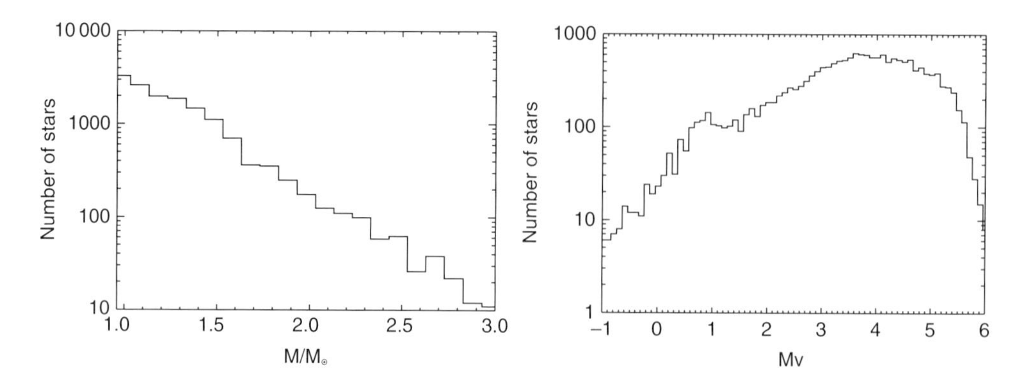

Figure 7.17 Luminosity function (left) and mass function (right) inferred from the overall distribution of Hipparcos stars in the HR diagram, constrained by the probabilistic estimation of masses and luminosities from stellar evolutionary models. From Allende Prieto & Lambert (1999, Figure 6).

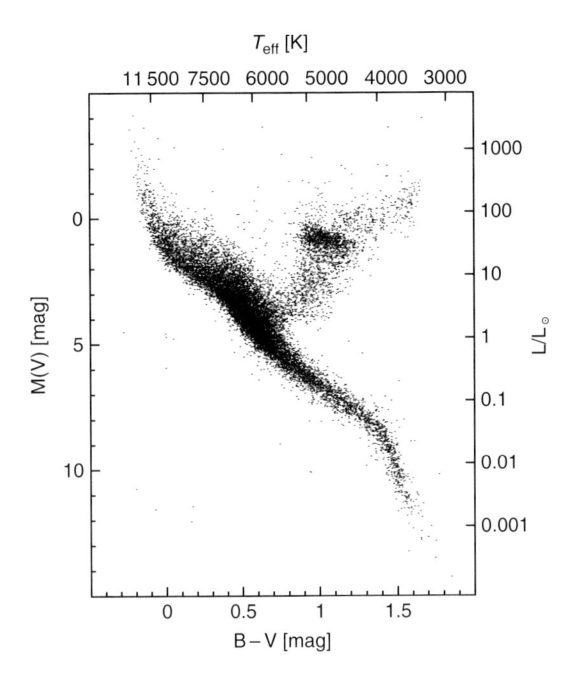

Figure 7.18 The observational Hertzsprung–Russell diagram, M_V versus $B-V$, for the 16 631 single stars with $\sigma_\pi/\pi < 0.1$, and with the additional constraint $\sigma_{B-V} < 0.025$ mag. A similar plot for the 20 853 stars before the elimination of binaries was given in Perryman et al. (1997, Figure 3). Note that while V and $B-V$ are listed in the catalogue, they are derived from the Hipparcos Hp and, usually, B_T and V_T. M_V takes no account of extinction or bolometric correction. Scales at top and right give approximate $T_{\rm eff}$ and L corresponding to main-sequence dwarfs.

method to 32 F–K dwarfs and subgiants based on high-quality echelle spectra, showing generally good agreement with the results of both Fuhrmann (1998a), and Hipparcos parallaxes interpreted in terms of evolutionary tracks. Takeda *et al.* (2005) extended the analysis to 160 mid-F through early-K stars, finding generally good agreement between parameters estimated purely from the spectroscopic results with those estimated from photometric colours, Hipparcos parallaxes, and evolutionary tracks.

Decin *et al.* (2003) used Hipparcos data combined with ISO SWS data for 11 G9–M2 giants to derive radii, luminosities and gravity-inferred masses, with CNO abundances derived in some cases.

Korn (2004) showed that a kinetic equilibrium or non-LTE approach for Fe I succeeds in fulfilling the Hipparcos trigonometric constraints if and only if inelastic collisions with hydrogen are properly accounted for. The model makes testable predictions for the gravities of metal-poor globular cluster giants in which non-LTE corrections are expected to reach up to +0.5 dex.

Da Silva *et al.* (2006) discussed the consistency of basic stellar parameters (age, mass, radius and gravity) for 72 evolved stars based on Hipparcos parallaxes. They derived the resulting age–metallicity relation in the solar neighbourhood based on field giants, rather than from clusters and field dwarfs. They found that the [Fe/H] dispersion of young stars (< 1 Gyr) is comparable to the observational errors, indicating that stars in the solar neighbourhood are formed from interstellar matter of rather homogeneous chemical composition.

7.5 Hipparcos results on stellar evolution

7.5.1 Nearby stars

The accurate distances for a large number of stars in the solar neighbourhood has allowed detailed studies of the fine structure of the HR diagram and related metallicity effects to be undertaken. As an illustration of the astrometric and photometric quality of the catalogue, Figure 7.18 gives the observational Hertzsprung–Russell diagram, M_V versus $(B-V)$, for the 20 853 stars for which distant determinations are better than 10% (i.e. $\sigma_\pi/\pi < 0.1$), and with the additional constraint $\sigma_{B-V} < 0.025$ mag. The general features of this diagram, based on the preliminary catalogue, were originally discussed by Perryman *et al.* (1995).

Among a sample of Hipparcos FGK stars closer than 25 pc, with error on parallax lower than 5%, Lebreton *et al.* (1997) selected stars with [Fe/H] in the range $[-1.0, +0.3]$ from detailed spectroscopic analysis, with $\sigma_{\rm [Fe/H]} \sim 0.10$ dex from Cayrel de Strobel *et al.* (1997b), with $F_{\rm bol}$ and $T_{\rm eff}$ from Alonso *et al.* (1995, 1996a) with $\sigma_{F_{\rm bol}}/F_{\rm bol} \sim 2\%$ and $\sigma_{T_{\rm eff}}/T_{\rm eff} \sim 1.5\%$, and not suspected to be unresolved binaries. Figure 7.19, left shows the HR diagram of the 34 selected stars: the error bars are the smallest obtained for stars in the solar neighbourhood ($\sigma_{M_{\rm bol}}$ are in the range 0.031–0.095 with an average value $\langle\sigma_{M_{\rm bol}}\rangle \sim 0.045$ mag). The sample is compared with theoretical isochrones derived from standard stellar models in Figure 7.19, centre and right. Models cover the entire [Fe/H]-range. They account for an α-element enhancement [α/Fe] = +0.4 dex for [Fe/H] ≤ -0.5 and, for non-solar [Fe/H], have a solar-scaled helium content, $Y = Y_{\rm p} + Z(\Delta Y/\Delta Z)_\odot$. The splitting of the sample into a solar metallicity sample and a moderately metal-deficient one (Figure 7.19, right pair) shows that the slope of the main sequence is well reproduced with the solar $\alpha_{\rm MLT}$, and that stars of solar metallicity and close to it occupy the theoretical band corresponding to their (LTE) metallicity range, while for moderately metal deficient stars there is a poor fit.

In general, stars have a tendency to lie on a theoretical isochrone corresponding to a higher metallicity than the spectroscopic (LTE) value. This trend was already noticed by Axer *et al.* (1995) but it is now even

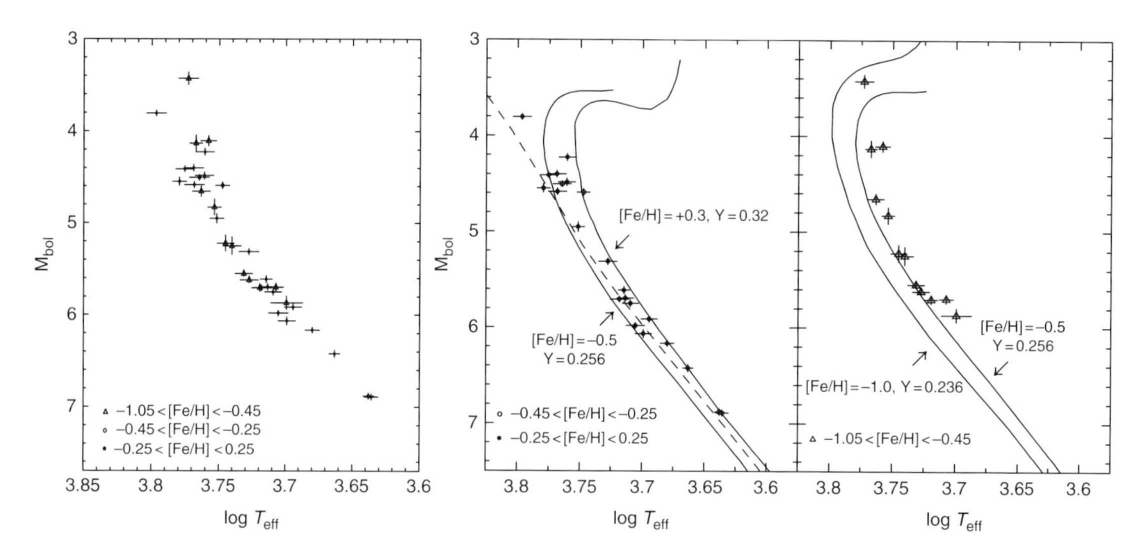

Figure 7.19 Left: the Hipparcos HR diagram of the 34 best-known nearby stars. The parallax accuracies σ_π/π are in the range 0.003–0.041. Bolometric fluxes and effective temperatures are available from the work of Alonso et al. (1995, 1996a), with $(\sigma_{F_{\rm bol}}/F_{\rm bol}) \sim 2\%$ and $(\sigma_{T_{\rm eff}}/T_{\rm eff}) \sim 1.5\%$. Resulting $\sigma_{M_{\rm bol}}$ are in the range 0.031–0.095 mag. Middle and right: the same sample split into two metallicity domains: (a) stars with [Fe/H] close to solar, $-0.45 \leq$ [Fe/H] +0.25. Theoretical isochrones are overlaid on the observational data. The lower isochrone (10 Gyr) is for [Fe/H] = −0.5, $Y = 0.256$ and [α/Fe] = 0.4; the upper isochrone (8 Gyr) is for [Fe/H] = +0.3, $Y = 0.32$ and [α/Fe] = 0.0; the dashed line is a solar ZAMS ($\alpha_{\rm MLT}$=1.65, $Y_\odot = 0.266$ and $Z_\odot = 0.0175$). The brightest star is the young star γ Lep. (b) Moderately metal-deficient stars, $-1.05 <$ [Fe/H] −0.45, with two 10 Gyr isochrones with [α/Fe] = 0.4: the lower is for [Fe/H] = −1.0, $Y = 0.236$ and the upper for [Fe/H] = −0.5, $Y = 0.256$. All stars but one are above the region defined by the isochrones. $(\Delta Y/\Delta Z)_\odot = 2.2$ is obtained for a primordial helium of $Y_{\rm p} = 0.227$ (Balbes et al., 1993). From Lebreton (2000b, Figures 1 and 2), reprinted with permission from the Annual Reviews of Astronomy and Astrophysics, Volume 38 ©2000 by Annual Reviews (www.annualreviews.org).

more apparent because of the high accuracy of the data. Helium content well below the primordial helium value would be required to resolve the conflict. This is exemplified by the star μ Cas A, the A-component of a well-known, moderately metal-deficient binary system that has a well-determined mass (error in mass of 8%). The standard model (Figure 7.20) is more than 200 K hotter than the observed point and is unable to reproduce the observed $T_{\rm eff}$ even if (reasonable) error bars are considered (Lebreton, 2000a). On the other hand, the mass–luminosity properties of the star are well reproduced if the helium abundance is chosen to be close to the primordial value, although the error bar in mass is somewhat too large to provide strong constraints. Several reasons can be invoked to explain the poor fit at low metallicities: (a) an erroneous temperature-scale: 3d model atmospheres could still change the $T_{\rm eff}$-scale as a function of metallicity (Gustafsson, 1998), but with the presently available one-dimensional models it seems difficult to increase the $T_{\rm eff}$ of Alonso *et al.* (1996a) by as much as 200–300 K. This scale is already higher than other photometric scales, by as much as 100 K (Nissen, 1998). Lebreton *et al.* (1999) verified that spectroscopic effective temperatures lead to a similar misfit; (b) erroneous metallicities: as discussed in Section 7.6, [Fe/H]-values inferred from model atmosphere analysis should be corrected for non-LTE effects. According to Thévenin & Idiart (1999) no correction is expected at solar metallicity, whereas for moderately metal-deficient stars the correction amounts to ~0.15 dex; (c) inappropriate interior models: in low-mass stars, microscopic diffusion by gravitational settling can make helium and heavy elements sink toward the centre, changing surface abundances as well as inner abundance profiles. In metal-deficient stars this process may be very efficient.

The two latter reasons are attractive because they qualitatively predict an increasing deviation from the standard case when metallicity decreases. As shown in Figure 7.20, a combination of microscopic diffusion effects with non-LTE [Fe/H] corrections could remove the discrepancy noted for μ Cas A: an increase of [Fe/H] by 0.15 dex produces a rightward shift of 80 K of the standard isochrone, representing about one third of the discrepancy. Additionally, adding microscopic diffusion effects, according to calculations by Morel & Baglin (1999), provides a match to the observed positions. Moreover, the general agreement for solar metallicity

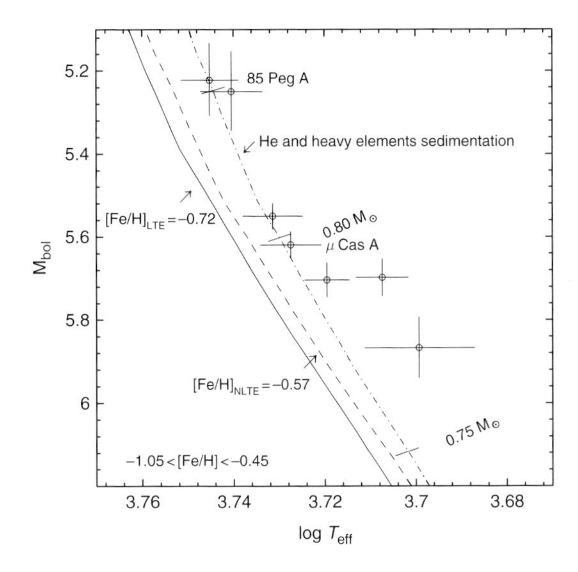

Figure 7.20 HR diagram for the unevolved moderately metal-deficient stars of Figure 7.19: mean LTE metallicity $[\mathrm{Fe/H}]_{\mathrm{LTE}} = -0.72$*, mean non-LTE value* $[\mathrm{Fe/H}]_{\mathrm{NLTE}} = -0.57$*. Full and dashed lines are standard isochrones (10 Gyr) computed with, respectively, the* $[\mathrm{Fe/H}]_{\mathrm{LTE}}$ *and* $[\mathrm{Fe/H}]_{\mathrm{NLTE}}$ *values. The dot-dashed isochrone (10 Gyr) includes He and heavy elements sedimentation: at the surface it has* $[\mathrm{Fe/H}]_{\mathrm{NLTE}} = -0.57$ *but the initial [Fe/H] was about −0.5. From Lebreton (2000b, Figure 3), reprinted with permission from the Annual Reviews of Astronomy and Astrophysics, Volume 38 ©2000 by Annual Reviews (www.annualreviews.org).*

stars (Figure 7.19, left) should remain: (1) at solar metallicities non-LTE corrections are found to be negligible, and (2) at ages of ~ 5 Gyr chosen as a mean age for those (expectedly) younger stars, diffusion effects are estimated to be smaller than the error bars on T_{eff} (Lebreton *et al.*, 1999). In summary, the high-level accuracy reached for a few tens of stars in the solar neighbourhood reveals imperfections in interior and atmosphere models. It casts doubts on abundances derived from model atmospheres in LTE, and favours models that include microscopic diffusion of helium and heavy elements toward the interior over standard models. Also, diffusion makes the surface [Fe/H]-ratio decrease by ~0.10 dex in 10 Gyr in a star like μ Cas (Morel & Baglin, 1999), which is rather small and hidden in the observational error bars. In very old, very deficient stars, the [Fe/H]-decrease is expected to be larger (Salaris *et al.*, 2000), which makes the relation between observed and initial abundances difficult to establish.

Complete HR diagrams of stars of the solar neighbourhood have been constructed by adopting different selection criteria, and have been compared to synthetic HR diagrams based on theoretical evolutionary tracks.

Schröder (1998a,b) proposed diagnostics of the degree of overshooting in main-sequence stars based on star counts in the different regions of the Hipparcos HR diagram for stars in the solar neighbourhood ($d < 50$–100 pc), employing a near solar chemical composition. In the mass range 1.2–$2M_{\odot}$, convective cores are small, and it is difficult to estimate the amount of overshooting from the shapes of isochrones. In contrast, the number of stars in the Hertzsprung gap, which is associated with the onset of H-shell burning, was shown to be a sensitive indicator of the extent of overshooting around $1.6M_{\odot}$; the greater the overshooting on the main sequence, the larger the He-burning core, leading to a more compact higher-gravity star, more rapid H-shell burning, and in turn a faster passage through the Hertzsprung gap. Actual star counts (Figure 7.5.1; with Schröder (1998b) also showing results for the $d < 100$ pc sample) indicate an onset of overshooting around $\sim 1.6M_{\odot}$ (no overshooting appears necessary below that mass), which is broadly consistent with other empirical calibrations based on the main sequence width or eclipsing binaries (e.g. Pols *et al.*, 1997). A model with overshooting for all stars, for example, is inconsistent with the number of stars observed in the Hertzsprung gap. It is noteworthy that Schröder (1998b) assumes a random distribution of stellar ages, without major bursts of star formation, and that this constant stellar birthrate (perhaps an observational consequence of diffusion and mixing of the stellar population in the course of Galactic rotation) reproduces the local HR diagram so well. The empirically-derived present-day mass function is well defined, at least between 1.6–$4M_{\odot}$. For a random age distribution this leads to an initial mass function close to that suggested by Scalo (1986), with exponent -1.7. The total time-independent stellar birth rate is dominated by stars below $1.6M_{\odot}$, and this region is poorly constrained, but for masses above this limit they establish a local birth rate of one such star within 50 pc per 6 Gyr.

Jimenez *et al.* (1998) compared the red envelope of Hipparcos subgiants ($\sigma_{\pi}/\pi < 0.15$, $\sigma_{(B-V)} < 0.02$ mag) with isochrones to determine a minimum age of the Galactic disk of 8 Gyr, which is broadly consistent with ages obtained with other methods (white-dwarf cooling curves, radioactive dating, or fits of various age-sensitive features in the HR diagram). The fit is still qualitative: the metallicities of subgiants are unknown because of the inadequacy of model atmospheres in that region. For this reason, Jimenez *et al.* (1998) investigated the isochrones in other regions, main sequence and clump (He core burning). They calculated the variations with mass of the clump position for a range of metallicities in the disk, and showed that stars with masses from 0.8–$1.3M_{\odot}$ (ages from 2–16 Gyr) all occupy a well-defined vertical branch, the red-edge of the clump. The

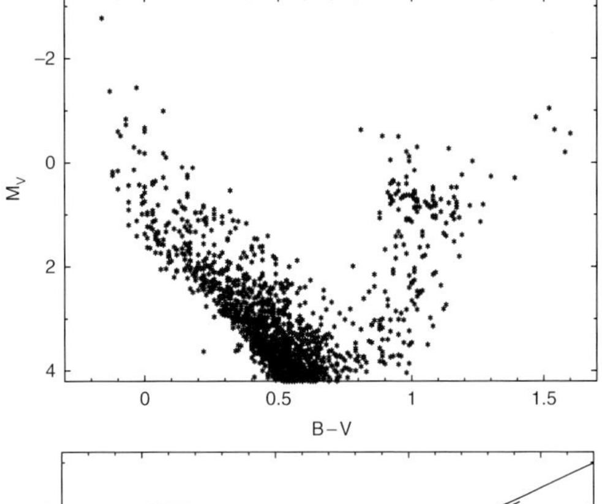

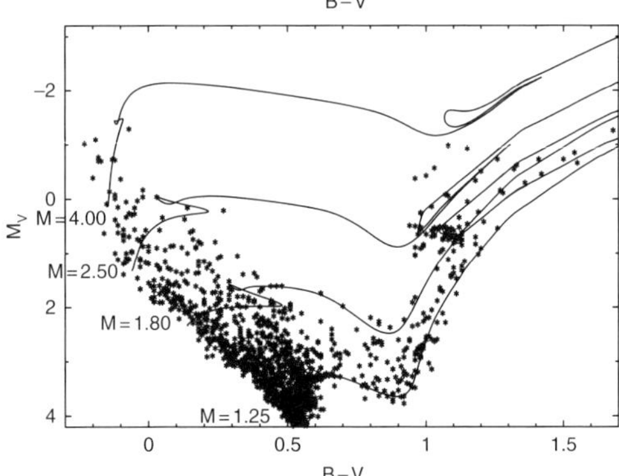

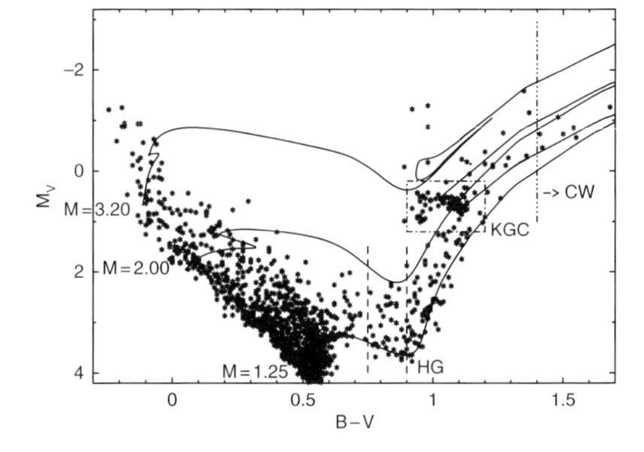

Figure 7.21 Left: the Hipparcos solar neighbourhood HR diagram for $d < 50$ pc. Below left: simulated HR diagram for $d < 50$ pc, with 1340 stars of random age with $1.15 < M < 10M_\odot$. The model (based on the Eggleton code) incorporates mass-dependent overshooting ($\alpha_{\rm ov} = 0$ at $1.6M_\odot$, 0.12 above $2.5M_\odot$, and 0.06 in between), reproducing the observed numbers in the various regions.Below right: simulation based on stellar evolution models without overshooting. The regions are: HG = Hertzsprung gap; KGC = K giant clump; CW = 'cool wind' region. From Schröder (1998b, Figures 1, 2, and 5).

colour of this border line is sensitive to metallicity, which makes it a good metallicity indicator in old metal-rich stars.

Ng & Bertelli (1998) revised the ages of stars of the solar neighbourhood and derived corresponding age–metallicity and age–mass relations. Fuhrmann (1998a) combined the [Mg/H]–[Fe/H] relation with age and kinematical information to distinguish thin and thick disk stars. Several features emerged from these studies: (1) no evident age–metallicity relation exists for the youngest (< 8 Gyr) thin-disk stars; some of them are rather metal-poor, and super metal-rich stars appear to have been formed early in the history of the thin disk; (2) there is an apparent lack of stars in the age-interval 10–12 Gyr which is interpreted by Fuhrmann (1998a) as a signature of the thin-disk formation; and (3) beyond 12 Gyr there is a slight decrease of metallicity with increasing age for stars of the thick disk; some of them are as old as halo stars.

Kotoneva *et al.* (2002b) showed that the luminosity of K dwarfs on the lower main sequence, $5.5 < M_V < 7.3$, is a simple function of metallicity, with $\Delta M_V = 0.04577$–0.84375 [Fe/H] where ΔM_V is the luminosity difference relative to a fiducial solar metallicity isochrone. They compared the data with a range of isochrones from the literature, finding none which fit the data at all metallicities, and concluding that metal-rich isochrones seem to be difficult to construct.

Böhm-Vitense gap In searching for observational evidence for a transition region between stars with convective rather than radiative upper layers, Böhm-Vitense (1970) suggested that stars with convective atmospheres would be approximately 0.08 mag redder in $B-V$ than stars with radiative upper layers for models with the same $T_{\rm eff}$, at least in the radiative–convective transition range. This implies that a population of main-sequence stars with a continuous range of effective temperatures could exhibit a 0.08 mag jump in $B-V$ at some critical transition temperature. This theoretical phenomenon has been referred to as 'the abrupt onset of convection', and the associated gap is referred to as the Böhm-Vitense gap. Although a gap was apparent in the main sequence of local field stars of Johnson & Morgan (1953), observations of the gap in the field population have generally been obtained using spectroscopically determined luminosity classes to select only main sequence objects. For field stars, the gap is typically in the $0.2 < B-V < 0.3$ range, consistent with the prediction of Böhm-Vitense (1970). Subsequent investigations, which appear to be equally divided between providing evidence for gaps, and for the absence of gaps, in field stars and clusters, are summarised by Newberg & Yanny (1998). These authors used Hipparcos parallaxes to investigate whether systematics of luminosity classification could be responsible for the gaps found by at least some authors. They constructed HR diagrams

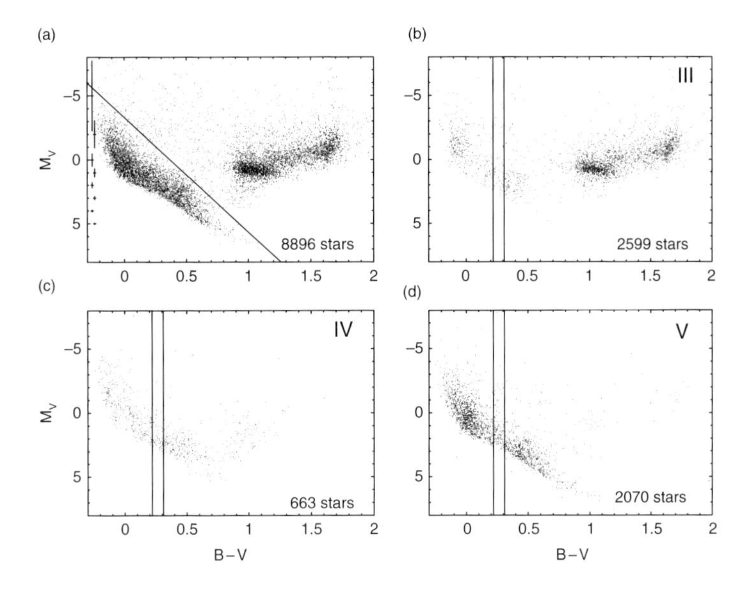

Figure 7.22 HR diagram of field stars using Hipparcos parallaxes, WBVR photometry, and luminosity classes from the Hipparcos Input Catalogue: (a) all stars with $V < 7$, with the line $M_V = 9.0(B - V) - 3.3$ shown for reference; (b) luminosity classes II–III, III and III–IV; (c) luminosity classes IV and IV–V; (d) luminosity class V. The limits of the Böhm-Vitense gap, $0.22 < B - V < 0.31$, as measured by Böhm-Vitense & Canterna (1974), are shown as vertical lines. Only about 50% of stars in the sample have luminosity class assignments in the Hipparcos Catalogue. From Newberg & Yanny (1998, Figure 1).

for two samples of bright stars using M_V from Hipparcos parallaxes (Figure 7.22). The first is magnitude-limited using photometry from the WBVR catalogue of northern sky bright stars ($\delta > -14°$), which shows a smooth distribution of stars along the main sequence, with no detectable gaps. The second contains all stars closer than 100 pc in the Hipparcos Catalogue with $\delta < -12°$. Comparison with the Michigan Spectral Survey shows that some stars which evidently lie on the main sequence in the HR diagram, particularly those in the $0.2 < B - V < 0.3$ region that has been labelled the Böhm-Vitense gap, are incorrectly classified as giants in the MK system. Other gaps that have been identified in the main sequence are also affected by such classification criteria. Their analysis finds no evidence for the existence of the Böhm-Vitense gap, and again implies that the standard identification of main-sequence stars with luminosity class V, and giants with luminosity class III, must be reconsidered for some spectral types. The true nature of the stars that lie on the main sequence in the HR diagram, but which do not have luminosity class V designations, remains to be investigated.

7.5.2 Zero-age main sequence

Hipparcos disk and halo stars span the whole range of Galactic metallicities. Lebreton (2000b) discussed the position of the zero-age main sequence (ZAMS) as a function of metallicity, and its implications for the unknown helium abundances. Figure 7.23 shows the non-evolved stars ($M_{\rm bol} > 5.5$) of Figures 7.19 and 7.24, along with standard isochrones of various metallicities and solar-scaled helium, $(\Delta Y/\Delta Z)_\odot = 2.2$. Lebreton (2000b) concluded that the observational and theoretical main-sequence widths are in reasonable agreement for $\Delta Y/\Delta Z = 2.2$. This qualitative agreement is broadly consistent with a $\Delta Y/\Delta Z$ ratio of $\sim 3 \pm 2$ derived from similar measures of the lower main-sequence width using Hipparcos data by Pagel & Portinari (1998), and the lower limit $\Delta Y/\Delta Z \gtrsim 2$ obtained by Fernandes *et al.* (1996) from the pre-Hipparcos main sequence. It also agrees with extragalactic determinations (Izotov *et al.*, 1997) and nucleosynthesis predictions.

The helium abundance at solar metallicities can be inferred from stellar evolutionary models. Figure 7.23 shows that there are four stars with Fe/H close to solar on the [Fe/H] = 0.3 isochrone. Non-LTE [Fe/H] corrections are negligible at solar metallicity. Microscopic diffusion may produce a shift in the HR diagram: for a $0.8M_\odot$ star of solar Fe/H at 5 Gyr the shift is small and comparable to the observational error bars, although it increases with age. These local disk stars are not expected to be very old, and the shift could instead indicate that their He-content is lower than the solar-scaled value. Calibration of individual objects and groups with metallicities close to solar indicate an increase of helium with metallicity corresponding to $\Delta Y/\Delta Z \sim 2.2$ from the Sun (Lebreton *et al.*, 1999) and $\Delta Y/\Delta Z \sim 2.3 \pm 1.5$

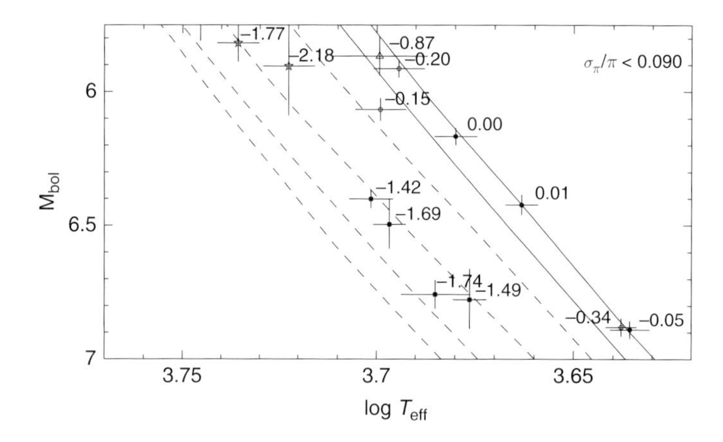

Figure 7.23 Hipparcos HR diagram of non-evolved stars with $\sigma_\pi/\pi < 0.10$. Each star is labelled with its [Fe/H]-value. Standard isochrones are plotted with, from left to right, [Fe/H] = −2.0, −1.5, −1.0, −0.5, 0.0, 0.3. From Lebreton (2000b, Figure 6), reprinted with permission from the Annual Reviews of Astronomy and Astrophysics, Volume 38 ©2000 by Annual Reviews (www.annualreviews.org).

from visual binaries (Fernandes *et al.*, 1998) but exceptions are found, such as in the (rather young) Hyades which, although metal-rich ([Fe/H] = 0.14), appear to have a solar or even slightly subsolar helium content with $\Delta Y/\Delta Z \sim 1.4$ (Perryman *et al.*, 1998).

Determination of the zero-age main sequence for Population II stars is problematic, because very few metal-deficient stars have accurate positions in the non-evolved part of the HR diagram: a gap appears for $-1.4 <$ [Fe/H] < -0.3 and only four subdwarfs are found below [Fe/H] ~ -1.4. The empirical dependence of the ZAMS location with metallicity is therefore impossible to establish for these stars, which are expected to have a practically primordial helium content. This adds to difficulties in estimating the distances of globular clusters (Eggen & Sandage, 1962; Sandage, 1970; Chaboyer, 1999).

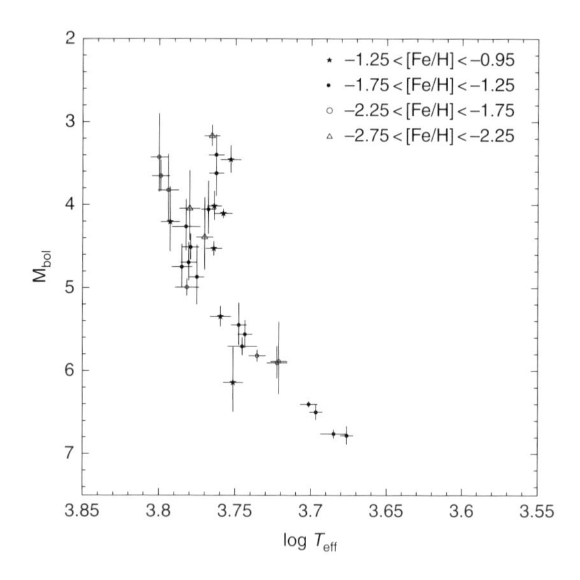

Figure 7.24 Hipparcos HR diagram of the 32 halo stars with $\sigma_\pi/\pi < 0.22$. Resulting $\sigma_{M_{bol}}$ are in the range 0.03–0.48 mag. A grouping of subgiants also follows an isochrone-like shape. From Cayrel et al. (1997b, Figure 3), as updated in Lebreton (2000b, Figure 4), reprinted with permission from the Annual Reviews of Astronomy and Astrophysics, Volume 38 ©2000 by Annual Reviews (www.annualreviews.org).

7.5.3 Subdwarfs and other Population II stars

Hipparcos provided the first high-quality parallaxes for a number of Population II, or halo population, stars (see review by Caputo, 1998). Such stars are identified by their low metallicity and, increasingly as a result of the Hipparcos astrometry, from their space motions. They can include main sequence (by definition, the low-metallicity, thick disk or halo Population II main sequence stars are also referred to as subdwarfs), turn-off, subgiant, giant, and horizontal branch stars. Age determinations of the local halo stars could be estimated from the shape of the resulting Population II main sequence and stars in the associated turn-off region, as well as providing comparisons with globular cluster sequences, a subject covered in more detail in Chapter 9.

Cayrel *et al.* (1997b) selected a sample of Population II Hipparcos halo stars, 32 having $\sigma_\pi/\pi < 0.22$. The theoretical HR diagram is shown in Figure 7.24, based on bolometric fluxes and effective temperatures from Alonso *et al.* (1995) and Alonso *et al.* (1996a); subdwarfs and a few subgiants are present, delineating an isochrone-like shape with a well-defined turn-off region. Bartkevičius *et al.* (1997) made a similar

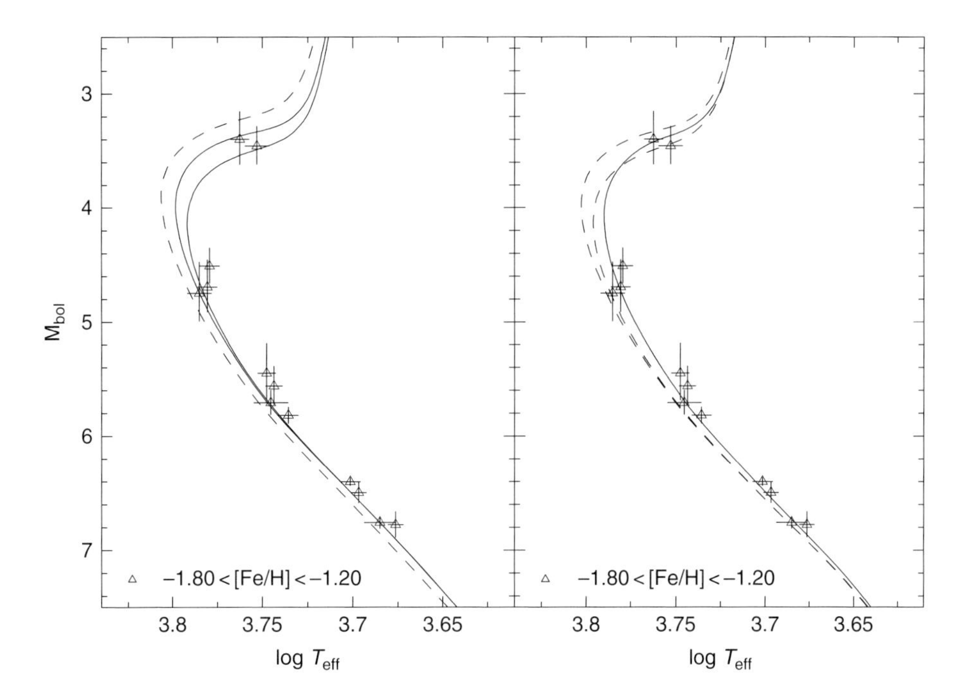

Figure 7.25 Hipparcos HR diagram of 13 halo stars with $[Fe/H]_{LTE} = -1.5 \pm 0.3$ *and* $\sigma_\pi/\pi < 0.12$ *(the parallax accuracies* σ_π/π *are in the range 0.01–0.12). Bolometric fluxes and effective temperatures are available from the work of Alonso et al. (1995) and Alonso et al. (1996a). Resulting* $\sigma_{M_{bol}}$ *are in the range 0.03–0.26 mag. Left: effect of a non-LTE correction of +0.2 dex in [Fe/H] as inferred from the models of Bergbusch & VandenBerg (2001): the dashed line is a standard isochrone of 12 Gyr ([α/Fe] = +0.3, Y ∼ 0.24) with [Fe/H] = −1.54 (LTE value), and the full lines are isochrones with [Fe/H] = −1.31 (non-LTE value) of 12 Gyr (upper line) and 14 Gyr (lower line). Right: effect of microscopic diffusion of He as inferred from the models of Proffitt & Vandenberg (1991): isochrones ([Fe/H] = −1.3 and [O/Fe] = 0.55), of 12 Gyr with (full line) and without (dashed-line; upper line 12 Gyr, lower 14 Gyr) microscopic diffusion are plotted. From Lebreton (2000b, Figure 5), reprinted with permission from the Annual Reviews of Astronomy and Astrophysics, Volume 38 ©2000 by Annual Reviews (www.annualreviews.org).*

study comprising some 150 Hipparcos Population II stars with $\sigma_\pi/\pi < 0.2$. They found that some 40% of stars previously classified as subdwarfs are, in fact, subgiants. The low-metallicity part of the main sequence is well-defined (their Figure 1), and they used theoretical isochrones and globular cluster fiducials to estimate an age spread of 7 ± 2 Gyr. Hipparcos-based kinematics of the 39 objects with [Fe/H] in the range -0.4 to -1.1 were consistent with the thick disk, with the 64 objects of lower metallicity being consistent with the halo population.

Cayrel *et al.* (1997b) and Pont *et al.* (1997) estimated the local halo to be 12–16 Gyr old on the basis of standard isochrones. Meillon *et al.* (1999) discussed an enlarged sample of subdwarf distances using the Hipparcos stars to calibrate photometric distance estimates for high-velocity stars in the sample of Carney *et al.* (1994). Viotti *et al.* (1997) discussed the properties of three hot subdwarfs, with $T_{eff} \sim 5 \times 10^4$ K and spanning a large range in visual luminosity, $M_V = 2.2$–4.6.

To estimate the age of the local halo, Cayrel *et al.* (1997b) kept 13 stars with the lowest error bars and spanning a narrow metallicity range, $[Fe/H] = -1.5 \pm 0.3$ (Figure 7.25). They found that halo stars, like disk stars, have lower T_{eff} than the theoretical isochrone corresponding to their metallicity. The discrepancy was also noted by Nissen *et al.* (1997) and Pont *et al.* (1997) in larger samples of halo stars, and amounts to 130–250 K depending on the metallicity, apparently independent of the model isochrones used. Again, non-LTE corrections leading to increased [Fe/H]-values (Δ[Fe/H] = +0.2 for [Fe/H] ~ -1.5 according to Thévenin & Idiart 1999), added to the effects of microscopic diffusion, can be invoked to reduce the misfit (Lebreton, 2000b). Figure 7.25, left compares the Cayrel *et al.* (1997b) sample with standard isochrones by Bergbusch & VandenBerg (2001), showing that the subdwarf main sequence cannot be reproduced by isochrones computed with the LTE [Fe/H]-value, but increasing the metallicity (to mimic non-LTE corrections) improves the fit. Figure 7.25, right compares the halo sequence with the Proffitt & Vandenberg (1991) isochrones that include He sedimentation. Microscopic diffusion makes the isochrones redder, modifies their shape, and predicts a lower turn-off

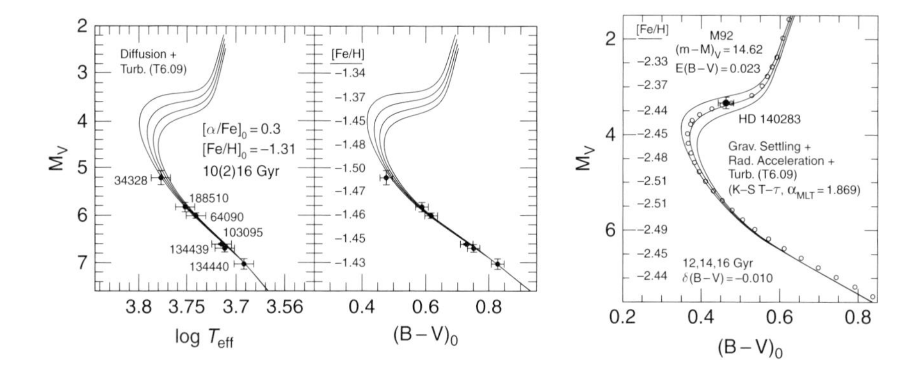

Figure 7.26 Left pair: comparison of the specific (their T6.09) isochrones from VandenBerg et al. (2002) with the [Fe/H] = −1.45 subdwarf sequence consisting of the six Hipparcos stars whose HD numbers are given in the left panel. The uncertainties in M_V correspond to uncertainties in the parallax data from Hipparcos. The predicted variation of [Fe/H] with M_V along these isochrones is approximately as given adjacent to the ordinate of the right panel. The isochrones reproduce the properties of the subdwarf calibrators in the colour–magnitude diagram, without adjustment of the colours or effective temperatures. Right: fit of the M92 fiducial (Stetson & Harris, 1988) to the field subgiant HD 140283; both have very similar metallicities and reddenings. Isochrones based on the T6.09KS tracks are compared with the observations. The predicted variation in [Fe/H] as a function of M_V along the 14 Gyr isochrone is indicated along the ordinate. From VandenBerg et al. (2002, Figures 5 and 9).

luminosity: the best fit with the observed sequence is achieved for an age smaller by 0.5–1.5 Gyr than that obtained without diffusion. Models by Castellani *et al.* (1997) show that, if sedimentation of metals is also taken into account, including its effects on the matter opacity, the isochrone shift is smaller than the shift obtained with He diffusion only. Morel & Baglin (1999) described further models in an attempt to explain the Hipparcos results, and showed that gravitational and thermal settling could produce shifts in excess of 100 K for Population II stars.

Salaris *et al.* (2000) also investigated the effects of atomic diffusion on the main sequence of metal-poor low-mass stars. Again, the general idea is that the spectroscopic metallicity determined for subdwarfs will be affected by metals and helium sinking below the boundary of the convective envelope, at least until the turn-off phase when the boundary of the convective envelope deepens. In effect, subdwarf models must be computed using a larger initial Z with respect to that observed, the exact value depending on age; the situation is different for globular cluster abundances, which are measured in red giants which have 'recovered' their initial abundances after the first dredge-up. They showed that their models with diffusion reproduce the Hipparcos HR diagram of subdwarfs with empirically-determined $T_{\rm eff}$ and spectroscopic [Fe/H] determinations. Accounting for diffusion yields ages lower by about 1 Gyr; thus absolute ages, age dispersion, the age–metallicity relation, and the helium enrichment ratio $\Delta Y/\Delta Z$ obtained from the width of the halo subdwarf main sequence, are all affected.

VandenBerg *et al.* (2002) constructed models of metal-poor stars accounting for both gravitational settling and radiative acceleration, in analogy with those required to match solar models based on helioseismic studies. Isochrones for ages between 12–18 Gyr were derived from evolutionary tracks for masses from 0.5–$1.0 M_\odot$ and specific initial chemical abundances, including $[\alpha/\mathrm{Fe}] = 0.3$, consistent with those noted previously. As found by Salaris *et al.* (2000), allowance for the additional diffusive processes leads to a 10–12% reduction in age at a given turn-off luminosity. However, in order for the diffusive models to satisfy the constraints from Li and Fe abundance data, and to reproduce the observed morphologies of globular cluster colour–magnitude diagrams in a straightforward way, extra mixing just below the boundary of the convective envelope seems to be necessary. Models with additional turbulent mixing appear to satisfy all of these constraints, as well as those provided by the colour–magnitude diagrams of the local Hipparcos subdwarfs (Figure 7.26, left). They went on to use the same models, combined with the position of the field subgiant HD 140283, auspiciously located in the HR diagram roughly halfway between the turn-off and the lower red giant branch (Grundahl *et al.*, 2000), to derive the distance of the globular cluster M92. Their results imply an age of 13.5 Gyr for M92, compared to an age of somewhat above 15 Gyr if non-diffusive models are used (Figure 7.26, right). Given that M92 is one of the most metal-deficient, and therefore presumably one of the oldest, of the Galaxy's globular clusters, this constraint on the age of the Universe is also in encouraging accord with the age of the Universe of 13.7 Gyr estimated from WMAP (Bennett *et al.*, 2003; Spergel *et al.*, 2003).

7.5.4 Subgiants

Subgiants are the next evolutionary stage after the main sequence, occupying the roughly 'horizontal' part of the evolutionary track between the main sequence and giant stages: the start of H-shell burning is accompanied by an increase in radius, and a movement away from the main sequence. After the main-sequence turn-off, the H-shell burning stars do not immediately evolve as fast as they do subsequently, as they rapidly cross the Hertzsprung gap. The subgiants therefore represent a mix of post-main-sequence stars: just above the main sequence and to the left of the Hertzsprung gap for the more massive stars, and on the slow rise up the lower giant branch for the low-mass stars. With $+2.5 \lesssim M_V < \lesssim 4$ and spectral types between G0 and K3, they are rare in the solar neighbourhood, being about 100 times rarer than subdwarfs: in the Hipparcos Catalogue there are only two subgiants with $\sigma_\pi/\pi < 12.5\%$, and none with $\sigma_\pi/\pi < 5\%$.

The history of the discovery of field subgiants, the role that they played in the development of the early understanding of stellar evolution, and their importance in the age dating of the Galactic disk, is reviewed by Sandage *et al.* (2003). The term 'subgiants' was first used by Strömberg (1930): their discovery did not fit the picture of stellar evolution developed by Russell during the years 1914–30, in which stars were believed to have been born as giants near $M_V = 0$, after which (it was believed) they contracted, becoming hotter, until they reached the main sequence. The M67 colour–magnitude diagram provided the key to the now-accepted explanation, and post-1960 terminology: that the giant-branch is the locus of stars on the 'first ascent' of post-main-sequence stars in a hydrogen shell burning phase, after core-hydrogen exhaustion. The helium flash at the top of the first-ascent branch is followed by the descent onto the core helium-burning phase, populated by the so-called 'clump' giants. The four Hyades giants were the local prototypes of the giant sequence, originally considered as first-ascent giants, but later suspected to be clump giants on the basis of spectroscopic isotope ratios combined with theories of nucleosynthesis. Sandage *et al.* (2003) commented that '*... when the colour magnitude diagram from the Hipparcos parallaxes became available (Perryman et al., 1995, Figure 6), it became immediately obvious that the Hertzsprung–Russell giants, especially the four in the Hyades central cluster, are clump stars rather than first-rise giants ... the Hipparcos colour–magnitude diagram provided the definitive proof because the entire clump sequence is directly revealed*'.

The closest and brightest subgiant is β Hyi (G2 IV, $V = 2.8$, $\pi_{\rm Hip} = 133.78 \pm 0.51$ mas). It provides an

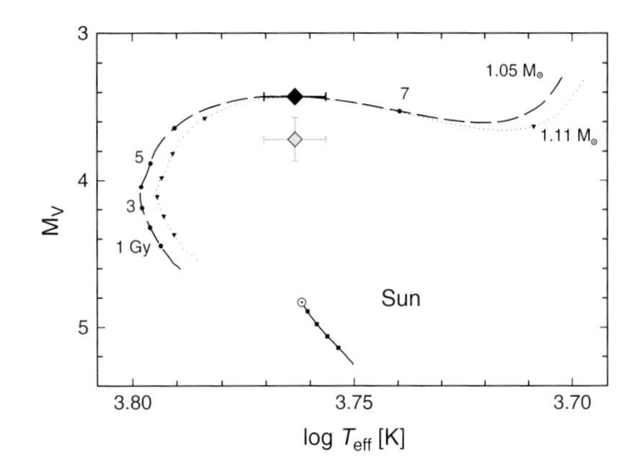

Figure 7.27 Stellar evolution, and the age of β Hyi. Post-main-sequence tracks show two representative models, passing through the post-Hipparcos position of the star in the M_V versus $T_{\rm eff}$ plane (◇). The position based on the old ground-based parallax is also marked in grey. At bottom, a track for a solar model (evolved until 4.6 Gy) demonstrates that the present solar luminosity and temperature are accurately reproduced for the adopted model parameters. From Dravins et al. (1998, Figure 1).

opportunity to study atmospheric and magnetic activity properties of stars like the Sun as they evolve off the main sequence. Accurate age determinations are possible for G subgiants that have evolved only slightly off the main sequence, since the evolutionary tracks in this region allow little scope for ambiguity. Detailed studies of β Hyi, its photospheric structure, chromospheric and coronal X-ray emission, make it one of the best-studied individual stars other than the Sun. It is typical of the old Galactic disk population, with weak chromospheric activity, an eccentric Galactic orbit, slightly metal poor at $[\rm Fe/H] = -0.20$, and with a temperature of $T_{\rm eff} = 5800 \pm 100$ K, close to that of the Sun. Pre-Hipparcos age estimates were in the range 8–10 Gyr, with the most recent pre-Hipparcos estimate of 9.5 Gyr, based on the ground-based parallax of 153 mas, given by Dravins *et al.* (1993). Dravins *et al.* (1998), see also Dravins *et al.* (1997), used evolutionary tracks for slightly different possible metallicities, and the same helium abundance as needed to fit the Sun, to determine $M_V = 3.43 \pm 0.01$, 0.3 mag brighter than previously believed, and with a significantly lower revised age of 6.7 Gyr (Figure 7.27). The fact that the appropriate stellar evolutionary tracks are running nearly horizontally means that uncertainty in $T_{\rm eff}$ has little effect on the star's inferred properties.

Thorén *et al.* (2004) determined detailed chemical abundances for 23 candidate subgiants. Ages and masses were derived from evolutionary models. Gravities were determined from the wings of strong, pressure-broadened metal lines, and compared with those from

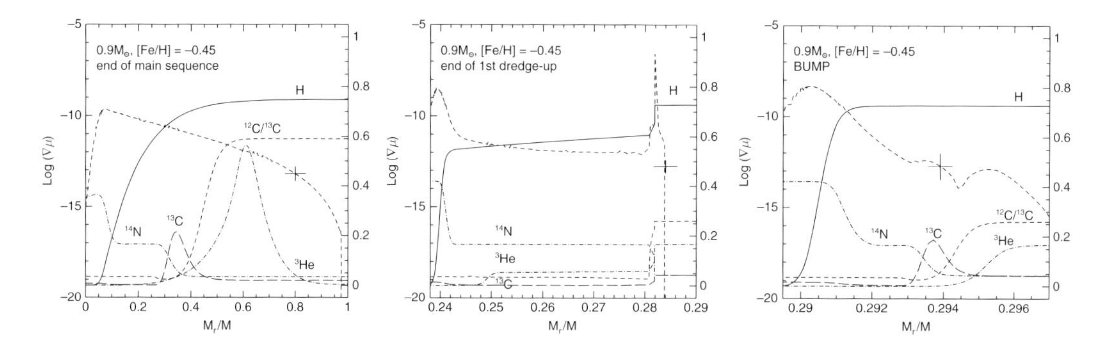

Figure 7.28 Left: molecular weight gradient (log scale, bold dashed line, left axis) and composition profiles (right axis) in a $0.9M_\odot$, [Fe/H] $= -0.45$ stellar model, at the end of the main sequence. The mass fractions are multiplied by 100 for ^{3}He and ^{14}N, by 1000 for ^{13}C; the ratio ^{12}C/^{13}C is divided by 100. The thick cross corresponds to $(\nabla \ln \mu)_c$, a critical molecular weight gradient discussed in their paper. Middle: same as left, after the completion of the first dredge-up, in the H-burning shell region. The base of the convection zone is located at $M_r/M \simeq 0.283$. Right: same, after the red giant branch bump, in the H-burning shell region. The base of the convection zone is located at $M_r/M \simeq 0.313$. From Charbonnel et al. (1998, Figures 4, 5 and 6).

Hipparcos parallaxes and evolutionary masses. Space velocities were derived from Hipparcos data, and as a result they confirmed only 12 objects as true subgiants. They found abundances corresponding closely to those of dwarfs, with the possible exceptions of lithium and carbon, implying that they show no chemical traces of post-main-sequence evolution, and that they are therefore valuable for studies of Galactic chemical evolution.

7.5.5 Giants

Giant stars are significantly more luminous than the Sun, but frequently much cooler, with their higher luminosity indicating a much larger diameter, although they are not necessarily more massive. Giants have exhausted core hydrogen burning, although they may still burn hydrogen in a surrounding shell, and may burn helium or even heavier elements in the core. The giant structure develops as increasing luminosity in the core leads to an expansion of the envelope, and thus to cooling, ion recombination, and an increasing opacity, which leads to further expansion which only ceases when the entire envelope becomes convective, thus transferring energy to the surface more efficiently. The locus of a convective envelope in the HR diagram is the Hayashi track, defining the position of the giant branch. Giants include both stars on the (red) giant branch and the asymptotic giant branch; these are only observationally separable in the HR diagram at relatively low luminosity, and it is generally with the guidance of theoretical models that the most luminous red giants and especially those with altered surface composition are assigned to the AGB and not to the red giant branch.

Charbonnel *et al.* (1998) used precise bolometric magnitudes derived from Hipparcos parallaxes to place five field giants with [Fe/H] ~ -0.5, including Arcturus,

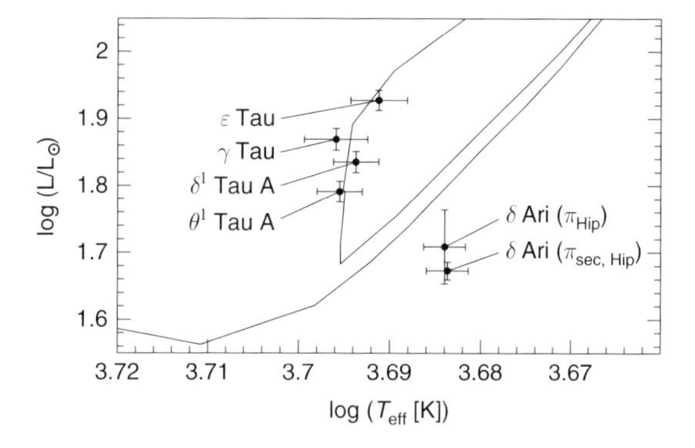

Figure 7.29 The Hyades red giant clump region based on the Hipparcos secular parallaxes from de Bruijne et al. (2001). The solid line is the $(Y, Z) = (0.273, 0.019)$ 631 Myr isochrone, with mass loss, from Girardi et al. (2000). The objects δ^1 and θ^1 Tau are spectroscopic binaries; their secular parallaxes and luminosities should be treated with some care. The suspected non-single K2III giant δ Ari is clearly not coeval with the Hyades, and therefore unlikely to be a cluster member. From de Bruijne et al. (2001, Figure 20).

into an evolutionary sequence along the ascent of the red giant branch. From studies of their C, N, O and Li abundances, and their ^{12}C/^{13}C data, they showed that the ^{12}C/^{13}C ratios drops from $\sim$ 20 to near 7 between M_{bol} = +0.9 and +0.5, while Li disappears. The C/N ratio is also affected, with log ^{12}C/^{14}N diminishing from +0.1 to -0.6 between $M_{bol} = +0.9$ and -2. Arcturus stands above the curve defined by the other stars by 0.4 dex. Their observations confirm the existence of an extra mixing process that becomes efficient on the red giant branch only when low-mass stars reach the so-called luminosity-function bump. Figure 7.28 shows the theoretical gradients of molecular weight,

$\nabla \ln \mu = d \ln \mu / dr$, and supported by these Hipparcos observations, from their standard stellar model of $0.9M_\odot$ with [Fe/H] $= -0.45$, at three different evolutionary stages: at the end of the main sequence, after the completion of the first dredge-up, and shortly after the red giant branch bump. Charbonnel *et al.* (1998) suggested that this extra mixing process could be related to rotation. Palacios *et al.* (2006) showed that this is probably not the case. The process has been subsequently attributed to a fundamental double diffusive instability originating from a molecular weight inversion created by the ^{3}He(^{3}He,2p)^{4}He reaction, referred to as thermohaline convection (Charbonnel & Zahn, 2007).

The red clump of the Hyades contains four giants: θ^1, δ^1, ϵ, and γ Tau (θ^1 and δ^1 Tau are themselves spectroscopic binaries); the determination of their effective temperatures and secular parallax-based luminosities is discussed by de Bruijne *et al.* (2001). The location of isochrones in the giant region depends quite sensitively on metallicity, on the mixing-length parameter $\alpha_{\rm MLT}$, and on the mass-loss history. All Hyades giants precisely follow the solar-metallicity 631 Myr isochrone of Girardi *et al.* (2000), which itself uses the Reimers (1975) empirical treatment of mass loss (Figure 7.29).

The Hyades membership list of Perryman *et al.* (1998) contains one additional evolved star of spectral type K2III, δ Ari which, however, has several characteristics of a non-member: it is 15.34 pc from the cluster centre, and has near-solar metallicity in contrast to the mean Hyades value. The isochrones show that δ Ari is not coeval with the classical giants, providing further evidence that the star is most likely a non-member.

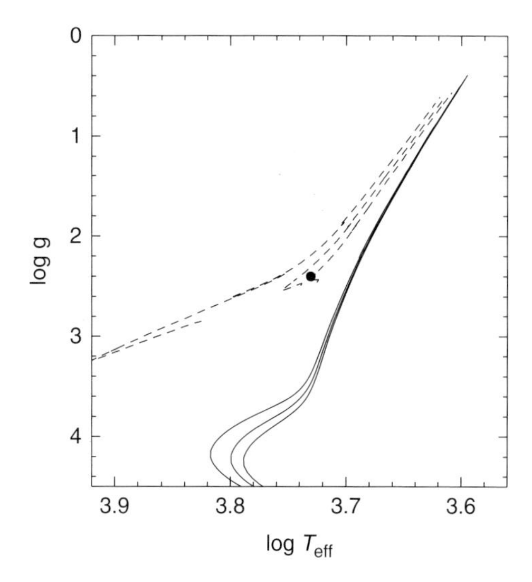

Figure 7.30 log $T_{\rm eff}$ *versus* log *g for the red horizontal branch star HD 17072. Solid lines are the Yale 1996 isochrones for* $Y = 0.23, Z = 0.001$, *[Fe/H] = –1.26 for ages of 10, 15 and 20 Gyr. Dashed lines are horizontal branch evolutionary tracks from Yi et al. (1997), also for* $Z = 0.001$. *HD 17072 falls on the more massive horizontal branch evolutionary track, rather than the red giant isochrones for any age or mass. From Carney et al. (1998, Figure 3).*

7.5.6 Horizontal branch

Main-sequence stars with initial masses between around 0.8 and $3M_\odot$ evolve through the red giant phase, losing mass through stellar winds, and adjusting their structure to evolve into horizontal branch stars. At the lower initial mass range, stars take some 10^{10} yr to evolve to the horizontal branch; they are thus old and therefore metal-poor. At the upper mass range, evolution to the horizontal branch takes some 10^9 yr; they are thus young, and would have formed in the disk probably with a solar-like composition. If the mass left in the hydrogen envelope is small, the star can be very blue and 'subluminous' (sdB/sdO, also known as extreme HB); higher envelope masses lead progressively to the blue HB stars (HB A and B, with spectra rather similar to those of main-sequence late B to early A stars), and to the red horizontal branch stars. Between the blue and red HB stars is the pulsational instability strip containing the RR Lyrae stars. In all cases, the mass in the He core is considered to be about $0.5M_\odot$, with envelope masses in the range < 0.04–$0.9M_\odot$, and total masses in the range 0.52 (HBB) to $0.6M_\odot$ (HBA). Evolution from the zero-age horizontal branch (ZAHB) leads to a small increasing luminosity (de Boer *et al.*, 1997b; de Boer, 1999).

The sdB (and sdO) classification arises from the fact that the bluest HB stars (in metal-poor globular clusters) can extend blueward of the Population I main sequence, and such a star can thus appear like a subdwarf of spectral-type B. That they cannot be subdwarfs in the sense of the term used for Population II main-sequence stars is evident from the short evolutionary time scales for B stars, with their implied ages thus inconsistent with the old ages characteristic of the low-metallicity Population II stars. Many sdB stars may have lost their hydrogen envelopes through interaction with a binary companion, and continue to reside in binary systems today (Morales-Rueda *et al.*, 2003).

A number of studies have used the Hipparcos data to refine estimates of absolute magnitudes, masses, and gravities of horizontal branch stars. These have been made either directly based on evolutionary models for nearby stars, or for horizontal branch stars in globular clusters based on globular cluster distances themselves derived from main-sequence fitting of local subdwarfs. The use of the the resulting absolute magnitudes of horizontal branch stars as distance indicators is covered further in Chapter 5.

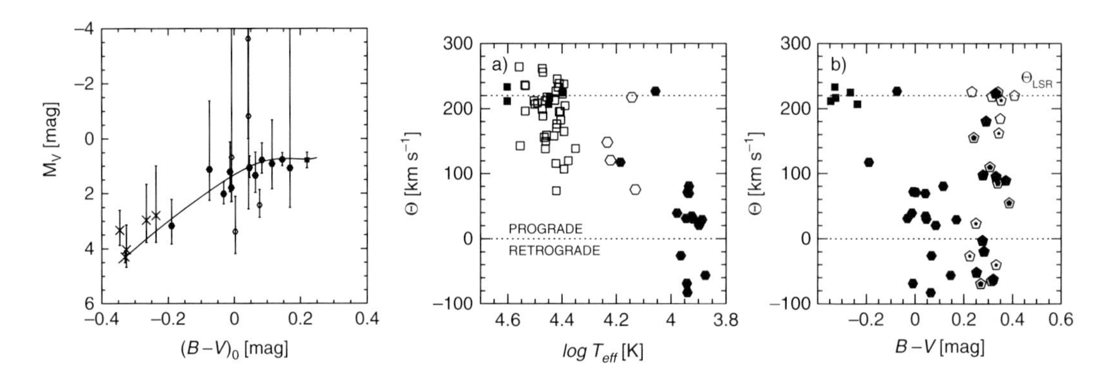

Figure 7.31 Left: colour–magnitude diagram showing the horizontal branch stars studied by Altmann & de Boer (2000), and the curve used as the shape of the field horizontal branch; hexagons: HBA/B stars (open symbols refer to stars not used for the fit); crosses: sdB/O stars; square: RR Lyrae. Right pair: kinematics of the field horizontal branch characterised by the orbital velocity, Θ. (a) Θ versus $T_{\rm eff}$; filled symbols show stars with Hipparcos data, and open symbols show results from previous studies; hexagons: HBA/B; squares: sdB/O. (b) Θ versus $B-V$, highlighting the cooler part of the horizontal branch and including RR Lyrae stars (pentagons), subdivided according to [Fe/H] < -1.6 (full symbols) through to [Fe/H] ≥ -0.9 (open symbols). From Altmann & de Boer (2000, Figures 1 and 3).

In the latter category, Heber *et al.* (1997) derived masses and gravities of blue horizontal branch stars for six globular clusters, for which previous mass estimates were in conflict with theoretical models in four cases. For M5 and NGC 6752 inferred masses were and remain in agreement with evolutionary theory. For M15 and M92, larger cluster distances led to larger masses, bringing them into agreement with theory. For NGC 288 and NGC 6397 the mass discrepancy was reduced, but discrepancies remained, suggesting that additional mixing processes need to be taken into account in the models.

Gratton (1998) used Hipparcos parallaxes for 22 metal-poor field horizontal branch stars with $V < 9$ to determine a mean value of $M_V = +0.69 \pm 0.10$ for [Fe/H] $= -1.41$, or $M_V = +0.60 \pm 0.12$ for [Fe/H] $= -1.51$ after eliminating HD 17072, which they considered could be on the first ascent of the giant branch rather than on the horizontal branch. Their values agreed with the determinations based on proper motions and application of the Baade–Wesselink method to field RR Lyraes.

The different conclusions regarding the nature and absolute magnitude of the HB stars was attributed by de Boer (1999) as a consequence of intrinsic differences due to variations in metallicity, total mass and core mass, and evolutionary routes.

For individual objects, de Boer *et al.* (1997b) used Hipparcos parallaxes to determine M_V for eight stars (five HBA and three sdB) with parallaxes in the range 3–6 mas. They used $T_{\rm eff}$ and $\log g$ from the literature to determine masses from evolutionary models, and they found $M = 0.38 \pm 0.07 M_\odot$ for $T_{\rm eff} = 7500$–9000 K, compared with theoretically-predicted masses near to $0.6 M_\odot$ (Renzini & Fusi Pecci, 1988).

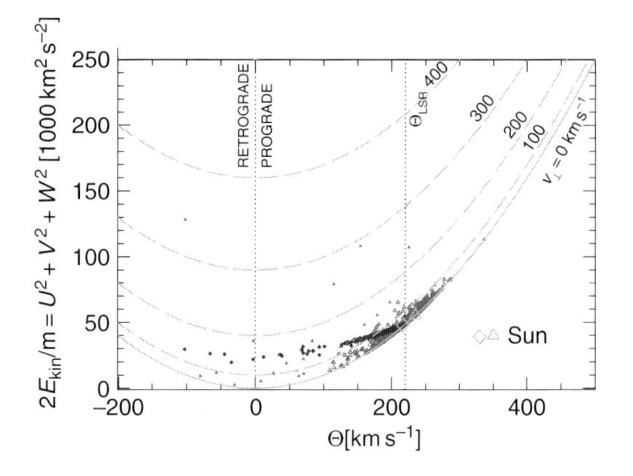

Figure 7.32 Diagram of Θ (grey triangles) and $\Theta_{\rm med}$ (black squares) against total kinetic energy and median kinetic energy, respectively. Filled symbols are stars of the halo population. Parabolae are isovelocity lines for $v_\perp$, which is orthogonal to the Θ coordinate. Median values lie on an almost straight line pointing from the LSR data point towards lower $E_{\rm kin,med}$ and $\Theta_{\rm med}$. A clear gap can be seen at $\Theta \sim 110\,{\rm km\,s^{-1}}$, below which there are only halo population stars. From Kaempf et al. (2005, Figure 6).

Carney *et al.* (1998) made a detailed study of the anomalous case HD 17072, which has the best-determined Hipparcos parallax, yielding $d = 132 \pm 8$ pc, and accordingly $M_V = 0.97 \pm 0.15$. Stellar models show convincingly that the metal-poor object is a genuine (thick disk population) red horizontal branch star (Figure 7.30). The faint luminosity confirmed for this and the other horizontal branch stars of Gratton (1998) was taken as indicating a conflict with results obtained from main-sequence fitting.

Since the atmospheres of most horizontal branch stars do not reflect their original metallicities, their kinematics provide important clues to their origins. Altmann & de Boer (2000) studied 21 field horizontal branch stars (14 HBA, 2 HBB, and 5 sdB/O), using Hipparcos parallaxes and proper motions, combined with ground-based radial velocities, to determine space motions (Figure 7.31). They found that HBA stars have low orbital velocities, with some even on retrograde orbits, with large eccentricities and in many cases reaching large distances above the Galactic plane, characteristics typical of the halo population. In contrast, the sdB/O stars show disk-like Galactic orbits. For a control sample of RR Lyrae stars, they found mixed kinematical behaviour, confirming previous studies, with some, mostly metal-rich, having disk-like orbits, but with the majority having halo-like orbits. The clear trend found in kinematics of stars along the horizontal branch, which is also a sequence in stellar mass, shows that the different kinds of field horizontal branch stars arise from stars having different origins in age and, e.g. metallicity or mass-loss rate.

A more extensive study of the kinematics of sdB stars, making incidental use of the Hipparcos data, was reported by Altmann *et al.* (2004). They found that sdB stars in the solar neighbourhood are divided into a disk population with a scale height of about 1 kpc, and a more extended population with a scale height of about 7 kpc. Kaempf *et al.* (2005) studied the spatial distribution and kinematics of the red horizontal branch stars, using the Hipparcos data both to distinguish them based on luminosity from the main sequence, subgiants and red clump stars, and to determine their kinematic properties and orbits from models of the Galaxy's gravitational potential. They showed that the red horizontal branch stars exhibit two populations, a disk population with a scale height of around 0.6 kpc, and a halo population with a scale height of about 4 kpc, similar to results found for the sdB stars (Figure 7.32).

Martin (2006) identified five blue horizontal branch stars in their study of B stars found far from the Galactic plane, finding a mix of halo and thick disk kinematics.

A search for new sdB stars from the FAUST catalogue of ultraviolet sources towards Ophiuchus was reported by Formiggini *et al.* (2002), using astrometric and photometric data from Hipparcos and Tycho to determine absolute magnitudes, and thus to identify 12 probably evolved hot stars. Spectroscopy of nine confirmed six as newly-discovered sdB stars, and two as newly-discovered white dwarfs. Discussions of the binary subluminous sdB star HD 188112, for which atmospheric parameters and the Hipparcos parallax yield a mass of only $0.24 M_\odot$, and its role as a test case for the double-degenerate scenario for progenitors of Type Ia supernovae comprising an sdB and white dwarf, can be found in Heber (2004).

7.5.7 Asymptotic giant branch

Towards the end of their lifetime, stars of low and intermediate mass ($< 8M_\odot$) evolve along the asymptotic giant branch stage. After core He-burning is exhausted, the core of carbon and oxygen is surrounded by a helium burning shell, the hydrogen burning shell and the intershell region in between. Evolved AGB stars undergo recurrent thermal instabilities of the He-burning shell (He-flash), referred to as thermal pulses. Helium-burning peak luminosities of $L_{\rm He} \sim 10^5$–$10^8 L_\odot$ cause complex convective mixing events. Subsequently, the bottom boundary of the convective envelope may engulf deeper regions where material previously synthesised by hydrogen and helium burning is present, the third dredge-up (Busso *et al.*, 1999; Herwig, 2000, 2004). Principal reaction products brought to the surface in this process are ^{12}C produced by the triple-alpha process of helium burning, thereby changing the elemental abundances of the stellar atmosphere from oxygen-rich into carbon-rich, and the heavier s-process elements. Many long-period variables (Miras, semi-regular and irregular variables) are considered to be on the AGB.

The presence of the TiO molecule is the defining characteristic of M-type spectra. The S stars are those in which the presence of TiO is matched and ultimately superseded by ZrO and other molecules containing elements produced by the s-process (the similar naming is coincidental).

Knauer *et al.* (2001) studied some 11 000 stars in common between IRAS and Hipparcos, calculating bolometric luminosities by integrating their spectral energy distributions from the B band to the far-infrared. They found a very narrow distribution in the luminosities of the AGB stars centred around $3000 L_\odot$. Based on an assumed bulge distance of 8 kpc, Jackson *et al.* (2002) found luminosities of $\sim 3500 L_\odot$ from a sample of around 10 000 candidate AGB stars selected from IRAS colours. The close relationship between the period–luminosity relation for large-amplitude AGB variables and that established for Mira variables in the solar neighbourhood, in Galactic globular clusters, and in the LMC, is discussed by Whitelock (2003).

Other aspects of AGB research include mass loss (Section 7.5.8), and stars with specific abundance anomalies including the technetium stars (Section 8.4.5).

7.5.8 Mass loss

An important process occurring in certain late stages of stellar evolution is that of mass loss, driven by the very low surface gravity and high luminosity. Observational evidence for expanding envelopes includes blue-shifted resonance lines in absorption, infrared dust emission, maser lines, and millimetre emission. These expanding envelopes indicate the existence of a stellar wind. Estimated mass-loss rates are of order 10^{-8}–$10^{-7} M_\odot \mathrm{yr}^{-1}$ for red giants, 10^{-7}–$10^{-6} M_\odot \mathrm{yr}^{-1}$ for red supergiants, and up to $10^{-5} M_\odot \mathrm{yr}^{-1}$ for Wolf–Rayet stars. These high mass-loss rates exceed the nuclear burning rate and cause the star to lose most of its hydrogen envelope. The mass-loss process is complex (Lamers, 1997; Sedlmayr & Winters, 1997) but qualitatively comprises a two-step process in which stellar pulsations drive strong shock waves into the photosphere, the shock waves then transporting mass to large radial distances where it condenses into dust particles from where radiation pressure accelerates the dust grains further outwards. Observed rates imply only brief such evolutionary stages, of order a few Myr, but a detailed understanding is important for studies of stellar evolution as well as the associated chemical enrichment of the interstellar medium. Stellar evolutionary models covering the relevant mass and evolutionary stages now frequently include the effects of mass loss explicitly (e.g. Schaerer *et al.*, 1993b; Claret, 1995; Jimenez & MacDonald, 1996; Girardi *et al.*, 2000).

The relevance of the Hipparcos data has been to supply more accurate distance estimates to specific stars or classes of stars, allowing the conversion of observed gas and/or dust mass-loss signatures, formulated in terms of mass loss per unit distance, into absolute mass loss rates, analogous to the importance of distances in transforming observed optical flux into absolute luminosity.

Groenewegen & de Jong (1998) studied 67 S-stars, deriving dust mass-loss rates per unit distance from IRAS 60 μm data, and gas mass-loss rates per unit distance from CO data. Absolute rates were then derived from estimates of distances and luminosities, of which 14 were directly determined from Hipparcos parallaxes. The S stars have mass-loss rates comparable to those found for carbon- and oxygen-rich Miras. Other studies of mass loss in Miras were made by Zijlstra & Bedding (1999). Olofsson *et al.* (2000) determined mass-loss rates from the carbon star TT Cyg based on its Hipparcos distance of 510 pc, from which they concluded that the star has recently gone through a period of strongly varying mass-loss properties. Ramdani (2003) also discussed the possible use of Hipparcos distances to determine mass-loss rates of Mira variables observed by ISO. Mass-loss rates for C-rich AGB stars based on mid-infrared observations, and calibrated using Hipparcos distances, were determined by Guandalini *et al.* (2004a,b). Bergeat & Chevallier (2005) derived mass-loss rates from millimetre observations of 119 C-rich giants, again using distances and luminosities from Hipparcos. They found

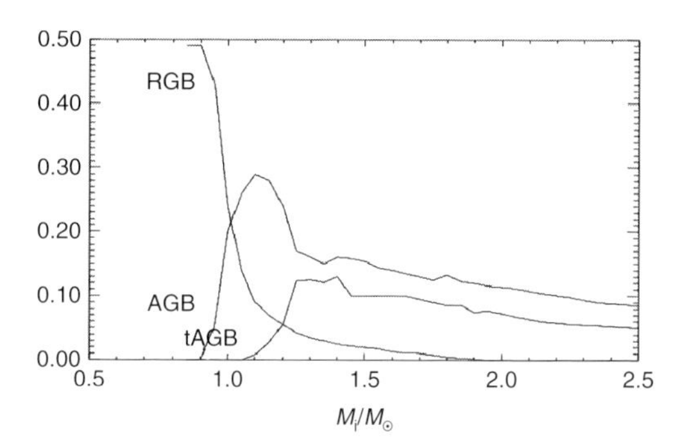

Figure 7.33 Relative contributions per mass interval (in units of $M_\odot$) to the total collective stellar mass re-injection from a sample of stars with an initial mass function dependence $\gamma = -1.8$. The respective fractions were obtained from stellar evolutionary models developed to match the observed Hipparcos HR diagram for single stars within 50 pc. From Schröder & Sedlmayr (2001, Figure 9).

fairly good agreement between predicted (Arndt *et al.*, 1997) and observed mass-loss rates as a function of T_{eff}. For most $M < 4M_\odot$ carbon-rich AGB stars over the range 2400–2900 K the mass-loss rate is rather independent of T_{eff}, with $\langle \dot{M} \rangle \simeq (5.8 \pm 4.4) \times 10^{-7} M_\odot \mathrm{yr}^{-1}$. At higher and lower temperatures the mass-loss rates increase with decreasing T_{eff}. Four stars with detached shells, evidence for strong episodic mass loss, and five cool infrared carbon-rich stars with optically-thick dust shells, have much larger mass-loss rates.

Schröder & Sedlmayr (2001) focused on the evolution of low-mass stars in the range $1-2.5M_\odot$, differentiating between the majority of the asymptotic giant branch stars which have moderate mass loss, and those which have reached the tip of the AGB, the short-lived but crucial evolutionary stage where the stellar atmosphere has turned carbon rich and where the mass loss is greatest. They adopted mass-loss rates from the empirical description of Reimers (1975) prior to the onset of C-rich winds, and from the dust-driven and pulsating wind models of Fleischer *et al.* (1992) for carbon-rich stars. The observed Hipparcos HR diagram for single stars within $d < 50$ pc was used to derive the single star initial mass function and star-formation rate of the solar neighbourhood by distributing synthetic stars on evolution tracks (from Pols *et al.*, 1998) to match the star counts in crucial regions of the observed HR diagram; a similar approach was used to constrain the model physics, including estimating the degree of overshooting for main-sequence stars, and described in Section 7.5.1. From the evolutionary models with consistent mass-loss descriptions, they computed the present-day sample of giants and tip-AGB stars, including their mass-loss rates, and derived collective yields of the resulting cool stellar winds. They argued that some 57% of the stellar mass is recycled; that stars below about $1.1M_\odot$ lose a significant amount of their mass already on the red giant branch due to the long time interval spent in that phase, and therefore contributing mainly oxygen-rich material; that pre-superwind AGB mass-loss returns some one third of all mass consumed by stars between 1 and $2.5M_\odot$; and that carbon-rich superwinds at the tip of the AGB should occur with stars more massive than $1.3M_\odot$, assuming that the dredge-up of carbon-rich material by thermal pulses operates for such giants (Figure 7.33).

Individual objects showing unusual or quantifiable mass-loss behaviour, where interpretation has been assisted by the Hipparcos data, include the oxygen-rich AGB star R Cas (Truong-Bach *et al.*, 1997), and the O star HD 191612, which shows two peculiar recurrent spectral states in the optical, varying between spectral types O6 and O8 (Walborn *et al.*, 2003). With the assistance of Hipparcos photometry the spectral variations are now considered to be periodic and predictable, perhaps indicating an eccentric binary with tidally-induced pulsations driving enhanced mass loss near periastron (Walborn, 2005).

7.5.9 Binary systems

Masses are available for a number of stars belonging to binary systems (see Chapter 3), allowing their calibration on the assumption that the stars have the same age and were born with the same chemical composition (Andersen, 1991; Noels *et al.*, 1991). A solution is sought which reproduces the observed positions in the HR diagram for both stars. Andersen (1991) argued that the only systems able to really constrain the internal structure theory are those with errors lower than 2% in mass, 1% in radius, 2% in $T_{\rm eff}$, and 25% in metallicity.

Possible variations of $\alpha_{\rm MLT}$ have been investigated through the simultaneous modelling of selected nearby visual binary systems (Fernandes *et al.*, 1998; Pourbaix *et al.*, 1999; Morel *et al.*, 2000a). Small variations of $\alpha_{\rm MLT}$ (not greater than ≈ 0.2) in the two components of α Cen (Pourbaix *et al.*, 1999) and ι Peg (Morel *et al.*, 2000a) have been suggested. Fernandes *et al.* (1998), who calibrated four systems and the Sun with the same program and input physics, found that $\alpha_{\rm MLT}$ is almost constant for [Fe/H] in the range $[{\rm Fe/H}]_\odot \pm 0.3$ dex and masses between 0.6 and $1.3M_\odot$. In this mass range, the sensitivity of models to $\alpha_{\rm MLT}$ increases with mass due to the increase with mass of the entropy jump across the superadiabatic layer, which makes the main-sequence slope vary with $\alpha_{\rm MLT}$. Lebreton (2000b) estimated that a change of $\alpha_{\rm MLT}$ of ± 0.15 around $1M_\odot$ translates into a change in $T_{\rm eff}$ of about 40–55 K depending on the metallicity. On the other hand, with the solar $\alpha_{\rm MLT}$ value the main-sequence slope of field stars and Hyades stars is well fitted. It is therefore reasonable to adopt the solar-$\alpha_{\rm MLT}$ value to model solar-type stars. For other stars, the situation is less clear. The calibration of $\alpha_{\rm MLT}$ depends on the external boundary condition applied to the model, itself sensitive to the low-temperature opacities, and on the colour transformation used for the comparison with observations. Chieffi *et al.* (1995) examined the main sequence and red giant branch in metal-deficient clusters and suggested a constancy of $\alpha_{\rm MLT}$ from main sequence to red giant branch and a decrease with decreasing Z. They found variations of $\alpha_{\rm MLT}$ with Z of $\approx 0.2 - 0.4$, but these are difficult to assess considering uncertainties in the observed and theoretical cluster sequences. On the other hand, calibration of $\alpha_{\rm MLT}$ with 2d simulations of convection gives complex results (Freytag & Salaris, 1999; Freytag *et al.*, 1999). In particular, for solar metallicity, $\alpha_{\rm MLT}$ is found to decrease when $T_{\rm eff}$ increases above solar $T_{\rm eff}$, and to increase slightly when stars move toward the red giant branch (by ≈ 0.10–0.15). At the same time, $\alpha_{\rm MLT}$ does not vary significantly when metallicity decreases at solar $T_{\rm eff}$.

Results for six binary systems and the Sun with the same program by Fernandes *et al.* (1998) and Morel *et al.* (2000a) show a general trend for Y to increase with Z : Y increases from 0.25 to 0.30 (± 0.02) when Z increases from 0.007 to 0.03 (± 0.002). However, the Hyades appear to depart from this tendency (see Section 6.3).

α Centauri AB The nearest visual binary system at just 1.3 pc, α Cen AB provides a powerful test for stellar evolutionary models. Masses, surface composition, luminosity and effective temperature are all rather well determined, and the component masses of around $1.1\,M_\odot$ and $0.9\,M_\odot$ bracket the mass of the Sun. In their early studies, Flannery & Ayres (1978) failed to find satisfactory models assuming a solar composition, but models were found corresponding to twice the solar metal abundance, consistent with there being a substantial dispersion in the metallicity of the interstellar medium at any one time. The system has been widely modelled subsequently (e.g. Noels *et al.*, 1991; Edmonds *et al.*, 1992; Lydon *et al.*, 1993; Fernandes & Neuforge, 1995). Fernandes & Neuforge (1995) used the system to test models of convection, in particular attempting to determine whether the two components require distinct mixing-length parameters. At that time, astrometric masses with an internal error of 1% were used, but the [Fe/H]-value was controversial, leading to various possibilities for the values of $\alpha_{\rm MLT}$. New radial velocity measurements have suggested masses higher than those derived from astrometry by about 6–7% (Pourbaix *et al.*, 1999). The higher masses imply a reduction in age by a factor of 2 and slightly different $\alpha_{\rm MLT}$-values for the two components. However, the orbital parallax corresponding to the high-mass option is smaller than and outside the error bars of both ground-based and Hipparcos parallax $\pi_{\rm Hip}$. Pourbaix *et al.* (1999) noted the lack of reliability of $\pi_{\rm Hip}$ given in the Hipparcos Catalogue, but since then it has been re-determined from Hipparcos intermediate data by Söderhjelm (1999).

Guenther & Demarque (2000) constructed detailed models using the most up-to-date stellar structure physics available, including helium and heavy element diffusion, OPAL equation-of-state, and OPAL and Alexander & Ferguson opacities. They constructed models for a range of likely metallicities, and for three different parallaxes: a Yale-based (ground-based) value of 750.6 ± 4.6 mas, the Hipparcos Catalogue value of 742.12 ± 1.40 mas, and the value revised by Söderhjelm (1999) using both ground and Hipparcos observations of 747.1 ± 1.2 mas; the revised value of Pourbaix *et al.* (1999), $737. \pm 2.6$ mas was not considered in their work. Figure 7.34 shows solutions for age and $\alpha_{\rm MLT}$ versus $Z_{\rm ZAMS}$, the inferred ZAMS metallicity, resulting from models which yield the same common age, $Y_{\rm ZAMS}$, and $Z_{\rm ZAMS}$ for the two stellar components, and for each of these three parallaxes. There is a jump that occurs in the common age of α Cen A and B between $Z_{\rm ZAMS} = 0.027$–0.030. This marks the transition from models that do not develop a convective core during their evolution, to the left, to models that do, to the right; core convection extending the main-sequence lifetime. Their models indicate that the age of α Cen AB is around 6–8 Gyr, with observational uncertainties, primarily in [Fe/H], limiting the accuracy. The unique evolutionary phase of the system is, however, such that the realistic uncertainty in composition spans the possibility $Z_{\rm ZAMS} \lesssim 0.03$, for which the absence of a convective core could push ages down to 6.8 ± 0.8 Gyr. Formally, they find a 10% difference in the mixing-length parameter for the two components, but in practice uncertainties in composition and $T_{\rm eff}$ exceed this difference. Adopting the Söderhjelm (1999) parallax yields self-consistent models of the two components (with $M_{\rm A} = 1.1015 \pm 0.008 M_\odot$, $M_{\rm B} = 0.9159 \pm 0.007 M_\odot$) that match the observed metallicity and which have $Y_{\rm ZAMS} = 0.28$, a composition consistent with expectations of Galactic nucleosynthesis. They go on to show how p-mode spacings would further constrain the models, in particular to test whether component A has a convective core.

Subsequent work includes further modelling (Morel *et al.*, 2000b); newly-observed oscillation frequencies which exclude the revised masses proposed by Pourbaix *et al.* (1999), and which yield similar values of $M_{\rm A} = 1.100 \pm 0.006 M_\odot$, $M_{\rm B} = 0.907 \pm 0.006 M_\odot$, $Y_{\rm ZAMS} = 0.300 \pm 0.008$ but a significantly smaller age of 4.85 ± 0.5 Gyr (Thévenin *et al.*, 2002); angular diameters of 8.314 ± 0.016 mas and 5.856 ± 0.027 mas yielding linear diameters and hence masses compatible with Thévenin *et al.* (2002), based on new VLTI–VINCI observations (Kervella *et al.*, 2003b); and an age of 6.52 ± 0.3 Gyr, $Y_{\rm ZAMS} = 0.275 \pm 0.010$, $Z_{\rm ZAMS} = 0.0434 \pm 0.0020$, and an 8% difference in $\alpha_{\rm MLT}$ for the two components, $\alpha_{\rm A} = 1.83 \pm 0.10$ and $\alpha_{\rm B} = 1.97 \pm 0.10$, with radii consistent with the VLTI–VINCI results, all based on detailed models and new seismology data (Eggenberger *et al.*, 2004). The discrepancy between non-seismic ages of 8.9 Gyr, and seismic constraints yielding 5.6–5.9 Gyr, however, continues (Yıldız, 2007).

Hyades binaries Lastennet *et al.* (1999) used three binary systems in the Hyades, 51 Tau, V818 Tau, and θ^2 Tau, with known metallicity and good Johnson photometric data, to test the validity of three sets of stellar evolutionary tracks (Geneva, Padova, and Granada groups, see Table 7.2), making use of the additional constraint that the metallicity should be the same and equal to the Hipparcos metallicity for all six components. They found consistent tracks for all three binary systems, except when using the orbital rather than the Hipparcos parallax for V818 Tau, which would lead to metallicities too large in comparison with the observed range. Comparisons between predicted and measured masses also showed good agreement. Although the Padova tracks predict correct masses for each component of V818 Tau, there is no Padova track that fits both components simultaneously in the mass–radius diagram. The approach was extended to 60 systems by Lastennet & Valls-Gabaud (2002).

Lebreton *et al.* (2001) used the Hyades binaries to estimate the helium abundance with smaller error bars than those obtained by Perryman *et al.* (1998) from their fit to the HR diagram (see also Section 7.6.3). They also noted the difficulties the models have in fitting the radii from eclipses in the vB 22 system, and confirmed, both from the HR diagram and from the vB 22 studies, that the mixing-length parameter could vary with mass. This had previously been suggested by Castellani *et al.* (2001) from their study of the Hyades HR diagram. Problems with the models for vB 22 were also reported by Pinsonneault *et al.* (2003). Yıldız *et al.* (2006) investigated the components of five Hyades binaries, and established a clear relation between the mixing-length parameter and the stellar mass. They proposed the functional dependence $\alpha_{\rm MLT} = 9.19(M/M_\odot - 0.74)^{0.053} - 6.65$ (their Figure 4) which is also in agreement with the solar value, and with models for α Cen.

Armstrong *et al.* (2006) combined the Hipparcos secular parallax of θ^2 Tau with visual orbit data derived from Mk III and NPOI interferometric observations to estimate a system mass of $4.03 \pm 0.20 M_\odot$. Individual masses were inferred from spectroscopic orbits. Despite uncertainties in luminosities and colours due to the rapid rotation and unknown rotational inclination, both components were found to be less massive and/or brighter than predicted from evolutionary models. The results provide an indication of still-unresolved conflicts between theory and observation even in the case of the Hyades.

7.5.10 Other results

Other related studies include: populations among high-velocity early-type stars (Royer, 1997); calibration of luminosity of hot stars (Nicolet, 1997); kinematics and M_V calibration of K and M dwarf stars (Upgren *et al.*, 1997); age and morphology of the main sequence from the Hipparcos HR diagram (Badiali *et al.*, 1997); the mass of horizontal-branch stars with Hipparcos (de Boer *et al.*, 1997a); Hipparcos and the theory of stellar interiors (Baglin, 1998); populating the HR diagram with the Hipparcos Catalogue (Casertano & Ratnatunga, 1999); evolutionary state and fundamental parameters of metallic A-F giants (North *et al.*, 1998); early-type stars and their origin (Royer, 1999); absolute magnitudes of Hipparcos M-type semi-regular variables (Aslan & Yeşilyaprak, 2000); luminosity calibrations with Hipparcos (Lebreton, 2002); the Hipparcos mission legacy for cool stars (Grenon, 2001); Hipparcos, IUE, and the stellar content of the solar neighbourhood (Allende Prieto, 2001); and mean stellar parameters from IUE and Hipparcos data (Niemczura *et al.*, 2003).

Related studies of individual objects include: atmospheric parameters for the giant stars MU Peg and λ Peg (Smith, 1998); non-LTE line-blanketed model for the B giant β CMa (Aufdenberg *et al.*, 1999); stellar evolution models for γ Per (Pourbaix, 1999); evolutionary status of HD 104237, ϵ Cha and HD 100546 (Shen & Hu, 1999); and UV Psc and the radiative flux scale (Vivekananda Rao & Radhika, 2003).

7.6 Abundances

An enormous literature exists on investigations of elemental abundances which impact, and have been affected by, the Hipparcos observations in one form or another. This part of the review is accordingly structured as follows. Since observational estimates of [Fe/H] are widely used as a proxy for overall metal content, in turn an essential ingredient for stellar evolutionary models, this section starts with a consideration of measurements and interpretation of [Fe/H] data which have been influenced by the availability of the Hipparcos results. This is followed by a similar overview of the α element abundances. Helium, although only directly measurable spectroscopically in hot stars, occupies a very special role in both its primordial content as generated in Big Bang nucleosynthesis, and the central role that its initial abundance plays in stellar evolutionary models. Lithium is generated in Big Bang nucleosynthesis, and also destroyed and created in specific evolutionary phases, and provides strong constraints on evolutionary models. Abundance studies of low-metallicity Population II stars, related to the investigations in Section 7.5.3, and metal-rich stars, are given short sections. Some aspects of the relationship between metallicity, age, and kinematics, also relevant to topics covered in Chapter 9, are included here. Finally, some of the many classes of chemically anomalous stars are noted; frequently arising from convective processes or mass transfer from a binary companion, under this category are sections on the class of 'chemically-peculiar' stars, technetium-rich stars (related to Section 7.5.7), and barium-rich stars.

7.6.1 [Fe/H]

Various compilations of [Fe/H] are found in the literature. Cayrel de Strobel *et al.* (1997b) presented the 5th edition of the catalogue of [Fe/H] determinations, containing 5946 determinations for 3247 stars with 700 bibliographic references. More recent work typically rests on Hipparcos-based samples. Haywood (2001a) used the Hipparcos Catalogue to define a sample of dwarfs in the solar neighbourhood, which they show to be centred on solar metallicity, with a 4% contribution of metal-poor stars, $[Fe/H] < -0.5$, in agreement with present estimates of the thick disk. Allende Prieto *et al.* (2004) reported a high-resolution spectroscopic survey for 118 stars within 14.5 pc and $M_V < +6.5$ mag. They found an inferred [Fe/H] distribution centred at about -0.1 dex, which they consider is representative of a larger volume of the local thin disk. Fuhrmann (2004) reported high-resolution spectroscopic observations of about 150 nearby stars. Luck & Heiter (2005) presented an abundance analysis for stars within 15 pc, $\delta > -30°$, and $M_V < +7.5$ mag, selected from the Hipparcos Catalogue and comprising 114 stars. They found the local mean metallicity to be $[Fe/H] = -0.07$ using all stars, and -0.04 when interlopers from the thick disk are eliminated (Figure 7.35). Jonsell *et al.* (2005) derived abundances for 43 metal-poor stars in the solar neighbourhood, $-3.0 < [Fe/H] < -0.4$, with $T_{\rm eff}$ estimated from Strömgren photometry and surface gravities from Hipparcos parallaxes.

Abundances estimated from model atmospheres in LTE neglect the perturbations of statistical equilibrium by the radiation field. Thévenin & Idiart (1999) found that in metal-deficient dwarfs and subgiants, the iron over-ionisation resulting from an enhanced ultraviolet flux modifies the line-widths. They obtained differential non-LTE/LTE abundance corrections increasing from 0.0 dex at [Fe/H] = 0.0 to +0.3 dex at $[Fe/H] = -3.0$. These corrections are supported by the agreement between

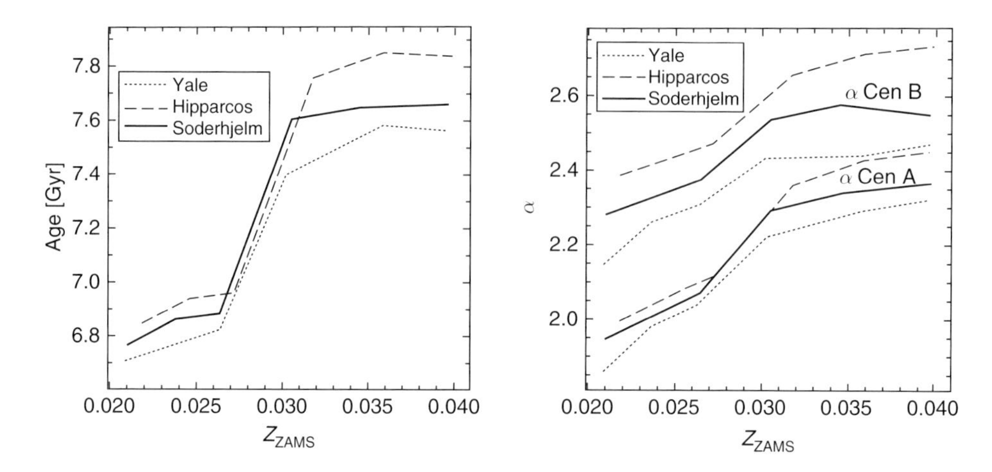

Figure 7.34 Age (left) and $\alpha_{\rm MLT}$ (right) versus $Z_{\rm ZAMS}$ for models yielding the same common age, $Y_{\rm ZAMS}$, and $Z_{\rm ZAMS}$ for the two stellar components of the binary α Cen. From Guenther & Demarque (2000, Figures 4 and 5).

spectroscopic gravities and those derived from the Hipparcos parallaxes. Korn *et al.* (2003) calculated non-LTE iron-line strengths and surface gravities under the influence of neutral hydrogen collisions in cool stars, demonstrating consistency between the resulting surface gravities and those derived from the Hipparcos parallaxes. Meléndez & Ramírez (2005) made an analysis of 25 Hipparcos metal-rich G and late F dwarfs, showing that assumptions of LTE do not simultaneously satisfy observational constraints imposed by the infrared flux method, and the Hipparcos parallaxes.

7.6.2 α-elements

Abundances of α-elements (O, Ne, Mg, Si, S, Ar, Ca, Ti) are important indicators of the star's enrichment history, since Type II supernovae create large amounts of the α elements with respect to iron, while Type Ia supernovae do not (see box on page 344). Until recently, metal-poor stars with $[\mathrm{Fe/H}] \lesssim -0.5$ dex have generally been considered to exhibit an $[\alpha/\mathrm{Fe}]$ enhancement with respect to the Sun quite independent of their metallicity (Wheeler *et al.*, 1989; McWilliam, 1997), but more recent results based on Hipparcos data show that metal-poor stars with lower α abundance ratios exist, perhaps with specific kinematic properties indicative of an accreted population (e.g. Hanson *et al.*, 1998). Other more recent α-element abundance studies have also been based on Hipparcos Catalogue samples, and their kinematic properties are summarised in Section 9.9. Clementini *et al.* (1999) determined $[\alpha/\mathrm{Fe}]$ for 99 metal-poor dwarfs with $-2.5 < [\mathrm{Fe/H}] < 0.2$ from high-resolution spectra, and found $[\alpha/\mathrm{Fe}] = +0.26\pm0.08$ dex. Gratton *et al.* (2003b) determined Fe and α-element abundances for about 150 field subdwarfs and early subgiants, using the Hipparcos astrometry to discuss their dependence on Galactic space velocity, population, and age. Bensby *et al.* (2005) performed a detailed abundance analysis for 102 F and G dwarfs, using the Hippacos astrometry to study the abundance trends in the thin and thick disk for 14 elements.

Theoretical models predict low-metallicity, α-enhanced stars at the earliest stages of galaxy evolution, when only Type II supernovae are responsible for the heavy elements. Later, as Type Ia supernovae occur, the $[\alpha/\mathrm{Fe}]$ abundance ratio decreases progressively as a result of the larger production of Fe. In broad terms, the picture has been summarised by Nissen (2000) as follows: in the thick disk, a fast initial star-formation rate resulted in a relatively high metallicity, $[\mathrm{Fe/H}] \sim -0.4$, before the iron elements produced in Type Ia supernovae decreased the value of $[\alpha/\mathrm{Fe}]$; in the thin disk, a slower initial star-formation rate led to decreasing $[\alpha/\mathrm{Fe}]$ at a lower metallicity, $[\mathrm{Fe/H}] \sim -0.6$. The overall abundance decreases with increasing Galactocentric distance, around $-0.07\,\mathrm{dex\,kpc^{-1}}$ out to 8.5 kpc, and slightly steeper decrease beyond, $-0.10\,\mathrm{dex\,kpc^{-1}}$; in the halo, most stars were formed at an even lower initial rate, and the decreasing $[\alpha/\mathrm{Fe}]$ occurs at $[\mathrm{Fe/H}] \sim -1.2$. Some halo stars have $[\alpha/\mathrm{Fe}]$ typical of the thick disk, consistent with a dual component formation model, in which an inner part formed with a fast formation rate, and an outer part with a low evolution or accreted from infall.

Halo stars are mainly metal-poor, typically in the range $-2.5 < [\mathrm{Fe/H}] < -1.0$, with a small fraction of very metal-poor, $[\mathrm{Fe/H}] < -2.5$. The thick disk population lies in the range $-1.2 < [\mathrm{Fe/H}] < -0.2$. For the metal-poor halo stars, $[\mathrm{Fe/H}] < -1.0$, $[\alpha/\mathrm{Fe}]$ is fairly constant at around 0.4 dex. Nissen (2000) argued that stars with low α abundances were formed in the outer part of the halo or were accreted from dwarf galaxies

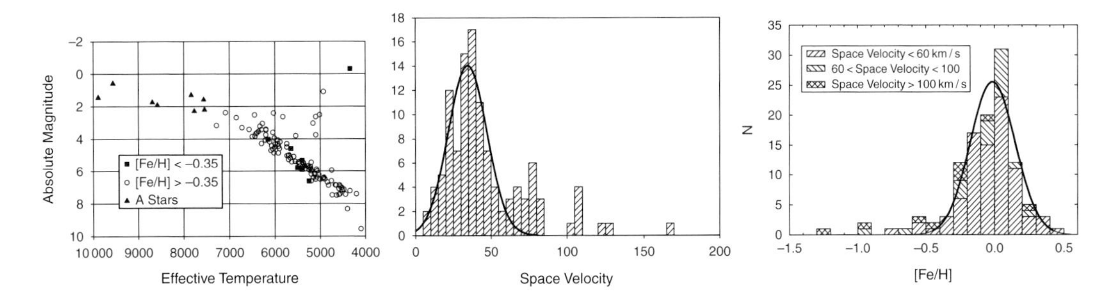

Figure 7.35 Northern stars within 15 pc of the Sun. Left: the HR diagram according to metallicity, with an additional sample of A stars with effective temperatures derived from Geneva photometry. Middle: space velocity distribution from Hipparcos astrometry and ground-based radial velocities: the maximum of the Gaussian fit lies at 34.5 km s^{-1}, with a standard deviation of 12.9 km s^{-1}, and there is an extended tail beginning at 60 km s^{-1}, or 2σ from the maximum. Right: metallicity distribution: the Gaussian fit yields a mean of [Fe/H] = −0.014 with a standard deviation of 0.016. From Luck & Heiter (2005, Figures 10, 11, 14).

like the Magellanic Clouds. For globular clusters, Carney (1996) found a fairly constant value of $[\alpha/\mathrm{Fe}] \sim +0.3$ for $[\mathrm{Fe/H}] < -0.6$, as for the halo field stars, but some have low $[\alpha/\mathrm{Fe}]$, closer to solar.

7.6.3 Helium

Helium and metal abundances both have a significant effect on the location of the stellar main sequence (e.g. Perrin *et al.*, 1977; Fernandes *et al.*, 1996; Pagel & Portinari, 1998), and these dependencies have consequences for all aspects of stellar astrophysics, the chemical evolution of galaxies, and cosmology. However, since helium lines are present in stellar spectra only at $T_{\mathrm{eff}} \gtrsim 20\,000$ K, helium content can typically only be determined indirectly, and contingently upon the correct interpretation of evolutionary models. The calibration of the solar model in luminosity and radius at solar age yields the initial helium content of the Sun (Christensen-Dalsgaard, 1982), while oscillation frequencies give access to the present value in the convection zone (Kosovichev *et al.*, 1992). For some binary systems of known mass and metallicity, the helium abundance can also be derived in an analogous manner; models have to satisfy the constraints on luminosity and effective temperature for two stars assumed to have the same age, metallicity, and initial helium content (see Section 7.5.9, and in particular the models for α Cen). In a few objects the helium abundance can be inferred from theoretical models with their associated caveats (see, e.g. the Hyades in Section 6.3). In single low-mass stars, however, neither mass nor age are known.

If the [Fe/H] ratio is known, it is common to use the scaling relation $Y - Y_{\mathrm{P}} = Z(\Delta Y/\Delta Z)$, which supposes that the helium abundance has grown with metallicity Z from the primordial value Y_{P} to its stellar birth value Y, where Y and Z represent abundances in mass fraction. Many attempts have been made to measure the enrichment factor, including pre- and post-Hipparcos measurements of the width of the lower main sequence in the solar neighbourhood (Section 7.5.2), and values in the range $\Delta Y/\Delta Z = 2$–6 have been reported (Fernandes *et al.*, 1996). The value is of interest for stellar evolution, the initial mass function, and the determination of the primordial helium abundance (Høg *et al.*, 1998; Pagel & Portinari, 1998; Fernandes, 2001). As already noted in Section 6.3 for the Hyades, models based on a helium content $Y \sim 0.28$ inferred from the $\Delta Y/\Delta Z$ enrichment law are not consistent with the observations, where a value $Y \sim 0.255 \pm 0.009$ was derived by Lebreton *et al.* (2001). For a primordial helium abundance $Y_{\mathrm{P}} = 0.235$, $\Delta Y/\Delta Z$ can again be derived from models reproducing the luminosity and radius of the Sun; thus for a solar model not including effects of microscopic diffusion, Lebreton *et al.* (2001) obtained $Y_\odot = 0.2674$, $Z_\odot = 0.0175$, and consequently $(\Delta Y/\Delta Z)_\odot = 1.9$. Jimenez *et al.* (2003) quoted $Y_\odot = 0.275$, $Z_\odot = 0.017$, and consequently $(\Delta Y/\Delta Z)_\odot = 2.3$.

Jimenez *et al.* (2003) estimated the cosmic production rate of helium relative to metals using 31 K dwarf stars in the Hipparcos Catalogue with accurate spectroscopic metallicities. K-dwarfs are very long lived, slowly evolving stars, so that their present day helium and metallicity is essentially the same as when they were born: they thus contain a fossil record of the amount of helium and metals which has been produced in successive stellar generations over the lifetime of the Galaxy. Their best-fitting value is $\Delta Y/\Delta Z = 2.1 \pm 0.4$ (Figure 7.36), in rather good agreement with determinations from H II regions (Peimbert *et al.*, 2002; Gruenwald *et al.*, 2002) and with theoretical predictions from stellar yields with standard assumptions for the initial mass function (Tsujimoto *et al.*, 1997), but significantly different from the value derived for the Hyades of less than one (Lebreton *et al.*, 2001). Flynn (2004) included

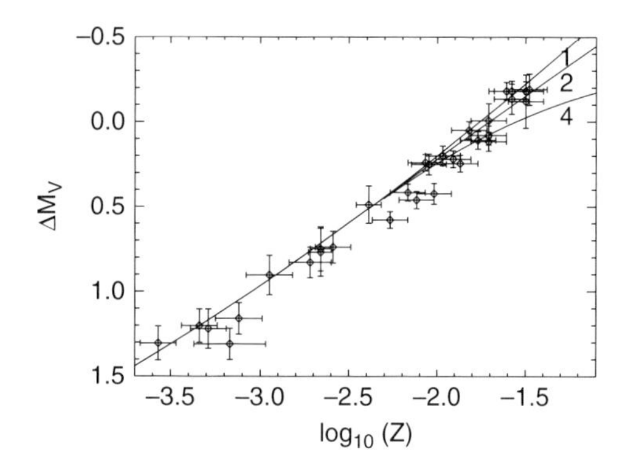

Figure 7.36 Determination of the cosmic production rate of helium, using 31 Hipparcos K dwarfs with spectroscopically determined metallicities. Solid lines correspond to ΔM_V, measured from the fiducial solar isochrone, from theoretical models for K dwarfs for different values of $\Delta Y/\Delta Z$ (1, 2 and 4 from top to bottom) as a function of metallicity. Errors in ΔM_V are determined by colour and parallax errors in about equal proportion. From Jimenez et al. (2003, Figure 2); reprinted with permission from AAAS.

recently measured K-dwarfs with supersolar metallicities, and obtained a further improvement on previous studies, finding a best-fitting value of $\Delta Y/\Delta Z = 2.4 \pm 0.4$. Casagrande *et al.* (2007) also used the Hipparcos data for K-dwarfs to derive $\Delta Y/\Delta Z = 2.1 \pm 0.9$.

At the time of the Hipparcos Catalogue publication, there had been a long-standing problem concerning the observed abundances of ^{3}He (Charbonnel & do Nascimento, 1998): briefly, its Galactic evolution was considered to be dominated by the processing of D to ^{3}He in low mass stars during the pre-main-sequence and main-sequence phases, engulfed in the convective envelope during the dredge-up on the lower red giant branch, and finally ejected into the interstellar medium during the final stages of evolution. In this standard view, the abundance of ^{3}He increases with time. The low observed abundance, of around $^3\mathrm{He/H} \sim 10^{-5}$, in the proto-solar nebula, the local interstellar medium, and in Galactic H II regions are in conflict with this model, and instead suggested that ^{3}He must be destroyed in some 70% of low-mass stars before becoming planetary nebulae. Rood *et al.* (1984) suggested that the destruction of ^{3}He could be related to the differences between predicted and observed ^{12}C/^{13}C ratios in low-mass red giants, pointing to an extra mixing process occurring at the point of the luminosity function bump. Charbonnel & do Nascimento (1998) studied 191 F–K giants, with Hipparcos parallaxes for the field giants providing luminosities and hence their evolutionary status. They determined that 96% of low-mass stars experience an extra mixing process on the red giant branch, revealed by the ^{12}C/^{13}C ratios, and are therefore expected to destroy their ^{3}He. While consistent non-standard stellar models were still needed to explain the various chemical anomalies in low-mass red giant branch stars in order to obtain reliable ^{3}He yields, they could already conclude that the very high percentage satisfies the Galactic requirements for the evolution of the ^{3}He abundance.

Understanding and quantifying the influence of the helium abundance assume an even greater importance with the discovery of colour bifurcation in the middle main sequence of the globular cluster ω Centauri (Bedin *et al.*, 2004), and a triple main sequence observed in the globular cluster NGC 2808 reported from HST observations by Piotto *et al.* (2007). The latter attribute the main-sequence branches, and the complexity of its horizontal branch, to successive episodes of star formation, with different helium abundances. The discovery perhaps invalidates one of the basic premises generally assumed in studies of globular and open clusters: that they host single stellar populations, coeval and originally chemically homogeneous.

7.6.4 Lithium

Lithium, amongst other elements, provides an important diagnostic of both primordial nucleosynthesis and of stellar evolutionary models. The disentangling of the two effects has proven to be a complex area of investigation, with many questions still unanswered. Essentially, ^{7}Li is produced in the Big Bang (along with D, ^{3}He, and ^{4}He), and its primordial abundance therefore provides an important test of Big Bang nucleosynthesis. ^{6}Li is produced primarily through cosmic-ray fusion reactions with the interstellar gas (e.g. Fields & Olive, 1999). Both isotopes are, however, destroyed at relatively low temperatures: at typical main-sequence densities, ^{7}Li survives only in the outer 2–3% of the stellar mass, where $T \lesssim 2.5 \times 10^6$ K. Otherwise, burning through the reaction ^{7}Li(p,α)^{4}He leads to Li depletion, which in turn provides an important test of evolutionary models (Pinsonneault *et al.*, 1992). High lithium abundance combined with high chromospheric activity appears to indicate stellar youth, although alone it is not considered to be a good tracer of age for solar-type stars (see, e.g. Pasquini *et al.*, 1994; Mallik, 1999). The picture is further complicated by the fact that abundances of lithium and other elements provide evidence for additional mixing processes (Pinsonneault, 1997), with depletion and additional creation processes now recognised. The Hipparcos data have been used to extend these investigations (e.g. Spite *et al.*, 1998; Cayrel *et al.*, 1999).

Lithium versus age In studies of lithium abundances versus age, Favata *et al.* (1997a) used the Hipparcos parallaxes of a number of high lithium abundance, high activity late-type dwarfs found from X-ray surveys with Einstein. The origin of these stars has been a matter of

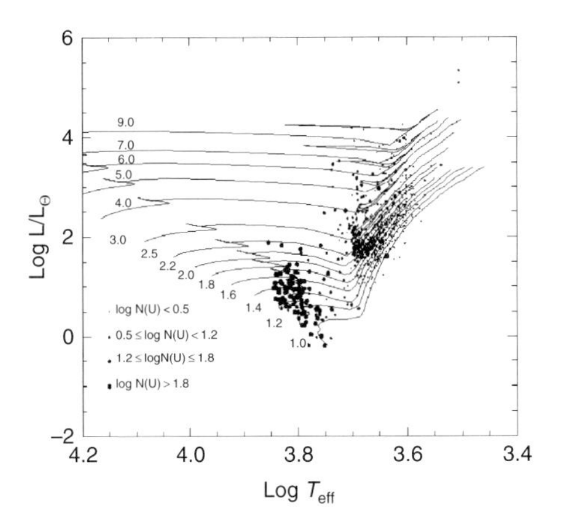

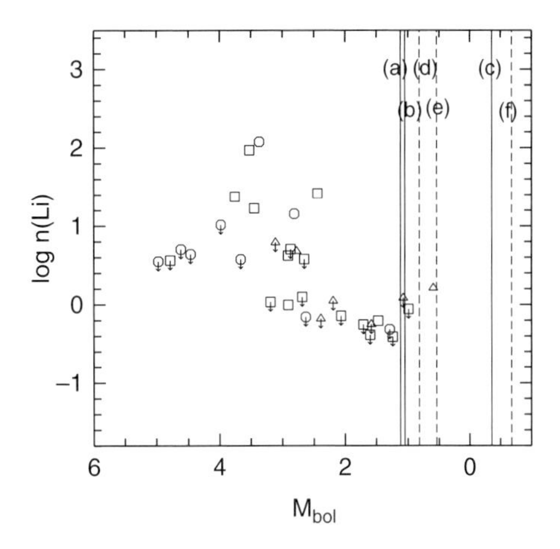

Figure 7.37 Left: HR diagram of the sample of subgiants, giants and supergiants studied by Mallik (1999) for which absolute visual magnitudes were determined from the Hipparcos data. Symbols of increasing size denote increasing Li abundances (see key). Theoretical evolutionary tracks are from Bressan et al. (1993) for initial masses ranging from $1-9M_\odot$. From Mallik (1999, Figure 3). Right: lithium abundance versus M_{bol} for subgiants with $B-V>0.7$ (i.e. stars that have completed the first-dredge up dilution), $M<1.4M_\odot$, and $M_V>1$. Symbols: △: [Fe/H] ≤ -0.3; □: $-0.3<$ [Fe/H] ≤ 0.05; ○: [Fe/H] > 0.05. Vertical lines represent evolutionary points where extra mixing should occur, according to Charbonnel (1994). Solid curves (a–c) refer to $Z=0.05, 0.02, 0.001$ respectively and to $M=1.0M_\odot$; dashed curves (d–f) refer to $M=1.25M_\odot$ for the same values of Z. From Randich et al. (1999, Figure 8a).

debate: if they were very young, with ages of a few million years and still on the Hayashi track, they would be inexplicably distant from obvious sites of star formation, while if they were close to the ZAMS, with ages of tens to hundreds of Myr, their origin would be less problematic. They found ages of $\sim$ 50 Myr, but with a spread in lithium abundance, X-ray emission, and rotational velocity. Pastori *et al.* (1999) reported further studies of lithium versus age for X-ray selected active cool stars, deriving X-ray luminosities from the Hipparcos parallaxes. Fischer *et al.* (1998) constructed a colour–magnitude diagram from a sample of lithium-rich nearby field stars. They showed that while most of the young field stars coincide with the distribution given by the Praesepe cluster, a few fall along the (more depressed) Pleiades sequence. This work is discussed further in Section 6.4.

Lithium depletion and evolutionary models Mallik (1999) measured Li abundances for 65 subgiants, giants and supergiants (spectral types F3–M3), combining the data with results for 802 similar objects for which Li abundances were already known, and for all of which Hipparcos parallaxes allowed the determination of absolute magnitudes and the construction of the corresponding HR diagram (Figure 7.37, left). Masses were estimated from evolutionary tracks. Some objects, especially of low mass, are located on or near the main sequence, with a few in the Hertzsprung gap, with others on the giant branch extending to $\log L/L_\odot = 4.0$ or higher. A pattern of decreasing Li with decreasing temperature is evident, with a large scatter at any given T_{eff}, and much larger depletions observed in the many of the higher mass ($> 2.5M_\odot$) cooler ($T_{eff} < 5000$ K) giants than predicted by standard evolutionary models. The results confirm an additional source of mixing at the end of the subgiant phase, or suggest significant mass loss even in the early parts of the giant phase.

Randich *et al.* (1999) made a similar analysis for 91 Population I subgiants, again using Hipparcos parallaxes to determine absolute magnitudes and evolutionary masses. Abundances are consistent with lithium depletion on the main sequence, along with two further episodes of depletion. From their lithium abundance versus $B-V$ colour index, they showed that the dilution of lithium in the first dredge-up is completed at around $B-V = 0.7$, or around $T_{eff} = 5600$–5700 K, independently of mass or metallicity. A second mixing episode on the red giant branch is predicted to occur when the hydrogen burning shell crosses the chemical discontinuity created by the convective envelope at its maximum extent, corrsponding to the so-called bump in the luminosity function, and being dependent on mass and metallicity (Charbonnel, 1994). Their dependence of lithium abundance versus M_{bol} for stars below $1.4M_\odot$ (Figure 7.37, right) demonstrates that $M_{bol} < 2$, or $\log L/L_\odot > 1.1$, providing an upper limit for the luminosity where extra mixing starts for these lower mass stars. This limit is significantly lower than predicted by the theoretical models of Charbonnel (1994).

Do Nascimento *et al.* (2000) studied lithium abundance and rotation for 120 subgiant stars, determining their evolutionary status from the Hipparcos parallaxes. They found good agreement with models of lithium dilution arising from the deepening of the convective envelope after the main-sequence turn-off, except for the more massive cooler stars. Chen *et al.* (2001) determined lithium abundances for 185 main-sequence field stars, deriving masses and ages from stellar evolution models and Hipparcos-based luminosities. They found a large gap in the lithium abundance versus T_{eff} plane, distinguishing the low abundance 'Li-dip' stars, in which Li depletion arises from some unknown process, perhaps the mass dependency of gravity waves (Talon & Charbonnel, 1998; Charbonnel & Talon, 2005), from others with a much higher abundance in which lithium depletion is attributed to normal convective dilution. They argue that Li depletion occurs early in the stellar life but with parameters other than mass and metallicity, e.g. initial rotation velocity or the rate of angular momentum loss, affecting the degree of dilution. Further analysis of the connection between rotation and lithium abundance, based on the evolutionary status determined from Hipparcos, was made for 121 subgiants by do Nascimento *et al.* (2003), for 127 F and G dwarfs and subgiants by Mallik *et al.* (2003), for 181 F and G dwarfs by Lambert & Reddy (2004), and for 56 FG giants by Jasniewicz *et al.* (2006).

Halo stars and the spite plateau One of the central lines of investigation related to lithium abundance is understanding the so-called 'Spite plateau': the discovery of a remarkably flat and constant Li abundance in Galactic halo dwarfs spanning a wide range of T_{eff} and metallicity (Spite & Spite, 1982a,b), and at a factor of 10 or so below the abundances found in young Population I objects. The two extreme interpretations are that either the halo population abundance represents the primordial Li abundance with the Galaxy having been enriched in Li by a factor of 10 since its birth, or that the value measured in Population I stars as well as in meteorites represents the primordial abundance with the halo stars having undergone a uniform depletion by a corresponding factor. The current paradigm, in part supported by the WMAP results on the baryon-to-photon ratio combined with standard Big Bang nucleosynthesis (Romano *et al.*, 2003), suggests that the primordial Li abundance lies somewhat midway between these two extremes (Charbonnel & Primas, 2005). The challenge is then to explain how the halo population has been so uniformally depleted by a factor of 3 over a large range in T_{eff} and metallicity (Pinsonneault *et al.*, 2000; Talon & Charbonnel, 2004).

Charbonnel & Primas (2005) studied 115 halo stars, deriving the evolutionary status from the Hipparcos parallaxes and evolutionary models, and thus rigorously excluding contamination from post main sequence stars. They found that the mean Li value as well as the dispersion appears to be marginally but significantly lower for the dwarfs than for the turn-off and subgiant stars. They concluded that the most massive halo stars have had a slightly different Li history than their less massive contemporaries, placing new constraints on possible depletion mechanisms for the halo stars.

Lithium abundance anomalies in G–K giants Another major area of investigation has been in the understanding of the unexpected occurrence of strong Li lines in about 1% of G–K giants (Wallerstein & Sneden, 1982), at levels much higher than any plausible initial composition as constrained by values found in the present interstellar medium (Balachandran *et al.*, 2000). Lithium is expected to be destroyed, at higher temperatures, in all but the outermost layers of a main-sequence star: observed depletion by some two orders of magnitude during the main-sequence phase is generally explained by slow rotationally-induced mixing, with some diffusion, and possible some early main-sequence mass loss. At the base of the red giant branch, a deepening convective envelope dilutes the remaining lithium during the first dredge-up, reducing the photospheric abundance by a further two orders of magnitude. G and K giants are therefore expected to have very low lithium abundances as a result of this overall convective dilution. The mechanism of 'hot bottom burning' used to explain the high lithium abundance in the most massive AGB stars (Sackmann & Boothroyd, 1999), cannot be invoked for the red giant branch and the less massive AGB stars.

To explain this phenomenon, de la Reza *et al.* (1997) proposed a model in which all ordinary, Li-poor, K giants become Li rich during a short time interval of order 10^5 yr, during which a thin expanding circumstellar shell triggered by an abrupt internal mixing mechanism was postulated to result in further surface ^{7}Li enrichment. Sackmann & Boothroyd (1999) predicted detailed results for a different mechanism, in which lithium could be created in low-mass red giants (via the Cameron–Fowler mechanism) in a process referred to as 'cool bottom processing', in which deep extra mixing transports envelope material into the outer wings of the hydrogen-burning shell, where it undergoes partial nuclear processing before being transported back to the envelope.

Jasniewicz *et al.* (1999) tested these hypotheses using 29 late-type giants with far infrared excess, finding no correlation between lithium abundance and infrared excess, and thus tending to preclude the first mechanism. Rather, they used their Hipparcos-derived positions in the HR diagram to argue that all stars in their sample have undergone mixing, and favoured instead the scenario in which additional lithium is created by extra-deep mixing.

Balachandran *et al.* (2000) made a detailed study of two K giants with high (supermeteoritic) lithium abundances, providing specific evidence for fresh ^{7}Li production, and excluding primordial lithium and planetary accretion as viable scenarios for the formation of Li-rich giants.

Charbonnel & Balachandran (2000) studied 20 lithium-rich giants where the Hipparcos data allowed accurate placing on the HR diagram (Figure 7.38). Five could be classified as lithium-rich because they have not completed the first dredge-up dilution, while three have abundances compatible with standard dilution; together these objects could be re-classified as lithium normal. For the remainder, they attribute the high lithium abundance to fresh synthesis occurring at two distinct episodes depending on mass: for low-mass red giants branch stars, lithium is generated at the phase referred to as the luminosity function bump, where the outwardly-moving hydrogen shell is burning through the mean molecular weight discontinuity created by the first dredge-up. Extra mixing can then connect the ^{3}He-rich envelope to the outer regions of the hydrogen-burning shell, enabling Li production by the Cameron & Fowler (1971) process. In intermediate-mass stars, the mean molecular weight gradient due to the first dredge-up persists until after core helium burning starts. The Li-rich phase in these stars occurs when the convective envelope deepens at the base of the AGB, permitting extra mixing to occur, and ceases when a strong mean molecular weight gradient is built up between the deepening convective envelope and the nuclear burning shell surrounding the inert CO core. Both episodes are very short lived.

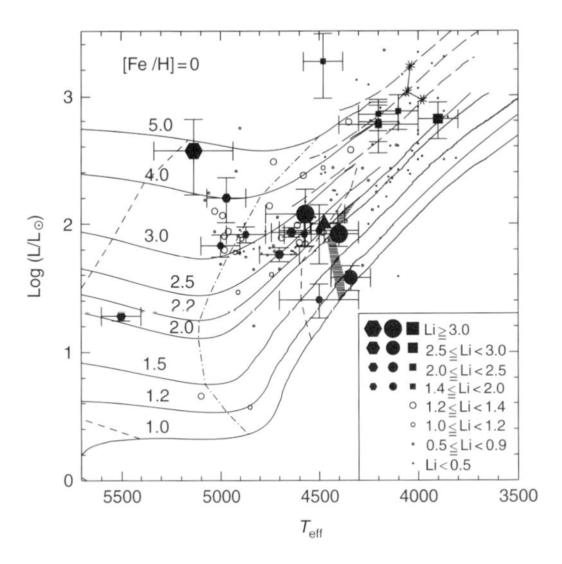

Figure 7.38 The HR diagram for Li-rich giants studied by Charbonnel & Balachandran (2000). Evolutionary tracks with [Fe/H] = 0 are labelled by mass. Two dashed lines delimit the first dredge-up; the higher temperature line marks the start of Li dilution, and the lower temperature line marks the deepest convective penetration. The dashed-dotted line indicates the start of the decrease in the carbon isotopic ratio due to dilution. The shaded region indicates the red giant branch bump. A solid line connects the asterisks where a molecular weight barrier appears due to the second dredge-up. Hexagons: Li-normal stars undergoing first dredge-up dilution; circles and triangle: stars at the red giant branch bump; squares: early-AGB stars; open circles: stars with Hipparcos parallaxes and $-0.1 \leq$ [Fe/H] $\leq +0.1$ from Brown et al. (1989). Symbol sizes indicate Li abundance. From Charbonnel & Balachandran (2000, Figure 1).

Studies of ^{6}Li The very different formation mechanisms of ^{6}Li and ^{7}Li can be investigated from the small isotopic shift of the ^{6}Li I 670.8 nm doublet of +0.0158 nm relative to the ^{7}Li doublet (Nissen *et al.*, 1999). Crifo *et al.* (1997, 1998) studied the Population II metal-poor star HD 84937, for which the pre-Hipparcos parallax had suggested it to be a dwarf. Significant ^{6}Li (at about 6% of ^{7}Li) is observed, but not predicted by standard models if it is indeed a dwarf of this colour and mass, for which convection should have depleted Li to levels below detectability. Chaboyer (1994) and Deliyannis & Malaney (1995) noted that the inconsistency would be removed if the star is a subgiant, and predicted that the star should have a parallax of about 11 mas. The Hipparcos parallax of 12.44 ± 1.06 mas indeed confirms that the star is a subgiant, with a ^{6}Li abundance in agreement with the standard models. Crifo *et al.* (1998) applied similar considerations to BD +26° 3578, HD 201891, and HD 140283. The latter had been a particularly perplexing case in terms of Li and Be abundances, with the ground-based parallax indicating that the star should be a dwarf, but the Hipparcos parallax showing that the star is a subgiant. Hobbs *et al.* (1999) used Hipparcos parallaxes to investigate the evolutionary states of halo stars with measured ^{6}Li/^{7}Li ratios, finding some consistent with predictions of cosmic ray production models, with others requiring significant ^{6}Li depletion mechanisms. Nissen *et al.* (1999) used the absolute magnitudes and evolutionary tracks of five metal-poor disk stars to show that their ^{6}Li abundances are in agreement with standard cosmic ray production in the Galactic disk (e.g. Fields & Olive, 1999) if combined with a moderate depletion for these stars.

7.6.5 Metal-poor stars

This section covers some of the detailed abundance studies of metal-poor stars influenced by the Hipparcos data availability. It is connected with the discussions of Section 7.5.3, but emphasises the consequences for studies of the chemical enrichment of the Galaxy. This is an extensive subject of investigation, and only some highlights are given.

The traditional explanation for the chemical evolution of the Galactic halo was formulated by Tinsley (1979), based on the differing products of the two main types of supernovae. Type Ia supernovae produce mainly Fe-group elements, while Type II supernovae produce lighter elements (including the α-elements), as well as some Fe-group and heavier elements. Since the time between star formation and explosion differs between the two types (Type II requiring 10^7 yr, while Type Ia require more than 10^9 yr), there is a time in which the enrichment is only from Type II events. The stars created out of the products of these early Type II supernovae will be relatively rich in the α- (and other light) elements, until enough Type Ia supernovae can explode to 'dilute' the light elements with Fe-group elements. This overall pattern is seen in observations of halo stars and clusters (e.g. McWilliam *et al.*, 1995) and indicates that element ratios can be used as an indicator of the history of a stellar population. More recent work has shown that the chemical evolution of the halo is more complicated: for example, Nissen & Schuster (1997) found that there were a number of stars on halo-like orbits that exhibited significantly lower [α/Fe] ratios than the disk stars at similar metallicities.

Hanson *et al.* (1998) studied the [Na/Fe] ratio in 68 metal-poor field giants with $-3 \lesssim$ [Fe/H] $\lesssim -1$, calculating Galactic velocity components from the Hipparcos data, and studying how abundances relate to kinematical structure in the halo. Na is considered to be an effective proxy for Mg and the α-elements, since it is considered to be created during C and Ne burning in massive stars, and subsequently ejected in Type II supernovae along with the other α elements. Their study is based on present evidence that [α/Fe] itself provides a kind of nuclear clock. Nine metal-poor (rather than very metal-poor) stars were assigned to the metal-weak thick disk with prograde orbits. The most metal-poor stars, [Fe/H] < -1.89, show a wider dispersion of [Na/Fe] ratios than do the less metal-poor stars; the difference is most striking for stars on retrograde Galactic orbits. Some 20% of the giants on retrograde orbits have [Na/Fe] ≤ -0.35, and may therefore be significantly younger than the oldest halo objects.

Various measurement programmes have been directed to determining detailed abundances for various metal-poor stars from the Hipparcos Catalogue, many of the objects common to the various programmes. Gratton *et al.* (1997) and Clementini *et al.* (1999) reported homogeneous photometric data (Johnson V, $B-V$, $V-K$, Cousins $V-I$ and Strömgren $b-y$), radial velocities, and abundances of Fe, O, Mg, Si, Ca, Ti, Cr and Ni, for 99 metal-poor Hipparcos stars in the range $-2.5 \lesssim$ [Fe/H] $\lesssim -0.2$. Their main results were that the equilibrium of ionisation of Fe is well satisfied in late F-early K dwarfs, and that oxygen and the α–elements are over-abundant by ~ 0.3 dex. Fulbright (2000) determined abundances for 168 metal-poor dwarfs. Element-to-iron ratios were derived for various α-, odd-Z, Fe-peak, and r- and s-process elements. Effects of non-LTE on the analysis of Fe I lines were shown to be very small on average, and spectroscopically determined surface gravities were rather close to those obtained from the Hipparcos parallaxes. They also concluded that $T_{\rm eff}$ can be determined from the abundance versus excitation potential for Fe I, that $\log g$ can be determined by matching iron abundances determined by the neutral and singly-ionised lines, and that the microturbulent velocity can be determined by the abundance versus line-strength plot for Fe I. Oxygen abundances were derived from metal-poor Hipparcos stars, and variously interpreted in terms of the Hipparcos-based gravities, by García López *et al.* (2001), Israelian *et al.* (2001), and Meléndez *et al.* (2001).

Gratton *et al.* (2003a) determined spectroscopic abundances for 150 Hipparcos metal-poor subdwarfs and subgiants, belonging to both the thick disk and the halo, and with $-2 <$ [Fe/H] < -0.6. Abundances were estimated for Fe, O, Na, Mg, Si, Ca, Ti, Sc, V, Cr, Mn, Ni, and Zn; the α-element abundance was taken as the average of Mg, Si, Ca, and Ti. Effective temperatures, surface gravities, and microturbulent velocities were estimated from evolutionary models and the Hipparcos-based luminosities, with population assigned according to Galactic orbit derived from the Hipparcos proper motions. Gratton *et al.* (2003a) identified three stellar components in their chemical and dynamical analysis: (a) a 'dissipative component', being a rotating population and probably produced during the dissipative collapse of the bulk of the early Galaxy, and identified with the population first identified by Eggen *et al.* (1962). It includes part of what is usually called the halo as well as the thick disk; their choice is governed by the fact that they could discern no clear discontinuity between the properties of the rotating part of the halo, and those of the thick disk; (b) an 'accretion component' comprising non-rotating and counter-rotating stars; it includes the remaining part of what is usually called the halo, is substantially identified with the accretion population proposed by Searle & Zinn (1978), and has a distinct chemical composition from that of the dissipative component, most likely due to a different origin; (c) the thin disk, which also has a distinct chemical composition from the dissipative population, likely due to a phase of low star formation that occurred during the early evolution of the Galaxy.

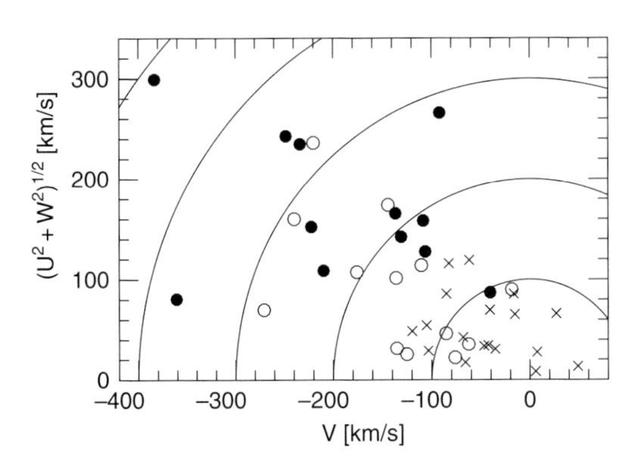

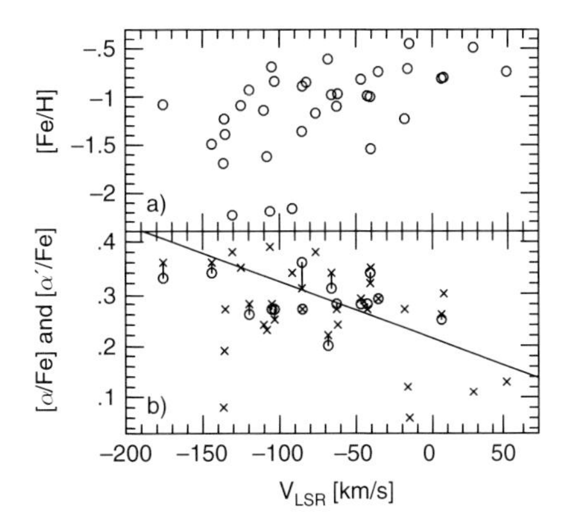

Figure 7.39 Left: Toomre diagram from the study of Jonsell et al. (2005), where symbols indicate overall stellar metallicity: $-1.0 \leq$ [Fe/H] (×), $-1.5 \leq$ [Fe/H] < -1.0 (∘) and [Fe/H] < -1.5 (•). Curves connect loci with identical total velocities relative to the LSR, and the diagram shows an obvious anticorrelation between overall metallicity and velocity relative to the LSR. Right: correlations of iron abundances and [α/Fe] with $V_{\rm LSR}$ for stars of the 'dissipative collapse component' of Gratton et al. (2003b). In the lower panel the line shows their Equation 3; ∘ [α/Fe] = $\frac{1}{4}$([Mg/Fe]+[Si/Fe]+[Ca/Fe]+[Ti/Fe]); × [α′/Fe] = $\frac{1}{3}$([Mg/Fe]+[Si/Fe]+[Ca/Fe]). From Jonsell et al. (2005, Figures 2 and 18).

Gehren *et al.* (2004) determined abundances of the light neutron-rich elements ^{23}Na and ^{27}Al in a small sample of moderately metal-poor stars with surface gravities derived from the Hipparcos data. They found that [Na/Mg] and [Al/Mg], in combination with [Mg/Fe], space velocities and stellar evolutionary ages, provide a discrimination between thick disk and halo stars, with the defining gap at [Al/Mg] ~ -0.15 and [Fe/H] ~ -1.0, and with an absolute but preliminary age of 14 Gyr, derived from the models of VandenBerg *et al.* (2000), i.e. without accounting for the effects of diffusion. While the absolute age boundary will therefore be lowered once diffusion effects are included, they take their results to indicate a need to revise current models of chemical evolution and/or stellar nucleosynthesis to allow for an adequate production of neutron-rich species in early stellar generations. Jonsell *et al.* (2005) determined O, Na, Mg, Al, Si, Ca, Sc, Ti, V, Cr, Fe, Ni, and Ba for 43 Hipparcos metal-poor field stars in the solar neighbourhood, most of them subgiants or turn-off-point stars with iron abundances $-3.0 <$ [Fe/H] < -0.4. They found some indication that [α/Fe] varies, and that some of the scatter around the trends in abundances relative to iron may be real, supporting the idea that the formation of halo stars occurred in smaller systems with different star-formation rates. They verified the finding by Gratton *et al.* (2003a) that stars that do not participate in the rotation of the Galactic disk show a lower mean and larger spread in [α/Fe] than stars participating in the general rotation, with the latter stars also showing some correlation between [α/Fe] and rotation speed (Figure 7.39). Zhang & Zhao (2005) made abundance determinations of Fe, O, Na, Mg, Al, Si, K, Ca, Sc, Ti, V, Cr, Mn, Ni, Cu and Ba, for 32 very metal-poor Hipparcos stars. Abundances were determined from LTE analyses using gravities determined from Hipparcos parallaxes and evolutionary tracks. Based on kinematics, the stars were also separated into dissipative collapse and accretion components of the halo population, confirming the trends observed in the earlier studies.

In conclusion, considerable data on the detailed chemical abundances of the metal-poor Hipparcos stars now exists. These can be tied to stellar masses, ages, and Galactic motions through use of the Hipparcos data combined with evolutionary models in which diffusion forms a necessary ingredient. There is a current tendency to assign the low metallicity stars to one of the three components proposed by Gratton *et al.* (2003a), within which framework the complex chemical evolution history of the Galaxy is now being routinely interpreted.

7.6.6 Super metal-rich stars

The term 'super-metal-rich' was introduced in the 1960s to describe a population where photometric results indicated over-abundances up to a factor 10 with respect to the Sun, itself considered as metal-rich. These very high over-abundance were not confirmed by spectroscopy, but the term continues to be used for (G and K) stars with metallicities equal to or higher than those of the Hyades, i.e. [Fe/H] $\sim 0.14 \pm 0.5$ (Cayrel de Strobel *et al.*, 1999). Metallicities reach [Fe/H] ~ 0.55 in the solar neighbourhood, or a little higher in the bulge, significantly exceeding the maximum reached now in the solar neighbourhood in recently formed stars. They are rather rare in the solar vicinity, representing some 3% of stars in the [Fe/H] Catalogue of Cayrel de Strobel *et al.* (1997b), and may include bulge objects at their orbital apocentres. They are important for determining the upper limit of metal enrichment in stars, and for understanding nucleosynthesis products in supernovae and AGB stars (Cayrel de Strobel, 1987; Barbuy & Grenon, 1990; Feltzing & Gustafsson, 1998).

Cayrel de Strobel *et al.* (1997a) selected stars from the [Fe/H] Catalogue of Cayrel de Strobel *et al.* (1997b), and used the Hipparcos data to identify two groups: one comprising G and K dwarfs and subgiants and reflecting their initial chemical composition, the other comprising giants in which at least part of the enrichment could reflect their own nucleosynthesis products. They used the parallaxes to define a turn-off age of the slightly evolved subgiants, confirming that the metal-rich stars are on average older than those of solar metallicity, implying a more chemically uniform present-day interstellar medium compared with the older, more active, interstellar medium.

Cayrel de Strobel *et al.* (1999) studied about 100 nearby G and K subgiants and dwarfs with metallicities with respect to solar of between 0.14–0.55 dex, having convective zones sufficiently developed to ensure that the atmospheric abundances reflect their initial chemical composition. Among those with well-determined ages, about 80% have intermediate ages of 2–5 Gyr, but only 20% have ages of 8 Gyr or more. They showed that the super-metal-rich phenomenon has existed from the beginning of the thin and thick disk populations, but with insufficent data to establish how the phenomenon has varied with time.

Grenon (1999) determined ages and kinematics for 270 Hipparcos super-metal-rich objects. He attributed part of the metal-rich to super-metal-rich stars, of intermediate age, as representing objects formed a few kpc inside the solar orbit from slowly-enriched gas, with some metal-rich bulge-like stars probably scattered out from the inner Galaxy. The oldest super-metal-rich objects were born earlier, closer to the Galactic centre where metal enrichment had been more rapid. Their eccentric orbits and outward migration can be explained by pertubations due to the Galactic bar.

7.6.7 Chemical enrichment of the Galaxy

Chemical enrichment of the Galaxy is a substantial area of research that has been affected by the Hipparcos data. A large body of literature relates to the enrichment products of successive generations of star formation, and the resulting inter-relationships between various abundance signatures, kinematics, and age (see, e.g. Pagel, 1997). The topic relates to stellar evolution through the products of nucleosynthesis returned to the interstellar medium at the various stages of stellar evolution, and to studies of the structure, kinematics, and age of the various stellar populations in the Galaxy. This section is restricted to a summary of some of the studies that have been made based on the Hipparcos data which impact on knowledge of the Galaxy's chemical enrichment. Aspects more closely connected to the spatial structure and kinematics of the various Galactic components are considered in Section 9.5.

The basic objectives are to derive the observed metallicity distribution in the solar neighbourhood, the observed age–metallicity relation, and to establish how these compare with theoretical models. The observational basis for these investigations include extensive surveys of metallicity distribution of solar neighbourhood dwarfs (e.g. Wyse & Gilmore, 1995; Rocha Pinto & Maciel, 1996; Favata *et al.*, 1997b; Flynn & Morell, 1997). Another central result came from the spectroscopic survey of F and G stars within 50 pc by Edvardsson *et al.* (1993). This supplied metallicity estimates as well as estimated ages from evolutionary tracks, and revealed a slowly increasing gradient with decreasing age in the age–metallicity relation, but with a larger scatter in metallicity than expected at a particular age (a review is given by Rocha Pinto *et al.*, 2000). The metallicity surveys had repeatedly pointed out a deficit of metal-poor stars relative to the simple paradigm given by the 'closed-box model' of the solar neighbourhood, in which no matter flows in or out, where the gas is initially free of metals, where the initial mass function is constant, and where the interstellar medium is homogeneous and well-mixed at all times. A related issue is the 'G dwarf problem', essentially the lack of long-lived metal-poor stars in the vicinity of the Sun relative to the simple closed-box model. A more detailed introduction to the observational framework is given by Haywood (2001a), with further discussion of the role of the thick disk (its contribution to the observed metallicity distribution, and whether it should be taken into account in the G dwarf problem), the fact that G-dwarfs are inherently biased against metal-rich stars, and the conversion between star counts and mass fraction. Some considerations on the specific effects of binary stars in the study of chemical evolution are reported by Vanbeveren & de Donder (2003).

Models proposed to explain the observed metallicity distribution, and the Edvardsson *et al.* (1993) scatter in the local age–metallicity relation, essentially involve relaxing one or more of the assumptions underlying the closed-box model; they include inhomogeneities in the interstellar medium (Malinie *et al.*, 1993), radial diffusion of stellar orbits (Grenon, 1987; Francois & Matteucci, 1993), episodic infall of metal-poor gas (Pilyugin & Edmunds, 1996a,b), and sequential or stochastic enrichment by stellar populations (van den Hoek & de Jong, 1997).

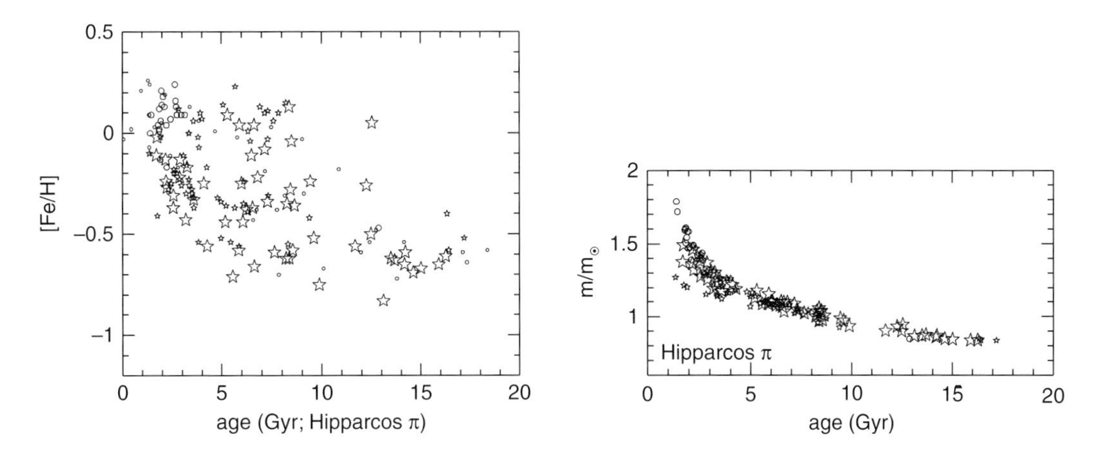

Figure 7.40 Left: the age–metallicity relation for F and G stars from the Edvardsson et al. (1993) sample, re-analysed using the Hipparcos data. Symbols represent main sequence (○) and subgiants (⋆). Right: the age–mass relation for the main sequence and subgiant branch stars having an uncertainty in log (age) $<$ *0.05. From Ng & Bertelli (1998, Figures 6 and 8).*

The Hipparcos contributions to these questions include a better definition of complete samples of survey stars based on distance, as well as their photometric and binary star statistics; improvement in estimating evolutionary ages through improved luminosities and other fundamental parameters; and the provisional of improved Galactic kinematic data for population studies, i.e. identifying halo, thin, or thick disk stars through the proper motions (as well as ground-based radial velocities).

Haywood *et al.* (1997a,b) constructed the local metallicity distribution of a sample of G and K dwarfs, using their positions in the HR diagram to select 243 stars still on the main sequence. Compared to the distribution obtained by Wyse & Gilmore (1995), both peak at the same metallicity of -0.1 dex; however, they found about 6% with [M/H] > 0.25 dex, while these are non-existent in the Wyse & Gilmore (1995) sample; a similar proportion have [M/H] < -0.5 dex, while they represent 19% in the earlier sample. Their results were an early confirmation from Hipparcos of the relative deficit of metal-poor stars in the solar neighbourhood.

Ng & Bertelli (1998) computed new ages, partly Hipparcos-based, for the sample of F and G stars within 50 pc studied by Edvardsson *et al.* (1993). Their objective was to re-examine whether the scatter in metallicity at a given age is real, or whether it resides in the earlier age estimates. Ng & Bertelli (1998) derived ages using the evolutionary models of Bertelli *et al.* (1994), with luminosities from the Hipparcos parallaxes. The resulting age–metallicity relation for the stars on the subgiant branch has a small but distinct slope, of ~ 0.07 dex Gyr^{-1}. They also derived the corresponding age–mass relation for the main sequence and subgiant branch (Figure 7.40).

Caloi *et al.* (1999) studied the relation between age, kinematics, and abundances for nearly 10 000 stars in the solar neighbourhood. They found that samples with space velocity $V > -30$ km s^{-1}, although including stars as old as 10^{10} yr, have a very young component (10^7–10^8 yr), and have a metallicity close to solar. Most with $V < -40$ km s^{-1} have a minimum age of about 2 Gyr. Stars with $V \lesssim -80$ km s^{-1} share a common age of about 10 Gyr. The distribution in [Fe/H] of these oldest objects suggests a decrease in metal content with increasing $|V|$ up to [Fe/H] < -2. For $V < -180$ km s^{-1} only the metal-poor component ([Fe/H] < -0.7) is found. The common age for large space velocity suggests that no substantial age spread exists in the inner halo, at least in the local sample.

Chen *et al.* (2000) obtained high-resolution spectra of 90 F and G disk dwarfs, covering the metallicity range $-1.0 <$ [Fe/H] $< +0.1$, and analysed in a similar way to the work of Edvardsson *et al.* (1993). They determined accurate evolutionary ages using Hipparcos-based luminosities, effective temperatures derived using the infrared flux method and the Hipparcos-based gravities, and Galactic velocity components from the Hipparcos astrometry. Some stars in the range $-1.0 <$ [Fe/H] < -0.6 having a small mean Galactocentric distance in the stellar orbits were shown to be older than the other disk stars and probably belong to the thick disk. Excluding these stars, a slight decreasing trend of [Fe/H] with increasing Galactocentric distance and age was found, but a large scatter in [Fe/H] (up to 0.5 dex) is present at a given age and Galactocentric distance. Abundance ratios with respect to Fe showed no significant scatter at a given [Fe/H]. Trends of O, Mg, Si, Ca, Ti, Ni and Ba as a function of [Fe/H] agreed rather well with those of Edvardsson *et al.* (1993), but the over-abundance of Na and Al for metal-poor stars found in their work was not confirmed.

Garnett & Kobulnicky (2000) also examined the age–metallicity relation based on the Edvardsson *et al.* (1993) data along with Hipparcos parallaxes to define distance-basd samples, and making use of the revised age estimates made by Ng & Bertelli (1998). They found that the scatter in the age–metallicity relation depends on distance, such that stars within 30 pc of the Sun show significantly less scatter in [Fe/H]. Stars of intermediate age at distances 30–80 pc from the Sun are systematically more metal poor than those nearby. They also found that the slope of the apparent age–metallicity relation is different for stars within 30 pc from that of more distant stars, probably attributable to selection biases in the original star sample. They concluded that the intrinsic dispersion in metallicity is much smaller than that measured at fixed age by Edvardsson *et al.* (1993), being less than 0.15 dex for field stars in the solar neighbourhood, and consistent with less than 0.1 dex for Galactic open star clusters and the interstellar medium.

Haywood (2001a), see also Haywood (2001b), updated the analysis of Haywood *et al.* (1997a), defining a sample of 393 stars within 25 pc for which metallicity estimates were derived mainly from calibrated Geneva and Strömgren photometry. From a detailed study, and a review of the earlier related surveys, he made the following conclusions: that most previous metallicity distributions have been biased against metal-rich stars; that his revised distribution, corrected for biases, is centred on solar metallicity; that the percentage of thick disk metallicity stars with $[Fe/H] < -0.5$ is about 4%, compatible with other estimates of the local density of this population; and that the observed distribution is well reproduced by a simple closed-box model. The rather radical conclusion is that the most widely-accepted mechanism for alleviating the previously-perceived deficit of metal-poor stars, the infall model, appears to be unnecessary for describing the solar vicinity metallicity distribution. If it can be demonstrated that old stars (older than about 8 Gyr), with disk-like kinematics and solar abundances, exist in significant proportions, it would imply that part of the old disk has formed through an evolutionary path that differs from the simple closed-box model. Further discussions are given by Haywood (2006) and Holmberg *et al.* (2007).

Kotoneva *et al.* (2002a, 2003) determined metallicities for 431 single K dwarfs selected from the Hipparcos Catalogue. The fact that G dwarfs are sufficiently massive that some have started to evolve away from the main sequence, requiring evolutionary corrections when determining space densities and metallicities, makes the longer-lived K dwarfs more suitable for studies of the local metallicity distribution; their intrinsically lower luminosity has made accurate spectroscopic abundances more problematic until recently, and as a result it is now known that the 'G dwarf problem' extends to the K dwarfs (Flynn & Morell, 1997), and even to the M dwarfs (Mould, 1976). Kotoneva *et al.* (2002a) used evolutionary isochrones to determine masses, selecting a subset of 220 stars complete within a narrow mass interval. The data were then adjusted to a model of the chemical evolution of the solar cylinder. They found that only a modest cosmic scatter was required to fit their age–metallicity relation, based on two main infall episodes for the formation of the halo-thick disk and thin disk respectively. Their study confirms that the solar neighbourhood formed on a long time scale of order 7 Gyr.

Reddy *et al.* (2003) measured abundances for 27 elements from C to Eu, including α-elements, in 181 F and G dwarfs; $T_{\rm eff}$ were adopted from an infrared flux calibration of Strömgren photometry, $\log g$ and ages from evolutionary tracks based on Hipparcos luminosities, and Galactic velocities (showing that most are thin disk population) from Hipparcos parallaxes. Relative abundances generally confirm previous published results: $[\alpha/Fe]$ increases slightly with decreasing [Fe/H]. Heavy elements with dominant contributions at solar metallicity from the s-process show [s/Fe] to decrease slightly with decreasing [Fe/H]. Scatter in [X/Fe] at a fixed [Fe/H] is entirely attributable to measurement errors, after excluding the few thick disk stars and the s-process-enriched CH subgiants. Tight limits are set on the remaining cosmic scatter. If a weak trend with [Fe/H] is taken into account, the composition of a thin disk star expressed as [X/Fe] is independent of the star's age and birthplace for elements contributed in different proportions by massive stars (Type II supernovae), exploding white dwarfs (Type Ia supernovae) and asymptotic red giant branch stars. Their results, taken in combination with other published studies, extend previous ideas about composition differences between the thin and thick disk: the thick disk stars are primarily identified by their velocity with respect to the LSR in the range -40 to $-100\,{\rm km\,s^{-1}}$, representing very old stars with origins in the inner Galaxy, and with metallicities $[Fe/H] \leq -0.4$. At the same [Fe/H], the thin disk stars have velocities with respect to the LSR close to zero, and are generally younger with a birthplace at about the Sun's Galactocentric distance. In the range $-0.35 \geq [Fe/H] \geq -0.70$, well represented by present thin and thick disk samples, [X/Fe] of the thick disk stars is greater than that of thin disk stars for Mg, Al, Si, Ca, Ti and Eu. [X/Fe] is very similar for the thin and thick disk for Na and the iron-group elements. [Ba/Fe] may be under-abundant in thick relative to thin disk stars (Figure 7.41). Zhang & Zhao (2006) made a similar study of abundances in 32 mildly metal-poor Hipparcos stars from the thin and thick disk. Reddy *et al.* (2006) extended the analysis to

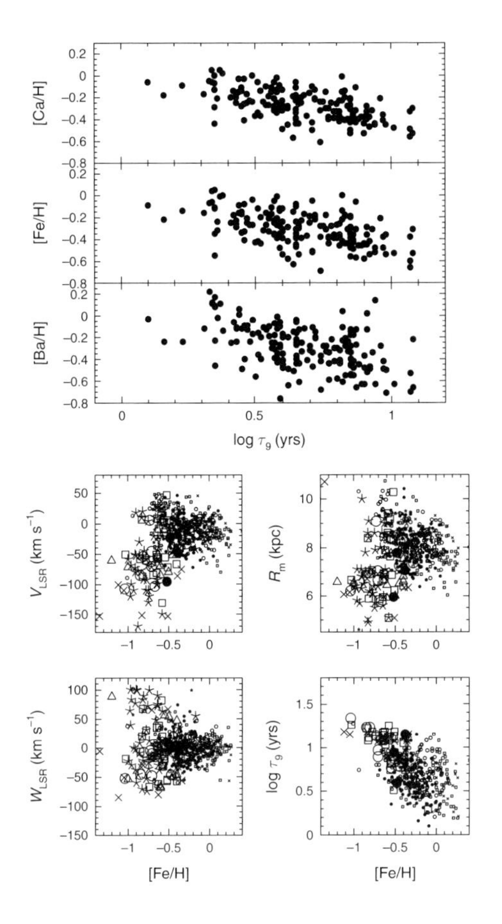

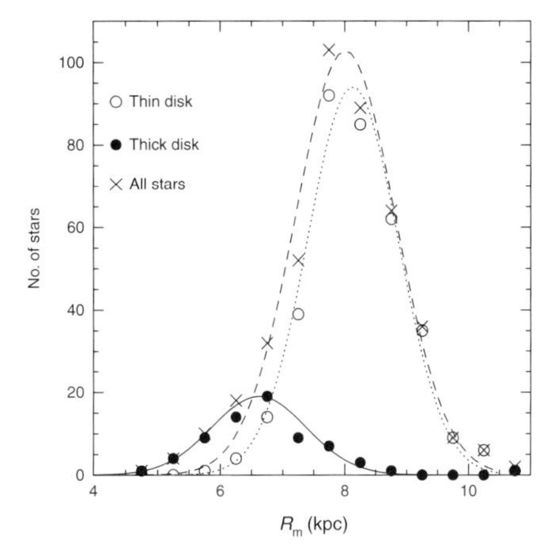

Figure 7.41 Left: relative abundances of Ca, Fe, and Ba shown versus age. Below left: V_{LSR}, W_{LSR}, R_m, and age versus [Fe/H], where R_m is the mean of the peri- and apo-Galactic distances, taken as a measure of the star's birthplace (Grenon, 1987). Values determined by Reddy et al. (2003) (•) are compared with earlier studies: ○: Chen et al. (2000), □: Edvardsson et al. (1993), ×: Fuhrmann (1998a), ⋆: Fulbright (2000), and △: Prochaska et al. (2000). Larger symbols represent [Mg/Fe] ≥ 0.2. Below right: distribution of around 500 disk stars in R_m, constructed from: Edvardsson et al. (1993), Chen et al. (2000), and Fulbright (2000), in addition to the sample of Reddy et al. (2003). The Gaussian fit to the entire sample (×) is asymmetric (dashed), while fits are symmetric for thin (dotted), and thick (solid) disk populations. From Reddy et al. (2003, Figures 7, 16, and 17).

176 nearby, $d < 150$ pc, thick disk candidate F and G dwarfs, segregating them into thick disk, thin disk, and halo populations.

Ibukiyama (2004) derived age–metallicity relations and orbits for 1658 solar neighbourhood stars with Hipparcos distances, comprising 1382 thin disk, 229 thick disk, and 47 halo stars. They concluded that the thin disk age–metallicity relation has a scatter which is an essential feature in the formation and evolution of the Galaxy, while that for the thick disk shows that star formation terminated there 8 Gyr ago. They confirmed that thick disk stars are more Ca-rich than thin disk stars with the same [Fe/H], and show a vertical abundance gradient. These results taken together support a model in which the thick disk formed as a result of monolithic collapse and/or the accretion of satellite dwarf galaxies.

Other related abundance studies making use of the Hipparcos data include studies of the Galactic evolution of Be and B stars (King, 2001); abundance analysis of two G dwarfs of the thick disk (Pettinger *et al.*, 2001); metallicity determination from ultraviolet–visual spectrophotometry (Morossi *et al.*, 2002); oxygen abundances in bulge-like dwarfs showing a turn-off at 10–11 Gyr (Pompéia *et al.*, 2002a,b); oxygen abundances in the halo giant HD 122563 (Barbuy *et al.*, 2003); oxygen abundance in 13 metal-poor subgiants (García Pérez *et al.*, 2006); zinc abundances in 38 metal-poor stars (Saito *et al.*, 2006); age–metallicity relation from (the revised photometric calibration of) the Geneva–Copenhagen survey of the solar neighbourhood (Holmberg *et al.*, 2007); and oxygen abundances in a sample of 523 nearby FGK disk and halo stars (Ramírez *et al.*, 2007).

7.7 Other stellar properties

7.7.1 Rotation

As stars form from clouds of interstellar gas and dust, they inherit their rotation from the effects of turbulent motion and magnetic fields of the parent cloud, and their subsequent rotation is governed by conservation of angular momentum as they collapse, magnetic braking, and mass loss. The detailed relationship between rotation rate and magnetic phenomena, mass, and age, is a subject of considerable observational and theoretical investigation. In terms of observational effects in the HR diagram, the effect of rotation on stellar

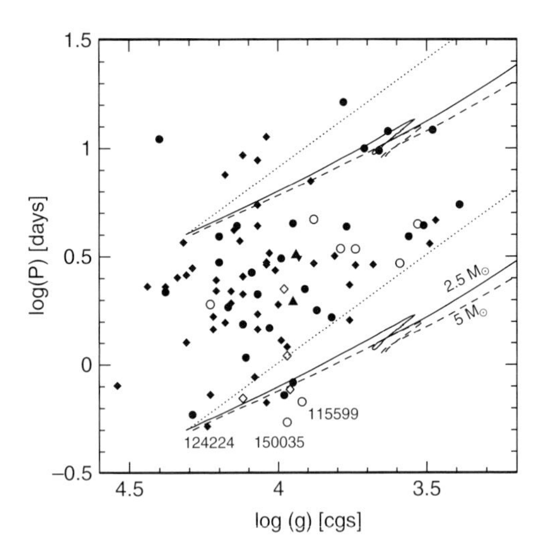

Figure 7.42 Rotational period versus surface gravity. Full symbols represent stars with a reliable period, open symbols are for possibly ambiguous periods. Circles (and triangles) represent stars with a spectroscopic value of $\log g$*, while diamonds are for stars with* $\log g$ *determined from Hipparcos. The three triangles are stars with a rotational period newly determined from Hipparcos photometry. Continuous and broken lines represent the evolution of period predicted from the moment of inertia, assuming rigid-body rotation, and for initial periods of 0.5 and 4 days. The dotted lines show the ideal case of conservation of angular momentum in independent spherical shells. From North (1998, Figure 4, as published in the erratum).*

evolutionary tracks was studied for example by Langer (1992), who found that rotation-induced mixing can increase the luminosity during core H burning by about 0.1 dex for the same value of M and $T_{\rm eff}$. Effects of rotation on M_V and spectral type was reviewed by Cassinelli (1987): fast rotators have lower gravities and smaller $T_{\rm eff}$ at their equators (gravity-darkening, or the von Zeipel effect, von Zeipel, 1924). A review of the main physical effects to be considered in the building of evolutionary models of rotating stars on the upper main sequence has been given by Maeder & Meynet (2000).

Lamers *et al.* (1997) determined the absolute visual magnitudes of 14 Hipparcos OB stars of luminosity classes III–V. They found that M_V can differ by up to 1.5 mag from the M_V spectral type calibration, with slowly-rotating stars, with $v \sin i < 100\,{\rm km\,s^{-1}}$, being generally fainter than the fast rotators with $v \sin i > 100\,{\rm km\,s^{-1}}$ (Figure 7.43 below). They argued that the effect is due to the influence of rotation on the assignment of spectral types and luminosity class. The study was extended to more stars, with similar result, by Schröder *et al.* (2004).

Fitzpatrick & Massa (2005) used the Hipparcos data for 45 B and early A stars to demonstrate that spectroscopic-based surface gravity determinations (from Strömgren photometry or line-profile analysis) may systematically underestimate the actual Newtonian gravity, GM/R^2 in the presence of even moderate rotation, $V \gtrsim 200\,{\rm km\,s^{-1}}$ (see Section 5.4).

North (1998) studied a sample of chemically-peculiar (CP) Ap and Bp stars, selecting about 100 Si stars, and constructing the $\log P_{\rm rot}$ versus $\log g$ diagram for stars having both a rotational period in the literature, and a reliable surface gravity either from spectroscopy (40 stars) or from Hipparcos parallaxes and theoretical evolutionary tracks (56 stars). The diagram (Figure 7.42) includes an unusual form of the evolutionary tracks, in which the models of Schaller *et al.* (1992) have been supplemented by code to determine how the moment of inertia evolves, assuming a solid body rotation with no loss of angular momentum. He concludes that field Si stars do not undergo significant magnetic braking during their life on the main sequence, and therefore that their slow rotation must be a property acquired before they arrive on the ZAMS. Based on the hypothesis that normally rotating A stars must be slowed down by at least a factor of four before spectral peculiarities can develop (Abt & Morrell, 1995), and that rotational velocity is the only parameter which determines whether a star will have a normal or peculiar spectrum, Hubrig *et al.* (2000b) argued that a much larger sample than that provided by Hipparcos would be needed to prove the existence of slowly-rotating Ap star progenitors.

Gondoin (1999) used the Hipparcos parallaxes for a sample of GK giants to examine the dependence of magnetic field on rotation, using coronal X-ray emission as a magnetic field diagnostic. These 2.5–$5M_\odot$ giants have rapidly rotating A and late B type progenitors on the main sequence with no outer convection zones. As they evolve off the main sequence, in the shell H burning stage, they develop thin outer convection zones, then rapidly traverse the consequently low-density F and G spectral type zone of the HR diagram known as the Hertzsprung gap. During this rapid evolution, their internal structure changes substantially, with rapidly deepening convection zones at mid-late G. The increasing convection zone combined with fast rotation is considered to trigger dynamo processes which generate the magnetic fields that, by analogy with the Sun, cause the X-ray emission of the outer stellar atmospheres. A large fraction of the yellow giants are able to maintain their high rotation until mid-G, at which point rotation velocities indicate strong rotational braking during their further evolution across the Hertzsprung gap. Gondoin (1999) used rotational velocities from the literature, X-ray luminosities from Einstein and ROSAT, parallaxes from Hipparcos, and evolutionary tracks from Schaller *et al.* (1992). He confirmed that there is a sharp decrease of X-ray emission for spectral type K1, and showed that

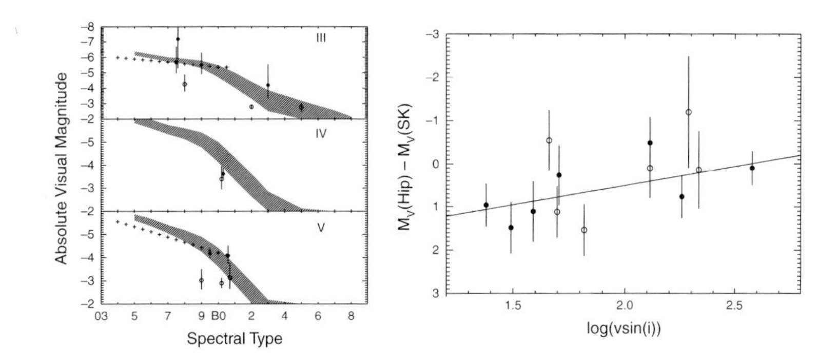

Figure 7.43 Left: M_V (Hip) versus spectral type for three luminosity classes. Error bars correspond to a 1σ parallax uncertainty. Calibrations by Schmidt–Kaler (in Schaifers & Voigt, 1982, Table 13), with the uncertainty (shaded regions) and Vacca et al. (1996) (+) are shown for comparison. Slow rotators, with $v \sin i < 100\,\mathrm{km\,s^{-1}}$ (○) are generally fainter than the fast rotators with $v \sin i > 100\,\mathrm{km\,s^{-1}}$ (●) and than the standard calibration. Right: difference between M_V from Hipparcos and the standard calibration as a function of $v \sin i$. Open symbols: class III; closed symbols: classes IV and V. Solid line: weighted least-squares fit. From Lamers et al. (1997, Figures 1 and 2).

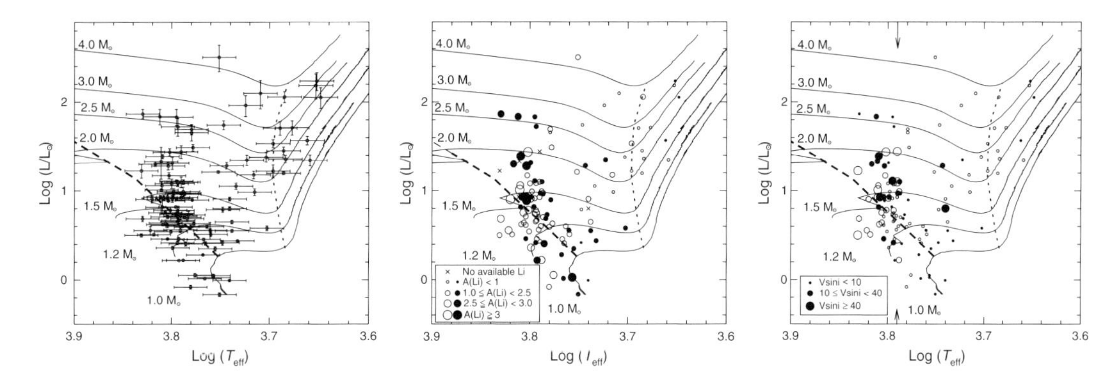

Figure 7.44 Left: HR diagram for 120 subgiants with luminosities derived from Hipparcos parallaxes. Evolutionary tracks at [Fe/H] = 0 are shown for masses 1–4$M_\odot$. The turn-off and the beginning of the ascent on the red giant branch are indicated by the dashed and dotted lines respectively in order to discriminate dwarfs, subgiants and giants. Middle: distribution of Li abundances, for single (○) and binary (●) stars, where symbol sizes are proportional to Li abundances; ×: stars without available Li abundance. Evolutionary tracks are as in the left figure. Right: rotation in the same plane, where the symbol size is proportional to $v \sin i$ in $\mathrm{km\,s^{-1}}$. The rotational discontinuity on the subgiant branch is indicated by the two arrows. From do Nascimento et al. (2000, Figure 1, 2 and 6).

the rotational velocities reach a minimum at the same location. No tight relation between rotational velocity and X-ray luminosity was evident, suggesting that differential rotation may influence their level of coronal emission. Further work was reported by Gondoin (2005, 2007).

Pasquini *et al.* (2000) used observed rotational velocities and evolutionary tracks from Hipparcos parallaxes for a sample of evolved F–K giants to study the evolution of angular momentum in the range 1–5$M_\odot$. For moment of inertia I and angular rotation Ω they found that the observed rotational velocities can be represented by simple $I\Omega^\beta$ = constant laws. For the mass range 1.6–3$M_\odot$ they found results consistent with angular momentum conservation, $\beta = 1$, while for higher and lower mass, $\beta = 2$ seemed more appropriate.

Do Nascimento *et al.* (2000), see also Canto Martins *et al.* (2003), studied the lithium abundance and rotation for a sample of 120 subgiants, for which masses and evolutionary status were estimated from the Hipparcos data (Figure 7.44). They showed that low-mass stars leave the main sequence with a low rotation rate, while more massive stars are slowed only when reaching the subgiant branch. A slight increase of the depth of the convective envelope seems to be sufficient for magnetic braking to occur at this point. The observed lithium discontinuity, which occurs when the convective envelope starts to deepen after the turn-off and reaches the inner

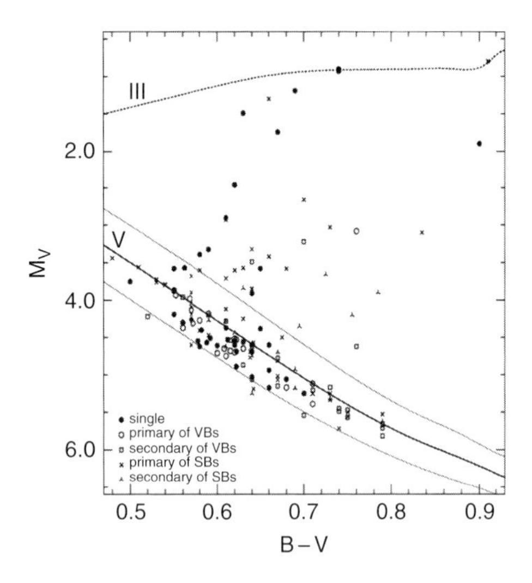

Figure 7.45 The B − V versus M_V diagram for the stars in the sample of Cutispoto et al. (2002). The single stars, the components of visual binaries (VB) and of spectroscopic binaries (SB) are indicated with different symbols. The continuous line and the long-dashed lines are the main sequence and the class III giant sequences, respectively, from Hipparcos data; the short-dashed lines indicate the limits of the dispersion of main-sequence stars from Hipparcos data. From Cutispoto et al. (2002, Figure 4).

lithium-free layers, is at roughly the same location as the rotational discontinuity.

Cutispoto *et al.* (2002) studied the lithium abundances and X-ray luminosities for 129 late F and G-type dwarfs selected on the basis of their fairly large rotational velocity, in a programme extending the study of the relationship between age, lithium abundance, chromospheric activity, and X-ray emission. The Coravel observations were used to determine radial velocities, $v \sin i$, and lithium abundances from the 670.78 nm doublet. Their results (Figure 7.45) showed a large fraction (62%) of binaries, with 30 new spectroscopic binaries, and nine pre-main-sequence or zero-age main-sequence objects.

Other Hipparcos-based studies related to chromospheric activity and rotation include studies of individual chromospherically active stars (e.g. Fekel, 1997); activity in late-type giants (Dorado & Montesinos, 1998); the age of active main-sequence stars in the solar neighbourhood (Catalano *et al.*, 2000); and Mg II chromospheric radiative loss rates in cool stars (Cardini, 2005; Cardini & Cassatella, 2007); the double-lined late-type binary HD 19485 (Fekel & Bolton, 2007).

Compilations or measurements of rotational velocities directly related to Hipparcos stars include: radial velocities and rotational velocities for 186 chemically peculiar stars (Levato *et al.*, 1996); rotational velocities of 250 A-type stars (Royer *et al.*, 1998); rotational and radial velocities for ~ 2000 FGK evolved stars (de Medeiros & Mayor, 1999); measurement of $v \sin i$ for 525 B8–F2 stars with $V < 8$ in the southern hemisphere (Royer *et al.*, 2002a); measurement of $v \sin i$ for 249 B8–F2 stars with $V < 7$ in the northern hemisphere (Royer *et al.*, 2002b); rotational and radial velocities for 231 FGK evolved supergiants (de Medeiros *et al.*, 2002); rotational and radial velocities for 78 double-lined binary stars (de Medeiros *et al.*, 2004); and rotational and radial velocities for 100 evolved metal-poor stars (de Medeiros *et al.*, 2006).

7.7.2 Magnetic field

Estimates of magnetic field strengths can be made from photoelectric polarimetry, for example by measuring the fractional circular polarisation in the wings of Hβ (Landstreet, 1980). Measurement of magnetic fields in unpolarised light can be based on spectral lines resolved into magnetically-split components, or from lines with differential magnetic broadening. In the magnetic chemically peculiar (CP2) stars, determination of stellar magnetic field geometries combined with surface chemical abundance distributions can place constraints on the manner in which these effects interact.

The General Catalogue of Ap and Am stars of Renson *et al.* (1991) includes 6684 objects showing abnormal enhancements of one or several elements in their atmospheres. The Hipparcos Catalogue includes some 940 Ap and Bp stars, with around 370 having $\sigma_\pi/\pi < 0.2$. Romanyuk & Kudryavtsev (2001) listed 149 Ap stars with reliably-measured longitudinal fields ranging from hundreds of Gauss to tens of kG, although more have subsequently been measured (e.g. Hubrig *et al.*, 2005).

Although it is well known that Ap and Bp stars belong to the main sequence, the evolutionary state of those with a measured magnetic field has remained more controversial, with some work suggesting that magnetic Ap and Bp stars are distributed across the width of the main sequence, while other suggesting that they are at the end of their main-sequence life (e.g. Stępień, 1994). The question is tied to the origin of magnetic fields in Ap and Bp stars, which is still a matter of debate: stars may have acquired their field at the time of their formation or early in their evolution (what is currently observed then being a fossil field), or the field may have been generated and maintained by a contemporary dynamo operating in the convective core (Moss, 1989; North, 1998).

Hubrig *et al.* (2000a) used Hipparcos data for 23 out of 42 then-known stars with resolved magnetically-split lines, and for 10 out of 14 then-known stars with differentially-broadened lines. The comparison sample comprised 416 single main-sequence B7–F2 stars with

$d < 100$ pc. Both from the usual HR diagram (L versus $T_{\rm eff}$), and from the 'astrometry-based luminosity' $a_V = 10^{0.2M_V} = \pi 10^{0.2m_v-2}$ of Arenou & Luri (1999, see Section 5.2.4), they showed that the positions in the HR diagram of magnetic Ap stars with masses below $3M_\odot$ differs significantly from that of normal stars in the same temperature range (Figure 7.46). The magnetic stars appear concentrated toward the centre of the main-sequence band, with magnetic fields appearing only in stars that have completed some 30% of their main-sequence lifetime. A marginal trend of the magnetic flux to be lower in more slowly rotating stars may suggest a dynamo origin for the field. No correlation existed between the rotation period and the fraction of the main-sequence lifetime completed, indicating that the slow rotation in these stars must already have been achieved before they became observably magnetic, perhaps the consequence of braking during the Hayashi phase, rather than the more general assumption that magnetic stars are slow rotators because of magnetic braking (see also Ryabchikova, 2004). Hubrig *et al.* (2005) enhanced the statistics of the study by making longitudinal magnetic field measurements of additional stars. Their results (Figure 7.47) confirmed that magnetic fields for masses below $3M_\odot$ only become observable after they have spent some 30% or more of their life on the main sequence, and showed that stronger magnetic fields tend to be found in hotter, younger, and more massive stars. A further 136 Hipparcos stars were measured by Hubrig *et al.* (2006). The distribution of magnetic field geometry across the main sequence was studied in further detail by Hubrig *et al.* (2007) for 90 Ap and Bp stars with accurate Hipparcos parallaxes and well-determined longitudinal magnetic fields. They confirmed a significant difference between the distribution of high-mass and low-mass magnetic stars in the HR diagram, with the longest rotation periods, of up to 70 years, found in the range 1.8–$3M_\odot$ and for stars already older than 40% of their main-sequence lifetime.

Pöhnl *et al.* (2005) studied 182 magnetic CP stars within 200 pc, divided into the three main chemical subgroups: 61 Si, 34 SiCr, and 87 SrCrEu objects. From the resulting HR diagram and comparison with relevant evolutionary tracks (Figure 7.48), they found that the magnetic CP phenomenon already occurs at the ZAMS, with some 16% of their objects having relative ages (compared to the main-sequence lifetime) below 20%. The effective temperature and magnetic field strength are the main parameters that determine the chemical peculiarity. Si, Cr, Sr, and Eu dominate sequentially as temperatures fall below 10 000 K, with the incidence of magnetic CP stars with $M < 2.25M_\odot$ decreasing to almost zero for $T_{\rm eff} < 8000$ K. Kochukhov & Bagnulo (2006) reported a related statistical study of 150 Hipparcos magnetic CP stars, concluding that the mechanism that originates and sustains the magnetic field in the upper main-sequence stars may be different in CP stars of different mass.

There have been various studies of individual magnetic stars. For the Bp star 84 UMa, for which the dipole magnetic field strength and magnetic obliquity were determined from spectropolarimetry, the Hipparcos photometric data were used by Wade *et al.* (1998) to confirm the rotational period of 1.385 76 day. The parallax was used to determine, with the aid of evolutionary tracks, estimates of the radius, mass, surface gravity, and age. The results place it close to the ZAMS, a result inconsistent with earlier suggestions that the magnetic Ap stars may be close to the end of their main-sequence life.

The G8-giant HR 1362 (EK Eri) is a slowly rotating ($P = 300$ d) and unusually magnetically active star, given the empirical relation for rotation versus activity normally found for cool giants. Strassmeier *et al.* (1999b) examined 20 years of V-band photometry, and used the Hipparcos parallax to derive a mass of $1.85M_\odot$ and radius $4.7 \pm 0.3R_\odot$. They conclude that the high magnetic field is not internally dynamo driven, but rather an unusual example of a relic of perhaps one in 10^2–10^3 main-sequence stars possessing a strong (several kG) primordial magnetic dipole field frozen in and surviving due to slow Ohmic decay through to the giant phase (Stępień, 1993).

Leone *et al.* (2000a) and Leone & Catanzaro (2001) made spectropolarimetric measurements of the mean longitudinal magnetic field of seven chemically peculiar stars, using Hipparcos photometry to confirm the rotational periods, and the parallaxes to determine the absolute magnitudes, hence the radii and, on the hypothesis of a rigid rotator, the inclination of the rotational axis with respect to the line-of-sight. They related the photometric variability to the non-homogeneous distribution of elements on the stellar surface, in which high metallicity regions block the emerging ultraviolet flux.

Other studies making use of the Hipparcos data include that of the magnetic chemically peculiar star ν For (Leone *et al.*, 2000b); studies of the alignment of the magnetic dipole with respect to the rotational axis in the case of the two slowly-rotating Ap/Bp stars HD 12288 and HD 14437 (Wade *et al.*, 2000); the stars 36 Aur, HR 2722, 13 And, and HD 220147 (Adelman, 2005); the magnetic Bp star 36 Lyn (Wade *et al.*, 2006); and studies of the co-existence of chemically-peculiar Bp stars, slowly-pulsating B stars, and constant B stars in the same part of the HR diagram, suggesting that magnetic field strength is an important factor in B-type stars becoming chemically peculiar (Briquet *et al.*, 2007).

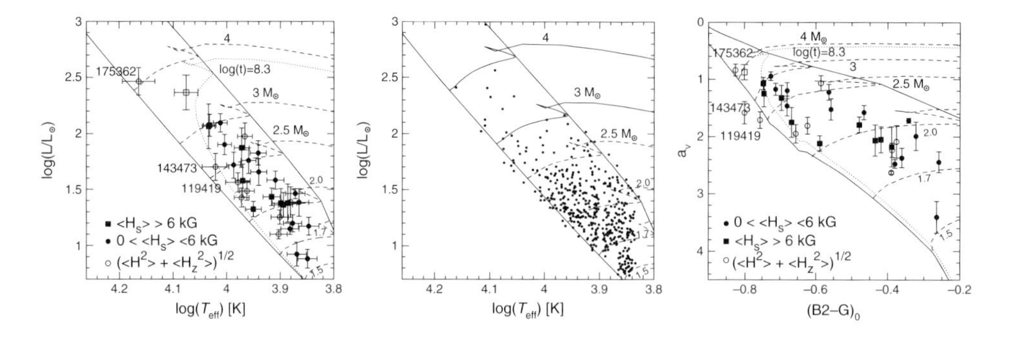

Figure 7.46 Left: HR diagram for Ap stars with resolved magnetically-split lines (filled symbols) and for Ap stars whose mean quadratic magnetic field has been determined (open symbols). Filled squares: stars with a mean magnetic field exceeding 6 kG. The open square is HD 137509, which has an extremely strong field. Other hot stars whose effective temperature has not been derived from photometry are identified. Evolutionary tracks are from Schaller et al. (1992). Middle: HR diagram for normal dwarf B7–F2 stars within 100 pc. Relative precision on parallax is generally better than 10%. Right: alternative HR diagram, showing the astrometry-based luminosity a_V versus the Geneva $(B2–G)_0$ index. Symbols are otherwise as for the figure at left. From Hubrig et al. (2000a, Figures 1, 2 and 5).

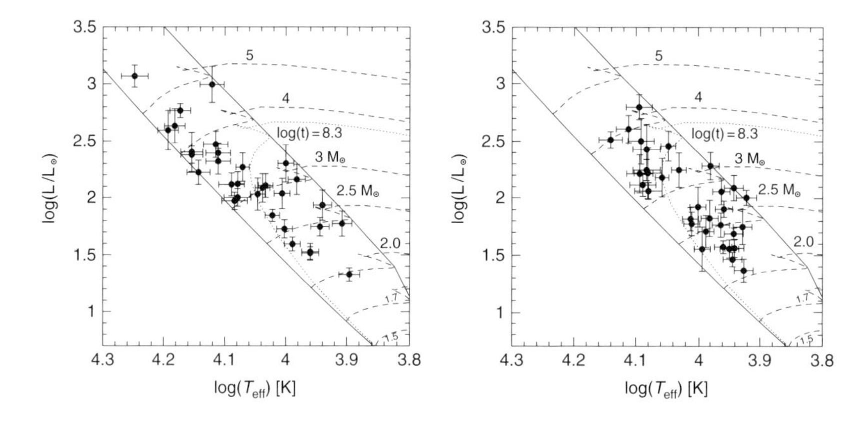

Figure 7.47 Left: HR diagram for magnetic stars with already-known mean longitudinal fields. Right: HR diagram for magnetic stars with mean longitudinal fields measured with VLT–FORS 1 by Hubrig et al. (2005). Evolutionary tracks are from Schaller et al. (1992). From Hubrig et al. (2005, Figure 1).

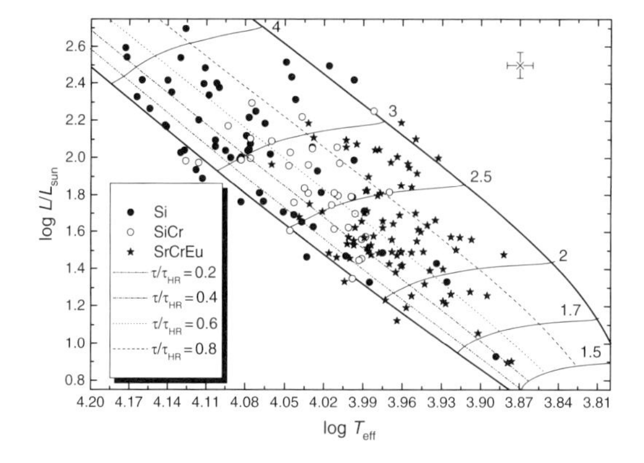

Figure 7.48 HR diagram for the magnetic CP stars studied by Pöhnl et al. (2005): indicating Si (•), SiCr (○), and SrCrEu (⋆) objects. Evolutionary tracks from 1.5 to $4M_\odot$, lines of equal τ/τ_{HR}, and the ZAMS and TAMS, are for $[Z] = 0.016$ taken from Schaller et al. (1992) and Schaerer et al. (1993b). Typical errors are shown in the upper right corner. From Pöhnl et al. (2005, Figure 2).

7.7.3 Imaging of surface structure

Magnetic chemically peculiar (CP) A and B stars are characterised by abnormal spectral line strengths of certain chemical elements and possess strong globally-organised magnetic fields. Periodic variations are interpreted within the 'oblique rotator model' (Stibbs, 1950), in which periodic variations are attributed to rotational modulation due to inhomogeneous surface abundances and magnetic fields. The abnormal chemical composition is considered to be limited to the outer envelopes, and produced by chemical diffusion (Michaud, 1970), now attributed to the competing processes of gravitational settling, radiative acceleration acting on individual elements, and other processes such as convection, magnetic fields, or strong mass loss (Michaud & Proffitt, 1993).

It is possible to recover the surface structure of some stars by transforming the time-varying spectroscopic line profiles due to rotating surface features into geometric images. The method, frequently referred to as Doppler imaging, has its origins in the work of Deutsch (1958), and is described and illustrated by Rice (1996). Adapting the method to more complex geometries (e.g. binary systems) and to spectropolarimetry (magnetic or Zeeman Doppler imaging) have been introduced in the last few years. Rice (1996) gives examples of studies undertaken for Ap stars (where the objective is to correlate the magnetic field structure with the abundance patterns on the stellar surface), for RS CVn stars (where the objective is to establish characteristics and evolution of the surface spots), for young main sequence stars (where spot features appear as extended areas of temperature depression of a few hundred K), and for T Tauri stars.

The selection of suitable objects for detailed study has been assisted by the Hipparcos Catalogue. Strassmeier *et al.* (2000) used it as a basis for a spectroscopic Ca II H and K survey of 1058 late-type stars, using the presence of emission in the core of these lines as a diagnostic of magnetic activity in their chromospheres. Other activity indicators were compiled, including Balmer Hα and Li lines. They derived rotational velocities, and space motions from the Hipparcos astrometry and radial velocities, and a final list of 21 stars considered as most suitable candidates for Doppler imaging based on stellar activity, rotational broadening of spectral lines, and knowledge of the precise rotation period.

For individual Doppler imaging targets, the Hipparcos data have been routinely used to derive absolute stellar properties, such as mass and age, based on appropriate stellar evolutionary models. Examples include the G8 giant HD 51066 (CM Cam) whose progenitor was inferred to be a very rapidly rotating Bp star with a magnetic field of several kG (Strassmeier *et al.*, 1998); the FK Comae-type star HD 199178 (V1794 Cyg) with $M = 1.65M_\odot$ and $L = 11L_\odot$, with low-latitude spots having lifetimes as short as 1 month, and the polar spot having a lifetime of order 10 years or 1000 stellar rotations (Strassmeier *et al.*, 1999a); the K0 giant HD 12545, with a mass of around $1.8M_\odot$, with the largest star spots ever observed with dimensions of order $12 \times 20R_\odot^2$ (Strassmeier, 1999); and the abundance distribution and magnetic field geometry of α^2 CVn, for which the Hipparcos parallax leads to a mass of about $4M_\odot$ and an age of $10^{8.145}$ Gyr (Kochukhov *et al.*, 2002).

Other determinations of surface structure in which the Hipparcos astrometric data have been invoked include the origin of the 1667 MHz OH maser in U Her (van Langevelde *et al.*, 2000); the positional coincidence between the thermal jet and CO bipolar outflow in Z CMa (Velázquez & Rodríguez, 2001); phase coherence of spots on UZ Lib (Oláh *et al.*, 2002); studies of the circumstellar 22 GHz water masers in U Her (Vlemmings *et al.*, 2002); and evidence for a polar spot on SV Cam (Jeffers *et al.*, 2005).

7.8 Asteroseismology

Asteroseismology uses stellar oscillations, typically excited by convection or by the κ mechanism, to study the internal structure, dynamics, and evolutionary state of stars (Christensen-Dalsgaard, 1984). A major goal of helioseismology (see review by Gough & Toomre, 1991), and of asteroseismology as applied to other stars (see the review by Brown & Gilliland, 1994), is to refine theories of stellar structure and evolution. Non-radial oscillations were invoked more than 50 years ago to explain puzzling spectral characteristics of β Cephei stars, leading to the search for and discovery of the 5-min oscillations in the Sun. The Sun comprises a core of radius $\sim 0.25\,R_\odot$ representing half the mass and some 98% of the energy generation, a stable radiative zone where the energy is transported outward by radiation, and a lower temperature region beyond $\sim 0.7\,R_\odot$ where convection dominates the energy transport. Standing pressure (or acoustic) waves are trapped between the density decrease toward the surface, and the increasing sound speed toward the centre which refracts the downward propagating wave back to the surface.

The normal modes of oscillation of a spherical star have amplitudes proportional to spherical harmonics $Y_l^m(\theta,\phi)e^{i\sigma t}$, with frequency σ, harmonic degree $l = 0, 1, 2, \ldots$ ($l = 0$ are radial modes, of order n), and azimuthal number $-l \le m \le l$. For a non-rotating star the frequency is independent of m, although this degeneracy is broken in the presence of rotation or magnetic

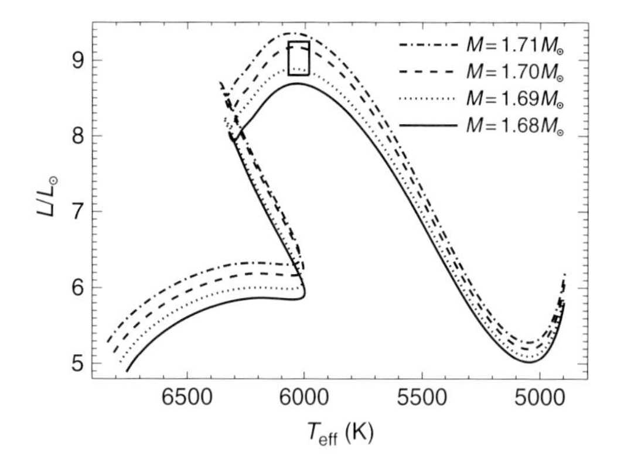

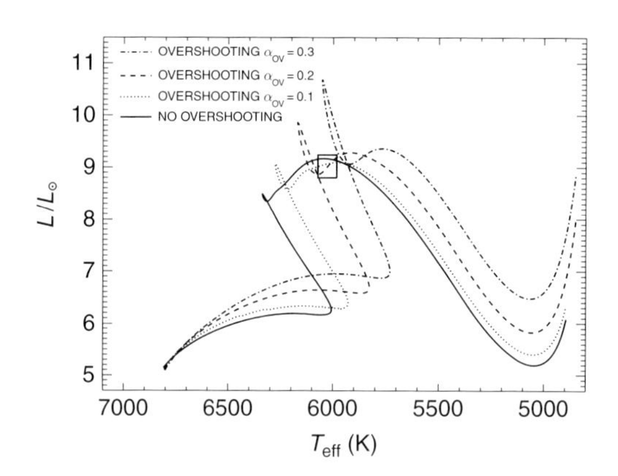

Figure 7.49 Asteroseismology results for η Boo. Left: evolutionary tracks plotted in an HR diagram for various assumed masses and the observed metallicity ($Z = 0.04$). The rectangle defines the 1σ error box for the inferred effective temperature and the Hipparcos-derived luminosity. Right: the effect of convective overshooting, for various assumed values for its extent from the convective stellar core, for the case $M = 1.7 M_\odot$ and $Z = 0.04$. From Di Mauro et al. (2003, Figures 1 and 3).

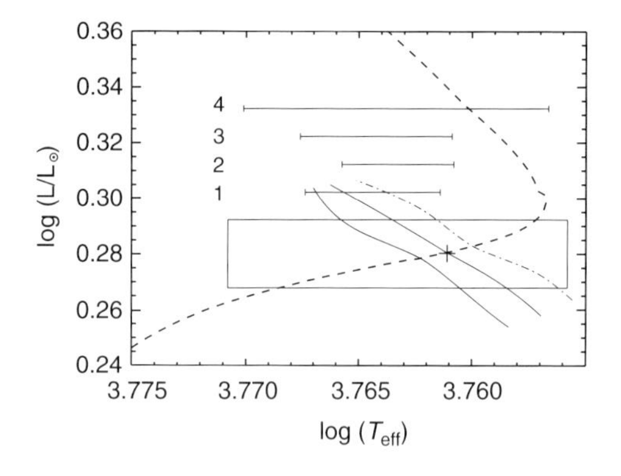

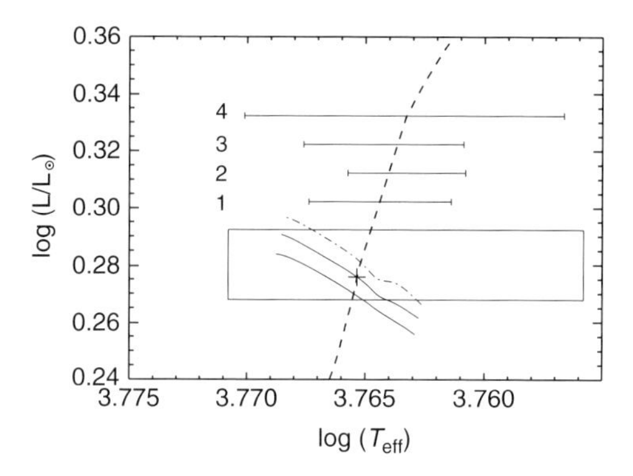

Figure 7.50 Asteroseismology results for exoplanet host μ Arae. Example evolutionary tracks (dashed lines) for assumed 'over-metallic' (left) and 'accretion' (right) scenarios. The three oblique lines correspond to asteroseismic separations of $90\,\mu$Hz (as observed, solid line), as well as $89\,\mu$Hz (higher curve) and $91\,\mu$Hz (lower curve). The error boxes correspond to the Hipparcos-based luminosity and adopted $T_{\rm eff}$, while the other horizontal lines correspond to other temperature determinations. The cross corresponds to the model described in detail in their paper. From Bazot et al. (2005, Figures 1 and 2).

field. The precise mode frequencies are detectable by precision photometry (intensity changes) or spectroscopy (radial velocity or equivalent width changes). Unlike the case for the Sun, the surface of stars are essentially unresolved, so that asteroseismology is limited to modes of low degree, which involve the sound travel time across the stellar diameter, and conditions near the stellar surface and the convective boundary layer. For non-radial modes, problems include detection of the tiny signals, identifying the modes, and explaining why particular modes are stimulated. The class of oscillation is determined by the dominant restoring force, either pressure (p-mode) or gravity (g-mode), and these waves propagate (or are damped) depending on certain local characteristic frequencies (see, e.g. Schatzman & Praderie, 1993, p294ff).

The main contribution of Hipparcos has been to provide precise parallaxes (and hence luminosities) which in turn provide important constraints on the stellar models which predict the star's luminosity from the observed oscillation frequencies. For variable stars, for example, a different evolutionary stage may give a drastically different eigenmode spectrum, and in turn may change the mode identification and asteroseismology analysis (Liu *et al.*, 1997).

There are four classes of non-degenerate star (i.e. excluding white dwarfs) for which the Hipparcos parallaxes have so far been particularly relevant (Eyer & Grenon, 1998; Favata, 1998; Kjeldsen & Bedding, 2001; Bedding, 2002): solar-like objects, high-amplitude δ Scuti variables, β Cephei variables, and rapidly oscillating Ap stars.

(a) In solar-like objects, oscillations are excited stochastically by turbulent convection, and are expected for all stars with significant convection in their outer regions, i.e. those on the cool side of the δ Scuti instability strip. The last 10 years have seen significant evidence for detections in η Boo, Procyon, ζ Her, α Cen A, and β Hyi (Bedding & Kjeldsen, 2006).

For the bright G0 subgiant, η Boo, detections of p-mode oscillations at the level of a few ppm were reported by Kjeldsen *et al.* (1995), at which time the observed frequencies were consistent with a luminosity implied by the ground-based parallax of 85.8 ± 2.3 mas, thus quantifying the evolutionary state and an age of about 2.3 Gyr (Christensen-Dalsgaard *et al.*, 1995; Guenther & Demarque, 1996): it is in the post-main-sequence phase with a He-core and H-burning shell, in which its central H is exhausted: the near-surface convection excites modes to the observed amplitudes. Hipparcos gave a factor of 3 improvement in parallax precision of 88.17 ± 0.75 mas ($d = 11.3$ pc), maintaining agreement with the models (Bedding *et al.*, 1998). Di Mauro *et al.* (2003) refined the modelling of η Boo to include convective overshooting (Figure 7.49). During the main-sequence phase, overshooting produces a chemical and thermal mixing in the region where the turbulent motions penetrate, outward from the edge of the convective core, making more H available for nuclear burning, and increasing the main-sequence lifetime, by about 15% for $\alpha_{\rm ov} = 0.2$. Models with $\alpha_{\rm ov} > 0.25$ were precluded for all plausible values of M and [Fe/H]. Improved luminosities, along with improved $T_{\rm eff}$ and [Fe/H], will allow the effects of overshooting to be further constrained in future.

Application of asteroseismology to exoplanet formation and evolution has been presented by Bazot *et al.* (2005) in the case of μ Arae, which hosts a triple planetary system including one of low-mass, $\sim 14\,M_{\rm Earth}$. In common with many exoplanet host stars, μ Arae shows a metallicity excess compared to stars without detected planets. The excess may be a bulk property of the star, tied to an originally metal-rich proto-planetary cloud in which planet condensation was facilitated by excess solid material in the protoplanetary disks. Alternatively, the high metallicity could be a surface phenomenon due to capture of metal-rich disk material or inwardly migrating proto-planets during the early phases of planetary formation, an explanation only plausible for main-sequence stars with a relatively small outer convection zone and therefore limited dilutional mixing. The luminosity is constrained by the Hipparcos parallax of 65.5 ± 0.8 mas. The metal-rich models have a convective core, while the accreting models do not, leading to a specific predicted signature for the $l = 0$ and $l = 2$ modes. At present, of the 43 detected oscillation modes with harmonic degrees $l = 0$–3, the crucial frequency region from 2.5–2.7 mHz has only low signal-to-noise, and the conclusions are currently ambiguous (Figure 7.50).

(b) High-amplitude δ Scuti (or dwarf Cepheids) variables occupy the few hundred Kelvin wide instability strip, and like the Cepheids and RR Lyrae stars, undergo self-excited radial pulsations. They oscillate in a large number of modes with periods of minutes to hours, predominantly in modes of low degree l and low radial overtone $n \sim l$, i.e. with simple spherical harmonic patterns on their surface and with frequencies close to their fundamental resonance. While the number of detected and accurately-known frequencies is large, some unknown selection mechanism excites only certain ones to observable amplitudes, making mode identification (necessary to compare with model frequencies) problematic. Only fairly rapidly rotating stars show multiple modes, and the effects of stellar rotation remain incompletely modelled. Studies of these objects using the Hipparcos data are described in Section 4.9.1. Some 70% of λ Bootis stars within the instability strip show δ Scuti-type pulsations, and in 1995 the Central Asian Network selected 29 Cyg as a key object for detailed pulsation studies. Results of the multi-site photometric and spectroscopic campaigns carried out in 1995–96 were reported by Mkrtichian *et al.* (2007). From the rich multi-periodic spectrum, they derived an asteroseismic luminosity from the observed frequency spacings of $\log L/L_\odot = 1.12$, in good agreement with the Hipparcos value of $\log L/L_\odot = 1.16$.

(c) The class of β Cephei variables are massive near-main-sequence stars between B0–B2 with well-developed convective cores, and are also good candidates for the study of convective properties through asteroseismology. Specific studies of these objects using the Hipparcos data are described in Section 4.9.4.

(d) Although mode identification is in general problematic, prior mode identification is not essential in the case of pulsations of low degree l and high radial overtone, $n \gg l$, for which asymptotic pulsation theory predicts that the observed eigenfrequencies should be nearly equally spaced in frequency for the p-modes. This is the case for rapidly oscillating Ap stars (roAp stars, a class numbering about 35 known objects), with periods of a few minutes. Specific studies of these objects using the Hipparcos data are described in Section 4.9.2.

Atmospheric effects and limited window functions from ground-based observations are being superseded by space-based asteroseismology studies underway by MOST (Matthews, 2005) and COROT (Baglin, 2003; Baglin *et al.*, 2006), where the use of the Hipparcos parallaxes to assist target selection has been described by

Bruntt *et al.* (2004). Meanwhile, the fact that asteroseismology predictions are consistent with luminosity measurements for a variety of stars with a range of masses and evolutionary states is an encouraging validation of the models.

References

Abbett WP, Beaver M, Davids B, *et al.*, 1997, Solar convection: comparison of numerical simulations and mixing length theory. *ApJ*, 480, 395–399 {**345**}

Abt HA, Morrell NI, 1995, The relation between rotational velocities and spectral peculiarities among A-type stars. *ApJS*, 99, 135–172 {**390**}

Adams WS, Joy AH, 1922, A spectroscopic method of determining the absolute magnitudes of A-type stars and the parallaxes of 544 stars. *ApJ*, 56, 242–264 {**354**}

Adelman SJ, 2005, uvby FCAPT photometry of the magnetic chemically-peculiar stars 36 Aur, HR 2722, 13 And, and HD 220147. *PASP*, 117, 476–482 {**393**}

Alexander DR, Ferguson JW, 1994, Low-temperature Rosseland opacities. *ApJ*, 437, 879–891 {**344, 346, 349**}

Allard F, Hauschildt PH, Alexander DR, *et al.*, 1997, Model atmospheres of very low-mass stars and brown dwarfs. *ARA&A*, 35, 137–177 {**345**}

Allende Prieto C, 2001, Hipparcos, IUE, and the stellar content of the solar neighbourhood. *Revista Mexicana de Astronomia y Astrofisica Conf. Ser.*, volume 10, 205–208 {**377**}

Allende Prieto C, Barklem PS, Lambert DL, *et al.*, 2004, A spectroscopic survey of stars in the solar neighbourhood. The nearest 15 pc. *A&A*, 420, 183–205 {**377**}

Allende Prieto C, García López RJ, Lambert DL, *et al.*, 1999, A consistency test of spectroscopic gravities for late-type stars. *ApJ*, 527, 879–892 {**354, 355**}

Allende Prieto C, Lambert DL, 1999, Fundamental parameters of nearby stars from the comparison with evolutionary calculations: masses, radii and effective temperatures. *A&A*, 352, 555–562 {**359, 360**}

Alonso A, Arribas S, Martinez-Roger C, 1995, Determination of bolometric fluxes for F, G and K subdwarfs. *A&A*, 297, 197–215 {**341, 361, 362, 366, 367**}

—, 1996a, Determination of effective temperatures for an extended sample of dwarfs and subdwarfs (F0–K5). *A&AS*, 117, 227–254 {**342, 349, 361, 362, 366, 367**}

—, 1996b, The empirical scale of temperatures of the low main sequence (F0V–K5V). *A&A*, 313, 873–890 {**342**}

Alonso A, Arribas S, Martínez-Roger C, 1999, The effective temperature scale of giant stars (F0–K5). I. The effective temperature determination by means of the infrared flux method. *A&AS*, 139, 335–358 {**352, 353**}

Alonso A, Salaris M, Arribas S, *et al.*, 2000, The effective temperature scale of giant stars (F0–K5). III. Stellar radii and the calibration of convection. *A&A*, 355, 1060–1072 {**352, 354**}

Altmann M, de Boer KS, 2000, Kinematical trends among the field horizontal branch stars. *A&A*, 353, 135–146 {**372, 373**}

Altmann M, Edelmann H, de Boer KS, 2004, Studying the populations of our Galaxy using the kinematics of sdB stars. *A&A*, 414, 181–201 {**373**}

Anders E, Grevesse N, 1989, Abundances of the elements: meteoritic and solar. *Geochim. Cosmochim. Acta*, 53, 197–214 {**339**}

An D, Terndrup DM, Pinsonneault MH, *et al.*, 2007, The distances to open clusters from main-sequence fitting. III. Improved accuracy with empirically calibrated isochrones. *ApJ*, 655, 233–260 {**350**}

Andersen J, 1991, Accurate masses and radii of normal stars. *A&A Rev.*, 3, 91–126 {**346, 352, 356, 358–360, 375**}

Andersen J (ed.), 1999, *Proceedings of the Twenty-third General Assembly*, volume 23 {**341**}

Antoniucci S, Paresce F, Wittkowski M, 2005, VLTI–VINCI measurements of HR 4049: the physical size of the circumbinary envelope. *A&A*, 429, L1–L4 {**358**}

Arenou F, Luri X, 1999, Distances and absolute magnitudes from trigonometric parallaxes. *ASP Conf. Ser. 167: Harmonizing Cosmic Distance Scales in a Post-Hipparcos Era*, 13–32 {**393**}

Armstrong JT, Mozurkewich D, Hajian AR, *et al.*, 2006, The Hyades binary θ^2 Tau: confronting evolutionary models with optical interferometry. *AJ*, 131, 2643–2651 {**376**}

Armstrong JT, Nordgren TE, Germain ME, *et al.*, 2001, Diameters of δ Cep and η Aql measured with the Navy Prototype Optical Interferometer. *AJ*, 121, 476–481 {**358**}

Arndt TU, Fleischer AJ, Sedlmayr E, 1997, Circumstellar dust shells around long-period variables. VI. An approximative formula for the mass loss rate of C-rich stars. *A&A*, 327, 614–619 {**374**}

Aslan Z, Yeşilyaprak C, 2000, Absolute magnitudes of M-type semi-regular variables in the Hipparcos Catalogue. *NATO ASIC Proc. 544: Variable Stars as Essential Astrophysical Tools*, 503–510 {**377**}

Asplund M, Grevesse N, Sauval AJ, 2005, The solar chemical composition. *Cosmic Abundances as Records of Stellar Evolution and Nucleosynthesis* (eds. Barnes TG, Bash FN), ASP Conf. Ser. 336, 25–38 {**339**}

Aufdenberg JP, Hauschildt PH, Baron E, 1999, A non-local thermodynamic equilibrium spherical line-blanketed stellar atmosphere model of the early B giant β CMa. *MNRAS*, 302, 599–611 {**377**}

Axer M, Fuhrmann K, Gehren T, 1994, Spectroscopic analyses of metal-poor stars. I. Basic data and stellar parameters. *A&A*, 291, 895–909 {**343**}

—, 1995, Spectroscopic analyses of metal-poor stars. II. The evolutionary stage of subdwarfs. *A&A*, 300, 751–768 {**355, 361**}

Badiali M, Cardini D, Emanuele A, *et al.*, 1997, Age and morphology of the main sequence from Hipparcos HR diagram data. *ESA SP–402: Hipparcos, Venice '97*, 661–664 {**377**}

Badnell NR, Bautista MA, Butler K, *et al.*, 2005, Updated opacities from the Opacity Project. *MNRAS*, 360, 458–464 {**344**}

Baglin A, 1998, Hipparcos and theory of stellar interiors. *Highlights in Astronomy*, 11, 555–557 {**377**}

—, 2003, COROT: a minisat for pioneer science, asteroseismology and planet finding. *Advances in Space Research*, 31, 345–349 {**397**}

Baglin A, Michel E, Auvergne M, *et al.*, 2006, The seismology programme of the CoRoT space mission. *Beyond the Spherical Sun: Proceedings of SOHO 18/GONG 2006/HELAS I; ESA SP-624*, volume 18, 34–41 {**397**}

Bahcall JN, Pinsonneault MH, 1992, Standard solar models, with and without helium diffusion, and the solar neutrino problem. *Reviews of Modern Physics*, 64, 885–926 {**349**}

Balachandran SC, Fekel FC, Henry GW, *et al.*, 2000, Two K giants with supermeteoritic lithium abundances: HDE 233517 and HD 9746. *ApJ*, 542, 978–988 {**382, 383**}

Balbes MJ, Boyd RN, Mathews GJ, 1993, The primordial helium abundance as determined from chemical evolution of irregular galaxies. *ApJ*, 418, 229–234 {**362**}

Barbuy B, Grenon M, 1990, Oxygen in bulge-like super-metal-rich stars. *Workshop on Bulges of Galaxies* (eds. Jarvis BJ, Terndrup DM), 83–86 {**386**}

Barbuy B, Meléndez J, Spite M, *et al.*, 2003, Oxygen abundance in the template halo giant HD 122563. *ApJ*, 588, 1072–1081 {**389**}

Barnes TG, Evans DS, Moffett TJ, 1978, Stellar angular diameters and visual surface brightness. III. An improved definition of the relationship. *MNRAS*, 183, 285–304 {**351**}

Bartkevičius A, Bartašiute S, Lazauskaite R, 1997, The HR diagram of metal-deficient stars. *ESA SP–402: Hipparcos, Venice '97*, 343–346 {**366**}

Basu S, Christensen-Dalsgaard J, Chaplin WJ, *et al.*, 1997, Solar internal sound speed as inferred from combined BiSON and LOWL oscillation frequencies. *MNRAS*, 292, 243–251 {**339**}

Bazot M, Vauclair S, Bouchy F, *et al.*, 2005, Seismic analysis of the planet-hosting star μ Ara. *A&A*, 440, 615–621 {**396, 397**}

Bedding TR, 2002, Hipparcos luminosities and asteroseismology. *Highlights in Astronomy*, 12, 694–695 {**396**}

Bedding TR, Kjeldsen H, Christensen-Dalsgaard J, 1998, Hipparcos parallaxes for η Boo and κ Boo: two successes for asteroseismology. *ASP Conf. Ser. 154: Cool Stars, Stellar Systems, and the Sun*, 741–744 {**397**}

Bedding TR, Kjeldsen H, 2006, Observing solar-like oscillations: recent results. *Memorie della Societa Astronomica Italiana*, 77, 384–388 {**397**}

Bedin LR, Piotto G, Anderson J, *et al.*, 2004, ω Centauri: the population puzzle goes deeper. *ApJ*, 605, L125–L128 {**380**}

Bennett CL, Halpern M, Hinshaw G, *et al.*, 2003, First-year Wilkinson Microwave Anisotropy Probe (WMAP) observations: preliminary maps and basic results. *ApJS*, 148, 1–27 {**368**}

Bensby T, Feltzing S, Lundström I, *et al.*, 2005, α-, r-, and s-process element trends in the Galactic thin and thick disks. *A&A*, 433, 185–203 {**378**}

Bergbusch PA, VandenBerg DA, 2001, Models for old, metal-poor stars with enhanced α-element abundances. III. Isochrones and isochrone population functions. *ApJ*, 556, 322–339 {**367**}

Bergeat J, Chevallier L, 2005, The mass loss of C-rich giants. *A&A*, 429, 235–246 {**374**}

Bernasconi PA, 1996, Grids of pre-main sequence stellar models. The accretion scenario at $Z = 0.001$ and $Z = 0.020$. *A&AS*, 120, 57–61 {**350**}

Bernkopf J, 1998, Unified stellar models and convection in cool stars. *A&A*, 332, 127–134 {**345**}

Bertelli G, Bressan A, Chiosi C, *et al.*, 1994, Theoretical isochrones from models with new radiative opacities. *A&AS*, 106, 275–302 {**350, 359, 387**}

Bessell MS, 1998, Cool star empirical temperature scales. *IAU Symp. 189: Fundamental Stellar Properties*, 127–136 {**342**}

Bessell MS, Castelli F, Plez B, 1998, Model atmospheres broad-band colours, bolometric corrections and temperature calibrations for O–M stars. *A&A*, 333, 231–250 {**341, 345, 349, 352**}

Bi SL, Di Mauro MP, Christensen-Dalsgaard J, 2000, An improved equation of state under solar interior conditions. *A&A*, 364, 879–886 {**345**}

Blackwell DE, Lynas-Gray AE, 1994, Stellar effective temperatures and angular diameters determined by the infrared flux method: revisions using improved Kurucz LTE stellar atmospheres. *A&A*, 282, 899–910 {**349**}

—, 1997, Determination of the temperatures of selected ISO flux calibration stars using the infrared flux method. *Technical Report, OUAST/98/1 Astrophysics, Nuclear Physics Lab.*, 98, 69702 {**349**}

—, 1998, Determination of the temperatures of selected ISO flux calibration stars using the infrared flux method. *A&AS*, 129, 505–515 {**349, 351**}

Blackwell DE, Petford AD, Arribas S, *et al.*, 1990, Determination of temperatures and angular diameters of 114 F–M stars using the infrared flux method. *A&A*, 232, 396–410 {**342**}

Blackwell DE, Petford AD, Shallis MJ, 1980, Use of the infrared flux method for determining stellar effective temperatures and angular diameters: the stellar temperature scale. *A&A*, 82, 249–252 {**342**}

Blackwell DE, Shallis MJ, 1977, Stellar angular diameters from infrared photometry: application to Arcturus and other stars with effective temperatures. *MNRAS*, 180, 177–191 {**342**}

Blackwell DE, Willis RB, 1977, Stellar gravities from metallic line profiles, with application to Arcturus. *MNRAS*, 180, 169–176 {**343**}

Blöcker T, 1995a, Stellar evolution of low and intermediate-mass stars. I. Mass loss on the asymptotic giant branch and its consequences for stellar evolution. *A&A*, 297, 727–738 {**350**}

—, 1995b, Stellar evolution of low- and intermediate-mass stars. II. Post-asymptotic giant branch evolution. *A&A*, 299, 755–769 {**348, 350**}

Böhm-Vitense E, 1953, Die Wasserstoffkonvektionszone der Sonne. *Zeitschrift fur Astrophysik*, 32, 135–164 {**345**}

—, 1958, Über die Wasserstoffkonvektionszone in Sternen verschiedener Effektivtemperaturen und Leuchtkräfte. *Zeitschrift fur Astrophysik*, 46, 108 {**345**}

—, 1970, The UBVr colours of main sequence stars. *A&A*, 8, 283–298 {**364**}

Böhm-Vitense E, Canterna R, 1974, The gap in the two-colour diagram of main sequence stars. *ApJ*, 194, 629–636 {**365**}

Bonneau D, Labeyrie A, 1973, Speckle interferometry: colour-dependent limb darkening evidenced on α Ori and o Cet. *ApJ*, 181, L1–L4 {**356**}

Bressan A, Chiosi C, Bertelli G, 1981, Mass loss and overshooting in massive stars. *A&A*, 102, 25–30 {**346**}

Bressan A, Fagotto F, Bertelli G, *et al.*, 1993, Evolutionary sequences of stellar models with new radiative opacities. II. $Z = 0.02$. *A&AS*, 100, 647–664 {**381**}

Brett JM, 1995, Opacity sampling model photospheres for M dwarfs. I. Computations, sensitivities and comparisons. *A&A*, 295, 736–754 {**345**}

Briquet M, Hubrig S, de Cat P, *et al.*, 2007, On the co-existence of chemically peculiar Bp stars, slowly pulsating B stars and constant B stars in the same part of the HR diagram. *A&A*, 466, 269–276 {**393**}

Brown JA, Sneden C, Lambert DL, *et al.*, 1989, A search for lithium-rich giant stars. *ApJS*, 71, 293–322 {**383**}

Brown TM, Christensen-Dalsgaard J, Weibel-Mihalas B, *et al.*, 1994, The effectiveness of oscillation frequencies in constraining stellar model parameters. *ApJ*, 427, 1013–1034 {**346**}

Brown TM, Gilliland RL, 1994, Asteroseismology. *ARA&A*, 32, 37–82 {**395**}

Bruntt H, Bikmaev IF, Catala C, *et al.*, 2004, Abundance analysis of targets for the COROT/MONS asteroseismology missions. II. Abundance analysis of the COROT main targets. *A&A*, 425, 683–695 {**398**}

Busso M, Gallino R, Wasserburg GJ, 1999, Nucleosynthesis in asymptotic giant branch stars: relevance for Galactic enrichment and Solar System formation. *ARA&A*, 37, 239–309 {**373**}

Caloi V, Cardini D, D'Antona F, *et al.*, 1999, Kinematics and age of stellar populations in the solar neighbourhood from Hipparcos data. *A&A*, 351, 925–936 {**387**}

Cameron AGW, Fowler WA, 1971, Lithium and the s-process in red giant stars. *ApJ*, 164, 111–114 {**383**}

Canto Martins BL, do Nascimento JD, Melo CHF, *et al.*, 2003, On the link between rotation, Ca II emission flux and Rossby number in subgiant stars. *Bulletin of the Astronomical Society of Brazil*, 23, 96–97 {**391**}

Caputo F, 1998, Evolution of Population II stars. *A&A Rev.*, 9, 33–61 {**366**}

Carbon DF, 1979, Model atmospheres for intermediate- and late-type stars. *ARA&A*, 17, 513–549 {**345**}

Cardini D, 2005, Mg II chromospheric radiative loss rates in cool active and quiet stars. *A&A*, 430, 303–311 {**392**}

Cardini D, Cassatella A, 2007, Colour, rotation, age, and chromospheric activity correlations in late-type main-sequence stars. *ApJ*, 666, 393–402 {**392**}

Carney BW, 1996, The constancy of [α/Fe] in globular clusters of differing [Fe/H] and age. *PASP*, 108, 900–910 {**379**}

Carney BW, Latham DW, Laird JB, *et al.*, 1994, A survey of proper motion stars. XII. An expanded sample. *AJ*, 107, 2240–2289 {**367**}

Carney BW, Lee J, Habgood MJ, 1998, The red horizontal-branch star HD 17072. *AJ*, 116, 424–428 {**371, 372**}

Casagrande L, Flynn C, Portinari L, *et al.*, 2007, The helium abundance and $\Delta Y/\Delta Z$ in lower main-sequence stars. *MNRAS*, 382, 1516–1540 {**380**}

Casertano S, Ratnatunga KU, 1999, Populating the HR diagram with the Hipparcos Catalogue. *ASP Conf. Ser. 167: Harmonizing Cosmic Distance Scales in a Post-Hipparcos Era*, 101–106 {**377**}

Cassinelli JP, 1987, Rotating stellar atmospheres. *IAU Colloq. 92: Physics of Be Stars* (eds. Slettebak A, Snow TP), 106–121 {**390**}

Castellani V, Ciacio F, Degl'Innocenti S, *et al.*, 1997, Heavy element diffusion and globular cluster ages. *A&A*, 322, 801–806 {**368**}

Castellani V, Degl'Innocenti S, Prada Moroni PG, 2001, Stellar models and Hyades: the Hipparcos test. *MNRAS*, 320, 66–72 {**376**}

Castelli F, Kurucz RL, 1994, Model atmospheres for Vega. *A&A*, 281, 817–832 {**352**}

Catalano S, Frasca A, Marilli E, *et al.*, 2000, Ca II H and K fluxes and luminosities after Hipparcos. The age of main sequence stars in the solar neighbourhood from the Ca II chromospheric emission. *ASP Conf. Ser. 198: Stellar Clusters and Associations: Convection, Rotation, and Dynamos*, 439–442 {**392**}

Cayrel de Strobel G, Crifo F, Lebreton Y, *et al.*, 1997a, The contribution of Hipparcos to the study of the stellar metal-rich population in the solar neighbourhood. *The First Results of Hipparcos and Tycho, IAU Joint Discussion 14*, 35–37 {**386**}

Cayrel de Strobel G, Lebreton Y, Soubiran C, *et al.*, 1999, Old, low-mass, metal-rich stars. *Ap&SS*, 265, 345–352 {**385, 386**}

Cayrel de Strobel G, Soubiran C, Friel ED, *et al.*, 1997b, A catalogue of [Fe/H] determinations: 1996 edition. *A&AS*, 124, 299–305 {**342, 343, 361, 377, 385, 386**}

Cayrel de Strobel G, 1969, On the magnesium green triplet lines and the sodium D lines as temperature and gravity indicators in late-type stars. *Proceedings of the 3rd Harvard–Smithsonian Conference on Stellar Atmospheres* (ed. Gingerich O), 35–40 {**343**}

—, 1987, On a possible metallicity gradient in super-metal-rich stars. *Journal of Astrophysics and Astronomy*, 8, 141–145 {**386**}

—, 1996, Stars resembling the Sun. *A&A Rev.*, 7, 243–288 {**341**}

Cayrel R, Castelli F, Katz D, *et al.*, 1997a, The bolometric correction $M_{\rm bol}$–*Hp*. *ESA SP–402: Hipparcos, Venice '97*, 433–436 {**349**}

Cayrel R, Faurobert-Scholl M, Feautrier N, *et al.*, 1996, On the use of Ca I triplet lines as luminosity indicators. *A&A*, 312, 549–552 {**343**}

Cayrel R, Lebreton Y, Morel P, 1999, Survival of ^{6}Li and ^{7}Li in metal-poor stars. *Galaxy Evolution: Connecting the Distant Universe with the Local Fossil Record* (ed. Spite M), 87–94 {**380**}

Cayrel R, Lebreton Y, Perrin MN, *et al.*, 1997b, The HR diagram in the plane $T_{\rm eff}$, $M_{\rm bol}$ of Population II stars with Hipparcos parallaxes. *ESA SP–402: Hipparcos, Venice '97*, 219–224 {**366, 367**}

Chaboyer BC, 1994, The primordial abundance of ^{6}Li and ^{9}Be. *ApJ*, 432, L47–L50 {**383**}

—, 1998, The age of the universe. *Phys. Rep.*, 307, 23–30 {**339**}

—, 1999, Globular cluster distance determinations. *ASSL Vol. 237: Post-Hipparcos Cosmic Candles*, 111–124 {**366**}

Chabrier G, Baraffe I, Allard F, *et al.*, 2000, Evolutionary models for very low-mass stars and brown dwarfs with dusty atmospheres. *ApJ*, 542, 464–472 {**350**}

Chabrier G, Baraffe I, 1997, Structure and evolution of low-mass stars. *A&A*, 327, 1039–1053 {**345, 350**}

Charbonnel C, 1994, Clues for non-standard mixing on the red giant branch from ^{12}C/^{13}C and ^{12}C/^{14}N ratios in evolved stars. *A&A*, 282, 811–820 {**381**}

Charbonnel C, Balachandran SC, 2000, The nature of the lithium-rich giants. Mixing episodes on the red giant branch and early-asymptotic giant branch. *A&A*, 359, 563–572 {**383**}

Charbonnel C, Brown JA, Wallerstein G, 1998, Mixing processes during the evolution of red giants with moderate metal deficiencies: the role of molecular-weight barriers. *A&A*, 332, 204–214 {**370, 371**}

Charbonnel C, do Nascimento JD, 1998, How many low-mass stars destroy ^{3}He? *A&A*, 336, 915–919 {**380**}

Charbonnel C, Meynet G, Maeder A, *et al.*, 1993, Grids of stellar models. III. From 0.8–120 $M_\odot$ at $Z = 0.004$. *A&AS*, 101, 415–419 {**350**}

—, 1996, Grids of stellar models. VI. Horizontal branch and early asymptotic giant branch for low-mass stars ($Z = 0.020, 0.001$). *A&AS*, 115, 339–344 {**350**}

Charbonnel C, Primas F, 2005, The lithium content of the Galactic halo stars. *A&A*, 442, 961–992 {**382**}

Charbonnel C, Talon S, 2005, New generation stellar models with rotation and internal gravity waves: the Li dip revisited. *EAS Publications Series 17* (eds. Alecian G, Richard O, Vauclair S), 167–176 {**382**}

Charbonnel C, Zahn JP, 2007, Thermohaline mixing: a physical mechanism governing the photospheric composition of low-mass giants. *A&A*, 467, L15–L18 {**371**}

Chen Y, Nissen PE, Benoni T, *et al.*, 2001, Lithium abundances for 185 main sequence stars: Galactic evolution and stellar depletion of lithium. *A&A*, 371, 943–951 {**382**}

Chen Y, Nissen PE, Zhao G, *et al.*, 2000, Chemical composition of 90 F and G disk dwarfs. *A&AS*, 141, 491–506 {**387, 389**}

Chesneau O, Meilland A, Rivinius T, *et al.*, 2005, First VLTI–MIDI observations of a Be star: α Ara. *A&A*, 435, 275–287 {**358**}

Chieffi A, Limongi M, Straniero O, 1998, The evolution of a $25M_\odot$ star from the main sequence up to the onset of the iron core collapse. *ApJ*, 502, 737–762 {**350**}

Chieffi A, Straniero O, Salaris M, 1995, Calibration of stellar models. *ApJ*, 445, L39–L42 {**375**}

Chieffi A, Straniero O, 1989, Isochrones for hydrogen-burning globular cluster stars. I. [Fe/H] from -2 to -1. *ApJS*, 71, 47–87 {**350**}

Chollet F, Sinceac V, 1999, Analysis of solar radius determination obtained by the modern CCD astrolabe of the Calern Observatory: a new approach of the solar limb definition. *A&AS*, 139, 219–229 {**356**}

Christensen-Dalsgaard J, 1982, On solar models and their periods of oscillation. *MNRAS*, 199, 735–761 {**379**}

—, 1984, What will asteroseismology teach us? *Space Research in Stellar Activity and Variability* (eds. Mangeney A, Praderie F), 11–18 {**395**}

Christensen-Dalsgaard J, Bedding TR, Kjeldsen H, 1995, Modeling solar-like oscillations in η Boo. *ApJ*, 443, L29–L32 {**397**}

Christensen-Dalsgaard J, Däppen W, 1992, Solar oscillations and the equation of state. *A&A Rev.*, 4, 267–361 {**345**}

Christensen-Dalsgaard J, Duvall TL, Gough DO, *et al.*, 1985, Speed of sound in the solar interior. *Nature*, 315, 378–382 {**343**}

Christensen-Dalsgaard J, Proffitt CR, Thompson MJ, 1993, Effects of diffusion on solar models and their oscillation frequencies. *ApJ*, 403, L75–L78 {**346**}

Claret A, 1995, Stellar models for a wide range of initial chemical compositions until helium burning. I. From $X = 0.60$ to $X = 0.80$ for $Z = 0.02$. *A&AS*, 109, 441–446 {**374**}

—, 2004, New grids of stellar models including tidal-evolution constants up to carbon burning. I. From $0.8-125M_\odot$ at $Z = 0.02$. *A&A*, 424, 919–925 {**350**}

—, 2005, New grids of stellar models including tidal-evolution constants up to carbon burning. II. From $0.8-125M_\odot$: the Small Magellanic Cloud ($Z = 0.002 - 0.004$). *A&A*, 440, 647–651 {**350**}

—, 2006, New grids of stellar models including tidal-evolution constants up to carbon burning. III. From $0.8-125M_\odot$: the Large Magellanic Cloud ($Z = 0.007 - 0.01$). *A&A*, 453, 769–771 {**350**}

—, 2007, New grids of stellar models including tidal-evolution constants up to carbon burning. IV. From $0.8-125M_\odot$: high metallicities ($Z = 0.04 - 0.10$). *A&A*, 467, 1389–1396 {**350**}

Claret A, Gimenez A, 1998, Stellar models for a wide range of initial chemical compositions until helium burning IV. From $X = 0.65$ to $X = 0.80$, for $Z = 0.004$. *A&AS*, 133, 123–127 {**350**}

Clementini G, Gratton RG, Carretta E, *et al.*, 1999, Homogeneous photometry and metal abundances for a large sample of Hipparcos metal-poor stars. *MNRAS*, 302, 22–36 {**353, 378, 384**}

Clem JL, VandenBerg DA, Grundahl F, *et al.*, 2004, Empirically constrained colour-temperature relations. II. uvby. *AJ*, 127, 1227–1256 {**353**}

Code AD, Bless RC, Davis J, *et al.*, 1976, Empirical effective temperatures and bolometric corrections for early-type stars. *ApJ*, 203, 417–434 {**341, 342, 358**}

Cohen ER, Taylor BN, 1986, *Codata Bulletin 63*. Pergamon, New York {**339**}

Cordier D, Pietrinferni A, Cassisi S, *et al.*, 2007, A large stellar evolution data base for population synthesis studies. III. Inclusion of the full asymptotic giant branch phase and web tools for stellar population analyses. *AJ*, 133, 468–478 {**350**}

Cox AN, 2000, *Allen's Astrophysical Quantities*, 4th Edition. Springer, New York {**341, 356**}

Cox JP, Giuli RT, 1968, *Principles of Stellar Structure*. Gordon and Breach, New York {**339**}

Crifo F, Spite F, Spite M, 1997, A first Hipparcos contribution to the lithium problem. *ESA SP–402: Hipparcos, Venice '97*, 351–354 {**383**}

Crifo F, Spite M, Spite F, 1998, A first Hipparcos contribution to the light elements problem. *A&A*, 330, L25–L28 {**383**}

Currie DG, Knapp SL, Liewer KM, 1974, Four stellar-diameter measurements by a new technique: amplitude interferometry. *ApJ*, 187, 131–134 {**356**}

Cutispoto G, Pastori L, Pasquini L, *et al.*, 2002, Fast-rotating nearby solar-type stars, Li abundances and X-ray luminosities. I. Spectral classification, $v \sin i$, Li abundances and X-ray luminosities. *A&A*, 384, 491–503 {**392**}

D'Antona F, Mazzitelli I, 1994, New pre-main sequence tracks for $M \leq 2.5M_\odot$ as tests of opacities and convection model. *ApJS*, 90, 467–500 {**350**}

—, 1997, Evolution of low-mass stars. *Memorie della Societa Astronomica Italiana*, 68, 807–822 {**350**}

Däppen W, 2006, The equation of state for the solar interior. *Journal of Physics A Mathematical General*, 39, 4441–4446 {**344**}

da Silva L, Girardi L, Pasquini L, *et al.*, 2006, Basic physical parameters of a selected sample of evolved stars. *A&A*, 458, 609–623 {**361**}

de Boer KS, Geffert M, Tucholke HJ, *et al.*, 1997a, Calculating the mass of horizontal-branch stars with Hipparcos. *ESA SP–402: Hipparcos, Venice '97*, 331–334 {**377**}

de Boer KS, Tucholke HJ, Schmidt JHK, 1997b, Calibrating horizontal-branch stars with Hipparcos. *A&A*, 317, L23–L26 {**371, 372**}

de Boer KS, 1999, Horizontal-branch stars: their nature and their absolute magnitude. *ASP Conf. Ser. 167: Harmonizing Cosmic Distance Scales in a Post-Hipparcos Era*, 129–139 {**371, 372**}

de Bruijne JHJ, Hoogerwerf R, de Zeeuw PT, 2001, A Hipparcos study of the Hyades open cluster. Improved colour-absolute magnitude and HR diagrams. *A&A*, 367, 111–147 {**354, 370, 371**}

de la Reza R, Drake NA, da Silva L, *et al.*, 1997, On a rapid lithium enrichment and depletion of K giant stars. *ApJ*, 482, L77–L80 {**382**}

de Medeiros JR, Mayor M, 1999, A catalogue of rotational and radial velocities for evolved stars. *A&AS*, 139, 433–460 {**392**}

de Medeiros JR, Silva JRP, do Nascimento JD, *et al.*, 2006, A catalogue of rotational and radial velocities for evolved stars. IV. Metal-poor stars. *A&A*, 458, 895–898 {**392**}
de Medeiros JR, Udry S, Burki G, *et al.*, 2002, A catalogue of rotational and radial velocities for evolved stars. II. Ib supergiant stars. *A&A*, 395, 97–98 {**392**}
de Medeiros JR, Udry S, Mayor M, 2004, A catalogue of rotational and radial velocities for evolved stars. III. Double-lined binary systems. *A&A*, 427, 313–317 {**392**}
Decin L, Vandenbussche B, Waelkens C, *et al.*, 2003, ISO-SWS calibration and the accurate modeling of cool-star atmospheres. IV. G9 to M2 stars. *A&A*, 400, 709–727 {**361**}
Deliyannis CP, Malaney RA, 1995, Flare production of ^{6}Li in Population II stars. *ApJ*, 453, 810–812 {**383**}
Demarque P, Woo JH, Kim YC, *et al.*, 2004, Yonsei–Yale (Y^2) isochrones with an improved core overshoot treatment. *ApJS*, 155, 667–674 {**350**}
Deutsch AJ, 1958, Harmonic analysis of the periodic spectrum variables. *Electromagnetic Phenomena in Cosmical Physics* (ed. Lehnert B), IAU Symp. 6, 209–221 {**395**}
Di Benedetto GP, 1998, Towards a fundamental calibration of stellar parameters of A, F, G, K dwarfs and giants. *A&A*, 339, 858–871 {**351**}
Di Benedetto GP, Rabbia Y, 1987, Accurate angular diameters and effective temperatures for eleven giants cooler than K0 by Michelson interferometry. *A&A*, 188, 114–124 {**341, 352**}
Di Mauro MP, Christensen-Dalsgaard J, Kjeldsen H, *et al.*, 2003, Convective overshooting in the evolution and seismology of η Boo. *A&A*, 404, 341–353 {**396, 397**}
Dorado M, Montesinos B, 1998, Convective parameters and activity in late-type giants. *ASP Conf. Ser. 154: Cool Stars, Stellar Systems, and the Sun*, 827–832 {**392**}
do Nascimento JD, Canto Martins BL, Melo CHF, *et al.*, 2003, On the link between rotation, chromospheric activity and Li abundance in subgiant stars. *A&A*, 405, 723–731 {**382**}
do Nascimento JD, Charbonnel C, Lèbre A, *et al.*, 2000, Lithium and rotation on the subgiant branch. II. Theoretical analysis of observations. *A&A*, 357, 931–937 {**382, 391**}
Dravins D, Lindegren L, Nordlund A, *et al.*, 1993, The distant future of solar activity: a case study of β Hyi. I. Stellar evolution, lithium abundance, and photospheric structure. *ApJ*, 403, 385–395 {**369**}
Dravins D, Lindegren L, VandenBerg DA, 1997, β Hyi (G2 IV): a revised age for the closest subgiant. *ESA SP–402: Hipparcos, Venice '97*, 397–400 {**369**}
—, 1998, β Hyi (G2 IV): a revised age for the closest subgiant. *A&A*, 330, 1077–1079 {**369**}
Dumm T, Schild H, 1998, Stellar radii of M giants. *New Astronomy*, 3, 137–156 {**356, 357**}
Dyck HM, Benson JA, van Belle GT, *et al.*, 1996, Radii and effective temperatures for K and M giants and supergiants. *AJ*, 111, 1705–1712 {**341**}
Dyck HM, van Belle GT, Thompson RR, 1998, Radii and effective temperatures for K and M giants and supergiants. II. *AJ*, 116, 981–986 {**352, 357**}
Edmonds P, Cram L, Demarque P, *et al.*, 1992, Evolutionary models and the p-mode oscillation spectrum of α Cen AB. *ApJ*, 394, 313–319 {**375**}
Edvardsson B, 1988, Spectroscopic surface gravities and chemical compositions for 8 nearby single sub-giants. *A&A*, 190, 148–166 {**343**}
Edvardsson B, Andersen J, Gustafsson B, *et al.*, 1993, The chemical evolution of the Galactic disk. I. Analysis and results. *A&A*, 275, 101–152 {**386–389**}
Eggen OJ, Lynden-Bell D, Sandage AR, 1962, Evidence from the motions of old stars that the Galaxy collapsed. *ApJ*, 136, 748–767 {**384**}
Eggen OJ, Sandage AR, 1962, On the existence of subdwarfs in the $M_{\rm bol}$, log $T_{\rm eff}$ plane. II. *ApJ*, 136, 735–747 {**366**}
Eggenberger P, Charbonnel C, Talon S, *et al.*, 2004, Analysis of α Cen AB including seismic constraints. *A&A*, 417, 235–246 {**376**}
Eggleton PP, Faulkner J, Flannery BP, 1973, An approximate equation of state for stellar material. *A&A*, 23, 325–330 {**343, 345**}
Emilio M, Leister NV, 2005, Solar diameter measurements at São Paulo Observatory. *MNRAS*, 361, 1005–1011 {**356**}
Evans TL, 2002, Red giant stars. *Encyclopedia of Astronomy and Astrophysics;* (ed. Murdin, P.) {**347**}
Eyer L, Grenon M, 1998, Results from Hipparcos mission on stellar seismology. *IAU Symp. 185: New Eyes to See Inside the Sun and Stars*, 291–294 {**396**}
Favata F, 1998, The impact of high-accuracy astrometry on asteroseismology. *The First MONS Workshop: Science with a Small Space Telescope* (eds. Kjeldsen H, Bedding TR), 89–97 {**396**}
Favata F, Micela G, Sciortino S, *et al.*, 1997a, The evolutionary status of high-lithium, high-activity cool dwarfs. *ESA SP–402: Hipparcos, Venice '97*, 347–350 {**380**}
Favata F, Micela G, Sciortino S, 1997b, The [Fe/H] distribution of a volume limited sample of solar-type stars and its implications for Galactic chemical evolution. *A&A*, 323, 809–818 {**386**}
Fekel FC, Bolton CT, 2007, Chromospherically-active stars. XXVI. The double-lined late-type binary HD 19485 = WZ Ari. *AJ*, 134, 2079–2085 {**392**}
Fekel FC, 1997, Chromospherically-active stars. XVI. The double-lined binary 42 Cap. *AJ*, 114, 2747–2752 {**392**}
Feltzing S, Gustafsson B, 1998, Abundances in metal-rich stars. Detailed abundance analysis of 47 G and K dwarf stars with $[Me/H] > 0.10$ dex. *A&AS*, 129, 237–266 {**386**}
Fernandes J, 2001, Helium-to-metal enrichment ratio: determinations on the HR diagram after Hipparcos. *Astronomy and Astrophysics: Recent Developments*, 149–154 {**379**}
Fernandes J, Lebreton Y, Baglin A, *et al.*, 1998, Fundamental stellar parameters for nearby visual binary stars: η Cas, XI Boo, 70 Oph and 85 Peg. Helium abundance, age and mixing length parameter for low-mass stars. *A&A*, 338, 455–464 {**366, 375**}
Fernandes J, Lebreton Y, Baglin A, 1996, On the width of the theoretical lower main sequence. Consequences for the determination of the $\Delta Y/\Delta Z$ ratio in the solar neighbourhood. *A&A*, 311, 127–134 {**365, 379**}
Fernandes J, Neuforge C, 1995, α Cen and convection theories. *A&A*, 295, 678–684 {**375**}
Fields BD, Olive KA, 1999, The revival of Galactic cosmic-ray nucleosynthesis? *ApJ*, 516, 797–810 {**380, 383**}
Fischer DA, Stauffer JR, Jones B, 1998, Hipparcos-based colour–magnitude diagram for Li-rich late-type field stars: a comparison with the Pleiades and Praesepe clusters. *ASP Conf. Ser. 154: Cool Stars, Stellar Systems, and the Sun*, 2097–2100 {**381**}

Fitzpatrick EL, Massa D, 2005, Determining the physical properties of B stars. II. Calibration of synthetic photometry. *AJ*, 129, 1642–1662 {**390**}
Flannery BP, Ayres TR, 1978, Evolution of the α Cen system. *ApJ*, 221, 175–185 {**375**}
Fleischer AJ, Gauger A, Sedlmayr E, 1992, Circumstellar dust shells around long-period variables. I. Dynamical models of C-stars including dust formation, growth and evaporation. *A&A*, 266, 321–339 {**374**}
Flower PJ, 1996, Transformations from theoretical HR diagrams to colour–magnitude diagrams: effective temperatures, $B-V$ colours, and bolometric corrections. *ApJ*, 469, 355–365 {**341**}
Flynn C, 2004, Cosmic helium production. *Publications of the Astronomical Society of Australia*, 21, 126–128 {**379**}
Flynn C, Morell O, 1997, Metallicities and kinematics of G and K dwarfs. *MNRAS*, 286, 617–625 {**386, 388**}
Formiggini L, Brosch N, Almoznino E, *et al.*, 2002, Hidden sub-luminous stars among the FAUST ultraviolet sources towards Ophiuchus. *MNRAS*, 332, 441–455 {**373**}
Francois P, Matteucci F, 1993, On the abundance spread in solar neighbourhood stars. *A&A*, 280, 136–140 {**386**}
Freytag B, Ludwig HG, Steffen M, 1999, A calibration of the mixing-length for solar-type stars based on hydrodynamical models of stellar surface convection. *Stellar Structure: Theory and Test of Connective Energy Transport* (eds. Gimenez A, Guinan EF, Montesinos B), ASP Conf. Ser. 173, 225–228 {**375**}
Freytag B, Salaris M, 1999, Stellar envelope convection calibrated by radiation hydrodynamics simulations: influence on globular cluster isochrones. *ApJ*, 513, L49–L52 {**375**}
Fuhrmann K, 1998a, Nearby stars of the Galactic disk and halo. *A&A*, 338, 161–183 {**361, 364, 389**}
—, 1998b, Surface gravities of very metal-poor stars from Hipparcos parallaxes. *A&A*, 330, 626–630 {**342, 343, 353**}
—, 2004, Nearby stars of the Galactic disk and halo. III. *Astronomische Nachrichten*, 325, 3–80 {**377**}
Fuhrmann K, Pfeiffer MJ, Frank C, *et al.*, 1997, The surface gravities of cool dwarf stars revisited. *A&A*, 323, 909–922 {**343**}
Fulbright JP, 2000, Abundances and kinematics of field halo and disk stars. I. Observational data and abundance analysis. *AJ*, 120, 1841–1852 {**384, 389**}
García López RJ, Israelian G, Rebolo R, *et al.*, 2001, Oxygen abundances derived in unevolved very metal-poor stars. *New Astronomy Review*, 45, 519–523 {**384**}
García Pérez AE, Asplund M, Primas F, *et al.*, 2006, Oxygen abundances in metal-poor subgiants as determined from [O I], O I and OH lines. *A&A*, 451, 621–642 {**389**}
Garnett DR, Kobulnicky HA, 2000, Distance dependence in the solar neighbourhood age–metallicity relation. *ApJ*, 532, 1192–1196 {**388**}
Gautschy A, 1987, On the Baade–Wesselink method. *Vistas in Astronomy*, 30, 197–241 {**356**}
Gehren T, Liang YC, Shi J, *et al.*, 2004, Abundances of Na, Mg and Al in nearby metal-poor stars. *A&A*, 413, 1045–1063 {**385**}
Gerbaldi M, Faraggiana R, Burnage R, *et al.*, 1998, Determination of $T_{\rm eff}$ and log g for A0 type stars and its impact on the interpretation of the HR diagram with the Hipparcos results. *Contributions of the Astronomical Observatory Skalnate Pleso*, 27, 197–204 {**352**}
—, 1999, Search for reference A0 dwarf stars: masses and luminosities revisited with Hipparcos parallaxes. *A&AS*, 137, 273–292 {**352**}
Gezari DY, Labeyrie A, Stachnik RV, 1972, Speckle interferometry: diffraction-limited measurements of nine stars with the 200-inch telescope. *ApJ*, 173, L1–L5 {**356**}
Girardi L, Bressan A, Bertelli G, *et al.*, 2000, Evolutionary tracks and isochrones for low- and intermediate-mass stars: from $0.15-7\,M_\odot$ at $Z=0.0004-0.03$. *A&AS*, 141, 371–383 {**350, 370, 371, 374**}
Girardi L, Bressan A, Chiosi C, *et al.*, 1996, Evolutionary sequences of stellar models with new radiative opacities. VI. $Z=0.0001$. *A&AS*, 117, 113–125 {**350**}
Girardi L, Castelli F, Bertelli G, *et al.*, 2007, On the effect of helium enhancement on bolometric corrections and $T_{\rm eff}$–colour relations. *A&A*, 468, 657–662 {**349**}
Gondoin P, 1999, Evolution of X-ray activity and rotation on G-K giants. *A&A*, 352, 217–227 {**390**}
—, 2005, The relation between X-ray activity and rotation in intermediate-mass G giants. *A&A*, 444, 531–538 {**391**}
—, 2007, The rotation-activity correlation among G and K giants in binary systems. *A&A*, 464, 1101–1106 {**391**}
Gough DO, Toomre J, 1991, Seismic observations of the solar interior. *ARA&A*, 29, 627–685 {**395**}
Gratton RG, 1998, The absolute magnitude of field metal-poor horizontal branch stars. *MNRAS*, 296, 739–745 {**372**}
Gratton RG, Carretta E, Claudi R, *et al.*, 2003a, Abundances for metal-poor stars with accurate parallaxes. I. Basic data. *A&A*, 404, 187–210 {**384, 385**}
Gratton RG, Carretta E, Clementini G, *et al.*, 1997, Metal abundances of one hundred Hipparcos dwarfs. *ESA SP–402: Hipparcos, Venice '97*, 339–342 {**384**}
Gratton RG, Carretta E, Desidera S, *et al.*, 2003b, Abundances for metal-poor stars with accurate parallaxes. II. α-elements in the halo. *A&A*, 406, 131–140 {**378, 385**}
Gray DF, 1967, Photometric determination of stellar radii. *ApJ*, 149, 317–343 {**356**}
Gray RO, 2000, Effective temperature scale and bolometric corrections. *Encyclopedia of Astronomy and Astrophysics* (ed. Murdin, P.) {**341**}
Green EM, Demarque P, 1996, Revised Yale isochrones and luminosity functions. *VizieR Online Data Catalogue*, 6040 {**350**}
Grenon M, 1987, Past and present metal abundance gradient in the Galactic disk. *Journal of Astrophysics and Astronomy*, 8, 123–139 {**386, 389**}
—, 1999, The kinematics and origin of super metal-rich stars. *Ap&SS*, 265, 331–336 {**386**}
—, 2001, The Hipparcos mission legacy for cool stars. *ASP Conf. Ser. 223: 11th Cambridge Workshop on Cool Stars, Stellar Systems and the Sun*, 359–367 {**377**}
Grevesse N, Asplund M, Sauval AJ, 2007, The solar chemical composition. *Space Science Reviews*, 105–111 {**339**}
Grevesse N, Noels A, 1993, Cosmic abundances of the elements. *Origin and Evolution of the Elements. Cambridge University Press* (eds. Prantzos N, Vangioni-Flam E, Casse M), 14–25 {**349**}
Grevesse N, Sauval AJ, 1999, The solar abundance of iron and the photospheric model. *A&A*, 347, 348–354 {**343**}
Groenewegen MAT, de Jong T, 1998, CO observations and mass loss of main sequence and S-stars. *A&A*, 337, 797–807 {**374**}

Gruenwald R, Steigman G, Viegas SM, 2002, The evolution of He and H ionisation corrections as H II regions age. *ApJ*, 567, 931–939 {**379**}

Grundahl F, VandenBerg DA, Bell RA, *et al.*, 2000, A distance-independent age for the globular cluster M92. *AJ*, 120, 1884–1891 {**368**}

Guandalini R, Ciprini S, Busso M, *et al.*, 2004a, Mass-losing asymptotic giant branch stars: infrared observations and evolutionary implications. *Memorie della Societa Astronomica Italiana*, 75, 617 {**374**}

—, 2004b, Mass loss in asymptotic giant branch stars from infrared colours. *Memorie della Societa Astronomica Italiana Supplement*, 5, 119 {**374**}

Guenther DB, Demarque P, Kim YC, *et al.*, 1992, Standard solar model. *ApJ*, 387, 372–393 {**339**}

Guenther DB, Demarque P, 1996, Seismology of η Boo. *ApJ*, 456, 798–810 {**397**}

—, 1997, Seismic tests of the Sun's interior structure, composition, and age, and implications for solar neutrinos. *ApJ*, 484, 937–959 {**339**}

—, 2000, α Cen AB. *ApJ*, 531, 503–520 {**339, 376, 378**}

Gustafsson B, 1998, What do we do when models don't fit? On model atmospheres and real stellar spectra. *Fundamental Stellar Properties* (eds. Bedding TR, Booth AJ, Davis J), IAU Symp. 189, 261–276 {**362**}

Gustafsson B, Bell RA, Eriksson K, *et al.*, 1975, A grid of model atmospheres for metal-deficient giant stars. I. *A&A*, 42, 407–432 {**345**}

Guzik JA, 1998, Solar structure: models and inferences from helioseismology. *Structure and Dynamics of the Interior of the Sun and Sun-like Stars SOHO 6/GONG 98 Workshop* (ed. Korzennik S), volume 6, 417–424 {**339**}

Hanbury Brown R, 1974, *The Intensity Interferometer: its Application to Astronomy*. Taylor and Francis, New York {**356**}

Hanbury Brown R, Davis J, Allen LR, *et al.*, 1967, The stellar interferometer at Narrabri Observatory. II. The angular diameters of 15 stars. *MNRAS*, 137, 393–402 {**358**}

Hanbury Brown R, Davis J, Allen LR, 1974, The angular diameters of 32 stars. *MNRAS*, 167, 121–136 {**356, 358**}

Haniff CA, Scholz M, Tuthill PG, 1995, New diameter measurements of 10 Mira variables: implications for effective temperatures atmospheric structure and pulsation modes. *MNRAS*, 276, 640–650 {**357**}

Hanson RB, Sneden C, Kraft RP, *et al.*, 1998, On the use of [Na/Fe] and [α/Fe] ratios and Hipparcos-based velocities as age indicators among low-metallicity halo field giants. *AJ*, 116, 1286–1294 {**378, 384**}

Hauschildt PH, Baron E, Allard F, 1997, Parallel implementation of the PHOENIX generalized stellar atmosphere program. *ApJ*, 483, 390–398 {**345**}

Hauschildt PH, Lowenthal DK, Baron E, 2001, Parallel implementation of the PHOENIX generalized stellar atmosphere program. III. A parallel algorithm for direct opacity sampling. *ApJS*, 134, 323–329 {**345**}

Haywood M, 2001a, A revision of the solar neighbourhood metallicity distribution. *MNRAS*, 325, 1365–1382 {**377, 386, 388**}

—, 2001b, A revision of the solar neighbourhood metallicity distribution. *ASP Conf. Ser. 230: Galaxy Disks and Disk Galaxies*, 15–16 {**388**}

—, 2006, Revisiting two local constraints of the Galactic chemical evolution. *MNRAS*, 371, 1760–1776 {**388**}

Haywood M, Palasi J, Gómez AE, *et al.*, 1997a, The metallicity distribution of late-type dwarfs and the Hipparcos HR diagram. *ESA SP–402: Hipparcos, Venice '97*, 489–494 {**349, 387, 388**}

Haywood M, Palasi J, Gómez AE, 1997b, The Hipparcos HR diagram and the metallicity distribution of late-type stars. *The First Results of Hipparcos and Tycho, IAU Joint Discussion 14*, 13–15 {**387**}

Heber U, 2004, The nature of the compact companion of the sdB star HD 188112. *FUSE Proposal*, 37 {**373**}

Heber U, Moehler S, Reid IN, 1997, Masses and gravities of blue horizontal branch stars revisited. *ESA SP–402: Hipparcos, Venice '97*, 461–464 {**372**}

Henyey LG, Wilets L, Böhm KH, *et al.*, 1959, A method for atomic computation of stellar evolution. *ApJ*, 129, 628–636 {**339**}

Herwig F, 2000, The evolution of asymptotic giant branch stars with convective overshoot. *A&A*, 360, 952–968 {**350, 373**}

—, 2004, Dredge-up and envelope burning in intermediate-mass giants of very low metallicity. *ApJ*, 605, 425–435 {**350, 373**}

Hestroffer D, Mignard F, 1997a, Determination of limb-darkening from Hipparcos observations. *ESA SP–402: Hipparcos, Venice '97*, 173–176 {**357**}

—, 1997b, Photometry with a periodic grid. I. A new method to derive angular diameters and brightness distribution. *A&A*, 325, 1253–1258 {**357**}

Hobbs LM, Thorburn JA, Rebull LM, 1999, Lithium isotope ratios in halo stars. III. *ApJ*, 523, 797–804 {**383**}

Høg E, Pagel BEJ, Portinari L, *et al.*, 1998, Primordial helium and $\Delta Y/\Delta Z$ from H II regions and from fine structure in the main sequence based on Hipparcos parallaxes. *Space Science Reviews*, 84, 115–126 {**379**}

Holmberg J, Flynn C, Portinari L, 2006, The colours of the Sun. *MNRAS*, 367, 449–453 {**339**}

Holmberg J, Nordström B, Andersen J, 2007, The Geneva–Copenhagen survey of the Solar neighbourhood. II. New uvby calibrations and rediscussion of stellar ages, the G dwarf problem, age–metallicity diagram, and heating mechanisms of the disk. *A&A*, 475, 519–537 {**388, 389**}

Hubrig S, North P, Mathys G, 2000a, Magnetic AP stars in the HR diagram. *ApJ*, 539, 352–363 {**392, 394**}

Hubrig S, North P, Medici A, 2000b, Rotation and evolution of A stars: looking for progenitors of cool Ap stars. *A&A*, 359, 306–310 {**390**}

Hubrig S, North P, Schoeller M, 2007, Evolution of magnetic fields in stars across the upper main sequence. II. Observed distribution of the magnetic field geometry. *Astronomische Nachrichten*, 328, 475–490 {**393**}

Hubrig S, North P, Schöller M, *et al.*, 2006, Evolution of magnetic fields in stars across the upper main sequence. I. Catalogue of magnetic field measurements with FORS 1 at the VLT. *Astronomische Nachrichten*, 327, 289–297 {**393**}

Hubrig S, North P, Szeifert T, 2005, Evolution of magnetic fields in stars across the upper main sequence: results from recent measurements with FORS 1 at the VLT. *ASP Conf. Ser. 343* (eds. Adamson A, Aspin C, Davis C), 374–378 {**392–394**}

Huebner WF, Merts AL, Magee NH, *et al.*, 1977, *Los Alamos Sci. Rep.* LA-6760-M {**343**}

Hutter DJ, Johnston KJ, Mozurkewich D, *et al.*, 1989, Angular diameter measurements of 24 giant and supergiant stars from the Mark III optical interferometer. *ApJ*, 340, 1103–1111 {**352**}

Iben I, 1967, Stellar evolution within and off the main sequence. *ARA&A*, 5, 571–626 {**346**}

Iben I, Renzini A, 1983, Asymptotic giant branch evolution and beyond. *ARA&A*, 21, 271–342 {**346**}

Ibukiyama A, 2004, Solar neighbourhood age–metallicity relation based on Hipparcos data. *Publications of the Astronomical Society of Australia*, 21, 121–125 {**389**}

Iglesias CA, Rogers FJ, Wilson BG, 1992, Spin–orbit interaction effects on the Rosseland mean opacity. *ApJ*, 397, 717–728 {**346**}

Iglesias CA, Rogers FJ, 1996, Updated Opal opacities. *ApJ*, 464, 943–953 {**344, 346, 349**}

Israelian G, Rebolo R, García López RJ, *et al.*, 2001, Oxygen in the very early Galaxy. *ApJ*, 551, 833–851 {**384**}

Itoh N, Adachi T, Nakagawa M, *et al.*, 1989, Neutrino energy loss in stellar interiors. III. Pair, photo-, plasma, and bremsstrahlung processes. *ApJ*, 339, 354–364 {**349**}

Izotov YI, Thuan TX, Lipovetsky VA, 1997, The primordial helium abundance: systematic effects and a new determination. *ApJS*, 108, 1–39 {**365**}

Jackson T, Ivezić Ž, Knapp GR, 2002, The Galactic distribution of asymptotic giant branch stars. *MNRAS*, 337, 749–767 {**373**}

Jasniewicz G, Parthasarathy M, de Laverny P, *et al.*, 1999, Late-type giants with infrared excess. I. Lithium abundances. *A&A*, 342, 831–838 {**382**}

Jasniewicz G, Recio Blanco A, de Laverny P, *et al.*, 2006, Lithium abundances for early F stars: new observational constraints for the Li dilution. *A&A*, 453, 717–722 {**382**}

Jeffers SV, Cameron AC, Barnes JR, *et al.*, 2005, Direct evidence for a polar spot on SV Cam. *ApJ*, 621, 425–431 {**395**}

Jerzykiewicz M, Molenda-Zakowicz J, 2000, Empirical luminosities and radii of early-type stars after Hipparcos. *Acta Astronomica*, 50, 369–380 {**357, 358**}

Jimenez R, Flynn C, Kotoneva E, 1998, Hipparcos and the age of the Galactic disk. *MNRAS*, 299, 515–519 {**363**}

Jimenez R, Flynn C, MacDonald J, *et al.*, 2003, The cosmic production of helium. *Science*, 299, 1552–1555 {**379, 380**}

Jimenez R, MacDonald J, Dunlop JS, *et al.*, 2004, Synthetic stellar populations: single stellar populations, stellar interior models and primordial protogalaxies. *MNRAS*, 349, 240–254 {**346, 350**}

Jimenez R, MacDonald J, 1996, Stellar evolutionary tracks for low-mass stars. *MNRAS*, 283, 721–732 {**350, 374**}

Johnson HL, Morgan WW, 1953, Fundamental stellar photometry for standards of spectral type on the revised system of the Yerkes spectral atlas. *ApJ*, 117, 313–352 {**364**}

Jonsell K, Edvardsson B, Gustafsson B, *et al.*, 2005, Chemical abundances in 43 metal-poor stars. *A&A*, 440, 321–343 {**377, 385**}

Kaempf TA, de Boer KS, Altmann M, 2005, Kinematics of red horizontal branch stars to trace the structure of the Galaxy. *A&A*, 432, 879–888 {**372, 373**}

Kervella P, 2006, Cepheid distances from interferometry. *Memorie della Societa Astronomica Italiana*, 77, 227–230 {**358**}

Kervella P, Coudé du Foresto V, Traub WA, *et al.*, 1999, Interferometric Observations of the Cepheid ζ Gem with IOTA/FLUOR. *ASP Conf. Ser. 194: Working on the Fringe: Optical and Infrared Interferometry from Ground and Space*, 22–27 {**358**}

Kervella P, Nardetto N, Bersier D, *et al.*, 2004a, Cepheid distances from infrared long-baseline interferometry. I. VLTI–VINCI observations of seven Galactic Cepheids. *A&A*, 416, 941–953 {**358**}

Kervella P, Thévenin F, Morel P, *et al.*, 2003a, The interferometric diameter and internal structure of Sirius A. *A&A*, 408, 681–688 {**358, 359**}

—, 2004b, The diameter and evolutionary state of Procyon A. Multi-technique modeling using asteroseismic and interferometric constraints. *A&A*, 413, 251–256 {**358, 359**}

Kervella P, Thévenin F, Ségransan D, *et al.*, 2003b, The diameters of α Cen A and B. A comparison of the asteroseismic and VLTI–VINCI views. *A&A*, 404, 1087–1097 {**376**}

King JR, 2001, The Galactic evolution of beryllium and boron revisited. *AJ*, 122, 3115–3135 {**389**}

Kjeldsen H, Bedding TR, Viskum M, *et al.*, 1995, Solar-like oscillations in η Boo. *AJ*, 109, 1313–1319 {**397**}

Kjeldsen H, Bedding TR, 2001, Current status of asteroseismology. *ESA SP–464: Helio- and Asteroseismology at the Dawn of the Millennium* (eds. Wilson A, Pallé PL), 361–366 {**396**}

Knauer TG, Ivezić Ž, Knapp GR, 2001, Analysis of stars common to the IRAS and Hipparcos surveys. *ApJ*, 552, 787–792 {**373**}

Kochukhov O, Bagnulo S, 2006, Evolutionary state of magnetic chemically peculiar stars. *A&A*, 450, 763–775 {**393**}

Kochukhov O, Piskunov N, Ilyin I, *et al.*, 2002, Doppler imaging of stellar magnetic fields. III. Abundance distribution and magnetic field geometry of α^2 CVn. *A&A*, 389, 420–438 {**395**}

Korn AJ, 2004, Unbiased stellar parameters. *Origin and Evolution of the Elements* (eds. McWilliam A, Rauch M), 32–32 {**361**}

Korn AJ, Shi J, Gehren T, 2003, Kinetic equilibrium of iron in the atmospheres of cool stars. III. The ionisation equilibrium of selected reference stars. *A&A*, 407, 691–703 {**378**}

Kosovichev AG, Christensen-Dalsgaard J, Däppen W, *et al.*, 1992, Sources of uncertainty in direct seismological measurements of the solar helium abundance. *MNRAS*, 259, 536–558 {**379**}

Kotoneva E, Flynn C, Chiappini C, *et al.*, 2002a, K dwarfs and the chemical evolution of the solar cylinder. *MNRAS*, 336, 879–891 {**388**}

—, 2003, K dwarf metallicity indicators and chemical evolution of the Milky Way. *Revista Mexicana de Astronomia y Astrofisica Conf. Ser.*, volume 17, 94–94 {**388**}

Kotoneva E, Flynn C, Jimenez R, 2002b, Luminosity–metallicity relation for stars on the lower main sequence. *MNRAS*, 335, 1147–1157 {**346, 364**}

Künzli M, North P, Kurucz RL, *et al.*, 1997, A calibration of Geneva photometry for B to G stars in terms of $T_{\rm eff}$, log g and [M/H]. *A&AS*, 122, 51–77 {**342**}

Kurucz RL, 1979, Model atmospheres for G, F, A, B, and O stars. *ApJS*, 40, 1–340 {**345**}

—, 1991, New opacity calculations. *NATO ASIC Proc. 341: Stellar Atmospheres – Beyond Classical Models* (eds. Crivellari L, Hubeny I, Hummer DG), 441–445 {**344, 346**}

—, 1993a, ATLAS9 stellar atmosphere programs and 2 km s^{-1} grid. *CD No. 13, Smithsonian Astrophysical Observatory* {**352**}

—, 1993b, A new opacity-sampling model atmosphere program for arbitrary abundances. *IAU Colloq. 138: Peculiar versus Normal Phenomena in A-type and Related Stars* (eds. Dworetsky MM, Castelli F, Faraggiana R), ASP Conf. Ser. 44, 87–97 {**345**}

Lambert DL, Reddy BE, 2004, Lithium abundances of the local thin disk stars. *MNRAS*, 349, 757–767 {**382**}

Lamers HJGLM, 1997, Stellar wind theories. *Stellar Atmospheres: Theory and Observations* (eds. de Greve JP, Blomme R, Hensberge H), volume 497 of *Lecture Notes in Physics, Springer–Verlag, Berlin*, 69–88 {**374**}

Lamers HJGLM, Harzevoort JMAG, Schrijver H, *et al.*, 1997, The effect of rotation on the absolute visual magnitudes of OB stars measured with Hipparcos. *A&A*, 325, L25–L28 {**390, 391**}

Landstreet JD, 1980, The measurement of magnetic fields in stars. *AJ*, 85, 611–620 {**392**}

Lane BF, Kuchner MJ, Boden AF, *et al.*, 2000, Direct detection of pulsations of the Cepheid star ζ Gem and an independent calibration of the period–luminosity relation. *Nature*, 407, 485–487 {**358**}

Langer N, 1992, Helium enrichment in massive early-type stars. *A&A*, 265, L17–L20 {**390**}

Lastennet E, Valls-Gabaud D, Lejeune T, *et al.*, 1999, Consequences of Hipparcos parallaxes for stellar evolutionary models. Three Hyades binaries: V818 Tau, 51 Tau, and θ^2 Tau. *A&A*, 349, 485–494 {**376**}

Lastennet E, Valls-Gabaud D, 2002, Detached double-lined eclipsing binaries as critical tests of stellar evolution: age and metallicity determinations from the HR diagram. *A&A*, 396, 551–580 {**376**}

Lebreton Y, 2000a, Science results for stellar structure and evolution from Hipparcos. *Unsolved Problems in Stellar Evolution* (ed. Livio M), 107–125 {**362**}

—, 2000b, Stellar structure and evolution: deductions from Hipparcos. *ARA&A*, 38, 35–77 {**340, 349, 362, 363, 365–367, 375**}

—, 2002, Luminosity calibrations with Hipparcos: theoretical point of view. *Highlights in Astronomy*, 12, 669–669 {**377**}

Lebreton Y, Fernandes J, Lejeune T, 2001, The helium content and age of the Hyades: constraints from five binary systems and Hipparcos parallaxes. *A&A*, 374, 540–553 {**376, 379**}

Lebreton Y, Michel E, Goupil MJ, *et al.*, 1995, Accurate parallaxes and stellar ages determinations. *Astronomical and Astrophysical Objectives of Sub-Milliarcsecond Optical Astrometry* (eds. Høg E, Seidelmann PK), IAU Symp. 166, 135–142 {**346**}

Lebreton Y, Perrin MN, Cayrel R, *et al.*, 1999, The Hipparcos HR diagram of nearby stars in the metallicity range $-1.0 <$[Fe/H]< 0.3. A new constraint on the theory of stellar interiors and model atmospheres. *A&A*, 350, 587–597 {**362, 363, 365**}

Lebreton Y, Perrin MN, Fernandes J, *et al.*, 1997, The HR diagram for late-type nearby stars as a function of helium and metallicity. *ESA SP–402: Hipparcos, Venice '97*, 379–382 {**361**}

Leone F, Catanzaro G, Catalano S, 2000a, Spectropolarimetric measurements of the mean longitudinal magnetic field of chemically peculiar stars. On the light, spectral and magnetic variability. *A&A*, 355, 315–326 {**393**}

Leone F, Catanzaro G, Malaroda S, 2000b, A spectroscopic study of the magnetic chemically peculiar star ν For. *A&A*, 359, 635–638 {**393**}

Leone F, Catanzaro G, 2001, Spectropolarimetric measurements of the mean longitudinal magnetic field of chemically peculiar stars. II. Phase relating the magnetic and luminosity variabilities. *A&A*, 365, 118–127 {**393**}

Levato H, Malaroda S, Morrell NI, *et al.*, 1996, Radial velocities and axial rotation for a sample of chemically peculiar stars. *A&AS*, 118, 231–238 {**392**}

Liu YY, Baglin A, Auvergne M, *et al.*, 1997, The lower part of the instability strip: evolutionary status of δ Scuti stars. *ESA SP–402: Hipparcos, Venice '97*, 363–366 {**396**}

Livio M, 2000, *Unsolved Problems in Stellar Evolution*. Cambridge University Press {**340**}

Luck RE, Heiter U, 2005, Stars within 15 pc: abundances for a northern sample. *AJ*, 129, 1063–1083 {**377, 379**}

Ludwig HG, Freytag B, Steffen M, 1999, A calibration of the mixing-length for solar-type stars based on hydrodynamical simulations. I. Methodical aspects and results for solar metallicity. *A&A*, 346, 111–124 {**345**}

Lydon TJ, Fox PA, Sofia S, 1993, A formulation of convection for stellar structure and evolution calculations without the mixing-length theory approximations. II. Application to α Cen AB. *ApJ*, 413, 390–400 {**375**}

Lyubimkov LS, Rachkovskaya TM, Rostopchin SI, *et al.*, 2002, Surface abundances of light elements for a large sample of early B-type stars. II. Basic parameters of 107 stars. *MNRAS*, 333, 9–26 {**359**}

Maeder A, Mermilliod JC, 1981, The extent of mixing in stellar interiors. Evolutionary models and tests based on the HR diagrams of 34 open clusters. *A&A*, 93, 136–149 {**346**}

Maeder A, Meynet G, 2000, The evolution of rotating stars. *ARA&A*, 38, 143–190 {**390**}

—, 2001, Stellar evolution with rotation. VII. Low metallicity models and the blue-to-red supergiant ratio in the SMC. *A&A*, 373, 555–571 {**350**}

Magain P, 1989, The chemical composition of the extreme halo stars. I. Blue spectra of 20 dwarfs. *A&A*, 209, 211–225 {**355**}

Malinie G, Hartmann DH, Clayton DD, *et al.*, 1993, Inhomogeneous chemical evolution of the Galactic disk. *ApJ*, 413, 633–640 {**386**}

Mallik SV, 1999, Lithium abundance and mass. *A&A*, 352, 495–507 {**380, 381**}

Mallik SV, Parthasarathy M, Pati AK, 2003, Lithium and rotation in F and G dwarfs and subgiants. *A&A*, 409, 251–261 {**382**}

Marigo P, Chiosi C, Kudritzki RP, 2003, Zero-metallicity stars. II. Evolution of very massive objects with mass loss. *A&A*, 399, 617–630 {**350**}

Marigo P, Girardi L, Chiosi C, *et al.*, 2001, Zero-metallicity stars. I. Evolution at constant mass. *A&A*, 371, 152–173 {**350**}

Martin JC, 2006, The origins and evolutionary status of B stars found far from the Galactic plane. II. Kinematics and full sample analysis. *AJ*, 131, 3047–3068, erratum: 133, 755 {**373**}

Masana E, Jordi C, Ribas I, 2006, Effective temperature scale and bolometric corrections from 2MASS photometry. *A&A*, 450, 735–746 {**349**}

Matthews JM, 2005, Canada's little space telescope that could: another year of scientific surprises from the MOST microsatellite. *JRASC*, 99, 131–131 {**397**}

Mazzitelli I, 1989, The core mass at the helium flash: influence of numerical and physical inputs. *ApJ*, 340, 249–255 {**350**}

Mazzitelli I, D'Antona F, Caloi V, 1995, Globular cluster ages with updated input physics. *A&A*, 302, 382–400 {**350**}

McWilliam A, Preston GW, Sneden C, *et al.*, 1995, Spectroscopic analysis of 33 of the most metal poor stars. *AJ*, 109, 2757–2799 {**384**}

McWilliam A, 1997, Abundance ratios and Galactic chemical evolution. *ARA&A*, 35, 503–556 {**378**}
Mégessier C, 1995, Accuracy of the astrophysical absolute flux calibrations: visible and near-infrared. *A&A*, 296, 771–778 {**342**}
—, 2000, Accuracy of reference star locations in the HR diagram. *Hipparcos and the Luminosity Calibration of the Nearer Stars, IAU Joint Discussion 13*, 26–26 {**349**}
Meillon L, Crifo F, Cayrel R, *et al.*, 1999, Calibration of photometric absolute magnitudes for subdwarfs with Hipparcos. *ASP Conf. Ser. 167: Harmonizing Cosmic Distance Scales in a Post-Hipparcos Era*, 284–287 {**367**}
Meléndez J, Barbuy B, Spite F, 2001, Oxygen abundances in metal-poor stars ($-2.2 <$ [Fe/H] < -1.2) from infrared OH lines. *ApJ*, 556, 858–871 {**384**}
Meléndez J, Ramírez I, 2005, Spectroscopic equilibrium of iron in metal-rich dwarfs. *ASP Conf. Ser. 336: Cosmic Abundances as Records of Stellar Evolution and Nucleosynthesis* (eds. Barnes TG, Bash FN), 343–346 {**378**}
Meynet G, Maeder A, Schaller G, *et al.*, 1994, Grids of massive stars with high mass-loss rates. V. From $12-120\,M_\odot$ at $Z = 0.001, 0.004, 0.008, 0.020, 0.040$. *A&AS*, 103, 97–105 {**350**}
Meynet G, Maeder A, 2002, Stellar evolution with rotation. VIII. Models at $Z = 10^{-5}$ and CNO yields for early Galactic evolution. *A&A*, 390, 561–583 {**348, 350**}
Michaud G, 1970, Diffusion processes in peculiar A stars. *ApJ*, 160, 641–658 {**395**}
Michaud G, Proffitt CR, 1993, Mechanisms of separation of elements. *IAU Colloq. 138: Peculiar versus Normal Phenomena in A-type and Related Stars* (eds. Dworetsky MM, Castelli F, Faraggiana R), ASP Conf. Ser. 44, 439–449 {**395**}
Michaud G, Richard O, Richer J, *et al.*, 2004, Models for solar abundance stars with gravitational settling and radiative accelerations: application to M67 and NGC 188. *ApJ*, 606, 452–465 {**340, 350**}
Mihalas D, Binney JJ, 1981, *Galactic Astronomy: Structure and Kinematics*, 2nd edition. W.H. Freeman, San Francisco {**347**}
Mihalas D, Dappen W, Hummer DG, 1988, The equation-of-state for stellar envelopes. II. Algorithm and selected results. *ApJ*, 331, 815–825 {**344**}
Missoulis V, Crockett H, Oliver S, *et al.*, 1999, Calibration of the ISOCAM LW2 and LW3 filters using ELAIS stars. *ESA SP–427: The Universe as Seen by ISO*, 85 {**351**}
Mkrtichian DE, Kusakin AV, López de Coca P, *et al.*, 2007, Multi-mode pulsations of the λ Bootis star 29 Cyg: the 1995 and 1996 multisite campaigns. *AJ*, 134, 1713–1727 {**397**}
Mondal S, Chandrasekhar T, 2005, Angular diameter measurements of evolved variables by lunar occultations at 2.2 and 3.8 μm. *AJ*, 130, 842–852 {**341**}
Moon TT, Dworetsky MM, 1985, Grids for the determination of effective temperature and surface gravity of B, A and F stars using uvbyβ photometry. *MNRAS*, 217, 305–315 {**342**}
Morales-Rueda L, Maxted PFL, Marsh TR, *et al.*, 2003, Orbital periods of 22 subdwarf B (sdB) stars. *MNRAS*, 338, 752–764 {**371**}
Morel P, 1997, CESAM: a code for stellar evolution calculations. *A&AS*, 124, 597–614 {**339, 350, 352, 358**}
Morel P, Baglin A, 1999, Microscopic diffusion and subdwarfs. *A&A*, 345, 156–162 {**362, 363, 368**}
Morel P, Morel C, Provost J, *et al.*, 2000a, Calibration of ι Peg system. *A&A*, 354, 636–644 {**375**}
Morel P, Provost J, Lebreton Y, *et al.*, 2000b, Calibrations of α Cen AB. *A&A*, 363, 675–691 {**376**}
Morel P, van't Veer C, Provost J, *et al.*, 1994, Incorporating the atmosphere in stellar structure models: the solar case. *A&A*, 286, 91–102 {**345**}
Morossi C, di Marcantonio P, Franchini M, *et al.*, 2002, Metallicity determinations from ultraviolet-visual spectrophotometry. I. The test sample. *ApJ*, 577, 377–388 {**389**}
Moss D, 1989, The origin and internal structure of the magnetic fields of the chemically peculiar stars. *MNRAS*, 236, 629–644 {**392**}
Mould JR, 1976, Accretion and the abundance distribution of late-type dwarfs. *MNRAS*, 177, 47P–52P {**388**}
Mourard D, Bonneau D, Koechlin L, *et al.*, 1997, The mean angular diameter of δ Cep measured by optical long-baseline interferometry. *A&A*, 317, 789–792 {**358**}
Nather RE, Evans DS, 1970, Photoelectric measurement of lunar occultations. I. The process. *AJ*, 75, 575–582 {**356**}
Newberg HJ, Yanny B, 1998, An absence of gaps in the main sequence population of field stars. *ApJ*, 499, L57–L60 {**364, 365**}
Ng YK, Bertelli G, 1998, Revised ages for stars in the solar neighbourhood. *A&A*, 329, 943–950 {**364, 387, 388**}
Nicolet B, 1997, Calibration of luminosity of hot stars: preliminary results. *ESA SP–402: Hipparcos, Venice '97*, 307–310 {**377**}
Niemczura E, Daszyńska J, Cugier H, 2003, The mean stellar parameters from IUE and Hipparcos data. *Advances in Space Research*, 31, 399–404 {**377**}
Nissen PE, 1998, Problems in modeling low-metallicity F and G stars. *The First MONS Workshop: Science with a Small Space Telescope* (eds. Kjeldsen H, Bedding TR), 99–104 {**362**}
—, 2000, Chemical composition of mildly metal-poor stars. *The Galactic Halo: from Globular Clusters to Field Stars* (eds. Noels A, Magain P, Caro D, *et al.*), 125–134 {**378**}
Nissen PE, Høg E, Schuster WJ, 1997, Surface gravities of metal-poor stars derived from Hipparcos parallaxes. *ESA SP–402: Hipparcos, Venice '97*, 225–230 {**342, 353–355, 367**}
Nissen PE, Lambert DL, Primas F, *et al.*, 1999, Isotopic lithium abundances in five metal-poor disk stars. *A&A*, 348, 211–221 {**383**}
Nissen PE, Schuster WJ, 1997, Chemical composition of halo and disk stars with overlapping metallicities. *A&A*, 326, 751–762 {**384**}
Noël F, 2002, On solar radius measurements with Danjon astrolabes. *A&A*, 396, 667–672 {**356**}
Noels A, Fraipont-Caro D, Gabriel M, *et al.* (eds.), 1995, *Stellar Evolution: What Should be Done* {**340**}
Noels A, Grevesse N, Magain P, *et al.*, 1991, Calibration of the α Cen system: metallicity and age. *A&A*, 247, 91–94 {**375**}
Nordgren TE, Armstrong JT, Germain ME, *et al.*, 2000, Astrophysical quantities of Cepheid variables measured with the Navy Prototype Optical Interferometer. *ApJ*, 543, 972–978 {**358**}
North P, 1998, Do Si stars undergo any rotational braking? *A&A*, 334, 181–187, erratum: 336, 1072 {**390, 392**}
North P, Erspamer D, Künzli M, 1998, The evolutionary state and fundamental parameters of metallic A-F giants. *Contributions of the Astronomical Observatory Skalnate Pleso*, 27, 252–254 {**377**}

Oláh K, Strassmeier KG, Granzer T, 2002, Time series photometric spot modeling V. Phase coherence of spots on UZ Lib. *Astronomische Nachrichten*, 323, 453–461 {**395**}

Olofsson H, Bergman P, Lucas R, *et al.*, 2000, A high-resolution study of episodic mass loss from the carbon star TT Cyg. *A&A*, 353, 583–597 {**374**}

Pagel BEJ, 1997, *Nucleosynthesis and Chemical Evolution of Galaxies*. Cambridge University Press {**386**}

Pagel BEJ, Portinari L, 1998, $\Delta Y/\Delta Z$ from fine structure in the main sequence based on Hipparcos parallaxes. *MNRAS*, 298, 747–752 {**365, 379**}

Palacios A, Charbonnel C, Talon S, *et al.*, 2006, Rotational mixing in low-mass stars. II. Self-consistent models of Population II red giant branch stars. *A&A*, 453, 261–278 {**371**}

Palla F, Stahler SW, 1993, The pre-main sequence evolution of intermediate-mass stars. *ApJ*, 418, 414–425 {**350**}

—, 1999, Star formation in the Orion nebula cluster. *ApJ*, 525, 772–783 {**348, 350**}

Pasquini L, de Medeiros JR, Girardi L, 2000, Ca II activity and rotation in F-K evolved stars. *A&A*, 361, 1011–1022 {**391**}

Pasquini L, Liu Q, Pallavicini R, 1994, Lithium abundances of nearby solar-like stars. *A&A*, 287, 191–205 {**380**}

Pastori L, Tagliaferri G, Pasinetti Fracassini LE, 1999, Lithium in X-ray selected active cool stars. *Galaxy Evolution: Connecting the Distant Universe with the Local Fossil Record*, 443 {**381**}

Pease FG, 1931, *Erg. Naturwiss.* 10,84 {**356**}

Peimbert A, Peimbert M, Luridiana V, 2002, Temperature bias and the primordial helium abundance determination. *ApJ*, 565, 668–680 {**379**}

Perrin G, Coude Du Foresto V, Ridgway ST, *et al.*, 1998, Extension of the effective temperature scale of giants to types later than M6. *A&A*, 331, 619–626 {**352**}

Perrin MN, Cayrel de Strobel G, Cayrel R, *et al.*, 1977, Fine structure of the HR diagram for 138 stars in the solar neighbourhood. *A&A*, 54, 779–795 {**379**}

Perryman MAC, Brown AGA, Lebreton Y, *et al.*, 1998, The Hyades: distance, structure, dynamics, and age. *A&A*, 331, 81–120 {**366, 371, 376**}

Perryman MAC, Lindegren L, Kovalevsky J, *et al.*, 1995, Parallaxes and the HR diagram for the preliminary Hipparcos solution H30. *A&A*, 304, 69–81 {**361, 369**}

—, 1997, The Hipparcos Catalogue. *A&A*, 323, L49–L52 {**361**}

Pettinger MM, Bernkopf J, Fuhrmann K, *et al.*, 2001, Stellar abundances of the Galactic thick disk. *Astronomische Gesellschaft Meeting Abstracts*, 166 {**389**}

Pietrinferni A, Cassisi S, Salaris M, *et al.*, 2004, A large stellar evolution data base for population synthesis studies. I. Scaled solar models and isochrones. *ApJ*, 612, 168–190 {**346, 350**}

—, 2006, A large stellar evolution data base for population synthesis studies. II. Stellar models and isochrones for an α-enhanced metal distribution. *ApJ*, 642, 797–812 {**350**}

Pilyugin LS, Edmunds MG, 1996b, Chemical evolution of the Milky Way Galaxy. II. On the origin of scatter in the age–metallicity relation. *A&A*, 313, 792–802 {**386**}

—, 1996a, Chemical evolution of the Milky Way Galaxy. I. On the infall model of Galactic chemical evolution. *A&A*, 313, 783–791 {**386**}

Pinsonneault MH, 1997, Mixing in stars. *ARA&A*, 35, 557–605 {**346, 380**}

Pinsonneault MH, Charbonnel C, Deliyannis CP, 2000, Sinks of light elements in stars. *The Light Elements and their Evolution* (eds. da Silva L, de Medeiros R, Spite M), IAU Symp. 198, 74–86 {**382**}

Pinsonneault MH, Deliyannis CP, Demarque P, 1992, Evolutionary models of halo stars with rotation. II. Effects of metallicity on lithium depletion, and possible implications for the primordial lithium abundance. *ApJS*, 78, 179–203 {**380**}

Pinsonneault MH, Terndrup DM, Hanson RB, *et al.*, 2003, The distances to open clusters from main-sequence fitting. I. New models and a comparison with the properties of the Hyades eclipsing binary VB 22. *ApJ*, 598, 588–596 {**376**}

Piotto G, Bedin LR, Anderson J, *et al.*, 2007, A triple main sequence in the globular cluster NGC 2808. *ApJ*, 661, L53–L56 {**380**}

Pöhnl H, Paunzen E, Maitzen HM, 2005, On the formation and evolution of magnetic chemically peculiar stars in the solar neighbourhood. *A&A*, 441, 1111–1116 {**393, 394**}

Pols OR, Schröder KP, Hurley JR, *et al.*, 1998, Stellar evolution models for $Z = 0.0001 - 0.03$. *MNRAS*, 298, 525–536 {**350, 374**}

Pols OR, Tout CA, Eggleton PP, *et al.*, 1995, Approximate input physics for stellar modeling. *MNRAS*, 274, 964–974 {**343, 350**}

Pols OR, Tout CA, Schröder KP, *et al.*, 1997, Further critical tests of stellar evolution by means of double-lined eclipsing binaries. *MNRAS*, 289, 869–881 {**363**}

Pompéia L, Barbuy B, Grenon M, 2002a, Detailed analysis of nearby bulge-like dwarf stars. I. Stellar parameters, kinematics, and oxygen abundances. *ApJ*, 566, 845–856 {**389**}

—, 2002b, Oxygen abundances in bulge-like dwarf stars. *IAU Symp. 207*, 122–125 {**389**}

Pont F, Charbonnel C, Lebreton Y, *et al.*, 1997, Hipparcos subdwarfs and globular clusters: towards reliable absolute ages. *ESA SP–402: Hipparcos, Venice '97*, 699–704 {**367**}

Popper DM, 1998, Hipparcos parallaxes of eclipsing binaries and the radiative flux scale. *PASP*, 110, 919–922 {**351**}

Pourbaix D, 1999, γ Per: a challenge for stellar evolution models. *A&A*, 348, 127–132 {**377**}

Pourbaix D, Neuforge-Verheecke C, Noels A, 1999, Revised masses of α Cen. *A&A*, 344, 172–176 {**375, 376**}

Prochaska JX, Naumov SO, Carney BW, *et al.*, 2000, The Galactic thick disk stellar abundances. *AJ*, 120, 2513–2549 {**389**}

Proffitt CR, Vandenberg DA, 1991, Implications of helium diffusion for globular cluster isochrones and luminosity functions. *ApJS*, 77, 473–514 {**367**}

Ramdani A, 2003, Low mass loss in late-type stars. *ESA SP–511: Exploiting the ISO Data Archive: Infrared Astronomy in the Internet Age* (eds. Gry C, Peschke S, Matagne J, *et al.*), 145–148 {**374**}

Ramírez I, Allende Prieto C, Lambert DL, 2007, Oxygen abundances in nearby stars: clues to the formation and evolution of the Galactic disk. *A&A*, 465, 271–289 {**389**}

Ramírez I, Meléndez J, 2005b, The effective temperature scale of FGK stars. II. $T_{\rm eff}$–colour–[Fe/H] calibrations. *ApJ*, 626, 465–485 {**353**}

—, 2005a, The effective temperature scale of FGK stars. I. Determination of temperatures and angular diameters with the infrared flux method. *ApJ*, 626, 446–464 {**353**}

Randich S, Gratton RG, Pallavicini R, *et al.*, 1999, Lithium in Population I subgiants. *A&A*, 348, 487–500 {**381**}

Reddy BE, Lambert DL, Allende Prieto C, 2006, Elemental abundance survey of the Galactic thick disk. *MNRAS*, 367, 1329–1366 {**388**}

Reddy BE, Tomkin J, Lambert DL, *et al.*, 2003, The chemical compositions of Galactic disk F and G dwarfs. *MNRAS*, 340, 304–340 {**388, 389**}

Reimers D, 1975, Circumstellar absorption lines and mass loss from red giants. *Mem. Soc. Roy. Sciences de Liège*, 8, 369–382 {**371, 374**}

Reis Neto E, Andrei AH, Penna JL, *et al.*, 2003, Observed Variations of the solar diameter in 1998–2000. *Sol. Phys.*, 212, 7–21 {**356**}

Renson P, Gerbaldi M, Catalano FA, 1991, General catalogue of Ap and Am stars. *A&AS*, 89, 429–434 {**392**}

Renzini A, Fusi Pecci F, 1988, Tests of evolutionary sequences using colour–magnitude diagrams of globular clusters. *ARA&A*, 26, 199–244 {**372**}

Ribas I, Gimenez A, Torra J, *et al.*, 1998, Effective temperature of detached eclipsing binaries from Hipparcos parallax. *A&A*, 330, 600–604 {**351**}

Rice JB, 1996, Doppler imaging of stellar surfaces. *Stellar Surface Structure* (eds. Strassmeier KG, Linsky JL), volume 176 of *IAU Symp. 176*, 19–33 {**395**}

Richard O, Vauclair S, Charbonnel C, *et al.*, 1996, New solar models including helioseismological constraints and light-element depletion. *A&A*, 312, 1000–1011 {**339, 346**}

Richichi A, Ragland S, Stecklum B, *et al.*, 1998, Infrared high angular resolution measurements of stellar sources. IV. Angular diameters and effective temperatures of fifteen late-type stars. *A&A*, 338, 527–534 {**341**}

Ridgway ST, Jacoby GH, Joyce RR, *et al.*, 1980, Angular diameters by the lunar occultation technique. III. *AJ*, 85, 1496–1504 {**341, 352**}

Rocha Pinto HJ, Maciel WJ, Scalo JM, *et al.*, 2000, Chemical enrichment and star formation in the Milky Way disk. I. Sample description and chromospheric age–metallicity relation. *A&A*, 358, 850–868 {**386**}

Rocha Pinto HJ, Maciel WJ, 1996, The metallicity distribution of G dwarfs in the solar neighbourhood. *MNRAS*, 279, 447–458 {**386**}

Rogers FJ, Iglesias CA, 1992, Radiative atomic Rosseland mean opacity tables. *ApJS*, 79, 507–568 {**343, 346**}

Rogers FJ, Swenson FJ, Iglesias CA, 1996, OPAL equation-of-state tables for astrophysical applications. *ApJ*, 456, 902–908 {**344, 349**}

Romano D, Tosi M, Matteucci F, *et al.*, 2003, Light element evolution resulting from WMAP data. *MNRAS*, 346, 295–303 {**382**}

Romanyuk II, Kudryavtsev D, 2001, Catalogue of magnetic chemically peculiar stars: some preliminary analysis. *Magnetic Fields Across the Hertzsprung–Russell Diagram* (eds. Mathys G, Solanki SK, Wickramasinghe DT), ASP Conf. Ser. 248, 299–304 {**392**}

Rood RT, Bania TM, Wilson TL, 1984, The 8.7 GHz hyperfine line of $^3He^+$ in Galactic H II regions. *ApJ*, 280, 629–647 {**380**}

Roxburgh I, 1997, Convective overshooting and mixing. *ASSL Vol. 225: Solar Convection and Oscillations and their Relationship* (eds. Pijpers FP, Christensen-Dalsgaard J, Rosenthal CS), 23–50 {**346**}

Royer F, 1997, Populations among high-velocity early-type stars. *ESA SP–402: Hipparcos, Venice '97*, 595–598 {**377**}

—, 1999, Study of early-type stars populations and of their origin. *Galaxy Evolution: Connecting the Distant Universe with the Local Fossil Record*, 433–434 {**377**}

Royer F, Gerbaldi M, Faraggiana R, *et al.*, 2002a, Rotational velocities of A-type stars. I. Measurement of $v \sin i$ in the southern hemisphere. *A&A*, 381, 105–121 {**392**}

Royer F, Gómez AE, Zorec J, *et al.*, 1998, On the rotational velocities distribution of A-type stars observed by Hipparcos. *IAU Symp. 189: Fundamental Stellar Properties*, 92–95 {**392**}

Royer F, Grenier S, Baylac MO, *et al.*, 2002b, Rotational velocities of A-type stars in the northern hemisphere. II. Measurement of $v \sin i$. *A&A*, 393, 897–911 {**392**}

Ryabchikova T, 2004, Observations of magnetic chemically peculiar stars. *IAU Symp. 224* (eds. Zverko J, Ziznovsky J, Adelman SJ, *et al.*), 283–290 {**393**}

Sackmann IJ, Boothroyd AI, 1999, Creation of ^{7}Li and destruction of ^{3}He, ^{9}Be, ^{10}B, and ^{11}B in low-mass red giants due to deep circulation. *ApJ*, 510, 217–231 {**382**}

Saha MN, 1920, *Philos. Mag.* 40, 72 {**343**}

Saito YJ, Takada-Hidai M, Takeda Y, *et al.*, 2006, Zinc abundances in metal-poor stars. *Origin of Matter and Evolution of Galaxies* (eds. Kubono S, Aoki W, Kajino T, *et al.*), American Institute of Physics Conf. Ser. 847, 464–466 {**389**}

Salaris M, Groenewegen MAT, Weiss A, 2000, Atomic diffusion in metal-poor stars. The influence on the main sequence fitting distance scale, subdwarfs ages and the value of $\Delta Y/\Delta Z$. *A&A*, 355, 299–307 {**346, 363, 368**}

Salaris M, Weiss A, 1998, Metal-rich globular clusters in the Galactic disk: new age determinations and the relation to halo clusters. *A&A*, 335, 943–953 {**353, 354**}

Salasnich B, Girardi L, Weiss A, *et al.*, 2000, Evolutionary tracks and isochrones for α-enhanced stars. *A&A*, 361, 1023–1035 {**348, 350**}

Sandage AR, 1970, Main sequence photometry, colour-magnitude diagrams, and ages for the globular clusters M3, M13, M15, and M92. *ApJ*, 162, 841–870 {**366**}

Sandage AR, Lubin LM, VandenBerg DA, 2003, The age of the oldest stars in the local Galactic disk from Hipparcos parallaxes of G and K subgiants. *PASP*, 115, 1187–1206 {**369**}

Saumon D, Chabrier G, 1991, Fluid H at high density: pressure dissociation. *Phys. Rev. A*, 44, 5122–5141 {**345**}

Scalo JM, 1986, The stellar initial mass function. *Fundamentals of Cosmic Physics*, 11, 1–278 {**363**}

Schaerer D, Charbonnel C, Meynet G, *et al.*, 1993a, Grids of stellar models. IV. From $0.8–120\,M_\odot$ at $Z = 0.040$. *A&AS*, 102, 339–342 {**350**}

Schaerer D, Meynet G, Maeder A, *et al.*, 1993b, Grids of stellar models. II. From $0.8–120\,M_\odot$ at $Z = 0.008$. *A&AS*, 98, 523–527 {**350, 374, 394**}

Schaifers K, Voigt HH (eds.), 1982, *Landolt-Börnstein: Numerical Data and Functional Relationships in Science and Technology; New Series, Group 6 Astronomy and Astrophysics, Volume 2*. Springer–Verlag, Berlin {**391**}

Schaller G, Schaerer D, Meynet G, *et al.*, 1992, New grids of stellar models. I. From $0.8–120\,M_\odot$ at $Z = 0.020, 0.001$. *A&AS*, 96, 269–331 {**350, 352, 358, 390, 394**}

Schatzman EL, Praderie F, 1993, *The Stars*. Springer–Verlag, Berlin {**396**}

Schröder KP, 1998a, Core-overshooting and the distribution of giants in the HR diagram. *ASP Conf. Ser. 154: Cool Stars, Stellar Systems, and the Sun*, 856–864 {**363**}

—, 1998b, The solar neighbourhood HR diagram as a quantitative test for evolutionary time scales. *A&A*, 334, 901–910 {**363, 364**}

Schröder KP, Sedlmayr E, 2001, The Galactic mass injection from cool stellar winds of the $1-2.5M_{\odot}$ stars in the solar neighbourhood. *A&A*, 366, 913–922 {**374**}

Schröder SE, Kaper L, Lamers HJGLM, *et al.*, 2004, On the Hipparcos parallaxes of O stars. *A&A*, 428, 149–157 {**390**}

Schwarzschild M, 1958, *Structure and Evolution of the Stars.* Princeton University Press {**339, 343**}

Searle L, Zinn R, 1978, Compositions of halo clusters and the formation of the Galactic halo. *ApJ*, 225, 357–379 {**384**}

Seaton MJ, Yan Y, Mihalas D, *et al.*, 1994, Opacities for stellar envelopes. *MNRAS*, 266, 805–828 {**343, 344**}

Sedlmayr E, Winters JM, 1997, Cool star winds and mass loss: theory. *Stellar Atmospheres: Theory and Observations* (eds. de Greve JP, Blomme R, Hensberge H), volume 497 of *Lecture Notes in Physics, Springer–Verlag, Berlin*, 89–132 {**374**}

Shen C, Hu J, 1999, The evolutionary status of HD 104237, ϵ Cha and HD 100546. *Chinese Astronomy and Astrophysics*, 23, 493–497 {**377**}

Siess L, Dufour E, Forestini M, 2000, An internet server for pre-main sequence tracks of low- and intermediate-mass stars. *A&A*, 358, 593–599 {**350**}

Siess L, Forestini M, Dougados C, 1997, Synthetic HR diagrams of open clusters. *A&A*, 324, 556–565 {**350**}

Sills A, Pinsonneault MH, Terndrup DM, 2000, The angular momentum evolution of very low-mass stars. *ApJ*, 534, 335–347 {**350**}

Simon NR, 1982, A plea for reexamining heavy element opacities in stars. *ApJ*, 260, L87–L90 {**343**}

Smith G, 1998, Stellar atmospheric parameters for the giant stars MU Peg and λ Peg. *A&A*, 339, 531–536 {**377**}

Söderhjelm S, 1999, Visual binary orbits and masses post Hipparcos. *A&A*, 341, 121–140 {**376**}

Soubiran C, Katz D, Cayrel R, 1998, On-line determination of stellar atmospheric parameters $T_{\rm eff}$, log g, [Fe/H] from Elodie echelle spectra. II. The library of F5 to K7 stars. *A&AS*, 133, 221–226 {**359**}

Spergel DN, Verde L, Peiris HV, *et al.*, 2003, First-Year Wilkinson Microwave Anisotropy Probe (WMAP) observations: determination of cosmological parameters. *ApJS*, 148, 175–194 {**368**}

Spite F, Spite M, Hill V, 1998, Lithium abundance in Population II stars: a post-Hipparcos discussion. *Space Science Reviews*, 84, 155–160 {**380**}

Spite F, Spite M, 1982a, Abundance of lithium in unevolved halo stars and old disk stars: interpretation and consequences. *A&A*, 115, 357–366 {**382**}

Spite M, Spite F, 1982b, Lithium abundance at the formation of the Galaxy. *Nature*, 297, 483–485 {**382**}

Stępień K, 1993, HR 1362: the evolved 53 Cam. *ApJ*, 416, 368–371 {**393**}

—, 1994, Properties and position of chemically peculiar stars on the HR diagram. *Chemically Peculiar and Magnetic Stars* (eds. Zverko J, Ziznovsky J), 8 {**392**}

Stein RF, Nordlund A, 1998, Simulations of solar granulation. I. General properties. *ApJ*, 499, 914–933 {**345**}

Stetson PB, Harris WE, 1988, CCD photometry of the globular cluster M92. *AJ*, 96, 909–975 {**368**}

Stibbs DWN, 1950, A study of the spectrum and magnetic variable star HD 125248. *MNRAS*, 110, 395–404 {**395**}

Strassmeier KG, 1999, Doppler imaging of stellar surface structure. XI. The super starspots on the K0 giant HD 12545: larger than the entire Sun. *A&A*, 347, 225–234 {**395**}

Strassmeier KG, Bartus J, Kovari Z, *et al.*, 1998, Doppler imaging of stellar surface structure. VIII. The effectively single and rapidly-rotating G8-giant HD 51066. *A&A*, 336, 587–603 {**395**}

Strassmeier KG, Lupinek S, Dempsey RC, *et al.*, 1999a, Doppler imaging of stellar surface structure. X. The FK Comae-type star HD 199178 = V1794 Cyg. *A&A*, 347, 212–224 {**395**}

Strassmeier KG, Stępień K, Henry GW, *et al.*, 1999b, Evolved, single, slowly rotating, but magnetically active. The G8-giant HR 1362 = EK Eri revisited. *A&A*, 343, 175–182 {**393**}

Strassmeier KG, Washuettl A, Granzer T, *et al.*, 2000, The Vienna-KPNO search for Doppler-imaging candidate stars. I. A catalogue of stellar-activity indicators for 1058 late-type Hipparcos stars. *A&AS*, 142, 275–311 {**395**}

Strömberg G, 1930, The distribution of absolute magnitudes among K and M stars brighter than the sixth apparent magnitude as determined from peculiar velocities. *ApJ*, 71, 175–190 {**347, 369**}

Takeda Y, Ohkubo M, Sadakane K, 2002a, Spectroscopic determination of atmospheric parameters of solar-type stars: description of the method and application to the Sun. *PASJ*, 54, 451–462 {**359**}

Takeda Y, Ohkubo M, Sato B, *et al.*, 2005, Spectroscopic study on the atmospheric parameters of nearby F–K dwarfs and subgiants. *PASJ*, 57, 27–43 {**361**}

Takeda Y, Sato B, Kambe E, *et al.*, 2002b, Spectroscopic determination of stellar atmospheric parameters: application to mid-F through early-K dwarfs and subgiants. *PASJ*, 54, 1041–1056 {**359**}

Talon S, 2008, Transport processes in stars: diffusion, rotation, magnetic fields and internal waves. *Stellar Nucleosynthesis: 50 Years after B2FH, EAS Publications Series* (eds. Charbonnel C, Zahn JP), in press {**346**}

Talon S, Charbonnel C, 1998, The Li dip: a probe of angular momentum transport in low-mass stars. *A&A*, 335, 959–968 {**382**}

—, 2004, Angular momentum transport by internal gravity waves. II. Population II stars from the Li plateau to the horizontal branch. *A&A*, 418, 1051–1060 {**382**}

Tej A, Chandrasekhar T, 2000, Angular diameter and effective temperature of a sample of 15 M giants at 2.2 μm from lunar occultation observations. *MNRAS*, 317, 687–696 {**358**}

Thévenin F, Idiart TP, 1999, Stellar iron abundances: non-LTE effects. *ApJ*, 521, 753–763 {**343, 355, 362, 367, 377**}

Thévenin F, Provost J, Morel P, *et al.*, 2002, Asteroseismology and calibration of α Cen binary system. *A&A*, 392, L9–L12 {**376**}

Thielemann F, 2002, Nucleosynthesis. *Encyclopedia of Astronomy and Astrophysics;* (ed. Murdin, P.) {**344**}

Thorén P, Edvardsson B, Gustafsson B, 2004, Subgiants as probes of Galactic chemical evolution. *A&A*, 425, 187–206 {**369**}

Thoul AA, Bahcall JN, Loeb A, 1994, Element diffusion in the solar interior. *ApJ*, 421, 828–842 {**349**}

Tilton GR, 1988, *Age of the Solar System*, 259–275. Meteorites and the Early Solar System {**340**}

Tinsley BM, 1979, Stellar life times and abundance ratios in chemical evolution. *ApJ*, 229, 1046–1056 {**384**}

Tomkin J, Lemke M, Lambert DL, *et al.*, 1992, The carbon-to-oxygen ratio in halo dwarfs. *AJ*, 104, 1568–1584 {**355**}

Trampedach R, Däppen W, Baturin VA, 2006, A synoptic comparison of the Mihalas–Hummer–Däppen and OPAL equations of state. *ApJ*, 646, 560–578 {**344**}

Truong-Bach, Nguyen-Q-Rieu, Sylvester RJ, *et al.*, 1997, ISO LWS observations of H_2O from R Cas: a consistent model for its circumstellar envelope. *Ap&SS*, 255, 325–328 {**375**}

Tsujimoto T, Yoshii Y, Nomoto K, *et al.*, 1997, A new approach to determine the initial mass function in the solar neighbourhood. *ApJ*, 483, 228–234 {**379**}

Upgren AR, Ratnatunga KU, Casertano S, *et al.*, 1997, Kinematics and M_v calibration of K and M dwarf stars. *ESA SP–402: Hipparcos, Venice '97*, 583–586 {**377**}

Vacca WD, Garmany CD, Shull JM, 1996, The Lyman-continuum fluxes and stellar parameters of O and early B-type stars. *ApJ*, 460, 914–931 {**391**}

Vanbeveren D, de Donder E, 2003, The chemical evolution of the solar neighbourhood. *Astrophysics and Space Science Library* (eds. Cheng KS, Leung KC, Li TP), volume 298, 99–110 {**386**}

VandenBerg DA, Bell RA, 1985, Theoretical isochrones for globular clusters with predicted BVRI and Strömgren photometry. *ApJS*, 58, 561–621 {**350**}

VandenBerg DA, Bergbusch PA, Dowler PD, 2006, The Victoria-Regina stellar models: evolutionary tracks and isochrones for a wide range in mass and metallicity that allow for empirically constrained amounts of convective core overshooting. *ApJS*, 162, 375–387 {**350**}

VandenBerg DA, Clem JL, 2003, Empirically constrained colour-temperature relations. I. $BV(RI)_C$. *AJ*, 126, 778–802 {**353, 354**}

VandenBerg DA, Richard O, Michaud G, *et al.*, 2002, Models of metal-poor stars with gravitational settling and radiative accelerations. II. The age of the oldest stars. *ApJ*, 571, 487–500 {**368**}

VandenBerg DA, Swenson FJ, Rogers FJ, *et al.*, 2000, Models for old, metal-poor stars with enhanced α-element abundances. I. Evolutionary tracks and ZAHB loci; observational constraints. *ApJ*, 532, 430–452 {**349, 350, 385**}

van Belle GT, Ciardi DR, ten Brummelaar TA, *et al.*, 2006, First results from the CHARA array. III. Oblateness, rotational velocity, and gravity darkening of Alderamin. *ApJ*, 637, 494–505 {**358**}

van Belle GT, Lane BF, Thompson RR, *et al.*, 1999, Radii and effective temperatures for G, K, and M giants and supergiants. *AJ*, 117, 521–533 {**352, 357**}

van Belle GT, Thompson RR, 1998, Evolved star sizes, temperatures as directly measured with interferometry. *IAU Symp. 191*, 225 {**357**}

van den Hoek LB, de Jong T, 1997, Inhomogeneous chemical evolution of the Galactic disk: evidence for sequential stellar enrichment? *A&A*, 318, 231–251 {**386**}

van Langevelde HJ, Vlemmings WHT, Diamond PJ, *et al.*, 2000, VLBI astrometry of the stellar image of U Her, amplified by the 1667 MHz OH maser. *A&A*, 357, 945–950 {**395**}

van't Veer-Menneret C, Katz D, Cayrel R, *et al.*, 1999, A grid of metal-poor model stellar atmospheres for stars born in the early galaxy. *Galaxy Evolution: Connecting the Distant universe with the Local Fossil Record*, 257–258 {**349**}

Velázquez PF, Rodríguez LF, 2001, VLA observations of Z CMa: the orientation and origin of the thermal jet. *Revista Mexicana de Astronomia y Astrofisica*, 37, 261–267 {**395**}

Ventura P, Zeppieri A, Mazzitelli I, *et al.*, 1998, Full spectrum of turbulence convective mixing. I. Theoretical main sequences and turn-off for $0.6–15M_\odot$. *A&A*, 334, 953–968 {**350**}

Viotti R, Cardini D, Emanuele A, *et al.*, 1997, The luminosity and kinematics of a sample of hot subdwarfs. *ESA SP–402: Hipparcos, Venice '97*, 395–396 {**367**}

Vivekananda Rao P, Radhika P, 2003, UV Psc and the radiative flux scale. *Ap&SS*, 283, 225–231 {**377**}

Vlemmings WHT, van Langevelde HJ, Diamond PJ, 2002, Astrometry of the stellar image of U Her amplified by the circumstellar 22 GHz water masers. *A&A*, 393, L33–L36 {**395**}

von Zeipel H, 1924, The radiative equilibrium of a rotating system of gaseous masses. *MNRAS*, 84, 665–683 {**390**}

Wade GA, Hill GM, Adelman SJ, *et al.*, 1998, Magnetic field models for A and B stars. V. The magnetic field and photometric variation of 84 UMa. *A&A*, 335, 973–978 {**393**}

Wade GA, Kudryavtsev D, Romanyuk II, *et al.*, 2000, Magnetic field geometries of two slowly rotating Ap/Bp stars: HD 12288 and HD 14437. *A&A*, 355, 1080–1086 {**393**}

Wade GA, Smith MA, Bohlender DA, *et al.*, 2006, The magnetic Bp star 36 Lyn. I. Magnetic and photospheric properties. *A&A*, 458, 569–580 {**393**}

Walborn NR, Howarth ID, Herrero A, *et al.*, 2003, The remarkable alternating spectra of the Of?p star HD 191612. *ApJ*, 588, 1025–1038 {**375**}

Walborn NR, 2005, Stellar-wind variations in the spectrum alternator HD 191612. *FUSE Proposal*, 88 {**375**}

Wallerstein G, Sneden C, 1982, A K giant with an unusually high abundance of lithium: HD 112127. *ApJ*, 255, 577–584 {**382**}

Weigelt G, Beckmann U, Berger JP, *et al.*, 2003, JHK band spectro-interferometry of T Cep with the IOTA interferometer. *Interferometry for Optical Astronomy II. Proc. SPIE, Vol. 4838* (ed. Traub WA), 181–184 {**358**}

Weiss A, Hillebrandt W, Thomas HC, *et al.*, 2004, *Cox and Giuli's Principles of Stellar Structure*. Princeton Publishing Associates {**339**}

Wesselink AJ, Paranya K, de Vorkin K, 1972, Catalogue of stellar dimensions. *A&AS*, 7, 257–289 {**356**}

Wesselink AJ, 1969, Surface brightnesses in the UBV system with applications of M_V and dimensions of stars. *MNRAS*, 144, 297–311 {**351**}

Wheeler JC, Sneden C, Truran JW, 1989, Abundance ratios as a function of metallicity. *ARA&A*, 27, 279–349 {**378**}

Whitelock PA, 2003, Luminosities of asymptotic giant branch variables. *ASSL Vol. 283: Mass-Losing Pulsating Stars and their Circumstellar Matter* (eds. Nakada Y, Honma M, Seki M), 19–26 {**373**}

Whitelock PA, van Leeuwen F, Feast MW, 1997, The luminosities and diameters of Mira variables from Hipparcos parallaxes. *ESA SP–402: Hipparcos, Venice '97*, 213–218 {**356, 357**}

Wittkowski M, Aufdenberg JP, Driebe T, *et al.*, 2006a, Tests of stellar model atmospheres by optical interferometry. IV. VINCI interferometry and UVES spectroscopy of Menkar. *A&A*, 460, 855–864 {**358**}

Wittkowski M, Aufdenberg JP, Kervella P, 2004, Tests of stellar model atmospheres by optical interferometry. VLTI–VINCI limb-darkening measurements of the M4 giant ψ Phe. *A&A*, 413, 711–723 {**358**}

Wittkowski M, Hummel CA, Aufdenberg JP, *et al.*, 2006b, Tests of stellar model atmospheres by optical interferometry.

III. NPOI and VINCI interferometry of the M0 giant γ Sge covering 0.5–2.2 μm. *A&A*, 460, 843–853 {**358**}
Wittkowski M, Hummel CA, Johnston KJ, *et al.*, 2001, Direct multi-wavelength limb-darkening measurements of three late-type giants with the Navy Prototype Optical Interferometer. *A&A*, 377, 981–993 {**358**}
Worden SP, 1976, Digital analysis of speckle photographs: the angular diameter of Arcturus. *PASP*, 88, 69–72 {**356**}
Wyse RFG, Gilmore G, 1995, Chemistry and kinematics in the solar neighbourhood: implications for stellar populations and for galaxy evolution. *AJ*, 110, 2771–2787 {**386, 387**}
Yıldız M, 2007, Models of α Cen A and B with and without seismic constraints: time dependence of the mixing-length parameter. *MNRAS*, 374, 1264–1270 {**376**}
Yıldız M, Yakut K, Bakış H, *et al.*, 2006, Modeling the components of binaries in the Hyades: the dependence of the mixing-length parameter on stellar mass. *MNRAS*, 368, 1941–1948 {**376**}
Yi S, Demarque P, Kim YC, *et al.*, 2001, Toward better age estimates for stellar populations: the Yonsei–Yale (Y^2) isochrones for solar mixture. *ApJS*, 136, 417–437 {**350**}
Yi S, Demarque P, Kim YC, 1997, On the ultraviolet-bright phase of metal-rich horizontal-branch stars. *ApJ*, 482, 677–684 {**371**}
Yi S, Kim YC, Demarque P, 2003, The Yonsei–Yale (Y^2) stellar evolutionary tracks. *ApJS*, 144, 259–261 {**348–350**}
Zahn JP, 1992, Circulation and turbulence in rotating stars. *A&A*, 265, 115–132 {**346**}
Zhang HW, Zhao G, 2005, Chemical abundances of very metal-poor stars. *MNRAS*, 364, 712–724 {**385**}
—, 2006, Chemical abundances of 32 mildly metal-poor stars. *A&A*, 449, 127–134 {**388**}
Zhao G, Gehren T, 2000, Non-LTE analysis of neutral magnesium in cool stars. *A&A*, 362, 1077–1082 {**343**}
Zijlstra AA, Bedding TR, 1999, Miras and mass loss in the local group. *IAU Symp. 192*, 348–355 {**374**}

8

Specific stellar types and the ISM

This chapter presents results for certain specific stellar types, divided into pre-main-sequence stars, main-sequence stars, X-ray sources in general, and post-main-sequence evolutionary phases. Additionally, it covers the use of the Hipparcos and Tycho stars as probes of the local interstellar medium.

8.1 Pre-main-sequence stars

8.1.1 Introduction

According to the theory of Hayashi (1961), a star begins its pre-main-sequence life with a very large radius, well in excess of the value that it finally assumes when it reaches the main sequence. The large surface area implies a high luminosity, more than can be transported through radiation alone. Thus, all sufficiently massive pre-main-sequence objects were predicted to be highly convective, spanning a broad region of the HR diagram above the main sequence, and evolving to the main sequence over the Kelvin–Helmholtz time scale. Subsequent refinements regarding the ignition of light elements and final approach to the main sequence were added by Iben (1965) and others. Detailed models of pre-main-sequence evolution have subsequently been developed by, e.g. D'Antona & Mazzitelli (1994, 1997), Palla & Stahler (1993, 1999), amongst others.

The starting radius is now considered to follow a certain locus, or 'birthline', in the HR diagram, where an object first becomes visible. Subsequently, detailed evolution depends on mass, protostellar mass accretion rate, mass loss due to winds, internal rotation, magnetic field, etc. Typical models, however, consider the pre-main-sequence star to be a non-rotating sphere of constant mass, contracting from the birthline to the ZAMS, with fusion dominated by hydrogen with contributions from deuterium and, at ages close to 10^7 yr, pre-main-sequence lithium burning (for further details see, e.g. D'Antona & Mazzitelli 1997 and Palla & Stahler 1999).

In practice, pre-main-sequence stars are characterised by active phenomena such as winds, jets, and outflows, and strong interaction with the circumstellar environment in which they are embedded. Most have strong infrared and/or ultraviolet excess, and show photometric and spectroscopic variability on time scales from minutes to years, attributed to variable extinction due to circumstellar dust, accretion, chromospheric activity, and pulsational instability if the object is lying in the instability strip.

As described in the box on page 414, pre-main-sequence stars are conventionally divided into two main classes: the lower-mass pre-main-sequence stars, or T Tauri stars; and the intermediate-mass pre-main-sequence stars, or Herbig Ae/Be stars. The distinction is retained because of their different spectral signatures, just as it is for stars on the main sequence.

8.1.2 T Tauri stars

T Tauri stars are newly-formed (age $\lesssim$ 10 Myr), non-pulsating, low-mass ($M < 2M_\odot$) stars that have recently become optically visible. They were discovered by Joy (1945), and named after the brightest member in the Taurus–Auriga cloud. They are of late spectral type, show apparently normal photospheres with additional continuum and line-emission characteristic of a hotter $\sim$ 10 000 K envelope, are often found in connection with groups of OB stars, show strong infrared and/or ultraviolet excess, and have luminosities which place them above the main sequence. They are subdivided, according to increasing evolutionary age, into classical T Tauri, weak-line T Tauri, and post-T Tauri.

Martín (1997, 1998) offered quantitative spectroscopic criteria for the classification of (classical,

Pre-main-sequence stars – some nomenclature: Pre-main-sequence stars form and arrive on the main sequence with the full range of main-sequence masses. The nomenclature is based on historical phenomenology: pre-main-sequence stars of low-mass, $< 2M_\odot$, the youngest visible (F)GKM spectral type stars, are referred to as T Tauri stars; pre-main-sequence stars of intermediate mass, 2–$8M_\odot$, i.e. A and B type pre-main-sequence stars, are referred to as Herbig Ae/Be stars; more massive stars, $\gtrsim 8M_\odot$, in the pre-main-sequence stage are not observed because they evolve very quickly: when they become visible through the surrounding circumstellar gas and dust cloud, core H-burning is already underway, and they are main-sequence objects. In the HR diagram pre-main-sequence stars are located to the above right of the main sequence: they have higher luminosities for a given $T_{\rm eff}$ than their main-sequence counterparts, because of their larger radii. In the absence of distance, and hence luminosity information, this can be recognised by the lower surface gravity.

Protostars form under self-gravity from the denser parts of molecular clouds, whose typical total mass is around $10^4 M_\odot$. The protostar starts with a small fraction of its final mass, but continues to accrete infalling material until, after a few million years, thermonuclear fusion begins in its core, and a strong stellar wind develops which stops further infall. The protostar is now considered a young star since its mass is fixed, and its future evolution largely determined.

T Tauri stars are the low-mass pre-main-sequence stars named after their prototype, T Tauri. They are found near molecular clouds and identified by their optical variability and strong chromospheric lines. Their surface temperatures are similar to those of main sequence stars of the same mass, but they are significantly more luminous because their radii are larger. Their central temperatures are too low for hydrogen burning, being powered by gravitational energy released as the star contracts towards the main sequence, which they reach after about 10^8 yr. Most T Tauri stars are in binary systems. They typically rotate with a period between one and twelve days, compared to a month for the Sun, and are very active and variable. There is evidence of large areas of star spot coverage, and they have intense and variable X-ray and radio emissions (approximately 1000 times that of the Sun). Many have extremely powerful stellar winds. Another source of brightness variability are clumps in the disk surrounding the T Tauri star.

Their spectra show higher lithium abundance than the Sun and other main-sequence stars because it has not yet been destroyed. Lithium depletion, which occurs above $\sim 2.5 \times 10^6$ K, varies strongly with size, and during the last highly convective and unstable stages of pre-main-sequence contraction contributes to the energy source. Rapid rotation increases the transport of lithium into deeper layers where it is destroyed. T Tauri stars generally increase their rotation rates as they age, through contraction and spin-up, as they conserve angular momentum. This causes an increased rate of lithium loss with age. Lithium burning lasts for ~ 100 Myr.

Circumstellar disks: roughly half of T Tauri stars are known to have circumstellar disks, which in this phase are called protoplanetary disks because they are the progenitors of planetary systems. Circumstellar disks are estimated to dissipate on time scales of up to 10^7 yr.

Herbig Ae/Be stars: Herbig (1960) was the first to realise that Be and Ae stars 'associated with nebulosity' are in fact intermediate-mass pre-main-sequence stars: young (< 10 Myr) stars of spectral types A and B, which have lost most of their envelope of infalling gas and dust, but are not yet fusing H into He, and whose main energy source is gravitational contraction. H and Ca emission lines are observed in their spectra, and they may be surrounded by circumstellar disks. Defining characteristics are now taken to be (Thé *et al.*, 1994): spectral type earlier than F8, presence of emission lines, excess infrared radiation due to circumstellar dust, and location in or near a probable star-formation region. Objects isolated from obvious regions of star formation are, however, now known, and these criteria may also select evolved massive stars (Davies *et al.*, 1990). Some authors also include stars without strong emission lines, although this makes the distinction between Vega-type stars (Section 10.7.5) less distinct. Thé *et al.* (1994) published a catalogue of 287 confirmed or possible candidates. Herbig Be stars are distinct from the classical Be stars, which are usually main-sequence objects in which emission arises from a gaseous disk around the star, originating partly from rotation (Section 8.2.1).

Isolated pre-main-sequence stars: pre-main-sequence stars are generally found in or close to active sites of star formation. Isolated objects of both T Tauri and Herbig Ae/Be type are now recognised, and frequently referred to as isolated young stellar objects. They may be isolated because they have been ejected from their birth locations, or because their parent cloud has since dispersed.

Herbig–Haro objects: are small regions of nebulosity associated with newly-born stars. The radiation, caused by shocks, results from gas ejected by young stars colliding with clouds of nearby gas and dust at speeds of several hundred kilometres per second. Herbig–Haro objects are ubiquitous in star-forming regions, and several are often seen around a single star, aligned along its rotation axis. The phenomenon is transient, lasting only a few thousand years. They can evolve visibly over quite short time scales as they move rapidly away from their parent star into the surrounding gas clouds.

weak-line, and post-) T Tauri stars: (1) spectral type in the range K0–M6, and either (2a) strong emission lines and 1 ultraviolet-optical-near infrared continuum excesses, or (2b) weak emission lines and a photospheric Li I λ670.8 nm absorption feature with a minimum equivalent width which depends on spectral type. Classical T Tauri stars meet criteria (1) and (2a), while weak-line T Tauri stars meet criteria (1) and (2b). Herbig (1978) noted that the T Tauri phase represents only about 10% of the pre-main-sequence evolution of solar-type stars, and used the term 'post-T Tauri' for a pre-main-sequence star which has lost its T Tauri properties, showing only some remnants of line emission and infrared excess, but still marked by the presence of the strong Li 670.8 nm resonance line, which is subsequently destroyed through convective mixing. T Tauri stars occupy a different region in the $T_{\rm eff}$ versus Li equivalent width diagram than the low-mass members of young open clusters, with the post-T Tauri stars intermediate between them.

Rotation is another important property of pre-main-sequence objects. Theories of angular momentum evolution in young solar-type stars predict that they will experience their maximum rotational velocity on the ZAMS, with the pre-ZAMS spin-up being due to contraction, and main sequence spin-down due to the braking effect of magnetic field interactions with the stellar wind. The classical T Tauri stars generally rotate slower than coeval weak-line T Tauri stars, in which the braking effect of the disk is missing.

While such stars are generally closely associated with sites of active star formation, objects more isolated from such regions are now known, such as the isolated classical T Tauri star TW Hya. The small group of known isolated (classical, weak-line, or post-) T Tauri stars now includes CD $-29°8887$, Hen 3–600, CD $-33°7795$, and HD 98800, all of which lie within a $10° \times 10°$ field, at least 13° from the nearest known dark cloud (Kastner *et al.*, 1997). The discovery of numerous candidate weak-line T Tauri stars up to 10–50 pc away from known molecular cloud cores of nearby star-forming regions prompted several scenarios to explain their location: the ejection model where star formation takes place in the cloud cores and the stars are subsequently ejected (Sterzik & Durisen, 1995); the model in which star formation takes place in small cloudlets which subsequently disperse (Feigelson, 1996); and triggered star formation by means of supernova explosions or the impacts of high-velocity clouds with the Galactic plane (Lépine & Duvert, 1994).

Hipparcos contributions An important Hipparcos contribution has been to provide direct distance estimates to star-forming regions and individual pre-main-sequence objects, to further constrain theoretical models. The luminosity, adjusted as always for bolometric correction and interstellar extinction, combined with the effective temperature, can also immediately confirm the pre-main-sequence nature if it lies above the main sequence. In a number of these studies, X-ray emission is used as an important diagnostic of activity and, frequently but not necessarily, of a low age. Additionally, proper motions allow the origin of an object to be traced back in time to the site of its putative formation. A catalogue of proper motions of 1250 pre-main-sequence objects associated with star-forming regions over the entire sky was given by Ducourant *et al.* (2005), updating the work of Teixeira *et al.* (2000). Astrometric material was from various sources, reaching $V \leq 16.5$ mag, with all proper motions reduced to the Tycho 2 Catalogue.

Wichmann *et al.* (1997), subsequently described in Wichmann *et al.* (1998), made an early analysis of the Hipparcos data of pre-main-sequence stars, comparing with pre-Hipparcos distance estimates for nearby star-forming regions mainly based on a few T Tauri stars in each. They found: for the Taurus–Auriga cloud, five stars gave $\langle\pi_{\rm Hip}\rangle = 7.06 \pm 0.71$ mas, or 142 ± 14 pc, compared with, for example, 140 ± 20 pc from Elias (1978); for Chamaeleon I, four stars gave $\langle\pi_{\rm Hip}\rangle = 6.25 \pm 0.68$ mas, or 160 ± 17 pc, compared with, for example, 140–150 pc from Schwartz (1992); for Lupus, five stars gave $\langle\pi_{\rm Hip}\rangle = 5.26 \pm 0.75$ mas, or 190 ± 27 pc, compared with, for example, 140 ± 20 pc from Hughes *et al.* (1993). They discussed the consequences for the upward adjustment of 0.25 mag in the resulting birthline of Lupus, although still much lower than that of the other two regions. They also discussed the cases of AB Dor and TW Hya, objects which had been considered as possible isolated pre-main-sequence objects, for which Hipparcos confirmed the pre-main-sequence nature only of TW Hya, but without identifying a possible birth place.

Frink *et al.* (1998), see also Frink (1999), examined the proper motions of 45 stars in the cores of the Chamaeleon clouds: two early-type stars, six classical and six weak-line T Tauri stars, eight pre-main-sequence, and 23 presumably older (dubious pre-main-sequence) stars discovered with ROSAT. For 12 stars, Hipparcos parallaxes were used to derive constraints on the distance distribution of the other stars. Proper motions, not all from the Hipparcos or Tycho Catalogues, were nevertheless all on the Hipparcos/ICRS system. Their sample divided into several subgroups, which they argued are consistent with both the high-velocity cloud impact model of Lépine & Duvert (1994) and the cloudlet model of Feigelson (1996), but inconsistent with any kind of a dynamical ejection model.

Hoff *et al.* (1998) searched for young stellar objects around two isolated T Tauri stars: the classical T Tauri TW Hya and the weak-line T Tauri CD $-29°8887$. From ROSAT observations detecting 107 X-ray sources in the joint fields, they found no other X-ray emitting young stellar objects around them. Hipparcos parallaxes for TW Hya and HD 98800 place them at distances of 56 and 46 pc respectively, making them the closest known T Tauri stars with dusty circumstellar disks. The space velocities of the two objects, from Hipparcos proper motions, are 3–5 km s^{-1}, with evolutionary ages from their position in the HR diagram of about 10^7 yr. They concluded that these two objects could not have travelled far from their original birthplace, suggesting that their parent molecular cloud must have dispersed.

Jensen *et al.* (1998) made an X-ray survey of the isolated T Tauri stars HD 98800 and CD $-33°7795$, and also failed to find other pre-main-sequence candidates within a 40 arcmin radius. Additionally, they searched the Hipparcos Catalogue for stars within a 10 pc volume centred on all four isolated stars, and identified

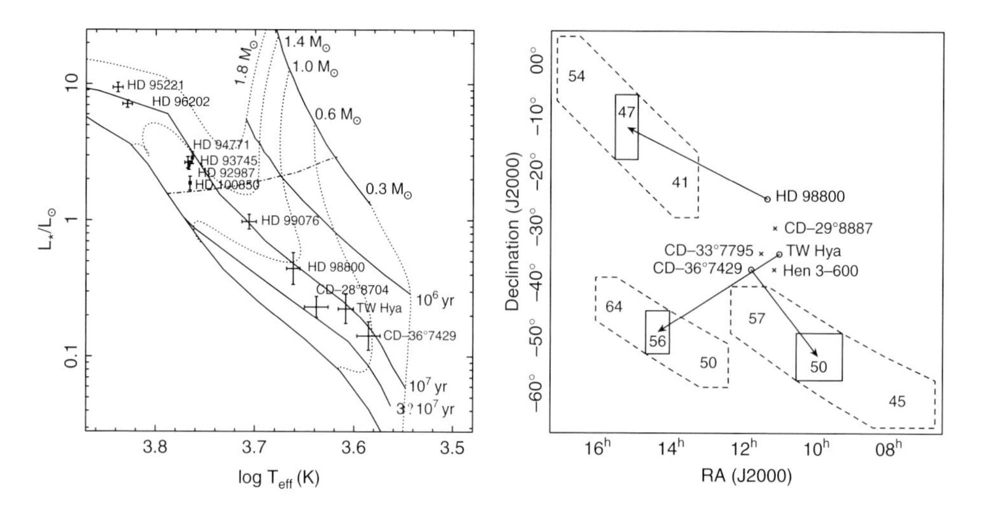

Figure 8.1 Left: HR diagram of candidate young stars. The evolutionary tracks and isochrones are from D'Antona & Mazzitelli (1994). The dashed line shows $V = 7.8$ at a distance of 50 pc, with bolometric correction included. Right: proper motions of confirmed young stars, projected back in time from the current positions. The arrow tips show positions on the sky 10^7 yr ago, corresponding to the estimated ages of TW Hya and HD 98800; the boxes show the uncertainty due to the proper motion. If the stars formed together, the arrows should point towards a common origin, which they do not. Young stars with unknown proper motions are shown with crosses. Dashed lines show the effect of 1σ distance uncertainties on the subtracted solar motion. Numbers represent the current distances in pc used to calculate the motions. From Jensen et al. (1998, Figures 2 and 3).

one other candidate, CD −36°7429, with a low space velocity, X-ray emission comparable to that of HD 98800, and Li absorption. They showed that the current positions and proper motions of TW Hya, HD 98800, and CD −36°7429 are inconsistent with them having formed as a group (Figure 8.1).

Martín (1998) made spectroscopic observations of six proposed T Tauri stars around B-type members of the Sco–Cen OB association, and compiled a list of secure pre-main-sequence stars in the region, identifying two classical, 18 weak-line, and 10 post-T Tauri stars. The presence of a mixture of T Tauri and post-T Tauri stars implies that previous results based on isochrone fitting that indicated an extremely young age for the Sco–Cen pre-main-sequence low-mass population were incorrect. The Hipparcos-derived distance of about 125 pc for the average of 6 B-type stars in the Sco–Cen field, instead of the 160 pc used in previous works, would lead to older evolutionary ages. Taking into account that post-T Tauri stars are generally fainter and harder to identify than weak-line T Tauri stars, he argued that the ratio of weak-line T Tauri to post-T Tauri objects in Sco–Cen may be of order unity, which would in turn suggest that the low-mass stars of the OB association span an age range similar to the B-type members (5–15 Myr), such that the low- and high-mass star populations are essentially coeval.

Over the elapsed period of several million years, the low-mass stars have had time to disperse around the OB association, which would explain why only a few such stars are found in modest area surveys, why the weak and post-T Tauri stars of Sco–Cen are widely dispersed in the sky, and might account for a significant number of the active lithium-rich stellar sources discovered with the ROSAT All-Sky Survey far from molecular clouds (e.g. Magazzù *et al.*, 1997). This study was developed by Guillout *et al.* (1998a), who detected a late-type stellar population in the direction of the Gould Belt by cross-correlating the ROSAT All-Sky Survey with the Tycho Catalogue (see Sections 6.10.5 and 8.3).

Soderblom *et al.* (1998) made a detailed study of HD 98800, a system comprising four post-T Tauri stars, and including an infrared excess attributed to a dust system. The system comprises two visible objects, HD 98800A and B, separated by about 0.8 arcsec, with both components being spectroscopic binaries, thus Aa+Ab, Ba+Bb. The former pair has an orbit with $P = 262$ d and $e = 0.484$, while the latter has $P = 315$ d and $e = 0.781$. The visual binary orbit has $P \gtrsim 10^5$ yr with $e \sim 0.993$; it is inferred to be marginally bound and currently near periastron. Improved estimates of $T_{\rm eff}$ combined with the Hipparcos distance of 47 pc, confirms that the component stars are too luminous to be on the main sequence (Figure 8.2). They found that HD 98800 is most probably about 10 Myr old, although it may lie in the range 5–20 Myr, confirming it as a member of the post-T Tauri class. Tracing back its orbit based on the Hipparcos proper motion, they showed that it may have formed in the Centaurus star-forming region, but it remains enigmatic in being so young and yet so far from where it was born. It provides evidence that the 'missing'

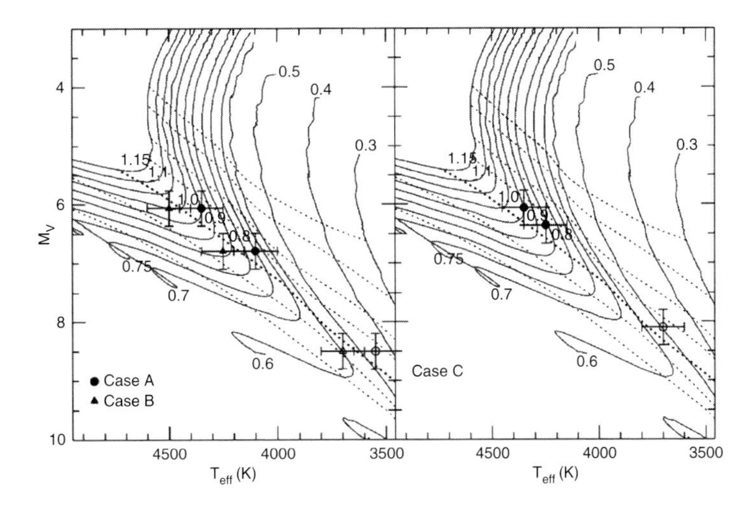

Figure 8.2 HR diagram for the components of HD 98800. Solid lines: evolutionary tracks. Dotted: isochrones, from top to bottom, ages of 1, 1.5, 3, 5, 10 (heavy dotted line), 20, and 40 Myr. Data points are for different T_{eff} estimates. Left: cases A (•) and B (△). Right: case C; open points are used for HD 98800Bb because their position assumes that Bb has the same age as Aa or Ba. For any one case the stars are in the order Aa, Ba, and Bb from hottest to coolest. From Soderblom et al. (1998, Figure 3).

post-T Tauri stars exist, but may be far removed from star-forming regions, making them difficult to find.

Neuhäuser & Brandner (1998) considered the lithium-rich ROSAT counterparts, which had been presumed to be low-mass pre-main-sequence stars. Of approximately 500 classified at the time, 21 stars were observed by Hipparcos. For seven out of 10 Taurus and Lupus stars in their sample, proper motions and parallaxes were not inconsistent with membership of these associations, while most of the stars in the Chamaeleon and Scorpius regions were inferred to be young foreground stars. Combined with ground-based photometry and spectroscopy, 15 stars could be placed in the HR diagram; all lie above the ZAMS and thus are indeed pre-main-sequence stars with ages from 1–15 Myr. Only two of the stars are located on the Hayashi-tracks, whereas the other 13 are post-T Tauri stars located on radiative tracks. Although containing only 3% of the total sample of lithium-rich ROSAT counterparts, the results did not confirm predictions made by some authors that the majority of ROSAT sources represented a dispersed population with ages in the range from 20–100 Myr – such stars were not found. They considered that the foreground pre-main-sequence stars may have been ejected toward us, or they could belong to the Gould Belt system.

Favata *et al.* (1998) studied the evolutionary status of a larger sample of late F to M0, 'solar-type stars', selected on the basis of their activity, mainly from Einstein-based X-ray surveys. The parallaxes were also used to place the objects in the HR diagram, with ages then determined by comparison with theoretical evolutionary tracks from D'Antona & Mazzitelli (1997), and computed for solar metallicity. They compared these evolutionary ages with those estimated from lithium abundance, activity level and the presence of circumstellar disks. To complement the sample at the young end, they used Hipparcos distances of a sample of optically-selected pre-main-sequence stars, mostly classical T Tauri stars. Some of the latter were found to be nearer than the pre-Hipparcos estimates, and far from their putative parent cloud. This implies a significantly larger age, thus providing observational evidence for the existence of long-lived T Tauri disks which could, in turn, produce slow rotators on the ZAMS. They argued that none of the age proxies (lithium, activity, and disks) appears to reliably and unambiguously select very young stars in the range of spectral types considered, with some apparently very young objects effectively lying on or very close to the main sequence. Assignment of ages to young solar-type stars on the basis of any of the standard proxies may thus significantly under- or over-estimate the evolutionary age (Figure 8.3, left and centre). Their data show that solar-type stars with very similar masses and evolutionary states can have very different properties. HD 17925 and HD 36705 (AB Dor) are on the same evolutionary track, with age estimates overlapping at the 1σ level, yet they differ widely in terms of X-ray emission, rotation, and lithium abundance. Primordial lithium abundance, high activity, and very high rotation are present in both HD 36705 (AB Dor) and HD 174429 (PZ Tel), yet their evolutionary ages are quite different, with PZ Tel approaching the main sequence on the radiative track and with an age of ~20 Myr, while AB Dor appears to be already on the main sequence, on the nominal 100 My isochrone, although only a lower

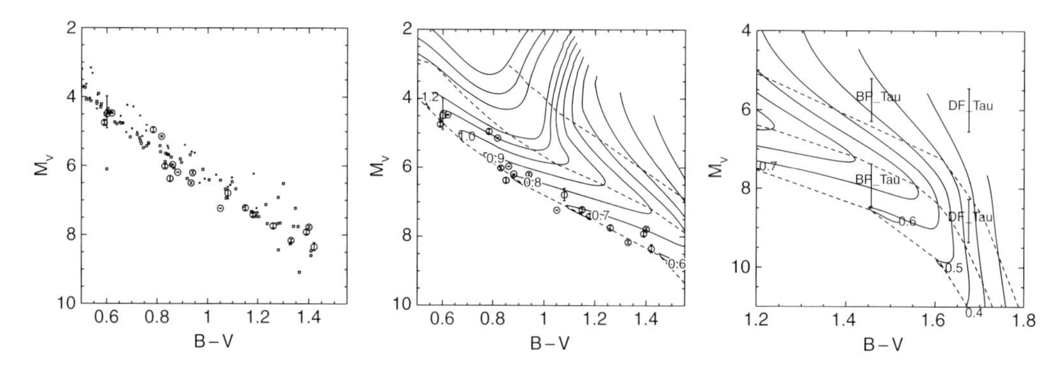

Figure 8.3 Left: evolutionary HR diagram for the X-ray selected sample of Favata et al. (1998), plotted as open circles, with error bars showing uncertainties due to distance. Also shown is the observed Hyades main sequence of Perryman et al. (1998), with open squares signifying single stars, and filled squares showing binaries. Middle: the same sample, with the evolutionary tracks of D'Antona & Mazzitelli (1997), transformed to M_V and $B-V$. Dashed lines are the 10^6, 10^7, 10^8 yr isochrones (the 10^9 yr isochrone is essentially indistinguishable from the latter). The mass is indicated by the number at the end of each track near the 10^8 yr isochrone. Right: the position of BP Tau and DF Tau on the evolutionary tracks of D'Antona & Mazzitelli (1997), using the stellar parameters of Kenyon & Hartmann (1995) scaled to their estimated distance of 140 pc (upper points), and also for distances corresponding to the Hipparcos parallaxes (lower points). Also indicated are the 10^6, 10^7, 10^8 yr isochrones. Tracks are spaced by $0.1M_\odot$, with mass indicated at the end of each track near the 10^8 yr isochrone. From Favata et al. (1998, Figures 1 and 2).

limit of ~35 Myr can be put on its age. BP Tau and DF Tau move from the distance of 140 ± 10 pc estimated by Kenyon & Hartmann (1995), to much closer distances of 42–70 pc and 31–52 pc respectively. While the Hipparcos results thus place them, unexpectedly, well in front of the Taurus–Auriga star-forming region, at the same time their luminosities would decrease to values more consistent with their presumed evolutionary status (Figure 8.3, right).

Bertout *et al.* (1999) re-examined the Hipparcos data for pre-main-sequence stars in the light of the unexpected result of Favata *et al.* (1998), that Hipparcos distances of certain classical T Tauri stars would place them far from known star-forming regions. They provided mean astrometric distances for groups of T Tauri stars in various star-forming regions. For the Taurus–Auriga complex, they obtained a mean parallax of 7.21 ± 0.49 mas, corresponding to a distance of 139^{+10}_{-9} pc, in agreement with the pre-Hipparcos estimate of Kenyon *et al.* (1994). Using only single stars, they found mean distances of three subgroups of 125^{+21}_{-16}, 140^{+16}_{-13} and 168^{+42}_{-28} pc, which might reflect a real depth effect. For the Orion A and B clouds, for which the Hipparcos Catalogue includes five of the eight distance indicators of Racine (1968), they determined a distance of 381^{+86}_{-59} pc, compared with the value of 600 ± 50 pc derived by Racine (1968). For the Chamaeleon, Lupus, and Scorpius star-forming clouds, their space velocities are all close to that of the Sco OB2 (Scorpius–Centaurus–Lupus–Crux) association, around $(U, V, W) = (0, -12, 0) \pm (5, 5, 5)$ km s^{-1}, such that their histories are considered connected, even if their present ages, motions and distances are not exactly the same. The Hipparcos data are limited and somewhat open to interpretation, but yield mean distances of around 170 pc, 140 pc, and 145 pc respectively (see also Section 6.10). In summary, they concluded that young stellar objects are located in their associated molecular clouds, as anticipated by a large body of previous work. A careful discussion of the effects of binarity for several of the objects was made, but failed to resolve the distance uncertainties in the discrepant cases of BP Tau and DF Tau. As a by-product, they confirmed the general validity of the Hipparcos Variability-Induced Mover (VIM) binary star solutions (however, see Section 3.4.1), citing the Hipparcos detection of binarity in RY Tau as a long-awaited but somewhat unexpected result. The mean distance to a sample of Li-rich late-type weak-line T Tauri stars in the Lupus star-forming region was determined by Wichmann *et al.* (1999), and shown to be in good agreement with the Hipparcos-based distance.

Gerbaldi *et al.* (2001) used the Hipparcos data to identify and date suspected post-T Tauri secondaries in binary systems. This search approach for elusive field post-T Tauri stars is based on an analysis of visual binaries with early-type primaries and late-type secondaries, first discussed by Murphy (1969) and carried out by Gahm *et al.* (1983). The idea is that if the main-sequence lifetime of the high-mass component is short, and comparable to the contraction time scale of a solar-type star, then the late-type secondaries should still be contracting to the ZAMS, or have recently arrived on it. Lindroos (1985, 1986) identified 78 such visual systems, determined ages based on Strömgren photometry and low-resolution spectroscopy, and discussed the coherence of the resulting ages using isochrones available at the

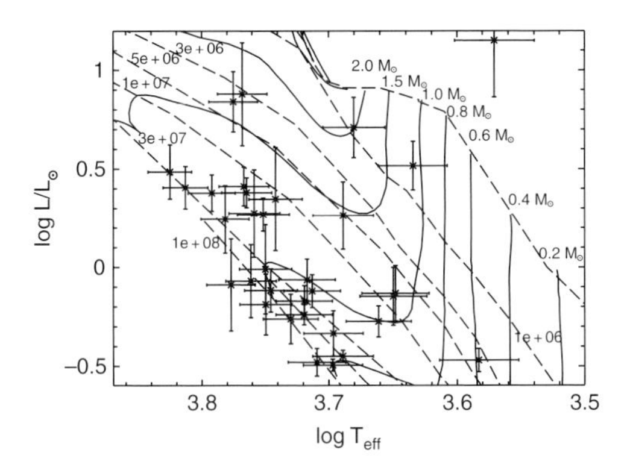

Figure 8.4 log $L/L_\odot$ versus log $T_{\rm eff}$ for the secondary stars with the pre-main-sequence tracks (thin line) and the isochrones (thick dashed line) from the models of Palla & Stahler (1999). Ages are given in years. Their location above the ZAMS demonstrates the pre-main-sequence evolutionary phase of many of these binary components. From Gerbaldi et al. (2001, Figure 5).

time. Spectroscopic studies of the sample were subsequently made by Pallavicini *et al.* (1992) and Martín *et al.* (1992) to identify genuine physical systems with post-T Tauri components. Gerbaldi *et al.* (2001) re-examined their ages, using estimates of $T_{\rm eff}$ from Strömgren and Geneva photometry, along with Hipparcos parallaxes of the early-type component, and by forcing the late-type secondary companion to be at the same distance as the primary. Ages and masses of each component were then derived from evolutionary tracks in the HR diagram for pre- and post-main-sequence stars (Palla & Stahler, 1999). Results for the secondary stars are shown in Figure 8.4, confirming the pre-main-sequence nature of many of the secondaries, especially those considered by Pallavicini *et al.* (1992) to be almost certain or possible physical systems.

Wichmann *et al.* (2003a) made an all-sky survey of nearby stars of spectral type F8 or later in a systematic search for young, zero-age main-sequence objects. The sample was derived by cross-correlating the ROSAT All-Sky Survey and the Tycho Catalogue, yielding 754 candidates distributed more-or-less randomly over the sky. Follow-up spectroscopy on 748 of them revealed a tight kinematic group of ten stars with very high Li equivalent widths, that are presumably younger than the Pleiades, but again distributed rather uniformly over the sky. About 43% of the sample had detectable levels of Li, indicating relatively young objects with ages not significantly above those of the Pleiades.

Other work on low-mass pre-main-sequence stars making use of the Hipparcos data include: positions and motions of Hα emission-line objects in the Chamaeleon star-forming regions (Relke & Pfau, 1998); radio emission from young stellar objects near LkHα 101 (Stine & O'Neal, 1998); the rapidly-rotating radio and X-ray F0 V star 47 Cas (Güdel *et al.*, 1998); two run-away T Tauri stars RW Aur and RK Ser (Trimble & Kundu, 1998); a search for star formation in the translucent clouds MBM 7 and MBM 55 (Hearty *et al.*, 1999); optical identification and study of the ROSAT-selected weak-line T Tauri candidates around Taurus–Auriga (Li & Hu, 1999); proper motions of faint ROSAT weak-line T Tauri stars in the Chamaeleon region (Terranegra *et al.*, 1999); ROSAT observations of T Tauri stars in MBM12, the nearest known molecular cloud with recent star formation (Hearty *et al.*, 2000b); proper motions of 223 pre-main-sequence stars in Chamaeleon, Lupus, Upper Scorpius, and Ophiuchus, not contained in the Hipparcos Catalogue but with proper motions on the same system (Ducourant *et al.*, 2000); motion and formation places of young stars in the solar neighbourhood (Palouš, 2000); multiplicity of X-ray-selected T Tauri stars in Chamaeleon (Köhler, 2001); post-T Tauri stars in the Scorpius–Centaurus OB association (Mamajek *et al.*, 2002); the pre-main-sequence spectroscopic binary AK Sco (Alencar *et al.*, 2003); new spectroscopic binaries among nearby stars (Wichmann *et al.*, 2003b); ROSAT X-ray active young stars toward Taurus–Auriga (Li, 2004); proper motion study of the ROSAT-selected weak-line T Tauri candidates around Taurus–Auriga (Li, 2005); the nature of the radio source T Tauri S (Johnston *et al.*, 2004); high-resolution spectroscopy of 13 candidate post-T Tauri stars (Bubar *et al.*, 2007); and the common proper motion companion to HIP 115147 (Makarov *et al.*, 2007).

8.1.3 Herbig Ae/Be stars

Herbig Ae/Be stars are pre-main-sequence stars of intermediate mass, $\sim$ 2–10$M_\odot$, in radiative contraction onto the main sequence, with a near- and/or far-infrared excess, surrounding nebulosity, often associated with the presence of circumstellar disks, and with luminosities which place them above the main sequence (see box on page 414). A few objects have now had their disks directly imaged using coronographic techniques, for example, HD 100546 (Augereau *et al.*, 2001). Hipparcos provided distance measurements which allow absolute luminosities to be determined, the pre-main-sequence nature of candidate stars to be confirmed, and their physical characteristics to be inferred.

Hipparcos contributions Van den Ancker *et al.* (1997) derived fundamental physical parameters (temperature, luminosity, mass, and age) of 10 Herbig Ae/Be candidates, and three non-emission-line A and B stars in star-forming regions. Pre-main-sequence evolutionary tracks, with different accretion rates, were taken from Palla & Stahler (1993). They found that the genuine Herbig stars are located between the birthline (where

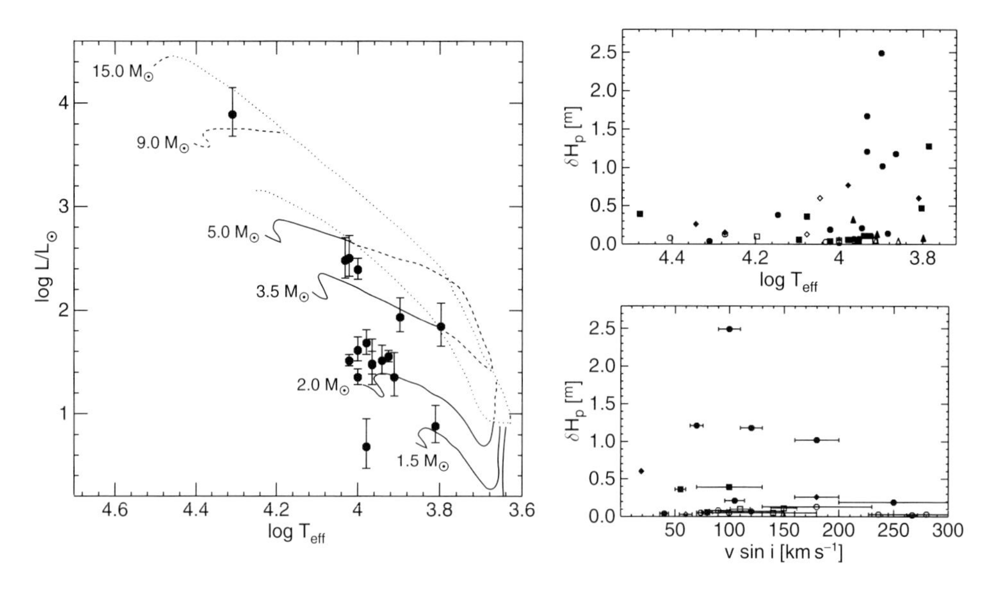

Figure 8.5 Left: HR diagram of Herbig Ae/Be stars with parallaxes measured by Hipparcos. Also shown are the theoretical pre-main-sequence evolutionary tracks (solid lines and dashed lines) and the birthlines for mass-loss rates of 10^{-4} (upper dotted line) and $10^{-5} M_\odot\, yr^{-1}$ (lower dotted line) from Palla & Stahler (1993). Top right: amplitude of the variations found in Herbig Ae/Be stars as a function of stellar effective temperature. •: stars in which the Hα line is predominantly double-peaked; ■: stars in which Hα shows a P Cygni profile; ▲: stars with a single-peaked Hα line in emission; ♦: stars in which Hα is in emission, but for which no information on its shape is available. Open symbols indicate stars for which the measured range in variability could be due to the uncertainty in the individual photometric measurements. Bottom right: amplitude of the variations found in Herbig Ae/Be stars as a function of $v \sin i$. From van den Ancker et al. (1998, Figures 1–3).

the star first becomes optically visible in its evolution) and the zero-age main sequence in the HR diagram, in agreement with what is expected for pre-main-sequence stars. The region close to the birthline is relatively devoid of stars as compared with the region closer to the ZAMS, in agreement with the expected evolutionary time scales. The Herbig Ae/Be stars not associated with star-forming regions were also found to be located close to the ZAMS. A fair number of stars clustered around the $2.5 M_\odot$ evolutionary track, with ages and energy distributions consistent with the evolutionary scenario of Waelkens *et al.* (1994), in which a broad dip around 10 μm in the energy distribution develops with time.

Van den Ancker *et al.* (1998), see also van den Ancker (1999), extended the work to 44 Herbig Ae/Be candidates, also making use of the Hipparcos photometric data to study their variability. Their conclusions (see also Figure 8.5) were as follows: (a) more than 65% show photometric variations with an amplitude larger than 0.05 mag; (b) those earlier than A0 show only moderate (< 0.5 mag) variations, whereas those of later spectral type may show variations of more than 2.5 mag, explained as stars with lower masses becoming optically visible while still contracting towards the ZAMS, whereas their more massive counterparts only become optically visible after having reached the ZAMS; (c) those with the smallest infrared excesses do not show large photometric variations, explicable if they are more evolved objects; (d) no correlation between photometric variability and $v \sin i$ was found: if large photometric variations are due to variable amounts of extinction by dust clouds in the equatorial plane of the system, evolutionary effects probably disturb the expected correlation between the two. Wichmann *et al.* (1998) also showed that their 11 Herbig Be stars lie closer to the ZAMS than their sample of eight Herbig Ae stars, in agreement with the models of Palla & Stahler (1993) which place the earlier spectral types closer to the ZAMS when the circumstellar material disperses and the star first becomes visible. The Hipparcos data have thus demonstrated that the majority of candidate Herbig Ae/Be stars are indeed pre-main-sequence objects.

Vieira *et al.* (2003) investigated 131 Herbig Ae/Be stars: 108 new candidates identified in the Pico dos Dias Survey (Gregorio-Hetem *et al.*, 1992), together with 19 previously known, and four from the IRAS Faint Source Catalogue. In addition to investigating emission-line properties, they combined distance estimates with details of the interstellar medium distribution, to find that 84 candidates can be associated with some of the more conspicuous star-forming regions, being in the

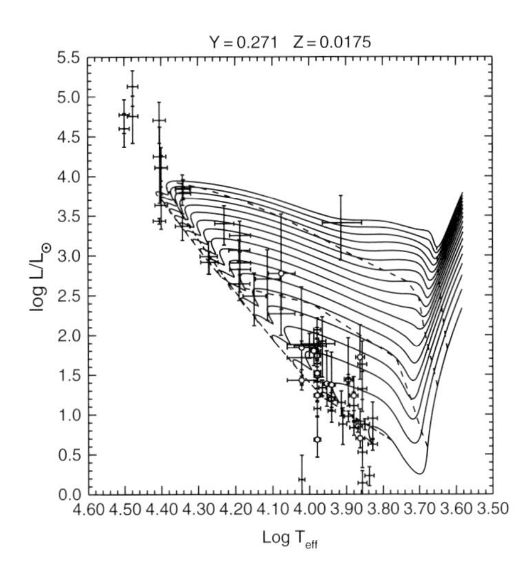

Figure 8.6 Evolutionary tracks from the ATON2.0 models of Ventura et al. (1998) from 1.5 to 10$M_\odot$ (solid lines) and isochrones corresponding to 10^3, 10^4, 10^5, 10^6 yr (dashed lines) and 10^7 yr (dash-dotted line), respectively. Symbols: ○: stars with Hipparcos distances; •: stars with distances estimated from star-forming regions. From Vieira et al. (2003, Figure 13).

right direction and at a compatible distance. As a further means of checking their properties, as well as their present evolutionary status, derived luminosities and effective temperatures of stars with possible association with star-forming regions and/or Hipparcos distances were plotted together with a set of pre-main-sequence evolutionary tracks on the HR diagram (Figure 8.6). Most of their candidates are located within the expected mass range ($2-10M_\odot$), and are also near the ZAMS. Stars with lower masses, between 1.5 and $2M_\odot$, were also expected given the adopted selection criteria. Considering the error bars, 14 stars have luminosities and temperatures outside the location expected for Herbig Ae/Be objects.

Hernández *et al.* (2005) studied early-type (BAF) stars in the nearby (< 500 pc) OB associations with the aim of determining the fraction of stars that belong to the Herbig Ae/Be class. They selected stars from the Upper Scorpius, Perseus OB2, Lacerta OB1, and Orion OB1 associations, with membership determined from Hipparcos data, spanning an age range of $\sim 3-16$ Myr. They also included the early-type stars in the Trumpler 37 cluster, part of the Cep OB2 association. They obtained spectra for 440 Hipparcos stars in these associations from which, in combination with the Hipparcos photometry, they determined spectral types, visual extinctions, effective temperatures, luminosities and masses. Using colours corrected for reddening, they found that the Herbig Ae/Be stars and the classical Be stars occupy clearly different regions in the *JHK* diagram. They showed that the Herbig Ae/Be stars constitute a small fraction of the early-type stellar population even in the younger associations, and showed that the time scales for disk evolution are much shorter in the intermediate-mass stars, which could be a consequence of more efficient mechanisms of inner disk dispersal.

Studies of individual objects making use of the Hipparcos data include: a study of the Herbig Ae/Vega-type binary HD 35187, revealing evidence for 0.4 mag of grey circumstellar extinction (Dunkin & Crawford, 1998); a study of the properties and spectral energy distributions of the Herbig Ae/Be stars HD 34282 and HD 141569 (Merín *et al.*, 2004); the parameters and evolutionary state of the Herbig Ae star candidate HD 35929 (Miroshnichenko *et al.*, 2004); and radio continuum sources associated with AB Aur (Rodríguez *et al.*, 2007).

8.2 Main-sequence evolutionary phases

This section covers certain main-sequence phases not considered in Chapter 7.

8.2.1 Be stars

Classical Be stars were originally defined as non-supergiant B-type stars, i.e. with $T_{\rm eff} \sim 10\,000$–$30\,000$ K, whose spectra have, or have had at some time, one or more Balmer lines in emission (Jaschek *et al.*, 1981; Collins, 1987). Since this definition in practice encompassed more than one type of star, the term has become increasingly used to signify B-type stars close to the main sequence that exhibit line emission in addition to the photospheric spectrum, but excluding Herbig Ae/Be stars or Algol systems (Porter & Rivinius, 2003; Smith *et al.*, 2000). The emission lines are attributed to a circumstellar gaseous component, believed to have been ejected by the star itself, and commonly accepted to be in the form of a disk, although the disk-formation processes are not fully understood (Bjorkman, 2000). Line emission is attributed to re-processing of the stellar ultraviolet light in the gaseous disk, while infrared excess and polarisation result from the scattering of stellar light in the disk. In conjunction with the fundamental property of rapid stellar rotation, with a mean rotation rate of some 70–80% of the critical velocity, physical mechanisms like non-radial pulsations, magnetic activity, or binarity are invoked to explain the circumstellar envelope formation. The Be nature is transient: Be stars exhibit a normal B-type spectrum at times, and normal B stars may become Be stars, with short-term periodic variability restricted to the earlier types. Extreme variability episodes may result from the complete loss of the disk, and its eventual rebuilding. Be stars are important for studies of asymmetric mass-loss processes, evolution of angular momentum distribution, asteroseismology, and magnetic field evolution.

There are three hypothesised origins for the very rapid rotation rates of the central B star: they were born as rapid rotators which somehow were able to avoid a spin-down by magnetic, wind loss, or binary tidal effects (Abt & Cardona, 1984); they were spun up at the end of the main-sequence phase as a result of core contraction (Meynet & Maeder, 2000; Heger & Langer, 2000); or they were spun up by mass and angular momentum transfer in an interacting binary system (Gies, 2000). Which of these processes dominate the formation of Be stars is presently unknown.

Hipparcos astrometry The location of Be stars in the HR diagram is important for understanding the luminosity of the central star, the characteristics of the stellar envelope, and their evolutionary path. B stars, and even more so Be stars, are scarce in the solar neighbourhood, and the pre-Hipparcos absolute magnitudes were determined only indirectly, through membership of open clusters, associations, or binary systems, or statistically through their spatial and kinematic distributions. There was nevertheless a consensus that Be stars were brighter, by about 0.5–1 mag, than B stars without emission.

Briot & Robichon (2000), see also Briot & Robichon (1998), studied the positions of 353 Be stars, and 5718 B stars without emission, in the Hipparcos HR diagram. Taking into account Malmquist and Lutz–Kelker biases, they showed that B stars are fainter than previous estimations by about 0.5 mag on average, that on average Be stars are brighter than B stars of the same spectral type, and that this over-luminosity increases with spectral type, possibly explained by the fact that the rotational velocities of the late Be stars is near the critical rotation velocity. Wegner (2000) found the same over-luminosity effect from his sample of 1207 unreddened and 441 reddened OB stars, compared with a sample of 90 unreddened and 25 reddened Be stars. Zhang *et al.* (2006) reported similar results for 457 Hipparcos Be stars, while also ruling out a correlation between near infrared excess and absolute visual magnitude.

One mechanism originally suggested for the creation of disks around Be stars, and to explain their rapid rotation, was interaction with a binary companion. In this model, matter overflows from the Roche lobe of the massive companion, then settles into an accretion disk around the Be star. The Be star becomes more massive, and the companion evolves off the main sequence as it exhausts its nuclear fuel. The Be star enters a phase of mass loss, and radiation pressure and rotation provide support for the massive disk remaining. This hypothesis had been largely dismissed, because the majority of Be stars have no observed companion, for example in the form of eclipsing or interacting binaries. Rinehart (2000) argued that this difficulty can be circumvented if single Be stars are actually ejected from binary systems, for example due to the supernova of the companion, the same mechanism put forward to explain the incidence of OB runaways and the observed properties of neutron stars (see Section 8.4.2). This explanation would predict that the velocity distribution of Be stars should be wider than that of normal B stars. Rinehart (2000) used the Hipparcos distances and proper motions to derive and compare the space velocity distributions of 129 Be stars with those of 5756 normal B stars. The distributions were found to be consistent, and thus not supporting the formation of single Be stars through binary breakup.

Berger & Gies (2001) made a similar search on a similar sample, but came to a different conclusion, probably because they were looking for a less pronounced difference between the two velocity distributions. The evolutionary path of a massive close binary system depends on what happens when the first star explodes as a supernova. As described in Section 8.4.2, an explosion can unbind the system, and the surviving companion can emerge moving with the orbital velocity at the time of the explosion, perhaps appearing as a high-velocity OB runaway star. On the other hand, if the total mass ejected in the supernova event is less than half the total system mass then the binary remains bound (van den Heuvel, 1978). The descendants of these systems containing neutron star and black hole companions are observed as high-mass X-ray binaries. The systemic velocity of the surviving binary will also have a runaway velocity related to the pre-supernova orbital velocity, and given by Nelemans *et al.* (1999) as

$$v_{\rm sys} = 213 \left(\frac{\Delta M}{M_\odot}\right) \left(\frac{m}{M_\odot}\right) \left(\frac{P_{\rm recirc}}{\rm day}\right)^{-1/3} \left(\frac{M_{\rm rem} + m}{M_\odot}\right)^{-5/3} \ {\rm km\,s^{-1}} \tag{8.1}$$

where ΔM is the mass lost during the supernova, m is the mass of the surviving companion, $M_{\rm rem}$ is the mass of the neutron star or black hole remnant, and $P_{\rm recirc}$ is the orbital period after subsequent re-circularisation of the orbit. Evidence that the massive supergiant X-ray binaries do indeed have unusually large space velocities is discussed in Section 8.4.2. Equation 8.1 provides an indication of the runaway velocities expected for the Be X-ray binaries. Typical values of $m/M_\odot = 10$, $M_{\rm rem}/M_\odot = 1.4$, and $P_{\rm recirc} = 100$ d, lead to $v_{\rm sys} = 8(\Delta M/M_\odot)\,{\rm km\,s^{-1}}$. Most models predict considerable mass loss prior to the supernova event, so that $\Delta M/M_\odot$ will be of order unity, and therefore, large runaway velocities are not expected in general. The results of Rinehart (2000) indeed suggest that the majority of Be stars are not post-supernova binaries for which $\Delta M/M_\odot$ was large. However, the possibility remains that many Be stars have evolved companions that avoided becoming supernovae and, thus, did not receive a runaway velocity boost. Berger & Gies (2001) constructed a sample of 344 Hipparcos Be stars, and derived their peculiar space velocities (after subtracting the Sun's motion and the motion of the LSR), and also their height above the Galactic plane, using $Z_0 = 20.5$ pc (Humphreys & Larsen 1995; although see also Section 9.2.2). In so doing, they confirmed the large distances from the Galactic plane of the three targets in common with the high Galactic latitude group identified by Slettebak *et al.* (1997): CD$-30°850$ = HD 14850; BD$-17°631$ = HD 20340; BD$+7°678$ = HD 29441. The distribution of peculiar velocities is shown in Figure 8.7, along with a Maxwellian velocity distribution fitted to the low-velocity portion, $V_{\rm Sp} < 30\,{\rm km\,s^{-1}}$. The mean velocity dispersion of $14.3\,{\rm km\,s^{-1}}$ is typical for massive young stars in the solar neighbourhood (Dehnen & Binney, 1998). Although some apparently high-velocity objects are probably due to large radial velocity errors, there are a significant number of stars that have very large peculiar velocities in one or both components. Those with $V_{\rm Sp} > 40\,{\rm km\,s^{-1}}$ represent only a modest 3–7% of the total sample of Be stars, but the existence of these runaway Be stars is an indication that at least some fraction of the Be stars probably formed in binaries. Even the general similarity of the Be and B star kinematics probably does not rule out the binary formation hypothesis, since there are other ways of forming a Be star in binaries that do not produce a runaway velocity.

Hipparcos photometry Be stars are known to exhibit different types of variability, often present simultaneously, characterised by time scales between a few minutes and several years (see, for example, Percy & Bakos, 2001). Some of these variations are periodic or quasi-periodic, and may be induced by the presence of a companion, by rotation, or pulsation of the Be star, or by inhomogeneities or density waves within its rotating disk. Binarity is usually invoked to explain the mid-term periodic variability, $P \sim 3-500$ d, while more rapid variability, < 3.5 d, of both B and Be stars, is explained as a consequence of non-radial pulsations or rotational modulation. The actual numbers of Be stars with companions in advanced evolutionary stages remains largely unknown because they are difficult to detect (Gies, 2000), and photometry offers one way of searching for binarity.

Hubert *et al.* (1997) searched the Hipparcos photometric data for short-period variability in a bright sample, $V < 5$ mag, of 26 B and 23 Be stars. Among B stars listed in the NSV (New Suspected Variable) compilation, 14 were confirmed as microvariables. Almost all the Be stars of the sample show rapid variability superimposed on long-term changes, with total amplitude of the long-term variability ranging between about 0.05–0.3 mag. The total amplitude of the short-term variations was slightly higher for Be stars than for B stars. Some Be stars showed recurrent outbursts with a variable strength and/or fading events. Short periodicities, with $P \leq 3.5$ d, were suspected or confirmed in more than half of Be stars and in a few B stars. Floquet *et al.* (1998) reported a similar analysis of the Tycho Catalogue photometry.

Hubert & Floquet (1998) extended the analysis to 273 Hipparcos Be stars; from their starting sample of 289 stars, 23 were listed in the catalogue as periodic, 157 as unsolved, 16 as microvariables with amplitude below 0.03 mag, 39 with a probability > 0.5 of being constant, and 54 without a specific variability indicator. The high accuracy of individual measurements of the *Hp* magnitude allowed a number of interesting light-curves to be analysed (Figures 8.8). They found that the degree of variability is highly dependent on temperature. Recurrent short-lived outbursts were generally not detected in late (B6–B9) Be stars, but were mainly detected in early Be stars with rather low to moderate $v \sin i$. Temporary fading events were preferentially seen in stars with large $v \sin i$. They detected a number of rapid periodic variables, as well as confirming the existence of short periodicities extending over a long duration, previously seen in intensive photometric monitoring over several days. A list of Be stars having rapid variations of large amplitude, $\Delta Hp \sim 0.1$ mag, were proposed as candidates for a search for multi-periodicity, i.e. as non-radial pulsators. Some showed quasi-periodic oscillations, ≤ 200 d, analogous to those discovered in Be stars by Mennickent *et al.* (1994). Estimates of cycles of long-term variations, $\gtrsim 1$ yr, could be made for some early spectral types. Light increases or decreases associated with transition phases from a quasi-normal B phase to a Be or Be shell phase have time scales dependent on the temperature of the star.

Hubert *et al.* (2000) extended the analysis after discovering that the B stars whose 'unsolved' light-curves are reported in the Hipparcos Catalogue are mainly Be stars. Short-lived outbursts were seen in 38 B and Be stars, and long-lived outbursts in 32 cases. They used the data to propose a model for the light-curves of the long-lived outbursts: they assumed that the star ejects a massive layer whose mass is given by $\Delta M/M_\odot \sim 8 \times 10^{-11}(R_\star/R_\odot)^2 e\tau_e$ where τ_e is the initial radial electron scattering opacity of the ejected layer, and e its ellipticity. In one scenario, the ejected layer goes on expanding, conserving its total mass and ellipticity; in the other, the mass is conserved, but its ellipticity changes as $e = R_\star/R$ where R is the equatorial radius of the expanding ellipsoidal layer. During the expansion, the opacity starts high, and the photometric evolution is determined by the increase of the layer radius R; in a subsequent phase,

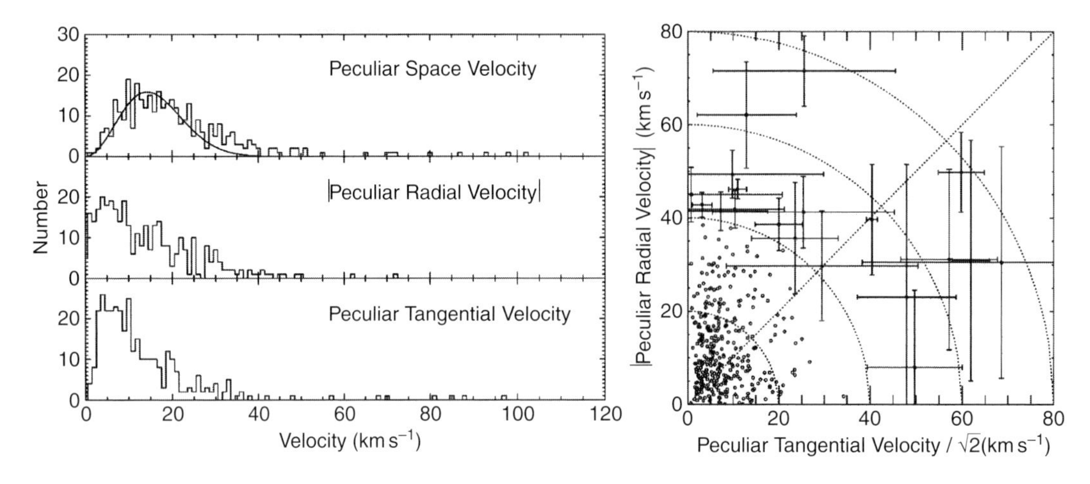

Figure 8.7 Left: histograms of peculiar tangential, radial, and space velocities (bottom to top) for the Be star sample studied by Berger & Gies (2001). The solid line in the upper panel represents a Maxwellian distribution fitted to the low-velocity portion of the sample. Right: distribution of peculiar tangential and radial velocities. The peculiar tangential velocities which sample two spatial dimensions are reduced by a factor of $\sqrt{2}$ for direct comparison with radial velocity (one spatial dimension). The individual high-velocity stars are plotted with error bars. From Berger & Gies (2001, Figures 1 and 2).

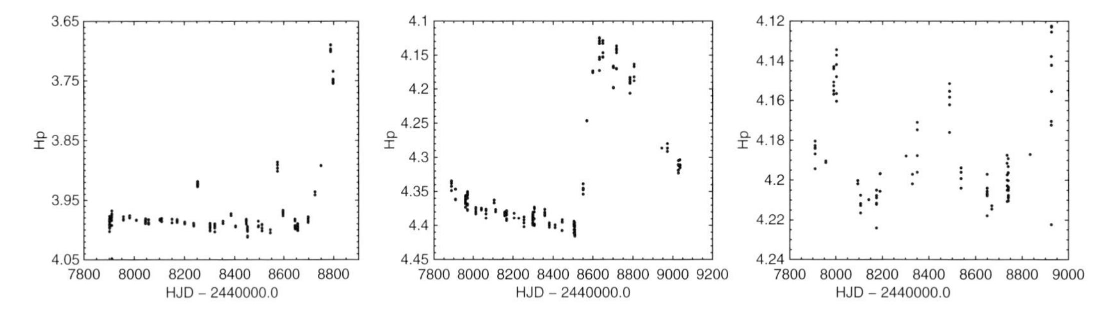

Figure 8.8 Some example Hipparcos light-curves, in Hp, for classical Be stars. Left: short-lived outbursts in ω CMa. Middle: strong long-lived outburst in υ Cyg. Right: outbursts in λ Eri, for which Hα outbursts from ground-based observations have also been observed during the intervals HJD 2447900–8000 and 2448400–8500. From Hubert & Floquet (1998, Figures 8–10).

the layer becomes progressively transparent. The light-curves corresponding to the first scenario can explain the light curves seen by Hipparcos.

Carrier *et al.* (2002a) studied four Be stars which have photometric variations according to Hipparcos, Tycho, or Geneva photometry, which can be linked to the presence of a companion: HR 1960, HR 2968, HR 3237 and HR 3642. For HR 1960, the Hipparcos photometric data show an amplitude of just 3 mmag in V, with $P = 395.48$ d, and for which the hypothesis of binarity is confirmed by radial velocity data. For HR 3237, the Hipparcos photometry gave a periodicity of $P = 11.546$ d claimed by Hubert & Floquet (1998) but unlikely to be real. For HR 3642, the Hipparcos photometry yields an amplitude of 0.07 mag and a period of $P = 137.99$ d (Figure 8.9), interpreted as interactions in a binary system.

Mennickent *et al.* (2002) extracted a sample of ~1000 Be candidates from the OGLE experiment light-curves and colours for about two millions stars in the Small Magellanic Cloud. They reported outbursts in 139 objects (13%), high and low states in 154 objects (15%), periodic variations in 78 (7%), and the more usual stochastic or quasi-periodic variations seen in Galactic Be stars in 685 (65%). They associated the outbursts with the Galactic Be star outbursts found by Hubert & Floquet (1998) from their analysis of Hipparcos photometry. They could correspond to Be stars with accreting white dwarf companions or alternatively, blue pre-main-sequence stars surrounded by thermally-unstable accretion disks.

Percy *et al.* (2002) used the Hipparcos epoch photometry and a form of autocorrelation analysis to investigate the amplitude and time scale of the short-period variability of 82 Be stars, including 46 Be stars analysed

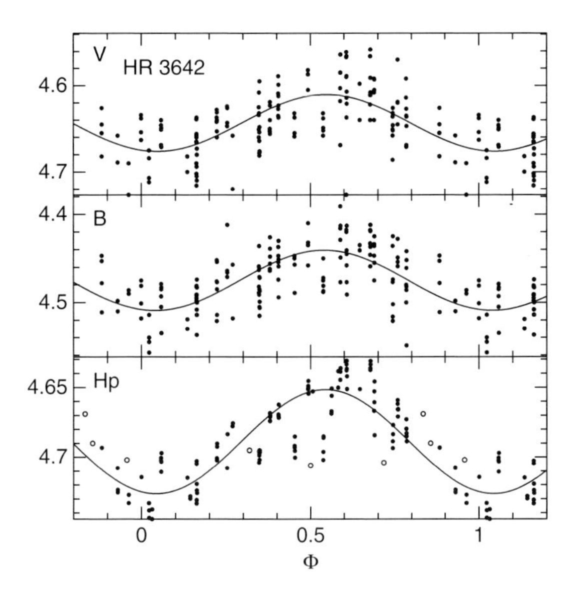

Figure 8.9 Photometric variability of the binary Be star HR 3642. Top: Tycho V_T magnitude folded at the period $P = 137.99$ d. Middle: same for Tycho B_T. Bottom: the Hipparcos Hp magnitude (•) and Geneva V magnitude (○). From Carrier et al. (2002a, Figure 12).

by Hubert & Floquet (1998), and 36 other Be stars suspected of short-period variability. Their algorithm takes, for all pairs of measurements, the difference in magnitude and the difference in time, plotting Δmag against Δt, and binning the Δmag to give a few values in each Δt bin. Their analysis gave useful information for about 84% of the stars, either confirming or refining the earlier analysis; for the rest, the time distribution of the Hipparcos epoch photometry limits its capability. Percy *et al.* (2004) applied the same approach to a larger sample of 277 Hipparcos Be stars, searching for short-period (0.2–2 d) photometric variability. They reported the discovery, time scale, and short-period variability amplitude for about 20 objects (Figure 8.10). They found that short-period variability is more prevalent in early Be stars than in late Be stars, consistent with earlier photometric results by Stagg (1987), and with high-resolution echelle spectroscopy of 27 early Be stars by Rivinius *et al.* (2003). Based on the conclusions of the latter work, that the short-term periodic line profile variability is due to non-radial pulsation, most notably with $l = m = +2$, the Hipparcos results presumably reflect the dependence of the pulsation mechanism on temperature. The idea that Be stars are non-radial pulsators, is reminiscent of the class of β Cephei stars (Section 4.9.4), which are early B stars that pulsate radially with $P \sim 0.1$–0.3 d. Aside from the emission lines, a substantial fraction of the Be stars have spectral types similar to those of the β Cephei stars, viz. B0.5–2 III–IV. β Cephei stars have also been identified through self-correlation analysis (Percy *et al.*, 2003), but why no Be stars show β Cephei-like variability, at least above 0.01 mag, is unclear.

Individual Be stars Carrier *et al.* (1999) monitored the Be star HR 2968 from 1978–1998 in Geneva photometry, overlapping the Hipparcos photometric observations between November 1989 to March 1993, during which period the two datasets showed good agreement. The mean luminosity, which was stable since 1978 (the normal B-star phase), increased from 1990 to 1995 (the Be phase), and then decreased until 1998. Also, in 1990 a periodic light variation phase started, with a period of 371 d. Five periods of this mid-term light variation have been observed. They propose a model to explain this periodic variability: the Be star is the main component of a binary system having an eccentric orbit of period 371 d; from 1990, the Be star was surrounded by matter expelled in its equatorial plane and, at each periastron passage, the companion star interacts gravitationally and/or radiatively with the disk.

Donati *et al.* (2001) used spectropolarimetric observations of β Cep, and the Hipparcos parallax-derived parameters ($M = 12M_\odot$, $R = 7R_\odot$, $T_{\rm eff} = 26\,000$ K and an age of 12 Myr, viewed with an inclination of the rotation axis of about 60°) to propose a consistent model of the large-scale magnetic field and of the associated magnetically-confined wind and circumstellar environment. The magnetic field is strong enough to magnetically confine the wind up to a distance of about 8–9$R_\star$. Both the X-ray luminosity and variability can be explained within the framework of the magnetically-confined wind-shock model of Babel & Montmerle (1997), in which the stellar-wind streams from both magnetic hemispheres collide with each other in the magnetic equatorial plane, producing a strong shock, an extended post-shock region and a high-density cooling disk. Field lines can support the increasing disk weight for no more than a month before they become significantly elongated in an effort to balance the gravitational plus centrifugal force, thereby generating strong field gradients across the disk. The associated current sheet eventually tears, forcing the field to reconnect through resistive diffusion, and causing the disk plasma to collapse towards the star. They proposed that this collapse is the cause for the recurrent Be episodes of β Cep, and discussed the applicability of the model to classical Be and normal non-supergiant B stars.

The underlying characteristic that Be stars rotate rapidly is elegantly supported by interferometric measurements of the rotational distortion of Achernar (α Eri, HD 10144), the brightest Be star. Domiciano de Souza *et al.* (2003) reported repeated measurements carried out during a two-month period in 2002, using Earth-rotation synthesis at the VLTI. The resulting angular

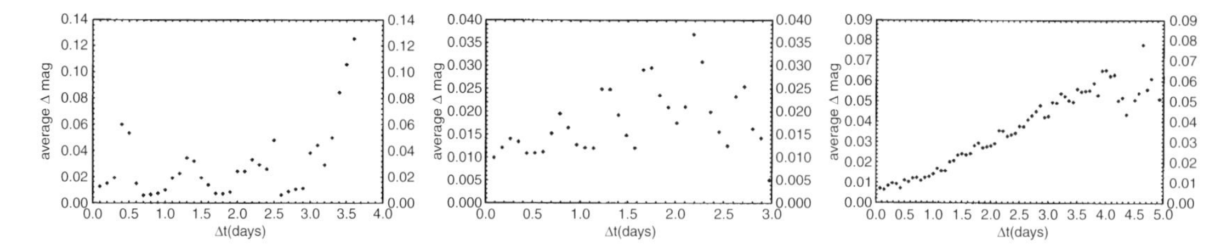

Figure 8.10 Self-correlation diagram of the Hipparcos Hp epoch photometry for three Be stars analysed by Percy et al. (2004). Left: HD 36408 (HIP 25950): minima occur at multiples of 0.9 d, defining the characteristic variability time scale. The points at $\Delta t > 3$ d are not significant (too few data points). Middle: HD 40978 (HIP 28783, V447 Aur): minima occur at multiples of 0.5 d. The points at $\Delta t > 2.5$ d are not significant. The rising trend in the maxima may be due to additional variability time scales of several days or more. Right: HD 58343 (HIP 35951, FW CMa): the rising trend is almost entirely due to one four-day monitoring of the star by Hipparcos, during which time the star varied significantly, suggesting an active star. The points at $\Delta t > 4$ d are not significant. From Percy et al. (2004, Figures 1–3).

diameters were combined with the Hipparcos distance ($d = 44.1 \pm 1.1$ pc), to derive the equatorial and apparent polar radii $R_{\rm eq} = 12.0 \pm 0.4 R_\odot$ and $R_{\rm pol} = 7.7 \pm 0.2 R_\odot$, respectively. Differentially rotating stellar models were developed by Jackson *et al.* (2004) to interpret this inferred rotational flattening.

Other studies of individual of Be stars making use of the Hipparcos data include: HD 6226: a possibly unrecognised Be star (Božić & Harmanec, 1998); photometry of HR 1960, showing periodicity at extremely small amplitudes (Burki, 1999); time-resolved spectroscopy of the peculiar Hα-variable HD 76534 (Oudmaijer & Drew, 1999); variability of NW Ser (Percy *et al.*, 1999); spectral and light variability of 60 Cyg (Koubský *et al.*, 2000); outbursts in HR 2501 (Carrier & Burki, 2003); long-term light and spectral variations of HD 6226 (Božić *et al.*, 2004); long-term variations and properties of κ Dra (Saad *et al.*, 2004); outbursts in HD 6226 (Slechta & Skoda, 2005); and the critically-rotating binary V360 Lac.

8.2.2 Shell stars

Be-shell stars are characterised by narrow absorption cores in addition to the broad photospheric absorption lines, and are now considered to be Be stars viewed equator-on (Rivinius *et al.*, 2006).

A-shell stars are A stars similarly characterised by the co-existence of two types of line profiles in their spectra: one originating in the photosphere of the rapidly rotating star, and the other originating in a cooler shell and comprising sharp low-ionisation absorption features in transitions such as Ca II or the Balmer series, especially Hα. Historically, β Pic was classified as an A-shell star. The work of Grady *et al.* (1996) suggests that field A-shell stars are the evolutionary successors of the pre-main-sequence Herbig Ae/Be stars, favourably oriented, with β Pic intermediate in age between pre-main-sequence and main sequence. Hauck & Jaschek (2000) used the Hipparcos data for 62 of their compilation of 76 A-shell stars. From their position in the HR diagram, they determined the luminosity class of each object, finding good agreement with that derived on the basis of Geneva photometry, and also finding that the majority are evolved (luminosity class I–III) rather than unevolved (luminosity class V) objects. Some 40% of the sample were binaries, and they pointed out similarities with the circumstellar components noted in some λ Bootis stars.

8.2.3 Chemically peculiar (Ap/Bp/Am stars)

Various studies of Ap stars have been made with the Hipparcos data in addition to those related directly to the rotation characteristics (Section 7.7.1) or magnetic field properties (Section 7.7.2).

North *et al.* (1997b) used the Hipparcos parallaxes to demonstrate that their absolute magnitudes are similar to those in clusters, and scattered across the whole width of the main sequence, except for the roAp stars. North *et al.* (1997a) estimated surface gravities from the corresponding evolutionary masses. Figueras *et al.* (1998) evaluated luminosity calibration based on Strömgren colour indices, deriving a new absolute magnitude calibration for late A-type main-sequence stars, taking into account the effects of evolution, metallicity, and rotation. Carrier *et al.* (2002b) derived orbits and masses for 16 cool magnetic Ap stars, finding similar mass ratios as for normal G dwarfs, but longer orbital periods.

Gómez *et al.* (1998a,b) made use of the full photometric and kinematic data (using the LM method, see page 213) for some 1000 Bp–Ap stars, to show that they belong to the young disk population (ages $\lesssim$ 1 Gyr), with evidence for the kinematic presence of the Pleiades, Sirius, and Hyades moving groups, with some found to be high-velocity objects. A search for the A stars progenitors of cool Ap stars was made by Hubrig *et al.* (2000, see Section 7.7.1).

Chemically peculiar (Ap and Bp) stars: There is a significant division in main-sequence properties around early-F stars. Cooler stars generate energy by the p-p cycle, have radiative cores, deep surface convection zones, rotate relatively slowly, and have generally weak magnetic fields. Hotter stars generate energy by the CNO cycle, have convective cores, little or no surface convection, and rotate relatively rapidly, and frequently have strong large-scale magnetic fields.

On and near the main sequence for the hotter stars are a large variety of spectroscopically or chemically peculiar (CP) stars, with a profusion of characteristics and names: Ap, Bp, Am, roAp, noAp, Si, SrTi, SrCrEu, HgMn, λ Boo, δ Scuti, ρ Pup, γ Dor, SPB, β Cephei, γ Cas, λ Eri, α Cyg, and more.

Strongly magnetic stars are known mostly as Ap (A peculiar) stars, although many of the Ap stars are B stars, sometimes referred to as Bp stars. The non-magnetic peculiar stars are known as Am (A metallic-lined) stars, whose sub-type classification can be strongly affected by the classification criterion: Balmer lines, Ca II K line, or metal lines. The Ap stars are variously classified (Preston, 1974) as He-weak, HgMn, Si, or SrCrEu stars, or as the subsets CP1 (Am); CP2 (magnetic Ap or Bp); CP3 (HgMn); and CP4 (He-weak B stars).

Models to explain these chemical peculiarities (which must be confined to a thin atmospheric layer) involve nucleosynthesis and dredging, surface spallation, magnetic accretion, planetesimal accretion, binary mass transfer, and diffusion. The latter invokes layers which are stable against turbulent mixing, in which heavy elements preferentially sink unless they have absorption lines intercepting the outward flux such that they are radiatively driven to the surface. Peculiar abundances arise as a result of these two competing effects. While this model has its weaknesses, it has been invoked to explain, amongst others: over-abundances of the Fe peak, rare-Earth, and lanthanide elements; under-abundances of Ca, Sc, C, and He in Am stars; isotope ratios of Hg in HgMn stars; slow rotation in Am and Ap stars; the binary nature of Am stars; and the fact that only a few Am and Ap stars pulsate (Kurtz, 2000).

Pöhnl *et al.* (2003) showed that the magnetic chemically peculiar stars of the upper main sequence already appear at very early stages of stellar evolution, before reaching 30% of their main sequence lifetime. For the 13 objects found in nearby open clusters, their derived ages of only 10–140 Myr are in agreement with those of the parent clusters.

Various photometric studies have also been reported. Adelman (1998) detected possible photometric variability in some objects, which was confirmed by Paunzen & Maitzen (1998), who attributed most of the newly-identified variability to an apparent magnetic field coupled with stellar rotation (the oblique rotator model). Variations of confirmed Am–Fm stars are explained by eclipsing binary systems. Catanzaro & Leone (2003) showed that variability in the equivalent width of the He I 587.6 nm lines is sometimes in phase, and sometimes in anti-phase, with the Hipparcos light-curves.

Studies of individual Am/Ap stars include: multiplicity in the Ap stars HD 8441 and HD 137909, and the Am stars HD 43478 and HD 96391 (North *et al.*, 1998); discovery of pulsations in the Am star HD 13079 (Martínez *et al.*, 1999); the equal-mass SB2 binary system 66 Eri (Yushchenko *et al.*, 1999); metal abundances of field A and Am stars (Hui-Bon-Hoa, 2000); Hβ photometry of southern chemically peculiar stars Maitzen *et al.* (2000); luminosities and Hβ calibration in Ap stars (Martínez *et al.*, 2000); the Ap binary HD 81009 (Wade *et al.*, 2000); abundances and chemical stratification in HD 204411 (Ryabchikova *et al.*, 2005); and photometry of HD 16545, HD 93226, HR 7575, and HR 8206 (Adelman, 2007).

A short review of the status of peculiar Am star research after Hipparcos was given by Gerbaldi (2003).

8.2.4 Flare stars

Flare stars are defined in the General Catalogue of Variable Stars as dwarf stars of spectral classes dM3e–dM6e characterised by rare and very short flares with amplitudes from 1 to 6 mag. Maximum brightness is attained in a few or several tens of seconds, the total duration of the flare being about 10–50 minutes. According to Gershberg *et al.* (1999), flare stars are stars on the lower part of the main sequence which show phenomena inherent to solar activity: sporadic flares, dark spots, variable emission from chromospheres and coronae, radio, X-ray and ultraviolet bursts. Gershberg *et al.* (1999) provide a catalogue of 463 flare stars, containing astrometric, spectral and photometric data as well as information on the infrared, radio and X-ray properties and general stellar parameters. The most well-known, UV Cet, was discovered in 1948 and, in consequence, flare stars are generally known as UV Ceti variables. Tens are known in the solar vicinity (including Proxima Centauri, and probably Barnard's star), and hundreds in the nearest stellar clusters. It has been suggested that all late-type stars go through a flare stage during their early evolution (König *et al.*, 2003), and flare stars may be weak-line, or post-T Tauri stars, or young zero-age main-sequence stars, the flare rates declining along the age sequence.

Saar (1998) used the Hipparcos parallax to argue that PZ Mon, long considered as a UV Ceti flare star, is actually a distant active giant. König *et al.* (2003) studied the TW Hya association, one of the closest known associations of young stars at about 60 pc. At the time of the study there were 19 known and confirmed members of the association, TWA 1–19. Hipparcos observed five confirmed members, along with three young flare stars located in the same region of sky (HIP 57269,

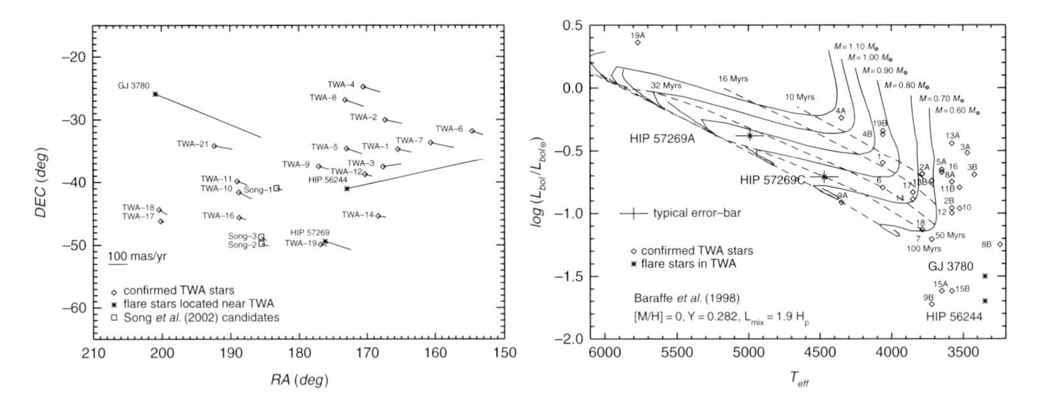

Figure 8.11 Left: proper motions of the known TW Hya association members and the three flare stars discussed by König et al. (2003): HIP 57269, GJ 3780 and HIP 56244. Proper motions were from Torres et al. (2003) for association members, from Song et al. (2002) for other candidates, and from the Hipparcos Catalogue or from the Simbad database for the flare stars. Song 1 corresponds to TYC 7760–0835–1, Song 2 to TYC 8238–1462–1, and Song 3 to TYC 8234–2856–1. Right: HR diagram of the three flare stars (HIP 57269 A, HIP 56244, and GJ 3780) and HIP 57269 C (⋆) together with the confirmed members of the TW Hya association. Theoretical tracks and isochrones are from Baraffe et al. (1998). Both HIP 57269 A and C lie above the main sequence, and on the same isochrone within the errors. $T_{\rm eff}$ are from Kenyon & Hartmann (1995), photometry from various sources, and distances to the individual objects was taken from Hipparcos or from Frink (2001) when possible; for the other stars a mean distance of 61.5 pc was used. GJ 3780 and HIP 56244 are unrelated foreground stars. From König et al. (2003, Figures 2 and 4). Figures copyright Wiley–VCH Verlag GmbH & Co. KGaA, and reproduced with permission.

HIP 56244, and GJ 3780), with distances around 46.7–103.9 pc. HIP 57269 shows strong lithium absorption with spectral type K1/K2V and a high level of chromospheric and coronal activity, and is located at a distance of 48.7 ± 6.3 pc. It shares the same proper motion as the association member HIP 57268, and was therefore proposed as a new association member candidate. It has a close companion, B, resolved visually by Tycho, and a wide companion, C, which also shows lithium absorption. HIP 57269 A and C lie above the main sequence, and are clearly pre-main-sequence stars, although the *UVW*-space velocity is more consistent with the star system being a Pleiades super cluster member (Figures 8.11). König *et al.* (2006) determined the properties of a further eight F-, G-, and K-type flare stars from the catalogue of Gershberg *et al.* (1999), estimating ages from their Hipparcos positions in the HR diagram. Expecting the flare stars to be young, they found instead that one of their sample was a young post-main-sequence giant, while all the others (including EK Dra and HN Peg) were main-sequence stars. They attributed the flare activity to interaction between differential rotation and the stellar magnetic field.

8.2.5 λ Bootis stars

Introduction The λ Bootis stars are a small class around spectral type A with anomalous abundances: moderate to extreme (up to factor 100) surface underabundances of most iron-peak elements (e.g. Mg, Ca, and Fe, yet with solar abundances of lighter elements (C, N, O, and S). Although the phenomenon now appears to be related to accretion, the details are incomplete, with the source of accreting material still to be unambiguously identified, and with the superposition of binary component spectra complicating the observational data.

For the prototype, λ Boo, the overall metal weakness compared to the normal H-line strength was first noted in the spectral classification survey by Morgan *et al.* (1943), and the first quantitative identification of significant metal deficiency was made by Burbidge & Burbidge (1956). The postulate that nuclear reactions had destroyed most of the heavier elements in the outer atmosphere (Sargent, 1965), was contradicted by its normal oxygen abundance (Kodaira, 1967). The λ Boo class are now considered to be Population I, typically late B to early F, with masses in the range 1.5–2.6 $M_\odot$, and moderate to high (projected) rotational velocities. Assignment to Population I is soundly based on trigonometric parallaxes which place them on or near the main sequence, proper motions and radial velocities which reflect the kinematics of young disk stars, and colours distinct from those of the more luminous, less massive A stars of Population II (Hauck & Slettebak, 1983; Venn & Lambert, 1990). That the abundance pattern is restricted to a surface phenomenon is inferred from the fact that they would fall below the relevant Population II ZAMS if they were metal-poor throughout. An initial absence of λ Bootis stars observed in open clusters was countered by the discovery of objects in the young cluster NGC 2264 and in the Orion OB1 association (Paunzen, 2001). The overall representation among stars in the spectral range B8–F4 is now considered to be about 2% in both the Galactic field as well as in open clusters.

In the last 10–15 years, three models have been advanced to explain the phenomenon (see summaries in Paunzen *et al.*, 2003; Paunzen, 2004): diffusion accompanied by mass-loss, diffusion accompanied by accretion, or various binary models.

The diffusion model has been successful in explaining the Am/Fm stars, Population I non-magnetic main-sequence stars with under-abundances of Ca and Sc, but large over-abundances of most heavier elements. Gravitational settling of He leads to the disappearance of the outer convection zone associated with He ionisation. However, even moderate equatorial rotation of $\sim$ 50 km s^{-1} probably eliminates under-abundance patterns due to meridional circulation in the stellar atmosphere.

The diffusion/accretion model explains surface abundance anomalies by accretion of metal-depleted interstellar or circumstellar gas. It is a characteristic of interstellar gas that different metals are depleted by different amounts due to their condensation in dust grains (Spitzer, 1998), and the model was proposed following the observed similarity between the λ Bootis abundances and the depletion pattern of the interstellar medium. Venn & Lambert (1990, e.g. their Figure 6) show how accretion from a mixture of interstellar and solar-abundance gas can explain why C, N, O, and S have similar and near normal abundances, while the metals can be under-abundant by a nearly uniform factor that varies from star to star. Their Figure 7 shows how the abundances correlate with the elemental first ionisation potential. Accretion could be from a surrounding disk or shell, with required accretion rates of $\gtrsim 10^{-14} M_\odot$ yr^{-1}, or from a diffuse interstellar cloud, with accretion rates of 10^{-14}–10^{-10} $M_\odot$ yr^{-1} depending on cloud density and relative impact velocity (Kamp & Paunzen, 2002). In either case, ongoing accretion is implied by the short mixing time scales of $\lesssim 10^6$ yr. For stars less massive than A-type, the deeper convective envelopes will quickly dilute any accreted material.

In the binary-based models, spectral peculiarities arise in one of two ways. They could be a result of merging through dynamical evolution of W UMa contact binary systems, before both stars finish their main-sequence lifetimes, and with progenitors each having masses between 0.8–1.5 $M_\odot$. Alternatively, they could be artifacts of undetected spectroscopic binaries (Faraggiana & Bonifacio, 1999; Gerbaldi *et al.*, 2003): the composite spectrum of two normal, solar-abundant, stars with different effective temperatures and gravities

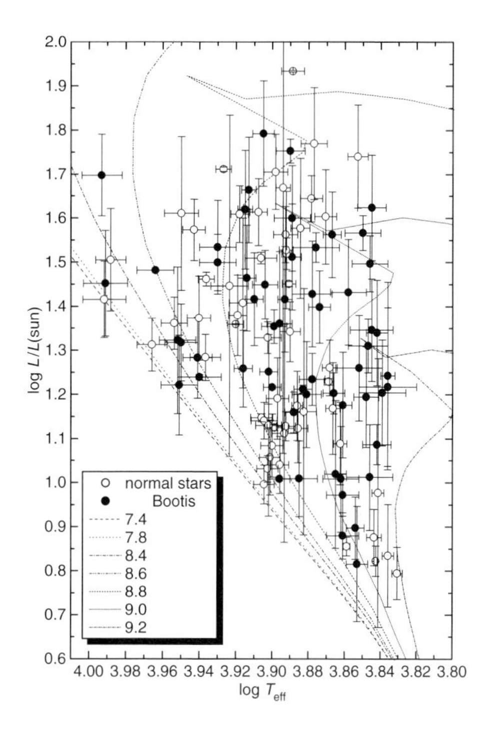

Figure 8.12 λ Bootis stars in the HR diagram, showing the location of the λ Bootis stars (filled circles) and a sample of normal stars (open circles). From the ages of post-evolutionary isochrones (the legend indicates log t in years), no object lies below the ZAMS, providing confidence in the models, while the λ Bootis stars are found at evolutionary stages throughout the main sequence with a peak at about 1 Gyr. The temperature range depicted roughly corresponds to the borders of the instability strip. From Paunzen et al. (2002b, Figure 1).

will have a metal-weak character, further amplified for components with different rotational velocities, although the mimicking of apparent under-abundances of up to −2 dex at the appropriate location in the HR diagram has not been demonstrated.

Hipparcos contributions The Hipparcos data have been applied in a variety of ways, using the photometric and astrometric data as indicators of binarity and/or pulsation, distance and proper motion data to constrain space velocities and hence population, and distance (and photometry) for absolute magnitude, mass and age (using bolometric corrections and theoretical evolutionary tracks), since mass loss coupled with diffusion and accretion predict widely different stellar age distributions. The composite spectra of binary pairs, partly identifiable using the Hipparcos double star data, complicates the classification of the phenomenon, and may explain at least part of the observed abundance anomalies (Faraggiana & Bonifacio, 1999; Gerbaldi *et al.*, 2003; Gerbaldi & Faraggiana, 2007).

In a study soon after the catalogue release, the absolute magnitude and evolutionary status of a subset of 38 Hipparcos objects placed them very close to the main sequence (Paunzen, 1997; Heiter *et al.*, 1998; Paunzen *et al.*, 1998a) leading to the suggestion that all the other stars were also in their pre-main-sequence phase, although very close to the main sequence. This was revised by subsequent analyses, which now support the view that Galactic field λ Boo objects are found at all evolutionary stages (Figure 8.12), with age ranges between 0.2–1 Gyr (Faraggiana & Bonifacio, 1999; Paunzen *et al.*, 2002b). This would seem to discount an earlier picture in which material is from accretion disk remnants (Waters *et al.*, 1992), which although present during the pre-main-sequence phase, would thereafter disappear within about 400 Myr. A fraction of pre-main-sequence objects is not, however, excluded.

Since most of the members are located within the classical instability strip, knowledge of their pulsational properties could provide information on their overall occurrence and evolutionary status, using asteroseismology to determine their masses and ages, and thereby confronting competing theories for their origin (Paunzen *et al.*, 1998b,c). Paunzen *et al.* (1998b) summarised the photometric monitoring of 52 such stars, including λ Bootis itself, classifying 22 as pulsating and 30 as non-variable, of which a total of 42 have Hipparcos parallaxes. Four were discovered as newly-pulsating, with magnitudes in the range 4.2–8.9 mag, further evidence that the photometric properties of even the brightest stars in the sky remain incompletely explored. The 17 pulsating stars have periods in the range 0.5–3.8 hr and amplitudes in the range 2–38 mmag, yielding a ratio of at least 50% for variable to non-variable members within the instability strip. Paunzen *et al.* (1998b, Figures 3 and 4) show the location of the stars in the observational (M_V versus $b-y$) and the theoretical ($T_{\rm eff}$ versus $L/L_\odot$) HR diagrams, along with the corresponding borders of the instability strip, the ZAMS, and evolutionary tracks corresponding to 1.5–2.5 $M_\odot$. Such analyses are possible only because of the availability of accurate distances which in turn yield accurate absolute magnitudes. The location of the pulsating stars in the log P versus log $\overline{\rho}/\overline{\rho}_\odot$ (period–density) diagram (their Figure 5), compared with the log Q values for the different radial modes from Breger (1979), is consistent with the (solar-abundant) δ Scuti stars, and thus supports the hypothesis that the λ Bootis abundance pattern is restricted to the stellar surface. Excited modes range from the fundamental in a few possible cases, up to high overtones, such that the pulsation properties of these two groups also appear indistinguishable, this similarity extending to the period–luminosity–colour relation (Paunzen *et al.*, 2002a). Variability, perhaps related to non-radial pulsations, was reported in the Hipparcos epoch photometry for HD 111604 by Jordan *et al.* (1997) and for HD 84948 by Iliev *et al.* (2002).

Paunzen *et al.* (2002b) examined the kinematics of 36 nearby ($d < 220$ pc) λ Bootis stars using Hipparcos proper motions and parallaxes, radial velocities from the literature, and the resulting heliocentric Galactic space-velocity components (U, V, W), corrected for solar motion. Comparison with the relationship between kinematics, age and heavy-element content for the solar neighbourhood from Caloi *et al.* (1999), also based on the Hipparcos data, confirms that the space velocities are typical of Population I objects. A similar analysis by Faraggiana & Bonifacio (1999), using a different sample, and restricted to the parameter $(U^2 + W^2)^{1/2}$, a measure of the kinematic energy not associated with Galactic rotation, also implied that all of their stars were members of the disk population.

Details of the λ Bootis phenomenon remain uncertain. Small in number, the class is valuable for observational tests of accretion, diffusion, mass-loss and asteroseismology. Knowledge of their evolutionary state and kinematical distributions have been expanded.

8.3 X-ray sources

Introduction X-ray emission from late-type stars is generally attributed to a magnetically-heated stellar corona. In the case of the Sun, the hot ($> 10^6$ K) emitting plasma is confined by coronal magnetic fields, which originate from the interaction between rotation and outer convection zones through the dynamo mechanism. Its X-ray luminosity varies between $\sim$ 3–100 $\times$ 10^{25} erg s^{-1} during the solar cycle, and is known to be strongly correlated with other magnetic activity indicators, notably sunspot numbers, flare frequency, and chromospheric Ca II emission. In principle, any late-type star should be able to sustain a corona and, thus, be an X-ray source.

The X-ray emission level and temperature of the coronal plasma vary with evolutionary changes in the star's rotation and internal structure. The current picture indicates that intermediate-mass 2–3$M_\odot$ G giants in the Hertzsprung gap may reach high X-ray emission levels 10^2–10^3 times higher than a solar-like G dwarf, while lower-mass evolved stars show only weak magnetic activity during the shell-hydrogen burning phases, eventually fading out as X-ray sources in the red giant phase. Evolved stars do not follow the activity-rotation relationship observed for late-type dwarfs. Although their coronal activity appears mass-dependent, this dependency is less than that expected from theoretical models of convective zone evolution in the framework of a solar-type magnetic dynamo.

The early Einstein observations revealed that stars of almost all spectral types were X-ray emitters, with luminosities in the range 10^{26}–10^{34} erg s^{-1} (Vaiana *et al.*, 1981). X-ray surveys have since been recognised as a powerful method of selecting young active late-type stars (Feigelson *et al.*, 1987). Those found near to star-forming regions, from the Einstein Medium Sensitivity Survey and the Extended Medium Sensitivity Survey, were mostly classified as weak-line T Tauri stars, i.e. still in the pre-main-sequence phase, but with optical characteristics less extreme than those of the classical T Tauri (Section 8.1.2). X-ray selected stars detected in the high-latitude surveys from Einstein, away from known star-forming regions, have mostly been classified as young stars already on the main sequence, with ages of the same order as the Pleiades, although their precise evolutionary state was difficult to assess due to the lack of reliable distance information. Similar conclusions were drawn from the High Latitude X-ray EXOSAT survey (Tagliaferri *et al.*, 1994), and from the ROSAT Wide Field Camera Bright Source Catalogue (Hodgkin & Pye, 1994), but it was unclear whether the two stellar populations really represented distinct classes of stars.

ROSAT also carried out the sensitive ROSAT All-Sky Survey, providing a flux-limited but otherwise unbiased sample of X-ray sources (Voges, 1992). About one third of the $\sim 1.5 \times 10^5$ detected sources are considered to be coronal, with optically faint counterparts, while a smaller number have bright optical counterparts. Studies have shown the existence of a large number of young stars located both in or near to known sites of star formation (e.g. Alcalá *et al.*, 1995; Sterzik *et al.*, 1995; Wichmann *et al.*, 1996), but they have also been detected in more extended areas of the sky. These stars were considered to be weak-line T Tauri stars ejected from the parent star-formation region but, due to their large distances, they posed a challenge to the understanding of star-formation processes (Neuhäuser *et al.*, 1995; Gorti & Bhatt, 1996). Subsequently Briceño *et al.* (1997) argued that the more distant objects were instead young main-sequence stars of ages up to $\sim 10^8$ yr, with little evidence from the All-Sky Survey results for the existence of a post-T Tauri population.

The main problem in assessing the nature of the X-ray selected stars is that the weak-line T Tauri stars have characteristics, both in X-ray and in optical, very similar to those of the young main-sequence stars. The most important characteristic which can be used to segregate the two populations is their evolutionary status given by their position in the HR diagram. The following discussion of the Hipparcos contribution to the study of X-ray sources is organised along a series of broad themes, but with some overlap between them.

Surveys and young stars Some specific Hipparcos studies related to young stars, including some related to X-ray emission, are considered in Section 8.1.2.

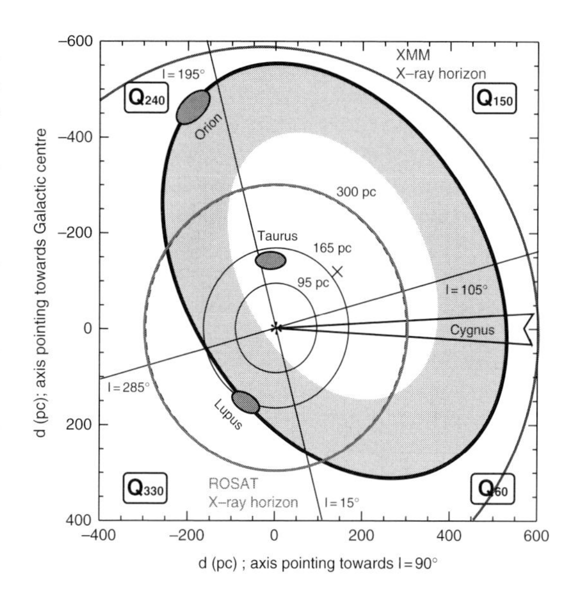

Figure 8.13 Geometry of the Gould Belt projected onto the Galactic plane. The ellipsoidal ring has semi-major and minor axes of 500 and 340 pc respectively. The Sun (⋆) is displaced from the structure's centre (×), which is located at about 200 pc towards $l = 130°$. Gould Belt members are assumed to be located near the outer edge of the belt (solid thick curve). The concentric circles of radius 95, 165 and 300 pc centred on the Sun show the X-ray horizon at a PSPC count-rate of 0.03 cts s^{-1} for X-ray luminosities of $\log L_X = 29.5, 30.0, 30.5$ erg s^{-1} respectively. 160 pc is also the optical horizon of a G5 ZAMS star at the Tycho limit of 10.5 mag. The grey shaded area sketches an alternative picture of a 'Gould Disk' in which members are spatially distributed between the inner and outer rings. Typical ROSAT and XMM horizons are included. From Guillout et al. (1998a, Figure 2).

Micela *et al.* (1997) used the Hipparcos parallaxes for a sample of 84 X-ray selected stars from the Einstein Extended Medium Sensitivity Survey, all at Galactic latitude $|b| > 20°$, generally far from known star-forming regions, and typically with the smallest ratio of X-ray to optical flux. They found that the majority are on the main sequence, with some 20% giants, and an absence of true pre-main-sequence stars. The fact that several showed lithium abundance and activity levels typical of very young stars was consistent with them being young main-sequence objects with ages comparable to the Pleiades, of around 60 Myr. The X-ray luminosities of the main-sequence stars, for the first time based on the accurate Hipparcos distances, were found to be very similar or slightly lower than those of Pleiades stars of similar spectral type.

Guillout *et al.* (1998b) studied the large-scale distribution of a sample of the more active X-ray stars from the ROSAT All-Sky Survey. At the typical limiting flux of $\sim 2 \times 10^{-13}$ erg cm^{-2} s^{-1}, the Sun could be detected at a

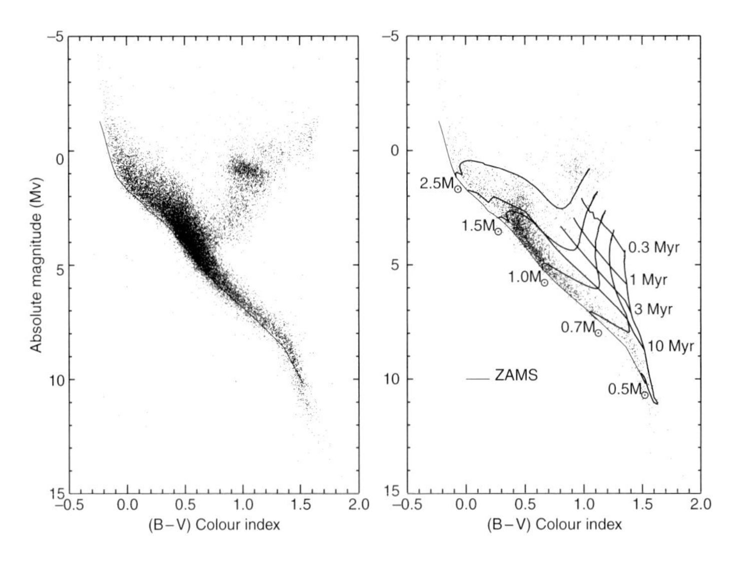

Figure 8.14 Observational HR diagram of the 19 350 Hipparcos stars (left) and 3407 ROSAT All-Sky Survey–Hipparcos stars with accurate distances, $\sigma_\pi/\pi \leq 0.1$, and colours, $\sigma_{B-V} \leq 0.025$ (right). The ZAMS is shown in both panels, while the right panel also includes pre-main-sequence evolutionary tracks (0.5–2.5$M_\odot$) and isochrones (0.3–10 Myr), all from Siess et al. (1997). From Guillout et al. (1999, Figure 5).

distance of about 10 pc, while highly-active stars with X-ray luminosities a factor of 1000 higher can be detected to about 300 pc. Such objects are of typically low mass, $< 2M_\odot$, rapidly rotating, and of young age, $\lesssim 10^7$ yr. From 8593 X-ray emitting stars cross-correlated with the Tycho Catalogue, they detected a density gradient decreasing from the Galactic plane to the Galactic pole, which they attributed to the scale height of the young late-type star population of the Galactic disk. They also detected a low Galactic latitude density enhancement with respect to the mean plane density, for which they derived a best-fit inclination $i = 27.5 \pm 1^\circ$ and an ascending node $l_\Omega = 282 \pm 3^\circ$ with respect to the Galactic plane. They discussed the Gould Belt as a possible explanation for the observed enhancement. Although the nature of the Gould Belt has long been debated (see page 325), studies at X-ray wavelengths can enhance the young stellar structures which are less pronounced at optical wavelengths. Guillout *et al.* (1998a) extended the analysis by taking into account the distance information of some 6200 stars contained in the Hipparcos Catalogue. They found that the excess stars can be understood as part of a coherent structure filling a large fraction of the solar vicinity, geometrically in agreement with a standard but somewhat modified picture of the Gould Belt (Figure 8.13): extending from the solar vicinity up to about 300 pc towards a quadrant centred on $l = 240^\circ$, but with a distribution extending to only 180 pc towards $l = 330^\circ$. The structure could also be understood as a disk-like arrangement of stars having the same inclination towards the Galactic plane as the Gould Belt, and extending to its outer boundary. They suggested that these stars are the residuals of original associations, and interpreted them as the late-type stellar component of the Gould Belt.

Hünsch *et al.* (1998a) presented data for all main sequence and subgiant stars of spectral types A, F, G, and K and luminosity classes IV and V listed in the Bright Star Catalogue (Hoffleit & Jaschek, 1991) that have been detected as X-ray sources in the ROSAT All-Sky Survey. The catalogue contains 980 entries, yielding an average detection rate of 32%, providing count rates and X-ray fluxes, source detection parameters, hardness ratios (an X-ray 'colour' influenced by the plasma temperature and the hydrogen column density), and X-ray luminosities derived from Hipparcos distances. Similar analyses based on the Bright Star Catalogue, but without the use of the Hipparcos parallaxes, had been previously made for OB stars (Berghöfer *et al.*, 1996), and late-type giants and supergiants (Hünsch *et al.*, 1998b).

Guillout *et al.* (1999) constructed the cross-correlation of the ROSAT All-Sky Survey with the Hipparcos and Tycho Catalogues, yielding the ROSAT All-Sky Survey–Hipparcos sample of 6200 matches, and the ROSAT All-Sky Survey–Tycho sample of 13 875 matches. While the ROSAT survey provides the opportunity to study young stars in the solar neighbourhood, cross-correlation with the Hipparcos and Tycho Catalogues provides access to many of the detailed properties of these X-ray emitters. The X-ray survey probes distances up to about 200 pc for F–G stars younger than 100 Myr, and to 80 pc or less for older stars. The

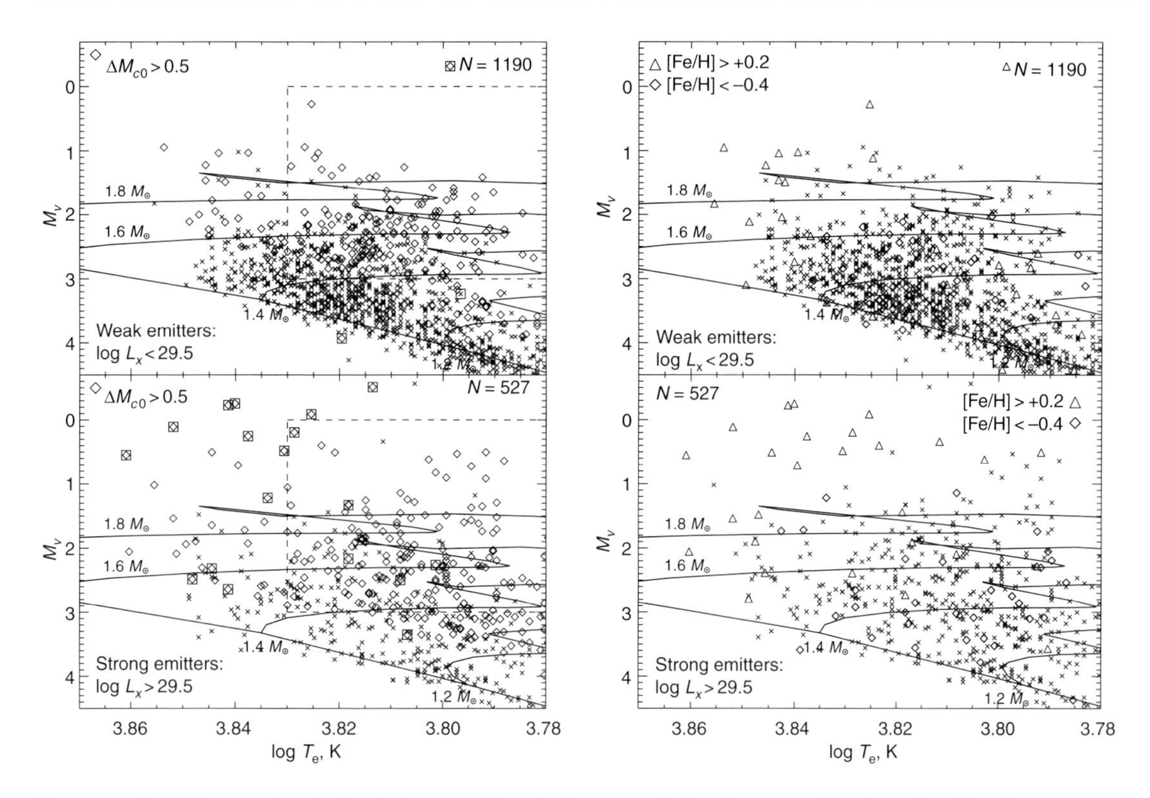

Figure 8.15 Left: M_V versus log $T_{\rm eff}$ for single stars, with evolutionary tracks for solar composition from Girardi et al. (2000). Boxed symbols indicate stars with reddening $E(b - y) \geq 0.04$. The fraction of strong X-ray emitters (bottom) among stars high above the ZAMS is substantially higher than that of moderate/weak X-ray emitters (top). Over-luminous stars are more conspicuous in the bottom panel, with many reddened stars at high temperatures and luminosities. These properties are consistent with the hypothesis that most of the over-luminous strong X-ray emitters at bright absolute magnitudes are either pre-main-sequence or anomalously-evolving stars. Right: as left, but illustrating metallicity effects. Metal-poor stars among strong X-ray emitters are mostly old and well evolved, illustrating that X-ray emission does not necessarily decline with age. From Suchkov et al. (2003, Figures 8–9).

magnitude limit of the optical catalogues determines the horizon for K–M stars which are sampled only within about 50 pc or less of the Sun, independent of age. The Hipparcos and ROSAT All-Sky Survey–Hipparcos HR diagrams are shown in Figure 8.14. X-ray selection strengthens delineation of the ZAMS, but evolved stars are also detected. Detection rate, mean $F_{\rm X}/F_{\rm opt}$, and X-ray luminosity could be computed with unprecedented colour resolution for regions both on the main sequence (i.e. between the ZAMS and the TAMS, luminosity class V) and off the main sequence (i.e. above the TAMS, luminosity class III). Corrected for $F_{\rm X}/F_{\rm opt}$ bias, the detection rate is remarkably constant for G-M main-sequence stars, but reveals a peak of detection for F-type stars. This may indicate that the heating of the coronae is a very sensitive mechanism, and that both internal structure and physical properties of main-sequence F stars strongly favour the efficiency of the dynamo. Detection rates for A-type stars, where effective temperatures are too low for radiatively driven winds but too high for deep envelope convection, are compatible with those computed for F-M stars, as expected if a late-type companion is responsible for the X-ray emission. High-mass stars evolving along the post-main-sequence evolutionary tracks are clearly detected in the main-sequence turn-off, and in the blue part of the K giant 'clump', while no significant detection arises on the cool side. They noted that the database has considerable potential for studies of star formation rate, of the initial mass function below $1M_\odot$, of the scale height of young stars, and of the distribution and dynamics of young stars in the solar neighbourhood.

Suchkov *et al.* (2003) explored the relationship with X-ray emission detected in the ROSAT All-Sky Survey. For 11 900 Hipparcos F stars with Strömgren *uvby* photometry, they identified 1980 X-ray counterparts (17%) in the ROSAT All-Sky Survey. The small fraction of highest X-ray luminosity sources, $\log L_{\rm X} > 30.4$, includes a variety of stellar types, dominated by very young objects, but including RS CVn stars and other active binaries.

The sample as a whole consists of at least three distinct groups of F stars that differ in X-ray luminosity, and in the evolution of their X-ray emission. The first, roughly half of the sample, is represented by normal zero-age main sequence and post-ZAMS stars whose X-ray emission declines with age, as in the conventional picture of stellar X-ray evolution. The second group consists of strong X-ray emitters that appear to be very young, possibly pre-main-sequence stars, characterised by high metallicity, high temperature, and high optical luminosity, many of which are located in the distance range 130–200 pc, near to known regions of ongoing star formation or OB associations. The third group also consists of strong X-ray emitters which are extremely over-luminous in the optical. They have a higher mean tangential velocity and velocity dispersion (derived from the Hipparcos astrometry), implying that these stars are on average the oldest rather than the youngest, most having evolved significantly away from the ZAMS. Their existence challenges the idea that X-ray emission continuously declines with age. They speculated that this old population of optically over-luminous strong X-ray emitters develops an extensive outer convective zone as they evolve, capable of maintaining a high level of coronal activity, and hence X-ray emission, despite decaying rotation velocity (Figure 8.15).

Cutispoto *et al.* (1999) presented high-precision $UBV(RI)_C$ photometric observations and spectroscopic radial velocity measurements for 51 cool stars detected in the extreme ultraviolet by the ROSAT Wide Field Camera. They used Hipparcos data to infer spectral types, as well as the single or binary nature of the sample stars. Similarly, Cutispoto *et al.* (2000) presented photometric and spectroscopic studies of 32 cool (northern) stars discovered in EXOSAT X-ray images. Boese (2004) estimated the positional precision of the ROSAT All-Sky Bright Source Catalogue by cross-correlation with the Tycho Catalogue.

Radial velocity studies of X-ray emitters from the ROSAT All-Sky Survey–Tycho sample have been reported by Frasca *et al.* (2006b) and Frasca *et al.* (2006a).

Post-main-sequence phases Schröder *et al.* (1998) studied a volume-limited sample of 36 single giants within 35 pc of the Sun, complete to $M_V \lesssim 3.0$ and for X-ray luminosities $L_X \gtrsim 1.5 \times 10^{28}$ erg s^{-1}. They used ROSAT data to determine stellar activity, Hipparcos parallaxes to place stars in the HR diagram, and evolutionary models of Pols *et al.* (1998) to derive individual masses and ages (Figure 8.16). They confirmed the suggestion by Hünsch & Schröder (1996), that stellar activity evolution is strongly coupled to stellar mass, and that X-ray emission is a very common feature among giants with $M \gtrsim 1.3M_\odot$. Most pointed ROSAT observations for objects on the giant branch and in the K giant clump,

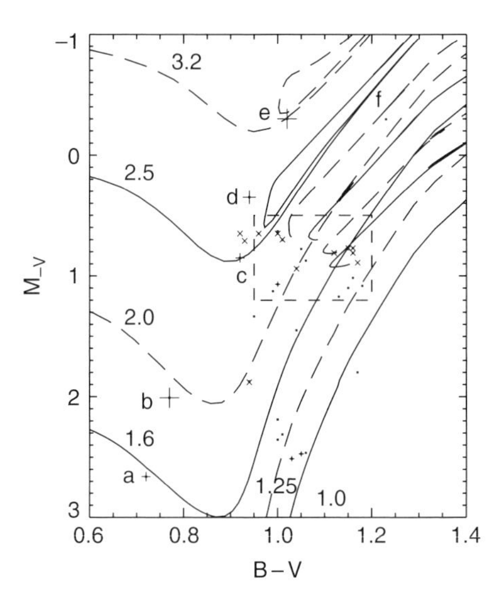

Figure 8.16 ROSAT detections of late-type giants, with evolutionary tracks. The K giant clump region is indicated and certain stars of interest are labelled: a: κ Del, b: 24 UMa, c: η Her, d: ϵ Vir, e: β Cet, f: α Boo. +: ROSAT All-Sky Survey detections, ×: pointed observations, and the symbol sizes indicate X-ray luminosity; dots are X-ray non-detections. From Schröder et al. (1998, Figure 1).

with masses between about 1.3 and $2.3M_\odot$, resulted in detections at typically solar levels. This indicates that for $M \gtrsim 1.3M_\odot$ magnetic activity mostly survives the He-flash, and possibly also persists onto the asymptotic giant branch. The more massive stars, $M \gtrsim 3M_\odot$, show even larger levels of activity in their advanced evolutionary stages.

Pizzolato *et al.* (1998, 2000) investigated the variation of coronal X-ray emission with age during early post-main-sequence phases for a sample of 120 late-type stars within 100 pc. Changes are expected to be relatively rapid in the evolutionary phases across the Hertzsprung gap and on the giant branch, accompanying the development of a deep convection zone, the rapid expansion of the stellar radius with the associated increase of moment of inertia and decrease of surface gravity, the significant drop of surface rotational velocity for intermediate-mass stars, and the drop in photospheric temperature. They selected a stellar sample with estimated masses in the range 1–$3M_\odot$, based on Hipparcos parallaxes and recent evolutionary models. These stars were observed with ROSAT/PSPC pointed observations, providing a sensitivity limit some two orders of magnitude fainter than observations in the All-Sky Survey. Their sample suggests a trend of decreasing X-ray emission with age for the lower mass stars, $M < 1.5M_\odot$, which have spent most of their main-sequence lives as X-ray emitters, but an increasing X-ray emission level

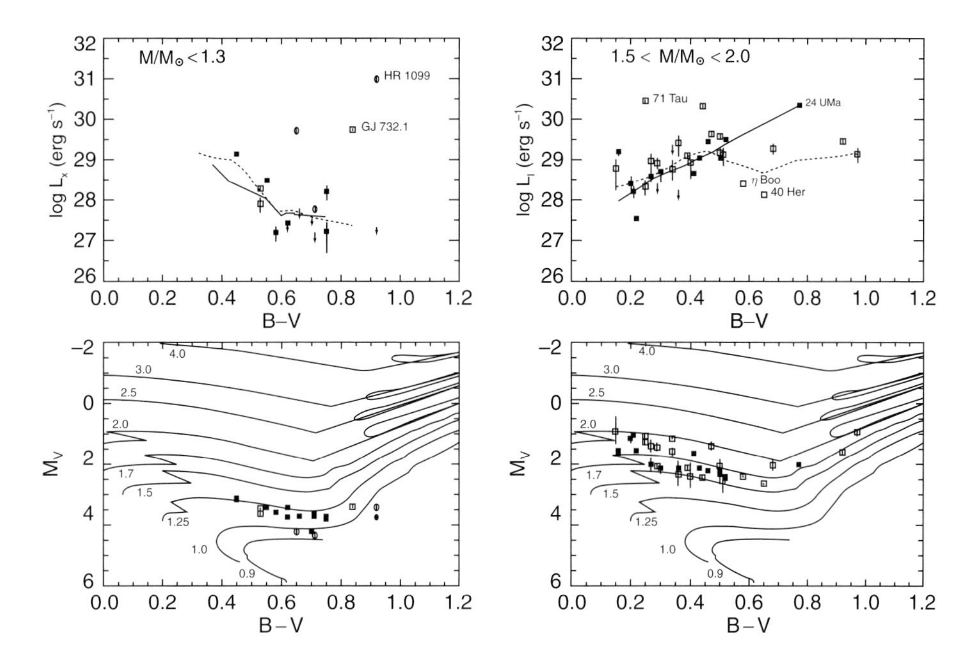

Figure 8.17 Left: X-ray luminosities versus B–V colour for single and resolved binary stars (filled symbols) and unresolved binaries (open symbols), for $M < 1.3M_\odot$; ◇: stars with estimated masses $< 1M_\odot$ included for comparison (errors on the X-ray luminosities are often smaller than the symbol size). Regression lines are for single (solid) and binary (dotted) stars. The bottom panel shows the corresponding HR diagram, including the evolutionary tracks from Schaller et al. (1992). Right: similar plots for masses in the range 1.5–2$M_\odot$. From Pizzolato et al. (2000, Figures 8–9).

with age for stars with masses $M > 1.5M_\odot$, which have been inactive A-type stars on the main sequence (Figure 8.17). A similar behaviour holds for the average coronal temperature, which follows a power-law correlation with the X-ray luminosity, independent of mass and evolutionary state. Their results are interpreted in terms of a magnetic dynamo whose efficiency depends on the stellar evolutionary state through the mass-dependent changes of the stellar internal structure, including the properties of envelope convection and the internal rotation profile.

Gondoin (1999) used the Hipparcos data in combination with published Einstein fluxes and rotational velocities for 88 apparently single G-K giants. He confirmed the existence of a sharp decrease of X-ray emission at spectral type K1 for giants with masses between 2.5 and $5M_\odot$, showing that their rotational velocity reaches a minimum at the same location in the HR diagram (see Section 7.7.1). He suggested that the results could support the importance of differential rotation in determining the level of coronal emission among $M > 2.5M_\odot$ G and K giants; the reversion to rigid rotation at the bottom of the red giant branch could prevent the maintenance of large-scale magnetic fields, thus explaining the sharp decrease of coronal X-ray emission at spectral type K1.

Maggio *et al.* (1998) reported a specific study of β Cet, the single giant with the highest X-ray luminosity in the solar neighbourhood, and making use of the Hipparcos distance of $29.38^{+0.63}_{-0.69}$ pc. Considered as a probable He-burning clump giant, its relatively high X-ray activity level is poorly understood. Smith & Shetrone (2000) obtained spectra covering the Ca II K line for 64 field red giants with X-ray emission detected by ROSAT. The majority of non-interacting giants in the sample have $M_V > -2.0$, as determined from the Hipparcos parallaxes. The X-ray and Ca II K-line data indicate that the giants of spectral types G and early K have coronae and chromospheres seemingly analogous to those of the Sun.

Binary fraction A large rate of visual binaries among strong X-ray sources has been reported based on the Tycho 2 Catalogue correlation for both the Hyades (Makarov, 2000) and the Pleiades (Makarov & Robichon, 2001). In the Hyades, 15 out of 32 Tycho 2 stars with $L_X > 1.3 \times 10^{29}$ erg s^{-1} are long-period binaries with separations between 5 and 50 AU, yielding a fraction of resolved components of at least 0.64. The Pleiades contain relatively fewer visual binaries, but the correlation is rather pronounced: 14 binary and multiple systems out of 17 were detected by ROSAT, with a resulting fraction of resolved components of 0.55 at $\log L_X > 29.4$, or 0.71 after including known spectroscopic binaries.

Makarov (2002) compared the occurrence of resolved visual doubles among the brightest X-ray stars from the ROSAT All-Sky Survey Bright Source Catalogue, and in the Hipparcos and Tycho Catalogues. A strong correlation between visual binarity and brightness in X-rays was found, with the rate of binarity, quantified as the fraction of resolved components, being seven times higher for the ROSAT Bright Source Catalogue stars than generally for all Tycho 2 stars. For a distance-limited sample of Hipparcos stars within 50 pc, the fraction of visual and unresolved binaries among the X-ray emitters is 2.4 times higher than among the fainter (or non-emitting) X-ray stars. This correlation between binarity and X-ray luminosity for field stars tends to confirm the results found for the Hyades and for the Pleiades. Since the rate of binarity is known to correlate well with average age (e.g. Reipurth & Zinnecker, 1993; Padgett *et al.*, 1997), the results may be ascribed to an underlying binarity–age or binarity–rotational velocity dependency, due to the fact that binaries are more frequent among young stars, while young stars are generally stronger X-ray sources, probably because they rotate faster. This binarity–age relation in which unresolved binary stars are significantly younger than single stars, was also found for F stars by Suchkov (2000). The HR diagram of components of nearby visual binaries (Figure 8.18) also indicates that the red clump giants luminous in X-rays are shifted blue-wards with respect to the average population of the red clump: they may be unresolved binaries, or giants younger than the other clump stars.

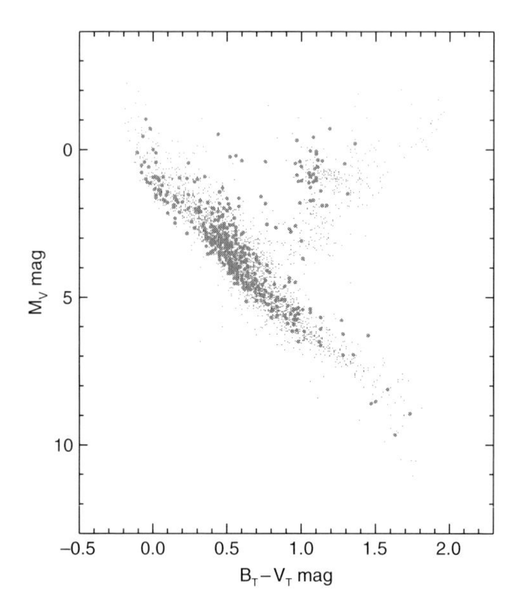

Figure 8.18 The HR diagram for the components of double stars resolved in the Tycho Double Star Catalogue, with Hipparcos parallaxes $\pi > 20$ mas and $\sigma_\pi < 1.8$ mas (small points). Components of binaries detected by ROSAT with $\log L_X > 29.3$ erg s^{-1} are indicated as larger filled circles. The majority of the resolved X-ray components follow the main sequence rather closely. X-ray luminous red giants are shifted bluewards with respect to the average population of the red clump. From Makarov (2002, Figure 2).

Nearby stars Hünsch *et al.* (1999) made a compilation of data for the ROSAT All-Sky Survey sources contained within the Third Catalogue of Nearby Stars (CNS3, Gliese & Jahreiß, 1991). The resulting catalogue contains 1252 entries, yielding an average X-ray detection rate of around 33%.

Makarov (2003) used the Hipparcos and Tycho 2 Catalogues to define the 100 most luminous X-ray stars within 50 pc of the Sun. The stars are mostly well studied, allowing the characteristics of different categories of emitters to be established, based on X-ray luminosity, kinematics, and binarity. In particular, the objective was to find out if very young stars in the solar neighbourhood can be selected in the absence of spectroscopic observations. He classified the stars into the following categories, notwithstanding the somewhat arbitrary nature of some of the boundaries: short-period spectroscopic binaries of RS CVn type (mostly stars with rapid rotation synchronous with the orbital motion, often with at least one evolved, giant or subgiant, component); other short-period spectroscopic or eclipsing binaries (typically with both components on the main sequence, including binary stars of BY Dra type, semi-detached binaries or Algols, and eclipsing binaries of β Lyr type); pre-main-sequence stars, post-T Tauri stars and very young main-sequence stars, typically younger than the Pleiades ($\sim$ 60 Myr); evolved (luminosity class IV and III) single or long-period binary stars; single or long-period variable stars of BY Dra type, rapidly rotating, and chromospherically active; contact binaries of WU UMa type; and emitters of unknown nature.

Makarov (2003) found that some 40% of stars with $L_X \gtrsim 10^{30}$ erg s^{-1} are short-period, mainly spectroscopic binaries. Their activity is probably underpinned by the fast rotation maintained by the transfer of orbital momentum. Some of the active short-period binaries include evolved stars, such as the RS CVn type objects, apparently kinematically very old. Around 10% are single or long-period evolved stars (giants and subgiants), often with fairly slow rotation, and whose activity cannot be explained by dynamical interaction between components in binary systems. Giants are significantly harder X-ray sources than main-sequence stars. Young stars, whether single or binary, comprise only a third of the X-ray population and, confirming other findings, cannot be identified purely on the basis of X-ray brightness.

Main-sequence active binaries also have very similar X-ray properties to young stars. In contrast, the kinematic properties may provide a more robust age diagnostic: young stars within 50 pc have very orderly patterns of motion at moderate velocities with respect to the LSR. Moreover, the majority of nearby stars seem to be members of the Pleiades stream, with the mean velocity of $(-9.6, -21.8, -7.7)$ km s^{-1} (cf. Montes *et al.*, 2001), and only a few may be members of other kinematic groups identified in the more extended solar neighbourhood. Most of the nearest young stars belong to sparse, probably expanding mini-associations.

The study confirmed the high rate of binaries among young stars, but the cause of the empirical relation between the rate of well-separated, long-period binarity and X-ray activity is far from obvious. It also found a large number of hierarchical multiple systems within the RS CVn category. These have somehow attained high space velocities with respect to the LSR, whilst surviving several Gyr of dynamical evolution. At larger distances, $\sim 50-200$ pc, a larger fraction of very young stars will appear due to the presence of massive OB associations and star-forming regions in the near part of the Gould Belt. More distant X-ray stars seem to be considerably harder, including T Tauri stars in the vicinity of active star-forming regions. At still larger distances, other types, barely represented in the current sample, e.g. thermally-emitting hot white dwarfs or supermassive wind-generating O stars, will also contribute.

Specific Regions The 543 pointed ROSAT High Resolution Imager observations covering a $10° \times 10°$ field of the Large Magellanic Cloud, and carried out between 1990–98, were compiled into a catalogue of 397 discrete X-ray sources. Cross-correlation with the Tycho Catalogue and the SIMBAD database was reported by Sasaki *et al.* (2000a). The results were used to classify the sources into foreground stars, supernova remnants, supersoft sources, X-ray binaries, and background active galactic nuclei. A similar analysis was made by Sasaki *et al.* (2000b) for the 71 pointed observations made of the Small Magellanic Cloud, covering a field of $5° \times 5°$, and containing 121 discrete X-ray sources.

Morley *et al.* (2001) carried out a medium-sensitivity survey using pointed observations with ROSAT, covering nine limited regions of the Galactic plane in the interval of Galactic longitude $l = 180$–$280°$. These observations, each of about 10^4 s, reached a factor of 5 better sensitivity than previous Galactic plane surveys made by Einstein and ROSAT. In total, 93 sources were detected, which were cross-correlated with objects in the Tycho 2 and USNO A2.0 Catalogues to find optical counterparts brighter than 19 mag. Amongst these, the Tycho 2 search yielded all 13 stellar matches made in SIMBAD, and six further stellar candidates. The majority of the optical counterparts are consistent with them being late-type main-sequence stars, as previous studies had proposed from incompletely identified surveys. The data were used to rule out models of stellar X-ray source counts in which the young stellar population is characterised either by a large scale-height and constant star-formation rate, or a small scale-height and an increased star-formation rate over the last billion years.

High-mass X-ray binaries High-mass X-ray binaries consist of a massive OB star and a compact X-ray source, a neutron star or a black hole. They are powered by wind-driven or Roche-lobe overflow material via an accretion disk. The role of the wind accretion means that, in practice, only OB supergiants or Be-star companions have a strong-enough wind to result in observable X-ray emission, only visible in the latter case when the neutron star moves through the dense Be disk at periastron passage.

The Hipparcos Input Catalogue included eight of the then-known high-mass X-ray binaries. Since that time (1982) the number of optically identified systems has increased, and a cross-correlation by Chevalier & Ilovaisky (1998) between the catalogue of van Paradijs (1995) and the Hipparcos Catalogue revealed nine additional X-ray sources (all Be systems) which were observed by Hipparcos as part of its magnitude-limited survey. Among the 17 O or B stars in the Hipparcos Catalogue optically identified with high-mass X-ray binaries as of 1997, 13 were classified as Be stars (γ Cas, LSI+61°303, X Per, HD 34921, V725 Tau, HD 63666, HD 65663, HD 91188, V801 Cen, BZ Cru, HD 109857, μ^2 Cru, and SAO 51568), and four as OB supergiants (V662 Cas, GP Vel, V884 Sco, and V1357 Cyg = Cyg X–1). Chevalier & Ilovaisky (1998) used the Hipparcos parallaxes (where significant) and proper motions to infer that the 13 Be systems have normal space velocities, $\overline{v_t} = 11.3 \pm 6.7$ km s^{-1}, while the four OB supergiant systems are high-velocity objects, with an average transverse velocity $\overline{v_t} > 60$ km s^{-1}. This supports a different formation mechanism for the two subgroups: the Be systems appear as normal low-velocity stars, consistent with them being systems with white dwarf companions (Section 8.2.1), while the OB supergiant systems appear to have originated in massive binaries which have survived a supernova explosion (Section 8.4.2).

Their results for two systems, LSI+61°303 (A0236+610) with $\pi_{\rm Hip} = 5.65 \pm 2.28$ mas and V725 Tau (A0535+262) with $\pi_{\rm Hip} = 3.00 \pm 1.72$ mas, suggested that they were some factor of 10 closer than previously believed. Steele *et al.* (1998) used CCD spectroscopy to show that their spectral types, reddening and absolute magnitudes are inconsistent with the closer distances, and that their 'traditional' distances of order 2 kpc are to be preferred. These are still consistent with

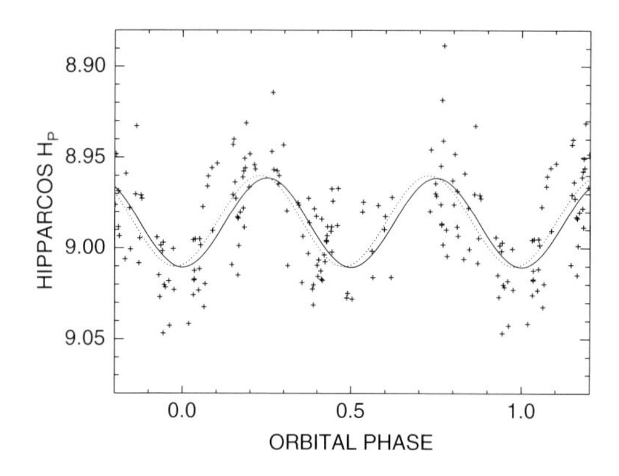

Figure 8.19 Hipparcos photometry of the high-mass X-ray binary Cyg X–1 as a function of orbital phase, according to the ephemeris of Voloshina et al. (1997). The solid line shows a double-wave sinusoidal fit of the tidal 'ellipsoidal' variation. The dotted line shows the best fit when the epoch of minimum light is a free parameter. From Sowers et al. (1998, Figure 1).

the Hipparcos estimates at 2σ, although they also have the worst goodness-of-fit values in the Hipparcos Catalogue of all objects in the Chevalier & Ilovaisky (1998) sample. Clark & Dolan (1999) made a similar analysis for eight of the Hipparcos X-ray sources, arguing that their parallaxes were consistent with zero in all but two cases.

Sowers *et al.* (1998) measured radial velocities, and made a tomographic mapping of the tidal stream, for Cyg X–1 (HIP 98298), based on observations made in 1985 and 1986. They used the resulting orbital solution together with a Hipparcos light-curve of the ellipsoidal variation to arrive at a revised orbital period of $P = 5.599\,77 \pm 0.000\,02$ d (Figure 8.19).

Kinematic aspects of the Hipparcos high-mass X-ray binaries are considered further in the section on runaway stars (Section 8.4.2).

8.4 Late stages of stellar evolution

8.4.1 Wolf–Rayet stars

General The Wolf–Rayet phenomenon represents the final He-burning phase in the evolution of massive O stars of initial mass $\gtrsim 25M_\odot$ (Maeder, 1996); the class is named after their discovery by two members of the Paris Observatory in 1867. In addition to the Doppler-broadened He lines visible in the stellar winds, the WN/WC/WO subtypes are dominated by N, C, and O lines respectively, representing successive late-stage fusion products. After the end of core-He burning, the Wolf–Rayet star explodes as a supernova, resulting in the expulsion of the outer shell of reaction products to the interstellar medium, and the accompanying collapse of the central core to a neutron star or black hole. Evolutionary considerations were first discussed by Chandrasekhar (1934), who later noted that *'continued and unrestricted contraction is possible, in theory'.* Beneath the dense winds that often hide the stellar surface, Wolf–Rayet stars have $T_{\rm eff} \sim 30\,000$–$150\,000$ K, $R \sim 1$–$15R_\odot$, $L \sim 10^5$–$10^6 L_\odot$, high mass-loss rates of ~ 1–$10 \times 10^{-5} M_\odot\,{\rm yr}^{-1}$, with terminal outflow velocities of $v_\infty \sim 1000$–$3000\,{\rm km\,s}^{-1}$. Although only a relatively small number have been detected in the Galaxy (van der Hucht, 2006, lists 298, of which 24 are in the open cluster Westerlund 1, and 60 are in open clusters near the Galactic Centre), their characteristic spectra are easily recognisable to large distances, making them potential tracers of Galactic structure.

Soon after the Hipparcos Catalogue release, two studies of Wolf–Rayet stars were published using the data: an astrometric/kinematic study by Moffat *et al.* (1998), and a photometric study by Marchenko *et al.* (1998). These represented the combined efforts of three groups who had independently proposed the inclusion of Wolf–Rayet stars in the Hipparcos Input Catalogue in the early 1980s. Early results were also reported by Moffat *et al.* (1997), and in a subsequent synthesis by Seggewiss *et al.* (1999).

The emphasis of the former study was on the kinematic nature of the Wolf–Rayet and O star runaways, and is considered further in Section 8.4.2. The photometric study of these 141 O stars, Wolf–Rayet stars, and high-mass X-ray binaries by Marchenko *et al.* (1998) included a wealth of new information. For the high-mass X-ray binaries, they identified optical outbursts in HD 102567, coinciding with periastron passages, and drastic changes in the light-curve of HD 153919. Other results included previously unknown long-term variability of HD 39680 (O6V) and WR 46; unusual major flares at irregular intervals of HDE 308399 (O9V); ellipsoidal variations of HD 64315, HD 115071 and HD 160641; rotational modulation in HD 66811(ζ Pup, O4) and HD 210839 (λ Cep, O); and a major dust formation episode in WR 121. The incidence of variability is slightly higher among the Wolf–Rayet stars than for the O stars, which might be explained by the higher percentage of known binary systems. Among the presumably single Wolf–Rayet stars, the candidate runaways appear to be more variable than the others (Figure 8.20).

Three Wolf–Rayet stars were detected as Hipparcos binaries: two new, WR 31 and WR 66, and the previously-known WR 86, with separations of 0.635, 0.396, and 0.230 arcsec respectively.

γ^2 Velorum Only one Wolf–Rayet star, the brightest known, γ^2 Vel (WR 11, HD 68273, HIP 39953), has a significant parallax, $\pi_{\rm Hip} = 3.88 \pm 0.53$ mas. At just 7σ, it corresponds to a distance of $d = 258^{+41}_{-31}$ pc, compared with the previous ground-based estimated distance of

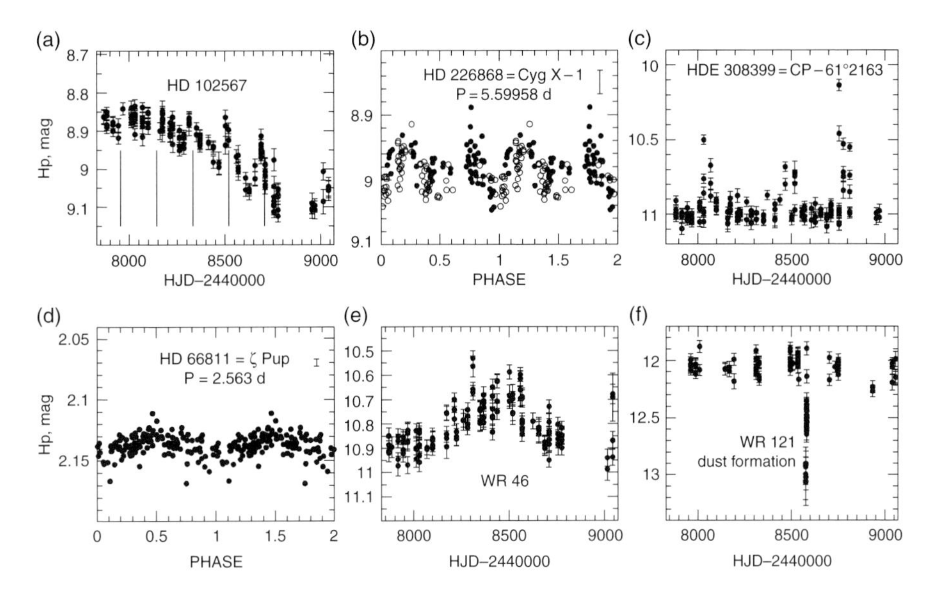

Figure 8.20 A selection of light-curves from a study of Wolf–Rayet stars, O stars, and high-mass X-ray binaries. From top left to bottom right: (a) HD 102567, a high-mass X-ray binary showing X-ray outbursts accompanied by significant growth of the optical flux: vertical lines mark the times of periastron passage; (b) Cyg X-1, a high-mass X-ray binary, and with the light-curve folded at the well-established orbital period of $P = 5.599\,85$ d: various features are discussed by the authors; (c) HDE 308399, an O star, with the most unusual light-curve of the objects studied, with large flares occurring at irregular intervals; (d) ζ Pup, an O star, with periodicity attributed to rotation. The pre-Hipparcos period was taken as $P = 5.21$ d, while the Hipparcos photometry gives half this value, $P = 2.563 \pm 0.005$ d, suggesting a large-scale inhomogeneity of the stellar surface modulating the structure and kinematics of the stellar wind; (e) WR 46, a Wolf–Rayet star, with significant long-term variations in the Hipparcos photometry; (f) WR 121, a Wolf–Rayet star, showing a very deep and narrow R CrB-like light minimum (see Section 8.4.4), in addition to a previously-reported dust formation episode. From Marchenko et al. (1998, Figures 2–7). The compilation of the selected figures was kindly provided by Sergey Marchenko.

450–460 pc, which had been considered as rather secure. The system is a spectroscopic binary, WC8+O9 I, with $P_{\rm orb} = 78.5$ d, and a projected semi-major axis of about 4 mas. A re-reduction of the Hipparcos Intermediate Astrometric Data, taking account of the orbit, confirmed the published parallax. An interferometric observation by Hanbury Brown *et al.* (1970), combined with a from the orbital period, gave $d = 350 \pm 50$ pc. Its visual companion γ^1 Vel was not measured by Hipparcos.

Discussions of the unexpected parallax of γ^2 Vel are given by van der Hucht *et al.* (1997) and Schaerer *et al.* (1997). Adoption of the Hipparcos parallax distance would assist in solving the problem of the mass-loss rate derived from the radio data, where the radio rate would be scaled down to $\dot{M} = 3.5 \times 10^{-5} M_\odot\,{\rm yr}^{-1}$, and thus in better agreement with the value derived from polarisation measurements (St.-Louis *et al.*, 1988). Considered as somewhat more problematic is the fact that, at the Hipparcos distance, γ^2 Vel would no longer be associated with the Vela OB2 association, in contrast to its well-studied neighbour, ζ Pup which, with $d_{\rm Hip} = 429^{+120}_{-77}$ pc, is in agreement with its earlier canonical distance. It could, however, still be an ionising source of the Gum nebula (van der Hucht *et al.*, 1997).

Schaerer *et al.* (1997) derived a total mass from its parallax, the angular size of the semi-major axis as measured with intensity interferometry by Hanbury Brown *et al.* (1970), and the period, finding $M({\rm WR+O}) = 29.5 \pm 15.9 M_\odot$. The stellar parameters for the O star companion derived from line blanketed non-LTE atmosphere models are $T_{\rm eff} = 34\,000 \pm 1500$ K, $\log L/L_\odot = 5.3 \pm 0.15$, from which an evolutionary mass of $M = 29 \pm 4 M_\odot$ and an age of $4.0^{+0.8}_{-0.5}$ Myr was obtained from single star evolutionary models. With non-LTE model calculations including He and C, they derived a luminosity $\log L/L_\odot \sim 4.7 \pm 0.2$ for the Wolf–Rayet star. Combined with the mass–luminosity relation for hydrogen-free Wolf–Rayet stars, they derived $M({\rm WR}) \sim 5 \pm 1.5 M_\odot$.

Oberlack *et al.* (2000) used the Hipparcos distance to place limits on the 1.809 MeV γ-ray line from the radioactive decay of ^{26}Al from γ^2 Vel, using COMPTEL data. They derived an upper limit of $6.3^{+2.1}_{-1.4} \times 10^{-5} M_\odot$ on the ^{26}Al yield, providing an additional constraint on theories of nucleosynthesis in Wolf–Rayet stars. Pozzo *et al.* (2000) reported the discovery of a low-mass, pre-main-sequence stellar association around γ^2 Vel, using it to argue that both are associated and approximately coeval, and that both are at a distance of 360–490 pc, disagreeing at the 2σ level with the Hipparcos parallax.

North *et al.* (2007) derived a complete orbital solution for the double-lined spectroscopic binary, using measurements from the Sydney University Stellar Interferometer (SUSI). They derived $d = 336^{+8}_{-7}$ pc, at roughly 2σ more distant than the Hipparcos parallax estimate, and yielding a distance which would reinstate membership of the Vela OB2 association. They determined the O-star primary component parameters to be $R = 17 \pm 2R_\odot$ and $M = 28.5 \pm 1.1M_\odot$, in good agreement with Schaerer *et al.* (1997), and the parameters of the Wolf–Rayet component to be $M = 9.0 \pm 0.6M_\odot$, significantly different from Schaerer *et al.* (1997). Based on some theoretical modelling, Millour *et al.* (2007) derived $d = 368^{+38}_{-13}$ pc from VLTI–AMBER observations.

The large-scale re-reduction of the Hipparcos data by van Leeuwen (2007) gives a revised parallax of $\pi = 2.99 \pm 0.32$, or $d = 334^{+40}_{-32}$ pc, in close agreement with the SUSI-based value of North *et al.* (2007). Accordingly, γ^2 Vel appears to be an example of the tail of $2-3\sigma$ error distribution to be expected in the Hipparcos parallaxes.

Other Wolf–Rayet stars Other studies of individual Wolf–Rayet stars making use of the Hipparcos data include study of a large H I shell around HD 191765 (Gervais & St-Louis, 1999); long-term photometry of the Wolf–Rayet stars WR 137, WR 140, WR 148, and WR 153 (Panov *et al.*, 2000); detection of molecular gas associated with the Wolf–Rayet ring nebula NGC 3199, in which the earlier hypothesis of nebula formation via a bow shock was shown to be unlikely because of the direction of the Hipparcos proper motion (Marston, 2001); a detailed photometric study of WR 46 confirming the brightening detected by Hipparcos (Veen *et al.*, 2002); and a distance to WR 134 of around 1.17 kpc, based on the use of interstellar Ca II H and K lines (Czart & Strobel, 2006). A study of the interstellar medium local to the binary WR 140, using the Hipparcos proper motion to estimate the epoch at which the binary acquired its large space velocity, is reported in Section 8.5.

8.4.2 Runaway stars

Introduction The study of runaway stars, stars with such extremely high space velocities that they must have been imparted by a particular formation process, somewhat distinct from the overall velocity distribution of stars in the Galaxy, is related to the study of the escape velocity in the vicinity of the Sun, treated in Section 9.4.2. While typical space velocities of OB stars are some 10 km s^{-1}, OB runaways can reach velocities of some 100 km s^{-1} or more. Although not uniquely the products of late stages of stellar evolution, some runaways are, and others are by-products of a binary supernova ejection; they are treated together here.

Many O stars and Wolf–Rayet stars lie well outside the limits of their likely birth places in open clusters and OB associations, and can manifest themselves by their large separations from the Galactic plane, or by their rapid motion. Blaauw (1961) recognised 19 O-type runaways, i.e. O stars having space motions greater than his adopted threshold of 40 km s^{-1}, some of whose velocity vectors point back to their presumed origin in recognised clusters or associations. This study was significantly extended by Gies & Bolton (1986) and Gies (1987).

The observed runaway nature of Wolf–Rayet stars was examined by Moffat & Isserstedt (1980). The following introduction to the topic is largely taken from Moffat *et al.* (1998), where additional background details are given, along with the selection criteria used for their Hipparcos-based studies of the kinematic/runaway nature of a sample of O stars, Wolf–Rayet stars, and high-mass X-ray binaries.

Two plausible theories for the origin of runaways exist, and both may be operating: the binary-supernova scenario (Blaauw, 1961), and the cluster-ejection scenario (Poveda *et al.*, 1967), in which a star is ejected via dynamical interaction between stars in a young, compact cluster.

A complete scenario of massive binary evolution was described by van den Heuvel (1973) and Tutukov & Yungelson (1973), and summarised as

$$\underset{(1)}{\mathrm{O+O}} \rightarrow \underset{(2)}{\mathrm{WR+O}} \rightarrow \underset{(3)}{\mathrm{c+O}} \rightarrow \underset{(4)}{\mathrm{c+WR}} \rightarrow \underset{(5)}{\mathrm{c(+)c}} \tag{8.2}$$

in which c stands for a 'compact' companion, a neutron star or black hole, left after the supernova explosion of its progenitor. In this scenario, it is the more massive star that evolves faster at first. Wind-driven mass-loss, possibly assisted by Roche-lobe overflow in the closest massive binaries, makes O stars evolve into lower-mass Wolf–Rayet stars, which also evolve along a sequence from cool to hot sub-types within the WN and then WC sequences. This occurs for each component in turn, i.e. at (1) → (2) and (3) → (4). At the end of the Wolf–Rayet phase, it is assumed that the star explodes as a supernova, at both (2) → (3) and (4) → (5). If the first supernova explosion is symmetric, the binary system will remain bound, since the less massive star explodes, leading to the class of high-mass X-ray binaries. If the first supernova is asymmetric, the binary system may disrupt, depending on the magnitude and direction of the extra kick velocity (de Cuyper, 1982). In either case, the supernova explosion duration is very short compared to the orbital period, so that the star receives a recoil velocity, and becomes a runaway, with velocities reaching 200 km s^{-1} for the closest, most massive pre-supernova binaries. In the case of the second supernova, it is the more massive star that explodes, so that the system, if it has not already separated after the first supernova, will normally become unbound, producing two high-velocity, single pulsars. In rare cases, the binary can survive this second supernova, producing a binary pulsar (de Cuyper, 1985). Detailed observational evidence for these various interaction products is described by Moffat *et al.* (1998).

The observed low but non-zero O+O binary frequency among runaways, as well as runaway pairs like AE Aur+μ Col (see below), slightly favours the cluster-ejection scenario. The binary supernova scenario appears to do better in accounting for observed space frequencies, low-mass cut-off, velocity-mass correlation, kinematical ages and presence of high-mass X-ray binaries among the OB runaways, and also accounts better for the high frequency of fast rotators and abundance anomalies among O runaways, compared to low-velocity O stars. Both scenarios can lead to the large range of runaway space velocities observed.

Pre-Hipparcos studies of runaway stars were based primarily on radial velocities. Proper motions were known for the brighter stars, but even then with a generally inadequate precision. The extremely broad emission lines corresponding to the high outflow velocities mean, however, that the determination of radial velocities by classical Doppler techniques is severely restricted.

Hipparcos studies Moffat *et al.* (1998) studied a sample comprising all 67 Galactic Wolf–Rayet stars observed by Hipparcos, including almost all down to $V = 12$ mag, and representing some one third of those known at the time. A total of 66 O stars with peculiar radial velocity components exceeding 30 km s^{-1} were also included, along with the seven high-mass X-ray binaries known at the time (γ Cas, X Per, 4U 05335+26, Vela X–1, 4U 1145–61, V861 Sco, 4U 1700–37, Cyg X–1). Selection details are given by Moffat *et al.* (1998).

Only four of the O stars, and one of the Wolf–Rayet, have reliable parallaxes, and for the remainder photometric distances were adopted. For a precision of 1 mas yr^{-1}, the detection of runaways with transverse velocities $v_t \sim 100$ km s^{-1}can then be made out to $d \sim 7$ kpc at the 3σ level. Moffat *et al.* (1998) derived Galactic velocity components, being the sum of basic solar motion (for which they adopted $UVW = 9, 11, 6$ km s^{-1}, from Delhaye 1965), Galactic rotation (for which they adopted a flat rotation curve with $\Theta_0 = 220$ km s^{-1} and $R_0 = 8.5$ kpc), and each object's additional or 'peculiar' motion. Figure 8.21, top shows net proper motions, after removal of the solar motion, in Galactic coordinates b and l. For the l component, the sinusoidal lines show the expected motions due to Galactic angular rotation with the above parameters, and for three different distances. The double-wave trend in μ_l versus l, with minima in absolute value at $l = 0°$ and $180°$, are as expected from the rotation model. The perfect double sine-wave for zero distance becomes increasingly distorted with larger distance. The weak dependence of the model curves on distance allows stars deviating significantly from the trend of Galactic rotation to be identified, but conversely makes proper motions essentially useless for rotational parallaxes.

They then identified possible runaways based on an object's peculiar tangential motion (Figure 8.21, bottom), adopting the criterion $v_{t,\rm pec} > 42 + \sigma_{v_t,\rm pec}$, consistent with the 30 km s^{-1} criterion previously adopted for radial velocities (Cruz González *et al.*, 1974) when allowing a factor $\sqrt{2}$ to account for the two components of transverse motion. This led to the selection of 19 potential runaway stars, six Wolf–Rayet and 13 O-type, with no significant difference between them in their frequency of runaways. They suggested that the cluster-ejection scenario is rather compatible with the observations, using the following line of argument. A kinetic age can be obtained for each star from the distance to their origin, assumed to be the Galactic mid-plane, divided by the runaway velocity: for the supernova binary scenario it can be argued that $\tau_{\rm kin} \simeq 0.3\tau_{\rm nuc}$, and for the cluster-ejection scenario $\tau_{\rm kin} \simeq \tau_{\rm nuc}$, where $\tau_{\rm nuc}$ is the current nuclear age of the observed star. O stars have $\tau_{\rm tot} \sim 2$–10 Myr, while Wolf–Rayet have $\tau_{\rm tot} \sim 2$–7 Myr for solar metallicity. In the solar neighbourhood, typical mean values are in the range $\tau_{\rm tot} \sim 8$–9 Myr for the average O star, and $\tau_{\rm tot} \sim 5$–6 Myr for the average Wolf–Rayet star. In a random sample, given that the Wolf–Rayet He-burning phase is very short, $\tau_{\rm nuc} \simeq \tau_{\rm tot}/2$ for O stars, and $\tau_{\rm nuc} \simeq \tau_{\rm tot}$ for Wolf–Rayet stars. Accordingly, $\tau_{\rm nuc} \simeq 5$ Myr represents an overall average value for the combined sample. Accounting for the K_z force law perpendicular to the Galactic plane (Section 9.4.1), they then estimate $\tau_{\rm kin} = 9^{+7}_{-3}$ Myr, a value consequently compatible with the cluster-ejection model, $\tau_{\rm kin} \simeq \tau_{\rm nuc}$, at the 1σ level.

Other studies Mammano *et al.* (1997) discussed the luminosities and velocities of some 100 N-rich OB stars, or OBN stars, observed by Hipparcos. These are believed to be subluminous OB stars in binary systems in which mass transfer (Bolton & Rogers, 1978), perhaps coupled with internal tidally-induced mixing (Leushin, 1988), brings N to the surface. Mammano *et al.* (1997) reported three runaways: HIP 18614, 70574, and 92198.

Lindblad *et al.* (1997) studied a sample of Hipparcos stars with complete Strömgren $uvby\beta$ photometry and belonging to Strömgren's 'early group'. In addition to distance and kinematic information from Hipparcos, ages were estimated from the evolutionary tracks of Maeder

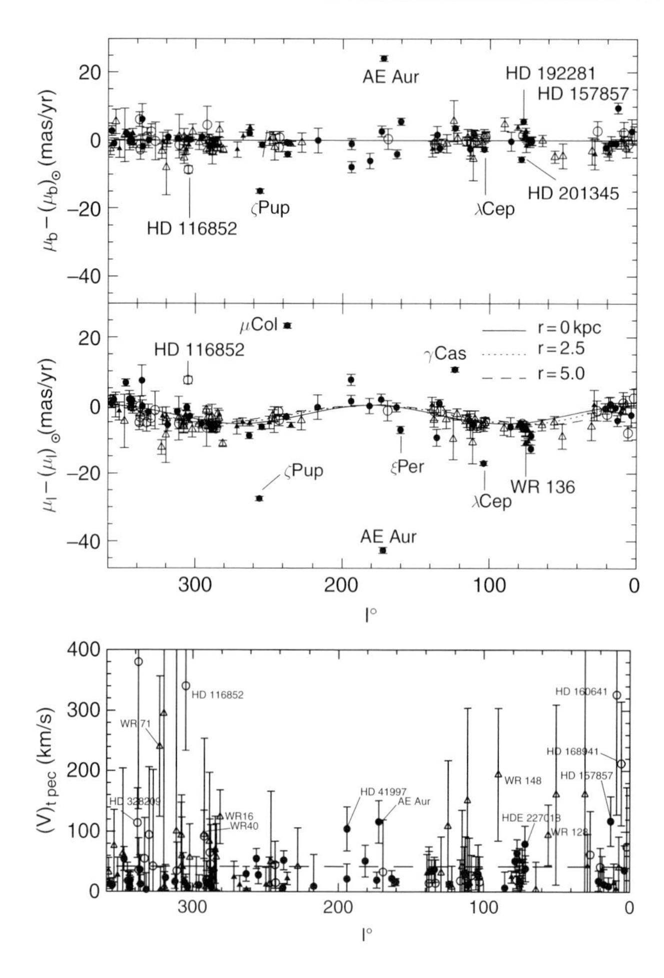

Figure 8.21 Top: net proper motion $\mu - \mu_\odot$ versus Galactic longitude for the b (upper) and l (lower) components of Galactic motion (after removal of solar motion), showing O stars (○), and Wolf–Rayet (△). Filled symbols are for $d \leq 2.5$ kpc, open symbols $d > 2.5$ kpc. Flat rotation curves ($\Theta_0 = 220$ km s^{-1}, $R_0 = 8.5$ kpc) are shown for the l component: solid, dotted, and dashed curves are for $d = 0, 2.5, 5.0$ kpc respectively. Stars that deviate by at least $10\sigma_\mu$ in either l or b are identified. Bottom: peculiar tangential motion of all O and Wolf–Rayet stars studied versus Galactic longitude (same symbols). The dashed horizontal line is the 42 km s^{-1} limit for selecting runaways. Stars with $v_{t,\rm pec} > 42 + \sigma_{v_{t,\rm pec}}$ are identified. From Moffat et al. (1998, Figures 1 and 2); revised figures corresponding to the cited erratum were kindly provided by Sergey Marchenko.

& Meynet (1988). From the distribution of the W velocity component perpendicular to the Galactic plane, they identified 24 runaway stars younger than 40 Myr, being high-W-velocity stars either at large z distances moving outward from the plane, or moving towards the plane, the latter being a phenomenon only observed at small z distances since few of these young stars would have had time to begin their gravitationally-driven return to the plane. They discussed the applicability of the sample for studies of the K_z method of determining the force law perpendicular to the Galactic plane.

Maitzen *et al.* (1998) used the Hipparcos astrometry and their own measured radial velocity, already known to be extremely high and around +230 km s^{-1}, to identify HIP 60350 as an extreme, young, B-type runaway star, ejected from the Galactic plane some 20 Myr ago, probably from a spiral arm region. Both its mass, $\sim 5M_\odot$, and its velocity with respect to the LSR of 417 km s^{-1}, suggest dynamical cluster ejection rather than a supernova origin. In contrast with the expected upper limit to the ejection speed of some 150 km s^{-1} in the case of the supernova ejection scenario (Stone, 1982), this velocity compares favourably with the upper limit for a low-mass companion to an early O-type star ejected from a cluster of around 1400 km s^{-1} (Leonard & Tremaine, 1990), and more particularly the upper limit reasonably expected for a sample of a few hundred runaways, of around 300–400 km s^{-1} (Leonard, 1993). Tenjes *et al.* (2001) followed up the analysis of the possible birthplace of HIP 60350 by integrating the orbit backwards in time,

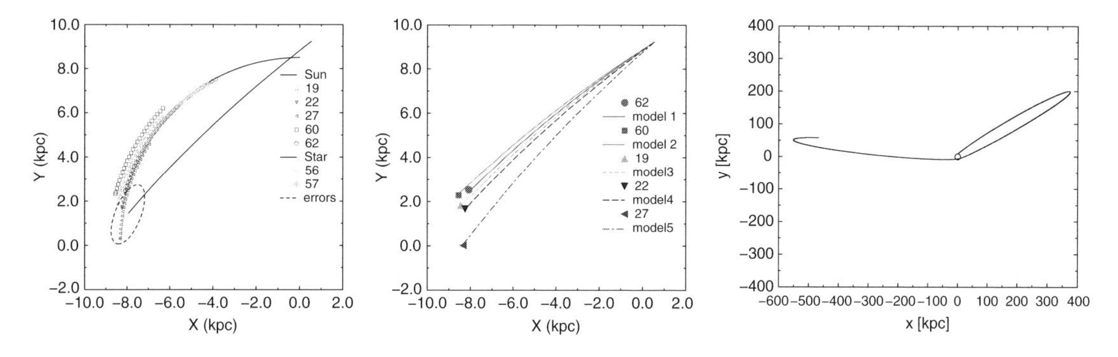

Figure 8.22 Left: motion of the H II regions Georgelin 56, 57, 60 and 62 (= NGC 3603), the molecular gas clouds Grabelsky 19, 22, and 27, the Sun, and the extreme runaway HIP 60350. Motions are in, or projected onto, the Galactic plane during the last 20.4 Myr. Middle: the positions of the H II regions and gas clouds at the times of star ejection. Corresponding orbits of the ejected star are labelled according to the different models studied. Right: orbit of HIP 60350 for the next 13.5 Gyr, projected onto the Galactic plane. The circle with centre coordinates (0, 0) *is situated at the Galactic centre and has a radius 10 kpc. The X, Y axes are directed toward the direction of rotation (at the solar position) and away from the centre, respectively. From Tenjes et al. (2001, Figures 3–5).*

using the Galactic potential model of Haud & Einasto (1989), to its intersection with the plane of the Galaxy. Its high velocity takes it far from the plane and towards the outer halo. Their inferred birthplace (Figure 8.22) lies near to the propagated position of the young open cluster NGC 3603, in which the initial mass function is known to extend up to some $120 M_\odot$ (Drissen *et al.*, 1995). However the ejection event, timed at 20 Myr ago, would be in contradiction to the cluster mean age of 3–4 Myr. They therefore suggest that it occurred at an earlier phase in the sequential star formation in that region, in which massive stars from a first stage of star formation ejected HIP 60350 by dynamical interaction, then exploded as supernovae triggering star formation in the adjacent parts of the cloud. This two-step process might be similar to the scenario outlined by Preibisch & Zinnecker (1999) for the Sco–Cen OB association, and as suggested by the triple main sequence observed in NGC 2808 by Piotto *et al.* (2007). The Hipparcos runaway HIP 60350 thus provides further circumstantial evidence against the commonly-assumed hypothesis that all stars in an open cluster are coeval.

Hoogerwerf *et al.* (2000, 2001) used parallaxes and proper motions from Hipparcos to retrace the orbits of 56 runaway O and B stars with distances less than 700 pc, in order to identify the parent stellar group. They started with all 1118 O–B5 stars in the Hipparcos Catalogue with radial velocities listed in the Hipparcos Input Catalogue, which included 153 of the 162 proposed runaway candidates known in 1982 to have large radial velocities. They included in their study nine compact radio sources, comprising eight radio pulsars with known distances and proper motions, and Geminga. They were able to deduce the specific formation scenario with near certainty for two cases. In the first, (Figure 8.23, left) the runaway star ζ Oph and the pulsar PSR J1932+1059 originated about 1 Myr ago in a supernova explosion in a binary in the Upper Scorpius subgroup of the Sco OB2 association. The pulsar received a kick velocity of $\sim$350 km s^{-1}, which dissociated the binary, and gave ζ Oph its large space velocity. In the second (Figure 8.23, right), they were able to study in considerable detail the runaway pair AE Aur and μ Col, some 70° apart on the sky, for which Blaauw & Morgan (1954) and Gies & Bolton (1986) had already postulated a common origin, possibly involving the massive highly-eccentric binary ι Ori, based on their almost equal and opposite velocities of $\sim$100 km s^{-1}. From the Hipparcos distances ($d_{\rm AE\ Aur} = 446^{+220}_{-111}$ pc, $d_{\mu\ \rm Col} = 397^{+110}_{-71}$ pc, $d_{\iota\ \rm Ori} = 406^{+185}_{-96}$ pc) and space motions, they were able to demonstrate that these three objects indeed occupied a very small volume $\sim$2.5 Myr ago, and showing that they were thus ejected from the nascent Trapezium cluster, with ι Ori being the surviving binary of a binary–binary collision that ejected AE Aur and μ Col. They also identified the parent group for two more pulsars: both likely originate in the $\sim$50 Myr old association Per OB3, which contains the open cluster α Per.

At least 21 of the 56 runaway stars in the sample of Hoogerwerf *et al.* (2000, 2001) can be linked to nearby associations and young open clusters, including the classical runaways 53 Ari (Ori OB1), ξ Per (Per OB2), and λ Cep (Cep OB3), and 15 new identifications. Amongst these are a pair of runaway stars analogous to AE Aur+μ Col, also travelling in opposite directions with similar space velocities: 86.5 km s^{-1} for HIP 22061 and 63.0 km s^{-1} for HIP 29678. Tracing their orbits back, they were close together some 1.1 Myr ago, in a region coinciding with the λ Ori cluster. They suggested that the neutron star Geminga may also have

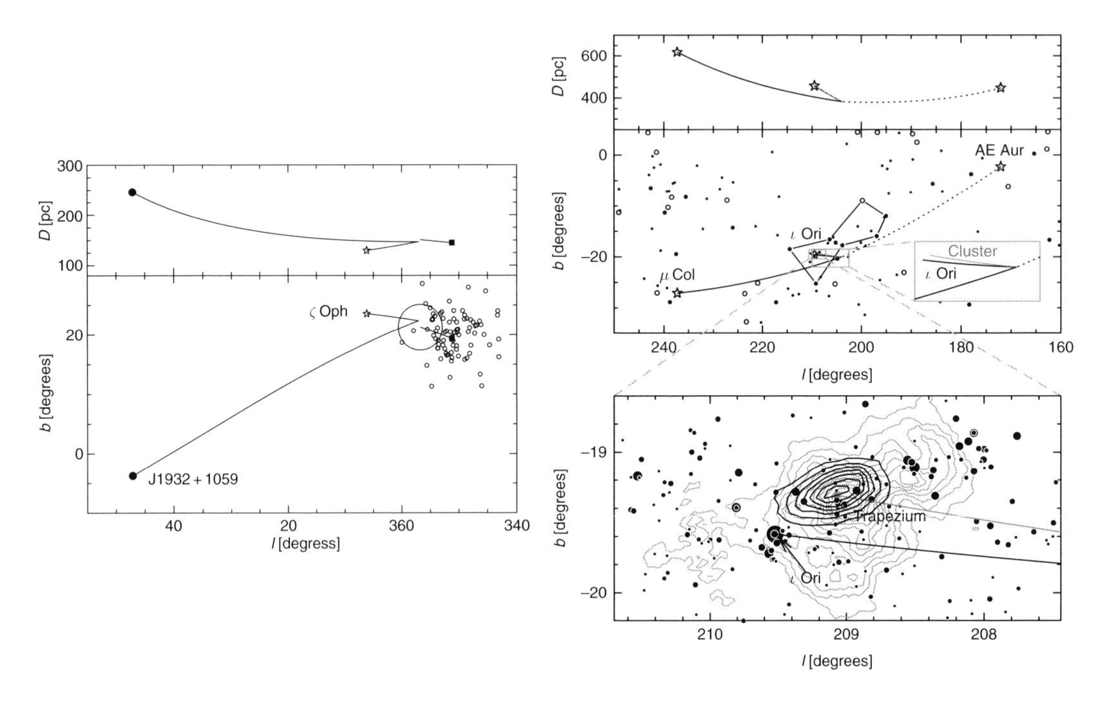

Figure 8.23 Left: orbits of ζ Oph, PSR J1932+1059, and Upper Scorpius. Present positions are denoted: ⋆: runaway, •: pulsar, ▪: association. Top: distance versus Galactic longitude. Bottom: orbits projected on the sky in Galactic coordinates. The small open circles in the bottom panel denote the present-day positions of the O, B, and A-type members of Upper Scorpius, from de Zeeuw et al. (1999). The large circle, of 10 pc radius, is the position of the association at the time of the supernova explosion. Right: top and middle: orbits, calculated back in time, of the runaways AE Aur (dotted) and μ Col (solid) and the binary ι Ori for a specific Monte Carlo simulation. Top and middle panels are as at left. The starred symbols depict the present position of the three stars, which met ∼2.5 Myr ago. Using conservation of linear momentum, the orbit of the parent cluster (grey solid line, see blow up) is calculated from the time of the assumed encounter to the present. Large circles denote all Hipparcos stars brighter than $V = 3.5$; •: O and B stars, ∘: other spectral type. Small circles denote the O and B type stars with $3.5 \leq V \leq 5$. Bottom: predicted position of the parent cluster (black contours) together with all Tycho stars to $V = 12.4$. Symbol size scales with magnitude; the brightest star is ι Ori. Black and dark grey lines are the past orbits of ι Ori and the Trapezium, respectively. The triangle denotes the predicted present-day position of the parent cluster for this particular simulation. The grey contours are the IRAS 100 μm flux map, and mainly outline the Orion Nebula. From Hoogerwerf et al. (2001, Figures 6 and 11).

been ejected from this region (Frisch, 1993; Smith *et al.*, 1994), a possibility supported by the fact that the age of Geminga, ∼350 000 yr, agrees well with the epoch of the supernova explosion which created the λ Orionis ring. The possibility was later rejected by Pellizza *et al.* (2005). Other currently nearby runaways and pulsars originated beyond 700 pc, where knowledge of the parent groups remains very incomplete.

Comerón & Pasquali (2007) used the Tycho 2 Catalogue proper motion of the O star BD +43°3654, along with an estimated distance of 1450 pc, to infer its space velocity in the plane of the sky of $39.8 \pm 9.8\,\mathrm{km\,s^{-1}}$. This is several times the estimated local sound speed, and consistent with the nearby bow shock observed in the infrared (Figure 8.28, right below). Their estimated mass of $70 \pm 15 M_\odot$ is similar to that of ζ Pup and λ Cep as estimated by Hoogerwerf *et al.* (2001). Its estimated age of ∼1.6 Myr, and its space motion, would place it near the centre of the Cygnus OB2 association at the time of its birth.

Two other insights into the nature of the runaway phenomenon were investigated by Hoogerwerf *et al.* (2001). They constructed a subset of all their O and B stars with known rotational velocities (Penny, 1996) and He abundances (Kudritzki & Hummer, 1990; Herrero *et al.*, 1992), along with an assessment of whether individual stars were runaways. The resulting correlation (Figure 8.24), still however based on a statistically limited and incomplete sample, demonstrates a provisional quantitative confirmation of the suggestion by Blaauw (1993), that massive runaways predominantly have high He abundances and large rotational velocities, suggesting in turn that they are indeed formed mainly by the binary-supernova scenario.

One of the characteristics of runaway stars produced according to the binary-supernova scenario is that they

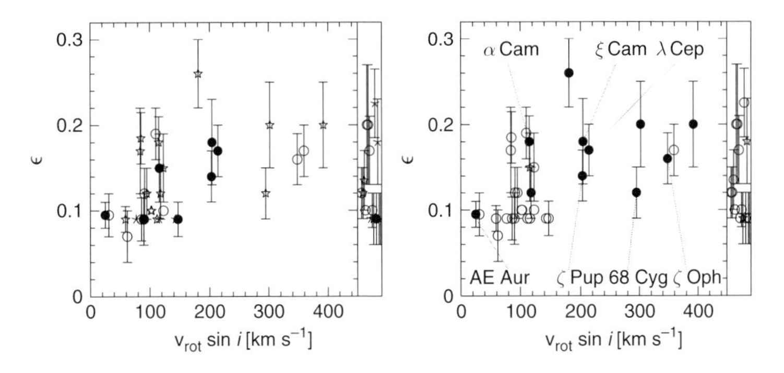

Figure 8.24 He/H abundance (ϵ) versus rotational velocity for O stars. ○: non-runaway stars; •: runaway stars; ⋆: doubtful runaways. The right of each panel display the He abundances of stars with an unknown rotational velocity. From Hoogerwerf et al. (2001, Figure 16a).

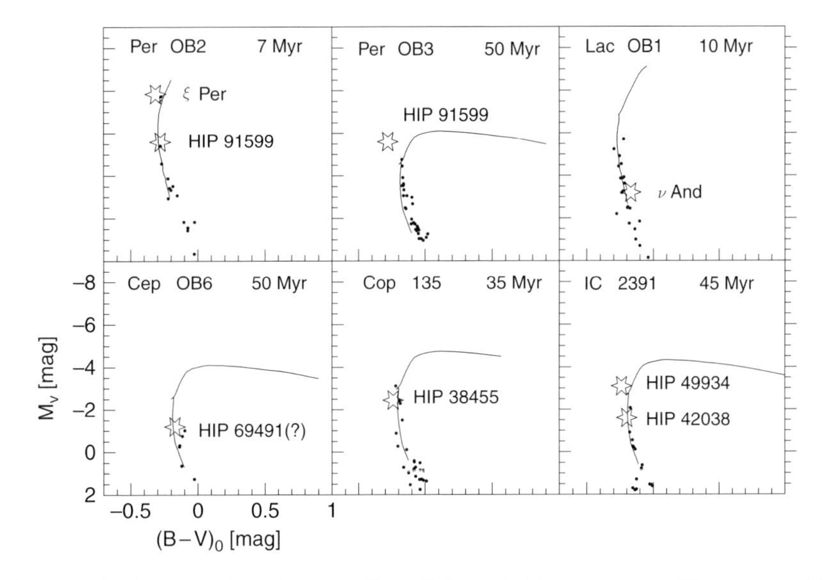

Figure 8.25 Absolute magnitude versus colour, determined from their spectral types, for some of the runaways (⋆) and their parent association or cluster (•). Of those shown, blue stragglers are ξ Per; possible stragglers are HIP 49934, and HIP 91599. Isochrones are from Schaller et al. (1992). Association ages are given at the top right of each panel. From Hoogerwerf et al. (2001, Figure 17).

are expected to be blue stragglers: the mass transfer in close binary systems from the primary to the secondary prior to the supernova explosion deposits a large amount of H onto the future runaway. This new fuel makes the runaway appear younger than the association or cluster in which it was born. The de-reddened absolute magnitude versus colour diagrams of the parent clusters discussed by Hoogerwerf *et al.* (2001) are shown in Figure 8.25. Three stars are identified as blue stragglers: HIP 38518, ξ Per, and λ Cep; and three others could be blue stragglers depending on the correct identification of the parent association: ζ Pup, HIP 49934, and HIP 91599. The blue straggler nature confirms their origin through the binary supernova scenario. In contrast, those produced by the dynamical ejection scenario are expected to follow the main sequence of the parent group. The stars which they identified securely as dynamical ejection runaways (the pairs AE Aur+μ Col, HIP 22061+HIP 29678) indeed fall on the main sequence of their parents: the Trapezium and the λ Ori cluster.

Mdzinarishvili & Chargeishvili (2005), updated from Mdzinarishvili (2004), used the same (peculiar) transverse velocity threshold limit as that adopted by Moffat *et al.* (1998), i.e. $v_{t,\mathrm{pec}} > 42 + \sigma_{v_t,\mathrm{pec}}$, to search for further OB runaways in the Hipparcos Catalogue. Adopting photometric distance estimates they first derived 61 candidate runaways of which 8 had been previously recognised by Blaauw (1961), Moffat *et al.* (1998), or Hoogerwerf *et al.* (2001). After comparing photometric and parallax distance estimates, noting that in all cases the former are larger than the latter, and therefore redefining the peculiar velocity threshold on the basis of

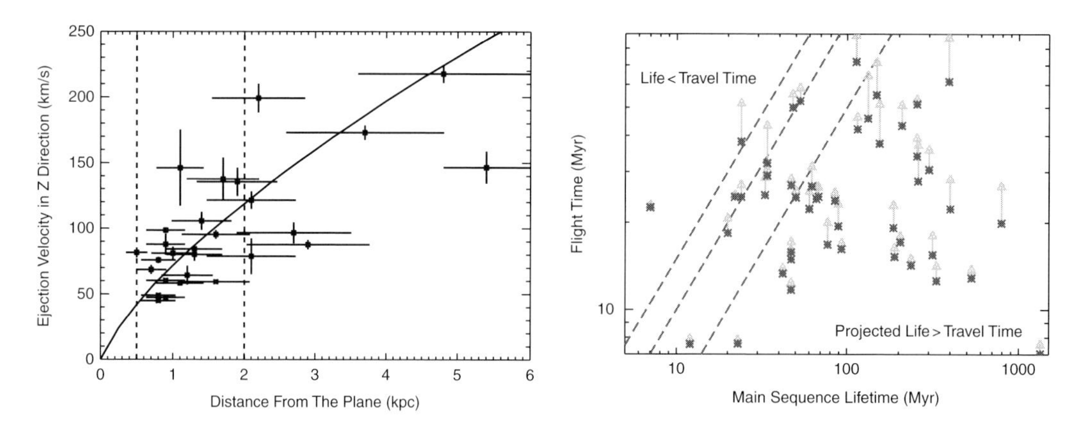

Figure 8.26 Left: the Population I runaway stars in a sample of B stars found far from the Galactic plane. The solid curve is the theoretical maximum Z-distance which a star will reach with a given vertical component of its ejection velocity. The dashed vertical lines mark $0.5 \leq Z \leq 2.0$ kpc. Right: the flight time of the star versus its estimated main sequence lifetime. ⋆ and △ denote the flight times calculated with the different Galactic potential models, the latter using a lower disk surface density, and thus increasing the actual flight time. Between the outer dotted lines, the main sequence lifetime is within ±50% of the flight time; stars to the lower right have flight times significantly shorter than the star's main-sequence lifetime, and can therefore be explained as runaways. The one star to the left, HD 140543, could have a large error in its estimated lifetime. From Martin (2006, Figures 1 and 14).

Hipparcos parallaxes accurate to better than 50%, their total number of runaway detections dropped to eight, with just five new candidates: HIP 29678, 36243, 57870, 66524, and 106917. The conclusion is that most significant OB runaways, with accurate astrometry from the Hipparcos Catalogue, have now been detected.

Martin (2003, 2004, 2006) made a further detailed study of the origins and evolutionary status of a sample of B stars found far from the Galactic plane. Faint blue stars at high Galactic latitudes were first discovered in the survey by Humason & Zwicky (1947), and are continuing to be found within the Galaxy (Saffer *et al.*, 1997; Rolleston *et al.*, 1999) and perhaps in the halo of M31 (Smoker *et al.*, 2000). There are currently three explanations for B stars found far from the plane: they could be normal massive Population I stars ejected from the disk (as discussed above), older evolved stars (subluminous blue horizontal branch or post-asymptotic giant branch stars) or, somewhat speculatively, young massive stars formed *in situ* in the Galactic halo. Martin (2004) analysed the abundances of a sample of these B stars, and identified several which are chemically similar to Population I stars, while others were more like an evolved population. No classical Be stars were found. Martin (2006) combined the abundance data with the kinematic motions of the stars derived from Hipparcos proper motions to estimate the maximum Z-distance a star will reach with a given ejection velocity, and the flight time versus the estimated main sequence lifetime (Figure 8.26). The sample was shown to contain 31 Population I runaways, fifteen old evolved stars (including five blue horizontal branch stars, three post-horizontal branch stars, a pulsating helium dwarf, and six stars of ambiguous classification), one F-dwarf, and two unclassified stars, probably extreme Population I runaways. No star in the sample unambiguously shows the characteristics of a young massive star formed *in situ* in the halo. The low binary frequency and rotational velocity distribution of the Population I runaways imply that most were ejected from dense star clusters by the dynamical ejection scenario. The absence of runaway classical Be stars stands unexplained: it could be related to the lack of binary companions, to the ejection mechanism, or to the cluster environment.

Studies of the Hipparcos photometry of two runaways revealed periodic variations: with $P = 0.89$ d in the case of HD 218915 (Barannikov, 2006b, with the same periodicity also measured in radial velocity), and $P = 67.385$ d in the case of HD 192281 (Barannikov, 2006a).

Trapezium systems Trapezium systems, named after the Trapezium in the Orion Nebula Cluster, are physical systems of three or more stars with roughly equal separations. Ambartsumian (1954) defined them to be a multiple system in which at least three of the component separations are of the same order of magnitude, i.e. their ratio lies between 0.33 and 3.3 (in contrast, hierarchical systems have factors of at least 10 between the largest and the smallest separations, typically consisting of a close pair and a distant third star, or a close pair and distant close pair). Since the orbits of the component stars in such systems are not closed, they should be dynamically unstable, either evolving towards hierarchical configurations in a few million years or less, or dissolving by successive ejections until only a close pair remains (Allen & Poveda, 1974; Allen *et al.*, 2004). The occurrence of projected 'pseudo-Trapezium' systems was studied by Abt & Corbally (2000).

Allen *et al.* (2006) studied runaway stars associated with 44 Trapezium systems. Making limited use of the Hipparcos distance information for some, the fastest components among them could be classified as runaway stars. The relatively high number of runaways results from the dynamical interactions that occur in the unresolved sub-Trapezium systems which, in analogy with the Orion trapezium, likely constitute the brightest components of their sample.

High-mass X-Ray binaries According to the binary supernova ejection scenario, all high-mass X-ray binaries (Section 8.3) should be runaways. This was not necessarily supported by pre-Hipparcos radial velocity observations in all cases (Gies & Bolton, 1986), although van Oijen (1989) found strong indications that they are high-velocity objects, a conclusion supported by the suggestion that Vela X–1 originated in the Vela OB1 association (van Rensbergen *et al.*, 1996), and the discovery of the wind bow-shock around it (Kaper *et al.*, 1997).

In addition to the study of the space motions of the Hipparcos massive X-ray binaries by Moffat *et al.* (1998), Chevalier & Ilovaisky (1998) used the Hipparcos data to demonstrate that OB supergiant X-ray binaries do indeed have relatively large runaway velocities. They studied the four supergiant high-mass X-ray binaries for which proper motions are available, 4U 0114+65, 4U 0900–40 (Vela X–1), 4U 1700–37, and Cyg X–1, to determine their space velocities, and to demonstrate their larger mean peculiar transverse velocity of $\sim 42 \pm 14\,\mathrm{km\,s^{-1}}$, compared to that of Be X-ray binaries which is $\sim 15 \pm 6\,\mathrm{km\,s^{-1}}$.

Van den Heuvel *et al.* (2000) demonstrated that these mean velocities are in good agreement with the predicted runaway velocities of these two types of system on the basis of simple conservative evolutionary models (van den Heuvel, 1994, and references). The difference results from the variation of the fractional He core mass as a function of stellar mass, in combination with the conservation of orbital angular momentum during the mass-transfer phase that preceded the formation of the compact object. This combination results in systematically narrower pre-supernova orbits in the OB-supergiant systems than in the Be-systems, and a larger fractional mass ejected in the supernova in high-mass systems relative to systems of lower mass. Regardless of possible kick velocities imparted to neutron stars at birth, this leads to a considerable difference in average runaway velocity between these two groups, a difference

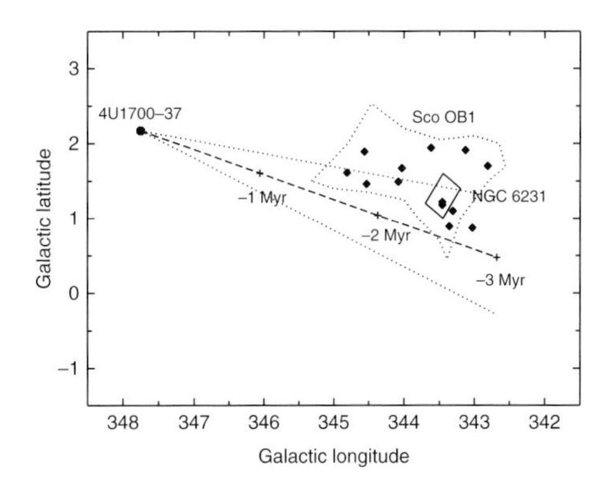

Figure 8.27 The reconstructed path of the runaway high-mass X-ray binary 4U 1700–37 intersects with the location of Sco OB1; the error cone is indicated by the dotted straight lines. ♦: association members confirmed by Hipparcos. The dotted line marks the region studied by Perry et al. (1991), including the young open cluster NGC 6231 (box). The corresponding kinematic age of 4U 1700–37 is 2 ± 0.5 Myr. The current angular separation between 4U 1700–37 and NGC 6231 (at 2 kpc) corresponds to a distance of about 150 pc. From Ankay et al. (2001, Figure 3).

even more pronounced when non-conservative mass transfer is taken into account (Portegies Zwart, 2000). The observed low runaway velocities of the Be X-ray binaries confirms that in most cases not more than 1–$2M_\odot$ was ejected in the supernovae that produced their neutron stars. In combination with their large mean orbital eccentricities, this indicates that their neutron stars must have received a kick velocity in the range 60–$250\,\mathrm{km\,s^{-1}}$ at birth. The considerable (transverse) runaway velocity of Cyg X–1, of $\sim 50 \pm 15\,\mathrm{km\,s^{-1}}$, shows that significant mass ejection also takes place with the formation of a black hole. A short review of the field following these Hipparcos results is given by Kaper *et al.* (2004).

Ankay *et al.* (2001) studied the origin of the runaway high-mass X-ray binary 4U 1700–37. Based on its Hipparcos proper motion, they proposed that it originated in the OB association Sco OB1 (Figure 8.27). At a distance of 1.9 kpc its space velocity with respect to Sco OB1 is $75\,\mathrm{km\,s^{-1}}$. This runaway velocity indicates that the progenitor of the compact X-ray source lost about $7M_\odot$ during the (assumed symmetric) supernova explosion. The system's kinematic age is about 2 ± 0.5 Myr, which marks the date of the progenitor supernova explosion. The present age of Sco OB1 is $\lesssim 8$ Myr, while its suggested core, NGC 6231, seems to be somewhat younger, at about 5 Myr. If 4U 1700–37 was born as a member of Sco OB1, this implies that the initially most massive star in the system terminated its evolution within $\lesssim 6$ Myr, corresponding to an initial mass of $\gtrsim 30M_\odot$.

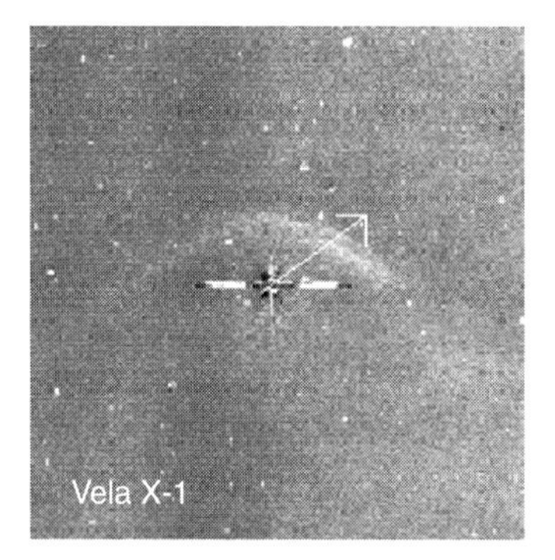

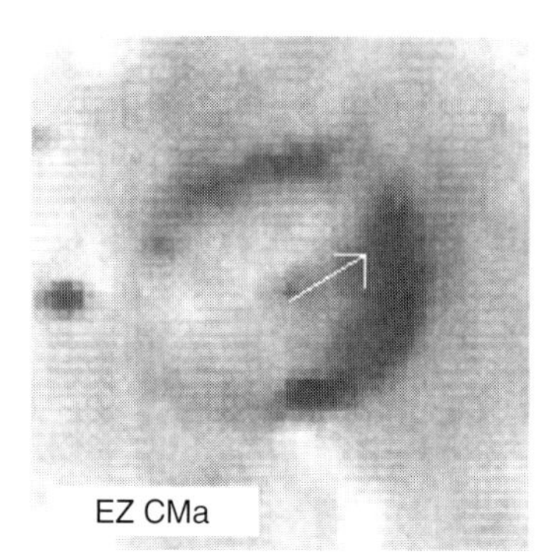

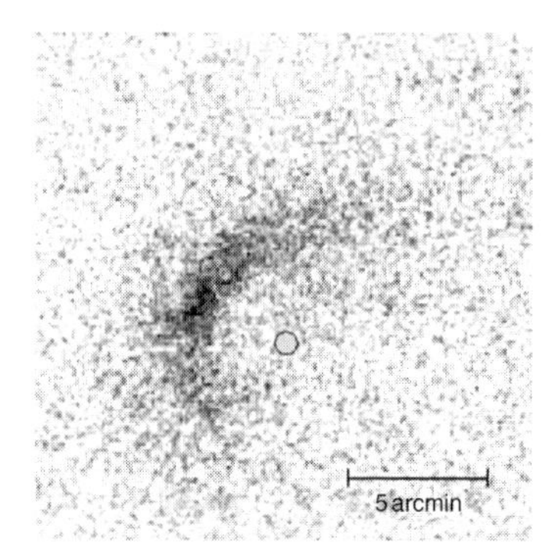

Figure 8.28 Left pair: proper motions superposed on interstellar medium emission maps for HD 77581 = Vela X–1 in which the star is moving directly towards the bow-shock maximum (left), and HD 50896 = EZ CMa = WR6 where the star is moving towards the brightest part of the ring nebula (right). Maps from Kaper et al. (1997) and van Buren et al. (1995), respectively. From Moffat et al. (1998, Figure 5). Right: Midcourse Space Experiment (MSX) infrared image of the region around the O star runaway BD $+43^\circ$ 3654. From Comerón & Pasquali (2007, Figure 3).

Runaway Be stars The possibility that some Be stars have formed in binary systems but have since been disrupted, with a consequently enhanced space velocity, is discussed in Section 8.2.1.

Runaways and bow shocks The high space velocity and powerful stellar wind of an OB runaway can give rise to a strong interaction with the ambient interstellar medium. In the case of supersonic motion, a wind bow-shock is formed (Baranov *et al.*, 1971). Thus, if the peculiar velocity exceeds the local sound speed, being some 0.7–6 km s^{-1} in H I regions and some 10 km s^{-1} in H II regions (Scheffler & Elsaesser, 1987), then regions of enhanced emission will occur in the direction of motion. The accumulated gas and dust is heated by the ultraviolet radiation field of the OB star, the heated dust radiates at infrared wavelengths, and the shocked gas emits in strong optical emission lines such as Hα and [O III]. OB stars, but not the Wolf–Rayet stars, had been previously searched for systematically by van Buren & McCray (1988) and van Buren *et al.* (1995) using IRAS 60 μm maps. These authors reported excess infrared emission for about 30% of the 188 OB stars studied. For a significant fraction, prominent arc-like features can be seen in high-resolution IRAS maps (Noriega-Crespo *et al.*, 1997), providing constraints on the stellar wind parameters and the density of the ambient medium.

In their study of Wolf–Rayet, O stars, and high-mass X-ray binaries, Moffat *et al.* (1998) also examined the correlation of proper motions with known bow-shock regions of the interstellar medium, using the tangential velocity vector, i.e. the projection of the space velocity on the sky, and searched the literature for known bow-shocks. Five such bow-shocks could be identified with well-established runaways (Figure 8.28, left pair).

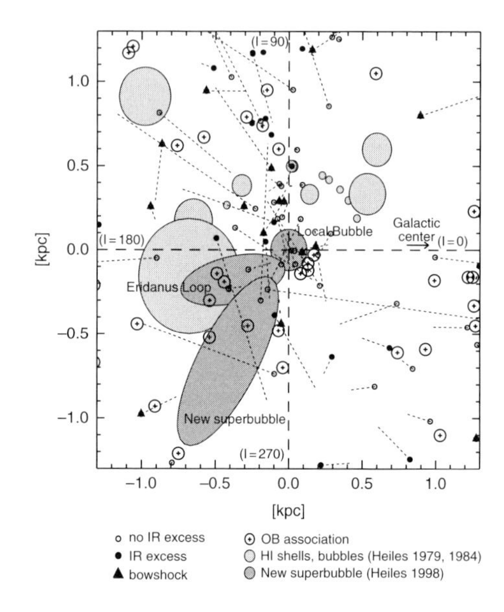

Figure 8.29 The distribution of H I shells in the Galactic plane, in the range $10^\circ \leq l \leq 250^\circ$ and $-10^\circ \leq b \leq 10^\circ$ according to Heiles (1979, 1987), shown as shaded regions. The Sun is in the middle; the direction of the Galactic centre is indicated. ⊕: OB associations from Mel'Nik & Efremov (1995). Also shown are the OB runaways and their distances travelled during the last 10 Myr. The large concentration of OB associations in the fourth quadrant suggests the presence of hot bubbles, although this region was not included in the study. Most OB runaways producing a bow-shock do not seem to be contained inside a hot bubble. From Huthoff & Kaper (2002, Figure 10).

Comerón & Pasquali (2007) reported a similar result for the O star BD $+43^\circ$3654 (Figure 8.28, right).

Huthoff & Kaper (2002) also used IRAS maps to search for bow-shocks around high-mass X-ray binaries, whose high space velocities were confirmed by the Hipparcos proper motions. Apart from the already-known bow-shock around Vela X–1 (Kaper *et al.*, 1997, see also Figure 8.28, left), they did not find convincing evidence for shocks around any other high-mass X-ray binaries. Also in the case of (supposedly single) OB-runaway stars, only a minority appears to be associated with a bow-shock (van Buren *et al.*, 1995). They proposed that in a hot tenuous medium, like that pervading the inside of Galactic superbubbles, the sound speed is sufficiently high, around 100 km s^{-1}, that many runaways are still only moving at a subsonic velocity, and thus are not creating an observable bow-shock. These super-bubbles are expected, and observed, around OB associations, where the OB-runaway stars were once born (Figure 8.29). Conversely, this implies that the presence or absence of wind bow-shocks around OB runaways effectively probes the physical conditions of the interstellar medium in the solar neighbourhood.

Hypervelocity stars The class of even higher velocity, hypervelocity stars, is a field not touched by the Hipparcos observations. Predicted to result from disruption of a close but Newtonian encounter between a tightly-bound binary and a $\sim 10^6 M_\odot$ black hole, with ejection velocities of up to 4000 km s^{-1}, the discovery of even one such hypervelocity star coming from the Galactic centre was considered as providing nearly definitive evidence for a massive black hole (Hills, 1988; Yu & Tremaine, 2003). The first such star, J090945.0+024507, was discovered from the SDSS survey, travelling with a heliocentric radial velocity of 853 ± 12 km s^{-1} in a direction 174° from the Galactic centre, at a distance of 55 ± 16 kpc (at which its proper motion would be $\sim$0.3 mas yr^{-1}), and at more than twice the local escape velocity (Brown *et al.*, 2005). A number of other such stars are now known (Brown *et al.*, 2006a,b, 2007b,a).

8.4.3 Carbon stars

Introduction Carbon stars, recognised amongst the first examples of chemically peculiar stars in the Harvard spectroscopic classification for the Henry Draper Catalogue, are recognised by the presence of absorption bands of C_2 at 438.3 and 473.7 nm, and the 421.6 nm band of CN (Shane, 1928). Their early classification had shown that they were subdivided into the 'warm' R stars and the 'cool' N stars (both giants), which are now recognised as two distinct classes rather than as representing a continuum of properties. A further subdivision was introduced by Keenan (1942), who noted that some members of the R class had unusually strong CH molecular bands.

Chemically, carbon stars are characterised by the ratio C/O > 1. Since C/O < 1 everywhere in the interstellar medium, and in most stars including the Sun, their excess carbon is attributed to stellar nucleosynthesis, either in the star itself, or as a result of mass transfer from a carbon-enriched binary companion. In either case, it signals production during the late stages of stellar evolution, in red giants or in asymptotic giant branch stars. Thus C/O > 1 occurs in some Miras, and in some semi-irregular and irregular variables, where they exhibit moderate (0.6 mag) variations on time scales of tens to thousands of days.

This section covers the contribution made by Hipparcos to the red carbon stars as defined within the revised MK system by Keenan (1993). These are signified by C-Rn, C-Nn, and C-Hn, where n is a homogeneous temperature index for three sequences, replacing the previously-designated R, N, and CH types, and corresponding to the G-K-M temperature sequence in normal 'oxygen' stars. Other types of carbon stars are also recognised: there are carbon-rich Cepheids, carbon-rich Wolf–Rayet (Section 8.4.1), the H-deficient carbon stars including the R CrB stars (Section 8.4.4), barium stars (Section 8.4.6), carbon white dwarfs (Section 8.4.8), and dwarf carbon stars (Dearborn *et al.*, 1986; Green *et al.*, 1992), themselves perhaps originating in a binary system by mass transfer from a cool white dwarf.

The pre-Hipparcos understanding of these various objects is summarised by Wallerstein & Knapp (1998) and Knapp *et al.* (2001). The N stars are very luminous red giants undergoing shell H and He burning, with extended cool envelopes. The outer convective zone penetrates deep into the core, bringing carbon produced in the core through the triple-alpha process, and sometimes other heavy elements including s-process elements, to the surface, and hence into the interstellar medium by their intense stellar winds. Theory suggests that all stars with initial masses in the range $1.5 - 4 M_\odot$ should go through the carbon star phase, lasting around 300 000 yr, before ending as a white dwarf. Their importance as a class is related to their very high luminosity, where they can contribute significantly to the entire energy output of galaxies in the infrared. Their chemical composition (Wallerstein & Knapp, 1998) shows nearly solar C/H, N/H, and $^{12}C/^{13}C$ ratios, indicating that much of the C and N in the Galaxy came from mass-losing carbon stars. In general, the N stars are long-period variables, either irregular, semi-regular, or Mira types.

Mass loss from the asymptotic giant branch carbon stars, at rates up to several times $10^{-5} M_\odot$ yr^{-1}, contributes about half of the total mass return to the interstellar medium (Wallerstein & Knapp, 1998). Mass-loss rates for Miras are about 10 times higher than for semi-regular and irregular stars, whose properties are similar enough to suggest that they belong to the same population. The distribution of carbon star mass-loss rates peaks at about $10^{-7} M_\odot$ yr^{-1}, close to the rate of growth of the core mass, and demonstrating the close relationship between mass loss and evolution. Detached shells are seen around some stars; they appear to form on the time scales of the helium shell flashes and to be a normal occurrence in carbon star evolution.

Other carbon stars with peculiar surface abundances, such as the CH, barium, and some technetium stars, originate by mass transfer from a binary companion. The hydrogen-deficient carbon stars, including the R CrB stars, arise from envelope loss.

The R stars are also red giants but with luminosities, determined by statistical parallaxes, too low for shell He burning, and do not display such prominent mass loss. Their O abundances are similar to that of the Sun, they do not have an over-abundance of s-process elements, and none are known in binary systems. Consequently, it has been suggested that the R stars are coalesced binary systems.

Hipparcos contributions Knapp *et al.* (2001) identified about 320 carbon stars in the Hipparcos Catalogue. They illustrated the complexity of their classification with two examples: HIP 12028 (HD 161115) had been variously classified in the past as R, CH, and as J, while HIP 85750 had been variously classified as R2, CH, and as N. They found that the Hipparcos Catalogue is reasonably complete in carbon stars to $Hp \sim 9.5$, or to $K \sim 7$. From the $(J-H, H-K)$ colour–colour diagram (Figure 8.30, left) it is apparent that R stars occupy a specific region, which is also populated by CH stars. A few R stars appear to occupy the same region as N-type carbon stars, which may also imply that most if not all late R-type carbon stars may actually be N stars.

The Hipparcos Catalogue gives $\sigma_\pi/\pi < 0.5$ for just 17 R stars, of which 15 are of types R0–R2. Resulting absolute magnitudes fall in the range $-2.8 < M_V < +4.1$. However, the estimated parallax uncertainties for many of the stars are larger, and sometimes very much larger, than those expected for stars of these magnitudes and at these ecliptic latitudes, probably due to the use of a mean and slightly erroneous colour index for these red and very variable stars (Section 4.10). Knapp *et al.* (2001) therefore reprocessed the Hipparcos Intermediate Astrometric Data for a subset of 83 of the class R carbon stars, solving directly for the distance modulus while rejecting outliers. They identified two distinct populations: the late-type R stars, perhaps identical to the N stars, which are mostly long-period variables of the semi-regular or Mira type, often with infrared emission due to dust and indicative of mass loss, and located at the bottom of the asymptotic giant branch. The other population comprises the early R type, mostly populating the red clump region (Figure 8.30, centre), and with a Gaussian distribution of absolute magnitudes, $M_K \sim -2.0 \pm 1.0$. They are non-variable and, as implied by the absence of an infrared excess, appear not to be undergoing strong mass loss. Their location in the red clump provides strong support for the formation scenario suggested by Dominy (1984), in which the He core flash mixes carbon to the surface of the star, perhaps rotationally-induced as the star is spun up by the accretion of a former companion. This scenario also has the attraction of accounting for the lack of binary systems among R stars. Izzard *et al.* (2007) examined binary merger models, constrained by the number of Hipparcos R stars observed, to propose the most likely formation scenario as resulting from the merger of a He white dwarf with a H-burning red giant branch star during a common envelope phase, followed by a He flash in a rotating core which mixes carbon to the surface.

Knapp *et al.* (2001) estimated the space density of the R stars by assigning to them an absolute magnitude of $M_K^0 = -2.0$, assuming a uniform density in the Galactic plane, n_0, a scale height of z_0, and fitting to the observed cumulative frequency distribution as a function of distance (Figure 8.30, right). They found a good fit to this uniform/exponential distribution out to about 600 pc for $z_0 = 300$ pc and $n_0 \sim 45\,\mathrm{kpc}^{-3}$. The sample of 284 red clump stars with Hipparcos parallaxes better than 5% from Alves (2000), roughly complete to ~80 pc, yields a volume density of $\sim 1.14 \times 10^5\,\mathrm{kpc}^{-3}$, such that R stars form only about 0.05% of the total number of red clump stars. In other words, only a very small fraction of clump stars in the solar neighbourhood is carbon rich.

A number of other related studies were published before and after this work, but without re-analysis of the Hipparcos Intermediate Astrometric Data. Alksnis *et al.* (1998) determined absolute visual and bolometric magnitudes for about 40 Hipparcos carbon stars, largely confirming previous absolute magnitude estimates, especially those from statistical parallaxes. They compared evolutionary tracks for initial masses $1 \le M/M_\odot \le 4$, from Charbonnel *et al.* (1996) and from Schaller *et al.* (1992), and concluded that the majority of CH- and R-stars are on the giant and subgiant branches, but that N-stars occupy a region $-4 < M_V < -1$ and $1.6 < (B-V)_0 < 3.6$ and which corresponds to an advanced stage of thermally-pulsing asymptotic branch giants. One object, HD 97578 (HIP 54806), appears to be of particularly low luminosity, and a potential carbon dwarf, although its Hipparcos parallax is particularly uncertain, $\pi = 31.2 \pm 14.1$ mas, and no re-processed solution was found by Knapp *et al.* (2001).

A series of detailed studies of the Hipparcos carbon stars was undertaken by the Lyon group. They studied the space distribution of Hipparcos carbon stars (Knapik *et al.*, 1998) finding a very similar local space density to Knapp *et al.* (2001) of $n_0 \sim 40$–$70\,\mathrm{kpc}^{-3}$, and with no clear spatial correlation with the local spiral arm structure or interstellar extinction. Hipparcos distances

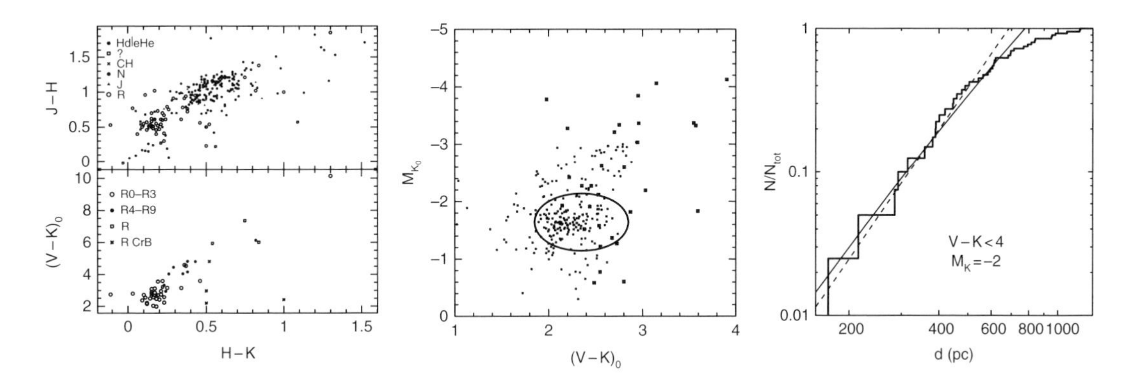

Figure 8.30 Left top: $(J - H)$ versus $(H - K)$ for the carbon stars from the Hipparcos Catalogue in the study by Knapp et al. (2001); J, H and K are from the literature. Symbols are listed in the figure, with Hd = H-deficient and eHe = extreme-He carbon stars. Left bottom: $(V - K)_0$ versus $(H - K)$ for the sample of R stars. Middle: M_K versus $V - K$ for early carbon R stars with $K \leq 8$ (⋆), and red clump giants (■) from Alves (2000), in which the ellipse indicates the typical location of the clump giants. Right: cumulative frequency distribution of the distances of early R stars (thick line), as compared to predictions for a stellar sample uniformly distributed in the plane and with an exponential distribution perpendicular to the plane (thin line). A scale height of 300 pc and an absolute magnitude $M_K^0 = -2.0$ have been adopted. The thin dashed line corresponds to the predictions for a spherically-symmetric homogeneous sample. From Knapp et al. (2001, Figures 1, 5 and 6).

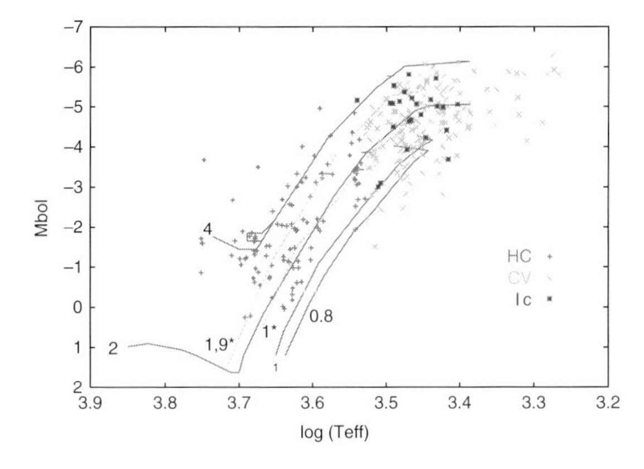

Figure 8.31 The HR diagram of about 300 carbon stars. The nearly vertical limit between hot carbon (HC) stars on the left (+), and carbon variables (CV) on the right (×), is evident. The Tc-rich stars (⋆) are located towards the bright locus of carbon variables, which are thermal-pulse asymptotic giants branch objects. Most stars populate a curved strip which corresponds to theoretical evolutionary tracks of red giant branch and asymptotic giants branch with initial masses 0.8–4 $M_\odot$. For $Z = 0.02$, theoretical tracks were adapted from Bressan et al. (1993) and Marigo (1998). For $Z = 0.008$, tracks for 1 and 1.9 $M_\odot$, labelled with asterisks, were adapted from Fagotto et al. (1994) and Marigo (1998), which illustrate the resulting leftward shift. From Bergeat et al. (2002a, Figure 9).

were also used to determine the degree of dust extinction and intrinsic spectral energy distribution of the hot carbon stars and related objects, the R CrB, barium, and helium-deficient carbon stars (Bergeat *et al.*, 1999), as well as the carbon Miras and very cool non-Miras (Knapik *et al.*, 1999). Using data shown to be consistent with the Hipparcos distance estimates, as well as angular diameters for 52 stars, they then derived a new scale of homogeneous effective temperatures for all carbon giants and related objects (Bergeat *et al.*, 2001). Using these temperatures, and bolometric corrections, Bergeat *et al.* (2002a) constructed the HR diagram for 370 objects located in the red giant branch and asymptotic giant branch regions (Figure 8.31). Mean luminosities increase with increasing radii and decreasing $T_{\rm eff}$, while theoretical tracks show that initial masses range from ~ 0.8 to $4.0 M_\odot$, with a range of metallicities likely. The luminosity function of Galactic carbon giants shows two maxima corresponding to the hot carbon stars (their HC class) of the thick disk, and to the carbon variables (their CV class) of the old thin disk respectively. While theoretical models give consistent results up to the early-asymptotic giant branch phase, and are at least partly consistent with evolved models of low- and intermediate-mass stars (Iben & Renzini, 1983; Busso *et al.*, 1999), they diverge in the thermal pulsing/third dredge-up phase. The number of thermal pulses required for C/O to exceed unity, for example, remains a complex function of initial mass, metallicity, dredge-up efficiency, mixing-length and convective overshoot parameters. Furthermore, the low-luminosity carbon stars are not yet explained by standard models (Marigo *et al.*, 1996, 1999). Bergeat *et al.* (2002c) made further investigations of their space and velocity distributions. They identified the hot carbon stars as a component of the thick disk contaminated by the spheroidal-component CH stars, and with initial masses $\lesssim 1.15 M_\odot$. They identified the carbon variable stars as components of the old (thin) disk, with a likely age spread from a few 10^8 Myr up to 8–12 Gyr, typically with higher initial masses of up to a few $M_\odot$.

Photometric studies of carbon stars, at least partly making use of the Hipparcos data, include: far infrared mapping of extended dust shells in Y CVn and U Ant (Izumiura & Hashimoto, 1997); photometric characteristics of Hipparcos carbon stars (Balklavs *et al.*, 1998); episodic mass loss of TT Cyg (Olofsson *et al.*, 1998, 2000); pulsation modes of long-period variables analysed in the period-radius diagram (Bergeat *et al.*, 2002b); near infrared photometry for 20 Hipparcos carbon stars (Chen *et al.*, 2003b,a); light-curve analysis for seven carbon stars (Dušek *et al.*, 2003); analysis of light-curves of 26 periodic variable carbon stars (Mikulášek & Gráf, 2005); and near-infrared observations of TU Gem and SS Vir (Richichi & Chandrasekhar, 2006).

8.4.4 Hydrogen-deficient carbon-rich stars

The class of hydrogen-deficient carbon stars have unusual abundances with very little H. They include (see Figure 8.30, left) the H-deficient carbon stars, the R Coronae Borealis (R CrB) stars (similar objects, but with dust obscuration episodes), and extreme helium stars (similar to but hotter than the R CrB stars).

The R CrB stars are distinguished from other H-deficient objects by their spectacular dust formation episodes (Clayton, 1996). They were discovered by Pigott & Englefield (1797) who observed that the star of the same name, previously observed to be about 6 mag, was missing from the sky. They may decline by up to 8 mag in a few weeks, revealing a rich emission-line spectrum. Their atmospheres have unusual abundances with very little H, and an over-abundance of C and N. They are thought to be the product of a final He-shell flash (or possibly the coalescence of a binary white-dwarf system), in which dust may form in non-equilibrium conditions created behind shocks caused by atmospheric pulsations. Many aspects of the R CrB phenomenon remain unclear, including details of the dust formation mechanism, their evolutionary status, and the nature of their emission-line regions. The R CrB stars represent a rare, or short-lived, stage of stellar evolution, since the number of known R CrB stars is small, with just 40 listed by Milone (1990). None is close enough to have a meaningful ground-based parallax, and information about their luminosities and masses originally came from three objects in the Large Magellanic Cloud (Feast, 1972), discovered from the Harvard Observatory survey plates during the early 1900s, for which $M_{\rm bol} = -4$ to -5. Others have been found more recently, notably in the Large Magellanic Cloud (Alcock *et al.*, 2001) and in the Galactic bulge (Zaniewski *et al.*, 2005) by the MACHO project, and in the Small Magellanic Cloud from Spitzer observations (Kraemer *et al.*, 2005).

Trimble & Kundu (1997) examined the brightest seven out of the 12 R CrB stars in the Hipparcos Catalogue: R CrB, RY Sgr, XX Cam, S Aps, UW Cen, V CrA, and SU Tau. None have parallaxes statistically different from zero, though most of the proper motions are. They considered two, somewhat extreme, assumptions. If all seven stars are at the same distance of 1200 pc, the resulting velocity ellipsoid is defined by $\langle U^2\rangle^{1/2}, \langle V^2\rangle^{1/2}, \langle W^2\rangle^{1/2} = 41, 30, 35\,{\rm km\,s^{-1}}$, not inconsistent with values typical of old disk populations, such as carbon stars, long-period variables, and planetary nebulae. If, in contrast, all stars have $M_V = -4.5$ as for those in the Large Magellanic Cloud, this leads to $\langle U^2\rangle^{1/2}, \langle V^2\rangle^{1/2}, \langle W^2\rangle^{1/2} = 52, 60, 104\,{\rm km\,s^{-1}}$, very different from either the disk or halo.

Cottrell & Lawson (1998) extended the analysis to the 21 H-deficient carbon stars observed by Hipparcos, including most of the brighter R CrB variables, other cool H-deficient carbon stars, and several higher-temperature extreme helium stars. Again, most have either negative or statistically insignificant parallaxes, indicating that they lie beyond the Hipparcos measurement limits. Although the distances to the Galactic H-deficient carbon stars thus remain unknown, the Hipparcos data confirm that they must have high luminosities like those in the Large Magellanic Cloud. Based on the Hipparcos proper motions, they derived *UVW* velocities for all objects, assuming two different values of $M_{\rm bol} = -3$ and -5. Their *UW*-velocity dispersion of the R CrB and H-deficient carbon stars is similar to that already reported for the extreme helium stars by Drilling (1986), further supporting the latter's finding of a predominantly bulge-like distribution. UW Cen may be an example of a halo R CrB star currently crossing the Galactic plane.

Other studies include variability in seven extreme He stars (Sahin & Jeffery, 2007), and fundamental parameters of He-weak and He-strong stars (Cidale *et al.*, 2007).

8.4.5 Technetium stars

Technetium-rich stars originate in the asymptotic giant branch phase (Section 7.5.7). Van Eck *et al.* (1998), see also Van Eck *et al.* (1997), compared the location of Tc-rich and Tc-poor S stars from the Hipparcos Catalogue in the HR diagram: ^{99}Tc has a half-life of only 2×10^5 yr, making it a reliable indicator of the third dredge-up, and likely to be detectable at the surface after only a few thermal pulses (Goriely & Mowlavi, 2000). They found that Tc-rich S stars are cooler and intrinsically brighter than Tc-poor S stars (Figure 8.32). Comparison with the Geneva evolutionary tracks reveals that the line marking the onset of thermal pulses on the asymptotic

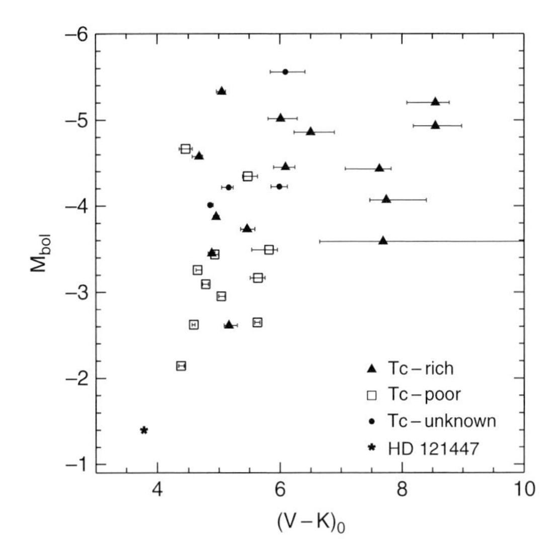

Figure 8.32 HR diagram for S stars with $\sigma_\pi/\pi < 0.85$; ▲: Tc-rich S stars; □: Tc-poor S stars; •: S stars with unknown Tc. HD 121447, the boundary case between Ba and S stars, is represented by ⋆. From Van Eck et al. (1998, Figure 2).

giant branch matches well the observed limit between Tc-poor and Tc-rich S stars. Tc-rich S stars are, as expected, identified with thermally-pulsing asymptotic giant branch stars of low and intermediate masses, whereas Tc-poor S stars are mostly low-mass stars either on the red giant branch or on the early asymptotic giant branch. Like barium stars, Tc-poor S stars are known to belong exclusively to binary systems, and their location in the HR diagram is consistent with the average mass of $1.6 \pm 0.2 M_\odot$ derived from their orbital mass-function distribution.

Lebzelter & Hron (2003) studied the occurrence of Tc as a function of luminosity for a sample of long-period variables selected according to the Hipparcos parallaxes. They found that a large number of asymptotic giant branch stars above the estimated theoretical limit for the third dredge-up do not show Tc. They confirmed previous findings that only a small fraction of the semi-regular variables show Tc lines, while they found a significant number of Miras without Tc. They estimated that a large fraction of their sample have current masses $< 1.5 M_\odot$ which, combined with evolutionary models, implies that the fraction of time a star is observed as a semi-regular variable or as a Mira is dependent on its mass.

8.4.6 Barium stars

Ba stars were defined as a group by Bidelman & Keenan (1951). They are S stars, mainly of spectral type G and K, that show enhancement of many s-process elements, particularly Ba II. Many belong to binary systems. The abundance anomalies are explained by mass transfer onto a 'normal' star, from an initially more massive companion already enriched in s-process elements (thermally pulsating asymptotic giant branch stars), either through Roche lobe overflow or wind accretion, and in which the primary has become a white dwarf (McClure *et al.*, 1980).

Various Hipparcos-based studies of Ba stars have been undertaken. Mennessier *et al.* (1997) and Luri *et al.* (1997) used parallaxes and proper motions of 297 objects to classify them according to luminosity and kinematics, confirming that the Ba stars form an inhomogeneous group with five distinct classes: halo stars, as well as dwarfs, subgiants, giants, and supergiants from the disk population. Confirmed or suspected multiplicity supports the binary star enrichment scenario. Bergeat & Knapik (1997) came to similar conclusions based on an analysis of 52 Ba stars with the highest parallax accuracies. Although only 21 out of 121 were detected by Hipparcos as binaries, the low proportion can be explained by the large magnitude difference and close angular separation of their white dwarf companions. North (1998) derived Hipparcos-based evolutionary masses of Ba dwarfs by interpolation in evolutionary tracks, and proposed a new statistical determination of the mass of the average white dwarf companion using mass functions determined from radial velocity measurements. Mashonkina *et al.* (1999) made a detailed spectral analysis of 11 dwarf Ba stars using gravities determined from the Hipparcos parallaxes, studying the relative importance of the r- and s-processes and the resulting odd-to-even isotope ratio from the Ba II lines, and concluding that the s-process indeed dominates Ba production in these stars. Jorissen *et al.* (2005) used the binary information from Hipparcos to probe the relation between Ba stars and yellow symbiotic stars.

8.4.7 Planetary nebulae

Planetary nebulae are the end stage of stellar evolution for stars other than those with masses more than $\sim 8 M_\odot$ which terminate in a supernova explosion. During the red giant phase, the He core is contracting and heating, and the outer layers are expanding and cooling, the process continuing until core temperatures reach $\sim 10^8$ K, initiating core He fusion, and leading to an inert C and O core with a He and a H burning shell. He fusion reaction rates are extremely temperature sensitive, $\propto T^{40}$, leading to an unstable pulsational feedback. This successively ejects more of the remaining stellar atmosphere, progressively exposing deeper and higher temperature regions of the core. At exposed temperatures of $T_{\rm eff} \sim 30\,000$ K, the dense ultraviolet photon flux ionises the

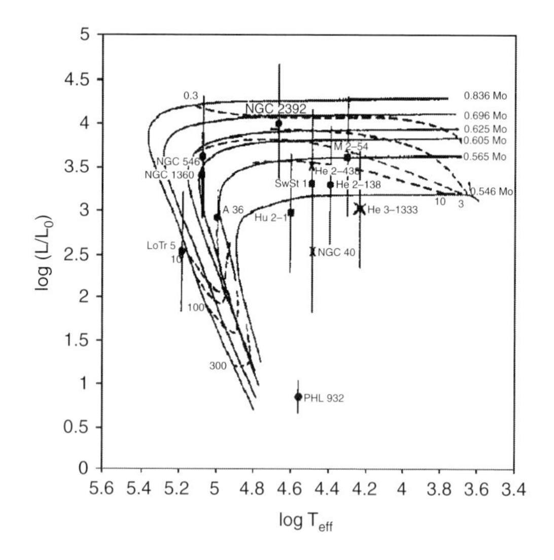

Figure 8.33 HR diagram of the central stars of planetary nebulae, based on adopted distances from both Hipparcos and other methods. Theoretical curves were taken from the post-asymptotic giant branch evolutionary sequences of Blöcker (1995). Lines of constant age are marked in units of 10^3 yr. Small nebulae are identified by filled squares. Negative Hipparcos parallaxes, for which other distance estimates are used, are identified by crosses. From Acker et al. (1998, Figure 2).

ejected atmosphere, illuminating the spectacular planetary nebula. The first discovered, M27, was by Messier in 1764.

Planetary nebula distances are typically very large, and difficult to determine by trigonometric parallax, although progress from the USNO ground-based parallax programme has been reported by Harris *et al.* (1997, 2007). Accordingly, various other methods have been used to estimate distances, including extinction-based methods, use of the interstellar Na D lines, spectroscopic distances for the cases where the central star shows an absorption-line spectrum, methods based on the nebula expansion, and the Schlovsky method which assigns a fixed value to the ionised nebula mass. A pre-Hipparcos review by Pottasch (1997) gives parallax distances for nine objects (S216, PW1, NGC 6853, A21, NGC 6720, A31, A24, NGC 7293, and A74), with distances ranging between 130–752 pc. The central star magnitudes range from $V = 12.3$ to 16.6 mag, and typical inferred stellar properties for this subgroup are $T_{\rm eff} \sim 10^5$ K, $\log g \sim 7$–7.5, and $L \sim 10$–$200 L_\odot$. Gravities, luminosities, and spectra indicate that at least these central stars are white dwarfs, or very close to the white dwarf stage, for which nuclear burning has ceased, and the objects are on the cooling tracks. Other objects with lower surface gravities, $\log g \sim 4$–6, and higher luminosities, $L \sim 10^3$–$10^4 L_\odot$, are found for less evolved stars with estimated expansion. Of the various challenges facing planetary nebula research was the fact that the predicted ages of low-luminosity objects were much higher than the kinematic ages (although these are now considered of little value due to the complicated hydrodynamics), and that the high-luminosity objects were predicted to evolve more rapidly than observed. Studies of planetary nebulae within the Magellanic Clouds, for which a reasonable distance estimate is known, do not entirely solve these problems because the central star properties are largely inferred from modelling of the nebula spectra.

Hipparcos measured the parallaxes of a small number of central stars of planetary nebulae. Pottasch & Acker (1998) considered the results for three objects for which spectroscopic distances had also been determined from stellar gravities, in turn derived from line-profile fitting of selected H and He lines. These objects (PHL 932 = HIP 4666, NGC 1360 = HIP 16566, A36 = HIP 66732) are close to the limiting magnitude of the Hipparcos observations, with parallaxes of 9.12 ± 2.79, 2.86 ± 2.12, and 4.12 ± 2.47 mas respectively. The Hipparcos distances are all considerably smaller than the spectroscopic distances, implying that the surface gravities are at least a factor of 3 higher than that found from the Balmer line profile fitting (for example, for PHL 932, $\log g_{\rm Hip} \sim 6.80$, $\log g_{\rm spect} \sim 5.5$). Nebular masses, calculated by assuming them to be homogeneous spheres, gave values of 0.0013, 0.32, and $0.0057 M_\odot$ respectively, with nebular expansion ages of 3.9, 15.0, and 6.5×10^3 yr respectively. Though considerably less than their typical theoretical ages of 10^5 yr, the conclusions are evidently based on a small number of rather marginally significant parallaxes.

Acker *et al.* (1998) extended the study to 19 planetary nebulae observed by Hipparcos. They discussed selection effects in the original samples, and suggested the possibility that they represent just two extreme evolutionary phases: either very young and compact, or very old and extended. All are distant, in the probable range 100–2000 pc, making individual distance determinations rather unreliable, but collectively indicating that previous distances to planetary nebulae had been overestimated. For very compact nebulae, they noted that the parallaxes could also be biased by the nebular emission. Positions in the HR diagram, based on Hipparcos distances, with other methods for the more compact objects, were not significantly different from previous studies, with the exception of PHL 932 and SaSt 2–12 (Figure 8.33).

Napiwotzki (2001) made a spectroscopic investigation of old planetary nebulae, and compared distance scales resulting from the Hipparcos and ground-based parallaxes (Harris *et al.*, 1997), with estimates derived from the interstellar Na D lines, and the Shlovsky method.

Space motions The most comprehensive pre-Hipparcos study of proper motions of planetary nebulae was made for 51 objects by Cudworth (1974), who used them to calculate distances using statistical parallaxes.

Napiwotzki *et al.* (2001b) derived abundance patterns in the central star of the planetary nebula BD+33°2642, and used the measured mean radial velocity of this binary system, combined with the Hipparcos proper motion, to derive a Galactic orbit providing strong evidence that the object is a member of the halo population.

Kerber *et al.* (2004) determined proper motions for four central stars of planetary nebulae: NGC 7293, Sh 2–216, IC 4593, and Sh 2–174. None are in the Hipparcos Catalogue, and only two (Sh 2–216 and IC 4593) appear in Tycho 2. However, all appear in the UCAC 2 and USNO B Catalogues, which have themselves been reduced to the Hipparcos/ICRS reference frame. They used these with published radial velocities to determine their Galactic orbits, traced back in time using the simplified model of the Galactic gravitational potential described by Pauli *et al.* (2003). They classified NGC 7293 and Sh 2–216 as thin disk objects, and IC 4593 and Sh 2–174 as thick disk objects (Figure 8.34). Kerber *et al.* (2002) had already shown how the high (ground-based) proper motion of Sh 2–68, of 53.2 ± 5.5 mas yr^{-1}, can account for the interaction between the planetary nebula and its ambient interstellar medium, where data from the SHASSA Hα survey (Gaustad *et al.*, 2001) have revealed a tail extending over 45 arcmin which is likely formed by matter stripped off the main nebula in the course of the planetary nebula trajectory along its Galactic orbit. The new orbital knowledge allows some further diagnostics to be made: for Sh 2–216, their orbit confirms that the object is in a mild but advanced interaction with the surrounding interstellar medium, with the central star starting to leave the quasi-static planetary nebula shell some 45 000 years ago. For low-velocity interactions in the dense interstellar medium close to the Galactic plane, the shell fossilises, diffuses slowly, and thus returns processed matter back into the interstellar medium. For IC 4593, the object crosses the Galactic plane with a large relative velocity and at very large angles, leading to the substantial bow shock observed. Sh 2–174 is an extreme case with the central star located outside of the main nebula; the proper motion of the thick disk orbit is consistent with the observed displacement over about 10 000 years, and they conclude that the interstellar medium stopped the expansion of the nebula at that time. Kerber *et al.* (2008) enlarged their sample to provide proper motion estimates for a total of 234 planetary nebulae based on various all-sky astrometric catalogues reduced to the Hipparcos system.

Other studies of planetary nebulae making use of the Hipparcos data include a photometric study of the central star of M2–54 (Handler, 1999); and the white dwarf binary central star of the planetary nebula A35 (Herald & Bianchi, 2002).

8.4.8 White dwarfs

Introduction Stellar evolution theory indicates that some 90% of stars will end their lives as white dwarfs, with different routes leading to the same fate. Theory predicts that most white dwarfs, which are of intermediate mass, have an internal composition dominated by C or C–O cores and a narrow mass distribution which peaks around $0.58M_\odot$ (e.g. Koester *et al.*, 1979; Bergeron *et al.*, 1992; Finley *et al.*, 1997). The canonical mass for the ignition of He in the cores of red giants (the helium flash) is around 0.46–$0.48M_\odot$, so that white dwarfs with lower masses are inferred to have He cores. The lower-mass objects are not believed to be a result of standard single star evolution since the main-sequence lifetime of their progenitors exceeds the Hubble time. Instead, they are considered to have been formed by Roche lobe overflow in a binary star system, in which the low-mass He core is exposed as a result of mass transfer to the secondary before the white dwarf progenitor reaches the tip of the red giant branch. For the high-mass tail, theory predicts interiors dominated by Ne and Mg (Panei *et al.*, 2000a).

White dwarfs span a vast temperature range, with the H-rich white dwarfs extending from 170 000 K to 4500 K, and with the temperature providing a direct luminosity and age indicator. The classification scheme DA, DB, etc. is according to the dominant spectral features (McCook & Sion, 1999, Table 1): thus DA dwarfs, subdivided into DA.25–DA13 according to $T_{\rm eff}$ (where the temperature index was originally defined as $50\,400/T_{\rm eff}$), have an atmosphere dominated by H I; DB are dominated by He I; DC have a continuous spectrum without deep lines; DO are dominated by He II; DZ by metal lines; and DQ by C features. The temperature ranges intersect instability regions analogous to those seen around the main sequence, and span the DA dwarfs (including the very common pulsating ZZ Ceti stars), the DB dwarfs, and the DO dwarfs (including the pulsating GW Vir stars). The relationship between $T_{\rm eff}$, abundances, and the instability strip, is illustrated in Vauclair & Vauclair (1982, Figure 1).

For a fully-degenerate electron gas at matter densities of 10^6–10^8 gm cm^{-3}, the equation-of-state leads to a mass–radius relation first derived by Chandrasekhar (1931). The relation was refined by Hamada & Salpeter (1961), who calculated zero-temperature (fully degenerate) white dwarf models of different chemical composition (He, C, Mg, Si, S, and Fe). Wood (1990, 1995)

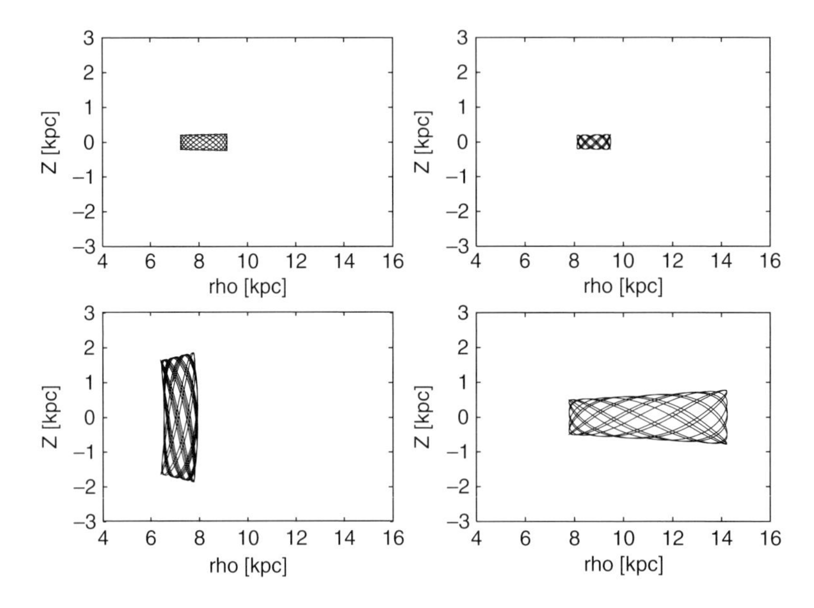

Figure 8.34 The simulated Galactic orbits of four planetary nebulae, based on the Hipparcos/ICRS proper motion reference frame. From upper left to lower right: NGC 7293, Sh 2–216, IC 4593 and Sh 2–174. NGC 7293 and Sh 2–216 (top pair) are classified as thin disk objects, and IC 4593 and Sh 2–174 (bottom pair) as thick disk objects. From Kerber et al. (2004, Figure 1).

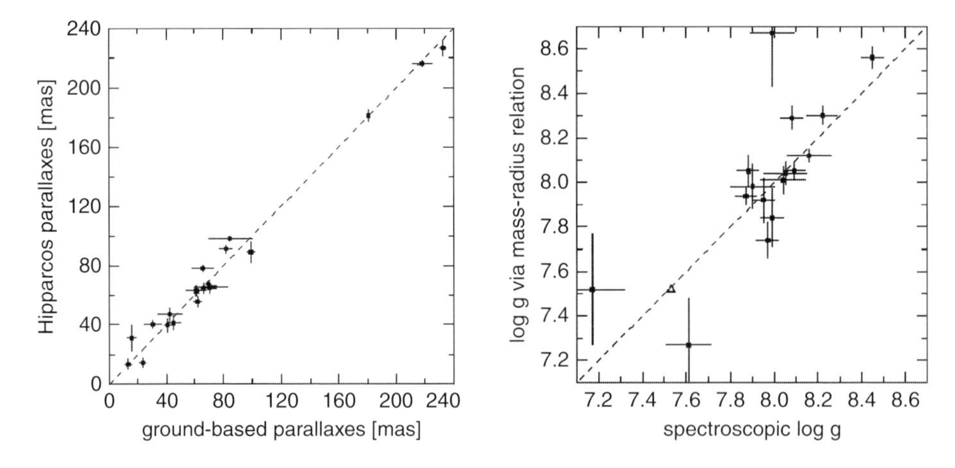

Figure 8.35 Left: comparison between the ground-based and Hipparcos parallax estimates for white dwarfs. The general agreement is very satisfactory, with the error bars being significantly reduced with the Hipparcos results in a number of cases. Right: comparison between log g determined spectroscopically with its value estimated from the theoretical mass–radius relation. From Vauclair et al. (1997b, Figures 1 and 2).

calculated models with C cores and different configurations of H and/or He layers, showing that a white dwarf with an envelope of reasonable thickness (10^{-4}–$10^{-7} M_\odot$) and non-zero temperature will depart from the Hamada–Salpeter relation.

The mass–radius relation remains a largely theoretical construct, but forms a basic underlying assumption in most studies of white dwarf properties. It enters the determination of their mass, and in turn their mass distribution and luminosity function, and thus a range of astrophysical applications including distance calibration to globular clusters (Renzini *et al.*, 1996; Zoccali *et al.*, 2001), estimation of the age of Galactic disk and halo by means of white dwarf cooling sequences (Winget *et al.*, 1987; D'Antona & Mazzitelli, 1990), dark matter investigations (Tamanaha *et al.*, 1990; Ibata *et al.*, 2000), and establishing constraints on possible variations of fundamental physical constants, notably $\dot{G}/G$ (García-Berro *et al.*, 1995, 2007), although see also Biesiada & Malec (2004). A precise observational determination of the mass–radius relation constrains theoretical models of their interiors, including tests of the inner chemical composition, the thickness of the H envelope of DA white dwarfs, and the characterization of their inner magnetic fields.

Empirical confirmation of the mass–radius relation, and hence a direct observational confirmation of stellar degeneracy, is challenging because of the very few white dwarfs with accurately-determined masses and radii, and because their intrinsic mass distribution is concentrated in a rather small interval. After publication of the theory by Chandrasekhar (1931), the observational situation was summarised 50 years later by Shipman & Sass (1980) as not very satisfactory: only three stars were available to test the relation, and two of these were 1.5σ from the predicted relation. Since then, other white dwarfs have been added to the test, but agreement between theory and observation has remained poor.

Various steps can be followed (and combined) in determining the empirical relation (Schmidt, 1996): (a) Surface brightness method, making use of the fundamental relation $f_\lambda = 4\pi(R^2/d^2)H_\lambda(T, \log g)$, which relates the observed monochromatic flux, f_λ, to the (Eddington) flux at the stellar surface H_λ, observed through solid angle R^2/d^2, where d is the distance and R the radius: if $T_{\rm eff}$ and $\log\, g$ are determined, generally from spectroscopy, then model atmospheres allow calculation of the energy flux at the stellar surface which, when compared to the flux on Earth, yields the angular diameter. The radius, R, is then obtained from the parallax and angular diameter, while the mass can be deduced from the spectroscopic gravity, $M = gR^2/G$. The method is the only one that can be applied to field white dwarfs, but yields masses only indirectly by appeal to model atmospheres. (b) Gravitational redshift: the strong gravitational field at the surface of a white dwarf causes a redshift of the spectral lines, the size of which depends on the gravitational velocity $v_{\rm grs} = GM/Rc$, where c is the speed of light. If $v_{\rm grs}$ can be measured and the gravity is known, then M and R can be obtained independently of the parallax. $v_{\rm grs}$ can be measured in members of binary systems, common proper motion pairs, or clusters, for which (in the case of bright white dwarfs which show narrow non-LTE line cores) the common component of the radial velocity allows separation of the gravitational redshift from the Doppler line shift due to the radial component of space motion. It has also been determined for single objects with very high quality spectroscopy, by comparing line shifts with respect to theoretical atmosphere models (e.g. in the case of Sirius B by Barstow *et al.*, 2005). (c) White dwarfs in visual binary systems: mass limits may be derived from radial velocities and orbital parameters through Kepler's third law if the parallax is known (Shipman, 1979). Radii are then derived from the knowledge of $T_{\rm eff}$ and distance.

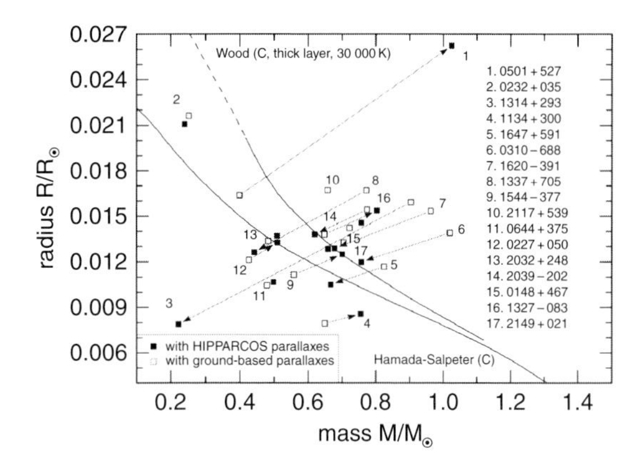

Figure 8.36 Comparison of masses and radii obtained with ground-based and Hipparcos parallaxes for the 17 DA white dwarfs. Masses are calculated via the surface gravities. Arrows show the change in parameters in going from the ground-based parallax to the Hipparcos parallax. From Vauclair et al. (1997b, Figure 5).

White dwarfs are very common, but with a faint intrinsic luminosity. As a result, reasonably bright dwarfs are relatively nearby, such that pre-Hipparcos ground-based parallaxes were already of reasonably high relative accuracy. The pre-Hipparcos comparison between theory and observation is described by Schmidt (1996), who argued that although errors in M and R were consistent with theory, the observational scatter on both quantities was too large to demonstrate a clear correlation.

Hipparcos contributions Hipparcos observed only 22 white dwarfs (11 field white dwarf, four known white dwarfs in visual binaries, and seven in common proper motion systems) among which the majority are of spectral type DA, but with one each of the spectral types DB, DC, DQ, DZ (Vauclair *et al.*, 1997b). The original object selection for the Hipparcos Input Catalogue was based on the white dwarf catalogue of 1279 objects of McCook & Sion (1987); the catalogue has subsequently been updated to include 2249 objects in the fourth edition (McCook & Sion, 1999), and more than 10 000 in the on-line version in 2007.

Vauclair *et al.* (1997a,b) and Provencal *et al.* (1997a, 1998) studied the complete sample of white dwarfs observed by Hipparcos. They are close to the faint magnitude limit of the Hipparcos observations, with a mean parallax accuracy of $\sigma_\pi \simeq 3.6$ mas. The mass–radius relation is nevertheless significantly narrower than the pre-Hipparcos situation (Figures 8.35 and 8.36). Most points are within 1σ of evolutionary models of white dwarfs with C cores and H surface layers (Provencal *et al.*, 1997a). The theoretical shape is still difficult to confirm, because of the small number of objects, the absence of objects in regions of either high or low mass, and because of different possible evolutionary sequences due to internal composition and unknown thickness of the H envelope (Vauclair *et al.*, 1997b), except for some particular stars. Other effects such as alterations due to strong internal magnetic fields (Suh & Mathews, 2000) are not yet testable.

Provencal *et al.* (1997a, 1998) developed the analysis, examining Hipparcos parallaxes and resulting radii for all 22 Hipparcos objects. For all seven common proper motion systems, the white dwarf has a gravitational redshift velocity determination (Reid, 1996), typically in the range 20–40 $\mathrm{km\,s^{-1}}$, as compared with $\sim$0.635 $\mathrm{km\,s^{-1}}$ in the case of the Sun. Some of the relevant results are shown in Figures 8.37 and 8.38. For the field white dwarfs, the Hipparcos and ground-based parallaxes generally agree to within 1σ, with the Hipparcos parallaxes being some 1.8 times more accurate than the ground-based values.

For the four white dwarfs in binary systems, Sirius B was the only star whose pre-Hipparcos parallax implied a location close to the expected theoretical mass–radius relations; the others (Procyon B, 40 Eridani B and Stein 2051) were at least 1.5σ below the theoretical position (Figure 8.37). After Hipparcos, the error on the radius is dominated by errors on flux and $T_{\rm eff}$, but no longer by the parallax. The parallax error still dominates the uncertainty on mass, except for Procyon where the error on the component separation dominates. The parallax of Stein 2051 was not improved by the Hipparcos observations. For the remainder, results (expressed in solar units) are: Sirius B ($\rho \sim 7$ arcsec): $M = 1.000 \pm 0.016$, $R = 0.0084 \pm 0.0002$; 40 Eri B ($\rho \sim 6943$ arcsec): $M = 0.501 \pm 0.011$, $R = 0.0136 \pm 0.0002$; Procyon B ($\rho \sim 4$ arcsec): $M = 0.604 \pm 0.018$, $R = 0.0096 \pm 0.0004$.

Sirius B is now more precisely located on the Wood (1995) mass–radius relation for DA white dwarfs of the observed $T_{\rm eff}$ with a thick H layer and C core (Holberg *et al.*, 1998). V471 Tau is also compatible with the thick H layer models. The estimated mass of 40 Eri B has been increased by 14%, placing the star back on the Hamada & Salpeter (1961) mass–radius relation for C cores, making it compatible with single star evolution without a thick H layer, and removing the disagreement between observation and theory which has been debated for many years. Sirius B (the most massive known white dwarf) and 40 Eri B (one of the less massive) fix the high-mass and low-mass limits of the mass–radius relation. The results provide evidence that not all DA white dwarfs have the same H surface thickness layer. A similar result was found for the specific case of 40 Eri B, using the same Hipparcos data, by Shipman *et al.* (1997).

Iron or strange cores Provencal *et al.* (1998) drew attention to the fact that three objects, the visual binary Procyon B, and two field dwarfs, GD 140 and EG 50, have radii much smaller than predicted by their inferred masses based on models with C cores, and would be better accounted for with Fe or Fe-rich core models (Provencal *et al.*, 1997b; Panei *et al.*, 2000a). For Procyon B, a reduction in the error of $T_{\rm eff}$ from 200 to 60 K would allow still clearer confirmation of this possibility. The results for Procyon B were indeed later revised by Hubble Space Telescope observations, as noted at the end of this section.

Although a Fe core composition had previously been considered as unexpected, and its reality is still debated, Isern *et al.* (1991) had earlier predicted that an explosive ignition of electron degenerate O/Ne/Mg cores may give rise to the formation of neutron stars, supernovae or iron white dwarfs. Extremely compact 'strange dwarfs', comprising quark matter at densities of $5 \times 10^{14}\ \mathrm{gm\,cm^{-3}}$, could also have a much smaller radius than a standard white dwarf, while evolving in a broadly similar way (Glendenning *et al.*, 1995a,b; Benvenuto & Althaus, 1996a,b).

Panei *et al.* (2000a) accordingly presented detailed mass–radius relations for models with various cores, with a He layer containing 1% of the stellar mass, and with and without a H envelope, considering masses from 0.15–$0.5M_\odot$ for a He core, from 0.45–$1.2M_\odot$ for C, O, and Si cores, and from 0.45–$1.0M_\odot$ for Fe cores. They also explored the effects of gravitational, chemical and thermal diffusion on low-mass He models with H and He envelopes. They found that for 40 Eri B, the observed mass, radius and $T_{\rm eff}$ are consistent with models having a C, O, or Si interior and a thin H envelope. The other objects all fall below these standard composition sequences, indicating a denser interior: there is little difference between the mass–radius relation for He, Ca, O, or Si cores because their mean molecular weight per electron is very similar, $\mu_e \simeq 2.00$. If GD 140 and Procyon B were assumed to have an Fe core (but see the comment at the end of this section), resulting in a very different sequence given $\mu_e \simeq 2.15$, they not only fell on a plausible evolutionary sequence, but one for which the model $T_{\rm eff}$ is consistent with the observed value. Tantalisingly, the radius of EG 50 is smaller than that predicted even for an Fe core object for the observed $T_{\rm eff}$, and thus is seemingly even denser than an Fe white dwarf (Figure 8.39).

Panei *et al.* (2000b) investigated further details of the evolution of an Fe-core white dwarf, determining the neutrino luminosity, surface gravity, crystallisation, internal luminosity profile and ages. They found that their evolution is markedly different from that of their C–O counterparts: cooling is strongly accelerated, by up to a factor of 5 for models with a pure

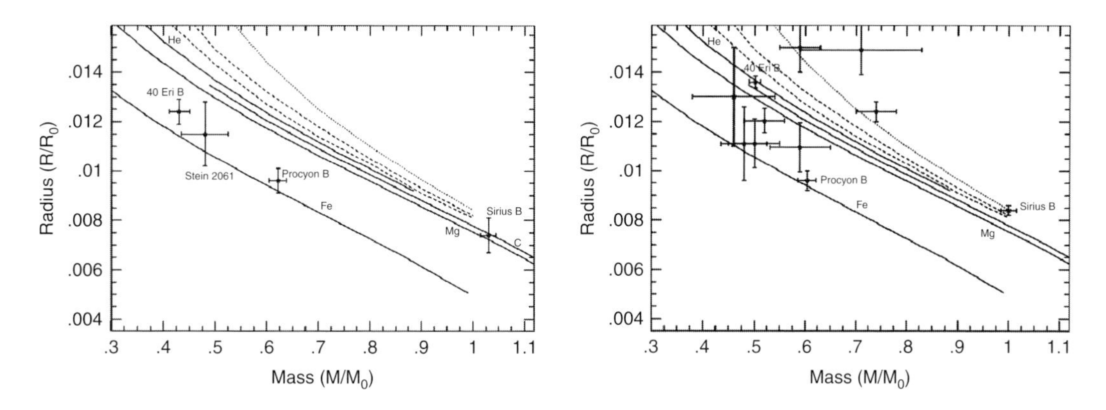

Figure 8.37 Left: the white dwarf mass–radius relation based on ground-based parallaxes prior to Hipparcos. The solid lines labelled He, C, Mg, and Fe denote the zero-temperature mass–radius relation of Hamada & Salpeter (1961). The dotted line is a 30 000 K H atmosphere relation from Wood (1995), and the dashed lines are those for 15 000 and 8000 K H-surface white dwarfs. Right: the white dwarf mass–radius relation after Hipparcos, showing revised positions for the visual binaries and including results from the common proper-motion systems. From Provencal et al. (1998, Figures 1 and 2).

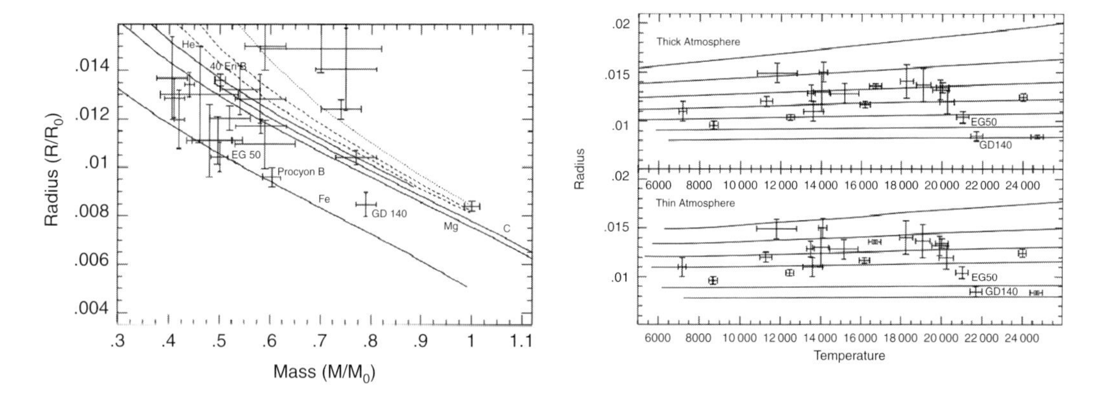

Figure 8.38 Left: the white dwarf mass–radius relation, showing the positions of the visual binaries, common proper-motion systems, and field white dwarfs. The field white dwarf masses were derived using published surface gravity measurements and radii based on Hipparcos parallaxes. Right: predicted masses for white dwarfs based on the model of Provencal et al. (1998). The top panel uses models with thick, log $q(H) = -4$, surface layers, and the second has log $q(H) = 0$. The solid lines are white dwarf cooling curves at constant mass, beginning at $0.4M_\odot$; and increasing by steps of 0.1 sequentially downward. From Provencal et al. (1998, Figures 3 and 4).

Fe composition. However, if iron white dwarfs were very numerous, some of them would have had time enough to evolve to lower luminosities than that corresponding to the fall-off in the observed luminosity function.

Mathews *et al.* (2006b) and Mathews *et al.* (2006a) provided further arguments that GD 140, EG 50, and six other white dwarfs have deduced masses and radii marginally consistent with strange-matter cores, with the alternative interpretation of Fe cores difficult to reproduce from stellar evolution models (Iben & Renzini, 1983; Woosley & Weaver, 1986, 1995). They also discussed the possible observational consequences of colour superconductivity for the strange cores, either in the colour-flavour locked (CFL) or two-flavour superconductor (2SC) states.

For Procyon B, subsequent Hubble Space Telescope STIS observations by Provencal *et al.* (2002) in fact yielded a revised $T_{\rm eff} = 7740 \pm 50$ K; the lower temperature implies a larger radius of $R = 0.01234 \pm 0.00032 R_\odot$. This places it back on the locus of C core white dwarfs, without needing to invoke an iron core. Iron or strange cores remain a non 'main stream' interpretation for the observations, and future improved observations will be required to advance the debate.

Sirius Sirius is the prototype of the class of Sirius-like binaries, comprising a main sequence or evolved star,

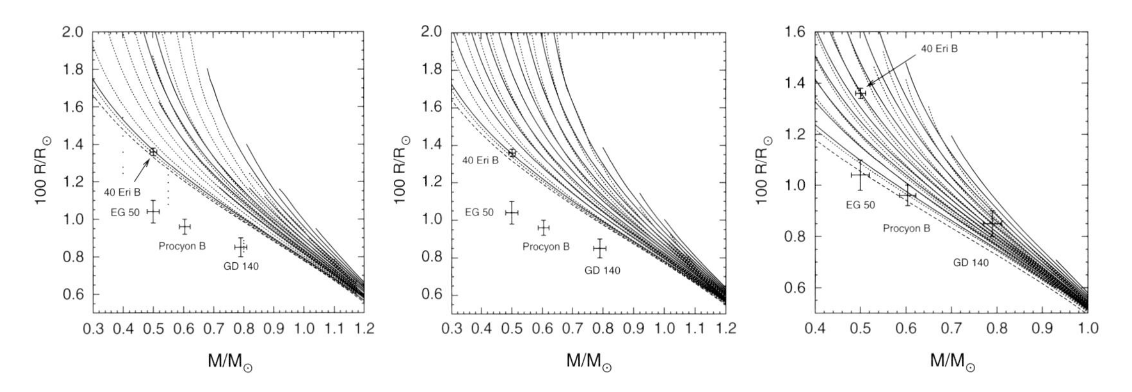

Figure 8.39 Left: mass–radius relation for white dwarfs with a C core surrounded by a He layer with a thickness of $10^{-2}M_\odot$. Solid and short dashed lines are for models with and without an outermost H envelope. The medium-dashed line corresponds to the mass–radius relation for homogeneous C models from Hamada & Salpeter (1961). $T_{\rm eff}$ ranges from $5 - 145 \times 10^3$ K, and the figure includes 'strange dwarf' models with $T_{\rm eff}$ ranging from $10 - 50 \times 10^3$ K. Middle: same, but for an O core. Right: same, but for an Fe core (note change in vertical scale). Data corresponding to 40 Eri B, EG 50, Procyon B and GD 140 are included, as taken from Provencal et al. (1998). Note that the position of Procyon B has since moved back to the C core locus with the results of Provencal et al. (2002). From Panei et al. (2000a, Figures 4a, 5 and 7).

and a non-interacting white dwarf secondary. Assuming that the pair have never interacted, the white dwarf is inferred to have evolved from an object more massive than the current primary. Such systems can be used to investigate the initial-to-final mass relation. In addition, if the two components are resolved and an astrometric mass determined for the degenerate star, they can be used to investigate the mass–radius relation.

Sirius B itself is in a wide visual orbit around the A0 V star, Sirius A, with an orbital period of about 50 yr. Holberg *et al.* (1998) published a detailed study of Sirius B, using the Hipparcos distance, and a revised $T_{\rm eff} = 24\,790 \pm 100$ K and $\log g = 8.57 \pm 0.06$ from IUE (International Ultraviolet Explorer) and EUVE (Extreme Ultraviolet Explorer) observations. Combining the spectroscopic results with the gravitational redshift of $89 \pm 16\,{\rm km\,s^{-1}}$ (Greenstein *et al.*, 1971) to derive a mass, and combining this with the existing astrometric mass, yields $M = 1.034 \pm 0.026 M_\odot$ and $R = 0.0084 \pm 0.00025 R_\odot$, consistent with the theoretical mass–radius relation for a C-core white dwarf. The EUVE spectrum provides an upper limit of ${\rm He/H} = 1.8 \times 10^{-5}$ in the photosphere.

Barstow *et al.* (2005) also used the Hipparcos parallax and Hubble Space Telescope STIS observations of the hydrogen Balmer lines, to derive $T_{\rm eff} = 25\,193 \pm 37$ K, $\log g = 8.556 \pm 0.010$, and $v_{\rm grs} = 80.42 \pm 4.83\,{\rm km\,s^{-1}}$, the latter from the shift of the narrow Hα core with respect to a model profile for the observed temperature and gravity. Their best estimates of the white dwarf mass and radius are $M = 0.978 \pm 0.005 M_\odot$ and $R = 0.00864 \pm 0.00012 R_\odot$. From various arguments, Liebert *et al.* (2005) derived a progenitor mass for Sirius B of $5.056^{+0.374}_{-0.276} M_\odot$.

White dwarfs in the Hyades The Hyades cluster contains a dozen known white dwarfs (e.g. Humason & Zwicky, 1947; Luyten, 1954, 1956; Böhm-Vitense, 1995). They typically have $V \gtrsim 14$ mag, too faint to have been observed directly by Hipparcos. However, several are contained in multiple systems (e.g. Lanning & Pesch, 1981), the primary components of which were observed. The present-day luminosity function of the Hyades predicts that the cluster contains about 25–30 white dwarfs (Chin & Stothers, 1971). The discrepancy between the observed and predicted numbers is possibly explained by evaporation from the cluster (e.g. Weidemann *et al.*, 1992; Eggen, 1993a,b). One possible example of an escaped white dwarf is HIP 12031 (Perryman *et al.*, 1998), located at more than 40 pc from the cluster centre.

V471 Tau (HIP 17962) is a post-common-envelope detached eclipsing binary. It is composed of a DA white dwarf and a coronally active K2V star. Hosting the hottest and youngest Hyades white dwarf, it has been studied extensively (e.g. Nelson & Young, 1970; Guinan & Sion, 1984; Clemens *et al.*, 1992; Shipman *et al.*, 1995; Marsh *et al.*, 1997). The system is a pre-cataclysmic variable: the K star does not yet fill its Roche lobe. Periodic optical and X-ray variations are related to material from the K-star wind being accreted onto the magnetic poles of the rotating white dwarf (e.g. Jensen *et al.*, 1986; Barstow *et al.*, 1992). The effective temperature and surface gravity were determined by fitting synthetic spectral energy distributions to observed spectra of the hydrogen Lyman lines using far-ultraviolet observations made with the first ORFEUS mission by both Barstow *et al.* (1997), who estimated $T_{\rm eff} = 32\,400^{+270}_{-800}$ K, and by Werner & Rauch (1997), who estimated $T_{\rm eff} = 35\,125$ K. The relatively large uncertainty in the best-fit $\log g$ values prevents an accurate mass determination based on the surface gravity. It has therefore been common practice to infer the surface gravity of the white dwarf from its astrometric mass, obtained from the orbital elements of the binary, and its radius, obtained from its observed and modelled flux combined with its Hipparcos parallax. A more precise estimate of the white dwarf radius was made using its secular parallax determined by de Bruijne *et al.* (2001): $\pi_{\rm sec,Hip} = 21.00 \pm 0.40$ mas compared with $\pi_{\rm Hip} = 21.37 \pm 1.62$ mas. The long time-baseline Tycho 2 secular parallax, which might be preferred over the Hipparcos secular parallax in view of the binary nature of the system, although with only $P_{\rm orb} = 0.521$ d (Stefanik & Latham, 1992), places the object at a slightly larger distance, with $\pi_{\rm sec,Tycho\,2} = 20.56 \pm 0.33$ mas. The Hipparcos secular parallax fits the Werner & Rauch (1997) model for a radius of $R = 0.0098 \pm 0.0011 R_\odot$; the corresponding surface gravity is $\log g = 8.34 \pm 0.10$ for $M = 0.759 \pm 0.020 M_\odot$. These values are consistent with but more precise than the results of Werner & Rauch (1997) and Barstow *et al.* (1997).

Böhm-Vitense (1993) reported the serendipitous discovery of a DA white dwarf companion around the close F6V/F6V binary HD 27483 (HIP 20284). She interpreted her spectra using the unblanketed white dwarf models of Wesemael *et al.* (1980), and the Hamada & Salpeter (1961) mass radius relation, assuming a distance to the system of 47.6 pc, according to the secular parallax derived by Schwan (1991). Burleigh *et al.* (1998) presented an analysis of the object, based on updated atmosphere and evolutionary models, using its Hipparcos parallax of $\pi_{\rm Hip} = 21.80 \pm 0.85$ mas. The orbital motion of the binary, with $P_{\rm orb} = 3.05$ d (Mayor & Mazeh, 1987), has not perturbed the Hipparcos measurements: its secular parallax, $\pi_{\rm sec,Hip} = 20.59 \pm 0.35$ mas, is well defined. De Bruijne *et al.* (2001) used cubic spline interpolation in Table 5 of Burleigh *et al.* (1998) to derive $\log g$, $T_{\rm eff}$, M, and R for both the Hipparcos trigonometric and secular parallaxes. The secular parallax white dwarf mass, $M = 0.70 \pm 0.04 M_\odot$, is significantly smaller than the value of $M = 0.94 M_\odot$ derived by Burleigh *et al.* (1998). This new estimate thus resolves the problem identified by Burleigh *et al.* (1998), that the sum of the cooling age of a $M = 0.94 M_\odot$ white dwarf and the evolutionary age of its progenitor is significantly shorter than the nuclear age of the Hyades.

Other individual white dwarfs Burleigh *et al.* (2001) obtained far ultraviolet spectroscopy from FUSE and EUVE to determine the mass of the white dwarf companion of the 4 mag A1 III star β Crt of $0.43 M_\odot$. The orbital period is highly uncertain, although the Hipparcos data suggest a binary period of about 10 yr. A low-mass white dwarf in a system with such a long period would be difficult to explain by the conventional model for low-mass white dwarfs, which explains their existence in terms of Roche lobe overflow in a binary system. The system could instead be a remnant of Algol-type evolution. In an Algol-type binary, a cool F–K III–IV secondary star fills its Roche lobe, and transfers mass to a hot B–A V primary. These systems represented a problem for a long time, because the less massive secondary is more evolved than its companion. The paradox was resolved when it was realised that the current secondary was originally the more massive star, and that the current configuration of the binary was the result of extensive mass transfer, resulting in a reversal of the mass ratio of the system. A remnant of such an Algol-type binary would then consist of an early-type intermediate-mass star with a low-mass white dwarf companion. Although Algols are the most numerous of the known eclipsing binaries, their post-mass transfer counterparts have eluded discovery because of their bright companions. If β Crt is the first confirmed post-Algol system, its currently uncertain orbital period would be expected to be some tens of days or less, rather than the longer period hinted at by the Hipparcos data.

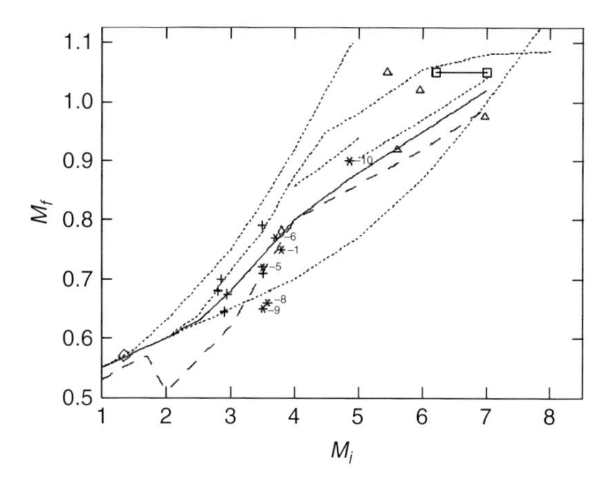

Figure 8.40 Initial-to-final mass relations. From top to bottom, relations are from Girardi et al. (2000), Herwig (1995), Marigo (1998), Dominguez et al. (1999), Weidemann (1987). Long dashed line: first thermal pulse relation. Full line: revised initial-to-final mass relation as given in Table 3 of Weidemann (2000), and partly based on the Hipparcos-derived age of the Hyades cluster. From Weidemann (2000, Figure 2).

Makarov (2004) analysed the Hipparcos astrometric solution for the nearest white dwarf, van Maanen 2 (HIP 3829), at a distance of just 4.4 pc. It is listed in the Hipparcos Double and Multiple Systems Annex as an 'acceleration solution', with $\dot\mu_{\alpha*} = 33.87 \pm 9.71$,

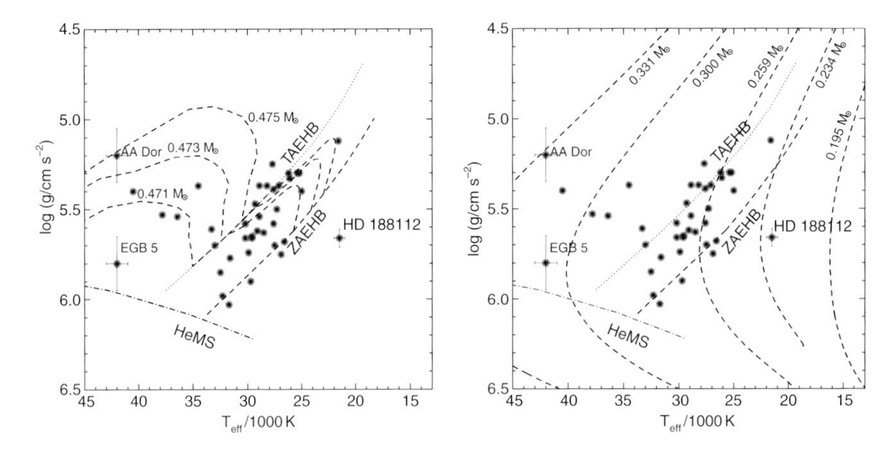

Figure 8.41 Left: HD 188112 in the $T_{\rm eff}$, log g plane. Other points are sdB stars in close binary systems with known periods (Morales-Rueda et al., 2003). Inclined labelled lines show the zero-age and terminal-age extreme horizontal branch. Other dashed lines indicate models for post-extreme horizontal branch evolution for masses of 0.471, 0.473, and 0.475$M_\odot$ from Dorman et al. (1993). Right: the data points are as left, but now showing evolutionary models for post red giant branch evolution from Driebe et al. (1998). From Heber et al. (2003, Figures 3 and 4).

$\dot{\mu}_\delta = 6.41 \pm 7.36$ mas yr^{-2}. After eliminating perspective acceleration as a possible explanation, Makarov (2004) argued that the acceleration is caused by a companion, citing investigations of the AC 2000.2 Catalogue as confirming the long-term proper motion determined by Hipparcos. From an analysis of the Hipparcos Intermediate Astrometric Data, he determined an orbital fit with $a_0 = 27$ mas, $P = 1.57$ yr, $i = 89°$, $\omega = 95°$, $\Omega = 92°$, and $T_0 = 1991.26$. Assuming a mass of $0.83\,M_\odot$ for the primary white dwarf leads to a mass of the secondary of $0.06 \pm 0.02 M_\odot$. The Hipparcos results thus give further evidence for an astrometric companion to van Maanen 2, which had long been suspected (van de Kamp, 1971; Gatewood & Russell, 1974).

Initial-to-final mass relation The existence of a relation between initial main-sequence stellar mass, and the final white dwarf mass, has been discussed by various authors. Weidemann (1977) investigated the Hyades white dwarfs, concluding that the data favoured relatively strong mass loss during the stellar lifetime, and indicating a possible upper limit to the initial mass of around $5M_\odot$. Weidemann & Koester (1983) presented a detailed method to obtain progenitor masses for individual white dwarfs: determining the cooling age from spectroscopically derived $\log g$ and $T_{\rm eff}$ by comparison with theoretical white dwarf cooling tracks, followed by subtraction of the cooling age from the cluster age to obtain the total pre-white dwarf evolutionary age, which is then fitted to theoretical stellar evolution model calculations of total ages as a function of initial mass and composition. An upper mass limit around $8M_\odot$ for white dwarf production was established, agreeing with the then theoretical upper mass limit for the development of a degenerate C–O core of around 8–9$M_\odot$ (Iben & Renzini, 1983). It was thus predicted that all intermediate-mass stars undergo heavy mass loss in order to reach the relatively small final white dwarf masses, an idea seemingly confirmed by observations of asymptotic giant branch stars which produce circumstellar shells and planetary nebulae. Further support comes from recent stellar evolution models which predict core masses at the beginning of the thermally pulsing asymptotic giant branch which practically coincide with final white dwarf masses derived for NGC 3532 and the binary white dwarf PG 0922+162 (Weidemann & Koester, 1983).

Weidemann (2000) revised his earlier initial-to-final mass relation (Weidemann, 1987), based on various new observational data, including a revision of the relation for the Hyades white dwarfs. These objects are now relocated at higher initial masses, not because of any change in the estimated distance to the cluster, but because the Hipparcos data lead to a significantly reduced cluster age of 625 ± 50 Myr (Perryman *et al.*, 1998), thus moving them closer temporally to the first thermal pulse. A comparison with theoretically predicted initial-to-final mass relations supports evolutionary models with exponential diffusive overshoot, which undergo core mass reduction by a strong third dredge-up. The upper mass limit for C–O white dwarf production was estimated to be smaller than previously assumed, at around $6.5M_\odot$, yielding resulting white dwarf masses below $1M_\odot$. More massive white dwarfs must have Ne–O cores after off-centre C burning, or are due to mergers or rotational lifting. Differential mass loss seems to be small, justifying the use of an universal initial-to-final mass

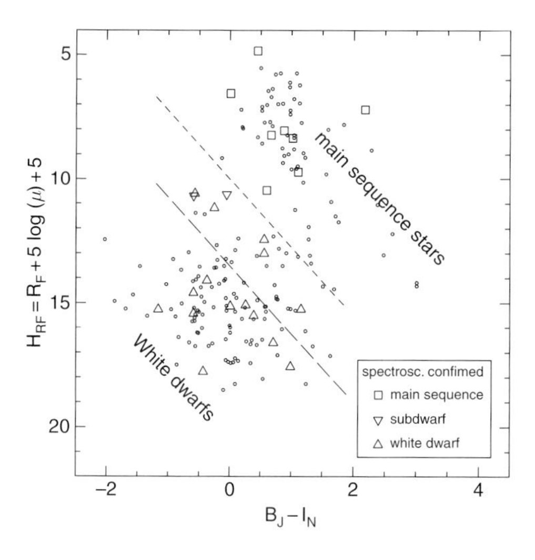

Figure 8.42 Reduced proper motion diagram of ultraviolet-bright sources with large proper motions studied by Lanning & Lépine (2006). The sources are distributed around two separate loci, depending on their spectral class: white dwarfs populate the lower left of the diagram, while main-sequence stars are found above and to the right of the white dwarf locus. Stars located in between the two main groups are possibly subdwarfs. This segregation is confirmed by the location of stars which have formal spectroscopic classifications (given by the large symbols identified in the legend). From Lanning & Lépine (2006, Figure 2).

relation for calculations of stellar and Galactic evolution (Figure 8.40).

Heber *et al.* (2003) reported a Hipparcos object which appears to be a rather convincing He-core white dwarf progenitor. At a distance of 80 pc, $\pi_{\rm Hip} = 12.33 \pm 1.70$ mas, HD 188112 (HIP 97962) is a $V = 10.2$ mag sdB star (subluminous B star), a class usually identified with the core He burning extreme horizontal branch with masses around $0.5 M_\odot$ (Section 7.5.6). Such stars evolve directly to the white dwarf cooling sequence omitting the asymptotic giant branch phase. Their spectral analysis places it below the extreme horizontal branch, with $T_{\rm eff} = 21\,500 \pm 500$ K, $\log g = 5.66 \pm 0.05$, and in a close binary system with $P = 0.606\,585$ d. From the atmospheric parameters and the Hipparcos parallax, they concluded that the sdB star is of low mass, $\sim 0.24^{+0.10}_{-0.07} M_\odot$, too low to sustain core He burning, and thus now evolving into a He core white dwarf with an unseen companion. Encouragingly, its inferred mass is fully consistent with the appropriate evolutionary tracks (Figure 8.41). It could be a pre-supernova Type Ia system (sdB + massive white dwarf) if the total binary mass is above the Chandrasekhar limit of $1.4 M_\odot$, or a post-supernova system (sdB + neutron star) if the companion mass alone is above that limit.

Other white dwarf studies Schröder *et al.* (2005) used their model of the local stellar population, which matches the number counts of Hipparcos stars within 100 pc in specific regions of the HR diagram (see Section 7.5.1), to predict the distribution of white dwarfs in the solar neighbourhood as a function of mass and effective temperature. They found good agreement between their model and the number of white dwarfs contained in the SPY (Supernova Ia Progenitor Survey) observing campaign for double-degenerate binaries, undertaken at the ESO VLT, and containing 1078 prospective objects (Napiwotzki *et al.*, 2001a; Koester *et al.*, 2001).

Pauli *et al.* (2003) presented kinematics of a sample of 107 DA white dwarfs from the SPY project. They used their own radial velocity and spectroscopic distance measurements combined with proper motions to determine space motions, and hence Galactic orbits. They used the kinematic criteria for assigning population membership deduced from a sample of F and G stars taken from the literature for which chemical criteria can be used to distinguish between thin disk, thick disk and halo. Candidates for thick disk and halo members were selected in a first step from the classical UV-velocity diagram, and final assignment of population membership was based on orbits and position in the (angular momentum) J_z-eccentricity diagram. They found four halo and twelve thick disk white dwarfs. The study is included here as a first study of the population-dependent kinematics of white dwarfs. While not drawing directly on the Hipparcos data, both the proper motion reference frame used for the space velocity studies, and the development of chemical criteria to distinguish between thin disk, thick disk, and halo population studies, are themselves based on Hipparcos kinematic data.

Zoccali *et al.* (2001) used eight local white dwarfs with known trigonometric parallaxes, including three measured by Hipparcos, to match the local white dwarf cooling sequence with that observed in the globular cluster 47 Tuc. They derived a distance modulus of $(m-M)_V = 13.27 \pm 0.14$ and, from the apparent magnitude of the main sequence turn-off based on Hubble Space Telescope data, derived an age of 13 ± 2.5 Gyr.

Lanning & Lépine (2006) examined 572 ultraviolet-bright sources from the Sandage two-colour survey of the Galactic plane. They used reduced proper motions (Section 5.2.5) based on proper motions from the USNO B1.0 and Tycho 2 Catalogues to argue that some two thirds of the high proper motion sources are likely to be previously-unidentified white dwarfs (Figure 8.42). Kilic *et al.* (2005) reported the discovery of 282 white dwarfs from the SDSS data release 2 imaging area, again using on reduced proper motions based on USNO plate astrometry.

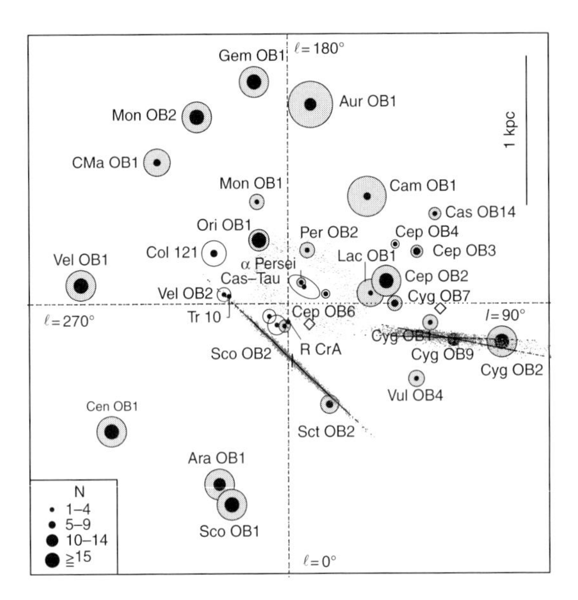

Figure 8.43 Monte Carlo estimates of the birthplaces for PSR B1929+10 (= PSR J1932+1059), as also studied by Hoogerwerf et al. (2001), and PSR B2021+51 (= PSR J2022+5154) with respect to the OB associations in the solar neighbourhood from de Zeeuw et al. (1999, see also Figure 6.26). Circle sizes represent projected association dimensions, enlarged by a factor 2 with respect to the distance scale. The size of the central dots indicates the degree of current or recent star formation activity, as given by the number N of stars more luminous than absolute magnitude $M_V \sim -5$. ♦: current VLBI pulsar positions. The motion is traced back for the duration of the pulsar spin-down age (the Hipparcos connection is in tracing the motion of the OB associations backwards in time). Since this work was published, the pulsar astrometry has improved further: the parallax uncertainties have decreased from ∼ 0.2 mas to about 0.08 mas, and the proper motions to about 0.4 mas yr^{-1} (Brisken et al., 2002). From Campbell (2001, Figure 1).

8.4.9 Supernovae, pulsars, and neutron stars

Reynolds *et al.* (1995) aligned the optical and radio images of SN 1987A, using the newly-available Hipparcos reference frame, and showed that the centres of the optical and radio emission coincide more closely than previously thought, indicating that the radio emission is due to interaction between an expanding shock front and the surrounding medium.

Caraveo *et al.* (1998), see also Caraveo *et al.* (1997), used the Hipparcos reference frame to locate the optical position of the radio-quiet isolated neutron star, Geminga (radio emission from Geminga was eventually reported by Malofeev & Malov 1997). At $V = 25.5$ mag, they bridged the magnitude range of the Hipparcos reference frame using a five-step transfer procedure, starting with astrometric plates taken in Torino covering 19 Hipparcos/Tycho reference stars, and leading to CCD images obtained with the Hubble Space Telescope. The final error with respect to the ICRS was 40 mas per coordinate, representing a 25-fold improvement in the previous absolute source position. An accurate absolute position is necessary for accurate pulsar timing, which is normally provided by radio observations for the radio pulsars. This offered the prospects of phase-referencing some 20 years of γ-ray photons (from SAS-2, COS-B, and EGRET), pulsed at the ∼ 237 ms spin period determined from ROSAT X-ray data. Such phase stability, over the corresponding $\sim 2.5 \times 10^9$ spin periods, was needed to determine the second derivative of the spin period, and hence the braking index, $n = \ddot{\nu}\nu/\dot{\nu}^2$. For a canonical braking index of $n = 3$, $\ddot{\nu} \simeq 2.7 \times 10^{-26}\,\mathrm{s}^{-3}$ was expected. Subsequent interpretation was complicated by a glitch in 1996, while the EGRET ephemeris yielded a pre-glitch value of $\ddot{\nu} \simeq 1.49(3) \times 10^{-25}\,\mathrm{s}^{-3}$ (Jackson & Halpern, 2005). Related discussions of the space motion and birthplace are given by Pellizza *et al.* (2005) and De Luca *et al.* (2006). The possibility that Geminga may have been ejected from the λ Orionis association (Hoogerwerf *et al.*, 2001, see Section 8.4.2) was rejected by Pellizza *et al.* (2005).

Campbell (2001) traced the motion of the pulsars PSR B1929+10 (PSR J1932+1059) and PSR B2021+51 (PSR J2022+5154) backwards in time for the duration of their spin-down age (Figure 8.43), and compared their inferred birthplaces with the Hipparcos astrometry of the OB associations in the solar neighbourhood from de Zeeuw *et al.* (1999). For the former, they found similar results to those given by Hoogerwerf *et al.* (2001).

Freire *et al.* (2001) derived the mean common proper motion, from individual timing solutions, for five of the known millisecond pulsars in the globular cluster 47 Tuc, obtaining $\mu_{\alpha*} = 6.6 \pm 1.9$ mas yr^{-1} and $\mu_\delta = -3.4 \pm 0.6$ mas yr^{-1}. This is in good agreement with that derived optically by Odenkirchen *et al.* (1997), based on Hipparcos data, of $\mu_{\alpha*} = 7.0 \pm 1.0$ mas yr^{-1} and $\mu_\delta = -5.3 \pm 1.0$ mas yr^{-1} (see Section 9.11.5).

8.5 Local interstellar medium

This section deals with the use of Hipparcos stars as tracers of the local interstellar medium. It might more logically have been placed within (the already rather lengthy) Chapter 9.

8.5.1 Local bubble

Introduction The Sun is thought to be moving just within the boundaries of a warm ($T \sim 7000$ K) low-density ($\sim 0.1\,\mathrm{cm}^{-3}$), and partially-ionised interstellar cloud. The heliosphere results from the expanding high-velocity solar wind within this cloud, and extends from the Sun to about 100–150 AU in the direction of the solar

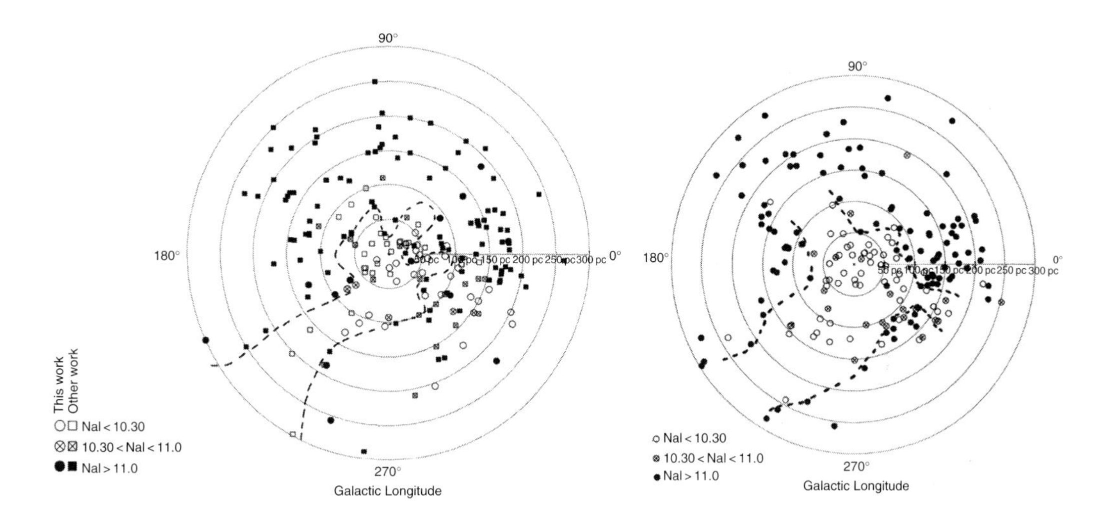

Figure 8.44 Polar projection of the line-of-sight Na I column densities as a function of distance from the Sun and Galactic longitude. Only stars with Galactic latitude between $-35° < b < +35°$ are plotted. The dotted line delineates a possible neutral absorption boundary to the Local Bubble. Left: using pre-Hipparcos distance estimates, from Welsh et al. (1994, Figure 3). Right: using Hipparcos distances, from Welsh et al. (1998, Figure 2).

motion; the termination shock, preceding the transition to the ambient interstellar medium, was crossed by Voyager 1 in December 2004. This and other nearby clouds are themselves located in a low-density region of the interstellar medium referred to as the local cavity. This is partially filled with hot ($\sim 10^6$ K), low number density ($n_{\rm H}$ $\sim$0.005 cm^{-3}) coronal gas, of approximate dimensions 100 pc; the hot component of the local cavity is referred to as the Local Bubble (Cox & Reynolds, 1987; Frisch, 1995). This Local Bubble is detectable in soft X-rays, and is perhaps the result of one or more nearby supernova explosions within the last 10 Myr, and/or the interaction of the stellar winds from OB stars in the nearby Sco–Cen association (Breitschwerdt, 2001). Although most of the volume in the Local Bubble appears to be filled with hot gas, most of the mass is contained within either warm and partially-ionised plasma, or several diffuse interstellar clouds of very low neutral gas density.

The exact size and shape of the tenuous Local Bubble has been much debated. Its dimensions can be determined directly by measuring the extent of the hot emitting bubble gas, or more indirectly by tracing the absence of absorption by neutral interstellar gas. Direct absorption measurements of the interstellar neutral hydrogen column density out to 100 pc at the required levels of N(H I)$<$ 10^{18} cm^{-2} have generally been problematic, and other indirect means of determining its extent have therefore been used. Paresce (1984) used both direct and indirect estimates of N(H I) absorption towards 82 stars closer than 250 pc to infer the absence of dense neutral gas out to $\sim$100 pc in most Galactic directions. Observations of the diffuse soft X-ray background radiation were modelled by Snowden *et al.* (1990) to reproduce the observed negative correlation between X-ray intensity and neutral interstellar hydrogen column density out to 300 pc. Diamond *et al.* (1995) used ROSAT Wide-Field Camera observations of extreme ultraviolet sources to model the distribution of inferred neutral hydrogen column densities as a function of distance out to 150 pc. Snowden *et al.* (1998) used ROSAT data to determine a typical radius of 50 pc for the Local Bubble.

Based on the fact that the two interstellar Na I D1 and D2 absorption lines at 589 nm are believed to be a good tracer of relatively cold, T < 1000 K, neutral gas clouds in the general interstellar medium, Welsh *et al.* (1994) used the observed distribution of interstellar Na I absorption within 250 pc of the Sun to infer the contours of neutral gas absorption in the local interstellar medium.

All of the above representations of the Local Bubble are dependent on knowledge of the distances towards the respective line-of-sight stellar targets towards which the level of absorption is measured. For stars closer than 20 pc, these distances, mostly derived from ground-based parallax measurements, were generally accurate to within 20%, whereas distance estimates for stars beyond 20 pc were of far greater uncertainty. The new distance determination for stars within 300 pc by Hipparcos allow the Local Bubble absorption characteristics to be defined with much greater certainty.

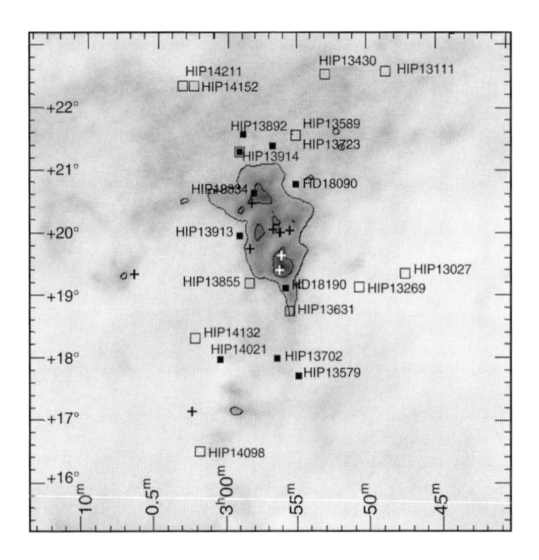

Figure 8.45 Linear grey-scale image showing the extent of the IRAS 100 μm emission from the star-forming cloud MBM 12, and the Hipparcos stars for which the presence of interstellar sodium absorption was used to constrain the distance. ■: stars previously observed to determine the cloud distance; □: stars observed as part of the Hipparcos-based study; black '+': known T Tauri stars; white '+': the two lines-of-sight towards the cataclysmic variable H 0253+193 and the G9 star DC48 which are known to have A_V > 5 mag. From Hearty et al. (2000a, Figure 1).

Interstellar medium morphology from absorption line measurements The study of interstellar cloud morphology is a relatively new research topic for which Linsky *et al.* (2000) proposed the name 'astronephography'. They used principally Hubble Space Telescope measurements of the H I column density from the spectra of hot white dwarfs and B-type stars, and the assumption that the clouds have a constant density. Another approach is to use Na I lines due to local interstellar gas observed in stars within and beyond these extended clouds.

Welsh *et al.* (1998) reconstructed the Na I absorption data originally presented by Welsh *et al.* (1994) using Hipparcos distances to investigate the distribution of neutral gas in the local interstellar medium. The new data (Figure 8.44) show that the picture of the detailed Galactic distribution of Na I absorption has changed significantly from that originally determined from ground-based distance estimates, and that the Local Bubble is some 50% larger than previously believed.

Sfeir *et al.* (1999) and Lallement *et al.* (2003) explored the structure of the Local Bubble by combining equivalent widths of the interstellar Na I D-line doublet with stellar distances from the Hipparcos parallaxes to determine 3d absorption maps of the local distribution of neutral gas towards 1005 lines-of-sight out to ~350 pc. The maps reveal 'interstellar tunnels' of different widths which connect the Local Bubble to surrounding cavities, supporting the model of Cox & Smith (1974) in which expanding supernova-driven bubbles interact and merge to form large-scale interstellar cavities. The neutral gas boundary to the Local Bubble is consequently determined to an accuracy of around 20 pc in many directions. Specifically, it is extended in a direction perpendicular to the plane of the Gould Belt (Section 6.10.5). Scenarios for the origin of the Gould Belt are various: the oblique impact of a high-velocity cloud with the Galactic disk, highly-energetic supernovae, or a gamma ray burst (Perrot & Grenier, 2003). Lallement *et al.* (2003) speculate that the Local Bubble could be the remnant of the central region of the burst, a scenario which might also explain why the Local Bubble 'chimney' (Welsh *et al.*, 1999; Crawford *et al.*, 2002) opens towards more intermediate- and high-velocity clouds compared to other regions of the halo, if these clouds represent ejected burst material now falling back onto the disk.

Magnani *et al.* (1985, MBM) made a CO survey of high-latitude molecular gas, and presented a catalogue of 57 clouds in 35 complexes. The mean distance to the clouds is about 100 pc; they are the nearest molecular clouds to the Sun, and may contain the nearest regions of star formation. The two nearest are MBM 12 and MBM 20. They include several T Tauri stars, and were considered to lie within or at the edge of the Local Bubble. Distances were estimated by Hearty *et al.* (2000a) using spectroscopic observations of Hipparcos stars, of known distance, along the line-of-sight to both clouds. Essentially, stars showing narrow interstellar Na I absorption are presumed to lie behind the cloud, and *vice versa*, thus constraining the distances without needing to assume a spectral type or luminosity as needed to interpret earlier spectroscopic parallax distance estimates. They found the distance to MBM 12 and MBM 20 to be $d = 58$–90 pc and $d = 112$–161 pc respectively (Figure 8.45).

Grant & Burrows (1999), see also Grant (1999), reported distance determinations for the two high-Galactic latitude cloud complexes, G192–67 and MBM 23–24. Thirty-four early-type stars were observed towards the two clouds, more than half of which have Hipparcos parallaxes. Interstellar Na I D absorption lines were detected towards some of the stars, which enabled estimates of the distances to the clouds of 109 ± 14 pc for G192–67 and 139 ± 33 pc for MBM 23–24. The relationship of the clouds to the Local Bubble and the local neutral hydrogen cavity is discussed.

Various other related studies have been undertaken using the Hipparcos distances. Cha *et al.* (2000), see also Cha (2000), studied the local interstellar medium in the direction of Puppis–Vela ($245° < l < 275°$, $-15° < b < +5°$, $d < 200$ pc), using observations of the

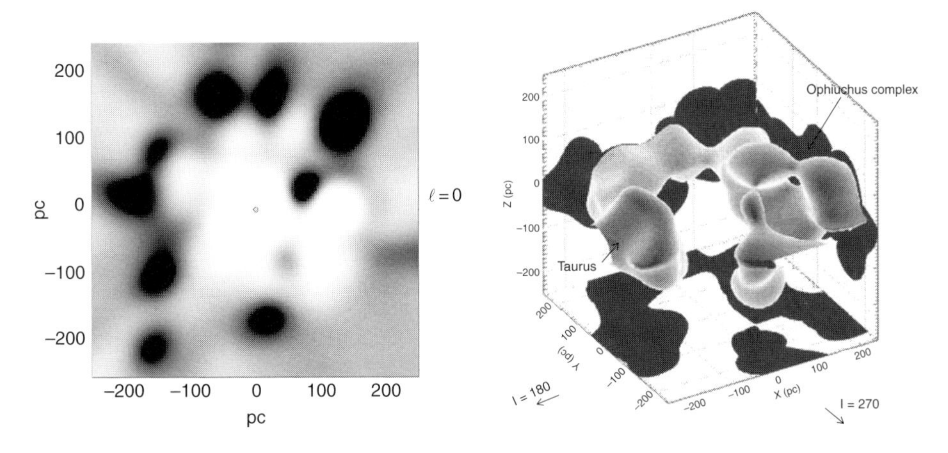

Figure 8.46 Left: cross-section of the 3d interstellar density distribution in the Galactic plane, showing the Na I clouds for a smoothing length of 40 pc. The Sun is in the centre, with the Galactic centre at the right. The grey scale runs from 0 (white) to 3×10^{-9} cm^{-3} (black). Right: 3d distribution (grey scale) of the Na I clouds around the Sun for the contour level 1×10^{-9} cm^{-3}. The structures are projected in black onto the three fundamental planes. From Vergely et al. (2001, Figures 4 and 8).

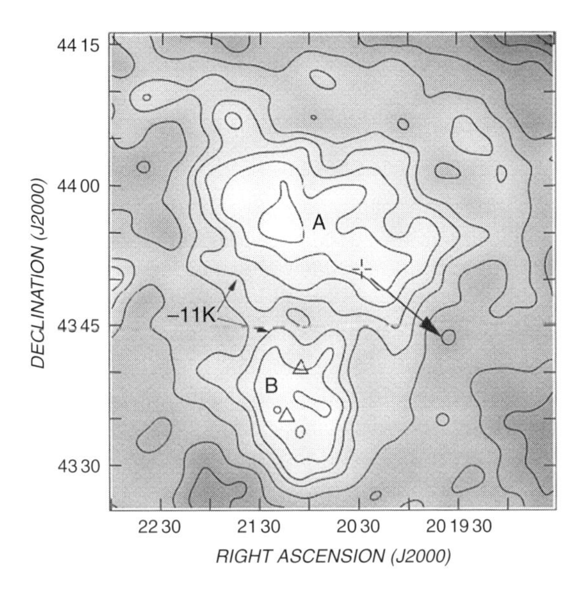

Figure 8.47 Neutral hydrogen in the neighbourhood of the Wolf–Rayet binary WR 140 (marked by a cross). The arrow indicates the proper motion direction of WR 140 as determined by Hipparcos. Open triangles indicate the position of the stellar objects seen in projection against a second minimum, B. Regions of low hydrogen emissivity are in light grey. Contour lines from −17 to −9 K are in steps of 2 K; and from −5 to +15 K are in steps of 4 K. There are two well-developed inner minima within the main H I cavity. From Arnal (2001, Figure 5).

Na I D lines of 11 stars. The stars flank the region of the apparent extension of the Local Bubble referred to as the β CMa tunnel, and the measurements were used to modify earlier estimates of the extent of the tunnel.

Crawford *et al.* (2000) made a study of the interstellar Na I D lines towards 29 stars in the general direction of the Lupus molecular clouds ($330° < l < 350°$, $0° < b < +25°$). Based on the minimum distance at which strong interstellar Na I lines appear in the spectra, he obtained a distance of $\sim 150 \pm 10$ pc to the Lupus complex.

Vergely *et al.* (2001) applied a robust inversion method, similar to tomographic mapping, to the Na I and H I measurements from García (1991), Welsh *et al.* (1994), and Sfeir *et al.* (1999), combined with Hipparcos parallaxes of some 1000 stars, to determine the 3d density distribution of the interstellar matter in the solar neighbourhood. They showed that the neutral interstellar matter is distributed in compact clouds or in cloud complexes with cavities between them. They were able to distinguish the Local Bubble and the Loop I cavities, and also two tunnels linking the Local Bubble to the outer regions of the Galaxy, away from the Galactic plane (Figure 8.46). Better accuracy could be achieved for the Na I data, where a larger number of lines-of-sight and target stars are available than for H I, and a rather detailed 3d density distribution was obtained with a 40 pc smoothing length. They showed that extended high-density regions in the Na I and H I maps are correlated.

Frisch *et al.* (2002) studied the velocity distribution of the nearby interstellar gas, determining the bulk flow velocity for the cluster of interstellar cloudlets within about 30 pc of the Sun. From 60 stars, they determined a streaming velocity through the local standard of rest of 17.0 ± 4.6 km s^{-1}, with an upstream direction of $l = 2.°3$, $b = -5.°2$, when making use of the Hipparcos values for the solar apex motion from Dehnen & Binney (1998). They estimated that the Sun may consequently

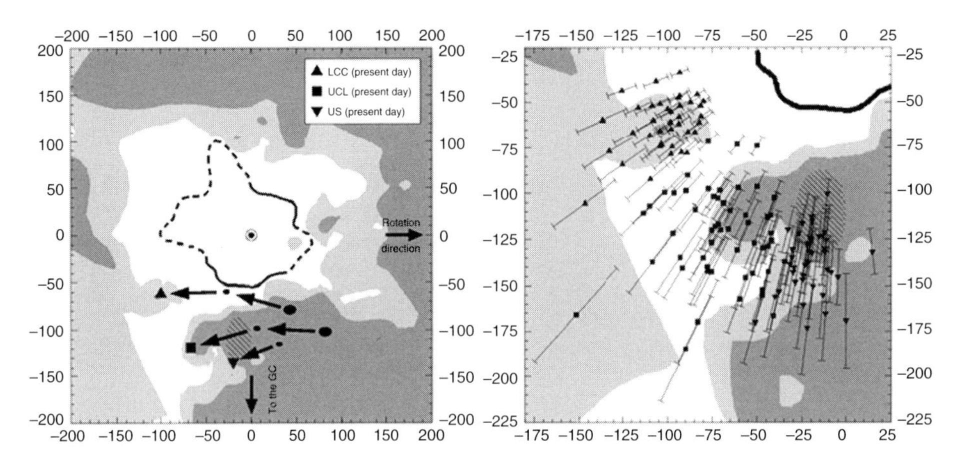

Figure 8.48 Left: the local cavity and Local Bubble in the plane of the Galactic equator. The filled contours show the Na I distribution, with white used for low-density regions and dark grey for high density regions. The black contour shows the present size of the Local Bubble as determined from X-ray data, with the dashed lines indicating areas where its limits cannot be accurately determined. The diagonal-line-filled ellipse shows the approximate position of the Ophiuchus molecular cloud. The present and past x, y coordinates, in pc, of the centre of the three subgroups of the Sco–Cen association are shown. For Lower Centaurus Crux (LCC) and Upper Centaurus Lupus (UCL) the past positions shown are those of 5 Myr and 10 Myr ago while for Upper Scorpius (US) only the position of 5 Myr ago is shown. The dimensions of the solid-filled ellipses indicate the uncertainties in the past positions. Right: enlargement of the left figure with the present positions of the OB stars in each of the three subgroups. Only those stars with accurately determined positions are shown. The symbol used in each case indicates the subgroup membership as coded in the left panel. From Maíz Apellániz (2001, Figure 1).

emerge from the surrounding gas patch within several thousand years. Such significant changes in ambient medium may also contribute to the kind of long-term climatic changes on Earth described in Section 10.6.

Arnal (2001) made a high-resolution H I 21-cm study of the region surrounding the Wolf–Rayet binary HD 193793 (WR 140), to look for evidence of an interaction between the star and its local interstellar medium. Based on the H I velocity structure, and the Hipparcos proper motion measurement indicating a space velocity of $28 \pm 3\,\mathrm{km\,s^{-1}}$ along the major axis of one of the H I minima attributed to the winds of the binary system, they concluded that the large tangential velocity was acquired by the system only some 1.3×10^5 yr ago (Figure 8.47). This could indicate that the binary was set in motion at this time, perhaps by a process such as the disruption of a hierarchical triple star system. IRAS maps show a corresponding large-scale feature attributed to dust grains heated by the stellar continuum.

Origin of the local bubble Various attempts have been made to trace the origin of the supernovae explosions, if indeed they are responsible for excavation of the Local Bubble. However, since the number density of early-type stars in the immediate solar neighbourhood is very small, with only three O–B5 stars within 67 pc of the Sun, it appears unlikely that several nearby isolated massive stars would have exploded within such a short time period. Furthermore, no OB associations of the right age currently exist within 100 pc of the Sun, so the identity of the supernovae progenitors which produced the Local Bubble has remained uncertain. Maíz Apellániz (2001) analysed the motions of the nearby OB associations determined by de Zeeuw *et al.* (1999, see also Section 6.10). The nearest of these is the Scorpius–Centaurus (Sco–Cen) OB association, which is itself divided into three subgroups: Lower Centaurus Crux, Upper Centaurus Lupus, and Upper Scorpius. Although currently located at ~130 pc from the Sun, extrapolating backwards in time to 5–10 Myr both the Sun's position, and those of these various association subgroups and their members, all using the Hipparcos astrometry, shows that the Sco–Cen association, and especially the Lower Centaurus Crux subgroup, was closer to the Sun's position 5 Myr ago. It therefore makes a likely source of the few supernovae needed to produce the Local Bubble (Figure 8.48). This conclusion is supported both by evolutionary synthesis calculations, and by the detection of 4–5 runaway stars escaping from it (Hoogerwerf *et al.*, 2000, 2001). Berghöfer & Breitschwerdt (2002) calculated the trajectory of the Pleiades subgroup B1 backwards in time, and found that 19 supernovae could have exploded between 10–20 Myr ago close to the region that is occupied by the Local Bubble, and that this scenario is in good agreement with the size of the Local Bubble and the present soft X-ray emissivity. A similar approach was followed by Fuchs *et al.* (2006) who extended their

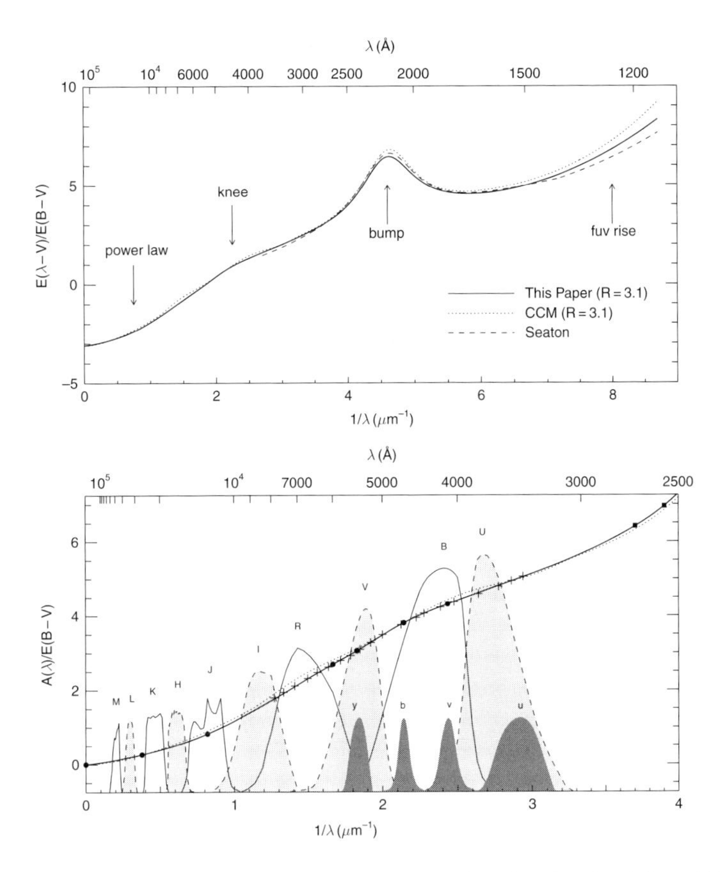

Figure 8.49 Top: normalised interstellar extinction curves from the far-infrared to the ultraviolet. Solid and dotted curves are estimates for the case $R \equiv A_V/E(B-V) = 3.1$ derived by Fitzpatrick (1999) and by Cardelli et al. (1989), respectively. The dashed curve shows the average Galactic ultraviolet extinction derived from OAO2, Copernicus and TD1 satellite data by Seaton (1979). Bottom: wavelength dependence of extinction in the infrared/optical region for the case $R = 3.1$ (thick curve). Arbitrarily scaled profiles of the Johnson UBVRIJHKLM and Strömgren uvby filters are shown for comparison. Neither figure is based on Hipparcos data, but extinction and reddening are crucial for interpretation of the Hipparcos data, and they are included here for reference. From Fitzpatrick (1999, Figures 1 and 6).

analysis beyond the presently-known association subgroups, and concluded that the Local Bubble was excavated around 14 Myr ago. Benítez *et al.* (2002) estimated that the association has generated some 20 supernova explosions during the last 11 Myr, some of them probably as close as 40 pc. Knie *et al.* (2004) reported rather direct evidence for such an event, on Earth, in the form of a significant enhancement of the ^{60}Fe concentration in a deep-sea ferromanganese crust at far above natural terrestrial levels. Produced in Type II supernovae, with a half-life of several million years, and detectable at ratios as low as $^{60}\text{Fe}/\text{Fe} \sim 10^{-16}$, ^{60}Fe appears to be a powerful indicator for the detection of supernova debris on Earth. Specifically they report an event occurring 2.8 Myr ago, at concentrations consistent with supernova ejecta at a distance of a few tens of parsec, and hence supporting the hypothesis of a supernova explosion from the Sco–Cen association. This epoch coincides with the onset and duration of an enhanced cosmic-ray flux and an African climate shift towards more arid conditions, attributed to the onset of a northern-hemisphere glacial event, perhaps provoking or contributing to the Pliocene–Pleistocene boundary marine extinction (see Section 10.6).

8.5.2 Extinction and reddening

Interstellar extinction, and its wavelength dependence, is attributed to absorption and scattering by dust particles located between the star and the Earth. Early studies were made by Trumpler (1930), van Rhijn (1949), Schatzman (1950), Münch (1952), Chandrasekhar & Münch (1952), and many others.

The details and spatial variability of interstellar extinction depend on the optical properties of dust grains along a line-of-sight, and measurements convey information about the composition and size distribution of these grains. Knowledge of the wavelength dependence of extinction is itself required to correct the effects of dust obscuration from observed energy distributions, since most objects are viewed through at least some small amount of interstellar dust. Determination of the wavelength dependence of the mean Galactic extinction from the infrared to the ultraviolet has been made by various authors.

Figure 8.49 illustrates the determination by Fitzpatrick (1999), with the commonly used normalisation $E(\lambda - V)/E(B - V)$, plotted against inverse wavelength. The parameter $R \equiv A_V/E(B - V)$ characterises the ratio of total to selective extinction at V. It ranges between about 2.2 and 5.8 for lines-of-sight along which ultraviolet extinction has been measured, with a mean value of ~ 3.1 for the diffuse interstellar medium (3.15 is given by Straižys, 1992). Accordingly, the relationship $A_V = 3.1\, E(B-V)$ is frequently used. More recent details are given by Fitzpatrick & Massa (2007) and references therein.

Spatial variations determine the accuracy with which energy distributions can be corrected for the effects of extinction, a process referred to as 'dereddening' given its wavelength dependence. The effects are more pronounced close to the Galactic plane. Pre-Hipparcos models of interstellar extinction over the celestial sphere include the work of Bahcall & Soneira (1981), who used the extinction law of Sandage (1972) in their Galaxy number count model, finding that it gave good agreement with star counts at $|b| > 10°$, but that at lower Galactic latitudes the absorption is patchy and poorly known. Neckel *et al.* (1980) derived extinctions and distances for more than 11 000 stars, investigating the spatial distribution of interstellar extinction for Galactic latitudes $|b| < 7\overset{\circ}{.}6$.

Arenou *et al.* (1992) used a sample of 17 000 stars with MK spectral types and photoelectric photometry to construct an extinction model in which the sky was divided into 199 cells, with a quadratic distance-dependency $A_V = \alpha r + \beta r^2$ adopted for each cell. Hakkila *et al.* (1997) merged several published studies to develop a model of large-scale visual extinction.

Hipparcos contributions The Hipparcos data have contributed to the determination of these spatial variations. Vergely *et al.* (1997) used extinctions determined from Strömgren photometry and Hipparcos parallaxes for 3700 stars, in an inverse method used to construct the 3d distribution of absorbing interstellar matter in the solar neighbourhood. The average opacity was estimated as 1.5 mag kpc^{-1} in A_V, with a scale height of 70 pc; the correlation functions give some indication of the cloud sizes.

Vergely *et al.* (1998) extended the study to 11 837 Hipparcos stars with $r < 400$ pc for which individual extinctions could be derived from Strömgren photometry. They defined four groups covering spectral types B0–A0, A0–A3, A3–F0, and F0–G2, each including stars of luminosity classes III–V. The colour excess is given by $E(b-y) = (b-y) - (b-y)_0$, where $(b-y)_0$ is the unreddened colour index defined statistically for each group through specific calibration of the Strömgren indices $(b-y), (u-b), m_1, c_1, \beta$; with $E(b-y) \sim 0.74E(B-V)$ according to Knude & Høg (1999). Mean extinction and its dispersion as a function of distance is shown in Figure 8.50a–c; it stays more-or-less constant and close to zero out to 70 pc, and then increases linearly out to ~ 260 pc. The lack of extinction out to 70 pc seems to confirm the existence of an almost empty Local Bubble surrounding the Sun. Figure 8.50d shows the derived extinction in the region of the ρ Ophiuchus cloud. To estimate the dependency on Galactic latitude, they considered two models. For a homogeneous interstellar medium disk of uniform density and height h_0, they derived a fit to a cosecant law as a function of distance r

$$\begin{aligned} E(r,b) &= E_0\, r \qquad\qquad \text{for } r < \frac{h_0}{|\sin b|} \\ &= E_0\, h_0 \operatorname{cosec} b \;\; \text{for } r > \frac{h_0}{|\sin b|} \end{aligned} \tag{8.3}$$

with $h_0 = 55$ pc and $E_0 = 2.5 \times 10^{-4}$. For an exponential decreasing density with height above the Galactic plane they derived

$$E(r,b) = \frac{E_0\, h_0}{|\sin b|}\left(1 - \exp\left(\frac{-r|\sin b|}{h_0}\right)\right) \tag{8.4}$$

with $h_0 = 35$ pc and $E_0 = 4.0 \times 10^{-4}$. Both fits are shown in Figure 8.50e. They went on to consider that the extinction is caused by a single type of interstellar cloud characterised by a certain cloud density, extinction cross-section, and mean extinction per cloud, demonstrating that this model gives a tolerable match to the observed dispersion in extinction for two different Galactic latitude ranges (Figure 8.50f–g). Allowing for more than one type of cloud, distributed homogeneously and following an exponential density law perpendicular to the Galactic plane, they found best fits to the extinction histograms for a model with faint clouds with $E_1 = 0.012$, medium clouds with $E_2 = 0.05$, and darker clouds with $E_3 > 0.1$. They also found that the angular correlation function (Figure 8.50h) could be well fitted, although only weakly constrained, by medium-extinction clouds with $E_2 = 0.05$ and diameters $\Phi \lesssim 5$ pc, combined with dark clouds with $E_3 = 0.15$ and $\Phi \sim 10$–60 pc, the low-extinction clouds having a negligible contribution. In comparing their work with previous similar parameterisations (Ambartsumian & Gordeladse, 1938; Schatzman, 1950; Chandrasekhar & Münch, 1952; Scheffler, 1966; Knude, 1979; Davidson *et al.*, 1987), they noted that the medium-extinction clouds have been detected by all previous studies, characterised by $\langle E \rangle = 0.03$–0.07,

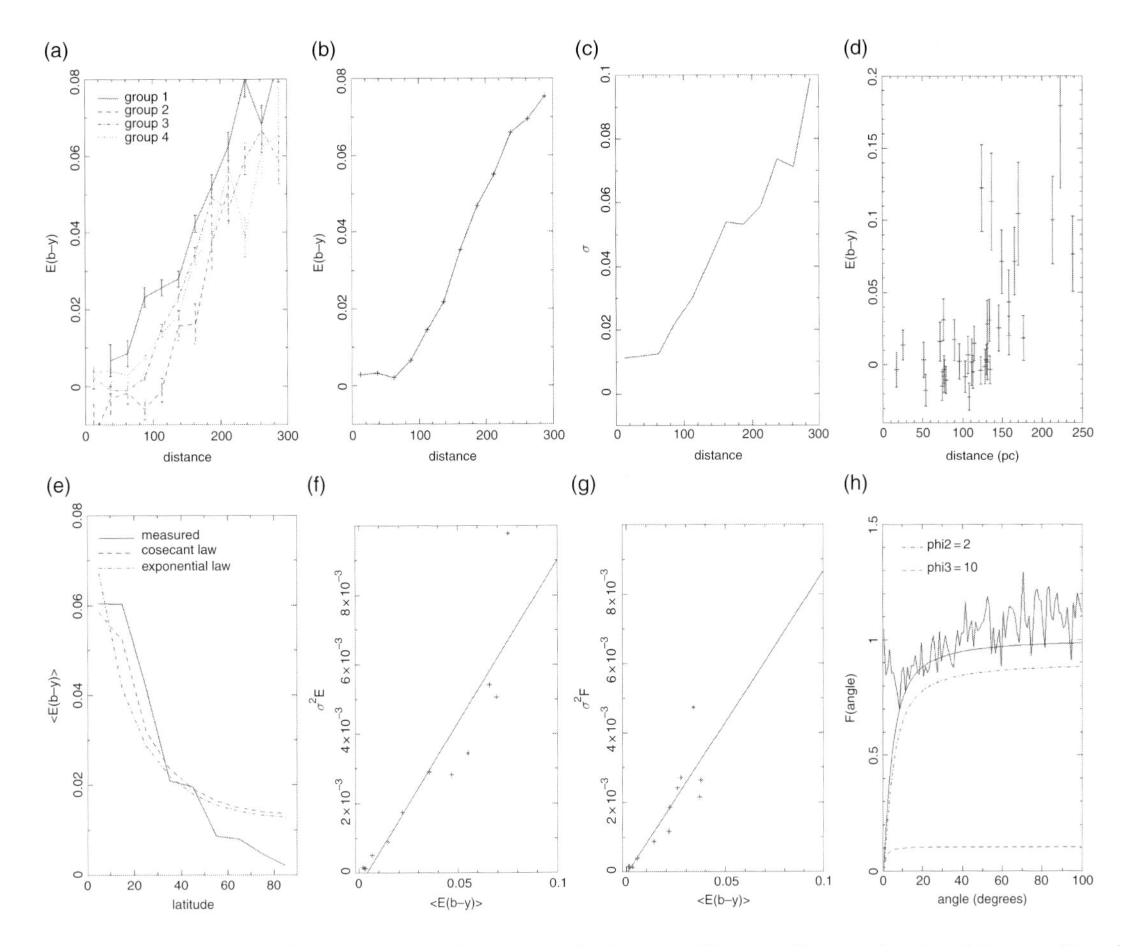

Figure 8.50 Top row, left-to-right: (a) mean extinction for stars of each group within $|z| < 40$ pc, as a function of distance; (b) total mean extinction as a function of distance, for $|z| < 40$ pc; (c) dispersion in extinction as a function of distance, for $|z| < 40$ pc; (d) extinction for single stars in the ρ Ophiuchi region: $0° < l < 10°$, $0° < b < 10°$, and $\sigma_\pi/\pi < 0.3$. Bottom row, left-to-right: (e) observed mean extinction as a function of Galactic latitude compared to cosecant and exponential laws in the distance range $R = 200$–250 pc; (f) σ_E^2 versus $\langle E \rangle$ from observations for different values of R, for stars within $|z| < 25$ pc; (g) σ_E^2 versus $\langle E \rangle$ for different values of R, for stars with $20° < |b| < 40°$; (h) observed angular correlation function $F_{\rm obs}(\theta)$ for a distance 200–400 pc. The line represents a fit with two kinds of clouds: $(E_2, \Phi_2) = (0.05, 2.0)$ and $(E_3, \Phi_3) = (0.15, 10.0)$, where E is the extinction and Φ the cloud diameter in pc. From Vergely et al. (1998, Figures 3, 4, 6, 8, 11, 12, 13, 15).

while the low-extinction clouds had only been inferred by Chandrasekhar & Münch (1952) based on brightness fluctuations on the sky.

Chen *et al.* (1998) used distances from open clusters determined by Hipparcos to construct the extinction map in the Galactic plane, for $|b| < 10°$, also using an inverse method and based on the assumption that the extinction is a linear function of distance. They provided an analytic expression for the interstellar extinction as a function of Galactic longitude and distance with a quartic expression for the extinction out to 1 kpc provided for each 10° interval of Galactic longitude (Figure 8.51). They compared their model with that of Arenou *et al.* (1992), by comparing its predictions with the Hipparcos observations for the distributions in V, $B - V$, distance r, and the reduced proper motion (Section 5.2.5). Joshi (2005) used 772 open clusters from the updated compilation by Dias *et al.* (2002) to study extinction, obtaining a scale height of 53 ± 5 pc for the distribution of open clusters, 186 ± 25 pc for the scale height of reddening material, and a distance of the Sun above the reddening plane of 22.8 ± 3.3 pc.

Knude & Høg (1998) studied the local distribution of interstellar reddening from the combination of Hipparcos parallaxes, the Tycho Catalogue colour indices $B_T - V_T$, and spectral and luminosity classification compiled from the literature. They used some 30 000 lines-of-sight for stars of luminosity classes V and III, primarily for negative declinations where most of the revised MK spectral classifications are available (see page 214). For $V_T < 9$, the median standard error of V_T and $(B_T - V_T)$ is better than 0.014 mag and 0.025 mag respectively, so that the data essentially allow the computation of a unique distance versus $E(B - V)$ dataset. They

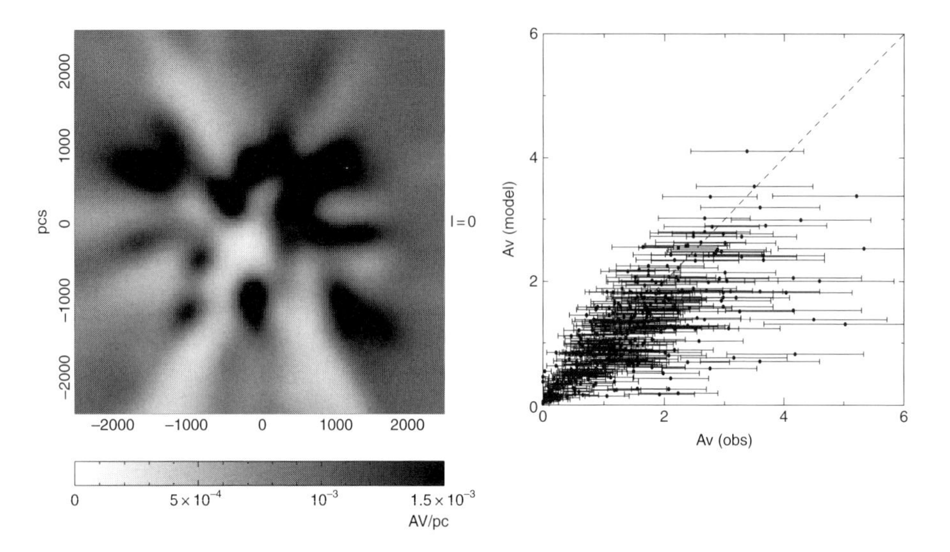

Figure 8.51 Left: extinction dependency determined in the Galactic plane. Right: extinction derived from the model of Chen et al. (1998) compared with the observations. From Chen et al. (1998, Figures 5 and 8).

determined a distance to the Southern Coalsack molecular cloud of 100–150 pc. For four of the five star-forming molecular clouds estimated to lie within $\sim$ 200 pc, not counting the high-latitude MBM clouds of Magnani *et al.* (1985), they determined distances of 150 pc for Chamaeleon, 100 pc for Lupus, 170 pc for Corona Australis, and 120 pc for ρ Ophiuchus. Certain extinction features were also noted in these regions. Their distance to Lupus is significantly less than the 150 ± 10 pc estimated by Crawford *et al.* (2000), which the latter attributes to a biased interstellar extinction value due to lower-density foreground material flowing outwards from the Sco–Cen OB association.

Knude & Høg (1999) used a similar approach to study dust features towards the north Galactic pole, and to correlate them with the positions of known high-velocity (< -90 km s^{-1}) and intermediate-velocity (-50 to -90 km s^{-1}) clouds. The existence of these anomalous negative-velocity H I clouds over much of the sky in the first and second quadrants of Galactic longitude, especially in the northern hemisphere, remains largely unexplained. They have been attributed by Verschuur (1993) to a vast supershell with an elliptical cross-section normal to the Galactic disk, and with its origin in the Perseus spiral arm. Knude & Høg (1999) noted a positional coincidence between the H I gas in narrow velocity intervals and the $E(B-V)$ versus distance features extracted independently from the Hipparcos and Tycho Catalogues. These dust counterparts include the 'intermediate velocity arch', a single complex of neutral gas with intermediate line-of-sight velocities that covers nearly one third of the northern Galactic hemisphere, and which has a z distance considered to be between 450–1700 pc. Interpretation is complicated by the fact that grain destruction in interstellar shocks may reduce the dust-to-gas ratio in high-velocity and intermediate-velocity gas, but they found an indication for a shell-like structure between 30–50 pc in a region where the proposed Perseus super-bubble is expected to be located between 50–80 pc.

The same approach was used to determine distances and absorption features in the cometary globules (interstellar clouds with a comet-like appearance) CG 30/CG 31/CG 38 (Knude *et al.*, 1999; Knude & Nielsen, 2000); and for distances to absorbing features in the direction of the star-forming molecular cloud LDN 1622 (Knude *et al.*, 2002).

Other studies of interstellar extinction making use of the Hipparcos data include: *uvbyβ* photometry in three extreme ultraviolet shadow directions (Knude, 1998); reclassification of spectra and calculation of interstellar extinction for objects in the Hipparcos Catalogue (Kilpio, 1998); reference stars for 12 low-extinction fields towards the Galactic bulge, including and extending the Tycho Catalogue stars (Dominici *et al.*, 1999); interstellar extinction along the Camelopardalis, Perseus and Cassiopeia border (Zdanavičius & Zdanavičius, 2002; Zdanavičius *et al.*, 2002); interstellar extinction towards the interaction zone between the local and Loop I bubbles (Corradi *et al.*, 2003); and 2MASS wide-field extinction maps and distance to the Pipe Nebula (Lombardi *et al.*, 2006).

Grey extinction The interstellar dust model of Li & Greenberg (1997) identifies 3–4 grain size populations responsible for the shape of the extinction curve in the ultraviolet, while in the visual a single population of

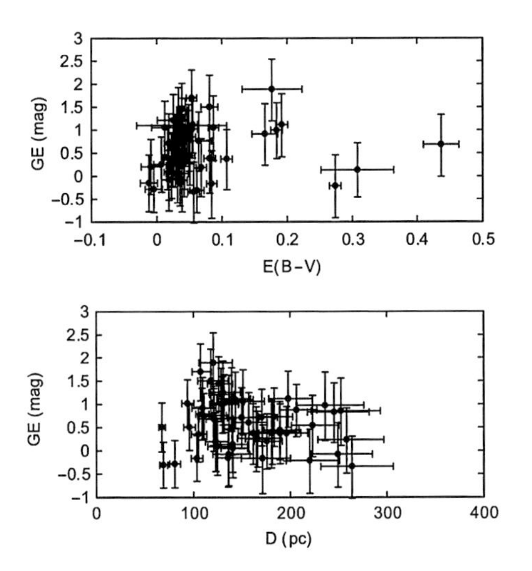

Figure 8.52 The GE (grey extinction) versus E(B − V) (top) and Hipparcos distances (bottom) for 56 O and B stars with parallax errors less than 15%. From Skórzyński et al. (2003, Figure 5).

large particles, of size $\sim 10^{-8}$ m, was postulated. These large particles cause selective extinction at visual wavelengths, but appear to be neutral or 'grey' in the ultraviolet. Very large particles would be expected to cause grey extinction at both visual and ultraviolet wavelengths.

Since the work of Trumpler (1930), it has generally been assumed that the total interstellar extinction is proportional to the selective extinction, and that if there is any non-selective extinction component, then its contribution is also strictly proportional to the selective component. This has led to the commonly-accepted direct proportionality of the total-to-selective interstellar extinction, characterised by $R = A_V/E(B-V)$, with R being, at least for a given direction in the sky, constant. While small particles, responsible for selective extinction, probably dominate in the typical interstellar medium, some regions may have conditions which allow creation or survival of very large particles as well. Various lines of evidence for these large particles have been summarised by Skórzyński *et al.* (2003): it includes the position of the lower main sequence in the open clusters NGC 2264 and the Pleiades, condensation around η Carina, sub-mm fluxes from various binary stars, the X-ray halo around Nova Cygni 1992, and grains detected by the space probes Ulysses and Galileo. Evidence for grey extinction from an optical to X-ray spectral analysis of the afterglow of the gamma-ray burst GRB 020405 is given by Stratta *et al.* (2005).

The Hipparcos data have provided further evidence for the existence of such grey extinction. Dunkin & Crawford (1998) reported spatially-resolved optical spectra of the individual stars in the young binary system HD 35187. From their Hipparcos-derived position in the HR diagram, they concluded that HD 35187B

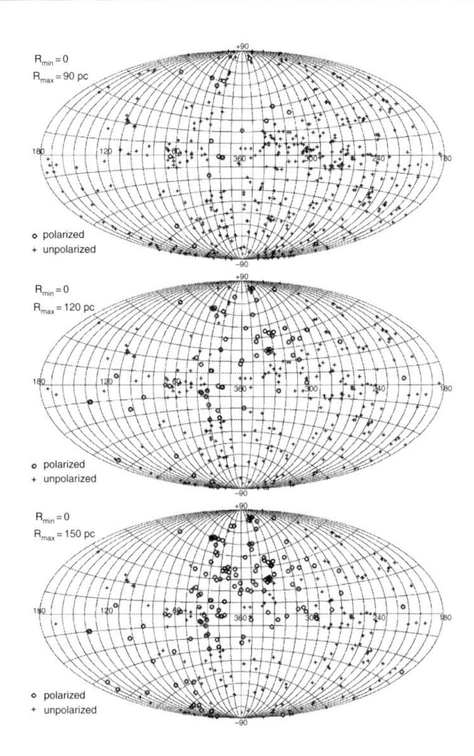

Figure 8.53 Lines-of-sight to Hipparcos stars for which polarisation measurements are available. Small circles correspond to lines-of-sight to stars showing polarisation of at least 0.10%. Crosses are lines-of-sight to unpolarised stars; they thus delineate regions free of interstellar polarisation. From top to bottom, distance limits are 90 pc, 120 pc, and 150 pc. From Leroy (1999, Figure 1).

is attenuated by about 0.4 mag of grey circumstellar extinction.

Skórzyński *et al.* (2003) used the Hipparcos astrometry and photometry, along with the spectral type and luminosity class and the intrinsic absolute magnitudes from Schmidt–Kaler (in Schaifers & Voigt, 1982, Table 13), to calculate the contribution of the grey extinction, GE, defined as the difference in absolute magnitude with and without the standard selective extinction

$$\mathrm{GE} = V - 5\log d + 5 - 3.1E(B-V) - M_V(\text{intrinsic}) \quad (8.5)$$

The results for 56 O and B stars with parallax precision better than 15% versus $E(B-V)$ and versus Hipparcos distances suggest that the effect does not depend on $E(B-V)$, but reaches a maximum at 110–150 pc (Figure 8.52). They concluded that matter responsible for both components of interstellar extinction is distributed inhomogeneously in the solar neighbourhood. They argued that the interpretation is supported by a correlation of the inferred GE measure, and the total

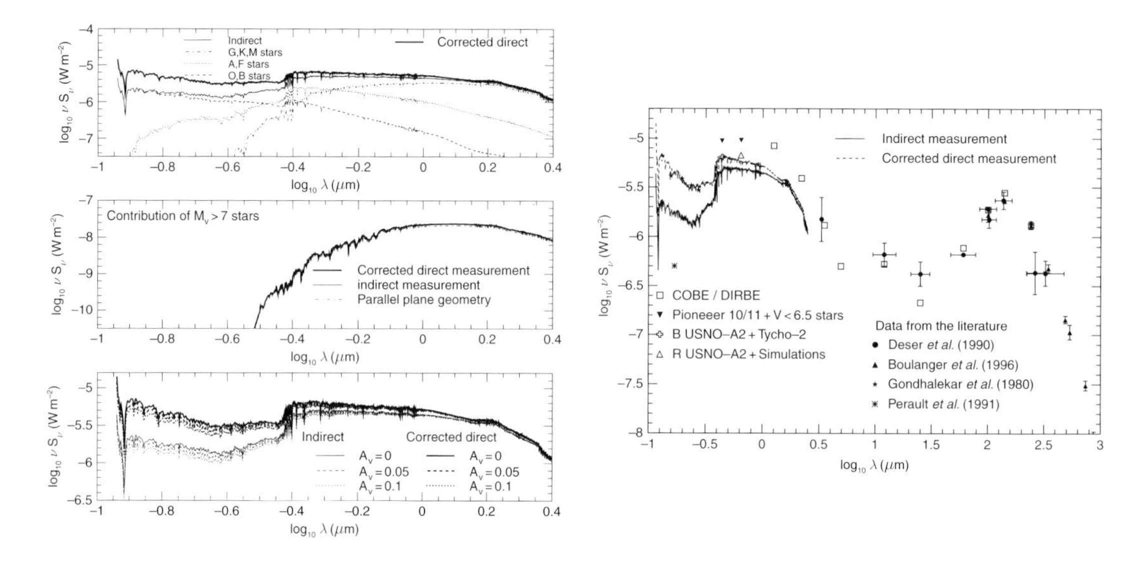

Figure 8.54 Left: direct and indirect synthetic estimates of the surface brightness of the Galaxy at the solar neighbourhood. Top panel: based on the $V < 7.3$ Hipparcos stars. Middle panel: simulated contribution from $M_V > 7$ stars, with $\pm 1\sigma$ luminosity function uncertainties. These dim stars mainly contribute to the near-infrared spectrum, but still at an order of magnitude below their bright counterparts. Bottom panel: influence of the extinction, perpendicular to the Galactic plane, which might have a marginal influence in the ultraviolet. A fixed extinction A_V is assumed for the whole population. Right: summary of the measured contributions to the radiation field. Data points with no error bars have been computed with a cosecant law applied at given wavelengths (COBE/DIRBE, Pioneer 10/11, USNO A2 and Tycho 2). The Pioneer 10/11 and COBE/DIRBE 1.25 μm and 2.2 μm data are corrected for extinction. Dashed and full curves correspond to the stellar synthetic estimates based on Hipparcos. The COBE/DIRBE estimates are compatible with Désert et al. (1990), Boulanger et al. (1996) (symbols with error bars). In the ultraviolet, uncertainties are large due to inhomogeneities of the OB associations. The indirect estimate is compatible with the all-sky average of Gondhalekar et al. (1980). From Melchior et al. (2007, Figures 9 and 16).

amount of interstellar matter indicated by the column density of krypton, some diffuse interstellar bands, and the degree of interstellar polarisation. They raised the possibility that, in some regions, neutral extinction could reach 1 mag or more at distances of only 400 pc. Some further Hipparcos evidence for grey extinction was provided by Patriarchi *et al.* (2003) in their study of interstellar reddening towards 185 O stars using *JHK* photometry from 2MASS; the few stars in their sample with a reliable Hipparcos parallax were found to have distances systematically smaller than those estimated from spectroscopic parallaxes.

8.5.3 Polarisation

Polarimetry yields information on the intrinsic polarisation of some stars due, for example, to circumstellar envelopes or to large-scale magnetic fields (e.g. Leroy, 2000). It is also sensitive to even small amounts of interstellar dust, at least for the dust component consisting of anisotropic particles likely to be aligned by the ambient magnetic field (Appenzeller, 1975; Tinbergen, 1982). Tinbergen (1982) reported almost no polarisation in the light of stars closer than 35 pc, demonstrating that the local interstellar medium has a very low dust content. A similar result was obtained by Leroy (1993a,b) who extended the analysis to 1000 stars within 50 pc. At the time of the Hipparcos Catalogue availability, the closest dust clouds were considered to be beyond 50 pc, perhaps beyond the radius of the Local Bubble (Snowden *et al.*, 1998). Detailed interpretation of the collective polarimetric data was primarily limited by the inaccuracy in the individual stellar distances.

Leroy (1999) used published polarimetric data combined with the Hipparcos parallaxes to determine the distribution of dust in the solar neighbourhood and, in particular, to search for the presence of a 'dust wall' which may lie at the boundary of the Local Bubble (Snowden *et al.*, 1998). Polarisation catalogues used were those of Behr (1959), Appenzeller (1966, 1968), Axon & Ellis (1976), Schröder (1976), Piirola (1977), Krautter (1980), Tinbergen (1982), Korhonen & Reiz (1986), Leroy (1993b), Berdyugin *et al.* (1995), and Reiz & Franco (1998). Taking account of the distance resolution of the Hipparcos parallaxes, measured stars were divided into three distance intervals: 60–90 pc, 90–120 pc, and 120–150 pc (Figure 8.53). The existing polarisation data, albeit inhomogeneous and with only poor sampling for some sky areas, show that significant dust clouds appear only at 70–80 pc, slightly beyond the Local Bubble boundary defined by X-ray data. Traces of polarisation appear at high as well as low Galactic latitudes. Almost no polarisation is seen, even at 150 pc, around $l = 240°$, especially in the southern hemisphere; this behaviour had been noted previously, and is attributed either to an extension of the Local Bubble in this direction, or to a possible connection with another nearby bubble (Heiles, 1998).

Harjunpää *et al.* (1999) reported linear polarisation measurements in three filamentary molecular clouds: the L1400 complex, L204, and MBM 25. They derived a new distance of ~145 pc from the Hipparcos parallaxes of two stars surrounded by bright nebulosities, on the assumption that the stars are physically associated with the cloud. Silva & Magalhães (2003) studied the magnetic field structure in the diffuse interstellar medium towards three high Galactic latitude DIRBE molecular clouds: DIR 292−37, DIR 314−37, and DIR 349−46, using the Hipparcos stars HIP 16850, HIP 114678, and HIP 106445 respectively.

Alves & Franco (2006) obtained *B*-band imaging linear polarimetry data for 131 Hipparcos stars within 250 pc, to re-examine the distribution of the local interstellar medium towards the IRAS 100 μm emission void in the Lupus dark clouds. For Lupus 1, they obtained a distance between 130–150 pc, with a low column density region coincident with the observed infrared void. They detected absorbing material closer than 60–100 pc, perhaps associated with the interface boundary between the Local Bubble and its neighbourhood Loop I superbubble. Alves & Franco (2007) used similar data for 82 Hipparcos stars to derive a distance of 145 ± 16 pc to the Pipe Nebula, placing it within the Ophiuchus dark cloud complex. They reported evidence that the largest filament in the nebula has collapsed along magnetic field lines, indicating that magnetic pressure plays an important role in the cloud's evolution.

8.5.4 Interstellar radiation field

The physics of the local interstellar medium is strongly influenced by the density and spectrum of the local interstellar radiation field, particularly in the ionising ultraviolet. In addition, the local radiation field provides a zero-point for models of the multi-wavelength continuum emission of high-redshift galaxies, used in studies of the formation and evolution of galaxies (see, e.g. Melchior *et al.*, 2007).

Henry (1977) estimated the brightness of the local interstellar radiation field by integrating the flux expected from stars in the Yale Bright Star Catalogue; the integration converged, demonstrating that the catalogue used was deep enough to provide a complete result. However, that analysis did not include possible contributions from any scattered light component of the interstellar ultraviolet radiation field. This aspect was addressed by Henry (2002) using the Hipparcos Input Catalogue, together with Kurucz model stellar atmospheres and information on interstellar extinction, to create a model of the expected intensity and spectral distribution of the local interstellar ultraviolet radiation field, under various assumptions concerning the albedo of the interstellar grains. In this analysis, the stellar distances are irrelevant (in the sense that they are already reflected in the star's apparent magnitude); the local radiation field in the ultraviolet is the integral of the individual contributions, determined by each star's apparent magnitude, its luminosity class, and the extinction at each wavelength of interest, but taking into account scattering determined by interstellar grains of given albedo. He concluded that the albedo in the far-ultraviolet is very low, perhaps $a = 0.1$, arguing that this determination is more reliable than any of the many previous, and conflicting, ultraviolet interstellar grain albedo determinations. He also confirmed that the ultraviolet background radiation observed at high, and even moderate, Galactic latitudes is extragalactic in origin, as it cannot be backscatter of the interstellar radiation field. The model, taking the corresponding data from the Hipparcos Catalogue, was also described by Sujatha *et al.* (2004).

Melchior *et al.* (2007), see also Melchior *et al.* (2001), made a detailed determination of the local interstellar radiation field, thereby determining the surface brightness of the Galaxy at the solar neighbourhood as observed from outside the Galaxy. They adopted two approaches: a 'synthetic' determination using the Hipparcos dataset, and an associated stellar spectral library from Pickles (1998), to estimate the optical spectrum of the Galaxy's surface brightness (Figure 8.54, left). They also compiled various datasets from the ultraviolet to the far infrared using, in the optical, the Tycho 2 Catalogue data, as well as data from the USNO A2 Catalogue and from Pioneer 10/11 (Figure 8.54, right). The latter is systematically larger than the synthetic local estimate based on Hipparcos data. They interpret this disagreement as due to the presence of a local minimum of the stellar density compatible with the Gould Belt. Their results suggest that the global luminosity of the Milky Way should follow the Tully–Fisher relation established for external galaxies (Tully & Fisher, 1977). Their on-line data include a series of images and animations of the distribution of the Hipparcos data, for different spectral types.

References

Abt HA, Cardona O, 1984, Be stars in binaries. *ApJ*, 285, 190–194 {**422**}

Abt HA, Corbally CJ, 2000, The maximum age of trapezium systems. *ApJ*, 541, 841–848 {**447**}

Acker A, Fresneau A, Pottasch SR, *et al.*, 1998, A sample of planetary nebulae observed by Hipparcos. *A&A*, 337, 253–260 {**454**}

Adelman SJ, 1998, On the Hipparcos photometry of chemically peculiar B, A, and F stars. *A&AS*, 132, 93–97 {**427**}

—, 2007, FCAPT uvby photometry of the magnetic CP stars HD 16545, HD 93226, HR 7575, and HR 8206. *PASP*, 119, 980–985 {**427**}

Alcalá JM, Krautter J, Schmitt JHMM, *et al.*, 1995, A study of the Chamaeleon star-forming region from the ROSAT All-Sky Survey. I. X-ray observations and optical identifications. *A&AS*, 114, 109–134 {**431**}

Alcock C, Allsman RA, Alves DR, *et al.*, 2001, The MACHO project LMC variable star inventory. X. The R CrB stars. *ApJ*, 554, 298–315 {**452**}

Alencar SHP, Melo CHF, Dullemond CP, *et al.*, 2003, The pre-main sequence spectroscopic binary AK Sco revisited. *A&A*, 409, 1037–1053 {**419**}

Alksnis A, Balklavs A, Dzervitis U, *et al.*, 1998, Absolute magnitudes of carbon stars from Hipparcos parallaxes. *A&A*, 338, 209–216 {**450**}

Allen C, Poveda A, Hernández-Alcántara A, 2004, Internal motions of trapezium systems. *Revista Mexicana de Astronomia y Astrofisica Conf. Ser.*, volume 21, 195–199 {**447**}

—, 2006, Runaway stars, trapezia, and subtrapezia. *Revista Mexicana de Astronomia y Astrofisica Conf. Ser.*, volume 25, 13–15 {**447**}

Allen C, Poveda A, 1974, The dynamical evolution of trapezium systems. *Stability of the Solar System and of Small Stellar Systems* (ed. Kozai Y), IAU Symp. 62, 239–246 {**447**}

Alves DR, 2000, K-band calibration of the red clump luminosity. *ApJ*, 539, 732–741 {**450, 451**}

Alves FO, Franco GAP, 2006, The infrared void in the Lupus dark clouds revisited: a polarimetric approach. *MNRAS*, 366, 238–246 {**475**}

—, 2007, An accurate determination of the distance to the Pipe nebula. *A&A*, 470, 597–603 {**475**}

Ambartsumian VA, 1954, *Contrib. Byurakan Obs.* 15, 3 {**446**}

Ambartsumian VA, Gordeladse SG, 1938, *Contrib. Byurakan Obs.* 2, 37 {**470**}

Ankay A, Kaper L, de Bruijne JHJ, *et al.*, 2001, The origin of the runaway high-mass X-ray binary HD 153919 (4U 1700–37). *A&A*, 370, 170–175 {**447**}

Appenzeller I, 1966, Polarimetrische, photometrische und spektroskopische Beobachtungen von Sternen im Cygnus und Orion. *Zeitschrift fur Astrophysik*, 64, 269–295 {**474**}

—, 1968, Polarimetric observations of nearby stars in the directions of the Galactic poles and the Galactic plane. *ApJ*, 151, 907–918 {**474**}

—, 1975, Lower limits for the interstellar reddening at the Galactic poles. *A&A*, 38, 313–314 {**474**}

Arenou F, Grenon M, Gómez AE, 1992, A tridimensional model of the Galactic interstellar extinction. *A&A*, 258, 104–111 {**470, 471**}

Arnal EM, 2001, A high-resolution H I study of the interstellar medium local to HD 193793. *AJ*, 121, 413–425 {**467, 468**}

Augereau JC, Lagrange AM, Mouillet D, *et al.*, 2001, HST–NICMOS2 coronographic observations of the circumstellar environment of three old pre-main sequence stars: HD 100546, SAO 206462 and MWC 480. *A&A*, 365, 78–89 {**419**}

Axon DJ, Ellis RS, 1976, A catalogue of linear polarisation measurements for 5070 stars. *MNRAS*, 177, 499–511 {**474**}

Babel J, Montmerle T, 1997, X-ray emission from Ap–Bp stars: a magnetically confined wind-shock model for IQ Aur. *A&A*, 323, 121–138 {**425**}

Bahcall JN, Soneira RM, 1981, Predicted star counts in selected fields and photometric bands: applications to Galactic structure, the disk luminosity function, and the detection of a massive halo. *ApJS*, 47, 357–403 {**470**}

Balklavs A, Dzervitis U, Eglitis I, 1998, Photometric characteristics of Hipparcos carbon stars sample. *Modern Problems in Stellar Evolution*, 273 {**452**}

Baraffe I, Chabrier G, Allard F, *et al.*, 1998, Evolutionary models for solar metallicity low-mass stars: mass-magnitude relationships and colour–magnitude diagrams. *A&A*, 337, 403–412 {**428**}

Barannikov AA, 2006a, Autocorrelation and Fourier analysis of the Crimean and Hipparcos photometry of the runaway star HD 192281. *Astronomical and Astrophysical Transactions*, 25, 101–103 {**446**}

—, 2006b, Search for the periodicity in brightness and radial velocity variations of the runaway star HD 218915. *Astronomical and Astrophysical Transactions*, 25, 95–100 {**446**}

Baranov VB, Krasnobaev KV, Kulikovskii AG, 1971, A model of the interaction of the solar wind with the interstellar medium. *Soviet Physics Doklady*, 15, 791–799 {**448**}

Barstow MA, Bond HE, Holberg JB, *et al.*, 2005, Hubble Space Telescope spectroscopy of the Balmer lines in Sirius B. *MNRAS*, 362, 1134–1142 {**457, 460**}

Barstow MA, Holberg JB, Cruise AM, *et al.*, 1997, The mass, temperature and distance of the white dwarf in V471 Tau. *MNRAS*, 290, 505–514 {**461**}

Barstow MA, Schmitt JHMM, Clemens JC, *et al.*, 1992, ROSAT sky survey observations of the eclipsing binary V471 Tau. *MNRAS*, 255, 369–378 {**460**}

Behr A, 1959, Beobachtungen zur Wellenlängenabhängigkeit der interstellaren Polarisation. *Zeitschrift fur Astrophysik*, 47, 54–58 {**474**}

Benítez N, Maíz Apellániz J, Canelles M, 2002, Evidence for nearby supernova explosions. *Physical Review Letters*, 88(8), 1101.1–4 {**469**}

Benvenuto OG, Althaus LG, 1996a, Luminosity evolution of strange dwarf stars. *Phys. Rev. D*, 53, 635–638 {**458**}

—, 1996b, The structure and thermal evolution of strange dwarf stars. *ApJ*, 462, 364–375 {**458**}

Berdyugin AV, Snare MO, Teerikorpi P, 1995, Interstellar polarisation at high Galactic latitudes from distant stars. II. First results for $Z \leq 600$ pc. *A&A*, 294, 568–574 {**474**}

Bergeat J, Knapik A, Rutily B, 1999, Dust extinction and intrinsic spectral energy distributions of carbon-rich stars. II. The hot carbon stars. *A&A*, 342, 773–784 {**451**}

—, 2001, The effective temperatures of carbon-rich stars. *A&A*, 369, 178–209 {**451**}

—, 2002a, Carbon-rich giants in the HR diagram and their luminosity function. *A&A*, 390, 967–986 {**451**}

—, 2002b, The pulsation modes and masses of carbon-rich long-period variables. *A&A*, 390, 987–999 {**452**}

—, 2002c, Thick disk and old disk carbon-rich giants in the Sun's vicinity. *A&A*, 385, 94–110 {**451**}

Bergeat J, Knapik A, 1997, Barium stars in the HR diagram. *A&A*, 321, L9–L1997 {**453**}

Berger DH, Gies DR, 2001, A search for high-velocity Be stars. *ApJ*, 555, 364–367 {**422–424**}

Bergeron P, Saffer RA, Liebert J, 1992, A spectroscopic determination of the mass distribution of DA white dwarfs. *ApJ*, 394, 228–247 {**455**}

Berghöfer TW, Breitschwerdt D, 2002, The origin of the young stellar population in the solar neighbourhood – a link to the formation of the Local Bubble? *A&A*, 390, 299–306 {**468**}

Berghöfer TW, Schmitt JHMM, Cassinelli JP, 1996, The ROSAT All-Sky Survey Catalogue of optically bright OB-type stars. *A&AS*, 118, 481–494 {**432**}

Bertout C, Robichon N, Arenou F, 1999, Revisiting Hipparcos data for pre-main sequence stars. *A&A*, 352, 574–586 {**418**}

Bidelman WP, Keenan PC, 1951, The Ba II stars. *ApJ*, 114, 473–476 {**453**}

Biesiada M, Malec B, 2004, A new white dwarf constraint on the rate of change of the gravitational constant. *MNRAS*, 350, 644–648 {**456**}

Bjorkman JE, 2000, The formation and structure of circumstellar disks. *IAU Colloq. 175: The Be Phenomenon in Early-Type Stars* (eds. Smith MA, Henrichs HF, Fabregat J), ASP Conf. Ser. 214, 435–446 {**421**}

Blaauw A, 1961, On the origin of the O- and B-type stars with high velocities (the runaway stars), and some related problems. *Bull. Astron. Inst. Netherlands*, 15, 265–290 {**440, 445**}

—, 1993, Massive runaway stars. *Massive Stars: Their Lives in the Interstellar Medium* (eds. Cassinelli JP, Churchwell EB), ASP Conf. Ser. 35, 207–219 {**444**}

Blaauw A, Morgan WW, 1954, The space motions of AE Aur and μ Col with respect to the Orion Nebula. *ApJ*, 119, 625–630 {**443**}

Blöcker T, 1995, Stellar evolution of low- and intermediate-mass stars. II. Post-asymptotic giant branch evolution. *A&A*, 299, 755–769 {**454**}

Bocse FG, 2004, On the precision of X-ray source parameters estimated from ROSAT data. *A&A*, 426, 1119–1134 {**434**}

Böhm-Vitense E, 1993, Detection of a white dwarf companion to the Hyades star HD 27483. *AJ*, 106, 1113–1117 {**461**}

—, 1995, White dwarf companions to Hyades F stars. *AJ*, 110, 228–231 {**460**}

Bolton CT, Rogers GL, 1978, The binary frequency of the OBN and OBC stars. *ApJ*, 222, 234–245 {**441**}

Boulanger F, Abergel A, Bernard JP, *et al.*, 1996, The dust/gas correlation at high Galactic latitude. *A&A*, 312, 256–262 {**474**}

Božić H, Harmanec P, Yang S, *et al.*, 2004, Properties and nature of Be stars. XXII. Long-term light and spectral variations of the new bright Be star HD 6226. *A&A*, 416, 669–676 {**426**}

Božić H, Harmanec P, 1998, HD 6226: a new bright B variable with occasional brightenings. Is it an unrecognized Be star? *A&A*, 330, 222–224 {**426**}

Breger M, 1979, δ Scuti and related stars. *PASP*, 91, 5–26 {**430**}

Breitschwerdt D, 2001, Modeling the local interstellar medium. *Ap&SS*, 276, 163–176 {**465**}

Bressan A, Fagotto F, Bertelli G, *et al.*, 1993, Evolutionary sequences of stellar models with new radiative opacities. II. $Z = 0.02$. *A&AS*, 100, 647–664 {**451**}

Briceño C, Hartmann LW, Stauffer JR, *et al.*, 1997, X-ray surveys and the post-T Tauri problem. *AJ*, 113, 740–752 {**431**}

Briot D, Robichon N, 1998, Absolute magnitudes for some B[e] stars from Hipparcos. *ASSL Vol. 233: B[e] stars*, 47 {**422**}

—, 2000, Hipparcos positions of B/Be stars in the HR diagram. *ASP Conf. Ser. 214: IAU Colloq. 175: The Be Phenomenon in Early-Type Stars*, 117 {**422**}

Brisken WF, Benson JM, Goss WM, *et al.*, 2002, Very Long Baseline Array measurement of nine pulsar parallaxes. *ApJ*, 571, 906–917 {**464**}

Brown WR, Geller MJ, Kenyon SJ, *et al.*, 2005, Discovery of an unbound hypervelocity star in the Milky Way halo. *ApJ*, 622, L33–L36 {**449**}

—, 2006a, A successful targeted search for hypervelocity stars. *ApJ*, 640, L35–L38 {**449**}

—, 2006b, Hypervelocity stars. I. The spectroscopic survey. *ApJ*, 647, 303–311 {**449**}

—, 2007a, Hypervelocity stars. III. The space density and ejection history of main sequence stars from the Galactic centre. *ApJ*, 671, 1708–1716 {**449**}

—, 2007b, Hypervelocity stars. II. The bound population. *ApJ*, 660, 311–318 {**449**}

Bubar EJ, King JR, Soderblom DR, *et al.*, 2007, Keck HIRES spectroscopy of candidate post-T Tauri stars. *AJ*, 134, 2328–2339 {**419**}

Burbidge EM, Burbidge GR, 1956, The chemical compositions of five stars which show some of the characteristics of Population II. *ApJ*, 124, 116 {**428**}

Burki G, 1999, Geneva photometry of the Be star HR 1960: periodicity and extremely small amplitude. *A&A*, 346, 134–138 {**426**}

Burleigh MR, Barstow MA, Holberg JB, 1998, A search for hidden white dwarfs in the ROSAT extreme ultraviolet survey. II. Discovery of a distant DA+F6/7V binary system in a direction of low-density neutral H. *MNRAS*, 300, 511–527 {**461**}

Burleigh MR, Barstow MA, Schenker KJ, *et al.*, 2001, The low-mass white dwarf companion to β Crt. *MNRAS*, 327, 1158–1164 {**461**}

Busso M, Gallino R, Wasserburg GJ, 1999, Nucleosynthesis in asymptotic giant branch stars: relevance for Galactic enrichment and Solar System formation. *ARA&A*, 37, 239–309 {**451**}

Caloi V, Cardini D, D'Antona F, *et al.*, 1999, Kinematics and age of stellar populations in the solar neighbourhood from Hipparcos data. *A&A*, 351, 925–936 {**430**}

Campbell RM, 2001, VLBI pulsar astrometry. *IAU Symp. 205*, 404 {**464**}

Caraveo PA, Lattanzi MG, Massone G, *et al.*, 1997, The importance of positioning isolated neutron stars: the case of Geminga. *ESA SP–402: Hipparcos, Venice '97*, 709–714 {**464**}

—, 1998, Hipparcos positioning of Geminga: how and why. *A&A*, 329, L1–L4 {**464**}

Cardelli JA, Clayton GC, Mathis JS, 1989, The relationship between infrared, optical, and ultraviolet extinction. *ApJ*, 345, 245–256 {**469**}

Carrier F, Burki G, Burnet M, 2002a, Search for duplicity in periodic variable Be stars. *A&A*, 385, 488–502 {**424, 425**}

Carrier F, Burki G, Richard C, 1999, Geneva photometry of the open cluster NGC 2451 and its exceptional Be star HR 2968 satellite. *A&A*, 341, 469–479 {**425**}

Carrier F, Burki G, 2003, Outbursts in the Be star HR 2501. *A&A*, 401, 271–279 {**426**}

Carrier F, North P, Udry S, *et al.*, 2002b, Multiplicity among chemically peculiar stars. II. Cool magnetic Ap stars. *A&A*, 394, 151–169 {**426**}

Catanzaro G, Leone F, 2003, Variability of the He I 587.6 nm line in early-type chemically peculiar stars. II. *Astronomische Nachrichten*, 324, 445–453 {**427**}

Cha AN, 2000, Probing the interstellar medium in Puppis–Vela through optical absorption line spectroscopy. Ph.D. Thesis {**466**}

Cha AN, Sahu MS, Moos HW, *et al.*, 2000, The local interstellar medium in Puppis–Vela. *ApJS*, 129, 281–294 {**466**}

Chandrasekhar S, 1931, The highly collapsed configurations of a stellar mass. *MNRAS*, 91, 456–466 {**455, 457**}

—, 1934, On the hypothesis of the radial ejection of high-speed atoms for the Wolf–Rayet stars and the novae. *MNRAS*, 94, 522–538 {**438**}

Chandrasekhar S, Münch G, 1952, The theory of the fluctuations in brightness of the Milky Way. *ApJ*, 115, 103–123 {**469–471**}

Charbonnel C, Meynet G, Maeder A, *et al.*, 1996, Grids of stellar models. VI. Horizontal branch and early asymptotic giant branch for low-mass stars ($Z = 0.020, 0.001$). *A&AS*, 115, 339–344 {**450**}

Chen B, Vergely JL, Valette B, *et al.*, 1998, Comparison of two different extinction laws with Hipparcos observations. *A&A*, 336, 137–149 {**471, 472**}

Chen P, Yang X, Wang X, 2003a, Near-infrared photometry of 20 Hipparcos carbon stars. *Chinese Astronomy and Astrophysics*, 27, 285–291 {**452**}

Chen P, Yan X, Wang X, 2003b, Near infrared photometry for 20 Hipparcos carbon stars. *Acta Astronomica Sinica*, 44, 110–115 {**452**}

Chevalier C, Ilovaisky SA, 1998, Hipparcos results on massive X-ray binaries. *A&A*, 330, 201–205 {**437, 438, 447**}

Chin CW, Stothers R, 1971, Low-mass white dwarfs and the cooling sequences in the Hyades cluster. *ApJ*, 163, 555–565 {**460**}

Cidale LS, Arias ML, Torres AF, *et al.*, 2007, Fundamental parameters of He-weak and He-strong stars. *A&A*, 468, 263–272 {**452**}

Clark LL, Dolan JF, 1999, The distance to eight X-ray sources derived from Hipparcos observations. *A&A*, 350, 1085–1088 {**438**}

Clayton GC, 1996, The R Coronae Borealis stars. *PASP*, 108, 225–241 {**452**}

Clemens JC, Nather RE, Winget DE, *et al.*, 1992, Whole Earth Telescope observations of V471 Tau: the nature of the white dwarf variations. *ApJ*, 391, 773–783 {**460**}

Collins GW, 1987, The use of terms and definitions in the study of Be stars. *IAU Colloq. 92: Physics of Be Stars* (eds. Slettebak A, Snow TP), 3–19 {**421**}

Comerón F, Pasquali A, 2007, A very massive runaway star from Cygnus OB2. *A&A*, 467, L23–L27 {**444, 448**}

Corradi WJB, Guimaraes MM, Vieira SLA, 2003, Interstellar extinction towards the interaction zone between the local and Loop I bubbles. *Astrophysics of Dust*, 9–9 {**472**}

Cottrell PL, Lawson WA, 1998, Hipparcos observations of H-deficient carbon stars. *Publications of the Astronomical Society of Australia*, 15, 179–182 {**452**}

Cox DP, Reynolds RJ, 1987, The local interstellar medium. *ARA&A*, 25, 303–344 {**465**}

Cox DP, Smith BW, 1974, Large-scale effects of supernova remnants on the Galaxy: generation and maintenance of a hot network of tunnels. *ApJ*, 189, L105–L108 {**466**}

Crawford IA, 2000, A study of interstellar Na I D absorption lines towards the Lupus molecular clouds. *MNRAS*, 317, 996–1004 {**467, 472**}

Crawford IA, Lallement R, Price RJ, *et al.*, 2002, High-resolution observations of interstellar Na I and Ca II towards the southern opening of the 'Local Interstellar Chimney': probing the disk-halo connection. *MNRAS*, 337, 720–730 {**466**}

Cruz González C, Recillas Cruz E, Costero R, *et al.*, 1974, A catalogue of Galactic O stars and the ionisation of the low density interstellar medium by runaway stars. *Revista Mexicana de Astronomia y Astrofisica*, 1, 211–259 {**441**}

Cudworth KM, 1974, New proper motions, statistical parallaxes, and kinematics of planetary nebulae. *AJ*, 79, 1384–1395 {**455**}

Cutispoto G, Pastori L, Guerrero A, *et al.*, 2000, Photometric and spectroscopic studies of cool stars discovered in EXOSAT X-ray images. IV. The northern hemisphere sample. *A&A*, 364, 205–216 {**434**}

Cutispoto G, Pastori L, Tagliaferri G, *et al.*, 1999, Classification of extreme ultraviolet stellar sources detected by the ROSAT WFC. I. Photometric and radial velocity studies. *A&AS*, 138, 87–99 {**434**}

Czart K, Strobel A, 2006, Distance to Wolf–Rayet star WR 134. *The Ultraviolet Universe: Stars from Birth to Death, IAU 26, JD 4*, 4 {**440**}

D'Antona F, Mazzitelli I, 1990, Cooling of white dwarfs. *ARA&A*, 28, 139–181 {**456**}

—, 1994, New pre-main sequence tracks for $M \leq 2.5M_{\odot}$ as tests of opacities and convection model. *ApJS*, 90, 467–500 {**413, 416**}

—, 1997, Evolution of low-mass stars. *Memorie della Societa Astronomica Italiana*, 68, 807–822 {**413, 417, 418**}

Davidson GT, Claflin ES, Haisch BM, 1987, On the $B - V$ colours of the bright stars. *AJ*, 94, 771–791 {**470**}

Davies JK, Evans A, Bode MF, *et al.*, 1990, Photometric monitoring of pre-main sequence stars. III. Variability of Herbig Be stars. *MNRAS*, 247, 517–522 {**414**}

Dearborn DSP, Liebert J, Aaronson M, *et al.*, 1986, On the nature of the dwarf carbon star G77–61. *ApJ*, 300, 314–324 {**449**}

de Bruijne JHJ, Hoogerwerf R, de Zeeuw PT, 2001, A Hipparcos study of the Hyades open cluster. Improved colour-absolute magnitude and HR diagrams. *A&A*, 367, 111–147 {**461**}

de Cuyper JP, 1982, Supernovae in binary systems: production of runaway stars and pulsars. *Astrophysics and Space Science Library 98* (eds. Kopal Z, Rahe J), 417–443 {**440**}

—, 1985, Progenitor systems of two neutron-star binaries. *Astrophysics and Space Science Library 120* (eds. Boland W, van Woerden H), 207–210 {**441**}

Dehnen W, Binney JJ, 1998, Local stellar kinematics from Hipparcos data. *MNRAS*, 298, 387–394 {**423, 467**}

Delhaye J, 1965, Solar motion and velocity distribution of common stars. *Galactic Structure* (eds. Blaauw A, Schmidt M), 61–84 {**441**}

De Luca A, Caraveo PA, Mattana F, *et al.*, 2006, On the complex X-ray structure tracing the motion of Geminga. *A&A*, 445, L9–L13 {**464**}

Désert FX, Boulanger F, Puget JL, 1990, Interstellar dust models for extinction and emission. *A&A*, 237, 215–236 {**474**}

de Zeeuw PT, Hoogerwerf R, de Bruijne JHJ, *et al.*, 1999, A Hipparcos census of the nearby OB associations. *AJ*, 117, 354–399 {**444, 464, 468**}

Diamond CJ, Jewell SJ, Ponman TJ, 1995, ROSAT extreme ultraviolet observations of DA white dwarfs and late-type stars and the structure of the local interstellar medium. *MNRAS*, 274, 589–601 {**465**}

Dias WS, Alessi BS, Moitinho A, *et al.*, 2002, New catalogue of optically visible open clusters and candidates. *A&A*, 389, 871–873 {**471**}

Domiciano de Souza A, Kervella P, Jankov S, *et al.*, 2003, The spinning-top Be star Achernar from VLTI–VINCI. *A&A*, 407, L47–L50 {**425**}

Dominguez I, Chieffi A, Limongi M, *et al.*, 1999, Intermediate-mass stars: updated models. *ApJ*, 524, 226–241 {**461**}

Dominici TP, Teixeira R, Horvath JE, *et al.*, 1999, Extension of Tycho catalogue for low-extinction windows in the Galactic bulge. *A&AS*, 136, 261–267 {**472**}

Dominy JF, 1984, The chemical composition and evolutionary state of the early R stars. *ApJS*, 55, 27–43 {**450**}

Donati JF, Wade GA, Babel J, *et al.*, 2001, The magnetic field and wind confinement of β Cep: new clues for interpreting the Be phenomenon? *MNRAS*, 326, 1265–1278 {**425**}

Dorman B, Rood RT, O'Connell RW, 1993, Ultraviolet radiation from evolved stellar populations. I. Models. *ApJ*, 419, 596–614 {**462**}

Driebe T, Schoenberner D, Blöcker T, *et al.*, 1998, The evolution of helium white dwarfs. I. The companion of the millisecond pulsar PSR J1012+5307. *A&A*, 339, 123–133 {**462**}

Drilling JS, 1986, Basic data on hydrogen-deficient stars. *IAU Colloq. 87: Hydrogen Deficient Stars and Related Objects* (eds. Hunger K, Schoenberner D, Kameswara Rao N), ASSL Vol. 128, 9–20 {**452**}

Drissen L, Moffat AFJ, Walborn NR, *et al.*, 1995, The dense Galactic starburst NGC 3603. I. HST–FOS spectroscopy of individual stars in the core and the source of ionisation and kinetic energy. *AJ*, 110, 2235–2241 {**443**}

Ducourant C, Teixeira R, Périé JP, *et al.*, 2005, Pre-main sequence star proper motion catalogue. *A&A*, 438, 769–778 {**415**}

Ducourant C, Teixeira R, Sartori MJ, *et al.*, 2000, Proper motions of pre-main sequence stars in southern star-forming regions. *Hipparcos and the Luminosity Calibration of the Nearer Stars, IAU Joint Discussion 13*, 41–44 {**419**}

Dunkin SK, Crawford IA, 1998, Spatially resolved optical spectroscopy of the Herbig Ae/Vega-like binary star HD 35187. *MNRAS*, 298, 275–284 {**421, 473**}

Dušek J, Mikulášek Z, Papoušek J, 2003, Preliminary analysis of light curves of seven carbon stars. *Contributions of the Astronomical Observatory Skalnate Pleso*, 33, 119–133 {**452**}

Eggen OJ, 1993a, Degenerate stars in the Hyades supercluster. *AJ*, 106, 642–649 {**460**}

—, 1993b, The low-mass Hyades and the evaporation of clusters. *AJ*, 106, 1885–1905 {**460**}

Elias JH, 1978, A study of the Taurus dark cloud complex. *ApJ*, 224, 857–872 {**415**}

Fagotto F, Bressan A, Bertelli G, *et al.*, 1994, Evolutionary sequences of stellar models with very high metallicity. V. $Z = 0.1$. *A&AS*, 105, 39–45 {**451**}

Faraggiana R, Bonifacio P, 1999, How many λ Bootis stars are binaries? *A&A*, 349, 521–531 {**429, 430**}

Favata F, Micela G, Sciortino S, *et al.*, 1998, The evolutionary status of activity-selected solar-type stars and of T Tauri stars as derived from Hipparcos parallaxes: evidence for long-lived T Tauri disks? *A&A*, 335, 218–226 {**417, 418**}

Feast MW, 1972, The R CrB variables in the Large Magellanic Cloud. *MNRAS*, 158, 11P–13P {**452**}

Feigelson ED, Jackson JM, Mathieu RD, *et al.*, 1987, An X-ray survey for pre-main sequence stars in the Taurus–Auriga and Perseus molecular cloud complexes. *AJ*, 94, 1251–1259 {**431**}

Feigelson ED, 1996, Dispersed T Tauri stars and Galactic star formation. *ApJ*, 468, 306–322 {**415**}

Figueras F, Luri X, Gómez AE, *et al.*, 1998, Chemically peculiar stars: photometric calibrations of luminosity using Hipparcos data. *Contributions of the Astronomical Observatory Skalnate Pleso*, 27, 184–191 {**426**}

Finley DS, Koester D, Basri G, 1997, The temperature scale and mass distribution of hot DA white dwarfs. *ApJ*, 488, 375–396 {**455**}

Fitzpatrick EL, Massa D, 2007, An analysis of the shapes of interstellar extinction curves. V. The infrared through ultraviolet curve morphology. *ApJ*, 663, 320–341 {**470**}

Fitzpatrick EL, 1999, Correcting for the effects of interstellar extinction. *PASP*, 111, 63–75 {**469, 470**}

Floquet M, Halbwachs JL, Hubert AM, 1998, Tycho photometry of some B[e] stars. *ASSL Vol. 233: B[e] stars*, 53 {**423**}

Frasca A, Guillout P, Marilli E, *et al.*, 2006a, Newly-discovered active binaries in the Rosat All-Sky Survey–Tycho sample of stellar X-ray sources. I. Orbital and physical parameters of six new binaries. *A&A*, 454, 301–309 {**434**}

Frasca A, Marilli E, Guillout P, *et al.*, 2006b, Late-type X-ray emitting binaries in the solar neighbourhood and in star-forming regions. *Ap&SS*, 304, 17–20 {**434**}

Freire PC, Camilo F, Lorimer DR, *et al.*, 2001, Timing the millisecond pulsars in 47 Tucanae. *MNRAS*, 326, 901–915 {**464**}

Frink S, 1999, Kinematics of T Tauri stars in nearby star-forming regions. *Ph.D. Thesis* {**415**}

—, 2001, Kinematic distances to TW Hya members stars via the convergent point method. *Young Stars Near Earth: Progress and Prospects* (eds. Jayawardhana R, Greene T), ASP Conf. Ser. 244, 16–20 {**428**}

Frink S, Röser S, Alcalá JM, *et al.*, 1998, Kinematics of T Tauri stars in Chamaeleon. *A&A*, 338, 442 {**415**}

Frisch PC, Grodnicki L, Welty DE, 2002, The velocity distribution of the nearest interstellar gas. *ApJ*, 574, 834–846 {**467**}

Frisch PC, 1993, Whence Geminga. *Nature*, 364, 395–396 {**444**}

—, 1995, Characteristics of nearby interstellar matter. *Space Science Reviews*, 72, 499–592 {**465**}

Fuchs B, Breitschwerdt D, de Avillez MA, *et al.*, 2006, The search for the origin of the Local Bubble redivivus. *MNRAS*, 373, 993–1003 {**468**}

Gahm GF, Ahlin P, Lindroos KP, 1983, A study of visual double stars with early-type primaries. I. Spectroscopic results. *A&AS*, 51, 143–159 {**418**}

García-Berro E, Hernanz M, Isern J, *et al.*, 1995, The rate of change of the gravitational constant and the cooling of white dwarfs. *MNRAS*, 277, 801–810 {**456**}

García-Berro E, Isern J, Kubyshin YA, 2007, Astronomical measurements and constraints on the variability of fundamental constants. *A&A Rev.*, 14, 113–170 {**456**}

García B, 1991, The interstellar lines catalogue. *A&AS*, 89, 469–527 {**467**}

Gatewood G, Russell J, 1974, Astrometric determination of the gravitational redshift of van Maanen 2. *AJ*, 79, 815–818 {**462**}

Gaustad JE, McCullough PR, Rosing W, *et al.*, 2001, A robotic wide-angle Hα survey of the southern sky. *PASP*, 113, 1326–1348 {**455**}

Gerbaldi M, 2003, Peculiar Am stars after Hipparcos. *ASSL Vol. 298: Stellar Astrophysics: A Tribute to Helmut A. Abt* (eds. Cheng KS, Leung KC, Li TP), 159–163 {**427**}

Gerbaldi M, Faraggiana R, Balin N, 2001, Binary systems with post-T Tauri secondaries. *A&A*, 379, 162–184 {**418, 419**}

Gerbaldi M, Faraggiana R, Lai O, 2003, The heterogeneous class of λ Bootis stars. *A&A*, 412, 447–464 {**429**}

Gerbaldi M, Faraggiana R, 2007, Duplicity among λ Bootis stars: the new case of HD 204041. *Binary Stars as Critical Tools and Tests in Contemporary Astrophysics, IAU Symp. 240*, 120–123 {**429**}

Gershberg RE, Katsova MM, Lovkaya MN, *et al.*, 1999, Catalogue and bibliography of the UV Ceti-type flare stars and related objects in the solar vicinity. *A&AS*, 139, 555–558 {**427, 428**}

Gervais S, St-Louis N, 1999, A large H I shell surrounding the Wolf–Rayet star HD 191765. *AJ*, 118, 2394–2408 {**440**}

Gies DR, 1987, The kinematical and binary properties of association and field O stars. *ApJS*, 64, 545–563 {**440**}

—, 2000, Glimpses of Be binary evolution. *IAU Colloq. 175: The Be Phenomenon in Early-Type Stars* (eds. Smith MA, Henrichs HF, Fabregat J), ASP Conf. Ser. 214, 668–680 {**422, 423**}

Gies DR, Bolton CT, 1986, The binary frequency and origin of the OB runaway stars. *ApJS*, 61, 419–454 {**440, 443, 447**}

Girardi L, Bressan A, Bertelli G, *et al.*, 2000, Evolutionary tracks and isochrones for low- and intermediate-mass stars: from $0.15–7\,M_\odot$ at $Z = 0.0004–0.03$. *A&AS*, 141, 371–383 {**433, 461**}

Glendenning NK, Kettner C, Weber F, 1995a, From strange stars to strange dwarfs. *ApJ*, 450, 253–261 {**458**}

—, 1995b, Possible new class of dense white dwarfs. *Physical Review Letters*, 74, 3519–3522 {**458**}

Gliese W, Jahreiß H, 1991, *Preliminary version of the Third Catalogue of Nearby Stars*. Astronomischen Rechen-Instituts Heidelberg. On: The Astronomical Data Centre CD-ROM: Selected Astronomical Cataloguess, Vol. I, NASA/ADC/GSFC {**436**}

Gómez AE, Luri X, Grenier S, *et al.*, 1998a, The HR diagram from Hipparcos data. Absolute magnitudes and kinematics of Bp–Ap stars. *A&A*, 336, 953–959 {**426**}

Gómez AE, Luri X, Sabas V, *et al.*, 1998b, Absolute magnitudes and kinematics of chemically peculiar stars from Hipparcos data. *Contributions of the Astronomical Observatory Skalnate Pleso*, 27, 171–178 {**426**}

Gondhalekar PM, Phillips AP, Wilson R, 1980, Observations of the interstellar ultraviolet radiation field from the S2/68 sky-survey telescope. *A&A*, 85, 272–280 {**474**}

Gondoin P, 1999, Evolution of X-ray activity and rotation on G-K giants. *A&A*, 352, 217–227 {**435**}

Goriely S, Mowlavi N, 2000, Neutron-capture nucleosynthesis in asymptotic giant branch stars. *A&A*, 362, 599–614 {**452**}

Gorti U, Bhatt HC, 1996, Dynamics of embedded protostar clusters in clouds. *MNRAS*, 278, 611–616 {**431**}

Grady CA, Perez MR, Talavera A, *et al.*, 1996, The β Pic phenomenon in A-shell stars: detection of accreting gas. *ApJ*, 471, L49–L52 {**426**}

Grant CE, 1999, The three-dimensional structure of the hot interstellar medium. Ph.D. thesis, Pennsylvania State University {**466**}

Grant CE, Burrows DN, 1999, Distances to the high Galactic latitude molecular clouds G192–67 and MBM 23–24. *ApJ*, 516, 243–249 {**466**}

Greenstein JL, Oke JB, Shipman HL, 1971, Effective temperature, radius, and gravitational redshift of Sirius B. *ApJ*, 169, 563–566 {**460**}

Green PJ, Margon B, Anderson SF, *et al.*, 1992, Carbon star luminosity indicators. *ApJ*, 400, 659–664 {**449**}

Gregorio-Hetem J, Lépine JRD, Quast GR, *et al.*, 1992, A search for T Tauri stars based on the IRAS point source catalogue. *AJ*, 103, 549–563 {**420**}

Güdel M, Guinan EF, Etzel PB, *et al.*, 1998, Assembling the pieces of the puzzle: a nearby, rapidly rotating young sun in 47 Cas? *ASP Conf. Ser. 154: Cool Stars, Stellar Systems, and the Sun*, 1247–1256 {**419**}

Guillout P, Schmitt JHMM, Egret D, *et al.*, 1999, The stellar content of soft X-ray surveys. II. Cross correlation of the ROSAT All-Sky Survey with the Tycho and Hipparcos Catalogues. *A&A*, 351, 1003–1015 {**432**}

Guillout P, Sterzik MF, Schmitt JHMM, *et al.*, 1998a, Discovery of a late-type stellar population associated with the Gould Belt. *A&A*, 337, 113–124 {**416, 431, 432**}

—, 1998b, The large-scale distribution of X-ray active stars. *A&A*, 334, 540–544 {**431**}

Guinan EF, Sion EM, 1984, IUE spectroscopy of the degenerate components in the Hyades close binaries V471 Tau and HZ 9. *AJ*, 89, 1252–1255 {**460**}

Hakkila J, Myers JM, Stidham BJ, *et al.*, 1997, A computerized model of large-scale visual interstellar extinction. *AJ*, 114, 2043–2053 {**470**}

Hamada T, Salpeter EE, 1961, Models for zero-temperature stars. *ApJ*, 134, 683–698 {**455, 458–461**}

Hanbury Brown R, Davis J, Herbison-Evans D, *et al.*, 1970, A study of γ^2 Vel with a stellar intensity interferometer. *MNRAS*, 148, 103–117 {**439**}

Handler G, 1999, Variable central stars of young planetary nebulae. A photometric study of the central star of M2-54. *A&AS*, 135, 493–498 {**455**}

Harjunpää P, Kaas AA, Carlqvist P, *et al.*, 1999, Linear polarisation and molecular filamentary clouds. *A&A*, 349, 912–926 {**475**}

Harris HC, Dahn CC, Canzian B, *et al.*, 2007, Trigonometric parallaxes of central stars of planetary nebulae. *AJ*, 133, 631–638 {**454**}

Harris HC, Dahn CC, Monet DG, *et al.*, 1997, Trigonometric parallaxes of planetary nebulae. *Planetary Nebulae* (eds. Habing HJ, Lamers HJGLM), volume 180 of *IAU Symp. 180*, 40–45 {**454**}

Hauck B, Jaschek C, 2000, A-shell stars in the Geneva system. *A&A*, 354, 157–162 {**426**}

Hauck B, Slettebak A, 1983, The λ Bootis stars: a reappraisal. *A&A*, 127, 231–234 {**428**}

Haud U, Einasto J, 1989, Galactic models with massive corona. II. Galaxy. *A&A*, 223, 95–106 {**443**}

Hayashi C, 1961, Stellar evolution in early phases of gravitational contraction. *PASJ*, 13, 450–452 {**413**}

Hearty T, Fernández M, Alcalá JM, *et al.*, 2000a, The distance to the nearest star-forming clouds: MBM 12 and MBM 20. *A&A*, 357, 681–685 {**466**}

Hearty T, Magnani L, Caillault JP, *et al.*, 1999, A search for star formation in the translucent clouds MBM 7 and MBM 55. *A&A*, 341, 163–173 {**419**}

Hearty T, Neuhäuser R, Stelzer B, *et al.*, 2000b, ROSAT PSPC observations of T Tauri stars in MBM 12. *A&A*, 353, 1044–1054 {**419**}

Heber U, Edelmann H, Lisker T, *et al.*, 2003, Discovery of a helium-core white dwarf progenitor. *A&A*, 411, L477–L480 {**462, 463**}

Heger A, Langer N, 2000, Pre-supernova evolution of rotating massive stars. II. Evolution of the surface properties. *ApJ*, 544, 1016–1035 {**422**}

Heiles C, 1979, H I shells and supershells. *ApJ*, 229, 533–537 {**448**}

—, 1987, Supernovae versus models of the interstellar medium and the gaseous halo. *ApJ*, 315, 555–566 {**448**}

—, 1998, Whence the Local Bubble, Gum, Orion? GSH 238+00+09, a nearby major superbubble toward Galactic longitude 238°. *ApJ*, 498, 689–703 {**475**}

Heiter U, Kupka F, Paunzen E, *et al.*, 1998, Abundance analysis of the λ Bootis stars HD 192640, HD 183324, and HD 84123. *A&A*, 335, 1009–1017 {**430**}

Henry RC, 1977, Far-ultraviolet studies. I. Predicted far-ultraviolet interstellar radiation field. *ApJS*, 33, 451–458 {**475**}

—, 2002, The local interstellar ultraviolet radiation field. *ApJ*, 570, 697–707 {**475**}

Herald JE, Bianchi L, 2002, The binary central star of the planetary nebula A35. *ApJ*, 580, 434–446 {**455**}

Herbig GH, 1960, The spectra of Be- and Ae-type stars associated with nebulosity. *ApJS*, 4, 337–368 {**414**}

—, 1978, Can post-T Tauri stars be found? *Problems of Physics and Evolution of the Universe*, 171–179 {**414**}

Hernández J, Calvet N, Hartmann L, *et al.*, 2005, Herbig Ae/Be stars in nearby OB associations. *AJ*, 129, 856–871 {**421**}

Herrero A, Kudritzki RP, Vilchez JM, *et al.*, 1992, Intrinsic parameters of Galactic luminous OB stars. *A&A*, 261, 209–234 {**444**}

Herwig F, 1995, The impact of improved theoretical and observational data on the initial-to-final mass relation. *Liège International Astrophysical Colloquia 32* (eds. Noels A, Fraipont-Caro D, Gabriel M, *et al.*), 441–446 {**461**}

Hills JG, 1988, Hypervelocity and tidal stars from binaries disrupted by a massive Galactic black hole. *Nature*, 331, 687–689 {**449**}

Hodgkin ST, Pye JP, 1994, ROSAT extreme ultraviolet luminosity functions of nearby late-type stars. *MNRAS*, 267, 840–870 {**431**}

Hoffleit D, Jaschek C, 1991, *The Bright Star Catalogue*. Yale University Observatory, 5th edition {**432**}

Hoff W, Henning T, Pfau W, 1998, The nature of isolated T Tauri stars. *A&A*, 336, 242–250 {**415**}

Holberg JB, Barstow MA, Bruhweiler FC, *et al.*, 1998, Sirius B: a new, more accurate view. *ApJ*, 497, 935 {**458, 460**}

Hoogerwerf R, de Bruijne JHJ, de Zeeuw PT, 2000, The origin of runaway stars. *ApJ*, 544, L133–L136 {**443, 468**}

—, 2001, On the origin of the O and B-type stars with high velocities. II. Runaway stars and pulsars ejected from the nearby young stellar groups. *A&A*, 365, 49–77 {**443–445, 464, 468**}

Hubert AM, Floquet M, Gómez AE, *et al.*, 1997, Photometric variability of B and Be stars. *ESA SP–402: Hipparcos, Venice '97*, 315–318 {**423**}

Hubert AM, Floquet M, Zorec J, 2000, Short-lived and long-lived outbursts in B and Be stars from Hipparcos photometry and modeling. *ASP Conf. Ser. 214: IAU Colloq. 175: The Be Phenomenon in Early-Type Stars*, 348 {**423**}

Hubert AM, Floquet M, 1998, Investigation of the variability of bright Be stars using Hipparcos photometry. *A&A*, 335, 565–572 {**423–425**}

Hubrig S, North P, Medici A, 2000, Rotation and evolution of A stars: looking for progenitors of cool Ap stars. *A&A*, 359, 306–310 {**426**}

Hughes J, Hartigan P, Clampitt L, 1993, The distance to the Lupus star-formation region. *AJ*, 105, 571–575 {**415**}

Hui-Bon-Hoa A, 2000, Metal abundances of field A and Am stars. *A&AS*, 144, 203–209 {**427**}

Humason ML, Zwicky F, 1947, A search for faint blue stars. *ApJ*, 105, 85–91 {**446, 460**}

Humphreys RM, Larsen JA, 1995, The Sun's distance above the Galactic plane. *AJ*, 110, 2183–2188 {**423**}

Hünsch M, Schmitt JHMM, Sterzik MF, *et al.*, 1999, The ROSAT All-Sky Survey Catalogue of the nearby stars. *A&AS*, 135, 319–338 {**436**}

Hünsch M, Schmitt JHMM, Voges W, 1998a, The ROSAT All-Sky Survey Catalogue of optically bright main sequence stars and subgiant stars. *A&AS*, 132, 155–171 {**432**}

—, 1998b, The ROSAT All-Sky Survey Catalogue of optically bright late-type giants and supergiants. *A&AS*, 127, 251–255 {**432**}

Hünsch M, Schröder KP, 1996, The revised X-ray dividing line: new light on late stellar activity. *A&A*, 309, L51–L54 {**434**}

Huthoff F, Kaper L, 2002, On the absence of wind bow-shocks around OB-runaway stars: probing the physical conditions of the interstellar medium. *A&A*, 383, 999–1010 {**448**}

Ibata RA, Irwin MJ, Bienaymé O, *et al.*, 2000, Discovery of high proper motion ancient white dwarfs: nearby massive compact halo objects? *ApJ*, 532, L41–L45 {**456**}

Iben I, Renzini A, 1983, Asymptotic giant branch evolution and beyond. *ARA&A*, 21, 271–342 {**451, 459, 462**}

Iben I, 1965, Stellar evolution. I. The approach to the main sequence. *ApJ*, 141, 993–1018 {**413**}

Iliev IK, Paunzen E, Barzova IS, *et al.*, 2002, First orbital elements for the λ Bootis spectroscopic binary systems HD 84948 and HD 171948: implications for the origin of the λ Bootis stars. *A&A*, 381, 914–922 {**430**}

Isern J, Canal R, Labay J, 1991, The outcome of explosive ignition of O/Ne/Mg cores: supernovae, neutron stars, or 'iron' white dwarfs? *ApJ*, 372, L83–L86 {**458**}

Izumiura H, Hashimoto O, 1997, ISOPHOT mapping observations of carbon stars. *Ap&SS*, 255, 341–347 {**452**}

Izzard RG, Jeffery CS, Lattanzio J, 2007, Origin of the early-type R stars: a binary-merger solution to a century-old problem? *A&A*, 470, 661–673 {**450**}

Jackson MS, Halpern JP, 2005, A refined ephemeris and phase-resolved X-ray spectroscopy of the Geminga pulsar. *ApJ*, 633, 1114–1125 {**464**}

Jackson S, MacGregor KB, Skumanich A, 2004, Models for the rapidly rotating Be star Achernar. *ApJ*, 606, 1196–1199 {**426**}

Jaschek M, Slettebak A, Jaschek C, 1981, Be star terminology. *Be Star Newsletter 4*, 9–11 {**421**}

Jensen ELN, Cohen DH, Neuhäuser R, 1998, ROSAT and Hipparcos observations of isolated pre-main sequence stars near HD 98800. *AJ*, 116, 414–423 {**415, 416**}

Jensen KA, Swank JH, Petre R, *et al.*, 1986, EXOSAT observations of V471 Tau: a 9.25 minute white dwarf pulsation and orbital phase dependent X-ray dips. *ApJ*, 309, L27–L31 {**460**}

Johnston KJ, Fey AL, Gaume RA, *et al.*, 2004, The enigmatic radio source T Tauri S. *ApJ*, 604, L65–L68 {**419**}

Jordan J, Dukes RJ, Adelman SJ, 1997, A search for light variations in the λ Bootis star HD 111604. *Bulletin of the American Astronomical Society*, 29, 1275 {**430**}

Jorissen A, Začs L, Udry S, *et al.*, 2005, On metal-deficient barium stars and their link with yellow symbiotic stars. *A&A*, 441, 1135–1148 {**453**}

Joshi YC, 2005, Interstellar extinction towards open clusters and Galactic structure. *MNRAS*, 362, 1259–1266 {**471**}

Joy AH, 1945, T Tauri variable stars. *ApJ*, 102, 168–195 {**413**}

Kamp I, Paunzen E, 2002, The λ Bootis phenomenon: interaction between a star and a diffuse interstellar cloud. *MNRAS*, 335, L45–L49 {**429**}

Kaper L, van der Meer A, Tijani AH, 2004, High-mass X-ray binaries and OB runaway stars. *Revista Mexicana de Astronomia y Astrofisica Conf. Ser.*, volume 21, 128–131 {**447**}

Kaper L, van Loon JT, Augusteijn T, *et al.*, 1997, Discovery of a bow shock around Vela X–1. *ApJ*, 475, L37–L40 {**447–449**}

Kastner JH, Zuckerman B, Weintraub DA, *et al.*, 1997, X-ray and molecular emission from the nearest region of recent star formation. *Science*, 277, 67–71 {**415**}

Keenan PC, 1942, The spectra of CH stars. *ApJ*, 96, 101–105 {**449**}

—, 1993, Revised MK spectral classification of the red carbon stars. *PASP*, 105, 905–910 {**449**}

Kenyon SJ, Dobrzycka D, Hartmann L, 1994, A new optical extinction law and distance estimate for the Taurus–Auriga molecular cloud. *AJ*, 108, 1872–1880 {**418**}

Kenyon SJ, Hartmann L, 1995, Pre-main sequence evolution in the Taurus–Auriga molecular cloud. *ApJS*, 101, 117–166 {**418, 428**}

Kerber F, Guglielmetti F, Mignani R, *et al.*, 2002, Proper motion of the central star of the planetary nebula Sh 2–68. *A&A*, 381, L9–L12 {**455**}

Kerber F, Mignani R, Pauli EM, *et al.*, 2004, Galactic orbits of planetary nebulae unveil thin and thick disk populations and cast light on interaction with the interstellar medium. *A&A*, 420, 207–211 {**455, 456**}

Kerber F, Mignani R, Smart RL, *et al.*, 2008, Galactic planetary and their central stars. II Proper motions. *A&A*, 479, 155–160 {**455**}

Kilic M, Munn JA, Harris HC, *et al.*, 2005, Discovery of 282 cool white dwarfs in the Sloan Digital Sky Survey. *14th European Workshop on White Dwarfs* (eds. Koester D, Moehler S), volume 334 of *ASP Conf. Ser. 334*, 131–134 {**463**}

Kilpio E, 1998, Reclassification of spectra and calculation of interstellar extinction for the objects of Hipparcos Catalogue. *Modern Problems in Stellar Evolution*, 268–272 {**472**}

Knapik A, Bergeat J, Rutily B, 1998, The space distribution of Hipparcos carbon stars. *A&A*, 334, 545–551 {**450**}

—, 1999, Dust extinction and intrinsic spectral energy distributions of carbon-rich stars. III. The Miras, CS, and SC stars. *A&A*, 344, 263–276 {**451**}

Knapp GR, Pourbaix D, Jorissen A, 2001, Reprocessing the Hipparcos data for evolved giant stars. II. Absolute magnitudes for the R-type carbon stars. *A&A*, 371, 222–232 {**449–451**}

Knie K, Korschinek G, Faestermann T, *et al.*, 2004, ^{60}Fe anomaly in a deep-sea manganese crust and implications for a nearby supernova source. *Physical Review Letters*, 93(17), 1103.1–4 {**469**}

Knude J, Fabricius C, Høg E, *et al.*, 2002, Distances of absorbing features in the LDN 1622 direction. An application of Tycho 2 photometry and Michigan Classification. *A&A*, 392, 1069–1079 {**472**}

Knude J, Høg E, 1998, Interstellar reddening from the Hipparcos and Tycho catalogues. I. Distances to nearby molecular clouds and star-forming regions. *A&A*, 338, 897–904 {**471**}

—, 1999, Interstellar reddening from the Hipparcos and Tycho catalogues. II. Nearby dust features at the NGP associated with approaching H I gas. *A&A*, 341, 451–457 {**470, 472**}

Knude J, Jønch-Sørensen H, Nielsen AS, 1999, Distance and absorption of the tails in the CG30/CG31/CG38 complex. an application of a $(V-I)_0$–M_V main sequence relation derived from the Hipparcos and Tycho Catalogues. *A&A*, 350, 985–996 {**472**}

Knude J, Nielsen AS, 2000, Distance and absorption features in the CG 30/CG 31/CG 38 complex. *A&A*, 362, 1138–1142 {**472**}

Knude J, 1979, A catalogue of low-mass clouds in the solar vicinity: results from a photometric survey of 84 volumes. *A&AS*, 38, 407–421 {**470**}

—, 1998, Stellar uvbyβ photometry in three extreme ultraviolet shadow directions. *A&AS*, 130, 477–484 {**472**}

Kodaira K, 1967, The abundance ratio of oxygen to magnesium in the atmosphere of λ Bootis-type stars. *PASJ*, 19, 556 {**428**}

Koester D, Napiwotzki R, Christlieb N, *et al.*, 2001, High-resolution VLT–UVES spectra of white dwarfs observed for the ESO supernova Type Ia progenitor survey (SPY). *A&A*, 378, 556–568 {**463**}

Koester D, Schulz H, Weidemann V, 1979, Atmospheric parameters and mass distribution of DA white dwarfs. *A&A*, 76, 262–275 {**455**}

Köhler R, 2001, Multiplicity of X-ray-selected T Tauri stars in Chamaeleon. *AJ*, 122, 3325–3334 {**419**}

König B, Guenther EW, Esposito M, *et al.*, 2006, Spectral synthesis analysis and radial velocity study of the northern F-, G- and K-type flare stars. *MNRAS*, 365, 1050–1056 {**428**}

König B, Neuhäuser R, Guenther EW, *et al.*, 2003, Flare stars in the TW Hya association: the HIP 57269 system. *Astronomische Nachrichten*, 324, 516–522 {**427, 428**}

Korhonen T, Reiz A, 1986, Interstellar polarisation from observations of A and F stars in high and intermediate Galactic latitudes, and from stars in the Mathewson and Ford polarisation catalogue. *A&AS*, 64, 487–494 {**474**}

Koubský P, Harmanec P, Hubert AM, *et al.*, 2000, Properties and nature of Be stars. XIX. Spectral and light variability of 60 Cyg. *A&A*, 356, 913–928 {**426**}

Kraemer KE, Sloan GC, Wood PR, *et al.*, 2005, R CrB candidates in the Small Magellanic Cloud: observations of cold, featureless dust with the Spitzer Infrared Spectrograph. *ApJ*, 631, L147–L150 {**452**}

Krautter J, 1980, Polarisation measurements of 313 nearby stars. *A&AS*, 39, 167–172 {**474**}

Kudritzki RP, Hummer DG, 1990, Quantitative spectroscopy of hot stars. *ARA&A*, 28, 303–345 {**444**}

Kurtz DW, 2000, Pulsating and chemically peculiar upper main sequence stars. *Encyclopedia of Astronomy and Astrophysics;* (ed. Murdin, P.) {**427**}

Lallement R, Welsh BY, Vergely JL, *et al.*, 2003, 3d mapping of the dense interstellar gas around the Local Bubble. *A&A*, 411, 447–464 {**466**}

Lanning HH, Lépine S, 2006, Proper motions of faint ultraviolet-bright sources in the Sandage two-colour survey of the Galactic plane. *PASP*, 118, 1639–1647 {**463**}

Lanning HH, Pesch P, 1981, HZ 9: a white dwarf/red dwarf spectroscopic binary in the Hyades. *ApJ*, 244, 280–285 {**460**}

Lebzelter T, Hron J, 2003, Technetium and the third dredge up in asymptotic giant branch stars. I. Field stars. *A&A*, 411, 533–542 {**453**}

Leonard PJT, 1993, Mechanisms for ejecting stars from the Galactic plane. *Luminous High-Latitude Stars* (ed. Sasselov DD), ASP Conf. Ser. 45, 360–364 {**442**}

Leonard PJT, Tremaine S, 1990, The local Galactic escape speed. *ApJ*, 353, 486–493 {**442**}

Lépine JRD, Duvert G, 1994, Star formation by infall of high-velocity clouds on the Galactic disk. *A&A*, 286, 60–71 {**415**}

Leroy JL, 1993a, A polarimetric investigation on interstellar dust within 50 pc from the Sun. *A&A*, 274, 203–213 {**474**}

—, 1993b, Optical polarisation of 1000 stars within 50 pc from the Sun. *A&AS*, 101, 551–562 {**474**}

—, 1999, Interstellar dust and magnetic field at the boundaries of the Local Bubble. Analysis of polarimetric data in the light of Hipparcos parallaxes. *A&A*, 346, 955–960 {**473, 474**}

—, 2000, *Polarisation of Light and Astronomical Observation*. Gordon & Breach Science (Advances in Astronomy and Astrophysics) {**474**}

Leushin VV, 1988, Nitrogen abundance in the atmospheres of single and binary stars. *Soviet Astronomy*, 32, 430–435 {**441**}

Liebert J, Young PA, Arnett D, *et al.*, 2005, The age and progenitor mass of Sirius B. *ApJ*, 630, L69–L72 {**460**}

Lindblad PO, Loden K, Palouš J, *et al.*, 1997, Runaway stars and the force perpendicular to the Galactic plane. *ESA SP–402: Hipparcos, Venice '97*, 665–668 {**441**}

Lindroos KP, 1985, A study of visual double stars with early-type primaries. IV. Astrophysical data. *A&AS*, 60, 183–211 {**418**}

—, 1986, A study of visual double stars with early-type primaries. V. Post-T Tauri secondaries. *A&A*, 156, 223–233 {**418**}

Linsky JL, Redfield S, Wood BE, *et al.*, 2000, The three-dimensional structure of the warm local interstellar medium. I. Methodology. *ApJ*, 528, 756–766 {**466**}

Li A, Greenberg JM, 1997, A unified model of interstellar dust. *A&A*, 323, 566–584 {**472**}

Li JZ, 2004, The status of ROSAT X-ray active young stars toward Taurus–Auriga. *Chinese Journal of Astronony and Astrophysics*, 4, 258–266 {**419**}

—, 2005, Proper motion study of the ROSAT selected weak-line T Tauri candidates around Taurus–Auriga. *Ap&SS*, 298, 525–535 {**419**}

Li JZ, Hu J, 1999, Optical identification and study of the ROSAT selected weak-line T Tauri candidates around Taurus–Auriga. *Star Formation* (ed. Nakamoto T), 385–386 {**419**}

Lombardi M, Alves J, Lada CJ, 2006, 2MASS wide field extinction maps. I. The Pipe nebula. *A&A*, 454, 781–796 {**472**}

Luri X, Gómez AE, Mennessier MO, *et al.*, 1997, Barium stars: luminosity and kinematics from Hipparcos data. *ESA SP–402: Hipparcos, Venice '97*, 355–358 {**453**}

Luyten WJ, 1954, A search for faint blue stars. II. The Hyades and the south Galactic polar region. *AJ*, 59, 224–227 {**460**}

—, 1956, The search for faint blue stars. IV. More blue stars in the Hyades region. *AJ*, 61, 261–262 {**460**}

Maeder A, 1996, The Conti scenario for forming Wolf–Rayet stars: past, present and future. *Liège International Astrophysical Colloquia 33* (eds. Vreux JM, Detal A, Fraipont-Caro D, *et al.*), 39–54 {**438**}

Maeder A, Meynet G, 1988, Tables of evolutionary star models from $0.85-120M_\odot$ with overshooting and mass loss. *A&AS*, 76, 411–425 {**442**}

Magazzù A, Martín EL, Sterzik MF, *et al.*, 1997, Search for young low-mass stars in a ROSAT selected sample south of the Taurus–Auriga molecular clouds. *A&AS*, 124, 449–467 {**416**}

Maggio A, Favata F, Peres G, *et al.*, 1998, X-ray spectroscopy of the active giant β Cet: the SAX LECS view. *A&A*, 330, 139–144 {**435**}

Magnani L, Blitz L, Mundy L, 1985, Molecular gas at high Galactic latitudes. *ApJ*, 295, 402–421 {**466, 472**}

Maitzen HM, Paunzen E, Pressberger R, *et al.*, 1998, HIP 60350: an extreme runaway star. *A&A*, 339, 782–786 {**442**}

Maitzen HM, Paunzen E, Vogt N, *et al.*, 2000, Hβ photometry of southern chemically peculiar stars: is the uvbyβ luminosity calibration also valid for peculiar stars? *A&A*, 355, 1003–1008 {**427**}

Maíz Apellániz J, 2001, The origin of the Local Bubble. *ApJ*, 560, L83–L86 {**468**}

Makarov VV, 2000, Kinematics versus X-ray luminosity segregation in the Hyades. *A&A*, 358, L63–L66 {**435**}

—, 2002, The rate of visual binaries among the brightest X-ray stars. *ApJ*, 576, L61–L64 {**436**}

—, 2003, The 100 brightest X-ray stars within 50 pc of the Sun. *AJ*, 126, 1996–2008 {**436**}

—, 2004, A sub-stellar companion to van Maanen 2. *ApJ*, 600, L71–L73 {**461, 462**}

Makarov VV, Robichon N, 2001, Internal kinematics and binarity of X-ray stars in the Pleiades open cluster. *A&A*, 368, 873–879 {**435**}

Makarov VV, Zacharias N, Hennessy GS, *et al.*, 2007, The nearby young visual binary HIP 115147 and its common proper motion companion LSPM J2322+7847. *ApJ*, 668, L155–L158 {**419**}

Malofeev VM, Malov OI, 1997, Detection of Geminga as a radio pulsar. *Nature*, 389, 697–699 {**464**}

Mamajek EE, Meyer MR, Liebert J, 2002, Post-T Tauri stars in the nearest OB association. *AJ*, 124, 1670–1694, erratum: 131, 2360 {**419**}

Mammano M, Margoni R, Oliva G, *et al.*, 1997, Luminosities and velocities of OBN stars from Hipparcos. *ESA SP–402: Hipparcos, Venice '97*, 303–306 {**441**}

Marchenko SV, Moffat AFJ, van der Hucht KA, *et al.*, 1998, Wolf–Rayet stars and O-star runaways with Hipparcos. II. Photometry. *A&A*, 331, 1022–1036 {**438, 439**}

Marigo P, 1998, Envelope burning over-luminosity: a challenge to synthetic thermal pulse asymptotic giant branch models. *A&A*, 340, 463–475 {**451, 461**}

Marigo P, Bressan A, Chiosi C, 1996, The thermally-pulsing asymptotic giant branch phase: a new model. *A&A*, 313, 545–564 {**451**}

Marigo P, Girardi L, Bressan A, 1999, The third dredge-up and the carbon star luminosity functions in the Magellanic Clouds. *A&A*, 344, 123–142 {**451**}

Marsh MC, Barstow MA, Buckley DA, *et al.*, 1997, An extreme ultraviolet-selected sample of DA white dwarfs from the

ROSAT All-Sky Survey. I. Optically derived stellar parameters. *MNRAS*, 286, 369–383 {**460**}

Marston AP, 2001, First detections of molecular gas associated with the Wolf–Rayet ring nebula NGC 3199. *ApJ*, 563, 875–882 {**440**}

Martínez P, Kurtz DW, Ashoka BN, *et al.*, 1999, Discovery of pulsations in the Am star HD 13079. *MNRAS*, 309, 871–874 {**427**}

Martínez P, Matthews JM, Kurtz DW, 2000, Hipparcos luminosities and the Hβ calibration in Ap stars. *Hipparcos and the Luminosity Calibration of the Nearer Stars, IAU Joint Discussion 13*, 25 {**427**}

Martín EL, 1997, Quantitative spectroscopic criteria for the classification of pre-main sequence low-mass stars. *A&A*, 321, 492–496 {**413**}

—, 1998, Weak and post-T Tauri stars around B-type members of the Scorpius–Centaurus OB association. *AJ*, 115, 351–357 {**413, 416**}

Martín EL, Magazzù A, Rebolo R, 1992, On the post-T Tauri nature of late-type visual companions to B-type stars. *A&A*, 257, 186–198 {**419**}

Martin JC, 2003, The origins and evolutionary status of B stars found far from the Galactic plane. Ph.D. Thesis {**446**}

—, 2004, The origins and evolutionary status of B stars found far from the Galactic plane. I. Composition and spectral features. *AJ*, 128, 2474–2500 {**446**}

—, 2006, The origins and evolutionary status of B stars found far from the Galactic plane. II. Kinematics and full sample analysis. *AJ*, 131, 3047–3068, erratum: 133, 755 {**446**}

Mashonkina L, Gehren T, Bikmaev IF, 1999, Barium abundances in cool dwarf stars as a constraint to s- and r-process nucleosynthesis. *A&A*, 343, 519–530 {**453**}

Mathews GJ, Lan NQ, Suh IS, *et al.*, 2006a, White dwarfs with strange matter cores: an analysis of candidates. *The Ultraviolet Universe: Stars from Birth to Death, IAU 26, JD4*, 4 {**459**}

Mathews GJ, Suh IS, O'Gorman B, *et al.*, 2006b, Analysis of white dwarfs with strange-matter cores. *Journal of Physics G Nuclear Physics*, 32, 747–759 {**459**}

Mayor M, Mazeh T, 1987, The frequency of triple and multiple stellar systems. *A&A*, 171, 157–177 {**461**}

McClure RD, Fletcher JM, Nemec JM, 1980, The binary nature of the barium stars. *ApJ*, 238, L35–L38 {**453**}

McCook GP, Sion EM, 1987, A catalogue of spectroscopically identified white dwarfs. *ApJS*, 65, 603–671 {**457**}

—, 1999, A catalogue of spectroscopically identified white dwarfs. *ApJS*, 121, 1–130 {**455, 457**}

Mdzinarishvili TG, 2004, New runaway O stars based on data from Hipparcos. *Astrophysics*, 47, 155–161 {**445**}

Mdzinarishvili TG, Chargeishvili KB, 2005, New runaway OB stars with Hipparcos. *A&A*, 431, L1–L4 {**445**}

Melchior AL, Combes F, Gould A, 2001, A new determination of the local interstellar radiation field. *SF2A-2001: Semaine de l'Astrophysique Française*, 63 {**475**}

—, 2007, The surface brightness of the Galaxy at the solar neighbourhood. *A&A*, 462, 965–976 {**474, 475**}

Mel'Nik AM, Efremov YN, 1995, New list of OB associations of our Galaxy. *Pis ma Astronomicheskii Zhurnal*, 21, 13–30 {**448**}

Mennessier MO, Luri X, Figueras F, *et al.*, 1997, Barium stars, Galactic populations and evolution. *A&A*, 326, 722–730 {**453**}

Mennickent RE, Pietrzyński G, Gieren W, *et al.*, 2002, On Be star candidates and possible blue pre-main sequence objects in the Small Magellanic Cloud. *A&A*, 393, 887–896 {**424**}

Mennickent RE, Vogt N, Sterken C, 1994, Long-term photometry of Be stars. I. Fading events and variations on time-scales of years. *A&AS*, 108, 237–250 {**423**}

Merín B, Montesinos B, Eiroa C, *et al.*, 2004, Study of the properties and spectral energy distributions of the Herbig Ae–Be stars HD 34282 and HD 141569. *A&A*, 419, 301–318 {**421**}

Meynet G, Maeder A, 2000, Stellar evolution with rotation. V. Changes in all the outputs of massive star models. *A&A*, 361, 101–120 {**422**}

Micela G, Favata F, Sciortino S, 1997, Hipparcos distances of X-ray selected stars: implications on their nature as stellar population. *A&A*, 326, 221–227 {**431**}

Mikulášek Z, Gráf T, 2005, Preliminary analysis of *V* and *Hp* light curves of 26 carbon Miras. *Ap&SS*, 296, 157–160 {**452**}

Millour F, Petrov RG, Chesneau O, *et al.*, 2007, Direct constraint on the distance of γ^2 Vel from VLTI–AMBER observations. *A&A*, 464, 107–118 {**440**}

Milone LA, 1990, Identification charts for southern R CrB stars. *Ap&SS*, 172, 263–271 {**452**}

Miroshnichenko AS, Gray RO, Klochkova VG, *et al.*, 2004, Fundamental parameters and evolutionary state of the Herbig Ae star candidate HD 35929. *A&A*, 427, 937–944 {**421**}

Moffat AFJ, Isserstedt J, 1980, The nature of single-line Population I Wolf–Rayet stars: evidence for high space velocity. *A&A*, 85, 201–207 {**440**}

Moffat AFJ, Marchenko SV, Seggewiss W, *et al.*, 1997, Searching for Wolf–Rayet and O-star runaways using Hipparcos proper motions. *ESA SP–402: Hipparcos, Venice '97*, 237–238 {**438**}

—, 1998, Wolf–Rayet stars and O-star runaways with Hipparcos. I. Kinematics. *A&A*, 331, 949–958, erratum affecting figures and tables: 345, 321-322 {**438, 440–442, 445, 447, 448**}

Montes D, López-Santiago J, Gálvez MC, *et al.*, 2001, Late-type members of young stellar kinematic groups. I. Single stars. *MNRAS*, 328, 45–63 {**437**}

Morales-Rueda L, Maxted PFL, Marsh TR, *et al.*, 2003, Orbital periods of 22 subdwarf B (sdB) stars. *MNRAS*, 338, 752–764 {**462**}

Morgan WW, Keenan PC, Kellman E, 1943, *An Atlas of Stellar Spectra, with an Outline of Spectral Classification*. University of Chicago Press {**428**}

Morley JE, Briggs KR, Pye JP, *et al.*, 2001, A ROSAT medium-sensitivity Galactic plane survey at $180^\circ < l < 280^\circ$. *MNRAS*, 326, 1161–1182 {**437**}

Münch IG, 1952, Statistics of stellar colour excesses. *ApJ*, 116, 575–586 {**469**}

Murphy RE, 1969, A spectroscopic investigation of visual binaries with B-type primaries. *AJ*, 74, 1082–1094 {**418**}

Napiwotzki R, 2001, Spectroscopic investigation of old planetaries. V. Distance scales. *A&A*, 367, 973–982 {**454**}

Napiwotzki R, Christlieb N, Drechsel H, *et al.*, 2001a, Search for progenitors of supernovae Type Ia with SPY. *Astronomische Nachrichten*, 322, 411–418 {**463**}

Napiwotzki R, Herrmann M, Heber U, *et al.*, 2001b, BD+33°2642: abundance patterns in the central star of a halo planetary nebula. *Post-Asymptotic Giant Branch Objects as a Phase of Stellar Evolution*, 277 {**455**}

Neckel T, Klare G, Sarcander M, 1980, The spatial distribution of the interstellar extinction. *A&AS*, 42, 251–281 {**470**}

Nelemans G, Tauris TM, van den Heuvel EPJ, 1999, Constraints on mass ejection in black hole formation derived from black hole X-ray binaries. *A&A*, 352, L87–L90 {**422**}

Nelson B, Young A, 1970, A new eclipsing binary containing a very hot white dwarf. *PASP*, 82, 699–706 {**460**}

Neuhäuser R, Brandner W, 1998, Hipparcos results for ROSAT-discovered young stars. *A&A*, 330, L29–L32 {**417**}

Neuhäuser R, Sterzik MF, Torres G, *et al.*, 1995, Weak-line T Tauri stars south of Taurus. *A&A*, 299, L13–LL16 {**431**}

Noriega-Crespo A, van Buren D, Dgani R, 1997, Bow shocks around runaway stars. III. The high-resolution maps. *AJ*, 113, 780–786 {**448**}

North JR, Tuthill PG, Tango WJ, *et al.*, 2007, γ^2 Vel: orbital solution and fundamental parameter determination with SUSI. *MNRAS*, 377, 415–424 {**440**}

North P, 1998, Fundamental parameters of BA dwarfs and CH subgiants. *IAU Symp. 191*, 517–517 {**453**}

North P, Carquillat JM, Ginestet N, *et al.*, 1998, Multiplicity among peculiar A stars. I. The Ap stars HD 8441 and HD 137909, and the Am stars HD 43478 and HD 96391. *A&AS*, 130, 223–232 {**427**}

North P, Erspamer D, Babel J, 1997a, Fundamental parameters of chemically peculiar stars: observational aspects. *Spectroscopy with Large Telescopes of Chemically Peculiar Stars, IAU Joint Discussion 16*, 105–105 {**426**}

North P, Jaschek C, Hauck B, *et al.*, 1997b, Absolute magnitudes of chemically peculiar stars. *ESA SP–402: Hipparcos, Venice '97*, 239–244 {**426**}

Oberlack U, Wessolowski U, Diehl R, *et al.*, 2000, COMPTEL limits on ^{26}Al 1.809 MeV line emission from γ^2 Vel. *A&A*, 353, 715–721 {**439**}

Odenkirchen M, Brosche P, Geffert M, *et al.*, 1997, Globular cluster orbits based on Hipparcos proper motions. *New Astronomy*, 2, 477–499 {**464**}

Olofsson H, Bergman P, Lucas R, *et al.*, 1998, Episodic mass loss of the carbon star TT Cyg. *IAU Symp. 191*, 415 {**452**}

—, 2000, A high-resolution study of episodic mass loss from the carbon star TT Cyg. *A&A*, 353, 583–597 {**452**}

Oudmaijer RD, Drew JE, 1999, Time-resolved spectroscopy of the peculiar Hα variable Be star HD 76534. *A&A*, 350, 485–490 {**426**}

Padgett DL, Strom SE, Ghez A, 1997, Hubble Space Telescope WFPC2 observations of the binary fraction among pre-main sequence cluster stars in Orion. *ApJ*, 477, 705–710 {**436**}

Pallavicini R, Pasquini L, Randich S, 1992, Optical spectroscopy of post-T Tauri star candidates. *A&A*, 261, 245–254 {**419**}

Palla F, Stahler SW, 1993, The pre-main sequence evolution of intermediate-mass stars. *ApJ*, 418, 414–425 {**413, 419, 420**}

—, 1999, Star formation in the Orion nebula cluster. *ApJ*, 525, 772–783 {**413, 419**}

Palouš J, 2000, Motion and formation places of young stars in the solar neighbourhood. *Astronomische Gesellschaft Meeting Abstracts*, 74 {**419**}

Panei JA, Althaus LG, Benvenuto OG, 2000a, Mass-radius relations for white dwarfs of different internal compositions. *A&A*, 353, 970–977 {**455, 458, 460**}

—, 2000b, The evolution of iron-core white dwarfs. *MNRAS*, 312, 531–539 {**458**}

Panov KP, Altmann M, Seggewiss W, 2000, Long-term photometry of the Wolf–Rayet stars WR 137, WR 140, WR 148, and WR 153. *A&A*, 355, 607–616 {**440**}

Paresce F, 1984, On the distribution of interstellar matter around the Sun. *AJ*, 89, 1022–1037 {**465**}

Patriarchi P, Morbidelli L, Perinotto M, 2003, A study of R_V in Galactic O stars from the 2MASS catalogue. *A&A*, 410, 905–909 {**474**}

Pauli EM, Napiwotzki R, Altmann M, *et al.*, 2003, 3d kinematics of white dwarfs from the SPY project. *A&A*, 400, 877–890 {**455, 463**}

Paunzen E, 1997, On the evolutionary status of λ Bootis stars using Hipparcos data. *A&A*, 326, L29–L32 {**430**}

—, 2001, A spectroscopic survey for λ Bootis stars. III. Final results. *A&A*, 373, 633–640 {**429**}

—, 2004, The λ Bootis stars. *IAU Symp. 224* (eds. Zverko J, Ziznovsky J, Adelman SJ, *et al.*), 443–450 {**429**}

Paunzen E, Handler G, Weiss WW, *et al.*, 2002a, On the period-luminosity–colour–metallicity relation and the pulsational characteristics of λ Bootis type stars. *A&A*, 392, 515–528 {**430**}

Paunzen E, Heiter U, Handler G, *et al.*, 1998a, On the new λ Bootis-type spectroscopic binary systems HD 84948 and HD 171948. *A&A*, 329, 155–160 {**430**}

Paunzen E, Iliev IK, Kamp I, *et al.*, 2002b, The status of Galactic field λ Bootis stars in the post-Hipparcos era. *MNRAS*, 336, 1030–1042 {**429, 430**}

Paunzen E, Kamp I, Weiss WW, *et al.*, 2003, A study of λ Bootis type stars in the wavelength region beyond 700 nm. *A&A*, 404, 579–591 {**429**}

Paunzen E, Maitzen HM, 1998, New variable chemically peculiar stars identified in the Hipparcos archive. *A&AS*, 133, 1–6 {**427**}

Paunzen E, Weiss WW, Kuschnig R, *et al.*, 1998b, Pulsation in λ Bootis stars. *A&A*, 335, 533–538 {**430**}

Paunzen E, Weiss WW, Martínez P, *et al.*, 1998c, Pulsation of the λ Bootis stars HD 111786 and HD 142994. *A&A*, 330, 605–611 {**430**}

Pellizza LJ, Mignani R, Grenier IA, *et al.*, 2005, On the local birth place of Geminga. *A&A*, 435, 625–630 {**444, 464**}

Penny LR, 1996, Projected rotational velocities of O-type stars. *ApJ*, 463, 737–746 {**444**}

Percy JR, Bakos AG, 2001, Photometric monitoring of bright Be stars. IV. 1996–1999. *PASP*, 113, 748–753 {**423**}

Percy JR, Coulter M, Mohammed F, 2003, Hunting for β Cephei stars using self-correlation analysis. *Be Star Newsletter 36*, 12–15 {**425**}

Percy JR, Harlow CDW, Wu APS, 2004, Short-period variable Be stars discovered or confirmed through self-correlation analysis of Hipparcos epoch photometry. *PASP*, 116, 178–183 {**425, 426**}

Percy JR, Hosick J, Kincaide H, *et al.*, 2002, Autocorrelation analysis of Hipparcos photometry of short-period Be stars. *PASP*, 114, 551–558 {**424**}

Percy JR, Marinova MM, Božić H, *et al.*, 1999, The photometric variability of the Be star NW Ser. *A&A*, 348, 553–556 {**426**}

Perrot CA, Grenier IA, 2003, 3d dynamical evolution of the interstellar gas in the Gould Belt. *A&A*, 404, 519–531 {**466**}

Perry CL, Hill G, Christodoulou DM, 1991, A study of Sco OB1 and NGC 6231. II. A new analysis. *A&AS*, 90, 195–223 {**447**}

Perryman MAC, Brown AGA, Lebreton Y, *et al.*, 1998, The Hyades: distance, structure, dynamics, and age. *A&A*, 331, 81–120 {**418, 460, 462**}

Pickles AJ, 1998, A stellar spectral flux library: 115–2500 nm. *PASP*, 110, 863–878 {**475**}

Pigott E, Englefield HC, 1797, On the periodical changes of brightness of two fixed stars. *Royal Society of London Philosophical Transactions Series I*, 87, 133–141 {**452**}

Piirola V, 1977, Polarisation observations of 77 stars within 25 pc from the Sun. *A&AS*, 30, 213–216 {**474**}

Piotto G, Bedin LR, Anderson J, *et al.*, 2007, A triple main sequence in the globular cluster NGC 2808. *ApJ*, 661, L53–L56 {**443**}

Pizzolato N, Maggio A, Sciortino S, 1998, Evolution of X-ray activity of $1-3\,M_\odot$ late-type stars in early post-main sequence phases. *ASP Conf. Ser. 154: Cool Stars, Stellar Systems, and the Sun*, 1146 {**434**}

—, 2000, Evolution of X-ray activity of $1-3M_\odot$ late-type stars in early post-main sequence phases. *A&A*, 361, 614–628 {**434, 435**}

Pöhnl H, Maitzen HM, Paunzen E, 2003, On the evolutionary status of chemically peculiar stars of the upper main sequence. *A&A*, 402, 247–252 {**427**}

Pols OR, Schröder KP, Hurley JR, *et al.*, 1998, Stellar evolution models for $Z = 0.0001-0.03$. *MNRAS*, 298, 525–536 {**434**}

Portegies Zwart SF, 2000, The characteristics of high-velocity O and B stars which are ejected from supernovae in binary systems. *ApJ*, 544, 437–442 {**447**}

Porter JM, Rivinius T, 2003, Classical Be stars. *PASP*, 115, 1153–1170 {**421**}

Pottasch SR, 1997, Comments on planetary nebula evolution. *Planetary Nebulae* (eds. Habing HJ, Lamers HJGLM), IAU Symp. 180, 483–492 {**454**}

Pottasch SR, Acker A, 1998, A comparison of Hipparcos parallaxes with planetary nebulae spectroscopic distances. *A&A*, 329, L5–L8 {**454**}

Poveda A, Ruiz J, Allen C, 1967, Runaway stars as the result of the gravitational collapse of proto-stellar clusters. *Boletin de los Observatorios Tonantzintla y Tacubaya*, 4, 86–90 {**440**}

Pozzo M, Jeffries RD, Naylor T, *et al.*, 2000, The discovery of a low-mass, pre-main sequence stellar association around γ^2 Vel. *MNRAS*, 313, L23–L27 {**439**}

Preibisch T, Zinnecker H, 1999, The history of low-mass star formation in the Upper Scorpius OB association. *AJ*, 117, 2381–2397 {**443**}

Preston GW, 1974, The chemically peculiar stars of the upper main sequence. *ARA&A*, 12, 257–277 {**427**}

Provencal JL, Shipman HL, Høg E, *et al.*, 1997a, Testing the white dwarf mass–radius relation with Hipparcos. *ESA SP–402: Hipparcos, Venice '97*, 375–378 {**457, 458**}

—, 1998, Testing the white dwarf mass–radius relation with Hipparcos. *ApJ*, 494, 759 {**457–460**}

Provencal JL, Shipman HL, Koester D, *et al.*, 2002, Procyon B: outside the iron box. *ApJ*, 568, 324–334 {**459, 460**}

Provencal JL, Shipman HL, Wesemael F, *et al.*, 1997b, Wide Field Planetary Camera 2 photometry of the bright, mysterious white dwarf Procyon B. *ApJ*, 480, 777–783 {**458**}

Racine R, 1968, Stars in reflection nebulae. *AJ*, 73, 233–245 {**418**}

Reid IN, 1996, White dwarf masses-gravitational redshifts revisited. *AJ*, 111, 2000–2016 {**458**}

Reipurth B, Zinnecker H, 1993, Visual binaries among pre-main sequence stars. *A&A*, 278, 81–108 {**436**}

Reiz A, Franco GAP, 1998, UBV polarimetry of 361 A- and F-type stars in selected areas. *A&AS*, 130, 133–140 {**474**}

Relke H, Pfau W, 1998, Accurate positions and proper motions of Hα emission-line objects in the Chamaeleon star-forming region. *Applied and Computational Harmonic Analysis*, 3, 196–197 {**419**}

Renzini A, Bragaglia A, Ferraro FR, *et al.*, 1996, The white dwarf distance to the globular cluster NGC 6752 (and its age) with the Hubble Space Telescope. *ApJ*, 465, L23–L26 {**456**}

Reynolds JE, Jauncey DL, Staveley-Smith L, *et al.*, 1995, Accurate registration of radio and optical images of SN 1987A. *A&A*, 304, 116–120 {**464**}

Richichi A, Chandrasekhar T, 2006, Near-infrared observations of the carbon stars TU Gem and SS Vir at milliarcsecond resolution. *A&A*, 451, 1041–1044 {**452**}

Rinehart SA, 2000, Single Be stars as binary ejecta? *MNRAS*, 312, 429–432 {**422**}

Rivinius T, Baade D, Stefl S, 2003, Non-radially pulsating Be stars. *A&A*, 411, 229–247 {**425**}

Rivinius T, Stefl S, Baade D, 2006, Bright Be-shell stars. *A&A*, 459, 137–145 {**426**}

Rodríguez LF, Zapata L, Ho PTP, 2007, Radio continuum sources associated with AB Aur. *Revista Mexicana de Astronomia y Astrofisica*, 43, 149–154 {**421**}

Rolleston WRJ, Hambly NC, Keenan FP, *et al.*, 1999, Early-type stars in the Galactic halo from the Palomar-Green Survey. II. A sample of distant, apparently young Population I stars. *A&A*, 347, 69–76 {**446**}

Ryabchikova T, Leone F, Kochukhov O, 2005, Abundances and chemical stratification analysis in the atmosphere of Cr-type Ap star HD 204411. *A&A*, 438, 973–985 {**427**}

Saad SM, Kubát J, Koubský P, *et al.*, 2004, Properties and nature of Be stars. XXIII. Long-term variations and physical properties of κ Dra. *A&A*, 419, 607–621 {**426**}

Saar SH, 1998, PZ Mon: an active evolved star. *Informational Bulletin on Variable Stars*, 4580 {**427**}

Saffer RA, Keenan FP, Hambly NC, *et al.*, 1997, A large-scale spectroscopic survey of early-type stars at high Galactic latitudes. *ApJ*, 491, 172–180 {**446**}

Sahin T, Jeffery CS, 2007, Variability and evolution in various classes of post-AGB stars. *Astronomische Nachrichten*, 328, 848–851 {**452**}

Sandage AR, 1972, The redshift-distance relation. II. The Hubble diagram and its scatter for first-ranked cluster galaxies: a formal value for q_0. *ApJ*, 178, 1–24 {**470**}

Sargent WLW, 1965, A possible relationship between the peculiar A stars and the λ Bootis stars. *ApJ*, 142, 787–790 {**428**}

Sasaki M, Haberl F, Pietsch W, 2000a, ROSAT HRI catalogue of X-ray sources in the Large Magellanic Cloud region. *A&AS*, 143, 391–403 {**437**}

—, 2000b, ROSAT HRI catalogue of X-ray sources in the Small Magellanic Cloud region. *A&AS*, 147, 75–91 {**437**}

Schaerer D, Schmutz W, Grenon M, 1997, Fundamental stellar parameters of γ^2 Vel from Hipparcos data. *ApJ*, 484, L153–L156 {**439, 440**}

Schaifers K, Voigt HH (eds.), 1982, *Landolt-Börnstein: Numerical Data and Functional Relationships in Science and Technology; New Series, Group 6 Astronomy and Astrophysics, Volume 2*. Springer–Verlag, Berlin {**473**}

Schaller G, Schaerer D, Meynet G, *et al.*, 1992, New grids of stellar models. I. From $0.8-120\,M_\odot$ at $Z = 0.020, 0.001$. *A&AS*, 96, 269–331 {**435, 445, 450**}

Schatzman EL, 1950, Sur l'abondance des grands nuages de matière interstellaire. *Annales d'Astrophysique*, 13, 367–383 {**469, 470**}

Scheffler H, Elsaesser H, 1987, *Physics of the Galaxy and Interstellar Matter*. Springer-Verlag, Berlin {**448**}
Scheffler H, 1966, Interstellar absorption II. *Zeitschrift fur Astrophysik*, 63, 267–281 {**470**}
Schmidt H, 1996, The empirical white dwarf mass–radius relation and its possible improvement by Hipparcos. *A&A*, 311, 852–857 {**457**}
Schröder KP, Hünsch M, Schmitt JHMM, 1998, X-ray activity and evolutionary status of late-type giants. *A&A*, 335, 591–595 {**434**}
Schröder KP, Napiwotzki R, Pauli EM, 2005, A model of the local white dwarf population. *ASP Conf. Ser. 334: 14th European Workshop on White Dwarfs* (eds. Koester D, Moehler S), 93 {**463**}
Schröder R, 1976, Optical polarisation of stars of Galactic latitude $b < -45°$. *A&AS*, 23, 125–137 {**474**}
Schwan H, 1991, The distance and main sequence of the Hyades cluster based on 145 stars with highly accurate proper motions obtained from work on the catalogues FK5 and PPM. *A&A*, 243, 386–400 {**461**}
Schwartz RD, 1992, The Chamaeleon dark clouds and T-associations. *Low-Mass Star Formation in Southern Molecular Clouds* (ed. Reipurth B), 93–118 {**415**}
Seaton MJ, 1979, Interstellar extinction in the ultraviolet. *MNRAS*, 187, 73P–76P {**469**}
Seggewiss W, Moffat AFJ, van der Hucht KA, *et al.*, 1999, Wolf–Rayet stars before and after Hipparcos. *Revista Mexicana de Astronomia y Astrofisica Conf. Ser.*, volume 8, 33–39 {**438**}
Sfeir DM, Lallement R, Crifo F, *et al.*, 1999, Mapping the contours of the Local Bubble: preliminary results. *A&A*, 346, 785–797 {**466, 467**}
Shane CD, 1928, The spectra of the carbon stars. *Lick Observatory Bulletin*, 13, 123–129 {**449**}
Shipman HL, 1979, Masses and radii of white dwarfs. III. Results for 110 H-rich and 28 He-rich stars. *ApJ*, 228, 240–256 {**457**}
Shipman HL, Provencal JL, Høg E, *et al.*, 1997, The mass and radius of 40 Eri B from Hipparcos: an accurate test of stellar interior theory. *ApJ*, 488, L43 {**458**}
Shipman HL, Provencal JL, Roby SW, *et al.*, 1995, Photospheric, circumstellar, and interstellar features of He, C, N, O, and Si in the HST spectra of four hot white dwarfs. *AJ*, 109, 1220–1230 {**460**}
Shipman HL, Sass CA, 1980, Masses and radii of white dwarfs. IV. The two-colour diagram. *ApJ*, 235, 177–185 {**457**}
Siess L, Forestini M, Dougados C, 1997, Synthetic HR diagrams of open clusters. *A&A*, 324, 556–565 {**432**}
Silva FN, Magalhães AM, 2003, The magnetic field structure towards NGC 6755 and three high latitude molecular clouds. *Bulletin of the Astronomical Society of Brazil*, 23, 223–223 {**475**}
Skórzyński W, Strobel A, Galazutdinov G, 2003, Grey extinction in the solar neighbourhood? *A&A*, 408, 297–304 {**473**}
Slechta M, Skoda P, 2005, An outburst detected in the spectrum of HD 6226. *Ap&SS*, 296, 179–182 {**426**}
Slettebak A, Wagner RM, Bertram R, 1997, Spectroscopic observations of some Be/B stars at high Galactic latitudes. *PASP*, 109, 1–8 {**423**}
Smith GH, Shetrone MD, 2000, Ca II K emission-line asymmetry among red giants detected by the ROSAT satellite. *PASP*, 112, 1320–1329 {**435**}
Smith MA, Henrichs HF, Fabregat J (eds.), 2000, *The Be Phenomenon in Early-Type Stars*, ASP Conf. Ser. 214 {**421**}
Smith VV, Cunha K, Plez B, 1994, Is Geminga a runaway member of the Orion association? *A&A*, 281, L41–L44 {**444**}
Smoker JV, Keenan FP, Marcha MJ, *et al.*, 2000, Radio continuum observations of possible B-type stars in the halo of M 31. *A&A*, 361, 60–62 {**446**}
Snowden SL, Cox DP, McCammon D, *et al.*, 1990, A model for the distribution of material generating the soft X-ray background. *ApJ*, 354, 211–219 {**465**}
Snowden SL, Egger R, Finkbeiner DP, *et al.*, 1998, Progress on establishing the spatial distribution of material responsible for the 0.25 keV soft X-ray diffuse background local and halo components. *ApJ*, 493, 715–729 {**465, 474**}
Soderblom DR, King JR, Siess L, *et al.*, 1998, HD 98800: a unique stellar system of post-T Tauri stars. *ApJ*, 498, 385 {**416, 417**}
Song I, Bessell MS, Zuckerman B, 2002, Additional TW Hya association members? Spectroscopic verification of kinematically selected candidates. *A&A*, 385, 862–866 {**428**}
Sowers JW, Gies DR, Bagnuolo WG, *et al.*, 1998, Tomographic analysis of Hα profiles in HDE 226868 (Cygnus X-1). *ApJ*, 506, 424–430 {**438**}
Spitzer LJ, 1998, *Physical Processes in the Interstellar Medium*. Wiley {**429**}
Stagg C, 1987, A photometric survey of the bright southern Be stars. *MNRAS*, 227, 213–240 {**425**}
Steele IA, Negueruela I, Coe MJ, *et al.*, 1998, The distances to the X-ray binaries LSI +61°303 and A0535+262. *MNRAS*, 297, L5 {**437**}
Stefanik RP, Latham DW, 1992, Binaries in the Hyades. *IAU Colloq. 135: Complementary Approaches to Double and Multiple Star Research* (eds. McAlister HA, Hartkopf WI), ASP Conf. Ser. 32, 173–175 {**461**}
Sterzik MF, Durisen RH, 1995, Escape of T Tauri stars from young stellar systems. *A&A*, 304, L9–L12 {**415**}
Sterzik MF, Alcalá JM, Neuhauser R, *et al.*, 1995, The spatial distribution of X-ray selected T-Tauri stars. I. Orion. *A&A*, 297, 418–426 {**431**}
Stine PC, O'Neal D, 1998, Radio emission from young stellar objects near LkHα 101. *AJ*, 116, 890–894 {**419**}
Stone RC, 1982, The velocity-mass correlation of the O-type stars: model results. *AJ*, 87, 90–97 {**442**}
Straižys V, 1992, *Multicolour Stellar Photometry*. Pachart Pub. House, Tucson {**470**}
Stratta G, Perna R, Lazzati D, *et al.*, 2005, Extinction properties of the X-ray bright/optically faint afterglow of GRB 020405. *A&A*, 441, 83–88 {**473**}
St-Louis N, Moffat AFJ, Drissen L, *et al.*, 1988, Polarisation variability among Wolf–Rayet stars. III. A new way to derive mass-loss rates for Wolf–Rayet stars in binary systems. *ApJ*, 330, 286–304 {**439**}
Suchkov AA, 2000, Age difference between the populations of binary and single F stars revealed from Hipparcos data. *ApJ*, 535, L107–L110, erratum: 539, L75 {**436**}
Suchkov AA, Makarov VV, Voges W, 2003, ROSAT view of Hipparcos F stars. *ApJ*, 595, 1206–1221 {**433**}
Suh IS, Mathews GJ, 2000, Mass-radius relation for magnetic white dwarfs. *ApJ*, 530, 949–954 {**458**}
Sujatha NV, Chakraborty P, Murthy J, *et al.*, 2004, A model of the stellar radiation field in the ultraviolet. *Bulletin of the Astronomical Society of India*, 32, 151 {**475**}
Tagliaferri G, Cutispoto G, Pallavicini R, *et al.*, 1994, Photometric and spectroscopic studies of cool stars discovered

in EXOSAT X-ray images II. Lithium abundances. *A&A*, 285, 272–284 {**431**}

Tamanaha CM, Silk J, Wood MA, *et al.*, 1990, The white dwarf luminosity function: a possible probe of the Galactic halo. *ApJ*, 358, 164–169 {**456**}

Teixeira R, Ducourant C, Sartori MJ, *et al.*, 2000, Proper motions of pre-main sequence stars in southern star-forming regions. *A&A*, 361, 1143–1151 {**415**}

Tenjes P, Einasto J, Maitzen HM, *et al.*, 2001, Origin and possible birthplace of the extreme runaway star HIP 60350. *A&A*, 369, 530–536 {**442, 443**}

Terranegra L, Morale F, Spagna A, *et al.*, 1999, Proper motions of faint ROSAT weak-line T Tauri stars in the Chamaeleon region. *A&A*, 341, L79–L83 {**419**}

Thé PS, de Winter D, Perez MR, 1994, A new catalogue of members and candidate members of the Herbig Ae/Be stellar group. *A&AS*, 104, 315–339 {**414**}

Tinbergen J, 1982, Interstellar polarisation in the immediate solar neighbourhood. *A&A*, 105, 53–64 {**474**}

Torres G, Guenther EW, Marschall LA, *et al.*, 2003, Radial velocity survey of members and candidate members of the TW Hya association. *AJ*, 125, 825–841 {**428**}

Trimble V, Kundu A, 1997, Parallaxes and proper motions of prototypes of astrophysically interesting classes of stars. I. R CrB variables. *PASP*, 109, 1089–1092 {**452**}

—, 1998, Parallaxes and proper motions of prototypes of astrophysically interesting classes of stars. *AJ*, 115, 358–360 {**419**}

Trumpler RJ, 1930, Absorption of light in the Galactic system. *PASP*, 42, 214–227 {**469, 473**}

Tully RB, Fisher J, 1977, A new method of determining distances to galaxies. *A&A*, 54, 661–673 {**475**}

Tutukov A, Yungelson L, 1973, Evolution of close binary with relativistic component. *Nauchnye Informatsii*, 27, 86–88 {**440**}

Vaiana GS, Cassinelli JP, Fabbiano G, *et al.*, 1981, Results from an extensive Einstein stellar survey. *ApJ*, 245, 163–182 {**431**}

van Buren D, McCray R, 1988, Bow shocks and bubbles are seen around hot stars by IRAS. *ApJ*, 329, L93–L96 {**448**}

van Buren D, Noriega-Crespo A, Dgani R, 1995, An IRAS/ISSA survey of bow shocks around runaway stars. *AJ*, 110, 2914–2925 {**448, 449**}

van den Ancker ME, 1999, Circumstellar material in young stellar objects. Ph.D. Thesis {**420**}

van den Ancker ME, de Winter D, Tjin A Djie HRE, 1998, Hipparcos photometry of Herbig Ae/Be stars. *A&A*, 330, 145–154 {**420**}

van den Ancker ME, Thé PS, Tjin A Djie HRE, *et al.*, 1997, Hipparcos data on Herbig Ae/Be stars: an evolutionary scenario. *A&A*, 324, L33–L36 {**419**}

van den Heuvel EPJ, 1973, X-ray stars: evolved from Wolf–Rayet binaries? *Nature*, 242, 71–73 {**440**}

—, 1978, Evolution of close binaries, with possible applications to massive X-ray binaries. *Physics of Neutron Stars and Black Holes*, 828–871 {**422**}

—, 1994, The binary pulsar PSR J2145–0750: a system originating from a low or intermediate mass X-ray binary with a donor star on the asymptotic giant branch? *A&A*, 291, L39–L42 {**447**}

van den Heuvel EPJ, Portegies Zwart SF, Bhattacharya D, *et al.*, 2000, On the origin of the difference between the runaway velocities of the OB-supergiant X-ray binaries and the Be/X-ray binaries. *A&A*, 364, 563–572 {**447**}

van der Hucht KA, 2006, New Galactic Wolf–Rayet stars, and candidates: an annex to the VIIth Catalogue of Galactic Wolf–Rayet stars. *A&A*, 458, 453–459 {**438**}

van der Hucht KA, Schrijver H, Stenholm B, *et al.*, 1997, The Hipparcos distance determination of the Wolf–Rayet system γ^2 Vel and its ramifications. *New Astronomy*, 2, 245–250 {**439**}

van de Kamp P, 1971, An astrometric study of van Maanen's star. *White Dwarfs* (ed. Luyten WJ), IAU Symp. 42, 32–34 {**462**}

Van Eck S, Jorissen A, Udry S, *et al.*, 1997, The Hipparcos HR diagram of S stars: probing nucleosynthesis and dredge-up. *ESA SP–402: Hipparcos, Venice '97*, 327–330 {**452**}

—, 1998, The Hipparcos HR diagram of S stars: probing nucleosynthesis and dredge-up. *A&A*, 329, 971–985 {**452, 453**}

van Leeuwen F, 2007, *Hipparcos, the New Reduction of the Raw Data*. Springer, Dordrecht {**440**}

van Oijen JGJ, 1989, Are massive X-ray binaries runaway stars? *A&A*, 217, 115–126 {**447**}

van Paradijs J, 1995, Catalogue of X-ray binaries. *X-Ray Binaries* (eds. Lewin WHG, van Paradijs J, van den Heuvel EPJ), 536–540, Cambridge Astrophysics Series {**437**}

van Rensbergen W, Vanbeveren D, de Loore C, 1996, OB runaways as a result of massive star evolution. *A&A*, 305, 825–834 {**447**}

van Rhijn PJ, 1949, On the distribution of interstellar grains. *Publications of the Kapteyn Astronomical Laboratory Groningen*, 53, 1–44 {**469**}

Vauclair G, Schmidt H, Koester D, *et al.*, 1997a, White dwarfs observed by Hipparcos. *ESA SP–402: Hipparcos, Venice '97*, 371–374 {**457**}

—, 1997b, White dwarfs observed by the Hipparcos satellite. *A&A*, 325, 1055–1062 {**456–458**}

Vauclair S, Vauclair G, 1982, Element segregation in stellar outer layers. *ARA&A*, 20, 37–60 {**455**}

Veen PM, van Genderen AM, van der Hucht KA, *et al.*, 2002, The enigmatic WR 46: A binary or a pulsator in disguise. I. The photometry. *A&A*, 385, 585–599 {**440**}

Venn KA, Lambert DL, 1990, The chemical composition of three λ Bootis stars. *ApJ*, 363, 234–244 {**428, 429**}

Ventura P, Zeppieri A, Mazzitelli I, *et al.*, 1998, Full spectrum of turbulence convective mixing. I. Theoretical main sequences and turn-off for $0.6-15 M_\odot$. *A&A*, 334, 953–968 {**421**}

Vergely JL, Egret D, Freire Ferrero R, *et al.*, 1997, The extinction in the solar neighbourhood from the Hipparcos data. *ESA SP–402: Hipparcos, Venice '97*, 603–606 {**470**}

Vergely JL, Ferrero RF, Egret D, *et al.*, 1998, The interstellar extinction in the solar neighbourhood. I. Statistical approach. *A&A*, 340, 543–555 {**470, 471**}

Vergely JL, Freire Ferrero R, Siebert A, *et al.*, 2001, Na I and H I 3d density distribution in the solar neighbourhood. *A&A*, 366, 1016–1034 {**467**}

Verschuur GL, 1993, A supershell model for the high- and intermediate-velocity neutral hydrogen clouds. *ApJ*, 409, 205–233 {**472**}

Vieira SLA, Corradi WJB, Alencar SHP, *et al.*, 2003, Investigation of 131 Herbig Ae/Be candidate stars. *AJ*, 126, 2971–2987 {**420, 421**}

Voges W, 1992, The ROSAT All-Sky X-ray survey. *Environment Observation and Climate Modeling Through International Space Projects* (eds. Guyenne TD, Hunt JJ), 9–19 {**431**}

Voloshina IB, Lyutyi VM, Tarasov AE, 1997, Photometric behavior of the binary system Cyg X-1 (V1357 Cyg) during the 1996 X-ray outburst. *Astronomy Letters*, 23, 293–298 {**438**}

Wade GA, Debernardi Y, Mathys G, *et al.*, 2000, An analysis of the Ap binary HD 81009. *A&A*, 361, 991–1000 {**427**}

Waelkens C, Bogaert E, Waters LBFM, 1994, The spectral evolution of Herbig Ae/Be stars. *The Nature and Evolutionary Status of Herbig Ae/Be Stars* (eds. Thé PS, Perez MR, van den Heuvel EPJ), ASP Conf. Ser. 62, 405–408 {**420**}

Wallerstein G, Knapp GR, 1998, Carbon stars. *ARA&A*, 36, 369–434 {**449**}

Waters LBFM, Trams NR, Waelkens C, 1992, A scenario for the selective depletion of stellar atmospheres. *A&A*, 262, L37–L40 {**430**}

Wegner W, 2000, Absolute magnitudes of OB and Be stars based on Hipparcos parallaxes. I. *MNRAS*, 319, 771–776 {**422**}

Weidemann V, 1977, Mass loss towards the white dwarf stage. *A&A*, 59, 411–418 {**462**}

—, 1987, The initial-to-final mass relation: Galactic disk and Magellanic Clouds. *A&A*, 188, 74–84 {**461, 462**}

—, 2000, Revision of the initial-to-final mass relation. *A&A*, 363, 647–656 {**461, 462**}

Weidemann V, Jordan S, Iben I, *et al.*, 1992, White dwarfs in the halo of the Hyades cluster: the case of the missing white dwarfs. *AJ*, 104, 1876–1891 {**460**}

Weidemann V, Koester D, 1983, The upper mass limit for white dwarf progenitors and the initial-to-final mass relation for low- and intermediate-mass stars. *A&A*, 121, 77–84 {**462**}

Welsh BY, Craig N, Vedder PW, *et al.*, 1994, The local distribution of Na I interstellar gas. *ApJ*, 437, 638–657 {**465–467**}

Welsh BY, Crifo F, Lallement R, 1998, A new determination of the local distribution of interstellar Na I. *A&A*, 333, 101–105 {**465, 466**}

Welsh BY, Sfeir DM, Sirk MM, *et al.*, 1999, EUV mapping of the local interstellar medium: the Local Chimney revealed? *A&A*, 352, 308–316 {**466**}

Werner K, Rauch T, 1997, Mass and radius of the white dwarf in the binary V471 Tau from Orfeus and Hipparcos observations. *A&A*, 324, L25–L28 {**461**}

Wesemael F, van Horn HM, Savedoff MP, *et al.*, 1980, Atmospheres for hot, high-gravity stars. I. Pure H models. *ApJS*, 43, 159–303 {**461**}

Wichmann R, Bastian U, Krautter J, *et al.*, 1997, Hipparcos observations of pre-main sequence stars. *ESA SP–402: Hipparcos, Venice '97*, 359–362 {**415**}

—, 1998, Hipparcos observations of pre-main sequence stars. *MNRAS*, 301, L39 {**415, 420**}

Wichmann R, Covino E, Alcalá JM, *et al.*, 1999, High-resolution spectroscopy of ROSAT-discovered weak-line T Tauri stars near Lupus. *MNRAS*, 307, 909–918 {**418**}

Wichmann R, Krautter J, Schmitt JHMM, *et al.*, 1996, New weak-line T Tauri stars in Taurus–Auriga. *A&A*, 312, 439–454 {**431**}

Wichmann R, Schmitt JHMM, Hubrig S, 2003a, Nearby young stars. *A&A*, 399, 983–994 {**419**}

—, 2003b, New spectroscopic binaries among nearby stars. *A&A*, 400, 293–296 {**419**}

Winget DE, Hansen CJ, Liebert J, *et al.*, 1987, An independent method for determining the age of the universe. *ApJ*, 315, L77–L81 {**456**}

Wood MA, 1990, Astero-archaeology: reading the Galactic history recorded in the white dwarfs. *PASP*, 102, 954, abstract of Ph.D. Thesis {**455**}

—, 1995, Theoretical white dwarf luminosity functions: DA models. *White Dwarfs* (eds. Koester D, Werner K), volume 443 of *Lecture Notes in Physics, Springer–Verlag, Berlin*, 41–45 {**455, 458, 459**}

Woosley SE, Weaver TA, 1986, The physics of supernova explosions. *ARA&A*, 24, 205–253 {**459**}

—, 1995, The evolution and explosion of massive stars. II. Explosive hydrodynamics and nucleosynthesis. *ApJS*, 101, 181–235 {**459**}

Yushchenko AV, Gopka VF, Khokhlova VL, *et al.*, 1999, Atmospheric chemical composition of the twin components of equal mass in the chemically peculiar SB2 system 66 Eri. *Astronomy Letters*, 25, 453–466 {**427**}

Yu Q, Tremaine S, 2003, Ejection of hypervelocity stars by the (binary) black hole in the Galactic centre. *ApJ*, 599, 1129–1138 {**449**}

Zaniewski A, Clayton GC, Welch DL, *et al.*, 2005, Discovery of five new R CrB stars in the MACHO Galactic bulge data base. *AJ*, 130, 2293–2302 {**452**}

Zdanavičius J, Straižys V, Corbally CJ, 2002, Interstellar extinction law near the Galactic equator along the Camelopardalis, Perseus and Cassiopeia border. *A&A*, 392, 295–300 {**472**}

Zdanavičius J, Zdanavičius K, 2002, Interstellar extinction along the Camelopardalis and Perseus border. *Baltic Astronomy*, 11, 441–463 {**472**}

Zhang P, Liu CQ, Chen PS, 2006, Absolute magnitudes of Be stars based on Hipparcos parallaxes. *Ap&SS*, 306, 113–128 {**422**}

Zoccali M, Renzini A, Ortolani S, *et al.*, 2001, The white dwarf distance to the globular cluster 47 Tucanae and its age. *ApJ*, 553, 733–743 {**456, 463**}

9

Structure of the Galaxy

9.1 Introduction

9.1.1 Overall structure of the Galaxy

Billions of stars, as well as planets, interstellar gas (predominantly atomic and molecular hydrogen), interstellar dust, and dark matter are gravitationally bound to form the Galaxy – a magnificent disk spiral system, supported by rotation. Most of the visible stars in the Galaxy, some 10^{11} objects of all types, masses and ages, lie in a flattened disk, a roughly axisymmetric structure whose constituent stars have formed at a fairly steady rate throughout the history of the Galaxy. They have a relatively high metal content compared to primordial abundances, inherited from the interstellar gas from which they formed, itself comprising the metal-rich debris of exploding supernovae. The Sun is located at about 7.5–8.5 kpc from the Galactic centre, and moves around it in a roughly circular orbit in a period of about 250 million years. The disk appears to be segregated into 'thin' and 'thick' components with different scale heights, angular rotation, ages, and metallicity. It shows significant spatial and kinematic structure, notably spiral arms, a central bar, and a warping of the outer plane.

The inner kpc of the disk also contains the bulge, which is less flattened, and consists mostly of fairly old stars. At its centre lies a super-massive black hole of $\sim 3\times10^6 M_\odot$. The stellar bulge is a major element of galaxy classification schemes, and may be relatively small (like the Milky Way), or large and luminous. It includes an old population traced by bulge RR Lyrae stars, but probably includes stars with a wide range of ages. There is also a wide spread in abundances, but generally closer to that of the old disk stars than to the old metal poor stars in the halo. The bulge in the Milky Way is triaxial, indicating the presence of a bar.

The bulge and bar are too distant to be observed directly by Hipparcos (although dynamical imprints of the bar in the local velocity structure are probably detected), and so will receive little further attention in this review. The disk appears to have a 'flared' structure in the outer parts, manifested as an increasing scale height with distance beginning well inside the solar circle (Gyuk *et al.*, 1999; López Corredoira *et al.*, 2002), but this will also receive no further attention here.

The disk and bulge are surrounded by a halo of about 10^9 old and metal-poor stars, as well as around 150 globular clusters and a small number of satellite dwarf galaxies. The visible stars have a roughly spheroidal distribution, and their chemical composition and kinematics, and presumably their formation history, are quite distinct from those of the disk, with low metal content, little or no systematic rotation, but large individual velocities. The relative importance of the disk and spheroid accounts for much of the variety in galaxy morphologies. The entire baryon system is believed to be embedded in a massive halo of dark material of unknown composition and poorly known spatial distribution, the primary evidence for whose existence is the flat rotation curve of the disk in the outer parts of the Galaxy.

The disk and central bulge together represent some 90% of the visible light, but only some 5% of the mass of the Galaxy. The stellar halo and globular clusters together represent some 0.2% of the halo mass, while the unseen dark halo is considered to account for some 95% of the mass of the Galaxy.

Determination of the overall size and structure of the Galaxy, based on the distribution of globular clusters (Shapley, 1918) and star counts (Kapteyn & van Rhijn, 1920), was one of the highlights of early twentieth century astronomical research. Radial velocities and proper motions were used, along with the assumption of hydrostatic equilibrium perpendicular to the Galactic plane, to make the first reasonable estimate of its mass (Kapteyn, 1922). Detailed studies of Galactic rotation were initiated by Lindblad (1927) and Oort (1927b), who developed a model of differential rotation, assuming an axisymmetric Galaxy, circular motion, and with rotation rate depending only on the Galactocentric distance (the Oort–Lindblad model). Such a hypothesis was sufficient to account for the observed double sine-wave variation of radial velocities with Galactic longitude. Using proper motions and radial velocities, Oort made the first determinations of the two constants A and B to describe the observable consequences of differential Galactic rotation. His underlying model, that the Galaxy is axisymmetric, was later generalised to non-axisymmetry by Ogorodnikov (1932) and Milne (1935).

Models used to describe the stellar kinematics of the Galactic disk have become ever more sophisticated as the quality of the observations improved, starting from the ellipsoidal and Gaussian model for the velocity distribution in the solar neighbourhood (Schwarzschild, 1908). However, it was known early on that this picture is too simple: for example, Kapteyn (1905) and others pointed out that the distribution of stellar velocities in the solar neighbourhood is not smooth but clumpy, in particular for early-type stars, with these clumps now hypothesised to be the debris of stellar associations and open clusters (see Chapter 6).

Any stellar system with metallicity, age and kinematic properties similar to the disk is traditionally called a Population I system, or disk population. Any stellar system with properties resembling the halo (low heavy element abundances, old stars, and little or no rotation) is traditionally called a Population II system, or halo population. This nomenclature is still widely and conveniently used, and corresponds to major events in the formation of the Galaxy, although stars of intermediate ages, metallicity, or kinematics do also exist.

9.1.2 Hipparcos contributions

Hipparcos has measured parallaxes and proper motions in the solar neighbourhood for large stellar samples free of kinematic biases, which in the past plagued studies of local stellar kinematics. Used as elements of the space velocities, these data have revealed that the local stellar velocity distribution is indeed much more structured and complex than Schwarzschild's simple model: there is substantial substructure even for late-type, and hence on average old, stars. The interpretation of these structures is still incomplete, but it seems clear already that the bar in the inner Galaxy, the spiral arms in the solar neighbourhood, and the stellar warp outside the solar circle, have all left their imprints.

One of the main lines of research using the Hipparcos data has been to use the improved distances and proper motions to refine our knowledge of the detailed dynamical state of the observed stars through determination and interpretation of the phase-space distribution function (positions and velocities). This also leads to estimates of the various numerical quantities describing the Sun's motion around the centre of the Galaxy, its distance to the Galactic centre, R_0, parameters describing the rotational characteristics of the Galaxy, notably the Oort constants A and B, and properties of the local velocity distribution. It is useful to recall that stellar motions are governed by the collective gravitational potential of all stars and other (including invisible) matter. As will be seen in Section 9.1.3, A can be derived from radial velocities for stars of known distance, while A and B can be derived from proper motions, independent of a distance scale. However, given the small numerical effects, $\mu_l \sim 0.003\cos 2l - 0.002$ arcsec yr^{-1} (Equation 9.4b, using $\mu = v_T/4.74d$, where μ is in arcsec yr^{-1}, v_T in km s^{-1}, and d in pc) the importance of the accuracy and absence of systematics in the Hipparcos data is evident.

The local velocity distribution of disk stars in the solar neighbourhood shows several well-established properties: an increase in velocity dispersion with age, attributed to the gravitational effects of spiral arms or massive compact objects like molecular clouds; the tendency of young stars to appear in moving groups or streams, attributed to dissolving ensembles of stars born at the same place and time; and the asymmetric drift and 'vertex deviation' of classes of stars (see box on page 305) attributed to non-equilibrium effects (for young stars) and/or to departures from an axisymmetric potential (e.g. the Galactic bar).

Previous investigations into the bulk motions of the local stellar velocity field have also been hampered by the non-inertiality of the reference frame within which ground-based proper motions have been referred. An important attribute of the Hipparcos Catalogue in this respect, in addition to the high accuracy and homogeneity of the proper motions, is the fact that the observations are referred to a quasi-inertial reference frame (see Section 1.4). The residual uncertainty of ± 0.25 mas yr^{-1} in the rate of rotation translates into a systematic effect of the same magnitude in the Hipparcos proper motion system with respect to an inertial frame, and corresponds to an uncertainty of 1.2 km s^{-1} kpc^{-1} on the rotational components of the velocity field, and on the Oort constant B.

A recurrent feature of many of the kinematic and dynamic studies reported from the Hipparcos data is the absence of the third Cartesian component of the space velocity vector, the radial velocity. This leads to various limitations in the exploitation of the data.

9.1.3 Concepts and definitions

Some of the main concepts underlying the discussions of this chapter are collected here by way of introduction,

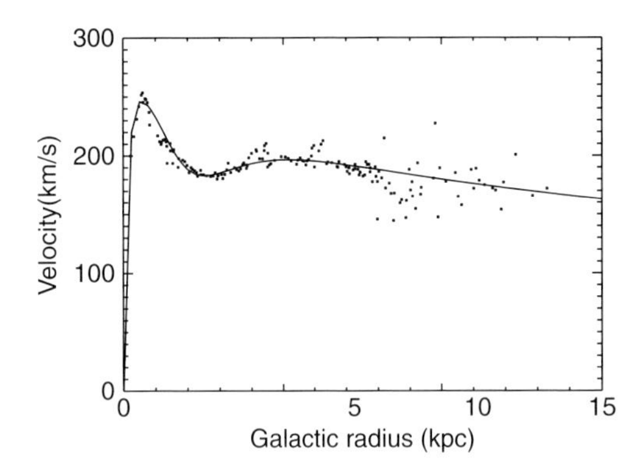

Figure 9.1 CO-based rotation curve of the Galaxy, with data points from Clemens (1985). The analytical fit, normalised to $R_0 = 7.5$ kpc and $\Theta_0 = 190\,\text{km s}^{-1}$, is represented by Equation 9.1. From Dias & Lépine (2005, Figure 7).

while reference should be made to more extensive texts for further details (e.g. Binney & Tremaine, 1987; Binney & Merrifield, 1998):

(a) Distance of the Sun from the Galactic centre, R_0: it is generally assumed that the centre of the Galaxy is defined by a black hole, or other central mass concentration, which coincides with the Galaxy's barycentre. For studies of internal Galactic dynamics, it is a working hypothesis that the Galactic centre defines the origin of an inertial coordinate system. This is clearly only an approximation for dynamical motions related to the barycentre of the local group of galaxies.

(b) Circular velocity, $\Theta(R)$ (also denoted as $v_c(R)$, and usually measured in km s^{-1}): this is the velocity of an object moving in a circle of radius R, in the Galactic plane and about the Galactic centre, for which centrifugal force balances the Galaxy's gravity. The definition assumes an axisymmetric galaxy, but can be extended to conditions of non-axisymmetry by constructing an azimuthal average over a small range of radii to obtain a regional average.

(c) The 'solar neighbourhood': this is a loose concept considered to be a volume centred on the Sun much smaller than the overall size of the Galaxy, but containing a statistically representative subset of its population, and thus with a somewhat arbitrary size dependent on the objects under investigation. It may range, for example, from a sphere of typical radius 10 pc for the faint, common white dwarfs or M dwarfs, 25 pc for the Catalogue of Nearby Stars (Section 9.3), 100 pc for the volume sampled to roughly 10% parallax precision by Hipparcos, and out to 1 kpc or more for the brighter and rarer O and B stars.

(d) The 'local standard of rest', or LSR: this is the velocity of a hypothetical group of stars in strictly circular orbits at the solar position. Its practical definition is complicated by the choice of stars representing it. Studies continue to be reported with reference to a wide range of stellar types, although it is now recognised that young stars are not in dynamical equilibrium and should not be used as a basis for the LSR definition.

(e) The 'solar motion': this can be determined with respect to a range of stellar and interstellar constituents of the Galaxy. Most frequently, this motion is estimated with respect to the LSR, $\mathbf{v}_\odot = (u_\odot, v_\odot, w_\odot)$, being the difference between the Sun's velocity and that of the reference system which, by definition, moves around the Galaxy with circular velocity $\Theta_0 \equiv \Theta(R_0)$. When estimated with respect to the mean velocity of specific stars in the solar neighbourhood, it is again a somewhat loose concept since it depends (for example) on the spectral class of star used as kinematic reference. Concepts such as 'standard solar motion' relative to commonly observed bright stars, and 'basic solar motion' with respect to commonly measured velocities, are encountered in the historical literature, but have limited physical relevance and will not be considered further.

[For completeness, it should be noted that some studies have discussed the possibility that an additional outward component of 5–7 km s^{-1} is required to transform the local standard of rest to a purely rotating reference system. This amounts to asserting that stars in the solar neighbourhood have a systematic radial component of motion, extending out to some unknown distance, perhaps of order 1–2 kpc. The original evidence was from asymmetries in 21-cm (Kerr, 1962) and CO surveys (Shuter, 1982); the latter identified a motion of the LSR, with respect to a true rotational standard of rest, both radially outward with a velocity of 4–10 km s^{-1} and azimuthally ahead of the direction of circular Galactic rotation with a velocity of $\sim$8 km s^{-1}; part of these effects are attributable to the use of an old solar motion value. It is now accepted that young objects and interstellar gas may be moving relative to the old stars at a velocity of a few km s^{-1} in both the radial and circular directions, depending on the spiral model adopted. But also in bar-driven resonances (Section 9.6), a small periodic radial oscillation of the LSR may result (Kotoneva *et al.*, 2005). Errors in the azimuthal component of this motion will have an effect on the dynamical determinations of R_0 (e.g Nikiforov, 1999), while the effect of a local radial systematic motion is important, for example, in establishing the Sun's Galactic orbit. Thus, if the LSR is also the local rotational standard of rest, then the Sun is close to its Galactic pericentre and with an apocentre at $R \simeq 1.14R_0$ (Stothers, 1985). If, however, the LSR is moving outwards at 7 km s^{-1}, then its true orbit would be more nearly circular.]

(f) Rotation curve: the disk of the Galaxy is in a state of differential rotation around an axis through the Galactic centre. The rotation curve of the Galaxy describes the observed rotation of some tracer population, $\Theta(R)$ (and may be distinguished from the 'circular-speed curve' giving the theoretical speed of circular orbits in an axisymmetric potential). It has an innermost part $R \lesssim 3$ kpc in almost solid body rotation with a rotation velocity which rises outwards, is roughly constant at $R \sim R_0$, and is fairly flat or with a slow decline at still larger radii, implying the presence of invisible or dark matter in the outer parts. The determination of $\Theta(R)$ from stars in the Galactic disk is limited by interstellar absorption to $R \simeq R_0$, while the determination of the rotation curve of the Galaxy at large distances from the Sun is based rather on 21 cm radio observations of the hyperfine transition of atomic hydrogen, or CO observations. Figure 9.1 shows the rotation curve derived from the CO observations of Clemens (1985). The curve is well represented by the following expression derived by Dias & Lépine (2005), which they have normalised to $R_0 = 7.5$ kpc and $\Theta_0 = 190\,\mathrm{km\,s^{-1}}$

$$\begin{aligned} V = & \,228\,\exp\left(-R/50 - (3.6/R)^2\right) \\ & +350\,\exp\left(-R/3.25 - 0.1/R\right) \end{aligned} \tag{9.1}$$

(g) Oort constants: description of the rotation law in terms of the Oort constants rests on the assumption of circular motion around the Galactic centre. Under these conditions, general expressions for radial and transverse velocities relative to the Sun, v_R and v_T, are derived by e.g. Mihalas & Binney (1981, Equation 8–4, 8–8) as

$$v_R = (\Omega - \Omega_0)\, R_0 \sin l \tag{9.2a}$$

$$v_T = (\Omega - \Omega_0)\, R_0 \cos l - \Omega d \tag{9.2b}$$

where $\Omega = \Theta/R$ and $\Omega_0 = \Theta_0/R_0$ are the angular rotation rate of the Galaxy at radius R and R_0, d is the distance from the Sun to the object at radius R from the Galactic centre, and l is the Galactic longitude of the observed object. The transverse space velocity, $v_{\rm T}$, is determined from the proper motions as

$$v_{\rm T} = \frac{A_v \mu}{\pi} \tag{9.3}$$

where $A_v = 4.74047\ldots$ equals the astronomical unit expressed in $\mathrm{km\,yr\,s^{-1}}$ when v_T is in $\mathrm{km\,s^{-1}}$, μ is in $\mathrm{arcsec\,yr^{-1}}$, and π is in arcsec (see Table 1.3).

Equations valid for $d \ll R_0$ and thus characterising differential galactic rotation in the vicinity of the Sun are given in terms of the Oort constants A and B (Mihalas & Binney, 1981, Equation 8–15,16 and 8–19,20)

$$v_R = d\,(A \sin 2l); \qquad A \equiv +\frac{1}{2}\left(\frac{\Theta}{R} - \frac{\mathrm{d}\Theta}{\mathrm{d}R}\right)_{R_0} \tag{9.4a}$$

$$v_T = d\,(A \cos 2l + B); \qquad B \equiv -\frac{1}{2}\left(\frac{\Theta}{R} + \frac{\mathrm{d}\Theta}{\mathrm{d}R}\right)_{R_0} \tag{9.4b}$$

where Θ and Θ_0 are the linear rotation velocities of the Galaxy at radii R and R_0, usually expressed in $\mathrm{km\,s^{-1}}$. The Oort 'constants' have units of frequency, and are usually expressed in $\mathrm{km\,s^{-1}\,kpc^{-1}}$, or in $\mathrm{mas\,yr^{-1}}$ (with $1\,\mathrm{km\,s^{-1}\,kpc^{-1}} \sim 0.211\,\mathrm{mas\,yr^{-1}}$). They are normally evaluated at $R = R_0$ (and at the solar azimuth) but could in principle describe the (circular) rotation curve at any R through the Oort functions, $A(R)$ and $B(R)$. For a flat rotation curve, A and B would be inversely proportional to R, with smaller A and larger B at larger Galactocentric distances. Local values of the angular rotation rate and its local derivative can be expressed directly in terms of local values of A and B

$$\left(\frac{\Theta}{R}\right)_{R_0} \equiv \Omega_0 = A - B \tag{9.5a}$$

$$-\left(\frac{\mathrm{d}\Theta}{\mathrm{d}R}\right)_{R_0} = A + B \tag{9.5b}$$

Physically, and in analogy with fluid dynamics, A describes the azimuthal shear of the velocity field, while B describes its vorticity.

(h) More general expression for the velocity field: if the assumptions of strictly circular motion and axisymmetry are relaxed, but are still restricted to motions in the plane, a more general expression for the velocity field at some point $\mathbf{x}$ is

$$\mathbf{V} = -\mathbf{v}_\odot + \mathbf{D}\cdot\mathbf{x} + \mathcal{O}(\mathbf{x}^2) \tag{9.6a}$$

$$\begin{aligned} \mathbf{D} = & \begin{pmatrix} \partial v_x/\partial x & \partial v_x/\partial y \\ \partial v_y/\partial x & \partial v_y/\partial y \end{pmatrix}_{\mathbf{x}=0} \\ \equiv & \begin{pmatrix} K + C & A - B \\ A + B & K - C \end{pmatrix} \end{aligned} \tag{9.6b}$$

where $\mathbf{v}_\odot$ is the Sun's velocity with respect to the local 'streaming' (or average) velocity. These four (Oort) constants measure the local divergence (K), vorticity (B), and azimuthal (A) and radial (C) shear of the velocity field. Expressed in cylindrical coordinates (R, φ) with the Sun at $(R_0, 0)$ these are (Chandrasekhar, 1942; Olling & Dehnen, 2003)

$$A = \frac{1}{2}\left(+\frac{v_\varphi}{R} - \frac{\partial v_\varphi}{\partial R} - \frac{1}{R}\frac{\partial v_R}{\partial \varphi}\right) \tag{9.7a}$$

$$B = \frac{1}{2}\left(-\frac{v_\varphi}{R} - \frac{\partial v_\varphi}{\partial R} + \frac{1}{R}\frac{\partial v_R}{\partial \varphi}\right) \tag{9.7b}$$

$$C = \frac{1}{2}\left(-\frac{v_R}{R} + \frac{\partial v_R}{\partial R} - \frac{1}{R}\frac{\partial v_\varphi}{\partial \varphi}\right) \tag{9.7c}$$

$$K = \frac{1}{2}\left(+\frac{v_R}{R} + \frac{\partial v_R}{\partial R} + \frac{1}{R}\frac{\partial v_\varphi}{\partial \varphi}\right) \tag{9.7d}$$

Corresponding equations for v_R and v_T with respect to the LSR, are (Olling & Dehnen, 2003, Equation 7)

$$v_R = d\,(K + A \sin 2l + C \cos 2l) \tag{9.8a}$$

$$v_T = d\,(B + A \cos 2l - C \sin 2l) \tag{9.8b}$$

In the axisymmetric limit and with no radial expansion or contraction, $C = K = 0$, and the expressions for A and B reduce to those given by Oort (Equation 9.4). Expressions for the full deformation tensor, i.e. including motions out of the plane, is discussed in Section 9.2.7.

(i) Motions of stars in the solar neighbourhood: stellar motions can be interpreted as a streaming (average) velocity plus random motions. In disk galaxies, systematic motions dominate over the random, and such stellar systems are said to be dynamically 'cold'. In the solar neighbourhood, the velocity dispersion in the plane is $\sim 45\,\mathrm{km\,s^{-1}}$ for the old stellar disk and $\sim 18\,\mathrm{km\,s^{-1}}$ for early-type stars, while the streaming motion is of the order of $200\,\mathrm{km\,s^{-1}}$. In the cold limit of vanishing random motions, the streaming is along the closed orbits supported by the gravitational potential.

(j) Orbits of stars in the solar neighbourhood: stellar orbits can be described by epicyclic motions due to their random or coherent motions with respect to the LSR superimposed on a circular orbit; in describing the coherent motion of a population, the term 'asymmetric drift' (see box on page 305) describes the difference between the local standard of rest and the mean rotational velocity of the population. In the epicycle approximation in which higher-order potential terms are ignored, the equations of motion become (Binney & Tremaine, 1987, Equation 3–57)

$$\ddot{x} = -\kappa^2 x \quad \text{and} \quad \ddot{z} = -\nu^2 z \tag{9.9}$$

defining the (local) epicycle frequency, κ, and vertical frequency, ν; while the approximation in x is reasonably accurate and hence quite useful, that in z is too inaccurate to be of practical value due to the fact that the vertical density structure of the disk is (roughly) exponential rather than constant with height (see Section 9.4.1). The period of radial motion in a near circular orbit is then $2\pi/\kappa$, with (Binney & Tremaine, 1987, Equation 3–62)

$$\kappa^2(R) = 4|B(R)|\Omega(R) = 4B(R)[B(R) - A(R)]$$
$$\text{with } \kappa_0^2 = 4B[B - A] \text{ at the Sun} \tag{9.10}$$

Using values of A and B from, e.g. Kerr & Lynden-Bell (1986) gives $\kappa_0 = 36 \pm 10\,\mathrm{km\,s^{-1}\,kpc^{-1}}$ and a ratio

$$\frac{\kappa_0}{\Omega_0} = 2\sqrt{\frac{-B}{A - B}} \simeq 1.3 \tag{9.11}$$

so that the Sun executes some 1.3 radial oscillations for each orbit around the Galactic centre. The semi-major axis of the epicycle in the radial direction depends on the velocity in the radial direction u as $a = (u/2)\sqrt{-B(A - B)}$, which is of order 350, 700, and 1200 pc for Cepheids, early-type stars, and late-type stars respectively, thus introducing radial smearing of kinematic properties over these length scales (Olling & Merrifield, 1998). Equation 9.11 holds for an equilibrium distribution in the 'cold' limit, i.e. when A and B are representative of circular orbits, and the asymmetric drift vanishes (it cannot be applied to OB stars, for example, which are dynamically 'cold', but not in equilibrium).

(k) Velocity dispersion: it has been known for more than 50 years that velocity dispersion increases with increasing $B - V$, due to the correlation of mean colour with age combined with scattering processes which increase the random motions with time (Section 6.9). Above $B - V \sim 0.61$ (Dehnen & Binney, 1998a), known as Parenago's discontinuity (Parenago, 1950), the dispersion is constant because subsamples of main-sequence stars have approximately the same mean age. At the same time, younger groups have larger mean Galactic rotation velocities than older groups. For small asymmetric drifts, i.e. in the epicycle approximation, the dispersions of peculiar velocities in the solar neighbourhood in the radial and tangential directions σ_u and σ_v are given by (Binney & Tremaine, 1987, Equation 3–76)

$$\frac{\sigma_v}{\sigma_u} = \left[\frac{B}{B - A}\right]^{1/2} \tag{9.12}$$

This relationship has been used in the past to indirectly estimate B given A and the velocity dispersions for different types of stars, and especially for the later-type stars which should be well-mixed dynamically. Although subject to the same conditions as for Equation 9.11, it introduces errors of only $\sim 10\%$ for typical samples with $\sigma_v \sim 30\,\mathrm{km\,s^{-1}}$ (e.g. Dehnen, 1999a, Figure 5).

(l) Orbital resonances: two resonances, the corotation resonance and the Lindblad family of resonances, are important in the study of bar and spiral structure instabilities which appear in rotating stellar systems. They appear in the treatment of orbits in non-axisymmetric potentials, in which the figure of the potential rotates rigidly at some 'pattern' speed Ω_p ($\equiv \Omega_\mathrm{b}$ for the bar, or Ω_s for the spiral, usually expressed in $\mathrm{km\,s^{-1}\,kpc^{-1}}$). The general stellar orbit is then represented as a combination of the circular motion of some 'guiding centre' and small oscillations about it (Binney & Tremaine, 1987, Equation 3–107 to 3–122). In a polar coordinate system (R, φ) in a frame rotating with the potential, and for a potential of the form $\Phi_1(R, \varphi) = \Phi_b(R)\cos(m\varphi)$, orbital solutions can be shown to have singularities when

$$\Omega_0 = \Omega_\mathrm{p} \tag{9.13a}$$
$$\text{or} \quad m(\Omega_0 - \Omega_\mathrm{p}) = \pm\kappa_0 \tag{9.13b}$$

where κ_0 is the epicycle frequency of the radial motion. These singularities correspond to resonance conditions which arise between the forcing frequency seen by the star, $m(\Omega_0 - \Omega_\mathrm{p})$, and the two natural frequencies

0 and κ_0. The former is known as the co-rotation resonance: in this case $\dot{\varphi}_0 = 0$, and consequently the guiding centre co-rotates with the rotating potential.

The latter are known as the Lindblad resonances, of which a galaxy may have 0, 1, 2, or more, depending on the value of m ($m = 2$ for a bar). At these locations, a star encounters successive peaks of potential, in phase with the frequency of its radial oscillation, overtaking the potential for $+\kappa_0$ (an inner Lindblad resonance) and conversely for $-\kappa_0$ (an outer Lindblad resonance).

9.2 The Sun within the Galaxy

This section covers studies of a number of the basic parameters defining the location of the Sun within the Galaxy (notably the distance to the Galactic centre, the distance from the Galactic plane, and the local solar motion), and the characterisation of the associated bulk motions of the local stellar population (notably the detailed description of the rotational properties of the disk). Kerr & Lynden-Bell (1986) provided a review of some of the major Galactic constants 10 years before the Hipparcos data became available, while a compilation of updated reference quantities including a summary of relevant Hipparcos results is given in Appendix A.

Despite the quantity and quality of the Hipparcos data, many basic quantities remain poorly constrained or modelled. Findings remain confusing in many areas; to a large extent, this review simply aims to collate the various investigations that have been carried out in the different areas.

9.2.1 Distance to the Galactic centre

The distance to the Galactic centre arguably represents the most fundamental distance scale entering into discussions of the Galaxy's structure and dynamics. The contributions made by Hipparcos are largely indirect and certainly inconclusive, and the overall status is described for completeness.

At around 7.5–8.5 kpc, the Galactic centre is too distant for direct trigonometric parallax determinations, and the large and variable extinction complicates the use of secondary indicators to derive its distance. Distance estimates were originally made from studies of the space density of objects believed to be distributed symmetrically around the Galactic centre, notably globular clusters, RR Lyrae stars and Mira variables. From a straight mean of some 25 determinations between 1974–86 in the range 6.7–10.5 kpc, Kerr & Lynden-Bell (1986) estimated $R_0 = 8.54 \pm 1.1$ kpc. The more recent pre-Hipparcos status is described by Reid (1993) and Huterer *et al.* (1995). The former reviewed the various techniques then available to determine R_0, combining different measurements to derive an estimate of $R_0 = 8.0 \pm 0.5$ kpc. Table 9.1 provides a selection of pre-Hipparcos distance estimates, along with more recent estimates based on Hipparcos and other observations.

Based on Hipparcos observations, the kinematic analysis of Hipparcos Cepheids has (indirectly) provided $R_0 = 8.5 \pm 0.5$ kpc (Feast & Whitelock, 1997), while an estimate based on the RR Lyrae period–luminosity relation and K-band observations of M5 has been revised using Hipparcos data to 9.3 ± 0.7 kpc (Reid, 1998). While large compared with other current estimates, revising the Walker & Terndrup (1991) result to take account of the Reid (1998) globular cluster distances would also raise the distance to 9.6 kpc. RR Lyrae-based methods assume that the luminosities of globular cluster and field RR Lyrae stars are identical. Other estimates of R_0 based on different absolute magnitudes for RR Lyrae stars are also found in the literature (e.g. Dambis & Rastorguev, 2001).

The most direct method relying partly on Hipparcos measurements is that based on red clump giants in Baade's window detected by OGLE: Paczyński & Stanek (1998) used some 600 red clump stars from Hipparcos with I-band photometry, out of some 2000 such stars in the Hipparcos database, to calibrate their absolute magnitudes and hence determine R_0. The same approach with an improved treatment of local interstellar extinction was used by Stanek & Garnavich (1998) to derive $R_0 = 8.2 \pm 0.15 \pm 0.15$ kpc (statistical plus systematic error), and Alves (2000) in the K-band to derive $R_0 = 8.24 \pm 0.42$ kpc, and correcting for evolutionary effects by Girardi & Salaris (2001) to derive $R_0 = 7.8 \pm 0.2$ kpc, although with possibly large systematics.

Estimates of R_0 can, in principle, also be derived from the Oort constants A and B according to Equation 9.5a, but this requires an independent estimate of Θ_0. Other observations of the inner regions of the Galaxy can also provide values of Ω_0 (e.g. Kalirai *et al.*, 2004).

Dynamical studies of the Galactic centre based on high-resolution adaptive optics imaging in the infrared appear to offer the most direct distance determination method currently available. Genzel *et al.* (2000) used statistical parallaxes of more than 100 Galactic centre stars to estimate $R_0 = 7.8 - 8.2(\pm 0.9)$ kpc. Eisenhauer *et al.* (2003) reported an orbital parallax determination of 7.9 ± 0.4 kpc. Ghez *et al.* (2008) estimated 8.30 ± 0.31 kpc.

In conclusion, estimates for R_0 still lie in the rather broad range 7.5–8.5 kpc. In most of the studies described here, authors assign an assumed value typically in this range, although values as small as 7.1 kpc, as derived by Olling & Merrifield (1998), are still discussed. A value of $R_0 = 8.2$ kpc, taking into account the red clump and Galactic centre results, is suggested as reference, and forms part of the constants given in Appendix A.

Table 9.1 Estimates of the distance to the Galactic centre, including recent Hipparcos-based determinations. Pre-Hipparcos RR Lyrae estimates should be revised to take account of any subsequent revision in the RR Lyrae absolute magnitudes, e.g. as derived from Hipparcos.

Method	Reference	R_0 (kpc)
Independent of Hipparcos:		
RR Lyrae in the Galactic bulge	Oort & Plaut (1975)	8.0 ± 0.6
Photometry and radial velocity of Cepheids	Caldwell & Coulson (1987)	7.8 ± 0.7
H_2O masers in Sgr B2	Reid *et al.* (1988)	7.1 ± 1.5
RR Lyrae in Baade's window	Walker & Terndrup (1991)	8.2 ± 1.0
H_2O masers W49	Gwinn *et al.* (1992)	8.1 ± 1.1
Compilation of various measurements	Reid (1993)	8.0 ± 0.5
RR Lyrae at infrared wavelengths	Carney *et al.* (1995)	7.8 ± 0.4
Galaxy rotation and Oort constants	Olling & Merrifield (1998)	7.1 ± 0.4
Statistical parallaxes of Galactic centre stars	Genzel *et al.* (2000)	8.0 ± 0.9
Orbital parallax of the Galactic centre star S2	Eisenhauer *et al.* (2003)	7.9 ± 0.4
Orbital parallax of the Galactic centre star S2	Ghez *et al.* (2008)	8.3 ± 0.3
Hipparcos-based:		
Cepheids	Feast & Whitelock (1997)	8.5 ± 0.5
RR Lyrae re-calibrated from globular clusters	Reid (1998)	9.3 ± 0.7
Red clump giants	Paczyński & Stanek (1998)	8.4 ± 0.4
Red clump giants with improved extinction	Stanek & Garnavich (1998)	8.2 ± 0.2
Red clump giants in K-band	Alves (2000)	8.2 ± 0.4
Red clump giants including evolution	Girardi & Salaris (2001)	7.8 ± 0.2

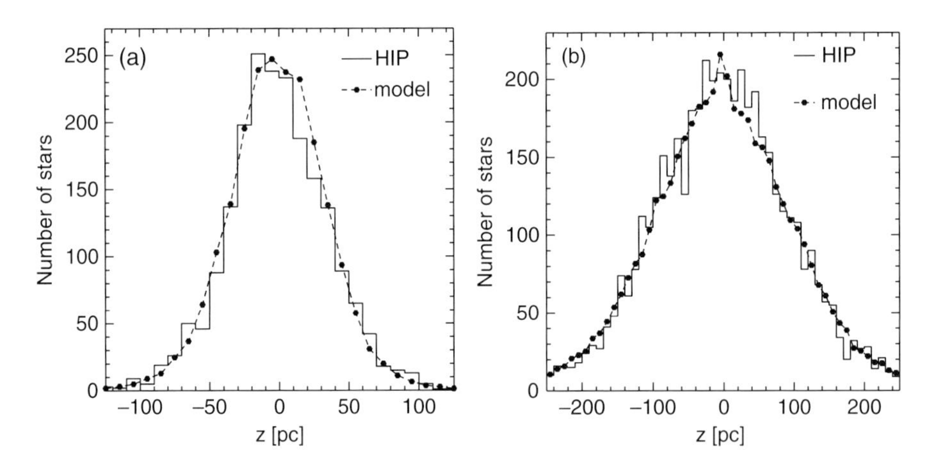

Figure 9.2 Determination of Z_0 from the distribution of stars as a function of distance from the Galactic mid-plane: (a) for stars with $0.3 < B - V < 0.6$ and $V < 7$ (roughly F stars) within a radial distance of 100 pc from the Sun; (b) for stars with $B - V > 0.8$ and $V < 7$ (almost exclusively red giants) within a radial distance of 200 pc from the Sun. The diagrams give counts from the observed and simulated catalogues, taking account of various selection effects, and therefore do not represent the true distribution of stars. From Holmberg et al. (1997, Figure 3).

9.2.2 Distance from the Galactic plane

It has been known for some time that the Sun is not exactly in the mid-plane of the Galactic disk as defined by the distribution of neutral hydrogen. The equatorial plane, $b = 0°$, of the Galactic coordinate system, as adopted by the IAU in 1958, passes through the Sun by definition. If the Sun were exactly in the mid-plane, the origin of the Galactic coordinate system should coincide with the Galactic centre. However the coordinates of Sgr A* imply that the Sun is slightly above the mid-plane. Knowledge of this displacement, Z_0, is important in studies relying on asymmetries in star counts or dust emission for their interpretation, for example, studies of the Galaxy warp from the DIRBE (Diffuse Infrared Background Experiment) on COBE (Freudenreich *et al.*, 1994; Porcel *et al.*, 1997). A value for Z_0 combined with the Sun's vertical velocity component $w_\odot$ also provides a phase reference for the Galactic environment of the Earth during its evolutionary history (Section 10.6).

Pre-Hipparcos estimates of Z_0 placed the Sun at a distance from the mid-plane ranging from 10 pc from interstellar dust measurements, 15 pc from IRAS source counts and COBE data, 37 pc from Cepheids, to as much as 42 pc from some classical star counts (Humphreys & Larsen, 1995). As examples, a value of $Z_0 = 20.5 \pm 3.5$ pc was derived by Humphreys & Larsen (1995) based on star counts at the Galactic poles, an assumed scale height for the disk, and a disk mid-plane defined by neutral hydrogen. From star counts for several thousand OB stars within 4 kpc of the Sun, Reed (1997) derived $Z_0 \sim 10-12$ pc, with possible effects due to the warp, the Gould Belt, and extinction noted (see also Reed, 2005).

Pham (1997) used a sample of F-stars from Hipparcos to estimate $Z_0 = 9 \pm 4$ pc, based on a maximum likelihood estimate of the scale height, and the Sun's distance from the mid-plane, as part of a determination of the local mass density (Section 9.4.1). Holmberg *et al.* (1997) estimated $Z_0 = 8 \pm 4$ pc using a similar approach for both F stars and red giants (Figure 9.2).

Higher values are still obtained from other investigations. Chen *et al.* (2001) used data from the Sloan Digital Sky Survey to derive $Z_0 = 27 \pm 4$ pc. Joshi (2005) found $Z_0 = 22.8 \pm 3.3$ pc with respect to the reddening plane defined by open clusters. Maíz Apellániz (2001) used Hipparcos parallaxes for 3531 O–B5 stars with $|b| > 5°$ to derive $Z_0 = 24.2 \pm 1.7 \pm 0.4$ pc (random + systematic), modelling the distribution as a self-gravitating isothermal profile with a 5% contribution from a 'parabolic' halo distribution (Figure 9.3); they argue that the early-type stars are well suited to the problem since they have a scale height comparable to the value of Z_0, and they identify extinction, the presence of spiral arms, and the presence of the Gould Belt as complicating factors. The fact that these stars are not yet in dynamical equilibrium may, however, make their use rather suspect. Branham (2003) used 93 106 Hipparcos parallaxes with a range of spectral and luminosity classes, but excluding O and B stars belonging to the Gould Belt, to derive $Z_0 = 34.56 \pm 0.56$ pc, and the corresponding coordinates of the Galactic pole, $l_g = 0.^\circ004 \pm 0.^\circ039$, $b_g = 89.^\circ427 \pm 0.^\circ035$.

There is evidently no clear consensus from these various Hipparcos studies, but a value of $Z_0 = 20$ pc is somewhat arbitrarily offered as a reference value.

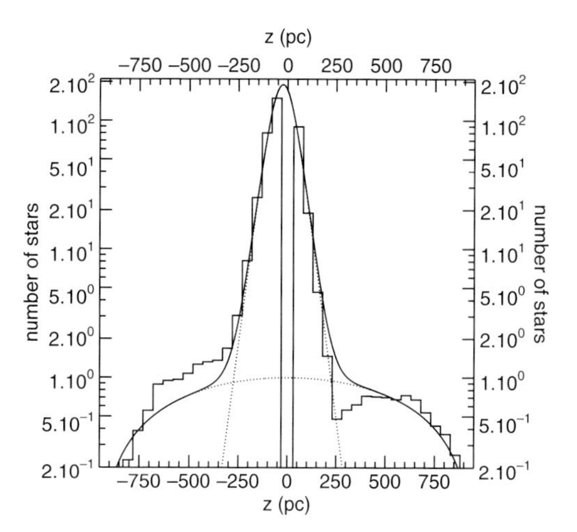

Figure 9.3 Determination of Z_0 from observed (histogram) and modelled frequency distributions as a function of the vertical coordinate for O–B5 stars, using a self-gravitating isothermal disk plus parabolic halo distribution. Dotted lines represent the individual components, and the continuous line is the sum of the two. The region immediately surrounding $z = 0$ is not considered for the fit, as it corresponds to the space between two semi-infinite cylinders. Large residuals are evident for $|z| >$ 250 pc. From Maíz Apellániz (2001, Figure 2).

9.2.3 Velocity dispersion and vertex deviation

Dehnen & Binney (1998a) used a sample of 11 865 Hipparcos stars with $\sigma_\pi/\pi < 0.1$ and no kinematic bias, and binned in $B-V$ to explore the dependence of solar motion, $\mathbf{v}_\odot = -\langle \mathbf{v} \rangle$, and S, a measure of the total velocity dispersion (Figures 9.4 and 9.5). A number of conclusions were drawn from this early Hipparcos study: (1) extrapolating to zero dispersion (Figure 9.4b) yields the velocity of the Sun with respect to the LSR as given in Table 9.2; (2) Parenago's discontinuity is clearly visible as an abrupt flattening of S (also seen in v) redward of $B-V \sim 0.61$ (Figure 9.4, right); with velocity dispersion increasing with age, the discontinuity should occur at the colour for which the main-sequence lifetime equals the age of the Galactic disk; (3) the vertex deviation, defined by Equation 9.52, shows a strong dependence on star colour blueward of the discontinuity, of about 10° for old disk stars and reaching about 20° for early-type stars (Figure 9.5); commonly used to parameterise deviation from dynamical symmetry, its persistence to late-type stars implies that the Galactic potential is significantly asymmetric at the solar radius; (4) redder than $B-V = 0.1$ (younger stars presumably being less well mixed) the ratios of the principal velocity dispersion components are $\sigma_1 : \sigma_2 : \sigma_3 \simeq 2.2 : 1.4 : 1$. They derive an estimate of the disk scale length, R_d, of $R_0/R_d \simeq 3-3.5$, consistent with recent infrared studies of the Galaxy.

9.2.4 Solar motion with respect to the local standard of rest

Various estimates of the solar motion have been made based on the Hipparcos data, and employing a wide variety of spectral types as reference. Table 9.2 shows various results for the Galactic components of $\mathbf{v}_\odot$. The component towards the North Galactic Pole is fairly

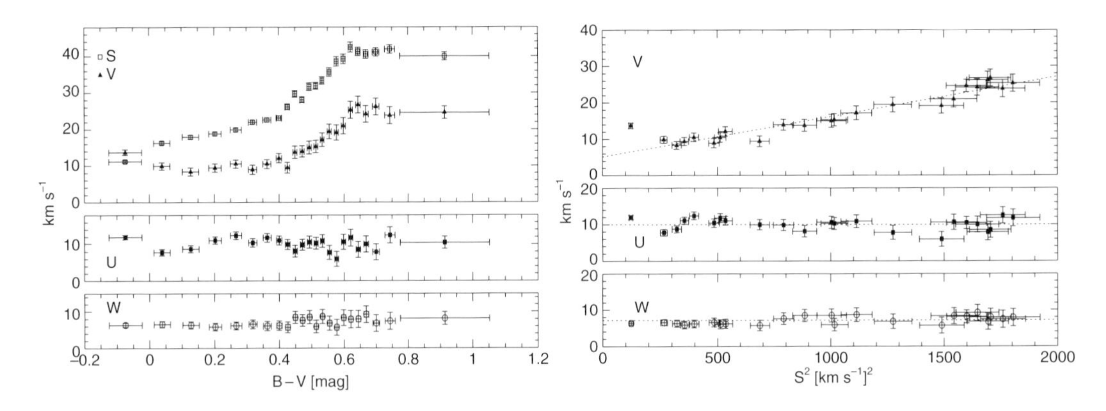

Figure 9.4 Left: The components U, V, W of the solar motion for stars with different colour $B - V$. Also shown is the variation of the dispersion S with colour (upper points in top panel), with Parenago's discontinuity visible at $B - V \sim 0.61$. Right: The dependence of U, V, W on S^2. Dotted lines correspond to the linear relation fitted (V) or mean values (U and W) for stars bluer than $B - V = 0$. From Dehnen & Binney (1998a, Figures 3–4).

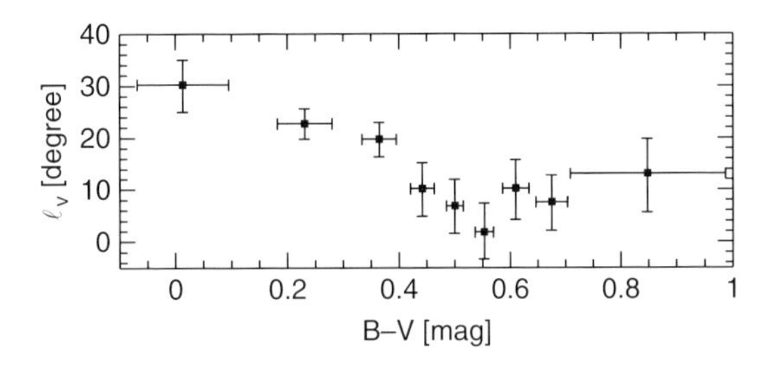

Figure 9.5 The vertex deviation l_v versus $B - V$ colour. Error bars are 1σ. From Dehnen & Binney (1998a, Figure 6).

constant with spectral type, has a weighted average of $w_\odot = 7.1 \pm 0.2\,\mathrm{km\,s^{-1}}$ (i.e. is moving upward from the Galactic plane), and is consistent with the most recent pre-Hipparcos results (e.g. Comerón *et al.*, 1994). Hipparcos results for the K giants, for example, are consistent with Hipparcos values found by Feast & Whitelock (1997) for Cepheids, and by Dehnen & Binney (1998a) for early- and late-type dwarfs. The other two components of motion are less precisely defined, with a sharp transition between the young disk dwarfs and the late-type (mainly giant) objects. The study of Fehrenbach *et al.* (2001) used 11 978 stars in the range $V = 7$–11, spectral types between B5–K8, and radial velocities to complete the space motions, then divided according to distance rather than spectral type; no convincing explanation was given for the small values of $w_\odot$. Hogg *et al.* (2005) modelled the three-dimensional velocity distribution function of 11 865 main-sequence Hipparcos stars, in a statistical formulation designed to handle the case of 'missing data', specifically the absence of radial velocities, and non-Gaussian velocity distributions. Using 20 colour-selected subsamples, they found that the local velocity dispersion is, again, a strong function of colour, which can be extrapolated to zero velocity dispersion to give the motion of the LSR (Figure 9.6).

Overall, the results typically provide better formal accuracies than pre-Hipparcos determinations, e.g. that of Evans & Irwin (1995) from APM proper motions, but still show a heterogeneity with spectral type and distance reflecting the complex velocity distributions in the solar neighbourhood. The main differences, notably in $v_\odot$, appear to arise from differences in interpretation of the asymmetric drift relation, which should be used to correct to zero velocity dispersion (i.e. to the case of closed orbits). From the existing literature, several pitfalls can be identified: (1) Several older studies used kinematically-biased samples, notably samples based on proper motion data for which the asymmetric drift relation will not apply. (2) A number of authors have used samples of early-type stars, considering that they are most representative of closed orbits because their velocity dispersion is low. The fact that the OB stars are not in dynamical equilibrium, however, again means that the asymmetric drift relation cannot be applied. (3) The asymmetric drift relation applies only to the mean and dispersion of a sample of an equilibrium population, and using a Gaussian fit or similar may give systematic errors. Dehnen & Binney (1998a) explicitly treated these aspects, using a kinematically-unbiased sample to infer the mean and dispersion, deriving the asymmetric drift relation directly from the data, and excluding early-type stars which did not follow the asymmetric drift relation of the rest. For these reasons, until a more critical review is published, their determination of $u_\odot, v_\odot, w_\odot = 10.0 \pm 0.36, 5.25 \pm 0.62, 7.17 \pm 0.38\,\mathrm{km\,s^{-1}}$ is suggested as a reference.

Table 9.2 Solar motion with respect to the LSR as represented by specific spectral types or distance classes. For the Hipparcos-based derivations: (1) boundaries for these types are listed in Table 9.4; (2) the in-plane components were estimated as part of studies of the spiral-wave density pattern; (3) from samples of spectral types B5–K8, organised into distance ranges 0–100 pc, 100–300 pc, and 300–500 pc; (4) derived within a statistical formulation to take account of the absence of radial velocity data.

		Solar motion wrt LSR ($\mathrm{km\,s^{-1}}$)			Total
Reference	Class	$u_\odot$	$v_\odot$	$w_\odot$	$V_\odot$
Pre-Hipparcos					
Mihalas & Binney (1981)	Compilation	9	12	7	16.5
Evans & Irwin (1995)	APM-based	7.3 ± 1.5	13.9 ± 2.3	8.8 ± 2.2	18.0
Hipparcos – Oort–Lindblad:					
Feast & Whitelock (1997)	Cepheids	9.3	11.2	7.61 ± 0.64	16.4
Miyamoto & Zhu (1998)	Cepheids	10.62 ± 1.20	16.06 ± 1.14	8.60 ± 1.02	21.1
Hipparcos – Ogorodnikov–Milne:					
Miyamoto & Zhu (1998)	O–B5 stars	11.59 ± 0.49	13.39 ± 0.48	7.12 ± 0.44	19.1
"	Cepheids	10.46 ± 1.19	15.95 ± 1.14	8.96 ± 1.03	21.1
Mignard (2000)[1]	A0–A5 dwarfs	9.92 ± 0.25	10.71 ± 0.26	6.96 ± 0.21	16.2
"	A5–F0 dwarfs	11.58 ± 0.32	10.37 ± 0.33	7.19 ± 0.31	17.1
"	F0–F5 dwarfs	11.46 ± 0.37	11.16 ± 0.37	7.02 ± 0.41	17.5
"	K0–K5 giants	7.99 ± 0.35	14.97 ± 0.36	7.39 ± 0.40	18.5
"	K5–M0 giants	8.72 ± 0.49	19.71 ± 0.51	7.28 ± 0.55	22.7
"	M0–M5 giants	7.37 ± 0.61	20.29 ± 0.63	6.85 ± 0.66	22.6
Branham (2000)	all Hipparcos	10.30 ± 0.06	19.13 ± 0.05	7.09 ± 0.04	22.8
Branham (2002)	OB stars	14.49 ± 0.12	19.68 ± 0.09	2.81 ± 0.07	24.6
Branham (2006) and priv. comm.	OB stars	7.76 ± 0.83	10.15 ± 0.89	5.29 ± 0.71	13.8
Hipparcos – vectorial harmonics:					
Vityazev & Shuksto (2004)	113 646 stars	–	–	–	23.3
Makarov & Murphy (2007)	non-binary	9.9 ± 0.2	15.6 ± 0.2	6.9 ± 0.2	19.7
Hipparcos – spiral-density wave:					
Mishurov & Zenina (1999b)[2]	Cepheids	7.8 ± 1.3	13.6 ± 1.4	–	–
Lépine *et al.* (2001)[2]	Cepheids	8.8 ± 1.0	11.9 ± 1.1	–	–
Hipparcos – other:					
Dehnen & Binney (1998a)	Dwarfs	10.0 ± 0.36	5.25 ± 0.62	7.17 ± 0.38	13.4
Brosche *et al.* (2001)	K0–K5 giants	9.0 ± 0.5	21.0 ± 0.5	7.7 ± 0.4	24.1
Fehrenbach *et al.* (2001)[3]	$\overline{d} = 46$ pc	9.79	13.20	3.25	16.7
"	$\overline{d} = 195$ pc	8.24	11.58	5.97	15.4
"	$\overline{d} = 378$ pc	2.93	10.36	4.79	11.8
Hogg *et al.* (2005)[4]	Dwarfs	10.1 ± 0.5	4.0 ± 0.8	6.7 ± 0.2	12.8

9.2.5 Rotation speed of the disk

The most direct method of determining the circular velocity of the LSR around the Galactic centre, Θ_0, is to determine the Sun's velocity with respect to a group of objects considered to be at rest relative to the Galactic centre, and correcting for the Sun's peculiar velocity with respect to the LSR. Such estimates have been made in the past based on three different approaches (e.g. Mihalas & Binney, 1981, Section 6.5): measurements with respect to halo population objects; measurements with respect to Local Group galaxies; and analysis of the escape speed of high-velocity stars. The principles, but not the complexities or results, of these three approaches are summarised here.

Measurements with respect to halo objects assume that the fundamental standard of rest is the Galactic centre and, furthermore, that since the halo system itself is not strongly flattened, it does not rotate rapidly about the Galactic centre. The halo globular clusters and metal-poor RR Lyrae stars thus represent, to first order, a non-rotating reference frame, with respect to which the solar motion reflects the rotation speed of the LSR. While this ignores the possibility that the halo rotates with some non-zero velocity, values of

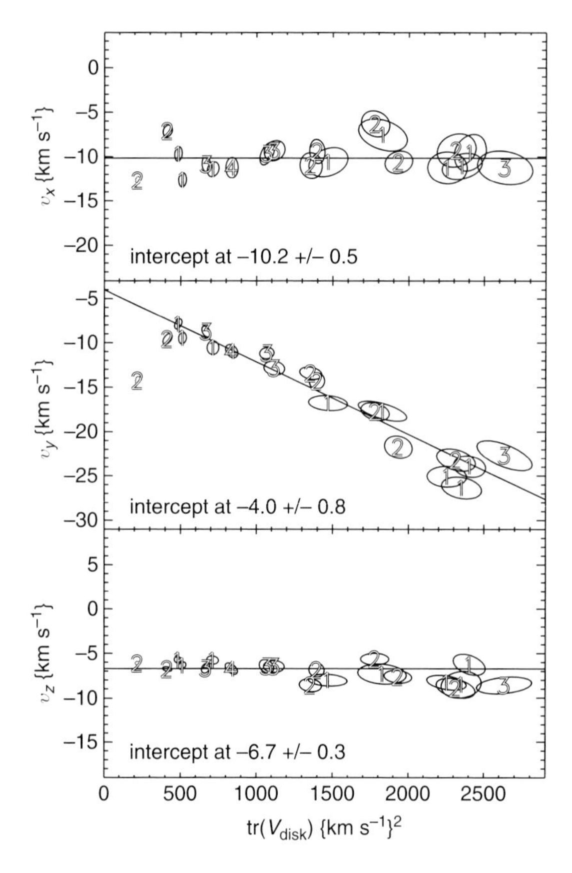

Figure 9.6 The mean velocity as a function of total velocity variance for 20 colour-selected subsamples for the determination of the local standard of rest (LSR). In this generalised version of the model, the disk velocity dispersion is fit with a mixture of K Gaussian ellipsoids with a common mean (this number is indicated within each ellipsoid), in order to accommodate non-Gaussian velocity distributions. In each panel, the ellipses indicate the 1σ uncertainty regions. From Hogg et al. (2005, Figure 6).

$\Theta_0 = 220–250\,\mathrm{km\,s^{-1}}$ also result in roughly equal numbers of halo RR Lyrae stars lying on prograde (direct) and retrograde orbits.

Measurements with respect to Local Group galaxies are based on their radial velocities providing the solar motion with respect to the velocity centroid of the Local Group. Such analyses must account for the fact that the observed motion includes not only the velocity arising from the motion of the LSR around the Galactic centre, and the peculiar velocity of the Sun with respect to the LSR, but also the velocity of the centre of mass of the Galaxy relative to the centre of mass of the Local Group.

Measurements of the kinematics of halo stars with the most extreme space velocities provide indirect estimates of the circular velocity, as well as the escape speed, v_{esc}, at which stars in the solar neighbourhood have enough energy to escape from the Galaxy's gravitational field. Such a cutoff in the stellar velocity distribution should occur first for stars moving in the sense of the Galactic rotation, with the observed value of v_{crit} satisfying $(\Theta_0 + v_{crit})^2 = v_{esc}^2$. Isobe (1974), for example, obtained $\Theta_0 = 275 \pm 20\,\mathrm{km\,s^{-1}}$ and $v_{crit} \simeq 110\,\mathrm{km\,s^{-1}}$, implying $v_{esc} \simeq 385\,\mathrm{km\,s^{-1}}$, although the latter is now considered to be an underestimate (see Section 9.4.2).

A somewhat independent possibility is to assume that the proper motion component μ_l of the radio source associated with the Galactic centre, Sgr A*, is entirely due to the reflex motion of our own rotation around the Galaxy, corrected for solar motion. Using weighted values from Backer & Sramek (1999) and Reid *et al.* (1999), Dehnen (1999b) obtained $\Omega_0 = 29.0 \pm 1 - v_\odot/R_0\,\mathrm{km\,s^{-1}\,kpc^{-1}}$, where the first term is from the proper motion, and the second corrects for the solar motion with respect to the LSR. For values of $v_\odot$ in the range 5–10 km s^{-1} (Table 9.2) and R_0 in the range 7.5–8.5 kpc, resulting values for Θ_0 are in the plausible range 210–240 km s^{-1}. Complications in the interpretation of μ_l, linked to the interpretation of the value of the μ_b component, were discussed by Gould & Ramirez (1998). Reid & Brunthaler (2004) used VLBA observations of Sgr A* to obtain $\Omega_0 = 27.19 \pm 0.87\,\mathrm{km\,s^{-1}\,kpc^{-1}}$, after correcting for the solar motion in longitude as determined by Dehnen & Binney (1998a, as given in Table 9.2). Their residual motion of Sgr A* perpendicular to the Galactic plane, also based on the solar motion of Dehnen & Binney (1998a), is $-0.4 \pm 0.9\,\mathrm{km\,s^{-1}}$.

Less direct estimates can also be made via Equation 9.5a, i.e. $\Theta_0 = (A - B)\,R_0$, based on an assumed value for R_0. Essentially the complex pattern of proper motions in the solar neighbourhood is used to derive the angular rotation of the Galaxy based on assumptions of the overall motion. In 1964, the IAU adopted a specific set of values (Table 9.3) which were representative, and which were self-consistent in the sense that $\Omega_0 \equiv \Theta_0/R_0$ derived from $A - B$ is the same as that obtained from the values of Θ_0 and R_0 (Equation 9.5a). An alternative self-consistent set ($A = 16, B = -11$) was given by Mihalas & Binney (1981), where A was estimated from a variety of radial velocity measurements for O and B stars, Cepheid variables, galactic clusters, etc.; B from proper motion data from fundamental star catalogues, proper motions relative to external galaxies, and velocity dispersion data; R_0 from space densities of globular clusters and RR Lyrae stars etc.; and Θ_0 from radial velocities of globular clusters, spheroidal components stars, and external galaxies in the Local Group.

Pre-Hipparcos values were variously estimated in the range $\Theta_0 = 220–250\,\mathrm{km\,s^{-1}}$, and $R_0 = 7–8.5\,\mathrm{kpc}$, compatible with observations of rotation curves in other galaxies of similar morphological type. Kerr & Lynden-Bell (1986) derived a straight mean of $\Theta_0 = 222 \pm 20\,\mathrm{km\,s^{-1}}$ from about 20 different determinations over the period 1974–85, with a brief discussion of systematic effects arising from kinematic or spatial asymmetries, or the perturbing effect of spiral density waves on the circular velocity.

As evident in the following two sections, values for $A - B$ remain rather weakly constrained even with the Hipparcos results. Taking a range of 26–30 $\mathrm{km\,s^{-1}\,kpc^{-1}}$ (from Table 9.3), in combination with estimates of R_0 in the range 7.5–8.5 kpc, yield Θ_0 lying in the still broad range of 195–255 $\mathrm{km\,s^{-1}}$. Within this range, the value of $\Theta_0 = 220\,\mathrm{km\,s^{-1}}$ is often adopted, but its uncertainty should be stressed.

The values of Θ_0 and R_0 considered by Mihalas & Binney (1981), for example, indicate that the Sun revolves around the centre of the Galaxy in a period of $P = 2\pi R_0/\Theta_0 = 2.2 \times 10^8$ yr. Their values of A and B imply that $(d\Theta/dR)_{R_0} = -(A + B) = -5\,\mathrm{km\,s^{-1}\,kpc^{-1}}$, such that the rotation curve of the Galaxy would be slightly declining in the solar neighbourhood. The fact that $A \simeq -B$ implies that the inferred form of the rotation curve at the solar circle (flat, slightly rising, or slightly declining) is sensitive to the derived quantities A and B. The Hipparcos results have not led to a clear conclusion on this point either, and if an estimate of the local behaviour is needed it seems preferable to be guided by the 21 cm or CO-based observations such as Figure 9.1, suggesting that the rotation curve at the solar circle is slightly decreasing with R, i.e. that $-(A+B) < 0$. This is also supported by the vectorial harmonic analysis presented by Makarov & Murphy (2007), discussed in Section 9.2.8.

9.2.6 Stellar kinematics in the Oort–Lindblad model

Since Oort's pioneering efforts, numerous studies have been made to determine A and B and the related values of Θ_0 and R_0, generally using a mixture of stellar types with intrinsically different kinematics. This effort has continued with the availability of the Hipparcos results, where the effects of rotation are seen even at large distances, notably for the Cepheids (Figure 5.15) and for the Wolf–Rayet stars (Figure 8.21).

Dehnen & Binney (1998b) discussed the current models for the mass distribution of the Galaxy, and showed that the free parameters of their model of the disk, bulge, and halo are constrained by the the tangent velocities at $R < R_0$, rotation velocities at $R > R_0$, the Oort constants, and the local surface density. They argued that models of the mass distribution are limited by poor knowledge of the vertical mass distribution, with small changes in $\Theta(R)$ associated with dramatic changes in the distribution of mass between the different components. They showed the difficulty of reconciling existing data with a dynamically plausible density profile for the halo, raising the question of whether a physically distinct dark halo is indicated, rather than a strongly evolving mass-to-light ratio of the disk or bulge with radius.

Olling & Merrifield (1998) calculated mass models for the Milky Way that include the density of interstellar gas, which they showed varied non-monotonically with radius, and so contributes significantly to the local gradient of the rotation curve. The corresponding radial variation in the Oort constants was estimated (Figure 9.8, left), and compared with values for tracers with different distance horizons. Between 0.9 and $1.2\,R_0$ the Oort functions $A(R)$ and $B(R)$ differ significantly from the general $\sim \Theta_0/R$ dependence, due to the local contributions from the interstellar gas (e.g. the ring of H I just beyond the solar circle. A consistent picture only emerged by adopting a (small) value of $R_0 = 7.1 \pm 0.4$ kpc which, however, agrees with the direct determination of $R_0 = 7.2 \pm 0.7$ kpc from H_2O masers (Reid, 1993). They also derived a small value of $\Theta_0 = 184 \pm 8\,\mathrm{km\,s^{-1}}$. With these Galactic constants the rotation curve of the Milky Way declines slowly in the outer Galaxy, with $\Theta_{20\,\mathrm{kpc}} = 166\,\mathrm{km\,s^{-1}}$. Although these models are not based on Hipparcos data, they showed that the resulting model radial velocities and proper motions are consistent with the radial velocities of Cepheids and the Hipparcos measurements of their proper motions (Figure 9.8, right). The various determinations of A (see Table 9.3) may show broadly the predicted dependency on location of the tracer: thus dwarfs at typical distances of 1 kpc yield smaller values of $A \sim 11\,\mathrm{km\,s^{-1}\,kpc^{-1}}$ (Mignard, 2000). Olling & Merrifield (1998) also suggested that the larger values of A generally found for the Cepheids is consistent with their preferential location in the inner Galaxy, although such a tendency is not strongly evident for the Hipparcos Cepheids (Figure 9.28). They explained the good fits of their model to the Cepheid radial velocity and proper motion data, at the same time conflicting values of Θ_0 and R_0 estimated by Feast & Whitelock (1997), as due to a wide range of Galactic and Oort constants being compatible with the primary observable, the rotation curve. On the possible effects of non-circular motions, they also conclude that if streaming motions are present to a significant degree, the stellar orbits will become more complex, and so the Oort constraints derived from simple stellar kinematic measurements will also be compromised. Finally, while most studies assume that the rotation curve is rather flat at R_0, there remains no clear consensus on whether it is slightly rising or falling in the immediate solar vicinity, with perhaps evidence for a marginal rise of around $+2 \pm 1\,\mathrm{km\,s^{-1}\,kpc^{-1}}$

Table 9.3 Determination of the Oort constants of Galactic rotation, A and B, and associated quantities. Units of A, B, A − B, A + B are kms^{-1} kpc^{-1}. Derivations based on: (1) μ and V_R; (2) adoption of conventional values; (3) compilation of methods; (4) proper motions in the outer Galaxy out to ∼ 1 kpc; (5) mass models including interstellar gas; (6–7) generalised 2d formulation using proper motions from Tycho 2/ACT (corrected for 'mode mixing') for young main-sequence stars extrapolated to zero asymmetric drift, and red giants respectively; (8–9) generalised 3d formulation for A0–A5 dwarfs and more distant giants respectively (in both cases, A has been reconstructed from the quoted values of $\overline{A}$ and ϕ); (10) generalised 3d formulation for ∼100 000 stars including some radial velocities. A value of R_0 and the resulting value of Θ_0 (from A − B) are given in the case of Feast & Whitelock (1997). For the other Hipparcos determinations, no value of R_0 was estimated. Estimates of Θ_0 can be derived for any assumed value of R_0. For the Hipparcos-based derivations, the uncertainty on the inertiality of the Hipparcos proper motion system, ±0.25 mas yr^{-1} in the rate of rotation, corresponds to an explicit additional uncertainty on the Oort constant B of 1.2 kms^{-1} kpc^{-1}.

Reference	A	$-B$	$A-B=$ Θ_0/R_0	$-(A+B)=$ $(d\Theta/dR)_{R_0}$	R_0 (kpc)	Θ_0 (km s^{-1})
Pre-Hipparcos:						
Oort (1927a)[1]	19	24	43	+5		
IAU (1964) standard[2]	15	10	25	−5	10	250
Kerr & Lynden-Bell (1986)[3]	14.4 ± 1.2	12.0 ± 2.8	26.4 ± 1.6	−2.5 ± 3.1	8.5 ± 1.1	220 ± 20
Hanson (1987)[4]	11.3 ± 1.1	13.9 ± 0.9	25.2 ± 1.9	+2.6 ± 1.4		
Olling & Merrifield (1998)[5]	11.3 ± 1.1	13.9 ± 0.9	25.2 ± 1.9	+2.6 ± 1.4	7.1 ± 0.4	184 ± 8
Hipparcos–Oort–Lindblad:						
Feast & Whitelock (1997) Cepheids	14.8 ± 0.8	12.4 ± 0.6	27.2 ± 1.0	−2.4 ± 1.0	8.5 ± 0.5	231 ± 15
Olling & Dehnen (2003) dwarfs[6]	9.6 ± 0.5	11.6 ± 0.5	21.1 ± 0.5	+2.0 ± 0.5		
Olling & Dehnen (2003) giants[7]	15.9 ± 1.2	16.9 ± 1.2	32.8 ± 1.2	+1.0 ± 1.2		
Liu & Ma (1999) O–B[14]	17.6 ± 0.2	14.6 ± 0.2	32.2 ± 0.3	−3.0 ± 0.3		
Hipparcos – Ogorodnikov–Milne:						
Miyamoto & Zhu (1998) Cepheids	16.5 ± 1.1	12.1 ± 0.9	28.6 ± 1.4	−4.4 ± 1.4		
Miyamoto & Zhu (1998) O–B5	16.1 ± 1.1	15.5 ± 0.9	31.6 ± 1.4	−0.6 ± 1.4		
Mignard (2000) dwarfs[8]	10.9 ± 0.8	13.3 ± 0.6	24.2 ± 1.1	+2.4 ± 1.1		
Mignard (2000) giants[9]	13.0 ± 1.0	11.4 ± 1.0	24.4 ± 1.4	−1.6 ± 1.4		
Branham (2000) all[10]	10.8 ± 0.5	11.0 ± 0.5	21.8 ± 0.7	+0.2 ± 0.7		
Branham (2002) O–B	14.9 ± 0.8	15.4 ± 0.7	30.3 ± 1.1	+0.5 ± 1.1		
Branham (2006) O–B	16.1 ± 0.7	10.7 ± 0.6	26.8 ± 1.0	−5.3 ± 1.0		
Hipparcos – vectorial harmonics:						
Vityazev & Shuksto (2004) 113 646 stars	13.5 ± 2.0	12.6 ± 1.6	26.1 ± 2.6	−0.9 ± 2.6		
Makarov & Murphy (2007) non-binary	13.8 ± 1.4	13.4 ± 1.2	27.1 ± 1.8	−0.4 ± 1.8		
Hipparcos – spiral-density waves:						
Lépine *et al.* (2001) Cepheids	17.5 ± 0.8	8.8 ± 1.5	26.3 ± 1.7	−8.7 ± 1.7		

from the studies of both Olling & Merrifield (1998) and Mignard (2000).

Olling & Dehnen (2003) argued that a hitherto overlooked source of systematic error in the determination of the Oort constants is the longitudinal variations of the mean stellar parallax. Caused by intrinsic density inhomogeneities and interstellar extinction, this leads to contributions to mean longitudinal proper motions $\mu_l(l)$ indistinguishable from the Oort constants at $\lesssim 20\%$ of their amplitude, an effect they refer to as 'mode mixing'. This implies a corruption of the Oort constants due to the harmonics of the underlying star distribution. They corrected for this using the latitudinal proper motions $\mu_b(l)$ of some 10^6 stars from the ACT/Tycho 2 Catalogues, brighter than 11 mag and with median $\sigma_\mu \sim$ 3 mas yr^{-1}. They found colour-dependent variations of A and B, correlating linearly with the asymmetric drift, thus deviating from values expected for an axisymmetric potential. Selecting red giants, with $B-V \gtrsim 1.2$, old enough to be in equilibrium but distant enough to be unaffected by local anomalies, and correcting for inhomogeneities and asymmetric drift, they found (in km s^{-1} kpc^{-1}) $A = 15.9 \pm 1.2$ and $B = -16.9 \pm 1.2$ ($A - B = 32.8$) (Table 9.3), with a significant value of $C = 9.8 \pm 1.2$ indicating non-axisymmetry. The

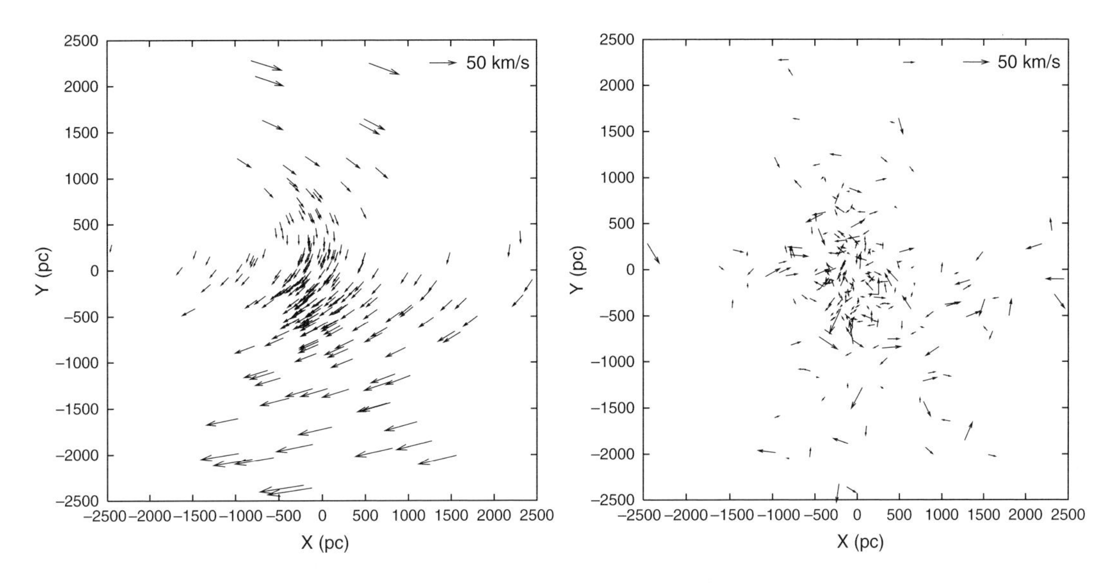

Figure 9.7 Velocity field in the Galactic plane for the 243 O–B5 stars studied by Uemura et al. (2000) showing the systematic velocity field modelled by their solution A (left) and the residual vectors (right). From Uemura et al. (2000, Figures 4 and 5).

corresponding values for young main-sequence stars, extrapolated to zero asymmetric drift, are $A = 9.6 \pm 0.5$ and $B = -11.6 \pm 0.5$ ($A - B = 21.1$), with $C = 0.4 \pm 0.5$.

Famaey *et al.* (2005b,a) presented a kinematic analysis of 5311 K and 719 M giants in the solar neighbourhood, including radial velocity data from a large Coravel survey, and proper motions from the Tycho 2 Catalogue. The UV-plane shows the rich small-scale structure discussed in Chapter 6, with several clumps corresponding to the Hercules stream, the Sirius moving group, and the Hyades and Pleiades superclusters, which they attribute to dynamical effects of the bar (in the case of the Hercules stream) or otherwise to perturbations by transient spiral waves (as recently modelled by De Simone *et al.*, 2004). This structure again raises the question of how to derive the solar motion in the presence of dynamical perturbations altering the kinematics of the solar neighbourhood: whether there exists in the solar neighbourhood a subset of stars having no net radial motion which can be used as a reference against which to measure the solar motion.

Numerous other studies related to understanding the characteristics of Galactic rotation, based on the Oort–Lindblad model, have been carried out using the Hipparcos data. These include the study of 1011 OB stars by Liu & Ma (1999), of Cepheids and young open clusters by Zabolotskikh *et al.* (2002), for distances between 10 and 100 pc according to colour index by Cepeda (2006), determination of the mean radial velocity in the solar neighbourhood by Cubarsi & Alcobé (2006), for 22 000 stars divided into 72 equal area regions by Teixeira & de Souza (2006) and de Souza & Teixeira (2007), for OB stars in the solar neighbourhood by Elias *et al.* (2006), for studies of the rotational vector by Tsvetkov (2006), and for 1400 KM dwarfs by Upgren *et al.* (2006).

As for the results on solar motion, the heterogeneity of results for the Oort constants is very apparent. Any reliable analysis must properly account for kinematic bias, for the fact that the very early-type (OB) stars are too young to have settled into an equilibrium distribution, and should correct for the effects of 'mode mixing'. The analysis by Olling & Dehnen (2003) considered all of these effects, and is suggested as a reference.

9.2.7 Stellar kinematics in the Ogorodnikov–Milne model

The usual derivation of the Oort constants is based on the assumption of a strictly axisymmetric potential. However, complications in deriving and interpreting A and B occur in a non-axisymmetric potential, specifically in the presence of spiral density waves and the central bar, in which case the Oort constants will vary with azimuth, and the numerical values obtained will depend on the distance of the tracers adopted. Several analyses have therefore discussed the Hipparcos space velocities in terms of a more general expression for the velocity field in the vicinity of the Sun, assuming only that it can be represented by a continuous smooth flow; for comparison, the more conventional Oort constants can then be derived from these solutions

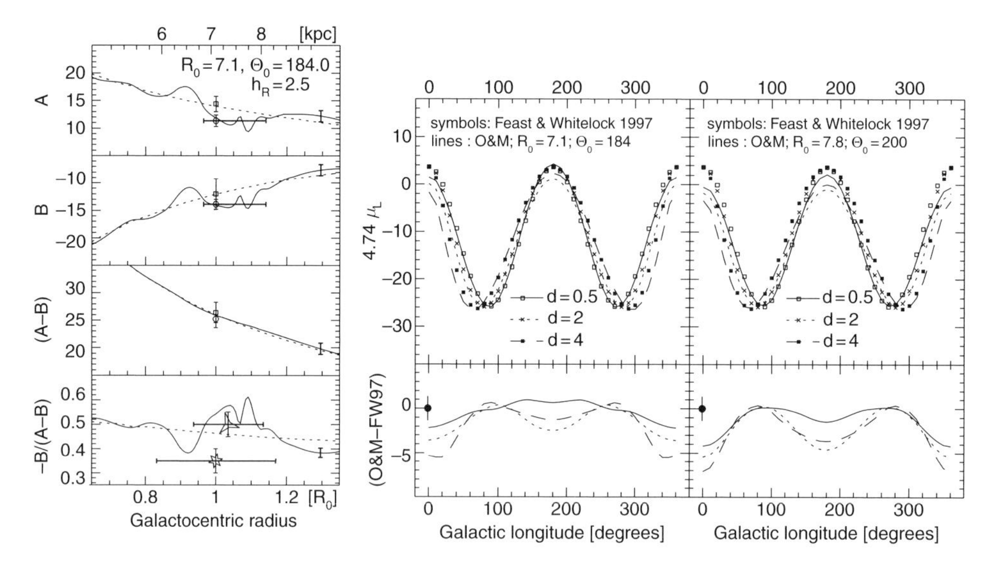

Figure 9.8 From the study of Olling & Merrifield (1998), parts of their Figures 3 (left) and 5 (right). Left: the Oort functions, $A(R)$ and $B(R)$, with $(A-B)$ and $-B/(A-B)$ ($\mathrm{km\,s^{-1}\,kpc^{-1}}$). Solid lines show the Oort functions for the best-fitting full mass model, while ignoring the gas component results in the dashed lines. Error bars at the right of each panel shows the uncertainty due to other model parameters, including stellar mass-to-light ratio and disk scale length $h(R)$. Observational estimates are shown, with the horizontal error bars indicating the relevant radial range. The upper panel shows values from Kerr & Lynden-Bell (1986) (open squares) and Hanson (1987) (open circles). Bottom panel shows $-B/(A-B)$ predicted by the velocity dispersions of M dwarfs and G–M giants (triangles) and G–K dwarfs (hexagons). Right: proper motions (top row) as calculated using models, and with the Oort and Galactic constants derived by Feast & Whitelock (1997) (using $R_0 = 8.5$ kpc). Lower panel shows the difference between the proper motions predicted by the O&M models and those predicted by Feast & Whitelock (1997). Units for vertical axes are $\mathrm{km\,s^{-1}\,kpc^{-1}}$. Leftmost panel shows the model which best fits the Oort constants constraints ($R_0 = 7.1$ kpc, $\Theta_0 = 184\,\mathrm{km\,s^{-1}}$); the other panels were calculated for the Galactic constants from Sackett (1997). d are the various adopted distance limits (kpc).

under appropriate assumptions on the form of the velocities. This more general analysis was formulated as a first-order Taylor-series expansion by Ogorodnikov (1932) and Milne (1935), and applied later in similar studies in particular by Clube (1972, 1973), Du Mont (1977) and Miyamoto & Soma (1993). Clube (1973), for example, showed that the local proper motions are not adequately described by the Oort–Lindblad differential rotation model, finding a significant expansion of the solar neighbourhood, away from the Galactic centre, of $75\pm27\,\mathrm{km\,s^{-1}}$, which he attributed qualitatively to spiral density waves and, within 300 pc or so, to the burst of recent star formation which gave rise to the Gould Belt and to a significant systematic motion in addition to the normal Galactic rotation.

Ogorodnikov (1965) considered this approximation to be valid out to distances of around 1 kpc, rather well matched to the Hipparcos trigonometric parallax accuracies of around 1 mas. Extension of the model to second-order was presented by Edmondson (1937).[1]

[1] Some didactic considerations appear in Milne (1935): *'It will be shown that any spatio-velocity stellar distribution-function whatever which possesses certain differential coefficients is accompanied by, in general, a K-term [an expansion] and a double wave in radial velocities, and a systematic drift and an associated double wave in the transverse motions, in any arbitrary plane. The result is capable of immediate specialisation to the particular cases of galactic rotation and galactic expansion. Instead, however, of proceeding by modifying the analyses of Oort and his successors as they stand, I here apply to stellar motions the method first used by Stokes, in a classical memoir of 1845, for discussing differential displacements in a strained elastic solid and differential motions in a (possibly viscous and compressible) fluid. Stokes showed that any displacement of an elastic solid, measured relative to a given particle of the solid in the immediate neighbourhood of that particle, could be resolved into a dilatation (positive or negative), a shear and a rotation; and that any motion of a fluid . . . could be resolved into a rate of dilatation, a rate of shear and a spin. Oort's analysis is in effect Stokes's method applied to the particular case of galactic rotation. It thus appears that the existence of a double wave in radial velocities is by itself no evidence for galactic rotation, though this is not to deny the reality of galactic rotation on other evidence. A by-product of the investigation is that . . . in order that the results may have simple physical interpretations, it is necessary to use the solar motion as given by the stars immediately surrounding the Sun, and not that derived from a reduction of the motions of the stars under consideration'.*

Generalising Equation 9.6a gives

$$\mathbf{V} = -\mathbf{v}_\odot + \mathbf{D}\cdot\mathbf{x} + \mathcal{O}(\mathbf{x}^2) \tag{9.14}$$

$$\mathbf{D} = \begin{pmatrix} \partial v_x/\partial x & \partial v_x/\partial y & \partial v_x/\partial z \\ \partial v_y/\partial x & \partial v_y/\partial y & \partial v_y/\partial z \\ \partial v_z/\partial x & \partial v_z/\partial y & \partial v_z/\partial z \end{pmatrix}_{\mathbf{x}=0} \equiv \begin{pmatrix} u_x & u_y & u_z \\ v_x & v_y & v_z \\ w_x & w_y & w_z \end{pmatrix} \tag{9.15}$$

which involves 12 unknowns: the three components of the solar reflex motion, and the nine components of the displacement tensor, **D**. The latter can itself be decomposed into the sum of a symmetric (strain) tensor **S** and an anti-symmetric (rotation) tensor Ω

$$\mathbf{S} = \begin{pmatrix} S_{11} & S_{12} & S_{13} \\ S_{21} & S_{22} & S_{23} \\ S_{31} & S_{32} & S_{33} \end{pmatrix} = \begin{pmatrix} +u_x & +\frac{1}{2}(u_y+v_x) & +\frac{1}{2}(u_z+w_x) \\ +\frac{1}{2}(u_y+v_x) & +v_y & +\frac{1}{2}(w_y+v_z) \\ +\frac{1}{2}(u_z+w_x) & +\frac{1}{2}(w_y+v_z) & +w_z \end{pmatrix} \tag{9.16}$$

$$\Omega = \begin{pmatrix} 0 & -\omega_z & +\omega_y \\ +\omega_z & 0 & -\omega_x \\ -\omega_y & +\omega_x & 0 \end{pmatrix} = \begin{pmatrix} 0 & +\frac{1}{2}(u_y-v_x) & +\frac{1}{2}(u_z-w_x) \\ -\frac{1}{2}(u_y-v_x) & 0 & +\frac{1}{2}(v_z-w_y) \\ -\frac{1}{2}(u_z-w_x) & -\frac{1}{2}(v_z-w_y) & 0 \end{pmatrix} \tag{9.17}$$

Mignard (2000) used a Hipparcos sample of about 20 000 single stars lying between about 0.1–2 kpc to study simultaneously, and for several categories of stars ranging from early-type dwarfs to K and M giants, three specific components of motion: the relative systematic velocity field due to Galactic rotation, the motion of the Sun relative to the LSR, and the associated velocity dispersion for each category. The proper motion in Galactic longitude and latitude is then modelled as (Mignard, 2000, Equation 33–34)

$$\mu_l \cos b = \Phi(S_{kl}, \omega_k, u_\odot, v_\odot, w_\odot) + \epsilon_l \tag{9.18a}$$

$$\mu_b = \Psi(S_{kl}, \omega_k, u_\odot, v_\odot, w_\odot) + \epsilon_b \tag{9.18b}$$

in which S_{kl} and ω_k represent the symmetric and anti-symmetric tensor components, $\mathbf{v}_\odot = (u_\odot, v_\odot, w_\odot)$ the components of the solar motion with respect to the LSR, and the random terms ϵ_l and ϵ_b include the random errors of the observed proper motion components as well as the contribution of the peculiar velocity of each star. The Hipparcos positions and proper motion components in equatorial coordinates were transformed into Galactic components using the full covariance matrix given in the catalogue for each star, star selection made so as to avoid any *a priori* statistical bias resulting from the input catalogue definition, and observations suitably-weighted. While the least-squares fit was performed on the S_{kl}, the following nonlinear condition equations then describe the systematic velocity field in the solar neighbourhood (Mignard, 2000, Equation 27–29)

$$\mu_l \cos b = +\overline{A}\cos b \cos(2l - 2\phi) + A' \sin b \cos(l - \psi) + B\cos b - B' \sin b \cos(l - \chi) \tag{9.19a}$$

$$\mu_b = -\frac{\overline{A}}{2}\sin 2b \sin(2l - 2\phi) + A' \cos 2b \sin(l - \psi) - \frac{K}{2}\sin 2b + B' \sin(l - \chi) \tag{9.19b}$$

$$v_R = +\overline{A}\cos^2 b \sin(2l - 2\phi) + A' \sin 2b \sin(l - \psi) + K\cos^2 b + Z \tag{9.19c}$$

where $A = \overline{A}\cos 2\phi$ and $C = \overline{A}\sin 2\phi$. The equation for the radial velocity was not used in view of the absence of homogeneous compilations of radial velocity observations. It is these eight 'generalised Oort constants', amplitudes $\overline{A}, B, K, A', B'$, and phases ϕ, ψ, χ, which were determined as shown in Table 9.4 (one further constant, Z, is only accessible through the radial velocities). Again, in terms of physical interpretation, the terms describe a local divergence (K), vorticity (B), and azimuthal (A) and radial (C) shear of the velocity field. For motion in a plane, $b = \mu_b = 0$ and the expressions for μ_l and v_R reduce to those given in Equation 9.8a. For strictly circular motion, $\phi = 0$, in which case $A = \overline{A}$ and $C = 0$.

The straight mean of $\overline{A}$ over all spectral types yields 14.8 ± 1.0 km s^{-1} kpc^{-1} (the mean of $A = \overline{A}\cos 2\phi$ yields 13.8 km s^{-1} kpc^{-1}), with the lower value of 11.5 km s^{-1} kpc^{-1} for the more distant A0–A5 stars perhaps being significant. The review by Kerr & Lynden-Bell (1986) gave an average 14.4 ± 0.7 km s^{-1} kpc^{-1}, but a range can be found in the literature (all in units of km s^{-1} kpc^{-1}): 11.3 ± 1.06 from faint stars in the Lick Northern Proper Motion programme (Hanson, 1987); 16 ± 1.5 from radial velocities of 272 supergiants out to 3 kpc (Dubath *et al.*, 1988); $12{-}13 \pm 0.6$ from 30 000 K–M giants (Miyamoto & Soma, 1993); $10{-}11.3 \pm 1$ from O and B stars in the Hipparcos Input Catalogue (Comerón *et al.*, 1994), etc. Interestingly, the Feast & Whitelock (1997) value of 14.8 ± 0.8 is similar to the result for the giants but, because the Cepheids are faint, only a small part of the two samples overlap.

For the Oort constant B (again, all in units of km s^{-1} kpc^{-1}), Mignard (2000) gives -13.2 ± 0.5 for early-type dwarfs and -11.5 ± 1.0 for distant giants, consistent with the Kerr & Lynden-Bell (1986) average value

Table 9.4 Determination of the generalised Oort constants, from Mignard (2000, Table 4). N_* are the numbers of stars in the selected spectral type interval: A0–F5 are dwarfs with $B-V=0.00, 0.15, 0.30, 0.45$; K0–M5 are giants with $B-V=0.85, 1.15, 1.40, 1.60$. Amplitudes are in km s^{-1} kpc^{-1}, and angles (ϕ, ψ, χ) are in degrees. $\sigma_v/\sigma_u = [B/B-A]^{1/2}$ is discussed in Section 9.1.3.

Class	N_*	$\overline{A}$	ϕ	B	K	A'	ψ	B'	χ	σ_v/σ_u
A0–A5	3936	11.5 ± 0.8	9.6 ± 2.1	-13.3 ± 0.6	-0.9 ± 2.4	5.3 ± 1.2	106 ± 13	3.1 ± 1.2	49 ± 23	0.733
A5–F0	3163	14.3 ± 1.3	0.2 ± 2.8	-13.2 ± 1.0	2.7 ± 3.3	5.5 ± 1.6	-43 ± 17	5.7 ± 1.6	-124 ± 16	0.693
F0–F5	3098	18.9 ± 2.0	4.9 ± 3.0	-13.0 ± 1.5	-1.6 ± 4.3	9.6 ± 2.0	-50 ± 12	9.6 ± 2.0	-119 ± 12	0.638
K0–K5	6246	13.4 ± 1.9	9.2 ± 4.0	-10.9 ± 1.5	-0.2 ± 3.0	2.0 ± 1.7	112 ± 47	2.9 ± 1.7	40 ± 34	0.669
K5–M0	3459	16.4 ± 2.1	13.9 ± 3.6	-12.4 ± 1.7	-3.1 ± 3.2	4.1 ± 1.9	-24 ± 28	1.3 ± 2.0	-23 ± 87	0.656
M0–M5	2492	14.1 ± 2.0	16.4 ± 4.0	-10.9 ± 1.6	-4.1 ± 3.1	1.8 ± 1.9	102 ± 59	1.5 ± 1.9	-26 ± 74	0.660

of -12.0 ± 0.7. Hanson (1987) gave -13.91 ± 0.92. Feast & Whitelock (1997) gave -12.3 ± 0.6 from the Cepheids.

Averaging over dwarfs and giants gives $\Omega_0 = A - B = 25.1\pm0.8$ km s^{-1} kpc^{-1}, close to the Feast & Whitelock (1997) value of 27.2 ± 0.9 km s^{-1} kpc^{-1} from the Hipparcos Cepheids, and yielding, with their $R_0 = 8.5\pm0.2$ kpc the rotation speed $\Theta_0 = 213\pm7$ km s^{-1}.

For the evolved K–M giants, these three classes yield a very uniform result of $\sigma_v/\sigma_u = 0.662\pm0.004$, close to the estimate of 0.63 derived by Kerr & Lynden-Bell (1986, see also Section 9.1.3).

Branham (2000, 2002, 2006) developed an analysis of the Ogorodnikov–Milne model in terms of total least-squares, in contrast to least-squares in linear regression where all errors are attributed to the dependent variables. Branham (2000) retained 98 269 Hipparcos stars with proper motions, after excluding double stars and stars believed to be members of the Gould Belt, and complemented these with 8613 published radial velocities. His condition equations corresponding to Equation 9.19a are given in Branham (2000, Equations 10–12). In addition to providing the solution for the 12 unknowns (including the solar motion and rotation terms), he also provided a restricted solution assuming motion only in the Galactic plane, and also assuming that $z=0$ is a plane of symmetry, in which case expressions for the Oort constants A and B plus a K-term follow from (Ogorodnikov, 1965)

$$A' = (+u_y + v_x)/2 \tag{9.20}$$

$$C' = (+u_x - v_y)/2 \tag{9.21}$$

$$B = (-u_y + v_x)/2 \tag{9.22}$$

$$K = (+u_x + v_y)/2 \tag{9.23}$$

and defining $A' = A\cos 2l_1$, $C' = -A\sin 2l_1$, yields the Oort constant, A, and the direction defining the longitude of the Galactic centre, l_1

$$A = \sqrt{A'^2 + C'^2} \tag{9.24}$$

$$l_1 = \frac{1}{2}\arctan\left(-\frac{C'}{A'}\right) \tag{9.25}$$

Derived values and mean errors, including the total solar velocity, $V_\odot$, are (Branham, 2000, Table 4)

$$\begin{aligned}
V_\odot &= 22.850\pm0.055 \text{ km s}^{-1}\\
A &= 10.788\pm0.495 \text{ km s}^{-1}\text{ kpc}^{-1}\\
B &= -10.973\pm0.488 \text{ km s}^{-1}\text{ kpc}^{-1}\\
l_1 &= -3.571\pm0.805 \text{ degrees}\\
K &= 6.351\pm0.809 \text{ km s}^{-1}\text{ kpc}^{-1}\\
\omega_x &= -0.118\pm0.104 \text{ mas yr}^{-1}\\
\omega_y &= -1.124\pm0.141 \text{ mas yr}^{-1}\\
\omega_z &= -2.314\pm0.092 \text{ mas yr}^{-1}
\end{aligned} \tag{9.26}$$

The components of the solar motion are included in Table 9.2, where it is evident that reasonable consistency with other determinations is found, and notably with the results of Mignard (2000). The direction to the Galactic centre, l_1, is close to the accepted value. The near equality of A and B implies a rather flat rotation curve at the solar circle. Together they imply $\Theta_0 = 185$ km s^{-1}for $R_0 = 8.5$ kpc, or $\Theta_0 = 155$ km s^{-1} for $R_0 = 7.1$ kpc. The K-component is significant. The component w_y appears significant, and exceeds the value of 0.25 mas yr^{-1} corresponding to the expected inertiality of the Hipparcos reference frame, but was not confirmed in the work of Branham (2006). Presumably it reflects the complex velocity structure of the solar neighbourhood in the same way as the Oort constant B greatly exceeds the inertial reference frame spin component of the Hipparcos Catalogue as a whole.

Branham (2002) applied the same method to 1817 O–B5 stars of which 1655 stars out to 1 kpc are from Hipparcos, and the remainder between 1 and 3 kpc are from Westin (1985). He used the second-order expansion of Edmondson (1937), restricted the analysis to Hipparcos O–B5 giants, and excluded those OB stars representing the Gould Belt. Branham (2006) applied the method to 290 non-Gould Belt O–B5 giant stars, also deriving the velocity dispersion and vertex deviation for the sample. Resulting rotation terms in mas yr^{-1}, which may be compared with those in Equation 9.26, are

$$\omega_{x,y,z} = -1.16 \pm 0.29, -1.08 \pm 0.40, -3.24 \pm 0.13 \quad \text{Branham (2002)} \quad (9.27)$$

$$= -0.01 \pm 0.17, +1.23 \pm 0.18, -2.27 \pm 0.13 \quad \text{Branham (2006)} \quad (9.28)$$

The resulting solar motion and, as constrained above, values of the Oort constants, are given in Tables 9.2 and 9.3 respectively.

Based on some further investigations of these results (Branham, 2007, priv. comm), the following complex picture emerges: (1) using 6705 Hipparcos O–B9 stars, with Gould Belt stars eliminated, 8% having radial velocities, no evidence is found for an asymmetry in the LSR for northern hemisphere (actually Galactic longitude 0–180°) versus southern hemisphere ($l = 180 - 360°$). In both instances, the solar velocity is ~18 km s^{-1}; (2) using 6686 O–A0 stars, the majority from Hipparcos, with some 3% of stars in the 1–3 kpc range from Westin (1985), again with Gould Belt stars eliminated, and of which about 18% have radial velocities, the results show a significant difference in the LSR for an 'interior solution' ($R < R_0$), and for an 'exterior solution' ($R > R_0$): the solar velocity changes from 27.7 km s^{-1} for the inner to 18.8 km s^{-1} for the outer solution; (3) dividing the O–A0 stars into four quadrants and three distance groups from the Sun (~200 pc, ~350 pc, and > 600 pc), shows a significantly higher solar velocity, of about 40 km s^{-1}, for the fourth quadrant ($l = 270$–$360°$) for the near and intermediate groups, and for the second quadrant ($l = 90$–$180°$) for the intermediate group, compared to about 20 km s^{-1} or lower for all of the other groups. This may indicate a bulk motion of the nearer stars towards the second and fourth quadrants.

Miyamoto & Zhu (1998) analysed Hipparcos proper motions in the same framework for two classes of stars: for 1352 O-B5 non-Gould Belt stars, and for 170 Hipparcos classical Cepheids. These are expected to be components of the same disk population, and should share the same kinematics. Their young age, and large mass, should result in the lowest velocity dispersion of the disk population of around 15 km s^{-1}. and should therefore best represent circular Galactic motion according to the epicycle approximation. On the other hand, they may have additional components of radial and azimuthal motion due to their recent formation in spiral arms, and the youngest stellar population appears to show the characteristics of the Galactic warp (Section 9.8). However, results for the two groups are very different. For the O-B5 stars the Ogorodnikov–Milne formulation reveals a clear stellar warping motion amounting to a rotation of 3.8 ± 1.1 km s^{-1} kpc^{-1} about an axis pointing to the Galactic centre (Section 9.8), and a large Galactic rotation of $\Theta_0 = 268.7 \pm 11.9$ km s^{-1} for $R_0 = 8.5$ kpc. In contrast, the Cepheids display no evidence for rotation or shear other than that described by the Oort differential rotation, and yield $\Theta_0 = 243.3 \pm 12.0$ km s^{-1} for $R_0 = 8.5$ kpc. Associated solar motion components are shown in Table 9.2. Branham (2000) has made the plausible suggestion that the anomolously large Galactic rotation for the O–B5 stars may arise from the use of only a first-order Taylor expansion out to distances of ~3 kpc.

Uemura *et al.* (2000) adopted a similar 3d formulation, using Hipparcos parallaxes for about 240 O–B5 stars. They derive a Galactic rotation in the solar vicinity of $\Theta_0 = 255.52 \pm 8.33$ km s^{-1} assuming $R_0 = 8.5$ kpc, and a local slope of the Galactic rotation curve of -1.98 ± 0.98 km s^{-1} kpc^{-1}. They found neither contraction nor expansion with respect to the LSR, and a compelling fit to the modelled systematic velocity field (Figure 9.7). Zhu (2000a) made a similar analysis using a larger sample of 1523 O–B5 stars, subtracting the Gould Belt contribution, and employing spectroscopic distances for those beyond the range of the Hipparcos parallax measurements. He also found a large value of $\Theta_0 = 269 \pm 12$ km s^{-1} assuming $R_0 = 8.5$ kpc, which he considered as more reliable than Cepheid-based determinations which may be contaminated by moving groups, and which would then imply a Galaxy mass interior to the solar circle increased by about 50%, to some $1.4 \times 10^{11} M_\odot$. He also identified a clear warping motion described as a systematic rotation $(\partial V_z/\partial\theta)/R = -3.79 \pm 1.05$ km s^{-1} kpc^{-1} around the axis pointing to the Galactic centre (Section 9.8).

Various other Hipparcos analyses based on the Ogorodnikov–Milne formulation have been reported: for Cepheids by Zhu (1999, 2000b), for OB stars by Zhu (2006), for dwarfs and giants by Drobitko & Vityazev (2003), for 70 000 dwarfs within 2 kpc using the Tycho 2 Catalogue data by Rybka (2004), and for 25 000 stars in the Orion spiral arm.

The range of results that emerges from these various analyses again suggests that a more critical evaluation of the underlying assumptions is required. As for the Oort–Lindblad model, any reliable analysis must properly account for kinematic bias, and the fact that the very early-type (OB) stars are presumably too young to have settled into an equilibrium distribution. Mode-mixing, discussed by Olling & Dehnen (2003) in the context of the Oort–Lindblad model, refers to the reflex of the Solar motion together with non-uniformity of the stellar distances producing features identical to those due to shear, divergence, and rotation of the local velocity field. The effect is presumably more problematic for any three-dimensional analysis and has not, as yet, been taken into account. In addition, the underlying assumption of the Ogorodnikov–Milne model, that of a smooth continuous flow of the stellar velocity field in the solar neighbourhood, may also be invalid.

9.2.8 Stellar kinematics and vector harmonics

Mignard & Morando (1990), subsequently developed by Vityazev & Shuksto (2004) and Makarov & Murphy (2007), introduced an interesting approach to describing the systematic field of tangential velocities derived from Hipparcos proper motions and parallaxes, based on vector spherical harmonics. The underlying idea is that for every spherical harmonic function on the sphere, S_n^m, one vector field can be constructed from ∇S_n^m, and another from $\mathbf{n} \times \nabla S_n^m$, where $\mathbf{n}$ is the unit normal. Just as any function on the sphere can be expanded in terms of S_n^m, any vector field on the sphere can be expanded in terms of these vector spherical harmonics, which then satisfy the same sort of orthogonality and completeness conditions.

While a new approach in this field, the methodology is an established mathematical technique (e.g. Hill, 1954; Arfken, 1975; Weisstein, 2007), and used in areas such as magnetostatics and electrodynamics (e.g. Jackson, 1975; Scanio, 1977; Barrera *et al.*, 1985; Carrascal *et al.*, 1991; Matute, 2006), and in general relativity, gravitational-wave, and cosmic microwave background studies, for example in analysing the cosmic microwave background polarisation induced by Thomson scattering and the stochastic gravitational wave background (e.g. Zaldarriaga, 1998; Pritchard & Kamionkowski, 2005). In these applications, the formulation is usually described in terms of 'spin-weighted spherical harmonics' (spin = 2), which are generalisations of the usual (spin = 0) spherical harmonics. For heritage reasons, the vector harmonics are then referred to as 'magnetic' and 'electric' components.

Following the notation of Makarov & Murphy (2007), the global proper motion field is represented by

$$\boldsymbol{\mu}(l, b) = \sum_{n=1}^{\infty} \sum_{m=-n}^{n} [h_n^m \mathbf{H}_n^m(l, b) + e_n^m \mathbf{E}_n^m(l, b)] \qquad (9.29)$$

where l, b are the Galactic longitudes and latitudes, and $\mathbf{H}_n^m$ and $\mathbf{E}_n^m$ are the orthogonal (magnetic and electric) vector harmonics. These vector harmonics are derived from the corresponding scalar spherical harmonics, S_n^m, over angular coordinates

$$\mathbf{H}_n^m(l, b) = \left[\frac{\partial S_n^m(l, b)}{\partial b} \boldsymbol{\tau}_l - \frac{1}{\cos b} \frac{\partial S_n^m(l, b)}{\partial l} \boldsymbol{\tau}_b \right]$$

$$\mathbf{E}_\mathbf{n}^\mathbf{m}(l, b) = \left[\frac{1}{\cos b} \frac{\partial S_n^m(l, b)}{\partial l} \boldsymbol{\tau}_l + \frac{\partial S_n^m(l, b)}{\partial b} \boldsymbol{\tau}_b \right] \qquad (9.30)$$

where $\boldsymbol{\tau}_l$ and $\boldsymbol{\tau}_b$ define the tangential coordinate directions toward increasing Galactic longitude and the north pole, respectively. They then constitute an orthogonal basis for the space of continuous vector functions on the unit sphere. When sampled over a large and sufficiently uniform set of points, the discretised functions remain nearly orthogonal, which makes the fitting algorithm both stable and accurate.

Mignard & Morando (1990) outlined the main properties of this basis, and illustrated the method applied to a preliminary solution of the Hipparcos sphere with 3500 stars.

Vityazev & Shuksto (2004, Equations 8–11) detailed the connection between the kinematic parameters of the Ogorodnikov–Milne model and the low-degree vector harmonic functions; consequently, the physical meaning and interpretation of these low-degree terms remain straightforward. The harmonics beyond those appearing in the Ogorodnikov–Milne model can then be used to identify more complex systematic motions not accommodated by that model. They used a total of 113 646 Hipparcos stars to estimate the classical contributions to the proper motions, deriving estimates both of the solar motion and of Galactic rotation, and identifying additional higher-order terms comparable to and sometimes even exceeding the contribution of previously-identified terms.

Makarov & Murphy (2007) developed this approach to analyse the local field of stellar tangential velocities for a sample of 42 339 non-binary Hipparcos stars, 50% within 112 pc and 75% within 160 pc. They also presented simple relations between the parameters of the classical linear (Ogorodnikov–Milne) model of the local systemic field and low-degree terms of the general vector harmonic decomposition. Accordingly, they determined the solar velocity with respect to a sample of local stars of $u_\odot, v_\odot, w_\odot = (10.5, 18.5, 7.3) \pm 0.1$ km s^{-1}, although not corrected for the asymmetric drift with respect to the LSR. With respect to stars beyond 100 pc, the peculiar solar motion is $(9.9, 15.6, 6.9) \pm 0.2$ km s^{-1}. The components $u_\odot$ and $w_\odot$ are very close to those determined by Dehnen & Binney (1998a), while $v_\odot$ differs by more than 10 km s^{-1}, presumably due to the residual asymmetric drift. The Oort constants determined by a least-squares adjustment in vector spherical harmonics, are $A = 14.0 \pm 1.4$, $B = -13.1 \pm 1.2$, $K = 1.1 \pm 1.8$, and $C = -2.9 \pm 1.4$ km s^{-1} kpc^{-1}.

Like Vityazev & Shuksto (2004), they found a few statistically significant higher-degree harmonic terms which do not correspond to any parameters in the classical linear model. One of them, a third-degree 'electric' harmonic, is tentatively explained as the response to a negative linear gradient of rotation velocity with distance from the Galactic plane, which they estimate at ~ -20 km s^{-1} kpc^{-1}. A similar vertical gradient of rotation velocity was detected for more distant stars representing the thick disk ($z > 1$ kpc), but possibly detected within the thin disk at $z < 200$ pc. The most unexpected and unexplained term within the Ogorodnikov–Milne model is the first-degree 'magnetic' harmonic representing a rigid rotation of the stellar field about the axis $-Y$

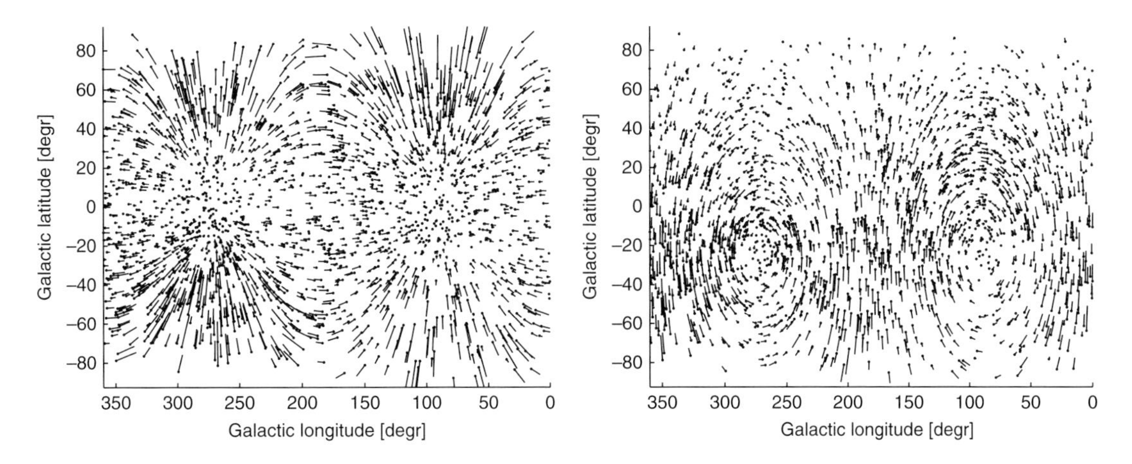

Figure 9.9 Left: the velocity field of Hipparcos stars generated by the vertical gradient of rotational velocity. Right: the velocity field of Hipparcos stars generated by the two unexpected 'magnetic' vector harmonics, $\boldsymbol{H}_1^{-1}$ and $\boldsymbol{H}_2^{-1}$. From Makarov & Murphy (2007, Figures 2 and 3).

pointing opposite to the direction of rotation. This harmonic is detected with a statistically robust coefficient $6.2\pm0.9\,\mathrm{km\,s^{-1}\,kpc^{-1}}$, and is also present in the velocity field of more distant stars. The ensuing upward vertical motion of stars in the general direction of the Galactic centre, and the downward motion in the anti-centre direction, are opposite to the vector field expected from the stationary Galactic warp model (Figure 9.9).

The approach is in its infancy, and appears to offer potential for further understanding the complex stellar velocity patterns in the solar neighbourhood. Whether these analyses suffer from kinematic bias, or the effects of 'mode mixing', remains to be demonstrated.

9.3 Census of nearby stars

The solar neighbourhood provides an important laboratory for models of the Galactic disk. Star counts provide an estimate of the luminous mass density of the Galactic disk near to the plane. The age distribution reflects the star-formation history of the disk. Chemical abundances as a function of age provide evidence of the chemical evolution and enrichment history of the disk. Space motions and Galactic orbits constrain the dynamical evolution of the Galaxy and the mixing of its stellar populations. Detailed knowledge of the nearby star population down to low-mass (and hence low-luminosity and faint-magnitude) limits is important for studies of the stellar luminosity function, for studies of the initial mass function and star-formation rate, and for categorising the different origins and evolutionary histories of the more 'exotic' populations which appear.

It remains a difficult task to establish a complete census of stars within the immediate solar neighbourhood, even out to distances of only 10–20 pc. Ground-based parallax surveys have been very successful in identifying nearby bright stars, but problems exist at the faint end of the luminosity function, $M_V \gtrsim 15$, where a complete parallax survey even out to only 10 pc remains impossible. Surveys searching for high-proper motion stars (e.g. Luyten, 1979) have been efficient at detecting nearby candidate stars which were then added to parallax programmes (including the Hipparcos Input Catalogue), but they implied a strong bias towards high-velocity halo objects. For this reason, the early nearby star compilations used spectroscopic and photometric distance estimates to identify additional nearby candidates. The advent of accurate all-sky multi-colour surveys has further facilitated the search for nearby, low-luminosity stars.

The *'Catalogue of Stars within Twenty-Five Parsecs of the Sun'* (Woolley, 1970) was one of the first attempts to compile a census of known stars in the solar neighbourhood, largely based on trigonometric parallaxes.

An evolving compilation has been maintained by the Astronomisches Rechen-Institut in Heidelberg over the last 50 years. Gliese (1957) published the *'Katalog der Sterne näher als 20 Parsek für 1950.0'*, containing 915 single stars and systems within 20 pc (1094 components altogether), with probable parallax errors of 9.2 mas. Gliese (1969) published the updated *'Catalogue of Nearby Stars'*, or CNS2, with a slightly enlarged distance limit of 22.5 pc ($\pi \geq 0.045$ arcsec). It contained 1049 stars or systems within 20 pc, and the probable errors were estimated as 7.6 mas. In both compilations trigonometric, photometric, and spectroscopic parallax estimates were employed.

The *'Third Catalogue of Nearby Stars'*, or CNS3, was only published in preliminary form, the formal reference being Gliese & Jahreiß (1991, where the accompanying file includes a short description of the content), but see also Jahreiß (1993) and Jahreiß & Gliese (1993). This extended the census to some 1700 stars nearer or apparently nearer than 25 pc (the trigonometric parallax limit is actually 0.0390 arcsec), and was based on the latest edition of the General Catalogue of Trigonometric Stellar Parallaxes. Information includes spectral types from a variety of sources, broadband *UBVRI* photometric data, photometric parallaxes, parallaxes based on luminosity and space-velocity components, spectral type–luminosity and colour–luminosity relations, positions, and proper motions. Contrary to the CNS2, trigonometric parallaxes and photometric or spectroscopic parallaxes were not combined. The resulting parallax is the trigonometric parallax if $\sigma_\pi/\pi < 0.14$, or the photometric or spectroscopic parallax if the trigonometric parallax was less accurate or not available.

The *'Fourth Catalogue of Nearby Stars'*, or CNS4, incorporates data from the Hipparcos Catalogue, and provided a major development in the comprehensive inventory of the solar neighbourhood up to a distance of 25 pc from the Sun. Although not yet published as of late 2007, Table 9.5 and Figure 9.10, from Jahreiß & Wielen (1997) and Jahreiß *et al.* (1998), illustrate the resulting impact that the Hipparcos measurements made on the knowledge of nearby stars. Although the number of stars within the chosen distance limit of 25 pc remained largely the same, some important details changed: the Hipparcos measurements identified 119 'new' nearby stars of which the closest was a high-proper motion star (from the NLTT survey) at 5.5 pc. More significantly, the results implied a considerable shift to larger typical distances, with associated implications for the local stellar mass density. CNS4 can be regarded as complete, at least in a statistical sense, for stars $M_V < 9$ mag, allowing more reliable studies of the stellar content of the solar neighbourhood. Makarov (1997) found just six stars in the CNS3 included in the Tycho Catalogue but not in the Hipparcos Catalogue, and a further 29 candidates for inclusion within the 25 pc distance limit, confirming that nearby stars in the Hipparcos Catalogue are complete to $M_V \lesssim 8$ mag. The binary star content of CNS4 is discussed by Jahreiß & Wielen (2000).

Northern Arizona University 'NStars Database' also maintains a compilation of all stellar systems within 25 pc, and at the end of 2006 listed 2029 systems comprising 2633 objects. Within the SIM preparatory science programme, they are also obtaining spectra, spectral types, and basic parameters ($T_{\rm eff}$, $\log g$, [M/H], and chromospheric activity) for some 3600 Hipparcos dwarf and giant stars earlier than spectral type M0 within 40 pc. Gray *et al.* (2003) reported results for the first 664 stars in the northern hemisphere, and Gray *et al.* (2006) gave results for 1676 stars in the southern hemisphere.

Table 9.5 The statistics of the nearby stars catalogues CNS1–CNS4, from Jahreiß & Wielen (1997, Table 1).

r (pc)	CNS1 1957	CNS2 1969	CNS3 1993	CNS4 1997	Stars pc^{-3}
0– 5	52	54	65	61	0.116
5–10	179	207	268	257	0.070
10–20	863	918	1593	1552	0.053
20–25			949	1126	0.035
0–20	1094	1179	1926	1870	0.056

Andronova (2000) has separately compiled a catalogue of about 5000 stellar systems considered to be within 25 pc, containing data from the Hipparcos, Tycho 2, Washington Double Star, MSC, digital sky surveys and other sources.

Georgia State University's 'Research Consortium on Nearby Stars' (RECONS, Henry *et al.*, 2006, and references) aims to discover and characterise 'missing' stars within 10 pc, via astrometric, photometric, and spectroscopic techniques. In mid-1998, their compilation listed 234 stars within 10 pc, compared with 182 Hipparcos entries out to 10 pc, and they estimated that there were still some 130 'missing' stars within this distance limit. At the end of 2006, they listed 100 stars within 6.7 pc, compared with 100 stars within 7.7 pc in Hipparcos.

Additional nearby candidates are emerging from the large-scale high proper motion surveys, themselves benefiting considerably from an astrometric re-calibration of multi-epoch Schmidt plates (Section 2.10.1). As described there, Lépine (2005) has established, in the northern hemisphere, a list of 539 new candidate stars within 25 pc of the Sun, including 63 estimated to be within only 15 pc. He estimates that some 18% of nuclear-burning stars within 25 pc of the Sun remain to be located.

9.4 Derived characteristics

9.4.1 Mass density in the solar neighbourhood

Background Various approaches have been taken to determining the total mass in the solar neighbourhood. Since visible matter (stars, gas, etc.) is presumably only a part of this total mass, estimates are also made of the total gravitating matter (including dark matter), expressed either in mass per unit volume, $M_\odot$ pc^{-3}, or as the column density out to a given z distance, expressed in $M_\odot$ pc^{-2}. For example, mass models of the Galaxy, constrained by observed star counts, can be used to infer the amount of matter in the disk (e.g. Bienaymé *et al.*, 1987).

Measurement of the dynamical effects of the local mass density, based on the vertical motions of stars in

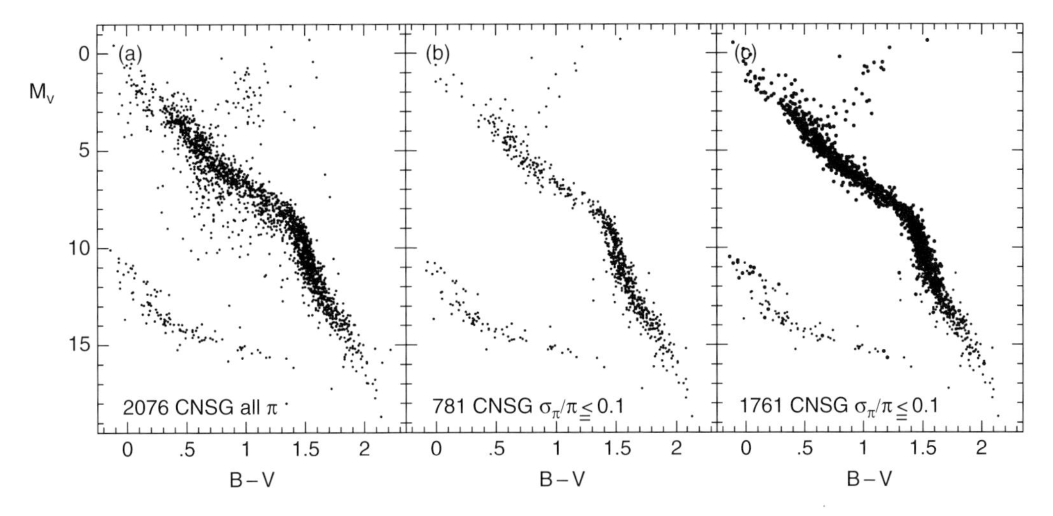

Figure 9.10 Colour–magnitude diagrams for nearby stars having $r \leq 25$ pc: (a) all CNSG stars (CNS-ground, i.e. CNS3) having trigonometric parallaxes. The large spread in the upper part shows clearly how many poor parallaxes were involved; (b) all CNSG stars with relative parallaxes better than 10%. This restriction provides a much cleaner picture, but leaves the upper part of the diagram sparsely populated; (c) all CNSH stars (CNS-Hipparcos, i.e. CNS4) with relative parallaxes better than 10%. Introducing the Hipparcos parallaxes left only 350 stars from (b), such that the upper part of (c) is exclusively occupied by the 1411 Hipparcos stars. From Jahreiß & Wielen (1997, Figure 3).

the disk, is potentially more robust. Analysis of the density and velocity distribution of a tracer sample of stars can provide estimates of the local volume density or the local column density of matter in the disk. This is often referred to as the K_z problem in Galactic dynamics, aiming to quantify the force law perpendicular to the Galactic plane. K_z is defined to be the component of the Galaxy's gravitational acceleration towards the Galactic plane in the solar neighbourhood. In a plane-stratified approximation out to a few kpc from the plane, K_z increases monotically with z. Studies use suitable tracer populations whose number density and velocities can be determined as a function of height, along with the Poisson and Boltzmann equations in various forms, i.e. in which matter produces the potential (as described by Poisson's equation), and is at the same time influenced by it (as described by Boltzmann's equation, or the Jeans' equation).

Under certain conditions, and if the vertical motion can be separated from the radial and azimuthal motions, this vertical force K_z can be described as a function of only the vertical velocity dispersion and the vertical density profile, which itself depends on the scale height h_z and the Sun's height above the Galactic plane Z_0. Near the plane of a highly-flattened system, for example, the vertical motions can be approximated by (Binney & Tremaine, 1987, Equation 4.38)

$$\frac{\partial}{\partial z}\left[\frac{1}{\nu}\frac{\partial(\nu\overline{v_z^2})}{\partial z}\right] = -4\pi G\rho \qquad (9.31)$$

This can be used to estimate the local mass density ρ (usually defined as the mass density on the Galactic plane in the neighbourhood of the Sun) if the number density ν and the mean-square vertical velocity $\overline{v_z^2}$ of any population of stars in the solar neighbourhood can be measured as a function of height from the plane. Estimates of ρ using several different stellar populations were used by Oort (1932, 1965) to conclude that, at the solar radius, $\rho_0 \equiv \rho(R_0, z=0) \simeq 0.15\,M_\odot\,\mathrm{pc}^{-3}$; ρ_0 is generally referred to as the Oort limit. Bahcall (1984a,b) derived $\rho_0 = 0.18 \pm 0.03\,M_\odot\,\mathrm{pc}^{-3}$ by considering the overall non-isothermal stellar disk as a linear combination of self-consistent isothermal distributions of different sub-populations.

The column density out to a specific distance can be determined from (Binney & Tremaine, 1987, Equation 4.39)

$$\Sigma(z) \equiv \int_{-z}^{+z} \rho(z')\mathrm{d}z' = -\frac{1}{2\pi G\nu}\frac{\partial(\nu\overline{v_z^2})}{\partial z} \qquad (9.32)$$

with Oort finding $\Sigma(700\,\mathrm{pc}) \simeq 90\,M_\odot\,\mathrm{pc}^{-2}$, and Bahcall (1984b) deriving $\Sigma(200\,\mathrm{pc}) \simeq 40\,M_\odot\,\mathrm{pc}^{-2}$ for F stars, and $\Sigma(700\,\mathrm{pc}) \simeq 75\,M_\odot\,\mathrm{pc}^{-2}$ for K giants. The more luminous K giants provide an effective tracer population visible to large distances.

Estimates of the amount of gravitating matter obtained in this way can then be compared to the total amount of visible disk matter observed in the solar neighbourhood, with corresponding limits then placed on any dark matter. The total density of visible matter

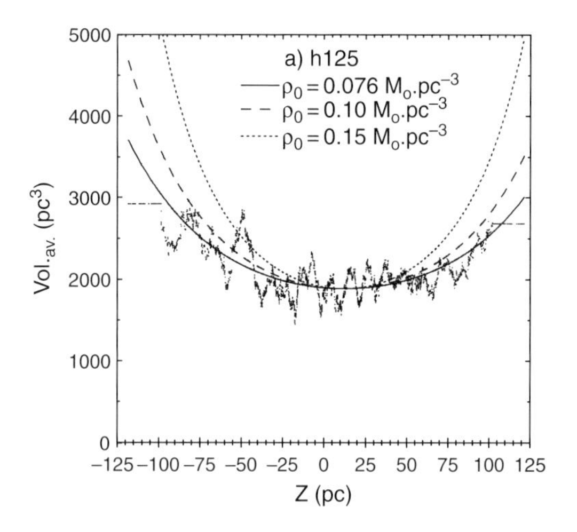

Figure 9.11 For the sample of Hipparcos A stars complete to 125 pc, the figure shows the inverse density profile calculated under three different assumed values of the local mass density; from bottom to top: 0.076 $M_\odot$ pc^{-3} (best estimate), 0.10 $M_\odot$ pc^{-3}, and 0.15 $M_\odot$ pc^{-3}. Moving averages are over 101 stars. From Crézé et al. (1998, Figure 12a).

near the Sun was summarised by Binney & Tremaine (1987, Equation 4.45) as

$$\begin{aligned}\rho_{\rm vis} &= \rho_{\rm stars} + \rho_{\rm stellar\ remnants} + \rho_{\rm ISM}\\ &\simeq 0.044 + 0.028 + 0.042\\ &= 0.114\, M_\odot\,{\rm pc}^{-3}\end{aligned} \tag{9.33}$$

where the contribution of stellar remnants is considered to be dominated by white dwarfs, and the interstellar medium (ISM) contribution includes H I and H_2 gas.

Discrepant though these dynamical and visible mass estimates were, a still larger surface density is required to maintain the circular rotation speed of the Galaxy at 220 km s^{-1} near $R_0 = 8.5$ kpc (Binney & Tremaine, 1987, Equation 4.44)

$$\Sigma_{\rm rot} = \frac{v_c^2}{2\pi G R_0} \simeq 210\, M_\odot\,{\rm pc}^{-2} \tag{9.34}$$

It thus appears that at least half the mass in a column near the Sun is in a component with a total scale height much greater than 700 pc, generally identified with the approximately spherical halo. For more details of the pre-Hipparcos situation see, for example, Kerr & Lynden-Bell (1986), Kuijken & Gilmore (1991), Kuijken (1995) and Stothers (1998).

Hipparcos contribution The Hipparcos data brought a number of observational improvements to this problem: primarily an improved accuracy on distances and space velocities, but also an improved size of the tracer sample, and an improved measurement of the local stellar luminosity function such that the amount of visible disk matter can be better estimated.

Crézé *et al.* (1998) used the Hipparcos data and a formulation for the local dynamical density (von Hoerner, 1960; Fuchs & Wielen, 1993) in which the density law $\nu(z)$, the potential $\Phi(z)$ and the velocity distribution at plane crossing $f(w_0)$ are connected with the local dynamical density ρ_0 as

$$\nu(\Phi) = 2\int_{\sqrt{2\Phi}}^{\infty} \frac{f(|w_0|)\, w_0\, {\rm d}w_0}{\sqrt{w_0^2 - 2\Phi}} \tag{9.35}$$

$$\rho_0 = \frac{1}{4\pi G} \left.\frac{{\rm d}^2\Phi}{{\rm d}z^2}\right|_{z=0} \tag{9.36}$$

Only rather general conditions of smoothness of f and Φ apply, although the method again assumes that the vertical motions of stars in the disk are decoupled from their motions in the plane (see Statler, 1989, for a discussion). Since estimates using this approach are dependent on the second derivative of the distribution of test stars close to the disk mid-plane, careful consideration of effects such as local extinction is required. Crézé *et al.* (1998) used A-type stars more luminous than $M_V = 2.5$ and within 125 pc of the Sun, stars bright enough to be seen at statistically meaningful distances, but (presumably) not so young as to be influenced by velocity inhomogeneities as a result of recent star formation. They demonstrated that any inhomogeneities in the phase-space distribution of this volume-limited and absolute magnitude-limited sample due to clumped stars (clusters and moving groups) is rather small (Chereul *et al.*, 1998). They then used the Hipparcos proper motions in Galactic latitude, combined with the distances, to derive the distribution of the vertical velocity components near the Galactic plane. Using a quadratic approximation to the potential, they derived a local dynamical density of $\rho_0 = 0.076 \pm 0.015\, M_\odot\,{\rm pc}^{-3}$ (Figure 9.11). Their revised assessment of the visible matter density was

$$\begin{aligned}\rho_{\rm vis} &= \rho_{\rm stars} + \rho_{\rm stellar\ remnants} + \rho_{\rm ISM}\\ &\simeq (0.030 \pm 0.003) + (0.015 \pm 0.005) + 0.04^{+0.04}_{-0.02}\\ &= 0.085\, M_\odot\,{\rm pc}^{-3}\end{aligned} \tag{9.37}$$

These dynamical and visible estimates being broadly compatible, Crézé *et al.* (1998) argued that a dark massive halo is still required to explain the rotation curve, that the halo must be largely spherical for its density to remain within the range permitted by the visible and dynamical density estimates, and that a (significant) contribution of dark matter in the form of a disk is precluded. A similar conclusion was reached by Bienaymé (1999) based on the local stellar velocity distribution, and the dependence of the vertical and horizontal velocity components formulated in the framework of Stäckel potentials.

Pham (1997) used a sample of about 10 000 Hipparcos F stars, determining their scale height and velocity dispersion assuming an isothermal distribution, and deriving a higher value of $\rho_0 = 0.11 \pm 0.01 \, M_\odot \, \mathrm{pc}^{-3}$.

Holmberg & Flynn (2000) used a slightly different implementation in which the velocity distribution at the mid-plane is integrated in a pre-established model of the local disk potential based on various subcomponents to yield its density fall-off in the vertical direction. This is then compared with the fall-off in density measured by Hipparcos. Their disk mass model includes the (improved) Hipparcos-based main-sequence stellar luminosity function from Holmberg *et al.* (1997), and revised contributions of M dwarfs, white dwarfs, brown dwarfs, and interstellar matter. It results in

$$\begin{aligned} \rho_{\rm vis} &= \rho_{\rm stars} + \rho_{\rm stellar\ remnants} + \rho_{\rm ISM} \\ &\simeq 0.038 + 0.006 + 0.050 \\ &= 0.094 \, M_\odot \, \mathrm{pc}^{-3} \end{aligned} \qquad (9.38)$$

and $\Sigma \simeq 48 \, M_\odot \, \mathrm{pc}^{-2}$. For the local dynamical mass density in the solar neighbourhood, they derived $0.103 \pm 0.006 \, M_\odot \, \mathrm{pc}^{-3}$ for 2026 A-stars within 200 pc, $0.094 \pm 0.017 \, M_\odot \, \mathrm{pc}^{-3}$ for 3080 F-stars within 100 pc, and $0.102 \pm 0.006 \, M_\odot \, \mathrm{pc}^{-3}$ for the combined sample. They were able to (partly) explain the lower value of Crézé *et al.* (1998) and the higher value of Pham (1997) by the forms of the adopted velocity distribution function and vertical acceleration respectively. They also conclude that there is no compelling evidence for significant amounts of dark matter in the disk.

Korchagin *et al.* (2003) estimated the mass surface density, using an approach similar to that used by Kuijken & Gilmore (1989a,b,c). The method is based on the assumption that the phase-space distribution function of any tracer population depends only on $E_z = \Phi(z) + \frac{1}{2} v_z^2$ where $-\mathrm{d}\Phi(z)/\mathrm{d}z = K_z(z)$. They used a sample of 1476 red giants inferred to be older than $\sim$3 Gyr (mainly first-ascent giants), about 93% complete within about ± 0.4 kpc from the Sun, with parallaxes and proper motions from Hipparcos, and radial velocities from Barbier-Brossat & Figon (2000). They determined the integral surface density of the disk based on the vertical velocity dispersion and the vertical scale height, $(1/\rho)(\mathrm{d}\rho/\mathrm{d}z)$, itself determined from the z distribution. They found a surface density of the disk of $10.5 \pm 0.5 \, M_\odot \, \mathrm{pc}^{-2}$ within ± 50 pc, and $42 \pm 6 \, M_\odot \, \mathrm{pc}^{-2}$ within ± 350 pc. The former, more closely approximating the local situation, leads to a volume density of gravitating matter of about $0.105 \pm 0.005 \, M_\odot \, \mathrm{pc}^{-3}$ under the assumption that the gravitating matter over this distance is distributed homogeneously with z. While this volume density is less robust than their estimate of the surface density, it is in reasonable agreement with the value obtained by Holmberg & Flynn (2000).

Soubiran *et al.* (2003) and Siebert *et al.* (2003) extended this kinematic analysis to measurements of the disk potential at larger distances from the plane, extending to 0.3–1 kpc, using a homogeneous and complete sample of nearly 400 red clump giants selected from the Tycho 2 Catalogue in the direction of the north Galactic pole. As well as extending the dynamical analysis to larger distances, analysis based on K giants or K dwarfs has the merit that these tracers are considerably older on average than the A and F stars. They are therefore well mixed dynamically, and thus better representing the overall Galactic potential. Absolute magnitudes and space velocities were derived from the Tycho 2 data combined with high-resolution spectroscopy. They determine $\Sigma(800\,\mathrm{pc}) \simeq 76 \, M_\odot \, \mathrm{pc}^{-2}$ or, removing the dark halo contribution of $\sim 0.01 \, M_\odot \, \mathrm{pc}^{-3}$, $\Sigma \simeq 67 \, M_\odot \, \mathrm{pc}^{-2}$. The local volume mass density is poorly constrained by this sample of tracers which, due to its large velocity dispersion, results in a rather uniform density distribution within, say, 125 pc, such that the changing vertical density distribution with height cannot easily be measured. They inferred a (dynamical) scale height of the total disk of 390^{+330}_{-120} pc.

Bienaymé *et al.* (2006) reported an extension of these studies, using 203 nearby (Hipparcos) and 523 (Tycho 2) red clump K giants, with high-resolution spectroscopy confirming classification and providing radial velocities. A two-parameter adjustment gave a dark halo contribution in the range $0\text{–}0.021 \, M_\odot \, \mathrm{pc}^{-3}$, and a total surface mass density $\Sigma(800\,\mathrm{pc}) \simeq 57\text{–}66 \, M_\odot \, \mathrm{pc}^{-2}$, or $\Sigma(1100\,\mathrm{pc}) \simeq 57\text{–}79 \, M_\odot \, \mathrm{pc}^{-2}$. Further studies of the age–metallicity and age–velocity relation for 891 Hipparcos stars, mostly clump giants, have been reported by Soubiran *et al.* (2008).

Holmberg & Flynn (2004) also used the surface density of K giants, this time in a cone towards the south Galactic pole, using Hipparcos parallaxes for the nearby stars, supplemented by distances inferred from intermediate-band photometry calibrated through the Hipparcos absolute magnitudes for more distant stars. Their disk model is similar to that constructed by Holmberg & Flynn (2000), but slightly revised to give a local mass density of $\rho_0 = 0.102 \, M_\odot \, \mathrm{pc}^{-3}$ (disk + dark halo), a total local mass surface density within 1100 pc of the Galactic mid-plane of $\Sigma(1100\,\mathrm{pc}) \simeq 70.6 \, M_\odot \, \mathrm{pc}^{-2}$ (disk + dark halo), and a local disk mass surface density (i.e. correcting for the implied halo contribution) of $\Sigma = 52.8 \, M_\odot \, \mathrm{pc}^{-2}$. They found a dynamical disk mass surface density of $\Sigma = 56 \pm 6 \, M_\odot \, \mathrm{pc}^{-2}$ in close agreement with the estimate in visible matter. Their model also yields $\Sigma(350\,\mathrm{pc}) = 41 \, M_\odot \, \mathrm{pc}^{-2}$, close to that reported by Korchagin *et al.* (2003).

From these various studies, a consistent picture now emerges of the dynamical disk mass surface density, the local volume mass density, and the local mass density identified with visible matter. This confirms a picture in which disk matter is well accounted for, and in which dark matter in the Galaxy is distributed in the form of the halo, with little if any concentrated in the form of the disk.

Vertical frequency For moderate excursions above and below the Galactic plane, for example in the case of the Sun, the restoring force is approximately linear with z, and the resulting vertical oscillation frequency is then

$$\omega_0^2 = -\frac{\partial K_z}{\partial z} = 4\pi G\rho_0 \tag{9.39}$$

which yields a vertical oscillation period of

$$P_0 = \sqrt{\frac{\pi}{G\rho_0}} = 26.42\left(\frac{\rho_0}{M_\odot\,\mathrm{pc}^{-3}}\right)^{-1/2}\ \mathrm{Myr} \tag{9.40}$$

Accordingly, a mass density in the range $\rho_0 = 0.076\text{–}0.102\,M_\odot\,\mathrm{pc}^{-3}$ results in a vertical oscillation period of between 96–83 Myr respectively. Taking the local volume mass density to be $\rho_0 = 0.10\,M_\odot\,\mathrm{pc}^{-3}$, yields a half-period of the vertical oscillations of the Sun through the Galactic plane of 42 ± 2 Myr. Strictly, the vertical motion is not harmonic due to the vertical mass density gradient, and the validity of these estimates may be questioned.

However, such determinations would be in conflict with the conclusions of Stothers (1998). He considered 26 pre-Hipparcos and two Hipparcos determinations of ρ_0 which span the range $0.07\text{–}0.26\,M_\odot\,\mathrm{pc}^{-3}$, and suggested that they follow a Gaussian distribution with a mean of $0.15\pm0.01\,M_\odot\,\mathrm{pc}^{-3}$, yielding $P_0 = 68\pm2$ Myr. He argued that such a value is consistent with a mean periodicity of 36 ± 1 Myr (corresponding to half the oscillation period) derived from cometary impact cratering records over the past 600 Myr, although noting that this period is of marginal statistical significance. Similar cratering periodicities have been cited by Moon *et al.* (2003, who gave 16.1 and 34.7 Myr), and Yabushita (2004, who gave 37.5 Myr). Arguments against such periodicities are presented by Jetsu & Pelt (2000). In any case, it seems premature to argue that long-term cratering records could tightly constrain the local value of ρ_0 given that, over the Sun's 250 Myr Galactic rotation period, the average vertical oscillation period might be larger than its present instantaneous value due to an elliptical component of the Sun's Galactic orbit combined with the radial density profile of the disk, even if azimuthal variations at the solar circle are ignored (Stothers, 1985). This is a further application of the need to characterise the motion of the Sun with respect to the Local Standard of Rest.

9.4.2 Escape velocity

The escape velocity from the solar neighbourhood has provided a fundamental constraint on galaxy models since the early studies of Kapteyn and Oort. Most dynamical measurements are rather insensitive to the mass distribution in the outer parts of the halo, since such mass exerts no net gravitational force on a nearby test particle so long as the halo is spherical or ellipsoidal. Local measurements can, however, provide a dynamical constraint on the halo extent, since the escape speed from the Galaxy should exceed the largest speed of any star observed in the solar neighbourhood. There are two caveats: (a) stars may be bound to the Local Group of galaxies, and just passing through the Galaxy; although this cannot be ruled out on mass arguments, it may be unlikely given the apparent absence of unbound giant stars in the solar neighbourhood (Leonard & Tremaine, 1990); (b) dynamical interactions involving binary stars in clusters can lead to ejected high-velocity runaway stars with velocities beyond the escape velocity.

In a highly simplified model in which the Galaxy is spherical, with a circular speed Θ_c constant out to some maximum radius r_*, the mass within radius r is

$$M(r) = \begin{cases} \Theta_c^2 r/G & \text{at } r < r_* \\ \Theta_c^2 r_*/G \equiv M_* & \text{at } r > r_* \end{cases} \tag{9.41}$$

from which (Binney & Tremaine, 1987, Equation 2–192)

$$v_{\mathrm{esc}}^2 = \begin{cases} 2\Theta_c^2[1+\ln(r_*/r)] & \text{at } r < r_* \\ 2\Theta_c^2 r_*/r & \text{at } r > r_* \end{cases} \tag{9.42}$$

A number of high-velocity stars near the Sun have velocities, with respect to an inertial frame, of around $500\,\mathrm{km\,s^{-1}}$, and various determinations of the escape speed over the period 1980–90 put it in the range $v_{\mathrm{esc}} = 400\text{–}640\,\mathrm{km\,s^{-1}}$. Using $v_{\mathrm{esc}} = 500\,\mathrm{km\,s^{-1}}$, $\Theta_c = 220\,\mathrm{km\,s^{-1}}$ and $R_0 = 8.5$ kpc, for example, yields an outer halo limit of $r_* \simeq 4.9R_0 \simeq 41$ kpc, and a total Galaxy mass of $M_* \simeq 4.6\times10^{11}\,M_\odot$.

Leonard & Tremaine (1990) developed a statistical technique for estimating the escape speed from the high-velocity tail of a uniform sample of stars. Estimates were strongly correlated with the shape of the high-velocity tail of the velocity distribution, and estimates based on space velocities were very sensitive to errors on distances and proper motions. Estimates from radial velocities were in the range $450\text{–}650\,\mathrm{km\,s^{-1}}$, with the most secure limit of $v_{\mathrm{esc}} > 430\,\mathrm{km\,s^{-1}}$ based on the radial velocity of the star G166–37. They argued that it is impossible to estimate the mass of the Galaxy using the local escape speed without detailed knowledge of how mass exterior to the solar circle is distributed.

The same statistical technique was applied by Meillon *et al.* (1997) to a set of 5307 F–M stars, including the known subdwarfs and high-velocity stars, using space velocities derived from the Hipparcos parallaxes and proper motions combined with radial velocities from Coravel. Estimates based on the total space velocities were in the range v_{esc} = 440–490 km s^{-1}, slightly smaller than that derived by Leonard & Tremaine (1990), probably due to the absence from the Hipparcos Catalogue of the (fainter) highest velocity stars known from the compilation of Carney *et al.* (1994). Meillon (1999) identified only 10 Hipparcos stars with $V_{tot} >$ 350 km s^{-1}. They were, however, able to use the intersection of the two samples (770 stars) to re-calibrate the absolute magnitudes and hence derive improved distances of about 20 non-Hipparcos stars. Together they found 98 stars with $|V_{radial}| \geq$ 250 km s^{-1}, 33 with $|V_{tangential}| \geq$ 300 km s^{-1}, and 24 with $|V_{total}| \geq$ 350 km s^{-1}. Four have $|V_{total}| \geq$ 400 km s^{-1}, and the fastest has $|V_{total}| = 458$ km s^{-1}. They concluded that the escape velocity is unlikely to exceed 530 km s^{-1}.

It has already been noted (Section 9.2.5) that measurements of the kinematic properties of halo-population stars with the most extreme space velocities provide indirect estimates of the Galaxy's circular rotation velocity, as well as the escape speed. Such a cutoff in the stellar velocity distribution should occur first for stars moving in the sense of the Galactic rotation, with the observed value of v_{crit} satisfying $(\Theta_0 + v_{crit})^2 = v_{esc}^2$. Isobe (1974), for example, obtained $\Theta_0 = 275 \pm 20$ km s^{-1} and $v_{crit} \simeq 110$ km s^{-1}, implying $v_{esc} \simeq$ 385 km s^{-1}.

A median likelihood of $v_{esc} \simeq 544$ km s^{-1} has been given by the RAVE survey (Smith *et al.*, 2007). Implications for modified Newtonian dynamics (MOND) on galactic scales are discussed by Famaey *et al.* (2007).

Runaway and hypervelocity stars, with extremely high space velocities imparted by a particular formation process, are considered separately in Section 8.4.2.

9.4.3 Initial mass function

The observed mass function describes the relative number of main sequence stars in a given volume as a function of mass at the present time. It is found to be reasonably well described (and is therefore usually expressed) in power-law form (Salpeter, 1955; Scalo, 1986)

$$\phi(M) = \frac{dN}{dM} \propto M^{-\alpha} \tag{9.43}$$

$$\text{or} \quad \xi(M) = \frac{dN}{d\log M} \propto M^{-\alpha+1} \tag{9.44}$$

or equivalently using $\Gamma = \alpha - 1$. A more fundamental quantity, directly connected to the problem of star formation, is the initial mass function (IMF). This describes the relative number of stars forming as a function of mass, $\Phi(M)$ or (in logarithmic units) $\Xi(M)$.

Deriving the initial mass function from the observed mass function requires a consideration of stellar evolution requiring a knowledge of the age of the population(s), star-formation rates, the local mix of stellar populations, and the evolution of the disk population due to, e.g. disk heating. The IMF is generally found to be well approximated by $\alpha \simeq 2.35$ for stars with $m \gtrsim 1\,M_\odot$, corresponding to the so-called Salpeter initial mass function. Other analyses (e.g. Scalo, 1986) give a steeper slope of $\alpha \simeq 2.7$, while Scalo (1998) and Kroupa (2001) have argued that there are probably systematic variations in the IMF, especially between that in the field and in clusters. Reviews of the pre-Hipparcos status are provided by Miller & Scalo (1979) and Scalo (1986).

Determining the initial mass function is motivated by attempts to understand and model the star-formation process in general, and whether and how the star-formation rate has changed with time, within the Galaxy and elsewhere (see, e.g. Kroupa, 2002). Both are important ingredients of many aspects of galaxy models, for example describing chemical and luminosity evolution, and estimating the visible contribution to the local matter density. Complications include the degeneracy of the star-formation rate and the low end of the initial mass function due to the increasing mean age of stars on the lower main sequence, as well as dynamical heating of the Galactic disk leading to the depletion of stars in the Galactic plane with time. The problem is made more complex for studies of the Galaxy compared with external galaxies because of the large fractional differences in distance for nearby samples. Classical studies of the IMF and star-formation rate have used column integrals to compensate for the disk heating, but have been consequently constrained by limited knowledge of space-density scale heights as a function of stellar age. Deriving the IMF in the disk has essentially been based on the following steps (e.g. Schröder & Pagel, 2003): (i) derive the absolute magnitude distribution from the apparent magnitude distribution and associated scale height; (ii) convert the absolute magnitude distribution into the luminosity function for main-sequence stars; (iii) use a theoretical mass–luminosity relation to obtain the present day mass function; and (iv) model the initial mass function on the present day mass function for a given star-formation rate history.

The availability of large numbers of accurate stellar distances embracing relatively large volume-limited samples has allowed a more direct approach to determining both the IMF and the star-formation rate.

A series of studies have used the Hipparcos distances to follow a more direct approach to establishing the IMF in a well-defined local volume (Bertelli *et al.*, 1997; Sabas, 1997; Binney *et al.*, 2000; Bertelli & Nasi, 2001; Schröder & Pagel, 2003; Girardi *et al.*, 2005; Ninkovic & Trajkovska, 2006). Schröder & Pagel (2003), for example, created synthetic stars randomly in mass and time, following a given IMF and an adopted star-formation rate in the given volume. An evolutionary grid was used to place the stars on the HR diagram, and their numbers were compared with star counts in specific regions of the HR diagram, seven located along the main sequence and four representing different evolved stages. They used single stars from Hipparcos with $d < 100$ pc and $M_V < 4.0$, further constrained by $|z| < 25$ pc, and using an offset of the Sun from the plane of $Z_0 = 15$ pc. They adopted scale heights directly from the corresponding age groups of Hipparcos stars, also finding good agreement with the thin disk scale height as a function of population and age as follows (see Schröder & Pagel, 2003, for details), where a reference is included only for those observations based on Hipparcos data:

- local interstellar medium and star-forming clouds (Vergely *et al.*, 1998): 35–55 pc
- young massive stars (OB and Wolf–Rayet): 25–65 pc; 34.2 ± 0.8 pc from Maíz Apellániz (2001)
- F-type stars within 50 pc: 160 pc
- thin disk K-dwarfs for $z > 300$ pc: 249 pc
- Galactic disk subgiants: 250 pc
- K and M (O-rich) variable AGBs and Miras, $P =$ 300–400 d: 250 pc; $P =$ 200–300 d: 500 pc
- RR Lyrae stars with moderate metal abundance: 700 pc

The evolved star counts remove the degeneracy found in models restricted to the main sequence, and their best fitting models with appropriate diffusion to larger scale heights lead to

$$\alpha = 2.70 \pm 0.15 \qquad 1.1 < M < 1.6\,M_\odot \tag{9.45}$$

$$= 3.10 \pm 0.15 \qquad 1.6 < M < 4.0\,M_\odot \tag{9.46}$$

similar to that considered by Scalo (1998). Comparison of the observed main-sequence star counts with different models of core overshooting (see Chapter 7) implies that the onset of overshooting occurs at $M = 1.50\,M_\odot$. They derived a locally constant star-formation rate, for $M > 0.9\,M_\odot$, of 2.0 ± 0.15 stars formed per 1000 yr and per kpc^3, and a column-integrated (non-local) thin disk star-formation rate, reasonable constant over time, for $M > 0.9\,M_\odot$, of 0.82 stars pc^{-2} Gyr^{-1}.

The form at the low-mass end, $M < 1\,M_\odot$, was investigated by Chabrier (2001), using the nearby Hipparcos-based luminosity function within 8 pc from Delfosse *et al.* (2000), and normalising their various functional forms of the initial mass function (power-law, log-normal, and exponential) to a Hipparcos-based value of $(dN/dM) = 2.8 \pm 0.2 \times 10^{-2}\,M_\odot\,\text{pc}^{-3}$ at $0.8\,M_\odot$.

Reid *et al.* (2002) studied the solar neighbourhood luminosity function and initial mass function using a sample of stars extending from A to M within 25 pc. They selected M dwarfs from the CNS3 with specific attention to possible kinematic bias, and used Hipparcos data to define a volume-limited sample of 558 main-sequence stars in 448 systems in the range $8 < M_V < +17$. With a local density of ~ 0.07 pc^{-3}, the M dwarfs account for the majority of stars in the Galaxy, and represent excellent tracers of many properties of the disk. Combined with a Hipparcos-based sample of more than 800 AFGK dwarfs within 25 pc, they constructed the solar neighbourhood luminosity function for stars with $-1 < M_V < +17$ mag. They excluded stars lying both above (evolved giants) and below (halo subdwarfs and white dwarfs) the main sequence, adjusted the photometry for known multiplicity, and made further adjustments to allow for undetected binaries. The resulting composite luminosity function, $\Phi(M_V)$, is shown in Figure 9.12. They found systematically lower space densities compared with those given by Wielen *et al.* (1983), with an overall space density of 0.112 stars pc^{-3}. To transform the stellar luminosity function to a mass function requires use of the mass–luminosity relation which they constructed from a number of studies covering the entire main sequence (Figure 9.13). They derived a present-day mass function consistent with a two-component power-law, with $\alpha \sim 1.2$ for $M < 0.6\,M_\odot$ and $\alpha \sim 5$ for $M > 1.1\,M_\odot$, and indicating that the main-sequence stars contribute 0.030–$0.033\,M_\odot\,\text{pc}^{-3}$ to the total mass density in the solar neighbourhood. Allowing for stellar evolution at high masses, integrating the density distribution perpendicular to the plane, and allowing for a 10% contribution from the thick disk, they derived a thin disk initial mass function close to the Salpeter value at high mass, $\alpha \sim 2.5-2.8$, but much flatter at low masses, $\alpha \sim 1.1-1.3$. The contribution of the thick disk stars is seen in a two-component Gaussian distribution of velocities, with the W velocity component of the local disk dwarfs represented by $\sigma_{W,\text{thin}} \sim 16$ km s^{-1} and a fraction $f = 0.12$ of high-velocity stars represented by $\sigma_{W,\text{thick}} \sim 36$ km s^{-1}.

Luminosity functions derived from *UBV* and *RGU* photometry in various selected areas (SA 141, 133, 51, and the direction of M5) were compared with those derived from Hipparcos by Karataš *et al.* (2001), Karataš *et al.* (2004), Bilir *et al.* (2004), and Karaali *et al.* (2004) respectively.

Initial mass function estimates have been made for only a few open clusters including, most recently, the Pleiades (Hambly *et al.*, 1999) and Praesepe (Pinfield *et al.*, 1999). Sanner & Geffert (2001) used the Tycho 2 proper motions to select stars in nine nearby open star clusters with ages ranging from 10–650 Myr (Blanco 1, Stock 2, α Per, Pleiades, NGC 2451A, IC 2391, Praesepe, IC 2602, NGC 7092). They derived ages, distance moduli and reddenings from the Tycho 2 photometry, then estimated the initial mass function from the main-sequence turn-off down to about $1\,M_\odot$. They found α in the range 1.69 ± 0.63 for NGC 2451A to 3.27 ± 0.70 for Blanco 1, with an average $\alpha = 2.65$. With the notable exception of NGC 7092, individual estimates are in agreement with those given by Tarrab (1982), and the slopes are within the range given by, e.g. Scalo (1998).

9.4.4 Star-formation rate

Star-formation rate as a function of time, combined with an appropriate stellar initial mass function, constrain cosmological models, galaxy formation scenarios, galactic evolution models and star-formation theories in general. The objective is to recover the star-formation rate as a function of time that gave rise to a given HR diagram. Even assuming a known initial mass function and initial chemical composition, uncertainties in the theories of stellar formation and evolution, as well as degeneracy in the observational parameters of the star between age and metallicity, combined with observational errors and uncertain distance and reddening corrections, make the problem a complex one. Some studies assume that the star formation is rather constant with time, others assume that it first increases then decreases to the present time, while others have found that the local rate has been rather irregular, with periods of enhancement and quiescence (see, e.g. Bertelli & Nasi, 2001).

The general approach is to vary the star-formation rate from a range of plausible models until a good match between the synthesised and observed HR diagrams is obtained. This may be, for example, by modelling the star-formation history as a series of bursts, and solving for the time, duration, and amplitude of each burst (see, e.g. Hernández *et al.*, 1999). Hernández *et al.* (2000) have accordingly derived the recent star-formation history in the solar neighbourhood, using Hipparcos data and the maximum-likelihood approach described by Hernández *et al.* (1999). The method aims to avoid assumptions on the underlying functional form of the star-formation rate, but rather iterates to find a time-dependent function which yields a vanishing first derivative for the likelihood, assuming that the relevant initial mass function and metallicity are known. They worked with a complete volume-limited sample defined by $M_V < 3.15$, $m_v < 7.25$, and $\sigma_\pi/\pi < 0.2$, which limits the age range over which the star-formation rate can be recovered to the interval 0–3 Gyr. They used the Padova isochrones (Bertelli *et al.*, 1994; Girardi *et al.*, 1996) assuming solar metallicity for all stars, and the initial mass function from Kroupa *et al.* (1993)

$$\rho(m) \propto \begin{cases} m^{-1.3} & 0.08M_\odot < m \le 0.5M_\odot \\ m^{-2.2} & 0.50M_\odot < m \le 1.0M_\odot \\ m^{-2.7} & 1.00M_\odot < m \end{cases} \tag{9.47}$$

The observed colour–magnitude diagram and the resulting star-formation rate are shown in Figure 9.14. The dotted envelope encloses alternative reconstructions using different cuts in M_V, giving an indication

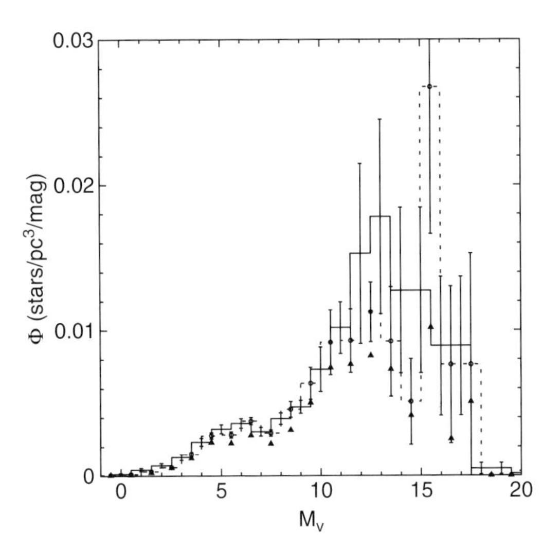

Figure 9.12 The nearby star luminosity function derived by combining data for the Hipparcos 25 pc sample of AFGK dwarfs with the Palomar/MSU volume-limited sample of M dwarfs. Triangles: single stars and primaries; circles and dashed histogram: space densities once companions are included. The luminosity function has been adjusted to include suspected 'missing' binary components by doubling the contribution from known companions to the Hipparcos 25 pc sample. Error bars reflect statistical uncertainties only. The solid histogram shows the luminosity function $\Phi(M_V)$ from Wielen et al. (1983). From Reid et al. (2002, Figure 10).

of the probable errors which can be seen to increase with time. The results suggest a certain level of rather constant star-formation activity, with a number of additional strong quasi-periodic components with a period close to 0.5 Gyr. They suggest various possibilities to explain this inferred periodicity: (a) Due to the passage of spiral waves as a result of the difference between the spiral pattern speed and the local circular velocity. For a spiral pattern angular frequency Ω_s equal to twice the circular frequency at the Sun's position, Ω_0, the time interval between spiral arm encounters is

$$\begin{aligned}\Delta t &= \frac{2\pi}{m\,|\Omega_0 - \Omega_s|} \\ &= \frac{0.22\,\text{Gyr}}{m}\left(\frac{\Omega_0}{29\,\text{km}\,\text{s}^{-1}\,\text{kpc}^{-1}}\right)^{-1}\left|\frac{\Omega_s}{\Omega_0} - 1\right|^{-1}\end{aligned} \tag{9.48}$$

where m is the number of arms in the spiral pattern. A value $\Omega_s = 0.5\,\Omega_0 \approx 14.5\,\text{km}\,\text{s}^{-1}\,\text{kpc}^{-1}$ would imply that the interaction with a single arm ($m = 1$) would be enough to account for the observed regularity in the recent star-formation history. However, more recent determinations suggest larger values $\Omega_s \sim 24\,\text{km}\,\text{s}^{-1}\,\text{kpc}^{-1}$, more consistent with a two-armed spiral. (b) Alternatively, if the solar neighbourhood

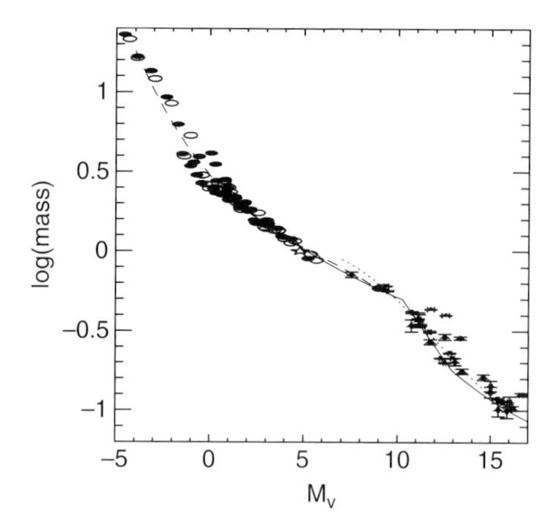

Figure 9.13 The mass–luminosity relation for main-sequence stars. Filled and open circles: primary and secondary stars, respectively, from the compilation of eclipsing binaries by Andersen (1991); triangles: lower main-sequence binaries from Ségransan et al. (2000). The five-pointed star marks the Sun. Dashed line: empirical fit to the upper main-sequence; dotted line: mass–luminosity relation from Delfosse et al. (2000); solid line: the three-component fit from Henry & McCarthy (1993). From Reid et al. (2002, Figure 11).

is close to the co-rotation radius, the Galactic bar could have triggered star formation in the solar neighbourhood with episodes separated by about 0.5 Gyr if the pattern speed of the bar is larger than about $40\,\mathrm{km\,s^{-1}\,kpc^{-1}}$. (c) The cloud formation, collision and stellar feedback models of Vazquez & Scalo (1989) predict a phase of oscillatory star formation as a result of self-regulation. (d) Close encounters with the Magellanic Clouds have also been suggested to explain the intermittent nature of the star-formation rate on longer time scales (Rocha Pinto *et al.*, 2000). The latter used chromospheric ages to argue that the disk has experienced enhanced episodes of star formation at 0–1 Gyr, 2–5 Gyr and, perhaps, 7–9 Gyr ago.

In a study of nearly 12 000 main sequence and subgiant stars aimed at determining the age of the solar neighbourhood, Binney *et al.* (2000) were also able to place some constraints on the star-formation rate, although with the caveat that the IMF near $1\,M_\odot$ is degenerate with the rate at which star formation declines. If the slope of the IMF is close to the Salpeter value of -2.35, then the star formation was inferred to be nearly constant with time.

Bertelli & Nasi (2001) used a sample of 1844 stars within 50 pc of the Sun corresponding to a completeness limit of $M_V = 4.5$, or $M \geq 0.9 - 1.0\,M_\odot$. They assume a form of the star-formation rate, an initial mass function initially following the classical Salpeter form but adjusted subsequently to a two-slope form, and appropriate representation of chemical composition and multiplicity, matching to the observed colour–magnitude diagram for main sequence and evolved stars. All their accepted solutions support a star-formation rate that was broadly increasing from the beginning, around 10 Gyr ago, up to the present time. Volume mass densities are strongly influenced by the form of the IMF at the low-mass end of the main sequence. A particular difficulty emerges for the ratio of the number of stars in the He burning phase to that in the main-sequence phase. The theoretical value is uncertain because of the uncertainty in the input physics, but the models produce an inadequate match to the observations, suggesting that stellar models in the mass range $1–1.6\,M_\odot$ are inadequate. These results are in contrast to the satisfactory match obtained by Schröder & Pagel (2003), for a specific onset of core overshooting. Vallenari *et al.* (2000) argued that colour–magnitude diagrams to $V \sim 23.5$ in four low Galactic fields do not support the idea of a constant or strongly increasing star-formation rate with time.

Schröder & Sedlmayr (2001) examined a further level of complexity in constraining the IMF and star-formation rate by examining the contribution of the heavy element-enriched interstellar medium resulting from stellar winds in the late stages of stellar evolution. This required detailed mass-loss models for stars on the red giant and asymptotic giant branches and, in particular, including dust-driven pulsating wind models for carbon-rich stars at the tip of the asymptotic giant branch. Stars with initial masses in the range $1–2.5\,M_\odot$ make a significant contribution to this process since they have had time to evolve into the red giant or asymptotic giant branches where they contribute to the cool-wind stellar mass loss; less massive stars evolve too slowly, and in the higher mass regime where supernovae are important the steeply increasing IMF strongly reduces stellar numbers. The Hipparcos data were used to define a sample of 1340 single stars complete to $d = 50$ pc and $M_V \leq 4$. Using evolutionary tracks, they then derived an IMF and star-formation rate matched to the observed solar neighbourhood HR diagram, from which they estimated present-day mass-loss rates. The best-fitting synthetic stellar sample was computed with an IMF of $\alpha = 2.9$ for $M > 1.8\,M_\odot$, and $\alpha = 2.7$ for lower masses, and with a SFR(t) of $1.14 \times 10^{-6} \cdot \mathrm{e}^{-t/6.3\,\mathrm{Gyr}}$ stars per year for single stars within 50 pc exceeding $0.9\,M_\odot$. SFR(0) is the present star-formation rate, equivalent to 2.2×10^{-3} stars $\mathrm{kpc^{-3}\,yr^{-1}}$. Together, this consumes a total mass of $4.5 \times 10^{-3}\,M_\odot$ per year in a volume of $1\,\mathrm{kpc^3}$.

Other studies of the star-formation rate derived from the Hipparcos data of the solar neighbourhood are described by Vergely *et al.* (2002); Cignoni & Shore (2006) and Cignoni *et al.* (2006).

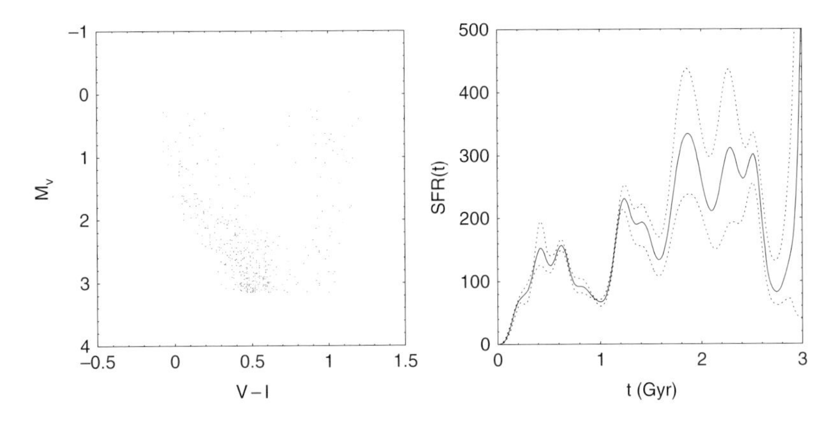

Figure 9.14 Left: colour–magnitude diagram of the volume-limited sample complete to $M_V < 3.15$ from Hipparcos. Right: inferred star-formation rate, SFR(t), from an inversion of this dataset. Only stars bluewards of $V - I = 0.7$ were considered in the inversion procedure, representative of the last 3 Gyr. From Hernández et al. (2000, Figure 4).

9.5 Properties of the disk

Background This section covers Hipparcos studies of the characteristics of the local stellar population. While dominated by thin disk stars, the local population contains approximately 5–10% of thick disk stars, and perhaps 0.1–0.5% of halo stars, such that studies of the disk population are intimately connected with the determination and segregation of the ages, chemical composition, and velocity structures of the various component populations. Subsequent sections are devoted to Hipparcos studies of specific large-scale phase-space features of the disk: the bar, the spiral arms, and the stellar warp. Significant substructure in the local disk is also present in the form of open clusters, moving groups, and associations, including the Gould Belt; this large topic is covered separately in Chapter 6. Further details of the chemical evolution of the disk, including the age–metallicity relation, are covered in Section 7.6.

The recognition of the existence of stellar populations differing in age, chemical composition, spatial distribution, and kinematic properties, has represented a breakthrough in the knowledge of Galactic structure, and the basis for recent models of galaxy formation and evolution. An extensive and complex literature now exists on the topic of the Galaxy's disk: its separation into thin and thick disk components, whether they represent discrete or continuous populations, their respective scale heights, their kinematic, metallicity, and age properties, the relationship between them, and their origin. The aim of this section is to convey a picture of this complexity, and the areas in which the Hipparcos data have contributed to an understanding.

The vertical distribution of the different populations is frequently described either in terms of a 'characteristic thickness', defined as the ratio of the surface density (integrated over disk thickness) to its volume density at the Galactic plane, or in terms of a scale height z_h, defined by $\exp(-z/z_h)$ for an exponential distribution, or equivalently for other functional forms such as sech^2. The thin disk has a characteristic thickness of $\sim$180–200 pc, while the thick disk has a characteristic thickness of $\sim$700–1000 pc, with the interstellar medium having a scale height of about 40 pc (Dehnen & Binney, 1998b). Even for the thin disk, however, its scale height is different for different classes of stars, with old stars found at greater distances from the plane partly as a result of disk heating (Section 6.9) in which the irregular gravitational field of spiral arms and molecular clouds gradually increases their random velocities over time.

The thick disk was first discussed as a possibly discrete population by Gilmore & Reid (1983) using star counts: the number density of stars between 1–5 kpc above the plane was significantly higher than expected from the existing disk/halo model. They described their excess population as having a scale height of $\sim$1500 pc and a local density of $\sim$2% of the (thin) disk; subsequent analyses favoured a somewhat higher local normalised density and a smaller scale height: for example, Robin *et al.* (1996) estimated a scale height of 760±50 pc, a local density of 5.6 ± 1% of the thin disk, and a scale length of 2.8 ± 0.8 kpc comparable to that of the thin disk of 2.5 ± 0.3 kpc. The relatively red colour of the main-sequence turn-off at heights of $z \sim 1$ kpc indicated a relatively old population. Most studies have converged on an average [Fe/H] in the range -0.5 to -0.7 with a significant low-metallicity tail extending to at least $\mathrm{[Fe/H]} \sim -1.6$, with no evidence for a significant abundance gradient with height. Reid (1998), but see also Schuster *et al.* (1993) and references therein for more details of this complex field, summarised the understanding at that time of the metallicity of the local thin disk, thick disk, and halo populations as follows:

$$\begin{aligned} \text{thin disk}: & \langle[\mathrm{Fe/H}]\rangle = -0.11 \; \sigma_{[\mathrm{Fe/H}]} = 0.16 \\ \text{thick disk}: & \langle[\mathrm{Fe/H}]\rangle = -0.50 \; \sigma_{[\mathrm{Fe/H}]} = 0.25 \\ \text{halo}: & \langle[\mathrm{Fe/H}]\rangle = -1.30 \; \sigma_{[\mathrm{Fe/H}]} = 0.60 \end{aligned} \tag{9.49}$$

There nevertheless remained debate as to whether these thin and thick disks are separate populations or some form of continuum distribution. The origins of the thick disk also remained unclear (Gilmore *et al.*, 1989): whether it was created during the slow formation of the disk, or as a result of the rapid formation of the thin disk followed by successive heating, or through a merger early in the formation history of the Galaxy.

F- and G-type dwarfs recur as important tracer populations for the history of the disk. They are relatively numerous; sufficiently long-lived to have survived from the formation of the disk, their convective atmospheres reflect their initial chemical composition; and their ages can be estimated using stellar evolutionary models at least for the more evolved stars.

Component populations and ages Estimates of the age of the Galactic disk have been obtained from the cooling rates of white dwarfs, from isochrone fits to evolved F and G stars, from nucleosynthesis models and radioactive decay rates, from open cluster main-sequence fitting, from the lower locus of the red giant branch, and from isochrone fitting to the subgiant branch. Pre-Hipparcos lower limits on the local disk age ranged upwards of 6 Gyr, with some estimates as high as 11 Gyr (Jimenez *et al.*, 1998).

Jimenez *et al.* (1998) and Flynn *et al.* (1999) determined the age of the Galactic disk from a fit to the red envelope of field subgiants in the Hipparcos colour–magnitude diagram with synthetic isochrones covering the range of observed disk metal abundance observed in both the G and K clump giants and dwarfs. They used horizontal branch stars to check the isochrones, and showed that the colour of the clump can be used to estimate the population metallicity. They derived a minimum age of 8 Gyr (Figure 9.15).

Ng & Bertelli (1998) made a re-assessment of the study by Edvardsson *et al.* (1993), who had obtained ages of up to 12 Gyr for disk stars with $[\mathrm{Fe/H}] > -0.5$. Ng & Bertelli (1998) used Hipparcos parallaxes to re-determine ages, and demonstrated that stars previously found to be older than 12 Gyr actually have $[\mathrm{Fe/H}] < -0.5$, and therefore belong to the thick disk.

Liu & Chaboyer (2000) selected 21 Hipparcos field stars with good parallaxes, and with [Fe/H] within 0.07 dex of solar to ensure likely membership of the thin disk, then compared the region of the main-sequence turn-off with solar metallicity isochrones to conclude that the oldest solar metallicity stars of the thin disk have an age of 7.5 ± 0.7 Gyr. Given that the local thin disk extends down to metallicities of around $[\mathrm{Fe/H}] \simeq -0.20$ (Edvardsson *et al.*, 1993), they also studied an extended sample of more metal-poor stars, also using the space motion to distinguish thin and thick disk stars. They concluded that the metal-poor thin disk stars in the solar neighbourhood have an age of 9.7 ± 0.6 Gyr; age estimates of their few thick disk stars were inconclusive, but range between 6.6–11.7 Gyr, with the oldest similar to that of the thick disk globular cluster 47 Tuc, and suggestive of a significant age spread of the thick disk. Binney *et al.* (2000) meanwhile derived an age of the solar neighbourhood of 11.2 ± 0.75 Gyr from a sample of nearly 12 000 main-sequence and subgiant stars together with the Padova isochrones, although a younger age of down to $\sim$9 Gyr was not entirely precluded.

Bernkopf *et al.* (2001) studied a volume-limited sample of about 360 FGK stars brighter than $M_V \sim 6.0$ within 25 pc and with $\delta > -15^\circ$. They identified only $\sim$18 stars belonging to the thick disk population, with the remaining 95% belonging to the thin disk, and no stars belonging to the halo. From the Hipparcos absolute magnitudes and detailed evolutionary tracks, extended to subgiants out to 50 pc, they identified a significant gap of 3–5 Gyr between star formation in the thick and thin disk populations: the thin disk has a mean age that is approximately solar and does not exceed about 8 Gyr, while the thick disk is very old, at around 14 Gyr. At such an age, the thick disk would be 'remnant dominated' implying that the solar neighbourhood would be embedded in a large population of dead stars, perhaps accounting for many of the massive compact halo objects and blue white dwarfs (e.g. Hansen, 1998; Harris *et al.*, 2001).

Jørgensen (2000) re-examined the 'G dwarf problem', a long-standing problem in Galactic astronomy. Models for Galactic chemical evolution in which the halo stars, with low metallicity, formed first out of gas which then collapsed dissipatively to form the disk (including the solar neighbourhood) predict too many low-metallicity stars amongst the longest-lived stars. Although various theoretical models have been proposed to resolve the problem, the observations have been of inadequate quality to discriminate between them. Based on a final sample of 253 G dwarfs in the mass interval 0.7–$1.0\,M_\odot$, Jørgensen (2000) argued that the 'infall models', following the accretion model of the halo (Searle & Zinn, 1978), such as those by Lynden-Bell (1975) and Pagel & Tautvaišienė (1995), can explain the G dwarf problem, with the G dwarf metallicity distribution providing a constraint on galaxy formation models.

Early microlensing surveys indicated the presence of massive compact halo objects (MACHOs) with masses of order 0.3–$0.8\,M_\odot$ (Alcock *et al.*, 1996). Following

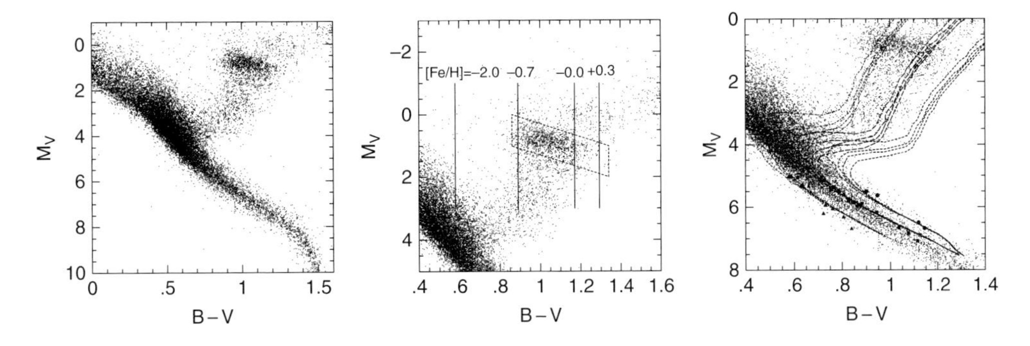

Figure 9.15 Left: colour–magnitude diagram for the Hipparcos stars with $\sigma_\pi/\pi < 0.15$, and with B_T, V_T colour errors less than 0.02 mag. Middle: The colour–magnitude diagram of the giant branch region: vertical lines show the expected position of the red edge of the clump defined by the theoretical ZAHB models for various indicated metallicities; the dashed box is used to isolate the clump stars. Right: colour–magnitude diagram overlaid by isochrone fits: G and K stars for which abundances were available are plotted in three metallicity ranges: $-0.60 <$ [Fe/H] < -0.40 (triangles and leftmost isochrones); $-0.05 <$ [Fe/H] $< +0.05$ (squares and middle isochrones); and $+0.25 <$ [Fe/H] $< +0.35$ (circles and rightmost isochrones). From Jimenez et al. (1998, Figures 1, 4, 6).

this announcement, halo dwarfs were initially considered as potential candidates for these elusive MACHOs, although deep star counts from HST data had already excluded them from serious consideration (e.g. Flynn *et al.*, 1996). A confirmation of this was possible with the Hipparcos data. Fuchs & Jahreiß (1998) searched the CNS4 for halo stars in the immediate solar neighbourhood, based on position on the subdwarf main sequence in the colour–magnitude diagram, their high space velocities, and their low metallicity (which however overlaps with the much more abundant thick disk stars), identifying 15 subdwarfs and a high-velocity white dwarf (Gliese 699.1). Of these, nine have Hipparcos parallaxes while the others were too faint or simply omitted from the Hipparcos Input Catalogue compilation. They derived a resulting halo dwarf mass density of $1\text{–}1.5 \times 10^{-4}\,M_\odot$. Compared with the local density of the dark halo of about $90 \times 10^{-4}\,M_\odot$ (Gates *et al.*, 1996) this implies that the local density of halo stars only is some 1.7% of the halo density, or some 3% of the local mass density of MACHOs implied by the associated microlensing surveys. The non-detection of such analogues in the solar neighbourhood also implied an upper limit to the MACHO luminosities of $M_B > 21$ mag.

Fuhrmann (2002) re-analysed the CNS4 sample, and accepted only seven of the 16 objects as halo candidates, the others being assigned to the thick disk, based on their location in the 'Toomre diagram' of the Galactic space velocity components $\sqrt{U^2 + W^2}$ versus V. This would correspond to a revised local stellar halo mass density of only $0.315 \times 10^{-4}\,M_\odot$. He considered even this to be an upper limit since, as inferred from the [Fe/H] and Fe/Mg ratios, part of the remaining halo stars, including possibly Groombridge 1830, could be accreted objects. He concluded that as a little as $0.15 \times 10^{-4}\,M_\odot$ might be associated with the classical halo. He summarised the disk parameters as follows (Figure 9.16): a local stellar mass density $\rho_0 = 3.82 \times 10^{-2}\,M_\odot\,\text{pc}^{-3}$; a halo mass density in the range $6\text{–}15 \times 10^{-5}\,M_\odot\,\text{pc}^{-3}$; a local normalisation $\rho_{0,\text{thick}}/\rho_{0,\text{thin}} = 0.06\text{–}0.18$; a stellar halo density varying as $\rho \sim r^{-3.5}$; and an exponential thin disk scale height of between 200–325 pc.

Sandage *et al.* (2003) studied the form of the main-sequence, subgiant and giant star distributions in the Hipparcos HR diagram for all stars with $\sigma_\pi/\pi \leq 0.010$. Isochrones for [Fe/H] = +0.37 provide the best fit to the reddest giants between $0 \leq M_V \leq +3$ as well as to the redmost envelope of the reddest main sequence stars at $M_V \geq 5$, while the red edge of the densest part of the field giants is best matched by [Fe/H] = +0.23. These high metallicities are also seen in the old thick disk cluster NGC 6791. Corresponding isochrone ages of the field stars in the solar neighbourhood are between $(7.4 - 7.9) \pm 0.7$ Gyr (the former accounting for elemental diffusion).

Fuhrmann (2004) combined new high-resolution spectroscopy with earlier data (Fuhrmann, 1998) to study ∼250 mid-F to early-K stars within 25 pc, with a few stars of the thick disk and halo at larger distances. Starting with the physical parameters of the star derived from the spectroscopic measurements (T_{eff}, $\log g$, etc.) the absolute magnitude derived from stellar models provides the 'spectroscopic parallax' which can be compared with the Hipparcos parallax. Agreement between the two indicates a 1σ spectroscopic parallax error of some 5%, allowing discordant results to be identified with spectroscopic binaries or with inconsistent Hipparcos parallaxes due to astrometric binaries. Figure 9.17 illustrate some of the spectroscopic results as a function of T_{eff} and $\log g$.

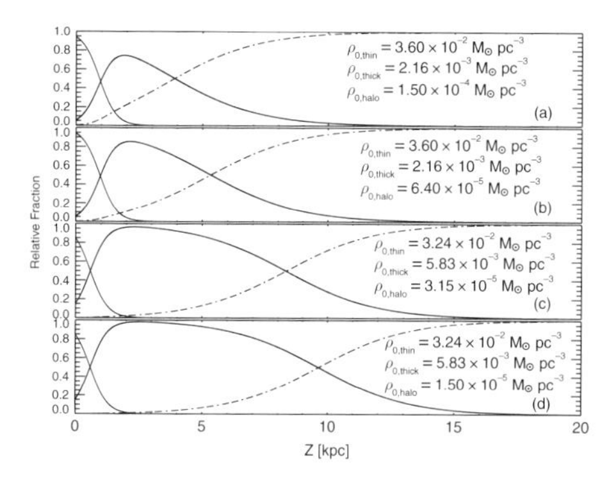

Figure 9.16 Relative fraction of thin disk (thin lines), thick disk (thick lines) and halo (dash–dotted lines) as a function of height z above the plane, at a Galactocentric distance R_0 = 8.0 kpc. Panels (a) and (b) refer to a 6% local normalization of thick disk versus thin disk, and with halo densities from Fuchs & Jahreiß (1998) and Gould et al. (1998) respectively. (c) 18% local normalization of the thick disk as derived from local long-lived FGK stars, and a halo mass density in which the local thick disk outweighs the halo by about a factor 200. (d) Same as (c), but with half of the stellar halo assumed to be accreted. From Fuhrmann (2002, Figure 2), reproduced with the permission of Elsevier.

Allende Prieto *et al.* (2004) reported a high-resolution spectroscopic survey, at $R \simeq 50\,000$ and between 362–921 nm, of 118 Hipparcos FGK stars with $M_V < 6.5$ mag and within the corresponding catalogue completeness limit (at $V = 7.3$ mag) of 14.5 pc from the Sun. With [Fe/H] centred at -0.1 dex and with a standard deviation of 0.2 dex, the sample is considered representative of the local thin disk. A number of very old, metal-rich, K stars are found. In agreement with earlier studies, they found that the abundance ratios of Si, Sc, Ti, Co, and Zn become smaller as the iron abundance increases to approximately solar, with the trend reversing for higher iron abundances. At a given metallicity, stars with a low Galactic rotational velocity tend to have high abundances of Mg, Si, Ca, Sc, Ti, Co, Zn, and Eu, but low abundances of Ba, Ce, and Nd. The Sun appears roughly 0.1 dex deficient in O, Si, Ca, Sc, Ti, Y, Ce, Nd, and Eu compared to its immediate neighbours with similar iron abundances. This could be explained by the Sun being somewhat older than most other dwarfs with similar spectral types in the thin disk. The precise iron abundance for local thin disk stars is, however, a matter of continued dispute, especially when derived from photometric surveys (e.g. Haywood, 2001; Kotoneva *et al.*, 2002; Fuhrmann, 2004).

Component populations and kinematics Kinematic information using Hipparcos distances and proper motions (and preferably radially velocities) is now widely and rather routinely used to discriminate between thin disk, thick disk, and halo stars.

Bonifacio *et al.* (1999) studied objects previously considered as metal-weak thick disk members (with [Fe/H] reaching -2.9), re-assigning part to the halo population, but confirming others as true thick disk members, and thus confirming the existence of a metal-poor tail of the thick disk. Caloi *et al.* (1999) studied kinematic signatures versus age. Fuchs *et al.* (1999) studied kinematics of subdwarfs versus metallicity, finding that most such nearby objects are thick disk stars, but that a subset are extreme metal-poor halo stars with no systematic rotation, while some of intermediate metallicity have a mean rotation speed of about 100 km s^{-1}. Naoumov (1999) studied the kinematics, age, and metallicity of thick disk stars. Beers *et al.* (2000) defined a sample of thick disk and halo stars used to study the halo dynamics (see Section 9.9).

Feltzing *et al.* (2001) investigated the age–metallicity relation in the solar neighbourhood for 5828 Hipparcos dwarfs and subdwarfs, deriving ages from evolutionary tracks, and metallicities from Strömgren photometry (Figure 9.18). They showed that (a) the age–metallicity diagram is well populated at all ages and especially that old, metal-rich stars do exist; (b) the scatter in metallicity at any given age is larger than the observational errors; (c) the exclusion of cooler dwarf stars from an age–metallicity sample preferentially excludes old, metal-rich stars, depleting the upper right-hand corner of the age–metallicity diagram; (d) the distance dependence found, also on the basis of Hipparcos data, by Garnett & Kobulnicky (2000) is an artifact of selection biases in the sample of Edvardsson *et al.* (1993); (e) a large part of the observed age–metallicity scatter is intrinsic to the formation processes of stars, with only part being attributable to stellar migration in the Galactic disk. Similar analyses were made by Chen *et al.* (2000b) and Chen *et al.* (2002b) using 90 F and G disk dwarfs.

Ibukiyama & Arimoto (2002) made a similar investigation for 1658 nearby Hipparcos stars, comprising 1382 thin disk, 229 thick disk, and 47 halo stars according to their orbital parameters. They confirmed the significant scatter of the thin disk age–metallicity relation, even for stars with similar orbits. The relation for the thick disk stars indicates that star formation terminated about 8 Gyr ago in the thick disk. As also reported by Gratton *et al.* (2000) and Prochaska *et al.* (2000) the thick disk stars are more Ca-rich than the thin disk stars at the same [Fe/H]. The orbital components were used to derive the maximum distance from the Galactic plane achieved for each star, $z_{\rm max}$, and to show that, while there is no metallicity gradient for the thin disk stars, [Fe/H] decreases with increasing $z_{\rm max}$ for thick disk stars (Figure 9.19).

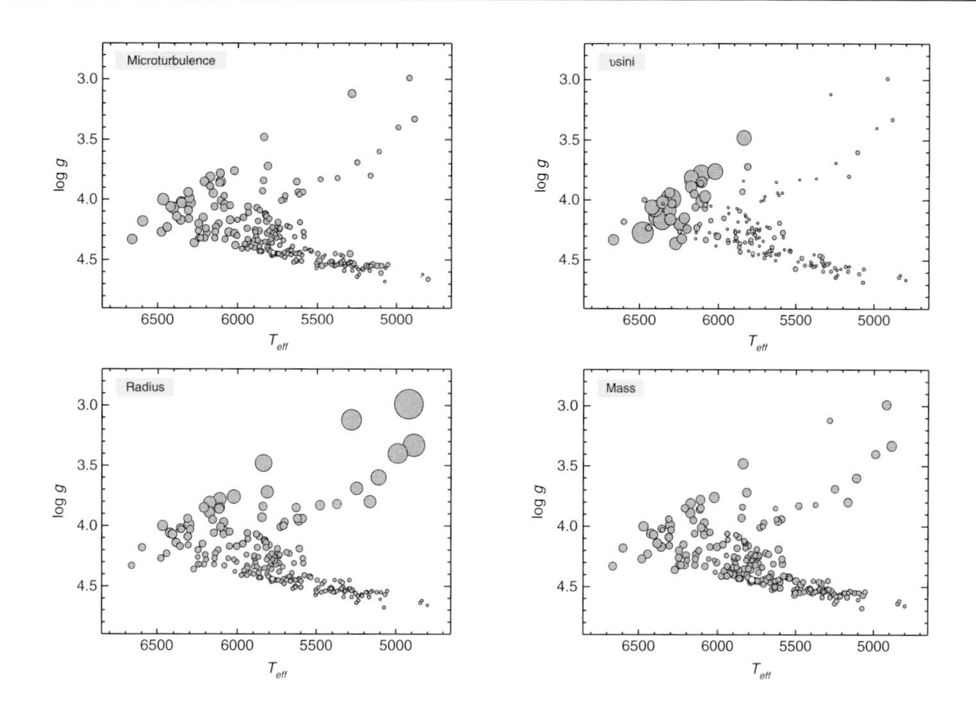

Figure 9.17 Spectroscopic results for the nearby FGK stars, all as a function of $T_{\rm eff}$ and logg: (a) atmospheric microturbulence, with circle diameters proportional to ξ_t. A smooth correlation along the main sequence and towards evolved stars is noticeable; (b) projected rotational velocities, confirming the well-known transition to high $v \sin i$ values for F-type stars. The largest circle corresponds to $v \sin i = 19\,{\rm km s^{-1}}$. Young and chromospherically active stars are omitted; (c) stellar radii, illustrating the larger radii for early main sequence and subgiants; for most objects $\sigma_R/R \sim 5\%$; (d) stellar masses, in which the partly inhomogeneous appearance is a direct reflection of the various metal enrichment levels. Although these astrophysical parameters are determined spectroscopically, it illustrates a dataset defined the Hipparcos parallaxes. From Fuhrmann (2004, Figures 4, 7, 8, 9). Figures copyright Wiley–VCH Verlag GmbH & Co. KGaA, and reproduced with permission.

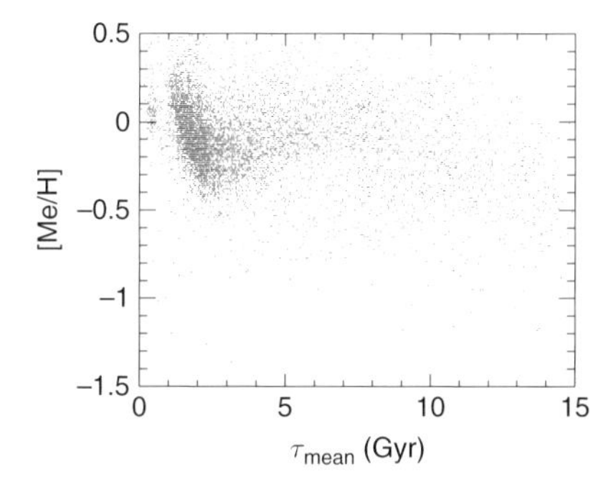

Figure 9.18 Age–metallicity diagram for 5828 Hipparcos stars with $\sigma_\pi/\pi < 0.25$ and well-defined isochrone ages satisfying $\sigma_\tau/\tau \leq 0.5$ and $M_V < 4.4$. Metallicities are derived from Strömgren photometry, and ages from theoretical isochrones. From Feltzing et al. (2001, Figure 10).

Quillen & Garnett (2001) re-examined the age–velocity dispersion relation in the solar neighbourhood using improved stellar age estimates based on Hipparcos parallaxes and theoretical isochrones. The results (Figure 9.20) show that the disk was relatively quiescent, suffering little heating or dispersion increase between 3 and 9 Gyr. However, at an age of 9 Gyr there is an abrupt increase, by almost a factor 2, in the stellar velocity dispersions. They propose that the Milky Way suffered a minor merger 9 Gyr ago that created the thick disk. The quiescent phase is consistent with modest heating caused by scattering from tightly wound transient spiral structure.

Rafikov (2001) used the parameters of the solar neighbourhood estimated from the Hipparcos data by Holmberg & Flynn (2000) to assess the dynamical stability of the disk, showing that it is stable against local axisymmetric density waves propagating through it.

Beers *et al.* (2002) used space velocities derived from Hipparcos and Tycho proper motions (and radial velocities from dedicated spectroscopy) for 39 giant candidates. These were selected in a search for nearby metal-deficient giant stars, whose discovery is complicated if they have the kinematic properties of the (thick) disk rather than of the halo. Their rotational velocities indicate that within 1 kpc from the Sun some 30–40% of metal-poor stars, below [Fe/H] = −1.0 and possibly extending below −1.6, might be associated with the

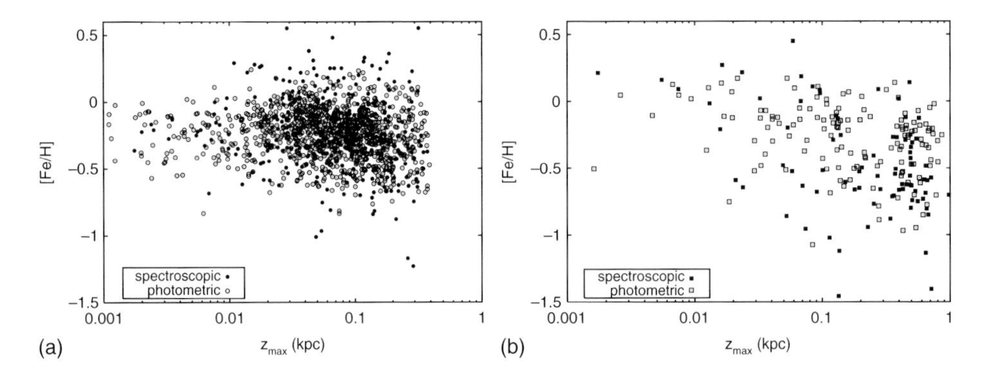

Figure 9.19 The [Fe/H] versus $z_{\max}$ relation for (a) thin disk stars, and (b) thick disk stars. [Fe/H] values are from the catalogue of Cayrel de Strobel et al. (2001), or derived from published uvbyβ photometry. Space motions and the resulting $z_{\max}$ are derived from the Hipparcos data and published radial velocities. From Ibukiyama & Arimoto (2002, Figures 12 and 14).

metal-weak thick disk. The lowest metallicity among these stars is [Fe/H] = −2.35. Such objects are relevant in models in which the metal-weak thick disk arises from the merging of small proto-Galactic fragments.

Nordström *et al.* (2004b), also reported in Nordström *et al.* (2004a) and Andersen *et al.* (2005), describe the extensive Geneva–Copenhagen photometric and radial velocity survey of the solar neighbourhood; consolidated photometric calibration and its consequences are described by Holmberg *et al.* (2007). It provides Strömgren *uvbyβ* photometry for ∼30 000 F and G dwarfs complete to ∼40 pc. The dedicated radial velocity survey was based on around 1000 nights of Coravel observations at the 1-m Observatoire de Haute Provence and 1.5-m Danish telescope at La Silla, and complemented by observations of several hundred fast rotators using the digital spectrometers of the Harvard–Smithsonian Center for Astrophysics, in total comprising ∼63 000 new radial velocity observations for nearly 13 500 stars. Combined with Hipparcos data this provides an all-sky, magnitude-limited, and kinematically-unbiased sample of ∼14 000 F and G dwarfs. It yields: binary identification from the Hipparcos astrometry and radial velocity measurements; metallicity, effective temperatures, interstellar reddening, and rotational velocities from the spectroscopy; isochrone ages and masses from stellar models; and kinematic properties including approximate Galactic orbits from the distances, proper motions, and radial velocities as a result of orbit integration within a simplified potential. The sample was designed to alleviate some of the selection biases in the much-cited study of 189 stars by Edvardsson *et al.* (1993), which excluded known binaries, fast-rotators, unevolved stars, young metal-poor and old metal-rich stars.

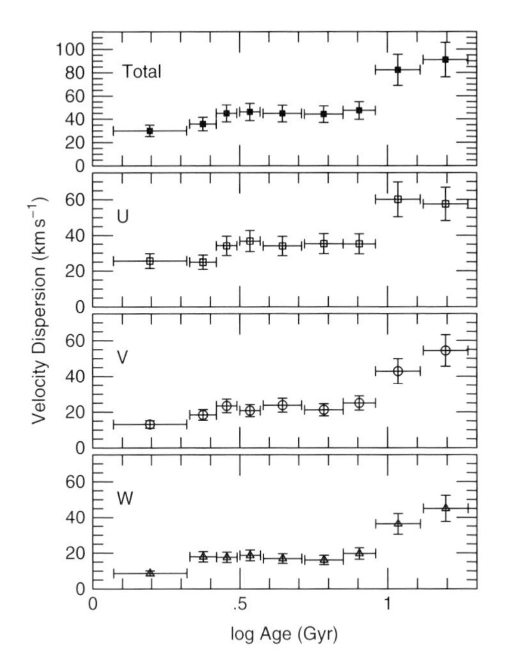

Figure 9.20 Velocity dispersion versus age for the stars from Edvardsson et al. (1993). Velocities are from the Hipparcos data supplemented by radial velocities from the literature; the ages are from Ng & Bertelli (1998). Vertical error bars represent statistical uncertainties; horizontal error bars represent the width of each age bin. From Quillen & Garnett (2001, Figure 1).

Figure 9.21 shows the resulting space velocities in Galactic *UVW* space: the $U - W$ and $V - W$ diagrams show a rather smooth distribution, with the only clearly discernible structure being due to the Hyades. In $U - V$ the abundant structure related to the classic moving groups or stellar streams has been discussed in Section 6.9. Galactic orbits were computed based on a specific Galactic potential, and their catalogue provides the present radial and vertical positions of each star, the computed mean peri- and apo-Galactic orbital distances, the orbital eccentricity, and the maximum distance from the Galactic plane. Their complete sample

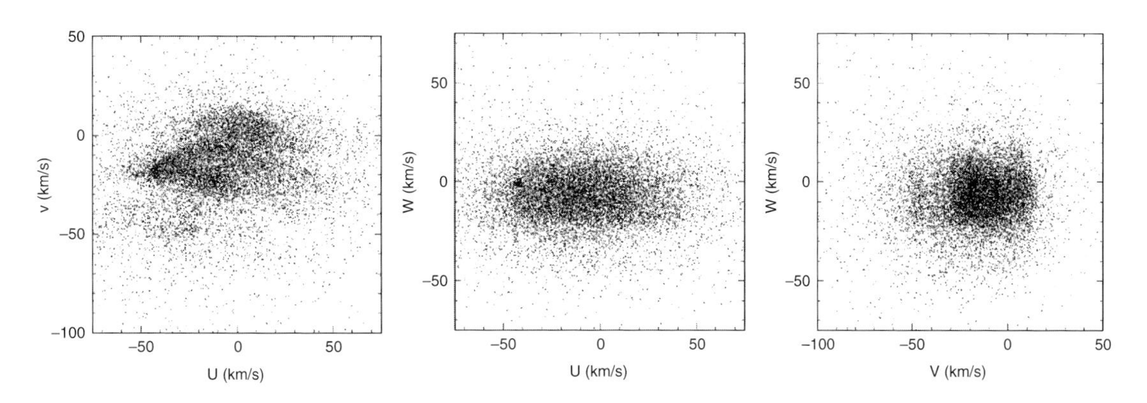

Figure 9.21 U − V, U − W, and V − W diagrams for the sample of ∼14 000 F and G stars from Nordström et al. (2004b). The obvious structure in U − V is related to the classic moving groups or stellar streams, discussed in Section 6.9. In U − W and V − W the only clearly discernible structure is due to the Hyades open cluster. From Nordström et al. (2004b, Figure 20).

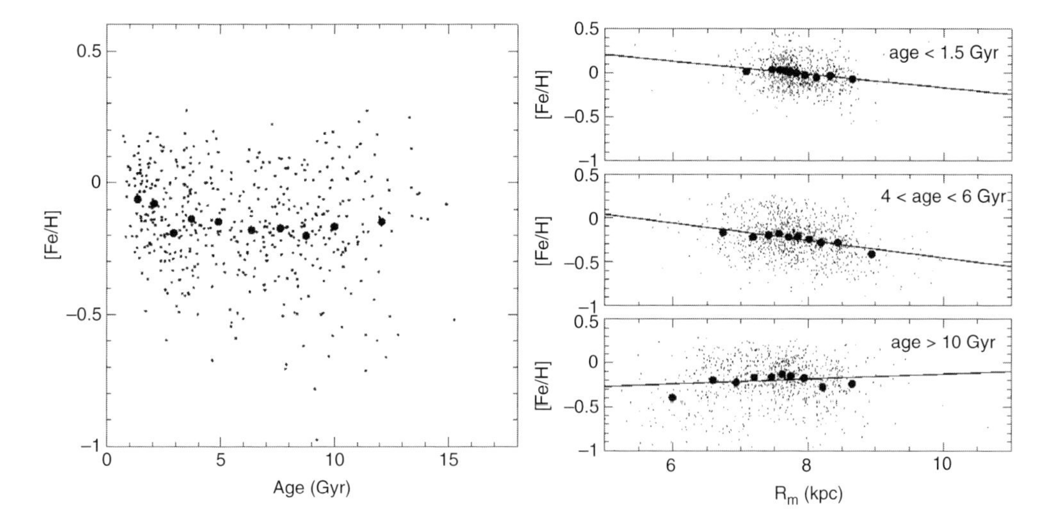

Figure 9.22 Left: age–metallicity diagram for the 462 single stars in the volume-limited subsample within 40 pc with well-defined ages; large dots show mean ages and metallicities in 10 bins with equal numbers of stars. Right: radial metallicity gradient for single stars in three age ranges, where R_m is the mean radius of the stellar orbits constructed from R_{min} and R_{max}. The slopes of the fitted lines are -0.076 ± 0.014, -0.099 ± 0.011, and $+0.028 \pm 0.036$ dex kpc^{-1}. From Nordström et al. (2004b, Figures 28 and 29).

contains 34% binary stars (21% visual double stars and 19% spectroscopic binaries, with some visual binaries also being spectroscopic binaries). This corresponds well to the 32% of binary systems of all types with $P < 10^5$ days found for G dwarfs by Duquennoy & Mayor (1991). Of the 11 060 stars not known to be double, 7817 have measured radial velocities consistent with their being single stars. Their results include: (a) a clearer picture of the age–metallicity relation for field stars, indicating little if any variation of mean metallicity with age, but with a real scatter significantly above that caused by observational errors (Figure 9.22, left); (b) clear signatures of a radial metallicity gradient, except for the oldest stars, with a mild steepening of the radial gradient with age (Figure 9.22, right); (c) an age–velocity relation reflecting the slow increase of the random velocities with age attributed to heating of the disk by massive objects such as spiral arms or giant molecular clouds (see Figure 9.23, and Section 6.9).

Ibukiyama (2004) constructed age–metallicity relations and orbits for 1658 solar neighbourhood stars for which accurate Hipparcos distances are available. Orbital information was used to classify the objects into 1382 thin disk, 229 thick disk, and 47 halo stars. For the thin disk there is a scatter in the age–metallicity relation, even for stars with similar orbits, suggesting that this scatter is an essential feature in the formation and evolution of the Galaxy. The thick disk age–metallicity relation indicated that star formation terminated some 8 Gyr ago, that there is a vertical abundance gradient; and that they are more Ca-rich than thin disk stars with the same [Fe/H].

Cubarsi & Alcobé (2004) and Alcobé & Cubarsi (2005) developed a maximum-entropy approach to search for subpopulations in phase space. They applied the method to a sample of 13 678 Hipparcos disk stars within 300 pc for which full space velocities could be constructed. They identified two discrete components corresponding to the thin and thick disks with the following properties: a thin disk comprising 91% of the sample, with velocity dispersions in cylindrical heliocentric coordinates ($\sigma_{1,2,3} = 28\pm1, 16\pm2, 13\pm1$) km s^{-1}, and vertex deviation $10 \pm 2°$; and a thick disk comprising the remaining 9% of the sample, a velocity dispersion ($65 \pm 2, 39 \pm 9, 41 \pm 2$) km s^{-1}, and vertex deviation $7 \pm 3°$. The thin disk structure is further resolved into three sub-components with a roughly constant velocity dispersion in the vertical direction: 37% assigned to an early-type star population with smaller velocity dispersion, $\sigma_1 = 12 \pm 31$ km s^{-1}, slower rotation and a clear radial expansion which could be identified with the Gould Belt; 38% assigned to young disk stars, with $\sigma_1 = 16 \pm 22$ km s^{-1}, and probably dominated by the moving groups; and the remaining 16% assigned to a continuous old disk population merging with the kinematics of the younger disk stars, reasonably consistent with the results of Famaey *et al.* (2005b). Similar objectives motivated the kinematic analysis described by Abad & Vieira (2005).

Bensby *et al.* (2004) investigated the age–metallicity relation for the thick disk stars using space velocities derived from the Hipparcos data, published radial velocities, and various restrictions in $B-V$ and M_V. They then used kinematical selection criteria to estimate the probability of a given star belonging to the thin disk, thick disk, or halo, using the properties shown in Table 9.6. They finally selected 295 stars as likely belonging to the thick disk. They found an age–metallicity relation in the thick disk, with the median age decreasing by about 5–7 Gyr when going from [Fe/H] ~ -0.8 to [Fe/H] ~ -0.1. Using α-element trends for a local sample of thick disk stars, they inferred that the time scale for the peak rate of supernova Type Ia extends for about 3–4 Gyr. Together this suggests that star formation was an ongoing process for several billion years, strengthening the hypothesis that the thick disk originated from a merger event with a companion galaxy that inflated the pre-existing thin disk.

Arifyanto *et al.* (2005) examined the space motions of 742 subdwarfs, the majority of them thick disk stars as diagnosed from their space velocities, based on the sample of Carney *et al.* (1994), and using Hipparcos and Tycho 2 parallaxes and proper motions, and radial

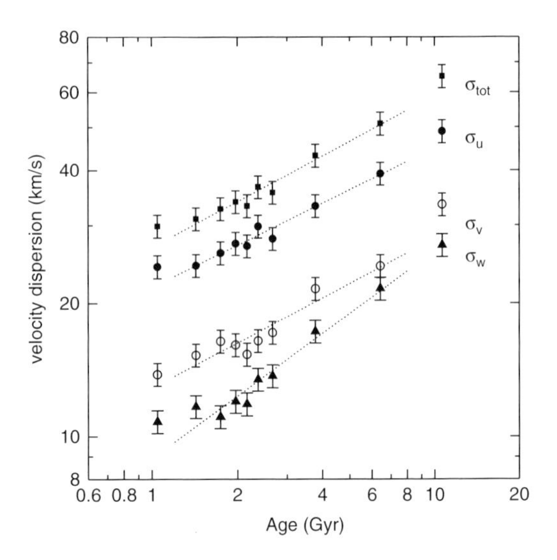

Figure 9.23 Velocity dispersions for single stars with relative age errors <25% as functions of age. From top to bottom, the total, U, V, and W velocity dispersion are plotted in 10 bins with equal numbers of stars. The lines show fitted power laws. The youngest and oldest age bins were excluded from the fits to avoid biases due to unrelaxed young structures and thick disk stars, respectively. From Nordström et al. (2004b, Figure 31). Similar results for their revised photometric calibration are given in Holmberg et al. (2007, Figure 33).

velocities and metallicities from Carney *et al.* (1994). The majority have [Fe/H] > -1 and represent the thick disk population. The halo component, with [Fe/H] < -1.6, is characterised by a low mean rotation velocity and a radially elongated velocity ellipsoid. In the intermediate metallicity range, [Fe/H] = -1 to -1.6, they found a significant number of subdwarfs with disk-like kinematics, which they interpret as a metal-weak thick disk population generated from the disrupted accreted merging galaxy. In the model of an accretion event, the kinematics of the halo depend on the kinematics of the infalling satellites, whereas the kinematics of the thick disk are determined by the heating of the rotating thin disk. The kinematic trace is then visible as the mean orbital rotation velocity of the thick disk stars.

Borkova & Marsakov (2004) considered 77 main-sequence F–G stars in the solar neighbourhood, mainly from the lists by Fuhrmann (1998, 2004), and containing representatives of all Galactic subsystems except the bulge. They compiled iron, magnesium, and europium abundances from published high-dispersion spectra, ages from theoretical isochrones, and space velocities derived using Hipparcos data. They identified a subset of retrograde orbits, similar to those of globular clusters considered to have been accreted by the Galaxy in

Table 9.6 Characteristic velocity dispersions (σ_u, σ_v, and σ_w) for the thin disk, thick disk, and stellar halo. X is the observed fraction of stars for the populations in the solar neighbourhood and v_{asym} is the asymmetric drift. From Bensby et al. (2004, Table 1). Similar values based on Hipparcos data are given elsewhere, e.g. by Reid (1998): $\langle V_{rot} \rangle = 205, 180 \pm 50, 20$ and $\sigma_{uvw} = 20,50,100$ for the thin, thick and halo components respectively; for the thick disk Chiba & Beers (2000) obtain $\langle V_{rot} \rangle = 200$ and $\sigma_{uvw} = 46,50,35$; while Soubiran et al. (2003) obtain $v_{asym} = -51 \pm 5$ and $\sigma_{uvw} = 63 \pm 6, 39 \pm 4, 39 \pm 4$.

	X	σ_u (km s^{-1})	σ_v (km s^{-1})	σ_w (km s^{-1})	v_{asym} (km s^{-1})
Thin disk	0.90	35	20	16	−15
Thick disk	0.10	67	38	35	−46
Halo	0.0015	160	90	90	−220

the past, as individual stars of extragalactic origin, with distinct abundance ratios of r-process and α-elements, and in particular a significant Eu over-abundance relative to Mg, attributed to high-mass Type II supernovae.

Soubiran & Girard (2005) compiled a catalogue of metallicities and abundance ratios from the literature in order to investigate abundance trends of several α and Fe-peak elements in the thin disk and the thick disk. The catalogue includes 743 stars with abundances of Fe, O, Mg, Ca, Ti, Si, Na, Ni and Al in the metallicity range $-1.30 <$ [Fe/H] $< +0.50$. Galactic orbits were constructed from the Hipparcos data and published radial velocities for 639 stars, and ages derived for 322 from isochrone fitting. Two samples kinematically representative of the thin and thick disks were selected, taking into account the Hercules stream which is intermediate in kinematics. Their results show that the two disks are chemically well separated, although overlapping in metallicity, and both show parallel decreasing α elements with increasing metallicity in the interval $-0.8 <$ [Fe/H] < -0.3. The Mg enhancement with respect to Fe of the thick disk is measured to be 0.14 dex. An even larger enhancement is observed for Al. The thick disk is clearly older than the thin disk with tentative evidence of an age–metallicity relation over 2–3 Gyr and a hiatus in star formation before the formation of the thin disk. They do not, however, observe a vertical gradient in the metallicity of the thick disk. The Hercules stream has properties similar to that of the thin disk, with a wider range of metallicity. Metal-rich stars assigned to the thick disk and super-metal-rich stars assigned to the thin disk appear as outliers in all their properties.

Schuster *et al.* (2006) presented a database of 1533 high-velocity and metal-poor stars, all with $uvby\beta$ photometry and complete kinematic data. Again, the [Fe/H] versus Galactic orbital velocities provide separation between the old thin disk, the thick disk and the halo. Based on the isochrones of Bergbusch & VandenBerg (2001), the most metal-poor halo stars have a mean age of 13.0 ± 0.2 Gyr, consistent with current age estimates of the Universe of 13.7 ± 0.2 Gyr from WMAP and subsequent time scales of 0.8–1.0 Gyr for structure formation. Their thick disk stars have age structures in the range 10–12.5 Gyr, consistent with the ΛCDM hierarchical-clustering scheme, where star bursts are triggered by merger and accretion events (e.g. Abadi *et al.*, 2003).

9.6 Properties of the bar

Theoretical models of rotating stellar systems (e.g. Maclaurin or Kalnajs disks, and Maclaurin spheroids) show that there are inherently dynamically unstable modes, the most important being the fundamental $l = m = 2$ ellipsoidal or 'bar' mode, which sets in when the ratio of kinetic energy in rotation to potential energy introduced by the deformation exceeds ~ 0.14 (Binney & Tremaine, 1987, Chapters 5–6). Widespread instabilities seen in rapidly rotating stellar systems are taken as evidence that this unstable mode provides an explanation for bars observed in both S0 and spiral galaxies, including the Milky Way. Given that bars are straight, the bar pattern is presumed to rotate rigidly at some rotation rate or 'pattern speed'. Yet this simple picture is complicated by the presence of differential rather than uniform Galaxy rotation, and by the possibility of other (secular) instabilities due to processes such as tidal heating and mergers. Manifestations of the bar extend to the solar neighbourhood, where stars with larger velocities relative to the Sun frequently move on elliptical orbits taking them to sufficiently small Galactocentric radii that they are profoundly influenced by the non-axisymmetric potential of the bar.

In our own galaxy the shape, orientation, and scale length of the bulge, and the presence of an additional bar-like structure in the disk plane, remain matters of theoretical and observational investigation, complicated partly because the Sun is located near to the Galactic plane from where the non-axisymmetric structure is less visible, and partly due to its obscuration by dust (Gerhard, 2002). Evidence for a bar comes from observations of motions of atomic and molecular gas, near infrared photometry, IRAS and clump star counts, and stellar kinematics (Gerhard, 1996). It has a semi-major axis $a = 3.1$–3.5 kpc (around half the Sun's distance from the Galactic centre); an axis ratio 10 : 3–4 : 3; an orientation angle between semi-major axis and Sun–Galaxy centre line of $\phi_b = 15 - 35°$; a pattern speed $\Omega_b = 50$–$60\,\mathrm{km\,s^{-1}\,kpc^{-1}}$ with a corresponding co-rotation radius of 3.3–5 kpc; a mass of $\sim 10^{10}\,M_\odot$; and an age of up to 6–7 Gyr (Gerhard, 2002).

While there is evidence that bars are confined to regions of the galaxy within the co-rotation resonance, their dynamical influence may extend to the outer parts through resonant effects: asymmetries in the potential may impart specific signatures in the velocity distribution in the solar neighbourhood. Even though the non-axisymmetric component of the bar-induced forces fall off steeply, as r^{-4} for the quadrupole component, reaching only about 1% of the total force at R_0, its otherwise restricted influence at the solar radius may be observable for orbits that are nearly in resonance with it.

Hipparcos has demonstrated rather convincing evidence for such an effect, until now seen as an unexplained anomaly in the local stellar velocity distribution. It is seen as a broad stream of low angular momentum and mainly outward moving stars ($u < 0$) with a mean heliocentric asymmetric drift of $\sim 45\,\mathrm{km\,s^{-1}}$, typical of the thick disk (Gilmore *et al.*, 1989). Classified by Eggen as his moving group ζ Hercules, but referred to also as the u-anomaly, it was also recognised by Mayor (1972). Clearest evidence for the stream comes from the Hipparcos local velocity distribution (u, v) data, already noted in the studies of moving groups (see Section 6.9), and from the proper motions combined with radial velocities from which dataset the hypothesis that the effect could be ascribed to the bar was developed (Raboud *et al.*, 1998; Martinet & Raboud, 1999).

Dehnen (1999b) focused attention on the (u, v) distribution for 6018 late-type stars from Hipparcos (Figure 9.24). In addition to the low-velocity region (solid ellipse) comprising several peaks associated with the moving groups and discussed in Section 6.9, is an intermediate-velocity structure (dashed ellipse) containing 15% of the late-type stars but few early-type stars, and only vaguely recognisable in pre-Hipparcos data. The mean outward motion rules out axisymmetric equilibrium, while their clear separation from the low-velocity stars suggests an orbital resonance.

Apart from the co-rotation resonance, a rotating bisymmetric bar creates, from Equation 9.13a with $m = 2$ for the bar mode, an inner and outer Lindblad resonance, the latter occurring when $\Omega_b - \Omega_0 = \kappa_0/2$. Dehnen (1999b) showed that the bi-modality so clearly evident in Figure 9.24 can be explained by the bar's outer Lindblad resonance, provided that the Sun lies

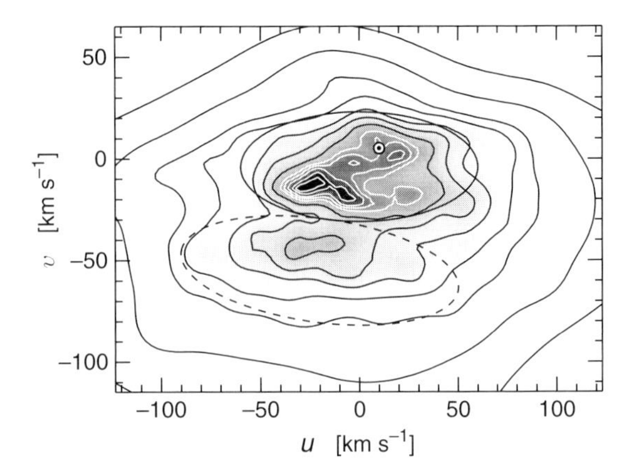

Figure 9.24 The velocity distribution $f(u, v)$ in the solar neighbourhood from Hipparcos data for 3527 main-sequence stars with $B - V \geq 0.6$ and 2491 mainly late-type non-main-sequence stars. The symbol $\odot$ indicates the solar velocity. Samples of early-type stars contribute almost exclusively to the low-velocity region (solid ellipse), which contains the most prominent moving groups, and is referred to as the 'LSR mode'. Intermediate velocities (broken ellipse) are mainly represented by late-type stars, of which $\sim$ 15% fall into this region. Interpreted as arising from the bar's outer Lindblad resonance, it is referred to as the 'OLR mode'. From Dehnen (1999b, Figure 1).

outside it.[2] The observed saddle point between the two stable modes yields $v_{\rm OLR} = -31 \pm 3\,\mathrm{km\,s^{-1}}$ and thus $R_0 - R_{\rm OLR} \simeq |v_{\rm OLR}/\Theta_0| R_0 \simeq 1$ kpc for a flat rotation curve at R_0. Thus, only stars with epicycle amplitudes $\gtrsim 1$ kpc, i.e. predominantly late-type stars, can visit the solar neighbourhood from inside the outer Lindblad resonance, consistent with the absence of early-type stars from this velocity component. Such orbits are elongated perpendicular to the bar's major axis, and move on average outward for bar angles $0° < \phi < 90°$. Evidence from the form of the bar compared to that seen in other galaxies, as well as infrared photometry, suggests that $\phi_b \sim 45°$.

Dehnen (1999b) used $v_\odot = 5.25 \pm 0.62\,\mathrm{km\,s^{-1}}$ and $R_0 \simeq 8$ kpc to estimate a pattern speed of the bar of

$$\Omega_b = 53 \pm 3\ \mathrm{km\,s^{-1}\,kpc^{-1}} \tag{9.50}$$

The estimate is rather insensitive to the choice of R_0, with the largest uncertainty arising from the proper

[2] Closed or quasi-periodic orbits in a barred potential are elongated either parallel or perpendicular to the bar, referred to as $x_1(1)$ or $x_1(2)$ orbits respectively, with the orientation changing at each of the fundamental resonances: anti-aligned inside the inner Lindblad resonance, aligned between the inner Lindblad resonance and the co-rotation resonance, anti-aligned between the co-rotation resonance and the outer Lindblad resonance, and aligned again outside of the outer Lindblad resonance.

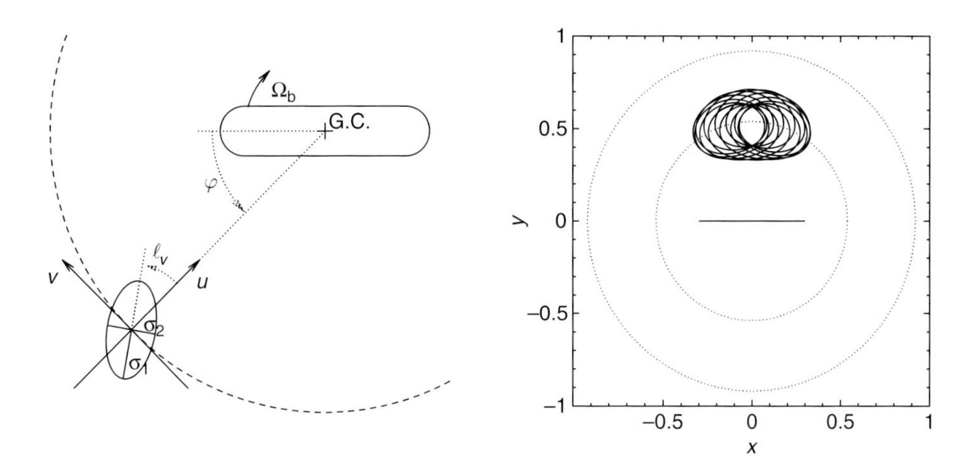

Figure 9.25 Left: modelling of the bar, showing the bar and its rotation with respect to the Galactic centre, and a resulting velocity dispersion ellipsoid with its principal components σ_1 and σ_2 and positive vertex deviation l_v. Right: the same study found that many stellar orbits are trapped near co-rotation, with the L_4 Lagrange point circulating in a retrograde sense (anticlockwise for a clockwise rotating bar). The bar is shown by a line, while the dotted circles are the co-rotation and outer Lindblad resonances. From Mühlbauer & Dehnen (2003, Figures 1 and 5).

motion of Sgr A* and its interpretation as the reflex motion of the Sun. He also derived a co-rotation radius $R_{\rm CR} \sim 0.5{-}0.6R_0$ which, combined with infrared photometry estimates of a bar length $R_b \sim 0.45R_0$, yields $R_{\rm CR}/R_b \sim 1.25 \pm 0.2$, in agreement with estimates from external galaxies.

Dehnen (2000) made detailed numerical orbital simulations depending on bar angle, shape of the rotation curve, etc., which revealed other features in common with the observed velocity distribution of Figure 9.24. Some features close to the location of the known moving groups arise from stars on orbits that are nearly closed, originating from near the outer Lindblad resonance and with large deviations from circularity such that they nevertheless visit the solar neighbourhood. This study leaves a clear picture of the main velocity peak (or 'LSR' mode' in his terminology) resulting from the bar elongated orbits, and the Hercules stream, or 'OLR mode', resulting from the anti-bar elongated orbits. The valley between the two modes corresponds to stars on unstable outer Lindblad resonance orbits.

Fux (2001) investigated further how the barred potential divides the disk's phase space into regions of regular and chaotic motion using three-dimensional N-body simulations, introducing refinements such as a variation of the bar pattern speed with time, and the change in effective potential for stars moving outside of the plane. He also concluded that the splitting of the (LSR) orbits into the Pleiades, Coma Ber, Sirius and other streams is unlikely to be related to the bar.

While moving groups can explain the high vertex deviation of young populations, the old populations should also show such a vertex deviation if the potential deviates from axisymmetry. Mühlbauer & Dehnen (2003) modelled the dynamical effects of the bar using a restricted N-body numerical method, equivalent to first-order perturbation theory, in which the self-gravity due to the wake induced by the bar is neglected (Figure 9.25). Orbits were integrated and velocity moments computed from the phase-space positions, and their simulation results were in broad agreement with analytic expressions for mean velocities under a non-axisymmetric perturbation of multipole order m, for which $\overline{u} \propto \sin 2\phi$ and $\overline{v} \propto \cos 2\phi$ for a flat rotation curve, with only limited sensitivity to its precise gradient at R_0. They determined the resulting radial motion of the LSR, and the resulting vertex deviation and velocity dispersion. The ratio of the principal axes of the velocity dispersion ellipsoid (the eigenvalues $\sigma^2_{1,2}$ of the tensor σ^2) for the undisturbed case (for which $\sigma_1 = \sigma_{uu}$ and $\sigma_2 = \sigma_{uv}$) is

$$\lim_{\sigma\to 0} \frac{\sigma_2^2}{\sigma_1^2} = \frac{\kappa^2}{4\Omega^2} \equiv \frac{1}{2}\left(1 + \frac{{\rm d}\ln\Theta}{{\rm d}\ln R}\right) \tag{9.51}$$

which, for a flat rotation curve, is expected to be 0.5 (Oort's relation). The vertex deviation

$$l_v = \frac{1}{2}\arctan\frac{2\sigma_{uv}^2}{\sigma_{uu}^2 - \sigma_{uv}^2} \tag{9.52}$$

is the angle between the direction of σ_1 and the line to the Galactic centre, with $l_v = 0$ for axisymmetric models. They assumed that the Sun lags the bar major axis by $\sim 20°$ (Dehnen, 2000); this value is uncertain but mostly quoted to lie between 15–45°.

Their simulations showed a radial motion of the LSR frequently reaching magnitudes of order $0.02\Theta_0$,

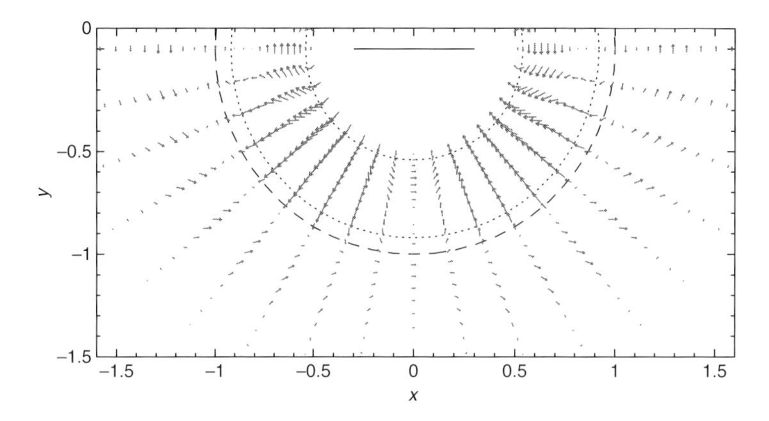

Figure 9.26 Bar-induced deviations of the mean velocity from the unperturbed state up to $m = 2$ for $\sigma_0 = 0.2v\Theta_0$. The bar is aligned with the horizontal axis, and rotates clockwise. The solar circle is dashed, with dotted circles again showing the co-rotation resonance and the outer Lindblad resonance. From Mühlbauer & Dehnen (2003, Figure 8).

or $\simeq 5\,\mathrm{km\,s^{-1}}$ for the Milky Way (for a circular velocity $\Theta_0 \sim 250\,\mathrm{km\,s^{-1}}$), and maximal at $\phi = 45°$, close to the proposed position of the Sun. However, they also found that $\overline{u}$ swings through zero shortly outside the outer Lindblad resonance, and if the Sun is close to this point, the bar-induced radial motion of the LSR is likely to be small (Figure 9.26), consistent with the value of $\overline{u} = -1.4 \pm 2.2\,\mathrm{km\,s^{-1}}$ for the mean radial motion of a sample of halo subdwarfs with respect to the LSR (Gould, 2003). While the traditional interpretation of vertex deviations is that young stars deviate from equilibrium, they also found a vertex deviation of order that observed, $\sim 10°$, arises in equilibrium in the case of a bar with orientation, strength, and pattern speed consistent with other data, suggesting that the vertex deviation in the solar neighbourhood could be predominantly caused by deviations from axisymmetry. Finally, they found that the ratio of σ_2^2/σ_1^2 is clearly affected by the bar, with values less than 0.5 (Oort's value for a flat rotation curve and an unperturbed disk) being possible, consistent with $\sigma_2^2/\sigma_1^2 \sim 0.42$ for the old stellar disk found by Dehnen & Binney (1998a). In summary, they argued that the kinematics of the solar neighbourhood are in agreement with the lowest-order kinematic deviations from axisymmetric equilibrium attributable entirely to the influence of the Galactic bar.

In their Hipparcos study of space velocities of Mira variables, Feast & Whitelock (2000) showed a dependence of kinematics on period, with the most extreme behaviour in the range 145–200 days. These 18 stars showed mean motions of $u = -73 \pm 17, v = -97 \pm 20, w = -11 \pm 11\,\mathrm{km\,s^{-1}}$, i.e. in a compact region to the bottom left of Figure 9.24. Again, this strong radially outward motion was taken to indicate Galactic axial asymmetry, and they interpreted the orbital asymmetry as indicating an extension of the bar out to beyond the solar circle, with the orbits of some local Miras probably penetrating into the bulge. The interpretation was questioned by Kharchenko *et al.* (2002) using a larger sample.

9.7 Properties of the spiral arms

Spiral arms are the primary sites of ongoing star formation in the Galaxy, and their behaviour plays a central role in understanding its dynamical and chemical evolution. They are the subject of great theoretical and observational interest, and remain poorly understood. The Lin–Shu hypothesis (Lin & Shu, 1964), that spiral structure consists of a density wave maintained in a steady state over many rotation periods, is frequently adopted as a paradigm. Specifically, it is the fact that the wave pattern is considered to be in the rotating gravitational potential, rather than locked into specific stars, that theory avoids an otherwise excessively tight 'winding' of the spiral arms after $\sim 10^{10}$ yr of differential Galactic rotation. Interstellar gas passes through the spiral potentials, is periodically compressed, leading to rapid star formation lighting up the quasi-instantaneous location of the spiral arms. In this model, the density-wave pattern extends only between the inner and outer Lindblad resonances (Equation 9.13a, in which m is the number of spiral arms).

The spiral pattern may simply originate as an unstable normal mode of a galaxy disk, with an amplitude building up until increased damping caused by energy dissipation in the interstellar medium leads to a stable finite amplitude perturbation. Nevertheless it has proved difficult to determine the basic model parameters such as the pattern speed (the angular speed of the spiral pattern rotation viewed from an inertial frame), or the location and number of the Lindblad resonances. Thus other models in which, for example, the spiral structure is a transitory even if recurrent instability, possibly tidally-induced, are also considered.

In some galaxies, and in some models, a central bar appears to provide the source of spiral waves that propagate outward from the inner Lindblad resonance. In general, however, the pattern speeds of the bar and of the spiral waves (if both exist in the same galaxy) have no unique relation: the rotating pattern of spiral density-wave perturbations is associated with its own co-rotation circle, at which the rotation velocity of the galactic disk coincides with that of the spiral pattern.

A summary of what is known about spiral density waves propagating in the solar neighbourhood is given by Quillen & Minchev (2005): the disk may contain a dominant two-armed, or tightly-wound four-armed structure, or may be a superposition of the two (this confusion is not surprising given results from numerical simulations such as those seen in Figure 9.27, left). Values of the pitch angles of the arms range from 5–27° (e.g. Elmegreen, 1985). Of the less ambiguous structures, the Carina–Sagittarius arm is 0.9 kpc from the Sun in the direction of the Galactic centre, and the Perseus arm is toward the anti-centre, and separated by about 2.5 kpc from the former (see Figure 10.11 below).

More controversial are estimates of the pattern speed. Various methods have been adopted, using age gradients of young objects, the birth place of intermediate-age open clusters, and velocity matching to spiral density waves. A compilation of estimates has been presented by Shaviv (2003, Table 3). These fall into two main groups. The first group clusters around a pattern speed of $\Omega_s \sim$ 13–16 km s^{-1} kpc^{-1} which, taking into account the flat rotation curve in the outer parts of the Galaxy, places the co-rotation radius at around 16 kpc, i.e. at the outer edge of the disk (Lin *et al.*, 1969; Yuan, 1969a,b; Palouš *et al.*, 1977; Gordon, 1978; Grivnev, 1983; Ivanov, 1983; Comerón & Torra, 1991). Other estimates cluster around $\Omega_s \sim 20$ km s^{-1} kpc^{-1}, placing the co-rotation radius closer to the solar circle (Marochnik *et al.*, 1972; Crézé & Mennessier, 1973; Palouš *et al.*, 1977; Nelson & Matsuda, 1977; Mishurov *et al.*, 1979; Grivnev, 1981; Efremov, 1983; Amaral & Lépine, 1997). Avedisova (1989) observed the age gradient across the Carina–Sagittarius spiral arm to derive an even larger value of $\Omega_s \sim$ 28–30 km s^{-1} kpc^{-1}.

Mishurov & Zenina (1999a,b) used the Hipparcos observations of Galactic Cepheids, combined with existing radial velocity measurements, to estimate the parameters of the spiral pattern. Cepheids are bright enough to be seen at large distances from the Sun, comparable to the inter-arm spacing, with well-determined proper motions from Hipparcos, and distances which can be estimated from the period–luminosity relation. Various selection criteria led to the use of 131 radial and 117 transverse velocities of short-period Cepheids ($P < 9$ d), with objects at distances of typically $\lesssim 4$ kpc. The total gravitational potential of the Galaxy was represented by

$$\Phi_G = \Phi_0 + \Phi_S = \Phi_0 + A\cos\chi \tag{9.53a}$$

$$\chi = m(\cot i \log(R/R_0) - \varphi) + \chi_0 \tag{9.53b}$$

where Φ_0 is the axisymmetric steady-state potential, Φ_S the perturbation due to the spiral-density wave in cylindrical coordinates (R, φ), m is the number of spiral arms, and i is the pitch angle, i.e. the angle at any radius r between the tangent to the arm and the circle r = constant. Radial and azimuthal stellar velocity components are represented as $\tilde{u} = f_u \cos\chi$ and $\tilde{v} = f_v \sin\chi$, where f_u and f_v are the amplitudes of the perturbed velocities. The Cepheid observations are used to estimate values for the nine model parameters, $\Omega_0, A, R_0\Omega_0'', f_u, f_v, i, \chi_\odot, u_\odot, v_\odot$; from which the difference between the angular rotation velocity of the spiral pattern and that of the Sun is determined, $\Delta\Omega = \Omega_s - \Omega_0$, a hence the displacement of the Sun relative to the co-rotation radius. For $m = 2$ and a sample restricted to 2 kpc they found $i = 6°$, $\chi_\odot = 322 \pm 9°$, along with the values of $u_\odot, v_\odot$ shown in Table 9.2. While unable to discriminate between models with two or four spiral arms, they concluded that the effects of the spiral arm perturbations are significant and, based on an assumed $R_0 =$ 7.5 kpc, found $R_c = 7.4$ kpc, placing the Sun within 0.1 kpc of the co-rotation resonance, and an associated spiral pattern speed of $\Omega_s = 27.7 \pm 3$ km s^{-1} kpc^{-1}.

Lépine *et al.* (2001) extended the analysis, again using the Hipparcos Cepheid observations, to suggest that the Galaxy's structure is well-represented by a superposition of two- and four-armed patterns. They also adopted $R_0 = 7.5$ kpc, then demonstrated good consistency with the self-sustained spiral-wave model in which superimposed two-armed and four-armed modes have different pitch angles (6° and 12° respectively). N-particle simulations were run to make visible the structure of the potential 'visible' and to construct the corresponding $l - v$ diagram (Figure 9.27). The distribution of a sample of Galactic H II regions gave support to the model. The model confirms the suggestion that the Sun lies very close to the co-rotation circle, and also provides estimates of the Oort constants (Table 9.3) and solar motion values (Table 9.2).

Investigation of the agreement between the observed u-anomaly, or Hercules stream, and its association with the bar's outer Lindblad resonance was developed by Quillen (2003), who investigated dynamical evolution under the combined potential of both bar and spiral. The bar pattern was adjusted in speed to match the location of the u-anomaly ($\Omega_b/\Omega_0 = 1.90$, equivalent to that given by Equation 9.50), and the number of spiral arms (two or four) and the spiral pattern speed were

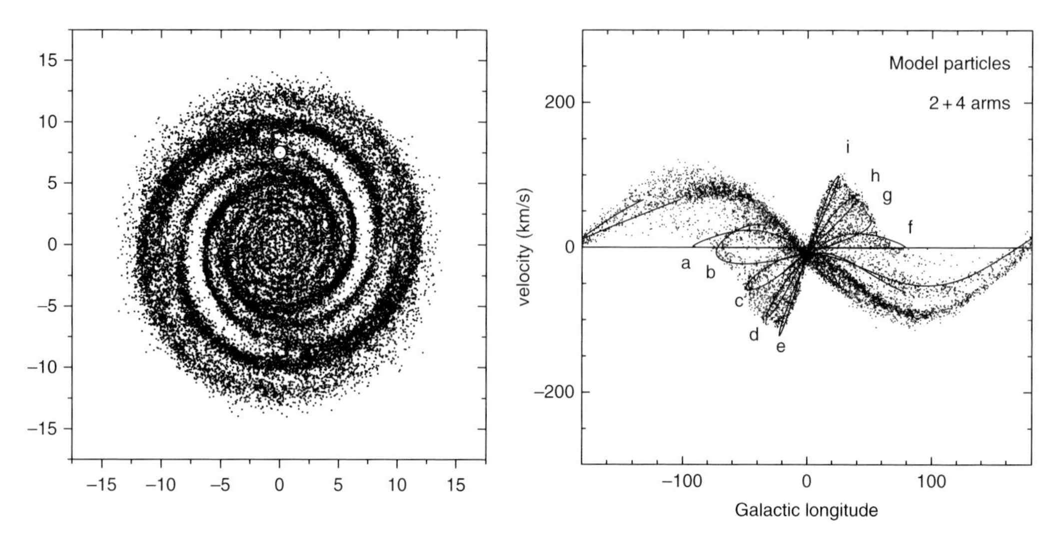

Figure 9.27 Left: visible structure of the Galaxy, constructed from cloud-particle simulations based on the spiral-wave model derived by Lépine et al. (2001, Figure 3), using the Hipparcos Cepheid observations fitted to a superposed 2+4 self-sustained wave harmonic model of spiral structure. Scale is in kpc, with the location of the Sun indicated. Right: the reconstructed $l - v$ diagram (velocity component in Galactic longitude versus Galactic longitude) for the same model, computed from cloud-particle simulations, and with the lines indicating the fits to the observed H II regions, from Lépine et al. (2001, Figure 7).

variables. Since the bar and spiral structure in general have different pattern speeds, the system resembles a forced pendulum, with chaotic orbits consequently occupying larger regions than those found by Fux (2001). For the streams seen in the local velocity distribution to persist, widespread chaos due to resonance overlap must be avoided, constraining the spiral pattern speed passing through the bar's outer Lindblad resonance to be $\Omega_s/\Omega_0 \gtrsim 0.45$ if the spiral structure is two-armed, or $\Omega_s/\Omega_0 \gtrsim 0.75$ if four-armed. Boundaries of the u-anomaly region could be identified with the onset of resulting chaotic orbits.

Quillen & Minchev (2005) extended the analysis in an attempt to explain the observed structure of the moving groups, i.e. the structure in the u, v diagram closer to the circular orbits (or LSR mode), as a direct result of spiral density wave resonance, rather than as the dissolution of large agglomerations associated with spiral arms (Asiain *et al.*, 1999) or, to account for the existence of older stars, as the superposition of moving groups of different ages (Chereul *et al.*, 1998), or to irregularities in the Galactic potential (De Simone *et al.*, 2004). They searched for orbits that are nearly closed or periodic in a frame moving with the spiral pattern. A moderately strong two-armed perturbation leads to a very narrow range of acceptable pattern speeds of $\Omega_s = 18.1 \pm 0.8\,\mathrm{km\,s^{-1}\,kpc^{-1}}$, or $\Omega_s/\Omega_0 = 0.66 \pm 0.3$ for the assumed $R_0 = 8$ kpc and $\Theta_0 = 220\,\mathrm{km\,s^{-1}}$, a value consistent with other recent estimates although not with that of Mishurov & Zenina (1999b). This places the Sun near the 4:1 Lindblad resonance, resulting in two regions near the origin in the (u, v) plane which exhibit nearly closed orbits, and hence resonance 'islands' populated by stars spanning a range of ages, as are the moving groups. For a specific angular offset of the spiral pattern $\gamma_2 = 15°$, nearly closed orbits can be matched to the location of existing moving groups. One family, with diamond-shaped orbits with peaks on top of the two dominant stellar arms and mean radii within R_0, corresponds to the Hyades/Pleiades moving group; the model predicts that the two groups are kinematically related, as indicated by Nordström *et al.* (2004b) and Famaey *et al.* (2005b). The other, with orbits 45° out of phase and with guiding radii outside R_0, correspond to the Coma Berenices moving group. The authors postulate that the Sirius/Ursa Major moving group could be related to a higher-order Lindblad resonance. They concluded that if the Sun is indeed located near a Lindblad resonance causing large epicyclic amplitudes, both the velocity of the LSR, and estimates of the Oort constants, are likely to be biased.

Fernández *et al.* (2001) set out to establish whether stars which make up the spiral patterns actually follow velocities predicted by density wave theory. They studied two Hipparcos datasets, one containing 448 O- and B-type stars (307 with radial velocities) and another composed of 186 Cepheids (165 with radial velocities). The velocity field of these young stars, beyond the influence of the Gould Belt at $R < 0.6$ kpc (Figure 9.28), is inadequately explained by a Galactic model which includes only solar motion and differential Galactic rotation. An improved fit is obtained

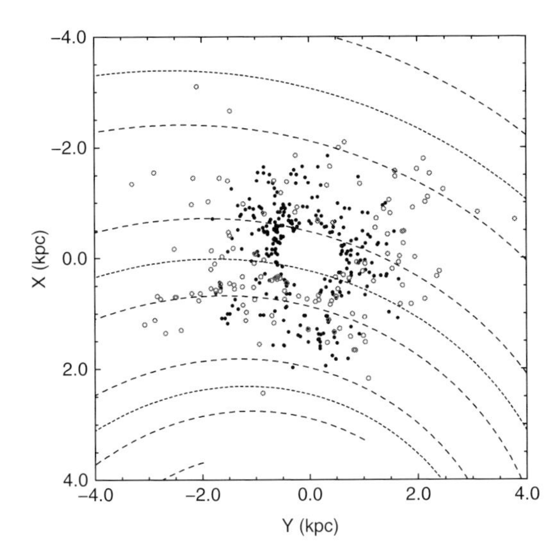

Figure 9.28 Star distribution in the X − Y galactic plane for O and B stars (filled circles) with 0.6 < R < 2 kpc and Cepheids (open circles) with 0.6 < R < 4 kpc (short cosmic scale distances). Spiral arm positions assume $\psi_\odot = 330°$. Dotted lines show the centre of the spiral arms as given by the minimum of the spiral potential ($\psi = 0°$), and dashed lines draw their approximate edges ($\psi = \pm 90°$). Note that the position of optical tracers of the spiral arms are in general displaced from $\psi = 0°$. From Fernández et al. (2001, Figure 04).

including effects of spiral arm kinematics, with the number of arms, the pitch angle, and the Sun's phase within the potential as free parameters. The results show a Galactic rotation curve with a classical value of the A Oort constant for the O- and B-star sample ($A = 13.7$–$13.8\,\mathrm{km\,s^{-1}\,kpc^{-1}}$), but a higher value for Cepheids ($A = 14.9$–$16.9\,\mathrm{km\,s^{-1}\,kpc^{-1}}$, depending on the cosmic distance scale adopted). The second-order term was found to be small, compatible with zero, but a K-term (indicating a compression, but with unidentified physical origin) out to 4 kpc, of $K = -(1-3)\,\mathrm{km\,s^{-1}\,kpc^{-1}}$. The results obtained for the spiral structure from O and B stars and Cepheids show good agreement. The Sun is located relatively near the minimum of the spiral perturbation potential ($\psi_\odot = 284$–$380°$, placing it between the centre and outer edge of a spiral arm) and very near the co-rotation circle. The angular rotation velocity of the spiral pattern was found to be $\Omega_s \sim 30\,\mathrm{km\,s^{-1}\,kpc^{-1}}$.

Mel'Nik *et al.* (1998) and Sitnik & Mel'Nik (1999) studied the detailed velocity distribution of Cepheids, at distances of 1–2 kpc, in the local Carina–Sagittarius and Cygnus–Orion spiral arms, the latter traced by three giant star-gas complexes including the 'local group' complexes. They used radial velocities and Hipparcos proper motions to derive a mean velocity for each association, which they then demonstrated consistency with the predictions of density wave theory, which leads to systematic motions of stars and gas along and across the arm and inter-arm regions (Figure 9.29). They used $R_0 = 7.1$ kpc, and deduced the Sun's velocity with respect to the LSR ($u_\odot, v_\odot, w_\odot = 10.0, 13.0, 7\,\mathrm{km\,s^{-1}}$) from the Cepheids such that the residual velocities are in the frame of reference that rotates in a circular orbit with the rotation velocity of young disk stars. In the Cygnus–Orion arm, for example, the residual radial velocity component is directed toward the Galactic centre with magnitude decreasing from 10–23 $\mathrm{km\,s^{-1}}$ (in Cyg OB1, OB3, OB8, and OB9) to 2–8 $\mathrm{km\,s^{-1}}$ (in Cyg OB7, Cep OB2 and OB3). The residual azimuthal component changes direction from opposite to that of Galactic rotation ($-16\ldots-3\,\mathrm{km\,s^{-1}}$) for the former associations, to that coinciding with the sense of Galactic rotation ($0\ldots+7\,\mathrm{km\,s^{-1}}$) for the latter. Per OB2, Ori OB1, Mon OB1, and Coll 121 are located in the inter-arm region, with corresponding radial and azimuthal components. They conclude that the Cygnus–Orion arm lies within the co-rotation radius. They also identified a concentration of Cepheids towards the outer edge of the arm, and an approximately two-fold increase of density in the arm within 1–2 kpc of the Sun. The position of the Cygnus–Orion arm is then given by pitch angle $i = 7°$ and $\chi_0 = 5°$, and for the Carina–Sagittarius arm $i = 7°$ and $\chi_0 = 260°$, in reasonable agreement with parameters for the two-armed model of Mishurov & Zenina (1999b) with $i = 6°$ and $\chi_0 = 322 \pm 9°$.

A rather different conclusion was drawn by Yano *et al.* (2002), who studied 78 Hipparcos Cepheids lying between 0.5–4 kpc, to see whether the internal motions within the spiral arms obeyed the linear correlation between positional phase and their epicyclic motions expected in density wave theory. They used $R_0 = 8.3$ kpc, $i = -12°$, $A = 14.5\,\mathrm{km\,s^{-1}\,kpc^{-1}}$ and $\Theta = 220\,\mathrm{km\,s^{-1}}$. They concluded that no such correlation was present for the Cepheids, probably due to their inaccurate space motions. Similarly inconclusive results were found for a sample of O–B5 stars. Evidently, the space motion accuracies and sample sizes are at the limits of what is needed to establish such velocity patterns, and we are left unclear whether the spiral wave parameters derived by some workers are spurious, or whether the additional structural terms considered fixed in the work of Yano *et al.* (2002) have masked the faint signatures of the spiral patterns detected by some workers.

Dias & Lépine (2005) estimated the spiral pattern rotation velocity by determining the birthplaces of some 600 open clusters in the Galactic disk as a function of age, either by assuming that their orbits are circular, or by integrating their Galactic orbits over time, in which the Hipparcos and Tycho data were used for estimates of the proper motions, distances, and ages.

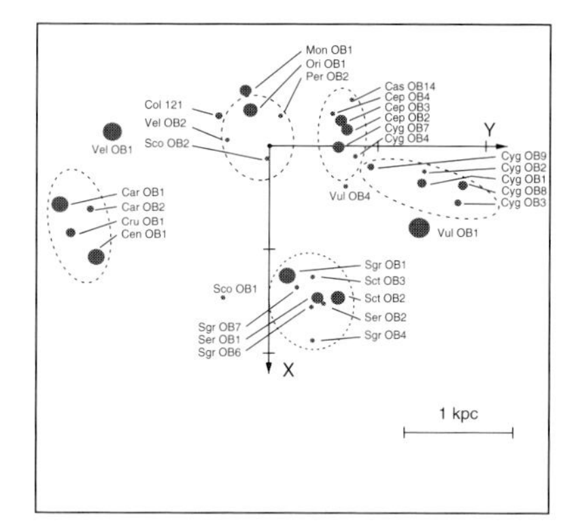

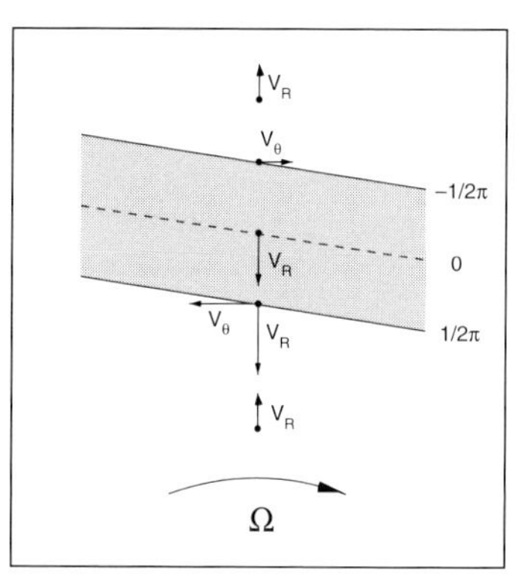

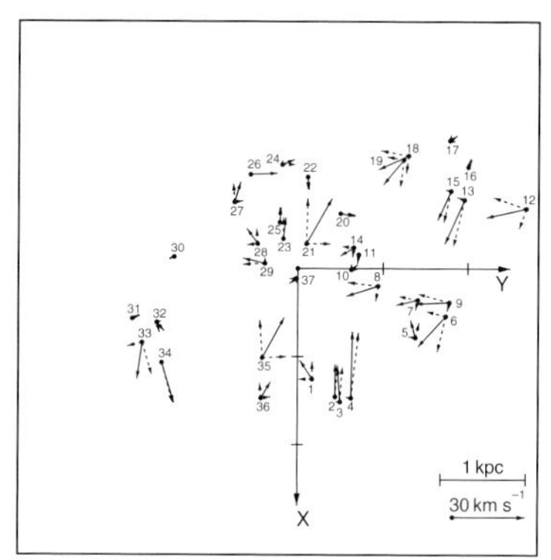

Figure 9.29 Left: the distribution of associations in the Cygnus–Orion (upper three dashed regions) and the Carina–Sagittarius arms (lower two-dashed regions), projected on the Galactic plane, with X toward the Galactic centre, the Sun lying at the coordinate origin, and the solid circles corresponding to the association size. Middle: the expected residual velocity field in the spiral density wave within the co-rotation radius. Vectors V_R and V_θ indicate the radial and azimuthal velocity perturbations, and Ω indicates the sense of Galactic rotation. Right: the observed residual velocity field of the associations in the two arms. Associated with each association is shown the space vector (solid line) and the radial and azimuthal components. Dotted lines mark the possible location of the middle of the two arms (phase $\chi = 0^\circ$). From Sitnik & Mel'Nik (1999, Figures. 1b, 5a) and Sitnik (2003, Figure 1a) respectively (the latter updates Figure 5b of the former), reproduced with kind permission of Springer Science and Business Media.

Their results (Figure 9.30) demonstrate that a dominant fraction of open clusters are formed in spiral arms, and that the spiral arms rotate like a rigid body, with the three main arms in the solar neighbourhood showing the same rotation velocity. Numerical results depend on the choice of R_0, Θ_0 and the form of the rotation curve at the solar circle. Adopting the rotation curve derived from CO observations, they found support for a small value of $R_0 \sim 7.5$ kpc (as, e.g. Olling & Merrifield, 1998) and $\Theta_0 \sim 190\,\mathrm{km\,s^{-1}}$, yielding a pattern speed of $\Omega_s \sim 24\,\mathrm{km\,s^{-1}\,kpc^{-1}}$, and implying that the Sun is indeed close to the co-rotation radius of $R_c \sim 7.9$ kpc. It also follows that the inner and outer Lindblad resonances for the two-arm spiral pattern lie at 2.5 and 11.5 kpc, with those for a four-arm mode at about 5 and 10 kpc.

The analysis was further extended by Lépine *et al.* (2006) and Lépine *et al.* (2008). They used the same method to derive the initial cluster velocities, and the angle between the initial velocity perturbation and the direction of circular motion. They found that the initial velocities are organised in preferential directions with respect to the spiral arms, from which they measured the epicycle frequency directly, that the preferential initial velocities extend for more than 100 Myr, and that the observations directly support the model of star formation originating from spiral density waves (see also Section 6.6).

The Hipparcos data have provided some valuable insights into, but not as yet a definitive value for, the spiral arm pattern speed. An application where an unambiguous value is desirable is discussed in Section 10.6.2, where considerations of long-term ice age patterns on Earth make a tantalising prediction of $\Omega_\odot - \Omega_s = 11.9 \pm 0.7\,\mathrm{km\,s^{-1}\,kpc^{-1}}$, suggesting $\Omega_s = 14.4\,\mathrm{km\,s^{-1}\,kpc^{-1}}$ for $\Omega_\odot = 26.3\,\mathrm{km\,s^{-1}\,kpc^{-1}}$. The weight of evidence, however, seems to point to $\Omega_s = 24$–$28\,\mathrm{km\,s^{-1}\,kpc^{-1}}$ in which case the Sun indeed lies close to the co-rotation circle. This conclusion is of interest for habitability considerations, where such a preferential location minimises peaks in the local star formation, the influence of supernovae explosions, and of gravitational perturbations to the Oort cloud (Section 10.5). Equally, the conclusion would be of importance for understanding the age–metallicity relation in the solar neighbourhood, since a prolonged minimum in star formation should correspond to a minimum in metallicity.

The duration of coherent spiral patterns remains an open question. There is some evidence that long-lived spiral patterns may be more prevalent in galaxies with a central bar. Numerical simulations of the evolution of barred spirals by Rautiainen & Salo (1999) suggest that spiral patterns may last upwards of 1 Gyr. Some theories and N-body simulations argue for short-lived patterns, with fresh spirals appearing in rapid succession, driven by structural features in the disk's velocity distribution function, and developing through 'swing amplification' of long-wavelength disturbances. Sellwood (2000) used the kinematically unbiased subsample of Hipparcos stars selected by Dehnen & Binney (1998a) to construct an energy–momentum diagram (Figure 9.31). If the ridge-like features in the plot are due to scattering at an inner Lindblad resonance, and not the scattering

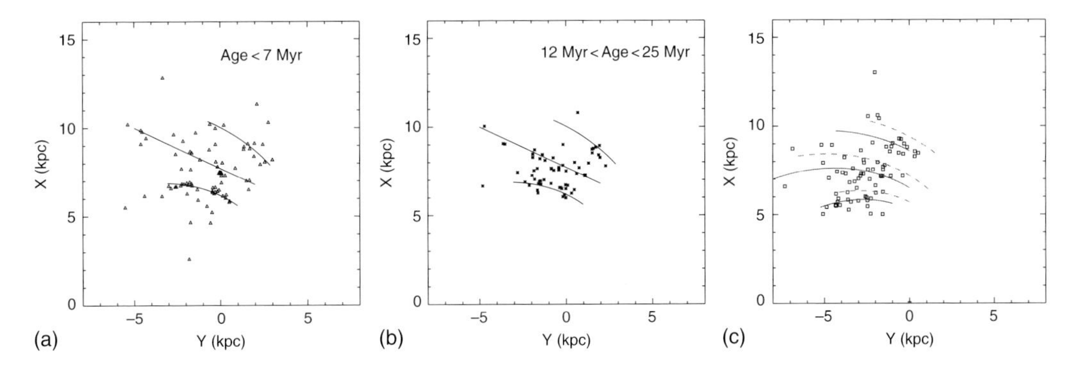

Figure 9.30 Determination of the spiral arm pattern speed based on Galactic open clusters in the solar neighbourhood. (a) Open clusters with age < 7 Myr; XY represents the usual Galactic coordinate system, with the Sun at (0,7.5). (b) Open clusters with ages in the range 12–25 Myr, showing current positions. (c) Birthplaces of clusters with ages in the range 9–15 Myr: dashed lines were fitted to a younger sample with ages in the range 5–8 Myr (not shown); solid lines are the same arms rotated by a best-fit angle of 10° around the Galactic centre. From Dias & Lépine (2005, Figures 1, 3, and 4).

at the bar's outer Lindblad resonance considered by Dehnen (2000), it would support the idea that the spiral arms are transient in nature. Independent of the explanation, Sellwood (2000) commented that *'the Hipparcos data provide the first observational confirmation that it is wrong to assume that the distribution function of a disk galaxy is smooth'*. Radial velocities subsequently made available from the Geneva–Copenhagen survey (Nordström *et al.*, 2004b) have improved the data somewhat, and an assessment of the implications is planned (Sellwood 2006, priv. comm.).

9.8 Properties of the stellar warp

A warped galaxy has an outer component whose plane is twisted upward at one side and down at the other. While many galaxies display such a warped structure including probably our own, with at least half of all spirals warped both in H I and in stars, there is only a limited understanding of their origin and dynamics (Binney, 1992). Warps are most apparent in the H I disk, and generally begin around the radius at which the optical disk ends. For most galaxies it is not known whether the optical disk (i.e. the stars) participates with the interstellar gas with the warp. Various explanations have been proposed, for example tidal forcing through a close encounter with a companion galaxy, infall of intergalactic material, or magnetic or hydrodynamic forces acting on the gas. The existence of many isolated warped galaxies is generally taken to imply misalignment of the symmetry axes of the disk and a flattened halo, such that gas in the outer parts of the disk settles into the symmetry plane of the former rather than of the latter; such a tilted disk embedded in a massive halo would precess. The occurrence of major ice ages at intervals of roughly half the Galactic year have even been used to postulate a connection between Earth's climate and the presence of a warp in our Galaxy (Section 10.6).

Previous studies of the Galactic warp have largely focused on its spatial structure, because its kinematic signature is mainly evident in the velocity component tangential to the line-of-sight viewed from in the plane, the component which cannot be measured directly from radio observations. Most searches for a stellar warp have been confined to counts of bright and mostly young objects, such as Cepheids and OB stars, Wolf–Rayet stars, open clusters, and supernova remnants, although studies of older objects have confirmed the feature (mostly using asymptotic giant branch stars from IRAS, and results from 2MASS) albeit of smaller amplitude (Djorgovski & Sosin, 1989; López Corredoira *et al.*, 2002).

Figure 9.32, left shows the type of structure under investigation. Miyamoto *et al.* (1993) had already detected a warping motion, based on the proper motions of about 2000 young stars in the heliocentric distance interval 0.5–3.0 kpc. They used the 3d Ogorodnikov–Milne model (see Section 9.2.7) to detect additional shear and rotation components around the two mutually orthogonal axes in the Galactic plane, in addition to the (classical) components around the axis perpendicular to the Galactic plane. The results indicated that the young stars are streaming around the Galactic centre in a tilted sheet (the warp) with a velocity of 225 km s^{-1}, and that the sheet itself is also rotating around the nodal line of the warp with an angular velocity of 4 km s^{-1} kpc^{-1}.

The Hipparcos data provide proper motions with systematic and zonal errors significantly smaller than the systematic motions expected. Studies have confirmed the spatial structure of the warp, and revealed interesting dynamical signatures (Miyamoto & Zhu, 1998). Figure 9.32, right shows the mean vertical motion of stars in the local solar neighbourhood as a function of

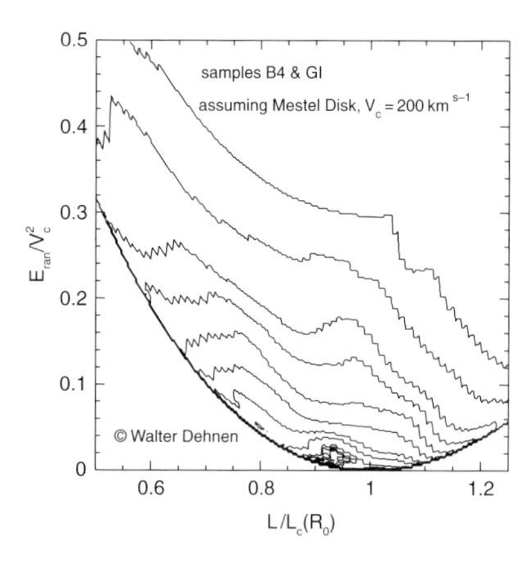

Figure 9.31 The density of solar neighbourhood stars in energy–momentum space for ~ 6000 stars from Hipparcos. L is the specific angular momentum, and $E_{\rm ran}$ is the excess of energy above that of a circulate orbit at this angular momentum. The skew to lower L is caused by the asymmetric drift. A smooth distribution function would have produced a featureless plot. From Sellwood (2000, Figure 5), constructed by Walter Dehnen, and reproduced with kind permission of Springer Science and Business Media.

azimuthal velocity from the inferred velocity distribution derived by Dehnen (1998) and discussed in Section 6.9. It shows a skewed distribution in which, at positive V, more stars are moving upward with respect to the LSR than downward. The upward curve for $V \gtrsim 10\,{\rm km\,s^{-1}}$ can be explained by the Galactic warp: the Sun lies roughly at the inner edge of the warp on the line of nodes such that nearby stars that participate in the warp have $W > 0$. Qualitatively, for such stars to enter the very local sample, they must be near their orbital pericentre, and hence have $V > 0$. The dashed line is for a warp starting well inside the solar circle, while better fits (solid lines) are found for an inner edge closer to the solar radius.

Smart & Lattanzi (1996) used simulations to demonstrate that the warp should be pronounced in the Galactic latitude component of the Hipparcos proper motions, but unobservable in the longitudinal component. Smart *et al.* (1998) used a sample of 2422 distant OB stars ($\pi < 2$ mas) in the range $70° < l < 290°$, i.e. towards the anti-centre. The choice of young stars assumes that they trace the motions of the gaseous component from which they were born, and thus more likely to trace the warp; the distance criterion was adopted to avoid local small-scale structural confusion arising from features such as the Gould Belt. The warp was modelled in cylindrical coordinates (R, φ, z) in which z is perpendicular to the plane of the inner Galaxy, and the Sun lies at $R_0 = 8$ kpc on the line $\varphi = z = 0$. The Galactic disk is flat to approximately R_0, then turns up to the north in the direction of Cygnus at $l \simeq 90°$, and south in the direction of Vela at $l \simeq 270°$. Consistent with this model, the Milky Way appears as a sinusoidal variation in the distribution of their star sample on the sky (Figure 9.33, left). The average height of their model disk was assumed to vary as

$$z = h(R)\,\sin(\varphi - \varphi_w + \omega_p t) \qquad (9.54)$$

where φ_w is the phase of the warp, assumed to be 0 (consistent with radio observations showing that the Sun lies close to the line of nodes); the warp rotates with angular frequency ω_p, opposite to that of Galactic rotation for $\omega_p > 0$. Central to the model is the assumption that the warp is long-lived, i.e. lasting for more than a few Galactic rotation periods, with $h(R)$ being time-independent. The variation of $\langle z \rangle$ with R shows the expected behaviour, increasing with R, and with the warp beginning within the orbit of the Sun. They found a height function, consistent with that from neutral hydrogen maps, of

$$h(R) = 0.067\,(R - 6.5)^2 \quad {\rm kpc,\ for}\ R > 6.5\ {\rm kpc} \qquad (9.55)$$

At any time t the azimuth of a star is given by $\varphi = \varphi_0 + \Omega(R)\,t$, where $\Omega(R)$ is the circular frequency. The predicted vertical velocity is then given by

$$V_z(R) \equiv {\rm d}z/{\rm d}t = [\Omega(R) + \omega_p]h(R)\cos\varphi \qquad (9.56)$$

From the observed proper motions, estimated vertical velocities were compared with the model, corrected for solar motion assuming $v_\odot = (9, 5, 7)\,{\rm km\,s^{-1}}$, and for Galactic rotation assuming $\Theta_0 = 220\,{\rm km\,s^{-1}}$ and with $(d\Theta/dR)_0 = -3\,{\rm km\,s^{-1}\,kpc^{-1}}$. While current theories, assuming an oblate halo, predict that $\omega_p > 0$ (Binney, 1992), this yields an upward trend in the diagram contrary to the observations. They concluded that while the spatial distribution of stars was consistent with previous studies using neutral hydrogen, the velocity distribution has the opposite sign to that expected (Figure 9.33). Thus the stellar kinematics do not follow the expected signature of a long-lived warp, whether precessing or not. Drimmel *et al.* (2000) (see also Zhu & Zhang, 2005) confirmed these results using an enlarged Hipparcos sample of 4538 OB stars with $\pi \leq 2$ mas covering all Galactic longitudes, and distances from a weighted mean of trigonometric and photometric estimates. They concluded that either the warp evolves on a time scale of the Galactic rotation period or shorter (possibly originating from an impulsive interaction such as with the Sagittarius dwarf galaxy), or that other large but unidentified systematic motions (vertical oscillations) are present within the disk. The latter may be indicated by the larger warp amplitude for the OB stars, of 0.4–0.8 kpc at $r = 10$ kpc, compared to that found at similar distances for the gas and dust of 0.3–0.4 kpc (Porcel *et al.*,

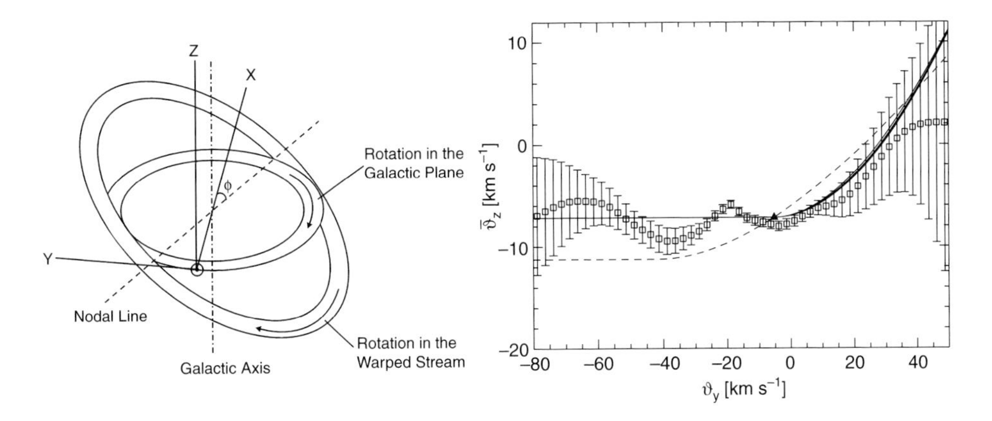

Figure 9.32 Left: the geometry of the modelled warp, with the Sun indicated by $\odot$*, from Smart & Lattanzi (1996). Right: the mean vertical motion,* $\bar{v}_z \equiv W$*, as a function of azimuthal velocity,* $\bar{v}_y \equiv V$*, for* $\sim$ *14 000 Hipparcos stars. Error bars are 1σ values, coupled by the adopted smoothing. Small-scale structure is considered as due to the moving groups. The solid triangle indicates the velocity of the Local Standard of Rest, with the various lines derived from models for the Galactic warp. From Dehnen (1998, Figure 6).*

1997), or at larger distances for the Cepheids (Miyamoto & Zhu, 1998).

Ideta *et al.* (2000) proposed an explanation for these results, by comparing the evolution of galactic warps contained within prolate and oblate spheroidal halos.[3] The halos were approximated as fixed potentials, with embedded self-gravitating disks represented by N-body particles, but ignoring dynamical friction between the two. They found that kinematic warps in oblate halos, in which the precession rate is a function of radius, wind up and disappear within a few dynamical times. In contrast, winding does not occur in prolate halos, and warps persist with continued alignment of the line of nodes, due to the fact that the precession rate of the outer disk increases when the precession of the outer disk recedes from that of the inner disk, and vice versa. Their results suggest that prolate halos could sustain galactic warps, and that warps persist due to the torques between the halo and disk, and between the inner and outer regions of the disk (see also Nelson & Tremaine, 1995).

Cosmological CDM (cold dark matter) simulations also suggest that dark matter halos surrounding individual galaxies are highly triaxial, and that the fraction of prolate and oblate halos is roughly equal (Dubinski & Carlberg, 1991). Although this is not necessarily in contradiction with the findings that the metal-weak halo tracers form a flattened distribution, at least in the inner halo (Section 9.9), it appears to be in conflict with the findings from the Sagittarius stream that the dark halo is rather round (Johnston *et al.*, 2005).

Zhu (2000a) used a sample of 1523 O–B5 stars to model the full Galaxy rotation in the 3d Ogorodnikov–Milne model framework, subtracting the Gould Belt contribution, and employing spectroscopic distances for those beyond the range of the Hipparcos parallax measurements. He again identified a clear warping motion described as a systematic rotation $(\partial V_z/\partial\theta)/R = -3.79 \pm 1.05\,\mathrm{km\,s^{-1}\,kpc^{-1}}$ around the axis pointing in the direction of the Galactic centre, in which the negative value corresponds to an increase in the inclination of the H I warp.

Peñarrubia *et al.* (2005) used absolute proper motions based on the Hipparcos reference system to determine the space motion of the postulated Canis Major dwarf galaxy, itself considered as a likely source of the Monoceros tidal stream (Section 5.10). The resulting prograde, low-inclination orbit of low eccentricity may have led to a strong coupling with the existing disk motion, with the interaction perhaps responsible for the observed warp.

Makarov & Murphy (2007) made a comprehensive analysis of the local stellar velocity field using vector spherical harmonics (Section 9.2.8). The most unexpected and unexplained term within their model is a first-degree 'magnetic' harmonic representing a rigid rotation of the stellar field about the axis $-Y$ pointing opposite to the direction of Galactic rotation. This harmonic was detected with a statistically robust coefficient $6.2\pm0.9\,\mathrm{km\,s^{-1}\,kpc^{-1}}$, and is also present in the velocity field of more distant stars. The ensuing upward vertical motion of stars in the general direction of the Galactic centre, and the downward motion in the anti-centre direction, are opposite to the vector field expected from

[3] An ellipsoid is represented by $x^2/a^2 + y^2/b^2 + z^2/c^2 = 1$. A spheroid is an ellipsoid with two equal semi-axes: $(x^2 + y^2)/a^2 + z^2/c^2 = 1$. An oblate spheroid has $a > c$, like a pancake or disk, and is generated by rotating an ellipse about its minor axis. A prolate spheroid has $a < c$, like a rugby ball, and is generated by rotating an ellipse about its major axis.

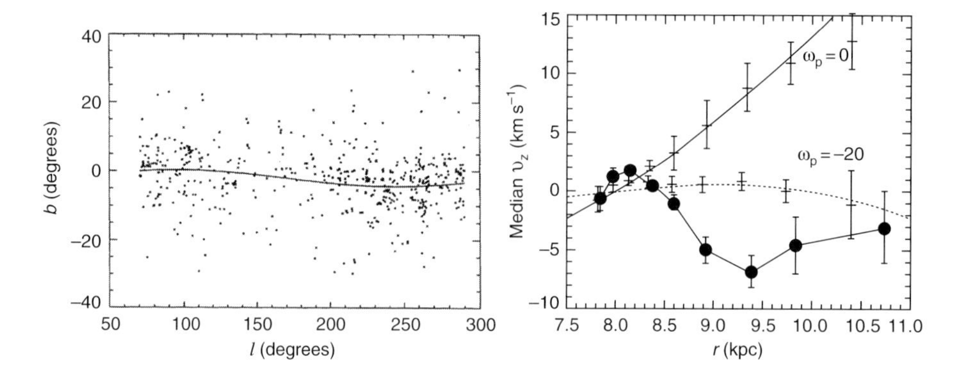

Figure 9.33 Left: distribution of OB stars in Galactic coordinates to the sample completeness limit, $V < 7.5$ mag. The curve is a sine fit. The shape results from stars in the first and fourth quadrants being on average closer to the Galactic centre, and therefore lower in the warp structure. Right: systematic vertical velocities as a function of Galactocentric distance r. Filled circles show medians of the observed vertical velocity, with error bars. Curves show predicted $V_z(r) - V_z(r_\odot)$ using Equation 9.56, for $\Omega_p = 0$ (full line) and $\Omega_p = -20$ kms^{-1} kpc^{-1} (dashed line). Error bars represent one standard deviation of medians from 30 simulated catalogues for each set of warp parameters. From Smart et al. (1998, Figures 2 and 3a), reprinted by permission from Macmillan Publishers Ltd: Nature ©1998.

the stationary Galactic warp model. They consider that a precessing line of nodes is conceivable, but that it is difficult to reconcile the observed pattern of vertical motion with a plausible precession model. They suggest, instead, that the line of nodes is stationary, but that the shape of the warp reverse every 50 Myr or so, curling one way and then the other.

9.9 The stellar halo

9.9.1 Mass and extent

The stellar halo was originally detected through the presence of high-velocity low-metallicity stars in the solar neighbourhood (Roman, 1955), moving on highly eccentric or highly inclined orbits, and typified by the high proper motion star Groombridge 1830 (Eggen & Sandage, 1959). Current knowledge of the stellar halo is reviewed by Bland-Hawthorn & Freeman (2000) and Helmi (2008). Its structure is closely linked to how the Galaxy formed (Freeman & Bland-Hawthorn, 2002), with different formation scenarios predicting different shapes (Navarro *et al.*, 1996), and different dependencies on age and metallicity.

The entire baryon system is believed to be embedded in a massive halo of dark material of unknown composition and poorly known spatial distribution, the primary evidence for whose existence is the flat rotation curve of the disk in the outer parts of the Galaxy (e.g. Merrifield, 2005). In cold-dark matter dominated cosmological models the structure of the stellar and dark halo do not necessarily correspond: while the stellar halo has a half-light radius of the smooth component lying within the solar circle (Freeman, 1996), the dark matter halo appears to be much more extended, with a half-mass radius of ~150 kpc (Klypin *et al.*, 2002).

Studies of the kinematics of various stellar populations in the Galaxy, especially for the halo, have long been limited by the absence of large samples with accurate distances, space motions, and metallicities, and the best measurements of halo kinematics are still obtained from samples of stars located in the solar neighbourhood. Searches for halo stars include proper motion surveys which can identify high-velocity stars out to a few hundred pc, and surveys for RR Lyrae, metal-poor giants using objective prisms, blue horizontal branch, and carbon stars. The fact that the stellar sample down to about 12 mag observed by Hipparcos is dominated by the disk population (the halo population constitutes only ~ 0.1% of local stars), combined with the large distances and hence individually insignificant parallaxes of typical halo stars, limits the impact of Hipparcos on kinematic and dynamic studies of the halo population. Nevertheless some interesting contributions have been made, based on the nearby well-measured stars.

The flat rotation curve of the Galaxy out to at least twice the radius of the solar circle implies that the total gravitational mass density behaves like an isothermal gas sphere, while the luminous matter density decreases exponentially with Galactocentric distance. Various investigations have tried to probe how far this massive halo, now generally believed to be dominated by dark matter, extends. In pre-Hipparcos studies, Carney *et al.* (1988) estimated the local escape velocity of stars passing through the solar neighbourhood to be about 500 km s^{-1}, leading to an estimated boundary radius of about 45 kpc and a total Galaxy mass of $\sim 5 \times 10^{11}\, M_\odot$.

Miyamoto & Tsujimoto (1997) used the space velocities of RR Lyrae stars from Hipparcos as well as other sources (in a programme limited by the availability of radial velocities) to identify maximum space velocities of 420–480 km s^{-1}, allowing a redetermination of the boundary radius of the halo. Compared with the Carney *et al.* (1988) condition for trapped orbits, $V_r^2 + V_t^2 \leq 2\Phi(r)$, where V_r and V_t denote the radial and tangential velocity components, Miyamoto & Tsujimoto (1997) derived the confinement condition within the Galactic boundary at $r = r_*$ as

$$V_r^2 + \left(1 - \frac{r^2}{r_*^2}\right) V_t^2 \leq 2[\Phi(r) - \Phi(r_*)] = 2\Theta_0^2 \ln\left(\frac{r_*}{R_0}\right) \tag{9.57}$$

giving a family of marginal velocity ellipses for a given set of R_0 and Θ_0. The maximum angular momentum attainable at $r = R_0$ is given by setting $V_r = 0$. Using $R_0 = 8.5$ kpc and $\Theta_0 = 220$ km s^{-1} (Figure 9.34) leads to an estimated Galactic boundary radius in the range $r_* = 50$–100 kpc, and a total Galaxy mass in the range $M_* = 5.5$–$11 \times 10^{11}\, M_\odot$.

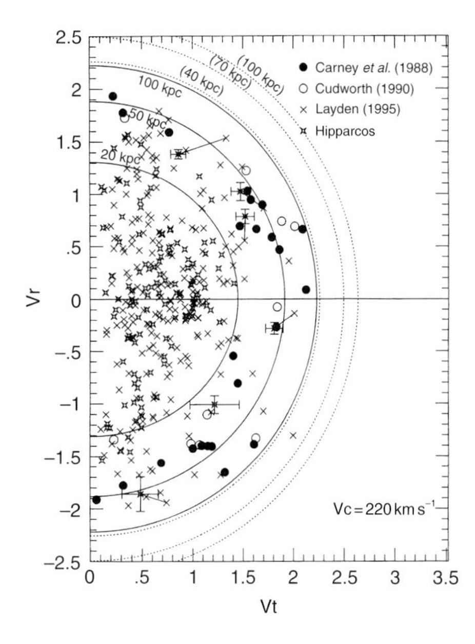

Figure 9.34 The velocity space (V_r, V_t), in units of $V_c = \Theta_0$, at the Sun, $r = R_0$. Marginal velocities given by Equation 9.57 are shown by solid ellipses, the usual escape velocities by the dotted circles for the indicated boundary radii r_*. Velocity sources are indicated. Error estimates are shown for the the six extreme velocity cases, with lines connecting the crosses indicating the effect of improved proper motions. From Miyamoto & Tsujimoto (1997, Figure 1).

9.9.2 Rotation, shape and velocity dispersion

Estimates of the rotation velocity of the stellar halo vary widely. Observationally, nearby halo population samples with good space velocities are statistically limited, more distant samples are limited by the accuracy of parallaxes and proper motions, and measurements of the circular speed of the disk and the rotation velocity of the halo are observationally coupled (Section 9.5). Many different estimates have been reported, but representative pre-Hipparcos values range from an absolute prograde rotation of 25 ± 15 km s^{-1} from metal-weak halo giants in the solar neighbourhood (Morrison *et al.*, 1990) to a retrograde motion of -55 ± 16 km s^{-1} from faint stars at the north Galactic pole (Majewski, 1992). Independently of the precise rotation, the existence of a chemically well-defined subcomponent with small rotation in an otherwise rapidly rotating Galaxy, is striking.

Chen (1998) used a sample of high proper motion stars (therefore with some kinematic bias) selected by Carney *et al.* (1994) with proper motions exceeding 0.18 arcsec yr^{-1} (from the New Luyten Two-Tenths Catalogue) for which detailed photometry and spectroscopy had been obtained. Improving the proper motion errors from typically 20–25 mas yr^{-1} to some 1 mas yr^{-1}, and limiting errors on the space velocities by restricting the sample to relative parallax errors $\sigma_\pi/\pi < 0.25$, provided a sample of 552 stars. The component of the space velocity in the direction of Galactic rotation, V, corrected for differential Galactic rotation, is shown as a function of metallicity, [Fe/H], in Figure 9.35. The cluster analysis identifies two populations: 511 disk component stars characterised by a mean [Fe/H] -0.37 ± 0.01 and mean circular velocity $\langle V \rangle = -51 \pm 2$ km s^{-1}, and 41 halo population stars characterised by [Fe/H] -1.73 ± 0.08 and $\langle V \rangle = -221 \pm 11$ km s^{-1} (the U and W velocity components are indistinguishable). The halo $\langle V \rangle$ velocity with respect to the LSR implies a rotation velocity of 1 ± 11 km s^{-1}. Figure 9.35 provides support for the disk and halo being distinct kinematic structures with different formation histories. Their halo sample is well represented by the 16 Gyr isochrones of Bergbusch & VandenBerg (1992). Their absence of a significant rotation of the halo component is confirmed by results from a different sample of red giants and RR Lyrae stars studied by Chiba & Yoshii (1998).

Beers *et al.* (2000) constructed an enlarged catalogue of 2106 stars selected to minimise kinematic bias, and with available radial velocities, distance estimates, and metal abundancies in the range $-4.0 \leq$ [Fe/H] ≤ 0.0. As a significant source of halo star kinematics, more than half the stars have high accuracy proper motions from Hipparcos and other sources. A derived dataset of 1203 stars with [Fe/H] ≤ -0.6 was used by Chiba & Beers (2000) to study details of the halo kinematics.

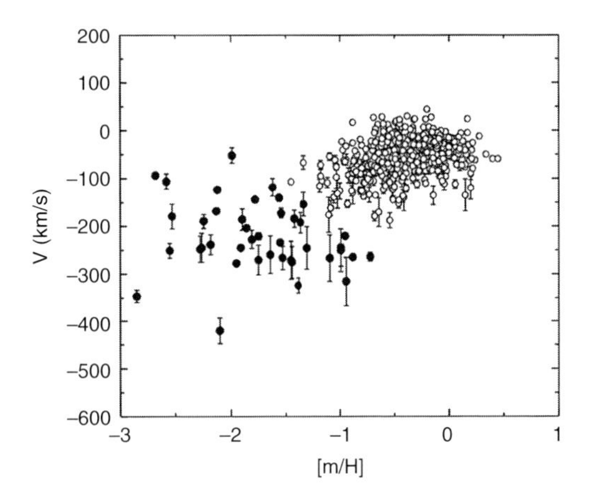

Figure 9.35 The space velocity component, V, versus metallicity, [Fe/H], for 552 stars. Filled circles indicate objects classified as halo stars by a cluster analysis. From Chen (1998, Figure 1).

For the rotation of the halo system as a whole, the values reported are

$$\langle U\rangle, \langle V\rangle, \langle W\rangle = -20 \pm 27, -221 \pm 11, -\ 9 \pm 11 \text{ km s}^{-1}$$

Chen (1998)

$$= -16 \pm 18, -217 \pm 21, -10 \pm 12 \text{ km s}^{-1}$$

Chiba & Yoshii (1998)

where the large negative $\langle V\rangle$, with respect to the LSR, added to Θ_0, indicates a non-rotating population.

Evidence for more detailed structure became evident in the enlarged sample of Chiba & Beers (2000), from which they inferred a nearly spherical outer halo (beyond 15–20 kpc) and highly flattened in the inner region. At first sight, there appears to be a conflict between the flattened, oblate, inner halo indicated by this work, and the prolate halo suggested by Ideta *et al.* (2000) to explain the development of the Galactic warp. In fact, the work of Chiba & Beers (2000) is relevant for the stellar halo, not that of the dark halo (i.e. the total mass) which is more relevant for the conclusions of Chiba & Beers (2000). In general, the density distribution of the halo tracer stars can be different from that of a background source of gravitational potential. So while the flattened stellar halo may not conflict with conclusions about the warp, more problematic is the finding from the Sagittarius stream that the dark halo is rather round (Johnston *et al.*, 2005).

Working in cylindrical coordinates, Chiba & Beers (2000) also reported a small prograde rotation for $|Z| < 1$ kpc (and for [Fe/H] ≤ -2.2 and assuming $\Theta_0 = 220 \text{ km s}^{-1}$), with a decrease in rotation rate with increasing distance from the Galactic plane

$$\langle V_\phi\rangle = 30 - 50 \text{ km s}^{-1} \tag{9.58a}$$

$$\Delta\langle V_\phi\rangle/\Delta|Z| = -52 \pm 6 \text{ km s}^{-1}\text{ kpc}^{-1} \tag{9.58b}$$

which may represent the signature of a dissipatively-formed flattened inner halo. V_ϕ decreases with $|Z|$, with no distinct boundary between any inner and outer halo. The outer halo, represented by local stars on orbits reaching more than 5 kpc from the plane, exhibits no systematic rotation; previous evidence for a counter-rotating halo at high $|Z|$ is not supported. The density distribution of the outer halo is nearly spherical and exhibits a power-law profile described as $\rho \propto R^{-3.55\pm0.13}$.

All three studies derived a radially-elongated velocity ellipsoid of

$$\sigma_U, \sigma_V, \sigma_W = 174 \pm 19,\ \ 72 \pm 8,\ \ 67 \pm 7 \text{ km s}^{-1}$$

Chen (1998)

$$= 161 \pm 10, 115 \pm 7, 108 \pm 7 \text{ km s}^{-1}$$

Chiba & Yoshii (1998)

$$= 141 \pm 11, 106 \pm 9,\ \ 94 \pm 8 \text{ km s}^{-1}$$

Chiba & Beers (2000)

with little dependency on Z (Chiba & Beers, 2000). All indicate a more anisotropic distribution than that reported by Morrison *et al.* (1990).

At higher metallicities, stars from Chiba & Beers (2000) exhibit disk-like kinematics and higher mean rotation: for $-0.7 \leq$ [Fe/H] ≤ -0.6, they found $\sigma_U, \sigma_V, \sigma_W = 46 \pm 4, 50 \pm 4, 35 \pm 3 \text{ km s}^{-1}$ and $\langle V_\phi\rangle = 200 \text{ km s}^{-1}$, more characteristic of the thick disk. Previous results indicating a sharp discontinuity of the rotational properties of the Galaxy at [Fe/H] $\simeq -1.7$ are confirmed (Figure 9.36).

9.9.3 Formation

Orbital time scales in the outer parts of the Galaxy are several billion years, such that the halo retains kinematic evidence of the surviving remnants of accretion, as well as chemical diagnostics of early low-mass stars as a result of their very long evolutionary lifetimes.

While the Galactic disk is thought to have been formed by the rapid collapse of a rotating galaxy-sized gas cloud, billions of years ago, seminal papers by Eggen *et al.* (1962) and Searle & Zinn (1978) proposed contrasting scenarios for the formation of the halo. Eggen *et al.* (1962) demonstrated an inverse correlation between metallicity and orbital eccentricity (from *UVW* velocity vectors) for 221 dwarfs, which led them to hypothesise that the oldest stars were formed out of gas falling toward the Galactic centre in the radial direction, and collapsing rapidly in a free-fall time of $\sim 2 \times 10^8$ yr from the halo onto the plane. Searle & Zinn (1978) studied 177 red giants in 19 globular clusters at typical distances beyond 8 kpc, and suggested that the halo clusters originated within transient protogalactic fragments that gradually lost gas while undergoing chemical evolution

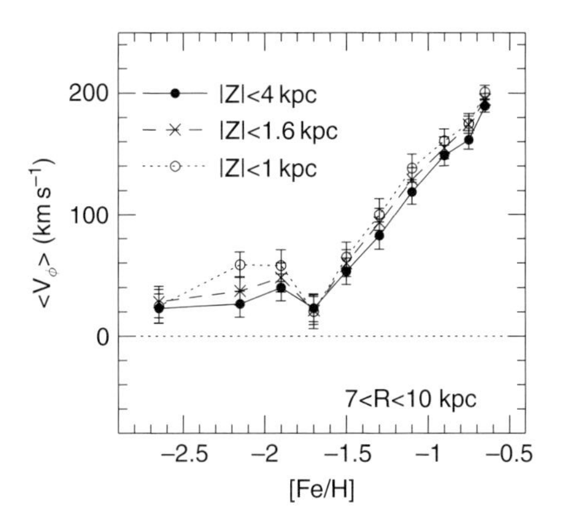

Figure 9.36 Distribution of mean rotational velocities around the Galactic centre $\langle V_\phi \rangle$ versus [Fe/H], based on complete proper motions and radial velocities. Values assume an LSR rotation velocity of 220 kms^{-1}, so that velocities close to 0 are non-rotating with respect to the Galactic centre. The halo population is considered to be defined by [Fe/H] < -2.2. The sharp discontinuity at [Fe/H] $\simeq -1.7$ represents the onset of contributions from the thick disk. Note that V_ϕ decreases with increasing $|Z|$. From Chiba & Beers (2000, Figure 3a).

and continued to fall into the Galaxy after the collapse of its central regions had been completed. Present understanding of kinematic and abundance patterns suggest a halo built up over billions of years from infalling low-mass objects ($10^7-10^8\,M_\odot$) such as dwarf galaxies: evidence includes the wide range of globular cluster metallicities independent of Galactocentric radius, a wide age spread of halo field stars and globular clusters of 2–3 Gyr, a subset of intermediate abundance globular clusters with retrograde mean motions, metal-poor halo stars of intermediate age, and metal-rich halo A stars (Bland-Hawthorn & Freeman, 2000).

While formation through progressive accretion, even at the present time, is now generally favoured, there has been a continued debate as to whether the correlation discovered by Eggen *et al.* (1962) was an artifact arising from the selection of metal-poor stars with high eccentricities (Chiba & Yoshii, 1997). Chiba & Yoshii (1997, 1998) used a nearby sample ($\lesssim 2$ kpc) of 122 metal-poor red giants and 124 RR Lyrae stars, Hipparcos proper motions, radial velocities, and both a two-dimensional and more realistic three-dimensional Galactic potential, to establish the three-dimensional orbital motions as a function of metallicity. Figure 9.37 shows the resulting Bottlinger diagram constructed from the two-dimensional planar potential given by Eggen *et al.* (1962)

$$\Phi_{\rm ELS}(R) = -\frac{GM}{b + (R^2 + b^2)^{1/2}} \qquad (9.59)$$

(and in equivalent form as the isochrone potential by Hénon 1959) where R is Galactocentric distance, M the mass of the disk, and b the scale length. Values of b and M can be derived from the Oort constants (A, B) combined with values of R_0 and Θ_0, or using the escape velocity in the vicinity of the Sun (Chiba & Yoshii, 1998, Appendix). The distribution of eccentricities can then be constructed from the measured space velocities. Confirming pre-Hipparcos studies, they concluded that orbital eccentricities of metal-poor halo stars are not correlated with their metallicities, in conflict with the Eggen *et al.* (1962) results. The enlarged sample of 1203 stars with [Fe/H] ≤ -0.6 was also used by Chiba & Beers (2000) to confirm the absence of a strong correlation between metallicity and orbital eccentricity of the halo stars.

Although the Searle & Zinn (1978) accretion scenario is consistent with this specific result, Chiba & Yoshii (1998) argued that the detailed form of the velocity ellipsoid as a function of Galactocentric distance is not. In the more recent cold dark matter (CDM) scenario of galaxy formation in which initial density fluctuations in the early Universe have larger amplitudes on smaller scales, overdense regions that end up with giant galaxies like our own contain larger density fluctuations on subgalactic scales, which develop into numerous self-interacting fragments in the collapsing protogalaxy. Bekki & Chiba (2000) performed a numerical simulations of structure formation from redshifts $z = 25$ to $z = 0$, using 'standard' CDM initial conditions, combined with star formation modelled by the conversion of the collisional gas into collisionless stars, and subsequent chemical and dynamical evolution. They demonstrated that this hierarchical merging of CDM clumps yields no significant correlation between [Fe/H] and eccentricity for [Fe/H] ≤ -0.6, and that the existence of low-metallicity ([Fe/H] ≤ -1) low-eccentricity ($e < 0.4$) stars is successfully reproduced. In this picture, the orbital eccentricities of metal-poor stars, once formed, are not greatly influenced by the change of an overall gravitational potential. At the same time, most of the metal-poor stars have been confined within massive clumps, whose orbits are gradually circularised due to dynamical friction and dissipative merging with smaller clumps.

Further studies by Chiba & Beers (2001) and simulations by Bekki & Chiba (2001) provide a picture in which the present-day stellar halo is characterised by a highly flattened inner halo and a nearly spherical outer halo. This could be a consequence of an initially two-component density distribution of the halo (itself perhaps a signature of dissipative halo formation through the merging of small subgalactic clumps) and of the subsequent flattening of the inner part via three different mechanisms: dissipative merging between larger and more massive clumps, adiabatic contraction due to the

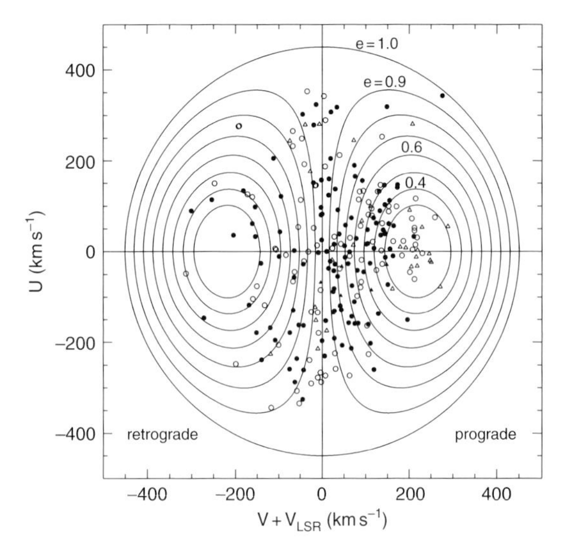

Figure 9.37 Bottlinger diagram for the halo stars. Filled and open circles represent red giants and RR Lyrae stars respectively, with the triangles for objects with larger proper motion errors. Abscissa and ordinate give UV space velocity components. Stars to the right of the vertical line travel on prograde (direct) orbits; stars to the left travel on retrograde orbits. Curves represent loci of constant eccentricity derived from a specific model Galaxy potential, in this case the two-dimensional planar potential given by Eggen et al. (1962), $\Phi_{\rm ELS}$, such that the eccentricity distribution can be derived from the measured velocity distribution. From Chiba & Yoshii (1998, Figure 11).

growing Galactic disk, and gaseous accretion onto the equatorial plane.

9.9.4 Halo substructure

If the Galaxy halo was indeed assembled from the merging or accretion of subgalactic clumps, as argued by Searle & Zinn (1978), signatures of those events might still be present in the form of kinematic substructures since the mixing of phase space for such stars is expected to be incomplete (Helmi & White, 1999). Evidence for this includes the Sagittarius dwarf galaxy which may be an ongoing merging event at the current epoch (Ibata *et al.*, 1994), the Magellanic Clouds which may ultimately follow a similar fate as they lose energy via dynamical friction (Lynden-Bell & Lynden-Bell, 1995), and signatures of past merging events in the halo which have been reported over the last few years, mainly from SDSS data (e.g. Majewski *et al.*, 1994, 2000; Newberg *et al.*, 2002; Odenkirchen *et al.*, 2003; Majewski, 2004; Majewski *et al.*, 2004; Martin *et al.*, 2004; Law *et al.*, 2005; Belokurov *et al.*, 2006; Rocha Pinto *et al.*, 2006).

Clear evidence in the solar neighbourhood for a previous merger of the Galaxy with what may have been a 'Searle & Zinn fragment' was discovered by Helmi *et al.* (1999). They examined a subset of the metal-deficient red giants and RR Lyrae stars from Chiba & Yoshii (1998), and identified a statistically significant clumping of stars in angular momentum space J_z versus $J_\perp = (J_x^2 + J_y^2)^{1/2}$. The underlying idea is that coherent streams of relatively rare objects can be identified from clumping in the space of adiabatic invariants, since stars from the same progenitor should have similar integrals of motion, even though they may be widely and sparsely distributed over the sky. Estimation of $J_\perp$ requires radial velocities, distances (estimated from photometry), and accurate proper motions provided by Hipparcos. The proposed substructure consists of just seven stars out of 97 with $[{\rm Fe/H}] \leq -1.6$ and $d < 1$ kpc, or 12 stars out of 275 with $[{\rm Fe/H}] \leq -1$ and $d < 2.5$ kpc (Figure 9.38a). They inferred that roughly 10% of the halo stars outside the solar radius may have arisen from a single coherent object with a total mass of about $10^8 M_\odot$, disrupted during the process of halo formation.

Chiba & Beers (2000) confirmed the presence of this clump in their sample (Figure 9.38b), although representing a smaller fraction of the total halo sample. In addition, they identified a possible trail in angular momentum space, which appears to connect the clump and the high J_z region, and which is most clearly evident in the higher abundance stars shown in the bottom panel. Although $J_\perp$ is not an exact integral of motion in an axisymmetric potential, similar features are also seen in the conserved 'third integral', I_3, for which no general analytic expression exists, but which can be estimated (along with other integrals such as the orbital energy E and J_z) using the Stäckel form of the potential (de Zeeuw, 1985; Chiba & Beers, 2000). This, in turn, implies that the relevant potential is nearly spherical, consistent with the relevant stars spending most of their orbits far from the plane where the effect of the disk potential is modest. Fiorentin *et al.* (2005) used the same survey sample to identify three stars with similar kinematic and metallicity characteristics using a two-point velocity correlation analysis.

Helmi *et al.* (2006) used the catalogue of Nordström *et al.* (2004b), along with Hipparcos astrometry, to search for further accretion events amongst 13 240 nearby stars, focusing on correlations between orbital parameters notably apocentre, pericentre, and z-component angular momentum, L_z. They identified a wealth of structure, much linked to the dynamical pertubations induced by spiral arms and the Galactic bar (see Section 6.9). They also found a significant excess of stars on orbits of common (moderate) eccentricity, analogous to the pattern expected for merger debris. Besides being dynamically peculiar, the 274 stars in these substructures have very distinct metallicity and age distributions, providing further evidence of their extragalactic provenance. They identified three coherent

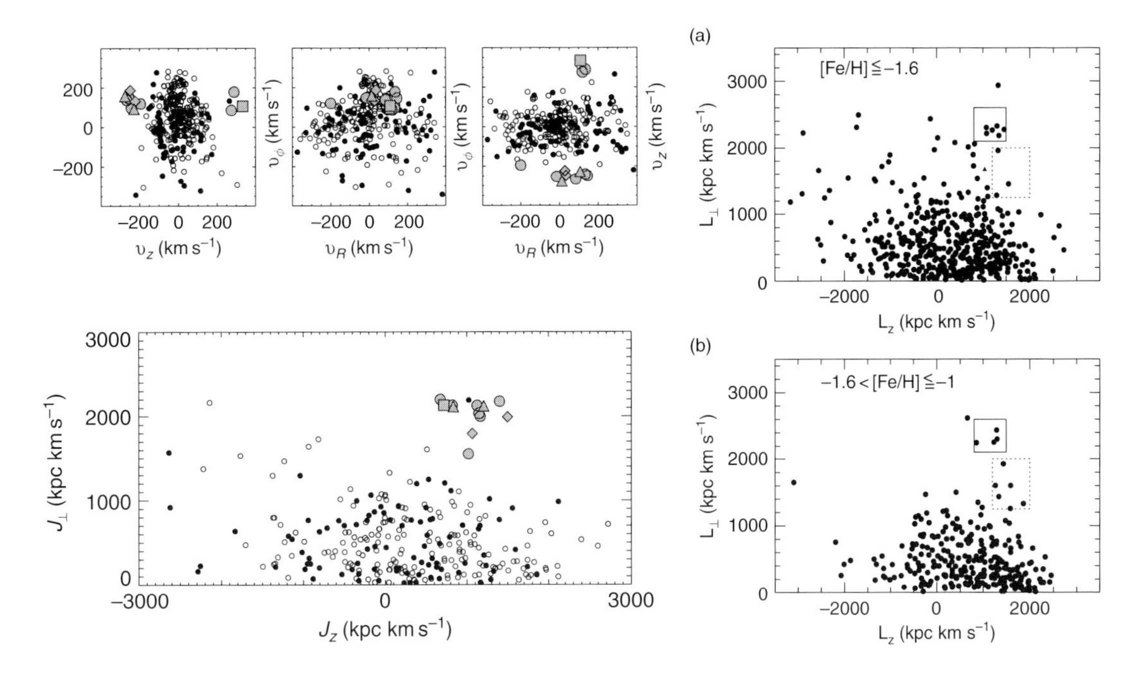

Figure 9.38 The distribution of nearby halo stars in velocity space (top three, left) and in the $J_z - J_\perp$ plane (bottom, left), from Helmi et al. (1999, Figure 2), reprinted by permission from Macmillan Publishers Ltd: Nature ©1999. Data are shown for the 1 kpc limited sample (filled circles) and for the extended sample of more metal-rich and more distant giants (open circles). Candidates for the detected substructure are in grey: triangles: more metal-rich giant stars beyond 1 kpc; diamonds: more metal-rich giants within 1 kpc; squares: metal-poor giants beyond 1 kpc; circles: metal-poor giants within 1 kpc. For $J_\perp \leq 1000$ kpc kms^{-1} and $|J_z| \leq$ 1000 kpc kms^{-1}, the observed distribution is fairly smooth; stars have relatively low angular momentum and at all inclinations. In contrast, for $J_\perp \geq 1000$ kpc kms^{-1}, there are a few stars moving on retrograde low-inclination orbits, an absence of stars on polar orbits, and an apparent clump on a prograde high-inclination orbit. Right pair: comparable plots (symbol L in place of J) from Chiba & Beers (2000, Figure 15) for halo stars within 2.5 kpc and two different metallicity ranges. Solid and dotted boxes denote the regions of the clump and trail noted in the text.

groups: the most metal-rich group, [Fe/H] > -0.45 dex, has 120 stars distributed into two stellar populations of ages ~8 Gyr (33%) and ~12 Gyr (67%); the second group, with [Fe/H] ≥ -0.6 dex, has 86 stars and shows evidence of three populations of ages 8 Gyr (15%), 12 Gyr (36%) and 16 Gyr (49%); the third group has 68 stars, with typical metallicity around -0.8 dex and a single age of ~14 Gyr. The identification of substantial amounts of debris in the Galactic disc whose origin can be traced back to more than one satellite galaxy, provides further evidence of the hierarchical formation of the Milky Way.

Dettbarn *et al.* (2007) analysed the phase space distribution of about 900 non-kinematically selected low-metallicity stars in the solar vicinity from the sample of Chiba & Beers (2000), all with Hipparcos astrometry. The search was conducted in angular momentum–eccentricity space. In addition to recovering all well-known star streams in the thick disk, they isolated four statistically-significant phase-space overdensities amongst halo stars. One is associated with the halo star stream discovered by Helmi *et al.* (1999) and confirmed by Chiba & Beers (2000), but they consider three of them as new halo stream structures.

Kotoneva *et al.* (2005) reported a detailed study of Kapteyn's star, a nearby M dwarf with the second largest known proper motion. Combined with its large radial velocity, it yields one of the few well-established retrograde Galactic orbits, at least amongst the nearest solar neighbourhood stars. They used the Hipparcos astrometry to derive a *UVW* Galactic space velocity of $21.1 \pm 0.3, -287.8 \pm 0.3, -52.6 \pm 0.3$ km s^{-1}, with representative Galactic orbits resulting from empirical bar/axisymmetric distortions. They suggest that the star may have formed in a dwarf satellite of the Milky Way, which was then dragged into a low-energy orbit, losing most of its mass in the process. They speculate that the remnant may be the globular cluster ω Centauri, whose retrograde orbit is similar to that of Kapteyn's star, and which includes among its established members stars with similar metallicity.

Prospects for mapping the substructure in the Galactic halo in greater detail with the next generation of astrometric satellites has been discussed by Helmi & de Zeeuw (2000) using different functional forms for the dark halo, disk, and bulge contributions to the total Galactic potential.

9.10 Models of the various Galaxy components

General considerations Much of the knowledge of large-scale Galaxy structure comes from a determination of star counts as a function of magnitude, colour, and coordinate direction. In the absence of distance information, the luminosity function and density distribution are related by (e.g. Mihalas & Binney, 1981, Section 4.2)

$$A_\lambda(m, l, b) = \int_0^\infty \Phi_\lambda(M)\rho(r, l, b)r^2\,\mathrm{d}r \tag{9.60}$$

where $A_\lambda(m, l, b)\mathrm{d}m\,\mathrm{d}l\,\mathrm{d}(\sin b)$ is the number of stars that have apparent magnitude in the range $[m, m + \mathrm{d}m]$, $\Phi_\lambda(M)$ is the luminosity function, which depends on M and the colour-band λ, and $\rho(r, l, b)$ is the density at radius r in Galactic coordinate direction l, b. In general this cannot be solved or inverted, although the luminosity function can be determined if a density distribution is assumed (e.g. Reid *et al.*, 1996). Progress can be made if it is assumed that different populations have the same luminosity function (e.g. Robin & Crézé, 1986), or by inferring the luminosity function using isochrone fitting (e.g. Haywood *et al.*, 1997). Pichon *et al.* (2002) show that the degeneracy is also lifted if proper motions are used in addition to the star counts in apparent magnitude, giving access to both the vertical density law of each stellar population and their luminosity functions, e.g. splitting the thin disk, the thick disk, and the halo components. Additional assumptions could provide a non-parametric determination of the local luminosity function, but complicated by interstellar reddening, complementary to that obtained by evolutionary track fitting with an assumed IMF and star-formation rate. That the proper motion data (and hence transverse velocities, v_perp) alone contain a wealth of information on the population is conveyed in the v_perp colour–magnitude diagram constructed by Gould (2004) directly from the Hipparcos Catalogue using distance and proper motion data only (Figure 9.39). For example, early main sequence stars, being young, tend to be moving more slowly than later main-sequence stars, which are predominantly older. Halo subdwarfs, which lie below the main-sequence due to their lower metallicities, also tend to travel quite fast relative to the Sun. Thick disk stars are intermediate in both metallicity and kinematics.

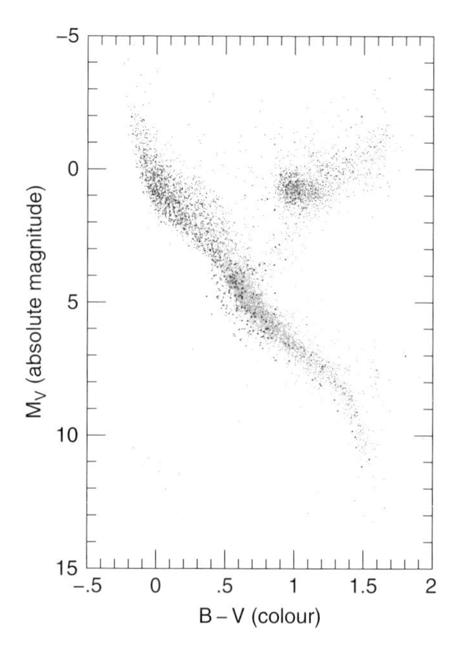

Figure 9.39 The colour–magnitude diagram, encoded according to $v_\perp$. The sample was constructed from the Hipparcos Catalogue using $\sigma_\pi/\pi < 0.2$, $V \le 7.3$, $\pi \ge 20$ mas, and proper motion $\mu \ge 200$ mas yr^{-1}. The original figure is in colour, in which the colour-coding is (in km s^{-1}): black: $v_\perp < 10$ km s^{-1}; red: $10 < v_\perp < 20$; yellow: $20 < v_\perp < 50$; green: $50 < v_\perp < 100$; cyan: $100 < v_\perp < 150$; blue: 150 $< v_\perp < 200$; magenta: $v_\perp > 200$. From Gould (2004).

Component contributions from Hipparcos Holmberg *et al.* (1997) constructed a rather simple model of the thin disk population, in which the radial density distribution is modelled with a scale length of 3.5 kpc and a z distribution of the form

$$\rho(Z) = \rho(0)\,\mathrm{sech}^{2/n}(nZ/2Z_e) \tag{9.61}$$

where $\rho(0)$ is the density in the plane, Z_e is the exponential scale height, and where the form was introduced by van der Kruit (1988) to provide a representation varying between isothermal ($n = 1$) and exponential ($n \to \infty$). Holmberg *et al.* (1997) found that $n = 3$ gave the best representation of the stellar distribution in the Hipparcos Catalogue for the 'survey' component (Figure 9.40), at the same time providing estimates of the Sun's height from the Galaxy mid-plane, Z_0 (Section 9.2.2).

Table 9.7 summarises component disk properties from the subsequent Hipparcos-based work of Chabrier (2001), although the thick disk mass has been criticised by Fuhrmann (2004).

Robin *et al.* (2003) described the Besançon Galaxy 'population synthesis' model constrained by source counts in the visible and near-infrared, and constructed to be dynamically self-consistent through the Galactic potential and the velocity dispersion of the different populations. The potential, comprising the bulge,

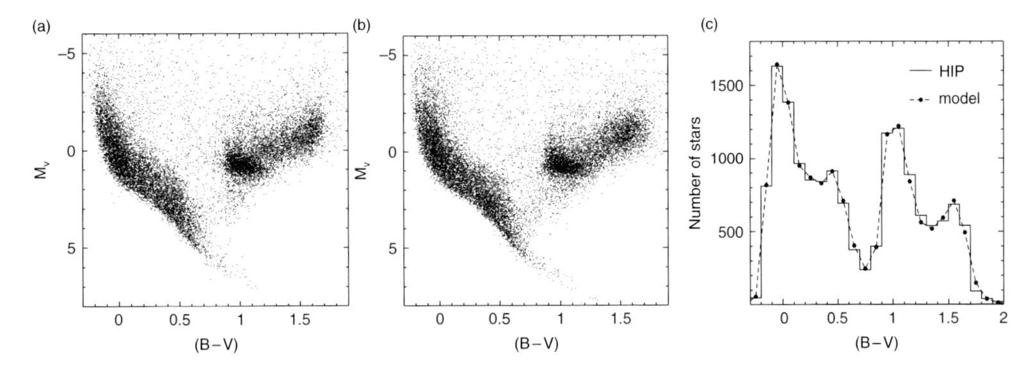

Figure 9.40 The simple form of the thin disk density distribution, Equation 9.61, provides a very reasonable match between the observed HR diagram for $V < 7$ from the Hipparcos Catalogue (left), and the simulated model catalogue (middle). An excellent fit was also obtained (right) for the total number of stars with $V < 7$ in the catalogue (solid line) and in the model (dashed line). From Holmberg et al. (1997, Figures 1 and 5).

the warped and flared disk, the thick disk, and the spheroidal populations, is constrained to be in agreement with the observed Galaxy rotation curve, and the various relevant Hipparcos results: the revised luminosity function and an associated two-component IMF with $\alpha = 3$ at high mass and $\alpha = 1.6$ at low mass, the age–velocity dispersion relation, and the local mass density. The total local mass density of $0.0759\,M_\odot\,\mathrm{pc}^{-3}$, comprised as shown in Table 9.8, was constructed to be in agreement with the value obtained by Crézé *et al.* (1998) from the K_z relation, which may be an underestimate according to the discussions in Section 9.4.1. The thin disk is basically modelled as a component of age 10 Gyr with constant star formation, the thick disk as a single burst of age 11 Gyr, and the halo as a single burst of age 14 Gyr. Resulting component density profiles and scale lengths are given in their Table 3.

Galactic model parameters, including disk and halo scale heights, estimated by Ak *et al.* (2007) from SDSS data for two fields at $l = 180^\circ, b = \pm 45^\circ$, made use of the Hipparcos data to constrain the corresponding local space densities.

Models of the potential Although the Hipparcos data provide some insight into the overall potential of the Galaxy, component masses and scale lengths have not been fundamentally revised, meaning that model potentials have not been significantly affected. Since various dynamical studies do employ models of the potential, two examples which have been used for the determination of orbital parameters are included here for reference. Of the various models in the literature, Table 9.9 reproduces the two used and conveniently summarised by Dinescu *et al.* (1999a): from Johnston *et al.* (1995) and Paczyński (1990) respectively. Both consist of axisymmetric potentials with three components: bulge, disk, and dark halo. The disks are the Miyamoto & Nagai (1975) potentials with different coefficients (these are highly simplistic, and cannot reproduce orbital properties for certain types of orbit). The dark halo is modelled in both cases as a logarithmic potential, which generates a flat rotation curve but is unphysical in the sense that it implies an infinite mass. The bulge is modelled as a Hernquist (1990) potential in the former (more massive and less centrally concentrated), and as a Plummer (1915) potential in the latter. The disks are of comparable mass, with the former being more flattened.

Famaey & Dejonghe (2003) derived a set of axisymmetric Stäckel potentials for the thin disk, the thick disk, and the halo (these are the most general set of non-rotating potentials for which the Hamilton–Jacobi equation is separable, which contain one free function, for which three exact integrals of motion are known, and which are considered relevant as models for a global Galaxy potential; Stäckel 1890). Together they satisfied the various constraints on the adopted Galaxy parameters: Galactocentric radius, flat rotation curve, Oort constants, local circular speed, and local dynamical mass.

9.11 Globular clusters

9.11.1 Introduction

Globular clusters are tight groups of $\sim 10^4$–10^6 stars with central densities of $\sim 10^3$–$10^4\,M_\odot\,\mathrm{pc}^{-3}$. Structurally they are described in terms of various characteristic radii, such as the core radius, R_c, reflecting the compactness of the cluster; the tidal radius, R_t, delimiting all stars bound to the cluster; and the half-mass radius, related to the global properties; with the cluster concentration $c \equiv \log(R_t/R_c)$.

Most galaxies appear to contain globular clusters. The Galaxy contains about 150 known clusters, with 20% within a few kpc of the centre, and extending out to 30–40 kpc. Although they represent a negligible

Table 9.7 Properties of the thin and thick disk, with separate contributions from the lower main sequence, main sequence, and stellar remnants (white dwarfs, neutron stars, and red giants). The columns give the number density, n_*, the matter density, ρ_*, and the disk local surface density, $\Sigma_\odot$. The scale height of the more massive main-sequence stars is mass dependent. From Chabrier (2001, Table 2).

	scale height (pc)	$n_\star$ (pc^{-3})	$\rho_\star$ (M$_\odot$ pc^{-3})	$\Sigma_\odot$ (M$_\odot$ pc^{-2}
Thin disk:				
lower main sequence ($\leq 1.0\,M_\odot$)	250	0.12 ± 0.02	$3.10 \pm 0.3 \times 10^{-2}$	15.5 ± 2
main sequence ($> 1.0\,M_\odot$)	Scalo (1986)	0.43×10^{-2}	0.6×10^{-2}	2.3
stellar remnants	250	0.7×10^{-2}	0.6×10^{-2}	2.8
all stars		0.13 ± 0.02	$4.30 \pm 0.3 \times 10^{-2}$	20.6 ± 2
Thick disk:	760–1000			
all stars			$\approx 0.22 \pm 0.02 \times 10^{-2}$	$\approx 3.2 - 4.3$
Total:		0.13 ± 0.02	$4.50 \pm 0.3 \times 10^{-2}$	24.4 ± 2.5

Table 9.8 Local mass density ρ_0 of the stellar components, the dark matter halo, and the interstellar medium (ISM); σ_W is the velocity dispersion used for the dynamical self-consistency, and ϵ are the resulting disk axis ratios. The white dwarf (WD) mass density assumes a white dwarf mass of 0.6 M$_\odot$. From Robin et al. (2003, Table 2).

	Age (Gyr)	ρ_0 ($M_\odot$ pc^{-3})	σ_W (km s^{-1})	ϵ
Disk	0-0.15	4.0×10^{-3}	6	0.0140
	0.15-1	7.9×10^{-3}	8	0.0268
	1-2	6.2×10^{-3}	10	0.0375
	2-3	4.0×10^{-3}	13.2	0.0551
	3-5	5.8×10^{-3}	15.8	0.0696
	5-7	4.9×10^{-3}	17.4	0.0785
	7-10	6.6×10^{-3}	17.5	0.0791
	WD	3.96×10^{-3}		
Thick disk	11	1.34×10^{-3}		
	WD	3.04×10^{-4}		
Stellar halo	14	9.32×10^{-6}		0.76
Dark halo		9.9×10^{-3}		1.
ISM		2.1×10^{-2}		

Table 9.9 Two Galactic potential models reproduced from Table 4 of Dinescu et al. (1999a). These are considered as realistic but relatively simple analytical models of the Galaxy's potential, well suited for orbital calculations based on the conservation of total energy, and of the z-component of the angular momentum.

Parameter	Johnston *et al.* (1995)	Paczyński (1990)
Φ_b (bulge)	$-\frac{GM_b}{r+c}$	$-\frac{GM_b}{\sqrt{R^2+(a_b+\sqrt{z^2+b_b^2})^2}}$
	$M_b = 3.4 \times 10^{10}\,M_\odot$, $c = 0.7$ kpc	$M_b = 1.12 \times 10^{10}\,M_\odot$, $a_b = 0.0$ kpc, $b_b = 0.277$ kpc
Φ_d (disk)	$-\frac{GM_d}{\sqrt{R^2+(a_d+\sqrt{z^2+b_d^2})^2}}$	$-\frac{GM_d}{\sqrt{R^2+(a_d+\sqrt{z^2+b_d^2})^2}}$
	$M_d = 10^{11}\,M_\odot$, $a_d = 6.5$ kpc, $b_d = 0.26$ kpc	$M_d = 8.07 \times 10^{10}\,M_\odot$, $a_d = 3.7$ kpc, $b_d = 0.20$ kpc
Φ_h (halo)	$v_o^2 \ln\left(1+\frac{r^2}{d^2}\right)$	$\frac{GM_h}{d}\left[\frac{1}{2}\ln\left(1+\frac{r^2}{d^2}\right)+\frac{d}{r}\arctan\frac{r}{d}\right]$
	$v_0 = 128$ km s^{-1}, $d = 12.0$ kpc	$M_h = 5 \times 10^{10}\,M_\odot$, $d = 6.0$ kpc
Φ_0 (km s^{-1})^2	-5.2×10^4	-12.3×10^4
$P_{\rm LSR}$ (10^6 yr)	218.0	223.0
Θ_0 (km s^{-1})	222.5	220.1

fraction of the light and mass of the stellar halo, they are important tracers of the age and dynamics of the Galaxy. The globular cluster population as a whole appears to be representative of the Galaxy's overall stellar population: in addition to a metal-poor $[Fe/H] < -0.8$ slowly-rotating halo-like population, there is a flattened metal-rich population in rapid rotation resembling the disk (Zinn, 1985), a bulge-like distribution in the inner 3 kpc region (Minniti, 1995), and a concentration reflecting the properties of the bar (Burkert & Smith, 1997).

Ages of the oldest globular clusters are of the same order as estimates of the age of the Universe derived from its expansion rate, and thus suggest that at least some were among the first objects to form when the Galaxy formed. The oldest clusters in the halo, the Large Magellanic Cloud, and nearby dwarf spheroidal galaxies, appear to be coeval, showing that the onset of globular cluster formation was synchronised to a precision of about 1 Gyr over a volume of radius $>$ 100 kpc. However, they also appear to span an age range of several Gyr, such that some are likely to be remnants of an early accretion phase of the Galaxy, having been captured from other galaxies during mergers or collisions, or formed during these interactions.

Absolute age determination requires a distance estimate to compare luminosities of the main-sequence turn-off point with theoretical models of stellar evolution. Like their open cluster counterparts, it is assumed that the constituent stars have roughly the same age and initial chemical composition (there are a few well-known exceptions to this, including ω Centauri and M22), but with a range of masses. As the cluster ages, the most massive stars move away from the main sequence, with a turn-off point which moves down the main sequence as the cluster ages (Figure 9.41). The absolute magnitude of the turn-off is considered to be an excellent age indicator, a basic evolutionary 'clock' for globular clusters. The main limitation in determining absolute ages has been in deriving sufficiently accurate distances: the scarcity of good parallax data for nearby calibrating subdwarfs, as well as the practical difficulties in acquiring accurate and uniform photometry and metallicity, have hindered application of the most direct distance determination method, namely main-sequence fitting of cluster colour–magnitude diagrams to nearby field subdwarfs of known distance.

This is important because the comparison between their colour–magnitude diagrams and the predictions of stellar evolutionary models probes the star-formation history of the early Universe and can set a lower limit on its age. Establishing the distribution of ages of globular clusters and open clusters is important in establishing whether they overlap in age, and whether there is a continuity of properties between the oldest open clusters, such as Berkeley 17 at around 10 Gyr (Krusberg & Chaboyer, 2006), and the youngest globular clusters, such as Terzan 7 at around 9 Gyr, and the transition between the halo/thick disk to the thin disk (Phelps, 1997).

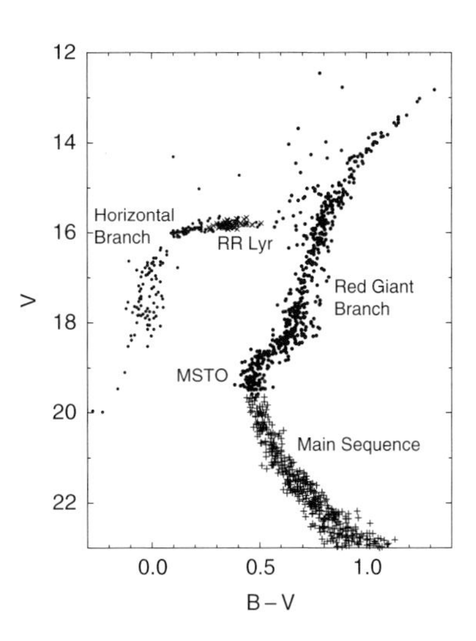

Figure 9.41 Colour–magnitude diagram of the globular cluster M15, based on data from Durrell & Harris (1993). Most stars are on the main sequence, in which core hydrogen fusion occurs (only some 10% of the main-sequence stars are shown for clarity); slightly higher mass stars have exhausted their supply of hydrogen in the core and 'turn off' at the main-sequence turn-off region (MSTO); subsequently, the stars quickly expand and become brighter on the red giant branch during a shell-hydrogen burning phase; still higher mass stars undergo core helium fusion on the horizontal branch; stars on this branch unstable to radial pulsations are the RR Lyrae stars. From Chaboyer (1998, Figure 1), reproduced with the permission of Elsevier.

9.11.2 Ages

The 'age paradox' was the focus of a number of Hipparcos and other studies around 1997–98. Before the availability of the Hipparcos results, ages of the oldest globular clusters were estimated to be around 15 Gyr (VandenBerg *et al.*, 1996), and possibly as much as 15.8 ± 2.1 Gyr (Bolte & Hogan, 1995), or even 16–17 Gyr in the case of NGC 6541 (Alcaino *et al.*, 1997).

At the same time, some estimates of the Hubble constant, including those from preliminary HST observations of Cepheids out to the Virgo cluster, indicated a high value for the Hubble constant of $H_0 = 80 \pm 17\,\mathrm{km\,s^{-1}\,Mpc^{-1}}$, and a resulting Universe expansion age of some 8 Gyr (Freedman *et al.*, 1994). The disturbing implication was that globular cluster ages might actually exceed the 'expansion age' of the Universe determined from the recessional velocities of high-redshift galaxies. Subsequent HST Cepheid observations converged on the value $H_0 = 72 \pm 8\,\mathrm{km\,s^{-1}\,Mpc^{-1}}$ (Freedman *et al.*, 2001), revising the expansion age to around 12 Gyr. Other determinations of H_0 appear to be converging on approximately the same value (see Section 5.7.1).

The first indications of a possible solution to the 'age paradox' came, however, from revised trigonometric parallaxes for the nearest Hipparcos Cepheids (Feast & Catchpole, 1997). The inference that the Large Magellanic Cloud Cepheids were about 10% further away than previously estimated, and thus brighter, led to the conclusion that the overall distance scale had to be revised by the same amount, with the implication that globular clusters themselves were more distant than previously thought, that their luminosities were larger, and that the turn-off point thus indicates younger ages than previously thought. Cepheids are, however, young and short-lived, and not typically found in the old globular clusters.

More direct results for the ages of globular clusters are based on subdwarf main-sequence fitting. Given the typically very large distances of globular clusters ($\gtrsim 10$ kpc) compared with the distance horizon of Hipparcos trigonometric parallaxes accurate to around 10% (< 100 pc), distance and age estimation based on the Hipparcos data are both somewhat indirect. The related problem of globular cluster distance determination has already been discussed in Chapter 5. To summarise, the Hipparcos contribution to globular cluster distance estimates comes mainly through the observation of nearby subdwarfs (metal-poor halo stars) which happen to be passing close to the Sun at this time. Since they also occur in globular clusters, careful luminosity calibration as a function of metallicity allows them to be used in globular cluster distance determination through main-sequence fitting. Independent distance estimates can be made from calibration of the RR Lyrae luminosities, since these objects also occur in both the field and in globular clusters. However, even the nearest RR Lyrae are distant and at the limits of Hipparcos parallaxes, and their use also rests on the assumption that the field and cluster RR Lyrae stars have identical luminosities.

Main-sequence fitting brought estimated ages for the oldest clusters down to around 11.5 ± 1.3 Gyr (Chaboyer, 1998) or 12.9 ± 2.9 Gyr (Carretta *et al.*, 2000). At around 12 Gyr, or some 3 Gyr younger than pre-Hipparcos estimates, these results were widely heralded as alleviating the original age paradox. Although individual estimates varied, broadly some 1 Gyr of the decrease can be attributed to the longer distance scale to globulars resulting from Hipparcos, while some 2 Gyr arose from improvements in the theoretical treatment of convection, helium diffusion, the equation-of-state, and a more realistic representation of the Population II elemental abundances, which were independently suggesting younger ages (Mazzitelli *et al.*, 1995; D'Antona *et al.*, 1997; Salaris *et al.*, 1997).

Hipparcos estimates of globular cluster ages have nevertheless continued to vary over the last few years. Estimated ages of the oldest globular clusters have generally risen slightly after 2000 as a result of improved control of observational errors and improved evolutionary models, with estimates for the oldest globular clusters of $13.4 \pm 0.8 \pm 0.6$ Gyr being given by Gratton *et al.* (2003).

9.11.3 Independent age estimates of the oldest halo objects

In addition to the main-sequence turn-off ages based on stellar evolution models, two other estimates of the minimum age of the Universe can be obtained for metal-poor stars in the halo for consistency checks (Chaboyer, 1998): nucleochronology, and white dwarf cooling curves. For the former, results for the thorium abundance of CS 22892 (Sneden *et al.*, 1996) gave an age of 15.2 ± 3.7 Gyr (thorium half-life is 14.05 Gyr). Hill *et al.* (2002) used the ^{238}U/^{232}Th ratio in the extreme halo giant CS 31082–001 to propose an age of 14.0 ± 2.4 Gyr.

White dwarfs are the end stage of stellar evolution for stars below about $8\,M_\odot$. Due to their low luminosities, the brightest can only be seen out to small distances, but none are observed with temperatures lower than about 4000 K. This implies that even the oldest white dwarfs have not had time to cool further, and thus provides a constraint on the age of the Universe, or at least on the age of the disk, although results are sensitive to the star-formation rate as a function of time. Oswalt *et al.* (1996) found a local disk age of $9.5^{+1.1}_{-0.8}$ Gyr. Adding 2 Gyr for the time taken for the Galaxy to collapse and the disk to form, yields a lower limit to the age of the Universe of about 11–12 Gyr. Hansen *et al.* (2002) estimated 12.7 ± 0.7 Gyr from a deep HST exposure of M4, but with substantial uncertainty arising from the evolutionary models.

The compilation of Chaboyer (1998) includes a variety of distance indicators and a detailed study of error sources, leading to a minimum age of the Universe of 9.5 Gyr at 95% confidence.

The age estimate for the oldest globular clusters of $13.4 \pm 0.8 \pm 0.6$ Gyr (Gratton *et al.*, 2003) agrees rather well with that of CS 31082–001 of 14.0 ± 2.4 Gyr, and with that of the age of the Universe from the WMAP group of 13.7 ± 0.2 Gyr (Bennett *et al.*, 2003; Spergel *et al.*, 2003), in which acceleration in the expansion of the early Universe is accounted for.

9.11.4 Consequences of globular cluster ages

Although the age paradox has therefore largely disappeared, accurate ages of globular clusters can still be

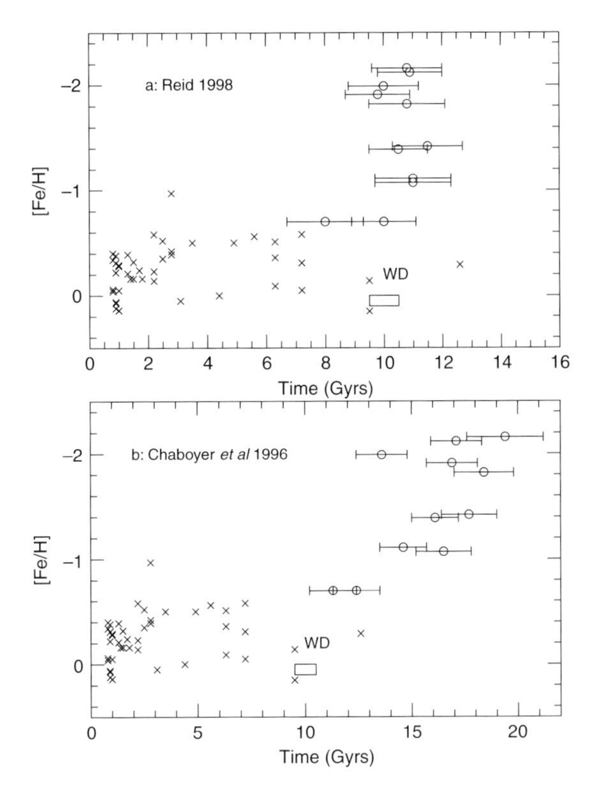

Figure 9.42 Star-formation history of the Galaxy, from Reid (1998, Figure 10). (a) is based on their globular ages, and (b) is that from Chaboyer et al. (1996b) based on their (pre-Hipparcos) globular cluster distances and ages derived using one of their M_V(HB) versus [Fe/H] relations. Crosses are ages of Galactic open clusters (from Friel, 1995), and the box 'WD' marks the age estimated for the Galactic disk from analysis of the white dwarf luminosity function (Oswalt et al., 1996).

used to constrain the epoch of formation of the earliest stellar populations in the Galaxy, and to determine a value of Ω_M independently of observations of Type Ia supernovae. Again, the topic is much broader than that implied by the use of the Hipparcos data alone, although this section again focuses only on the immediate Hipparcos implications.

First attempts to use the new age estimates from the Hipparcos data to map out the onset of star formation within the disk and halo populations were made by Reid (1998), who compared his results with pre-Hipparcos estimates as shown in Figure 9.42a–b. He argued that the pre-Hipparcos ages required a progressive formation of halo clusters with increasing metallicity over some 7 Gyr, with some 8 Gyr elapsing between the start of the halo and the first star formation in the disk, in turn requiring an unidentified process to inhibit collapse and star formation within the rotating disk over these times. His revised globular cluster ages implied a gap of at most 1.5 Gyr between the onset of star formation in the halo and significant star formation occurring within the disk.

Salaris & Weiss (1998) estimated ages of around 9.2 Gyr for the disk clusters 47 Tuc and M71 from the turn-off versus ZAHB, or turn-off using red giant branch photometry, and their larger globular cluster sample shows some correlation between age and metallicity. From 34 clusters, Rosenberg *et al.* (1999) concluded that there is no evidence for an age spread for clusters with Fe/H< -1.2, but clear evidence of age dispersion for higher metallicities. The resulting model suggests that the oldest globular clusters formed at the same age throughout the halo, with metal-rich clusters formed at a later time, and significantly younger halo clusters arising from later mergers of dwarf galaxies.

The age for 47 Tuc, as well as the relative age estimates by Rosenberg *et al.* (1999) and Salaris & Weiss (2002), indicate that the epoch of formation of globular clusters lasted about 2.6 Gyr, ending about 10.8 ± 1.4 Gyr ago, which corresponds to a redshift of $z > 1.3$. The formation of the thick disk and its associated population of globular clusters like 47 Tuc would then have occurred before the formation of the thin disk. If subsequent violent accretion events required for the formation of younger globular clusters had occurred, this would probably have disrupted or heated the thin disk. Most such young objects, like Pal 12, are probably connected to the Sagittarius dwarf galaxy (Dinescu *et al.*, 2000). In the standard ΛCDM the end of the phase of formation of globular clusters in the Galaxy corresponds to a redshift of $z > 1.3$, and suggests a close link between the epoch of formation of the Milky Way spheroid, and that of the early spheroids in high redshift galaxies.

Carretta *et al.* (2000) evaluated the consequences of his revised ages as follows. In Figure 9.43a, the relationships are constructed for the pre-Hipparcos, or 'short', distance scale. In this case M_V(HB) ~ 0.75 at [Fe/H] $\simeq -1.5$, globular cluster absolute ages would be around 16 Gyr and the derived distance modulus for the LMC of ~18.25, implies H_0 in the range 65–85 km s^{-1} Mpc^{-1}. In Figure 9.43b, the relationships are constructed for the post-Hipparcos, or 'long', distance scale. In this case M_V(HB) ~ 0.5 at [Fe/H] $\simeq -1.5$, globular cluster absolute ages are around 13 Gyr and the derived distance modulus for the LMC of ~18.5 is consistent with H_0 in the range 55–75 km s^{-1} Mpc^{-1}. Shaded regions show the permitted areas, where H_0 is consistent with the distance modulus used to derive the ages for the globular clusters. They concluded that Galactic globular clusters formed at $z > 1$.

Gratton *et al.* (2003) combined ages of NGC 6397, NGC 6752, and 47 Tuc with the precise estimates of the age and geometry of the standard ΛCDM Universe from WMAP and the HST key-project results (Spergel *et al.*, 2003), for which the age of the Universe is estimated

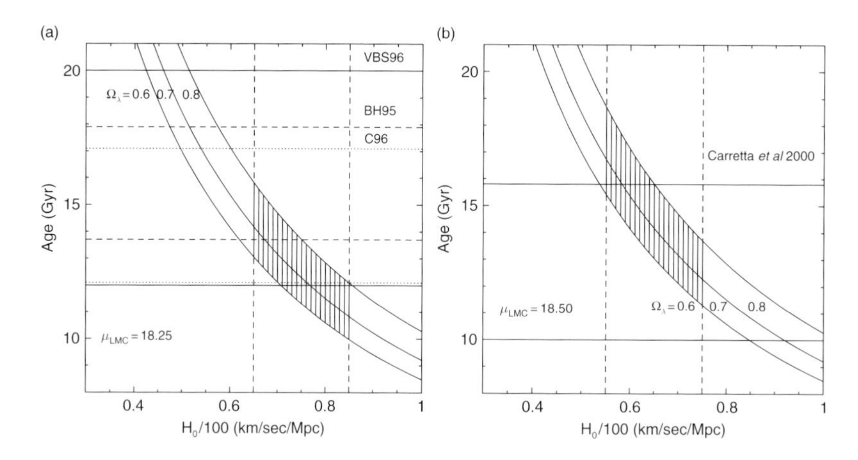

Figure 9.43 Age versus H_0 relationship, from Carretta et al. (2000, Figure 1), for various cosmological models of flat Universe and for different values of $\Omega_\Lambda = 1-\Omega_m$, with Ω_m in the range suggested by recent Type Ia supernovae data ($\Omega_m \sim 0.3\pm0.1$). The shaded area is the permitted region according to values of H_0 consistent with the distance moduli used to derive ages for the globular clusters. Left: pre-Hipparcos parameter space, with 1σ upper limits $t = 15.8 \pm 2.1$ Gyr (Bolte & Hogan, 1995), $t = 14.6 \pm 2.5$ Gyr (Chaboyer et al., 1996a), t=15^{+5}_{-3} Gyr (VandenBerg et al., 1996). Right: after Hipparcos, with the age constraint derived by Carretta et al. (2000).

as 13.7 ± 0.2 Gyr; the Hubble constant as $H_0 = 71 \pm 4$ km s^{-1} Mpc^{-1}; and the values of Ω_M and Ω_Λ are fixed at 0.27 ± 0.04 and 0.73 ± 0.04 respectively. Coupled with age estimates for the oldest Galactic globular clusters of $13.4 \pm 0.8 \pm 0.6$ Gyr, this implies that they formed within 1.7 Gyr of the Big Bang at the 1σ confidence level, corresponding to a redshift of $z > 2.5$. Alternatively, assuming that the Universe is 0.3 Gyr older than the oldest globular clusters, Gratton *et al.* (2003) showed that the value of Ω_M in a flat Universe is constrained to $\Omega_M < 0.57$ and $\Omega_M < 0.75$ at the 95% and 99% confidence levels respectively. This suggests the need for a dark energy contribution, independently of results from high-redshift Type Ia supernovae (Perlmutter *et al.*, 1999), and from galaxy clusters (Rosati *et al.*, 2002).

9.11.5 Kinematics and dynamics

The kinematic properties of Galactic globular clusters reflect the gravitational potential and thus mass distribution of the Galaxy. The combination of kinematics, metallicities, and ages provide important constraints on theories of formation of the Galaxy, including when and how the halo and disk formed, and constraints on the age of the oldest populations supported by rotation.

Establishing and understanding the nature of globular cluster orbits has been a long-standing problem, and a review of the early advances is given by Frenk & White (1980). Studies in the 1950s suggested that cluster orbits within the Galaxy were predominantly radial, although subsequent analyses provided conflicting conclusions. Frenk & White (1980) themselves found a systemic rotation for a sample of 66 clusters, a velocity dispersion increasing with Galactocentric distance, and a Galactic rotation curve supporting the existence of a massive Galaxy halo extending out to at least 33 kpc. Lynden-Bell & Lynden-Bell (1995) identified possible streams of clusters likely to trace the orbits of satellites that may have long since merged with the Galaxy, including streams associated with the Magellanic Clouds, Fornax, and the Sagittarius dwarf. They also listed proper motions of 22 clusters predicted to exceed 1 mas yr^{-1}, or 4.74 km s^{-1} kpc^{-1}.

Further advances in understanding cluster orbits have been limited by the fact that their space velocities have generally been incompletely known; only four kinematic quantities were generally measurable: the three-dimensional position and the line-of-sight velocity. In contrast, a complete orbit determination requires the full phase-space information (i.e. the two other, transverse, Cartesian components of the space velocity vector), along with the form of the potential in which the cluster moves. As an indication of the pre-Hipparcos observational status, Tucholke *et al.* (1996) tied the mean proper motion of M92 to an inertial system, represented by external galaxies, with a formal accuracy of about 1.0 mas yr^{-1}.

Since individual globular cluster stars are fainter than the Hipparcos Catalogue limit, and since the overall catalogue stellar density on the sky is only 2.5 stars per square degree, part of the construction of the Hipparcos Input Catalogue involved including a number of bright reference stars lying close to 13 selected globular clusters (NGC 4147, 5024, 5272, 5466, 5904, 6205, 6218, 6254, 6341, 6779, 6934, 7078, and 7089), specifically to

Table 9.10 Space velocities and orbital parameters for the representative case of the globular cluster NGC 7078, taken from Dinescu et al. (1999a, Tables 2 and 3), for which [Fe/H] = −2.17, $d_\odot = 9.5$ kpc, $R_{GC} = 9.8$ kpc, and $V_{rad} = -106.6 \pm 0.6$ km s^{-1}. The adopted absolute proper motion components were $\mu_\alpha \cos\delta = -0.95 \pm 0.51$ and $\mu_\delta = -5.63 \pm 0.50$ mas yr^{-1}, partly based on Hipparcos results. Listed are the derived space velocity components (U, V, W) with respect to the Galactic standard of rest, using $R_0 = 8.0$ kpc and $\Theta_0 = 220$ km s^{-1}. Apo-Galactic and peri-Galactic distances, R_a and R_p, maximum height above the Galactic plane, z_{max}, orbital period, P, and eccentricity, e, were calculated using axisymmetric Galaxy potentials with bulge, disk, and dark halo components. The table is given an example of the orbital information that can be reconstructed from high-quality astrometric data.

U km s^{-1}	V km s^{-1}	W km s^{-1}	R_a kpc	R_p kpc	z_{max} kpc	P 10^6 yr	e
-148 ± 28	-219 ± 14	-58 ± 24	10.3 ± 0.7	5.4 ± 1.1	4.9 ± 0.8	242 ± 25	0.32 ± 0.05

provide an inertial reference frame for globular cluster proper motion studies. Geffert *et al.* (1997) used some 4–9 Hipparcos stars in the fields of 10 globular clusters to derive their absolute proper motions in the inertial system of Hipparcos; differences with respect to pre-Hipparcos 'absolute' proper motions amounted to as much as ±4 mas yr^{-1}. Resulting space velocity components with respect to the Galactic coordinate system UVW, as well as peri- and apo-Galactic distances, eccentricities and z-components of angular momentum, were calculated using a simple logarithmic potential. Apo-Galactic distances thus derived were typically smaller than those estimated in pre-Hipparcos studies, and more consistent with the observed distribution extending out to only R = 30–40 kpc (Zinn, 1985). A significant fraction have retrograde motions, with the mean value of the angular momentum $\langle I_z \rangle = -130 \pm 340$ kpc km s^{-1}, corresponding to a rotational velocity at R_0 of $+15 \pm 40$ km s^{-1}. A similar study by Odenkirchen *et al.* (1997) for 15 globular clusters indicated that the more metal-rich clusters are concentrated towards the Galactic centre. Those with significant retrograde motion appear relatively homogeneous in metallicity, with $-2.0 <$ [Fe/H] < -1.5, with a subset of the young halo clusters orbiting with a small retrograde rotation of −9 km s^{-1}. Observed cluster radii are consistent with the tidal limits imposed by orbital motion in the Galactic field. It is encouraging that their proper motion for 47 Tuc, for example, of $\mu_\alpha = 7.0 \pm 1.0$ mas yr^{-1} and $\mu_\delta = -5.3 \pm 1.0$ mas yr^{-1}, is in good agreement with that derived by Freire *et al.* (2001) from the mean common proper motion (from timing solutions) for five millisecond pulsars within the cluster, of $\mu_\alpha = 6.6 \pm 1.9$ mas yr^{-1} and $\mu_\delta = -3.4 \pm 0.6$ mas yr^{-1} (see Section 8.4.9).

Kutuzov & Ossipkov (2001) reported reconstructed orbits for 24 clusters, 15 based on these Hipparcos reductions. Geffert (1998) used the same procedure as Geffert *et al.* (1997) to derive space motions for M3 and M92, also using the improved positions for optical identification of X-ray and radio sources in the two fields. Their formal uncertainties are around 0.8 mas yr^{-1} in each proper motion component, around 20–30 km s^{-1} in UVW, and around 0.3–1.4 kpc in apo- and peri-Galactic distances. Wu & Wang (1999) used 14 plates over 81 yr, reduced to the Hipparcos system, to derive consistent proper motions for M3, but with a significantly improved standard error of about 0.2 mas yr^{-1} in each component. Dinescu *et al.* (1999b) derived absolute proper motions and space velocities for 10 clusters, one based on Hipparcos reference stars, the others using galaxies. Space velocities for a further six clusters, also on the Hipparcos system, were reported by Casetti-Dinescu *et al.* (2007); these included a metal-rich thick disk cluster (NGC 5927), two unexpected 'pairs' of dynamically-associated clusters (NGC 2808 and NGC 4372; NGC 4833 and NGC 5986), and one with a highly-retrograde orbit with an apocentric distance of 22 kpc (NGC 3201). Similar analyses were made by Chen *et al.* (2000a) for M10; Wang *et al.* (2000) for NGC 4147; Wang *et al.* (2001) and Chen *et al.* (2002a) for M13; and Wu *et al.* (2002a,b) for M3.

Van Leeuwen *et al.* (2000) made an extensive study of the internal kinematics of the globular cluster ω Centauri based on ~8000 cluster members reaching 16.0–16.5 mag, using 100 plates spanning 1931–1983 to derive individual proper motion accuracies of 0.1–0.65 mas yr^{-1}. From three Hipparcos and 53 additional Tycho 2 Catalogue reference stars, they derived global proper motion components of $\mu_x = -3.97 \pm 0.41$ and $\mu_y = -4.38 \pm 0.41$ mas yr^{-1}, significantly different from the pre-Hipparcos value of $\mu_x = +0.5 \pm 0.6$ and $\mu_y = -7.7 \pm 0.5$ mas yr^{-1} from Murray *et al.* (1965), implying a total tangential space motion of 142 ± 19 km s^{-1} at 5.1 kpc, and a (non-significant) global rotational of 0.6 ± 0.7 mas yr^{-1} R_c^{-1} ($R_c \sim 3$ arcmin).

Halo clusters Dinescu *et al.* (1999a) analysed the global results for 38 halo clusters. Adopted proper motions were weighted combinations of all recent high-quality estimates, including but not restricted to reductions with respect to the Hipparcos reference frame. Orbits and orbital parameters were calculated using axisymmetric Galaxy potentials with bulge, disk,

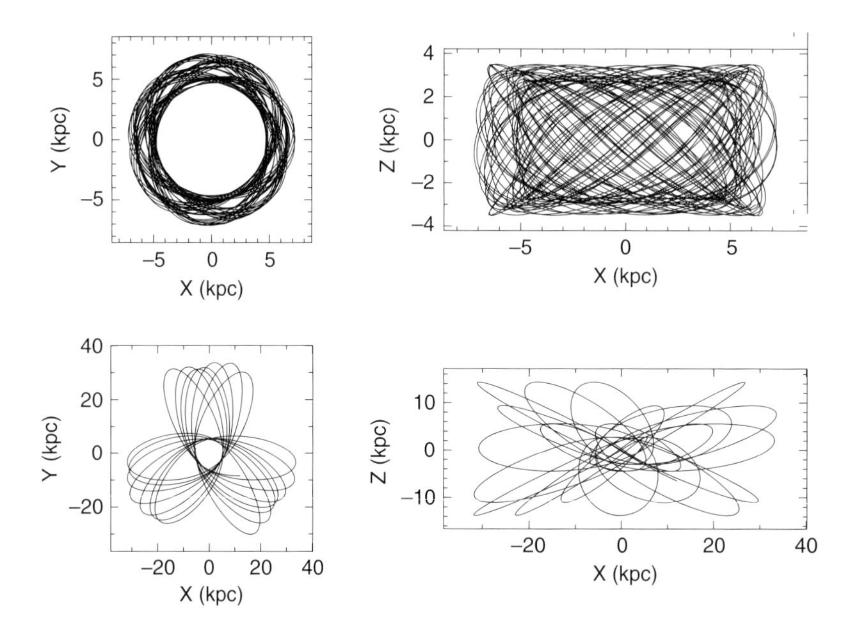

Figure 9.44 Galaxy orbits for the globular clusters NGC 104 (top pair) and NGC 1851 (bottom pair), with components in the Galactic plane (left) and perpendicular to it (right). Orbits were derived from the six components of position and velocity, including proper motions from Hipparcos, combined with an analytic potential of the Galaxy, and are shown for a time interval of 10^{10} yr. From Dinescu et al. (1999a, Figure 2).

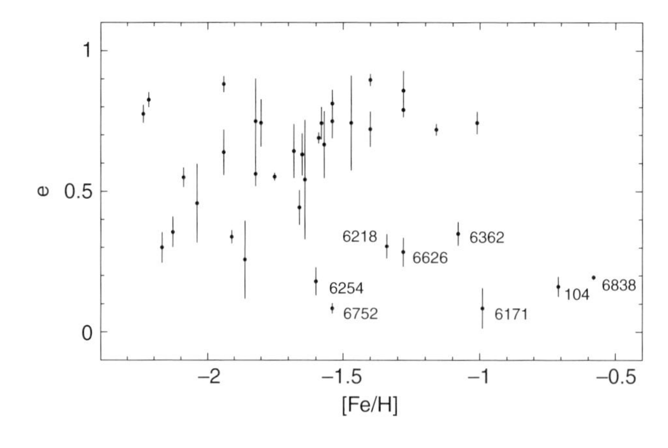

Figure 9.45 Eccentricity as a function of metallicty for the globular cluster orbits derived by Dinescu et al. (1999a, Figure 5). Higher eccentricities are expected for a hot halo population, while the distribution at low metallicities also extends to the lower eccentricities typical of standard metal-rich thick-disk clusters like 47 Tuc.

and dark halo components. Table 9.10 shows orbital parameters for a representative cluster, NGC 7078, from Dinescu *et al.* (1999a), while Figure 9.44 shows two representative examples of the resulting Galactic orbits over a time interval of 10^{10} yr. Results were reasonably insensitive to the potential adopted, while contribution to the potential from the bar was ignored, as was possible chaotic orbital behaviour perhaps relevant for peri-Galactic distances within about 1 kpc of the Galactic centre.

Dinescu *et al.* (1999a) identified three metal-poor clusters (NGC 6254, 6626, and 6752) with orbits similar to those of the metal-rich disk clusters, like 47 Tuc and NGC 6838, i.e. with low-eccentricity, low-inclination, and rotation velocities close to that of the thick disk. This suggests that part of the inner metal-poor halo is the low-metallicity tail of the thick disk, with comparable ages (see Figure 9.45). Similar behaviour was found for the Hipparcos sample of red giants and RR Lyrae stars by Chiba & Yoshii (1997), where some 20% of the low-metallicity objects had much lower eccentricities than expected for a hot halo population. In contrast, the young halo or red horizontal branch clusters with a radially anisotropic velocity distribution, high orbital energy, apocentric radii larger than 10 kpc, and high eccentricity, may represent accreted components. These clusters could have been produced in satellite galaxies whose orbits were modified and circularised due to dynamical friction before they were completely disrupted, or formed under suitable pressure and density conditions from a disk-like structure fairly early on in the history of the Galaxy. They also used the orbital data to search for streams of globular clusters that might have originated in the same accreted satellite, largely invalidating such a hypothesis for three members of an association suggested by Rodgers & Paltoglou (1984) on the basis of metallicity and significant retrograde motion, but finding two possible pairs with common values of $E_{\rm tot}$ and L_z.

Inner Galaxy clusters Dinescu *et al.* (2003) measured the absolute proper motion of four low-latitude inner Galaxy clusters (NGC 6266, 6304, 6316 and 6723), using the Hipparcos reference system due to the absence of extragalactic reference objects in these high-extinction regions. Along with three other bulge clusters, whose space motions were determined with respect to bulge stars and are consequently somewhat less secure, they found space velocities all smaller than the escape velocity of the bulge and thus confined to the bulge region (Figure 9.46). Of the three metal-poor systems, $[{\rm Fe/H}] < -1.0$, two have halo-like orbits and one has a disk-like orbit. Of the four metal-rich clusters, two are consistent with a disk-like system. One, NGC 6528, itself one of the most metal-rich Galactic globular clusters, has a space velocity consistent with membership of the bar, having a very low vertical velocity, and a transverse velocity in agreement with the rotation of a solid body having the Galactic bar's pattern speed of $\Omega_{\rm b} = 60\,{\rm km\,s^{-1}\,kpc^{-1}}$ (Section 9.6). It was presumably trapped into a bar-like orbit from an initially disk-like configuration. Despite the small numbers, these clusters represent an important dynamical tracer of the bulge population in terms of distance, metallicity, age, radial velocity, and now proper motion, in the most complex region of the Galaxy where a variety of stellar populations are densely superimposed.

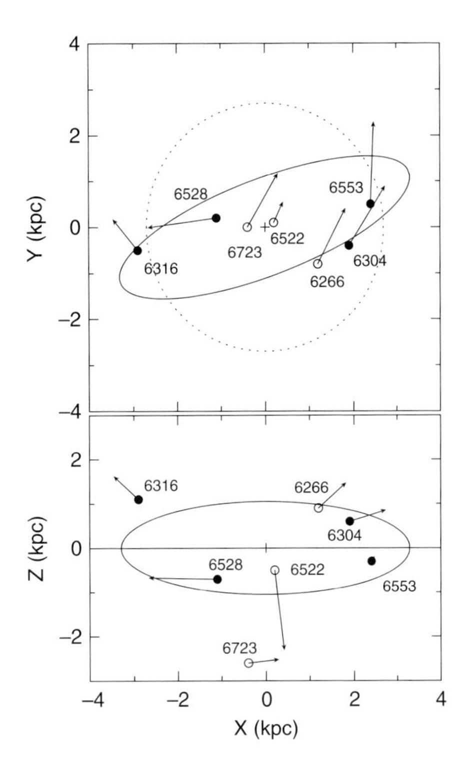

Figure 9.46 Components in the Galactic plane (top) and perpendicular to it (bottom) of positions and velocities of the seven inner Galaxy globular clusters, three with absolute proper motions from Hipparcos. Open symbols: metal-poor; solid symbols: metal-rich. The cross is the Galactic centre, the ellipse represents the Galactic bar, and the dotted circle (top panel) indicates the bulge region of radius ~ 2.7 kpc. Absolute proper motion were based on Hipparcos for NGC 6266, 6304, 6316, and 6723. From Dinescu et al. (2003, Figure 4), who classified them as halo-like: 6522, 6723; disk-like: 6266, 6304, 6553; bar-like: 6528; hotter than the bar: 6316.

9.11.6 Cluster disruption

Globular cluster kinematics also provide important clues about their past and ongoing disruption due to passages through the disk or close to the bulge. Globular clusters in the inner Galaxy are particularly affected by externally-driven destruction processes, with a consequent decrease of their system velocity dispersion with time. Improved orbits allow quantitative constraints to be placed on the destruction mechanisms. An introduction to the dynamical evolution of clusters, and the importance of accurate Galactic orbits in this context, is given by Gnedin *et al.* (1999b) and references therein.

Internal to the cluster, destruction processes include effects of stellar evolution, and two-body relaxation coupled with cluster evaporation in which loosely-bound stars acquire velocities above the escape velocity and evaporate from the cluster. Core collapse, or gravothermal catastrophe (Binney & Tremaine, 1987, Section 8.2), is a runaway phenomenon driven by the negative heat capacity of the stellar dynamical system dominated by gravity (similar considerations govern the stability of nuclear burning in stellar cores), and the transport of kinetic energy from the inner to the outer parts. It can be reversed through heating in the core via binary formation and stellar mergers (e.g. Hut *et al.*, 1992). External tidal effects also play an important role in the destruction of globular clusters (Gnedin & Ostriker, 1997, 1999). The stellar distribution being essentially truncated at the point where the ambient Galactic density exceeds that of the cluster stars, they have shown that clusters dissolve in $\sim N/200$ Galactic orbits, where N is the initial number of cluster stars. Furthermore, a net cluster rotation may develop as stars in prograde orbits are preferentially ejected compared with those on retrograde orbits. Time-varying tidal forces cause gravitational shocks as the cluster passes through the Galaxy disk or comes near to the bulge, with subsequent relaxation accelerating cluster evolution (Weinberg, 1994; Gnedin *et al.*, 1999a).

Gnedin *et al.* (1999b) included all of these effects in their Fokker–Planck code, calibrated against the intrinsically more accurate but computationally more complex N-body code. The Fokker–Planck development follows the gradual evolution of stellar systems driven by encounters by truncating terms above second-order in incremental phase space (Binney & Tremaine, 1987, Section 8.3). They used the Hipparcos-based three-dimensional velocity components for NGC 6254 to calculate its orbit in an analytic Galaxy potential. Like the examples in Figure 9.44, the inferred orbit is of the rosette type, with a small eccentricity, $e = 0.21$, moving around the Galaxy with a period of about 1.4×10^8 yr, during which it passes close to the Galactic centre every 9.5×10^7 yr, and crosses the Galactic disk every 5.3×10^7 yr. The amplitude of the tidal shocks varies approximately as a Gaussian function of time, thus satisfying the conditions under which the adiabatic corrections were calculated. These shocks accelerate the dynamical evolution of the cluster, stripping stars in the outer regions and adding energy dispersion in the core, although as the clusters lose mass and become more compact, two-body relaxation effects increase in relative importance. Including the effects of shocks implied by the inferred orbit of NGC 6254 reduces the destruction time from 24 to 18 Gyr, which demonstrates that the effects of tidal shocks and two-body relaxation are roughly comparable in importance at present, and that tidal shocks must have been more important in the past. It follows that a large fraction of the initial population of globular clusters might already have been destroyed as a result of their Galactic orbits.

Dinescu *et al.* (1999a) also examined destruction rates along similar lines, finding that tidal shocks due to the resulting bulge and disk interactions are significant compared to internal relaxation only for about eight of the 38 systems which they studied. Allen *et al.* (2006) studied the effects of a barred potential on the Galactic orbits of 48 clusters for which absolute proper motions are now known. They also determined destruction rates due to bulge and disk shocking, for cases with and without a bar.

De Marchi *et al.* (2006) found that the Hipparcos-based Galactic orbit of NGC 6218 determined by Odenkirchen *et al.* (1997), in contrast with pre-Hipparcos orbits, is now in line with its unusually flat stellar mass function measured to $V \sim 25$ mag. They attributed this not to some anomalous form of the initial mass function, but to severe tidal stripping due to its Galactic orbit. Conversely, they argue, if the orbit of a cluster is known, the slope of its observed stellar mass function can assist in determining the Galactic potential.

Meusinger *et al.* (2001) reported a (non-Hipparcos based) proper motion search for stars escaping with high velocities from the globular cluster M3 as a result of binary hardening. At the distance of 10 kpc, the escape velocity of 20 km s^{-1} corresponds to $\mu = 0.4$ mas yr^{-1}. Some 20 candidates out of 10^4 stars were identified for follow-up studies.

9.11.7 Tidal streams and the mass of the Galaxy

Murali & Dubinski (1999) described the use of absolute proper motions, again calibrated using the Hipparcos astrometry, for a number of globular clusters, as probes of the mass distribution of the Galaxy. While various methods are available for mass determination for the inner Galaxy (including observable material in the disk, rotation curves, and the K_z relation), estimates for the outer Galaxy typically rely on the dynamics of satellites, and are subject to uncertainties as to their gravitationally-bound nature, or their overall equilibrium given their long orbital time scales. At intermediate halo distances $R \sim 20$ kpc, far above the Galactic disk, the Hipparcos globular cluster sample (Pal 5, NGC 4147, NGC 5024, and NGC 5466) should provide information on the profile and shape of the inner halo, regions for which the mass distribution is relatively poorly known. Although the path of the cluster through the Galaxy cannot be followed directly, traces of their tidal streams do remain. Murali & Dubinski (1999) describe two methods to use evaporating stars, which reach the inner and outer Lagrange points nearly at rest, to determine the mass and potential of the Galaxy. The first requires complete phase-space data for the stream stars, assuming that the tidal stream follows a streamline in the Galactic potential. This is a generalisation of rotation curve measurements to non-circular orbits and, in principle, provides direct measurement of the local gravitational acceleration and density field (see their Figure 1). Even for nearby clusters at $d \lesssim 5$ kpc, this approach will require considerably improved astrometric data on individual stars, e.g. from Gaia or SIM. The second approach involves fitting a model stream curve to the complete position and velocity vector for the cluster, while knowing only the projected position and radial velocities for the stream stars. This reflects the status based on the Hipparcos cluster space motion combined with current ground-based observational capabilities for individual stream stars. Their estimation of mass determination as a function of, for example, proper motion errors or mass-loss rates, suggests that these four clusters could already provide information on the mass distribution at intermediate Galactocentric distances. The other 11 'Hipparcos' clusters lie closer to the disk and to the Sun, and should be good candidates for probing the mass distribution close to the disk using a similar approach.

9.11.8 Individual globular clusters

In addition to the studies of ages, distances, colour–magnitude diagrams and bulk motions, the Hipparcos data have little direct application to globular clusters. An example of the use of improved registration of the optical and radio reference frames is found in the study of Verbunt & Johnston (2000), who used the Hipparcos and Tycho Catalogue positions to improve the identification of ROSAT HRI X-ray sources in the globular clusters ω Centauri, NGC 6397 and NGC 6752, reducing their systematic errors some around 5 arcsec to around 1–2 arcsec.

References

Abad C, Vieira K, 2005, Systematic motions in the Galactic plane found in the Hipparcos Catalogue using Herschel's method. *A&A*, 442, 745–755 {**526**}

Abadi MG, Navarro JF, Steinmetz M, *et al.*, 2003, Simulations of galaxy formation in a Λ Cold Dark Matter Universe. II. The fine structure of simulated galactic disks. *ApJ*, 597, 21–34 {**527**}

Ak S, Bilir S, Karaali S, *et al.*, 2007, Estimation of Galactic model parameters with the Sloan Digital Sky Survey and the metallicity distribution in two fields in the anti-centre direction of the Galaxy. *Astronomische Nachrichten*, 328, 169–177 {**545**}

Alcaino G, Liller W, Alvarado R, *et al.*, 1997, Multicolour CCD photometry of the poorly-studied globular cluster NGC 6541. *AJ*, 114, 2638–2643 {**547**}

Alcobé S, Cubarsi R, 2005, Disk populations from Hipparcos kinematic data. Discontinuities in the local velocity distribution. *A&A*, 442, 929–946 {**526**}

Alcock C, Allsman RA, Axelrod TS, *et al.*, 1996, The MACHO project first-year Large Magellanic Cloud results: the microlensing rate and the nature of the Galactic dark halo. *ApJ*, 461, 84–103 {**520**}

Allen C, Moreno E, Pichardo B, 2006, The orbits of 48 globular clusters in a Milky Way-like barred galaxy. *ApJ*, 652, 1150–1169 {**554**}

Allende Prieto C, Barklem PS, Lambert DL, *et al.*, 2004, A spectroscopic survey of stars in the solar neighbourhood. The nearest 15 pc. *A&A*, 420, 183–205 {**522**}

Alves DR, 2000, *K*-band calibration of the red clump luminosity. *ApJ*, 539, 732–741 {**495, 496**}

Amaral LH, Lépine JRD, 1997, A self-consistent model of the spiral structure of the Galaxy. *MNRAS*, 286, 885–894 {**531**}

Andersen J, 1991, Accurate masses and radii of normal stars. *A&A Rev.*, 3, 91–126 {**518**}

Andersen J, Nordström B, Mayor M, 2005, The solar neighbourhood in four dimensions. *ASP Conf. Ser. 336: Cosmic Abundances as Records of Stellar Evolution and Nucleosynthesis* (eds. Barnes TG, Bash FN), 305–308 {**524**}

Andronova AA, 2000, The catalogue of the nearest stellar systems – NESSY. *Hipparcos and the Luminosity Calibration of the Nearer Stars, IAU Joint Discussion 13*, 12–14 {**510**}

Arfken G, 1975, *Vector Spherical Harmonics, in Mathematical Methods for Physicists*. Academic Press, Orlando {**508**}

Arifyanto MI, Fuchs B, Jahreiß H, *et al.*, 2005, Kinematics of nearby subdwarf stars. *A&A*, 433, 911–916 {**526**}

Asiain R, Figueras F, Torra J, 1999, On the evolution of moving groups: an application to the Pleiades moving group. *A&A*, 350, 434–446 {**532**}

Avedisova VS, 1989, Structure of the Sagittarius–Carina spiral arm and parameters of the spiral pattern. *Astrophysics*, 30, 83–90 {**531**}

Backer DC, Sramek RA, 1999, Proper motion of the compact, nonthermal radio source in the Galactic centre, Sgr A*. *ApJ*, 524, 805–815 {**500**}

Bahcall JN, 1984a, Self-consistent determinations of the total amount of matter near the Sun. *ApJ*, 276, 169–181 {**511**}

—, 1984b, The distribution of stars perpendicular to Galactic disk. *ApJ*, 276, 156–168 {**511**}

Barbier-Brossat M, Figon P, 2000, Mean radial velocities catalogue of Galactic stars. *A&AS*, 142, 217–223 {**513**}

Barrera RG, Estevez GA, Giraldo J, 1985, Vector spherical harmonics and their application to magnetostatics . *European Journal of Physics*, 6, 287–294 {**508**}

Beers TC, Chiba M, Yoshii Y, *et al.*, 2000, Kinematics of metal-poor stars in the Galaxy. II. Proper motions for a large non-kinematically selected sample. *AJ*, 119, 2866–2881 {**522, 539**}

Beers TC, Drilling JS, Rossi S, *et al.*, 2002, Metal abundances and kinematics of bright metal-poor giants selected from the LSE survey: implications for the metal-weak thick disk. *AJ*, 124, 931–948 {**523**}

Bekki K, Chiba M, 2000, Formation of the Galactic stellar halo: origin of the metallicity–eccentricity relation. *ApJ*, 534, L89–L92 {**541**}

—, 2001, Formation of the Galactic stellar halo. I. Structure and kinematics. *ApJ*, 558, 666–686 {**541**}

Belokurov V, Zucker DB, Evans NW, *et al.*, 2006, The field of streams: Sagittarius and its siblings. *ApJ*, 642, L137–L140 {**542**}

Bennett CL, Halpern M, Hinshaw G, *et al.*, 2003, First-year Wilkinson Microwave Anisotropy Probe (WMAP) observations: preliminary maps and basic results. *ApJS*, 148, 1–27 {**548**}

Bensby T, Feltzing S, Lundström I, 2004, A possible age–metallicity relation in the Galactic thick disk? *A&A*, 421, 969–976 {**526, 527**}

Bergbusch PA, VandenBerg DA, 1992, Oxygen-enhanced models for globular cluster stars. II. Isochrones and luminosity functions. *ApJS*, 81, 163–220 {**539**}

—, 2001, Models for old, metal-poor stars with enhanced α-element abundances. III. Isochrones and isochrone population functions. *ApJ*, 556, 322–339 {**527**}

Bernkopf J, Fidler A, Fuhrmann K, 2001, The dark side of the Milky Way. *ASP Conf. Ser. 245: Astrophysical Ages and Times Scales* (eds. von Hippel T, Simpson C, Manset N), 207–215 {**520**}

Bertelli G, Bressan A, Chiosi C, *et al.*, 1994, Theoretical isochrones from models with new radiative opacities. *A&AS*, 106, 275–302 {**517**}

Bertelli G, Nasi E, Bressan A, *et al.*, 1997, Synthetic diagrams for the interpretation of star-formation rate and initial mass function. *ESA SP–402: Hipparcos, Venice '97*, 501–506 {**515**}

Bertelli G, Nasi E, 2001, Star-formation history in the solar vicinity. *AJ*, 121, 1013–1023 {**515, 517, 518**}

Bienaymé O, 1999, The local stellar velocity distribution of the Galaxy. Galactic structure and potential. *A&A*, 341, 86–97 {**512**}

Bienaymé O, Robin AC, Crézé M, 1987, The mass density in our Galaxy. *A&A*, 180, 94–110, erratum: 186, 359 {**510**}

Bienaymé O, Soubiran C, Mishenina TV, *et al.*, 2006, Vertical distribution of Galactic disk stars. *A&A*, 446, 933–942 {**513**}

Bilir S, Karaali S, Buser R, 2004, Analysis of RGU photometry in selected area 51. *Turkish Journal of Physics*, 28, 289–299 {**516**}

Binney JJ, 1992, Warps. *ARA&A*, 30, 51–74 {**535, 536**}

Binney JJ, Dehnen W, Bertelli G, 2000, The age of the solar neighbourhood. *MNRAS*, 318, 658–664 {**515, 518, 520**}

Binney JJ, Tremaine S, 1987, *Galactic Dynamics*. Princeton University Press {**492, 494, 511, 512, 514, 527, 553, 554**}

Binney J, Merrifield M, 1998, *Galactic Astronomy*. Princeton University Press {**492**}

Bland-Hawthorn J, Freeman KC, 2000, The baryon halo of the Milky Way. *Science*, 287, 79–84 {**538, 541**}

Bolte M, Hogan CJ, 1995, Conflict over the age of the universe. *Nature*, 376, 399–402 {**547, 550**}

Bonifacio P, Centurion M, Molaro P, 1999, Abundances of metal-weak thick-disk candidates. *MNRAS*, 309, 533–542 {**522**}

Borkova TV, Marsakov VA, 2004, Stars of extragalactic origin in the solar neighbourhood. *Astronomy Letters*, 30, 148–158 {**526**}

Branham RL, 2000, The parameters of Galactic kinematics determined from total least squares. *Revista Mexicana de Astronomia y Astrofisica*, 36, 97–112 {**499, 502, 506, 507**}

—, 2002, Kinematics of OB stars. *ApJ*, 570, 190–197 {**499, 502, 506, 507**}

—, 2003, The Sun's distance from the Galactic plane. *Ap&SS*, 288, 417–419 {**497**}

—, 2006, Kinematics of O–B5 giants. *MNRAS*, 370, 1393–1400 {**499, 502, 506, 507**}

Brosche P, Schwan H, Schwarz O, 2001, The Galactic motion field of K0-5 giants from Hipparcos data. *Astronomische Nachrichten*, 322, 15–42 {**499**}

Burkert A, Smith GH, 1997, Sub-structure in the globular cluster system of the Milky Way: the highest metallicity clusters. *ApJ*, 474, L15–L18 {**547**}

Caldwell JAR, Coulson IM, 1987, Milky Way rotation and the distance to the Galactic centre from Cepheid variables. *AJ*, 93, 1090–1105 {**496**}

Caloi V, Cardini D, D'Antona F, *et al.*, 1999, Kinematics and age of stellar populations in the solar neighbourhood from Hipparcos data. *A&A*, 351, 925–936 {**522**}

Carney BW, Fulbright JP, Terndrup DM, *et al.*, 1995, The distance to the Galactic centre obtained by infrared photometry of RR Lyrae variables. *AJ*, 110, 1674–1685 {**496**}

Carney BW, Laird JB, Latham DW, 1988, A survey of proper motion stars. V. Extreme-velocity stars and the local Galactic escape velocity. *AJ*, 96, 560–566 {**538, 539**}

Carney BW, Latham DW, Laird JB, *et al.*, 1994, A survey of proper motion stars. XII. An expanded sample. *AJ*, 107, 2240–2289 {**515, 526, 539**}

Carrascal B, Estevez GA, Lee P, *et al.*, 1991, Vector spherical harmonics and their application to classical electrodynamics. *European Journal of Physics*, 12, 184–191 {**508**}

Carretta E, Gratton RG, Clementini G, *et al.*, 2000, Distances, ages, and epoch of formation of globular clusters. *ApJ*, 533, 215–235 {**548–550**}

Casetti-Dinescu DI, Girard TM, Herrera D, *et al.*, 2007, Space velocities of southern globular clusters. V. A low Galactic latitude sample. *AJ*, 134, 195–204 {**551**}

Cayrel de Strobel G, Soubiran C, Ralite N, 2001, Catalogue of [Fe/H] determinations for FGK stars: 2001 edition. *A&A*, 373, 159–163 {**524**}

Cepeda W, 2006, Comparison of solar motion relative to stars of different colour index in the solar neighbourhood. *Revista Mexicana de Astronomia y Astrofisica Conf. Ser.*, volume 25, 63–64 {**503**}

Chaboyer BC, 1998, The age of the universe. *Phys. Rep.*, 307, 23–30 {**547, 548**}

Chaboyer BC, Demarque P, Kernan PJ, *et al.*, 1996a, A lower limit on the age of the universe. *Science*, 271, 957–961 {**550**}

Chaboyer BC, Demarque P, Sarajedini A, 1996b, Globular cluster ages and the formation of the Galactic halo. *ApJ*, 459, 558–569 {**549**}

Chabrier G, 2001, The Galactic disk mass budget. I. Stellar mass function and density. *ApJ*, 554, 1274–1281 {**516, 544, 546**}

Chandrasekhar S, 1942, *Principles of Stellar Dynamics*. University of Chicago Press {**493**}

Chen B, 1998, Kinematic sub-structure of halo stars from Hipparcos observations. *ApJ*, 495, L1, erratum: 507, L79 {**539, 540**}

Chen B, Stoughton C, Smith JA, *et al.*, 2001, Stellar population studies with the SDSS. I. The vertical distribution of stars in the Milky Way. *ApJ*, 553, 184–197 {**497**}

Chen L, Geffert M, Wang J, *et al.*, 2000a, A proper motion study of the globular cluster M10. *A&AS*, 145, 223–228 {**551**}

Chen L, Wang J, Zhao J, 2002a, A space motion study of the globular cluster M13. *IAU Symp. 207*, 119–121 {**551**}

Chen Y, Nissen PE, Zhao G, *et al.*, 2000b, Chemical composition of 90 F and G disk dwarfs. *A&AS*, 141, 491–506 {**522**}

Chen Y, Zhao G, Shi J, 2002b, Information from the kinematics of F and G stars in the solar neighbourhood. *Chinese Journal of Astronony and Astrophysics*, 2, 419–428 {**522**}

Chereul E, Crézé M, Bienaymé O, 1998, The distribution of nearby stars in phase space mapped by Hipparcos. II. Inhomogeneities among A-F type stars. *A&A*, 340, 384–396 {**512, 532**}

Chiba M, Beers TC, 2000, Kinematics of metal-poor stars in the Galaxy. III. Formation of the stellar halo and thick disk as revealed from a large sample of non-kinematically selected stars. *AJ*, 119, 2843–2865 {**527, 539–543**}

—, 2001, Structure of the Galactic stellar halo prior to disk formation. *ApJ*, 549, 325–336 {**541**}

Chiba M, Yoshii Y, 1997, Three-dimensional orbits of metal-poor halo stars and the formation of the Galaxy. *ApJ*, 490, L73–L76 {**541, 552**}

—, 1998, Early evolution of the Galactic halo revealed from Hipparcos observations of metal-poor stars. *AJ*, 115, 168–192 {**539–542**}

Cignoni M, Degl'Innocenti S, Prada Moroni PG, *et al.*, 2006, Recovering the star-formation rate in the solar neighbourhood. *A&A*, 459, 783–796 {**518**}

Cignoni M, Shore SN, 2006, Restoring colour-magnitude diagrams with the Richardson–Lucy algorithm. *A&A*, 454, 511–516 {**518**}

Clemens DP, 1985, Massachusetts–Stony Brook Galactic plane CO survey: the Galactic disk rotation curve. *ApJ*, 295, 422–428 {**492, 493**}

Clube SVM, 1972, Galactic rotation and the precession constant. *MNRAS*, 159, 289–314 {**504**}

—, 1973, Another look at the absolute proper motions obtained from the Lick pilot programme. *MNRAS*, 161, 445–463 {**504**}

Comerón F, Torra J, Gómez AE, 1994, On the characteristics and origin of the expansion of the local system of young objects. *A&A*, 286, 789–798 {**498, 505**}

Comerón F, Torra J, 1991, A study on the kinematics of the local system of young stars. *A&A*, 241, 57–61 {**531**}

Crézé M, Chereul E, Bienaymé O, *et al.*, 1998, The distribution of nearby stars in phase space mapped by Hipparcos. I. The potential well and local dynamical mass. *A&A*, 329, 920–936 {**512, 513, 545**}

Crézé M, Mennessier MO, 1973, An attempt to interpret the mean properties of the velocity field of young stars in terms of Lin's theory of spiral waves. *A&A*, 27, 281–289 {**531**}

Cubarsi R, Alcobé S, 2004, Cumulants and symmetries in a trivariate normal mixture. A qualitative study of the local velocity distribution. *A&A*, 427, 131–144 {**526**}

—, 2006, Radial mean velocity in the solar neighbourhood. *A&A*, 457, 537–540 {**503**}

Dambis AK, Rastorguev AS, 2001, Absolute magnitudes and kinematic parameters of the subsystem of RR Lyrae variables. *Astronomy Letters*, 27, 108–117 {**495**}

D'Antona F, Caloi V, Mazzitelli I, 1997, The universe and globular clusters: an age conflict? *ApJ*, 477, 519–534 {**548**}

Dehnen W, 1998, The distribution of nearby stars in velocity space inferred from Hipparcos data. *AJ*, 115, 2384–2396 {**536, 537**}

—, 1999a, Simple distribution functions for stellar disks. *AJ*, 118, 1201–1208 {**494**}

—, 1999b, The pattern speed of the Galactic bar. *ApJ*, 524, L35–L38 {**500, 528**}

—, 2000, The effect of the outer Lindblad resonance of the Galactic bar on the local stellar velocity distribution. *AJ*, 119, 800–812 {**529, 535**}

Dehnen W, Binney JJ, 1998a, Local stellar kinematics from Hipparcos data. *MNRAS*, 298, 387–394 {**494, 497–500, 508, 530, 534**}

—, 1998b, Mass models of the Milky Way. *MNRAS*, 294, 429–438 {**501, 519**}

Delfosse X, Forveille T, Ségransan D, *et al.*, 2000, Accurate masses of very low-mass stars. IV. Improved mass–luminosity relations. *A&A*, 364, 217–224 {**516, 518**}

Dettbarn C, Fuchs B, Flynn C, *et al.*, 2007, Signatures of star streams in the phase space distribution of nearby halo stars. *A&A*, 474, 857–861 {**543**}

De Marchi G, Pulone L, Paresce F, 2006, Why is the mass function of NGC 6218 flat? *A&A*, 449, 161–170 {**554**}

De Simone R, Wu X, Tremaine S, 2004, The stellar velocity distribution in the solar neighbourhood. *MNRAS*, 350, 627–643 {**503, 532**}

de Souza RE, Teixeira R, 2007, Kinematic segregation of nearby disk stars from the Hipparcos database. *A&A*, 471, 475–484 {**503**}

de Zeeuw PT, 1985, Elliptical galaxies with separable potentials. *MNRAS*, 216, 273–334 {**542**}

Dias WS, Lépine JRD, 2005, Direct determination of the spiral pattern rotation speed of the Galaxy. *ApJ*, 629, 825–831 {**492, 493, 533, 535**}

Dinescu DI, Girard TM, van Altena WF, *et al.*, 2003, Space velocities of southern globular clusters. IV. First results for inner galaxy clusters. *AJ*, 125, 1373–1382 {**553**}

Dinescu DI, Girard TM, van Altena WF, 1999a, Space velocities of globular clusters. III. Cluster orbits and halo substructure. *AJ*, 117, 1792–1815 {**545, 546, 551, 552, 554**}

Dinescu DI, Majewski SR, Girard TM, *et al.*, 2000, The absolute proper motion of Palomar 12: a case for tidal capture from the Sagittarius dwarf spheroidal galaxy. *AJ*, 120, 1892–1905 {**549**}

Dinescu DI, van Altena WF, Girard TM, *et al.*, 1999b, Space velocities of southern globular clusters. II. New results for 10 clusters. *AJ*, 117, 277–285 {**551**}

Djorgovski S, Sosin C, 1989, The warp of the Galactic stellar disk detected in IRAS source counts. *ApJ*, 341, L13–L16 {**535**}

Drimmel R, Smart RL, Lattanzi MG, 2000, The Galactic warp in OB stars from Hipparcos. *A&A*, 354, 67–76 {**536**}

Drobitko EV, Vityazev VV, 2003, Kinematic analysis of near and far stars in the Hipparcos Catalogue. *Astrophysics*, 46, 224–233 {**507**}

Dubath P, Mayor M, Burki G, 1988, The kinematics of late-type supergiants. *A&A*, 205, 77–85 {**505**}

Dubinski J, Carlberg RG, 1991, The structure of cold dark matter halos. *ApJ*, 378, 496–503 {**537**}

Duquennoy A, Mayor M, 1991, Multiplicity among solar-type stars in the solar neighbourhood. II. Distribution of the orbital elements in an unbiased sample. *A&A*, 248, 485–524 {**525**}

Durrell PR, Harris WE, 1993, A colour-magnitude study of the globular cluster M15. *AJ*, 105, 1420–1440 {**547**}

Du Mont B, 1977, A three-dimensional analysis of the kinematics of 512 FK4/FK4 Sup stars. *A&A*, 61, 127–132 {**504**}

Edmondson FK, 1937, Stellar kinematics and mean parallaxes. *MNRAS*, 97, 473–485 {**504, 506**}

Edvardsson B, Andersen J, Gustafsson B, *et al.*, 1993, The chemical evolution of the Galactic disk. I. Analysis and results. *A&A*, 275, 101–152 {**520, 522, 524**}

Efremov YN, 1983, Cepheids and spiral structure. *Soviet Astronomy Letters*, 9, 51–55 {**531**}

Eggen OJ, Lynden-Bell D, Sandage AR, 1962, Evidence from the motions of old stars that the Galaxy collapsed. *ApJ*, 136, 748–767 {**540–542**}

Eggen OJ, Sandage AR, 1959, Stellar groups. IV. The Groombridge 1830 group of high-velocity stars and its relation to the globular clusters. *MNRAS*, 119, 255–277 {**538**}

Eisenhauer F, Schödel R, Genzel R, *et al.*, 2003, A geometric determination of the distance to the Galactic centre. *ApJ*, 597, L121–L124 {**495, 496**}

Elias F, Alfaro EJ, Cabrera Caño J, 2006, OB stars in the solar neighbourhood. II. Kinematics. *AJ*, 132, 1052–1060 {**503**}

Elmegreen DM, 1985, Spiral structure of the Milky Way and external galaxies. *IAU Symp. 106: The Milky Way Galaxy* (eds. van Woerden H, Allen RJ, Burton WB), 255–270 {**531**}

Evans DW, Irwin MJ, 1995, The APM proper motion project. II. The lower proper motion stars. *MNRAS*, 277, 820–844 {**498, 499**}

Famaey B, Bruneton JP, Zhao H, 2007, Escaping from modified Newtonian dynamics. *MNRAS*, 377, L79–L82 {**515**}

Famaey B, Dejonghe H, 2003, Three-component Stäckel potentials satisfying recent estimates of Milky Way parameters. *MNRAS*, 340, 752–762 {**545**}

Famaey B, Jorissen A, Luri X, *et al.*, 2005a, Dynamical streams in the solar neighbourhood. *ESA SP–576: The Three-Dimensional Universe with Gaia* (eds. Turon C, O'Flaherty KS, Perryman MAC), 129–133 {**503**}

—, 2005b, Local kinematics of K and M giants from Coravel, Hipparcos, and Tycho 2 data. Revisiting the concept of superclusters. *A&A*, 430, 165–186 {**503, 526, 532**}

Feast MW, Catchpole RM, 1997, The Cepheid period–luminosity zero-point from Hipparcos trigonometrical parallaxes. *MNRAS*, 286, L1–L5 {**548**}

Feast MW, Whitelock PA, 1997, Galactic kinematics of Cepheids from Hipparcos proper motions. *MNRAS*, 291, 683–693 {**495, 496, 498, 499, 501, 502, 504–506**}

—, 2000, Mira kinematics from Hipparcos data: a Galactic bar to beyond the solar circle. *MNRAS*, 317, 460–487 {**530**}

Fehrenbach C, Duflot M, Burnage R, 2001, New determination of the solar apex. *A&A*, 369, 65–73 {**498, 499**}

Feltzing S, Holmberg J, Hurley JR, 2001, The solar neighbourhood age–metallicity relation: does it exist? *A&A*, 377, 911–924 {**522, 523**}

Fernández D, Figueras F, Torra J, 2001, Kinematics of young stars. II. Galactic spiral structure. *A&A*, 372, 833–850 {**532, 533**}

Fiorentin PR, Helmi A, Lattanzi MG, *et al.*, 2005, Structure in the motions of the fastest halo stars. *A&A*, 439, 551–558 {**542**}

Flynn C, Gould A, Bahcall JN, 1996, Hubble Deep Field constraint on baryonic dark matter. *ApJ*, 466, L55–L58 {**521**}

Flynn C, Jimenez R, Kotoneva E, 1999, Eight Gyr minimum disk age from Hipparcos. *Ap&SS*, 265, 243–244 {**520**}

Freedman WL, Madore BF, Gibson BK, *et al.*, 2001, Final results from the Hubble Space Telescope key project to measure the Hubble Constant. *ApJ*, 553, 47–72 {**548**}

Freedman WL, Madore BF, Mould JR, *et al.*, 1994, Distance to the Virgo cluster galaxy M100 from Hubble Space Telescope observations of Cepheids. *Nature*, 371, 757–762 {**547**}

Freeman KC, 1996, The importance of galactic halos to understanding galaxy formation. *Formation of the Galactic Halo...Inside and Out* (eds. Morrison HL, Sarajedini A), ASP Conf. Ser. 92, 3–13 {**538**}

Freeman K, Bland-Hawthorn J, 2002, The new Galaxy: signatures of its formation. *ARA&A*, 40, 487–537 {**538**}

Freire PC, Camilo F, Lorimer DR, *et al.*, 2001, Timing the millisecond pulsars in 47 Tucanae. *MNRAS*, 326, 901–915 {**551**}

Frenk CS, White SDM, 1980, The kinematics and dynamics of the Galactic globular cluster system. *MNRAS*, 193, 295–311 {**550**}

Freudenreich HT, Berriman GB, Dwek E, *et al.*, 1994, DIRBE evidence for a warp in the interstellar dust layer and stellar disk of the galaxy. *ApJ*, 429, L69–L72 {**496**}

Friel ED, 1995, The old open clusters of the Milky Way. *ARA&A*, 33, 381–414 {**549**}

Fuchs B, Jahreiß H, Wielen R, 1999, Kinematics of nearby subdwarfs. *Ap&SS*, 265, 175–178 {**522**}

Fuchs B, Jahreiß H, 1998, Halo stars in the immediate solar neighbourhood. *A&A*, 329, 81–86 {**521, 522**}

Fuchs B, Wielen R, 1993, Kinematical constraints on the dynamically determined local mass density of the Galaxy. *AIP Conf. Proc. 278: Back to the Galaxy* (eds. Holt SS, Verter F), 580–583 {**512**}

Fuhrmann K, 1998, Nearby stars of the Galactic disk and halo. *A&A*, 338, 161–183 {**521, 526**}

—, 2002, Where are the halo stars? *New Astronomy*, 7, 161–169 {**521, 522**}

—, 2004, Nearby stars of the Galactic disk and halo. III. *Astronomische Nachrichten*, 325, 3–80 {**521–523, 526, 544**}

Fux R, 2001, Order and chaos in the local disk stellar kinematics induced by the Galactic bar. *A&A*, 373, 511–535 {**529, 532**}

Garnett DR, Kobulnicky HA, 2000, Distance dependence in the solar neighbourhood age–metallicity relation. *ApJ*, 532, 1192–1196 {**522**}

Gates EI, Gyuk G, Turner MS, 1996, Gravitational microlensing and the Galactic halo. *Phys. Rev. D*, 53, 4138–4176 {**521**}

Geffert M, 1998, Hipparcos based astrometric analysis of M3 and M92 fields: optical identification of X-ray and radio sources, space motions of globular clusters M3 and M92 and a Galactic orbit of the sdB star PG 1716+426. *A&A*, 340, 305–308 {**551**}

Geffert M, Hiesgen M, Colin J, *et al.*, 1997, Absolute proper and space motions of globular clusters. *ESA SP–402: Hipparcos, Venice '97*, 579–582 {**551**}

Genzel R, Pichon C, Eckart A, *et al.*, 2000, Stellar dynamics in the Galactic centre: proper motions and anisotropy. *MNRAS*, 317, 348–374 {**495, 496**}

Gerhard O, 1996, Dynamics of the bar at the Galactic centre. *IAU Symp. 169: Unsolved Problems of the Milky Way* (eds. Blitz L, Teuben PJ), 79–91 {**528**}

—, 2002, The Galactic bar. *ASP Conf. Ser. 273: The Dynamics, Structure and History of Galaxies* (eds. Da Costa GS, Jerjen H), 73–84 {**527, 528**}

Ghez A, Salim S, Weinberg N, *et al.*, The distance to the Galactic centre. *ApJ*, in press {**495, 496**}

Gilmore G, Reid IN, 1983, New light on faint stars. III. Galactic structure towards the south pole and the Galactic thick disk. *MNRAS*, 202, 1025–1047 {**519**}

Gilmore G, Wyse RFG, Kuijken K, 1989, Kinematics, chemistry, and structure of the Galaxy. *ARA&A*, 27, 555–627 {**520, 528**}

Girardi L, Bressan A, Chiosi C, *et al.*, 1996, Evolutionary sequences of stellar models with new radiative opacities. VI. $Z = 0.0001$. *A&AS*, 117, 113–125 {**517**}

Girardi L, Groenewegen MAT, Hatziminaoglou E, *et al.*, 2005, Star counts in the Galaxy: simulating from very deep to very shallow photometric surveys with the trilegal code. *A&A*, 436, 895–915 {**515**}

Girardi L, Salaris M, 2001, Population effects on the red giant clump absolute magnitude, and distance determinations to nearby galaxies. *MNRAS*, 323, 109–129 {**495, 496**}

Gliese W, 1957, Katalog der Sterne näher als 20 Parsek für 1950.0. *Astron. Rechen-Institut, Heidelberg*, 8, 1–89 {**509**}

—, 1969, Catalogue of Nearby Stars. *Veroeffentlichungen des Astronomischen Rechen-Instituts Heidelberg*, 22, 1–117 {**509**}

Gliese W, Jahreiß H, 1991, *Preliminary version of the Third Catalogue of Nearby Stars*. Astronomischen Rechen-Instituts Heidelberg. On: The Astronomical Data Centre CD-ROM: Selected Astronomical Catalogues, Vol. I, NASA/ADC/GSFC {**510**}

Gnedin OY, Hernquist L, Ostriker JP, 1999a, Tidal shocking by extended mass distributions. *ApJ*, 514, 109–118 {**554**}

Gnedin OY, Lee HM, Ostriker JP, 1999b, Effects of tidal shocks on the evolution of globular clusters. *ApJ*, 522, 935–949 {**553, 554**}

Gnedin OY, Ostriker JP, 1997, Destruction of the Galactic globular cluster system. *ApJ*, 474, 223–255 {**554**}

—, 1999, On the self-consistent response of stellar systems to gravitational shocks. *ApJ*, 513, 626–637 {**554**}

Gordon MA, 1978, Determination of the spiral pattern speed of the galaxy. *ApJ*, 222, 100–102 {**531**}

Gould A, 2003, Stellar halo parameters from 4588 subdwarfs. *ApJ*, 583, 765–775 {**530**}

—, 2004, The $v_\perp$ colour-magnitude diagram. *ArXiv Astrophysics e-prints*, only published as astro-ph/0403506 {**544**}

Gould A, Flynn C, Bahcall JN, 1998, Spheroid luminosity and mass functions from Hubble Space Telescope star counts. *ApJ*, 503, 798–808 {**522**}

Gould A, Ramirez SV, 1998, Non-acceleration of Sgr A*: implications for Galactic structure. *ApJ*, 497, 713–716 {**500**}

Gratton RG, Bragaglia A, Carretta E, *et al.*, 2003, Distances and ages of NGC 6397, NGC 6752 and 47 Tuc. *A&A*, 408, 529–543 {**548–550**}

Gratton RG, Carretta E, Matteucci F, *et al.*, 2000, Abundances of light elements in metal-poor stars. IV. [Fe/O] and [Fe/Mg] ratios and the history of star formation in the solar neighbourhood. *A&A*, 358, 671–681 {**522**}

Gray RO, Corbally CJ, Garrison RF, *et al.*, 2003, Contributions to the nearby stars (NStars) project: spectroscopy of stars earlier than M0 within 40 pc: The northern sample. *AJ*, 126, 2048–2059 {**510**}

—, 2006, Contributions to the nearby stars project: spectroscopy of stars earlier than M0 within 40 pc: the southern sample. *AJ*, 132, 161–170 {**510**}

Grivnev EM, 1981, Galactic spiral structure and the kinematics of H II regions. *Soviet Astronomy Letters*, 7, 303–305 {**531**}

—, 1983, Birthplaces of 55 classical Cepheids. *Soviet Astronomy Letters*, 9, 287–289 {**531**}

Gwinn CR, Moran JM, Reid MJ, 1992, Distance and kinematics of the W49N H_2O maser outflow. *ApJ*, 393, 149–164 {**496**}

Gyuk G, Flynn C, Evans NW, 1999, Star counts and the warped and flaring Milky Way disk. *ApJ*, 521, 190–193 {**490**}

Hambly NC, Hodgkin ST, Cossburn MR, *et al.*, 1999, Brown dwarfs in the Pleiades and the initial mass function across the stellar/substellar boundary. *MNRAS*, 303, 835–844 {**516**}

Hansen BMS, Brewer J, Fahlman GG, *et al.*, 2002, The white dwarf cooling sequence of the globular cluster Messier 4. *ApJ*, 574, L155–L158 {**548**}

Hansen BMS, 1998, Old and blue white dwarfs as a detectable source of microlensing events. *Nature*, 394, 860–862 {**520**}

Hanson RB, 1987, Lick northern proper motion programme. II. Solar motion and Galactic rotation. *AJ*, 94, 409–415 {**502, 504–506**}

Harris HC, Hansen BMS, Liebert J, *et al.*, 2001, A new very cool white dwarf discovered by the Sloan Digital Sky Survey. *ApJ*, 549, L109–L113 {**520**}

Haywood M, Robin AC, Crézé M, 1997, The evolution of the Milky Way disk. II. Constraints from star counts at the Galactic poles. *A&A*, 320, 440–459 {**544**}

Haywood M, 2001, A revision of the solar neighbourhood metallicity distribution. *MNRAS*, 325, 1365–1382 {**522**}

Helmi A, 2008, The stellar halo of the Galaxy. *A&A Rev.*, 15, 145–188 {**538**}

Helmi A, de Zeeuw PT, 2000, Mapping the sub-structure in the Galactic halo with the next generation of astrometric satellites. *MNRAS*, 319, 657–665 {**544**}

Helmi A, Navarro JF, Nordström B, *et al.*, 2006, Pieces of the puzzle: ancient sub-structure in the Galactic disc. *MNRAS*, 365, 1309–1323 {**542**}

Helmi A, White SDM, de Zeeuw PT, *et al.*, 1999, Debris streams in the solar neighbourhood as relics from the formation of the Milky Way. *Nature*, 402, 53–55 {**542, 543**}

Helmi A, White SDM, 1999, Building up the stellar halo of the Galaxy. *MNRAS*, 307, 495–517 {**542**}

Hénon M, 1959, L'amas isochrone. *Annales d'Astrophysique*, 22, 126–139 {**541**}

Henry TJ, Jao WC, Subasavage JP, *et al.*, 2006, The solar neighbourhood. XVII. Parallax results from the CTIOPI 0.9-m programme: 20 new members of the RECONS 10 pc sample. *AJ*, 132, 2360–2371 {**510**}

Henry TJ, McCarthy DW, 1993, The mass–luminosity relation for stars of $1.0-0.08\,M_\odot$. *AJ*, 106, 773–789 {**518**}

Hernández X, Valls-Gabaud D, Gilmore G, 1999, Deriving star-formation histories: inverting HR diagrams through a variational calculus maximum likelihood method. *MNRAS*, 304, 705–719 {**517**}

—, 2000, The recent star-formation history of the Hipparcos solar neighbourhood. *MNRAS*, 316, 605–612 {**517, 519**}

Hernquist L, 1990, An analytical model for spherical galaxies and bulges. *ApJ*, 356, 359–364 {**545**}

Hill EL, 1954, The theory of vector spherical harmonics. *American Journal of Physics*, 22, 211–214 {**508**}

Hill V, Plez B, Cayrel R, *et al.*, 2002, First stars. I. The extreme r-element rich, iron-poor halo giant CS 31082–001. Implications for the r-process site(s) and radioactive cosmochronology. *A&A*, 387, 560–579 {**548**}

Hogg DW, Blanton MR, Roweis ST, *et al.*, 2005, Modeling complete distributions with incomplete observations: the velocity ellipsoid from Hipparcos data. *ApJ*, 629, 268–275 {**498–500**}

Holmberg J, Flynn C, Lindegren L, 1997, Towards an improved model of the Galaxy. *ESA SP-402: Hipparcos, Venice '97*, 721–726 {**496, 497, 513, 544, 545**}

Holmberg J, Flynn C, 2000, The local density of matter mapped by Hipparcos. *MNRAS*, 313, 209–216 {**513, 523**}

—, 2004, The local surface density of disk matter mapped by Hipparcos. *MNRAS*, 352, 440–446 {**513**}

Holmberg J, Nordström B, Andersen J, 2007, The Geneva–Copenhagen survey of the Solar neighbourhood. II. New uvby calibrations and rediscussion of stellar ages, the G dwarf problem, age–metallicity diagram, and heating mechanisms of the disk. *A&A*, 475, 519–537 {**524, 526**}

Humphreys RM, Larsen JA, 1995, The Sun's distance above the Galactic plane. *AJ*, 110, 2183–2188 {**497**}

Huterer D, Sasselov D, Schechter PL, 1995, Distances to nearby galaxies: combining fragmentary data using four different methods. *AJ*, 110, 2705–2714 {**495**}

Hut P, McMillan S, Goodman J, *et al.*, 1992, Binaries in globular clusters. *PASP*, 104, 981–1034 {**553**}

Ibata RA, Gilmore G, Irwin MJ, 1994, A dwarf satellite galaxy in Sagittarius. *Nature*, 370, 194–196 {**542**}

Ibukiyama A, 2004, Solar neighbourhood age–metallicity relation based on Hipparcos data. *Publications of the Astronomical Society of Australia*, 21, 121–125 {**525**}

Ibukiyama A, Arimoto N, 2002, Hipparcos age–metallicity relation of the solar neighbourhood disk stars. *A&A*, 394, 927–941 {**522, 524**}

Ideta M, Hozumi S, Tsuchiya T, *et al.*, 2000, Time evolution of Galactic warps in prolate halos. *MNRAS*, 311, 733–740 {**537, 540**}

Isobe S, 1974, The local velocity of rotation in the Galaxy. *A&A*, 36, 327–332 {**500, 515**}

Ivanov GR, 1983, The aging of open clusters across Galactic spiral arms. *Soviet Astronomy Letters*, 9, 107–109 {**531**}

Jackson JD, 1975, *Classical Electrodynamics, Second Edition.* Wiley, New York {**508**}

Jahreiß H, 1993, The Third Catalogue of Nearby Stars: last edition before Hipparcos. *Hipparcos: une nouvelle donne pour l'astronomie. Ecole d'Astrophysique de Goutelas* (eds. Benest D, Froeschlé C), 593–598 {**510**}

Jahreiß H, Gliese W, 1993, The Third Catalogue of Nearby Stars: results and conclusions. *IAU Symp. 156: Developments in Astrometry and their Impact on Astrophysics and Geodynamics* (eds. Mueller II, Kolaczek B), 107–112 {**510**}

Jahreiß H, Wielen R, Fuchs B, 1998, The impact of Hipparcos on our knowledge of nearby stars. *Applied and Computational Harmonic Analysis*, 3, 171–180 {**510**}

Jahreiß H, Wielen R, 1997, The impact of Hipparcos on the Catalogue of Nearby Stars. The stellar luminosity function and local kinematics. *ESA SP–402: Hipparcos, Venice '97*, 675–680 {**510, 511**}

—, 2000, The census of nearby stars binaries. *IAU Symp. 200*, 129 {**510**}

Jetsu L, Pelt J, 2000, Spurious periods in the terrestrial impact crater record. *A&A*, 353, 409–418 {**514**}

Jimenez R, Flynn C, Kotoneva E, 1998, Hipparcos and the age of the Galactic disk. *MNRAS*, 299, 515–519 {**520, 521**}

Johnston KV, Law DR, Majewski SR, 2005, A Two Micron All-Sky Survey view of the Sagittarius dwarf galaxy. III. Constraints on the flattening of the Galactic halo. *ApJ*, 619, 800–806 {**537, 540**}

Johnston KV, Spergel DN, Hernquist L, 1995, The disruption of the Sagittarius dwarf galaxy. *ApJ*, 451, 598–606 {**545, 546**}

Jørgensen BR, 2000, The G dwarf problem: analysis of a new data set. *A&A*, 363, 947–957 {**520**}

Joshi YC, 2005, Interstellar extinction towards open clusters and Galactic structure. *MNRAS*, 362, 1259–1266 {**497**}

Kalirai JS, Richer HB, Hansen BMS, *et al.*, 2004, The Galactic inner halo: searching for white dwarfs and measuring the fundamental galactic constant, Θ_0/R_0. *ApJ*, 601, 277–288 {**495**}

Kapteyn JC, van Rhijn PJ, 1920, On the distribution of the stars in space especially in the high Galactic latitudes. *ApJ*, 52, 23–38 {**490**}

Kapteyn JC, 1905, Star streaming. *Brit. Assoc. Adv. Sci. Rep.*, 1905, 257–265 {**491**}

—, 1922, First attempt at a theory of the arrangement and motion of the sidereal system. *ApJ*, 55, 302–328 {**490**}

Karaali S, Bilir S, Buser R, 2004, Comprehensive analysis of RGU photometry in the direction to M5. *Publications of the Astronomical Society of Australia*, 21, 275–283 {**516**}

Karataš Y, Bilir S, Karaali S, *et al.*, 2004, Analysis of UBV photometry in selected area 133. *Astronomische Nachrichten*, 325, 726–732 {**516**}

Karataš Y, Karaali S, Buser R, 2001, Analysis of RGU photometry in selected area 141. *A&A*, 373, 895–898 {**516**}

Kerr FJ, 1962, Galactic velocity models and the interpretation of 21-cm surveys. *MNRAS*, 123, 327–345 {**492**}

Kerr FJ, Lynden-Bell D, 1986, Review of Galactic constants. *MNRAS*, 221, 1023–1038 {**494, 495, 501, 502, 504–506, 512**}

Kharchenko NV, Kilpio E, Malkov O, *et al.*, 2002, Mira kinematics in the post-Hipparcos era. *A&A*, 384, 925–936 {**530**}

Klypin A, Zhao H, Somerville RS, 2002, ΛCDM-based models for the Milky Way and M31. I. Dynamical models. *ApJ*, 573, 597–613 {**538**}

Korchagin VI, Girard TM, Borkova TV, *et al.*, 2003, Local surface density of the Galactic disk from a three-dimensional stellar velocity sample. *AJ*, 126, 2896–2909 {**513**}

Kotoneva E, Flynn C, Chiappini C, *et al.*, 2002, K dwarfs and the chemical evolution of the solar cylinder. *MNRAS*, 336, 879–891 {**522**}

Kotoneva E, Innanen K, Dawson PC, *et al.*, 2005, A study of Kapteyn's star. *A&A*, 438, 957–962 {**492, 543**}

Kroupa P, 2001, On the variation of the initial mass function. *MNRAS*, 322, 231–246 {**515**}

—, 2002, The initial mass function of stars: evidence for uniformity in variable systems. *Science*, 295, 82–91 {**515**}

Kroupa P, Tout CA, Gilmore G, 1993, The distribution of low-mass stars in the Galactic disk. *MNRAS*, 262, 545–587 {**517**}

Krusberg ZAC, Chaboyer BC, 2006, UBVI CCD photometry of the old open cluster Berkeley 17. *AJ*, 131, 1565–1573 {**547**}

Kuijken K, 1995, Dark matter in the Milky Way. *IAU Symp. 164: Stellar Populations* (eds. van der Kruit PC, Gilmore G), 195–204 {**512**}

Kuijken K, Gilmore G, 1989a, The mass distribution in the Galactic disk. III. The local volume mass density. *MNRAS*, 239, 651–664 {**513**}

—, 1989b, The mass distribution in the Galactic disk. II. Determination of the surface mass density of the Galactic disk near the Sun. *MNRAS*, 239, 605–649 {**513**}

—, 1989c, The mass distribution in the Galactic disk. I. A technique to determine the integral surface mass density of the disk near the Sun. *MNRAS*, 239, 571–603 {**513**}

—, 1991, The Galactic disk surface mass density and the Galactic force at $Z = 1.1$ kpc. *ApJ*, 367, L9–L13 {**512**}

Kutuzov SA, Ossipkov LP, 2001, Galactic orbits of globular clusters. *ASP Conf. Ser. 228: Dynamics of Star Clusters and the Milky Way* (eds. Deiters S, Fuchs B, Just A, *et al.*), 500–502 {**551**}

Law DR, Johnston KV, Majewski SR, 2005, A Two Micron All-Sky Survey view of the Sagittarius dwarf galaxy. IV. Modeling the Sagittarius tidal tails. *ApJ*, 619, 807–823 {**542**}

Leonard PJT, Tremaine S, 1990, The local Galactic escape speed. *ApJ*, 353, 486–493 {**514, 515**}

Lépine JRD, Dias WS, Mishurov YN, *et al.*, 2006, Statistics of initial velocities of Galactic clusters. *IAU Symp. 237*, 154–158 {**534**}

Lépine JRD, Dias WS, Mishurov YN, 2008, Direct determination of the epicycle frequency in the Galactic disk, and the derived rotation velocity. MNRAS, 386, 2081–2090 {**534**}

Lépine JRD, Mishurov YN, Dedikov SY, 2001, A new model for the spiral structure of the Galaxy: superposition of 2- and 4-armed patterns. *ApJ*, 546, 234–247 {**499, 502, 531, 532**}

Lépine S, 2005, Nearby Stars from the LSPM-North Proper Motion Catalogue. I. Main sequence dwarfs and giants within 33 pc of the Sun. *AJ*, 130, 1680–1692 {**510**}

Lin CC, Shu FH, 1964, On the spiral structure of disk galaxies. *ApJ*, 140, 646–655 {**530**}

Lin CC, Yuan C, Shu FH, 1969, On the spiral structure of disk galaxies. III. Comparison with observations. *ApJ*, 155, 721–746 {**531**}

Lindblad B, 1927, On the state of motion in the Galactic system. *MNRAS*, 87, 553–564 {**491**}

Liu WM, Chaboyer BC, 2000, The relative age of the thin and thick Galactic disks. *ApJ*, 544, 818–829 {**520**}

Liu Y, Ma W, 1999, A kinematic study of the Galaxy. *Chinese Astronomy and Astrophysics*, 23, 395–400 {**502, 503**}

López Corredoira M, Cabrera Lavers A, Garzón F, *et al.*, 2002, Old stellar Galactic disk in near-plane regions according to 2MASS: scales, cut-off, flare and warp. *A&A*, 394, 883–899 {**490, 535**}

Luyten WJ, 1979, *The LHS Catalogue. A Catalogue of Stars with Proper Motions Exceeding 0.5 arcsec Annually.* University of Minnesota, Minneapolis {**509**}

Lynden-Bell D, Lynden-Bell RM, 1995, Ghostly streams from the formation of the Galaxy's halo. *MNRAS*, 275, 429–442 {**542, 550**}

Lynden-Bell D, 1975, The chemical evolution of galaxies. *Vistas in Astronomy*, 19, 299–316 {**520**}

Maíz Apellániz J, 2001, The spatial distribution of O–B5 stars in the solar neighbourhood as measured by Hipparcos. *AJ*, 121, 2737–2742 {**497, 516**}

Majewski SR, 1992, A complete, multicolour survey of absolute proper motions to $B \sim 22.5$: Galactic structure and kinematics at the north Galactic pole. *ApJS*, 78, 87–152 {**539**}

—, 2004, Sub-structure in the Galactic halo. *Publications of the Astronomical Society of Australia*, 21, 197–202 {**542**}

Majewski SR, Munn JA, Hawley SL, 1994, Absolute proper motions to $B \simeq 22.5$: evidence for kimematical sub-structure in halo field stars. *ApJ*, 427, L37–L41 {**542**}

Majewski SR, Ostheimer JC, Patterson RJ, *et al.*, 2000, Exploring halo sub-structure with giant stars. II. Mapping the extended structure of the Carina dwarf spheroidal galaxy. *AJ*, 119, 760–776 {**542**}

Majewski SR, Ostheimer JC, Rocha Pinto HJ, *et al.*, 2004, Detection of the main-sequence turnoff of a newly-discovered Milky Way halo structure in the Triangulum–Andromeda region. *ApJ*, 615, 738–743 {**542**}

Makarov VV, 1997, Candidates to the Catalogue of Nearby Stars from Tycho astrometry and photometry. *ESA SP–402: Hipparcos, Venice '97*, 541–544 {**510**}

Makarov VV, Murphy DW, 2007, The local stellar velocity field via vector spherical harmonics. *AJ*, 134, 367–375 {**499, 501, 502, 508, 509, 537**}

Marochnik LS, Mishurov YN, Suchkov AA, 1972, On the spiral structure of our Galaxy. *Ap&SS*, 19, 285–292 {**531**}

Martin NF, Ibata RA, Bellazzini M, *et al.*, 2004, A dwarf galaxy remnant in Canis Major: the fossil of an in-plane accretion on to the Milky Way. *MNRAS*, 348, 12–23 {**542**}

Martinet L, Raboud D, 1999, Chemo-dynamical effects of bars in Galactic evolution. *Galaxy Evolution: Connecting the Distant Universe with the Local Fossil Record*, 371–374 {**528**}

Matute EA, 2006, On the vector solutions of Maxwell equations with the spin-weighted spherical harmonics. *Rev. Mex. Fis. E*, 51, 116–117 {**508**}

Mayor M, 1972, On the vertex deviation. *A&A*, 18, 97 {**528**}

Mazzitelli I, D'Antona F, Caloi V, 1995, Globular cluster ages with updated input physics. *A&A*, 302, 382–400 {**548**}

Meillon L, 1999, An estimation of the Galactic escape velocity from high-velocity metal-poor stars. *Ap&SS*, 265, 179–180 {**515**}

Meillon L, Crifo F, Gómez AE, *et al.*, 1997, First steps toward the determination of the escape velocity in the solar neighbourhood. *ESA SP–402: Hipparcos, Venice '97*, 591–594 {**514**}

Mel'Nik AM, Sitnik TG, Dambis AK, *et al.*, 1998, Kinematic evidence for the wave nature of the Carina–Sagittarius arm. *Astronomy Letters*, 24, 594–602 {**533**}

Merrifield MR, 2005, Dark matter on Galactic scales (or the lack thereof). *The Identification of Dark Matter* (eds. Spooner NJC, Kudryavtsev V), 49–58 {**538**}

Meusinger H, Scholz RD, Irwin MJ, 2001, A proper motion search for stars escaping from a globular cluster with high velocity. *ASP Conf. Ser. 228: Dynamics of Star Clusters and the Milky Way* (eds. Deiters S, Fuchs B, Just A, *et al.*), 520–522 {**554**}

Mignard F, 2000, Local Galactic kinematics from Hipparcos proper motions. *A&A*, 354, 522–536 {**499, 501, 502, 505, 506**}

Mignard F, Morando B, 1990, Analyse de catalogues stellaires au moyen des harmoniques vectorielles. *Journées 1990: Systèmes de Référence Spatio-Temporels. Colloque André Danjon*, 151–158 {**508**}

Mihalas D, Binney JJ, 1981, *Galactic Astronomy: Structure and Kinematics*, 2nd edition. W.H. Freeman, San Francisco {**493, 499–501, 544**}

Miller GE, Scalo JM, 1979, The initial mass function and stellar birthrate in the solar neighbourhood. *ApJS*, 41, 513–547 {**515**}

Milne EA, 1935, Stellar kinematics and the K-effect. *MNRAS*, 95, 560–561 {**491, 504**}

Minniti D, 1995, Metal-rich globular clusters with $R < 3$ kpc: disk or bulge clusters. *AJ*, 109, 1663–1669 {**547**}

Mishurov YN, Pavlovskaya ED, Suchkov AA, 1979, Galactic spiral structure parameters derived from stellar kinematics. *Soviet Astronomy*, 23, 147–152 {**531**}

Mishurov YN, Zenina IA, 1999a, Parameters of the Galactic rotation curve and spiral pattern from Cepheid kinematics. *Astronomy Reports*, 43, 487–493 {**531**}

—, 1999b, Yes, the Sun is located near the corotation circle. *A&A*, 341, 81–85 {**499, 531–533**}

Miyamoto M, Nagai R, 1975, Three-dimensional models for the distribution of mass in galaxies. *PASJ*, 27, 533–543 {**545**}

Miyamoto M, Soma M, Yoshizawa M, 1993, Is the vorticity vector of the Galaxy perpendicular to the Galactic plane? II. Kinematics of the Galactic warp. *AJ*, 105, 2138–2147 {**535**}

Miyamoto M, Soma M, 1993, Is the vorticity vector of the Galaxy perpendicular to the Galactic plane? I. Precessional correction and equinoctial motion correction to the FK5 system. *AJ*, 105, 691–701 {**504, 505**}

Miyamoto M, Tsujimoto T, 1997, Where does the dark halo of the Galaxy end? *ESA SP–402: Hipparcos, Venice '97*, 537–540 {**539**}

Miyamoto M, Zhu Z, 1998, Galactic interior motions derived from Hipparcos proper motions. I. Young disk population. *AJ*, 115, 1483–1491 {**499, 502, 507, 535, 537**}

Moon HK, Min BH, Kim SL, 2003, Terrestrial impact cratering chronology. II: Periodicity analysis with the 2002 database. *Journal of Astronomy and Space Sciences*, 20, 269–282 {**514**}

Morrison HL, Flynn C, Freeman KC, 1990, Where does the disk stop and the halo begin? Kinematics in a rotation field. *AJ*, 100, 1191–1222 {**539, 540**}

Mühlbauer G, Dehnen W, 2003, Kinematic response of the outer stellar disk to a central bar. *A&A*, 401, 975–984 {**529, 530**}

Murali C, Dubinski J, 1999, Determining the Galactic mass distribution using tidal streams from globular clusters. *AJ*, 118, 911–919 {**554**}

Murray CA, Candy MP, Jones DHP, 1965, Studies of the globular cluster ω Centauri. *Royal Greenwich Observatory Bulletin*, 100, 81–98 {**551**}

Naoumov SO, 1999, Origins of the Galactic thick disk: two populations or one? Ph.D. Thesis {**522**}

Navarro JF, Frenk CS, White SDM, 1996, The structure of cold dark matter halos. *ApJ*, 462, 563–575 {**538**}

Nelson AH, Matsuda T, 1977, On one-dimensional Galactic spiral shocks. *MNRAS*, 179, 663–670 {**531**}

Nelson RW, Tremaine S, 1995, The damping and excitation of Galactic warps by dynamical friction. *MNRAS*, 275, 897–920 {**537**}

Newberg HJ, Yanny B, Rockosi C, *et al.*, 2002, The ghost of Sagittarius and lumps in the halo of the Milky Way. *ApJ*, 569, 245–274 {**542**}

Ng YK, Bertelli G, 1998, Revised ages for stars in the solar neighbourhood. *A&A*, 329, 943–950 {**520, 524**}

Nikiforov II, 1999, Modeling the rotation curve of the plane subsystem and determination of the distance to the Galactic centre: analysis of data for gas complexes. *Astronomy Reports*, 43, 345–360 {**492**}

Ninkovic S, Trajkovska V, 2006, On the mass distribution of stars in the solar neighbourhood. *Serbian Astronomical Journal*, 172, 17–20 {**515**}

Nordström B, Andersen J, Holmberg J, *et al.*, 2004a, The Geneva–Copenhagen survey of the solar neighbourhood. *Publications of the Astronomical Society of Australia*, 21, 129–133 {**524**}

Nordström B, Mayor M, Andersen J, *et al.*, 2004b, The Geneva–Copenhagen survey of the solar neighbourhood: ages, metallicities, and kinematic properties of ~14 000 F and G dwarfs. *A&A*, 418, 989–1019 {**524–526, 532, 535, 542**}

Odenkirchen M, Brosche P, Geffert M, *et al.*, 1997, Globular cluster orbits based on Hipparcos proper motions. *New Astronomy*, 2, 477–499 {**551, 554**}

Odenkirchen M, Grebel EK, Dehnen W, *et al.*, 2003, The extended tails of Palomar 5: a 10° arc of globular cluster tidal debris. *AJ*, 126, 2385–2407 {**542**}

Ogorodnikov KF, 1932, *Astron. Zhur.* 4,190 {**491, 504**}

Ogorodnikov KF, 1965, *Dynamics of Stellar Systems*. Pergamon, Oxford {**504, 506**}

Olling RP, Dehnen W, 2003, The Oort constants measured from proper motions. *ApJ*, 599, 275–296 {**493, 502, 503, 507**}

Olling RP, Merrifield MR, 1998, Refining the Oort and Galactic constants. *MNRAS*, 297, 943–952 {**494–496, 501, 502, 504, 534**}

Oort JH, 1927a, Investigations concerning the rotational motion of the Galactic system together with new determinations of secular parallaxes, precession and motion of the equinox. *Bull. Astron. Inst. Netherlands*, 4, 79–89 {**502**}

—, 1927b, Observational evidence confirming Lindblad's hypothesis of a rotation of the Galactic system. *Bull. Astron. Inst. Netherlands*, 3, 275–282 {**491**}

—, 1932, The force exerted by the stellar system in the direction perpendicular to the Galactic plane and some related problems. *Bull. Astron. Inst. Netherlands*, 6, 249–287 {**511**}

—, 1965, Stellar dynamics. *Galactic Structure* (eds. Blaauw A, Schmidt M), 455–512 {**511**}

Oort JH, Plaut L, 1975, The distance to the Galactic centre derived from RR Lyrae variables, the distribution of these variables in the Galaxy's inner region and halo, and a rediscussion of the Galactic rotation constants. *A&A*, 41, 71–86 {**496**}

Oswalt TD, Smith JA, Wood MA, *et al.*, 1996, A lower limit of 9.5 Gyr on the age of the Galactic disk from the oldest white dwarfs. *Nature*, 382, 692–694 {**548, 549**}

Paczyński B, 1990, A test of the Galactic origin of gamma-ray bursts. *ApJ*, 348, 485–494 {**545, 546**}

Paczyński B, Stanek KZ, 1998, Galactocentric distance with the Optical Gravitational Lensing Experiment and Hipparcos red clump stars. *ApJ*, 494, L219–L222 {**495, 496**}

Pagel BEJ, Tautvaišien e G, 1995, Chemical evolution of primary elements in the Galactic disk: an analytical model. *MNRAS*, 276, 505–514 {**520**}

Palouš J, Ruprecht J, Dluzhnevskaia OB, *et al.*, 1977, Places of formation of 24 open clusters. *A&A*, 61, 27–37 {**531**}

Parenago PP, 1950, *Astron. Zhur.* 27, 150-158 {**494**}

Perlmutter S, Aldering G, Goldhaber G, *et al.*, 1999, Measurements of Ω and Λ from 42 high-redshift supernovae. *ApJ*, 517, 565–586 {**550**}

Peñarrubia J, Martínez-Delgado D, Rix HW, *et al.*, 2005, A comprehensive model for the Monoceros tidal stream. *ApJ*, 626, 128–144 {**537**}

Pham HA, 1997, Estimation of the local mass density from an F-star sample observed by Hipparcos. *ESA SP–402: Hipparcos, Venice '97*, 559–562 {**497, 513**}

Phelps RL, 1997, Berkeley 17: the oldest open cluster? *ApJ*, 483, 826–836 {**547**}

Pichon C, Siebert A, Bienaymé O, 2002, On the kinematic deconvolution of the local neighbourhood luminosity function. *MNRAS*, 329, 181–194 {**544**}

Pinfield DJ, Hodgkin ST, Jameson RF, *et al.*, 1999, Praesepe deep survey work: towards the bottom of the stellar mass function. *Irish Astronomical Journal*, 26, 94–98 {**516**}

Plummer HC, 1915, The distribution of stars in globular clusters. *MNRAS*, 76, 107–121 {**545**}

Porcel C, Battaner E, Jimenez-Vicente J, 1997, Geometric differences between the gaseous and stellar warps in the Milky Way. *A&A*, 322, 103–108 {**496, 537**}

Pritchard JR, Kamionkowski M, 2005, Cosmic microwave background fluctuations from gravitational waves: an analytic approach. *Annals of Physics*, 318, 2–36 {**508**}

Prochaska JX, Naumov SO, Carney BW, *et al.*, 2000, The Galactic thick disk stellar abundances. *AJ*, 120, 2513–2549 {**522**}

Quillen AC, Garnett DR, 2001, The saturation of disk heating in the solar neighbourhood and evidence for a merger 9 Gyr

ago. *ASP Conf. Ser. 230: Galaxy Disks and Disk Galaxies*, 87–88 {**523, 524**}
Quillen AC, 2003, Chaos caused by resonance overlap in the solar neighbourhood: spiral structure at the bar's outer Lindblad resonance. *AJ*, 125, 785–793 {**531**}
Quillen AC, Minchev I, 2005, The effect of spiral structure on the stellar velocity distribution in the solar neighbourhood. *AJ*, 130, 576–585 {**531, 532**}
Raboud D, Grenon M, Martinet L, *et al.*, 1998, Evidence for a signature of the Galactic bar in the solar neighbourhood. *A&A*, 335, L61–L64 {**528**}
Rafikov RR, 2001, The local axisymmetric instability criterion in a thin, rotating, multicomponent disk. *MNRAS*, 323, 445–452 {**523**}
Rautiainen P, Salo H, 1999, Multiple pattern speeds in barred galaxies. I. Two-dimensional models. *A&A*, 348, 737–754 {**534**}
Reed BC, 1997, The Sun's displacement from the Galactic plane: limits from the distribution of OB-star latitudes. *PASP*, 109, 1145–1148 {**497**}
—, 2005, New estimates of the solar-neighbourhood massive star birthrate and the Galactic supernova rate. *AJ*, 130, 1652–1657 {**497**}
Reid IN, 1998, Hipparcos subdwarf parallaxes: metal-rich clusters and the thick disk. *AJ*, 115, 204–228 {**495, 496, 519, 527, 549**}
Reid IN, Gizis JE, Hawley SL, 2002, The Palomar/MSU nearby star spectroscopic survey. IV. The luminosity function in the solar neighbourhood and M dwarf kinematics. *AJ*, 124, 2721–2738 {**516–518**}
Reid IN, Yan L, Majewski SR, *et al.*, 1996, Starcounts redivivus. II. Deep star counts with Keck and HST and the luminosity function of the Galactic halo. *AJ*, 112, 1472–1486 {**544**}
Reid MJ, 1993, The distance to the centre of the Galaxy. *ARA&A*, 31, 345–372 {**495, 496, 501**}
Reid MJ, Brunthaler A, 2004, The proper motion of Sgr A*. II. The mass of Sgr A*. *ApJ*, 616, 872–884 {**500**}
Reid MJ, Readhead ACS, Vermeulen RC, *et al.*, 1999, The proper motion of Sgr A*. I. First VLBA results. *ApJ*, 524, 816–823 {**500**}
Reid MJ, Schneps MH, Moran JM, *et al.*, 1988, The distance to the centre of the Galaxy – H_2O maser proper motions in Sgr B2(N). *ApJ*, 330, 809–816 {**496**}
Robin AC, Crézé M, 1986, Stellar populations in the Milky Way: comparisons of a synthetic model with star counts in nine fields. *A&AS*, 64, 53–64 {**544**}
Robin AC, Haywood M, Crézé M, *et al.*, 1996, The thick disk of the Galaxy: sequel of a merging event. *A&A*, 305, 125–134 {**519**}
Robin AC, Reylé C, Derrière S, *et al.*, 2003, A synthetic view on structure and evolution of the Milky Way. *A&A*, 409, 523–540, erratum: 416, 157 {**544, 546**}
Rocha Pinto HJ, Majewski SR, Skrutskie MF, 2006, Argo and other tidal structures around the Milky Way. *Revista Mexicana de Astronomia y Astrofisica Conference Series*, volume 26, 84–85 {**542**}
Rocha Pinto HJ, Scalo JM, Maciel WJ, *et al.*, 2000, Chemical enrichment and star formation in the Milky Way disk. II. Star-formation history. *A&A*, 358, 869–885 {**518**}
Rodgers AW, Paltoglou G, 1984, Kinematics of Galactic globular clusters. *ApJ*, 283, L5–L7 {**553**}
Roman NG, 1955, A catalogue of high-velocity stars. *ApJS*, 2, 195–195 {**538**}
Rosati P, Borgani S, Norman C, 2002, The evolution of X-ray clusters of galaxies. *ARA&A*, 40, 539–577 {**550**}
Rosenberg A, Saviane I, Piotto G, *et al.*, 1999, Galactic globular cluster relative ages. *AJ*, 118, 2306–2320 {**549**}
Rybka SP, 2004, Local kinematics of dwarfs from the Tycho 2 data. *Kinematika i Fizika Nebesnykh Tel*, 20, 133–141 {**507**}
Sabas V, 1997, Determination of the initial mass function in the solar neighbourhood between 1.2–4 $M_\odot$. *ESA SP–402: Hipparcos, Venice '97*, 563–566 {**515**}
Sackett PD, 1997, Does the Milky Way have a maximal disk? *ApJ*, 483, 103–110 {**504**}
Salaris M, Degl'Innocenti S, Weiss A, 1997, The age of the oldest globular clusters. *ApJ*, 479, 665–672 {**548**}
Salaris M, Weiss A, 1998, Metal-rich globular clusters in the Galactic disk: new age determinations and the relation to halo clusters. *A&A*, 335, 943–953 {**549**}
—, 2002, Homogeneous age dating of 55 Galactic globular clusters. Clues to the Galaxy formation mechanisms. *A&A*, 388, 492–503 {**549**}
Salpeter EE, 1955, The luminosity function and stellar evolution. *ApJ*, 121, 161–167 {**515**}
Sandage AR, Lubin LM, VandenBerg DA, 2003, The age of the oldest stars in the local Galactic disk from Hipparcos parallaxes of G and K subgiants. *PASP*, 115, 1187–1206 {**521**}
Sanner J, Geffert M, 2001, The initial mass function of open star clusters with Tycho 2. *A&A*, 370, 87–99 {**516**}
Scalo JM, 1986, The stellar initial mass function. *Fundamentals of Cosmic Physics*, 11, 1–278 {**515, 546**}
—, 1998, The inital mass function revisited: a case for variations. *ASP Conf. Ser. 142: The Stellar Initial Mass Function* (eds. Gilmore G, Howell D), 201–236 {**515–517**}
Scanio JJG, 1977, Spin-weighted spherical harmonics and electromagnetic multipole expansions. *American Journal of Physics*, 45, 173–178 {**508**}
Schröder KP, Pagel BEJ, 2003, Galactic archaeology: initial mass function and depletion in the thin disk. *MNRAS*, 343, 1231–1240 {**515, 516, 518**}
Schröder KP, Sedlmayr E, 2001, The Galactic mass injection from cool stellar winds of the $1-2.5M_\odot$ stars in the solar neighbourhood. *A&A*, 366, 913–922 {**518**}
Schuster WJ, Moitinho A, Márquez A, *et al.*, 2006, uvbyβ photometry of high-velocity and metal-poor stars. XI. Ages of halo and old disk stars. *A&A*, 445, 939–958 {**527**}
Schuster WJ, Parrao L, Contreras ME, 1993, uvbyβ photometry of high-velocity and metal-poor stars. VI. A second catalogue, and stellar populations of the Galaxy. *A&AS*, 97, 951–983 {**519**}
Schwarzschild K, 1908, *Göttingen Nachrichten*. 191 {**491**}
Searle L, Zinn R, 1978, Compositions of halo clusters and the formation of the Galactic halo. *ApJ*, 225, 357–379 {**520, 540–542**}
Ségransan D, Delfosse X, Forveille T, *et al.*, 2000, Accurate masses of very low-mass stars. III. 16 new or improved masses. *A&A*, 364, 665–673 {**518**}
Sellwood JA, 2000, Spiral structure as a recurrent instability. *Ap&SS*, 272, 31–43 {**534–536**}
Shapley H, 1918, Globular clusters and the structure of the Galactic system. *PASP*, 30, 42–54 {**490**}
Shaviv NJ, 2003, The spiral structure of the Milky Way, cosmic rays, and ice age epochs on Earth. *New Astronomy*, 8, 39–77 {**531**}

Shuter WLH, 1982, A rotational standard of rest. *MNRAS*, 199, 109–113 {**492**}

Siebert A, Bienaymé O, Soubiran C, 2003, Vertical distribution of Galactic disk stars. II. The surface mass density in the Galactic plane. *A&A*, 399, 531–541 {**513**}

Sitnik TG, 2003, The line-of-sight velocities of OB associations and molecular clouds in a wide solar neighbourhood: the streaming motions of stars and gas in the Perseus arm. *Astronomy Letters*, 29, 311–320 {**534**}

Sitnik TG, Mel'Nik AM, 1999, The density-wave nature of the Cygnus–Orion arm. *Astronomy Letters*, 25, 156–168 {**533, 534**}

Smart RL, Drimmel R, Lattanzi MG, *et al.*, 1998, Unexpected stellar velocity distribution in the warped Galactic disk. *Nature*, 392, 471–473 {**536, 538**}

Smart RL, Lattanzi MG, 1996, The Galactic warp in the proper motions of young stars. *A&A*, 314, 104–107 {**536, 537**}

Smith MC, Ruchti GR, Helmi A, *et al.*, 2007, The RAVE survey: constraining the local Galactic escape speed. *MNRAS*, 379, 755–772 {**515**}

Sneden C, McWilliam A, Preston GW, *et al.*, 1996, The ultra-metal-poor, neutron-capture-rich giant star CS 22892–052. *ApJ*, 467, 819–840 {**548**}

Soubiran C, Bienayme O, Mishenina TV, *et al.*, 2008, Vertical distribution of Galactic disk stars. IV. Age–metallicity and age–velocity relations from clump giants. *A&A*, 480, 91–101 {**513**}

Soubiran C, Bienaymé O, Siebert A, 2003, Vertical distribution of Galactic disk stars. I. Kinematics and metallicity. *A&A*, 398, 141–151 {**513, 527**}

Soubiran C, Girard P, 2005, Abundance trends in kinematical groups of the Milky Way's disk. *A&A*, 438, 139–151 {**527**}

Spergel DN, Verde L, Peiris HV, *et al.*, 2003, First-Year Wilkinson Microwave Anisotropy Probe (WMAP) observations: determination of cosmological parameters. *ApJS*, 148, 175–194 {**548, 549**}

Stäckel P, 1890, *Math. Ann.* 35, 91-101 {**545**}

Stanek KZ, Garnavich PM, 1998, Distance to M31 with the Hubble Space Telescope and Hipparcos red clump stars. *ApJ*, 503, L131–L134 {**495, 496**}

Statler TS, 1989, Problems in determining the surface density of the Galactic disk. *ApJ*, 344, 217–231 {**512**}

Stothers RB, 1985, Terrestrial record of the Solar System's oscillation about the Galactic plane. *Nature*, 317, 338–341 {**492, 514**}

—, 1998, Galactic disk dark matter, terrestrial impact cratering and the law of large numbers. *MNRAS*, 300, 1098–1104 {**512, 514**}

Tarrab I, 1982, The initial mass function for young open clusters. *A&A*, 109, 285–288 {**517**}

Teixeira R, de Souza RE, 2006, Kinematics of nearby disk stars from Hipparcos database. *Memorie della Societa Astronomica Italiana*, 77, 1183–1187 {**503**}

Tsvetkov AS, 2006, The rotational vector of the local stellar system. *Astronomical and Astrophysical Transactions*, 25, 165–169 {**503**}

Tucholke HJ, Scholz RD, Brosche P, 1996, Proper motion study of the globular cluster M92. *A&A*, 312, 74–79 {**550**}

Uemura M, Ohashi H, Hayakawa T, *et al.*, 2000, Galactic rotation derived from OB stars using Hipparcos proper motions and radial velocities. *PASJ*, 52, 143–151 {**503, 507**}

Upgren AR, Boyle RP, Sperauskas J, *et al.*, 2006, Kinematics of nearby K-M dwarfs: first results. *Memorie della Societa Astronomica Italiana*, 77, 1168–1171 {**503**}

Vallenari A, Bertelli G, Schmidtobreick L, 2000, The Galactic disk: study of four low latitude Galactic fields. *A&A*, 361, 73–84 {**518**}

VandenBerg DA, Stetson PB, Bolte M, 1996, The age of the Galactic globular cluster system. *ARA&A*, 34, 461–510 {**547, 550**}

van der Kruit PC, 1988, The three-dimensional distribution of light and mass in disks of spiral galaxies. *A&A*, 192, 117–127 {**544**}

van Leeuwen F, Le Poole RS, Reijns RA, *et al.*, 2000, A proper motion study of the globular cluster ω Centauri. *A&A*, 360, 472–498 {**551**}

Vazquez EC, Scalo JM, 1989, Evolution of the star-formation rate in galaxies with increasing densities. *ApJ*, 343, 644–658 {**518**}

Verbunt F, Johnston HM, 2000, Multiple and variable X-ray sources in the globular clusters ω Centauri, NGC 6397, NGC 6752, and Liller 1. *A&A*, 358, 910–922 {**555**}

Vergely JL, Ferrero RF, Egret D, *et al.*, 1998, The interstellar extinction in the solar neighbourhood. I. Statistical approach. *A&A*, 340, 543–555 {**516**}

Vergely JL, Köppen J, Egret D, *et al.*, 2002, An inverse method to interpret colour-magnitude diagrams. *A&A*, 390, 917–929 {**518**}

Vityazev VV, Shuksto A, 2004, Stellar kinematics by vectorial harmonics. *Order and Chaos in Stellar and Planetary Systems* (eds. Byrd GG, Kholshevnikov KV, Myllri AA, *et al.*), ASP Conf. Ser. 316, 230–233 {**499, 502, 508**}

von Hoerner S, 1960, Die zeitliche Rate der Sternentstehung. *Fortschritte der Physik VIII*, 8, 191–244 {**512**}

Walker AR, Terndrup DM, 1991, The metallicity of RR Lyrae stars in Baade's window. *ApJ*, 378, 119–126 {**495, 496**}

Wang J, Chen L, Chen D, 2001, Absolute proper motions of 264 stars in the region of the globular cluster M13. *Shanghai Observatory Annals*, 22, 63–74 {**551**}

Wang J, Chen L, Wu Z, *et al.*, 2000, Kinematics and colour–magnitude diagram of the globular cluster NGC 4147. *A&AS*, 142, 373–387 {**551**}

Weinberg MD, 1994, Adiabatic invariants in stellar dynamics. II. Gravitational shocking. *AJ*, 108, 1403–1413 {**554**}

Weisstein EW, 2007, Vector spherical harmonics. *MathWorld–A Wolfram Web Resource, http://mathworld.wolfram.com/VectorSphericalHarmonic.html* {**508**}

Westin TNG, 1985, The local system of early-type stars: spatial extent and kinematics. *A&AS*, 60, 99–134 {**506, 507**}

Wielen R, Jahreiß H, Krüger R, 1983, The determination of the luminosity function of nearby stars. *IAU Colloq. 76: Nearby Stars and the Stellar Luminosity Function* (eds. Philip AGD, Upgren AR), 163–170 {**516, 517**}

Woolley RvdR, 1970, Catalogue of stars within twenty-five parsecs of the Sun. *Annals of the Royal Greenwich Observatory*, 5, 1–227 {**509**}

Wu Z, Wang J, Chen L, 2002a, Determination of the proper motions and membership of the globular cluster M3 and of its orbit in the Galaxy. *Chinese Journal of Astronony and Astrophysics*, 2, 216–225 {**551**}

Wu Z, Wang J, Li C, 2002b, Absolute proper motions of 534 stars in the region of globular cluster M3. *Shanghai Observatory Annals*, 23, 61–82 {**551**}

Wu Z, Wang J, 1999, The absolute proper motion of M3 and its Galactic orbit. *Observational Astrophysics in Asia and its Future*, 327–331 {**551**}

Yabushita S, 2004, A spectral analysis of the periodicity hypothesis in cratering records. *MNRAS*, 355, 51–56 {**514**}

Yano T, Chiba M, Gouda N, 2002, Kinematic analysis of spiral structures in the local disk. *A&A*, 389, 143–148 {**533**}

Yuan C, 1969a, Application of the density-wave theory to the spiral structure of the Milky Way system. I. Systematic motion of neutral H. *ApJ*, 158, 871–888 {**531**}

—, 1969b, Application of the density-wave theory to the spiral structure of the Milky Way system. II. Migration of stars. *ApJ*, 158, 889–898 {**531**}

Zabolotskikh MV, Rastorguev AS, Dambis AK, 2002, Kinematic parameters of young subsystems and the Galactic rotation curve. *Astronomy Letters*, 28, 454–464 {**503**}

Zaldarriaga M, 1998, Cosmic microwave background polarization experiments. *ApJ*, 503, 1–15 {**508**}

Zhu Z, 1999, Galactic kinematics derived from classical Cepheids. *Chinese Astronomy and Astrophysics*, 23, 445–453 {**507**}

—, 2000a, Kinematics of the Galaxy from Hipparcos proper motions. *PASJ*, 52, 1133–1139 {**507, 537**}

—, 2000b, Local kinematics of classical Cepheids. *Ap&SS*, 271, 353–363 {**507**}

—, 2006, Velocity space of Galactic O–B stars. *Chinese Journal of Astronomy and Astrophysics*, 6, 363–371 {**507**}

Zhu Z, Zhang H, 2005, Galactic warping motion from Hipparcos proper motions and radial velocities. *Fundamental Astronomy: New Concepts and Models for High Accuracy Observations* (ed. Capitaine N), 244–245 {**536**}

Zinn R, 1985, The globular cluster system of the galaxy. IV. The halo and disk subsystems. *ApJ*, 293, 424–444 {**547, 551**}

10

Solar System and exoplanets

This chapter covers the use of the Hipparcos data in two loosely connected areas: observations and interpretation related to the Solar System, and observations related to exoplanets.

10.1 Hipparcos Solar System objects

Astrometric and photometric measurements of a number of Solar System objects were performed either by the Hipparcos main instrument (and thus appear as the Hipparcos Catalogue Solar System Annex) or by the star mapper instrument (and thus appear in the Tycho Catalogue Solar System Annex). A summary is provided in Table 10.1. The results concern mainly asteroids, but also the planetary satellites Europa, Ganymede, Callisto, Titan and Iapetus, and the major planets Uranus and Neptune. Reductions were accurate to the mas level for the main mission. A detailed description of the resulting catalogue contents is given in ESA (1997), Volume 1, Section 2.7. Volume 3, Chapter 15 gives details of the specific data analysis aspects. Hestroffer *et al.* (1998a) also summarised the measurements and reductions relevant for Solar System objects, the objects observed, and the presentation of results.

There were two main objectives for the inclusion of Solar System objects within the observing programme. The first was to establish the relationship between the resulting dynamical reference system and the stellar reference frame (ICRF). To date, the theoretical positional precision of asteroids had never been met observationally, and they only entered into the FK5 solution, for example, with relatively modest weight. The second objective was to acquire improved positional data at a series of epochs to enable dynamical and physical studies of these objects. In the case of high-accuracy astrometry of very close encounters, for example, masses of some asteroids can be obtained. Observations of planetary satellites relative to the background stars also yield, in an indirect manner, accurate positions of the gravitating major planet's centre of mass (Morrison *et al.*, 1997; Fienga *et al.*, 1997; Fienga, 1998).

Observations and Catalogue considerations Solar System objects could only be included in the Hipparcos observation programme if they satisfied a number of conditions: (a) if their apparent magnitudes were sufficiently bright; (b) the angular diameter of each observed object had to be sufficiently small for the depth of modulation on the focal plane grid to be sufficiently high; (c) planetary satellites had to satisfy constraints on proximity and magnitude difference with respect to the bright major parent planet to avoid scattered light masking the image modulation (from which positions were inferred). These combined restrictions resulted in only a few observable planets and planetary satellites, and 48 asteroids (out of the 63 on the initial observing programme) observed by the main mission and appearing in the Hipparcos Catalogue, and/or in the Tycho Catalogue.

As for the stars, objects observed with the Hipparcos main instrument required *a priori* positions known to about 1 arcsec at each observation epoch, a condition which demanded extensive ground-based observations pre-launch, and the use of a consistent Solar System ephemeris for the satellite observations.

Analysis of the data related to Solar System objects formed part of the whole Hipparcos reduction procedure described in Chapter 1. The velocity for the Earth made use of the VSOP 82 theory for the motion of the planets (Bretagnon, 1982), and the lunar ephemeris ELP 2000 (Chapront-Touzé & Chapront, 1983). These ephemerides have since been superseded, most recently by the JPL development ephemeris solution DE414 (Konopliv *et al.*, 2006), and the numerical planetary ephemeris developed at the IMCCE–Observatoire de

Planets, dwarf planets, and asteroids: Resolution B5 of the IAU General Assembly XXVI in 2006 divided planets and other bodies of the Solar System, except satellites, into three distinct categories according to the following definitions: (1) a planet is a celestial body that is in orbit around the Sun, has sufficient mass for its self-gravity to overcome rigid body forces so that it assumes a hydrostatic equilibrium (nearly round) shape, and has cleared the neighbourhood around its orbit; (2) a 'dwarf planet' is a celestial body that is in orbit around the Sun, has sufficient mass for its self-gravity to overcome rigid body forces so that it assumes a hydrostatic equilibrium (nearly round) shape, has not cleared the neighbourhood around its orbit, and is not a satellite; (3) all other objects, except satellites, orbiting the Sun, shall be referred to collectively as 'small Solar System bodies'. The latter category includes most of the Solar System asteroids, most trans-Neptunian objects, comets, and other small bodies. These definitions imply that there are accordingly eight 'classical' planets, while Pluto, Ceres, Charon, 2003 UB_{313}, and potentially other objects become 'dwarf' planets (Pluto is actually recognised as the prototype of a new category of trans-Neptunian objects). In this new system of IAU definitions, the term 'minor planet' is no longer used, such an object being referred to as an 'asteroid' or, more generally, a 'small Solar System body'. The IAU www pages give further details. Early Hipparcos literature, including the catalogue publication in 1997, followed a previous IAU recommendation to use the term 'minor planet' instead of 'asteroid'. The nomenclature used here follows the 2006 IAU recommendations.

Paris, INPOP06 (Fienga *et al.*, 2008). The geocentric Hipparcos satellite ephemeris was provided by the satellite operation centre (ESOC) with an accuracy of ~1.5 km in position and ~0.2 m s^{-1} in velocity. The apparent, or 'proper', instantaneous satellitocentric directions were corrected for stellar aberration due to the satellite's barycentric velocity expanded to second order in v/c, and for general relativistic gravitational light bending due to the Sun's spherical gravitational field, based on a knowledge of the heliocentric direction of the object. The positions and epoch of observation were referred to the geocentre, and the position was corrected for the satellite's parallax based on its known geocentric position. Final adjustment was made to the ICRS, as for the main catalogue. Results were published as one-dimensional positions per field transit, projected onto the tangent plane, and with respect to the appropriate reference great circle. Standard errors refer to the uncertainty of the abscissa parallel to the reference great circle, and passing through the chosen reference point. Phase, shape or albedo corrections were not taken into account. The given position corresponds to the 'photocentre' for the smallest objects (although this position is specified differently in the case of the NDAC and FAST analyses). For Uranus, Neptune and to a lesser extent the two Jovian satellites, whose angular diameters are larger than the star mapper (Tycho experiment) slit width, the position on the surface of the body depends on its albedo distribution and the scanning geometry.

Two estimators of the apparent magnitude were derived, $Hp_{\rm dc}$ giving the mean intensity corrected for background noise, and $Hp_{\rm ac}$ derived from the amplitude of the modulation. Due to the rather random observation epochs of the objects, the asteroid data rarely yield magnitudes over the object's rotation period, nor representative light-curves. They can nevertheless be used for deriving magnitudes reduced to unit heliocentric and geocentric distances (accurate to about 0.03 mag) over a large range of solar phase angles, rotational phase, aspect and obliquity. The Hipparcos Solar System objects photometric catalogue was completed, for convenience, with some additional calculated aspect data: the distance to the Sun, the distance to the satellite, and the solar phase angle. The resulting photometric observations of asteroids provide information about their rotational properties (spin-vector orientation and rotation period) and in some cases their shape, and the scattering properties of their surface, in turn yielding insight into their collisional evolution and conditions in the early Solar System.

From ground-based measurements, direct measurements of the equatorial coordinates of the outer planets, such as Jupiter or Saturn, are complicated by phase effects due to illumination and scattering effects, making the determination of the centre of mass with respect to the photocentre very difficult, and limiting the external mean error of such observations to typically 200–500 mas (e.g. Pascu & Schmidt, 1990). Nonetheless, accurate positions are required for constructing planetary ephemerides, which are used for a variety of purposes, including supporting spacecraft missions to these planets and minor Solar System bodies (NEAR, DS1, Cassini–Huygens, Rosetta, Pluto Express), for computing dynamical masses for asteroids, for accurately predicting occultation events used to investigate the atmospheres of planetary satellites, and for investigating the dynamics of asteroids and planetary satellites. With the Hipparcos Catalogue, and catalogues of higher star density reduced to the same reference frame, differential CCD reductions providing good positional accuracies are now routinely achieved. Stone (1999a) reported that more than 30 000 observations were taken with the Flagstaff Astrometric Scanning Transit Telescope (FASTT) in 1999, with accuracies reaching 60 mas in each coordinate, and an observing programme comprising Uranus, Neptune and Pluto, 18 satellites of Jupiter, Saturn, Uranus and Neptune, and over 2000 asteroids.

Table 10.1 Solar System objects observed by the satellite and contained in either of the Hipparcos or Tycho Catalogues. The total of 48 asteroids observed by Hipparcos are listed in Table 10.2. As of 2006, Ceres is classified by the IAU as a 'dwarf planet'.

	Hipparcos		Tycho	
Object	Astrometry	Photometry	Astrometry	Photometry
Classical planets:				
Uranus	–	–	√	–
Neptune	–	–	√	–
Satellites:				
J 2 Europa	√	–	–	–
J 3 Ganymede	–	–	√	–
J 4 Callisto	–	–	√	–
S 6 Titan	√	–	√	√
S 8 Iapetus	√	–	–	–
Asteroids:				
(1) Ceres	√	√	√	√
(2) Pallas	√	√	√	√
(4) Vesta	√	√	√	√
(6) Hebe	√	√	√	√
(7) Iris	√	√	√	√
...43 others	√	√	–	–

10.2 Asteroids: Masses and orbits

10.2.1 Mass determination

Large asteroids induce non-negligible and occasionally strong gravitational perturbations on the orbits of a great number of Solar System objects, both main belt asteroids and some planets. The DE403 ephemerides (Standish *et al.*, 1995), for example, takes into account the perturbation of 300 asteroids. However, masses are generally only poorly known. Over the last 40 years, several mass determinations of large asteroids have been made, starting with an attempt to determine the mass of (4) Vesta from its perturbations on the orbit of (197) Arete (Hertz, 1966). Usually repeated, moderately close encounters of objects with the largest asteroids are used in order to accumulate their perturbing effects in the orbits of the test asteroids. For this purpose, the usual time span taken into account is typically 50–100 years (e.g. Viateau, 2000). Due to their accuracy, Hipparcos data of asteroids allow the possibility of mass determinations over a much shorter time interval of 3.3 years. Originally, mass determination used the effect of one object on another, with the simultaneous correction of the six orbital elements for each. Several perturbed objects can be simultaneously used to calculate the mass of the perturbing object, reducing the correlations between parameters by simultaneously calculating the mass of the perturber and the orbits of all perturbed objects. Easier to apply, but less rigorous, is to make one mass determination for each perturbed asteroid, and calculate the weighted mean of all (Viateau & Rapaport, 1998).

Determination of potential close encounters for all 48 Hipparcos asteroids with expected effects on the asteroid trajectories larger than 5 mas were made by Bange (1998), who took into account not only the minimal distance, but also the encounter velocity and the estimated mass of the perturbing body. Perturbations from planets Mercury to Pluto were taken into account, as well as perturbations by the asteroid/dwarf planet (1) Ceres, and asteroids (2) Pallas, (4) Vesta and (10) Hygiea. Significant results were predicted in the case of the (20) Massalia/(44) Nysa encounter, a possibility identified during the Input Catalogue compilation by Scholl and Schmadel. For this pair, a succession of three close encounters occurred just before or during the Hipparcos mission, with minimal distances between 0.037–0.116 AU, and particularly low encounter velocities of around 1.65 km s^{-1}, compared to an average encounter velocity between asteroids of about 5 km s^{-1}. The diameters are 151 km for Massalia and 73.3 km for Nysa, implying a mass ratio of about 9, very favourable for this type of application. Bange (1998) derived $M = 2.42 \pm 0.41 \times 10^{-12} M_{\odot}$ for Massalia, leading to a density of 2.67 ± 1.06 g cm^{-3}. Due to another close approach with the large asteroid (4) Vesta, the mass obtained is somewhat dependent on the assumed mass of Vesta.

Ceres is the largest asteroid (now classified as a dwarf planet) with a mass about half that of the entire main asteroid belt, and a best pre-Hipparcos mass estimate of $(4.71 \pm 0.05) \times 10^{-10} M_{\odot}$. Viateau & Rapaport (1998) made a re-determination from its gravitational perturbations on the orbits of nine asteroids, of which four were observed by Hipparcos, and the five others

Table 10.2 Taxonomic classification, poles, and shapes for the Hipparcos asteroids. Classification is from Tholen (1989). Poles and shapes are from the Asteroids II database and updated version cited. Such information is relevant, for example, for interpretation of the Hipparcos epoch photometry. From ESA (1997, Volume 3, Chapter 15).

		Magnusson (1989)			Magnusson *et al.* (1994)		
IAU Number and Name	Tholen class	database	pole	shape	database	pole	shape
(1) Ceres	G	√	√	√	√	√	√
(2) Pallas	B	√	√	√	√	√	√
(3) Juno	S	√	√	√	√	√	√
(4) Vesta	V	√	√	√	√	√	√
(5) Astraea	S	√	√	√	√	√	√
(6) Hebe	S	√	√	√	√	√	√
(7) Iris	S	√	√	√	√	√	√
(8) Flora	S	√	√	–	√	√	√
(9) Metis	S	√	√	√	√	√	√
(10) Hygiea	C	√	–	–	√	√	√
(11) Parthenope	S	–	–	–	–	–	–
(12) Victoria	S	√	√	–	√	√	√
(13) Egeria	G	–	–	–	–	–	–
(14) Irene	S	–	–	–	–	–	–
(15) Eunomia	S	√	√	√	√	√	√
(16) Psyche	M	√	√	√	√	√	√
(18) Melpomene	S	–	–	–	√	√	–
(19) Fortuna	G	√	√	√	√	√	√
(20) Massalia	S	√	√	√	√	√	√
(22) Kalliope	M	√	√	√	√	√	√
(23) Thalia	S	–	–	–	√	√	√
(27) Euterpe	S	–	–	–	–	–	–
(28) Bellona	S	√	√	√	√	√	√
(29) Amphitrite	S	√	√	√	√	√	√
(30) Urania	S	–	–	–	–	–	–
(31) Euphrosyne	C	√	√	√	√	√	√
(37) Fides	S	√	√	√	√	√	√
(39) Laetitia	S	√	√	√	√	√	√
(40) Harmonia	S	–	–	–	√	√	√
(42) Isis	S	–	–	–	–	–	–
(44) Nysa	E	√	√	√	√	√	√
(51) Nemausa	CU	–	–	–	√	√	√
(63) Ausonia	S	√	√	√	√	√	√
(88) Thisbe	CF	√	√	√	√	√	√
(115) Thyra	S	–	–	–	√	√	√
(129) Antigone	M	√	√	√	√	√	√
(192) Nausikaa	S	√	√	–	√	√	√
(196) Philomela	S	–	–	–	√	√	√
(216) Kleopatra	M	√	√	√	√	√	√
(230) Athamantis	S	–	–	–	–	–	–
(324) Bamberga	CP	–	–	–	–	–	–
(349) Dembowska	R	√	√	√	√	√	√
(354) Eleonora	S	√	√	√	√	√	√
(451) Patientia	CU	√	√	√	√	√	√
(471) Papagena	S	–	–	–	–	–	–
(511) Davida	C	√	√	√	√	√	√
(532) Herculina	S	√	√	√	√	√	√
(704) Interamnia	F	√	√	–	√	√	√

by ground-based meridian circles in a reference frame reduced to that of Hipparcos. Their results gave a mass of $(4.759 \pm 0.023) \times 10^{-10}\,M_\odot$. Assuming a mean diameter from star occultations of 932.6 ± 5.2 km led to a bulk density of $2.23 \pm 0.05\,\mathrm{g\,cm^{-3}}$, some 20% higher than the mean density of C-class asteroids obtained in the determination of the DE403/LE403 ephemerides. The result confirms that the mass previously recommended by the IAU, $5.0 \times 10^{-10}\,M_\odot$, is too large by about 5%, with non-negligible consequences on the calculation of the orbits of Mars and various other asteroids. A subsequent determination from the perturbation

Osculating elements: The orbits of Solar System objects such as planets, satellites, asteroids, and comets, are continuously modified by the gravitational effects of all other sufficiently massive objects, and by other perturbative effects such as planet oblateness, non-gravitational forces, and relativistic effects. Their orbit representation is therefore by means of 'osculating elements', which themselves change with time. Six elements, and their associated epoch, are usually used to describe the orbit: the time of perihelion passage T or, alternatively, the corresponding angular measure called the 'mean anomaly' M; the semi-major axis a (in the case of elliptical comet or asteroid orbits, or the perihelion distance q in the case of both elliptic and hyperbolic orbits); the eccentricity of the orbit e; and three angles (and an associated mean equinox): the argument of perihelion ω, the longitude of the ascending node Ω, and the orbital inclination with respect to the ecliptic i.

on 25 asteroids, but not making use of the Hipparcos data, yielded $(4.70 \pm 0.04) \times 10^{-10} M_\odot$ for Ceres, as well as estimates for the highly-inclined (2) Pallas and (4) Vesta (Michalak, 2000).

These determinations refer to the estimates made from the Hipparcos observations themselves. Estimates also benefit, significantly if indirectly, from the improvement in the ground-based measurements transformed to the uniform (Hipparcos-based) ICRF: for example, the mass of (16) Psyche determined from its close approach at about 3.1 km s^{-1} with (94) Aurora in 1937, and the mass of (121) Hermione from its close approach at about 2.9 km s^{-1} with (278) Paulina in 1944 (Viateau, 2000). Other examples can be found in Hilton (1999), Michalak (2001), Vitagliano & Stoss (2006), and Aslan *et al.* (2006). Nevertheless in general, whether from the short time-span and small number of Hipparcos asteroids, or from the more extended database of lower-accuracy ground-based observations, the number of accurate asteroid mass determinations remains small; close encounters are rather infrequent, and most of these have relatively small effects on the perturbed objects.

10.2.2 Orbits and photometry

The use of CCDs and a dense optical reference frame based on the Hipparcos ICRF has led to improved prospects in astrometry and orbit determination of minor Solar System bodies (e.g. Fienga *et al.*, 1997; Carpino, 1998). Although not routinely referenced in such work, the Hipparcos and Tycho Catalogues now provide a routine but important framework for the astrometric analysis of Solar System bodies, in particular over large fields of view for the re-reduction of positions of planetary satellites, and ongoing surveys for small bodies of the Solar System, including the LINEAR, LONEOS, Catalina and Kitt Peak surveys.

Observations at a single apparition of newly-discovered asteroids, comets or Kuiper-belt objects, rarely yield an accurate ephemeris because the observation times are too short compared to the sidereal period of the celestial bodies. The main uncertainties arise in the determinations of the mean anomaly, semi-major axis, and eccentricity, where a and e are highly correlated over short arcs. Such poorly constrained orbits may result in the loss of a newly-discovered object, whereas older and somewhat less precise data can provide useful constraints on the future path. Hestroffer *et al.* (1998b) showed how to combine the accurate, short-interval observations from Hipparcos with lower-accuracy but often much longer interval observations from the ground, based on an analysis of the variance/covariance matrix in the least-squares solutions of the satellite observations. Taken with the ground-based observations used to prepare the satellite observations, they applied the method to (2) Pallas (67 Hipparcos and 985 ground-based observations, dating back to 1802) and (324) Bamberga (75 Hipparcos and 1211 ground-based observations, dating back to 1890). They showed that the Hipparcos observations substantially constrain the parameters of the osculating trajectory, with a precision of a few 10^{-6} degrees, but not to a significant improvement of the semi-major axis. The method developed could be applied to all 48 asteroids observed by Hipparcos, and to other asteroids when combining datasets at different epochs with very different accuracies. Errors in the mass of Ceres for the calculation of its perturbation, or relativistic precession of the perihelion, were not taken into account.

Systematic effects in the residuals can be partly attributed to phase effects in the photocentre due to oblique illumination by the Sun, especially since the Hipparcos observations were obtained near quadrature (i.e. at large solar phase angles), in contrast to the situation on the ground where observations are generally concentrated around opposition. While the photocentric offset can be large, and dependent on the solar phase angle, the apparent size of the object, its shape, and its surface properties, these offsets could not be determined during the construction of the catalogue; as a result the published observations refer to the photocentre, while ephemerides refer to the centre of mass.

Hestroffer (1998) derived a limb-darkening parameterization for the Hipparcos asteroids, and the resulting improvement of ephemerides for the largest objects (1) Ceres, (2) Pallas, (4) Vesta, and (7) Iris. He derived a photocentre offset over a given Hipparcos reference great circle $\Delta v \propto C(i)$, with

$$C(i) = 0.670 + 0.045i + \mathcal{O}(i^2) \quad i \leq 0.2 \qquad (10.1)$$

$$= 0.686 + 0.037i + \mathcal{O}(i^2) \quad i > 0.2 \qquad (10.2)$$

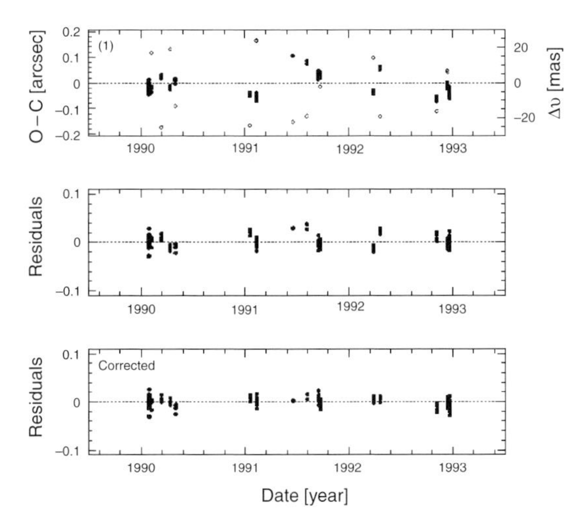

Figure 10.1 Improvement of the residuals for (1) Ceres when the correction for the photocentre offset is applied. Top: initial residuals where the ephemerides are calculated from the osculating elements for the year 1996. The calculated photocentre offsets are shown as squares, whose values are indicated on the right-hand axis. Middle: residuals after correction of the six osculating elements, but without taking the phase effect into account. Bottom: with the residuals corrected for the phase effect. From Hestroffer (1998, Figure 2).

The best result was obtained for (4) Vesta where the rms of the residuals dropped from 14.0 to 7.2 mas, close to the mean standard deviation of a single observation for this asteroid (Figure 10.1). Investigation of light scattering and photocentre offsets for the Hipparcos observations was also made by Batrakov *et al.* (1999), who compared results using various scattering laws: those of Lambert, Lommel–Seeliger, and Buratti–Veverka, as well as that proposed by Hestroffer (1998).

The Hipparcos modulated photometry data were used for more detailed modelling of the limb-darkening parameters for J2 Europa and S6 Titan (Hestroffer, 2003), information which cannot be obtained from conventional disk-integrated photometry. This yielded a brightness distribution characterised by the Minnaert parameter $k = 0.58 \pm 0.01$ and $k = 0.90 \pm 0.02$ respectively, in good agreement with independent methods including lunar occultation and spacecraft fly-bys. The form of limb darkening significantly affects the determination of fundamental parameters such as stellar radius with stellar interferometry, or the determination of the position and motion of Solar System bodies showing phase effects. Europa, for example, is known to have two significantly different hemispheres: the leading one visible at eastern elongation is bright and covered with ice, while the trailing hemisphere is dark and the ice still covered by dust, leading to a significant variation of magnitude with orbital phase.

Examples of the analysis of historical photographic plate archives of Solar System bodies have been reported: Di Sisto & Orellana (1999) described the classification of more than 2000 photographic plates of asteroids, and 156 plates of comets, taken between 1948–86 at the La Plata Observatory, a pilot group of which had already by that time been reduced with the Hipparcos Catalogue as reference. Rizvanov & Nefedjev (2005) described the database of about 3000 photographic observations of planets obtained at the high-altitude Zelenchuck station of the Engelhardt astronomical observatory in Tatarstan, and now also reduced to the Hipparcos/Tycho 2 reference frame.

10.3 Planets, satellites, occultations and appulses

The outer planets Fienga (1998) used the Hipparcos data for the determination of astrometric positions of Jupiter, based on 41 photographic plates containing Hipparcos stars and more than two Galilean satellites, taken between 1967–74. These were calibrated using the Hipparcos star positions to yield accurate planetary satellite positions transferred to the equatorial coordinates of Jupiter using the G5 (Arlot, 1982) and JPL DE403 ephemerides. Resulting positions had standard errors of 42 mas in right ascension, and 38 mas in declination, some 5–10 times better than previous techniques.

While more accurate methods, namely radar ranging and interferometry, can be applied at the smaller distances of the inner planets, optical positions for the outer planets still provide the main source of data for improving their ephemerides. Various papers have reported compilations of the positions of the outer planets and their satellites, now reduced to the Hipparcos-based ICRF, generally identifying only small systematics offsets with respect to the dynamical reference frame ephemerides predictions: 250 plates with measurements of Pluto during 1930–94 with 25–40 Hipparcos references stars on each from Pulkovo, Latvia and Arkhyz (Ryl'Kov *et al.*, 1997); positions for Uranus, Neptune and Pluto, as well as for 17 satellites of Jupiter and Neptune, obtained in 1996–2001 with the Flagstaff Astrometric Scanning Transit Telescope with accuracies of around ± 0.06 arcsec or better in each coordinate (Stone, 2001, and the series of preceding papers referenced therein); CCD observations at the 1.2-m OHP of the faint satellites of Jupiter J6–J13 (Arlot *et al.*, 1999); and 1758 observations of Uranus and its satellites Miranda, Ariel, Umbriel, Titania, and Oberon, from Brazil between 1982–98 (Veiga *et al.*, 2003).

Occultations by minor Solar System bodies Occultations by Solar System objects are important in providing fundamental knowledge about their shapes, sizes,

Orbital resonances: Orbital resonance occurs when two orbiting bodies exert a regular, periodic gravitational influence on each other, usually due to their orbital periods being related by a ratio of two small integers (a mean-motion resonance). Orbital resonances greatly enhance the mutual gravitational influence of the bodies. In most cases, this results in an unstable interaction, in which the bodies exchange momentum and shift orbits until the resonance no longer exists. Examples include: (a) the series of almost empty lanes in the asteroid belt, the Kirkwood gaps, corresponding to mean-motion resonances with Jupiter, in which almost all asteroids have been ejected by repeated perturbations; (b) resonances between Saturn and its inner moons, giving rise to gaps in the rings of Saturn; (c) the 1:1 resonances between bodies with similar orbital radii causing large Solar System bodies to 'clear out' the region around their orbits by ejecting nearly everything else around them, an effect incorporated into the IAU 2006 definition of a planet. Under some circumstances, a resonant system can be stable and self correcting, so that the bodies remain in resonance. An orbital resonance may involve any combination of the orbit parameters, such as eccentricity versus semi-major axis, or eccentricity versus orbit inclination.

A mean motion resonance occurs when two bodies have revolution periods in a simple integer ratio, which may either stabilize or destabilize the orbit. Stabilization occurs when the two bodies move such that they never closely approach. Examples in the Solar System include: (a) The stable orbit of Pluto and the Plutinos, despite crossing the orbit of the much larger Neptune, because of a 3:2 resonance. Other, more numerous, Neptune-crossing bodies that were not in resonance were ejected from that region by strong perturbations due to Neptune. (b) The Trojan asteroids, protected by a 1:1 resonance with Jupiter. A Laplace resonance occurs when three or more orbiting bodies have a simple integer ratio between their orbital periods. For example, Jupiter's inner moons Ganymede, Europa, and Io are in a 1:2:4 orbital resonance, i.e. Ganymede completes 1 orbit in the time that Europa makes 2, and Io makes 4.

A secular resonance occurs when the precession of two orbits is synchronised, usually a precession of the perihelion or ascending node. A small body in secular resonance with a much larger one (e.g. a planet) will precess at the same rate as the large body. Over long times, of order 1 Myr, a secular resonance will change the eccentricity and inclination of the small body. An example is the secular resonance between asteroids and Saturn. Asteroids which approach it have their eccentricity slowly increased until they become Mars-crossers, at which point they are usually ejected from the asteroid belt due to a close pass to Mars. This resonance forms the inner and lateral boundaries of the main asteroid belt around 2 AU, and at inclinations of about 20°. Secular resonances can also occur for a combination of three or more objects, notably Jupiter, Saturn, and an asteroid.

A Kozai resonance occurs when the inclination and eccentricity of a perturbed orbit oscillate synchronously, i.e. increasing eccentricity while decreasing inclination and vice versa. This resonance applies only to bodies on highly inclined orbits. One of the consequences of this resonance is the lack of bodies on highly-inclined orbits, as the growing eccentricity would result in small pericentres, typically leading to a collision or destruction by tidal forces for large moons.

Detailed behaviour over long periods is very sensitive to the initial conditions and masses. Early examples of the application of accurate positions can be seen in the long-term numerical integration of the planetary trajectories to study stability and chaos in the Solar System, for example, over the 100 Myr of the LONGSTOP project (long-term gravitational stability test for the outer planets, e.g. Nobili, 1988; Roy *et al.*, 1988; Milani *et al.*, 1989), the 200 Myr of the MIT project (Applegate *et al.*, 1986), and up to 5 Gyr in more global terms (Laskar, 1996). LONGSTOP, for example, identified long-periodic variations in the semi-major axes of the outer planets with periods of the order of 1 million years (Nobili, 1988). In the case of Pluto, Milani *et al.* (1989) used a 100 Myr integration to confirm the 3:2 mean-motion resonance with Neptune, the 19 900-year longitude libration (defined as a very slow oscillation, real or apparent, of a satellite as viewed from the larger celestial body around which it revolves), and the 3.78 Myr libration of the argument of the pericentre of Pluto. In addition, they reported a third resonance with a libration period of 34.5 Myr.

albedos and atmospheres if present. More than 300 occultations of stars by asteroids have been observed since the first one reported in 1958 involving (3) Juno. Until the Hipparcos results became available, the comparatively low accuracy of astrometric measurements made it difficult to predict occultations with a high level of confidence and, as a result, there had been limited success in observing these events (Stone, 1999b). The combination of accurate ephemerides and Hipparcos/Tycho star positions now allows the prediction of close conjunctions or occultations between Hipparcos/Tycho stars defined by the ICRF and planets or satellites defined by appropriate Solar System ephemerides. Differential reductions using reference stars taken from those catalogues and CCD observations now enable the positions of planetary bodies and the background occultation stars to be determined with accuracies of around ±60 mas, resulting in occultation predictions accurate to ±50 mas or better. As a consequence, occultations can be predicted now with much improved accuracy, and a large number have been observed since the Hipparcos Catalogue was released; typically some 30 events are now observed each year, triple the previous number, with ellipse parameters now determined for many (Dunham *et al.*, 2002). The International Occultation Timing Association (IOTA), as a collaboration between professional and amateur observers, has also provided more accurate timings based on the Hipparcos data, and led to the first confirmed observations of lunar meteor impacts in November 1999 (Dunham & Timerson, 2000). Predictions of stellar occultations by satellites of asteroids, specifically for (22) Kalliope, (121) Hermione, (45) Eugenia and (90) Antiope, to determine the size of the binary companion and mass and density of the primary, were reported by Berthier *et al.* (2004). Occultations of Tycho Catalogue stars have been reported for TYC 1310–2435–1 by Tethys in December 2002, for TYC 1396–00214 by Jupiter in April 2003, and TYC 5806–696–1 by Titania in August 2003.

Soma *et al.* (2006) reported the first asteroid satellite occultation for a satellite discovered previously by other means: that of the 9.1 mag star TYC 1886–01206–1 (SAO 78190) by asteroid (22) Kalliope and its satellite Linus in November 2006. Both were observed by eight stations within the Japanese Occultation Information Network, and gave a projected size of $(209 \pm 40) \times (136 \pm 26)$ km^2 for Kalliope and 33 ± 3 km for Linus.

Occultations of objects such as Centaurs and Kuiper Belt Objects, can also now be planned with improved chances for success. With more chords now observable with each occultation event, more information is becoming available about the shapes of planetary objects.

Over 800 trans-Neptunian objects have been discovered since 1992. No particularly reliable angular size estimates have been made, although indirect size information has been derived for about 15 of them from Spitzer radiometric data (Brucker *et al.*, 2007; Stansberry *et al.*, 2008). The largest, Quaoar, is estimated at about 0.04 ± 0.01 arcsec, at the resolution limit of the Hubble Space Telescope. The only direct method of estimating sizes is through observation of stellar occultations. At 30–50 AU, corresponding to a parallax of 0.2–0.3 arcsec, an occultation will only happen somewhere on Earth if the geocentric path of the object passes within 0.2–0.3 arcsec from the star. Their very small angular motion, 0.1–1 mas s^{-1}, makes the rate of stellar occultations by them about three orders of magnitude smaller than by main belt asteroids, and about 100 times less than the Jovian Trojans. Denissenko (2004) estimated that any given trans-Neptunian object will occult a 15 mag or brighter star every 4–5 years. For 10^5 Hipparcos stars of 10–11 mag this results in one expected occultation in 10 years for the 20 largest objects and only one event per century for 10^4 stars brighter than 6.5 mag. They computed expected occultations by Hipparcos and associated catalogue stars brighter than 15 mag by the 17 largest trans-Neptunian and four known Kuiper Belt objects between 2004–14, including the rare occultation of a 6.5 mag star (HIP 111398) by the double asteroid (66652) 1999 RZ253 on 2007 October 4. Searches for small Kuiper Belt objects are being made along similar lines using fast photometric observations of target stars at the Pic du Midi and Haute Provence Observatories (Roques *et al.*, 2001), from which a 3σ detection due to a 120-m size Kuiper Belt Object is predicted roughly once a day.

Unvisited by spacecraft, Pluto's tenuous nitrogen atmosphere was first detected by an occultation in 1985 and studied more extensively by a second event in 1988. Combined with ground-based spectroscopy, atmospheric modelling led to the conclusions that Pluto's atmosphere is composed primarily of N_2 in vapour-pressure equilibrium with surface N_2 frost. Above the surface, the atmospheric temperature is maintained at about 100 K by a small amount of CH_4 absorbing solar radiation in the 2.3 and 3.3 μm bands, and emitting in the 7.8 μm band. The first occultations since 1988 were observed in July and August 2002 by Sicardy *et al.* (2003), revealing a twofold increase in Pluto's atmospheric pressure over 14 years. This observation is consistent with nitrogen cycle models despite Pluto's increasing heliocentric distance: sublimation of the south polar N_2 frost cap increases as it goes into sunlight following equinox on Pluto in 1987. Pluto is crossing the plane of the Milky Way over the last few years, and other occultations (not involving Hipparcos or Tycho stars) have been reported in 2006 and 2007, and also an important occultation of Charon by a 14 mag star on 11 July 2005 (Gulbis *et al.*, 2006; Sicardy *et al.*, 2006; Person *et al.*, 2006).

A few examples of various other related results are given. The north polar region of Jupiter occulted the bright star HIP 9369 in October 1999 (Raynaud *et al.*, 2001). Light curves with strong peaks in ingress and egress were interpreted as temperature profile fluctuations possibly caused by gravity wave propagation in the planet atmosphere; from time lags between events measured from different observatories, the direction of propagation of several gravity wave packets propagating in the atmosphere were established. Uranus and HIP 106829 had a close separation of 5.90 arcsec in September 2001, and 2.08 arcsec between the star and the brightest Uranian satellites (Peng, 2001). Titan occulted two bright Tycho stars in November 2003, with data gathered by both professional and amateur astronomers from visible to the near infrared (Sicardy *et al.*, 2004). Inversion of the light-curves provides sub-km resolution of the density and temperature profiles of Titan's upper stratosphere at 220–550 km; a well-confined inversion layer, with a temperature increase of more than 15 K in 6 km, lies at an altitude of about 510 km, attributed to a localised heating source or to an unidentified dynamical processes. Equatorial positions of ~ 0.05 arcsec in each coordinate were obtained for the Galilean satellite Callisto with respect to HIP 104297 in November 1997, thus demonstrating good agreement between the dynamical reference system from DE405 and the ICRF defined by Hipparcos (Peng & Li, 2002).

Appulses Much more common than occultations are appulses, the close approach of one celestial object to another as seen from a third body, specifically of interest here being those in which objects such as asteroids, Kuiper Belt Objects, and other Solar System objects pass

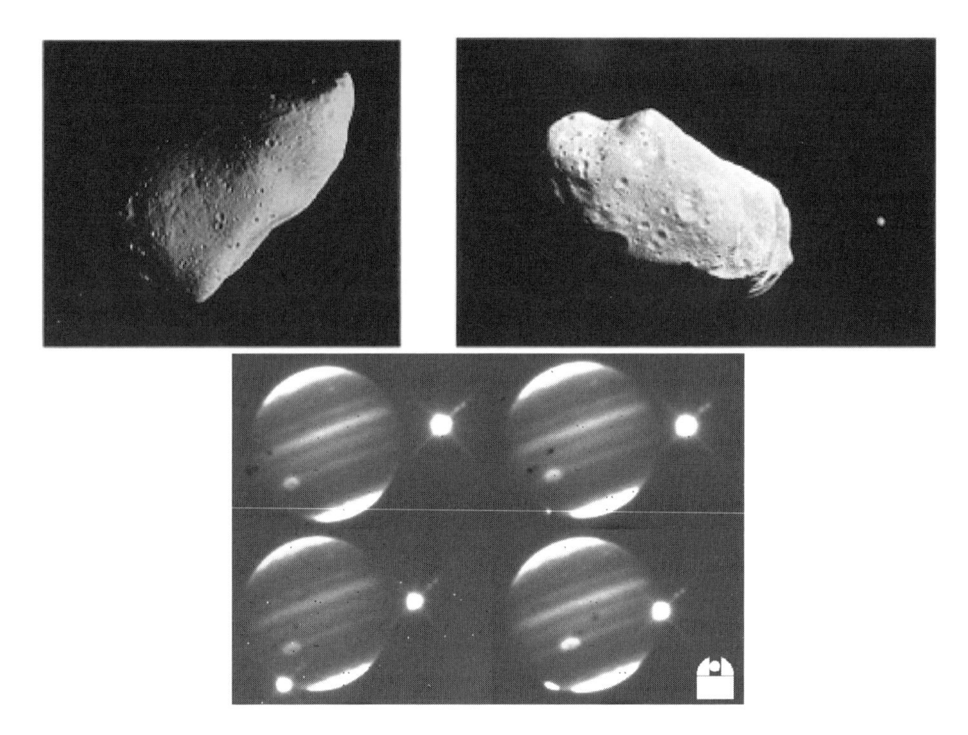

Figure 10.2 Top left: Gaspra from 1600 km; Top right: Ida from 3400 km; Ida's moon, Dactyl, is to the right. Courtesy: NASA. Above: the first impact fragment of Comet Shoemaker–Levy 9 colliding with Jupiter. Top left: just before impact (bright object at right is Io; fainter oval structure at bottom is the Great Red Spot). Top right and bottom left: the impact appears above the lower left limb. Bottom right: approximately 20 min after impact. Images taken at Calar Alto, with the near-IR camera of MPIA, Heidelberg. Courtesy: Calar Alto Observatory.

close to bright stars. These appulse events have occasionally been used for high-angular resolution observations of faint objects using adaptive optics systems to yield diffraction-limited resolution. Use of the Tycho 2 Catalogue for groups of natural guide stars for Multi-Conjugate Adaptive Optics (MCAO) on 8–10 m telescopes is discussed by Bello *et al.* (2002) and Marchetti *et al.* (2002). Berthier & Marchis (2001) calculated events out to 65 arcsec angular separation, roughly corresponding to the field of view of the majority of adaptive optics systems, and using the Tycho 2 Catalogue which, limited to stars brighter than 13–14 mag, also corresponds to the limit of the wavefront sensor for systems such as those at ESO and Keck. While such a highly-accurate reference frame for this application might seem unnecessary, observations are in practice limited by the precision of the ephemeris, which influences the number of points used to fit the orbital parameters and the duration of the observations. High-resolution images of trans-Neptunian objects provide better sampled light-curves in the infrared *JHK* bands, giving direct information on shape, rotational state and albedo distribution. On the basis of 480 objects with semi-major axis greater than 20 AU, they predict around 100 events observable every month.

Appulses were calculated over the interval 2000–09 for the Pluto–Charon system by McDonald & Elliot (2000a), and for Neptune's moon Triton by McDonald & Elliot (2000b), based on CCD observations aligned to the USNO–A2 Catalogue, itself re-calibrated with respect to the Hipparcos reference frame. Using the DE405 ephemeris, 486 appulses within 1 arcsec were predicted for Pluto and 479 for Charon. Over this decade, the mean apparent radius of Earth from Pluto–Charon is 0.285 arcsec, while the mean apparent radii of Pluto and Charon from Earth are 0.054 and 0.027 arcsec respectively. Thus a Pluto appulse with a minimum separation less than about 0.34 arcsec or a Charon appulse with less than about 0.31 arcsec separation will be visible as an occultation somewhere on Earth.

Evidently, the use of laser guide stars instead of natural guide stars (e.g. as used on Keck, Gemini, and VLT), makes the technique less important.

Navigational aspects Two early applications of the provisional Hipparcos positions (see Figure 10.2) were their provision to the NASA/JPL team involved in the planning for the fly-by of asteroids Gaspra (in 1992) and Ida (1993) by the spacecraft Galileo on its way to Jupiter. Galileo passed twice through the asteroid belt. Accurate predictions of the asteroid positions were made with respect to the stellar reference frame, allowing correct orientation of the Galileo cameras during the fly-by. Early stellar positions were also supplied to ESO, an organisation involved in the observing campaign accompanying the impact of the break-up fragments of comet Shoemaker–Levy 9 (1993e) with the surface of Jupiter (West *et al.*, 1994).

10.4 Dynamical reference system

10.4.1 Constraining precession

Background Since the Hipparcos observations were made from space, the effects of the Earth's precession, nutation and polar motion are not relevant for the derived network of Hipparcos positions, parallaxes and proper motions. Understanding and quantifying these complex effects nevertheless remains of relevance for a number of reasons: for comparing astronomical observations made from the ground at different epochs, for understanding the detailed observational effects and their causes (specifically in the context of the information that this provides about the Earth's interior), and because the determination of precession is intimately connected with establishing an inertial frame of reference. While current modern precession–nutation models are of appropriate quality for the transformation of optical ground-based observations, and while satellite observations provide accurate measurements of Earth orientation parameters, ground-based optical observations were for a long-time used to discuss the reality or otherwise of the 'non-precessional (or fictitious) motion of the equinox'. To understand the literature related to this phenomenon, some further background is needed.

Precession has been traditionally modelled as two components: the dominant luni-solar precession due to the Sun and Moon, and amounting to about 50.40 arcsec yr^{-1}, or 1° every 71.6 years. The smaller component of planetary precession is due to the combined gravitational effects of the other planets, and amounts to about 0.12 arcsec yr^{-1}, but in the opposite direction. The sum of the two effects, about 50.2877 arcsec yr^{-1}, is called general precession.

In the formulation of Lieske *et al.* (1977), for example, the speed of precession was derived from stellar motions relative to a coordinate frame rotating due to the precession of the Earth's axis, equivalent to the determination of the absolute speed of rotation of the reference frame used for the observations. In this case the true stellar proper motions are given by

$$\mu_\alpha = \mu'_\alpha - (m + n \sin\alpha \tan\delta) \qquad (10.3)$$

$$\mu_\delta = \mu'_\delta - n \cos\alpha \qquad (10.4)$$

where μ'_α, μ'_δ are the stellar motions derived from absolute observations covering some interval of time, and referring to a rotating coordinate frame defined by the mean equator and equinox at the observation epochs, and m, n are the rates of general precession in right ascension and declination respectively

$$m = p \cos\epsilon - \lambda \qquad (10.5)$$

$$n = p \sin\epsilon \qquad (10.6)$$

Here, ϵ is the instantaneous obliquity of the ecliptic, p the luni-solar precession, and λ is the planetary precession; note that p and λ are denoted as ϕ and χ in the cited literature, and that this formulation is equivalent to that given in Equation 1.26 (with $\Delta e = 0$). Strictly, the astrometrically determined quantity is $p = p_0 - p_g$, where p_0 is the luni-solar precession, and p_g is the geodesic precession, a relativistic effect caused by the slow rotation of the coordinate frame in the neighbourhood of the Earth with respect to the barycentric inertial frame. A model for this was originally given by de Sitter & Brouwer (1938), and the value of $p_g \equiv \phi_g = 1.92$ arcsec per Julian century was adopted by Lieske *et al.* (1977).

'Non-precessional motion' (or fictitious motion) of the equinox was a term given to any component of the long-term precessional motion not accounted for by these specific dynamical effects. The associated correction to λ is then described by $\Delta\lambda + \Delta e$, where Δe characterises this non-precessional term (cf. Equation 1.26). Evidently, if all dynamical terms are correctly accounted for, if a stellar reference frame is truly inertial, and if a non-rotating cosmological frame along the lines of Mach's principle exists, then the dynamical and stellar reference frames should coincide, and non-precessional motion should be zero. The existence or otherwise of such a term has been long debated.

Determination of luni-solar precession can in principle be made in a variety of ways (Walter & Sovers, 2000). Estimates from theoretical models of lunar and solar torques are in practice limited by knowledge of the Earth's mechanical ellipticity, and no longer compete with empirical estimates; in practice, the value of precession is used to estimate the dynamic ellipticity of the Earth with, for example, $e_{\rm dyn} = 0.003\,284\,547\,9(12)$ given by Mathews *et al.* (2002). Empirical methods are based on estimates from (a) stellar proper motions, an approach with numerous difficulties, as summarised by Walter & Sovers (2000, Section 1.6.3); Fricke (1967), for example, determined a correction to Newcomb's value of luni-solar precession of $\Delta p = +1.20 \pm 0.11$ arcsec per Julian century; (b) dynamical determinations: Laubscher (1976a,b) used observations of the Sun, Mercury, Venus, and Mars to determine a similar (but less precise) correction of +1.12 arcsec per Julian century; and (c) estimates from the positions of extragalactic objects:

Precession, nutation and polar motion: The Earth's rotation axis is not perpendicular to the ecliptic plane, but inclined to it by about 23°.5. Furthermore, it does not remain fixed in space, but moves as a result of various external and internal forces. The overall spin axis motion, and the accompanying motion of the Earth's equator, is conventionally decomposed into three components: precession, nutation, and polar motion. Precession and nutation are caused by the gravitational perturbations from other bodies in the Solar System on the Earth's equatorial bulge. The combined effect causes the spin axis to precess, tracing a circle of radius equal to the Earth's inclination axis, with a period of about 25 700 years. Precession also makes the tropical year, measured with respect to the equinoxes, about 20 minutes shorter than the sidereal year, measured with respect to the stars. The effect was discovered by Hipparchus in around 130 BC, after adding his own observations to those of Babylonian astronomers in the preceding centuries (for further historical details see, e.g. Kokott, 1998; Débarbat, 2001; Savoie, 2001). The combination of effects means that the obliquity of the ecliptic, the angle between the planes of the Earth's equator and the ecliptic, oscillates between about 22°.0–24°.6, with a period of about 41 000 years. The eccentricity of the Earth's orbit also fluctuates with a period of about 100 000 years, and is close to the minimum of its cycle at present.

Nutation is the collective term given to a set of smaller periodic oscillations of the rotation axis also caused by gravitational effects of the Sun and Moon. Lunar nutation is the principal term, caused by the 5° inclination of the Moon's orbit to the ecliptic. It has a period 18.6 years and an amplitude of around 17 arcsec in longitude, and was discovered in 1728 by James Bradley. Polar motion is the term used to describe the small and irregular movements of the Earth's geographic poles relative to the crust, originating from the misalignment between the rotation axis and the symmetry axis. The dominant term is a 435-day period in the rotation axis (the Chandler period), thought to be due to seasonal changes in the Earth's mass distribution. Polar motion has an amplitude of ~ 0.3 arcsec, and has to be taken into account when interpreting positional measurements made with transit instruments or similar from the ground.

The present distinction between the long-period terms expressed as a polynomial development in time (precession) and the shorter, periodic components of motion (nutation) follows from the form of the semi-analytical development of the expressions for the luni-solar-planetary torques acting on the Earth, and from the form of the precession–nutation quantities obtained by integrating these equations. Similarly, nutation and periodic polar motion terms are currently distinguished according to whether the periods are longer or shorter than 2 days.

Early estimates of the general precession in longitude were given by Bessel (1830), Peters (1842) and Struve (1843). The value calculated by Newcomb (1898), of $p = 5025.64$ arcsec per tropical century, remained that generally accepted for many years. Subsequent contributions were made by Morgan & Oort (1951), Lieske *et al.* (1977), Bretagnon & Chapront (1981), Laskar (1986), Bretagnon & Francou (1988), Simon *et al.* (1994), Williams (1994), Bretagnon *et al.* (1997), Bretagnon *et al.* (1998), Mathews *et al.* (2002), Bretagnon *et al.* (2003), Fukushima (2003), Capitaine *et al.* (2003) and others. The development of Lieske *et al.* (1977), giving the formula for precession in arcsec per Julian century, was adopted as the IAU 1976 model. The nutation model of Wahr (1981) was adopted as the IAU 1980 model.

combining data from VLBI and lunar laser ranging led Charlot *et al.* (1995) to their correction, to the IAU 1976 value, of $\Delta p = -0.30 \pm 0.02$ arcsec per Julian century.

Newcomb concluded from observations of the eighteenth and nineteenth centuries that any precessional motion of the equinox not accounted for by existing dynamical models was negligible. However, determination of the motion of the equinox from stellar proper motions have tended to give slightly different results from dynamical determinations: the study by Oort (1943) based on the FK3 motions yielded a correction $\Delta\lambda + \Delta e = 1.19 \pm 0.10$ arcsec per century, while that by Fricke (1967) based on the FK4 motions gave $\Delta\lambda + \Delta e = 1.20 \pm 0.11$ arcsec per century. The discrepancy was tackled by Fricke (1982), who determined the equinox motion dynamically, after rejecting all nineteenth century determinations. With the transition to the IAU 1976 system, the FK4 proper motions have been corrected by 1.27 arcsec per century, and it is now known from comparison of the FK5 with Hipparcos that any residual rotation of the ground-based motions is within 0.1 arcsec per century in absolute value.

Hipparcos contributions The first idea of using asteroids to define the dynamical frame of reference has been attributed to Dyson (1928). Bougeard *et al.* (1997) investigated methods to quantify any residual rotation between ICRS–Hipparcos, and the dynamical reference frame as represented by the nearly 3000 Hipparcos measurements of 48 asteroids. Unambiguous results were not obtained due to outlier behaviour.

Batrakov *et al.* (1999) also aimed to determine any rotational correction to the Hipparcos Catalogue from observations of 12 asteroids obtained at the Mykolaiv (formerly Nikolayev) Observatory, Ukraine, during 1961–95, both separately (which yielded poor results) and in combination with the Hipparcos observations of all 48 asteroids. Photocentre offsets were taken into account. For their preferred solutions they found insignificant spin terms, e.g. using the FAST data, $\omega_{x,y,z} = +0.3 \pm 0.3, -0.6 \pm 0.3, -0.8 \pm 0.6$ mas yr^{-1} (and similarly for NDAC), while for the orientation, at epoch 1988 October 1 (that of DE200/LE200), they determined $\epsilon_{x,y,z} = +3.0 \pm 1.9, -9.4 \pm 2.3, -1.4 \pm 3.6$ mas (and similarly for NDAC), compared with similar results from Folkner *et al.* (1994) of $\epsilon_{x,y,z} = -2 \pm 2, -12 \pm 3, -6 \pm 3$ mas.

Yagudina (2001) collected 13 000 radar and optical observations of 24 near-Earth asteroids and main belt asteroids, and used them to determine precise orbits and the FK5 Catalogue orientation parameters within

IAU 2000 precession–nutation: Since 1979, the variable orientation of the Earth's spin axis in space has been observed with very high precision by VLBI, currently at accuracy levels of about 0.2 mas or better. Since 1994, the Earth's spin rate has also been measured continuously by the Global Positioning System, GPS, which measures the orientation of the Earth with respect to satellite orbits, whose tie to the ICRF is itself subject to many gravitational and non-gravitational effects.

These observations have led to a new model of precession, adopted by the IAU in 2000 (Capitaine *et al.*, 2003). The new model, P03, is based on the theory of rotation of the rigid Earth (Souchay *et al.*, 1999), and matches the IAU 2000 nutation model (Mathews *et al.*, 2002). This contains terms exerted by the Moon, Sun and planets, and a transfer function for a non-rigid Earth model, expressing frequency-dependent modifications of the amplitudes and phases of individual nutation terms due to its non-rigidity. The Earth model consists of a visco-elastic mantle, a fluid outer and solid inner core, and includes inner core anelasticity, ocean tide effects, electromagnetic couplings of the mantle and the solid inner core with the fluid outer core, annual atmospheric tide and the relativistic contributions of geodesic precession and nutation. Seven basic parameters of the model have been estimated by a fit to VLBI observations of precession–nutation during the past two decades or so, in which extragalactic sources are used as reference points, such that the Earth orientation parameters derived from VLBI are directly referred to the ICRF.

The IAU 2000 precession–nutation model very closely represents the real motion of the Earth's spin axis in space, such that the celestial pole offsets measured by VLBI are typically below 1 mas. However, the nutation model does not contain free motions that cannot be reliably predicted. Specifically, it excludes various significant effects at eigenperiods of the Earth model (Capitaine *et al.*, 2005), due to the fact that the outer fluid core and inner solid core can rotate around axes that are not coincident with the spin axis of the mantle: Chandler wobble, with a terrestrial period of about 435 days, 'retrograde free core nutation', with a celestial period of about 430 days, 'prograde free core nutation', with a celestial period of about 1020 days, and 'inner core wobble', with a terrestrial period of about 2400 days. Further details of the IAU 2000 resolutions on astronomical reference systems, time scales, and Earth rotation models, are given by Kaplan (2005). Practical implementation is discussed by Wallace & Capitaine (2006).

the Hipparcos system. Again, the main motivation was to place limits on any possible non-precessional motion of the dynamical equinox over the interval 1750–2000, via the apparent change in the zero-point of right ascension from the Newcomb catalogue and through FK3, FK4, and FK5. It is still debated whether this evolution represents a true kinematic effect, or simply systematic errors in the old observations. Their results, along with most other modern determinations, give results close to zero, and suggest that the effect can be attributed to the lower quality of the old observations, although future observational monitoring is still advocated.

Although optical observations are no longer the method of choice for positional measurements and ephemerides improvement of the inner planets, an apparent disagreement between the two approaches is still discussed (see Kolesnik & Masreliez, 2004). The latter compiled some 240 000 worldwide optical observations of the Sun, Mercury, and Venus, accumulated over 250 years – from James Bradley to the present, and estimated by the authors as comprising around 90% of all angular positional measurements made during this interval. Evidently, the data vary considerably in accuracy: the typical internal precision of eighteenth century observations of daytime objects (Bradley, Maskelyne, Hornsby) was estimated at 2 arcsec in right ascension and 1.5 arcsec in declination. In the early nineteenth century, Bessel observed with 1.3 arcsec accuracy in both right ascension and declination, while the best instruments of the second half of the century (Pulkovo, USNO) reached some 0.5–0.8 arcsec, with 0.5 arcsec remaining typical for twentieth century observations. Even such (relatively) low-precision measurements can give estimates of the secular trends comparable in accuracy to those provided by modern methods of observations made over a limited time span. Kolesnik & Masreliez (2004) investigated the secular variations (i.e. slowly changing components roughly proportional to time) of the longitudes of Mercury, Venus, and Earth by reducing all observations to the Hipparcos-based reference frame using a uniform reduction scheme for all objects, and compared with the ICRS-based numerical ephemeris DE405. This involved treatment of the tidal acceleration of the Moon (first estimated by Clemence (1948) who used the results of Spencer Jones (1939) on the apparent accelerations of longitudes of the Sun, Mercury and Venus observed with respect to Universal Time), of systematic differences between the various proper motion systems, and the adjustment of astronomical constants used over the last 250 years. Their resulting equinox corrections are shown in Figure 10.3. The linear fits suggest systematic errors in right ascension in the nineteenth century, as shown by observations of the Sun and both planets, and are not explicable by a simple rotation of the Hipparcos system. While this supports the conclusions of Yagudina (2001), that non-precessional motion may be non-existent, and an artifact of the lower quality of the older observations, they argued that some aspects of the problem, including the fact that the tidal acceleration of the Moon estimated in the study is about 4 arcsec century^{-2} larger than that determined from lunar laser ranging, remain unsettled. This has led Masreliez (1999) to indicate how such differences can be explained in terms of his 'scale expanding cosmos' model.

the difference of the zenith distance, provided that the star's apparent position and geographic position of the observatory are known. These three types of measured quantities lead to three different types of observation equation, all expressed in arcsec, given in the accompanying box as an aid to understanding the modelling involved.

Amongst the corrections applied were the following known effects of geophysical origin: (a) plate tectonic motions after the geophysical model NNR–NUVEL1 (Argus & Gordon, 1991), replaced by values based on recent space geodetic data in case of observatories close to plate boundaries: plate motions used, typically some 0.01–0.05 arcsec per century, are listed for each observatory by Vondrák *et al.* (1997c); (b) oceanic tide-locking variations of local verticals, using the Schwiderski ocean-tide model for its long-term contribution and the Le Provost model for the diurnal and semi-diurnal contributions (Shum *et al.*, 1997); (c) short-period zonal tide variations for periods shorter than 35 days (Yoder *et al.*, 1981).

Figure 10.5 shows the polar motion components x, y, the celestial pole offsets $\Delta\varepsilon$, $\Delta\psi \sin\varepsilon$, and the length-of-day variations, all at five-day intervals. The polar motion (Figure 10.5, top) displays a clear secular trend in both coordinates x, y of 0.68 mas and 3.32 mas yr^{-1} respectively, i.e. a total of 3.39 mas yr^{-1}. The beat period between the Chandler wobble ($\sim$14 months) and the annual component, of about six years is also visible. Figure 10.6 represents the polar motion in three dimensions, as presented by Vondrák & Ron (2000b). Again visible is the beat period, of about six years, between the annual and 'Chandler' term.

The celestial pole offsets (Figure 10.5, lower top) reflect the deviation of the spin axis of the Earth from the adopted precession–nutation model, in this case the IAU 1976 precession and the IAU 1980 nutation models. There is an obvious trend in $\Delta\psi \sin\varepsilon$ of -0.60 ± 0.01 mas yr^{-1} and a smaller trend in $\Delta\varepsilon$ of -0.10 ± 0.01 mas yr^{-1}. The Hipparcos Catalogue being nominally in the same reference system as the positions of the radio sources observed by VLBI, these trends should correspond to the error in the precession constant detected from VLBI observations. However, the value obtained for $\Delta\psi \sin\varepsilon$ was initially much smaller than that reported from VLBI observations by Souchay *et al.* (1995) of -1.28 ± 0.01 mas yr^{-1}, or the one adopted in the IERS Conventions (McCarthy, 1996) of -1.19 mas yr^{-1}. The inconsistency became smaller in subsequent solutions (Vondrák & Ron, 2000a). Also visible are quasi-periodic changes in both components, corresponding to errors in nutation; resulting corrections to selected nutation terms with periods ranging from 14 days to 18.6 years, are formally estimated with a precision of about 0.5 ms due to the very long period of observation (Vondrák & Ron, 1997a).

The five-day values of UT1R–TAI were converted into the length-of-day changes (Figure 10.5, middle), with short-period tidal variations removed. A slight increase in the length-of-day over the entire interval corresponds to a slow deceleration of the Earth's rotation, mostly due to tidal friction. An impressive feature of the graph is the long-term quasi-periodic 'decadal' variation ascribed to core–mantle interaction. The annual and semi-annual periodic variations due to the interaction of the rotating Earth with the atmosphere are also visible.

The same data can be used to determine the time-dependent orientation of the Hipparcos reference system with respect to the ICRS (e.g. Poma *et al.*, 1997). Making use only of the celestial pole offset yields just four of the orientation and rotation terms, i.e. those defining the orientation and rotation of the catalogue around the x and y axes (Vondrák, 1996; Vondrák & Ron, 1997b; Vondrák *et al.*, 1997c). It was one of the methods used in the link of the Hipparcos intermediate catalogue to the ICRS (Kovalevsky *et al.*, 1997), entering the solution with due weight. It is qualitatively different to the other methods used, based on the rotating Earth as an intermediate reference frame. It produced results marginally discordant with the final solution, with respect to which, at epoch J199.25, gave

$$\epsilon_{0x} = -3.5 \pm 1.9 \text{ mas} \tag{10.7}$$

$$\epsilon_{0y} = +5.8 \pm 1.9 \text{ mas} \tag{10.8}$$

$$\omega_x = -0.96 \pm 0.28 \text{ mas yr}^{-1} \tag{10.9}$$

$$\omega_y = -0.32 \pm 0.28 \text{ mas yr}^{-1} \tag{10.10}$$

The orientation and spin terms have not been re-determined using these data from the subsequent solutions of the Earth orientation parameters.

Various other studies have reported re-reduction of ground-based Earth orientation observations based on the Hipparcos reference frame. These include re-reduction of ILS latitude observations (Poma & Uras, 1996); long-term proper motions using ground-based observations of the Northern Photographic Zenith Tube (NPZT) Catalogue stars (Yoshizawa, 1997); correlation between inter-annual variations of latitude positions and geophysical phenomena on the Earth, due to variations in the vertical at the level of 0.01–0.02 arcsec (Li, 1997); re-reduction of the ILS latitude observations in both the Melchior–Dejaiffe and (Boss) General Catalogues (Poma *et al.*, 1997); re-reduction of the Belgrade zenith telescope observations obtained over the period 1949–1985 (Damljanović, 1997); analysis of PIP (Punta Indio) and Belgrade latitude data (Damljanović, 2000); reduction of data from the Ondřejov (Prague) photographic zenith tube (Ron

Equations used in the analysis of polar motion: The detailed equations used in the analysis of polar motion by Vondrák (1999) were as follows

$$v_\varphi = (\varphi - \varphi_0) - (1 - 0.0042\cos 2\varphi_0)(x\cos\lambda_0 - y\sin\lambda_0) + \Delta\varepsilon\sin\alpha + \Delta\psi\sin\varepsilon\cos\alpha$$
$$- (A + A_1 T + B\sin 2\pi t + C\cos 2\pi t + D\sin 4\pi t + E\cos 4\pi t) - \Lambda D_\varphi$$
$$v_T = 15.041\cos\varphi_0\,[(\mathrm{UT0{-}TAI}) - (\mathrm{UT1{-}TAI})]$$
$$- 1.0042\sin\varphi_0(x\sin\lambda_0 + y\cos\lambda_0) - \cos\varphi_0\tan\delta(\Delta\varepsilon\cos\alpha - \Delta\psi\sin\varepsilon\sin\alpha)$$
$$- 15\cos\varphi_0(A' + A_1' T + B'\sin 2\pi t + C'\cos 2\pi t + D'\sin 4\pi t + E'\cos 4\pi t) - 15\Lambda D_\lambda\cos\varphi_0$$
$$v_h = -\delta h + 15.041\cos\varphi_0\sin a(\mathrm{UT1{-}TAI})$$
$$+ x\left[(1 - 0.0042\cos 2\varphi_0)\cos\lambda_0\cos a + 1.0042\sin\varphi_0\sin\lambda_0\sin a\right]$$
$$- y\left[(1 - 0.0042\cos 2\varphi_0)\sin\lambda_0\cos a - 1.0042\sin\varphi_0\cos\lambda_0\sin a\right]$$
$$+ \Delta\varepsilon(\sin q\sin\delta\cos\alpha - \cos q\sin\alpha) - \Delta\psi\sin\varepsilon(\sin q\sin\delta\sin\alpha + \cos q\cos\alpha)$$
$$+ (A + A_1 T + B\sin 2\pi t + C\cos 2\pi t + D\sin 4\pi t + E\cos 4\pi t)\cos a$$
$$+ 15\cos\varphi_0(A' + A_1' T + B'\sin 2\pi t + C'\cos 2\pi t + D'\sin 4\pi t + E'\cos 4\pi t)\sin a$$
$$+ \Lambda(D_\varphi\cos a + 15 D_\lambda\cos\varphi_0\sin a)$$

where φ_0, λ_0 are the adopted (constant) mean geographic coordinates of the instrument, α, δ apparent equatorial coordinates of the observed star, a, q its azimuth and paralactic angle and D_φ, D_λ theoretical tidal variations of the vertical, calculated for the rigid Earth. T is measured in centuries from the mean epoch of observation of each instrument and t in years from the beginning of the current Besselian year. The equations are used in a least-squares estimation of: coordinates of the pole in terrestrial reference frame x, y; universal time differences UT1–TAI after 1955; celestial pole offsets $\Delta\varepsilon, \Delta\psi\sin\varepsilon$; and, for each instrument, constant, linear, annual and semi-annual deviations in latitude (A, A_1, B, C, D, E); constant, linear, annual and semi-annual deviation in universal time $(A', A_1', B', C', D', E')$; and rheological parameter $\Lambda = 1 + k - l$ governing the tidal variations of the local vertical. The matrix of normal equations is large (with about 30 000 unknowns) and very sparse, and solved by a modified Cholesky decomposition. In the process, old observations had to be reduced to the appropriate system of astronomical constants.

& Vondrák, 2003); photographic zenith tubes observations to improve Hipparcos proper motion in declination (Damljanović, 2005; Damljanović & Vondrák, 2005); observations with the zenith telescopes and photoelectric transit instruments of the Pulkovo Observatory made during the twentieth century (Gorshkov *et al.*, 2005); and further improvements in star positions from a combination of the Hipparcos and ground-based latitude data (Damljanović & Pejovic, 2006a,b; Damljanović & Vondrák, 2006; Damljanović *et al.*, 2006).

Binary stars All of the above solutions were worked out in the frame of the Hipparcos Catalogue, although some 20% of the positions and proper motions were corrected in the process due to systematic displacements arising from their binary nature. Subsequently, Vondrák (2003) and Vondrák & Ron (2003) used the same basic ideas as Wielen *et al.* (1999) to identify which of the Hipparcos stars are less suitable for positional extrapolation backwards in time over the long periods of this study, due to the perturbing effects of long-period binaries on the 'instantaneous' proper motions measured by Hipparcos (see Section 1.11). Their first Earth Orientation Catalogue (EOC–1), constructed to be used as a catalogue reference for long-term Earth rotation studies, used the local meridian circle observations combined with ARIHIP, Tycho 2, and some other catalogues. Subsequent inclusion of astrolabe data resulted in EOC–2 containing 4418 objects (Vondrák, 2004; Vondrák & Ron, 2005a). Periodic residuals for certain stars in the subsequent solution for the Earth orientation terms (Vondrák & Ron, 2005b) led to the construction of EOC–3 (Stefka & Vondrák, 2006; Vondrák & Stefka, 2007), and the identification of 585 stars with significant periodic motions (Figure 10.7). The process is essentially an iterative one: the Hipparcos proper motions, already improved and expanded by Tycho 2 and ARIHIP, have been further improved according to constraints from the Earth orientation observations, leading to EOC–3 with median standard errors of 0.70 and 0.35 mas yr^{-1} in right ascension and declination respectively.

Combined with the most recent solution of the Earth orientation parameters, the new Earth Orientation catalogue EOC–3, and new IAU 2000 model of precession–nutation (Mathews *et al.*, 2002) has led to a re-analysis of the great changes observed in the amplitude and phase of the Chandler wobble over the period 1923–40 (Vondrák & Ron, 2005). They propose an explanation for this event based on a very small circular excitation, presumably by atmospheric winds or ocean currents, at the Chandler frequency and in phase with it. A revised solution of Earth orientation parameters, using EOC–3 as star catalogue, and based on the IAU 2000 model

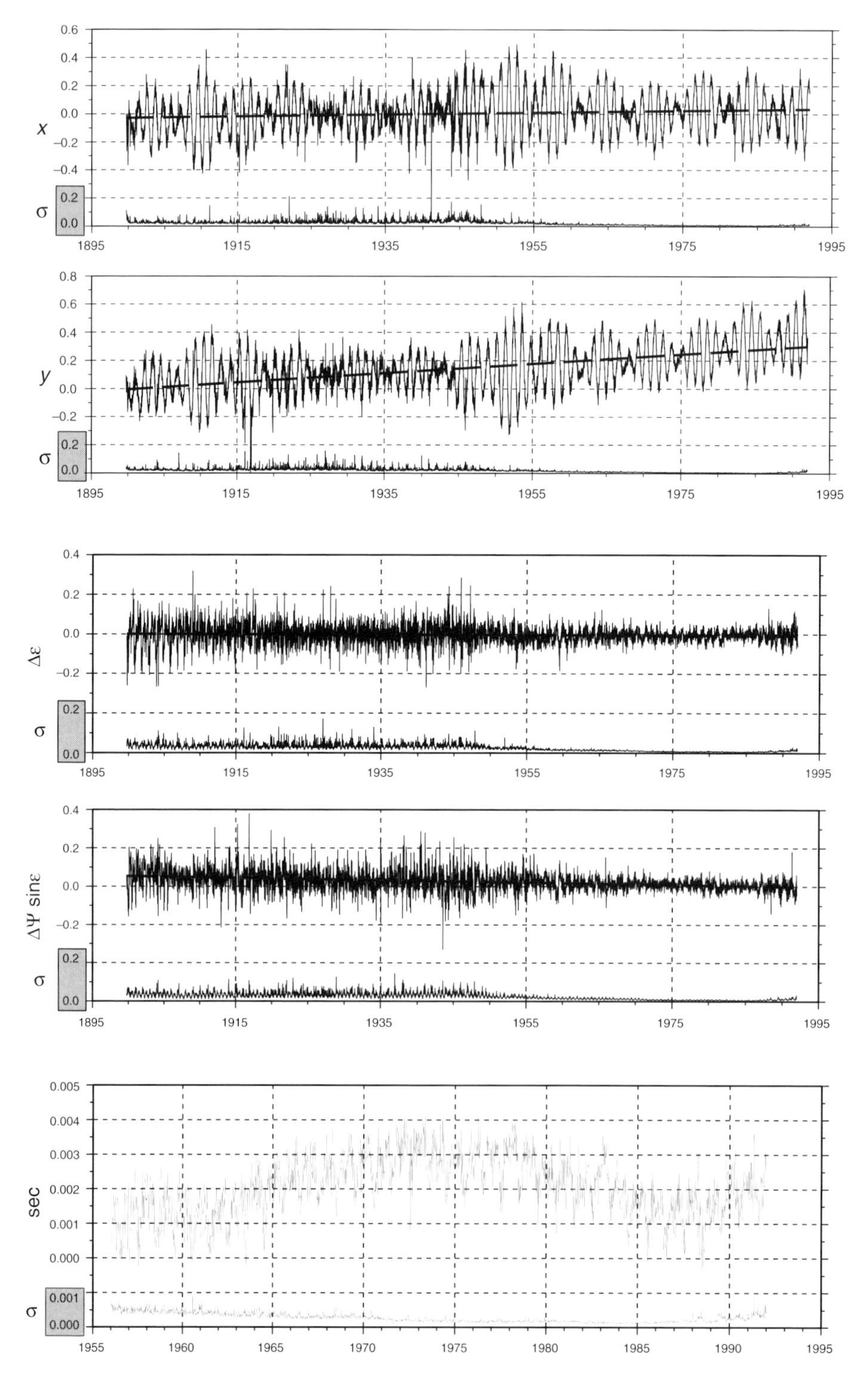

Figure 10.5 Results from a large-scale re-determination of the Earth orientation parameters based on the Hipparcos reference frame. Top pair: polar motion components x, y; middle pair: celestial pole offsets $\Delta\varepsilon$, $\Delta\psi \sin\varepsilon$; bottom: excess of the length-of-day compared with the nominal value of 86 400 s, with short-period tidal variations, $P < 35$ days, removed; this only extends back to the introduction of TAI in 1955. All results are given at five-day intervals, with standard errors σ below the main curves. From Vondrák et al. (1997a, Figures 3–5).

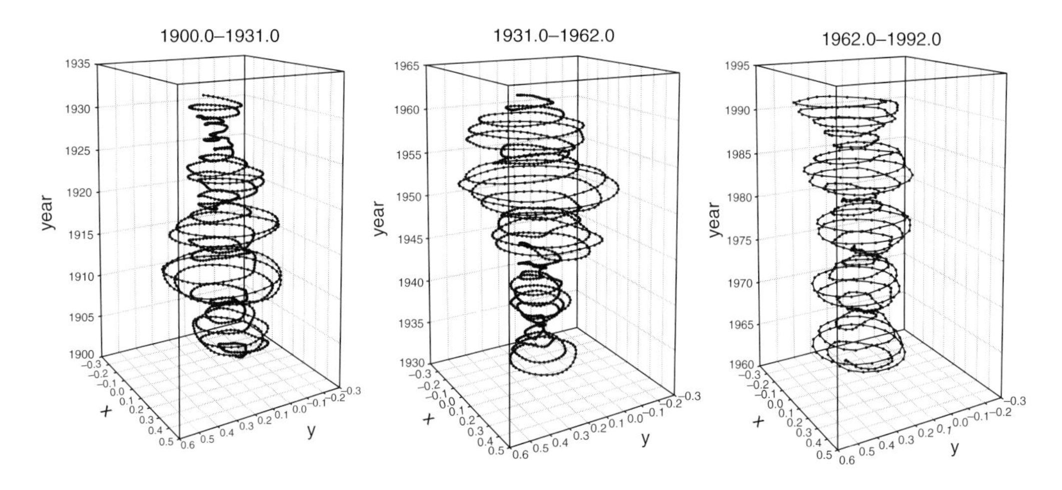

Figure 10.6 Polar motion in x, y as a function of time (running upwards). The data were filtered with a smoothing coefficient $\epsilon = 0.18 \times 10^{-6}$ day^{-6}. Visible is the beat period, of about 6 years, between the annual and 'Chandler' term. From Vondrák & Ron (2000b, Figure 7).

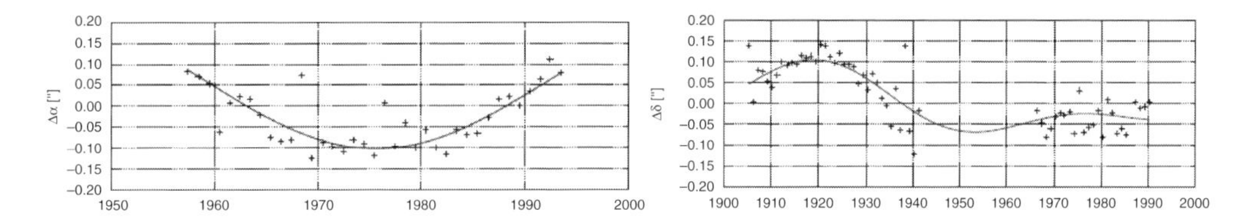

Figure 10.7 Examples of long-term binaries from the EOC–3, identified from their periodic residuals in the Earth orientation parameters. Left: HIP 86974 observed over 36 years in both coordinates. The residuals revealed a period, identified as a binary with P = 65 years in the Sixth Catalogue of Orbits of Visual Binary Stars. Right: HIP 63503 observed over a period of 85 years in declination, and identified in the Sixth Catalogue of Orbits of Visual Binary Stars as a binary with P = 106.7 years. From Vondrák & Stefka (2007, Figures 3 and 5).

of precession and nutation, was reported by Vondrák & Ron (2005b).

10.5 Passage of nearby stars

The mean-free time between collisions between the Sun and other stars in the solar neighbourhood is estimated to be of order 10^{13} years, and thus very much longer than the present age of the Galaxy. Stars in the solar neighbourhood therefore behave essentially as a collisionless gas. Nonetheless, measurable effects on Earth due to nearby star passages may occur.

An extensive study of close approaches of stars to the Solar System, and the possible consequences of stellar encounters with the Oort Cloud, was made by García Sánchez *et al.* (1997, 1999, 2001) and García Sánchez (2000). The basic idea is to combine Hipparcos proper motion and parallax data for nearby stars with ground-based radial velocity measurements to determine accurate space velocities with respect to the Sun, and hence identify stars which may have passed (or will pass) close to the Sun. Close stellar encounters could deflect large numbers of comets from the Oort Cloud into the inner Solar System, with possibly consequences for an increased impact hazard on Earth.

Some work has been done in the past to search for stellar perturbers of the cometary cloud. Mullari & Orlov (1996) used ground-based data to predict close encounters with the Sun by stars contained in a preliminary version of the Third Catalogue of Nearby Stars (CNS3 Gliese & Jahreiß, 1991). They found that in the past, three stars may have had encounters with the Sun within 2 pc, and that in the discernible future, 22 may have them. Matthews (1994) made a similar study, limited to stars within about 5 pc, and listing close approach distances for six stars within the next 50 000 yr. However, pre-Hipparcos, the accuracy of ground-based parallax and proper motion measurements imposed a limit on the accuracy of predictions of past or future close stellar passages.

In their initial study, García Sánchez *et al.* (1997, 1999) used the Hipparcos data combined with published and newly-acquired radial velocities to identify 1194

Close encounters with the Oort Cloud: The Solar System is inferred to be surrounded by the Oort Cloud, a vast body of $\sim 10^{12}$–10^{13} comets with orbits extending to interstellar distances, and with a total estimated mass of some tens of Earth masses (Oort, 1950); a review is given by Weissman (1996b). The boundary of stable cometary orbits, which is the outer dimension of the Oort Cloud, is a prolate spheroid with the long axis oriented toward the Galactic centre, and with maximum semi-major axes of about 10^5 AU for direct orbits of comets oriented along the Galactic radius vector, about 8×10^4 AU for orbits perpendicular to the radius vector, and about 1.2×10^5 AU for retrograde orbits, i.e. those opposite to the direction of Galactic rotation (Antonov & Latyshev, 1972; Smoluchowski & Torbett, 1984). These cometary orbits are perturbed by random passing stars, by giant molecular clouds, by the Galactic gravitational field, and by supernova shock waves. In particular, close or penetrating passages of stars through the Oort Cloud can deflect large numbers of comets into the inner planetary region (Hills, 1981; Weissman, 1996a), initiating Earth-crossing cometary showers and possible collisions with Earth, in addition to those brought by small bodies closer to the Sun. Sufficiently large impacts or multiple impacts closely spaced in time could result in biological extinction events. Some terrestrial impact craters and stratigraphic records of impact and extinction events suggest that such showers may have occurred in the past, specifically in the late Eocene (Farley *et al.*, 1998). Hut *et al.* (1987) reported four weak peaks in the age of distribution of impact craters over the past 100 Myr, as well as two compact clusters of ages of impact glass broadly coincident with crater-age peaks. Recent paleontological observations indicate a stepwise character for some well-documented mass extinctions in the past 100 Myr, which roughly coincide with three of the four peaks in crater ages and which have a duration compatible with comet shower predictions of about 2–3 Myr (Hut *et al.*, 1987; Fernandez & Ip, 1987).

Evidence of the dynamical influence of close stellar passages on the Oort Cloud might be found in the distribution of cometary aphelion directions. Although the distribution of long-period (10^6–10^7 yr) comet aphelia is largely isotropic on the sky, some non-random clusters of orbits exist, and it has been suggested that these groupings record the tracks of recent stellar passages close to the Solar System (Biermann *et al.*, 1983). Weissman (1992) showed that it would be difficult to detect a cometary shower in the orbital element distributions of the comets, except for the inverse semi-major axis ($1/a_0$) energy distribution, and that there is currently no evidence of a cometary shower in this distribution.

stars that could pass close enough to significantly perturb the Oort Cloud, i.e. within 2–3 pc. They assumed both a simple rectilinear motion model, and one based on dynamical integrations of the motion of the stars in the Galactic potential, with relatively good agreement between both. The effect of stellar interactions on the rectilinear motion is small: a star passing 1 pc from a $1\,M_\odot$ star with a relative velocity of $20\,\mathrm{km\,s^{-1}}$ results in an angular deflection of only 4.5 arcsec. Even over a path length of 100 pc, the rms deflection due to such encounters (assuming a local stellar density of $0.1\,\mathrm{pc^{-3}}$) is less than 1 arcmin. This deflection at 100 pc initial distance would change the impact parameter by less than 0.03 pc. They tabulate 147 stars with closest approach distances within 5 pc, contained within a time interval of ± 10 Myr, with a roughly similar number of approaches in the past and in the future (see Figure 10.8). One object, Gliese 710, has a predicted closest approach of less than 10^5 AU, or ~ 0.5 pc, although its possible binary nature and consequent effects on its inferred space velocity remain unclear. Several other stars come within about 1 pc during an interval of ± 8.5 Myr. In most cases, the uncertainty in closest approach distance was dominated either by uncertainties in the published radial velocity measurements, or uncertainties in the barycentric motion of binary systems.

Proxima Centauri (HIP 70890) is currently the nearest star to the Sun. Based on its proximity on the plane of the sky and similar distance, it is commonly thought to be a third component of the binary system α Cen A/B (HIP 71683 and 71681), although kinematic data do not unambiguously signify a bound orbit. Matthews (1994) used a radial velocity of $-22.37\,\mathrm{km\,s^{-1}}$ for Proxima Centauri, required to account for a bound orbit, and found a closest approach distance to the Sun of 0.941 pc, 26.7×10^3 yr from now. For the α Cen A/B system, he found a closest approach distance of 0.957 pc in about 28.0×10^3 yr. García Sánchez *et al.* (1999) found results consistent with these predictions. Barnard's star (HIP 87937) will have its closest approach to the Sun 9.7×10^3 yr from now at a distance of 1.143 pc. Algol had a close encounter of about 2.4 pc some 6.9 Myr ago, in agreement with the estimate made by Lestrade *et al.* (1999) using VLBI astrometry.

To assess the dynamical effect on the Oort Cloud, García Sánchez *et al.* (1999) used the dynamical model of Weissman (1996a), which uses the impulse approximation to estimate the velocity perturbations on the Sun and on hypothetical comets, and thus changes in cometary orbits in a modelled Oort Cloud. Based on simulations containing 10^8 hypothetical comets, they found that the maximum effect occurs for the encounters with Gliese 710, which results in a minor shower with $\sim 4 \times 10^{-7}$ of the Oort Cloud population being thrown into Earth-crossing orbits. Assuming an estimated Oort Cloud population of 6×10^{12} comets, this predicts a total excess flux of about 2.4×10^6 Earth-crossing comets in each shower. Because the arrival times of the comets are spread over about 2×10^6 yr, the net increase in the Earth-crossing cometary flux is only about one new comet per year, compared with the estimated steady-state flux of perhaps two dynamically new long-period comets per year entering the planetary system directly from the Oort Cloud. Thus, the net increase

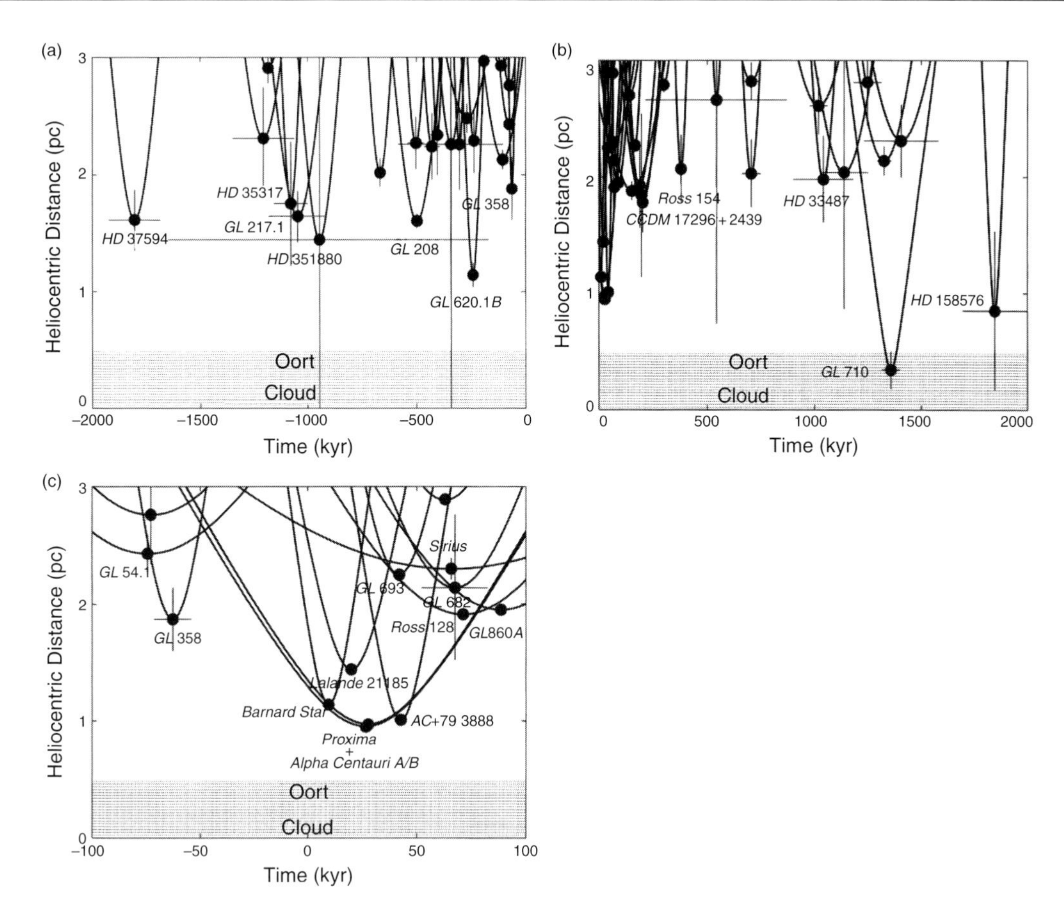

Figure 10.8 Closest predicted stellar passages (a) within the past 2 Myr; (b) up to 2 Myr in the future: Gliese 710 has the most plausible passage through the Oort Cloud in the sample; (c) for $\pm 10^5$ yr. Error bars in time and 'miss distance' are plotted at the closest approach. Several close passages are predicted over the next tens of thousands of years. From García Sánchez et al. (1999, Figures 4–6).

in the cometary flux is about 50%. Since long-period comets probably account for only ~10% of the steady-state flux at Earth, the net increase in the cratering rate is about 5%, unlikely to be detectable given the stochastic nature of comet and asteroid impacts.

In a subsequent paper, García Sánchez *et al.* (2001) included more detailed models of the Galactic potential (local, global, and perturbative potentials, the latter including effects of the spiral arms, with all three giving almost identical results in the case of Gliese 710, for example), confirming the validity of the method within time periods of ±10 Myr, and yielding a frequency of stellar encounters within 1 pc of the Sun of $2.3 \pm 0.2\,\mathrm{Myr}^{-1}$. They also demonstrated that the Hipparcos data is observationally incomplete, a fact implicit in the catalogue completeness limit as a function of magnitude. Correcting for the fact that only about one-fifth of the stars or stellar systems out to 50 pc are present in the catalogue (Figure 10.9), they obtained a corrected frequency of $11.7 \pm 1.3\,\mathrm{Myr}^{-1}$ for stellar encounters within 1 pc. Gaia will significantly improve this type of study, by extending the completeness census, by improving the determination of impact parameters, and by extending the temporal baseline, for example, extending to the possible cometary shower during the late Eocene, some 36 Myr ago. For a close-up view of exoplanetary systems, HD 217107 is a G7V star with a planetary companion of mass $M \sin i = 1.27\,M_\mathrm{J}$, presently located at 19.7 pc, and which will pass at a heliocentric distance of 2.3 pc in 1.4 Myr.

Frogel & Gould (1998) argued that a star passing within 10^4 AU would trigger a comet shower that would reach the inner Solar System some 0.18 Myr later. With an *a priori* probability of 0.4% that a star has passed this close to the Sun but that the comet shower has not yet reached the Earth, they searched the Hipparcos Catalogue for such a close-encounter candidate and, in agreement with García Sánchez *et al.* (1999), found

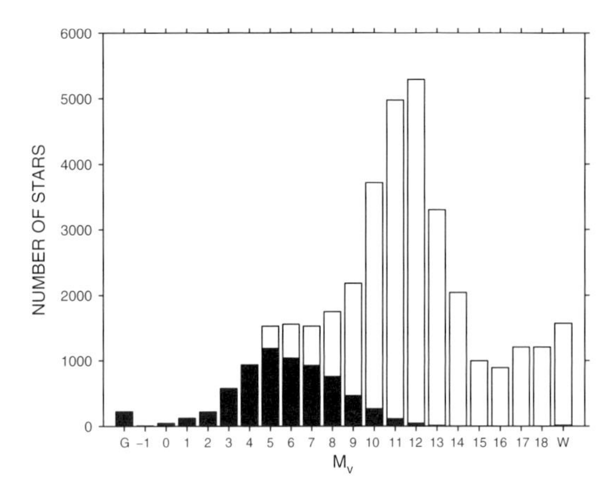

Figure 10.9 Related to the study of the passage of nearby stars, the figure shows the total number of star systems within a heliocentric distance of 50 pc, in 1 mag bins, derived from knowledge of the overall stellar luminosity function (upper histogram). The black part of each bar is the subset of star systems contained in the Hipparcos Catalogue, and hence available for this study. The bar to the left (label G) represents giant stars, and the bar to the right (label W) represents white dwarfs. From García Sánchez et al. (2001, Figure 13).

none. They also confirmed Gliese 710 as the candidate with the closest approach in the future. More quantitatively, they assess the sensitivity to high-damage encounters. The outer planets progressively deplete cometary phase space of all orbits with perihelia within the orbital radius of Neptune, $a_{\rm Nep} = 30\,{\rm AU}$. The most severe encounter possible is one where this loss cone is suddenly filled, increasing the new comet rate by a factor $\sim$40. They show this is possible only for slowly-moving stars of order the mass of the Sun, and that it is generally the encounters with very little effect that would have escaped notice. An investigation with similar objectives was made by Serafin & Grothues (2002).

10.6 Earth's climate

10.6.1 Maunder minimum

Interestingly, the relationship between solar activity, sunspot cycle, and Earth's climate (see box on page 587) can be probed, albeit indirectly, using the Hipparcos data. Maunder-type minima have been examined both as a probe of the nature of the solar dynamo, and of underlying assumptions in climate models (Soon & Yaskell, 2004; Langematz *et al.*, 2005). Whether the present cycle of solar activity is unusual or transitory, or whether Maunder-type minima are common phenomena in other stars like the Sun, can be investigated through surveys of chromospheric emission. Baliunas & Jastrow (1990) noted that 30% of Sun-like stars appeared to be in these low-activity states, a figure subsequently revised to 10–15% by Saar & Baliunas (1992). Henry *et al.* (1996) observed more than 800 stars within 50 pc for chromospheric emission in Ca H and K, and also concluded that some 10% of these stars were very inactive; they inferred that if the observations are considered to be a sequence of snapshots of the Sun during its life, then the Sun will spend about 10% of the remainder of its main-sequence life in Maunder minimum-type phases.

Wright (2004, 2006) used Hipparcos parallaxes for stars within 60 pc to improve knowledge of their location with respect to the main sequence which, from stellar evolutionary models, provides an estimate of their age (Chapter 7). The accurate Hipparcos parallaxes, and hence luminosities, are important because the luminosity evolution is relatively small over the main-sequence lifetime. He showed that nearly all stars previously classified as Maunder-minimum candidates, i.e. very inactive stars with $\log R'_{\rm HK} < -5.1$ where $R'_{\rm HK}$ is a measure of the fraction of a star's total luminosity emitted in the Ca II H and K line cores excluding the photospheric component, are evolved stars, previously mistaken for Sun-like stars in extraordinarily low states of activity. The $R'_{\rm HK}$ metric is, in other words, uncalibrated for the effects of metallicity and gravity, which in turn implies that it is not obvious how to identify a star in a Maunder-minimum state from a single observation. However, from long-term monitoring programmes (since 1966 in the case of Baliunas *et al.*, 1995) it appears that Sun-like stars exhibit a variety of activity behaviour, including solar-like cycles and flat-activity states. If the Sun's Maunder Minimum was a transition between these two states, then continued monitoring should detect some stars switching between them.

10.6.2 Sun's orbit and the spiral arms

Since its formation about 4.5 billion years ago, the Sun has made about 20 orbits around the Galaxy, and in the process has made many passages through the spiral arms. There is a growing interest in determining how these passages may have affected Earth's environment. Terrestrial fossil records show that the rapid rise in biodiversity since the Precambrian period has been punctuated by large extinctions, at intervals of 40–140 Myr, representing extremes over a background of smaller events and the natural process of species extinction. Leitch & Vasisht (1998) pointed out that the non-terrestrial phenomena proposed to explain these events, such as boloidal impacts (a candidate for the Cretaceous–Tertiary or K/T boundary extinction some 65 million years ago) and nearby supernovae, are collectively far more effective during the Solar System's

Astronomy and the Earth's climate: The Earth's climate is linked to several well-known, and some not so well-known, astronomical phenomena. At the most basic level are three dominant cycles, related to the Earth's motion, and sometimes referred to as the Milankovitch cycles: (1) the ellipticity of its orbit, (2) the inclination of its spin axis with respect to the ecliptic plane, or 'obliquity of the ecliptic', and (3) long-term precession (and shorter-term nutation) of its spin axis (see box on page 577). Together they create seasonal variations of the solar radiation reaching the Earth's surface, and explain in large part the episodic nature of the Earth's glacial and interglacial periods within the present ice age, i.e. over the last two million years. The orbital eccentricity fluctuates on a cycle of about 100 000 years, between 0–5%, a variation of prime importance to the glacial cycles. Currently about 3%, this leads to a 6% increase in received solar energy in January compared to July. The axial tilt oscillates between about 22°.0 – 24°.5 with a period of about 41 000 years. Currently about 23°.5, it largely accounts for the seasons. Smaller tilts lead to less seasonal change in solar irradiation, but greater disparity between the equatorial and polar regions probably promoting the growth of ice sheets. Spin axis precession has a period of about 25 700 years. Climatic variations are enhanced when the solstices coincide with orbital aphelion and perihelion. Shorter-term disturbance of the Earth's rotation, called polar motion, is influenced by less predictable effects as ocean currents, wind systems, and motions in the Earth's core.

A link between solar irradiance variability and climate has been suspected for more than 200 years since the work of William Herschel, and the current status is summarised by Shaviv (2003). While recent anthropogenic contributions to climate change are now uncontested, and effects ranging from El Niño to volcanic eruptions play their part, other processes correlated with solar activity seem to contribute (Friis-Christensen & Lassen, 1991), notably at levels larger than are expected from the typical 0.1% changes in solar irradiance. A particularly cold period in Europe which caused great hardship, the so-called 'Maunder Minimum' of 1645–1715, was apparently correlated with the virtual disappearance of sunspots, and coincided with the coldest excursion of the 'Little Ice Age' (Eddy, 1976; Schaefer, 1997). This and other cold episodes in Europe including the Spörer Minimum (1460–1550) correlate with peaks in the ^{14}C flux, as observed in tree rings. Contrasting warmer periods have also existed, notably the 'Medieval Maximum' around 1000 AD which was accompanied by the migration of the Vikings. An increased solar wind when the Sun is active more effectively reduces the Galactic cosmic ray flux which reaches Earth (e.g. Lockwood, 2001; Rouillard & Lockwood, 2004), and produces more ^{14}C, and *vice versa*. There are various studies which, furthermore, demonstrate a correlation between cosmic ray and climate variations (Carslaw *et al.*, 2002), possibly linked through atmospheric ionisation and Earth's cloud cover, in which the cosmic ray flux leads to ionisation of tropospheric aerosols required for the condensation of cloud droplets (Marsh & Svensmark, 2000, e.g.). Other mechanisms have been proposed to link the Galactic environment with climate variability: via encounters with an interstellar cloud through increasing solar luminosity as a result of accretion (Hoyle & Lyttleton, 1939); through shrinking of the heliosphere while crossing interstellar clouds (Begelman & Rees, 1976); and through perturbation of the Oort Cloud and injection of comets into the inner Solar System (Napier & Clube, 1979). These mechanisms all predict ice age epochs synchronised with spiral arm crossings. Still other mechanisms may be related to the revolution period around the Galaxy, which may introduce a phase lag if connected to star formation, for example for passages through the spiral arms: Williams (1975) noted that the major ice ages on Earth are at intervals of $\sim$ 150 Myr, roughly half the Galactic year, and therefore hypothesised some mechanism related to a tidally warped disk. The Hipparcos observations provide some further insight into some of these mechanisms.

journey through the spiral arms. Using the best available data on the location and kinematics of the Galactic spiral structure, including distance scale and kinematic uncertainties, they presented evidence that arm crossings provide a viable explanation for the timing of the large extinctions. Although terrestrial impact crater records also suggest variations on a time scale shorter than the inter-arm crossings, these may be related to the Sun's oscillations perpendicular to the disk (e.g. Stothers, 1998).

Shaviv (2002, 2003) argued that there is a correlation between extended cold periods on Earth and its exposure to a varying cosmic ray flux: the flux varies not only from changes in solar wind strength, but also as the Sun moves through Galactic spiral arms, where enhanced star formation and increased supernova rates create more intense exposure to cosmic rays. Such a periodic variation is confirmed by the variable cosmic ray flux recorded in iron meteorites. The driving mechanism proposed is that the cosmic ray flux experienced by Earth affects the atmospheric ionisation rate and, in turn, the formation of charged aerosols that promote cloud condensation nuclei. A close correlation has been demonstrated between the cosmic ray flux and low-altitude cloud cover over a 15 year time span (Marsh & Svensmark, 2000). The inference is that extended periods of high cosmic-ray flux may lead to increased cloud cover and surface cooling that could result in ice ages extending for millions of years. Shaviv (2002) demonstrated a correlation between ice age epochs and spiral arm crossings, agreeing in both period and phase. Shaviv (2003) also show that apparent peaks in the star formation rate history coincides with particularly icy epochs, while the long period of 1–2 Gyr before present, during which no glaciations are known to have occurred, coincides with a significant paucity in the past star formation rate (Figure 10.10). In related studies, Yeghikyan & Fahr (2004a,b) suggested that during some spiral passages, the Earth may encounter interstellar clouds of sufficient density to alter the chemistry of the upper atmosphere, decreasing the ozone concentration in the

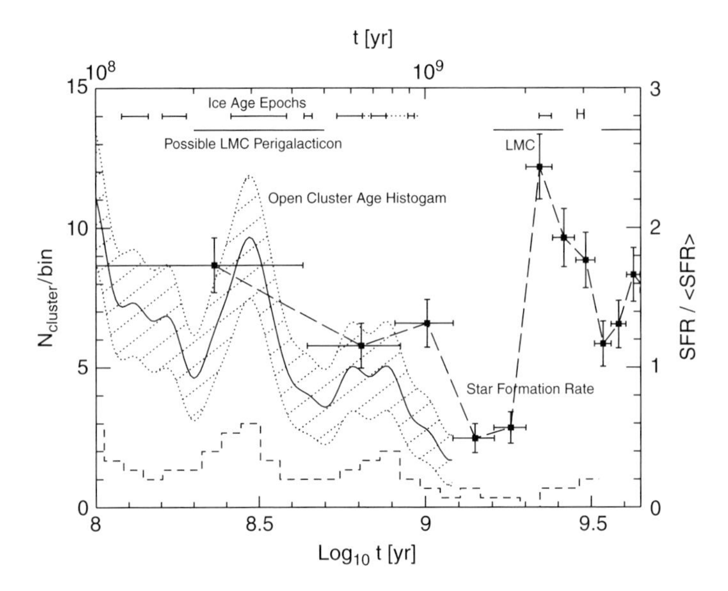

Figure 10.10 Star formation rate (SFR, right scale) based on chromospheric ages of nearby stars (Rocha Pinto et al., 2000, black squares). The hatched region is a histogram from ages of nearby open clusters (left scale). There is a minimum in the star-formation rate between 1–2 Gyr ago (9–9.3 in log t), and two peaks around 0.3 and 2.2 Gyr (8.5 and 9.35 in log t, all ages before present). At top: glaciation periods on Earth: the late Archean (3 Gyr) and mid-Proterozoic (2.2–2.4 Gyr) correlate with the previous Large Magellanic Cloud (LMC) perigalacticon passage, and a possible consequent peak in star-formation rate in the Milky Way and the LMC. Lack of glaciations at 1–2 Gyr correlates with a lower star-formation rate in the Milky Way. The long Carboniferous–Permian glaciation correlates with the peak at 300 Myr and the last LMC perigalacticon. The late Neo-Proterozoic ice ages correlate with a smaller peak around 500–900 Myr. From Shaviv (2003, Figure 2), reproduced with the permission of Elsevier.

mesosphere, and perhaps triggering an ice age of long duration.

The geological record of glacial deposits and the distribution of ice rafted debris (Frakes *et al.*, 1992; Crowell, 1999) indicate that the Earth has experienced periods of extended cold ('icehouses') and hot temperatures ('greenhouses') lasting tens of million years. The long periods of cold may be punctuated by much more rapid episodes of ice age advances and declines. The climate variations indicated by the geological evidence of glaciation are confirmed by measurements of ancient tropical sea temperatures through oxygen isotope levels in biochemical sediments (Veizer *et al.*, 2000). These studies lead to a generally coherent picture in which four periods of extended cold have occurred over the last 500 Myr. The midpoints of these ice age epochs are summarised in Table 10.3 (Shaviv, 2003).

A comparison of the geological record of temperature variations with estimates of the Sun's position relative to the spiral arms of the Galaxy is difficult for a number of reasons. First, the Solar System's location within the disk makes it hard to discern the spiral structure of the Galaxy, particularly in more distant regions. Nevertheless, there is now some evidence that a four-arm spiral pattern is successful in explaining the emissions from the star-forming complexes of the Galaxy. Second, the angular rotation speed of the Galactic spiral pattern is still poorly known, with estimates ranging between 11.5 and 30 km s^{-1} kpc^{-1} (see Section 9.7). Finally, the Sun's orbit in the Galaxy is not circular. Therefore, accounting for its variation in distance from the Galactic centre, and for its orbital speed, is needed to make a prediction of its position at epochs extending significantly into the past.

Gies & Helsel (2005) calculated the Sun's motion through the Galaxy over the last 500 million years, based on estimates of the Sun's current position and space velocity from Hipparcos, and using a realistic model for the Galactic gravitational potential (Figure 10.11). Times of the Sun's past spiral arm crossings were estimated for a range of assumed values for the spiral pattern angular speed. Their calculation of the Sun's motion in the Galaxy is consistent with the suggestion that ice age epochs occur around the times of spiral arm passages as long as the spiral pattern speed is close to Ω_s = 14–17 km s^{-1} kpc^{-1}, at least consistent with a number of the more direct estimates discussed in Section 9.7. Specifically, for a difference between the mean solar and pattern speed of $\Omega_\odot - \Omega_s = 11.9 \pm 0.7\,\mathrm{km\,s^{-1}\,kpc^{-1}}$ (Figure 10.11b) the Sun has traversed four spiral arms at times that appear to correspond well with long-duration cold periods on Earth, indicated by thick line segments in each of Figures 10.11a–c. Similar results, with a similar

Table 10.3 Mid-points of ice age epochs from the various sources compiled by Shaviv (2003, Table 2, who also discuss the errors, typically 10–15 Myr), and also summarised by Gies & Helsel (2005, Table 1). The final column gives the latter's estimates of the spiral arm crossing epochs based on their proposed smaller spiral pattern speed (Figure 10.11b). BP = before present. The correspondence between ice-age epoch, spiral arm crossing, and geological period is (Svensmark, 2007): (1) Carina–Sagittarius arm (Miocene), leading almost immediately in geological terms to the Orion spur (Pliocene to Pleistocene); (2) Scutum–Crux arm (Jurassic to early Cretaceous); (3) Norma–Cygnus arm (Carboniferous); (4) Perseus arm (Ordovician to Silurian).

Ice Age Epoch	Crowell (1999) (Myr BP)	Frakes *et al.* (1992) (Myr BP)	Veizer *et al.* (2000) (Myr BP)	Shaviv (2003) (Myr BP)	Arm Crossing (Myr BP)
1	< 22	< 28	30	20	80
2	155	144	180	160	156
3	319	293	310	310	310
4	437	440	450	446	446

relative pattern speed of $12.3 \pm 1.4\,\mathrm{km\,s^{-1}\,kpc^{-1}}$, were found in the analysis by Svensmark (2006b, 2007). Shaviv (2003) estimated that the mid-point of resulting ice ages may occur some 21–35 Myr after the spiral arm crossing, due to the difference in the stellar and spiral pattern speeds and to the time delay between stellar birth and the cosmic ray generation accompanying the ensuing supernovae events. Collectively, these results have been considered as supporting the idea that extended exposure to the higher cosmic ray flux associated with spiral arm passages can lead to increased cloud cover and long ice age epochs on Earth.

Further aspects of this topic have been considered by Svensmark (2006a, 2007). According to stellar evolution models, the Sun's luminosity around four billion years ago was less than 75% of its present value, yet there is mineral evidence for liquid water some 4.4 billion years ago, and for life itself in ancient sea sediments some 3.8 billion years ago. The long-standing discussion as to why the Earth was not frozen at that time could have an associated explanation in terms of a more vigorous solar wind, a corresponding absence of cosmic rays, and a resulting limited cloud cover compensating for the lower solar irradiance.

Additional evidence of a connection between Earth's climate and supernovae events during a closer passage of the Sco–Cen OB star association, and in particular the signature of a specific event some 2.8 Myr ago indicated by enhanced ^{60}Fe concentration in the deep-sea ferromanganese crust, is discussed in Section 8.5.1.

While intriguing, the influence of cosmic rays on long-term glaciation has been challenged, specifically being considered as secondary compared with changes in atmospheric CO_2 and oceanic pH (Royer *et al.*, 2004; Rahmstorf *et al.*, 2004). On time scales of centuries, solar effects on the Earth's climate are generally attributed to the changing magnetic field that emerges from the Sun, but the breakdown according to direct cosmic ray effects, total solar irradiance, and solar ultraviolet irradiance is presently unclear (Lockwood, 2006; Lockwood & Fröhlich, 2007).

10.6.3 Sun's orbit and Galactic plane passages

Fuhrmann (2004) argued that the Sun is a very typical thin-disk star. But the high multiplicity fraction amongst nearby stars at least raises a possibility that the Sun was born in a binary system within its parent association. That its equatorial plane is inclined at about 8° with respect to the plane of the planets may support this idea (Heller, 1993), or may instead indicate the consequences of a nearby stellar passage early in the formation of the Solar System. There exists a faint possibility that a distant brown dwarf companion remains gravitationally bound but still undetected, or that a formerly wide companion was lost from the Sun during one of its 150 or so Galactic plane crossings.

The small peculiar velocity of the Sun with respect to the Local Standard of Rest (Section 9.2.4) is noteworthy. Its low relative U and V velocities compared to the local surroundings are more characteristic of a young disk object, decreasing its collision probability while crossing the Galactic plane. Its typical W velocity, in contrast, minimises its mid-plane crossing time and hence minimises encounters, enhancing the environmental stability for the orbiting planets with infrequent collisions induced from stellar encounters. Estimates of the most recent mid-plane crossing time relies on values of $Z_0 \simeq 8-27$ pc (Section 9.2.2) and $W_0 \simeq 7\,\mathrm{km\,s^{-1}}$, away from the plane (Section 9.2.4), i.e. between 1–4 Myr ago. Previous and future passages require knowledge of the Sun's vertical oscillation period, generally considered to be around 60–66 Myr, although the revised estimates of $\rho_0 \sim 0.102\,M_\odot\,\mathrm{pc^{-3}}$ (Section 9.4.1) leads to $P \sim 83$ Myr.

Svensmark (2006b) used δ^{18}O proxy data from the Phanerozoic database over the last 500 Myr, isolating a ~140 Myr period identified with the Sun's passage through the spiral arms, and a ~30 Myr period which could be related to the crossing of the Galactic plane through the vertical oscillatory motion. Their fitting

Astrometric planet detection in context: As of the end of 2007, more than 250 exoplanets are known, including more than 25 multiple planet systems. The majority have been discovered using the radial velocity technique: some 25 others have been detected (and 20 discovered) from photometric transits and a few others by microlensing. Radial velocity experiments using high-resolution spectrographs currently reach accuracies of $\sim$1 m s^{-1} in the most favourable cases, leading to planets of five Earth masses being detectable around nearby M-dwarfs. The number of radial velocity detections should increase steadily over the next few years using currently-available instrumentation, including more multiple systems with longer-period outer planets. Further advances are expected with the advent of precision infrared radial velocities and substantial improvements in long-term instrument stability. For example, Espresso is a third-generation ESO VLT instrument concept targetting accuracies of 0.05–0.1 m s^{-1} on large numbers of nearby stars: for comparison, the Earth's orbit causes the Sun to move with a reflex velocity of 0.09 m s^{-1}.

Transit surveys are likely to yield a significant number of new detections as the temporal observing baseline builds up, such that periodic signals can be detected in the data. On the ground, wide-field monitoring will yield transits of gas giant planets around the brightest, nearest stars most amenable to follow-up radial velocity and other studies, allowing characterisation of the planetary mass, radius, and spectral features. There is also a move towards near-infrared transit surveys in order to study M-dwarfs, whose lower luminosity will make it possible to detect smaller transiting planets into the terrestrial-mass range. Higher photometric precision is possible from space, and the French-led COROT mission, launched in December 2006, has reported its first planet detections, with many hundreds plausible over its operational lifetime. NASA's Kepler mission will be launched in 2009 and is designed to detect terrestrial-mass planets, although Earth-mass planets in Earth-like orbits are unlikely to be confirmed before 2012 (i.e. launch plus three orbits). Most detections from COROT and Kepler will be around relatively faint stars (12–15 mag) at distances of tens to hundreds of parsecs, making ground-based physical characterisation problematic. The PLATO mission, selected in 2007 for competitive study under the ESA Cosmic Vision programme for possible launch in 2017, would provide both transit and asteroseismology data for some 10 000 stars, yielding planet and host star data for each target.

Astrometric detections will only become realistic, in large numbers, with the launch of ESA's Gaia in 2011: more than 1000 exoplanets with Jupiter mass or greater should be discovered. NASA's SIM PlanetQuest should provide astrometric detections of Earth-mass planets through careful target selection. Astrometric discovery may also be feasible, but highly challenging, using ground-based interferometry. Experiments are planned for the ESO VLTI PRIMA infrastructure, where a planet-search programme (ESPRI) targeting 10 microarcsec accuracy will start in 2009: a somewhat lower-precision programme is planned for the Keck interferometer.

Microlensing exoplanet candidates should continue to be discovered in small numbers from the ground over the coming years using dedicated networks of small telescopes to conduct detailed monitoring of stellar lensing events. Earth-mass planets are potentially detectable. The most significant result to date was the detection of a five Earth-mass planet, OGLE–2005–BLG–390Lb. Significantly larger samples would be valuable in clarifying the statistical distribution of orbits and masses, and thus providing data for theories of planetary formation. Small networks of dedicated ground-based telescopes would be required to make significant advances; the alternative would be a dedicated space mission capable of imaging high-density regions without confusion, although there are no plans currently for such a mission.

Direct imaging detection and atmospheric characterisation through spectroscopy will be very challenging, particularly for terrestrial-mass planets. Ultimately, the latter will require space observations such as the long-baseline infrared nulling interferometry underlying ESA's Darwin and NASA's TPF-I concepts. Attempts will also be made for direct detection from the ground where, with the current generation of 8–10 m-class telescopes, direct imaging of young (and thus self-luminous) gas giants should be possible. The ESO VLT second-generation instrument SPHERE, for example, aims to image Jupiter-mass exoplanets at distances of 1–100 AU from the host star using extreme adaptive optics, integral field spectroscopy, and imaging polarimetry. A roughly equivalent, albeit technologically-different extreme-adaptive optics direct-imaging instrument, GPI, is under construction for Gemini south.

Other major planet detection facilities will include large approved facilities such as ALMA on the ground and JWST in space, and longer-term initiatives still at the study phase, such as the extremely large (30–40 m) telescopes, and optical and infrared telescopes on the high Antarctic plateau.

spectroscopic measurements, d from the star's parallax, and if M_* can be estimated from its spectral type or from evolutionary models, then the astrometric displacement yields M_p directly. A single measurement of the angular separation at one epoch from ground-based interferometric astrometry would also provide orbital constraints on $\sin i$. The principles are the same as those adopted for higher mass secondaries, as discussed in Chapter 3, and as described by the relations given in the box on page 112. For multi-planet systems, astrometric measurements could in principle determine their relative orbital inclinations (i.e. whether the planets are coplanar), an important ingredient for formation theories and dynamical stability analyses.

Discussions of ground-based, pre-Hipparcos, optical observations related to planet detection are given by Black & Scargle (1982) and Gatewood (1987). Measurement of submilliarcsec displacements from ground has been largely impossible to date because of atmospheric effects, although accuracies for narrow-angle ground-based measurements are improving, with prospects for interferometry at the tens of microarcsec.

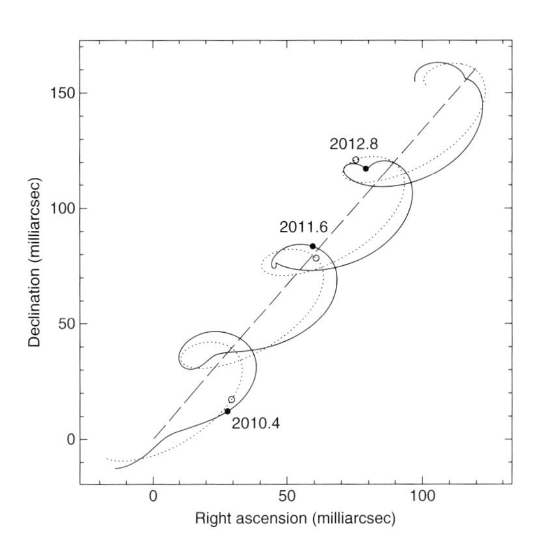

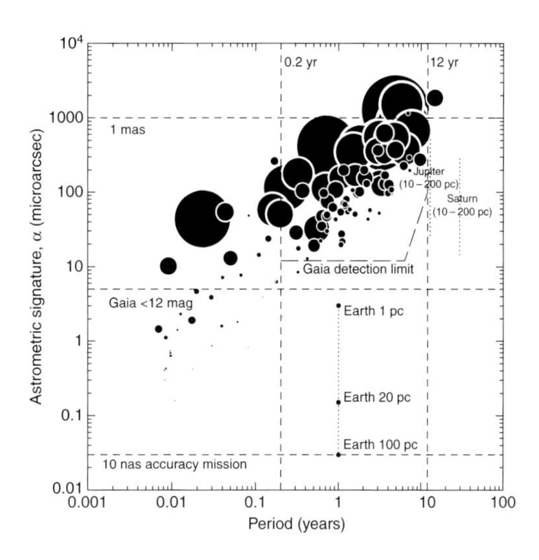

Figure 10.13 Left: the simulated path on the sky of a star at a distance of 50 pc, with a proper motion of 50 mas yr^{-1}, and orbited by a planet of mass $M_p = 15\,M_J$, eccentricity $e = 0.2$, and semi-major axis $a = 0.6$ AU. The straight dashed line shows the path of the system's barycentric motion viewed from the Solar System barycentre. The dotted line shows the additional effect of parallax (the Earth's orbital motion around the Sun, with a period of 1 year). The solid line shows the apparent motion of the star as a result of the planet, the additional perturbation being magnified by ×30 for visibility. Labels indicate (arbitrary) times in years. Right: astrometric signature, α (Equation 10.11), induced on the parent star for the planetary systems known in 2005, as a function of orbital period. Circles are shown with a radius proportional to $M_p \sin i$. Astrometry at the milliarcsec level has marginal power in detecting these systems, while the situation changes dramatically for microarcsec measurements. Also shown are the corresponding effects of Earth, Jupiter, and Saturn at the distances indicated.

Hipparcos astrometry: Extra-solar planets Although some inferences about the properties of systems discovered by radial velocity measurements have been possible from the Hipparcos results, it is evident from Equation 10.11 (and Figure 10.13, right) that milliarcsec astrometry can contribute only rather marginally to extra-solar planet detection.

For known planetary systems, the Hipparcos data have, however, provided some constraints on planetary masses. The principle is that periodic motion of a star due to an invisible companion may still be detectable at levels below which the star is clearly identifiable as double, typically $\rho > 0.1$ arcsec and $\Delta m \leq 3$–4 mag. While milliarcsec astrometry does not reach the sensitivity of radial velocity measurements for inclined orbits and for secondaries of order 1–10 M_J, smaller inclinations lead to smaller radial velocities, so that the astrometric signature due to the primary's reflex motion is present at all inclinations. Perryman *et al.* (1996) derived weak upper limits on M_p for 47 UMa ($< 7\,M_J$), 70 Vir ($< 38\,M_J$), and 51 Peg ($< 500\,M_J$) based on adjustment of the orbital elements using a large number of trial periods, including those of the known planets. Comparable upper limits for 47 UMa were later given by Zucker & Mazeh (2001) and, following the discovery of a second planet orbiting the same system, by Fischer *et al.* (2002).

Mazeh *et al.* (1999) used a similar approach, but using the Hipparcos intermediate astrometric data measured on 27 independent reference great circles, to derive a mass for the outer companion of the triple planetary system υ And. The stellar orbit associated with the outermost planet has a period of 1269 ± 9 days, and a minimum semi-major axis of 0.6 mas. The procedure was based on taking the five spectroscopically-derived orbital elements (period P, epoch of periastron passage T_0, eccentricity e, longitude of periastron ω, and radial-velocity amplitude K) and the 12 astrometric values (the five astrometric parameters of the barycentre, i.e. two components of position, two of proper motion, and the parallax, combined with the seven astrometric orbital elements, i.e. in addition to P, T_0, e, ω, the semi-major axis a_1, the inclination i, and the longitude of nodes Ω), fixing P, T_0, e, ω to their spectroscopic values, and then solving for the remaining eight parameters. They did this by choosing a dense grid in the a_1, i plane, and finding values of the five regular astrometric parameters and Ω that minimises the χ^2 statistics for each pair of (a_1, i). This led to a pronounced value at $a_1 = 1.4$ mas indicating the detection of an astrometric motion.

Using the radial velocity amplitude K constrains the product

$$a_1 \sin i = 0.56 \pm 0.02 \left(\frac{P}{1269\ \text{days}}\right)\left(\frac{K}{69.5\ \text{m s}^{-1}}\right)\left(\frac{\sqrt{1-e^2}}{0.95}\right)\left(\frac{\pi}{74.25\ \text{mas}}\right)\ \text{mas} \quad (10.12)$$

where the terms come from the spectroscopic orbit (P, K, e) or Hipparcos parallax (π). They found a clear minimum of $i = 156°.0$, and a 1σ range of $131°.4$–$163°.9$, yielding an implied planet mass of $M_p = 10.1^{+4.7}_{-4.6}\,M_J$, compared with an $M_p \sin i$ from radial velocity measurements of $4.1\,M_J$. They also derived estimates of the mass of the two inner planets on the assumption that the orbits of all three are co-aligned, which yielded $1.8 \pm 0.8\,M_J$ and $4.9 \pm 2.3\,M_J$ respectively. While this assumption is unproven, a large relative value of the inclination angles can be excluded on dynamical stability grounds. Similar arguments (Zucker & Mazeh, 2000) have been used to demonstrate that the companion of HD 10697 is, in contrast, probably a brown dwarf (see Section 10.7.2).

A discussion of possible bias in mass estimates obtained from the Hipparcos astrometry has been given by Arenou & Palasi (2000). In the case of ρ CrB, Gatewood *et al.* (2001) combined the 28 independent Hipparcos reference great circle data with Multichannel Astrometric Photometer data to derive $a_1 = 1.66 \pm 0.35$ mas and $M_p = 0.14 \pm 0.05\,M_\odot$, implying a mass some 100 times larger than the minimum mass from radial velocity studies, corresponding to that of an M dwarf star. The question of the general distribution of planetary masses derived from the Hipparcos data subsequently became the subject of some debate. Han *et al.* (2001) used the Hipparcos intermediate astrometric data for all known planetary candidates with $P > 10$ days to draw the provisional conclusion that spectroscopic programmes can be biased to small values of sin *i*, leading in turn to masses much in excess of the associated minimum masses, a possibility loosely supported by certain other lines of evidence (see, e.g. McGrath *et al.*, 2002). However, subsequent work negated this suggestion. Pourbaix (2001) and Pourbaix & Arenou (2001) re-analysed the same intermediate astrometric data, and argued that the trend to low inclinations is an artifact of the adopted reduction procedure, and that the astrometric data are not accurate enough to allow the conclusion that a significant fraction of the radial velocity companions have stellar masses. This conclusion was supported by the work of McGrath *et al.* (2002, 2003), who used HST Fine Guidance Sensor observations, formally at the level of 0.3 mas, from 14 orbits over a 30-day interval, to examine the motion of ρ^1 Cnc (or 55 Cnc). They placed 3σ upper limits of 0.3 mas on the semi-major axis of the reflex motion due to the 14.65-day $M \sin i = 0.88\,M_J$ planetary candidate, ruling out the 1.15 mas perturbation proposed by Han *et al.* (2001), and placing an upper limit of about 30 M_J on the planetary mass.

Other limits on masses using the Hipparcos data have been reported for various planetary candidates discovered through radial velocity observations: for HD 179949 and HD 164427 (Tinney *et al.*, 2001), for the K2 giant ι Dras (Frink *et al.*, 2002), and for the intermediate-mass giant HD 11977 (Setiawan *et al.*, 2005). Zucker & Mazeh (2001) analysed the Hipparcos astrometry for 47 planetary and 14 brown dwarf secondary candidates, finding that the lowest derived upper limit is for 47 UMa at $0.014\,M_\odot$, confirming the planetary nature of the unseen companion, and similar to the limits given by Perryman *et al.* (1996). For 13 other planet candidates, the upper limits exclude a stellar companion although brown dwarf secondaries are still an option, again negating the idea that most extra-solar planets are disguised stellar secondaries. The Hipparcos Intermediate Astrometric Data were used to constrain, or in three cases to solve for, the orbital inclinations and masses for a further ten low-mass companions from the Keck precision radial velocity survey (Vogt *et al.*, 2002).

Unconfirmed reports of small long-period astrometric displacements, consistent with planetary bodies, were made more than 20 years ago for Barnard's star. Observations over many years yielded two proposed planetary mass bodies (0.7 and 0.5 M_J) with periods of 12 and 20 years respectively (van de Kamp, 1963, 1977, 1982). This was refuted by the ground-based astrometric studies of Gatewood (1995), and by HST FGS studies by Benedict *et al.* (1999), who placed limits of 2.1–$0.37\,M_J$ for orbital periods in the range 50–600 days. From 2.5 years of radial velocity data, Kürster *et al.* (2003) placed upper limits of $M \sin i = 0.12-0.86\,M_J$ in the separation range 0.017–0.98 AU, and $M \sin i = 7.5\,M_{\rm Earth} - 3.1\,M_{\rm Neptune}$ throughout the habitable zone 0.034–0.082 AU. They were able to measure the secular acceleration of around 2.97–5.15 m s^{-1} yr^{-1}, in good agreement with the predicted value of 4.50 m s^{-1} yr^{-1} based on the Hipparcos proper motion and parallax combined with the known absolute radial velocity. Kürster *et al.* (2006) extended the monitoring to more than five years, clearly confirming the secular trend of 4.50 m s^{-1} yr^{-1}, with residual fluctuations attributed to stellar activity (see Section 1.13.2 and Figure 1.20).

An astrometric planet detection for Lalande 21185 (HD 95735), with $M_p = 0.9\,M_J$ and $P = 5.8$ years was claimed from ground-based observations by Gatewood (1996), but no Hipparcos or radial velocity confirmation has subsequently been published.

Combination of radial velocity and astrometry Hauser & Marcy (1999) illustrate the use of combined radial velocity and astrometry analysis in the case of the highly

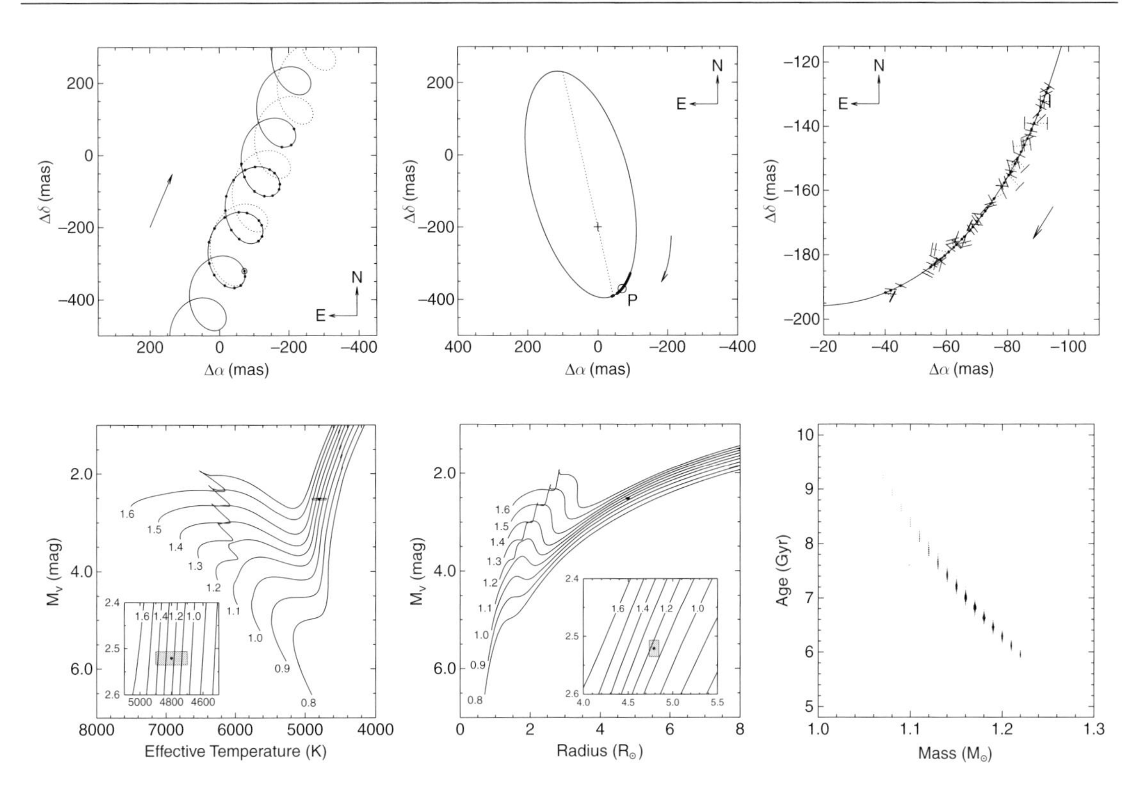

Figure 10.14 Results from the astrometric–spectroscopy study of the planet-hosting star γ Cep by Torres (2007). Top left: path of γ Cep A on the sky resulting from the combined effects of proper motion (arrow), orbital motion, and parallactic motion (solid curve). •: Hipparcos observations shown at their predicted locations (the actual measurements are 1d). Dotted curve: star path in the absence of orbital motion; Top middle: computed orbit of γ Cep A around the system barycentre (indicated '+'). The retrograde motion is indicated by the arrow, and the dotted line is the line of nodes. The Hipparcos observations bracket periastron passage (○, labelled 'P'); Top right: enlargement of previous, showing the individual Hipparcos observations, and a graphical representation of the 1d measurements; Bottom left: evolutionary tracks from Yi et al. (2001) and Demarque et al. (2004) in the M_V versus $T_{\rm eff}$ plane. Masses are in solar units; the dot with the shaded error box represents the measurements (enlargement in inset); Bottom middle: as previous, but in the M_V versus R plane; Bottom right: theoretical mass and age combinations consistent with the measured properties of $T_{\rm eff}$, [Fe/H], M_V, and R. The best fit is for $M_{\rm Aa} = 1.18^{+0.04}_{-0.11} M_\odot$ and an age of $6.6^{+2.6}_{-0.7}$ Gyr. Larger point sizes indicate a closer match. From Torres (2007, Figures 3–5 and 9–11).

eccentric orbit of the planet around 16 Cyg B which, it was hypothesised, may have been induced by the companion star, 16 Cyg A; but only if the stellar binary has a sufficiently small periastron distance. The long period of the stellar binary, $\sim 3 \times 10^4$ yr, implies that less than 1% of the orbit has been observed since its first astrometric measurements in 1830. They therefore computed the orbit from the measured instantaneous velocity and position vectors, based on new precise Doppler and astrometric data, making use of the Hipparcos parallax to constrain the linear separation.

The only unknown parameter is then the separation between the two stars along the line-of-sight, which can be constrained by the condition that the orbit is bound, which led to possible orbits with $P = 1.82 \times 10^4 - 1.3 \times 10^6$ yr, $a =$ 877–15 180 AU, and $e = 0.54$–0.96. The uncertainty on these specific parameters is still dominated by the uncertainty on the (mean) parallax, $\pi_{\rm Hip} = 46.475 \pm 0.50$ mas, while the stellar binary orbit remains consistent with the possibility that perturbations from 16 Cyg A cause the eccentricity in the planet around 16 Cyg B.

Torres (2007) used a combination of radial velocity measurements and Hipparcos astrometry in a detailed study of the orbit of the bright K1 III–IV star γ Cep. This had been reported to have a substellar companion in a 2.5 yr orbit, as well as an unseen stellar companion at a larger separation. Torres (2007) used the Hipparcos intermediate astrometric data, as well as ground-based positional observations going back more than a century, to establish that the orbit of the secondary star is eccentric, $e = 0.4085 \pm 0.0065$, with a period $P = 66.8 \pm 1.4$ yr. Evolutionary models imply that the primary star is on the first ascent of the giant branch and has $M = 1.18 \pm$

$0.11M_\odot$, $T_{\rm eff} = 4800 \pm 100$ K, and an age ~6.6 Gyr for an assumed metallicity [Fe/H] = $+0.01 \pm 0.05$. The unseen secondary star was found to be an M4 dwarf with a mass of $0.362 \pm 0.022M_\odot$ and is therefore predicted to be ~6.4 mag fainter than the primary in the K-band. The minimum mass of the putative planetary companion is $M_p \sin i = 1.43 \pm 0.13M_{\rm J}$. The Hipparcos data place a dynamical upper limit on this mass of $13.3M_{\rm J}$ at the 95% confidence level, thus confirming that it is indeed substellar in nature. The orbit of this object is only 9.8 times smaller than the orbit of the secondary star, and thus the smallest ratio among exoplanet host stars in multiple systems, but its orbit would be stable if coplanar with the binary (Figures 10.14).

Hipparcos astrometry: Brown dwarfs Brown dwarfs (Burrows & Liebert, 1993; Basri, 2000; Reid & Hawley, 2000) occupy the mass range of about 12–80 $M_{\rm J}$ (0.01–$0.08M_\odot$), bridging the gap between the higher mass cool M stars and the lower mass giant Jupiter-like planets. They are not massive enough to ignite stable H burning, which occurs above about $0.08M_\odot$, although lithium fusion contributes to their luminosity above about $65M_{\rm J}$, and deuterium fusion above about 12–13 $M_{\rm J}$. Below this limit, which is sensitive to chemical composition, structural uniformity, nuclear processes, and the role of dust (Tinney, 1999), objects should retain essentially their entire deuterium complement of around the protosolar value, $D/H \sim 2 \times 10^{-5}$, and derive no luminosity from thermonuclear fusion at any stage in their evolutionary lifetime (Burrows *et al.*, 1993; Saumon *et al.*, 1996).

The first identification of a low-mass dwarf cooler than the M dwarfs (classified as having strong molecular bands of TiO and VO) was the discovery of the first L dwarf (classified as having strong lines of neutral alkali elements and hydrides), a very red companion to the white dwarf GD 165, by Becklin & Zuckerman (1988). The first lower temperature T dwarf (having strong methane bands at near-infrared wavelengths), a faint companion to the bright M1V star Gliese 229, was discovered by Nakajima *et al.* (1995). Rapid developments in the field have followed from the identification of many new L and T dwarfs from the survey programs DeNIS, 2MASS, and SDSS. By early 2000, the number of L dwarfs was nearly 100 (Kirkpatrick *et al.*, 2000), and by 2002 the number of T dwarfs was around 30 (Burgasser *et al.*, 2002). By 2005, the numbers were around 400 and 60, respectively. Thus, while the early Hipparcos studies were based on blind searches of catalogue stars for previously-unknown low-mass astrometric binary components, later work, such as that reported below by Dahn *et al.* (2002), used *a priori* knowledge of the existence of these companions.

Bernstein (1997), see also Bernstein (2003), reported the first analysis of the Hipparcos astrometric data in a blind search for brown dwarfs, showing that the data do permit the detection of brown dwarf components in astrometric binaries. The method used was an extension of the classical description of orbital motion using the Thiele–Innes constants for the description of the orbit in space (see box on page 112), leading to a linear model with the sine and cosine of the mean anomaly as coefficients in a linearised observation equation. No *a priori* orbital information is required, and an optimum solution is found by varying the search period within reasonable limits. A systematic search of low-mass companions among M and K stars in the solar neighbourhood (within 50 pc) resulted in the six candidates shown in Table 10.4.

Guirado *et al.* (1997) used the Hipparcos astrometric data in combination with radio VLBI data to detect an orbital motion of the $0.76M_\odot$ star AB Dor, consistent with a dynamical mass of the secondary of $M_2 = 0.08$–$0.11M_\odot$.

Halbwachs *et al.* (2000) combined the Hipparcos astrometric data with the spectroscopic orbital elements of 11 spectroscopic binaries with brown dwarf candidate components, i.e. with $M_2 \sin i \sim 0.01-0.08M_\odot$, with the Hipparcos observations in order to derive astrometric orbits including the masses of the secondary components (Figure 10.15). They found that seven objects have secondary masses more than 1–2σ above the planetary limit of $0.08M_\odot$; 1 brown dwarf was accepted with low confidence, leaving only three viable brown dwarf candidates. They developed a statistical approach, based on the relation between the semi-major axes of the photocentric orbit and the frequency distribution of the mass ratios, q, to conclude that a minimum in the q distribution exists for M_2 in the range 0.01–$0.1M_\odot$, for companions of solar-type stars. This feature could correspond to the transition between giant planets and stellar companions. Due to the relatively large frequency of single brown dwarfs found in open clusters, it was concluded that the mass distribution of the secondary components in binary systems does not correspond to the initial mass function, at least for masses below the H-ignition limit.

Zucker & Mazeh (2000) used similar arguments to derive the mass of the companion to the star HD 10697. From the radial velocity data (Vogt *et al.*, 2000), the stellar orbit has a period of about 3 yr, $M_2 \sin i \sim 6.35M_{\rm J}$, and a minimum semi-major axis of 0.36 mas. Using the Hipparcos data together with the spectroscopic elements, Zucker & Mazeh (2000) found a semi-major axis of 2.1 ± 0.7 mas, implying a mass of $38 \pm 13M_{\rm J}$ for the unseen companion (Figure 10.16). They concluded that the secondary of HD 10697 is probably a brown dwarf, orbiting its parent star at a distance of 2 AU.

Table 10.4 Candidate brown dwarfs from the astrometric analysis by Bernstein (1997).

Star	V	M_1 ($M_\odot$)	M_2 ($M_\odot$)	P (d)
Gliese 176 (HIP 21932)	9.9	0.39	0.07 ± 0.03	463.8 ± 39.1
Gliese 375 (HIP 48904)	11.3	0.22	0.07 ± 0.03	294.3 ± 9.5
Gliese 408 (HIP 53767)	10.0	0.33	0.04 ± 0.02	250.1 ± 11.4
G 206–40 (HIP 91699)	11.3	0.27	0.07 ± 0.03	258.6 ± 8.9
Gliese 570 (HIP 73184)	5.7	0.70	0.06 ± 0.02	33.3 ± 0.3
Gliese 433 (HIP 56528)	9.8	0.50	0.03 ± 0.01	523.6 ± 36.9

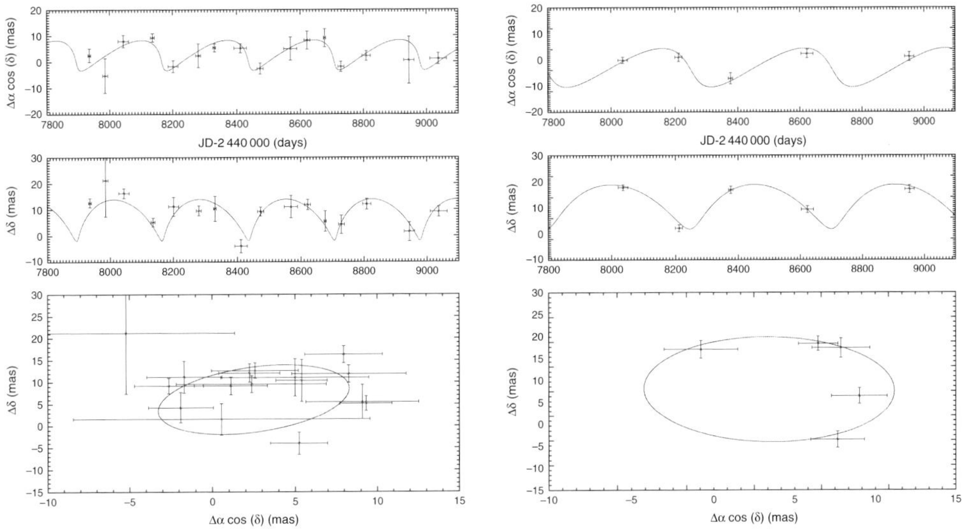

Figure 10.15 Astrometric orbits of two binary systems, derived from the Hipparcos observations taking the orbital elements of the spectroscopic binary into account. Left: HIP 62145, for which $P = 271.16$ d, $M_2 = 0.137 \pm 0.011 M_\odot$. Right: HIP 113718, for which $P = 454.66$ d, $M_2 = 0.161 \pm 0.013 M_\odot$. The results show that both secondaries are too massive to be brown dwarfs. From Halbwachs et al. (2000, Figures 3 and 4).

Zucker & Mazeh (2001) included 14 brown dwarf candidates in their study of low-mass companions. They confirmed the results of Halbwachs *et al.* (2000), deriving astrometric orbits for six systems that imply secondaries with stellar masses, supporting their conclusion about a possible 'brown dwarf' desert separating planet and stellar secondaries (Figure 10.17). Similar results were given by Frink (2003). Reffert & Quirrenbach (2006) used the Hipparcos intermediate astrometry data to complete the orbital information for the two brown dwarf companions HD 38529 and HD 168443. They found best-fit solutions for the orbital inclination and ascending nodes implying masses of $37^{+36}_{-19} M_{\rm J}$ and $34 \pm 12 M_{\rm J}$, respectively. As well as providing the masses directly, the results provide the orbital orientation in space, which should be valuable in future attempts at directly imaging.

Dahn *et al.* (2002) presented trigonometric parallaxes from the USNO CCD faint star parallax program for 28 late-type dwarfs and brown dwarfs: eight M dwarfs, 17 L dwarfs, and three T dwarfs. Broadband CCD photometry (*VRIz**) and near-infrared photometry (*JHK*) were obtained for these and for 24 additional late-type dwarfs. These data were supplemented with various astrometric and photometric data from the literature, including ten L and two T dwarfs whose parallaxes were established by association with bright Hipparcos primaries (GJ 1048, Gl 229, Gl 337, HD 89744, Gl 417, HD 130948, Gl 569, Gl 570, Gl 584, Gl 618.1). Colours were used to determine bolometric corrections, and models used to estimate stellar radii and hence effective temperatures, with the coolest L dwarfs having $T_{\rm eff} \sim 1360$ K. Distances were used to place the objects in the HR diagram (Figure 10.18). The results were used to show that the $I - J$ colour is a good predictor of absolute magnitude for late-M and L dwarfs; M_J becomes monotonically fainter with $I - J$ colour and with spectral type through late-L dwarfs, then brightens for early-T dwarfs. The combination of z^*JK colours alone can be used to classify late-M, early-L, and T dwarfs accurately, and to

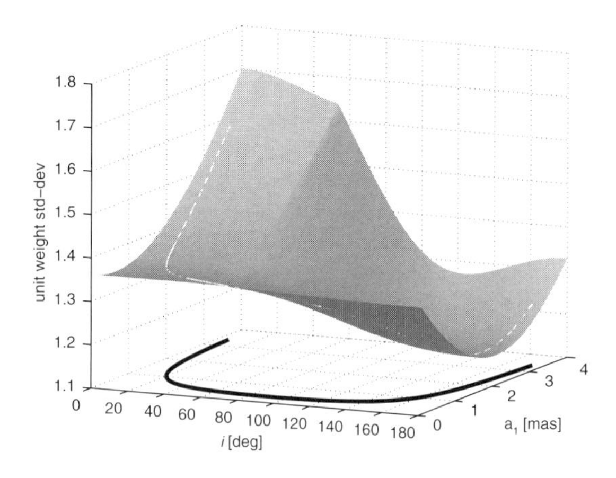

Figure 10.16 Analysis of the Hipparcos measurements of the brown dwarf secondary in HD 10697, showing the minimum square-root normalised χ^2 statistics as a function of the semi-major axis, a_1, and the orbital inclination i. The continuous line is the $a_1 \sin i = 0.36$ mas constraint. From this, the authors determined a semi-major axis of 2.1 ± 0.7 mas, implying a mass of $38 \pm 13 M_J$ for the unseen companion. From Zucker & Mazeh (2000, Figure 1).

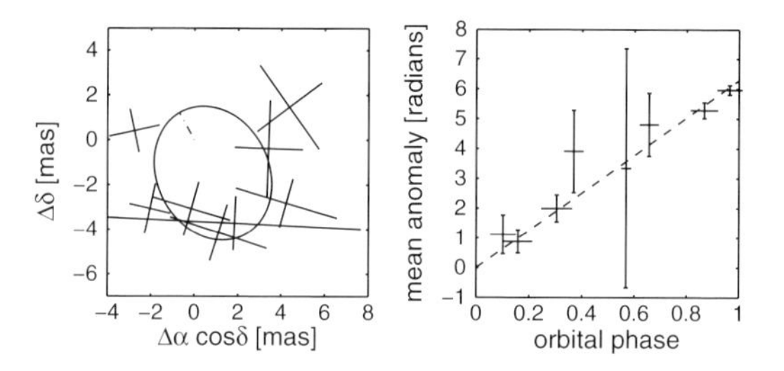

Figure 10.17 Left: eight 'two-dimensionally averaged' Hipparcos data points for the brown dwarf candidate HD 164427, showing the derived astrometric orbit (solid line), and the direction of periastron (dot-dashed line). Right: the mean anomaly as a function of orbital phase. The dashed line is the expected dependency. From Zucker & Mazeh (2001, Figure 4).

predict their absolute magnitudes, but is less effective at disentangling the scatter among mid- and late-L dwarfs. The mean tangential velocity was found to be slightly less than that for M dwarfs in the solar neighbourhood, consistent with a sample with a mean age of several Gyr.

Hipparcos data were also used in determining an orbit for the low-mass binary Gliese 22 AC from speckle interferometry (Woitas *et al.*, 2003), in characterising the companion around the intermediate-mass giant HD 11977 (Setiawan *et al.*, 2005), and in the first directly imaged brown dwarf companion of an exoplanet host star, the $H = 16.5$ mag HD 3651B at a distance of 11 pc (Mugrauer *et al.*, 2006). A proposed search for brown dwarf companions around Hipparcos subgiant stars, to establish the present-day mass function and brown dwarf formation history using benchmark brown dwarfs whose age can be determined independently, was discussed by Pinfield *et al.* (2006).

10.7.3 Photometric transits

Aside from orbital data and corresponding lower mass limits from radial velocity measurements, specific planetary characteristics, in the absence of direct imaging or spectroscopy, are presently very limited. However, the detection of photometric transits provides additional important information.

The first object for which transits were observed was detected through radial velocity measurements HD 209458 (Charbonneau *et al.*, 2000). The precise shape of the transit curve is determined by five parameters: the planetary and stellar radii, the stellar mass, the orbital inclination i, and the limb-darkening parameter. For assumed values of R_p and i, the relative flux change can be calculated at each phase, by integrating the flux occulted by a planet of given radius at the correct projected location on the limb-darkened disk. Best-fit parameters yielded (Charbonneau *et al.*, 2000) $R_p = 1.27 \pm 0.02\, R_J$ and $i = 87\overset{\circ}{.}1 \pm 0\overset{\circ}{.}2$ which, in combination with $M_p \sin i = 0.63\, M_J$ from the radial velocity solution, yields $M_p = 0.63\, M_J$ ($\sin i \sim 1$). The radius agrees with predictions for a hydrogen-dominated gas giant, with Charbonneau *et al.* (2000) estimating $\rho \sim 0.38\,\mathrm{g\,cm^{-3}}$, significantly less dense than Saturn, the least dense of the Solar System gas giants, and a surface gravity of $g \sim 9.7\,\mathrm{m\,s^{-2}}$. Subsequent observations have led to improved stellar parameters, and detection and corresponding physical diagnostics in other spectral lines. Transits from a few other planets have also been discovered. However, the smallness of the effect (a photometric signature of around 1% for a Jupiter-type planet, 0.01% for an Earth – an amplitude which can only be detected above the Earth's atmosphere), and the geometrical alignment probability for such a transit to be seen on Earth, means that transits for known radial velocity detected transits are uncommon. The long observation time needed, not only to detect such an event but to establish and confirm the periodicity, also means that, despite the large number of transit programmes underway, candidates discovered by this method are rare.

The photometric transit signal of HD 209458 was confirmed *a posteriori* in the Hipparcos epoch photometry data by Söderhjelm *et al.* (1999) and by Robichon & Arenou (2000a). Hipparcos observed the star on 89 occasions, of which five corresponded to epochs of the planetary transits (Figure 10.19): the orbital period of about 3.5 days and a transit duration of about 0.1 day implies a 3% probability of observing a transit at any given epoch. The Hipparcos median magnitude for this

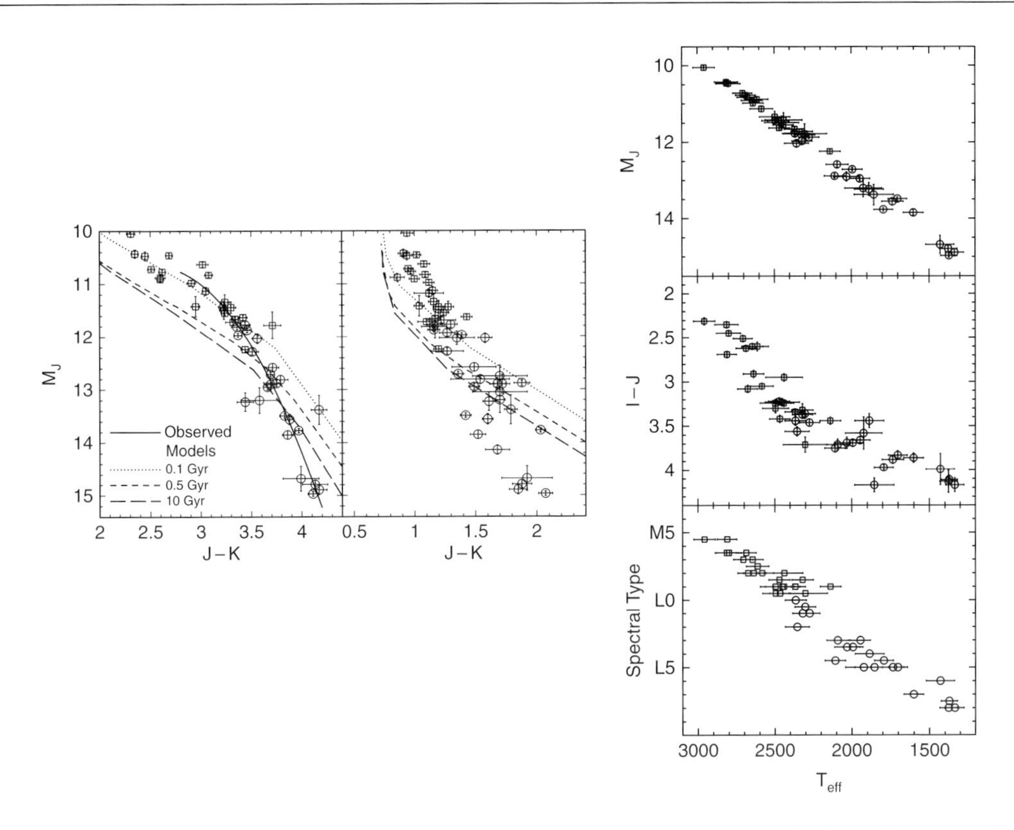

Figure 10.18 Above: M and L dwarfs compared to the AMES DUSTY models (Chabrier et al., 2000) for three different ages. The solid line shows a fit to the observed M6.5–L8 dwarfs. Squares and circles are M and L dwarfs, respectively. Right: T_{eff} calculated from M_J, with K-band bolometric corrections from Leggett et al. (2001), and $R/R_\odot$ from models by Burrows et al. (1997) and Chabrier et al. (2000). Most of the parallaxes are from the USNO ground-based CCD faint star parallax program, but 12 parallaxes (for 10 L and two T dwarfs) were inferred from their brighter binary companions observed by Hipparcos. From Dahn et al. (2002, Figures 6 and 7).

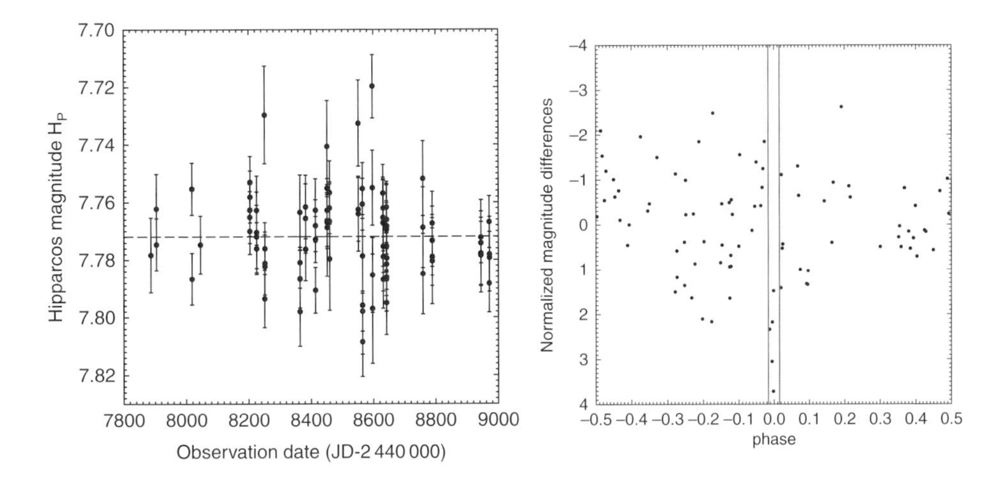

Figure 10.19 Left: Hipparcos individual photometric observations of HD 209458 as a function of observation epoch, in barycentric Julian Date. Right: normalised magnitude residuals versus phase, using the period obtained by the authors. The transit duration is indicated. From Robichon & Arenou (2000a, Figures 1 and 3).

star is $Hp = 7.7719 \pm 0.002$, and the transits resulted in a $2.3 \pm 0.4\%$ mean decrease in flux in the Hp band. As a result of the long temporal baseline of more than eight years, or nearly 1000 periods, between the Hipparcos observations in the early 1990s and the first ground-based detection in the late 1990s, the period could be improved some 20-fold, from 3.524 47 days as determined on ground (Mazeh *et al.*, 2000) to $3.524\,739 \pm 0.000\,014$ days when combined with the Hipparcos data (Robichon & Arenou, 2000a). Castellano *et al.* (2000) assessed the likelihood of these 'transit-like' events occurring in the Hipparcos data by chance, in view of the sparse sampling of the light-curve and the non-Gaussian distribution of the data points, and derived a probability of the signals occurring with the time and depth observed as 2.1×10^{-5}. Their arguments indicate that hundreds or thousands of transiting planets of this type may be discoverable from the ground, and that sparsely-sampled datasets may not be a large impediment to their discovery. Period improvement using the Hipparcos data will remain a possibility for transits discovered independently in the future. Castellano *et al.* (2000) assumed that the planet's orbital eccentricity is zero as indicated by the radial velocity data. In the case of non-zero eccentricity (a value of $e = 0.04$, for example, also being consistent with the data), the observed period is the sum of the Keplerian period and relativistic precession with a time scale of order $(v_p/c)^{-2}$ orbits, or about 0.06 s for $e = 0.04$, which is much smaller than the error bars.

This transit event was only detected in the Hipparcos epoch photometry data because a period and transit epoch was already known from the ground-based radial velocity and transit data. Presumably, more transit events remain buried in the Hipparcos epoch photometry database, although they remain difficult to identify in part because of the computationally-intensive processes required to search for them, but also because of the noise characteristics of the dataset. Robichon & Arenou (2000b) estimated that ~25 such candidates exist. Castellano (2001) put the figure at 6–24, assuming that 1 in 20 FGK dwarfs possess short-period planets, and described statistical and catalogue-based methods for discriminating transits from intrinsic stellar variability and eclipses due to stellar companions. Koen & Lombard (2002) also described refined procedures for identifying transit events in photometric time series, applying them to 10 820 bright ($Hp < 7$ mag) non-variable Hipparcos Catalogue stars, and illustrating their method with examples from the Hipparcos database (Figure 10.20). Laughlin (2000) defined a sample of 206 metal-rich stars of spectral type FGK which have an enhanced probability of harbouring short-period planets, and searched the Hipparcos epoch photometry database for associated transits: although the quality

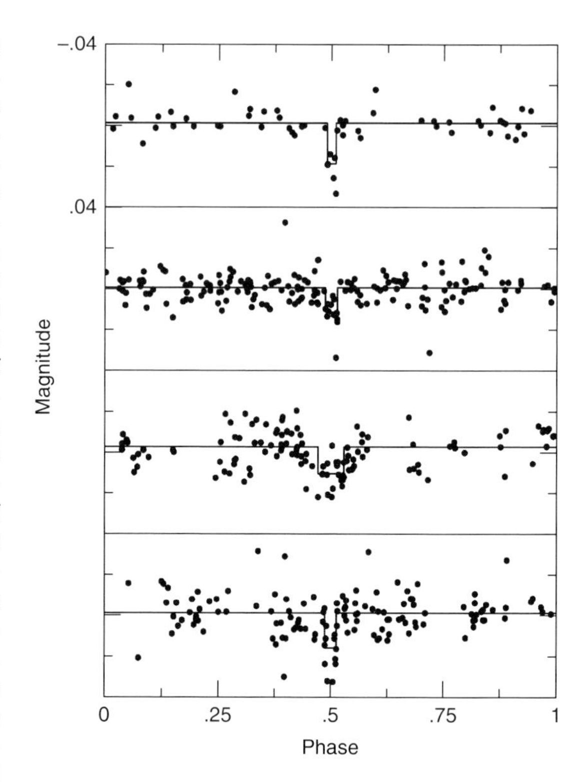

Figure 10.20 Four examples from the Hipparcos epoch photometry database which satisfy certain statistical criteria to be classified as planet transit candidates. From top to bottom: HIP 3559, P = 6.25 d; HIP 5021, P = 2.04 d; HIP 48223, P = 7.82 d; HIP 93042, P = 4.06 d. The signals are not necessarily due to planetary transits, and could be due to, e.g. stellar eclipses. From Koen & Lombard (2002, Figure 1).

was not adequate to permit unambiguous transit detections, various candidate transit periods were identified and listed in their Table 2. Jenkins *et al.* (2002) presented empirical methods for setting appropriate detection thresholds and for establishing the confidence level in planetary candidates obtained from transit photometry, including the sparse photometric data provided by Hipparcos.

Hébrard *et al.* (2006) systematically searched the Hipparcos photometry data for periods compatible with expected planetary transits, and constructed a ranked list of candidates for follow-up measurements. Radial velocities of 194 of the candidates were measured using HARPS, but the observations did not identify new transiting hot Jupiters, instead preferentially selecting active stars. They showed that the second transiting star bright enough to appear in the Hipparcos Catalogue, HD 149026 (Sato *et al.*, 2005), does not have transits deep enough to be detected, even *a posteriori*, in the Hipparcos photometric data.

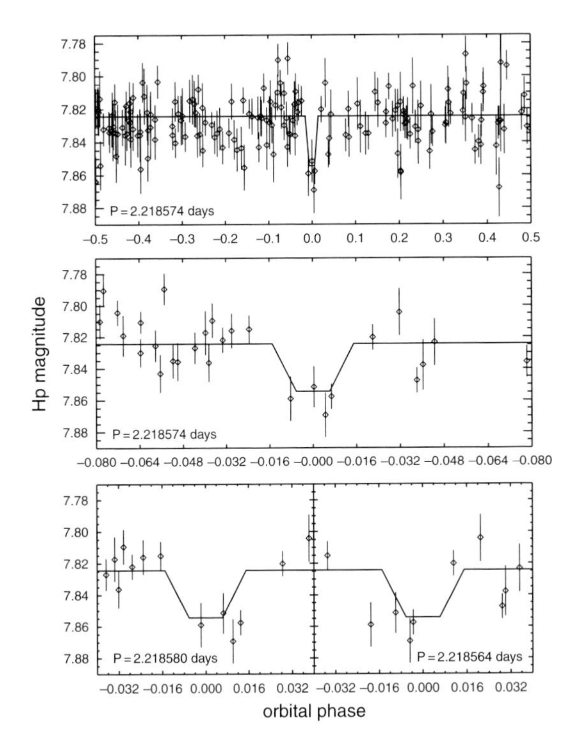

Figure 10.21 Top: Hipparcos photometric measurements for HD 189733, folded with a period $P_{Hip} = 2.218\,574$ d, determined from a χ^2 analysis of the Hipparcos measurements. It agrees with the value reported by Bouchy et al. (2005). The approximation of the transit curve used for the χ^2 computation is shown as the solid line. Middle: zoom of the measurements around the transit (phase 0). Bottom: same plots, but for the two extreme values of the error bar on P_{Hip}. From Hébrard & Lecavelier des Etangs (2006, Figure 3).

Bouchy *et al.* (2005) and Hébrard & Lecavelier des Etangs (2006) both reported transits in the Hipparcos photometric data of the third transiting extra-solar planet bright enough to be in the Hipparcos Catalogue, HD 189733 (Bouchy *et al.*, 2005), the ninth transiting planet to be discovered. Also detected through radial velocity measurements, its orbital period is 2.219 d, one of the shortest known, and it currently has the largest photometric transit depth of around 3%, with a transit duration of around 1.6 hr. Estimated planetary parameters from the combined ground-based data (radial velocity and transits) are $M_p = 1.15 \pm 0.04 M_J$ and $R_p = 1.26 \pm 0.03 R_J$. Hébrard & Lecavelier des Etangs (2006) showed that, with the ratio of transit duration to orbital period, around 3% of randomly chosen observations would be expected to fall during a transit, such that out of the 176 available Hipparcos observations, some five corresponding to transit periods would be expected. Folding of the Hipparcos light-curve indeed shows that such a number of data points fall at the time of transits (Figure 10.21). Using the zero phase from Bouchy *et al.* (2005), the combination of Hipparcos and ground-based data span a total measurement interval of 15 years, from which their χ^2 analysis yields a considerably more accurate period of $P_{Hip} = 2.218\,574^{+0.000\,006}_{-0.000\,010}$ d, and a corresponding accuracy in the orbital period of ~1 s. Baines *et al.* (2007) measured the interferometric angular diameter of the host star to be $\theta = 0.377 \pm 0.024$ mas using the CHARA array, combining this with the Hipparcos parallax to derive a linear radius of the host star of $0.779 \pm 0.052 R_\odot$, and a radius of the planet of $1.19 \pm 0.08 R_J$.

Absence of a pronounced photometric signal in the Hipparcos data was used to confirm the planetary nature of the companion to γ Cep (Hatzes *et al.*, 2003), and also to β Gem (Hatzes *et al.*, 2006). The Hipparcos photometry shows the lowest photometric variances amongst the FG dwarfs (Eyer & Grenon, 1997), an ingredient used by Batalha *et al.* (2002) in their assessment of the Kepler mission's planet detection capabilities. Since real transits can be mimicked by various non-planetary phenomena (e.g. transits by brown dwarfs, late-M dwarfs, or evolved stars, and grazing eclipses by ordinary stars) Gould & Morgan (2003) used the Hipparcos, Tycho 2, and 2MASS data (the latter to supplement the degraded Tycho photometry at fainter magnitudes) to design criteria to exclude giant stars too large to permit straightforward detection of planets, based on the reduced proper motion in the V band (Section 5.2.5).

10.7.4 Host star properties

Radial velocity searches, which represent the principal method for planet discovery to date, have concentrated on F–K stars. Stars earlier than F5 have fast rotation, making precision radial velocity measurements impossible, while M dwarfs have lower luminosities and relatively few have been surveyed (the discontinuity in rotation may itself be in part associated with planetary formation). Of the latter, Gliese 876 is the nearest known system, at 4.7 pc. Results suggest that planet formation is not restricted to more massive stars, and given that M dwarfs outnumber G dwarfs by a factor of 10, most planets in our Galaxy may orbit stars whose luminosity and mass are significantly lower than that of the Sun.

The Hipparcos data, notably the distances and derived luminosities, have been extensively used in establishing target lists and host star properties for radial velocity and other planetary search programmes, for example, for the choice of targets in the Keck planet search programme (Vogt *et al.*, 2002); for the Darwin infrared interferometry mission (Kaltenegger *et al.*, 2006); and more generally for studies of multiplicity of known exoplanetary systems (Raghavan *et al.*, 2006). An estimate of temperatures and metallicities for more than 100 000 FGK dwarfs from the Tycho 2 Catalogue, used to

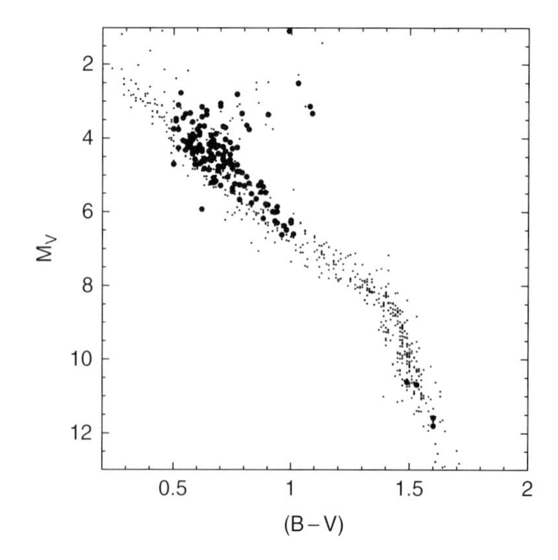

Figure 10.22 The Hipparcos HR diagram for nearby stars, with systems identified with planetary companions shown as filled circles. The reference stars have a distance limit of 25 pc, and correspond to the sample discussed by Reid et al. (2002). The original figure appeared in Hawley & Reid (2003, Figure 3). The planet hosting stars were updated to reflect the status as of mid-2007, with the updated figure kindly prepared by Neill Reid.

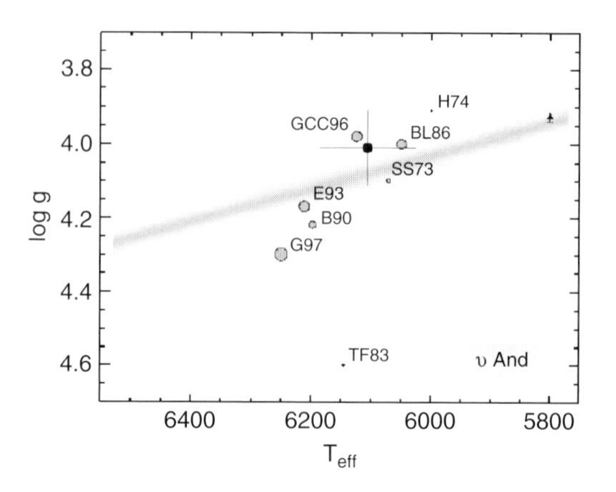

Figure 10.23 The HR diagram for υ And. The black dot marks the spectroscopically derived value for the effective temperature and surface gravity with error bars. The upwards tilted greyscale bar displays the most probable $T_{\rm eff}$, log g parameter space, as essentially prescribed by the accurate Hipparcos distance scale. The small circles represent comparisons with previous work from the literature, with diameters proportional to the derived metallicity. A systematic shift of the bar as a result of deviating metallicity scales from different analyses is indicated by the vertical arrow for a decrease in [Fe/H] by 0.1 dex. From Fuhrmann et al. (1998, Figure 2).

identify targets for N2K, a large-scale radial velocity survey of 2000 stars for hot Jupiters (Fischer *et al.*, 2005), was reported by Ammons *et al.* (2006).

Ecuvillon *et al.* (2007) studied the kinematics of metal-rich stars with and without planets, using Hipparcos astrometry and radial velocities from the CORALIE database, and then examined their relation to the Hyades, Sirius, and Hercules dynamical streams in the solar neighbourhood (see Section 6.9). Showing that the planet host targets have a kinematic behaviour similar to that of the metal-rich comparison subsample, and appealing to the scenarios proposed for the origin of the dynamical streams, they argued that systems with giant planets could have formed more easily in the metal-rich inner Galaxy, and then been brought into the solar neighbourhood by these dynamical streams.

Evolutionary state The host stars cover a mass range of $M_* = 0.8–1.2\,M_\odot$ (and lower for the K–M dwarfs), and a broad age range. Surveyed stars are typically bright and rather nearby (out to distances of ~50 pc), so that distances are well determined from Hipparcos (Figure 10.22). In most cases, detailed spectroscopic analyses of the parent stars have been carried out. Valenti & Fischer (2005), for example, present a compilation of 1040 FGK dwarfs from the Keck, Lick, and AAT planet search programmes, with estimates of $T_{\rm eff}$, $\log g$, projected $V_{\rm rot}$, [Fe/H], and abundances of Na, Si, Ti, Fe, and Ni. The Hipparcos parallaxes, combined with V-band photometry and bolometric corrections, have been used to determine stellar luminosities. Interpolating Yonsei–Yale isochrones with respect to luminosity, effective temperature, metallicity, and α-element enhancement of each star then yields a theoretical mass, radius, gravity, and age range for most stars in the catalogue. Estimated precision was 44 K in $T_{\rm eff}$, 0.03 dex in metallicity, 0.06 dex in $\log g$, and 0.5 km s^{-1} in $V_{\rm rot}$. Ages from isochrones nevertheless have a precision that varies significantly according to location in the HR diagram.

In a related application that has become routine in the field of extra-solar planets, and for which only a couple of examples are therefore given (see, e.g. Gonzalez, 1998), Fuhrmann *et al.* (1997) used the Hipparcos parallaxes to estimate absolute magnitudes, and hence masses and ages from stellar evolutionary tracks, deriving masses of $1.12\pm0.06\,M_\odot$ and $1.03\pm0.05\,M_\odot$ for 51 Peg and 47 UMa respectively, and ages of 4.0 ± 2.5 Gyr (not significantly different from the age of the Sun) and 7.3 ± 1.9 Gyr respectively. Fuhrmann *et al.* (1998) made a similar analysis for other F and G-type stars with planetary companions: υ And, ρ^1 Cnc, τ Boo, 16 Cyg and ρ CrB (Figure 10.23). The solid age determinations not only provide information on the stars' evolutionary state, but also provide constraints on the efficiency of tidal interactions for orbital synchronisation and circularisation.

Reid *et al.* (2007) used the Hipparcos Catalogue to argue that most stars now known to have gas giant planetary companions are younger than 5 Gyr, while stars with planetary companions within 0.4 AU have a significantly flatter age distribution, indicating stability on time scales of many Gyr. Furthermore, if the frequency of terrestrial planets is indeed correlated with stellar metallicity, then the median age of such planetary systems is likely to be around 3 Gyr.

Hipparcos constraints applied to asteroseismology models for the planet-hosting star μ Ara were obtained by Bazot *et al.* (2005), and discussed in Section 7.8. Other discussions of asteroseismology for host stars, making use of the Hipparcos data, are given for the COROT primary target HD 52265 by Soriano *et al.* (2007), and for ι Hor by Laymand & Vauclair (2007).

Effects of metallicity It was surmised from some of the earliest studies of planet-hosting stars that, on average, stars hosting planets have significantly higher metal content, compared to the average solar-type star in the solar neighbourhood, although some are metal poor (Gonzalez, 1999a). For the subset of short-period giants, Gonzalez & Laws (2000) concluded that the parent stars are all metal rich; values of [Fe/H] = +0.45 for ρ^1 55 Cnc and 14 Her place them amongst the most metal-rich stars in the solar neighbourhood. This suggests that their physical parameters are affected by the process that formed the planet.

There are two hypotheses to explain the connection between high metallicity and the presence of planets: high metallicity may favour the formation of rocky planets (and large rocky cores of gas giants) as a result of excess solid material in the protoplanetary disk: essentially the metals make condensation easier. Alternatively, as originally formulated to explain the high Li content of certain stars (Li being rapidly destroyed even at relatively low temperatures), the high metallicity could be due to the capture of metal-rich disk material by the star during its early history, and possibly even to capture of the planet itself as a result of dynamical friction (e.g. Siess & Livio, 1999; Sandquist *et al.*, 2002; Israelian *et al.*, 2001, 2003). Detailed studies, using the Hipparcos Catalogue as a volume-limited sample of 486 FGK stars (Reid, 2002), or to derive masses, ages, and surface gravities for additional systems (Laws *et al.*, 2003; Santos *et al.*, 2004), have continued to confirm this picture. Indeed, Fuhrmann (2004) notes the probable accretion of a giant planet or brown dwarf onto 59 Vir in the recent past, while additional suggestions for recent planet consumptions by dwarf stars have been inferred by the presence of ^{6}Li in the otherwise metal-rich dwarfs HD 219542 (Gratton *et al.*, 2001) and HD 82943 (Israelian *et al.*, 2001).

The relationship between metallicity and the presence of short-period planets was used by Laughlin (2000) to construct a sample of 206 metal-rich FGK stars from the Hipparcos database combined with Strömgren *uvby* photometry which have an enhanced probability of harbouring short-period planets, and which would therefore be good candidates for radial velocity surveys. More recently, Fuhrmann (2004) used the Hipparcos trigonometric distances compared with the spectroscopic distances from the nearby exoplanet sample of Santos *et al.* (2001) to show that the latter's surface gravity scale is systematically offset.

Suchkov & Schultz (2001) reported a correlation between *Hp* variability and age for a sample of planet-bearing F stars (their Figure 3), although this may be an artifact caused by the parameters used in stellar age estimation (Suchkov 2006, priv. comm.). Reid (2002) used a control sample of 486 Hipparcos FGK stars to show that the planetary host stars exhibit a velocity distribution well matched to a Gaussian in each component, but with lower dispersions than in the field star sample, suggesting that the average age is only some 60% that of a representative subset of the disk, perhaps reflecting the higher proportion of metal-rich stars in the planet host sample. A comparison with IRAS-based infrared colours, deriving the evolutionary status from the Hipparcos parallaxes and model evolutionary tracks, was made by do Nascimento & de Medeiros (2003). Cayrel de Strobel (2005) studied 99 known host stars with reliable Hipparcos parallaxes and hence absolute magnitudes. For many of them, their space velocities, (U, V, W), are also known. The effective temperatures, spectroscopic gravities and chemical compositions were derived from detailed spectroscopic analyses in the literature. Taking into account the metallicity of each star, the ages of the sufficiently evolved host stars were estimated based on isochrones obtained from metal-poor, solar-metal-normal, and metal-rich stellar models. The resulting age-distribution of the evolved planet hosts is bimodal, reflecting the mean age of the thin disk (4–6 Gyr) and the presence of an older-disk population of about 10–12 Gyr.

An accretion event? Dall *et al.* (2005) argued that the common proper motion system FH Leo, a wide visual binary of separation 8.31 arcsec and observed together by Hipparcos as HIP 54268, and classified in the catalogue as a nova-like variable due to an outburst observed in the Hipparcos photometry (Figure 10.24), cannot be a nova. Further analysis by Vogt (2006) has suggested that it still may be a dwarf nova in a triple system (see Section 3.7), but the debate is still open, and the accretion possibility will be considered further here.

From spectroscopy, and a study of the elemental abundances including Li and α-elements, Dall *et al.* (2005) concluded that the component stars, HD 96273

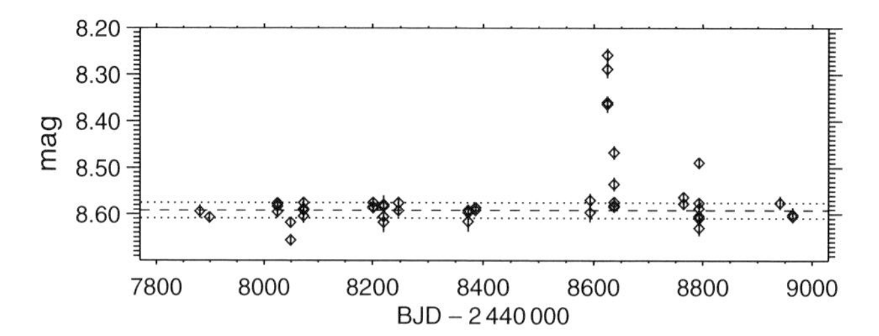

Figure 10.24 Full Hipparcos light-curve for FH Leo. The dashed line is the mean magnitude (not including the outburst); dotted lines are the 1σ standard deviations. Error bars are the Hipparcos intrinsic errors. The authors' favoured explanation for this outburst is a planetary accretion event. From Dall et al. (2005, Figure 1).

and BD+07° 2411 B, do constitute a physical binary, but are of normal late-F and early-G type. At a distance $d = 117$ pc, the lower limit for the physical separation is 936 AU. They studied the Hipparcos light-curve, and concluded that the rise-time might have been very fast, while the decay probably lasted at least 13 days with a possible second event about 170 days later. They discarded instrumental errors, because the duration of the event and the gradual fading argues against erroneous measurements over an extended period of about two weeks, without all other measurements during that period being affected. They examined several possible explanations for the outburst: a transient background or foreground object including the proximity of Jupiter or a background supernova, magnetic interaction with unseen companion, a planetary accretion event, or a microlensing event. Their favoured explanation is a planetary accretion event leaving evidence in the form of an energy outburst and a polluted stellar atmosphere. From the magnitude rise, they estimated a mass of the accreted matter as 5×10^{20} kg, about the mass of a large asteroid like Pallas or Vesta. A scenario where BD+07° 2411 B accreted such a companion could explain both the Hipparcos outbursts, the possible slight over-abundances found in BD+07° 2411 B with respect to HD 96273, and the presence of lithium in BD+07° 2411 B. A test of this explanation would be to measure the ^{6}Li/^{7}Li ratio, and the abundance of beryllium. There has been circumstantial evidence for such accretion events affecting other stars based, for example, on enhancement in the ^{6}Li abundance of the accreting body (e.g. Israelian *et al.*, 2001), but never a conclusive photometric signature of such a rare event.

As well as accretion onto the host star by an inwardly spiraling planet, planet–planet collisions during the early stages of planetary formation are also expected to give rise to measurable extreme ultraviolet/soft X-ray flashes lasting for hours, and a bright infrared afterglow lasting for thousands of years (Zhang & Sigurdsson, 2003). This scenario was not, however, considered.

Slower, less dramatic, but more persistent collisions between asteroids in the Solar System generate a tenuous cloud of dust known as the zodiacal light. In the young Solar System, dust production rates should have been many times larger. Yet copious dust in the zodiacal region around stars much younger than the Sun has rarely been found (cold dust from a Kuiper-belt analogous region out beyond the orbit of Neptune is known around several hundred main-sequence stars). Song *et al.* (2005) reported a large amount of warm, small, silicate dust particles around the solar-type star BD+20 307 (HIP 8920). The composition and quantity of dust could be explained by recent frequent or large collisions between asteroids or other planetesimals whose orbits are being perturbed by a nearby planet.

10.7.5 Proto-planetary disks

During a calibration scan of the photometric standard star α Lyr (Vega), the Infrared Astronomical Satellite (IRAS) found that the flux densities at 25, 60 and 100 μm were in excess of those expected from a blackbody at the temperature of the star (Aumann *et al.*, 1984). During the course of the mission, three other bright main-sequence stars were found to exhibit this infrared excess: α PsA (Fomalhaut), ϵ Eri and β Pic which, together with Vega, have become known as prototypes of the 'Vega-type' phenomenon. The infrared excess emission from these main-sequence stars has been attributed to circumstellar material in the form of a disk or ring. Coronagraphic (including HST), infrared, and submillimetre imaging of these and other Vega-type stars have confirmed that the dust is indeed located in a disk or torus, and have shown that significant clumps of dust can exist in the disks.

There is an intimate connection between protoplanetary disks and the planets which are formed from them. In contrast to the observational difficulties for planets, it is now relatively easy to observe disks. Not only are they extremely large – a typical disk extends to of order 1000 AU from the star – but the surface area of the small particles which make up the disk is many orders of magnitude larger than that of a planet. They emit and reflect light very well, and can be seen to relatively large distances from the central star. Disks also appear to be long-lived, $\sim 10^{6} - 3 \times 10^{7}$ years, and robust against disruption by the events that commonly accompany early stellar evolution. They are remarkably common throughout the Galaxy, and their properties appear quite similar to the picture of our primitive solar nebula. They are particularly evident due to their strong emission at infrared red wavelengths, between about 2 μm and 1 mm, with a spectrum much broader than any single-temperature black body, originating from thermal emission over a wide variation of temperatures, from ∼1000 K very close to the stars to ∼30 K near the outer edges of the disks at several hundred AU from the central star.

The star β Pic (HIP 27321) is a prototype normal main-sequence star surrounded by a circumstellar dust disk seen nearly edge-on (Artymowicz, 1997). The innermost parts of the disk appear to be devoid of gas and dust, and an absence of diffuse material suggests the presence of planets. In the outer disk, silicate dust and sand occur in quantities orders of magnitude larger than those in the present Solar System, but are consistent with a young Solar System, of age ∼100 Myr, in the clearing stage. Cometary-like bodies, perturbations of a planet on the disk, and light variations possibly associated with planets have all been reported for this system.

The Hipparcos parallax placed the star at 19.28 ± 0.19 pc, a value close to the upper limit of previous estimates. This removed speculation that the star was underluminous, and brought models requiring only a very low extinction back into line with the observations (Crifo *et al.*, 1997). The star is located very close to the zero-age main sequence, or on it, suggesting that it is old enough to possess a well-formed planetary system, and implying an age of at least 8 Myr. Barrado y Navascués *et al.* (1999) derived an age estimate in a very different way: they used proper motions from Hipparcos to identify two M dwarfs with almost identical space motions, to within 1 km s^{-1}. Based on a colour–magnitude diagram from accurate photometry and Hipparcos parallaxes, they derived ages of ∼20 Myr by comparison with theoretical evolutionary tracks. They inferred that this small common motion group is likely to be real, and hence derived an age of 20 ± 10 Myr for β Pic.

Kalas *et al.* (2001) studied star passages which have occurred close to β Pic in the past, using similar considerations used for the study of star passages close to the Earth (Section 10.5), to test the hypothesis that its planetesimal disk has been disrupted by one or more recent (10^5–10^6 yr) close stellar encounters, perhaps resulting in the asymmetric disk structure observed at large radii. Using Hipparcos astrometry and supplementary radial velocities in a star sample that they consider is about 20% complete, they traced the space motions of 21 497 stars and identified 18 that have passed within 5 pc in the past 1 Myr, with four probably having penetrated

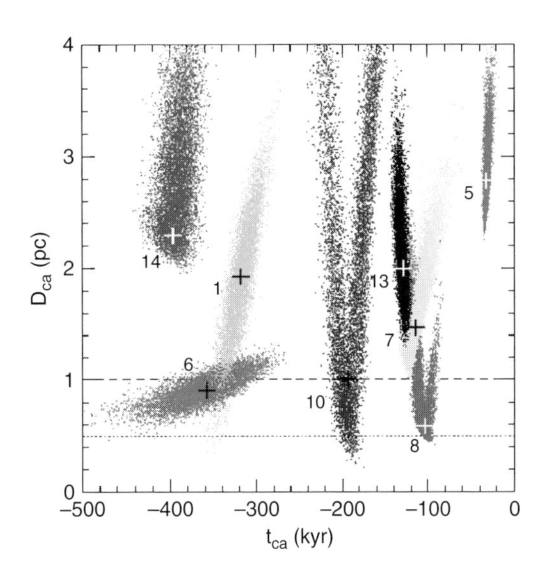

Figure 10.25 Positions in the closest approach plane (D_{ca}, t_{ca}) for eight of the 18 candidate perturbers of β Pic with $D_{ca} < 3$ pc. Each numbered 'cloud' corresponds to possible approaches for a different star passage: the individual points are from a Monte Carlo analysis, where the spread in the distribution of points reflects the initial uncertainties on the stellar proper motions, parallaxes, and radial velocities. Crosses mark the maxima of the probability distributions. The horizontal dashed line marks $D = 1.0$ pc (the Roche radius of β Pic set by the Galactic tidal field) and the dotted line marks $D = 0.5$ pc. From Kalas et al. (2001, Figure 1, original in colour).

the Roche radius, set by the Galactic tidal field, in the past 1 Myr (Figure 10.25). Its closest encounter was with a K2 III star at about 0.6 pc, but dynamically the most important encounter was with an F7 V star at about 0.9 pc some 0.3 Myr in the past. From the velocity and eccentricity changes induced by the 18 perturbations, they inferred that they are dynamically significant if planetesimals exist in a β Pic Oort-type cloud, concluding that the primary role of such stellar perturbations would be to help build a comet cloud rather than to destroy a pre-existing structure. The probability that encounters penetrated within 0.01 pc is negligible, meaning that the inferred encounters cannot account for the observed large-scale disk asymmetries.

Deltorn & Kalas (2001) used the same Hipparcos-based space motions of 21 497 stars, and the same methodology, to search for close stellar encounters over the past 1 Myr with three other nearby stars with asymmetric debris disks: Vega, ϵ Eri and Fomalhaut. Thus ϵ Eri has apparently experienced at least three < 2 pc encounters over the past 100 000 yr, with an ongoing close encounter with Kapteyn's star and a 42% probability that the closest approach distance is < 1 pc. Vega and Fomalhaut experienced respectively four and six

$<$ 2 pc encounters over the past 1 Myr. These encounter rates are comparable to those found for the Sun (Section 10.5). Each had one encounter with a roughly 2% probability that the closest approach distance is less than 0.5 pc. These encounters will not directly influence the debris disks observed around Vega, ϵ Eri and Fomalhaut, but they may pass through hypothetical Oort clouds surrounding them. They found that two other Vega-type stars, HD 17848 and HD 20010, experienced $<$0.1 pc stellar encounters that are more likely to directly perturb their circumstellar disks. The greatest limitation in conducting these studies was with the incompleteness or inaccuracy of the radial velocity catalogues.

In contrast to β Pic, most disks occur around young stars, in particular around T Tauri stars which lie close to star-forming clouds, and with an excess of infrared emission most likely arising from dusty material in orbit around, and heated by, the central star. Calculations imply that the disk phase of planetary formation lasts for only a relatively small fraction of a system's total lifetime, although some stars may maintain disks for a billion years or more, perhaps never forming any planets.

Many disks have been detected through their infrared, sub-mm, or radio emission, frequently showing a roughly flattened shape, with Doppler measurements indicating rotation. Dust disks have been observed around some 100 main-sequence stars within 50 pc of the Sun, for example, around the binary BD +31°643; ϵ Eri; HR 4796; the pre-main-sequence binary HK Tau, and around HD 98800, a very young bright planetary debris system bearing a strong similarity to the zodiacal dust bands in our own Solar System. Observations of comets have been made in HD 100546. A number of disk systems have been imaged using the Hubble Space Telescope, including β Pic, and HD 141569.

Hipparcos-derived distances, and hence luminosities, have been used to argue that some 20% of all A-type stars pass through an early phase where they possess an amount of circumstellar dust comparable to that found around HR 4796 or β Pic (Jura *et al.*, 1998). Hipparcos distances have been used in discussions of the physical characteristics of various disks, for example SCUBA photometry (Sylvester *et al.*, 2001); in a search for Vega-type stars using a correlation of the Hipparcos and IRAS Faint Source Catalogue (Song *et al.*, 2002); for studies of the M III star HD 97048 (Piétu *et al.*, 2003); and in a study of their kinematics, lifetimes and temporal evolution (Manoj & Bhatt, 2005). Rhee *et al.* (2007) correlated the Hipparcos distances with IRAS luminosities, and confirmed that the dust population decreases with age. The parallaxes have also been used to estimate the number of nearby potential targets for a search for exo-zodiacal dust emission with the next generation of millimeter-wave telescopes (Simon, 1998).

Hipparcos-based kinematic studies have led to the discovery or confirmation of various young moving groups in the solar neighbourhood, with ages in the range ~10–30 Myr, notably the β Pic, AB Dor, Tucana–Horologium and TW Hya moving groups (Section 6.10.4). These provide a sample of nearby stars suitable for the study of ongoing formation of planetary systems (e.g. Zuckerman *et al.*, 2004). For one candidate exoplanet companion, to the young 2MASS brown dwarf 2M 1207, a moving cluster distance estimate of 53 ± 6 pc was derived by Mamajek (2005) and a trigonomeric parallax distance of 63 ± 7.5 pc was derived by Biller & Close (2007).

10.7.6 Habitability and related issues

Assessment of the suitability of a planet for supporting life, or habitability, is based on our knowledge of life on Earth. The habitable zone is consequently presently defined by the range of distances from a star where liquid water can exist on the planet's surface. For a given stellar luminosity, this is primarily controlled by the star–planet separation, but is affected by factors such as planet rotation combined with atmospheric convection. For Earth-like planets orbiting main-sequence stars, the inner edge is bounded by water loss and the runaway greenhouse effect, as exemplified by the CO_2-rich atmosphere and resulting temperature of Venus. The outer boundary is determined by CO_2 condensation and runaway glaciation, but it may be extended outwards by factors such as internal heat sources including long-lived radionuclides (U^{235}, U^{238}, K^{40}, etc., as on Earth), tidal heating due to gravitational interactions (as in the case of Jupiter's moon Io), and pressure-induced far-infrared opacity of H_2, since even for effective temperatures as low as 30 K, atmospheric basal temperatures can exceed the melting point of water. These considerations result, for a 1 $M_\odot$ star, in an inner habitability boundary at about 0.7 AU and an outer boundary at around 1.5 AU or beyond. The habitable zone evolves outwards with time because of the increasing Sun's luminosity with age, resulting in a narrower width of the continuously habitable zone over ~4 Gyr of around 0.95–1.15 AU. Positive feedback due to the greenhouse effect and planetary albedo variations, and negative feedback due to the link between atmospheric CO_2 level and surface temperature may limit these boundaries further.

Within the ~1 AU habitability zone, Earth 'class' planets can be considered as those with masses between about 0.5–10 $M_{\rm Earth}$ or, equivalently, radii between 0.8–2.2 $R_{\rm Earth}$. Planets below this mass in the habitable zone are likely to lose their life-supporting atmospheres because of their low gravity and lack of plate tectonics, while more massive systems are unlikely to be habitable because they can attract a hydrogen–helium atmosphere and become gas giants.

Habitability is also likely to be governed by the range of stellar types for which life has enough time to evolve, i.e. stars not more massive than spectral type A. However, even F stars have narrower continuously habitable zones because they evolve more rapidly, while late K and M stars may not host habitable planets because they can become trapped in synchronous rotation due to tidal damping. Mid- to early-K and G stars may therefore be optimal for the development of life. Other effects complicate considerations of habitability: in the absence of our Moon, thought to be an accident of accretion, an Earth-like planet would undergo large-amplitude chaotic fluctuations due to secular resonances (spin-axis/orbit axis or spin-axis precession/perihelion precession) on time scales of order 10 Myr, depending also on the planet's land–sea distribution.

Habitability may be further confined within a narrow range of [Fe/H] of the parent star (e.g. Gonzalez, 1999b, 2005). If the occurrence of gas giants decreases at lower metallicities, their shielding of inner planets in the habitable zone from frequent cometary impacts, as occurs in the Solar System, would also be diminished. At higher metallicity, asteroid and cometary debris left over from planetary formation may be more plentiful, enhancing impact probabilities.

Gonzalez (1999a, and references therein) has investigated whether the anomalously small motion of the Sun with respect to the local standard of rest, both in terms of its pseudo-elliptical component within the Galactic plane, and its vertical excursion with respect to the mid-plane, may be explicable in anthropic terms. Such an orbit could provide effective shielding from damage to the Earth's ozone layer from high-energy ionising photons and cosmic rays from nearby supernovae, from the diffuse X-ray background by neutral hydrogen in the Galactic plane, and from increases in the perturbed Oort cloud comet impact rate due to Galactic tidal forces and z tides as the Sun oscillates perpendicular to the plane. In this sense, both the very small height of the Sun from the Galactic mid-plane (Section 9.2.2), and its apparent proximity to the spiral arm co-rotation circle (Section 9.7; the further it is from co-rotation, the more frequently it crosses the spiral arms, with the possible consequences described in Section 10.6.2) may endow the Sun with more stable and benign conditions necessary for the evolution of life. This discussion moves in the direction of the Weak Anthropic Principle, formulated by Barrow & Tipler (1986, p16) as *'The observed values of all physical and cosmological quantities are not equally probable but they take on values restricted by the requirement that there exist sites where carbon-based life can evolve and by the requirement that the Universe be old enough for it to have already done so'*, and discussed in this context by Gonzalez (1999a) and Reid (2002). The study of Barbieri & Gratton (2002), relying on Hipparcos-based space velocities, suggests that the Galactic orbits of stars with and without planets are not kinematically different, although the same conclusion is not necessarily true for planets harbouring intelligent life. Because of the general increase of metallicity with perigalactic distance, this in turn was taken to imply that a higher metallicity of the parent protostellar nebula does not significantly enhance the probability of planet formation, lending weight to scenarios where the the presence of planets is the cause of the higher metallicity, rather than *vice versa*.

The detection of habitable-zone Earth-mass planets around solar-type stars will be a challenge for radial velocity measurements, where a precision of $1\,\mathrm{m\,s^{-1}}$ corresponds to a planetary mass of around $10\,M_{\rm Earth}$ for a planet in a 1 AU orbit. While transit searches are more sensitive to habitable planets around late-type stars, Gould *et al.* (2003) evaluated astrometric detection of habitable-zone planets around early-type stars. Being relatively massive is generally unfavourable for astrometric detection since the astrometric signal falls off inversely with the stellar mass, and being relatively rare and hence typically more distant, the astrometric signal again falls inversely with distance for a given orbital separation. However, their higher luminosity means that their habitable zones are at larger orbital distances. Gould *et al.* (2003) used the Hipparcos Catalogue to show that F and A stars represent the majority of viable targets for astrometric searches with SIM, for planets with semi-major axes currently in the habitable zone, $a \sim 1\,\mathrm{AU}\,\sqrt{L/L_\odot}$. For planetary periods $P < 5$ yr and planets of mass $M_p < 3M_{\rm Earth}$, they find 19 Hipparcos candidate stars with $M < M_\odot$ and 30 with $M > M_\odot$.

10.7.7 Solar twins and solar analogues

Solar twins are, by definition, non-binary stars (currently) identical to the Sun in all astrophysical parameters: mass, age, luminosity, chemical composition, temperature, surface gravity, magnetic field, rotation velocity, and chromospheric activity. Such stars are considered as being those most likely to possess planetary systems similar to our own, and best-suited to host life forms based on carbon chemistry and water oceans. Solar analogues are those that looked in the past, or will look in the future, very similar to the Sun, thus providing a look at the Sun at some other point in its evolution.

The systematic search for and study of solar twins and analogues started with the work of Hardorp (1978), who surveyed the near-ultraviolet (360–410 nm) spectra of 77 solar-type stars in parts of the northern and southern hemispheres, finding no G2V star which matches the properties of the Sun. The pre-Hipparcos status is reviewed by Cayrel de Strobel (1996). Starting with 109 photometric solar-like candidates from various authors, they were also unable to identify a 'perfect' twin although, interestingly, two of the first three exoplanetary systems discovered, 51 Peg and 47 UMa, were on the starting list. The G2 star HD 146233 (HR 6060 = HIP 79672 = 18 Sco) at 14 pc comes very close to being such a twin, although with slightly higher luminosity and age as indicated by the Hipparcos parallax (Porto de Mello & da Silva, 1997; Cayrel de Strobel & Friel, 1998). The two G components of the binary 16 Cyg A/B were considered to be the next closest twins.

Solar twins Porto de Mello *et al.* (2000) reported a survey of solar twins within 50 pc using Hipparcos parallaxes. Pinho *et al.* (2003) and Pinho & Porto de Mello (2003) studied the 50 solar-type stars (FGK dwarfs or subgiants) within 10 pc of the Sun, selected from the Hipparcos Catalogue with $0.4 < (B - V) < 1.15$ and $2 < M_V < 8$. A detailed analysis of their evolutionary state, atmospheric parameters, chemical composition, multiplicity, and chromospheric activity indicates that only 80% of these stars should be retained in lists of truly interesting targets for exobiology. Porto de Mello *et al.* (2006) updated their study with detailed consideration of the presence of a suitable continuously habitable zone, based on Hipparcos parallaxes, derived Galactic orbits, and variability data. They identified 33 candidates as astrobiologically interesting stars within 10 pc, suitable for study by space infrared interferometry, headed by HD 1581, 109358, and 115617 which most closely resemble the Sun.

Galeev *et al.* (2004) made a spectroscopic analysis of 15 stars that are close in photometric properties to the Sun, with abundances determined for 33 elements ranging from Li to Eu. Effective temperatures and surface gravities were derived from published photometric indices and Hipparcos parallaxes. They showed that stars photometrically similar to the Sun can be divided into three groups according to their elemental abundances: six have solar chemical composition, four have abundance excesses, and five have some abundance deficiencies. The sample contains two metal-deficient subgiants (HD 133002 and HD 225239), demonstrating that photometric similarity is not a sufficient criterion to consider a star as a solar twin. When several criteria, including chemical composition, are simultaneously taken into account, only four stars from the sample can be considered truly solar like: HD 10307, HD 34411, HD 146233, and HD 186427 (16 Cyg B). The results confirm the previous suggestions that HR 6060 (18 Sco) is the most probable twin of the Sun: essentially all the parameters of the two stars coincide within the errors.

King *et al.* (2005) reported Keck spectroscopy of four solar twin candidates from their Hipparcos-based Ca II H and K survey: HIP 71813 (HD 129357), HIP 76114 (HD 138573), HIP 77718 (HD 142093), and HIP 78399 (HD 143436). Hipparcos distances were used to derive absolute luminosities and hence ages by comparison with evolutionary models. They also derived the Galactic kinematics from the Hipparcos proper motions. While evidently solar-type stars, HIP 76114 and 77718 are a few percent less massive, significantly older, and more metal-poor than the Sun; they are neither good solar twin candidates nor solar analogues. HIP 71813 appears to be an excellent solar analogue of age ~8 Gyr. Their results for HIP 78399 suggest that it may be as close a solar twin as HR 6060.

As part of their spectroscopic survey of nearly 3600 Hipparcos dwarf and giant stars earlier than M0 within 40 pc, Gray *et al.* (2006) have provided a list of 61 dwarf stars that have spectral types between G0–G5, $\log g > 4.20$, $[M/H] > -0.10$, and are single or members of wide doubles, distinguishing stars that have spectral types, basic physical parameters, and activity levels that are close, or very close, to those of the Sun. Their 14 objects in the latter class are considered as solar twin candidates. They also indicate those that have known exoplanets and those that are currently on the Keck, Lick, and AAT Doppler planet-search programme. Only HD 186427 appears in the above-mentioned lists of twin candidates, and they consider this as being close, rather than very close, to the properties of the Sun.

Solar analogues As noted above, a solar analogue provides a look at the Sun at some other point in its evolutionary phase. Gaidos (1998) identified 38 young solar analogues within 25 pc, selecting G and early K stars from the Hipparcos Catalogue, based on lack of known stellar companions within 800 AU, bolometric luminosities close to that of the zero-age Sun and consistent with the zero-age main sequence, and ROSAT X-ray luminosities commensurate with the higher rotation rate and level of dynamo-driven activity in solar-mass stars less than 0.8 Gyr old.

Fuhrmann (2004) discuss the young solar analogue HD 63433. With $T_{\rm eff} = 5650$ K and $[Fe/H] = -0.02$, the star has a mass of about $1\,M_\odot$ with an age of 200 Myr. It represents what the Sun could have been like 4.3 billion years ago. According to García Sánchez *et al.* (2001) the star, currently at 21.8 pc, will pass the Solar System in 1.3 Myr at a distance of only 2.1 pc, providing *'good prospects to study the young Sun in detail'*. The nearest subgiant, β Hyi (Dravins *et al.*, 1998), also appears to be a close solar analogue of age ~ 6.7 Gyr (Section 7.5.4). King *et al.* (2005) noted that HIP 71813 appears to be an excellent solar analogue of age ~ 8 Gyr.

SETI target list: The creation of the Catalogue of Habitable Stellar Systems was motivated by a need for an expanded target list for use in the search for extraterrestrial intelligence by Project Phoenix of the SETI Institute. Project Phoenix is a privately funded continuation of NASA's High Resolution Microwave Survey, a programme to search for continuous and pulsed radio signals generated by extra-solar technological civilizations. This consisted of an all-sky survey at 1–10 GHz, as well as a targeted search of 1000 nearby stars at higher spectral resolution and sensitivity in the 1–3 GHz range. Although Congress terminated the survey in 1993, the SETI Institute raised private funds to continue the targeted portion of the search as Project Phoenix, which now carries out observations at the Arecibo Observatory in conjunction with simultaneous observations from the Lovell Telescope at Jodrell Bank, UK. The project uses a total of three weeks of telescope time per year and observes around 200 stars per year. In a joint effort by the SETI Institute and the University of California at Berkeley, the Allen Telescope Array is currently being constructed at Hat Creek Observatory in northern California. The array consists of 350 dishes, each 6.1 m in diameter, a bandwidth covering 0.5–11 GHz, and a capability of observing some 10 000 target stars per year. Hence the observing list for Project Phoenix had to be greatly expanded from its original scope of about 2000 of the nearest and most Sun-like stars.

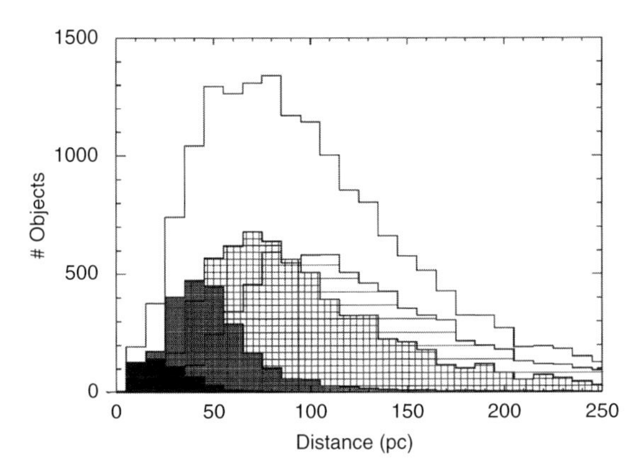

Figure 10.26 Number of potentially habitable stars as a function of distance for M stars (solid histogram), K stars (dark-hatched histogram), G stars (light-hatched histogram), F stars (horizontal-lined histogram), and all stars (open histogram). The farthest star in their Catalogue of Nearby Habitable Systems is at about 300 pc, and transmitted power comparable to the Arecibo planetary radar will be detectable by the Allen Telescope Array at this distance. From Turnbull & Tarter (2003b, Figure 11).

10.7.8 Search for extraterrestrial intelligence

Two papers made extensive use of the Hipparcos and Tycho Catalogues in defining target catalogues for use in the search for extraterrestrial intelligence (SETI). In

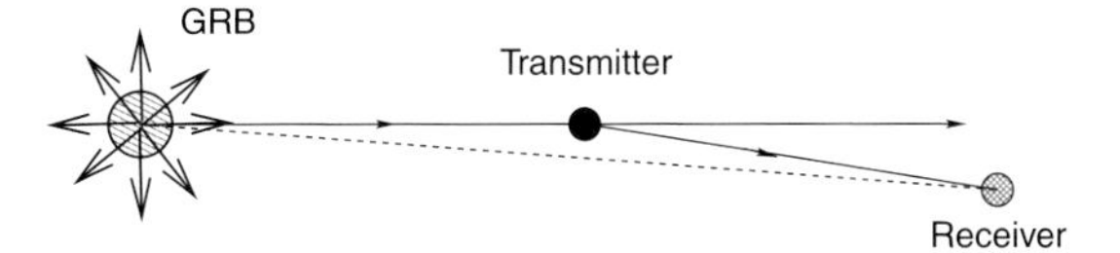

Figure 10.27 Geometry of the SETI search configuration described by Corbet (1999). On receipt of a gamma-ray burst pulse (GRB), the transmitting civilisation despatches a signal 180° away on the sky. A receiving civilisation focuses its detection strategy on candidate habitable stars in the direction of the gamma-ray burst. Although the path difference GRB–Transmitter–Receiver compared with GRB–Receiver can be days or months, the time of the signalling event can be predicted with a precision of hours or minutes depending on the knowledge of the astrometric position of the GRB with respect to the transmitter.

preparation for their Allen Telescope Array, the SETI Institute expanded its former list of about 2000 targets compiled for Project Phoenix. Turnbull & Tarter (2003b) constructed a catalogue of stellar systems that are potentially habitable to complex life forms (including intelligent life), which comprises the largest portion of the new SETI target list. Habitability is defined in their work as a stellar system in which an Earth-like planet could have formed and supported liquid water throughout the last three billion years. Their Catalogue of Nearby Habitable Systems was created from the Hipparcos Catalogue using the information on distances (for signal propagation), variability (for climate stability), multiplicity (for orbital stability), kinematics (as a metallicity indicator), and spectral classification (as a criterion for various habitability conditions) for the 118 218 catalogue stars, complemented by data from several other catalogues containing data on X-ray luminosity, Ca II H and K activity, rotation, spectral types, kinematics, metallicity, and Strömgren photometry. Combined with theoretical studies on habitable zones, evolutionary tracks, and three-body orbital stability, these data were used to remove unsuitable stars, leaving a residue of stars that, according to present knowledge, are potentially habitable hosts for complex life. The analysis resulted in 17 129 Hipparcos habitable star candidates near the Sun, of which 75% lie within 140 pc, and some 2200 of which are known or suspected to be members of binary or triple systems (Figure 10.26). Turnbull & Tarter (2003a) presented the full target list and prioritization algorithm developed for use by the microwave search for technological signals at the SETI Institute. They included the Catalogue of Nearby Habitable Systems, all of the nearest 100 stars (out to about 7 pc, see Section 9.3) and 14 old open clusters, and further augmented by a subset of the Tycho 2 Catalogue based on reduced proper motions. This larger catalogue should routinely provide at least three target stars, each acting as an off-source reference for the others, within the large primary field of view of the Allen Telescope Array. The algorithm for prioritising objects in the full target list includes scoring based on the category of each target, its distance, and its proximity to the Sun in the colour–magnitude diagram.

A novel application of the Hipparcos data is the use of the accurate parallaxes for a specific type of SETI search described by Corbet (1999). Detecting a civilisation's signal over a large distance is clearly most efficient when transmitting and receiving a brief and beamed signal, rather than a continuous omnidirectional transmission. For such a scheme to be feasible, the transmitter and the recipient, both unknown to the other, must find a strategy that will enable the transmitter and receiver to transmit and observe at an appropriate time and in an appropriate direction. Strategies to achieve this synchronisation have been developed based on natural astronomical events (e.g. Pace & Walker, 1975) and in particular using as temporal and spatial marker the frequent, powerful, and short-duration gamma-ray bursts (see Corbet, 1999, and references therein). In this approach, the search on Earth is concentrated in the direction of a candidate habitable star which has been selected (from amongst a presumably large number of candidates distributed over the sky) to lie closely aligned, to within an angle θ, with a specific celestial burst event; the signalling event despatched by an alien civilisation has been correspondingly transmitted in a direction roughly anti-aligned with the same celestial event, i.e. the gamma-ray burst, the alien civilisation, and the Earth, are roughly co-aligned and in that sequence (Figure 10.27). Accurate knowledge of the distance to the target planet-hosting star allows the time delay between the gamma-ray burst event (the time marker), and the signalling event being sought, to be estimated from simple trigonometry. Ignoring the uncertainties associated with the gamma-ray burst position, a planet-hosting star at $d = 20$ pc from Earth, and at an angle of 1° from the direction of a gamma-ray burst event seen from Earth, will transmit an event which will be detected 3.63 days later, but one whose arrival time on Earth can be predicted with a precision of ± 1.8 hours (their Table 1). The timing uncertainty would drop to just 1 minute in the case of 10 μas parallaxes from SIM or Gaia. A strategy based on observations at opposition, i.e. in the anti-Sun direction, and benefiting from Hipparcos-level accuracy proper motions or better, is described by Corbet (2003).

References

Ammons SM, Robinson SE, Strader J, *et al.*, 2006, The N2K Consortium. IV. New temperatures and metallicities for more than 100 000 FGK dwarfs. *ApJ*, 638, 1004–1017 {**602**}

Andrei AH, Penna JL, Benevides-Soares P, *et al.*, 1997, Astrolabe observations of the Sun on the Hipparcos reference frame. *ESA SP–402: Hipparcos, Venice '97*, 161–164 {**579**}

Antonov VA, Latyshev IN, 1972, Determination of the form of the Oort Cloud as the Hill surface in the Galactic field. *IAU Symp. 45: The Motion, Evolution of Orbits, and Origin of Comets* (eds. Chebotarev GA, Kazimirchak-Polonskaia EI, Marsden BG), 341–345 {**584**}

Applegate JH, Douglas MR, Gursel Y, *et al.*, 1986, The outer Solar System for 200 million years. *AJ*, 92, 176–194 {**572**}

Arenou F, Palasi J, 2000, Upper limit masses from Hipparcos astrometry. *IAU Symp. 202*, 35–35 {**594**}

Argus DF, Gordon RG, 1991, No-net-rotation model of current plate velocities incorporating plate motion model NUVEL-1. *Geophys. Res. Lett.*, 18, 2039–2042 {**580**}

Arlot JE, Thuillot W, Fienga A, *et al.*, 1999, Observations of the natural planetary satellites for dynamical and physical purpose. *AAS/Division for Planetary Sciences Meeting*, 31 {**571**}

Arlot JE, 1982, New constants for Sampson–Lieske theory of the Galilean satellites of Jupiter. *A&A*, 107, 305–310 {**571**}

Artymowicz P, 1997, β Pic: an early Solar System? *Annual Review of Earth and Planetary Sciences*, 25, 175–219 {**605**}

Aslan Z, Gumerov RI, Hudkova LA, *et al.*, 2006, Preliminary results in asteroid mass determination. *Nomenclature, Precession and New Models in Fundamental Astronomy, IAU Joint Discussion 16*, 16, 67–70 {**570**}

Aumann HH, Beichman CA, Gillett FC, *et al.*, 1984, Discovery of a shell around α Lyr. *ApJ*, 278, L23–L27 {**604**}

Baines EK, van Belle GT, ten Brummelaar TA, *et al.*, 2007, Direct measurement of the radius and density of the transiting exoplanet HD 189733b with the CHARA array. *ApJ*, 661, L195–L198 {**601**}

Baliunas SL, Donahue RA, Soon WH, *et al.*, 1995, Chromospheric variations in main sequence stars. *ApJ*, 438, 269–287 {**586**}

Baliunas SL, Jastrow R, 1990, Evidence for long-term brightness changes of solar-type stars. *Nature*, 348, 520–523 {**586**}

Bange JF, 1998, An estimation of the mass of asteroid (20) Massalia derived from the Hipparcos minor planets data. *A&A*, 340, L1–L4 {**568**}

Barbieri M, Gratton RG, 2002, Galactic orbits of stars with planets. *A&A*, 384, 879–883 {**607**}

Barrado y Navascués D, Stauffer JR, Song I, *et al.*, 1999, The age of β Pic. *ApJ*, 520, L123–L126 {**605**}

Barrow JD, Tipler FJ, 1986, *The Anthropic Cosmological Principle*. Clarendon Press, Oxford {**607**}

Basri G, 2000, The discovery of brown dwarfs. *Scientific American*, 282, 57–63 {**596**}

Batalha NM, Jenkins J, Basri G, *et al.*, 2002, Stellar variability and its implications for photometric planet detection with Kepler. *ESA SP–485: First Eddington Workshop on Stellar Structure and Habitable Planet Finding* (eds. Favata F, Roxburgh IW, Galadi D), 35–40 {**601**}

Batrakov YV, Chernetenko YA, Gorel GK, *et al.*, 1999, Hipparcos Catalogue orientation as obtained from observations of minor planets. *A&A*, 352, 703–711 {**571, 576**}

Bazot M, Vauclair S, Bouchy F, *et al.*, 2005, Seismic analysis of the planet-hosting star μ Ara. *A&A*, 440, 615–621 {**603**}

Becklin EE, Zuckerman B, 1988, A low-temperature companion to a white dwarf star. *Nature*, 336, 656–658 {**596**}

Begelman MC, Rees MJ, 1976, Can cosmic clouds cause climatic catastrophes? *Nature*, 261, 298–299 {**587**}

Bello D, Leroux CR, Le Roux B, *et al.*, 2002, Performances of a NGS-based MCAO demonstrator: the NGC 3366 and NGC 2346 simulations. *Beyond Conventional Adaptive Optics. ESO Proceedings Vol. 58* (eds. Vernet E, Ragazzoni R, Esposito S, *et al.*), 231 {**574**}

Benedict GF, McArthur B, Chappell DW, *et al.*, 1999, Interferometric astrometry of Proxima Centauri and Barnard's star using Hubble Space Telescope FGS3: detection limits for substellar companions. *AJ*, 118, 1086–1100 {**594**}

Bernstein HH, 1997, Astrometric indications of brown dwarfs based on Hipparcos data. *ESA SP–402: Hipparcos, Venice '97*, 705–708 {**596, 597**}

—, 2003, Detection of brown dwarfs with astrometric satellites. *IAU Symp. 211* (ed. Martín E), 505 {**596**}

Berthier J, Marchis F, Descamps P, *et al.*, 2004, Prediction of stellar occultations by satellite of asteroids. *Bulletin of the American Astronomical Society*, 1142 {**573**}

Berthier J, Marchis F, 2001, A tool for observations of Centaurs and Kuiper Belt Objects with adaptive optics systems. *Bulletin of the American Astronomical Society*, 1049 {**574**}

Bessel FW, 1830, *Tabulae Regiomontanae reductionum observationum astronomicarum ab anno 1750 ad 1850.* Regiomonti Prussorum: Borntraeger {**576**}

Biermann L, Huebner WF, Lust R, 1983, Aphelion clustering of 'new' comets: star tracks through Oort's Cloud. *Proceedings of the National Academy of Science*, 80, 5151–5155 {**584**}

Biller BA, Close LM, 2007, A direct distance and luminosity determination for a self-luminous giant exoplanet: the trigonometric parallax to 2MASSW J1207334-393254Ab. *ApJ*, 669, L41–L44 {**606**}

Black DC, Scargle JD, 1982, On the detection of other planetary systems by astrometric techniques. *ApJ*, 263, 854–869 {**592**}

Bouchy F, Udry S, Mayor M, *et al.*, 2005, ELODIE metallicity-biased search for transiting hot Jupiters. II. A very hot Jupiter transiting the bright K star HD 189733. *A&A*, 444, L15–L19 {**601**}

Bougeard ML, Bange JF, Caquineau C, *et al.*, 1997, Robust estimation with application to Hipparcos minor planet data. *ESA SP–402: Hipparcos, Venice '97*, 165–168 {**576**}

Bretagnon P, 1982, Theory for the motion of all the planets: the VSOP82 solution. *A&A*, 114, 278–288 {**566**}

Bretagnon P, Chapront J, 1981, A note on the numerical expressions for precession calculations. *A&A*, 103, 103–107 {**576**}

Bretagnon P, Fienga A, Simon JL, 2003, Expressions for precession consistent with the IAU 2000A model. Considerations about the ecliptic and the Earth orientation parameters. *A&A*, 400, 785–790 {**576**}

Bretagnon P, Francou G, Rocher P, *et al.*, 1998, SMART97: a new solution for the rotation of the rigid Earth. *A&A*, 329, 329–338 {**576**}

Bretagnon P, Francou G, 1988, Planetary theories in rectangular and spherical variables: VSOP87 solutions. *A&A*, 202, 309–315 {**576**}

Bretagnon P, Rocher P, Simon JL, 1997, Theory of the rotation of the rigid Earth. *A&A*, 319, 305–317 {**576**}

Brucker M, Grundy WM, Stansberry J, *et al.*, 2007, Determining the radii of sixteen trans-Neptunian bodies through thermal modeling. *AAS/Division for Planetary Sciences Meeting Abstracts*, volume 39, 52.05 {**573**}

Burgasser AJ, Kirkpatrick JD, Brown ME, *et al.*, 2002, The spectra of T dwarfs. I. Near-infrared data and spectral classification. *ApJ*, 564, 421–451 {**596**}

Burrows A, Hubbard WB, Saumon D, *et al.*, 1993, An expanded set of brown dwarf and very low-mass star models. *ApJ*, 406, 158–171 {**596**}

Burrows A, Liebert J, 1993, The science of brown dwarfs. *Reviews of Modern Physics*, 65, 301–336 {**596**}

Burrows A, Marley M, Hubbard WB, *et al.*, 1997, A non-gray theory of extrasolar giant planets and brown dwarfs. *ApJ*, 491, 856–875 {**599**}

Capitaine N, Wallace PT, Chapront J, 2003, Expressions for IAU 2000 precession quantities. *A&A*, 412, 567–586 {**576, 577**}

—, 2005, Improvement of the IAU 2000 precession model. *A&A*, 432, 355–367 {**577**}

Carpino M, 1998, Prospects in astrometry and orbital determination of minor bodies. *Planet. Space Sci.*, 47, 29–34 {**570**}

Carslaw KS, Harrison RG, Kirkby J, 2002, Cosmic rays, clouds, and climate. *Science*, 298, 1732–1737 {**587**}

Castellano TP, 2001, Discovering transits of HD 209458-type planets with Hipparcos and FAME photometry. Ph.D. Thesis {**600**}

Castellano TP, Jenkins J, Trilling DE, *et al.*, 2000, Detection of planetary transits of the star HD 209458 in the Hipparcos data set. *ApJ*, 532, L51–L53 {**600**}

Cayrel de Strobel G, Friel ED, 1998, Hipparcos results for solar analogues. *Solar Analogues: Characteristics and Optimum Candidates*, 93–104 {**608**}

Cayrel de Strobel G, 1996, Stars resembling the Sun. *A&A Rev.*, 7, 243–288 {**608**}

—, 2005, The study of stars with planets. *ESA SP–576: The Three-Dimensional Universe with Gaia* (eds. Turon C, O'Flaherty KS, Perryman MAC), 267–270 {**603**}

Chabrier G, Baraffe I, Allard F, *et al.*, 2000, Evolutionary models for very low-mass stars and brown dwarfs with dusty atmospheres. *ApJ*, 542, 464–472 {**599**}

Chandler SC, 1891, On the variation of latitude. I. *AJ*, 11, 59–61 {**578**}

Chapront-Touzé M, Chapront J, 1983, The lunar ephemeris ELP 2000. *A&A*, 124, 50–62 {**566**}

Charbonneau D, Brown TM, Latham DW, *et al.*, 2000, Detection of planetary transits across a Sun-like star. *ApJ*, 529, L45–L48 {**598**}

Charlot P, Sovers OJ, Williams JG, *et al.*, 1995, Precession and nutation from joint analysis of radio interferometric and lunar laser ranging observations. *AJ*, 109, 418–427 {**576**}

Clemence GM, 1948, On the system of astronomical constants. *AJ*, 53, 169–179 {**577**}

Corbet RHD, 1999, The use of gamma-ray bursts as direction and time markers in SETI strategies. *PASP*, 111, 881–885 {**609, 610**}

—, 2003, Synchronized SETI: the case for 'opposition'. *Astrobiology*, 3, 305–315 {**610**}

Crifo F, Vidal-Madjar A, Lallement R, *et al.*, 1997, β Pic revisited by Hipparcos. *A&A*, 320, L29–L32 {**605**}

Crowell JC, 1999, *Pre-Mesozoic Ice Ages.* Geological Society of America, Boulder {**588, 589**}

Dahn CC, Harris HC, Vrba FJ, *et al.*, 2002, Astrometry and photometry for cool dwarfs and brown dwarfs. *AJ*, 124, 1170–1189 {**596, 597, 599**}

Dall TH, Schmidtobreick L, Santos NC, *et al.*, 2005, Outbursts on normal stars. FH Leo misclassified as a nova-like variable. *A&A*, 438, 317–324 {**603, 604**}

Damljanović G, 1996, The Hipparcos mission and the re-reduction of Belgrade zenith-telescope observations. *IAU Symp. 172: Dynamics, Ephemerides, and Astrometry of the Solar System*, 489–490 {**579**}

—, 1997, Some long-period variations of BLZ latitude observations. *Joint European and National Astronomical Meeting*, 227 {**580**}

—, 2000, The analysis of PIP and BLZ latitude data and the free diurnal nutation. *Models and Constants for Sub-microarcsecond Astrometry, IAU Joint Discussion 2*, 24–27 {**579, 580**}

—, 2005, Photographic zenith tubes observations to improve Hipparcos proper motion in declination of some stars. *Serbian Astronomical Journal*, 170, 127–132 {**581**}

Damljanović G, Pejovic N, Jovanovic B, 2006, Improvement of Hipparcos proper motions in declination. *Serbian Astronomical Journal*, 172, 41–51 {**581**}

Damljanović G, Pejovic N, 2006a, Corrections of proper motions in declination by using ILS data. *Serbian Astronomical Journal*, 173, 95–99 {**581**}

—, 2006b, Ground-based latitude data and the improved reference frame. *Publications de l'Observatoire Astronomique de Beograd*, 80, 325–325 {**581**}

Damljanović G, Vondrák J, 2005, Improved proper motions in declination of Hipparcos stars derived from observations of latitude. *Fundamental Astronomy: New Concepts and Models for High Accuracy Observations* (ed. Capitaine N), 230–231 {**581**}

—, 2006, Better accuracy of Hipparcos proper motions in declination for stars observed with 10 photographic zenith tubes. *Nomenclature, Precession and New Models in Fundamental Astronomy, IAU 26, JD16*, 16 {**581**}

Débarbat S, 2001, The evolution of the precession and nutation constants over two centuries. *J2000, a Fundamental Epoch for Origins of Reference Systems and Astronomical Models* (ed. Capitaine N), 153–157 {**576**}

Deltorn JM, Kalas P, 2001, Search for Nemesis encounters with Vega, ϵ Eri, and Fomalhaut. *ASP Conf. Ser. 244: Young Stars Near Earth: Progress and Prospects* (eds. Jayawardhana R, Greene T), 227–231 {**605**}

Demarque P, Woo JH, Kim YC, *et al.*, 2004, Yonsei–Yale (Y^2) isochrones with an improved core overshoot treatment. *ApJS*, 155, 667–674 {**595**}

Denissenko DV, 2004, Occultations of stars brighter than 15 mag by the largest trans-Neptunian objects in 2004–2014. *Astronomy Letters*, 30, 630–633 {**573**}

de Sitter W, Brouwer D, 1938, On the system of astronomical constants. *Bull. Astron. Inst. Netherlands*, 8, 213–217 {**575**}

Di Sisto RP, Orellana RB, 1999, Determinación de posiciones de asteroides utilizando el catálogo Hipparcos. *Boletín de la Asociación Argentina de Astronomia La Plata Argentina*, 43, 8–13 {**571**}

do Nascimento JD, de Medeiros JR, 2003, IRAS colours of exoplanet host stars. *Bulletin of the Astronomical Society of Brazil*, 23, 11–12 {**603**}

Dravins D, Lindegren L, VandenBerg DA, 1998, β Hyi (G2 IV): a revised age for the closest subgiant. *A&A*, 330, 1077–1079 {**608**}

Dunham DW, Goffin E, Manek J, *et al.*, 2002, Asteroid occultation results multiply helped by Hipparcos. *Memorie della Societa Astronomica Italiana*, 73, 662–665 {**572**}

Dunham DW, Timerson B, 2000, The International Occultation Timing Association's long history of amateur-professional collaboration. *American Astronomical Society Meeting*, 196 {**572**}

Dyson FW, 1928, *IAU Transactions*. 3, 227-231 {**576**}

Ecuvillon A, Israelian G, Pont F, *et al.*, 2007, Kinematics of planet-host stars and their relation to dynamical streams in the solar neighbourhood. *A&A*, 461, 171–182 {**602**}

Eddy JA, 1976, The Maunder Minimum. *Science*, 192, 1189–1202 {**587**}

ESA, 1997, *The Hipparcos and Tycho Catalogues. Astrometric and Photometric Star Catalogues derived from the ESA Hipparcos Space Astrometry Mission, ESA SP–1200* (17 volumes including 6 CDs). European Space Agency, Noordwijk, also: VizieR Online Data Catalogue {**566, 569**}

Eyer L, Grenon M, 1997, Photometric variability in the HR diagram. *ESA SP–402: Hipparcos, Venice '97*, 467–472 {**601**}

Farley KA, Montanari A, Shoemaker EM, *et al.*, 1998, Geochemical evidence for a comet shower in the late Eocene. *Science*, 280, 1250–1253 {**584**}

Fernandez JA, Ip WH, 1987, Time-dependent injection of Oort Cloud comets into Earth-crossing orbits. *Icarus*, 71, 46–56 {**584**}

Fienga A, 1998, On the use of Hipparcos data for the determination of astrometric positions of outer planets. *A&A*, 335, 1111–1116 {**566, 571**}

Fienga A, Arlot JE, Pascu D, 1997, Impact of Hipparcos data on astrometric reduction of Solar System bodies. *ESA SP–402: Hipparcos, Venice '97*, 157–160 {**566, 570**}

Fienga A, Manche H, Laskar J, *et al.*, 2008, INPOP06: a new numerical planetary ephemeris. *A&A*, 477, 315–327 {**567**}

Fischer DA, Laughlin G, Butler RP, *et al.*, 2005, The N2K Consortium. I. A hot Saturn planet orbiting HD 88133. *ApJ*, 620, 481–486 {**602**}

Fischer DA, Marcy GW, Butler RP, *et al.*, 2002, A second planet orbiting 47 UMa. *ApJ*, 564, 1028–1034 {**593**}

Folkner WM, Charlot P, Finger MH, *et al.*, 1994, Determination of the extragalactic-planetary frame tie from joint analysis of radio interferometric and lunar laser ranging measurements. *A&A*, 287, 279–289 {**576**}

Frakes LA, Francis JE, Syktus JI, 1992, *Climate Modes of the Phanerozoic*. Cambridge Monographs in Physics {**588, 589**}

Fricke W, 1967, Precession and Galactic rotation derived from fundamental proper motions of distant stars. *AJ*, 72, 1368–1379 {**575, 576**}

—, 1982, Determination of the equinox and equator of the FK5. *A&A*, 107, L13–L16 {**576**}

Friis-Christensen E, Lassen K, 1991, Length of the solar cycle: an indicator of solar activity closely associated with climate. *Science*, 254, 698–700 {**587**}

Frink S, Mitchell DS, Quirrenbach A, *et al.*, 2002, Discovery of a sub-stellar companion to the K2 III giant ι Dra. *ApJ*, 576, 478–484 {**594**}

Frink S, 2003, Upper mass limits for known radial velocity planets from Hipparcos intermediate astrometric data. *ESA SP–539: Earths: Darwin/TPF and the Search for Extrasolar Terrestrial Planets* (eds. Fridlund M, Henning T, Lacoste H), 413–418 {**597**}

Frogel JA, Gould A, 1998, No death star – for now. *ApJ*, 499, L219–L222 {**585**}

Fuhrmann K, Pfeiffer MJ, Bernkopf J, 1997, Solar-type stars with planetary companions: 51 Peg and 47 UMa. *A&A*, 326, 1081–1089 {**602**}

—, 1998, F- and G-type stars with planetary companions: υ And, ρ^1 Cnc, τ Boo, 16 Cyg and ρ CrB. *A&A*, 336, 942–952 {**602**}

Fuhrmann K, 2004, Nearby stars of the Galactic disk and halo. III. *Astronomische Nachrichten*, 325, 3–80 {**589, 603, 608**}

Fukushima T, 2003, A new precession formula. *AJ*, 126, 494–534 {**576**}

Gaidos EJ, 1998, Nearby young solar analogues. I. Catalogue and stellar characteristics. *PASP*, 110, 1259–1276 {**608**}

Galeev AI, Bikmaev IF, Musaev FA, *et al.*, 2004, Chemical composition of 15 photometric analogues of the Sun. *Astronomy Reports*, 48, 492–510 {**608**}

García Sánchez J, 2000, Close approaches of stars to the Solar System based on Hipparcos data. *PASP*, 112, 422–422 {**583**}

García Sánchez J, Preston RA, Jones DL, *et al.*, 1997, A search for stars passing close to the Sun. *ESA SP–402: Hipparcos, Venice '97*, 617–620 {**583**}

—, 1999, Stellar encounters with the Oort Cloud based on Hipparcos data. *AJ*, 117, 1042–1055, erratum: 118, 600 {**583–585**}

García Sánchez J, Weissman PR, Preston RA, *et al.*, 2001, Stellar encounters with the Solar System. *A&A*, 379, 634–659 {**583, 585, 586, 608**}

Gatewood G, Han I, Black DC, 2001, A combined Hipparcos and Multichannel Astrometric Photometer study of the proposed planetary system of ρ CrB. *ApJ*, 548, L61–L63 {**594**}

Gatewood G, 1987, The multichannel astrometric photometer and atmospheric limitations in the measurement of relative positions. *AJ*, 94, 213–224 {**592**}

—, 1995, A study of the astrometric motion of Barnard's star. *Ap&SS*, 223, 91–98 {**594**}

—, 1996, Lalande 21185. *Bulletin of the American Astronomical Society*, 885 {**594**}

Gies DR, Helsel JW, 2005, Ice age epochs and the Sun's path through the Galaxy. *ApJ*, 626, 844–848 {**588–590**}

Gliese W, Jahreiß H, 1991, *Preliminary Version of the Third Catalogue of Nearby Stars.* Astronomischen Rechen-Instituts Heidelberg. On: The Astronomical Data Centre CD-ROM: Selected Astronomical Cataloguess, Vol. I, NASA/ADC/GSFC {**583**}

Gonzalez G, 1998, Spectroscopic analyses of the parent stars of extrasolar planetary system candidates. *A&A*, 334, 221–238 {**602**}

—, 1999a, Are stars with planets anomalous? *MNRAS*, 308, 447–458 {**603, 607**}

—, 1999b, Is the Sun anomalous? *Astronomy and Geophysics*, 40(5), 25–29 {**607**}

—, 2005, Habitable zones in the Universe. *Origins of Life and Evolution of the Biosphere*, 35, 555–606 {**607**}

Gonzalez G, Laws C, 2000, Parent stars of extrasolar planets. V. HD 75289. *AJ*, 119, 390–396 {**603**}

Gorshkov V, Naumov V, Prudnikova E, *et al.*, 2005, Pulkovo coordinates from astrooptical observations. *Kinematika i Fizika Nebesnykh Tel Supplement*, 5, 311–315 {**581**}

Gould A, Ford EB, Fischer DA, 2003, Early-type stars: most favourable targets for astrometrically detectable planets in the habitable zone. *ApJ*, 591, L155–L158 {**607**}

Gould A, Morgan CW, 2003, Transit target selection using reduced proper motions. *ApJ*, 585, 1056–1061 {**601**}

Gratton RG, Bonanno G, Claudi R, *et al.*, 2001, Non-interacting main sequence binaries with different chemical compositions: evidences of infall of rocky material? *A&A*, 377, 123–131 {**603**}

Gray RO, Corbally CJ, Garrison RF, *et al.*, 2006, Contributions to the nearby stars project: spectroscopy of stars earlier than M0 within 40 pc: the southern sample. *AJ*, 132, 161–170 {**608**}

Guirado JC, Reynolds JE, Lestrade JF, *et al.*, 1997, Astrometric detection of a low-mass companion orbiting the star AB Dor. *ApJ*, 490, 835–839 {**596**}

Gulbis AAS, Elliot JL, Person MJ, *et al.*, 2006, Charon's radius and atmospheric constraints from observations of a stellar occultation. *Nature*, 439, 48–51 {**573**}

Halbwachs JL, Arenou F, Mayor M, *et al.*, 2000, Exploring the brown dwarf desert with Hipparcos. *A&A*, 355, 581–594 {**596, 597**}

Han I, Black DC, Gatewood G, 2001, Preliminary astrometric masses for proposed extrasolar planetary companions. *ApJ*, 548, L57–L60 {**594**}

Hardorp J, 1978, The Sun among the stars. I. A search for solar spectral analogues. *A&A*, 63, 383–390 {**607**}

Hatzes AP, Cochran WD, Endl M, *et al.*, 2003, A planetary companion to γ Cep A. *ApJ*, 599, 1383–1394 {**601**}

—, 2006, Confirmation of the planet hypothesis for the long-period radial velocity variations of β Gem. *A&A*, 457, 335–341 {**601**}

Hauser HM, Marcy GW, 1999, The orbit of 16 Cyg AB. *PASP*, 111, 321–334 {**594**}

Hawley SL, Reid IN, 2003, An outsiders view of extrasolar planets. *The Future of Cool-Star Astrophysics: 12th Cambridge Workshop on Cool Stars , Stellar Systems, and the Sun* (eds. Brown A, Harper GM, Ayres TR), volume 12, 128–140 {**602**}

Hébrard G, Lecavelier des Etangs A, 2006, A posteriori detection of the planetary transit of HD 189733b in the Hipparcos photometry. *A&A*, 445, 341–346 {**601**}

Hébrard G, Robichon N, Pont F, *et al.*, 2006, Search for transiting planets in the Hipparcos database. *Tenth Anniversary of 51 Peg b: Status of and Prospects for Hot Jupiter Studies* (eds. Arnold L, Bouchy F, Moutou C), 193–195 {**600**}

Heller CH, 1993, Encounters with protostellar disks. I. Disk tilt and the nonzero solar obliquity. *ApJ*, 408, 337–346 {**589**}

Henry TJ, Soderblom DR, Donahue RA, *et al.*, 1996, A survey of Ca II H and K chromospheric emission in southern solar-type stars. *AJ*, 111, 439–465 {**586**}

Hertz HG, 1966, The mass of Vesta. *IAU Circ.*, 1983, 3 {**568**}

Hestroffer D, 1998, Photocentre displacement of minor planets: analysis of Hipparcos astrometry. *A&A*, 336, 776–781 {**570, 571**}

—, 2003, Photometry with a periodic grid. II. Results for J2 Europa and S6 Titan. *A&A*, 403, 749–756 {**571**}

Hestroffer D, Morando B, Høg E, *et al.*, 1998a, The Hipparcos Solar System objects catalogues. *A&A*, 334, 325–336 {**566**}

Hestroffer D, Viateau B, Rapaport M, 1998b, Minor planets ephemerides improvement from joint analysis of Hipparcos and ground-based observations. *A&A*, 331, 1113–1118 {**570**}

Hills JG, 1981, Comet showers and the steady-state infall of comets from the Oort Cloud. *AJ*, 86, 1730–1740 {**584**}

Hilton JL, 1999, US Naval Observatory ephemerides of the largest asteroids. *AJ*, 117, 1077–1086 {**570**}

Holmberg J, Flynn C, 2004, The local surface density of disk matter mapped by Hipparcos. *MNRAS*, 352, 440–446 {**591**}

Hoyle F, Lyttleton RA, 1939, The effect of interstellar matter on climatic variation. *Proceedings of the Cambridge Philisophical Society*, 405–409 {**587**}

Hut P, Alvarez W, Elder WP, *et al.*, 1987, Comet showers as a cause of mass extinction. *Nature*, 329, 118–126 {**584**}

Israelian G, Santos NC, Mayor M, *et al.*, 2001, Evidence for planet engulfment by the star HD 82943. *Nature*, 411, 163–166 {**603, 604**}

—, 2003, New measurement of the ^{6}Li/^{7}Li isotopic ratio in the extra-solar planet host star HD 82943 and line blending in the Li region. *A&A*, 405, 753–762 {**603**}

Jenkins J, Caldwell DA, Borucki WJ, 2002, Some tests to establish confidence in planets discovered by transit photometry. *ApJ*, 564, 495–507 {**600**}

Jetsu L, Pelt J, 2000, Spurious periods in the terrestrial impact crater record. *A&A*, 353, 409–418 {**590**}

Jura M, Malkan M, White R, *et al.*, 1998, A protocometary cloud around HR 4796A? *ApJ*, 505, 897–902 {**606**}

Kalas P, Deltorn JM, Larwood J, 2001, Stellar encounters with the β Pic planetesimal system. *ApJ*, 553, 410–420 {**605**}

Kaltenegger L, Eiroa C, Stankov A, *et al.*, 2006, Target star catalogue for Darwin: nearby habitable star systems. *IAU Colloq. 200: Direct Imaging of Exoplanets: Science and Techniques* (eds. Aime C, Vakili F), 89–92 {**601**}

Kaplan GH, 2005, The IAU resolutions on astronomical reference systems, time scales, and earth rotation models: explanation and implementation. *US Naval Observatory Circulars*, 179 {**577**}

King JR, Boesgaard AM, Schuler SC, 2005, Keck HIRES spectroscopy of four candidate solar twins. *AJ*, 130, 2318–2325 {**608**}

Kirkpatrick JD, Reid IN, Liebert J, *et al.*, 2000, 67 additional L dwarfs discovered by the Two Micron All Sky Survey. *AJ*, 120, 447–472 {**596**}

Koen C, Lombard F, 2002, Testing photometry of stars for planetary transits. *ESA SP–485: First Eddington Workshop on Stellar Structure and Habitable Planet Finding* (eds. Favata F, Roxburgh IW, Galadi D), 159–161 {**600**}

Kokott W, 1998, Variations of a constant – on the history of precession. *Astronomische Gesellschaft Meeting Abstracts*, 7 {**576**}

Kolesnik YB, Masreliez CJ, 2004, Secular trends in the mean longitudes of planets derived from optical observations. *AJ*, 128, 878–888 {**577, 578**}

Konopliv AS, Yoder CF, Standish EM, *et al.*, 2006, A global solution for the Mars static and seasonal gravity, Mars orientation, Phobos and Deimos masses, and Mars ephemeris. *Icarus*, 182, 23–50 {**566**}

Korsun A, Kurbasova G, 2000, New version of the polar motion in 1846–1900. *Models and Constants for Submicroarcsecond Astrometry, IAU Joint Discussion 2*, 27–32 {**579**}

Kovalevsky J, Lindegren L, Perryman MAC, *et al.*, 1997, The Hipparcos Catalogue as a realisation of the extragalactic reference system. *A&A*, 323, 620–633 {**580**}

Kürster M, Endl M, Rodler F, 2006, In search of terrestrial planets in the habitable zone of M dwarfs. *The Messenger*, 123, 17–20 {**594**}

Kürster M, Endl M, Rouesnel F, *et al.*, 2003, The low-level radial velocity variability in Barnard's star: secular acceleration, indications for convective redshift, and planet mass limits. *A&A*, 403, 1077–1087 {**594**}

Langematz U, Claussnitzer A, Matthes K, *et al.*, 2005, The climate during the Maunder Minimum: a simulation with the Freie Universität Berlin Climate Middle Atmosphere Model. *Journal of Atmospheric and Terrestrial Physics*, 67, 55–69 {**586**}

Laskar J, 1986, Secular terms of classical planetary theories using the results of general theory. *A&A*, 157, 59–70 {**576**}

—, 1996, Marginal stability and chaos in the Solar System. *IAU Symp. 172: Dynamics, Ephemerides, and Astrometry of the Solar System* (eds. Ferraz-Mello S, Morando B, Arlot JE), 75–88 {**572**}

Laubscher RE, 1976b, Dynamical determinations of the general precession in longitude. II. A determination from observations of Mars. *A&A*, 51, 13–20 {**575**}

—, 1976a, Dynamical determinations of the general precession in longitude. I. A rediscussion of previous determinations. *A&A*, 51, 9–20 {**575**}

Laughlin G, 2000, Mining the metal-rich stars for planets. *ApJ*, 545, 1064–1073 {**600, 603**}

Laws C, Gonzalez G, Walker KM, *et al.*, 2003, Parent stars of extrasolar planets. VII. New abundance analyses of 30 systems. *AJ*, 125, 2664–2677 {**603**}

Laymand M, Vauclair S, 2007, Asteroseismology of exoplanets host stars: the special case of ι Hor (HD 17051). *A&A*, 463, 657–662 {**603**}

Leggett SK, Allard F, Geballe TR, *et al.*, 2001, Infrared spectra and spectral energy distributions of late M and L dwarfs. *ApJ*, 548, 908–918 {**599**}

Leitch EM, Vasisht G, 1998, Mass extinctions and the Sun's encounters with spiral arms. *New Astronomy*, 3, 51–56 {**586**}

Lestrade JF, Preston RA, Jones DL, *et al.*, 1999, High-precision VLBI astrometry of radio-emitting stars. *A&A*, 344, 1014–1026 {**584**}

Lieske JH, Lederle T, Fricke W, *et al.*, 1977, Expressions for the precession quantities based upon the IAU 1976 system of astronomical constants. *A&A*, 58, 1–2 {**575, 576**}

Li ZX, 1997, Inter-annual variations of the vertical and their possible influence on the star catalogues derived from ground-based astrometric observations. *The New International Celestial Reference Frame, IAU Joint Discussion 7*, 20 {**580**}

Lockwood M, 2001, Long-term variations in the magnetic fields of the Sun and the heliosphere: their origin, effects, and implications. *J. Geophys. Res.*, 106, 16021–16038 {**587**}

—, 2006, What do cosmogenic isotopes tell us about past solar forcing of climate? *Space Science Reviews*, 125, 95–109 {**589**}

Lockwood M, Fröhlich C, 2007, Recent oppositely directed trends in solar climate forcings and the global mean surface air temperature. *Proc. R. Soc. A.*, 463, 2447–2460 {**589**}

Magnusson P, Lagerkvist CI, Dahlgren M, *et al.*, 1994, The Uppsala Asteroid Data Base. *IAU Symp. 160: Asteroids, Comets, Meteors* (eds. Milani A, di Martino M, Cellino A), 471–476 {**569**}

Magnusson P, 1989, Pole determinations of asteroids. *Asteroids II* (eds. Binzel RP, Gehrels T, Matthews MS), 1180–1190 {**569**}

Mamajek EE, 2005, A moving cluster distance to the exoplanet 2M 1207b in the TW Hya association. *ApJ*, 634, 1385–1394 {**606**}

Manoj P, Bhatt HC, 2005, Kinematics of Vega-like stars: lifetimes and temporal evolution of circumstellar dust disks. *A&A*, 429, 525–530 {**606**}

Marchetti E, Falomo R, Bello D, *et al.*, 2002, A search for star asterisms for natural guide star based MCAO correction. *Beyond Conventional Adaptive Optics. ESO Proceedings Vol. 58* (eds. Vernet E, Ragazzoni R, Esposito S, *et al.*), 403 {**574**}

Marsh ND, Svensmark H, 2000, Low cloud properties influenced by cosmic rays. *Physical Review Letters*, 85, 5004–5007 {**587**}

Masreliez CJ, 1999, The scale expanding cosmos. *Ap&SS*, 266, 399–447 {**577**}

Mathews PM, Herring TA, Buffett BA, 2002, Modeling of nutation and precession: new nutation series for non-rigid Earth and insights into the Earth's interior. *Journal of Geophysical Research*, 107(B4), 3.1–3.26 {**575–577, 581**}

Matthews RAJ, 1994, The close approach of stars in the solar neighbourhood. *QJRAS*, 35, 1–9 {**583, 584**}

Mayor M, Queloz D, 1995, A Jupiter-mass companion to a solar-type star. *Nature*, 378, 355–359 {**591**}

Mazeh T, Naef D, Torres G, *et al.*, 2000, The spectroscopic orbit of the planetary companion transiting HD 209458. *ApJ*, 532, L55–L58 {**600**}

Mazeh T, Zucker S, dalla Torre A, *et al.*, 1999, Analysis of the Hipparcos measurements of υ And: a mass estimate of its outermost known planetary companion. *ApJ*, 522, L149–L151 {**593**}

McCarthy DD, 1995, Modern use of astronomical observations of Earth orientation since 1600. *Bulletin of the American Astronomical Society*, 27, 1286 {**579**}

McCarthy DD, 1996, *IERS conventions, IERS Technical Notes 21*. Observatoire de Paris {**580**}

McDonald SW, Elliot JL, 2000a, Pluto–Charon stellar occultation candidates: 2000–2009. *AJ*, 119, 1999–2007, erratum: 120, 1599 {**574**}

—, 2000b, Triton stellar occultation candidates: 2000–2009. *AJ*, 119, 936–944 {**574**}

McGrath MA, Nelan E, Black DC, *et al.*, 2002, An upper limit to the mass of the radial velocity companion to ρ^1 Cnc. *ApJ*, 564, L27–L30 {**594**}

McGrath MA, Nelan E, Noll K, *et al.*, 2003, An upper limit to the mass of the radial velocity companion to ρ^1 Cnc: an update. *ASP Conf. Ser. 294: Scientific Frontiers in Research on Extrasolar Planets* (eds. Deming D, Seager S), 145–150 {**594**}

Michalak G, 2000, Determination of asteroid masses. I. (1) Ceres, (2) Pallas and (4) Vesta. *A&A*, 360, 363–374 {**570**}

—, 2001, Determination of asteroid masses. II. (6) Hebe, (10) Hygiea, (15) Eunomia, (52) Europa, (88) Thisbe, (444) Gyptis, (511) Davida and (704) Interamnia. *A&A*, 374, 703–711 {**570**}

Milani A, Nobili AM, Carpino M, 1989, Dynamics of Pluto. *Icarus*, 82, 200–217 {**572**}

Morgan HR, Oort JH, 1951, A new determination of the precession and the constants of Galactic rotation. *Bull. Astron. Inst. Netherlands*, 11, 379–386 {**576**}

Morrison LV, Hestroffer D, Taylor DB, *et al.*, 1997, Check on JPL ephemerides DExxx using Hipparcos and Tycho observations. *ESA SP–402: Hipparcos, Venice '97*, 149–152 {**566**}

Mugrauer M, Seifahrt A, Neuhäuser R, *et al.*, 2006, HD 3651B: the first directly imaged brown dwarf companion of an exoplanet host star. *MNRAS*, 373, L31–L35 {**598**}

Mullari AA, Orlov VV, 1996, Encounters of the Sun with nearby stars in the past and future. *Earth, Moon and Planets*, 72, 19–23 {**583**}

Nakajima T, Oppenheimer BR, Kulkarni SR, *et al.*, 1995, Discovery of a cool brown dwarf. *Nature*, 378, 463–465 {**596**}

Napier WM, Clube SVM, 1979, A theory of terrestrial catastrophism. *Nature*, 282, 455–459 {**587**}

Newcomb S, 1898, A new determination of the precessional constant with the resulting motions. *Astron. Papers Amer. Ephem. and Nautical Almanac*, 8, 1–76 {**576**}

Nobili AM, 1988, Long-term dynamics of the outer Solar System review of the LONGSTOP project. *ASSL Vol. 140: IAU Colloq. 96: The Few Body Problem* (ed. Valtonen MJ), 147–163 {**572**}

Oort JH, 1943, Tentative corrections to the FK3 and GC systems of declinations. *Bull. Astron. Inst. Netherlands*, 9, 423–428 {**576**}

—, 1950, The structure of the cloud of comets surrounding the Solar System and a hypothesis concerning its origin. *Bull. Astron. Inst. Netherlands*, 11, 91–110 {**584**}

Pace GW, Walker JCG, 1975, Time markers in interstellar communication. *Nature*, 254, 400–401 {**610**}

Pascu D, Schmidt RE, 1990, Photographic positional observations of Saturn. *AJ*, 99, 1974–1984 {**567**}

Peng QY, Li ZL, 2002, CCD observations for the Hipparcos star HIP 104297 measured with respect to the Galilean satellite Callisto. *Acta Astronomica Sinica*, 43, 84–89 {**573**}

Peng QY, 2001, A very close conjunction between Uranus and a Hipparcos star. *Publications of the Yunnan Observatory*, 86, 55–58 {**573**}

Perryman MAC, 2000, Extra-solar planets. *Reports of Progress in Physics*, 63, 1209–1272 {**591**}

Perryman MAC, Lindegren L, Arenou F, *et al.*, 1996, Hipparcos distances and mass limits for the planetary candidates: 47 UMa, 70 Vir, and 51 Peg. *A&A*, 310, L21–L24 {**593, 594**}

Person MJ, Elliot JL, Gulbis AAS, *et al.*, 2006, Charon's radius and density from the combined data sets of the 2005 July 11 occultation. *AJ*, 132, 1575–1580 {**573**}

Peters CAF, 1842, *Numerus Constans Nutationis*. Academy of Sciences, St Petersburg {**576**}

Piétu V, Dutrey A, Kahane C, 2003, A Keplerian disk around the Herbig Ae star HD 34282. *A&A*, 398, 565–569 {**606**}

Pinfield DJ, Jones HRA, Lucas PW, *et al.*, 2006, Finding benchmark brown dwarfs to probe the substellar initial mass function as a function of time. *MNRAS*, 368, 1281–1295 {**598**}

Pinho LGF, Porto de Mello GF, de Medeiros JR, *et al.*, 2003, The Sol project: the sun in time. *Bulletin of the Astronomical Society of Brazil*, 23, 126–126 {**608**}

Pinho LGF, Porto de Mello GF, 2003, Astrobiologically interesting stars in the solar neighbourhood. *Bulletin of the Astronomical Society of Brazil*, 23, 128–128 {**608**}

Poma A, Uras S, Pannunzio R, 1997, Comparison between the ILS-MD and Hipparcos Catalogues. *ESA SP–402: Hipparcos, Venice '97*, 113–116 {**580**}

Poma A, Uras S, 1996, Re-reduction of the ILS latitude observations using the Hipparcos data. *Proc. 3rd International Workshop on Positional Astronomy and Celestial Mechanics*, 455–460 {**580**}

Porto de Mello GF, da Silva L, 1997, HR 6060: the closest ever solar twin? *ApJ*, 482, L89–L92 {**608**}

Porto de Mello GF, da Silva R, da Silva L, 2000, A survey of solar twin stars within 50 pc of the Sun. *ASP Conf. Ser. 213: Bioastronomy 99*, 73 {**608**}

Porto de Mello GF, del Peloso EF, Ghezzi L, 2006, Astrobiologically interesting stars within 10 pc of the Sun. *Astrobiology*, 6, 308–331 {**608**}

Pourbaix D, Arenou F, 2001, Screening the Hipparcos-based astrometric orbits of sub-stellar objects. *A&A*, 372, 935–944 {**594**}

Pourbaix D, 2001, The Hipparcos observations and the mass of sub-stellar objects. *A&A*, 369, L22–L25 {**594**}

Raghavan D, Henry TJ, Mason BD, *et al.*, 2006, Two suns in the sky: stellar multiplicity in exoplanet systems. *ApJ*, 646, 523–542 {**601**}

Rahmstorf S, Archer D, Ebel DS, *et al.*, 2004, Cosmic rays, carbon dioxide, and climate. *EOS Transactions*, 85(4), 38–41 {**589**}

Raynaud E, Drossart P, Sicardy B, *et al.*, 2001, Scale-time and correlation analysis of light curves from the HIP 9369 occultation by the northern polar region of Jupiter. *AAS/Division for Planetary Sciences Meeting*, 33 {**573**}

Reffert S, Quirrenbach A, 2006, Hipparcos astrometric orbits for two brown dwarf companions: HD 38529 and HD 168443. *A&A*, 449, 699–702 {**597**}

Reid IN, 2002, On the nature of stars with planets. *PASP*, 114, 306–329 {**603, 607**}

Reid IN, Gizis JE, Hawley SL, 2002, The Palomar/MSU nearby star spectroscopic survey. IV. The luminosity function in the solar neighbourhood and M dwarf kinematics. *AJ*, 124, 2721–2738 {**602**}

Reid IN, Hawley SL (eds.), 2000, *New Light on Dark Stars: Red Dwarfs, Low-Mass Stars, Brown Dwarfs* {**596**}

Reid IN, Turner EL, Turnbull MC, *et al.*, 2007, Searching for Earth analogues around the nearest stars: the disk age–metallicity relation and the age distribution in the solar neighbourhood. *ApJ*, 665, 767–784 {**603**}

Rhee J, Song I, Zuckerman B, *et al.*, 2007, Characterization of dusty debris disks: the IRAS and Hipparcos Catalogues. *ApJ*, 660, 1556–1571 {**606**}

Rizvanov N, Nefedjev J, 2005, Photographic observations of Solar System bodies at the Engelhardt astronomical observatory. *A&A*, 444, 625–627 {**571**}

Robichon N, Arenou F, 2000a, HD 209458 planetary transits from Hipparcos photometry. *A&A*, 355, 295–298 {**598–600**}

—, 2000b, Planetary transits in space-borne missions. *IAU Symp. 202* {**600**}

Rocha Pinto HJ, Scalo JM, Maciel WJ, *et al.*, 2000, An intermittent star-formation history in a normal disk galaxy: the Milky Way. *ApJ*, 531, L115–L118 {**588**}

Ron C, Vondrák J, 2001, On the celestial pole offsets from optical astrometry in 1899–1992. *J2000, a Fundamental Epoch for Origins of Reference Systems and Astronomical Models* (ed. Capitaine N), 201–202 {**579**}

—, 2003, An improved star catalogue for Ondřejov PZT. *Astrometry from Ground and from Space* (eds. Capitaine N, Stavinschi M), 191–195 {**581**}

Roques F, Lavillonnière N, Auvergne M, *et al.*, 2001, Research of small Kuiper Belt Objects by stellar occultations. *Bulletin of the American Astronomical Society*, 1031 {**573**}

Rouillard A, Lockwood M, 2004, Oscillations in the open solar magnetic flux with a period of 1.68 years: imprint on galactic cosmic rays and implications for heliospheric shielding. *Annales Geophysicae*, 22, 4381–4395 {**587**}

Roy AE, Walker IW, MacDonald AJ, *et al.*, 1988, Project LONGSTOP. *Vistas in Astronomy*, 32, 95–116 {**572**}

Royer DL, Berner RA, Montañez IP, *et al.*, 2004, CO_2 as a primary driver of Phanerozoic climate. *GSA Today, Geol. Soc. Amer.*, 14(3), 4–10 {**589**}

Ryl'Kov VP, Dement'eva AA, Narizhnaja NV, 1997, The Pulkovo catalogue of 284 positions of Pluto in 1930–1994 based on observations from three observatories. *Baltic Astronomy*, 6, 349–350 {**571**}

Saar SH, Baliunas SL, 1992, Recent advances in stellar cycle research. *ASP Conf. Ser. 27: The Solar Cycle* (ed. Harvey KL), 150–167 {**586**}

Sandquist EL, Dokter JJ, Lin DNC, *et al.*, 2002, A critical examination of Li pollution and giant-planet consumption by a host star. *ApJ*, 572, 1012–1023 {**603**}

Santos NC, Israelian G, Mayor M, 2001, The metal-rich nature of stars with planets. *A&A*, 373, 1019–1031 {**603**}

—, 2004, Spectroscopic [Fe/H] for 98 extrasolar planet-host stars: exploring the probability of planet formation. *A&A*, 415, 1153–1166 {**603**}

Sato B, Fischer DA, Henry GW, *et al.*, 2005, A transiting hot Saturn around HD 149026 with a large dense core. *ApJ*, 633, 465–473 {**600**}

Saumon D, Hubbard WB, Burrows A, *et al.*, 1996, A theory of extrasolar giant planets. *ApJ*, 460, 993–1018 {**596**}

Savoie D, 2001, The precession of the equinoxes from Hipparchus to Tycho Brahe. *J2000, a Fundamental Epoch for Origins of Reference Systems and Astronomical Models* (ed. Capitaine N), 125–130 {**576**}

Schaefer BE, 1997, Sunspots that changed the world. *S&T*, 93(4), 34–36 {**587**}

Serafin RA, Grothues HG, 2002, On stellar encounters and their effect on cometary orbits in the Oort Cloud. *Astronomische Nachrichten*, 323, 37–48 {**586**}

Setiawan J, Rodmann J, da Silva L, *et al.*, 2005, A sub-stellar companion around the intermediate-mass giant star HD 11977. *A&A*, 437, L31–L34 {**594, 598**}

Shaviv NJ, 2002, Cosmic ray diffusion from the Galactic spiral arms, iron meteorites, and a possible climatic connection. *Physical Review Letters*, 89(5), 1102–1105 {**587**}

—, 2003, The spiral structure of the Milky Way, cosmic rays, and ice age epochs on Earth. *New Astronomy*, 8, 39–77 {**587–589**}

Shum CK, Woodworth PL, Andersen OB, *et al.*, 1997, Accuracy assessment of recent ocean tide models. *J. Geophys. Res.*, 102(C11), 25173–25194 {**580**}

Sicardy B, Bellucci A, Gendron E, *et al.*, 2006, Charon's size and an upper limit on its atmosphere from a stellar occultation. *Nature*, 439, 52–54 {**573**}

Sicardy B, Colas F, Widemann T, *et al.*, 2004, The two stellar occultations of 14 November 2003: revealing Titan's stratosphere at sub-km resolution. *AAS/Division for Planetary Sciences Meeting Abstracts*, 36 {**573**}

Sicardy B, Widemann T, Lellouch E, *et al.*, 2003, Large changes in Pluto's atmosphere as revealed by recent stellar occultations. *Nature*, 424, 168–170 {**573**}

Siess L, Livio M, 1999, The accretion of brown dwarfs and planets by giant stars. II. Solar-mass stars on the red giant branch. *MNRAS*, 308, 1133–1149 {**603**}

Simon JL, Bretagnon P, Chapront J, *et al.*, 1994, Numerical expressions for precession formulae and mean elements for the Moon and the planets. *A&A*, 282, 663–683 {**576**}

Simon RS, 1998, Getting the dirt on exo-zodis with the MMA. *Exo-Zodiacal Dust Workshop*, 199 {**606**}

Smoluchowski R, Torbett M, 1984, The boundary of the Solar System. *Nature*, 311, 38–39 {**584**}

Söderhjelm S, Robichon N, Arenou F, 1999, HD 209458. *IAU Circ.*, 7323, 3 {**598**}

Soma M, Hayamizu T, Berthier J, *et al.*, 2006, (22) Kalliope and (22) Kalliope I. *Central Bureau Electronic Telegrams*, 732, 1 {**573**}

Song I, Weinberger AJ, Becklin EE, *et al.*, 2002, M-type Vega-like stars. *AJ*, 124, 514–518 {**606**}

Song I, Zuckerman B, Weinberger AJ, *et al.*, 2005, Extreme collisions between planetesimals as the origin of warm dust around a Sun-like star. *Nature*, 436, 363–365 {**604**}

Soon WW, Yaskell SH, 2004, *Maunder Minimum and the Sun–Earth Connection*. World Scientific {**586**}

Soriano M, Vauclair S, Vauclair G, *et al.*, 2007, The COROT primary target HD 52265: models and seismic tests. *A&A*, 471, 885–892 {**603**}

Souchay J, Feissel M, Bizouard C, *et al.*, 1995, Precession and nutation for a non-rigid Earth: comparison between theory and VLBI observations. *A&A*, 299, 277–287 {**580**}

Souchay J, Loysel B, Kinoshita H, *et al.*, 1999, Corrections and new developments in rigid earth nutation theory. III. Final tables 'REN-2000' including crossed-nutation and spin-orbit coupling effects. *A&AS*, 135, 111–131 {**577**}

Spencer Jones H, 1939, The rotation of the Earth, and the secular accelerations of the Sun, Moon and planets. *MNRAS*, 99, 541–558 {**577**}

Standish EM, Newhall JG, Williams WF, *et al.*, 1995, JPL planetary and lunar ephemerides DE403/LE403. *JPL IOM*, 314, 10 {**568**}

Stansberry J, Grundy W, Brown M, *et al.*, 2008, Physical properties of Kuiper Belt and Centaur objects: constraints from Spitzer Space Telescope. *The Solar System beyond Neptune*, 161–179 {**573**}

Stefka V, Vondrák J, 2006, Earth orientation catalogue EOC-3: an improved optical reference frame. *Nomenclature, Precession and New Models in Fundamental Astronomy, IAU 26, JD16*, 16 {**581**}

Stone RC, 1999a, CCD Solar System astrometry reduced to the Hipparcos reference frame. *Bulletin of the American Astronomical Society*, 31, 849 {**567**}

—, 1999b, Improved predictions for planetary occultations. *Southern Skies*, 38, 148–159 {**572**}

—, 2001, Positions for the outer planets and many of their satellites. V. FASTT observations taken in 2000–2001. *AJ*, 122, 2723–2733 {**571**}

Stothers RB, 1998, Galactic disk dark matter, terrestrial impact cratering and the law of large numbers. *MNRAS*, 300, 1098–1104 {**587, 590**}

Struve O, 1843, *Astronomische Nachrichten*. 21, 65-68 {**576**}

Suchkov AA, Schultz AB, 2001, Evidence for a very young age of F stars with extrasolar planets. *ApJ*, 549, L237–L240 {**603**}

Svensmark H, 2006a, Cosmic rays and the biosphere over 4 billion years. *Astronomische Nachrichten*, 327, 871–875 {**589**}

—, 2006b, Imprint of Galactic dynamics on Earth's climate. *Astronomische Nachrichten*, 327, 866–870 {**589, 591**}

—, 2007, Cosmoclimatology: a new theory emerges. *Astronomy and Geophysics*, 48(1), 18–24 {**589**}

Sylvester RJ, Dunkin SK, Barlow MJ, 2001, SCUBA photometry of candidate Vega-like sources. *MNRAS*, 327, 133–140 {**606**}

Tholen DJ, 1989, Asteroid taxonomic classifications. *Asteroids II* (eds. Binzel RP, Gehrels T, Matthews MS), 1139–1150 {**569**}

Tinney CG, Butler RP, Marcy GW, *et al.*, 2001, First results from the Anglo–Australian planet search: a brown dwarf candidate and a 51 Peg-like planet. *ApJ*, 551, 507–511 {**594**}

Tinney CG, 1999, Brown dwarfs: the stars that failed. *Nature*, 397, 37–40 {**596**}

Torres G, 2007, The planet host star γ Cep: physical properties, the binary orbit, and the mass of the substellar companion. *ApJ*, 654, 1095–1109 {**595**}

Turnbull MC, Tarter JC, 2003a, Target selection for SETI. II. Tycho 2 dwarfs, old open clusters, and the nearest 100 stars. *ApJS*, 149, 423–436 {**609**}

—, 2003b, Target selection for SETI. I. A catalogue of nearby habitable stellar systems. *ApJS*, 145, 181–198 {**609**}

Valenti JA, Fischer DA, 2005, Spectroscopic properties of cool stars. I. 1040 F, G, and K dwarfs from Keck, Lick, and AAT planet search programmes. *ApJS*, 159, 141–166 {**602**}

van de Kamp P, 1963, Astrometric study of Barnard's star from plates taken with the 24-inch Sproul refractor. *AJ*, 68, 515–521 {**594**}

—, 1977, Barnard's star 1916–1976: a sexagintennial report. *Vistas in Astronomy*, 20, 501–521 {**594**}

—, 1982, The planetary system of Barnard's star. *Vistas in Astronomy*, 26, 141–157 {**594**}

Veiga CH, Vieira Martins R, Andrei AH, 2003, Positions of Uranus and its main satellites. *AJ*, 125, 2714–2720 {**571**}

Veizer J, Godderis Y, François LM, 2000, Evidence for decoupling of atmospheric CO_2 and global climate during the Phanerozoic eon. *Nature*, 408, 698–701 {**588, 589**}

Viateau B, 2000, Mass and density of asteroids (16) Psyche and (121) Hermione. *A&A*, 354, 725–731 {**568, 570**}

Viateau B, Rapaport M, 1998, The mass of (1) Ceres from its gravitational perturbations on the orbits of 9 asteroids. *A&A*, 334, 729–735 {**568**}

Vitagliano A, Stoss RM, 2006, New mass determination of (15) Eunomia based on a very close encounter with (50278) 2000CZ12. *A&A*, 455, L29–L31 {**570**}

Vityazev VV, 2001, Irregular variations in Earth rotation: the singular spectrum analysis and wavelets. *J2000, a Fundamental Epoch for Origins of Reference Systems and Astronomical Models* (ed. Capitaine N), 287–291 {**579**}

Vogt N, 2006, FH Leo, the first dwarf nova member of a multiple star system? *A&A*, 452, 985–986 {**603**}

Vogt SS, Butler RP, Marcy GW, *et al.*, 2002, Ten low-mass companions from the Keck precision velocity survey. *ApJ*, 568, 352–362 {**594, 601**}

Vogt SS, Marcy GW, Butler RP, *et al.*, 2000, Six new planets from the Keck precision velocity survey. *ApJ*, 536, 902–914 {**596**}

Vondrák J, 1996, Indirect linking of the Hipparcos Catalogue to extragalactic reference frame via Earth orientation parameters. *IAU Symp. 172: Dynamics, Ephemerides, and Astrometry of the Solar System*, 491–496 {**580**}

—, 1999, Earth rotation parameters 1899.7–1992.0 after reanalysis within the Hipparcos frame. *Surveys in Geophysics*, 20, 169–195 {**579, 581**}

—, 2003, Earth Orientation Catalogue: an improved reference frame. *The International Celestial Reference System: Maintenance and Future Realization, IAU Joint Discussion 16*, 17–20 {**581**}

—, 2004, Astrometric star catalogues as combination of Hipparcos/Tycho Catalogues with ground-based observations. *Serbian Astronomical Journal*, 168, 1–8 {**581**}

Vondrák J, Feissel M, Essaifi N, 1992, Expected accuracy of the 1900–1990 Earth orientation parameters in the Hipparcos reference frame. *A&A*, 262, 329–340 {**579**}

Vondrák J, Ron C, Pešek I, *et al.*, 1995, New global solution of Earth orientation parameters from optical astrometry in 1900–1990. *A&A*, 297, 899–906 {**579**}

—, 1997a, The Hipparcos Catalogue: a reference frame for Earth orientation in 1899.7–1992.0. *ESA SP–402: Hipparcos, Venice '97*, 95–100 {**579, 582**}

Vondrák J, Ron C, Pešek I, 1997b, Earth rotation in the Hipparcos reference frame. *Celestial Mechanics and Dynamical Astronomy*, 66, 115–122 {**579**}

—, 1997c, Using the Earth orientation parameters to link the Hipparcos and VLBI reference frames. *A&A*, 319, 1020–1024 {**580**}

Vondrák J, Ron C, 1997a, Long-periodic nutation terms as determined by optical astrometry in 1899.7–1992.0. *Precession–Nutation and Astronomical Constants for the Dawn of the 21st Century, IAU Joint Discussion 3*, 17–20 {**580**}

—, 1997b, The orientation of the Hipparcos celestial system with respect to ICRS as induced from the Earth orientation parameters. *The New International Celestial Reference Frame, IAU Joint Discussion 7*, 39–42 {**580**}

—, 2000a, Precession–nutation estimates from optical astrometry 1899.7–19920.0 and comparison with VLBI results. *IAU Colloq. 180: Towards Models and Constants for Sub-Microarcsecond Astrometry* (eds. Johnston KJ, McCarthy DD, Luzum BJ, *et al.*), 248–253 {**579, 580**}

—, 2000b, Survey of observational techniques and Hipparcos reanalysis. *ASP Conf. Ser. 208: IAU Colloq. 178: Polar Motion: Historical and Scientific Problems*, 239–250 {**579, 580, 583**}

—, 2003, An improved optical reference frame for long-term Earth rotation studies. *Astrometry from Ground and from Space* (eds. Capitaine N, Stavinschi M), 49–55 {**581**}

—, 2005a, Combined astrometric catalogue EOC-2: an improved reference frame for long-term Earth rotation studies. *Fundamental Astronomy: New Concepts and Models for High Accuracy Observations* (ed. Capitaine N), 210–215 {**581**}

—, 2005b, Solution of Earth orientation parameters in the frame of the new Earth Orientation Catalogue. *Kinematika i Fizika Nebesnykh Tel Supplement*, 5, 305–310 {**581, 583**}

Vondrák J, Ron C, 2005, The great Chandler wobble change in 1923–1940 re-visited. *Forcing of Polar Motion in the Chandler Frequency Band: A Contribution to Understanding Inter-Annual Climate Variations, CCEGS 24* (eds. Plag HP, Chao B, Gross R, *et al.*), 39–47 {**581**}

Vondrák J, Stefka V, 2007, Combined astrometric catalogue EOC-3. An improved reference frame for long-term Earth rotation studies. *A&A*, 463, 783–788 {**581, 583**}

Wahr JM, 1981, The forced nutations of an elliptical, rotating, elastic and oceanless Earth. *GJRAS*, 64, 705–727 {**576**}

Wallace PT, Capitaine N, 2006, Precession–nutation procedures consistent with IAU 2006 resolutions. *A&A*, 459, 981–985 {**577**}

Walter HG, Sovers OJ, 2000, *Astrometry of Fundamental Catalogues: the Evolution from Optical to Radio Reference Frames*. Springer–Verlag, Berlin {**575**}

Weissman PR, 1992, The effects of star passages through the Oort Cloud. *Bulletin of the American Astronomical Society*, 1063 {**584**}

—, 1996a, Star passages through the Oort Cloud. *Earth Moon and Planets*, 72, 25–30 {**584**}

—, 1996b, The Oort Cloud. *ASP Conf. Ser. 107: Completing the Inventory of the Solar System* (eds. Rettig T, Hahn JM), 265–288 {**584**}

West RM, Hainaut O, Schulz R, *et al.*, 1994, Periodic comet Shoemaker–Levy 9 (1993e). *IAU Circ.*, 6017, 1 {**575**}

Wielen R, Dettbarn C, Jahreiß H, *et al.*, 1999, Indications on the binary nature of individual stars derived from a comparison of their Hipparcos proper motions with ground-based data. I. Basic principles. *A&A*, 346, 675–685 {**581**}

Williams JG, 1994, Contributions to the Earth's obliquity rate, precession, and nutation. *AJ*, 108, 711–724 {**576**}

Williams GE, 1975, Possible relation between periodic glaciation and the flexure of the Galaxy. *Earth and Planetary Science Letters*, 26, 361–369 {**587**}

Woitas J, Tamazian VS, Docobo JA, *et al.*, 2003, Visual orbit for the low-mass binary Gliese 22 AC from speckle interferometry. *A&A*, 406, 293–298 {**598**}

Wright JT, 2004, Do we know of any Maunder minimum stars? *AJ*, 128, 1273–1278, erratum: 128, 1273 {**586**}

—, 2006, Maunder minimum stars revisited. *Solar and Stellar Activity Cycles, IAU Joint Discussion 8*, 28–31 {**586**}

Yagudina EI, 2001, The use of radar observations of near-Earth asteroids in the determination of the dynamical equinox. *Celestial Mechanics and Dynamical Astronomy*, 80, 195–203 {**576, 577**}

Yeghikyan A, Fahr H, 2004a, Effects induced by the passage of the Sun through dense molecular clouds. I. Flow outside of the compressed heliosphere. *A&A*, 415, 763–770 {**587**}

—, 2004b, Terrestrial atmospheric effects induced by counterstreaming dense interstellar cloud material. *A&A*, 425, 1113–1118 {**587**}

Yi S, Demarque P, Kim YC, *et al.*, 2001, Toward better age estimates for stellar populations: the Yonsei-Yale (Y^2) isochrones for solar mixture. *ApJS*, 136, 417–437 {**595**}

Yoder CF, Williams JG, Parke ME, 1981, Tidal variations of Earth rotation. *J. Geophys. Res.*, 86, 881–891 {**580**}

Yoshizawa M, 1997, Hipparcos and ground-based proper motions of NPZT stars. *ESA SP–402: Hipparcos, Venice '97*, 109–112 {**580**}

Zhang B, Sigurdsson S, 2003, Electromagnetic signals from planetary collisions. *ApJ*, 596, L95–L98 {**604**}

Zhu Z, 2007, Precession constant correction and proper motion systems of FK5 and Hipparcos. *Chinese Astronomy and Astrophysics*, 31, 296–307 {**578**}

Zuckerman B, Song I, Bessell MS, 2004, The AB Dor moving group. *ApJ*, 613, L65–L68 {**606**}

Zucker S, Mazeh T, 2000, Analysis of the Hipparcos measurements of HD 10697: a mass determination of a brown dwarf secondary. *ApJ*, 531, L67–L69 {**594, 596, 598**}

—, 2001, Analysis of the Hipparcos observations of the extrasolar planets and the brown dwarf candidates. *ApJ*, 562, 549–557 {**593, 594, 597, 598**}

Appendix A

Numerical quantities

Numerical quantities

This compilation of numerical quantities includes both fundamental defining quantities, and certain relevant derived and associated reference quantities. The system of astronomical units relevant for astrometry is essentially defined by four numbers: the length of the day, d; the mass of the Sun, $M_\odot$, or in practice $GM_\odot$; the astronomical unit, A_m; and the Gaussian constant of gravitation, k (a discussion of the complexities and limitations can be found in Klioner 2008). Many of the quantities here have been compiled for the Gaia mission's parameter database by J. de Bruijne. As a result of changing definitions, some of the physical constants used in the construction of the Hipparcos and Tycho Catalogues (Table 1.3, notably A_m, $GM_\odot$, $GM_\oplus$ and ϵ) have been updated compared to those listed on page 9. Other comments are as follows:

- CODATA06: these parameters are '2006 CODATA recommended values', from the CODATA Task Group on Fundamental Constants (see http://www.codata.org and http://physics.nist.gov/cuu/Constants).
- INPOP06: self-consistent (TCB-compatible) Solar System quantities from the numerical planetary ephemeris developed at the IMCCE–Observatoire de Paris (Fienga *et al.*, 2008). They provide an alternative to the JPL development ephemeris solutions DE405 and the latest version DE414 (Konopliv *et al.*, 2006).
- IAU(1976): the IAU (1976) system of astronomical constants.

(1) The astronomical unit (in m) is a defining constant in INPOP06, consistent with the combination of the light travel time and the speed of light used in the JPL DE solutions.

(2) The value here is derived from the other defining quantities (CODATA06 gives $5.670\,400(40) \times 10^{-8}$).

(3) Because of the limited accuracy in the knowledge of G, planetary ephemerides, and associated masses and linear dimensions, are still calculated in terms of the Gaussian constant of gravitation, $k = 4\pi^2 A_\mathrm{m}^3/(P_\oplus^2(M_\odot + M_\oplus))$. Gauss (1857) used a numerical value of k which was subsequently incorporated into the IAU (1976) system of astronomical constants, and remains that used as a defining constant in, e.g. INPOP06.

(4) The Julian Year is an IAU definition, entering directly in the definition of the unit of proper motion, mas yr^{-1}.

(5) In high-accuracy applications, care is needed in representing quantities dependent on π due to rounding errors. Double-precision floating-point numbers following IEEE standards have 64 bits (16 significant digits) yielding, to appropriate accuracy:

- $\pi = 3.141\,592\,653\,589\,793\,238$
- 1 degree in radians: $\pi/180 = 1.745\,329\,251\,994\,33 \times 10^{-2}$ rad
- 1 mas in radians: $\pi/(180 \times 3600 \times 1000) = 4.848\,136\,811\,095\,36 \times 10^{-9}$ rad

(6) $\gamma = 1$ is fixed in INPOP06. Other terms given here can be derived numerically, e.g. Klioner (2003). $\delta_\mathrm{pN}(\perp)$ is the post-Newtonian deflection angle for an observer at 1 AU from the Sun, of a light ray arriving at right angles to the solar direction due to the spherically symmetric part of the gravitational field of the Sun. $\delta_\mathrm{pN}(R_\odot)$ is the post-Newtonian deflection angle for an observer at 1 AU from the Sun, of a solar-limb grazing light ray due to the spherically symmetric part of the gravitational field of the Sun.

(7) Accurate barycentric arrival time determinations require corrections for both geometrical (Rømer) delay associated with the observer's motion around the barycentre, and relativistic (Shapiro) delay caused by the gravitational bodies in the Solar System; see, e.g. Lindegren & Dravins (2003, Section 4.3), and Will (2003, Equation 7).

(8) The value in INPOP06 is itself derived from other defining constants. Although $M_\odot$ is currently considered as a constant, the physical mass of the Sun decreases at $\sim 10^{-13} M_\odot\,\mathrm{yr}^{-1}$ carried by solar radiation (Krasinsky & Brumberg, 2004). Secular acceleration in the mean longitudes of the inner planets currently places a limit of $\dot{G}/G = -2 \pm 5 \times 10^{-14}\,\mathrm{yr}^{-1}$ (Pitjeva, 2005), equivalent to the same limit on the determination of $\dot{M}/M$ (Klioner, 2008).

(9) The solar apparent radius at 1 AU is that derived from astrolabe observations by Chollet & Sinceac (1999). At this accuracy level, solar oblateness is irrelevant.

(10) See Section 7.2.1.

(11) Average of the three measurements cited.

(12) Grevesse & Noels (1993) give $Y_\odot$ and $Z_\odot$, from which $X_\odot = 1 - Y_\odot - Z_\odot$. Associated parameters relevant for the equation-of-state are the mean molecular weight $\mu_\odot = 1/(2X_\odot + 3Y_\odot/4 + Z_\odot/2) = 0.6092$ and the mean molecular weight per free electron $\mu_{e,\odot} = 2/(X_\odot + 1) = 1.1651$. These derivations assume complete ionisation, and that metals give ~0.5 particles per m_H (e.g. Karttunen, 1987, Section 11.2). See Section 7.1 for recent developments.

(13) Here, 'Earth' includes the Earth's atmosphere, but excludes the Moon. 'Earth-system' includes the Moon. Similarly, the 'Jupiter-system' includes the contribution from its moons.

(14) Mean value at J2000.0 from a 250-year fit of the JPL DE200 ephemeris. The orbital period is derived from mean longitude rates. The resulting quoted orbital semi-major axis of the Sun's orbit around the hypothetical barycentre of the Sun/Jupiter-system is the major component of many contributions to the Sun's orbit around the solar system barycentre.

(15) Hipparcos-derived values from Dehnen & Binney (1998). For a discussion of alternatives, and errors, see Section 9.2.4.

(16) For concepts and definitions, see Section 9.1.3. For a discussion of R_0, see Section 9.2.1. Hipparcos-derived values of$A$, B (and derived quantities) are from Feast & Whitelock (1997). For a discussion of alternatives, and errors, see Section 9.2.5.

(17) See Kovalevsky (2003, Equation 4).

(18) Rounded mean of the various Hipparcos values.

Symbol	Meaning	Source/Derivation	Value
	Various constants:		
c	Speed of light in vacuum (exact)	CODATA06	$299\,792\,458\,\mathrm{m\,s^{-1}}$
A_m	Astronomical unit (in m)[1]	INPOP06	$1.495\,978\,706\,910\,000 \times 10^{11}\,\mathrm{m}$
G	Newton's constant of gravitation	CODATA06	$6.674\,28(67) \times 10^{-11}\,\mathrm{m^3\,kg^{-1}\,s^{-2}}$
h	Planck constant	CODATA06	$6.626\,068\,96(33) \times 10^{-34}\,\mathrm{J\,s}$
σ	Stefan–Boltzmann constant[2]	$2\pi^5 k^4/(15c^2h^3)$	$5.670\,399 \times 10^{-8}\,\mathrm{W\,m^{-2}\,K^{-4}}$
k	Gaussian gravitational constant (exact)[3]	IAU (1976)	$0.017\,202\,098\,95\,\mathrm{AU^{3/2}\,day^{-1}}\,M_\odot^{-1/2}$
	Time scales (Section 1.4.1):		
d	Day in s (any time scale)	Klioner (2008)	86 400 (exactly)
y	Julian Year in days[4]		365.25 (= 31 557 600 s)
J2000	Julian date of standard epoch, J2000.0	Section 1.4.1	2 451 545. 0 JD
	Derived quantities[5]:		
A_v	Proper motion constant (Equation 1.5)	$A_\mathrm{m}/(y \times d \times 1000)$	$4.740\,470\,463\,\mathrm{km\,yr\,s^{-1}}$
$\mathrm{pc_m}$	Parsec in m	$A_\mathrm{m} \times 180 \times 3600/\pi$	$3.085\,677\,581\,305\,729 \times 10^{16}\,\mathrm{m}$
$\mathrm{pc_{AU}}$	Parsec in AU	$180 \times 3600/\pi$	$2.062\,648\,06 \times 10^5\,\mathrm{AU}$
$\mathrm{ly_m}$	Light-year in m	$y \times d \times c$	$9.460\,730\,473 \times 10^{15}\,\mathrm{m}$
$\mathrm{ly_{AU}}$	Light-year in AU	$y \times d \times c/A_\mathrm{m}$	$6.324\,107\,708\,8 \times 10^4\,\mathrm{AU}$
$\mathrm{ly_{pc}}$	Light-year in pc	$y \times d \times c/\mathrm{pc_m}$	0.306 601 pc
	General Relativity:		
γ	General relativistic PPN parameter[6]	Current assumption	1
$\delta_\mathrm{pN}(\perp)$	Deflection angle at 1 AU perpendicular to ecliptic[6]	Klioner (2003)	4.072×10^{-3} arcsec
$\delta_\mathrm{pN}(R_\odot)$	Deflection angle at 1 AU at solar limb[6]	Klioner (2003)	1.750 453 arcsec
Δ	Shapiro time-delay constant[7]	$(1+\gamma)(GM_\odot)/c^3$	$9.851 \times 10^{-6}\,\mathrm{s}$

Symbol	Quantity	Source	Value
	Sun:		
$GM_\odot$	Heliocentric gravitational constant[8]	INPOP06	$1.327\,124\,420\,76 \times 10^{20}$ m^3 s^{-2}
$M_\odot$	Mass[8]	$(GM_\odot)/G$	$1.988\,4 \times 10^{30}$ kg
$\theta_\odot$	Apparent radius in arcsec[9]	Chollet & Sinceac (1999)	959.63(8) arcsec
$R_\odot$	Apparent radius in m	$A_\mathrm{m} \times \pi\theta_\odot/(3600 \times 180)$	$6.959\,917\,56 \times 10^8$ m
$\rho_\odot$	Mean mass density	$M_\odot/((4/3)\pi R_\odot^3)$	1.408 g cm^{-3}
$g_\odot$	Surface gravity	$(GM_\odot)/R_\odot^2$	274.0 m s^{-2}
$J_{2,\odot}$	Oblateness	INPOP06	$1.95(55) \times 10^{-7}$
$M_{V,\odot}$	Johnson absolute V magnitude[10]	Cayrel de Strobel (1996)	+4.83 mag
$M_{\mathrm{bol},\odot}$	Bolometric magnitude[10]	Cayrel de Strobel (1996)	+4.75 mag
$\mathrm{BC}_{V,\odot}$	Bolometric correction, Johnson V band[10]	Cayrel de Strobel (1996)	−0.08 mag
$J_\odot$	Energy flux at 1 AU[11]	Duncan *et al.* (1982)	1371 W m^{-2}
$L_\odot$	Luminosity	$4\pi A_\mathrm{m}^2 J_\odot$	3.856×10^{26} W
$T_{\mathrm{eff},\odot}$	Effective (black-body) temperature	$(L_\odot/4\pi\sigma R_\odot^2)^{1/4}$	5781 K
$X_\odot, Y_\odot, Z_\odot$	H, He, and metal abundances by mass[12]	Grevesse & Noels (1993)	0.716 6, 0.265 9, 0.017 5
$\alpha_{\mathrm{MLT},\odot}$	Mixing-length parameter	Girardi *et al.* (2000)	1.68
	Other Solar System values:		
$GM_\oplus$	Geocentric gravitational constant[13]	INPOP06	$3.986\,004\,390\,77 \times 10^{14}$ m^3 s^{-2}
$M_\oplus$	Earth mass[13]	$(GM_\oplus)/G$	$5.972\,2 \times 10^{24}$ kg
$M_\odot/M_\oplus$	Sun/Earth mass ratio[13]	INPOP06	$3.329\,460\,508\,95 \times 10^5$
	Sun/Earth-system mass ratio[13]	INPOP06	$3.289\,005\,614\,00 \times 10^5$
$M_\odot/M_\mathrm{J}$	Sun/Jupiter-system mass ratio[13]	INPOP06	1047.348 6
M_J	Jupiter-system mass[13]	$M_\odot/(M_\odot/M_\mathrm{J})$	1.899×10^{27} kg $\sim 9.5 \times 10^{-4} M_\odot$
a_J	Orbital semi-major axis of Jupiter-system[14]	Seidelmann (1992)	5.203 363 01 AU
P_J	Jupiter sidereal orbital period[14]	Seidelmann (1992)	11.862 615 yr
$a_{\odot,\mathrm{J}}$	Sun's orbital semi-major axis wrt Jupiter-system[14]	$a_\mathrm{J}A_\mathrm{m}/(1000 \times (M_\odot/M_\mathrm{J}))$	743 222 km
α_J	Sun's astrometric signature from Jupiter at 10 pc	$a_\mathrm{J}/(10(M_\odot/M_\mathrm{J}))$	497×10^{-6} arcsec
	Transformations:		
ϵ	Obliquity of ecliptic (J2000.0)	Chapront *et al.* (2002)	$23^\circ\,26'\,21.411'' = 84\,381.411$ arcsec
$\alpha_G, \delta_G, l_\Omega$	Galactic coordinates within ICRS	Section 1.4.5	$\alpha_G = 192.^\circ 859\,48$, $\delta_G = +27.^\circ 128\,25$, $l_\Omega = 32.^\circ 931\,92$
$\mathbf{A}'_\mathrm{K}$	ICRS to ecliptic coordinates	Section 1.4.5	see Equation 1.15
$\mathbf{A}'_\mathrm{G}$	ICRS to Galactic coordinates	Section 1.4.5	see Equation 1.16
	Solar motion with respect to LSR:[15]		
$U_\odot$	Component towards Galactic centre	Dehnen & Binney (1998)	+10.00 km s^{-1}
$V_\odot$	Component in direction of Galactic rotation	Dehnen & Binney (1998)	+ 5.25 km s^{-1}
$W_\odot$	Component towards north Galactic pole	Dehnen & Binney (1998)	+ 7.17 km s^{-1}
V_LSR	Total velocity $(U_\odot^2 + V_\odot^2 + W_\odot^2)^{1/2}$	Dehnen & Binney (1998)	13.38 km s^{-1}
	Local Galactic constants:[16]		
A	Oort constant, A	Feast & Whitelock (1997)	+14.82 km s^{-1} kpc^{-1}
B	Oort constant, B	Feast & Whitelock (1997)	−12.37 km s^{-1} kpc^{-1}
Ω_0	Angular velocity	$A - B$	+27.19 km s^{-1} kpc^{-1}
R_0	Galactocentric radius of the Sun	Section 9.2.1	8.2 kpc
V_0	Circular velocity at R_0	$R_0\,\Omega_0$	223 km s^{-1}
P_rot	Galactic rotation period	$(2\pi/\Omega_0)(\mathrm{pc_m}/y \times d)$	2.26×10^8 yr
$\Delta\mu$	Maximum aberration due to Galactic rotation[17]	$V_0^2/(R_0 c)$ rad s^{-1} (V_0, R_0 in SI)	4.3×10^{-6} arcsec yr^{-1}
κ_0	Epicycle frequency	$(-4B(A-B))^{0.5}$	36.7 km s^{-1} kpc^{-1}
Z_0	Sun's height from Galactic mid-plane[18]	Section 9.2.2	+20 pc
ρ_0	Oort limit (disk+dark halo; Section 9.4.1)	Holmberg & Flynn (2000)	$0.102\,M_\odot$ pc^{-3}
$P_\perp$	Vertical oscillation period (Section 9.4.1)	$(\pi/G\rho_0)^{0.5}$ s (ρ_0 in SI)	8.2×10^7 yr

References

Cayrel de Strobel G, 1996, Stars resembling the Sun. *A&A Rev.*, 7, 243–288 {**621**}

Chapront J, Chapront-Touzé M, Francou G, 2002, A new determination of lunar orbital parameters, precession constant and tidal acceleration from lunar laser ranging measurements. *A&A*, 387, 700–709 {**621**}

Chollet F, Sinceac V, 1999, Analysis of solar radius determination obtained by the modern CCD astrolabe of the Calern Observatory: a new approach of the solar limb definition. *A&AS*, 139, 219–229 {**620, 621**}

Dehnen W, Binney JJ, 1998, Local stellar kinematics from Hipparcos data. *MNRAS*, 298, 387–394 {**620, 621**}

Duncan CH, Willson RC, Kendall JM, *et al.*, 1982, Latest rocket measurements of the solar constant. *Solar Energy*, 28, 385–387 {**621**}

Feast MW, Whitelock PA, 1997, Galactic kinematics of Cepheids from Hipparcos proper motions. *MNRAS*, 291, 683–693 {**620, 621**}

Fienga A, Manche H, Laskar J, *et al.*, 2008, INPOP06: a new numerical planetary ephemeris. *A&A*, 477, 315–327 {**619**}

Girardi L, Bressan A, Bertelli G, *et al.*, 2000, Evolutionary tracks and isochrones for low- and intermediate-mass stars: from $0.15-7\,M_\odot$ at $Z = 0.0004-0.03$. *A&AS*, 141, 371–383 {**621**}

Grevesse N, Noels A, 1993, Cosmic abundances of the elements. *Origin and Evolution of the Elements. Cambridge University Press* (eds. Prantzos N, Vangioni-Flam E, Casse M), 14–25 {**620, 621**}

Holmberg J, Flynn C, 2000, The local density of matter mapped by Hipparcos. *MNRAS*, 313, 209–216 {**621**}

Karttunen H, 1987, *Fundamental Astronomy*. Springer–Verlag, Berlin {**620**}

Klioner SA, 2003, A practical relativistic model for microarcsecond astrometry in space. *AJ*, 125, 1580–1597 {**619, 620**}

—, 2008, Relativistic scaling of astronomical quantities and the system of astronomical units. *A&A*, 478, 951–958 {**619, 620**}

Konopliv AS, Yoder CF, Standish EM, *et al.*, 2006, A global solution for the Mars static and seasonal gravity, Mars orientation, Phobos and Deimos masses, and Mars ephemeris. *Icarus*, 182, 23–50 {**619**}

Kovalevsky J, 2003, Aberration in proper motions. *A&A*, 404, 743–747 {**620**}

Krasinsky GA, Brumberg VA, 2004, Secular increase of astronomical unit from analysis of the major planet motions, and its interpretation. *Celestial Mechanics and Dynamical Astronomy*, 90, 267–288 {**620**}

Lindegren L, Dravins D, 2003, The fundamental definition of radial velocity. *A&A*, 401, 1185–1201 {**620**}

Pitjeva EV, 2005, Relativistic effects and solar oblateness from radar observations of planets and spacecraft. *Astronomy Letters*, 31, 340–349 {**620**}

Seidelmann PK, 1992, *Explanatory Supplement to the Astronomical Almanac*. University Science Books, New York {**621**}

Will CM, 2003, Propagation speed of gravity and the relativistic time delay. *ApJ*, 590, 683–690 {**620**}

Appendix B

Acronyms

- 2MASS Two-Micron All Sky Survey
- AAO Anglo–Australian Observatory
- AAT Anglo–Australian Telescope
- AAVSO Amateur Association of Variable Star Observers
- ACRS Astrographic Catalogue Reference Stars (catalogue)
- ACS Advanced Camera for Surveys (instrument on HST)
- ACT Astrographic Catalogue plus Tycho (reference catalogue)
- ADS Aitken Double Star (catalogue)
- ADS Astrophysics Data System
- AFOEV Association Française des Observateurs d'Etoiles Variables
- AGB Asymptotic Giant Branch (stars)
- AMEX Astrometric Mapping Explorer (satellite concept)
- APM Automatic Plate Measuring (machine)
- APS Automated Plate Scanner
- ARI Astronomisches Rechen-Institut (Heidelberg)
- ASAS All Sky Automated Survey
- ASCC All-Sky Compiled Catalogue
- ATLAS Program for calculating model stellar atmospheres
- AXAF X-ray Astrophysics Facility (satellite)
- BCRS Barycentric Celestial Reference System
- BD Bonner Durchmusterung (catalogue)
- BIH Bureau International de l'Heure
- BJD Barycentric Julian Date
- BPM Bruce Proper Motion (catalogue)
- BSC Bright Star Catalogue
- BSS Bright Star Supplement (to UCAC catalogue)
- BTA Big Telescope Alt-azimuthal (SAO 6-m telescope)
- CCD Charge-Coupled Device
- CCDM Catalogue of Components of Double and Multiple Stars
- CD Córdoba Durchmusterung (catalogue)
- CDM Cold Dark Matter
- CDS Centre de Données Astronomique de Strasbourg
- CERGA Centre de Recherche en Géodynamique et Astrométrie
- CESAM Stellar evolution model code
- CFH/CFHT Canada–France–Hawaii telescope
- CHARA Centre for High Angular Resolution Astronomy
- CMC Carlsberg Meridian Catalogue
- CNO Carbon–Nitrogen–Oxygen (cycle)
- CNS Catalogue of Nearby Stars
- COBE Cosmic Background Explorer (satellite)
- COMPTEL Compton Telescope (instrument on GRO)
- CORAVEL Radial velocity instrument
- COROT COnvection ROtation and planetary Transits (satellite)
- COS-B Gamma ray observatory (satellite)

CPC Cape Photographic Catalogue
CPD Cape Durchmusterung (catalogue)
CPIRSS Catalogue of Positions of Infrared Stellar Sources
CRVAD Catalogue of Radial Velocities with Astrometric Data
CTIO Cerro Tololo Inter-American Observatory
CUO Copenhagen University Observatory

DDO David Dunlap Observatory
DENIS Deep Near Infrared Survey of the Southern Sky
DIB Diffuse Interstellar Band
DIRBE Diffuse Infrared Background Experiment (on COBE)
DIVA Double Interferometer for Visual Astrometry (satellite concept)
DM Durchmusterung (catalogue)
DMSA Double and Multiple Systems Annex (Hipparcos)
DSS Digitized Sky Survey

EFF Eggleton–Faulkner–Flannery (equation-of-state)
EGRET Energetic Gamma Ray Experiment Telescope (on GRO)
ELODIE Radial velocity instrument
ESA European Space Agency
ESO European Southern Observatory
ESOC European Space Operations Centre
EUV Extreme Ultraviolet
EUVE Extreme Ultraviolet Explorer (satellite)
EXOSAT X-ray satellite

FAME Full-Sky Astrometric Mapping Explorer (satellite concept)
FAST Fundamental Astrometry by Space Techniques (Hipparcos consortium)
FASTT Flagstaff Astrometric Scanning Transit Telescope
FAUST Experiment on ATLAS-1 Shuttle mission
FC Fundamental Catalog
FGS Fine Guidance Sensor (instrument on HST)
FLS Fornax–Leo I and II–Sculptor stream
FLUOR Fiber Linked Unit for Recombination (on IOTA)
FRANEC Stellar evolution model code
FUSE Far Ultraviolet Spectroscopic Explorer (satellite)
FUV Far Ultraviolet
FWHM Full Width at Half Maximum

GALAXY Photographic plate measuring machine
GC General Catalogue
GCRS Geocentric Celestial Reference System
GCTP General Catalogue of Trigonometric Parallaxes
GCVS General Catalogue of Variable Stars
GJ Gliese–Jahreiß (catalogue)
GL Gliese (catalogue)
GOMOS Global Ozone Monitoring by Occultation of Stars (on ENVISAT)
GPS Global Positioning System
GRB Gamma Ray Burst
GRO Gamma Ray Observatory (satellite)
GSC Guide Star Catalogue
GSFC Goddard Space Flight Centre

HARPS High Accuracy Radial velocity Planet Searcher (at ESO)
HB Horizontal Branch (stars)
HCRF Hipparcos Celestial Reference Frame (ICRF materialised by Hipparcos)
HD Henry Draper (catalogue)
HDE Henry Draper Extension (catalogue)
HEPA Hipparcos Epoch Photometry Annex
HEPAE Hipparcos Epoch Photometry Annex Extension
HIC Hipparcos Input Catalogue (identifier number)
HIP Hipparcos Catalogue (identifier number)
HRI High Resolution Imager (instrument on ROSAT)
HR Hertzsprung–Russell (diagram)
HST Hubble Space Telescope (satellite)

IAU International Astronomical Union
ICDB Identified Counts Data Base (Tycho 2 Catalogue)
ICRF International Celestial Reference Frame
ICRS International Celestial Reference System
IERS International Earth Rotation Service
ILR Inner Lindblad Resonance
ILS International Latitude Service
IMC Institut de Mécanique Céleste (Paris)
IMF Initial Mass Function
INAG Institut National d'Astronomie et de Géophysique (predecessor of INSU)
INCA Input Catalogue Consortium (Hipparcos consortium)
INSU Institut National des Sciences
IOTA Infrared Optical Telescope Array
IPMS International Polar Motion Service
IR Infrared
IRAS Infrared Astronomical Satellite
IRC Infrared Catalogue
IRS International Reference Stars (catalogue)
ISM Interstellar Medium
ISO Infrared Space Observatory (satellite)

ISO	International Standards Organization
IUE	International Ultraviolet Explorer (satellite)
JASMINE	Japan Astrometry Satellite Mission for INfrared Exploration (satellite concept)
JD	Julian Date
JMSTAR	stellar evolution model code
JPL	Jet Propulsion Laboratory
JWST	James Webb Space Telescope (satellite)
KPNO	Kitt Peak National Observatory
LBT	Large Binocular Telescope
LCC	Lower Centaurus Crux (star association)
LHS	Luyten Half-Second (catalogue)
LIDA	Light Interferometric Device for Astrometry (satellite concept)
LM	Maximum likelihood luminosity calibration algorithm
LMC	Large Magellanic Cloud
LONGSTOP	Long-Term Gravitational Stability Test for the Outer Planets
LPM	Luyten Proper Motion (catalogue)
LPV	Long-Period Variable
LSPM	Lépine–Shara Proper Motion (catalogue)
LSR	Local Standard of Rest
LSST	Large Synoptic Survey Telescope
LTE	Local Thermodynamic Equilibrium
LTT	Luyten Two-Tenths (catalogue)
LTT	Light Travel Time
MACHO	Massive Compact Halo Object
MAMA	Multianode Microchannel Array (measuring machine)
MAP	Multichannel Astrometric Photometer
MAPS	Milli-Arcsecond Pathfinder Survey (satellite concept)
MARCS	Program for calculating model stellar atmospheres
MBM	Magnani–Blitz–Mundy (molecular gas clouds)
MCAO	Multi-Conjugate Adaptive Optics
MDI	Michelson Doppler Imager (instrument on SOHO)
MEGA	Proper motion catalogue
MERIT	Monitor Earth Rotation and Intercompare Techniques
MERLIN	Multi-Element Radio Linked Interferometer Network
MHD	Magnetohydrodynamic
MIRA	Mitaka Optical and Infrared Array
MIT	Massachusetts Institute of Technology
MK	Morgan–Keenan (spectral classification system)
MKK	Morgan–Keenan–Kellman (spectral classification system)
MLT	Mixing Length Theory
MOST	Microvariability and Oscillations of Stars (satellite)
MPDC	Minor Planet Data Centre
MPIA	Max-Planck-Institut für Astronomie
MS	Main Sequence
MSC	Multiple Star Catalogue
MSTO	Main Sequence Turn-Off
MSX	Mid-Course Experiment (satellite)
NACO	NAOS-CONICA (VLT instrument)
NCP	North Celestial Pole
NDAC	Northern Data Analysis Consortium (Hipparcos)
NEAR	Near Earth Asteroid Rendezvous Mission (satellite)
NFK	Neuer Fundamental Katalog (catalogue)
NGC	New General Catalogue
NLTE	Non-Local Thermodynamical Equilibrium
NLTT	New Luyten Two-Tenths (catalogue)
NMARCS	Program for calculating model stellar atmospheres
NOAO	National Optical Astronomy Observatory
NOMAD	Naval Observatory Merged Astrometric Dataset
NPM	Northern Proper Motion (catalogue)
NPOI	Navy Prototype Optical Interferometer
NPZT	Northern Photographic Zenith Tube (catalogue)
NSV	New and Suspected Variable Stars (catalogue)
OATO	Osservatorio Astronomico di Torino
OBN	N-rich OB stars
OBSS	Origins Billion Star Survey (satellite concept)
OCA	Observatoire de la Côte d'Azur (Nice)
ODF	Opacity Distribution Function
OGLE	Optical Gravitational Lensing Experiment
OHP	Observatoire de Haute Provence
OLR	Outer Lindblad Resonance
OPAL	Opacity Code
ORFEUS	Orbiting Retrievable Far and Extreme Ultraviolet Spectrometer (instrument)
OSIRIS	Ohio State InfraRed Imager/Spectrometer
PC	Period-Colour (relation)
PCRS	Pointing Calibration and Reference Sensor (instrument on SPITZER)
PCRV	Pulkovo Compilation of Radial Velocities (catalogue)
PGC	Principal Galaxy Catalogue
PISCO	Speckle Camera (at Pic Du Midi Observatory)
PK	Perek–Kohoutek (catalogue)

PLC Period–Luminosity–Colour (relation)
PLCZ Period–Luminosity–Colour–Metallicity (relation)
PMFS Proper Motions of Fundamental Stars (catalogue)
PMM Precision Measuring Machine
PMS Pre-Main Sequence
POSS Palomar Observatory Sky Survey
POSS-I Palomar Observatory Sky Survey (first epoch)
POSS-II Palomar Observatory Sky Survey (second epoch)
PPARC Particle Physics and Astronomy Research Council (predecessor of STFC)
PPM Positions and Proper Motions (catalogue)
PPN Parameterized Post-Newtonian
PPO Petit Prisme Objectif
PRAVELO Pulkovo RAdial VELOcities (catalogue)
PRIMA Phase Referenced Imaging and Micro-arcsecond Astrometry (VLTI instrument)
PSC Point Source Catalogue (IRAS)
PSPC Position Sensitive Proportional Counter (instrument on ROSAT)
PTI Palomar Testbed Interferometer
PZT Photographic Zenith Tube

RASS ROSAT All-Sky Survey (catalogue)
RAVE Radial Velocity Experiment (at AAO)
RECONS Research Consortium on Nearby Stars
RGB Red Giant Branch
RGO Royal Greenwich Observatory
ROSAT Röntgen Satellite
RPM Reduced Proper Motion
RYTSI RIT-Yale Tip-Tilt Speckle Imager

SAAO South African Astronomical Observatory
SAO Smithsonian Astrophysical Observatory (catalogue)
SCP South Celestial Pole
SCUBA Submillimetre Common-User Bolometer Array (instrument on JCMT)
SDSS Sloan Digital Sky Survey
SERC Science and Engineering Research Council (predecessor of STFC)
SES Second Epoch Survey
SETI Search for Extraterrestrial Intelligence
SFR Star Formation Rate/Star Formation Region
SHASSA Southern H-Alpha Sky Survey Atlas
SI Système International d'Unités (International System (of Units))
SIM Space Interferometry Mission (satellite concept)
SIMBAD Astronomical Database (at CDS)
SIRTF Space Infrared Telescope Facility (satellite)
SMC Small Magellanic Cloud
SMM Solar Maximum Mission (satellite)
SN Supernova
SOHO Solar and Heliospheric Observatory (satellite)
SPB Slowly-Pulsating B (star)
SPM Southern Proper Motion (catalogue)
SPO Schmidt Prisme Objectif (Schmidt Objective Prism)
SPY Supernova Ia Progenitor Survey
SR Semi-Regular (variable)
SRS Southern Reference System (catalogue)
SSSC Sydney Southern Star Catalogue
STFC Science and Technology Facilities Council
STIS Space Telescope Imaging Spectrograph (instrument on HST)
SUSI Sydney University Stellar Interferometer
SWS Short Wavelength Spectrometer (instrument on ISO)

TAC Twin Astrographic Catalogue
TAI Temps Atomique International (International Atomic Time)
TAMS Terminal Age Main Sequence
TCB Temps Coordonné Barycentrique (Barycentric Coordinate Time)
TDAC Tycho Data Analysis Consortium (Hipparcos)
TDB Temps Dynamique Barycentrique (Barycentric Dynamical Time)
TDSC Tycho Double Star Catalogue
TDT Temps Dynamique Terrestre (Terrestrial Dynamical Time)
TOPBASE Opacity Project on-line atomic database
TRC Tycho Reference Catalogue
TT Temps Terrestre (Terrestrial Time)
TYC Tycho Catalogue (identifier number)
TYC2 Tycho 2 (catalogue)

UCAC USNO CCD Astrograph Catalogue
UCL Upper Centaurus Lupus (star association)
USNO United States Naval Observatory
UTC Temps Universel Coordonné (Coordinated Universal Time)
UV Ultraviolet
UVES UV-Visual Echelle Spectrograph (VLT instrument)

VBLUW Colours in Walraven photometric system
VIM Variability Induced Mover (type of binary star)
VLA Very Large Array

VLBA Very Long Baseline Array
VLBI Very Long Baseline Interferometry
VLT Very Large Telescope (ESO)
VLTI Very Large Telescope Interferometer
VSOP Variations Séculaires des Orbites Planétaires

WCCD Washington Comprehensive Catalogue Database
WD White Dwarf
WDS Washington Double Star (catalogue)
WEB Wilson–Evans–Batten (radial velocity catalogue)
WIYN Wisconsin, Indiana, Yale, NOAO (observatory)
WMAP Wilkinson Microwave Anisotropy Probe (satellite)
WR Wolf–Rayet (star)

XMM X-ray Multi-Mirror (satellite)

YREC Yale Rotating Evolutionary Code

ZAHB Zero-Age Horizontal Branch (star)
ZAMS Zero Age Main Sequence (star)

Appendix C

Author gallery

Carlos Abad
Mérida, Venezuela

Connie Aerts
Leuven

Luis A. Aguilar
UNAM, Mexico

Christine Allen
UNAM, Mexico

Angel Alonso
IAC, Tenerife

Frédéric Arenou
Obs. de Paris

Pierre Bacchus‡
Lille

Yuri Balega
SAO, Russia

Ulrich Bastian
ARI, Heidelberg

Holger Baumgardt
Univ. Bonn

Tim Bedding
Sydney

Jacques Bergeat
Lyon

Pierluigi Bernacca
Univ. Padova

Hans-Heinrich Bernstein
ARI, Heidelberg

Mike Bessell
Mt Stromlo Observatory

Olivier Bienaymé
Strasbourg

James Binney
Oxford

Adriaan Blaauw
Groningen

Richard Branham
Mendoza, Argentina

Danielle Briot
Obs. de Paris

Anthony Brown
Leiden

Beatricia Bucciarelli
Torino

Alexey Butkevich
Dresden

Nicole Capitaine
Obs. de Paris

Bruce W. Carney
Univ. of North Carolina

Eugenio Carretta
OA, Bologna

Fabien Carrier
Leuven

Dana Casetti-Dinescu
Wesleyan Univ.

Angelo Cassatella
INAF, Roma

Giusa Cayrel de Strobel
Obs. de Paris

Roger Cayrel
Obs. de Paris

Brian Chaboyer
Dartmouth

Julio Chanamé
STScI, Baltimore

Corinne Charbonnel
Genève

Masashi Chiba
Tohoku, Japan

Fernando Comerón
ESO

Alan Cousins‡
SAAO, South Africa

Noel Cramer
Genève

Michel Crézé
APC, Université Paris 7

Giuseppe Cutispoto
INAF, Catania

Thomas H. Dall
Gemini

Andrei Dambis
SAI, Moscow

Klaas de Boer
Univ. Bonn

Jos de Bruijne
ESA–ESTEC

Walter Dehnen
Leicester

Jean Delhaye‡
Obs. de Paris

Tim de Zeeuw
Leiden/ESO

Maria Pia Di Mauro
IASF, Roma

Jean Dommanget
ORB, Bruxelles

José Dias do Nascimento
UFRN, Brazil

Francesco Donati
Torino

Dainis Dravins
Lund

Ronald Drimmel
Torino

Jiri Dusek
Brno

Olin Eggen[‡]
CTIO

Daniel Egret
Strasbourg/Paris

Dafydd Wyn Evans
Cambridge

Laurent Eyer
Genève

Claus Fabricius
Barcelona

Benoit Famaey
ULB, Bruxelles

Fabio Favata
ESA–ESTEC

Michael Feast
Cape Town

Sofia Feltzing
Lund

David Fernández
Barcelona

Francesca Figueras
Barcelona

Debra Fischer
San Francisco

Chris Flynn
Tuorla Observatory

Adam Frankowski
ULB, Bruxelles

Walter Fricke[‡]
ARI, Heidelberg

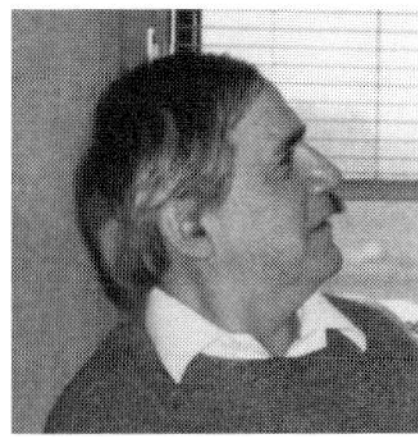

Michel Froeschlé
CERGA, Grasse

Joan García Sanchez
Barcelona

Michele Gerbaldi
IAP, Paris

Douglas Gies
Georgia State

Terry Girard
Yale

Léo Girardi
Padova

Wilhelm Gliese[‡]
ARI, Heidelberg

Ana Gómez
Obs. de Paris

Philippe Gondoin
ESA–ESTEC

George Gontcharov
Pulkovo Observatory

Andrew Gould
Ohio State

Rafaele Gratton
Padova

Michel Grenon
Genève

Michael Grewing
IRAM, Grenoble

Frank Grundahl
Aarhus

David Guenther
St Mary's Univ., Halifax

Patrick Guillout
Strasbourg

Jean-Louis Halbwachs
Strasbourg

Nigel Hambly
Edinburgh

Gerald Handler
Vienna

Ulrich Heber
Erlangen

Guillaume Hébrard
IAP, Paris

Amina Helmi
Groningen

Jesús Hernández
CIDA, Venezuela

Xavier Hernández
UNAM, Mexico

Daniel Hestroffer
IMCCE, Paris

Erik Høg
Copenhagen

David W. Hogg
New York

Ronnie Hoogerwerf
ISC, Inc

Elliott Horch
SCSU

Anne-Marie Hubert
Obs. de Paris

Swetlana Hubrig
ESO

Hartmut Jahreiß
ARI, Heidelberg

Carlos Jaschek[‡]
Strasbourg

Eric Jensen
Swarthmore

Raul Jimenez
ICE, Barcelona

Carme Jordi
Barcelona

Alain Jorissen
ULB, Bruxelles

Torsten Kaempf
AIFI, Bonn

Paul Kalas
UC Berkeley

Philip Keenan[‡]
Ohio State

Florian Kerber
ESO

Pierre Kervella
Obs. de Paris

Nina Kharchenko
MAO, Kiev

Sergei Klioner
Dresden

Jill Knapp
Princeton University

Jens Knude
Copenhagen

Chris Koen
Univ. Western Cape

Brigitte König
Munich

Jean Kovalevsky
CERGA, Grasse

Pavel Kroupa
Bonn

Martin Kürster
MPIA, Heidelberg

Pierre Lacroute[‡]
Strasbourg

Rosine Lallement
Verrières-le-Buisson

Henny Lamers
Utrecht

Patricia Lampens
ORB, Bruxelles

Yveline Lebreton
Obs. de Paris

Jacques Lépine
São Paulo

Sebastien Lépine
AMNH, New York

Rudolf Le Poole
Leiden

Jean-François Lestrade
Obs. de Paris

Per Olof Lindblad
Stockholm

Lennart Lindegren
Lund

R. Earle Luck
CWRU, Ohio

Xavier Luri
Barcelona

Chopo Ma
NASA GSFC

Søren Madsen
Lund

Gisela Maintz
Bonn

Hans-Michael Maitzen
Vienna

Jesús Maíz Apellániz
IAA, Granada

Eric Mamajek
Harvard–Smithsonian CfA

Eduardo Martín
IAC, Tenerife

John C. Martin
Univ. of Illinois

Brian Mason
US Naval Observatory

Janet Mattei[‡]
AAVSO, Cambridge MA

Jaymie Matthews
Univ. of British Columbia

Tsevi Mazeh
Tel Aviv

Marie-Odile Mennessier[‡]
Montpellier

François Mignard
Nice

Anthony Moffat
Univ. Montréal

David Monet
US Naval Observatory

Leslie Morrison
RGO

Andrew Murray
RGO

Heidi Jo Newberg
Rensselaer, New York

Poul Erik Nissen
Aarhus

Birgitta Nordström
Copenhagen

Pierre North
EPF, Lausanne

Edouard Oblak
Besançon

Rob Olling
Univ. of Maryland

Sebastian Otero
CEA, Argentina

René Oudmaijer
Leeds

Ferhat Fikri Özeren
Kayseri, Turkey

Bohdan Paczyński‡
Princeton

Jan Palouš
Prague

Alexey Pamyatnykh
Warsaw/Moscow

Ernst Paunzen
Vienna

Susan Percival
Liverpool John Moores

John R. Percy
Toronto

Michael Perryman
ESA–ESTEC

Jeff Pier
USNO

Anatoly Piskunov
INASAN Moscow

Imants Platais
Johns Hopkins

Frédéric Pont
Genève

Gustavo Porto de Mello
Rio de Janeiro

Stuart Pottasch
Groningen

Dimitri Pourbaix
ULB, Bruxelles

Thomas Preibisch
MPIFR Bonn

Theodor Pribulla
Tatranská Lomnica, Slovakia

Jean-Louis Prieur
OMP Toulouse

Judith Provencal
Univ. of Delaware

Alice Quillen
Univ. of Rochester

Bacham Eswar Reddy
IIA, Bangalore

Sabine Reffert
LSW, Heidelberg

Mark Reid
Harvard–Smithsonian CfA

Yves Réquième
Bordeaux

Ignasi Ribas
CSIC–IEEC Barcelona

Noël Robichon
Obs. de Paris

Slavek Rucinski
Toronto

Maurizio Salaris
Liverpool John Moores

Allan Sandage
Obs. of Carnegie Inst.

Marília J.Sartori
LNA, Brazil

Hans Schrijver
Utrecht

Klaus–Peter Schröder
Guanajuato, Mexico

Selim Selam
AUO, Ankara

Harry Shipman
Univ. Delaware

Roger Sinnott
Sky & Telescope

Tanya Sitnik
Sternberg, Moscow

Jovan Skuljan
New Zealand

Ricky Smart
Torino

David Soderblom
STScI, Baltimore

Staffan Söderhjelm
Lund

Inseok Song
Caltech

Caroline Soubiran
Bordeaux

John Southworth
Warwick

James Sowell
Georgia Tech.

Krzysztof (Kris) Stanek
Ohio State

Klaus Strassmeier
Potsdam

Anatoly Suchkov
STScI, Baltimore

Henrik Svensmark
Danish Space Centre

Peeter Tenjes
Tartu

Wil Tirion
Capelle a/d IJssel, NL

Andrei Tokovinin
CTIO, Chile

Jordi Torra
Barcelona

Guillermo Torres
Harvard–Smithsonian CfA

Catherine Turon
Obs. de Paris

Makoto Uemura
Hiroshima

Sean Urban
US Naval Observatory

William van Altena
Yale

Gerard van Belle
ESO

Mario van den Ancker
ESO

Hans van der Marel
TU, Delft

Sophie Van Eck
ULB, Bruxelles

Floor van Leeuwen
Cambridge

Gérard Vauclair
Toulouse

Sergio Vieira
Minas Gerais, Brazil

Jan Vondrák
Prague

Christoffel Waelkens
Leuven

Walter Wegner
Bydgoszcz, Poland

Barry Welsh
UC Berkeley

Patricia Whitelock
SAAO, South Africa

Roland Wielen
ARI, Heidelberg

Jason Wright
Berkeley

Norbert Zacharias
US Naval Observatory

Zi Zhu
Nanjing

Shay Zucker
Tel Aviv

Ben Zuckerman
UC Los Angeles

Index of first authors

Subject index